2018 International Power Electronics Conference (IPEC-Niigata 2018 –ECCE Asia-)

Niigata, Japan
20-24 May 2018

Pages 759-1459

IEEE Catalog Number: CFP1854I-POD
ISBN: 978-1-5386-4190-3

Copyright © 2018, IEEJ Industry Applications Society
All Rights Reserved

*** This is a print representation of what appears in the IEEE Digital Library. Some format issues inherent in the e-media version may also appear in this print version.*

IEEE Catalog Number: CFP1854I-POD
ISBN (Print-On-Demand):   978-1-5386-4190-3
ISBN (Online):   978-4-88686-405-5

Additional Copies of This Publication Are Available From:

Curran Associates, Inc
57 Morehouse Lane
Red Hook, NY 12571 USA
Phone: (845) 758-0400
Fax: (845) 758-2633
E-mail: curran@proceedings.com
Web: www.proceedings.com

TABLE OF CONTENTS

THREE-PHASE INDUCTIVE POWER TRANSFER SYSTEM WITH 12 COILS FOR RADIATION NOISE REDUCTION 69
Keisuke Kusaka ; Jun-Ichi Itoh

SECONDARY-SIDE-ONLY CONTROL FOR SMOOTH VOLTAGE STABILIZATION IN WIRELESS POWER TRANSFER SYSTEMS WITH CONSTANT POWER LOAD 77
Giorgio Lovison ; Takehiro Imura ; Hiroshi Fujimoto ; Yoichi Hori

CONSTANT CURRENT CHARGING AND THE MAXIMUM SYSTEM EFFICIENCY TRACKING FOR WIRELESS CHARGING SYSTEMS EMPLOYING DUAL-SIDE CONTROL 84
Zhenjie Li ; Xiaoliang Huang ; Kai Song ; Jinhai Jiang ; Chunbo Zhu ; Zhijiang Du

ELECTRIC FIELD COUPLING TYPE HIGH POWER WIRELESS POWER TRANSFER WITH LEAKAGE ELECTRIC FIELD STRUCURE 88
Mitsuru Masuda

TRANSFER POWER ANALYSIS OF CAPACITIVELY ISOLATED OUTLET AND PLUG (CAPISOP) USING SERIES RESONANCE 94
Hirohito Funato ; Koki Amano ; Takuya Hatsumi ; Junnosuke Haruna

WIDE VOLTAGE GAIN RANGE LLC DC/DC TOPOLOGIES: STATE-OF-THE-ART 100
Qi Cao ; Zhiqing Li ; Haoyu Wang

DUAL HALF-BRIDGE LLC RESONANT CONVERTER WITH HYBRID-SECONDARY-RECTIFIER (HSR) FOR WIDE-OUPUT-VOLTAGE APPLICATIONS 108
Jae-Il Baek ; Chong-Eun Kim ; Keon-Woo Kim ; Min-Su Lee ; Gun-Woo Moon

A STUDY ON THE ANALYSIS AND CONTROL OF NO-LOAD CHARACTERISTICS OF LLC RESONANT CONVERTER FOR PLASMA PROCESS 114
Min-Jun Kwon ; Woo-Cheol Lee

MECHANISM OF CURRENT IMBALANCE IN LLC RESONANT CONVERTER WITH CENTER TAPPED TRANSFORMER 118
Mitsuru Sato ; Shingo Nagaoka ; Takeshi Uematsu ; Toshiyuki Zaitsu

PERFORMANCE STUDY OF HIGH-POWER HALF-BRIDGE INTERLEAVED LLC CONVERTER 123
Hung-I Hsieh ; Hui-Lung Chiu ; Guan-Chyun Hsieh

MULTI-CHIP SIC MOSFET POWER MODULES FOR STANDARD MANUFACTURING, MOUNTING AND COOLING 130
Alberto Castellazzi ; Asad Fayyaz ; Emre Gurpinar ; Abdallah Hussein ; Jianfeng Li ; Bassem Mouawad

AN ALTERNATIVE METHOD TO ACCURATELY DETERMINE THE THERMAL RESISTANCE OF SIC MOSFET STRUCTURES WITH DISCRETE DIODES 137
Andras Vass-Varnai ; Young Joon Cho ; Gabor Farkas ; Marta Rencz

HEAT-RESISTANT PACKAGING TECHNOLOGY FOR WIDE BANDGAP POWER DEVICES AND THERMAL RELIABILITY TESTING 142
K. Suganuma ; H. Zhang ; S. Nagao ; C. Chen ; T. Sugahara ; A. Shimoyama ; A. Suetake

VERIFICATION OF IDENTIFICATION ACCURACY OF LOSS CALCULATED BY INVERSE THERMAL ANALYSIS 148
Yuki Ikari ; Kazushige Nakao

PACKAGING ARCHITECTURES FOR SILICON CARBIDE POWER ELECTRONIC MODULES 153
H. Alan Mantooth ; Simon S. Ang

DEVELOPMENT OF A HOMO-POLAR BEARINGLESS MOTOR WITH CONCENTRATED WINDING FOR HIGH SPEED APPLICATIONS 157
Dai Suzuki ; Takaaki Oiwa

HIGH-SPEED SLOTLESS PERMANENT MAGNET MACHINES: MODELLING AND DESIGN FRAMEWORKS 161
S. Jumayev ; K.O. Boynov ; E.A. Lomonova ; J. Pyrhonen

DEVELOPMENT AND PERFORMANCE OF HIGH-SPEED SPM SYNCHRONOUS MACHINE 169
Kota Kawanishi ; Keisuke Matsuo ; Takayuki Mizuno ; Koji Yamada ; Takashi Okitsu ; Kouki Matsuse

1.2KW 100,000RPM HIGH SPEED MOTOR FOR AIRCRAFT 177
Takehiro Jikumaru ; Gen Kuwata

COMPARATIVE EVALUATION OF Y-INVERTER AGAINST THREE-PHASE TWO-STAGE BUCK-BOOST DC-AC CONVERTER SYSTEMS 181
Michael Antivachis ; Dominik Bortis ; David Menzi ; Johann W. Kolar

DC-POWERED OFFICE BUILDINGS AND DATA CENTRES : THE FIRST 380 VDC MICRO GRID IN A COMMERCIAL BUILDING IN GERMANY 190
Tilo Pueschel

RECENT TREND IN POWER ELECTRONICS FOR ICT SYSTEMS 196
Hiroshi Nakao ; Yu Yonezawa ; Yoshiyasu Nakashima

GREEN BASE STATION USING ROBUST SOLAR SYSTEM AND HIGH PERFORMANCE LITHIUM ION BATTERY FOR NEXT GENERATION WIRELESS NETWORK (5G) AND AGAINST MEGA DISASTER 201
M. Nakamura ; K. Takeno

OPTIMIZATION OF MAINTENANCE BY FAILURE PREDICTION CONSIDERING INSTANTANEOUS AND CUMULATIVE EFFECTS OF EXTERNAL ENVIRONMENTS 207
Kaisei Kanetani ; Masahiro Yamazaki ; Tadatoshi Babasaki ; Hideaki Kim ; Tatsushi Matsubayashi

HYBRID CONVERTERS WITH REDUCED INDUCTOR LOSS FOR INTEGRATABLE POWER CONVERSION 213
Gab-Su Seo ; Hanh-Phuc Le

ENERGY SAVING SYSTEM TREND FOR HARBOR CRANE WITH LITHIUM ION BATTERY 219
Hidemasa Yoshihara

INVERTER DRIVE OF DYNAMOMETERS FORAUTOMOTIVE EVALUATION SYSTEM 227
Shizunori Hamada ; Toshimichi Takahashi ; Nobutaka Kezuka ; Masaju Kouketsu ; Shingo Ishigaki

EXPERIMENTAL INVESTIGATION OF PROTOTYPE ALL-SIC CONVERTER FOR ULTRA-HIGH-SPEED ELEVATOR 233
Kazuhisa Mori ; Kaoru Katoh ; Yohei Matsumoto ; Tatsushi Yabuuchi ; Naoto Ohnuma

HIGH-VOLTAGE, LARGE-CAPACITY CONVERTER TECHNOLOGIES AND THEIR APPLICATIONS 238
Daisuke Yoshizawa ; Paul Bixel ; Masahiko Tsukakoshi

HIGHER RADIAL SUSPENSION FORCE OF MAGNETIC BEARING ON CENTRIFUGAL COMPRESSOR FOR HVAC 244
Yuji Nakazawa ; Yusuke Irino ; Atsushi Sakawaki ; Kazunobu Ohyama

NOVEL SWITCHING CONTROL METHOD FOR FULL-BRIDGE DC-DC CONVERTERS FOR IMPROVING LIGHT-LOAD EFFICIENCY USING REVERSE RECOVERY CURRENT 250
Fumihiro Sato ; Takae Shimada ; Takayuki Ouchi

A 800V/14V SOFT-SWITCHED CONVERTER WITH LOW-VOLTAGE RATING OF SWITCH FOR XEV APPLICATIONS 256
Byeongwoo Kim ; Kangsan Kim ; Sewan Choi

HIGH SPEED CONTROL METHOD FOR SUPERPOSING HIGH-FREQUENCY-HIGH-SINUSOIDAL-CURRENT WITH DC CURRENT TO ANALYZE BATTERY AC IMPEDANCE 261
Jin Xu ; Toshihiko Kishimoto ; Noboru Shimosato

EV BMS WITH TIME-SHARED ISOLATED CONVERTERS FOR ACTIVE BALANCING AND AUXILIARY BUS REGULATION 267
Z. Gong ; B.A.C. Van De Ven ; Y. Lu ; Y. Luo ; K. Gupta ; C. Da Silva ; H.J. Bergveld ; O. Trescases

A DRIVING CIRCUIT WITH PARTIAL POWER REGULATION FOR RGB LED LAMPS 275
You-Chun Huang ; Yu-Jen Chen ; Yong-Jyun Li ; Chin-Sien Moo

FPGA-BASED DYNAMIC DUTY CYCLE AND FREQUENCY CONTROLLER FOR A CLASS-E2DC-DC CONVERTER 282
Sanghyeon Park ; Juan Rivas-Davila

DESIGN METHODOLOGY OF 3 KW INDUCTION HEATING SYSTEM FOR BOTH LOW RESISTANCE AND HIGH RESISTANCE CONTAINERS IN A SINGLE BURNER 289
Si-Hoon Jeong ; Hwa-Pyeong Park ; Jee-Hoon Jung

MULTI-RESONANT INVERTER REALIZING DOWNSIZING AND LOSS REDUCTION FOR ALL-METALLIC IH COOKTOP 296
Takayuki Hirokawa ; Makoto Imai ; Atsushi Fujita

TEMPERATURE ESTIMATION OF ALUMINUM ELECTROLYTIC CAPACITOR UNDER ACTUAL CIRCUIT OPERATION 302
Kazuki Urata ; Toshihisa Shimizu

DESIGN AND EVALUATION OF CURRENT DISTRIBUTION IN POWER MODULE 309
Takaaki Ibuchi ; Eisuke Masuda ; Tsuyoshi Funaki

DEVELOPMENT OF IMPEDANCE-SOURCE INVERTER USING SIC-MOSFET 313
Ryuji IIjima ; Thilak Senanayake ; Takanori Isobe ; Hiroshi Tadano

CONTROL METHODOLOGY FOR REALIZATION OF 100KW HEECS CHOPPER WITH 99.5% EFFICIENCY 318
Yukinori Tsuruta ; Atsuo Kawamura

IRON LOSS REDUCTION IN THE CORES OF INDUCTION HEATING COILS FOR SMALL-FOREIGN-METAL PARTICLE DETECTOR WITH A 400-KHZ SIC-MOSFETS HIGH-FREQUENCY INVERTER 324

Takuya Shijo ; Yuki Uchino ; Yujiro Noda ; Hiroaki Yamada ; Toshihiko Tanaka

FREQUENCY TRACKING BURST-MODE PDM-CONTROLLED CLASS-D ZERO VOLTAGE SOFT-SWITCHING RESONANT CONVERTER FOR INDUCTIVE POWER TRANSFER APPLICATIONS 329

Yoichiro Tabata ; Tomokazu Mishima ; Tatsuya Kido

REDUCED-ORDER DYNAMICAL MODELS OF TUNED WIRELESS POWER TRANSFER SYSTEMS 337

Hongchang Li ; Jingyang Fang ; Yi Tang

DYNAMIC MODELLING AND CLOSED LOOP CONTROL OF TRANSMITTER PARALLEL AND RECEIVER SERIES COMPENSATED IPT TOPOLOGY FOR EV APPLICATIONS 342

Suvendu Samanta ; Akshay Kumar Rathore

DEVELOPMENT OF INDUCTIVE POWER TRANSFER SYSTEM FOR EXCAVATOR UNDER LARGE LOAD FLUCTUATION : CONSIDERATION OF RELATIONSHIP BETWEEN LOAD VOLTAGE AND RESONANCE PARAMETER 348

Jun-Ichi Itoh ; Kent Inoue ; Keisuke Kusaka

WIRELESS POWER TRANSFER SYSTEM USING THREE-PHASE TO SINGLE-PHASE MATRIX CONVERTER 356

Yuji Hayashi ; Hiromasa Motoyama ; Takaharu Takeshita

DESIGN OF A REDUCED-ORDER OBSERVER FOR SENSORLESS CONTROL OF DUAL-ACTIVE-BRIDGE CONVERTER 363

Nguyen Duy Dinh ; Goro Fujita

IMPROVED LOAD TRANSIENT RESPONSE OF A DUAL-ACTIVE-BRIDGE CONVERTER 370

Sheng-Zhi Zhou ; Chuan Sun ; Song Hu ; Guo Chen ; Xiaodong Li

MODULATION AND ACTIVE MIDPOINT CONTROL OF A THREE-LEVEL THREE-PHASE DUAL-ACTIVE BRIDGE DC-DC CONVERTER UNDER NON-SYMMETRICAL LOAD 375

Philipp Joebges ; Anton Gorodnichev ; Rik W. De Doncker

A NOVEL SWITCHING ALGORITHM TO IMPROVE EFFICIENCY AT LIGHT LOAD CONDITIONS FOR THREE-PHASE DAB CONVERTER IN LVDC APPLICATION 383

Hyun-Jun Choi ; Si-Hoon Jung ; Jee-Hoon Jung

DESIGN OF A HIGH-FREQUENCY DUAL-ACTIVE BRIDGE CONVERTER WITH GAN DEVICES FOR AN OUTPUT POWER OF 3.7 KW 388

Philipp Schülting ; Christian Winter ; Rik W. De Doncker

EXPLORATION OF THE DESIGN AND PERFORMANCE SPACE OF A HIGH FREQUENCY 166 KW/10 KV SIC SOLID-STATE AIR-CORE TRANSFORMER 396

Piotr Czyz ; Thomas Guillod ; Florian Krismer ; Johann W. Kolar

NOVEL CALCULATION METHOD OF IRON LOSS OF GAPPED INDUCTORS USING LOSS MAP 404

Yoshihiro Miwa ; Toshihisa Shimizu

VERIFICATION OF THE REDUCTION OF THE COPPER LOSS BY THE THIN COIL STRUCTURE FOR INDUCTION COOKERS 410

Morimasa Hataya ; Koki Kamaeguchi ; Eiji Hiraki ; Kazuhiro Umetani ; Takayuki Hirokawa ; Makoto Imai ; Hideki Sadakata

CONDITION MONITORING OF ELECTROLYTIC CAPACITOR BASED ON ESR ESTIMATION AND THERMAL IMPEDANCE MODEL USING IMPROVED POWER LOSS COMPUTATION 416

Sundararajan Prasanth ; Mohamed Halick ; Mohamed Sathik ; Firman Sasongko ; Tan Chuan Seng ; Peng Yaxin ; Rejeki Simanjorang

TEST SETUP FOR CHARACTERISATION OF BIASED MAGNETIC HYSTERESIS LOOPS IN POWER ELECTRONIC APPLICATIONS 422

Min Luo ; Drazen Dujic ; Jost Allmeling

A FAST OPEN-CIRCUIT FAULT DIAGNOSIS SCHEME FOR MODULAR MULTILEVEL CONVERTERS WITH MODEL PREDICTIVE CONTROL 428

Dehong Zhou ; Shunfeng Yang ; Yi Tang

AN ONLINE OPEN-CIRCUIT FAULT DIAGNOSIS AND FAULT TOLERANT SCHEME FOR THREE-PHASE AC-DC CONVERTERS WITH MODEL PREDICTIVE CONTROL 434

Dehong Zhou ; Yi Tang

THE LIFETIME ASSESSMENT OF A MICRO-INVERTER FOR PV APPLICATIONS 439

Tohihiro Shimao ; Koji Kato ; Youichi Ito ; Akio Iwabuchi ; Yongheng Yang ; Frede Blaabjerg

ONLINE HEALTH MONITORING OF MULTIPLE MOSFETS IN A GRID-TIED PV INVERTER USING SPREAD SPECTRUM TIME DOMAIN REFLECTOMETRY (SSTDR) 446
Sourov Roy ; Faisal Khan

AN IMPROVED EQUIVALENT MODEL FOR A LONG PV STRING UNDER PARTIAL SHADING CONDITIONS 453
Xiaoyang Wang ; Huiqing Wen ; Xingshuo Li

OPTIMIZED FLUX-WEAKENING CONTROL OF INDUCTION MOTOR FOR TORQUE ENHANCEMENT IN VOLTAGE EXTENSION REGION 459
Zhen Dong ; Yong Yu ; Bo Wang ; Qinghua Dong ; Dianguo Xu

IMPROVED PERFORMANCE OF CFTC-BASED DIRECT TORQUE CONTROL OF INDUCTION MACHINES BY INCREASING TORQUE LOOP BANDWIDTH 466
Ibrahim Mohd Alsofyani ; June-Hee Lee ; Byung-Moon Han ; Kyo-Beum Lee

µ-ANALYSIS EVALUATION OF A NOVEL COMBINED CURRENT-AND-SPEED CONTROL FOR INDUCTION MOTORS VIA ILQ DESIGN METHOD 471
Shuto Omori ; Hiroshi Takami ; Masashi Nakamura

LOSS MINIMIZATION CONTROL OF SENSORLESS SCALAR-CONTROLLED INDUCTION MOTOR DRIVES CONSIDERING IRON LOSS 478
Nguyen Anh Tan ; Dong-Choon Lee

TUNING OF INDUCTION MOTOR DRIVE WITH TORQUE SENSOR 483
Hajime Kubo ; Yugo Tadano

QUASI-TWO-LEVEL CONVERTER FOR OVERVOLTAGE MITIGATION IN MEDIUM VOLTAGE DRIVES 488
F. Bertoldi ; M. Pathmanathan ; R. S. Kanchan ; K. Spiliotis ; J. Driesen

A MEDIUM-VOLTAGE THREE-PHASE AC-DC CONVERTER CONSISTING OF CASCADED THREE-LEVEL BOOST-TYPE RECTIFIERS AND AN OPEN-END WINDING TRANSFORMER ... 495
Ryoji Tsuruta ; Hiromitsu Suzuki ; Ritaka Nakamura

A FAULT TOLERANT CONTROL STRATEGY FOR THE DELTA-CONNECTED CASCADED CONVERTER 503
Ping-Heng Wu ; Po-Tai Cheng

COOLING PERFORMANCE IMPROVEMENT OF HEAT SINK BY OSCILLATING HEAT PIPE ADDITION AND DESIGN FOR ENVIRONMENT OF OSCILLATING HEAT PIPE REFRIGERANT 511
Kuan-Chung Tey ; Kenichiro Suzuki

COMPACT LARGE CAPACITY GAS TURBINE STATIC STARTER 517
Hironori Kawaguchi ; Shigeyuki Nakabayashi ; Akinobu Ando ; Hiroshi Ogino ; Yasuaki Matsumoto ; Ikuto Udagawa ; Takahiro Ohta

VOLTAGE REFERENCE MODIFICATION SCHEME FOR RESONANCE SUPPRESSION IN LCL-FILTERED INVERTERS WITH DISCONTINUOUS PWM METHOD 521
Hyeon-Sik Kim ; Seung-Ki Sul

PARAMETRIC ROBUSTNESS ANALYSIS FOR PARALLEL FEEDFORWARD COMPENSATION BASED ACTIVE DAMPING OF LCL GRID CONNECTED INVERTER 528
Muhammad Talib Faiz ; Muhammad Mansoor Khan ; Xu Jianming ; Muhammad Ali ; Houjun Tang

OPEN-LOOP-BASED ISLAND-MODE VOLTAGE CONTROL METHOD FOR SINGLE-PHASE GRID-TIED INVERTER WITH MINIMIZED LC FILTER 534
Satoshi Nagai ; Jun-Ichi Itoh

EXPERIMENTAL VALIDATION OF ADAPTIVE CURRENT INJECTING METHOD FOR GRID-SYNCHRONIZATION IMPROVEMENT OF GRID-TIED REGS DURING SHORT-CIRCUIT FAULT 542
Shaokang Ma ; Hua Geng ; Geng Yang ; Bo Liu

ADAPTIVE CONTROL OF GRID-VOLTAGE FEEDFORWARD FOR GRID-CONNECTED INVERTERS BASED ON REAL-TIME IDENTIFICATION OF GRID IMPEDANCE 547
Roni Luhtala ; Tuomas Messo ; Tomi Roinila

MODEL BASED TUNING OF PROPORTIONAL RESONANT CONTROLLERS FOR VOLTAGE SOURCE INVERTERS 555
Stefan Almér ; Thomas Besselmann ; Mario Schweizer

AN SOC-BASED PLATFORM FOR INTEGRATED MULTI-AXIS MOTION CONTROL AND MOTOR DRIVE 560
Yongping Sun ; Ming Yang ; Yangyang Chen ; Wangpin He ; Dianguo Xu

VARIABLE SWITCHING FREQUENCY STRATEGY FOR ENHANCED SETTLING PERFORMANCE OF POSITION CONTROL WITHIN INVERTER LOSS LIMIT 565
Choongin Lee ; Jung-Ik Ha

TWO-WHEEL CANE FOR WALKING ASSISTANCE...571
Phi Van Lam ; Yasutaka Fujimoto

FALL PREVENTION AND VIBRATION SUPPRESSION OF WHEELCHAIR USING RIDER MOTION STATE...575
Isseki Takahashi ; Toshiyuki Murakami

STABILIZATION METHOD FOR RESIDENTIAL DC SYSTEM BASED ON PASSIVITY CRITERION...583
Hiroaki Kakigano

A NOVEL CONTROL APPROACH TO MULTI-TERMINAL POWER FLOW CONTROLLER FOR NEXT-GENERATION DC POWER NETWORK...588
Kenji Natori ; Yuta Nakao ; Yukihiko Sato

DC MICROGRID FOR TELECOMMUNICATIONS SERVICE AND RELATED APPLICATION...593
Keiichi Hirose

MVDC DISTRIBUTION GRIDS FOR ELECTRIC VEHICLE FAST-CHARGING INFRASTRUCTURE...598
Marco Stieneker ; Benedict J. Mortimer ; Arne Hinz ; Adolf Müller-Hellmann ; Rik W. De Doncker

REVIEW OF RESONANT GATE DRIVER IN POWER CONVERSION...607
Bainan Sun ; Zhe Zhang ; Michael A.E. Andersen

A LOW PROFILE HIGH FREQUENCY LED DRIVING SYSTEM BASED ON AIRCORE PLANAR INDUCTOR...614
Yueshi Guan ; Xihong Hu ; Shu Zhang ; Yijie Wang ; Dianguo Xu ; Wei Wang

ANALYSIS AND COMPENSATION OF DEAD-TIME EFFECT IN SIC-DEVICE-BASED HIGH-SWITCHING-FREQUENCY INVERTERS...619
Qingzeng Yan ; Xibo Yuan ; Xiaojie Wu ; Yiwen Geng

CONTROL AND PERFORMANCE OF NEW ASYMMETRICAL OPERATION FOR SWITCHED-CAPACITOR-BASED RESONANT CONVERTERS...626
Hadi Setiadi ; Hideaki Fujita

HIGH-FREQUENCY RESONANT CONVERTER WITH SYNCHRONOUS RECTIFICATION FOR HIGH CONVERSION RATIO AND VARIABLE LOAD OPERATION...632
Lei Gu ; Kawin Surakitbovorn ; Juan Rivas-Davila

SMART PV INVERTERS FOR SMART GRID APPLICATIONS...639
Cheng-Jhen Yang ; Terng-Wei Tsai ; Yi-Chan Li ; Cheng-Yu Tang ; Yaow-Ming Chen ; Yung-Ruei Chang

HIGH-VOLTAGE BI-DIRECTIONAL HALF-BRIDGE THREE-LEVEL SERIES RESONANT CONVERTER WITH FREQUENCY MODULATION CONTROL...645
Lee Sih-Yi ; Jhang Jynu-Jhe ; Lin Jing-Yuan ; Hsieh Yao-Ching ; Chiu Haung-Jen

A CONTROL STRATEGY FOR FLYING-START OF SHAFT SENSORLESS PERMANENT MAGNET SYNCHRONOUS MACHINE DRIVE...651
Zih-Cing You ; Sheng-Ming Yang

CONTACTLESS EV POWER TRACK SYSTEM WITH SEGMENT-EXCITED INDUCTIVELY COUPLED STRUCTURE...657
Jia-You Lee ; Yu-Chi Wang ; Chih-Yi Liao

DRIVING TEST EVALUATION OF SENSORLESS VEHICLE DETECTION METHOD FOR IN-MOTION WIRELESS POWER TRANSFER...663
Katsuhiro Hata ; Kensuke Hanajiri ; Takehiro Imura ; Hiroshi Fujimoto ; Yoichi Hori ; Motoki Sato ; Daisuke Gunji

A SYSTEM DESIGN METHOD OF HIGH-FREQUENCY CLASS-D INVERTER FOR WIDEBAND CURRENT CONTROL...669
Hiroki Kurumatani ; Seiichiro Katsura

ANALYSIS OF INTERIOR PERMANENT MAGNET TWO DEGREES OF FREEDOM MOTOR BASED ON CROSS-COUPLED STRUCTURE...675
Yoshiyuki Hatta ; Tomoyuki Shimono

STUDY COMPARISON BETWEEN FIREFLY ALGORITHM AND PARTICLE SWARM OPTIMIZATION FOR SLAM PROBLEMS...681
Mounia Janah ; Yasutaka Fujimoto

BANDWIDTH LIMITATIONS IN FORCE CONTROL OF A SERIES ELASTIC ACTUATOR WITH BACKLASH AND QUANTIZATION...688
Hanul Jung ; Chan Lee ; Sehoon Oh

ROTOR SHAPE OPTIMIZATION OF INTERIOR PERMANENT MAGNET SYNCHRONOUS MOTORS WITH CONCENTRATED WINDINGS BY CONSIDERING END-LEAKAGE FLUX...693
Katsumi Yamazaki ; Hiroki Narushima

LOSS ANALYSIS OF PERMANENT-MAGNET SYNCHRONOUS MACHINES CONSIDERING IN-PLANE EDDY CURRENT IN ELECTRICAL STEEL SHEETS 699
Hideki Ohguchi ; Satoshi Imamori ; Katsumi Yamazaki ; Haiyan Yui ; Masao Shuto

STUDY ON INFLUENCE OF DIFFERENCE IN STRUCTURE OF CONCENTRATED WINDING IPMSMS OBTAINED BY AUTOMATIC DESIGN 704
A. Ura ; M. Sanada ; S. Morimoto ; Y. Inoue

CARRIER HARMONIC LOSS REDUCTION TECHNIQUE ON DUAL THREE-PHASE PERMANENT-MAGNET SYNCHRONOUS MOTORS WITH PHASE-SHIFT PWM 711
Yoshihiro Miyama ; Haruyuki Kometani ; Kan Akatsu

FLUX INTENSIFYING PM-MOTOR WITH VARIABLE LEAKAGE MAGNETIC FLUX TECHNIQUE 718
Masahiro Aoyama ; Toshihiko Noguchi

CONTINUOUS OPERATION CONTROL OF PMSM IN THE CASE OF DC POWER SUPPLY LOSS 726
Jongwon Heo ; Keiichiro Kondo

MODEL PREDICTIVE CONTROL FOR MULTIPHASE MOTOR DRIVES – A TECHNOLOGY STATUS REVIEW 732
A. Tenconi ; S. Rubino ; R. Bojoi

INFLUENCE OF FAST SWITCHING SEMICONDUCTORS ON THE WINDING INSULATION SYSTEM OF ELECTRICAL MACHINES 740
Kay Hameyer ; Andreas Ruf ; Florian Pauli

CENTRALIZED CONTROL OF MODULAR MULTI RECTIFIER FOR MOTOR DRIVE APPLICATIONS UNDER UNBALANCED GRID 746
Yipeng Song ; Pooya Davari ; Frede Blaabjerg

VECTOR CONTROL OF MAGNETICALLY MODULATED MOTOR FOR POWER SPLITTING OF HEV APPLICATION 753
Toshihiko Noguchi ; Sawanth Krishna Machavolu ; Masahiro Aoyama ; Yuto Motohashi

IMPEDANCE-BASED STABILITY EVALUATION OF VIRTUAL SYNCHRONOUS MACHINE IMPLEMENTATIONS IN CONVERTER CONTROLLERS 759
Eneko Unamuno ; Atle Rygg ; Mohammad Amin ; Marta Molinas ; Jon Andoni Barrena

STABLE POWER SUPPLY METHOD FOR HOUSEHOLD APPLIANCES VIA VIRTUAL SYNCHRONOUS GENERATOR IN SINGLE-PHASE THREE-WIRE MICROGRID 767
Yuko Hirase ; Hidehiko Nakagawa ; Eiji Yoshimura ; Shogo Katsura ; Kensho Abe ; Osamu Noro ; Kazushige Sugimoto ; Kenichi Sakimoto

A NOVEL OSCILLATION DAMPING METHOD OF VIRTUAL SYNCHRONOUS GENERATOR CONTROL WITHOUT PLL USING POLE PLACEMENT 775
Jia Liu ; Yushi Miura ; Toshifumi Ise

OPERATION OF A MODULAR MULTILEVEL CONVERTER CONTROLLED AS A VIRTUAL SYNCHRONOUS MACHINE 782
Salvatore D'arco ; Giuseppe Guidi ; Jon Are Suul

ASSESSMENT OF VIRTUAL SYNCHRONOUS MACHINE BASED CONTROL IN GRID-TIED POWER CONVERTERS 790
Chi Li ; Igor Cvetkovic ; Rolando Burgos ; Dushan Boroyevich

RESEARCH ON THE BLOCKCHAIN-BASED INTEGRATED DEMAND RESPONSE RESOURCES TRANSACTION SCHEME 795
Shengnan Zhao ; Yang Li ; Beibei Wang ; Huiling Su

INDIRECT CURRENT CONTROL FOR SEAMLESS TRANSFER OF UTILITY INTERACTIVE INVERTER 803
Kyungbae Lim ; Injong Song ; Jaeho Choi

STUDY OF AC POWER INTERCHANGE AND DC POWER INTERCHANGE FOR MICRO GRID SYSTEMS 809
Kazuto Yukita ; Daiki Owaki ; Shunsuke Horie ; Toshiro Matsumura ; Yasuyuki Goto

STABILITY ENHANCEMENT STRATEGY FOR ISLANDING MICROGRID WITH MULTI-TYPE INVERTERS BASED ON HYBRID IMPEDANCE MODELLING 815
Meiqin Mao ; Yong Ding ; Yatao Shen ; Liuchen Chang

DC POWERED DATA CENTER WITH 200 KW PV PANELS 822
Keiichi Hirose

INFLUENCES OF DETERIORATION IN CAPACITOR AND INDUCTOR ON CURRENT SENSORLESS STATIC MODEL DC-DC CONVERTER 826
Fujio Kurokawa ; Masashi Taguchi ; Jizhe Wang ; Hidenori Maruta ; Nobumasa Matsui

CAPACITIVE DIVIDER BASED PASSIVE START-UP METHODS FOR FLYING CAPACITOR STEP-DOWN DC-DC CONVERTER TOPOLOGIES 831

Michael Halamicek ; Tom Moiannou ; Nenad Vukadinovic ; Aleksandar Prodic

HIGH VOLTAGE GAIN INTERLEAVED ACTIVE-CLAMP FORWARD (IACF) CONVERTER HAVING REDUCED PRIMARY CONDUCTION LOSS 838

Yeonho Jeong ; Mu-Hyun Park ; Gun-Woo Kim ; Byoung-Hee Lee ; Gun-Woo Moon

CONTROL OF SWITCHING-CAPACITOR BASED BUCK-BOOST CONVERTER 845

M. Veerachary ; Vasudha Khubchandani

IMPROVEMENT OF UPLOAD TRANSIENT RESPONSES FOR ULTRA HIGH STEP-DOWN CONVERTER 851

Y.T. Yan ; K.I. Hwu

POWER ELECTRONICS AND CONTROL TECHNOLOGIES FOR HOUSEHOLD WASHER 856

Toru Niki

DEVELOPMENT OF ROOM AIR CONDITIONER WITH TWIN-PROPELLER FANS 860

Takamasa Uemura ; Tomoya Fukui ; Kenichi Sakoda

ELECTROLYTIC CAPACITOR-LESS SINGLE-PHASE TO THREE-PHASE INVERTER WITH HARMONICS SUPPRESSION CONTROL FOR AIR CONDITIONER 866

Nobuo Hayashi ; Takuro Ogawa ; Tomoisa Taniguchi ; Morimitsu Sekimoto

LATEST DEVELOPMENT OF SIC POWER MODULE-BASED SINGLE-STAGE AC-AC RESONANT CONVERTER FOR HIGH-FREQUENCY INDUCTION HEATING APPLICATIONS 872

Tomokazu Mishima

AN OPTIMIZED CONTROL STRATEGY TO IMPROVE THE CURRENT ZERO-CROSSING DISTORTION IN BIDIRECTIONAL AC/DC CONVERTER BASED ON V2G CONCEPT 878

Lei Jing ; Xiaoqing Wang ; Bodong Li ; Maohang Qiu ; Bo Liu ; Min Chen

PER-PHASE CONTROL STRATEGY OF THE THREE-PHASE FOUR-WIRE INVERTER 883

Yi-Chan Li ; Terng-Wei Tsai ; Cheng-Jhen Yang ; Yaow-Ming Chen ; Yung-Ruei Chang

OPPORTUNITIES FOR PERFORMANCE IMPROVEMENT OF SINGLE-PHASE POWER CONVERTERS THROUGH ENHANCED AUTOMATIC-POWER-DECOUPLING CONTROL 889

Huawei Yuan ; Sinan Li ; Wenlong Qi ; Siew-Chong Tan ; S. Y. Ron Hui

ZERO VOLTAGE SWITCHING SCHEME FOR FLYBACK CONVERTER TO ENSURE COMPATIBILITY WITH ACTIVE POWER DECOUPLING CAPABILITY 896

Hiroki Watanabe ; Jun-Ichi Itoh

MODEL PREDICTIVE FAULT TOLERANT CONTROL OF BIDIRECTIONAL AC/DC CONVERTER WITH VOLTAGE BALANCE OF SPLIT CAPACITOR 904

Nan Jin ; Chongyan Zhao ; Leilei Guo

PWM STRATEGY FOR PARALLEL OPERATION OF THREE PHASE CONVERTERS TIED TO GRID 911

Hyun-Sam Jung ; Seung-Ki Sul

PRACTICAL ISSUES AND IMPLEMENTATION CIRCUITS OF THE DIGITAL-ANALOG HYBRID FULL FEED-FORWARD METHOD WITH UNIPOLAR AND BIPOLAR MODULATIONS 917

Xin Zhang ; Henry S. H. Chung ; Zhixun Ma

AN AC-DC POWER CONVERTER FOR ELECTROLYTIC CAPACITOR-LESS LED DRIVER WITH HIGH LUMINOUS EFFICACY 922

Kwon-Sik Park ; Byuong-Jun Seo ; Kyoung-Suk Kang ; Eui-Cheol Nho

AN IMPROVED CASCADED DUAL-BUCK INVERTER 927

Usman Ali Khan ; Honnyong Cha ; Ashraf Ali Khan ; Heung-Geun Kim ; Wilson Eberle ; Liwei Wang

A SINGLE-SWITCH INTEGRATED-STAGE LED DRIVER BASED ON CUK AND CLASS-E CONVERTER 934

Shu Zhang ; Yijie Wang ; Xiaosheng Liu ; Yan Zhou ; Dianguo Xu

A FAULT-TOLERANT PARALLEL INVERTER APPLIED TO MICRO-GRID 939

Xiangyue Shi ; Jinjie Peng ; Zhifeng Qiu ; Wei Xiong

STABILITY ANALYSIS OF GRID-CONNECTED CONVERTERS WITH ADD-ON VOLTAGE SUPPORT FUNCTIONALITY USING REPETITIVE CONTROL 946

Y. Zhang ; M. G. L. Roes ; M. A. M. Hendrix ; J. L. Duarte

ADAPTIVE SERIES STABILIZER MODULE FOR THE GRID CONNECTED INVERTER UNDER VARIABLE GRID CONDITIONS 953

Xin Zhang

AN IMPROVED DROOP CONTROL BASED SMOOTH TRANSFER CONTROL STRATEGY 957

Xin Meng ; Jinjun Liu ; Zeng Liu ; Ronghui An

FREQUENCY RESPONSE ANALYSIS OF LOAD EFFECT ON DYNAMICS OF GRID-FORMING INVERTER 963
Matias Berg ; Tuomas Messo ; Teuvo Suntio

A NEW CONTROL METHOD FOR TRIPLE-ACTIVE BRIDGE CONVERTER WITH FEED FORWARD CONTROL 971
Takanobu Ohno ; Nobukazu Hoshi

ANALYSIS OF PFM OPERATION MODEL FOR CAPACITOR CHARGER RESONANT TOPOLOGY WITH ENERGY DOSAGE 977
Pengyu Jia ; Yiqin Yuan ; Shengwen Fan ; Zhenyu Shan

AN ACTIVE-CLAMPED CURRENT-FED HALF-BRIDGE DC-DC CONVERTER WITH THREE SWITCHES 982
Truong-Duy Duong ; Minh-Khai Nguyen ; Young-Cheol Lim ; Joon-Ho Choi

A HIGH GAIN QUASI SINGLE STAGE LLC RESONANT DC/DC CONVERTER WITH COUPLED INDUCTOR AND PARTIAL ACTIVE CLAMP 987
Chongcan Huo ; Xiaogao Xie ; Shuai Jiang ; Hanjing Dong

SUPPRESSION OF RIPPLE CURRENT IN HIGH STEP-UP DC-DC CONVERTER UTILIZING COCKCROFT-WALTON CIRCUIT WITH INDUCTOR 992
Takumi Yasuda ; Masataka Minami ; Shin-Ichi Motegi ; Masakazu Michihira

AN OPTIMAL DESIGN METHOD CONSIDERING TRANSFORMER PARASITIC CAPACITANCE OF LLC RESONANT CONVERTERS 998
Naizeng Wang ; Xu Yang ; Mofan Tian ; Haiyang Jia ; Guangzhao Xu ; Zhenwei Li

COMPARISON OF HARMONIC LINEARIZATION AND HARMONIC STATE SPACE METHODS FOR IMPEDANCE MODELING OF MODULAR MULTILEVEL CONVERTER 1004
Jing Lyu ; Xin Zhang ; Jingjing Huang ; Jianwen Zhang ; Xu Cai

AN IMPROVED PHASE-SHIFTED PWM FOR A FIVE-LEVEL HYBRID-CLAMPED CONVERTER 1010
Kui Wang ; Nianzhou Liu ; Zedong Zheng ; Yongdong Li

INTEGRATED CONTROL METHODS FOR ASYMMETRICAL CASCADED H-BRIDGE RECTIFIER 1015
Wenjing Dai ; Jie Chen ; Xin Chen ; Chunying Gong

TRANSIENT VOLTAGE STRESS MODELING FOR SUBMODULES OF MODULAR MULTILEVEL CONVERTERS UNDER GRID VOLTAGE SAGS 1021
Zhijian Yin ; Yongheng Yang ; Huai Wang

SVPWM STRATEGY BASED ON MULTILEVEL 3LNPC-CR 1027
Xiaoqiong He ; Pengcheng Han ; Xiaolan Lin ; Yi Wang ; Xu Peng

THE MULTIPLE DEGREE OF FREEDOM BASED NEUTRAL POINT POTENTIAL CONTROL OF THREE LEVEL NEUTRAL POINT CLAMPED CONVERTERS 1032
Bo Guan ; Shinji Doki

A MODIFIED PHASE-SHIFTED PWM TECHNIQUE FOR THE GRID-CONNECTED HYBRID CASCADED CONVERTER 1038
Yu-Chen Su ; Po-Tai Cheng

NOVEL T-TYPE DUAL-BUCK INVERTER WITH MINIMUM NUMBER OF INDUCTORS 1046
Tien-The Nguyen ; Honnyong Cha ; Bang Le-Huy Nguyen ; Heung-Geun Kim

CONTROL OF DIRECT AC/AC MODULAR MULTILEVEL CONVERTER IN RAILWAY POWER SUPPLY SYSTEM 1051
Shuguang Song ; Jinjun Liu ; Shaodi Ouyang ; Xingxing Chen ; Baojin Liu

WIRELESS POWER TRANSFER: CRITICAL REVIEW OF RELATED STANDARDS 1062
Mohamad Abou Houran ; Xu Yang ; Wenjie Chen ; Mehdi Samizadeh

COMPARATIVE STUDY OF SINGLE-PHASE FUNDAMENTAL COMPONENT FREQUENCY ESTIMATION SCHEMES UNDER TIME-VARYING HARMONIC DISTORTION OPERATION 1067
E. B. Kapisch ; J. L. Duarte ; C. A. Duque

A COMPREHENSIVE DEAD-TIME COMPENSATION METHOD FOR A THREE-PHASE DUAL-ACTIVE BRIDGE CONVERTER WITH HYBRID MODULATION SCHEMES 1073
Jingxin Hu ; Zhiqing Yang ; Rik W. De Doncker

EVALUATION OF A HIGH-FREQUENCY REACTOR WITH A NEW WIRE GUIDE FOR A TOROIDAL CORE 1080
Hideki Ayano ; Akira Fujimura ; Yoshihiro Matsui

CORE LOSS EVALUATION IN POWDER CORES: A COMPARATIVE COMPARISON BETWEEN ELECTRICAL AND CALORIMETRIC METHODS 1087
Yuki Ishikura ; Jun Imaoka ; Mostafa Noah ; Masayoshi Yamamoto

MODELING, MAGNETIC DESIGN, AND SIMULATION METHODS CONSIDERING DC SUPERIMPOSITION CHARACTERISTIC OF POWDER CORES USED IN POWER CONVERTERS 1095
Jun Imaoka ; Kenkichiro Okamoto ; Masahito Shoyama ; Yuki Ishikura ; Mostafa Noah ; Masayoshi Yamamoto

MODELLING AND DESIGN OF A MEDIUM FREQUENCY TRANSFORMER FOR HIGH POWER DC-DC CONVERTERS 1103
Miloš Stojadinovic ; Jürgen Biela

EVALUATION OF INDUCTOR LOSSES ON Z-SOURCE INVERTER CONSIDERING AC AND DC COMPONENTS 1111
Ryuji Iljima ; Naoki Kamoshida ; Rene Alexander Barrera Cardenas ; Takanori Isobe ; Hiroshi Tadano

AN INTEGRATING STRUCTURE OF OUTPUT FILTER FOR GRID CONNECTED INVERTER BASED ON FMLF TECHNIQUE 1118
Jie Ma ; Yenan Chen ; Pingping Chen ; Wenxing Zhong ; Dehong Xu

NEW SCREENING METHOD FOR IMPROVING TRANSIENT CURRENT SHARING OF PARALLELED SIC MOSFETS 1125
Junji Ke ; Zhibin Zhao ; Peng Sun ; Huazhen Huang ; James Abuogo ; Xiang Cui

PSPICE MODELING AND APPLICATION FOR SIC POWER MOSFET TO EVALUATE THE POWER LOSS IN FULL-BRIDGE CONVERTER 1131
Juan Wei ; Fei Lin ; Zhongping Yang ; Xianjin Huang ; Chanjuan Xiao ; Hao Zhang ; Wencai Liang

ALL-SIC MODULE PACKAGING TECHNOLOGY 1137
Kento Shirata ; Norihiro Nashida ; Hideyo Nakamura ; Yoshitaka Nishimura

A NEW SMALLEST 1200V INTELLIGENT POWER MODULE FOR THREE PHASE MOTOR DRIVES 1141
Minsub Lee ; Miran Baek ; Junbae Lee ; Daewoong Chung

DESIGN AND ENHANCEMENT OF ESD RELIABILITY IN CIRCULAR UHV 300-V NLDMOS POWER COMPONENTS 1145
Shen-Li Chen ; Yi-Hao Chao ; Chih-Ying Yen ; Jen-Hao Lo ; Chun-Ting Kuo ; Yu-Lin Lin ; Yi-Hao Chiu ; Pei-Lin Wu ; Yu-Lin Jhou

A TECHNOLOGY ANALYSIS OF VOLTAGE SHARING IN SERIES CONNECTED POWER DEVICES 1149
Z Davletzhanova ; O Alatise ; R Bonyadi ; J Ortiz-Gonzalez ; T Dai ; M Jennings ; L Ran ; P Mawby

FAILURE MECHANISM ANALYSIS AND PHYSICS-OF-FAILURE LIFETIME PREDICTION METHOD FOR PRESS-PACK THYRISTOR OF CONVERTER VALVE 1157
Ning Liang ; Zhigang Zhang ; Yating Gou ; Cuicui Liu ; Zebin Yang ; Jiangnan Chen ; Fang Zhuo ; Feng Wang

SURGE VOLTAGE ABSORPTION BY A SILICON CARBIDE AVALANCHE-DIODE WITH P-N STRUCTURE 1162
K. Koseki ; Y. Tanaka

CALCULATION OF THYRISTOR RELIABILITY PARAMETER OF UHVDC CONVERTER VALVE IN HEMP ENVIRONMENT 1167
Zhigang Zhang ; Yating Gou ; Cuicui Liu ; Zebin Yang ; Xiaotong Du ; Jiangnan Chen ; Fang Zhuo ; Feng Wang ; Yuanliang Lan ; Caiwang Sheng

GENERALIZED STACKELBERG GAME-THEORETIC APPROACH FOR JOINTED ENERGY AND RESERVE COORDINATION OF ELECTRIC VEHICLES 1172
Tianyang Zhao ; Xuewei Pan ; Lei Li ; Fei Zhao ; Can Wang

IMPEDANCE INFLUENCE ANALYSIS OF PHASE-LOCKED LOOPS ON THREE-PHASE GRID-CONNECTED INVERTERS 1177
Yuncheng Wang ; Xin Chen ; Yang Zhang ; Jie Chen ; Chunying Gong

PULSE-INJECTION-BASED SENSORLESS CONTROL METHOD WITH IMPROVED DYNAMIC CURRENT RESPONSE FOR PMSM 1183
Hechao Wang ; Kaiyuan Lu ; Dong Wang ; Frede Blaabjerg

INFLUENCE OF PARAMETER VARIATIONS ON OPERATING CHARACTERISTICS OF MTPF CONTROL FOR DTC-BASED PMSM DRIVE SYSTEM 1189
Keisuke Fujii ; Yukinori Inoue ; Shigeo Morimoto ; Masayuki Sanada

A QUIET POSITION SENSORLESS CONTROL FOR AN IPMSM BASED ON EXTENDED EMF AND VOLTAGE INJECTION SYNCHRONIZED WITH PWM CARRIER 1196
Yuki Ishii ; Hiroki Yamashita ; Hisao Kubota

STUDY OF TORQUE RIPPLE REDUCTION AND TORQUE BOOST BY MODIFIED TRAPEZOIDAL MODULATION 1202
Satoshi Joryo ; Kazuto Tatsumi ; Toshimitsu Morizane ; Katsunori Taniguchi ; Noriyuki Kimura ; Hideki Omori

FAULT DIAGNOSIS METHOD OF CURRENT SENSOR FOR PERMANENT MAGNET SYNCHRONOUS MOTOR DRIVES 1206
Guoqiang Zhang ; Guoxin Wang ; Gaolin Wang ; Junya Huo ; Lianghong Zhu ; Dianguo Xu

SENSORLESS SPEED CONTROL OF DIESEL-GENERATOR SYSTEMS BASED ON MULTIPLE SOGI-FLLS 1212
  Ngoc Dat Dao ; Dong-Choon Lee ; Dae-Sik Lim

ROBUSTNESS OF SIMPLIFIED SPEED-SENSORLESS VECTOR CONTROL FOR INDUCTION MOTOR 1217
  Naoki Akao ; Mineo Tsuji ; Shin-Ichi Hamasaki

MAXIMUM TORQUE CONTROL REFERENCE FRAME BASED ON A TORQUE MAP FOR IPMSMS WITH LARGE INDUCTANCE VARIATION 1223
  Kazuki Ohta ; Takumi Ohnuma ; Shinji Doki

PMSM MODEL DISCRETIZATION IN CONSIDERATION OF PARK TRANSFORMATION FOR CURRENT CONTROL SYSTEM 1228
  Masamichi Inoue ; Shinji Doki

PSEUDO-RANDOM HIGH-FREQUENCY SINUSOIDAL VOLTAGE INJECTION BASED SENSORLESS CONTROL FOR IPMSM DRIVES 1234
  Guoqiang Zhang ; Huiying Wang ; Gaolin Wang ; Junya Huo ; Lianghong Zhu ; Dianguo Xu

AT-NPC 3-LEVEL INVERTER-FED INDUCTION MOTOR VECTOR CONTROL WITH NEUTRAL POINT VOLTAGE CONTROL 1240
  K. Sudo ; M. Tsuji ; S. Hamasaki ; T. Fukuoka ; H. Ichinose

INVESTIGATION OF VARIOUS POSITION ESTIMATION ACCURACY ISSUES IN PULSE-INJECTION-BASED SENSORLESS DRIVES 1246
  Hechao Wang ; Kaiyuan Lu ; Dong Wang ; Frede Blaabjerg

POSITION SENSORLESS CONTROL OF SWITCHED RELUCTANCE MOTOR USING ESTIMATED PWM PHASE VOLTAGE 1253
  Y. Nakazawa ; K. Ohyama ; H. Fujii ; H. Uehara ; Y. Hyakutake

EXPERIMENTAL CONFIRMATION OF THRUST AND ATTRACTIVE FORCE CONTROL OF LINEAR INDUCTION MOTOR BY TWO DIFFERENT FREQUENCY COMPONENTS 1259
  Kenta Sannomiya ; Toshimitsu Morizane ; Noriyuki Kimura ; Hideki Omori

GA BASED OPTIMIZED TRAJECTORIES OF ROTATING SPEED AND D-Q AXIS CURRENTS FOR AN IPMSM 1264
  Shuta Kumagai ; Kaoru Inoue ; Toshiji Kato

2-DEGREE-OF-FREEDOM DEADBEAT CONTROL WITH DISTURBANCE COMPENSATION FOR PMSM DRIVE SYSTEM USING FPGA 1270
  Arata Takahashi ; Shotaro Takakura ; Tomoki Yokoyama

EXTENDED EMF-BASED SIMPLE IPMSM SENSORLESS VECTOR CONTROL USING COMPENSATED CURRENT CONTROLLER 1276
  Takatoshi Inoue ; Yasumasa Hamabe ; Mineo Tsuji ; Shin-Ichi Hamasaki

FULL-BAND OUTPUT IMPEDANCE MODEL OF VIRTUAL SYNCHRONOUS GENERATOR IN DQ FRAMEWORK 1282
  Li Wenbing ; Wang Jianhua ; Song Jingyu ; Luo Fangfang ; Gao Shang ; Wu Zaijun

AN MTPA CONTROL METHOD OF A PMSM AND A SYNRM BASED ON A DTC IN THE STATOR FLUX LINKAGE SYNCHRONOUS FRAME 1289
  Gimpei Itoh ; Yukinori Inoue ; Shigeo Morimoto ; Masayuki Sanada

EEMFS EXCITED BY SIGNAL INJECTION FOR POSITION SENSORLESS CONTROL OF PMSMS AND THEIR PERFORMANCE COMPARISON BY USING IMAGINARY ELECTROMOTIVE FORCE 1295
  Takumi Nimura ; Shota Kondo ; Shinji Doki ; Mutuwo Tomita

HARMONIC CURRENT CANCELLATION METHOD FOR PMSM DRIVE SYSTEM USING RESONANT CONTROLLERS 1301
  Dongsheng Li ; Yoshitaka Iwaji ; Yasuo Notohara ; Ken Kishita

ESTIMATION ERROR ANALYSIS OF STATOR FLUX OBSERVER FOR DTC-BASED PMSM DRIVES 1308
  Atsushi Shinohara ; Kichiro Yamamoto

APPLICATION OF FICTITIOUS REFERENCE ITERATIVE TUNING TO CONTROLLER DESIGN FOR VARIOUS MACHINES 1315
  Hidehiro Ikeda ; Kazuya Goto ; Feili Zhang ; Kazuya Kayashima ; Tsuyoshi Hanamoto

HIGH EFFICIENCY CONTROL FOR PERMANENT MAGNET MOTOR DRIVE SYSTEM WITH FUEL CELLS CONNECTED IN SERIES WITH ELECTRIC DOUBLE-LAYER CAPACITORS 1322
  Kichiro Yamamoto ; Fumiya Ohdera ; Atsushi Shinohara

COMPARATIVE STUDY OF SPEED RIPPLE REDUCTION BY VARIOUS CONTROL METHODS IN PMSM DRIVE SYSTEMS WITH PULSATING LOAD 1329
  Yuma Komaru ; Yukinori Inoue ; Shigeo Morimoto ; Masayuki Sanada

ESTIMATION OF THE PARAMETERS OF THE SERVO DRIVE SYSTEM USING PARTICLE SWARM OPTIMIZATION ALGORITHM 1336

Helin Zhu ; Jae Hyuk Choi ; Sang Uk Park ; Jusuk Lee ; Hyong Gun Lee ; Hyung Soo Mok

A PROGRAMMABLE BATTERY TEST SYSTEM WITH ENERGY RECYCLING FEATURE BASED ON SINUSOIDAL LOADING TECHNIQUE 1341

Chang-Hua Lin ; Guan-Jung Chen ; Hwa-Dong Liu ; Kun-Feng Chen

DEVELOPMENT OF LARGE-CAPACITY CONVERTER FOR BATTERY ENERGY STORAGE SYSTEMS 1346

Hiroyoshi Komatsu ; Tatsuji Katayama ; Noriko Kawakami

ANALYSIS AND COMPARISON OF DC/DC TOPOLOGIES IN PARTIAL POWER PROCESSING CONFIGURATION FOR ENERGY STORAGE SYSTEMS 1351

Maria C. Mira ; Zhe Zhang ; A. E. Michael Andersen

TWO-STAGE PROTECTION FOR MULTI-CHANNEL POWER ELECTRONIC CONVERTERS FED LARGE ASYNCHRONOUS HYDRO-GENERATING UNIT 1358

R. R. Semwal ; Anto Joseph

CURRENT SHARING CONTROL FOR SERIES-PARALLEL CHANGEOVER USING BATTERY AND ELECTRIC DOUBLE-LAYER CAPACITOR BANK 1364

Taisei Nishino ; Keisaku Isozaki ; Naoki Kogai ; Kyungmin Sung

CONTROL METHOD OF ENERGY STORAGE SYSTEM TO IMPROVE OUTPUT POWER OF PCS 1370

Mikiya Ishibashi ; Hitoshi Haga ; Kenji Arimatsu ; Koji Kato

A CONTROL STRATEGY OF MMC BATTERY ENERGY STORAGE SYSTEM BASED ON ARM CURRENT CONTROL 1376

Liu Danqing ; Wang Guangzhu ; Ou Zhujian ; Liu Jiaxing

EQUIVALENT RESISTANCE CONTROL FOR MAXIMUM POWER TRANSFER METHOD OF PIEZOELECTRIC ELEMENT IN VIBRATION POWER GENERATION 1381

Kenya Takamura ; Hiroaki Yamada ; Toshihiko Tanaka ; Tomoharu Yada ; Hajime Fujiwara

DC BUS VOLTAGE STABILIZATION FOR CASCADED POWER CONVERTER BY INTEGRATING AN EXTRA PORT INTO LOAD SIDE PSFB 1386

Jiang You ; Weiyan Fan ; Mengyan Liao

COMMON MODE CURRENT REDUCTION OF THREE-PHASE CASCADED MULTILEVEL TRANSFORMERLESS INVERTER FOR PV SYSTEM 1391

Wenjie Wang ; Ke Chen ; Lijun Hang ; Anping Tong ; Yiliang Gan

CURRENT SHARING/VOLTAGE SHARING CONTROL STRATEGY FOR CASCADED DC/DC CONVERTER IN PHOTOVOLTAIC DC COLLECTION SYSTEM 1397

Bo Chen ; Yi Wang ; Yanjun Tian ; Shilei Wei

PCC VOLTAGE COMPENSATION OF PV INVERTER WITH ACTIVE POWER DECOUPLING CIRCUIT 1403

Duck-Hwan Hwang ; Jung-Yong Lee ; Younghoon Cho

A NOVEL PARTIAL SHADING DETECTION ALGORITHM UTILIZING POWER LEVEL MONITORING OF PHOTOVOLTAIC PANELS 1409

Thusitha Randima Wellawatta ; Sung-Jin Choi

BOOST INTEGRATED THREE-PHASE SOLAR INVERTER USING CURRENT UNFOLDING AND ACTIVE DAMPING METHODS 1414

N. Ha Pham ; Tomoyuki Mannen ; Keiji Wada

LINEAR ACTIVE DISTURBANCE REJECTION CONTROL FOR ISOLATED THREE-PORT CONVERTER 1421

Jiang You ; Mengyan Liao ; Weiyan Fan

STABILITY CONSTRAINED GAIN OPTIMIZATION OF DROOP CONTROLLED CONVERTERS IN DC NANOGRIDS 1426

Soumya Bandyopadhyay ; Laura Ramirez-Elizondo ; Pavol Bauer

SIC BASED SSPC FOR HIGH VOLTAGE SPACE APPLICATIONS 1435

D. Marroquí ; A. Garrigós ; José M. Blanes ; R. Gutiérrez

AN IMPROVED VOLTAGE-TYPE GRID-CONNECTED CONTROL STRATEGY FOR COMPENSATING UNBALANCED VOLTAGE 1442

Liu Hongpeng ; Zhou Jiajie ; Wang Wei

DUAL TWO-STAGE ISOLATED BIDIRECTIONAL DC-DC CONVERTER FOR DC GRID STORAGE 1447

Gabriel Tibola ; Jorge L. Duarte

MODULAR MULTILEVEL CONVERTER WITH CAPACITOR VOLTAGE SELF-BALANCING USING REDUCED NUMBER OF VOLTAGE SENSORS 1455

Taiyuan Yin ; Yue Wang ; Xiaolei Wang ; Shiyuan Yin ; Shumin Sun ; Guanglei Li

PLUG AND OUTLET IN HOUSEHOLD DC LOW VOLTAGE MICRO-GRID POWER DISTRIBUTION..1460
Worapong Pairindra ; Surin Khomfoi

PERFORMANCE PROGRAMMING TECHNIQUE FOR MULTI-STAGE DC POWER DISTRIBUTION SYSTEMS..1465
Syam Kumar Pidaparthy ; Hansang Kim ; Yeonjung Kim ; Byungcho Choi

COORDINATION CONTROL FOR PARALLELED INVERTERS BASED ON VSG FOR PV/BATTERY MICROGRID..1472
Meiqin Mao ; Cheng Qian ; Liuchen Chang ; Yan Du

ADAPTIVE VOLTAGE CONTROL SCHEME FOR DAB BASED MODULAR CASCADED SST IN PV APPLICATION..1478
Tao Liu ; Yang Xuan ; Xu Yang ; Peng Xu ; Yang Li ; Lang Huang ; Xiang Hao

SIX-STEP MMC-BASED HIGH POWER DC-DC CONVERTER..1484
Stefan Milovanovic ; Dražen Dujic

COMBINED DC POWER FLOW CONTROLLER FOR DC GRID..1491
Yongning Chi ; Xizhou Du ; Siqi Liu ; Xu Cai

AN APPROACH FOR THE EMULATION OF DC GRID ADMITTANCES: IMPLEMENTATION ON A BUCK CONVERTER..1498
Enrique Rodriguez-Diaz ; Fracisco D. Freijedo ; Drazen Dujic ; Juan C. Vasquez ; Josep M. Guerrero

A COMPOUND CONTROLLER FOR POWER FLOW AND SHORT-CIRCUIT FAULT IN DC GRID..1504
Han Ye ; Wu Chen ; Pengpeng Pan ; Xiaokun He

DESIGN PROCEDURE AND CONTROL OF A HYBRID CIRCUIT BREAKER WITH ADAPTABLE PULSE CURRENT INJECTION..1509
Andreas Jehle ; Jürgen Biela

A PRAGMATIC SOH AND SOC CO-ESTIMATOR FOR LITHIUM-ION BATTERIES IN SMART GRID APPLICATIONS..1517
Kaiyuan Li ; King Jet Tseng ; Feng Wei ; Boon-Hee Soong

MODELING AND STABILITY ANALYSIS OF PARALLEL DROOP-CONTROLLED AND CURRENT-CONTROLLED INVERTERS..1524
Shike Wang ; Zeng Liu ; Jinjun Liu ; Ronghui An

DIRECT WIRELESS BATTERY CHARGING SYSTEM..1530
Woo-Seok Lee ; Jin-Hak Kim ; Shin-Young Cho ; Il-Oun Lee

AN IMPROVED PWM SCHEME TO ACHIEVE ZERO-VOLTAGE SWITCHING FOR ALL DEVICES IN THREE-PHASE ISOLATED MATRIX RECTIFIER..1537
Xuerui Lin ; Yunwei Ryan Li ; Jahangir Afsharian ; Dewei David Xu

FIXED-FREQUENCY HF GATE DRIVER BY A PUSH-PULL SELF-EXCITATION LC OSCILLATOR HAVING A CAPACITANCE TRANSISTOR..1543
Naoyuki Ishibashi ; Takuya Mizushima ; Masahiko Hirokawa ; Akihiko Katsuki

A FLEXIBLE REDUCED CAPACITOR VOLTAGES STRATEGY FOR VARIABLE-SPEED DRIVES WITH MODULAR MULTILEVEL CONVERTER..1549
Fangzhou Zhao ; Guochun Xiao ; Daoshu Yang ; Zhiqian Wu ; Xin Meng

A LEAKAGE FLUX CANCELLATION TECHNIQUE FOR SERIES-PARALLEL COMBINED RESONANT CIRCUITS WITH ASYMMETRIC ROTARY TRANSFORMERS USED FOR ULTRASONIC SPINDLE DRIVE..1554
Jun Imaoka ; Masahito Shoyama

A NOVEL STRUCTURAL HEALTH MONITORING SYSTEM WITH WIRELESS POWER AND BI-DIRECTIONAL DATA TRANSFER..1562
Yujin Jangs ; Keon-Woo Kim ; Moo-Hyun Park ; Nayoung Lee ; Gun-Woo Moon

CONTROL STRATEGY FOR STARTER GENERATOR IN UAV WITH MICRO JET ENGINE..1567
Jun-Ichi Itoh ; Kazuki Kawamura ; Hiroyuki Koshikizawa ; Kazuyuki Abe

STUDY ON THE INFLUENCE OF VOLTAGE VARIATIONS FOR NON-INTRUSIVE LOAD IDENTIFICATIONS..1575
Yu-Hsiu Lin ; Shun-Kang Hung ; Men-Shen Tsai

BASIC EXPERIMENT OF A MAGLEV SYSTEM FOR A FLEXIBLE STEEL PLATE WITH CURVATURE: FUNDAMENTAL CONSIDERATION ON LEVITATION STABILITY UNDER DISTURBANCE..1580
Makoto Tada ; Kazuki Ogawa ; Takayoshi Narita ; Hideaki Kato ; Hiroyuki Moriyama

PERFORMANCE OF HYBRID MAGNETIC LEVITATION CONTROL SYSTEM FOR THIN STEEL PLATE BY EMS AND PMS: EXPERIMENTAL EVALUATION OF APPLYING OPTIMAL GAP AND ARRANGEMENT OF PMS.. 1586

Yasuaki Ito ; Yoshiho Oda ; Kengo Okuno ; Toshiki Suzuki ; Masahiro Kida ; Takayoshi Narita ; Hideaki Kato ; Hiroyuki Moriyama

A PRACTICAL LITHIUM-ION BATTERY MODEL BASED ON THE BUTLER-VOLMER EQUATION 1592

Kaiyuan Li ; King Jet Tseng ; Feng Wei ; Boon-Hee Soong

BONDING TECHNOLOGY USING COLD-ROLLED AG SHEET IN DIE-ATTACHMENT APPLICATIONS 1598

Seungjun Noh ; Chanyang Choe ; Chuantong Chen ; Hao Zhang ; Katsuaki Suganuma

HIGH-FREQUENCY SELF-DRIVEN SYNCHRONOUS RECTIFIER CONTROLLER FOR WPT SYSTEMS 1602

Akihiro Konishi ; Kazuhiro Umetani ; Eiji Hiraki

AUTOMATIC RESONANCE FREQUENCY TUNING METHOD FOR REPEATER IN RESONANT INDUCTIVE COUPLING WIRELESS POWER TRANSFER SYSTEMS 1610

Masataka Ishihara ; Kazuhiro Umetani ; Eiji Hiraki

INDUCTIVE POWER TRANSFER FOR T5 FLUORESCENT LAMP LIGHTING SYSTEM 1617

Chung-Chuan Hou ; Tang-Jung Chen ; Ching-Chen Chen ; Chen-Wei Chang ; Po-Wei Wang

AN IMPLEMENT 1.5 MHZ OF INDUCTION HEATING FOR ALUMINUM BASED ON VACUUM TUBE OSCILLATOR CIRCUIT 1622

A. Bilsalam ; P. Chanmontree ; S. Supanyapong ; V. Chunkag

SINGLE-INDUCTOR MULTIPLE-OUTPUTS DIMMABLE LED DRIVER WITH BUCK CONVERTER 1626

Ta-Wei Huang ; Wei-Jing Tseng ; Jun-Xian Huang

A SOFT-SWITCHED THREE-LEVEL T-TYPE INVERTER WITH AUXILIARY COMMUTATED POLES 1634

Apollo Charalambous ; Xibo Yuan

CARRIER-BASED REALIZATION OF ARBITRARY SPACE-VECTOR PWM METHODS FOR THREE-LEVEL INVERTERS 1642

Somboon Sangwongwanich ; Supakorn Paiboon

MULTI-LEVEL TOPOLOGY BASED LINEAR AMPLIFIER FAMILY FOR REALIZATION OF NOISE-LESS INVERTERS 1649

Hidemine Obara ; Tatsuki Ohno ; Atsuo Kawamura

A NEW ZERO-VOLTAGE SWITCHING THREE-LEVEL CONVERTER WITH REDUCED RECTIFIER VOLTAGE STRESS 1655

Keon-Woo Kim ; Cheon-Yong Lim ; Dong-Kwan Kim ; Yu-Jin Jang ; Gun-Woo Moon

MODEL PREDICTIVE CONTROL OF A THREE-LEVEL NPC RECTIFIER WITH A SLIDING MANIFOLD TERM 1661

Xiaonan Gao ; Wei Tian ; Xicai Liu ; Zhenbin Zhang ; Ralph Kennel

H∞ CONTROL-BASED VIBRATION SUPPRESSION IN ROBOT ARM WITH STRAIN WAVE GEARING 1666

Tran Vu Trung ; Makoto Iwasaki

FINE FORCE SENSORLESS FORCE CONTROL BASED ON FRICTION-FREE DISTURBANCE OBSERVER 1673

Ohishi Kiyoshi ; Naoki Kamiya ; Toshimasa Miyazaki ; Yuki Yokokura

KINEMATICS AND TRACKING CONTROL OF A FOUR AXIS ANTENNA FOR SATCOM ON THE MOVE 1680

Oguz Kaan Hancioglu ; Mustafa Celik ; Ugur Tumerdem

POSITION SENSORLESS POSITION CONTROL FOR DUAL SOLENOID ACTUATOR 1687

Sakahisa Nagai ; Atsuo Kawamura

CAE TECHNOLOGY APPLICATION TREND FOR LARGE-CAPACITY POWER ELECTRONICS DEVELOPMENT 1692

Teruo Yoshino ; Kuniaki Nagasaka ; Shigeaki Nakabayashi ; Ikuto Udagawa ; Isamu Tominaga ; Junya Konno

XILINX SYSTEM GENERATOR BASED MODELLING OF FINITE STATE MPC 1698

Vijay Kumar Singh ; Ravi Nath Tripathi ; Tsuyoshi Hanamoto

POWER HARDWARE-IN-THE-LOOP SETUP FOR STABILITY STUDIES OF GRID-CONNECTED POWER CONVERTERS 1704

Tommi Reinikka ; Henrik Alenius ; Tomi Roinila ; Tuomas Messo

PASSIVITY-BASED LCL FILTER DESIGN OF GRID-CONNECTED VSCS WITH CONVERTER SIDE CURRENT FEEDBACK 1711

Shih-Feng Chou ; Xiongfei Wang ; Frede Blaabjerg

ADAPTIVE CONTROL OF DC POWER DISTRIBUTION SYSTEMS: APPLYING PSEUDO-RANDOM SEQUENCES AND FOURIER TECHNIQUES............1719
Tomi Roinila ; Hessamaldin Abdollahi ; Silvia Arrua ; Enrico Santi

AN IMPROVED FINITE-SET MODEL PREDICTIVE TORQUE CONTROL FOR INTERIOR PERMANENT MAGNET SYNCHRONOUS MOTOR DRIVES............1724
Xinan Zhang ; Gilbert Foo ; Tung Ngo

PREDICTIVE TORQUE CONTROL FOR FIVE PHASE INDUCTION MOTOR DRIVE WITH COMMON MODE VOLTAGE REDUCTION............1730
Apekshit Bhowate ; Mohan Aware ; Sohit Sharma ; Yogesh Tatte

INDIRECT MATRIX CONVERTER FOR PERMANENT-MAGNET-SYNCHRONOUS-MOTOR DRIVES BY IMPROVED TORQUE PREDICTIVE CONTROL............1736
Yun Jang ; Yeongsu Bak ; Kyo-Beum Lee

PREDICTIVE DC-LINK CURRENT CONTROL BASED ON IPMSM DISCRETE STATE EQUATION FOR INVERTER WITHOUT INDUCTOR OR ELECTROLYTIC CAPACITOR............1741
Yousuke Akama ; Kodai Abe ; Kiyoshi Ohishi ; Yuki Yokokura ; Koji Kobayashi ; Tatsuki Kashihara

NEW SEARCH ALGORITHM OF MODEL PREDICTIVE CONTROL TO REDUCING CALCULATION AMOUNT FOR IMPROVING STEADY CURRENT CONTROL PERFORMANCE............1747
Masahiro Shimaoka ; Shinji Doki

DISTRIBUTED POWER SHARING STRATEGY FOR ISLANDED MICROGRIDS WITHOUT FREQUENCY AND VOLTAGE DEVIATIONS............1752
Tuan V. Hoang ; Hong-Hee Lee

LIFETIME-ORIENTED DROOP CONTROL STRATEGY FOR AC ISLANDED MICROGRIDS............1758
Yanbo Wang ; Dong Liu ; Fujin Deng ; Dao Zhou ; Zhe Chen

EXPERIMENT ON HIERARCHICAL CONTROL BASED POWER QUALITY ENHANCEMENT FOR STANDALONE MICROGRID............1764
Darith Leng ; Sompob Polmai ; Kittichot Soontorntaweesub

A DISTRIBUTED PREDICTIVE CONTROL STRATEGY BASED ON STATE ESTIMATOR FOR ISLANDED MICROGRID............1771
Mi Dong ; Li Li ; Xiaoyu Tian

MAXIMUM POWER POINT TRACKING METHOD FOR PV MODULE UNDER WIDE RANGE VARYING IRRADIANCE LEVELS............1777
Hwa-Dong Liu ; Chang-Hua Lin

DUAL MPPT CONTROL AND FIELD TESTING FOR SWITCHED CAPACITOR-BASED CELL-LEVEL POWER BALANCING UTILIZING DIFFUSION CAPACITANCE OF PHOTOVOLTAIC CELLS............1782
Masatoshi Uno ; Yota Saito ; Masaya Yamamoto ; Shinichi Urabe

SERIES RESONANT DC-DC CONVERTER WITH DUAL-MODE RECTIFIER FOR PV MICROINVERTERS............1788
Yanfeng Shen ; Huai Wang ; Zhan Shen ; Yongheng Yang ; Frede Blaabjerg

VOLTAGE-REFERENCE ACTIVE POWER DECOUPLING BASED ON BOOST CONVERTER FOR SINGLE-PHASE BRIDGE INVERTER............1793
Shuang Xu ; Meiqin Mao ; Riming Shao ; Liuchen Chang

A SINGLE-PHASE COMMON GROUND BOOST INVERTER FOR PHOTOVOLTAIC APPLICATIONS............1799
Tan-Tai Tran ; Minh-Khai Nguyen ; Young-Cheol Lim ; Joon-Ho Choi

STUDY FOR FURTHER INTRODUCTION OF THE ELECTRONIC FREQUENCY CONVERTERS TO THE TOKAIDO SHINKANSEN............1803
Toshimasa Shimizu ; Ken Kunomura ; Masahiko Kai ; Hiroki Miyajima ; Teruhisa Matsui

COUNTERMEASURE FOR PARTIAL TURN-OFF OF THYRISTOR CHANGEOVER SWITCH INTRODUCED TO TOHOKU SHINKANSEN SHIN-YONO SECTIONING POST............1810
Yuki Mizumoto ; Nobuhito Kurosawa

HARDWARE–IN–THE–LOOP REAL–TIME SIMULATION EXPERIMENT PLATFORM FOR TRACTION POWER SUPPLY SYSTEM BASED ON DSPACE-XSIM............1816
Runze Zhang ; Fei Lin ; Zhongping Yang ; Hu Cao ; Yuping Liu

EVALUATING THE NON-SINUSOIDAL AND NON-SYMMETRIC REGIMES FROM A RAILWAY SUPPLYING SUBSTATION............1822
Ileana-Diana Nicolae ; Petre-Marian Nicolae ; Radu-Florin Marinescu

A FUNDAMENTAL TRAIN RUNNING EXPERIMENT FOR A BASIC PERFORMANCE VERIFICATION OF A TRAIN POWER DEMAND CONTROL SYSTEM BY DECENTRALIZED CONTROL ALGORITHM............1828
Yusuke Oki ; Tomoyuki Ogawa ; Yoko Takeuchi ; Tatsuhito Saito ; Jun'ichiro Kawaguchi

VERIFICATION OF SIC BASED MODULAR MULTILEVEL CASCADE CONVERTER (MMCC) FOR HVDC TRANSMISSION SYSTEMS 1834
Y. Ishii ; T. Jimichi

CONTROL OF A 6.6-KV TRANSFORMERLESS STATCOM BASED ON THE MMCC-SDBC USING SIC MOSFETS 1840
Laxman Maharjan ; Toshihisa Tajyuta ; Hiroshi Shinohara ; Akio Suzuki ; Akio Toba

ISOLATED THREE–PHASE AC/DC CONVERTER USING A SOFT–SWITCHING TECHNIQUE FOR BATTERY CHARGER 1847
Yuto Matsui ; Kazuma Suzuki ; Takaharu Takeshita ; Wataru Kitagawa

IMPLEMENTATION OF A MINIATURIZED SIC INVERTER 1854
Hideaki Fujita ; Cristian Andres Garces Guajardo

DESIGN CONSIDERATION OF FLYING CAPACITOR MULTILEVEL INVERTERS USING SIC MOSFETS 1860
Yukihiko Sato ; Kenji Natori

A CONTROL METHOD OF OVERVOLTAGE SUPPRESSION ACROSS THE DC CAPACITOR IN A GRID-CONNECTION CONVERTER USING LEG SHORT-CIRCUIT OF POWER MOSFETS DURING THE INITIAL CHARGE 1866
Tomoyuki Mannen ; Keiji Wada

THE ESSENTIAL RELATIONSHIP BETWEEN DEADBEAT PREDICTIVE CONTROL AND CONTINUOUS-CONTROL-SET MODEL PREDICTIVE CONTROL FOR PWM CONVERTERS 1872
Bi Liu ; Tao Chen ; Wensheng Song

DEADBEAT CONTROL FOR MULTI-LEVEL INVERTER USING 1MHZ MULTISAMPLING METHOD FOR UTILITY INTERACTIVE SYSTEM 1877
Ryosuke Kikuchi ; Ryunosuke Araumi ; Tomoki Yokoyama

1MHZ MULTISAMPLING DEADBEAT CONTROL WITH DISTURBANCE COMPENSATION METHOD FOR THREE PHASE PWM INVERTER 1883
Hiroaki Ueta ; Tomoki Yokoyama

MODULAR MULTILEVEL CONVERTER REPLACED ONE MODULE WITH HIGH VOLTAGE IGBT 1890
Kazunobu Oi ; Kenta Takasho ; Yugo Tadano

INCREASED EFFICIENCY AND REDUCED REALIZATION EFFORT OF DSBC AND DSCC MODULAR MULTILEVEL CONVERTERS (MMCS) 1896
A. Hillers ; J. Biela

COMMON-MODE VOLTAGE INJECTION TECHNIQUES FOR QUASI TWO-LEVEL PWM-OPERATED MODULAR MULTILEVEL CONVERTERS 1904
Jakub Kucka ; Axel Mertens

CURRENT TRACKING AND CELL-VOLTAGE LIMITATIONS OF MODULAR MULTILEVEL CONVERTERS WITH DIRECT DIGITAL CONTROL 1912
T.-F. Wu ; T.-C. Chou ; K.-E. Lin ; T.-Y. Li

SWITCHING LOSS ANALYSIS OF SIC-MOSFET BASED ON STRAY INDUCTANCE SCALING 1919
Keiji Wada ; Masato Ando

MODELING AND OPTIMIZATION OF DISPLACEMENT WINDINGS FOR TRANSFORMERS IN DUAL ACTIVE BRIDGE CONVERTERS 1925
Zhan Shen ; Yanfeng Shen ; Zian Qin ; Huai Wang

OPTIMIZED SELECTION AND UTILIZATION OF DC-LINK CAPACITOR IN A SINGLE-PHASE PV GRID INVERTER SYSTEM 1931
Caspar Collins ; Li Ran

AN EVALUATION CIRCUIT FOR DC-LINK CAPACITORS USED IN A HIGH-POWER THREE-PHASE INVERTER WITH CONDITION MONITORING 1938
Kazunori Hasegawa ; Ichiro Omura ; Shin-Ichi Nishizawa

RECENT MARKET AND TECHNICAL TRENDS IN COPPER ROTORS FOR HIGH-EFFICIENCY INDUCTION MOTORS 1943
Daniel Liang ; Victor Zhou

OVERVIEW OF THE LATEST RESEARCH AND DEVELOPMENT FOR COPPER DIE-CAST SQUIRREL-CAGE ROTORS 1949
Shu Yamamoto

A NOVEL HEAT-RESISTANT INSULATION-PROCESSING AGENT APPLICABLE TO COPPER DIE-CAST SQUIRREL-CAGE ROTORS 1955
Junichi Uchida ; Yuki Sueuchi ; Naosumi Kamiyama

INSULATION-PROCESSING OF COPPER DIE-CAST SQUIRREL-CAGE ROTOR ON MOTOR EFFICIENCY IN HIGH-SPEED OPERATION OVER 10,000 R/MIN 1960
Hideaki Hirahara ; Akira Tanaka ; Shu Yamamoto

HIGH-PRECISION ROTOR POSITION ESTIMATION FOR HIGH-SPEED SPMSM DRIVE BASED ON STATE OBSERVER AND HARMONIC ELIMINATION 1966
Peng Yang ; Xi Xiao ; Meng Zhang ; Shkodyrev Vyacheslav

HARMONIC LOSS REDUCTION IN HIGH SPEED MOTOR DRIVE SYSTEMS BY FLYING CAPACITOR MULTILEVEL INVERTER 1972
Anudari Tumurbaatar ; Sae Mochidate ; Koji Yamaguchi ; Tomohiro Matsuda ; Yukihiko Sato

CURRENT SOURCE TYPE PMSG WIND TURBINE SYSTEM WITH THREE-PHASE THREE-SWITCH BUCK-TYPE RECTIFIER FOR MACHINE-SIDE CONVERTER 1977
Beomseok Chae ; Tahyun Kang ; Yongsug Suh

A STUDY OF 10MW LOAD COMMUTATED INVERTER FOR GAS-TURBINE START-UP 1985
An Hyunsung ; Cha Hanju

PROTOTYPING OF 500 KVA MEDIUM FREQUENCY TRANSFORMER FOR OFFSHORE DIRECT-CURRENT COLLECTION GRID 1991
Tomoyuki Hatakeyama ; Naoyuki Kurita ; Mamoru Kimura

PSCAD/EMTDC AND RTDS SIMULATION ANALYSIS OF MULTIVENDOR MULTI-TERMINAL HVDC SYSTEM CONNECTED TO OFFSHORE WINDFARMS 1997
Hiroshi Suwa ; Takuro Arai ; Takahiro Ishiguro ; Tohru Yoshihara ; Mamoru Kimura ; Tsuneshisa Wachi ; Takahiro Horikoshi ; Tatsuhito Nakajima

INTEROPERABILITY OF MODULAR MULTILEVEL CONVERTERS AND 2-LEVEL VOLTAGE SOURCE CONVERTERS IN A LABORATORY-SCALE MULTI-TERMINAL DC GRID 2003
Salvatore D'arco ; Atsede G. Endegnanew ; Giuseppe Guidi ; Jon Are Suul

PRINCIPLE EXPERIMENT OF CURRENT COMMUTATED HYBRID DCCB FOR HVDC TRANSMISSION SYSTEMS 2011
Ryuta Hasegawa ; Kazuhisa Kanaya ; Yushi Koyama ; Toshiaki Matsumoto ; Takahiro Ishiguro

A THREE-INPUT CENTRAL CAPACITOR DC/DC CONVERTER 2016
Jiaxin Liu ; Feng Gao

SERIES/PARALLEL SWITCHING CIRCUITS USING POWER MOSFETS FOR PHOTOVOLTAIC MODULES 2022
Masamichi Tanemo ; Koki Matsudate ; Shinichi Nomura

MODULARIZED EQUALIZATION ARCHITECTURE BASED ON SWITCHED CAPACITOR CONVERTER TO VIRTUALLY UNIFY MISMATCHED PHOTOVOLTAIC PANEL CHARACTERISTICS 2030
Masatoshi Uno ; Masaya Yamamoto

BUCK-BOOST TYPE MPPT CIRCUIT SUITABLE FOR PHOTOVOLTAIC GENERATION OF VEHICLE INSTALLATION 2036
Fumihisa Kano ; Yuji Kasai ; Hideki Kimura ; Kouhei Sagawa ; Junnosuke Haruna ; Hirohito Funato

VERIFICATION TEST OF ENERGY-EFFICIENT OPERATIONS AND SCHEDULING UTILIZING AUTOMATIC TRAIN OPERATION SYSTEM 2042
Shoichiro Watanabe ; Yasuhiro Sato ; Takafumi Koseki ; Eisuke Isobe ; Jun Kawashita

THE DIRECT BENEFIT OF SIC POWER SEMICONDUCTOR DEVICES FOR RAILWAY VEHICLE TRACTION INVERTERS 2047
Shingo Makishima ; Kazuki Fujimoto ; Keiichiro Kondo

THE LOSS CHARACTERISTICS OF PSFB ZVS DC-DC CONVERTER APPLIED TO THE AUXILIARY POWER SYSTEM 2051
Xianjin Huang ; Juan Zhao ; Fei Lin

SURVEY ON ELECTROMAGNETIC INTERFERENCE ANALYSIS FOR TRACTION CONVERTERS IN RAILWAY VEHICLES 2058
Zhichang Yang ; Hong Li ; Chao Feng ; Yanfeng Jiang ; Fei Lin ; Zhongping Yang

DEVELOPMENT OF TRACTION MOTOR FOR NEW ZERO - EMISSION VEHICLE 2066
Akinobu Iwai ; Satoshi Honjo ; Hirofumi Suzumori ; Toshio Okazawa

EMC DESIGN AND DEVELOPMENT METHODOLOGY FOR TRACTION POWER INVERTERS OF ELECTRIC VEHICLES 2073
Isao Hoda ; Jia Li ; Hiroki Funato

SIMULATION-DRIVEN DESIGN OPTIMIZATION OF A MULTILAYER EMC INPUT FILTER 2078
Fatou Diouf ; Nadim Sakr ; Anna Gheonjian

EV TRACTION INVERTER EMPLOYING DOUBLE-SIDED DIRECT-COOLING TECHNOLOGY WITH SIC POWER DEVICE 2082
Takashi Hirao ; Masami Onishi ; Yusuke Yasuda ; Akihiro Namba ; Kinya Nakatsu

AN OVERVIEW OF STABILITY IMPROVEMENT METHODS FOR WIDE-OPERATION-RANGE FLYBACK CONVERTER WITH VARIABLE FREQUENCY PEAK-CURRENT-MODE CONTROL 2086
Ching-Hsiang Cheng ; Ching-Jan Chen ; Shinn-Shyong Wang

DESIGN AND IMPLEMENTATION OF A HIGH POWER DENSITY ACTIVE-CLAMPED FLYBACK CONVERTER 2092
Yu-Chen Liu ; Bing-Siang Huang ; Cheng-Hung Lin ; Katherine A. Kim ; Huang-Jen Chiu

OPTIMIZED VARIABLE ON-TIME CONTROL FOR LED LIGHTING DRIVER 2097
Jizhe Wang ; Haruhi Eto ; Fujio Kurokawa

DESIGN OF MULTIMODE BATTERY CHARGER WITH DYNAMIC VOLTAGE TRACKING CONTROL 2102
Pang-Jung Liu ; Lin-Hao Chien ; Song-Kai Lee ; Ang-Tung Chen

DUAL-SLOT POWER-PICKUP STRUCTURE FOR CONTACTLESS STRIP INDUCTIVE POWER TRACK SYSTEM 2107
Jia-You Lee ; I-Lin Chen ; Chien-Tzu Ko

DISCONTINUOUS SVM TECHNIQUE FOR THREE-LEG VSI FED BALANCED/UNBALANCED TWO-PHASE LOADS 2113
Supanut Charoensuksirikul ; Yuttana Kumsuwan

REDUCTION OF POWER LOSSES BASED ON GENERALIZED TWO-LEVEL PWM ALGORITHM FOR A NINE-SWITCH VSI 2121
Neerakorn Jarutus ; Yuttana Kumsuwan

SIC-BASED THREE-PHASE QUASI-Z-SOURCE INVERTER VERSUS THE TWO-STAGE TOPOLOGY - A COMPARISON 2129
Kornel Wolski ; Mariusz Zdanowski ; Jacek Rabkowski

DC-SIDE CIRCUIT IMPLEMENTATION OF A THREE-PHASE INVERTER FOR BALANCING PHASE-LEG CAPACITOR CURRENTS 2137
Takashi Hirao ; Keiji Wada ; Toshihisa Shimizu

A THREE-PHASE HYBRID SWITCHED-BOOST INVERTER 2145
Minh-Khai Nguyen ; Tan-Tai Tran ; Hoan-Tien Luong ; Kyoung-Won Lee ; Youn-Ok Choi ; Geum-Bae Cho

THE EFFECT OF BUILT-IN CR SNUBBER CAPACITOR INTO THE POWER MODULE 2149
Ryotaro Hata ; Shigeki Nishiyama

EVALUATION OF NOVEL HYBRID PROTECTION BASED ON PYROSWITCH AND FUSE TECHNOLOGIES 2153
Tomokazu Sakuraba ; Rémy Ouaida ; Song Chen ; Thibaut Chailloux

OPTIMAL DESIGN OF A MAGNETICALLY COUPLED FILTER FOR HIGH EFFICIENCY, LOW COST AND LOW VOLUME DC-DC BATTERY STORAGE CONVERTER 2158
Timothé Delaforge ; Robert Pasterczyk ; Mickaël Robert ; Hervé Chazal ; Jean-Luc Schanen ; Sébastien Mariethoz

HIGH POWER/CURRENT INDUCTOR LOSS MEASUREMENT WITH SHUNT RESISTOR CURRENT-SENSING METHOD 2165
Pin Yu Huang ; Toshihisa Shimizu

SENSITIVITY ANALYSIS OF MEDIUM FREQUENCY TRANSFORMER DESIGN 2170
Marko Mogorovic ; Drazen Dujic

STANDARD MODELS FOR POWER ELECTRONIC SYSTEM SIMULATION 2176
Koichi Shigematsu ; Hiroki Ishikawa ; Taku Noda ; Kentarou Fukushima ; Yoichi Sekiba ; Yusuke Kouno ; Takashi Abe ; Takayuki Sekisue ; Shinji Katoh

MODELING AND MODEL PARAMETER EXTRACTION OF WIDE BANDGAP POWER SEMICONDUCTOR DEVICE, PACKAGE, AND CIRCUIT FOR SIMULATING FAST SWITCHING BEHAVIOR 2181
Tsuyoshi Funaki

STABILITY ANALYSIS METHODS OF A GRID-CONNECTED INVERTER IN TIME AND FREQUENCY DOMAINS 2186
Toshiji Kato ; Kaoru Inoue ; Taiki Sakiyama

FINITE ELEMENT METHODS FOR MULTI-OBJECTIVE OPTIMIZATION OF A HIGH STEP-UP INTERLEAVED BOOST CONVERTER 2193
Wilmar Martinez ; Camilo Cortes ; Ahmad Bilal ; Jorma Kyyra

HIGH FIDELITY REAL-TIME SIMULATION OF MULTI-LEVEL CONVERTERS 2199
Jost Allmeling ; Niklaus Felderer ; Min Luo

AN ENHANCED HIGH FREQUENCY PULSATING VOLTAGE INJECTION METHOD BASED ON IMMUNE ALGORITHM FOR SENSORLESS IPMSM DRIVES 2204
Yanping Zhang ; Zhonggang Yin ; Chao Du ; Youyun Wang ; Xiangdong Sun

POSITION ESTIMATION ACCURACY IMPROVEMENT FOR MAGNETIC SALIENCY BASED SENSORLESS CONTROL INCLUDING CROSS-COUPLING FACTOR 2210
Keita Shimamoto ; Shinya Morimoto

SENSORLESS DRIVE IN THE LOW SPEED REGION AND AUTO-TUNING METHOD FOR PERMANENT MAGNET SYNCHRONOUS MOTORS 2216
Naofumi Nomura ; Shinichi Higuchi

HIGH STABILITY V/F CONTROL OF PMSM USING STATE FEEDBACK CONTROL BASED ON N-T COORDINATE SYSTEM 2224
Yosuke Matsuki ; Shinji Doki

STABILIZATION METHOD USING EQUIVALENT RESISTANCE GAIN BASED ON V/F CONTROL FOR IPMSM WITH LONG ELECTRICAL TIME CONSTANT 2229
Jun-Ichi Itoh ; Takato Toi ; Koroku Nishizawa

SINGLE-PHASE SOLID-STATE TRANSFORMER USING MULTI-CELL WITH AUTOMATIC CAPACITOR VOLTAGE BALANCE CAPABILITY 2237
Jun-Ichi Itoh ; Kazuki Aoyagi ; Keisuke Kusaka ; Masakazu Adachi

A DEVELOPED DUAL MMC ISOLATED DC SOLID STATE TRANSFORMER AND ITS MODULATION STRATEGY 2245
Yan Li ; Chao Liu ; Xu Cai

DC FAULT RIDE-THROUGH OF A THREE-PHASE DUAL-ACTIVE BRIDGE CONVERTER FOR DC GRIDS 2250
Jingxin Hu ; Shenghui Cui ; Rik W. De Doncker

A COMPOUND 10KV DVR SYSTEM BASED ON SOLID STATE TRANSFORMER STRUCTURE 2262
Yaqian Zhang ; Jianzhong Zhang ; Xing Hu ; Zakiud Din

A DUAL-ENERGY-SOURCE UNINTERRUPTIBLE POWER SUPPLY (UPS) 2270
Hao Wang ; Dehong Xu ; Binci Xu ; Haijin Li ; Ye Zhu

INFLUENCE OF WIND POWER FORECASTS ON EQUITABLE DISTRIBUTION METHOD OF WIND POWER CURTAILMENT 2278
Daisuke IIoka ; Hiroumi Saitoh

COMPARISON OF OPTIMIZED DEMAND OF EGS FOR MINIMIZING FUEL CONSUMPTION AND EGS MODEL WITH POWER GRID FREQUENCY USING A HPSPITAL LOAD WITH PV 2283
Yuji Mizuno ; Teppei Baba ; Fujio Kurokawa ; Nobumasa Matsui

COORDINATED DFIG WIND TURBINES AND SOLAR PV GENERATORS FOR INTER-AREA OSCILLATION DAMPING 2287
Tossaporn Surinkaew ; Issarachai Ngamroo

ENERGY MANAGEMENT USING A QUICK CHARGER WITH STORAGE BATTERIES FOR ELECTRIC VEHICLES 2292
Taku Ishibashi ; Toyonari Shimakage ; Norikazu Takeuchi ; Takaaki Kikuchi ; Midori Nonogaki

A METHOD FOR JUNCTION TEMPERATURE ESTIMATION UTILIZING TURN-ON SATURATION CURRENT FOR SIC MOSFET 2296
Hui-Chen Yang ; Rejeki Simanjorang ; Kye Yak See

FIELD BUS FOR DATA EXCHANGE AND CONTROL OF MODULAR POWER ELECTRONIC SYSTEMS WITH HIGH SYNCHRONISATION ACCURACY 2301
Stefan Rietmann ; Simon Fuchs ; André Hillers ; Jürgen Biela

ANALYTICAL INVESTIGATION ON ASYMMETRIC LCC COMPENSATION CIRCUIT FOR TRADE-OFF BETWEEN HIGH EFFICIENCY AND POWER 2309
Kodai Takeda ; Takafumi Koseki

PROBABILISTIC PCA-SUPPORT VECTOR MACHINE BASED FAULT DIAGNOSIS OF SINGLE PHASE 5-LEVEL CASCADED H-BRIDGE MLI 2317
Nagendra Vara Prasad Kuraku ; Yigang He ; Murad Ali

A STUDY ON EDGE SUPPORTED ELECTROMAGNETIC LEVITATION SYSTEM: FUNDAMENTAL CONSIDERATION ON LEVITATION PERFORMANCE OF THIN STEEL PLATE 2324
Yoshiho Oda ; Yasuaki Ito ; Kengo Okuno ; Masahiro Kida ; Toshiki Suzuki ; Takayoshi Narita ; Hideaki Kato ; Hiroyuki Moriyama

APPLICATION OF FACTS DEVICES FOR A DYNAMIC POWER SYSTEM WITHIN THE USA 2329
Jan Paramalingam ; Fuminori Nakamura ; Akihiro Matsuda ; Daisuke Yamanaka ; Taichiro Tsuchiya

CAPACITOR VOLTAGE BALANCING IN SEMI-FULL-BRIDGE SUBMODULE WITH DIFFERENTIAL-MODE CHOKE : (INVITEDPAPER) 2335
Kalle Ilves ; Yuhei Okazaki ; Nan Chen ; Muhammad Nawaz ; Antonios Antonopoulos

RESEARCH ON KEY TECHNOLOGY AND EQUIPMENT FOR ZHANGBEI 500KV DC GRID 2343
Hui Pang ; Xiaoguang Wei

WHAT LED TO SUCCESS IN ACADEMIC RESEARCH ON THE FAMILY OF MODULAR MULTILEVEL CASCADE CONVERTERS? 2352
Hirofumi Akagi

OPERATING PRINCIPLE OF CURRENT RESONANT CONVERTER USING AIR CORE TRANSFORMER FOR ISOLATED POWER SUPPLY ON CHIP 2360
Seiya Abe ; Hikaru Kaishakuji ; Satoshi Matsumoto

ANALYSIS FOR HIGH-FREQUENCY LLC RESONANT CONVERTER WITH PLANAR TRANSFORMER AT LIGHT-LOAD CONDITION 2365
Keon-Woo Kim ; Jae-Il Baek ; Yeonho Jeong ; Ki-Mok Kim ; Gun-Woo Moon

A NOVEL FULL DIGITAL CONTROL H-BRIDGE DC-DC CONVERTER FOR POWER SUPPLY ON CHIP APPLICATIONS 2370
Shigeki Nakano ; Toshiomi Oka ; Seiya Abe ; Satoshi Matsumoto

A HIGH-EFFICIENCY POWER SUPPLY FROM MAGNETIC ENERGY HARVESTERS 2376
Cheon-Yong Lim ; Yeonho Jeong ; Keon-Woo Kim ; Feel-Soon Kang ; Gun-Woo Moon

OPPORTUNITIES FOR LEVERAGING LOW-VOLTAGE GAN DEVICES IN MODULAR MULTI-LEVEL CONVERTERS FOR ELECTRIC-VEHICLE CHARGING APPLICATIONS 2380
Mojtaba Ashourloo ; Mohammad Shawkat Zaman ; Miad Nasr ; Olivier Trescases

A NEW CONTROL STRATEGY FOR MODULAR MULTILEVEL CONVERTER OPERATING IN QUASI TWO-LEVEL PWM MODE 2386
Chao Wang ; Kui Wang ; Zedong Zheng ; Yongdong Li

A CURRENT-SOURCE TYPE MMC WITH DELTA-CONNECTED ARMS FOR SMES 2393
Yushi Miura ; Toshifumi Ise

NEW MODULE WITH ISOLATED HALF BRIDGE OR ISOLATED FULL BRIDGE FOR MODULAR MEDIUM VOLTAGE CONVERTER 2400
Yunpeng Si ; Yifu Liu ; Qin Lei

DEVELOPMENT OF A 700-V-CLASS REVERSE-BLOCKING IGBT FOR ADVANCED T-TYPE NEUTRAL POINT-CLAMPED POWER CONVERSION SYSTEM 2404
Hiroki Wakimoto ; Haruo Nakazawa ; David H. Lu ; Takashi Matsumoto ; Yoichi Nabetani

CERAMIC EMBEDDING AS PACKAGING SOLUTION FOR FUTURE POWER ELECTRONIC APPLICATIONS 2410
Hoang Linh Bach ; Tobias Maximilian Endres ; Daniel Dirksen ; Sigrid Zischler ; Christoph Friedrich Bayer ; Andreas Schletz ; Martin März

MICROELECTROMECHANICAL SYSTEM (MEMS) RESONATOR: A NEW ELEMENT IN POWER CONVERTER CIRCUITS FEATURING REDUCED EMI 2416
A N M Wasekul Azad ; Sourov Roy ; Abu Saleh Imtiaz ; Faisal Khan

A LUMPED THERMAL MODEL INCLUDING THERMAL COUPLING EFFECTS AND BOUNDARY CONDITIONS FOR CAPACITOR BANKS 2421
Qiusheng Wang

HYSTERESIS MODELING OF MAGNETIC DEVICES BASED ON RELUCTANCE NETWORK ANALYSIS 2426
Yoshiki Hane ; Kenji Nakamura

OPTIMAL SIZING AND PLACEMENT OF SOLAR POWERED CHARGING STATION UNDER EV LOADS PENETRATION USING ARTIFICIAL BEE COLONY TECHNIQUE 2430
Yuttana Kongjeen ; Kulsomsup Yenchamchalit ; Krischonme Bhumkittipich

A COMPARISON OF AVERAGE MODEL, SAMPLED-DATA MODEL AND MULTI-FREQUENCY MODEL BASED ON DC/DC CONVERTERS 2435
Xiangpeng Cheng ; Jinjun Liu ; Zeng Liu ; Yiming Tu ; Danhong Xue

SMALL-SIGNAL DISCRETE-TIME MODELING AND DIGITAL CONTROL OF THE BI-DIRECTIONAL DC/DC CONVERTERS 2441
Jia Yaoqin ; Xu Yingchun ; Hou Yijie

ENERGY MANAGEMENT OF HYDROGEN-STORAGE PHOTOVOLTAIC GENERATION SYSTEM WITH A FUNCTION OF SUPPRESSING SHORT-PERIOD COMPONENTS 2449
Yuuki Machida ; Akihisa Goto ; Akiko Takahashi ; Shigeyuki Funabiki

A DYNAMIC BATTERY CHARGING APPROACH FOR ENERGY TRADING IN THE SMART GRID 2456
Avinash Sharma ; Akshay Kumar Rathore ; Rajesh Kumar

A FORCED COMMUTATION METHOD OF THE SOLID-STATE TRANSFER SWITCH IN THE UNINTERRUPTED POWER SUPPLY APPLICATIONS 2462
Meng-Jiang Tsai ; Jiuyang Zhou ; Po-Tai Cheng

ONLINE INTERNAL IMPEDANCE MEASUREMENTS OF LI-ION BATTERY USING PRBS BROADBAND EXCITATION AND FOURIER TECHNIQUES: METHODS AND INJECTION DESIGN ... 2470
Jussi Sihvo ; Tuomas Messo ; Tomi Roinila ; Roni Luhtala

A DC CURRENT FLOW CONTROLLER FOR MESHED HVDC GRIDS ... 2476
Viktor Hofmann ; Mark-M. Bakran

AN ISOLATED SOFT-SWITCHING HYBRID-SOURCE DC-DC CONVERTER FOR DC OFFSHORE WIND FARMS ... 2484
Shenghui Cui ; Jingxin Hu ; Marco Stieneker ; Rik W. De Doncker

A TRANSFORMERLESS MULTI-CELL SOLID-STATE FAULT CURRENT LIMITER FOR MEDIUM VOLTAGE POWER SYSTEM ... 2490
Pantarote Techama ; Sompob Polmai ; Chanin Bunlaksananusorn

A NOVEL DC POWER FLOW CONTROLLER FOR HVDC GRIDS WITH DIFFERENT VOLTAGE LEVELS ... 2496
Ya'nan Wu ; Han Ye ; Wu Chen ; Xiaokun He

DESIGN AND CONTROL OF SINGLE-PHASE GRID-CONNECTED PHOTOVOLTAIC MICROINVERTER WITH REACTIVE POWER SUPPORT CAPABILITY ... 2500
Geon-Hong Min ; Kyung-Hwan Lee ; Jung-Ik Ha ; Myong Hwan Kim

OPTIMAL SIZE AND MULTI-OBJECTIVE CONTROL OF BATTERY ENERGY STORAGES IN DISTRIBUTION SYSTEM WITH HIGH PENETRATION OF DISTRIBUTED PV GENERATORS ... 2505
Meiqin Mao ; Lei Zhou ; Yangyang Wang ; Liuchen Chang

MISSION PROFILE-ORIENTED CONTROL FOR RELIABILITY AND LIFETIME OF PHOTOVOLTAIC INVERTERS ... 2512
Ariya Sangwongwanich ; Yongheng Yang ; Dezso Sera ; Frede Blaabjerg

DISCONTINUOUS CURRENT MODE CONTROL FOR MINIMIZATION OF THREE-PHASE GRID-TIED INVERTER IN PHOTOVOLTAIC SYSTEM ... 2519
Hoai Nam Le ; Jun-Ichi Itoh

A THEORETICAL ANALYSIS ON STATIC CHARACTERISTICS OF VOLTAGE BASED CONTROL METHOD AND CURRENT BASED CONTROL METHOD FOR THE WAYSIDE ENERGY STORAGE SYSTEM IN DC-ELECTRIFIED RAILWAY ... 2527
Hiroyasu Kobayashi ; Keiichiro Kondo ; Diego Iannuzzi

IMPROVEMENT OF A DC ELECTRICAL RAILWAY SIMULATOR USING ARTIFICIAL INTELLIGENCE ... 2534
Alvaro J. Lopez-Lopez ; Ramon R. Pecharroman ; Antonio Fernandez-Cardador ; Asuncion P. Cucala

FEEDING-LOSS REDUCTION BY HIGHER-VOLTAGE DC RAILWAY FEEDING SYSTEM WITH DC-TO-DC CONVERTER ... 2540
Hidenori Shigeeda ; Hiroaki Morimoto ; Kazuhiko Ito ; Toshiyuki Fujii ; Naoki Morishima

MODELING AND SIMULATION OF NOVEL RAILWAY POWER SUPPLY SYSTEM BASED ON POWER CONVERSION TECHNOLOGY ... 2547
Minwu Chen ; Ruofei Liu ; Shaofeng Xie ; Xiaofang Zhang ; Yimin Zhou

COMPARATIVE STUDY ON FRONT-END PARAMETER IDENTIFICATION METHODS FOR WIRELESS POWER TRANSFER WITHOUT WIRELESS COMMUNICATION SYSTEMS ... 2552
Sinan Li ; S. Y. Ron Hui

A NEW TYPE OF WIRELESS V2X SYSTEM WITH A DUAL-ACTIVE BIDIRECTIONAL SINGLE-ENDED CONVERTER AND OPTIMIZED SIC-MOSFET ... 2558
Hideki Omori ; Aoto Yamamoto ; Naoki Mukaiyama ; Masahito Tsuno ; Kenji Fukuda ; Hisato Michikoshi ; Noriyuki Kimura ; Toshimitsu Morizane

METAL OBJECT DETECTION SYSTEM WITH PARALLEL-MISTUNED RESONANT CIRCUITS AND NULLIFYING INDUCED VOLTAGE FOR WIRELESS EV CHARGERS ... 2564
Seog Y. Jeong ; Van X. Thai ; Jun H. Park ; Chun T. Rim

WIRELESS EV CHARGING SYSTEM WITHOUT AIR-GAP AND MISALIGNMENT ... 2569
Wenxing Zhong ; Dehong Xu

FIXED SLOPE CARRIER PWM FOR INDIRECT MATRIX CONVERTER ... 2576
Tzung-Lin Lee ; Chun-Yao Hung ; Yen-Wen Chen ; Wen-Mei Huang

CARRIER-BASED OVERMODULATION STRATEGY FOR MATRIX CONVERTERS ... 2581
Paiboon Kiatsookkanatorn ; Somboon Sangwongwanich

THREE-PHASE TO HIGH-FREQUENCY SINGLE-PHASE MATRIX CONVERTER : A FREQUENCY CONTROL SUITABLE FOR SOFT SWITCHING ... 2589
Wataru Kodaka ; Satoshi Ogasawara ; Koji Orikawa ; Masatsugu Takemoto ; Takashi Hyodo ; Hiroyuki Tokusaki

TWO-STEP COMMUTATION FOR ISOLATED DC-AC CONVERTER WITH MATRIX CONVERTER ... 2596
Shunsuke Takuma ; Jun-Ichi Itoh

A DC-LINK CAPACITOR VOLTAGE OSCILLATION REDUCTION METHOD FOR A MODULAR MULTILEVEL CASCADE CONVERTER WITH SINGLE DELTA BRIDGE CELLS (MMCC-SDBC) 2604

Takaaki Tanaka ; Huai Wang ; Frede Blaabjerg

OPTIMIZED DECOUPLING CONTROL OF FLYING CAPACITOR IN ANPC FIVE-LEVEL INVERTER 2611

Fusheng Wang ; Deyou Zheng ; Jianing Wang ; Fei Li ; Fang Liu ; Shuying Yang ; Zhen Xie

CASCADED DUAL-BUCK AC-AC CONVERTER USING COUPLED INDUCTORS 2619

Sanghun Kim ; Duekjin Jang ; Heung-Geun Kim ; Honnyong Cha

INSTANTANEOUS POWER LOSS CALCULATION FOR MMC BASED ON VIRTUAL ARM MATHEMATICAL MODEL 2625

Yin Shiyuan ; Wang Yue ; Yin Taiyuan ; Nie Cheng ; Duan Guozhao ; Wang Zhang

COMPARISON OF CURRENT CONTROL STRATEGIES IN MODULAR MULTILEVEL CONVERTER 2630

Jianzhao Wei ; Anirudh Budnar Acharya ; Lars Norum ; Pavol Bauer

MODEL PREDICTIVE CONTROL OF A MODULAR MULTILEVEL CONVERTER WITH AN IMPROVED CAPACITOR BALANCING METHOD 2638

Shichong Zhang ; Baodong Bai ; Dezhi Chen

HIGH STEP-UP DC-DC CONVERTER BASED ON MULTI-CELL COUPLED INDUCTOR DIODE-CAPACITOR NETWORK 2646

Xinying Li ; Yan Zhang ; Jinjun Liu ; Pengxiang Zeng

NOVEL ACTIVE CLAMPING STEP-DOWN DC-DC CONVERTER WITH LOWER VOLTAGE STRESS 2653

Chi-Hsuan Hsu ; Jun-Min Jian ; Jiann-Fuh Chen ; Hsuan Liao

DESIGN AND EVALUATION OF A MAGNETICALLY-LOOSELY-COUPLED INDUCTOR FOR A FOUR-PHASE INTERLEAVED BOOST CHOPPER 2660

Hiroki Kowatari ; Toshinori Kitamura ; Nobukazu Hoshi

A SYNCHRONOUS-REFERENCE-FRAME I-V DROOP CONTROL METHOD FOR PARALLEL-CONNECTED INVERTERS 2668

Mingshen Li ; Yonghao Gui ; Zheming Jin ; Yajuan Guan ; Josep M. Guerrero

TRANSIENT STABILITY IMPACT OF THE PHASE-LOCKED LOOP ON GRID-CONNECTED VOLTAGE SOURCE CONVERTERS 2673

Heng Wu ; Xiongfei Wang

COMPREHENSIVE ANALYSIS OF VIRTUAL IMPEDANCE-BASED ACTIVE DAMPING FOR LCL RESONANCE IN GRID-CONNECTED INVERTERS 2681

Teng Liu ; Zeng Liu ; Jinjun Liu ; Yiming Tu ; Zipeng Liu

A COMPARATIVE STUDY OF THE TRADITIONAL FS-MPC AND THE PROPOSED CSF-PCC FOR THE THREE-PHASE GRID-CONNECTED INVERTERS 2688

Zhixun Ma ; Xin Zhang ; Jingjing Huang

CONSTANT SWITCHING-FREQUENCY PREDICTIVE- CURRENT-CONTROL METHOD WITH A DICHOTOMY SOLUTION FOR THE GRID-TIED INVERTERS 2692

Zhixun Ma ; Xin Zhang ; Jingjing Huang ; Zhao Bin ; Lyu Jing

OBSERVER-BASED ACTIVE DAMPING FOR GRID-CONNECTED CONVERTERS WITH LCL FILTER 2697

Y. Zhang ; M. G. L. Roes ; M. A. M. Hendrix ; J. L. Duarte

CONDUCTION LOSS ANALYSIS AND OPTIMIZATION DESIGN OF FULL BRIDGE LLC RESONANT CONVERTER 2703

Yugang Yang ; Lifei Zhang ; Tianshu Ma

FULL-BRIDGE T-TYPE ISOLATED DC/DC CONVERTER WITH WIDE INPUT VOLTAGE RANGE 2708

Dong Liu ; Yanbo Wang ; Fujin Deng ; Zhe Chen

RESEARCH ON HIGH EFFICIENCY LLC DC-DC CONVERTER BASED ON SIC MOSFET 2714

Pengcheng Han ; Xiaoqiong He ; Haijun Ren ; Zhiqing Zhao ; Xu Peng

AN IMPROVED DUAL PHASE SHIFT CONTROL STRATEGY FOR DUAL ACTIVE BRIDGE DC-DC CONVERTER WITH SOFT SWITCHING 2718

Miao Hong ; Gao Xuanjie ; Zeng Chengbi ; Duan Shujiang

DEVELOPMENT OF AN SIC HIGH-FREQUENCY PWM INVERTER USING A THICK MULTILAYER PCB TO MINIMIZE STRAY INDUCTANCE 2725

Kohsuke Ishikawa ; Satoshi Ogasawara ; Masatsugu Takemoto ; Koji Orikawa

FAST SWITCHING PLANAR POWER MODULE WITH SIC MOSFETS AND ULTRA-LOW PARASITIC INDUCTANCE 2732

Arash Edvin Risseh ; Hans-Peter Nee ; Konstantin Kostov

EXPERIMENTAL EVALUATION OF INVERTER SYSTEM CONSISTING OF 4-PARALLEL GAN DEVICES UNIT..2738

Yoshiya Ohnuma ; Satoshi Miyawaki ; Fumiya Hattori ; Masayoshi Yamamoto

IMPACT OF THE THERMAL-INTERFACE-MATERIAL THICKNESS ON IGBT MODULE RELIABILITY IN THE MODULAR MULTILEVEL CONVERTER..2743

Yi Zhang ; Huai Wang ; Zhongxu Wang ; Yongheng Yang ; Frede Blaabjerg

NANOSCALE INVESTIGATION OF THE POWER MOSFET BY THE AFM/KFM/SCFM..............2750

Mizuki Nakajima ; Yuuki Uchida ; Nobuo Satoh ; Hidekazu Yamamoto

SIMULATION ANALYSIS OF OPTIMUM GATE DRIVING CONDITIONS OF IGBTS..................2756

Satoshi Sugahara ; Masaki Kawakami ; Kousuke Kamakura

IMPROVEMENT OF THE I2T CAPABILITY FOR XEV ACTIVE SHORT CIRCUIT PROTECTION BY COMBINATION OF RC-IGBT AND LEADFRAME TECHNOLOGIES.............2764

Keiichi Higuchi ; Hayato Nakano ; Akihiro Osawa ; Akio Kitamura ; Shunji Takenoiri ; Daisuke Inoue ; Souichi Yoshida ; Hiromichi Gohara

INVESTIGATION OF SWITCHING BEHAVIOR OF AN IGBT UNDER SOFT TURN-OFF IN APPLICATION FOR DUAL-ACTIVE BRIDGE CONVERTERS..2768

Eri Ogawa ; Yuichi Onozawa ; Rik W. De Doncker

600 V HIGH VOLTAGE GATE DRIVER IC (HVIC) WITH 1.0 MHZ HIGH FREQUENCY OPERATION FOR LLC CURRENT RESONANT POWER SUPPLY...2774

Masaharu Yamaji ; Masashi Akahane ; Takahide Tanaka ; Akihiro Jonishi ; Hidetomo Ohashi ; Masahiro Sasaki ; Hitoshi Sumida

AN INTEGRATED VOLTAGE AND CURRENT BALANCING STRATEGY OF SERIES-PARALLEL CONNECTED IGBTS...2780

Xiaotong Du ; Fang Zhuo ; Haotian Sun ; Hao Yi ; Yanlin Zhu

THERMAL DESIGN AND ANALYSIS OF A CABLE CHARGER USED FOR PORTABLE ELECTRONICS..2785

Mofan Tian ; Xu Yang ; Naizeng Wang ; Yang Chen ; Laili Wang

PARASITIC INDUCTANCE DESIGN CONSIDERATIONS TO SUPPRESS GATE VOLTAGE OSCILLATION OF FAST SWITCHING POWER SEMICONDUCTOR DEVICES.....................2789

Yusuke Sugihara ; Kimihiro Nanamori ; Masayoshi Yamamoto ; Yasuki Kanazawa

THE EXAMINATION OF INCREASING OPERATION SPEED OF CONSEQUENT POLE TYPE AXIAL GAP MOTOR FOR HIGHER OUTPUT POWER DENSITY..2796

Toru Ogawa ; Tomohira Takahashi ; Masatsugu Takemoto ; Satoshi Ogasawara ; Hideaki Arita ; Akihiro Daikoku

BASIC STUDY OF PMASYNRM WITH BONDED MAGNETS FOR TRACTION APPLICATIONS.............2802

Marika Kobayashi ; Shigeo Morimoto ; Masayuki Sanada ; Yukinori Inoue

STUDY ON ROTOR STRUCTURE SUITABLE FOR IMPROVING POWER DENSITY AND EFFICIENCY IN IPMSMS FOR AUTOMOTIVE APPLICATIONS..2808

R. Imoto ; M. Sanada ; S. Morimoto ; Y. Inoue

EXAMINATION OF THE DEMAGNETIZATION SUPPRESSION EFFECT OF PLACING FLUX BARRIERS IN AN IPMSM USING RARE-EARTH BONDED MAGNETS..............................2814

Takashi Umeda ; Masayuki Sanada ; Shigeo Morimoto ; Yukinori Inoue

A NOVEL POLE-CHANGING METHOD WITH A MULTIPLE THREE-PHASE INVERTER.............2820

Yuki Hidaka ; Taiga Komatsu ; Hideaki Arita

STARTING CHARACTERISTICS OF AN ULTRA-LIGHTWEIGHT MOTOR USING MAGNETIC RESONANCE COUPLING..2826

Kenta Takishima ; Kazuto Sakai

DESIGN AND BASIC CHARACTERISTICS ANALYSIS OF TOROIDAL WINDING AXIAL GAP INDUCTION MOTOR...2832

Ryosuke Sakai ; Yukihiro Yoshida ; Katsubumi Tajima

MAGNET ARRANGEMENT SUITABLE FOR LARGE AIR GAP LENGTH IN LINEAR PM VERNIER MOTOR...2836

Tatsuya Ninomiya ; Abdulaziz Gasim ; Shoji Shimomura

MICRO ELECTROMAGNETIC VIBRATION ENERGY HARVESTER WITH MECHANICAL SPRING AND IRON FRAME FOR LOW FREQUENCY OPERATION..................................2842

Yecheng Shen ; Kaiyuan Lu ; Yongming Xia

MEASUREMENT OF TWO-LEVEL INVERTER INDUCED CURRENT SLOPES AT HIGH SWITCHING FREQUENCIES FOR CONTROL AND IDENTIFICATION ALGORITHMS OF ELECTRICAL MACHINES...2848

Simon Decker ; Andreas Liske ; Daniel Schweiker ; Johannes Kolb ; Michael Braun

A NEW TOPOLOGY OF SWITCHED-CAPACITOR MULTILEVEL INVERTER FOR SINGLE-PHASE GRID-CONNECTED WITH ELIMINATING LEAKAGE CURRENT......2854

Mehdi Samizadeh ; Xu Yang ; Bagher Karami ; Wenjie Chen ; Mohamad Abou Houran ; Adib Abrishamifar ; Abdolreza Rahmati

AN INTERLEAVED BUCK-CASCADED BUCK-BOOST INVERTER FOR PV GRID-CONNECTION APPLICATIONS......2860

Chien-Hsuan Chang ; Chun-An Cheng ; Hung-Liang Cheng

A NOVEL PV ARRAY CONNECTION STRATEGY WITH PV-BUCK MODULE TO IMPROVE SYSTEM EFFICIENCY......2866

Chi Shao ; Wenjie Wang ; Lijun Hang ; Anping Tong ; Shitao Wang

A COMMON-MODE VOLTAGE REDUCTION FOR TWO-STAGE THREE-PHASE TRANSFORMERLESS PV INVERTERS......2871

Adisak Promyoo ; Surapong Suwankawin

A GRID-CONNECTED PV-ENERGY STORAGE SYSTEM WITH SYNCHRONOUS GENERATOR CHARACTERISTICS......2877

Huadian Xu ; Jianhui Su ; Ning Liu ; Yong Shi ; Yan Du

A TRANSFORMERLESS BIDIRECTIONAL DC-DC CONVERTER BASED ON POWER UNITS WITH UNIPOLAR AND BIPOLAR STRUCTURE FOR MVDC INTERCONNECTION......2882

Lejia Sun ; Fang Zhuo ; Feng Wang ; Hao Yi ; Baohui Ma

NEW MODULATION CONTROL OF CONVERTER SYSTEM APPLIED FOR OFFSHORE WIND FARMS......2887

Naoki Kawabata ; Noriyuki Kimura ; Toshimitsu Morizane ; Hideki Omori

SPHERE DECODING BASED LONG-HORIZON PREDICTIVE CONTROL OF THREE-LEVEL NPC BACK-TO-BACK PMSG WIND TURBINE SYSTEMS......2895

Ferdinand Grimm ; Zhenbin Zhang ; Ralph Kennel

BASED ON PCHD AND HPSO SLIDING MODE CONTROL OF D-PMSG WIND POWER SYSTEM......2901

Lijun Hou ; Xuemei Zheng ; Chao Wang ; Yangman Li ; Haoyu Li

ESTABLISHMENT AND DYNAMIC CONTROL OF WIND INDUCTION GENERATOR......2907

M. Z. Lu ; V. K. Ganisetti ; C. M. Liaw

MIDDLE FREQUENCY SOLID STATE TRANSFORMER FOR HVDC TRANSMISSION FROM OFFSHORE WINDFARM......2914

Noriyuki Kimura ; Toshimitsu Morizane ; Isao Iyoda ; Kazushige Nakao ; Tomoki Yokoyama

SIMULATION OF WIND POWER GENERATION SYSTEM USING SWITCHED RELUCTANCE GENERATOR AND CAPACITOR-LESS AC-AC CONVERTER......2921

Guyuan Ji ; Kazuhiro Ohyama

VARIABLE FREQUENCY CONTROL AND FILTER DESIGN FOR OPTIMUM ENERGY EXTRACTION FROM A SIC WIND INVERTER......2932

Abdallah Hussein ; Alberto Castellazzi

EXPERIMENTAL VERIFICATIONS OF UPFC USING DEADBEAT CONTROL WITH 3-PHASE UNBALANCED COMPENSATION......2938

Shin-Ichi Hamasaki ; Hiroto Fukuda ; Syohei Tokumaru ; Mineo Tsuji

A CONTROL METHOD FOR TWO TYPES OF THREE-PHASE TRANSFORMERLESS UNIFIED POWER QUALITY CONDITIONER......2944

Fujian Li ; Guochun Xiao ; Fangzhou Zhao ; Shuai Zhang ; Baojin Liu

DESIGN OF CUSTOMER-END CONVERTER SYSTEMS FOR LOW VOLTAGE DC DISTRIBUTION FROM A LIFE CYCLE COST PERSPECTIVE......2948

A. Mattsson ; P. Nuutinen ; T. Kaipia ; P. Peltoniemi ; J. Karppanen ; V. Tikka ; A. Lana ; P. Pinomaa ; P. Silventoinen ; J. Partanen

A CONTROL METHOD OF DC CAPACITOR VOLTAGE IN MMC FOR HVDC SYSTEM USING NEGATIVE SEQUENCE CURRENT......2956

Hanis Afiqah Binti Jaffar ; Ahmad Arif Bin Abd Rahman ; Hiroaki Kakigano

A COORDINATE AND DISTRIBUTED CONTROL SCHEME FOR MULTILEVEL AND MULTI-STAGE MEDIUM VOLTAGE SOLID STATE TRANSFORMER......2963

Jintong Nie ; Liqiang Yuan ; Qing Gu ; Jianning Sun ; Zhengming Zhao

AN IMPROVED HARMONIC POWER SHARING SCHEME OF PARALLELED INVERTER SYSTEM......2969

Liu Hongpeng ; Liu Xiaoxi ; Zhang Wei ; Wang Wei

THE GRID IMPEDANCE ADAPTATION DUAL MODE CONTROL STRATEGY IN WEAK GRID......2973

Ming Li ; Xing Zhang ; Ying Yang ; Pengpeng Cao

TRANSMISSION POWER ANALYSIS AND CONTROL OF THE DC TRANSFORMER IN HYBRID AC/DC MICROGRID 2980
Jingjin Huang ; Xin Zhang ; Tengfei Zhang

A NOVEL FLEXIBLE INTERCONNECTION SCHEME FOR MICROGRID TO OPTIMIZE THE CAPACITY OF ENERGY STORAGE SYSTEM (ESS) 2986
Zhou Jianqiao ; Zhang Jianwen ; Cai Xu ; Li Zhuyong ; Wang Jiacheng ; Zang Jiajie

VSC CONTROL AND PARAMETERS DESIGN BASED ON VIRTUAL SYNCHRONOUS GENERATOR 2992
Fang Liu ; Meng Wang ; Zhen Xie ; Fusheng Wang ; Jinxin Deng ; Xing Zhang

MULTI-TARGET VIRTUAL RESISTANCE CONTROL STRATEGY IN A 400 HZ LOW VOLTAGE MICROGRID 2997
Yuze Li ; Xuejun Pei ; Zhi Chen ; Hanyu Wang ; Yong Kang

AN ADAPTIVE POWER COMPENSATION STRATEGY FOR THE VOLTAGE STABILIZATION OF LCL-VSC BASED MICROGRIDS 3002
Sheng Xu ; Wu Cao ; Dongchen Fan ; Jianfeng Zhao ; Shunyu Wang

RESONANCE DETECTION STRATEGY FOR MULTIPLE GRID-CONNECTED INVERTERS-BASED SYSTEM USING CASCADED SECOND-ORDER GENERALIZED INTEGRATOR 3010
Wu Cao ; Dongchen Fan ; Kangli Liu ; Jianfeng Zhao ; Liheng Ruan ; Xiaojun Wu

HARMONIC STABILITY ASSESSMENT BASED ON GLOBAL ADMITTANCE FOR MULTI-PARALLELED GRID-CONNECTED VSIS USING MODIFIED NYQUIST CRITERION 3015
Wu Cao ; Dongchen Fan ; Kangli Liu ; Jianfeng Zhao ; Liheng Ruan ; Xiaojun Wu

THE AC TRACTION POWER SUPPLY SYSTEM FOR URBAN RAIL TRANSIT BASED ON NEGATIVE SEQUENCE CURRENT COMPENSATOR 3020
Tianshu Zhao ; Xu Peng

GRID CONNECTED POWER GENERATION CONTROL METHOD FOR Z-SOURCE INTEGRATED BIDIRECTIONAL CHARGING SYSTEM 3025
Xu Jia ; Guoming Chuai ; Haonan Niu ; Qianfan Zhang

AN ISOLATED PFC CONVERTER WITH HARMONIC MODULATION TECHNIQUE FOR EV CHARGERS 3030
Byung-Kwon Lee ; Jun-Young Lee ; Dong-Hun Kang

HIGHLY DYNAMIC SWITCHING FREQUENCY-BASED CALCULATION OF POWER QUANTITIES, FUNDAMENTAL WAVEFORMS, AND RMS VALUES OF INVERTER-FED ELECTRICAL MACHINES 3034
Alexander Stock ; Johannes Teigelkötter ; Johannes Büdel

DESIGN AND ANALYSIS OF HIGH VOLTAGE POWER SUPPLY FOR INDUSTRIAL ELECTROSTATIC PRECIPITATORS 3040
Shengwen Fan ; Yiqin Yuan ; Pengyu Jia ; Zhigang Chen ; Haisi Li

LOAD SHARING OPERATION IN N+1 UPS SYSTEM BY USING HARMONIC SHARING CONTROL METHOD 3046
Prashant Patel ; Sagar Naina ; Utsav Patel ; Premal Patwa

RESEARCH ON CAPACITY OPTIMIZATION OF PV-WIND-DIESEL-BATTERY HYBRID GENERATION SYSTEM 3052
Cailing Zhu ; Furong Liu ; Sheng Hu ; Shu Liu

A NUMERICAL ANALYSIS AND IMPROVEMENT OF OUTPUT CHARACTERISTICS IN DIFFERENT PASSIVE RECTIFIERS BASED ON VIBRATION GENERATORS 3058
Tomoki Sakabe ; Masataka Minami ; Shin-Ichi Motegi ; Masakazu Michihira

CIRCUIT MODELING APPROACH FOR ANALYZING TRIBOELECTRIC NANOGENERATORS FOR ENERGY HARVESTING 3063
Bo-Kyung Yoon ; Jeong Min Baik ; Katherine A. Kim

GENERAL POWER ELECTRIC CONVERTER MODEL 3069
Jingwen Xie

A MODULAR CONVERTER- AND SIGNAL-PROCESSING-PLATFORM FOR ACADEMIC RESEARCH IN THE FIELD OF POWER ELECTRONICS 3074
Rüdiger Schwendemann ; Simon Decker ; Marc Hiller ; Michael Braun

CONTROL IC FOR BOOST-FLYBACK CONVERTER FOR ENERGY HARVESTING APPLICATIONS 3081
Jhih-Sian Li ; Kai-Hui Chen ; Jui-Hung Lai ; Jun-Xian Huang

NEW CONCEPT OF THE DC-DC CONVERTER CIRCUIT APPLIED FOR THE SMALL CAPACITY UNINTERRUPTIBLE POWER SUPPLY 3086
Dang Minh Huynh ; Yoichi Ito ; Shinji Aso ; Koji Kato ; Kenji Teraoka

COMPARATIVE STUDY ON THE PERFORMANCE OF DUAL-PHASE TAPPED-INDUCTOR BOOST CONVERTER AND INTERLEAVED BOOST PARALLEL-INPUT SERIES-OUTPUT CONVERTER IN 40 TO 400V APPLICATIONS 3092

Niño Christopher Ramos ; Tsuyoshi Funaki

A NEW STANDBY STRUCTURE INTEGRATED WITH BOOST PFC CONVERTER FOR SERVER POWER SUPPLY 3100

Jae-Il Baek ; Jae-Kuk Kim ; Jae-Bum Lee ; Moo-Hyun Park ; Gun-Woo Moon

NONISOLATED TWO-CHANNEL LED DRIVER WITH SIMPLE SNUBBER 3107

Jong-Woo Kim ; Jung-Kyu Han ; Jih-Sheng Lai

DESIGN AND IMPLEMENTATION OF SINGLE-PHASE ASYMMETRIC MULTILEVEL STATCOM 3112

Hao Chen ; Yang Han ; Ping Yang ; Congling Wang ; Josep M. Guerrero

SUBMODULE VOLTAGE BALANCING AND LOSS EQUALISATION IN ALTERNATE ARM CONVERTERS BASED ON VIRTUAL VOLTAGES 3117

Georgios Konstantinou ; Harith R. Wickramasinghe ; Salvador Ceballos ; Josep Pou

BALANCED CONDUCTION LOSS DISTRIBUTION AMONG SMS IN MODULAR MULTILEVEL CONVERTERS 3123

Zhongxu Wang ; Huai Wang ; Yi Zhang ; Frede Blaabjerg

SIMPLIFICATION OF MODEL PREDICTIVE CONTROL FOR MODULAR MULTILEVEL CONVERTER THROUGH DIRECT VOLTAGE LEVEL SELECTION 3129

Xingxing Chen ; Jinjun Liu ; Shaodi Ouyang ; Shuguang Song ; Rui Luo

FAMILY OF INTEGRATED MULTI-INPUT MULTI-OUTPUT DC-DC POWER CONVERTERS 3134

Bang Le-Huy Nguyen ; Honnyong Cha ; Tien-The Nguyen ; Heung-Geun Kim

LOW-COMPLEXITY STATE-SPACE BASED SYSTEM IDENTIFICATION AND CONTROLLER AUTO-TUNING METHOD FOR MULTI-PHASE DC-DC CONVERTERS 3140

Marc Kanzian ; Harald Gietler ; Christoph Unterrieder ; Matteo Agostinelli ; Michael Lunglmayr ; Mario Huemer

A PHASE-SHIFT DOUBLE FULL-BRIDGE (PSDB) CONVERTER WITH THREE SHARED LEADING-LEGS 3145

Junjie Zhu ; Qinsong Qian ; Shengli Lu ; Weifeng Sun ; Le Zhang

DUAL ACTIVE BRIDGE SYNCHRONOUS RECTIFIED STEP-DOWN CONVERTER 3151

Chien-Chun Huang ; Chang-Lin Tsai ; Tsung-Lin Tsai ; Yao-Ching Hsieh ; Huang-Jen Chiu ; Jing-Yuan Lin

ACCURATE IMPEDANCE MODEL OF GRID-CONNECTED INVERTER FOR SMALL-SIGNAL STABILITY ASSESSMENT IN HIGH-IMPEDANCE GRIDS 3156

Tuomas Messo ; Roni Luhtala ; Aapo Aapro ; Tomi Roinila

MODELING OF UNBALANCED THREE-PHASE GRID-CONNECTED CONVERTERS WITH DECOUPLED TRANSFER FUNCTIONS 3164

Wei Liu ; Xiongfei Wang ; Frede Blaabjerg

PREDICTING VOLTAGE CHARACTERISTIC OF CHARGING MODEL FOR LI-ION BATTERY WITH ANN FOR REAL TIME DIAGNOSIS 3170

Minella Bezha ; Naoto Nagaoka

IMPEDANCE MODELING AND STABILITY ANALYSIS OF THE CASCADED THREE-PHASE SYMMETRIC SYSTEMS USING COMPLEX TRANSFER FUNCTIONS 3176

Teng Liu ; Zeng Liu ; Jinjun Liu ; Yiming Tu ; Zipeng Liu

ACOUSTIC NOISE REDUCTION OF 12/8 POLES SRM WITHOUT EFFICIENCY DROP USING SIMPLE CURRENT WAVEFORMS 3182

Kyohei Kiyota ; Kenji Amei ; Takahisa Ohji ; Jun Jisaki ; Masanobu Nakai

STUDY OF SWITCHED RELUCTANCE MOTOR DIRECTLY DRIVEN BY COMMERCIAL THREE-PHASE POWER SUPPLY 3186

Masaki Takahashi ; Kohei Aiso ; Kan Akatsu

DOUBLE STATOR AXIAL-FLUX SWITCHED RELUCTANCE MOTOR FOR ELECTRIC CITY COMMUTERS 3192

Hiroki Goto

TORQUE RIPPLE REDUCTION USING ASYMMETRIC FLUX BARRIERS IN SYNCHRONOUS RELUCTANCE MOTOR 3197

Yuuto Yamamoto ; Shigeo Morimoto ; Masayuki Sanada ; Yukinori Inoue

ON-BOARD SINGLE-PHASE ELECTRIC VEHICLE CHARGER WITH ACTIVE FRONT END 3203

Theodore Soong ; Peter W. Lehn

A BIDIRECTIONAL BUFFERED CHARGING UNIT FOR EV'S (BBCU) 3209

Gabriel Fernandez

RECONFIGURABLE CONVERTER WITH MULTIPLE-VOLTAGE MULTIPLE-POWER FOR E-MOBILITY CHARGING 3215

Mohamed S A Dahidah ; He Liu ; Vassilios G. Agelidis

DEVELOPMENT OF A SERIES HYBRID ELECTRIC VEHICLE LABORATORY TEST BENCH WITH HARDWARE-IN-THE-LOOP CAPABILITIES...3223
Poria Fajri ; Nima Lotfi ; Mehdi Ferdowsi

NEW THREE-PHASE STATIC TRANSFER SWITCH USING AC SSCB......................................3229
Seung-Min Song ; Jin-Young Kim ; In-Dong Kim

HARMONICS COMPENSATION IN HIGH FREQUENCY RANGE OF ACTIVE POWER FILTER WITH SIC-MOSFET INVERTER IN DIGITAL CONTROL SYSTEM...3237
Shin-Ichi Hamasaki ; Kengo Nakahara ; Mineo Tuji

CONTROL OF BUCK-BOOST DIRECT MATRIX CONVERTER WITH LOW VOLTAGE RIDE-THROUGH CAPABILITY 3243
Nico Remus ; Martin Leubner ; Wilfried Hofmann

AN IMPROVED PLL BASED SEAMLESS TRANSFER CONTROL STRATEGY 3251
Xin Meng ; Jinjun Liu ; Zeng Liu ; Ronghui An

EFFICIENT URBAN RAILWAY DESIGN INTEGRATING TRAIN SCHEDULING, ONBOARD ENERGY STORAGE, AND TRACTION POWER MANAGEMENT 3257
Warayut Kampeerawar ; Takafumi Koseki ; Fulin Zhou

OPTIMAL CONTROL METHOD OF AN ENERGY STORAGE SYSTEM FOR ENERGY SAVING 3265
Yoko Takeuchi ; Tomoyuki Ogawa ; Keisuke Sato ; Hiroaki Morimoto ; Tatsuhito Saito

START-UP AND TRANSIENT OPERATION OF A BIDIRECTIONAL CHOPPER WITH AN AUXILIARY CONVERTER...3273
Hamzeh J. Ahmad ; Haruna Ohnishi ; Makoto Hagiwara

EXPERIMENTAL RESULTS OF QUASI-OPTIMAL CHARGING CURRENT PATTERNS TO REDUCE THE INTERNAL HEAT GENERATION OF THE LITHIUM-ION BATTERY 3280
Yoshiaki Taguchi ; Gaku Yoshikawa

DEVELOPMENT OF TEST METHODS AND EVALUATION RESULTS FOR 500KV HVDC CONVERTER...3286
Keisuke Hattori ; Asuka Ohtake ; Takayoshi Kamejima ; Haruhisa Wada

DISSIPATION LOOP FOR SHOOT-THROUGH FAULTS IN HVDC CONVERTER CELLS................3292
Keijo Jacobs ; Staffan Norrga ; Hans-Peter Nee

A SUPPRESSION METHOD OF HARMONIC INSTABILITY IN LINE-COMMUTATED CONVERTERS APPLYING ACTIVE HARMONIC FILTERS..3299
Kenichiro Sano ; Toshiaki Kikuma ; Tatsuhito Nakajima ; Junya Kanno

EXPERIMENT OF SEMICONDUCTOR BREAKER USING SERIES-CONNECTED IEGTS FOR HYBRID DCCB..3304
Kazuyasu Takimoto ; Hiroshi Takenaka ; Toshiaki Matsumoto ; Takahiro Ishiguro

STUDY OF EMI CAUSED BY BUCK CONVERTER ON CONTROLLER AREA NETWORK.........3309
Ryo Shirai ; Toshihisa Shimizu

A STUDY ON REDUCTION TECHNIQUES OF A WIDEBAND COMMON-MODE VOLTAGE PRODUCED BY A PWM INVERTER 3315
Shotaro Takahashi ; Satoshi Ogasawara ; Masatsugu Takemoto ; Koji Orikawa ; Michio Tamate

A MODIFIED DISCONTINUOUS PWM FOR COMMON-MODE VOLTAGE ELIMINATION IN 3-LEVEL 4-LEG PWM CONVERTER SYSTEM..3323
Seon-Ik Hwang ; Jun-Hyung Jung ; In-Ho Cho ; Jang-Mok Kim ; Yung-Deug Son

EMI ANALYSIS OF FULL-SIC INTEGRATED POWER MODULE 3329
Xiliang Chen ; Wenjie Chen ; Yu Ren ; Liang Qiao ; Yilin Sha ; Xu Yang

EXPERIMENTAL VERIFICATION OF COUPLING EFFECT AND POWER TRANSFER CAPABILITY OF DYNAMIC WIRELESS POWER TRANSFER 3332
Chan Anyapo ; Nithiphat Teerakawanich ; Chowarit Mitsantisuk ; Kiyoshi Ohishi

NEIGHBORING EFFECTS ON THE DEACTIVATED INVERTER IN A SEGMENTED DYNAMIC WIRELESS EV CHARGING SYSTEM...3338
Qingwei Zhu ; Yanjie Guo ; Lifang Wang ; Shufan Li ; Chenglin Liao

MULTIPLE EXCITING VOLTAGE CONTROL FOR MAXIMIZATION OF MULTI-HOP WIRELESS POWER TRANSFER EFFICIENCY 3344
Masato Sasaki ; Masayoshi Yamamoto

GENERAL ANALYTICAL MODEL FOR INDUCTIVE POWER TRANSFER SYSTEM WITH EMF CANCELING COILS..3349
Keita Furukawa ; Keisuke Kusaka ; Jun-Ichi Itoh

STABILITY INFLUENCE OF FILTER COMPONENTS PARASITIC RESISTANCE ON LCL-FILTERED GRID CONVERTERS..3357
Hiroaki Matsumori ; Toshihisa Shimizu ; Frede Blaabjerg ; Xiongfei Wang ; Dongsheng Yang

REAL-TIME ESTIMATION CONTROL OF INDUCTANCE PARAMETERS USING DUST CORE MATERIALS FOR PWM INVERTER 3363
Kazu Imai ; Takuma Yoshino ; Ohasi Shunsuke ; Tomoki Yokoyama

CONTROL DESIGN OF OUTPUT-STAGE FILTERLESS SINUSOIDAL-WAVE INVERTER 3369
Shinichi Hiroshige ; Kenji Yamanaka ; Masahide Hojo

SERIES REACTIVE POWER COMPENSATOR WITH REDUCED CAPACITANCE FOR HYBRID TRANSFORMER 3375
Yuki Takahashi ; Takanori Isobe ; Hiroshi Tadano

AN INSIGHT INTO THE VOLTAGE RISING BEHAVIOR DURING TURN-OFF PROCESS OF SERIES CONNECTED SIC MOSFETS ON CIRCUIT LEVEL 3383
Panrui Wang ; Feng Gao ; Yang Jing ; Yufeng Chen ; Lei Zhang

PARALLELING SIX 320A 1200V ALL-SIC HALF-BRIDGE MODULES FOR A LARGE CAPACITY POWER STACK 3390
David Hongfei Lu ; Hiromu Takubo ; Sho Takano ; Yuhei Suzuki

3.3KV ALL-SIC MODULE FOR ELECTRIC DISTRIBUTION EQUIPMENT 3396
Ryohei Takayanagi ; Katsumi Taniguchi ; Satoshi Kaneko ; Naoyuki Kanai ; Keishirou Kumada ; Motohito Hori ;
Yoshinari Ikeda ; Kouji Maruyama ; Itsuo Kawamura

PRESENT STATUS OF SIC BASED POWER CONVERTERS AND GATE DRIVERS – A REVIEW 3401
Abhijit Choudhury

METHOD OF APPLYING FORCE DISTRIBUTION FUNCTION FOR LINEAR SWITCHED RELUCTANCE MOTOR DRIVEN BY CURRENT SOURCE INVERTER 3406
Tadashi Hirayama ; Shuma Kawabata

A NOVEL DRIVE CIRCUIT FOR SWITCHED RELUCTANCE MOTORS WITH BIPOLAR CURRENT DRIVE 3412
Hiroki Ishikawa ; Yuma Uesugi ; Seiya Sakurai

TORQUE RIPPLE MINIMIZATION CONTROL OF SRM BASED ON NOVEL MOTOR MODEL CONSIDERING MUTUAL COUPLING EFFECT 3418
Sungyong Shin ; Naruse Hikaru ; Takashi Kosaka ; Nobuyuki Matsui

COMPARISON OF HIGH FREQUENCY VOLTAGE INJECTION METHODS FOR SHAFT SENSORLESS CONTROL OF WOUND-FIELD FLUX SWITCHING MACHINE 3426
Hong-Quan Nguyen ; Sheng-Ming Yang

DESIGN AND EXPERIMENTAL VERIFICATION OF A DAB MEDIUM FREQUENCY TRANSFORMER FOR A 6.6KV/200V SOLID STATE TRANSFORMER 3431
Rene Barrera-Cardenas ; Takanori Isobe ; Terazono Katsushi ; Tadano Hiroshi

RESEARCH ON THE UNBALANCED COMPENSATION RANGE OF DELTA-CONNECTED CASCADED H-BRIDGE MULTILEVEL SVG 3439
Rui Luo ; Yingjie He ; Yiming Tu ; Xingxing Chen ; Jinjun Liu

STATIC SYNCHRONOUS COMPENSATOR TO STABILIZE GRID VOLTAGE FOR WIND AND PHOTOVOLTAIC POWER PLANT 3450
Ryota Okuyama ; Naoki Morishima ; Yusuke Ashizaki ; Yohei Itaya

LARGE EQUALIZATION CURRENT CONTROL STRATEGY FOR SERIES CONNECTED BATTERY PACKS BASED ON BUCK-BOOST CONVERTER 3455
Xinbo Liu ; Zhuo Gao ; Xuehao Huang ; Yaohan Zou

A MULTI-PORT BIDIRECTIONAL POWER CONVERSION SYSTEM FOR REVERSIBLE SOLID OXIDE FUEL CELL APPLICATIONS 3460
Xiang Lin ; Kai Sun ; Jin Lin ; Zhe Zhang ; Wei Kong

SELF-PREHEATING METHOD FOR LI-ION BATTERY USING BATTERY IMPEDANCE ESTIMATOR 3466
Dong-Kwan Kim ; Young-Dal Lee ; Sang-Hyun Ha ; Yu-Jin Jang ; Gun-Woo Moon

ACTIVE ANTI-ISLANDING TECHNIQUE WITH REDUCED NON-DETECTION ZONE FOR CENTRALIZED INVERTERS 3471
Prashant Jain ; Vivek Agarwal ; Bishnu Prasad Muni ; Eswar Rao ; Deepak Gehlot ; S. Gautam Kumar

DEVELOPMENT OF SIC APPLIED TRACTION SYSTEM FOR SHINKANSEN HIGH-SPEED TRAIN 3478
Kenji Sato ; Hirokazu Kato ; Takafumi Fukushima

DEVELOPMENT OF A HIGH POWER DENSITY AUXILIARY CONVERTER BASED ON 1700V 225A SIC MOSFET FOR TRAMS 3484
Liu Hao ; Fei Lin ; Zhongping Yang ; Hu Cao ; Meng Xia

EXPERIMENTAL TESTS RESULTS OF DAMPING CONTROL WITH OVER VOLTAGE RESISTOR FOR REGENERATIVE BRAKE CONTROL OF RAILWAY VEHICLE 3490
Natsuki Kawagoe ; Febry Pandu Wijaya ; Hiroyasu Kobayashi ; Keiichiro Kondo ; Tetsuya Iwasaki ; Akihiko
Tsumura ; Takumi Nagashima ; Yoshinori Yamashita ; Ryota Gondo

COILS LAYOUT OPTIMIZATION OF DYNAMIC WIRELESS POWER TRANSFER SYSTEM TO REALIZE OUTPUT VOLTAGE STABLE..............3495
Yi Wang ; Fei Lin ; Zhongping Yang ; Panpan Cai ; Zhiyuan Liu

QUICK CHARGER FOR A BATTERY USING MODULAR MATRIX CONVERTER (MMXC)..............3501
Kazuma Suzuki ; Takaharu Takeshita

VARIABLE OUTPUT VOLTAGE CONTROL OF AN ISOLATED BI-DIRECTIONAL AC/DC CONVERTER WITH A SOFT-SWITCHING TECHNIQUE..............3507
Takumi Hamaguchi ; Kazuma Suzuki ; Wataru Kitagawa ; Takaharu Takeshita

A NEW MODULATION METHOD APPLYING OPTIMAL DUTY CYCLE AND PHASE SHIFT FOR BIDIRECTIONAL ISOLATED THREE-PHASE AC/DC CONVERTER BASED ON MATRIX CONVERTER..............3514
Koji Shigeuchi ; Jin Xu ; Noboru Shimosato ; Yukihiko Sato

DECOUPLING CONTROL METHOD FOR ELIMINATING DC BIAS FLUX OF HIGH FREQUENCY TRANSFORMER IN A BIDIRECTIONAL ISOLATED AC/DC CONVERTER..............3522
Kensuke Sakuma ; Koji Shigeuchi ; Jin Xu ; Noboru Shimosato ; Yukihiko Sato

INTERLEAVED VOLTAGE-DOUBLER BOOST CONVERTER FOR POWER FACTOR CORRECTION..............3528
Bo-Jia Huang

ZVS INTERLEAVED TOTEM-POLE BRIDGELESS PFC CONVERTER WITH PHASE-SHIFTING CONTROL..............3533
Moo-Hyun Park ; Jae-Il Baek ; Jung-Kyu Han ; Cheon-Yong Lim ; Gun-Woo Moon

A ZERO-VOLTAGE-SWITCHING TOTEM-POLE BRIDGELESS BOOST POWER FACTOR CORRECTION RECTIFIER HAVING MINIMIZED CONDUCTION LOSSES..............3538
Young-Dal Lee ; Chong-Eun Kim ; Jae-Il Baek ; Dong-Kwan Kim ; Gun-Woo Moon

POWER-FACTOR-CORRECTION WITH POWER DECOUPLING FOR AC-TO-DC CONVERTER..............3544
Wan-Jung Chen ; Tsung-Hsi Wu ; Yao-Ching Hsieh ; Chin-Sien Moo ; Po-Hsiang Wen

DESIGN AND ANALYSIS OF THE DISTRIBUTED CONTROLLER FOR THE MODULAR MULTILEVEL CASCADED CONVERTER..............3549
Ping-Heng Wu ; Yu-Chen Su ; Po-Tai Cheng

ASYMMETRIC MIXED MODULAR MULTILEVEL CONVERTER TOPOLOGY IN HYBRID BIPOLAR HVDC TRANSMISSION SYSTEMS..............3557
Joon-Hee Lee ; Jae-Jung Jung ; Seung-Ki Sul

HIGH POWER MEDIUM VOLTAGE 10 KV SIC MOSFET BASED BIDIRECTIONAL ISOLATED MODULAR DC–DC CONVERTER..............3564
Sayan Acharya ; Ritwik Chattopadhyay ; Anup Anurag ; Satish Rengarajan ; Yos Prabowo ; Subhashish Bhattacharya

MULTI-LEVEL POWER CONVERTER USING SERIES-CONNECTED SOLID-STATE TRANSFORMERS..............3572
Yuichi Mabuchi ; Yuki Kawaguchi ; Kimihisa Furukawa ; Mitsuhiro Kadota ; Mizuki Nakahara ; Akihiko Kanoda

CAPACITOR VOLTAGE CONTROL OF MMC-STATCOM DURING UNBALANCED AC SYSTEM FAULT..............3578
Kaho Nada ; Takeshi Kikuchi ; Tsuguhiro Takuno ; Toshiyuki Fujii ; Ryosuke Uda ; Takashi Sugiyama

SIC BASED POWER SEMICONDUCTOR IN APPLICATIONS - ASPECTS AND PROSPECTS..............3584
Peter Friedrichs

ELECTROMAGNETIC MODELING APPROACHES TOWARDS VIRTUAL PROTOTYPING OF WBG POWER ELECTRONICS..............3588
Ivana Kovacevic-Badstübner ; Daniele Romano ; Giulio Antonini ; Jonas Ekman ; Ulrike Grossner

SILICON BASED DEVICES FOR DEMANDING HIGH POWER APPLICATIONS..............3596
A. Kopta ; J. Vobecky ; M. Rahimo ; T. Wikström ; U. Vemulapati ; C. Papadopoulos ; C. Corvasce ; M. Andenna ; F. Dugal ; F. Fischer ; S. Hartmann

RECENT PROGRESS IN HIGH TO ULTRA-HIGH-VOLTAGE SIC POWER DEVICES: DEVELOPMENT AND APPLICATION..............3603
Y. Yonezawa

DYNAMIC DRIFT EFFECTS IN GAN POWER TRANSISTORS: CORRELATION TO DEVICE TECHNOLOGY AND MISSION PROFILE..............3607
Joachim Würfl ; Eldad Bahat-Treidel ; Oliver Hilt ; Maria Troppenz ; Mihaela Wolf ; Jan Böcker ; Carsten Kuring ; Sibylle Dieckerhoff

COMPENSATION METHOD OF RADIAL UNBALANCE FORCE AT FAILURE OF A MOTOR SECTION IN A D-Q AXIS CURRENT CONTROL BEARINGLESS MOTOR..............3613
Masahide Ooshima

A BEARINGLESS SYNCHRONOUS RELUCTANCE SLICE MOTOR WITH ROTOR FLUX BARRIERS 3619

Thomas Holenstein ; Thomas Nussbaumer ; Johann W. Kolar

PARAMETER IDENTIFICATIONS OF CURRENT-FORCE FACTOR AND TORQUE CONSTANT IN SINGLE-DRIVE BEARINGLESS MOTORS 3627

Hiroya Sugimoto ; Akira Chiba

DAMPENING OF AXIAL VIBRATIONS IN A BEARINGLESS FLUX-SWITCHING SLICE MOTOR BY FIELD CURRENT REGULATION 3632

Bianca Klammer ; Karlo Radman ; Wolfgang Gruber

ANALYSIS AND DESIGN OF A BEARINGLESS AXIAL-FORCE/TORQUE MOTOR WITH FLEX-PCB WINDINGS 3640

Nobuyuki Kurita ; Walter Bauer ; Gerald Jungmayr ; Wolfgang Gruber ; Wolfgang Amrhein

A PLOTTER-BASED AUTOMATIC MEASUREMENT AND STATISTICAL CHARACTERIZATION OF MULTIPLE DISCRETE POWER DEVICES 3644

Michihiro Shintani ; Benjamin Dauphin ; Kazuki Oishi ; Masayuki Hiromoto ; Takashi Sato

A NOVEL HIGH-SPEED SIC MOSFET DRIVER WITH A LOW SWITCH-VOLTAGE STRESS 3650

Xiuqin Wei ; Yuchong Sun ; Hiroo Sekiya

ENHANCEMENT OF DRIVING CAPABILITY OF GATE DRIVER USING GAN HEMTS FOR HIGH-SPEED HARD SWITCHING OF SIC POWER MOSFETS 3654

Takafumi Okuda ; Takashi Hikihara

DESIGN AND EXPERIMENTAL VERIFICATION OF ROBOT ARM OPERATION FOR POWER PACKET DISPATCHING SYSTEM 3658

Tomoki Yokoyama ; Ryunosuke Araumi ; Kazunori Asada ; Takashi Ando

A RESOURCE SHARING MODEL IN A POWER PACKET DISTRIBUTION NETWORK 3665

H. Ando ; R. Takahashi ; S. Azuma ; M. Hasegawa ; T. Yokoyama ; T. Hikihara

DECOUPLED DSOGI-PLL FOR IMPROVED THREE PHASE GRID SYNCHRONISATION 3670

A. A. Nazib ; D. G. Holmes ; B. P. Mcgrath

A DEVIATION ELIMINATION CONTROL BASED ON AUTONOMOUS CURRENT-SHARING CONTROLLER FOR THE PARALLEL-CONNECTED INVERTERS IN AC MICROGRIDS 3678

Yajuan Guan ; Wei Feng ; Baoze Wei ; Wenzhao Liu ; Mingshen Li ; C. Juan Vasquez ; M. Josep Guerrero

SISO TRANSFER FUNCTIONS FOR STABILITY ANALYSIS OF GRID-CONNECTED VOLTAGE-SOURCE CONVERTERS 3684

Hongyang Zhang ; Lennart Harnefors ; Xiongfei Wang ; Jean-Philippe Hasler ; Hans-Peter Nee

A COMMUNICATION-INDEPENDENT REACTIVE POWER SHARING SCHEME WITH ADAPTIVE VIRTUAL IMPEDANCE FOR PARALLEL CONNECTED INVERTERS 3692

Ronghui An ; Zeng Liu ; Jinjun Liu ; Shike Wang

DESIGN AND INTEGRATION OF THE BI-DIRECTIONAL ELECTRIC VEHICLE CHARGER INTO THE MICROGRID AS EMERGENCY POWER SUPPLY 3698

Yang Song ; Pengcheng Li ; Yuanliang Zhao ; Shuai Lu

STABILITY IMPACT OF PV INVERTER GENERATION ON MEDIUM VOLTAGE DISTRIBUTION SYSTEMS 3705

Ye Tang ; Rolando Burgos ; Chi Li ; Dushan Boroyevich

1MW POWER CONDITIONING SYSTEM WITH MULTIPLE DC INPUTS FOR PVS AND BATTERIES 3711

Yasuaki Furusho ; Yasuyuki Noto ; Kansuke Fujii

A ROBUST AND FLEXIBLE DC-LINKED 3-PHASE ENERGY MANAGEMENT SYSTEM WITH ADAPTIVE DROOP CONTROL STRATEGY 3717

Yue Ma ; Yuki Ishikura ; Hitoshi Tsuji ; Kazuaki Mino

MAXIMUM POWER POINT TRACKING CONTROL FOR SMALL HYDROELECTRIC GENERATION 3723

Kazuya Azegami ; Masashi Takiguchi ; Junya Yano ; Hirohiko Tsutsumi ; Toshitake Masuko

DESIGN AND EXPERIMENTAL VERIFICATION OF A THREE-PHASE DUAL-ACTIVE BRIDGE CONVERTER FOR OFFSHORE WIND TURBINES 3729

Takushi Jimichi ; Murat Kaymak ; Rik W. De Doncker

OPTIMIZED BIDIRECTIONAL PFC RECTIFIERS & INVERTERS - SI VS. SIC VS. GAN IN 2L AND 3L TOPOLOGIES - 3734

Jonas Wyss ; Jürgen Biela

A STANDARD BLOCK OF "SERIES CONNECTED SIC MOSFET" FOR MEDIUM/HIGH VOLTAGE CONVERTER 3742

Qin Lei ; Chunhui Liu ; Yunpeng Si ; Yifu Liu

DESIGN AND TESTING OF 1 KV H-BRIDGE POWER ELECTRONICS BUILDING BLOCK BASED ON 1.7 KV SIC MOSFET MODULE 3749
Jun Wang ; Rolando Burgos ; Dushan Boroyevich ; Zeng Liu

A FLYBACK CONVERTER WITH SIC POWER MOSFET OPERATING AT 10 MHZ: REDUCING LEAKAGE INDUCTANCE FOR IMPROVEMENT OF SWITCHING BEHAVIORS 3757
Kazuki Hashimoto ; Takafumi Okuda ; Takashi Hikihara

A STUDY ON LOAD FLUCTUATION OF ISOLATED DC-DC CONVERTER WITH CLASS PHI-2 INVERTER USING GAN-HFET 3762
Yuta Yanagisawa ; Yushi Miura ; Hiroyuki Handa ; Tetsuzo Ueda ; Toshifumi Ise

SINGLE-INDUCTOR MULTIPLE-OUTPUT CURRENT-SOURCE CONVERTER WITH IMPROVED CROSS REGULATION AND SIMPLE CONTROL STRATEGY 3768
Zheng Dong ; Xiaolu Lucia Li ; Chi K. Tse

LIMIT OPERATING FREQUENCY OF PEAK CURRENT-MODE CONTROL DC-DC CONVERTER CONSIDERING TURN-OFF DELAY TIME 3773
Ryo Ute ; Kazuya Fujiwara ; Jun Imaoka ; Masahito Shoyama

A NOVEL SINGLE SWITCH HIGH FREQUENCY DC/DC CONVERTER AND ITS MATHEMATIC MODEL 3780
Yueshi Guan ; Xihong Hu ; Shu Zhang ; Yijie Wang ; Dianguo Xu ; Wei Wang

ANALYSIS OF CLOSED LOOP OPERATION OF AN ISOLATED BIDIRECTIONAL DAB DC-DC CONVERTER WITH LC COUPLING 3785
Bruno Yukio Enomoto ; Kelly C. M. Carvalho ; Lourenço Matakas Junior ; Wilson Komatsu

ISOLATED AC/DC CONVERTER USING SIMPLE PWM STRATEGY 3791
Naoki Hirose ; Yuto Matsui ; Takaharu Takeshita

ANALYSIS OF ONE PHASE LOSS OPERATION OF THREE-PHASE ISOLATED BUCK MATRIX-TYPE RECTIFIER WITH EIGHT-SEGMENT PWM SCHEME 3797
Jahangir Afsharian ; Dewei David Xu ; Bin Wu ; Bing Gong ; Zhihua Yang ; Jun-Ichi Itoh

NOVEL ISOLATED BIDIRECTIONAL INTEGRATED DUAL THREE-PHASE ACTIVE BRIDGE (D3AB) PFC RECTIFIER 3805
F. Krismer ; E. Hatipoglu ; J. W. Kolar

LOAD VOLTAGE REGULATION METHOD FOR AN ISOLATED AC-DC CONVERTER WITH POWER DECOUPLING OPERATION 3813
Shohei Komeda ; Hideaki Fujita

OPTIMAL DESIGN OF A LOW COST 20KW 99.1% EFFICIENCY ACTIVE ZCS ISOLATED DC-DC CONVERTER 3820
Timothé Delaforge ; Sébastien Mariéthoz

SOFT-SWITCHING ANALYSIS AND PFM CONTROL METHOD OF BIDIRECTIONAL DC/DC CONVERTER TOPOLOGY 3825
Yijie Wang ; Haoyu Wang ; Hongyu Song ; Dianguo Xu

A FULLY SOFT-SWITCHED PWM DC-DC CONVERTER USING AN ACTIVE-SNUBBER-CELL 3833
Hai N. Tran ; Adhistira M. Naradhipa ; Sunju Kim ; Ali Tausif

FLYING CAPACITOR RESONANT POLE INVERTER WITH DIRECT INDUCTOR CURRENT FEEDBACK 3840
Sjef J. Settels ; Jorge L. Duarte ; Jeroen Van Duivenbode

DESIGN OF A GAN-BASED WIRELESS POWER TRANSFER SYSTEM AT 13.56 MHZ TO REPLACE CONVENTIONAL WIRED CONNECTION IN A VEHICLE 3848
Kawin Surakitbovorn ; Juan Rivas-Davila

EFFICIENCY MAXIMIZATION OF INDUCTIVE POWER TRANSFER SYSTEM BY IMPEDANCE AND SWITCHING FREQUENCY CONTROL IN SECONDARY-SIDE CONVERTER 3855
Ryosuke Ota ; Dannisworo S. Nugroho ; Nobukazu Hoshi

ANALYSIS OF OPTIMAL OPERATION FREQUENCY RANGE FOR BATTERY CHARGING IN WPT SYSTEM 3863
Yongbin Jiang ; Min Wu ; Junwen Liu ; Yue Wang ; Laili Wang ; Hailong Zhang

INITIAL CURRENT INJECTION METHOD OF A DIRECT THREE-PHASE TO SINGLE-PHASE AC/AC CONVERTER FOR INDUCTIVE CHARGER 3870
Ferdi Perdana Kusumah ; Jorma Kyyrä

MISSION PROFILE EMULATOR FOR PERMANENT MAGNET SYNCHRONOUS MACHINE BASED ON THREE-PHASE POWER ELECTRONIC CONVERTER 3877
Yubo Song ; Ran Cheng ; Ke Ma

A VARIABLE DC BUS VOLTAGE BASED POWER HARDWARE-IN-THE-LOOP EMULATION OF ELECTRIC MOTORS WITH WIDE VARIATION IN INTERFACE FILTER INDUCTANCE 3884
Tsai-Fu Wu ; Mitradatta Misra ; Ying-Yi Jhang ; Chang-Jun Yang ; Yin-Chi Xu

COPPER LOSS MINIMIZATION CONTROL AT ZERO OUTPUT VOLTAGE FOR ELECTROLYTIC CAPACITOR-LESS INVERTER 3890

Kodai Abe ; Haruya Kada ; Kiyoshi Ohishi ; Hitoshi Haga ; Yuki Yokokura

ARMATURE TEMPERATURE ESTIMATION INSENSITIVE TO ROTOR FLUX VARIATION FOR SPMSM 3896

Toshiki Sano ; Kiyoshi Ohishi ; Yuki Yokokura ; Hiroki Iwata ; Yuji Ide ; Daigo Kuraishi ; Akihiko Takahashi

VIRTUAL SYNCHRONOUS GENERATOR CONTROL WITH RELIABLE FAULT RIDE-THROUGH CAPABILITY BY ADOPTING MODEL PREDICTIVE CONTROL 3902

Jonggrist Jongudomkarn ; Jia Liu ; Toshifumi Ise

RESHAPING QUADRATURE-AXIS IMPEDANCE OF THREE-PHASE GRID-CONNECTED CONVERTERS FOR LOW-FREQUENCY STABILITY IMPROVEMENT 3910

Yi Tang ; Jingyang Fang ; Xiaoqiang Li ; Hongchang Li

COMPARISON BETWEEN TRADITIONAL DROOP AND A NEW AUTONOMOUS CONTROL SCHEME FOR PARALLEL INVERTERS 3916

Mohammad Bani Shamseh ; Teruo Yoshino ; Atsuo Kawamura

A NOVEL MICROGRID POWER SHARING SCHEME ENHANCED BY A NON-INTRUSIVE FEEDER IMPEDANCE ESTIMATION METHOD 3924

Baojin Liu ; Zeng Liu ; Jinjun Liu ; Ronghui An ; Shuguang Song

DEVELOPMENT OF A 3.2MW PHOTOVOLTAIC INVERTER FOR LARGE-SCALE PV POWER PLANTS 3929

Naoya Shibata ; Tsuguhiro Tanaka ; Masahiro Kinoshita

IMPEDANCE-BASED STABILITY ANALYSIS OF LARGE-SCALE PV STATION UNDER WEAK GRID CONDITION CONSIDERING SOLAR RADIATION FLUCTUATION 3934

Yiming Tu ; Jinjun Liu ; Teng Liu ; Xiangpeng Cheng

EXPERIMENTAL VERIFICATION OF GRID-CONNECTION OF A PV CONVERTER USING A SYMMETRICALLY CONNECTED BOOST CONVERTER FOR A HIGH-LEG DELTA TRANSFORMER 3940

Daiki Yamaguchi ; Hideaki Fujita

A NOVEL SINGLE- STAGE HIGH-FREQUENCY BOOST INVERTER CASCADED BY RECTIFIER-INVERTER SYSTEM FOR PV GRID-TIE APPLICATIONS 3945

Hamdy Radwan ; Mahmoud A. Sayed ; Takaharu Takeshita ; Adel A. Elbaset ; G. Shabib

NINE SWITCHES MATRIX CONVERTER USING BI-DIRECTIONAL GAN DEVICE 3952

Takashi Hirota ; Kentaro Inomata ; Daisuke Yoshimi ; Masato Higuchi

A MODEL PREDICTIVE DUAL CURRENT CONTROL METHOD FOR INDIRECT MATRIX CONVERTER FED INDUCTION MOTOR DRIVES 3958

Mei Yang ; Chen Lisha ; Liang Wang ; Yunwei Li

FAULT TOLERANT PREDICTIVE CONTROL OF THREE-LEVEL NEUTRAL-POINT-CLAMPED BACK-TO-BACK POWER CONVERTERS 3965

Zhenbin Zhang ; Xicai Liu ; Kejun Cai ; Feng Gao ; Ralph Kennel

TWO-STAGE OPTIMIZATION BASED PREDICTIVE TORQUE CONTROL WITH REDUCED COMPLEXITY FOR A THREE-LEVEL INVERTER DRIVEN INDUCTION MOTOR 3971

Ilham Osman ; Dan Xiao ; Faz Rahman

DESIGN CHALLENGES OF SIC DEVICES FOR LOW- AND MEDIUM-VOLTAGE DC-DC CONVERTERS 3979

Georges Engelmann ; Alexander Sewergin ; Markus Neubert ; Rik W. De Doncker

DESIGN AND TESTING OF 6 KV H-BRIDGE POWER ELECTRONICS BUILDING BLOCK BASED ON 10 KV SIC MOSFET MODULE 3985

Jun Wang ; Slavko Mocevic ; Jiewen Hu ; Yue Xu ; Christina Dimarino ; Igor Cvetkovic ; Rolando Burgos ; Dushan Boroyevich

HIGH POWER MEDIUM VOLTAGE CONVERTERS ENABLED BY HIGH VOLTAGE SIC POWER DEVICES 3993

Sanket Parashar ; Ashish Kumar ; Subhashish Bhattacharya

SOFT-SWITCHING – THE KEY TO HIGH POWER WBG CONVERTERS 4001

Deepak Divan ; Zheng An ; Prasad Kandula

SIC: TECHNOLOGY ENABLER FOR MV DC/DC GALVANICALLY INSULATED MODULAR CONVERTERS 4009

S. Alvarez ; M. Bellini ; U. Vemulapati ; F. Canales ; M. Rahimo

A BEARINGLESS SLICE MOTOR WITH A SOLID IRON ROTOR FOR DISPOSABLE CENTRIFUGAL BLOOD PUMP 4016

Tadahiko Shinshi ; Ryo Yamamoto ; Yoshiki Nagira ; Junichi Asama

REDUCED HARDWARE PARALLEL DRIVE FOR NO VOLTAGE BEARINGLESS MOTORS 4020

Eric L. Severson

DUAL FIELD-ORIENTED CONTROL OF BEARINGLESS MOTORS WITH COMBINED WINDING SYSTEM 4028
Wolfgang Gruber ; Siegfried Silber

OPEN-CIRCUIT FAULT TOLERANT STUDY OF BEARINGLESS MULTI-SECTOR PERMANENT MAGNET MACHINES 4034
G. Valente ; L. Papini ; A. Formentini ; C. Gerada ; P. Zanchetta

BALANCE CONTROL OF SPLIT CAPACITOR POTENTIAL FOR MAGNETICALLY LEVITATED MOTOR SYSTEM USING ZERO-PHASE CURRENT 4042
Takaaki Oiwa

ASYMMETRICAL HALF-BRIDGE CONVERTER WITH ZERO DC-OFFSET CURRENT IN TRANSFORMER USING NEW RECTIFIER STRUCTURE 4049
Jung-Kyu Han ; Jong-Woo Kim ; Seung-Hyun Choi ; Jih-Sheng Lai ; Gun-Woo Moon

CIRCULATING CURRENT-LESS PHASE-SHIFTED FULL-BRIDGE CONVERTER WITH NEW RECTIFIER STRUCTURE 4054
Jung-Kyu Han ; Gun-Woo Moon

A BI-DIRECTIONAL CURRENT DETECTION USING CURRENT TRANSFORMERS FOR BI-DIRECTIONAL DC-DC CONVERTER 4059
Seiji Iyasu ; Yuji Hahashi ; Yuuichi Handa ; Kimikazu Nakamura ; Keiji Wada

A 10 MHZ GANFET BASED ISOLATED HIGH STEP-DOWN DC-DC CONVERTER 4066
Prasanth Thummala ; Dorai Babu Yelaverthi ; Regan Zane ; Ziwei Ouyang ; Michael A. E. Andersen

ANALYSIS AND DESIGN OF A PARALLEL RESONANT CONVERTER FOR CONSTANT CURRENT INPUT TO CONSTANT VOLTAGE OUTPUT DC-DC CONVERTER OVER WIDE LOAD RANGE 4074
Tarak Saha ; Hongjie Wang ; Baljit Riar ; Regan Zane

NOVEL SINUSOIDAL INPUT CURRENT SINGLE-TO-THREE-PHASE Z-SOURCE BUCK+BOOST AC/AC CONVERTER 4080
M. Haider ; D. Bortis ; J. W. Kolar ; Y. Ono

SIMPLE PWM STRATEGY OF A MATRIX CONVERTER FOR MINIMIZING OUTPUT VOLTAGE HARMONICS 4088
Takuya Oshima ; Takaharu Takeshita

NOVEL THREE-LEVEL BACK-TO-BACK CONVERTERS: STRUCTURE, MODULATION METHOD, AND EXPERIMENT 4096
S. Sangwongwanich ; K. Niyomsatian ; S. Samermurn ; S. Nuchnoi ; S. Suwankawin

MODEL PREDICTIVE CONTROL USING SUBDIVIDED VOLTAGE VECTORS FOR CURRENT RIPPLE REDUCTION IN AN INDIRECT MATRIX CONVERTER 4104
Keon Young Kim ; Yeongsu Bak ; Jin-Hyuk Park ; Kyo-Beum Lee

DC-LINK RIPPLE CURRENT REDUCTION IN BACK-TO-BACK CONVERTERS WITH DPWM 4109
Anatolii Tcai ; Kyo-Beum Lee

AN ANALYSIS OF CLASS DE VOLTAGE-SOURCE PARALLEL RESONANT INVERTER 4114
Takeshi Kondo ; Tsuyoshi Inaba ; Yoshikazu Sakai ; Hirotaka Koizumi

AN IMPROVEMENT ON EXTENDED IMPEDANCE METHOD TOWARDS EFFICIENT STEADY-STATE ANALYSIS OF HIGH-FREQUENCY CLASS-E RESONANT INVERTERS 4122
Junrui Liang

OUTPUT POWER CAPABILITY COMPARISONS OF CLASS-E POWER AMPLIFIERS WITH HARMONIC RESONANCE 4127
Hiroo Sekiya ; Xiuqin Wei ; Yuchong Sun

A CLASS Φ2 RESONANT BUCK CONVERTER WITH RIPPLE INJECTION BURST CONTROL METHOD 4133
Min Lin ; Masahiko Hirokawa

PRACTICAL DESIGN TECHNIQUE FOR HIGH POWER DENSITY LLC RESONANT CONVERTER 4139
Shingo Nagaoka ; Hiroyuki Onishi ; Koji Takatori ; Toshiyuki Zaitsu ; Takeshi Uematsu

OPERATIONAL STUDY AND PROTECTION OF A SERIES RESONANT CONVERTER WITH DC CURRENT INPUT APPLIED IN DC CURRENT DISTRIBUTION SYSTEMS 4145
Hongjie Wang ; Tarak Saha ; Baljit Riar ; Regan Zane

A STUDY ON IMPROVEMENT OF POWER UTILIZATION RATE OF ENERGY SYSTEMS WITH PVS AND BATTERIES 4151
Hiroaki Endo ; Masakatsu Kurisaka ; Tsutomu Ueno ; Yusuke Yoshioka ; Kaoru Inoue ; Toshiji Kato

A NOVEL DC DISTRIBUTION NETWORK WITH MULTI-LEVEL BUS VOLTAGES AND ITS ENERGY MANAGEMENT SYSTEM DESIGN 4157
Jingjin Huang ; Xin Zhang ; Zhixun Ma ; Jianfang Xiao

A NOVEL DC-SIDE-PORT IMPEDANCE MODELING OF MODULAR MULTILEVEL CONVERTERS BASED ON HARMONIC STATE SPACE METHOD 4162

Jing Lyu ; Xin Zhang ; Zhixun Ma ; Xu Cai

AN IMPROVED MASTER-SLAVE CONTROL FOR THREE-PORT CONVERTER BASED DISTRIBUTED DC GRID-CONNECTED PV SYSTEM 4168

Siyue Jiang ; Kai Sun ; Hongfei Wu ; Haixu Shi ; Xiaofeng Dong ; Syed Muhammad Raza Kazmi

SENSORLESS POSITION ESTIMATION, PARAMETER IDENTIFICATION AND CONTROL INTEGRATION FOR PERMANENT MAGNET SYNCHRONOUS MACHINES USING CURRENT DERIVATIVE MEASUREMENTS 4174

M.X. Bui

DYNAMIC PERFORMANCE IMPROVEMENT OF BIDIRECTIONAL SWITCHED-CAPACITOR DC/DC CONVERTER BY RIGHT-HALF-PLANE ZERO ELIMINATION 4181

Ding Kaicheng ; Zhang Yan ; Liu Jinjun ; Zeng Pengxiang ; Zhang Jinshui

A MATRIX BASED ISOLATED BIDIRECTIONAL AC-DC CONVERTER WITH LCL TYPE INPUT FILTER FOR ENERGY STORAGE APPLICATION 4186

Prathamesh Pravin Deshpande ; Amit Kumar Singh ; Sanjib Kumar Panda

ON A STUDY OF VOLTAGE DIVIDING CLASS Φ AMPLIFIER 4193

Katsutoshi Hirayama ; Tadashi Suetsugu ; Yudai Furukawa ; Fujio Kurokawa

A DPWM BASED CONTROL STRATEGY TO INTEGRATE PHOTOVOLTAIC SYSTEM AND BATTERY STORAGE USING GRID CONNECTED THREE-LEVEL T-TYPE INVERTER 4198

Mohammad M. Hashempour ; Yue-Ting Tsai ; T. L. Lee

IMPEDANCE MEASUREMENT OF MEGAWATT-LEVEL RENEWABLE ENERGY INVERTERS USING GRID-FORMING AND GRID-PARALLEL CONVERTERS 4205

Matias Berg ; Tuomas Messo ; Tomi Roinila ; Henrik Alenius

IMPROVED VIRTUAL INDUCTANCE BASED CONTROL STRATEGY OF DFIG UNDER WEAK GRID CONDITION 4213

Ran Fang ; Wenjia Chen ; Xueguang Zhang ; Dianguo Xu

CONTROL OF VSC-HVDC FOR WIND FARM INTEGRATION WITH REAL-TIME FREQUENCY MIRRORING AND SELF-SYNCHRONIZING CAPABILITY 4220

Renxin Yang ; Chen Zhang ; Xu Cai ; Gang Shi ; Jing Lyu

A STUDY ON STEADY-STATE CHARACTERISTICS OF SERIES-CONNECTED WIND FARM USING AN EXPERIMENTAL SET OF LABORATORY SIZE 4227

Fujio Tatsuta ; Shoji Nishikata

A NOVEL ISLANDING DETECTION METHOD WITH TWO-PHASE MAGNIFICATION INSPECTION 4233

Jian-Tang Liao ; Shun-Hao Yeh ; Hong-Tzer Yang

Author Index

The 2018 International Power Electronics Conference

Impedance-Based Stability Evaluation of Virtual Synchronous Machine Implementations in Converter Controllers

Eneko Unamuno[1*], Atle Rygg[2], Mohammad Amin[3], Marta Molinas[2], Jon Andoni Barrena[1]

1 Electronics and Computer Science Department, Mondragon Unibertsitatea, Arrasate-Mondragón, Spain
2 Department of Engineering Cybernetics, Norwegian University of Science and Technology, Trondheim, Norway
3 Electrical and Computer Engineering, Illinois Institute of Technology, Chicago, United States
*E-mail: eunamuno@mondragon.edu

Abstract—This paper presents a stability evaluation of two Virtual Synchronous Machine (VSM) implementations in the control of Voltage Source Converters (VSCs): a current-controlled (CCVSM) and a voltage-controlled (VCVSM) version of the VSM. The performances of the two control implementations are analyzed under power reference and grid frequency variations and the evaluation of their stability properties is carried out by adopting the impedance-based stability analysis. The results of the *dq* impedance analysis of the two VSMs implementations reveal a clear differentiation between the CCVSM and the VCVSM, particularly in the very low frequency range, even if a close similarity is observed in the time domain responses. The *dq* impedances of both implementations show an *RL* behaviour in the medium frequency range, but when inspecting the stability margin on the Nyquist plot, the VCVSM appears to have a lower stability margin than the CCVSM from the intersection with the unit circle.

Index Terms—Frequency Control, Small-Signal Stability, Synchronous Machine Emulation, Swing Equation, Virtual Synchronous Machine.

I. INTRODUCTION

The increasing penetration of converter-interfaced generation, energy storage systems (ESS) and loads is changing the behavior of power systems that were originally dominated by synchronous machines (SMs) with high values of inertia [1]. The distributed nature of these devices is shifting the structure of electric grids from a classical centralized/top-down structure towards a more decentralized configuration [2], [3]. Although this transition provides several advantages—e.g. a reduction of the dependency on fossil fuels or an alternative to supply the increasing worldwide energy demands—it also poses several challenges to be addressed in order to ensure an adequate operation.

For instance, a higher amount of converter-interfaced systems on an electric grid may cause instabilities in the voltage or frequency because, unlike classical SMs, they do not provide any inherent inertia over power variations [1]. Therefore, these converters must be controlled by more advanced techniques in order to support the grid by providing this inertial behavior and primary reserve—i.e. becoming grid-forming systems instead of merely being grid-following devices [4]. In this context, in the last decades synchronous machine

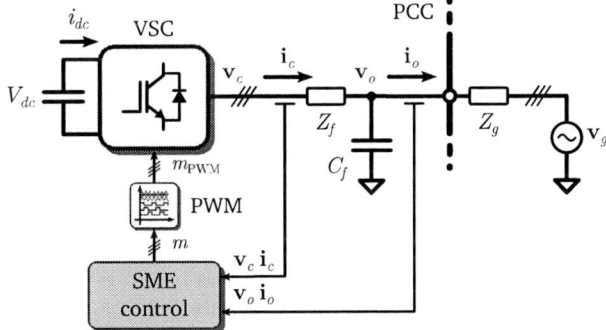

Fig. 1: Configuration of a VSC with synchronous machine emulation control.

emulation (SME) techniques have arisen as one of the most interesting alternatives. These techniques not only provide synthetic inertia to the grid, but they are also capable of operating in a distributed manner as classical SMs, which makes them suitable for both grid-connected or isolated systems. By employing only local measurements, SME controlled converters share power variations occurring in the grid without any extra communication network. Furthermore, some SME control types provide better flexibility when connecting to electric grids by eliminating the dedicated synchronization unit (phase-locked loop or PLL) [5], [6].

Fig. 1 shows a typical configuration of a SME-controlled system, which consists of a voltage-source converter (VSC) connected to the grid through a passive filter, and is controlled based on its local measurements.

In the literature there is a wide variety of SME techniques, and some of them have been already reviewed for instance in [7] and [8], respectively. In this case, we focus the analysis on the so-called virtual synchronous machines (VSMs), which are based on the explicit emulation of a synchronous machine swing equation and have been widely employed for different applications such as vehicle-to-grid applications [9] or microgrids [10]–[13]. Closed-loop VSM techniques can be classified in two main groups:

- Current-controlled virtual synchronous machines (CCVSM) [5], [10], [14]–[16]
- Voltage-controlled virtual synchronous machines (VCVSM) [9], [12], [13], [17]

The aim of this paper is to carry out a comparative evaluation in terms of performance and stability of non-augmented versions of these two types of VSM techniques controlling a grid-connected VSC as the one shown in Fig. 1. By non-augmented we refer to the basic structure of VSM controllers without any feed-forward terms or active damping loops.

In this context, in Section II we first revisit the structure of the two types of VSM techniques and their most relevant characteristics, discussing some of their advantages and drawbacks. Based on these configurations, in Section III we introduce the methodology followed to derive the small-signal state-space models in the dq-domain. Moreover, we highlight the importance of rotating impedance matrices so that analytical and measured impedances are compared in the same rotating reference frame.

Following the steps on this methodology, in Section IV we evaluate the performance of each VSM technique in time-domain simulations under different perturbations. Here, we analyze the response of the derived small-signal models and compare them to the non-linear systems, which allows us to verify the analytical models for subsequent analyses.

Section V covers the analysis and comparison of the stability of these systems to identify the advantages and drawbacks of each technique. Although there is a wide variety of methods for determining the stability of such systems, several challenges arise due to the non-linearities of the control strategies. In this work we adopt the impedance-based stability analysis, applying the generalized Nyquist criterion (GNC), which is one of the most used methods to determine the stability of power electronics-based power systems [18]–[22]. Finally, Section VI concludes with the most important remarks of the research.

II. BACKGROUND AND OVERVIEW OF VSM CONTROL STRATEGIES

VSM techniques are inspired by the concept of operation of classical synchronous machines, which are employed to regulate frequency and voltage of the electric grid. In this context, VSMs usually include an active power controller (APC) that is typically composed by a p/f droop regulator and an inertia-emulation part controlling the output frequency of the converter. Moreover, these techniques integrate a reactive power controller (RPC) that regulates the voltage amplitude through a q/v droop regulator. In the following sections we revisit the specific characteristics of the investigated CCVSM and VCVSM control strategies, showing their main differences.

A. Current-Controlled Virtual Synchronous Machine

From the two approaches, CCVSMs are the closest to classical SM-based systems because, in addition to the already mentioned active and reactive power controllers, they implement a virtual electrical model of a SM in the control strategy. The CCVSM technique analyzed in this paper is based on the recent study carried out by Mo et al. in [5], where two different SM electrical models are compared: a dynamic electrical model and a quasi-stationary model. For the sake of simplicity, in this case we have implemented the dynamic model shown in [5], which is illustrated in Fig. 2.

The active power controller is composed by a droop regulator and the emulation of inertia, which is done by integrating the swing equation of classical SMs. The difference of this APC compared to other approaches is that, instead of estimating the frequency of the grid with a PLL, the frequency is generated by filtering the value obtained in the inertia-emulation, assuming that it will vary very slowly. Another difference resides in the fact that the RPC includes a PI regulator to establish the voltage amplitude reference of the virtual SM in the d axis.

Apart from these controllers and the dynamic model of the SM, CCVSMs include a classical current control based on PI regulators with decoupling terms. As shown in Fig. 2, we divide the voltage references obtained in this current controller by the voltage of the dc bus. Thus, the dc-side dynamics can be effectively decoupled from the ac-side dynamics, and any dc-side control will not be considered in the following.

The main advantage of this control is that it increases the flexibility to adapt the behaviour of the converter by varying not only the virtual-inertia but also the equivalent stator impedance of the SM model. In addition, the integration of a current controller enables the direct limitation of the current, and there is no need to employ any synchronizing algorithm because the frequency is internally generated.

One of the drawbacks might be the fact that the SM dynamic model must be carefully designed to ensure an adequate dynamic behaviour of the converter.

B. Voltage-Controlled Virtual Synchronous Machine

As shown in Fig. 3, VCVSM techniques are more complex than the previous approach. They have been widely employed in the literature for different applications such as microgrids or electric vehicles [9], [12].

The configuration illustrated in Fig. 3 is composed of a cascaded current and voltage controller, a virtual RL impedance and similar active controller as in the previous technique.

The PLL, illustrated in Fig. 4, consists of a classical PI regulating the q axis term of the output voltage to zero.

One of the advantages of this technique compared to direct voltage control as commonly applied in [6], [8] is that it enables current limitation and the accurate control of the voltage at the output of the converter. In addition, the included virtual-impedance provides more degrees of freedom to adjust the dynamic behaviour of the converter.

However, this specific approach brings about some drawbacks. On the one hand, the high number of cascaded controllers significantly increases the complexity of the technique, making the tuning and configuration of parameters a challenging task. In addition, the strategy shown in Fig. 3, which is

The 2018 International Power Electronics Conference

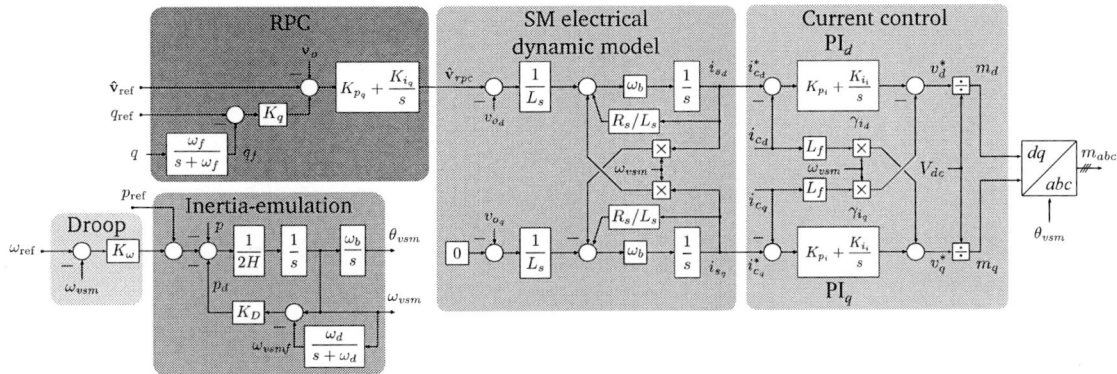

Fig. 2: Current-controlled virtual synchronous machine control diagram adapted from [5].

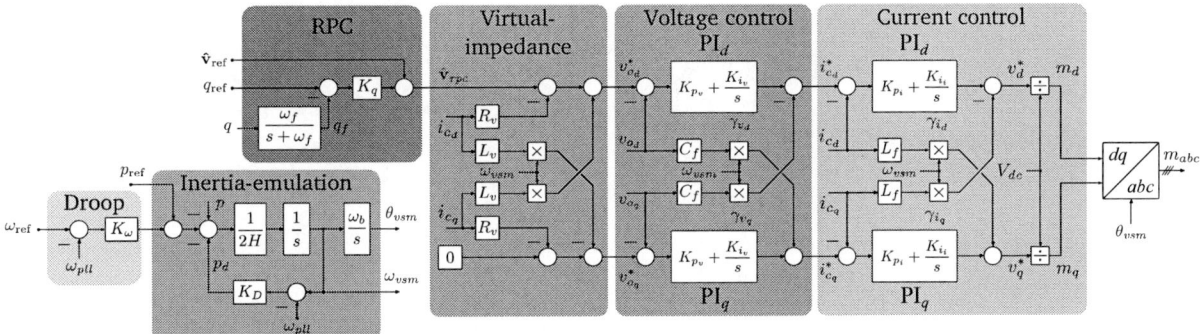

Fig. 3: Voltage-controlled virtual synchronous machine control diagram based on [12].

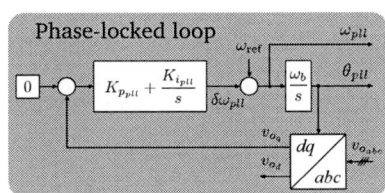

Fig. 4: Structure of the PLL employed for the VCVSM control strategy.

based on [12] employs a synchronizing algorithm to capture the frequency of the grid

In any case, we could avoid the use of a PLL by generating the frequency internally as in the CCVSM. As we are focusing this paper on previous studies of the literature, we leave this analysis for further research activities.

III. Methodology and Simulation Conditions

In order to carry out the performance and stability comparative evaluation, we have developed a methodology that is summarized in Fig. 5.

In this approach, we first derive the differential equations that define the dynamics of the analyzed system, including physical elements as well as control algorithms. Taking into account that some of the differential equations are non-linear,

we solve the system to obtain the linearization point $\bar{x}$. This is equal to the steady-state point of operation, and is computed by eliminating the derivative terms and solving the set of equations for specific input values. We then obtain the small-signal state-space model linearized over $\bar{x}$ employing Taylor series expansion, which enables the use of linear techniques for stability analyzes. In order to verify the small-signal model and observe its dynamic performance, we compare it with the non-linear model in time-domain simulations for different disturbances. This part is covered in detail in Section IV. Once the model is verified, we obtain the analytical and measured impedance of each system and analyze the stability applying the GNC.

In this paper all the equations employed to model the VSM-controlled converters are represented in rotating reference frames—i.e. in the dq domain—by applying the amplitude-invariant Park transform. In this sense, we can distinguish three main rotating reference frames:

1) Global reference frame (GRF): the d axis of the GRF is aligned with the grid voltage vector v_g, and the q axis, as in the rest of rotating reference frames, leads the d axis by 90°. The variables referenced to this rotating frame are denoted with superscript "G".

2) Local reference frame (LRF): the d axis of the LRF is aligned with the output voltage vector v_o. In this case,

761

The 2018 International Power Electronics Conference

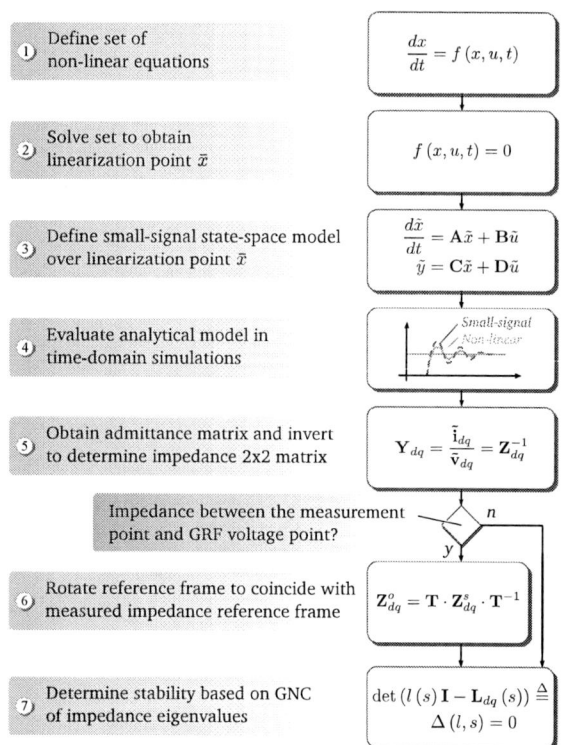

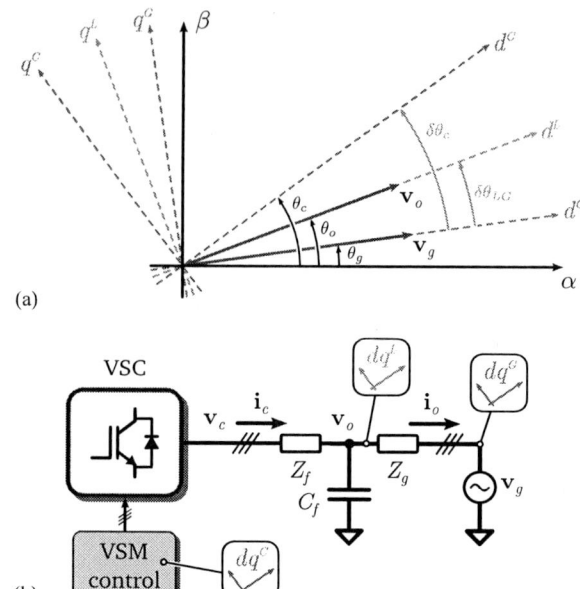

Fig. 5: Methodology for the analytical derivation of dq impedance 2×2 matrices and stability analyses based on GNC.

Fig. 6: Phasor diagram of VSM-controlled converter rotating reference frames.

this voltage corresponds to the voltage at the capacitor of the output LC filter of the converter.

3) Dynamic reference frame (DRF): the DRF is slightly different from the previous two and is aligned with the control frequency—so we denote variables referenced to this rotating frame with superscript "C"—and has an angle θ_c that corresponds to the angle generated by the inertia-emulation loop, θ_{vsm}. This reference frame can also be viewed as the local reference frame seen from the control system.

Fig. 6 represents the phasor diagram of the two VSM-controlled converters, where we illustrate the location where each rotating reference frame is aligned.

In relation to these rotating reference frames, special care must be taken when comparing the impedances obtained from the analytical equations and the simulation model. Usually, the latter is obtained by injecting a perturbation signal—either voltage or current—at the output of the converter and calculating the impedance matrix by dividing the voltage and current. These variables are measured in the abc-domain, and are transformed into the LRF through a fast Fourier transform (FFT). If there is no impedance between the analyzed system and the grid, or if the power transmission between these two systems is zero, the LRF and the GRF are completely aligned. This means that the impedances referenced to any of

these two rotating reference frames can be directly compared. However, when these two conditions are not met, the two rotating reference frames are not aligned any more due to the voltage drop in the impedance. As analytical impedances are most of the times referenced to the GRF, a rotation must be applied to one of the impedances so as to compare them in the same reference frame (step ⑥ in the proposed methodology). This is even more relevant when carrying out the impedance-based stability analysis of a system composed by multiple interconnected devices, as they must be all referenced to a common reference frame [23].

For the sake of simplicity and readability and due to space constraints we do not include the analytical development of the previously shown VSM-controlled converters. The reader is referred to [5] and [12] for the detailed equations of the CCVSM and the VCVSM, respectively. At this point it is worth mentioning that we slightly modified the technique proposed in [5] by employing the converter current instead of the output current in the controller.

The results shown in the next sections are obtained considering the following conditions:

1) The reactive power controllers are disabled by setting the reference q_{ref} and the droop gain K_q to zero in both cases.

2) No active damping loops are added in the control algorithms of converters, as they would influence their performance and stability characteristics. This means the analysis is done for non-augmented techniques, considering only their inherent behaviour.

3) All the analyzed systems begin the simulation from a steady-state point of operation of $p = 0.5$ p.u. to reproduce a more realistic point of operation.

The 2018 International Power Electronics Conference

4) We model the delay of the PWM as a second order transfer function following the Padé approximation proposed in [24], which takes into account the computation delay, the sampler and the zero order hold:

$$G_{pwm} = \frac{1 - 0.5T_s s}{(1 + 0.5T_s s)^2} \qquad (1)$$

In addition, we must mention that the parameters used for the simulations are gathered from [5] and [12].

IV. MODEL VALIDATION AND PERFORMANCE EVALUATION OF VSMS

In this section we evaluate the dynamic behaviour of VSM-controlled converters under different disturbances. These simulations serve not only to observe the response of each analyzed system but also to verify the analytical model by comparing it with the non-linear simulation model.

A. Active power reference variation

In this case we apply a step change in the active power reference, transitioning from $p_{ref} = 0.5$ p.u. to 0.6 p.u. The purpose of this simulation is to observe the response of the converter avoiding the effect of the frequency droop controller.

Fig. 7 shows the output power and frequency response of both cases for a 0.1 p.u. variation of the power reference $p_{m_{ref}}$.

The curves in this case show a very good accuracy between the non-linear and the developed small-signal models. Moreover, the CCVSM and VCVSM techniques show a similar behaviour; the difference resides in their overshooting and the time they require to reach the steady-state operation. However, this can be adapted by varying the emulated inertia and damping factor of the *swing equation* included in their control strategies.

B. Grid frequency variation

Here, the frequency of the grid begins at its rated value of 50 Hz and we apply a negative step variation of 0.2% of this rated frequency. This way we can study not only the dynamic behaviour but also the steady-state primary regulation of each system. The results are shown in Fig. 8.

In this case, the VCVSM technique has some overshooting and a small oscillation under the step variation, but it reaches the steady-state approximately in 0.5 s. On the other hand, the CCVSM has a very high overshoot ($\sim 30\%$) with no oscillation afterwards. Even though this can be problematic, CCVSM-controlled converters are very flexible and their response can be adapted by varying several control parameters. Moreover, despite its high overshooting, the CCVSM technique provides more inertial response compared to the VCVSM for the same H, which can be noticed by looking at the rate of change of frequency (RoCoF) of the curves shown at the right part of Fig. 8.

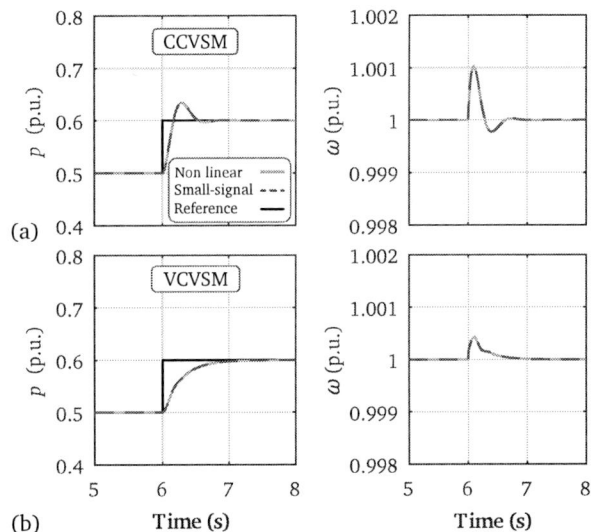

Fig. 7: Dynamic response of VSM-controlled converters under a power reference variation: (a) CCVSM and (b) VCVSM.

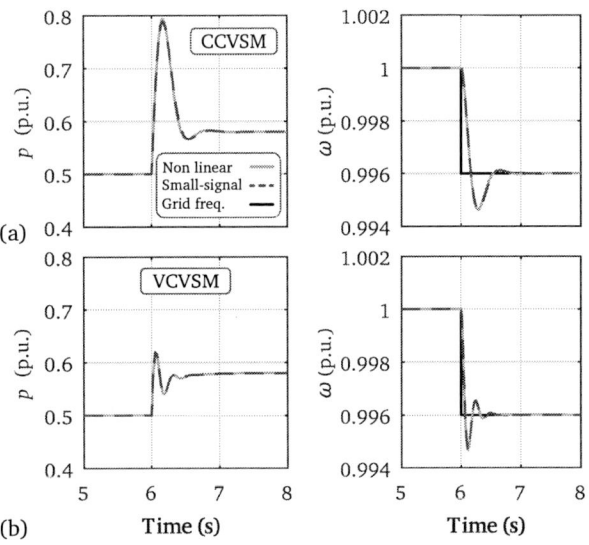

Fig. 8: Dynamic response of VSM-controlled converters under a grid frequency variation: (a) CCVSM and (b) VCVSM.

V. IMPEDANCE-BASED STABILITY ANALYSIS

Once we have verified the small-signal models in time-domain simulations and we have proven their performance under different types of perturbations, the next step is to study the stability of each system.

As we have already mentioned, in the literature there is a wide variety of methods to determine the stability of power electronic systems connected to ac grids. In this case we have decided to employ the generalized Nyquist criterion (GNC) following the process employed by Belkhayat in [20]

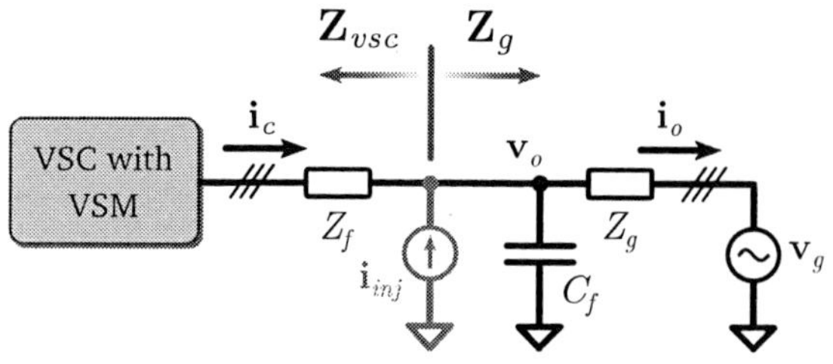

The 2018 International Power Electronics Conference

Fig. 9: Current injection point to measure the source and load impedances.

and by Burgos *et al.* in [21], which are based on the GNC theorem proposed by MacFarlane and Postlethwaite in [18] and [19]. With such method the stability of a closed-loop system can be determined by studying the location of the open-loop poles and encirclements of the $(-1+j0)$ point by the eigenvalues $\{\lambda_1, \lambda_2\}$ of the return-ratio matrix $\mathbf{L}_{dq}(s)$ (for the sake of simplicity the function of s is omitted in the notation, simplifying to $\mathbf{L}_{dq}$).

In order to obtain $\mathbf{L}_{dq}$, we need to extract the source ($\mathbf{Z}_{S_{dq}}$) and load impedances ($\mathbf{Z}_{L_{dq}}$) from the systems whose stability we want to analyze. In this context, we have assumed that the ideal voltage source, its series impedance, Z_g, and the capacitor C_f of the *LC* filter represent the source of the system and the VSC with its control and the output series Z_f impedance represent the load. This means that $\mathbf{Z}_{S_{dq}}$ corresponds to $\mathbf{Z}_g$ and $\mathbf{Z}_{L_{dq}}$ corresponds to $\mathbf{Z}_{vsc}$, as illustrated in Fig. 9.

We have derived the expressions of these impedances from the developed steady-state analytical models, splitting them into a source and load subsystem and calculating the transfer functions between the state variables and the inputs of the system.

Moreover, as we want to corroborate the correctness of these impedances prior to analyzing their stability, we have derived them not only by following the above-mentioned procedure, but also by measuring them in the simulations with the original non-linear systems. For the latter, we have employed a single-tone injection of current at the point where we split the source and the load (Fig. 9), following the procedure proposed by Rygg *et al.* in [25].

A. Output impedance comparative evaluation

Fig. 10 illustrates the frequency response of the impedances of the analyzed VSM techniques, as well as the equivalent grid impedance in black—which is equal in both cases.

The *dd* and *qq* terms of the CCVSM and VCVSM show an *RL* shape in most part of the frequency range, although the impedance of the VCVSM is more irregular than the CCVSM one. As we mentioned before, the VCVSM is more complex than the CCVSM, which has a direct impact on its impedance shape and stability characteristics. This can be clearly seen in the low frequency range (< 6 Hz), where the impedances of the VCVSM start showing a capacitive behaviour that could cause certain instabilities. This phenomenon might be influenced by

the use of a PLL for estimating the grid frequency, the voltage control loop or the differences in the reactive power controller.

We can say the impedances of the VSM techniques show a relatively clean shape in the entire frequency range. Moreover, the magnitude of their diagonal terms is higher than the grid impedance in most part of the frequency range, which is a desirable feature in terms of stability. One of the reasons for the higher magnitude impedances of these techniques is the virtual-impedances they integrate in the controllers. In the case of the CCVSM, for instance, these impedances could be adapted to emulate higher SM virtual windings in order to improve the stability characteristics of the overall system; however, this would also modify the dynamic response of the converter under disturbances, so we have to reach a compromise to ensure a correct operation.

In general, we can also see from Fig. 10 that at frequencies higher than 500 Hz the output filter of the converter is the dominating impedance, and the control techniques have less impact. In this range of frequencies the impedances of the converter interact with the grid side ones and their phase difference is greater than $100°$, which could lead to weak stability characteristics as will be shown in the next section.

B. Stability analysis based on the generalized Nyquist criterion

Based on the impedances obtained in the previous section, here we derive the return-ration matrix $\mathbf{L}_{dq}$ and the eigenvalues of each system in order to analyze their stability with the generalized Nyquist criterion.

Fig. 11 illustrates the Nyquist plots of the two analyzed VSM-controlled converters.

The results of the CCVSM technique (Fig. 11(a)) show very good characteristics in terms of stability compared to the VCVSM strategy. The eigenvalues do not encircle the $(-1+j0)$ point and the system has an infinite gain margin—as the eigenvalues do not cross the negative part of the real axis—and a phase margin of $45.24°$, which demonstrates that the system is very robust under parameter variations. In this case we can see that the frequency in which the eigenvalues cross the unity circle most near to the $(-1+j0)$ point is near 744 Hz. As previously mentioned this corresponds to the range close to the resonance of the filter capacitor and the grid side impedance.

On the other hand, the curves of the VCVSM (Fig. 11(b)) show that the system is very close to become unstable with a gain margin of 0.65 dB and a phase margin of $3.44°$, determined by the second eigenvalue. The frequency of the point in which the eigenvalues—in this case λ_2—cross the unity circle most near to the $(-1+j0)$ point is nearly 1230 Hz, which means that the converter is interacting with the grid in this range of frequencies. Looking back at Fig. 10, we can see that this frequency corresponds to the case where Z_{qq} crosses with the grid side impedance. As in this case the off-diagonal terms of $\mathbf{Z}_{vsc_{dq}}$ are very low compared to the diagonal terms, we can say that Z_{qq} is closely related to λ_2. At this crossing point we can observe that both impedances have a high phase

764

The 2018 International Power Electronics Conference

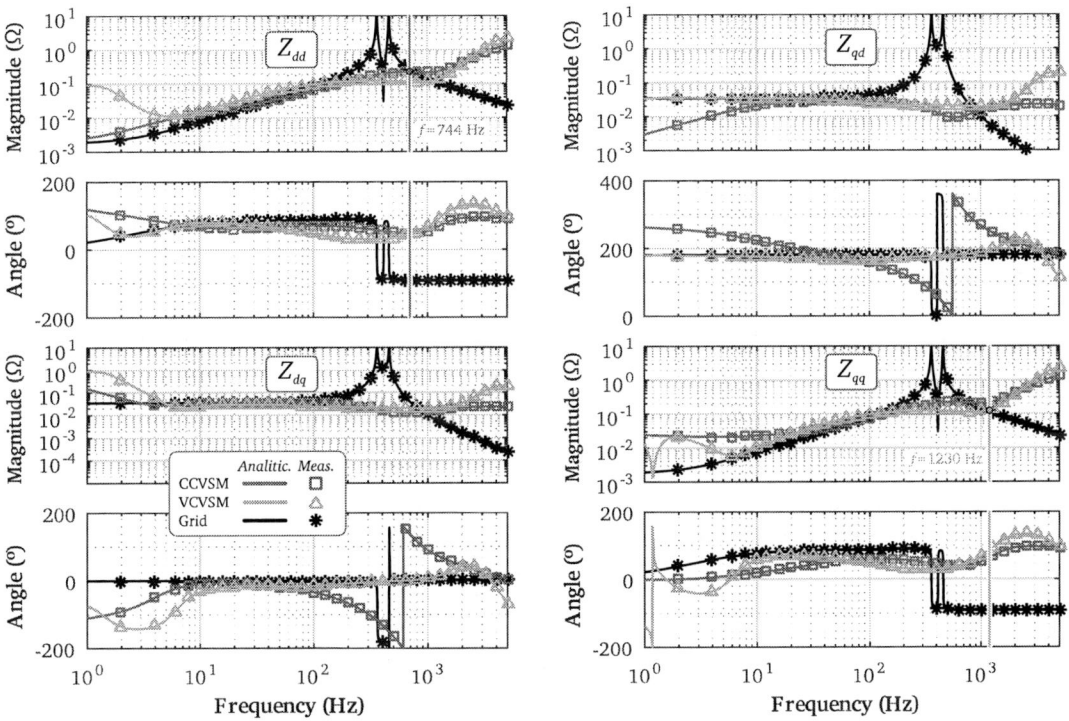

Fig. 10: *dq* domain output impedances of VSMs.

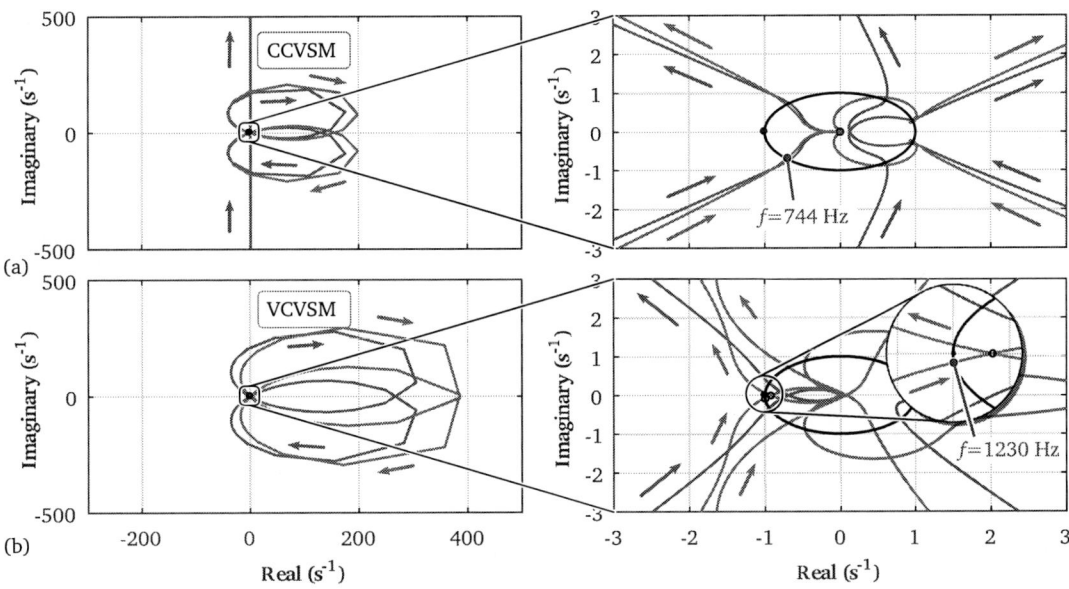

Fig. 11: Nyquist plots of VSM-controlled VSC eigenvalues: (a) Current-controlled VSM and (b) Voltage-controlled VSM.

difference, which is the source of these weak stability margins. Nevertheless, we must also take into account that even though in this case λ_1 shows better margins than λ_2, both eigenvalues have a similar shape and cross the unity circle at similar frequencies. This means that any of the two eigenvalues could

cause the system to become unstable for variations in system parameters.

Looking at these results, we can conclude that a method for improving the stability margins not only of the VCVSM but also of the CCVSM might be the integration of active

765

damping loops in the regulators of the converter to dampen the resonances of the output filter of the converter, as it is done in [12]. This would modify the shape of the impedances so that their interaction ranges are changed to reduce the phase differences and increase the stability margins.

VI. CONCLUSION

In this paper we have carried out a comparative evaluation of current-controlled and voltage-controlled virtual synchronous machines in terms of performance and impedance-based stability.

The results show that even though the CCVSM and VCVSM have a similar dynamic response under power reference and grid frequency perturbations, their inherent stability characteristics are quite different. In this sense, we have observed that the shape of the impedances of VCVSM techniques is more irregular than in the CCVSM, mainly in the low frequency range. This difference is mainly caused by how the frequency of the grid is estimated, because the bandwidth of the PLL integrated in the VCVSM technique makes the impedances capacitive at this frequency range, which could be the source of instabilities at certain conditions. In any case, the CCVSM as well as the VCVSM show an *RL* shape for most part of the analyzed frequency range, and we have seen that at frequencies over 500 Hz the control techniques have less impact and the passive devices of the system become the dominant elements. Moreover, from the obtained Nyquist curves we can conclude that the stability margins of both techniques are determined by the interactions of the impedances at this range of frequencies. In this context, the CCVSM shows very good inherent characteristics in terms of stability compared to the VCVSM. However, we also believe the integration of active damping loops could significantly improve the stability margins of the latter by adapting the shape of its impedances in the range where they interact with the side of the grid. We leave this analysis opened for further research activities.

In this paper we have also developed a methodology where we highlight the importance of taking into account the rotating reference frame to which the impedances are aligned. In order to compare impedances obtained at different points of the ac grid, we need to shift them to a common reference frame applying a simple rotation matrix because of the voltage drops across line impedances.

ACKNOWLEDGMENT

This work has been partially funded by a predoctoral grant of the Basque Government (PRE_2016_2_0241).

The authors would like to thank Dr. Jon Are Suul for his contributions and the valuable discussions about the topic.

REFERENCES

[1] B. Kroposki, B. Johnson, Y. Zhang, V. Gevorgian, P. Denholm, B.-M. Hodge, and B. Hannegan, "Achieving a 100% Renewable Grid: Operating Electric Power Systems with Extremely High Levels of Variable Renewable Energy," *IEEE Power Energy Mag.*, vol. 15, no. 2, pp. 61–73, mar 2017.

[2] E. Unamuno and J. A. Barrena, "Hybrid ac/dc microgrids—Part I: Review and classification of topologies," *Renew. Sustain. Energy Rev.*, vol. 52, pp. 1251–1259, dec 2015.

[3] ——, "Hybrid ac/dc microgrids—Part II: Review and classification of control strategies," *Renew. Sustain. Energy Rev.*, vol. 52, pp. 1123–1134, dec 2015.

[4] ——, "Hybrid AC/DC Microgrid Mode-Adaptive Controls," in *Dev. Integr. Microgrids*, W.-P. Cao and J. Yang, Eds. InTech, 2017, ch. 11, pp. 255–273.

[5] O. Mo, S. D'Arco, and J. A. Suul, "Evaluation of Virtual Synchronous Machines with Dynamic or Quasi-stationary Machine Models," *IEEE Trans. Ind. Electron.*, vol. 0046, no. c, pp. 1–1, 2016.

[6] Q.-C. Zhong, Z. Ma, W.-L. Ming, and G. C. Konstantopoulos, "Grid-friendly wind power systems based on the synchronverter technology," *Energy Convers. Manag.*, vol. 89, pp. 719–726, jan 2015.

[7] S. D'Arco and J. A. Suul, "Virtual synchronous machines— Classification of implementations and analysis of equivalence to droop controllers for microgrids," in *2013 IEEE Grenoble Conf.* IEEE, jun 2013, pp. 1–7.

[8] H. Bevrani, T. Ise, and Y. Miura, "Virtual synchronous generators: A survey and new perspectives," *Int. J. Electr. Power Energy Syst.*, vol. 54, pp. 244–254, jan 2014.

[9] J. A. Suul, S. DArco, and G. Guidi, "Virtual Synchronous Machine-Based Control of a Single-Phase Bi-Directional Battery Charger for Providing Vehicle-to-Grid Services," *IEEE Trans. Ind. Appl.*, vol. 52, no. 4, pp. 3234–3244, jul 2016.

[10] R. Hesse, D. Turschner, and H.-P. Beck, "Micro grid stabilization using the Virtual Synchronous Machine (VISMA)," *2009 Int. Conf. Renew. Energies Power Qual.*, 2009.

[11] P. F. Frack, P. E. Mercado, M. G. Molina, E. H. Watanabe, R. W. De Doncker, and H. Stagge, "Control Strategy for Frequency Control in Autonomous Microgrids," *IEEE J. Emerg. Sel. Top. Power Electron.*, vol. 3, no. 4, pp. 1046–1055, dec 2015.

[12] S. D'Arco, J. A. Suul, and O. B. Fosso, "A Virtual Synchronous Machine implementation for distributed control of power converters in SmartGrids," *Electr. Power Syst. Res.*, vol. 122, pp. 180–197, 2015.

[13] D. J. Hogan, F. Gonzalez-Espin, J. G. Hayes, G. Lightbody, L. Albiol-Tendillo, and R. Foley, "Virtual synchronous-machine control of voltage-source converters in a low-voltage microgrid," in *2016 18th Eur. Conf. Power Electron. Appl. (EPE'16 ECCE Eur.* IEEE, sep 2016, pp. 1–10.

[14] Y. Hirase, K. Abe, K. Sugimoto, and Y. Shindo, "A grid-connected inverter with virtual synchronous generator model of algebraic type," *Electr. Eng. Japan*, vol. 184, no. 4, pp. 10–21, sep 2013.

[15] Y. Chen, R. Hesse, D. Turschner, and H.-p. Beck, "Comparison of methods for implementing virtual synchronous machine on inverters," in *Int. Conf. Renew. Energies Power Qual.*, 2012, pp. 1–6.

[16] H.-P. Beck and R. Hesse, "Virtual synchronous machine," in *2007 9th Int. Conf. Electr. Power Qual. Util.* IEEE, oct 2007, pp. 1–6.

[17] S. D'Arco and J. A. Suul, "Equivalence of virtual synchronous machines and frequency-droops for converter-based Microgrids," *IEEE Trans. Smart Grid*, vol. 5, no. 1, pp. 394–395, 2014.

[18] A. G. MacFarlane and I. Postlethwaite, "The generalized Nyquist stability criterion and multivariable root loci," *Int. J. Control*, vol. 25, no. 1, pp. 81–127, jan 1977.

[19] A. G. MacFarlane, *Frequency-Response Methods in Control Systems.* IEEE Press, 1979, vol. 3, no. 1.

[20] M. Belkhayat, "Stability Criteria for AC Power Systems with Regulated Loads," Ph.D. dissertation, 1997.

[21] R. Burgos, D. Boroyevich, F. Wang, K. Karimi, and G. Francis, "On the Ac stability of high power factor three-phase rectifiers," in *2010 IEEE Energy Convers. Congr. Expo.* IEEE, sep 2010, pp. 2047–2054.

[22] M. Amin, M. Molinas, J. Lyu, and X. Cai, "Impact of Power Flow Direction on the Stability of VSC-HVDC Seen From the Impedance Nyquist Plot," *IEEE Trans. Power Electron.*, vol. 32, no. 10, pp. 8204–8217, oct 2017.

[23] A. Rygg, M. Molinas, E. Unamuno, C. Zhang, and X. Cai, "A simple method for shifting local dq impedance models to a global reference frame for stability analysis," pp. 1–5, jun 2017.

[24] J. L. Agorreta, M. Borrega, J. López, and L. Marroyo, "Modeling and Control of N-Paralleled Grid-Connected Inverters With LCL Filter Coupled Due to Grid Impedance in PV Plants," *IEEE Trans. Power Electron.*, vol. 26, no. 3, pp. 770–785, mar 2011.

[25] A. Rygg, M. Molinas, Chen Zhang, and Xu Cai, "Frequency-dependent source and load impedances in power systems based on power electronic

Stable Power Supply Method for Household Appliances via Virtual Synchronous Generator in Single-Phase Three-Wire Microgrid

Yuko Hirase, Hidehiko Nakagawa, Eiji Yoshimura,
Shogo Katsura, Kensho Abe, and Osamu Noro
Kawasaki Technology Co., Ltd.
1-1, Kawasaki-cho, Akashi 673-8666, Japan

Kazushige Sugimoto and Kenichi Sakimoto
Kawasaki Heavy Industries, Ltd.
1-1, Kawasaki-cho, Akashi 673-8666, Japan

Abstract—Distributed inverter-based power sources have become the main components of microgrids (MGs), and the virtual synchronous generator (VSG) concept is considered useful for improving the robustness of MGs. However, various problems such as the burden of unbalanced loads, parallel operation of multiple inverters, and harmonics of the system voltage received from household electric loads can arise when constructing an MG in an emergency using a domestic single-phase three-wire (1P3W) system. Therefore, we introduce a new VSG method for 1P3W systems to address these issues. This method is based on the previously proposed VSG methods for three-phase and single-phase two-wire systems. In this paper, we demonstrate that a small-capacity inverter with two VSGs can stably supply electrical power to 1P3W home appliances.

Keywords—Distributed Power Source, Harmonic Mitigation, Single-Phase Three-Wire, Virtual Synchronous Generator

I. INTRODUCTION

Integrating renewable energy sources (RESs) and distributed generators (DGs) into power systems has become an appealing countermeasure to overcome the challenges of potential power blackouts. Although distributed inverter-based power sources have become the main components of microgrids (MGs), dynamic stability, special protection schemes, and improved control strategies are important parameters for their implementation. In recent years, several studies have clarified that system stabilization becomes difficult because system inertia is reduced by the introduction of inverters equipped with RESs [1-5].

To solve this problem, numerous studies focused on the mechanism like virtual synchronous generator (VSG) control, which involves both frequency and voltage stabilizations [6-17]. In these studies, the dynamic behavior of a synchronous generator (SG) was simulated. Because the frequency/voltage control parameters can be tuned to enhance the dynamic stability of an MG, the VSG concept is considered to be useful for improving the robustness of MGs. Further, we previously proposed forerunner types of VSG control mechanisms that can be deployed in three-phase (3P) and single-phase two-wire (1P2W) systems, demonstrated their ease of realization and driving stability, and provided

mathematical analyses results [6-9, 13]. In this paper, we introduce a new VSG control method that can be applied to a single-phase three-wire (1P3W) system, and we demonstrate its stability in enabling continuous stress-free use of 1P3W household appliances even when they are disconnected from a commercial grid.

In countries with small land areas such as Japan, a delta-type distribution system is economically superior to a star distribution system. Therefore, in Japan, electricity delivered from a power plant by a delta-type distribution system is drawn into residences via pole transformers as a 1P3W system. In conventional 1P2W systems, the neutral line and voltage line of the star distribution system can be directly used for single-phase appliances. However, in the 1P3W system, by drawing two voltage lines (L1/L2) with a rated voltage of 100 V from both sides of the secondary (customer) side winding of the transformer, a power of 200 V (double the rated voltage) can be achieved. Simultaneously, each 100-V supply corresponding to N-L1/N-L2 can be obtained by drawing the neutral (N) line from the center of the windings. This approach is a feature of the 1P3W system wherein high-capacity equipment rated at 200 V can be used simultaneously with existing 100-V equipment. A 1P3W system can supply the same amount of power with half the current and one fourth the loss of a 1P2W system. Further, it is also advantageous that the same amount of power can be supplied with thinner wires, which helps save space.

In order to sufficiently exploit the abovementioned advantages of a 1P3W system, the loads to both voltage lines should be divided equally, as far as possible. Therefore, as described in the indoor wiring regulations specified by JEAC8001−a representative private standard in Japan−the equipment imbalance rate in 1P3W systems is limited to 40 %. However, unavoidable unbalanced load power needs to be supplied by the power supply; i.e., in order to construct an MG with an existing 1P3W residential system in an emergency (such as commercial power outage), the power supply in the MG has to also supply the unbalanced load power. Furthermore, even if unbalanced loads are arbitrarily connected to the L1 and L2 voltage lines, the frequency of each phase must be the same. This problem can be solved by utilizing the characteristics of a VSG (installed in each phase), which autonomously shares the load and cooperates with other SGs and VSGs to maintain a stable frequency. While many emergency 1P3W inverters use single-phase transformers to construct three-wire systems from two-wire systems, the VSG inverter proposed in this study realizes 1P3W MGs in a transformless manner; this is the novelty of our research.

The capacity of the VSG inverter as an emergency power supply is usually limited, because it is difficult to secure installation space within the typically small-sized residences in Japan. Given this circumstance, the system voltage output of the VSG is significantly affected by harmonics present in the driving current of household appliances, which form the main load of residences. Therefore, it is desirable that some type of harmonic mitigation be installed in the controller of the small power source. In our previously proposed 1P2W VSG control, we used a method called double decoupled synchronous reference frame (DDSRF) in order to apply the two-phase operation to the reference frame that is the same as in 3P VSG control, without any major changes [7]. By applying the DDSRF used for conversion from single phase to two phase, it is possible to decompose harmonics of arbitrary orders included in the original signal. We call this new harmonic decomposition method the multi decoupled synchronous reference frame (MDSRF). The harmonics on the voltage influenced by the load current is mitigated to a large degree, and stable power supply in an MG is enabled even with a small-capacity VSG inverter. In this regard, a method similar to DDSRF and MDSRF that decomposes the original signal into specific harmonic-order components has been introduced in [18, 19].

Very few studies have focused on the realization of an MG based on 1P3W inverters. Although single-phase MGs are the focus of certain studies [20-22], these are all 1P2W systems. Although 1P3W systems are not the most commonly used

commercial systems, overcoming their problems can lead to the development of efficient MGs for small areas.

The remainder of this paper is organized as follows. In Section II, we present the construction method of the VSG control in the 1P3W system. In Section III, the characteristics of a 1P3W system are explained based on the results of numerical simulations. The power supply to the unbalanced load and change in output according to power commands are evaluated under the scenarios of grid-connected operation, standalone operation, and parallel operation with multiple VSGs. In Section IV, we verify that the experimental test results agree with those of the simulations. In addition, the effect of the harmonic decomposition method is demonstrated via power supply tests to household appliances. Finally, we summarize our findings in Section V.

II. SINGLE-PHASE THREE-WIRED VSG CONTROL

A. System Configuration

Fig. 1 illustrates the system configuration of an inverter with 1P3W VSG control. The red boxes indicate voltage sensors and the blue circles indicate current sensors. The block arrows at these measurement points indicate positive signal flow directions. The measured signals are the AC voltages ($v_{\_1}$, $v_{\_2}$), AC currents ($i_{\_1}$, $i_{\_2}$), and the voltage of the

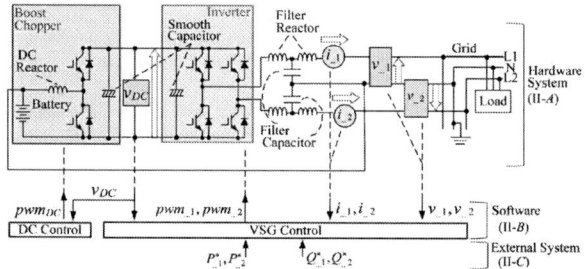

Fig. 1.　System configuration of a 1P3W inverter with virtual synchronous generator (VSG).

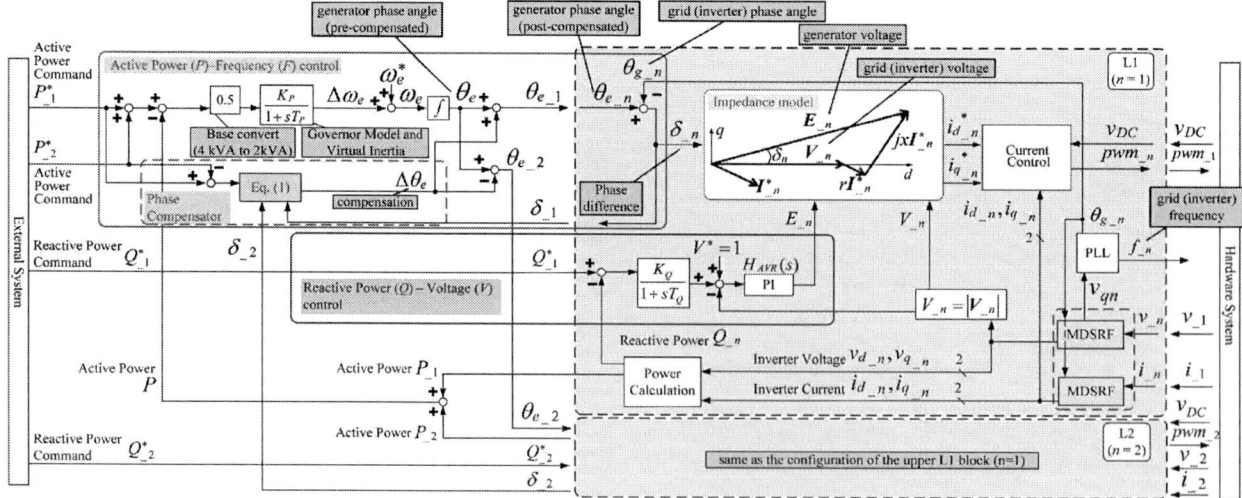

Fig. 2.　Virtual synchronous generator (VSG) control block diagram.

DC link (v_{DC}), and these values are sent to the control blocks (VSG control, DC control), which return the corresponding pulse-width modulator (PWM) command values (pwm_1, pwm_2, pwm_{DC}). These signals input to/output from the controllers are indicated by solid arrows. The VSG control receives power commands (P^*_1, P^*_2, Q^*_1, Q^*_2) from an external system.

The battery voltage is boosted up to the DC link voltage by the chopper. The N line is grounded as per legal requirements. As stated in the previous section, when loads connected to the L1 and L2 lines are unbalanced, the unbalanced current through the N line is compensated by the storage battery. In this study, we do not focus on the deterioration of the battery.

B. Design of Proposed 1P3W VSG Control

1) Overall Structure: Fig. 2 depicts the VSG control block diagram. The orange box regions indicate the control for each of the L1 and L2 phases. Hereafter, we call phases of the L1/L2 voltage lines as the L1/L2 phases, and subscript $\_n$ (n = 1 or 2) indicates the phase number. The input signals to each block are v_{DC}, v_n, and i_n, whereas the output signals from each block are pwm_n. The command values of the active and reactive powers (P^*_n, Q^*_n) are input from the external system. The pre-compensated generator phase angle (θ_e), post-compensated generator phase angles (θ_{e_n}), grid (inverter) phase angles (θ_{g_n}), and their difference angles (δ_n) are expressed in radians, and the other variables are expressed per unit (pu). The abbreviations PLL, PI, and AVR represent the phase-locked loop, proportional–integral, and automatic voltage regulator, respectively. The system frequencies (f_n) are generated in the PLL.

Grid (inverter) phase angles θ_{g_n} are also generated in the PLL, which works in combination with the MDSRFs (indicated by dashed purple line in Fig. 2.) The MDSRF is an advanced harmonic decomposition method based on the DDSRF. In this study, the MDSRF was adopted as the harmonic mitigation method.

Fig. 3 shows the structure of the DDSRF, which has been previously described in detail [7]. Here, $\boldsymbol{x}_{\alpha\beta}=[x_\alpha, x_\beta]^t$ denotes the two-phase vector of the original signal, wherein superscript t denotes the transpose. In the case of a 3P system, $\boldsymbol{x}_{\alpha\beta}$ denotes the Clarke transformed original signal,

and in the cases of 1P2W and 1P3W systems, x_α denotes the original signal and x_β is 0. The matrices $\boldsymbol{T}^{\pm1}$ denote the rotation matrices for rotation angles $\pm\theta$. Here, θ denotes the grid (inverter) phase angle. In the case of a 1P3W system (Fig. 2), $\theta=\theta_{g\_n}$. Further, $\boldsymbol{x}^\pm_{dq}$ denotes the signals on the reference frames, and the superscripts $+/-$ indicate if the signals are on the positive/negative synchronous reference frames, respectively. The cutoff frequency of the low-pass filter ($F(s)$) is empirically set to approximately 10 Hz, and its time constant coincides with that of the related integration. The signal input to the PLL is a q -axis signal on the positive synchronous reference frame (x^+_q). In this manner, the basic configuration of the DDSRF can separate the positive- and negative-phase components of the fundamental (primary) wave.

In the same manners, signals of orders including the fundamental (primary) wave can be separated by the MDSRF, as shown in Fig. 4. When $k=1$, the system function corresponds to the abovementioned DDSRF. When $k=0$, the MDSRF can be utilized to separate the DC component of the original signal, which cannot be eliminated before it is input to the CPU. When $k\geq2$, harmonics of arbitrary orders can be separated. In 1P2W and 1P3W systems, the order of harmonics is expressed as $k=4l+1$ (l denoting a positive integer.) In our system, the DC component ($k=0$) and the 3rd and 5th harmonics ($l=1$) are separated from the fundamental wave.

2) Active power (P)–Frequency (F) Control: The red solid line portion in Fig. 2 denotes the control section of the active power and frequency, wherein δ_n (n = 1 and 2) are generated. The active power (P) is related to the VSG torque and the angular velocity of the rotor (ω_e). The variation in ω_e ($\Delta\omega_e$) can be calculated using the governor model and virtual inertia. This calculation is expressed by the combination of the proportional control gain droop (K_P) and the first-order lag element of a time constant (T_P).

When replacing the 1P3W system with actual generators, it can be considered that two single-phase SGs are disposed in tandem around a single rotor. Based on this assumption, the load torque applied to the rotor is the sum of the active powers of the two phases. Therefore, in P–F control, the sum of the powers of the two phases is fed back. Because the frequencies of the L1 and L2 phases must be the same, ω_e is common in both phases. The generator phase angle θ_e denotes the

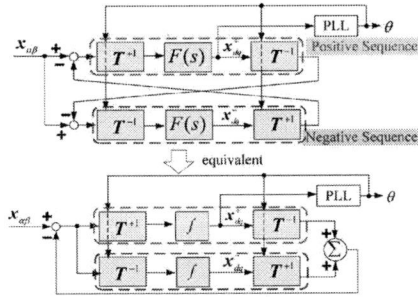

Fig. 3. Original structure of the double decoupled synchronous reference frame (DDSRF).

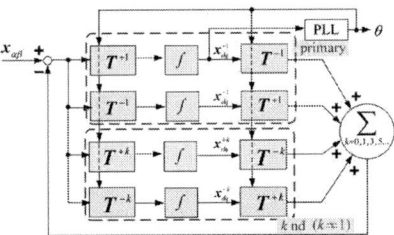

Fig. 4. Modified structure of the multi-decoupled synchronous reference frame (MDSRF).

rotation angle of the magnetic field. When equal active powers are output in both phases, θ_{e_n} in each phase is equal to θ_e. However, when a different active power is to be output for each phase, θ_{e_n} must be set accordingly. Otherwise, it becomes impossible to output unbalanced active powers in a grid-connected operation. Furthermore, when the generator phase angles of both phases are equal during the grid-disconnected operation, the inverter phase angle differs depending on the load in each phase. Therefore, a deviation of the line-to-line voltage between the L1 and L2 lines occurs, and synchronization to another grid becomes difficult. The red dashed-ine portion in Fig. 2 indicates one compensation method, which can solve the abovementioned issues. This method is based on the idea that the generator phase angles (θ_{e_n}) can be compensated by changing the relative location between the stator and rotor windings. Its compensation ($\Delta\theta_e$) is expressed as

$$\Delta\theta_e = \frac{(x\cos\delta_0 - r\sin\delta_0)^2}{2xV^2}(P^*_{\_1} - P^*_{\_2}) \qquad (1)$$

where $\delta_0 = (\delta_{\_1} + \delta_{\_2})/2$ and $V = V_{\_1} = V_{\_2}$, and x and r are the virtual impedances, which are described later in subsection (4). The derivation of Eq. (1) is presented in Appendix A.

3) Reactive power (Q)–Voltage (V) Control: The blue solid line portion in Fig. 2, which indicates the control portion of the reactive power and voltage, corresponds to the calculation of the induced electromotive force (EMF) ($\boldsymbol{E}_{\_n} = [e_{d\_n}, e_{q\_n}]^t$). The reactive power is related to the magnetic field of the VSG. The magnitude of $\boldsymbol{E}_{\_n}$ ($E_{\_n} = |\boldsymbol{E}_{\_n}|$) can be set using a proportional control gain droop (K_Q), the first-order lag element of a time constant (T_Q), and a PI compensator. Parameter K_Q denotes the droop constant of the reactive power and voltage, the first-order lag element indicates a measurement filter, and the PI compensator is a type of an AVR.

4) Impedance Model: The green solid line portion in Fig. 2 denotes the impedance model. The vector diagram shows the relationship among three vectors: $\boldsymbol{E}_{\_n}$, $\boldsymbol{V}_{\_n}$, and $\boldsymbol{I}^*_{\_n}$, where $\boldsymbol{V}_{\_n} = [v_{d\_n}, v_{q\_n}]^t$ denotes the grid (inverter) voltage vector, and $V_{\_n} = |\boldsymbol{V}_{\_n}|$ is its magnitude. Further, $\boldsymbol{I}^*_{\_n} = [i^*_{d\_n}, i^*_{q\_n}]^t$ denotes the inverter current command vector, and $i^*_{d\_n}$ and $i^*_{q\_n}$ are fed to the current control. The inverter current ($\boldsymbol{I}_{\_n} = [i_{d\_n}, i_{q\_n}]^t$) is regarded as the armature current. The vector diagram in the figure can be represented by

$$\begin{bmatrix} i^*_d \\ i^*_q \end{bmatrix} = \boldsymbol{Y}\left(\begin{bmatrix} e_d \\ e_q \end{bmatrix} - \begin{bmatrix} v_d \\ v_q \end{bmatrix}\right)$$

$$\boldsymbol{Y} = \frac{1}{r^2 + x^2}\begin{bmatrix} r & x \\ -x & r \end{bmatrix} = \begin{bmatrix} y_{dd} & y_{dq} \\ -y_{dq} & y_{dd} \end{bmatrix} . \qquad (2)$$

The $z = r + jx$ represents the virtual impedance (r : resistance, x : reactance), and r and x are set in the VSG control program. The VSG is an inverter that operates similar to a generator; it works on the assumption that virtual inertia and virtual impedance exist. In addition to the virtual impedance, the inverter has the actual impedance of the AC output filter. In the high-frequency range of the AC output filter, the actual impedance is dominant. Conversely, in the low-frequency range such as the fundamental frequency, the virtual impedance plays an important role for the transient behavior of the inverter [6].

When a sudden power outage occurs, the system frequency and voltage of the MG are maintained owing to the virtual inertia provided by the inverter. Detailed explanations and results of the demonstrations of VSG control are presented in [6, 7]. Precise analyses of active power (P)–frequency (F) control, reactive power (Q)–voltage (V) control, and their stabilities are provided in [8, 9]. The basic idea of VSG control in 3P and 1P2W systems described in these references also follows the 1P3W system.

C. Power Commands from External System

The power commands (P^*_n, Q^*_n) are input from the external system for various purposes: change in the power distribution with other power supplies, charge/discharge control of the storage battery, frequency/voltage control of the MG, and synchronization to other grids. All of these are

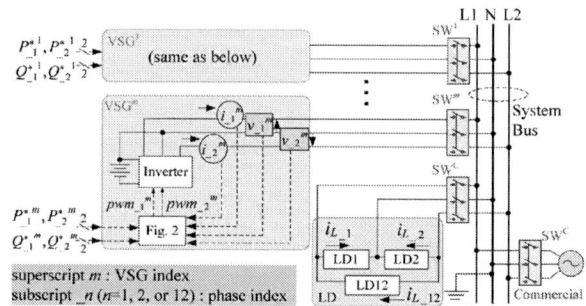

Fig. 5. Test configuration.

TABLE I
CONTROL PARAMETERS AND SYSTEM CONSTANTS OF VSGS

Droop Gain of P–F (K_P)	5	%
Time Constant of First-Order Lag Element (T_P)	0.12	s
Droop Gain of Q–V (K_Q)	20	%
Time Constant of Measurement Delay (T_Q)	0.005	s
Proportional Gain in PI compensator of AVR	2	(-)
Integral Time Constant of AVR	0.5	s
Virtual Resistance (r)	0.2	*pu*
Virtual Reactance (x)	0.4	*pu*
Rated Power VSG1/VSG2	4/6	kVA
Rated Current of Voltage Line VSG1/VSG2	20/30	A
AC Voltage of each phase	100	V
DC link Voltage	384	V
DC Reactor	2.9	mH
Smooth Capacitor in DC (inverter side)	1320	μF
Filter Reactor in AC (system side)	1.5	mH
Filter Reactor in AC (inverter side)	0.5	mH
Filter Capacitor in AC	3.9	μF

constructed as secondary controls, which require a longer processing time than the abovementioned transient (primary) control. The feature of the VSG is that transient stability can be achieved only by its primary control without any external control, and in this paper, we omit the detailed descriptions of these external systems.

III. CHARACTERIZATION BY NUMERICAL SIMULATION

In this section, we demonstrate the characteristics of the 1P3W VSG using numerical simulations. The test configuration is shown in Fig. 5. Superscript m indicates the index of a parallel-operated VSG. In this study, $m = 1, 2$. The orange and blue-filled portions indicate the VSG and loads (LDs), respectively. The LDs are connected/disconnected to the system bus via the switches indicated by the green-filled

boxes (SW^L). In the simulation, all LDs are considered as resistive loads. The VSGs are also connected/disconnected to the system bus via switches SW^m, and the test system can be connected to the commercial grid via switch SW^C. The control parameters and the system constants of the VSGs are listed in Table I. If the maximum current of VSG1/VSG2 exceeds 20/30 A owing to the input power commands or connected LDs, the VSGs stop at overload. If the current is within the above range, it is not necessary to equilibrate the loads to the L1/L2 phases.

Table II lists the test cases examined in the simulations. The case corresponding to Table II-1 demonstrates the grid-connected and standalone operations of VSG1, whereas the case corresponding to Table II-2 demonstrates the parallel operation of VSG1 and VSG2. Figs. 6 and 7 show the results

TABLE II

II-1 TEST CASE OF GRID-CONNECTED AND STANDALONE OPERATION OF VSG1

time	VSG1					VSG2					LD				Commercial
	SW^1	$P^{*\,1}_1$	$P^{*\,1}_2$	$Q^{*\,1}_1$	$Q^{*\,1}_2$	SW^2	$P^{*\,2}_1$	$P^{*\,2}_2$	$Q^{*\,2}_1$	$Q^{*\,2}_2$	SW^L	LD1	LD2	LD12	SW^C
(s)	close/open	(W)	(W)	(Var)	(Var)	close/open	(W)	(W)	(Var)	(Var)	close/open	(W)	(W)	(W)	close/open
0	close	0	0	0	0	open	-	-	-	-	close	200	700	600	close
5	close	500	1000	0	0	open	-	-	-	-	close	200	700	600	close
10	close	0	0	0	0	open	-	-	-	-	close	200	700	600	close
15	close	0	0	0	0	open	-	-	-	-	close	200	700	600	open
20	close	500	1000	0	0	open	-	-	-	-	close	200	700	600	open
25	close	0	0	0	0	open	-	-	-	-	close	200	700	600	open
30	close	0	0	0	0	open	-	-	-	-	open	-	-	-	open
35	close	0	0	0	0	open	-	-	-	-	close	200	700	600	open

II-2 TEST CASE OF PARALLEL OPERATION OF VSG1 AND VSG2

time	VSG1					VSG2					LD				Commercial
	SW^1	$P^{*\,1}_1$	$P^{*\,1}_2$	$Q^{*\,1}_1$	$Q^{*\,1}_2$	SW^2	$P^{*\,2}_1$	$P^{*\,2}_2$	$Q^{*\,2}_1$	$Q^{*\,2}_2$	SW^L	LD1	LD2	LD12	SW^C
(s)	close/open	(W)	(W)	(Var)	(Var)	close/open	(W)	(W)	(Var)	(Var)	close/open	(W)	(W)	(W)	close/open
0	close	0	0	0	0	close	0	0	0	0	close	200	700	600	open
5	close	500	1000	0	0	close	0	0	0	0	close	200	700	600	open
10	close	0	0	0	0	close	0	0	0	0	close	200	700	600	open
15	close	0	0	0	0	close	500	1000	0	0	close	200	700	600	open
20	close	0	0	0	0	close	0	0	0	0	close	200	700	600	open
25	close	0	0	0	0	close	0	0	0	0	open	-	-	-	open
30	close	0	0	0	0	close	0	0	0	0	close	200	700	600	open

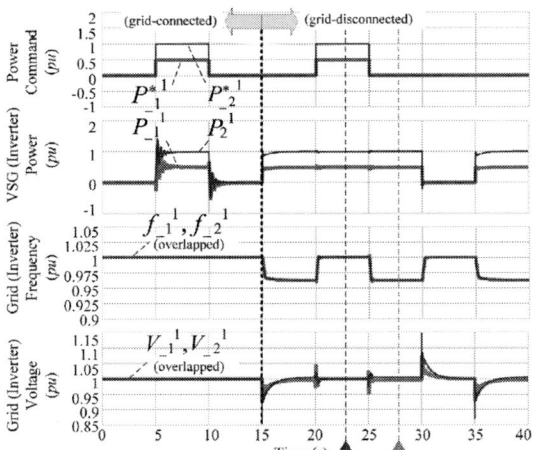

Fig. 6.   Grid-connected and standalone operation of VSG1 (Simulation test results).

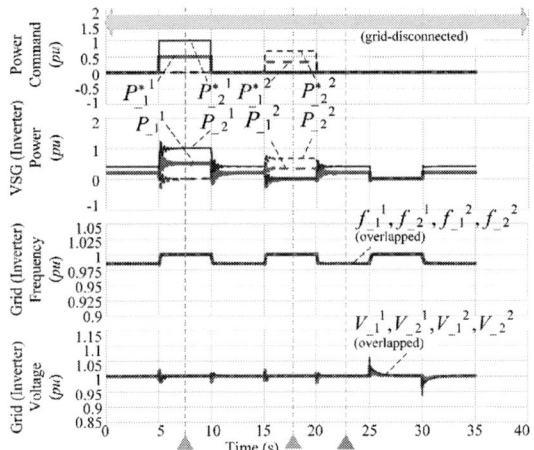

Fig. 7.   Parallel operation of VSG1 and VSG2 (Simulation test results).

of these simulations, respectively. In both figures, at the instant when the power command $P_{\_n}^m$ is input in a stepwise manner, the VSG powers fluctuate. The commercial power supply can be regarded as an SG with a droop constant of 0 and infinite capacity. That is, these fluctuations are caused by the resonance between multiple inertial generators (VSGs and an SG) [8]. When the power command from the external system is 0, the power distribution to the load is determined by the capacity of each generator, and the power distribution can be changed by the commands from the external system. In the case of the standalone operation, the system frequency can also be restored by the same power command.

Figs. 8 and 9 depict the vector diagrams corresponding to the impedance model of Fig. 2. The diagrams are shown at timings corresponding to the color-filled triangles in Figs. 6 and 7. The blue vector ($\hat{E}_{\_n}$) and light blue vector ($E_{\_n}$) represent the generator voltage vectors based on the pre-compensated generator phase angle (θ_e) and the post-compensated generator phase angles (θ_{e_n}), respectively. The coordinate axes are set such that the vector diagrams of the L1/L2 phases are symmetrical with respect to the origin. Therefore, the d/q axes of the L1 phases are rightward/upward, and the d/q axes of the L2 phases are leftward/downward; this coincides with the reference frame of $\hat{E}_{\_n}$. The red and green vectors denote $V_{\_n}$ and $I_{\_n}$, respectively. As described in Section II-B-2), using generator phase compensation, it is possible to output unbalanced power even if the inverter phases are equal to the L1/L2 phases. When the grid (inverter) voltage phases become equal by compensation, their voltage vectors $V_{\_n}$ (indicated by red arrows) lie in a nearly straight line as shown in Figs. 8 and 9. However, when the compensation is not used ($P_{\_n}^{*\,m}=0$), the phases of $V_{\_n}$ are not equal, and load powers are supplied using these different grid (inverter) phases. This mechanism is noteworthy in the 1P3W VSG.

IV. EXPERIMENTAL TEST RESULTS

The photograph of the VSG inverter is shown in Fig. 10. In the experimental tests, the circuit configuration was the same as that in Fig. 5, and the test scenarios are also the same as the simulation tests. The corresponding results are shown in Figs. 11 and 12, and they are in good agreement with the simulation results of Figs. 6 and 7, respectively. In the experimental tests (Fig. 11), the grid (inverter) voltage fluctuates when the command to the VSGs changes (5 s and 10 s.), whereas it is stable in the simulation tests. This is attributed to the fact that the commercial power source in the simulation is simulated with a constant voltage source, whereas, in the laboratory, the impedance of the distribution system is large.

Next, in order to confirm the effect of harmonic mitigation, we conducted a power supply test with a vacuum cleaner. Because household appliances such as a vacuum cleaner used in the experimental tests have a capacitor-input-type rectifier circuit at the output part, its current contains a large number of harmonic components. If the voltage of the VSG is distorted owing to this current, it may adversely affect other loads that are fed at the same time, or in the worst case, the other loads may undergo damage and stop working. Therefore, the aim of the harmonic mitigation method is not to distort the grid

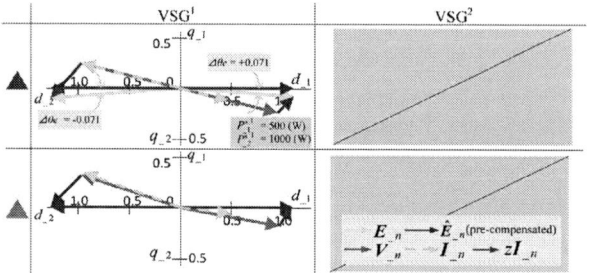

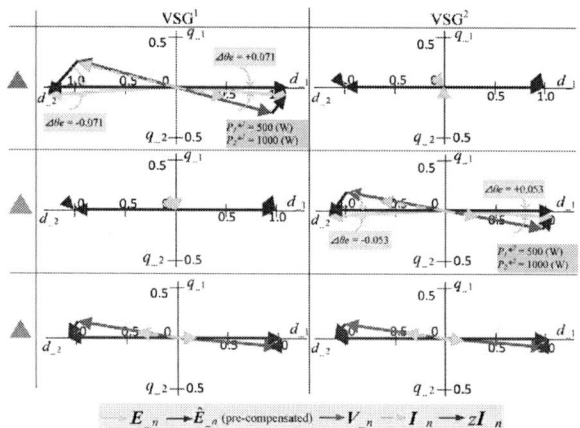

Fig. 8.   Vector diagram of virtual synchronous generator (VSG) in test case II-1.

Fig. 9.   Vector diagram of virtual synchronous generator (VSG) in test case II-2.

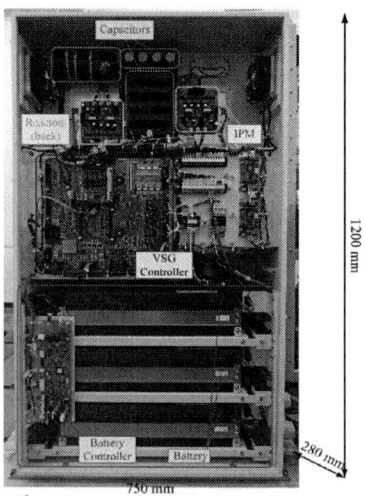

Fig. 10.   Photograph of VSG inverter.

772

The 2018 International Power Electronics Conference

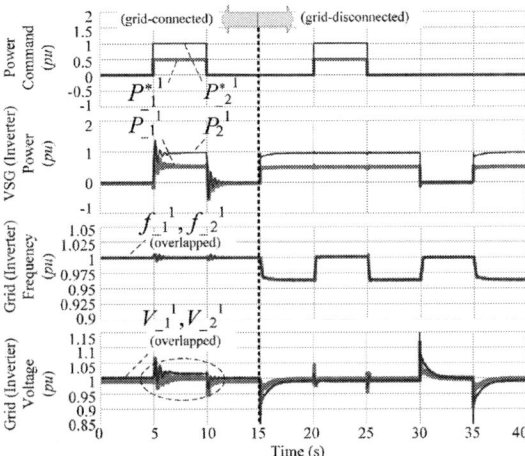

Fig. 11. Grid-connected and standalone operation of VSG1 (Experimental test results).

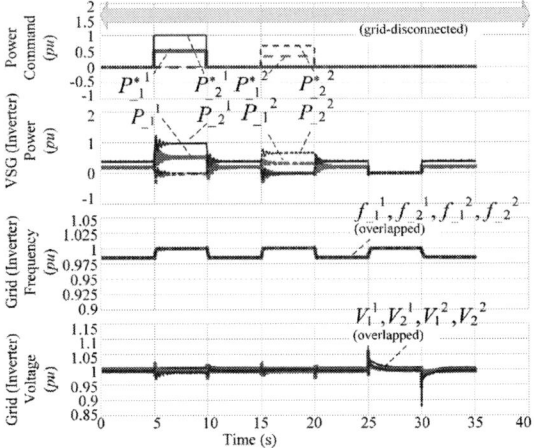

Fig. 12. Parallel operation of VSG1 and VSG2 (Experimental test results).

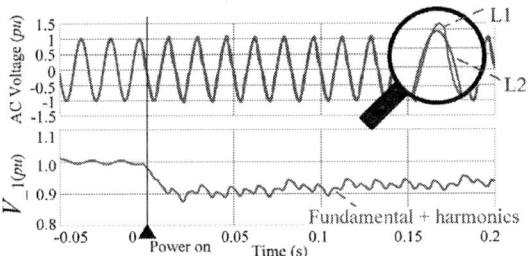

Fig. 13. Power supply to the vacuum cleaner (multi decoupled synchronous reference frame (MDSRF) is not adopted).

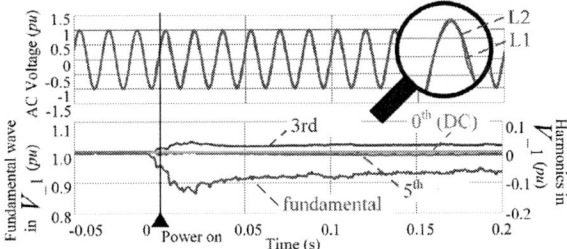

Fig. 14. Power supply to the vacuum cleaner (with adoption of multi decoupled synchronous reference frame (MDSRF)).

(inverter) voltage when power is supplied to the household appliances that generate harmonics (such as a vacuum cleaner).

Figs. 13 and 14 show the test results corresponding to the absence and presence of MDSRF, respectively. The nominal voltage of the vacuum cleaner was 100 V, and it was connected to L1-N. Here, the superscript indicating the VSG index is omitted. In each figure, the upper panel indicates the grid voltages measured at AC lines. We show both the L1 phase voltage (blue line) and the L2 phase voltage (red line) for comparison. The lower panel shows $V_{\_1} \cong v_{d\_1}$ of the L1 phase including the fundamental wave and extracted harmonics. The fundamental wave and the harmonics are represented by the left vertical axis and right vertical axis, respectively. At time instant zero, the vacuum cleaner was turned on.

As can be observed from Fig. 13, when there is no harmonic countermeasure, the AC voltage of the L1 line is distorted under the influence of harmonics included in the load current. However, in the case where the harmonic components are separated and removed by the MDSRF, the distortion of the grid voltage ($V_{\_1}$) is greatly reduced, as can be seen from Fig. 14. The lower panel in Fig. 14 indicates that the third-harmonic amplitude is large. After the period shown in Figs. 13 and 14, $V_{\_1}$ settled to approximately 0.95 pu, which is lower than the rated voltage of 1 pu. This value can be calculated from the reactive power load of the vacuum cleaner of about 0.25 pu and the Q–V droop gain (K_Q) of 20 %.

Because the capacity of the power supply constituting the MG in a small space is usually limited, the system voltage distortion caused by the harmonics included in the load current may become a serious problem. The application of MDSRF can be one effective countermeasure.

V. CONCLUSION

We proposed a new VSG method that is effective for constructing 1P3W MGs. The advantage of the 1P3W system is that both 100-V and 200-V loads can be supplied using the same distribution lines. By using VSGs in the 1P3W system, it is possible to construct an inexpensive and space-saving MG, which can make use of the abovementioned advantage. However, despite the fact that a balanced capacity load is not necessarily connected to the two voltage lines, the frequencies must be identical in both phases, which is difficult for the small distributed inverter with the conventional control method.

In this study, applying two VSG controls to both the L1 and L2 phases certainly proved that the abovementioned issues can be solved. Furthermore, by using the MDSRF even in the case where household appliances form the main loads within an MG, the inverter voltage can be output without being significantly affected by harmonics of the load current. That is, MDSRF can be an effective countermeasure.

Although there are very few countries that adopt 1P3W for commercial systems or use 100-V or 200-V household appliances simultaneously, the proposed concept can find applications in MGs such as those in aircraft and vessels connected with various voltage loads. Considering this potential application, our next challenge is to construct an MG with larger capacities of power supplies and loads, and to summarize the advantages and disadvantages of the VSG in 1P3W systems over 3P and 1P2W systems.

APPENDIX

A. Generator Phase Compensation

In the steady state, $\delta_{\_n}$ is set to satisfy Eq. (A1) according to the power command values. Here, the superscript indicating the VSG index is omitted, and $Q^*_{\_n} \cong 0$.

$$P^*_{\_n} = \frac{V}{r^2 + x^2}\left(r\left(E_{\_n}\cos\delta_{\_n} - V\right) + xE_{\_n}\sin\delta_{\_n}\right). \quad \text{(A1)}$$

Assuming that the system voltages between the L1 and L2 phases are balanced and $v_{d\_1} = v_{d\_2} = |V| = V$,

$$P^*_{\_n} = \frac{\tan\delta_{\_n}}{x - r\tan\delta_{\_n}}V^2 \quad \text{(A2)}$$

can be obtained from Eq. (A1). Let $\delta_1 = \delta_0 + \Delta\theta_e$ and $\delta_2 = \delta_0 - \Delta\theta_e$, where δ_0 is the average value of δ_1 and δ_2, and $\Delta\theta_e$ denotes the compensation of θ_{e_n}. Eq. (A2) is approximated by

$$P^*_1 = \left(\frac{\tan\delta_0}{x - r\tan\delta_0} + \frac{x}{\left(x\cos\delta_0 - r\sin\delta_0\right)^2}\Delta\theta_e\right)V^2$$
$$P^*_2 = \left(\frac{\tan\delta_0}{x - r\tan\delta_0} - \frac{x}{\left(x\cos\delta_0 - r\sin\delta_0\right)^2}\Delta\theta_e\right)V^2 \quad \text{(A3)}$$

From Eq. (A3), we can get

$$\Delta\theta_e = \frac{\left(x\cos\delta_0 - r\sin\delta_0\right)^2}{2xV^2}\left(P^*_1 - P^*_2\right). \quad \text{(A4)}$$

REFERENCES

[1] J. O'Sullivan, A. Rogers, D. Flynn, P. Smith, A. Mullane, and M. O. Malley, "Studying the maximum instantaneous non-synchronous generation in an island system—frequency stability challenges in Ireland," *IEEE Transactions on Power Systems*, vol. 29, no. 6, pp. 2943-2951, 2014.

[2] X. Tang, W. Deng, and Z. Qi, "Investigation of the dynamic stability of microgrid," *IEEE Transactions on Power Systems*, vol. 29, no. 2, pp. 698-706, 2014.

[3] F. Gonzalez-Longatt, A. Bonfiglio, R. Procopio, and D. Bogdanov, "Practical limit of synthetic inertia in full converter wind turbine generators: Simulation approach," in *Proceedings of 2016 19th*

International Symposium on Electrical Apparatus and Technologies (SIELA), May 2016, Bourgas, Bulgaria, pp. 1-5.

[4] P. Tielens and D. Van Hertem, "The relevance of inertia in power systems," *Renewable and Sustainable Energy Reviews*, vol. 55, pp. 999-1009, 2016.

[5] M. Dreidy, H. Mokhlis, and S. Mekhilef, "Inertia response and frequency control techniques for renewable energy sources: A review," *Renewable and Sustainable Energy Reviews*, vol. 69, pp. 144-155, 2017.

[6] Y. Hirase, K. Abe, K. Sugimoto, and Y. Shindo, "A grid-connected inverter with virtual synchronous generator model of algebraic type," *Electrical Engineering in Japan*, vol. 184, no. 4, pp. 10-21, 2013.

[7] Y. Hirase, O. Noro, E. Yoshimura, H. Nakagawa, K. Sakimoto, and Y. Shindo, "Virtual synchronous generator control with double decoupled synchronous reference frame for single-phase inverter," *IEEJ Journal of Industry Applications*, vol. 4, no. 3, pp. 143-151, 2015.

[8] Y. Hirase, K. Sugimoto, K. Sakimoto, and T. Ise, "Analysis of resonance in microgrids and effects of system frequency stabilization using a virtual synchronous generator," *IEEE Journal of Emerging and Selected Topics in Power Electronics*, vol. 4, no. 4, pp. 1287-1298, 2016.

[9] Y. Hirase, K. Abe, K. Sugimoto, K. Sakimoto, H. Bevrani, and T. Ise, "A novel control approach for virtual synchronous generators to suppress frequency and voltage fluctuations in microgrids," *Applied Energy*, vol. 210, pp. 699-710, 2018.

[10] H. Bevrani, T. Ise, and Y. Miura, "Virtual synchronous generators: A survey and new perspectives," *International Journal of Electrical Power & Energy Systems*, vol. 54, pp. 244-254, 2014.

[11] S. D'Arco, J. A. Suul, and O. B. Fosso, "A virtual synchronous machine implementation for distributed control of power converters in SmartGrids," *Electric Power Systems Research*, vol. 122, pp. 180-197, 2015.

[12] L.-Y. Lu and C.-C. Chu, "Consensus-based secondary frequency and voltage droop control of virtual synchronous generators for isolated AC micro-grids," *IEEE Journal on Emerging and Selected Topics in Circuits and Systems*, vol. 5, no. 3, pp. 443-455, 2015.

[13] K. Sakimoto, K. Sugimoto, Y. Shindo, and T. Ise, "Virtual synchronous generator without phase locked loop based on current controlled inverter and its parameter design," *IEEJ Transactions on Power and Energy*, vol. 135, no. 7, pp. 462-471, 2015.

[14] J. Liu, Y. Miura, H. Bevrani, and T. Ise, "Enhanced virtual synchronous generator control for parallel inverters in microgrids," *IEEE Transactions on Smart Grid*, vol. 8, no. 5, pp. 2268-2277, 2016.

[15] J. Liu, Y. Miura, and T. Ise, "Comparison of dynamic characteristics between virtual synchronous generator and droop control in inverter-based distributed generators," *IEEE Transactions on Power Electronics*, vol. 31, no. 5, pp. 3600-3611, 2016.

[16] R. Shi, X. Zhang, L. Fang, H. Xu, C. Hu, Y. Yu, and H. Ni, "Research on power compensation strategy for diesel generator system based on virtual synchronous generator," in *Proceedings of 2016 IEEE 8th International Power Electronics and Motion Control Conference (IPEMC-ECCE Asia)*, May 2016, Hefei, China, pp. 939-943.

[17] Q. C. Zhong, G. C. Konstantopoulos, B. Ren, and M. Krstic, "Improved synchronverters with bounded frequency and voltage for smart grid integration," *IEEE Transactions on Smart Grid*, (in press).

[18] S. Golestan, M. Monfared, and F. D. Freijedo, "Design-oriented study of advanced synchronous reference frame phase-locked loops," *IEEE Transactions on Power Electronics*, vol. 28, no. 2, pp. 765-778, 2013.

[19] J. Liu, Y. Miura, and T. Ise, "Model-predictive-control-based distributed control scheme for bus voltage unbalance and harmonics compensation in microgrids," in *Proceedings of 2017 IEEE Energy Conversion Congress and Exposition (ECCE)*, Oct. 2017, Cincinnatti, OH, USA, pp. 4439-4446.

[20] S. Li, X. Fu, M. Ramezani, Y. Sun, and H. Won, "A novel direct-current vector control technique for single-phase inverter with L, LC and LCL filters," *Electric Power Systems Research*, vol. 125, pp. 235-244, 2015.

[21] F. Shahnia and R. P. Chandrasena, "A three-phase community microgrid comprised of single-phase energy resources with an uneven scattering amongst phases," *International Journal of Electrical Power & Energy Systems*, vol. 84, pp. 267-283, 2017.

[22] V. Vásquez, L. M. Ortega, D. Romero, R. Ortega, O. Carranza, and J. J. Rodríguez, "Comparison of methods for controllers design of single phase inverter operating in island mode in a microgrid: Review," *Renewable and Sustainable Energy Reviews*, vol. 76, pp. 256-267, 2017.

The 2018 International Power Electronics Conference

A Novel Oscillation Damping Method of Virtual Synchronous Generator Control Without PLL Using Pole Placement

Jia Liu[1*], Yushi Miura[1] and Toshifumi Ise[1]

1 Division of Electrical, Electronic and Information Engineering, Osaka University, Suita, Japan
*E-mail: liu@eei.eng.osaka-u.ac.jp

Abstract— Recent years, in order to avoid probable inertia decrease of power system due to increasing inverter-interfaced distributed generators, virtual synchronous generator (VSG) control, an inverter control scheme providing virtual inertia, has been proposed. However, the resonant mode of basic VSG control results in transient oscillation damping. In this study, a state-space model of a synchronous generator connected to infinite bus is established to study the mechanism of this oscillation and to understand the damping effect of previous VSG control methods. Based on the state-space model, a novel oscillation damping method of VSG using pole placement is proposed. With a very simple state feedback control, the proposed method provides proper damping without any frequency measurement whereas keeps the inertial feature of VSG. Simulation results obtained with PSCAD/ EMTDC and experimental results obtained in a laboratorial microgrid testbed indicate that the proposed method outperforms previous VSG control methods with excellent transient performance.

Keywords— *Distributed generator, virtual synchronous generator, oscillation damping, state feedback control.*

I. INTRODUCTION

Over past two decades, stimulated by environmental policies, the penetration rate of renewable energy sources (RESs) in the power system increases rapidly all over the world. Different to the traditional centralized power plants where large synchronous generators (SG) are installed as the conversion interface between prime mover and the grid, most of these RESs are connected the grid in a distributed manner, and through inverters with much smaller power rating. Due to the absence of rotating mass in the inverters, these inverter-interfaced distributed generators (DGs) do not possess any intrinsic inertia. Therefore, they cannot provide frequency support to the power system like conventional SGs.

To stabilize the frequency of the future power system, which is expected to be highly penetrated by inverter-interfaced DGs, several control methods of inverters mimicking the inertial feature of SG are proposed [1]–[6]. These inverter control methods can be identified as virtual synchronous generator (VSG) control, as the

swing equation of the SGs are emulated in the control scheme [7]. It is demonstrated that the VSG-control-based DGs can contribute to the system inertia and thus restrain frequency fluctuation [7], [8]. Besides, as most VSG control methods are developed from droop control concept, the advantages of droop control, such as power sharing between multiple DGs, and smooth transition between grid-connected and islanded modes, are also inherited by VSG control [9].

However, the emulation of swing equation introduces a resonant mode, which causes low-frequency oscillation in the output active power and frequency of the inverter. A straightforward solution to this problem is to emulate the damping effect of SG, whereas the latter is realized by damper windings producing damping torque whenever the slip between electrical frequency and rotor frequency is nonzero [10]. However, a phase-locked loop (PLL) of grid frequency is needed to provide the information of electrical frequency. As PLL is nonlinear and sensitive to voltage unbalance and harmonics, its unexpected influence on VSG control is a matter of concern. To omit this PLL, the electrical frequency is set to nominal frequency in [3]–[6]; however, the damping ratio cannot be tuned independently due to this simplification. In [11], it is shown that this oscillation can be damped by alternating the moment of inertia of VSG; however, temporarily reduced inertia also leads to reduced contribution of frequency support. In [7], [12], it is proposed that the damping effect can be produced by the differential term of output active power; however, differential terms are sensitive to noises and ripples. In [9], it is demonstrated that increasing output reactance by virtual impedance control can increase the damping ratio of the VSG; however, the damping effect is insufficient and the transient response of VSG becomes slower.

In this paper, to understand the resonant mode and damping effect of SG and previous VSG control methods, a state-space model of a SG connected to infinite bus is established. Based on this model, a PLL-free oscillation damping method of VSG using pole placement is proposed. This method is verified by simulation executed in PSCAD/EMTDC and experiment carried out in a laboratorial microgrid testbed. It is shown that the proposed method outperforms previous VSG control methods with excellent transient performance without any frequency measurement.

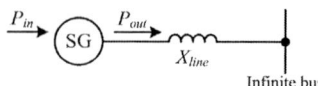

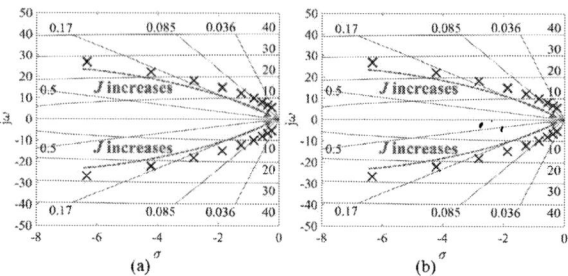

Fig. 1. Model of a SG connected to infinite bus.

II. STATE-SPACE MODEL OF SYNCHRONOUS GENERATOR

Fig. 1 illustrates a SG connected to infinite bus. The swing equation of this SG can be expressed as follows [13].

$$P_{in} - P_{out} = J\omega_m \frac{d\omega_m}{dt} + D(\omega_m - \omega_g) \tag{1}$$

where P_{in} is the shaft power, P_{out} is the output power, J is the moment of inertia, D is the damping factor, ω_m is the rotor frequency, and ω_g is the electrical frequency of infinite bus.

It is noteworthy the damping power $D(\omega_m - \omega_g)$ is produced by the damper windings of synchronous generators. As the principle of damper windings is similar to that of induction machine, the damping power is proportional to the slip frequency as it is shown in (1). In several previous researches on VSG, the damping power is formulated in the form of friction. However, this may be debatable as the damping of rotor motion by mechanical losses is small and negligible compared to the damping from damper windings [13].

Equation (2) shows the transient synchronizing power coefficient, which forces the SG to move to its stable equilibrium point in power-angle diagram [13].

$$\frac{\partial P_{out}}{\partial \delta'} = K \approx \frac{E'V_{bus}\cos\delta'_0}{X} \tag{2}$$

where δ' is the transient power angle and δ'_0 is its operating point, K is the transient synchronizing power coefficient, E' is the transient electromotive force, V_{bus} is the voltage of infinite bus, and $X = X'_d + X_{line}$ is the output reactance (X'_d is the d-axis transient reactance).

To simplify the calculation, when calculate the transient synchronizing power coefficient K, it can be assumed that $E' \approx V_{bus} \approx V_{base}$, $\delta'_0 \approx \sin^{-1}\frac{XS_{base}}{V_{base}^2}$, where V_{base} is the rated voltage and S_{base} is the rated power.

The relation between the power angle and frequencies can be expressed as

$$\int \delta' dt = \int \delta dt = \omega_m - \omega_g \tag{3}$$

where δ is the power angle.

The shaft power of SG is usually regulated by a governor as shown in (4).

$$P_{in} = P_0 - k_p(\omega_m - \omega_0) \tag{4}$$

where P_0 is the set value of active power, k_p is the droop coefficient, ω_0 is the nominal frequency. In a real synchronous generator, there is some mechanical delay in the governor system. However, it is demonstrated in previous researches that not mimicking this delay can improved the transient response of VSG control [7], [8]. As a result, this delay effect is omitted in (4).

Fig. 2. Eigenvalue loci of SG model when (a) J varies and (b) D varies.

TABLE I
PARAMETERS OF SYNCHRONOUS GENERATOR MODEL

Parameter	Value	Parameter	Value
V_{base}	200 V	M^*	8 s
S_{base}	5 kVA	D^*	0 pu
ω_0	376.99 rad/s	k_p^*	20 pu
		X^*	0.3 pu

Equations (1)–(4) can be linearized in small signal [7], and arranged in state-space form shown in (5).

$$\begin{cases} \dot{x} = Ax + Ew \\ y = x \end{cases} \tag{5}$$

where output vector y, state vector x, disturbance vector w, state-transition matrix A, disturbance-input matrix E are as follows.

$$y = x = [\Delta\omega_m \quad \Delta P_{out}]^T \tag{6}$$

$$w = [\Delta\omega_g \quad \Delta P_0]^T \tag{7}$$

$$A = \begin{bmatrix} -\dfrac{k_p + D}{J\omega_0} & -\dfrac{1}{J\omega_0} \\ K & 0 \end{bmatrix} \tag{8}$$

$$E = \begin{bmatrix} \dfrac{D}{J\omega_0} & \dfrac{1}{J\omega_0} \\ -K & 0 \end{bmatrix} \tag{9}$$

The eigenvalues of state-transition matrix A can be deduced as

$$\lambda = \sqrt{\frac{K}{J\omega_0}} \, e^{j(\pi \pm \cos^{-1}\frac{k_p + D}{2\sqrt{KJ\omega_0}})} \tag{10}$$

Therefore, the undamped natural frequency ω_n and damping ratio ξ can be derived as

$$\omega_n = \sqrt{\frac{K}{J\omega_0}} \tag{11}$$

$$\xi = \frac{k_p + D}{2\sqrt{KJ\omega_0}} \tag{12}$$

Similar equation to (11) and (12) is reported in [10]; however, the damping ratio is deduced as $\xi = \frac{D}{2\sqrt{KJ\omega_0}}$, because the damping effect of the governor is not considered in [10]. Therefore, when the "linearizing technique" method of [10] is applied in the following part of this paper, design of damping factor D is modified with (12). Besides, the non-decoupling effect of reactive power in the case of resistive line impedance is considered in [10]. In this paper, since the output impedance is controlled to be inductive by adding virtual stator inductance as discussed in Section III, this non-decoupling effect can be neglected.

From (11) and (12), it can be noticed that larger moment of inertia J results in smaller undamped natural frequency, which is not influenced by the damping factor D. Moreover, larger J results in smaller ξ, which makes the system more oscillatory; however, ξ can be increased by increasing D. These relations can also be confirmed from the eigenvalue loci plot shown in Fig. 2. The parameters used to plot Fig. 2 are shown in Table I. These parameters are shown in per unit value defined as follows.

$$M^* = \frac{J\omega_0^2}{S_{base}} \tag{13}$$

$$D^* = \frac{D\omega_0}{S_{base}} \tag{14}$$

$$k_p^* = \frac{k_p\omega_0}{S_{base}} \tag{15}$$

$$X^* = \frac{XS_{base}}{V_{base}^2} \tag{16}$$

where M^* is the inertia constant and subscript $*$ indicates per unit value. It is noteworthy that per unit values shown in this paper are based on respective power rating S_{base} of each DG.

From Fig. 2, it can be concluded that a well-damped VSG controller can be obtained by proper design of D as it is proposed in [10]. However, to emulate (1), a PLL to measure bus frequency ω_g is needed [2], [10]. As PLL is known for its nonlinearity and sensitivity to voltage unbalance and harmonics, it is preferred to not use PLL in a VSG controller. Besides, as the bus voltage ω_g may be difficult to be measured from inverter due to geographic restrain, ω_g is often measured from the output voltage of inverter. This approximation may also influence the transient performance of the VSG.

An alternative solution is to use nominal frequency ω_0 instead of ω_g [3]–[6]. In this case, (1) becomes

$$P_{in} - P_{out} = J\omega_m \frac{d\omega_m}{dt} + D(\omega_m - \omega_0) \tag{17}$$

By combining (17) with (4), (18) can be obtained.

$$P_{in} - P_{out} = J\omega_m \frac{d\omega_m}{dt} + (k_p + D)(\omega_m - \omega_0) \tag{18}$$

It can be noticed from (18) that D becomes identical to droop coefficient k_p. In fact, D is considered as droop coefficient and dedicated governor control is omitted in [3]–[6]. In this case, if D is changed in order to obtain desired damping ratio ξ, the operating point defined by droop relation will change consequently. Therefore, D is

determined by droop relation which is usually predefined by power rating and frequency tolerance, and thus cannot be tuned independently.

Contrarily, if (1) is used, as $\omega_m = \omega_g$ at steady state, D will not influence the operating point of droop relation. Therefore, to avoid ambiguity, the term $D(\omega_m - \omega_0)$ is defined as droop term and the term $D(\omega_m - \omega_g)$ is defined as dedicated damping term in this paper. Consequently, VSG control using (17) without other dedicated damping part can be classified as no-dedicated-damping VSG control.

In conclusion, well-damped VSG control without PLL is expected. However, as mentioned in the introduction, the other existing damping methods proposed in [7], [9], [11], [12] still have respective defects, and thus no ideal solution is available yet in the literature.

III. PROPOSED VSG CONTROL WITH POLE PLACEMENT

To realize ideal damping effect in VSG control without PLL, a control scheme using pole placement shown in Fig. 3 is proposed in this paper.

This control scheme is developed from [9]; therefore, most blocks are the same as those in [9] except the block "Governor + Damping" shown in Fig. 4. Besides, as the damping effect is produced by the damping power shown in Fig. 4, the damping term in (1) is neglected and (19) is emulated in the block "Swing Equation Function" instead. Therefore, the PLL to measure ω_g used in [9] is omitted.

$$P_{in} - P_{out} = J\omega_m \frac{d\omega_m}{dt} \tag{19}$$

In the block "Stator Impedance Adjuster" shown in Fig. 5, virtual impedance control is applied to adjust the total output reactance X of the inverter. As a result, from (2), transient synchronizing power coefficient K can be

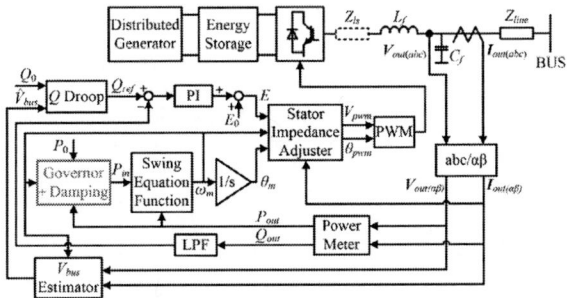

Fig. 3. Block diagram of the proposed VSG control.

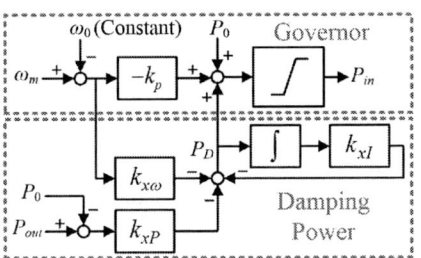

Fig. 4. Detail of the block "Governor + Damping".

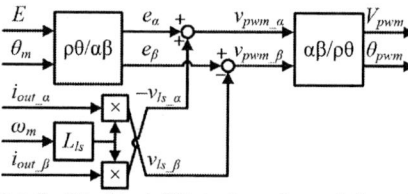

Fig. 5. Detail of the block "Stator Impedance Adjuster".

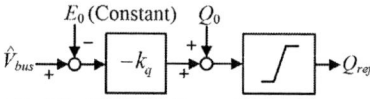

Fig. 6. Detail of the block "Q Droop".

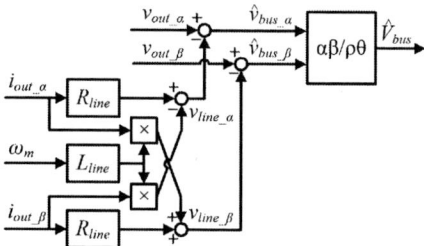

Fig. 7. Detail of the block "V_{bus} Estimator".

adjusted to better share transient active power between inverter. It is proposed in [9] to set X^* to 0.7 pu in order to increase the damping effect; however, large X^* slow down the transient dynamics of the inverter. Therefore, the recommended design of X^* is proposed to be 0.3 pu in this paper, because the damping effect provided by the proposed pole placement method is effective enough. As the output impedance should be controlled inductive to avoid non-decoupling of active and reactive power, it is not recommended to further reduce the value of X^*.

In the block "Q Droop" shown in Fig. 6, the droop relation between reactive power and voltage is applied. In the block "V_{bus} Estimator" shown in Fig. 7, the bus voltage is estimated from the measurement of output voltage and current, in order to provide a common reference to the Q–V droop control. The design details of these blocks are shown in [9] thus omitted in this paper.

In the upper half of the block diagram of "Governor + Damping" shown in Fig. 4, governor control similar to [9] is applied using the droop relation between frequency and active power. The proposed damping control using pole placement method is shown in the lower half of Fig. 4.

From Fig. 4, the governor control shown in (4) is replaced by

$$P_{in} = P_0 + P_D - k_p(\omega_m - \omega_0) \tag{20}$$

and the damping power P_D is obtained by

$$P_D = -k_{x\omega}(\omega_m - \omega_0) - k_{xP}(P_{out} - P_0) - k_{xI}\int P_D \tag{21}$$

Considering the damping power P_D, the state-space model show in (5) should be rewritten as

$$\begin{cases} \dot{\hat{x}} = \widehat{A}\hat{x} + \widehat{B}u + \widehat{E}w \\ \hat{y} = \hat{x} \end{cases} \tag{22}$$

where

$$\hat{y} = \hat{x} = \begin{bmatrix} x \\ \int \Delta P_D \end{bmatrix} \tag{23}$$

$$u = [\Delta P_D] \tag{24}$$

$$\widehat{A} = \begin{bmatrix} A & 0 \\ 0 & 0 \end{bmatrix} \tag{25}$$

$$\widehat{B} = \begin{bmatrix} \dfrac{1}{J\omega_0} & 0 & 1 \end{bmatrix}^T \tag{26}$$

$$\widehat{E} = \begin{bmatrix} E \\ 0 \end{bmatrix} \tag{27}$$

and the control input u is obtained through the state feedback control expressed in (28).

$$u = -K_x\hat{x} + k_{xP}\Delta P_0 \tag{28}$$

where feedback gain matrix $K_x = [k_{x\omega} \quad k_{xP} \quad k_{xI}]$.

The close-loop state-space model considering the state feedback control shown in (28) can be derived as

$$\begin{cases} \dot{\hat{x}} = (\widehat{A} - \widehat{B}K_x)\hat{x} + \widetilde{E}w \\ \hat{y} = \hat{x} \end{cases} \tag{29}$$

where

$$\widetilde{E} = \begin{bmatrix} \dfrac{D}{J\omega_0} & \dfrac{1 + k_{xP}}{J\omega_0} \\ -K & 0 \\ 0 & k_{xP} \end{bmatrix} \tag{30}$$

As known from pole placement method, if (22) is controllable, there must exist a K_x for any desired set of close-loop poles [14]. As shown in Fig. 8, in order to obtain a well-damped system, the damping ratio ξ of the desired close-loop poles is set to 0.9 and the undamped natural frequency ω_n is kept unchanged. Compared to Fig. 2, it can be noticed that the movement of poles is similar to that of increasing $D(\omega_m - \omega_g)$. This indicates that the oscillation is damped whereas the inertial feature of VSG is kept.

As an integral term of P_D appears in the state variables, P_D is forced to 0 at steady state, and thus the operating point of droop relation is not affected by P_D. However,

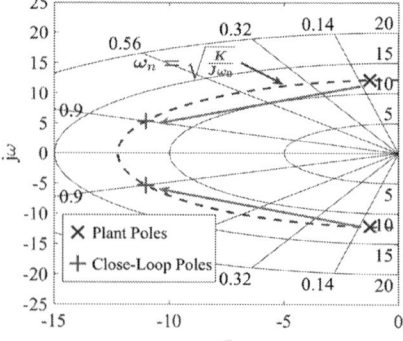

Fig. 8. Design of close-loop poles based on pole placement method (3rd pole not shown).

the integral term of P_D introduced a third pole to the system. This pole should be placed far away from the conjugate poles such as shown in (31), in order to make it negligible [15].

$$\lambda_3 = -10\xi\omega_n \qquad (31)$$

Generally, (22) is controllable. Therefore, the feedback gain matrix $\boldsymbol{K_x}$ can be computed from $\widehat{\boldsymbol{A}}, \widehat{\boldsymbol{B}}$ and the desired close-loop poles using Ackermann's formula.

As shown in Fig. 4, the inputs to calculate the proposed damping power are the virtual rotor frequency ω_m and the output active power P_{out}. Both are already available in the basic VSG control. Therefore, no additional measurement is required, and PLL is omitted in the proposed VSG control scheme.

IV. SIMULATION RESULTS

To verify the proposed VSG control and compare it to previous methods, simulations using the circuit shown

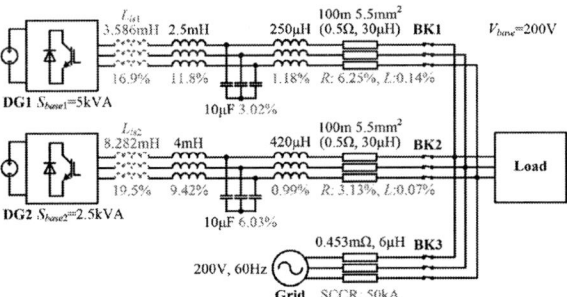

Fig. 9. Simulation circuit.

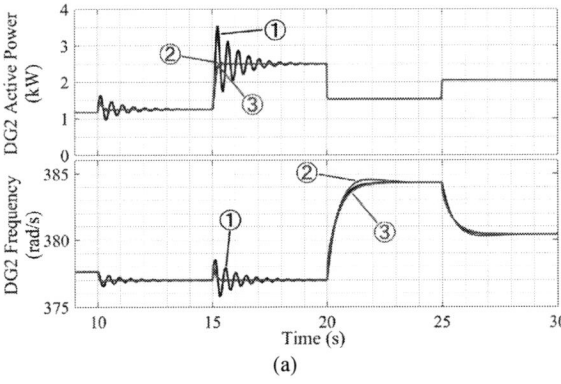

(a)

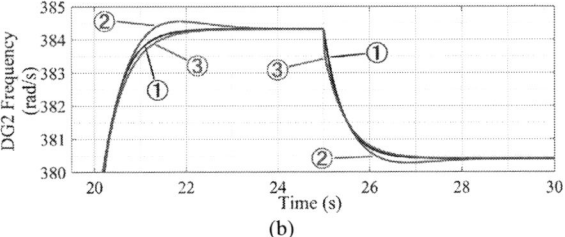

(b)

Fig. 10. (a) Simulation results of active power and frequency in the case of a single DG; (b) zoom of (a). ① no-dedicated-damping method; ② the "linearizing technique" method; ③ the proposed pole placement method.

in Fig. 9 are executed in PSCAD/ EMTDC. In the studied microgrid, two parallel inverters are connected to the bus of microgrid, to which the loads are connected. The bus is connected to the maingrid, which is modelled as infinite bus with an R-L type impedance equivalent to 50 kA short circuit current rating (SCCR). The main control parameters of the two DGs are the same as those shown in Table I, except that the power rating of DG2 is 2.5 kVA. Besides, droop coefficient of $Q\text{--}V$ droop k_q^* is 10 pu, and the parameters of reactive power PI controller are $K_{pq}^* = 0.005$ pu, $T_{iq} = 1.25 \times 10^{-4}$ s.

By using Ackermann's formula, the feedback gain of the proposed damping control are $k_{x\omega}^* = 156.27$ pu, $k_{xP}^* = 14.362$ pu, $k_{xI}^* = 110.17$ pu.

The simulation using the "linearizing technique" method [10] and the method without dedicated damping [3]–[6] are also executed to compare with the proposed method. In these methods, the proposed damping power P_D shown in Fig. 4 is set to 0. In the "Swing Equation Function" block, (1) is used for the "linearizing technique" method and (17) is used for the no-dedicated-damping method. The inertia moment J of all cases are set to the same value. The damping factor D of the "linearizing technique" method is set to 156.27 pu according to (12), and the damping factor D of the no-dedicated-damping method is equivalent to the droop coefficient k_p. Besides, the governor is omitted in the no-dedicated-damping method as it is discussed in Section II.

Normally, load frequency control (LFC) is applied for VSGs in islanded mode to restore the frequency deviation; however, in order to better understand the dynamic performance of the basic VSG control, it is not applied in this paper thus the VSGs operate in free-governor mode.

A. Single DG

Fig. 10 shows the simulation results where only DG2 is activated (BK1 off). The proposed VSG control is compared to the method using "linearizing technique" method [10] and the method without dedicated damping [3]–[6]. Initially, DG2 is connected to the maingrid (BK3 on), the grid frequency (the frequency of infinite bus) is 60. 1 Hz, the set value of active power P_0^* is 0.5 pu, and the load is 1.5 kW. At 10 s, grid frequency falls to 60 Hz to simulate a disturbance in the grid, and at 15 s, P_0^* is changed to 1.0 pu. Large oscillation in both transients can be found in the case of no-dedicated-damping method, whereas no oscillation occurs in the cases of the "linearizing technique" method and the proposed method. At 20 s, DG2 is disconnected from the grid (BK3 off), and the load is increased to 2 kW at 25 s. No oscillation occurs in all cases; however, frequency overshoots can be noticed in the case of the "linearizing technique" method. This unexpected overshoots may result from the nonlinearity of the PLL, which is only used in this method. Conversely, the proposed method shows excellent transient in both grid-connected and islanded modes.

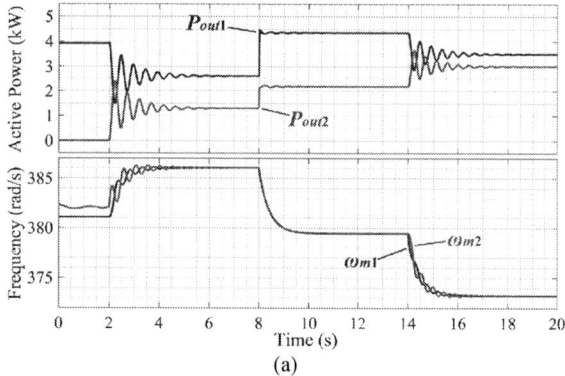

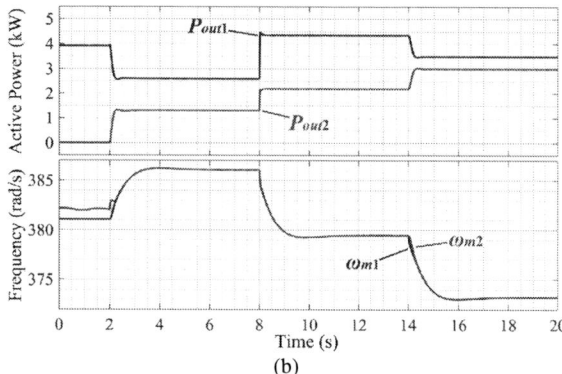

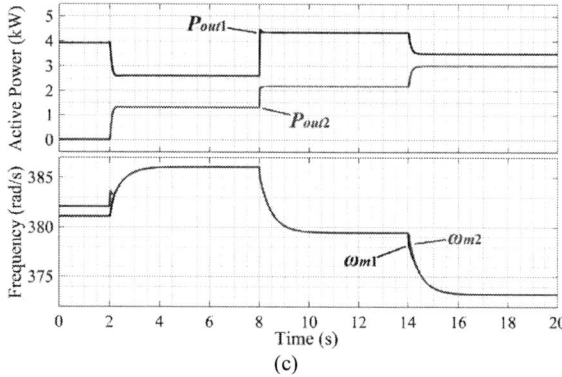

Fig. 11. Simulation results of active power and frequency in the case of two parallel DGs where DGs are controlled by (a) no-dedicated-damping method, (b) the "linearizing technique" method and (c) the proposed pole placement method.

B. Parallel DGs

Fig. 11 demonstrates the simulation results where two parallel DGs are connected. Initially, the load is 3.9 kW, 1.2 kvar, P_0^* of both DGs are set to 1.0 pu, and both grid and DG2 are not connected (BK2 and BK3 off). At around 2 s, DG2 is synchronized (BK2 on). Large oscillation can be noticed in the case of no-dedicated-damping method, whereas no oscillation is observed in the cases of the proposed method and the "linearizing technique" method. At 8 s, additional load of 2.6 kW, 0.4 kvar is connected. No oscillation occurs in all cases, because the output impedance of each DG is adjusted to

be the same in per unit value by the block "Stator Impedance Adjuster" shown in Fig. 5. According to [9], the oscillatory poles is cancelled by zeros during a loading transition as long as swing equation parameters and output impedance are the same in per unit value. At 14 s, P_0^* of DG 1 is changed to 0.5 pu to intentionally modify the power sharing ratio. Again, oscillation is observed in the case of no-dedicated-damping method, whereas well damped in the cases of the proposed method and the "linearizing technique" method. As the disturbance is no longer caused by loading transition, the abovementioned cancellation of oscillatory poles does not happen, thus the damping effect is the only solution to damp this kind of oscillation. Transient performance of intentional power sharing change is also studied in [9]; however, the proposed method in [9], i.e. increasing damping effect by increasing output impedance, is less effective than the proposed method as shown in respective simulation results. Moreover, although both the "linearizing technique" method and the proposed method show excellent performance regarding active power results, the frequency overshoots issue of the "linearizing technique" method mentioned in the single DG case occurs again in the parallel DGs case, whereas this kind of overshoots do not occur in the proposed method.

V. EXPERIMENTAL RESULTS

The simulation results of the parallel DGs case are further verified by experimental results in a laboratorial microgrid testbed shown in Fig. 12. The circuit of this testbed is the same as that of simulation shown in Fig. 9, and the control parameters and the experiment scenario are the same as those of simulation.

By comparing the experimental results shown in Fig. 13 to the simulation results shown in Fig. 11, it can be concluded that all comments on simulation results still stand for the experimental results, as they are almost the same. The only difference is the transient after synchronization of DG2 at 2 s, where active power overshoots in experimental results are much larger than

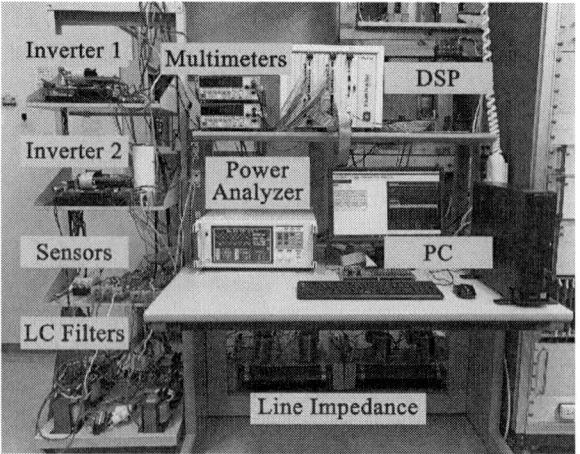

Fig. 12. Laboratorial microgrid testbed.

The 2018 International Power Electronics Conference

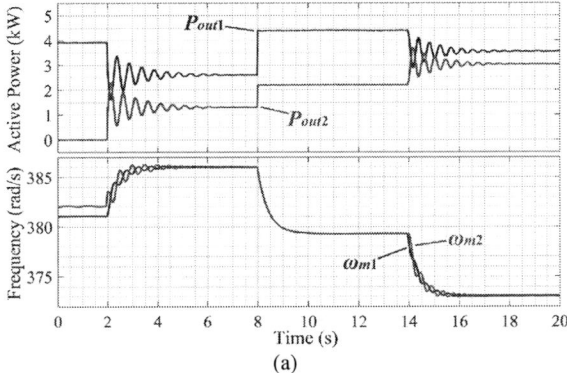

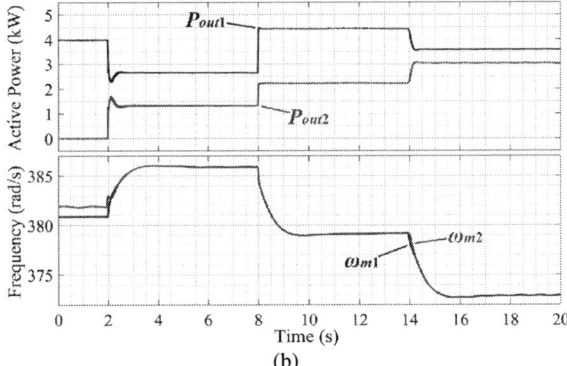

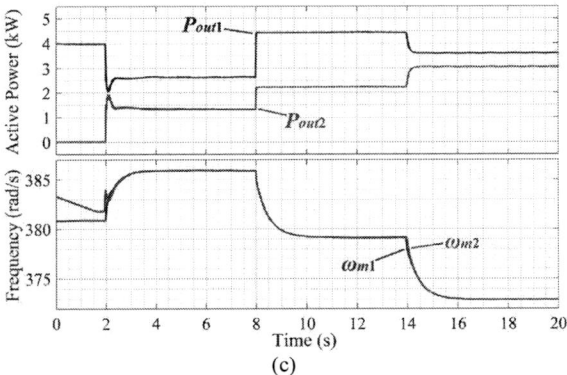

Fig. 13. Experimental results of active power and frequency in the case of two parallel DGs where DGs are controlled by (a) no-dedicated-damping method, (b) the "linearizing technique" method and (c) the proposed pole placement method.

those in simulation. Nevertheless, these overshoots are well damped by the proposed method and the "linearizing technique" method.

VI. CONCLUSIONS

In this paper, a state-space model of SG is presented to study the low-frequency oscillation of SG and VSG control. Based on the eigenvalue analyses of this model, a well-damped VSG control using pole placement is proposed. With a very simple state feedback control, the proposed method provides proper damping without PLL whereas keeps the inertial feature of VSG. Simulation

results and experimental results demonstrate that the proposed method outperforms previous methods with excellent transient performance of both active power and frequency.

ACKNOWLEDGMENT

This work was supported by JSPS KAKENHI Grant Number JP17K14643.

REFERENCES

[1] J. Driesen and K. Visscher "Virtual synchronous generators," in *Proc. IEEE Power Energy Soc. Gen. Meeting—Convers. Del. Elect. Energy 21st Century*, 2008, pp. 1–3.

[2] K. Sakimoto, Y. Miura, and T. Ise, "Stabilization of a power system including inverter type distributed generators by the virtual synchronous generator," *Electrical Engineering in Japan*, vol. 187, no. 3, pp. 7–17, May 2014 [*IEEJ Trans. Power and Energy*, vol. 132, no. 4, pp. 341–349, Apr. 2012].

[3] Q.-C. Zhong and G. Weiss, "Synchronverters: inverters that mimic synchronous generators," *IEEE Trans. Ind. Electron.*, vol. 58, no. 4, pp. 1259–1267, Apr. 2011.

[4] Q.-C. Zhong, P.-L. Nguyen, Z. Ma, and W. Sheng, "Self-synchronized synchronverters: inverters without a dedicated synchronization unit," *IEEE Trans. Power Electron.*, vol. 29, no. 2, pp. 617–630, Feb. 2014.

[5] M. Guan, W. Pan, J. Zhang, Q. Hao, J. Cheng, and X. Zheng, "Synchronous generator emulation control strategy for voltage source converter (VSC) stations," *IEEE Trans. Power Syst.*, vol. 30, no. 6, pp. 3093–3101, Nov. 2015.

[6] H. Wu, X. Ruan, D. Yang, X. Chen, W. Zhao, Z. Lv, and Q.-C. Zhong, "Small-signal modeling and parameters design for virtual synchronous generators," *IEEE Trans. Ind. Electron.*, vol. 63, no. 7, pp. 4292–4303, Jul. 2016.

[7] J. Liu, Y. Miura, and T. Ise, "Comparison of dynamic characteristics between virtual synchronous generator and droop control in inverter-based distributed generators," *IEEE Trans. Power Electron.*, vol. 31, no. 5, pp. 3600–3611, May 2016.

[8] Y. Hirase, K. Sugimoto, K. Sakimoto and T. Ise, "Analysis of resonance in microgrids and effects of system frequency stabilization using a virtual synchronous generator", *IEEE J. Emerg. Sel. Topics Power Electron.*, vol. 4, no. 4, pp. 1287–1298, Dec. 2016.

[9] J. Liu, Y. Miura, H. Bevrani, and T. Ise, "Enhanced virtual synchronous generator control for parallel inverters in microgrids," *IEEE Trans. Smart Grid*, vol. 8, no. 5, pp. 2268–2277, Sep. 2017.

[10] T. Shintai, Y. Miura, and T. Ise, "Oscillation damping of a distributed generator using a virtual synchronous generator," *IEEE Trans. Power Del.*, vol. 29, no. 2, pp. 668–676, Apr. 2014.

[11] J. Alipoor, Y. Miura, and T. Ise, "Power system stabilization using virtual synchronous generator with alternating moment of inertia," *IEEE J. Emerg. Sel. Topics Power Electron.*, vol. 3, no. 2, pp. 451–458, Jun. 2015.

[12] S. Dong and Y. C. Chen, "Adjusting synchronverter dynamic response speed via damping correction loop," *IEEE Trans. Energy Convers.*, vol. 32, no. 2, pp. 608–619, Jun. 2017.

[13] J. Machowski, J. Bialek, and J. R. Bumby, *Power System Dynamics and Stability*. New York, NJ, USA: John Wiley & Sons, 1997, pp. 141–182.

[14] K. Ogata, *Modern Control Engineering, 5th ed.* Upper Saddle River, NJ, USA: Prentice Hall, 2010, pp. 723–735.

[15] R. C. Dorf and R. H. Bishop, *Modern Control Systems, 12th ed.* Upper Saddle River, NJ, USA: Prentice Hall, 2011, pp. 314–317.

The 2018 International Power Electronics Conference

Operation of a Modular Multilevel Converter Controlled as a Virtual Synchronous Machine

Salvatore D'Arco[1]*, Giuseppe Guidi[1], and Jon Are Suul[1,2]
[1]SINTEF Energy Research, Trondheim, Norway
[2]Department of Engineering Cybernetics, Norwegian University of Science and Technology, Trondheim, Norway
E-mail: salvatore.darco@sintef.no, Giuseppe.Guidi@sintef.no, Jon.A.Suul@sintef.no

Abstract-This paper presents an implementation of a Virtual Synchronous Machine (VSM) based on a Modular Multilevel Converter (MMC). The control system relies on the internal simulation of an electromechanical swing equation, which emulates the inertia and damping effect of a synchronous machine and provides the frequency and phase angle needed for controlling the MMC to operate as a VSM. The ac-side control is based on cascaded voltage and current controllers in a synchronous reference frame defined by the virtual swing equation. The reference signals for the ac-side voltage control are generated from a reactive power control loop and a quasi-stationary virtual impedance which emulates the equivalent voltage drop over the stator windings of a synchronous machine. The control system also includes internal control loops for regulating the double frequency circulating currents of the MMC. The presented VSM implementation is tested in a laboratory environment using an MMC with 18 half-bridge sub-modules per arm. Experimental results demonstrate the operation of the VSM-controlled MMC when connected to an external ac grid and in islanded mode feeding a resistive load.

Keywords— HVDC Transmission, Modular Multilevel Converter, Virtual Inertia, Virtual Synchronous Machine

I. INTRODUCTION

The concept of Virtual Synchronous Machines (VSM), as first introduced in [1], allows for controlling power electronic converters so that they emulate the general characteristics of synchronous generators. This approach for control system design has been motivated by the expected decreasing equivalent inertia of power systems resulting from large-scale integration of distributed generation systems with power electronic grid interfaces and the corresponding decommissioning of traditional thermal power plants with synchronous generators [2]. Thus, the main purpose of VSM based control strategies for power electronic converters is to emulate the inertia and the damping of synchronous machines (SMs) [3].

Emulation of SM characteristics can be achieved by introducing an internal simulation of a virtual swing equation as part of the control system of a power electronic converter [1], [3]-[6]. In this case, the speed and phase angle of the virtual swing equation should be explicitly

represented as state variables in the simulation, and the control system should rely on a power-balance-based synchronization mechanism in the same way as a SM. Such approaches for SM emulation will ensure that the control system will be suitable for grid connected operation as well as islanding of systems dominated by power electronic converters [1], [3], [4] [7]. Alternatively, the power provided by a converter with a conventional control system could be controlled to emulate the inertial characteristics of a traditional generator [2], [8], [9]. However, this approach must rely on measurement of the grid frequency and its derivative, usually by a Phase Locked Loop, which implies that the control strategy is not suitable for islanded operation [3].

Since power system operation with very low equivalent physical inertia is already becoming a reality in microgrids or small islanded power systems, such applications have been studied already with the first concepts for VSM-based control that was introduced in the literature [1], [10], [11]. However, the expected future challenges of operating large-scale interconnected power systems with low inertia are already emerging [12], [13]. In this context, HVDC converter stations might be especially relevant for implementation of control strategies for providing virtual inertia, due to the high rating and correspondingly high impact on the characteristics of a power system. Thus, multiple studies of VSM-based control strategies applied to HVDC transmission systems have recently been presented [8], [14]-[19]. These previous publications have mainly considered the operation of traditional 2-level (2L) Voltage Source Converters (VSCs), even if the Modular Multilevel Converter introduced in [20] is becoming the preferred topology for VSC HVDC transmission schemes. The first examples of studying the utilization of MMC-based HVDC terminals for providing virtual inertia and frequency support have recently been published in [21], [22], based on numerical simulations.

In this paper, an implementation of the VSM concept on a MMC is presented and demonstrated with a laboratory-scale setup. The experiments are based on an MMC prototype with 18 half-bridge sub-modules per arm. The VSM-based control system is relying on a virtual swing equation, a quasi-stationary virtual impedance and

---

This work was supported by the project "HVDC Inertia Provision" (HVDC Pro), financed by the ENERGIX program of RCN with project number 268053/E20 and the industry partners; Statnett, Statoil, RTE and ELIA.

The 2018 International Power Electronics Conference

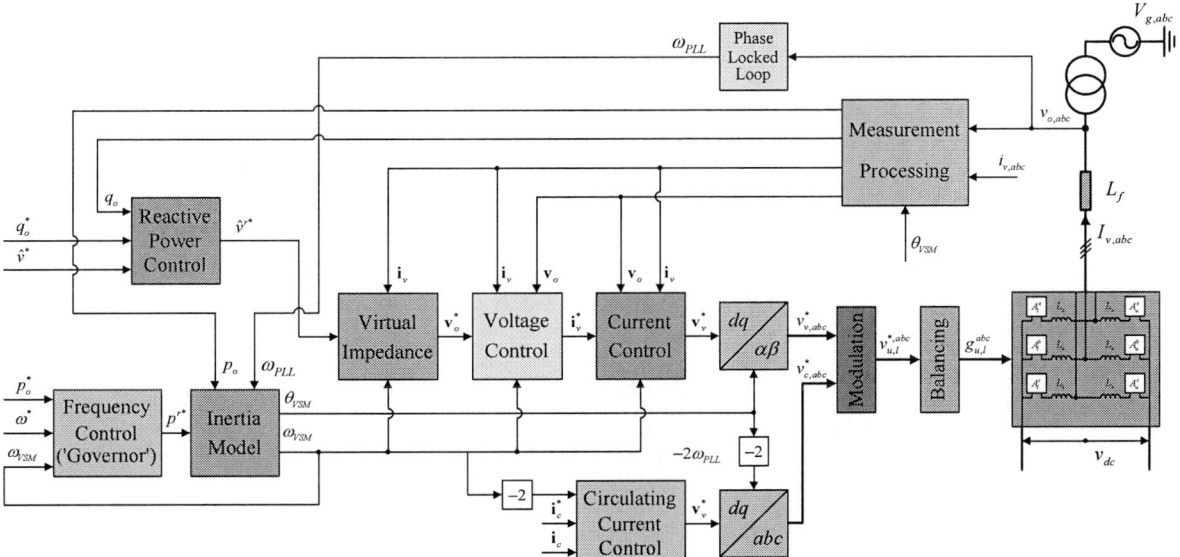

Fig. 1. Overview of VSM-based control system for a Modular Multilevel Converter

cascaded voltage and current controllers. Thus, the control of the ac-side of the MMC is derived from the VSM implementations presented in [7], [23]. The functionality of this VSM-based control scheme is integrated with the modulation strategy and the control of the circulating currents of the MMC. Experimental results demonstrate the intended operation and the expected dynamic performance in response to several external perturbations, both when connected to an external grid and when operated in islanded mode on a resistive load. Seamless transition between grid-connected and islanded mode following a sudden disconnection of the main grid is also demonstrated.

II. CONTROL STRATEGY FOR AN MMC CONTROLLED AS A VIRTUAL SYNCHRONOUS MACHINE

The control system implemented for the VSM-based operation of an MMC is outlined in this section. In the figures, upper case symbols indicate physical variables and parameters, while lower case symbols represent per unit quantities. Furthermore, bold symbols represent synchronous reference frame dq variables expressed as complex space vectors, i.e. as $\mathbf{x} = x_d + j\, x_q$.

A. VSM Control System Overview and Adaptations for Operation with an MMC

An overview of the implemented VSM-based control strategy for the MMC is shown in Fig. 1. For interpreting the structure of the control system, it should be considered that the MMC is controlled by two parallel control paths that are generating voltage references for driving the ac-side and circulating currents $i_{v,abc}$ and $i_{c,abc}$, respectively. Thus, the MMC has two independent sets of inner loop current controllers which are generating the ac-side and internal voltage references, v_v^* and v_c^*.

The individual modulation indices $n_{u,l}$ for controlling the upper (u) and lower (l) arms of each phase k of the MMC are calculated from the voltage references as:

$$n_{u,k} \approx \frac{-v_{v,k}^* + v_{c,k}^* + v_{dc}}{v_{dc}}, \quad n_{l,k} \approx \frac{v_{v,k}^* + v_{c,k}^* + v_{dc}}{v_{dc}} \quad (1)$$

$$for \quad k \in \{a, b, c\}$$

This calculation of the arm insertion indexes implies that the equivalent gain of the control system is compensated for any variations in the available dc voltage. However, this modulation strategy does not compensate for the periodic oscillations appearing in the equivalent arm capacitor voltages. Indeed, if the actual or estimated arm voltages were used in the calculation of the insertion indices, it would be necessary to introduce closed loop control of the equivalent voltage or energy balance of the MMC [24]-[26].

In the presented implementation, the explicit energy control is avoided and the total equivalent voltages of each arm of the MMC will settle naturally to an operating point according to the power flow and the parameters of the MMC topology. Thus, only a Circulating Current Suppression Controller (CCSC), implemented as a set of decoupled PI controllers in the double frequency negative sequence synchronous reference frame according to [27], is introduced to regulate the internal circulating currents of the MMC. The CCSC is utilized for reducing the peak arm currents, the losses associated with the double frequency circulating currents and for limiting the corresponding internal oscillations of the MMC capacitor voltage. Thus, the current references for the circulating current controller indicated in Fig. 1 are set to (i.e. $\mathbf{i}_c^* = 0 + j\,0$) [27].

It can be noted that a sum energy controller investigated in [28] could also be easily introduced in the presented control scheme. This would decouple the total energy stored in the MMC from the voltage at the dc terminals, without requiring explicit control of the energy balance within the topology.

As shown by Fig. 1, the rest of the VSM-based control system relies on a set of cascaded voltage and current controllers, as well as a virtual impedance, for controlling the ac-side variables of the MMC. Furthermore, the control

783

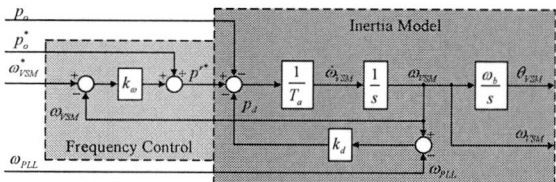

Fig. 2. Frequency droop control and the virtual swing equation implemented as the inertia model of the VSM

Fig. 3. Overview of laboratory setup and picture of circuit board with two half-bridge sub-modules

scheme includes a reactive power controller, the inertia model for emulating an SM swing equation, and a droop-based frequency control. These control functions can be almost identical to the VSM implementation presented in [7], [23] and are briefly summarized in the following.

B. Virtual Swing Equation and Frequency Control

The details of the inertia model and the frequency control from the lower left part of Fig. 1 are shown in Fig. 2 The implementation is based on [23], and, as seen from the figure, the emulated model includes the two main elements of the VSM, i.e. the inertia represented by the mechanical time constant T_a, and the mechanical damping power p_d. In this case, the damping of the swing equation is implemented by calculating the difference between the per unit speed of the virtual swing equation and the per unit grid frequency as detected by a PLL. Thus, under regular operation, the PLL in Fig. 1 does not have any other function than tracking the frequency, and is not involved in the grid synchronization of the control strategy.

Considering only the inertia emulation of the virtual swing equation in Fig. 2, the state equation of the virtual speed of the VSM can be defined by:

$$\frac{d\omega_{VSM}}{dt} = \frac{p^{r*}}{T_a} - \frac{p_o}{T_a} - \frac{k_d \cdot (\omega_{VSM} - \omega_{PLL})}{T_a} \quad (2)$$

However, the figure also shows that the input power reference to the inertia model, p^{r*}, is resulting from a simple power-frequency droop function according to:

$$p^{r*} = p_o^* + k_\omega \cdot (\omega_{VSM}^* - \omega_{VSM}) \quad (3)$$

From the control system overview in Fig. 1, it can be seen how the virtual per unit speed ω_{VSM} of the VSM and the phase angle θ_{VSM} are utilized in the control structure. It should be noted that the phase angle θ_{VSM}, resulting from integration of the virtual speed, is used for all dq-transformations in the system. Thus, the entire control system is implemented in synchronous reference frames defined by the virtual swing equation of the VSM.

TABLE I
MAIN PARAMETERS OF THE REDUCED-SCALE MMC PROTOTYPE

Converter parameters	Reference	18 HB model
Rated power	1059MVA	60 kVA
Rated DC voltage	640 kV DC	700V
Rated AC voltage|	333 kV	400V
Rated current	1836A	83A
Cells per arm	401 HB	18 HB
Nominal cell voltage	2 kV	50V
Arm inductance	50 mH	1,5 mH
Cell capacitance	10 mF	20 mF

C. Virtual Impedance

The structure of the outer loop reactive power droop control and cascaded voltage and current controllers shown in Fig. 1 can be identical for the MMC as for a 2L VSC. Thus, the implementation is based directly on [7], [23] and no further details are presented here. However, the virtual impedance emulating the voltage drop in the stator impedance of a SM has an important function since the current dependency of the voltage reference allows the cascaded voltage and current controllers to operate under strong grid conditions. For simplicity, a quasi-stationary implementation of this virtual impedance is preferred, and the resulting voltage reference is given by:

$$\mathbf{v}_o^* = \hat{v}^{r*} - (r_v + j \cdot \omega_{VSM} \cdot l_v) \cdot \mathbf{i}_o \quad (4)$$

III. HARDWARE IMPLEMENTATION OF MMC CONTROLLED AS A VSM

The VSM-based control system described in the previous section has been implemented on a reduced scale MMC prototype and tested in a laboratory environment with controllable ac- and dc-side voltage sources. This section provides a brief description of the hardware prototype and on the experimental setup.

A. Laboratory Scale MMC Prototype

The MMC converter prototype used in the experiments is a scaled-down version of a terminal from the HVDC transmission scheme described in [29]. The main specifications are reported in Table I. A picture with an overview of the laboratory setup and an image of the hardware for the sub-modules is shown in Fig. 3.

The control implementation for the MMC has a hierarchical and modular structure, as shown in Fig. 4. The system controller is located on the Opal-RT unit and is programmed in Simulink. It executes the VSM control and all the control loops of the MMC presented in the previous section, including the energy balancing and circulating current control. Outputs of the system controller are the six reference values for the converter arm voltages according to (1), which are sent to three individual leg controllers via a dedicated high-speed, full-duplex optical bus, operating at 5 Mbit/s. On the same bus, the leg controllers send back to the system controllers the measurements of the arm currents and the sum of the cell voltages in each arm, along with other monitoring information.

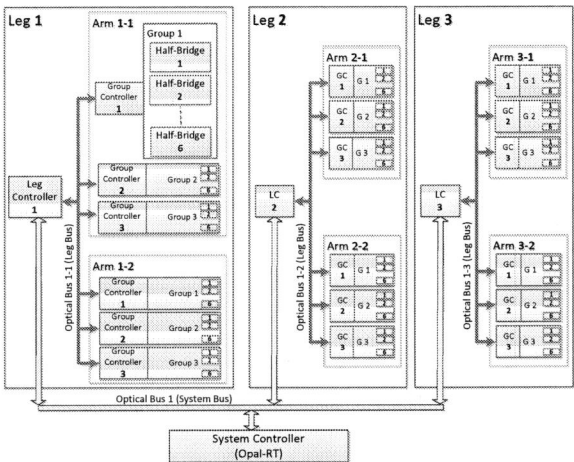

Fig. 4. MMC control and communication structure

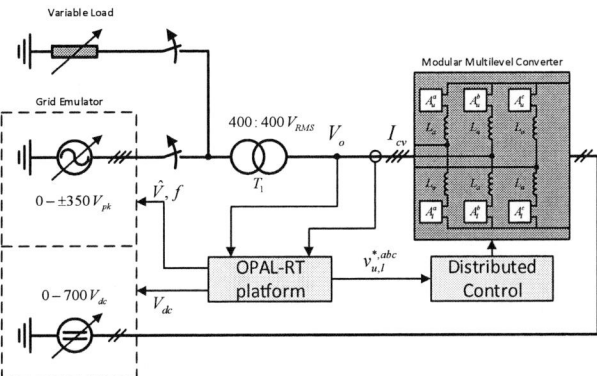

Fig. 5. System configuration for testing of MMC-based VSM

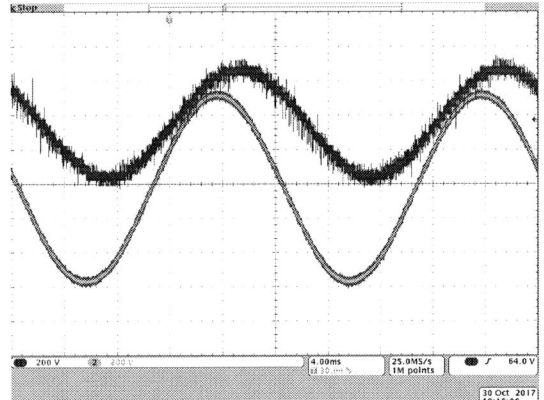

Fig. 6. MMC converter waveforms: Lower arm voltage of phase a and line-to-line voltage at output terminals

Each leg controller is responsible for the balancing of the dc-voltages of the cells within the connected arms. Information about the individual cell voltages is collected through a dedicated optical bus, operating at 3.5 Mbit/s, connecting the leg controller to all the group controllers of the leg. The group controllers are directly connected to the half-bridge switching cells and are responsible for translating the voltage commands for each cell received from the leg controller into suitable gate signals for the switching devices. Strict synchronization of all the switching cells is achieved by the custom-made time-triggered protocol implemented on the optical buses.

B. System Configuration

The system configuration for the experimental tests is shown in Fig. 5. The MMC is connected on both the ac side and the dc side to individual outputs of a grid emulator controlled in real time by the OPAL-RT platform. The grid emulator is rated for 200 kW in total, and provides the possibility to independently control six individual output terminals with a voltage control bandwidth up to 20 kHz. Three of the outputs have been grouped together to form a controlled ac grid at 380 V LL RMS, while two other outputs have been grouped to act as a dc voltage source at 690 V. An adjustable resistive load is also placed on the ac bus, acting as local load when the switch connecting the emulated ac bus to the rest of the system is open, resulting in islanded operation of the VSM-controlled MMC. A transformer with unity voltage ratio is connected between the MMC output and the rest of the ac bus to ensure galvanic insulation.

As shown in Fig. 4, the Opal-RT real time simulator controls the grid emulator by providing adjustable reference signals for the voltage amplitude and frequency. References for the grid emulator and the MMC are transferred via two different 5 Mbit/s optical buses.

The presented configuration offers a high degree of flexibility for testing the effect of changes of references or control parameters in the converter as well as the effect of external transients in the ac and dc grid by acting on the references for the grid emulator. Moreover, the possibility to disconnect the emulated ac grid from the rest of the ac

bus allows for testing the response of the VSM-controlled MMC in case of sudden islanding.

IV. EXPERIMENTAL RESULTS

This section presents an experimental validation of the VSM functionalities both in grid connected operation and in islanded mode. Moreover, the smooth transition between grid connected and islanded operation that characterizes the VSM is demonstrated.

For all the experiments, the VSM control has been configured with an equivalent inertia constant T_a equal to 2 s, a droop gain k_ω of 20 and a damping factor k_d of 200.

The oscilloscope screenshot in Fig. 6 shows the lower arm voltage of phase a measured before the filter and referred to the negative dc rail; the voltage waveform features 19 different levels, consistently with the number of cells per arm of the converter. The line-to-line voltage between phase a and b at the output terminals of the MMC after the arm inductors is also shown, clearly demonstrating the low harmonic distortion achieved even with the basic filter constituted by the arm inductors only.

A. Grid connected operation

The VSM is operated while connected to an emulated 50 Hz, 380V ac grid. The system is initially in steady state with active power reference to the MMC set to 20 kW, reactive power reference set to zero and internal VSM voltage setpoint equal to 380V.

785

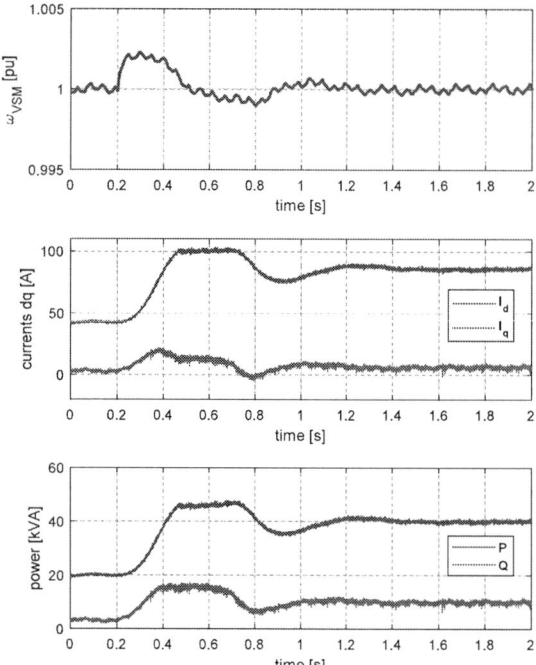

Fig. 7. Experimental results: VSM response to a step in power reference at nominal grid frequency

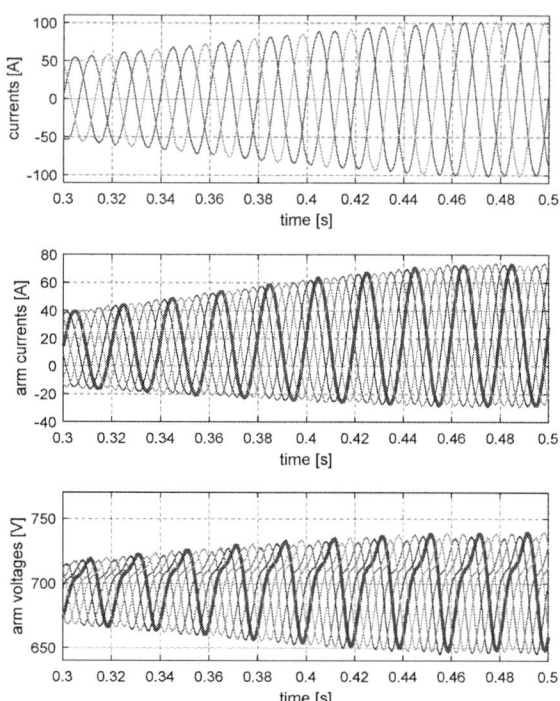

Fig. 8. Experimental results: VSM-MMC response to a step in power reference at nominal grid frequency. MMC output currents (top), arm currents (middle), arm voltages (bottom).

As a first test of dynamic response, the transient behavior of the system has been verified for a step change in the active power reference of the VSM from 20 kW to 40 kW, with key results plotted in Fig. 7. Since the grid emulator is maintaining the ac frequency at 50 Hz, the effect of the change in the active power reference directly translates into the same change in the power injected from the VSM to the grid. This is highlighted in the experimental results where the rotating speed of the VSM starts at 1.0 pu and returns to 1.0 pu after a damped transient as expected from the dynamics of the swing equation. The measured power displays a similar behavior with a damped second order transient before reaching the new steady state conditions. It should be noticed that the unlimited response of the VSM in this case would have presented an overshoot in the power and in the current. However, a current saturation set at 100 A limits the peak current during the transient, actively protecting the solid-state devices in the converter. Indeed, the ability to precisely control the currents during transients without altering the functional behavior of the VSM is a main feature of the proposed VSM implementation scheme.

The converter output currents in the stationary frame, together with the MMC arm voltages and currents are displayed in Fig. 8. Notably, arm voltages and currents remain well balanced throughout the transient, indicating that the VSM control is not interfering with the inner MMC control.

In the second test, the references of the VSM are maintained unaltered but the grid emulator is controlled to impose a step change in the frequency from 50 Hz to 49.8 Hz. The experiment aims at validating the inertia support offered by the VSM scheme and the effect of the droop in steady state. As shown by the results in Fig. 9, the VSM responds to a step decrease in the frequency as a synchronous generator, by a transient injection of power that in this case reaches a peak of approximately 35 kW. The active power and the virtual speed of the VSM settle to a new steady state condition after a brief oscillatory transient. The final steady state power is different from the initial operation, due to the presence of a droop characteristic in the VSM control. Indeed, the grid emulator imposes the frequency to a lower value, thus forcing the power injected from the VSM to a slightly higher value. This illustrates the double support contribution of the VSM to the receiving grid: a transient support action due to the response of the virtual inertia with an almost immediate power injection and a permanent support at steady state with an injection of power according to the droop characteristic. It should also be noticed that the VSM scheme offers the possibility to fine-tune these two contributions by acting individually on the inertia or the droop constant.

The response of the VSM to a change in the voltage amplitude of the grid voltage from 380 V to 345V is shown in Fig. 10. The transient indicates that the effect of the change on the active current component and on the VSM speed is relatively minor. However, due to the presence of a virtual impedance, the VSM reacts to the lowering grid voltage with an increased reactive current injection.

B. *Transition from grid-connected to islanded operation*

A main feature of the VSM is the capability of allowing for smooth and seamless transition from grid connected to islanded mode.

Fig. 11 shows the transient behavior of the system when the grid emulator is suddenly disconnected from the ac bus.

The 2018 International Power Electronics Conference

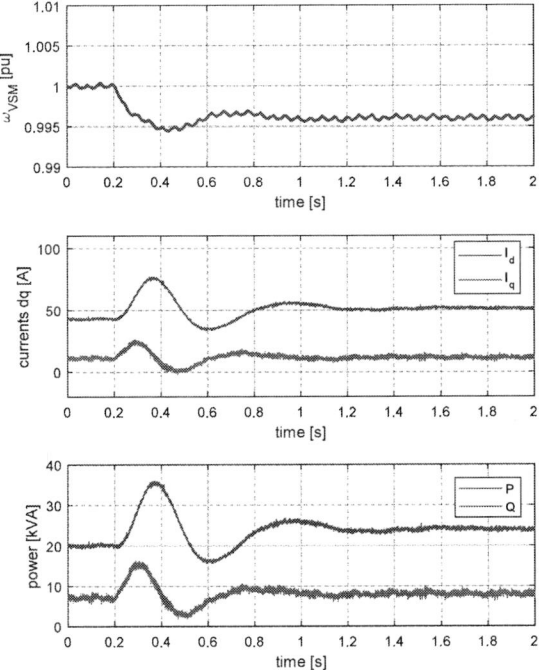

Fig. 9. Experimental results: VSM response to a step in the grid frequency

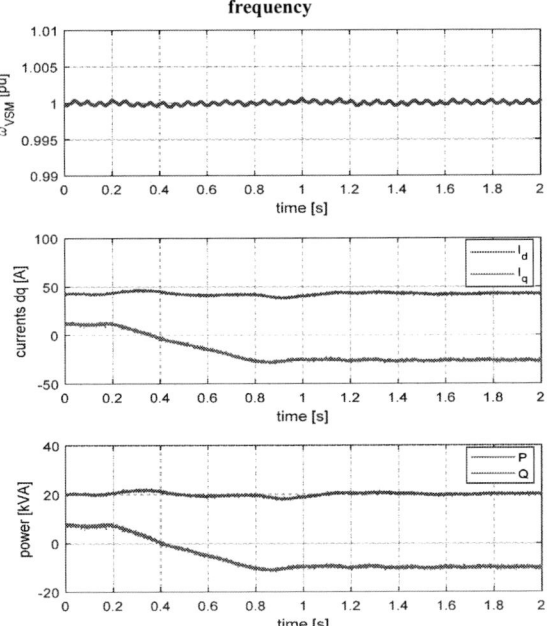

Fig. 10. Experimental results: VSM response to a variation in the grid voltage

The grid voltage feeding the local resistive load is controlled properly, with the voltage magnitude dropping less than 8% immediately after the islanding event and quickly recovering with a well-damped oscillation lasting for less than 300 ms. The transient on the grid frequency is properly damped by the virtual inertia, with oscillations extinguishing in a few seconds. The MMC operation is not disrupted by the islanding event, with arm voltages and

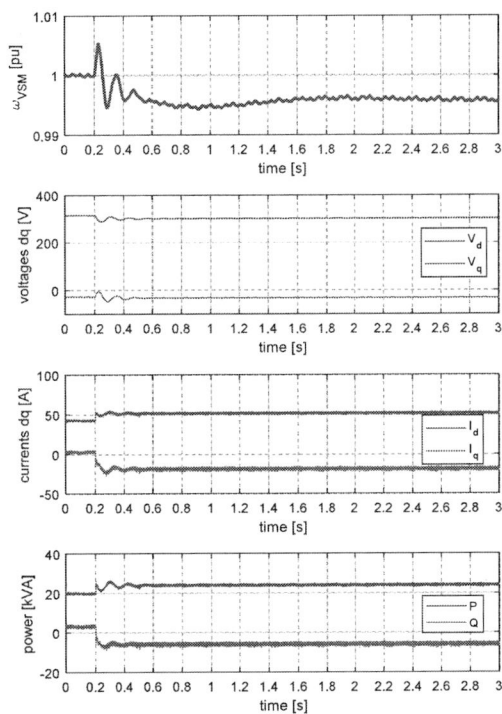

Fig. 11. Experimental results: VSM response to an islanding event

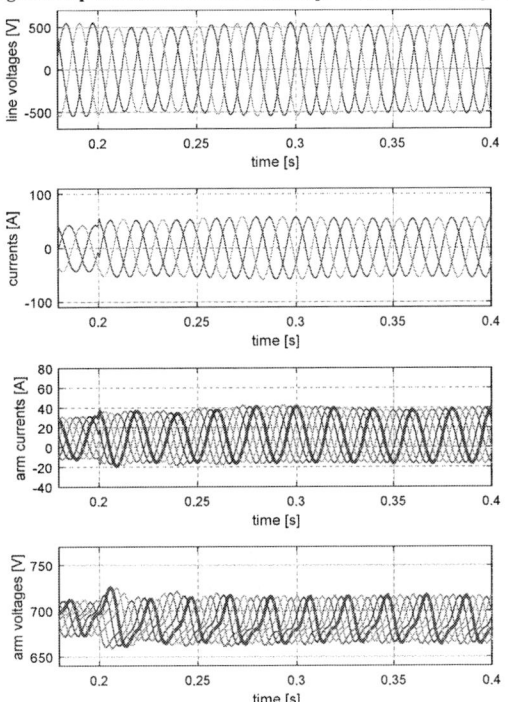

Fig. 12. Experimental results: VSM-MMC response to an islanding event; grid voltages and currents in stationary frame (top two graphs) and MMC arm currents and voltages (bottom two graphs)

currents remaining balanced and well under control throughout the transient, as shown in Fig. 12.

C. Islanded operation

As a last test the VSM is operated in islanded mode feeding a resistive load. The VSM is entirely responsible

The 2018 International Power Electronics Conference

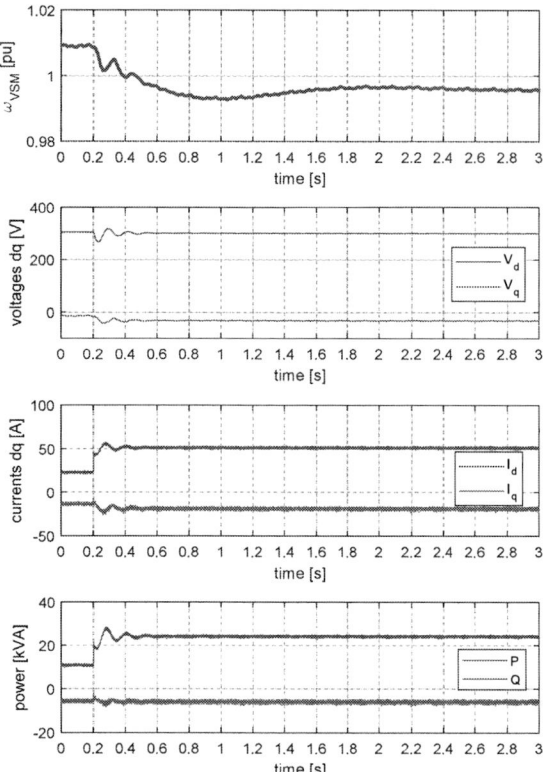

Fig. 13. Experimental results: VSM response to a load step while operating in islanded mode

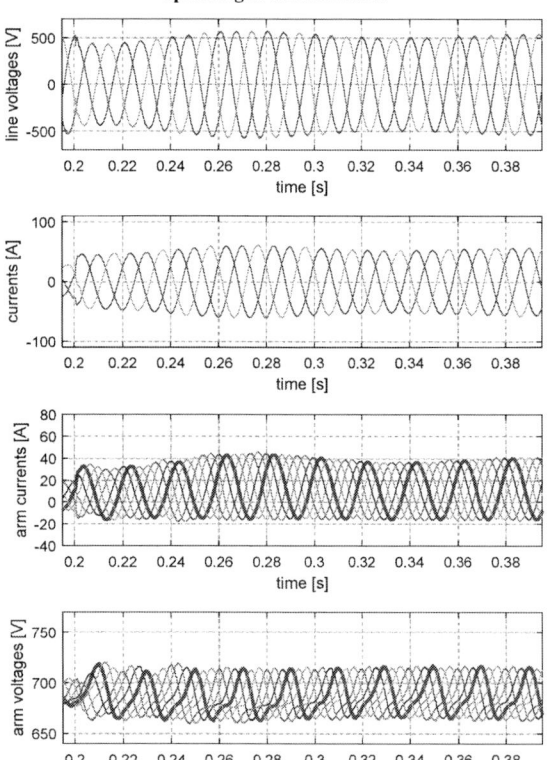

Fig. 14. Experimental results: VSM-MMC response to a load step while operating in islanded mode; grid voltages and currents in stationary frame (top two graphs) and MMC arm currents and voltages (bottom two graphs)

for the control of both voltage amplitude and frequency of the local grid. The effect of a step change in the load resistance from 11 Ohm to 5 Ohm is investigated.

Results in Fig. 13 show that the voltage regulation achieved by the VSM-controlled MMC is excellent, with hardly any dip observed in the grid voltage measured at the converter output terminals at t = 0.2 s, when the load is changed stepwise. The load step triggers small, well-damped oscillations with dynamics that are very similar to those observed after an islanding event.

The transient on the grid frequency, which in the islanded case coincides with the speed of the VSM, is also well damped, with the frequency settling to a lower value after the load is increased, as dictated by the droop coefficient in (3).

Robustness of the inner MMC control against grid disturbances is demonstrated by the results in Fig. 14, showing that arm currents and voltages remain balanced throughout the transient.

V.   CONCLUSION

Inertia support from power converters is assumed to become a critical element in future power systems where the presence of physical rotating inertia is expected to be gradually reduced. The concept of Virtual Synchronous Machines offer a convenient approach to integrate inertia support functionality in the control of power converters together with the possibility to seamlessly switch between grid-connected and islanded operation. This control approach has initially emerged in the context of distributed generation systems, but can also be implemented in HVDC transmission schemes for supporting the frequency regulation of the power system. This paper presented an implementation of a Virtual Synchronous Machine scheme for the control of a MMC and its experimental validation. Several tests have been performed on a reduced-scale, MMC prototype controlled as a VSM, demonstrating its performance both when connected to an external ac grid or when operating in islanded mode. It has been shown that the proposed VSM implementation can operate without interfering with the internal operation of the MMC converter itself, thus allowing a direct migration of the algorithms developed for conventional 2-level converters to the more complex multilevel converters used in state-of-the-art HVDC systems.

VI.   REFERENCES

[1] H.-P. Beck, R. Hesse, "Virtual Synchronous Machine," in Proceedings of the 9th International Conference on Electrical Power Quality and Utilisation, Barcelona, Spain, 9-11 October 2007, 6 pp.

[2] K. Visscher, S. W. H. De Haan, "Virtual Synchronous Machines (VSG's) for Frequency Stabilization in Future Grids with a Significant Share of Decentralized Generation," in CIRED seminar 2008: Smart Grids for Distribution, Frankfurt, Germany, 23-24 June 2008, 4 pp.

[3] S. D'Arco, J. A. Suul, "Virtual Synchronous Machines – Classification of Implementations and Analysis of Equivalence to Droop Controllers for Microgrids," in Proceedings of IEEE PowerTech Grenoble 2013, Grenoble, France, 16-20 June 2013, 7 pp.

[4] S. D'Arco, J. A. Suul, O. B. Fosso, "Small-signal modelling and parametric sensitivity of a Virtual Synchronous Machine in islanded operation," in *International Journal of Electric Power and Energy Systems*, Vol. 72, November 2015, pp. 3-15

[5] Q.-C. Zhong, G. Weiss, "Synchronverters: Inverters That Mimic Synchronous Generators," IEEE Trans. on Ind. Electronics, vol. 58, no. 4, pp. 1259-1267, April 2011

[6] P. Rodriguez, I. Candela, A. Luna, "Control of PV Generation Systems using the Synchronous Power Controller," in Proc. of the 2013 IEEE Energy Conversion Congress and Exposition, ECCE 2013, Denver, Colorado, USA, 15-19 September 2013, pp. 993-998

[7] S. D'Arco, J. A. Suul, "Small-Signal Analysis of an Isolated Power System controlled by a Virtual Synchronous Machine," in *Proceedings of the IEEE 17th International Conference on Power Electronics and Motion Control*, PEMC 2016, Varna, Bulgaria, 25-30 September 2016, pp. 462-469

[8] E. Rakhshani, P. Rodriguez, "Inertia Emulation in AC/DC Interconnected Power Systems Using Derivative Technique Considering Frequency Measurement Effects," in IEEE Transactions on Power Systems, Vol. 32, No. 5, September 2017, pp.3338-3351

[9] D. Duckwitz, B. Fisher, "Modeling and Design of df/dt-based Inertia Control for Power Converters," in IEEE Journal of Emerging and Selected Topics in Power Electronics, Vol. 5, No. 4, December 2017, pp.1553-1564

[10] R. Hesse, D. Turschner, H.-P. Beck, "Micro grid stabilization using the Virtual Synchronous Machine (VISMA)," in *Proceedings of the International Conference on Renewable Energies and Power Quality*, ICREPQ'09, Valencia, Spain, 15-17 April 2009, 6 pp.

[11] Y. Chen, R. Hesse, D. Turschner, H.-P. Beck, "Investigation of the Virtual Synchronous Machine in the Island Mode," in *Proceedings of the 2012 3rd IEEE Innovative Smart Grid Technologies Europe Conference*, Berlin, Germany, 15-17 October 2012, 6 pp.

[12] T. Ackerman, T. Prevost, V. Vittal, A. J. Roscoe, J. Matevosvan, N. Miller, "Paving the Way: A Future Without Inertia Is Closer Than You Think," in IEEE Power and Energy Magazine, Vol. 15, No. 6, November/December 2017, pp. 61-.69

[13] Y. Wang, V. Silva, M. Lopez-Botct-Zulueta, "Impact of high penetration of variable renewable generation on frequency dynamics in the continental Europe interconnected system," in IET Renewable Power Generation, Vol. 10, No. 1, January 2016, pp. 10-16

[14] J. Zhou, C. Booth, G. P. Adam, A. J. Roscoe, "Inertia Emulation Control of VSC-HVDC Transmission System," in Proceedings of the 2011 International Conference on Advanced Power System Automation and Protection, Beijing, China, 16-20 October 2011, 6 pp.

[15] J. Zhu, C. D. Booth, G. P. Adam, A. J. Roscoe, C. G. Bright, "Inertia Emulation Control Strategy for VSC-HVDC Transmission Systems" in IEEE Transactions on Power Systems, Vol. 28, No. 2, May 2013, pp. 1277-1287

[16] R. Aouini, B. Marinescu, K. B. Kilani, M. Elleuch, "Synchronverter-Based Emulation and Control of HVDC Transmission," in IEEE Transactions on Power Systems, Vol. 31, No. 1, January 2016, pp. 278-286

[17] M. Guan, W. Pan, J. Zhang, Q. Hao, J. Cheng, X. Zheng, "Synchronous Generator Emulation Control Strategy for Voltage Source Converter (VSC) Stations," in IEEE Transactions on Power Systems, Vol. 30, No. 6, November 2016, pp. 3093-3101

[18] W. Zhang, K. Rouzbehi, J. I. Candela, A. Luna, P. Rodriguez, "Control of VSC-HVDC with Electromechanical Characteristics and Unified Primary Strategy," in Proc. of the 2016 IEEE Energy Conversion Congress and Exposition, ECCE 2016, Milwaukee, Wisconsin, USA, 18-22 September 2016, 8 pp.

[19] E. Rakshani, D. Remon, A. M. Cantarellas, J. M. Garcia, P. Rodriguez, "Modeling and sensitivity analysis of VSP based virtual inertia controller in HVDC links of interconnected power systems," in Electric Power System Research, Vol. 141, Dec. 2016, pp. 246-263

[20] A. Lesnicar, R. Marquardt, "An innovative modular multilevel converter topology suitable for a wide power range, in Proceedings of the 2003 IEEE Bologna PowerTech Conference, Bologna, Italy, 23-26 June 2003, vol.3, pp. 272-277

[21] C. Verdugo, J. I. Candela, P. Rodriguez, "Grid Support Functionalities based on Modular Multilevel Converters with Synchronous Power Control," in Proceedings of the 5th International Conference on Renewable Energy Research and Applications, Birmingham, UK, 20-23 November 2016, pp. 572-577

[22] O. D. Adeuyi, M. Cheah-Mane, J. Liang, N. Jenkins, Y. Wu, Z. Li, X. Wu, "Frequency Support from Modular Multilevel Converter Based Multi-Terminal HVDC Schemes," in Proc. of the 2015 IEEE Power and Energy Society General Meeting, PESGM 2015, Denver, Colorado, USA, 26-30 July 2016, 5 pp.

[23] S. D'Arco, J. A. Suul, O. B. Fosso, "Small-Signal Modeling and Parametric Sensitivity of a Virtual Synchronous Machine," in *Proceedings of the 18th Power Systems Computation Conference*, PSCC 2014, Wrocław, Poland, 18-22 August 2014, 9 pp.

[24] A. Antonopoulos, L. Ängquist, H.-P. Nee, "On Dynamics and Voltage Control of the Modular Multilevel Converter," in *Proceedings of the 13th European Conference on Power Electronics and Applications*, EPE'09, Barcelona, Spain, 8-10 September 2009, 10 pp

[25] L. Harnefors, A. Antonopoulos, S. Norrga, L. Ängquist, H.-P Nee, "Dynamic Analysis of Modular Multilevel Converters," in *IEEE Transactions on Industrial Electronics*, vol. 60, no. 7, July 2013, pp. 2526-2537

[26] G. Bergna J. A. Suul, S. D'Arco, "State-Space Modeling of Modular Multilevel Converters for Constant Variables in Steady-State," in *Proceedings of the 17th IEEE Workshop on Control and Modeling for Power Electronics*, COMPEL 2016, Trondheim, Norway, 27-30 June 2016, 9 pp.

[27] Q. Tu, Z. Xu, L. Xu, "Reduced Switching-Frequency Modulation and Circulating Current Suppression for Modular Multilevel Converters," in *IEEE Transactions on Power Delivery*, Vol.26, No.3, pp.2009-2017, July 2011

[28] J. Freytes, G. Bergna, J. A. Suul, S. D'Arco, F. Gruson, F. Colas, H. Saad, X. Guillaud, "Improving Small-Signal Stability of an MMC with CCSC by Control of the Internally Stored Energy," in *IEEE Transactions on Power Delivery*, Vol. 33, No. 1, February 2018, pp. 429-439

[29] J. Peralta, H. Saad, S. Dennetiére, J. Mahseredjian and S. Nguefeu, "Detailed and Averaged Models for a 401-Level MMC-HVDC System," in IEEE Transactions on Power Delivery, vol. 27, No. 3, pp.1501-1508, July 2012

The 2018 International Power Electronics Conference

Assessment of Virtual Synchronous Machine based Control in Grid-Tied Power Converters

Chi Li[1*], Igor Cvetkovic[1], Rolando Burgos[1], Dushan Boroyevich[1]

1 Bradley Department of Electrical and Computer Engineering, Virginia Tech, Blacksburg, USA

E-mail: lichi@vt.edu

*Abstract-*This paper presents the assessment of virtual synchronous machine (VSM) control schemes applied to grid-tied power converters. Specifically, the paper analyzes first the merits of its usage in reactive power static compensators (STATCOM), and second in generic grid-tied inverters. VSM is compared against standard d-q frame vector control, and key operational aspects like synchronization and startup procedure are discussed in detail. The paper considers both VSM-based and inverter-synchronous machine duality based models for the development of control schemes. As a particularly important outcome of this duality study, it will be shown how converter synchronization can be, in general, realized by scaling down and integrating DC-link voltage of any voltage source converter. The concepts presented will be demonstrated experimentally with a 30 kVA 208 V ac synchronous generator and power converter test bed.

Keywords— virtual synchronous machine; STATCOM; modeling

I. INTRODUCTION

In the past decade, a control method named a virtual synchronous machine (VSM) has been proposed to improve the performance of grid-interfaced inverters, which controls the inverters to emulate the behavior of dominant existing conventional synchronous machines to obtain a more grid-friendly response due to the nature of synchronization mechanics in the swing equation [1-5]. Despite some differences in the details about how to model synchronous machines, the cardinal part of this kind of control method is to replace the D-Q frame phase-locked loops (PLL) with power balance-based synchronization, which eliminates the possible instability problems caused by PLLs. The VSM scheme makes inverters cooperate with each other inherently rather than fight and fall into instability, as may happen with multiple inverters connected together with PLLs. Moreover, with the virtual parameters such as virtual inertia and virtual impedance, one can easily program online to adjust to different conditions, getting rid of the slow responses of a real synchronous machine during transients.

Recent references [6-9] applied the VSM control method to various applications including STATCOMs, wind turbines and grid-tied converters for energy storage and compared against the traditional d-q frame vector control. Some advantages were shown in terms of better transient responses in the small-signal sense. Other, quite interesting work related to the virtual synchronous machine concept, can be found in [11]-[15], focusing on

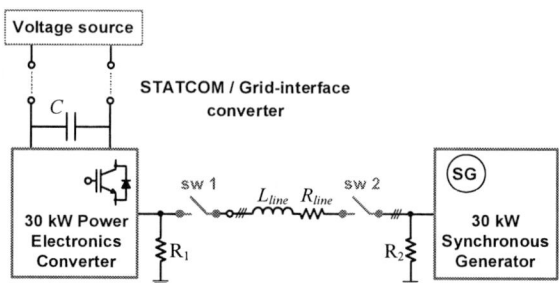

Fig. 1 A Single line diagram of the 3-phase test bed

damping of power oscillation [11, 12], power quality improvement [13], islanding [14], and system-level integration benefits [15].

This paper is organized as followings: Section II describes the testbed to be studied with renewable energy integration and STATCOM compensation; section III shows the advantages of applying VSM to STATCOM application against the standard d-q frame vector control; section IV gives some ideas on the start-up process for converters under VSM control; section V enables a new insight and ideas of how converter synchronization can be performed without the use of conventional synchronization methods, presenting an alternative to commonly used power-balance synchronization method. Conclusions are given in the section VI.

II. TESTBED DESCRIPTION

Fig. 1 shows the simplified test bed structure. It comprises 30 HP variable speed drive, 22 kW induction motor, and four-pole 27 kW synchronous generator with field and damper windings. It further contains passive components, and switches, as well as the power electronics converter of the same power-level used as both, STATCOM, and a grid-interface converter emulating synchronous machine. Fig. 2 shows the pictures of the testbed, without resistive loads and lines.

This power converter was built using IGBT-based integrated power module PM100CL1A060, accompanied custom designed high-speed digital controller featuring Texas Instruments TMS320F28343 Delfino MCU, and Lattice Complex Programmable Logic Device (CPLD) LCMX02-4000HC.

III. COMPARISON AGAINST D-Q FRAME VECTOR CONTROL

The advantage of the VSM-based controller is significantly shown in Fig. 3. The input power to the

790

The 2018 International Power Electronics Conference

Fig. 2 Synchronous generator and inverter under VSM control

generator varies by 0.25 pu and hence the frequency of the output voltage will fluctuate around the nominal value and finally come back to it. The settling time depends on the inertia or the time constant of the generator because the STATCOM has little capability of frequency regulation and the oscillation of frequency will not stop until there is no power imbalance applied in the rotating shaft. As observed, the VSM-based controller regulates the PCC bus voltage much better than the traditional d-q frame one. A minor reason is that the VSM-based controller acting like a synchronous machine can automatically participate in the frequency regulation using its inertia, which is actually the energy stored in the capacitor here, and thus absorb or provide some active power during the transients. But this effect is very tiny unless there is some energy storage device connected to the dc link because the energy of the dc capacitor itself is very small compared to the source active power rating. Although the direct influence of the effort to regulate the frequency is often negligible, the characteristic of the swinging axis of the VSM-based controller provides a better synchronization of the STATCOM to the grid and ensures exerting the compensating current in the proper phase with respect to the source voltage, regulating the PCC voltage magnitude better and facilitating the recovery process of the power imbalance of the source. Conversely, the phase information of the PLL-oriented d-q frame controller relies on the voltage it detects and the corresponding compensating current is injected to the grid regardless.

IV. DISCUSSION ON START-UP PROCESS

A practical consideration for converters under VSM control is the start-up process. Referring to synchronous condensers, there are typically two kinds of starting methods [10]. The first method is to connect a synchronous condenser to the grid via its transformer tapping down the voltage to speed up the torque, and then increase the secondary side voltage gradually to accelerate to the grid frequency. The second method is to use an additional motor to drive its torque to the grid frequency approximately, and then use a synchronizer to synchronize before closing the main circuit breaker.

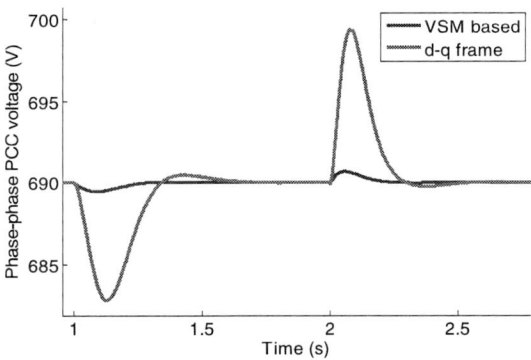

Fig. 3 Transient responses to frequency variance in the system

With these ideas, there are also two corresponding methods for VSM-controlled converters. The first one is to connect the converter to the grid and use the anti-parallel diodes of switching devices to charge the dc capacitor while in the meantime the controller begins to synchronize. When the dc bus voltage is stable and the synchronization completes, the converter starts switching with current loop and dc voltage loop running to charge the dc capacitor further to the nominal value before starting ac voltage loop or power loop to normal operation. This is suitable for converters with no start-up capability from the dc side, e.g. STATCOMs.

The second one is to use a separate dc source to charge the dc bus voltage to the nominal value through a dc pre-charging circuit without connection to the grid and mimic a synchronizer to synchronize, which can be achieved by a PLL only used during start-up process. After the dc bus is charged and synchronization is finished, a main circuit breaker is closed and all the loops except the ac voltage loop or power loop start working to ensure smooth connection. Finally, the outer ac voltage loop or power loop is turned on to function normally. When the grid-tied converters have the dc power input, this approach is straightforward and easy to implement.

The detailed start-up process is summarized in Table I for a STATCOM under VSM control in order to implement without addition of an extra dc power source and potential danger due to failure in mimicking synchronizer.

791

Table I Start-up process for VSM-STATCOMs

	Controller	Hardware
Step 1	• Start synchronization loop (virtual inertia)	DC capacitor charged by anti-parallel diodes.
If dc voltage charging completes & synchronization completes		
Step 2	• Start current loop • Start dc voltage loop	Bypass pre-charging resistor; Start PWM to charge dc bus further.
If dc voltage charging completes		
Step 3	• Start ac voltage loop	

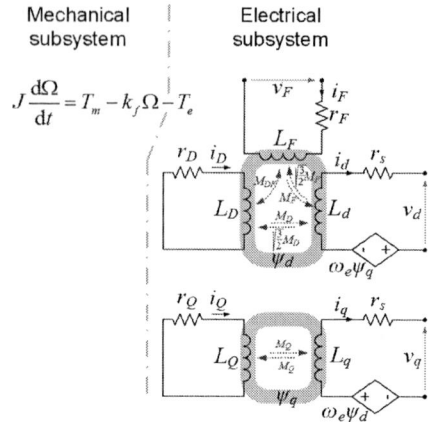

Fig. 3. Synchronous machine model [kundur]

V. MACHINE – CONVERTER DUAL MODEL

Another approach to modeling and control of power electronic converters as synchronous machines is to develop an electrical model of the machine completely dual to commonly used – electromechanical, and then restructure it mathematically to fully resemble generic average model of the power electronics converter. This would offer better understanding on which power converter parameters relate to which parameters of cylindrical or salient pole machine. One of the goals in the effort of implementing machine operation in the power electronics converter is to use us much as possible resources/components of the chosen converter, leaving only complementary parts to be implemented into the converter control algorithm. This could be achieved by, first, defining electrical parameters that feature physical/mathematical duality with corresponding mechanical parameters, and second, by mathematically restructuring synchronous machine model to resemble power converter models. Fig. 3 shows dynamic model of the salient pole synchronous machine with one damper winding in the q-axis, and one damper and one field winding in d-axis. The same figure also illustrates the machine fluxes in two axes, as well as the coupling coefficients. As this form does not offer directly observable relationships and equivalence with the power converter structure, model shown in Fig. 4 is derived from it offering a very evident similarity with the typical average model of the two-level voltage source converters widely used in the engineering practice.

By performing mechanical – electrical duality [], it can be shown that machine torque T corresponds to converter (dc-link) current I_{dc}, angular speed Ω corresponds to voltage v, moment of inertia J to dc-link capacitance C, and friction k_f to conductance G. The following equations describe torque and flux relationship (for $\omega_e = \Omega p$):

$$v_d = -r_s i_d - \omega_e \psi_q + \frac{d\psi_d}{d\theta_e}\frac{d\theta_e}{dt} = -r_s i_d + \Omega p(\frac{d\psi_d}{d\theta_e} - \psi_q)$$

$$v_q = -r_s i_q + \omega_e \psi_d + \frac{d\psi_q}{d\theta_e}\frac{d\theta_e}{dt} = -r_s i_q + \Omega p(\frac{d\psi_q}{d\theta_e} + \psi_d)$$

(1)

where

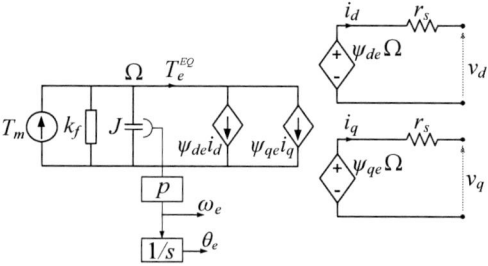

Fig. 4. Mathematically restructured synchronous machine model

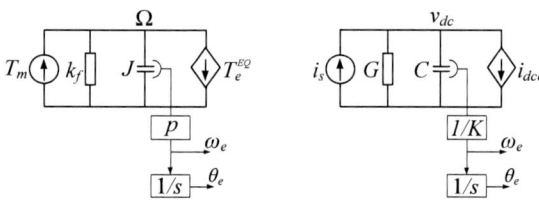

Fig. 5. Internal angle generation (left – machine, right – converter)

$$T_e^{EQ} = \frac{1}{\Omega}\left(i_d\frac{d\psi_d}{dt} + i_q\frac{d\psi_q}{dt}\right) + \frac{\omega_e}{\Omega}(\psi_d i_q - \psi_q i_d) =$$
$$= i_d\psi_{de} + i_q\psi_{qe}.$$

(2)

Another thing that is apparent from Fig. 4 is that directly reveals how converter duty cycles have to be controlled in order to achieve emulation of the synchronous machine. Relationships (5) show that:

$$\psi_d^{EQ} = p\left(\frac{d\psi_d}{d\theta_e} - \psi_q\right) \equiv d_d$$

$$\psi_q^{EQ} = p\left(\frac{d\psi_q}{d\theta_e} + \psi_d\right) \equiv d_q$$

(5)

where d_d, and d_q are converter duty cycles in d and q axis respectively. The machine emulation was performed experimentally using the test bed described earlier in this paper, and results are shown in Fig. 6 below. This figure shows the magnitudes of the terminal currents and

The 2018 International Power Electronics Conference

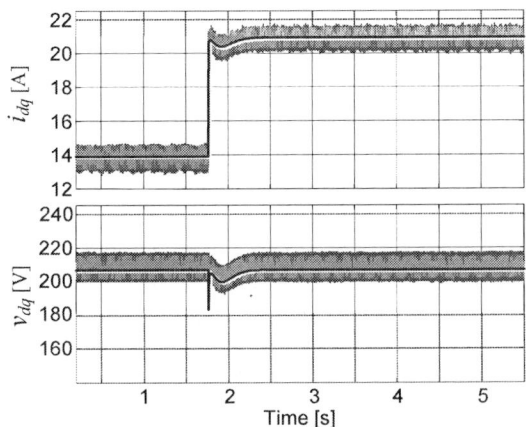

Fig. 6. Instantaneous current and voltage magnitude comparison; experiment machine - (green), experiment converter - (blue), simulation - (gray)

voltages (6) for synchronous machine, converter, and model from Fig. 4 overlaid. A very good matching is evident.

$$i_{dq} = \sqrt{i_d^2 + i_q^2}, \quad v_{dq} = \sqrt{v_d^2 + v_q^2}. \tag{6}$$

Once connected to the grid, operating at the particular operating point, synchronous machines self-synchronize to the system frequency, and precisely follow its changes with a zero steady-state error. This is their inherent feature due to the physics behind the power balance, and is described by Newton's motion law written for the mechanical subsystem shown in Fig. 3, and emphasized in Fig. 5 for both machine and converter behaving like the machine.

This feature can be emulated in the converters too, allowing them to "synchronize" their DC-link voltages with the system frequency. Hence, it is possible to obtain internal converter angle in the same way it is obtained in the synchronous machines – only by integrating converter's DC-link voltage according to (3):

$$v_{dc} \equiv K\omega_e \; \rightarrow \; \frac{d\theta_e}{dt} = \omega_e \; \rightarrow \; \theta_e = \int \frac{v_{dc}}{K} dt + \theta_o \tag{7}$$

Here K is the scaling constant – As DC-link voltage is directly proportional to the grid angular frequency, this constant can be used to maintain this voltage at the desired value.

This is experimentally verified, and shown in Fig. 7 where Dc-link dynamics has been recorded during the load step transient. Consequently, converter output voltages and currents feature frequency directly proportional to DC-link voltage (according to (7)).

VI. CONCLUSIONS

In this paper, virtual synchronous machine (VSM) control scheme was applied to STATCOMs and extended to generic grid-tied inverters. It was shown that STATCOMs under VSM control could enhance the transient responses of voltage regulation. Additionally, practical concerns about the VSM-controlled converters

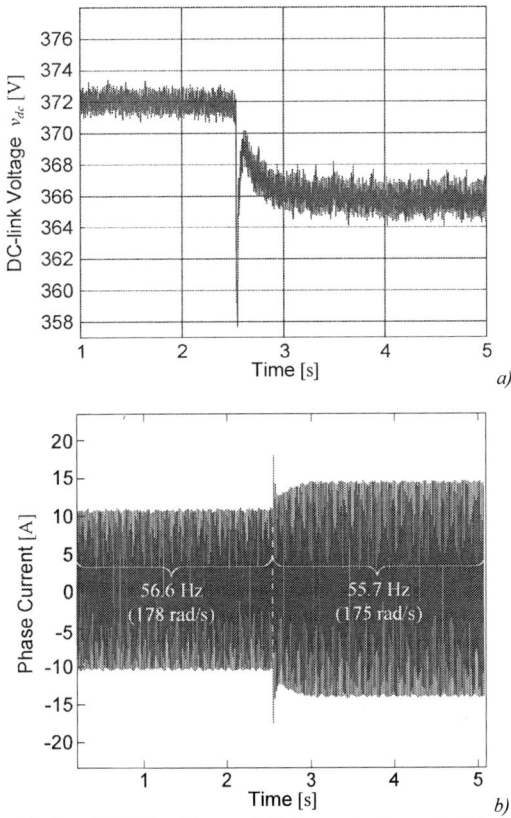

Fig. 7. *a)* DC-link voltage, and *b)* phase output current of the power converter during the load step

were discussed to conduct experiments. The digest then advances into the electromechanical duality between the electric machine and the power converter, showing restructured machine model derived in order to fully resemble power converter average model. Not only that this model allows for any type of machine to be accurately simulated, it also gives an insight of what needs to be done in order to successfully emulate machine with the power electronics converter. It is further shown how converter synchronization can be achieved by measuring and integrating its own DC-link voltage. Dynamic characterization of the particular machine is performed, and successful emulation demonstrated using power electronics converter prototype.

REFERENCES

[1] H. P. Beck and R. Hesse, "Virtual synchronous machine," in *Electrical Power Quality and Utilisation, 2007. EPQU 2007. 9th International Conference on*, 2007, pp. 1-6.

[2] J. Driesen and K. Visscher, "Virtual synchronous generators," in *Power and Energy Society General Meeting - Conversion and Delivery of Electrical Energy in the 21st Century, 2008 IEEE*, 2008, pp. 1-3.

[3] Z. Qing-Chang and G. Weiss, "Synchronverters: Inverters That Mimic Synchronous Generators," *Industrial Electronics, IEEE Transactions on*, vol. 58, pp. 1259-1267, 2011.

[4] P. Rodriguez, I. Candela, and A. Luna, "Control of PV generation systems using the synchronous power controller," in *Energy Conversion Congress and Exposition (ECCE), 2013 IEEE*, 2013, pp. 993-998.

[5] Z. Lidong, L. Harnefors, and H. P. Nee, "Power-Synchronization Control of Grid-Connected Voltage-Source Converters," *Power Systems, IEEE Transactions on,* vol. 25, pp. 809-820, 2010.

[6] L. Chi, R. Burgos, I. Cvetkovic, D. Boroyevich, L. Mili, and P. Rodriguez, "Analysis and design of virtual synchronous machine based STATCOM controller," in *Control and Modeling for Power Electronics (COMPEL), 2014 IEEE 15th Workshop on,* 2014, pp. 1-6.

[7] L. Chi, R. Burgos, I. Cvetkovic, D. Boroyevich, L. Mili, and P. Rodriguez, "Evaluation and control design of virtual-synchronous-machine-based STATCOM for grids with high penetration of renewable energy," in *Energy Conversion Congress and Exposition (ECCE), 2014 IEEE,* 2014, pp. 5652-5658.

[8] Y. Ma, W. Cao, L. Yang, F. Wang, and L. M. Tolbert, "Virtual Synchronous Generator Control of Full Converter Wind Turbines with Short Term Energy Storage," *IEEE Transactions on Industrial Electronics,* vol. PP, pp. 1-1, 2017.

[9] J. Liu, Y. Miura, and T. Ise, "Comparison of Dynamic Characteristics Between Virtual Synchronous Generator and Droop Control in Inverter-Based Distributed Generators," *IEEE Transactions on Power Electronics,* vol. 31, pp. 3600-3611, 2016.

[10] T. J. E. Miller, *Reactive power control in electric systems.* New York: Wiley, 1982.

[11] Y. Hirase, K. Sugimoto, K. Sakimoto, and T. Ise, "Analysis of Resonance in Microgrids and Effects of System Frequency Stabilization Using a Virtual Synchronous Generator," *IEEE Journal of Emerging and Selected Topics in Power Electronics,* vol. 4, pp. 1287-1298, 2016.

[12] K. Sakimoto, Y. Miura, and T. Ise, "Stabilization of a power system with a distributed generator by a Virtual Synchronous Generator function," in *Power Electronics and ECCE Asia (ICPE & ECCE),* 2011 IEEE 8th International Conference on, 2011, pp. 1498-1505.

[13] C. Yong, R. Hesse, D. Turschner, and H. P. Beck, "Improving the grid power quality using virtual synchronous machines," in *Power Engineering, Energy and Electrical Drives (POWERENG),* 2011 International Conference on, 2011, pp. 1-6.

[14] C. Yong, R. Hesse, D. Turschner, and H. P. Beck, "Investigation of the Virtual Synchronous Machine in the island mode," in *Innovative Smart Grid Technologies,* 2012 3rd IEEE PES International Conference and Exhibition on, 2012, pp. 1-6.

[15] D. Chen, Y. Xu, and A. Q. Huang, "Integration of DC Microgrids as Virtual Synchronous Machines into the AC Grid," *IEEE Transactions on Industrial Electronics,* vol. PP, pp. 1-1, 2017.

The 2018 International Power Electronics Conference

Research on the Blockchain-based Integrated Demand Response Resources Transaction Scheme

Shengnan Zhao[1], Yang Li[1*], Beibei Wang[1] and Huiling Su[2]
1 Southeast University, Nanjing, China
2 Jiangsu Electric Power Research Institute, Nanjing, China
*E-mail: li_yang@seu.edu.cn

Abstract—Under the background of rapid development of distributed renewable energy (DRE) and demand response (DR), the traditional DR will develop into integrated demand response (IDR). The current centralized trading of electricity market model is unable to meet the trading needs of scattered IDR resources. As the decentralized and distributed accounting mode, the blockchain technology fits the requirement of IDR resources to participate in energy market. The blockchain-based DRE transaction platform can support the credible transaction and settlement between the IDR resources, and promote the development of DER. Corresponding to the transaction principle, the frame of blockchain-based IDR resources transaction scheme was proposed. The transactions between DER and DR are taken for example to explain the detail trading process. Finally, the smart contracts of the transactions are designed and deployed on Ethereum private blockchain to prove the validity of the proposed transaction scheme.

Keywords—Blockchain, Decentralized market, Demand response, transaction scheme design

I. INTRODUCTION

With the rapid development of smart grid and the promotion of technologies of distributed renewable energy (DRE) generators and demand response (DR), the boundaries between the generators and the energy consumers are becoming blurred. Jeremy Rifkin puts forward the concept of the energy interconnection. He believes fundamental economic change occurs when new communication technologies converge with renewable electricity, which realizing the access of distributed energy and fair trade. Under this background, more and more other forms of demand side energy resources, such as heat and gas, will participate into the market. The traditional DR will develop into integrated demand response (IDR), which makes the transactions more complicated and the managements more difficult [1]. It is very important to design an efficient trading scheme and discover the value of IDR resources.

The managements of transactions can be divided into two main categories: the centralization and decentralization[2]. The centralized transactions can bring many problems such as high operation costs, long time consuming and security problems. The center has to collects all IDR information to arrange the scheduling instructions, which will lead to personal privacy concerns. What's more, when the number of IDR users increases dramatically, the data information will increase geometrically, and it will be more difficult for scheduling DR in real time. So some scholars have put forward the idea of applying decentralization to the demand side transactions[3].

The preconditions for IDR resources' involving in energy trading include the real-time information exchange, the self-optimizing strategy selection, the autonomic transaction settlement and decentralized trading platform for the regional IDR resources[4]. Considering the characteristics of being open, decentralized, transparent and non-tampering of block chain, it can effectively improve the efficiency of transactions and ensure the transaction security[5], which fit the needs of IDR trading. As the focus of current research, block chain has been thoroughly studied on its basic principles, characteristics, and applications in finance, energy and other fields[6-8]. The research on applications of block chain in electricity transactions can be divided into two aspects. One is feasibility analysis, which concentrates on the application prospect and the fitness with trading mechanism, such as distribution network transactions, ancillary services, direct power purchase[8-10]. The other is research on transaction process design and smart contracts design, such as automatic demand response, bilateral transactions in distribution network and so on[11-13]. However, the current research has not involved transaction process design and smart contracts design of multilateral trading among local IDR resources, which has broad prospects of application in the distribution network and micro-grid. DR users and distributed renewable energy generators, Additionally, the existing smart contract research focuses on the description of the contract implementation process and lacks engineering implementation.

This paper applies block chains to the transactions among local distributed IDR resources. Corresponding to the multilateral transaction principle, the scheme of blockchain-based decentralized trading was proposed. We took the typical DER and DR resource's trading for example, and designed the smart contracts of user's response strategy, the trading matchmaking, and the user's DR settlement. These contracts were deployed on the private chain of the Ethereum and the whole process are validated. In this way, the privacy information could be protected, and the coordination of local IDR resources is realized. The results can promote the development of block-chain based IDR trading protocol and the application of IDR transaction.

II. BLOCKCHAIN AND DECENTRALIZED TRANSACTIONS

A. Blockchain and the smart contract

Blockchain is a chain structure where all transaction data are packed into blocks, and the blocks are connected in chronological order[14]. With the technology of asymmetric encryption, Merkel tree, proof of work consensus mechanism and so on, the transaction data can be transparent, non-destructive and traceable. The essence of blockchain is a decentralized database. Compared with the traditional database, blockchain database is more difficult to tamper. However, if blockchain is only applied to data storage, its effect is limited. Therefore, combining blockchain with smart contracts is proposed to achieve more complex functions.

A smart contract is a set of programs defined in a digital form which prescribes the rights and obligations and is automatically executed by the computer system[4]. The characteristics of blockchain technology, such as programmability and decentralization, support the operation of the smart contract. With the aid of blockchain and smart contract technology, nodes in the network can participate in transactions' maintenance and management independently, realizing the safe funds transfer. If applying the decentralization to power transaction management, the cost can be reduced.

At the end of 2013, Vitalik Buterin, the founder of Ethereum, released the first edition of white paper. It indicates that Ethereum is a blockchain platform with the function of creating smart contracts. Ethereum has a fully developed Turing-complete programming language, which provides a programming environment for creating the smart contracts. Ethereum not only provides users with some pre-defined operations, but also allows users to create any complex operations. When an account executes a contract locally, the other accounts will update the contract's content to reach a new consensus. Thus, a credible transaction can be established without central supervision, which is very suitable for the multilateral transactions of distributed energy.

B. The development of Blockchain-based decentralized transactions

The blockchain has a profound impact on all walks of life, especially on the financial field. The most significant impact of blockchain on finance is asset tokenization[15] , which means trading with digital currency instead of legal currency. With the development of blockchain, the various tokens have come out, such as Bitcoin, Ether, Litecoin, which bring difficulties to the exchange of different tokens. At present, the trade of tokens is still done in centralized exchanges. Users deposit legal currencies or tokens into the exchange's bank account or blockchain account, and get other kind of tokens or legal currencies back. This mode makes the exchange face security problem, such as hacker attack, safekeeping risk and embezzlement of operators.

In order to solve the above problems, the open source community has carried out a series of attempts to build decentralized exchanges, such as EtherDelta, 0x project, KyberNetwork, etc. The 0x project is a decentralized transaction protocol which adopts the smart contract system in Ethereum[16]. It can be a shared infrastructure for the distributed application(Dapp) of the Ethereum and has wide application prospects. However, it cannot realize real-time order matching and complex order matching. Loopring, which is upcoming in April 2018, has also been paid much attention to. It has the identical design idea with 0x project. Users authorize the matchmaking contracts in Loopring to transfer the tokens to smart contracts. The exchange looks for a set of orders that can be matched and transfers the matching results to the smart contracts' address with its signature. The order can be matched out of chain while the transaction can be done on the chain[17].

C. Decentralized transactions of integrated demand response resources

The concept of blockchain corresponds to the need that the IDR resources participate in market transaction. On the one hand, the blockchain technology can support the massive peer-to-peer transactions among users and create a trusted distributed transaction environment for the P2P transaction among distributed energy resources. On the other hand, smart contracts provide a platform for the advanced transactions implementation by encapsulating the related protocols in the blockchain and executing them automatically[5]. In the process of transaction, the users' decision and transaction behaviors can be recorded as smart contracts and deployed in the blockchain. Thus, the decision can be made rapidly and the fairness of the transaction can be guaranteed.

Figure 1 shows a contrast of information exchange order between a centralized transaction scheme and a distributed transaction scheme[18]. Under the centralized transaction framework, users need to upload data information to the central processing system and make a scheduling decision of demand response resources by matchmaking or optimizing. Then users download the data information from the central processing system. Under the distributed transaction framework, the uploading of data is a broadcast process. After the contract is reached, the transaction's information is recorded in the blockchain and broadcasted. It can ensure the transparency of information and guarantee the privacy of users by anonymous accounts. In the transaction settlement process, under the centralized transaction framework, the central processing system needs to check users' information about energy, such as electricity, gas and heat and transfer funds to users through a third party such as banks. Under the distributed transaction framework, the tokens can be transferred according to the users' data and smart contracts without the central processing system.

There are two significant advantages of the decentralized electricity transactions. The first one is that blockchain technology allows the transactions to be carried on in a completely decentralized, trustless environment. It ensures the transparency of information and reduces the cost of transaction management. The other one is the adaptability to uncertainty of resources. For example, the users can go online and participate in the transaction at any time. During the transaction period, users can participate in the process of making contract only when they are online. The changes of schedulable IDR resources can be properly handled.

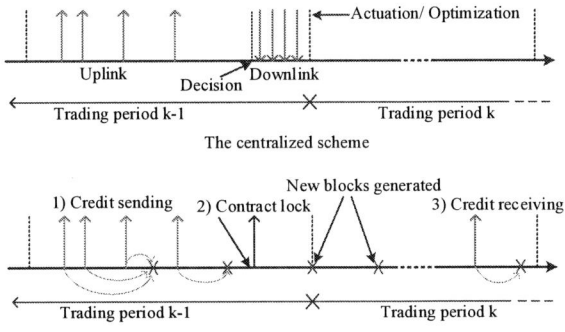

Fig. 1. Sequence of the communication exchanges in the centralized scheme and the blockchain-based scheme.

III. THE TRANSACTION FRAMEWORK OF THE DECENTRALIZED IDR TRADING

In the distributed IDR resource transaction framework designed in this paper, we assume that each user has a block-chain account with a pair of keys, called public key and private key respectively[19], and smart meters are available. All IDR users' available resources' amount need to be determined before the trading session. Inspired by the open source protocols of Loopring, the transaction process proposed in this paper is as follows:

Distributed trading participants need to authorize the matchmaking smart contracts to transfer a certain amount of token money from their block-chain account. The energy producers submit IDR resources orders, and each order contains the energy form, the amount and the IDR resources deliverable time. The energy consumers submit the upper limit of the resources he can provide in a certain period and his smart contract address of the response strategy. All participants send their own orders to the exchange with signature using their private keys. After the exchange received these separated orders, it will replace these orders into a corresponding order-book, while updating a new block and calculating each orders status to match the order set (the same energy form, the same deliverable period, etc.). Once the orders are successfully mix-matched, the exchange will send out a signature to the given matchmaking smart contract address. The matchmaking smart contract will calculate

and broadcast the difference between the IDR demand and the IDR supply based on the data interaction with the response strategy smart contracts. The matchmaking smart contract will send a fixed value (such as the retail price) as the initial IDR resources purchase price, the user response strategy smart contract will send back the available resources' amount. The price will be adjusted and broadcasted by the matchmaking smart contract according to the difference between the demand and supply, until to meet the target amount or reach the price ceiling. After the transaction agreement is reached, the settlement smart contracts will be generated. After the event, smart meters will automatically upload the data to the settlement smart contract and complete the transfer of token money.

We take the electricity transactions between DRE generators and DR users for example. The deliveries of information, funds and electricity in blockchain based market are given in Fig.2.

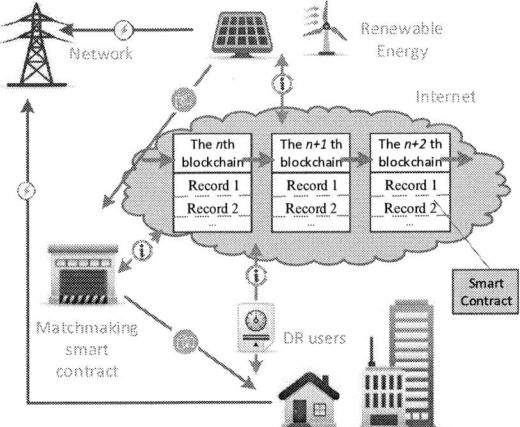

Fig. 2 The deliveries of information, funds and electricity

The DRE generators broadcast their DR resources demand in trading period T, which could be either positive or negative; and DR users also broadcast their resources available in period T.

The process of achieving the transaction consensus is as follows:

1) DRE generators and DR users authorize the matchmaking smart contract to transfer a certain number of tokens from their blockchain account. The user calls the response strategy smart contract and uploads the parameters of its cost function through the contract interface, and then broadcast to the whole blockchain network.

2) When a new generator or user participates into the decentralized market, a new node comes about. The nodes will generate their own orders and send it to the exchange with signature. Thus, all orders go into a medium and are ready to be mix-matched. The orders can be sent to multiple exchanges if there are some ways to avoid duplicate transactions.

3) The exchange replaces these orders into the electricity order-book, and monitors the order status (canceled, changed, etc.) from the blockchain. The

exchange matches the orders according to some principles, such as the same kind of energy form, the same deliverable time and DR oversupply principle.

4) The exchange will send the order set with a signature to the matchmaking smart contract address. The smart contract verifies all the signatures mentioned above and check the latest status of the orders.

5) First, the matchmaking smart contract calls the DR demand calculation smart contract to get the demand value and sends an initial DR price to DR users' response strategy smart contract. And then adjust the DR price according to the feedback data and formula (1) - (2). An empty block is built at set intervals (15s in Ethereum) to record the operation of the smart contracts.

6)After received new DR price, the response strategy smart contract modified the DR supply amount based on the cost function automatically, until meet the formula (3).

$$\sigma = \sum_{i=1}^{N_\sigma} P_{Gi} - \sum_{i=1}^{N_L} P_{Di} \qquad (1)$$

$$\lambda^{k+1} = \lambda^k + \rho \cdot (\sigma / \sum_{i=1}^{N_\sigma} P_{Gi}) \qquad (2)$$

$$\sigma \leq \varepsilon \qquad (3)$$

Where P_{Gi} is the DR demand of generator G_i; P_{Di} is the output of DR user D_i; and the term N_L, N_G are the number of DR users and DER generators, respectively. λ^k is the DR price in iteration k; ρ is a constant positive number; and σ is the difference between DR demand and supply.

The DR users change their supply according to their cost, which is calculated by the utility function [18]. The cost of positive output (consuming less power) and negative output (consuming more power) P_{Di} provided by users can be calculated by (4). The limit constraint is shown in (5).

$$f_{Di}(P_{Di}) = \begin{cases} \alpha_i P_{Di}^2 - (2\alpha_i P_i^0 - \beta_i)P_{Di} & P_{Di} \geq 0 \\ \alpha_i P_{Di}^2 & P_{Di} < 0 \end{cases} \qquad (4)$$

$$P_{Di}^{min} \leq P_{Di} \leq P_{Di}^{max} \qquad (5)$$

The marginal price of each DR user can be calculated by (6).

$$\lambda_i = \begin{cases} \beta_i - 2\alpha_i \cdot P_i^0 + 2\alpha_i \cdot P_{Di} & ,P_{Di} \geq 0, \forall i \in N_L \\ 2\alpha_i \cdot P_{Di} & ,P_{Di} < 0, \forall i \in N_L \end{cases} \qquad (6)$$

So, for each price, the DR user can provide a corresponding amount of DR resources.

7) The matchmaking smart contract creates settlement smart contracts when all the participants reach an agreement, which is executed when receiving the smart meters' data.

8) The exchange begins receiving new block and new data from the blockchain in order to update the order-book to mix-match new and existing orders.

The simplified transaction scheme is given in Fig.3.

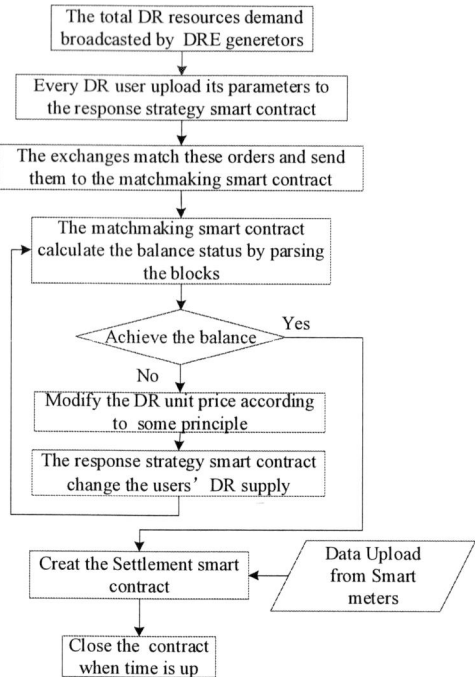

Fig. 3 Simplified flow chart of the proposed trading scheme

IV. THE FORMULATION OF SMART CONTRACTS

The order matching protocols of IDR resources' transaction should be developed on the basis of the widely accepted open source protocols. Considering the similarity of transaction matching, we will design this part after the Loopring comes out. This section focuses on the formulation of smart contracts for IDR transactions with the example of transactions between DRE generators and DR users.

Multilateral transactions can be divided into three major smart contracts in chronological order which are the DR demand calculation, the response strategy making and the transaction settlement. Once formulated, the smart contract will be broadcasted to all the accounts in the blockchain.

A. DR demand calculation contract

This paper uses Ethereum as the demonstration platform of smart contracts. When the renewable energy generators in distribution network send requests for response resources, they need to transfer a certain amount of ether as a margin to the smart contract address according to the historical price of DR resources. The smart contract will record all the requests, then calculates and publishes the total DR requirements. The basic elements of the contract are determined by Table I.

TABLE I BASIC ELEMENTS OF A DEMAND CALCULATION CONTRACT

Items	Type	Meaning
Buyer address	ADDRESS	Account address of buyer
DR amount	INT 256	Determined by the DRE generators
Delivery time	UINT	Determined by the DRE generators

B. Response strategy contract of DR users and the matchmaking contract

This paper divides the process of reaching consensus into two steps: submission of response amount and price updating. When the total demand respond amount meets the target or the time reaches the transaction deadline, the target response of each DR user can be obtained. The value of demand response and price will be used as reference for the formulation of settlement contract. Basic elements of the response strategy contract and the matchmaking contract are given in Table II.

TABLE II BASIC ELEMENTS OF A RESPONSE STRATEGY CONTRACT AND MATCHMAKING CONTRACT

Items	Type	Meaning
Buyer address	Address	Account address of buyer
Total DR amount	INT256	Determined by DR demand calculation contract
Unit price	INT256	Initial value is determined by historical data. Final value is determined after the transaction.
Delivery time	INT256	Determined by the order set
Consensus time	UINT256	Random values, such as 10MIN, but before delivery time

C. Settlement contract

When the number of DR users is very large, the settlement process will be very complex and hard to regulate. By means of smart contract, the funds can be automatically transferred according to the pre-agreed terms after the DR event.

The basic elements of the contract are shown in the Table III.

TABLE III BASIC ELEMENTS OF A SETTLEMENT CONTRACT

Items	Type	Meaning
Buyer address	ADDRESS	Account address of buyer
DR amount	INT 256	Determined by DR demand calculation contract
Unit price	UINT 256	Determined by matchmaking contract
Penalty charge	INT 256	Determined by market access agreement
Settlement time	UINT256	After the delivery time

The smart contract consists of DR amount, price, settlement time and the penalty charge which the users have to pay if their response amount is not equal to the contract value. We classify the response effectiveness into three types: over-response, under-response, and moderately response. Only moderately response type will not face penalties. The settlement rules should be clarified in the market access agreements.

Assume that smart ammeters can collect the data before and during the event. After the event period, the electric data will be transferred to the smart contract. The contract will be executed and funds transferred between accounts. Then, the transaction settlement completes.

Pseudocode for settlement smart contracts is as follows:

TABLE IV SMART CONTRACT DESIGN OF SETTLEMENT

M_i : *Each Smart Meter*
  Record energy consumption;
  Send day ahead and two hours ahead time-stamped & signed consumption to S1;
S1 : *Create settlement contract , on blockchain*
  Calculate baseline,
  Initial timer;
  Broadcast the new block which include the information of TABLE III. Notify all blockchain accounts;

M_i : *Each Smart Meter*
  Send time-stamped & signed consumption to S2;
S2: *Execute settlement contract, on blockchain*
  if timer ⩾ Time setting & the right account, then
    Compare baseline from S1 with the meter readings;
    Compute penalties, payments and charges
    Transfer tokens between accounts
  End
Broadcast

V. CASE STUDY

In order to verify the effectiveness of the transaction scheme proposed in this paper, the smart contracts on distributed transaction were deployed on the Ethereum for simulation test. There are four DR users and three DRE generators in this case.

A. Deploying a total DR demand calculation contract

The contract can be deployed on a public wallet, where DRE generators transfer a margin to before trading. This contract consists of the constructor, the DR demand uploading function, the amount query function and auxiliary functions. The constructor runs and get external variables when a contract created. DR demand uploading function is to record the delivery time and the demand amount. Amount query function provides interfaces for matchmaking contracts and the contract participants to query. DR demand uploading function is simulated in the virtual accounts of Remix. Suppose there are three DRE generator's accounts and the results are shown in the figure 4-5. The results of the DR demand uploading function are shown in Figure 4, which indicated the state of execution of the contract, the address of the contract, the address of the executor, the gas value of the execution of the contract, the contract's hash value, log and other information. Figure 5 shows the execution result of the query function. Considering that Solidity language does not support decimal numeric storage, in order to ensure the accuracy of the calculation results, the initial data are magnified. The data in figure 5 have been magnified 100 times and the actual DR demand is 5MW.

Each transaction needs to broadcast the hash value to other nodes for verification. If other nodes confirm the validity of the transactions which contained in the block, the block will be connected to the blockchain. The contract will be executed successfully.

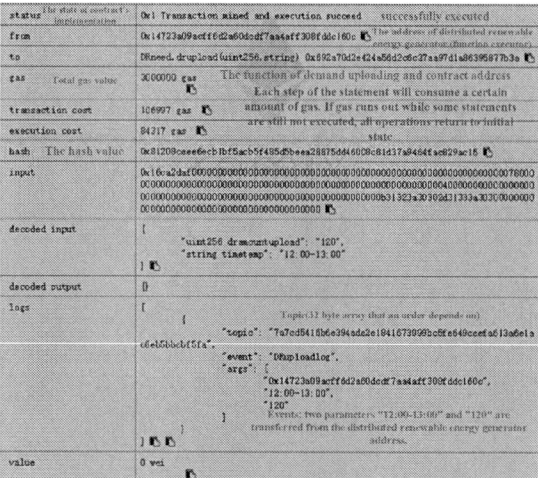

Fig. 4 The distributed renewable energy generator upload the demand of DR resource

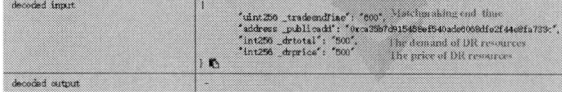

Fig. 5 The result of the total DR amount query function

B. Calculating users' response amount

The user's response calculation requires two contracts to coordinate with each other. One is the matchmaking smart contract (contract A) and the other is the response strategy smart contract (contract B).

(1) The matchmaking smart contract(contract A)

This contract realizes the adjustment of the DR price according to the data feedback date transferred from the response strategy smart contracts (contract B) of DR users. The number of users is set to 4, and the target value is 5MW.

The input of this contract is the amount of demand response submitted by users. The output are: the demand resources' new price(being sent back to the user's wallet when iterative termination criteria are not satisfied), the power of users(being output when iterative termination criteria are satisfied). This contract consists of a constructor, an input interface function, a price adjustment function, a result output function and auxiliary functions. It needs to be explained that if we do not update the users' output by means of the contract which calculating users' response amount, we can also refer to the classic quotation contract to count the users' output value which are submitted manually.

Assume that the initial value of constructor is 5ether/MW (ether is the unit of the virtual currency in Ethereum), which is magnified 100 times as figure 6.

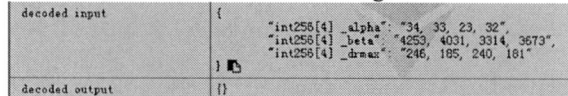

Fig. 6 Input of demand calculating function

During the first iteration, the function of data transferring transfers the price to contract B and the process is recorded in an event. In the figure 7, integer data 500 is transferred from the address of contract A to contract B.

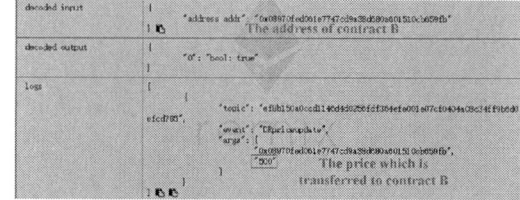

Fig. 7 Record of quoted price transferring

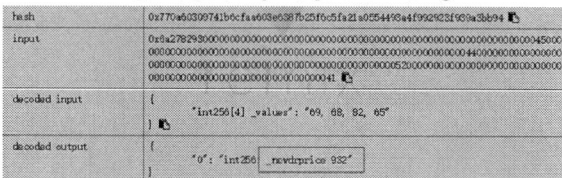

Fig. 8 Updating quoted price according to users' outputs

The output returned by contract B are [69,68,82,65]. Contract A adjusts prices according to the values and formula (1) - (2) and the update price is 9.32ether/MW.

The value will be transferred to contract B again and recorded by an event which contains the progress that an integer number 932 is transferred to contract B.

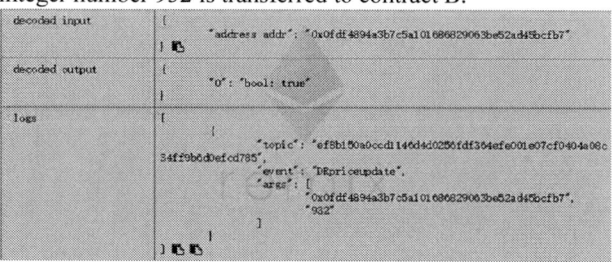

Fig. 9 Record of quoted price transferring

(2) The response strategy smart contract (contract B)

This contract consists of a constructor, a power calculation function, a data transfer function and some auxiliary functions. The inputs of constructor are characteristic parameters of the four DR users mentioned in contract A, which are shown in table 5. It needs to be explained that the contract B of four users can be create respectively. We puts all the user data into one contract for demonstration purpose.

TABLE V PARAMETERS OF DR USERS

No.	α_i	$2\alpha_i P_i^0 - \beta_i$	P_{Di}^{max} (MW)	P_{Di}^{min} (MW)
1	0.34	42.53	2.46	0
2	0.33	40.31	1.85	0
3	0.23	33.14	2.4	0
4	0.32	36.73	1.8	0

The values of users' parameters are magnified 100 times as the input of the constructor, and the operating results of the contract are shown in figure 10.

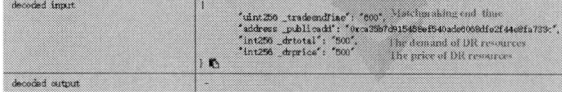

Fig. 10 Input of users' characteristic parameters

Contract B calculates the output based on the quoted

price according to the formula (4)-(6). As shown in figure 11, the outputs of four users under the initial quotation are 690,680,820,650KW separately.

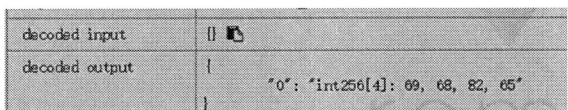

Fig. 11 Users' outputs under the initial quoted price

Invoking call function can transmit the output of each user (results got form contract B) to contract and get a Boolean parameter for return value. The process is shown in figure 12. Each user's power value was transferred to the contract A and recorded in the event DRmountTr: An array of 4 integer variables is transmitted from the address of contract B to the address of contract A.

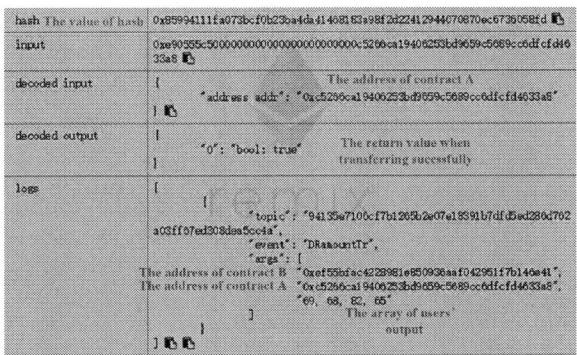

Fig. 12 The transferring of users' load under the initial quoted price

The function was convergent after six iterations and output the power of four users which were 1.16,1.16,1.51,1.14MW separately and the price was 36.64ether/MW. The output of each user during the last interaction in the remix was shown in Figure 13.

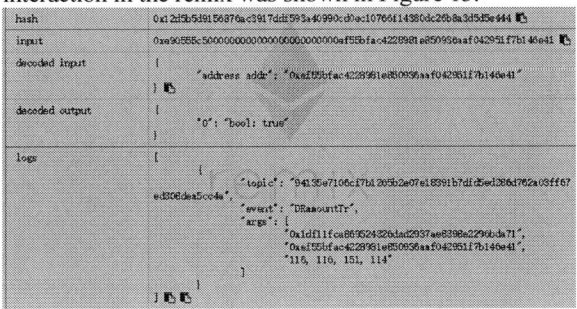

Fig. 13 Users' outputs when the iteration is converged

C. Transaction settlement contract

Transaction settlement contract mainly consists of constructor, baseline load calculation function, transaction amount calculation function, funds transferring function and other auxiliary functions. This contract is deployed by a public address, and the transfer of funds is confirmed by multiple signatures.

This paper takes the output of the first user as an example to illustrate the process of transaction settlement between the power users and the DER generators when the quotation has been determined.

(1) Constructor: this function has the same name with contract. With the aid of this function, the contract can be deployed preliminarily and initial values can be input which includes the address of user, the target output and the standard price. Suppose the input parameters of the contract are:
"0xb868ab9cf247345f586fa0f0750ce110c2202db3"(the address of user), 1160(the target output whose unit is KW), 3664(the standard price whose unit is 1013wei). The process of input is shown in figure 14.

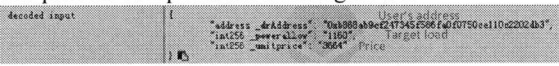

Fig. 14 The input of constructor

(2) Baseline load calculation function:

This function is to calculate users' baseline load value according to the historical load data. The calculation method refers to [20]. The inputs of this function include the users' day-before electricity data, two hours-before data and actual electricity data. The process of data input is shown in Figure 15.

(3) Calculation on transaction amount between two sides

Assume that the actual output of the user which within 10% of the target value belongs to moderate response when the contracts are settled at the standard price.

If the users' output belong over-response or under-response, the contracts are settled at 90% of the standard price. Then the formula for the amount of money paid to the user is as following:

Total amount=(Actual output-Baseline load)* Standard price* Penalty parameter

According to the first user's historical output data, the amount these generators should pay to the first user is shown in figure 16. It records the transfer of tokens between two addresses in the form of event. Converting the unit to Ether and we could obtain the amount of money is 1.58328Ether that paid to the user based on the value of the user's response.

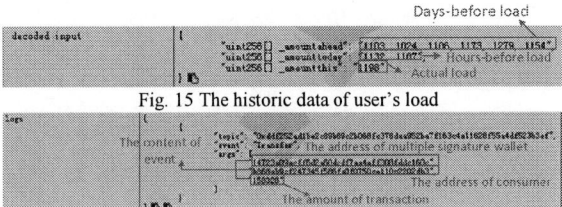

Fig. 15 The historic data of user's load

Fig. 16 The amount transfer between multiple signature wallet and the user's wallet

These contracts were deployed on metamask, which is an application that can allow users to run Ethereum dApps in browser without running a full Ethereum node. The event which recorded the transaction's process is shown in the figure 17. It shows that 1.58328Ether was transferred from the address of the multiple signature wallet to the user.

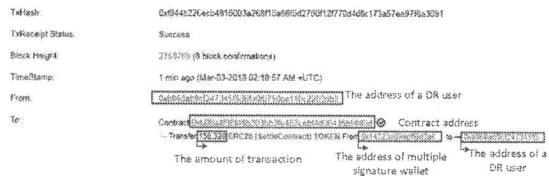

Fig. 17 The record of amount transfer in metamask

Similarly, the other three users received compensation for their response are 1.14348, 0.733, 0.70368Ether separately.

VI. CONCLUSIONS

The decentralized characteristic of the blockchain technology is highly compatible with the trading demand of the integrated demand response resources. This paper discussed the application of blockchain technology in the integrated demand response and proposed a decentralized transaction scheme and the interaction process of the smart contracts.

The trading of electricity demand response was taken for example to explain the scheme in detail. The DR users adjust their DR supply by monitoring the offer broadcasted in the blockchain, so that all the electricity generated by distributed renewable energy can be consumed. The article designed and deployed the trading smart contracts on the Ethereum and verified the trading process. The designed transaction scheme can be extended to transactions in other energy form.

REFERENCES

[1] A. Nieße, S. Lehnhoff, M. Tröschel, et al, "Market-based self-organized provision of active power and ancillary services: an agent-based approach for smart distribution grids," *Complexity in engineering (COM-PENG)*, pp. 1-5, 2012.

[2] Aitzhan, Nurzhan Zhumabekuly, and D. Svetinovic, "Security and Privacy in Decentralized Energy Trading through Multi-signatures, Blockchain and Anonymous Messaging Streams." *IEEE Transactions on Dependable & Secure Computing*, pp.99:1-14, 2016.

[3] Y.Yuan and F.Wang, "Blockchain: the state of the art and future trends," *Acta Automatica Sinica*, vol. 42, no 4, pp. 481-494, 2016(in Chinese).

[4] M.Zeng, J.Cheng, Y.Wang, et al, "Primarily Research for Multi Module Cooperative Autonomous Mode of Energy Internet Under Blockchain Framework," Proceedings of the CSEE, vol. 37, no 13, pp. 3672-3681, 2017(in Chinese).

[5] "Digital Point Based on Blockchain Technology," Goopal White Paper[EB/OL]. Goopal, 2015. http://goopal.online/Z-documents_white_paper.html.

[6] Deutsche Energie-Agentur GmbH (dena), "Blockchain in the energy transition: A survey among decision-makers in the German energy industry," *tech. rep.*, 2016.

[7] Al Kawasmi E, Arnautovic E, Svetinovic D, "Bitcoin-based decentralized carbon emissions trading infrastructure model," *SystEng*, vol. 18, no 2, pp. 115-130, 2015.

[8] J.Ping, S.Chen, N.Zhang, et al, "Decentralized transactive mechanism in distribution network based on smart contract," Proceedings of the CSEE, vol. 37, no 13, pp. 3682-3690, 2017(in Chinese).

[9] B.Li, W.Cao, B.Qi, et al, "Overview of Application of Block Chain Technology in Ancillary Service Market," *Power System Technology*, vol. 41, no 3, pp. 736-744, 2017(in Chinese).

[10] X.Ouyang, X.Zhu, L.Ye, et al, "Preliminary applications of blockchain technique in large consumers direct power trading,"

Proceedings of the CSEE, vol. 37, no 13, pp. 3737-3745, 2017(in Chinese).

[11] G.Wu, B.Zeng, R.Li, et al, "Research on the Application of Blockchain in the Integrated Demand Response Resource Transaction," Proceedings of the CSEE, vol. 37, no 13, pp. 3717-3728, 2017(in Chinese).

[12] W.She, Y.Hu, X.Yang, et al, "Virtual power plant operation and scheduling model based on energy blockchain network," Proceedings of the CSEE, vol. 37, no 13, pp. 3729-3736, 2017(in Chinese).

[13] X.Tai, H.Sun, and Q.Guo, "Electricity transactions and congestion management based on blockchain in energy internet," *Power System Technology*, vol. 40, no 12, pp.3630-3638, 2016(in Chinese).

[14] J. Mattila, T. Seppala, C. Naucler, et al, "Industrial Blockchain Platforms : An Exercise in Use Case Development in the Energy Industry," *tech. rep.*, *ETLA*, 2016.

[15] Christidis and M. Devetsikiotis, "Blockchains and Smart Contracts for the Internet of Things," *IEEE Access*, pp.4:2292-2303, 2016.

[16] Will Warren and Amir Bandeali, "0x: An open protocol for decentralized exchange on the ethereum blockchain," 2017, https://www.0xproject.com/pdfs/0x_white_paper.pdf.

[17] Loopring Project Ltd , "LOOPRING Decentralized Token Exchange Protocol v1.5,2017," https://github.com/Loopring/whitepaper/raw/master/en_whitepaper.pdf

[18] Koukoula, Despina I., N. D. Hatziargyriou, "Gossip Algorithms for Decentralized Congestion Management of Distribution Grids," *IEEE Transactions on Sustainable Energy*, vol. 7, no 3, pp. 1071-1080, 2016.

[19] Aitzhan N Z and Svetinovic D. "Security and Privacy in Decentralized Energy Trading through Multi-signatures, Blockchain and Anonymous Messaging Streams," *IEEE Transactions on Dependable & Secure Computing*, pp.99:1-1, 2016.

[20] W.Niu, L.Wang, Y.Li. "Calculation method and application of customer baseline load in demand response project," *Journal of Southeast University: Natural Science Edition*, vol. 44, no 3, pp. 556-560, 2014(in Chinese).

Indirect Current Control for Seamless Transfer of Utility Interactive Inverter

Kyungbae Lim[1], Injong Song[1], Jaeho Choi[1*]

1 School of Electrical Engineering, Chungbuk National University, Cheongju, Republic of Korea
*E-mail: choi@cbnu.ac.kr

Abstract— This paper describes the P+MR based indirect current control technique for seamless mode transfer of utility interactive inverter. In the utility interactive inverter operation, the inverter is operated as a current source or a sub power source for the utility interactive mode and also it operated as a voltage source in the islanded mode, respectively. Here, each single variable based control has been conventionally used in many researches, but it can cause the transient dynamic degradation due to the controller change at mode transfer. Recently, the indirect current control based inverter was proposed to improve the transient dynamic by controlling the output voltage consistently. In this paper, the P+MR based indirect current control topology was selected to improve the power quality at mode changes. Firstly, all design processes of P+MR based triple loop are described in detail with considering the system stability and transient dynamic. Secondly, the whole mode transfer technique which includes the unintentional islanding mode, is described based on the proposed indirect control. Finally, the validity of proposed method is verified by the PSiM simulation.

Keywords— *Indirect current control, seamless transfer, Unintentional Islanding, PR control.*

I. INTRODUCTION

A control of distributed generation (DG) using three-phase utility interactive inverter is one of the most popular issues in the microgrid research area.

In the utility interactive mode (grid-connected mode), the inverter should supply power to the grid or local critical loads with allowable power quality so a DG based inverter is operated as a current source or a sub power source for the utility interactive mode. If intentional islanding operation is needed or grid recloser is open due to the sudden grid fault, the inverter needs to be operated as a voltage source because it has to supply the full local load demand without the grid under islanded mode. On the contrary to this, the inverter needs to be transferred from islanded mode to utility interactive mode when the grid is recovered from fault or the intentional islanding is unnecessary anymore. During these above-mentioned transitions, the inverter needs to keep the power quality within allowable range to avoid the load damage due to the unrated local load voltage at mode transition.

There have been several researches that studied mode transition technique to improve the power quality at

mode transfer. In [1-3], the current control and the voltage control were used under the utility interactive mode and the islanded mode, respectively. Here, the load voltage quality might be deteriorated during the transient process due to the use of controller change at each mode transfer. Besides, the performance of load voltage regulation is quite sensitive to switch the operation process especially when the islanding occurs.

In [4], the droop based voltage control scheme was proposed. The droop based inverter was operated as a voltage source under both islanded and utility interactive mode. It was possible to guarantee the load voltage quality using proposed method during the operation mode transition, but the dynamic performance was little poor under the utility interactive mode.

In [5], a sine and cosine table based indirect current control was designed. This study contributed the correct voltage reference assignment with considering the correlation between both before and after voltage of the grid side inductor. Then local load voltage quality could be improved comparing with the case of which controller is just single loop. This method has two following drawbacks. Firstly, capacitor reference generation method is slightly complicated to adopt. Secondly, this method is hard to be extended to the non-linear load compensation due to its complexity when expected capacitor voltage reference is calculated by including the non-linearity value information.

In [6, 7], the synchronous frame control based indirect current control operation technique was proposed when the output local load is placed on output of LC filter. Its controller design includes the limiters for each d-q values which are outputs of grid current controller. Although it could cope with the control failure when unintentional islanding occurs due to the grid fault, its steady state output voltage quality is deteriorated because inverter output voltage is operated with threshold value of limiter [5-7].

In [8, 9], the PR (Proportional+ Resonant) control based indirect current control was used with same load location as in [6, 7]. If load voltage quality can be guaranteed using the proposed method under mode changes, this technique includes only the intentional islanding transfer for strategic islanding operation

The 2018 International Power Electronics Conference

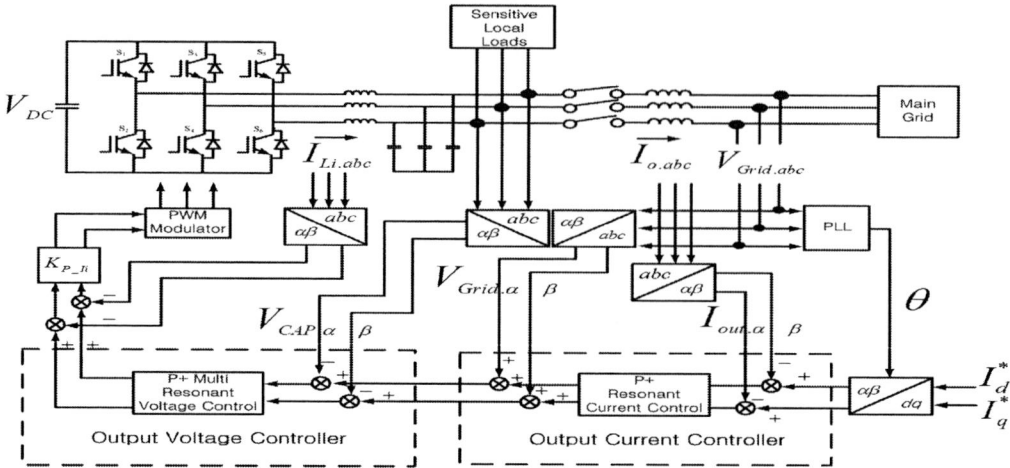

Fig. 1. Indirect current control method under both steady state islanded and utility interactive mode.

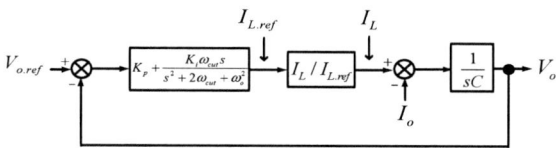

Fig. 2. Block diagram of middle voltage loop which includes inner current loop.

without the consideration about unintentional islanding event.

In this paper, P+MR (Proportional + Multi Resonant) based triple loop indirect current control has been adopted to realize seamless mode transfer. All the controller design processes of P+MR based triple loop are described in detail with considering the system stability and transient dynamics. The proposed method is based on the stationary reference frame and is basically designed by both the improved grid synchronization technique in [9] and the concept introduction of limiter in [5-7] to realize the seamless mode transfer under whole mode change events. It includes not only the intentional islanding but also the unintentional islanding. Comparing with [5-7], it is possible to transfer from unintentional islanding to steady state islanded mode. Then, it can contribute to maintain the voltage quality under islanded mode as a rated value.

Finally, the validity of proposed method is verified by the PSiM simulation.

II. CONTROLLER DESIGN

Figure 1 shows the steady state indirect current control scheme under both modes. As shown in this figure, it is triple loop controller which consists of output current control loop, middle capacitor voltage control loop and

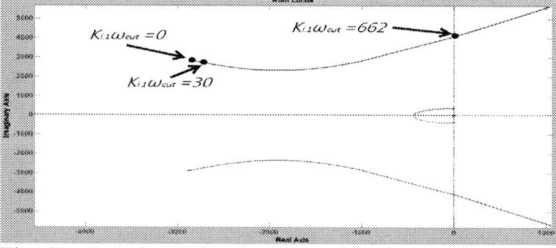

Fig. 3. Root locus for fundamental resonant gain $K_{i.1}\cdot\omega_{1.cut}$ variations under $K_p = 0.0577$.

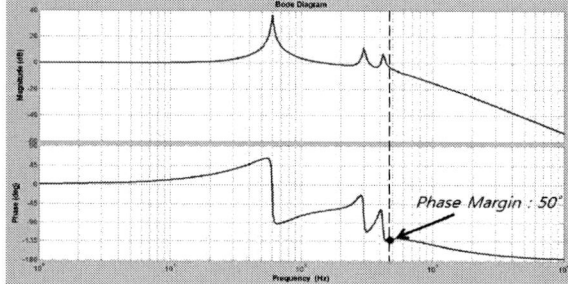

Fig. 4. Open loop bode plot of the voltage loop with harmonics compensation: $K_p = 0.0577$, $K_{i.1,5,7}\cdot\omega_{1,5,7.cut} = 30, 20, 15$

inner inverter side inductor current loop, respectively [8]. Therefore, output load voltage can be controlled consistently in spite of any mode changes, and that is why the seamless mode transition can be possible by this controller. The performance of this controller mainly depends on how robust the output voltage controller which includes inner P current control is designed under islanded mode. Based on robust middle voltage loop design, outer grid current control loop can be designed with the proper control bandwidth.

Figure 2 represents the block diagram of middle voltage loop which includes inner current loop. From Fig.

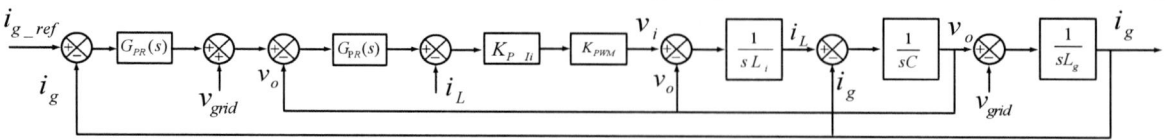

Fig. 5. Block diagram of outer grid current loop which includes middle capacitor voltage loop.

The 2018 International Power Electronics Conference

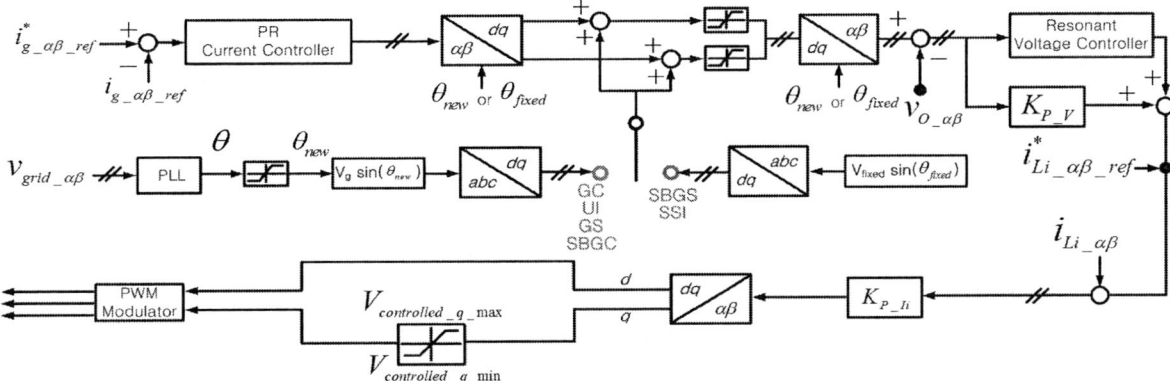

Fig. 8. Block diagram of proposed overall control method.

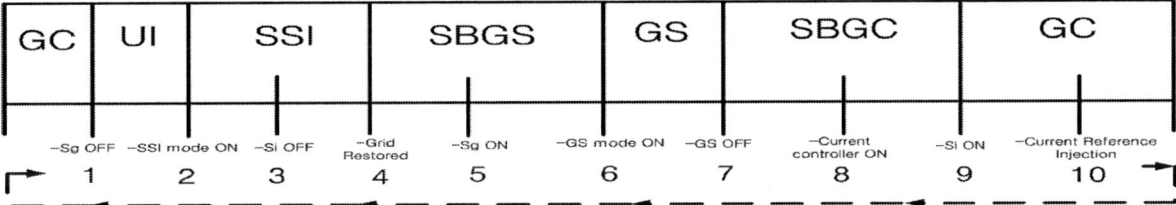

Fig. 9. Whole mode transfer sequence with considering Figs. 7 and 8.

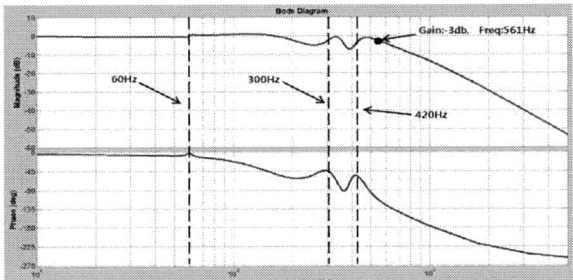

Fig. 6. Bode plot of the outer grid current closed loop transfer function.

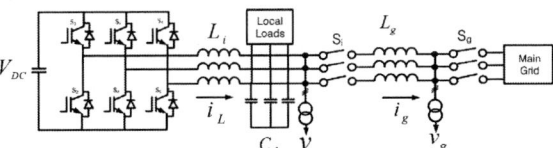

Fig. 7. Proposed system configuration with local loads.

2, the characteristic equation of the voltage loop can be obtained from its closed loop transfer function. Then, the proportional and resonant control gain can be obtained by assuming relative values as '0'.

Figure 3 shows the root locus for resonant gain $K_{i,1} \cdot \omega_{1\cdot cut}$ variations when K_p is 0.0577. As shown in this figure, the damping ratios are almost same even if they have different fundamental resonant gain. Hence, it can be checked that the transient dynamic of PR control is mainly affected by the proportional gain [10-13]. This fundamental component based resonant gain term can be extended to the harmonic resonant gain selection to compensate the degraded THD output voltage performance due to the non-linearity of local load.

Figure 4 shows the bode plot of the open loop P+MR based middle voltage controller. As shown in figure, it was designed to have large gain on the center frequency 60Hz and dominant harmonics frequencies 300, 420Hz with phase margin 50° which guarantees both the proper control stability and performance.

Figure 5 represents the block diagram of outer grid current loop which includes middle capacitor voltage

loop. Here, the closed loop transfer function can be obtained from Fig. 2 with predesigned middle voltage loop design. By using this closed loop transfer function, the bode plot of the closed outer grid current loop can be obtained as shown in Fig. 6. In this figure, the bandwidth of outer grid current loop is designed as 561Hz, which is about 1/10 times of 5kHz switching frequency. Here, it has to consider the tradeoff between the control margin and the control dynamic performance to select the outer current loop PR gain..

III. PROPOSED MODE TRANSFER METHOD

The mode transfer method using P+MR based indirect current control is described. As shown in Fig. 7, inverter supplies the local load through the LC filter and switch S_i is located between the local load and the grid-side inductor Lg. And then, this connects to the main grid through the recloser, S_g. This configuration of utility interactive inverter is referred from the configuration in [7]. Figure 8 presents the block diagram of the proposed overall control method. The mode transfer sequence in Fig. 8 is organized as the sequences circulation of GC (Grid Connected mode) → UI (Unintentional Islanded mode) → SSI (Steady State of Islanded mode) →SBGS (Standby of Grid Synchronization) → GS (Grid Synchronization) → SBGC (Standby of Grid Connected mode) → and GC again as shown in Fig. 9. The outer

805

grid current loop includes the well-designed the P+MR based multi loop voltage controller with inner proportional inverter side inductor current loop. The proposed method has two main design points.

Firstly, the main voltage reference is feedforwarded at the output of outer grid current loop to reduce the control action of the outer grid current loop and to return the overlarge voltage reference by the limiter threshold value to its rated value again. Secondly, the toggling switch design has been considered to change the feedforwarded voltage reference properly for seamless mode transfer. The mode transfer sequence is as following:

① Sudden grid fault occurs while the inverter operates under the utility interactive mode and Sg is opened to disconnect the DG based inverter with the main grid. Then, operation mode is automatically changed to the UI mode.

② In UI mode, the islanding detection signal is not received yet but the voltage reference can be protected by the limiter at the feedforward terms of d-q voltage references. In [7], magnitude of voltage reference is consistently maintained as this limiter threshold value. But this overlarge voltage cannot be used under the islanded mode fully because of the worsen quality of local load voltage regulation and also it becomes more serious if islanded mode is continued for a long time.

③ Islanding detection signal is received and inverter mode changes to the SSI mode by toggling the mode selection switch in Fig. 8 and Si can be opened at the same time and also outer grid loop is disabled to avoid the useless control action under SSI. In SSI mode, the fixed voltage magnitude and frequency are adopted for the feedforwarded voltage reference to improve the load voltage quality under SSI.

④ Grid has been restored from its pre-fault, then the inverter needs to change its mode from SSI to GC mode. SBGS and GS mode are needed between both modes. In SBGS mode, after detecting the grid restoration signal, Sg is closed and the grid voltage can be sensed by DSP. Then the GS mode should be included. For the seamless grid synchronization under the stationary reference frame, the GS mode reference modification method can be used referring from [9]. The limiter design of PLL was proposed with considering the compatibility between the limiter value and the allowable frequency range of fundamental resonant control.

⑤ After the GS mode is complete, mode needs to be changed to SBGC by toggling the mode selection switch. Here, the grid voltage can become directly the voltage reference. Before connecting to the main grid, the outer grid current loop is enabled again. (Note that it does not make any transition because of both the zero grid current, $I_{g\_\alpha\beta\_ref}$, and the initially zero assigned grid current reference, $I^*_{g\_\alpha\beta\_ref}$. In other words, output of outer current loop can be '0' due to the '0' outer current

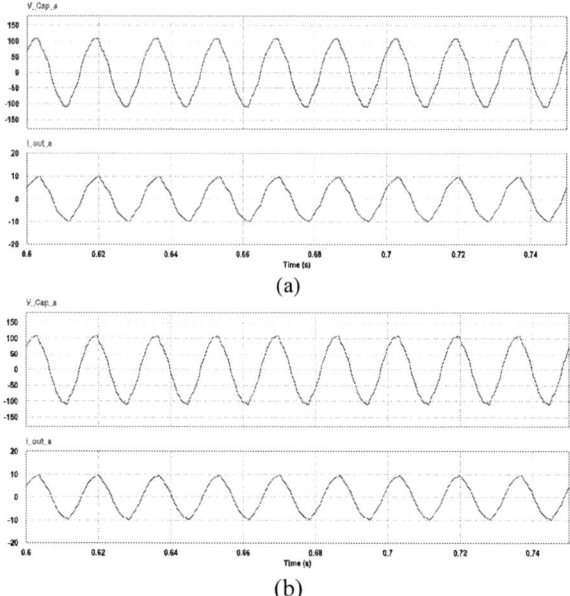

(a)

(b)

Fig. 10. Simulation result under GC mode according to existence of harmonics compensation: (a) Without harmonics compensation (THD: I_g(5.6%), V_o(5.1%)), (b) with harmonics compensation (THD: I_g(3.2%), V_o(3.7%)), (top) capacitor voltage (load voltage), (bot) grid current.

error value.)

⑥ For the mode change to the GC mode, the switch Si is closed then finally GC mode operation starts by changing the magnitude of grid current reference as its rated value smoothly with ramp function.

IV. SIMULATION RESULTS

Table 1 shows the simulation parameters for the utility interactive inverter. The system configuration and control scheme are based on Figs. 7 and 8. The Δ-Y transformer is located on the point between local load and the grid to synchronize inverter output voltage magnitude with grid voltage. The sampling frequency is 10kHz and switching frequency is 5kHz.

Figures 10 and 11 show the simulation results under SSI and GC modes, respectively. These figures are presented to verify the performance by gain selection criteria which include harmonics compensation in this paper. As shown in Fig. 10, THDs of both load voltage

TABLE I
SIMULATION PARAMETERS

Symbol	Meaning	Value
I_{Peak}	Peak grid current	9.5 A
V_{dc}	DC link voltage	400 V
f_{sw}	Switching frequency	5kHz
V_{rated}	Rated output voltage	105V
L_i, L_g	Filter inductance	3, 1mH
C_f	Filter capacitor	40(Y)
K_{p_Ili}	Proportional gain of I_{Li} Loop	17.3
K_{p_v}	Proportional gain of V_o Loop	0.0577
K_{p_Io}	Proportional gain of I_g Loop	2
K_{i_1}, ω_{i_1}	Resonant gain of V_o Loop	30, 20, 15
K_{ii_1}, ω_{ii_1}	Resonant gain of I_g Loop	100

806

The 2018 International Power Electronics Conference

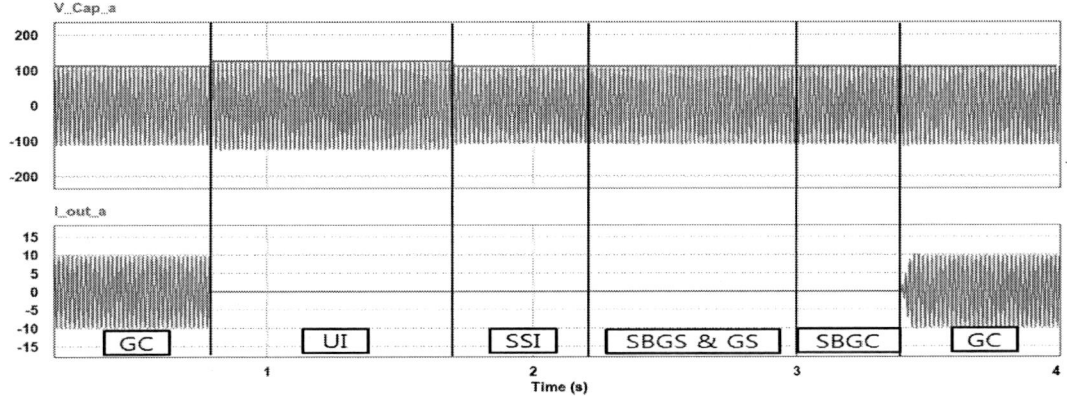

Fig. 12. Simulation result under whole operation modes.

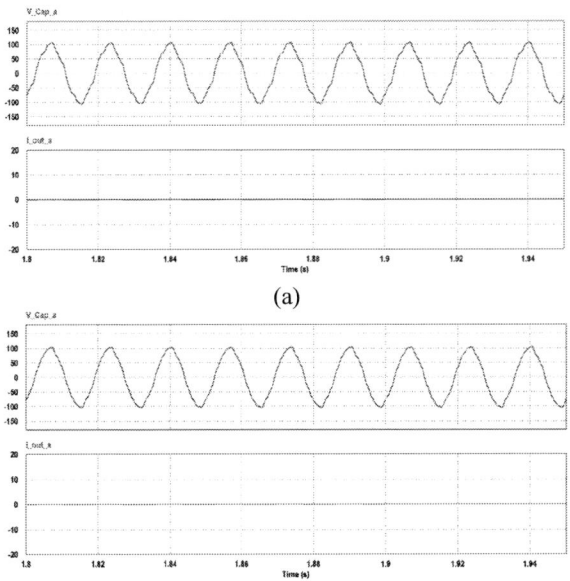

(a)

(b)

Fig. 11. Simulation result under SSI mode according to existence of harmonics compensation: (a) Without harmonics compensation (V_o_THD: 7.6%), (b) with harmonics compensation (V_o_THD: 3.8), (top) capacitor voltage (load voltage), (bot) grid current.

and grid current were improved from 5.1, 5.6% to 3.7, 3.2%, respectively. In Fig. 11, outer grid current is '0' because inverter operates under SSI mode without connection to the grid. THD of load voltage under SSI was improved from 7.6 to 3.8%. This THD compensation performance is better than that in Fig. 10 because inverter has to supply the full local load without the grid under SSI (Note that THD of load voltage under GC mode is not strongly affected by non-linearity of local load because robust main grid keeps the output load voltage stably comparing with SSI mode.)

Figure 12 shows the simulation result under whole operation modes. As shown in this figure, the inverter is operated under GC initially then sudden grid fault occurs. The mode changes to UI with keeping the magnitude as a threshold value of limiter. After islanding detection signal has been received, the inverter changes its mode to

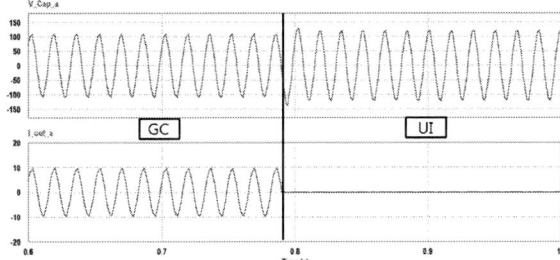

Fig. 13. Simulation result when mode changes from GC to UI : (top) capacitor voltage (load voltage), (bot) grid current.

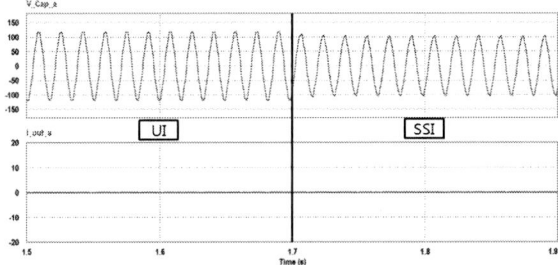

Fig. 14. Simulation result when mode changes from UI to SSI: (Top) capacitor voltage (load voltage), (bot) grid current.

SSI for better quality operation under islanded mode. After the grid has been restored, the mode changes to the SBGS and GS to track the phase of restored grid voltage. After the grid synchronization has been complete, the outer grid current loop is enabled to stand by the GC. After inverter switch Si is on, finally the inverter changes its mode to GC with assigning the rated grid current reference. Analysis for each mode can be provided with zoomed in waveforms at each mode transition event in Fig. 12.

Figure 13 shows the simulation result when mode changes from GC to UI. As mentioned before, mode changes from GC to UI at 0.79s by opening the grid recloser switch due to the sudden grid fault. But output load voltage is kept with the threshold value of limiter of voltage reference feedforwarded term.

Figure 14 shows the simulation result when mode changes from UI to SSI. As also mentioned before, it is hard to predict how long grid fault is continued so

807

The 2018 International Power Electronics Conference

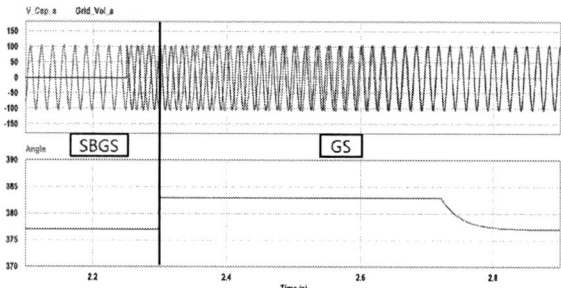

Fig. 15. Simulation result when mode changes from SBGS to GS: (Top) capacitor voltage (load voltage), (bot) phase angle of measured capacitor voltage.

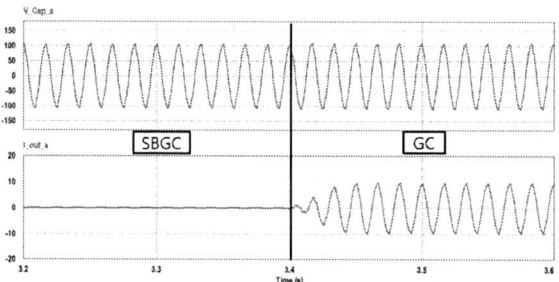

Fig. 16. Simulation result when mode changes from SBGC to GC: (Top) capacitor voltage (load voltage), (bot) grid current.

consistent UI mode operation is not preferred because output load voltage under UI mode is different from rated magnitude and frequency operation value under UI mode. Hence, SSI mode operation has been added to whole operation modes in this paper unlike the operation sequence in [7].

Figure 15 shows the simulation result when mode changes from SBGS to GS. At 2.1s, grid has been restored but sensed grid voltage is still '0' because grid recloser switch is open yet. After grid restoration was noticed, grid recloser switch Sg is on to sense the restored grid voltage. As soon as GS mode starts, feedforward voltage reference should be toggled as shown in Fig. 9. As shown in Fig. 10, grid synchronization is well performed with smooth transient.

Figure 16 shows the simulation result when mode changes from SBGC to GC. As soon as switch Si is closed to connect with the grid, rated grid current reference was assigned at 3.4s. After 3 cycles, inverter operates steady state GC mode operation well as shown in Fig. 16.

V. CONCLUSIONS

In this paper, the P+MR based indirect current control which considers the unintentional islanding due to the sudden grid fault, was investigated to realize the seamless mode transfer under whole inverter operation modes. The proposed scheme was designed to realize seamless mode transition at whole mode changes such as GC → UI, UI → SSI, SSI → GS, SBGC → GC and it has been verified that proposed scheme can achieve the improved performance than results in previous related researches through the PSiM simulation.

ACKNOWLEDGMENT

This work was supported by the Korea Institute of Energy Technology Evaluation and Planning (KETEP) and the Ministry of Trade, Industry & Energy (MOTIE) of the Republic of Korea (No. 20168530050030).

REFERENCES

[1] F.-S. Pai, "An improved utility interface for microturbine generation system with stand-alone operation capabilities," *IEEE Trans. Ind. Electron.*, vol. 53, no. 5, pp. 1529–1537, 2006.

[2] R. Teodorescu and F. Blaabjerg, "Flexible control of small wind turbines with grid failure detection operating in stand-alone and grid-connected mode," *IEEE Trans. on Power Electron.*, vol. 19, no. 5, pp. 1323 -1332, 2004.

[3] I. J. Balaguer, Q. Lei, S. Yang, U. Supatti, and F. Z. Peng, "Control for grid-connected and intentional islanding operations of distributed power generation," *IEEE Trans. Ind. Electron.*, vol. 58, no. 1, pp. 147–157, 2011.

[4] Y. Li, D. M. Vilathgamuwa, and P. C. Loh, "Design, analysis, and realtime testing of a controller for multibus microgrid system," *IEEE Trans. Power Electron.*, vol. 19, no. 5, pp. 1195–1204, 2004.

[5] J. Kwon, S. Yoon, and S. Choi, "Indirect current control for seamless transfer of three-phase utility interactive inverters," *IEEE Trans. Power Electron.*, vol. 27, no. 2, pp. 773–781, 2012.

[6] Z. Liu, J. Liu, and Y. Zhao, "A Unified control strategy for three-phase inverter in distributed federation," *IEEE Trans. Power Electron*, vol. 29, no. 3, pp. 1176-1190, 2014.

[7] Z. Liu and J. Liu, "Indirect current control based seamless transfer of three-phase inverter in distributed generation," *IEEE Trans. Power Electron.*, vol. 29, no. 7, pp.3368-3383, 2014.

[8] K. Lim and J. Choi, "PR based indirect current control for seamless transfer," *in Conf. Rec. of IPEMC'2016-ECCE Asia*, 2016.

[9] K. Lim and J. Choi, "Seamless grid synchronization of a proportional+resonant control-based voltage controller considering non-linear loads under islanded mode," *Energies 2017, 10*, 1514, 2017.

[10] D. N. Zmood and D. G. Holmes, "Stationary frame current regulation of PWM inverters with zero steady state error," *IEEE Trans. on Power Electron*, vol. 18, no. 2, pp. 814-822, 2003.

[11] D. N. Zmood, D. G. Holmes, and G. H. Bode, "Frequency-domain analysis of three-phase linear current regulators," *IEEE Trans. on Ind. Appl.*, vol. 37, no. 2, pp. 601-610, 2001.

[12] K. Lim, J. Jang, S. Moon, J. Kim, and J. Choi, "Output voltage regulation based on P plus resonant control in islanded mode of microgrids," *in Proc. of PEMC'2014.*,pp. 452-457, 2014.

[13] K. Lim, J. Choi, "Output voltage regulation for harmonic compensation under islanded mode of microgrid," *Journal of Power Electron.*, vol.17, no. 2, pp. 464-475, 2017.

The 2018 International Power Electronics Conference

Study of AC Power Interchange and DC Power Interchange for Micro Grid Systems

Kazuto Yukita[1*], Daiki Owaki[1], Shunsuke Horie[1] Toshiro Matsumura[1] and Yasuyuki Goto[1]

1 Aichi Institute of Technology, 1247 Yachigusa, Yakusa-cho Toyota Aichi, Japan

*E-mail: yukita@aitech.ac.jp

Abstract— The DC and AC interchanges are compared herein. Currently, an attempt has been made to prepare for the worsening of the power supply/demand balance arising from the spread of renewable energy by exchanging power between electricity companies. However, the synchronization of phases and frequencies in AC exchange is necessary. Therefore, a DC interchange is considered effective in simplifying the synchronization. Herein, we compare the DC and AC interchanges, and show the characteristics of each method from the results.

Keywords— power interchange UPFC PFC Distributed power supply

I. INTRODUCTION

In recent years, power generation facilities using renewable energies such as solar power and wind power have increased owing to global environmental and energy security problems. However, in these power generation facilities, the amount of power generation is unstable because it depends on weather conditions. Therefore, changes in frequency, voltage, etc. can adversely affect the electrical power quality. One of the countermeasures is to interchange electrical power. Power interchange is a method of exchanging power with other grids to compensate for the excess or deficiency of electrical power. Currently, electricity companies are further considering the implementation of power interchange using transmission and distribution lines. The method used currently is primarily electrical power interchange using AC. However, in Japan, the frequency used in the east is different from that in the west; therefore, frequency synchronization is necessary. To simplify the synchronization, power interchange using DC is considered an effective approach. The authors have studied AC power interchange, DC power interchange and various power interchanges.[1]~[5] Therefore, herein, we compared AC interchange with DC interchange.

II. TECHNICAL WORK PREPARATION

Power interchange refers to the sharing of surplus electrical power generated in a certain grid with other grids. Fig. 1 shows a schematic of power interchange. In this figure, for example, when surplus electrical power is generated in microgrid 1 (MG 1), it indicates that the excessive electrical power generated is sent to microgrid

2 (MG 2) and consumed by MG 2. If surplus power is generated in both MGs, it shall reverse the flow to the electrical power system. Table 1 presents an operation with power interchange. Further, Table 2 presents an operation without power interchange. Here, A shows that power is exchanged from MG1 to MG2; B shows that power is exchanged from MG2 to MG1; C shows that a reverse power flow from MG1 is performed; D shows that a reverse power flow from MG2 is performed; E shows that MG1 receives power from the power system; F shows that MG2 receives power from the power system; G shows no operation. In addition, PG shows the total amount of power generation, and PL shows the total load amount. These tables show that the electrical power received from the grid can be reduced using power interchange. Further, Table 3 gives the advantages and disadvantages of both the AC and DC power interchanges.

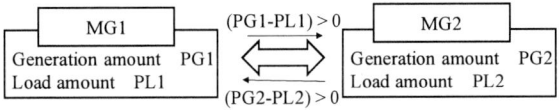

Fig.1. Schematic diagram of power interchange

TABLE I
Operation with power interchange

		MG2		
		PG>PL	PG<PL	PG=PL
MG1	PG>PL	C,D	A	C
	PG<PL	B	E,F	E
	PG=PL	D	F	G

TABLE II
Operation without power interchange

		MG2		
		PG>PL	PG<PL	PG=PL
MG1	PG>PL	C,D	C,F	C
	PG<PL	D,E	E,F	E
	PG=PL	D	F	G

809

TABLE III
Advantages and disadvantages of AC power interchange and
DC power interchange

	AC power interchange
Advantage	1. Reduce the amount of power storage device installed 2. Existing AC line can be used.
Disadvantage	DC power must be converted once to AC power before power interchange.
	DC power interchange
Advantage	1. Easy tidal current control 2. It is unnecessary to synchronize current, voltage phase, and frequency. 3. Solar power generation, high compatibility with storage batteries
Disadvantage	It is necessary to maintain a DC voltage such as a battery.

III. UPFC AND PFC STRUCTURES

Fig. 2 shows the unified power flow controller (UPFC) structure. The UPFC consists of two inverters: independent inverters on the series side and the parallel side. Each inverter is connected via a capacitor. The serial inverter independently controls the active power and reactive power. The parallel inverter controls the reactive power and active power generated by the series inverter.

Fig. 3 shows the power flow controller (PFC) structure, which is composed of two DC/DC converters. The control is divided at the master and slave. Using two DC/DC converters, the bidirectional power interchange control can be conducted. Fig. 4 shows a control block of the master and slave with PFC. Vin_ref is a DC voltage command value of between DC / DC converters , Vin is the DC voltage value of between DC / DC converters , Vdc_ref is a direct current system reference voltage value of the interconnection system, Vdc is the voltage value of the DC systems of interconnection systems. First of all, as shown in Fig. 2 (a), the control input of the master derives a voltage difference $\triangle$ Vin of between DC / DC converters .

$$\Delta V_{in\_m} = V_{in\_ref\_m} - V_{in\_m} \tag{1}$$

Then, enter the deviation, and performs PI control.

At this time, by using the output Iin_vin_ref_m, take the difference of DC current flowing through Idc_m in between the DC / DC converters, calculate the control input of the power m_chop_ref(master).

$$diin\_m = I_{in\_vin\_ref\_m} - I_{dc\_m} \tag{2}$$

Next, as shown in Fig. 2 (b), the control input of the slave calculates the deviation ΔVdc_s of the voltage value of the DC systems and the DC system reference voltage value of the interconnection systems.

$$\Delta V_{dc\_s} = V_{dc\_ref\_s} - V_{dc\_s} \tag{3}$$

Also, enter the deviation ΔVdc_s, and performs PI control. At this, just like control on the master, by using the output Idc_vdc_ref_s, take the difference between the current Idc of the DC system of the interconnection

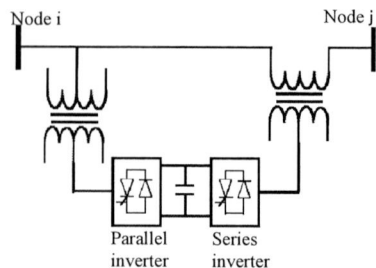

Fig.2. Configuration of UPFC

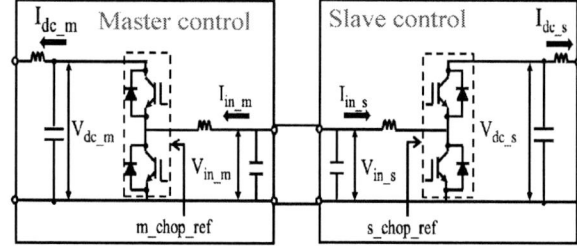

Fig.3. Configuration of PFC

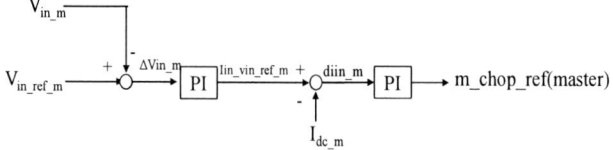

(a) Master control side control block

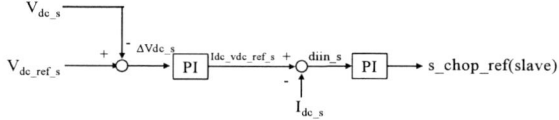

(b) Slave control side control block
Fig.4. Control block

systems, calculate the control input of the power s_chop_ref(slave).

IV. SYSTEM CONFIGURATION

Fig. 5 shows the system model for AC, in which the power interchange experiment was conducted. Further, Fig. 6 shows the system model for DC. The system model is composed of MG 1 and MG 2. Points A to G in the figure indicate the measurement points. Points A, B, C, and D in MG 1 are the measurement points for the received power from the power system, the generated power from the PV, the AC load power, and respectively. generation power of the PV, the AC load power and Similarly, points E, F, G, and H in MG 2 are the measurement points for the received power from the power system, the generated power from the PV, the AC load power, and the power storage battery power, respectively. The power interchange is measured at Point I. Here, the UPFC is used for the power interchange using AC, and PFC is used for the power interchange

810

using DC. As the power storage device is introduced via the bi-directional conversion device, the conversion from AC power to DC power, and vice versa, is possible. As shown in Table 4, the capacity of each device in this system is 10.0 kW for the PV power generation equipment, 10.0 kW for the PCS, 10.0 kW for the DC load, 3.0 kVA for the AC load, and 15.6 kWh for the power storage device. The bi-directional conversions of MG 1 and MG 2 are 20.0 kVA and 10 kVA, respectively.

V. EXPERIMENTAL METHOED

In this study, we conducted experiments on the AC power interchange and DC power interchange during the surplus PV power generation.

A. AC power interchange

Fig. 7 shows a control flowchart of AC power interchange. The flow of control in AC interchange detects the system power value of each MG. P1 indicates the power system on the MG1 side and P2 indicates the power system on the MG2 side. For each MG, if the power system value is positive or negative, no power interchange is made. Then, if one MG is positive and the other MG is negative, power interchange occurs according to the situation.

In the beginning of the system state, the generated power of PV on the MG 1 side is set to be smaller than the load power. The PV generated power on the MG 2 side is set to be larger than the load power. Specifically, on the MG 1 side, the PV output is approximately 2.5 kW and the load power is approximately 0.5 kW. Further, as the power storage device is fully charged, the surplus power is set to flow in the reverse direction to the power system. On the MG 2 side, the PV output is approximately 1.0 kW, the load power is approximately 2.0 kW, the power storage device is constantly charged at 0.5 kW, and power system is approximately 1.5 kW. From that state, the case where the load power on the MG 1 side and the PV generated power are made equal, the case in which the load power on the MG 2 side and the PV generated power are equalized were studied.

Under these conditions, power conversion using UPFC is performed.

B. DC power interchange

Fig. 8 and Fig. 9 show control flowcharts of DC power interchange. The flow of control in DC interchange detects the voltage of DC_bus in each MG. V1 indicates the DC voltage on the MG1 side and V2 indicates the DC voltage on the MG2 side. Also, Vbatt_max is the upper-limit value of the DC voltage in each MG. Vbatt_min is the lower-limit value of the DC voltage in each MG. First, if the DC voltage value of each MG is within the upper limit value and the lower limit value, no interchange is made. Then, if the DC voltage value of one MG is outside the set value and the DC voltage value of the other MG is within the set value, power interchange according to the situation is performed.

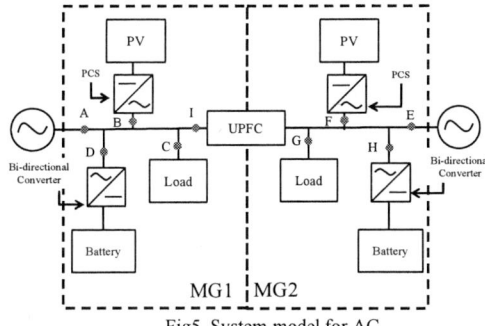

Fig5. System model for AC

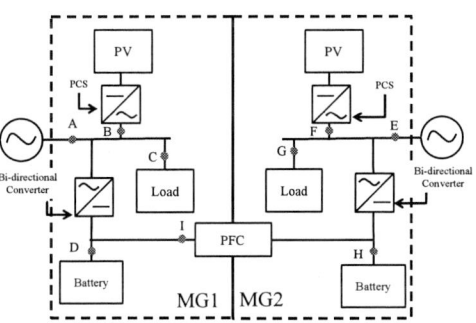

Fig.6. System model for DC

TABLE IV
System configuration

Device	Capacity
Photovoltaics system(PV)	10.0kW
Power Conditioning System(PCS)	10.0kW
AC load	3.0kVA
Bi-directional Converter(MG1)	20.0kVA
Bi-directional Converter(MG2)	10.0kVA
VRLA-Battery	15.6kWh
Unified power flow controller (UPFC)	1kVA
Power flow controller (PFC)	10kW

In both MG 1 and MG 2, the PV output power is set higher than the load power. Regarding the DC power interchange herein, control is performed using the DC voltage value of the power storage device in each MG. Table 5 gives the upper and lower limits of the DC voltage. The DC voltage upper-limit value of MG 1 is set to 336 V and that for MG 2 is set to 340 V. The DC voltage lower-limit value of MG 1 is set to 330 V and that for MG 2 is set to 320 V. In addition, the load power of MG 1 increased from 1.5 kW to 1.7 kW. The load power on the MG 2 side is increased gradually by 0.5 kW from 2.0 kW to 3.0 kW. The PV output increased gradually assuming a sunny day. Further, both MG 1 and MG 2 shall not receive power from the power system.

Under these conditions, power conversion using PFC is performed.

The 2018 International Power Electronics Conference

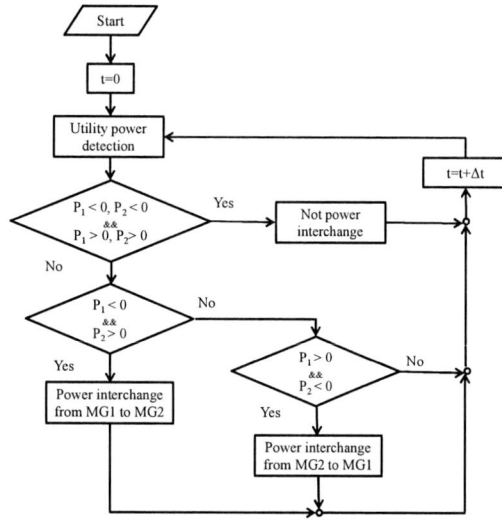

Fig.7. AC Power interchange flowchart

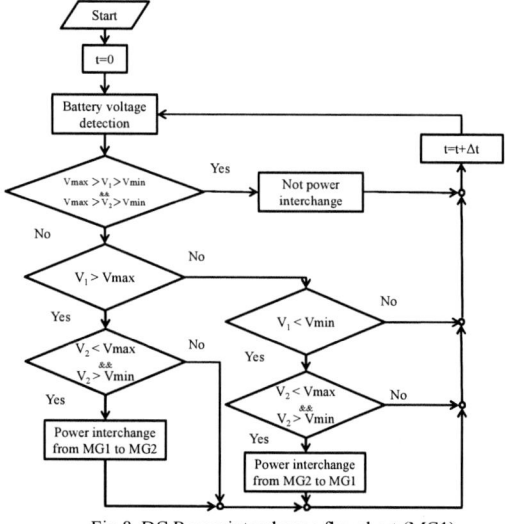

Fig.8. DC Power interchange flowchart (MG1)

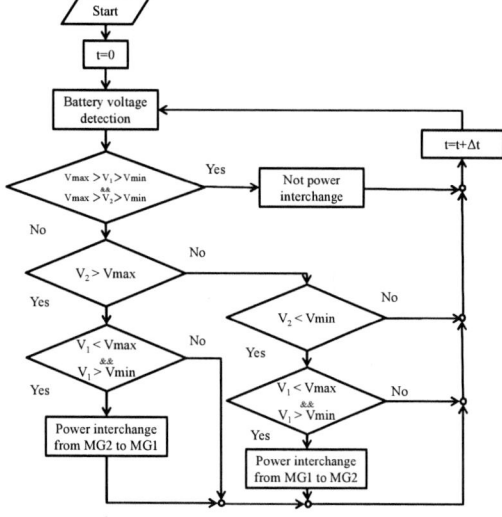

Fig.9. DC Power interchange flowchart (MG2)

TABLE V
System configuration

	Vbatt_max [V]	Vbatt_min [V]
MG1	336	330
MG2	340	320

VI. EXPERIMENTAL RESULTS

Here, the experimental results of the AC and DC power interchanges are presented. Describe characteristics and efficiency in each power interchange.

A. AC power interchange

Fig. 10 and 11 show the power characteristics of the AC power interchange in MG 1 and MG 2, respectively. The power characteristics of the power storage device are negative during charging, and positive during discharging. The power interchange characteristic is positive in that electrical power is flowing from MG 1 to MG 2. Conversely, it is negative when electrical power is flowing from MG 2 to MG 1. The PV surplus power is the reverse power flowing to the power system before MG 1 starts the power interchange. Moreover, MG 2 receives 1.7 kW from the power system. From Fig.6, it can be seen that on the MG 1 side, surplus power of PV is transmitted to the MG 2 side by the start of power interchange. Then, the load power on the MG 1 side was increased, and the generated power and PV generated power is set constant. The surplus electricity of PV ceased and the power interchange stopped. Then, the load power was reduced, surplus power of PV was generated, and the power interchange started again. Next, the load power on the MG 2 side was decreased, and the generated power and PV generated power are set constant. Then, the load power on the MG 2 side was decreased, and the generated power and PV generated power are set constant. Since there is no power to supply surplus power on the MG 1 side at the MG 2 side, the power interchange amount decreased. Then, as the load power increased, the power interchange amount again increased. In addition, from formula (1), the power conversion efficiency at the time of AC power interchange is calculated to be 86.1%.

From the above results, it can be seen that the AC power supply can be performed according to the flowchart, and it is understood that the PV surplus power can be power interchange.

B. DC power interchange

Fig. 12 and 13 show the power characteristics of the DC power interchange in MG 1 and MG 2, respectively. Fig. 14 shows the voltage value of DC_bus in each MG.

DC power interchange likewise, The power interchange characteristic is positive in that electrical power is flowing from MG 1 to MG 2. Conversely, it is negative when electrical power is flowing from MG 2 to MG 1. As shown, the surplus PV electrical power is charged in each storage battery of MG 1 and MG 2. Moreover, it is understood that no electrical power is supplied from the power system. Then, since the DC voltage of the MG 1 reaches the upper limit voltage 336 V and the DC voltage of the MG 2 is 334 V which is within the set value, the power exchange by the PFC is started. Subsequently, the amount of power interchange is increased every time the PV increases, and the charge amount of the power storage device is kept constant. Thereafter, the amount of power interchange is reduced owing to an increase in the load power on the MG 1 side. Further, when the load power on the MG 2 side increases, the shortage is covered by the discharging from the power storage device. Therefore, although the storage battery voltage of MG 1 gradually increases, MG 2 rises or decreases. In addition, from formula (1), the power conversion efficiency at the time of DC power interchange is calculated to be 87.6%.

From the results above, we found that even if PV surplus power is generated, the electrical power can be consumed in the MG by the DC power conversion and the charging and discharging of the power storage device.

$$\eta = \frac{P_{out}}{P_{in}} \times 100 \qquad (1)$$

η: Power conversion efficiency P_{in}: Input power

P_{out}: Output power

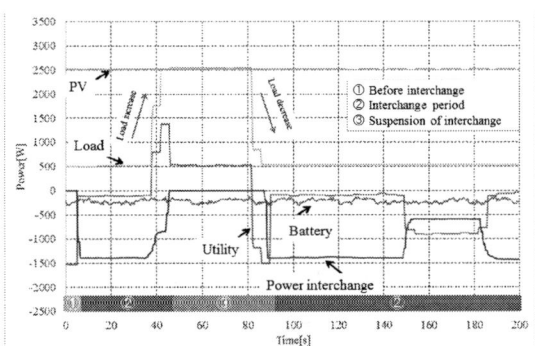

Fig.10. AC power interchange of MG 1

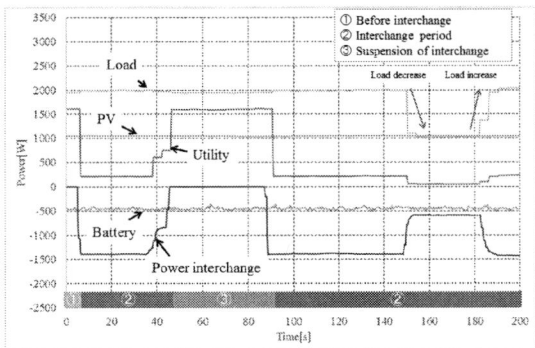

Fig11. AC power interchange of MG 2

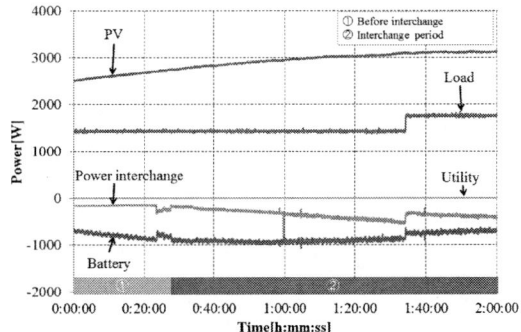

Fig.12. DC power interchange of MG 1

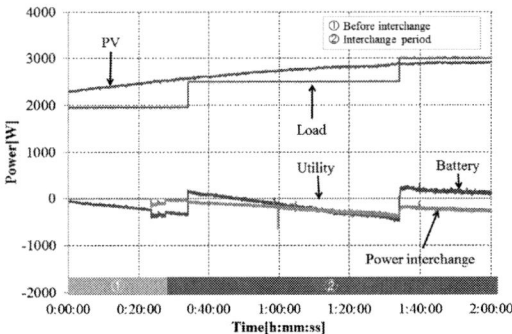

Fig.13. DC power interchange of MG 2

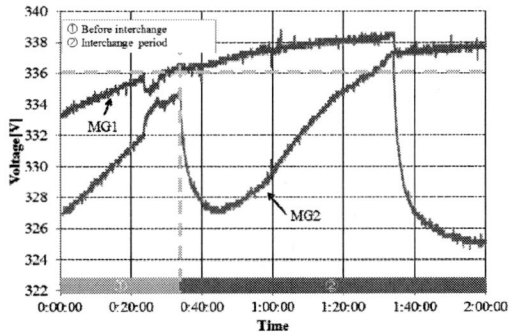

Fig.14. Power storage device voltage

TABLE VI
Efficiency comparison

	AC power interchange	DC power interchange	Difference
efficiency[%]	86.2	87.6	1.4

VII. CONCLUSION

This study investigated the AC and DC power interchanges. The results show a conversion efficiency of 86.2% for the AC power interchange. Further, the conversion efficiency of the DC power interchange is 87.6%. We found that the conversion efficiency of the DC power interchange is 1.6% better than that of the AC power interchange. Therefore, the effectiveness of the DC power interchange is confirmed.

References

[1] S. Tomoyasu, K. Yukita, Y. Goto, K. Ichiyanagi, T. Kinno, T. Takeda, K. Hirose, T. Ushirokawa, and T. Ota, "Power Interchange Using UPFC or PFC for Micro Grid" The 2013 Annual Conference of Power & Energy Society. IEE of Japan, vol.197

[2] S. Tomoyasu, T. Takeda, K. Yukita, Y. Goto, K. Ichiyanagi, and H. Morita, "Power Exchange Using PFC for Micro Grid" The 2014 International Power Electronics Conference, vol. 20F2-3.

[3] S. Tomoyasu, T. Takeda, K. Yukita, Y. Goto and K. Ichiyanagi, "Study of the power interchange considering microgrid characteristics" The papers of joint technical meeting on power engineering and power systems engineering vol PE-14-070 , vol. 20F2-3.

[4] H. Miyoshi, T. Takeda, K. Yukita, Y. Goto, K. Ichiyanagi, T. Ota, "A study of power interchange between micro-grids with distributed generations" The 2015 Annual Conference of Power & Energy Society. IEE of Japan, vol 219

[5] D. Owaki, K. Yukita, T. Matsumura and Y. Goto, "The power interchange using UPFC between power grids" The Institute of Electronics Information and Communication Engineers IEICE-117, no.240, EE2017-32, pp.7-12

The 2018 International Power Electronics Conference

Stability Enhancement Strategy for Islanding Microgrid with Multi-type Inverters Based on Hybrid Impedance Modelling

Meiqin Mao, Yong Ding, Yatao Shen, and Liuchen Chang
Hefei University of Technology
Hefei, China
mmqmail@163.com, yong_dsyuct@163.com, taoya_shen@163.com, lchang@unb.ca

Abstract—Because of the different output characteristic of deployed inverters including PQ-controlled and droop-controlled inverters, the system stability analysis and enhancement becomes a key issue for the islanding microgrid with multi-type inverters. This paper proposes an effective stability enhancement strategy for islanding microgrid with multi-type inverter based on the hybrid impedance modelling and deduced impedance-based stability analysis criterion. The proposed strategy provides different virtual impedance design methods for PQ and droop controlled inverters. For the droop-controlled inverter, the dual virtual impedance combining the inner virtual inductance and outer virtual inductance is adopted to ensure the inductive output impedance characteristic in the fundamental frequency to easily achieve the P/f and Q/V control and improve the resistive output impedance in the high frequency to enhance the system damping. For the PQ-controlled inverter, the inner virtual resistance control scheme is used to enhance the damping in the resonant frequency. An islanding microgrid model including multi-type inverters is established in Matlab/Simulink and the simulation results prove the effectiveness of the proposed strategy.

Keywords—Impedance-based model, Microgrid stability analysis, Inverter, Multiloop feedback system, virtual impedance.

I. INTRODUCTION

Islanding microgrids are localized power networks that incorporate with various distributed generators (DGs), energy storages, and local loads without connecting to the main power system [1]. Because of the different types and functions, the DGs in islanding microgrids are usually interfaced with the power electronic based inverters utilizing different control strategies, such as PQ-control, droop control and alike. The different output characteristic of deployed inverters makes the system stability analysis and enhancement become a key issue for the islanding microgrid with multi-type inverters.

Some modelling and analysis methods have been proposed to investigate the stability of microgrid, such as State Space Modelling (SSM) method [2], Dynamic Phasor (DP) method [3], Transfer Function Matrix (TFM) method and Impedance-based Modelling (IM) method [4].

The former three method mentioned above require detailed physical and internal control information of each inverter to derive the reference(input)-to-output relationship, which can be regarded as the "internal" stability and not convenient to be applied in practical applications. Compared with those former three methods, the IM method, which is firstly introduced for system stability analysis of DC system [5], can be seen as the "external" stability and focuses on the stability caused by the interconnections of system components. By taking the advantages of the terminal behaviors, the IM method does not need the internal information of each inverter, hence, has received wide attention in recent years [6-7].

There are several impedance-based stability criteria, most of which are based on the minor-loop concept [8-9], have been proposed to analyze the interactions of interconnected converters. [8] presents an impedance-based stability analysis method for systems consisting of paralleled voltage source converters by means of the Nyquist criterion for multi-loop systems. Grid-tied and islanded modes are both discussed in [8], however, the converters are represented by the Norton equivalent circuits in the grid-connected mode and the Thevenin equivalent circuits in the islanded mode respectively. A unified Impedance-based Stability Criterion (UIBSC) is proposed for paralleled grid-tied inverters in [9]. By modelling each inverter as an ideal current source in parallel with its output admittance, namely Norton equivalent circuits, a single grid-tied inverter can be derived as the equivalent stability analysis model for paralleled grid-tied inverters. Furthermore, a global minor loop gain (GMLG) can be obtained and checked to determine the system stability. Unfortunately, the converters applied in [8] and [9] used the same control method in each mode, the cases including the inverters with different control strategies have not been investigated.

For the Islanding Microgrid with Paralleled Inverters Controlled by Droop and PQ Strategies (IMG-PICDPS) system, there are few references have been published to analyze their stability problems. In previous work [4], a hybrid impedance modelling-based stability analysis method for IMG-PICDPS is initially proposed and two stability analysis criterions for IMG-PICDPS are obtained. However, stability enhancement strategies for IMG-PICDPS have not been further discussed in [4]. The main drive of this paper is to investigate the stability enhancement strategy for IMG-PICDPS based on the proposed hybrid impedance model.

Because of its impedance reshaping ability and performance improvement ability including damping the resonances, improving the output characteristics and the system stability, providing ancillary services including grid fault/disturbance ride-through, harmonic/unbalance compensation, the programmable impedances and improving the stability robustness of the converters against the different grid/load conditions, the virtual impedance-based control scheme have been receiving more and more attentions and thus is used in this paper to address the stability enhancement issue of IMG-PICDPS. The virtual impedance is in essence a lossless circuit-oriented control concept. Reference [10] presents an overview of the virtual impedance-based control strategies for voltage-source and current-source converters. And the proposed virtual impedance-based control strategies in literature can be divided into two categories, i.e., internal virtual impedance and external virtual impedance [11].Because the external virtual impedance is realized by modifying the inverter control reference, its control dynamics are limited by the inverter closed-loop control bandwidth. However, the internal virtual impedance is usually realized by introducing the feedback of state variables, so its control bandwidth can be high enough to damp the filter resonance.

In this paper, a virtual impedance-based stability enhancement strategy for the IMG-PICDPS is explored based on the hybrid impedance model and stability criterions proposed in [4]. In detail, two different virtual impedance control methods are proposed for DRoop-controlled Inverters (DR-Invs) and the PQ-controlled Inverters (PQ-Invs), respectively. For the DR-Invs, the double virtual impedance control scheme contains an inner virtual inductance and an external virtual inductance is adopted to guarantee that the output impedance of DR-Inv is inductive in the frequency near 50Hz and significantly resistive in the resonant frequency. For the PQ-Invs, the inner resistance control scheme is adopted to realize the goal of enhancing the damping of the PQ inverter at the resonance frequency.

The rest of the paper is arranged as follows. The structure and hybrid impedance model of the studied IMG-PICDPS are described in Section II. The proposed stability enhancement strategy is introduced in Section III. In Section IV, an IMG-PICDPS system is modelled in the Matlab/Simulink, and two cases are simulated to verify the effectiveness of the proposed stability analysis method, followed by the conclusions in Section V.

II. STRUCTURE AND HYBRID IMPENDENCE MODEL OF THE STUDIED IMG-PICDPS

A. Structure of the IMG-PICDPS

Fig.1 shows the power stage and control diagram of the studied IMG-PICDPS in this paper. The DR-Invs are connected to the energy storage systems (ESS), and the PQ-Invs are connected to DG units, such as photovoltaic and wind generators. The power loop in the impedance model can be neglected for that the inner loops have a bandwidth which is much greater than that of the low-pass filter of power loop. Fig.2 shows the control diagram of DR-Inv in α-β coordinate frame, where $G_u(s)$ and $G_c(s)$ are the transfer functions of the PR controller of voltage loop and the PI controller of current loop, respectively. G_{PWM} denotes the gain of PWM generator; Fig.3 is the control diagram of PQ-Inv, where $G_i(s)$ is the PI regulator of current loop, Kc is the capacitance current feedback coefficient for active damping.

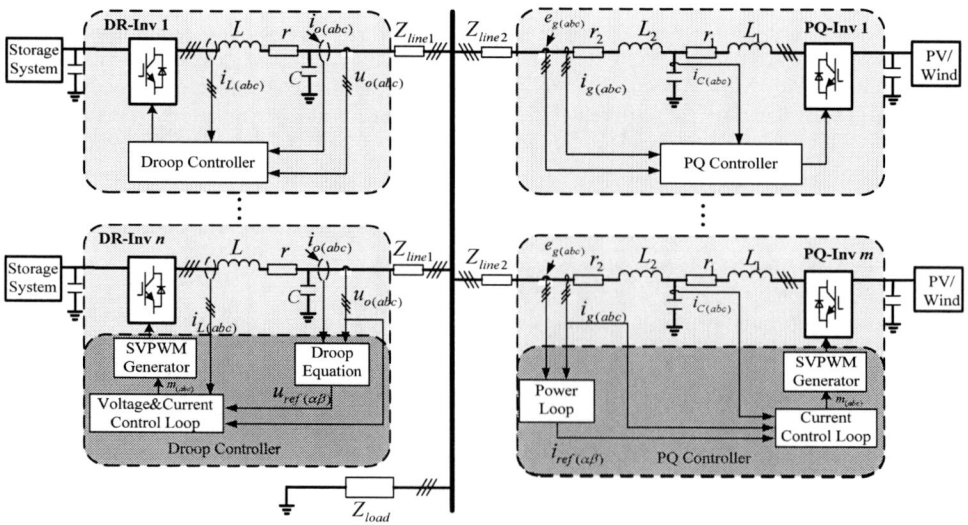

Fig. 1. Structure of the IMG-PICDPS

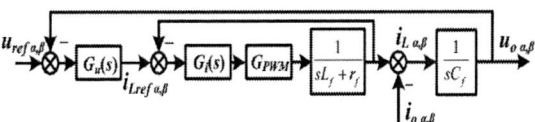

Fig. 2. Control diagram of DR-Inv

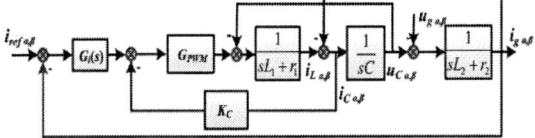

Fig. 3. Control diagram of PQ-Inv

B. Hybrid impedance model of IMG-PICDPS

The hybrid impedance model of the studied IMG-PICDPS is shown in Fig.4, where Z_{line1} and Z_{line2} denote the equivalent line impedances of PQ-Invs and DR-Invs, Z_L is the local load in the microgrid, subscript k and j denote the kth PQ-Inv and the jth DR-Inv, respectively. The detailed process of the derivation of impedance model can be found in [4].

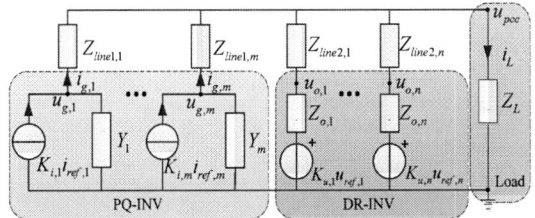

Fig. 4. Hybrid Impedance model of the IMG-PICDPS

Supposing that each PQ-Inv or each DR-Inv has the same parameters, that is, $Z_k = Z$ (or $Y_k = Y$); $K_{i,k} = K_i$, (where $Z_k = 1/Y_k$, $k =1, 2, \cdots, m$), $Z_{o,j} = Z_o$; $K_{u,j} = K_u$, (where $j =1, 2, \cdots, n$). Meanwhile, the line impedances of PQ-Invs and DR-Invs are all the same, that is, Z_{line}. According to the Fig.4, the node voltage can be derived as in equation (1). Since $i_L = Y_L u_{pcc}$, then equation (1) can be simplify to equation (2).

$$\left(Y_L + \frac{m}{Z+Z_{line}} + \frac{n}{Z_o+Z_{line}}\right)u_{pcc} = \frac{Z}{Z+Z_{line}}K_i\sum_{k=1}^{m}i_{ref,k}$$
$$+ \frac{1}{Z_o+Z_{line}}K_u\sum_{j=1}^{n}u_{ref,j} \quad (1)$$

$$u_{pcc} = \frac{1}{n}\frac{1}{1+\dfrac{m(Z_o+Z_{line})}{n(Z+Z_{line})}}[-(Z_o+Z_{line})i_L + K_u\sum_{j=1}^{n}u_{ref,j}$$
$$+ \frac{Z}{Z+Z_{line}}(Z_o+Z_{line})K_i\sum_{k=1}^{m}i_{ref,k}] \quad (2)$$

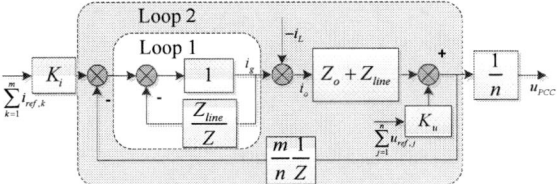

Fig. 5. Equivalent control diagram of IMG-PICDPS

According to the equation (2), the impedance model of the IMG-PICDPS can be equivalent to a double closed-loop control system as shown in Fig. 5.

C. Stability Analysis Criterion for IMG-PICDPS

Two stability analysis criterions which are expressed in equation (3) can be extracted from Fig.5 to predict the stability of the system. Previous work has demonstrated that the system stability of an IMG-PICDPS is not only related to the output impedance of PQ-Inv and the line impedance, but also affected by the output impedance of DR-Inv and the penetration of PQ-Inv.

$$C_1 : \frac{Z_{line}}{Z}$$
$$C_2 : \frac{m(Z_o+Z_{line})}{n(Z+Z_{line})} \quad (3)$$

III. STABILITY ENHANCEMENT STRATEGY

The proposed virtual impedance control schemes for DR-Inv and PQ-Inv are shown as Fig.6 and Fig.7, respectively.

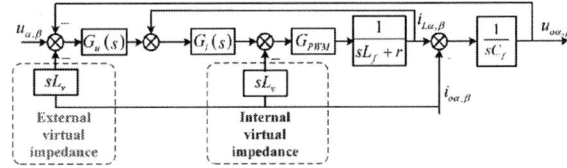

Fig. 6. Virtual impedance control schemes for DQ-Inv

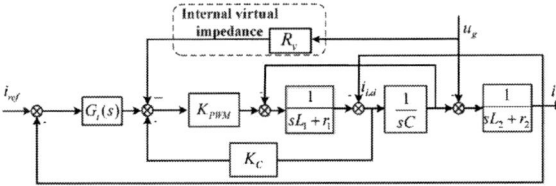

Fig. 7. Virtual impedance control schemes for PQ-Inv

A. Effect of virtual impedance on DR-Inv and PQ-Inv

According to Fig. 6 and Fig. 7, the output impedance $Z_o^{'}$ of DR-Inv and the output admittance $Y_i^{'}$ of PQ-Inv with virtual impedance are shown in equation (4) and (5). The Bode diagrams of the output impedance for DR-Inv with and without virtual inductance are shown in Fig.8. And the Fig.9 is the Bode diagrams of the output admittance for PQ-Inv with and without virtual resistance.

$$Z_o^{'} = \frac{sL_vG_{PWM}G_L(s)G_C(s)(1+G_u(s)G_i(s))+G_C(s)(1+G_i(s)G_{PWM}G_L(s))}{1+G_u(s)G_i(s)G_{PWM}G_L(s)G_C(s)+G_i(s)G_{PWM}G_L(s)} \quad (4)$$

$$Y_i^{'} = \frac{R_vG_{PWM}G_{L1}(s)G_C(s)G_{L2}(s)+G_{L2}(s)(1+G_{L1}(s)G_C(s))}{1+G_i(s)G_{PWM}G_{L1}(s)G_C(s)G_{L2}(s)+G_{L1}(s)G_C(s)+G_C(s)G_{L2}(s)} \quad (5)$$

In the Fig. 8, compared with the Bode diagram of Z_o, the virtual inductance makes the output impedance $Z_o^{'}$ of

DR-Inv is inductive in the frequency near 50Hz and significantly resistive in the resonant frequency. So it can facilitate the realization of P/f and Q/V droop control scheme, as well as enhance the damping of the system in the resonant frequency. For PQ-Inv., Fig. 9 shows that the resistive component of admittance for PQ-Inv is increased in the frequency near 1kHz by virtual resistance control scheme.

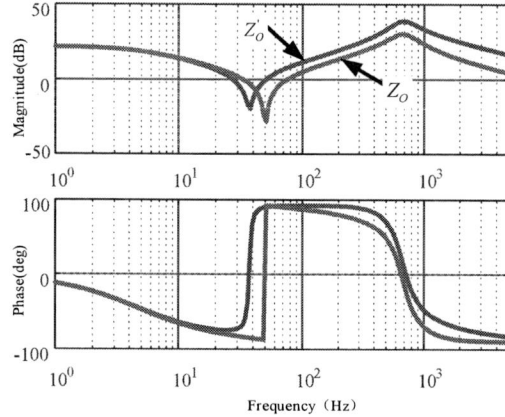

Fig. 8. Bode diagrams for the output impedance of DR-Inv with and without virtual impedance

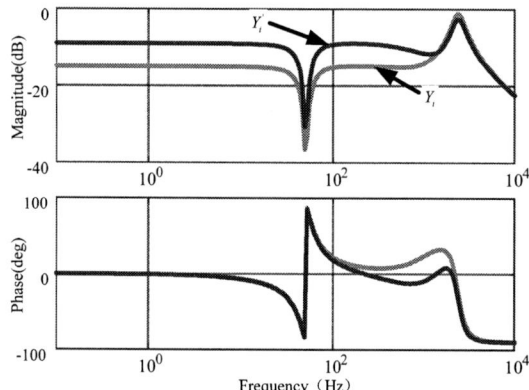

Fig. 9. Bode diagrams for the output admittance of PQ-Inv with and without virtual impedance

B. Stability verification

In this subsection, the effectiveness of the virtual impedance control scheme for PQ-Inv and DR-Inv will be validated by the proposed criterions C_1 and C_2. Let Z_i' $=1/Y_i'$, according to equation (4) and (5), the criterions C_1 and C_2 can be rewritten in equation (6). Fig. 10 is the Nyquist diagrams of C_1 and C_2 as line inductance increases with virtual impedance control both in PQ-Inv and DR-Inv and the Nyquist diagrams of C_1 and C_2 as the number of PQ-Inv increases with virtual impedance control both in PQ-Inv and DR-Inv are shown in Fig. 11, the Nyquist diagrams of C_1 and C_2 as line inductance increases without virtual impedance 1 and the Nyquist diagrams of C_1 and C_2 as the number of PQ-Inv increases without virtual impedance are also pictured in Fig.12 and Fig.13 for comparison and analysis.

Compared with Fig. 12 and Fig.13, it can be seen that the Nyquist diagrams of C_1 and C_2 in Fig. 10 and Fig.11

do not encircle the point (-1, j0) all the time. According to the Nyquist criterion for multiloop system, the system shown in Fig. 5 will be stable with the proposed virtual impedance control scheme.

$$C_1 : \frac{Z_{line}}{Z_i'} \qquad C_2 : \frac{m\left(Z_o' + Z_{line}\right)}{n\left(Z_i' + Z_{line}\right)} \qquad (6)$$

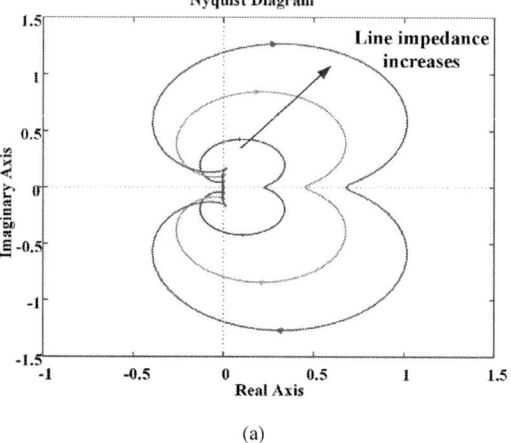

(a)

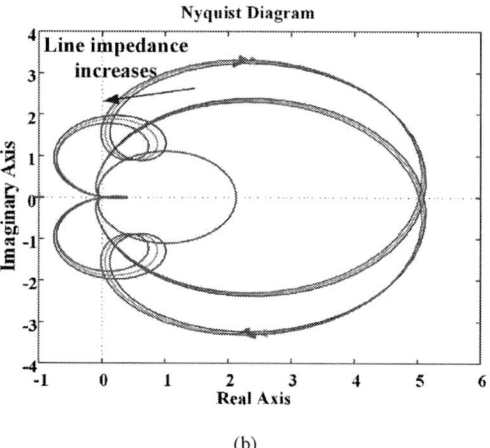

(b)

Fig. 10. Nyquist diagrams of C_1 and C_2 as line inductance increases with virtual impedance control both in PQ-Inv and DR-Inv ((a): C_1, (b): C_2)

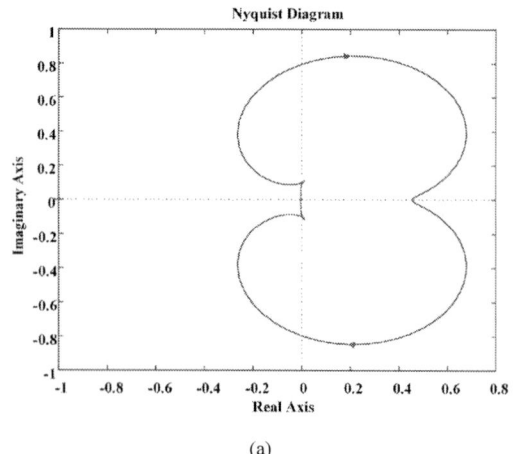

(a)

The 2018 International Power Electronics Conference

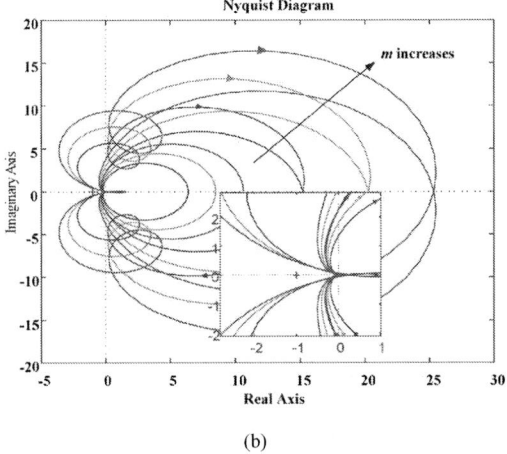

(b)

Fig. 11. Nyquist diagrams of C_1 and C_2 as the number of PQ-Inv increases with virtual impedance control both in PQ-Inv and DR-Inv ((a): C_1, (b): C_2)

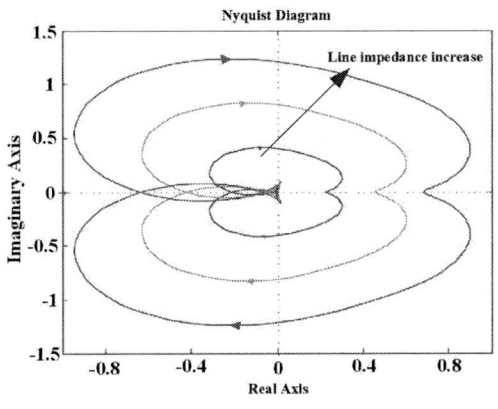

(a)

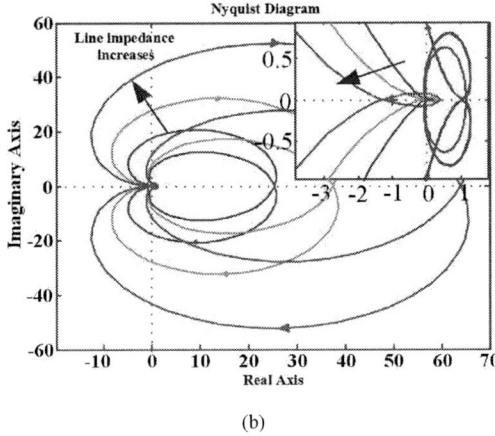

(b)

Fig. 12. Nyquist diagrams of C_1 and C_2 with the line inductance increases without virtual impedance control ((a): C_1, (b): C_2)

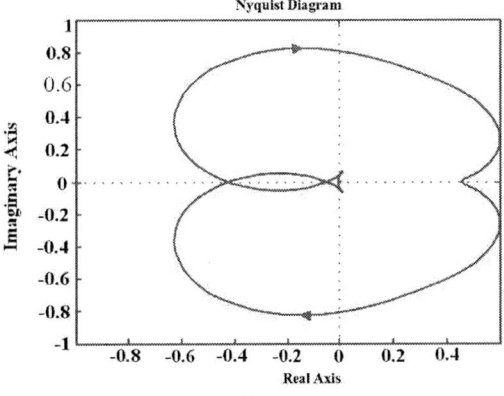

(a)

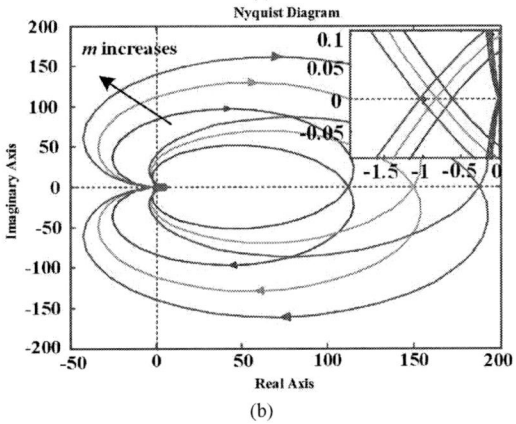

(b)

Fig. 13. Nyquist diagrams of C_1 and C_2 with the number of PQ-Inv increases without virtual impedance control ((a): C_1, (b): C_2)

IV. SIMULATION AND DISCUSSION

To validate the proposed virtual impedance control scheme in section II, two cases using Matlab/Simulink are performed. The parameters of the PQ-invs and DR-Invs in investigated IMG-PICDPS are shown in Table I.

TABLE I
PAPAMETERS OF THE PQ-INVS & DR-INVS IN IMG-PICDPS

	Parameter	Symbol	Value
PQ-Inv	Switching frequency	f_s(Hz)	10k
	LCL Filter Values	L_1(mH)	0.87
		C(uF)	22
		L_2(mH)	0.22
	PR controller	K_{ip}	5.6
		K_{ir}	1500
	Capacitive current feedback coefficient	K_C	0.7
DR-Inv	Switching frequency	f_s(Hz)	10k
	LC Filter Values	L(mH)	2
		C(uF)	20
	voltage loop controller	K_{up}	0.004
		K_{ur}	180
	current loop controller	K_{ip}	9
		K_{ti}	1.5

A. Case 1

For Case 1, one DR-Inv and one PQ-Inv are used. The initial value of line inductance for both inverters is 0.3mH and the line resistant is 0.1Ω. The rated PQ-Inv is 45kW and the load is resistive load of 100kW. The

819

simulation process consists of four steps as follows:

Step 1: The DR-Inv is put into operation at time t = 0;

Step 2: The PQ-Inv is put into operation at time t = 0.04s. At the same time, the active power reference is set to 45kW and the reactive power reference is 0;

Step 3: The line inductance is increased form 0.3mH to 1mH at 0.15s;

Step 4: The virtual impedance control schemes for DR-Inv and PQ-Inv are enabled at 0.3s, other conditions remain unchanged.

Fig. 14 shows the simulation results between 0.1s and 0.4s. From the Fig. 14, it can be seen that the two paralleled inverters can operate stably before increasing the line inductance at 0.15s. But serious oscillations of the PCC voltage and current will occur when the line inductance is increased to 1mH after 0.15s and before enabling the virtual impedance control at 0.3s. Then, the virtual impedance control schemes for PR-Inv and PQ-Inv are enabled at 0.3s. It can be seen that the oscillations of the PCC voltage and current are suppressed in Fig.14. The simulation results shown in Fig. 14 verify the theoretical analysis in section III.

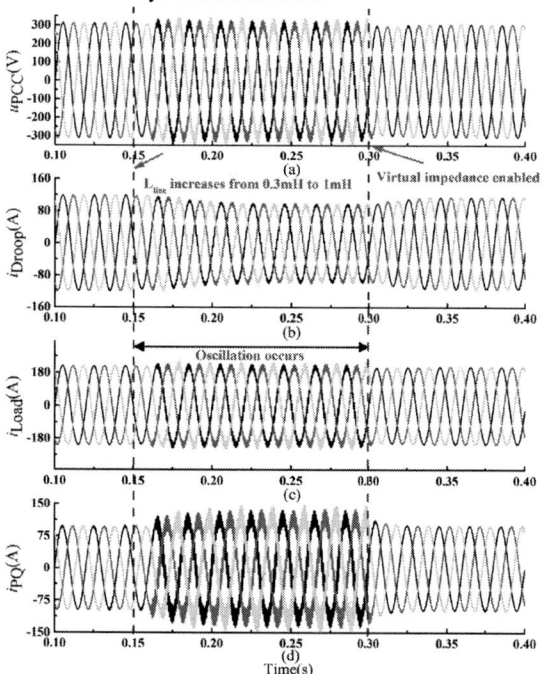

Fig. 14. Simulation results for changing the line inductance ((a): PCC voltage, (b):Load current, (c):DR-Inv current, (d)PQ-Inv current)

B. Case 2

In Case 2, there is one PQ-Inv and one DP-Inv, namely the PQ-Inv penetration is 50%, in the microgrid with the line inductance is 0.3mH and the line resistant is 0.1Ω before 0.14s. The load in this case is also resistive, but the capacity is 210kW. The number of PQ-Inv in the system increased to 2, 3 and 4, causing the PQ-Inv penetration increased to 66.6%, 75%, 80% at 0.14s, 0.2s and 0.26s, respectively. The virtual impedance control schemes for DR-Inv and PQ-Inv are enabled at 0.32s and the other conditions remain unchanged. The voltage of

the PCC is shown in Fig. 15(a), and the Fig. 15(b) and (c) show the current of load and DR-Inv, respectively.

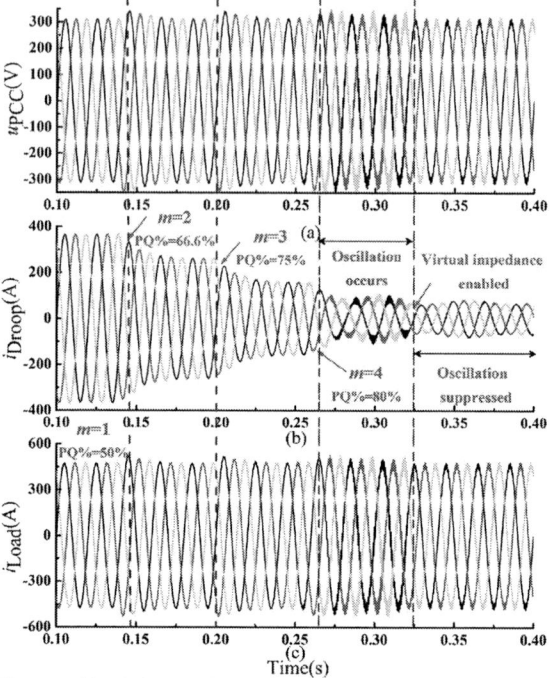

Fig. 15. Simulation results for changing the PQ-Inv penetration ((a): PCC voltage, (b): Load current, (c): DR-Inv current).

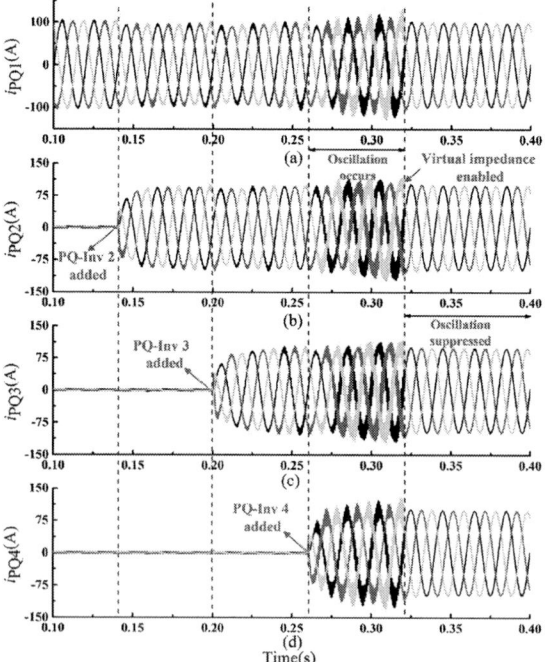

Fig. 16. PQ-Invs currents ((a):i_{PQ1}, (b): i_{PQ2}, (c): i_{PQ3}, (d): i_{PQ4}).

The output current of each PQ-Inv is shown in Fig. 16(a), (b), (c) and (d), respectively. From the Fig.15 and Fig.16, it can be seen that the system can operates stably when the number of PQ-Inv is less than four, i.e., the PQ-penetration is less than 80% before 0.26s. The oscillation occurs in the waveforms of voltage and the currents after the fourth PQ-Inv is added to the system at 0.26s. Then, the oscillations in the waveforms of voltage and the currents are suppressed when the virtual impedance control schemes for PR-Inv and PQ-Inv are enabled at 0.32s. These results imply that the stability of IMG-PICDPS will decrease as the PQ-Inv penetration increases, while the proposed stability enhancement strategy can effectively improve the system stability.

V. CONCLUSIONS

To realize the stability enhancement strategy for IMG-PICDPS, two different virtual impedance control methods are proposed for DR-Inv and PQ-Inv in IMG-PICDPS respectively. The double virtual impedance control scheme used for DR-Inv can facilitate the realization of P/f and Q/V droop control scheme, as well as enhance the damping of the system in the resonant frequency. Meanwhile, the inner resistance control scheme adopted for PQ-Inv can realize the damping enhancing at the resonance frequency, thus can do a favor in stability enhancement. Two cases are performed in Matlab/Simulink and the results verify that the stability of IMG-PICDPS will be decreased when the line inductance or the PQ-Inv penetration increases to a certain value in Case 1 and Case 2, respectively. However, when the virtual impedance control is enabled, the oscillations will both disappear and the IMG-PICDPS will recover to stable operation states in the two cases, which verifies the effectiveness of the proposed strategy.

ACKNOWLEDGMENT

This work was supported by NSFC (51577047); Foreign Science and Technology Cooperation Project of Anhui Province (1604b0602015).

REFERENCES

[1] M. Mao, P. Jin, N. D. Hatziargyriou, and L. Chang. "Multiagent-based hybrid energy management system for microgrids". *IEEE Transactions on Sustainable Energy*, vol.5, no.3, pp.938-946, 2014.

[2] Joel S. Modeling and control for microgrids[D]. Arizona State University, 2013.

[3] Mendoza-Araya, Patricio A., and Giri Venkataramanan. "Impedance matching based stability criteria for AC microgrids". in: *Energy Conversion Congress and Exposition (ECCE), 2014 IEEE*. IEEE, 2014. pp. 1558-1565.

[4] M. Mao, Y. Ding, Y. Shen, and L. Chang. "Hybrid Impedance-based Modelling and Stability Analysis of IMG-PICDPS". in *Energy Conversion Congress and Exposition (ECCE), 2017 IEEE, 2017*, pp.1-6.

[5] R. Middlebrook, "Input Filter Considerations in Design and Application of Switching Regulators", in *IEEE Industry Applications Society Annual Meeting*, pp. 366-382, 1976.

[6] Sun J. "Impedance-based stability criterion for grid-connected inverters". *Power Electronics, IEEE Transactions on*, 2011,26(11):3075-3078.

[7] B. Wen, D. Dong, D. Boroyevich, R. Burgos, P. Mattavelli, and Z. Shen, "Impedance-Based Analysis of GridSynchronization Stability for Three-Phase Paralleled Converters," *IEEE Transactions on Power Electronics*, vol.31, no. 1, pp. 26-38, Jan 2016.

[8] Wang, Xiongfei, Frede Blaabjerg, and Poh Chiang Loh. "An impedance-based stability analysis method for paralleled voltage source converters." In *Power Electronics Conference (IPEC-Hiroshima 2014-ECCE-ASIA), 2014 International*. IEEE, 2014.

[9] Ye, Qing, et al. "A unified impedance-based stability criterion (UIBSC) for paralleled grid-tied inverters using global minor loop gain (GMLG)." In *Energy Conversion Congress and Exposition (ECCE), 2015 IEEE*. IEEE, 2015.

[10] X. Wang, Y. Li, F. Blaabjerg, and P. C. Loh. "Virtual-Impedance-Based Control for Voltage-Source and Current-Source Converters". *IEEE Transactions on Power Electronics*, vol.30, no.12, pp.7019-7037, 2015.

[11] J. He, Y. Li. "Generalized Closed-Loop Control Schemes with Embedded Virtual Impedances for Voltage Source Converters with LC or LCL Filters". *IEEE Transactions on Power Electronics*, vol.27, no.4, pp.1850-1861, 2012.

DC powered data center with 200 kW PV panels

Keiichi Hirose

Data center business Headquarters, NTT FACILITIES, INC., Tokyo, Japan
E-mail: hirose36@ntt-f.co.jp

Abstract— Power consumption of ICT facilities and data centers has grown, and this has led to a need to improve energy efficiency of these facilities. DC power distribution systems employing 380VDC as the supply voltage is one promising approach to address this problem for countries around the world developing and deploying commercial services. The international team by the university of Texas, Austin, USA and NTT FACILITIEIS, Japan demonstrated a 380VDC power distribution system interconnected with a solar power generation system in Texas, USA. The purpose of this demonstration was to show that a 380VDC power supply system saves more energy than an AC power supply system, and to show how much carbon dioxide emissions can be reduced by integrating a solar power generation system. This demonstration resulted in an approximate 17% energy reduction compared with an AC power supply system having the same level of reliability. Also, an evaluation using Data center Performance Per Energy (DPPE) as a performance index of the efficiency of data centers was carried out. The results showed that Power Usage Effectiveness (PUE), one of the sub-metrics of DPPE, improved with the 380VDC power supply system compared with the AC power supply system.

Keywords— 380 VDC, Data center, PV panel, Litium Iion Battery.

I. INTRODUCTION

The amount of traffic flowing through communications infrastructure is increasing explosively. This is due to the increase in large-volume content as a result of cloud, AI, and IoT that comes with the popularization of mobile terminals such as smartphones, tablets, and many types of sensors. Along with this, the power consumption of data centers is growing annually, and measures for conserving energy are required. Fig. 1 describes various approaches for energy saving in data centers classified in four major areas: "Server Load/Computing Operations," "Cooling Equipment," "Power Conversion & Distribution," and "Alternative Power Generation/on site Generation" [1].

Under these circumstances, high voltage direct current (HVDC) power supply systems have been attracting attention as an innovative technique for energy conservation to improve power efficiency in ICT infrastructures, but the current situation is that high capacity HVDC power supply systems scaled up to cope with the increasing demands of data centers while delivering high energy savings performance have been limited worldwide.

Energy Efficiency Opportunities

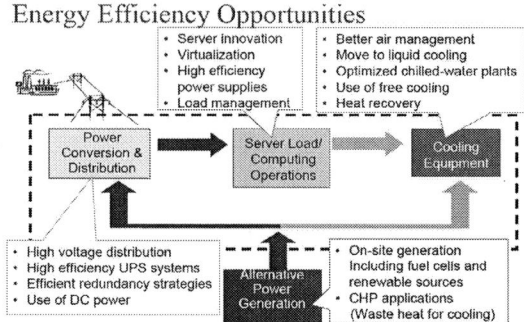

Fig. 1. Energy efficiency opportunities in data centers [1].

Previously, The New Energy and Industrial Technology Development Organization (NEDO) has demonstrated the practical use of high capacity (500 kW class) HVDC power supply system technology with high energy savings performance through the "Research and Development Project for Green Network System Technology" (FY2008-2012).

Against this background, NEDO has agreed to cooperate with the state of Texas in the United States to advance the spread of high voltage direct current (HVDC) power supply system technology, and a Memorandum of Understanding (MOU) has been signed. In this project, we will install the high capacity HVDC power supply system (500 kW class) NEDO cultivated through the aforementioned project, demonstrating the energy savings of the system while aiming to promote the use of HVDC power supply systems in the United States, which is the world's largest market in the ICT field.

More specifically, NTT Facilities, Inc. on the Japan side and the University of Texas at Austin on USA side participated in this project. The HVDC power supply system had be installed at the University of Texas at Austin's Texas Advanced Computing Center (TACC) and the energy savings of the system had been demonstrated. With regards to the HVDC power supply system, the direct current (DC) power supply system had been set up by combining high capacity HVDC power equipment with a solar power generation system and lithium-ion batteries, and the DPPE method has be used to measure the efficiency of the data center and evaluate the energy savings performance of the HVDC power supply system, as well as show the system's superiority over AC power supply systems. In addition to this, by connecting the system to renewable energy from solar power generation and tailoring the number of operating HVDC power supply units (module level on/off control) within the HVDC power supply system based on variations in solar power generated over the course of the day, we aim for further energy savings by optimizing the operating efficiency of the power supply system, with a target of 15% energy savings for the system as a whole compared to conventional systems. The project period is about one year and eight months from August 2015 to March 2017.

II. 380VDC POWER SUPPLY SYSTEM

Focusing on the perspective of power distribution, AC power supply systems are normally used to supply power in data centers, and AC-UPS systems are built in as backup power sources for use in the event of power outages.

Fig. 2 illustrates the configurations of a typical AC power distribution system, a conventional 48VDC power distribution system, and a 380VDC power distribution system. In an AC power distribution system, AC/DC conversions and DC/AC conversions are required inside the UPS in order to charge the embedded battery. Also, AC/DC conversion at the input of ICT equipment is necessary because the components in the ICT equipment (e.g., CPU) require DC power. Consequently, four stages of power conversion would be required from the input of the UPS to the input of the CPU inside the ICT equipment, which results in decrease of total efficiency due to the loss in each conversion stages.

A 380VDC power distribution system is a much simplified system with having only two stages of power conversion. It was originally proposed based on the architecture of the 48VDC power distribution system traditionally used in telecommunication buildings.

III. NEED OF DISTRIBUTED GENERTION FOR DATA CENTER

Most of the data centers in general receive electricity from electric utilities. There are several kinds of power generation type of utilities such as hydropower, thermal power, nuclear power, renewable energy etc. In Japan, on the occasion of the Great East Japan Great Earthquake that occurred on March 11, 2011, the dependence of thermal power generation on Increase. When relying on import for fuel for power generation, it is necessary to consider the supply chain to the data center sufficiently and in detail.

Figure 3 shows an example of supply chain of energy supply to the data center. When unit 1 is the energy of CPU or memory drive, primary energy of 5 to 6 times the total of power supply chain is required [2].

In some regions and countries in foreign countries, problems have occurred in the shortage of power transport capacity for DC of several tens to 100 MW level and stable supply. In order to lower the dependency rate of electricity supply from electric power companies and to improve energy efficiency and reliability, cases of

adopting distributed power sources as the main energy source of DC are increasing.

IV. DC POWER SYSTEM FOR DEMONSTRATION

A high efficiency power supply system should be installed at large-scale data centers [3]. Therefore, we developed a 300 kW-class 380 VDC power supply system. Figure 4 shows the main component of this power system is a 300-kW rectifier and its distribution panel, and Table I lists the main specifications. In general, the power supply is better when the load factor is high, making the operation highly efficient. Therefore, we developed the power supply system incorporating "adaptive power supply management scheme". "adaptive power supply management scheme" is a control method that always maximizes the efficiency of the power system by controlling the operating, standby, and halt statuses of the power module in response to the power consumption state of the load, such as the ICT equipment.

V. EVALUATION METHOD

The DPPE [4] is used as the system for evaluating data center efficiency quantitively. Fig. 5 shows the four guidelines shows as evaluation criteria, but since the DC power distribution system is ancillary equipment in the data center, we will apply Power Usage Effectiveness (PUE), which is a guideline for ancillary equipment operation, and Green Energy Coefficient (GEC), which is a guideline indicating green energy efficiency from the connection of a solar power generation system, and perform measurements.

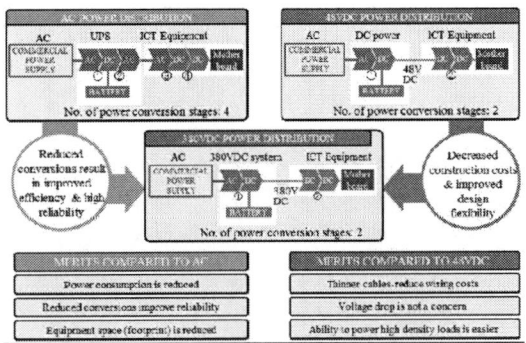

Fig. 2. 380 VDC power supply system.

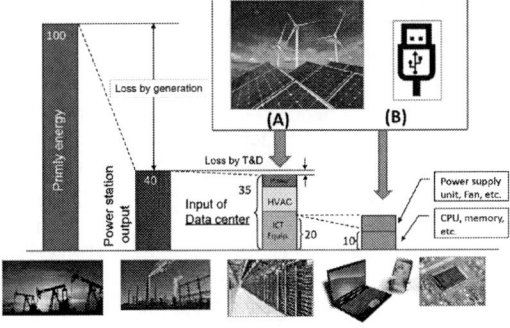

Fig. 3. Energy (Power) supply chain of data center.

The 2018 International Power Electronics Conference

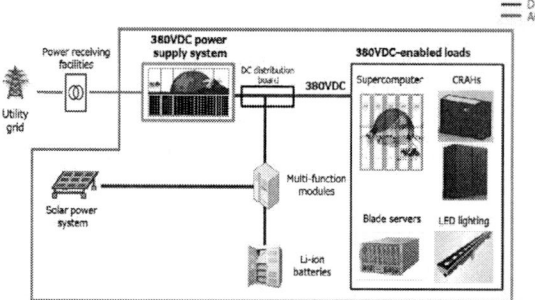

Fig. 4. Outline of basic system configuration.

TABLE I.
SPECIFICATION OF 380 VDC POWER SUPPLY

Item	Specification
Input voltage	174-235 VAC
Output voltage	380 VDC
Output capacity	300 kW
Maximum conversion efficiency	98%

Guideline name	Green Energy Efficiency (GEC)	Power Usage Effectiveness (PUE) of ancillary equipment	IT Equipment Energy Efficiency (ITEE)	IT Equipment Utilization (ITEU)
Measurement technique, amount of evaluation	Actual measurements • Proportion of green energy	Actual measurements • Overall energy consumption with respect to energy consumed by IT equipment	Aggregate catalog values • Efficiency values posed in IT equipment catalogs	Combination of aggregate catalog values and actual measured values • IT equipment utilization rate
Specific energy conservation example	Introduction of solar power generation system, etc.	Increased efficiency of air conditioning and power	Introduction of energy-saving IT equipment	Increase in IT equipment utilization rate, virtualization, etc.

Fig. 5. DPPE guidelines.

This demonstration shows that the 380 VDC power distribution system saves more energy than the AC power distribution system and shows how carbon dioxide emissions are reduced by integrating the solar power generation system.

A. Efficiency Comparison

The input and output powers of the 380 VDC power supply system were measured to compare the efficiencies of the 380 VDC power supply system and a conventional AC power supply system.

B. Energy Saving Effect of "Adaptive Power Supply Management Scheme"

Power consumption of general ICT equipment hardly changed. Electric power generated by the solar power system varies depending on the amount of solar radiation. Therefore, when the solar power system is connected to a 380 VDC power supply system but the power consumption of the ICT device is constant, the electric power generated by the photovoltaic power generation system varies and the output power of the DC power supply system fluctuates. Thus, by utilizing the adaptive power supply management scheme, the DC power supply system can be operated with high efficiency in accordance with the power generated by the photovoltaic power generation system. The input power and output power of the DC power supply system is measured when

it is not used and when using the adaptive power supply management scheme.

C. Effect of Integration of Solar Power Generation System

In addition, when a solar power generation system is connected to an DC power supply system, it can be connected to the output side of the power supply, as shown in Fig. 3, making it possible to reduce losses due to power conversion. For each 200-kW solar power generation system that is connected, the effect shown in Table II is likely, in comparison with connection to an AC power supply system.

VI. EVALUATION RESULTS

Monthly amounts of energy during demonstration period (kWh) are listed in Table II. In this demonstration, we aimed to show the superior performance of 380 VDC power supply system compared to AC power supply system, and show how much carbon dioxide emissions are reduced by integrating the solar power gene ration system.

The following shows findings:

A. Efficiency Comparison of 380 VDC and AC power Supply Systems

It is clear that in the case that the 380 VDC power supply system is used, compared to the case that AC power supply system is used, power loss is reduced by 36% on average. It is also clear that using the 380 VDC power supply system reduces commercial energy consumption by more than 4%. These reductions are possible because high conversion efficiency can be maintained, even during low-load periods, as a result of adaptive power supply management.

B. Evaluation of 380VDC Power Supply System by using DPPE

Fig. 6 shows the monthly PUE for 380 VDC power supply system and for AC power supply system. PUE for HVDC power supply system is lower for all months than that for AC power supply system. This result means that introducing the HVDC power supply system improved the energy efficiency of the facilities during the demonstration period.

TABLE II
MONTHLY AMOUNTS OF ENERGY DURING DEMONSTRATION PERIOD (KWH)

	Item	Sep-16	Oct-16	Nov-16	Dec-16	Jan-17	Feb-17	Total
380VDC	Apollo consumption power	53,580	51,624	54,810	65,253	65,822	44,437	315,626
	A/C device consumption power	19,372	28,018	29,307	33,880	34,317	23,871	169,265
	PV generated power	11,791	15,895	10,538	10,222	14,063	9,739	72,240
	Commercial power supply	47,527	70,789	80,501	96,012	93,522	64,493	452,844
	Power loss	5,962	6,295	6,193	6,391	6,709	2,229	35,893
	Total consumption power	79,318	87,454	91,041	106,734	107,525	74,735	575,044
	Compared with AC power supply system							
AC	Commercial power supply	49,717	73,944	84,007	100,367	97,895	67,141	473,371
	Power loss	7,252	9,480	9,696	10,748	11,072	8,207	56,419
	Total consumption power	61,508	89,820	94,537	118,589	111,958	77,130	545,611

824

Fig. 8 shows the monthly GEC for 380 VDC power supply system and for AC power supply system. GEC is about 4.5% higher for all months in the 380 VDC case. This result can be explained that solar power generation system is connected to 380 VDC power supply system directly without AC/DC power conversions, so it leads to more efficient system than AC power supply system.

VII. CONCLUSIONS

In cooperation with the university of Texas, Austin, we demonstrated the advantages of 380 VDC power supply system interconnected with a solar power generation system. This demonstration showed that the DC power distribution system saves more energy than the AC power distribution system and showed how much carbon dioxide emissions are reduced by integrating the solar power generation system.

Efficiencies of 380 VDC power supply and AC power supply systems were compared.

・ Power loss was reduced by 36% on average.

・ Reduction of commercial energy consumption was more than 4%.

Evaluation of 380 VDC power supply system by using the DPPE was conducted.

・ DPPE improved by 4.5% compared with AC power supply system.

Effect of the integration of a solar power generation system was evaluated.

・ Energy saving effect was 17%.

・ This was equivalent to 74.8t-CO2 reduction of greenhouse gases.

In the future, using the results of this demonstration as a foundation, we plan to promote the expanded deployment of 380 VDC power supply systems as one solution to the problem of reducing data center power consumption focusing on the U.S. but also looking more broadly towards countries around the world.

In this paper, a practical example of the challenges and measures in the energy surface of the data center is a key infrastructure of ICT services that are steadily year by year expanded to introduce some. The ICT infrastructure is a facility that has a nature that moves 24 hours a day, 356 days, and even if it improves 1 W, 1%, it should have a non-negligible effect over a long period of time. I hope to continue to contribute to the reliable and efficient operation of the data center and to the better society with wealth by research and technology development in the energy field.

REFERENCES

[1] P. Scheihing, U.S. Department of Energy, "DOE Data Center Energy Efficiency Program," February 2009.
[2] K. Hirose," Technical Trends on Energy Issues of Data Centers as ICT Infrastructure," Journal of IEICE, Vol.101 No.4pp.345-349, April 2018.
[3] J. Inamori, H. Hoshi, T. Tanaka, T. Babasaki, and K. Hirose, "380-VDC Power Distribution System for 4-MW-scale Cloud Facility", INTELEC, Vancouver, CS15-04, September 2014.
[4] Japan National Body/ Green IT Promotion Council, "DPPE: Hoplistic Framework for Data Center Energy Efficiency –KPIs for Infrastructure, IT Equipment, Operation (and Renewable Energy)-", August 2020.

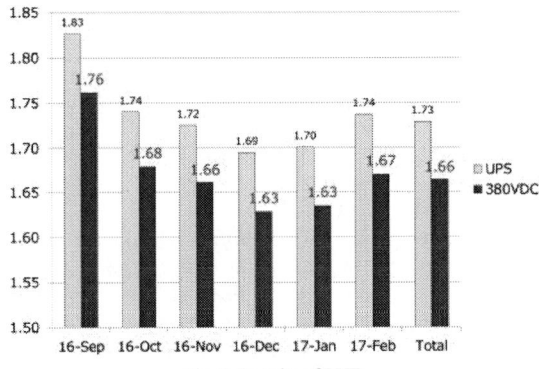

Fig. 7. Results of PUE.

Fig. 8. Results of GEC.

The 2018 International Power Electronics Conference

Influences of Deterioration in Capacitor and Inductor on Current Sensorless Static Model DC-DC Converter

Fujio Kurokawa[1*], Masashi Taguchi[2*], Jizhe Wang[2], Hidenori Maruta[2] and Nobumasa Matsui[1]

1 Institute for Innovative Science and Technology, Nagasaki Institute of Applied Science, Nagasaki, Japan

2 Graduate School of Engineering, Nagasaki University, Nagasaki, Japan

1*E-mail: kurokawa_fujio@nias.ac.jp

2*E-mail: bb52117220@ms.nagasaki-u.ac.jp

Abstract— This paper discussed the influences of deterioration on capacitor and inductor in the sensorless static model with the filter system. The effects of variations in input supply and output load to the proposed system have been confirmed in previous studies. Nonetheless, the influences of variations in the capacitor and inductor on the system have not been confirmed yet. Therefore, this paper focuses on discussing about what kind of impacts will be caused to the stability analysis of the filter and stable operation of the converter when the inductance and capacitance of converter are varied. The impacts have been confirmed by a 5 W, 20 V/5 V buck-type dc-dc converter via simulation.

Keywords— dc-dc converter, digital control, static model, sensorless model

I. INTRODUCTION

An increase of power consumption is expected in the future information society. An increase in demand of communication equipment is expected hereafter. A power supply is an important element in a communication system. The switching equipment and the transmission equipment are communication equipment. These are indispensable things for today's living. The stable power supply is required to operate them normally. A cellphone and internet are deeply involved in the society and our lives. Reliability of cellphone and internet are important problems especially. Therefore, the high reliability and the high accuracy are required for the power supply system for communication. The high efficiency and miniaturization are also required for the power supply. The communication system is indispensable things so that we live safely and comfortably. The improvement of communication system technology leads to an improvement of convenience. Our study on the digital control of switching power supply can meet the demand of modern power supply for communication. A digital control can make highly control compared with a conventional analog control power supply. The digital control can switch always optimal control. So, the digital control can be stabilized the output of power supply.

A buck type dc-dc converter is adopted for the switching power supply. The voltage mode control is the most basic method in the digital control of switching power supply. This method is superior to a disturbance removal. On the other hand, the responsiveness is not very good. The responsiveness of control is as important as the stability. The voltage mode control is the feedback control. The static model that sum current detection to the voltage mode control is proposed [1]. The static model realized the fast responsiveness against the load variation [2]-[5]. The current detection is required for the detection resistance of proposed static model. This detection resistance causes two problems. The problem is the increase of converter parts and power loss by the detection resistance. The sensorless static model excluding detection resistance is proposed in order to solve these problems [6]-[12]. The sensorless static model made the current prediction to obtain the information of output current instead of the current detection. Since the sensorless static model does not need the detection resistor, the efficiency reduction due to the detection resistor can be removed. In case of the current prediction, the prediction value of output current vibrates. A low pass filter to suppress vibration of the predicted value is introduced to the sensorless static model [13]. The filter is designed by the stability analysis. As a result, the sensorless static model including the filter can suppress the vibration of predicted value.

The influence of load and input voltage variation on the proposed method has been confirmed [3], [7], [8]. The stable dynamic characteristics of output voltage of converter were confirmed. Since the deterioration of capacitor and inductor cause variations of capacitance and inductance, thus it is assumed that the deterioration will give an effect on the stability of proposed method. The stability analysis of filter and stable operation of converter are affected when the value of the capacitance and inductance are changed. The circuit configuration of proposed method, the system description and the derivation of transfer function for analysis are shown in order. In this paper, the influences of deterioration to stability of proposed method is discussed. The influence is confirmed by indicating the results of simulation and analysis. Finally, a conclusion is indicated.

II. OPERATION PRINCIPLE

Figure 1 describes a proposed basic circuit structure of digital control dc-dc converter. In this figure, the input voltage e_i, and the output voltage e_O, are detected from the main circuit. The detecting resistor is eliminated because the output current is predicted by the proposed sensorless static model. Figure 2 shows the schematic diagram of digital control. The control circuit consists of the current prediction part, sensorless static model part, and PID control part. Detected values are sent to the pre-amplifier, and then they are converted to digital values $e_i[n]$, $e_o[n]$ via A-D converter. N indicates the nth value during the sampling period. The static model can derive an appropriate bias value by controlling with the detected input-output voltages and the output current. The static model in CCM is expressed as follows;

$$T_{on\_model\_CCM}[n] = \frac{NT_s}{(e_i[n])}\left(E_o^* + ri_{o\_pred}[n]\right) \quad (1)$$

where E_o^* is the reference value of output voltage, T_s indicates the switching period and N_{T_s} indicates its digital value, R is the total value of loss resistances in the circuit. Figure 3 shows $T_{on\_model}[n]$ for the output current. i_{oc} indicates the critical current. Equation (1) is derived from input-output characteristics of the buck-type dc-dc converter. In the conventional static model, the real detected value is used for the output current. However the predicted value $i_{o\_pred}[n]$ without the sensor resistor is used for the output current in the proposed sensorless static model. Equation (2) represents the predicted value of output current in the sensorless static model.

$$i_{o\_pred}[n] = \frac{T_{on}[n-1](e_i[n]) - N_{T_s}(e_o[n])}{rN_{T_s}} \quad (2)$$

Equation (2) is derived from the static model expression. In (2), the predicted value $i_{o\_pred}[n]$ of output current is calculated by $e_i[n]$, $e_o[n]$ and the on-time digital value $T_{on}[n-1]$ of previous period. The predicted value $i_{o\_pred}[n]$ may oscillate when a disturbance such as load fluctuation occurs. So the oscillation of $i_{o\_pred}[n]$ is suppressed by using a digital filter in the current prediction part.

The final on-time is determined by the sum of (1) and PID control.

$$T_{on}[n] = T_{on\_model}[n] + T_{on\_PID}[n] \quad (3)$$

Even if the output current and the input voltage changes, an appropriate bias is calculated for each switching period. As a result, a stable output voltage can be maintained.

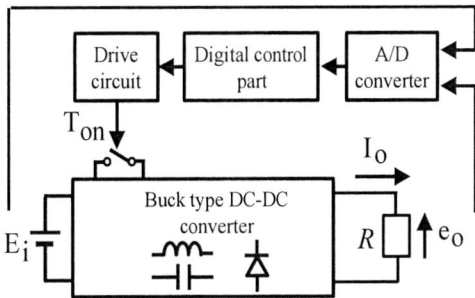

Fig. 1. Basic circuit structure.

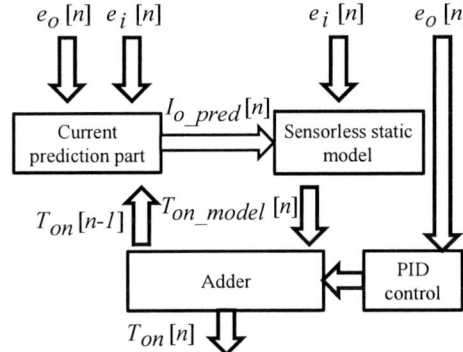

Fig. 2. Structure of digital control.

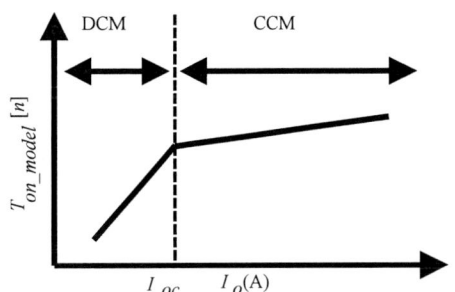

Fig. 3. ON time of static model.

III. ANALYSIS RESULTS

In this section, the design of sensorless static model including the filter is discussed. The small signal analysis is performed by deriving the transfer function related to e_o from the entire system including the main circuit and the control circuit. From the stability condition of e_o, the stability limit with the time constant T_c of the low pass filter is shown. The design index of filter coefficient that can suppress oscillation of $i_{o\_pred}[n]$ is analyzed. Based on the equivalent circuit model in a period of the main circuit, the transfer function of the main circuit, sensorless static model, current prediction, PID control and filter transfer functions are derived considering only infinitesimal change from each control expressions. The transfer functions of main circuit, static model, current prediction, PID control and low pass filter are shown below.

$$
\begin{cases}
\Delta e_o(s) = G(s)\left\{ E_i\dfrac{\Delta N\,T_{on}}{N\,T_s} + \dfrac{E_o(Ls+r)}{R}\dfrac{\Delta R(s)}{R} \right\} \\[2mm]
G(s) = \dfrac{\dfrac{1}{LC}}{s^2 + \left(\dfrac{1}{CR}+\dfrac{r}{L}\right)s + \dfrac{1}{LC}\left(1+\dfrac{r}{R}\right)}
\end{cases} \tag{4}
$$

$$
\Delta NT_{on\_model}(s) = \frac{1}{E_i}\left\{\left(1+\frac{r}{R_{pred}}\right)\Delta e_o(s)e^{-s\tau_2}\right. \\
\left. - \frac{E_o r}{R_{pred}^2}\Delta R_{pred}(s)C(s)\right\} \tag{5}
$$

$$
\Delta R_{pred}(s) = -\frac{R_{pred}^2}{E_o r N T_s}E_i N_{Ton}(s)e^{-s\tau_1} \\
+ \frac{R_{pred}}{E_o}\left(1+\frac{R_{pred}}{r}\right)\Delta e_o(s)e^{-s\tau_2} \tag{6}
$$

$$
\Delta NT_{on\_PID}(s) = -\left(H_p + sH_d + \frac{H_i}{s}\right)\Delta e_o(s)e^{-s\tau_2} \tag{7}
$$

$$
C(s) = \frac{1}{sT_c + 1} \tag{8}
$$

Figure 4 is created from these equations. Figure 4 is a block diagram of the entire system. $e^{-s\tau_1}$ and $e^{-s\tau_2}$ indicate the delay of control operation and the delay time of one cycle, respectively. From Fig. 4, the transfer function of output voltage regarding the load is derived.

$$
\frac{\Delta e_o(s)}{\dfrac{\Delta R(s)}{R}} = \frac{s^3\dfrac{E_o}{C}(\tau_1+T_c) + s^2\dfrac{E_o r}{LC}(\tau_1+T_c)}{a_4 s^4 + a_3 s^3 + a_2 s^2 + a_1 s + a_0} \tag{9}
$$

$$
\phi(s) = a_4 s^4 + a_3 s^3 + a_2 s^2 + a_1 s + a_0 \tag{10}
$$

$$
a_4 = R\left\{\tau_1 + T_c\left(1+\frac{E_i H_d \tau_2}{LC}\right)\right\} \tag{11}
$$

$$
a_3 = \frac{R}{LC}\left[E_i H_d \tau_2 + \frac{(L+rRC)}{R}\tau_1 \\
+ T_c\left\{\frac{L+rRC}{R} + E_i(H_d - \tau_2 H_p)\right\}\right] \tag{12}
$$

$$
a_2 = \frac{R}{LC}\left[E_i(H_d - \tau_2 H_p) + \frac{\tau_1(R+r)}{R} \\
+ T_c\left\{\frac{R+r}{R} + E_i(H_p - H_i\tau_2)\right\}\right] \tag{13}
$$

$$
a_1 = \frac{RE_i}{LC}\left(H_p - H_i\tau_2 + H_i T_c\right) \tag{14}
$$

$$
a_0 = \frac{RE_i H_i}{LC} \tag{15}
$$

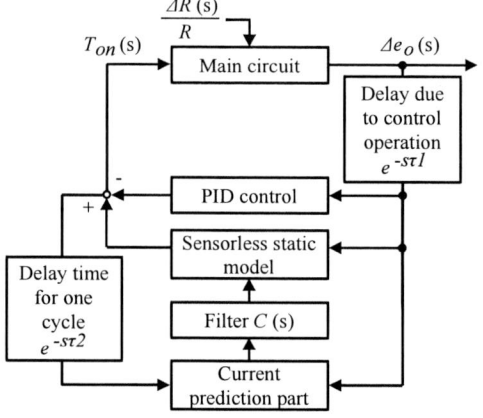

Fig. 4. Diagram of whole system.

From (9), coefficients of the characteristic equation $\varphi(s)$ can be represented by a_0 to a_4. Using the stability criterion of Hurwitz, its stability should satisfy the follow two conditions:

(i) The first condition is the coefficients a_0 to a_4 in (10) must be positive.

Since the rated output power is 5 W, the desired output voltage is 5 V, thus the parameter R is the 5 Ω. T_c is the time constant of the low-pass filter and $T_c = RC$ (s), it is a positive value. Each coefficient a_4 to a_0 in (10) is obtained by substituting the circuit parameters to (11)-(15). The filter coefficients a_4 to a_0 are shown in Table I. They must be positive values according to the Hurwitz submatrix.

(ii) The second condition is the eqation of Hurwitz submatrix (16) is positive:

$$
\begin{vmatrix}
a_3 & a_1 & 0 \\
a_4 & a_2 & a_0 \\
0 & a_3 & a_1
\end{vmatrix} = \alpha_3 T_c^3 + \alpha_2 T_c^2 + \alpha_1 T_c + \alpha_0 > 0 \tag{16}
$$

TABLE I
THE CONDITION OF STABILITY CRITERION OF HURWITZ (I)

$\varphi(s)$	Values after parameter substitution
a_4	$5.0000 \cdot 10^{-5} + 5.0115 \cdot T_C$
a_3	$1.2024 \cdot 10^{-2} + 5.7142 \cdot 10^3 \cdot T_C$
a_2	$296.62 + 1.4407 \cdot 10^8 \cdot T_C$
a_1	$1.1441 \cdot 10^8 + 1.1452 \cdot 10^{10} \cdot T_C$
a_0	$2.2905 \cdot 10^9$

The cubic inequality relating to T_C is derived from (16). The stability limit of T_C is derived by substituting each circuit parameters into the result of applying the Hurwitz stability criterion to (10). Therefore, the filter coefficient can be designed by using the stable T_C value. In case of digitizing the low pass filter and introducing it to the current prediction unit, the following equation is obtained.

$$i_{o\_pred}'[n] = K i_{o\_pred}'[n-1] + (1-K) i_{o\_pred}[n] \quad (17)$$

where $i_{o\_pred}'[n]$ and $i_{o\_pred}[n]$ indicate the predicted current values after passing through the filter and before passing through the filter, respectively. K represents the filter coefficient, and by adjusting this value, vibration suppression of the predicted current value is performed. K can be expressed by the following equation using the analog filter's time constant T_C.

$$K = \frac{T_c}{T_s} \bigg/ \left(1 + \frac{T_c}{T_s}\right) = \frac{T_c}{T_s + T_c} \quad (18)$$

The deterioration in capacitor and inductor cause the variations of capacitance C and inductance L. Normally, the C and L are decreased by 0 % ~ 20 % than real value due to the deterioration. The variations of C and L lead to the change in T_C and further influence the stability limits of K.

Figure 5 shows the relation between the stability limit of K and variations of C. In this study, the C is varied from the rated value 891 µF to 20 % decreased value 713 µF. It is shown that the stable region is extended by the decreasing of C.

Figure 6 shows the relation between the K and variations of L. The L is varied from the rated value 196 µH to 20 % decreased value 157 µF. It is found that the stable region of K is extended by the decreasing of L in whole load conditions.

IV. SIMULATION RESULTS

The effectiveness of proposed sensorless static model is verified by the simulator PSIM. It turned out to be stable the predicted current when it is designed more than $K = 0.990$ in the whole range of CCM as shown in Fig. 5

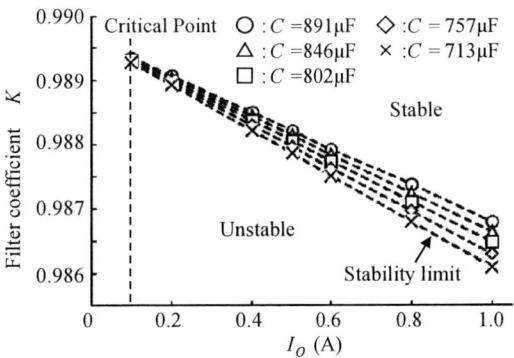

Fig. 5. Stability limits of proposed method due to variations of capacitance.

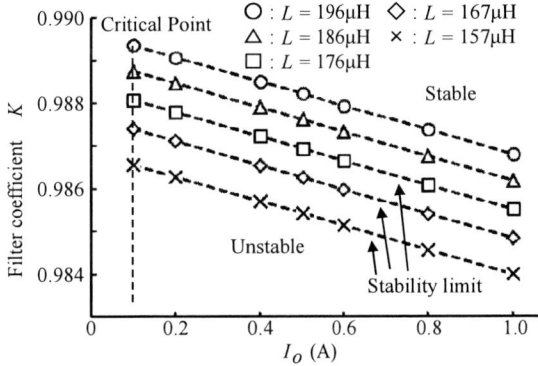

Fig. 6. Stability limits of proposed method due to variations of inductance.

and Fig. 6. Considering the modelling error such as the internal resistance, and perform operation check, the K is selected as 0.995. The transient response when the load is changed from 0.5 A to 1.0 A and from 1.0 A to 0.5 A is confirmed. Figure 7 to Fig. 9 show simulation results. The predicted current $i_{o\_pred}[n]$ follows the value of output current based on the step change, and its movements are represented by (2). The range of convergence condition is within ±1 % of the reference output voltage i.e. within 4.95 V to 5.05 V. From these figures, it is confirmed that the sensorless static model regulate the output voltage effectively in various conditions of C and L. Furthermore, based on the optimal value of filter coefficient K, the convergence time of the output voltage can be maintained.

V. CONCLUSION

The influence of deterioration in capacitor and inductor on current sensorless static model dc-dc converter is discussed in this paper. In the proposed method, the efficiency reduction caused by the sensing-resistor is eliminated. Based on the stability analysis, the value of

The 2018 International Power Electronics Conference

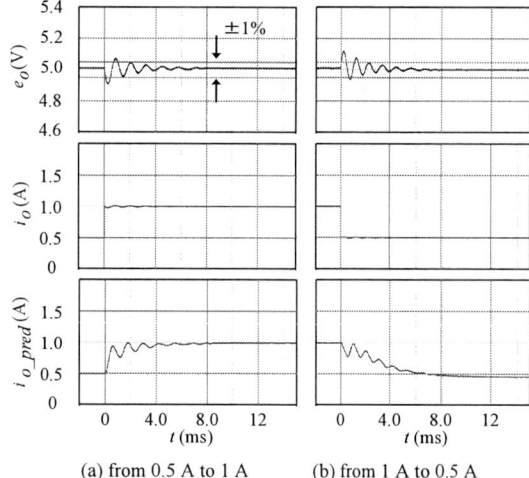

(a) from 0.5 A to 1 A   (b) from 1 A to 0.5 A

Fig. 7. Transient characteristics in case of $K = 0.995$ and $E_i = 20$ V(rated condition).

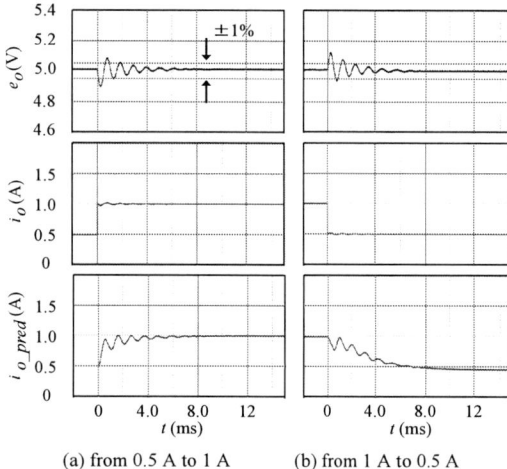

(a) from 0.5 A to 1 A   (b) from 1 A to 0.5 A

Fig. 8. Transient characteristics in case of $K = 0.995$ and $C = 713$ μF (-20 %).

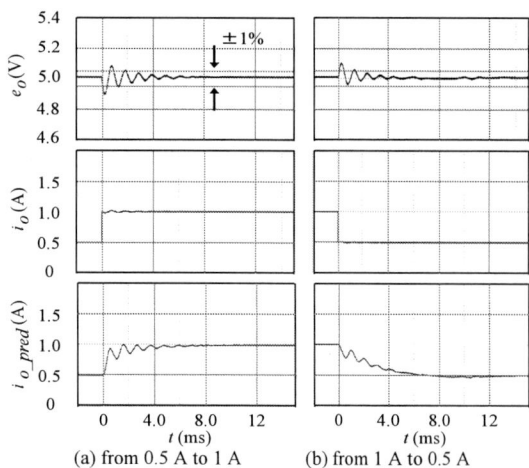

(a) from 0.5 A to 1 A   (b) from 1 A to 0.5 A

Fig. 9. Transient characteristics in case of $K = 0.995$ and $L = 157$ μH (-20 %).

filter coefficient K that satisfy stable condition of output voltage is determined as $K = 0.995$. It is confirmed that by using the design index of the filter, the stable output can be obtained even when the capacitance and inductance are verified. And the convergence time of the output voltage can be maintained when the K is optimized. The excellent transient response characteristics are obtained. Therefore it is revealed that the design index of proposed method is effectiveness to suppress the influence of deterioration in capacitor and inductor.

REFERENCES

[1] J. Liang and R. G. Harley, "Feed-forward transient compensation control for dfig wind generators during both balanced and unbalanced grid disturbances," Proc. of *IEEE Energy Conversion Congress and Exposition*, pp. 2389-2396, 2011.

[2] L. Jia, Z. Hu, Y. Liu and P. C. Sen, "A practical control strategy to improve unloading transient response performance for buck converters," Proc. of *IEEE Energy Conversion Congress and Exposition*, pp. 397-404. 2011.

[3] F. Kurokawa, J. Sakemi, A. Yamanishi, and H. Osuga, "A new STS model dc-dc converter," Proc. of *IEEE Energy Conversion Congress and Exposition*, pp. 680-684, 2011.

[4] P. Shangzhi; P. K. Jain, "A low-complexity dual-voltage-loop digital control architecture with dynamically varying voltage and current references, " *IEEE Trans. Power Electronics*, Vol. 29, No. 4, pp. 2049-2060, 2014.

[5] C. Wen, B. Fahimi, E. Cosoraba, Y. Fan, " Stability analysis and voltage control method based on virtual resistor and proportional voltage feedback loop for cascaded dc-dc converters," Proc. of *Energy Conversion Congress and Exposition*, pp. 3016-3022, 2014.

[6] S. Zhenyu, T. Siew-Chong and C. K. Tse, "Transient mitigation of dc-dc converters using an auxiliary switching circuit," Proc. of *IEEE ECCE*, pp. 1259-1264, 2011.

[7] F. Kurokawa, and S. Hirotaki, "A novel sensorless model control dc-dc converter," in Proc. *IEEE Renewable Energy Research and Application*, pp. 663 - 667, 2014.

[8] F. Kurokawa, and S. Hirotaki, "A new high performance dc-dc converter with sensoress model reference modification," Proc. of *International Telecommunications Energy Conference*, pp. 1-5, 2014.

[9] S. M. RakhtAla, M. Yasoubi and H. HosseinNia, "Design of second order sliding mode and sliding mode algorithms: a practical insight to dc-dc buck converter," *IEEE CAA Journal of Automatica Sinica*, vol. 4, pp. 483-497, 2017.

[10] Q. Ye, R. Mo and H. Li, "Low-Frequency Resonance Suppression of a Dual-Active-Bridge dc-dc converter Enabled DC Microgrid," *IEEE Trans. Power Electronics*, Vol. 5, pp.982-994, 2017.

[11] A. Radic, A. Straka and A. Prodic, "Synchronized Zero-Crossing-Based Self-Tuning Capacitor Time-Constant Estimator for Low-Power Digitally Controlled DC–DC Converters," *IEEE Trans. Power Electronics*, Vol. 29, pp. 5106-5110, 2014.

[12] N. Deng, P. Wang and X. P. Zhang, "A DC Current Flow Controller for Meshed Modular Multilevel Converter Multiterminal HVDC Grids," *CSEE Journal of Power and Energy Systems*, Vol. 1, pp. 43-51, 2015.

[13] F. Kurokawa, S. Watanabe, Y. Furukawa, H. Maruta, N. Matsui, and I. Colak, "Analysis of sensorless model control dc-dc converter in CCM," in Proc. *International Conference on Electrical Machines and Systems*, pp. 1-6, 2016.

The 2018 International Power Electronics Conference

Capacitive Divider Based Passive Start-up Methods for Flying Capacitor Step-down DC-DC Converter Topologies

Michael Halamicek[1*], Tom Moiannou[1], Nenad Vukadinović[1], and Aleksandar Prodić[1]

1 Laboratory of Power Management and Integrated SMPS, ECE Department, University of Toronto, Toronto, Canada
*E-mail: michael.halamicek@mail.utoronto.ca

Abstract— This paper introduces two methods for limiting the voltage stress during start-up across switches of multi-level flying capacitor (ML-FC) step-down dc-dc converters. For a general *N*-level converter, the presented methods reduce the voltage stress to (*N*-1) times lower value than that of a conventional buck, allowing lower voltage rating transistors with smaller specific on-resistances to be used. These methods require no active control of switches on initial start-up and rely on segmentation of the input filter capacitor or the utilization of flying capacitors as part of voltage dividers. The speed of these schemes is limited only by the size of the flying capacitor, parasitics in the conduction pathway, and the quality of the start-up diode and the low side switch body diodes. The methods have been verified on a 3-level buck 24V-to-5V, 20W prototype, with input voltage rise times of less than 1μs showing the effectiveness of the start-up circuits.

Keywords— *dc-dc converter, multi-level flying capacitor, passive start-up, step-down*

I. INTRODUCTION

Multi-level flying capacitor (ML-FC) converters are becoming an attractive alternative to the widely-used conventional buck in low power dc-dc applications processing power from a fraction of watt to several hundreds of watts, as they allow for a drastic reduction of the overall converter volume and improvement of power processing efficiency simultaneously [1] [2]. The advantages of ML-FC converters are obtained by reducing voltage swings across the inductors and switching components, allowing for smaller inductors to be used and lowering the voltage stress across switches during regular operation, resulting in reduced switching losses and smaller inductors. One of the drawbacks of ML-FC converters is the larger number of switches in the conduction path often cause higher conduction losses [3]. Theoretically, the lower voltage stress across the transistors during regular operation allows for the use of devices with lower blocking voltage and lower on-resistance, R_{on}, eliminating these higher conduction losses [4] [5] [6]. Furthermore, lower voltage silicon is more cost-effective while having much better figure of merit (FOM) [7], which represents the product of an R_{on} resistance and a device capacitance for a fixed silicon area.

However, start-up presents a challenge to fully exploiting the lower switch stresses offered by these topologies [8]. The example 3-level converter in Fig. 1 demonstrates this problem. Before power up, the initial voltage of the flying capacitor, C_{fly}, is zero, and is unable to provide the voltage division function as in regular operation. Consequently, during start-up, S_1 will be exposed to the full input voltage. Along with this immediate start-up problem, it also takes some time for the flying capacitor to charge to its steady state value once switching action begins. During this time, other switches may be exposed to larger than expected voltages. These issues force at least one of the transistors to be exposed to the full input voltage and prevents lower blocking voltage components from being used. As a result, ML-FC step-down converters usually still suffer from higher conduction losses compared to the conventional buck.

The goal of this paper is to introduce two sets of commutation cell based methods for limiting the voltage stress across switches of ML-FC step-down dc-dc converters during power ups. For a general *N*-level converter, the presented methods guarantee reduction in voltage stress to (*N*-1) times lower value than that of a conventional buck during start-up, enabling the use of lower voltage rated components. The presented solutions potentially allow for on-chip implementation of the entire power stage with lower voltage rated silicon components. In this paper, two methods based on the commutation cells shown in Figs. 2 and 3 are presented.

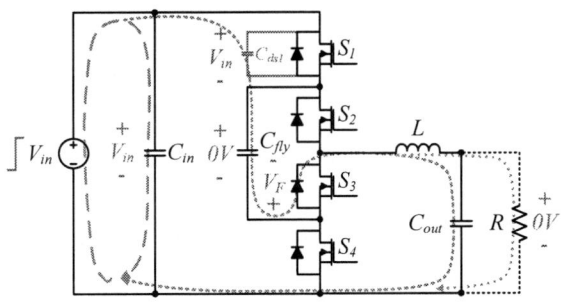

Fig. 1. 3L-FC buck during start-up. Initial inrush current path shown in orange. Initial voltage distribution showing full input voltage across S_1.

This work of the Laboratory for Power Management and Integrated SMPS is supported by Würth Elektronik eiSos GmbH & Co.

The 2018 International Power Electronics Conference

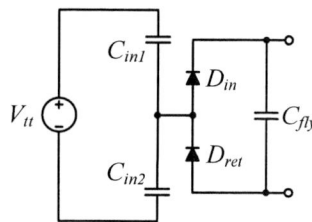

Fig. 2. Commutation cell for input capacitor divider start-up method

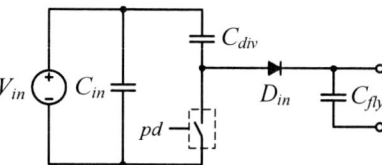

Fig. 3. Commutation cell for flying capacitor divider start-up method.

II. BACKGROUND

Fig. 1 shows the inrush current paths on start-up for a common flying capacitor topology, the three-level flying capacitor buck converter (3L-FC buck) [4]. In addition to the current charging the input filter capacitor, there is a small current that flows through the drain-source capacitance of S_1, C_{ds1}, through the flying capacitor, the body diode of S_3 and toward the relatively low impedance of the output filter. Upon initial ramp-up of the input voltage, C_{ds1} forms a capacitive divider with the flying and output capacitors, which due to the much larger impedance of C_{ds1} compared to C_{fly} and C_{out} results in nearly the entire input voltage being seen across S_1. Furthermore, once switching action begins, the flying capacitor takes many cycles to charge, during which time other switches may be exposed to the full input voltage. In the case of the 3L-FC buck, S_4 would also be exposed to the full input voltage upon initial switching.

Various solutions have been proposed to address these issues [9]-[14]. Some assume that the input voltage ramps up slowly enough that the flying capacitor can be pre-charged and the converter soft-started, which would still permit the use of half-rated switches [9]. These solutions assume input voltage ramp times of 100's of microseconds to a few milliseconds and low input voltage slew rates. Others propose shorter input voltage ramp times and higher slew rates of 1V/μs, but expect that prior to start-up, logic and gate drive supplies are already on and stable [10]. However, in the targeted low-power applications, these assumptions may not hold. Low power converters are usually exposed to much faster slew rates during input voltage ramp up due to the lower impedances of the reactive components at the converter input. To satisfy these criteria, some solutions propose adding a hot-swap circuit or some other series element in the conduction path at the converter input. One such approach involves adding two parallel fully rated switches at the converter input. One is a low impedance path that is enabled during normal operation, while the other path is a high impedance that serves to pre-charge the flying capacitor [11]. Similar hot-swapping structures can also be seen in power delivery

architectures for server applications [12]. While hot-swap events can be handled safely, the additional series switch serves no function as part of normal operation and generally degrades the power processing efficiency and/or increases silicon area, while also increasing complexity.

Instead of adding additional switches to the conduction pathway, one can also rate the main switch for the full input voltage. Pre-charging the flying capacitor can also be done in a more controlled manner through the action of two current sources placed in series with the flying capacitor. A feedback network forces the current sourced from the input voltage and the current discharged to ground to match [13] [14]. These solutions eliminate the need for having both S_1 and S_4 rated for the full input voltage, but still require the use of one high voltage rating transistor in the conduction path and thus do not provide the full utilization of the lower voltage rating advantages of multi-level topologies. It is suggested that half rated switches can still be used if the current sources are scaled up sufficiently thereby increasing the charge rate of the flying capacitor [13]. However, generating these current sources and their supply voltages, particularly for higher input voltages, may be a significant challenge. These supplies would also have to be stable before the input voltage ramp and may burn non-negligible static power. Effectiveness is limited in the presence of relatively high start-up inrush currents caused by high input voltage slew rates.

The previous discussion indicates that the ideal solution would be a start-up scheme that:

- requires no additional series elements to be placed in the conduction path,
- can respond to an infinitely fast input voltage ramp (i.e. hot-swap),
- is entirely passive and requires no additional supply voltages on start-up, and
- uses switches rated for steady state operation (e.g. $V_{in}/2$ plus some margin for the 3L-FC buck).

A converter that generally meets these criteria was recently proposed, which is functionally identical to the 2-phase series-capacitor buck [15]. The strategy involves moving the main switch, which is strictly in series with voltage source, down to the ground path and then adding an additional capacitor between the base of the flying capacitor and the base of the voltage source. The flying capacitor and additional capacitor are now in series and connected across the voltage source. Unfortunately, the lack of a common ground may make this solution unviable in a number of applications. Additionally, this strategy of moving the main switch to the ground side of the source only works for the one switch that is strictly in series with the source. Consequently, this strategy cannot be extended to higher level flying capacitor converters (e.g., 4L-FC buck or 3-phase series-capacitor buck).

In the following section, two start-up solutions that approach the ideal requirements and satisfy a large majority of practical applications today are presented. Both schemes rely on the general principle of ML-FC converters—the utilization of switching-capacitor network for obtaining voltage division.

832

III. PRINCIPLE OF OPERATION AND DESIGN REQUIREMENTS

A. Input Filter Capacitor Based Solution

This scheme can be described through the diagrams in Figs. 2 and 4. Fig. 2 shows the general commutation cell, where the voltage source V_{tt} is either the input voltage of the converter (for a 3L-FC buck converter, shown in Fig. 4) or the voltage between two neighboring taps of a capacitive divider for a general N-level case. For a 3L-FC buck, the single input capacitor of Fig. 1 is replaced with two stacked capacitors, with the center node connected to the terminals of the flying capacitor through two diodes. The cathode of the high current capacity inrush diode, D_{in}, is connected to the top of the flying capacitor, while the anode of the small return diode, D_{ret}, is connected to the base of the flying capacitor. Fig. 4 shows this method as implemented in the 3L-FC buck converter with the initial inrush current path and the subsequent inductor discharge path highlighted. On start-up, the inrush current flows through the input filter capacitors and splits the input voltage between them setting the center node voltage to approximately $V_{in}/2$. D_{in} and D_3 become forward biased and charge the flying capacitor. Once the flying capacitor voltage reaches $V_{in}/2$ at its peak current, D_{in} and D_3 will become reverse biased, and D_4 will begin to conduct until the inductor is fully discharged.

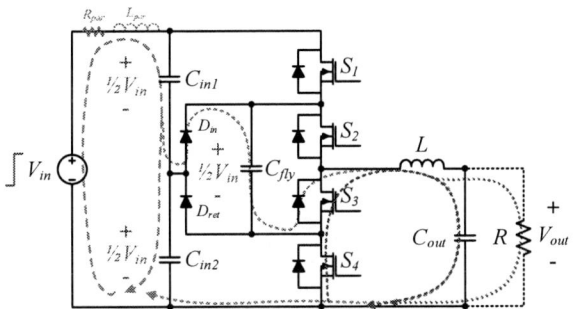

Fig. 4. 3L-FC buck with input capacitor divider start-up method. Initial inrush current path shown in orange, inductor discharge path shown in purple, along with initial voltage distribution.

Determining the final value of the flying capacitor and output capacitor voltages can be complicated by the multiple impedance paths and by the presence of the inductor. Due to the high initial inrush current, once the flying capacitor reaches its steady state voltage, the high instantaneous inductor current causes charge to continue to be forced to the output through the low side switch, S_4. Damping along these paths is also very low as reduction in series resistance is key to reducing conduction losses. The resultant capacitor voltages depend on the values of the input, output and flying capacitors, the inductor, the characteristics of the diodes in the conducting pathway and the parasitic resistances and inductances in the paths. However, if $C_{out} \gg C_{fly}$ and $C_{in1,2} \gg C_{fly}$, the output capacitor voltage is not greatly affected by the charging of the flying capacitor, which will charge to $V_{in}/2$.

With a few simplifications, we can derive expressions for the final values of the flying capacitor voltage, the output voltage, and the current through the inductor and diodes on start-up. To obtain a worst-case approximation of these values, we assume ideal diodes and negligible parasitics in the conduction pathway, giving the simplified circuit model for the 3L-FC buck on start-up shown in Fig. 5. Corresponding waveforms of the flying capacitor voltage, the output voltage and the charging currents on start-up are shown in Fig. 6.

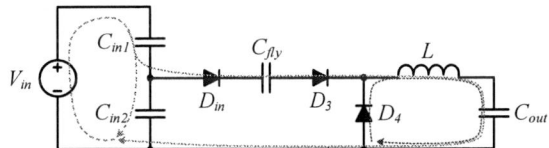

Fig. 5. Simplified ideal circuit model for the 3L-FC buck on start-up. Inrush current path shown in orange, inductor discharge path in purple.

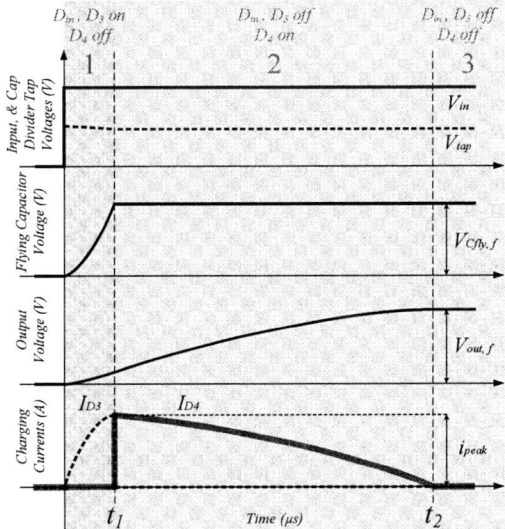

Fig. 6. Voltage and current waveforms corresponding to Fig. 5.

The equivalent capacitance of the flying and output capacitors is given by Eq. 1, where C_{eq} is the equivalent capacitance, C_{out} is the output capacitance, and C_{fly} is the value of the flying capacitor. The resonant frequency of the output filter during portion 1 when D_{in} and D_3 conduct is given by Eq. 2, and the resonant frequency of the output filter during portion 2 when D_4 conducts is given by Eq. 3, where ω_{o1} and ω_{o2} are the resonant frequencies during portions 1 and 2, respectively, and L is the value of the output inductor.

$$C_{eq} = \frac{C_{fly}C_{out}}{C_{fly}+C_{out}} \qquad (1)$$

$$\omega_{o1} = \frac{1}{\sqrt{LC_{eq}}} \qquad (2)$$

$$\omega_{o2} = \frac{1}{\sqrt{LC_{out}}} \qquad (3)$$

For an input voltage step, the inductor will initially be seen as a high impedance and the centre tap voltage of the input capacitors, V_{tap}, will initially be set by the voltage divider of C_{in1} and C_{in2}. This voltage will determine the maximum inductor charging current and consequently, the maximum output voltage. The final value of V_{tap} will be determined by a capacitive voltage divider of the input capacitors and the equivalent output capacitance, C_{eq}, which will affect the point at which the inrush diode turns off and thus the final voltage of the flying capacitor.

On start-up, the peak charging current, i_{peak}, through the inductor, the inrush diode and the body diodes of S_3 and S_4 is given by Eq. 4, where V_{in} is the input voltage, and C_{in1} and C_{in2} are the input capacitors. The maximum final output voltage, $V_{out,f}$, is given by Eq. 5. The final value of the flying capacitor voltage, $V_{Cfly,f}$, is given by Eq. 6.

$$i_{peak} = V_{in} \left(\frac{C_{in1}}{C_{in1}+C_{in2}} \right) \sqrt{\frac{C_{eq}}{L}} \tag{4}$$

$$v_{out_f} = V_{in} \left(\frac{C_{in1}}{C_{in1}+C_{in2}} \right) \frac{\sqrt{C_{eq}(C_{eq}+C_{out})}}{C_{out}} \tag{5}$$

$$v_{C_{fly_f}} = V_{in} \left(\frac{C_{in1}}{C_{in1}+C_{in2}+C_{eq}} \right) \tag{6}$$

The flying capacitor voltage settles at $t = t_1$. At $t = t_2$, the inductor finishes discharging fully, with the output voltage reaching its maximum. The expressions for these times are given in Eqs. 7 and 8, where t_1 and t_2 are the end times of portions 1 and 2 respectively.

$$t_1 = \frac{1}{\omega_{o1}} \frac{\pi}{2} \tag{7}$$

$$t_2 = \frac{1}{\omega_{o2}} \tan^{-1} \left(\sqrt{\frac{C_{out}}{C_{eq}}} \right) + t_1 \tag{8}$$

During regular converter operation, the flying capacitor voltage must be actively balanced at $V_{in}/2$ to limit the voltage stress across switches and maintain symmetrical current ripple, as well as overall stability of the system [16]. Consequently, since the positive and negative terminals of the flying capacitor will range between $V_{in}/2$ and V_{in}, and $0V$ and $V_{in}/2$ respectively, the inrush and return diodes, D_{in} and D_{ret}, keep V_{tap} fixed at approximately $V_{in}/2$. This allows the use of input capacitors rated for half the input voltage. Capacitor volume generally scales with the energy storage requirements of the capacitor [17]. Though the single input capacitor is replaced by two capacitors each with double the capacitance of the original, the rated voltage is halved and thus the total stored energy is the same. As a result, theoretically, the total volume of the stacked input capacitors, C_{in1} and C_{in2} should be comparable to that of the single input capacitor.

There is one limitation to the use of this input capacitor divider method. If the flying capacitor voltage is naturally lower than $V_{in}/2$, as is the case of the series-capacitor buck

operating in D^2 mode for conversion ratios above 0.25 [18], there will be current cycling through the diodes as the flying capacitor moves between the ground and input voltage rails during steady state. This cycling will reduce the efficiency of the system and disrupt its operation. In this case, the return diode can be omitted, only permitting unidirectional current flow. However, the stacked input capacitors will have to be rated for the full input voltage, increasing volume and negating some of the benefits of this solution. For converters operating with similar modulation schemes, the second start-up method may be more desirable as the start-up capacitor does not affect the steady state operation of the converter.

B. Flying Capacitor Divider Based Solution

The second start-up method also features capacitive division using additional passive circuitry but utilizes the flying capacitor itself as a part of the divider. Fig. 3 shows the general commutation cell and Fig. 7 shows implementation for the 3-level buck with the initial inrush current and inductor discharge paths highlighted. A start-up capacitor, C_{div}, of comparable size to the flying capacitor is connected between the input and the top of the flying capacitor forming a capacitor divider. A high current capacity inrush diode, D_{in}, in that path charges the flying capacitor and prevents reverse current flow during steady state operation. A pull-down switch pulls the diode anode to ground sometime after start-up and keeps it reversed biased during normal operation. In this method, the start-up capacitor, flying capacitor, and output capacitor are all in series during power-up.

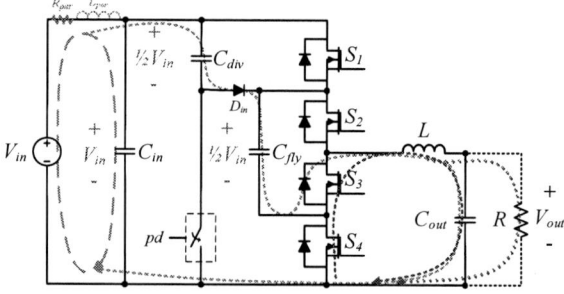

Fig. 7. 3L-FC buck with flying capacitor divider start-up method. Initial inrush current path shown in orange, inductor discharge path shown in purple, along with initial voltage distribution.

During steady state operation, the start-up capacitor is connected in parallel with the input capacitor but contributes little to input filtering as the high R_{on} of the small start-up switch makes the total impedance of the start-up capacitor plus switch much higher than that of the input capacitor at all frequencies. Since the start-up circuit does not interfere with steady state converter operation, this solution is better suited than the input capacitor divider solution to systems where modulation schemes may result in the flying capacitor voltage deviating from $V_{in}/2$, such as the series-capacitor buck operating in D^2 mode.

IV. EXTENSION TO OTHER MULTI-LEVEL FLYING CAPACITOR TOPOLOGIES

These methods can be extended to any N-level flying capacitor buck converter and reduce the switch voltage stresses during start-up to $(N\text{-}1)$ times lower value than that of a conventional buck. For both methods, extension to another 3-level topology, the 2-phase series-capacitor buck [19] [18], is shown in Figs. 8a and 8b.

These methods can also be used for start-up in higher order ML-FC topologies such as the 4L-FC buck [4] and the 3-phase series-capacitor buck [18]. These 4-level topologies have two flying capacitors, operating with steady state voltages of $2 \cdot V_{in}/3$ and $V_{in}/3$. For the input capacitor divider method, adding a third stacked input

capacitor provides two tap voltages, $2 \cdot V_{in}/3$ and $V_{in}/3$, from which two inrush diodes can be placed to charge the flying capacitors. There is some freedom in the placement of the return diodes, as long as the maximum voltage at the node connected to the return diode anode is exactly equal to the corresponding tap voltage. Implementations of this method with the 4L-FC buck and 3-phase series-capacitor buck are shown in Figs. 8c and 8e, respectively. Implementation of the flying capacitor divider method with 4-level topologies involves two additional capacitors, each one independently creating a divider with a single flying capacitor. Implementations of this method with the 4L-FC buck and 3-phase series-capacitor buck are shown in Figs. 8d and 8f, respectively.

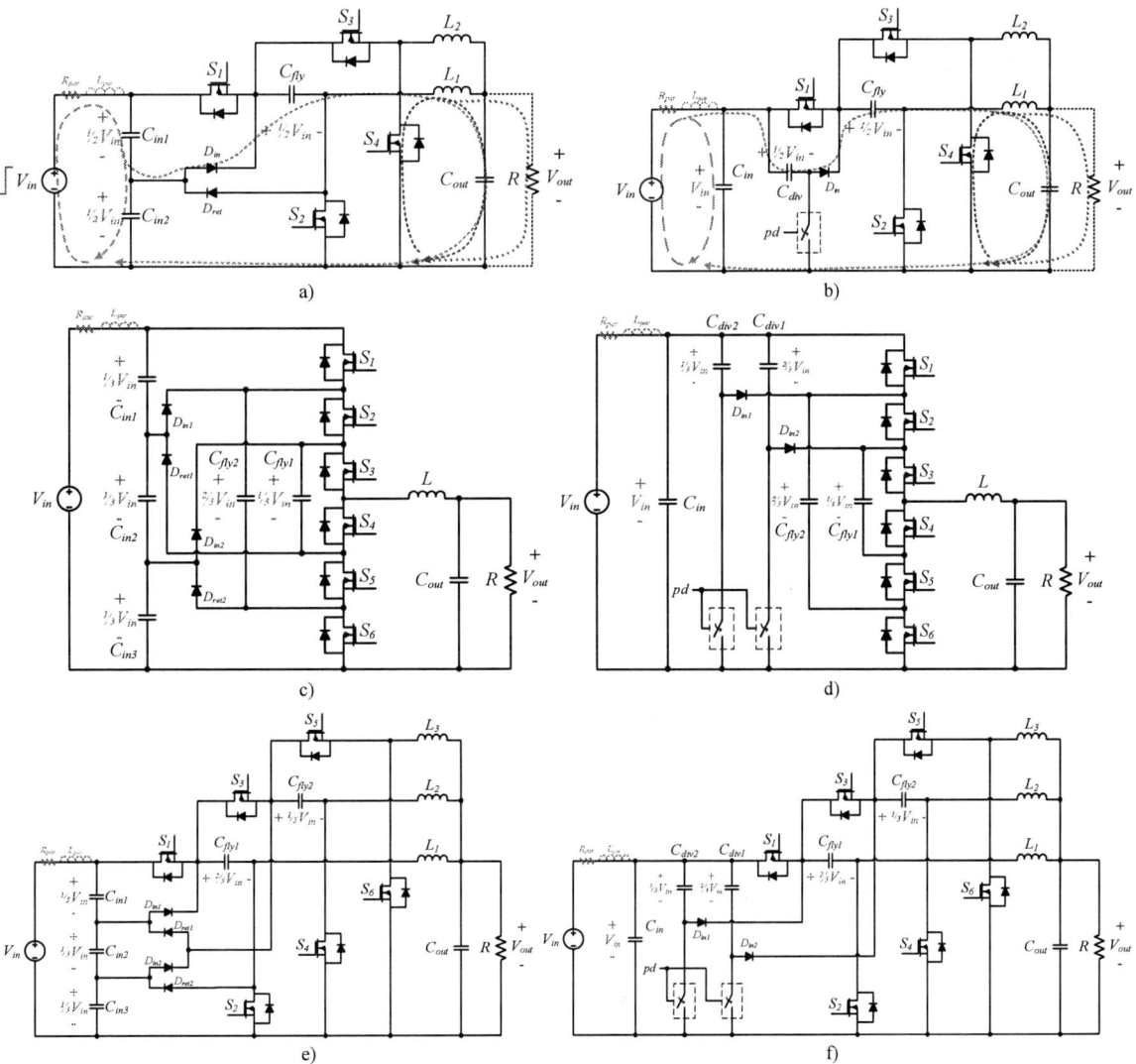

Fig. 8. Extension of start-up methods to different flying capacitor buck topologies. Figs 8a, 8c, and 8e show the split input capacitor divider start-up method with the 2-phase series-capacitor buck, 4L-FC buck, and the 3-phase series-capacitor buck, respectively. Figs 8b, 8d, and 8f show the flying capacitor divider start-up method with the 2-phase series-capacitor buck, 4L-FC buck, and the 3-phase series-capacitor buck, respectively.

V. EXPERIMENTAL RESULTS

A discrete PCB prototype of the 3L-FC buck converter was made to test the functionality of the start-up methods. The prototype operates at 1MHz switching frequency and is nominally rated for a conversion ratio of 24V-to-5V, 20W output. The component values corresponding to Figs. 4 and 7 are listed in Tables I and II.

The methods were tested with fast input voltage rise times of under 1μs. The results are detailed in Fig. 9. Both start-up methods guarantee that the voltage stress across each switch is limited to $V_{in}/2$ during start-up, and pre-charge the flying capacitor. The peak current, final output voltage and final flying capacitor voltage differ slightly from the approximations primarily due to diode non-idealities. The small ripple in the switch voltages at approximately 8μs following the input step for both start-up schemes indicates the point at which the body diode of S_4 ceases conducting and turns off.

TABLE I
EXPERIMENTAL PROTOTYPE DESIGN SPECIFICATIONS

V_{in}	24 V
V_{out}	5 V
P_{out}	20 W
f_{sw}	1 MHz

TABLE II
COMPONENT PARAMETERS FOR FIGS. 4 & 7

Parameters	Input Capacitor Divider	Flying Capacitor Divider
L_{par}	5 nH	5 nH
R_{par}	10 mΩ	80 mΩ
C_{in}	-	5 μF
C_{in1}, C_{in2}	10 μF	-
C_{fly}	1 μF	1 μF
C_{div}	-	1 μF
L	1.2 μH	1.2 μH
C_{out}	50 μF	50 μF

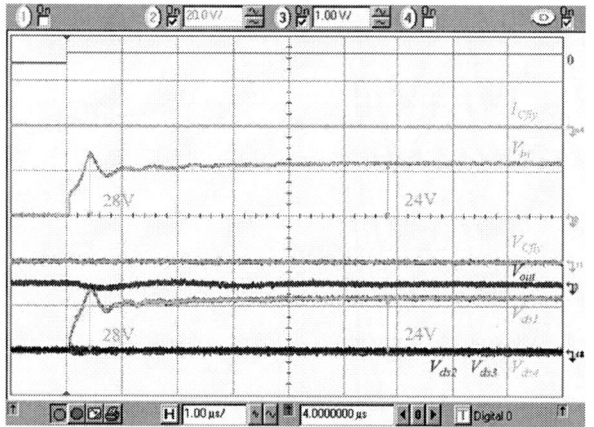

Fig. 9a. Fast input voltage ramp (~400ns) in the 3L-FC buck prototype showing the converter switch voltages on start-up in the absence of any start-up scheme, with the full input voltage seen across S_1 ($V_{ds1} = V_{in}$).

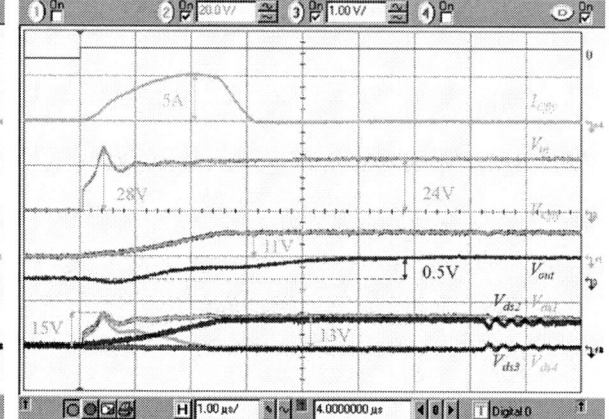

Fig. 9b. Fast input voltage ramp (~400ns) in the 3L-FC buck prototype using the input capacitor divider start-up method. All switch voltages are limited to ~$V_{in}/2$. The flying capacitor is also pre-charged to ~$V_{in}/2$.

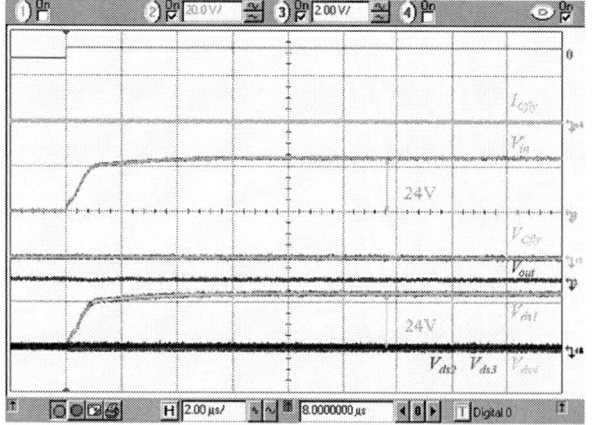

Fig. 9c. Fast input voltage ramp (~1μs) in the 3L-FC buck prototype showing the converter switch voltages on start-up in the absence of any start-up scheme, with the full input voltage seen across S_1 ($V_{ds1} = V_{in}$).

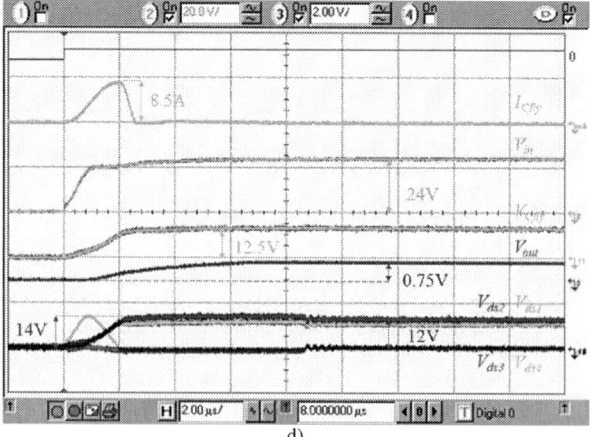

d)

Fig. 9d. Fast input voltage ramp (~1μs) in the 3L-FC buck prototype using the flying capacitor divider start-up method. All switch voltages are limited to ~$V_{in}/2$. The flying capacitor is also pre-charged to ~$V_{in}/2$.

VI. CONCLUSIONS

This paper introduces two sets of methods for limiting the voltage stress across switches of multi-level flying capacitor (ML-FC) step-down dc-dc converters during start-up. These methods split the input voltage across multiple capacitors and consequently across multiple switches. For a general N-level converter, in applications where the input voltage is expected to rise in a few microseconds or longer, the presented methods guarantee the voltage stress to be (N-1) times lower value than that of a conventional buck during start-up, allowing lower voltage rating transistors with smaller on-resistances to be used. The passive nature of these methods allows their use in applications where the gate drive and logic supply voltages are sourced from the converter input or an internal node. The methods are very simple and require only a limited number of discrete components. The speeds of these methods are limited only by the quality of the matching between high-frequency impedances of the components and can theoretically be pushed to respond to input voltage ramps of fractions of a microsecond. The input capacitor divider method also reduces the stress during input voltage transients. Experimental results demonstrate the functionality of these start-up methods.

REFERENCES

[1] V. Yousefzadeh, E. Alarcón and D. Maksimović, "Three-Level Buck Converter for Envelope Tracking Applications," *IEEE Transactions on Power Electronics*, vol. 21, no. 2, pp. 549 - 552, Mar. 2006.

[2] Y. Lei, W.-C. Liu and R. C. N. Pilawa-Podgurski, "An Analytical Method to Evaluate Flying Capacitor Multilevel Converters and Hybrid Switched-Capacitor Converters for Large Voltage Conversion Ratios," in *Proc. IEEE Control and Modeling for Power Electronics (COMPEL)*, Vancouver, 2015.

[3] A. Stupar, T. McRae, N. Vukadinovic, A. Prodic and J. A. Taylor, "Multi-Objective Optimization and Comparison of Multi-Level DC-DC Converters Using Convex Optimization Methods," in *Proc. IEEE Power Electronics and Applications (EPE'16 ECCE Europe)*, Karlsruhe, 2016.

[4] T. Meynard and H. Foch, "Multi-Level Conversion: High Voltage Choppers and Voltage-Source Inverters," in *Proc. IEEE Power Electronics Specialists Conference*, Toledo, 1992.

[5] M. Amato and V. Rumennik, "Comparison of Lateral and Vertical DMOS Specific On-Resistance," in *IEEE Proc. Electron Devices Meeting*, Washington, 1985.

[6] N. Vukadinović, A. Prodić, B. A. Miwa, C. B. Arnold and M. W. Baker, "Volume and Efficiency Comparison Between Multi-level Dc-Dc Converters and Buck Converter for Low-Power Mobile Applications," in *18th International Symposium on Power Electronics Ee 2015*, Novi Sad, 2015.

[7] S. M. Ahsanuzzaman, Y. Ma, A. A. Pathan and A. Prodić, "A Low-Volume Hybrid Step-Down DC-DC Converter Based on the Dual Use of Flying Capacitor," in *Proc. IEEE Applied Power Electronics Conference and Exposition (APEC)*, Long Beach, 2016.

[8] P. S. Shenoy, M. Amaro, J. Morroni and D. Freeman, "Comparison of a Buck Converter and a Series Capacitor Buck Converter for High-Frequency, High-Conversion-Ratio Voltage Regulators," *IEEE Transactions on Power Electronics*, vol. 31, no. 109, p. 7006–7015, Oct. 2016.

[9] G. Calabrese, M. Granato, G. Frattini and L. Capineri, "Integrated high step-down multiphase buck converter with high power

density," in *Proc. IEEE Power Electronics and Applications (EPE'14-ECCE Europe)*, Lappeenranta, 2014.

[10] J. Xue and H. Lee, "A 2 MHz 12–100 V 90% Efficiency Self-Balancing ZVS Reconfigurable Three-Level DC-DC Regulator With Constant-Frequency Adaptive-On-Time V^2 Control and Nanosecond-Scale ZVS Turn-On Delay," *IEEE Journal of Solid-State Circuits*, vol. 51, no. 12, p. 2854–2866, Dec. 2016.

[11] A. Stillwell and R. C. N. Pilawa-Podgurski, "A 5-level flying capacitor multi-level converter with integrated auxiliary power supply and start-up," in *Proc. IEEE Applied Power Electronics Conference and Exposition (APEC)*, Tampa, 2017.

[12] E. Candan, D. Heeger, P. S. Shenoy and R. C. N. Pilawa-Podgurski, "A series-stacked power delivery architecture with hot-swapping for high-efficiency data centers," in *Proc. IEEE Energy Conversion Congress and Exposition (ECCE)*, Montreal, 2015.

[13] D. Reusch, F. C. Lee and M. Xu, "Three level buck converter with control and soft startup," in *Proc. IEEE Energy Conversion Congress and Exposition*, San Jose, 2009.

[14] J. M. Khayat, S. Carlo-Rodriquez, M. G. Amaro, R. Ramani and P. S. Shenoy, "Series Capacitor Buck Converter Having Circuitry For Precharging The Series Capacitor". United States Patent US 2015/0311793 A1, 29 Oct. 2015.

[15] K. Kim, H. Cha, S. Park and I.-O. Lee, "A Modified Series-Capacitor High Conversion Ratio DC–DC Converter Eliminating Start-Up Voltage Stress Problem," *IEEE Transactions on Power Electronics*, vol. 33, no. 1, pp. 8-12, 2017.

[16] N. Vukadinović, A. Prodić, B. A. Miwa, C. B. Arnold and M. W. Baker, "Extended Wide-Load Range Model for Multi-Level DC-DC Converters and a Practical Dual-Mode Digital Controller," in *Proc. IEEE Applied Power Electronics Conference and Exposition (APEC)*, Long Beach, 2016.

[17] K. Raggl, T. Nussbaumer and J. W. Kolar, "Guideline for a Simplified Differential-Mode EMI Filter Design," *IEEE Transactions on Industrial Electronics*, vol. 57, no. 3, p. 1031–1040, Mar. 2010.

[18] Y. Jang, M. M. Jovanović and Y. Panov, "Multi-Phase Buck Converters with Extended Duty Cycle," in *Proc. IEEE Applied Power Electronics Conference and Exposition (APEC)*, Dallas, 2006.

[19] K. Nishijima, K. Harada, T. Nakano, T. Nabeshima and T. Sato, "Analysis of Double Step-Down Two-Phase Buck Converter for VRM," in *Proc. IEEE International Telecommunications Energy Conference*, Berlin, 2005.

The 2018 International Power Electronics Conference

High Voltage Gain Interleaved Active-Clamp Forward (IACF) Converter having Reduced Primary Conduction Loss

Yeonho Jeong[1], Mu-hyun Park[1], Gun-Woo Kim[1], Byoung-hee Lee[2], and Gun-Woo Moon[1]
1 Electrical Engineering, KAIST, Daejeon, Republic of Korea
2 Electrical Engineering, Han-bat University, Daejeon, Republic of Korea
*E-mail: gwmoon@kaist.ac.kr

Abstract— This paper proposes an interleaved active clamp forward (IACF) converter having small primary conduction loss. Based on the conventional IACF converter, the components on one ACF module is re-arranged to employ the increased input voltage for high voltage gain. By achieving the high voltage gain, the proposed converter can significantly reduce the primary condition loss with large turns-ratio and the expanded duty ratio at the nominal input voltage V_{nom}. In addition, since there is no additional components, a power density cannot be a burden in the proposed converter although high efficiency is achieved. Finally, the validity of the proposed converter is confirmed by the experimental results of a prototype converter with 36-72 VDC input and 480 W (12 V/40 A) output.

Keywords— *DC/DC power systems, high voltage gain, interleaved active-clamp forward converter, and wide range of low input voltage.*

I. INTRODUCTION

The DC/DC power supplies for 48V DC battery system has widely employed in many applications such as the battery charging/discharging systems, the server power systems in data centers, network power systems, and telecommunication applications [1]-[9]. Recently, the 48V DC/DC power system has expanded to small electrical vehicle such as electric bike [7] and neighborhood electric vehicles [8]-[9].

In those 48V DC/DC power systems, there are two important requirements. First, the power systems should operate with a wide input voltage range to cover the dynamic batter voltage, and the required voltage range is generally 36-72VDC [1]-[3]. In addition, since those power systems mostly operate at the nominal input voltage V_{nom}, the highest efficiency at V_{nom} is necessary. To satisfy two requirements, many DC/DC topologies have been considered. Among them, the phase-shifted full-bridged (PSFB) converter is widely applied due to low voltage stress on primary switches, zero-voltage-switching operation of the primary switches, and twice powering transmission during one switching cycle. However, to cover the wide input voltage range, power systems are designed at the minimum input voltage in

order to operate with the maximum duty ratio which is 0.5 in the PSFB converter. Therefore, the operating duty ratio at V_{nom} can be small. It causes large free-wheeling current which is circulating on the primary side. Moreover, due to the low input voltage of 48V DC/DC power systems, the transformer turns-ratio is also limited for regulating the output voltage V_O, and it causes a large reflected current on the primary side from the secondary side. As a result, the primary conduction loss becomes large due to large free-wheeling current by small duty ratio and small transformer turns-ratio. To reduce the large primary conduction loss of the PSFB converter, an interleaved active clamp forward (IACF) converter described in Fig. 1(a) is preferred. The IACF converter consists of two symmetric modules, and the each stage is independently operated with the interleaved gate signals, which normally defined as the 180-degree phase difference between two gate signals. From the interleaving operation, since the power can transfer to the load twice during one switching cycle, the primary conduction loss and the filter size can be reduced as the same with the PSFB converter. In addition, since the free-wheeling current on the primary side is noticeably reduced, the primary conduction loss can be reduced. However, since the duty ratio of IACF converter is also limited as 0.5, the small transformer turns-ratio and large reflected current can affect to the large primary conduction losses.

To solve the limitation of IACF converter having wide input voltage range and low input voltage, many papers have been studied [10]-[11]. In [10], IACF converter with parallel input and series-parallel output is proposed. According to the change of the input voltage, the secondary windings connection of the transformers is selected in series and parallel. Thus, the series-parallel output structure allow to operate over 0.5 of duty ratio, and the turns-ratio can be increased due to high voltage gain. It results in reducing primary conduction loss and switch turn-off loss. However, in the medium power applications with high output current, the synchronous rectifiers (SRs) should be employed instead of the rectifier diodes to reduce the conduction loss on the

The 2018 International Power Electronics Conference

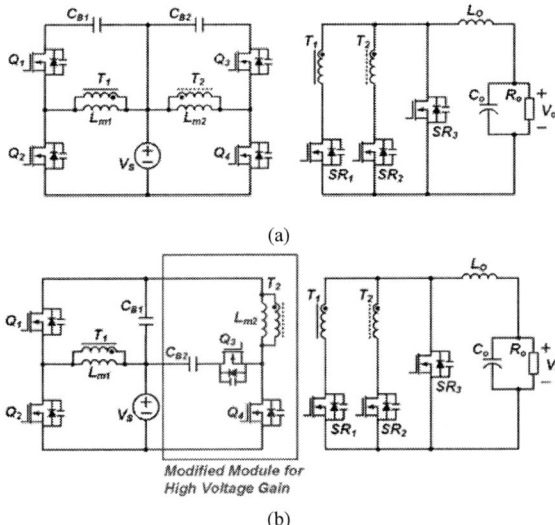

(a)

(b)

Fig. 1. Circuit diagram: (a) Conventional IACF converter and (b) proposed IACF converter.

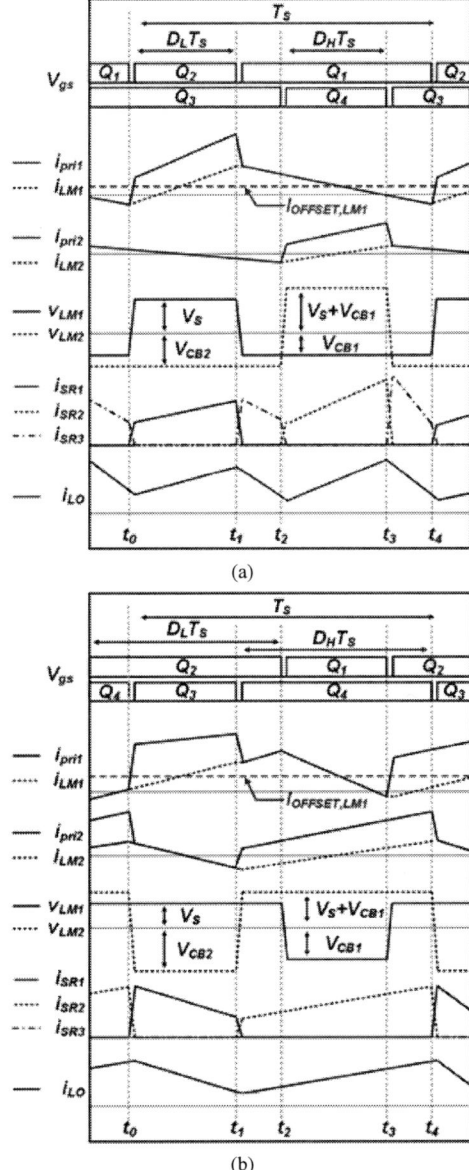

(a)

(b)

Fig. 2. Key waveforms of the proposed converter: (a) at D_L and $D_H \leq$ 0.5 and (b) at D_L and $D_H > 0.5$..

secondary side is large. However, in [10], it requests two floating gate drivers for SRs, which cause high circuit complexity and low power density. In [11], by adding windings of two transformers, which are connected in series on the secondary side, the duty ratio can be expanded over 0.5. However, because the current on the secondary side is large, the large space for the additional windings is required. Moreover, since the usage of the additional winding is restricted only when the duty ratio is over 0.5, the converter in [11] requires large transformer size and low power density.

To overcome abovementioned problems of previous approaches, the IACF converter having a new structure is proposed in this paper. By changing the components position of one module, the voltage gain can be increased with high input voltage of the modified module. In addition, since the proposed converter can operate when the duty ratio is over 0.5, the duty ratio at V_{nom} and the transformer turns-ratio can be increased by the high voltage gain. The primary conduction loss can be reduced, and it results in high efficiency at V_{nom}. Moreover, in the proposed converter, since there is no additional component, the power density can be kept as the conventional IACF converter. The analysis and design guideline for high efficiency are also presented, and the operation and performance of the proposed converter are verified by a prototype with 36-72 VDC input and 480 W (12 V/40 A) output.

II. DISCRIPTION OF PROPOSED CONVERTER

A. Derivation of Proposed Converter

The conventional IACF converter with synchronous rectifiers (SRs) on the secondary side, which is applying to reduce the secondary conduction losses, is presented as shown in Fig. 1(a). The conventional IACF converter consists of two ACF modules, and they operate with interleaved gate signals for achieving small primary

conduction loss and output filter. From the conventional IACF converter, the proposed converter is derived by modifying one ACF module. Basically, the input voltage V_S is applied to the transformer of ACF modules when low side switches are turned-on, and the voltage on the secondary side can have a role as the voltage source to transfer the power to the load. In the proposed converter, one module is re-arranged by changing the transformer T_2 and two components, one switch Q_3 and one clamping capacitor C_{B2}. From the change, the voltage across the transformer on the modified stage V_{LM2} is increased from V_S to the stacked voltage, which consists of V_S and the voltage across the clamping capacitor on the opposite side V_{CB1}. Moreover, since the duty ratio in the proposed converter can be increased over 0.5 by the modified

839

The 2018 International Power Electronics Conference

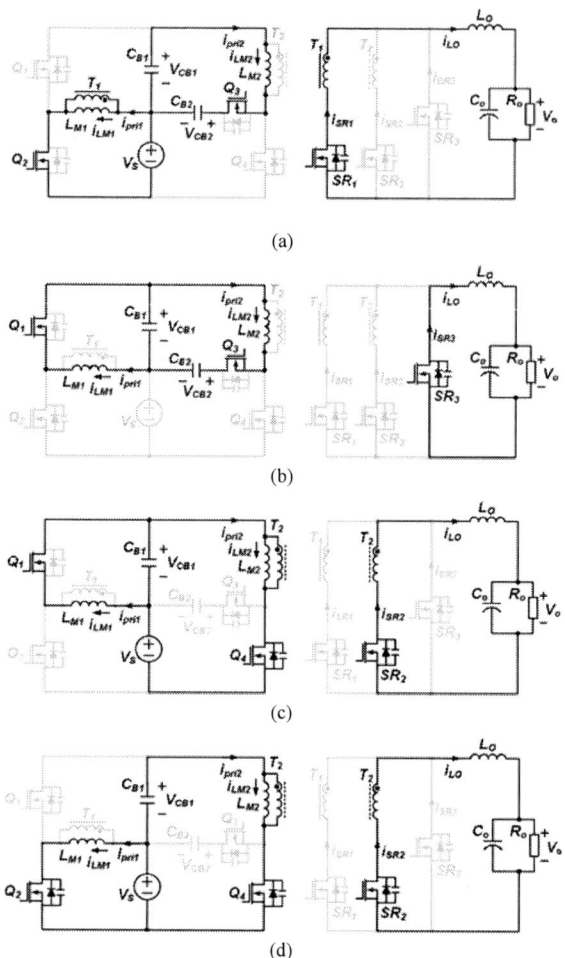

Fig. 3. Operating circuits of the proposed converter for each operational mode: (a) Mode 1 (t_0~t_1), (b) Mode 2 (t_1~t_2) and Mode 4 (t_3~t_4), (c) Mode 3 (t_2~t_3) during D_L and D_H are under 0.5, and (d) Mode 3 (t_2~t_3) during D_L and D_H are over 0.5..

module, the large duty ratio at V_{nom} can be designed. Those changes can allow to reduce the primary conduction losses.

B. Operational Principle

In the proposed converter, Q_2 and Q_4 are the powering switches, and Q_1 and Q_3 are the active clamping switches for the reset operation. And the duty ratio for Q_2 on the conventional ACF module is named as D_L, and the duty ratio for Q_4 on the modified ACF module is set as D_H. And the duty ratio for Q_1 and Q_3 are defined as 1-D_L and 1-D_H, respectively. For the high voltage gain, the proposed converter can have two operational modes that D_H and D_L are 1) under 0.5 and 2) over 0.5.

Firstly, to facilitate the explanation of the mode operation at D_L and $D_H \leq 0.5$, key waveforms and the operational description with the current path are described in Fig. 2 and Fig. 3, respectively. In steady state, the operation of the proposed converter is separated into four different modes, and the analysis for the mode operation is based on the assumptions as follows:

1) All active power devices are ideal switches with

parallel body diodes and parasitic capacitors C_{OSS}.

2) Two main transformers T_1 and T_2 have the turns-ratios of n_1 and n_2, the magnetizing inductance L_{M1} and L_{M2}, and the leakage inductances of them are named as L_{lkg1} and L_{lkg2}, respectively.

3) Clamping capacitors C_{B1} and C_{B2} are large enough to be constant.

Mode 1 [t_0~t_1]: At t_0, mode 1 begins with turning-on of Q_2. Because the voltage across L_{M1} of T_1 in a conventional module is the positive value as the input voltage V_S, and the inductor current $i_{LM1}(t)$ builds-up. The power is transferred through SR_1 on the secondary side. In the modified module, V_{LM2} defined as the voltage across L_{M2} of T_2 on the modified module is negative as -V_{CB2} (=$V_S D_H/(1-D_L)/(1-D_H)$), and the inductor current $i_{LM2}(t)$ is decreased. On the secondary side, the voltage across output inductor V_{LO} is defined as the difference of two voltages, V_S/n and the output voltage V_O. When the duty ratio is under 0.5, since the high input voltage is applied and the reflected voltage to the secondary side (=V_S/n) is larger than V_O, L_O is starting to increase.

Mode 2 [t_1~t_2]: At t_1, after Q_2 is turned off and Q_1 is turned-on with ZVS operation. The commutation voltage between Q_1 and Q_2 can be easily accomplished due to the large ZVS energy including the reflected output current. In the original module, the voltage across L_{M1} is changed to the negative voltage, which is the voltage across C_{B1}, -V_{CB1} (=$D_L V_S/(1-D_L)$). Although the modified module is keeping the operation in the previous mode, two modules are in a freewheeling mode, and the power is not transfer to the secondary-side. Therefore, L_O is in a freewheeling operation through SR_3. Mode 1 and mode 2 are the same with the conventional IACF converter, and two modules in the conventional IACF converter alternatively operates as the same with mode 1 and mode 2.

Mode 3 [t_2~t_3]: When Q_3 is turned-off and Q_4 is turned-on with ZVS operation, mode 3 is started. The original module is keeping the free-wheeling mode. In the modified module, the power is transferred through SR_2. For the voltage gain, because V_{LM2} is V_S+V_{CB1}, not V_S as the conventional converter, the proposed converter can have large voltage gain due to the increased input voltage.

Mode 4 [t_3~t_4]: Mode 4 begins when Q_4 is turned off and Q_3 is turned on, the inductor current of L_{M1}, L_{M2}, and L_O are in a freewheeling mode. The operations of two modules are the same with one of mode 2. In addition, because ZVS energy is not enough due to the offset current of L_{M1}, L_{M1} should be carefully designed to accomplish ZVS operation of Q_2 when mode 1(t_0~t_1) starts.

Meanwhile, in the conventional converter, when the duty ratio is over 0.5, the operation and the voltage gain cannot be increased compared with them at $D=0.5$. On the other hand, in the proposed converter, when D_H and D_L are over 0.5, the mode operation and the voltage gain can be changed. Therefore, the mode operation when D_H and D_L are over 0.5 should be described as shown in Fig. 2(b). There are two different operations compared with the operation when D_H and D_L are under 0.5. First, in the mode 1, because this operation with large duty ratio can

840

The 2018 International Power Electronics Conference

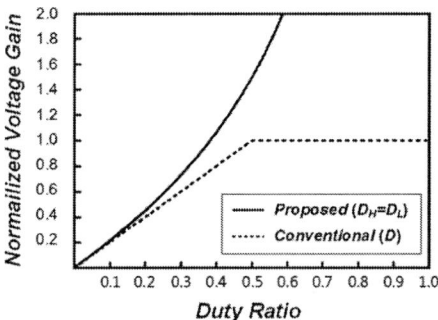

Fig. 4. Voltage gain comparison between conventional and proposed IACF converter.

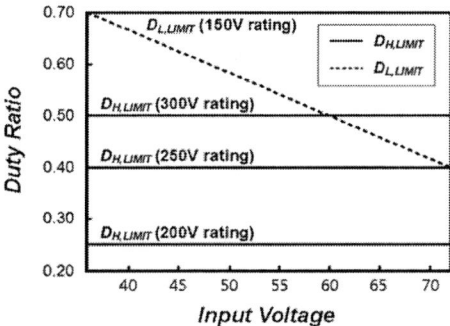

Fig. 5. Limitation of duty ratio according to the different voltage rating.

be shown at low V_S, V_S/n is lower than V_O. It means that the voltage across L_O is negative, not positive at D_L and $D_H < 0.5$. In addition, mode 2($t_1\sim t_2$) and mode 4($t_3\sim t_4$) are changed to overlap gate signals with Q_2 and Q_4, not Q_1 and Q_3. It notes that the power can be continuously transferred from the modified module to the output as shown in Fig. 3(d). Since V_{LM2} is V_S+V_{CB1} as the same with mode 3, the proposed converter can achieve high voltage gain.

III. ANALYSIS OF PROPOSED CONVERTER

A. DC Conversion Ratio

In the conventional IACF converter, the DC conversion ratio M_{CONV} is calculated as follows [10]-[11]:

$$M_{CONV}(D)$$
$$= \frac{V_O}{V_S} = \begin{cases} \dfrac{2D_{CONV}}{n_{CONV}}, & when\, D_{CONV} \leq 0.5. \\ 1, & when\, 0.5 < D_{CONV} \leq 1.0, \end{cases} \quad (1)$$

where D_{CONV} and n_{CONV} ($=N_P/N_S$) are an duty ratio and the transformer turns-ratio of the conventional converter, respectively.

On the other hand, the proposed converter has the increased voltage gain due to the re-arranged primary side structure. To calculate the voltage conversion ratio M_{PROP}, there are some assumptions: 1) The commutation period on the secondary-side is neglected, 2) all capacitors C_{B1}, C_{B2}, and C_O are enough large to have a constant voltage, and 3) the overlapped duty ratio, D_{over_L} and D_{over_H} ($=D_H$-0.5), are defined as the duty ratio over 0.5 on the conventional and modified stage. Based on the assumptions, M_{PROP} can be expressed as follows:

$$M_{PROP}(D_L, D_H)$$
$$= \frac{V_O}{V_S} = \frac{1}{n}\left[\left(D_L + \frac{D_H}{1-D_L} \right) - \left(D_{over\_L} + D_{over\_H} \right) \right], \quad (2)$$

Fig. 4 shows the normalized DC conversion ratio of the conventional and proposed converter. It notes that the proposed converter can have higher voltage gain. When the duty ratio is under 0.5, the proposed converter can show higher voltage gain due to the high input voltage on the modified structure. When the duty ratio is over 0.5, although the conventional converter operates with the same voltage gain at D=0.5, the proposed converter can transfer the power to the load through the modified module due to the higher input voltage of the modified structure. It causes the voltage gain can be extremely increased. Consequently, the proposed converter can have large turns-ratio of transformer, and it results in the reduced primary conduction loss with the small reflected current from the secondary-side.

B. Voltage Stress of Primary Switches

In the proposed converter, the higher voltage gain is achieved by modifying one module of IACF converter. In addition, since the duty ratio is also expanded, the voltage across capacitors C_{B1} and C_{B2} are increased. However, in general, to satisfy high voltage rating of switches, a thick and low-doped layer with highly resistive are required. Therefore, the channel resistance of switches which is $R_{DS(ON)}$ can be increased to satisfy the high voltage rating. It is possible to degrade efficiency by large conduction loss. Therefore, the maximum voltage stress of switches should be carefully designed to get the maximized efficiency. The maximum voltage stress on the conventional module $V_{DS(L),max}$ and the modified module $V_{DS(H),max}$ can be expressed as follows:

$$V_{DS(L),max} = \frac{V_S}{(1-D_L)}, \quad (3)$$

$$V_{DS(H),max} = \frac{V_S}{(1-D_L)(1-D_H)}. \quad (4)$$

From (3) and (4), it notes that the voltage stresses can be controlled by limiting of D_L and D_H, and they can be presented as follows:

$$D_{L,LIMIT} = 1 - \frac{V_S}{V_{DS(L),max}}, \quad (5)$$

$$D_{H,LIMIT} = 1 - \frac{V_S}{V_{DS(H),max}(1-D_L)} = 1 - \frac{V_{DS(L),max}}{V_{DS(H),max}}. \quad (6)$$

Based on (5) and (6), when 150V rating in the normal stage and the various rating in the modified stage with the enough stress margin are applied, the limit of D_L and D_H can be drawn as shown in Fig. 5. In the conventional module, the voltage stress is limited V_S, and the employed switches are the same with the conventional IACF converter. In the modified module, D_{H_LIMIT} is fixed

841

The 2018 International Power Electronics Conference

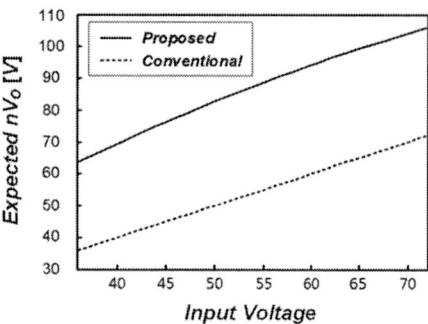

Fig. 6. Comparison on the maximum voltage gain (nV_O) according to V_S.

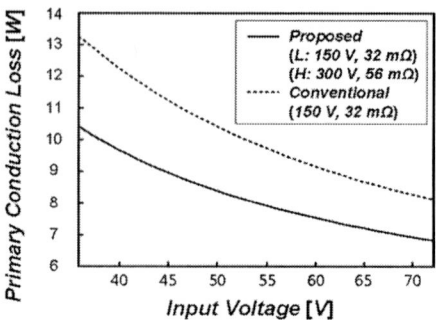

Fig. 7. Analysis of loss distribution comparing the conventional and proposed converter.

regardless of the change of V_S because the voltage stress in the original module is already controlled by D_L. In addition, to apply small $R_{DS(ON)}$ by considering efficiency, the voltage stress of switches in the modified stage should be minimized. However, if D_L and D_H are too small, V_O cannot be regulated due to the restricted voltage gain in (2). Therefore, the proposed converter for maximizing efficiency and satisfying the voltage regulation should be carefully designed.

C. DC Offset Current of Main Trnasformer

In the conventional IACF converter, the offset current of the main transformers can be calculated with the commutation time as follows:

$$I_{OFFSET,CONV} = \frac{-I_O}{2n(1-D_{CONV})} D_{COMM}, \quad (7)$$

where D_{COMM} is the commutation time the duration when SR_1 and SR_2 are turned-on/off and SR_3 is turned-off/on.

On the other hand, the proposed converter have the different offset current of two transformers. For T_2 in the modified stage, the offset current of L_{M2} named by $I_{LM2,OFFSET}$ has the same equation in (7), but it is smaller than $I_{OFFSET,CONV}$ due to the short commutation time with large voltage across T_2. In the conventional stage, the offset current of T_1 called as $I_{LM1,OFFSET}$ can be increased to satisfy the charge balance of C_{B1}, and it can be expressed as follows with neglecting the commutation time $D_{COMM,H2}$ and $D_{COMM,L2}$:

$$I_{LM1,OFFSET} = \frac{2D_H I_O}{n}, \quad (9)$$

Since the offset current can affect high RMS current on the primary switches and a transformer size, it should be minimized. To reduce $I_{LM1,OFFSET}$, D_H of the modified module, which is proportional on the power transfer capability, should be minimized, and it means the power capability of the original module is maximized. However, since the design for large D_L can be a burden on the voltage stress of switches, D_L and D_H should be considered with the offset current and voltage stress at the same time.

However, since the proposed converter can have high voltage gain based on the expected nV_O as shown in Fig. 6, the turns-ratio in the proposed converter can have doubled. Consequently, although the offset current of one

transformer in the proposed converter is increased, the transformer size is similar due to the increased turns-ratio by the high voltage gain.

D. Comparison on Primary Conduction Loss

To compare the efficiency of 48V DC/DC power systems, the primary conduction loss can be an important standard because the primary conduction loss is the largest part of all losses. In addition, since the highest efficiency should be achieved at V_{nom}, the primary conduction loss is also compared at the same conditions.

In the proposed converter, even though $I_{LM1,OFFSET}$ in the modified module is larger than one of the conventional converter, the primary current can be reduced by increasing the operating duty ratio at V_{nom} and reducing the reflected current by the large transformer turns-ratio. For the modified module, the primary current can be significantly reduced due to large turns-ratio by high voltage gain and no transformer offset current. Based on the design specification and components in Table I, Fig. 7 shows the comparison of the primary conduction losses including all switches and winding losses of transformer by the high voltage gain of the proposed converter.

E. Comparison on Power Density

To compare the power density of two transformers, the window-area product (A_P) of the magnetic core for the power capacity should be compared. For the conventional IACF converter, because two transformers operate as symmetric, two transformers has the same A_P value. From the equation of A_P, the window-area product for one transformer can be given as follows [12]:

$$A_{P,CONV} = W_a A_e = \frac{P_t(10^4)}{4\Delta B f_{SW} J K_u} = 10949[\text{cm}^4], \quad (10)$$

where W_a, A_e, J, and K_u are available window area of core in cm^2, the effective cross sectional area of core in cm^2, the current density in A/cm^2, and window utilization factor, respectively.

For the proposed converter, since two modules asymmetrically operate, A_P value of the transformer is separately considered. For the original module, A_P for L_{M1} can be calculated by (11):

842

The 2018 International Power Electronics Conference

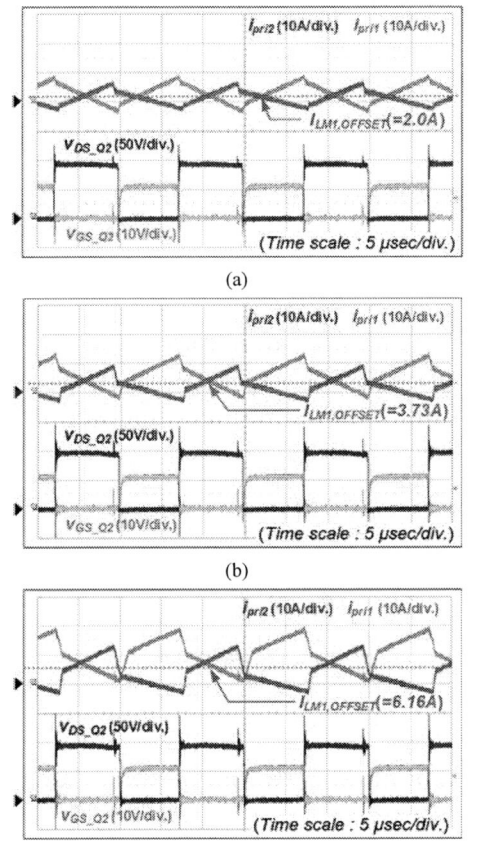

(a)

(b)

(c)

Fig. 8. Experimental waveforms of the proposed converter at nominal voltage (=48V_{DC}): (a) 20%, (b) 50%, and (c) 100% load conditions.

$$A_{P,T1} = W_a A_e = \frac{P_t(10^4)}{4\Delta B f_{SW} J K_u} = 10811[\text{cm}^4]. \qquad (11)$$

For the modified module, A_P value can be presented as follows:

$$A_{P,T2} = W_a A_e = \frac{P_t(10^4)}{4\Delta B f_{SW} J K_u} = 9906[\text{cm}^4]. \qquad (12)$$

For the total power density of two transformers, although two stages have the different A_P value, the proposed converter is the similar to one of the conventional converter from (10)-(12).

IV. EXPERIMENTAL RESULTS

To prove the feasibility of the proposed converter, 480W prototype for a DC/DC power system is utilized, and its design specifications are as follows: input voltage = 36-72 VDC (V_{nom}=48 V), output voltage = 12 V, and rated output power = 480 W (12 V/40 A). Table I shows the applied components and magnetics.

Fig. 8 shows the experimental waveforms at 48 V_{DC} nominal input voltage with 20%, 50% and 100% load conditions. The duty ratios at V_{nom} is set as close as 0.5 to minimize the RMS current. Although there is offset current of L_{M1}, the proposed converter can reduce the

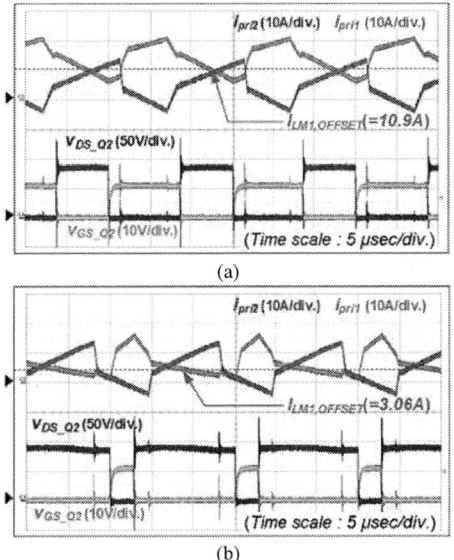

(a)

(b)

Fig. 9. Experimental waveforms of the proposed converter with 100% load conditions: (a) at $V_{S,min.}$ (=36 V_{DC}) and (b) at $V_{S,max.}$ (=72 V_{DC}).

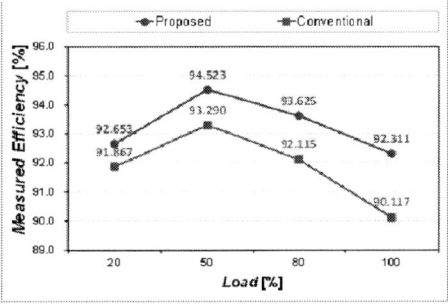

Fig. 10. Measured efficiency comparison of the conventional and proposed converter at nominal input voltage (=48 VDC)..

primary conduction loss by the increased voltage gain and small reflected current by large transformer turs ratio. Fig. 9 shows the operation with the maximum and minimum input voltage to check under the entire input voltage range. In Fig. 9(a), the operation with the minimum input voltage (=36 V_{DC}) can be presented, and Fig. 9(b) shows the operation with the maximum input voltage (=72 V_{DC}). Since only D_L is changed with the input voltage range, it can be noticed by the primary current of the modified module i_{pri1}. D_H is the same regardless of the variation of V_S. In Fig. 10, the efficiency comparison of proposed and conventional converter is described. Since the proposed converter can reduce the primary conduction loss, the efficiency at V_{nom} can be increased at the entire load range.

I. CONCLUSIONS

An IACF converter for 48V DC/DC power system having one modified module has been proposed in this paper. The proposed converter is derived only by re-arranging one ACF module of the conventional IACF converter, and the proposed converter achieve high voltage gain and reduce the primary conduction loss.

843

TABLE I. COMPONENTS LIST OF PROTOTYPE.

Components		Conventional IACF converter	Proposed
Switches (Q_1 and Q_2)		IRFB61N15D (150 V/60 A/32 mΩ)	IRFB61N15D (150 V/100 A/32 mΩ)
Switches (Q_3 and Q_4)			FDA59N30 (300 V/59 A/47 mΩ)
Synchronous Rectifiers (SR_1, SR_2, and SR_3)		IRFB3006 (60 V/195 A/2.1 mΩ)	
Capacitors (C_{B1} and C_{B2})		1 μF	
Main Trans.	T_1	L_M=40 μH Core : PQ3230 (Ae=161mm^2) Turns-ratio (N_P:N_S) = 4 : 2	L_{M1}=30 μH Core : PQ3230 (Ae=161mm^2) Turns-ratio (N_P:N_S) = 10 : 2
	T_2		L_{M2}=100 μH Core : PQ3230 (Ae=161mm^2) Turns-ratio (N_P:N_S) = 10 : 2
Output inductor (L_O)		L_O=5 μH Core : CH270125 (Ae=63.5mm^2), Wire : 1.5 mm x3, 6Ts	

Moreover, the analysis and the design guideline for the voltage stress on primary switches and offset current of transformer are briefly presented in this paper. The verification of the proposed converter is proved with the prototype with 36-72 V_{DC} input voltage and 480 W (12 V/40 A) output capability. As shown in the curve of measured efficiency, the proposed converter improve the efficiency under the entire load conditions by reducing large primary conduction loss. Therefore, the proposed converter is expected to be very attractive topology for a high efficiency of 48V DC/DC power systems having wide range of low input voltage.

ACKNOWLEDGMENT

This work was supported by the National Research Foundation of Korea (NRF) grant funded by the Korea government (MSIP) (No. 2016R1A2B2010328).

REFERENCES

[1] H. S. Kim, J. K. Kim, K. B. Park, H. W. Seong, G. W. Moon, and M. J. Youn, "On/off control of boost PFC converters to improve light load efficiency in paralleled power supply units for servers," *IEEE Trans. Ind. Electron.*, vol. 61, no. 3, pp. 1235–1242, Mar. 2014.

[2] S. Luo and I. Batarseh, "A review of distributed power systems Part I: DC distributed power system," *IEEE Aerosp. Electron. Syst. Mag.*, vol. 20, no. 8, pp. 5–16, Aug. 2005.

[3] W. Chen, X. Ruan, H. Yan, and C. K. Tse, "DC/DC conversion systems consisting of multiple converter modules: Stability, control, and experimental verifications," *IEEE Trans. Power Electron.*, vol. 24, no. 6, pp. 1463–1474, Jun. 2009.

[4] M. Murrill and B. Sonnenberg, "Evaluating the opportunity for DC power in the data center," *Emerson Network Power White Paper*, 2011.

[5] Y. Jang, M. M. Jovanovic, "A New Soft-Switched DCDC Front-End Converter for Applications with Wide-Range Input Voltage from Battery Power Sources" in *Proc. The 25th Int. Telecommunications Energy Conf.*, 2003, pp. 770-777.

[6] Y. Jeong, J. K. Kim, J. B. Lee, and G. W. Moon, "An Asymmetric Half-Bridge Resonant Converter Having a Reduced Conduction Loss for DC/DC Power Applications With a Wide Range of Low Input Voltage" *IEEE Trans. Power Electron.*, vol. 32, no. 10, pp. 7795-7804, Oct. 2017.

[7] W. Chen, S. Round, and R. Duke, "Design of an auxiliary power distribution network for an electric vehicle," in *Proc. The First IEEE International Workshop on Electronic Design, Test, and Applications*, 2002, pp. 257-161.

[8] F. Musavi, M. Craciun, D. S. Gautam, W. Eberle, and W. G. Dunford, "An LLC resonant DC–DC converter for wide output voltage range battery charging applications," *IEEE Trans. Power Electron.*, vol. 28, no. 12, pp. 5437–5445, Dec. 2013.

[9] I. O. Lee, "Hybrid DC–DC converter with phase-shift or frequency modulation for NEV battery charger," *IEEE Trans. Ind. Electron.*, vol. 63, no. 2, pp. 884–893, Feb. 2016.

[10] G. Zhang, X. Wu, W. Yuan, J. Zhang, and Z. Qian, "A new interleaved active-clamp forward converter with parallel input and series-parallel output," in *Proc. IEEE APEC'09 Conf.*, 2009, pp. 40-44.

[11] H. S. Youn, J. Il Baek, S. W. Jwa, J. K. Han, G. W. Moon, and J. K. Kim, "Interleaved active clamp forward converter with additional series-connected secondary windings for wide input and high current output applications," in *Proc. 2016 The 8th Int. Power Electron. Motion Control Conf. IPEMC-ECCE Asia*, 2016, pp. 3061–3065.

The 2018 International Power Electronics Conference

Control of Switching-Capacitor Based Buck-Boost Converter

M. Veerachary, Vasudha Khubchandani
Dept. of Electrical Engineering, Indian Institute of Technology Delhi, New Delhi, INDIA
E-mail: mveerachary@yahoo.co.in

Abstract— **Control of switching-capacitor based buck-boost converter for point of load applications is proposed in this paper. It provides buck-boost conversion with effective reduction in the source current ripples. The topological configuration of the proposed converter is such that it also eliminates the use of an additional input L-C filter and its associated damping network. The converter's performance, a detailed time-domain and steady-state analysis is presented. Voltage transformation characteristic features are established and equations defining L and C components are formulated in terms of ripple quantities. The state-space models are derived for two modes of operation, and small-signal analysis is performed to obtain the relevant transfer functions. Later, these are used in the loop-design procedure to obtain a suitable controller. The proposed circuit is able to perform bucking as well as boosting of the load voltage, similar to the conventional buck-boost converter and has no interaction issues as it has common ground. A 36 *Watt*, 100 *kHz* prototype point of load converter is built to supply power at constant load voltage of either 36 or 12 V. A 24 *V* battery source is connected in the prototype converter operation both in simulation and experimentation. The proposed point of load converters' effectiveness is demonstrated, both in simulation and experimentation, in terms of source current ripples reduction as well as in buck-boosting features.**

Keywords— **Buck-boost converter, Point of load converter, Two-switch buck-boost converter, State-space model.**

I. INTRODUCTION

With technological advancement, the requirement of power conversion at high frequencies is soaring and is dominant in applications requiring low power for their operation. Converters for point of load applications are being developed by the designers while laying special emphasis on achieving higher conversion efficiency at full-load, increased power density, and lower radiation. Use of several point of load converters is common in low power dc system wherein many design challenges [1]-[9] must be resolved by the application engineer so as to ensure reliable power distribution. Some of them are: (i) formulation of transformerless non-isolated topologies as minimization or elimination of transformer leakage inductance is a difficult task, (ii) achieving reduced ripples with minimal L, C component requirements, and (iii) reduction in size and weight of the filtering components which will result in enhanced power densities. Many dc-dc conversion circuits are present in the literature which produce stable voltages to drive the dc-loads and is briefly classified as: (i)

bucking based circuits, (ii) boosting circuits, and (iii) buck-boost and other higher-order or derived converter circuits. These converters find broad application in areas pertaining to controlled power such as: (i) customized low-power integrated circuits, (ii) powering compact and tiny automotive loads, (iii) sophisticated loads such as bio-medical equipment, (iv) internet, wide and local area network services, and (v) telecommunication power supply systems, on-board spacecraft power systems, and defence equipment, etc.

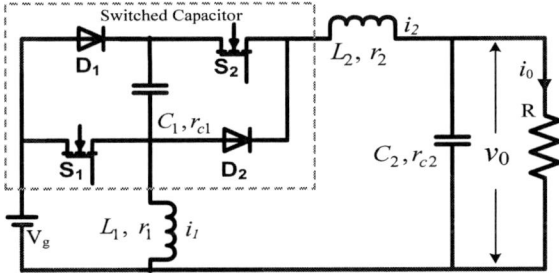

(a) Proposed topology: Switching-capacitor based buck-boost converter

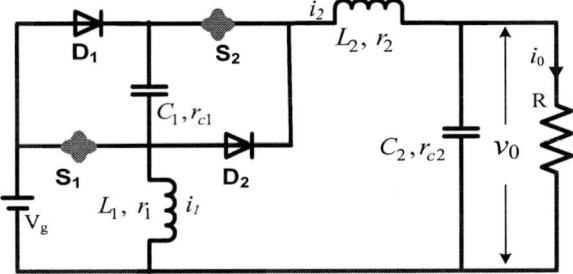

(b) Operation of proposed switching-capacitor based buck-boost converter (mode-1)

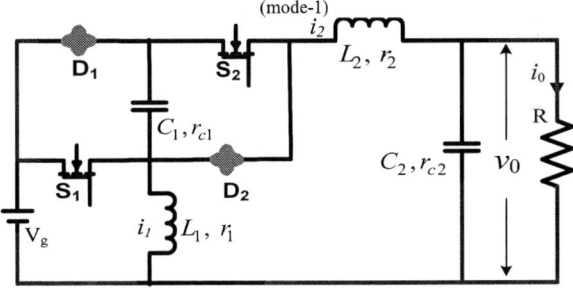

(c) Operation of proposed switching-capacitor based buck-boost converter (mode-2)

Fig. 1. Steady-state waveforms and switching-capacitor based buck-boost converter.

845

Generally, the use of buck-boost circuits is in back-end power processing in which wide variations occur in the battery voltage. They are also used in point of load converter back-end applications where low voltage dc batteries are used to drive high voltage rating loads or vice-versa. One such converter is the single switch buck-boost dc-dc converter (SSBBC) which is broadly used as it is capable of smooth changeover from buck to boost and vice-versa. However, this comes with the problem of higher ripples on source side which affects the battery life and reliability. The buck-boost converter with input filter (BBCIF), addresses the high ripple problem to some extent but at the expense of increased system order [7]. However, addition of only the input filter is not recommended because at times it induces unwanted oscillations and may force the system into instability [8]. The insertion of the damping network in the L-C filter may help in stabilizing the system by reducing the oscillations.

Some fourth-order topologies [1] like SEPIC and CUK converters are reported in the literature which limits the ripple content while ensuring buck-boost voltage transformation. However, they may pose control issues due to non-minimum phase behaviour. Two-switch buck-boost converter (TSBBC) [9], cascading of buck followed by boost, also exhibits identical voltage gain characteristics same as that of the SSBBC but with more flexibility in terms of control due to more number of controlled devices. From control point of view, TSBBC offers more flexibility but at the expense of high source and load side ripples. Several 4th-order dc-dc converters have also been reported [2]-[4] recently in voltage regulation applications. A fifth-order boost converter along with robust controller is reported in [7]-[8]. Features like lower source current ripples with voltage transformation ratio same as TSBBC is the main motive in the formulation of the proposed fourth-order buck-boost topology. This topology primarily uses a capacitor and its connection is realized through controlled switching. The following paragraph pay attention to the steady-state analysis, state-space modeling and control aspects. Discrete-time modelling formulations reported in [7] are adopted in the analysis and design of the closed-loop control of the proposed converter.

II. MODELING OF SWITCHING-CAPACITOR BASED BUCK-BOOST CONVERTER

A switching-capacitor based buck-boost converter (SBBC) is evolved in this paper and shown Fig. 1. It comprises of a bridge with active switches and a capacitance C_1. In the bridge there are two controllable switching devices (S_1, S_2) and two diodes (D_1, D_2). The circuit configuration, switching action of (S_1, S_2), permits the involvement of capacitor C_1 either in source side circuit or in the load side circuit. It is this switching action of the capacitor C_1 that is primarily responsible for energy transfer from the source to the load. When the switches (S_1, S_2) are ON then C_1 gets connected to load side, else (D_1, D_2 are in ON-state) gets connected to the source. Due to this interconnection of C_1 with L_1 and L_2 results in lower source current ripples. Further, the converter exhibits buck-boost voltage gain features of conventional SEPIC converter. The

reported TSBBC is a second-order, while the proposed SBBC is fourth-order and the increased order is responsible in mitigating the source current ripples. There is no change in its voltage gain transformation features. The number of L, C elements in this SBBC is the same as in the buck-boost converter with input filter and hence the factors responsible for dynamic performance are almost identical. Due to variations in the circuit connections (L, C elements connections) the proposed SBBC exhibits the following salient features: (i) no change in buck-boost features like conventional TSBBC, (ii) lesser source current ripples both in bucking and boosting operations, and (iii) positive output with common ground between the source and load. A comparison of SBBC component stress level with other buck-boost topologies is given in Table-I.

A. Steady-state Performance Analysis of Switching Capacitor Buck-Boost Converter

In order to determine the steady-state performance features, a time-domain analysis is presented in this section. It includes voltage boosting with voltage and current ripple along with L, C energy storage elements design expressions. The proposed SBBC has two controlling switches, and hence the circuit operating modes are decided by the control scheme adopted. According to the number of switches, the following are the possible gating schemes: (i) S_1 continuously ON while S_2 in pulse width modulation (PWM), (ii) S_2 continuously OFF while S_1 in PWM, (iii) S_1 gate signal is complementary to switch S_2, and (iv) S_1 and S_2 both in PWM simultaneously. Amongst them, the last controlling scheme i.e. simultaneous switching of S_1 and S_2 in PWM results in switching capacitor, is reported in this paper. There are two types of devices in this circuit operation, controlled switches and uncontrolled diodes. Due to the circuit configuration, the simultaneous switching of S_1 and S_2 automatically synchronize the complementary operation of diodes D_1, D_2. An application of volt-sec balance to the inductors L_1 and L_2 is useful in establishing buck-boost features of the proposed SBBC. Voltage across inductive components L_1, L_2 is given by (i) in mode-1: $v_{L1} = (V_g - v_{c1})$; $v_{L2} = V_g$, (ii) in mode-2: $v_{L1} = (V_g - v_{c1})$; $v_{L2} = (V_g - v_{c1} - v_0)$. Applying volt-sec balance to the inductor L_1 gives the identity defined by eqn. 1.

$$(D)(V_g - v_{c1}) + (1-D)(V_g - v_{c1}) = 0 \qquad (1)$$

Upon simplification, the above equation results in $v_{c1} = V_g$, which means that the capacitor C_1 charges to supply voltage. Hence, the input filter capacitor voltage rating is equal to the source voltage. Similar approach is extended to the inductor L_2 resulting in eqn. 2.

$$(D)(V_g) + (1-D)(V_g - v_{c1} - v_0) = 0 \qquad (2)$$

Incorporating the identity obtained from eqn. 1 and then simplification of eqn. 2 results in the voltage gain of the SBBC and it is given by eqn. 3. This voltage gain is identical to the TSBBC and SEPIC converter. Furthermore, the resultant voltage gain is the polarity inverted version of the conventional buck-boost converter. This polarity inversion is due to the switching effect of capacitor.

$$\frac{v_0}{V_g} = \frac{D}{(1-D)} \qquad (3)$$

For the given input/output specifications, the energy storage elements need to be properly designed to meet various ripple standards. The ripple currents and voltage expressions can easily be derived using network equations along with simple time-domain steady-state analysis. In steady-state the capacitor C_1 charges to source voltage, from eqn. 1, the ripple current in inductor L_1 is primarily decided by the non-ideal series resistances. The voltage impressed across L_2 is source voltage and hence its design expression in terms of ripple current is given by eqn. 5. Following the similar trend, the design equation for L_2 is given by eqn. 5. Likewise, eqns. 6 to 7 (C_1, and C_2 expressions) are obtained.

$$L_1 \ge D(V_g - v_{c1} - (r_1 + r_{c1})i_1)/(f_s \Delta i_1) \qquad (4)$$

$$L_2 > V_g D/(f_s \Delta i_2) \qquad (5)$$

$$C_1 \ge I_{L1} D/(f_s \Delta v_{c1}) \qquad (6)$$

$$C_2 > DI_0/(f_s \Delta v_0) \qquad (7)$$

B. Small-signal Analysis and Transfer Functions of the SBBC

Fig. 1b and 1c represents the SBBC operating modes wherein in first mode energy stored in the energy storage elements L_1, L_2 and C_1 while C_2 will supply the energy to the load. In the second mode, the stored energy is transferred to load via C_1, L_2 and also charge C_2. Using network mesh analysis, the behaviour of these two circuits can easily be analysed independently. Here, the circuit behaviour essentially defined by the inductor current and capacitive elements voltages, which leads to the formulation of a set of first-order differential equations through mesh analysis and then represented using matrix notation as stated in eqn. 8.

$$\left. \begin{aligned} [\dot{x}] &= [A_k][x] + [B_k][u] \\ [y] &= [E_k][x] + [F_k][u] \end{aligned} \right\rangle \quad t_k < t < t_{(k+1)} \qquad (8)$$

here $[A_k]$ is the system state matrix, $[B_k]$ the input matrix, $[E_k]$ the output matrix, $[x]$ the state vector, $[y]$ the output vector, and $[u]$ is the input forcing function vector.

$$[A_1] = \begin{bmatrix} \dfrac{(r_1+r_{c1})}{-L_1} & 0 & \dfrac{-1}{L_1} & 0 \\ 0 & \dfrac{-r_2}{L_2} & 0 & 0 \\ \dfrac{1}{C_1} & 0 & 0 & 0 \\ 0 & 0 & 0 & \dfrac{-b}{RC_2} \end{bmatrix}; \qquad (9)$$

$$B_1 = B_2 = [1/L_1 \quad 1/L_2 \quad 0 \quad 0]^T;$$

$$E_1 = [0 \quad 0 \quad 0 \quad b]$$

$$E_2 = [0 \quad a \quad 0 \quad b]$$

$$[A_2] = \begin{bmatrix} \dfrac{(r_1+r_{c1})}{-L_1} & \dfrac{r_{c1}}{L_1} & \dfrac{-1}{L_1} & 0 \\ \dfrac{r_{c1}}{L_2} & \dfrac{(r_1+r_{c1}+a)}{-L_2} & \dfrac{-1}{L_2} & \dfrac{-b}{L_2} \\ \dfrac{1}{C_1} & \dfrac{1}{C_1} & 0 & 0 \\ 0 & \dfrac{b}{C_2} & 0 & \dfrac{-b}{RC_2} \end{bmatrix}; \qquad (10)$$

The small-signal transfer functions, both in s and z-domain, can easily be obtained from eqn. 8 after its linearization. The state-space model in discrete-time domain, in terms of ϕ and Γ, is defined as:

$$\hat{x}[NT_s] = \phi \hat{x}[(N-1)T_s] + \Gamma \hat{d}[(N-1)T_s] \qquad (11)$$

where $\alpha = [(A_1 - A_2)X + (B_1 - B_2)V_g]$, $\phi_1 = e^{A_1 t_d}$, $\phi_2 = e^{A_2 D_2 T_s}$, $\phi = e^{A_1(DT_s - t_d)}\phi_1\phi_2$, $\Gamma = e^{A_1(DT_s - t_d)}\alpha\phi_2$. These 'phi' and 'gamma' matrices are used in formulating the z-transfer functions as listed in Table II.

TABLE -I. Component Stress Of SBBC Over The Reported Buck-Boost Converters

Quantity	SSBBC	TSBBC	Proposed SBBC
Voltage Gain	$\dfrac{-D}{(1-D)}$	$\dfrac{D}{(1-D)}$	$\dfrac{D}{(1-D)}$
PVS	$(v_0 - V_g)$	v_0	v_{C1}
PDS	$(v_0 - V_g)$	$(v_0 - V_g)$	v_{C1}
SCR	High	High	Low
OCPCS	Identical	Identical	Low

PVS: Switch peak voltage stress, PDS: Diode peak voltage stress, SCR: Source current ripple, OCPCS: Peak current stress in the output capacitor.

TABLE II. Z-Transfer Functions Formulation[8]

Transfer Function	Formulation
Control-to-Output	$\hat{v}_0(z) / \hat{d}(z) = E'(zI-\phi)^{-1}\Gamma$
Audio Susceptibility	$\hat{v}_0(z) / \hat{v}_g(z) = E'(zI-\phi)^{-1}\Gamma + F'$
Output Impedance	$\hat{v}_0(z) / \hat{i}_0(z) = [E'(zI-\phi)^{-1}\Gamma + J']$
Input Admittance	$\hat{i}_{in}(z) / \hat{v}_g(z) = P'[(zI-\phi)^{-1}\Gamma]$

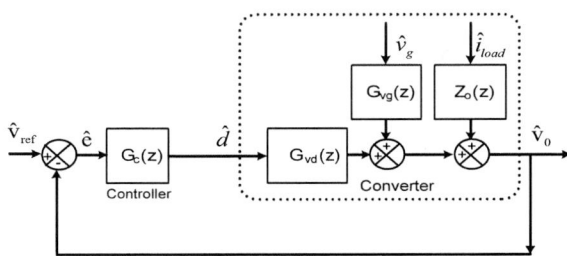

Fig. 2. Block-diagram: Closed-loop controlled SBBC.

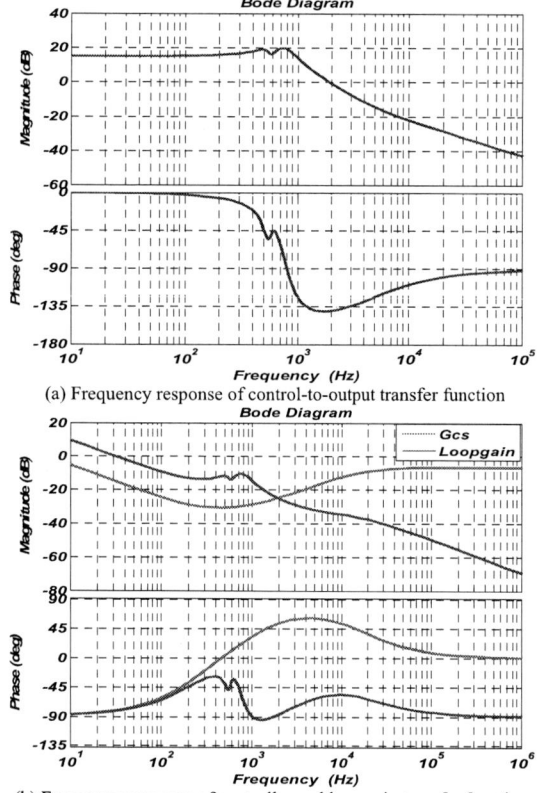

(a) Frequency response of control-to-output transfer function

(b) Frequency response of controller and loop gain transfer functions

Fig. 3. Frequency response plots.

III. SYSTEM DESIGN

Conventional linear control theory is well rooted in the literature [10]. It essentially uses the frequency domain models for controller design and then to predetermine the absolute/ relative stability aspects of closed-loop controlled converter systems (CLCS). The block-diagram representation of controlled SBBC is shown in Fig. 2. Here, the controller transfer function is in the feed-forward path along with the duty ratio -to- output load voltage transfer function. The remaining two transfer functions, output–to-input voltage and output impedance, reveal the impact of disturbances on the controlling quantity, in this case it is load voltage. In the design of controller the feed-forward path gain transfer function, indicated in Fig. 2, plays an important role. For the controller selection and design single-input single output tool of Matlab [10] is used. This tool offers placing of controller poles and zero in the desired locations such that the resulting feed-forward path gain transfer function exhibits absolute stability. Once it satisfies the absolute stability criterion, it becomes necessary to choose a controller which will ensure the relative stability specifications. In order to keep the closed-loop SBBC to be absolutely stable and also achieves sufficient disturbance rejection features, here it is fluctuations in the source voltage and load, the gain margin (GM) should be at least 6 dB (higher value is recommended), the phase margin (PM) somewhere in between (45°~75°) with sufficient bandwidth which is decided by a combination of controller and SBBC system. For

designing, firstly a pole is incorporated at the origin to reduce the steady-state error with placing of a second pole to attain sufficient disturbance rejection. Later, two zeros are added to the controller transfer function to achieve the relative stability margins. Furthermore, the load voltage is sensed and scaled-down to less than 5 V range. In view of this, controller gain is reduced and pole-zero locations get modified to meet the relative stability margins as discussed above. The final controller form in s-domain is

$$G_c(s) = \frac{k(s+a_1)(s+a_2)}{(s+b_1)(s+b_2)} \tag{12}$$

For experimentation, the controller defined in eqn. 12 is realized on a digital platform and the s-domain controller is then transformed into the digital domain. The digital equivalent of the s-domain controller is defined in eqn. 13.

$$G_c(z) = \frac{k_{11}(z+a_{11})(z+a_{22})}{(z+b_{11})(z+b_{22})} \tag{13}$$

Where gain $k_{11}=0.9563$, $a_{11}=-0.985$, $a_{22}=-0.949$, $b_{11}=-1.0$, $b_{22}=-0.19259$. Frequency response plots of the control-to-output and feed-forward path transfer functions are given in Fig. 3. From the frequency response of the feed-forward path transfer function it is clear that the SBBC converter system is stable and also has sufficient relative stability margins to maintain robustness against uncertainties. To reaffirm the correctness of the frequency responses generated/obtained, a step response in time-domain is obtained from MATLAB and it is depicted in Fig. 4. Although the step response has critically damped nature at starting, but over the time it reaches the step reference in nearly 30 ms. It is thus evident that the proposed closed-loop system is stable and also possesses sufficient disturbances rejection features.

TABLE III. PROPOSED SBBC PARAMETERS

Parameter	Value
Vg	24 V
Vo	15/ 36 V
L1	400 μH
L2	600 μH
C1, C2	47 μF
C3	100 μF
fs	100 kHz

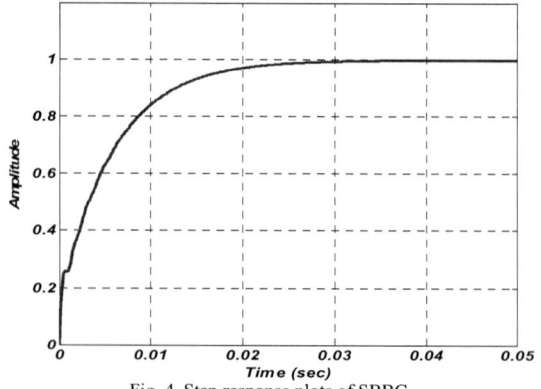

Fig. 4. Step response plots of SBBC.

848

The 2018 International Power Electronics Conference

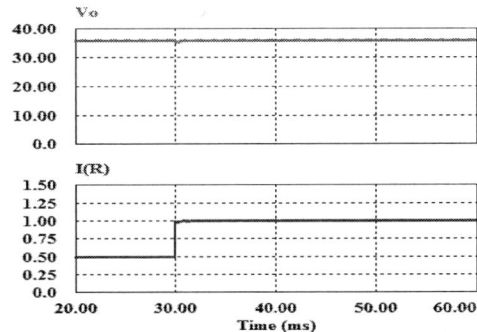

(b) Boost mode of operation (V$_{o\_ref}$: 36 V; Vg: 24 V; R: 72 to 36 Ω)
Fig. 6. Dynamic response of SBBC load voltage against load perturbation.

IV. EXPERIMENTAL RESULTS

A 36 Watt, 24 to 36 V SBBC prototype is built for testing its performance both in simulation and experiments. Table III is included with the parameters used in simulation as well as in experimentation. Steady-state and dynamic response simulations were performed in PSIM software simulator [11]. Initially, different reference values are chosen to demonstrate buck-boosting features. For a source voltage of 24 V, a reference load voltage of 15 V is chosen for bucking operation while a reference load voltage of 36 V is chosen for ensuring the boosting operation. Further, closed-loop simulations are performed with bucking followed by boosting operations and the load voltage response is shown in Fig. 5a. It is clear that up to first 30 ms, the load voltage is in bucking mode (D<0.5, 15 V), and for the next 20 ms, it is boosting mode (D>0.5, 36 V). Fig. 5c shows the source current (i_g) along with inductor currents. This source current is smooth and continuous alike inductor currents and hence its ripples are low. Load voltage regulation in buck and boosting operation is shown in Fig. 6, wherein the transient settling time is about 5.0 ms.

To demonstrate the proposed SBBC's merits, design and its operation principles with the controller, discussed in Section III, experiments [12] are performed. Parameters used for experimentation circuit are listed in Table III. The measured start-up and steady-state waveforms are plotted in Figs. 7 and 8. The measured starting response time in boost mode is close to 30 ms and this is in close agreement with the step-response simulation result shown in Fig. 4. Measured steady-state waveforms (Fig. 8) are also in close agreement with the waveforms generated through simulation (Fig. 5c). The dynamic response of SBBC load voltage against load and source perturbations is also measured and illustration of voltage regulation measurement, for both buck and boosting operations, is shown in Fig. 9. In bucking operation the SBBC is responding quickly (response time close to 5.0 ms) while in boosting operation it is a bit slow (response time close to 15.0 ms). The discrepancies and a slight mismatch in the waveforms, in the simulation and experiments, are due to: (i) nature of simulation platform, step size and simulation timings adopted, (ii) difficulty in estimating the time delays involved in the actual experimentation and their inclusion during simulation,

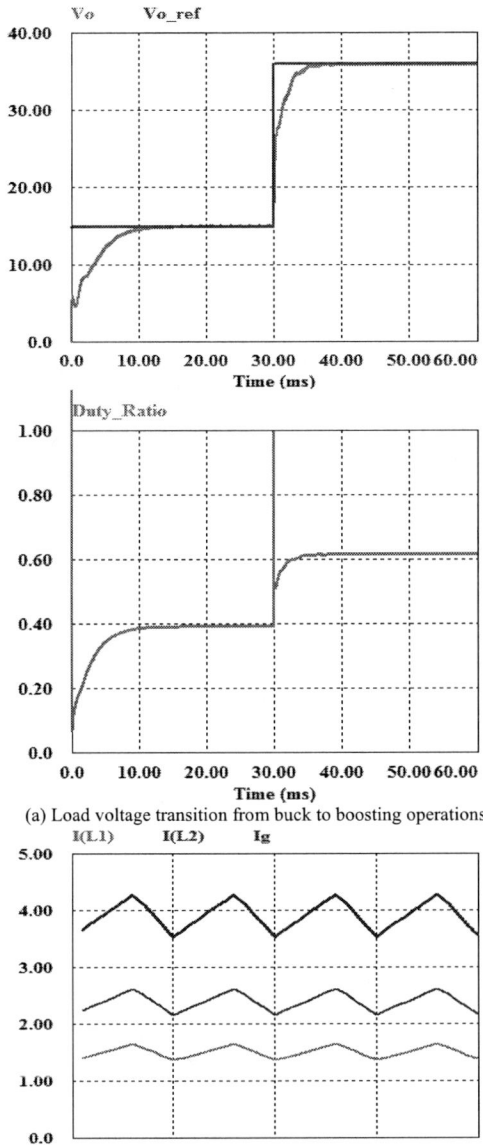

(a) Load voltage transition from buck to boosting operations

(b) Enlarged steady-state waveforms
Fig. 5. Step response and steady-state waveforms of SBBC.

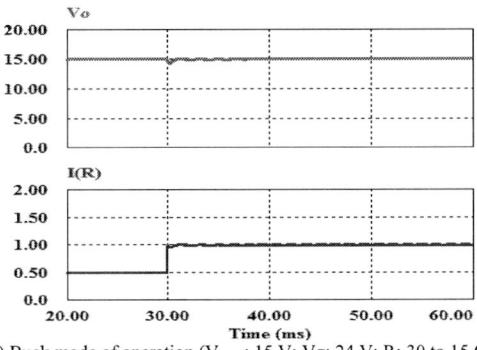

(a) Buck mode of operation (V$_{o\_ref}$: 15 V; Vg: 24 V; R: 30 to 15 Ω)

849

(iii) mismatch in accounting the circuit non-ideal parameters of measurement set-up while performing simulations, and (iv) adopted step size in the solution of differential equations in simulation platform.

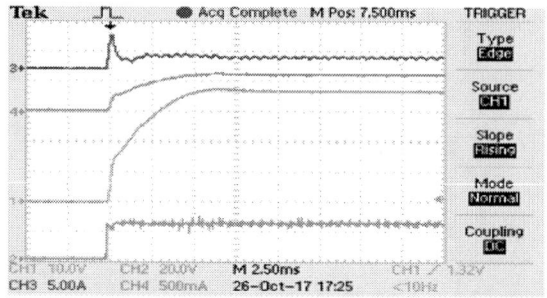

Fig. 7. Measured start-up response of SBBC (boost mode).

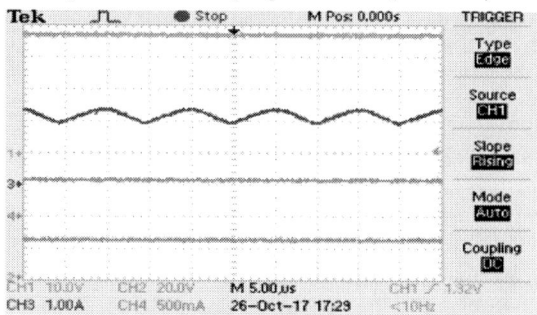

Ch-1:(V_0):10 V/div; Ch-2:(V_g):10 V/div; Ch-3:(i_g):1.0 A/div;
(a) Steady-state waveforms (V_{o_ref}: 36 V; Vg: 24 V; R: 60 Ω)

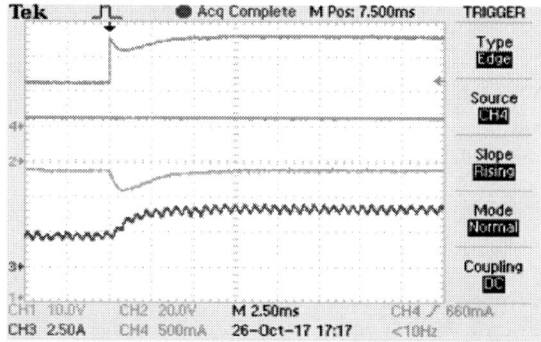

(b) Dynamic response (V_{o_ref}: 36 V; Vg: 24 V; R: 60 to 30 Ω)
Fig. 8. Measured waveforms of SBBC (boost mode).

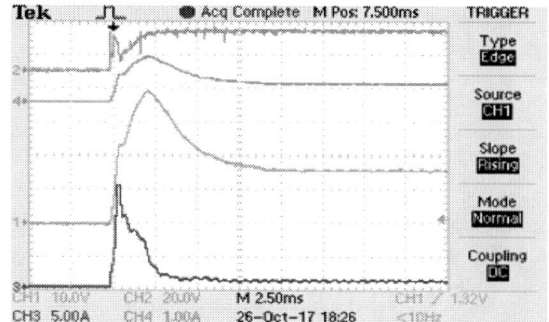

Ch-1:(V_0): 5 V/div; Ch-2:(V_g): 10 V/div; Ch-3:(i_g): 1.0 A/div; Ch-4:i_0: 0.5 A/div.
(a) Start-up response (V_{o_ref}:15 V; Vg: 24 V; R: 30 Ω)

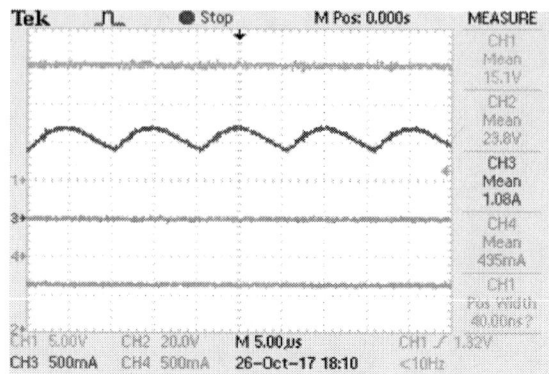

Ch-1:(V_0): 10 V/div; Ch-2:(V_g): 10 V/div; Ch-3:(i_g): 1.0 A/div; Ch-4:(i_0: 0.5 A/div
(b) Steady-state waveforms (V_{o_ref}: 15 V; Vg: 24 V; R: 30 Ω)
Fig. 9. Measured waveforms of SBBC (buck mode).

V. CONCLUSION

A switching-capacitor based buck-boost converter, which is a fourth-order topology, exhibiting identical voltage gain as that of conventional buck-boost/SEPIC converter was proposed. Even without the use of input damping filter, source current ripple was minimized which lead to effective EMI mitigation. Detailed small-signal analysis was established and accordingly a controller was designed. The effectiveness of the controller was demonstrated for both bucking as well as boosting operations. The analytics of the proposed SBBC, its effectiveness in terms of load voltage regulation were simulated and supported with experimental measurements. Simulation and measurement results were in close agreement.

REFERENCES

[1] K. C. Daly, "Ripple determination for switch-mode dc-dc converters," *IET Proc.*, vol. 129, Pt. G, no. 5, pp. 229-234, Oct. 1982.

[2] Jingquan Chen, D. Maksimovic, Robert Ericksom, "Buck-boost PWM converters having two independently controlled switches," Proc. of *IEEE APEC.*, vol. 2, 2001, pp. 736-741.

[3] Kerui Li, Andrian Ioinovici, "Large DC gain nonisolated converter based on a new L-C-D step-up switching cell," *Proc. of IEEE APCCAS*, 2014, pp. 284-287.

[4] Shiyu Zhang, Jianping Xu, Ping Yang, "A single-switch high-gain quadratic boost converter based on voltage-lift-technique", *Proc. of IEEE IPEC*, 2012, pp. 71-75.

[5] K. I. Hwu, Y. T. Yau, "A KY boost converter," *IEEE Trans. on Power Electron.*, vol. 25, no. 11, pp. 2699-2703, Nov. 2010.

[6] K. I. Hwu, T. J. Peng, "A novel buck-boost converter combining KY and buck converters," *IEEE Trans. on Power Electron.*, vol. 27, no. 5, pp. 2236-2241, May. 2012.

[7] M. U. Iftikhar, D. Sadarnac, C. Karimi, "Input filter damping design for control-loop stability of dc-dc converters," in *Proc. of IEEE, ISIE*, 2007, pp. 353-358.

[8] M. Veerachary, Anmol Ratna Saxena, "Design of Robust Digital Stabilizing Controller for Fourth-Order Boost DC–DC Converter: A Quantitative Feedback Theory Approach," *IEEE Trans. on Ind. Electron.*, vol. 59, no. 2, pp. 952-963, Feb. 2012.

[9] Chuan Yao, Xinbo Ruan, Weijie Cao, Peilin Chen, "A two-mode control scheme with input voltage feed-forward for the two-switch buck-boost dc-dc converter," *IEEE Trans. on Power Electron.*, vol. 29, no. 4, pp. 2037-2048, Apr. 2014.

[10] MATLAB, user manual, 2005.

[11] PSIM, user manual, 2005.

[12] dsPIC30f6010, user manual, 2010

The 2018 International Power Electronics Conference

Improvement of Upload Transient Responses for Ultra High Step-Down Converter

Y. T. Yau[1], Member, IEEE, IEEE, and K. I. Hwu[2], Member, IEEE
[1]Asian Power Devices Inc.
[2]Department of Electrical Engineering, National Taipei University of Technology, Taipei, Taiwan
[1]E-mail: pabloyau@apd.com.tw, [2]E-mail: eaglehwu@ntut.edu.tw

Abstract – In this paper, the enhancement of the upload transient response for the ultrahigh step-down converter is presented. By means of an external circuit containing one switch and one inductor, the voltage on the energy-transferring capacitor can be controlled so that upload transient responses can be improved.

Keywords—Buck, coupling inductor, step-down.

I. INTRODUCTION

For the ultrahigh step-down DC-DC converter, the traditional buck converter is adopted intuitively. This converter can transfer 48V to lower voltage [1]. However, if one or more components are destroyed or this converter is not under control, the input voltage of 48V will be directly passed to the load. Consequently, the costly CPU or RAM will be destroyed. In addition, the corresponding duty cycle will be too small so as to make the system hard to control. The literatures [2]-[7] propose no isolated floating output structures, which cannot be applied to the data center server board because many common-ground power supplies are needed on this board. The literature [8] uses overlap structure, which can transfer high voltage to low voltage. But, this structure has the demerit that many floating gate driving circuits and many magnetic components are used. The circuits shown in the literatures [9]-[14] have a small number of magnetizing components as well as the common ground between the input and the output, but possess a demerit of floating gate driving. On the other hand, for low output voltage applications, a synchronous switch is used to replace the diode, and hence an additional gate driving IC is required. The circuit shown in literature [15] uses a common half-bridge gate driving IC so as to facilitate driving switches as compared with floating gate driving, but possesses a demerit that the leakage inductance of the coupled inductor will result in a large voltage spike. Therefore, in the literature [16], an additional passive snubber is used to solve this problem, thereby increasing the number of components and power loss.

Based on the aforementioned, a non-isolated ultrahigh step-down converter with simple structure and high conversion efficiency [17] is presented. Such a converter has some advantages, such as a small number of components, easily driven switches, etc. Based on the circuit topology shown in [17], the purpose of this paper is to enhance upload transient responses. By means of the external circuit containing one switch and one inductor, the voltage on the energy-transferring capacitor C_R is controlled such that load transient responses is improved.

II. PROPOSED CIRCUIT CONFIGURATION

Fig. 1 shows the proposed topology, which contains three active switches Q_1, Q_2 and Q_3, one DC blocking capacitor C_R, and one coupling inductor composed of the winding N_1 and the winding N_2. The transient speedup circuit adds one diode D_B, one switch Q_4 and one inductor L_B to the ultrahigh step-down converter [17]. These three components are used to fast adjust the voltage on the capacitor C_R.

There are some symbols in Fig. 1 to be given as follows: (i) the currents direction of Q_1, Q_2 and Q_3 are signified by I_{ds1}, I_{ds2} and I_{ds3}, respectively; (ii) the voltage across C_R is denoted by v_{CB}; (iii) the currents flowing through N_1 and N_2 are defined by i_{N1} and i_{N2}, respectively; and (iv) the input and output voltages are represented by V_{in} and V_O.

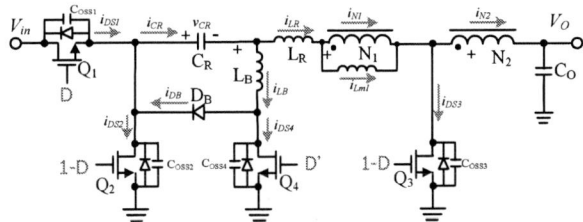

Fig. 1. Proposed high step-down circuit with fast load transient responses.

III. BASIC CIRCUIT OPERATING PRINCIPLES

The normal mode of the proposed circuit is first described, and then the accelerating mode follows. There are six operating states and eight operating states for the normal mode and the accelerating mode, respectively.

1) Normal Mode

a) State 1: As shown in the time interval between t_0 and t_1 in Fig. 2(a), the Q_1 is turned on, but the Q_2 and Q_3 are turned off. During this state, the input voltage V_{in} charges the C_R, and the leakage inductor L_R and magnetizing inductor L_m are magnetized. At the same time, the currents i_{N1} and i_{N2} are increasing, and the input voltage V_{in} is passing the energy to the load.

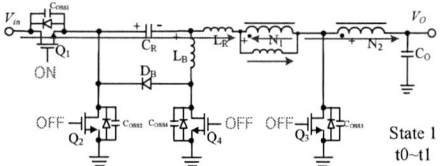

Fig. 2(a). Current direction in state 1 under normal mode.

b) *State 2*: As shown in the time interval between t_1 and t_2 in Fig. 2(b), the Q_1 is still turned off but the Q_2 and Q_3 are still turned off. This state belongs to one dead time zone. Because the current in the leakage inductor L_R is continuous, the body diodes of Q_2 and Q_3 are turned on.

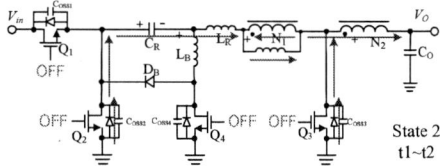

Fig. 2(b). Current direction in state 2 under normal mode.

c) *State 3*: As shown in the interval between t_2 and t_3 in Fig. 2(c), the Q_1 is still turned off but the Q_2 and Q_3 are turned on. Before this state, due to the continuous current in L_R, the currents i_{LR} and i_{N2} flow through the body diodes of Q_2 and Q_3. Therefore, the Q_2 and Q_3 are turned on with ZVS. As soon as the leakage current of the N_1 winding falls to zero, this state comes to the end, and the operation proceeds to state 4.

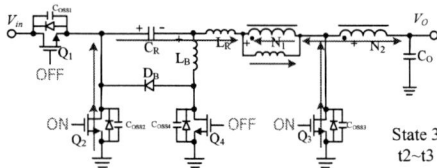

Fig. 2(c). Current direction in state 3 under normal mode.

d) *State 4*: As shown in the time interval between t_3 and t_4 in Fig. 2(d), the Q_1 is still turned off but the Q_2 and Q_3 are still turned on. During this state, the energy stored in C_R is released and magnetizing the N_1 winding in the opposite direction, thereby making the coupled inductor behave like the transformer. The corresponding energy is transferred to the load via the winding N_2, and hence the currents i_{N1} and i_{N2} are increased in the opposite direction.

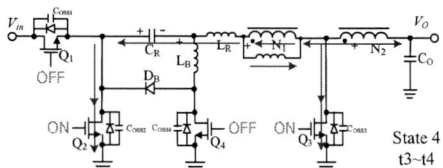

Fig. 2(d). Current direction in state 4 under normal mode.

e) *State 5*: As shown in the time interval between t_4 and t_5 in Fig. 2(e), the Q_1 is still turned off and the Q_2 and Q_3 are turned off. Due to the continuous current in L_R, the currents $-i_{N1}$ and $-i_{N2}$ flow through the body diodes of Q_1. This state belongs to the other dead time zone. During this dead time, the leakage

inductor L_R keeps demagnetized. At the same time, the energy stored in the leakage inductor can be passed to the input voltage. As the current i_{N1} falls to zero, the operation goes to state 5.

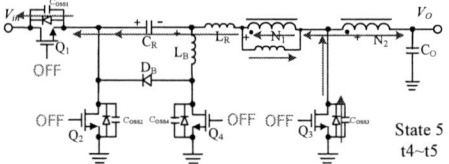

Fig. 2(e). Current direction in state 5 under normal mode.

f) *State 6*: As shown in the time interval between t_5 and t_0 in Fig. 2(f), the Q_1 is turned on with ZVS but the Q_2 and Q_3 are still turned off. At the same time, since the current in L_R is smaller than the current in the winding N_2, the body diode of the switch Q_3 is forward biased. As soon as the current i_{LR} is equal to the current i_{N2}, the current in the body of the switch Q_3 stops flowing, the operating state goes back to state 1, and the next cycle repeats.

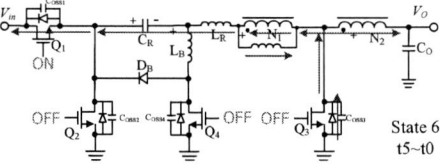

Fig. 2(f). Current direction in state 6 under normal mode.

2) Accelerating l Mode

This mode is activated by the Q_4 as the controller detects the output voltage abruptly dropped. Fig. 3 provides the illustrated simulated waveforms.

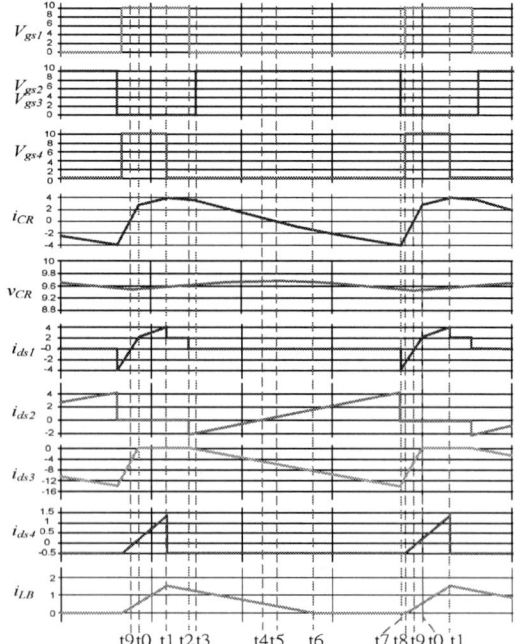

Fig. 3. Illustrated waveforms under the accelerating mode.

a) State 1: As shown in the time interval between t_0 and t_1 in Fig. 4(a), the Q_1 and Q_4 are turned on, but the Q_2 and Q_3 are turned off. During this state, the input voltage V_{in} charges the C_R. Due to the Q_4 being turned on, the inductor L_B is magnetized, thereby causing the current in L_B, named i_{LB}, to be increased. Because the current i_{CR} is the sum of the currents i_{LR} and i_{LB} and the inductor L_B is relatively small, the current i_{LB} is increasing abruptly, thereby making the voltage across C_R, called v_{CR}, fast increased. At the same time, the current i_{LR} is equal to the current i_{N2} and both of these two currents are increasing.

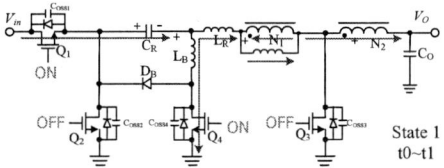

Fig. 4(a).Current direction in state 1 under accelerating mode.

b) State 2: As shown in the time interval between t_1 and t_2 in Fig. 4(b), the Q_4 is forced to be turned off, the other switches remain the same status shown in state 1. During this state, the inductor L_B is demagnetized due to the voltage across it being $-v_{CB}$, causing the current i_{LB} to be decreased and the diode D_B to be forward biased. At the same time, the capacitor C_R is charged.

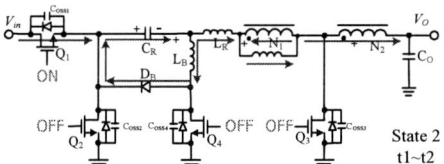

Fig. 4(b).Current direction in state 2 under accelerating mode.

c) State 3: As shown in the time interval between t_2 and t_3 in Fig. 4(c), the Q_1 are forced to be turned off and the Q_2, Q_3 and Q_4 still keep turned off. During this state, which is called one dead time zone, because the current in the leakage inductance is continuous, the currents i_{LR} and i_{N2} flow through the body diodes of Q_2 and Q_3. At the same time, the D_B is forward biased due to the current in L_B, called i_{LB}, being continuous. Therefore, the current i_{LB} is decreasing and hence the capacitor is charged.

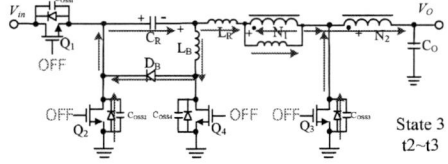

Fig. 4(c).Current direction in state 3 under accelerating mode.

d) State 4: As shown in the time interval between t_3 and t_4 in Fig. 4(d), the Q_1 and Q_4 keep turned off but the Q_2 and Q_3

are turned on. Because the currents i_{LR} and i_{N2} flow through the body diodes of Q_2 and Q_3 in state 3, the Q_2 and Q_3 are turned on with ZVS. At the same time, the D_B is still forward biased due to the current in L_B, called i_{LB}, being continuous. Therefore, the current i_{LB} is still decreasing and hence the C_R is still charged. Because the current in the leakage inductor is continuous, the energy stored in the C_R will magnetize the N_1 winding in the opposite direction, forcing the current i_{LR} to be increased in the opposite direction as well as transferring the energy to the output load via the N_2 winding based on the transformer behavior, so as to make the currents i_{LR} and i_{N2} increased.

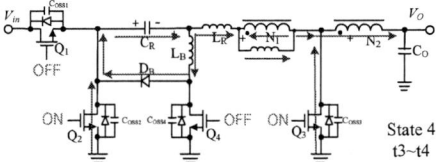

Fig. 4(d).Current direction in state 4 under accelerating mode.

e) State 5: As shown in the time interval between t_4 and t_5 in Fig. 4(e), the Q_1 and Q_4 are still turned off but the Q_2 and Q_3 are still turned on. During this state, the energy stored in the L_R charges the C_R, and this coupled inductor has the behavior of the transformer and passes the energy from the N_2 winding to the output load. As the current i_{CR} goes down to zero, the operating state goes to state 6.

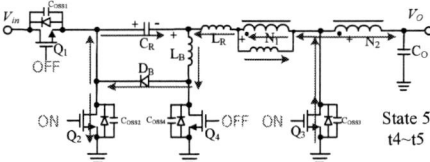

Fig. 4(e).Current direction in state 5 under accelerating mode.

f) State 6: As shown in the time interval between t_5 and t_6 in Fig. 4(f), the Q_1 and Q_4 are still turned off, but the are Q_2 and Q_3 are still turned on. During this state, the C_R still releases energy to the N_1 winding. On the other hand, this coupled inductor behaves like a transformer, transferring the energy from the N_2 winding to the output load. As the current i_{LB} goes down to zero, the operating state goes to state 7.

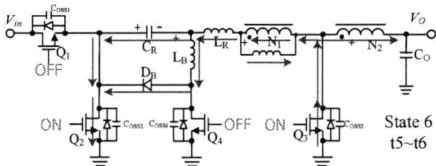

Fig. 4(e).Current direction in state 6 under accelerating mode.

g) State 7: As shown in the time interval between t_6 and t_7 in Fig. 4(g), the Q_1 and Q_4 are still turned off and the Q_2 and Q_3 are still turned on. During this state, since the current in the inductor L_B is zero, the D_B is reverse biased. As the Q_2 and Q_3 are turned off, the operating state proceeds to step 8.

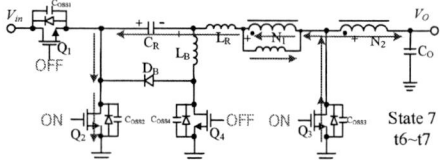

Fig. 4(f).Current direction in state 7 under accelerating mode.

h)   State 8: As shown in the time interval between t_7 and t_8 in Fig. 4(h), all the switches are still turned off. This time interval is the other dead time zone. During this state, the body diode of the Q_1 will be turned on due to the current in L_R being continuous. As the Q_1 and Q_4 are turned on, the operating state proceeds to step 9.

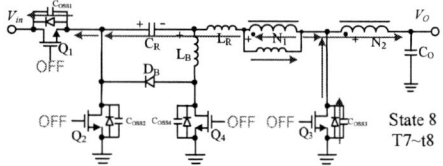

Fig. 4(h).Current direction in state 8 under accelerating mode.

i)   State 9: As shown in the time interval between t_8 and t_9 in Fig. 4(i), the Q_1 and Q_4 are turned on and the Q_2 and Q_3 are still turned off. Based on state 8, the Q_1 is turned on with ZVS. During this state, the energy stored in the leakage inductor L_R keeps transferred to the input. As soon as the L_R releases energy to zero, the operating state goes into state 10.

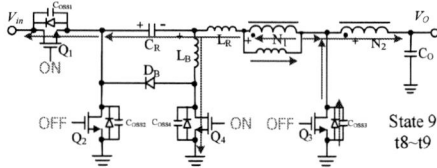

Fig. 4(f).Current direction in state 9 under accelerating mode.

j)   State 10: As shown in the time interval between t_9 and t_0 in Fig. 4(j), the Q_1 and Q_4 are still turned on and the Q_2 and Q_3 are still turned off. During this state, the L_R is magnetized due to the Q_1 being turned on, whereas the inductor L_B is also magnetized due to the Q_4 being turned on. During this state, the energy stored in the L_R keeps transferred to the input. As soon as the current in the body diode of the Q_3 is zero, the operating state proceeds to state 1. So, one cycle is completed and hence the next cycle is repeated.

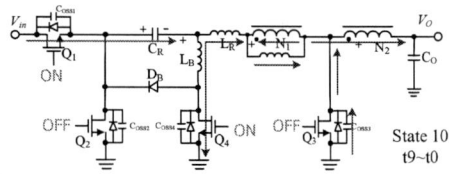

Fig. 4(j).Current direction in state 10 under accelerating mode.

IV.   CONTROL STRATEGY FOR THE ACCELERATING MODE

The Fig. 5 provides the procedure for the accelerating mode during the upload transient. In order to monitor the output voltage, two threshold voltages for the trigger-in point and the trigger-out point are preset. As the output voltage is lower than the trigger-in point, the accelerating mode is activated and hence the Q_4 is enabled, thereby causing the output voltage to be increased. Afterwards, as the output voltage reaches the trigger-out point, the circuit begins to break away from the accelerating mode. In order to avoid the unexpected effect on the circuit due to mode exchange, the fade out mode is inserted between these two modes. During the fade out mode, via gradually reducing the duty cycle of the PWM signal for the Q_4, the current i_{LB} will be decreased smoothly. In addition, because the response under the accelerating mode is faster than that under the normal mode, the controller parameters for these two modes are different. The system parameters and component specifications are tabulated in Table 1.

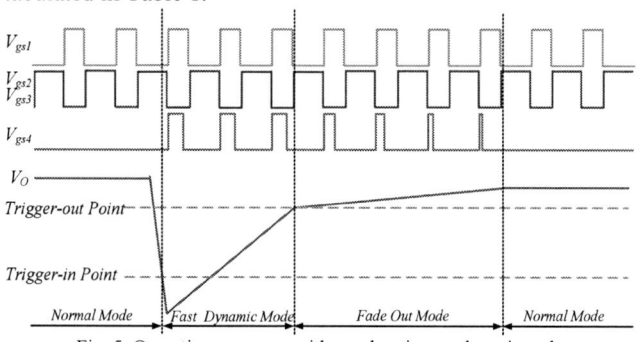

Fig. 5. Operating sequence with accelerating mode activated.

Table 1. System requirements and component specifications

Input voltage V_{in}	48V
Output voltage V_o and rated current I_o	3.3V/10A
Switching frequency f_s	160kHz
Capacitor C_R	20uF/50V TDK MLCC
Coupling inductor	N_1:N_2=24:8 with L_m=87.1uH and L_R=3.94uH
Inductor L_B	18uH
PID1parameters for normal mode	127-1-23
PID2parameters for accelerating mode	26-7-30

V.   EXPERIMENTAL RESULTS

Fig. 6 shows the gate driving signals for the Q_1, Q_2, Q_3 and Q_4, called v_{gs1}, v_{gs2}, v_{gs3} and v_{gs4}, respectively, in the steady state. From this figure, it can be seen that the switch Q_4 does not work. Fig. 7 shows the gate driving signals for the Q_1, Q_2, Q_3 and Q_4, called v_{gs1}, v_{gs2}, v_{gs3} and v_{gs4}, respectively, in the upload transient. From this figure, it can be seen that the Q_4 fades out before the output voltage goes into the normal mode. Figs. 8and9show the voltage across the C_R, called v_{CR}, the gate driving signal for the Q_4, called v_{gs4}, the current in the inductor L_B, called i_{LB}, and the output voltage V_o, due to step load change from 25% to 75% load without and with the accelerating mode, respectively. From these two figures, it can be seen that the latter has a better performance of load transient responses. Figs. 10 and 11 show the gate driving signal for the Q_4, called v_{gs4}, the voltage across the C_R, called v_{CR}, the current in the C_R, called i_{CR}, and the current in the inductor L_B, called i_{LB}, due to load current change from 25% to 75% without and with the accelerating mode, respectively. From these two figures, it can be seen that as compared with the former, the current i_{LB} in the latter relatively fast charges the C_R, thereby making the voltage v_{CR} relatively fast increased.

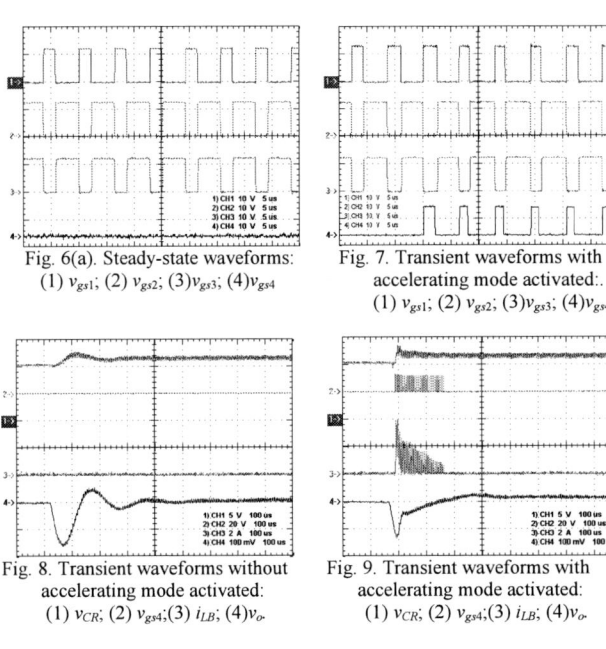

Fig. 6(a). Steady-state waveforms: (1) v_{gs1}; (2) v_{gs2}; (3)v_{gs3}; (4)v_{gs4}

Fig. 7. Transient waveforms with accelerating mode activated:. (1) v_{gs1}; (2) v_{gs2}; (3)v_{gs3}; (4)v_{gs4}

Fig. 8. Transient waveforms without accelerating mode activated: (1) v_{CR}; (2) v_{gs4};(3) i_{LB}; (4)v_o.

Fig. 9. Transient waveforms with accelerating mode activated: (1) v_{CR}; (2) v_{gs4};(3) i_{LB}; (4)v_o.

VII. CONCLUSION

This paper provides a control strategy so as to enhance the upload transient response for the ultrahigh step-down converter. Via an additional auxiliary circuit, the energy-transferring capacitor can be fast charged, thus making the voltage across this capacitor fast increased, and hence the fast upload transient response can be achieved. The proposed control strategy also can be applied to similar topology[18-21].

REFERENCES

[1] H. A. Mohamed, F. Chao, Fred C. Lee, and Q. Li, "48V voltage regulator module with PCB winding matrix transformer for future data centers," *IEEE Trans. Ind. Electron.*, Early Access, 2017.

[2] O. Pelan, N. Muntean, O. Cornea, and F. Blaabjerg, "High voltage conversion ratio, switched C & L cells, step-down DC-DC converter," *IEEE ECCE Conf.*, pp.5580-5585, 2013.

[3] Y. Chen, Z. H. Zhong, and Y. Kang, "Design and implementation of a transformerless single-stage single-switch double-buck converter with low DC-link voltage, high step-down, and constant input power factor features," *IEEE Trans. Power Electron.*, vol. 29 no.12, pp. 6660-6671, Aug. 2014.

[4] S.K. Ki and D. C. Lu, "A high step-down transformerless single-stage single-switch AC-DC converter," *IEEE Trans. Power Electron.*, vol. 28, no. 1, pp. 36-45, 2013.

[5] D. Cheshmdehmam, E. Adib and H. Farzanehfard, "Structure improvement of active-clamp to achieve high step-down conversion," *IEEE ICEE Conf.*, pp.670-675, 2016.

[6] C. F. Chuang, C. T. Pan and H. C. Cheng, "A novel transformer-less interleaved four-phase step-down DC converter with low switch voltage stress and automatic uniform current sharing characteristics," *IEEE Trans. Ind. Electron.*, vol. 31, no. 1, pp. 406-417, 2016.

[7] M. Esteki, B. Poorali, E. Adib, and H. Farzanehfard, "High step-down interleaved buck converter with low voltage stress," *IET Power Electronics*, vol. 8, no. 12, pp.2352 - 2360, 2015.

[8] G. Tibola and J. L. Duarte, and A. Blinov, "Multi-cell DC-DC converter with high step-down voltage ratio," *IEEE ECCE Conf.*, pp.2010-2016, 2015.

[9] K. Matsumoto, K. Nishijima, T. Sato, and T. Nabeshima, "A two-phase high step down coupled-inductor converter for next generation low voltage CPU," *IEEE ICPE-ECCE Asia Conf.*, pp.2813 - 2818, 2011.

[10] J. Zhao, "Non-isolation soft-switching buck converter for high-step-down conversion," *IEEE INTELEC Conf.*, pp.1-6, 2009.

[11] W. Martinez, J. Imaoka, Y. Itoh, M. Yamamoto, and K. Umetani, "A novel high step-down interleaved converter with coupled inductor," *IEEE INTELEC Conf.*, pp.1-6, 2015.

[12] Il-Oun Lee, S. Y. Cho, and G. W. Moon, "Interleaved buck converter having low switching losses and improved step down conversion ratio," *IEEE Trans. Power Electron.*, vol. 27 no.8 pp. 3664-3675, 2012.

[13] O. Kirshenboim and M. M. Peretz, "High efficiency non-isolated converter with very-high-step-down conversion ratio," *IEEE Trans. Power Electron.*, vol. 32, no. 5, pp. 3683-3690, 2017.

[14] K. I. Hwu, W. Z. Jiang, and P. Y. Wu, "An expandable four-phase interleaved high step-down converter with low switch voltage stress and automatic uniform current sharing," *IEEE Trans. Ind. Electron.*, vol. 63, no. 10, pp. 6064-6072, 2016.

[15] K. Yao, Y. Qiu, M. Xu, and F. C. Lee, "A novel winding-coupled buck converter for high-frequency, high-step-down DC-DC conversion," *IEEE Trans. Power Electron.*, vol. 20, no. 5, pp. 1017-1024, 2005.

[16] K. Yao, M. Ye, M. Xu, and F. C. Lee, "Tapped-inductor buck converter for high-step-down DC-DC conversion," *IEEE Trans. on Power Electron.*, vol. 20, no. 4, pp. 775-780, 2005.

[17] K. I. Hwu, W. Z. Jiang, and Y. T. Yau, "Ultrahigh step-down converter," *IEEE Trans. Power Electron.*, vol. 30, no. 6, pp.3262-3274, 2015.

[18] Y. T. Yau, W. Z. Jiang, and K. I. Hwu, "Analysis and design of a high-step-down ratio resonant converter," *IET Power Electronics*, vol. 9, no. 5, pp.864 - 873, 2016.

[19] Y. T. Yau, W. Z. Jiang, and K. I. Hwu, "Step-down converter with wide voltage conversion ratio," *IET Power Electronics*, vol. 8, no. 11, pp.2136 - 2144, 2015.

[20] Y. T. Yau, W. Z. Jiang, and K. I. Hwu, "Ultrahigh Step-Down Converter with Wide Input Voltage Range Based on Topology Exchange," *IEEE Transactions on Power Electronics*, vol. 32, no. 7, pp. 5341-5364, 2017.

[21] K. I. Hwu, W. Z. Jiang, and Y. T. Yau, "Nonisolated Coupled-Inductor-Based High Step-Down Converter With Zero DC Magnetizing Inductance Current and Nonpulsating Output Current," *IEEE Trans. Power Electronics*, vol. 31, no. 6, pp.4362 - 4377, 2016.

Power Electronics and Control Technologies for Household Washer

Toru Niki[1*]

1 Taga Home Appliance Works, Home & Echo Appliances Group, Hitachi Appliances, Inc., Ibaraki, Japan
*E-mail: toru.niki.pg@hitachi.com

Abstract— **This paper presents the motor control technology for household washers. Recently, customers have required not only basic functions (washing and drying performance), but also further advantages such as energy efficiency and saving water. Additionally, the customers require lower noise and vibration damping properties. To solve these requirements, we developed a multipole permanent magnet synchronous motor and control software with vector operation.**

Keywords— *household washer, parmanent magnet synchronous motor, position sensorless control, vector control*

I. INTRODUCTION

The technology of washing machines started from a single cistern type with wash and rinse functions, and a two cistern type with a fully automated process. More recent washing machines also have a drying function.

A fully automated washing machine has a single washing tub that works with washing, rinse, and spinning function. The tub consists of an outer tank and a rotatable inside tub. The washing tub is supported with a vibration absorption mechanism (suspension). In the case of a washer-dryer, the heating unit (heater and fan) and dehumidification unit are added with full automation.

The Japanese domestic demand for recent washing machines is around 4.5 million units per year, and more than 95% are the fully automated type. Furthermore, approximately 25% of the fully automated type are also combined dryer units.

The variation of washing machines are categorized as "top load washer-dryer" (figure 1) and "front load washer-dryer" (figure 2).

The top load washer-dryer version is developed for the Japanese market, and has become the most popular type in Japan. This type uses water flow caused by the rotation of the turbulator to wash clothes. The washing tub is set vertically in the machine, and the turbulator is integrated at the bottom of the washing tub. A clutch that is set between the motor and the turbulator exchanges mechanical energy transitioned at each process. For example, it connects the motor and the turbulator during washing and rinsing processes; it also connects the motor and the washing tub when spinning process.

This type can achieve high detergency, but uses a large amount of water and can tangle clothes. To avoid this issue, suitable rotation control of the turbulator and water circulation system is needed.

The front load washer-dryer type is common in Europe and other countries. It works by filling the bottom of the tub (drum) with a small amount of water and using the rotation of the tub and gravity to move the clothes through the water.

It can achieve low cloth entanglement, but detergency is lower than the top-load washer type. Mechanical vibration also often occurs via unbalanced cloth placement in the drum. To solve these problems, changes included

Fig. 1. Outline structure of top load washer-dryer

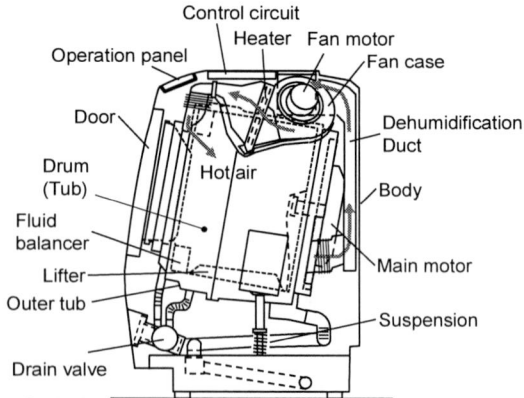

Fig. 2. Outline structure of front load washer-dryer

enlargement of the drum diameter, better circulation of high-water-flow and a higher concentration of water were effective. Also, a low-vibration rotating technique has been developed recently.

Washing machines not only require good washing and drying performance as a basic function, but also need energy saving characteristics for current markets. Even more so, due to the various changes in modern lifestyles and washing needs, advanced vibration control technology is required that can drive large-sized washing machines calmly and stably at all ranges of rotating speeds.

This paper describes the motor and control technologies of front-loading washer-dryers (figure 2).

II. VIBRATION AND NOISE REDUCTION OF THE MOTOR

The following is a description of the method to achieve low vibration and noise for the "main motor" of the washer-dryer.

The power transmission system from the motor to the washing tub is classified as follows: belt drive system, gear drive system, and direct drive system (Table 1). Brushless motors have good low-vibration, low-noise, controlled response, and energy saving characteristics. We adopted a direct drive system and a brushless motor to achieve low vibration and noise.

Brushless motors are categorized into varying types. One of the which is the permanent magnet synchronous motor (PMSM), one of the most applied versions recently used for washers because of low vibration and noise characteristics as well as efficient energy consumption. This trend is based on breakthroughs of motor drive design, such as vector control and increasing performance of power semiconductors and microcomputers.

Vibration generated from a motor reaches each part of the machine. When various parts match the resonance frequency of the motor, the parts will begin to oscillate. As a result, noise from the machine becomes noticeable.

To achieve low vibration and noise, reducing motor excitation power and resonance avoidance are required.

The multipolarization of the magnetic pole is effective to reduce excitation power of the motor. We designed 56 pole rotors to minimize cogging torque of the rotor, which is one of the sources of vibration (figure 3). We utilized magnetic field analysis simulation to determine

Table 1. Characteristics of power transmission system

	Belt-drive	Gear-drive	Direct-drive
Vibration and noise	Low	Medium	Low
Transmission efficiency	Low	Medium	High
Motor body size	Small	Medium	Big
	Tub Pulley Belt Motor Pulley	Tub Gear Motor	Tub Motor

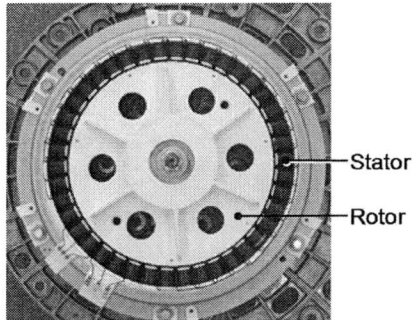

Fig. 3. Appearance of main motor (blushless motor)

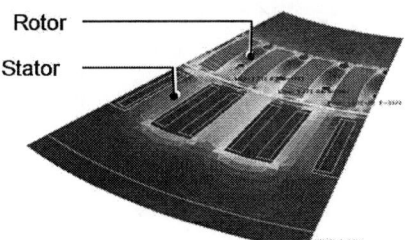

Fig. 4. Example of electromagnetic field finite element method

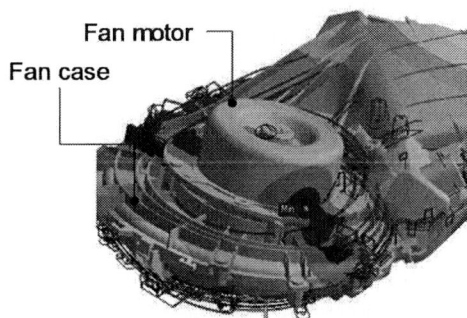

Fig. 5. Example of vibration system CAE analysis (fan motor for front load washer-dryer)

quantitative excitation power to decide the final specification of the motor [1]. An example of the electromagnetic field analysis using finite element method (FEM) is shown in figure 4.

In resonance avoidance, the resonance frequency of each component can be found using CAE (Computer Aided Engineering) analysis software. We utilize it at the initial stage of design to reduce development time and the testing process (figure 5).

III. MOTOR CONTROL

In recent years, it is common to use several motors which are controlled by inverters. For example, in the front load washer dryer, three inverter-controlled motors are used: a main motor driving the washing tub (drum), a fan motor which forces high-speed air (the high-tension force needed for drying wrinkles), and a circulating pump motor which circulates water in the tub to achieve high-cleaning performance and saving water.

The demanded performances are greatly varied for each motor. For example, the main motor requires both high torque in 30 - 50 min⁻¹ when washing, and a high rotating speed of 1900 min⁻¹ when spinning. In another instance, the fan motor requires both high speed (up to 16000min⁻¹) and minimized current consumption in order to achieve both reduction of drying out wrinkles and energy saving.

Other than basic performance of the motor control as mentioned above, clothing weight detection and clothing deflect detection in the washing tub (drum) are required. This demonstrates that the motor control can also be used as a sensor.

A system configuration diagram of the main motor control software is shown in figure 6.

The software is comprised of two sections: (A) motor control and, (B) washing process management.

(B) - The washing process management function manages all washing process and conditions. The software receives washing course commands (ex. "standard courses", "gentle course", etc.) from the user, senses various conditions, and then passes the target rotation speed to the motor control software.

(A) - The main motor control software has three major functions and these are operated in a sampling period of 33μs - 500μs; (A-1) vector calculation unit, (A-2) PWM waveform generation unit, and (A-3) rotational frequency measurement unit.

(A-1): The vector calculation unit calculates the output voltage to apply to the motor. The voltage comes from the voltage equation that models the motor. To solve the equation, a deviation of target rotation speed and current rotation speed of the motor are needed.

(A-2): The PWM waveform generation unit generates a pulse signal corresponding to the three-phase alternating current based on the voltage command value provided from the vector calculation unit. The inverter circuit is driven based on this pulse signal and applies voltage to the motor. As a result, the motor starts to turn.

(A-3): The rotor position of the motor is continuously measured by the rotational frequency measurement unit. The position information feeds back to other control functions.

The example of the current waveform of the main motor is shown in figure 7. In this case, a load (wet clothes) is in the washing tub, the amount is approximately 1/2 of washing tub capacity, and the position of the load changes at every moment. Therefore the motor current pulsates. This is a distinctive characteristic of the washing machine motor control and thus means a stationary state does not exist during most of the washing process. In other words, it is necessary to design the control system to function in a non-stationary state.

The initial spinning process is one of the most advanced motor control techniques (figure 8).

In this process, a high-level motor control technique is needed to achieve requirements as follows: reducing mechanical vibration and minimizing position bias of clothes in the drum, reducing mechanical vibration and

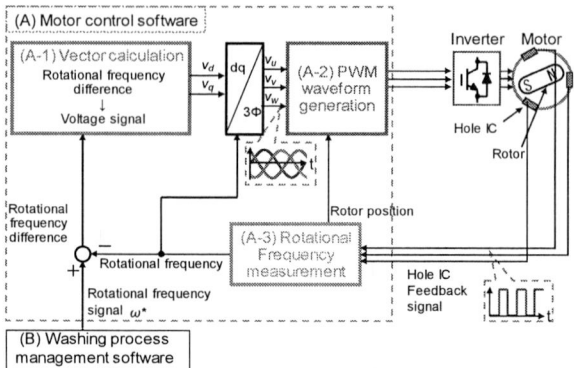

Fig. 6. Main motor control system

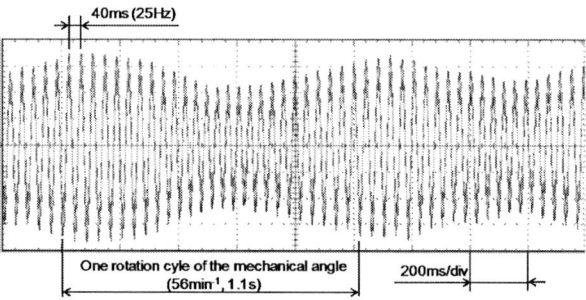

Fig. 7. Main motor current waveform

sound noise at the top speed when spinning.

General process for spinning is follows:

(1) Drain dirty water from the tub by opening the Drain valve (refer to figure 2). The clothes coalesce at the bottom of the tub. When the drum begins to rotate, the center of inertia is biased from center of the axis of the tub (unbalanced state). Serious mechanical vibration will thus occur.

(2) To solve this problem, some rotating and sensing techniques are used. At this step, the drum begins slowly. When the drum achieves a certain rotating speed, the clothes start to move frequently, arising from the effect of centrifugal force. Clothes are stick within the tub spread.

(3) Fix the rotating speed (N1), and begin monitoring variation of rotation. If the variation is larger than certain threshold, the rotate is stopped, and the spinning process is re-attempted. Otherwise the rotating speed will be accelerated.

(4) These process ((2) and (3)) are repeated to achieve the target speed (Nt).

With the recently enlarged washing machines, a number of large-sized blankets can be placed into the drum. Avoiding an unbalancing state will then become more difficult.

To solve this problem, we developed a drum spinning rotation speed detection method to optimize the spinning speed for unbalancing (figure 9).

(a) Clothes become adhered within the tub, using the effect of centrifugal force.

(b) Accelerate from scanning start speed (Ns) to scanning end speed (Ne). During acceleration, motor control software monitors variations of the rotation speed.

858

The 2018 International Power Electronics Conference

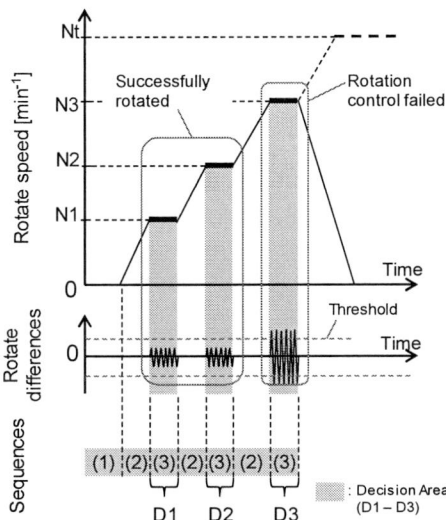

Fig. 8. Drum rotation control at spinning processs

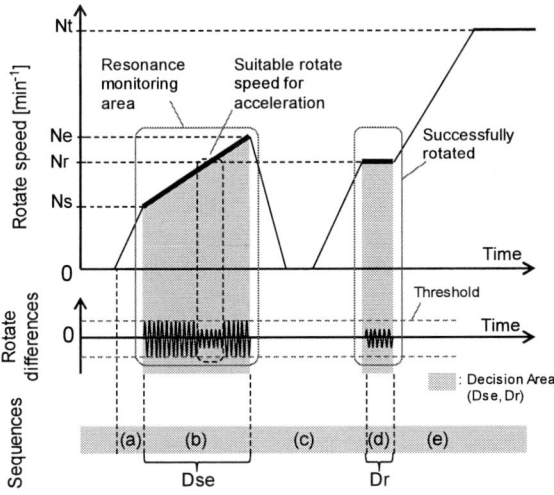

Fig. 9. Drum rotation control with new technique

After scanning, the software determine suitable speed to minimize mechanical vibration (Nr).

(c) Stop rotate.

(d) Restart rotating at the speed of Nr, determine the variation once again.

(e) If the variation is lower than threshold value, the drum will be accelerated to the top speed (Nt).

The fan motor control software is also comprised of the same two sections (figure 10); (a) motor control and, (b) washing process management. The fan motor is controlled by a "sensorless vector control technique" [2] – without position sensors and current sensors - because of space limitations of wiring electric cords, and a higher achievement of low energy consumption and noise characteristics.

(a) - The fan motor control software has five major functions and these are operated in sampling period of 33µs - 500µs; (a-1) vector calculation unit, (a-2) PWM

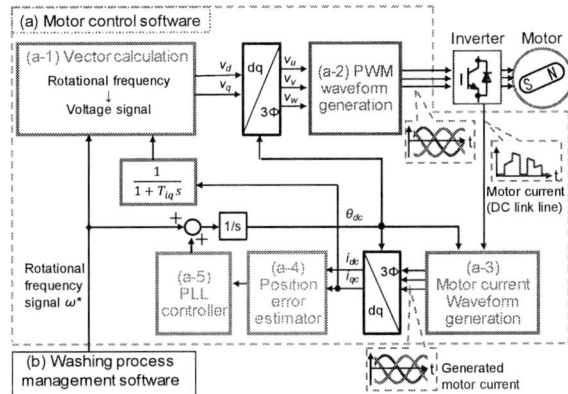

Fig. 10. Turbo fan motor control system

waveform generation unit, (a-3) motor current waveform generation unit, (a-4) position error estimator, and (a-5) PLL control unit.

(a-3): Initially, generate three phase alternative current wave from DC current of the shunt register on the DC link power stage of the inverter. The current wave is translated to the d-q axis current.

(a-4): The position of the rotor is estimated by a position error estimator. The position is estimated from the voltage command, current command values of the d and q axes, and the inverter rotational speed command.

(a-5): The PLL controller compares the estimated position error with the command value to calculate the inverter rotational speed command.

The vector calculation unit (a-1), and the PWM waveform generation unit are the same as the same section of the main motor control system ((A-1) and (A-2)).

Accordingly, these calculations achieved smoothed voltage applying to the motor, contributing to energy reduction performance with low noise.

IV. CONCLUSIONS

Washing machines are now ubiquitous, and also a maturated product. But cleaning performance and energy saving improvements are continuously expected. Consequently new technologies are still added to the products. In addition, we have to consider the growth of high-performance fiber for clothes and enriched type liquid detergents. We would like to create more evolved washing machines by applying new technologies to meet the customer's needs into the future.

REFERENCES

[1] K. Miyata, "Magnetic Field Analysis by the Edge Element FEM" *IEEJ Journal of Industry Applications*, vol. 124, no. 7, pp. 1404-1409, 2004.

[2] Y. Kawabata, T. Endo, Y. Takakura, "Study of Control for Position Sensorless and Motor Current Sensorless Permanent Magnet Synchronous Motor Drives", *IEEJ Transactions on Industry Applications*, vol.134, no.6, pp.579-587, 2014.

The 2018 International Power Electronics Conference

Development of Room Air Conditioner with Twin-Propeller Fans

Takamasa Uemura
Thermal and Fluid Systems Department
MITSUBISHI ELECTRIC CORPORATION
ADVANCED TECHNOLOGY
R&D CENTER
Hyogo, Japan
Uemura.Takamasa@dh.
MitsubishiElectric.co.jp

Tomoya Fukui
Thermal and Fluid Systems Department
MITSUBISHI ELECTRIC CORPORATION
ADVANCED TECHNOLOGY
R&D CENTER
Hyogo, Japan
Fukui.Tomoya@ea.
MitsubishiElectric.co.jp

Kenichi Sakoda
Thermal and Fluid Systems Department
MITSUBISHI ELECTRIC CORPORATION
ADVANCED TECHNOLOGY
R&D CENTER
Hyogo, Japan
Sakoda.Kenichi@cw.
MitsubishiElectric.co.jp

In order to realize a significant energy conservation and comfort of room air conditioners, we have developed a new structure indoor unit which drastically reviewed the internal structure for the first time in about half a century. This indoor unit improves energy saving performance by changing the shape of the air passage matching the air flow through the indoor unit from the upper side to the lower side, changing to the fan shape suitable for the air passage, and efficient arrangement of the heat exchanger. We reduced the power consumption of the indoor unit by 31% and improved the APF by 13.3% compared with the previous year. In addition, we realized an improvement in comfort by controlling the airflow making full use of the features of the new fan shape.

Keywords— Personal twin flow, Propeller fan, W shape heat exchanger

I.   INTRODUCTION

In order to protect the global environment, the energy saving law of "Top Runner Approach" has been enforced, and improvement of energy saving performance of room air conditioners which accounts for much of the power consumption at home is increasingly important. Room air conditioners development is also conducted linked with improvement of energy saving and comfort. It is not only improvement of APF (Annual Performance Factor) which is an indicator of energy saving performance, but also control of air flow by sensing the temperature distribution by sensors [1].

We drastically reviewed the composition of the room air conditioner indoor unit and developed a new type of indoor unit that combines high energy saving and comfort. In this paper, we introduce the features and the major technical developments performed to realize the indoor unit having a new structure.

1. Combination of thinner and higher efficiency of propeller fan
2. Reduction of aerodynamic noise
3. Combination of performance and reliability
4. High efficiency and high density mounting of heat exchanger

II.   REVIEW OF INDOOR UNIT STRUCTURE

A. Structure and problem of conventional indoor unit

Fig.1 shows the example of an internal structure of conventional indoor unit. In the conventional indoor unit, a cross-flow fan is used, and a heat exchanger is arranged in Λ shape to cover the upstream side of the cross-flow fan. The cross-flow fan is widely used in indoor units because it produces a wide sheet air flow.

Fig.2 shows the example of an vertical cross-sectional view of conventional indoor unit. As indicated by arrows in Fig.2, the cross-flow fan sucks air from the front side and exhausts it downward, and has a large axial length larger than the diameter of the blades.

Until around 2003, this type of air passage was suitable for cross-flow fans because indoor units were structured to take the air also from the front side and there were high demands in cooling applications of blowing out cooled air from the front side. However, as shown in Fig.2 (b), in recent years, the front side is used as a design surface and is not usable for air intake, requiring air intake to be from the top. Demands for blowing hot air downward are also increasing in heating applications. Therefore, it is necessary to configure air passage to allow the air intaking from the top side and blowing downward, rather than conventional air passage in which air intake is made from the front and air blowout also from the front. For that reason, we have been promoting high performance improvement by increasing the height and the thickness of the indoor unit, moving the front side panel, etc.

As described above, the structure of conventional indoor units cannot cope with the changes of market demands. The air stream meandering largely inside an indoor unit is one of the causes generating loss of blower power. In order to improve the energy saving performance drastically, not only making every component more efficient but also drastic reviews overturning traditional idea were needed for the indoor unit structure.

The 2018 International Power Electronics Conference

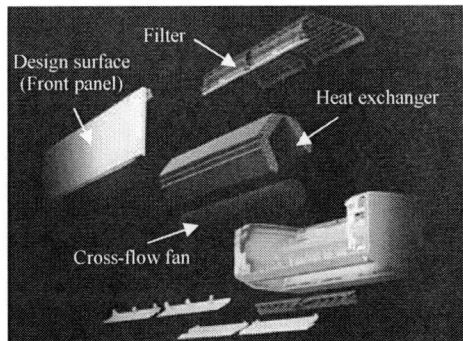

Fig. 1. Internal structure of conventional indoor unit

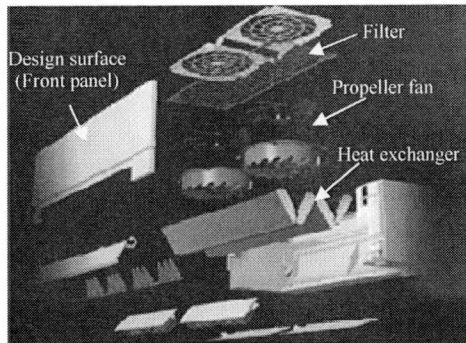

Fig. 3. Internal structure of new type indoor unit

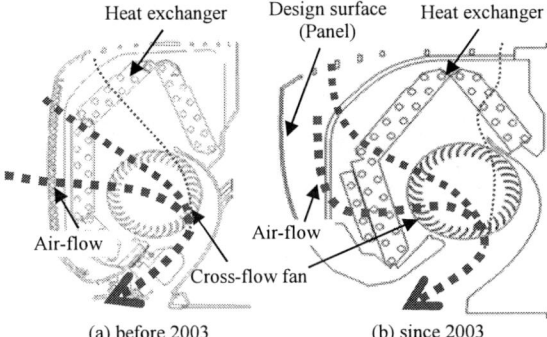

(a) before 2003　　　　(b) since 2003
Fig. 2. Vertical cross-sectional view of conventional indoor unit

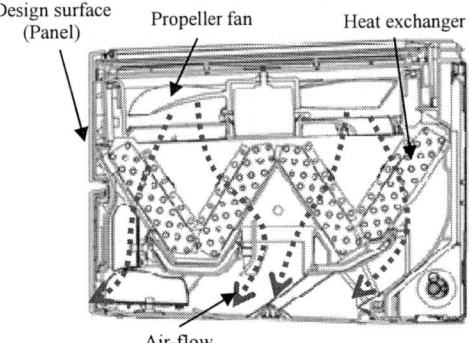

Fig. 4. Vertical cross-sectional view of new type indoor unit

B. Structure and technical developments of new type indoor unit

Among the components comprising a room air conditioner, the largest power consumer is a compressor. However, from the APF point of view, where APF (Annual Performance Factor) is the index for energy saving performance, the intermediate capacity is the one that greatly affects the annual power consumption. In the intermediate capacity, the power consumption of an indoor blower largely contributes to the total power consumption of an indoor unit. Therefore, to improve the APF, it is essential to reduce the power consumption by an indoor blower.

Fig.3 shows an internal structure of the indoor unit designed to significantly reduce the blower power, and Fig.4 shows a vertical cross sectional view. In this structure, a propeller fan was introduced as an indoor blower to suit air passage heading from the top to the bottom. The air flow of propeller fan is suitable for the current indoor unit. The propeller fan has the intake direction and exhaust direction matched, so air is sucked from above the indoor unit and exhausted downward. Furthermore, since the propeller fan generally has a higher fan efficiency than the cross-flow fan, we can reduce power consumption by installing it in the indoor unit. Further, by arranging the fan above the indoor unit, we can arrange the heat exchangers densely at the place where the cross-flow fan was conventionally arranged.

III. MAJOR TECHNICAL DEVELOPMENT

A. Combination of thinner and higher efficiency of propeller fan

Fig.5 shows a propeller fan developed for the indoor unit having the new structure as well as its configuration. Main components of the developed propeller fan are a multi-blade fan and a bell-mouth. While propeller fans for use in outdoor units or ventilation fans are being developed for higher efficiency and lower aerodynamic noise, those for indoor units need to be made much thinner [2].

Fig.6 shows a side view of the developed propeller fan. The height dimension of the propeller fan depends on the chord length and inclination angle of the blade to the rotating axis. To reduce a fan height, the chord length must be reduced. However, because the power produced by a blade is proportional to the chord length, the number of blades is generally increased in line with the reduction of a chord length. The dotted line in Fig.7 shows the change in fan efficiency versus the change in a height dimension. The fan efficiency is obtained by dividing the product of air volume and static pressure by the power given to a fan per unit time. When the above thinning method is used, the fan efficiency decreases with reduction of a height dimension. To solve this issue, the following approaches were taken.

Firstly, shapes of the blade and bell-mouth were investigated. In the investigation, the pressure

861

distribution and pressure variation were measured by small pressure sensors embedded in the bell-mouth's wall. The analysis was made to clarify the causes of degradation in the efficiency and aerodynamic noise. Fig.8 shows the located positions of the pressure sensors. Fig.9 shows the pressure distribution measured by the pressure sensors on the bell-mouth's wall. Fig.10 shows the variance of the pressure on the bell-mouth's wall. From the measurement data, we found that there is a correlation between the pressure variation at the outer peripheral side and the values of aerodynamic noise and efficiency. Then, we optimized the shapes of the blade and bell-mouth by changing parameters of bell-mouth's shape, the chord length at the outer peripheral side and the inclination angle. These optimizations improved the fan efficiency from approx. 37% to approx.46%.

Secondly, in order to reduce the pressure loss induced by the rotation-directional velocity component of the fan's blowout stream, we investigated stationary blades for the stays supporting a drive motor. The shape of the stationary blade was computed based on the blowout velocity distribution extracted from our fluid analysis made to the propeller fan. When stationary blades are arranged radially from the inner peripheral side towards the outer peripheral side, the space between the two blades becomes wider at the outer peripheral side, resulting in the loss of a static pressure recovery effect. To ensure pressure recovery, we prepared auxiliary blades that were extended from the bell-mouth wall between adjacent stationary blades, and relocated the stationary blades in order to have appropriate spaces at the inner peripheral side and the outer peripheral side respectively (See Fig.5). By implementing these optimizations, the fan efficiency was improved up to 51% as indicated by the mark ● in Fig.7, thereby we achieved the thinning of the propeller fan while establishing high efficiency. This is a significant improvement as compared with the efficiency (approx. 30%) of traditional cross-flow fans.

Another improvement was made to reduce the pressure loss caused by the heat exchanger located at the downstream side, by adding cutouts on the bell-mouth's outer peripheral edge of the blowout side. Depending on the point of operation, propeller fans increase the flow in the radial direction. If the flow in the radial direction is interrupted by the wall surface or the like of the bell-mouth, regions having high wind speeds are created on the outer peripheral side of the fan, generating a distribution of wind speeds that increases the pressure loss. Fig.11 shows the difference in flows with and without cutouts at the blowout side of the fan. By adding cutouts on the bell-mouth's outer peripheral edge of the blowout side that is the pressurized side of the stationary blades, the wind distribution for the heat exchanger was relaxed, thereby resulting in approx. 5% reduction in the pressure loss induced by the heat exchanger.

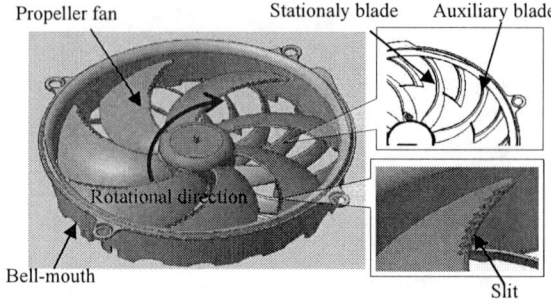

Fig. 5. Developed propeller fan

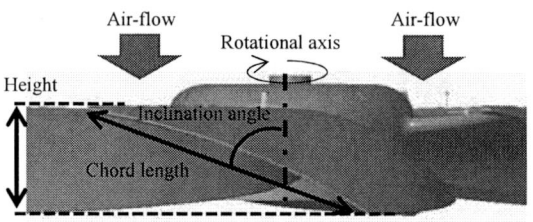

Fig. 6. Side view of the developed propeller fan

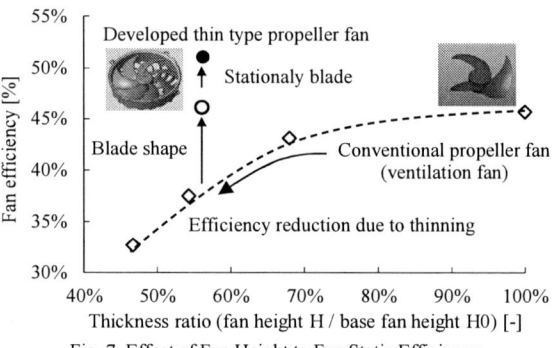

Fig. 7. Effect of Fan Height to Fan Static Efficiency

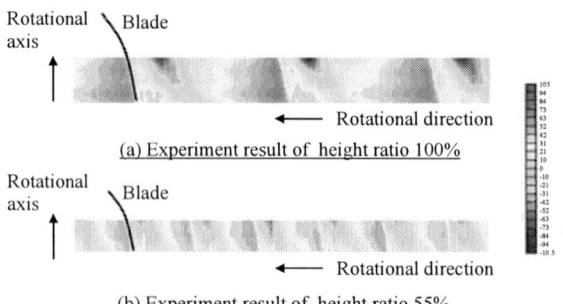

Fig. 8. Located positions of the pressure sensors

Fig. 9. Pressure distribution on the bell-mouth's wall

The 2018 International Power Electronics Conference

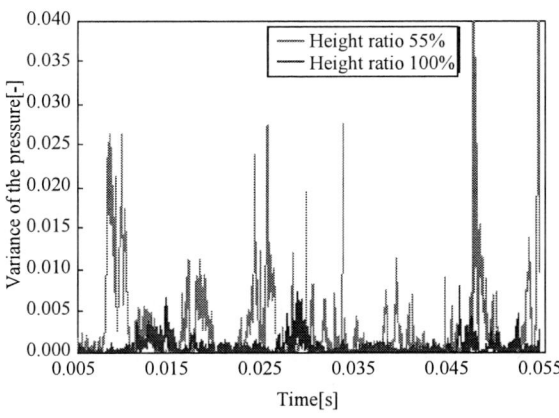

Fig. 10. Variance of the pressure on the bell-mouth's wall

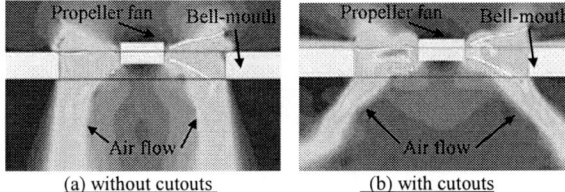

(a) without cutouts   (b) with cutouts

Fig. 11. Difference in flows with and without cutouts

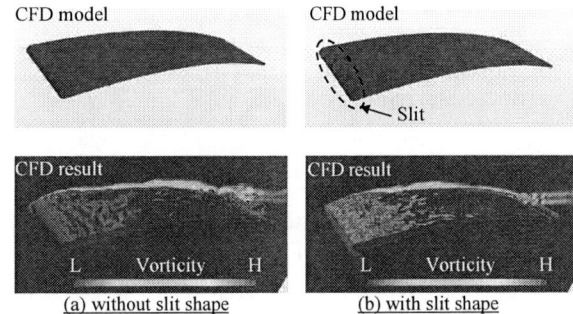

(a) without slit shape   (b) with slit shape

Fig. 12. Isosurface of vorticity on the negative pressure side of the blade

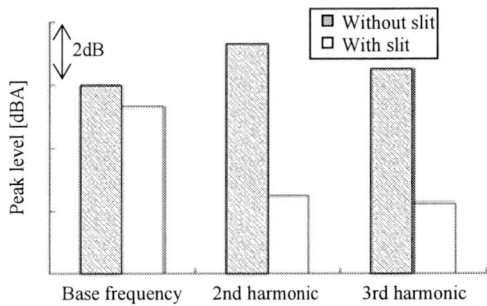

Fig. 13. Changes in peak levels of discrete frequency noise

B. Reduction of aerodynamic noise

The following presents the technology we developed to reduce aerodynamic noise. To protect customers from rotating fan blades, some structure such as a fan guard is provided at the air intake of the fan. When any structure is existing near the air intake, slipstream or drift is generated due to such structure, producing an abrupt lift variation against blades of a propeller fan. This lift variation generates discrete frequency noise at the fundamental frequency of the product of a fan rotation speed and number of blades. When the discrete frequency noise exceeds the sound pressure level at nearby frequencies, such noise is perceived as unpleasant noise in feeling.

To mitigate the abrupt lift variation produced by slipstream or drift due to the structure, we modified the blade to have a slit shape on the leading edge. Fig.12 illustrates the analyzed isosurface of vorticity on the negative pressure side of the blade with and without the slit shape. As seen, the large turbulence at the leading edge produced by streams flowing onto the blade is fragmented by the slit shape. By fragmenting the large turbulence, it is possible to mitigate the abrupt lift variation produced by slipstream or drift due to the structure. Fig.13 shows the changes in peak levels of discrete frequency noise with and without the slit shape. As seen, the slit shape at the blade leading edge has reduced discrete frequency noise effectively.

C. Combination of performance and reliability

For new-type indoor units, we investigated the reduction of indoor blower power as well as the issues relevant to reliability. Here we present the two improvements relating to thinning of the propeller fan as examples of our studies we have made to achieve both performance and reliability.

The first improvement was made to the shape of a stationary blade. The stationary blade structure for frames supporting a propeller fan and motor is often used in device cooling fans. In the case of our developed propeller fan, because the height of the fan is reduced, it is not possible to make the height of stationary blades higher. Furthermore, because the ratio of the boss diameter (Fig.14) holding the motor to the outer peripheral diameter is small, the space between the adjacent stationary blades becomes wide at the outer peripheral when radially locating stationary blades from the boss towards the outer peripheral. In such case, the proper chord-pitch ratio (Fig.14) cannot be ensured. When the proper chord-pitch ratio is not ensured, the pressure recovery effect becomes small as well as the stiffness is degraded. Therefore, it was necessary to ensure sufficient strength of the stationary blades and suppress their vibrations. To solve this issue, we developed a structure shown in Fig. 15 in which auxiliary blades are prepared on the outer peripheral where the space between the adjacent stationary blades becomes large, and the auxiliary blade and stationary blade are bridged. This structure has improved the stiffness as well as reduced the generation of vibration produced by propeller fans when they become unbalanced.

The second improvement relates to the lattice structure located in the upstream of the fan. This is the structural member to provide protection for fingers from touching the rotating blades and to hold an air purifying filter (refer to Fig.16). Since the slipstream generates discrete frequency noise, it is necessary to reduce such noise while ensuring strength of the lattice. The form of the lattice structure and that of the air-purifying filter are generally designed to have the same lattice shape to reduce the ventilation resistance and improve elegance in design. When the air intake is rectangular and the lattices are arranged in linear, the lattice structure shown in Fig.17 (a) is feasible to provide easiness of molding and to ensure the strength. However, this structure is likely to generate discrete frequency noise because:

- the timings at which the leading edge of a propeller fan blade crosses the slipstream behind the linear lattice become the same at the inner and outer peripheral portions of the fan blade, and

-drift is generated because the air intake at the lattice is rectangular whereas the air intake of the fan is circular.

To solve this issue, in addition to the provision of slit shape on the leading edge, we changed the air intake to the circular one as shown in Fig.17 (b), and arranged helical lattices in place of linear lattices so that the crossed-axis angle between the leading edge of a blade of the propeller fan and the lattices becomes almost more than 80 degrees. By doing this, it is now possible to shift the timing at which the leading edge of a propeller fan blade crosses the slipstream behind the helical lattice from the inner peripheral towards outer peripheral, thereby suppressing the generation of discrete frequency noise. In addition, each helical lattice is connected to circular rings to improve the strength of the lattices and easiness of molding.

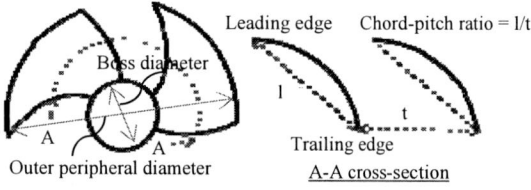

Fig. 14. Definition of propeller fan parameter

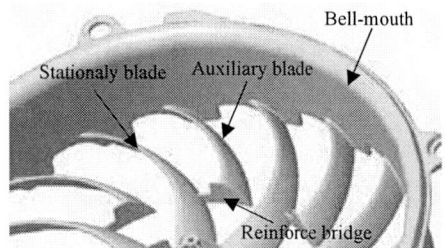

Fig. 15. Stationaly blade and auxiliary blade of bell-mouth

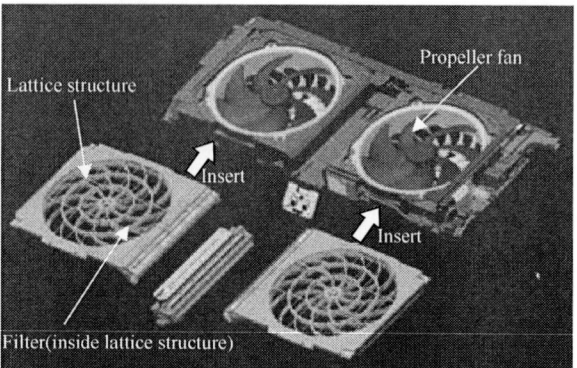

Fig. 16. Upstream structure of developed indoor unit

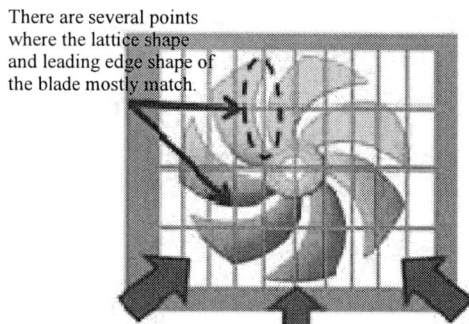

Drift is generated because the air intake of the indoor unit is ectangular whereas the air intake of the fan is circular.

(a) Linear lattice

The helical shaped lattices mitigate simultaneous crossover by the inner and outer peripheral portions of the blade on the slipstream.

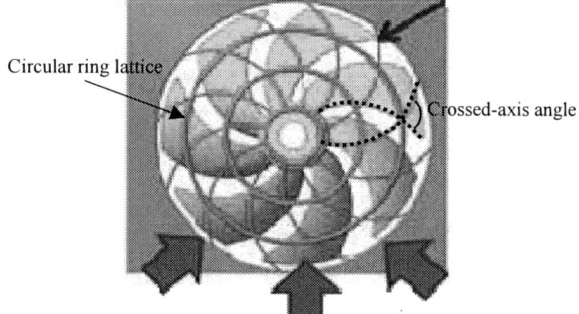

The circular (axial symmetry) air intake mitigates drift.

(b) Helical lattice

Fig. 17. Shape of lattice structure

D. High efficiency and high density mounting of heat exchanger

In the new type of indoor unit, since the thin propeller fan is arranged in the upper part of it, the space for disposing the heat exchanger is expanded compared to the conventional case, and the degree of freedom of arrangement is increased. In order to improve the performance of the indoor unit, it is necessary to arrange so as to maximize the temperature efficiency while minimizing the aerodynamic pressure loss of the heat

The 2018 International Power Electronics Conference

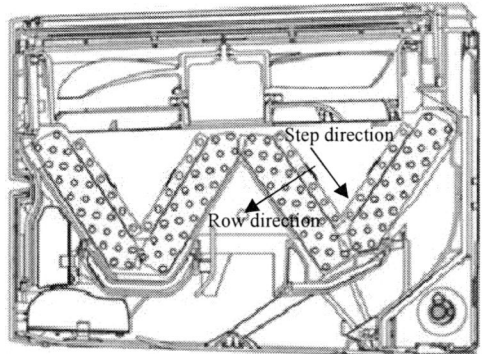

Fig. 18. Step direction and row direction of heat exchanger

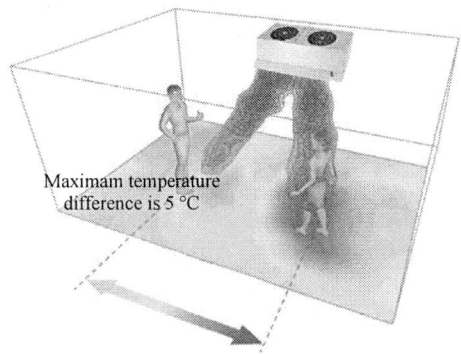

Fig. 19. Temperature Distribution of 5cm above the floor

Exchanger [3]. Therefore, by arranging the heat exchanger in a W shape and taking a long distance in the perpendicular direction to the air passage direction (step direction referred to Fig.18), the surface area of heat excahnger was enlarged and the aerodynamic pressure loss was reduced. In addition, by decreasing the distance in the air passage direction (row direction referred to Fig.18), the decrease of the temperature efficiency was suppressed. As a result, without increasing the height and the thickness of the indoor unit, the embarkation amount of the heat exchanger was increased up to 22%[*1].

*1 Comparison between 2016 model MSZ-ZW565S(Conventional) and MSZ-FZ5616S(New type)

IV. MAJOR TECHNICAL DEVELOPMENT

A. Improvement of energy saving

We reduce the power consumption of the indoor unit by 31%[*1] and improve the APF by 13.3%[*1], as shown in Table I, by the high efficient propeller fan and the efficient arrangement of heat exchangers. The developed new type of indoor unit realizes a dramatic improvement in energy saving performance. The followings are the major technical developments performed to realize the indoor unit having a new structure.

TABLE I
COMPARISON OF POWER CONSUMPTION AND APF

	Conventional	New type
Power consumption (Air Flow Late 18m³/min)	47.9 W	32.8 W
APF	6.0	6.8

*1 Comparison between 2016 model MSZ-ZW565S(Conventional) and MSZ-FZ5616S(New type)

B. Individual air conditioning by independent propulsion fan driving

Since the developed new type indoor unit has a plurality propeller fans, it is possible to independently control the rotation speed of each fan according to the sensing result of the temperature of the human body or the room. Therefore, it is possible to create different temperature spaces in the same room, by creating an air volume difference between the left and right of the indoor

unit. Fig.19 shows the temperature distribution at 5 cm above the floor when increasing the rotation speed of the fan on the left side and decreasing the rotation speed of the right side fan during the heating operation. With this left-right independent driving, the temperature difference of up to 5°C can be added to the left and right floor temperature.

V. CONCLUSIONS

We developed a new structure room air conditioner indoor unit which drastically reviewed the structure of conventional indoor unit and realized a drastic improvement in energy saving performance compared to the previous year model. In addition, we realized individual air conditioning by independently controlling a plurality of fans. We will continue to promote the development of the high efficiency air-conditioning equipment aiming for further energy saving and contribute to the protection of the global environment.

REFERENCES

[1] S. Miwa, S.Watanabe, T. Hirai and T. Matsumoto, "Air Conditioner Controlling Temperatures We Feel", *Journal of The Robotics Society of Japan*, Vol. 32, No. 3 (2014), pp. 218-221.

[2] S. Nakashima, S. Yamada, and K. Kise, "Experimental research into relation between propeller fan's flow fields and noise (relationships between difference of tip flow behavior in each operation point and its noise)", *Transactions of the Japan Society of Mechanical Engineers*, Series B, Vol. 76, No. 767 (2010), pp. 32-37.

[3] K. Kaga, S. Kotoh, T. Ogushi and H. Yoshida, "Prediction of Performance of a Heat Exchanger by Thermal Network Method : Improvement of an Evaporator's Performance and Restraining an Air Flow Bypass by Adjusting Refrigerant-Flow Path Pattern(Thermal Engineering)", *Transactions of the Japan Society of Mechanical Engineers Series B*, Vol. 75, No. 758 (2009), pp. 2049-2054.

The 2018 International Power Electronics Conference

Electrolytic Capacitor-Less Single-Phase to Three-Phase Inverter with Harmonics Suppression Control for Air Conditioner

Nobuo Hayashi[1*], Takuro Ogawa[1], Tomoisa Taniguchi[1] and Morimitsu Sekimoto[1]

1 Technology and Innovation Center, Daikin Industries, LTD. , Osaka, Japan

*E-mail: nobuo.hayashi@daikin.co.jp

Abstract— Global demand for highly energy-efficient inverter air conditioners is rising. However, the power converter must comply with harmonics regulations. This requires the converter to have a large inductor, PFC circuit and electrolytic capacitor. Therefore, the converter becomes expensive, which prevents inverter air conditioners from becoming popular around the world. In order to solve this problem, electrolytic-capacitor-less inverters were proposed.

This paper proposes a harmonics suppression control of an electrolytic capacitor-less inverter. As a result of applying this control, the inductance and the capacitance of power converter have been reduced to 1/6 and 1/50, respectively, compared with those for conventional power converters, which has resulted in the commercialization of low-cost inverter air conditioners.

Keywords— *electrolytic capacitor-less inverter, input current harmonics, repetitive control, spatial harmonics*

I. INTRODUCTION

In recent years, the penetration rate of air conditioners in Japan has reached 100%, and it is usual for each family to have several air conditioners at home. An air conditioner is characterized by its availability for both cooling and heating with clean air maintained, and highly efficient performance with the use of a heat pump. However, because of an increase in operating hours and an expansion of the number of air conditioners in use, the power consumption of air conditioners in each household has been occupying 25% of total household power

consumption. Accordingly, there has been an increasing demand for power saving on air conditioners from the viewpoint of reducing users' annual power consumption and preventing global warming.

An inverter air conditioner drives the compressor under revolution control according to the difference between the set temperature and room temperature, but the rotational speed of the compressor is less than half the rated value (under intermediate load) in most of the operating hours. Therefore, the power-saving effect of the inverter is great, and in combination with the permanent magnet motor used for the air conditioner, it is possible to reduce the annual power consumption of air conditioner by 30% to 40%.

As shown in Fig. 1, however, the current global diffusion rate of inverter air conditioners is still low. It is an extremely important and urgent task to spread inverter air conditioners worldwide from the viewpoint of global environmental protection. In order to spread them, low-cost inverter air conditioners, in particular, are required in emerging markets.

On the other hand, strict regulations are defined in overseas markets, including the European and Chinese markets, for power harmonics emitted from inverters. Topological measures, such as the use of large inductors and the addition of power factor correction (PFC) circuits, are used in response to the power harmonic regulations. However, these causes inverter cost

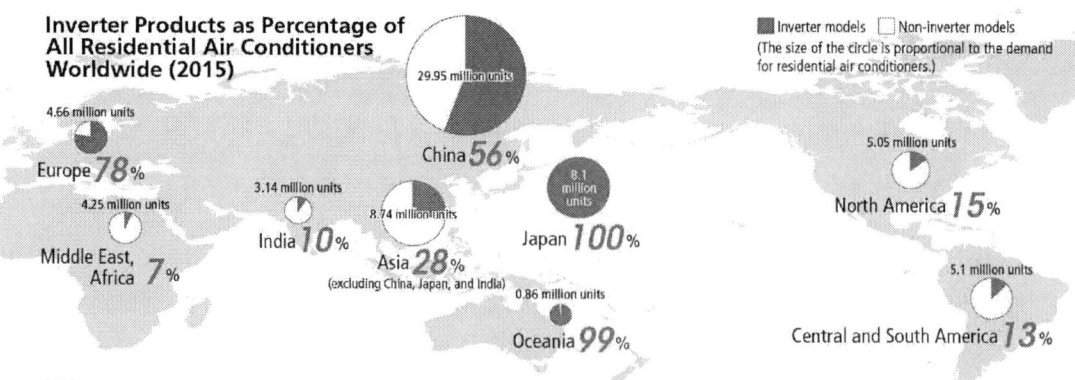

Fig. 1. Inverter Products as Percentage of All Residential Air Conditioners Worldwide (2015)
(Complied by Daikin Industries, LTD.)

increment.

Electrolytic capacitor-less inverters have been proposed to achieve cost reduction and to satisfy the power harmonic regulations [1]-[6]. The technology applied eliminates electrolytic capacitors, inductors, and PFC circuits, which are large and high-cost components, and makes inverters perform not only motor control but also input current control, thus satisfying the world's most strict power harmonic regulations.

This paper reports the control method for electrolytic capacitor-less inverters applied to room air conditioners and evaluates the method applied to commercialized room air conditioners.

II. COMPARISON OF VARIOUS TECHNIQUES FOR HARMONIC SUPPRESSION

A harmonic suppression circuit is selected for each air conditioner design from several types according to the performance capacity of the air conditioner. The main circuit configurations of typical harmonic suppression circuits are shown in Fig. 2. The outline of each circuit is shown below, and the features are summarized in Table I.

A. Diode rectifier

In this method, only an inductor is used for waveform improvements. Although this method is inexpensive compared to others, it results in an increase in power harmonics. Therefore, it is not possible to use this method for overseas equipment, and it is adopted only in small-capacity equipment in Japan.

B. Partial switching (single pulse)

In this method, a power supply short mode is set once in the half cycle of power supply to widen the conduction width of the input current. The method is adopted for overseas low-capacity equipment with a maximum input of 2 kW.

C. Partial switching (multiple pulse)

A power supply short mode is set a number of times in a half cycle. Therefore, power harmonics can be suppressed to a greater extent compared with the method specified in Partial switching (single pulse). This method is adopted for high-capacity equipment with an input current greater than 16 A (or 3.5 kW), where the power harmonic regulations are less strict. IEC 61000-3-12 applies to products with an input current in excess of 16A.

D. Boost chopper

The boost chopper can control the input current in a sinusoidal waveform. It can, therefore, significantly suppress harmonics. For this reason, it is applied to medium-capacity equipment with an input current of around 16 A, where the emission of power harmonics needs to be strictly controlled. The boost chopper in a high-carrier switching operation, however, involves the problem of both efficiency and noise characteristic deterioration.

Table I
COMPARATIVE TABLE OF EACH CIRCUIT

Circuit		Power	Cost	Harmonics	Noise
(A) Diode rectifier		Japan Only	Low	Large	Small
Partial Switch-ing	(B) Single pulse	below 2kW	Middle	Middle	Middle
	(C) Multiple pulse	above 3.5kW	Middle	Small	Middle
(D) Boost chopper		below 3.5kW	Very High	Very Small	Very Large
(E) Electrolytic capacitor-less		below 2kW	Very low	Small	Small

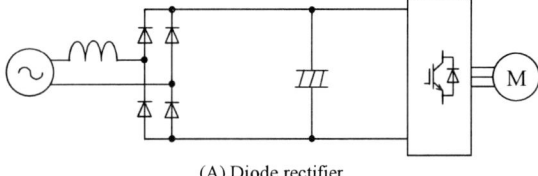

(A) Diode rectifier

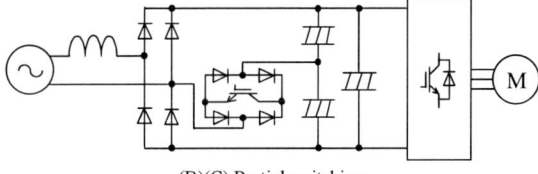

(B)(C) Partial switching

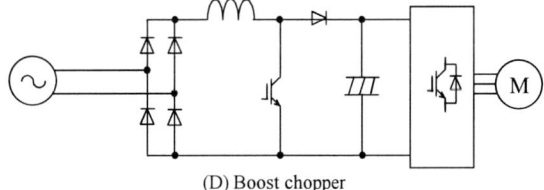

(D) Boost chopper

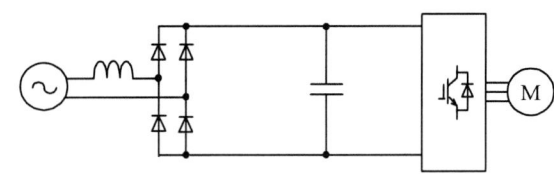

(E) Electrolytic-capacitor-less inverter
Fig. 2. Circuit configuration to improve harmonics

E. Electrolytic capacitor-less inverter

In this method, the reduction of harmonics is possible with inverter control modification without using an electrolytic capacitor, a large inductor, and a PFC circuit. This will reduce circuit components significantly and will result in a significant cost reduction. This method does, however, require complicated inverter control.

The 2018 International Power Electronics Conference

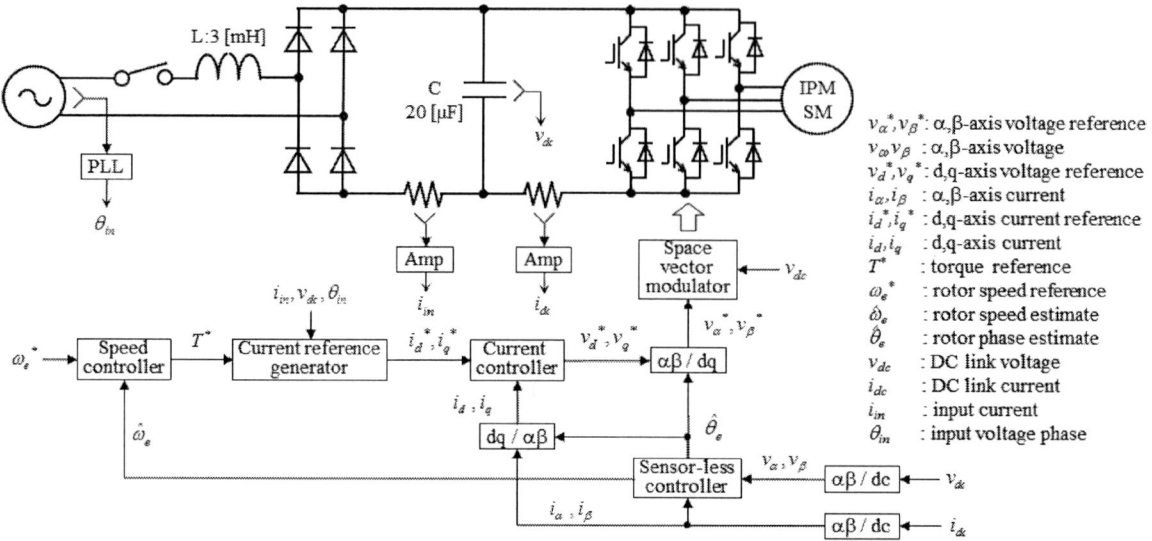

Fig. 3. System configuration of electrolytic capacitor-less inverter

v_α^*, v_β^* : α,β-axis voltage reference
v_α, v_β : α,β-axis voltage
v_d^*, v_q^* : d,q-axis voltage reference
i_α, i_β : α,β-axis current
i_d^*, i_q^* : d,q-axis current reference
i_d, i_q : d,q-axis current
T^* : torque reference
ω_e^* : rotor speed reference
$\hat{\omega}_e$: rotor speed estimate
$\hat{\theta}_e$: rotor phase estimate
v_{dc} : DC link voltage
i_{dc} : DC link current
i_{in} : input current
θ_{in} : input voltage phase

III. CONTROL METHOD OF ELECTROLYTIC-CAPACITOR-LESS INVERTER

The system configuration of an electrolytic capacitor-less inverter is shown in Fig. 3. DC link capacitor C is an extremely low-capacity (20 μF) film capacitor and has only a function of smoothing the switching ripples of the inverter.

Therefore, it is possible to control the DC link voltage with the interior permanent magnet synchronous motor (IPMSM) under field-weakening control. The DC link voltage, input voltage and input current of the converter are shown in Fig. 4 [1]. The DC link voltage is lowered greatly in step with the pulse of the input voltage as shown. The input current can then flow to a point where the instantaneous value of the input voltage is low, thereby improving the power factor of the converter.

The relationship between the power factor and ripple rate (v_M/v_m) of DC link voltage v_{dc}, as shown in Fig. 5, if the waveform shown in Fig. 4 occurs. This indicates that a power factor of 97% or more can be achieved by reducing the DC link voltage to less than one half of the input voltage under a field-weakening control.

In this circuit, the output energy nearly equals input energy, because there are no electrolytic capacitors or other elements that store input energy. Due to this characteristic, it is possible to reduce the power harmonics by instantaneously controlling the motor torque so that it pulsates at twice the power supply frequency.

However, the LC resonance and the spatial harmonics of the IPMSM causes control disturbances in the electrolytic capacitor-less method, thus increasing power harmonics. Therefore, it is necessary to control these disturbances.

First, this section describes the suppression control of LC resonance. LC resonance occurs because the method uses a low-capacity film capacitor and emits harmonics

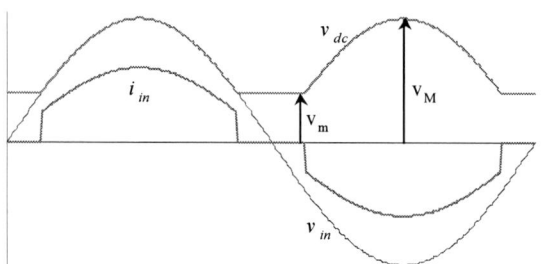

Fig. 4. Voltage and current waveform

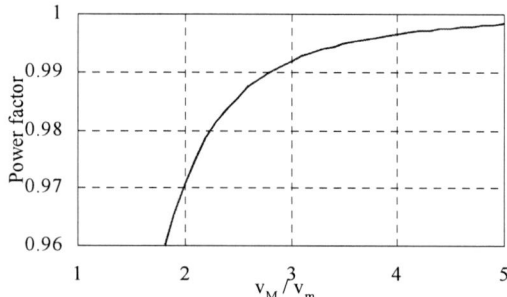

Fig. 5. Relation of power factor and ripple rate (v_M/v_m)

in the vicinity of the resonance frequency. LC resonance is generally suppressed under feedback control, such as proportional-integral (PI) control. However, this control circuit is composed of a general-purpose 32-bit RISC microcontroller with a carrier frequency of 5.9 kHz. Accordingly, PI control causes a phase delay, and as shown in Fig. 6, LC resonance cannot be sufficiently suppressed. LC resonance is generated every half cycle of the power supply. Therefore, it can be suppressed by repetitive controls with the feedback of the input current (as shown in Figure 7) to solve the phase delay.

Next, this section describes the suppression control of the spatial harmonics of the motor. The back electromotive force (EMF) of IPMSM is shown in Fig. 8.

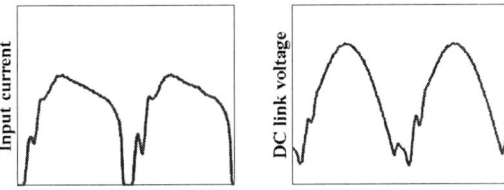

Fig. 6. The waveform of input current and DC link voltage
with LC resonance suppression by PI controller

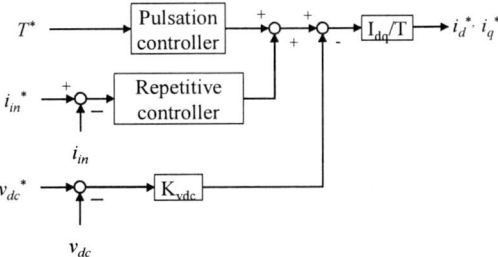

Fig. 7. Repetitive controller to suppress LC resonance
(Current reference generator)

The harmonic component of back EMF is shown in Fig. 9. This figure shows that the fifth harmonic wave is the highest. The fifth harmonic wave contained in the back EMF can be represented by the following equation in the dq coordinates.

$$
\begin{bmatrix} v_{d\_6th} \\ v_{q\_6th} \end{bmatrix} = [C] \begin{bmatrix} v_{u\_5th} \\ v_{v\_5th} \\ v_{w\_5th} \end{bmatrix} = \omega_e \psi_c [C] \begin{bmatrix} \sin(5\theta_e) \\ \sin\left(5\theta_e - \frac{2}{3}\pi\right) \\ \sin\left(5\theta_e + \frac{2}{3}\pi\right) \end{bmatrix}
$$

$$
= \sqrt{\frac{2}{3}}\, \omega_e \psi_c \begin{bmatrix} \sin(6\theta_e) \\ \cos(6\theta_e) \end{bmatrix} \tag{1}
$$

$$
[C] = \sqrt{\frac{2}{3}} \begin{bmatrix} \cos\theta_e & \cos\left(\theta_e - \frac{2}{3}\pi\right) & \cos\left(\theta_e + \frac{2}{3}\pi\right) \\ \sin\theta_e & \sin\left(\theta_e - \frac{2}{3}\pi\right) & \sin\left(\theta_e + \frac{2}{3}\pi\right) \end{bmatrix}
$$

where, ω_e is motor electric angular velocity [rad/s], θ_e is electric angle [rad], ψ_c is proportional factor of back EMF.

Therefore, if i_d and i_q are constant, the motor input power has a six-fold harmonic, which flows into the high-capacitance electrolytic capacitor of the DC link in the case of the conventional partial switching method or boost chopper method. Accordingly, this component does not appear as a power harmonic. However, in the case of the electrolytic capacitor-less method, the component leaks as a power harmonic because the electrolytic capacitor-less method uses a film capacitor.

In order to suppress power harmonics caused by spatial harmonics, v_d and v_q need to pulsate at six times the electric angular velocity of the motor. The maximum rotational speed of the motor in the operating area of an air conditioner exceeds 6000 min^{-1}. Therefore, for example, if the number of poles of the motor is six, v_d and v_q need to be controlled at 1.8 kHz or higher

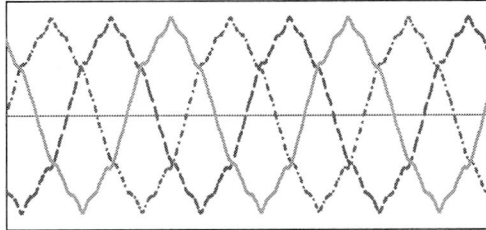

Fig. 8. Back EMF waveform of IPMSM

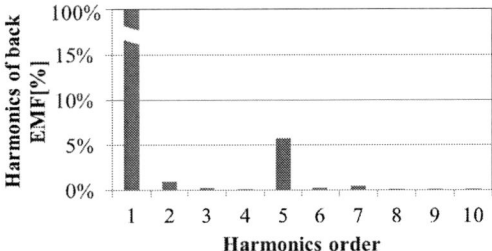

Fig. 9. Harmonics of back EMF of tested motor

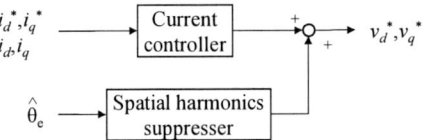

Fig. 10. Control block to suppress spatial harmonics

frequencies, and the suppression control of spatial harmonics requires responsiveness to high-frequency components. However, current control using the feedback has low responsiveness. Therefore, spatial harmonics need to be suppressed by directly correcting v_d^* and v_q^* as shown in Fig. 10.

IV. EXPERIMENTAL RESULTS

The evaluation results of the power harmonics, efficiency, and compressor vibration with the use of electrolytic capacitor-less inverters are important for the purposes of their commercialization. This section shows the evaluation results of these parameters.

First, this section describes the measurement results of power harmonics. Table II shows the equipment constants of the motor used for the experiment. The experiment was conducted at the maximum load point in the driving range with an input voltage of 220 V at 50 Hz, an input current of 7.1 A, an input power of 1,540 W, and a motor speed of 6540 min^{-1}. The waveform of the output current (motor phase currents), input current, and DC link voltage under the above operating conditions are shown in Fig. 11. The input current and the DC link voltage are controlled almost to be sinusoidal, from which almost no LC resonance occurs. The output currents pulsate at twice the power supply frequency because the motor torque pulsates. The measurement results of power harmonics are shown in Fig. 12. Yokogawa Electric's WT3000, which conforms to IEC 61000-4-7, was used for the

869

TABLE II
SPECIFICATION OF TESTED MOTOR

Item	Value
d-axis inductance	5 mH
q-axis inductance	8 mH
Magnet flux linkage	0.093 Wb
Number of poles	6

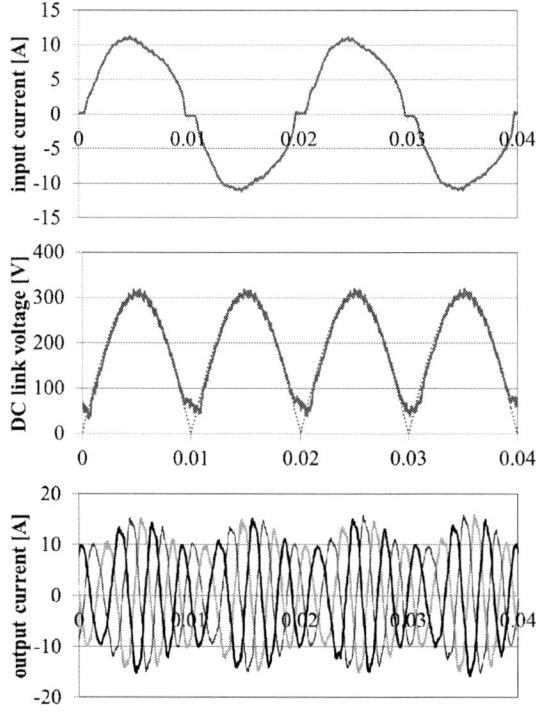

Fig. 11. Experimental result of waveform in maximum-load condition

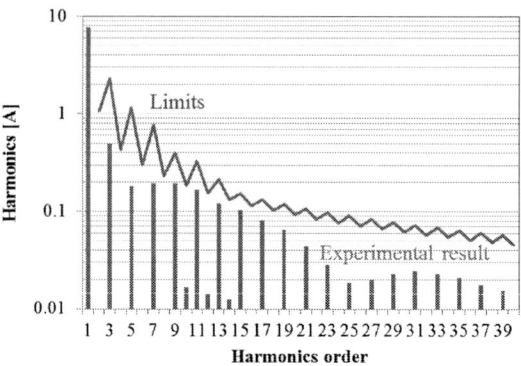

Fig. 12. Experimental result of harmonics in maximum-load condition

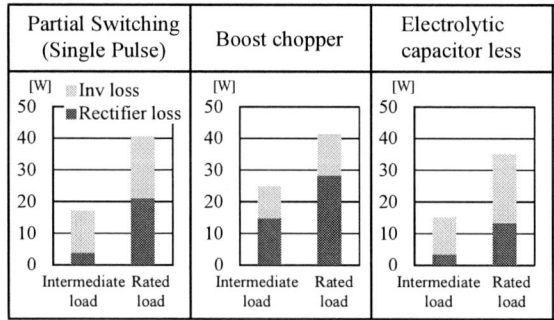

Fig. 13. Experimental result of loss of each converter

measurement of power harmonics. As shown in Fig. 12, each power harmonic order is less than 70% of the value regulated by harmonic current emissions class A, conforming to IEC 61000-3-2.

With regard to the efficiency of the inverter, Fig. 13 shows the loss of the main power converter. The partial switching method and the boost chopper method have switching elements and inductors in the rectifier, thus increasing the loss of the rectifier compared to the electrolytic capacitor-less method. On the other hand, the output current of the electrolytic capacitor-less method is larger than that of other methods, which causes increased inverter loss. As a result, the total of the rectifier loss and inverter loss of the electrolytic capacitor-less method is the smallest with both intermediate load and rated load.

Lastly, with regard to the vibration of the inverter, the load torque is in the shape shown in Fig. 14 when driving a single-cylinder compressor. Accordingly, for example, if the motor torque is constant, the deviation between the motor torque and the compressor load torque becomes excitation torque, thus vibrating the compressor. This vibration increases as the motor rotates at a lower speed. Furthermore, an increase in the vibration of the compressor causes a piping stress problem.

In order to suppress the vibration of the compressor, it is necessary to reduce the excitation torque by making the motor torque pulsate in synchronization with the basic component of the compressor load torque as shown in Fig. 14. Therefore, when the compressor is in low-speed operation, in particular, it is necessary to make the motor torque pulsate not only at double the frequency of the power supply frequency but also at the basic frequency of the compressor load torque in the case of the electrolytic capacitor-less method.

The input current, the DC link voltage and the motor current with the intermediate load are shown in Fig. 15. The rotational speed of the motor is 1560 min⁻¹, and the motor torque is controlled to suppress the vibration of the compressor. The levels of vibration with and without vibration suppression control are shown in Fig. 16. The vibration in the tangential direction of the compressor is measured by an acceleration pickup. By applying vibration suppression control as shown in the figure, the vibration of the compressor is reduced by 83%. The figure shows that the electrolytic capacitor-less method can suppress vibration sufficiently.

In the electrolytic capacitor-less method, the motor torque pulsated at a double frequency of the frequency of the power supply. It was confirmed that this did not cause a big problem in air conditioners operating with compressor loads.

The 2018 International Power Electronics Conference

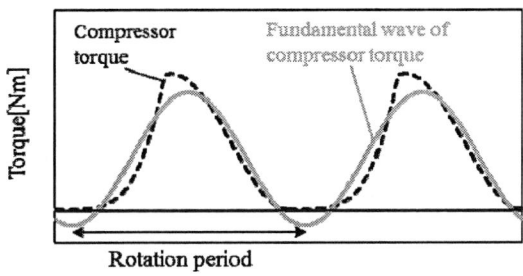

Fig. 14. The compressor torque and fundamental wave of the torque

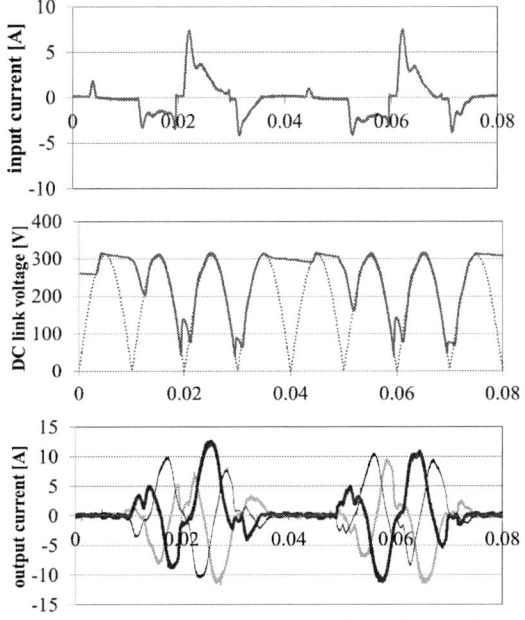

Fig. 15. Experimental result of waveform in half-load condition

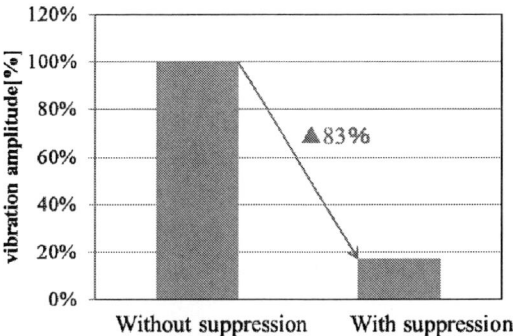

Fig. 16. Vibration amplitude of compressor

(a) Conventional (partial switching)

(b) Developed (electrolytic-capacitor-less)
Fig. 17. Printed circuit board

V. CONCLUSION

The electrolytic capacitor-less inverter applied to the room air conditioner was confirmed to satisfy the harmonic regulations of IEC 61000-3-2. A printed circuit board in the conventional partial switching method is shown in Fig. 17, along with a printed circuit board in the electrolytic capacitor-less method developed this time. Electrolytic capacitors are used in the printed circuit board developed this time, which are, however, used to absorb inductance energy when the inverter is stopped, and they do not affect the driving of the motor. The DC link capacitor used is a film capacitor in place of a conventional electrolytic capacitor, with the capacitance reduced to 1/50. The DC link in the partial switching method uses a film capacitor in place of an electrolytic capacitor. The inductance and weight of the inductor are reduced to 1/6 and 1/3, respectively. By applying the electrolytic capacitor-less method, it will be possible to deploy low-cost inverter air conditioners in the global market, and which will encourage the use of inverter air conditioners to spread and will greatly contribute to the reduction of annual power consumption.

REFERENCES

[1] I. Takahashi, "Improved Power Factor Rectifier Circuit for Inverter Controlled PM Motor," National Convention Record IEE Japan, pp.1591(2000-3) (in Japanese)

[2] I. Takahashi, H. Haga, "Power Factor Improvement of Single-Phase Diode Rectifier Circuit By Field Weakening of Inverter Driven IPM Motor," IEEJ Trans. on Industry Applications, vol.123-D, no.12, pp.1467-1473, 2003 (in Japanese)

[3] H. Haga, I. Takahashi, K. Ohishi, "IPM Motor Drive Method Using a New Inverter Having the Operation of High Power Factor Single-phase Diode Rectifier without Electrolytic Capacitor," IEEJ Trans. on Industry Application, vol.124-D, no.5, pp479-485, 2004 (in Japanese)

[4] T. Yokoyama, K. Ohishi, H. Haga, J. Shibata, "Control Method for achieving High Power Factor in Single-Phase to Three-Phase Converters without Electrolytic Capacitors," IEEJ Trans. on Industry Application, vol.129-D, no.5, pp.490-497, 2009 (in Japanese)

[5] A. Yoo, SK. Sul, H. Kim, KS. Kim: "Flux-weakening strategy of an induction machine driven by an electrolytic-capacitor-less inverter," IEEE Trans. on Industry Application, vol.47, no.3, pp.1328-1336, 2011

[6] K. Abe, H. Haga, K. Ohishi, Y. Yokokura, "Fine Current Harmonics Reduction Method for Electrolytic Capacitor-Less and Inductor-Less Inverter Based on Motor Torque Control and Fast Voltage Feedforward Control for IPMSM," IEEE Trans. on Industrial Electronics, vol.64, no.2, pp1071-1080, 2017

The 2018 International Power Electronics Conference

Latest Development of SiC Power Module-based Single-Stage AC-AC Resonant Converter for High-Frequency Induction Heating Applications

Tomokazu Mishima*

Dept. of Marine Engineering, Graduate School of Maritime Sciences, Kobe University

5-1-1, Higashinada, Kobe, Hyogo 658–0022, Japan

E-mail: mishima@maritime.kobe-u.ac.jp

Abstract—A new prototype of zero-voltage soft-switching (ZVS) single-stage ac-ac converter for high-frequency (HF) induction heating (IH) applications is presented in this paper. By adopting a Silicon Carbide (SiC)-MOSFET and SBD hybrid power module, named as "all SiC power module", the high-efficiency power conversion can be attained with ZVS under the condition of 100 kHz HF switching. The essential performances are demonstrated in experiment using a 2 kW-100 kHz prototype with a Si-IGBT module, ant the validity of wide-band-gap (WBG) - based power converter is discussed from the practical point of view in the field of industrial IH system.

Keywords—Inductive Heating (IH), Silicon Carbide Metal-Oxide-Field-Effect-Transistor (SiC-MOSFET), single-stage ac-ac converter, zero voltage soft-switching (ZVS).

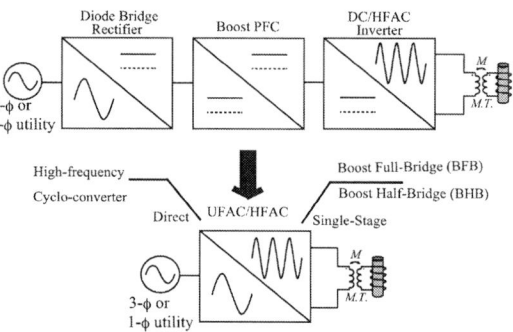

Fig. 1. Conceptual diagram of ac-ac converter and related circuit topologies.

I. INTRODUCTION

The resonant power converter technologies have been drawing much attention in the industrial induction heating system such as metal hardening. The emerging technology of wide-band-gap (WBG) power device, especially SiC MOSFET makes the higher switching and large-current conduction available in the power converters of IH applications. The advanced power converter topologies such as direct or single-stage ac-ac converters are useful and suitable for HF-IH power converters[1]-[4] as illustrated in Fig. 1, and higher power density switch-mode power converters are expected to the practical applications owing to the WBG power devices. The author of this paper has developed two types of single-stage power converters for IH applications; boost-half-bridge (BHB)[4] and boost full-bridge (BFB)[5]. The novel prototype of single-stage ac-ac converter with SiC power modules are originally developed in this paper for improving the power density of BFB topology, and its feasibility is investigated by experiment of 100 kHz prototype.

II. CIRCUIT TOPOLOGY AND OPERATION

The circuit topology of the proposed ac-ac converter is presented in Fig.2. The main power consists of the boost converter (totem-pole bridgeless boost rectifier) and full-bridge resonant inverter-integrated circuit topology "BFB" with the active switches Q_1–Q_4, while maintaining the bridgeless topology by Q_1–Q_2 and D_5–D_6. The inductor L_b contributes for boosting the utility voltage v_{in}

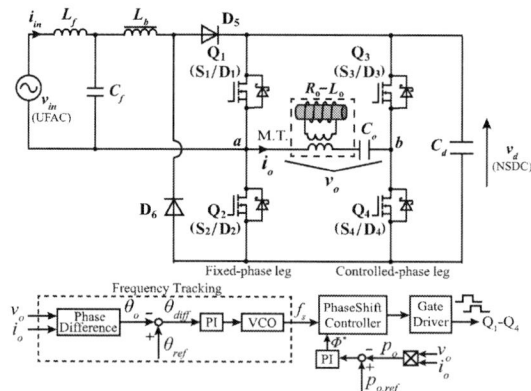

Fig. 2. A single-stage ac-ac converter based on boost full-bridge (BFB) circuit.

under the high frequency switching condition of Q_1–Q_4. The lossless snubber capacitors can be replaced by the parasitic capacitances of SiC-MOSFET, while are additionally connected in parallel with Q_1–Q_4 in case of Si-IGBT. Thereby, ZVS operation can achieve with the equivalent inductance of the IH load both at their turn-on and turn-off transitions. The switching frequency component is eliminated from the utility current i_{in} through the low-pass filter L_f-C_f.

The power controller is comprised of the resonant frequency tracking by Phase-Locked-Loop (PLL) and

872

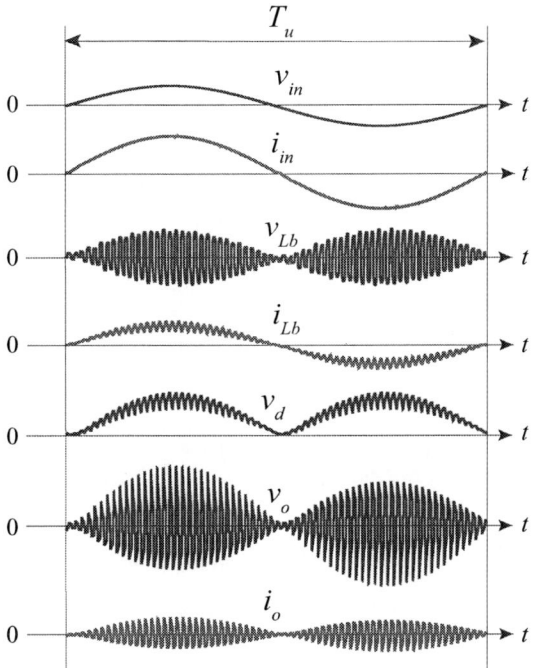

Fig. 3. Key voltage and current waveforms during the utility frequency ac cycle[5].

the phase-shift pulse-width-modulation (PS-PWM). The phase reference θ_{ref} is determined on the basis of ZVS range of the lagging phase switches Q_3 and Q_4.

Key waveforms of the proposed ac-ac converter are shown in Fig.3 for the utility frequency ac (UFAC). The twice UFAC component appears in the NSDC-link capacitor C_d for the sake of single stage power conversion. However, the related power pulsation does not have an impact on the performance of converter; no significant magnetic force appears as mechanical vibration due to the heavy weight of metal heated load. No electrolytic capacitor is necessary in the converter, which leads to a long lifetime and easy-to-maintenance implementation.

The relevant voltage and current waveforms during the HFAC cycle are illustrated in Fig. 5. The corresponding sub-mode transitions and equivalent circuits are depicted in Fig. 4. The outstanding feature of the converter mode transitions is the gate-signal pattern of Q_1–Q_4 is identical between the positive and negative cycles of v_{in}. Accordingly, the logic architecture of switch-gate pulse generator is very simple and cost-effective since a phase-shift PWM IC can be applied for the controller. The detailed descriptions of operation principle is presented in [5].

III. DESIGN CONSIDERATION OF CIRCUIT PARAMETERS

A. Load Resonant and Non-Smoothed DC-link Capacitors

The NSDC-link voltage v_d changes in accordance with v_{in} under the principle of single-stage frequency conversion and has its double frequency component. Since the ON-duty cycle of Q_1-Q_4 is fixed to 50%, the average value v_d for one HFAC cycle is two times as high as that of v_{in}. Fig. 6 depicts the relationship between the NSDC-link capacitor c_d, series resonant capacitor c_o and the peak voltage $v_{d,peak}$ at the NSDC link. It can be observed $v_{d.peak}$ depends on C_d rather than C_o. By giving the voltage ripple ratio $\gamma = \Delta v_{d,HF}/v_d$ so that the voltage stress is well under the specified voltage rating of power module, the NSDC-link capacitor can be determined.

The load quality factor Q of the proposed converter can be defined as

$$ Q = \frac{\omega_{r1} L_o}{R_o}, \quad \omega_{r1} = 2\pi f_{r1}. \tag{1}$$

The voltage v_{co} across C_o can be expressed by Q as

$$ v_{co} = Q v_d = Q\left(2|v_{in}| + \frac{\Delta v_{d,HF}}{2}\right). \tag{2}$$

The load quality factor Q can also be expressed by the voltage ratio as $Q = v_{co}/v_d$, as a result it should be determined in accordance with γ.

The proposed converter has two resonant frequencies f_{r1}, f_{r2} in accordance with the resonant load network as illustrated in Fig. 7:

$$ f_{r1} = \frac{1}{2\pi\sqrt{L_o C_r}}, \; f_{r2} = \frac{1}{2\pi\sqrt{L_o C_o}} \tag{3}$$

$$ C_r = \frac{C_d C_o}{C_d + C_o} = \frac{C_o}{1 + C_o/C_d}. \tag{4}$$

In order to keep a low distorted current at the output load, f_{r1} and f_{r2} should be close each other and less than switching frequency f_s for achieving ZVS.

B. Lossless Snubber Capacitors

The lossless snubber capacitors are related with the ZVS range in Q_1–Q_4. The capacitances can be expressed with the dead time interval T_d as

$$ C_x \le \frac{i_{Q,x}}{v_d} \cdot T_d \quad x \in 1, \ldots 4 \tag{5}$$

The current $i_{Q,x}$ in (5) corresponds with the load current i_o at their turn-off transitions, and the fixed phase switches Q_1, Q_2 are subject to the condition by

$$ \frac{1}{2}L_o i_o{}^2 > \frac{1}{2}(C_1 + C_2)v_d{}^2. \tag{6}$$

In the similar way, the controlled phase switches Q_3, Q_4 are subject to the condition by

$$ \frac{1}{2}L_o i_o{}^2 > \frac{1}{2}(C_{s3} + C_4)v_d{}^2. \tag{7}$$

The 2018 International Power Electronics Conference

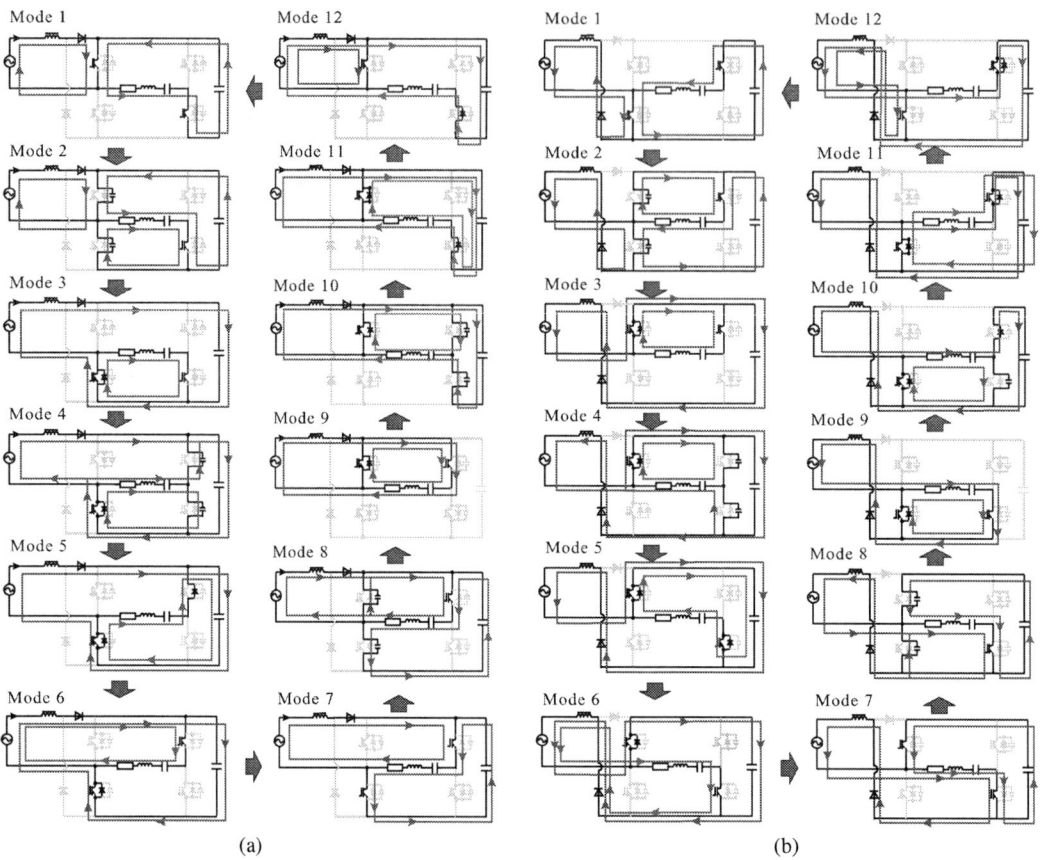

Fig. 4. Sub-mode transitions and equivalent circuits: (a) positive half cycle, and (b) negative half cycle.

Hence, the turn-off current of each switch are theoretically expressed as follows:

$$i_{Q1,off} > \frac{v_{Q2,off}}{Z_{r,fixed}}, \quad i_{Q2,off} > \frac{v_{Q1,off}}{Z_{r,fixed}} \quad (8)$$

$$i_{Q3,off} > \frac{v_{Q4,off}}{Z_{r,cont.}}, \quad i_{Q4off} > \frac{v_{Q3,off}}{Z_{r,cont.}} \quad (9)$$

where $Z_{r,fixed}$ and $Z_{r,cont.}$ represent the resonant characteristic impedances of the fixed and controlled phase legs of BFB as

$$Z_{r,fixed} = \sqrt{\frac{L_o}{2C_2}}, \quad C_2 = C_1 \quad (10)$$

$$Z_{r,cont.} = \sqrt{\frac{L_o}{2C_4}}, \quad C_4 = C_3. \quad (11)$$

The turn-off currents $i_{Q3,off}$ and $i_{Q4,off}$ are smaller than $i_{Q1,off}$ and $i_{Q2,off}$ while the voltages $v_{Q1,off}$– $v_{Q4,off}$ are almost equivalent, thereby $Z_{r,cont.}$ should be set smaller than $Z_{r,fixed}$.

IV. FREQUENCY TRACKING AND PHASE-SHIFT PWM

The power factor of IH load is expressed by

$$\cos\theta_o = \frac{V_{ou}I_{ou}\cos\theta_u + \sum_{h=1,3,5,...}^{\infty} V_{oh}I_{oh}\cos\theta_h}{V_{o,rms}I_{o,rms}} \quad (12)$$

$$V_{o,rms} = \sqrt{V_{o,h1}^2 + \sum_{3,5,...}^{\infty} V_{o,hn}^2} \quad (13)$$

$$I_{o,rms} = \sqrt{I_{o,h1}^2 + \sum_{3,5,...}^{\infty} I_{o,hn}^2} \quad (14)$$

The fundamental components of the HF voltage and current mainly contributes for heating energy. In order to maximize the fundamental components, the phase angle between v_{o1} and i_{o1} should be tracked in accordance with the impedance variation of the heat objective. Fig. 8 depicts the phasor diagram of the voltage and current at the output stage of BFB. The phase-angle reference θ_{ref} is determined with consideration for ZVS range. The resonant frequency of the heated load is tracked by the front-end part of the logic circuit in Fig. 2 under the principle of PLL.

Fig. 9 depicts the relationship between the switching frequency and the phase-angle reference by the assump-

874

The 2018 International Power Electronics Conference

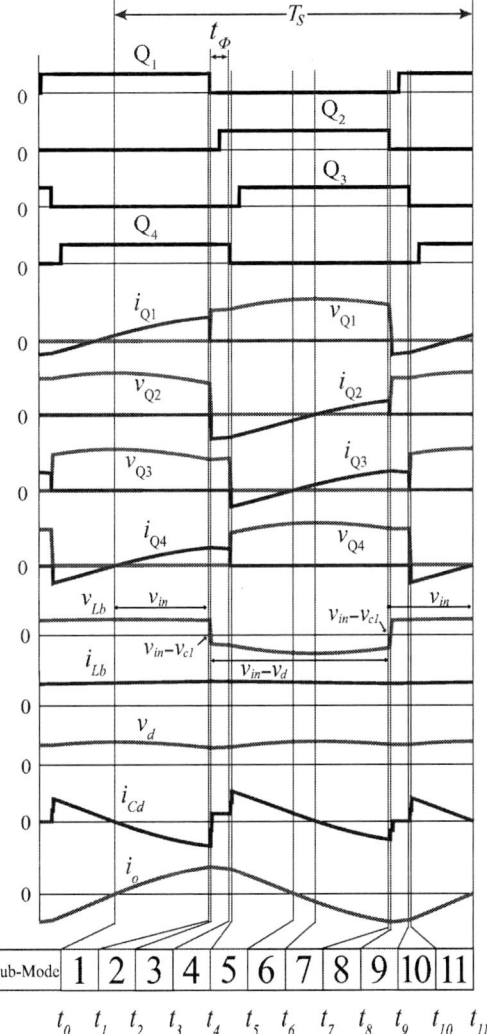

Fig. 5. Key voltage and current waveforms during the high frequency ac cycle[5].

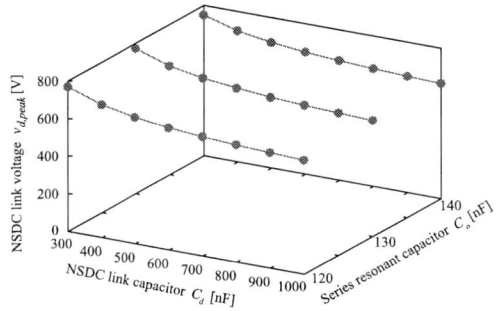

Fig. 6. NSDC-link voltage with variations of C_d and C_o.

tion; the resonant frequency is 78.4 kHz. It can be known the phase angle reference has an impact on the switching frequency and the load power factor as well as ZVS range. A computer-simulated result of the proposed frequency

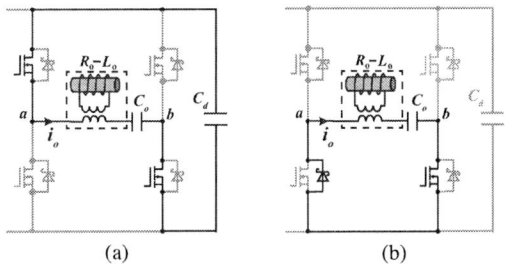

Fig. 7. Series load resonant networks: (a) power delivering state from NS-dc link, and (b) free-wheeling state.

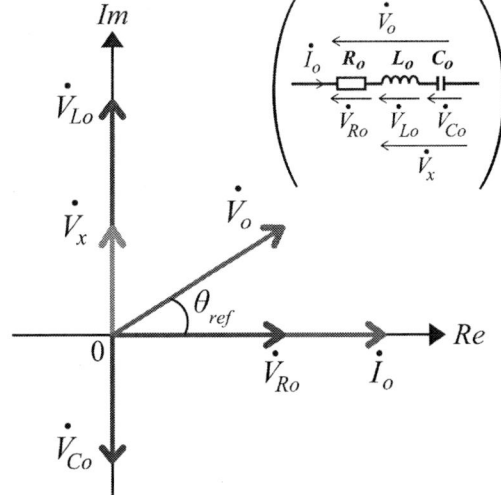

Fig. 8. Phasor diagram of output voltage and current in BFB.

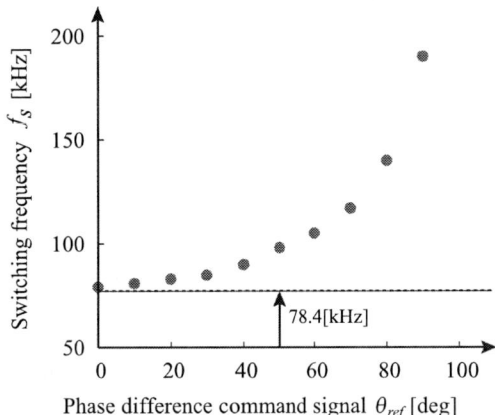

Fig. 9. Switching frequency versus phase angle characteristics in open loop frequency control.

controller is shown in Fig. 10, whereby the validity of controller logic is demonstrated.

V. EXPERIMENTAL VERIFICATION

A 2 kW-100 kHz prototype of the SiC power module-based proposed converter is developed and evaluated

The 2018 International Power Electronics Conference

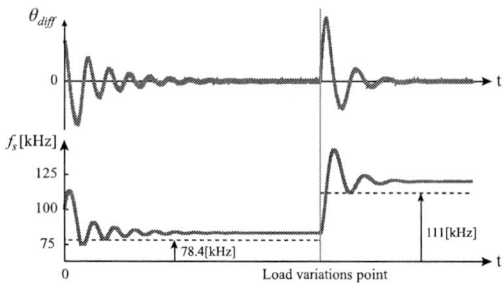

Fig. 10. Simulation results of frequency tracking closed loop control.

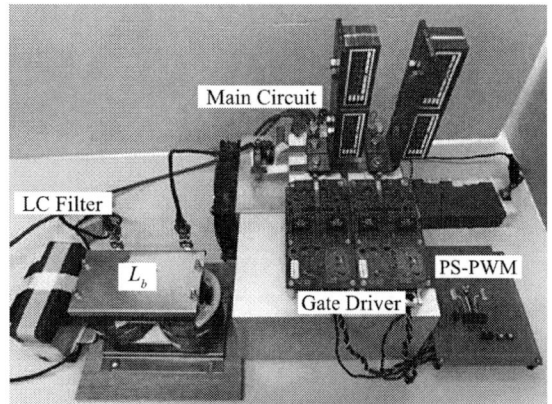

Fig. 11. Exterior appearance of the prototype converter.

TABLE I. EXPERIMENTAL CIRCUIT PARAMETERS

Item	Symbol	Value [unit]
Utility voltage	v_{in}	100 V
Output power rating	P_o	2 kW
Input UFAC frequency	f_u	60 Hz
Switching frequency	f_s	100 kHz
Input filter inductor	L_f	100 μH
Input filter capacitor	C_{in}	1 μF
Boost AC inductor	L_b	300 μH
Resonant capacitor	C_o	100 nF
Non-smoothed DC-link capacitor	C_d	2 μF
Equivalent resistance of IH load	R_o	6 Ω
Equivalent inductance of IH load	L_o	32 μH
M.T. winding turns ratio	w_1/w_2	5 / 5

by experiment. The exterior appearance of the main converter and the analog circuit controller is shown in Fig. 11. Two sets of a half-bridge all-SiC module (CAS120M12BM2, 1200 V, 193 A, 13 mΩ) with the mount-type gate driver (PT62SCMD12R02) are applied for the BFB circuit. The circuit parameters and specifications are displayed in TABLE I. The switching pattern signals are generated by a phase shift PWM controller (UCC3895).

The observed waveforms of the prototype are depicted in Figs. 12 and 13 for the 100 kHz HF switching condition. The ZVS commutation of all the active switches can be confirmed in the wide range of load power, thereby

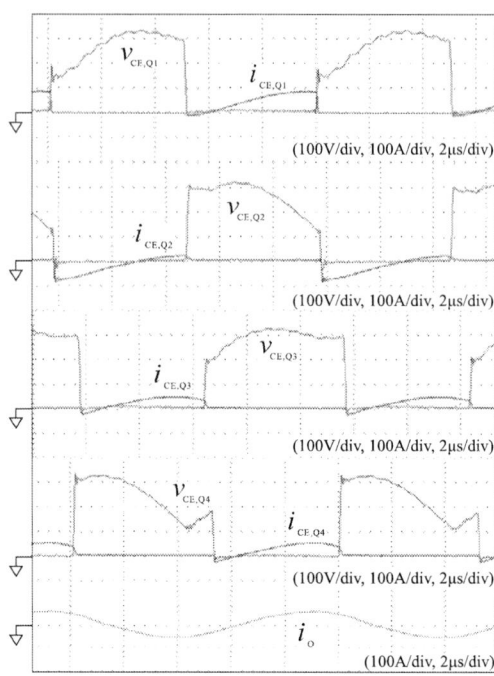

Fig. 12. Switching waveforms at the phase-shift command $\phi^* = 30°$ and $P_o = 2$ kW.

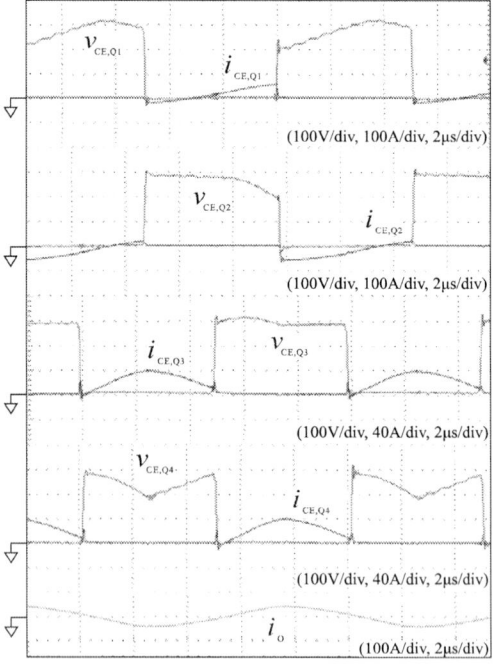

Fig. 13. Switching waveforms at the phase-shift angle waves at $\phi^* = 90°$ and $P_o = 1.2$ kW.

the validity of circuit parameter design is proven. The commutations of the controlled-leg switches Q_3 and Q_4 are susceptible to the capacitive (leading phase) current for the light load condition, then the ZVS operations

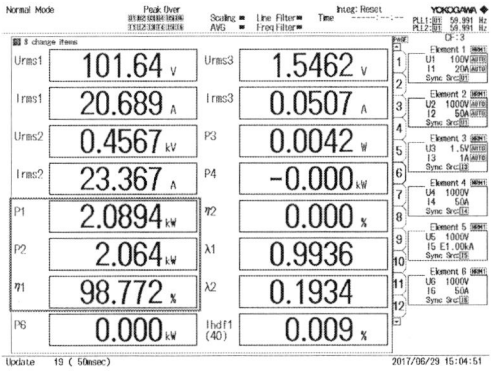

Fig. 14. Measured data of the SiC-MOSFET prototype at $P_o = 2\,\text{kW}$.

become critical as appears in Fig. 13. However, no significant surge and ringing phenomena emerge owing to the recovery free of SBD in the all SiC-Module.

The single-stage ac-ac conversion (frequency conversion) and ZVS operations are confirmed from the experimental result. The maximum efficiency 98.7 % is recorded at $P_o = 2\,\text{kW}$ for the SiC -based UFAC-HFAC conversion stage, as indicated in Fig. 14. The efficiency of the Si-IGBT prototype is measured by 91.3 % at $P_o = 2\,\text{kW}$ under the same operating condition. Thus, the efficiency of the single-stage ac-ac power conversion can improve drastically with application of SiC-MOSFET module.

VI. CONCLUSION

A new challenge of the SiC power module-based single-stage ac-ac converter for high-frequency induction heating applications has been presented in this paper. The design guideline with consideration of zero voltage soft switching and voltage stress of resonant tank has been described for assembly of prototype. The validity of analog circuit -based frequency tracking controller has been evaluated in computer simulation. The performance of the proposed converter has been evaluated by experiment with a SiC-MOSFET module-applied breadboard prototype, and the high efficiency of power conversion attains; the maximum efficiency is 98.8 % at 2 kW high frequency output and more than 7 % of efficiency improvement achieves.

ACKNOWLEDGEMENT

The authors acknowledges Fuji Electronics Industry, co., Ltd. for the technical advices and cooperations in experimental works.

REFERENCES

[1] N.A. Ahmed, "High-frequency soft-switching ac conversion circuit with dual-mode PWM/PDM control strategy for high-power IH applications," *IEEE Trans. Ind. Electron.*, vol.23, no.34, pp. 1440–1448, Apr. 2011.

[2] H. Sarnago, O. Lucía, A. Mediano, and J.M. Burdío, "Direct ac-ac resonant boost converter for efficient domestic induction heating applications," *IEEE Trans. Power. Electron.*, vol.29, no.3, pp.1128–1139, Mar. 2014.

[3] H. Sarnago, O. Lucía, M. Perez-Tarragona, and J.M. Burdío, "Dual-output boost resonant full-bridge topology and its modulation strategies for high-performance induction heating applications," *IEEE Trans. Ind. Electron.*, vol.63, no.6, pp.3554–3561, Jun. 2016.

[4] T. Mishima, Y. Nakagawa, and M. Nakaoka, "A bridgeless BHB ZVS-PWM ac-ac converter for high-frequency induction heating applications," *IEEE Trans. Ind. Appli.*, vol.51, No.4 pp.3304–3315, Jul./Aug, 2015.

[5] T. Mishima, S. Sakamoto, and C. Ide, "ZVS phase-shift PWM-controlled single-stage boost full-bridge ac-ac converter for high-frequency induction heating applications," *IEEE Trans. Ind. Electrons.*, vol.64, no.3, pp.2054-2061, Mar. 2017.

An Optimized Control Strategy to Improve the Current Zero-Crossing Distortion in Bidirectional AC/DC Converter based on V2G Concept

Lei Jing, Xiaoqing Wang, Bodong Li, Maohang Qiu, Bo Liu and Min Chen*
Department of Applied Electronics, ZheJiang University, Hang Zhou, China
*E-mail: calim@zju.edu.cn

Abstract— With the development of smart grid and new energy technology, V2G (vehicle-to-grid) has become a new research hotspot, and bidirectional converter plays an important role in it. In this paper, a bidirectional AC/DC converter using SiC device is developed, and unipolar PWM modulation is adopted in consideration of efficiency. Compared with other modulation methods that are widely used, unipolar modulation has the advantages of low switching loss, small current ripple and small common mode noise, however the current is distorted at zero-crossing. With the investigation on the characteristics of the current zero-crossing distortion based on the AC/DC converter prototype, an optimized control strategy which generates the switching signal for the line frequency bridge based on the control loop output reference (CLOR) instead of the grid voltage is proposed. The performance of the analyses and the proposed strategy is verified by experiments.

Keywords— *bidirectional AC/DC converter; current zero-crossing distortion; V2G (vehicle-to-grid); unipolar modulation; control loop output reference (CLOR)*

I. INTRODUCTION

As the global greenhouse effect intensifies, new energy technologies are becoming more and more important, V2G has been a research hotpot based on the development of smart grid and electric technology. The main idea of V2G lies in the interaction between electric vehicles and power grid, using a large number of batteries in electric vehicles to store energy and reducing the fluctuation of power grid introduced by distributed resource. With orderly control signal to charge or discharge EVs, V2G has the ability of peak shaving and valley filling for power grid and can optimize the operation of power grid [1]-[9].

Bidirectional AC/DC converter is the key component used to provide V2G capability for EVs. It is supposed to effectively control the power flow between the grid and the battery, the control system is responsible to ensure the quality of power as well as implementing power flow control [10]-[16]. Compared with other modulation

methods (e.g., bipolar modulation, unipolar frequency modulation, etc.) that are widely used in AC/DC converter, the unipolar modulation has the advantages of low switching loss, small current ripple and small common mode noise, therefore the unipolar modulation is preferred for single-phase AC/DC converter in consideration of the system efficiency. However, because of the phase difference between grid voltage and current, the discontinuity of modulation wave near the voltage zero crossing and so on, the grid-side current is distorted [17]-[23]. The current zero-crossing distortion will decrease the quality of power conversion and lead to deterioration in operation reliability. [24] - [25] proposed a hybrid modulation method to solve the problem of zero-crossing distortion. However, the effect of this method is obvious only when the power factor is low. Moreover, the control method is complex, and the advantages of low loss when using unipolar modulation is compromised

In order to improve the performance of the bidirectional AC/DC converter, this paper analyses the cause of current zero-crossing distortion when using unipolar modulation, and proposes an optimized control strategy to solve this problem. The improved control method can well improve the current zero-crossing distortion by generating the switching signal for the line frequency bridge based on the control loop output reference (CLOR) instead of the grid voltage. Experiment results from a 3-kW single-phase bidirectional converter are presented to verify the analysis and discussion.

II. THE CONTROL SYSTEM OF BIDIRECTIONAL AC/DC Converter

The topology of the bidirectional AC/DC converter mentioned in this paper is shown in Fig. 1. Full bridge topology is easier to achieve high efficiency with uncomplicated control strategy, so it is adopted here. The bidirectional converter operates in totem-pole bridgeless PFC mode when it is supposed to charge, and operates in grid-connected inverter mode when it is supposed to discharge. And unipolar modulation is also adopted in consideration of efficiency, T_1 and T_3 are SiC MOSFET working at switching frequency while T_2 and T_4 are Si

---

This project is supported by the National Natural Science Foundation of China (51477153)

MOSFET working at line frequency.

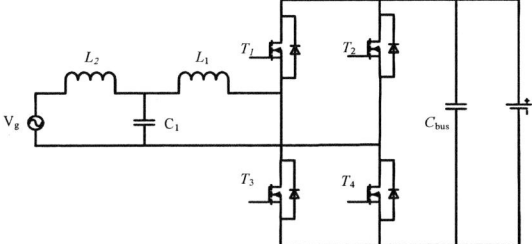

Fig. 1. The topology of bidirectional AC/DC converter.

The working modes of the converter are briefly described below, because the control strategy is unified and the operation is symmetrical when charging and discharging, only the positive half cycle of the PFC mode is presented. As shown in Fig. 2, when the converter works in the positive half cycle of the PFC mode, T_2 is always off while T_4 is always on, T_1 and T_3 works at high frequency PWM mode.

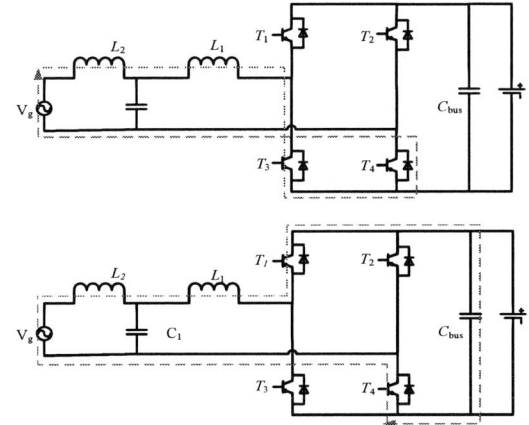

Fig. 2. Switching process when working in the positive half cycle of the PFC mode.

The control strategy of the bidirectional converter is shown in Fig. 3. The double-loop control structure is adopted in this paper, the voltage loop is used to stabilize the dc-bus voltage while the current loop is supposed to control the grid current.

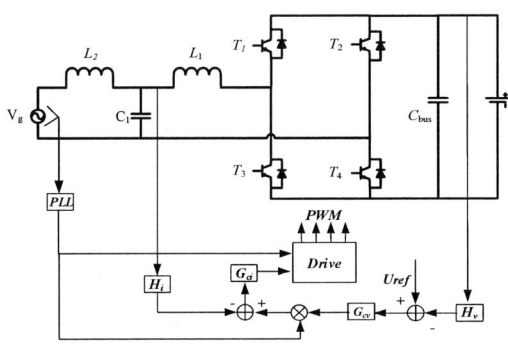

Fig. 3. Control system of the bidirectional AC/DC converter.

III. ANALYSIS OF THE CURRENT ZERO-CROSSING DISTORTION

According to the control system structure, we can get the control block shown in Fig. 4. Then the transfer function from grid voltage to current and the transfer function from reference current to grid current can be derived as equation (1) and equation (2).

$$G_{i_{L_1}\text{-}v_g} = \frac{1}{s^3 L_1 L_2 C + s L_1 + s L_2}\frac{1}{1+T} \tag{1}$$

$$G_{i_{L_1}\text{-}i_{ref}} = \frac{(s^2 L_2 C + 1)K_{PWM}G_{ci}}{s^3 L_1 L_2 C + s L_1 + s L_2}\frac{1}{1+T} \tag{2}$$

It can be seen from the transfer function that the grid current is not only depend on the reference current, but also influenced by the grid voltage. In order to eliminate the influence of the grid voltage and improve the quality of current waveform, grid voltage feedforward can be added to the control loop. However, the current waveform is not very good because the zero-crossing distortion still exists , the experiment waveform of the converter working in PFC mode is shown in Fig. 5 as an illustration.

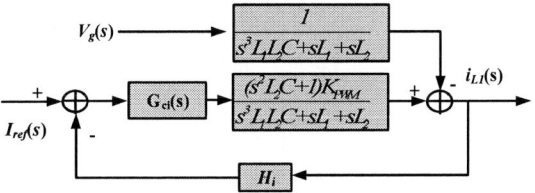

Fig. 4. The simplified control block of the bidirectional AC/DC converter.

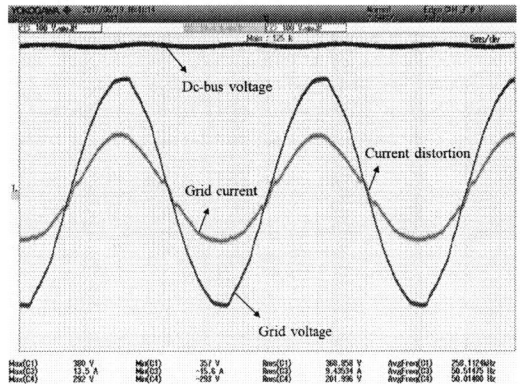

Fig. 5. Experiment waveform of the converter with the current distortion.

The cause of current zero-crossing distortion will be analyzed below. Same as the foregoing reason, only the analysis of the converter working in charge mode is presented.

When unipolar modulation is implemented in the

control strategy, the operation state of the converter can be analyzed as boost circuit in the positive half cycle and negative half cycle respectively, which can be seen in Fig. 2. When the grid voltage is positive and T_3 and T_4 are on, inductor stores energy; when T_3 is off and T_1 turns on, grids and inductors transfer energy to the DC side. If we set the duty ratio of T_1 as D_1, it is easy to know that D_1 will vary according to the sine curve during the positive half cycle, while the duty ratio of T_3 is complementary; symmetrically, the duty ratio of T_3 will vary according to the sine curve during the negative half cycle. For a clearer explanation, the variation trend of D_1 in a cycle is shown in Fig.6 as an example.

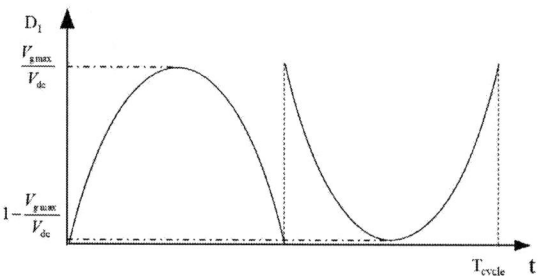

Fig. 6. The duty ratio variation trend of T3 during a power grid cycle.

As can be seen from the Fig. 6, the duty ratio is discontinuous between positive and negative cycles, which is inherent in unipolar modulation. If we cannot accurately control the synchronous switching of the discontinuous duty cycle and the line frequency switch (T_2 and T_4), there will be faults with waveform modulation. Actually, conventional control strategy detects the zero-crossing of the grid voltage to generate the switching signal for the line frequency bridge, and due to the calculation delay and modulation delay of digital control system, there will be phase diffidence between the control signal generated by the control loop and grid voltage. This leads to the inconsistency between the discontinuity point of the control signal and the switching point of T_2 and T_4. And the inconsistency will inevitably bring distortion of the modulation current at this point.

IV. PROPOSED OPTIMIZED CONTROL STRATEGY BASED ON CLOR

According to above analysis of the current zero-crossing distortion, this paper proposed an optimized control strategy to solve this problem. The new control strategy generates the switching signal for the line frequency bridge based on the control loop output reference (CLOR) instead of the grid voltage.

As analyzed in the previous section, the inconsistency between the discontinuity point of the control signal and the switching point of T_2 and T_4 will cause the current zero-crossing distortion. To synchronize the two parts, the new control strategy detects the control loop output, and use this as a reference signal to switch T_2 and T_4. In digital control systems, the CLOR, as shown in Fig. 7, is easy to obtain.

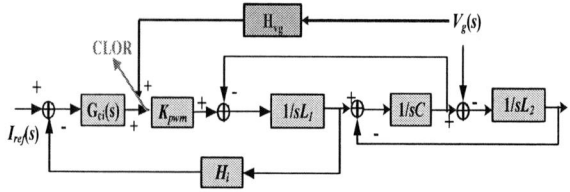

Fig. 7. The control loop block and the control loop output reference.

The proposed control strategy can get the value of CLOR during each switching cycle in digital control systems, this ensures the accuracy of detection and control. Then the converter will switch the MOSFETs which work at line frequency (T_2 and T_4) according to the CLOR. More specifically, we switch T_2 and T_4 at the discontinuity point of the CLOR jumping from 0 to 1(or 1 to 0). In this way, the circuit status and the control signal can be consistent. A flowchart of the optimized control strategy is shown in Fig. 8.

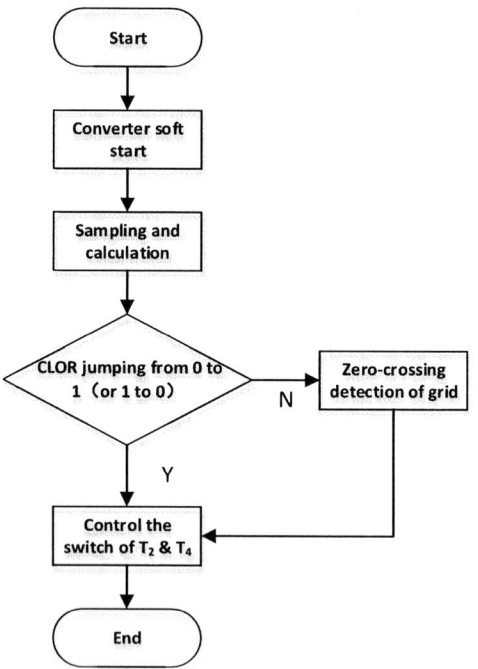

Fig. 8. The main flowchart of the optimized control strategy.

V. EXPERIMENT VERIFICATION

In order to examine the feasibility of the proposed control strategy and evaluate the performance of the converter, the optimized control strategy has been implemented using a 3kW experiment prototype. The single-phase bidirectional AC/DC converter specifications are presented in Table I .

880

TABLE I		
BIDIRECTIONAL AC/DC CONVERTER SPECIFICATIONS		
Symbol	Parameter	Value
P	Power	3 kW
V_{bus}	DC bus voltage	380 V
V_g	Grid voltage	220 V$_{rms}$
f_o	Line frequency	50 Hz
f_s	Switch frequency	50 kHz
L_{cov}	Converter side inductor	870 μH
C	Filter capacitor	2.2 μF
L_g	Grid side inductor	165 μH
C_{bus}	DC bus capacitor	2.8 mF

Fig. 9 shows the steady-state waveforms of the converter working in PFC mode, and Fig. 10 shows the steady-state waveforms of converter working in inverter mode. Table II shows the experiment data of PFC mode and Table III shows the experiment data of inverter mode. As can be seen from the figures, the current has almost no distortion near zero-crossing. And the overall quality of the current waveform is also good. Compared with the waveform before improvement, the THD is reduced by 2%.

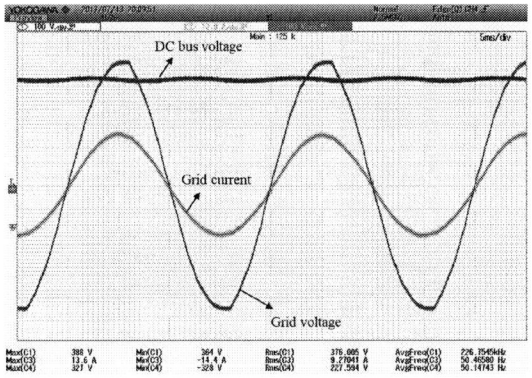

Fig. 9. The experiment waveform of the bidirectional AC/DC converter when operations in PFC mode.

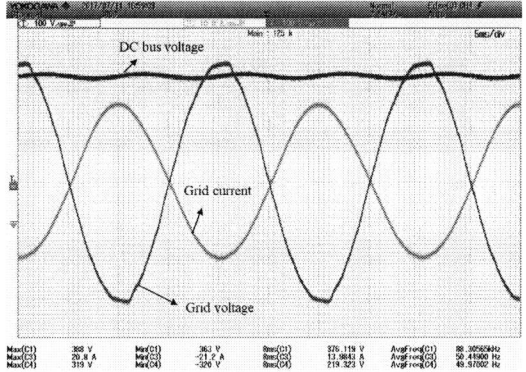

Fig. 10. The experiment waveform of the bidirectional AC/DC converter when operations in inverter mode.

TABLE II		
EXPERIMENT DATA IN PFC MODE		
Symbol	Parameter	Value
P_{in}	Input power	2992 W
P_{out}	Output power	2911 W
PF	Power factor	0.999
THD	Total harmonic distortion	3.03%
η	Efficiency	97.37%

TABLE III		
EXPERIMENT DATA IN INVERTER MODE		
Symbol	Parameter	Value
P_{in}	Input power	3039 W
P_{out}	Output power	2967 W
PF	Power factor	0.999
THD	Total harmonic distortion	3.24%
η	Efficiency	97.67%

VI. CONCLUSION

An optimized control strategy has been proposed in this paper, which is able to solve the problem of current zero-crossing distortion for the single-phase bidirectional AC/DC converter using unipolar modulation. The proposed control strategy generates the switching signal for the line frequency bridge based on the control loop output reference (CLOR) instead of conventional grid voltage reference, so the circuit status and the control signal can be consistent. Compared with other methods such as hybrid modulation, the proposed control strategy is easy to implement and has excellent performance. Experimental results confirm the superior performance of the proposed control strategy.

ACKNOWLEDGMENT

The author would like to thank TDK corporation for providing PCB board and materials for the research.

REFERENCES

[1] MH. Qiu, M. Chen, B. Liu and L. Jing, "A simplified control strategy to precisely control the reactive power through bi-directional switching in single phase bidirectional AC/DC converter for V2G techniques," *2017 IEEE Applied Power Electronics Conference and Exposition (APEC)*, Tampa, FL, 2017, pp. 1558-1562.

[2] S. Han, S. Han and K. Sezaki, "Development of an Optimal Vehicle-to-Grid Aggregator for Frequency Regulation," *IEEE Transactions on Smart Grid*, vol. 1, no. 1, pp. 65-72, 2010.

[3] E. Sortomme and M. A. El-Sharkawi, "Optimal Charging Strategies for Unidirectional Vehicle-to-Grid," *IEEE Transactions on Smart Grid*, vol. 2, no. 1, pp. 131-138, 2011.

[4] W. Su, H. Eichi, W. Zeng and M. Y. Chow, "A Survey on the Electrification of Transportation in a Smart Grid

Environment," *IEEE Transactions on Industrial Informatics*, vol. 8, no. 1, pp. 1-10, 2012.

[5] E. Sortomme and M. A. El-Sharkawi, "Optimal Scheduling of Vehicle-to-Grid Energy and Ancillary Services," *IEEE Transactions on Smart Grid*, vol. 3, no. 1, pp. 351-359, 2012.

[6] M. Yilmaz and P. T. Krein, "Review of the Impact of Vehicle-to-Grid Technologies on Distribution Systems and Utility Interfaces," *IEEE Transactions on Power Electronics*, vol. 28, no. 12, pp. 5673-5689, 2013.

[7] C. Pang, P. Dutta and M. Kezunovic, "BEVs/PHEVs as Dispersed Energy Storage for V2B Uses in the Smart Grid," *IEEE Transactions on Smart Grid*, vol. 3, no. 1, pp. 473-482, 2012.

[8] C. Liu, K. T. Chau, D. Wu and S. Gao, "Opportunities and Challenges of Vehicle-to-Home, Vehicle-to-Vehicle, and Vehicle-to-Grid Technologies," *Proceedings of the IEEE*, vol. 101, no. 11, pp. 2409-2427, 2013.

[9] J. R. Pillai and B. Bak-Jensen, "Integration of Vehicle-to-Grid in the Western Danish Power System," *IEEE Transactions on Sustainable Energy*, vol. 2, no. 1, pp. 12-19, 2011.

[10] M. Parvez Akter, S. Mekhilef, N. Mei Lin Tan and H. Akagi, "Modified Model Predictive Control of a Bidirectional AC–DC Converter Based on Lyapunov Function for Energy Storage Systems," *IEEE Transactions on Industrial Electronics*, vol. 63, no. 2, pp. 704-715, 2016.

[11] D. Dong, T. Thacker, R. Burgos, D. Boroyevich, F. Wang and B. Giewont, "Control design and experimental verification of a multi-function single-phase bidirectional PWM converter for renewable energy systems," *2009 13th European Conference on Power Electronics and Applications*, Barcelona, 2009, pp. 1-10.

[12] A. R. Vaz, J. B. Vieira, L. C. De Freitas, E. A. A. Coelho and V. J. Farias, "Bidirectional three-phase bridge converter driven to several functions," *2004 IEEE 35th Annual Power Electronics Specialists Conference (IEEE Cat. No.04CH37551)*, 2004, pp. 3932-3938 Vol.5.

[13] M. Yilmaz and P. T. Krein, "Review of Battery Charger Topologies, Charging Power Levels, and Infrastructure for Plug-In Electric and Hybrid Vehicles," *IEEE Transactions on Power Electronics*, vol. 28, no. 5, pp. 2151-2169, 2013.

[14] Y. Liu, H. Han, M. Su and Y. Sun, "A novel single-phase current-source-type bidirectional converter for V2G application," *Proceedings of the 33rd Chinese Control Conference*, Nanjing, 2014, pp. 7034-7039.

[15] P. Thomas, Sreekanth P K, Ganesh M and D. Nair, "Design, simulation and comparison of single phase bidirectional converters for V2G and G2V applications," *2015 Annual IEEE India Conference (INDICON)*, New Delhi, 2015, pp. 1-6.

[16] M. Yilmaz and P. T. Krein, "Review of the Impact of Vehicle-to-Grid Technologies on Distribution Systems and Utility Interfaces," *IEEE Transactions on Power Electronics*, vol. 28, no. 12, pp. 5673-5689, 2013.

[17] Y. Xia and R. Ayyanar, "Comprehensive comparison of THD and common mode leakage current of bipolar, unipolar and hybrid modulation schemes for single phase grid connected full bridge inverters," *2017 IEEE Applied Power Electronics Conference and Exposition (APEC)*, Tampa, FL, 2017, pp. 743-750.

[18] Z. Yao, L. Xiao and Chunying Gong, "Half-bridge-type inverter with hysteresis current control and unipolar modulation," *2010 5th IEEE Conference on Industrial Electronics and Applications*, Taichung, 2010, pp. 1113-1117.

[19] D. Marandi, T. N. Sowmya and B. C. Babu, "Comparative study between unipolar and bipolar switching scheme with LCL filter for single-phase grid connected inverter system," *Electrical, Electronics and Computer Science (SCEECS), 2012 IEEE Students' Conference on*, Bhopal, 2012, pp. 1-4.

[20] F. Wu, B. Sun, K. Zhao and L. Sun, "Analysis and Solution of Current Zero-Crossing Distortion With Unipolar Hysteresis Current Control in Grid-Connected Inverter," *IEEE Transactions on Industrial Electronics*, vol. 60, no. 10, pp. 4450-4457, 2013.

[21] M. Wang, S. Guo, Q. Huang, W. Yu and A. Q. Huang, "An Isolated Bidirectional Single-Stage DC–AC Converter Using Wide-Band-Gap Devices With a Novel Carrier-Based Unipolar Modulation Technique Under Synchronous Rectification," *IEEE Transactions on Power Electronics*, vol. 32, no. 3, pp. 1832-1843, 2017.

[22] J. Soomro, T. D. Memon and M. A. Shah, "Design and analysis of single phase voltage source inverter using Unipolar and Bipolar pulse width modulation techniques," *2016 International Conference on Advances Electrical, Electronic and Systems Engineering (ICAEES)*, Putrajaya, 2016, pp. 277-282.

[23] R. Sharma and J. A. R. Ball, "Unipolar switched inverter low-frequency harmonics caused by switching delay," *IET Power Electronics*, vol. 2, no. 5, pp. 508-516, Sept. 2009.

[24] A. Mesemanolis, D. Pontikidis and C. Demoulias, "A new modulation technique for reduced harmonic distortion of current in PV inverters," *2011 IEEE EUROCON - International Conference on Computer as a Tool*, Lisbon, 2011, pp. 1-4.

[25] T. F. Wu, C. L. Kuo, K. H. Sun and H. C. Hsieh, "Combined Unipolar and Bipolar PWM for Current Distortion Improvement During Power Compensation," *IEEE Transactions on Power Electronics*, vol. 29, no. 4, pp. 1702-1709, April 2014.

Per-Phase Control Strategy of the Three-Phase Four-Wire Inverter

Yi-Chan Li[1], Terng-Wei Tsai[1], Cheng-Jhen Yang[1], Yaow-Ming Chen[1] and Yung-Ruei Chang[2]

[1]Department of Electrical Engineering, National Taiwan University, Taipei, Taiwan
[2]Institute of Nuclear Energy Research, Atomic Energy Council, Taoyuan, Taiwan
E-mail: r05921020@ntu.edu.tw

Abstract—A per-phase control strategy of the three-phase four-wire inverter is proposed in this paper. The proposed control method is simple and does not require any complex transformation. The equivalent single-phase half-bridge inverter of the three-phase four-wire inverter is presented and used to derive the equations of the reference sinusoidal control signal of each phase. The active power and reactive power of each phase can be controlled, respectively. Computer simulation and experimental results for different operating cases are presented to verify the performance of the proposed per-phase control method.

Keywords— three-phase four-wire inverter, per-phase control.

I. INTRODUCTION

In recent years, in order to solve the issue of fossil fuel shortage and to utilize the renewable energy, the demand of microgrids have been increased rapidly [1]. The microgrid has many advantages, such as providing the electricity to the local load and reducing the power loss caused by long-range transmission. Furthermore, if a grid fault occurs, the microgrid with energy sources can be disconnected from the grid and operate on its own. Therefore, the load in the mircogrid is less affected by comparing to the conventional centralized grid.

Usually, the microgrid consists of the distributed generations (DGs), the distributed storages (DSs), the interconnection switches, the control systems, and the local loads [2]. In the microgrid, DGs and DSs are two indispensable and important components. They improve the stability and reliability of the microgrid.

The renewable energies, such as wind power and solar energy, have been developed for many years and are used as DGs. On the other hand, the battery bank which is a common type of DS has been is well developed. The DS can enhance the overall performance of microgrid system in many ways. First, since the renewable energy is often affected by the unpredictable weather and environmental factors, the power quality of the microgrid will be affected. This situation will cause the unbalance power between generators and loads. Thus, the DS plays an important role of providing the sufficient energy to meet the load demand.

Second, the DS provide the required power to balance the power system. If the power requirement of the loads is increased, the DS can deliver the required power to the loads. On the contrary, if the power requirement of the loads is reduced, the excess power generated by the DG can be transferred to the DS for future need. Third, the DS can enhance the power quality and improve the power factor by compensating the corresponding reactive power.

For the purpose of converting the energy of the DG and the DS into the grid-compatible ac power as well as to improve the stability and reliability of the microgrid, many power electronics interfaces, such as the DC-AC inverter, are necessary [3]-[7].

The DC-AC inverter can be classified to different types according to different circuit topologies including the single-phase, the three-phase three-wire, the three-phase four-wire type, and etc. Among them, the three-phase four-wire inverter can deal with the power system unbalanced problem is commonly used.

Also, many control methods have been developed and applied to the inverters [8]-[16]. However, some control strategies need complex calculation such as the dq0 transformation, which will increase the computing time of the digital signal processor (DSP). Therefore, the relatively simple per-phase control strategy is proposed.

This proposed per-phase control strategy can be directly implemented on the three-phase four-wire inverter which can be treated as three equivalent single-phase half-bridge circuits [13]. The sinusoidal reference voltage signals of the equivalent single-phase half-bridge circuit are derived and used to generate sinusoidal pulse-width modulation (SPWM) signals without using complex transformation.

The proposed per-phase control method can achieve the power control of each phase, independently. Also, the power transmission of each phase of the three-phase four-wire inverter can be bi-directional. Therefore, the proposed control strategy can be used to mitigate the problems of unbalanced loads. For example, the corresponding power can be transferred to each phase according to the different power requirements of the loads in each phase. On the other hand, according to different grid codes, the inverter can deliver different reactive power to different phases to compensate the ac grid variation.

In order to demonstrate the performance of the proposed per-phase control strategy, a 5kVA prototype circuit is implemented. The computer simulation and the experimental results will be presented in this paper to verify the feasibility of the proposed control method.

II. PER-PHASE CONTROL STRATEGY OF THE THREE-PHASE FOUR-WIRE INVERTER

Fig. 1 shows the circuit diagram of the three-phase four-wire inverter with a dc energy storage. The output of the inverter are connected to the ac mains. The inverter consists of the split-capacitors, three-leg switches and the output inductors. The center point of the split-capacitors and the neutral point of the three-phase ac mains are connected together with a neutral line.

In Fig. 1, the three-phase four-wire inverter can be viewed as three equivalent single-phase half-bridge inverters sharing the common dc source. The equivalent circuit for one single-phase inverter and its control loop are shown in Fig. 2. The subscript x can be the indication for phase r, s, or t of the three-phase inverter, while n is the neutral point.

Based on the output power requirements of phase x, the active power command $P_{cmd,x}$ and the reactive power command $Q_{cmd,x}$ are used to calculate the corresponding reference current $i_{acref,x}$, which can be expressed as (1). It should be mentioned that a variable with an arrow on the top is noted in phasor form. The amplitude $|\overrightarrow{i_{acref,x}}|$ and the phase angle $\angle\theta_{i_{acref,x}}$ of $i_{acref,x}$ in (1) are shown in equations (2) and (3), respectively, where $v_{ac,x,rms}$ and $\angle\theta_{v_{ac,x}}$ represent the root mean square (RMS) value and the phase angle of ac voltage of phase x.

$$i_{acref,x} = |\overrightarrow{i_{acref,x}}| \times cos(\omega t + \angle\theta_{i_{acref,x}}) \tag{1}$$

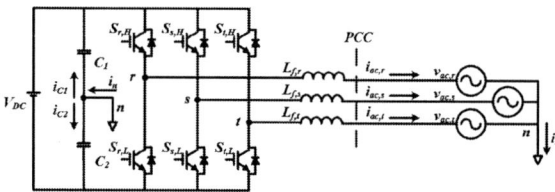

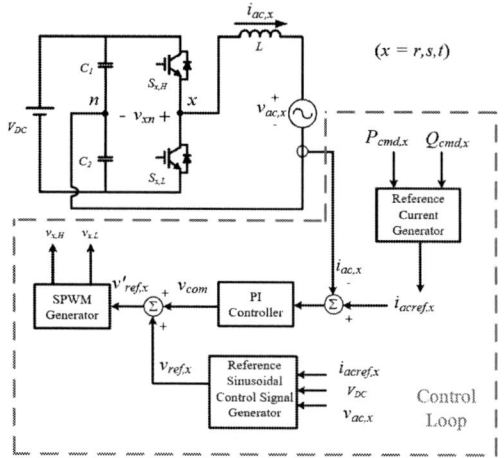

Fig. 2. Circuit diagram of the equivalent single-phase half-bridge inverter with its control loop.

Fig. 1. Circuit diagram of the grid-tied three-phase four-wire inverter with a dc energy storage.

$$|\overrightarrow{i_{acref,x}}| = \sqrt{2} \times \sqrt{\frac{P_{cmd,x}^2 + Q_{cmd,x}^2}{v_{ac,x,rms}^2}} \tag{2}$$

$$\angle\theta_{i_{acref,x}} = \angle\theta_{v_{ac,x}} - tan^{-1}\frac{Q_{cmd,x}}{P_{cmd,x}} \tag{3}$$

The conceptual modulation scheme of SPWM is shown in Fig. 3. The V_{Peak} and $-V_{Peak}$ are the maximum and minimum value of the triangular wave v_{tri}, respectively, and $v_{ref,x}$ is the reference sinusoidal control signal of phase x. The equation of $v_{ref,x}$ is expressed as (4), where $|\overrightarrow{v_{ref,x}}|$ and $\angle\theta_{v_{ref,x}}$ are the amplitude and phase angle of $v_{ref,x}$. Because the switching period of the switches is very small by comparing to the period of ac mains, the $v_{ref,x}$ can be regarded as a constant value within a switching period. Thus, the duty cycle D for switch $S_{x,H}$ can be obtained in equation (5). Since the two switches, $S_{x,H}$ and $S_{x,L}$, are operated with complementary gate drive signals in every switching period T_S, the duty cycle of switch $S_{x,L}$ is $(1-D)$.

$$v_{ref,x} = |\overrightarrow{v_{ref,x}}| \times cos(\omega t + \angle\theta_{v_{ref,x}}) \tag{4}$$

$$D = \frac{v_{ref,x} - (-V_{Peak})}{V_{Peak} - (-V_{Peak})} = \frac{v_{ref,x}}{2V_{Peak}} + \frac{1}{2} \tag{5}$$

According to Fig. 2, assuming that the voltages of the split-capacitors are balanced, the average voltages of capacitor C_1 and C_2 are $V_{DC}/2$, where V_{DC} is the voltage of dc energy storage. Hence, the average voltage V_{xn} across the node x and the neutral point n for a switching period T_S can be derived as (6).

$$V_{xn} = \frac{1}{T_S}\left[\frac{V_{DC}}{2}DT_S - \frac{V_{DC}}{2}(1-D)T_S\right]$$
$$= (2D-1)\frac{V_{DC}}{2} \tag{6}$$

By combining (5) and (6), the mathematical expression of V_{xn} in a switching cycle can be derived as (7).

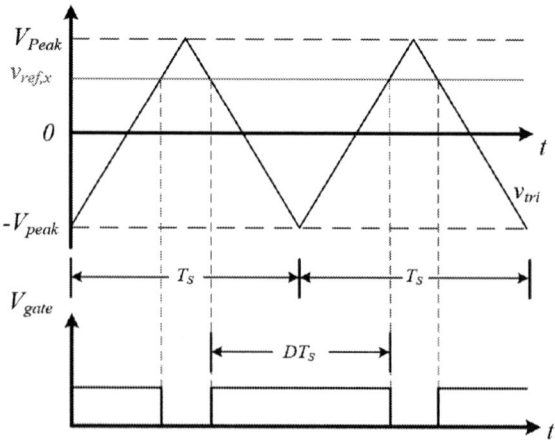

Fig. 3. The conceptual modulation scheme of SPWM.

$$V_{xn} = \frac{V_{DC}}{2V_{peak}} v_{ref,x} \tag{7}$$

Since the switching frequency is much higher than the ac mains frequency, the voltage across the node x and the neutral point n can be expressed:

$$v_{xn} = \frac{V_{DC}}{2V_{peak}} v_{ref,x} . \tag{8}$$

From Fig. 2, it can be found that the node x, the inductor the ac mains voltage, and the neutral point n form a loop. According to the Kirchhoff Voltage Law (KVL), equation (9) can be obtained. By combining (8) and (9), (10) can be derived.

$$v_{xn} - L\frac{di_{ac,x}}{dt} - v_{ac,x} = 0 \tag{9}$$

$$\frac{V_{DC}}{2V_{peak}} v_{ref,x} - L\frac{di_{ac,x}}{dt} - v_{ac,x} = 0 \tag{10}$$

In steady state, the time-domain variables can be expressed as phasor form. As the result, the phasor form of (10) can be rewritten as (11).

$$\frac{V_{DC}}{2V_{peak}} \overrightarrow{v_{ref,x}} - j\omega L \times \overrightarrow{i_{ac,x}} - \overrightarrow{v_{ac,x}} = 0 \tag{11}$$

From the equation (11), the phasor form of $v_{ref,x}$ can be derived as (12). The amplitude $|\overrightarrow{v_{ref,x}}|$ and the phase angle $\angle\theta_{v_{ref,x}}$ of $v_{ref,x}$ in (4) can be obtained as (13) and (14).

$$\overrightarrow{v_{ref,x}} = \frac{2V_{peak}}{V_{DC}} [\overrightarrow{v_{ac,x}} + j\omega L\overrightarrow{i_{ac,x}}] \tag{12}$$

$$|\overrightarrow{v_{ref,x}}| = \frac{2V_{peak}}{V_{DC}} [|\overrightarrow{v_{ac,x}}|^2 + \omega^2 L^2 |\overrightarrow{i_{ac,x}}|^2$$
$$-2\omega L|\overrightarrow{v_{ac,x}}||\overrightarrow{i_{ac,x}}|sin(\theta_{i_{ac,x}} - \theta_{v_{ac,x}})]^{\frac{1}{2}} \tag{13}$$

$$\angle\theta_{v_{ref,x}} = tan^{-1}\left[\frac{|\overrightarrow{v_{ac,x}}|\sin\theta_{v_{ac,x}} + \omega L|\overrightarrow{i_{ac,x}}|\cos\theta_{i_{ac,x}}}{|\overrightarrow{v_{ac,x}}|\cos\theta_{v_{ac,x}} - \omega L|\overrightarrow{i_{ac,x}}|\sin\theta_{i_{ac,x}}}\right] \tag{14}$$

If the output current is well-regulated, the output current $i_{ac,x}$ will be very close to the reference current $i_{acref,x}$, so the error between $i_{ac,x}$ and $i_{acref,x}$ will approximate to zero. Therefore, if the term $i_{ac,x}$ is replaced with $i_{acref,x}$, the reference sinusoidal control signal $v_{ref,x}$ will help the output current to achieve the required current more effectively.

With the above equations, the control loop for the single-phase half-bridge equivalent circuit of the three-phase four-wire inverter in Fig. 2 is constructed. The control loop includes the equivalent single-phase half-bridge inverter, Reference Current Generator, PI Controller, Reference Sinusoidal Control Signal Generator, and SPWM Generator. In Fig. 2, Reference Current Generator is realized by equations (2) and (3). Equation (13) and (14) will be performed in the Reference Sinusoidal Control Signal Generator. Gate signals are generated by the SPWM Generator according to (5). The term v_{com} represents the output compensation voltage of PI Controller. The reference sinusoidal control signal, $v'_{ref,x}$, is the sum of $v_{ref,x}$ and v_{com}. Variables $v_{x,H}$ and $v_{x,L}$ are the gate drive signals of the switches $S_{x,H}$ and $S_{x,L}$, respectively.

According to the control loop diagram shown in Fig. 2, it can be found that the control loop does not need any complex transformation. This control strategy can achieve the power control of each phase independently.

III. SIMULATION RESULTS

In order to verify the proposed per-phase control strategy, the grid-tied three-phase four-wire inverter with a dc energy storage circuit shown in Fig. 1 is simulated using Matlab/Simulink. The circuit specifications are displayed in Table I. Two cases are simulated to demonstrate the performance of the proposed control strategy. The circuit is simulated under the balanced three-phase ac voltages. The RMS value of the voltage of each phase is 220V. The output power commands of each phase of the two cases are shown in Table II.

The simulation results of case 1 in Table II are shown in Fig. 4. It demonstrates the balanced output of active power. Each of phase is required to inject 1500W and 0Var into the grid, respectively. According to the power demands and the ac voltage, the current's amplitude of each phase should be 9.64A. Besides, the voltage and current of each phase should be in phase.

Fig. 4 (a) shows the average output power of each phase, where Pr, Ps, Pt and Qr, Qs, Qt represent the active power and the reactive power of each phase, respectively. The corresponding legends of the output power of each phase are shown in the upper right corner of Fig. 4(a). The waveforms of three-phase ac voltages, $v_{ac,r}$, $v_{ac,s}$, $v_{ac,t}$ and the output currents, $i_{ac,r}$, $i_{ac,s}$, $i_{ac,t}$ are shown in Fig. 4(b). According to Fig. 4(a), it can be seen that the solid lines overlap together, and they are close to the value 1500W. It means that Pr, Ps and Pt are approximately equal to 1500W. On the other hand, the dashed lines overlap together at the value zero, so Qr, Qs and Qt are 0Var. From Fig. 4(b), it is found that the currents and voltages of phase-r, phase-s and phase-t are in phase. In addition, the peak values of $i_{ac,r}$, $i_{ac,s}$, $i_{ac,t}$ are equal, so the output currents are balanced. As a result, it can be concluded that the inverter only injects the balanced active power into the grid and the output power of each phase follows the power commands in Table II.

The simulation results of case 2 are shown in Fig. 5. Case 2 demonstrates that each phase can transmit different active power and reactive power to the grid. The corresponding output power commands of each phase are shown in Table II. According to the power demands and the ac voltage, the current's amplitude of each phase should be 9.64A, 6.43A, 3.21A, respectively, and the current of each phase should lags its corresponding voltage by 60°.

Fig. 5(a) shows the output active and reactive power of each phase. The corresponding legends of the power of each phase are in the upper right corner of Fig. 5(a). From Fig. 5(a), it is found that the average output active power and reactive power of each phase are approximately close to the power commands of case 2. According to the waveforms of three-phase ac voltages and output currents shown in Fig. 5(b), it can be found that the output currents are not balanced, and the peak value of the output current of each phase approximates to the one calculated from the power command and ac voltage. Moreover, the phase difference between the voltage and the current of each phase meets the calculation result of power commands in Table II. Consequently, the output power of each phase can be individually controlled.

From the above simulation results, the feasibility of the proposed control strategy is verified. This control strategy can indeed achieve the individual output power control of each phase.

IV. EXPERIMENTAL RESULTS

A 5kVA prototype three-phase four-wire inverter with the specifications shown in Table I is built and tested to verify the performance of the proposed control strategy. The DSP TMS320F28335 is used as the controller of the inverter. There are two cases to be tested, which are the same as the cases in Table II. These two cases are tested under the balanced three-phase voltages. The RMS value of the voltage of each phase is 220V. The experimental waveforms of the two cases are shown in Fig. 6 and Fig. 7, respectively. These waveforms include three-phase output currents and phase-r voltage in steady state.

TABLE I
THE CIRCUIT SPECIFICATIONS

Parameter	Value
Power rating	5kVA
Line-to-neutral voltage ($v_{ac,x}$)	220V$_{rms}$
Input DC voltage (V_{DC})	760V
Switching frequency (f_{SW})	30kHz
DC bus capacitor (C_1, C_2)	3.29mF
Output inductor (L)	2mH

TABLE II
THE OUTPUT POWER COMMANDS OF EACH PHASE OF THE TWO CASES

Case 1			
Phase	P_{cmd}(W)	Q_{cmd}(Var)	S_{cmd}(VA)
r	1500	0	1500
s	1500	0	1500
t	1500	0	1500
Case 2			
Phase	P_{cmd}(W)	Q_{cmd}(Var)	S_{cmd}(VA)
r	750	1300	1500
s	500	866	1000
t	250	433	500

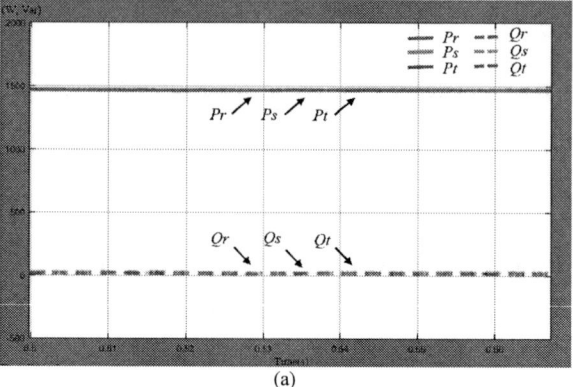

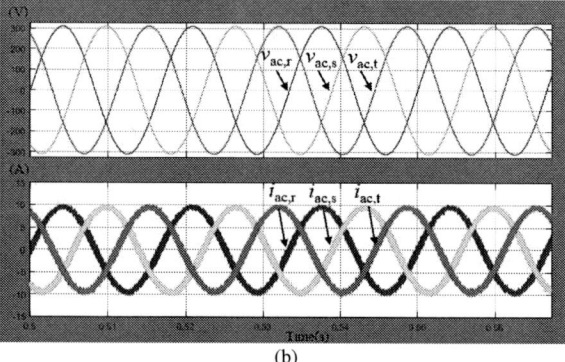

Fig. 4. The simulation results of case 1. (a) Average active power and reactive power, (b) Voltage and current waveforms.

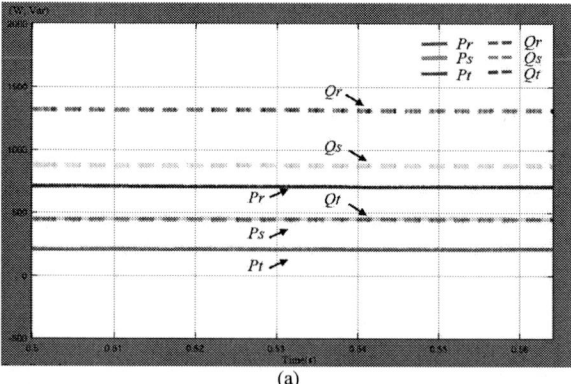

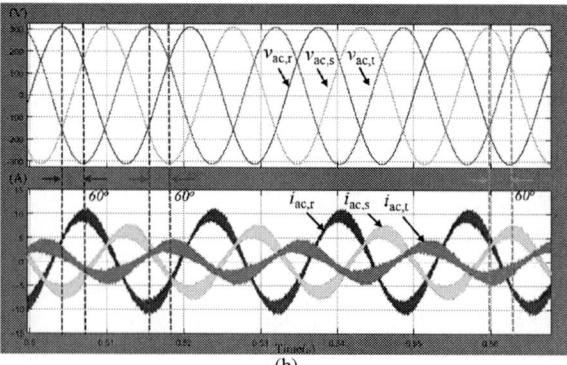

Fig. 5. The simulation results of case 2. (a) Average active power and reactive power, (b) Voltage and current waveforms.

Case 1 illustrates the balanced output of active power. The average output active power and reactive power of each phase are required to be 1500W and 0Var. According to the power demands and the ac voltage, the current's amplitude of each phase should be 9.64A. In addition, the voltage and current of each phase should be in phase. Fig. 6 shows the output waveforms of case 1. From Fig. 6, it is found that the output currents are well-regulated, and the peak values of the currents are equal. Moreover, the voltage and the current of phase-r are in phase. Since this case is tested under the balanced three-phase voltages, it can be expected that the voltages and the currents of phase-s and phase-t are in phase, too. Thus, the output currents are balanced. It can be found that the inverter only injects active power into the grid and the output power are balanced.

Case 2 demonstrates the unbalanced output of active power and reactive power. The required output power of each phase is shown in Table II. According to the power demands and the ac voltage, the current's amplitude of each phase should be 9.64A, 6.43A, 3.21A, respectively, and the current of each phase should lags the voltage of each phase by 60°. Fig. 7 shows the measured results of case 2. From Fig. 7, the current's amplitude of each phase is close to the one calculated from the power command and the ac voltage. Additionally, the voltage and the current of phase-r are 60° out of phase. The phase difference is accordance with the calculation result from power command of phase-r. Since this case is tested under the balanced three-phase voltages, and the output currents are 120° out of phase, it can be expected that the voltages and the currents of phase-s and phase-t are 60° out of phase. Thus, the current's amplitude the current's phase of each phase accurately follows the command. Fig. 7 proves that the output power of each phase can be controlled, respectively.

According to the experimental results, the output waveforms are consistent with the waveforms of the simulation results. Therefore, the feasibility of the proposed control method is verified.

V. CONCLUSIONS

This paper proposes a per-phase power control strategy for the three-phase four-wire inverter. The reference sinusoidal control signal of each phase is derived from the equivalent single-phase half-bridge inverter. The output power of each phase can be independently controlled by using the derived reference sinusoidal control signal. The proposed control strategy can achieve different power demands according to different conditions or requirements.

The simulation and experimental results are presented in this paper. The experimental results are consistent with the simulation results. Therefore, the feasibility of the proposed control method is verified.

ACKNOWLEDGMENT

This work is partially supported by a research grant from the Institute of Nuclear Energy Research, Taiwan.

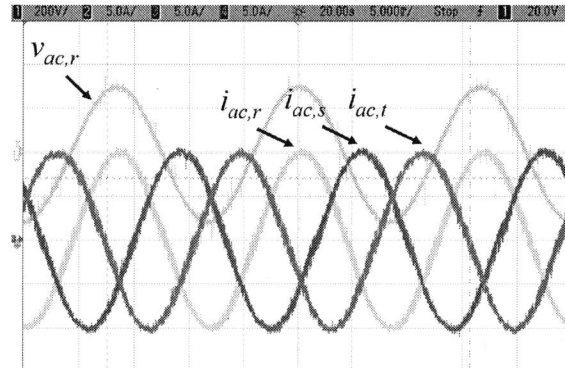

Fig. 6. Measured waveforms under balanced active power.

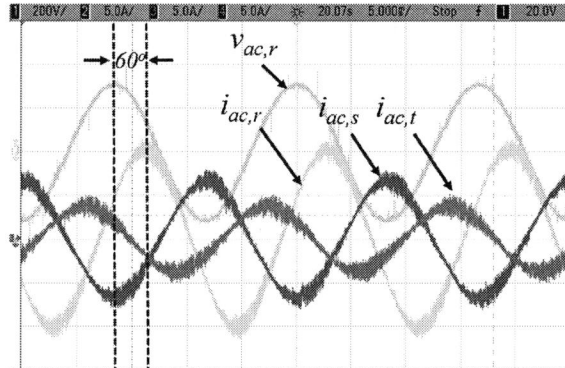

Fig. 7. Measured waveforms under unbalanced active power.

REFERENCES

[1] R. H. Lasseter, "MicroGrids," *2002 IEEE Power Engineering Society Winter Meeting. Conference Proceedings (Cat. No.02CH37309)*, 2002, pp. 305-308 vol.1.

[2] D. E. Olivares *et al.*, "Trends in Microgrid Control," in *IEEE Transactions on Smart Grid*, vol. 5, no. 4, pp. 1905-1919, July 2014.

[3] K. T. Tan, P. L. So, Y. C. Chu, and M. Z. Q. Chen, "Coordinated Control and Energy Management of Distributed Generation Inverters in a Microgrid," *IEEE Transactions on Power Delivery*, vol. 28, no. 2, pp. 704-713, April 2013.

[4] S. V. Araujo, P. Zacharias, and R. Mallwitz, "Highly Efficient Single-Phase Transformerless Inverters for Grid-Connected Photovoltaic Systems," *IEEE Transactions on Industrial Electronics*, vol. 57, no. 9, pp. 3118-3128, Sept. 2010.

[5] H. Qian, J. Zhang, J. S. Lai, and W. Yu, "A high-efficiency grid-tie battery energy storage system," *IEEE Transactions on Power Electronics*, vol. 26, no. 3, pp. 886-896, March 2011.

[6] M. Jang, M. Ciobotaru, and V. G. Agelidis, "A Single-Phase Grid-Connected Fuel Cell System Based on a Boost-Inverter," *IEEE Transactions on Power Electronics*, vol. 28, no. 1, pp. 279-288, Jan. 2013.

[7] J. Rocabert, A. Luna, F. Blaabjerg and P. Rodríguez, "Control of Power Converters in AC Microgrids," in *IEEE Transactions on Power Electronics*, vol. 27, no. 11, pp. 4734-4749, Nov. 2012.

[8] D. M. Brod and D. W. Novotny, "Current Control of VSI-PWM Inverters," *IEEE Transactions on Industry Applications*, vol. IA-21, no. 3, pp. 562-570, May 1985.

[9] M. P. Kazmierkowski and L. Malesani, "Current control techniques for three-phase voltage-source PWM converters: a survey," *IEEE Transactions on Industrial Electronics*, vol. 45, no. 5, pp. 691-703, Oct 1998.

[10] J. Rocabert, A. Luna, F. Blaabjerg, and P. Rodríguez, "Control of Power Converters in AC Microgrids," *IEEE Transactions on Power Electronics*, vol. 27, no. 11, pp. 4734-4749, Nov. 2012.

[11] M. F. Schonardie, A. Ruseler, R. F. Coelho, and D. C. Martins, "Three-phase grid-connected PV system with active and reactive

power control using dq0 transformation," *IEEE/IAS International Conference on Industry Applications,* 2010, pp. 1-6.

[12] W. T. Franke, C. Kürtz, and F. W. Fuchs, "Analysis of control strategies for a 3 phase 4 wire topology for transformerless solar inverters," *IEEE International Symposium on Industrial Electronics,* 2010, pp. 658-663.

[13] R. Ghosh and G. Narayanan, "Control of Three-Phase, Four-Wire PWM Rectifier," *IEEE Transactions on Power Electronics,* vol. 23, no. 1, pp. 96-106, Jan. 2008.

[14] A. Hintz, U. R. Prasanna, and K. Rajashekara, "Comparative Study of the Three-Phase Grid-Connected Inverter Sharing Unbalanced Three-Phase and/or Single-Phase systems," *IEEE Transactions on Industry Applications,* vol. 52, no. 6, pp. 5156-5164, Nov.-Dec. 2016.

[15] N. A. Ninad and L. A. C. Lopes, "Per-phase vector (dq) controlled three-phase grid-forming inverter for stand-alone systems," *IEEE International Symposium on Industrial Electronics,* 2011, pp. 1626-1631.

[16] N. A. Ninad and L. A. C. Lopes, "Per-phase DQ control of a three-phase battery inverter in a diesel hybrid mini-grid supplying single-phase loads," *IEEE International Conference on Industrial Technology,* 2011, pp. 204-209.

The 2018 International Power Electronics Conference

Opportunities for Performance Improvement of Single-Phase Power Converters through Enhanced Automatic-Power-Decoupling Control

Huawei Yuan[1*], Sinan Li[1], Wenlong Qi[1], Siew-Chong Tan[1], S. Y. (Ron) Hui[1,2]

1 Department of Electrical and Electronic Engineering, The University of Hong Kong, Hong Kong, China
2 Department of Electrical and Electronic Engineering, Imperial College London, London, U. K.
*E-mail: hwyuan@eee.hku.hk

Abstract— There is a growing demand for high power density, high efficiency, and high reliability (H³) single-phase power converters in power electronics applications. These H³ converters, which are equipped only with very small energy storage components, involve large-signal operation and exhibit highly coupled and nonlinear characteristics. Existing controller designs for these converters are based on small-signal models, and may not attain satisfactory dynamic performance and robustness. This paper presents a nonlinear control method based on the technique of input-output feedback linearization and the automatic-power-decoupling control strategy for the H³ converters. The controller not only achieves global system stability and enhanced dynamic performance over existing control solutions, but also enables us to exploit the full potential of H³ single-phase converters, thereby providing new application opportunities. Simulations and experiments are carried out to demonstrate the feasibilities and to verify the performance of the proposed control.

Keywords— *Automatic power decoupling, discontinuous current mode, feedback linearization, single-phase converter.*

I. INTRODUCTION

Single-phase power converters are commonly found in our lives and their applications are ranged from low power (e.g. LED drivers, laptop adapters, etc., that are typically <250 W) to medium power applications (e.g. micro-inverters, electric vehicle chargers, etc., that are typically in the kilowatt scale). Recently, there is a trend towards developing high power density, high efficiency and high reliability (H³) types of single-phase power converters mainly due to (i) the widening use of distributed renewable power generation which necessitates a prolonged operating lifespan and simple onsite deployment of the converters, and (ii) the advancement of power electronics technologies which enables a big leap in the attainable power density, efficiency and reliability of the converter hardware. The GOOGLE Little Box Challenge initiated in 2014 has also challenged worldwide professionals to achieve a most compact 2 kW single-phase solar inverter with the highest possible efficiency [1]. By incorporating advanced technologies, such as new power

semiconductor devices, circuit topologies and modulation concepts, a new generation of single-phase power converters, which concurrently achieve high efficiency (>98%) and high power density (>200 W/in³) that are far better than those of existing products in the market, have been devised [2]–[6].

Despite such improvements, there are clear performance limitations with the present types of H³ single-phase power converters from a control point-of-view. These limitations concern three aspects, namely, the (i) transient performance; (ii) startup, shutdown, and protection functionalities; and (iii) capabilities of providing ancillary services, as listed in Table I. H³ single-phase converters utilizing existing controllers (based on small-signal models and direct power decoupling control strategy) can easily suffer from poor dynamic performance and system failure [7], [8]. This makes them less competitive than existing products. The issues stated in Table I must be addressed before the H³ single-phase power converters can establish themselves as possible alternatives to industrial applications.

This paper aims at coping with the control challenges mentioned above, by proposing an enhanced-automatic-power-decoupling (E-APD) control, yielding fast dynamics, global stability and enhanced robustness ideally at all operating conditions. The paper also presents a comprehensive performance evaluation of the H³ single-phase power converters with the proposed E-APD controller to demonstrate the potential of the presented control.

TABLE I
PERFORMANCE LIMITATIONS OF H³ SINGLE-PHASE POWER CONVERTER WITH SMALL POWER PULSATION BUFFER

Aspects	Limitations
Transient Performance	1. Large-signal stability
	2. Fast transient response
	3. Low-cost control implementation
	4. Feed nonlinear load
Startup, Shutdown, and Protection	1. Fast and reliable startup
	2. Voltage holdup during ac mains outage
	3. Over-/under-voltage protection
	4. Over-current protection
Provision of Ancillary Services	1. Reactive power generation
	2. Bidirectional power flow
	3. Active power filtering

This work is supported by the Hong Kong Research Grant Council under GRF Project 17205817.

II. CHALLENGES OF H^3_i SINGLE-PHASE POWER CONVERTER CONTROL

A. Characteristics of H^3 Single-Phase Converters with Small Power Pulsation Buffers

To maintain a stable dc-port voltage, the single-phase power converter inherently requires a substantial amount of energy storage (also called power pulsation buffer or PPB) due to the double-line-frequency power difference between the ac-port and the dc-port of the system even during the steady state [9]. Conventionally, single-phase power converters utilize bulky dc-link capacitors to achieve power buffering in a passive way (see C_b in Fig. 1(a)), whereas H^3 single-phase power converters achieve buffering by active approaches (see C_b in Fig. 1(b)), leading to significantly reduced energy storage and thus a much more compact converter design.

Essentially, the converter depicted in Fig. 1(a) is a two-port passive network, while the converter depicted in Fig. 1(b) is with three-ports (with an additional ripple-port). Three key distinctions can be identified between the two converters regarding the (i) operation of the buffering capacitor C_b, (ii) the dynamics of the dc port, and (iii) the control model of the controller (see Table II). These differences are rarely discussed and should not be neglected. The unique characteristics of the H^3 single-phase converter impose critical challenges on its controller design. The converter experiences poor dynamic performances and suffer from instability issue when conventional linear control is employed.

B. Control Strategies

While the H^3 single-phase power converter is a three-port network, only the power of two ports needs to be controlled directly and independently. Below are all the possible control strategies for an H^3 single-phase converter:

Strategy A: direct control of ac- and ripple-port power;
Strategy B: direct control of dc- and ripple-port power;
Strategy C: direct control of ac- and dc-port power.

Strategy A and *B* involve the direct control of the ripple-port power and are called direct-power-decoupling (DPD) control. In contrast, no dedicated ripple-port power control is required with *Strategy C*. Such a control strategy is called automatic-power-decoupling (APD) control. In literature, *Strategy A* is adopted much more widely than the other two control strategies.

Among the three control strategies, DPD control is less robust in practice and possibly more complicated than APD control [8]. This is because DPD control requires the generation of a ripple-port voltage reference (e.g. $v_b{}^*$ in Fig. 1(b)) for its ripple-port power control. However, accurate prediction of $v_b{}^*$ is difficult especially when the system is in transient state, undergoing significant input-voltage/load/reference variations or other external disturbances. Therefore, the DPD control strategy cannot easily achieve perfect power buffering and easily fails to retain the desired operation of the converter. The APD control strategy, which does not involve the direct control of ripple-port power (and hence generation of $v_b{}^*$ is not needed) inherently avoids the above issue. Perfect buffering of the unbalanced power between the ac- and dc-ports is theoretically possible, which leads to improved dynamic performance and robustness against various kinds of disturbances.

III. TOWARDS ENHANCED AUTOMATIC-POWER-DECOUPLING CONTROL

Based on the discussions in Section II, a nonlinear controller based on a combination of input-output feedback linearization (FBL) [10] and APD control strategy, hereon known as enhanced APD (or E-APD) control, is proposed. The operating principle of the proposed control is illustrated on a typical H^3 single-phase power converter operating in the discontinuous conduction mode (DCM), as shown in Fig. 2. This topology is selected due to its high power density and high-quality design [11].

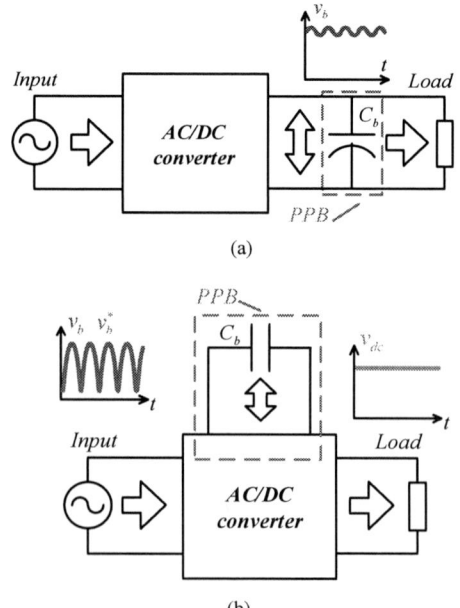

(a)

(b)

Fig. 1. Generic circuit model of (a) conventional single-phase converter with a large dc-link capacitor, and (b) H^3 single-phase converter with a small PPB.

TABLE II
KEY DISTINCTIONS BETWEEN CONVENTIONAL SINGLE-PHASE CONVERTERS AND H^3 SINGLE-PHASE CONVERTERS

Aspects	Fig. 1(a)	Fig. 1(b)
Operation of the Buffering Capacitor	Small-signal operation	Large-signal operation
Dynamic of Dc-Port	Almost decoupled from load and ac-port dynamics	Highly coupled to ac-port, ripple-port, and load dynamics
Control Model of Controller	single-input single-output	two-input two-output

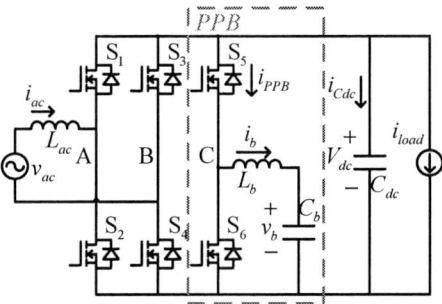

Fig. 2. Topology of the targeted H^3 single-phase power converter.

A. Modeling of the H^3 Single-Phase Power Converter

With the targeted H^3 converter (see Fig. 2) operating in DCM, the PPB must be operated in two modes, i.e., charging and discharging mode, depending on the instantaneous power flow directions of the PPB. The mathematical model of the overall system in each operating mode is derived as follows.

In the charging mode of the PPB, the state-space-averaged model of the converter can be obtained as

$$\dot{x} = f(x) + g(x) \cdot u \qquad (1)$$

where

$$x = \begin{bmatrix} x_1 \\ x_2 \\ x_3 \end{bmatrix} = \begin{bmatrix} i_{ac} \\ v_{dc} \\ v_b \end{bmatrix}, \quad u = \begin{bmatrix} u_1 \\ u_2 \end{bmatrix} = \begin{bmatrix} m \\ d_{PPB}^2 \end{bmatrix}, \qquad (2)$$

$$f(x) = \begin{bmatrix} \dfrac{v_{ac}}{L_{ac}} \\ -\dfrac{i_{load}}{C_{dc}} \\ 0 \end{bmatrix}, \qquad (3)$$

$$g(x) = \begin{bmatrix} -\dfrac{x_2}{L_{ac}} & 0 \\ \dfrac{x_1}{C_{dc}} & -\dfrac{(x_2 - x_3)}{cC_{dc}} \\ 0 & \dfrac{x_2(x_2 - x_3)}{cC_b x_3} \end{bmatrix}, \qquad (4)$$

where m is the modulation index of the full-bridge component, d_{PPB} is the duty cycle of the PPB (i.e. the duty cycle of S_5 during the charging mode and that of S_6 during the discharging mode), and $c = 2L_b/T_{sw}$.

Meanwhile, the state-space-averaged model in the discharging mode can be described by (1) but with a different $g(x)$ as

$$g(x) = \begin{bmatrix} -\dfrac{x_2}{L_{ac}} & 0 \\ \dfrac{x_1}{C_{dc}} & \dfrac{x_3^2}{cC_{dc}(x_2 - x_3)} \\ 0 & -\dfrac{x_2 x_3}{cC_b(x_2 - x_3)} \end{bmatrix}. \qquad (5)$$

Clearly, the targeted system is highly coupled and nonlinear, thereby justifying the adoption of the proposed E-APD control, which is a nonlinear controller.

B. Design of the Proposed E-APD Controller Based on Input-Output Feedback Linearization

According to the APD control strategy introduced in Section II-B, the ac-port current x_1 and the dc-port voltage x_2 are selected as the control outputs. For convenience of analysis, the gains of L_{ac} and C_{dc} are applied to the two control outputs respectively, i.e.

$$y = \begin{bmatrix} y_1 \\ y_2 \end{bmatrix} = \begin{bmatrix} L_{ac} x_1 \\ C_{dc} x_2 \end{bmatrix}. \qquad (6)$$

During the charging mode of the PPB, according to (1)–(4), the time differentiation of (6) leads to

$$\dot{y} = \begin{bmatrix} v_{ac} \\ -i_{load} \end{bmatrix} + \begin{bmatrix} -x_2 & 0 \\ x_1 & -\dfrac{(x_2 - x_3)}{c} \end{bmatrix} \begin{bmatrix} u_1 \\ u_2 \end{bmatrix}. \qquad (7)$$

By selecting a new control input v as

$$\begin{bmatrix} \dot{y}_1 \\ \dot{y}_2 \end{bmatrix} = \begin{bmatrix} v_1 \\ v_2 \end{bmatrix} = v, \qquad (8)$$

the original coupled and nonlinear system given in (7) can be transformed into two decoupled and linear first-order single-input-single-output (SISO) subsystems with respect to v. The solution of (7) and (8) leads to the decoupling control law of

$$u = \begin{bmatrix} \dfrac{v_{ac} - v_1}{x_2} \\ \dfrac{c\left[(v_{ac} - v_1)x_1 - (v_2 + i_{load})x_2\right]}{x_2(x_2 - x_3)} \end{bmatrix}. \qquad (9)$$

Similarly, during the discharging mode of the PPB, the first-order time derivative of the output y can be derived based on (1)–(3) and (5) as

$$\dot{y} = \begin{bmatrix} v_{ac} \\ -i_{load} \end{bmatrix} + \begin{bmatrix} -x_2 & 0 \\ x_1 & \dfrac{x_3^2}{(x_2 - x_3)c} \end{bmatrix} \begin{bmatrix} u_1 \\ u_2 \end{bmatrix}. \qquad (10)$$

Selection of v as (8) leads to a new decoupling control law of

$$u = \begin{bmatrix} \dfrac{v_{ac} - v_1}{x_2} \\ \dfrac{c\left[(v_{ac} - v_1)x_1 - (v_2 + i_{load})x_2\right]}{x_2 x_3^2}(x_2 - x_3) \end{bmatrix}. \qquad (11)$$

A block diagram of the H^3 single-phase converter controlled by the above-mentioned input-output FBL is depicted in Fig. 3(a). Its resultant equivalent block diagram is given in Fig. 3(b). Clearly, the equivalent block diagram is always linear and decoupled when the decoupling laws (9) and (11) are applied, regardless of the operation mode of the converter.

With a feedback law of

$$\begin{bmatrix} v_1 \\ v_2 \end{bmatrix} = \begin{bmatrix} \dot{y}_1^* + \alpha_1(y_1^* - y_1) \\ \dot{y}_2^* + \alpha_2(y_2^* - y_2) \end{bmatrix}, \qquad (12)$$

the error dynamics of the closed-loop system can be derived from (8) and (12) as

$$\begin{cases} \dot{e}_1 + \alpha_1 e_1 = 0 \\ \dot{e}_2 + \alpha_2 e_2 = 0 \end{cases}, \qquad (13)$$

where $e_i = x_i^* - x_i$, and $\tau_i = 1/\alpha_i$ is the time constant of the corresponding control loop. α_1 and α_2 directly determine the bandwidths of the closed-loop system. Asymptotic reference tracking with a desirable convergence rate can be attained if the bandwidths of the two control loops are properly designed.

With FBL control, the zero dynamics of the system, i.e., the dynamic of x_3 when e_1 and e_2 reach zero, determine the stability of the overall system. Following the procedures in [12], one can verify that the zero dynamic of the overall system is always stable in the sense of Lyapunov given the proposed control law of (9), (11) and (12).

Finally, the operation mode of the PPB can be determined from the sign of the PPB current i_{PPB} (as marked in Fig. 2) as follows:

· when $i_{PPB} > 0$, the PPB operates in the charging mode and absorbs power;

· when $i_{PPB} < 0$, the PPB operates in the discharging mode and releases power.

According to (9) and (11), the averaged i_{PPB} over one switching cycle T_{sw} can be derived as

$$i_{PPB} = \frac{(v_{ac} - v_1)x_1 - (v_2 + i_{load})x_2}{x_2}. \qquad (14)$$

Therefore, the operation mode of the PPB can be easily determined by checking the polarity of (14). A complete system block diagram taking account of input power factor correction and average PPB voltage (denoted as x_{3-0}) regulation is shown in Fig. 4.

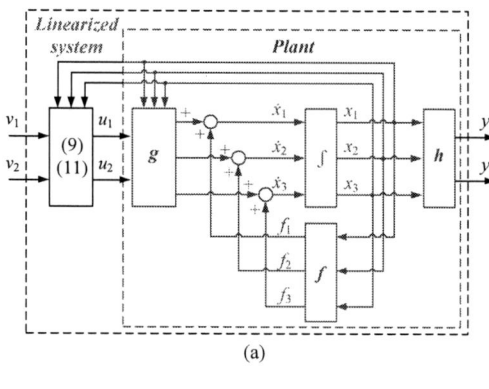

(a)

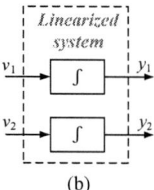

(b)

Fig. 3. (a) Block diagram of nonlinear and coupled model of the targeted H³ single-phase power converter with input-output FBL, and (b) its resultant equivalent block diagram.

IV. SIMULATION AND EXPERIMENTAL RESULTS

The proposed E-APD controller has been verified through processor-in-the-loop PSIM simulations, and some preliminary experimental results have also been obtained on a hardware prototype. The specifications of the tested H³ single-phase power converter are shown in Table III. The performance of the targeted converter with the proposed controller has been extensively examined with respect to its (i) steady-state operation, (ii) transient operation, (iii) startup and shutdown processes, and (vi) provision of active-power-filtering (APF) function.

A. Steady-State Operation

In this test, a resistive load of 1.53 kΩ is connected in parallel with the dc-port capacitor C_{dc}. The steady-state waveforms are displayed in Fig. 5. It can be observed that the ac-port current i_{ac} lines up well with the ac-port voltage v_{ac} and has a low THD of 1.14%, and that the dc-link voltage v_{dc} is regulated at 400 V with a peak-to-peak ripple of merely 2 V. On the other hand, the ripple-port voltage v_b fluctuates with a peak-to-peak ripple of 40 V. The large-signal variation of v_b implies the power-pulsation-buffering function of the PPB and thus constitutes the unique feature of the H³ single-phase power converter. The waveform of the ripple-port current i_{Lb} indicates the DCM operation of the PPB as i_{Lb} reaches zero in each switching cycle and remains zero until the next switching instant. The same test was also conducted on the hardware prototype given in Fig. 6 and the result conforms to the simulation result given in Fig. 5.

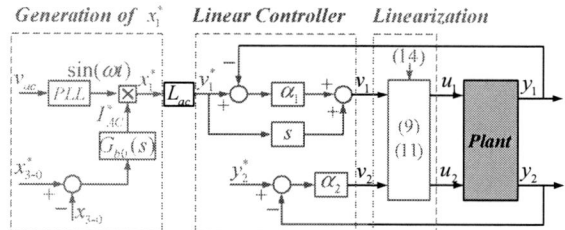

Fig. 4. Complete control block diagram of the proposed E-APD controller based on input-output FBL for the targeted H³ single-phase power converter.

TABLE III
SPECIFICATIONS OF THE H³ SINGLE-PHASE POWER CONVERTER

Parameter	Value
Rated power	100 W
Switching frequency	25 kHz
Input voltage v_{ac}	220 V (RMS)/ 50 Hz
Ac-port inductance L_{ac}	7 mH
Output voltage V_{dc}	400 V
Dc-link capacitance C_{dc}	10 μF
Ripple-port capacitance C_b	30 μF
Ripple-port inductance L_b	212 μH
Average ripple-port voltage v_{b0}	275 V
Time constant of current loop τ_1	80 μs
Time constant of voltage loop τ_2	250 μs

The 2018 International Power Electronics Conference

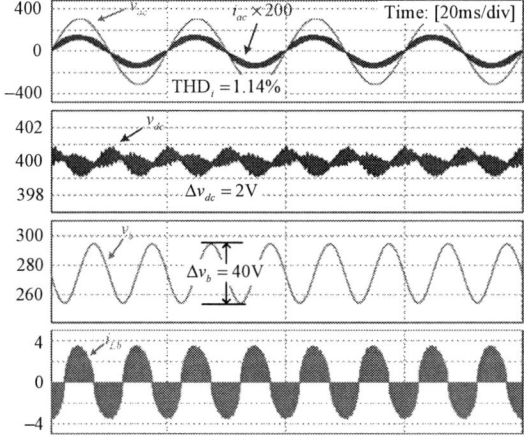

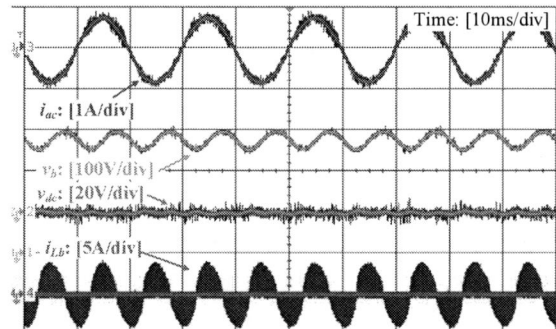

Fig. 5. Steady-state waveforms of the system.

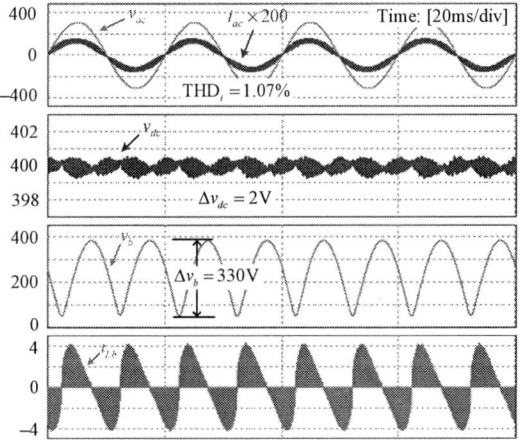

Fig. 7. Steady-state waveforms of the system with a reduced buffering capacitor of 4.6 μF.

Fig. 6. Steady-state waveforms of the system in the experiment (v_{dc} is ac-coupled to display the detail).

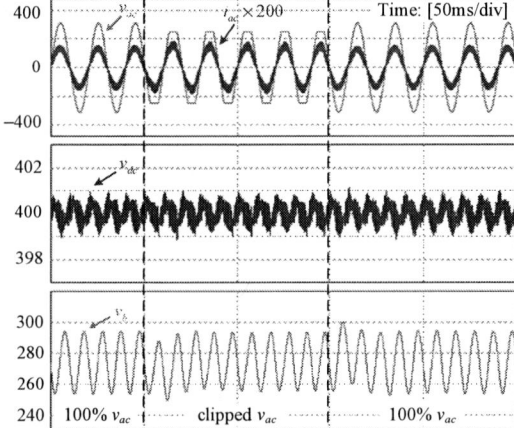

Fig. 8. Waveforms of the system under input-voltage clipping with a crest factor of 1.2.

As the proposed E-APD controller features large-signal stability, the buffering capacitor C_b can be further reduced to approach the physical limit of 3.98 μF, as analyzed in [9]. Fig. 7 displays the steady-state waveforms of the H³ single-phase power converter with a reduced C_b of 4.6 μF. Although the ripple-port voltage v_b swings greatly between 50 V and 380 V, the THD of i_{ac} and the peak-to-peak voltage ripple of v_{dc} remains unchanged.

In practice, the capacitance C_b cannot be too close to the physical limit in order to reserve adequate energy storage margins for transient intervals. Therefore, the buffering capacitor of 30 μF is used in all the other tests.

B. Transient Operation

The transient performance of the system is examined by imposing disturbances to input voltage, load, and output-voltage reference.

In the input-voltage-disturbance test, the input voltage v_{ac} is clipped with a crest factor of 1.2 and recovers after 100 ms as shown in Fig. 8. Although the clipped v_{ac} contains substantial harmonic components, the result shows that the output voltage v_{dc} is kept well-regulated in the whole transient process and that the ac-port current i_{ac} retains good alignment with v_{ac}.

Fig. 9 illustrates the dynamic responses of the power converter under the step changes of the dc-port load between 0 W and 100 W (full load). The dc-link voltage is shown to be robust against load disturbances. The imbalanced power between the ac-port and the dc-port introduced by the step change of the load is effectively buffered by the PPB, as can be observed in the waveform of v_b in Fig. 9.

The robustness of the proposed E-APD control is also examined by applying a time-varying nonlinear load at the dc-port. The nonlinear load comprises a resistive load (about 100 W) and a 37-Hz sinusoidal current sink (with a power fluctuation of 40 W). Fig. 10 shows that v_{dc} keeps tightly regulated at 400 V with a small variation of 2 V, despite the substantial and irregular load power fluctuation. The ac-port current i_{ac} is also in phase with the input voltage v_{ac} with a negligible deformation in the shape. The waveform of v_b implies an irregular unbalanced power difference between the ac- and dc-ports that is buffered. The waveforms in Fig. 10 demonstrate the enhanced robustness and decoupled characteristics with the proposed E-APD control.

The 2018 International Power Electronics Conference

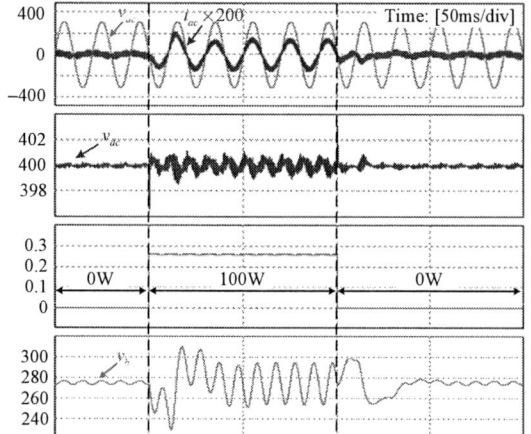

Fig. 9. Waveforms of the system under step change of load between 0 W and 100 W (full load).

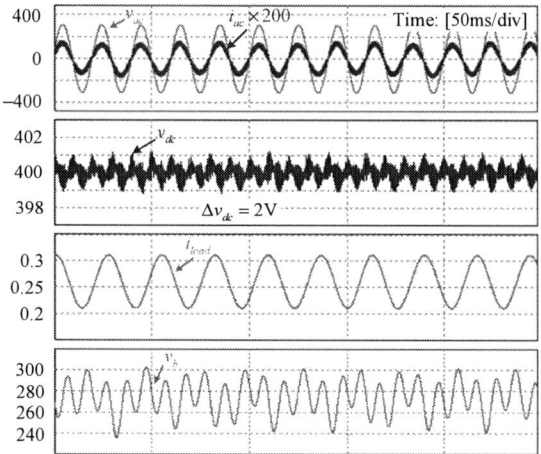

Fig. 10. Waveforms of the system with a time-varying nonlinear load.

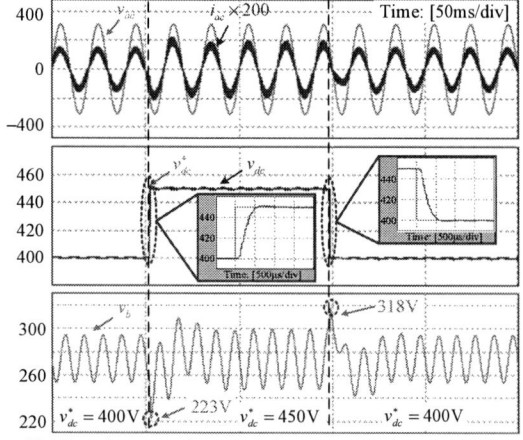

Fig. 11. Transient waveforms of the system under step change of output-voltage reference.

Fig. 11 illustrates the performance of the targeted H^3 single-phase power converter when the dc-port voltage reference v_{dc}^* is step changed between 400 V and 450 V. The waveforms of v_{dc} at the transient instants are enlarged, and exhibit two typical first-order responses

both with a settling time of around 1 ms. The dynamics of the output voltage v_{dc} confirm the theoretical analysis in (13) and match the desired time constant of 250 μs as listed in Table III.

C. Startup and Shutdown Processes

The startup and shutdown performances of the H^3 single-phase converter with the proposed E-APD control are also examined.

The startup process of the converter includes three phases (i.e. pre-charging phase, charging phase, and full-load phase) and the waveforms are illustrated in Fig. 12. In the pre-charging phase, the dc-link capacitor C_{dc} is charged to about 300 V through the body diodes of the switches S_1–S_4, and the ripple-port capacitor C_b is charged to about 220 V through switch S_5 by gradually increasing the duty cycle of S_5 in an open-loop manner. Since the PPB has been charged up, the system can switch into the charging phase and the output voltage can be quickly charged to the nominal value 400 V following the first-order curve as highlighted in Fig. 12. The waveforms suggest that the whole process of establishing the output voltage v_{dc} can be as short as 40 ms without any overshoot in v_{dc} and full load can be connected to the output immediately afterwards.

The shutdown waveforms of the converter are recorded in Fig. 13. In practice, output-voltage-hold-up of at least half to one line cycle is required for many consumer electronics applications (e.g PC power supply unit (PSU)). Due to its low energy storage, an H^3 single-phase power converter is generally regarded as being incapable of providing sufficient hold-up time. However, Fig. 13 shows that the output voltage v_{dc} remains regulated at 400 V for another 10 ms (i.e. half line cycle) after the input voltage v_{ac} is shut down. The output voltage starts to drop when the PPB is fully discharged. With the proposed E-APD control, the small energy storage in the converter can be fully utilized for supplying the load without compromising the hold-up time.

D. Provision of Active Power Filtering Function

The feasibility of providing the APF function with the H^3 single-phase converter is also studied. A single-phase diode-bridge rectifier with an RC load is used as a nonlinear load connecting to the main ac grid, whilst the H^3 single-phase converter is controlled to compensate the harmonic current of the nonlinear load. The results are depicted in Fig. 14, where i_g is the grid current which equals to the total current of the nonlinear current and the ac-port current of the H^3 single-phase converter. It can be observed that the THD of i_g greatly decreases from 101% to 5.42% after the APF function is activated. The output voltage v_{dc} has a small peak-to-peak variation of 5 V when the harmonics are compensated. The irregular waveform of v_b implies the unbalanced power between the ac- and dc-ports that is buffered.

894

The 2018 International Power Electronics Conference

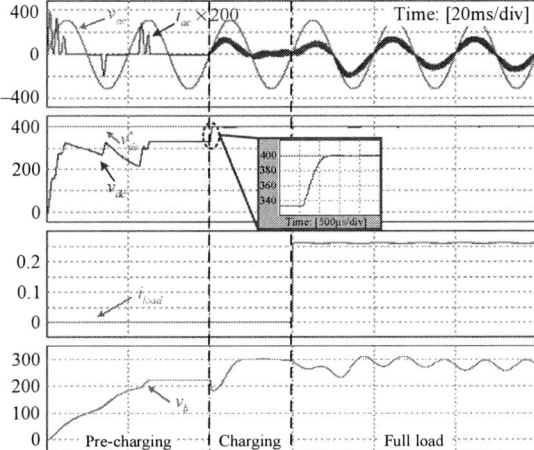

Fig. 12. Waveforms of the system during the start-up process.

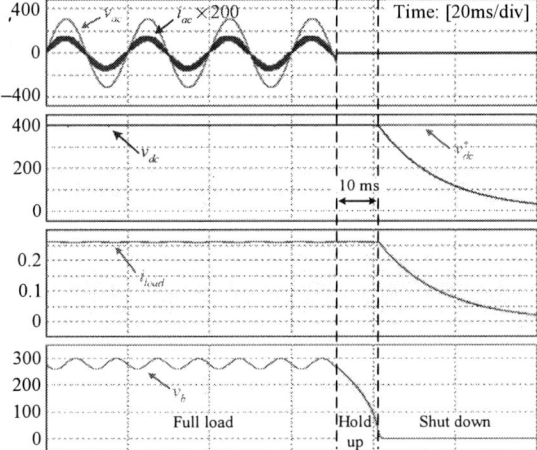

Fig. 13. Waveforms of the system during the shut-down process.

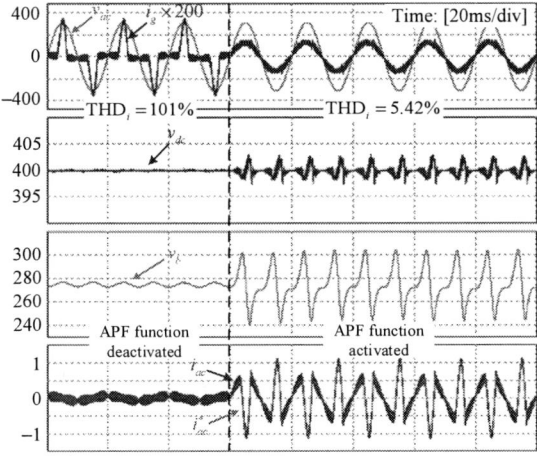

Fig. 14. Waveforms of the system in the active-power-filtering test.

power converters with small power-pulsation-buffering capacitors. Three key distinctions between H^3 single-phase power converters and their conventional counterparts are firstly identified through generic network modeling. The control challenges of H^3 single-phase power converters justify the adoption of the proposed nonlinear controller. A complementing APD control strategy is also applied to further enhance the dynamic performance and robustness of the system. The presented control has been verified through various simulation tests and some preliminary experimental tests. The results show that the proposed control gives faster dynamic responses and improved robustness as compared to conventional solutions.

REFERENCES

[1] GOOGLE, "Detailed inverter specifications, testing procedure, and technical approach and testing application requirements for the Little Box Challenge," Mountain View, Tech. Rep., 2015. [Online]. Available: https://www.littleboxchallenge.com/.

[2] D. Bortis, D. Neumayr, and J. W. Kolar, "$\eta\rho$-Pareto optimization and comparative evaluation of inverter concepts considered for the GOOGLE Little Box Challenge," in *IEEE 17th Workshop Contr. Modeling Power Electron.*, 2016, pp. 1–5.

[3] Y. Lei *et al.*, "A 2-kW single-phase seven-level flying capacitor multilevel inverter with an active energy buffer," *IEEE Trans. Power Electron.*, vol. 32, no. 11, pp. 8570–8581, Nov. 2017.

[4] Z. Liu, F. C. Lee, Q. Li, and Y. Yang, "Design of GaN-based MHz totem-pole PFC rectifier," *IEEE J. Emerg. Sel. Top. Power Electron.*, vol. 4, no. 3, pp. 799–807, Sep. 2016.

[5] A. S. Morsy, and P. N. Enjeti, "Comparison of active power decoupling methods for high-power-density single-phase inverters using wide-bandgap FETs for Google Little Box challenge," *IEEE J. Emerg. Sel. Top. Power Electron.*, vol. 4, no. 3, pp. 790–798, Sep. 2016.

[6] J. W. Kolar, D. Bortis, and D. Neumayr, "The ideal switch is not enough," in *28th Int. Symp. Power Semicond. Devices ICs*, 2016, pp. 15–22.

[7] Y. Tang, Z. Qin, F. Blaabjerg, and P. C. Loh, "A dual voltage control strategy for single-phase PWM converters with power decoupling function," *IEEE Trans. Power Electron.*, vol. 30, no. 12, pp. 7060–7071, Dec. 2015.

[8] S. Li, W. Qi, S. C. Tan, and S. Y. Hui, "Enhanced automatic-power-decoupling control method for single-phase ac-to-dc converters," *IEEE Trans. Power Electron.*, vol. 33, no. 2, pp. 1816-1828, Feb. 2018.

[9] P. T. Krein, R. S. Balog, and M. Mirjafari, "Minimum energy and capacitance requirements for single-phase inverters and rectifiers using a ripple port," *IEEE Trans. Power Electron.*, vol. 27, no. 11, pp. 4690–4698, Nov. 2012.

[10] J. E. Slotine, and W. Li, *Applied nonlinear control*. Englewood Cliffs, NJ: Prentice-Hall, 1991.

[11] D. Neumayr, D. Bortis, and J. W. Kolar, "Ultra-compact power pulsation buffer for single-phase dc/ac converter systems," in *IEEE 8th Int. Power Electron. Motion Contr. Conference*, 2016, pp. 2732–2741.

[12] T.-S. Lee, "Input-output linearization and zero-dynamics control of three-phase AC/DC voltage-source converters," *IEEE Trans. Power Electron.*, vol. 18, no. 1, pp. 11–22, Jan. 2003.

V. CONCLUSIONS

In this paper, a nonlinear control method is proposed for the performance enhancement of high power density, high efficiency, and high reliability (H^3) single-phase

The 2018 International Power Electronics Conference

Zero Voltage Switching Scheme for Flyback Converter to Ensure Compatibility with Active Power Decoupling Capability

Hiroki Watanabe[1*], Jun-ichi Itoh[1]

1 Department of Electrical, Electronics and Information Engineering, Nagaoka University of Technology, Nagaoka, Japan
*E-mail: hwatanabe@stn.nagaokaut.ac.jp

Abstract—In this paper, a novel ZVS method with synchronous rectifier is proposed in order to achieve the soft switching and the active power decoupling without additional component. The proposed method separates the ZVS operation and the snubber capability to keep the discontinuous current mode (DCM) for the power decoupling. In addition, an active clamp circuit which achieves soft switching operation is applied for the surge voltage suppression of a main switch. From the experimental results, 0.4% of a second-order harmonics on a DC input current is obtained in comparison with a DC component. In addition, it is confirmed that the surge voltage is reduced by 50% owing to the active clamp circuit, and the ZVS operation by the experiment.

Keywords— micro inverter; active power decoupling; zero voltage switching

I. INTRODUCTION

Recently, micro inverter systems for PV applications are actively researched for a sustainable power solution due to the attractive characteristics such as; flexibility, high-system efficiency, and low manufacturing cost [1]. The micro inverter promisingly becomes a trend for the future PV system instead of the large capacity inverters, and the micro inverter requires the high reliability because the large number of the converter units are adopted to the PV generation system. Generally, in the power converter for DC to single-phase AC grid system, an electrolytic capacitor is usually employed owing to the large capacitance for the double-line frequency power ripple compensation. These capacitors limit the life-time of the converter, which results in low reliability.

According to this problem, the active power decoupling techniques which have a film or ceramic capacitor are attracted as the one of this solution [2]-[9]. The active power decoupling circuit charges and discharges the decoupling capacitor in order to absorb the double-line frequency power ripple, which enables to small capacitance. As a result, large electrolytic capacitor does not necessary. However, a lot of conventional active power decoupling circuits require the additional components such as the switching devices and the additional inductor. Although the low cost is the one of the advantages in the micro inverter, these components

increase complexity of the circuit configuration and the cost. Therefore, the active power decoupling technique which utilize the discontinuous current mode (DCM) was already proposed by authors [10], and it reduces the DC link capacitance without the additional components.

The ZVS techniques for the flyback converter have been hot research topics in order to reduce the switching losses and the voltage stress due to the leakage energy of transformer [11]. The quasi-resonant (Q-R) operation is the one method, which adjusts the switching timing to the zero voltage by a pulse frequency modulation (PFM) [12].

The other technique utilizes an active clamp circuit for the ZVS operation. In this method, the magnetizing current becomes negative due to the series resonance between the leakage inductance and the clamp capacitor, and the parasitic capacitor is discharged before turn-on. As a result, the ZVS operation is obtained.

In the active power decoupling technique by authors, the flyback converter has to limit the operation condition because it is achieved under the constant switching frequency and the DCM operation. In this case, the conventional ZVS method as the Q-R operation and the active clamp circuit cannot apply to the proposed converter. This is because the Q-R operation needs the PFM, and conventional active clamp method has to operate under the continuous current mode (CCM).

This paper proposes a novel ZVS scheme to ensure compatibility with the active power decoupling circuit. For that, the synchronous rectifier is applied instead of the diode rectifier. In addition, the active clamp circuit is only used to reduce the turn-off surge voltage of the main switch. The originality of this paper is that the switching pattern is added to a synchronous rectifier for ZVS operation. Similar ZVS method with forward converter was already reported by Ref. [13], however, this method is also operated under the CCM condition. The differential point is that the proposed converter needs the zero current period of the magnetizing current for power decoupling. In addition, the magnetizing current should be negative to discharge the parasitic capacitor of the main switch. In order to satisfy these condition, the conduction modes for the synchronous rectifier switch is turned-on before the main switch is turned-on, and the active clamp circuit is turned-off when the released energy of the magnetizing inductance becomes zero to ensure the zero current period for DCM. Finally, the active power decoupling and the ZVS operation is confirmed by experiment.

II. CIRCUIT CONFIGURATION

Fig.1 shows a conventional micro inverter topology with the conventional active power decoupling circuit [14]. In order to reduce the buffer capacitance C_{buf}, the small decoupling capacitor C_{apd} and the auxiliary circuit are adopted on the transformer secondary side. The decoupling capacitor C_{apd} is charged and discharged in synchronized with the double-line frequency to compensate the power ripple with small capacitor. However, these additional components complex the circuit configuration, and the transformer has to compose the three winding transformer which has complicate configuration.

Fig. 2 shows the proposed converter which consists of the flyback converter, the voltage source inverter (VSI) and small buffer capacitor C_{buf}. The flyback converter isolates between the PV and the single-phase grid, and it has the high voltage gain to push up the DC input voltage. The proposed converter compensates the double-line frequency power ripple by the small DC link capacitor C_{buf} without additional components. In addition, the active clamp circuit is applied for the surge voltage suppression due to the leakage energy of the transformer. Moreover, the synchronous rectifier is implemented for the ZVS operation and the reduction of the conduction losses.

The hard switched flyback converters undergo substantial switching losses in the high switching frequency and high voltage stress on the each switching devices. This is because the parasitic capacitor of S_1 is quickly charged due to the leakage energy. Typically, The RCD snubber is connected to primary side in order to suppress the huge surge voltage. However, the reduction of the switching losses and the snubber losses is the important to improve the conversion efficiency. According to these reason, the active clamp circuit is considered. The leakage energy transfers to the clamp capacitor C_{clamp} when the active clamp switch S_{clamp} is turned-on. In order to reduce the peak surge voltage, the C_{clamp} should be large to delay the charge period in comparison with the parasitic capacitor.

Fig. 3 shows the principle of the power decoupling between the DC and single-phase AC sides. The proposed active power decoupling separates the DC component and double-line frequency component by the flyback converter. Thus, the DC input power P_{in} is always constant, and the DC link capacitor C_{buf} stores the differential power between input and output power.

When both the output voltage and current waveforms are sinusoidal, the instantaneous output power p_{out} is expressed as

$$p_{buf} = \frac{V_{acp} I_{acp}}{2}(1 - \cos 2\omega t) \qquad (1)$$

where V_{acp} is the peak voltage, I_{acp} is the peak current, and ω is the angular frequency of the output voltage. From (1),

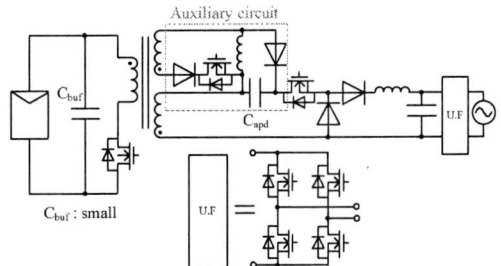

Fig 1. Conventional flyback micro inverter with active power decoupling circuit. It can reduce the capacitance of C_{buf}. However, auxiliary circuit is necessary.

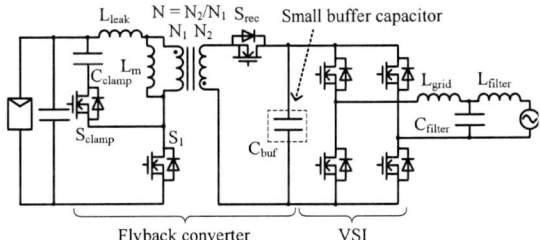

Fig 2. Circuit configuration of proposed flyback converter. Active clamp circuit suppress turn-off surge voltage. Synchronous rectifier utilizes ZVS operation for S_1.

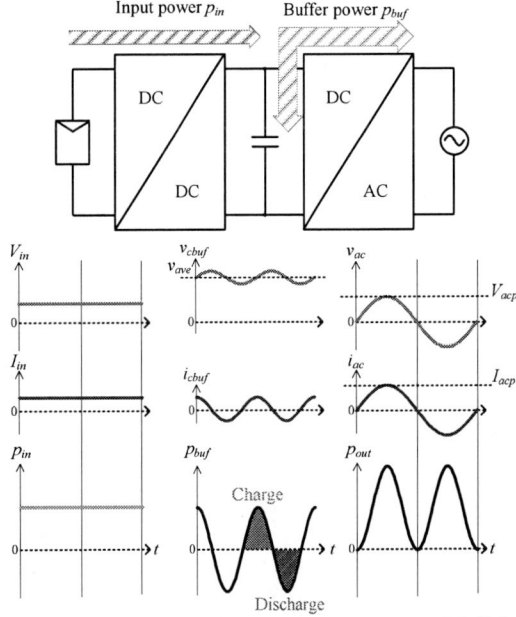

Fig 3. Relationship between input and output power. DC link capacitor receives differential power due to double-line frequency power ripple, and input power becomes constant.

the power ripple that contains double-line frequency of the power grid, appears at DC link.

In order to absorb the power ripple, the instantaneous power p_{buf} is expressed as

$$p_{buf} = \frac{1}{2} V_{acp} I_{acp} \cos 2\omega t \qquad (2)$$

897

where, the polarity of the p_{buf}, is defined as positive when the DC link capacitor C_{buf} discharges. Note that the active power of C_{buf} should be zero. Owing to the power decoupling capability, the input power is matched to the output power. Thus, the relationship between the input and output power is expressed as

$$p_{pv} = \frac{1}{2} V_{acp} I_{acp} = V_{in} I_{in} \qquad (3)$$

The proposed converter achieves the active power decoupling by the DC link capacitor C_{buf}. As a result, the DC link voltage v_{cbuf} fluctuates at the double-line frequency of the single-phase grid frequency as Fig. 3.

III. OPERATION MODES OF PROPOSED CONVERTER

A. Active power decoupling operation

Fig. 4 shows the drain current waveforms of the main switch S_1 without active clamp circuit. The proposed active power decoupling method is very simple, and the detail is following.

Firstly, the average primary current both CCM and DCM are expressed as

$$I_{ave\_CCM} = \frac{V_{dc}}{V_{in}} I_{dc} \qquad (4)$$

$$I_{ave\_DCM} = \frac{I_{peak}}{2} D_{on} \qquad (5)$$

$$I_{peak} = \frac{V_{in}}{L_m} D_{on} T_{sw} \qquad (6)$$

where, I_{ave_ccm} is the average current in CCM, V_{dc} is the DC link voltage, I_{dc} is the DC link current, V_{in} is the DC input voltage, L_m is the magnetizing inductance, D_{on} is the on duty ratio of the main switch S_1, T_{sw} is the switching period. Note that, the V_{dc} fluctuates at double-line frequency when the DC link capacitor is small. In the CCM condition, the PV side average current I_{ave_ccm} is decided from DC link parameters such as V_{dc} and I_{dc}. As a result, the power decoupling fails.

On the other hand, the PV side average current in DCM I_{ave_DCM} is only decided primary parameters such as DC input voltage, and the switching period and on duty. In order to achieve the active power decoupling, the proposed converter is operated by the constant on duty reference and the constant switching frequency. As a result, constant average current is obtained, and the double-line frequency power ripple does not occurs on PV side.

Fig.5 shows the current waveforms of the DC input current I_{in}, DC link current i_{dc}, and the inverter output current i_{ac}. The proposed converter obtains the constant input power by the flyback converter, and the DC link capacitor C_{buf} compensates the power ripple due to single-phase AC grid. In this case, the DC link current includes

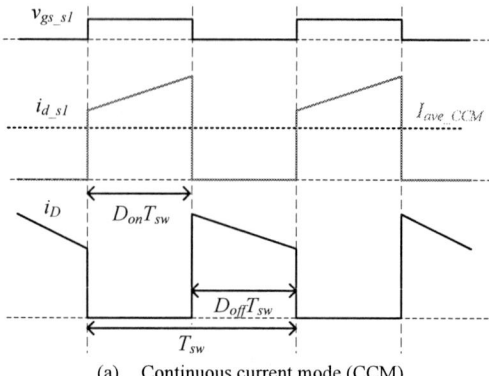

(a)   Continuous current mode (CCM)

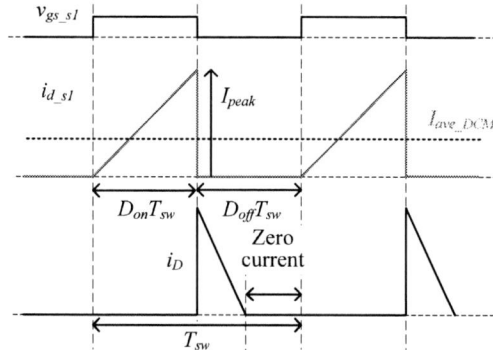

(b)   Discontinuous current mode (DCM)

Fig.4. Fundamental waveforms without active clamp circuit. In DCM, power decoupling can achieve under constant duty reference and switching frequency condition.

the charge and discharge current of C_{buf}, and it has the double-line frequency component as Fig. 5.

According to these method, the proposed converter decouples the DC components and the double-line frequency components. However, it limits the operation condition of the flyback converter to constant duty command and switching frequency. As a result, the conventional ZVS techniques cannot apply to the proposed converter. This is because the Q-R operation needs the PFM. In addition, the ZVS with the active clamp circuit is operated under the CCM condition. In order to solve this problem, the new ZVS method is proposed from the next chapter.

B. Operation modes of flyback converter

Fig. 6 shows the key ZVS switching waveforms of the conventional and proposed method. The large surge voltage occurs on main switch S_1 due to the resonance between leakage inductance L_{leak} and the parasitic capacitor of S_1 at turn-off transition when the snubber circuit does not places. The active clamp circuit suppresses the peak drain-source voltage V_{ds_s1} by the clamp capacitor C_{clamp}, and the clamp voltage is expressed as

$$V_{dS\_s1} = V_{in} + \frac{V_{dc}}{N} \qquad (7)$$

where, N is the turn ratio of the transformer.

According to Fig. 6 (a), the active clamp circuit is used for both the ZVS operation and the surge voltage suppression. The active clamp switch S_{clamp} and S_1 are switched to complementation through the dead-time period [15]. At the dead-time before turn-on of S_1, the magnetizing current i_{Lm} reaches to the negative current. In this state, the parallel diode is turned-on, and the parasitic capacitor discharged. As a result, ZVS is achieved at S_1. However, this method is only applied the continuous current mode, and it disturbs the proposed power decoupling operation as (4). Thus, the magnetizing current has to become discontinuous current for the proposed power decoupling.

Fig. 6 (b) shows the proposed ZVS method using synchronous rectifier. In the proposed method, the active clamp circuit is only used for the surge voltage suppression. Instead, synchronous rectifier switch S_{rec} has the turned-on state before turn-on of S_1 in order to generate the negative current for ZVS.

Fig.7 shows the operation modes of the proposed flyback converter, and the detail of each mode are described in the following.

Mode 1 (t_0 - t_1):

In mode 1, the main switch S_1 is turned-on, and the active clamp switch S_{clamp} is remaining off-state. The DC input voltage is applied to the transformer and each energy storage. The magnetizing current linearly increases until the end of the mode 1, and the energy is stored both leakage inductance and magnetizing inductance.

Mode 2 ($t_1 - t_2$, dead-time):

This mode starts when the main switch S_1 is turned-off. In this mode, the energy of the leakage inductance L_{leak} is transferred to the parasitic capacitor C_{oss}. In this mode, the synchronous rectifier switch S_{rec} and the parallel diode of S_{rec} is remaining off-state. Thus, the primary side composes the series resonance circuit with L_m, L_{leak} and C_{oss}. Note that C_{oss} is very small, and it is charged quickly. This mode is the end when the V_{ds_s1} reaches to (7).

Mode 3 ($t_2 - t_3$, dead-time):

This mode starts when the parallel diode of S_{rec} and S_{clamp} is turned-on, and the magnetizing inductor energy is released and delivered to the VSI side. L_{leak}, C_{clamp} are resonant during in this interval. This mode is the end when the S_{clamp} is turned-on.

Mode 4: ($t_3 - t_4$)

This mode starts when S_{rec} and S_{clamp} is turned-on. The period of this interval is set to the energy release period of the magnetizing inductance, and it is expressed as

$$T_{t3-t4} = I_{peak} \frac{NL_m}{V_{dc}} \qquad (8)$$

where, I_{peak} is the magnetizing peak current, N is the turn ratio of the transformer, L_m is the magnetizing inductance, and the V_{dc} is the DC link voltage. At the $t = t_3$, S_{clamp}

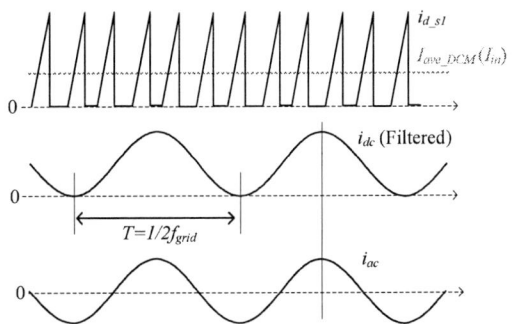

Fig.5. Current waveforms of PV input current I_{ave_DCM}, DC link current i_{dc} and inverter output current i_{ac}. DC link capacitor is charged and discharged due to double-line frequency in order to compensate the differential power during input and output power.

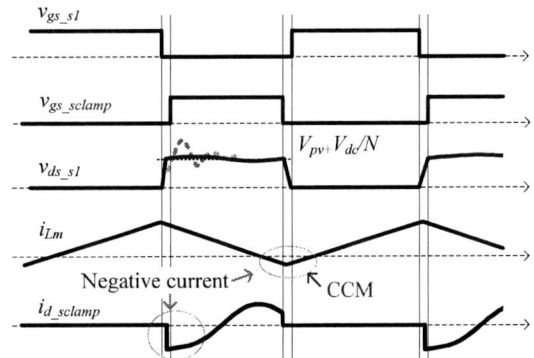

(a) Conventional ZVS method with active clamp circuit. This method, active clamp circuit and main circuit are switched to complementation.

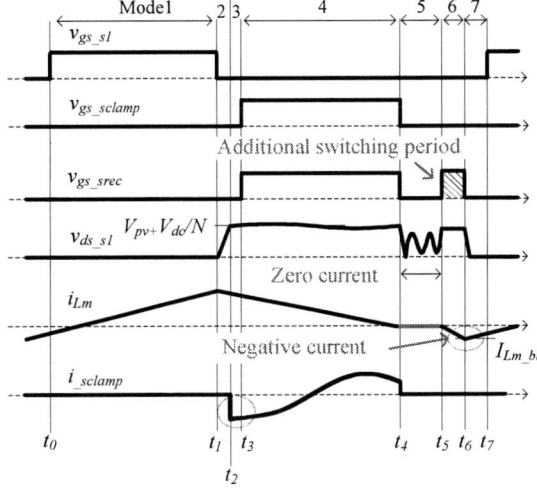

(b) Proposed ZVS method with active clamp circuit.

Fig.6. Key waveforms of flyback converter in switching cycle. In order to achieve ZVS, negative current period is necessary before S_1 is turned-on. In addition, zero current period is required for proposed active power decoupling capability.

achieves the ZVS because the parasitic capacitor of S_{clamp} is already discharged before the turn-on.

Mode 5: ($t_4 - t_5$)

This modes starts when the magnetizing current becomes zero. The all switches are tuned-off, and the magnetizing current remaining zero. The proposed active power decoupling method operates under DCM. Thus, this mode is important in order to achieve the active power decoupling capability without the additional components. In this interval, the primary side composes the series resonance circuit with L_m, L_{leak} and C_{oss}, and the voltage oscillation occurs. The resonance period is expressed as

$$T_{res} = 2\pi \sqrt{(L_{leak} + L_m)C_{oss}} \tag{9}$$

Mode 6: ($t_5 - t_6$)

This mode is for the ZVS operation of the main switch S_1. The negative voltage provides to the transformer, and the magnetizing current reaches negative when the synchronous rectifier S_{rec} is turned-on. The bottom current of the magnetizing inductance I_{Lm_bt} is expressed as

$$I_{Lm\_bt} = \frac{V_{dc}}{NL_m} T_{t5-t6} \tag{10}$$

Mode 7 ($t_6 - t_7$, dead-time):

At the $t = t_6$, the magnetizing current is remaining negative. In this interval, the parasitic capacitor C_{oss} is discharged, and the drain source voltage v_{ds-s1} is decreased to zero. After that, the parallel diode of S_1 is turned-on.

After this mode, S_1 is turned-on, and the ZVS is achieved.

The proposed converter is added the switching state of mode 6 and 7 for ZVS. Due to these modes, the parasitic capacitor C_{oss} is discharged before turn-on of S_1. In addition, the operation mode of the proposed converter has the zero current period as mode 5, and the magnetizing current becomes discontinuous current. As a result, the proposed power decoupling is also achieved.

C. Control block diagram

The control block diagrams of the flyback converter is shown in Fig. 7. The Flyback converter operates by the constant duty reference and switching frequency. In order to avoid the short circuit between S_1, S_{clamp}, and S_{rec}, the dead-time is implemented after modulation part. In addition, the dead-time of S_1 is separated to two steps for generation of the synchronous rectifier on-state.

According to (5), the PV side average current is decided by the peak current of the magnetizing current, and the on duty including the dead-time period is expressed as

$$D_{on\_s1} = D_{ref} - \frac{T_{dead}}{T_{sw}} \tag{11}$$

$$T_{dead} = T_{d1} + T_{d2} \tag{12}$$

where, T_{dead} is the dead-time period, and it is the sum of the dead-time T_{d1} and T_{d2}. The turn-on period of S_{clamp} equal to the energy release period of the magnetizing inductance, and it is expressed as

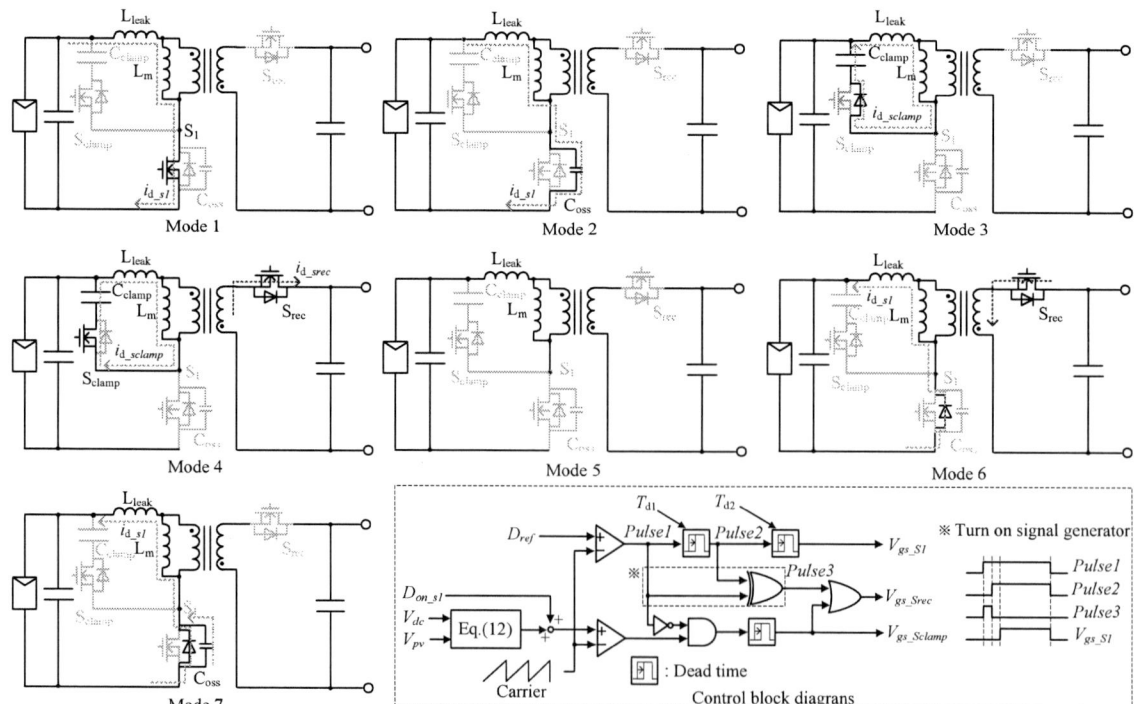

Fig.7. Operation modes and the control block diagrams of flyback converter. In order to achieve ZVS, mode 6 and 7 which discharge the parasitic capacitor are proposed.

900

$$D_{on\_sclamp} = \frac{NV_{pv}}{V_{dc}} D_{on\_s1} \qquad (13)$$

The gate signals for S_{rec} is the sum of the active clamp switch gate signals V_{gs_sclamp} and the turn-on signals for ZVS (e.g. *Pulse 3*). Note that, pulse period is set by the dead-time of T_{d1}.

IV. SIMULATION RESULT

Table. 1 shows the simulation parameters. In this simulation, the leakage inductance L_{leak} and the parasitic capacitor C_{oss} is adopted to evaluate the ZVS operation. In order to suppress the surge voltage, the clamp capacitor is set to 10μF.

Fig. 8 shows the switching waveforms with the diode rectification and the proposed method. Firstly, the surge voltage is suppressed by the Active clamp circuit both condition. However, according to Fig. 8 (a), the drain-source voltage v_{ds_s1} does not zero when the S_1 is turned-on, and the magnetizing current becomes positive value. It is meaning that the hard-switching condition occurs, and the parasitic capacitor cannot be discharged by the magnetizing current.

On the other hand, according to Fig. 8 (b), the v_{ds_s1} becomes zero before the turn-on of S_1. In addition, it is confirmed that the magnetizing current becomes negative value at dead-time period. From these result, the operation of the proposed method is confirmed. Note that, the conduction mode of the rectifier circuit is necessary in

order to achieve ZVS, and the conduction losses becomes increase on the synchronous rectifier circuit without the ZVS operation. However, the additional conductive period is short, and the secondary current is small in comparison with the primary current. Thus, the increase of conduction losses is small.

V. EXPERIMENTAL RESULT

Table. 2 shows the experimental parameters. This chapter presents experimental results using 300 W prototype drawn in Fig. 2 to confirm the active power decoupling operation and the proposed ZVS method. Note that VSI is operated under the open loop control with R-L load. Besides, the duty command compensation is applied to the VSI control. This is because the DC link voltage

Table.1 Simulation parameters

Symbol	Quantity	Value
V_{in}	Input voltage	50 V
P_{in}	Input power	300 W
f_{sw}	Switching frequency	50 kHz
L_m	Magnetizing inductance	10 μH
L_{leak}	Leakage inductance	100 nH
C_{oss}	Parasitic capacitor	1000 pF
C_{dc}	DC link capacitor	50 μF
C_{clamp}	Clamp capacitor	10 μF

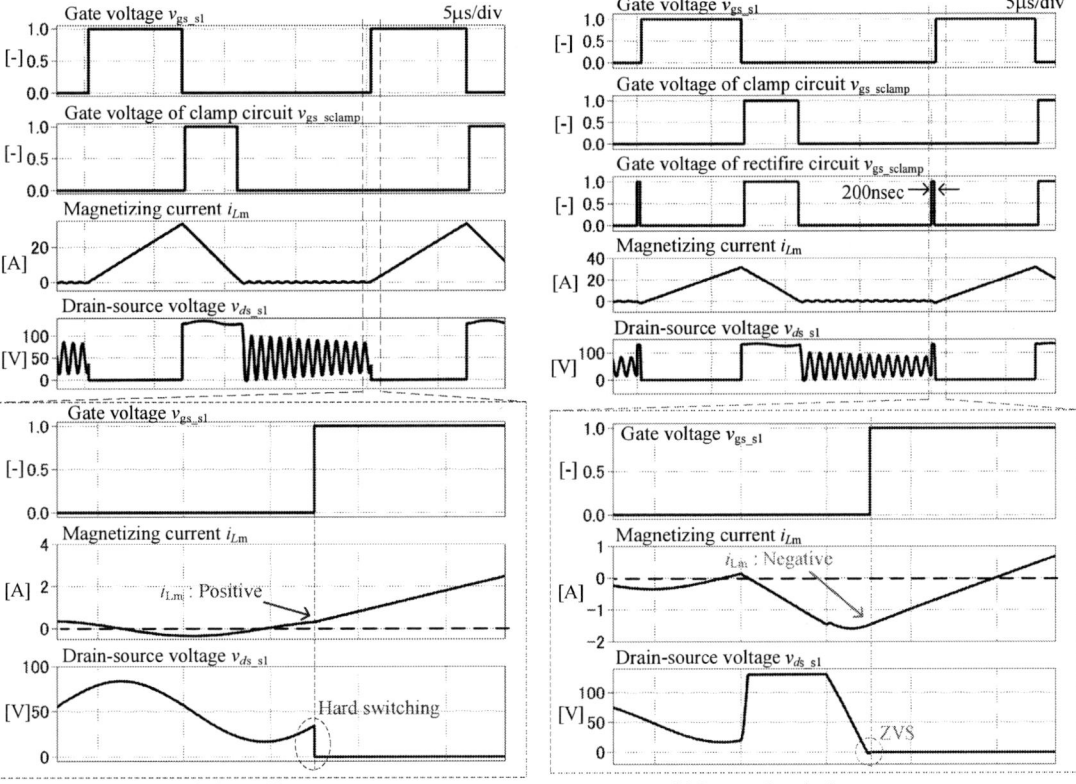

(a) Diode rectifier   (b) Proposed method

Fig.8. Simulation result with switching waveforms both diode rectifier and proposed method

fluctuates by the power decoupling operation. The switching frequency is set to 50 kHz, and the decoupling capacitor is selected to 33μF in order to apply the small film capacitor.

Fig. 9 shows the input and output waveforms with proposed method. According to Fig. 9 (a), the DC input current is regulated under the constant value, and the DC link capacitor voltage fluctuates at the double-line frequency as 100 Hz owing to compensate the single-phase AC power ripple. In addition, the sinusoidal waveforms of the inverter output voltage and current are obtained. According to Fig. 9 (b), the second order harmonics of 100 Hz is 0.4% by the DC components. From this result, the power decoupling operation with 33μF is confirmed by experiment.

Fig. 10 shows the experimental results which does not use the snubber circuit and the proposed method. Note that the evaluation of the surge voltage suppression and the ZVS operation is confirmed by light load condition because the huge surge voltage occurs by snubber less condition. As shown in Fig. 10, the surge voltage occurs when the S_1 is turned-off due to the leakage inductance of the transformer. In addition, the current oscillation also occurs at turn-on period by the hard switching operation. In order to improve these waveforms, the active clamp circuit and the proposed method is applied in Fig. 11.

Fig. 11 shows the switching forms with active clamp circuit and proposed method at light load condition. Note that, Fig. 11 (a) is the diode rectifier condition, and the synchronous rectifier is constantly turned-off. According to Fig.11 (a), the surge voltage is suppressed by 50% using the active clamp circuit. However, the current oscillation cannot improve due to the hard switching operation as shown in Fig. 11(c).

According to Fig. 11 (b), the surge voltage suppression and the ZVS are achieved by the active clamp circuit and the proposed method. As shown in Fig. 11 (d), the drain-source voltage of S_1 becomes zero before the S_1 is turned-on. From these results, the active power decoupling and the ZVS operation are confirmed by the experiment.

VI. CONCLUSION

This paper discussed ZVS operation method of the flyback converter with the active power decoupling capability. The proposed method converter is operated under constant frequency and duty command for the active power decoupling, and the synchronous rectifier is applied in order to achieve ZVS for the flyback converter. From the experimental result, the second-order harmonics in the DC input current becomes 0.4% of the DC average current. Finally, the surge voltage suppression and the ZVS operation are confirmed by the experiment.

In the future work, the circuit design and operation optimization will be considered in order to improve the conversion efficiency.

Table.2 Experimental parameters

Symbol	Quantity	value
V_{in}	Input voltage	50 V
P_{in}	Input power	300 W
f_{sw}	Switching frequency	50 kHz
D_{ref}	On duty command	0.5
L_m	Magnetizing inductance	11 μH
L_{leak}	Leakage inductance	250 nH
C_{oss}	Parasitic capacitor of S_1	1100 pF
C_{clamp}	Clamp capacitor	6 μF
C_{buf}	Decoupling capacitor (DC link cap.)	33 μF
R_{load}	Road	120 Ω

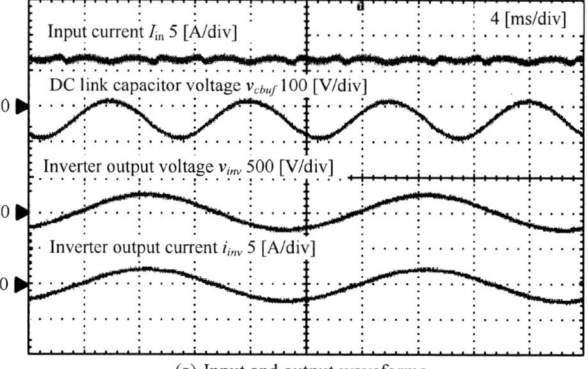

(a) Input and output waveforms

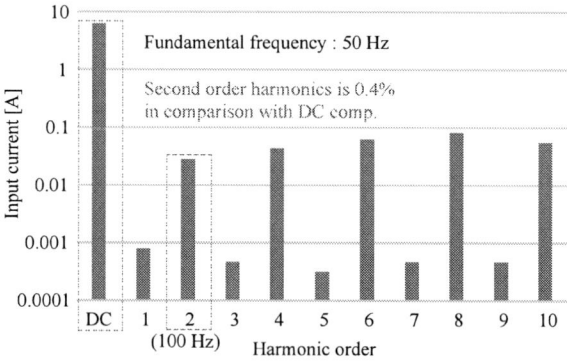

(b) DC input current harmonic analysis result

Fig.9. Experimental results with over all operation waveforms and DC input current harmonic analysis result.

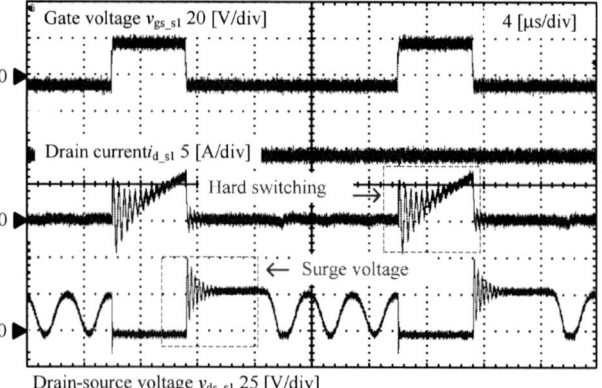

Fig.10. Experimental results with snubber less and diode rectifier condition.

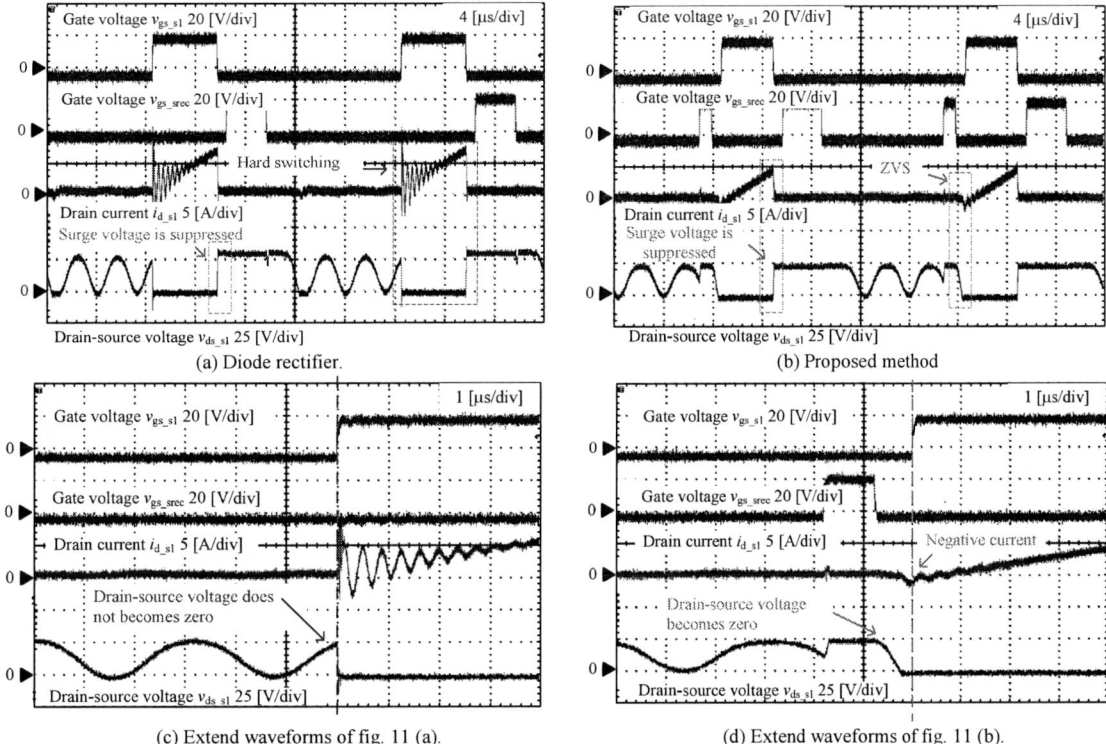

(a) Diode rectifier.

(b) Proposed method

(c) Extend waveforms of fig. 11 (a).

(d) Extend waveforms of fig. 11 (b).

Fig.11 Experimental result with switching waveforms both hard switching condition and ZVS condition.

VII. ACKNOWLAGEENT

This study was supported by New Energy and Industrial Technology Development Organization (NEDO) of Japan.

REFERENCES

[1] H-I Hsieh, and J. Hou, "Realization of Interleaved PV Microinverter by Quadrature-Phase-Shift SPWM Control", *IEEJ Journal of Industry Applications*, Vol.4, No.5, pp.643-649, (2015)

[2] C.-T. Lee; Y.-M. Chen; L.-C. Chen; P.-T. Cheng;" Efficiency improvement of a DCAC converter with the power decoupling capability" *IEEE Appl. Power Electron. Conf Expo.*,pp. 1462-1468, (2012)

[3] C. B. Barth; I. Moon; Y. Lei; S. Qin; C. N. Robert; Pilawa-Podgurski;" Experimental evaluation of capacitors for power buffering in single-phase power converters" *IEEE Energy Conversion Congr. Expo.*,pp. 6269-6276, (2015)

[4] W. Liu; K. Wang; H. S.-h. Chung; S. T.-h. Chuang;"Modeling and Design of Series Voltage Compensator for Reduction of DC-Link Capacitance in Grid-Tie Solar Inverter" *IEEE Trans. Power Electron.* Vol. 30, No.5, pp. 2534-2548, (2015)

[5] C. Y. Wu; C. H. Chen; J. W. Cao; M. T. Liu; " Power control and pulsation decoupling in a single-phase grid-connected voltage-source inverter" Tencon 2013., pp. 475-479, (2013)

[6] K. H. Chao, P.T. Cheng: "Power decoupling methods for single-phase three-poles AC/DC converters", *IEEE Energy Conversion Congr. Expo.*, pp.3742-3747, Sep. (2009)

[7] Shota Yamaguchi, and Toshihisa Shimizu,"Single-phase Power Conditioner with a Buck-boost-type Power Decoupling Circuit,"*IEEJ. Journal of Industry Applications*, vol. 5, no. 3, pp. 191-198, (2016)

[8] R. Wang, F. Wang, R. Lai, P. Ning, R. Burgos, D. Boroyevich : "Study of Energy Storage Capacitor Reduction for Single Phase PWM Rectifier" *2009 Twenty-Fourth Annu. IEEE Appl. Power Electron. Conf. Expo.*, pp1177-1183, Feb. (2009)

[9] C.-T. Lee, Y.-M. Chen, L.-C. Chen, P.-T. Cheng: "Efficiency Improvement of a DC/AC Converter with the Power Decoupling Capability", *2012 Twenty-Fourth Annu. IEEE Appl. Power Electron. Conf Expo.*, pp1462-1468, Feb. (2012)

[10] H. Watanabe, J. Itoh: "Highly-reliable Fly-back-based PV Micro inverter Applying Power Decoupling Capability without Additional Components", PCIM2017, pp. 681-688 (2017)

[11] M. A. Rezaei, K-J Lee, A, Huang "A High-Efficiency Flyback Micro-inverter With a New Adaptive Snubber for Photovoltaic Applications", *IEEE Trans. Power Electron.*, vol. 31, no. 1, pp. 318-327, (2016)

[12] X. Xie, J. Li, K. Peng, C. Zhao, Q. Lu: "Study on the Single-Stage Forward-Flyback PFC Converter With QR Control", *IEEE Trans. Power Electron.*, vol. 31, no. 1, pp. 430-442, (2016)

[13] T. Tanaka, K. Hirachi, J. itoh: "A novel control strategy for active clamping single ended forward DC/DC converter", IEEJ, Vol.36, pp.41-47 (2011) (Japanese).

[14] C.-Y. Liao, W.-S. Lin, C-Y- Chou: "A PV Micro-inverter with PV Current Decoupling Strategy", *IEEE Trans. Power Electron.*, vol. 32, no. 8, pp. 6544-6557 (2016)

[15] R. Hasan, S. Mekhilef: "A Resonant Double Stage Microinveters for PV Application", *IEEE Appl. Power Electron. Conf Expo.*, pp2051-2056, (2017)

The 2018 International Power Electronics Conference

Model Predictive Fault Tolerant Control of Bidirectional AC/DC Converter with Voltage Balance of Split Capacitor

Nan Jin[1], Chongyan Zhao[1] and Leilei Guo[1*]
1 Department of Electrical Engineering, Zhengzhou University of Light Industry, Zhengzhou, China
*E-mail: 2006guoleilei@163.com

Abstract—The bidirectional AC/DC converter is a key device for bidirectional power conversion between AC and DC side. Its fault tolerance performance is an important guarantee for the reliable and stable operation of the power conversion system. In order to improve the fault tolerant operation performance, a model predictive direct power control strategy with central point voltage balance and the switching frequency reduction is proposed. On this basis, the cost function with the central point voltage balance and the switching frequency reduction additional items is established, the two-step prediction model is adopted to realize the time delay compensation. The simulation results show that the three-phase four-switch bidirectional AC/DC converter has good steady state and dynamic performance as well as robustness. The effectiveness of the proposed control strategies is verified.

Keywords—Bidirectional AC/DC converter, three-phase four-switch, central point voltage balance, switching frequency reduction, model predictive control.

I. INTRODUCTION

The bidirectional AC/DC converter transfer the energy between DC and AC power supply, and has been widely application in micro-grids, motor control, energy storage systems and other fields [1-2]. When the bidirectional AC/DC converter operates in high voltage and high frequency conditions, power semiconductor devices are prone to damage and cause leg faults. Therefore, the fault phase is connected to the central point of the split capacitor at the DC side after the leg fault occurs, and the three-phase four-switch (TPFS) fault tolerant structure is constructed [3-4]. Three-phase four-switch fault tolerant converter has become one of the important research fields of power conversion reliability [5-8].

But, the fault phase current through the central point of the split capacitor will cause voltage fluctuation of the split capacitor, which will lead to unbalanced grid-connected current and influence the grid-connected power quality. On the other hand, the unbalance of the split capacitor voltage reduces the life-time of electrolytic capacitor for the over-voltage operation. Therefore, it is

This paper is supported by NSFC (National Natural Science Foundation of China) - Henan Province Joint Foundation (U1604136).

of great significance to solve the voltage unbalance problem of the split capacitor in the three-phase four-switch fault tolerant structure [9-13]. Model predictive control technology is introduced into the field of power electronics, which has the characteristics of simple structure, fast dynamic response and flexible control [14-15].

In [8], the topology of the three-phase four-switch fault tolerant structure is given, the fault tolerance model and implement method of three-phase four-switch converter are analyzed. In [9], the four-switch inverter space vector modulation (SVM) method is proposed, the situation of unbalance of DC side split capacitor voltage is analyzed. On this basis, the influence of the central point voltage fluctuation on the output current is suppressed by optimizing the time of the basic vector action. In [10], the three-phase four-switch structure is proposed. The method that adds DC side capacitor voltage control variables in fault phase output current command value is put forward to realize DC side split capacitor voltage balance.

In [11-13], the relationship between the DC side split capacitor current and the fault phase current is analyzed under the three-phase four-switch fault tolerant structure to eliminate the central point voltage fluctuation. At the same time, it is necessary to consider reducing the switching frequency owing to the high switching frequency. So that it can reduce the switching losses and obtain more system efficient.

Based on the above analysis of the working principle and voltage vector relationship of the three-phase four-switch fault tolerant structure, a model predictive direct power control strategy with central point voltage balance and switching frequency reduction is designed. The cost function with two step prediction model is established to realize the time delay compensation. In simulation part, three-phase four-switch fault tolerant structure is carried out to verify the steady-state and dynamic performance. The robustness of the system is verified by changing the AC side inductance parameter and the control algorithm calibration inductance parameter. Selecting the optimal weighting factors of the central point voltage balance and the switching frequency reduction additional items in the cost function to achieve a balanced state. Simulation results verify that the proposed control strategies are effective.

II. FAULT TOLERANT MODEL FOR BIDIRECTIONAL AC/DC CONVERTER

The fault tolerant structure of bidirectional AC/DC converter, such as Fig. 1(a), uses three bidirectional thyristors as the connection switches between three-phase and the central point of the split capacitor. In normal operation, the bidirectional thyristor is in the state of disconnection. When the short circuit or open circuit occurs in one phase leg (such as phase a leg), the fast fuse Fa is disconnected and the corresponding bidirectional thyristor TRa is triggered to reconstruct fault tolerance structure, to ensure continuous operation of the system.

Fig. 1(b) is reconstructed bidirectional AC/DC converter three-phase four-switch fault tolerant structure. Phase a leg is connected to the central point of the split capacitor at the DC side after the fault occurs. The four switching devices on phase b and phase c leg can still work. The AC side through the filter inductance L, line resistance R connected to the grid.

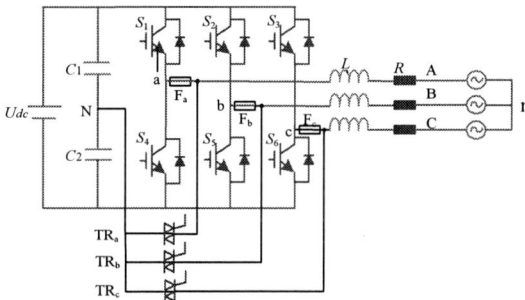

（a）Fault tolerant structure for bidirectional AC/DC converter

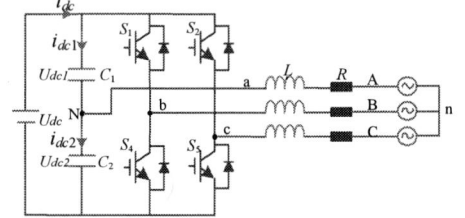

（b）Three-phase four-switch fault tolerant structure of bidirectional AC/DC converter

Fig. 1. Fault tolerant topology of bidirectional AC/DC converter

The switching state S_i ($i=b$, c) of the bidirectional AC/DC converter is defined as follows:

$$S_i = \begin{cases} 1 & \text{upper bridge of } i \text{ phase is on} \\ & \text{and lower bridge is off} \\ 0 & \text{upper bridge of } i \text{ phase is off} \\ & \text{and lower bridge is on} \end{cases} \quad (1)$$

The relationship between the AC side three-phase voltage and switching state of the three-phase four-switch converter can be obtained by (2):

$$\begin{cases} u_{an} = \dfrac{U_{dc1}}{3}\left(-S_b - S_c\right) + \dfrac{U_{dc2}}{3}\left(2 - S_b - S_c\right) \\ u_{bn} = \dfrac{U_{dc1}}{3}\left(2S_b - S_c\right) + \dfrac{U_{dc2}}{3}\left(2S_b - S_c - 1\right) \\ u_{cn} = \dfrac{U_{dc1}}{3}\left(2S_c - S_b\right) + \dfrac{U_{dc2}}{3}\left(2S_c - S_b - 1\right) \end{cases} \quad (2)$$

The formula (2) in the $\alpha\beta$ two phase stationary coordinate system can be obtained by:

$$\begin{bmatrix} u_\alpha \\ u_\beta \end{bmatrix} = \frac{1}{3}\begin{bmatrix} -S_b - S_c & 2 - S_b - S_c \\ \sqrt{3}\left(S_b - S_c\right) & \sqrt{3}\left(S_b - S_c\right) \end{bmatrix}\begin{bmatrix} u_{dc1} \\ u_{dc2} \end{bmatrix} \quad (3)$$

When the converter is in fault tolerant structure, due to the different switching state of phase b and phase c leg, converter can output voltage vector V_0, V_1, V_2, V_3, the relationship between voltage vector and switching state in the $\alpha\beta$ two phase stationary coordinate system is shown in Table I. According to different conditions of DC side split capacitor, such as $U_{dc1} = U_{dc2}$, $U_{dc1} < U_{dc2}$ and $U_{dc1} > U_{dc2}$, the basic voltage vectors in the $\alpha\beta$ coordinate system are different which are shown in Fig. 2.

TABLE I
VOLTAGE VECTORS OF TPFS CONVERTER WITH PHASE A FAULT

Vector	S_b	S_c	u_α	u_β
V_0	0	0	$2U_{dc2}/3$	0
V_1	0	1	$(U_{dc2}-U_{dc1})/3$	$-\sqrt{3}\left(U_{dc1}+U_{dc2}\right)/3$
V_2	1	0	$(U_{dc2}-U_{dc1})/3$	$\sqrt{3}\left(U_{dc1}+U_{dc2}\right)/3$
V_3	1	1	$-2U_{dc1}/3$	0

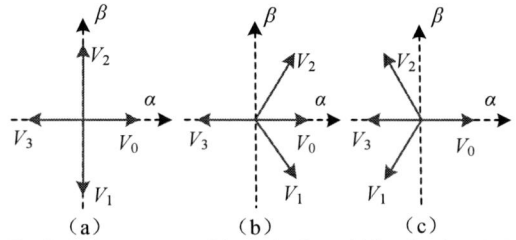

Fig. 2. Voltage vectors of the fault tolerant bidirectional AC/DC converter. (a) $U_{dc1} = U_{dc2}$ (b) $U_{dc1} < U_{dc2}$ (c) $U_{dc1} > U_{dc2}$

III. PREDICTIVE MODEL OF BIDIRECTIONAL AC/DC FAULT TOLERANT CONVERTER

A. Power Predictive Model for Bidirectional AC/DC Fault Tolerant Converter

According to Kirchhoff's law, the state equation of the system in the three-phase stationary coordinate system is obtained as shown in equation (4):

$$L\frac{d}{dt}\begin{bmatrix} i_a \\ i_b \\ i_c \end{bmatrix} = \begin{bmatrix} u_{an} \\ u_{bn} \\ u_{cn} \end{bmatrix} - R\begin{bmatrix} i_a \\ i_b \\ i_c \end{bmatrix} - \begin{bmatrix} e_a \\ e_b \\ e_c \end{bmatrix} \quad (4)$$

where e_a, e_b, e_c are power grid voltage, i_a, i_b, i_c are the converter output current, u_{an}, u_{bn}, u_{cn} are the converter output voltage.

In the $\alpha\beta$ two phase stationary coordinate system, the formula (4) can be obtained by:

$$L\frac{d}{dt}\begin{bmatrix}i_\alpha\\i_\beta\end{bmatrix}=\begin{bmatrix}u_\alpha\\u_\beta\end{bmatrix}-R\begin{bmatrix}i_\alpha\\i_\beta\end{bmatrix}-\begin{bmatrix}e_\alpha\\e_\beta\end{bmatrix} \tag{5}$$

where u_α, u_β, i_α, i_β, e_α, e_β are the $\alpha\beta$ coordinate system components of the converter output voltage, current and power grid voltage.

The discretization of formula (5) can be obtained by:

$$\frac{L}{T}\begin{bmatrix}i_\alpha(k+1)-i_\alpha(k)\\i_\beta(k+1)-i_\beta(k)\end{bmatrix}=\begin{bmatrix}u_\alpha(k)\\u_\beta(k)\end{bmatrix}-\begin{bmatrix}Ri_\alpha(k)\\Ri_\beta(k)\end{bmatrix}-\begin{bmatrix}e_\alpha\\e_\beta\end{bmatrix} \tag{6}$$

where $u_\alpha(k)$, $u_\beta(k)$, $i_\alpha(k)$, $i_\beta(k)$ are $\alpha\beta$ components of the converter output voltage and current at k instant. $i_\alpha(k+1)$, $i_\beta(k+1)$ are $\alpha\beta$ components of predictive current value at $k+1$ instant. T is sampling period.

The active power P and reactive power Q can be obtained by:

$$\begin{bmatrix}P\\Q\end{bmatrix}=\frac{3}{2}\begin{bmatrix}e_\alpha & e_\beta\\e_\beta & -e_\alpha\end{bmatrix}\begin{bmatrix}i_\alpha\\i_\beta\end{bmatrix} \tag{7}$$

The discretization of formula (7) can be obtained by:

$$\begin{bmatrix}P(k+1)-P(k)\\Q(k+1)-Q(k)\end{bmatrix}=\frac{3}{2}\begin{bmatrix}e_\alpha & e_\beta\\e_\beta & -e_\alpha\end{bmatrix}\begin{bmatrix}i_\alpha(k+1)-i_\alpha(k)\\i_\beta(k+1)-i_\beta(k)\end{bmatrix} \tag{8}$$

The power prediction equation for bidirectional AC/DC fault tolerant converter is obtained by:

$$\begin{bmatrix}P(k+1)\\Q(k+1)\end{bmatrix}=\frac{3T}{2L}\begin{bmatrix}e_\alpha & e_\beta\\e_\beta & -e_\alpha\end{bmatrix}\begin{bmatrix}u_\alpha(k)-Ri_\alpha(k)-e_\alpha\\u_\beta(k)-Ri_\beta(k)-e_\beta\end{bmatrix}+\begin{bmatrix}P(k)\\Q(k)\end{bmatrix} \tag{9}$$

B. Central Point Voltage Balance of Split Capacitor

As shown in Fig. 1 (b), the fault phase leg is connected to the central point of the split capacitors C_1 and C_2 at the DC side, the split capacitive currents i_{dc1}, i_{dc2} and i_a have the following relationship:

$$i_{dc1}=-i_{dc2}=\frac{1}{2}i_a \tag{10}$$

The relationship between the capacitor voltage and the current is as follows, where i_{dc} is the DC side power supply current, shown in Fig. 1.

$$\begin{bmatrix}C\dfrac{dU_{dc1}}{dt}\\C\dfrac{dU_{dc2}}{dt}\end{bmatrix}=\begin{bmatrix}i_{dc} & -i_b & -i_c\\i_{dc}+i_b+i_c & -i_b & -i_c\end{bmatrix}\begin{bmatrix}1\\S_b\\S_c\end{bmatrix} \tag{11}$$

The derivative of the voltage offset value of the split capacitor is obtained by:

$$C\frac{d(U_{dc1}-U_{dc2})}{dt}=-(i_c+i_b)=i_a \tag{12}$$

The voltage offset value of the split capacitor ΔU_{dc} is obtained by formula (13), where $U_{dc1}(0)$, $U_{dc2}(0)$ are the initial voltage of capacitor C_1 and C_2:

$$\Delta U_{dc}=\frac{1}{C}\int_0^t i_a+(U_{dc1}(0)-U_{dc2}(0)) \tag{13}$$

By discretization of formula (13), The predictive offset component at $k+1$ instant can be expressed as:

$$\Delta U_{dc}(k+1)=\Delta U_{dc}(k)+\frac{T}{C}i_a(k) \tag{14}$$

C. Switching Frequency Reduction

In the bidirectional power conversion system, in the case of lower switching frequency, system losses and switching losses generated by switching devices are smaller. It will get higher power conversion efficiency. Therefore, an additional item S that reduces the switching frequency is added into the cost function.

$$S=\sum_{i=b,c}\left|S_i^{k+1}-S_i^k\right| \tag{15}$$

S_i^k and S_i^{k+1} are switch status at current and next moment. By adding the additional item S in the cost function g, the purpose of switching frequency reduction is realized.

IV. Model Predictive Control of Bidirectional AC/DC Fault Tolerant Converter

In the model predictive control strategy, the cost function g is established to select the optimal switching vector, the voltage and current values of the system are collected at k instant, and the prediction power value of $k+1$ instant are obtained. Comparing the predictive value and the reference value by the cost function, the optimal voltage vector is selected to control the switching devices at k instant. In order to realize multiple control objectives, the additional items of central point voltage balance and switching frequency reduction are added into the cost function, as shown in formula (16).

$$\begin{aligned}g=&\left|P_{ref}-P(k+1)\right|+\left|Q_{ref}-Q(k+1)\right|\\&+\lambda_1\left|\Delta U_{dc}(k+1)\right|+\lambda_2 S\end{aligned} \tag{16}$$

where P_{ref}, Q_{ref} are reference values of active power and reactive power. $P(k+1)$, $Q(k+1)$ are predictive values of active power and reactive power at $k+1$ instant. $\Delta U_{dc}(k+1)$ is predictive compensation value of central point of the split capacitor at $k+1$ instant. λ_1 is the weighting factor of the central point voltage balance additional item, and λ_2 is the weighting factor of the switching frequency reduction additional item.

The model predictive direct power control algorithm is applied to the bidirectional AC/DC converter control system. When converter operates in high sampling frequency environments, there will be a time delay that influence system performance. In order to eliminate the influence, the two-step prediction method is used to compensate the time delay. The two-step prediction cost function is shown in formula (17).

$$g = \left| P_{ref} - P(k+2) \right| + \left| Q_{ref} - Q(k+2) \right| \\ + \lambda_1 \left| \Delta U_{dc}(k+2) \right| + \lambda_2 S \tag{17}$$

where $P(k+2)$, $Q(k+2)$ are predictive values of active power and reactive power at $k+2$ instant which are shown in formula (18). $\Delta U_{dc}(k+2)$ is predictive offset value of central point voltage of the split capacitor at $k+2$ instant which is shown in formula (19).

$$\begin{bmatrix} P(k+2) \\ Q(k+2) \end{bmatrix} = \frac{3T}{2L} \begin{bmatrix} e_\alpha & e_\beta \\ e_\beta & -e_\alpha \end{bmatrix} \begin{bmatrix} u_\alpha(k+1) - Ri_\alpha(k+1) - e_\alpha \\ u_\beta(k+1) - Ri_\beta(k+1) - e_\beta \end{bmatrix} \\ + \begin{bmatrix} P(k+1) \\ Q(k+1) \end{bmatrix} \tag{18}$$

$$\Delta U_{dc}(k+2) = \Delta U_{dc}(k) + \frac{T}{C} i_a(k) + \frac{T}{C} i_a(k+1) \tag{19}$$

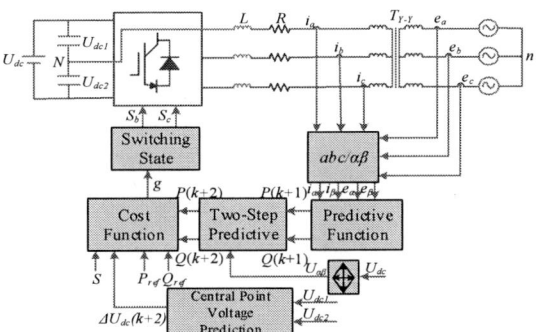

Fig. 3. Model predictive control structure diagram

The model predictive control structure is shown in Fig. 3, the grid voltage and converter output current are collected at k instant, the output voltage of the converter is calculated by the switching state and the DC side split capacitor voltage, the power prediction values of $k+2$ instant are obtained by two step prediction function, the additional items of the central point voltage balance and the switching frequency reduction are added into the cost function. The optimal switching state is selected to control the IGBT.

V. SIMULATION RESULTS AND ANALYSIS

The simulation model of bidirectional AC/DC converter is built in MATLAB/simulink, and the proposed control strategies are verified. The system parameters are shown in Table II.

TABLE II
SYSTEM PARAMETERS

Symbol	Meaning	Value
U_{dc}	DC side voltage	400V
C_1, C_2	Capacitance	1000μF
L	Filter inductance	20mH
e	AC side phase voltage	110V
f	AC side frequency	50Hz
f_{samp}	Sampling frequency	20KHz
T_{Y-Y}	AC transformer	Y-Y, 190V/75V, 5kVA

A. Fault Tolerant Grid Connected Control Simulation

When one phase leg fault occurs, the current and power waveforms are analyzed by giving the reference values of the active power and reactive power. The system can work normally under the fault tolerant structure. In Fig. 4(a), the converter works in the inverter mode with the reference active power is 1000W and reactive power is 0Var. The simulation results show that output current THD is 2.58% with good sinusoidal waveform, the output active power is stable at 1000W, reactive power is stable at 0Var. In Fig. 4(b), the converter works in the rectifier mode with the reference active power is -1000W and reactive power is 0Var. The simulation results show that output current THD is 2.52% with good sinusoidal waveform, the output active power is stable at -1000W, reactive power is stable at 0Var.

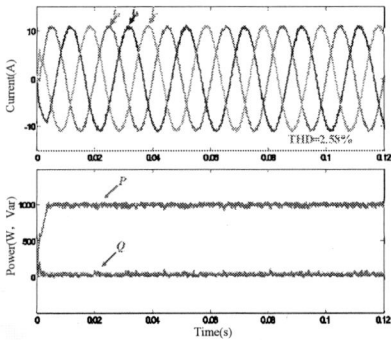

(a)Simulation results of fault tolerant control in inverter mode

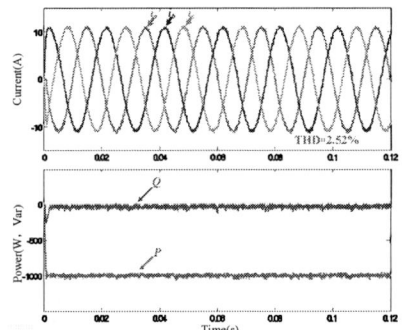

(b)Simulation results of fault tolerant control in rectifier mode

Fig. 4. Simulation results of fault tolerant control

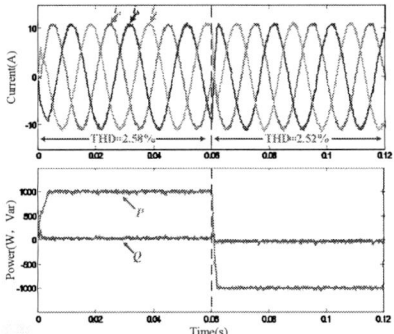

Fig. 5. Simulation results of active power change

In Fig. 5, the following simulation is designed to verify the dynamic performance of the converter under the proposed control strategies, the active power is changed from 1000W (inverter mode) to -1000W (rectifier mode) at 0.06s, and the reactive power is kept at 0Var. When the reference active power changes, the system has good dynamic performance without surge, spike and other transient shocks.

B. System Robustness Simulation

The influence of the change of the inductance parameter on the fault-tolerant operation of the converter is studied, two groups of simulation experiments are designed to lay the foundation for further study of the control method to improve the robustness.

(1) The inductance parameter L of the converter is changed from 80% to 120% of the rated value, and the inductance parameter L^* in the control algorithm is kept at the rated value 20mH. The system works in inverter and rectifier modes to verify the robustness. The simulation results are shown in Fig. 6 and Fig. 7.

Through the Fig. 6 and Fig. 7, it can be seen that the system has good robustness in the case of inductance parameters change, and the current THD reduces with the increase of inductance parameters.

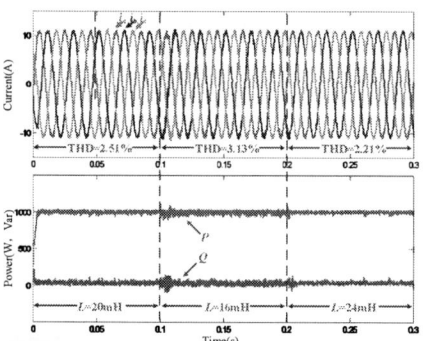

Fig. 6. Simulation results of inductance parameter L change in inverter mode

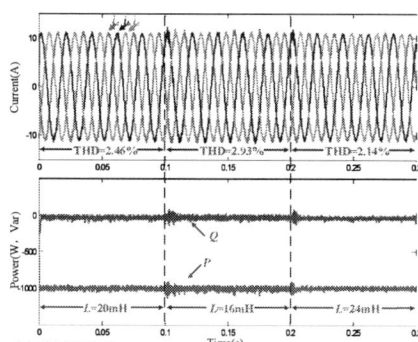

Fig. 7. Simulation results of inductance parameter L change in rectifier mode

(2) The reactor parameter L^* in the control algorithm follows the actual reactor parameter L changed from 80% to 120% of the rated value. The system works in inverter and rectifier modes to verify the robustness. The simulation results are shown in Fig. 8 and Fig. 9.

Through the Fig. 8 and Fig. 9, it can be seen that the system has good robustness in the case of inductance parameters L and L^* change simultaneously, and the

current THD reduces with the increase of inductance parameters.

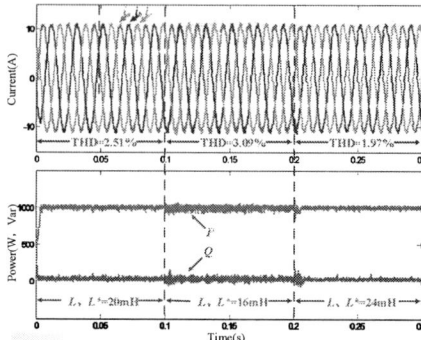

Fig. 8. Simulation results of inductance parameter L and L^* changing simultaneously in inverter mode

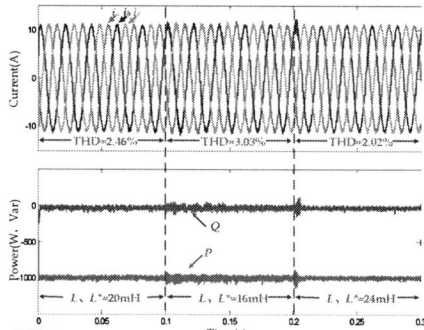

Fig. 9. Simulation results of inductance parameter L and L^* changing simultaneously in rectifier mode

C. Central Point Voltage Balance Simulation and Weighting Factor λ_1 Selection

In order to verify the effectiveness of the central point voltage balance control strategy and the influence of the weighting factor λ_1 on the central point voltage balance, the following simulation is designed, as shown in Fig. 10.

In the 0.05s time of $\lambda_1=0$, two split capacitor voltages are different and the central point voltage is unbalanced without the central point voltage balance control strategy. In Fig. 10(a), the λ_1 switches to 1300 at 0.05s, the central point voltage is balanced after 0.8s, the current THD=3.68%, the maximum power fluctuation is 250W. In Fig. 10(b), the λ_1 switches to 1500 at 0.05s, the central point voltage is balanced after 0.3s, the current THD=6.29%, the maximum power fluctuation is 400W.

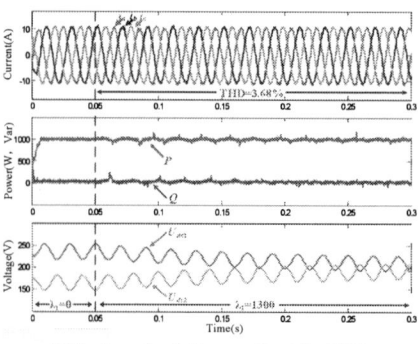

(a)System simulation results at $\lambda_1=1300$

908

The 2018 International Power Electronics Conference

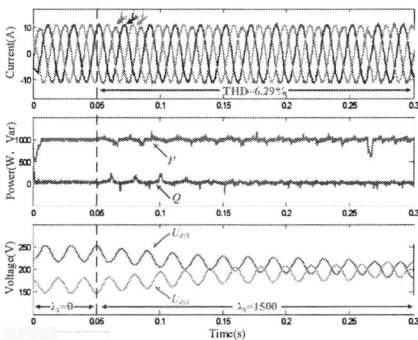

(b)System simulation results at λ_1=1500

Fig. 10. Simulation results of central point voltage balance control

It can be seen that the central point voltage balance response time can be effectively reduced with the increase of weighting factor λ_1, the power ripple and the current THD increase with the large λ_1. In this paper, in order to balance the system current THD, power fluctuation and central point voltage balance response time, λ_1=1300 is selected as the optimal weighting factor of central point voltage balance additional item.

D. The Switching Frequency Reduction Simulation and Weighting Factor λ_2 Selection

The following simulation is designed to verify the effectiveness of the switching frequency reduction control strategy. In the case of the optimal value λ_1=1300, the influences of the change of weighting factor λ_2 in the cost function on the current THD, the central point voltage deviation value V_{dc} and the switching frequency average $f_{sw(av)}$ are shown in Fig. 11.

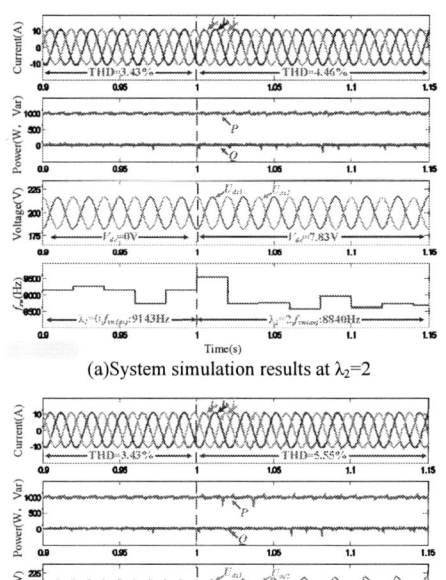

(a)System simulation results at λ_2=2

(b)System simulation results at λ_2=3

Fig. 11. Simulation results of switching frequency reduction

When the λ_2=0 before 1s, the output current THD is 3.43%, the central point voltage deviation value V_{dc} is 0V, and the switching frequency average $f_{sw(av)}$ is 9143Hz. In Fig. 11(a), the λ_2 is switched to 2 at 1s, the current THD=3.23%, the central point voltage deviation value V_{dc} is 7.83V, and the switching frequency average $f_{sw(av)}$ is 8840Hz. In Fig. 11(b), the λ_2 is switched to 3 at 1s, the current THD=5.55%, the central point voltage deviation value V_{dc} is 12.71V, and the switching frequency average $f_{sw(av)}$ is 8516Hz.

It can be seen that the switching frequency can be effectively reduced with the increase of the weighting factor λ_2, the split capacitor voltage deviation value V_{dc} and the current THD increase with the larger λ_2, but the growth range is small. In order to ensure the current THD and the central point voltage balance of the split capacitor have a good performance, and effectiveness of switching frequency reduction control strategy at the same time. λ_2=2 is selected as the optimal weighting factor of the switching frequency reduction additional item.

VI. CONCLUSIONS

The model predictive direct power control method for three-phase four-switch fault tolerant structure of the bidirectional AC/DC converter with the central point voltage balance and the switching frequency reduction is proposed. The proposed control strategies offer a higher reliability for the bidirectional AC/DC converter. The main contributions of the paper are mainly reflected in the following aspects:

1. The working principle and voltage vector relationship of the three-phase four-switch fault tolerant converter are analyzed in detail.

2. The model predictive direct power control is designed with central point voltage balance of the DC side split capacitor. The central point voltage balance control of the split capacitor optimizes the quality of grid connected power.

3. When the reference power changes, the system operation modes can be switched smoothly from the inverter mode to the rectifier mode with a good dynamic performance. The good robustness of the system is verified by the variation of inductance parameters. The additional item S in the cost function can effectively reduce the switching frequency.

REFERENCES

[1] C. Cecati, A. Tommaso and F. Genduso, "Comprehensive modeling and experimental testing of fault detection and management of a nonredundant fault-tolerant VSI," *IEEE Trans on Industrial Electronics*, vol. 62, no. 6, pp. 3945-3954, 2015.

[2] X. Zheng, L. Xiao and Y. Tian, "A control strategy of bidirectional three-phase AC/DC converter without PLL," *Proceeding of the CSEE*, vol. 33, no. 36, pp. 79-87+12, 2013.

[3] W. Zheng, Z. Zeng and R. Zhao, "Modeling and modulation of the fault-tolerant grid-connected three-phase PWM converter," *Proceeding of the CSEE*, vol. 36, no. 8, pp. 2202-2212, 2016.

[4] D. Xu, X. Liu and Y. Yu, "A survey on fault diagnosis and tolerant control of inverters," *Trans of China Electrotechnical Society*, vol. 30, no. 21, pp. 1-12, 2015.

[5] B. A. Welchko, T. A. Lipo and T. M. Jahns, "Fault tolerant three-phase AC motor drive topologies: a comparison of features, cost, andlimitations," *IEEE Transactions on Power Electronics*, vol. 19, no. 4, pp. 1108 -1116, 2004.

[6] D. Sun, Z. He and Y. He, "Four-switch inverter fed PMSM DTC with SVM approach for fault tolerant operation," *IEEE International Electric Machines and Drives Conference*, vol. 56, no. 12, pp. 295-299, 2007.

[7] F. W. Fuchs, "Some diagnosis methods for voltage source inverters in variable speed drives with induction machines a survey," *The 29th Annual Conference of the IEEE Industrial Electronics Society*, vol. 12, no. 8, pp. 1378-1385, 2003.

[8] K. Zhao, Q. An, "Fault tolerant inverter permanent magnet synchronous motor position sensorless control system," *Electric Machines and Control*, vol. 14, no. 4, pp. 25-30, 2010.

[9] C. Zhu, Z. Zeng and R. Zhao, "Carrier-based modulation algorithm of fault-tolerant three-phase four-switch inverter for better current performance," *Electric Power Automation Equipment*, vol. 5, no. 37, pp. 40-47, 2017.

[10] W. Wang, A. Luo and Y. Li, "Control Method of Three-Phase Four-Switch Shunt Active Power Filter," *Transactions of China Electrotechnical Society*, vol. 29, no. 10, pp. 183-190, 2014.

[11] S. Dasgupta, S. N. Mohan and S. K. Sahoo, "Application of four-switch-based three-phase grid-connected inverter to connect renewable energy source to a generalized unbalanced microgrid system," *IEEE Trans. Ind. Electron*, vol. 60, no. 6, pp. 1204-1215, 2013.

[12] R. Wang, J. Zhao and Y. Liu, "A comprehensive investigation of four-switch three-phase voltage source inverter based on double fourier integral analysis," *IEEE Trans. Power Electron*, vol. 26, no. 6, pp. 2774-2787, 2011.

[13] D. Zhou, J. Zhao and Y. Liu, "Predictive torque control scheme for three-phase four-switch inverter-fed induction motor drives with dc-link voltages offset suppression," *IEEE Trans. Power Electron*, vol. 30, no. 6, pp. 3309-3318, 2015.

[14] K. Shen, J. Zhang and L. Wang, "Model predictive control of three-phase voltage source inverter," *Transactions of China Electrotechnical Society*, vol. 28, no. 12, pp. 283-289, 2013.

[15] J. Hu, J. Zhu and D. Dorrell, "Model predictive direct power control of doubly-fed induction generators under unbalanced grid voltage conditions in wind energy applications," *IET Renewable Power Generation*, vol. 8, no. 6, pp. 687-695, 2015.

The 2018 International Power Electronics Conference

PWM Strategy for Parallel Operation of Three Phase Converters Tied to Grid

Hyun-Sam Jung[1] and Seung-Ki Sul[1]
* Seoul National University
1 School of Electrical Engineering & Computer Science, Seoul National University, 1 Gwanak-ro, Gwanak-gu,
Seoul, Korea

Abstract- **When three phase converters operate, tied to the grid, interleaved PWM with coupled inductor can be used to enhance current quality. As the results, the harmonics of the output current are reduced significantly because of the increased equivalent switching frequency due to interleaving operation. However, the circulating current increases between converters. The coupled inductor is employed to suppress this circulating current. In this system, optimization of PWM pulse pattern by interleaved carrier would be the best strategy to enhance harmonics characteristics of output voltage and current. In this paper, based on harmonics analysis of PWM, a new PWM strategy for interleaved PWM is proposed and an offset voltage of each converter is optimized in terms of the harmonics of output voltage and current. As the results, the grid side inductor can be significantly reduced by decreasing voltage and current harmonics of the system to the grid. The validity, feasibility and effectiveness of the proposed method have been confirmed by computer simulation and experimental results.**

Keywords— Parallel operations, Pulse Width Modulations, Interleaved PWM and Harmonics reduction.

I. INTRODUCTION

Parallel operation of converter of the same power capacity is widely used in Power Conditioning System (PCS) to interface renewable energy source to the grid. Because, the converters in parallel can be designed as a modular structure, which has significant advantages on reducing production and maintenance cost of the system. Besides, even if one of the converters, which are in parallel, fails, PCS can be operated under reduced power capacity. Furthermore, when the power generated from the renewable source is reduced, and then some converters can be turn-off to minimize the loss of overall PCS. In this way, the efficiency of the system at light load condition can be enhanced remarkably. In other words, parallel operation could increase system reliability, efficiency and reduce the cost simultaneously. Due to these advantages, parallel operation of PWM converters has been investigated actively for several decades [1]-[13].

When converters are operated in parallel, there are two types of PWM methods. One is the synchronized PWM, which uses identical carrier [4]-[8]. This kind of PWM methods minimizes the size of the sharing inductors between converters. The other is interleaved PWM, which employs interleaved carrier. When interleaved carrier is used in parallel operation of PWM converters, some harmonics of each PWM converter are cancelled each other and harmonic characteristics can be improved conspicuously at output node, where output of converters

are connected. [9]-[11]. However, interleaved PWM invokes circulating current unavoidably. To reduce the circulating current and minimize inductor size, a coupled inductor as the sharing inductor is employed in the parallel path [12].

In this paper, interleaved PWM with the coupled inductor is considered for parallel operation of PWM converter tied to the grid. For harmonics analysis of PWM, at first, conventional PWM method is analyzed. Next, a new PWM method is proposed based on the analysis. By using the proposed PWM strategy, harmonics of output current and voltage of system, consisted with multiple PWM converters and coupled inductor as sharing inductors, can be conspicuously reduced. Feasibility and validity of the proposed method are proven by computer simulation and experimental results.

II. HARMONIC ANALYSIS OF CONVENTIONAL INTERLEAVED PWM

Output voltage of a converter with PWM can be represented as Fourier series as (1) [14].

$$
\begin{aligned}
f(x,y) = \frac{A_{00}}{4} &+ \sum_{n=1}^{\infty} [A_{0n} \cos ny_0 + B_{0n} \sin ny_0] \\
&+ \sum_{m=1}^{\infty} [A_{m0} \cos mx_c + B_{m0} \sin mx_c] \\
&+ \sum_{m=1}^{\infty} \sum_{\substack{n=-\infty \\ (n!=0)}}^{\infty} [A_{mn} \cos(mx_c + ny_0) + B_{mn} \sin(mx_c + ny_0)]
\end{aligned}
\tag{1}
$$

where, $x_c(t) = \omega_c t + \theta_c$, $y_0(t) = \omega_0 t + \theta_0$

In this equation, the fourth term in right hand side is defined as sideband harmonics, which is the residual components except integer multiples of the fundamental frequency, $n \times \omega_0$, and that of the carrier frequency, $m \times \omega_c$. If C_{mn} is defined as (2), then the term can be rewritten as (3), which represents the characteristic of sideband harmonics and is denoted as $v_{sideband}$ as in (3).

$$
C_{mn} = A_{mn} + jB_{mn} = |C_{mn}| e^{j\phi_{C_{mn}}}
\tag{2}
$$

where, $|C_{mn}| = \sqrt{A_{mn}^2 + B_{mn}^2}$ *and* $\phi_{C_{mn}} = \tan^{-1}\left(\frac{B_{mn}}{A_{mn}}\right)$

$$
\begin{aligned}
v_{sideband} &= \sum_{m=1}^{\infty} \sum_{\substack{n=-\infty \\ (n!=0)}}^{\infty} [A_{mn} \cos(mx + ny) + B_{mn} \sin(mx + ny)] \\
&= \sum_{m=1}^{\infty} \sum_{\substack{n=-\infty \\ (n!=0)}}^{\infty} |C_{mn}| \cos(mx + ny + \phi_{C_{mn}})
\end{aligned}
\tag{3}
$$

This equation can be transformed from time domain to phasor domain as (4), where *x* is a certain phase among a, b, and c, and *k* is *k*th converter in parallel. If the same phase of each converter in parallel is considered, (4) can be simplified to (5), because $n \times \theta_0$ is a constant. These

The 2018 International Power Electronics Conference

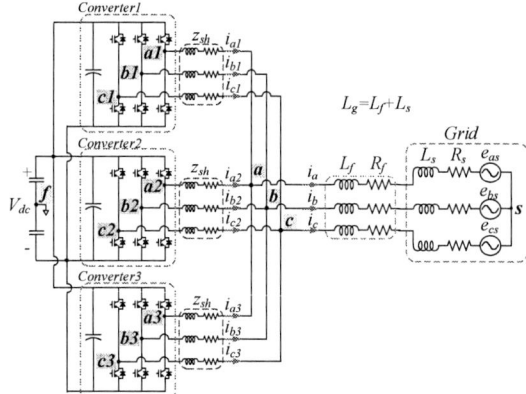

Fig. 1: Circuit of 3-converters in parallel operation

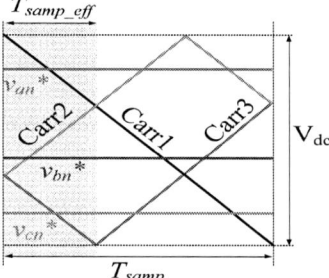

Fig. 2: Phase shifted carrier in case of 3-parallel operation

equations are employed to develop a new PWM method in this paper.

$$V_{xnk}(m,n) = C_{mn} e^{j(m\theta_c + n\theta_0)} . \tag{4}$$

$$V_{xnk}(m,n) = C_{mn}' e^{j(m\theta_c)}, \text{ where } C_{mn}' = C_{mn} e^{j(n\theta_0)} . \tag{5}$$

In this paper, for convenience of explanation, parallel operation of three converters tied to grid is considered as a representative example as shown in Fig. 1, where 'xk' denotes x-phase of kth converter, 'x' is output node of x-phase and 'f' is neutral point of DC link. In this figure, z_{sh} is a sharing inductor. A coupled inductor is employed as the sharing inductor. L_f and L_s are a filter inductor and a grid inductor, respectively. Sum of L_f and L_s is defined as L_g. In this case, PWM pulses are generally interleaved with angle 0, $2/3\pi$ and $4/3\pi$, respectively, because all carriers are shifted as shown in Fig. 2, where Carr1, Carr2 and Carr3 are carriers for converter1, converter2 and converter3, respectively.

Harmonics in phasor form represented as (5) can be divided into two groups based on whether it has an effect on total current, i_x or not. The output voltage, V_{xf}, which denotes voltage between the node 'x' and the neutral node 'f' has an influence on total current, i_x. It can be deduced as (6). The other is named as the circulating voltage, V_{xn_cir}, which has an effect only on the circulating current. These harmonics are defined as (7). All of m can be represented in $3l$, $3l+1$ and $3l+2$, where l is an integer. Output voltage, v_{xf}, is related only to $3l$ components and circulating voltage

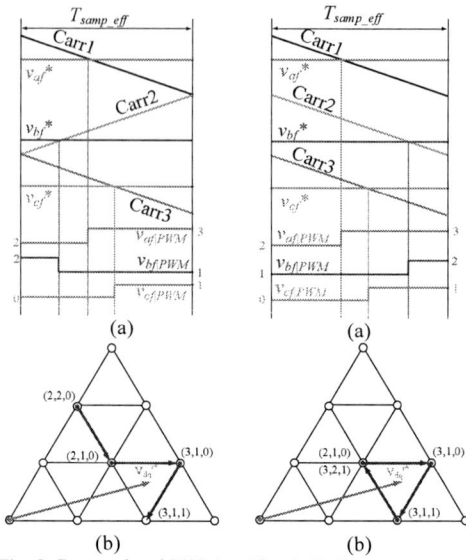

Fig. 3: Conventional PWM   **Fig. 4:** EPDPWM

is related only to $3l+1$ components and $3l+2$ components according to (6) and (7). It means that $3l$ components should be minimized in order to improve output current and reduce the size of filter inductor, L_f, when three converters operate in parallel.

$$V_{xf}(m,n) = C_{mn} \frac{1 + 2\cos(m\frac{2}{3}\pi)}{3} . \tag{6}$$

$$V_{xf\_cir}(m,n) = C_{mn} \frac{2}{3}\left(1 - \cos(m\frac{2\pi}{3})\right). \tag{7}$$

In order to reduce the $3l$ components of PWM harmonics, conventional PWM should be analyzed and modified to be suitable for the case of p-parallel operation, where p is the number of paralleled converters. When each PWM pulse is interleaved with $2\pi/p$, output voltage, v_{xf}, can be generally defined as (8).

In case of 3-parallel operation, carrier waves and reference voltages in Fig. 2 can be analyzed like multi-level converter as shown in Fig. 3 (a). According to Fig. 3 (a), voltage vectors are selected by the sequence as shown in Fig. 3 (b). It is the well-known that three voltage vectors adjacent to a voltage reference, V_{dq}^{r*} can synthesize the reference with the least switching harmonics of voltage and current [14]. However, as shown in Fig. 3 (b), the conventional interleaved PWM could not synthesize the reference voltage with the adjacent vectors. It makes the harmonics of current and inductance of the filter inductor, L_f, increase. Therefore, there is the possibility to improve the harmonics by modifying PWM.

III. PROPOSED PWM STRATEGY

As shown in Fig. 3 (b), conventional method is optimized not to total current, i_x, but to current of each converter, i_{xk}. However, according to the harmonic

$$v_{xf}|_{ILPWM} = V_{dc} + V_{dc}MI\cos(\omega_0 t) + \frac{1}{p}\sum_{k=0}^{p-1}\sum_{n=-\infty}^{\infty}\frac{4V_{dc}}{\pi}\sum_{m=1}^{\infty}\frac{1}{m}J_n\left(m\frac{\pi}{2}MI\right)\sin\left(\left[m+n\right]\frac{\pi}{2}\right)\cos(m(\omega_c t + k\frac{2\pi}{p}) + n(\omega_0 t + \theta_o)) . \tag{8}$$

912

The 2018 International Power Electronics Conference

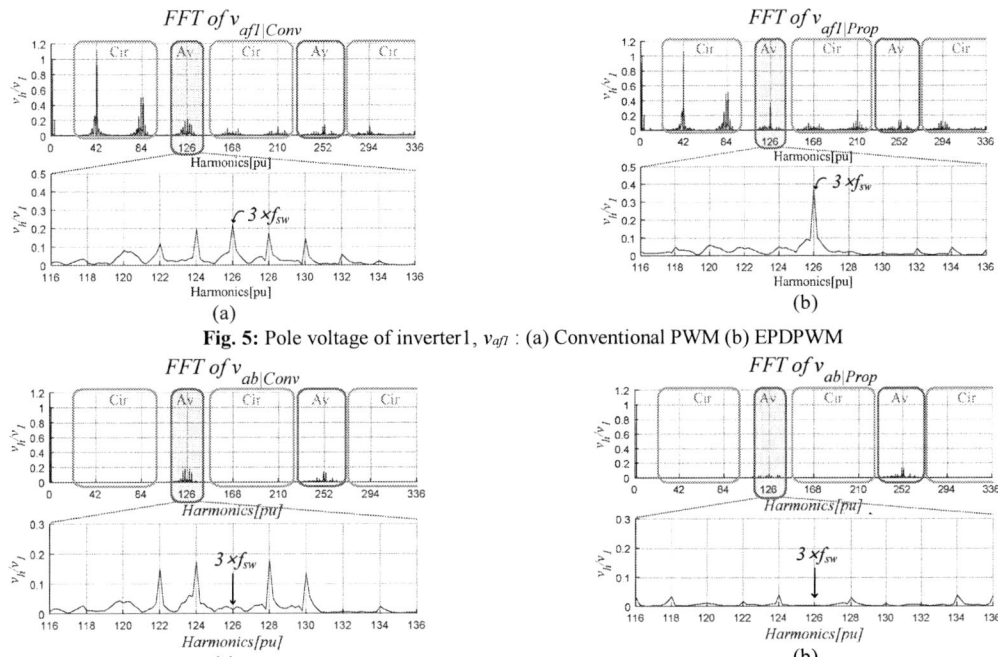

Fig. 5: Pole voltage of inverter1, v_{af1} : (a) Conventional PWM (b) EPDPWM

Fig. 6: Line to line voltages of output node, v_{ab} : (a) Conventional PWM (b) EPDPWM

analysis of PWM in section II, because interleaved PWM can be analyzed as a form of multi-level PWM, PWM can be devised with consideration of output voltage, v_{xf}, and output current, i_x. If PWM is changed from Fig. 3 (a) to Fig. 4 (a) by modifying carrier, it can make V_{dq}^{r*} be synthesized by the adjacent vectors as shown Fig. 4 (b). In this paper, this PWM is named as Effective Phase Disposition PWM, EPDPWM. In (4), which represents the harmonics spectrum of x-phase pole voltage of kth converter, if it is assumed that n is an integer multiple of the number of

phase, $3 \times u$, and m is an integer multiple of the number of parallel, $p \times l$, then these pole voltages can be expressed as

$$v_{af1}(3u, pl) = v_{bf1}(3u, pl) = v_{cf1}(3u, pl) = C_{mn}$$
$$\vdots$$
$$v_{afq}(3u, pl) = v_{bfq}(3u, pl) = v_{cfq}(3u, pl) = C_{mn} \quad . \qquad (9)$$
$$\vdots$$
$$v_{afN}(3u, pl) = v_{bfN}(3u, pl) = v_{cfN}(3u, pl) = C_{mn}$$

It means that these components are not included in line to line voltages of each converter. Therefore, they will be called as Zero Sequence Voltage Harmonics, ZSVH, in this paper. In other words, increasing the ZSVH would have no influence on circulating current and output current. Concentrating more voltages on the ZSVH makes the magnitude of the other harmonics in the sideband decrease because it is a common sense that the RMS value of AC voltage of PWM of each converter is not affected by PWM method itself. EPDPWM has tendency of increasing the

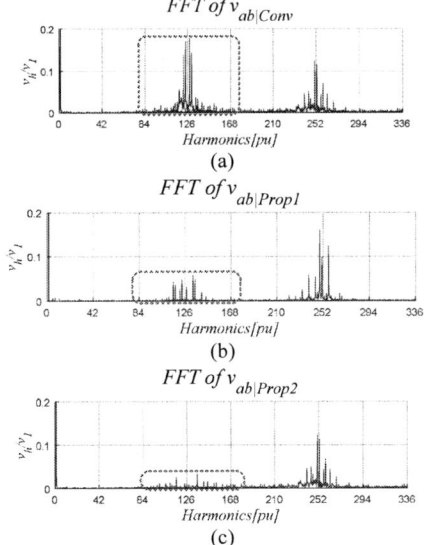

Fig. 7: Line to line voltage of output node:
(a) Conventional PWM (b) EPDPWM (c) EPDPWM with suboptimal offset voltage

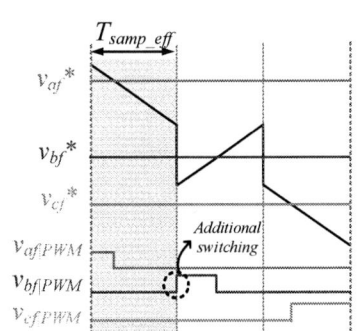

Fig. 8: Modifying carrier of kth converter:

913

The 2018 International Power Electronics Conference

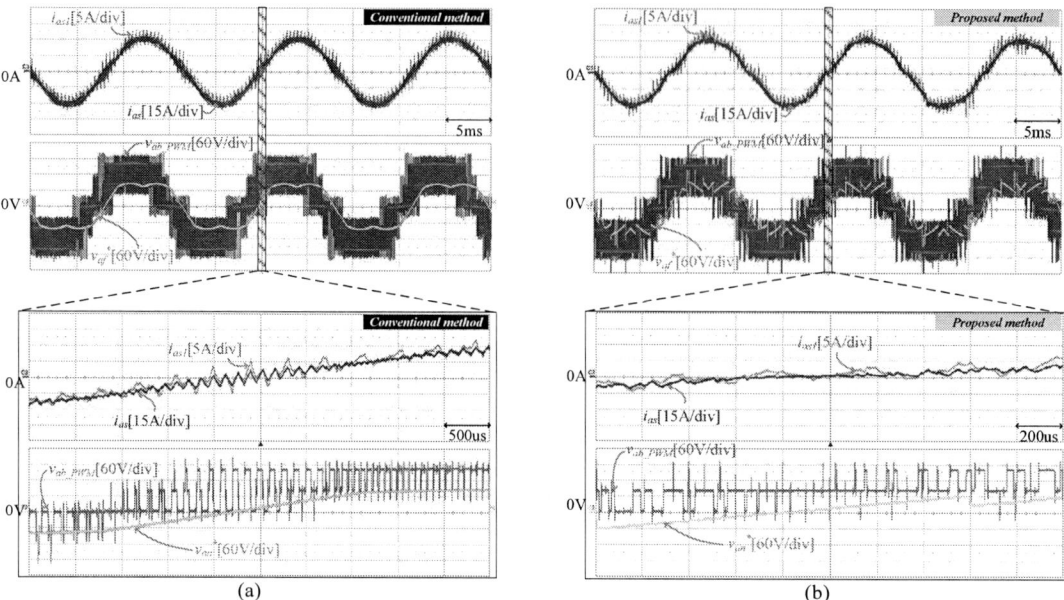

Fig. 9: Experimental results: (a) Conventional PWM (b) EPDPWM with suboptimal offset voltage

ZSVH in 3rd sideband. Therefore, if PWM pulse is modified from conventional PWM to EPDPWM, it is expected that harmonics of voltage and current at node, where PWM converters are connected, would be reduced. In order to prove the effect of modification, simple simulation in per unit basis has been carried out. Simulation conditions are as follows: V_{dc} is 1 pu., switching frequency is 41.67 pu, which means that the switching frequency is 41.67 times of the fundamental frequency, and Modulation Index (MI) is 0.65. The number of paralleled converters, p, is 3. The results of this simulation are shown in Fig. 5 and Fig. 6, where Av and Cir denote harmonics of v_{xf} and v_{xf_cir}, respectively. Especially, 3rd sideband harmonics are expanded in the figure for explanation. From the expanded figure, it can be seen that with the proposed method harmonic component at $3 \times f_{sw}$ in pole voltage in Fig. 5 increases and the other components in the 3rd sideband decrease. Because of this, the harmonics of the line to line voltages of grid side are conspicuously reduced as shown in Fig.6. It can be said that the harmonics of the grid side currents are also reduced.

An offset voltage, v_{sf}, can be easily optimized, because this PWM is developed based on concept of multi-level PWM. In case of two level converter, the suboptimal PWM is when dwelling times of zero vectors are equally distributed in a switching period, in terms of switching ripple component [16]. This tendency is guaranteed, when voltage reference is synthesized by three voltage vectors adjacent to voltage reference, even in the case of multilevel converter where the number of the level of output voltage is high. EPDPWM satisfies the condition according to Fig. 4 (b). Therefore, it can be noted that equal dwelling time of redundant vectors of pivot is suboptimal. Fig. 7 (c) shows the simulation result showing this suboptimal case. By only changing the offset voltage, v_{sf}, 3rd sideband harmonics in Fig. 7 (c) are conspicuously reduced

compared to the result of conventional PWM, Fig. 7 (a), and that of EPDPWM, Fig. 7 (b).

As aforementioned, the carrier of EPDPWM is not the same as that of conventional Phase Disposition PWM (PDPWM). Fig. 8 shows the modifying carrier of the kth converter for the EPDPWM. This carrier can be implemented by using simple triangular carriers and logic circuits in FPGA. When EPDPWM is applied to system, there would be additional switching in a switching period of PWM. The additional switching is shown in a black dashed line in the Fig. 8. Therefore, it is necessary to prove that the proposed method is more cost-effective compared to conventional method, although switching frequency may increase. In order to prove this, Cost Index, C.I., is defined as (10). $H_{max}|3^{rd}$ is the maximum harmonic in 3rd sideband harmonics in this equation. In the case of three converters in parallel, the 3rd sideband harmonics is the first sideband harmonics of line to line voltage of output as shown in Fig.5 (b) and Fig.6 (b). It is well known fact that the largest harmonic in the first side band is strongly related to the size of filter inductance to keep harmonic regulation of grid such as IEEE 519 [15]. Therefore, the maximum harmonic in 3rd sideband harmonics, $H_{max}|3^{rd}$, is considered in the C.I., as in

$$C.I. = \frac{\left(H_{max} \mid 3^{rd} \times f_{sw}\right)_{Proposed}}{\left(H_{max} \mid 3^{rd} \times f_{sw}\right)_{Conventional}} . \tag{10}$$

In other words, decreasing the harmonic in this side band is the most important to cut down the system size and cost. When MI is over 0.5, C.I. goes down below 0.4. That means that the proposed PWM is at least 2 times superior in higher modulation index, where usually converters tied to grid operate. When it is considered both switching frequency and inductance of grid side inductor, it would be said that the proposed method could save the size of filter inductor at least by 60% compared to the conventional PWM.

The 2018 International Power Electronics Conference

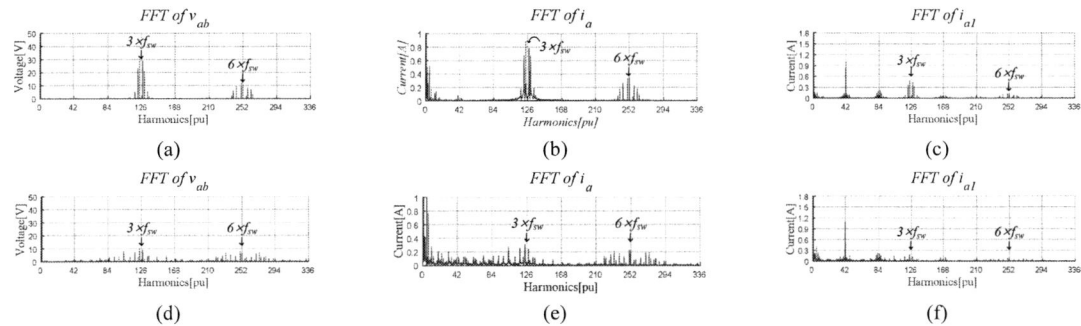

Fig. 10: Experimental results: (a), (b), (c) : Conventional PWM, (d), (e), (f): EPDPWM with suboptimal offset voltage

IV. EXPERIMENT RESULTS

Experimental set was implemented as Fig. 1. Line to line voltage of grid side is 110V$_{rms}$. DC link is kept as 300 V by using a regulated DC supply whose maximum current is 45 A$_{peak}$. A q-axis current reference on the synchronous reference frame of each converter is 10A$_{peak}$, and output current is 30A$_{peak}$. Output voltage of system is connected to the grid through a filter inductor, L_f, whose inductance is 200 uH, which is 0.025pu. Switching frequency and MI of each converter is 2.5kHz, 0.65, respectively. The z_{sh} is the coupled inductor whose inductance is 3.2mH at switching frequency.

Fig. 9 shows experimental results. Upper figure shows a-phase current of 1st converter, i_{a1}, and a-phase output current, i_a. Lower figure shows the a-phase voltage reference, $v_{af}{}^*$, and line to line voltage, v_{ab}, between node 'a' and 'b' in Fig. 1. In the conventional method, offset voltage is optimized according to two level PWM. In the proposed method, that is optimized under the consideration of parallel operation like multilevel converter. In the expanded waveform in Fig. 9, in conventional method, line to line voltage cannot be synthesized by the adjacent voltage levels in a switching period. However, in the proposed method, line to line voltage is synthesized by the adjacent voltage levels in the period. These can be explained by Fig. 3 and Fig. 4. This difference of the voltage synthesis is reflected on FFT result of line to line voltage, v_{ab}, and output current, i_a, in Fig 10. Fig 10 (a) and (d) show the line to line voltage of conventional PWM and that of proposed PWM, respectively. In this figure, there are harmonics, whose frequency are three times and six times of switching frequency. Harmonics of output current are also improved in conjunction with the line to line voltage as shown in Fig. 10 (b) and (e). The largest harmonic in this sideband harmonics of i_a, which is the main concern in designing filter inductance, L_f, is cut down by 66% compared to that of conventional PWM. All of the sideband components are included in a phase current of converter1, i_{a1} as shown in Fig. 10 (c) and (f). Among these components, the residual harmonics except harmonics of i_a flow between converters as circulating current.

V. CONCLUSION

In this paper, the conventional interleaved PWM is analyzed by using Fourier series. Based on the analysis, it

has been found that the Zero Sequence Voltage Harmonics, ZSVH, have no influence on line to line voltage of each converters and output voltage of paralleled converters. Based on this observation, PWM pattern has been modified to increase the ZSVH, intentionally. It makes the other components in its sideband harmonics reduced. Besides, offset voltage is optimized based on the proposed PWM. Finally, the validity, feasibility and effectiveness of the proposed method have been proven by computer simulation and experimental results. The experimental results show that the largest harmonic in sideband harmonics of i_a, which is the main concern in designing grid side inductance, L_f, of the proposed PWM is cut down by 66% compared to that of conventional PWM.

REFERENCES

[1] F. Bovolini and H. Pinheiro, "Flexible arrangement of static converters for grid connected wind energy conversion systems," IEEE Trans. Ind. Electron., vol. 61, no. 9, pp. 4707–4721, Sep. 2014.

[2] Z. Ye, D. Boroyevich, J. Y. Choi, and F. C. Lee, "Control of circulating current in two parallel three-phase boost rectifiers," IEEE Trans. Power Electron., vol. 17, no. 5, pp. 609–615, 2002.

[3] T. Kawabata and S. Higashino, "Parallel operation of voltage source inverters," IEEE Trans. Ind. Appl., vol. 24, no. 2, pp. 281–287,Mar./Apr. 1988.

[4] B. Shi and G. Venkataramanan, "Parallel operation of voltage source inverters with minimal intermodule reactors," in Conf. Rec. IEEE IASAnnu. Meeting, Oct. 2004, pp. 156–162.

[5] H. S. Jung, J. M. Yoo, S. K. Sul, H. J. Lee and C. Hong, "Suppression of circulating current in paralleled inverters with isolated DC-link," 2016 IEEE Energy Conversion Congress and Exposition (ECCE), Milwaukee, WI, 2016, pp. 1-8.

[6] H. S. Jung, J. M. Yoo, S. K. Sul, H. J. Lee and C. Hong, "Parallel Operation of Inverters With Isolated DC Link for Minimizing Sharing Inductor," in IEEE Transactions on Industry Applications, vol. 53, no. 5, pp. 4450-4459, Sept.-Oct. 2017.

[7] H. S. Jung and S. K. Sul, "A design of circulating current controller for paralleled inverter with non-isolated dc-link," *2017 IEEE 3rd International Future Energy Electronics Conference and ECCE Asia (IFEEC 2017 - ECCE Asia)*, Kaohsiung, 2017, pp. 1913-1919.

[8] M. Hashii, K. Kousaka, and M. Kaimoto, "New approach to a high-power GTO PWMinverter for AC motor drives," IEEE Trans. Ind. Appl.,vol.IA-23, no. 2, pp. 263–269, Mar./Apr. 1987.

[9] K. Xing, F. C. Lee, D. Borojevic, Y. Zhihong, and S. Mazumder, "Interleaved PWM with discontinuous space-vector modulation," IEEE Trans.Power Electron., vol. 14, no. 5, pp. 906–917, Sep. 1999.

[10] T. Geyer and S. Schr¨oder, "Reliability Considerations and Fault-Handling Strategies for Multi-MW Modular Drive Systems," IEEETransactions on Industry Applications, vol. 46, no. 6, pp. 2442–2451,November/December 2010.

[11] D. Zhang, F. Wang, R. Burgos, and D. Boroyevich, "Common-Mode Circulating Current Control of Paralleled Interleaved Three-Phase Two-Level Voltage-Source ConvertersWith Discontinuous

Space-Vector Modulation," IEEE Trans. Power Electron., vol. 26, no. 12, pp. 3925–3935, 2011.

[12] S. Ohn, H. S. Jung and S. K. Sul, "A novel filter structure to suppress harmonic currents based on the sequence of sideband harmonics," *2017 IEEE Energy Conversion Congress and Exposition (ECCE)*, Cincinnati, OH, 2017, pp. 2795-2802.

[13] J. L. Agorreta, M. Borrega, J. Lopez, and L.Marroyo, "Modeling and control of N-paralleled grid-connected inverters with LCL filter coupled due to grid impedance in PV plants," IEEE Trans. Power Electron., vol. 26, no. 3, pp. 770–785, Mar. 2011 .

[14] D.G. Holmes, T.A. Lipo, *Pulse Width Modulation For Power Converters*. USA: Wiley, 2003.

[15] B. G. Cho and S. K. Sul, "Non-iterative LCL filter design for three-phase two-level voltage-source PWM converters," 2014 Int. Power Electron. Conf. IPEC-Hiroshima - ECCE Asia 2014, pp. 2802–2809, 2014.

[16] V. Blasko, "Analysis of a hybrid PWM based on modified space-vectorand triangle-comparison methods," IEEE Trans. Ind. Appl., vol. 33, no.3, pp. 756–764, May/Jun. 1997.

Practical Issues and Implementation Circuits of the Digital-Analog Hybrid Full Feed-Forward Method with Unipolar and Bipolar Modulations

Xin Zhang[1*], Henry S. H. Chung[2], ZhiXun Ma[1]

1. School of Electrical and Electronic Engineering, Nanyang Technological University, Singapore
2. Department of Electronic Engineering, City University of Hong Kong, Hong Kong.
E-mail: jackzhang@ntu.edu.sg*

Abstract— **Full Feed-forward (FF) method has been proved to be an effective method to suppress the output current distortion of the grid connected inverter (GCI) which caused by grid voltage harmonics. Nowadays, most of the FF methods are realized through digital solutions. However, the digital delay of the digital control based FF (DFF) method reduces the suppression ability of the GCI current distortion and may cause the the system instability problem at weak grid case. In order to solve this problem, the concept of the digtal-analog hybrid full feed-forward (DAH-FF) approach has been firstly proposed in the recently reported work of the authors. In this paper, the DAH-FF method will be further discussed, which more focuses on the practical issues of implementaion circuits of the DAH-FF method with different modulation strategies, such as unipolar modulation and bipolar modulation. Furthermore, the selection principle of the digital part and analog part of the DAH-FF method is also discussed. Finally, the experimental results verify the correctness of the above theorectical analysis.**

I. INTRODUCTION

Nowadays, the rapid increase of renewable energy systems utilization has rendered power systems as one of the most promising areas of research in the last decades. Under this background, the grid-connected inverters (GCI) which connect the gird and the renewable energy sources play more and more important roles [1].

However, the harmonics in the grid voltage may bring distortion to the output current of the GCI. full feed-forward (FF) method is usually treated as one of the effective solutions to reject the harmonic components from the grid to the GCI current via theoretically increasing the out impedance of the GCI to infinity [2~4]. Since the original control of the GCI is realized by digital controller, the digital controller is naturally utilized as the first implementation option for the existing FF method. Unfortunately, the digital control-based FF (DFF) method has delay problem, which not only reduce the current suppression ability of the GCI, but also may cause instability problem at the weak grid scenario [5].

The straightforward method to solve the above DFF problem is to use a pure analogue circuit to realize all of the control parts of the FF method. However, the analogue circuit is inadvisable to realize some complex controller of the GCI, such as the phase-locked loop, proportional - resonant controller, etc. As a result, a novel concept of the digtal-analog hybrid full feed-forward (DAH-FF) approach has been proposed in the recently presented work of the authors [6]. It points out that, only the analogue based feedforward control parts of the FF method is needed to totally remove the delay problem of the

DFF method. However, though the DAH-FF concept is very clear, it still suffers many practical issues, such as how to combine digital control parts and analogue control circuits perfectly? How to guarantee the necessary deadtime of the upper & lower swithces in the same switching leg with bipolar and unipolar modulations? etc.

In this paper, the DAH-FF approach has been further analyzed with explaining how to select its digital part and analogue part. The practical issues of the DAH-FF method with bipolar and unipolar modulations have also been discussed respectively. As series of detailed implementation circuits of the DAH-FF have been developed. Finally, the experimental work demonstrates conclusively that the implementaion cicuits are very effective and efficient in practice.

II. FURTHER ANALYSIS OF THE DAH-FF METHOD AND ITS SELECTION PRINCIPLES

In this section, the original purpose of the FF method and the delay problem of the DFF method has been reviewed. Then, the selection principle of the digital & analogue parts for the DAH-FF method are carefully discussed.

A. The Original Purpose of the FF Method

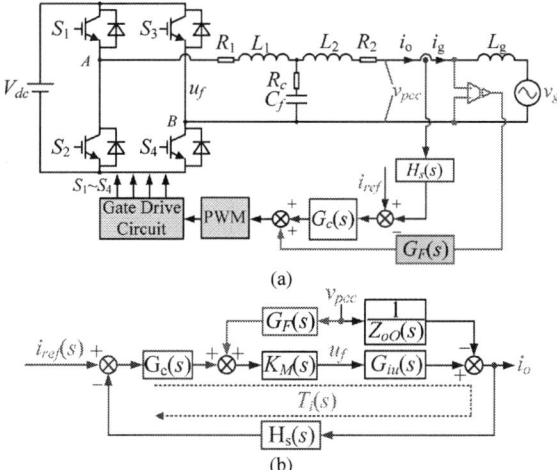

(a)

(b)

Fig. 1 A typical GCI with FF method: (a) circuit; (b) control block.

According to Fig. 1, the expression of the output current of the GCI, i_o, can be expressed as

$$i_o(s) = I_{inv}(s) - \left(\frac{1}{Z_{inv}(s)} + \frac{1}{Z_{oF}(s)} \right) \cdot v_{pcc}(s) \quad (1)$$

TABLE I
DEFINITIONS OF THE TRANSFER FUNCTIONS IN FIG. 1

Transfer function	Definition
$G_F(s)$	Feedforward transfer function
$G_c(s)$	Transfer function of the current regulator
$H_s(s)$	Sampling coefficient of the GCI output current
$K_M(s)$	V_{dc}/V_m (V_{dc}: GCI input voltage; V_m: GCI carrier voltage)
$Z_{oO}(s)$	Open loop output impedance of the GCI [6]
$G_{iu}(s)$	Open loop u_f to i_o transfer function [6]

where

$$I_{inv}(s) = \frac{T_i(s)}{1+T_i(s)} \cdot \frac{i_{ref}(s)}{H_s(s)} \qquad (2)$$

$$Z_{inv}(s) = Z_{oO}(s) \cdot [1+T_i(s)] \qquad (3)$$

$$Z_{oF}(s) = \frac{-(1+T_i(s))}{G_F(s)K_M(s)G_{iu}(s)} \qquad (4)$$

here, $T_i(s)$ is the loop gain of the GCI and can be expressed as

$$T_i(s) = G_c(s) \cdot K_M(s) \cdot G_{iu}(s) \cdot H_s(s) \qquad (5)$$

According to (1), it can be seen that if $Z_{oF}(s)=-Z_{inv}(s)$, the expression of i_o can be simplified as:

$$i_o(s)\big|_{Z_{inv}(s)=-Z_{oF}(s)} = I_{inv}(s) \qquad (6)$$

According to (4) and (6), if a suitable feedforward transfer function $G_F(s)$ can be found/designed to meet (6), the output current of the GCI will act as an ideal current source $I_{inv}(s)$ and hence, it will not suffer any power quality or instability problems. **This is the original purpose of the FF method, which can be visual presented in Fig. 2.** In addition, by (4) and (6), the suitable $G_F(s)$ can be expressed as

$$G_F(s) = \frac{1}{Z_{oO}(s) \cdot K_M(s) \cdot G_{iu}(s)} \qquad (7)$$

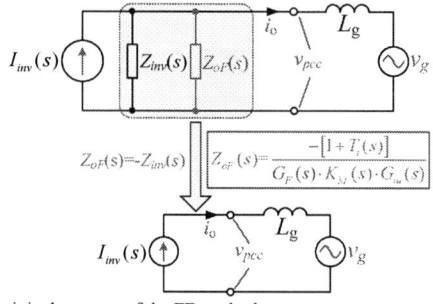

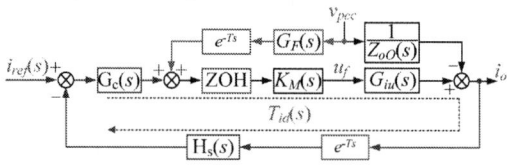

Fig. 2 The original purpose of the FF method.

B. The Delay Problem of the DFF Method

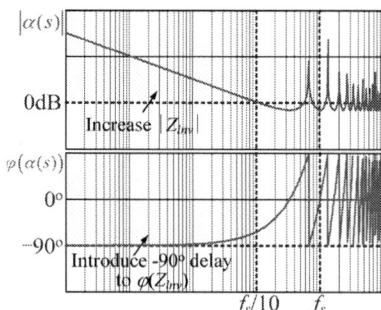

Fig. 3 Control block of the GCI with the DFF method.

Currently, since the original feedback control of the GCI is usually realized via digital controller, the digital controller is naturally utilized as the first implementation method for the FF method. The control block of the DFF is depicted at Fig. 3. As seen, there two types of delay: the computation delay and the PWM delay [7]. The computation delay is one sampling period in the commonly synchronous sampling scheme, which can be

modeled as $e^{-T_s S}$. The PWM delay is caused by the zero-order hold effect, which can be expressed as $(1- e^{-T_s S})/s \approx T_s e^{-0.5T_s S}$. Here, T_s is the sampling time of the digital control system.

According to Fig. 3, the i_o expression with DFF method can be derived as

$$i_o(s) = I_{invd}(s) - \left(\frac{1}{Z_{invd}(s)} - \frac{1}{Z_{oFd}(s)}\right) \cdot v_{pcc}(s) \qquad (8)$$

where

$$I_{invd}(s) = \frac{T_{id}(s)}{1+T_{id}(s)} \cdot \frac{i_{ref}(s)}{H_s(s)} \qquad (9)$$

$$Z_{invd}(s) = Z_{oO}(s) \cdot [1+T_{id}(s)] \qquad (10)$$

$$Z_{oFd}(s) = -\frac{1+T_{id}(s)}{e^{-1.5T_s} \cdot G_F(s) \cdot K_M(s) \cdot G_{iu}(s)} \qquad (11)$$

here, $T_{id}(s)$ is the loop gain of the GCI with the DFF method and can be expressed as

$$T_{id}(s) = e^{-1.5T_s} \cdot T_i(s) \qquad (12)$$

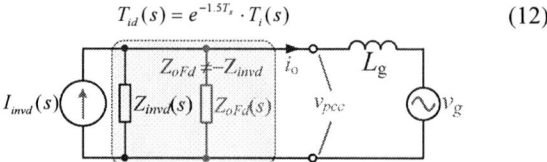

Fig. 4 The real impedance model of the GCI with the DFF method.

By (8), the real impedance model of the GCI with the DFF method has been also presented at Fig. 4. Subsisting (7) to (8), the real i_o expression with DFF method can be further derived as

$$i_o(s) = I_{invd}(s) - \underbrace{\frac{1}{1/\left(1-e^{-1.5T_s}\right)}}_{\alpha(s)} \cdot \frac{1}{Z_{invd}(s)} \cdot v_{pcc}(s) \qquad (13)$$

According to (13), it is shown that the DFF method lost the original purpose of the FF method, i.e., the output current of the GCI cannot act as an ideal current source anymore. As shown in (13), the impact of the digital delay of the DFF method can be evaluated via $\alpha(s)$. For the sake of analysis, the bode plot of $\alpha(s)$ has been depicted in Fig. 5.

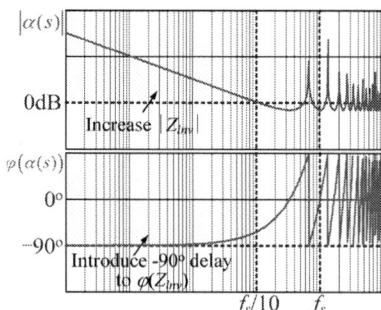

Fig. 5 Bode plots of $\alpha(s)$ --- the impact of the digital delay of the DFF method.

a) The delay problem of the DFF method on the GCI output current harmonic suppression: the current distortion suppression function of the DFF method only effective below the 1/10 sampling frequency. Besides, with the DFF method, the GCI current distortion may become more serious out of the 1/10 sampling frequency.

b) The delay problem of the DFF method on the system stability when the GCI connect to the weak grid: According to the existing stability criterion [8] and (13), the system loop gain with the DFF method can be expressed as

918

$$T_m(s) = \frac{Z_g}{\alpha(s) \cdot Z_{invd}(s)} \qquad (14)$$

According to (14), the system phase margin (PM) with the DFF method can be derived as

$$PM = \underbrace{180^0 - \varphi(Z_g) + \varphi(Z_{invd}(s))}_{original\ system\ PM} + \varphi(\alpha(s)) \qquad (15)$$

According to Fig. 5, $\varphi(\alpha(s)) = -90^\circ$ within $f_s/10$. Therefore, the DFF method reduces the system PM within 1/10 sampling frequency, which may bring negative impact to the system stability when GCI connects to the weak grid.

C. Selection Principles of the DAH-FF Approach

As shown in Fig. 6, there are two types of DAH-FF approach: a) digital feedforward + analogue feedback; b) analogue feedforward + digital feedback. This section will explain how to select the right type of the DAH-FF method.

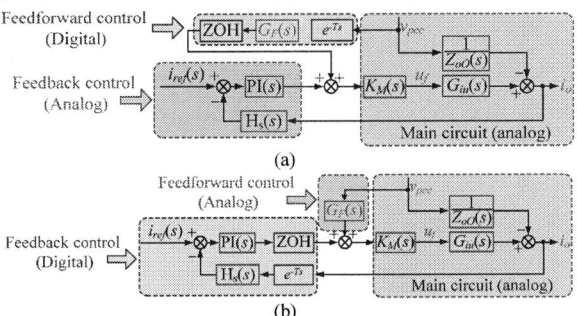

(a)

(b)

Fig. 6 Two types of the DAH-FF method: a) digital feedforward + analogue feedback; b) analogue feedforward + digital feedback.

According to Fig. 6(a), if the DAH-FF utilizes the type of 'digital feedforward + analogue feedback', the output current of the GCI can be derived as

$$i_o(s) = I_{inv}(s) - \frac{1}{\alpha(s)} \cdot \frac{1}{Z_{inv}(s)} \cdot v_{pcc}(s) \qquad (16)$$

Compare (16) and (13), it is obvious that, the 'digital feedforward + analogue feedback' type DAH-FF has the same delay problem with the DFF method. It does not solve anything.

According to Fig. 6(b), if the DAH-FF utilizes the type of 'analogue feedforward + digital feedback', the output current of the GCI can be derived as

$$i_o(s) = I_{invd}(s) - \left(\frac{1}{Z_{invd}(s)} - \frac{1}{Z_{invd}(s)} \right) \cdot v_{pcc}(s) = I_{invd}(s) \qquad (17)$$

According the (17), the 'analogue feedforward + digital feedback' type DAH-FF can guarantee the output current of the GCI act as an ideal current source. It realizes the original purpose of the FF method.

Thus, though the DAH-FF approach has two realization types, only the 'analogue feedforward + digital feedback' type DAH-FF is effective. This is the DAH-FF selection principle.

III. PRACTICAL ISSUES OF THE DAH-FF METHOD AND ITS IMPLEMENTATION SOLUTIONS

A. Practical Issues of the DAH-FF Approach

According to the study of the authors, there are still 4 practical issues of the DAH-FF method need to be solved:

- How to realize the analogue feedforward controller?
- How to realize the DAH-FF approach when the GCI utilizing unipolar modulation with guaranteeing the deadtime function?
- How to realize the DAH-FF approach when the GCI utilizing bipolar modulation with guaranteeing the deadtime function?

B. The Analogue Implementation Circuit of the Feedforward Controller

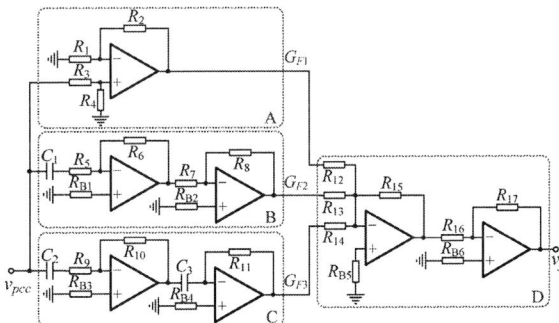

Fig. 7 The analogue implementation circuit of the feedforward controller.

C. DAH-FF Realization when the GCI utilizing unipolar modulation with guaranteeing the deadtime function

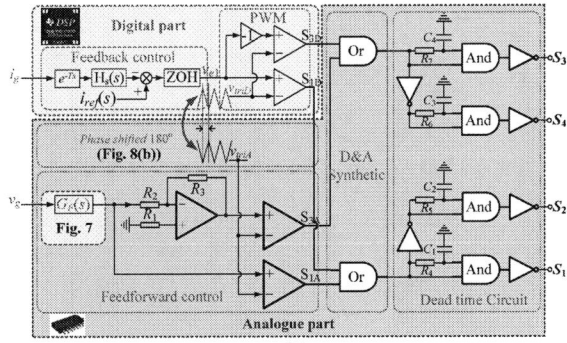

(a)

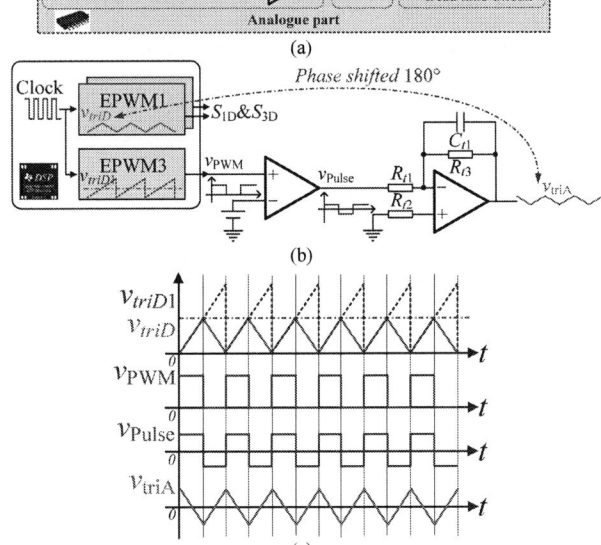

(b)

(c)

Fig. 8 The implementation circuit of the DAH-FF method for the unipolar modulation: a) implementation circuit; b) carrier phase shifted circuit; c) key waveforms in the carrier phase shifted circuit.

919

According to Fig. 1 and (7), the expression of the feedforward controller $G_F(s)$ can be derived as:

$$G_F(s) = \underbrace{\frac{V_{triA}}{V_{dc}}}_{G_{F1}} + \underbrace{\frac{sC_f r_1}{1+sC_f r_c} \cdot \frac{V_{tri}}{V_{dc}}}_{G_{F2}} + \underbrace{\frac{sC_f V_{tri} L_1 / V_{dc}}{1+sC_f r_c} \cdot s}_{G_{F3}} \quad (18)$$

According to (18), the analogue implementation circuit of the feedforward controller has been depicted at Fig. 7.

The implementation circuit of the DAH-FF method for the unipolar modulation has been depicted in Fig. 8(a). As seen, it contains an additional deadtime circuit to guarantee the deadtime function. In addition, in order to combine the digital gate signal and the analogue gate signal perfectly, it required a 180° phase shifted between the digital carrier and the analogue carriers. Figs. 8(b) and (c) gives the detailed circuit and key waveforms of the phase shifted circuit.

D. DAH-FF Realization when the GCI utilizing bipolar modulation with guaranteeing the deadtime function

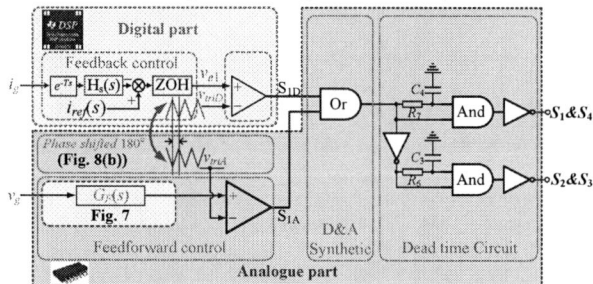

Fig. 9 The implementation circuit DAH-FF method for the bipolar modulation.

Similarly, the implementation circuit of the DAH-FF method for the bipolar modulation has been depicted in Fig. 9. As seen, it also contains an additional deadtime circuit to guarantee the deadtime function. Its phase shifted circuit is the same with the circuit at Figs. 8 (b)&(c) and not repeated here.

Fig. 10 PCB of the analogue part of the DAH-FF method with both unipolar and bipolar modulations.

Therefore, if the engineer wants to realize the DAH-FF method for the unipolar modulation in practice, it can refer to the implementation circuits shown in Figs. 7, 8(a) and 8(b); if the engineer wants to realize the DAH-FF method for the bipolar modulation in practice, it can refer to the implementation circuits shown in Figs. 7, 8(a) and 9.

For the sake of application, the detailed PCB of the analogue part of the DAH-FF method has been presented in Fig. 10, which contains both unipolar and bipolar modulations.

IV. EXPERIMENTAL VERIFICATION

In this section, a 600 W experimental prototype is fabricated and operated with 115V/50Hz grid to verify the real effects of the proposed realization circuits for the DAH-FF method.

The GCI is connected to a weak grid (Z_g = 6 mH). The experimental results of the unipolar modulation based GCI with the DFF and the DAH-FF methods are shown in Fig. 11. The experimental results of the bipolar modulation based GCI with the DFF and the DAH-FF methods are shown in Fig. 12.

According to Figs. 12 and 13, the experimental results verify the correctness of the implementation circuits of the DAH-FF method with both unipolar and bipolar modulations.

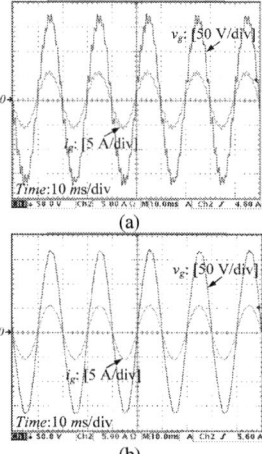

Fig. 11 Experimental results of the unipolar modulation based GCI: a) with DFF method; b) with the proposed realization circuit of the DAH-FF method.

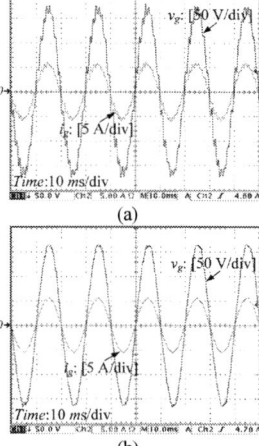

Fig. 12 Experimental results of the bipolar modulation based GCI: a) with DFF method; b) with the proposed realization circuit of the DAH-FF method.

V. CONCLUSIONS

The proposed DAH-FF method has been further analyzed in this paper. It is pointed out that, though the DAH-FF approach has two types of realization, only the 'analogue feedforward + digital feedback' type DAH-FF is effective. In addition, the practical issues of the DAH-FF method with unipolar and bipolar modulations have been considered, such as how to guarantee the deadtime function? how to combine the

The 2018 International Power Electronics Conference

digital and analogue control parts perfectly? etc. Furthermore, detailed implementation circuits of the DAH-FF methods with unipolar & bipolar modulations have been designed carefully. Finally, a 600 W GCI has been built to experimental verify the correctness of the proposed implementation solutions of the DAH-FF method.

REFERENCES

[1] K. Jalili and S. Bernet, "Design of LCL filter of active-front-end two-level voltage-source converters," IEEE Trans. Ind. Electron., vol. 56, no. 5, pp. 1674–1689, May 2009.

[2] T. Abeyasekera, C. M. Johnson, D. J. Atkinson, and M. Armstrong, "Suppression of line voltage related distortion in current controlled grid connected inverters," IEEE Trans. Power Electron., vol. 20, no. 6, pp. 1393– 1401, Nov. 2005.

[3] S. Y. Park, C. L. Chen, J. S. Lai, and S. R. Moon, "Admittance compensation in current loop control for a grid tie LCL fuel cell converter," IEEE Trans. Power Electron., vol. 23, no. 4, pp. 1716–1723, 2008.

[4] X. Wang, X. Ruan, S. Liu and C. K. Tse, "Full Feedforward of Grid Voltage for Grid-Connected Inverter With LCL Filter to Suppress Current Distortion Due to Grid Voltage Harmonics," *IEEE Trans. Power Electron.*, vol. 25, no. 12, pp. 3119–3127, Jan. 2010.

[5] S. Buso and P.Mattavelli, Digital Control in Power Electronics. Seattle,WA, USA: Morgan & Claypool, 2006, pp. 17–64.

[6] X. Zhang, Henry Shu-hung Chung, Y. He, C. Lai and W. Wu, "DAH-FF Approach to Improve the Current Quality and Stability of the LCL Type Grid-Connected Inverter," in IEEE ECCE, Oct. 2017, USA.

[7] D. Holmes, T. Lipo, B. McGrath, and W. Kong, "Optimized design of stationary frame three phase ac current regulators," IEEE Trans. Power Electron., vol. 24, no. 11, pp. 2417–2426, Nov. 2009.

[8] J. Sun, "Impedance-based stability criterion for grid-connected inverters," *IEEE Trans on. Power Electronics Letters*, vol. 26, no. 11, pp. 3075-3078, 2011.

The 2018 International Power Electronics Conference

An AC-DC Power Converter for Electrolytic Capacitor-less LED Driver with High Luminous Efficacy

Kwon-Sik Park, Byuong-Jun Seo, Kyoung-Suk Kang and Eui-Cheol Nho[*]
Dept. of Electrical Engineering, Pukyong National University, Busan, Korea
*E-mail: nhoec@pknu.ac.kr

Abstract— Operating principle and output filter analysis are described about a new topology for the electrolytic capacitor-less LED driving AC-DC power converter. Since the luminous flux of LED increases with the average value of the LED current, pulsating current is available for the LED drive. However, the peak current should be minimized to obtain high luminous efficacy with minimized driving power and high reliability of the LED. The proposed scheme is applied to 60 [W] LED driver. The input power factor of the power supply is 0.945 and the peak-to-average ratio of the LED driving current is 1.469.

Keywords—AC-DC converter, electrolytic capacitor-less, LED, power factor

I. INTRODUCTION

High-brightness light-emitting diode (LED) has good feature such as high efficiency and long lifetime. In order to provide dc current for the LED drive, various power converters have been reported. In most case the power converter utilizes the commercial AC voltage source, therefore, AC-DC power converter is necessary to convert the AC source to a regulated DC source. One of the important index of the converter is input power factor higher than 0.9 for most commercial luminaries as required by regulation standard. Therefore, the converter should be designed to have power factor correction (PFC) function.

Some AC-DC converters adopt two-stage power conversion consisting with AC-DC PFC and DC-DC regulation stages to provide good operating performance. However, this scheme has demerits of high cost and large sized volume due to many components. To achieve reliable and cost-effective LED drive power supply many research works have been done on single-stage PFC converters. Since the AC input current of the PFC converter tracks the AC input voltage, the instantaneous input power has two components of DC and AC, where, the frequency of the AC component is twice the AC source frequency, and the peak value of the AC power is the same with DC power.

The most beneficial power for the LED drive is the pure DC power which is corresponding to the average of the input power. To obtain the pure DC power many topologies have been proposed. In order to filter the pulsating power large capacitance capacitors like electrolytic capacitors are widely used as energy storage component. However, the lifetime of electrolytic capacitor is very short compared with that of LED. Therefore, electrolytic capacitor-less AC-DC LED drivers have been suggested. The proposed scheme in [1] provide perfect DC current to the LEDs, however, two additional active switches, inductor, and capacitor should be added to the output side to absorb the pulsating power.

By the way, the luminous flux of the LED increases with the average value of the LED drive current, therefore, pulsating driving current is available [2], however, the LED flux tends to saturate at high current [3], and the flux decreases with the junction temperature of the LED [4, 5]. Therefore, it is necessary for the pulsating high peak current to be reduced as small as possible to increase the luminous efficacy (lumen/watt) and decrease the junction temperature. The LED driving current of the topology in [6, 7] is pulsating, however, the peak-to-average ratio of the current is reduced by the injection of third harmonic current component into the input current.

This paper proposes a new topology for a LED driver to provide minimized peak-to-average ratio driving current. Circuit topology and operating principle are described with output filter analysis, and the simulation results show the usefulness of the proposed scheme.

II. PROPOSED CIRCUIT DIAGRAM

Fig. 1 shows the proposed circuit diagram. The magnetizing inductance L_m of the transformer used in flyback converter is 600 [μH], and the turns ratio of the transformer is $N_1:N_2$=4:1. The switch S_1 operates in discontinuous current mode (DCM) with a switching frequency of 50 [kHz], providing power factor correction as well as output power regulation. The capacitance of the output filter C_o is around 10 [μF], which is very small compared with that of conventional electrolytic capacitor used for the reduction of 120 [Hz] ripple component of the output voltage. The inductor L_o inserted at the anode side of the series connected LEDs reduces the output current ripple component, and the inductance value is about 30 [μH].

A series connected S_2 and C_i are connected between positive and negative rails of the diode rectifier output, playing the role as an auxiliary circuit to reduce the peak-to-average ratio of the LED driving current. The capacitance of C_i is 1 [μF] which is 1/10 of that of C_o. The charging current i_{S2} flows thorough the body diode of the switch S_2 at the increasing region of the diode rectified voltage. The charged energy is discharged at the valley region of the rectified voltage v_1 by turn-on of S_2. The discharged energy enables the power converter to have reduced peak-to-average ratio of the LED driving current. The average value of charging or discharging current is very small compared with that of the main current i_{S1}, therefore, the power rating of S_2 is small.

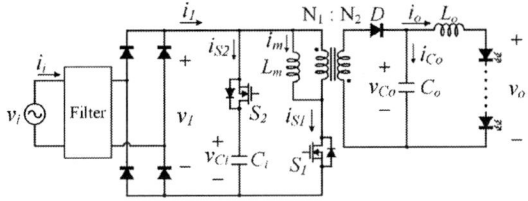

Fig. 1. Proposed circuit diagram for LED drive

III. OPERATING PRINCIPLE OF THE PROPOSED CIRCUIT

Fig. 2 shows waveforms of v_i, i_i, v_1, i_1, i_{S2}, v_{Ci}, i_{S1}, i_{Co}, v_{Co}, and i_o. There are 4 operating modes according to the power transfer mechanism, and the current path in each mode is shown in Fig. 3. The circuit operation is described as follows.

1) Mode - I ($t_0 \leq t < t_1$)
The input current i_i increases as the conventional one as shown in Fig. 2. In other words, the current i_i corresponds to the average value of the switch current i_{S1}, and the current i_{S1} increases when the switch S_1 is turned-on and decreases when S_1 is turned-off with the following equation.

$$
i_{S1}(t) = \begin{cases} \dfrac{v_i(t)}{L_m} t, & S_1 : ON \\[3mm] i_1(t_p) - \dfrac{\dfrac{N_1}{N_2} v_{Co}(t)}{L_m}, & S_1 : OFF \end{cases} \quad (1)
$$

Where, t_p means the time when i_{S1} reaches the top value, and the current decreases to zero by turn-off of S_1. This operation repeats as the switching is carried out in DCM as shown the waveform of i_{S1} in Fig. 2. The red lines in Fig. 3(a) show the current path during the switch S_1 is turned-on. Besides the basic DCM operation, the switch S_1 is turned-on and off also according to the program to obtain high power factor above 0.9 and to inject the third harmonics into the input current to reduce the peak-to-average ratio of the LED driving current. During this mode the load power is provided from the AC source, therefore, the waveform of load current i_o resembles that of source

current i_i. There is no role of the switch S_2 and capacitor C_i in this mode.

2) Mode - II ($t_1 \leq t < t_2$)
As soon as the input voltage v_1 reaches the capacitor voltage v_{Ci} at $t=t_1$, the body diode of S_2 begins to conduct to charge the capacitor C_i, and the charging continues up to the peak voltage of v_1. In this mode the AC input power is delivered to the capacitor C_i as well as the LED load. Therefore, the input current has two components of capacitor charging current and a current corresponding load power. The effect of the charging current on the input power factor and input current THD is insignificant because the magnitude of charging current component is small. The switch S_1 is operated under the same purpose described in Mode - I. Each current path is shown in red color in Fig. 3(b). The current waveform of i_1 in Fig. 2 is the sum of i_{S2} and i_{S1}. The charging current i_{S2} is

$$
i_{S2}(t) = C_i \frac{dv_{ci}(t)}{dt} \quad (2)
$$

3) Mode - III ($t_2 \leq t < t_3$)
Basically, the circuit operation in this mode is almost the same with that in Mode - I except the capacitor voltage level of v_{Ci} as shown in Fig. 2. The capacitor voltage is maintained at the peak value of the input source voltage, while the voltage level is kept constant below the peak value in Mode - I, and the level is determined according to the capacitance value of C_i, the magnitude of $i_o(t_o)$, and the interval between t_3 and t_4. Furthermore, input power factor is also influenced by those values. Therefore, sufficient consideration is needed in the design of the parameters.

4) Mode - IV ($t_3 \leq t < t_4$)
In this mode the magnitude of the rectified voltage v_1 is small, therefore, it is inevitable for the load current of the conventional LED drive power converter to be as small as zero. The small current in this mode results in increased peak-to-average ratio of the LED driving current in the conventional scheme. The proposed topology in this paper can provide much higher load current even in this mode to reduce the peak-to-average ratio. The stored energy in capacitor C_i begins to discharge by turn-on the switch S_2, and the turned-on state of the switch S_2 maintains during this Mode - IV as shown in Fig. 2. Therefore, the switching loss of S_2 is negligible. Once the switch S_1 is turned-on the switch current i_{S1} increases as follows.

$$
i_{S1}(t) = \frac{v_{ci}(t)}{L_m} t \quad (3)
$$

The capacitor voltage v_{Ci} begins to decrease as the switch current i_{S1} flows as follows.

$$
v_{ci}(t) = v_{ci}(t_3) - \frac{1}{C_i} \int_0^t i_{S1}(t)\, dt \quad (4)
$$

The pulsating frequency of i_{S1} is 50 [kHz], therefore, the discharging rate is determined by the average value of i_{S1}.

As the voltage v_{Ci} decreases with time, the duty ratio of the switch S_1 should be increased continuously to provide constant load current during this mode. The discharging current path is shown in red lines in Fig. 3(c). If the period of this mode increases, the peak-to-average ratio can be reduced more, however, input power factor decreases, and input current THD increases, therefore, the period should be set properly.

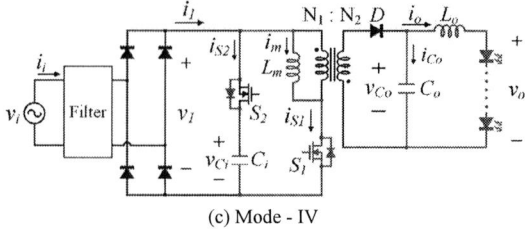

(c) Mode - IV

Fig. 3. Current path in each operating mode

IV. SIMULATION REULTS

Simulations are carried out with the circuit diagram of Fig. 1 and parameters used in simulations are listed in Table I.

TABLE 1
SIMULATION PARAMETERS

Symbol	Value
v_i	220 [V], 60 [Hz]
C_i	1 [μF]
L_m	600 [μH]
$N_1 : N_2$	4 : 1
C_o	5~30 [μF]
L_o	5~50 [μH]
P_o	60 [W]

Fig. 4(a) shows the waveforms of i_{Co}, v_{Co}, and i_o in case of C_o=5 [μF] and L_o=5 [μH]. The rms current of i_{Co} is 2.8 [A], and the rated current of 5 [μF] film capacitor is 6.7 [A], therefore, it is sufficient for the capacitor to be used as an output filter in the viewpoint of current rating. However, the voltage ripple component of v_{Co} is 6.2 [V$_{peak-to-peak}$], which requires large value of L_o. In case of L_o=5 [μH] the LED current ripple component is 3.0 [A$_{peak-to-peak}$] as shown in Fig. 4(a).

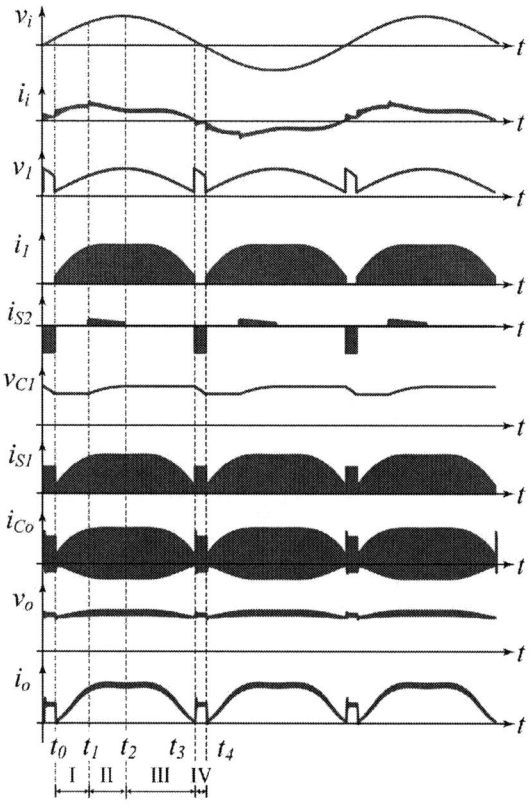

Fig. 2. Waveforms of v_i, i_i, v_1, i_1, i_{S2}, v_{C1}, i_{S1}, i_{Co}, v_o, i_o, and operating mode

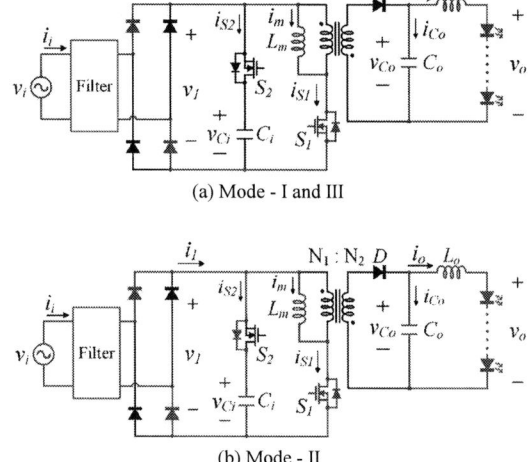

(a) Mode - I and III

(b) Mode - II

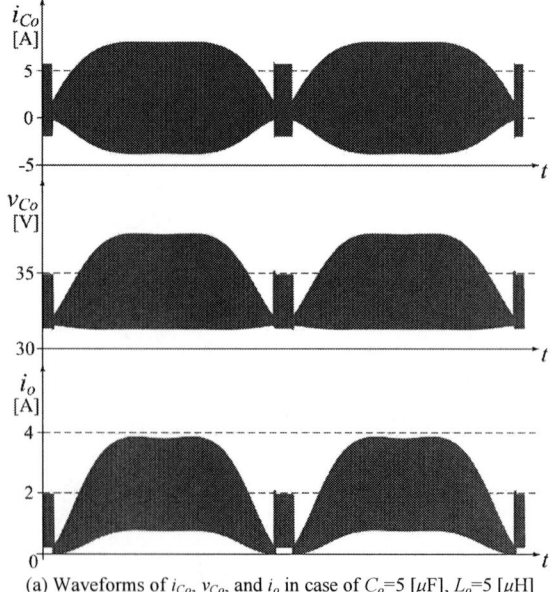

(a) Waveforms of i_{Co}, v_{Co}, and i_o in case of C_o=5 [μF], L_o=5 [μH]

The 2018 International Power Electronics Conference

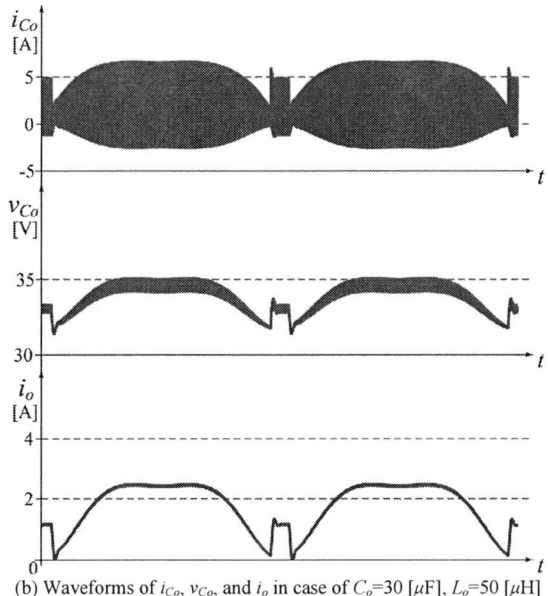

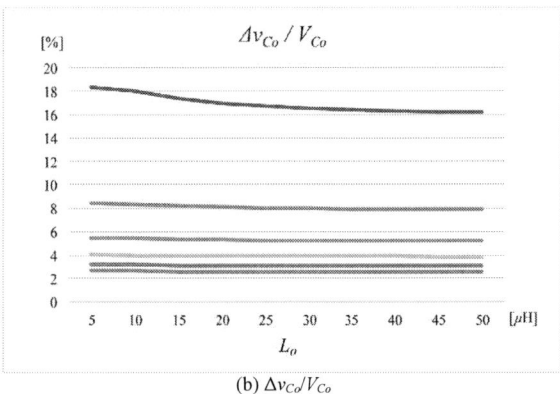

(b) $\Delta v_{Co}/V_{Co}$

Fig. 5. $\Delta i_o/I_o$ and $\Delta v_{Co}/V_{Co}$ with the variation of C_o and L_o

(b) Waveforms of i_{Co}, v_{Co}, and i_o in case of C_o=30 [μF], L_o=50 [μH]

Fig. 4. Waveforms of i_{Co}, v_{Co}, and i_o

Fig. 4(b) shows the waveforms in case of C_o=30 [μF] and L_o=50 [μH]. The voltage ripple component of v_{Co} is 0.87 [$V_{\text{peak-to-peak}}$], which is 14 [%] of that in Fig. 4(a). The LED current ripple component is 0.05 [$A_{\text{peak-to-peak}}$] that is very small compared to that in Fig. 4(a).

Fig. 5 shows the ripple component of the LED driving current and the ripple voltage of the v_{Co} with the variation of C_o and L_o. Table II shows the combination of C_o and L_o that guarantees the ripple component within 15 [%] of the average value of the LED driving current.

The input power factor is 0.945 and the peak-to-average ratio of the LED drive current is 1.469 regardless of the combination of the C_o and L_o.

TABLE II
COMBINATION OF C_o AND L_o TO MAKE Δi_o WITHIN 15% OF i_o

C_o [μF]	L_o [μH]
10	30
15	20
20	15
25	15
30	10

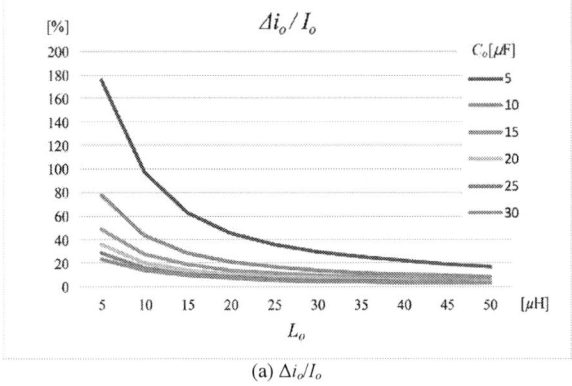

(a) $\Delta i_o/I_o$

V. CONCLUSION

Circuit operating principle and output filter characteristics of a new power converter for electrolytic capacitor-less LED driver is described in this paper. In order to minimize the peak-to-average ratio of the LED driving current, proper current is added to the LED driving current near the current valley area as well as the third harmonic component injection into the input current. The added current is provided by the auxiliary circuit connected between the positive and negative rails of the diode rectifier output. The power rating of the auxiliary switch is as small as 1/20 of the main switch, and the capacitance of the auxiliary capacitor is also 1/20 of the output capacitor.

Simulation results for 60 [W] LED driver show the usefulness of the proposed circuit. High input power factor of 0.945 and low peak-to-average ratio LED driving current of 1.469 can be obtained.

It is expected that the proposed topology can be used for high luminous efficacy with low input power and high reliability in LED driving system.

ACKNOWLEDGMENT

This work was supported by the Korea Institute of Energy Technology Evaluation and Planning(KETEP) and the Ministry of Trade, Industry & Energy(MOTIE) of the Republic of Korea (No. 20174030201490).

REFERENCES

[1] K. Lee, Y. Hsieh, T. Liang, "A Current Ripple Cancellation Circuit for Electrolytic Capacitor-less AC-DC LED Driver," *in Proc. IEEE Appl. Power Electron. Conf. (APEC)*, 2013, pp. 1058-1061.

[2] J. Wei, Z. Yi, L. Wang, L. Liu, H. Wu, G. Wang, B. Zhang, "White LED Light Emission as a Function of Current and Junction Temperature," *In Proceedings of the 10th China International Forum on Solid State Lighting (ChinaSSL)*, Beijing, China, 10–12 November 2013; pp. 166–169.

[3] S.C. Hsia, M.H. Sheu, and S. Y. Lai," Chip Implementation of High-Efficient LED Dimming Driver for High-Power LED Lighting System," *IET Power Electronics*, vol. 8, no. 6, pp.1043-1051, 2015.

[4] S. Y. Hui and Y. X. Qin, "A general photo-electro-thermal theory for light emitting diode (LED) systems," *IEEE Trans. on Power Electronics*, vol. 24, no. 8, pp. 1967–1976, 2009.

[5] S. Buso, G. Spiazzi, M. Meneghini, and G. Meneghesso, "Performance degradation of high-brightness light emitting diodes under DC and pulsed bias," *IEEE Trans. Device Mater. Rel.*, vol. 8, no. 2, pp. 312–322, 2008.

[6] B. Wang, X. Ruan, K. Yao, and M. Xu, "A Method of Reducing the Peak-to-Average Ratio of LED Current for Electrolytic Capacitor-Less AC–DC Drivers," *IEEE Trans. on Power Electronics*, vol. 25, no. 3, pp. 592-601, 2010.

[7] X. Ruan, B. Wang, K. Yao, and S. Wang, "Optimum injected current harmonics to minimize peak-to-average ratio of led current for electrolytic capacitor-less ac–dc drivers," *IEEE Trans. on Power Electronics*, vol. 26, no. 7, pp. 1820-1825, 2011.

The 2018 International Power Electronics Conference

An Improved Cascaded Dual-Buck Inverter

Usman Ali Khan[1*], Honnyong Cha[1], Ashraf Ali Khan[2], Heung-Geun Kim[3], Wilson Eberle[2] and Liwei Wang[2]
1 School of Energy Engineering, Kyoungpook National University, Daegu, Korea
2 School of Engineering, The University of British Columbia, Okanagan, Canada
3 Department of Electrical Engineering, Kyoungpook National University, Daegu, Korea
*Email: Usman_705@ymail.com

Abstract— The cascaded H-bridge inverter plays an important role for achieving a high output voltage using standard low voltage rating semiconductor devices in dc-ac power conversion. It possesses shoot-through problem, and requires dead-time in switching signals to solve it. The dead-time reduces practical voltage gain and output waveforms quality. To overcome these inconveniences, an improved cascaded dual-buck inverter having no shoot-through problem is presented in this paper. It can be operated without using dead-times in switching signals. Additionally, the proposed inverter is capable of using power MOSFETs without reverse recovery issues of their body diodes to boost efficiency and increase switching frequencies. Compared to the conventional dual-buck dc-ac inverter, the proposed inverter reduces the inverter volume and cost by using fewer inductors. To validate the feasibility of the proposed inverter simulation and experimental results are presented.

Keywords— *Dual-buck inverter, inductor, MOSFET, power density.*

I. INTRODUCTION

The H-bridge inverter is a famous topology in dc-ac power conversion. Practically, it is operated with dead-time between the switches to avoid shoot-through occurrence. The dead-time reduces practical voltage gain and output waveforms quality. Additionally, in H-bridge inverter at high bus voltage, the body diode of MOSFET can generate more reverse recovery issues [1-2]. To limit the reverse recovery issues, the H-bridge inverter is normally designed with IGBT. Compared to MOSFET, IGBT has fixed voltage drop in conduction, low switching speed and high switching loss.

As an alternative to the H-bridge inverter, the dual-buck inverter [3] shown in Fig. 1 is presented. It offers the following advantages.

- The shoot-through risk is eliminated by inserting inductors L_1 and L_2 between switch S_1 and S_2, and L_3 and L_4 between switch S_3 and S_4. Owing to no shoot-through risk, PWM dead-times can be eliminated.

- MOSFETs can be used without conducting their body diodes and reverse recovery issues to boost efficiency.

The dual-buck inverter is a two-level having switch voltage stress of input voltage.

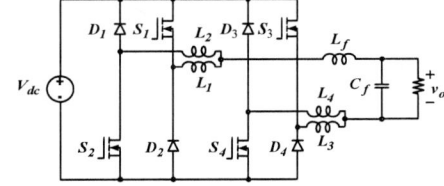

Fig. 1. Full-bridge dual-buck inverter [3].

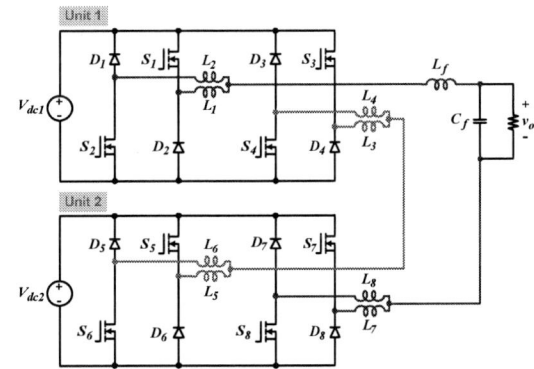

Fig. 2. Conventional cascaded dual-buck inverter [5].

Its peak output voltage cannot be greater than the input voltage. Due to limited voltage handling capability of semiconductor switches ($S_1 - S_4$), dual-buck inverter cannot be connected to a high input dc voltage to generate a high output ac voltage.

For high voltage and power operations, multilevel inverters including flying capacitor [8], cascaded [7, 14, 16] and neutral point clamped multilevel converters [6] have been advanced [4-9, 17]. Compared to the neutral point clamped and flying capacitor multilevel inverters, the cascaded inverter can produce a higher output voltage. The traditional cascaded H-bridge inverter inherits the shortcomings of H-bridge inverter in each cascaded module.

A cascaded dual-buck inverter shown in Fig. 2 is presented in [5]. Similar to the cascaded H-bridge inverter, it can produce a high output voltage using standard low voltage rating switches. Additionally, it has no shoot-through risk and eliminates dead-times in switching signals. Further, it can use MOSFETs as switching devices without reverse recovery issues of their body diodes.

927

The 2018 International Power Electronics Conference

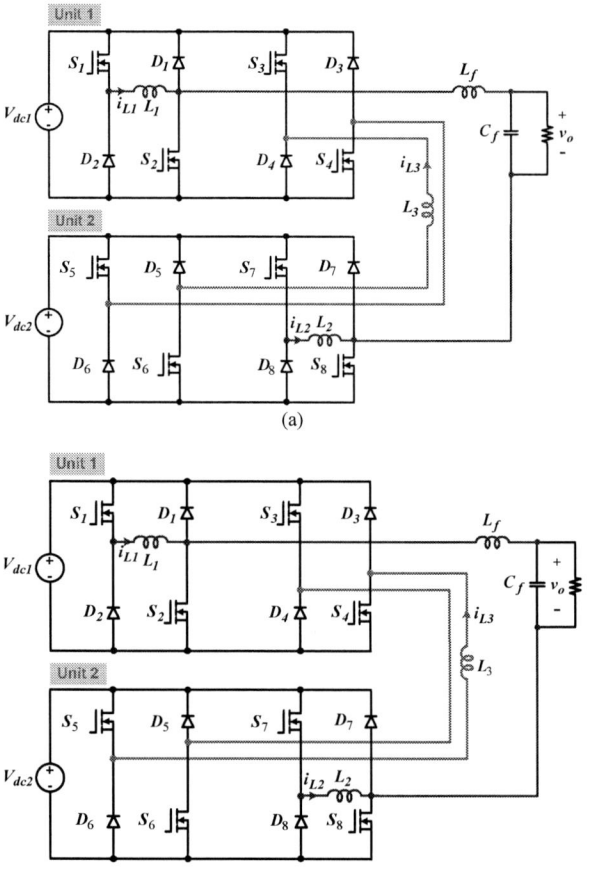

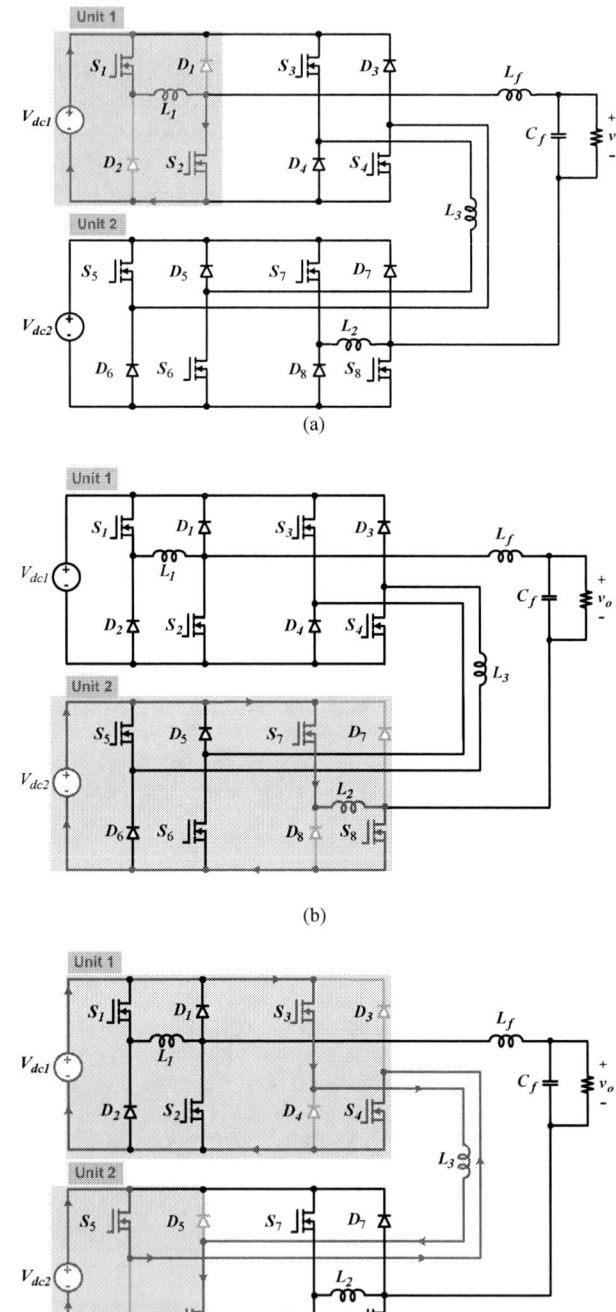

(a)

(b)

Fig. 3. Proposed 2-unit cascaded dual-buck inverters.

Nevertheless, this inverter requires four current limiting inductors in each cascaded unit, which increase the magnetic volume, cost and complexity of circuit layout.

In this paper, an improved cascaded dual-buck inverter requiring fewer current limiting inductors is proposed to reduce inverter volume, cost and circuit complexity. Additionally, it has no shoot-through risk, eliminating dead-times in PWM signals, and producing a high output voltage using standard low voltage rating devices.

Further, the cascaded modules of the proposed inverter can be phase-shifted to increase switching frequency of inductors and capacitors without increasing switching frequency of semiconductor devices. The experimental and simulation results obtained validate the feasibility of the proposed inverter.

II. Proposed Cascaded Inverter

The schematic diagrams of the proposed inverters are depicted in Fig. 3. In the proposed inverter, in each leg only one active switch is connected.

(c)

Fig. 4. Opposition to shoot-through during overlap between the switches. (a) S_1 and S_2 are ON. (b) S_7 and S_8 are ON. (c) $S_3 - S_6$ are ON.

Therefore, similar to the dual-buck converters [10-16], the proposed inverter has no shoot-through concerns, and the switches can be switched without dead-times.

928

As shown in Fig. 3, the proposed inverter requires fewer current limiting inductors than of the inverter in Fig. 2. The proposed 2, 3 and 4-unit cascaded dual-buck inverters require three, four and five current limiting inductors, respectively. Whereas, the conventional 2, 3 and 4-unit cascaded dual-buck inverters require eight, twelve, and sixteen current limiting inductors, respectively. As the number of cascaded unit increases, more current limiting inductors can be saved with the proposed inverter.

In the 2-unit inverter, L_1 protects shoot-through between S_1 and S_2 (see Fig. 4(a)), L_2 protects shoot-through between S_7 and S_8 (see Fig. 4(b)), and the inductor L_3 protects shoot-through when overlap occurs between (S_3 and S_4) or (S_5 and S_6) or ($S_3 - S_6$) (see Fig. 4(c)).

The inductors $L_1 - L_3$ also work as output filtering inductors. By designing $L_1 - L_3$ with required inductance, the output filter inductor L_f can be removed. The body diodes of MOSFETs do not conduct, and external fast recovery diodes are used for current freewheeling.

III. OPERATION OF THE 2-UNIT CASCADED INVERTER USING HYBRID BIPOLAR PWM SWITCHING

In this section, the operation of the proposed 2-unit cascaded inverter using hybrid-bipolar PWM switching (HBPS) is discussed. The block diagram and gate signals of HBPS are shown in Fig. 5. The duty cycle D of S_1 varies between 0.5 and 1. For $v_{ref} > 0$, $S_1, S_4, S_5,$ and S_8 receive high frequency signals, while $S_2, S_3, S_6,$ and S_7 are OFF. For $v_{ref} < 0$, $S_2, S_3, S_6,$ and S_7 receive high frequency signals, while $S_1, S_4, S_5,$ and S_8 are OFF. In a switching cycle the converter has two operating states.

1. Mode 1
As shown in Fig. 6(a), $S_1, S_4, S_5,$ and S_8 are ON, and all diodes $D_1 - D_8$ are reversed biased. The slope of inductor current in terms of V_{dc} and v_o can be expressed as,

$$\frac{di_{L1}}{dt} = \frac{2V_{dc} - v_o}{2L + L_f} \tag{1}$$

where L is the inductance of L_1 and L_3.

2. Mode 2
It is shown in Fig. 6(b). $S_1 - S_8$ are OFF, D_1, D_4, D_5 D_8 are reverse biased, and D_2, D_3, D_6, and D_7 freewheel the inductor current.

$$\frac{di_{L1}}{dt} = \frac{-2V_{dc} - v_o}{2L + L_f} \tag{2}$$

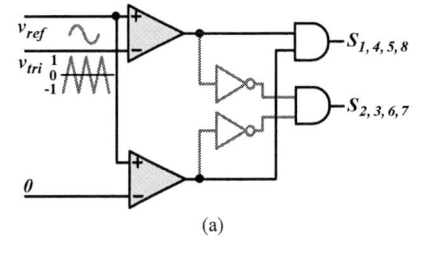

(a)

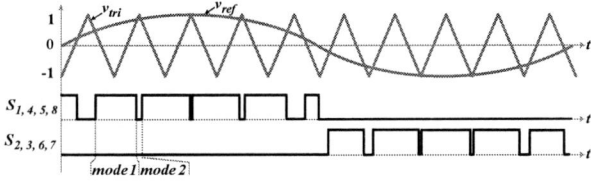

(b)

Fig. 5. HBPS. (a) Block diagram (b). Gate signal generation.

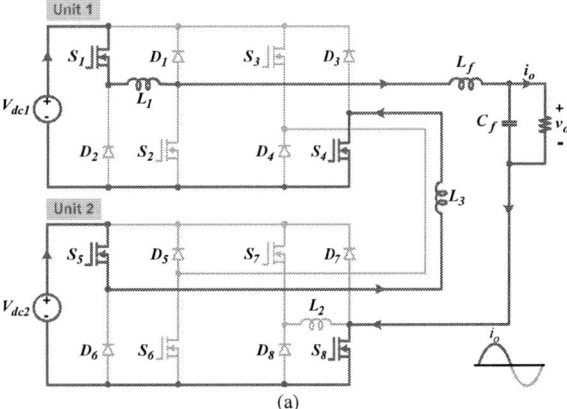

(a)

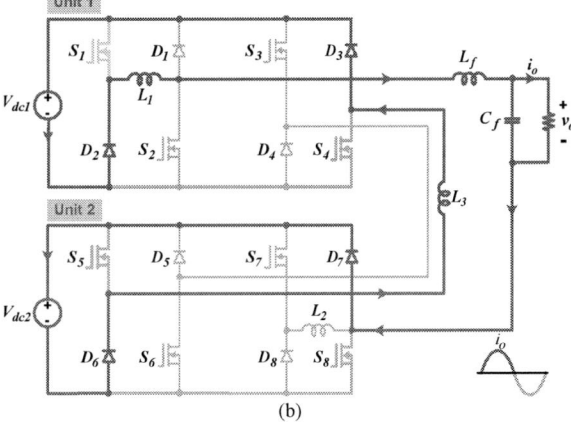

(b)

Fig. 6. Operation with HBPS. (a) Mode 1. (b) Mode 2.

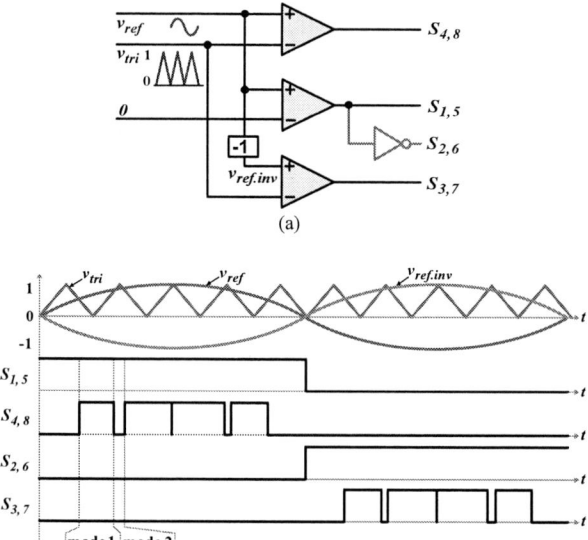

(a)

(b)

Fig. 7. HUPS. (a) Block diagram. (b) Gate signal generation.

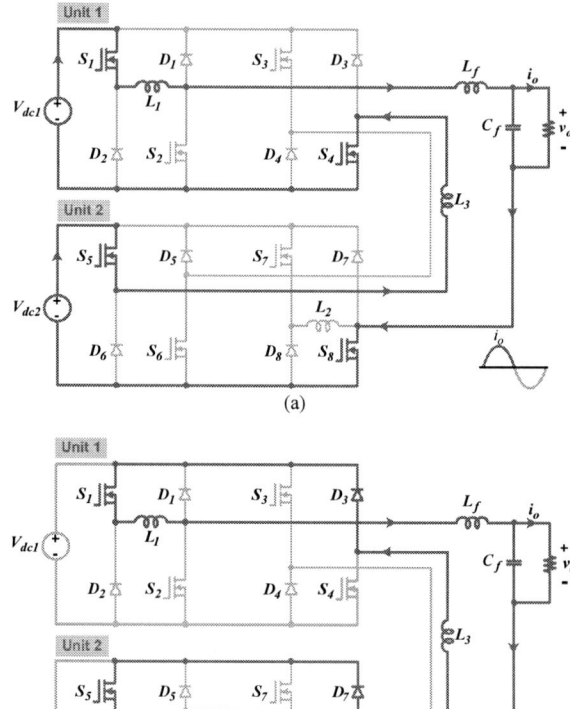

(a)

(b)

Fig. 8. Operation with HUPS. (a) Mode 1. (b) Mode 2.

IV. OPERATION OF THE 2-UNIT CASCADED INVERTER USING HYBRID UNIPOLAR PWM SWITCHING

The gate signal generation and block diagram of hybrid unipolar PWM switching (HUPS) are shown in Fig. 7. In this switching strategy, D the duty cycle of S_4 varies between 0 and 1. Using this strategy, four switches receive low frequency signals. In each cascaded unit at a time, only one switch works at high frequency. As shown in Fig. 7, for $v_{ref} > 0$, S_1 and S_5 are always ON, S_2, S_6, S_3 and S_7 are always OFF, and S_4 and S_8 are switching at high frequency. Similarly, for $v_{ref} < 0$, S_2 and S_6 are always ON, S_1, S_5, S_4 and S_8 are always OFF, and S_3 and S_7 are switching at high frequency.

In a switching cycle the converter has two operating states as discussed below for $v_{ref} > 0$.

1. Mode 1

It is shown in Fig. 8(a). $S_1, S_4, S_5,$ and S_8 are ON, while S_2, S_3, S_6 and S_7 are OFF, and diodes $D_1 - D_8$ are reversed biased. The slope of inductor current in terms of V_{dc} and v_o can be expressed as

$$\frac{di_{L1}}{dt} = \frac{2V_{dc} - v_o}{2L + L_f} \qquad (3)$$

2. Mode 2

It is shown in Fig. 8(b). S_1 and S_5 are ON while $S_2 - S_4$ and $S_6 - S_8$ are OFF, and D_3 and D_7 freewheel the current. The slope of inductor current can be obtained as

$$\frac{di_{L1}}{dt} = \frac{-v_o}{2L + L_f} \qquad (4)$$

V. SIMULATION RESULTS OF THE PROPOSED 2-UNIT CASCADED INVERTER

The input voltages V_{dc1} and V_{dc2} are 310 V_{dc}, output voltage (v_o) is 420 V_{rms}, modulation index is 0.95, output power is 2 kW, switching frequency is 40 kHz, the inductors ($L_1 - L_3$) are of 0.1 mH, and output filter inductor (L_f) is 1 mH. Fig. 9 shows the simulation results of the 2-unit cascaded inverter using HBPS. The input voltages V_{dc1}, V_{dc2} and output voltage (v_o) are shown in Fig. 9(a). Fig. 9(b) shows inductor currents ($i_{L1} - i_{L3}$). Fig. 9(c) shows output current i_o, and drain-source voltages V_{DS1} and V_{DS7} of switches S_1 and S_7, respectively.

Fig. 10 shows simulation results of 2-unit cascaded inverter using HUPS. Fig. 10(a) shows input and output voltages. Fig. 10(b) shows inductor currents. Fig. 10(c) shows output current, and drain-source voltages. The inductor current ripples with HUPS is smaller.

930

The 2018 International Power Electronics Conference

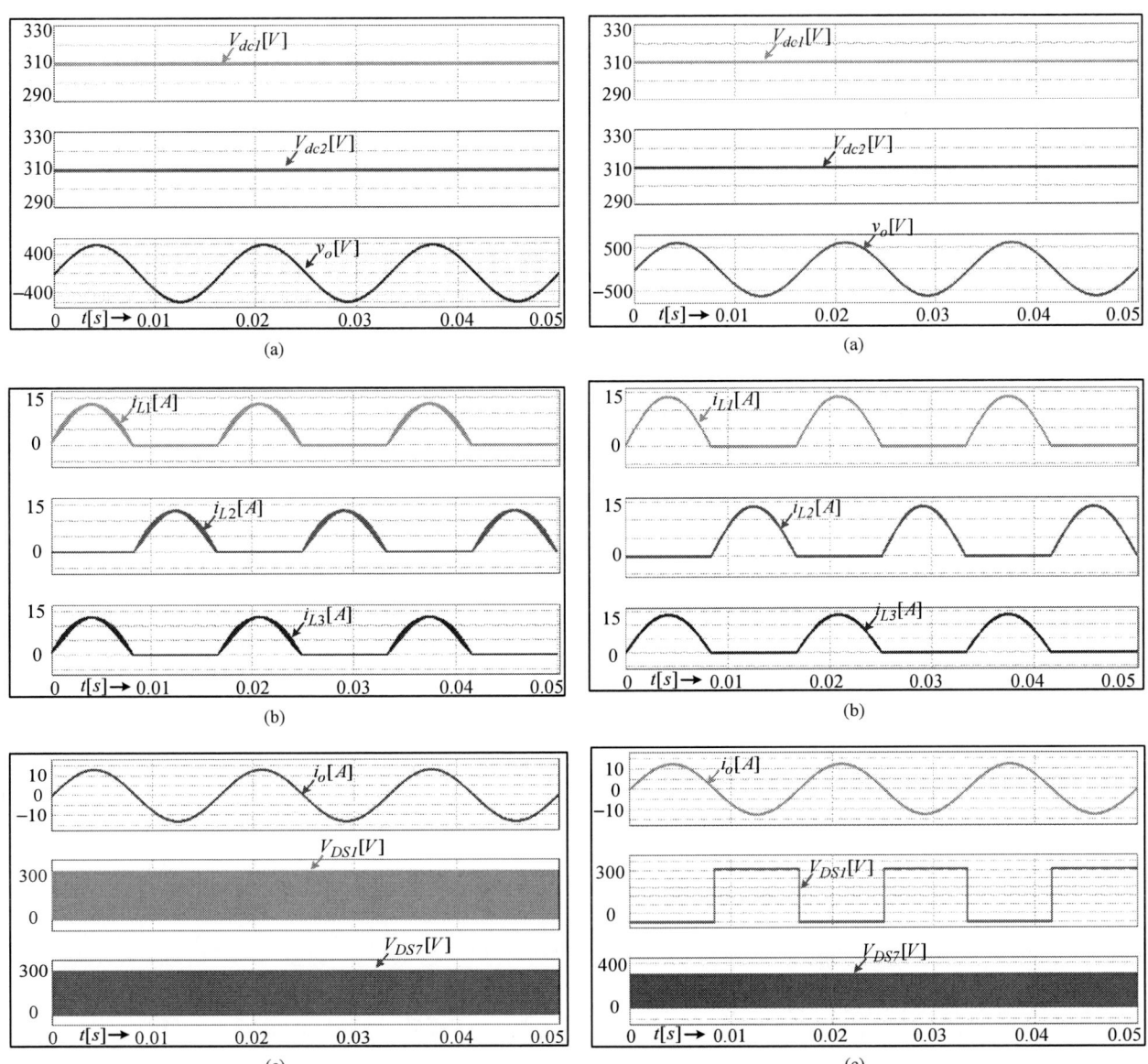

Fig. 9. Simulation results of 2-unit cascaded dual-buck inverter using HBPS. (a) Input and output voltages. (b) Inductor currents. (c) Output current and drain-source voltages.

Fig. 10. Simulation results of 2-unit cascaded dual-buck inverter using HUPS. (a) Input and output voltages. (b) Inductor currents. (c) Output current and drain-source voltages.

VI. EXPERIMENTAL RESULTS

A 2 kW hardware prototype of the proposed 2-unit cascaded inverter is constructed and tested using hybrid bipolar and unipolar PWM switching strategies. Phase-shift control is used, and gate signals of cascaded units are phase-shifted by 180°. The electrical specifications of the inverter are given in Table I. The cascaded units are supplied with separate DC voltage sources, each of 310 Vdc.

TABLE I. ELECTRICAL SPECIFICATIONS

Output voltage	420 V_{rms}/60 Hz
Input voltage of each unit	310 V_{dc}
Output power	2 kW
Switching frequency	40 kHz
MOSFET	47N60CFD
Diode	RHRG3060
Inductors ($L_1 - L_3$)	0.1 mH
Inductor (L_f)	1 mH
Output capacitor (C_f)	4.5 µF

The 2018 International Power Electronics Conference

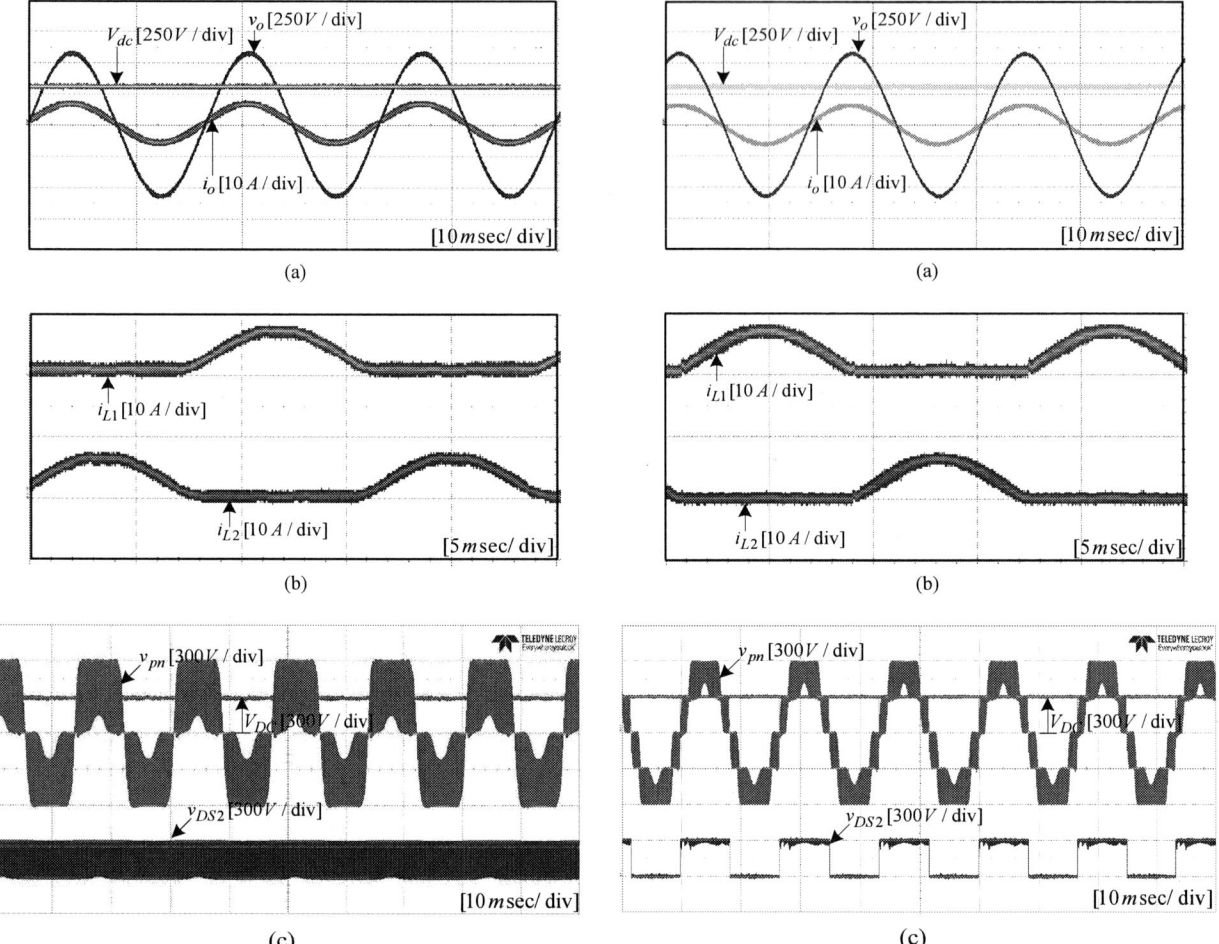

(a)

(b)

(c)

Fig. 11. Experimental results with HBPS.

(a)

(b)

(c)

Fig. 12. Experimental results with HUPS.

Fig. 11 shows the experimental results using hybrid bipolar PWM switching scheme. Fig. 11(a) shows input voltage V_{dc}, output voltage v_o and output current i_o. Fig. 11(b) shows inductor currents i_{L_1} and i_{L_2}. As shown, the inductor current ripple is greatly reduced due to phase-shift control. Fig. 11 (c) shows the drain-source voltage of S_2, input voltage V_{dc} and voltage v_{pn} between the source of S_1 and drain of S_8. v_{pn} has three levels, $2V_{dc}$, 0 and $-2V_{dc}$.

Fig. 12 shows the experimental results using hybrid unipolar PWM switching. Fig. 12(a) shows V_{dc}, v_o and i_o. Fig. 12(b) shows i_{L_1} and i_{L_2}. 12 (c) shows the drain-source voltage of S_2, input voltage V_{dc} and voltage v_{pn} between the source of S_1 and drain of S_8. v_{pn} has five levels, $2V_{dc}, V_{dc}, 0, -V_{dc}$ and $-2V_{dc}$.

The output voltage waveforms are more sinusoidal owing to absence of dead-times in switching signals.

The unidirectional positive inductor currents in Figs. 11(b) and 12(b) indicates, the body diodes of MOSFETs are not conducting, thus eliminating the reverse recovery issues of MOSFET body diodes.

VII. CONCLUSION

In this paper, an improved cascaded dual-buck inverter requiring fewer inductors is presented to reduce the volume and cost of the inverter. The main features of the proposed inverter are given below.

- A high output voltage can be obtained using low voltage rating semiconductor devices.
- The inverter can be operated with phase-shift control to shrink the volume of passive components.
- The shoot-through risk is eliminated, and dead-times in switching signals are not required.

- The reverse recovery issues are minimized by disabling body diodes of MOSFETs and using external fast recovery diodes.
- The inverter can be switched at higher switching frequencies due to adoption of MOSFETs and fast recovery diodes.

The feasibility of the proposed inverter is validated on a 2 kW hardware prototype using phase-shifted hybrid bipolar and unipolar PWM switching controls.

ACKNOWLEDGMENT

This work was supported by the Korea Institute of Energy Technology Evaluation and Planning (KETEP) and the Ministry of Trade, Industry & Energy (MOTIE) of the Republic of Korea (No. 20174030201490) and was also supported by Basic Science Research Program through the National Research Foundation of Korea (NRF) funded by the Ministry of Education (NRF-2016R1D1A1B03934577).

REFERENCES

[1] A. Fiel and T.Wu, "MOSFET failure modes in the zero-voltage-switched full-bridge switching mode power supply applications," in Proc. IEEE Appl. Power Electron. Conf., 2001, pp. 1247–1252.

[2] Q. Zhao and G. Stojcic, "Characterization of cdv/dt induced power loss in synchronous buck dc–dc converters," IEEE Trans. Power Electron., vol. 22, no. 4, pp. 1508–1513, Jul. 2007.

[3] D. Garabandic, "Method and apparatus for reducing switching losses in a switching circuit," U.S. Patent 6 847 196, Aug. 28, 2002.

[4] C. Liu, P. Sun, J.-S. Lai, Y. Ji, M. Wang, C.-L. Chen, and G. Cai, "Cascaded dual-boost/buck active-front end converter for intelligent universal transfer," IEEE Trans. Ind. Electron., vol. 59, no. 12, pp. 4671–4680, Dec. 2012.

[5] P. W. Sun, C. Liu, J.-S. Lai, and C.-L. Chen, "Cascade dual buck inverter with phase-shift control," IEEE Trans. Power Electron., vol. 27, no. 4, Apr. 2012.

[6] A. Nabae, I. Takahashi, and H. Akagi, "A new neutral-point-clamped PWM inverter," IEEE Trans. Ind. Appl., vol. IA-17, no. 5, pp. 518–523, Sep./Oct. 1981.

[7] P. W. Hammond, "A new approach to enhance power quality for medium voltage AC drives," IEEE Trans. Ind. Appl., vol. 33, no. 1, pp. 202–208, Jan./Feb. 1997.

[8] J. Wen and K. M. Smedley, "Synthesis of multilevel converters based on single- and/or three-phase converter building block," IEEE Trans. Power Electron., vol. 23, no. 3, pp. 1247-1256, May 2008.

[9] P. Lezana and G. Ortiz, "Extended operation of cascade multicell converters under fault condition," IEEE Trans. Ind. Electron., vol. 56, no. 7, pp. 2697-2703, Jul. 2009.

[10] P. Sun, C. Liu, J.-S. Lai, C.-L. Chen, and N. Kees, "Three-phase dual-buck inverter with unified pulse width modulation," IEEE Trans. Power Electron., vol. 27, no. 3, pp. 1159–1167, Mar. 2012.

[11] Z. Yao, L. Xiao, and Y. Yan, "Dual-buck full-bridge inverter with hysteresis current control," IEEE Trans. Ind. Electron., vol. 56, no. 8, pp. 3153–3160, Aug. 2009.

[12] A. A. Khan, H. Cha, and H.-G. Kim, "Three-phase three-limb coupled inductor for three-phase direct PWM AC-AC converters solving commutation problem," IEEE Trans. Ind. Electron, vol. 63, no. 1, pp. 189–201, Jan. 2016.

[13] A. A. Khan, H. Cha, and H. F. Ahmed, "High efficiency single-phase AC-AC converters without commutation problem," IEEE Trans. Power Electron., vol. 31, no. 8, pp. 5655–5665, Aug. 2016.

[14] A. A. Khan, H. Cha, and J. Lai, "Cascaded dual-buck inverter with reduced number of inductors," IEEE Trans. Power Electron, vol. 33, no. 4, pp. 2847-2856, April 2018.

[15] A. Khan, and H. Cha, "Dual-buck structured high reliability and high efficiency single-stage buck-boost inverters," IEEE Trans. Ind. Electron, vol. 65, no. 4, pp. 3176-3187, April 2018.

[16] A. A. Khan, H. Cha, J. Baek, J. Kim, and J. Cho, "Cascaded dual-buck AC–AC converter with reduced number of inductors," IEEE Trans. Power Electron, vol. 32, no. 10, pp. 7509-7520, Oct. 2017.

[17] M. Aly, E. Ahmed, and M. Shoyama, "A new single phase five-level inverter topology for single and multiple switches fault tolerance," IEEE Trans. Power Electron, 10.1109/TPEL.2018.2792146.

The 2018 International Power Electronics Conference

A single-switch integrated-stage LED driver based on Cuk and Class-E converter

Shu Zhang[1], Yijie Wang[1], Xiaosheng Liu[1], Yan Zhou[1], Dianguo Xu[1]

1. School of Electrical Engineering & Automation, Harbin Institute of Technology, Harbin, China
wangyijie@hit.edu.cn

Abstract-**This paper presents a single-switch integrated-stage LED driver based on Cuk and Class-E converter. For the LED driver, a cost-effective characteristic is of main concern in general lighting application. Therefore, a single-switch integrated-stage structure is adopted in the paper. The Cuk converter works in discontinuous current mode (DCM) to realize the power factor correction (PFC) function. Meanwhile, Cuk converter has the continuous input and output current characteristic which can simply filter circuit design. The Class-E converter is operated in resonant working condition to transform power to the LED loads as the DC-DC converter cell. By integrating Cuk and Class-E converter, the cost of system can be decreased and power density can also be increased. At rated condition, the power factor (PF) can reach 0.995 and the efficiency can get 90%. A 100W prototype is presented to verify the validity of the presented LED driver.**

I. INTRODUCTION

Owing to the advantages of lighting emitting diodes (LEDs), the traditional lighting sources, such as incandescent lamps, tungsten lamps and compact fluorescent lamps are gradually replaced in the near future. The LEDs show higher luminous efficacy, pure color and environmentally friendly. According to the I-V curve of LEDs, a constant control gear is necessary, that is the usually called LED driver[1-3].

For the off-line LED driver, in order to avoid polluting the grid, Energy Star and IEC61000-3-2 C standard demands a higher PF and low total harmonic distortion (THD). Therefore a PFC cell is needed to add in the LED driver. In traditional topology of LED driver, a two-stage structure is adopted to support the LED loads depicted in Fig.1 (a). The PFC cell and DC-DC converter cell are separately controlled by independent control circuits. Therefore, the dynamic response is satisfying and the high performance can be achieved. But this structure has also some disadvantages. For example, the cost of system is higher, especially in general lighting application. And the whole system power density decreases due to the additional components of control circuits. In order to decrease the cost and increase the power density, an integrated-stage topology is presented, shown in Fig.1 (b). The PFC cell and DC-DC converter cell have the common active switch. Only one active switch should be operated to control the LED driver which decreases the cost of system sharply. In the street lighting application,

integrated-stage LED driver is a good scheme.

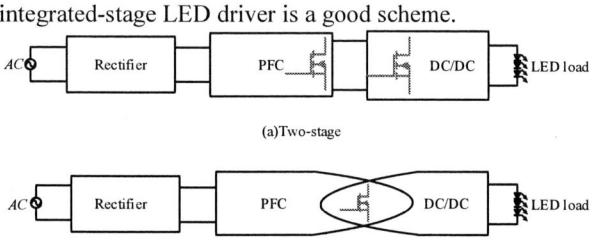

(a)Two-stage

(b) Integrated-stage
Fig.1 The topology of two-stage and integrated-stage of LED driver

Usually there are some basic converters to be used as PFC cell, such as Buck, Boost, Buck-boost, SEPIC and so on[4-7]. When these converters work in DCM, the PFC function can be achieved which can cancel relevant inner current-loop control circuit. Therefore, only an outer current loop control circuit is needed to regulate the output current. For the Buck converter, it has the step-down characteristic. When the output bus voltage is higher than input voltage, the dead zone of output voltage will result in decreasing PF. And Buck converter has poor performance in THD. Although Buck converter has simple structure, it is not optimal converter used as PFC cell. For the Boost converter, the input inductor can smooth input current which can filter the noise from the grid. Therefore Boost converter is usually adopted as the PFC cell. But Boost converter has the step-up characteristic which increases the voltage stress of downstream circuit, especially across the bus capacitor. The selection of E-cap is more or less difficult to balance the size and the value. For the step-up and step-down converters, such as, Buck-boost, SEPIC, Cuk, Zeta, have been widely used as the PFC cells. For the Cuk converter, it has the similar working principle to the Buck-boost converter, while it also has more advantages over Buck-boost converter. Firstly, the Cuk converter has the connected-to-ground switch which can be drive easily. Secondly, the source current from grid and output current through the load (E-cap in the integrated-stage topology) is continuous which simply the filter circuit design. Therefore, Cuk converter has some advantages over the Buck-boost converter. In this paper, Cuk converter is adopted as the PFC cell.

For the DC-DC converters, they can divided into two

934

main categories: non-resonant converter and resonant converter. As discussed above, these PFC converters can belong to the non-resonant converter. They are commonly used in medium and low power applications. By applying resonant techniques into these common DC-DC converters, these topologies can also belong to resonant converters[8-9]. Therefore, the category is not strict. For the resonant converter, LLC converter is the typical topology. Because of the high efficiency and wide regulation range of output load, LLC is widely used in lighting applications. In high power level application, LLC is optimal scheme with relative mature control techniques. For the medium and low lighting applications, two actives switch in LLC circuit still increase the cost of system. In the radio frequency amplifier applications, Class-E converter is usually adopted. In reference[10], Class-E converter can also operate as DC-DC converter which has a potential scheme to apply for the medium and low power applications. Therefore, Class-E becomes a potential topology to operate as the LED driver. Moreover, only one active switch is needed to be controlled which restricts the cost of system. Adopting pulse-frequency-modulation (PFM) with constant duty ratio, the Class-E converter can work on nominal working condition. In this paper, Class-E converter is chosen as the DC-DC converter cell.

For the integrated-stage LED driver, integrating PFC cell and DC-DC converter cell, the cost of system can be decreased and power density is also increasing. In order to balance the ripple power between input power and output power, an E-cap is usually adopted. Because E-cap decreases the reliability, a few techniques are going to solve the problems. One of E-cap elimination techniques is harmonic injection which can decrease the ripple of output power. But it will degrade the PF which limits the lighting application area. Another one of E-cap elimination techniques is adopting bidirectional converter. Adopting store capacitor, compensates the change of output power to eliminate the E-cap. While, additional components are needed to realize the function, so the algorithm complexity and cost will increase sharply. Therefore, elimination of E-cap is also trade-off between cost and reliability of system. In this paper, an E-cap is still adopted. If ignoring the cost of system, a multi-layer capacitor can replace the E-cap which can also increase the span life of LED driver system. So the E-cap less research can be done in the future.

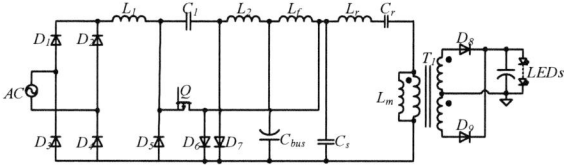

Fig.2 The proposed LED driver based on Cuk and Class-E converter

The paper presents a 100-W prototype based on integrated-stage Cuk and Class-E converter shown in Fig.2. The paper will be organized as follows: In Section II, the working principle will be discussed. The design of system is presented in Section III. In Section IV, the experimental results will be shown. A conclusion is discussed in Section V.

II. WORKING PRINCIPLE OF PROPOSED LED DRIVER

A. Cuk PFC cell

The Cuk converter works in DCM to realize the PFC function. Cuk converter has several features such as low-noise level, natural protection against inrush current and high overall conversion efficiency.

The input voltage $v_{ac}(t)$ is shown in Eq.(1).

$$v_{ac}(t) = V_m \sin(2\pi f_L t) \tag{1}$$

Where, V_m is the amplitude of input voltage, f_L is the line frequency.

After the diode bridge rectifier, the input voltage can be rewritten as Eq.(2).

$$V_{in}(t) = |V_m \sin(2\pi f_L t)| \tag{2}$$

Where $||$ represents the modulus function.

The current i_{L1} through the inductor L_1 will be increased linearly when the switch is turned on. The peak current $i_{L1,pk}(t)$ is shown in Eq.(3). Then the average current $i_{L1,avg}(t)$ can be obtained in Eq.(4) which is equal to the value of input current $i_{ac}(t)$.

$$i_{L1,pk}(t) = \frac{V_{in}(t)}{L_1} DT_s \tag{3}$$

$$i_{L1,avg}(t) = i_{ac}(t) = \frac{1}{2} i_{L1,pk}(t) DT_s = \frac{1}{2} \frac{V_{in}(t)}{L_1} D^2 T_s^2 \tag{4}$$

Where, D is the duty ratio of drive signal, T_s is the working period, L_1 is the value of inductor L_1.

From the Eq.(4) shows, the input current $i_{ac}(t)$ follows the input voltage $V_{in}(t)$ ($v_{ac}(t)$). So when Cuk converter works in DCM, the PFC function can be achieved.

B. Class-E DC-DC converter cell

The Class-E converter works in resonant working condition to transform the energy to the output LED loads. For the Class-E converter, it has simple structure instead of complex resonant loop which has been a promising program to be used in DC-DC converter, especially in high frequency applications.

In [11-12], a resonant topology like Class-E converter is proposed in details. The main difference between the resonant circuit and Class-E converter is the choke inductor which is considered to join the resonance. Although the analysis provides a concise overview about the resonant circuit which only has two inductors and two capacitors, the scheme is adjusted to the small power application. Because the choke inductor is replaced by resonant inductor which improves the current to a higher level to result in a higher loss. Therefore, the traditional Class-E converter still has the advantage in medium application.

For the Class-E converter, in order to improve the efficiency, the reactive power should be depressed sharply. The compensation components are adopted which will transform more active power to LED loads. The equivalent model of Class-E converter is presented in Fig.3.

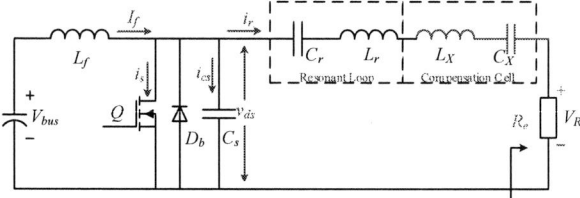

Fig.3 The equivalent model of proposed Class-E DC-DC converter cell

The Resonant Loop contains inductor L_r and capacitor C_r which are working at fundamental switching frequency. In order depress the reactive power, compensation capacitor C_X is adopted to compensate the equivalent inductor L_X. The L_X contains the primary-side inductor Lm and parasitic inductance.

For the Class-E DC-DC converter, the parameter y is assumed as the half period of switch turn-off period. The expression is shown as follows.

$$D = \frac{\pi - y}{\pi} \quad (1)$$

Where, D is the duty ratio of switch.

As the model of Class-E converter shown in Fig.2, the voltage v_R across the equivalent resistor R_e is shown as Eq.(2).

$$v_R(\omega t) = V_R \cdot \sin(\omega t + \varphi) \quad (2)$$

Where, φ is the initial phase, ω is the switch working angle frequency.

Another crucial voltage is the drain-source voltage v_s, which is shown in Eq.(3). According to fundamental wave analysis, the fundamental component of voltage v_s can be calculated in Eq.(4).

$$v_s(\omega t) = \frac{1}{\omega C_s}\int_{\frac{\pi}{2}-y}^{\omega t}(i_1 - i_r)d(\omega t) = \frac{1}{\omega C_s}\int_{\frac{\pi}{2}-y}^{\omega t}[I_1 - \frac{V_R}{R}\sin(\omega t + \varphi)]d(\omega t) \quad (3)$$

$$= \frac{I_1}{\omega C_s}(\omega t + y - \frac{\pi}{2}) + \frac{V_R}{\omega C_s R}[\cos(\omega t + \varphi) + \sin(\varphi - y)]$$

$$v_{s1}(\omega t) = V_R \cdot \sin(\omega t + \varphi) + (\omega L - \frac{1}{\omega C_r}) \cdot \frac{V_R}{R} \cdot \cos(\omega t + \varphi) \quad (4)$$

$$= V_{s1m} \cdot \sin(\omega t + \varphi_1)$$

Where,

$$\varphi_1 = \varphi + \tan^{-1}(\frac{\omega L - 1/\omega C_r}{R}) = \varphi + arc\tan(\frac{X}{R}) \quad (5)$$

$$V_{s1m} = V_R \cdot \sqrt{1 + \frac{X^2}{R^2}} = V_R \cdot \rho \quad (6)$$

The main working diagram is shown in Fig.4.

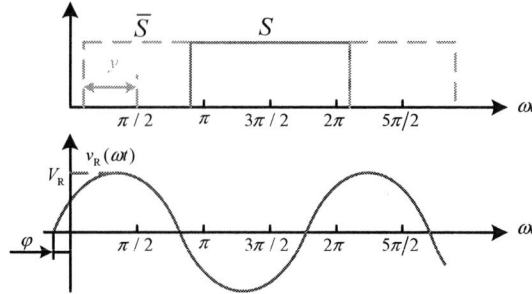

Fig.4 Main working waveforms of fundamental components

According to the Fourier expansion, the fundamental component of switch voltage can be rewritten in Eq.(7).

$$V_{s1m} = -2[\frac{I_1}{\pi \omega C_s}(\frac{\pi}{2} - y + \varphi_1) + \frac{V_R}{\pi \omega C_s R}\sin(y - \varphi)]\cos\varphi_1 \sin y + \frac{I_1}{\omega C_s}[-2\sin\varphi_1 \sin y +$$

$$(\pi + 2\varphi_1)\cos\varphi_1 \sin y + 2y\sin\varphi_1 \cos y] - \frac{V_R}{2\pi \omega C_s R}[\sin(2\varphi + \psi)\sin(2y) - 2y\sin\psi]$$

$$(7)$$

Then submitting Eq.(6) into Eq.(7), the amplitude of output voltage v_R can be obtained in Eq.(8). Then the initial phase φ is shown in Eq.(9).

$$V_R = I_1 R \frac{2y\sin\varphi_1 \sin y - 2y\cos\varphi_1 \cos y + 2\cos\varphi_1 \sin y}{-2\sin(\varphi - y)\sin\varphi_1 \sin y - 0.5\cos(2\varphi + \psi)\sin(2y) + y\cos\psi} \quad (8)$$

$$= I_1 R \cdot g(\varphi, \psi, y)$$

$$\varphi = \arctan(\frac{q_1 s_1 + q_2 s_2 + q_2 r_1 - q_1 r_2}{-q_1 s_2 - q_2 r_2}) \quad (9)$$

Where,

$$q_1 = 2y\sin y\cos\psi + 2y\cos y\sin\psi - 2\sin y\sin\psi$$

$$q_2 = 2y\sin y\sin\psi - 2y\cos y\cos\psi + 2\sin y\cos\psi$$

$$s_1 = -2\sin\psi\sin^2 y$$

$$s_2 = \omega\pi C_s R\rho - y\sin\psi + \sin\psi\sin y\cos y$$

$$r_1 = 2\sin^2 y\cos\psi$$

$$r_2 = y\cos\psi - \sin y\cos y\cos\psi$$

For the choke inductor L_f, in a whole switching period, the average voltage is zero. So the bus voltage V_{bus}, across the bus capacitor C_{bus}, is calculated in (10).

$$V_{bus} = \frac{1}{2\pi}\int_0^{2\pi} v_s d(\omega t) = \frac{I_1}{2\pi\omega C_s}[2y^2 + 2y\cdot g\cdot\sin(\varphi - y) - 2g\cdot\sin\varphi\sin y] \quad (10)$$

Then the gain of system can be obtained as the depicted Eq.(9) and (10). The expression M_A is shown in (11).

$$M_A = \frac{2\pi\cdot\omega\cdot C_s\cdot R\cdot g}{2y^2 + 2y\cdot g\cdot\sin(\varphi - y) - 2g\cdot\sin\varphi\cdot\sin y} \quad (11)$$

According to the Eq.(11), the waveform of system gain is depicted in Fig.5. From the Fig.5, when the working frequency equals to the fundamental frequency of resonant loop (L_r, C_r), the different quality factor Q will not influence the system gain M_A.

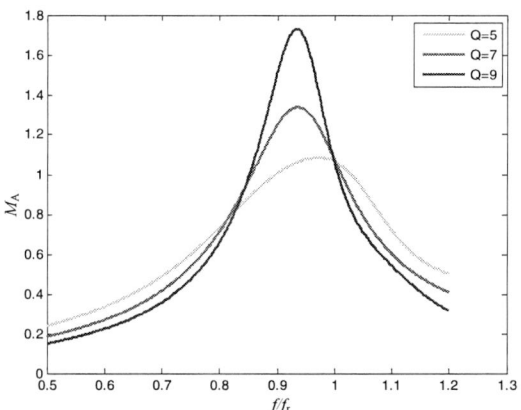

Fig.5 Waveform of system gain M_A and quality factor Q

III. DESIGN OF PROPOSED LED DRIVER

A 100W LED driver is designed to apply for the street lighting application. The main characteristics are shown as follows:

1) Input voltage: $v_{in}(t) = 110\sqrt{2}\sin(2\pi \cdot 50t)$ V;

2) Working frequency: $f_s = 100$ kHz;

3) Output voltage: $V_o = 50$ V;

4) System efficiency: $\eta = 85\%$.

For the Cuk converter, the discontinuous inductor current mode (DICM) is usually adopted to integrate with the backward DC-DC converter. The main Equations are shown as follows.

$$L_{eq} = \frac{L_1 L_2}{L_1 + L_2} \qquad (12)$$

$$D = \sqrt{2} M \sqrt{K_a} \qquad (13)$$

$$M = \frac{V_{bus}}{V_m} \qquad (14)$$

$$K_a < K_{a,crit} = \frac{2L_{eq}}{R_{Class-E} T_s} \qquad (15)$$

$$C_1 = \frac{1}{\omega_r^2 (L_1 + L_2)} \qquad (16)$$

Where, L_{eq} is the equivalent inductor, $R_{Class-E}$ is the equivalent load of Class-E converter, M is the gain of Cuk converter, K_a is the relative coefficient.

The values of main components are depicted in Table I.

TABLE I
MAIN COMPONENTS OF LED DRIVER

Component	Device Model	Value
L_1		869.8μH
L_2		400μH
C_1		4.7nF
Cs		3.6nF
Cr		2.8nF
$Cbus$		100μF
Co		47μF
Lf		3.5mH
Lr		1mH
Q	10N80C	

IV. EXPERIMENTAL RESULTS

The experimental results are shown as follows.

Fig.6 is the input voltage and current. From the Fig.6, the input current follows the input voltage. So the PFC function of Cuk converter has been realized when it works in DICM.

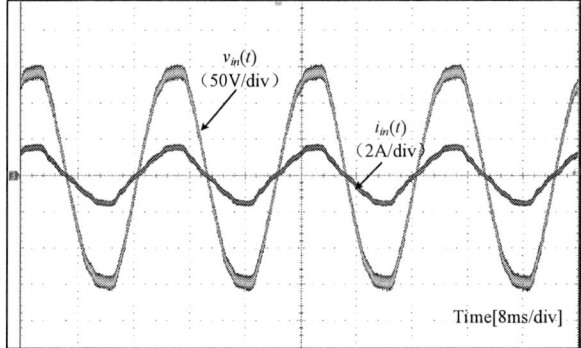

Fig.6 Waveforms of input current $i_{in}(t)$ and input voltage $v_{in}(t)$

Fig.7 shows the waveforms of current trough the output diodes D_8 and D_9. From the Fig.7, the ZCS condition can be achieved at the rated working condition.

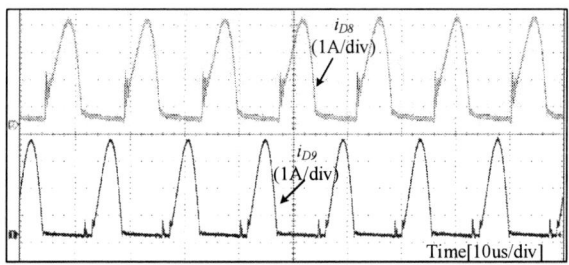

Fig.7 Waveforms of output current through diodes D_8, D_9

Fig.8 shows the waveforms of bus voltage V_{bus} and resonant current i_r. The bus voltage is constant and below the rated voltage of capacitance. And the resonant current i_r is near sinusoidal which means an adequate resonant condition.

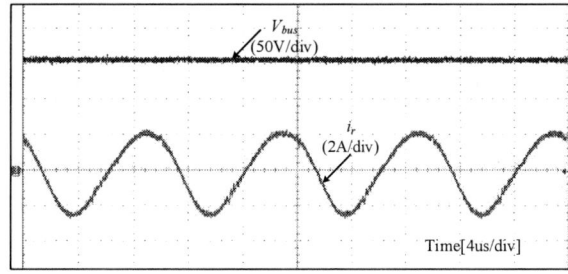

Fig.8 Waveforms of bus voltage V_{bus} and resonant current i_r

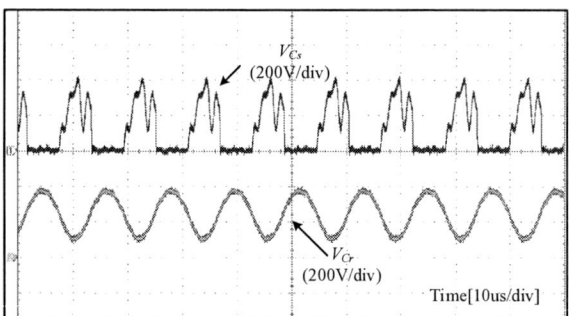

Fig.9 Waveforms of voltage across capacitors C_s and C_r

Fig.9 shows the waveforms of voltage across the capacitors C_s and C_r. The voltages V_{Cr} and V_{Cs} are below the 1000Vdc which is the rated voltage.

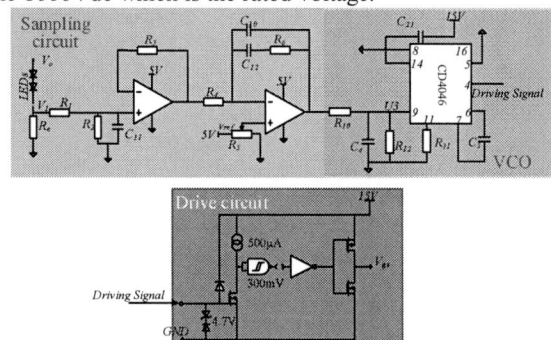

Fig.10 Control circuit of proposed LED driver

Fig.10 is the control circuit of proposed LED driver. It consists of sampling circuit, voltage control oscillating (VCO) cell and drive circuit.

V. CONCLUSION

An integrated LED driver is proposed to apply in the street lighting application. This topology can decrease the cost of system and Class-E can provide a high efficiency at rated working condition. Cuk converter can realize the PFC function and Class-E converter support the LED loads at resonant working condition. The experimental results verify the validity of proposed LED driver.

ACKNOWLEDGEMENT

This work is supported in part by the National Natural Science Foundation of China under Grant 51407044, and in part by the National Key Research and Development Program of China under Grant 2017YFB0402800.

REFERENCES

[1] Y. Wang, J. M. Alonso and X. Ruan, "A Review of LED Drivers and Related Technologies," *IEEE Trans. on Indu. Electron.*, vol. 64, no. 7, pp. 5754-5765, July 2017.

[2] K. P. Reshma, R. Sreenath and N. Pai, "Design and implementation of an isolated switched-mode power supply for led application," in *Proc. 2016 Inter. Conf. on Computation of Power, Energy Information and Commuincation (ICCPEIC)*, Chennai, 2016, pp. 459-461.

[3] J. He, X. Ruan and L. Zhang, "Adaptive Voltage Control for Bidirectional Converter in Flicker-Free Electrolytic Capacitor-Less AC–DC LED Driver," *IEEE Trans. on Indu. Electron.*, vol. 64, no. 1, pp. 320-324, Jan. 2017.

[4] Y.-Z Xu, W.-M. Lin, Y.-C. Xu and Y.-J. Shao, "Inductor optimize design for BCM BUCK-PFC in LED driver," in Proc. *2011 Inter. Conf. on Electric Information and Control Engineering*, Wuhan, 2011, pp. 2264-2267.

[5] Y. Hu, L. Huber and M. M. Jovanović, "Single-Stage, Universal-Input AC/DC LED Driver With Current-Controlled Variable PFC Boost Inductor," *IEEE Trans. on Power Electron.*, vol. 27, no. 3, pp. 1579-1588, March 2012.

[6] U. Ramanjaneya Reddy and B. L. Narasimharaju, "Single-stage electrolytic capacitor less non-inverting buck-boost PFC based AC–DC ripple free LED driver," *IET Power Electron.*, vol. 10, no. 1, pp. 38-46, 1 20 2017.

[7] B. Poorali and E. Adib, "Analysis of the Integrated SEPIC-Flyback Converter as a Single-Stage Single-Switch Power-Factor-Correction LED Driver," *IEEE Trans. on Indu. Electron.*, vol. 63, no. 6, pp. 3562-3570, June 2016.

[8] Y. Guan, Q. Bian, Y. Wang, X. Hu, B. Liu, W. Wang, D. Xu, "Analysis and Design of High Frequency Converter with Resistive Matching Network and Spiral Inductor," *IEEE Trans. on Power Electron.*, vol.PP, no.99, pp.1-1.

[9] Y. Guan, Y. Wang, Q. Bian, X. Hu, W. Wang, D. Xu, "High Efficiency Self-Driven Circuit with Parallel Branch for High Frequency Converters," *IEEE Trans. on Power Electron.*, vol.PP, no.99, pp.1-1.

[10] M. Hayati, S. Roshani, M. K. Kazimierczuk and H. Sekiya, "A Class-E Power Amplifier Design Considering MOSFET Nonlinear Drain-to-Source and Nonlinear Gate-to-Drain Capacitances at Any Grading Coefficient," *IEEE Trans. on Power Electron.*, vol. 31, no. 11, pp. 7770-7779, Nov. 2016.

[11] K. H. Lee, E. Chung, Y. Han and J. I. Ha, "A Family of High-Frequency Single-Switch DC–DC Converters With Low Switch Voltage Stress Based on Impedance Networks," *IEEE Trans. on Power Electron.*, vol. 32, no. 4, pp. 2913-2924, April 2017.

[12] Y. Guan; Y. Wang; W. Wang; D. Xu, "Analysis and Design of High Frequency DC/DC Con-verter Based on Resonant Rectifier," *IEEE Trans. on Indu. Electron.*, vol.PP, no.99, pp.1-1

The 2018 International Power Electronics Conference

A Fault-tolerant parallel inverter applied to micro-grid

Yan Li *member IEEE*, Xiangyue Shi, Jinjie Peng*, Zhifeng Qiu and Wei Xiong
School of information science and engineering, Central south university, Changsha, China
*E-mail: pjj2770111@126.com

Abstract— A topology structure of the fault-tolerant parallel inverter applied to micro-grid has been proposed in this paper. In normal operation state, two parallel inverters run stably. The structure of the fault-tolerant inverter is reconstructed after failure. The reconfiguration of the fault-tolerant inverter results in the phase voltage deviation and the circulating current between the two parallel inverters. The circulating current between the parallel inverters is reduced greatly by increasing the inverter filter capacitor value to support the DC side voltage. Simulation results verify the validity and effectiveness of the proposed topology structure.

Keywords— fault-tolerant, inverter, loop current, micro-grid

I. INTRODUCTION

With the development of human society and the growing demand of electric power consumption, energy and environment problems have become increasingly prominent and attracted more and more attention. Distributed generation has the advantages of less pollution, higher energy utilization and flexible installation site[1]. Owning to the development of the power electronics and modern control theory, the concept of micro-grid has emerged. Photovoltaic, wind power, energy storage and other devices in micro-grid connect the public power grid through the power electric inverter circuit. Therefore, the inverter is the crucial component in micro-grid[2-4].

The fault-tolerant inverter technology has been proposed by many scholars. The inverter circuit is reconstructed after failure, then it combines with the suitable control strategy to maintain the stability of the system and restore the original performance as much as possible[5-6]. Literatures [7-9] proposed the redundant fault-tolerant inverter. A bridge arm in the inverter was cut off when it is faulty, then the forth redundant bridge arm was connected. The reconstructed inverter topology structure is as same as the normal inverter. The three-phase four-switch inverter is widely considered as the fault-tolerant topology of the conventional three-phase six-switch inverter[10-11]. Most of the literatures focus on the fault-tolerant control technology of one-level inverter, literature [12] proposed the two-level and multi-level fault-tolerant inverters. The inverter topology structure relied on the hardware and software reconfigurations.

The fault-tolerant technology has a wide range of applications, such as aerospace, robot, traction, ship drive and so on. However, few articles study the fault-tolerant parallel inverters applied in the micro-grid.

A topology structure of the fault-tolerant parallel inverters applied to micro-grid is proposed in this paper. The fault-tolerant inverter topology structure is reconstructed after failure, which results in the phase voltage deviation and the circulating current between the two parallel inverters. Increasing the inverter output side filter capacitor value can support the DC side voltage, then the circulating current is reduced greatly. The fault-tolerant parallel inverter proposed in the paper can operate stability in both islanded and grid-connected mode. The stability and reliability of the micro-grid is greatly improved. The operation cost of the system is saved.

This paper is arranged as follows. The micro-grid topology structure with the fault-tolerant parallel inverter is proposed in the second section. The third section analyzes the working principle of the inverters, which demonstrates the cause of the circulating current between the two parallel inverters. The forth section conducts the simulations to verify the validity and effectiveness of the fault-tolerant parallel inverter. The fifth section is the conclusion.

II. THE MICRO-GRID TOPOLOGY STRUCTURE WITH THE FAULT-TOLERANT PARALLEL INVERTER

The equivalent micro-grid topology structure with the fault-tolerant parallel inverters is shown in Fig. 1. It can operate in both islanded and grid-connected mode. Inverter I is the fault-tolerant one and inverter II is the normal one. Inverter I and inverter II are connected in parallel at PCC. $V_1 \sim V_6$ are the power semiconductor devices IGBT. They constitute the three-phase six-switch structure of inverter I. $F_1 \sim F_9$ are the fast fuses. T_{oa}, T_{ob}, T_{oc}, T_{ab}, T_{bc} and T_{ac} are six TRIACs. They constitute the fault-tolerant inverter topology structure together. The output side of inverter I and inverter II both adopt a set of LC filters to suppress higher order harmonics. We assume that the line parameters of inverter I and inverter II are the same.

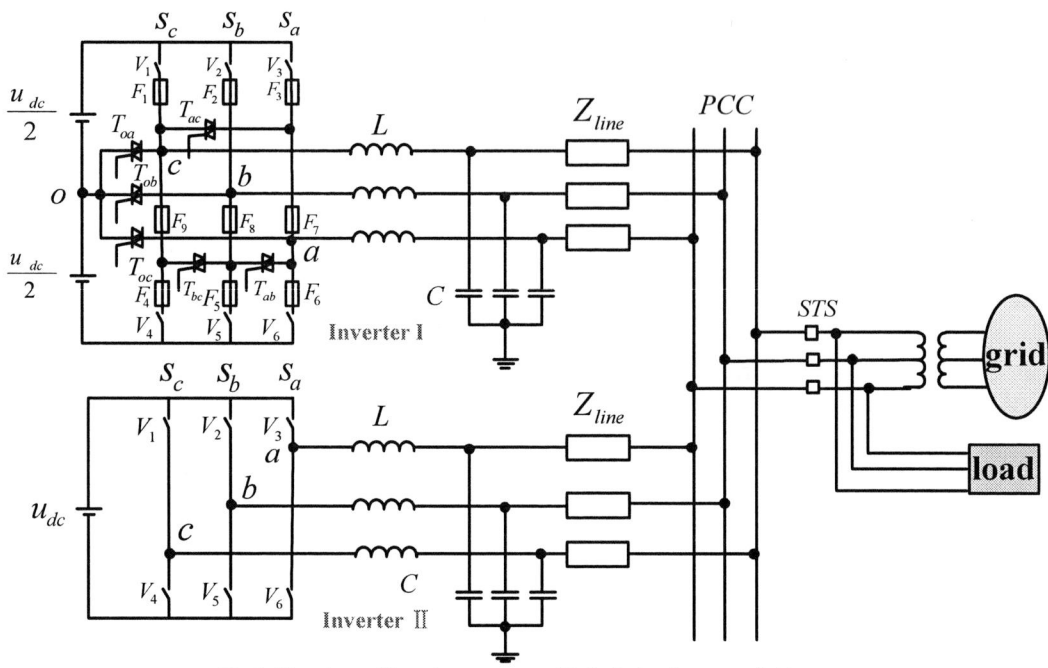

Fig. 1. The micro-grid topology structure with the fault-tolerant parallel inverter

In Fig. 1, when inverter I is operating normally, its equivalent circuit is three-phase six-switch. Take faulty c-phase as example. When the switches in c-phase of inverter I occur open-circuit fault or short-circuit fault, the corresponding fuses are fused and then the relevant bridge arm is removed, and turn on the corresponding TRIACs, then the three-phase four-switch inverter circuit is constituted. The point o is the midpoint c of c-phase in original normal state. There always are four switch on-off states after any phase of inverter I is faulty. The four on-off states are listed in Table I.

TABLE I
THE ON-OFF STATES OF INVERTER I WHEN c-PHASE IS FAULTY

Case	On	Off
1	V_3 , V_5	V_2 , V_6
2	V_5 , V_6	V_2 , V_3
3	V_2 , V_3	V_5 , V_6
4	V_2, V_6	V_3 , V_5

III. WORKING PRINCIPLE OF THE INVERTERS

A. The on-off States Analysis of Inverter II

There are six on-off states of inverter II which are listed in Table II.

TABLE II
THE ON-OFF STATES OF INVERTER II

Case	On	Off
1	V_1, V_5, V_6	V_4, V_2, V_3
2	V_1, V_5, V_3	V_4, V_2, V_6
3	V_1, V_2, V_6	V_4, V_5, V_3
4	V_4, V_5, V_3	V_1, V_2, V_6
5	V_4, V_2, V_6	V_1, V_5, V_3
6	V_4, V_2, V_3	V_1, V_5, V_6

Through the analysis of Table I and Table II, we can know that the topology structure of the fault-tolerant parallel inverters have 72 switch on-off states, when the fault phase and the switch on-off states of inverter I are determined, the fault-tolerant parallel inverter circuit has six different circulating current paths. In order to analyze the working principle of the fault-tolerant parallel inverters, we need to study the inner circulating current of inverter I, inverter II and the circulating current between inverter I and inverter II under the six cases shown in Table II. Similarly, when the other two phases in inverter I are faulty, the working principle and the analysis method are as same as above-mentioned, so it is sufficient to analyze only one on-off state of inverter I shown in Table I.

B. The Circulating Current Analysis Between the Inverters

When the power switches in the c-phase of inverter I are faulty, the c-phase need to be removed, so it remains four power switches V_2, V_3, V_5 and V_6 according to the six on-off states shown in Table II. The circulating current condition of the equivalent fault-tolerant parallel inverters under case 1 is shown in Fig. 2.

940

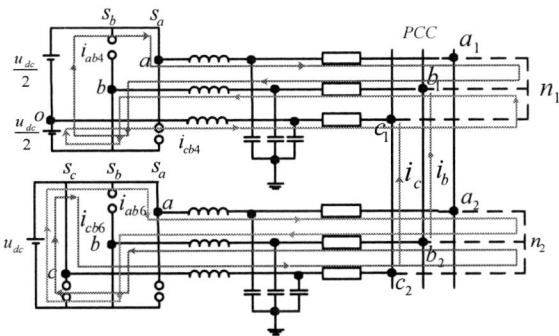

Fig. 2. The circulating current circuit of the equivalent fault-tolerant parallel inverters

In Fig. 2, the voltage equations of inverter I are:

$$\begin{cases} u_{on1} &= u_{an1} + u_{oa1} \\ u_{on1} &= u_{bn1} + u_{ob1} \\ u_{on1} &= u_{on1} \end{cases} \tag{1}$$

Define the circulating current equation as $i_{loop} = \dfrac{i_{x2} - i_{x1}}{2}$. x represents the a, b, c phase, respectively. Set the current direction from inverter II to inverter I is the positive direction. In order to simplify the numerical calculation, we set the equivalent output impedance of inverter I and inverter II is Z_a, Z_b, Z_c, and $Z_a = Z_b = Z_c = Z$.

According to Fig. 2, it is assumed that the inverter I and inverter II are running independently. There are two inner circulating current paths in inverter I (i_{ab4}, i_{cb4}) and two inner circulating current in inverter II (i_{ab6}, i_{cb6}), so it can be calculated that the inner circulating current of inverter I and inverter II is:

$$i_{ab4} = \frac{u_{dc}}{Z_a + Z_b} = \frac{u_{dc}}{2Z} \tag{2}$$

$$i_{cb4} = \frac{\dfrac{u_{dc}}{2}}{Z_b + Z_c} = \frac{u_{dc}}{4Z} \tag{3}$$

$$i_{cb6} = i_{ab6} = \frac{u_{dc}}{Z_b + Z_c} = \frac{u_{dc}}{2Z} \tag{4}$$

The subscript 4, 6 represent inverter1 and inverter II, respectively.

When inverter I and inverter II are running in parallel at PCC, the equivalent line impedance between the output side of inverter I and inverter II is zero. The intersections a_1, b_1, c_1 are correspondingly equal with the intersections a_2, b_2, c_2 on physical properties. In Fig. 2, the current through the point a_1 is $i_{a1} = i_{ab4} = \dfrac{u_{dc}}{2Z}$. The current through the point a_2 is $i_{a2} = i_{ab6} = \dfrac{u_{dc}}{2Z}$. i_{a1} and i_{a2} shunt to each other through the points a_1, a_2, respectively. i_{a1} and i_{a2} are equal and opposite. i_{ab6} offsets against i_{ab4}. So

there is no circulating current of a-phase between inverter I and inverter II.

Analogy with the circulating current analysis method of a-phase, the b-phase circulating current is

$$i_{loop} = i_b = \frac{i_{b2} - i_{b1}}{2} = \frac{u_{dc}}{8Z} \tag{5}$$

The c-phase circulating current is

$$i_{loop} = i_c = \frac{i_{c2} - i_{c1}}{2} = \frac{u_{dc}}{8Z} \tag{6}$$

Combined with the aforementioned circulating current analysis process and the six on-off states shown in Table II, the circulating current of a, b, c phase under the remaining five on-off conditions are listed in Table III.

TABLE III
THE CIRCULATING CURRENT OF A, B, C PHASE

Case	a-phase	b-phase	c-phase
2	$i_H = 0$	$i_H = -\dfrac{u_{dc}}{8Z}$	$i_H = \dfrac{3u_{dc}}{8Z}$
3	$i_H = \dfrac{u_{dc}}{4Z}$	$i_H = -\dfrac{u_{dc}}{8Z}$	$i_H = \dfrac{u_{dc}}{8Z}$
4	$i_H = \dfrac{u_{dc}}{4Z}$	$i_H = -\dfrac{u_{dc}}{8Z}$	$i_H = \dfrac{u_{dc}}{8Z}$
5	$i_H = 0$	$i_H = \dfrac{u_{dc}}{8Z}$	$i_H = \dfrac{u_{dc}}{8Z}$
6	$i_H = 0$	$i_H = -\dfrac{u_{dc}}{8Z}$	$i_H = \dfrac{3u_{dc}}{8Z}$

In general, the main causes of the circulating current between the parallel inverters are the output voltage deviation and the output impedance mismatch. Through the analysis of this section, we can know that the DC side voltage is divided equally when inverter I is reconstructed after failure, so the phase voltage of inverter I decreases to half of the original voltage. The source voltage of inverter II remains unchanged, so the phase voltage deviation between inverter I and inverter II is formed. It demonstrates that the primary cause of the circulating current is the output phase voltage deviation between inverter I and inverter II at PCC.

The circulating current will not only affect the load distribution ratio and decrease the transmission efficiency, but also causes a large number of power electronic equipment overheating, loss, life expectancy and safety performance degradation in the system. In this paper, in order to eliminate the circulating current, the two parallel inverters output side filter capacitors are all high-capacity filter capacitance. Increasing the filter capacitor value can make up the shortage of the load capacity of the power supply unit, so it can decrease the phase voltage deviation and the circulating current between inverter I and inverter II.

IV. SIMULATIONS AND ANALYSIS

In order to verify the validity and effectiveness of the micro-grid topology structure with the fault-tolerant parallel inverter proposed in this paper, the simulation model is built on the MATLAB/Simulink platform. The main simulation parameters are listed in Table 4.Conclusions are one of the most important parts of a paper. Please give careful consideration to this section.

TABLE IV
MAIN PARAMETERS OF THE SIMULATION MODEL

Parameters of inverter I , inverter II	Value
DC side voltage u_{dc} /V	800
Frequency f/Hz	50
Filter inductance L/mH	3e-3
Filter capacitance C/uF	150e-6
Line resistance R/Ω	0.01
High-capacity filter capacitance C/uF	1500e-6

A. Simulations of the Micro-grid in Islanded Mode

The simulation process takes 1s. During the period of 0s~0.5s, inverter I is in normal state. During the period of 0.5s~1s, inverter I is in fault state. The simulation results are shown in Fig. 3 and Fig. 4.

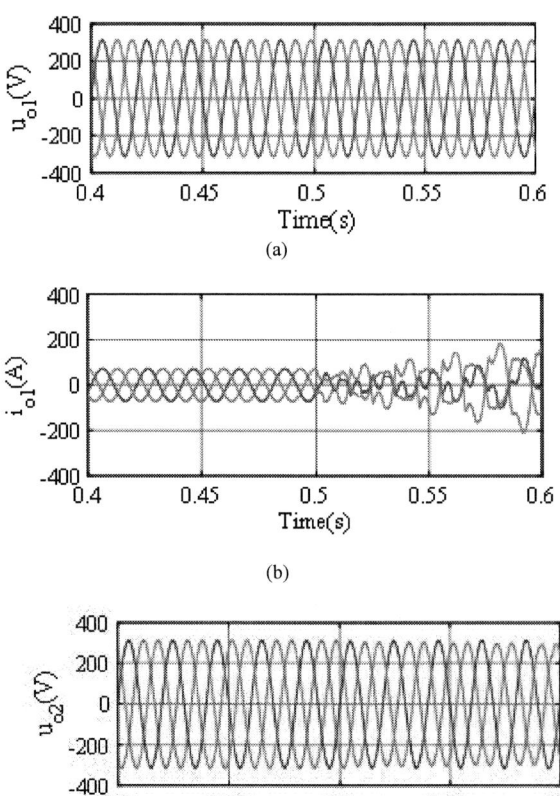

(a)

(b)

(c)

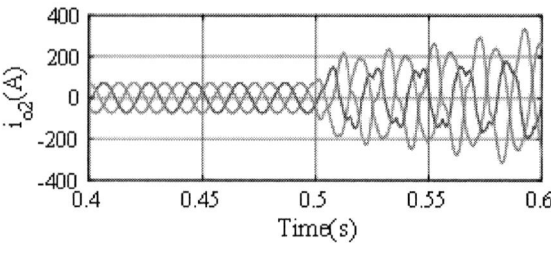

(d)

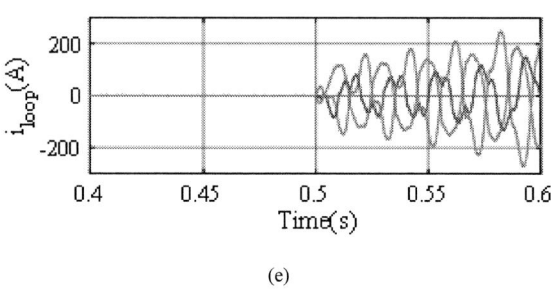

(e)

Fig. 3. Simulation waveforms of the micro-grid in islanded mode(C=150e-6uF) (a) The output voltage of inverter I (b) The output current of inverter I (c) The output voltage of inverter II (d) The output current of inverter II (e) The circulating current between inverter I and inverter II

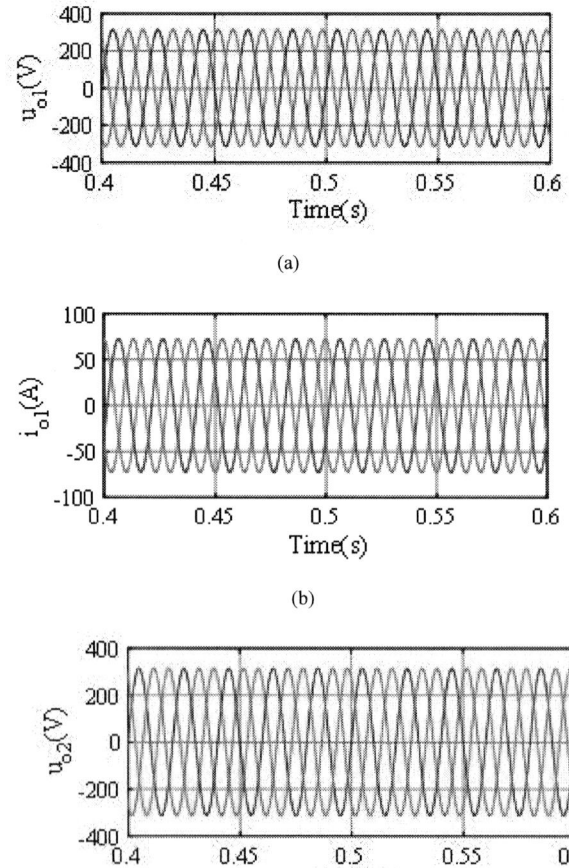

(a)

(b)

The 2018 International Power Electronics Conference

(c)

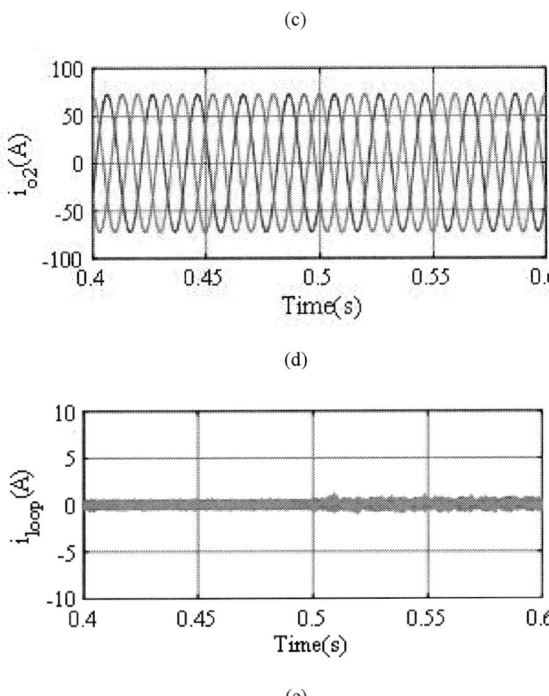

(d)

(e)

Fig. 4. Simulation waveforms of the micro-grid in islanded mode (C=1500e-6uF) (a) The output voltage of inverter Ⅰ (b) The output current of inverter Ⅰ (c) The output voltage of inverter Ⅱ (d) The output current of inverter Ⅱ (e) The circulating current between inverter Ⅰ and inverter Ⅱ

According to Fig. 3.(a)~(e), the output voltage and output current waveforms of inverter Ⅰ and inverter Ⅱ are smooth and stable sinusoid when inverter Ⅰ is in normal state, the circulating current is small and can be ignored. The output voltage and output current waveforms of inverter Ⅰ and inverter Ⅱ are distorted and three-phase imbalanced when inverter Ⅰ is faulty, the circulating current has increased largely and the stability of the system is destroyed seriously. In Fig. 4.(a)~(e), the output voltage and output current waveforms of inverter Ⅰ and inverter Ⅱ are smooth and stable sinusoidal during the whole simulation process, the circulating current is small and maintains around 1A.

B. Simulations of Micro-grid in Grid-connected Mode

The simulation is in grid-connected mode. The simulation results are shown in Fig. 5 and Fig. 6.

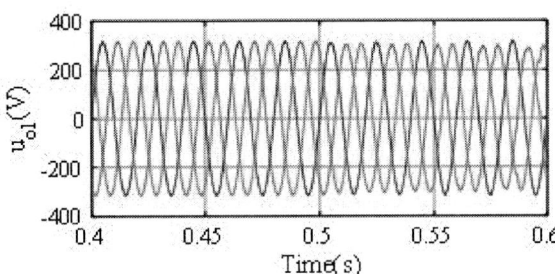

(a)

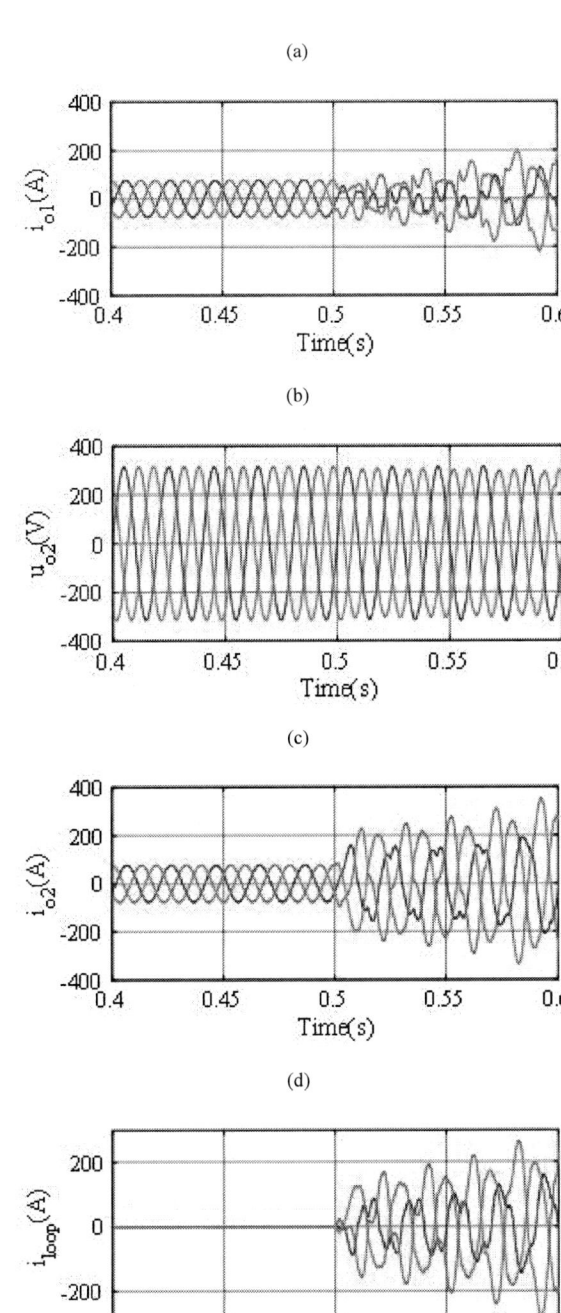

(b)

(c)

(d)

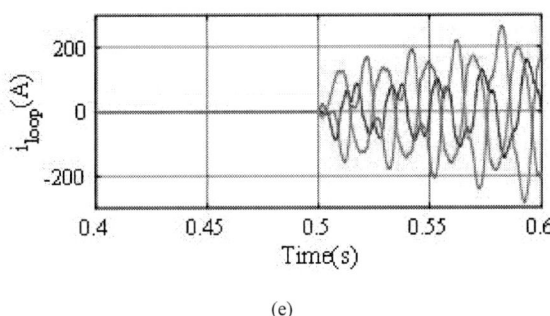

(e)

Fig. 5. Simulation waveforms of the micro-grid in grid-connected mode (C=150e-6uF) (a) The output voltage of inverter I (b) The output current of inverter I (c) The output voltage of inverter Ⅱ (d) The output current of inverter Ⅱ (e) The circulating current between inverter I and inverter Ⅱ

943

The 2018 International Power Electronics Conference

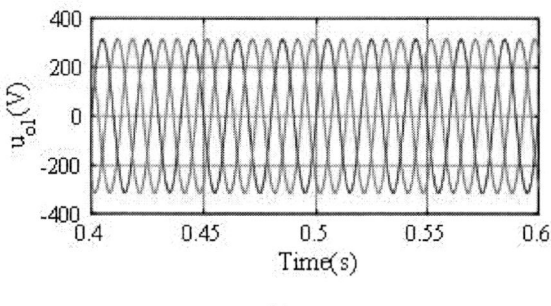

(a)

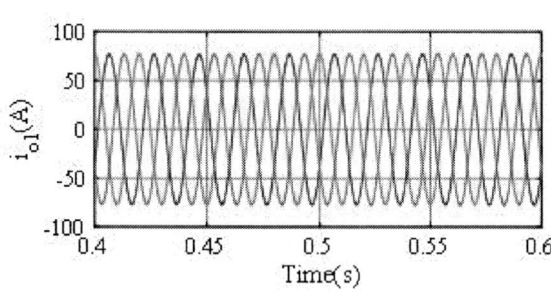

(b)

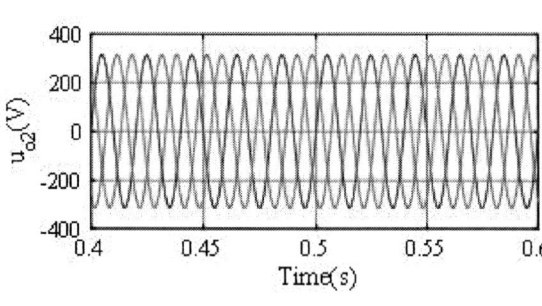

(c)

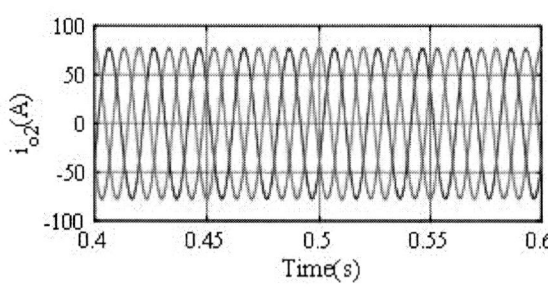

(d)

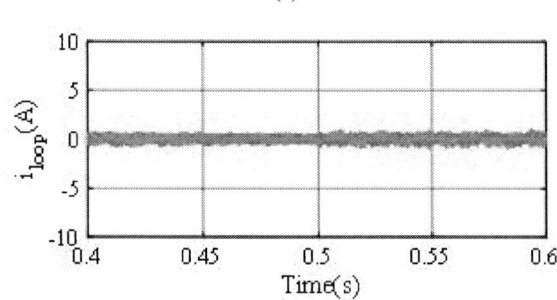

(e)

Fig. 6. Simulation waveforms of the micro-grid in grid-connected mode (C=1500e-6uF) (a) The output voltage of inverter I (b) The output current of inverter I (c) The output voltage of inverter II (d) The output current of inverter II (e) The circulating current between inverter I and inverter II

According to Fig. 5.(a)~(e), the output voltage and output current waveforms of inverter I and inverter II are all standard sinusoid when inverter I is in normal state. The circulating current is small and can be ignored. The output voltage and output current waveforms of inverter I and inverter II are distorted and three-phase imbalanced when inverter I is faulty. The circulating current is increased largely and the stability and reliability of the system is destroyed. In Fig. 6.(a)~(e), the output voltage and output current waveforms of inverter I and inverter II are stable and smooth sinusoid during the whole simulation process. The circulating current is small and remains around 1A.

Simulation results verify that the proposed structure is feasibility. The reliability of the micro-grid is improved.

V. CONCLUSION

A topology structure of the fault-tolerant parallel inverter applied to the micro-grid was proposed in this paper. The fault-tolerant parallel inverter can run stably in normal state. The structure of the fault-tolerant inverter was reconstructed after failure. The fault unit was removed to keep the fault-tolerant parallel inverter operating normally. Moreover, the reconfiguration of the fault-tolerant inverer resulted in the phase voltage deviation and the circulating current between the two parallel inverters. The inverter filter capacitance was increased to compensate the load capacity of the DC side power and supplied the DC side power with voltage support. The circulating current between the two parallel inverters was reduced greatly. At last, simulation results verified the validity and effectiveness of the proposed topology structure.

ACKNOWLEDGMENT

This work is supported by National Nature Science Foundations under Grant 51507192 and 61403429, China.

REFERENCES

[1] J. M. Carrasco, L. G. Franquelo, J. T. Bialasiewicz, et al, "Power-electronic systems for the grid integration of renewable energy sources: A survey," *IEEE Trans. on Industry Electronics*, vol. 33, no.4, pp. 1002-1016, Aug. 2006.

[2] Guerrero J M, Matas J, De-Vicuna L G, et al. "Decentralized control for parallel operation of distributed generation inverters using resistive output impedance," *IEEE Trans. on Industry Electronics*, vol. 54, no. 2, pp. 994-1004, April 2007.

[3] Y. W. Li and C. N. Kao. "An accurate power control strategy for power electronics- interfaced distributed generation units operating in a low voltage multibus micro-grid," *IEEE Trans. on Industry Electronics*, vol. 24, no.12, pp.2977-2988, December 2009.

[4] F. Blaabjerg, R. Teodorescu, M. Liserre, et al. "Overview of control and grid synchronization for distributed power generation systems," *IEEE Trans. on Power Electronics*, vol.53, no.5, pp.1398-1407, October 2006.

[5] B. A. Welchko, T. A. Lipo, T. M. Jahns et al, "Fault tolerant three-phase AC motor drive topologies: a comparison of features, cost,and limitations," *IEEE Trans. on Power Electronics*, vol. 19, no. 4, pp.1108-1116, July 2004.

[6] J. Bennett, G. Atkinson, B. Mecrow, and D. Atkinson, "Fault-tolerant design considerations and control strategies for aerospace drives," *IEEE Trans. on Industry Electronics*, vol. 59, no. 5, pp. 2049–2058, May 2012.

[7] Haitao Li, Wenzhuo Li, and Hongliang Ren, "Fault-tolerant inverter for high-speed low-inductance BLDC Drives in aerospace applications," *IEEE Trans. on Power Electronics*, vol. 32, no.3, pp. 2452-2463, March 2017.

[8] RIBEIRO R L A,JACOBINA C B,SILVA E R C, et al. "Fault-tolerant voltage-fed PWM inverter ac motor drive system," *IEEE Trans. on Industrial Electronics*, vol.32, no.3, pp. 439-446, April 2004.

[9] Rammohan Rao Errabelli , Peter Mutschler, "Fault-Tolerant Voltage Source Inverter for Permanent Magnet Drives," *IEEE Trans. on Power Electronics*, vol. 27, no. 2, pp. 500-508, February 2012.

[10] AN Qun-tao, SUN Xing-tao, ZHAO ke, et al. "Control Strategy for Fault –tolerant Three-phase Four-switch Inverters," *Proceeding of the CSEE*, vol. 30, no. 3, pp. 14-30, January, 2010 .

[11] MENDES A M S,CARDOSO A J M. "Fault-tolerant operating strategies applied to three-phase induction-motor drives," *IEEE Trans. on Induatrial Electronics*, vol. 53, no. 6, pp. 1807-1817, December 2006.

[12] Behrooz Mirafzal, "Survey of fault–tolerance techniques for three-phase voltage source inverters," *IEEE Transa. on Induatrial Electronics*, vol. 61, no. 10, pp. 5192-5201 October 2014.

The 2018 International Power Electronics Conference

Stability Analysis of Grid-connected Converters with Add-on Voltage Support Functionality using Repetitive Control

Y. Zhang*, M. G. L. Roes, M. A. M. Hendrix and J. L. Duarte
Department of Electrical Engineering
Eindhoven University of Technology
P.O. Box 513, 5600MB Eindhoven, The Netherlands
*E-mail: ya.zhang@tue.nl

Abstract—**The aim of this paper is to analyse the stability of grid-connected converters using repetitive control for voltage harmonic compensation. Supplementary to an earlier proposed control strategy to upgrade grid-connected converters for add-on voltage support, the stability analysis of the overall system is presented. A straightforward procedure is proposed and a couple of stability conditions are derived. The analysis is broken down into two subparts and therefore, the Nyquist's criterion can be easily applied. Simulations and experiments support the stability model analysis. The proposed method is not limited to the application with voltage support but also applicable for other converter systems based on repetitive-control algorithms.**

Keywords—*grid-connected converter, repetitive-control, stability, voltage support*

I. INTRODUCTION

Power electronic converters are widely applied to interface sustainable energy sources with the public grid and local loads [1]–[4]. Figure 1 shows the typical diagram of a single-phase grid-connected converter with local loads. In such a converter the local voltage is conventionally measured for grid synchronization and the converter output current is measured for active power regulation [5]–[9]. This type of converter generally experiences difficulties to compensate (the local grid voltage) for harmonic distortions that are generated by the local load currents, unless an additional sensor of the non-linear load current [10]–[13], or the grid current [14], or the grid voltage [15] is applied. In the application where multiple non-linear loads are connected to the point of common connection (PCC), more than one additional sensor would become necessary, which adds up to the total cost of the converter system.

As demonstrated in [16], [17], a single-phase grid-connected inverter can be incorporated with an added functionality to sink the non-linear load current harmonics without using an additional sensor. This is achieved by employing a dedicated control strategy, where the PCC voltage is used for more than grid synchronization, e.g. for harmonic compensation to improve the local voltage quality. This is advantageous for the local grid, especially for a grid where the non-linear loads are the main contributor to the harmonics. Since the harmonics are locally

countered, the distortion of the current to the grid can be diminished as a consequence.

Compared to regular current controllers applied in grid-connected converters [5], [18]–[20], the control strategy in [17] exhibits distinctive features, e.g. two references and two control loops in parallel, which makes a single-input single-output (SISO) feedback control strategy not applicable. In addition to that, since both proportional-resonance (PR) and repetitive-control (RC) algorithms were applied, the stability analysis of the overall system is non-trivial. A straightforward procedure is proposed and a couple of stability conditions are derived in this paper. The proposed stability analysis strategy is also applicable for a SISO converter system using repetitive-control, since a SISO system can be considered as an extreme case of a single-input dual-output servo.

The structure of the paper is as follows. Firstly, the voltage support control strategy for a single-phase grid-connected converter system is recalled. Then the procedure to derive the stability conditions of the overall system is presented. In Section IV considerations of the control design are discussed. This is followed by simulation results in Section V and experimental results in Section VI. Finally, conclusions are drawn in Section VII.

II. RECALL OF THE CONTROL STRATEGY WITH ADD-ON VOLTAGE SUPPORT FUNCTIONALITY

Figure 2 shows the structure of the control strategy. The control strategy for additional voltage harmonic compensation includes two loops [16], one for active power regulation and another for harmonic compensation. The RC algorithm is applied in the second loop. There the discrete-Fourier-transformation (DFT) filter is applied to selectively compensate for harmonics. The transfer function of the DFT filter in the z-domain is

$$C_{DFT}(z) = \frac{2}{N} \sum_{i=0}^{N-1} \left(\sum_{h \in N_h} \cos\left[\frac{2\pi}{N} hi\right] \right) z^{-i}, \quad (1)$$

where N_h and N are the set of harmonics to be compensated and the number of samples per grid period, respectively. The transfer function of the proportional-

946

The 2018 International Power Electronics Conference

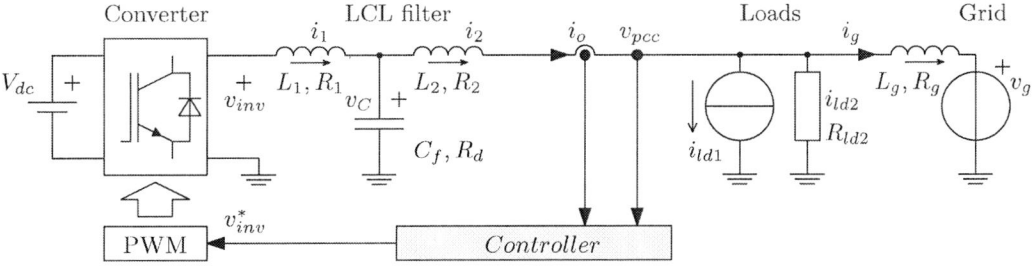

Fig. 1. System diagram of a single-phase grid-connected inverter with LCL filter and local loads

resonance filter in the s-domain is

$$G_{pr}(s) = K_p + K_r \frac{s}{s^2 + \omega_o^2}, \tag{2}$$

where K_p is the proportional gain, K_r the resonance gain, and ω_0 the central resonance frequency which is designed to be equal to the grid angular frequency.

III. SYSTEM MODELLING AND STABILITY CONDITION DERIVATION

This section presents a procedure to derive a closed-loop dynamic model of the grid-connected inverter of Fig. 1 with the controller of Fig. 2. Subsequently, based on the derived model a couple of stability conditions are formulated.

A. Modelling of the inverter and delay of PWM

The open-loop transfer function from the desired inverter output voltage v_{inv}^* to the inverter's output voltage v_{inv} (refer to Fig. 1) is denoted as $H_0(s)$. Taking into account the calculation time and the delays caused by pulse-width-modulation (PWM), we approximate $H_0(s)$ as

$$H_0(s) = e^{-sT_s}, \tag{3}$$

where T_s is the sampling period. The sampling frequency $f_s = \dfrac{1}{T_s}$ is identical to the inverter's switching frequency.

B. LCL filter and grid

The current source i_{ld1} in Fig. 1 denotes the non-linear load and the voltage source v_g represents the grid. Both i_{ld1} and v_g are regarded as external disturbances and therefore the former is regarded as an open-circuit and another as a short-circuit during small-signal modelling. Therefore, the open loop dynamics from the inverter output voltage to the converter output current and the PCC voltage can be described by a 4th order state-space model, written as

$$\frac{\mathrm{d}}{\mathrm{d}t}x(t) = A_c x(t) + B_c u(t)$$
$$y(t) = C_c x(t), \tag{4}$$

where the input, states and outputs correspond to

$$u(t) = v_{inv}(t), \quad x(t) = \begin{bmatrix} i_1(t) \\ i_2(t) \\ i_g(t) \\ v_C(t) \end{bmatrix}, \quad y(t) = \begin{bmatrix} i_o(t) \\ v_{pcc}(t) \end{bmatrix}. \tag{5}$$

The matrices in (4) are found to be

$$A_c = \tag{6}$$
$$\begin{bmatrix} -\dfrac{R_1 + R_d}{L_1} & \dfrac{R_d}{L_1} & 0 & -\dfrac{1}{L_1} \\ \dfrac{R_d}{L_2} & -\dfrac{R_d + R_{ld2} + R_2}{L_2} & \dfrac{R_{ld2}}{L_2} & \dfrac{1}{L_2} \\ 0 & \dfrac{R_{ld2}}{L_g} & -\dfrac{R_d + R_{ld2}}{L_g} & 0 \\ \dfrac{1}{C_f} & -\dfrac{1}{C_f} & 0 & 0 \end{bmatrix},$$
$$B_c = \begin{bmatrix} \dfrac{1}{L_1} & 0 & 0 & 0 \end{bmatrix}^T, C_c = \begin{bmatrix} 0 & 1 & 0 & 0 \\ 0 & R_{ld2} & -R_{ld2} & 0 \end{bmatrix}.$$

The transfer functions from the input to the outputs can be obtained as follows:

$$\begin{bmatrix} H_1(s) \\ H_2(s) \end{bmatrix} = C_c(sI - A_c)^{-1} B_c. \tag{7}$$

We can see that $H_1(s)$ and $H_2(s)$ are the transfer functions from $v_{inv}(s)$ to $i_o(s)$ and to $v_{pcc}(s)$, respectively. The *zero-order-hold* method is used for the discretization of $H_0(s)$, $H_1(s)$ and $H_2(s)$, which results in $H_0(z)$, $H_1(s)$ and $H_2(z)$, respectively.

C. Closed-loop model

Figure 3 shows the closed loop diagram where the plant is given by Fig. 1 and the controller by Fig. 2. The two references r_1 and r_2 in Fig. 3 correspond to the desired fundamental inverter output current $i_{o,1}^*$ and the desired pcc voltage harmonic content $v_{pcc,h}^*$ (zero under normal circumstances) in Fig. 2; the two outputs y_1 and y_2 correspond to the inverter output current i_o and the pcc voltage v_{pcc} in Fig. 1, respectively. The system in Fig. 3 is a single-input dual-output feedback control system, and

947

The 2018 International Power Electronics Conference

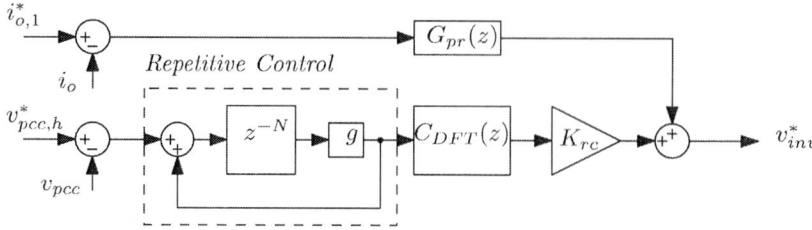

Fig. 2. Structure of the control strategy using repetitive-control for add-on voltage harmonic compensation

we have

Controller: $c_1(z) = G_{pr}(z)$

$$c_2(z) = \frac{gz^{-N}}{1 - gz^{-N}} C_{DFT}(z) K_{rc}; \quad (8)$$

Plant: $p_1(z) = H_0(z) H_1(z)$

$$p_2(z) = H_0(z) H_2(z). \quad (9)$$

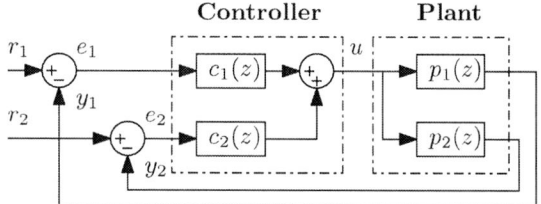

Fig. 3. Diagram of a single input dual output feedback control system.

Analysing the closed-loop system model in Fig. 3 yields a transfer function matrix, given by

$$\begin{bmatrix} y_1(z) \\ y_2(z) \end{bmatrix} = \begin{bmatrix} \dfrac{p_1(z)c_1(z)}{1+T(z)} & \dfrac{p_1(z)c_2(z)}{1+T(z)} \\ \dfrac{p_2(z)c_1(z)}{1+T(z)} & \dfrac{p_2(z)c_2(z)}{1+T(z)} \end{bmatrix} \begin{bmatrix} r_1(z) \\ r_2(z) \end{bmatrix}, \quad (10)$$

where $T(z)$ is

$$T(z) = p_1(z)c_1(z) + p_2(z)c_2(z). \quad (11)$$

It can be seen from (10) that all the entries in the closed-loop transfer function matrix share the same denominator, and that $T(z)$ is the loop gain transfer function of the system. Therefore, $T(z)$ can be analysed to derive the stability conditions of the system in Fig. 3.

D. Stability condition derivation

The closed-loop characteristic polynomial of the system in Fig. 3 is

$$D(z) = 1 + T(z). \quad (12)$$

From substituting (8) into (11) and (12) follows

$$D(z) = 1 + p_1(z)c_1(z) + \frac{p_2(z)gz^{-N}C_{DFT}(z)K_{rc}}{1 - gz^{-N}}. \quad (13)$$

Observing (13) indicates that the closed-loop characteristic polynomial $D(z)$ is of quite high order, making a direct application of Nyquist's Stability Criterion not straightforward. By decomposition of $D(z)$, however, it is possible to reach a set of stability conditions. Denoting

$$\bar{T}_1(z) = p_1(z)c_1(z) \quad (14)$$

$$\bar{T}_2(z) = -gz^{-N}(1 - \mathcal{P}(z)C_{DFT}(z)K_{rc}(z)) \quad (15)$$

$$\mathcal{P}(z) = \frac{p_2(z)}{1 + p_1(z)c_1(z)}, \quad (16)$$

the polynomial in (13) is rewritten as

$$D(z) = \frac{(1 + \bar{T}_1(z))(1 + \bar{T}_2(z))}{1 - gz^{-N}}. \quad (17)$$

Since $\bar{T}_1(z)$ is stable and g is chosen smaller than one, a set of sufficient stability conditions of the system in Fig. 3 is therefore derived as follows:

> **Condition 1** The equation $1 + \bar{T}_1(z) = 0$ has no unstable roots;

> **Condition 2** The equation $1 + \bar{T}_2(z) = 0$ has no unstable roots.

As a result, with the proposed analysis procedure in this section, the overall system stability problem is broken down into two explicit sub-problems.

IV. CONTROL DESIGN CONSIDERATIONS

This section discusses how to properly tune the control parameters to satisfy the stability conditions derived in Section III-D. Table I lists the parameters of the grid-connected converter that are used for the examples in this section (with reference to Fig. 1). The controller parameters are set according to Table II unless mentioned otherwise.

It can be observed from the expressions of $\bar{T}_1(z)$ and $\bar{T}_2(z)$ that weather condition **1** holds is determined by $c_1(z)$ alone; for condition **2** this is determined by both $c_1(z)$ and $c_2(z)$. Therefore, the stability strategy is to satisfy condition **1** first and thereafter satisfy condition **2**.

A. Transfer function $\bar{T}_1(z)$

Using the control parameters in Table II, the Nyquist plot of $\bar{T}_1(z)$ is depicted in Fig. 4. It can be concluded that the sufficient stability condition **1** in Section III-D

948

The 2018 International Power Electronics Conference

TABLE I. GRID-CONNECTED CONVERTER PARAMETERS

Description	Symbol	Value
DC power supply voltage	V_{dc}	400 V
Converter switching frequency	f_{sw}	10 kHz
LCL filter	L_1, R_1	5.22 mH, 0.2 Ω
	C_f, R_d	2.82 μF, 0.2 Ω
	L_2, R_2	5.22 mH, 0.2 Ω
Linear load resistance	R_{ld2}	94 Ω
Grid rms voltage	V_g	220 V
Nominal grid frequency	f_g	50 Hz
Grid impedance	L_g	10.44 mH
	R_g	0.4 Ω

TABLE II. INITIAL CONTROL PARAMETERS

Parameters	Symbol	Value
Sampling rate	f_s	10 kHz
PR filter	K_p	30
	K_r	6000
	ω_0	$2\pi f_g$
RC filter	N	200
	N_h	$\{3, 5, 7\}$
	K_{rc}	1.2
	g	0.99

is satisfied. Moreover, it can be observed from the expression of $c_1(z)$ in (8) and (2) that decreasing K_p and K_r simultaneously will make condition **1** easier to be satisfied.

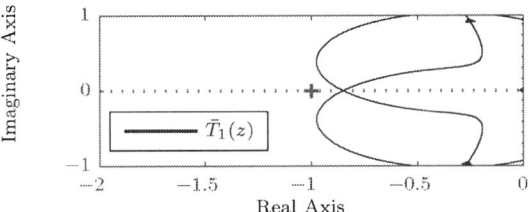

Fig. 4. Nyquist plot of $\bar{T}_1(z)$.

B. Transfer function $\bar{T}_2(z)$

Using the control parameters in Table II, the Nyquist plot of $\bar{T}_2(z)$ is depicted in Fig. 5 with two different values of g. It can be concluded that the sufficient stability condition **2** in Section III-D is satisfied in both cases. Clearly, decreasing g helps to increase the system gain margin, according to the expression of $\bar{T}_2$ in (15) and its Nyquist plot in Fig. 5.

V. SIMULATION RESULTS

A series of simulations were performed in *Matlab/Simulink/Plecs* to investigate the stability discussion in Section IV. An H-bridge is chosen for the converter indicated in Fig. 1. Since a weak grid, namely a grid with relatively large grid impedance, experiences more difficulties in maintaining the local voltage quality in the presence of local non-linear loads, a weak grid

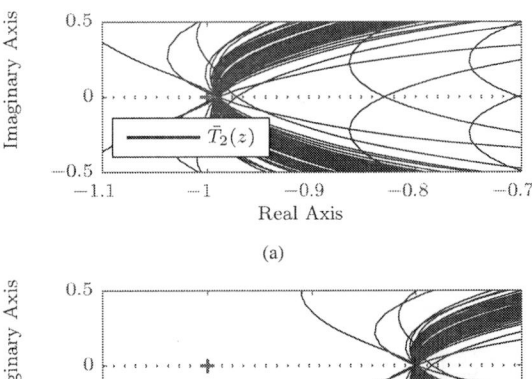

(a)

(b)

Fig. 5. Nyquist plot of $\bar{T}_2(z)$: (a) with $g = 0.99$; (b) with $g = 0.8$.

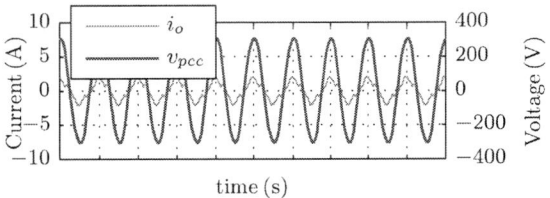

Fig. 6. Simulation results. The RC loop in Fig. 2 disabled resulting in high PCC voltage distortion (THD(v_{pcc})=3.82%).

is assumed in this paper. The parameters of the grid-connected converter system is shown in Table I. In the simulation the sampling process is synchronized with respect to the converter switching behaviour according to the considerations in [21], and therefore, the sampling rate coincides with the switching frequency.

Figure 6 and Figure 7 show the simulation results when the RC loop is disabled ($c_2 = 0$) and enabled. It can be observed that compared to when the RC loop is disabled in Fig. 6, the PCC voltage waveform is less distorted when the RC loop is enabled in Fig. 7. A further analysis indicates that the total harmonic distortion (THD) of the PCC voltage is 3.82% in Fig. 6 and 1.07% in Fig. 7, which numerically validates the improvement in the PCC

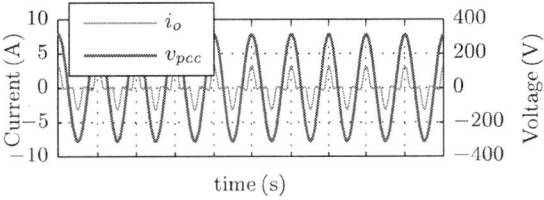

Fig. 7. Simulation results. The RC loop in Fig. 2 enabled resulting in low PCC voltage distortion, $g = 0.99$ (THD(v_{pcc})=1.07%).

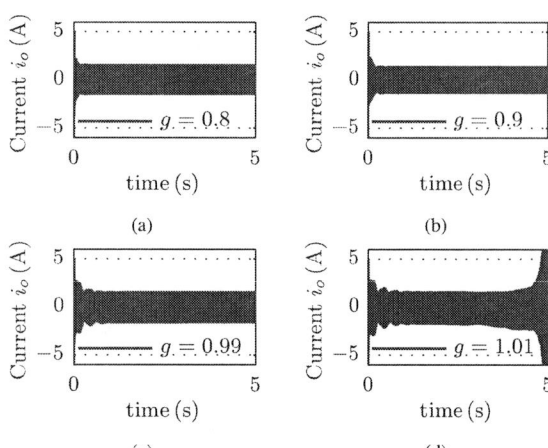

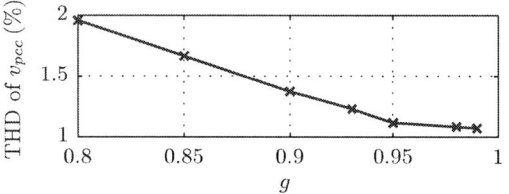

Fig. 8. Simulated waveform of i_o with different values of g.

Fig. 9. Simulation results. Total harmonic distortion of the pcc voltage with different values of g.

voltage quality.

Figure 8 shows the start-up transient dynamics of the converter with different values of g. It can be seen that decreasing g accelerates the stabilization of the system; however, consequently, the PCC voltage THD increases, as shown in Fig. 9. In a deliberately even worse case (in Fig. 8(d)) the system becomes unstable when g is bigger than one, which is in agreement with the stability discussions in Section IV, Fig. 5.

VI. EXPERIMENTAL RESULTS

This section considers the overall system performance in experiments, including steady-state and transient behaviours. A laboratory setup was realised in accordance with the system diagram shown in Fig. 1. It consists of: a DC power supply, a single-phase H-bridge converter, an LCL filter, a linear load, a programmable load, a *Spitzenberger* type DM 3000 grid emulator and a *dSPACE DS1104* real-time control system. The nominal parameters of the converter system are shown in Table I; the control parameters are set according to Table II.

A. Steady-state performance

When the pulse-width-modulation of the inverter is disabled, Figure 10 and Figure 11 show the respective experimental results when the non-linear load is disconnected from and connected to the PCC. Figure 12 and Figure 13 show the respective experimental results when the RC loop is disabled and enabled. Comparing Fig.

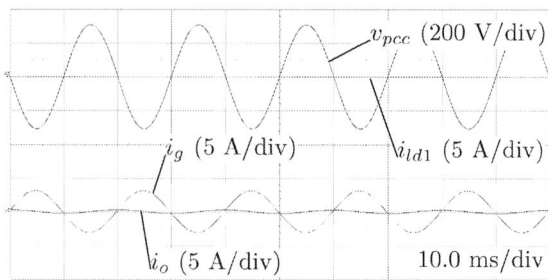

Fig. 10. Experimental results. The non-linear load is disconnected, resulting in low PCC voltage distortion.

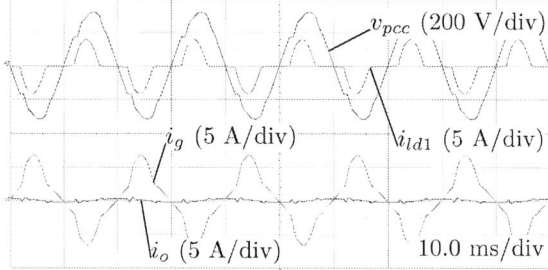

Fig. 11. Experimental results. The non-linear load is connected, resulting in high PCC voltage distortion (THD(v_{pcc})=6.65%).

10 and Fig. 11 indicates that the non-linear load can distort the local PCC voltage. However, using only the conventional PR current controller, the harmonics are poorly suppressed, as demonstrated in Fig. 12. When the RC loop is enabled, the local PCC voltage distortion is reduced, as shown in Fig. 13, which is in accordance with the simulation results in Section V.

Figure 14 shows the results when g is set to 0.8. It can be seen that the grid current in Fig. 14 is more distorted than in the situation in Fig. 13. The experimental THD of the PCC voltage against g is plotted in Fig. 15. It can be seen that increasing g results in lower THD of the local PCC voltage, which is in agreement with the simulation results in Section V. Nevertheless, it is noticeable that the experiments (Fig. 15) show more total harmonic distortion than the simulation (Fig. 9), which could be explained by two main factors. Firstly, the non-linear load in simulation

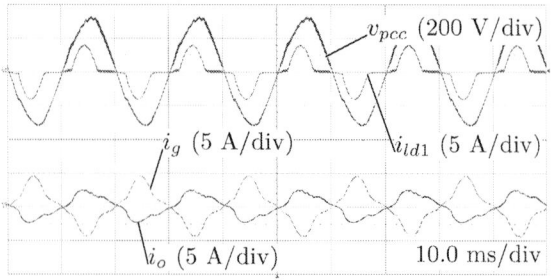

Fig. 12. Experimental results. The RC loop in Fig. 2 disabled resulting in high PCC voltage distortion (THD(v_{pcc})=4.45%).

The 2018 International Power Electronics Conference

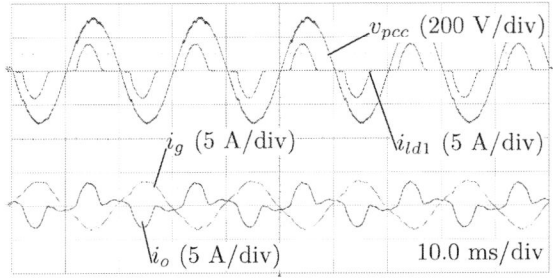

Fig. 13. Experimental results. The RC loop in Fig. 2 enabled resulting in reduced PCC voltage distortion, $g = 0.99$ (THD(v_{pcc})=2.28%).

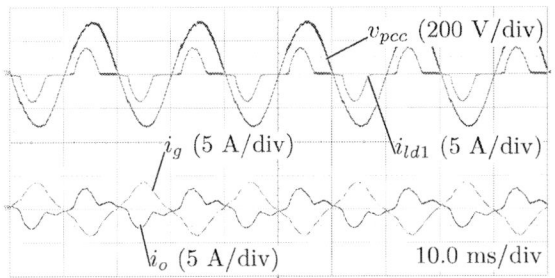

Fig. 14. Experimental results. The RC loop in Fig. 2 enabled and $g = 0.80$.

is not exactly the same as in experiments. The former one is created by two Zener diodes anti-series-connected. In experiments the non-linear load is programmed to have a root mean square value of 2 A, a Crest factor of 2 and a power factor of 0.8. Secondly, the performance of the repetitive-control algorithm is highly dependent on the grid frequency estimation accuracy. In simulation the grid frequency is perfectly stabilized at a set value; however, in practice the grid frequency was between 49.93 Hz and 49.97 Hz.

B. Transient performance

The experimental transient dynamics of the converter when changing the value of g are shown in Fig. 16. Consistent to the prediction from the model analysis and the observation in simulation, the system keeps stable when g is changed from 1 to 0.99 and unstable when

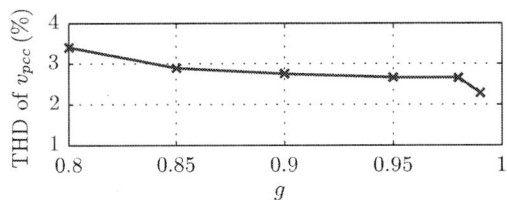

Fig. 15. Experimental results. Total harmonic distortion of the pcc voltage with different values of g.

g is changed from 1 to 1.01.

As mentioned earlier, a smaller g increases the overall system stability gain margins at the cost of reduced harmonic attenuation. In the extreme case, $g = 0$, the voltage support functionality is disabled (with reference to the control strategy structure in Fig. 2); in this case the stability condition **2** in Section III-D is always satisfied.

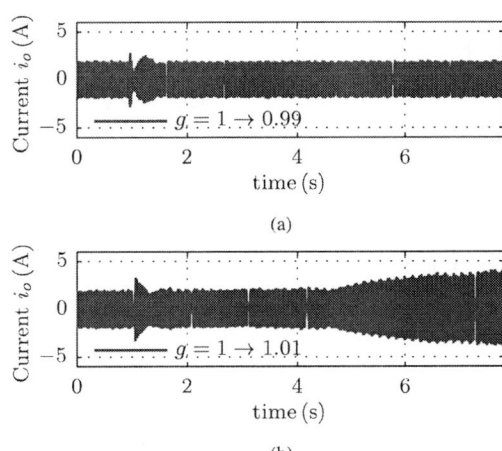

Fig. 16. Experimental results. Waveform of i_o when changing $g = 1$ to (a) 0.99 at around 1s; (b) 1.01 at around 1s.

VII. CONCLUSION

As an extension of the proposed control strategy of grid-connected converters with add-on voltage support in [17], this paper presents the stability analysis. It introduces a method to quantify the stability margins of the proposed strategy thus improving the robustness of the system.

Theoretically, when the feedback gain in the RC is unity, zero-error disturbance rejection is achievable; however, it is at the cost of bringing the system operating at boundary stability condition. A gain smaller than one results in a system that stabilizes faster. However, the disturbance rejection ability is less; namely, reduced performance in voltage harmonic attenuation. As demonstrated by simulation and experimental results in this paper, there is a trade-off between stability margins and the performance of harmonic disturbance rejection.

ACKNOWLEDGEMENT

This project has received funding from the Electronic Components and Systems for European Leadership Joint Undertaking under grant agreement No 737434. This Joint Undertaking receives support from the European Unions Horizon 2020 research and innovation programme and Germany, Slovakia, Netherlands, Spain, Italy.

REFERENCES

[1] Y. Yang, K. Zhou, H. Wang, F. Blaabjerg, D. Wang, and B. Zhang, "Frequency adaptive selective harmonic control for grid-connected inverters," *IEEE Transactions on Power Electronics*, vol. 30, no. 7, pp. 3912–3924, Jul. 2015.

[2] Y. Yang, K. Zhou, and F. Blaabjerg, "Enhancing the frequency adaptability of periodic current controllers with a fixed sampling rate for grid-connected power converters," *IEEE Transactions on Power Electronics*, vol. 31, no. 10, pp. 7273–7285, Oct. 2016.

[3] M. Castilla, J. Miret, J. Matas, L. G. d. Vicuna, and J. M. Guerrero, "Linear current control scheme with series resonant harmonic compensator for single-phase grid-connected photovoltaic inverters," *IEEE Transactions on Industrial Electronics*, vol. 55, no. 7, pp. 2724–2733, Jul. 2008.

[4] P. M. d. Almeida, J. L. Duarte, P. F. Ribeiro, and P. G. Barbosa, "Repetitive controller for improving grid-connected photovoltaic systems," *IET Power Electronics*, vol. 7, no. 6, pp. 1466–1474, Jun. 2014.

[5] M. Castilla, J. Miret, J. Matas, L. G. d. Vicuna, and J. M. Guerrero, "Control design guidelines for single-phase grid-connected photovoltaic inverters with damped resonant harmonic compensators," *IEEE Transactions on Industrial Electronics*, vol. 56, no. 11, pp. 4492–4501, Nov. 2009.

[6] J. Xu, Q. Qian, S. Xie, and B. Zhang, "Grid-voltage feedforward based control for grid-connected LCL-filtered inverter with high robustness and low grid current distortion in weak grid," in *IEEE Applied Power Electronics Conference and Exposition (APEC)*, Mar. 2016, pp. 1919–1925.

[7] S. H. Lee, W. J. Cha, B. H. Kwon, and M. Kim, "Discrete-time repetitive control of flyback CCM inverter for PV power applications," *IEEE Transactions on Industrial Electronics*, vol. 63, no. 2, pp. 976–984, Feb. 2016.

[8] X. Li, J. Fang, Y. Tang, X. Wu, and Y. Geng, "Capacitor-voltage feedforward with full delay compensation to improve weak grids adaptability of LCL-filtered grid-connected converters for distributed generation systems," *IEEE Transactions on Power Electronics*, vol. 33, no. 1, pp. 749–764, Jan. 2018.

[9] Q. Zhao, Y. Ye, G. Xu, and M. Zhu, "Improved repetitive control scheme for grid-connected inverter with frequency adaptation," *IET Power Electronics*, vol. 9, no. 5, pp. 883–890, Apr. 2016.

[10] R. I. Bojoi, L. R. Limongi, D. Roiu, and A. Tenconi, "Enhanced power quality control strategy for single-phase inverters in distributed generation systems," *IEEE Transactions on Power Electronics*, vol. 26, no. 3, pp. 798–806, Mar. 2011.

[11] M. Rashed, C. Klumpner, and G. Asher, "Repetitive and resonant control for a single-phase grid-connected hybrid cascaded multilevel converter," *IEEE Transactions on Power Electronics*, vol. 28, no. 5, pp. 2224–2234, May 2013.

[12] Z. Yao and L. Xiao, "Control of single-phase grid-connected inverters with nonlinear loads," *IEEE Transactions on Industrial Electronics*, vol. 60, no. 4, pp. 1384–1389, Apr. 2013.

[13] R. R. Chilipi, N. A. Sayari, A. R. Beig, and K. A. Hosani, "A multitasking control algorithm for grid-connected inverters in distributed generation applications using adaptive noise cancellation filters," *IEEE Transactions on Energy Conversion*, vol. 31, no. 2, pp. 714–727, Jun. 2016.

[14] J. Miret, M. Castilla, J. Matas, J. M. Guerrero, and J. C. Vasquez, "Selective harmonic-compensation control for single-phase active power filter with high harmonic rejection," *IEEE Transactions on Industrial Electronics*, vol. 56, no. 8, pp. 3117–3127, Aug. 2009.

[15] R. A. Mastromauro, M. Liserre, T. Kerekes, and A. Dell'Aquila, "A single-phase grid-connected photovoltaic systems with power quality conditioner functionality," *IEEE Transactions on Industrial Electronics*, vol. 56, no. 11, pp. 4436–4444, Nov. 2009.

[16] Y. Zhang, M. A. M. Hendrix, M. G. L. Roes, J. L. Duarte, and E. A. Lomonova, "Voltage harmonics suppression add-on functionality without additional sensors for existing grid-connected inverters," in *19th European Conference on Power Electronics and Applications (EPE'17 ECCE Europe)*, Sep. 2017, pp. 1–10.

[17] Y. Zhang, M. A. M. Hendrix, M. G. L. Roes, and J. L. Duarte, "Grid-connected converters with voltage support using only local measurements," in *IEEE 8th International Symposium on Power Electronics for Distributed Generation Systems (PEDG)*, Apr. 2017, pp. 1–6.

[18] M. Liserre, R. Teodorescu, and F. Blaabjerg, "Stability of photovoltaic and wind turbine grid-connected inverters for a large set of grid impedance values," *IEEE Transactions on Power Electronics*, vol. 21, no. 1, pp. 263–272, Jan. 2006.

[19] Y. Yang, K. Zhou, and F. Blaabjerg, "Current harmonics from single-phase grid-connected inverters - examination and suppression," *IEEE Journal of Emerging and Selected Topics in Power Electronics*, vol. 4, no. 1, pp. 221–233, Mar. 2016.

[20] S. Golestan, E. Ebrahimzadeh, J. M. Guerrero, and J. C. Vasquez, "An adaptive resonant regulator for single-phase grid-tied VSCs," *IEEE Transactions on Power Electronics*, vol. 33, no. 3, pp. 1867–1873, Mar. 2018.

[21] L. Corradini, D. Maksimovic, P. Mattavelli, and R. Zane, *The Digital Control Loop*. Wiley-IEEE Press, 2015, pp. 51–78.

The 2018 International Power Electronics Conference

Adaptive Series Stabilizer Module for the Grid Connected Inverter under Variable Grid Conditions

Xin Zhang[1*]

1. School of Electrical and Electronic Engineering, Nanyang Technological University, Singapore
E-mail: jackzhang@ntu.edu.sg*

Abstract— **This paper proposes an adaptive series stabilizer module (AS²M) to automatically shape the grid impedance to guarantee the stability of the GCI with different grid condition. Since the AS²M is in series with the GCI and the grid, it only needs to burden very small power. In addition, only a very simple current control loop is employed by the AS²M without any additional phase locked loop (PLL) or voltage controller. Furthermore, the AS²M can be treated as an independent stabilizer module and do not need to get any informtion from the grid or the GCI. Finally, an LCL type GCI is tested to verify the correctness of the proposed AS²M.**

I. Introduction

To offer a high degree of autonomic, modularity, scalability and maintainability, lots of distributed power generation units have been expected to deliver energy to the grid via grid connected inverters (GCI) [1]. GCI are generally voltage source converters with different output filters, such as the popular *LCL* filter and the classical *L* filter [2]. However, no matter for the LCL type GCI or the L type GCI, the ac output current regulation faces a critical challenge of filter resonance. Such resonant frequencies may further interact with the GCI control loop, get worse with the grid impedance, leading to harmonic instability phenomena in a wide frequency range [3].

In order to address the stability issue, many prior arts have been proposed. In [4, 5], both passive and active damping techniques have been reported to tackle the filter resonance. Among them, active damping techniques are preferable, as they do not cause extra power dissipation. However, the filter resonant frequency is actually affected by the equivalent grid impedance, which is not constant and always determined by the line impedance & power transformer leakage impedance [6]. As a result, it is a challenge to develop effective damping methods that can guarantee the stability of the GCI at different grid conditions. Thus, continues active damping research efforts have been made to further improve the robustness of the damping methods against the changing of the grid impedance, such as hybrid damping [7], adaptive cascaded notch filer [8] and active damper connected in shunt at the point-of-common-coupling (PLL) [9], etc. In addition, grid-voltage feedforward mechanism is also popularly integrated into the GCI control to alleviate the effect of variable grid impedance and suppress the grid harmonic distortion [10, 11]. However, due to the digital delay and the noise magnification of the differentiator, it is pointed out that the grid voltage feedforward method may make the GCI be unstable at weak grid case.

It is very interesting to find out that all the existing damping schemes are only focused on how to shape the output impedance of the GCI instead of the grid impedance. Certainly, this phenomenon is also well understood: compared to the unknown grid impedance, the GCI information is easier to be

obtained and hence, the GCI impedance is also easier to be modified. However, since shaping the GCI output impedance needs to know the internal information of the GCI, the existing damping schemes blocks the modularity development of the power electronic systems. For instance, even the damping method is designed very well for a special GCI to meet the grid impedance, it should be designed again if the GCI is changed. Unfortunately, this will increase the research and development cycle and the time costs.

In order to solve the above limitations of the shaping GCI output impedance methods, an adaptive series stabilizer module (AS²M) has been proposed in this paper. Different with the existing damping methods, the AS²M can shape the grid impedance without knowing any information of the GCI or the grid. If the GCI is oscillated with the grid impedance, the AS²M will adaptively detect the resonant frequency and add a large virtual resistor in series with the grid impedance to fix the stability problem. If the GCI is working very well with the grid impedance, the AS²M will automatically change to a traditional series active power filter (SAPF) to improve the power quality of the GCI. In addition, since the AS²M is in series with the GCI and the grid impedance, only small power loss will be generated.

In the rest of this paper, it is arranged as follows: In section II, the stability problem of between the GCI and the grid impedance is reviewed; Section III introduces the proposed AS²M; A detailed LCL type GCI has been built to verify the function of the AS²M in Section IV; Finally, Section V concludes this paper.

II. Reviewer of the Stability Problem of the GCI with the Weak Grid

A. The Impedance Model of the GCI

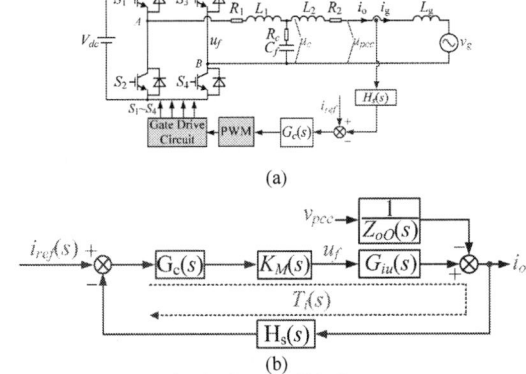

Fig. 1 A typical GCI: (a) circuit; (b) control block.

As shown in Fig. 1, a LCL type GCI was taken as an example to review the impedance model of the GCI. The topology of the GCI is presented at Fig. 1(a). Its control block is depicted at Fig. 1(b). Table I shows the definition of the transfer functions in Fig. 1(b).

TABLE I
DEFINITIONS OF THE TRANSFER FUNCTIONS IN FIG. 1(B)

Transfer function	Definition
$G_c(s)$	Transfer function of the current regulator
$H_s(s)$	Sampling coefficient of the GCI output current
$K_M(s)$	V_{dc}/V_m (V_{dc}: GCI input voltage; V_m: GCI carrier voltage)
$Z_{oO}(s)$	Open loop output impedance of the GCI
$G_{iu}(s)$	Open loop u_f to i_o transfer function

$Z_{oO}(s)$ and $G_{iu}(s)$ can be expressed as

$$G_{iu}(s) = \frac{\hat{i}_o}{\hat{u}_f}\bigg|_{\hat{v}_{pcc}=0} = \frac{1/sC_f + R_c}{\begin{bmatrix}(sL_1+R_1)\cdot(1/sC_f + R_c)\\+(sL_1+R_1)\cdot(sL_2+R_2)+\\(1/sC_f + R_c)\cdot(sL_2+R_2)\end{bmatrix}} \quad (1)$$

$$Z_{oO}(s) = -\frac{\hat{v}_{pcc}}{\hat{i}_o}\bigg|_{\hat{u}_f=0} = \frac{\begin{bmatrix}(sL_1+R_1)\cdot(1/sC_f + R_c)+\\(sL_1+R_1)\cdot(sL_2+R_2)+\\(1/sC_f + R_c)\cdot(sL_2+R_2)\end{bmatrix}}{(sL_1+R_1)+(1/sC_f + R_c)} \quad (2)$$

According to Fig. 1, the expression of the output current of the GCI, i_o, can be expressed as

$$i_o(s) = I_{inv}(s) - \left(\frac{1}{Z_{inv}(s)} + \frac{1}{Z_{oF}(s)}\right)\cdot v_{pcc}(s) \quad (3)$$

where

$$I_{inv}(s) = \frac{T_i(s)}{1+T_i(s)}\cdot\frac{i_{ref}(s)}{H_s(s)} \quad (4)$$

$$Z_{inv}(s) = Z_{oO}(s)\cdot[1+T_i(s)] \quad (5)$$

$$Z_{oF}(s) = \frac{-(1+T_i(s))}{G_F(s)K_M(s)G_{iu}(s)} \quad (6)$$

here, $T_i(s)$ is the loop gain of the GCI and can be expressed as

$$T_i(s) = G_c(s)\cdot K_M(s)\cdot G_{iu}(s)\cdot H_s(s) \quad (7)$$

By (3), the GCI can be presented by an impedance based current source model as shown in the virtual box part in Fig. 2.

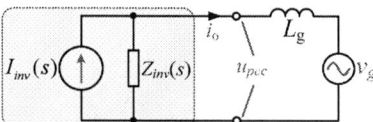

Fig. 2 Impedance model of the GCI with the weak grid.

B. Stability Problem of the GCI with the Weak Grid

According to the existing stability criterion of the GCI with the weak grid [12], the system equivalent loop gain can be expressed as

$$T_m(s) = \frac{sL_g}{Z_{inv}(s)} \quad (8)$$

If the $T_m(s)$ in (8) meets the Nyquist criterion, the system will be stable, otherwise, the system will be unstable. Therefore, even a GCI is designed very well individually, it may suffer instability issues when its output impedance mismatches with the grid impedance. This is the stability problem of the GCI with the weak grid.

In order to show the stability problem clearly, a detailed

LCL type GCI with a real weak grid has been tested here. The parameters of the tested LCL type GCI and the weak grid has been depicted in Table II.

TABLE II
PARAMETERS OF THE TEST GCI AND THE WEAK GRID

Para.	Value	Para.	Value	Para.	Value
V_{dc1}	405 V	R_{11}	0.01 Ω	R_{c1}	10 Ω
f_{sw1}	8 kHz	L_{21}	1.2 mH	H_{s1}	0.55
P_{o1}	1.4 kW	R_{21}	0.01 Ω	K_{p1}	35
L_{11}	7.2 mH	C_{f1}	6 μH	K_{i1}	60000
Grid impedance: 3 mH					

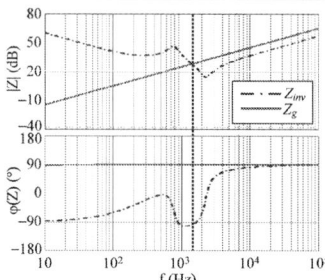

Fig. 3 Bode plot of $Z_{inv}(s)$ and $Z_g(s)$.

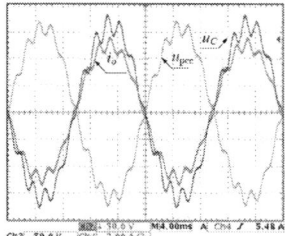

Fig. 4 Experimental waveforms of the test system at unstable case.

The bode plots of the $Z_{inv}(s)$ and the Z_g (sL_g) of the test system have been presented in Fig. 3. As seen, the test GCI output impedance not only being intersected with the grid impedance, their phase difference is also larger than 180°. Therefore, by the Nyquist criterion and (8), this system is unstable. Fig. 4 gives the experimental results of the test system, which presents the oscillated waveforms caused by the instability issues.

III. THE PROPSOED ADAPTIVE SERIES STABILIZER MODULE (AS²M)

A. The Position of the Proposed AS²M

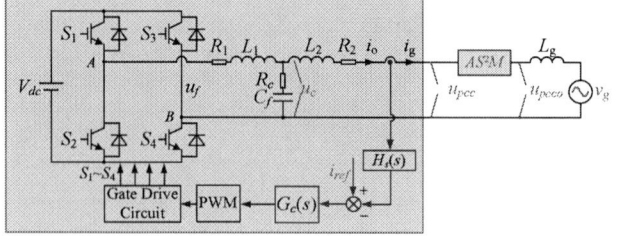

Fig.5 The position of the proposed AS²M.

Fig. 5 shows the position of the proposed AS²M. As seen, it is in series with the weak grid and the GCI, which located at the grid side of the Point of the common coupling (PCC). It treats the GCI as a block box and does not need to know its internal information.

954

The 2018 International Power Electronics Conference

B. Main Circuit and Control Block of the Proposed AS²M

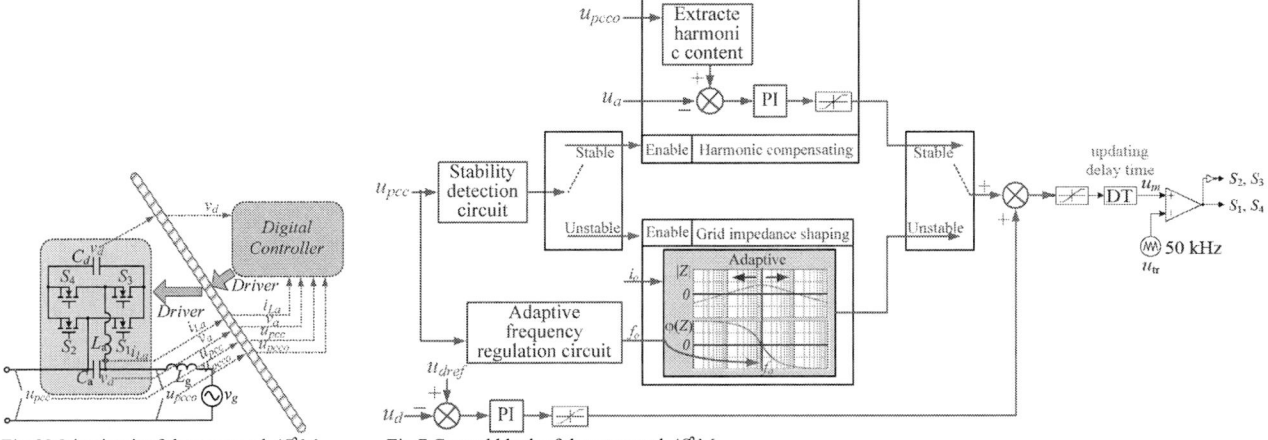

Fig.6 Main circuit of the proposed AS²M.　　Fig.7 Control block of the proposed AS²M.

The main circuit of the proposed AS²M has been shown in Fig. 6. As seen, the topology of the AS²M is the same with the traditional SAPF.

The control block of the proposed AS²M has been shown in Fig. 7. As seen, it contains two control modes:

- If the GCI is stable with the grid impedance, the AS²M can work as the traditional SAPF to compensate the harmonic.
- If the GCI is unstable with the grid impedance, the AS²M can automatically add a virtual resistor in series with the grid impedance to shape the grid impedance. As a result, the GCI will be stable again with the proposed AS²M.

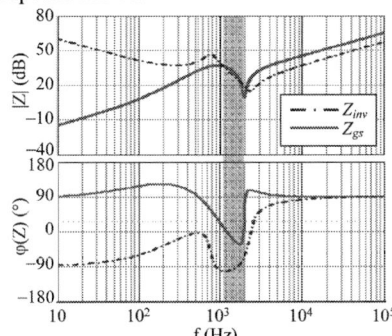

Fig.8 Bode plot of $Z_{inv}(s)$ and $Z_{gs}(s)$ with the proposed AS²M.

Fig. 8 shows the Bode plots of $Z_{inv}(s)$ and $Z_{gs}(s)$ with the proposed AS²M, where $Z_{gs}(s)$ is the shaped grid impedance with the AS²M. As seen, the AS²M can make the grid impedance acts similar to a resistor at the intersection area of the amplititude of the $Z_{inv}(s)$ and $Z_{gs}(s)$. Therefore, the AS²M can guarantee the GCI work stable with the shaped grid impedance. This is the stabilization function of the proposed AS²M.

IV. EXPERIMETNAL RESULTS

According to Fig 1 and table II, a real GCI was built at the lab. The proposed AS²M was tested as well.

A. Verify the traditional SAPF function of the Proposed AS²M

As shown in Fig. 9, if the grid is a still grid. The original GCI is stable with the grid. Then, the AS²M works as a good SAPF to improve the power quality.

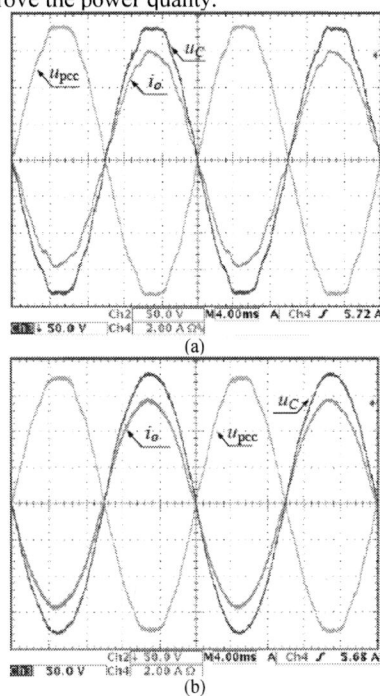

Fig.9 Experimental waveforms when L_g=0: a) without the proposed AS²M; b) with the propsoed AS²M.

B. Verify the stabilizaiton function of the Proposed AS²M

As shown in Fig. 10, if the grid is a weak grid. The original GCI is unstable with the grid. Then, the AS²M works as a good stabilizer to make the system become stable.

C. Verify the dynamic performance of the Proposed AS²M

As shown in Fig. 11, the Proposed AS²M enjoys a very good dynamic performance.

The 2018 International Power Electronics Conference

In summary, the experimental results verify the correctness of the function of Proposed AS²M.

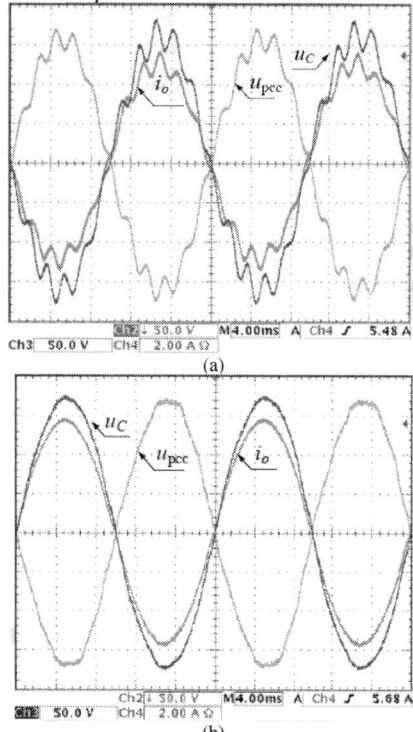

(a)

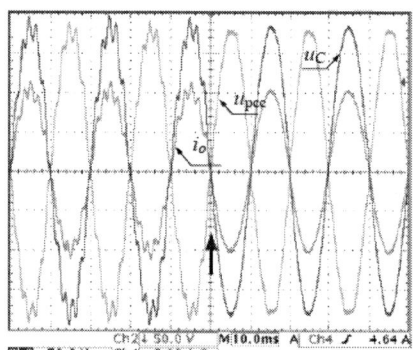

(b)

Fig.10 Experimental waveforms when L_g=3mH (the cases at table II): a) without the proposed AS^2M, b) with the propsoed AS^2M.

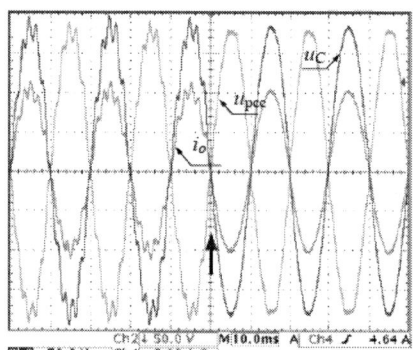

Fig.11 Dynamic experimental waveforms of the propsoed AS^2M when L_g=3mH (the cases at table II).

V. CONCLUSIONS

In this paper, an adaptive series stabilizer module (AS²M) is propsoed to automatically shape the grid impedance to guarantee the stability of the GCI with different grid condition. It contains two operation modes: a) if the GCI is stable with the grid impedance, the AS²M can work as the traditional SAPF to compensate the harmonic; b) if the GCI is unstable with the grid impedance, the AS²M can automatically add a virtual resistor in series with the grid impedance to shape the grid impedance. As a result, the GCI will be stable again with the proposed AS²M. In addition, since the AS²M is

in series with the GCI and the grid, it only suffers a little power just like the traditional SAPF. Finally, the experimental results verify the correctness of the function of Proposed AS²M

REFERENCES

[1] F. Blaabjerg, R. Teodorescu, M. Liserre, and A.V. Timbus, "Overview of control and grid synchronization for distributed power generation systems," IEEE Trans. Ind. Electron., vol. 53, no. 5, pp. 1398-1409, Oct. 2006.

[2] M. Liserre, F. Blaabjerg, and S. Hansen, "Design and control of an LCL-filter-based three-phase active rectifier," IEEE Trans. Ind. Appl., vol. 41, no. 5, pp. 1281-1291, Sept./Oct. 2005

[3] P. Brogan, "The stability of multiple, high power, active front end voltage sourced converters when connected to wind farm collector system," in Proc. EPE WECS 2010, pp. 1-6.

[4] S.G. Parker, B.P. McGrath, and D.G. Holmes, "Regions of Active Damping Control for LCL Filters," IEEE Trans. Ind. Appl., vol. 50, no. 1, pp. 424-432, Jan. /Feb. 2014.

[5] K. Jalili and S. Bernet, "Design of LCL filter of active-front-end two-level voltage-source converters," IEEE Trans. Ind. Electron., vol. 56, no. 5, pp. 1674–1689, May 2009.

[6] J.L. Agorreta, M. Borrega, J. Lopez, and L. Marroyo, "Modeling and Control of NParalleled Grid-Connected Inverters with LCL Filter Coupled Due to Grid Impedance in PV Plants," IEEE Trans. Power Electron., vol. 26, no. 3, pp. 770-785, Mar. 2011.

[7] Y. Lei, W. Xu, C. Mu, Z. Zhao, H. Li, and Z. Li, "New Hybrid Damping Strategy for Grid-Connected Photovoltaic Inverter With LCL Filter," IEEE Trans. Appl. Superconduct., vol. 24, no. 5, Oct. 2014.

[8] R.P. Alzola, M. Liserre, F. Blaabjerg, R. Sebasti´an, and T. Kerekes. "Selfcommissioning Notch Filter for Active Damping in Three Phase LCL-filter Based Grid Inverters," IEEE Trans. Power Electron., vol. 29, no. 12, pp. 6754-6761, Dec. 2014.

[9] X. Wang, Y. Pang, P.C. Loh, and F. Blaabjerg, "A Series-LC-Filtered Active Damper with Grid Disturbance Rejection for AC Power-Electronics-Based Power Systems," IEEE Trans. Power Electron., vol. 30, no. 8, pp. 4037-4041, Aug. 2015.

[10] X. Wang, X. Ruan, S. Liu and C. K. Tse, "Full Feedforward of Grid Voltage for Grid-Connected Inverter With LCL Filter to Suppress Current Distortion Due to Grid Voltage Harmonics," *IEEE Trans. Power Electron.*, vol. 25, no. 12, pp. 3119–3127, Jan. 2010.

[11] D. Holmes, T. Lipo, B. McGrath, and W. Kong, "Optimized design of stationary frame three phase ac current regulators," IEEE Trans. Power Electron., vol. 24, no. 11, pp. 2417–2426, Nov. 2009.

[12] J. Sun, "Impedance-based stability criterion for grid-connected inverters," *IEEE Trans on. Power Electronics Letters*, vol. 26, no. 11, pp. 3075-3078, 2011.

The 2018 International Power Electronics Conference

An Improved Droop Control Based Smooth Transfer Control Strategy

Xin Meng, Jinjun Liu, Zeng Liu, Ronghui An
State Key Lab of Electrical Insulation and Power Equipment
School of Electrical Engineering, Xi'an Jiaotong University
Xi'an China
Email: mengxinstar@stu.xjtu.edu.cn

Abstract—This paper proposes a smooth transfer control method, which can transfer from the current-source control in the grid connected (GC) mode to droop control in the stand alone (SA) mode automatically. When the grid is normal, the DG unit is based on current-source control. Upon the occurrence of utility outage, the power and voltage control loop are automatically activated to regulate the load voltage. The advantages of this proposed strategy are as follows. Firstly, compared with droop control, the proposed control strategy can improve the grid current quality and output constant power when the grid voltage has oscillation in GC mode. Secondly, compared with the hybrid voltage and current control strategy, the quality of critical load voltage can be maintained during the transferring process, and the reliability of microgrid in SA mode can be improved. The simulation results are provided to verify the effectiveness of the proposed control method.

Keywords—smooth transfer; improved droop control; grid connected mode; stand alone mode

I. INTRODUCTION

The distributed generation (DG) should be able to operate in both GC mode and SA mode, providing a smooth transfer between these two modes when sensitive and critical loads exist [1]. The existing smooth transfer control methods can be classified into three categories.

The first category named indirect current control is proposed in [1-4]. The grid current is regulated by controlling the capacitor voltage of LC filter indirectly. When islanding happens, the DG unit can transfer from controlled current source to controlled voltage source automatically.

In the second category, the DG unit is based on droop control in both GC and SA modes [5-7]. Therefore, the voltage quality can be guaranteed during the transition process. However, if the grid voltage has oscillation in GC mode, the grid current and inverter output power will have oscillation, too.

In the third category, which named as hybrid voltage and current control strategy, the DG unit is controlled as a current source in GC mode by one control system and as a voltage source in SA mode by the other control system [8-11]. The main issues of this control method are as follows. On the one hand, the load voltage quality heavily relies on the speed and accuracy of the islanding detection algorithm. On the other

hand, this kind of control method is based on master-slave microgrid structure. One DG unit acts as master unit, which function is to provide voltage support in SA mode. Other DG units act as slave units. The master DG unit adopts the hybrid voltage and current control. The reliability of microgrid system in SA mode is poor, because the whole system cannot operate normally once the master DG unit fails.

This paper proposes an improved droop control based smooth transfer control strategy, which can transfer from the current-source control to voltage-source control automatically. In GC mode, the power and voltage control loops are inactivated, and only the inductor current loop is utilized to control the DG unit acting as a current source. Upon the occurrence of utility outage, the power and voltage control loop will be automatically activated to regulate the load voltage. In SA mode, the DG unit is based on droop control.

Compared with existing smooth transfer control method, the proposed method has two advantages. Firstly, compared with droop control method, this proposed strategy can improve the grid current quality and output constant power when the grid voltage has oscillation in GC mode. Secondly, compared with the hybrid voltage and current control method, the quality of critical load voltage can be maintained during the transferring process, and the reliability of microgrid in SA mode can be improved.

This paper is organized as follows. Section II introduces the power stage of DG units and the proposed control method. Section III verifies the proposed seamless transfer control method by simulation. Finally, the conclusions are given in Section IV.

II. PROPOSED SMOOTH TRANSFER CONTROL METHOD

The power stage of DG unit is shown in Fig.1. There is a static switch S_i between DG and the utility, which is controlled by the DG unit. The utility protection switch S_u is governed by the utility. When the grid is normal, both S_i and S_u are closed. Upon the occurrence of grid outage, S_u turns off immediately, and then islanding is formed. After the islanding state is confirmed by islanding detection algorithm of DG, the transfer switch S_i turns off.

The control block is composed of three cascaded feedback loops, i.e., power control loop, capacitor voltage loop and inductor current loop, as shown in Fig.2 and 3. In GC mode,

This work was supported by the National Natural Science Foundation of China under Grant 51437007, and the Power Electronics Science and Education Development Program of Delta Environmental & Educational Foundation under Grant DREM2014002.

The 2018 International Power Electronics Conference

the power control loop and capacitor voltage loop are inactivated, and only the inductor current loop is utilized to control the DG unit acting as a current source, which will be analyzed later.

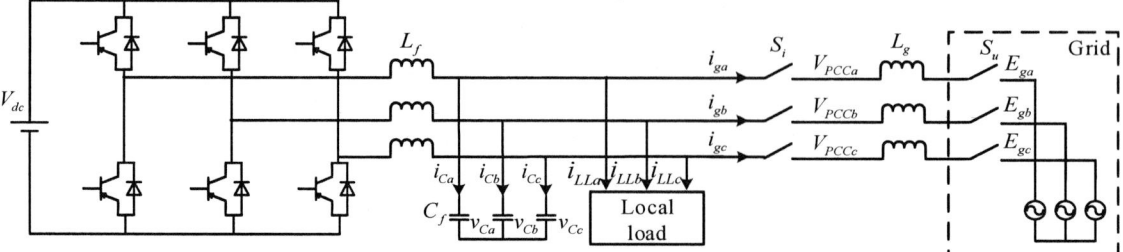

Fig. 1. Power stage of the DG unit

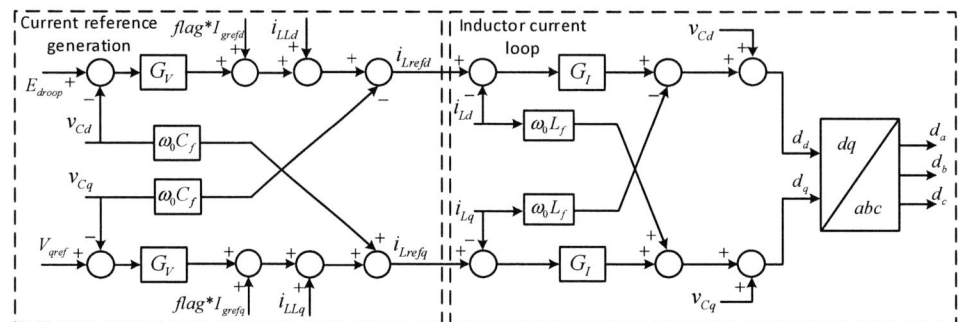

Fig. 2. The voltage and current control block for the proposed control method

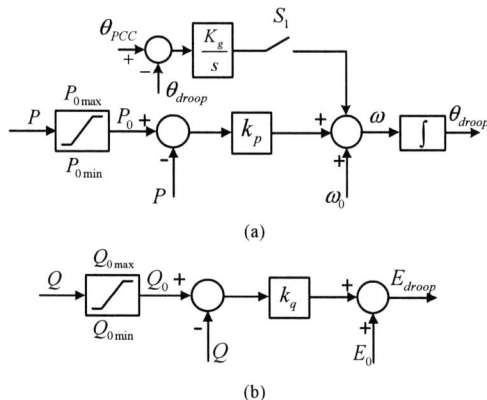

(a)

(b)

Fig. 3. The improved droop control block

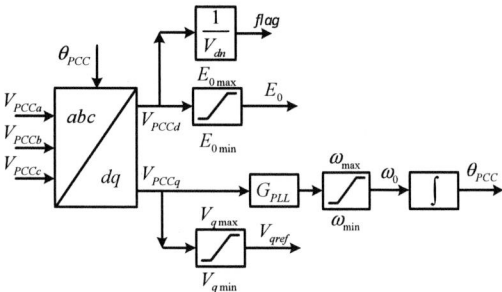

Fig. 4. The improved PLL block

Another important part is the PLL block, as shown in Fig.4, which produces the voltage set value E_0 in the reactive power control loop, the Q-axis voltage reference V_{qref}, the phase angle of PCC voltage θ_{PCC} and the control signal "*flag*".

There are totally four states for the DG, including the GC mode, the transition from GC mode to SA mode, the SA mode and the transition from SA mode to GC mode.

1) GC mode:

When the utility is normal, both Si and Su are closed. The DG unit is controlled as a current source to supply given active and reactive power by the inductor current loop.

Before the inverter output power reaches steady state, the limiter in the power loop are not activated. When the inverter output power reaches steady state, the limiter in the power loop are activated. The limiting values of limiter in the active and reactive power loop are set as

$$P_{0\max} = 1.1 * P_{nom}, P_{0\min} = 0.9 * P_{nom} \quad (1)$$

$$Q_{0\max} = 1.1 * Q_{nom}, Q_{0\min} = 0.9 * Q_{nom} \quad (2)$$

where P_{nom} and Q_{nom} are the inverter output active and reactive power in steady state respectively, which are decided by I_{grefd} and I_{grefq}. There exists $P_0=P_{nom}=P$ and $Q_0=Q_{nom}=Q$ in Fig.3. The control switch S_1 is open in GC mode; therefore, the output of active power control loop ω equals to ω_0. Similarly, the output of reactive power control loop E_{droop} equals to E_0. ω_0 and E_0 are produced by the improved PLL block. Therefore, the active and reactive power control loops don't work.

The voltage control loop doesn't work in GC mode either. The voltage controller G_v adopts proportional (P) regulator.

958

The reason for utilizing P regulator in voltage loop is to avoid the integral effect, making sure the output of G_v is zero when the input of G_v is zero. The voltage of point of common coupling (PCC) equals to the LC filter capacitor voltage. The limiter of D-axis and Q-axis in the PLL block will not function; therefore, the relationship $E_{droop}=v_{cd}$ and $V_{qref}=v_{cq}$ are established, and the voltage control loop is out of function in GC mode.

V_{dn} in PLL block is the nominal voltage value of PCC. V_{PCCq} is regulated to zero by the PLL, so V_{PCCd} equals to the amplitude of PCC voltage V_{PCC}. Therefore, $flag = V_{PCCd} / V_{dn} \approx 1$.

Based on above analysis, in the D-axis, the inductor current reference i_{Lrefd} can be expressed as (3)

$$i_{Lrefd} = I_{grefd} + i_{LLd} - \omega_0 C v_{Cq} \qquad (3)$$

The first part I_{grefd} is a given current reference. The second part is the load current of D-axis i_{LLd}, which is determined by the characteristic of the local load. The third part $-\omega_0 C v_{Cq}$ equals to zero due to $v_{Cq} = 0$. Consequently, the current reference i_{Lrefd} is imposed by the given current reference I_{grefd} and the load current i_{LLd}. In the Q-axis, the inductor current reference i_{Lrefq} is similar with which of D-axis.

The control diagram in GC mode can be simplified as Fig.5 and Fig.6, and the inverter is controlled as a current source by the inductor current loop.

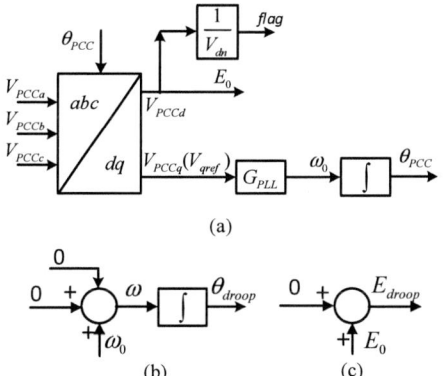

(a)

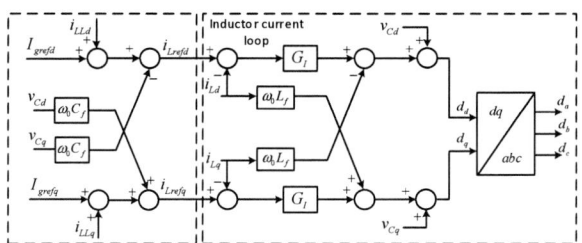

(b)   (c)

Fig. 5.   Simplified PLL and droop control block in GC mode

Fig. 6.   Simplified PLL and droop control block in GC mode

2) Transition from GC mode to SA mode:

When the utility switch S_u opens, the islanding is formed, and the amplitude and frequency of load voltage will drift because of the power mismatch between the DG output and the load demand. The transition can be divided into two parts as shown in Fig.7. The first time interval starts from the instant of opening S_u to the instant of opening S_i when the islanding is detected. The second time interval starts from the instant of opening S_i.

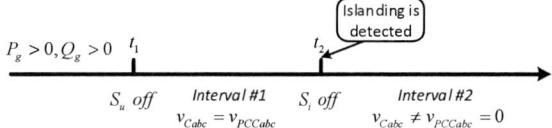

Fig. 7.   Operation sequence from GC mode to SA mode

It is assumed that the DG unit injects active power and reactive power into the utility in GC mode. Therefore, when islanding happens, the local load must absorb the extra power injected to the grid, as the output power of inverter is not changed instantaneously. The magnitude of load voltage will rise with the increase of load active power. At the same time, the angular frequency will drop to consume more reactive power. The PCC voltage V_{PCCabc} is still the same with the load voltage v_{Cabc} because the switch S_i is in ON state. When the amplitude of PCC voltage V_{PCCabc} increases, the D-axis output of PLL block will also increases. Because of the existence of D-axis limiter in Fig.4, if the increase of V_{PCCabc} is large, E_0 will equal to E_{0max}. Therefore, $E_0 \neq v_{cd}$, the voltage control loop will be activated.

The local load cannot absorb too much power, otherwise, the load voltage is out of the allowable range. Therefore, the inverter output active power and reactive power will change to adjust the load voltage within permissible operation range. The limiter in the active and reactive power control loop, as shown in Fig.3, will function, and there exists $P_0 \neq P$ and $Q_0 \neq Q$. Therefore, the active and reactive power control loop will be activated.

The second time interval of the transition begins from the instant when the switch S_i is open after the islanding has been confirmed. If the switch S_i opens, the PCC voltage V_{PCCabc} will decrease to zero, and the input voltage of PLL block is zero. The voltage set value E_0 equals to the lower limit V_{dmin}, and V_{qref} equals to zero. The control signal "$flag$" changes to zero when the switch S_i opens, which will make the grid current reference out of function. The power control loop and voltage control loop will function, and the load voltage is always controlled within allowed range.

3) SA mode:

In SA mode, switching S_i and S_u are both in OFF state. The control signal "$flag$" is zero. The DG unit is controlled as a voltage source based on droop control. The reliability of microgrid system can be improved in SA mode because that multiple DG units transfer from current source to voltage source to regulate the voltage of microgrid when islanding happens.

4) Transition from SA mode to GC mode:

Firstly, if the grid is restored and S_u is closed, the PCC voltage equals to the grid voltage because that the switch S_i is open, and the voltage drop on grid inductor equals to zero. The control signal "*flag*" equals to one. The PLL block will track the phase of grid voltage automatically. Then the switch S_l in Fig.3 is closed to synchronize θ_{droop} with θ_{PCC}.

Secondly, the output of the PLL block V_{PCCd} and V_{PCCq} will be adjusted within the upper and lower value of the limiter. The amplitude of load voltage will be regulated to follow the grid voltage. As a result, the synchronization of load voltage and grid voltage is finished.

Thirdly, the switch S_i turns on, and the DG unit operates in GC mode again. The switch S_l in Fig.3 turns off.

III. SIMULATION RESULTS

PSCAD simulations are conducted to verify the proposed method. The simulation circuit is shown in Fig.8. Master1 and Master2 are controlled by this proposed control strategy. Slave1 and Slave2 are controlled as current sources based on PQ control. The simulation parameters are shown in Table I.

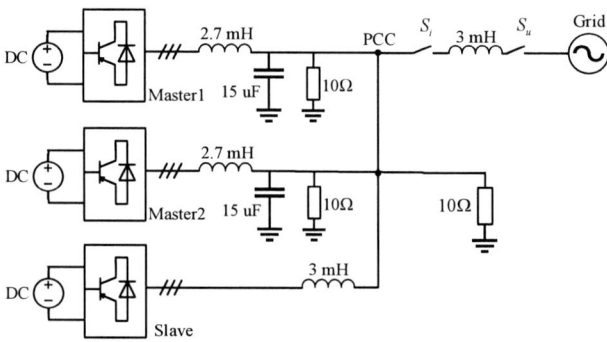

Fig. 8.   The simulation circuit

Table I Simulation parameters

Parameters	Value
Grid voltage amplitude	162.6V
Grid current reference value I_{grefd}	15A
D-axis limiting value V_{dmax}	177V
D-axis limiting value V_{dmin}	146.5V
Q-axis limiting value V_{qmax}	16V
Q-axis limiting value V_{qmin}	-16V
P_{0max}, P_{0min}	8415W,6885W
Q_{0max}, Q_{0min}	1000Var,-1000Var

1) At first, the utility is normal, and the microgrid is connected with the grid.

2) At 2s, the grid is broken and S_u turns off.

3) At 2.02s, the islanding is confirmed and S_i turns off.

4) During 2.02s-5s, the microgrid operates in SA mode.

5) At 3.2s, the grid is restored and S_u turns on.

6) At 5s, S_i turns on. The synchronization of load voltage and grid voltage is completed during 3.2s-5s.

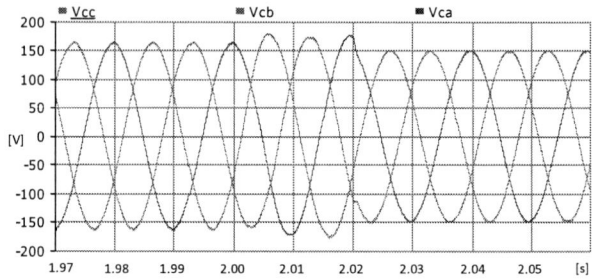

(a)

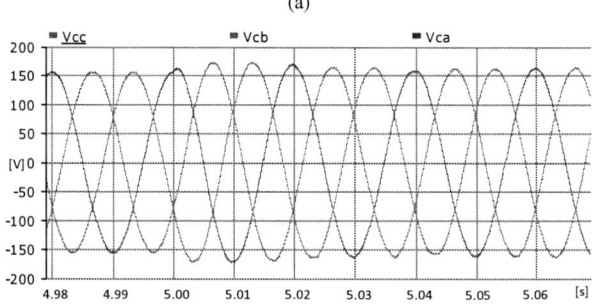

(b)

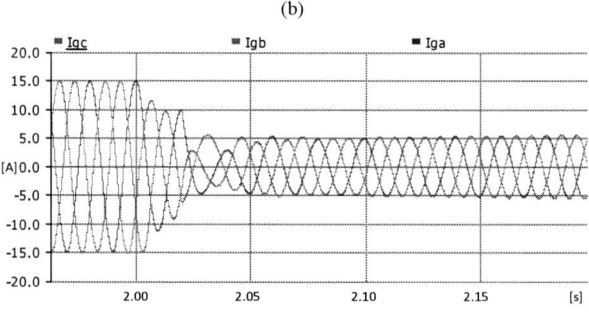

(c)

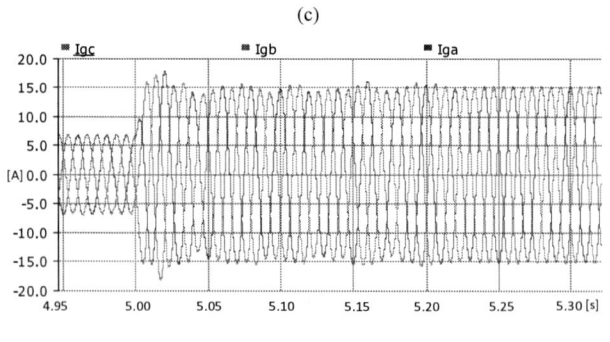

(d)

Fig. 9. Simulation results of load voltage and grid current

Fig.9 (a) and (c) shows the load voltage near the capacitor of LC filter and grid current when the microgrid transfers from GC mode to SA mode. Fig.9 (b) and (d) shows the load voltage and grid current when the microgrid transfers from SA mode to GC mode. It can be seen from Fig.9 (a) and (b), there is no voltage distortion during the transferring process, and the change of load voltage is smooth. In Fig.9 (c) and (d), the grid current during the transferring process is also smooth, and there is no apparent inrush current.

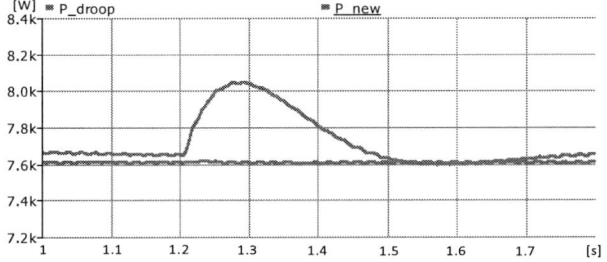

Fig. 10. Simulation results of inverter output power

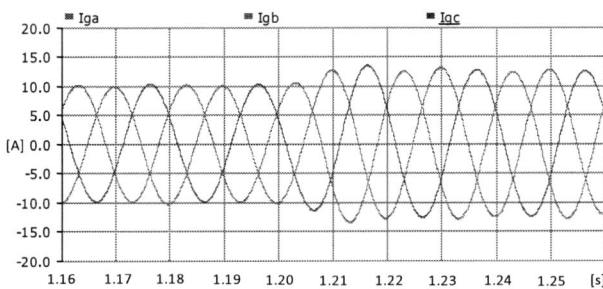

(a)

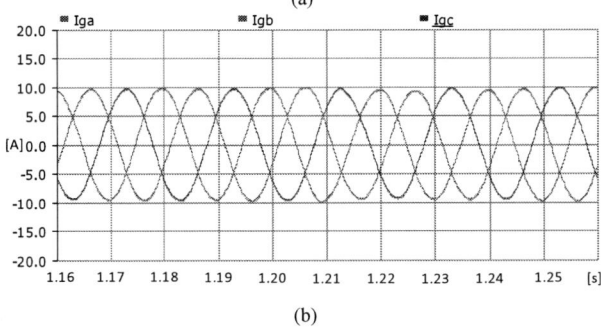

(b)

Fig. 11. Simulation results of grid current with (a) Conventional droop control (b) Proposed control method

The gird frequency drops from 50Hz to 49.5Hz at 1.2s and returns to 50Hz at 1.21s. As shown in Fig.10 (the red line) and Fig.11 (a), the inverter output power and the grid current change a lot with conventional droop control method. However, with proposed control method, the inverter output power and the grid current have no change, as shown in Fig.10 (the blue line) and Fig.11 (b). It verifies that the proposed control strategy can output constant power and improve the grid

current quality when the grid voltage has oscillation in GC mode.

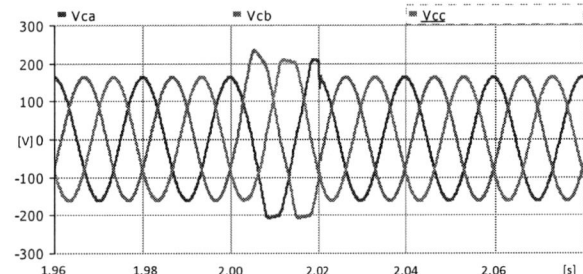

Fig. 12. Simulation results of load voltage with hybrid voltage and current control method

In Fig.12, the load voltage quality will be worsened during the transferring process (2s-2.02s) from GC mode to SA mode with hybrid voltage and current control method. However, the quality of load voltage can be maintained with proposed control method, as shown in Fig.9 (a).

IV. CONCLUSIONS

This paper proposes an improved droop control based seamless transfer strategy. In GC mode, the power and voltage control loop are inactivated, and the inductor current loop is utilized to control the DG unit to act as a current source. Upon the occurrence of utility outage, the power and voltage control loop will be automatically activated to regulate the load voltage. Compared with existing smooth transfer control method, the proposed method has two main contributions. Firstly, compared with droop control method, it can improve the grid current quality and output constant power when the grid voltage has oscillation in GC mode. Secondly, compared with the hybrid voltage and current control method, the quality of critical load voltage can be maintained during the transferring process, and the reliability of microgrid in SA mode can be improved.

ACKNOWLEDGMENT

This work was supported by the National Natural Science Foundation of China under Grant 51437007, and the Power Electronics Science and Education Development Program of Delta Environmental & Educational Foundation under Grant DREM2014002.

REFERENCES

[1] Z. Liu and J. Liu, "Indirect Current Control Based Seamless Transfer of Three-phase Inverter in Distributed Generation," *IEEE Transactions on Power Electronics*, vol. 29, no. 7, pp. 3368-3383, Jul. 2014.

[2] T. Yu, S. Choi, and H. Kim, "Indirect current control algorithm for utility interactive inverters for seamless transfer," in *Power Electronics Specialists Conference, 2006. PESC '06. 37th IEEE*, 2006, pp. 1–6.

[3] J. Kwon, S. Yoon and S. Choi, "Indirect Current Control for Seamless Transfer of Three-Phase Utility Interactive Inverters," *IEEE Transactions on Power Electronics*, vol. 27, no. 2, pp. 773-781, Feb. 2012.

[4] S. Yoon, H. Oh, and S. Choi, "Controller design and implementation of indirect current control based utility interactive inverter system," in *2011 IEEE Energy Conversion Congress and Exposition*, 2011, pp. 955–960.

[5] Y. Shi and J. Su, "A Seamless Mode Transfer Method for Microgrid Based on Mode Adaptive Droop Control," in *TENCON 2013 - 2013 IEEE Region 10 Conference (31194)*,2013, pp.1-5.

[6] Y. Jia, D. Liu and J. Liu, "A Novel Seamless Transfer Method for a Microgrid Based on Droop Characteristic Adjustment," in *Power Electronics and Motion Control Conference (IPEMC), 2012 7th International* , 2012, pp.362-367.

[7] S. Hu, C. Kuo, T. Lee and J. Guerrero, "Droop-Controlled Inverters with Seamless Transition between Islanding and Grid-Connected Operations," in *2011 IEEE Energy Conversion Congress and Exposition*, 2011, pp.2196-2201.

[8] T.Hwang and S.Park, "A Seamless Control Strategy of Distributed Generation Inverter For Critical Load Safety under Strict Grid Disturbance," in *2012 Twenty-Seventh Annual IEEE Applied Power Electronics Conference and Exposition (APEC) Conference*, 2012, pp.254-261.

[9] I. Balaguer, Q. Lei, S. Yang, U. Supatti and F. Peng, "Control for Grid-Connected and Intentional Islanding Operations of Distributed Power Generation," *IEEE Transactions on Industrial Electronics*,vol.58, no.1, pp.147-157, Jan. 2011.

[10] D. Ochs, B. Mirafzal and P. Sotoodeh, "A Method of Seamless Transitions Between Grid-Tied and Stand-Alone Modes of Operation for Utility-Interactive Three-Phase Inverters," *IEEE Transactions on Industrial Applicayions*,vol. 50, no. 3, pp. 1934-1941, May. 2014.

[11] Y. Wang and G. Zhao, "Direct Current Control Strategy for Seamless Transfer of Voltage Source Inverter in Distributed Generation Systems," in *International Conference on Renewable Power Generation (RPG 2015)*, 2015, pp.1-5.

The 2018 International Power Electronics Conference

Frequency Response Analysis of Load Effect on Dynamics of Grid-Forming Inverter

Matias Berg*, Tuomas Messo, Teuvo Suntio
Laboratory of Electrical Energy Engineering, Tampere University of Technology, Tampere, Finland
*E-mail: matias.berg@tut.fi

Abstract—The grid-forming mode of the voltage source inverters (VSI) is applied in uninterruptible power supplies and micro-grids to improve the reliability of electricity distribution. During the intentional islanding of an inverter-based micro-grid, the grid-forming inverters (GFI) are responsible for voltage control, similarly as in the case of uninterruptible power supplies (UPS). The unterminated model of GFI can be developed by considering the load as an ideal current sink. Thus, the load impedance always affects the dynamic behavior of the GFI. This paper proposes a method, to analyze how the dynamics of GFI and the controller design are affected by the load. Particularly, how the frequency response of the voltage loop gain changes according to the load and, how it can be used to the predict time-domain step response. The frequency responses that are measured from a hardware-in-the-loop simulator are used to verify and illustrate explicitly the load effect.

Keywords—grid-forming inverter, dynamics, dq-domain, load effect

I. INTRODUCTION

The recent years have witnessed a huge growth in the number of installed distributed photovoltaic generation systems. Distributed generation with an energy storage system in a micro-grid enables the intentional islanding of the micro-grid during a failure in the utility network [1], [2]. If there are no rotating generators in the micro-grid, the inverters that normally operate in the grid-feeding mode, have to form the grid during the intentional islanding [3]. The dynamics of the grid-forming inverter (GFI) differ from the dynamics of the grid-feeding inverter. The grid-forming inverter is a voltage-output converter and the grid-feeding inverter is a current-output converter [4]. An ideal current sink as the load of GFI is the basis fo the dynamic analysis, but the load effect has to be taken into account.

The importance of modeling the output impedance of power-electronics-based systems has been widely addressed [5]–[7]. In order to derive the output impedance, the output current has to be considered an input variable. The output impedance has been derived this way in [5]. The output impedance modeled in [5] is verified by frequency response measurements, but the other transfer functions are not measured. Dynamics of an LC-filter has been included to the input admittance of an active rectifier for the purposes of impedance-based analysis in [8]. Passive loads have been modeled as a part of system consisting of grid-connected solar inverter and an active rectifier in [9]. However, the analysis is focused on the frequency responses of impedances and the effect of

the filter on control-to-output transfer functions was not analyzed in [8], [9].

The control-related transfer functions change if the load is changed from a current sink to a passive or active load. The output impedance of the grid-forming inverter has been derived also in [10] and the output current is considered as an input variable. The unterminated dynamics are analyzed when the controllers are tuned. However, the simulation and practical tests are done with passive and non-linear loads without analyzing the load effect to the control loops. In [11] the output current of a single-phase system is considered as an input variable and the unterminated model is used to derive the transfer functions. The time-domain behavior is tested under a resistive load and a non-linear load. However, the load effect on the loop gains is not shown.

A load-affected transfer function is directly derived in phase domain in [1]. However, no frequency-response verification is presented. A dynamic model of a passive load is derived in [12], but it is not used for frequency response analysis of the system. In [2] the load is analyzed in the dq-domain and included in the system model, but frequency response analysis is missing. A passive load has been addressed also in [13] and [14], but no frequency responses are analyzed.

This paper proposes a method, that can be used to analyze the load effect on the unterminated dynamics of GFI in the frequency domain. The rest of the paper is organized as follows: Section II introduces the modeling of the unterminated dynamics of the three-phase grid-forming inverter in dq-domain. Section III examines the load effect on the dynamics of GFI. Frequency response analysis of the load effect is used to tune the controllers and to predict the time-domain response in Section IV. The conclusions are finally presented in Section V.

II. UNTERMINATED SMALL-SIGNAL MODEL

The used averaging and linearizing method originates from the work of Middlebrook [15]. Figure 1 shows the circuit diagram of a three-phase grid-forming inverter. The load is assumed to be an ideal three phase current sink in the dynamic analysis. Thus, the grid inductance or load side inductors of the LCL-filter cannot be included in the unterminated models due to violation of Kirchoff's law. Output impedance of the grid-forming inverter and the other input-to-output transfer function can be derived by analyzing the power stage in Fig. 1. The input variables are input voltage, duty-ratios and output currents. The

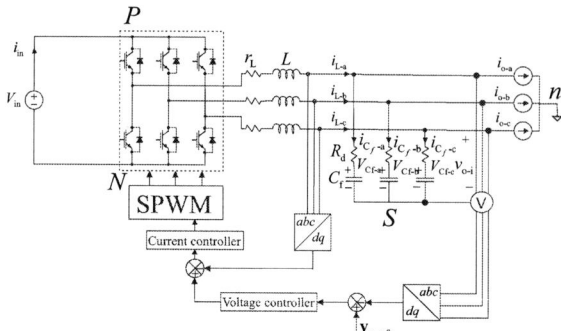

Figure 1. Circuit diagram of the grid-forming inverter including a simplified control system.

output variables are input current, inductor currents and output voltages. The inductor currents are chosen as output variables, because they are commonly needed in the cascaded control of the output voltage.

A state-space model of the grid-forming inverter is derived. The capacitor voltages and inductor currents are chosen as the state variables. Modeling in the synchronous reference frame is applied. For brevity, the equations are shown directly in the synchronous reference frame (DQ-frame). In the following equations, subscripts d and q denote whether the corresponding variable is either the direct or quadrature component. i_L is the inductor current, i_o the output current, d the duty ratio, v_{in} the input voltage, v_{Cf} the filter capacitor voltage, i_{in} the input current. Angle brackets around the variables in (1)–(7) denote that equations are averaged over one switching period. Thus, on and off-time equations are not shown separately.

$$\langle i_{in} \rangle = \frac{3}{2} \left(d_\mathrm{d} \langle i_\mathrm{Ld} \rangle + d_\mathrm{q} \langle i_\mathrm{Lq} \rangle \right) \tag{1}$$

$$\frac{d \langle i_\mathrm{Ld} \rangle}{dt} = \frac{1}{L} \big[d_\mathrm{d} \langle v_\mathrm{in} \rangle - (r_\mathrm{L} + r_\mathrm{sw} + R_\mathrm{d}) \langle i_\mathrm{Ld} \rangle \\ + \omega_s i_\mathrm{Lq} + R_\mathrm{d} \langle i_\mathrm{od} \rangle - \langle v_\mathrm{Cfd} \rangle \big] \tag{2}$$

$$\frac{d \langle i_\mathrm{Lq} \rangle}{dt} = \frac{1}{L} \big[d_\mathrm{q} \langle v_\mathrm{in} \rangle - (r_\mathrm{L} + r_\mathrm{sw} + R_\mathrm{d}) \langle i_\mathrm{Lq} \rangle \\ - \omega_s i_\mathrm{Ld} + R_\mathrm{d} \langle i_\mathrm{oq} \rangle - \langle v_\mathrm{Cfq} \rangle \big] \tag{3}$$

$$\frac{d \langle v_\mathrm{Cfd} \rangle}{dt} = \frac{1}{C_f} \big[\langle i_\mathrm{Ld} \rangle + \omega_s v_\mathrm{Cfq} - \langle i_\mathrm{od} \rangle \big] \tag{4}$$

$$\frac{d \langle v_\mathrm{Cfq} \rangle}{dt} = \frac{1}{C_\mathrm{f}} \big[\langle i_\mathrm{Lq} \rangle - \omega_s v_\mathrm{Cfd} - \langle i_\mathrm{oq} \rangle \big] \tag{5}$$

$$\langle v_\mathrm{od} \rangle = \langle v_\mathrm{Cfd} \rangle + R_\mathrm{d} \langle i_\mathrm{Ld} \rangle - R_\mathrm{d} \langle i_\mathrm{od} \rangle \tag{6}$$

$$\langle v_\mathrm{oq} \rangle = \langle v_\mathrm{Cfq} \rangle + R_\mathrm{d} \langle i_\mathrm{Lq} \rangle - R_\mathrm{d} \langle i_\mathrm{oq} \rangle , \tag{7}$$

where C_f, L, d, r_{sw} and ω_s denote filter capacitor, filter inductor, duty ratio, parasitic resistance of a switch, grid angular frequency, respectively. r_L is the equivalent series resistance of the filter inductor. The damping resistance that includes the filter capacitor equivalent series resistance is denoted by R_D. Capital letters denote steady-state values at the operating point.

Equations (1)–(7) are linearized at the steady-state operation point and transformed into the frequency domain. The linearized equations are expressed by coefficient matrices **A**, **B**, **C** and **D** , input variable vector **U**, output variable vector **Y** and state variable vector **X**. Equation (8) shows the state space after transformation to frequency domain using the Laplace variable 's' and the output and input variable vectors are shown in (9) and (10), respectively.

$$s\mathbf{X}(s) = \mathbf{A}\mathbf{X}(s) + \mathbf{B}\mathbf{U}(s)$$
$$\mathbf{Y}(s) = \mathbf{C}\mathbf{X}(s) + \mathbf{D}\mathbf{U}(s) \tag{8}$$

The coefficient matrices are shown in (11)–(14).

$$\mathbf{Y} = \begin{bmatrix} \hat{i}_\mathrm{in} & \hat{i}_\mathrm{Ld} & \hat{i}_\mathrm{Lq} & \hat{v}_\mathrm{od} & \hat{v}_\mathrm{oq} \end{bmatrix}^T \tag{9}$$

$$\mathbf{U} = \begin{bmatrix} \hat{v}_\mathrm{in} & \hat{i}_\mathrm{od} & \hat{i}_\mathrm{oq} & \hat{d}_\mathrm{d} & \hat{d}_\mathrm{q} \end{bmatrix}^T \tag{10}$$

$$\mathbf{A} = \begin{bmatrix} -\frac{r_\mathrm{eq}}{L} & \omega_\mathrm{s} & -\frac{1}{L} & 0 \\ -\omega_\mathrm{s} & -\frac{r_\mathrm{eq}}{L} & 0 & -\frac{1}{L} \\ \frac{1}{C_\mathrm{f}} & 0 & 0 & \omega_\mathrm{s} \\ 0 & \frac{1}{C_\mathrm{f}} & -\omega_\mathrm{s} & 0 \end{bmatrix} \tag{11}$$

$$\mathbf{B} = \begin{bmatrix} \frac{D_\mathrm{d}}{L} & \frac{R_\mathrm{d}}{L} & 0 & \frac{V_\mathrm{in}}{L} & 0 \\ \frac{D_\mathrm{q}}{L} & 0 & \frac{R_\mathrm{d}}{L} & 0 & \frac{V_\mathrm{in}}{L} \\ 0 & -\frac{1}{C_\mathrm{f}} & 0 & 0 & 0 \\ 0 & 0 & -\frac{1}{C_\mathrm{f}} & 0 & 0 \end{bmatrix} \tag{12}$$

$$\mathbf{C} = \begin{bmatrix} \frac{3D_\mathrm{d}}{2} & \frac{3D_\mathrm{q}}{2} & 0 & 0 \\ 1 & 0 & 0 & 0 \\ 0 & 1 & 0 & 0 \\ R_\mathrm{d} & 0 & 1 & 0 \\ 0 & R_\mathrm{d} & 0 & 1 \end{bmatrix} \tag{13}$$

$$\mathbf{D} = \begin{bmatrix} 0 & 0 & 0 & \frac{3I_\mathrm{Ld}}{2} & \frac{3I_\mathrm{Lq}}{2} \\ 0 & 0 & 0 & 0 & 0 \\ 0 & 0 & 0 & 0 & 0 \\ 0 & -R_\mathrm{d} & 0 & 0 & 0 \\ 0 & 0 & -R_\mathrm{d} & 0 & 0 \end{bmatrix}, \tag{14}$$

where r_eq denotes $r_\mathrm{L} + r_\mathrm{sw} + R_\mathrm{d}$. The transfer functions from the inputs to the outputs can be solved as shown in (15).

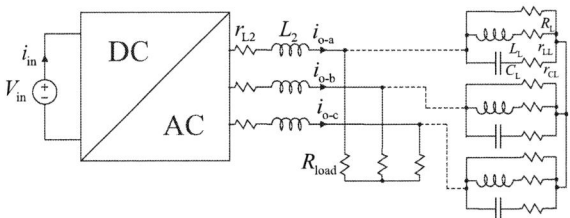

Figure 2. Circuit diagram of the grid-forming inverter including the load-side inductor and a resistive load or alternatively a RLC-load.

$$\mathbf{Y}(s) = \overbrace{\left(\mathbf{C}(s\mathbf{I} - \mathbf{A})^{-1}\mathbf{B} + \mathbf{D}\right)}^{\mathbf{G}} \mathbf{U}(s), \quad (15)$$

where matrix $\mathbf{G}$ contains the transfer functions. Different transfer functions can be collected from the matrix as shown in (16).

$$\begin{bmatrix} Y_{\text{in}} & T_{\text{oid}} & T_{\text{oiq}} & G_{\text{cid}} & G_{\text{ciq}} \\ G_{\text{ioLd}} & G_{\text{oLd}} & G_{\text{oLqd}} & G_{\text{cLd}} & G_{\text{cLqd}} \\ G_{\text{ioLq}} & G_{\text{oLdq}} & G_{\text{oLq}} & G_{\text{cLdq}} & G_{\text{cLq}} \\ G_{\text{iod}} & -Z_{\text{od}} & -Z_{\text{oqd}} & G_{\text{cod}} & G_{\text{coqd}} \\ G_{\text{ioq}} & -Z_{\text{odq}} & -Z_{\text{oq}} & G_{\text{codq}} & G_{\text{coq}} \end{bmatrix} \quad (16)$$

In this paper the transfer functions are merged into transfer matrices [5], [16]. Equation (17) shows the transfer matrices that were solved in (15) and the corresponding input and output variables. Hats over the input and output variables denote small-signal variables. Input voltage and input current are scalar variables and their small signal dependency is denoted by Y_{in}. The input and output variables that are collected into 2-by-1 vectors are shown in (18).

$$\begin{bmatrix} \hat{i}_{\text{in}} \\ \hat{\mathbf{i}}_{\text{L}} \\ \hat{\mathbf{v}}_{\text{o}} \end{bmatrix} = \begin{bmatrix} Y_{\text{in}} & \mathbf{T}_{\text{oi}} & \mathbf{G}_{\text{ci}} \\ \mathbf{G}_{\text{iL}} & \mathbf{G}_{\text{oL}} & \mathbf{G}_{\text{cL}} \\ \mathbf{G}_{\text{io}} & -\mathbf{Z}_{\text{o}} & \mathbf{G}_{\text{co}} \end{bmatrix} \begin{bmatrix} \hat{v}_{\text{in}} \\ \hat{\mathbf{i}}_{\text{o}} \\ \hat{\mathbf{d}} \end{bmatrix} \quad (17)$$

$$\hat{\mathbf{i}}_{\text{L}} = \begin{bmatrix} \hat{i}_{\text{Ld}} & \hat{i}_{\text{Lq}} \end{bmatrix}^T \hat{\mathbf{v}}_{\text{o}} = \begin{bmatrix} \hat{v}_{\text{od}} & \hat{v}_{\text{oq}} \end{bmatrix}^T$$
$$\hat{\mathbf{i}}_{\text{o}} = \begin{bmatrix} \hat{i}_{\text{od}} & \hat{i}_{\text{oq}} \end{bmatrix}^T \hat{\mathbf{d}} = \begin{bmatrix} \hat{d}_{\text{d}} & \hat{d}_{\text{q}}^T \end{bmatrix} \quad (18)$$

III. LOAD EFFECT

Grid-feeding inverters are commonly equipped with an LCL-filter. It is assumed that an inverter that is used in the grid-feeding mode will be used also in the grid-forming mode. In the case of grid-feeding inverters, load impedances have been included in the model in [17] and analytical equations for generalized source and load interactions are shown in [4]. The effect of load dynamics on the unterminated dynamics has been analyzed in the case of DC-DC converters in [18]. Fig. 2 shows a circuit diagram of the grid-forming inverter, where the load is a resistor or alternatively a parallel RLC-load (as depicted

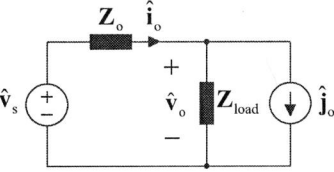

Figure 3. An equivalent small-signal circuit that has been widely used to analyze to impedance based stability.

using dashed lines). The load-side inductor is taken into account in the model. In the unterminated model in Fig. 1 the load-side inductor is not included, because the series connection of an inductor and current sink is inconsistent according to circuit theory.

Fig. 3 shows an equivalent small-signal circuit of two interconnected systems. Very similar circuits have been widely used in the literature for impedance-based stability analysis [7], [9], [19], [20]. Variables $\hat{v}_{\text{s}}$ and $\hat{j}_{\text{o}}$ denote small-signal source voltage and load current, respectively. However, they do not indicate, how the voltage and current are dependent on the inverter input parameters. In following, the general voltage source is replace by the control-to-output transfer function matrices so that the load-affected transfer functions can be solved.

The output dynamics of the grid-forming inverter are shown as an equivalent linear circuit in Fig. 4(a) which corresponds to the equation of $\hat{v}_{\text{o}}$ in (17) that is developed from the case the load is an ideal current sink in Fig. 1. However, the load effect of the load-side inductor, its ESR and the load resistor in Fig. 2 must be taken into account. Figure 4(b) shows the output dynamics model, where the load impedance $\mathbf{Z}_{\text{load}}$ and the impedance of the load-side inductor $\mathbf{Z}_{\text{L2}}$ are included. The transfer functions for the load and the inductor impedance are derived similarly as the unterminated model. Appendix A shows the state-space coefficient matrices that are used to solve as the admittance matrix of the grid-side inductor. The inverse of the admittance matrix is $\mathbf{Z}_{\text{L2}}$. Appendix B shows the coefficient matrices for the RLC-load. Equations for the small-signal output voltage $\hat{v}_{\text{o}}$ are written for the both circuits in Figs. 4(a) and 4(b). The equations are shown in (19) and (20), respectively.

$$\hat{\mathbf{v}}_{\text{o}} = \mathbf{G}_{\text{io}}\hat{v}_{\text{in}} - \mathbf{Z}_{\text{o}}\hat{\mathbf{i}}_{\text{o}} + \mathbf{G}_{\text{co}}\hat{\mathbf{d}} \quad (19)$$

$$\hat{\mathbf{v}}_{\text{o}} = \mathbf{Z}_{\text{L2}}\hat{\mathbf{i}}_{\text{o}} + \mathbf{Z}_{\text{load}}\hat{\mathbf{i}}_{\text{o}} - \mathbf{Z}_{\text{load}}\hat{\mathbf{j}}_{\text{o}} \quad (20)$$

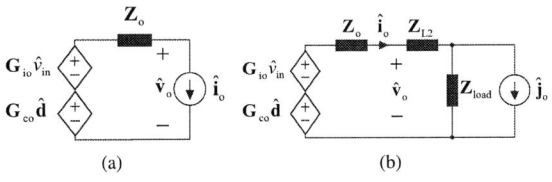

Figure 4. a) Output dynamics and b) load-affected output dynamics.

Small-signal output current vector $\hat{\mathbf{i}}_o$ is solved from (20). The solution is shown in (21).

$$\hat{\mathbf{i}}_o =(\mathbf{Z}_{L2} + \mathbf{Z}_{\text{load}})^{-1}\hat{\mathbf{v}}_o \\ + (\mathbf{Z}_{L2} + \mathbf{Z}_{\text{load}})^{-1}\mathbf{Z}_{\text{load}}\hat{\mathbf{j}}_o \tag{21}$$

Eq. (22) shows the result when $\hat{\mathbf{i}}_o$ in (21) is substituted to (19). The $(\mathbf{I} + \mathbf{Z}_o(\mathbf{Z}_{L2} + \mathbf{Z}_{\text{load}})^{-1})^{-1}$ is the common factor in all of the equations. The load-affected control-to-output transfer functions are collected from the bottom row of the matrix in (27). The transfer functions from $\hat{\mathbf{j}}_o$ to $\hat{\mathbf{v}}_o$ could be also manipulated to (23). This format shows that the small-signal current $\hat{\mathbf{i}}_o$ is solved analogously to circuit theory by dividing $\hat{\mathbf{j}}_o$ according to the impedances and the multiplying by $-\mathbf{Z}_o$ to solve the output voltage $\hat{\mathbf{v}}_o$.

$$\hat{\mathbf{v}}_o =(\mathbf{I} + \mathbf{Z}_o(\mathbf{Z}_{L2} + \mathbf{Z}_{\text{load}})^{-1})^{-1} \\ (\mathbf{G}_{io}\hat{v}_{in} - \mathbf{Z}_o(\mathbf{Z}_{L2} + \mathbf{Z}_{\text{load}})^{-1}\mathbf{Z}_{\text{load}}\hat{\mathbf{j}}_o \\ + \mathbf{G}_{co}\hat{\mathbf{d}}) \tag{22}$$

$$\hat{\mathbf{v}}_o = -(\mathbf{Z}_o + \mathbf{Z}_{L2} + \mathbf{Z}_{\text{load}})^{-1}\mathbf{Z}_{\text{load}}\mathbf{Z}_o\hat{\mathbf{j}}_o \tag{23}$$

The control-to-output voltage transfer function $\mathbf{G}_{co}^{L}$ can be solved also directly from the load affected output dynamics diagram in Fig. 4(b). The load affected circuit can be understood as a voltage divider, which divides the small-signal voltage caused by $\mathbf{G}_{co}$ or $\mathbf{G}_{io}$ over the impedances $\mathbf{Z}_o$, $\mathbf{Z}_{L2}$ and $\mathbf{Z}_{\text{load}}$. A very similar equations has been analyzed in [19], [20]. However, in [19], [20] the equations are derived in the case of an arbitrary voltage source as in Fig. 3 – not in the case of input-output dynamics of the converter.

The remaining load-affected transfer functions in (27) are solved by substituting $\hat{\mathbf{i}}_o$ in (17) by (21) as shown in (24) and then substituting $\hat{\mathbf{v}}_o$ by (22). Solving for $\hat{i}_{in}$ and $\hat{\mathbf{i}}_L$ as a function of $\hat{v}_{in}$, $\hat{\mathbf{j}}_o$ and $\hat{\mathbf{d}}$ gives the load affected transfer functions. Equation (25) shows the result in the case of inductor current. Load-affected output transfer functions $\mathbf{G}_{io}^{L}$, $\mathbf{G}_o^{L}$ and $\mathbf{G}_{co}^{L}$ are used for brevity in (25) instead of using the expression in (22). The load affected input current dynamics (26) can be solved similarly as the load affected inductor current dynamics.

$$\hat{\mathbf{i}}_L =\mathbf{G}_{iL}\hat{v}_{in} + \mathbf{G}_{oL}((\mathbf{Z}_{L2} + \mathbf{Z}_{\text{load}})^{-1}\hat{\mathbf{v}}_o \\ + (\mathbf{Z}_{L2} + \mathbf{Z}_{\text{load}})^{-1}\mathbf{Z}_{\text{load}}\hat{\mathbf{j}}_o) + \mathbf{G}_{cL}\hat{\mathbf{d}} \tag{24}$$

$$\hat{\mathbf{i}}_L =(\mathbf{G}_{iL} + \mathbf{G}_{oL}(\mathbf{Z}_{L2} + \mathbf{Z}_{\text{load}})^{-1}\mathbf{G}_{io}^{L})\hat{v}_{in} \\ + (\mathbf{G}_{cL} + \mathbf{G}_{oL}(\mathbf{Z}_{L2} + \mathbf{Z}_{\text{load}})^{-1}\mathbf{G}_{co}^{L})\hat{\mathbf{d}} \\ + \mathbf{G}_{oL}((\mathbf{Z}_{L2} + \mathbf{Z}_{\text{load}})^{-1}\mathbf{Z}_{\text{load}} \\ - ((\mathbf{Z}_{L2} + \mathbf{Z}_{\text{load}})^{-1}(-\mathbf{G}_o^{L})))\hat{\mathbf{j}}_o \tag{25}$$

$$\hat{i}_{in} =(Y_{in} + \mathbf{T}_{oi}(\mathbf{Z}_{L2} + \mathbf{Z}_{\text{load}})^{-1}\mathbf{G}_{io}^{L})\hat{v}_{in} \\ + (\mathbf{G}_{ci} + \mathbf{T}_{oi}(\mathbf{Z}_{L2} + \mathbf{Z}_{\text{load}})^{-1}\mathbf{G}_{co}^{L}))\hat{\mathbf{d}} \\ + \mathbf{T}_{oi}((\mathbf{Z}_{L2} + \mathbf{Z}_{\text{load}})^{-1}\mathbf{Z}_{\text{load}}) \\ - ((\mathbf{Z}_{L2} + \mathbf{Z}_{\text{load}})^{-1}(-\mathbf{G}_o^{L})))\hat{\mathbf{j}}_o \tag{26}$$

$$\begin{bmatrix} \hat{i}_{in} \\ \hat{\mathbf{i}}_L \\ \hat{\mathbf{v}}_o \end{bmatrix} = \begin{bmatrix} Y_{in}{}^{L} & \mathbf{T}_{oi}^{L} & \mathbf{G}_{ci}^{L} \\ \mathbf{G}_{iL}^{L} & \mathbf{G}_{oL}^{L} & \mathbf{G}_{cL}^{L} \\ \mathbf{G}_{io}^{L} & -\mathbf{G}_o^{L} & \mathbf{G}_{co}^{L} \end{bmatrix} \begin{bmatrix} \hat{v}_{in} \\ \hat{\mathbf{j}}_o \\ \hat{\mathbf{d}} \end{bmatrix} \tag{27}$$

The resulting load-affected dynamics can be expressed as shown in (27), where superscript L denotes that the transfer functions are affected by the load impedance. It should be noted that $\hat{\mathbf{j}}_o$ replaces $\hat{\mathbf{i}}_o$ as an input variable as it can be seen from Fig. 4(b). Since $\hat{\mathbf{j}}_o$ and $\hat{\mathbf{v}}_o$ are not defined at an interface according to the definition of an impedance, a transfer function matrix $\mathbf{G}_o^{L}$ is used instead of an impedance matrix.

References [21] and [22] have pointed out that the impedances of interconnected three-phase systems should be shifted to a global reference frame to enable impedance-based stability analysis. However, the load impedance matrix of the pure resistive load and the RLC-load analyzed in this paper are symmetric, which means that no impedance shifting is required.

IV. FREQUENCY RESPONSE ANALYSIS

The parameters and the operating point values of the grid-forming inverter are shown in Tables I and II. The resistive load in Fig. 2 is considered first. Fig. 6 shows both the frequency response given by analytical model G_{cod}^{L} and the frequency response measured from a hardware-in-the-loop simulator. G_{cod}^{L} is the transfer function from the duty ratio d-component to the output voltage d-component, which has major importance for control design. The resistive load R_{load} is chosen according to (28) so that nominal operation point is maintained. Eq. (29) shows, how the load impedance matrix is defined in the case of the resistive load.

$$R_{\text{load}} = \frac{V_{od}}{I_{od}} - r_{L2} \tag{28}$$

$$\mathbf{Z}_{\text{load}} = \begin{bmatrix} R_{\text{load}} & 0 \\ 0 & R_{\text{load}} \end{bmatrix} \tag{29}$$

Table I. INVERTER PARAMETER VALUES

Parameter	Value	Parameter	Value
L	1.4 mH	r_L	25 mΩ
L_2	0.47 mH	r_{L2}	22 mΩ
C_f	10 μF	R_d	1.96 Ω
f_s	10 kHz	r_{sw}	10 mΩ
ω_s	2π60 Hz		

Table II. OPERATING POINT VALUES

Parameter	Value	Parameter	Value
V_{od}	169.7 V	V_{oq}	0.0000 V
I_{od}	19.64 A	I_{oq}	0.0000 A
I_{Ld}	19.65 A	I_{Lq}	0.6397 A
V_{Cfd}	169.7 V	V_{Cfq}	-1.254 V
V_{in}	416.0 V	I_{in}	16.93 A
D_{d}	0.4088	D_{q}	0.0250

The correlation between the measured and predicted based frequency responses in Fig. 6 confirms that the proposed model is correct. The mainly resistive load damps the resonance caused by the LC-filter. The instruments utilized in the measurements were Typhoon HIL -real time simulator, Boombox control platform from Imperix and Venable frequency response analyzer. A photograph of the HIL simulation setup is shown in Fig. 5. An oscilloscope was used additionally.

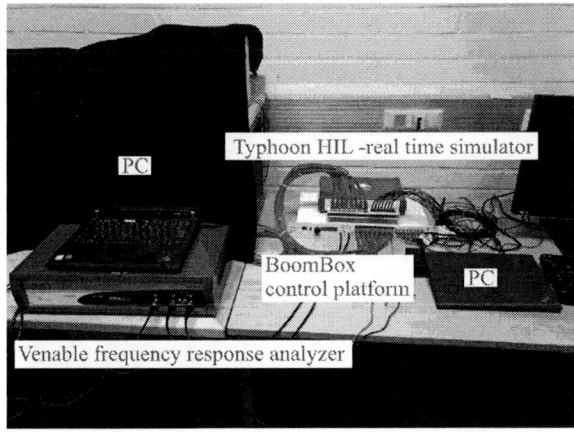

Figure 5. Real-time simulation setup: PC, Venable frequency response analyzer, Boombox control platform, and Typhoon HIL -real time simulator.

Taking advantage of the steps to derive the load-affected transfer function, the load effect can be also removed from the frequency response. Eq. (30) shows, how $\mathbf{G}_{\text{co}}$ can be calculated if the load-affected transfer function matrix, $\mathbf{G}_{\text{co}}^{\text{L}}$ is known from measurements, i.e., the unterminated dynamic model can be solved even if the load is not an ideal current sink.

$$\mathbf{G}_{\text{co}} = (\mathbf{I} + \mathbf{Z}_{\text{o}}(\mathbf{Z}_{\text{L2}} + \mathbf{Z}_{\text{load}})^{-1})\mathbf{G}_{\text{co}}^{\text{L}} \qquad (30)$$

Fig. 6 shows also a comparison between the derived G_{cod} in (17) and the transfer function calculated according to (30). G_{cod} corresponds to the situation of Fig. 1, where the load is an ideal current sink. Thus, the ideal transfer functions can be illustrated even though, the converter is affected by the load impedance. Assuming that the impedance matrices $\mathbf{Z}_{\text{o}}$, $\mathbf{Z}_{\text{L2}}$ and $\mathbf{Z}_{\text{Load}}$ are known.

A cascaded controller is commonly used to control the output voltage of the grid-forming inverter [5], [23], [24]. The controller consists of the inner inductor current loop and outer output voltage loop. The controller is tuned according to the control-to-inductor-current and control-to-output voltage transfer functions affected by the R-load. Fig. 7 shows the measured and model-based frequency response of the $G_{\text{cLd}}^{\text{L}}$. The unterminated control-to-current d-component G_{cLd} is also shown in Fig. 7. The resistive load clearly damps the resonance and increases the low-frequency gain, which greatly simplifies the tuning of the current controller. The current controller G_{cc} is a PI-controller. Consisting of an integrator, a zero at 1 kHz and a gain of 36.8 dB.

Fig. 8 shows the frequency response of the full-order current loop gain $L_{\text{outCd}}^{\text{FO}}$. The phase margin is 65.4 °at 551 Hz. The gain margin is 8.51 dB. The full-order current loop gain includes the cross-coupling between d and q-components. The loop gain is given in (31) and it has been derived in [4]. The delay caused by sampling and PWM is 1.5 $1/f_{\text{s}}$ and it is modeled by a third order Padé approximation [16]. The delay transfer function is omitted for brevity from (31), but it is shown in (32) and in completed block diagram of the system in Fig. 11.

$$L_{\text{outCd}}^{\text{FO}} = G_{\text{cLd}}^{\text{L}}G_{\text{cc}} - \frac{G_{\text{cLqd}}^{\text{L}}G_{\text{cLqd}}^{\text{L}}}{1 + G_{\text{cLq}}^{\text{L}}G_{\text{cc}}}G_{\text{cc}}G_{\text{cc}} \qquad (31)$$

The matrix current loop gain is shown in (32).

$$\mathbf{L}_{\text{outC}} = \mathbf{G}_{\text{cL}}\mathbf{G}_{\text{del}}\mathbf{G}_{\text{cc}}\mathbf{G}_{\text{seC}} \qquad (32)$$

The current loop is closed in (33) and (34) shows, how the inductor current reference-to-output voltage transfer function $\mathbf{G}_{\text{co}}^{\text{sec}}$ is calculated. Superscript 'sec' denotes secondary and means that the secondary control loop (i.e. current loop) is closed. $G_{\text{cod}}^{\text{sec}}$ is used to tune the voltage controller. Fig. 11 shows the control block diagram of the complete system. The block diagram can be used to calculate also other closed-loop transfer functions and loop gains.

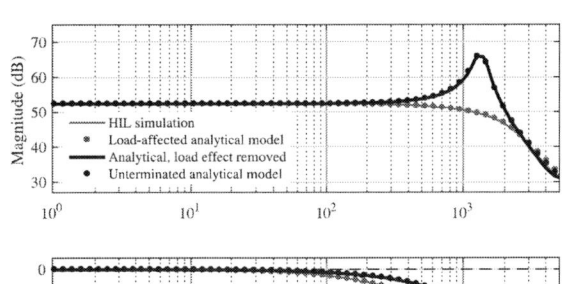

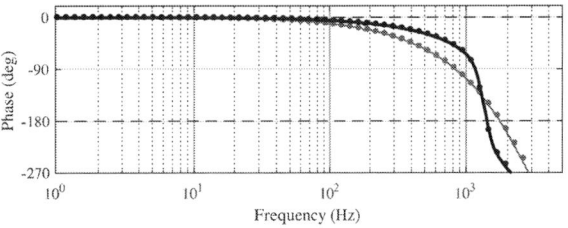

Figure 6. Measured and model-based frequency responses of $G_{\text{cod}}^{\text{L}}$ (resistive load).

The 2018 International Power Electronics Conference

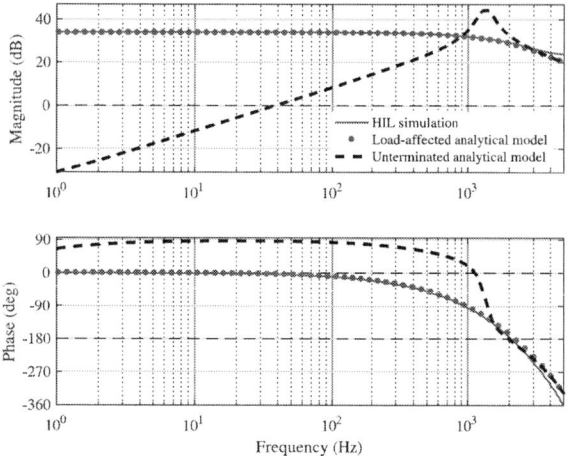

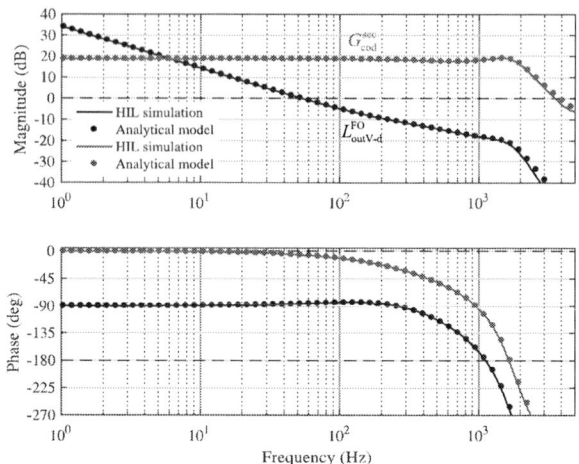

Figure 7. Bode plot of HIL-simulated and derived G_{cLd}^L for the R-load and the unterminated G_{cLd}.

Figure 9. Bode plot of simulated and derived voltage controller loop gain L_{outVd}^{FO} and the current loop affected control-to-output dynamics G_{cod}^{Lsec}.

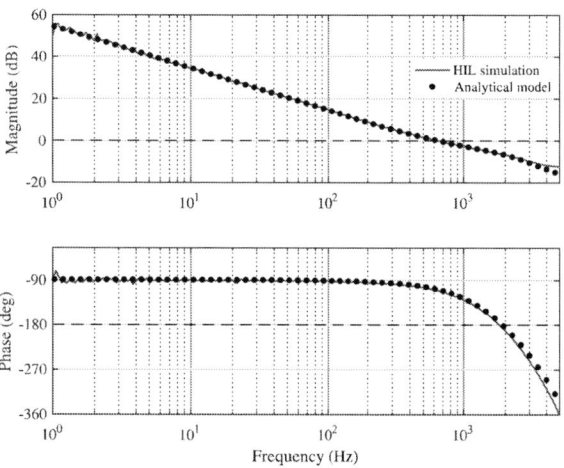

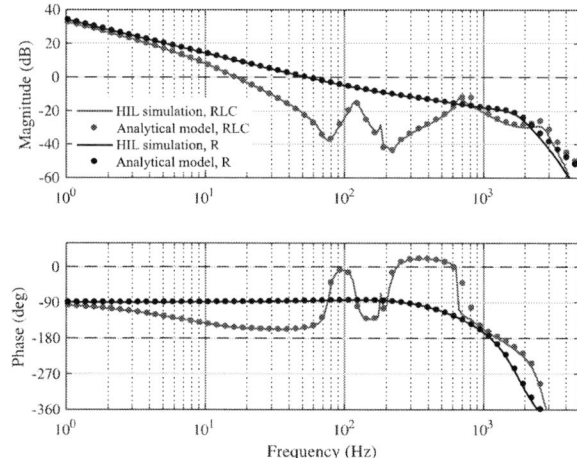

Figure 8. Bode plot of measured and derived current loop gain L_{outCd}^{FO} for the R-load.

Figure 10. Bode plot of measured and derived voltage controller loop gain L_{outVd}^{FO} for the R-load and for the RLC-load.

$$\mathbf{G}_{cL}^{sec} = (\mathbf{I} + \mathbf{L}_{outC})^{-1} \mathbf{G}_{cL}^{L} \mathbf{G}_{del} \mathbf{G}_{cc} \tag{33}$$

$$\mathbf{G}_{co}^{sec} = \mathbf{G}_{co}^{L} \mathbf{G}_{del} \mathbf{G}_{cc} - \mathbf{G}_{co}^{L} \mathbf{G}_{cL}^{-1} \mathbf{L}_{outC} \mathbf{G}_{cL}^{sec} \tag{34}$$

$$L_{outV-d}^{FO} = G_{cod}^{L} \mathbf{G}_{vc} - \frac{G_{coqd}^{L} G_{coqd}^{L}}{1 + G_{coq}^{L} \mathbf{G}_{vc}} \mathbf{G}_{vc} \mathbf{G}_{vc} \tag{35}$$

Fig. 9 shows the simulated and the frequency response of the analytic voltage loop gain L_{outVd}^{FO} (35) and the model of $\mathbf{G}_{cod}^{sec}$. The crossover frequency the voltage loop is 53.9 Hz and the phase margin is 93.5 °when the load is pure resistance. The voltage controller $\mathbf{G}_{vc}$ consists of an integrator, a zero at 200 Hz, a pole at 600 Hz and a gain of 31.6 dB.

The cascaded controller is kept unchanged, but the

load is changed to the RLC-load that is similar to the RLC-load used in [13]. L_L and C_L are 4.584 mH and 1.535 mF, respectively. The resonance frequency is at around 60 Hz as in [13]. 30 mΩ resistances r_{LL} and r_{CL} are connected in series with the parallel capacitance and inductance, respectively. R_L of the parallel load equals R_{load}, the load of the first case. Figure 10 shows also the frequency response of the voltage loop in this case. It can be seen that the crossover frequency is 16.5 Hz and the phase margin is reduced to 26.7 °. The low phase margin indicates that there will be oscillation in the step response of the system.

A step response comparison by using the R and RLC-loads is done. The system is simulated in Typhoon HIL as in the case of frequency-domain measurement. An oscilloscope is connected to the analog outputs of Imperix Boombox to analyze the response in detail in dq-domain. Fig. 12 shows the output voltage response to a step

968

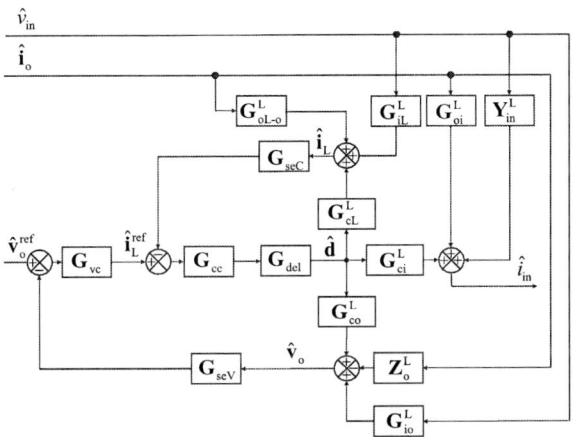

Figure 11. Control block diagram of the closed-loop system.

change in the voltage reference d-component as the R-load is used. The step is from 155 V to the nominal amplitude 169,7 V. There is no overshoot or oscillation in the response. The response with the RLC-load as the controller remains unchanged is also shown in Fig. 12. A significant overshoot and decaying oscillation is present in the response.

The model is used to retune the controllers so that a proper step-response is achieved with the RLC-load. The current controller pole location is changed to 100 Hz and the new gain is 24.8 dB. The voltage controller zero is moved to 5 Hz, two poles are located at 60 Hz and the new gain is 24.1 dB. Fig. 13 shows the predicted and HIL-simulated voltage loop gains. The phase margin is 58.2 ° at 20.6 Hz. The gain margin is 14 dB at 129 Hz. The step response in Fig. 14 is good, as the higher phase margin than with previous controller tuning implies.

The previous analysis shows that the proposed model can be used to analyze the effect of different loads on the control dynamics. Frequency responses of the load-affected transfer functions can be used to predict the time-domain behavior and to design the controllers according

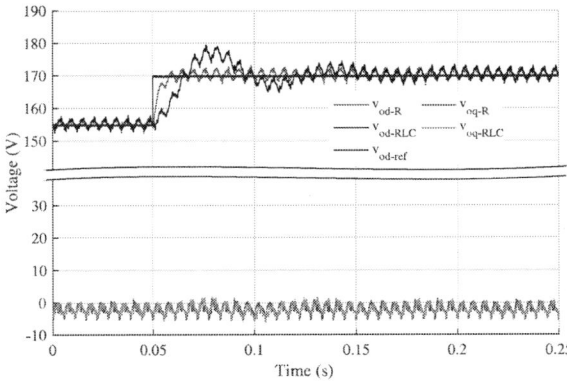

Figure 12. Typhoon HIL simulation of output voltage d and q-components step response to a reference step for the R-load and for the RLC-load.

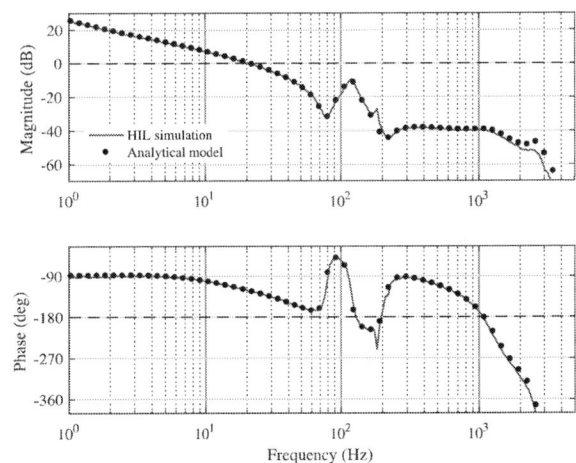

Figure 13. Bode plot of simulated and predicted voltage loop gain with the controller that is tuned for the RLC-load.

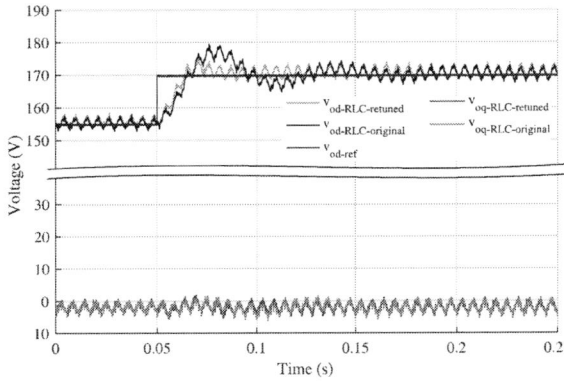

Figure 14. Typhoon HIL simulation of output voltage d and q-component step responses to a reference step with the RLC-load with the original controller and the controller retuned for RLC-load.

to a specific load or a worst-case scenario.

V. CONCLUSION

This paper proposes a method to model unterminated dynamics of a grid-forming inverter. The effect of a non-ideal load is included in the model by calculating, how the load impedance affects the output dynamics. Furthermore, the unterminated model of the grid-forming inverter includes the output impedance that is required to calculate the load-affected model and an important tool in the stability analysis of interconnected systems.

The HIL measurements provided in this paper confirm that the frequency response analysis is a powerful tool for predicting the time-domain response of the grid-forming inverter under distinct loads. One possible application of the proposed modeling technique is to tune the controller according to a specific load so that a desired time-domain response is achieved. The model can be also used to examine worst-case load conditions. The load effect can be also removed from the measured load affected frequency response and the unterminated model can be verified. Future work will concentrate on the on the load-affected dynamics in the case of an active load, such as an active rectifier.

APPENDIX A

Eq. (36) shows the grid-side inductor admittance state-space coefficient matrices.

$$\mathbf{A}_{\text{L2}} = \begin{bmatrix} \frac{-r_{\text{L2}}}{L_2} & \omega_{\text{s}} \\ -\omega_{\text{s}} & \frac{-r_{\text{L2}}}{L_2} \end{bmatrix} \quad \mathbf{B}_{\text{L2}} = \begin{bmatrix} \frac{1}{L_2} & 0 \\ 0 & \frac{1}{L_2} \end{bmatrix}$$

$$\mathbf{C}_{\text{L2}} = \begin{bmatrix} 1 & 0 \\ 0 & 1 \end{bmatrix} \quad \mathbf{D}_{\text{L2}} = \begin{bmatrix} 0 & 0 \\ 0 & 0 \end{bmatrix}$$

(36)

APPENDIX B

Eq. (37) shows the RLC-load admittance state-space coefficient matrices.

$$\mathbf{A}_{\text{RLC}} = \begin{bmatrix} \frac{-r_{\text{LL}}}{L_{\text{L}}} & \omega_{\text{s}} & 0 & 0 \\ -\omega_{\text{s}} & \frac{-r_{\text{LL}}}{L_{\text{L}}} & 0 & 0 \\ 0 & 0 & \frac{-1}{C_{\text{L}}r_{\text{CL}}} & \omega_{\text{s}} \\ 0 & 0 & -\omega_{\text{s}} & \frac{-1}{C_{\text{L}}r_{\text{CL}}} \end{bmatrix}$$

$$\mathbf{B}_{\text{RLC}} = \begin{bmatrix} \frac{1}{L_{\text{L}}} & 0 \\ 0 & \frac{1}{L_{\text{L}}} \\ \frac{1}{C_{\text{L}}} & 0 \\ 0 & \frac{1}{C_{\text{L}}} \end{bmatrix} \quad \mathbf{C}_{\text{RLC}} = \begin{bmatrix} 1 & 0 & \frac{-1}{r_{\text{CL}}} & 0 \\ 0 & 1 & 0 & \frac{-1}{r_{\text{CL}}} \end{bmatrix}$$

$$\mathbf{D}_{\text{RLC}} = \begin{bmatrix} \frac{1}{R_{\text{L}}} + \frac{1}{r_{\text{CL}}} & 0 \\ 0 & \frac{1}{R_{\text{L}}} + \frac{1}{r_{\text{CL}}} \end{bmatrix}$$

(37)

REFERENCES

[1] I. J. o. Balaguer, "Control for grid-connected and intentional islanding operations of distributed power generation," *IEEE Trans. Ind. Electron.*, vol. 58, no. 1, pp. 147–157, 2011.

[2] M. Rasheduzzaman, J. A. Mueller, and J. W. Kimball, "An accurate small-signal model of inverter- dominated islanded microgrids using dq reference frame," *IEEE Trans. Emerg. Sel. Topics Power Electron.*, vol. 2, no. 4, pp. 1070–1080, 2014.

[3] J. Rocabert et al., "Control of power converters in AC microgrids," *IEEE Trans. Power Electron.*, vol. 27, no. 11, pp. 4734–4749, nov 2012.

[4] T. Suntio, T. Messo, and J. Puukko, *Power Electronic Converters: Dynamics and Control in Conventional and Renewable Energy Applications*. Wiley VCH, 2017.

[5] B. Wen et al., "Modeling the output impedance of three-phase uninterruptible power supply in D-Q frame," in *2014 IEEE Energy Conversion Congress and Exposition (ECCE)*. IEEE, 2014, pp. 163–169.

[6] S. Lissandron et al., "Experimental validation for impedance-based small-signal stability analysis of single-phase interconnected power systems with grid-feeding inverters," *IEEE Journal of Emerging and Selected Topics in Power Electronics*, vol. 4, no. 1, pp. 103–115, 2016.

[7] T. Roinila, T. Messo, and E. Santi, "Mimo-identification techniques for rapid impedance-based stability assessment of three phase systems in DQ domain," *IEEE Trans. Power Electron.*, vol. 33, no. 5, pp. 1–1, 2017.

[8] B. Wen et al., "AC stability analysis and dq frame impedance specifications in power-electronics-based distributed power systems," *IEEE Trans. Emerg. Sel. Topics Power Electron.*, vol. 5, no. 4, pp. 1455–1465, dec 2017.

[9] B. Wen and othe, "Impedance-based analysis of grid-synchronization stability for three-phase paralleled converters," *IEEE Trans. Power Electron.*, vol. 31, no. 1, pp. 26–38, 2016.

[10] M. Ramezani, S. Li, and S. Golestan, "Analysis and controller design for stand-alone vsis in synchronous reference frame," *IET Power Electron.*, vol. 10, no. 9, pp. 1003–1012, 2017.

[11] J. M. Guerrero et al., "Output impedance design of parallel-connected UPS inverters with wireless load-sharing control," *IEEE Trans. Ind. Electron.*, vol. 52, no. 4, pp. 1126–1135, 2005.

[12] A. Yazdani, "Control of an islanded distributed energy resource unit with load compensating feed-forward," in *2008 IEEE Power and Energy Society General Meeting - Conversion and Delivery of Electrical Energy in the 21st Century*. IEEE, 2008, pp. 1–7.

[13] R. J. Vijayan, S. Ch, and R. Roy, "Dynamic modeling of microgrid for grid connected and intentional islanding operation," in *2012 International Conference on Advances in Power Conversion and Energy Technologies (APCET)*. IEEE, aug 2012, pp. 1–6.

[14] T. Vandoorn et al., "Theoretical analysis and experimental validation of single-phase direct versus cascade voltage control in islanded microgrids," *IEEE Trans. Ind. Electron.*, vol. 60, no. 2, pp. 789–798, 2013.

[15] R. D. Middlebrook, "Small-signal modeling of pulse-width modulated switched-mode power converters," *Proc. IEEE*, vol. 76, no. 4, pp. 343–354, 1988.

[16] A. Aapro et al., "Effect of active damping on output impedance of three-phase grid-connected converter," *IEEE Trans. Ind. Electron.*, vol. PP, no. 99, pp. 1–1, 2017.

[17] J. Puukko and T. Suntio, "Modelling the effect of non-ideal load in three-phase converter dynamics," *Electronics Letters*, vol. 48, no. 7, p. 402, 2012.

[18] T. Suntio, *Dynamic Profile of Switched-Mode Converter: Modeling, Analysis and Control*. Weinheim, Germany: Wiley, 2009.

[19] B. Wen and other, "D-Q impedance specification for balanced three-phase AC distributed power system," in *2015 IEEE Applied Power Electronics Conference and Exposition (APEC)*. IEEE, 2015, pp. 2757–2771.

[20] R. Burgos et al., "On the ac stability of high power factor three-phase rectifiers," in *2010 IEEE Energy Conversion Congress and Exposition*. IEEE, 2010, pp. 2047–2054.

[21] A. Rygg et al., "On the equivalence and impact on stability of impedance modelling of power electronic converters in different domains," *IEEE Trans. Emerg. Sel. Topics Power Electron.*, no. 4, pp. 1–1, 2017.

[22] J. Chen and J. Chen, "Stability analysis and parameters optimization of islanded microgrid with both ideal and dynamic constant power loads," *IEEE Trans. Ind. Electron.*, vol. 65, no. 4, pp. 1–1, 2017.

[23] Poh Chiang Loh et al., "A comparative analysis of multiloop voltage regulation strategies for single and three-phase UPS systems," *IEEE Trans. Power Electron.*, vol. 18, no. 5, pp. 1176–1185, 2003.

[24] P. Loh and D. Holmes, "Analysis of multiloop control strategies for LC/CL/LCL-filtered voltage-source and current-source inverters," *IEEE Trans. Ind. Appl.*, vol. 41, no. 2, pp. 644–654, 2005.

The 2018 International Power Electronics Conference

A New Control Method for Triple-Active Bridge Converter with Feed Forward Control

Takanobu OHNO[1]*, Nobukazu HOSHI
Electrical Engineering, Tokyo University of Science, Chiba, Japan
*E-mail: aglioolio26@gmail.com

Abstract—**Triple Active Bridge (TAB) DC/DC converter contributes to the realization of micro grid by controlling the power flow among distributed power systems and load. The control system of the TAB converter has characteristics of a 2-input 2-output MIMO structure and strong nonlinearity. There is a trade-off problem between the accuracy of the decoupler and the simplicity of the controller structure. Hence, this paper proposes the new control method that enables the decoupler to operate in a wide range while keeping the simplicity of the controller structure. Also, this paper describes experimental results of 1 kW prototype TAB converter in order to verify the validity of the theoretical considerations. As a result, it was confirmed that the interference is suppressed by the proposed controller.**

Keywords—Bidirectional converter; decoupling control; feed forward control; triple active bridge;

I. INTRODUCTION

In recent years, the demands for distributed power generation systems have increased due to environmental protection. In the systems, it is required to make an efficient use of various power sources such as lithium ion batteries, solar cells, and other renewable energy power supplies. From the viewpoint of such energy resource allocation, the management based on the smart grid concept has been proposed [1][2]. In such situations, equipment that can control power flow with high performance is required. Triple Active Bridge (TAB) converter shown in

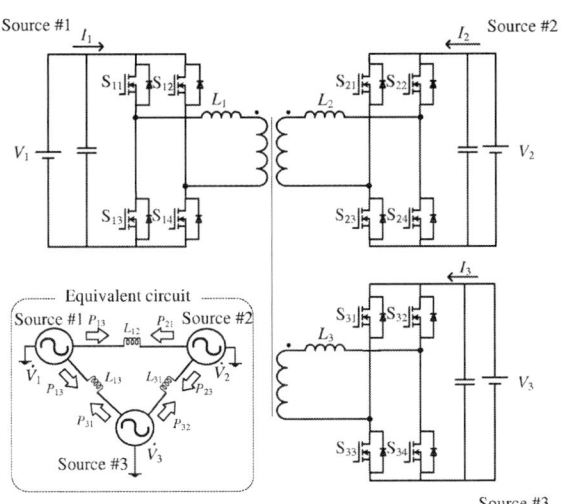

Fig. 1. A circuit configuration of triple active bridge DC/DC converter.

Fig. 1 has ideal properties which are able to satisfy these requirements such as bidirectionality, electrical insulation, and wide voltage range by transformer winding ratio[3]-[9]. However, TAB converter has difficult property regards to controller design including the dependencies of the system parameters for operating point and interaction among the operation variables which are caused by Multi-Input-Multi-Output (MIMO) system[10]. In a previous research, a control method using a diagonalization decoupling compensator derived after a linear modeling of TAB converter has been studied[11]. However, its operation range is limited because the decoupling compensator depends on the operating point from its nonlinearity. This paper presents a control method that satisfied both of the decoupling capability in a wide operation range and the simplicity of the controller configuration. This paper is organized as follows: the next section describes a circuit configuration and the modeling of TAB; control theory for the converter discusses in section III; followed by the controller design in section IV; section V are prototype implementations and experimental results; and at last Section VI concludes the paper.

II. THE MODELING AND THE CIRCUIT CONFIGURATION OF TAB CONVERTER

A circuit configuration and the equivalent circuit of TAB converter are shown in Fig. 1. TAB converter is composed of three full-bridge inverters that have independently DC sources V_1, V_2 and V_3, a 3-winding transformer, and external inductors L_1, L_2 and L_3. Each parameter in the equivalent circuit is calculated as follows by approximating the square wave output of the inverter to the fundamental wave.

$$\dot{V}_i = \frac{4V_i}{\pi} \angle \phi_i \ (i = 1, 2, 3) \tag{1}$$

$$L_{ij} = \frac{L_i L_j + L_j L_k + L_k L_i}{L_k} \ (i, j, k = 1, 2, 3) \tag{2}$$

$$P_{ij} = \frac{4V_i \cos \delta_i V_j \cos \delta_j}{\pi^3 f_{sw} L_{ij}} \sin (\phi_i - \phi_j) \tag{3}$$

Where $(\phi_i - \phi_j)$ represents the phase shift amount of the control signals between i-th and j-th source, f_{sw} represents the switching frequency, and δ_i represents the duty cycle of i-th inverter. From equations (1)~(3), the transmission power to the circuit of each port is expressed

971

by the following equation with the phase of source #1 as a reference i.e. $\phi_1 = 0$.

$$P_2 = K_{12}\sin\phi_2 + K_{23}\sin(\phi_2 - \phi_3) \quad (4)$$
$$P_3 = K_{13}\sin\phi_3 + K_{23}\sin(\phi_3 - \phi_2) \quad (5)$$
$$0 = P_1 + P_2 + P_3 \quad (6)$$

Where K_{ij} is described by the following equation.

$$K_{ij} = \frac{4V_iV_j\cos\delta_i\cos\delta_j}{\pi^3 f_{sw}L_{ij}} \quad (i,j = 1,2,3) \quad (7)$$

Under the condition that the bus voltages do not change locally, equations (4)~(6) can be transformed as follows.

$$I_2 = \frac{K_{12}}{V_2}\sin\phi_2 + \frac{K_{23}}{V_2}\sin(\phi_2 - \phi_3) \quad (8)$$
$$I_3 = \frac{K_{13}}{V_3}\sin\phi_3 - \frac{K_{23}}{V_3}\sin(\phi_2 - \phi_3) \quad (9)$$
$$0 = I_1V_1 + I_2V_2 + I_3V_3 \quad (10)$$

TAB converter is a 2-input 2-output system from the fact that each port currents I_2, I_3 are determined by the phase shift angle ϕ_2, ϕ_3 in (8) and (9). Also it is indicated that the system has nonlinearity as a consequence of including the sine function. In order to design the controller, linearization of the system described by the following equation is necessary.

$$\begin{bmatrix} I_2 \\ I_3 \end{bmatrix} = \begin{bmatrix} G_{11} & G_{12} \\ G_{21} & G_{22} \end{bmatrix} \begin{bmatrix} \phi_2 \\ \phi_3 \end{bmatrix} \quad (11)$$

Where $G_{ij}(i,j = 1,2)$ is the coefficient of Taylar expansion near the operating point, and is expressed as the following equations.

$$G_{11} = K_{12}\cos\phi_2 + K_{23}\cos(\phi_2 - \phi_3) \quad (12)$$
$$G_{12} = -K_{23}\cos(\phi_2 - \phi_3) \quad (13)$$
$$G_{21} = -K_{23}\cos(\phi_2 - \phi_3) \quad (14)$$
$$G_{22} = K_{13}\cos\phi_3 + K_{23}\cos(\phi_2 - \phi_3) \quad (15)$$

Therefore, the matrix $\mathbf{G}$ varies depending on the operating points (ϕ_2, ϕ_3) as a consequence of nonlinearity.

III. CONTROL THEORY IN TAB CONVERTER

In this section, the stability and the decoupling method with compensator are described for controller design.

A. Phase restriction for stability

The combination of the operation and the output variables in the multi-loop control such as 2-input 2-output system is called 'pairing'; and the influence from the variables other than the pairing variable is called 'interference'; e.g. in (11), (I_2, ϕ_2) and (I_3, ϕ_3) are paired independently then G_{12} and G_{21} represent interference. It is known that the performance of pairing and the degree of the interference can be evaluated by Niederlinski Index (NI), which is expressed as the following equation[12][13].

$$\text{NI} = \frac{\det(\mathbf{G}(0))}{\prod_{i=1}^{n} G_{ii}(0)} \quad (16)$$

Where $\mathbf{G}(0)$ is a system in a steady state. On the assumption that each controller in the multi-loop system includes integrating operation and each control loops are stable when any other loops are opened, if and only if NI is positive then the multi loop system is stable. Furthermore, as the NI is closer to 1, the response of the paring variable is better and the interference is smaller; on the other hand, the system becomes uncontrollable when the index is less than 0. The NI of the TAB converter is expressed as the following equation.

$$\frac{1}{\text{NI}} = 1 + \frac{1}{\alpha\beta\frac{\cos\phi_2}{\cos(\phi_2-\phi_3)} + \alpha\frac{\cos\phi_2}{\cos(\phi_2-\phi_3)} + \beta\frac{\cos\phi_3}{\cos(\phi_2-\phi_3)}} \quad (17)$$

Where α and β are expressed as the following equations.

$$\alpha = \frac{K_{12}}{K_{23}}, \quad \beta = \frac{K_{13}}{K_{23}} \quad (18)$$

Fig. 2 shows the region where NI is negative under the assumption that all the full-bridge inverters have same variables, i.e. $V_1 = V_2 = V_3$ and $\delta_1 = \delta_2 = \delta_3$. To maintain a stable operation range with a margin, this paper proposes the limiter by the following equation.

$$|\phi_2 - \phi_3| < \frac{\pi}{2} \quad (19)$$

B. Decoupling control

In order to deconstruct a multi variable control system into a series of independent single-loop subsystem, a novel decoupling control has been proposed. Fig. 3 shows the block diagram of a control system with a decoupling network. The pre-compensator $\mathbf{H}$ of the system $\mathbf{G}$ mathematically diagonalizes the system matrix and plays the role of decoupler. In the case of the 2-input 2-output

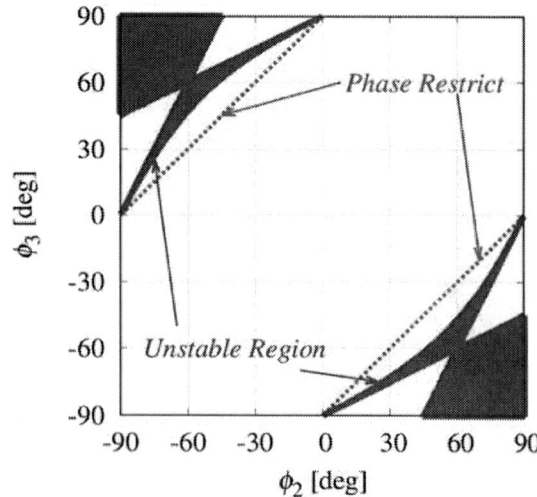

Fig. 2. Unstable operation range derived from (17) and phase restriction for safety operation range derived from (19).

system, $\mathbf{H}$ can be designed as

$$\mathbf{H} = \mathbf{G}^{-1} = \frac{1}{G_{11}G_{22} - G_{12}G_{21}} \begin{bmatrix} G_{22} & -G_{12} \\ -G_{21} & G_{11} \end{bmatrix}. \tag{20}$$

As described in section III, since $\mathbf{G}$ depends on the operating point, $\mathbf{H}$ also depends on the operating point. Hence, precalculation of the look up table is necessary to design the compensator.

IV. CONTROL STRATEGY

In this section, a controller that is capable of decoupling without precalculation over the entire operation range is proposed. The main idea of this method is that a feed forward controller suppresses the interference regarded as the disturbance on the closed loop, after whole system is divide into closed and open loop system. Firstly, one closed loop of the multi loop in TAB system is converted to an open loop. Control variable ϕ_3 and its output variable I_2 are expressed as follows by the formula transformation of (8) and (9).

$$I_2 = \frac{1}{V_2}\sqrt{(K_{12} + K_{23}\cos\phi_3)^2 + (K_{23}\cos\phi_3)^2}$$
$$\times \sin\left(\phi_2 - \arctan\left(\frac{K_{23}\cos\phi_3}{K_{12} + K_{23}\cos\phi_3}\right)\right) \tag{21}$$

$$\phi_3 = \arcsin\left(\frac{-V_3 I_3}{\sqrt{(K_{13} + K_{23}\cos\phi_2)^2 + (K_{23}\sin\phi_2)^2}}\right)$$
$$+ \arctan\left(\frac{K_{23}\sin\phi_2}{K_{13} + K_{23}\cos\phi_2}\right) \tag{22}$$

In that case, I_2-ϕ_2 system is treated as closed loop control and I_3-ϕ_3 system is treated as open loop control. From the difference of the response speed between the open loop and closed loop systems, the interference arises mainly at the closed loop system as a stepwise disturbance from the open loop. Therefore, arctangent function in sine function of (21) mainly induces interference. In Fig. 4, the interference block indicates the influence of I_2-ϕ_2 closed loop system from ϕ_3 as shown in (21); and the subsystem block indicates I_3-ϕ_3 open loop system as shown in (22). Feed forward block is expressed as the following equation.

$$y = \arctan\left(\frac{K_{23}\cos\phi_3}{K_{12} + K_{23}\cos\phi_3}\right) \tag{23}$$

Where y is the output of the feed forward block.

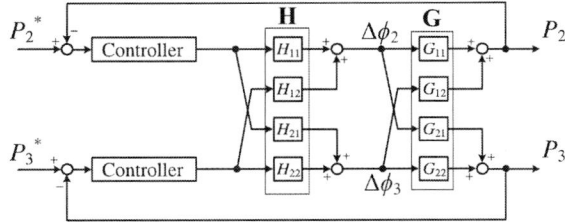

Fig. 3. Block diagram of a control system with a decoupling network.

V. EXPERIMENTAL RESULTS

This section describes the implementation of a prototype TAB converter and the experimental results.

A. Controller implementation

An appearance and the system diagram of prototype TAB converter are shown in Figs. 5 and 6 respectively. Whole system of prototype is comprised of an FPGA board as a controller, sensor circuits, triple active bridge, external inductors, and 3-winding transformer. Table I lists its specifications. The controller implemented using the method is described in section IV with FPGA and softcore CPU Nios-II. In order to avoid forming a closed loop by the feed forward and, to operate the feed forward at appropriate timing, delay of the control variables is required from the view point of digital control. Fig. 7 illustrates the block diagram of signal delay block and the selector which is switched in accordance with the condition for feed forward control. The feed forward control is operated if the deviation in feed back control is larger than threshold value which is determined by the magnitude of the current ripple in the steady state after estimating the feed forward amount by delaying the command signal. The phase restrictor in the final output is what was proposed in section III-A.

B. Experimental results in transient state.

Transient state waveforms in the prototype TAB converter were measured. Fig. 8 illustrates the experimental results of DC bus currents I_1, I_2 and I_3 when the command I_2^* was changed from 0 to 5 A with closed loop control. In the case that the open loop is affected by the change of the closed loop's command value, the constant value control by the open loop calculation is

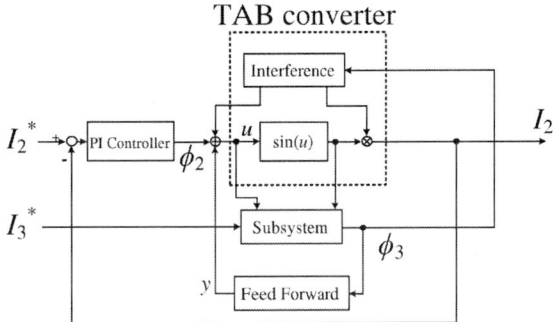

Fig. 4. Block diagram of the control loops using feed forward in the TAB converter system. Subsystem block denotes open loop control of ϕ_3 derived from (22) and Feed Forward block indicates (23).

TABLE I. SPECIFICATIONS OF PROTOTYPE TAB CONVERTER.

Controller		FPGA Cyclone V
Switching device		Power MOSFET IXFN140N30P
Switching frequency	f_{sw}	20 [kHz]
DC bus voltage	V	100 [V]
Transformer winding ratio		1:1:1
External inductance	L_1, L_2, L_3	18.98, 20.10, 19.57 [μH]
Resolution of phase control		0.36 [degree/bit]

realized due to the change of ϕ_2 in higher-order time delay system of the closed loop by the PI controller. Fig. 9 shows the experimental results of DC bus currents I_2, I_3 when the command I_3^* was changed from 0 to 5 A with open loop control while keeping I_2^* constant 2 A with closed loop control. In the case that the closed loop is

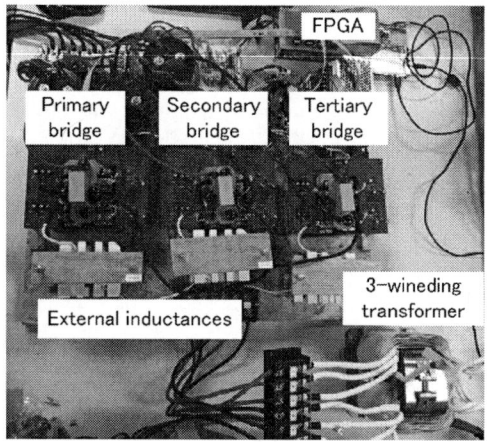

Fig. 5. Appearance of prototype TAB converter.

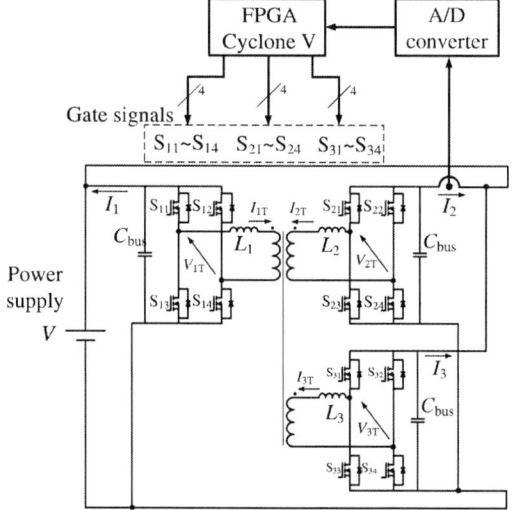

Fig. 6. System diagram of prototype TAB converter in the experiment.

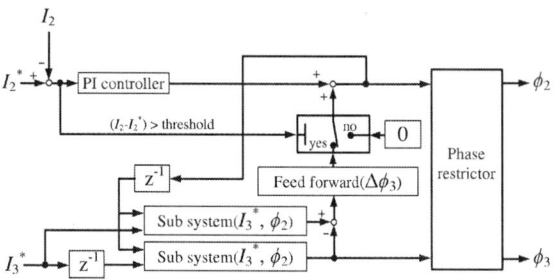

Fig. 7. Block diagram of control with the delay and selector of the feed forward in order to avoid forming a closed loop.

affected by the change of the open loop's command value, the constant value control is difficult to apply. This is caused by ϕ_3, which is the operating value of the open loop system, varies stepwise, and interference is applied as a stepwise disturbance on the open loop system. Also Fig. 9 shows the comparison between the transient states with and without feed forward controller. It is confirmed that the interference is suppressed by the feed forward controller.

C. Experimental results in steady state.

Steady state waveforms of the prototype TAB converter were measured. Fig. 10 illustrates that the experimental results of the DC bus currents I_i ($i = 1 \sim 3$) and the inductor currents I_{iT} ($i = 1 \sim 3$) when the DC current command values I_2^*, I_3^* are 2, 6 A, respectively. From Fig. 11(a), it was confirmed that both I_2 and I_3 follow the command value and the controllers performs the operation as designed. Fig. 11(b)~11(d) shows the inductor currents; and these waveforms demonstrated that the TAB converter operated adequately. Also Fig. 11 illustrates the experimental results at another operating point $(I_2^*, I_3^*) = (8, -8)$ A. From these results, it was confirmed that TAB converter can be operated in a wide operation range by the proposed control method.

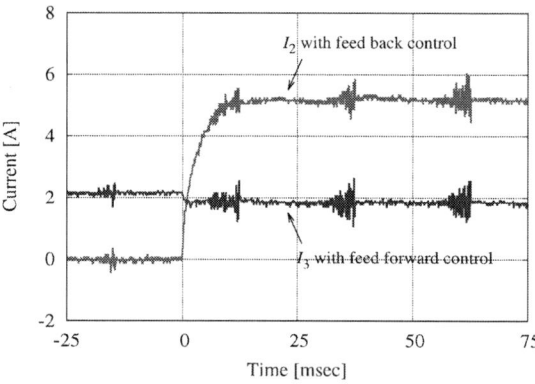

Fig. 8. A step response with the proposed control scheme.

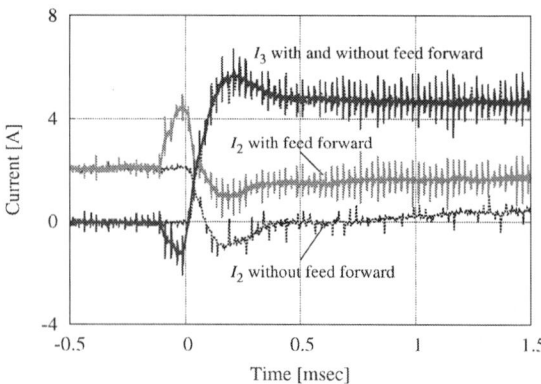

Fig. 9. A comparison between with and without feed forward controller.

The 2018 International Power Electronics Conference

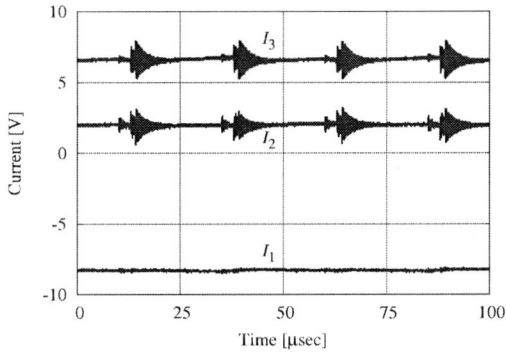

(a) DC bus currents in each port.

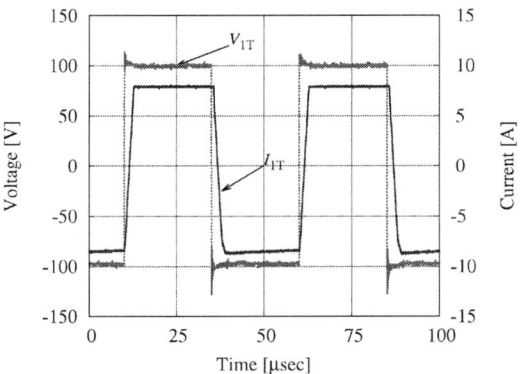

(b) Primary bridge inductor current and voltage waveforms.

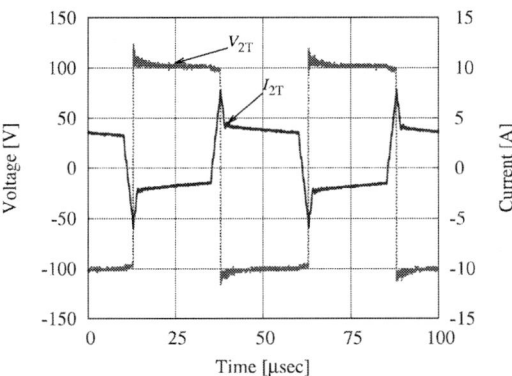

(c) Secondary bridge inductor current and voltage waveforms.

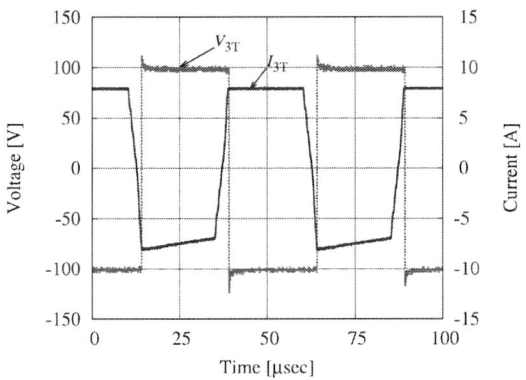

(d) Tertiary bridge inductor current and voltage waveforms.

Fig. 10. Steady state currents and voltages waveforms at the power transmission from port 1 to port 2 and port 3.

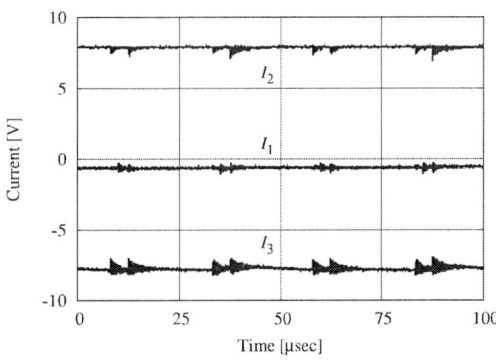

(a) DC bus currents in each port

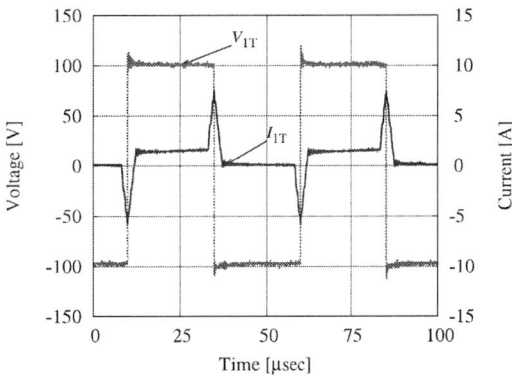

(b) Primary bridge inductor current and voltage waveforms.

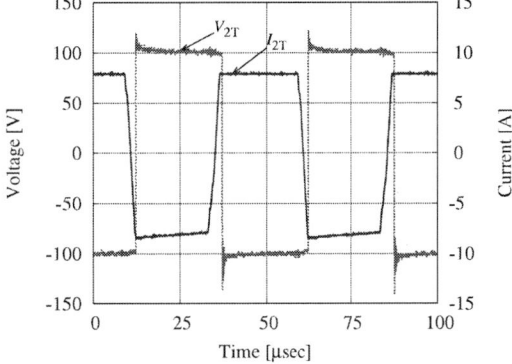

(c) Secondary bridge inductor current and voltage waveforms.

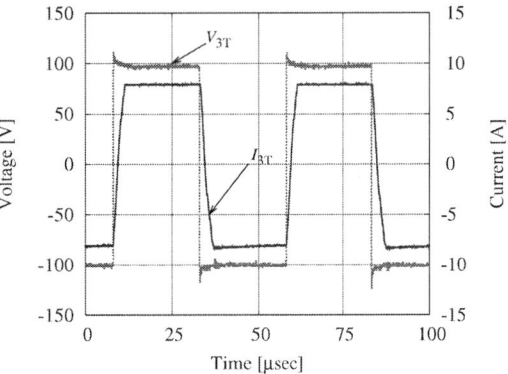

(d) Tertiary bridge inductor current and voltage waveforms.

Fig. 11. Steady state currents and voltage waveforms at the power transmission from port 3 to port 2.

975

VI. CONCLUSION

This paper proposed a control method of TAB converter which can decouple the interference in all stable operation range without precaluculations. The proposed controller realized decoupling function without using a decoupler depending on the operating points by a feed forward control. Hence, the structure of the controller is not complex. It was confirmed that each DC bus current follows the command value as designed in the experiment. Also, the experiment indicated that interference in the transient response can be suppressed by feed forward control in the transient response where the command value in closed loop is changed. When the proposed control method is applied, since the transient characteristics of each port are different, it is necessary to consider which control is applied to which source. This consideration will be conducted in a future work.

REFERENCES

[1] Rikiya Abe, Hisao Taoka, David McQuilkin "Digital Grid: Communicative Electrical Grids of the Future", *Proc. of IEEE Transactions on Smart Grid*, Vol.2, Issue.2, pp.399-410, 2011.

[2] Alex Q. Huang, Mariesa L. Crow, Gerald Thomas Heydt, Jim P. Zheng, Steiner J. Dale "The Future Renewable Electric Energy Delivery and Management (FREEDM) System: The Energy Internet", *Proc. of the IEEE*, Vol.9, Issue.1, pp.133-148, 2011.

[3] Mohammad Jafari, Zahra Malekjamshidi, Glenn Platt, Jian Guo Zhu, D.G.Dorrell "A Multi-Port Converter Based Renewable Energy System for Residential Consumers of Smart Grid", *Industrial Electronics Society, IECON 2015 - 41st Annual Conference of the IEEE*, pp.5168-5173, 2015.

[4] MEI Qiang, Xu Zhen-lin WU Wei-yang "A Novel Multi-Port DC-DC Converter for Hybrid Renewable Energy Distributed Generation Systems Connected to Power Grid", *IEEE International Conference on Industrial Technology*, August, 2008.

[5] XIE Junm, ZHANG Xing, ZHANG Chong-wei, LIU Sheng-yong, "A Novel Three-port Bi-directional DC-DC Converter", *Proc. of 2010 2nd IEEE international Symposium on Power Electronics for Distributed Generation Systems*, 2010.

[6] Walbermark M. dos Santos, Marcio S. Ortmann, Romulo Schweutzer, Samir A. Mussa, Denizar C. Martins, "DESIGN AND EXPERIMENTAL RESULTS OF THE TAB CONVERTER WITH PV POWER INJECTION", *Proc. of Power Electronics Conference*, 2011.

[7] Shota Nakagawa, Junichi Arai, Ryosuke Kasashima, Koya Nishimoto, Yuichi Kado, Keiji Wada "Dynamic Performance of Triple-Active Bridge Converter rated at 400 V, 10 kW, and 20 kHz", *Proc. of Future Energy Electronics Conference and ECCE Asia*, pp.1090-1094, 2017.

[8] Kreigo Katagiri, Shota Nakagawa, Kento Kurosawa, Junichi Arai, Yuichi Kado, Keiji Wada "Power Flow Control of Triple Active Bridge Converter Equipped with AC/DC Converter for Constucting Autonomous Hybrid AC/DC Microgrid Systems",*Proc. of Industrial Electronics Society, IECON 2017*, pp.1442-1446, 2017.

[9] J. L. Duarte, M. Hendrix, and M. G. Simoes, "Three-port bidirectional converter for hybrid fuel cell systems" *IEEE Transactions on Power Electron*, vol.22, no.2, pp.480-487, 2007.

[10] Chuanhong Zhao, Simon D. Round "An Isolated Three-Port Bidirectional DC-DC Converter With Decoupled Power Flow Management", *IEEE Transaction on Power Electronics*, vol.23, no.5, pp.2443-2453, 2008.

[11] Koya Nishimoto, Yuichi Kado, Ryosuke Kasashima, Shota Nakagawa, Keiji Wada "Decoupling Power Flow Control System in Triple Active Bridge Converter Rated at 400 V, 10 kW, and 20 kHz", *Proc. of Power Electronics for Distributed Generation Systems*, 2017.

[12] A. Niederlinski, "A Heuristic Approach to the Design of Linear Multivariable Interacting Control Systems ", *Automatica*, vol.7, no.2, pp.691-701, 1971.

[13] Min-Sen Chiu, Yaman Arkun, "A new result on Relative Gain Array, Niederlinski Index and decentralized stability condition: 2×2 plant cases", *Automatica*, vol.27, no.2, pp.419-421, 1991.

Analysis of PFM Operation Model for Capacitor Charger Resonant Topology with Energy Dosage

Pengyu Jia, Yiqin Yuan, Shengwen Fan, Zhenyu Shan

College of Electrical and Control Engineering, North China University of Technology

Abstract- **Capacitor charger resonant topology has been widely used in high-power and high-voltage filed due to the high efficiency and low switching loss. However, the steady-state and dynamic model is not clearly proposed in the past few literatures. This paper shows the derivation process for the topology and presents the operation model under PFM control. The PFM control model is presented and analytical solution is obtained. Experiment results are proposed to verify the theoretical conclusions.**

I. INTRODUCTION

In high-power and high voltage field, power supply design process is always a challenge due to the consideration of efficiency and device stress. Traditional hard-switching topology is hard to realize high-voltage-gain due to the switching loss and poor efficiency. To reduce power loss, resonant converters are widely researched in many literatures [1-4]. Resonant converter can utilize parasitic parameters to achieve soft-switching process. Under certain control methods, zero-current-switching (ZCS) or zero-voltage-switching (ZVS) can be realized in order to reduce the switching loss. A capacitor charger resonant topology with energy dosage have been presented in [5, 6], which can realize soft-switching during both switch-on and switch-off transitions, but the relationship with traditional resonant converter and the analytical solution for steady-state model under different control modes is not clearly proposed.

This paper presents the derivation of the capacitor charger resonant topology (CCRT) and gives the full derivation process for the steady-state model under pulse-frequency-mode (PFM) control method. A prototype is designed to verify the conclusions.

II. DERIVATION OF CCRT

The schematic of main circuit for CCRT is showed in Fig.1. It consists of two active power switches and two diodes at primary side of transformer. Full bridge rectifier with large capacitor is connected with the secondary side of transformer. Capacitor C_1 is equal to C_2 and both work as the resonant capacitors. L_1 is the resonant inductor which can be realized by the leakage inductance of

transformer. C_o indicates the output capacitor and R_{Load} indicates the load.

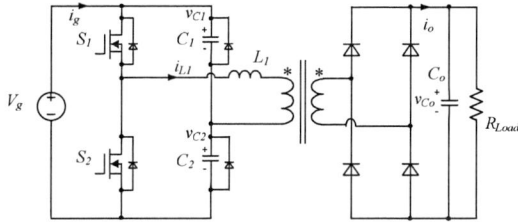

Fig.1 Schematic of main circuit for CCRT

CCRT can be considered as the derivation from LC series resonant converter, which is showed in Fig.2.

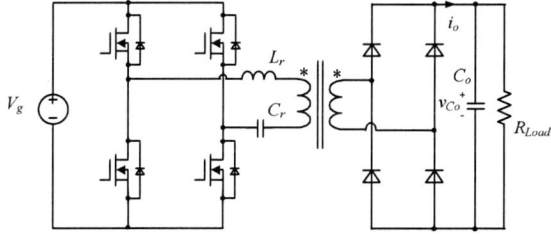

(a) LC series resonant converter

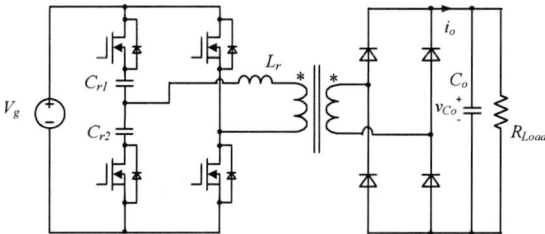

(b) Position change the resonant capacitor

Fig.2 Derivation of CCRT from LC series resonant converter

LC series resonant converter has variety operation modes, which can be divided into odd/even k discontinuous conduction modes and odd/even k continuous conduction modes, where k is integer depending on the switching frequency, resonant frequency and main circuit parameters. So operation mode needs to be designed carefully to ensure k is controllable [7, 8].

Different from LC series resonant converter, CCRT have certain limited operation modes and which can be easily designed. Thanks to the diodes in parallel with C_1

and C_2, the resonant capacitor voltage must change monotonically when power switch S_1 or S_2 turns on. When S_1 or S_2 turns off, the capacitor voltage changes reversely and also monotonically. During the whole switching period, the voltage of capacitor is always positive. The inductor current i_{L1} must change from zero when the switch S_1 or S_2 turns on.

III. ANALYTICAL SOLUTION FOR THE MODEL OF CCRT

Here f_r indicates resonant frequency and f_s indicates switching frequency. If $f_r < f_s$, then only zero-current-switching-on is obtained. There is still some power loss caused by switching-off process. To realize both zero-current-switching-on and zero-current-switching-off results, f_s can be designed less than f_r. The output capacitor C_o is designed sufficiently large as filter.

The theoretical waveforms of different stage are given in Fig.3 and Fig.4, respectively, corresponding to the conditions that whether the energy stored in resonant capacitors is entirely transferred to load.

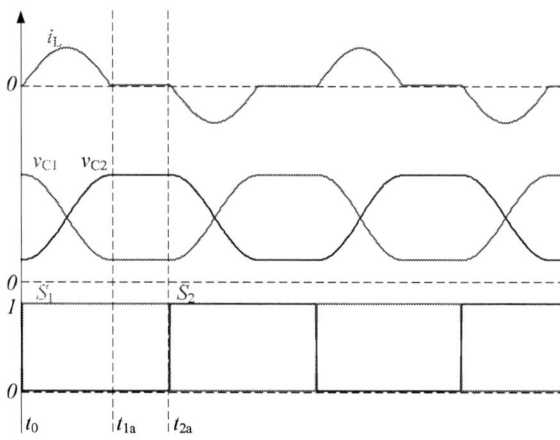

Fig.3 Energy of resonant capacitors is partly transferred into load

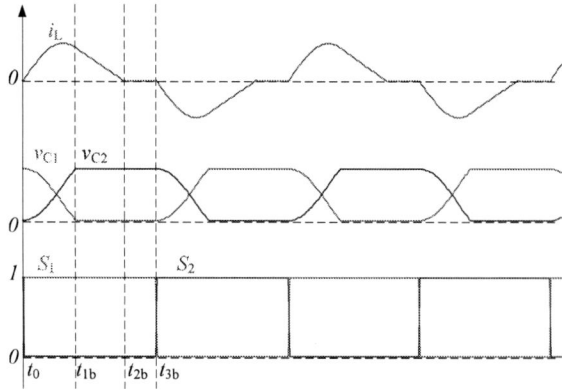

Fig.4 Energy of resonant capacitors is entirely transferred into load

Here operation mode I and II are used respectively to describe the conditions in Fig.3 and Fig.4. To simplify the computation, parameters in secondary side are converted into primary side with 'sec' in subscript. Therefore the effect of transformer can be reduced.

Operation mode I: This mode is corresponding to the condition that energy in C_1 and C_2 are partly transferred to load. From t_0 to t_{1a}, the status-equations can be described as follows.

$$\begin{cases} L_1 \dfrac{di_{L1}}{dt} = v_{C1} - v_{o\,sec} \\ C_1 \dfrac{dv_{C1}}{dt} = -\dfrac{i_L}{2} \qquad (1\sim3) \\ C_2 \dfrac{dv_{C2}}{dt} = \dfrac{i_L}{2} \end{cases}$$

$$v_{C1}(0) = V_1 \qquad (4)$$

$$v_{C2}(0) = V_g - V_1 \quad (5)$$

$$i_{L1}(0) = 0 \qquad (6)$$

Because $C_1 = C_2 = C$, according to the equations, the analytical solutions have been found as follows.

$$\begin{cases} v_{C1} = V_{o\,sec} + (V_1 - V_{o\,sec}) \cos(\omega t) \\ v_{C2} = V_g - V_{o\,sec} + (V_{o\,sec} - V_1) \cos(\omega t) \qquad (7\sim9) \\ i_{L1} = \dfrac{V_1 - V_{o\,sec}}{R_0} \sin(\omega t) \end{cases}$$

where

$$V_{o\,sec} = \frac{V_g}{2} \quad (10)$$

$$V_1 = V_g \left(\frac{1}{16 C R_{Load\,sec} f_s} + \frac{1}{2} \right) \quad (11)$$

$$\omega = \frac{1}{\sqrt{2LC}} \quad (12)$$

$$R_0 = \sqrt{\frac{L}{2C}} \quad (13)$$

If the transformer ratio is 1: N, therefore

$$V_o = \frac{N V_g}{2} \quad (14)$$

Operation mode II: This mode is corresponding to the condition that energy in C_1 and C_2 is entirely transferred to load. From t_0 to t_{1b}, the circuit is in the process of resonance. At the time point t_{1b}, $v_{c1}(t_{1b})$ decrease to zero and the inductor current i_{L1} decrease linearly under the effect of V_o.

From t_0 to t_{1b}, the status-equations can be described as $(1\sim3)$ but the initial conditions are in the following.

$$v_{C1}(0) = V_g \qquad (15)$$

$$v_{C2}(0) = 0 \qquad (16)$$

$$i_{L1}(0) = 0 \qquad (17)$$

At time point t_{1b}, the status variable equations can be derived as follows.

$$v_{C1}(t_{1b}) = V_{o\,sec} + (V_g - V_{o\,sec})\cos(\omega t_{1b}) = 0 \quad (18)$$

$$v_{C2}(t_{1b}) = V_g - V_{o\,sec} + (V_{o\,sec} - V_g)\cos(\omega t_{1b}) = V_g \quad (19)$$

$$i_{L1}(t_{1b}) = \frac{V_g - V_{o\,sec}}{R_0}\sin(\omega t_{1b}) \quad (20)$$

From t_{1b} to t_{2b}, i_{L1} decreases to zero by voltage V_o.

$$i_{L1}(t_{1b}) - \frac{V_{o\,sec}}{L}(t_{2b} - t_{1b}) = 0 \quad (21)$$

To get the analytical solutions of v_o, t_{1b} and t_{2b}, the relation between i_{L1} and v_o is found here,

$$<i_o>_{AVG\,sec} R_{Load\,sec} = V_{o\,sec} \quad (22)$$

where $<i_o>_{AVG}$ means the average current for output. Here it is transferred to primary side in the computation. By integration (23) can be derived.

$$<i_o>_{AVG\,sec} = 2f_s \left[\begin{array}{c} 2\omega(V_g - V_{o\,sec})(1 - \cos \omega t_{1b}) + \\ \dfrac{t_{2b} - t_{1b}}{2} \bullet \dfrac{V_g - V_{o\,sec}}{R_0}\sin \omega t_{1b} \end{array} \right] \quad (23)$$

By the same integration method the average current for input can also be derived, here $<i_g>_{AVG}$ is used to represent it.

$$<i_g>_{AVG} = f_s \left[2\omega(V_g - V_{o\,sec})(1 - \cos \omega t_{1b}) \right] \quad (24)$$

According to the law of energy conservation, (25) can be derived.

$$V_g <i_g>_{AVG} = V_{o\,sec} <i_o>_{AVG\,sec} \quad (25)$$

The voltage ratio between v_g and v_o can be derived by (21~25) and listed in the following.

$$\frac{V_o}{V_g} = \sqrt{2f_s C R_{Load}} \quad (26)$$

From operation mode I and II we can conclude that in operation mode I, the voltage can't be controlled through PFM method and always depends on input and transformer ratio. In operation mode II, PFM is valid to regulate the output and the maximum output voltage is equal to $NV_g/2$. The transform condition from mode I to mode II is that all energy stored in C_1 and C_2 is entirely consumed by load resistor in the duration of $T_s/2$. Thus the capacitance value C is computed by (27).

$$\frac{V_o^2}{R_{Load}} \bullet \frac{T_s}{2} = \frac{1}{2} \bullet 2C \bullet V_g^2 \quad (27)$$

Where V_o can be substitute by $NV_g/2$, so the constraints for switching frequency f_s, R_{Load} and C can be derived as follows if PFM method needs to be applied to converter.

$$C \leq \frac{N^2}{8f_s R_{Load}} \quad (28)$$

It is important to be pointed out that $f_s < f_r$ is a necessary but not sufficient condition to guarantee the converter operating in mode II. From Fig.4, it is clear that t_{3b} must larger than t_{2b}. To satisfy this constraint, the following conditions can be derived.

$$\frac{T_s}{2} \geq \frac{1}{w}\frac{NV_g}{V_o}\sqrt{1 - \frac{2V_o}{NV_g}} + \frac{1}{w}\arccos\sqrt{\frac{V_o}{V_o - NV_g}} \quad (29)$$

$$f_s \leq \frac{1}{\dfrac{2}{w}\dfrac{NV_g}{V_o}\sqrt{1 - \dfrac{2V_o}{NV_g}} + \dfrac{2}{w}\arccos\sqrt{\dfrac{V_o}{V_o - NV_g}}} \quad (30)$$

According to Fig.4 it is clear that the active switches and rectifier diodes can realize ZCS during both turn-on and turn-off process. Clamp diodes can realize ZVS turn-on and ZCS turn-off process.

From the analysis above, it is clear that CCRT can be considered as a constant power source if the switching frequency is fixed, according to (27). That means output voltage is only related with switching frequency, resonant capacitor and load resistor. Usually for the consideration of saving cost, voltage-double capacitors can be placed instead of two rectifier diodes. The transformer ratio is designed to ensure equivalent voltage V_{osec} must be less than half of the input voltage.

IV. INPUT-PARALLEL-OUTPUT-SERIES CONFIGURATION AND EXPERIMENT RESULTS

To improve the power level of single converter and achieve high-voltage-gain, Input-Parallel-Output-Series (IPOS) configuration is always applied. IPOS configuration of CCRT is proposed in Fig.5.

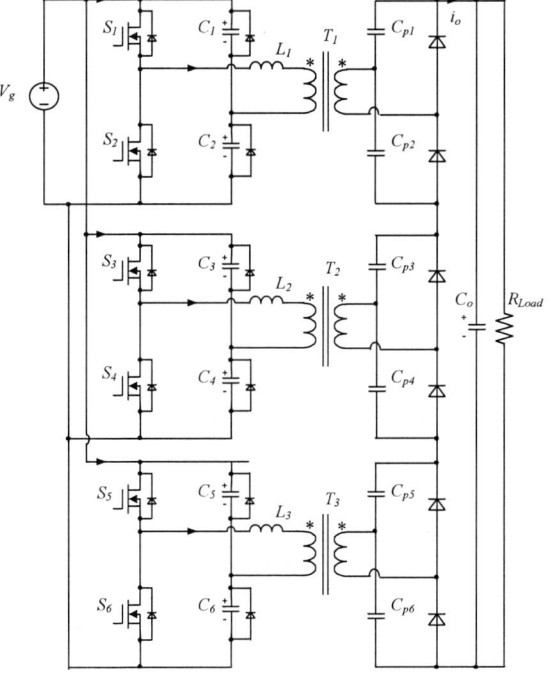

Fig.5 Schematic of experiment platform

Corresponding to Fig.5, (26) and (27) should be changed into (31) and (32).

$$\frac{V_o}{V_g} = \sqrt{6f_s C R_{Load}} \quad (31)$$

$$\frac{V_o^2}{R_{Load}} \cdot \frac{T_s}{2} = \frac{3}{2} \cdot 2C \cdot V_g^2 \quad (32)$$

Since the voltage can't be regulated in mode I, here the transfer function G_{vf} from frequency to output in mode II is presented in (33). It is obvious that in mode II the plant can be considered as a capacitive load and the frequency f_s controls the current source injecting to the load so to form voltage output.

$$G_{vf}(s) = V_g \sqrt{\frac{2C}{f_s R_{Load}}} \frac{s6R_{Load}(6C_o + C_p) + 36}{s^2 R_{Load} C_p (C_p + 7C_o) + s7C_p} \quad (33)$$

A 1kW power supply platform with 22kV voltage output is built. The main circuit schematic is shown in Fig.5 and component parameters are listed in Table I.

TABLE I MAIN CIRCUIT PARAMETERS

Symbol	Meaning	Value
Vg	Input voltage	V_g=200V
$C_1 \sim C_6$	Energy dosage capacitor	C=1uF
$C_{p1} \sim C_{p6}$	Multiplying-circuit capacitor	C_p=10nF
C_o	Output capacitor	C_o=0.33nF
L_1, L_2, L_3	Leakage inductance	L=2uH
R_{Load}	Load resistor	R_{Load}=500k
$1:N$	Transformer ratio	$1:N$=1/40

According to (33), the bode plot of G_{vf} with 1kW power point is presented in Fig.(6). It is clear that the plant is a minimum phase system and this brings much convenient to design control system.

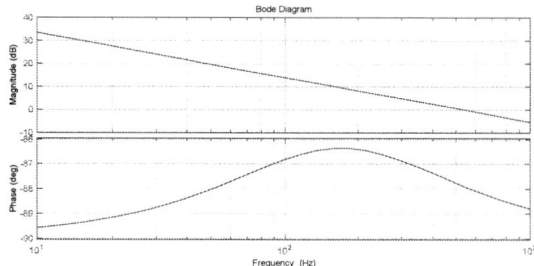

Fig.6 Bode plot of G_{vf} (f_s=4.8kHz) in mode II

Because of the parasitic parameters such as conduction resistance, wire resistance, etc. Switching frequency f_s is always higher than the theoretical computation result. In operation mode II, on the basis of (26), the input to output voltage ratio can be described as show in (31) and the maximum gain is 120. Here the experiment results under operation mode I and mode II are presented as shown in Fig.7 and Fig.8. Three probe

channels are corresponding to S_1 driver pulse (20V/div), current of transformer T1 (25A/div), output voltage v_o with the ratio of 1:10k (2V/div), respectively.

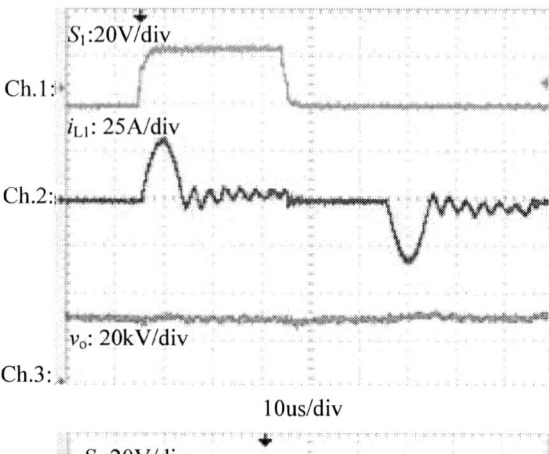

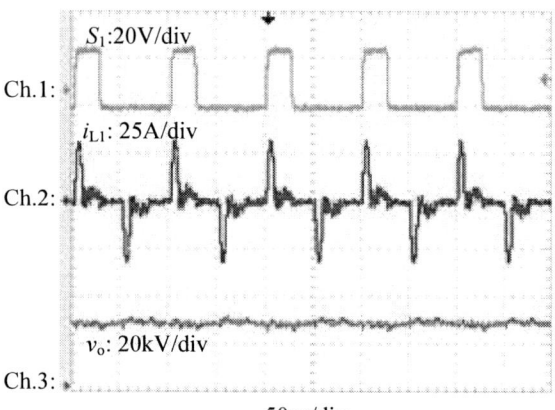

Fig.7 Waveforms of inductor current and output (f_s=10kHz, v_o=24kV, mode I)

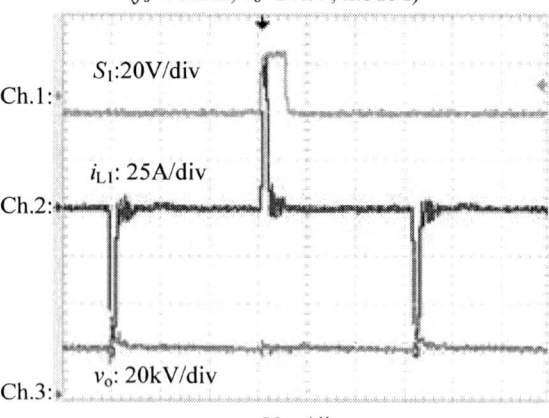

(a) Waveforms of inductor current and output (f_s=3kHz, v_o=19kV, mode II)

The 2018 International Power Electronics Conference

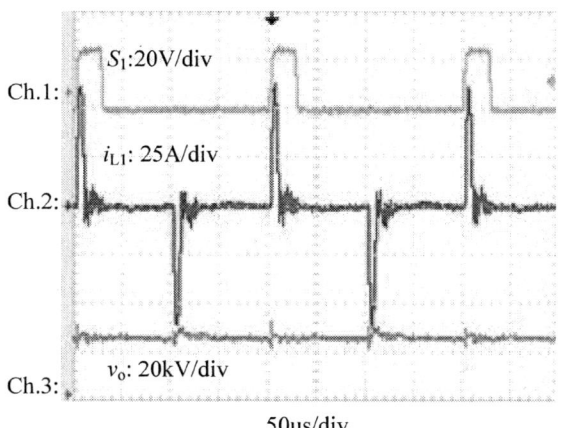

(b) Waveforms of inductor current and output (f_s=4.8kHz, v_o=22kV, mode II)

Fig.8 experiment waveforms in mode II

The experiment results show that the output voltage has a monotonic relationship with switching frequency until the output reaches 24kV, then the output voltage is constant even if the switching frequency still increases.

V. CONCLUSION

The derivation of CCRT is presented in this paper and the basic operation mode is analyzed. By setting the converter in DCM operation mode, the active switches and rectifier stage can achieve soft-switching status during both turn-on and turn-off process. Clamp diodes can achieve ZVS turn-on and ZCS turn-off. The converter can be considered as a constant power source and the output voltage has a monotonic relationship with switching frequency. Besides, the transfer function for the dynamic model indicates CCRT is a minimum phase plant so the control system can be easily designed. Two operation modes for CCRT is described explicitly and the analytical solutions have been presented. A prototype with 1kW power level is built to verify the theoretical analysis.

REFERENCES

[1] C. q. Lee and K. Siri, "Analysis and Design of Series Resonant Converter by State-Plane Diagram," in IEEE Transactions on Aerospace and Electronic Systems, vol. AES-22, no. 6, pp. 757-763, Nov. 1986.

[2] A. F. Wittulski and R. W. Erickson, "Steady-State Analysis of the Series Resonant Converter," in IEEE Transactions on Aerospace and Electronic Systems, vol. AES-21, no. 6, pp. 791-799, Nov. 1985.

[3] J. A. Martin-Ramos, J. Diaz, A. M. Pernia, J. M. Lopera and F. Nuno, "Dynamic and Steady-State Models for the PRC-LCC Resonant Topology With a Capacitor as Output Filter," in IEEE Transactions on Industrial Electronics, vol. 54, no. 4, pp. 2262-2275, Aug. 2007.

[4] Y. Chen, J. Xu, J. Cao, L. Lin and H. Ma, "PWM–PFM hybrid controlled LCC resonant converter with wide ZVS range and narrow switching frequency variation," in

Electronics Letters, vol. 53, no. 17, pp. 1218-1220, 8 17 2017.

[5] A. Pokryvailo, C. Carp and C. Scapellati, "High-Power High-Performance Low-Cost Capacitor Charger Concept and Implementation," in IEEE Transactions on Plasma Science, vol. 38, no. 10, pp. 2734-2745, Oct. 2010.

[6] M. Wolf and A. Pokryvailo, "High Voltage Resonant Modular Capacitor Charger Systems With Energy Dosage," 2005 IEEE Pulsed Power Conference, Monterey, CA, 2005, pp. 1029-1032.

[7] V. Vorperian and S. Cuk, "A complete DC analysis of the series resonant converter," 1982 IEEE Power Electronics Specialists conference, Cambridge, MA, USA, 1982, pp. 85-100.

[8] A. F. Wittulski and R. W. Erickson, "Steady-State Analysis of the Series Resonant Converter," in IEEE Transactions on Aerospace and Electronic Systems, vol. AES-21, no. 6, pp. 791-799, Nov. 1985.

The 2018 International Power Electronics Conference

An Active-Clamped Current-Fed Half-bridge DC-DC Converter With Three Switches

Truong-Duy Duong[1], Minh-Khai Nguyen[2*], Young-Cheol Lim[1] and Joon-Ho Choi[1]

1 Department of Electrical Engineering, Chonnam National University, Gwangju 500-757, Korea.
2 Department of Electrical Engineering, Chosun University, Gwangju 61452, Korea.
*E-mail: khaibk@ieee.org

Abstract—An active-clamped current-fed half-bridge DC-DC converter with three switches is proposed in this paper. The characteristics of the proposed converter have following as the input current is continuous with low ripple; decreasing one active switch; and achieving wide range zero-voltage switching (ZVS) on two switches. Which leads to high efficiency, reduce size and cost. The operating principles and parameters selection are discussed. A 250 W prototype with PI controller is built to validate the proposed converter.

Keywords—DC-DC conversion, active snubber, current-fed half-bridge (CFHB) converter, soft switching.

I. INTRODUCTION

Since the green energy has become essential when the industrialization is quickly developing. Thus, the using of green energy sources are increased and many subjects have been researched with applications such as uninterruptible power supplies (UPS), electric vehicles, photovoltaic and fuel cell systems, where the DC voltage is needed to step up high DC voltage [1]-[2]. To use the low input voltage and provide the continuous power with characteristics: low DC voltage, high current with low ripple, the high boost DC-DC converter should be applied. Until now, many proposed topologies are found and developed with non-isolated [1]-[5] and isolated topology [6]-[17].

The isolated DC-DC converter can step up the low voltage to the high voltage with large ratio conversion and the soft-switching technique is needful. The isolated topology for the fuel cell applications can include current-fed and voltage-fed topology. The comparison of advantages and disadvantages between current-fed and voltage-fed topology has been illustrated as [10]-[11]. The current-fed topology has a low input current ripple, low voltage switches stress and low power losses which is good for efficiency converter. Thus, the current-fed isolated converters are suitable for the fuel cell application systems, are proposed [10]-[16]. Many topologies of isolated current-fed have proposed based on full-bridge or half-bridge topologies. The current-fed half-bridge (CFFB) is better the current-fed full-bridge (CFFB) topology for many characteristics: the input current ripple is lower; the utilization of high-frequency

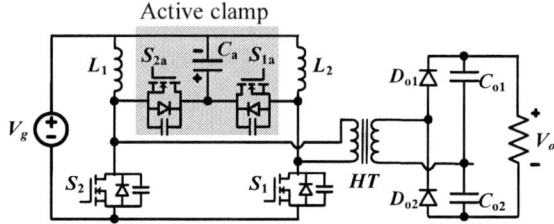

Fig. 1. An active clamped ZVS CFHB DC-DC converter.

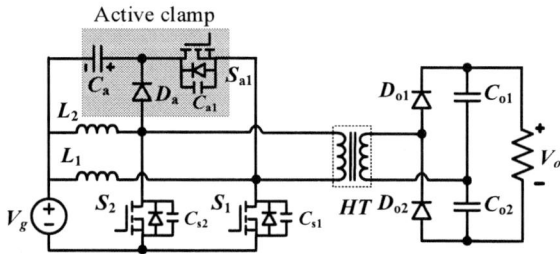

Fig. 2. An active clamped ZVS CFHB DC-DC converter.

transformer is higher because the half-bridge topology can generate a zero voltage at the primary side of the transformer, the half-bridge converter has the voltage rating of the primary side is doubled and the current rating of the primary side drop one-half while the turns ratio of the transformer is halved, the CFHB is limited zero voltage switching (ZVS) range. Hence, the half-bridge is more suitable to the fuel cell application which requires high current, low ripple current and low voltage. However, the large voltage rating and high voltage spike on the switches because of the leakage inductance of the transformer is a problem for the CFHB topology. To solve these problem, many previous topologies were proposed [12]-[16], the active clamping circuit was used and introduced in [12]-[13] with the ZVS of all power switches, the CFHB quasi-resonant with an active-clamp circuit was added, is addressed in [14], the active-clamped L-L type current-fed isolated converter was presented in [15]-[16], by using the clamping circuit, the voltage spike was reduced and all switches were achieved ZVS.

The 2018 International Power Electronics Conference

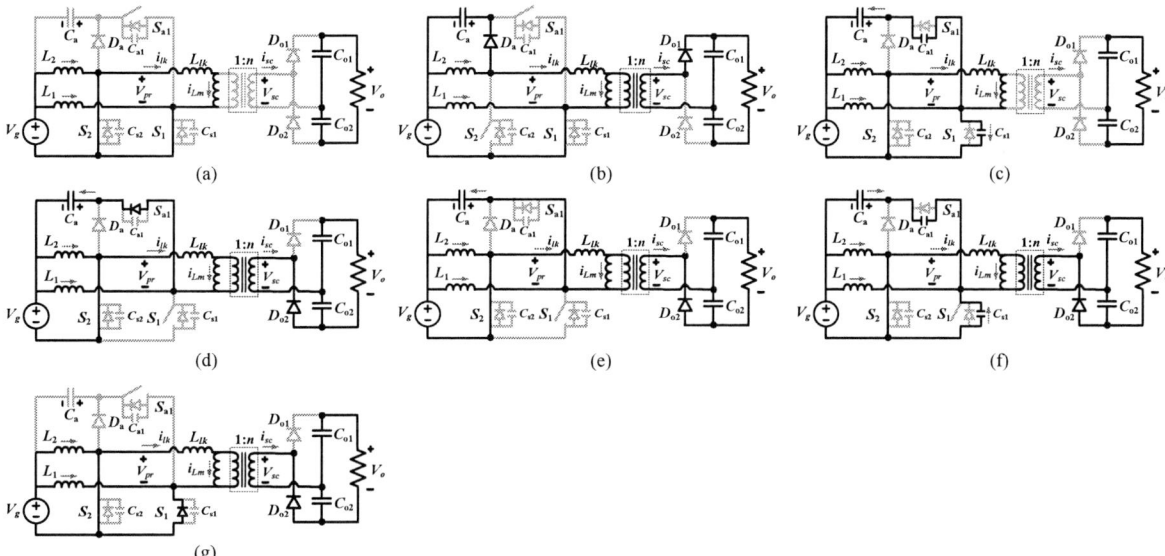

Fig. 3. Operation states of the proposed converter.

Similarly, an active-clamped ZVS CFHB DC-DC converter is addressed in [12]-[13] used two input inductors, two main switches, an active-clamping circuit (two auxiliary switches, one clamping capacitor), a high-frequency transformer and a voltage doubler rectifier (VDR) is included two diodes and two capacitors, shown as Fig. 1.

The active-clamped ZVS CFHB [12] have an active clamping snubber absorbs the switches voltage; achieve wide range ZVS on switches; does not need any clamp winding; PWM generation is simple; the input current ripple is low. This paper proposed a new active-clamped CFHB DC-DC converter with three switches which has some advantages of the active-clamped CFHB and used less than one auxiliary switch, as shown in Fig. 2. The cost and size are decrease and the efficiency is improved.

II. ANALYSIS MODES OF PROPOSED THREE-SWITCH ACTIVE-CLAMPED CURRENT-FED HALF-BRIDGE CONVERTER

As shown in Fig. 2, the proposed new active-clamped CFHB converter consists of two inductors, two main switches and active snubber circuit (such as one auxiliary switch, one diode, one snubber capacitor) at the primary side connected voltage doubler rectifier at the secondary side. The operation principle is illustrated base on the following conditions: two inductors L_1 and L_2 are large enough to be examined as constant current $I_{L1} = I_{L2} = I_{in}/2$; capacitors are large enough to maintain the constant capacitor voltage; switches and diodes are ideal; the high-frequency transformer is ideal, the leakage inductor current and the current flow to the windings of the transformer change linearly.

Interval 1–[t_0-t_1, Fig. 3(a)]: In this interval, two main switches S_1 and S_2 are turned on, the primary winding is

short-circuited. Two boost inductor is charged by the input voltage. We have

$$\begin{cases} L_1 \dfrac{di_{L1}}{dt} = L_2 \dfrac{di_{L2}}{dt} = V_g \\ L_{lk} \dfrac{di_{lk}}{dt} = L_m \dfrac{di_{Lm}}{dt} = 0, \end{cases} \tag{1}$$

Interval 2–[t_1-t_2, Fig. 3(b)]: Switch S_1 is turned on when switch S_2 and auxiliary S_{a1} are turned off. The primary voltage is charged. The primary voltage of the transformer is a positive voltage. We get:

$$\begin{cases} L_2 \dfrac{di_{L2}}{dt} = -V_{Ca} \\ L_1 \dfrac{di_{L1}}{dt} = V_g \end{cases} \text{and} \begin{cases} V_{pr} = V_g + V_{Ca} \\ L_m \dfrac{di_{Lm}}{dt} = \dfrac{V_o}{2n}, \end{cases} \tag{2}$$

Interval 3–[t_2-t_3, Fig. 3(a)]: two main switches S_1 and S_2 are turned on, when the auxiliary switch S_{a1} and diode D_1 are turned off, the primary voltage is equal to zero. So the diodes of the secondary side are turned off and the secondary voltage is equal to zero.

Interval 4–[t_3-t_4, Fig. 3(c)]: The inductor L_1 current charges C_{s1} and discharge C_{a1}.

Interval 5–[t_4-t_5, Fig. 3(d)]: The body diode of auxiliary switch S_{a1} is conducted and S_{a1} can turn on with ZVS.

$$\begin{cases} L_2 \dfrac{di_{L2}}{dt} = V_g \\ L_1 \dfrac{di_{L1}}{dt} = -V_{Ca} \\ V_{pr} = -V_g - V_{Ca}, \end{cases} \tag{3}$$

Interval 6– [t_5-t_7, Fig. 3(e)]: After S_{a1} is turn on with ZVS, the main switch S_2 and auxiliary switch S_{a1} keep on.

983

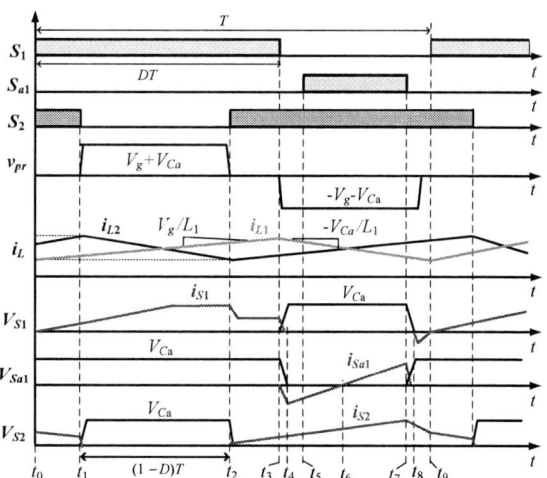

Fig. 4. Key waveforms of the proposed converter

The primary voltage is equal to $-V_g-V_{ca}$ and the secondary voltage is boost to $-nV_{ca}$. We have the equations

$$
\begin{cases}
L_2 \dfrac{di_{L2}}{dt} = V_g \\[2mm]
L_1 \dfrac{di_{L1}}{dt} = -V_{Ca} \\[2mm]
L_m \dfrac{di_{Lm}}{dt} = -\dfrac{V_o}{2n},
\end{cases}
\tag{4}
$$

Interval 7–[t_7-t_8, Fig. 3(f)]: In this case is the same with the interval 5. The leakage inductor and two capacitors C_1, C_{a1} are resonated

$$
f_r = \frac{1}{2\pi\sqrt{L_{lk}\left(C_1 + C_{a1}\right)}}.
\tag{5}
$$

Interval 8–[t_8-t_9, Fig. 3(h)]: In this interval, switch S_1 is turned on with ZVS.

If the dead time between the switch S_1 and switch S_{a1} is assumed to be zero and the average voltage across the inductor L_1 in period switching is equal to zero, we get

$$
\begin{cases}
V_{Ca} = \dfrac{D}{1-D}V_g \\[2mm]
V_o = \dfrac{2n}{1-D}V_g.
\end{cases}
\tag{6}
$$

Where D is the duty cycle of switch S_1.

III. DESIGN OF THE PROPOSED CONVERTER

A. Inductor Values

Assuming that two boost inductors L_1 and L_2 are large enough to be examined as constant current I_{L1} is equal to the current of I_{L2} and equal to one-half of the input current. The inductance of L_1 and L_2 are

$$
L_1 = L_2 = \frac{DV_g^2}{r_L\% fP_o}
\tag{7}
$$

where $r_L\%$ and f are the inductor current ripple of inductor L_1 or L_2 and the switching frequency, respectively.

B. Design of VDR circuit

At Fig. 3(a), two main switches are turned on, the secondary side current is zero and the two output capacitors are the output current, two output capacitors are given by

$$
C_{O1} = C_{O2} = \frac{(2D-1)}{2r_{C0}\% fR}
\tag{8}
$$

Where $r_{C0}\%$ and R are the output capacitor voltage ripple, resistance of load, respectively.

The current through two output diodes is one-half the output current and is calculated

$$
I_{Do1} = I_{Do2} = \frac{I_o}{2} = \frac{P_o}{2V_o},
\tag{9}
$$

C. ZVS Verification

To achieve the ZVS turn-on of main switch S_1 and auxiliary switch S_{a1}, the deadtime between S_1 and S_{a1} is given

$$
T_{dt} = \frac{D(1-D)(C_{S1} - C_{a1})R}{2n^2}
\tag{10}
$$

D. Power loss of MOSFET and Diode

The power loss of MOSFET can be calculated as

$$
P_M = P_{cM} + P_{swM} = R_{DSon} \cdot I_{Mrms}^2 + \left(E_{onM} + E_{offM}\right) \cdot f_{sw}.
\tag{11}
$$

where, R_{DSon}, I_{Mrms}, E_{onM}, E_{offM} and f_{sw} are the drain-source resistance, the RMS current, the switch-on, switch-off energy losses and switching frequency of the MOSFET, respectively.

The power losss of diode is calculated by

$$
P_D = P_{CD} + P_{swD} = u_D \cdot I_{Dav} + R_D \cdot I_{Drms}^2 + E_{onD} \cdot f_{sw},
\tag{12}
$$

where u_D, R_D are the on-state voltage and the on-state resistance, the average and RMS current of diode, the turn-on energy and the switching frequency, respectively.

IV. SIMULATION AND EXPERIMENT VERIFICATIONS

A. Simulation Verifications

The proposed converter was built with using PSIM 9.1 simulation when the input voltage is 32 V and the output voltage was set 400 V prototype. The converter was verified by operation condition of 260 W, as shown in Fig. 5. The simulation parameters of the converter is selected as the same Table I. Two switches S_1 and S_{a1} are

The 2018 International Power Electronics Conference

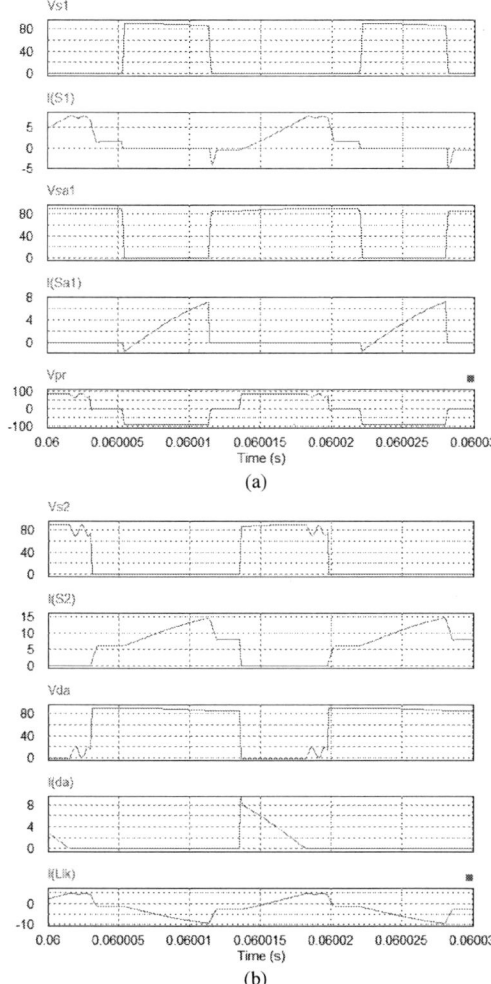

Fig. 5. Simulation results of the proposed converter.

TABLE I
EXPERIMENTAL PARAMETERS

Input voltage range		25- 42 V
Output voltage		400 V
Output power		250 W
Inductance of L_1 and L_2		370 μH
Transformer	Turn ratio	1:2.5
	Leakage inductance	6 μH
Capacitors	C_1	22 μF/ 305 V
	$C_{o1} = C_{o2}$	2.2 μF/ 630 V
	$C_{s1} = C_{sa1} = C_{s2}$	3.3 nF/ 300 V
Switching frequency		60 kHz

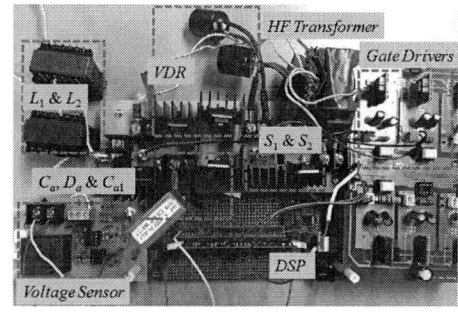

Fig. 6. The photograph of converter.

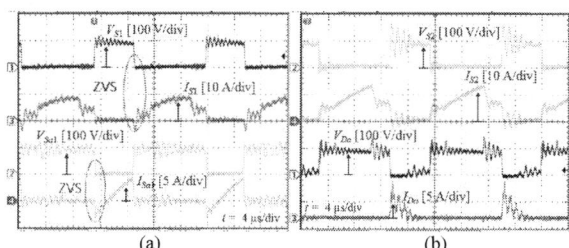

Fig. 7. Experimental waveforms of the proposed converter when $V_g = 32$ V and $P_o = 250$ W.

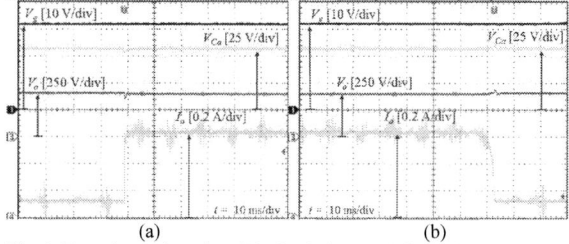

Fig. 8. Experimental results of the load change (a) from 50 W to 250 W, (b) from 250 W to 50 W. From top to bottom: input voltage, capacitor C_a voltage, output voltage and output current.

achieved ZVS with wide range load. The current of switches are negative show that the body diode of switches is conducting, the voltage on switches is equal to zero when there are turned on.

B. Experimental Verifications

The experimental results were tested at 250 W and used to kit TMS320F28335 DSP to verify the theoretical operation waveforms based on the same parameters, shown as Table. I. The main switches and auxiliary switch are used IRFP4668PbF MOSFETs ($R_{DSon} = 8$ mΩ); one STPS60SM200C Schottky diode; two DSEP30-12A; high-frequency transformer: PQ50/50 core; two boost inductors: BH E 30/30 core, $L_1 = L_2 = 370$ μH. The gating signal is generated by kit DSP through an isolated amplifier (TLP250).

The converter was performed for the input voltage is 32 V at the output power of 250 W. The output voltage is constant and equal to 400 V, the inductor L_1 current is continuous and interleaved inductor current. In Fig. 7(a), two switch S_1 and S_{a1} are turned off with ZVS; because

the body-diode of switches are conducted before the switches begin conducting. The switch S_2 is turned on with near-ZVS. Because of the parasitic effect in the devices, there is still ringing in the experimental results. The converter prototype was reconstructed as shown in Fig. 6.

To keep the output voltage at 400 V while the load change, a PI controller is applied, shown as Fig. 8. The

985

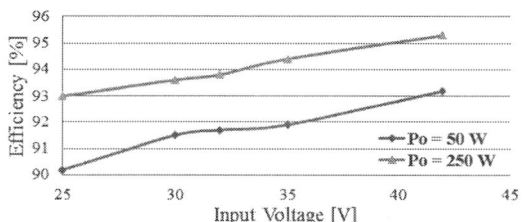

Fig. 9. Plot of efficiency.

compare signal between the feedback output voltage and the reference value of 400 V is crossed over the PI controller. The output signal PI controller is limited from 0 to 1.

The control signal of three switches are generated from the compare this signal with the high frequency triangle waveform. The PI controller experiment for the proposed converter was tested at V_g = 32 V, the output and capacitor C_a voltage are kept constant. The load change from 50 W to 250 W, shown as Fig. 8(a) and the load change from 250 W to 50 W, shown as Fig. 8(b).

Fig. 9 shows the efficiency of the proposed converter, it is measured when the input voltage changes from 25 V to 42 V. The maximum efficiency is 95.6% at P_o = 250 W and V_g = 42 V. Full load efficiency at V_g = 25 V and 42 V are 93% and 95.3%, respectively. The efficiency increase when the input voltage increases.

V. CONCLUSIONS

An active-clamped current-fed half-bridge DC-DC converter with three switches is presented in this paper. The proposed inverter achieves: save one gate driver for the switch, the two switches operate in ZVS with wider power range, the input current ripple is low. Compare to an active-clamped ZVS CFHB DC-DC converter, the proposed converter is limited one switch is operated in near- ZVS. The experimental and simulation results are shown to verify the analysis theory and design of the proposed converter. The proposed converter is suitable for fuel-cell applications.

REFERENCES

[1] D. Sha, F. You and X. Wang "A high-efficiency current-fed semi-dual-active bridge DC-DC converter for low input voltage applications," *IEEE Trans. Ind. Electron.*, vol. 63, no. 4, pp. 2155–2164, April. 2016.

[2] Y. P. Hsieh, J. F. Chen, L. S. Yang, C. Y. Wu, and W. S. Liu, "High-conversion-ratio bidirectional DC/DC converter with couple inductor," *IEEE Trans. Ind. Electron.*, vol. 61, no. 3, pp. 1311-1319, Mar. 2014.

[3] M. K. Nguyen, T. D. Duong, and L. C. Lim, "Switched-capacitor-based dual-switch high-boost dc-dc converter," *IEEE Trans. Power Electron.*, vol. 33, no. 5, pp. 4181–4189, May 2018.

[4] F. L. Tofoli, D. C. Pereira, W. J. Paula, and D. S. O. Junior, "Survey on non-isolated high-voltage step-up dc–dc topologies based on the boost converter," *IET Power Electron.*, vol. 8, no. 10, pp. 2044–2057, Oct. 2015.

[5] G. Wu, X. Ruan, and Z. Ye, "Nonisolated high step-up DC-DC converters adopting switched-capacitor cell," *IEEE Trans. Ind. Electron.*, vol. 62, no. 1, pp. 383–393, January 2015.

[6] P. J. Wolfs, "A current-sourced DC–DC converter derived via the duality principle from the half-bridge converter," *IEEE Trans. Power Electron.*, vol. 40, no. 1, pp. 139–144, Feb. 1993.

[7] M. K. Nguyen, L. C. Lim, J. H. Choi, and G. B. Cho "Isolated High Step-up DC-DC Converter Based on Quasi-Switched-Boost Network". *IEEE Trans. Ind. Electron*, vol. 63, no. 12, pp. 7553-7562, Dec. 2016.

[8] M. K. Nguyen, T. D. Duong, L. C. Lim, and Y. J. Kim, "Isolated boost dc-dc converter with three switches," *IEEE Trans. Power Electron.*, vol. 33, no. 2, pp. 1389–1398, Feb. 2018.

[9] U. R. Prasanna, and A. K. Rathore, "Extended Range ZVS Active-Clamped Current-Fed Full-Bridge Isolated DC/DC Converter for Fuel Cell Applications: Analysis, Design, and Experimental Results," *IEEE Trans. Ind. Electron.*, vol. 60, no. 7, pp. 2661-2672, July 2013.

[10] S. A. Teston, E. G. Carati, J. P Costa, R. Cardoso, and C. M. O. Stein, "Comparison of two connection possibilities of the clamp capacitor in the active-clamped ZVS current-fed half-bridge converter" *in IEEE 13th Brazilian Power Electronics Conference and 1 st Southern Power Electronics Conference (COBEP/SPEC)*, Nov. 2015.

[11] U. R. Prasanna, and A. K. Rathore "Current-fed interleaved phase-modulated single-phase unfolding inverter: analysis, design, and experimental results," *IEEE Trans. Ind. Electron.*, vol. 61, no. 1, pp. 310-319, Jan. 2014.

[12] S. K. Han, H. K. Yoon, G. W. Moon, M. J. Youn, Y. H. Kim, and K. H. Lee, "A new active clamping zero-voltage switching PWM current-fed half-bridge converter," *IEEE Trans. Power Electron.*, vol. 20, no. 6, pp. 1271–1279, Nov. 2005.

[13] S. J. Jang, C. Y. Won, B. K. Lee, and J. Hur, "Fuel cell generation system with a new active clamping current-fed half-bridge converter," *IEEE Trans. Energy Convers.*, vol. 22, no. 2, pp. 332–340, Jun. 2007.

[14] S. S. Dobakhshari, J. Milimonfared, M. Taheri, and H. Moradisizkoohi, "a quasi-resonant current-fed converter with minimum switching losses," *IEEE Trans. Power Electron.*, vol. 32, no. 1, pp. 353-362, Jan. 2017.

[15] A. K. Rathore, A. K. S. Bhat, and R. Oruganti, "Analysis, design and experimental results of wide range ZVS active-clamped L-L type current-fed dc/dc converter for fuel cells to utility interface," *IEEE Trans. Ind. Electron.*, vol. 59, no. 1, pp. 473–485, Jan. 2012.

[16] P. Xuewei, and A. K. Rathore, "Naturally clamped soft-switching current-fed three-phase bidirectional DC/DC converter," *IEEE Trans. Ind. Electron.*, vol. 62, no. 5, pp. 3316-3324, May 2015.

[17] J. Park, and S. Choi, "Design and control of a bidirectional resonant DC-DC converter for automotive engine/ battery hybrid power generators," *IEEE Trans. Power Electron.*, vol. 29, no. 7, pp. 3748-3757, July 2014.

The 2018 International Power Electronics Conference

A High Gain Quasi Single Stage LLC Resonant DC/DC Converter with Coupled Inductor and Partial Active Clamp

Chongcan Huo, Xiaogao Xie*, Shuai Jiang and Hanjing Dong
School of Automation, Hangzhou Dianzi University, Hangzhou, China
*E-mail: Xiexg@hdu.edu.cn

Abstract—For low DC input applications such as PV converters, low input voltage causes large conduction losses of the primary side of the isolated front stage DC/DC converter due to the high primary current, which makes the efficiency is difficult to improve. In order to improve the conversion efficiency and optimize the topological structure, a quasi single stage LLC converter with coupled inductor and partial active clamp has been proposed in this paper. This new DC/DC converter can achieve soft switching of all switches and secondary rectifiers. In addition, taking advantage of the coupling of the inductor, the DC voltage gain of the proposed converter has been doubled compared to the conventional full bridge LLC resonant converter voltage. A 250W experimental prototype with 24V-36VDC input and 400VDC output was designed and tested.

Keywords— LLC, High Gain, Coupled inductor, Partial active clamp.

I. INTRODUCTION

As the presence of energy crisis and the deteriorating of environmental problems, the renewable resource such as solar power, have attract more and more attention. The isolated solar conversion usually contains a former DC/DC converter and an afterward inverter [7]. For micro solar inverter, the input source is usually a single PV board with low input DC voltage and the output is typically AC 220V, which means the front-stage DC/DC requires a large voltage gain [1]. Therefore, a front stage DC/DC with high step-up voltage gain is adopted to improve the voltage gain in two-stage micro solar inverter. However, the low input voltage causes large conduction losses on the primary side if the single-stage LLC resonant converter is applied as the front stage DC/DC converter directly.

Usually, an additional step-up DC/DC converter is added into the front stage DC/DC converter. As an example, a two stage DC/DC structure formed by Boost and LLC is shown in Fig.2. Furthermore, some high gain step-up DC/DC topologies have been proposed to replace the conventional boost converter [1]-[3][10]. The main merit of this topology is that efficiencies of two stages can be optimized separately. However, these two stage DC/DC converters usually need two set of control circuit, and the increased number of components also increases the cost of the converter. Additionally, it is also difficult

for the two-stage structure to meet the requirement of high efficiency due to the two cascaded efficiency. As shown in Fig.3, the quasi single stage boost+LLC converter [5]-[7], which integrates the boost circuit and LLC resonant converter, can achieve double voltage gain compared to the conventional LLC resonant converter. However, the soft switching of the LLC switches, especially the bottom switch connected with the boost inductor is lost because the boost inductor current offsets the resonant current for soft switching [2]-[4], [8]-[9], as shown in Fig.4.

In this paper, a new quasi single stage LLC converter with coupled inductor and partial active clamp circuit has been proposed, as shown in Fig.5. The inductor current falls rapidly during the dead time between Q3 and Q4 so that ZVS of the bottom switches are not affected. Unlike conventional full bridge LLC resonant converter, the upper switches Q1 and Q2 are mainly used to absorb the energy of coupled inductor's leakage inductor and clamp the voltage stress of the bottom switches to be double of the input voltage. A partial active clamp strategy is applied so that the upper clamp switches only conduct very short time in a switching cycle and low size and low cost switches can be applied Moreover, ZVS and ZCS can be realized with optimal on-time of the clamp switches.

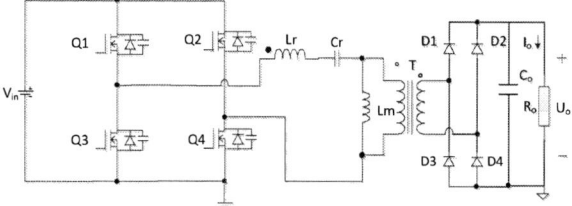

Fig.1. Single-stage LLC resonant converter

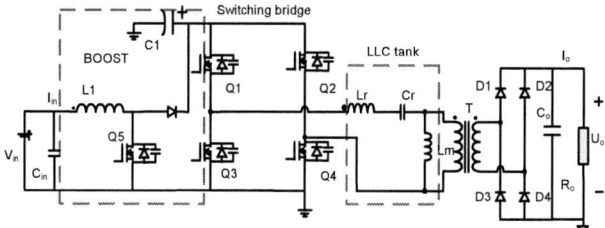

Fig.2. Two-stage Boost+LLC resonant converter

The 2018 International Power Electronics Conference

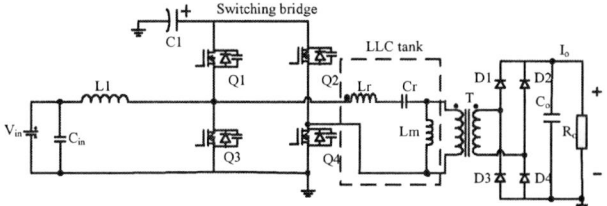

Fig.3. Quasi single stage Boost LLC resonant converter

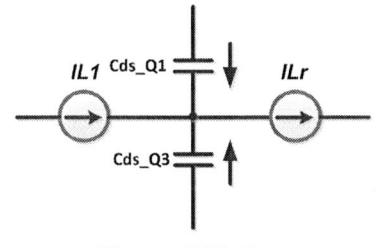

Q1 on-> off, Q3 off->on

Fig.4 Process during the dead time after Q1 off

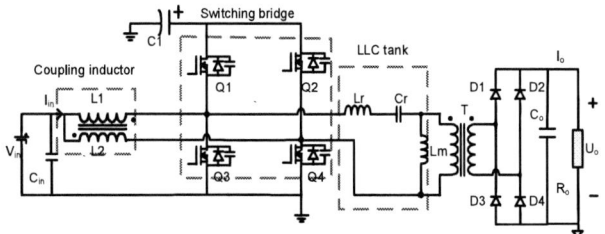

Fig.5. Proposed topology

II. PRINCIPLE OF OPERATION

In this part, the operation principle of the converter is introduced. The main operating waveform when operating frequency *fs* is equal to the resonant frequency *fr* is used to illustrate the principle of operation of the converter, as shown in Fig.6.

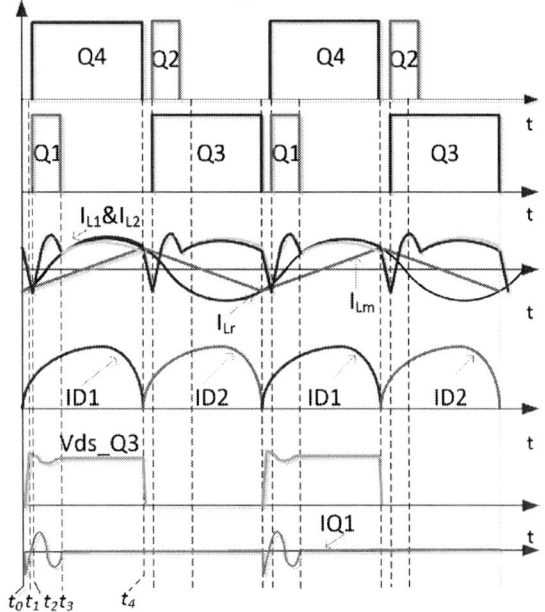

Fig.6 Operation waveforms of the proposed converter @ *fs=fr*

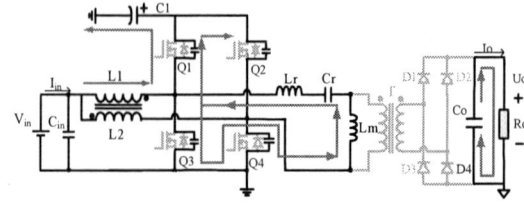

Mode 1

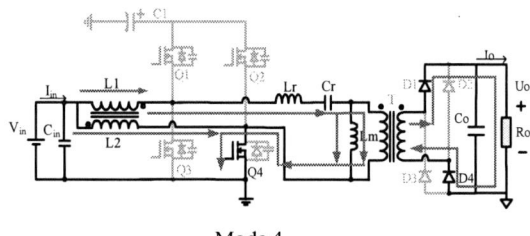

Mode 2

Mode 3

Mode 4

Fig.7 Operation modes

[1]Mode 1 (t0<t<t1): As shown in Fig.7a, at t0, Q3 turns off, the converter enters the dead zone and operates in Mode.1. The leakage inductors of coupled inductor resonant with the equivalent capacitors of Q1&Q3 and Q2&Q4 respectively, as shown in Fig.8. During this process, coupling inductor current drops rapidly. The difference value between coupling inductor current and the resonant current i_{Lr} discharges the parasitic capacitance of switch Q1 and switch Q4, while charges the parasitic capacitance of switch Q2 and switch Q3. At the same time, the secondary rectifier diodes D1 ~ D4 are is off.

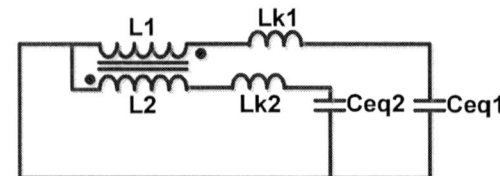

Fig.8 Equivalent resonant model under mode 1

988

[2]Mode 2 (t1<t<t2): In this interval, the converter will continue to work on dead zone. As shown in Fig.7b, at t1, the voltages across switches Q1 and Q4 have be discharged to zero, and the resonant current i_{Lr} continues to pass through the body diode of switch Q4. The voltage across the inductor L2 is clamped to the input voltage and the magnetizing inductor current of coupled inductor starts to rise linearly. The leakage inductor of the coupled inductor is in resonance with clamp capacitor C1 and the parasitic capacitor of Q3, the resonant current passes through the body diode of Q1. The Meanwhile, the output diode bridge does not work and the output capacitor Co provides the energy for the load consumption.

[3] Mode 3 (t2<t<t3): As shown in Fig.7c, at t2, switches Q1 and Q4 are zero voltage switched on (ZVS), the leakage inductor of the coupled inductor continues in resonance with clamp capacitor C1 and the parasitic capacitor of Q3. During this period. The secondary rectifier diodes D1 and D4 conduct and carry the output current.

[4] Mode 4 (t3<t<t4): As shown in Fig.7d, at t3, the reversed resonant current flowing through Q1 reaches zero, and Q1 is zero current switched off. At the same time, the current of iL1 and iL2 equal to the resonant current iLr. During this period, the voltage across Q3 keeps constant and equals to double of input voltage Vin. The LLC part operates like convention full bridge LLC converter with double input voltage.

The converter enters a second half cycle after t4, which is similar to the first half cycle and will not be described in detail here.

2.1 Converter voltage gain analysis

It can be found that the voltage across the LLC resonant tank has been increased due to the coupled inductor, which is doubled compared to conventional full-bridge LLC converter. Thus, the conduction losses of the resonant tank can be reduced. The DC gain of the proposed converter is,

$$G_{dc} = \frac{2 \cdot f^2 \cdot (K\text{-}1)}{\sqrt{(K \cdot f^2 - 1)^2 + f^2 (f^2 - 1)^2 \cdot (K\text{-}1)^2 \cdot Q^2}} \quad (1)$$

Here, L_r and C_r is resonant capacitor and resonant inductor of the resonant converter, "Q" refers to "quality factor" and $Q = \sqrt{L_r / C_r} / R_{ac}$, R_{ac} is the reflected load resistance and $R_{ac} = n^2 \cdot \frac{8}{\pi^2} \cdot \frac{V_o}{I_o} = n^2 \cdot \frac{8}{\pi^2} \cdot R_o$, f is the normalized switching and $f = f_s / f_r$, f_r is the resonant frequency and $f_r = 1 / 2 \cdot \pi \sqrt{L_r \cdot C_r}$, K is the ratio of total primary inductance to resonant inductance and $K = (L_r + L_m) / L_r$.

According to the (1), the calculated DC voltage gain G_{dc} curves of the proposed converter is shown in Fig.9b. For comparison, the DC voltage gain G_{dc} curves of conventional full bridge LLC converter is shown in Fig.9a. It can be seen that the voltage gain of the proposed converter is doubled.

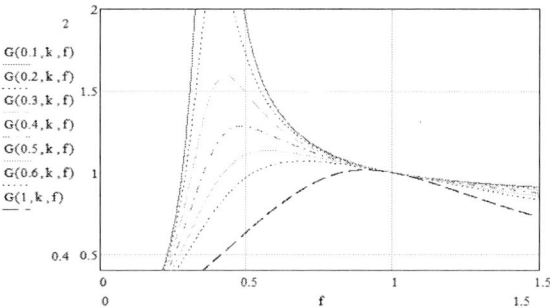

(a) DC voltage gain curves of conventional full bridge LLC

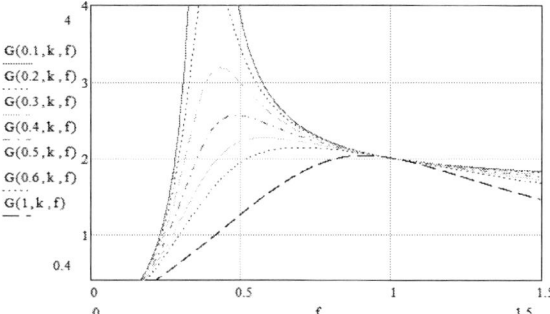

(b) DC voltage gain curves of proposed LLC converter
Fig.9 DC voltage gain curves

As everyone knows, a big advantage of the LLC converter is the ZVS characteristic of the switches. Because of the resonant process occurs after the main switch turns off, the current of the coupled inductor drops rapidly. Therefore, the boost stage of the quasi single stage converter will not affect the soft switching of the LLC resonant converter, which can not be achieved in conventional quasi single stage boost-LLC converter.

2.2 Design Consideration for Proposed Converter

For the proposed converter, the leakage inductance of the coupled inductor generates the resonance with the resonant capacitance C1 and the parasitic capacitance of switch. Then the zero current switching for the clamp switches can be achieved. Moreover, because the partial clamp control strategy is applied, the resonance period is short and the average current flows through the clam switch is much lower than that with complementary control. Correspondingly, the losses upper switches are reduced and the efficiency of proposed converter is improved.

Based on the characteristics of the selected switch, two main conditions should be satisfied. One is that the switch should be turned on completely, the other is the conduction time of clamp switch should approximately equal to the resonance period of the clamp process for reducing the conduction loss. The conduction time of clamp switch can be obtained according to (2).

$$T_{ra} = 2\pi \sqrt{L_k \cdot (C_1 // C_{oss\_d})} \quad (2)$$

Here, L_K is the leakage inductor in the coupled inductor, and C_{oss_d} is the parasitic capacitor of each bottom switch.

III. EXPERIMENTAL RESULTS

A 250 W lab-made prototype with 24-36VDC input and 400VDC output is built up to verify the proposed DC/DC resonant converter. The operating frequency range is designed to be 50 kHz to 120 kHz, and the resonant frequency is set to 100 kHz. The specification of the prototype example is shown in Table.I. For comparison, a 250W prototype of the traditional full bridge LLC resonant converter is also built up. Following the proposed procedure, the key circuit components of the prototype are listed in Table II. It can be seen that the MOSFET with small current stress are applied to upper switches.

Some tested waveforms of the proposed converter are shown in Fig.10~Fig.12. Fig.10 shows the currents of the windings of coupled inductor and LLC resonant current. The coupled inductor currents drop rapidly due to the resonant process between the clamp capacitor and the parasitic capacitor of the switch as the bottom switches turn off, which meets well with the theoretical analysis. Fig.11 shows the driving signal and switching waveform of bottom switch Q3. Q3 achieves ZVS very well. The switching waveform and current of upper switch Q1 are shown in Fig.12. The upper switch has achieved ZVS and ZCS.

Efficiency comparison between the proposed converter and the conventional full bridge LLC converter are shown in Fig.13. It can be seen that the proposed converter can achieve higher efficiency compared to the conventional full bridge LLC converter under different conditions.

Table I. Prototype specification

Input voltage	24V~36VDC
Output voltage	400VDC
Experimental specification	250W
Resonance frequency	100kHz

Table II. Parameters of main components

Components		Part name & Value
Full-Bridge MOSFET	Upper switches	IRF540
	Bottom switches	IPP045N10N3G
Rectifier diodes		MUR860
Resonant inductor (Lr)		PQ2620(7.36uH)
Magnetizing inductor(Lm)		PQ2625(30.9uH)
Coupled Inductor(L)		PQ2620(28uH)
Leakage Inductor		0.37uH
Resonant capacitor(Cr)		344.3nF
Turns ratio Np:Ns		10：58
Clamp capacitor C1		100nF

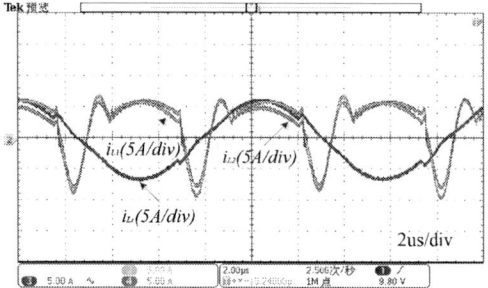

Fig.10 Measured current waveforms of the proposed converter @ 32VDC & full load

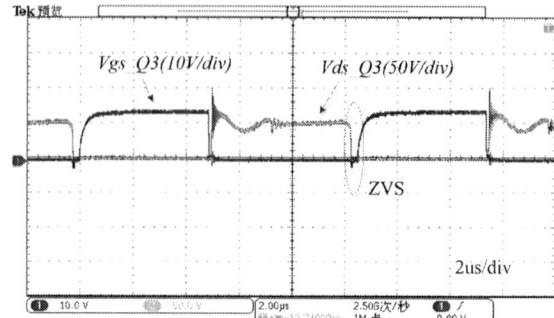

Fig.11 Measured waveforms of Q3 @ 32VDC & full load

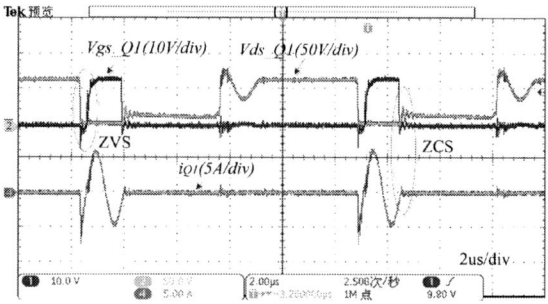

Fig.12 Vgs_Q1, Vgs_Q3 & i_{Q1} @ 32VDC & full load

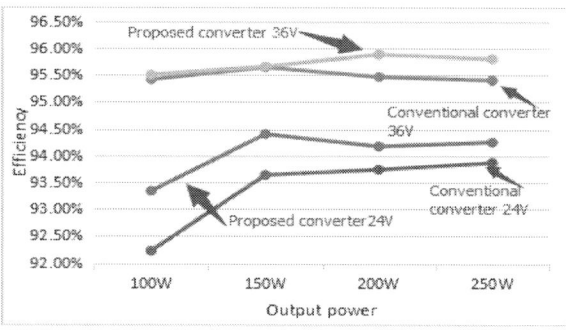

Fig.13 Efficiency comparison between the proposed converter and the conventional full bridge LLC converter

IV. CONCLUSION

In this paper, a high gain quasi single stage LLC resonant converter has been proposed for improving the conversion efficiency. The proposed converter can feature simple control strategy and simple structure. At the same time，this new proposed converter can achieve zero switching loss of the upper switches, ZVS of the bottom switches and ZCS of the secondary rectifiers. The total losses of the upper switches are reduced greatly and half-semiconductor components with small rated current can be applied. The proposed converter has been implemented to a 250 W prototype. Experimental results meet very well with theoretical analysis and the proposed converter can achieve higher efficiency than the conventional full bridge LLC converter. Thus, the proposed converter is a good candidate for renewable energy generation systems, such as photovoltaic, full cell and so on.

REFERENCES

[1] R. Beiranvand, B. Rashidian, M. R. Zolghadri, and S. M. H. Alavi,"Using LLC resonant converter for designing wide-range voltage Source," IEEE Trans. Ind. Electron, vol. 58, no.2,pp. 746-1756,May.2011

[2] B. Yang, F. C. Lee, A. J. Zhang, and G. Huang, "LLC resonant converter for front end dc–dc conversion," in Proc. APEC'02, 2002, pp.1108–1112.

[3] B. Chung, K. Yoon, S. Phum, E. Kim, and J. Won, "A novel LLC resonant converter for wide input voltage and load range,"in Proc. Power Electron.and ECCE Asia. Conf., May.2011, pp. 2825-2830.

[4] H. Guisong, G. Yilei, and Z. Jinfa, "LLC series resonant dc–dc converter," Power Supply China, vol. 1, no. 1, pp. 61–66, 2002.

[5] S.-Y. Chen, Z. R. Li, and C.-L. Chen, "Analysis and design of single-stage AC/DC LLC resonant converter," IEEE Trans. Ind. Electron., vol. 59, no. 3, pp. 1538–1544, Mar. 2012.

[6] C.-H. Chang, H.-Y. Chen, C.-T. Cho, and J.-Y. Chiu, "A novel single stage LLC resonant AC-DC converter with power factor correction feature," in Proc. 6th IEEE Conf. Ind. Electron. Appl., Jun. 2011, pp. 2191–2196.

[7] Lee. J. P, Min. B. D., Kim. T. J., "A Novel Topology for Photovoltaic DC/DC Full-Bridge Converter With Flat Efficiency Under Wide PV Module Voltage and Load Range". IEEE Trans. Ind Electron., vol. 55, June. 2008, pp. 2655-2663.

[8] Fang X., Hu H. B., Chen F., "Efficiency-Oriented Optimal Design of the LLC Resonant Converter Based on Peak Gain Placement". IEEE Trans. Power Electronics, vol.28, May. 2013, pp.2285-2296.

[9] Gu Y. L, Lu Z. Y, Qian Z. M., " Three-Level LLC Series Resonant DC/DC Converter". IEEE Trans. Power Electronics, vol 20 , April.2005, pp.781-789.

[10] Yang, Lung-Sheng, Liang, Tsorng-Juu, Chen, Jiann-Fuh," Transformerless DC-DC converters with high step-up voltage''.IEEE Trans. Industrial Electronics,vol.56,August. 2009, pp.3144-3152.

The 2018 International Power Electronics Conference

Suppression of Ripple Current in High Step-Up DC-DC Converter utilizing Cockcroft-Walton Circuit with Inductor

Takumi Yasuda[1], Masataka Minami[2*], Shin-ichi Motegi[2], and Masakazu Michihira[2]

1 Advanced Course of Electrical and Electronic Engineering, Kobe City College of Technology, Kobe, Hyogo, Japan
2 Department of Electrical Engineering, Kobe City College of Technology, Kobe, Hyogo, Japan
*E-mail: minami@kobe-kosen.ac.jp

Abstract—It has become necessary to reduce the volume of high step-up DC-DC converters for high-voltage DC power supplies. In this report, an isolated high step-up DC-DC converter is proposed for them. The proposed converter consists of the LLC converter and the Cockcroft-Walton (CW) circuit with an inductor. The proposed converter boosts the output voltage by double LC resonances of primary circuit and secondary circuit respectively. It is experimentally clarified that the inductor on the secondary side in the proposed converter plays the role of not only boosting the output voltage, but also suppressing ripples of transformer primary and secondary currents.

Keywords—Cockcroft-Walton circuit, LLC converter, double LC resonances

I. INTRODUCTION

At present, high-voltage DC power supplies are widely utilized in many apparatuses, such as X-ray system, electron beam systems, and so on [1]–[3]. However, such apparatuses are utilized in limited fields because of their large-size and high-cost. It has become necessary to reduce the volume of the high-voltage DC power supplies.

Non-isolated high step-up DC-DC converters, which proposed in the past, employs coupled-inductor [4], [5], voltage multiplier [6], [7], boost converter [8], [9], and combination of these techniques [10], [11] to obtain high voltage gain. These converters achieve more than 10 of boost ratio. Nevertheless, the non-isolated converters requires other method from electrical isolation to ensure safety. On the other hand, the DC-DC converters with transformer are capable of boosting the output voltage as well as electrical isolation. Thus, an isolated high step-up DC-DC converter is proposed in this report. Isolated high step-up DC-DC converters are often based on the topology of the boost converter [12], [13] and the full-bridge inverter [2] in the primary side of the transformer. Nevertheless, these converters are difficult to reduce their size because they contain a lot of elements such as plural transformers, inductors, and more than 3 switches.

This report proposes an isolated high step-up DC-DC converter shown in Fig. 1. The proposed converter adopts LLC converter utilizing leakage inductor of transformer as a resonance inductor as shown in Fig. 2 [14], [15]. The proposed converter is relatively small in size since

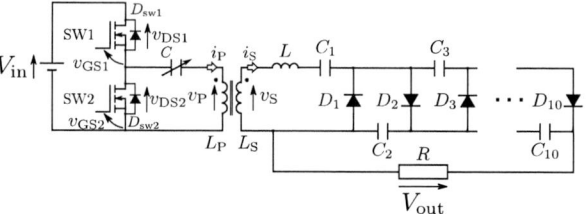

Fig. 1: Proposed isolated high step-up DC-DC converter.

the LLC converter has a small number of inductors and switching devices compared with the conventional converters. Furthermore, the proposed converter boosts the voltage by the primary circuit which utilizes characteristics of the LLC converter. In addition, because the LLC topology reduces switching loss with Zero Voltage Switching (ZVS), the proposed converter has small heat sinks. It is considered that the proposed converter achieves downsizing and high boost ratio.

The most isolated high step-up DC-DC converters adopt the voltage multiplying circuit or the Cockcroft-Walton (CW) circuit [16] as the rectifier [2], [12]. The CW circuit is well-known as a high step-up rectifier with a simple structure that composed only of passive elements. The proposed converter in this report employs the 5-stage CW circuit shown in Fig. 3. The ideal 5-stage CW circuit generates 5 times DC output voltage of the peak to peak (P-P) value of the input voltage [16]. However, the output voltage of an actual CW circuit decreases compared with the ideal CW circuit on account of load current [17], [18] and diode junction capacitor [19], [20]. In order to overcome these drawbacks, various methods are proposed such as high frequency operation, increasing the number of stages [17], parameter adjustment [18], [21], and novel circuit structure [19]. We proposed a novel CW circuit that inserted an inductor to input side [22]. And then, it is clarified that the inductor resonants with the diode junction capacitors in the CW circuit and the output voltage exceeds the ideal output voltage of the conventional CW circuit [22], [23]. It is considered that the CW circuit with the inductor is the parallel resonant converter [24] utilizing the diode junction capacitor. This result helps to downsize of high step-up DC-DC converters since the CW circuit with the inductor obtains same output voltage

992

The 2018 International Power Electronics Conference

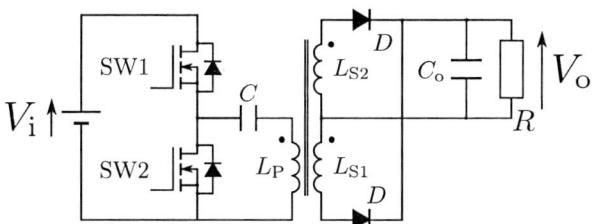

Fig. 2: Circuit diagram of LLC converter.

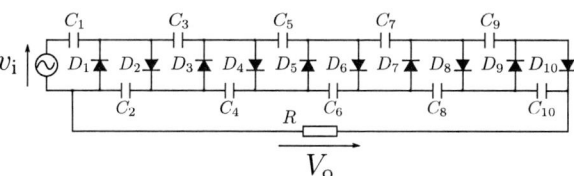

Fig. 3: Circuit diagram of 5-stage CW circuit.

of the conventional CW circuit in a smaller number of stages.

This report proposes the isolated high step-up DC-DC converter with the LLC converter and the novel CW circuit as shown in Fig. 1. The proposed converter boosts the output voltage by double LC resonances of the primary circuit and the secondary circuit respectively. The effect of the inductor in the CW circuit is investigated in this report. Firstly, this report presents frequency characteristics of the output voltage compared with the converter which has a different resonance frequency of primary circuit. Then, voltage and current waveforms of the transformer of the proposed converter is compared with the converter without inductor. As a result, the ripples of transformer primary and secondary currents are generated owing to the resonance between the diode junction capacitors of the CW circuit and leakage inductor of the transformer if there is no inductor in the input side of the CW circuit. On the other hand, the inductor in the proposed converter suppresses the ripples of the transformer primary and secondary currents. In addition, the output voltage increases due to the resonance between the diode junction capacitors in the CW circuit and the inserted inductor.

II. CONSTRUCTION OF PROPOSED CONVERTER

The proposed converter is composed of the LLC topology and the CW circuit with an inductor. This section describes the topology of the LLC converter and the CW circuit with the inductor respectively. Then, the construction of the proposed converter and its characteristics are concisely presented.

A. LLC Converter

The proposed converter utilizes the LLC topology on the primary side. Fig. 2 depicts the circuit diagram of the LLC converter. Two switches of the LLC converter are operated in 50 % duty cycle and variable frequency [25]. The LLC converter has several advantages, such as simple circuit configuration, large output voltage range, and ZVS in all load range [26]. In addition, the voltage gain of the LLC converter increasing thanks to the resonance between a resonance capacitor C and a mutual inductor L_M of the transformer. Meanwhile, the switching frequency range of the LLC converter has to be higher than the resonance frequency between resonance capacitor C and the mutual inductor L_M of the

transformer to achieve ZVS [26]. Therefore, the proposed converter, which utilizes the LLC converter, also operates the frequency above the resonance frequency between the resonance capacitor C and the mutual inductor L_M of the transformer.

B. Cockcroft-Walton circuit with inductor

The CW circuit consists of the capacitors and the diodes connected as shown in Fig. 3. When the input voltage v_i in Fig. 3 is downward, the diodes numbered odd turn on and capacitors numbered odd are charged. On the other hand, when the input voltage v_i is upward, the diodes numbered even turn on and capacitors numbered even are charged. As a result, capacitors numbered even of the ideal unloaded CW circuit are charged DC voltage which is equal to the P-P value of the input voltage v_i. Therefore, the ideal output voltage V_o of unloaded 5-stage CW circuit in Fig. 3 is 5 times P-P value of the input voltage. However, the output voltage of the CW circuit decreases on account of load current [17], [18]. Furthermore, the ON priod of the diodes in the CW circuit with light load is a very short. All diodes are regarded as junction capacitors C_T at the OFF period. Thus, the output voltage V_o of the CW circuit is divided to the junction capacitors C_T and decreased [19].

Therefore, an inductor was inserted in series with the CW circuit in previous reports [22], [23]. In almost cases, the capacitance of a diode junction capacitor is much smaller than the capacitance C_1, ..., and C_{10} in the CW circuit. Thus, the CW circuit in Fig. 3 is regard as a capacitive load which 10 diode junction capacitors C_T is connected in parallel. The total equivalent capacitor of CW circuit $10C_T$ is possible to increase the output voltage, resonating with the inductor inserted into the input side [22], [23]. It is considered that the CW circuit with the inductor is a parallel resonant converter [24] utilizing the diode junction capacitor. As a result, the output voltage V_o of the CW circuit with the inductor exceeds the ideal output voltage of the CW circuit without the inductor. This result helps to downsize the proposed converter since the CW circuit with the inductor obtains the same output voltage of the CW circuit without the inductor in a smaller number of stages.

C. Construction of proposed converter

The proposed converter utilizes the CW circuit with the inductor as a rectifier of the LLC converter. The two switches of the converter alternately operate in 50 % duty cycle in the same way as the LLC converter. The resonance capacitor C and the mutual inductor L_M

993

constitute a resonance circuit of the primary side. On the other hand, the inductor L and the equivalent capacitor of the CW circuit constitute a resonance circuit of the secondary side. The proposed converter boosts the output voltage with these double LC resonant circuits.

In order to downsize the circuit, all switching devices in the proposed converter are Silicon Carbide (SiC) devices. Recently, switching frequency tends to be higher with switching devices based on wide bandgap semiconductors [27], [28]. High switching frequency helps passive components to downsize. The switching devices based on SiC reduce switching losses and achieve higher frequency switching compared with Si devices [29]. Furthermore, since SiC devices are tough against heat, the volume of the heat sinks are possible to be reduced [28]. Moreover, SiC devices have an advantage in forward voltage compared with Si devices in high step-up DC-DC converters [28]. Therefore, all switching devices in the proposed converter are SiC devices.

III. DESIGN OF PROPOSED CONVERTER

This section presents the design of the transformer, the resonance capacitor C, and the inductor L in the proposed converter concisely.

The proposed converter boosts the output voltage by the double LC resonances of the primary circuit and the secondary circuit respectively. The output voltage of the proposed converter is anticipated to be the highest when the resonance frequency on the primary side is same with the resonance frequency on the secondary side. As mentioned last section, the LLC converter has to be operated with the switching frequency over the resonance frequency of the primary circuit to achieve ZVS. Hence, the proposed converter operates at the switching frequency f_{sw} which is little higher than the resonance frequency of the primary circuit.

Firstly, the design of the transformer is presented. To achieve ZVS, the limitation of mutual inductance L_M of LLC topology is written as follows:

$$L_M \leq \frac{t_{dt}}{16 C_S f_{sw,max}}, \quad (1)$$

where t_{dt} is the deadtime, C_S is the sum of the output capacitance of the MOSFETs and the junction capacitance of the anti-parallel diodes. And $f_{sw,max}$ is the maximum switching frequency [26]. The output capacitance C_{oss} of MOSFETs SW1 and SW2 (SCT2120AF) in this report are set to 250 pF, and the junction capacitances of anti-parallel diodes D_{sw1} and D_{sw2} (SCS205KG); C_T are 100 pF according to the datasheets. Also, the maximum switching frequency $f_{sw,max}$ is set to 1 MHz, and the deadtime of switching is 0.15 μs in this report. Therefore, calculated mutual inductance L_M is less than 17.8 μH. The mutual inductance L_M of the transformer is decided 12 μH in this report.

Secondly, the resonance frequency on the primary side is described. The transformer voltage is stepped up by the LC resonance between the resonance capacitor C

TABLE I: Experimental parameters

SiC MOSFET	SW1, SW2	650 V, 29 A
SiC SBD	D_{sw1}, D_{sw2}	1200 V, 5 A
	D_1, ..., and D_{10}	1200 V, 5 A
Capacitor	C	15 nF, 8.2 nF, 6.8 nF, 630 V
	C_1, ..., and C_{10}	4700 pF, 400 V
Transformer	L_P, L_S	12 μH, 1:1, $k = 0.98$
Inductor	L	70 μH
Load	R	1 MΩ, 3 W

and the mutual inductor L_M on the primary side in the proposed converter. Therefore, the resonance frequency on the primary side is written as

$$f_{r1} = \frac{1}{2\pi\sqrt{L_M C}}. \quad (2)$$

This is the same equation with the design equation of the LLC converter [26].

Finally, the selection of the inductance L on the secondary side is represented. The inductor L plays the role of boosting the output voltage by the LC resonance with the equivalent capacitor of the CW circuit $10C_T$ as described in Sec. II. B. Since the secondary resonance circuit is considered to be a parallel resonance circuit which consists of the inductor L and the equivalent capacitor of the CW circuit, the inductance L is written as

$$L = \frac{1}{(2\pi \cdot f_{r2})^2 10 C_T}, \quad (3)$$

where the diode junction capacitance (SCS205KG) C_T is 100 pF as shown above. Therefore, the equivalent capacitance of the CW circuit $10C_T$ is 1000 pF. The resonance frequency on the secondary side of the proposed converter is set at 600 kHz. Therefore, the inductance L calculated is 70.4 μH. The inductance L is 70 μH in this report.

IV. EXPERIMENTAL ANALYSIS

This section experimentally investigates the proposed converter. At first, experimental conditions are described. Then, the switching frequency characteristics of the output voltage V_{out} depending on the presence or absence of the inductor L are represented. Thirdly, the differences of the output voltage V_{out} due to the resonance frequency on the primary side are compared. Finally, the voltage and current waveforms of each element depending on the presence or absence of the inductor L are investigated.

A. Experimental conditions

Table I lists experimental conditions. In experiments, the resonance frequency on the primary side is changed with the resonance capacitor C, which is 15 nF, 8.2 nF, 6.8 nF. The calculated resonance frequency on the primary side is respectively 375 kHz, 507 kHz, and 557 kHz. Changing the resonance frequency on the primary side, the output voltages V_{out} are compared with the gaps of the resonance frequency of the primary circuit and the resonance frequency of the secondary circuit. It is anticipated that the output voltage V_{out} is the highest when the resonance capacitance C is 6.8 nF since the

The 2018 International Power Electronics Conference

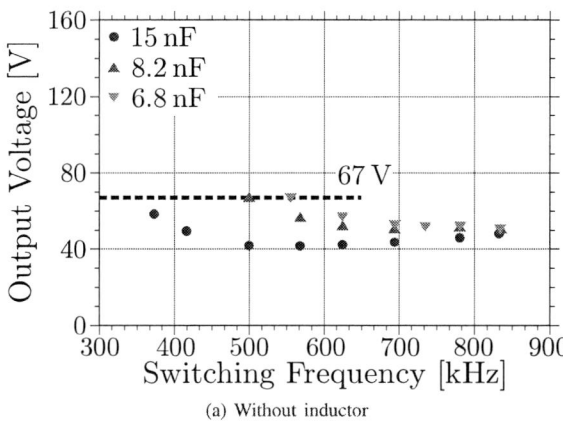

(a) Without inductor

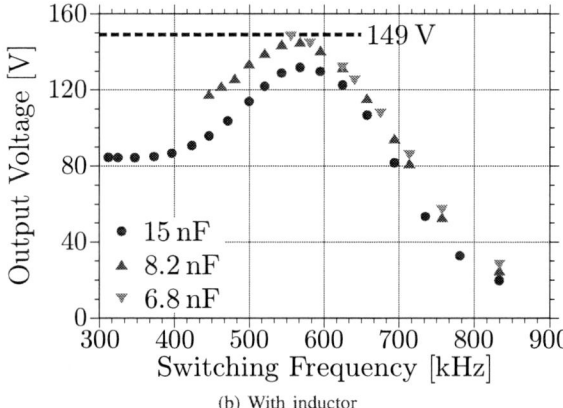

(b) With inductor

Fig. 4: Frequency characteristics of output voltage in the proposed converter depending on the presence or absence of the inductor.

resonance frequency on the primary side is the nearest to the resonance frequency of the secondary circuit in the three capacitors C. The turn ratio of the transformer is 1:1. The effect of stepping up without the transformer is investigated in this report. In addition, coupling coefficient k of the transformer is equal to 0.98.

B. Frequency characteristics of output voltage

Fig. 4 shows the frequency characteristics of the output voltage V_{out}. The keys in Fig. 4 mean the resonance capacitance C. In Fig. 4a, which is the result without the inductor L and $C = 6.8\,\text{nF}$, the output voltage V_{out} increases and reaches 67 V as the switching frequency f_{sw} decreases. It is considered that the output voltage V_{out} increases because lower switching frequency is close to the resonance frequency of the primary circuit.

Whereas in Fig. 4b, which is the result with the inductor L, the frequency characteristics of the output voltage V_{out} becomes a peak value at 550 kHz regardless of the resonance capacitance C. Because the switching frequency f_{sw} where the output voltage V_{out} becomes peak value is near the resonance frequency on the secondary side 600 kHz, it is considered that the peak output voltage V_{out} produced by the LC resonance of the

secondary circuit. In Fig. 4b, the peak value of the output voltage $V_{\text{out}} = 149\,\text{V}$, where the resonance capacitance C is 6.8 nF, is little higher than the peak value of the output voltage V_{out} where the resonance capacitance C are 8.2 nF and 15 nF. These differences suppose to occur by the resonance of the primary circuit. Nevertheless, the output voltages V_{out} in the different value of the resonance capacitance C are almost same. Thus, it is suggested that stepping up in the proposed converter depends on the resonance of the secondary circuit. In other words, it is not necessary to input high voltage to the transformer of the proposed converter since the secondary circuit steps up enough. Thus, a transformer with low boost ratio could be adopted to the proposed converter. As a result, the isolation between the wire and the core of the transformer in the proposed converter becomes uncomplicated and the proposed converter must downsize.

C. Voltage and current waveforms

Fig. 5 demonstrates voltage and current waveforms of each element depending on the presence or absence of the inductor. The waveforms of Fig. 5 is measured under the condition that the resonance capacitance C is 15 nF and the switching frequency f_{sw} is 500 kHz. In Fig. 5a, which is measured under no secondary inductor L, the transformer currents i_{P} and i_{S} contain about 7.5 MHz ripple. The frequency of the current ripples agrees with the resonance frequency between the leakage inductance of the transformer ($0.48\,\mu\text{H}$) and the equivalent capacitance of the CW circuit ($10C_{\text{T}}=1000\,\text{pF}$). Therefore, it is considered that the ripples are caused by the resonance between the leakage inductor of the transformer and the equivalent capacitor of the CW circuit. The ripples must cause high frequency noise on other elements and the gate-driver of itself. Therefore, it is necessary to suppress these ripples. On the other hand, the current ripples are suppressed in Fig. 5b, which measured under the condition that the converter inserted the inductor L. It is considered that the inserted inductor L suppresses the current ripples. Consequently, the inductor L in the secondary side plays the role of not only boosting the output voltage, but also suppressing the current ripples.

V. CONCLUSION

This report proposes the isolated high step-up DC-DC converter for X-ray systems, electron beam systems, and so on. The proposed converter consists of the LLC converter and the CW circuit with an inductor. The proposed converter steps up by the double LC resonances on the primary side and the secondary side respectively. This report clarified that the secondary inductor plays the role of not only boosting the output voltage, but also suppressing the ripples of the transformer currents caused by the resonance between the leakage inductor of the transformer and the equivalent capacitor of the CW circuit.

In future works, the turn ratio of the transformer will be changed. In addition, the primary side of the converter

The 2018 International Power Electronics Conference

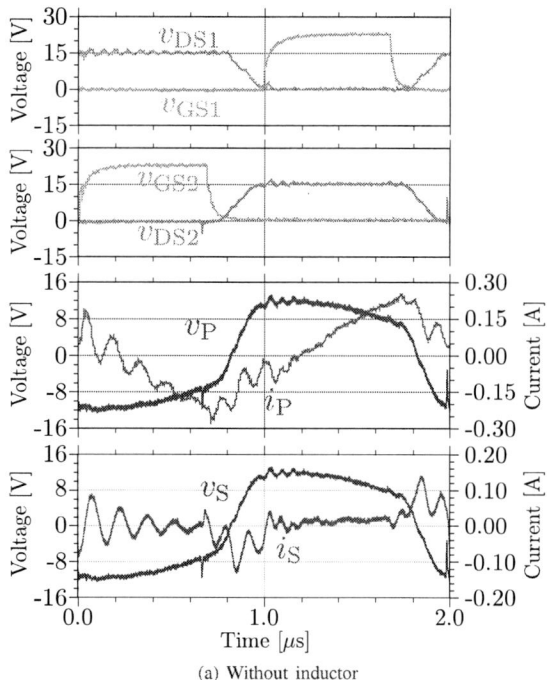

(a) Without inductor

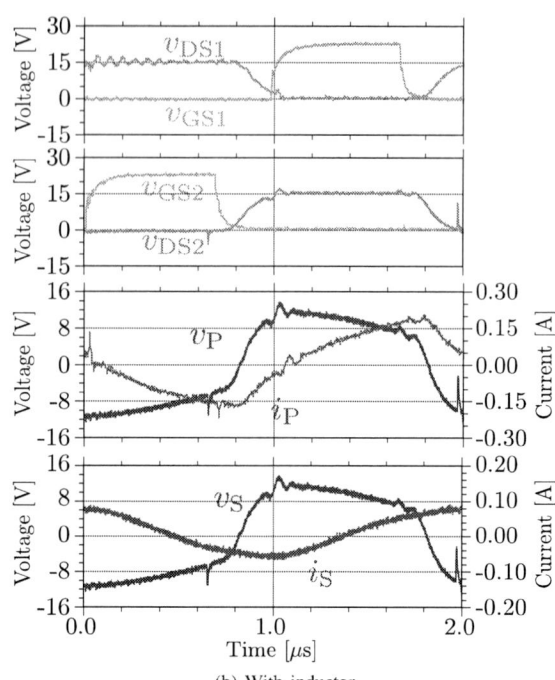

(b) With inductor

Fig. 5: Voltage and current waveforms at the transformer in the proposed converter at $C = 15\,\mathrm{nF}$ and $f_{\mathrm{sw}} = 500\,\mathrm{kHz}$.

will be improved to obtain a higher output voltage than this reports.

ACKNOWLEDGEMENT

This research is partially supported by the Kyoto Technoscience Center in Japan.

REFERENCES

[1] Z. Ghazali, K. Z. Dahlan, S. H. Aiasah, A. B. Khomsaton, and T. M. Ting, "Status of Electron Beam Processing Technology in Malaysia," in *Proceedings of the FNCA 2002 workshop on application of electron accelerator. Radiation system for liquid samples*, 2003.

[2] W. C. Hsu, J. F. Chen, Y. P. Hsieh, and Y. M. Wu, "Design and Steady-State Analysis of Parallel Resonant DC-DC Converter for High-Voltage Power Generator," *IEEE Transactions on Power Electronics*, vol. 32, no. 2, pp. 957–966, 2017.

[3] A. Pernía, M. J. Prieto, P. J. Villegas, J. Díaz, and J. A. Martín-Ramos, "LCC Resonant Multilevel Converter for X-ray Applications," *Energies*, vol. 10, no. 10, p. 1573, 2017.

[4] Q. Zhao and F. C. Lee, "High-Efficiency, High Step-Up DC-DC Converters," *IEEE Transactions on Power Electronics*, vol. 18, no. 1, pp. 65–73, 2003.

[5] S. Dwari and L. Parsa, "An Efficient High-Step-Up Interleaved DC-DC Converter with a Common Active Clamp," *IEEE Transactions on Power Electronics*, vol. 26, no. 1, pp. 66–78, 2011.

[6] C. M. Young, M. H. Chen, T. A. Chang, C. C. Ko, and K. K. Jen, "Cascade Cockcroft–Walton Voltage Multiplier Applied to Transformerless High Step-Up DC-DC Converter," *IEEE Transactions on Industrial Electronics*, vol. 60, no. 2, pp. 523–537, 2013.

[7] L. H. Barreto, P. P. Praça, G. A. Henn, R. N. Silva, and D. S. Oliveira, "Single Stage High Voltage Gain Boost Converter with Voltage Multiplier Cells for Battery Charging Using Photovoltaic Panels," in *Applied Power Electronics Conference and Exposition*

(APEC), 2012 Twenty-Seventh Annual IEEE. IEEE, 2012, pp. 364–368.

[8] K. Hwu and Y. Yau, "High Step-Up Converter Based on Charge Pump and Boost Converter," *IEEE Transactions on Power Electronics*, vol. 27, no. 5, pp. 2484–2494, 2012.

[9] Y. Tang, D. Fu, T. Wang, and Z. Xu, "Hybrid Switched-Inductor Converters for High Step-Up Conversion," *IEEE Transactions on Industrial Electronics*, vol. 62, no. 3, pp. 1480–1490, 2015.

[10] F. L. Tofoli, D. de Souza Oliveira, R. P. Torrico-Bascopé, and Y. J. A. Alcazar, "Novel Nonisolated High-Voltage Gain DC-DC Converters Based on 3SSC and VMC," *IEEE Transactions on Power Electronics*, vol. 27, no. 9, pp. 3897–3907, 2012.

[11] Y. P. Hsieh, J. F. Chen, T. J. P. Liang, and L. S. Yang, "Novel High Step-Up DC-DC Converter with Coupled-Inductor and Switched-Capacitor Techniques for a Sustainable Energy System," *IEEE Transactions on Power Electronics*, vol. 26, no. 12, pp. 3481–3490, 2011.

[12] T. J. Liang, J. H. Lee, S. M. Chen, J. F. Chen, and L. S. Yang, "Novel Isolated High-step-up DC-DC Converter with Voltage Lift," *IEEE Transactions on Industrial Electronics*, vol. 60, no. 4, pp. 1483–1491, 2013.

[13] F. Evran and M. T. Aydemir, "Isolated High Step-Up DC-DC Converter with Low Voltage Stress," *IEEE Transactions on Power Electronics*, vol. 29, no. 7, pp. 3591–3603, 2014.

[14] B. Yang, R. Chen, and F. C. Lee, "Integrated Magnetic for LLC Resonant Converter," in *Applied Power Electronics Conference and Exposition APEC 2002. Seventeenth Annual IEEE*, vol. 1. IEEE, 2002, pp. 346–351.

[15] J. Y. Lee, Y. S. Jeong, and B. M. Han, "An Isolated DC/DC Converter Using High-Frequency Unregulated *LLC* Resonant Converter for Fuel Cell Applications," *IEEE Transactions on Industrial Electronics*, vol. 58, no. 7, pp. 2926–2934, 2011.

[16] J. D. Cockcroft and E. T. S. Walton, "Experiments with High Velocity Positive Ions," *Proceedings of the Royal Society of London. Series A*, vol. 129, no. 811, pp. 477–489, 1930.

[17] A. Iijima, T. Hakamada, S. Matsui, and K. Kanaya, "DC High-Voltage Power Supply Molded by Epoxy Resin for Electron Beam

Equipment," *Bulletin of the Electrotechnical Laboratory*, vol. 33, no. 11, pp. 1351–1365, 1969.

[18] I. C. Kobougias and E. C. Tatakis, "Optimal Design of a Half-Wave Cockcroft–Walton Voltage Multiplier with Minimum Total Capacitance," *IEEE Transactions on Power Electronics*, vol. 25, no. 9, pp. 2460–2468, 2010.

[19] E. Everhart and P. Lorrain, "The Cockcroft-Walton Voltage Multiplying Circuit," *Review of Scientific Instruments*, vol. 24, no. 3, pp. 221–226, 1953.

[20] M. Minami, T. Ito, S. Motegi, and M. Michihira, "Theoretical Analysis of Decreased Boost Ratio in Unloaded Cockcroft-Walton Circuit (in Japanese)," *IEEJ Transaction on Industrial Application*, vol. 136, no. 3, pp. 246–247, 2016.

[21] T. Fukuyama and K. Sugihara, "Study on Operating Principle of Cockcroft-Walton Circuit to Produce Plasmas Using High-Voltage Discharge," *Plasma and Fusion Research*, vol. 11, pp. 2 401 008–2 401 008, 2016.

[22] M. Minami, T. Ito, S. Motegi, and M. Michihira, "Boost Ratio and Power Factor Improvement in Cockcroft-Walton Circuit with Diode Junction Capacitor (in Japanese)," *IEEJ Transactions on Industry Applications*, vol. 136, no. 12, pp. 991–996, 2016.

[23] ——, "An Experimental Analysis of Output Voltage Characteristics in Cockcroft-Walton Circuit with SiC Diode and Input Inductor," in *Electrical Machines and Systems (ICEMS), 2016 19th International Conference on*. IEEE, 2016, pp. 1–4.

[24] G. Ivensky, A. Kats, and S. Ben-Yaakov, "An RC Load Model of Parallel and Series-Parallel Resonant DC-DC Converters with Capacitive Output Filter," *IEEE Transactions on Power Electronics*, vol. 14, no. 3, pp. 515–521, 1999.

[25] J. Zhang, J. Liao, J. Wang, and Z. Qian, "A Current-Driving Synchronous Rectifier for an LLC Resonant Converter with Voltage-Doubler Rectifier Structure," *IEEE Transactions on Power Electronics*, vol. 27, no. 4, pp. 1894–1904, 2012.

[26] J. Jung and J. Kwon, "Theoretical Analysis and Optimal Design of LLC Resonant Converter," in *Power Electronics and Applications, 2007 European Conference on*. IEEE, 2007, pp. 1–10.

[27] H. Ohashi, I. Omura, S. Matsumoto, Y. Sato, H. Tadano, and I. Ishii, "Power Electronics Innovation with Next Generation Advanced Power Devices," *IEICE Transactions on communications*, vol. 87, no. 12, pp. 3422–3429, 2004.

[28] J. Biela, M. Schweizer, S. Waffler, and J. W. Kolar, "SiC versus Si—Evaluation of Potentials for Performance Improvement of Inverter and DC-DC Converter Systems by SiC Power Semiconductors," *IEEE Transactions on Industrial Electronics*, vol. 58, no. 7, pp. 2872–2882, 2011.

[29] T. Friedli, S. Round, and J. W. Kolar, "A 100 kHz SiC Sparse Matrix Converter," in *Power Electronics Specialists Conference, 2007. PESC 2007. IEEE*. IEEE, 2007, pp. 2148–2154.

An Optimal Design Method Considering Transformer Parasitic Capacitance of LLC Resonant Converters

Naizeng Wang, Xu Yang, Mofan Tian, Haiyang Jia, Guangzhao Xu, Zhenwei Li
State Key Laboratory of Electrical Insulation and Power Equipment, Xi'an Jiaotong University
Xi'an, People's Republic of China
E-mail: wangnaizeng3878212@163.com

Abstract— **LLC resonant converters have been widely used in DC-DC conversion applications due to the advantages of high efficiency and high power density. There are many design methods of LLC resonant converters in former research. However, none of them take transformer parasitic capacitance into consideration. This paper proposes an optimal design method considering transformer parasitic capacitance. Firstly, a brief summary of design process is introduced. Next, a wire-wound transformer is designed and the transformer parasitic capacitance is measured. After that, the resonant parameters are selected aiming at reducing device conduction loss and satisfying gain requirement. Finally, experiments are conducted on a 400V-12V 200W GaN-based LLC resonant converter. Experimental results show that the optimal design method has good performance and the efficiency of the prototype is up to 94.2%.**

Keywords— ***Dead Time, LLC, Optimal Design Method, Transformer Parasitic Capacitance***

I. INTRODUCTION

LLC resonant converters are attracting increasing attention because it can achieve both high efficiency and high power density. LLC resonant converters have many advantages over conventional resonant converters. For example, it can regulate the output voltage over wide line and load variations with a relatively small variation of switching frequency. It can also achieve zero voltage switching (ZVS) over the entire operating range. Another advantage of LLC resonant converters is that the two physical inductors can be integrated into one physical component, including both the series resonant inductance and the magnetizing inductance [1-4].

Several design methods of LLC resonant converters have been developed in the past, which can be summarized into three categories: 1) fundamental harmonic approximation (FHA), 2) FHA with time domain correction, 3) time-domain analysis. LLC resonant tank has two degrees of freedom, the quality factor Q and the inductance ratio λ. Accordingly, previous design methods usually use the peak gain requirement and additional conditions to complete a design. In [5,6], a universal approach to optimally design LLC converters is proposed, the core of which is an accurate algorithm that can find all possible designs satisfying the peak gain requirement. Designers can conveniently evaluate several design results and find the optimal one for their respective applications. In [7], a semi-empirical design methodology is proposed by fixing maximum switching frequency and dead time and estimating efficiency. In [8], when LLC resonant converter is used as a wide output range voltage source, maximum switching frequency and output voltage ripple are predefined to start the design flowchart. In [9], the selection of quality factor aims at limiting start-up current. The only way to limit the start-up current is to enlarge quality factor in design. In [10], the inductance ratio is selected empirically by observing the gain curve to sustain stable gain for control. In [11], the selection of quality factor and inductance ratio is limited by the maximum voltage or AC stress of resonant capacitor. In [12], nine operation modes are analyzed. Resonant frequency, maximum switching frequency and dead time are swept to attain the optimal design.

However, few design methods considering transformer parasitic capacitance were proposed in former research. When LLC resonant converter operates at a high frequency, the parasitic capacitance presents in power transformers can't be neglected any more [13,14]. In order to guarantee ZVS of power devices, sufficient magnetizing current and dead time are needed to ensure that all the parasitic capacitances have been discharged, including the output capacitances of the primary side devices and secondary side devices as well as the transformer parasitic capacitance.

This paper proposes an optimal design method, which takes transformer parasitic capacitance into account. In section II, several concise design steps are listed. In section III, six lumped capacitances model is built, which can be reduced to one lumped capacitance model. After that, a wire-wound transformer is designed and its parasitic capacitance is measured experimentally. In section IV, the resonant parameters are determined aiming at reducing device conduction loss and satisfying gain requirement. In section V, experiments are conducted on a 400V-12V 200W LLC prototype. Experimental results show that the optimal design method has good performance and the efficiency of the prototype is up to 94.2%.

II. DESIGN PROCESS

Fig. 1 shows the LLC design flowchart. The first step is to determine the transformer turns ratio according to given specifications. The second step is to design a transformer preliminarily, including structure design, winding design and core design. After that, the transformer parasitic capacitance can be measured. Next, according to the relationship between dead time and device conduction loss, optimal dead time can be attained. Corresponding magnetizing inductance can be calculated. Then there will be a one-to-one relationship between inductance ratio λ and quality factor Q. Simulation is necessary in order to judge which group of λ and Q can satisfy the gain requirement. If there are no groups of λ and Q can satisfy the gain requirement, then the dead time and magnetizing inductance should be adjusted slightly. "Trial and error" is inevitable in the design process. Finally, detailed design of transformer is indispensable to get proper series inductance and magnetizing inductance.

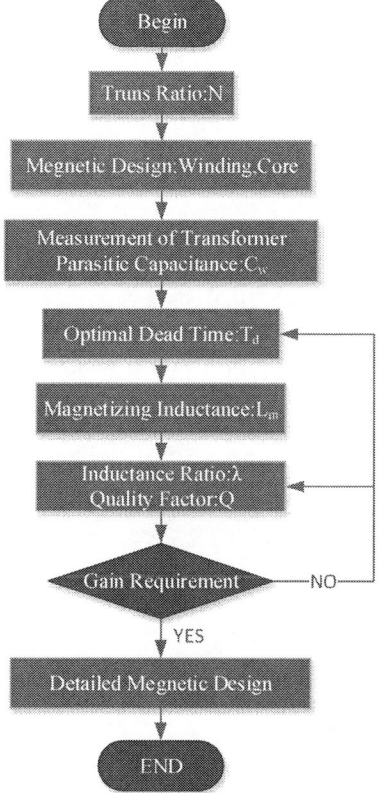

Fig. 1. LLC design flowchart

Fig. 2 shows the topology of half bridge LLC resonant converter. The specifications are listed in Table I. Detailed design will be explained in following sections. In order to explain the design process clearly, some symbols and meanings are listed in Table II.

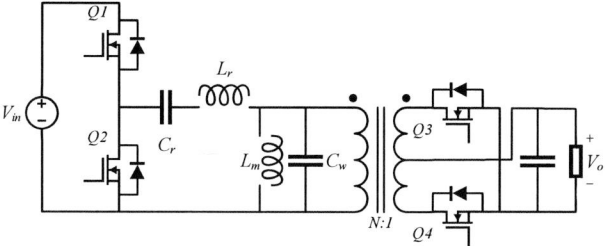

Fig. 2. Half bridge LLC resonant converter

TABLE I
SPECIFICATIONS OF THE LLC CONVERTER

Symbol	Meaning	Value
V_{in}	Input voltage	300V~400V
V_{in_nom}	Nominal input voltage	400V
P_o	Output power rating	200W
f_r	Resonant frequency	1MHz
V_o	Output voltage	12V

TABLE II
PARAMETERS OF THE LLC CONVERTER

Symbol	Meaning
N	Turns ratio
C_w	Equivalent transformer parasitic capacitance
I_{mp}	Peak magnetizing current
T_d	Dead time
n	Number of secondary side devices in parallel
T_s	Switching period
R_L	Load resistance
I_p	RMS current of primary side devices
I_s	RMS current of secondary side devices
$R_{ds(on)\_p}$	On-resistance of primary side devices
$R_{ds(on)\_s}$	On-resistance of secondary side devices
C_{po}	Output capacitance of primary side devices
C_{so}	Output capacitance of secondary side devices

III. MODELING AND DESIGN OF TRANSFORMER

A. Six Lumped Capacitances Model

Transformer parasitic capacitance consists of intra-winding capacitance and inter-winding capacitance. The former includes the self-capacitance of primary winding and secondary winding, and the latter is the mutual capacitance between primary and secondary winding [19]. In order to analyze inter-winding capacitance between primary winding and secondary winding, a three-port-network of six lumped capacitances is shown in Fig. 3 [20].

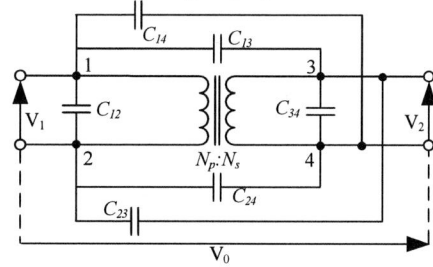

Fig. 3. Three-port-network of six lumped capacitances

The total electrostatic energy stored in the transformer is

$$W = \frac{1}{2}C_{12}V_1^2 + \frac{1}{2}C_{34}V_2^2 + \frac{1}{2}C_{24}V_0^2 + \frac{1}{2}C_{14}(V_1 - V_0)^2$$
$$+ \frac{1}{2}C_{23}(V_2 + V_0)^2 + \frac{1}{2}C_{13}(V_1 - V_0 - V_2)^2 \tag{1}$$

Rearrange (1) based on voltages,

$$W = \frac{1}{2}(C_{12} + C_{14} + C_{13})V_1^2 + \frac{1}{2}(C_{34} + C_{23} + C_{13})V_2^2$$
$$+ \frac{1}{2}(C_{24} + C_{14} + C_{23} + C_{13})V_0^2 + (-C_{14} - C_{13})V_1V_0$$
$$+ (C_{23} + C_{13})V_2V_0 + (-C_{13})V_1V_2 \tag{2}$$

A wire-wound component consisting of one primary layer

and one secondary layer can be modeled as a plate capacitor with a height H, a length L and a separation D shown in Fig. 4.

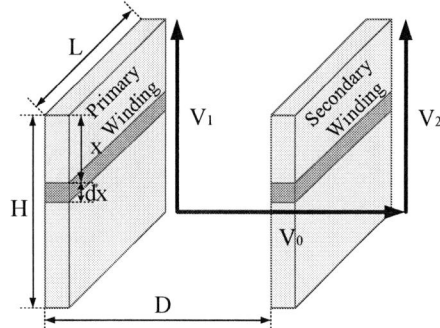

Fig. 4. Capacitance between primary and secondary winding

The static capacitance between primary winding and secondary winding can be calculated by (3) in case of H and L being much larger than D.

$$C_0 = \varepsilon_0 \varepsilon_r \frac{HL}{D} \tag{3}$$

Suppose the voltage V_1 and V_2 increased linearly. Therefore, at the coordinate x the voltage V_x between the primary and secondary winding is

$$V_x = V_0 + (1 - \frac{x}{H})V_2 - (1 - \frac{x}{H})V_1 \tag{4}$$

The electrostatic energy stored in the small capacitor (the shadow area) is

$$dW = \frac{1}{2}\varepsilon_0 \varepsilon_r \frac{Ldx}{D} V_x^2 \tag{5}$$

Integration of (5) between $x = 0$ and $x = H$ results in the total energy stored between the two layers,

$$W = \int_0^H \frac{1}{2}\varepsilon_0 \varepsilon_r \frac{LV_x^2}{D} dx \tag{6}$$

$$W = \frac{1}{6}C_0 V_1^2 + \frac{1}{6}C_0 V_2^2 + \frac{1}{2}C_0 V_0^2$$
$$- \frac{1}{2}C_0 V_1 V_0 + \frac{1}{2}C_0 V_2 V_0 - \frac{1}{3}C_0 V_1 V_2 \tag{7}$$

Compare (2) with (7), corresponding coefficients should be equal. The capacitances are

$$C_{12} = -\frac{1}{6}C_0 \quad C_{14} = \frac{1}{6}C_0 \quad C_{24} = \frac{1}{3}C_0$$
$$C_{34} = -\frac{1}{6}C_0 \quad C_{23} = \frac{1}{6}C_0 \quad C_{13} = \frac{1}{3}C_0 \tag{8}$$

The overall effect of six parasitic capacitances can be modeled as an equivalent capacitance shown in Fig. 5, which can be expressed as follows [21],

$$C_{eq} = C_{12} + k^2 C_{34} + 2k \frac{C_{14}C_{23} - C_{13}C_{24}}{C_{13} + C_{14} + C_{23} + C_{24}}$$
$$+ \frac{(C_{14} + C_{13})(C_{23} + C_{24}) + k^2(C_{13} + C_{23})(C_{14} + C_{24})}{C_{13} + C_{14} + C_{23} + C_{24}} \tag{9}$$

Fig. 5. Equivalent capacitance model

B. Design a Wire-Wound Transformer

In order to design a wire-wound transformer, the first step is to determine its turns ratio, which is

$$N = \frac{V_{in\_nom}}{2V_o} \tag{10}$$

Next, the transformer is preliminarily designed using AP method. The structure of the transformer is shown in Fig. 6. The magnetic core is EQ20/R-3F36. The primary winding is wound into a ring. The secondary winding adopts interleaving parallel copper sheets, which can minimize resistance and make the leakage inductance symmetrical. 3D print technology is used here to build the ABS skeleton, which can improve utility ratio of the core window areas and avoid large eddy loss created by gap.

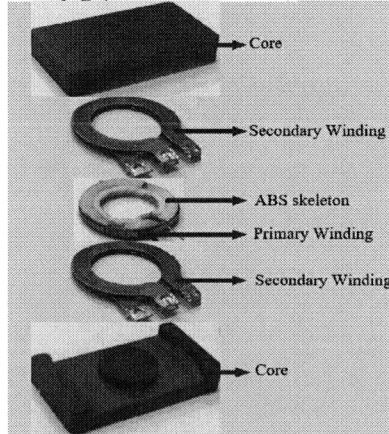

Fig. 6. Structure of the transformer

C. Measurment of Transformer Parasitic Capacitance

For wire-wound transformers, the parasitic capacitance is difficult to calculate theoretically. In order to measure the equivalent transformer parasitic capacitance, experiments are conducted on an open-loop LLC converter under no load condition. Waveforms of gate-source voltage V_{GS} and drain-source voltage V_{DS} of Q_2 are shown in Fig. 7. The input voltage is 200V and the driving voltage is 6V. The switching frequency is about 1MHz. The driving voltage doesn't rise up until the drain-source voltage drops to zero. Then zero voltage turn-on can be achieved.

During the dead time, the magnetizing current charges both the device output capacitances and the transformer parasitic capacitance,

$$I_{mp}T_d = 2C_{po}V_{in} + C_{eq}V_{in} + \frac{2V_o}{N}2nC_{so} \qquad (11)$$

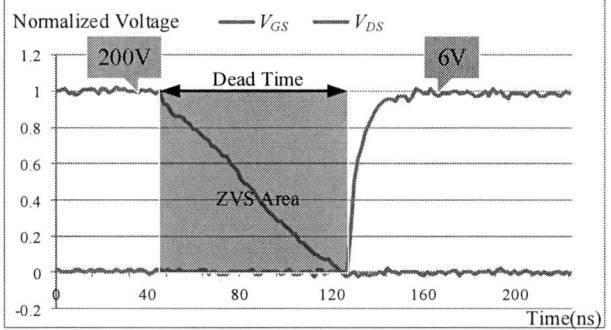

Fig. 7. Zero voltage switching during dead time

I_{mp}, T_d, and V_o can be read directly from oscilloscope. C_{po} and C_{so} can be calculated according to their respective C-V curve. Finally, the equivalent transformer capacitance can be obtained using (11), which is about 20pF.

IV. DESIGN OF RESONANT TANK PARAMETERS

A. Relationship between Dead Time and Conduction Loss

For the LLC topology shown in Fig. 2, the waveforms at the resonance are shown in Fig. 8.

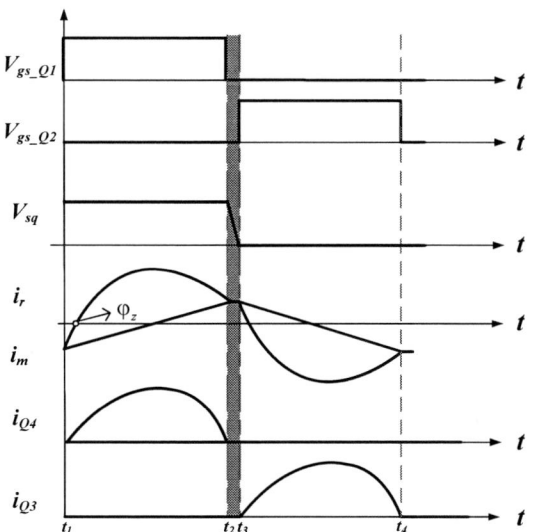

Fig. 8 Waveforms at the resonance

Considering t_1 to t_2 in Fig. 8, the peak magnetizing current can be expressed as

$$I_{mp} = \frac{NV_o}{2L_m}(\frac{T_s}{2} - T_d) \qquad (12)$$

Combining (10), (11), and (12), the magnetizing inductance is

$$L_m = \frac{(\frac{T_s}{2} - T_d)T_d}{4(2C_{po} + C_w + \frac{1}{N^2}2nC_{so})} \qquad (13)$$

The resonant current can be seen as a sine wave during t_1 to t_2 and a constant during t_2 to t_3 approximately.

$$i_p = \begin{cases} \sqrt{2}I_{rms\_p}\sin(w_0 t - \varphi_z), & 0 \le t \le \frac{T_s}{2} - T_d \\ \\ I_{mp}, & \frac{T_s}{2} - T_d \le t \le \frac{T_s}{2} \end{cases} \qquad (14)$$

The angular frequency of the sine wave is

$$w_0 = \frac{2\pi}{T_s - 2T_d} \qquad (15)$$

The magnetizing current can be seen as a triangular wave during t_1 to t_2 and a constant during t_2 to t_3 approximately.

$$i_m = \begin{cases} -I_{mp} + \frac{NV_o}{L_m}t, & 0 \le t \le \frac{T_s}{2} - T_d \\ \\ I_{mp}, & \frac{T_s}{2} - T_d \le t \le \frac{T_s}{2} \end{cases} \qquad (16)$$

The average output current is

$$N\frac{2}{T_s}\int_0^{\frac{T_s}{2}-T_d}(i_p - i_m)dt = \frac{V_o}{R_L} \qquad (17)$$

Then the RMS current during t_1 to t_2, I_{rms_p} can be derived,

$$I_{rms\_p} = \sqrt{\frac{V_o^2\pi^2T_s^2}{8N^2R_L^2(T_s - 2T_d)^2} + \frac{1}{2}I_{mp}^2} \qquad (18)$$

The primary RMS resonant current during t_1 to t_3, I_p, can be derived using (19). Detailed formula is deduced in (23) by replacing I_{rms_p} and I_{mp}.

$$I_p = \sqrt{\frac{2}{T_s}[I_{rms\_p}^2(\frac{T_s}{2} - T_d) + I_{mp}^2T_d]} \qquad (19)$$

The secondary RMS current during t_1 to t_3, I_s, is derived using (20). Detailed formula is deduced in (24) by replacing I_{rms_p} and I_{mp}.

$$I_s = \sqrt{\frac{2N^2}{T_s}\int_0^{\frac{T_s}{2}-T_d}(i_p - i_m)^2 dt} \qquad (20)$$

Total conduction loss can be expressed in (21).

$$P = I_p^2 R_{ds(on)\_p} + I_s^2 \frac{R_{ds(on)\_s}}{n} \qquad (21)$$

The relationship between dead time and total conduction loss is shown in Fig. 9. Obviously, there is an optimal dead time to achieve the lowest conduction loss, which is about 60ns. Then L_m can be determined using (13). Then the relationship between inductance ratio λ and quality factor Q is

$$\lambda Q = \frac{2\pi f_r L_m}{R_e} \qquad (22)$$

The 2018 International Power Electronics Conference

$$I_p = \sqrt{\frac{V_o^2 \pi^2 T_s}{8N^2 R_L^2 (T_s - 2T_d)} + (\frac{1}{2} + \frac{T_d}{T_s})(\frac{2C_{po}V_{in} + C_w V_{in} + 4V_o nC_{so}/N}{T_d})^2} \quad (23)$$

$$I_s = \sqrt{N^2(1 - \frac{2T_d}{T_s})}\sqrt{\frac{V_o^2 \pi^2 T_s^2}{8N^2 R_L^2 (T_s - 2T_d)^2} + (\frac{5}{6} - \frac{8}{\pi^2})(\frac{2C_{po}V_{in} + C_w V_{in} + 4V_o nC_{so}/N}{T_d})^2} \quad (24)$$

There is only one degree of freedom left. Simulation is necessary so as to judge which group of λ and Q can satisfy the gain requirement. Finally, the L_m is 32.47µH, L_r is 4.22µH, C_r is 6.01nF. According to simulation results, above parameters can meet the gain requirement.

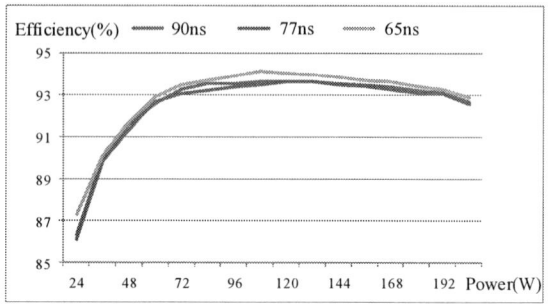

Fig. 12 Efficiency curves for different dead times

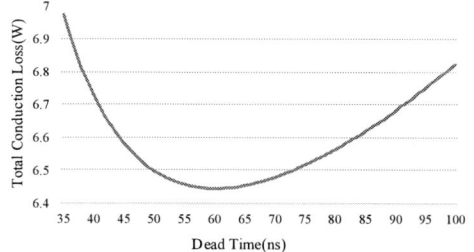

Fig. 9 Relationship between dead time and total conduction loss

V. Experimental Results

Experiments are conducted on a 1MHz 400V-12V 200W GaN-based LLC resonant converter shown in Fig. 10. The waveforms of synchronous driving signal and primary resonant current is shown in Fig. 11.

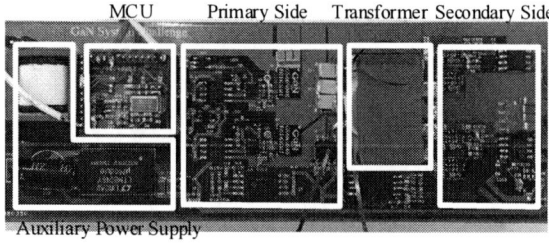

Fig. 10 LLC Prototype

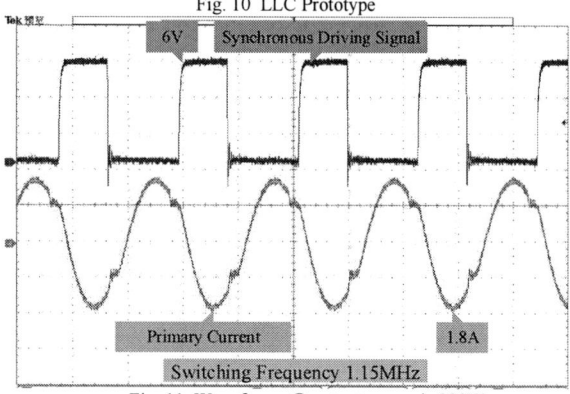

Fig. 11 Waveforms @ output power is 200W

Efficiency curves for different dead times are shown in Fig.12. When the dead time is 65ns, the LLC resonant converter can achieve the highest efficiency.

VI. Conclusions

This paper proposes an optimal design method considering transformer parasitic capacitance. After taking it into consideration, the optimal dead time will be longer. As a result, corresponding resonant parameters will change as well. According to the experimental results, this design method works well. Finally, the efficiency of the prototype is up to 94.2%.

Acknowledgment

At the point of finishing this paper, I'd like to express my sincere thanks to all those who have lent me hands in the course of my writing this paper. First of all, I'd like to take this opportunity to show my sincere gratitude to my supervisor, Mr. Yang, who has given me so much useful advice on my writing. Secondly, I'd like to express my gratitude to my teammates who offered me references and information on time. Last but not the least, I'd like to thank my girlfriend, Xiao Xiao, who is my inspiration and motivation. I love you forever.

References

[1] Bo Yang, F. C. Lee, A. J. Zhang and Guisong Huang, "LLC resonant converter for front end DC/DC conversion," *APEC. Seventeenth Annual IEEE Applied Power Electronics Conference and Exposition (Cat. No.02CH37335)*, Dallas, TX, 2002, pp. 1108-1112 vol.2.

[2] J. F. Lazar and R. Martinelli, "Steady-state analysis of the LLC series resonant converter," *APEC 2001. Sixteenth Annual IEEE Applied Power Electronics Conference and Exposition (Cat. No.01CH37181)*, Anaheim, CA, 2001, pp. 728-735 vol.2.

[3] T. Duerbaum, "First harmonic approximation including design constraints," *INTELEC - Twentieth International Telecommunications Energy Conference (Cat. No.98CH36263)*, San Francisco, CA, 1998, pp. 321-328.

[4] J. Marquart, S. Nigsch and K. Schenk, "Design Optimization for a High Power-Density, Wide Output, High Frequency LLC Resonant Converter for Lighting Applications," *PCIM Europe 2016; International Exhibition and Conference for Power Electronics, Intelligent Motion, Renewable Energy and Energy Management*, Nuremberg, Germany, 2016, pp. 1-9.

[5] Z. Hu, L. Wang, H. Wang, Y. F. Liu and P. C. Sen, "An Accurate Design Algorithm for LLC Resonant Converters–Part I," in *IEEE Transactions on Power Electronics*, vol. 31, no. 8, pp. 5435-5447, Aug. 2016.

[6] Z. Hu, L. Wang, Y. Qiu, Y. F. Liu and P. C. Sen, "An Accurate Design Algorithm for LLC Resonant Converters–Part II," in *IEEE Transactions on Power Electronics*, vol. 31, no. 8, pp. 5448-5460, Aug. 2016.

[7] C. Adragna, S. De Simone and C. Spini, "A design methodology for LLC resonant converters based on inspection of resonant tank currents.," *2008 Twenty-Third Annual IEEE Applied Power Electronics Conference and Exposition*, Austin, TX, 2008, pp. 1361-1367.

[8] R. Beiranvand, B. Rashidian, M. R. Zolghadri and S. M. Hossein Alavi, "A Design Procedure for Optimizing the LLC Resonant Converter as a Wide Output Range Voltage Source," in *IEEE Transactions on Power Electronics*, vol. 27, no. 8, pp. 3749-3763, Aug. 2012.

[9] T. Liu, Z. Zhou, A. Xiong, J. Zeng and J. Ying, "A Novel Precise Design Method for LLC Series Resonant Converter," *INTELEC 06 - Twenty-Eighth International Telecommunications Energy Conference*, Providence, RI, 2006, pp. 1-6.

[10] G. C. Hsieh, C. Y. Tsai and S. H. Hsieh, "Design Considerations for LLC Series-Resonant Converter in Two-Resonant Regions," *2007 IEEE Power Electronics Specialists Conference*, Orlando, FL, 2007, pp. 731-736.

[11] D. Funk, U. Schwalbe and T. Reimann, "Design Method for LLC Resonant Converter Considering Buck and Boost Mode with Limited Frequency Range for Wide Input Voltage Range," *PCIM Europe 2016; International Exhibition and Conference for Power Electronics, Intelligent Motion, Renewable Energy and Energy Management*, Nuremberg, Germany, 2016, pp. 1-8.

[12] Minghui Xu, Jing Ji, Zhan Li, Yuxi Wang, Jianhua Du and Hao Ma, "Design methodology of LLC converters based on simplified mode analysis for wide output range," *2016 IEEE 25th International Symposium on Industrial Electronics (ISIE)*, Santa Clara, CA, 2016, pp. 465-470.

[13] W. Zhang *et al.*, "Impact of planar transformer winding capacitance on Si-based and GaN-based LLC resonant converter," *2013 Twenty-Eighth Annual IEEE Applied Power Electronics Conference and Exposition (APEC)*, Long Beach, CA, 2013, pp. 1668-1674.

[14] W. Zhang, F. Wang, D. J. Costinett, L. M. Tolbert and B. J. Blalock, "Investigation of Gallium Nitride Devices in High-Frequency LLC Resonant Converters," in *IEEE Transactions on Power Electronics*, vol. 32, no. 1, pp. 571-583, Jan. 2017.

[15] Bing Lu, Wenduo Liu, Yan Liang, F. C. Lee and J. D. van Wyk, "Optimal design methodology for LLC resonant converter," Twenty-First Annual IEEE Applied Power Electronics Conference and Exposition, 2006. APEC '06., Dallas, TX, 2006, pp. 6 pp.-.

[16] N. Shafiei, M. Ordonez, S. R. Cove, M. Craciun and C. Botting, "Accurate modeling and design of LLC resonant converter with planar transformers," 2015 IEEE Energy Conversion Congress and Exposition (ECCE), Montreal, QC, 2015, pp. 5468-5473.

[17] T. Duerbaum and G. Sauerlaender, "Energy based capacitance model for magnetic devices," APEC 2001. Sixteenth Annual IEEE Applied Power Electronics Conference and Exposition (Cat. No.01CH37181), Anaheim, CA, 2001, pp. 109-115 vol.1.

[18] M. A. Saket, N. Shafiei and M. Ordonez, "LLC Converters With Planar Transformers: Issues and Mitigation," in IEEE Transactions on Power Electronics, vol. 32, no. 6, pp. 4524-4542, June 2017.

[19] N. Shafiei, M. Ordonez, S. R. Cove, M. Craciun and C. Botting, "Accurate modeling and design of LLC resonant converter with planar transformers," *2015 IEEE Energy Conversion Congress and Exposition (ECCE)*, Montreal, QC, 2015, pp. 5468-5473.

[20] T. Duerbaum and G. Sauerlaender, "Energy based capacitance model for magnetic devices," *APEC 2001. Sixteenth Annual IEEE*

Applied Power Electronics Conference and Exposition (Cat. No.01CH37181), Anaheim, CA, 2001, pp. 109-115 vol.1.

[21] M. A. Saket, N. Shafiei and M. Ordonez, "LLC Converters With Planar Transformers: Issues and Mitigation," in *IEEE Transactions on Power Electronics*, vol. 32, no. 6, pp. 4524-4542, June 2017.

The 2018 International Power Electronics Conference

Comparison of Harmonic Linearization and Harmonic State Space Methods for Impedance Modeling of Modular Multilevel Converter

Jing Lyu[1], Xin Zhang[2], JingJing Huang[2], Jianwen Zhang[1], and Xu Cai[1*]

1 Department of Electrical Engineering, Shanghai Jiao Tong University, Shanghai, China
2 School of Electrical and Electronic Engineering, Nanyang Technological University, Singapore
*E-mail: xucai@sjtu.edu.cn

Abstract— The impedance modeling of the modular multilevel converter (MMC) is the key for the resonance and stability analysis of MMC-based power systems. Especially, the internal dynamics of the MMC, which may have a great impact on the terminal characteristics of the MMC, need to be taken into account in the impedance modeling. To this end, two main methods have been reported in recent literatures to develop the MMC impedance models, i.e., harmonic linearization method and harmonic state space (HSS) method. So far, there is a lack of comparison between the two methods. Therefore, this paper presents a comparative study on the harmonic linearization and HSS methods in application of the MMC impedance modeling. The equivalence as well as the merits and demerits of these two methods are discussed.

Keywords— *Modular Multilevel Converter, impedance modeling, harmonic linearization, harmonic state space.*

I. INTRODUCTION

Modular multilevel converter (MMC) has been widely used in high-voltage direct current (HVDC) transmission and other high-voltage/high-power applications [1], [2], due to its advantages such as modularity, scalability, high efficiency, high performance, etc. However, due to its complex circuit structure, the modeling and control of the MMC becomes much more complicated than that of two-level voltage-source converters (VSCs) [3], [4]. Furthermore, the internal dynamics of the MMC may have a great impact on the operational stability of MMC-based power systems, e.g. MMC-HVDC connected wind farms [5], [6]. Hence, the internal dynamics such as harmonic circulating currents and capacitor voltage fluctuations are essential to be taken into account in MMC modeling.

The impedance-based analytical approach is preferred to be applied in stability analysis of power electronics interconnected systems due to its clear physical conception [7]. To do so, the impedance modeling of power converters is prerequisite. However, most of research has been focusing on the impedance modeling of two-level converters [8], [9] or MMCs without consideration of the internal dynamics [10]-[12], while only a few papers have so far investigated the MMC

impedance modeling with consideration of the internal dynamics [13]-[15]. In [13], the analytical derivation of the MMC sequence impedance by using the harmonic linearization method has been presented directly in the time-domain framework, where the effects of the internal dynamics can be included in the impedance modeling, but the lengthy algebra must be performed. In [14], the MMC sequence impedance model in the frequency-domain framework based on the harmonic state space (HSS) method has been derived, in which harmonics of state variables, inputs and outputs are posed separately in a state-space form and matrix operations are carried out. In addition, a method called as multi-harmonic linearization with matrix form is also presented in [15] to develop the MMC sequence impedance model, which is essentially equivalent to the HSS method in [14].

Though both the harmonic linearization method and HSS method are popular modeling approaches for the MMC, it will always be hard to judge the superiority of one method over another one without theoretical comparison. Therefore, this paper will carry out a comparative analysis of the harmonic linearization method and the HSS method in application of the MMC impedance modeling. For the sake of analysis, a brief derivation process of the MMC impedance modeling using these two methods is presented. The equivalence of these two methods in application of the impedance modeling is then revealed from impedance characteristics. Furthermore, the merits and demerits of these two methods are also discussed.

II. DERIVATION OF THE MMC IMPEDANCE USING HARMONIC LINEARIZATION

Fig. 1 shows the average-value model of the MMC (taking one phase leg for instance), where $C_{arm}=C_{SM}/N$, C_{SM} is the submodule (SM) capacitance, N is SM number per arm, L and R are the arm inductance and equivalent series resistance, i_u and i_l are the upper and lower arm currents, v_{cu}^{Σ} and v_{cl}^{Σ} are the sum capacitor voltages of the upper and lower arms, v_g and i_g are the ac-side phase voltage and current, i_c is the circulating current, n_u and n_l are the insertion indices of the upper and lower arms, and

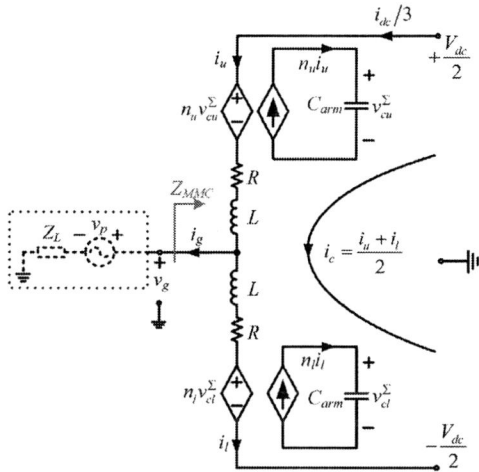

Fig. 1. Average-value model of one phase leg of MMC.

i_{dc} and V_{dc} are the dc-link voltage and current. In addition, $Z_L = R_L + j\omega_1 L_L$ is the ac-side equivalent load in order to determine the steady-state operating point, and v_p is the injected small perturbation voltage that is used to derive the ac-side small-signal impedance of the MMC. Furthermore, the dc-link voltage V_{dc} is assumed to be constant.

According to the harmonic linearization principle [13], by injecting a small sinusoidal perturbation voltage at frequency ω_p in the ac-side of the MMC and then calculating the resulting current response at the same frequency, the ac-side small-signal impedance of the MMC can thus be obtained by calculating the ratio of the resulting complex voltage to current at the perturbation frequency ω_p, which is defined as

$$Z_{MMC}(j\omega_p) = -\frac{V_{gp}}{I_{gp}} \quad (1)$$

where the bold capital letters V_{gp} and I_{gp} represent the complex phasors of the small perturbations $v_{gp}(t)$ and $i_{gp}(t)$ at frequency ω_p, respectively.

According to the operation characteristics of the MMC, the injected small perturbation voltage at frequency ω_p will lead to perturbations in all variables at frequencies that are listed as follows:

$$\omega_p, \omega_p \pm \omega_1, \omega_p \pm 2\omega_1, \cdots, \omega_p \pm h\omega_1 \quad (2)$$

The frequency components in (2) indicate that strong frequency couplings exist within MMC. This means that if only considering the frequency component at ω_p and ignoring other frequency components, the resulting MMC impedance model is probably inaccurate. In other words, the impedance model will be accurate if all harmonics can be included. However, it is impossible to include all harmonics in practice. Furthermore, it can be intuitively known that the higher the perturbation frequency in (2) is, the smaller the magnitude of the corresponding frequency component is [4]. Hence, the impedance model would be accurate enough if only several lower harmonics in (2) are considered.

In this section, the harmonic linearization is applied directly in the time-domain framework. For doing that, the specific frequency components in (2) must be predetermined. For instance, the frequency components at ω_p, ω_p-ω_1 and ω_p+ω_1 will be considered in this paper to derive the MMC impedance by using harmonic linearization. Hence, the ac-side phase current with a small perturbation can be expressed as

$$i_g(t) = I_{g1}\cos(\omega_1 t + \varphi_{i1}) + I_{gp}\cos(\omega_p t + \varphi_{ip}) \quad (3)$$

where I_{g1} and φ_{i1} are the magnitude and phase of the fundamental current at frequency ω_1, respectively, and I_{gp} and φ_{ip} are the magnitude and phase of the small perturbation current at frequency ω_p, respectively.

The resulting circulating current can then be expressed

$$i_c(t) = \frac{i_{dc}}{3} + I_{c2}\cos(2\omega_1 t + \varphi_{c2}) +$$
$$I_{cp0}\cos\left[(\omega_p - \omega_1)t + \varphi_{cp0}\right] + \quad (4)$$
$$I_{cp2}\cos\left[(\omega_p + \omega_1)t + \varphi_{cp2}\right]$$

where I_{c2} and φ_{c2} are the magnitude and phase of the second harmonic circulating current, I_{cp0} and φ_{cp0} are the magnitude and phase of the circulating current perturbation at frequency (ω_p-ω_1), and I_{cp2} and φ_{cp2} are the magnitude and phase of the circulating current perturbation at frequency (ω_p+ω_1).

Taking open-loop control for example, the analytical derivation of the MMC impedance by harmonic linearization will be presented as follows.

The SM capacitor voltages of the upper and lower arms can be obtained

$$v_{cu}^i = \frac{1}{C}\int n_u i_u dt = \frac{1}{C}\int \frac{1}{2}\left[1 - m\cos(\omega_1 t)\right]\left(i_c + \frac{i_g}{2}\right)dt \quad (5a)$$

$$v_{cl}^i = \frac{1}{C}\int n_l i_l dt = \frac{1}{C}\int \frac{1}{2}\left[1 + m\cos(\omega_1 t)\right]\left(i_c - \frac{i_g}{2}\right)dt \quad (5b)$$

where m is the fundamental modulation index.

Hence, the output voltages of the upper and lower arms can then be obtained

$$v_u = n_u v_{cu}^\Sigma = \frac{N}{2}\left[1 - m\cos(\omega_1 t)\right]v_{cu}^i \quad (6a)$$

$$v_l = n_l v_{cl}^\Sigma = \frac{N}{2}\left[1 + m\cos(\omega_1 t)\right]v_{cl}^i \quad (6b)$$

Furthermore, we have

$$L\frac{di_c}{dt} + Ri_c = \frac{V_{dc}}{2} - \frac{v_u + v_l}{2} \quad (7)$$

$$\frac{L}{2}\frac{di_g}{dt} + \frac{R}{2}i_g = \frac{v_l - v_u}{2} - v_g \quad (8)$$

Combining (3)-(8) and retaining only those perturbation components at frequency ω_p, ω_p-ω_1 and ω_p+ω_1 in the final expressions, the relationship between the ac-side small perturbation voltage and current at frequency ω_p can thus be obtained

$$-V_{gp}e^{j\varphi_{vp}} = B_{ip}I_{gp}e^{j\varphi_{ip}} + B_{cp0}I_{cp0}e^{j\varphi_{cp0}} + B_{cp2}I_{cp2}e^{j\varphi_{cp2}} \quad (9)$$

where $I_{cp0}e^{j\varphi_{cp0}}$ and $I_{cp2}e^{j\varphi_{cp2}}$ can be expressed as a function of $I_{gp}e^{j\varphi_{ip}}$, respectively. Additionally, B_{ip}, B_{cp0} and B_{cp2} are given in (10).

The 2018 International Power Electronics Conference

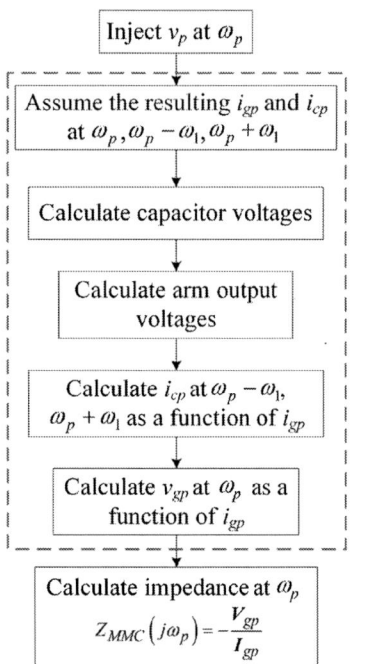

Fig. 2. Flowchart of the MMC impedance modeling by using harmonic linearization.

$$B_{ip} = \frac{j\omega_p L + R}{2} - j\frac{N}{32C}\left(\frac{4}{\omega_p} + \frac{m^2}{\omega_p + \omega_1} + \frac{m^2}{\omega_p - \omega_1}\right) \quad (10a)$$

$$B_{cp0} = j\frac{Nm}{8C}\left(\frac{1}{\omega_p} + \frac{1}{\omega_p - \omega_1}\right) \quad (10b)$$

$$B_{cp2} = j\frac{Nm}{8C}\left(\frac{1}{\omega_p} + \frac{1}{\omega_p + \omega_1}\right) \quad (10c)$$

Hence, the ac-side small-signal impedance of the MMC can be obtained by substituting $I_{cp0}e^{j\varphi_{cp0}}$ and $I_{cp2}e^{j\varphi_{cp2}}$ in (9) according to (1). The final expression of the MMC impedance can be given as (11). In addition, it can be observed from (9) that the terminal characteristics of the MMC are determined not only by the ac-side voltage and current dynamics but also by the circulating current dynamics.

$$Z_{MMC}\left(j\omega_p\right) = B_{ip} + B_{cp0}A_{cp0} + B_{cp2}A_{cp2} \quad (11)$$

where

$$A_{cp0} = \frac{a_{00}a_{21} - a_{01}z_{2p}}{z_{0p}z_{2p} - a_{00}a_{20}}, A_{cp2} = \frac{a_{20}a_{01} - a_{21}z_{0p}}{z_{0p}z_{2p} - a_{00}a_{20}} \quad (12a)$$

$$a_{00} = -j\frac{Nm^2}{16C\omega_p}, a_{01} = j\frac{Nm}{16C}\left(\frac{1}{\omega_p - \omega_1} + \frac{1}{\omega_p}\right)$$
$$ \quad (12b)$$
$$a_{20} = -j\frac{Nm^2}{16C\omega_p}, a_{21} = j\frac{Nm}{16C}\left(\frac{1}{\omega_p + \omega_1} + \frac{1}{\omega_p}\right)$$

$$z_{0p} = j\left(\omega_p - \omega_1\right)L + R - j\frac{N}{16C}\left(\frac{4}{\omega_p - \omega_1} + \frac{m^2}{\omega_p}\right)$$
$$ \quad (12c)$$
$$z_{2p} = j\left(\omega_p + \omega_1\right)L + R - j\frac{N}{16C}\left(\frac{4}{\omega_p + \omega_1} + \frac{m^2}{\omega_p}\right)$$

Fig. 2 shows the flowchart of the MMC impedance modeling process by using harmonic linearization. It is noted that every step in the red box, which involves lengthy algebra, needs to be performed by hand. The process is time-consuming and prone to error, especially for higher harmonics considered in the model. In this view, the harmonic linearization method applies to the small-signal modeling of power converters with less harmonics, e.g. two-level VSCs.

III.   DERIVATION OF THE MMC IMPEDANCE BASED ON HARMONIC STATE SPACE

According to Fig. 1, the time-domain state-space equation of one phase leg of the MMC can be expressed

$$\dot{x}(t) = A(t)x(t) + B(t)u(t) \quad (13)$$

where $x(t)$, $u(t)$, $A(t)$, and $B(t)$ are shown in (14)-(17).

$$x(t) = \begin{bmatrix} i_c & v_{cu}^\Sigma & v_{cl}^\Sigma & i_g \end{bmatrix}^T \quad (14)$$

$$u(t) = [V_{dc}] \quad (15)$$

$$A(t) = \begin{bmatrix} -\dfrac{R}{L} & -\dfrac{n_u}{2L} & -\dfrac{n_l}{2L} & 0 \\[2ex] \dfrac{n_u}{C_{arm}} & 0 & 0 & \dfrac{n_u}{2C_{arm}} \\[2ex] \dfrac{n_l}{C_{arm}} & 0 & 0 & -\dfrac{n_l}{2C_{arm}} \\[2ex] 0 & -\dfrac{n_u}{L} & \dfrac{n_l}{L} & -\dfrac{R+2Z_L}{L} \end{bmatrix} \quad (16)$$

$$B(t) = \begin{bmatrix} \dfrac{1}{2L} & 0 & 0 & 0 \end{bmatrix}^T \quad (17)$$

in which

$$\begin{cases} n_u(t) = \dfrac{1}{2}\left[1 - m\cos\left(\omega_1 t + \theta_{m1}\right)\right] \\[2ex] n_l(t) = \dfrac{1}{2}\left[1 + m\cos\left(\omega_1 t + \theta_{m1}\right)\right] \end{cases} \quad (18)$$

Open-loop control is assumed. Thereby, the small perturbation state-space expression of equation (13) can be written as

$$\dot{x}_p(t) = A_p(t)x_p(t) + B_p(t)u_p(t) \quad (19)$$

where $x_p(t)$, $u_p(t)$, and $B_p(t)$ are shown in (20)-(22), and $A_p(t)$ remains the same as (16).

$$x_p(t) = \begin{bmatrix} i_{cp} & v_{cup}^\Sigma & v_{clp}^\Sigma & i_{gp} \end{bmatrix}^T \quad (20)$$

$$u_p(t) = \begin{bmatrix} v_p \end{bmatrix} \quad (21)$$

$$B_p(t) = \begin{bmatrix} 0,0,0,-\dfrac{2}{L} \end{bmatrix}^T \quad (22)$$

Hence, by performing the HSS modeling to the small perturbation state-space equation in (19), the small perturbation HSS model of the MMC can be obtained

$$s\mathbf{X}_p = \left(\mathbf{A}_p - \mathbf{Q}_p\right)\mathbf{X}_p + \mathbf{B}_p\mathbf{U}_p \qquad (23)$$

where $\mathbf{X}_p$, $\mathbf{U}_p$, and $\mathbf{Q}_p$ are given in (24)~(26), in which the subscript "$p\pm h$" denotes the perturbation component at frequency "$\omega_p\pm h\omega_1$". I is a fourth-order identity matrix. Additionally, $\mathbf{A}_p$ and $\mathbf{B}_p$ are Toeplitz matrices, whose elements are given in (27) and (28). It is noted that the diagonal matrix $\mathbf{Q}_p$ contains all the perturbation frequencies defined in (2), which means that this model considers all the frequency coupling effects.

$$\mathbf{X}_p = \left[X_{p-h}, \cdots, X_p, \cdots, X_{p+h}\right]^T$$
$$X_{p\pm h} = \left[I_{cp\pm h}, V_{cup\pm h}^{\Sigma}, V_{clp\pm h}^{\Sigma}, I_{gp\pm h}\right] \qquad (24)$$

$$\mathbf{U}_p = \left[U_{p-h}, \cdots, U_p, \cdots, U_{p+h}\right]^T$$
$$U_p = \left[V_p\right], U_{p\pm h} = [0]\,(h \ge 1) \qquad (25)$$

$$\mathbf{Q}_p = \mathrm{diag}\left[j\left(\omega_p - h\omega_1\right)\cdot I, \ldots, j\omega_p\cdot I, \ldots, j\left(\omega_p + h\omega_1\right)\cdot I\right] \quad (26)$$

$$A_0 = \begin{bmatrix} -\dfrac{R}{L} & -\dfrac{1}{4L} & -\dfrac{1}{4L} & 0 \\[2mm] \dfrac{1}{2C_{arm}} & 0 & 0 & \dfrac{1}{4C_{arm}} \\[2mm] \dfrac{1}{2C_{arm}} & 0 & 0 & -\dfrac{1}{4C_{arm}} \\[2mm] 0 & -\dfrac{1}{2L} & \dfrac{1}{2L} & -\dfrac{R+2Z_L}{L} \end{bmatrix} \quad (27a)$$

$$A_{\pm 1} = \begin{bmatrix} 0 & \dfrac{me^{\pm j\theta_{m1}}}{8L} & -\dfrac{me^{\pm j\theta_{m1}}}{8L} & 0 \\[2mm] -\dfrac{me^{\pm j\theta_{m1}}}{4C_{arm}} & 0 & 0 & -\dfrac{me^{\pm j\theta_{m1}}}{8C_{arm}} \\[2mm] \dfrac{me^{\pm j\theta_{m1}}}{4C_{arm}} & 0 & 0 & -\dfrac{me^{\pm j\theta_{m1}}}{8C_{arm}} \\[2mm] 0 & \dfrac{me^{\pm j\theta_{m1}}}{4L} & \dfrac{me^{\pm j\theta_{m1}}}{4L} & 0 \end{bmatrix} \quad (27b)$$

$$A_{\pm h} = \begin{bmatrix} 0 & 0 & 0 & 0 \\ 0 & 0 & 0 & 0 \\ 0 & 0 & 0 & 0 \\ 0 & 0 & 0 & 0 \end{bmatrix}\,(h \ge 2) \qquad (27c)$$

$$B_0 = \left[\frac{1}{2L}, 0, 0, 0\right]^T \qquad (28a)$$

$$B_{\pm h} = [0,0,0,0]^T \,(h \ge 1) \qquad (28b)$$

Ignoring the transient behavior of the perturbation signals, the small perturbation components of the state variables at each perturbation frequency can be calculated by

$$\mathbf{X}_p = -\left(\mathbf{A}_p - \mathbf{Q}_p\right)^{-1}\left(\mathbf{B}_p\mathbf{U}_p\right) \qquad (29)$$

By solving (29), the perturbation components V_{gp} and I_{gp} of the ac phase voltage and current of the MMC at frequency ω_p can be expressed as a function of the injected small perturbation voltage V_p, respectively. Hence, the ac-side small-signal impedance of the MMC

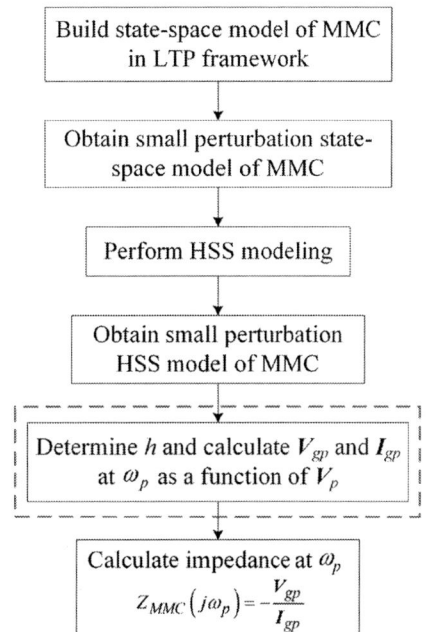

Fig. 3. Flowchart of the MMC impedance modeling by HSS.

can be obtained by substituting V_{gp} and I_{gp} into (1). It should be pointed out that the harmonic order h must be predetermined, which means how many harmonics need to be considered. Different values of h will generate different impedance expressions which means impedance models with different accuracies. For comparison with the harmonic linearization, the harmonic order is selected as 1, i.e. $h=1$, which means that only the three frequency components at ω_p, $\omega_p-\omega_1$ and $\omega_p+\omega_1$ are considered in the HSS modeling.

Fig. 3 shows the flowchart of the MMC impedance modeling process based on the HSS method, where the first four steps are very easy to be implemented by hand and the corresponding results can be directly obtained. The most computational task, which is indicated by the red box, can be readily implemented by computer based on matrix operation.

IV. COMPARATIVE ANALYSIS

As aforementioned, both the harmonic linearization and HSS methods can be used to derive the MMC impedance with consideration of the internal dynamics and frequency couplings. Intuitively, the harmonic linearization method is an algebraic method with a lengthy derivation process, especially for higher harmonics, while the HSS modeling method, which is based on matrix operation, is readily implemented by the help of computer and also easily extended to any number of harmonics. The comparison results between the harmonic linearization and HSS methods in application of MMC impedance modeling are presented in Table I.

Fig. 4 shows the ac-side small-signal impedance characteristics of the MMC derived by the harmonic linearization and HSS methods, respectively, where only those frequency components at ω_p, $\omega_p-\omega_1$ and $\omega_p+\omega_1$ are considered for both the two methods. As can be seen, the

The 2018 International Power Electronics Conference

Table I. Comparison between harmonic linearization and HSS in application of MMC impedance modeling

	Operation	Derivation process	Expansibility to higher harmonics	Model accuracy	Computer-aided
Harmonic linearization	Algebraic	Complex	Low	High	No
HSS	Matrix	Simple	High	High	Yes

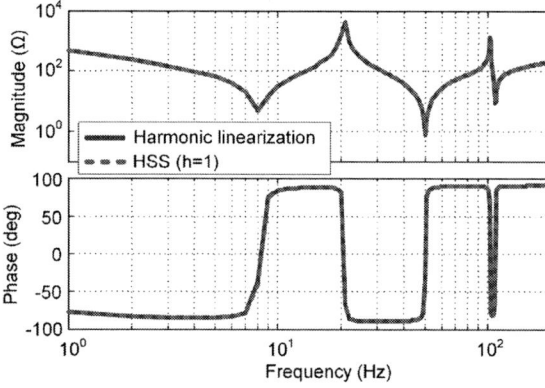

Fig. 4. Comparison between the analytical impedances obtained from harmonic linearization and HSS, respectively.

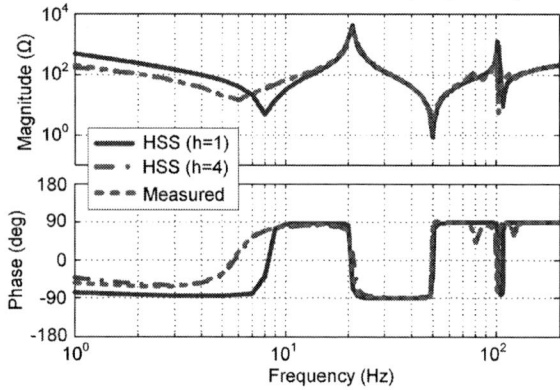

Fig. 5. Comparison between the measured impedance and the analytical ones with consideration of different number of harmonics.

two analytical impedances are completely overlapped, which indicates that both the two methods have the same model accuracy despite different derivation processes. In addition, the measured impedance based on a nonlinear time-domain simulation model of the MMC in Matlab/Simulink is also given to validate the analytical model, as shown in Fig. 5, where the impact of the number of harmonics considered in the HSS model on the accuracy of the analytical MMC impedance model is presented as well. As can be seen, the analytical impedance with $h=4$ is in good agreement with the measured one, which indicates that the MMC internal harmonics have great influence on the MMC impedance characteristics, especially at low frequencies. The main electrical parameters of the MMC are as follows: the fundamental frequency 50 Hz, converter-side voltage 166 kV, dc-link voltage 320 kV, SM number per arm 20, SM capacitance 140 μF, and arm inductance 360 mH. It is noted that the main electrical parameters of the MMC

used in this paper are identical to those of a real MMC system, but both the SM number and capacitance are one-tenth of the actual parameters, where keeping the arm equivalent capacitance C_{arm} unchanged.

V. CONCLUSION

This paper presents a comparative study of the harmonic linearization and HSS methods for the MMC impedance modeling. The comparative analysis indicates that the same impedance model accuracy can be achieved by both the two methods. Intuitively, the harmonic linearization method is an algebraic method while the HSS method is based on matrix operation. Both these two methods are able to include the MMC internal dynamics and frequency couplings. However, the derivation process will become much more time-consuming and prone to error when considering higher harmonics by using the harmonic linearization method, while the HSS method can be readily implemented with the aid of computer and also easily extended to any number of harmonics.

ACKNOWLEDGMENT

This work was partly supported by the National Key Research and Development Program under Grant 2016YFB0900901 and by Shanghai Science and Technology Committee Scientific Research Program under Grant 16DZ1203402.

REFERENCES

[1] S. Dehnath, J. Qin, B. Bahrani, M. Saeedifard, and P. Barbosa, "Operation, control, and applications of the modular multilevel converter: a review," *IEEE Trans. on Power Electronics*, vol. 30, no. 1, pp. 37-53, 2015.

[2] M. A. Perez, S. Bernet, J. Rodriguez, S. Kouro, and R. Lizana, "Circuit topologies, modeling, control schemes, and applications of modular multilevel converter," *IEEE Trans. on Power Electronics*, vol. 30, no. 1, pp. 4-17, 2015.

[3] L. Harnefors, A. Antonopoulos, S. Norrga, L. Ängquist, and H.P. Nee, "Dynamic analysis of modular multilevel converters," *IEEE Trans. on Industrial Electronics*, vol. 60, no. 7, pp. 2526-2537, 2013.

[4] K. Ilves, A. Antonopoulos, S. Norrga, and H.P. Nee, "Steady-state analysis of interaction between harmonic components of arm and line quantities of modular multilevel converters," *IEEE Trans. on Power Electronics*, vol. 27, no. 1, pp. 57-68, 2012.

[5] J. Lyu, X. Cai, and M. Molinas, "Frequency domain stability analysis of MMC-based HVdc for wind farm integration," *IEEE Journal of Emerging and Selected Topics in Power Electronics*, vol. 4, no. 1, pp. 141-151, 2016.

[6] J. Lyu, X. Cai, M. Amin, and M. Molinas, "Subsynchronous oscillation mechanism and its suppression in MMC-Based HVDC connected wind farms," *IET Generation, Transmission & Distribution*, vol. 12, no. 4, pp. 1021-1029, 2017.

[7] J. Sun, "Impedance-based stability criterion for grid-connected inverters," *IEEE Trans. on Power Electronics*, vol. 26, no. 11, pp. 3075-3078, 2011.

[8] M. Cespedes and J. Sun, "Impedance modeling and analysis of grid-connected voltage-source converters," *IEEE Trans. on Power Electronics*, vol. 29, no. 3, pp. 1254-1261, 2014.

[9] B. Wen, D. Boroyevich, R. Burgos, P. Mattavelli, and Z. Shen, "Analysis of D-Q small-signal impedance of grid-tied inverters," *IEEE Trans. on Power Electronics*, vol. 31, no. 1, pp. 675-687, 2016.

[10] J. Lyu, X. Cai, and M. Molinas, "Impedance modeling of modular multilevel converters," *IEEE IECON*, 2015, Yokohama, Japan.

[11] M. Beza, M. Bongiorno, and G. Stamatiou, "Analytical Derivation of the AC-Side Input Admittance of a Modular Multilevel Converter with Open- and Closed-Loop Control Strategies," *IEEE Trans. on Power Delivery*, vol. 33, no. 1, pp. 248-256, 2018.

[12] J. Khazaei, M. Beza, and M. Bongiorno, "Impedance analysis of modular multi-level converters connected to weak ac grids," *IEEE Trans. on Power Systems*, vol. PP, no. 99, pp. 1-1, 2017.

[13] J. Lyu, Q. Chen, and X. Cai, "Impedance modeling of modular multilevel converter by harmonic linearization," IEEE COMPEL 2016, Trondheim, Norway.

[14] Q. Chen, J. Lyu, R. Li, and X. Cai, "Impedance modeling of modular multilevel converter based on harmonic state space," IEEE COMPEL 2016, Trondheim, Norway.

[15] J. Sun and H. Chao, "Impedance modeling and analysis of modular multilevel converters," IEEE COMPEL 2016, Trondheim, Norway.

The 2018 International Power Electronics Conference

An Improved Phase-Shifted PWM for a Five-level Hybrid-Clamped Converter

Kui Wang[1*], Nianzhou Liu[2], Zedong Zheng[1] and Yongdong Li[1]

1 Department of Electrical Engineering, Tsinghua University, Beijing, China
2 Wuhan institute of marine electric propulsion, Wuhan, China
*E-mail: wangkui@tsinghua.edu.cn

Abstract—**Five-level hybrid-clamped (5L-HC) converter is a newly proposed topology which is suitable for medium-voltage high-power conversions without switches directly connected in series. Phase-shifted PWM (PSPWM) is commonly used to control this converter. However, the line voltages of PSPWM include many intervals that jump two voltage levels continuously, which will deteriorate the harmonic performance. This paper proposes an improved PSPWM method for the hybrid five-level converter, which can imitate the effect of phase disposition PWM (PDPWM) and improve the quality of the line voltages. Simulation and experimental results verify the validity of the proposed method.**

Keywords— Multilevel converter, Phase-shifted PWM, capacitor voltage balance, medium voltage drive.

I. INTRODUCTION

Multilevel converters are widely used in high voltage high power applications due to the advantages of low voltage stress on switches, low dv/dt and better harmonic performance [1]–[4]. Among all the topologies, neutral-point-clamped (NPC) and cascaded H-bridge (CHB) topologies are the most widely used and commercialized [3], [4]. However, they both have some drawbacks such as DC-link neutral-point voltage unbalance problem or complex input transformer. In order to overcome these problems, many new topologies have been studied, such as five-level stacked multi-cell (SMC) converter, five-level active neutral-point clamped (ANPC) converter, modular multilevel converter (MMC) and five-level nested neutral-point clamped (5L-NNPC) converter [5]–[14]. A novel five-level hybrid-clamped (5L-HC) converter is also proposed in [15], which is suitable for medium-voltage variable speed drive applications without switches directly connected in series, as shown in Fig. 1. Phase-shifted PWM (PSPWM) is proved to have natural voltage balancing ability and is suitable to control this converter [16]–[18]. However, the line voltages of PSPWM include many intervals that jump two voltage levels continuously, which will deteriorate the harmonic performance.

Therefore, this paper proposed an improved PSPWM technique for this 5L-HC converter to improve the line voltage quality. The phase voltage waveform of the proposed PSPWM is the same as phase-disposition PWM (PSPWM) while the carriers for the four switches are also

This work was supported by the National Natural Science Foundation of China (Grant No. 51407101).

phase shifted, hence the harmonic performance of line voltage can be improved and the voltage balancing method can be unchanged. The feasibility of this method is verified by both simulation and experimental results.

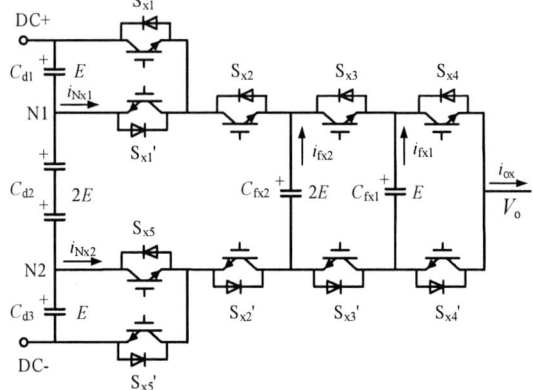

Fig. 1 The proposed five-level hybrid-clamped converter

II. OPERATING PRINCIPLES AND MODULATION METHOD

A single phase circuit of the 5L-HC inverter is shown in Fig. 1, which consists of 10 switches and 2 flying capacitors C_{fx1} and C_{fx2}, where x represents phase *a*, *b*, or *c*. The switches S_{x1}–S_{x5} and S_{x1}'–S_{x5}' are operated complementarily and S_{x1} and S_{x5} are operated synchronously. The nominal voltages of the upper, central and lower DC-link capacitors are E, $2E$ and E, and the nominal voltages of two flying capacitor C_{fx1} and C_{fx2} are E and $2E$, so five voltage levels can be obtained. Table I shows all the switching states.

If the switching functions of switches S_{x1}–S_{x5} are defined as S_{fx1}–S_{fx5}, according to Table I, the instantaneous output voltage level V_{ox} can be written as:

$$V_{ox} = (S_{fx1} + S_{fx2} + S_{fx3} + S_{fx4}) \cdot E \qquad (1)$$

The output voltage V_{ox} depends on the sum of S_{fx1}–S_{fx4}, which is the same as flying-capacitor multilevel converters. In order to simplify the control system and acquire the voltage natural balance ability, PS-PWM is a good choice for this converter.

TABLE I
SWITCHING STATES OF THE 5L-HC CONVERTER

S_{x1}	S_{x5}	S_{x2}	S_{x3}	S_{x4}	V_{ox}
0	0	0	0	0	0
0	0	0	0	1	E
0	0	0	1	0	E
0	0	1	0	0	E
1	1	0	0	0	E
0	0	0	1	1	$2E$
0	0	1	0	1	$2E$
0	0	1	1	0	$2E$
1	1	0	0	1	$2E$
1	1	0	1	0	$2E$
1	1	1	0	0	$2E$
0	0	1	1	1	$3E$
1	1	0	1	1	$3E$
1	1	1	0	1	$3E$
1	1	1	1	0	$3E$
1	1	1	1	1	$4E$

III. MODELING OF THE CAPACITOR CURRENTS AND VOLTAGES

The voltage balancing method under PSPWM of this topology is introduced in [15] and the diagram of traditional PS-PWM method is shown in Fig. 2(a). Although PSPWM has the advantage of flying capacitor voltage natural balancing ability in the flying-capacitor multilevel converters, its harmonic performance is much worse than PDPWM, which restrict its application in high power applications especially when the carrier frequency is very low.

In order to improve the harmonic performance of PSPWM and reserve the naturally balancing ability, an improved PSPWM is proposed, which reveals the same harmonic performance with phase-disposition PWM. As shown in Fig. 2(b), eight carriers are required and divided into two sets in the proposed PS-PWM. Each set of carriers includes four carriers and are phase shifted by π/2. The two sets are phase shifted by π/4. A carrier of the improved PSPWM method is shown in Fig. 3. The carrier set 1 is used when the reference voltage belongs to [0, 1] or [2, 3] and the carrier set 1 is used when the reference voltage belongs to [1, 2] or [3, 4].

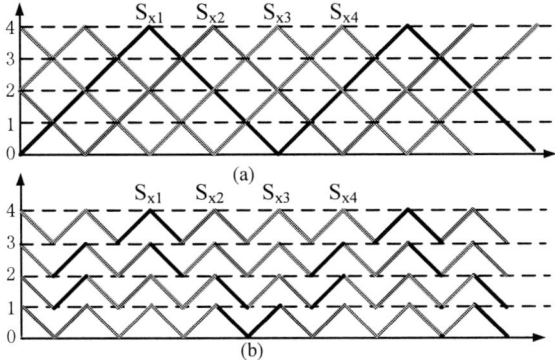

Fig. 2. (a) Traditional PSPWM and (b) the improved PSPWM method

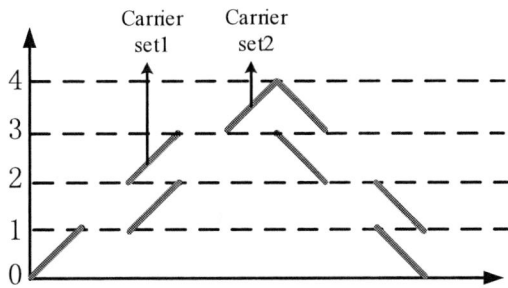

Fig. 3. A carrier of the improved PSPWM method

So the carrier wave of the proposed PSPWM is a broken line. By this way, the output phase voltage can be the same with PDPWM and the four switches are still controlled by PSPWM, only the carrier waves are changed.

IV. SIMULATION RESULTS

A three-phase 5L-HC inverter is simulated with MATLAB to demonstrate the performance of the proposed control method. Fig. 4 is the circuit structure and Table II shows the simulation parameters.

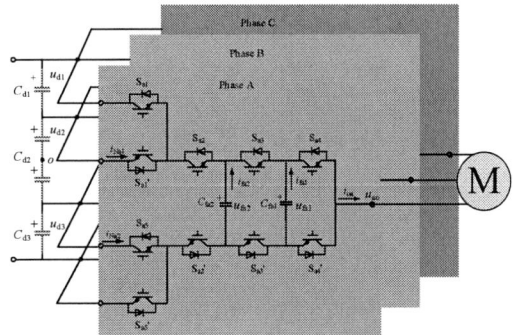

Fig. 4 Three-phase 5L-HC inverter.

TABLE II
SIMULATION PARAMETERS

Parameters	Value
Rated volume	1 MVA
Rated line voltage	6.6 kV
DC-link voltage	U_{dc} = 11200 V
DC-link capacitor	C_{d1} = C_{d3} = 500 μF, C_{d2} = 250 μF
Flying capacitor	C_{fx1} = 400 μF, C_{fx2} = 200 μF
Carrier frequency	f_c = 500 Hz
R-L Load	R_l = 40 Ω, L_l = 15 mH

Fig. 5 presents the phase voltage, line voltage and line current with traditional PSPWM and the proposed PSPWM respectively. It can be seen obviously that the line voltage of traditional PSPWM includes many intervals that jump two voltage levels continuously, which will deteriorate the harmonic performance.

The 2018 International Power Electronics Conference

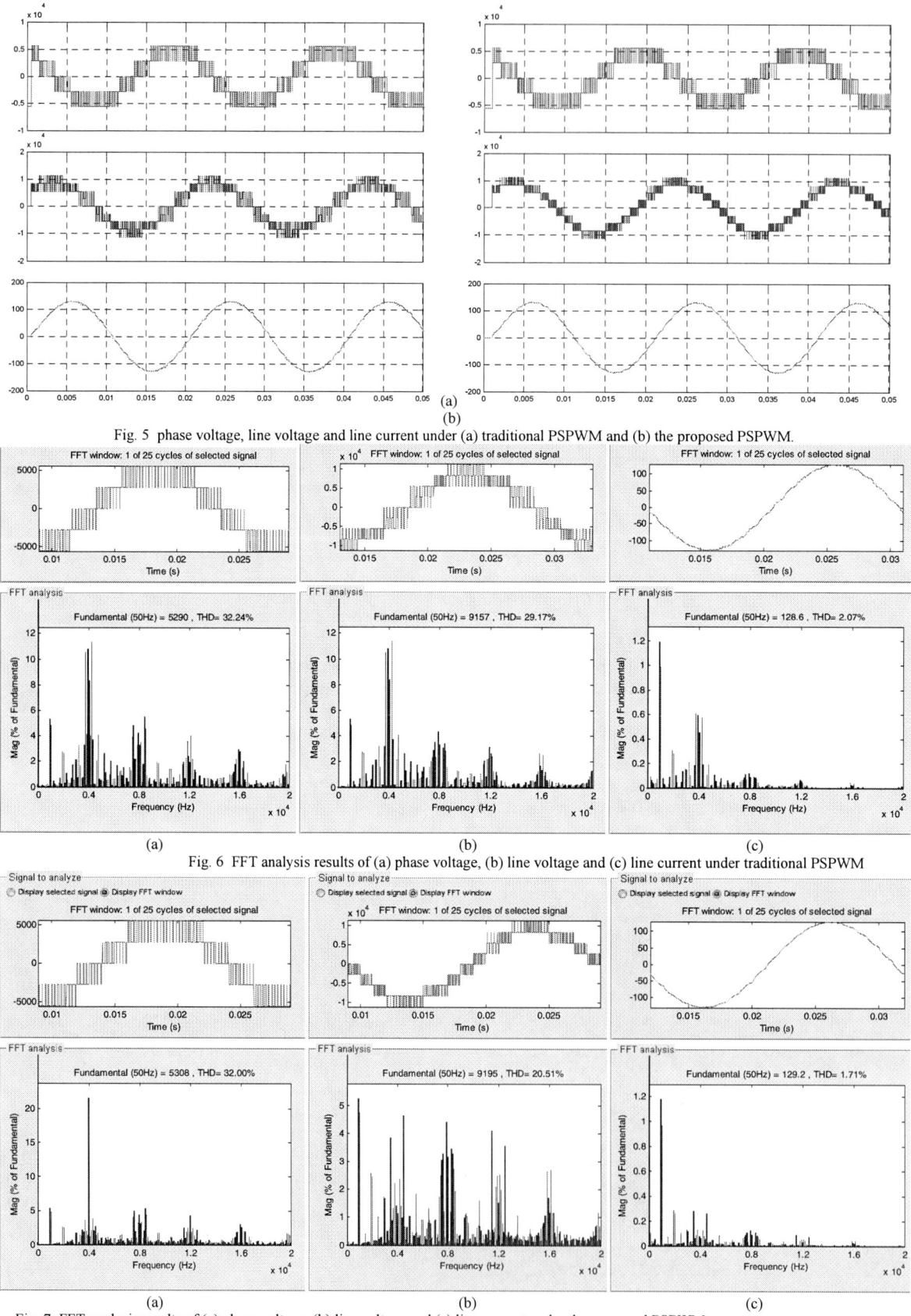

Fig. 5 phase voltage, line voltage and line current under (a) traditional PSPWM and (b) the proposed PSPWM.

(a)   (b)   (c)

Fig. 6 FFT analysis results of (a) phase voltage, (b) line voltage and (c) line current under traditional PSPWM

(a)   (b)   (c)

Fig. 7 FFT analysis results of (a) phase voltage, (b) line voltage and (c) line current under the proposed PSPWM

The 2018 International Power Electronics Conference

Fig. 6 and Fig. 7 are their FFT analysis results. It can be seen that the THDs of the phase voltage under different PSPWM methods are nearly the same while the THDs of line voltage and line current under the proposed PSPWM are much better than the traditional PSPWM, which demonstrates that the proposed modulation method is effective.

V. EXPERIMENTAL RESULTS

In order to demonstrate the validity of this modulation method, a three-phase 5L-HC inverter has been built, as shown in Fig. 8. The experimental results are shown in Table III.

TABLE III
EXPERIMENTAL PARAMETERS

Parameters	Value
DC-link voltage	U_{dc} =200 V
DC-link capacitor	$C_{d1} = C_{d3} = 2460 \ \mu F$, $C_{d2} = 1230 \ \mu F$
Flying capacitor	$C_{fx1} = C_{fx2} = 100 \ \mu F$
Carrier frequency	$f_c = 8$ kHz
R-L Load	$R_1 = 25 \ \Omega$, $L_1 = 5$ mH

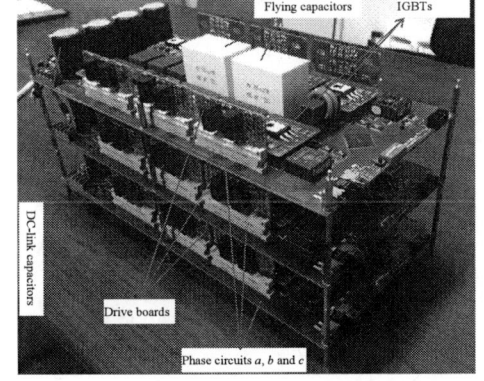

Fig. 8 5L-HC converter prototype.

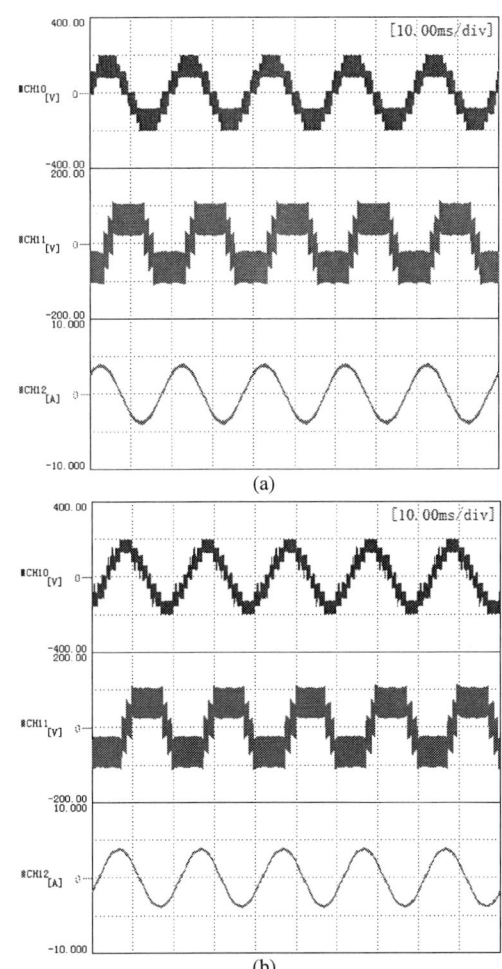

(a)

(b)

Fig. 9 Experimental results of phase voltage, line voltage and line current under (a) traditional PSPWM and (b) the proposed PSPWM.

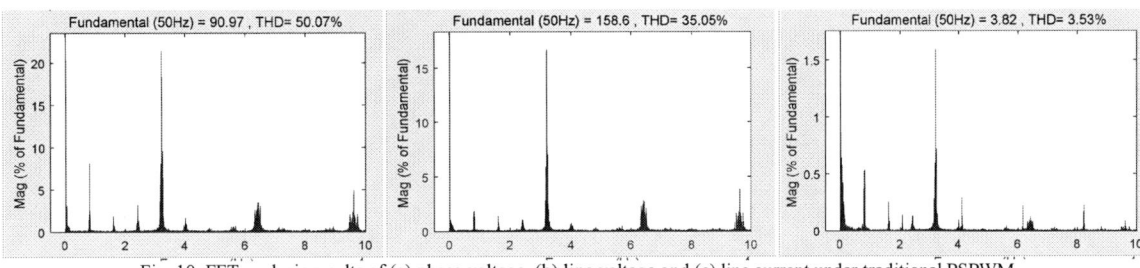

Fig. 10 FFT analysis results of (a) phase voltage, (b) line voltage and (c) line current under traditional PSPWM

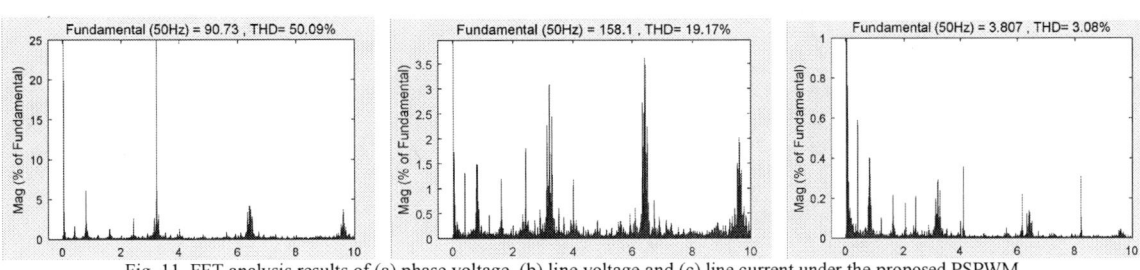

Fig. 11 FFT analysis results of (a) phase voltage, (b) line voltage and (c) line current under the proposed PSPWM

1013

Fig. 9 shows the phase voltage, line voltage and line current with traditional PSPWM and the proposed PSPWM respectively. The experimental results are similar with the simulation results. Fig. 10 and Fig. 11 are their FFT analysis results. The THDs of the phase voltage under different PSPWM method are nearly the same while the THD of line voltage under the proposed PSPWM is much better than the traditional PSPWM. The THD of line current is only a little better under the proposed PSPWM is because of the filter of load inductor. The simulation and experimental results demonstrate that the proposed modulation method is effective.

VI. CONCLUSIONS

5L-HC converter has great potential in high-power medium-voltage drive applications. This paper focuses on the modulation method of this converter and an improved PSPWM method is proposed to improve the harmonics performance, which can be seen as an equivalent to PDPWM. The validity of this method is verified by simulation and experimental results.

REFERENCES

[1] S. Kouro, M. Malinowski, K. Gopakumar, J. Pou, L. G. Franquelo and W. Bin, et al., "Recent Advances and Industrial Applications of Multilevel Converters," *IEEE Trans. Ind. Electron.*, vol.57, no. 8, pp. 2553-2580, Aug. 2010.

[2] H. Abu-Rub, J. Holtz, J. Rodriguez, and B. Ge, "Medium-Voltage Multilevel Converters—State of the art, challenges, and requirements in industrial applications," *IEEE Trans. Ind. Electron.*, vol.57, no. 8, pp. 2581-2596, Aug. 2010.

[3] J. Rodriguez, S. Bernet, P. Steimer, and I. Lizama, "A survey on neutral-point-clamped inverters," *IEEE Trans. Ind. Electron.*, Vol. 57, no. 7, pp. 2219–2230, July 2010.

[4] M. Malinowski, K. Gopakumar, J. Rodriguez, X. Pe and M. A. Rez, "A Survey on Cascaded Multilevel Inverters, " *IEEE Trans. Ind. Electron.*, Vol. 57, no. 7, pp. 2197-2206, July 2010.

[5] A. M. Y. M. Ghias, J. Pou and V. G. Agelidis, "Voltage-Balancing Method for Stacked Multicell Converters Using Phase-Disposition PWM," *IEEE Trans. Ind. Electron.*, vol. 62, no. 7, pp. 4001-4010, July 2015.

[6] T. A. Meynard, H. Foch, F. Forest, C. Turpin, F. Richardeau, L. Delmas, G. Gateau, and E. Lefeuvre, "Multicell converters:

Derived topologies," *IEEE Trans. Ind. Electron.*, vol. 49, no. 5, pp. 978–987, Oct. 2002.

[7] A. M. Y. M. Ghias, J. Pou and V. G. Agelidis, "An Active Voltage-Balancing Method Based on Phase-Shifted PWM for Stacked Multicell Converters," *IEEE Trans. Power Electron.*, vol. 31, no. 3, pp. 1921-1930, March 2016.

[8] P. Barbosa, P. Steimer, J. Steinke, L. Meysenc, M. Winkelnkemper, and N. Celanovic, "Active neutral-point-clamped (ANPC) multilevel converter technology," in *Proc. Conf. Rec. EPE*, 2005, CDROM.

[9] S. Pulikanti and V. Agelidis, "Hybrid Flying Capacitor Based Active-Neutral-Point-Clamped Five-Level Converter Operated with SHE-PWM," *IEEE Trans. Ind. Electron.*, vol. 58, no. 10, pp. 4643-4653, Oct. 2011.

[10] Z. Liu, Y. Wang, G. Tan, H. Li, Y. Zhang, "A Novel SVPWM Algorithm for Five-Level Active Neutral-Point-Clamped Converter," *IEEE Trans. Power Electron.*, vol.31, no. 5, pp. 3859-3866, May 2016.

[11] M. Narimani, B. Wu and N. Zargari, "A Novel Five-Level Voltage Source Inverter with Sinusoidal Pulse Width Modulator for Medium-Voltage Applications," *IEEE Trans. Power Electron.*, vol. 31, no. 3, pp. 1959-1967, March 2016.

[12] A. Lesnicar and R. Marquardt, "A new modular voltage source inverter topology," in *Proc. Conf. Rec. Eur. Conf. Power Electron. Appl.*, 2003, pp.1-10.

[13] M. A. Perez, S. Bernet, J. Rodriguez, S. Kouro, and R. Lizana, "Circuit Topologies, Modeling, Control Schemes, and Applications of Modular Multilevel Converters," *IEEE Trans. Power Electron.*, vol. 30, no. 1, pp. 4-17, Jan. 2015.

[14] S. Debnath, J. Qin, B. Bahrani, M. Saeedifard, and P. Barbosa, "Operation, Control, and Applications of the Modular Multilevel Converter: A Review," *IEEE Trans. Power Electron.*, vol. 30, no. 1, pp. 37-53, Jan. 2015.

[15] K. Wang, Z. Zheng, L. Xu, and Y. Li, "Topology and Control of a Five-Level Hybrid-Clamped Converter for Medium-Voltage High-Power Conversions," *IEEE Trans. Power Electron.*, vol. 33, no. 6, pp. 4690-4702, June 2018.

[16] C. Feng, J. Liang and V. G. Agelidis, "Modified Phase-Shifted PWM Control for Flying Capacitor Multilevel Converters," *IEEE Trans. Power Electron.*, vol. 22, no. 1, pp. 178-185, 2007.

[17] A. M. Y. M. Ghias, J. Pou, M. Ciobotaru, and V. G. Agelidis, "Voltage-Balancing Method Using Phase-Shifted PWM for the Flying Capacitor Multilevel Converter," *IEEE Trans. Power Electron.*, vol. 29, no. 9, pp. 4521-4531, Sep. 2014.

[18] S. R. Pulikanti, G. S. Konstantinou and V. G. Agelidis, "An n-level flying capacitor based active neutral-point-clamped converter," *2010 2nd IEEE International Symposium on Power Electronics for Distributed Generation Systems (PEDG)*, pp. 553-558.

The 2018 International Power Electronics Conference

Integrated Control methods for asymmetrical cascaded H-bridge rectifier

Wenjing Dai, Jie Chen, Xin Chen, Chunying Gong
Nanjing University of Aeronautics and Astronautics, Nanjing, P.R.China
E-mail: daiwenjing_nuaa@foxmail.com

Abstract- **The hybrid cascaded H-bridge rectifier (CHBR) with suitable modulation is superior in the performance for its ability to synthesize more levels at ac side. The structure of combing hybrid CHBR topology and the hybrid H-bridge inverter is applicable for AC/AC converter for aircraft applications, solid state transformer (SST), and so on. This paper proposes two kinds of modulation methods and corresponding control strategies applying to hybrid CHBR. First the control strategy for traditional hybrid modulation is proposed. Then a kind of quasi unipolar carrier disposition pulse width modulation (CD-PWM) is put forward to avoid the problem existing in hybrid modulation. Simulation verifies the adopted modulation and strategy.**

Keywords—hybrid cascaded H-bridge rectifier, carrier disposition pulse width modulation, control strategy of asymmetrical output.

I. Introduction

Among all of the multilevel converters, Cascaded H-bridge rectifier (CHBR) is popular for its simple lay out, enhanced input power quality, and great modularity [1].

Many modulation techniques have been proposed for CHBR, three key of which are Multi-carrier PWM, Multilevel Selective Harmonic Elimination (SHE) and Space Vector PWM (SVM-PWM) [2-5]. Among them, the Multi-carrier PWM is the most commonly used which can be classified into Carrier-Phase-Shifted PWM (CPS-PWM) and Hybrid-PWM.

CPS-PWM offers even power distribution at any modulation depth and the principle is simple, but each switch works at high frequency that causes great switching losses and only equal dc voltage values can be output in different cells. Additionally, it is found that for same device switching frequency, the overall performance of Hybrid-PWM technique in terms of voltage and current THDs is superior to those with CPS-PWM [6], because five levels voltage is synthesized at ac side of two cells CHBR using CPS-PWM while seven levels voltage is produced using Hybrid-PWM, which realizing the reduction of the current harmonics. Studies [7][8] also find that the hybrid H-bridge inverter topology with unequal dc bus voltages can improve electric performance and reduce system weight and volume significantly.

The CHBR topology with asymmetrical output and the hybrid H-bridge inverter can be combined to realize the "back to back" power structure, such as the AC/AC topology designed for aircraft applications shown in

fig.1, where the input voltage is 115V at variable frequency of 360-800Hz and the CHBR is a great substitute for the bulky ATRU. The DC/DC converters are used just for isolation effect.

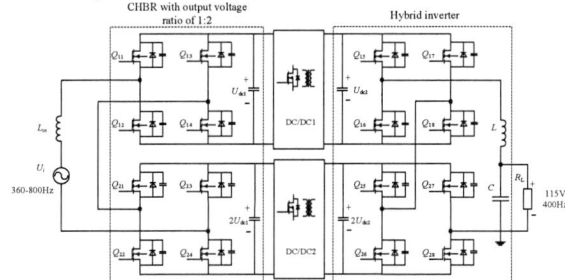

Fig.1. Topology of cascaded H-bridge multilevel AC/AC converter designed for aircraft application

In this paper, the modulation and control strategy of two cells hybrid CHBR with 1:2 output voltage ratio are considered. First the control strategy for traditional hybrid modulation is proposed and the power flow problem is analyzed in section II. After that, a kind of quasi unipolar Carrier Disposition Pulse Width Modulation (CD-PWM) which can avoid power flow is introduced, and corresponding strategy is proposed to maintain output voltage and guarantee the unity power factor in section III. Finally, the system performance was examined through a Matlab/ Simulink model during both transient and steady state conditions.

II. Traditional Control Method Of Hybrid Multilevel CHBR

The hybrid multilevel topology shown in fig.2 consists of two cascaded H-bridge units, in which the switches $Q_{11} \sim Q_{14}$ constitute the low-voltage DC output H-bridge unit with output u_{dc1} equal to E, and the switches $Q_{21} \sim Q_{24}$ constitute the high-voltage DC output H-bridge unit with output u_{dc2} equal to 2E. A kind of traditional hybrid multilevel modulation which has been used in hybrid inverter is introduced in the following paragraphs, and then the control strategy applying to the modulation is proposed in this paper. After that, the ac side voltage is analyzed theoretically and the deficiency of this kind of traditional hybrid modulation is revealed.

A. traditional hybrid modulation

The traditional hybrid modulation works as follows: The square wave modulation signal of high voltage unit shown in the fig.3(a) is obtained by comparing the sine-wave modulation signal v_{r1} with the threshold voltage v_{cmp} and $-v_{cmp}$, the ac side voltage v_H of high output unit

1015

shown in fig.3(c) is square wave voltage whose amplitude is 2E. The modulation signal v_{r2} of low voltage unit shown in the fig.3(b) is obtained by subtracting square wave modulation signal of high voltage unit from v_{r1}, compare v_{r2} with the high frequency carrier wave, then driving signal of the low output unit is got. The ac side voltage v_L of low DC output unit is shown as fig.3(d), when $v_{r1} > v_{c1}$, the ac side voltage is +E; when $v_{r1} < v_{c2}$, the ac side voltage is -E.

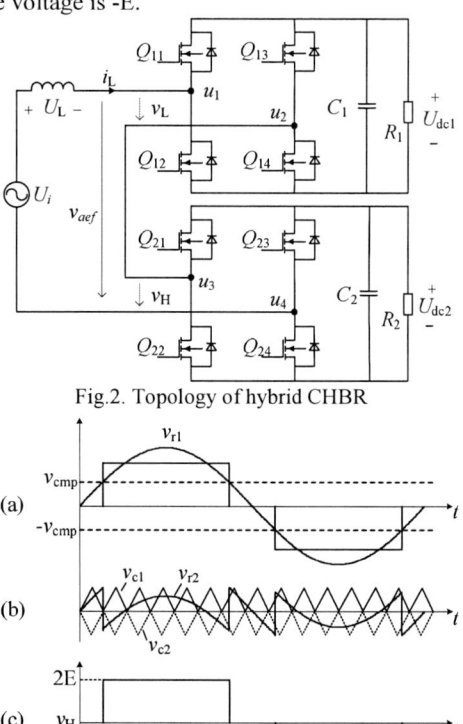

Fig.2. Topology of hybrid CHBR

Fig.3. Principle diagram of traditional hybrid modulation

As is shown in Fig.3, seven voltage levels can be synthesized at the ac side of the CHBR. Additionally, the switches in the high output voltage unit works at the basic frequency, which lead to less switching loss; The switches in the low output voltage unit bear less stress so low voltage switch devices can be chosen and the switching loss is also not large relatively although switches work at high frequency.

B. control strategy of traditional hybrid modulation

In order to achieve the output of CHBR with an output voltage ratio of 1:2, the control strategy suitable for the traditional hybrid modulation method is proposed as below.

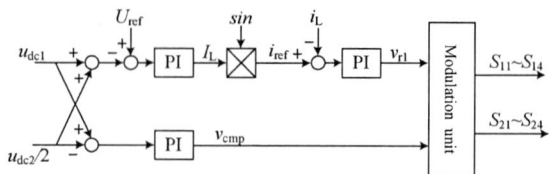

Fig.4. Proposed control diagram of traditional hybrid modulation in two cells CHBR

As is shown in fig.4, this controller includes an overall voltage control loop, a voltage distribution control loop and a current control loop. The sum of the DC-link voltage u_{dc1} and $u_{dc1}/2$ is controlled through the overall voltage controller which is a conventional PI controller, among which the overall reference voltage U_{ref} is 2E. The output of the controller multiply by synchronous sinusoidal signal of input voltage and then the command of the current i_{ref} is got. The actual current i_L is compared to the instantaneous current command i_{ref}, the error is sent to the PI current controller to produce the sinusoidal modulation wave v_{r1}, so the error between the reference current signals and the actual current can be eliminated; The voltage distribution control loop make the low-voltage output u_{dc1} compared to half of the high-voltage output $u_{dc1}/2$ and sending the error to PI controller so the threshold voltage v_{cmp} is produced. With the work of overall voltage control loop and voltage distribution control loop, the deviation between DC output voltage and its reference can be regulated, in other words, the DC output voltage ratio of 1:2 can be achieved.

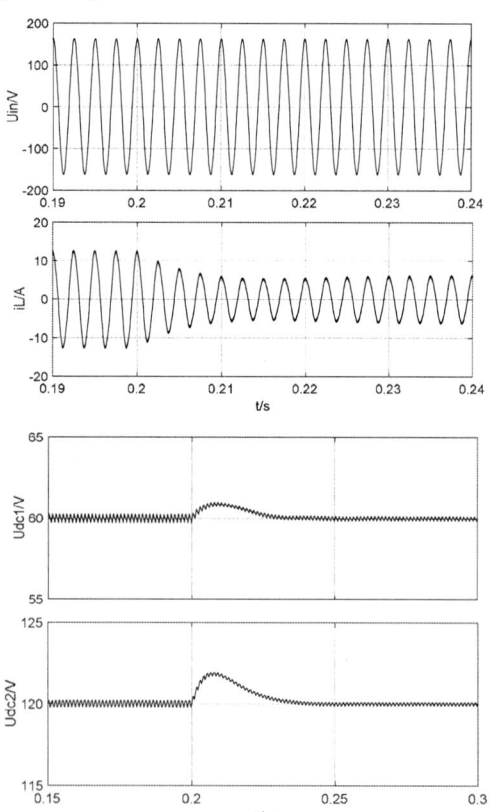

Fig.5 Simulation results of asymmetrical CHBR with hybrid modulation and corresponding control strategy.

Simulation of the asymmetrical output CHBR is carried out. At 0.2s，a load decrease of 50% is applied at the system. As is shown in fig.5, the validity of this kind of modulation and control strategy is verified.

C. The problem of power flow

Under the condition of achievement of the DC output voltage ratio as 1:2, status of synthesizing seven voltage levels at ac side with traditional hybrid modulation is list out at below.

TABLE I
STATUS OF SYNTHESIZING SEVEN VOLTAGE LEVELS AT AC SIDE WITH TRADITIONAL HYBRID MODULATION

v_{afe}	v_H	v_L
+3E	+2E	+E
+2E	+2E	0
+E	+2E	-E
	0	+E
0	0	0
-E	0	-E
	-2E	+E
-2E	-2E	0
-3E	-2E	-E

It is easy to see that, when the overall ac side voltage is +E (or–E), the high-voltage output unit and the low-voltage output unit are connected in series while their voltages are at the opposite polarity, which is shown in fig.6. Meanwhile, the polarity of the current flowing through the two cells is the same all the time. Therefore, the problem of energy exchange or power flow exist between two H-bridge with the traditional hybrid modulation, which may lead to the fluctuation of the DC voltage on the capacitor and the reduction of the system efficiency to some extent.

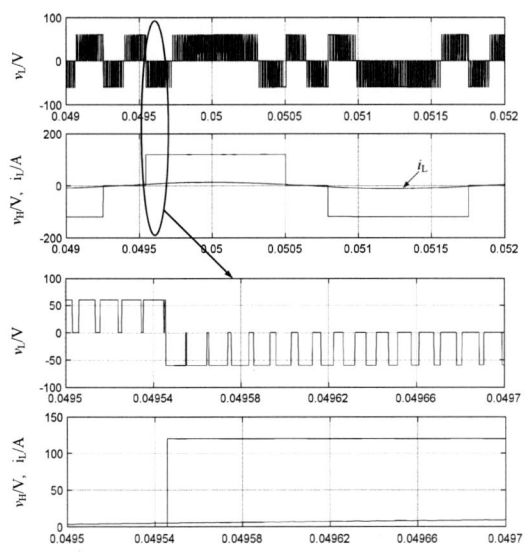

Fig.6. Simulation result displaying the power flow

Hence, in order to avoid the power flow between two H-bridge cells, the status of opposite polarity voltage must be eliminated. A kind of carrier disposition pulse width modulation (CD-PWM) is proposed to resolve the power flow problem in this paper.

III. CARRIER DISPOSITION PULSE WIDTH MODULATION

A. principle of CD-PWM

First of all, the way to synthesizing seven voltage levels at ac side as the operation model, is analyzed in the following paragraphs.

When the ac side voltage v_{afe} is in the [+2E, +3E] range as is shown in fig.7(a) and fig.7(b), the ac side voltage v_H in high-voltage output unit is a constant value of +2E while the ac side voltage v_L in low-voltage output unit changes between 0 and +E.

When the ac side voltage v_{afe} is in the [+E, +2E] range as is shown in fig.7(b) and fig.7(c), two cells complement each other, that means, when v_H is +2E, v_L is 0; Or when v_H is 0, v_L is +E.

When the ac side voltage v_{afe} is in the [0, +E] range as is shown in fig.7(c) and fig.7(d), v_H is a constant value of 0 while the v_L changes between 0 and +E.

When the ac side voltage v_{afe} is in the [0, -E] range as is shown in fig.7(e) and fig.7(f), v_H is a constant value of 0 while the v_L change between 0 and -E.

When the ac side voltage v_{afe} is in the [-E, -2E] range as is shown in fig.7(f) and fig.7(g), two cells complement each other, that means, when v_H is -2E, v_L is 0; Or when v_H is 0, v_L is -E.

When the ac side voltage v_{afe} is in the [-2E, -3E] range as is shown in fig.7(g) and fig.7(h), v_H is a constant value of -2E while the v_L changes between 0 and -E.

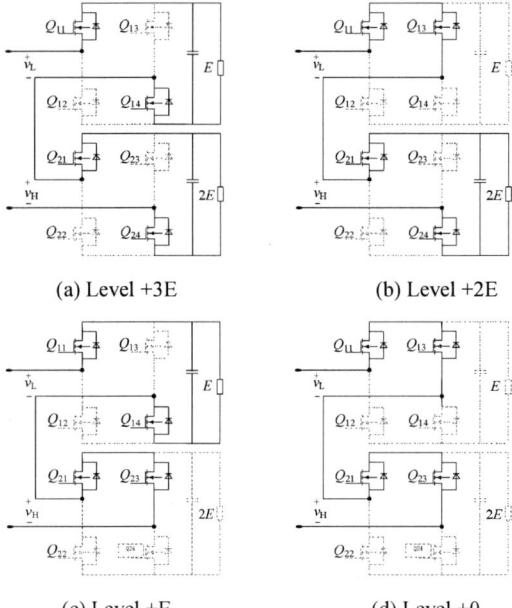

(a) Level +3E   (b) Level +2E

(c) Level +E   (d) Level +0

The 2018 International Power Electronics Conference

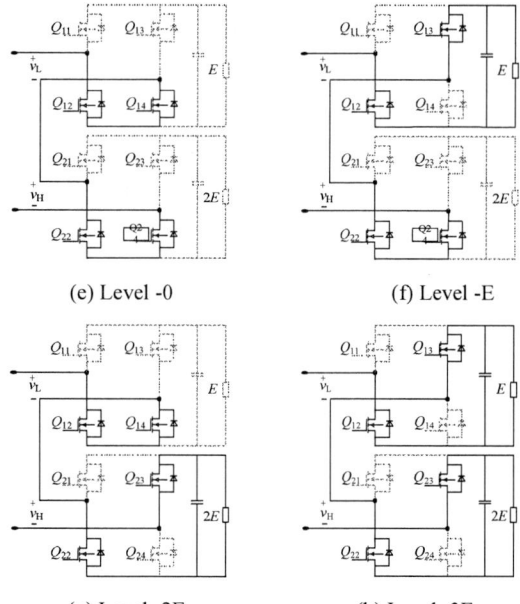

(e) Level -0   (f) Level -E

(g) Level -2E   (h) Level -3E

Fig.7 operating mode with CD-PWM

From analysis above we can see that, the ac side voltages of the two units are always at the same polarity, which avoid the power flow effectively. Moreover, while v_{afe} is positive, switches Q_{11} and Q_{21} are always on; While v_{afe} is negative, switches Q_{12} and Q_{22} are always on. That is to say, the status of the switches of two H-bridge left-bridge-arm is consistent, which depending on the positive or negative polarity of the modulated wave v_{ref}. So the CD-PWM proposed is a modulation in similar to unipolar modulation for each cascaded unit.

Quasi unipolar CD-PWM works as follows: First of all, the polarity signal D is obtained by the sinusoidal modulation wave v_{ref}, then the sinusoidal modulation wave v_{ref} and -v_{ref} are compared to three triangle wave v_a, v_b, v_c which are distributed above the zero reference line and arranged from top to bottom. Six logic signals A, B, C, A′, B′ and C′ are given by:

$$A = \begin{cases} 1 & v_{ref} > v_a \\ 0 & v_{ref} < v_a \end{cases} \qquad (1)$$

$$B = \begin{cases} 1 & v_{ref} > v_b \\ 0 & v_{ref} < v_b \end{cases} \qquad (2)$$

$$C = \begin{cases} 1 & v_{ref} > v_c \\ 0 & v_{ref} < v_c \end{cases} \qquad (3)$$

$$A' = \begin{cases} 1 & -v_{ref} > v_a \\ 0 & -v_{ref} < v_a \end{cases} \qquad (4)$$

$$B' = \begin{cases} 1 & -v_{ref} > v_b \\ 0 & -v_{ref} < v_b \end{cases} \qquad (5)$$

$$C' = \begin{cases} 1 & -v_{ref} > v_c \\ 0 & -v_{ref} < v_c \end{cases} \qquad (6)$$

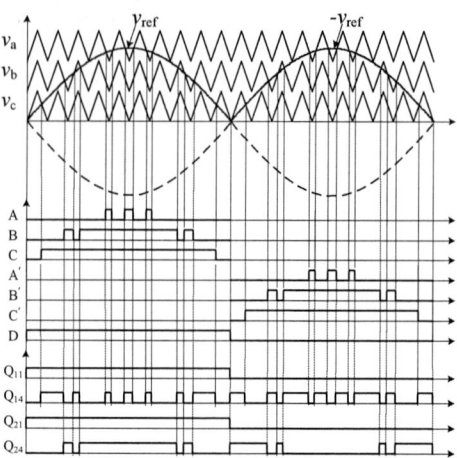

Fig.8 Schematic diagram of logic signal generation and driving signal of switches

In the positive half cycle of v_{ref}, the value of D is 1, as is shown in fig.8, so Q_{11}, Q_{21} are always on, Q_{12}, Q_{22} which is complement Q_{11}, Q_{21} stay the off state. When $v_c < v_{ref} < v_b$ or $v_{ref} > v_a$, Q_{14} open, otherwise Q_{13} open. When $v_{ref} > v_{trb}$, Q_{24} open, otherwise Q_{23} open. Therefore, during the positive half cycle of v_{ref}, driving signals of switch are given with logic equation as:

$$Q_{11} = D, \quad Q_{12} = \bar{D}, \quad Q_{14} = A + \bar{B}C, \quad Q_{13} = \bar{Q}_{14}$$

$$Q_{21} = D, \quad Q_{22} = \bar{D}, \quad Q_{24} = B, \quad Q_{23} = \bar{Q}_{24}$$

In the negative half cycle of v_{ref}, the value of D is 0, so Q_{12}, Q_{22} are always on and Q_{11}, Q_{21} stay the off state. When $v_c < -v_{ref} < v_b$ or $-v_{ref} > v_a$, Q_{13} open, otherwise Q_{14} open. When $-v_{ref} > v_{trb}$, Q_{23} open, otherwise Q_{24} open. Therefore, during the negative half cycle of v_{ref}, driving signals of switch are given with logic equation as:

$$Q_{11} = D \quad Q_{12} = \bar{D} \quad Q_{13} = A' + \bar{B}'C' \quad Q_{14} = \bar{Q}_{13}$$

$$Q_{21} = D \quad Q_{22} = \bar{D} \quad Q_{23} = B' \quad Q_{24} = \bar{Q}_{23}$$

Fig.9 Level status with quasi unipolar CD-PWM

Based on the analysis above, the logic expressions of the driving signals of each switch can be obtained in one cycle:

$$Q_{11} = D \quad Q_{12} = \bar{D}$$

$$Q_{14} = D(A + \bar{B}C) + \bar{D}\overline{(A' + \bar{B}'C')} \quad Q_{13} = \bar{Q}_{14}$$

1018

$$Q_{21} = D \quad Q_{22} = \overline{D} \quad Q_{24} = DB + \overline{D}\overline{B}' \quad Q_{23} = \overline{Q}_{24}$$

According to the driving signals of the switches, the seven voltage levels at the overall ac side can be synthesized which is shown in fig.9.

B. control strategy for asymmetrical output of CHBR

The overall voltage control loop and the current control loop proposed in this paper for the CD-PWM are similar to the strategy used for the traditional hybrid modulation above, which produce the modulation waveform v_m, as is shown in fig.10. Additionally, in order to achieve the output of CHBR with a voltage ratio of 1:2, a control strategy is proposed below to realize the asymmetrical output of two cells CHBR.

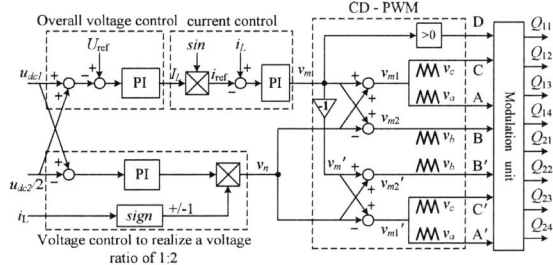

Fig.10 Proposed system control diagram of two cells CHBR with CD-PWM

Half of the output of the high-voltage output unit u_{dc2} is subtracted from output of the low-voltage output unit u_{dc1}, then the result is sent into a PI regulator. The output of the regulator multiply by the polarity signal of the input current i_L, after which the result called voltage ratio regulating signal v_n is obtained.

When v_m is positive, a new modulation wave v_{m1} is achieved by adding v_n to v_m while another new modulation wave v_{m2} is achieved by subtracting v_n from v_m. v_{m1} is compared to the triangle carrier wave v_a, v_c as the modulation wave of the low-voltage output unit, then the driving signal of the switch Q_{13} and Q_{14} is produced, while v_{m2} is compared to the triangle carrier wave v_b as the modulation wave of the high-voltage output unit, then the driving signal of the switch Q_{23} and Q_{24} is produced.

When v_m is negative, a new modulation wave v_{m2}' is achieved by adding v_n to $-v_m$ while another new modulation wave v_{m1}' is achieved by subtracting v_n from $-v_m$. v_{m1}' is compared to the triangle carrier wave v_a, v_c as the modulation wave of the low-voltage output unit, then the driving signal of the switch Q_{13} and Q_{14} is produced, while v_{m2}' is compared to the triangle carrier wave v_b as the modulation wave of the high-voltage output unit, then the driving signal of the switch Q_{23} and Q_{24} is produced.

The effect of the control strategy proposed above is analyzed below. First of all, assuming that u_{dc1} is bigger than $u_{dc2}/2$, so the modulation signal v_m is in the positive half cycle, which means that the switch Q_{11}, Q_{21} are on. The input current i_L is set as positive when it flows out from the left arm of low-voltage output unit.

When $i_L>0$, voltage ratio regulating signal $v_n>0$, leading to the amplitude of the new modulation signal v_{m1} is increased. The logic expressions of switch drive signal

analyzed above show that, at this point, time that the driving signal of Q_{14} is high is prolonged by comparing v_{m1} with the carrier wave, that is, the discharging time of the DC bus capacitor in the low voltage unit is prolonged. At the same time, the amplitude of the new modulation signal v_{m2} is decreased, the logic expressions of switch drive signal show that, time that the driving signal of Q_{24} is on is shortened, that is, the discharging time of the DC bus capacitor in the high voltage unit is shortened. The overall effect of the change of the charging or discharging time of the DC bus capacitors as is analyzed above is the reduction of u_{dc1} and the rise of $u_{dc2}/2$.

Similarly, others cases can be analyzed that, by using the control strategy shown in fig.10, the DC output voltages of the two cells are adjusted continuously until the balance of 1:2 output ratio is reached.

IV. SIMULATION RESULTS

A Matlab/Simulink model of single-phase two cells CHBR is built to verify the quasi unipolar CD-PWM and its control strategy proposed. The ac input is 115V400HZ. The target values of two cells output voltages are 60V and 120V respectively. The inductance at ac side is 80uH, the frequency of the triangle carrier is 100kHz.

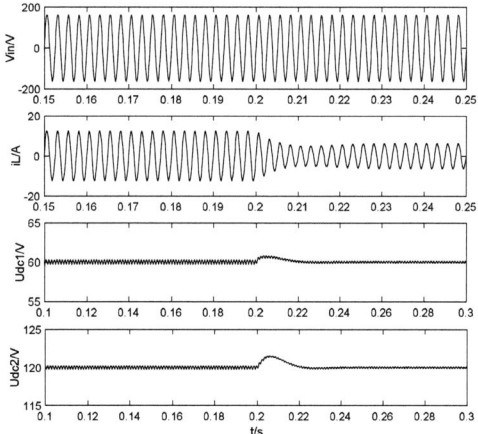

Fig.11 Simulation results of two cells CHBR with quasi unipolar CD-PWM

Input voltage and current at the ac side are shown below. When the system runs with full load, the input current is sinusoidal and THD is 3.49%. Output dc voltage u_{dc1} and u_{dc2} in fig.11 show that output voltage can also keep steady at 60V and 120V. At 0.2s a load decrease of 50% is applied at the system, the result shows that the system has a very fast dynamics.

V. CONCLUSIONS

Hybrid CHBR with asymmetrical output brings significant advantages in weight, volume and performance for its feasibility of synthesizing more levels at ac side. For two cells CHBR, the control strategy suitable for hybrid CHBR with traditional hybrid modulation is proposed in this paper. To avoid the power flow between two H-bridge cells in hybrid modulation, quasi unipolar CD-PWM is presented, and the corresponding control strategy is proposed. Simulation results are given for the verification of quasi unipolar CD-PWM and control strategy.

REFERENCES

[1] T. Xinghua, L. Yongdong and S. Min, "A Pi-based control scheme for primary cascaded H-bridge rectifier in transformerless traction converters," 2010 International Conference on Electrical Machines and Systems, Incheon, 2010, pp. 824-828.

[2] Y. Li, Y. Wang and B. Q. Li, "Generalized Theory of Phase-Shifted Carrier PWM for Cascaded H-Bridge Converters and Modular Multilevel Converters," in IEEE Journal of Emerging and Selected Topics in Power Electronics, vol. 4, no. 2, pp. 589-605, June 2016.

[3] A. Moeini, H. Zhao and S. Wang, "Improve Control to Output Dynamic Response and Extend Modulation Index Range With Hybrid Selective Harmonic Current Mitigation-PWM and Phase-Shift PWM for Four-Quadrant Cascaded H-Bridge Converters," in IEEE Transactions on Industrial Electronics, vol. 64, no. 9, pp. 6854-6863, Sept. 2017.

[4] H. Jiang-bo and F. Zhi-hong, "The simulation research of control modeling for three-phase voltage source SVPWM rectifier," 2016 IEEE International Conference on High Voltage Engineering and Application (ICHVE), Chengdu, 2016, pp. 1-5.

[5] H. Iman-Eini, S. Farhangi and J. L. Schanen, "A modular AC/DC rectifier based on cascaded H-bridge rectifier," 2008 13th International Power Electronics and Motion Control Conference, Poznan, 2008

[6] I. Sarkar and B. G. Fernandes, "Modified hybrid multi-carrier PWM technique for cascaded H-Bridge multilevel inverter," IECON 2014 - 40th Annual Conference of the IEEE Industrial Electronics Society, Dallas, TX, 2014, pp. 4318-4324.

[7] M. D. Manjrekar, P. K. Steimer, and T. A. Lipo, "Hybrid multilevel power conversion system: A competitive solution for high-power applications," IEEE Trans. Ind. Applicat., vol. 36, pp. 834–841, May/June 2000.

[8] B. P. McGrath and D. G. Holmes, "Multicarrier PWM strategies for multilevel inverters," in IEEE Transactions on Industrial Electronics, vol. 49, no. 4, pp. 858-867, Aug 2002.

[9] L. Xiang, W. Jian, Y. Xiaojie and W. Kun, "An improved proportional pulse compensation strategy for DC voltage balance of cascaded H-Bridge Rectifier," 2016 IEEE Energy Conversion Congress and Exposition (ECCE), Milwaukee, WI, 2016, pp. 1-6.

Transient Voltage Stress Modeling for Submodules of Modular Multilevel Converters under Grid Voltage Sags

Zhijian Yin, Yongheng Yang, and Huai Wang
Department of Energy Technology, Aalborg University, Aalborg, Denmark
E-mail: zyi@et.aau.dk, yoy@et.aau.dk, hwa@et.aau.dk

Abstract—The trend for the model-based design of modular multilevel converters (MMC) demands a better estimation of the submodule (SM) voltage stress. The transient voltage stresses of MMC SMs are important for the SM capacitor dimensioning and the robustness of power modules. This paper proposes an analytical model of the SM transient voltage stress under grid voltage sags. It is based on a simplified analysis of the MMC current control and circulating current control schemes. A case study reveals that a sufficient accuracy level can be achieved with the proposed model to estimate the transient voltage stress.

Index Terms—Submodulue voltage stress, modular multilevel converter, dynamic analysis, grid voltage sag.

I. INTRODUCTION

An intensive development of modular multilevel converters (MMC) has been acknowledged in the last decade. Compared with traditional line-commutated converters (LCC), the MMC has many merits in high voltage DC (HVDC) applications. For instance, MMC systems can operate with a low switching frequency, while maintaining high energy conversion efficiency, strong scalability, and low harmonic distortion [1]-[3]. In addition, the MMC has been widely used in various AC systems, such as static synchronous compensators (STATCOM) and motor drives [3], [4].

Due to the harsh operation environment in power transmission systems, the dimensioning of the MMC must meet the requirements for high reliability and strong fault tolerance. Among various failure modes of MMC systems, one critical mechanism is related to the over-voltage on submodules (SMs) in the case of unexpected grid faults [5], [6]. Thus, it is important to evaluate the maximum SM voltage stress under different operation circumstances to facilitate the component selection and then the installation for specific MMC-based projects.

To properly size the SM components, many efforts have been devoted to modeling the SM voltage stress [7]-[10]. For instance, the SM voltage variation is discussed in [7], seen from the prospective of energy storage requirements. The design parameters of this approach is then normalized and used for the SM capacitor selection in [8]. However, this method only considers the system operation in steady-state operating conditions (i.e. the grid voltage is stable, and there are no DC

voltage variations, etc.), which is not applicable to estimate the SM voltage stress during abnormal conditions. For example, a significant voltage oscillation may appear on the SM capacitor in response to grid dynamics, which may be caused by interaction of control loops. To address this issue, the steady-state SM voltage ripple in the case of unbalanced grid condition is analyzed in [9] considering the positive-negative-sequence grid voltage and the MMC arm current. The interaction between the circulating current control and the SM voltage stress in an imbalanced grid is proposed in [10], while the impact of the SM voltage balancing control is investigated in [11]. However, these provide a steady-state analytical solution for the SM voltage estimation without considering the transient system behaviors which, however, may induce catastrophic failures in operation, despite its necessity in MMC design.

Therefore, this paper investigates the transient SM voltage stress behaviors during grid voltage dips. The mathematical modeling and equivalent circuit of the MMC are discussed in Section II. The transient SM voltage stress analysis and the impact of typical MMC control loops are presented in Section III. The theoretical analysis is verified by a comparative study between simulation results and theoretical estimations in Section IV, followed by a conclusion.

II. MATHEMATICAL MODELING OF THE MMC

Fig. 1 illustrates a single-phase topology of MMC. Each arm of the MMC contains N identical half-bridge SMs and one series arm inductor L_a. The SM is formed by two insulated gate bipolar translators (IGBTs) and a parallel SM capacitor C_{sm}. In Fig. 1, v_{sm} represents the SM voltage. The DC-bus voltage and circulating current are denoted as V_{dc} and i_{cir}, respectively. The phase-grid voltage and injected current are represented as v_g and i_o. Moreover, L_{ac} is an equivalent series between the MMC and the grid. In the normal operation, the grid voltage can be expressed as

$$v_g = U \sin(\omega t) \tag{1}$$

where U is the amplitude of the phase voltage with ω being the grid fundamental frequency.

To simplify the circuit analysis, an average circuit model is derived as shown in Fig. 2(a) with the following assumptions:

The 2018 International Power Electronics Conference

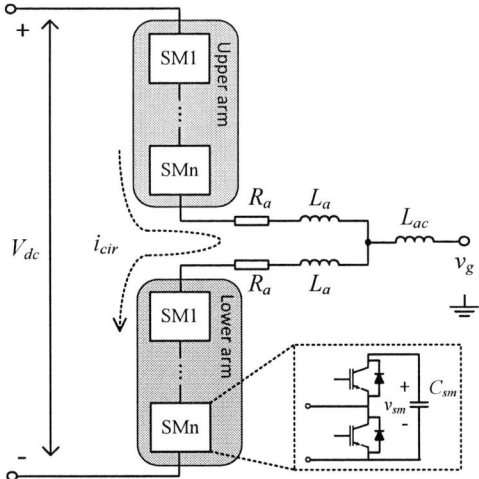

Fig. 1. Single-phase circuit topology of an MMC.

1) The voltage stress across each SM is balanced at any moment during the operation.

2) The switching frequency for each SM is much higher than the fundamental frequency.

In Fig. 2(a), all SMs are combined to a single one with an equivalent DC-link capacitor C_{eq} containing the total arm voltage stress $v_{c\Sigma}$. The capacitance of C_{eq} is equal to C_{sm}/N. Here, n_u and n_l are the insertion indices for the upper and lower arm; i_u and i_l represents the arm current for the upper and lower arms, respectively. Furthermore, i_o is the output current to the grid which can be calculated by

$$i_o = i_u - i_l \qquad (2)$$

According to the Kirchhoff's Voltage Law (KVL), the system dynamics can be represented by

$$-\frac{V_{dc}}{2} + \frac{n_u + n_l}{2} v_{c\Sigma} + L_a \frac{di_{cir}}{dt} = 0 \qquad (3)$$

$$\frac{n_u - n_l}{2} v_{c\Sigma} + (\frac{L_a}{2} + L_{ac}) \frac{di_o}{dt} + v_g = 0 \qquad (4)$$

with

$$i_{cir} = \frac{i_u + i_l}{2} \qquad (5)$$

According to (3) and (4), the dc-side and ac-side equivalent circuits can be derived as shown in Fig. 2(b) and Fig. 2(c) respectively, in which the system control variables d_{cir} and d_{out} are defined as

$$d_{cir} = \frac{n_u + n_l}{2} \qquad (6)$$

$$d_{out} = \frac{n_l - n_u}{2} \qquad (7)$$

Then the insertion indices for the upper and lower arms can be represented by

$$n_u = d_{cir} - d_{out} \qquad (8)$$

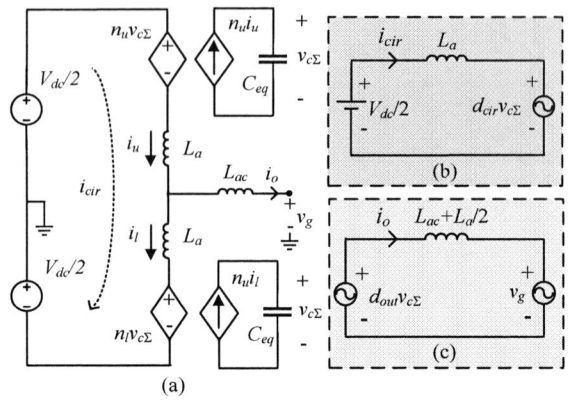

(a)

Fig. 2. Equivalent circuit model of the MMC. (a) Average model of the MMC. (b) DC-side equivalent circuit. (c) AC-side equivalent circuit.

$$n_l = d_{cir} + d_{out} \qquad (9)$$

The relationship among the arm current, output current, and circulating current is then obtained as

$$i_u = i_{cir} + \frac{i_o}{2} \qquad (10)$$

$$i_l = i_{cir} - \frac{i_o}{2} \qquad (11)$$

Noticing that the equivalent SM capacitor charging currents for the upper and lower arms are modeled as $n_u i_u$ and $n_l i_l$, respectively, in Fig. 2(a). Thus the SM capacitor voltage stress for each arm can be calculated as

$$v_{cu} = \frac{V_{dc}}{N} + \int \frac{n_u i_u}{C_{sm}} \qquad (12)$$

$$v_{cl} = \frac{V_{dc}}{N} + \int \frac{n_l i_l}{C_{sm}} \qquad (13)$$

III. TRANSIENT SM VOLTAGE STRESS ANALYSIS DURING GRID VOLTAGE SAGS

The grid voltage and output current of the MMC under grid voltage dips are defined as .

$$v_g' = k v_g \qquad (14)$$

$$i_o^* = I' \sin(\omega t - \varphi') \qquad (15)$$

where k indicates how deep the voltage sag is. The amplitude I' and phase angle φ' of the output current are determined by the reactive power control strategy, which is usually requested by the grid operators. The phase shifting of grid voltage during faults is neglected.

If a direct voltage control is applied, the typical structure of the MMC system can be set up as shown in Fig. 3(a), which includes an output current controller and a circulating current controller [12]. When the grid voltage drops, both controllers will affect the dynamics of the SM voltage. The P^* and Q^* are the reference value of active power and reactive power, respectively.

1022

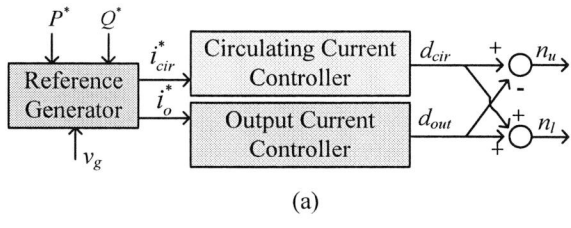

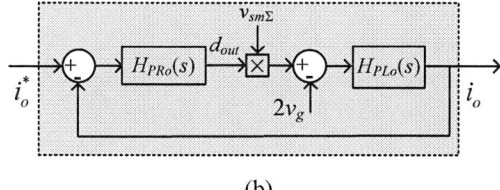

(a)

(b)

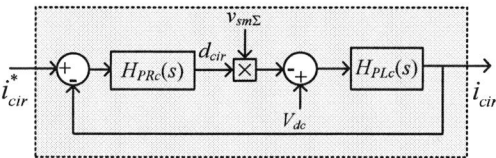

(c)

Fig. 3. Schematic diagrams for the MMC control system. (a) A typical structure of the MMC controller. (b) Output current control loop with a proportional resonant (PR) controller. (c) A typical structure of the circulating current control loop using a PR controller.

A. *Impact of the Output Current Control*

According to the AC equivalent circuit given in Fig. 2(b), an output current control loop using a Proportional Resonant (PR) controller is designed as shown in Fig. 3(b), where H_{PRo} is the transfer function of an PR controller with a resonant frequency of ω, H_{PLo} represents the plant transfer function and i_o^* is the reference value for the grid injected grid current. The transfer functions can be given as

$$H_{PRo}(s) = K_{po} + \frac{sK_{ro}}{s^2 + \omega^2} \tag{16}$$

$$H_{PLo}(s) = \frac{1}{sL_{eq}} \tag{17}$$

where K_{po} and K_{ro} are the gains of the PR controller for the output current loop, and L_{eq} is the equivalent output filter inductance that can be expressed as

$$L_{eq} = \frac{L_a}{2} + L_{ac} \tag{18}$$

According to (15), (16), and Fig. 3(b), the interaction among the grid voltage v_g, the output current reference i_o^*, the the output current i_o and d_{out} for a closed-loop MMC system can be derived as

$$\begin{bmatrix} i_o \\ d_{out} \end{bmatrix} = \begin{bmatrix} H_1(s) & H_2(s) \\ \frac{1}{v_{sm\Sigma}} H_3(s) & \frac{1}{v_{sm\Sigma}} H_4(s) \end{bmatrix} \begin{bmatrix} i_o^* \\ v_g \end{bmatrix} \tag{19}$$

in which $v_{sm\Sigma}$ represents the sum of the SM voltages in the upper and lower arms. The transfer functions $H_1(s)$ to $H_4(s)$ are expressed as

$$H_1(s) = \frac{H_{PRo}(s)H_{PLo}}{1 + H_{PRo}(s)H_{PLo}(s)} \tag{20}$$

$$H_2(s) = -\frac{2H_{PLo}(s)}{1 + H_{PRo}(s)H_{PLo}(s)} \tag{21}$$

$$H_3(s) = \frac{H_{PRo}(s)}{1 + H_{PRo}(s)H_{PLo}(s)} \tag{22}$$

$$H_4(s) = \frac{2H_{PRo}(s)H_{PLo}(s)}{1 + H_{PRo}(s)H_{PLo}(s)} \tag{23}$$

If the output current control is designed as that in [13], the system $H_1(s)$ is a low-pass filter with a relatively high bandwidth to keep the output current tracking with the reference i_o^*. In that case $H_3(s)$ behaves as a high-pass filter to enhance the dynamic response of d_{out} during the reference signal change. To investigate the transient SM capacitor voltage under a low frequency (i.e. the fundamental frequency and double-line frequency), the small signal response is then given as

$$\begin{bmatrix} \tilde{i}_o \\ \tilde{d}_{out} \end{bmatrix} = \begin{bmatrix} 1 & H_2(s) \\ 0 & \frac{2}{v_{sm\Sigma}} \end{bmatrix} \begin{bmatrix} \tilde{i}_o^* \\ \tilde{v}_g \end{bmatrix} \tag{24}$$

It should be noted that the output voltage of the MMC system is a coupled-term composed by d_{out} and $v_{sm\Sigma}$ in (22). To analyze the behavior of the total arm voltage during grid dynamics, d_{out} is considered as a constant value due to the high bandwidth of $H_3(s)$ and $H_4(s)$. Thus,

$$d_{out} = \frac{2v_g'}{v_{sm\Sigma}} = k\frac{U}{v_{sm\Sigma}} \sin \omega t \tag{25}$$

The transient output current i_o' is expressed as

$$i_o' = i_o^* - \mathcal{L}^{-1}\left[(1-k)U\frac{\omega}{s^2+\omega^2}H_2(s)\right] \tag{26}$$

B. *Impact of Circulating Current Control*

To reduce the SM voltage ripple and circulating energy losses, a circulating current controller needs to be applied to the MMC. Many different control strategies are proposed [14]-[17] to enhance the controller performance. A common circulating current suppression solution using a PR controller is taken as an example and analyzed.

The simplified model of the circulating current control is presented in Fig. 3(c), where H_{PRc} is the transfer function of the PR controller and H_{PLc} is the system plant transfer function. They can be given as

$$H_{PRc}(s) = K_{pc} + \frac{sK_{rc}}{s^2 + (2\omega)^2} \tag{27}$$

$$H_{PLc} = \frac{1}{2sL_a} \tag{28}$$

1023

where K_{pc} and K_{rc} are the gains of the PR controller for the circulating current loop.

According to (27), (28), and Fig. 3(c), the closed-loop transfer function for the circulating current control is given as

$$i_{cir} = \frac{H_{PRc}(s)H_{PLc}(s)}{1 + H_{PRc}(s)H_{PLc}(s)}\tilde{i}^*_{cir} \tag{29}$$

$$d_{cir} = \frac{V_{dc}}{v_{sm\Sigma}} \tag{30}$$

For a low frequency range, the gain of the circulating current control loop is close to 1. However, a low-pass filter H_{LPF} is needed to obtain the DC component of the circulating current. Due to its low bandwidth, the transient response of the circulating current i'_{cir} is dominated by this filter. According to the instantaneous power balancing theory,

$$V_{dc}i^*_{cir} = \frac{v'_g i^*_o}{2} \tag{31}$$

By substituting (1), (14), (15) into (31), the steady-state circulating current during the grid voltage sag should be

$$i^*_{cir} = \frac{kUI'}{2V_{dc}}\cos(\varphi') \tag{32}$$

Thus, the time-domain dynamic response of the circulating current during grid voltage drop can be represented as

$$i'_{cir} = \frac{kUI'}{2V_{dc}}\cos(\varphi') + \mathcal{L}^{-1}\left[H_{LPF}(s)\Delta i_{cir}\right] \tag{33}$$

$$\Delta i_{cir} = \frac{UI}{2V_{dc}}\cos(\varphi) - \frac{kUI'}{2V_{dc}}\cos(\varphi') \tag{34}$$

C. Transient SM Voltage Stress Estimation

According to the analysis, the transient behaviors of SM charging currents $n_u i_u$ and $n_l i_l$ can be estimated, which enables a mathematical evaluation of the SM voltage dynamics during grid voltage sags. By substituting (26) to (33) into (10), (11) and then inserting into (12) and (13), the transient voltages of the SMs in upper arm and lower arm are derived as

$$v_{cu}(t) = \frac{V_{dc}}{N} + \int_{T_i}^{T_i+t}\frac{[d_{cir} - d_{out}][i_{cir}'(t) + i_o'(t)]}{C_{sm}}dt \tag{35}$$

$$v_{cl}(t) = \frac{V_{dc}}{N} + \int_{T_i}^{T_i+t}\frac{[d_{cir} + d_{out}][i_{cir}'(t) - i_o'(t)]}{C_{sm}}dt \tag{36}$$

where T_i indicates the initial point on wave of the grid voltage dip.

Table I. Specifications of an MMC System.

Parameter	Value
Number of levels (N)	200
DC bus voltage (V_{dc})	±200 kV
AC grid voltage (RMS value)	110 kV (50 Hz)
AC grid current (RMS value)	500 A
SM capacitor (C_{sm})	6.3 mF
Arm inductance (L_a)	20 mH
AC inductance (L_{ac})	10 mH
Voltage dip ratio	$k = 0.5$
Point on wave of dip initiation	$T_i = 10$ s
Control system design	$K_{po} = 30$; $K_{ro} = 3000$
	$K_{pc} = 40$; $K_{rc} = 4000$

IV. CASE STUDY

A. System Specifications

A case study based on the Zhoushan VSC-HVDC project is simulated with the specifications shown in Table I [18].

During a voltage dip condition, one p.u. reactive current is injected into the grid. A second-order low-pass filter with a 20-Hz cut-off frequency is chosen to obtain the DC value of the circulating current. The transfer function of the low-pass filter is

$$H_{LFP}(s) = \frac{22000}{s^2 + 240s + 22000} \tag{37}$$

B. Simulation Results

According to the parameters in Table I, the system is simulated and the results are shown in Fig. 4. As the operation of each arm is symmetrical, only the waveforms of the upper arm are presented.

It can be seen in Fig. 4(b) that during the grid dip transient, the injected grid current can achieve a relatively fast response with a small overshoot. However, the adjustment for the circulating current is much more slower due to the low-pass filter in the loop, which is illustrated in Fig. 4(e). As shown in Fig. 4(f), this leads to a slow decrease in the arm current and an unbalanced energy flow between the system input and output. As a result, an over-voltage appears across each SM capacitor as shown in Fig. 4(g).

The discrepancies between the results from simulation results and the analytical model results are due to the neglecting of the coupling effect in the MMC controller and the system simplification. However, this method can still achieve a good accuracy to estimate the transient behaviors of the SM capacitor charging current, as shown in Fig. 4(h).

The 2018 International Power Electronics Conference

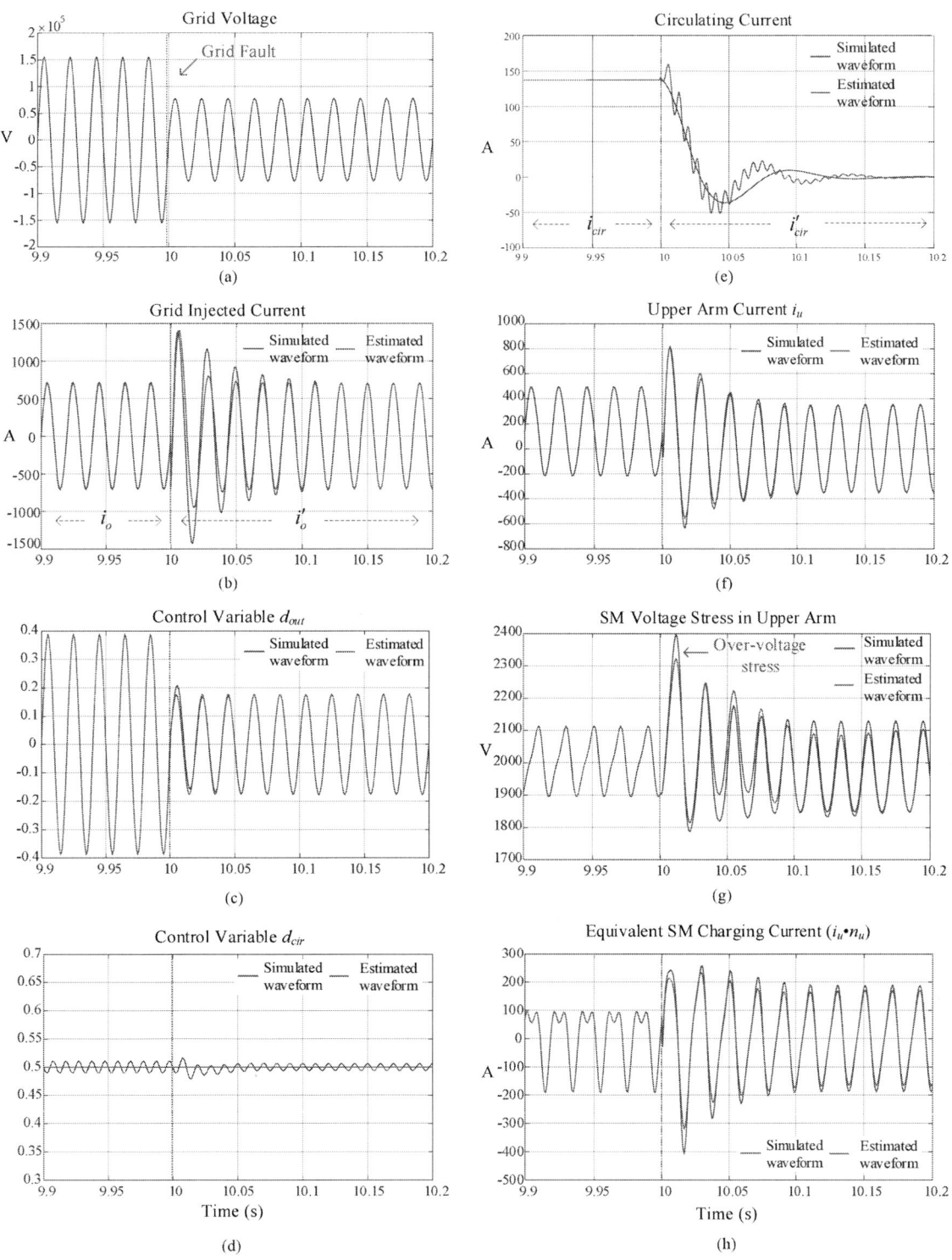

Fig. 4. Simulation results at $t = 10$ s. (a) Grid voltage. (b) Injected grid current. (c) Estimation of the control variable d_{out}. (d) Estimation of the control variable d_{cir}. (e) Comparison between the simulated and estimated circulating currents. (f) Comparison between the simulated and estimated upper arm currents. (g) Estimated transient SM voltages. (h) Equivalent SM capacitor charging currents.

1025

V. CONCLUSION

A closed-loop time-domain model was derived for the MMC SM transient voltage stress under grid faults. It enables a quantitative estimation of the peak voltage stresses of SM capacitors and power modules. The proposed model is a first step towards the model-based design of MMC systems to properly size key components without over-engineering or insufficient-robustness. In the presented case study of an MMC system, it achieved an error of 4% between the results from the proposed model and simulations, which is considered as sufficient to assist the MMC design.

REFERENCES

[1] M. A. Perez, S. Bernet, J. Rodriguez, S. Kouro, and R. Lizana, "Circuit Topologies, Modeling, Control Schemes, and Applications of Modular Multilevel Converters," *IEEE Trans. Power Electron.*, vol. 30, no. 1, pp. 4-17, Jan. 2015.

[2] A. Nami, J. Liang, F. Dijkhuizen, and G. D. Demetriades, "Modular Multilevel Converters for HVDC Applications: Review on Converter Cells and Functionalities," *IEEE Trans. Power Electron.*, vol. 30, no. 1, pp. 18-36, Jan. 2015.

[3] H. Akagi, "Classification, Terminology, and Application of the Modular Multilevel Cascade Converter (MMCC)," *IEEE Trans. Power Electron.*, vol. 26, no. 11, pp. 3119-3129, Nov. 2011.

[4] M. Hagiwara, K. Nishimura, and H. Akagi, "A medium-voltage motor drive with a modular multilevel PWM inverter," *IEEE Trans. Power Electron.*, vol. 25, no. 7, pp. 1786–1799, July 2010.

[5] B. Li, S. Shi, B. Wang, G. Wang, W. Wang, and D. Xu "Fault Diagnosis and Tolerant Control of Single IGBT Open-Circuit Failure in Modular Multilevel Converters," *IEEE Trans. Power Electron.*, vol. 31, no. 4, pp. 3165–3176, July 2015

[6] F. Deng, Y. Tian, R. Zhu, and Z. Chen, " Fault-Tolerant Approach for Modular Multilevel Converters Under Submodule Faults," *IEEE Trans. Ind. Electron.*, vol. 63, no. 11, pp. 7253 - 7263, Mar. 2016.

[7] K. Ilves, S. Norrga, L. Harnefors, and H. P. Nee, "On energy storage requirements in modular multilevel converters," *IEEE Trans. Power Electron.*, vol. 29, no. 1, pp. 77–88, Jan. 2014.

[8] Y. Tang, L. Ran, O. Alatise, and P. Mawby, "Capacitor Selection for Modular Multilevel Converter," *IEEE Trans. Ind. Appl.*, vol. 50, no. 3, pp. 1915–1923, May/Jun. 2014.

[9] X. Shi, Z. Wang, B. Liu, Y. Liu, L. M. Tolbert, and F. Wang, "Characteristic Investigation and Control of a Modular Multilevel Converter-Based HVDC System Under Single-Line-to-Ground Fault Conditions," *IEEE Trans. Power Electron.*, vol. 28, no. 8, pp. 3702-3713, Aug. 2013.

[10] J. Li, G. Konstantinou, H. R. Wickramasinghe, J. Pou, X. Wu, and X. Jin, "Impact of Circulating Current Control in Capacitor Voltage Ripples of Modular Multilevel Converters under Grid Imbalances," *IEEE Trans. Power Deli.*, vol. PP, no. 99, pp. 1–1, Aug. 2017.

[11] Y. Li, E. A. Jones, and F. Wang, "The Impact of Voltage-Balancing Control on Switching Frequency of the Modular Multilevel Converter," *IEEE Trans. Power Deli.*, vol. 31, no. 4, pp. 2829–2839, Apr. 2016.

[12] S. Debnath, J. Qin, B. Bahrani, M. Saeedifard, and P. Barbosa, "Operation, Control, and Applications of the Modular Multilevel Converter: A Review," *IEEE Trans. Power Electron.*, vol. 30, no. 1, pp. 37–53, Jan. 2015.

[13] K. Sharifabadi, L. Harnefors, H. Nee, S. Norrga, and R. Teodorescu, *Design, control, and application of modular multilevel converter for HVDC transmission systems*, J. Wiley & Sons, West Sussex, UK. pp. 157-164.

[14] G. Konstantinou. J. Pou, S. Ceballos, R. Picas, J. Zaragoza, and V. G. Agelidis, "Control of Circulating Currents in Modular Multilevel Converters Through Redundant Voltage Levels," *IEEE Trans. Power Electron.*, vol. 31, no. 1, pp. 7761 - 7769, Nov. 2016.

[15] L. He, K. Zhang, J. Xiong, and S. Fan, "A Repetitive Control Scheme for Harmonic Suppression of Circulating Current in Modular Multilevel Converters," *IEEE Trans. Power Electron.*, vol. 30, no. 1, pp. 471 - 481, Jan. 2015.

[16] L. Ben-Brahim, A. Gastli, M. Trabelsi, K. A. Ghazi, M. Houchati, and H. Abu-Rub, "Modular Multilevel Converter Circulating Current Reduction Using Model Predictive Control," *IEEE Trans. Ind. Electron.*, vol. 63, no. 6, pp. 3857 - 3866, Jun. 2015.

[17] B. Li, Z. Xu, S. Shi, D. Xu, and W. Wang, "Comparative Study of the Active and Passive Circulating Current Suppression Methods for Modular Multilevel Converters," *IEEE Trans. Power Electron.*, vol. 63, no. 6, pp. 1878 - 1883, Mar. 2018.

[18] G. Tang, Z. He, H. Pang, X. Huang, and X. Zhang, "Basic Topology and Key Devices of the Five-Terminal DC Grid," *CSEE Journal of Power and Energy System*, vol. 1, no. 2, pp. 22 - 35, Jun. 2015.

SVPWM Strategy Based on Multilevel 3LNPC-CR

Xiaoqiong He[1, 2*], Pengcheng Han[1], Xiaolan Lin[1], Yi Wang[1], Xu Peng[1]

1 School of Electrical Engineering, Southwest Jiaotong University, Chengdu, China
2 National Rail Transit Electrification and Automation Engineering Technique Research Center, Chengdu, China
*Email: hexq@home.swjtu.edu.cn

Abstract- **This paper proposed a new modulation strategy for three-level neutral point clamped cascaded rectifier to solve the DC-link voltage unbalance. The main idea is to reallocate the number of voltage levels generated by each module based on the space vector pulse width modulation. This proposed modulation strategy can reduce the switching frequency while keeping the mutual-module voltage balance. Firstly, the DC-link voltage unbalance problem is analyzed. Then, a new modulation strategy is introduced in detail. At the same time, the internal-module capacitor voltages are balanced by selecting the redundant vectors. The feasibility of this modulation strategy is verified by experiment.**

KEY WORDS: **Neutral point clamped cascaded rectifier (NPC-CR), voltage balancing, modulation, voltage level**

I. INTRODUCTION

The structure of multilevel convertor has the features of high capacity, high output voltage and low voltage rating requirement of power switches. Benefit from these features, the multilevel convertors are widely used in high power applications for medium or high voltage [1,2]. Nowadays, H-bridge cascaded rectifier has already been applied in the power electronic transformer, and has been put into service in the 15kV single-phase traction power supply system of Europe [2]. Compared with the H-bridge cascaded rectifier, more output voltage levels would be generated by the three-level neutral point clamped cascaded rectifier (3LNPC-CR), so that less number of the modules is required and the 3LNPC-CR has more advantages in high voltage and high power occasions. The unbalanced voltage inside the 3LNPC-CR module and between the modules is a central problem of the 3LNPC-CR topology [3].

The unbalanced DC-link capacitor voltages is an unavoidable problem of 3LNPC convertor. Both auxiliary voltage balancing circuit and modulation strategy can be adopted to solve the unbalanced problem of the DC-link capacitance voltage [4,5]. The method of using the auxiliary voltage balancing circuit is proved to be impactful to equalize the capacitance voltage of DC-link, while the problems of complex circuit, higher cost and higher losses are also caused. In order to control the capacitor voltage, the redundant vectors is used to adjust the charging or discharging state of the capacitor, which is the main idea of the modulation strategy [6]. The modulation strategy has already been widely applied in single-phase NPC convertor.

Besides the problem of the internal modules, when the loads of the NPC-CRs are different or the changes of loads occur, the difference would exist among the output power, and the unbalance voltage would appear among the modules in practical application. The control strategy, which is based on the proportional-integral (PI) controller, and the modulation strategy, which is based on the phase shift carrier (PSC), are the most general strategy for balancing the mutual-module voltage under the research and application [7]. The modulation strategy consists of the pulse width modulation (PWM) and space vector pulse width modulation (SVPWM). Especially, PSCPWM strategy has the advantages of the high power quality and is easy to implement the modular and distributed control, while the SVPWM strategy has the advantages of the clear physical meaning, the high voltage utilization rate, and is more adaptive digital realization [8].

PSC-SVPWM exists the problem that the input may change more than one level at a time [9]. More than one module would change its input voltage levels at the same time and the switching frequency would be higher. These could lead to the misfiring faults. For reducing the changes of the module input voltage level and making the level change to the adjacent levels, a novel SVPWM modulation strategy of multi-module 3LNPC-CR is proposed in this paper. While reducing the switching frequency, the strategy solve the unbalanced problem of internal-module and mutual-module. The simulation and experiment results verified the feasibility and validity of the proposed strategy.

II. THE STRUCTURE OF THE NPC-CR

The structure of the multi-module NPC-CR is shown as Fig.1. Each module is composed of two bridges, and every bridge could output three voltage level : +E、0、-E. E is defined as the capacitance voltage at the balanced case.

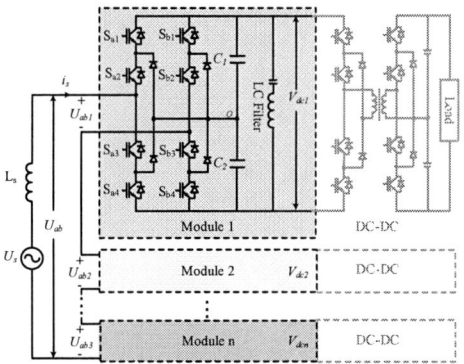

Fig.1 The structure of the 3LNPC-CR

In this paper, U_S is the sign of the input voltage, and V_a, V_b correspond to the output voltage level of the bridge a and bridge b respectively. U_{abx} (x=1,2,…,n) is the input voltage of the module x. U_s, i_s represent the voltage and current of the grid side. Table.1 exhibits the relationship between the switch states and the output voltage level of single module, when i_s>0.

TABLE I
THE VOLTAGE LEVEL STATE OF THE NPC RECTIFIER (i_s>0)

U_{abx}	Mode	V_a	V_b	C_1	C_2
2E	1	E	-E	Charge	Charge
E	2	E	0	Charge	No effect
	3	0	-E	No effect	Charge
0	4	0	0	No effect	No effect
-E	5	-E	0	No effect	Discharge
	6	0	E	Discharge	No effect
-2E	7	-E	0	Discharge	Discharge

III. THE MODULATION STRATEGY OF THE NPC-CR

3.1 working area determination

The new modulation scheme is introduced based on a 3 module 3LNPC-CR, which could generate 13 voltage levels in the overall input terminal. So, u_{ab}^* could be divided into 12 areas according to its instantaneous value. The mathematical expression of u_{ab}^* is given by formula (1), and the specific area determination rules are given in formula (2), (3).

$$u_{ab}^* = 6\,EM\sin(\omega t) \quad (1)$$

$$|area| = \begin{cases} 6, & 5E<\left|u_{ab}^*\right| \\ 5, & 4E<\left|u_{ab}^*\right| \le 5E \\ 4, & 3E<\left|u_{ab}^*\right| \le 4E \\ 3, & 2E<\left|u_{ab}^*\right| \le 3E \\ 2, & E<\left|u_{ab}^*\right| \le 2E \\ 1, & 0<\left|u_{ab}^*\right| \le 1E \end{cases} \quad (2)$$

$$area = \begin{cases} |area|, & u_{ab}^* \ge 0 \\ -|area|+1, & u_{ab}^* < 0 \end{cases} \quad (3)$$

3.2 Determination of Input Voltage Levels of each module

The output voltage will be different when the loads are unbalanced. Since the power flows in the rectifier direction. The main idea of this new modulation strategy is to adjust the power distribution by reallocating the input voltage levels of each power module. Specifically, when the grid current i_s>0, suppose the output voltage of module i is highest among the cascaded modules. When u$_{abi}$>0, the voltage level should be lowered so that there will be more power flow into module i. When u$_{abi}$<0, the voltage level should also be lowered so that the outflow power of module i will be increased. When the grid current i_s<0, suppose the output voltage of module k is lowest among the cascaded modules. When u$_{abk}$>0, the voltage level should be increased so that there will be more power flow into module k. when u$_{abk}$<0, the input voltage should also be increased so that the outflow power of module k will be decreased. In this way, the output voltages will be balanced by properly reallocating the input voltage levels of each module.

The reference voltage u_{ab}^* is composed by two nearest voltage vectors V_A [V_{A1},V_{A2},V_{A3}]and V_B[V_{B1},V_{B2},V_{B3}], which represents the input voltages of module1, module2 and module3. $V_{A,sum}$, $V_{B,sum}$ stands for the overall input voltage level when V_A, V_B is working. The reference voltage u_{ab}^* could be expressed by formula (1). The initial value of u_{ab}^* is zero. Thus, at the beginning, V_A and V_B are set to [0 0 0]. The relationship between V_A and V_B is shown in formula (4). When the slope of u_{ab}^* is positive, V_A represents the voltage level of the lower boundary in each area, while V_B represents the voltage level of the upper boundary. When the slope of u_{ab}^* is negative, V_A represents the voltage level of the upper boundary in each area, while V_B represents the voltage level of the lower boundary.

$$\begin{cases} V_{A,sum} - V_{B,sum} = -E, & u_{ab}^* \text{ is going up} \\ V_{A,sum} - V_{B,sum} = E, & u_{ab}^* \text{is going down} \end{cases} \quad (4)$$

When the grid current i_s>0, the procedure of this modulation strategy is shown in the flow chart.

1) Determine the area, in which u_{ab}^* is working.

2) When *area* changes, or u_{ab}^* reaches the minimum value and the maximum value of each period, the triggering signal T will be generated. And this signal will trigger V_A, V_B to plus or minus E.

3) Sort the V_{dc1}, V_{dc2}, V_{dc3} in descending order, and the result will be represented by $V_{dc,max}$, $V_{dc,mid}$, $V_{dc,min}$. M_{max}, M_{mid} and M_{min} represent the module with the maximum output voltage, the module with the medium voltage and the module with the minimum voltage respectively.

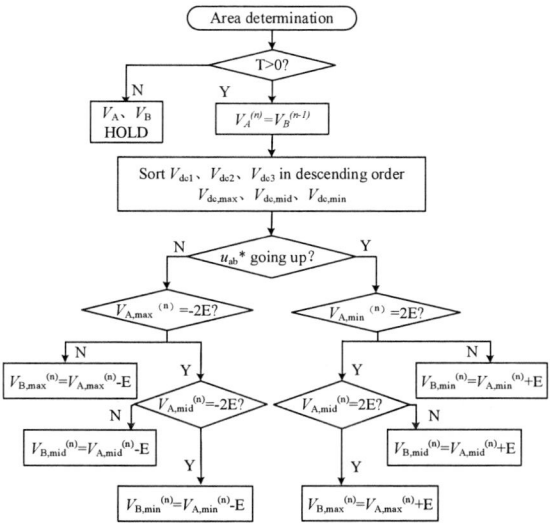

Fig.2 Flow chart of voltage vector distribution when $i_s>0$

4) Determine whether $u_{ab}*$ is in the tendency to go up. If yes, the input voltage of M_{min} will plus E. However, if the input voltage of M_{min} has already reached 2E and is not able to be increased anymore, then, the input voltage of M_{mid} will plus E if its input voltage has not reached 2E, otherwise, it has to be the input voltage of M_{min} to plus E. if $u_{ab}*$ is not in the tendency to go up, the input voltage of M_{max} will minus E. However, if the input voltage of M_{max} has already reached -2E and is not able to be decreased anymore, then, the input voltage of M_{mid} will minus E as the second priority. If the input voltage of M_{mid} has also reached -2E, the last choice will be the input voltage of M_{min} to minus E.

5) The input voltage vector before triggering signal T is assigned to V_A, while the input voltage vector after T is assigned to V_B. that is to say, V_B is obtained on the basis of V_A. So, prior to the before mentioned procession, V_B obtained in the previous triggering period should be assigned to V_A in the early stage of the current triggering period. So that the whole procession will be continuous.

The procession is similar when $i_s<0$, so it will not be introduced in detail.

3.3 On time calculation

T_A and T_B represent the on time of voltage vector V_A and V_B. In one carrier period T_s, the on time calculation is realized according to formula:

$$u_{ab}* = \begin{cases} V_{A,sum}T_A + V_{B,sum}T_B = u_{ab}*T_S \\ T_A + T_B = T_S \end{cases} \quad (6)$$

3.4 internal-module voltage balance strategy

The before-mentioned modulation procession aims to determine the voltage levels of each input terminal, so that the mutual-module voltage could be balanced. However, the modulation process should still be extended to cover voltage levels of each bridge leg, in this way, the capacitor voltages of each rectifier

module could be balanced. As shown in Table. I, even if with the same input voltage, the charging/discharging states of the dc-link capacitors could be different determined by redundant voltage vectors of each bridge leg. Considering the power flow direction and the difference between two capacitor voltages, these redundant voltage vectors are selected to provide charging/discharging path for the dc-link capacitors. The detailed selection rules are given in Table. II.

TABLE II
THE PRINCIPLE OF CHOOSING REDUNDANT VECTOR

	$V_{C1}>V_{C2},$ $i_s>0$	$V_{C1}>V_{C2},$ $i_s<0$	$V_{C1}<V_{C2},$ $i_s<0$	$V_{C1}<V_{C2},$ $i_s>0$
$U_{abx}=E$	$[V_a,V_b]=$ $[0,-E]$	$[V_a,V_b]=$ $[E,0]$	$[V_a,V_b]=$ $[0,-E]$	$[V_a,V_b]=$ $[E,0]$
$U_{abx}=-E$	$[V_a,V_b]=$ $[-E,0]$	$[V_a,V_b]=$ $[0,E]$	$[V_a,V_b]=$ $[-E,0]$	$[V_a,V_b]=$ $[0,E]$

IV. SIMULATION AND EXPERIMENT

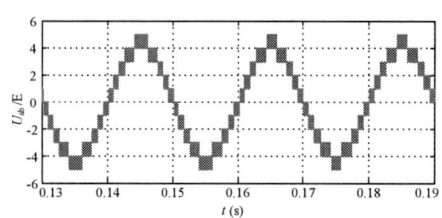

Fig. 3 Waveform of the input voltage levels of the 3 module cascaded NPC rectifier

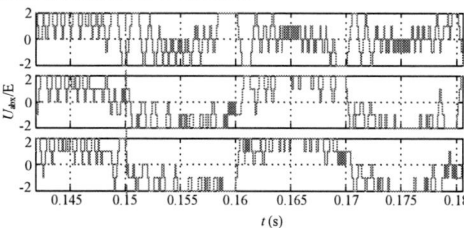

Fig. 4 Level-skip waveforms of module input voltages when the load of module 1 is cut off

TABLE III
THE PARAMETERS OF 3 MODULE 3LNPC-CR

Parameter	Value
Grid voltage(us)	75V
Grid inductor	2mH
load	30
Modulation index	0.7

In order to verify the validity of this new modulation technology, a simulation model is built in Simulink/MATLAB. The parameters are shown in table.3.At 0.15s, the load of module1 is cut off. Fig. 3 and Fig. 4 show the instantaneous waveforms of the overall input voltage and the input voltage of each rectifier module. As shown in Fig.3, since the modulation index is 0.7, when the reference voltage $u_{ab}*$ reaches its peak value, it works in the area 5 and is composed by 4E and 5E.even if the load of module 1 is cut off, the overall input voltage has not been affected. As shown in Fig. 4, since the load of module 1 is cut off at 0.15s, so the load voltage of module 1 becomes the highest among these three cascaded modules. It is obvious that when the grid current is positive, the input voltage of module 1 is lower than other modules, which means there will be less power flow into module 1; when the grid current is negative, the input voltage of module1 is higher than other modules, which means there will be more power flow out of module 1.

An experimental model is also built to verify the feasibility of the modulation scheme. The experimental parameters are identical to the simulation parameters.

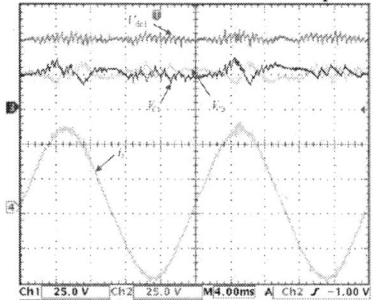

Fig. 5 Capacitor voltages of in the module

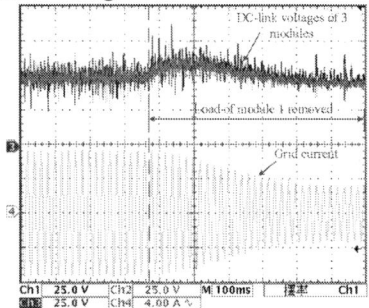

Fig. 6 voltage and current waveforms when the load of module 1 is removed

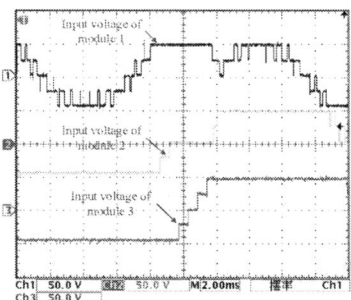

Fig.7 Level-skip waveforms of module input voltages when the load of module 1 is removed

It is shown in Fig.5 that the internal voltage of module 1 is balanced. To verify the dynamic properties, the load of module 1 is cut off. As shown in Fig.6, the output voltage of each module converge to the same value after approximately 0.26s. The grid current keeps steady except the reasonable change in its amplitude due to the load step.

As shown in Fig.7, after the load becomes unbalanced, the input voltage of module 1 changes rapidly to balance the output voltages. When the grid

current is positive, the input voltage of module 1 becomes lower, and when the grid current is negative, the input voltage of module 1 becomes higher, which exactly match with before-mentioned modulation rules.

V. CONCLUSION

Based on 3 module 3LNPC-CR, a new modulation strategy is proposed, which will successfully realize both internal-module and mutual-module voltage balance.

The properties of this new modulation strategy are as follows.

1) Only module changes its input voltage levels at one time. Which will effectively lower the frequency of the IGBTs.

2) When the loads are unbalanced, even if one of the loads are cut off, the dc-link voltage will converge to the same value rapidly.

3) At the same time, the capacitor voltage in each rectifier module is successfully balanced.

VI. ACKNOWLEDGE

This paper is supported by National Natural Science Foundation of China (Grant Nos.51477144).

REFERENCES

[1] D. Dujic et al., Power Electronic Traction Transformer Low Voltage Prototype[J], Power Electronics, IEEE Transactions, 2013, 28(12), pp. 5522-5534.

[2] Kouro, S.; Malinowski, M.; Gopakumar, K.; Pou, J.; Franquelo, L. G.; Wu, B.; et al., "Recent Advances and Industrial Applications of Multilevel Converters," IEEE Transactions on Industrial Electronics 2010, vol. 57, pp. 2553-2580.

[3] Z. Shu, Z. Kuang, S. Wang, X. Peng and X. He, Diode-clamped three-level multi-module cascaded converter based power electronic traction transformer[C], 2015 IEEE 2nd International Future Energy Electronics Conference (IFEEC), Taipei, 2015, pp.1-5.

[4] Zeliang, S.; Haifeng, Z.; Xiaoqiong, H.; Na D. and Yongzi, J. One-inductor-based auxiliary circuit for dc-link capacitor voltage equalisation of diode-clamped multilevel converter[J], IET Power Electronics, 2013, vol.6, no.7, pp.1339-1349.

[5] Zeliang, S.; Xiaoqiong, H.; Zhiyong, W.; Daqiang Q. and Yongzi, J. Voltage Balancing Approaches for Diode-Clamped Multilevel Converters Using Auxiliary Capacitor-Based Circuits[J], Power Electronics, IEEE transactions, 2013, vol.28, no.5, pp.2111-2124.

[6] Cecati, C.; Dell'Aquila, A.; Liserre M.; and Monopoli, V. G.; "Design of H-bridge multilevel active rectifier for traction systems," IEEE Transactions on Industry Applications 2003, vol. 39, no. 5, pp. 1541-1550.

[7] Dell'Aquila, A.; Liserre, M.; Monopoli V. G. and Rotondo, P. "Overview of PI-Based Solutions for the Control of DC Buses of a Single-Phase H-Bridge Multilevel Active Rectifier," IEEE Transactions on Industry Applications 2008, vol. 44, no. 3, pp. 857-866.

[8] Zeliang, S.; Na, D.; Jie, C.; Haifeng Z. and Xiaoqiong H. "Multilevel SVPWM With DC-Link Capacitor Voltage Balancing Control for Diode-Clamped Multilevel

Converter Based STATCOM," IEEE Trans. Ind. Electron 2013, vol.60, no.5, pp.1884-1896.

[9] Peng X, He X, Han P, et al. Smooth Switching Technique for Voltage Balance Management Based on Three-Level Neutral Point Clamped Cascaded Rectifier[J]. Energies, 2016, 9.

The Multiple Degree of Freedom based Neutral Point Potential Control of Three Level Neutral Point Clamped Converters

Bo Guan, Shinji Doki
Department of Information and Communication Engineering
Nagoya University
Aichi, Japan 464-8603
Email: guanbo_1989@nagoya-u.jp

Abstract-Although the virtual space PWM and the dual wave modulation (DWM) method can eliminate the low frequency neutral point potential (NPP) fluctuation, which appears in some NPP control based on the nearest three vector PWM essentially, they will increase the switching frequency and fade the control ability of the NPP drift problem. In this paper, a novel multiple degree of freedom based NPP control with the strong control ability of low frequency NPP fluctuation and NPP drift problem is proposed for three level converter. The traditional DWM method sacrifices many control domains of neutral point current (i_o) to keep i_o zero, by setting the zero sequence voltage to an appropriate value. On the contrary, the proposed method introduces three degree of freedom to control the NPP, which makes it have the largest NPP control domains. The comparison on the control domains of i_o for DMW method and proposed method is discussed at first. Then, some simulation are also carried out to verify the correctness and effectiveness of proposed method.

Index Terms—Dual wave modulation, neutral point potential, degree of freedom, control domains of neutral point current, three level converter.

I. INTRODUCTION

With the development of power electronics, pulse width modulation (PWM) and various types of converters become more and more important in modern industry applications. Especially, multilevel converters play an irreplaceable role for medium and high voltage high power applications [1, 2]. In this paper, some issues about three level neutral point clamped (TL-NPC) converters are discussed in detail. TL-NPC converters can generate a more sinusoidal output voltage, which is of benefit to get rid of the bulk output filters. In addition, since the rated voltage of switch devices can be reduced by half, TL-NPC converters can be utilized in a higher voltage application. Moreover, electromagnetic interference (EMI) problem can be improved due to the smaller du/dt.

Although TL-NPC converters have lots of advantages compared with two level converters, the neutral point potential (NPP) problem is very troublesome. Essentially, the NPP problem can be classified to 2 sorts: 1) the NPP drift problem; 2) the low frequency NPP fluctuation problem. The first one we concerned is the NPP drift problem, which is original from the dead time, the asymmetrical loads or the inconsistences of switch devices, etc [1-4]. In the past 20 years, many studies have been done against the NPP problem, since it may ruin the capacitors and switch devices. For example, we can control the NPP by adjusting the duty time of pairs of redundant vectors in three level space vector diagram [5, 6]. The same goal can also be realized via injecting a zero sequence voltages [7, 8]. However, the low frequency NPP fluctuation problem becomes more and more noticeable with further research. It limits the possibility to decrease the DC-link capacitors, which is directly related to the weight and volume of TL-NPC converters. It is inevitable to generate the low frequency NPP fluctuation essentially, when some modulation methods based on the nearest three vector PWM (NTV-PWM) are adopt in the system. Thus, some methods, such as the virtual space PWM and the dual wave modulation (DWM) method [9-11], are put forward to eliminate the low frequency NPP fluctuation by changing the modulation mode.

Unfortunately, even if the DWM method can solve the NPP drift and the low frequency NPP fluctuation problem at the same time, it also causes some issues, such as a higher switching frequency [9-11] and the weaker control ability of the NPP drift. This is because we have to fix the zero sequence voltage, when dual modulation waves are reconstructed based on the DMW method. As a result, the control domains of the neutral point current (i_o) is limited compared with the NTV-PWM. In order to reduce the switching frequency and expand the control domains of i_o, which is key to control the NPP drift, a novel multiple degree of freedom (M-DOF) based NPP control for TL-NPC converters is proposed in this paper. By introducing three degree of freedom (DOF), we have more choices to control the NPP. Therefore, the NPP control ability is enhanced substantially compared with the DMW method. It provides the strongest NPP control ability theoretically. In addition, at the moment that the two sorts of NPP problems are overcame, the switching frequency is also decreased.

The basic principles of NPP problems and the traditional DMW method with a compensator are introduced firstly in section II. Then, after defining the M-DOF, the NPP control theory based on the M-DOF is explained. In addition, the control domains of i_o under the traditional DMW method and the proposed method are also discussed in section III, In section IV, some simulation results are given to prove the correctness and effectiveness of proposed method.

II. THE NPP PROBLEMS AND THE TRADITIONAL DMW METHOD WITH A COMPENSATOR

A. The NPP problems for TL-NPC converters

TL-NPC converter is raised in 1981, which are consist of 4 full-controlled switch devices and 2 diodes in one phase, as shown in Fig. 1. It can be controlled based on the Table. I [7, 8].

Obviously, if we need to generate the output voltage $o(0)$, the NPP is directly linked to the loads. Thus, if i_o flows from the NPP to the loads, the upper capacitor (C_{up}) charges and the bottom capacitor (C_{down}) discharges. And if i_o flows into the NPP, C_{up} discharges and C_{down} charges. It is to say, the NPP problems are inevitable, if we try to use the output voltage $o(0)$ and can not keep i_o zero at the same time. Firstly, the NPP error (Δu_o) is defined as (1).

$$\begin{cases} \Delta u_o = \left(u_{cdown} - u_{cup}\right)/2 \\ u_{cdown} = u_{dc}/2 + \Delta u_o, u_{cup} = u_{dc}/2 - \Delta u_o \end{cases} \quad (1)$$

Therefore, i_o can be calculated from (1), as shown in (2).

$$\begin{cases} i_o = \dfrac{d\left(Cu_{cup}\right)}{dt} - \dfrac{d\left(Cu_{cdown}\right)}{dt} = -2C\dfrac{\Delta u_o}{T_s} \\ C = C_{up} = C_{down} \end{cases} \quad (2)$$

Here, T_s is the control period. A minus neutral point current (i_{oc}) is necessary to compensate i_o, as shown in (3).

$$i_{oc} = -i_o = 2C\frac{\Delta u_o}{T_s} \quad (3)$$

If the system contains the dead time, the asymmetrical loads or the inconsistent of switch devices, a drift i_o would appear. It leads to the NPP drift, which may damage the capacitors and switch devices. On the other hand, the modulation mode under NTV-PWM may also cause the NPP problem, since i_o can not keep zero at this time. A low frequency NPP fluctuation problem will appear. We have to change the modulation mode to solve the low frequency NPP fluctuation. And the i_{oc} derived from the modulation methods is also necessary to compensate the drift i_o.

B. The traditional DMW method with a compensator

Since the virtual space PWM is same as DMW method essentially, which can solve the NPP drift and the low frequency NPP fluctuation problem, the traditional DMW method is introduced here. Firstly, we should inject a zero-sequence voltage (u_z=-(u_{min}+u_{max})/2) into the original three phase reference voltage (u_u, u_v, u_w) to realize the maximum linear modulation, as shown in (4) [10, 11].

$$\begin{cases} u_{zu} = u_u + u_z \\ u_{zv} = u_v + u_z \\ u_{zw} = u_w + u_z \end{cases} \quad (4)$$
$$u_z = -\left(u_{max} + u_{min}\right)/2$$

Based on reference [10, 11], the reference voltage (u_{zi}, i=u,v,w) can be reconstructed into two modulation waves (u_{ip}, u_{in}), as shown in (5).

$$\begin{cases} u_{zu} = u_{up} + u_{un} \\ u_{zv} = u_{vp} + u_{vn} \\ u_{zw} = u_{wp} + u_{wn} \end{cases} \Leftrightarrow \begin{cases} u_{zi} = u_{ip} + u_{in} \\ 0 \le u_{ip} \le 1, -1 \le u_{in} \le 0 \\ i = u, v, w \end{cases} \quad (5)$$

Then, we use two carrier waves ($u^p_{carrier}$, $u^n_{carrier}$) to modulate u_{ip} and u_{in}, as shown in (6). The output voltage (u_{out}) can also be expressed as Fig. 2.

$$u_{out} = \begin{cases} 1, if \ u_{ip} > u^p_{carrier} \ \&\& u_{in} > u^n_{carrier} \\ -1, if \ u_{ip} < u^p_{carrier} \ \&\& v_{in} < u^n_{carrier} \\ 0, if \ u_{ip} > u^p_{carrier} \ \&\& u_{in} < u^n_{carrier} \\ \quad or \ u_{ip} < u^p_{carrier} \ \&\& u_{in} > u^n_{carrier} \end{cases} \quad (6)$$

Thus, i_o derived from DMW method is achieved based on the modulation rule of (6) [10, 11], as shown in (7).

$$i_o = \left|1 + u_{un} - u_{up}\right|i_u + \left|1 + u_{vn} - u_{vp}\right|i_v + \left|1 + u_{wn} - u_{wp}\right|i_w \quad (7)$$

If we can get the solutions, which are able to keep i_o zero in any time, the low frequency NPP fluctuation will be eliminated completely. For three phase three line system, there is a special solution, as (8).

$$\begin{aligned} &if \ \left|1 + u_x\right| = \left|1 + u_{un} - u_{up}\right| = \left|1 + u_{vn} - u_{vp}\right| = \left|1 + u_{wn} - u_{wp}\right|, \\ &then \ i_o = \left|1 + v_x\right|\left(i_u + i_v + i_w\right) = 0 \end{aligned} \quad (8)$$

Define the maximum, middle and minimum voltages of (u_u, u_v, u_w) as (u_{max}, u_{mid}, u_{min}). Define the corresponding currents of (u_{max}, u_{mid}, u_{min}) as (i_{max}, i_{mid}, i_{min}). Especially, if u_x is set to (u_{min}-u_{max})/2, three phase reference voltages can be rewritten as (9).

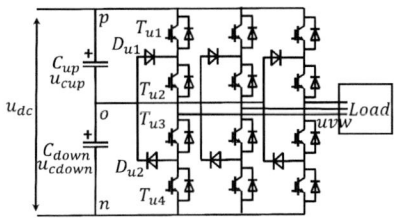

Fig. 1. Three-level NPC converter topology (IGBTs).

TABLE I
SWITCHING STATES AND OUTPUT LEVELS OF PHASE U

Output voltage	T_{u1}	T_{u2}	T_{u3}	T_{u4}
$p(u_{dc}/2)$	On	On	Off	Off
$o(0)$	Off	On	On	off
$n(-u_{dc}/2)$	Off	Off	On	On

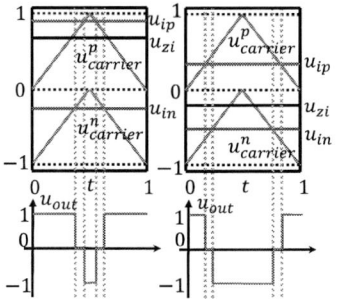

Fig. 2. The principle of the DMW method.

1033

$$\begin{cases} u_{ip} = \dfrac{u_i - u_{\min}}{2} \geq 0 \\[2mm] u_{in} = \dfrac{u_i - u_{\max}}{2} \leq 0 \end{cases}, \quad i = \{u, v, \mathrm{w}\} \quad (9)$$

Essentially, DMW method is just an open-loop way to solve the low frequency NPP fluctuation problem. In order to overcome the NPP drift problem, a compensator is also necessary. Reference [11] introduces an offset voltage to adjust the two reconstructed modulation waves, as (10).

$$\begin{cases} u_{ip} = u_{ip} + u_{ioff} \\ u_{in} = u_{in} - u_{ioff} \end{cases} \quad (10)$$

$$u_{ioff} = k_p \left| \Delta u_o \right| \cdot sign(\Delta u_o i_i) sign(1 + u_{in} - u_{ip})$$

On the one hand, there is no theoretical basis to determine the adjusting gain (k_p), which may cause the over-adjustment. And it is hard to obtain the direct relationship of the offset voltage, the NPP error, the DC-link capacitors and the phase current. One the other hand, although the compensator is adopt, the control ability of the NPP drift is still weak, since u_x has been limited to $(u_{min}-u_{max})/2$.

III. M-DOF BASED NPP CONTROL OF TL-NPC CONVERTERS

In order to suppress the low frequency NPP fluctuation, the traditional DMW method sacrifices many control domains of i_o to keep i_o zero in any time. As a result, the control ability of the NPP drift is weaken in some operation conditions, compared with the NPP control based on NTV-PWM [7, 8]. The comparison on the control domains of i_o between DMW method and NTV-PWM method is given as Fig. 3. We can control i_o from I_{max} to I_{min} freely. It is seen from Fig. 3(c)(e)(f) that the NTV-PWM is not able to keep i_o zero in the entire output period. Thus, the low frequency NPP fluctuation is generated. In contrast, the DMW method has a narrower control domains of i_o than that of NTV-PWM from Fig. 3(a)(b)(d).

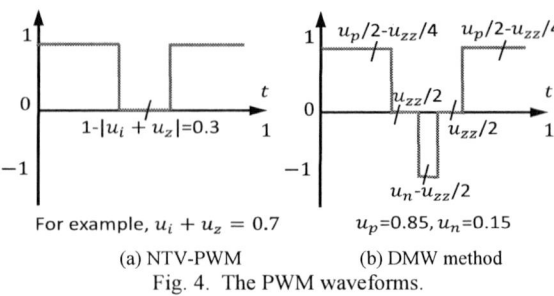

For example, $u_i + u_z = 0.7$ $u_p = 0.85, u_n = 0.15$

(a) NTV-PWM   (b) DMW method

Fig. 4. The PWM waveforms.

In this section, three DOF are introduced to control the NPP, which make the converter have the widest control domains of i_o to deal with the NPP drift.

A. The definition and limits of M-DOF

The first DOF is the zero sequence voltage (u_z). For the NPP control based on NTV-PWM, only u_z can be adjusted to amend i_o. Without changing the line voltages, it can revise the phase voltages, which affect the duty cycle of output voltage $o(0)$. The control range of u_z is given as (11)

$$-1 - u_{\min} \leq u_z \leq 1 - u_{\max} \quad (11)$$

The limit of (11) can ensure that the modulation does not exceed the maximum linear zone.

As analyzed in section II, in order to solve the low frequency NPP fluctuation, the modulation mode should be revised. We find from Fig. 4 that the PWM waveform of DMW method switches twice, while it switches one time for NTV-PWM. Although the special modulation mode increases the switching frequency, it also provides one DOF (u_{zz}) to amend the duty cycle of output voltage $o(0)$. For the traditional DMW method, it is necessary to fix u_x. As a result, u_{zz} is just the optimal value to keep i_o zero. In this paper, we will make an actively control of u_{zz} based on the NPP.

Also, the limit of u_{zz} should be determined at first. Actually, there are two extreme cases, which can be utilized to achieve the limit of u_{zz}. The first extreme case is shown as Fig. 5.

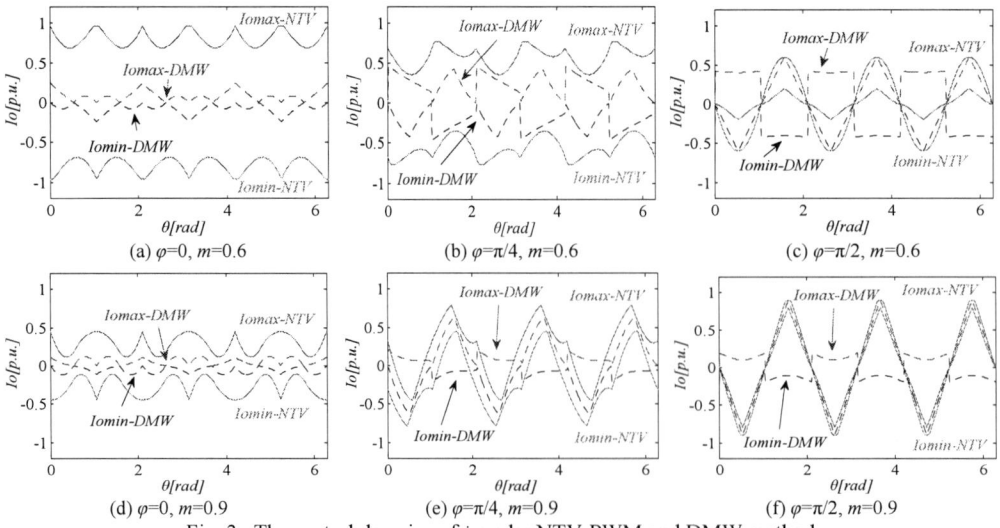

(a) $\varphi=0$, $m=0.6$ (b) $\varphi=\pi/4$, $m=0.6$ (c) $\varphi=\pi/2$, $m=0.6$

(d) $\varphi=0$, $m=0.9$ (e) $\varphi=\pi/4$, $m=0.9$ (f) $\varphi=\pi/2$, $m=0.9$

Fig. 3. The control domains of i_o under NTV-PWM and DMW method.

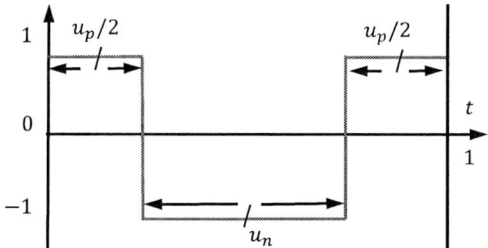

Fig. 5. The extreme case one.

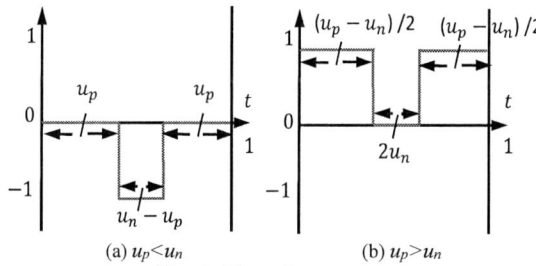

(a) $u_p < u_n$ (b) $u_p > u_n$

Fig. 6. The extreme case two.

For the extreme case one, the output voltage $o(0)$ is not employed. At this time, the three level modulation is degrade to two level modulation. It is not what we want, since it is harmful to improve the EMI problem. u_{zz} is decreased to zero, as shown in (12).

$$0 \le u_{zz} \qquad (12)$$

Here, the duty cycle (u_p, u_n) are calculate as (13).

$$\begin{cases} u_p = \left(1 + u_i + u_z\right)/2 \\ u_n = \left(1 - u_i - u_z\right)/2 \end{cases} \qquad (13)$$

$$i = \min \text{ or } mid \text{ or } \max$$

The other extreme case is depicted as Fig. 6. It is noted from Fig. 6 that the modulation changes from the special modulation as DMW method to NTV-PWM. It is of benefit to decrease the switching frequency compared with DMW method. u_{zz} is expanded as (14).

$$u_{zz} \le 2\min\left(u_p, u_n\right) \qquad (14)$$

Therefore, the limit range of u_{zz} is illustrated as (15).

$$0 \le u_{zz} \le 2\min\left(u_p, u_n\right) \qquad (15)$$

The last DOF is the phase chosen variable (phi). As mentioned above, the traditional DMW method fixes u_x to $(u_{min}\text{-}u_{max})/2$. This is because only u_{mid} is modulated via that special modulation mode at this time. Thus, the switching frequency increases by 4/3 not 2. Given the switching frequency, we also utilize the special modulation mode in one phase. However, it is harsh to know which one phase is the most beneficial to control the NPP, depending on the power factor (PF) of the loads. Therefore, the last DOF is defined as (16).

$$phi = \begin{cases} 0, if \ u_{\min} \ is \ modulated \ as \ Fig.6(b) \\ 1, if \ u_{mid} \ is \ modulated \ as \ Fig.6(b) \\ 2, if \ u_{\max} \ is \ modulated \ as \ Fig.6(b) \end{cases} \qquad (16)$$

B. The control principles of M-DOF

Next, it is necessary to take the knowledge of the principles of adjusting the M-DOF to control the NPP. The phase, which uses the special modulation mode, can generate an i_{o_phi}, as shown in (17).

$$i_{o\_phi} = \begin{cases} u_{zz}i_{\min}, if \ phi = 0 \\ u_{zz}i_{mid}, if \ phi = 1 \\ u_{zz}i_{\max}, if \ phi = 2 \end{cases} \qquad (17)$$

The other two phase voltages are modulated as NTV-PWM, so the i_{o_other} derived from the two phase voltages are calculated as,

$$i_{o\_other} = \begin{cases} \left(1-|u_{\max}+u_z|\right)i_{\max} + \left(1-|u_{mid}+u_z|\right)i_{mid}, if \ phi = 0 \\ \left(1-|u_{\max}+u_z|\right)i_{\max} + \left(1-|u_{\min}+u_z|\right)i_{\min}, if \ phi = 1 \\ \left(1-|u_{mid}+u_z|\right)i_{mid} + \left(1-|u_{\min}+u_z|\right)i_{\min}, if \ phi = 2 \end{cases} \qquad (18)$$

Therefore, the i_o generated from the modulation method is depicted as (19).

$$i_o = i_{o\_phi} + i_{o\_other} \qquad (19)$$

It is noted that the i_o in (19) can be changed via the defined three DOF. Moreover, it is difficult to achieve the explicit expressions of the three DOF and i_o. Thus, a search optimization method is adopt in this paper to obtain the optimal three DOF, of which the i_o is closest to the required i_{oc} in (3). It can be realized as following:

1) Compare the original three phase reference voltages (u_u, u_v, u_w) to get the maximum, middle, minimum voltage (u_{max}, u_{mid}, u_{min}). And the corresponding currents of (u_{max}, u_{mid}, u_{min}) are (i_{max}, i_{mid}, i_{min}).

2) Determine the limit range of the DOF(u_z) via (11).

3) Set the initial DOF(phi) to zero at first.

4) Set the initial u_z to -1-u_{min}.

5) Calculate the limit range of the DOF(u_{zz}) via (13)(15).

6) Set the initial u_{zz} to 0.

7) Predict the i_o derived from the modulation method as (19) and calculate the absolute value of neutral point current error ($|i_{o\_err}|$) between i_o and i_{oc}.

8) Go back to step 6) and increase u_{zz} until $2\min(u_p, u_n)$ by the step of $2\min(u_p, u_n)/n$.

9) Go back to step 4) and increase u_z until 1-u_{max} by the step of $(2-u_{max}+u_{min})/n$.

10) Go back to step 3) and increase phi until 2 by the step of 1.

11) Achieve the optimal three DOF, of which the $|i_{o\_err}|$ is the minimum.

12) Taking $phi=0$ as an example, three phase reference voltages are modulated by a unified way in (6). At this time, the duty cycle of the modulation waves is as shown in (20).

$$\begin{cases} u_{\max p} = \left(u_{\max} + u_z\right) \cdot \left[1 + floor\left(u_{\max} + u_z\right)\right] \\ u_{\max n} = -\left(u_{\max} + u_z\right) \cdot floor\left(u_{\max} + u_z\right) \end{cases}$$
$$\begin{cases} u_{mid p} = \left(u_{mid} + u_z\right) \cdot \left[1 + floor\left(u_{mid} + u_z\right)\right] \\ u_{mid n} = -\left(u_{mid} + u_z\right) \cdot floor\left(u_{mid} + u_z\right) \end{cases} \qquad (20)$$
$$u_{\min p} = u_p - u_{zz}/2, u_{\min n} = -\left(u_n - u_{zz}/2\right)$$

C. The comparison on the control domains of i_o for the DMW method and the proposed method

In this part, the comparison on the control domains of i_o for the DMW method and the proposed method is

1035

The 2018 International Power Electronics Conference

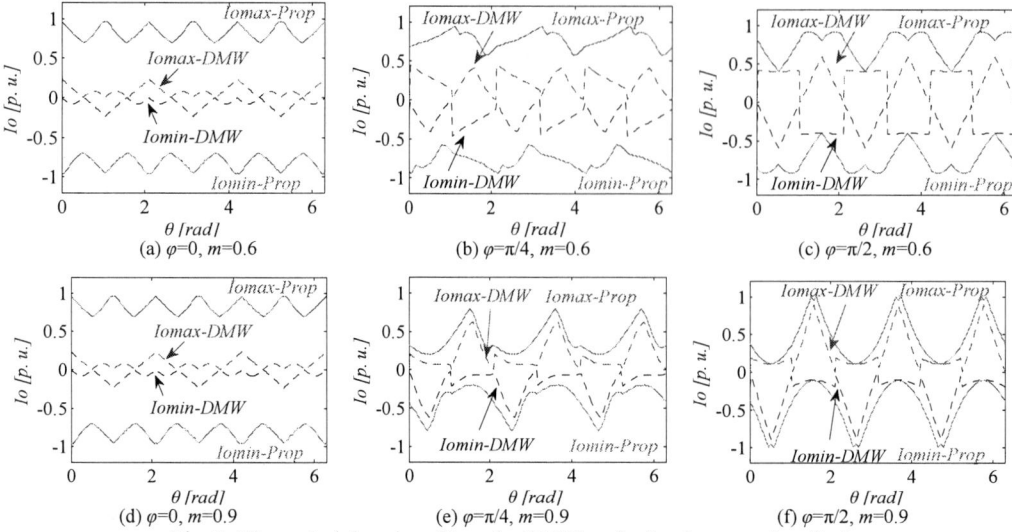

Fig. 7. The control domains of i_o under DMW method and proposed method.

discussed to explain the strong NPP control ability of proposed method, as shown in Fig. 7. Similarly to Fig. 3, i_o can only be revised from I_{max} to I_{min}. It is obviously seen from Fig. 7 that the proposed method has a wider control domains of i_o, compared with the DMW method in all the operation conditions. It is very vital to cope with the serious NPP drift problem. In addition, the proposed way can also keep i_o zero and overcome the low frequency NPP fluctuation.

IV. THE SIMULATION RESULTS

In order to proof the correctness and feasibility of proposed method, some simulation have be done here. The simulation is based on Matlab/Simulink software and the parameters are shown as Table. II. An initial NPP error (Δu_o) is set to 20V. And m is equal to 0.7.

Some simulation results of the traditional NTV-PWM, DMW method and proposed method are given in Fig. 8, Fig. 9 and Fig. 10. Firstly, it is known from Fig. 8(a) and Fig. 10(a) that the traditional NTV-PWM and proposed method can keep three level modulation. In contrast, the DMW method may cause three level modulation degrade to two level modulation sometimes, which is not want we want.

On the other hand, the traditional NTV-PWM can solve the NPP drift problem within 0.02s, as shown in Fig. 8(b). However, here is still a low frequency fluctuation in the NPP. It will limit the capacitor's reduction. It can be seen from Fig. 9(b) that the DMW does not only overcome the

NPP drift problem within 0.02s, but also suppress the low frequency NPP fluctuation. At the cost, the switching frequency increases from 4 kHz for the traditional NTV-PWM to 5.3 kHz, as shown in Fig. 8(c) and Fig. 9(c). Lastly, we can find from Fig. 10(b) that the proposed method deal with the low frequency NPP fluctuation and the NPP drift problem within 0.01s at the same time. It is noteworthy that the control speed of the NPP drift problem for proposed method is four times than that for the DMW method. It corresponds to the stronger control ability of proposed method to cope with the NPP drift problem. Moreover, as analyzed in section III, if the optimal u_{zz} is selected as $2\min(u_p, u_n)$, the modulation would come back to the NTV-PWM, which is good for reducing the switching frequency. Therefore, it is shown in Fig. 10(c) that the switching frequency of proposed method is about 4.7 kHz, which is smaller than that of the DMW method.

In addition, the FFT analysis of the line voltage and phase current for the traditional NTV-PWM, the DMW method and proposed method is also carried on, as shown in Table. III. It is noted that even if the proposed method changes the modulation mode for the NPP control, it does not lead to the deterioration of the output performance.

V. SUMMARIES

For TL-NPC converters, the NPP drift problem and the low frequency NPP fluctuation problem are the troublesome issues. The traditional NVT-PWM method can not deal with the low frequency NPP fluctuation problem. In contrast, the DMW method sacrifices many control domain of i_o to solve the low frequency NPP

TABLE II
THE SIMULATION PARAMETERS

Items	Parameters
DC-link voltage	220 V
C_{up} and C_{down}	1800 uF
Switching frequency	4 kHz
Output frequency	50 Hz
Loads	R-L(1Ω/25mH) PF=0.127

TABLE III
THE FFT ANALYSIS

Method	THD(U_{uv})	THD(I_u)
NTV-PWM	41.71%	0.43%
DMW method	49.94%	0.39%
Proposed method	48.78%	0.52%

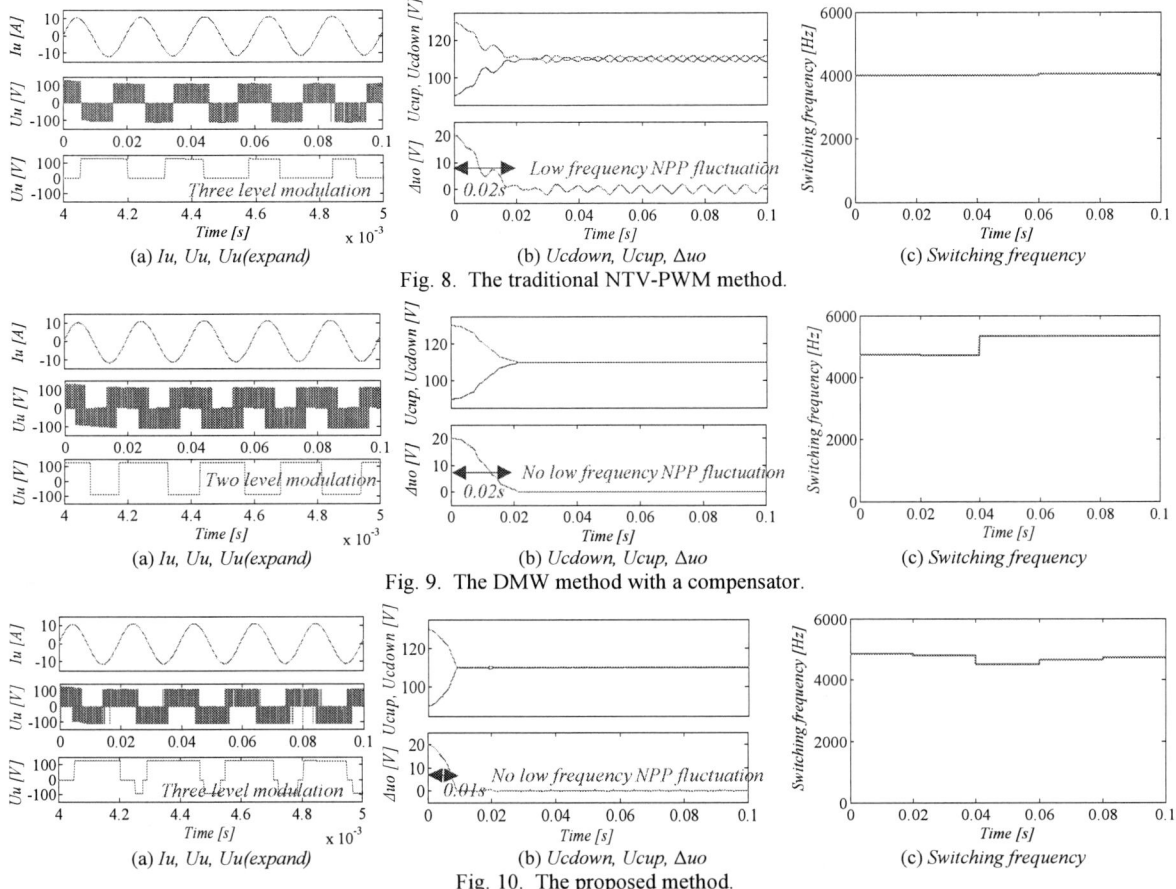

(a) Iu, Uu, Uu(expand)   (b) Ucdown, Ucup, Δuo   (c) Switching frequency

Fig. 8. The traditional NTV-PWM method.

(a) Iu, Uu, Uu(expand)   (b) Ucdown, Ucup, Δuo   (c) Switching frequency

Fig. 9. The DMW method with a compensator.

(a) Iu, Uu, Uu(expand)   (b) Ucdown, Ucup, Δuo   (c) Switching frequency

Fig. 10. The proposed method.

fluctuation problem. As a result, the switching frequency will increase to 4/3 and the control ability of NPP drift problem will be weaken. It is harmful to cope with the serious NPP drift problem, such as a low dead time or a strong asymmetrical load. In this paper, the M-DOF based NPP control is proposed, which can solve the two NPP problems and has a lower switching frequency. More importantly, the control ability of the NPP drift problem can be expand to the theoretical strongest.

REFERENCES

[1] A. Nabae, I. Takahashi and H. Akagi, "A new neutral-point-clamped PWM inverter," *IEEE Transactions on Industry Applications*, Vol. IA-17, No. 5, pp. 518-523, 1981.

[2] C. Wang and Y. Li, "A survey on topologies of multilevel converters and study of two novel topologies," *IEEE 6th International Power Electronics and Motion Control Conference*, pp. 860-865, 2009.

[3] B. Guan and S. Doki, "The compensation strategy on the drift voltage problem of dual modulation waves based neutral point potential control for three level converter," in *Proc. ISIE '17*, Edinburgh, UK, pp. 893-898, 2017.

[4] B. Guan and C. Wang, "A Narrow Pulse Compensation Method for Neutral-Point-Clamped Three-Level Converters Considering Neutral-Point Balance," in *Proc. ICPE-ECCE Asia '15*, Seoul, South Korea, pp. 2770-2775, 2015.

[5] A. Lewicki, Z. Krzeminski and H. Abu-Rub, "Space-vector pulsewidth modulation for three-level NPC converter with the neutral point voltage control," *IEEE Transactions on Industrial Electronics*, Vol. 58, No. 11, pp. 5076-5086, 2011.

[6] J. Zaragoza, J. Pou and S. Ceballos, "A comprehensive study of a hybrid modulation technique for the neutral-point-clamped converter," *IEEE Transactions on Industrial Electronics*, Vol. 58, No. 2, pp. 294-304, 2009.

[7] C. Wang and Y. Li, "Analysis and calculation of zero-sequence voltage considering neutral-point potential balancing in three-level NPC converters," *IEEE Transactions on Industrial Electronics*, Vol. 57, No. 7, pp. 2262-2271, 2010.

[8] J. Pou, J. Zaragoza and S. Ceballos, "A carrier-based PWM strategy with zero-sequence voltage injection for a three-level neutral-point-clamped converter," *IEEE Transactions on Power Electronics*, Vol. 27, No. 2, pp. 642-651, 2012.

[9] G. I. Orfanoudakis, M. A. Yuratich and S. M. Sharkh, "Nearest-vector modulation strategies with minimum amplitude of low-frequency neutral-point voltage oscillations for the neutral-point-clamped converter," *IEEE Transactions on Power Electronics*, Vol. 28, No. 10, pp. 4485-4499, 2013.

[10] J. Zaragoza, J. Pou and S. Ceballos, "Optimal voltage-balancing compensator in the modulation of a neutral-point-clamped converter," *IEEE International Symposium on Industrial Electronics*, pp. 719-724, 2007.

[11] J. Pou, J. Zaragoza and P. Rodriguez, "Fast-processing modulation strategy for the neutral-point-clamped converter with total elimination of low-frequency voltage oscillations in the neutral point," *IEEE Transactions on Industrial Electronics*, Vol. 54, No. 4, pp. 2288-2294, 2007.

A Modified Phase-Shifted PWM Technique for the Grid-Connected Hybrid Cascaded Converter

Yu-chen Su, *Student Member, IEEE*, and Po-tai Cheng, *Fellow, IEEE*
Center for Advanced Power Technologies, Department of Electrical Engineering
National Tsing Hua University, Hsinchu, Taiwan

Abstract—This paper investigates the harmonic spectrum of the grid-connected hybrid cascaded converter (HCC) with the phase-shifted PWM (PSPWM). Detailed mathematical analysis shows that the conventional PSPWM (CPSPWM) is unable to eliminate the lower order harmonic components, which leads to severe distortion in the output waveforms. In order to solve this problem, a modified PSPWM (MPSPWM) technique is proposed in this paper. The proposed MPSPWM technique is capable of achieving better output performance while maintaining the system efficiency, and it is confirmed by the simulation results and the laboratory test results.

Index Terms—Hybrid cascaded converter (HCC), phase-shifted PWM (PSPWM), harmonic spectrum.

I. INTRODUCTION

The modular multilevel cascaded converter with single-star bridge-cell (MMCC-SSBC) [1] has attracted great attention in the grid applications like renewable energy systems [2]–[5], static synchronous compensators (STATCOMs) [6]–[9], etc. However, the main disadvantage of the MMCC-SSBC is that a large number of power semiconductor devices are demanded, which increases the number of gate signals and gate drivers as well. By integrating three single-phase H-bridge cells into one three-phase two-level cell, the number of power semiconductor devices, gate signals, and gate drivers can be decreased, and the volume can also be reduced [10]. Consequently, the topology of the hybrid cascaded converter (HCC) is developed [11]–[14], as shown in Fig. 1.

For the operation of the HCC in the grid applications, a four-layer hierarchical control scheme has been presented in [14] to attain both the power control and the DC capacitor voltage balancing control no matter the system is balanced or unbalanced. Nevertheless, with the conventional PSPWM (CPSPWM) that has been generally implemented on the MMCC-SSBC [15]–[17], it seems that the output waveforms of the HCC behave poorly, which may not satisfy the power quality requirement of the utility grid.

In this paper, the harmonic spectrum of the grid-connected HCC with the CPSPWM is derived through the detailed mathematical analysis to demonstrate that the poor performance of the output waveforms results from the inability to eliminate the lower order harmonic components. In order to solve this problem, this paper presents a modified PSPWM (MPSPWM) technique for the grid-connected HCC to improve the harmonic spectrum, which makes the H-bridge cells fulfill their own PSPWM and then increases the carrier frequency of the two-level cell to match the H-bridge cells. The proposed

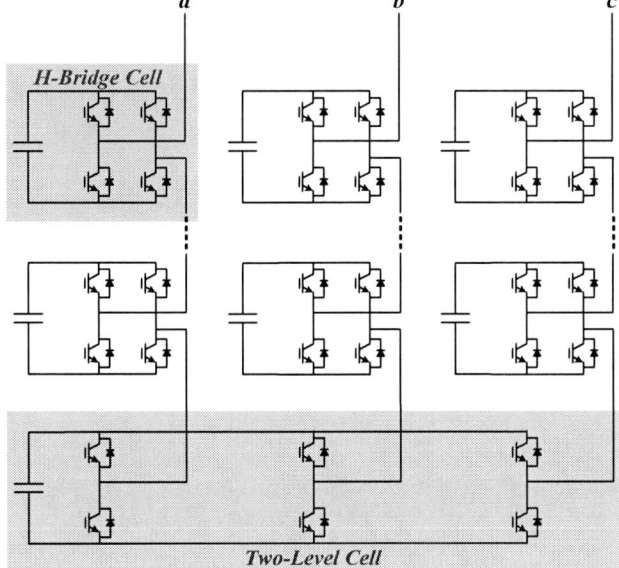

Fig. 1. Topology of the hybrid cascaded converter (HCC).

MPSPWM technique is further compared with the CPSPWM based on the same switching times to prove that better output performance can be achieved without affecting the system efficiency, and its effectiveness is also validated with the help of the simulation results and the laboratory test results.

II. SYSTEM CONFIGURATION

Fig. 2 shows the system configuration of the grid-connected HCC with the cascaded cell number $N=3$, which consists of two single-phase H-bridge cells at each phase and one three-phase two-level cell. The cascaded cell number can be adjusted depending on the system requirement. In this paper, Fig. 2 is taken as an example for the simplicity in explanation.

With the four-layer hierarchical control scheme presented in [14], the HCC in Fig. 2 works as a 1.5kVar STATCOM in this paper. The corresponding system parameters are given in TABLE I. Note that the DC voltage ratio between the H-bridge cells and the two-level cell is set as $1/2$, namely, $V_{dc3}=2V_{dcmn}$ ($m=a,b,c$ & $n=1,2$). This DC voltage ratio is selected so that each cell has the same DC voltage utilization to produce seven-level output voltage waveforms.

The 2018 International Power Electronics Conference

$$v_{o,h} = \frac{V_{dc,h}}{2} M \cos(\omega t) + \frac{2V_{dc,h}}{\pi} \sum_{m=1}^{\infty} \sum_{n=-\infty}^{\infty} \frac{1}{m} J_n(m\frac{\pi}{2}M) \sin([m+n]\frac{\pi}{2}) \cos(m\omega_c t + n\omega t)$$

$$v_{o,H} = V_{dc,H} M \cos(\omega t) + \frac{4V_{dc,H}}{\pi} \sum_{m=1}^{\infty} \sum_{n=-\infty}^{\infty} \frac{1}{2m} J_{2n-1}(m\pi M) \cos([m+n-1]\pi) \cos(2m\omega_c t + [2n-1]\omega t)$$

(1)

$$v_{o,CPSPWM} = 3V_{dc}^* M \cos(\omega t)$$
$$+ \frac{4V_{dc}^*}{\pi} \sum_{m=1}^{\infty} \sum_{n=-\infty}^{\infty} \frac{1}{2m} J_{2n-1}(3m\pi M) \cos([3m+n-1]\pi) \cos(6m\omega_c t + [2n-1]\omega t)$$
$$+ \frac{4V_{dc}^*}{\pi} \sum_{m=1}^{\infty} \sum_{n=-\infty}^{\infty} \frac{1}{2m-1} J_{2n}([2m-1]\frac{\pi}{2}M) \cos([m+n-1]\pi) \cos([2m-1][\omega_c t - \frac{2\pi}{3}] + 2n\omega t)$$

(2)

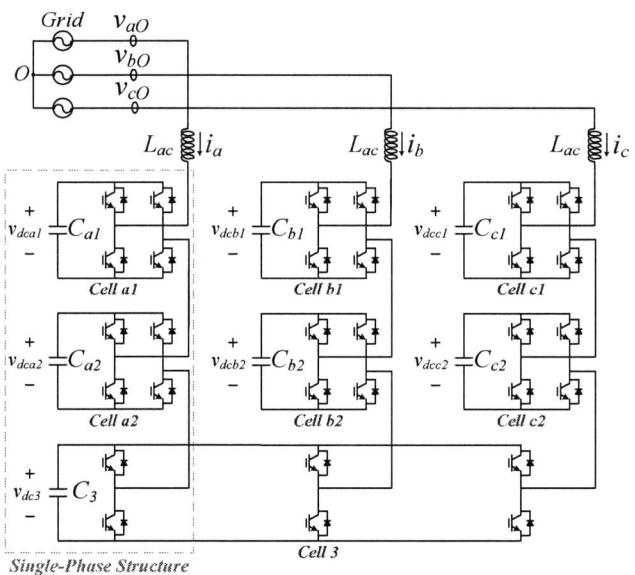

Fig. 2. System configuration of the grid-connected HCC.

TABLE I
SYSTEM PARAMETERS

	Symbol	Value
Grid voltage (L-L rms)	v_g	220(V)
Grid frequency	ω	60(Hz)
Rated reactive power	Q_R	1.5(kVar)
Cascaded cell number	N	3
AC filter inductor	L_{ac}	6.8(mH)
Nominal DC voltage of H-bridge cells	V_{dc}^*	80(V)
Nominal DC voltage of two-level cell	$2V_{dc}^*$	160(V)
DC capacitor of H-bridge cells	C_{dc}	840(μF)
DC capacitor of two-level cell	$3C_{dc}$	2520(μF)
Unit capacitance constant [18]	H	32.26(msec)

III. HARMONIC SPECTRUM OF THE HCC WITH CPSPWM

Following the system operation described in Section II, the harmonic spectrum of the HCC with the CPSPWM is evaluated in this section. It is evident that the phase voltage of the HCC is constructed from its single-phase structure including two H-bridges and one half-bridge, as shown in Fig. 2. Accordingly, it is necessary to establish the output voltage models of the half-bridge and the H-bridge first. Note that this paper adopts the sinusoidal PWM (SPWM) for the half-bridge and the unipolar SPWM for the H-bridge, which are the most well-known PWM methods for them [16].

By means of double Fourier integral analysis (DFIA) [19], the output voltages of the half-bridge ($v_{o,h}$) and the H-bridge ($v_{o,H}$) can be mathematically expressed as (1), where the first term denotes the fundamental component and the summation term denotes the harmonic components. M represents the modulation index; $V_{dc,h}$ and $V_{dc,H}$ represent their respective

DC voltages; ω and ω_c represent the reference frequency and the carrier frequency. From (1), it is found that the harmonic components of the half-bridge and the H-bridge are located around $1\omega_c, 2\omega_c, 3\omega_c...$ and $2\omega_c, 4\omega_c, 6\omega_c...$, respectively.

With the information in (1), the phase voltage of the HCC with the CPSPWM can be obtained. Fig. 3 illustrates the operational principle of the CPSPWM, in which the carriers of each bridge have the same frequency and $\pi/3$ phase shift. Then, taking into consideration the system DC voltage ratio, $V_{dc,h} = 2V_{dc,H} = 2V_{dc}^*$, the phase voltage of the HCC with the CPSPWM ($v_{o,CPSPWM}$) is derived by adding together all the bridge output voltages at the same phase, as shown in (2). The first term indicates the fundamental component, and the first summation term and the second summation term indicate the harmonic components located around the even multiples of ω_c and the odd multiples of ω_c, respectively.

1039

$$v_{o,MPSPWM} = 3V_{dc}^{*} M \cos(\omega t)$$
$$+ \frac{4V_{dc}^{*}}{\pi} \sum_{m=1}^{\infty} \sum_{n=-\infty}^{\infty} \frac{1}{2m} J_{2n-1}(2m\pi M) \cos([2m+n-1]\pi) \cos(4m\omega_c t + [2n-1]\omega t)$$
$$+ \frac{4V_{dc}^{*}}{\pi} \sum_{m=1}^{\infty} \sum_{n=-\infty}^{\infty} \frac{1}{m} J_n(m\frac{\pi}{2}M) \sin([m+n]\frac{\pi}{2}) \cos(4m\omega_c t + n\omega t) \tag{3}$$

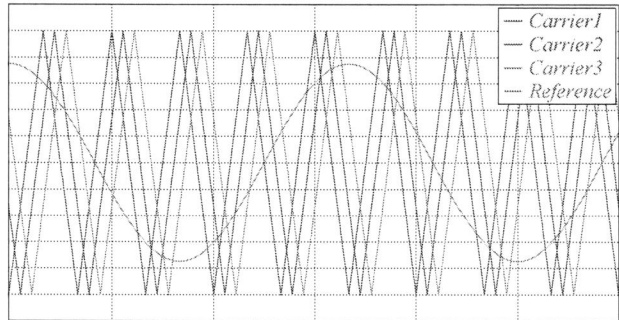

Fig. 3. Operational principle of the CPSPWM.

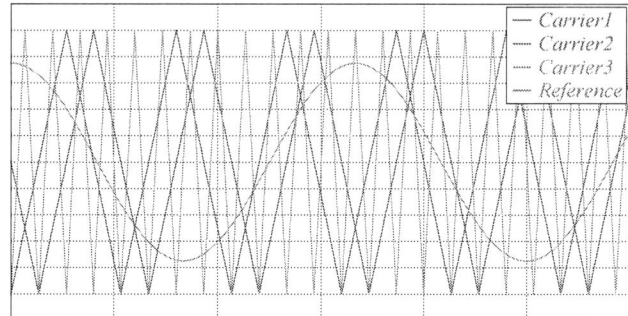

Fig. 4. Operational principle of the MPSPWM.

From (2), it is apparent that although the harmonic components located around the even multiples of ω_c are pushed to $6\omega_c, 12\omega_c, 18\omega_c...$, the harmonic components located around the odd multiples of ω_c remain at $1\omega_c, 3\omega_c, 5\omega_c....$ These lower order harmonic components are mainly produced by the two-level cell, and thus visible distortion may emerge from them.

IV. PROPOSED MPSPWM TECHNIQUE

As discussed in Section III, the CPSPWM fails in eliminating the lower order harmonic components. It is essentially because different harmonic components exist in the H-bridge cell and the two-level cell, and they are difficult to fully cancel. Hence, a MPSPWM technique is proposed in this paper to deal with this problem.

The target of the proposed MPSPWM technique is to optimize the harmonic distribution of the HCC. It is known from [17] that further harmonic cancellation can be accomplished by appropriately shifting the carrier phase of each series-connected H-bridge cell. Following this concept, the proposed MPSPWM technique makes the H-bridge cells carry out their own PSPWM and then increases the carrier frequency of the two-level cell to be in harmony with the H-bridge cells.

Fig. 4 illustrates the operational principle of the MPSPWM, in which the carriers of each H-bridge cell have the same frequency and $\pi/2$ phase shift, and the carrier frequency of the two-level cell is four times that of the H-bridge cells. Then, in the same way, the phase voltage of the HCC with the MPSPWM ($v_{o,MPSPWM}$) can be derived as (3). The first term indicates the fundamental component, and the first summation term and the second summation term indicate the harmonic components produced by the H-bridge cells and the two-level cell, respectively.

From (3), it can clearly be seen that the harmonic components produced by the H-bridge cells and the two-level cell are both located around $4\omega_c, 8\omega_c, 12\omega_c...$, which is significant improvement in the harmonic spectrum of the HCC.

V. COMPARISON BETWEEN CPSPWM AND MPSPWM

Although the MPSPWM can significantly improve the harmonic spectrum of the HCC, the increase of the carrier frequency of the two-level cell may result in more switching times and thereupon higher switching loss. Therefore, this section attempts to compare the MPSPWM with the CPSPWM under the condition of the same switching times for the fair comparison. For this purpose, the overall frequency of the MPSPWM needs to be rearranged.

Because the H-bridge switches four times per PWM cycle and the half-bridge switches twice per PWM cycle, the switching ratio of the CPSPWM to the MPSPWM is $10/16$ for the HCC with the cascaded cell number N=3. Accordingly, with the overall frequency of the MPSPWM multiplied by $10/16$, the comparison between the CPSPWM and the MPSPWM based on the same switching times can be realized, and the results are summarized in TABLE II. As can be seen, even under the condition of the same switching times, the MPSPWM is still superior to the CPSPWM in terms of the harmonic distribution, $2.5\omega_c, 5\omega_c....$

It is also interesting to note from TABLE II that the concept of the MPSPWM is to optimize the harmonic distribution of the HCC by reorganizing the carrier frequency of the H-bridge cells and the two-level cell while keeping the total switching times unchanged. It can thus be said that the proposed MPSPWM technique is able to achieve better output performance without affecting the system efficiency.

TABLE II
COMPARISON BETWEEN CPSPWM AND MPSPWM

	CPSPWM	MPSPWM
Carrier frequency of H-bridge cells	ω_c	$\frac{5}{8}\omega_c$
Carrier frequency of two-level cell	ω_c	$\frac{5}{2}\omega_c$
Harmonic distribution of the HCC	$1\omega_c, 3\omega_c...$	$2.5\omega_c, 5\omega_c...$

TABLE III
PARAMETERS FOR CPSPWM AND MPSPWM

	CPSPWM	MPSPWM
Carrier frequency of H-bridge cells	2000(Hz)	1250(Hz)
Carrier frequency of two-level cell	2000(Hz)	5000(Hz)

VI. SIMULATION RESULTS

In order to verify the proposed MPSPWM technique and compare it with the CPSPWM, the simulation results are provided in this section. The testbench is built in the PLECS simulation platform as the system configuration shown in Fig. 2 with the corresponding system parameters described in Section II. Besides, the parameters for the CPSPWM and the MPSPWM are designed according to TABLE II, and they are given in TABLE III. The simulation results of the STATCOM operation from the rated inductive 1.5kVar injection to the rated capacitive 1.5kVar injection are shown and illustrated as follows.

Fig. 5, Fig. 6, and Fig. 7 illustrate the simulation results of the DC capacitor voltages, the phase voltage and its harmonic spectrum, and the phase currents, respectively. As shown in Fig. 5(a) and Fig. 5(b), the DC capacitor voltages of the H-bridge cells ($80V$) and the two-level cell ($160V$) are well controlled no matter the CPSPWM or the MPSPWM is implemented. However, the phase voltages with the CPSPWM and the MPSPWM, as shown in Fig. 6(a) and Fig. 6(b), perform somewhat differently. By means of fast Fourier transform (FFT) analysis, their respective harmonic spectra are also obtained in Fig. 6(a) and Fig. 6(b). Obviously, there are lower order harmonic components existing in the harmonic spectrum of the phase voltage with the CPSPWM. In contrast, the harmonic spectrum of the phase voltage with the MPSPWM is significantly improved because the harmonic distribution is optimized by the MPSPWM. More importantly,

they agree with the results summarized in TABLE II. Due to the significant improvement in the harmonic spectrum of the phase voltage with the MPSPWM, it is obvious that the phase currents with the MPSPWM, as shown in Fig. 7(b), have more mitigated distortion than the phase currents with the CPSPWM, as shown in Fig. 7(a), in which the quality of the phase currents is greatly improved with 1.68% THD reduction for the rated inductive 1.5kVar operation and 1.81% THD reduction for the rated capacitive 1.5kVar operation.

In addition, the loss analysis between the CPSPWM and the MPSPWM is simulated in this section as well. Because the total loss of a power semiconductor device is basically composed of the conduction loss and the switching loss, the total loss of the HCC is analyzed by considering these two loss elements. According to the system operation, the 200V/34A power MOSFET (Infineon IPP320N20N3G) and the 560V/21A power MOSFET (Infineon SPP21N50C3) are employed in the H-bridge cells and the two-level cell, respectively, and their respective device parameters are established in the PLECS simulation platform for the comprehensive loss analysis.

The simulation results of the loss analysis between the CPSPWM and the MPSPWM are illustrated in Fig. 8(a) and Fig. 8(b), including the conduction loss of the H-bridge cells and the two-level cell, the switching loss of the H-bridge cells and the two-level cell, and the total loss of the HCC. Because of the cascaded system configuration, there is almost no difference between the conduction loss of the CPSPWM and that of the MPSPWM. Nevertheless, as mentioned in Section V, the carrier frequency of the H-bridge cells and the two-level cell is reorganized by the MPSPWM, so the switching loss of the MPSPWM is also redistributed compared with that of the CPSPWM. Despite this, the total loss of the MPSPWM is very close to that of the CPSPWM because the total switching times are kept the same, which once again proves that the implementation of the MPSPWM does not affect the system efficiency.

VII. LABORATORY TEST RESULTS

The laboratory test results are likewise provided for further verification in the practical implementation. The experimental hardware setup has been built in the laboratory following the system configuration shown in Fig. 2. Note that all the parameters are the same as those of the simulation. The system is then tested by implementing the CPSPWM and the MPSPWM on it, and the laboratory test results of the STATCOM operation with the rated capacitive 1.5kVar injection are shown and illustrated as follows.

Fig. 9 and Fig. 10 illustrate the laboratory test results of the DC capacitor voltages and the phase currents, respectively. Similar to the simulation results, Fig. 9(a) and Fig. 9(b) show that the DC capacitor voltages of the H-bridge cells ($80V$) and the two-level cell ($160V$) are both well controlled no matter the CPSPWM or the MPSPWM is implemented. Also, Fig. 10(a) and Fig. 10(b) show that the phase currents with the MPSPWM perform much better than the phase currents with the CPSPWM, as highlighted in this paper.

The 2018 International Power Electronics Conference

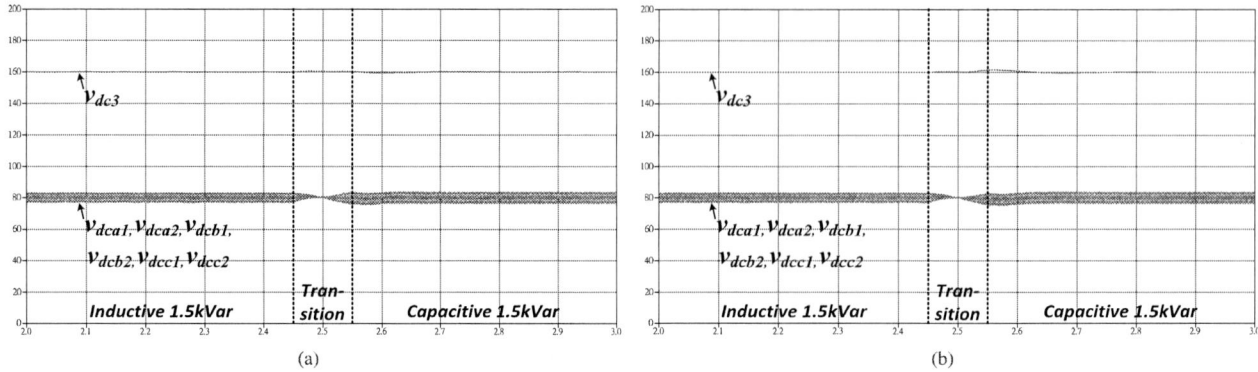

(a)   (b)

Fig. 5. Simulation results of the DC capacitor voltages. (a) CPSPWM. (b) MPSPWM.

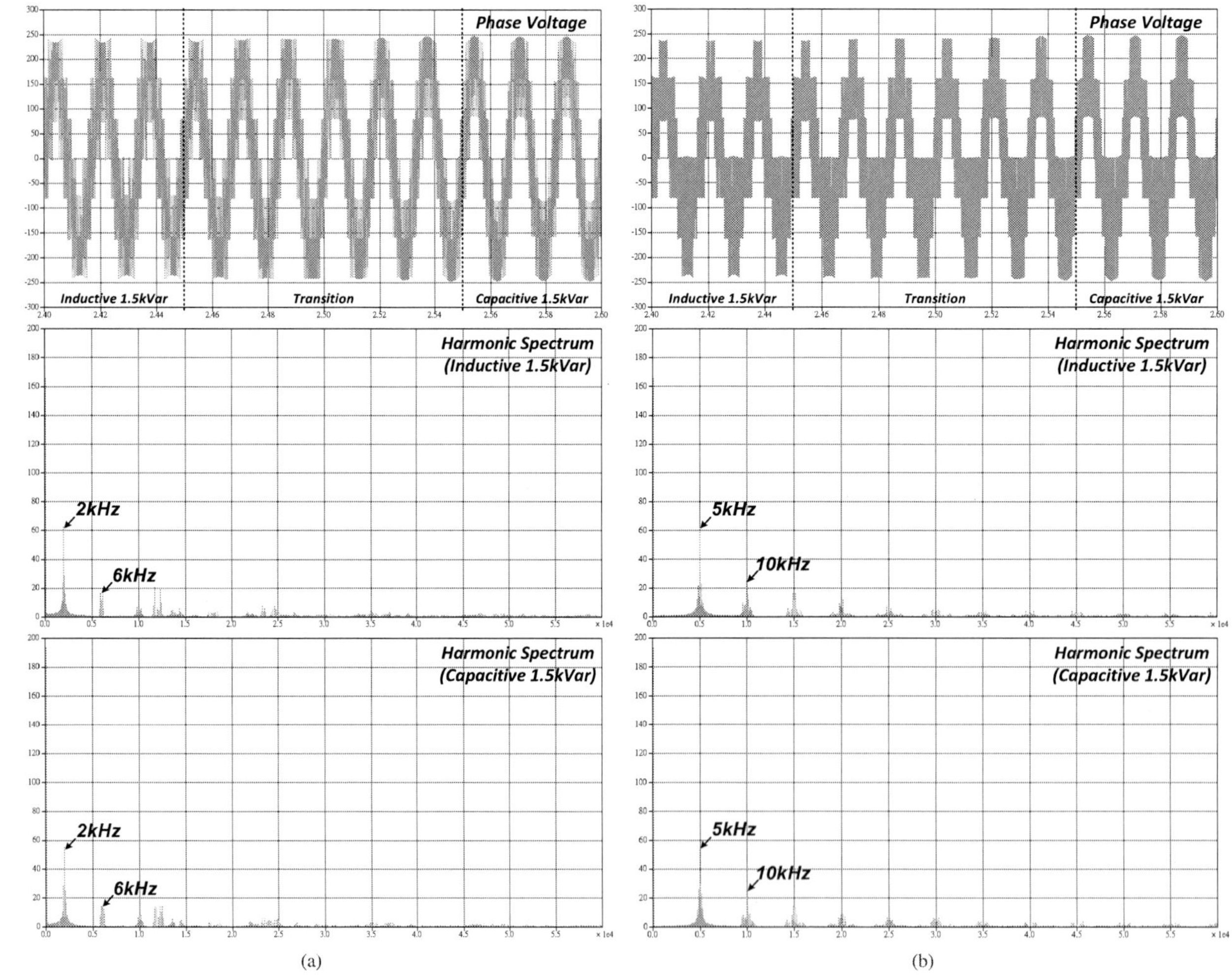

(a)   (b)

Fig. 6. Simulation results of the phase voltage and its harmonic spectrum. (a) CPSPWM. (b) MPSPWM.

1042

The 2018 International Power Electronics Conference

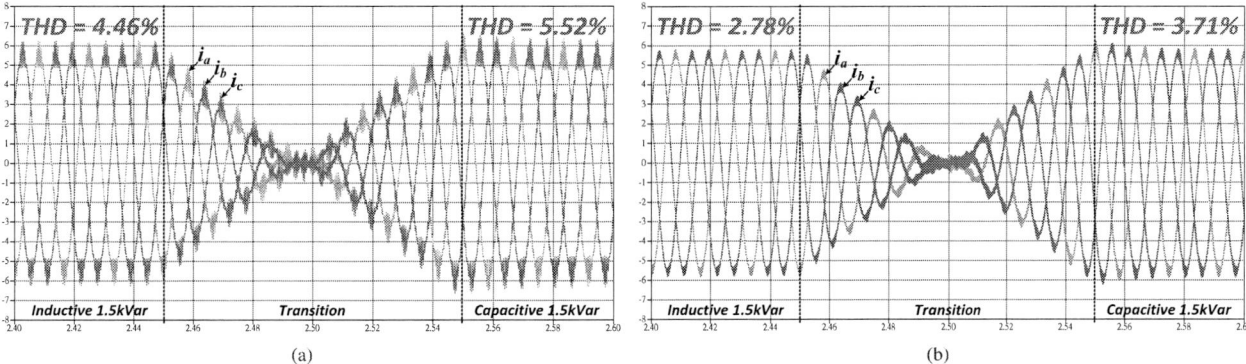

Fig. 7. Simulation results of the phase currents. (a) CPSPWM. (b) MPSPWM.

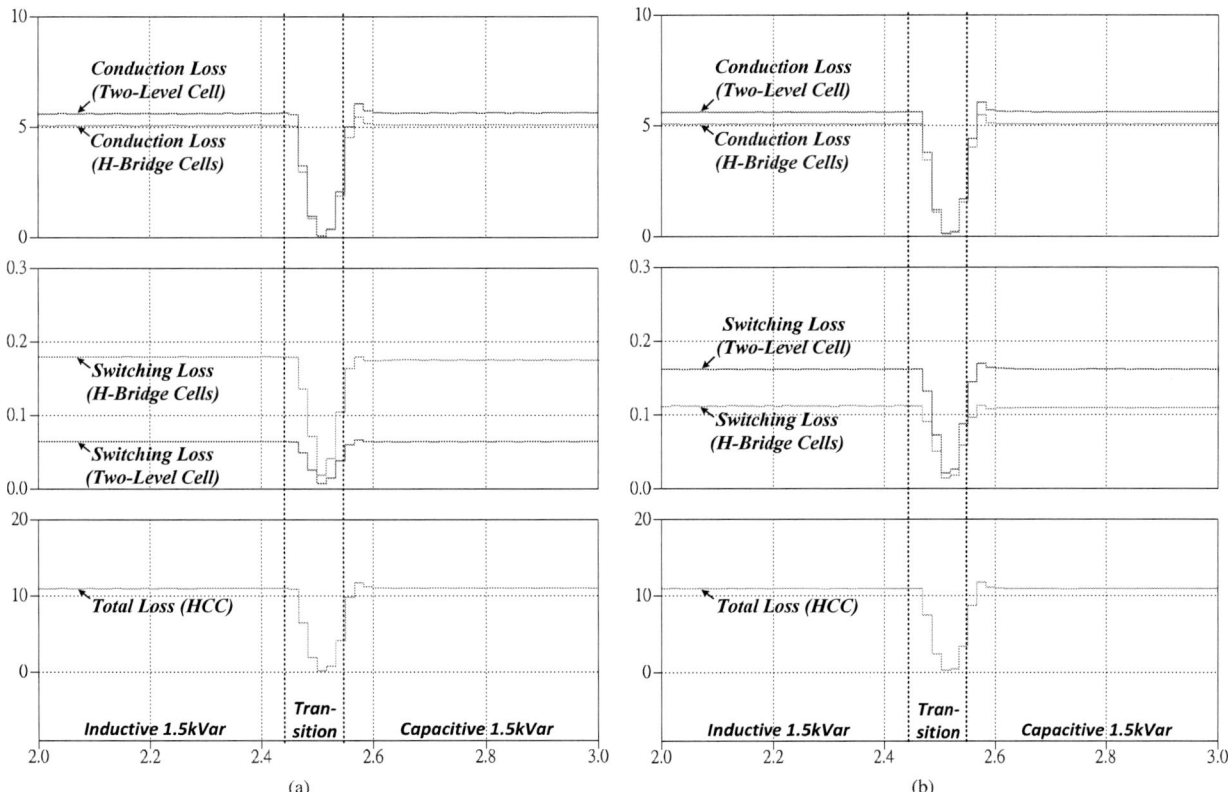

Fig. 8. Simulation results of the loss analysis. (a) CPSPWM. (b) MPSPWM.

VIII. CONCLUSION

In this paper, a MPSPWM technique is proposed for the grid-connected HCC to improve the harmonic spectrum. Through detailed mathematical analysis, the reason for the undesired lower order harmonic components in the output waveforms of the HCC when implementing the CPSPWM is clarified, and the proposed MPSPWM technique solves this problem by making the H-bridge cells complete their own PSPWM and then increasing the carrier frequency of the two-level cell to match the H-bridge cells. Furthermore, the comparison between the CPSPWM and the MPSPWM based on the same switching times shows that the proposed MPSPWM technique is capable of achieving better output performance while maintaining the system efficiency, which is also verified by the simulation results and the laboratory test results. Owing to this characteristic, the proposed MPSPWM technique is expected to be a promising modulation scheme for the grid-connected HCC.

REFERENCES

[1] H. Akagi. "Classification, terminology, and application of the modular multilevel cascade converter (mmcc)," *IEEE Transactions on Power Electronics*, vol. 26, no. 11, pp. 3119–3130, Nov 2011.

The 2018 International Power Electronics Conference

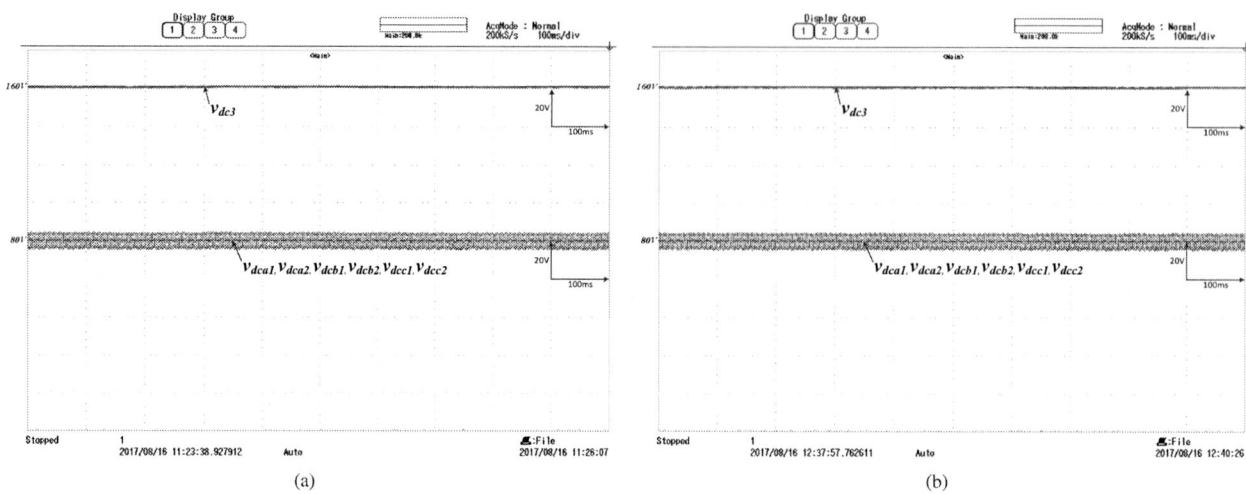

Fig. 9. Laboratory test results of the DC capacitor voltages (capacitive 1.5kVar operation). (a) CPSPWM. (b) MPSPWM.

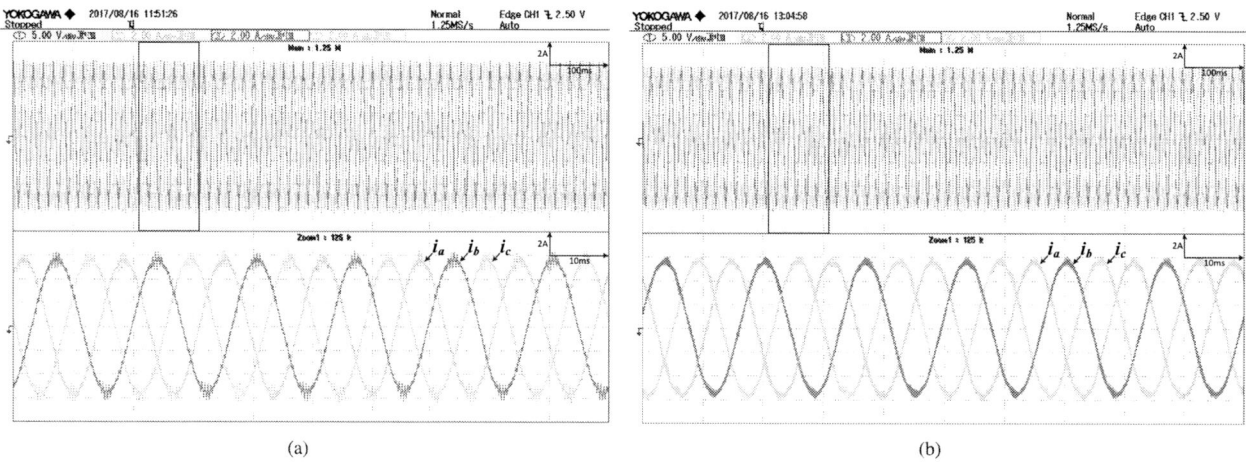

Fig. 10. Laboratory test results of the phase currents (capacitive 1.5kVar operation). (a) CPSPWM. (b) MPSPWM.

[2] S. Rivera, S. Kouro, B. Wu, J. I. Leon, J. Rodrguez, and L. G. Franquelo, "Cascaded h-bridge multilevel converter multistring topology for large scale photovoltaic systems," in *2011 IEEE International Symposium on Industrial Electronics*, June 2011, pp. 1837–1844.

[3] Y. Yu, G. Konstantinou, B. Hredzak, and V. G. Agelidis, "Power balance of cascaded h-bridge multilevel converters for large-scale photovoltaic integration," *IEEE Transactions on Power Electronics*, vol. 31, no. 1, pp. 292–303, Jan 2016.

[4] ——, "Power balance optimization of cascaded h-bridge multilevel converters for large-scale photovoltaic integration," *IEEE Transactions on Power Electronics*, vol. 31, no. 2, pp. 1108–1120, Feb 2016.

[5] B. Xiao, L. Hang, J. Mei, C. Riley, L. M. Tolbert, and B. Ozpineci, "Modular cascaded h-bridge multilevel pv inverter with distributed mppt for grid-connected applications," *IEEE Transactions on Industry Applications*, vol. 51, no. 2, pp. 1722–1731, March 2015.

[6] C. T. Lee, B. S. Wang, S. W. Chen, S. F. Chou, J. L. Huang, P. T. Cheng, H. Akagi, and P. Barbosa, "Average power balancing control of a statcom based on the cascaded h-bridge pwm converter with star configuration," *IEEE Transactions on Industry Applications*, vol. 50, no. 6, pp. 3893–3901, Nov 2014.

[7] K. Sano and M. Takasaki, "A transformerless d-statcom based on a multivoltage cascade converter requiring no dc sources," *IEEE Transactions on Power Electronics*, vol. 27, no. 6, pp. 2783–2795, June 2012.

[8] B. Gultekin and M. Ermis, "Cascaded multilevel converter-based transmission statcom: System design methodology and development of a 12

kv 12 mvar power stage," *IEEE Transactions on Power Electronics*, vol. 28, no. 11, pp. 4930–4950, Nov 2013.

[9] H. Akagi, S. Inoue, and T. Yoshii, "Control and performance of a transformerless cascade pwm statcom with star configuration," *IEEE Transactions on Industry Applications*, vol. 43, no. 4, pp. 1041–1049, July 2007.

[10] J. Wen and K. M. Smedley, "Synthesis of multilevel converters based on single- and/or three-phase converter building blocks," *IEEE Transactions on Power Electronics*, vol. 23, no. 3, pp. 1247–1256, May 2008.

[11] S. Mekhilef and M. N. A. Kadir, "Novel vector control method for three-stage hybrid cascaded multilevel inverter," *IEEE Transactions on Industrial Electronics*, vol. 58, no. 4, pp. 1339–1349, April 2011.

[12] S. Khomfoi, N. Praisuwanna, and L. M. Tolbert, "A hybrid cascaded multilevel inverter application for renewable energy resources including a reconfiguration technique," in *2010 IEEE Energy Conversion Congress and Exposition*, Sept 2010, pp. 3998–4005.

[13] Y. Zhang, G. Adam, S. Finney, and B. Williams, "Improved pulse-width modulation and capacitor voltage-balancing strategy for a scalable hybrid cascaded multilevel converter," *IET Power Electronics*, vol. 6, no. 4, pp. 783–797, April 2013.

[14] Y. C. Su, P. H. Wu, and P. T. Cheng, "Control of the hybrid cascaded converter under unbalanced conditions," in *2017 IEEE Energy Conversion Congress and Exposition (ECCE)*, Oct 2017, pp. 2858–2865.

[15] Y. Liang and C. O. Nwankpa, "A new type of statcom based on cascading voltage-source inverters with phase-shifted unipolar spwm,"

1044

IEEE Transactions on Industry Applications, vol. 35, no. 5, pp. 1118–1123, Sep 1999.

[16] Y. Li, Y. Wang, and B. Q. Li, "Generalized theory of phase-shifted carrier pwm for cascaded h-bridge converters and modular multilevel converters," *IEEE Journal of Emerging and Selected Topics in Power Electronics*, vol. 4, no. 2, pp. 589–605, June 2016.

[17] D. G. Holmes and T. A. Lipo, *CarrierBased PWM of Multilevel Inverters*. Wiley-IEEE Press, 2003, pp. 453–530. [Online]. Available: http://ieeexplore.ieee.org/xpl/articleDetails.jsp?arnumber=5311978

[18] H. Fujita, S. Tominaga, and H. Akagi, "Analysis and design of a dc voltage-controlled static var compensator using quad-series voltage-source inverters," *IEEE Transactions on Industry Applications*, vol. 32, no. 4, pp. 970–978, Jul 1996.

[19] D. G. Holmes and T. A. Lipo, *Modulation of One Inverter Phase Leg*. Wiley-IEEE Press, 2003, pp. 95–153. [Online]. Available: http://ieeexplore.ieee.org/xpl/articleDetails.jsp?arnumber=5311956

The 2018 International Power Electronics Conference

Novel T-type Dual-Buck Inverter with Minimum Number of Inductors

Tien-The Nguyen[1*], Honnyong Cha[1], Bang Le-Huy Nguyen[1] and Heung-Geun Kim[2]
1 School of Energy Engineering, Kyungpook National University, Daegu, Korea
2 Department of Electrical Engineering, Kyungpook National University, Daegu, Korea
*E-mail: nguyenthetien93@gmail.com

Abstract— It is well-known that the benefits of the dual-buck structure are no-reverse recovery problem of the body diodes, no shoot-through worry, and high efficiency. However, the conventional dual-buck structure requires two inductors for one-phase leg. This paper proposes a novel T-type dual-buck inverter with minimum number of inductors The proposed inverter contains only one inductor in each T-type dual-buck phase leg, thus, the total number of inductors is reduced by half. Therefore, the volume and the weight of inverter can be reduced. The operating principle and pulse width modulation method (PWM) is presented. A single-phase prototype hardware is built and tested, experimental results are provided to verify the theoretical analysis.

Keywords— *Dual-buck structure, five-level inverter, high reliability, single-stage, T-type NPC.*

I.   INTRODUCTION

Multilevel inverters are widely applied in high power system using the medium voltage level switches. They offer some benefits such as low voltage stress, low output voltage distortion, smaller common-mode voltage (CMV) and low total harmonic distortion (THD) compared to the two-level inverters [1]-[3]. Three well-known multilevel inverter topologies include neutral-point clamped (NPC), flying-capacitor, and cascade H-bridge inverters. The T-type inverter belongs to the family of NPC multilevel inverter. Compared to the conventional one, although two switches of the T-type inverters suffer from higher voltage stress, they benefit from lower conduction loss [4], [5]. Thus, higher efficiency can be achieved in low-voltage applications.

Power MOSFETs have some outstanding characteristics such as high switching frequency operation, low switching loss, low conduction loss at a certain level of current. Comparing to insulated-gate bipolar transistors (IGBTs), the fast transition feature of MOSFET increases the switching frequency of the switches. This results in reduced size of passive components. However, the body diode of MOSFET has poor reverse recovery characteristic, thus, it results in a huge reverse current, especially when the voltage excess 250 V. This reverse current could damage the device over time or kill them immediately. Thus, the converter reliability is reduced significantly. There are many

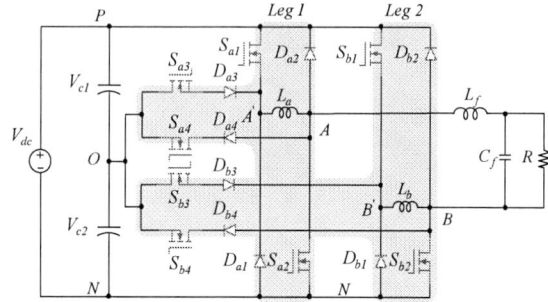

Fig. 1. Proposed five-level dual buck full-bridge inverter.

solutions introduced so far to solve this problem. One of them is the soft-switching feature of the resonant converter. However, auxiliary circuits are required, the switching frequency changes widely, and it is difficult to scale up to higher power level [14].

In conventional phase leg structure, the switches are connected in series. This suffers from short-circuit possibility. Then, the PWM dead-time is inevitably inserted to prevent the shoot-through. Nonetheless, the larger dead-time affects the quality of output waveforms and reduces energy transfer capability [6]-[8].

Dual-buck structure is introduced to overcome the reverse recovery of MOSFET's body diodes and PWM dead-time issues. There are no shoot-through worries owing to no shoot-through path. Thus, the dead-time can be minimized or eliminated. In addition, in dual-buck structure, the MOSFET's body diode does not have a chance to conduct. As a result, the reverse recovery issues can be relieved. Moreover, the external diodes in series with switches can be selected individually with optimal reverse recovery characteristics. Therefore, converter reliability and efficiency can be increased [10]-[11].

The family of the five-level dual-buck full-bridge inverter (DBFBI) is introduced in [12], [13], where, the different types of three-level dual-buck NPC leg are combined with a two-level dual-buck half-bridge to form the five-level inverters. In this case, four inductors are employed.

This paper presents a novel T-type dual-buck inverter with only one inductor for each phase leg. In Section II,

1046

the operation modes are analyzed and PWM technique is introduced. Simulation and experimental results are provided in Section III and IV. The paper is concluded at Section V.

II. OPERATING PRINCIPLE

A. Topological Configuration

The proposed single-phase five-level dual-buck inverter is illustrated in Fig. 1. It is the combination of two legs of the three-level T-type dual-buck structure. There are two cells in each phase leg: positive and negative cell.

For leg 1 in Fig. 1, the positive cell consists of switches (S_{a1}, S_{a3}) and diodes (D_{a1}, D_{a3}). Where, S_{a1}, D_{a1}, and S_{a3} in series with D_{a3} are connected to P, N, and O, respectively. Then, the current can only flow from P, N, O to the common point A' of this cell. The reverse direction is inhibited. On the other hand, in the negative cell, switches D_{a2}, S_{a2}, and S_{a4} in series with D_{a4} are connected to P, N, O, respectively. Unlike the positive cell, the current in the negative cell can only flow from the common point A to the P, N, O points. The inductor L_a is placed between A' and A to limit the shoot-through current. The similar configuration is used for leg 2 in Fig. 1.

Hence, when the output current is positive, the positive cell of the leg 1 and the negative cell of the leg 2 conduct the current through L_a, whereas other cells are inactive as depicted in Fig. 2(a). Reversely, when the output current is negative, the negative cell of the leg 1 and the positive cell of the leg 2 are involved in the operation with L_b as depicted in Fig. 2(b).

B. Operating Principle

This section illustrates the operating principle of the proposed converter. There are six operation modes per one phase leg as described in Fig. 3.

1) *Mode 1a*: Switch S_{a1} is ON while other switches are OFF. Assume that the output current direction of the leg 1 is positive which is shown in Fig. 3(a). The current flows from the point P through switch S_{a1} and inductor L_a. In this mode, the voltage $v_{AN} = V_{dc} - v_{La}$; neglecting the inductor voltage v_{La}, the voltage $v_{AN} = V_{dc}$.

2) *Mode 1b*: With the same condition of mode 1a, but the direction of current is negative, which is shown in Fig. 3(b). The current flows through the external diode D_{a2} instead of switch S_{a1}. The voltage $v_{AN} = V_{dc}$.

3) *Mode 2a*: Switch S_{a3} is ON while other switches are OFF. Assume that the output current of the leg 1 is positive as depicted in Fig. 3(c). The current flows from the middle point O through switch S_{a3}, diode D_{a3}, and inductor L_a. The voltage $v_{AN} = V_{dc}/2$ in this mode.

4) *Mode 2b*: Similar to mode 2a, switch S_{a3} and S_{a4} are ON state while other switches are OFF, but the direction of current is negative, which is shown in Fig. 3(d). The diode D_{a3} is reversed bias and blocks the current through switch S_{a3}. Hence, the current flows through the external diode D_{a4}, switch S_{a4} and returns to the middle point O. The voltage $v_{AN} = V_{dc}/2$ in this mode.

5) *Mode 3a*: S_{a2} is ON while other switches are OFF.

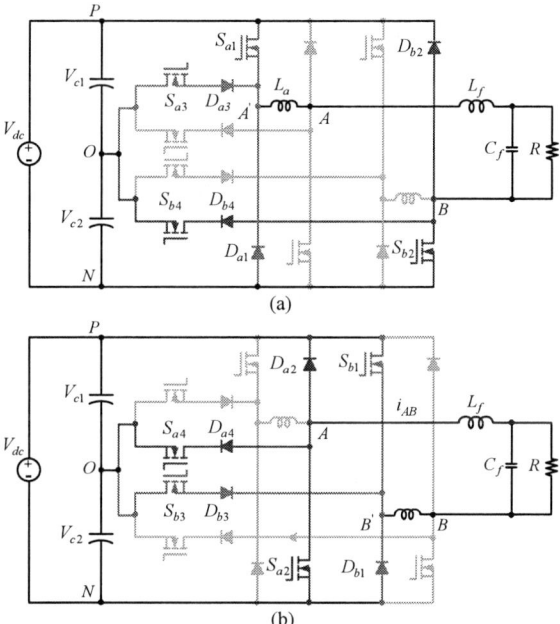

Fig. 2. Equivalent circuits of the proposed inverter when the output current is positive (a) and negative (b).

Assume that the current direction is positive as depicted in Fig. 3(e). The current flows from the point N through external diode D_{a3} and inductor L_a. In this mode, the voltage $v_{AN} = 0$.

6) *Mode 3b*: Similar to mode 3a, switch S_{a1} is ON while other switches are OFF, but the direction of current is negative, which is shown in Fig. 3(f). The diode D_{a3} is reversed bias. Hence, the current flows through the switch S_{a2} and returns to the point N. The voltage $v_{AN} = 0$ in this mode.

In the leg 1, the leg voltage v_{AN} equals to V_{dc} in mode (1a, 1b); $V_{dc}/2$ in mode (2a, 2b); and 0 in mode (3a, 3b). In the leg 2, similar to the analysis of the leg 1, the leg voltage v_{BN} also includes three voltage levels: $-V_{dc}$, $-V_{dc}/2$, and 0. The output voltage v_{AB} is calculated in the equation below:

$$v_{AB} = v_{AN} - v_{BN} \qquad (1)$$

In order to generate five-level voltage: $+V_{dc}$, $+V_{dc}/2$, 0, $-V_{dc}/2$ and $-V_{dc}$ at the output voltage v_{AB}, the six operating modes of the leg 1 and leg 2 are combined together. The combination modes between leg 1 and 2 produce 18 switching states which are shown in Table I for positive current ($i_{AB} > 0$) and in Table II for negative current ($i_{AB} < 0$). In case of positive current shown in Table I, the output current i_{AB} always flows from the A' point of the leg 1 to the leg 2. There are 9 switching states generating five voltage levels at the voltage $v_{A'B}$: $+V_{dc}$, $+V_{dc}/2$, 0, $-V_{dc}/2$, $-V_{dc}$. Similarly, in case of negative current shown in Table I, the output current i_{AB} always flows from the B' point of the leg 2 to the leg 1. In 18 switching states, there are some redundant states. Hence, the switching stated should be selected optimally.

Compared to the traditional T-type NPC inverter [4] and DBFBI in [13], although the number of passive components increases to 4 diodes and 1 inductor in each

1047

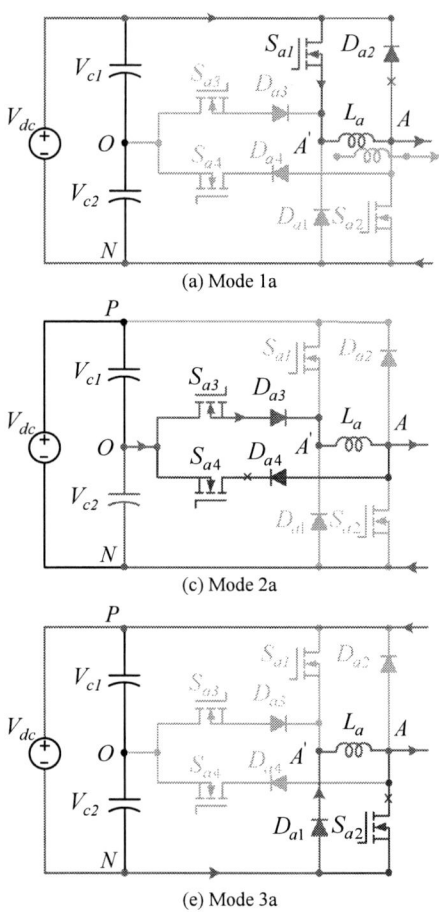
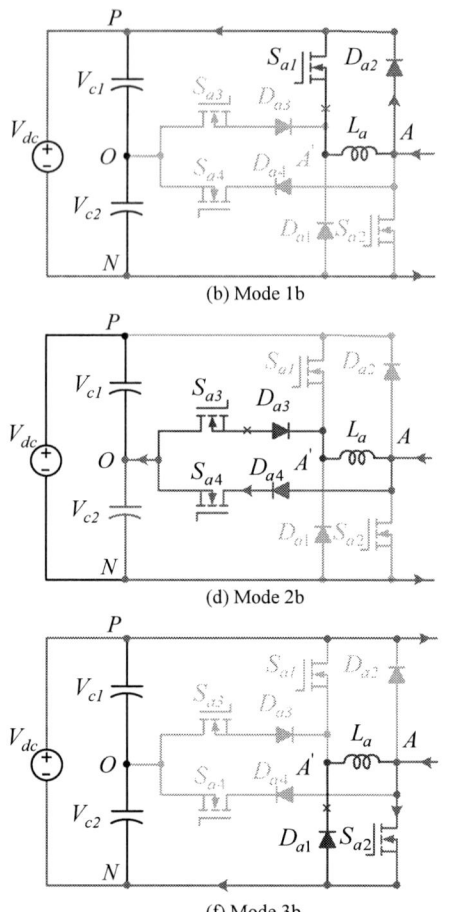

Fig. 3. Operation modes

phase leg, the power density of converter can be increased by combining two inductors of each phase leg into one magnetic core. The reliability of converter is also improved by adding small inductors L_a and L_b. When overlap time occurs between S_{a1} and S_{a2} due to the fault trigger caused by EMI issue, the shoot-through current is limited through inductor L_a. In addition, the volume of external filter inductor is reduced because L_a serves as filter inductor. The converter efficiency can be further improved by selecting power MOSFET and external diode separately. Fast reverse recovery diode is a good selection to optimize switching loss.

C. PWM Technique

The PWM gate signal is illustrated in Fig. 4. Two reference signals v_{AN} and v_{BN} are expressed in (2) and (3). The reference signal v_{BN} is shifted 180° in phase to the reference signal v_{AN}. The leg 1 and 2 employ the reference signals v_{AN} and v_{BN}, respectively. In the first half-cycle, switches S_{a1} and S_{b2} are gated complimentary of switches S_{a3} and S_{b4} respectively. Similarly, switches S_{a2} and S_{b1} are gated complimentary of switches S_{a4} and S_{b3} in the second half-cycle. Eight switching states are selected to generate PWM signals, states (1, 2, 3, 5) with $i_{AB} > 0$ and $i_{AB} < 0$.

$$\begin{cases} v_{AN} = m_a \sin(2\pi f_o t) & (i_{AB} > 0) \\ v_{AN} = m_a \sin(2\pi f_o t) + 1 & (i_{AB} < 0) \end{cases} \quad (2)$$

$$\begin{cases} v_{BN} = m_a \sin(2\pi f_o t + \pi) & (i_{AB} > 0) \\ v_{BN} = m_a \sin(2\pi f_o t + \pi) + 1 & (i_{AB} < 0) \end{cases} \quad (3)$$

Where m_a is modulation index, f_o is reference frequency (Hz). In this switching PWM scheme, four switches operate at a high frequency in each half-cycle while other switches are OFF.

TABLE I
SWITCHING STATES WITH POSITIVE CURRENT

State	Active switches		$v_{A'N}$	v_{BN}	$v_{A'B}$
	Leg 1	Leg 2			
1	S_{a1}	S_{b2}	V_{dc}	0	V_{dc}
2	S_{a1}	S_{b4}, D_{b4}	V_{dc}	$V_{dc}/2$	$V_{dc}/2$
3	S_{a3}, D_{a3}	S_{b2}	$V_{dc}/2$	0	$V_{dc}/2$
4	D_{a1}	S_{b2}	0	0	0
5	S_{a3}, D_{a3}	S_{b4}, D_{b4}	$V_{dc}/2$	$V_{dc}/2$	0
6	S_{a1}	D_{b2}	V_{dc}	V_{dc}	0
7	S_{a3}, D_{a3}	D_{b2}	$V_{dc}/2$	V_{dc}	$-V_{dc}/2$
8	D_{a1}	S_{b4}, D_{b4}	0	$V_{dc}/2$	$-V_{dc}/2$
9	D_{a1}	D_{b2}	0	V_{dc}	$-V_{dc}$

1048

TABLE II
SWITCHING STATES WITH NEGATIVE CURRENT

State	Active switches		v_{AN}	$v_{B'N}$	$v_{AB'}$
	Leg 1	Leg 2			
1	S_{b1}	S_{a2}	V_{dc}	0	V_{dc}
2	S_{b1}	S_{a4}, D_{a4}	V_{dc}	$V_{dc}/2$	$V_{dc}/2$
3	S_{b3}, D_{b3}	S_{a2}	$V_{dc}/2$	0	$V_{dc}/2$
4	D_{b1}	S_{a2}	0	0	0
5	S_{b3}, D_{b3}	S_{a4}, D_{a4}	$V_{dc}/2$	$V_{dc}/2$	0
6	S_{b1}	D_{a2}	V_{dc}	V_{dc}	0
7	S_{b3}, D_{b3}	D_{a2}	$V_{dc}/2$	V_{dc}	$-V_{dc}/2$
8	D_{b1}	S_{a4}, D_{a4}	0	$V_{dc}/2$	$-V_{dc}/2$
9	D_{b1}	D_{a2}	0	V_{dc}	$-V_{dc}$

TABLE III
PARAMETERS

Parameters	Value
Input voltage	100 V
Reference frequency	60 Hz
Switching frequency	20 kHz
Inductors La, Lb	100 μH
Filter inductor Lf	0.76 mH
Filter capacitor Cf	20 μF
Resistance R	50 Ω

III. SIMULATION RESULTS

This section shows simulation results of the proposed converter with parameters presented in Table III. The simulated waveforms are shown in Fig. 5. As can be seen, the inductor L_a and L_b conduct current in half positive and negative cycle respectively. The output voltage v_{AB} changes in step-level between 0 and ±50 V, ±50V and ±100 V. One of the drawbacks of this PWM technique is current zero-crossing distortion. In case of resistive load, the unit power factor is achieved. Therefore, current zero-crossing distortion issue does not happen. However, it will occur when the nonlinear load is applied. In order to eliminate current zero-crossing distortion, one of the methods is both switches S_{a1} and S_{a4} gating ON/OFF simultaneously near zero-crossing period, which is mentioned in [9].

In this structure, two inductors served as filter inductor, the number of inductors is reduced by half that increases the power density of the converter. Although the inductance value is still the same, this does not affect the converter significantly.

IV. EXPERIMENTAL RESULTS

The prototype of five-level dual-buck full-bridge is shown in Fig. 6 and the experimental parameters are listed in Table III. The experiment results are shown in Fig. 7. Output voltage v_{AB} and current i_{AB} waveforms are shown in Fig. 7(a). As can be seen, the output voltage v_{AB} has five voltage levels, which changes voltage level between $+V_{dc}$ and $+V_{dc}/2$; $V_{dc}/2$ and 0; 0 and $-V_{dc}/$; $-V_{dc}/2$ and $-V_{dc}$. The current waveforms of two inductors are presented in Fig. 7(b). The sum of two inductor currents is equal to the output current i_{AB}. A resistive load is used in the experiment, so the unit power factor is achieved. The load voltage waveform v_R is shown in Fig. 7(c), which is measured after the LC filter.

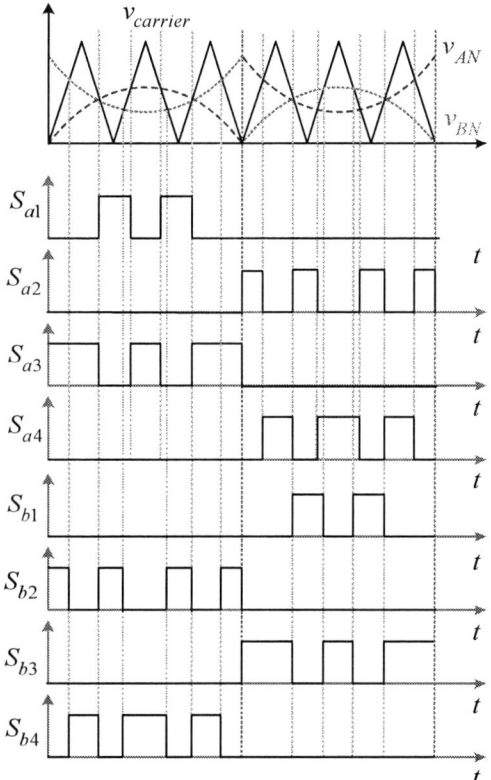

Fig. 4. Gate signal generation

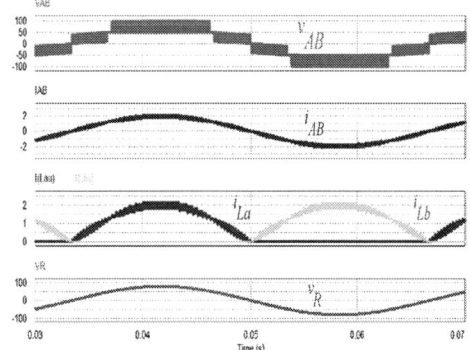

Fig. 5. Simulation results.

Fig. 6. Prototype picture.

1049

V. CONCLUSION

This paper derives the topology of the five-level dual-buck full-bridge inverter based on two T-type dual-buck inverters with only one inductor on each phase leg. All the switching states of the proposed inverter are presented. The eight optimal switching states are selected to control the converter efficiency. Based on the current direction, the PWM method is also introduced and analyzed. Four switches employ at high frequency in each half-cycle. The volume of the proposed inverter is also decreased by using only two in each phase leg, compared to 4 inductors in traditional one. Simulation and experimental results verify the analysis of proposed T-type dual-buck full-bridge inverter.

ACKNOWLEDGMENT

This work was supported by the Korea Institute of Energy Technology Evaluation and Planning (KETEP) and the Ministry of Trade, Industry & Energy (MOTIE) of the Republic of Korea (No. 20174030201490) and was also supported by Basic Science Research Program through the National Research Foundation of Korea (NRF) funded by the Ministry of Education (NRF-2016R1D1A1B03934577).

REFERENCES

[1] Rodriguez, J-S. Lai, and F. Peng, "Multilevel inverters: A survey of topologies, control, and applications," *IEEE Trans. Ind. Electron.*, vol. 49, no. 4, pp. 724-738, Aug. 2002.

[2] J. Rodriguez, S. Bernet, P. K. Steimer, and I. E. Lizama, "A survey on neutral point clamped inverters," *IEEE Trans. Ind. Electron.*, vol. 57, no. 7, pp. 2219-2230, Jul. 2010.

[3] S. J. Park, F. S. Kang, M. H. Lee, and C. U. Kim, "A new single-phase five-level PWM inverter employing a deadbeat control scheme," *IEEE Trans. Power Electron.*, vol. 18, no. 3, pp. 831-843, May. 2003.

[4] M. Schweizer and J. W. Kolar, "Design and implementation of a highly efficient three-level T-type converter for low-voltage applications," *IEEE Trans. Power Electron.*, vol. 28, no. 2, pp. 899–907, Feb. 2013.

[5] M. Schweizer, T. Friedly, and J. W. Kolar, "Comparative evaluation of advanced three-phase three-level inverter/converter topologies against two-level systems," *IEEE Trans. Ind. Electron.*, vol. 60, no. 12, pp. 5515– 5527, Dec. 2013.

[6] P. W. Sun, C. Liu, J.-S. Lai, and C.-L. Chen, "Cascade dual buck inverter with phase-shift control," *IEEE Trans. Power Electron.*, vol. 27, no. 4, pp. 2067-2077, Apr. 2012.

[7] Z. Yao, L. Xiao, and Y. Yan, "Dual-buck full bridge inverter with hysteresis current control," *IEEE Trans. Ind. Electron.*, vol. 56, no. 8, pp. 3153–3160, Aug. 2009.

[8] P. Sun, C. Liu, J-S. Lai, C-L. Chen, and N. Kees, "Three-phase dual buck inverter with unified pulse width modulation," *IEEE Trans. Power Electron.*, vol. 27, no. 3, pp. 1159–1167, Mar. 2012.

[9] P. Sun, C. Liu, J. S. Lai and C. L. Chen, "Cascade Dual Buck Inverter with Phase-Shift Control," *IEEE Trans. Power Electron.*, vol. 27, no. 4, pp. 2067-2077, April 2012.

[10] A. A. Khan and H. Cha, "Dual-Buck-Structured High-Reliability and High-Efficiency Single-Stage Buck–Boost Inverters," *IEEE Trans. Ind. Electron.*, vol. 65, no. 4, pp. 3176-3187, April 2018.

[11] A. A. Khan, H. Cha, H. F. Ahmed, J. Kim and J. Cho, "A Highly Reliable and High-Efficiency Quasi Single-Stage Buck–Boost Inverter," *IEEE Trans. Power Electron.*, vol. 32, no. 6, pp. 4185-4198, June 2017.

[12] B. Gu, J. Dominic, J. S. Lai, C. L. Chen, T. LaBella and B. Chen, "High Reliability and Efficiency Single-Phase Transformerless Inverter for Grid-Connected Photovoltaic Systems," *IEEE Trans. Power Electron.*, vol. 28, no. 5, pp. 2235-2245, May 2013.

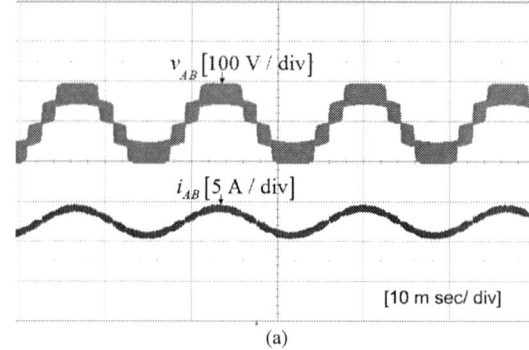

(a)

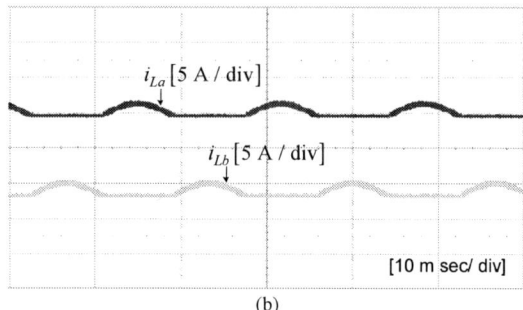

(b)

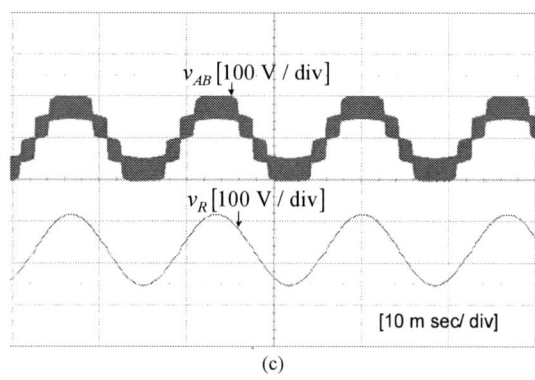

(c)

Fig. 7. Experimental results. (a) Output voltage v_{AB} and current i_{AB}. (b) Currents on L_a and L_b. (c) Output voltage v_{AB} and load voltage v_R.

[13] L. Zhang, K. Sun, Y. Xing, and J. Zhao, "A family of five-level dual-buck full-bridge inverters for grid-tied applications," *IEEE Trans. Power Electron.*, vol. 31, no. 10, pp. 7029–7042, Oct. 2016.

[14] Y. S. Lee and G. T. Cheng, "Quasi-Resonant Zero-Current-Switching Bidirectional Converter for Battery Equalization Applications," *IEEE Trans. Power Electron.*, vol. 21, no. 5, pp. 1213-1224, Sept. 2006.

The 2018 International Power Electronics Conference

Control of Direct AC/AC Modular Multilevel Converter in Railway Power Supply System

Shuguang SONG, Jinjun LIU, Shaodi OUYANG, Xingxing CHEN and Baojin LIU
School of Electrical Engineering, Xi'an Jiaotong University, Xi'an, China

Abstract— **In the railway power supply system, the direct three-to-single phase AC/AC Modular Multilevel Converter (MMC) has received lots of attention due to its salient features compared to the traditional traction transformers. Recent researches mainly focus on the single-phase side 16.7Hz situation, and the single-phase side 50Hz case has not been detailed discussed. In this paper, the authors firstly establish the steady-state model of MMC, and derive the expression of capacitor voltage variation in Submodules (SMs). The severe voltage imbalance between upper- and lower- arms is observed in the analysis when MMC operates at single-phase side 50Hz condition. To eliminate this threat and ensure safe operation of MMC, the high-frequency voltage and current injection method is applied, which effectively balance the voltage between upper- and lower-arms. The overall control and different levels of capacitor voltage control are presented in details. Simulation result in PSCAD/EMTDC software environment verifies the effectiveness of arm voltage balance method.**

Keywords— *AC/AC, Modular Multilevel Converter, Steady-state analysis, Voltage Regualtion.*

I. INTRODUCTION

In recently years, the high-speed train has achieved a fast development from all over the world. The high-speed train is single-phase load. Therefore, the corresponding power supply system needs to realize a three-to-single phase AC/AC power conversion since the utility grid is a three-phase system. Typically, balanced or unbalanced traction transformers including Scott transformers, Woodbridge transformers, three-phase V/V transformers, impedance-matching balance transformers, etc., are used [1]. The balanced transformers can eliminated the negative-sequence current when powers of two feeders are the same. However, the load condition in two feeders varies in real time, the power is rare to be balanced. Besides, the negative-sequence current in the unbalance transformers always exists. Nevertheless, duo to its simple structure, traction transformers are still widely used. In a word, the existing transformer solution will lead negative-sequence current to the utility grid.

The Modular Multilevel Converter (MMC) has gained lots of attention in middle/high power applications due to its salient features [2-6]. When using full-bridge cell as the Submodule (SM), the MMC can achieve direct AC/AC power conversion. Several papers have discussed the application of MMC as a direct AC/AC converter, 50Hz to 16.7Hz - one common traction supply standard in Europe

This work was supported by the State Key Laboratory of Electrical Insulation and Power Equipment under Grant EIPE14112.

[7-11]. Compared with the traction transformers, there will be no negative-sequence current in the utility grid. And this system can even operate under asymmetric grid conditions [12].

Except for the single-phase side 16.7Hz power supply system, the single-phase side 50Hz power supply system is implemented in China and Japan [13-14]. However, direct AC/AC power conversion with equal frequency may cause large voltage variation in MMC SMs, which will threaten the operation of whole system [15], [16]. In [17], the authors analyzed the arm energy under equal frequency situation and proposed a sept-branch modular multilevel converter. However, the behavior of SM capacitor voltage still need to be further discussed, and the cost of proposed topology is higher than typical MMC system. And so far, few other literatures discuss the MMC based direct three-to-single phase AC/AC power conversion under equal frequency situation.

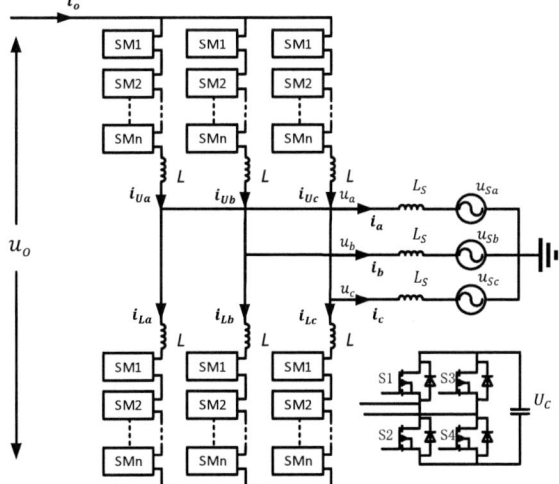

Fig. 1. Circuit configuration of direct AC/AC MMC and its submodule structure.

In this paper, the authors firstly establish the steady-state model of direct AC/AC MMC system. The expression of SM capacitor voltage is derived. Based on the analysis, voltage imbalance can be observed under equal frequency situation which coincides with the existing literatures. Then, the high-frequency voltage and current injection method is adopted to mitigate the voltage imbalance. Detailed voltage regulation control structure is presented. The rest of this paper is organized as follows:

1051

Section II describes the circuit configuration and the steady-state analysis of MMC and gives the expression of SM capacitor voltage. Section III presents the whole control structure which realizes power balance between upper- and lower- arms. Section IV shows simulation results in the PSCAD/EMTDC software environment, leading to Section V, which concludes the work.

II. CIRCUIT CONFIGURATION AND STEADY-STATE ANALYSIS

A. MMC Topology

The direct three-to-single phase AC/AC MMC topology is illustrated in Fig. 1. The topology resembles the existing MMC based DC/AC topology. The main differences are: 1) half-bridge SMs are replaced by full-bridge cell, shown in Fig. 1; 2) DC sides outputs single-phase AC voltage instead of a constant DC voltage.

B. Steady-state Analysis

Based on the MMC working principles, the mathematical expressions of MMC three-phase side output voltages and currents can be written as:

$$\begin{cases} u_a = U_a \sin(\omega t + \alpha) \\ u_b = U_b \sin\left(\omega t + \alpha - \frac{2\pi}{3}\right) \\ u_c = U_c \sin\left(\omega t + \alpha + \frac{2\pi}{3}\right) \end{cases} \tag{1}$$

$$\begin{cases} i_a = I_a \sin(\omega t + \varphi) \\ i_b = I_b \sin(\omega t + \varphi - \frac{2\pi}{3}) \\ i_c = I_c \sin(\omega t + \varphi + \frac{2\pi}{3}) \end{cases} \tag{2}$$

Similarly, the single-phase side variables can also be defined as:

$$\begin{cases} u_o = U_o \sin(\Omega t + \alpha_o) \\ i_o = I_o \sin(\Omega t + \varphi_o) \end{cases} \tag{3}$$

where Ω is the angular frequency of single-phase side.

Due to the symmetric structure of MMC, only the analysis of phase a is given below. The currents in the arms are composed of both single-phase side and three-phase side components if ignoring the harmonic circulating currents:

$$\begin{cases} i_{Ua} = \frac{1}{2}i_a + \frac{1}{3}i_o \\ i_{La} = -\frac{1}{2}i_a + \frac{1}{3}i_o \end{cases} \tag{4}$$

The voltage references for the arms can also be easily obtained:

$$\begin{cases} U_{Ua\_ref} = \frac{1}{2}u_o - u_a \\ U_{La\_ref} = \frac{1}{2}u_o + u_a \end{cases} \tag{5}$$

This will directly lead to the corresponding switching functions:

$$\begin{cases} S_{Ua} = \frac{U_{Ua\_ref}}{NU_c} = \frac{1}{2}M_o \sin(\Omega t + \alpha_o) - \frac{1}{2}M_a \sin(\omega t + \alpha) \\ S_{La} = \frac{U_{La\_ref}}{NU_c} = \frac{1}{2}M_o \sin(\Omega t + \alpha_o) + \frac{1}{2}M_a \sin(\omega t + \alpha) \end{cases} \tag{6}$$

The sum and difference of the arm capacitor voltages between upper and lower arms can be calculated by:

$$\Sigma = \frac{N}{C}\int (S_{Ua} \cdot i_{Ua} + S_{La} \cdot i_{La}) \, dt \tag{7}$$

$$\Delta = \frac{N}{C}\int (S_{Ua} \cdot i_{Ua} - S_{La} \cdot i_{La}) \, dt \tag{8}$$

Substituting (1) (2) and (3) into (7) (8), respectively, leads to (9) and (10).

Equation (9) shows the expression of the total capacitor voltage in phase a, and Equation (10) shows the voltage difference between upper- and lower- arms in phase a. The frequency of three-phase side ω and the frequency of single-phase side Ω will both have impact on them.

If the single-phase side frequency is equal to the three-phase side one, i.e., $\Omega = \omega$, (9) and (10) can be further simplified as (11) and (12). After the simplification, Terms that grow with time are observed in both (11) and (12).

$$\left[\frac{1}{6}M_o I_o \cos(\alpha_o - \varphi_o) - \frac{1}{4}M_a I_a \cos(\alpha - \varphi)\right]t \tag{13}$$

$$\left[\frac{1}{4}M_o I_a \cos(\alpha_o + \varphi) - \frac{1}{6}M_a I_o \cos(\alpha - \varphi_o)\right]t \tag{14}$$

The power balance between single-phase and three-phase side is presented by (13). In the normal operation, the power should always be balanced required by the energy conservation law, and this terms will be zero. Otherwise, the total energy stored in MMC system will not remain stable. Note that (13) exists regardless of the frequency.

While (14) indicates the interaction between voltages and currents of single-phase and three-phase sides. Different from (13), this term exists only under equal frequency situation. If no control is applied, it can hardly be zero, which in other words, causes the capacitor voltages keeping increasing in one arm and decreasing in the other. In this situation, the MMC system will fail.

$$\Sigma = \frac{N}{C}\int \frac{1}{4}M_a I_a \cos(2\omega t + \alpha + \varphi) - \frac{1}{4}M_a I_a \cos(\alpha - \varphi) - \frac{1}{6}M_o I_o \cos(2\Omega t + \alpha_o + \varphi_o) + \frac{1}{6}M_o I_o \cos(\alpha_o - \varphi_o) \, dt \tag{9}$$

$$\Delta = \frac{N}{C}\int -\frac{1}{4}M_o I_a \cos((\omega + \Omega)t + \alpha_o + \varphi) + \frac{1}{4}M_o I_a \cos((\omega - \Omega)t - \alpha_o + \varphi) + \frac{1}{6}M_a I_o \cos((\omega + \Omega)t + \alpha + \varphi_o) - \frac{1}{6}M_a I_o \cos((\omega - \Omega)t + \alpha - \varphi_o) \, dt \tag{10}$$

$$\Sigma = \frac{N}{C}\left\{\left[\frac{1}{6}M_o I_o \cos(\alpha_o - \varphi_o) - \frac{1}{4}M_a I_a \cos(\alpha - \varphi)\right]t + \frac{1}{8\omega}M_a I_a \sin(2\omega t + \alpha + \varphi) - \frac{1}{12\Omega}M_o I_o \sin(2\omega t + \alpha_o + \varphi_o) + 2CU_c\right\} \tag{11}$$

$$\Delta = \frac{N}{C}\left\{\left[\frac{1}{4}M_o I_a \cos(\alpha_o + \varphi) - \frac{1}{6}M_a I_o \cos(\alpha - \varphi_o)\right]t - \frac{1}{8\omega}M_o I_a \sin(2\omega t + \alpha_o + \varphi) + \frac{1}{12\omega}M_a I_o \sin(2\omega t + \alpha + \varphi_o)\right\} \tag{12}$$

III. CONTROL OF DIRECT AC/AC MMC

From the discussion in section II, when MMC operates at equal frequency condition, severe voltage imbalance between upper- and lower- arms occurs. This voltage imbalance is similar to the startup/low-speed operation of MMC based motor drive and the operation of MMC based DC-DC converter.

In the existing researches, the high-frequency voltage and current injection method has been proposed, which is capable of effectively balancing upper- and lower- arm voltages. The idea is that interaction of high-frequency components generates extra power flow between upper- and lower- arms, and with proper selection, this power will fully compensate the imbalance power. Detailed derivation can be found in [18-21], and will not be presented in this paper.

In spite of different cause for arm voltage imbalance, in the control of direct three-to-single phase AC/AC MMC, the high-frequency components injection method can also be applied. The overall control system is shown below.

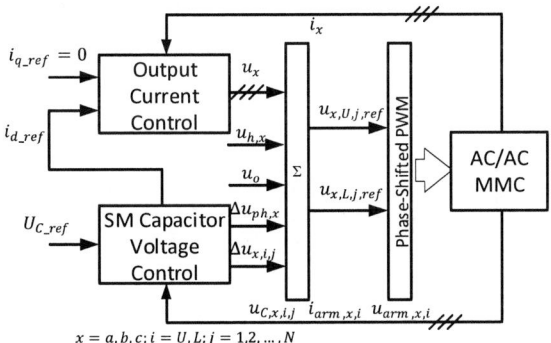

$$x = a, b, c; i = U, L; j = 1,2,...,N$$

Fig. 2. Overall control diagram of direct AC/AC MMC.

A. Overall Control

Fig. 2 shows the overall control diagram of direct three-to-single phase AC/AC MMC. In this paper, the single-phase side of MMC operates as a voltage source, providing stable 50Hz AC voltage u_o for the loads. The control of three-phase side is based on DQ rotating frame. The d-axis current reference i_{d_ref} is used to maintain the total capacitor voltage to be stable, while q-axis current reference i_{q_ref} is set to be zero for unity power factor operation. The details of the DQ frame output current control can be found in [22]. In addition, a high-frequency voltage component $u_{h,x}$ is added in the control signal to achieve arm voltage balance control.

$$u_{h,x} = U_h \sin(\omega_h t) \tag{15}$$

B. SM Capacitor Voltage Control

The SM capacitor voltage control structure consists of four levels: the total capacitor voltage control, the phase capacitor voltage control, arm capacitor voltage balance control and the individual capacitor voltage control.

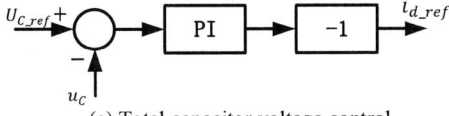

(a) Total capacitor voltage control

Fig. 3(a) shows the control block of total capacitor voltage control. Based on the SM capacitor voltage reference U_{C_ref} and the measured total average capacitor voltage u_C, the PI controller produces the d-axis current reference. This current reference will control the active power flow in and out the MMC to achieve power balance. The u_C can be calculated by

$$u_C = \frac{1}{6N} \sum_{x=a,b,c} \sum_{i=U,L} \sum_{j=1...N} u_{C,x,i,j} \tag{16}$$

The phase capacitor voltage control is presented in Fig. 3(b). The difference between measured total average capacitor voltage u_C and measured phase average voltage $u_{C,x}$, passes through a PI controller, generates the DC circulating current reference $i_{d,x,ref}$.

$$u_{C,x} = \frac{1}{2N} \sum_{i=U,L} \sum_{j=1...N} u_{C,x,i,j} \tag{17}$$

Fig. 3(c) draws the block of arm capacitor voltage balance control. It is based on the measured average capacitor voltage in upper- and lower- arms, i.e., $u_{C,x,U}$ and $u_{C,x,L}$. The difference of arm voltages times the united term of $\sin(\omega_h t)$ forming the final instantaneous high-frequency current reference $i_{h,x,ref}$. The high-frequency current interacts with the high-frequency voltage $u_{h,x}$, generates power flow, compensating term (14) in real time. The measured average arm capacitor voltage are calculated by

$$u_{C,x,U} = \frac{1}{N} \sum_{j=1...N} u_{C,x,U,j} \tag{18}$$

$$u_{C,x,L} = \frac{1}{N} \sum_{j=1...N} u_{C,x,L,j} \tag{19}$$

Fig. 3(d) shows the inner current loop which forces the

circulating current follows its reference value where the current reference is the sum of DC circulating current reference $i_{d,x,ref}$ and high-frequency circulating current reference $i_{h,x,ref}$.

Finally, Fig. 3(e) presents the individual capacitor voltage control. The reference signal $\Delta u_{x,i,j}$ is formed by the measured average capacitor voltage in one arm and the individual SM capacitor voltage $u_{C,x,i,j}$.

C. Modulation Signal Synthesis

Based on the output voltage components u_o, u_x and voltage control components u_h, $\Delta u_{ph,x}$, $\Delta u_{x,i,j}$, the final voltage reference for the SMs can be synthesized by:

$$\begin{cases} u_{x,U,j,ref} = \dfrac{u_o}{2N} - \dfrac{u_x + u_h}{N} + \Delta u_{ph,x} + \Delta u_{x,i,j} \\ u_{x,L,j,ref} = \dfrac{u_o}{2N} + \dfrac{u_x + u_h}{N} + \Delta u_{ph,x} + \Delta u_{x,i,j} \end{cases} \quad (20)$$

IV. SIMULAITON VERIFICATION

To verify the control method, a direct three-to-single phase AC/AC MMC model is built in PSCAD/EMTDC software environment.

The three-phase side MMC connects to the 50Hz utility grid, and the single-phase side provides AC voltage. The traction load here is emulated by the series RL loads.

The simulation results of single-phase side 16.7Hz situation is firstly presented in Fig. 4. In this part, no high-frequency components injection control is applied. The voltages between upper- and lower- arms still remain balanced. When changing the single-phase side frequency to 50Hz, from Fig. 5, it is obvious that the severe voltage imbalance appears. The whole system fails, which confirms the analysis in Section II.

In Fig. 6, the high-frequency components injection control is applied. The voltage of both upper- and lower-arm SMs remains their reference value 1.2kV. To test the control system, at t=2.00s, disable the high-frequency components injection control. Immediate voltage deviation occurs. The deviation shown in Fig.6 grows with time with constant slop. This result verifies the validity of the capacitor voltage analysis. At t=2.05s, enable the voltage control. The SM voltages return to the reference quickly, verifying the effectiveness of the control method.

Fig. 7 shows the corresponding MMC output currents. Without high-frequency components injection control, during 2.00 to 2.05s, large current distortion appears.

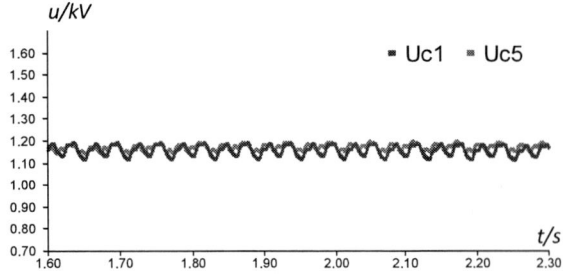

Fig. 4. Capacitor voltages at single-phase side 16.7Hz.

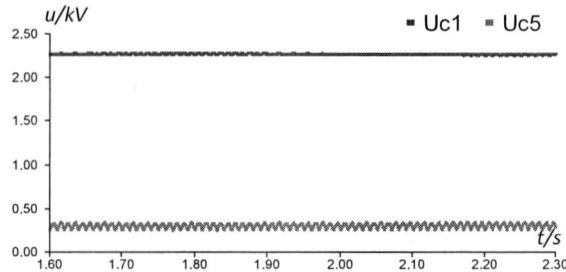

Fig. 5. Capacitor voltages at single-phase side 50Hz without control.

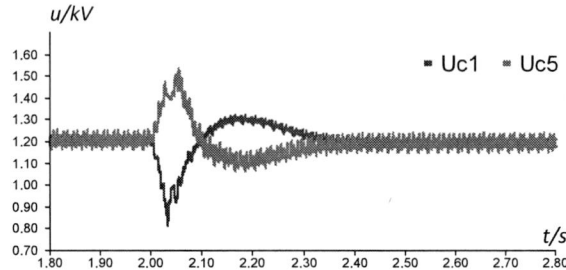

Fig. 6. Capacitor voltages at single-phase side 50Hz.

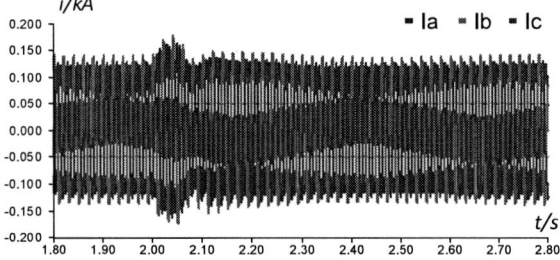

Fig. 7. Output currents at single-phase side 50Hz.

V. CONCLUSIONS

In this paper, control of direct three-to-single phase AC/AC MMC has been studied. Base on the steady-state analysis, severe voltage imbalance between upper- and lower- arm has been found in single-phase side 50Hz operation which threatens the stability of MMC system. The high-frequency components injection method is adopted in this paper, and the detailed voltage balance control structure is presented. Based on the proposed method, the capacitor voltages are well regulated which is verified by the simulation results.

REFERENCES

[1] Z.W. Zhang, B.Wu, J. S. Kang, and L. F. Luo, "A multi-purpose balanced transformer for railway traction applications," *IEEE Trans. Power Del.*, vol. 24, no. 2, pp. 711–718, Apr. 2009.

[2] N. Flourentzou, V. G. Agelidis, and G. D. Demetriades, "VSC-based HVDC power transmission systems: An overview," *IEEE Trans. Power Electron.*, vol. 24, no. 3, pp. 592–602, Mar. 2009.

[3] S. Debnath, J. Qin, B. Bahrani, M. Saeedifard, and P. Barbosa, "Operation, control, applications of the modular multilevel converter: A review," *IEEE Trans. Power Electron.*, vol. 30, no. 1, pp. 37–53, Jan. 2015.

[4] M. Saeedifard and R. Iravani, "Dynamic performance of a modular multilevel back-to-back HVDC system," *IEEE Trans. Power Del.*, vol. 25, no. 4, pp. 2903–2912, Oct. 2010.

[5] J. Peralta, H. Saad, S. Dennetiere, J. Mahseredjian, and S. Nguefeu, "Detailed and averaged models for a 401-level MMC–HVDC system," *IEEE Trans. Power Del.*, vol. 27, no. 3, pp. 1501–1508, Jul. 2012.

[6] H. Akagi, "Classification, terminology, and application of the modular multilevel cascade converter (MMCC)," *IEEE Trans. Power Electron.*, vol. 26, no. 11, pp. 3119–3130, Nov. 2011.

[7] M. Winkelnkemper, A. Korn, and P. Steimer, "A modular direct converter for transformerless rail interties," in *Proc. IEEE Int. Symp. Ind. Electron.*, pp. 562-567, July 4-7, 2010.

[8] T. Schrader, C. Heising, V. Staudt, and A. Steimel, "Multivariable control of MMC-based static converters for railway applications," in *Proc. Elect. Syst. Aircraft, Railway, Ship Propulsion (ESARS)*, pp. 1-6, Oct. 16-18, 2012.

[9] L. Ängquist, A. Haider, H.-P. Nee, and H. Jiang, "Open-loop approach to control a Modular Multilevel Frequency Converter," in *Proc. 14th Eur. Conf. Power Electron. Appl. (EPE)*, pp. 1-10, Aug. 30-Sept. 1, 2011.

[10] N. Thitichaiworakorn, M. Hagiwara, and H. Akagi, "A single-phase to three-phase direct modular multilevel cascade converter based on double-star bridge-cells (MMCC-DSBC)," in *Proc. 1st Int. Future Energy Electron. Conf.*, pp. 476-481, Nov. 3-6, 2013.

[11] L. Bessegato, L. Harnefors, K. Ilves, S. Norrga, and S. Östlund, "Control of Direct ACAC Modular Multilevel Converters Using Capacitor Voltage Estimation," in *Proc. EPE'16 ECCE Europe*, pp. 1-9, sept. 2016.

[12] M. Vasiladiotis, N. Cherix, and A. Rufer, "Operation and Control of Single-to-Three-Phase Direct AC/AC Modular Multilevel Converters under Asymmetric Grid Conditions," in *Proc. 9th Int. Conf. Power Electron. (ICPE)*, pp. 1-6, June 1-5, 2015.

[13] H. Morimoto, M. Ando, Y. Mochinaga, T. Kato, J. Yoshizawa, T. Gomi, T. Miyashita, S. Funahashi, M. Nishitoba, and S. Oozeki, "Development of railway static

power conditioner used at substation for Shinkansen," in *Proc. Power Conver. Conf.*, vol. 3, pp. 1108–1111, 2002.

[14] Z. He, H. Hu, Y. Zhang, and S. Gao, "Harmonic resonance assessment to traction power-supply system considering train model in china highspeed railway," *IEEE Trans. Power Del.*, vol. 29, no. 4, pp. 1735–1743, Aug. 2014.

[15] A. J. Korn, M. Winkelnkemper, P. Steimer and J. W. Kolar, "Direct modular multi-level converter for gearless low-speed drives," *Proceedings of the 2011 14th European Conference on Power Electronics and Applications*, Birmingham, 2011, pp. 1-7.

[16] K. Ilves, L. Bessegato and S. Norrga, "Comparison of cascaded multilevel converter topologies for AC/AC conversion," *International Power Electronics Conference (IPEC-Hiroshima 2014 - ECCE ASIA)*, Hiroshima, 2014, pp. 1087-1094.

[17] Ming. Lei, Y. Li, Z. Li, F. Xu, P. Wang and C. Zhao, "A Sept-Branch Modular Multilevel Converter for Three-Phase to Single-Phase Direct AC/AC Equal Frequency Conversion," in *Proc. IECON 2017*, pp. 2239-2244.

[18] A. Antonopoulos, L. Ängquist, S. Norrga, K. Ilves, L. Harnefors, and H. Nee, "Modular multilevel converter ac motor drives with constant torque from zero to nominal speed," *IEEE Trans. Ind. Appl.*, vol. 50, no. 3, pp. 1982–1993, May/Jun. 2014.

[19] M. Hagiwara, I. Hasegawa, and H. Akagi, "Start-up and low-speed operation of an electric motor driven by a modular multilevel cascade inverter," *IEEE Trans. Ind. Appl.*, vol. 49, no. 4, pp. 1556–1565, Jul./Aug. 2013.

[20] S. Debnath, J. Qin, and M. Saeedifard, "Control and stability analysis of modular multilevel converter under low-frequency operation," *IEEE Trans. Ind. Electron.*, vol. 62, no. 9, pp. 5329–5339, Sep. 2015.

[21] B. Li et al., "An improved circulating current injection method for modular multilevel converters in variable-speed drives," *IEEE Trans. Ind. Electron.*, vol. 63, no. 11, pp. 7215–7225, Nov. 2016.

[22] B. Wu. "High-Power Converters and AC Drives," New York/Piscataway: Wiley-IEEE Press, 2016.

Gap in pagination due to withheld paper.

Pages 1056-1061

Wireless Power Transfer: Critical Review of Related Standards

Mohamad Abou Houran[1], Xu Yang[*1], Wenjie Chen[1] and Mehdi Samizadeh[1]

1 School of Electrical Engineering, Xi'an Jiaotong University, Xi'an, China

[*]E-mail: yangxu@mail.xjtu.edu.cn

Abstract— **Wireless power transfer system (WPT) turned to be a reliable and an appropriate strategy, which found many applications. However, there is a growing concern about two safety issues. The first one is linked to the human exposure to electromagnetic fields (EMFs) that may cause potential health hazards. The second issue is the electromagnetic compatibility (EMC), where WPT systems may cause disturbance to electrical circuits. With the rapid development of WPT systems including the wide range of frequency and power, many leading countries in WPT industry have set many standards and regulations for safety usage of WPT. However, due to the complexity and variety of these standards, it is usually difficult to follow for both researchers and electrical engineers. This paper aims to provide a review for the recent standards and regulations concerning WPT industry and its applications. Therefore, a simple classification with a comprehensive comparison is presented.**

Keywords— *Electromagnetic compatibility, electromagnetic fields, EMI standards, wireless power transfer.*

I. INTRODUCTION

WIRELESS power transfer system (WPT) is a promising technology, which found its way as a successful strategy in many applications such as electric vehicles (EVs) [1], plug in hybrid electric vehicles (PHEVs) [2], [3], implantable medical devices (IMD) [4],[5], and home appliances [5]. The wireless power transfer system can be essentially divided into two categories. The first one is the far-field WPT. The second category is the near-field WPT or non-radiated, which classified into three sections. 1) Inductively coupled power transfer (ICPT) [7], [8]. 2) Capacitive power transfer (CPT) [9]. 3) Magnetically coupled resonances MCR WPT [10], [11]. ICPT is based on the law of electromagnetic induction between two coils, whereas CPT system is mostly designed for low-power applications. On the other hand, the MCR WPT used for a mid-range distance, which provides higher efficiency compared to MCI WPT in a longer transmission distance.

With the rapid development of WPT industry, WPT systems became widely used technology in many applications. Nowadays, there is a big concern that the generated electromagnetic fields (EMFs) by WPT systems cause a potential danger to the safety of consumers [12]. It has been reported that the EMF generated by WPT systems will stimulate current and heat in the body [13] that in turn cause nerve, tissue and muscle stimulation, change in the central nervous system as well [12]. On the other hand, since MCR WPT has a large air gap, especially in electric vehicle applications, as the high power being transferred to the vehicle wirelessly [14], it is unavoidable that strong EMF is generated. Consequently, the disturbance (EMI), which created by WPT system will affect other electrical circuits [15].

Based on the obvious EMI risks of using WPT technology in different applications, it is necessary to regulate the usage of WPT systems in order to mitigate the EMI disturbance. Therefore, several recommendations and standards were issued to protect against EMI and exposure to EMF. At the beginning of WPT industry, there were no specialized WPT standards, but ICNIRP and IEEE standards were applicable, such as IEEE standard C95.1-1991. After that, it was revised to IEEE standard C95.1™ in 2005. Moreover, safety levels with respect to the human exposure to RF EMFs in the range of 3 kHz-300 GHz were introduced [16], [17]. On the other hand, ICNIRP established guidelines for the protection of human body exposed to electric and magnetic fields in the low-frequency range of the EM spectrum [18]. In the last decade, with the growth of WPT technology, the industry group Qi or WPC (Wireless Power Consortium) introduced a wireless charging standard in 2008. In 2012, A4WP standards for MCR WPT were introduced. In addition, several leading countries and international organizations have published many standards. However, the standards are hard to track and usually make the search process so confusing for engineers and researchers. This paper aims to classify and simplify numerous standards. In this paper, these standards are reviewed and compared based on sub-standards, power and frequency ranges, applications as well as EMC Test Standards. Finally, the recommended limits for EMF exposure are displayed.

The remainder of the paper is organized as follows: Section II gives the fundamental principle of WPT. Section III describes the EMI standards, which related to WPT system. Finally, the conclusion and further research are reported in section IV.

II. BASIC PRINCIPLE of WPT

In this section, the MCR WPT, as shown in Fig. 1, will be discussed since it's the most common type of WPT recently, the transmitting coil and receiving coil are designed to resonate at the operating frequency, which set an efficient energy channel for energy transmission [11]. The efficiency depends on the characteristic parameters for each resonator and the coupling coefficient (κ).

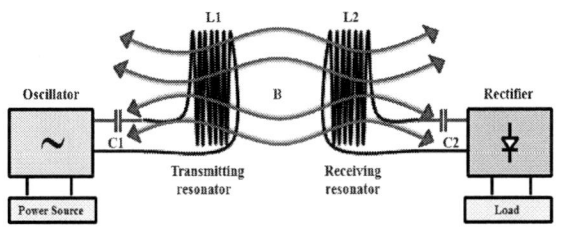

Fig. 1. Diagram of MCR WPT

Resonance involves energy oscillating between two modes; it can be described by two essential parameters, the resonant frequency and the intrinsic loss rate (Γ). The ratio of these two parameters is the quality factor (Q) of the resonator, where: $Q = (\frac{\omega_o}{2\Gamma}) = (\frac{\omega_o}{R} L)$ .If the two resonators are placed close to each other, there will be a coupling between them and an exchange of the energy. Moreover, the energy exchange efficiency depends on the characteristic parameters for each resonator and the coupling coefficient between them.

Fig. 2 shows an equivalent circuit of the WPT in a series-series mode (SS), where V_g is the amplitude of the voltage source with a resistance R_g. In addition, the transmitter and receiver resonators are represented by the inductances L_1 and L_2, respectively. They are coupled through M. Moreover, R_1 and R_2 are the parasitic resistances. Finally, C_1 and C_2 are the resonance capacitors.

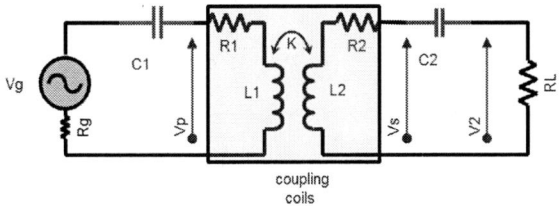

Fig. 2. Equivalent circuit

III. OVERVIEW of WPT RELATED STANDARDS

Fig. 3 shows the electromagnetic spectrum in the non-ionizing and the ionizing radiation sections. Non-ionizing means the energy of the waves is too low to ionize tissues. WPT products use electromagnetic waves within the non-ionizing marked area of the electromagnetic spectrum.

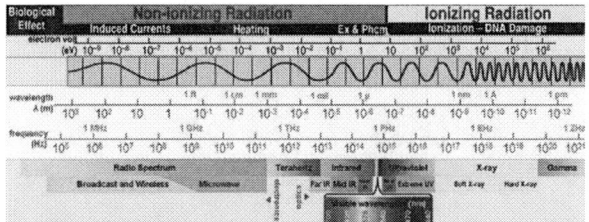

Fig. 3. The Electromagnetic Spectrum

WPT systems have a wide range in term of frequency, power, and transferring distance. If WPT system is placed next to the human body, an electromagnetic field EMF issue will be created. Therefore, to ensure the safety of human body, many exposure limits were issued. On the other hand, if WPT system is located beside other electrical circuits, an EMC issue will be created. Therefore, the EMI susceptibility test and EMI protection are needed to guarantee that emerging WPT system does not affect the operation of other systems [19]. TABLE I provides the latest EMI standards related to WPT considering different types of applications, test procedures in addition to the EMF and EMC safety issues, and interoperability requirements, etc. TABLE II presents WPT systems, the frequency and power bands under study in non-ISM bands for Japan and Korea.

The recent and ongoing efforts of research work related to WPT standards are fully covered. The current situation of WPT systems in Japan and South Korea (leading countries in WPT industry) is described. Many aspects such as power and frequency ranges, applications and WPT types are considered. Most electrical and electronic WPT products must comply with EMI emission, immunity, and the basic restrictions of specific absorption rate requirements. Finally, the standards differ from a country to another. For that reason, before the deployment of WPT system, it should be categorized regarding its frequency, power, application type, and the transferring distance. Thus, it can be treated as a normal equipment that should follow specific restrictions, or it can be treated as an important case and should be subjected to tighter restrictions.

The 2018 International Power Electronics Conference

TABLE I
WPT RELATED STANDARDS AND GUIDELINES

Standards	Substandard	Country	Frequency/power range	Application/Comments
FCC **Federal communications commission**	KDB 680106 Part 15B, Part 18	USA	Above 9 kHz are intentional radiators	RF exposure wireless charging apps (wireless chargers, and inductive chargers, wireless charging pads).
SAE **Society of Automotive Engineers**	SAE J2954/ J2836/6™ J2847/6-J2931/6		J2954™ EVs and PHEVs agreed on "85 kHz band	Wireless charging task force, specific use cases, and specific protocols.
ANSI **American national standards institute**	IEEE-ANSI C63.2 2016-1-1		10 Hz to 40 GHz	A standard for EM noise and the field strength instrumentation.
	IEEE-ANSI C63.14 2014-1-1		-	Standard dictionary for technologies of EMC.
ACGIH **The American Conference of Governmental Industrial Hygienists**	Threshold limit values (TLVs) Biological exposure indices (BEIs)		30 kHz and 300 GHz	RF and microwave radiation TLV documentation (2010).
CISPR **The International Special Committee on Radio Frequency Interference**	CISPR SC-B	International	from 9 kHz upwards, CISPR 11 range 150 kHz up to 1 GHz, or up to 18 GHz	Household appliances, ignition systems, and fluorescent lamps.
	CISPR 11:2015			Power electronics ISM RF equipment used in WPT.
IEC **International electro-technical commission**	IEC 61980 (IEC TC 69), IEC TC 100, IEC 62827-1:2016, IEC PAS 63095-1:2017(E)		For IEC PAS 63095-1:2017(E) baseline power profiles ($\leq$5W) and extended power profile ($\leq$15W)	<ul><li>IEC 61980-1: General requirements</li><li>IEC 61980-2: Communication</li><li>IEC 61980-3: Magnetic field power transfer</li></ul>
IEEE	IEC/IEEE 62704		30 MHz- 1 GHz	<ul><li>Specifies and provides the test for vehicle.</li><li>human body models and the general benchmark data for those models</li></ul>
	IEEE P62704-2 (DRAFT)		Extended up to 6 GHz	
ISO **International organization for standardization**	ISO PAS 19363 2017-1-1, (ISO/NP 19363 under development) ISO/IEC JTC 1 SC 6, ISO DIS 15118-8		close synchronization with IEC 61980 and SAE J2954	<ul><li>Magnetic field WPT, safety and interoperability requirements.</li><li>Requirement for wireless communication</li></ul>
ICNIRP **International commission on non-ionizing radiation protection**	ICNIRP 1998 ICNIRP 2003 ICNIRP 2009 ICNIRP 2010		EM Field: (1 Hz-100 kHz) /2003, 2010 (1 Hz- 300kHz) 1998	<ul><li>Guideline for limiting exposure to electric and magnetic fields, which vary by time.</li><li>Biological effects of exposure to low frequency EM fields (2003).</li><li>Some guidance is extended to 10 MHz to cover the nervous system effects (2009).</li><li>ICNIRP 2010 replaces the low-frequency part of the 1998 guidelines.</li></ul>
ETSI **European Telecommunication standards institute**	ETSI EN 303 417 V1.1.0 (Draft) 2016-12	EU	For WPT systems which use frequency other than RF beam investigate ranges [19 - 21 kHz, 59 - 61 kHz, 79 - 90 kHz, 100 - 300 kHz, 6 765 - 6 795 kHz]	Harmonized standard covering the essential requirements of article 3.2 of Directive 2014/53/EU.
CCSA **China Communication Standard Association**	CCSA TC9	China	part1: General; part2: Tightly Coupled; part3: Resonance wireless power	EMF evaluation methods for WPT and EMC limit and measurements.
	YD/T 2654-2013			Requirements and test methods of EMC of WPT equipment.

1064

BWF Broadband Wireless Forum ARIB Association of Radio Industries and Businesses	ARIB STD-T113 (2015)	Japan	6.78 MHz-band MCR WPT for mobile, 400 kHz-band CPT, EV/PHEV WPT spectrum (42 kHz ~ 48 kHz, 52 kHz ~ 58 kHz, 79 ~ 90 kHz, 140.91~148.5 kHz) with 3 kW and 7.7 kW	Study for WPT spectrum for all the applications and technologies. ■ Capacitive coupling WPT, ■ WPT using microwave two-dimensional waveguide sheet. ■ Magnetic resonance WPT using 6.78 MHz for mobile/portable devices. ■ Magnetic induction WPT for home appliances and office equipment. ■ WPT for EV/PHEV.
TTA Telecommunication Technology association	TTAR-06.162 (19/11/2015)	Korea	EV in 2011, OLEV (19 kHz ~ 21 kHz, 59 kHz ~ 61 kHz), Normal Power: 100 kW. 13.56MHz band is used for 3D glasses WPT	■ Efficiency measuring methods, wireless power transfer, and heavy duty EVs. ■ MCR WPT. ■ ICPT.
A4WP Alliance for Wireless Power	A4WP standards	Established 2012 (Samsung, Qualcomm and others)	6.78MHz for power transfer and 2.4GHz for the control signals	■ MCR WPT. ■ A4WP and PMA merge to form industry-leading organization for wireless charging standards.
Qi (WPC) Wireless Power Consortium	Qi standards Version 1.0. Version 1.1.	Industry group, since 2008	110 kHz – 205 kHz. Low power in the range: (0-5) W. Medium power: up to 120 W.	■ Details and specifics about the Qi WPC standards. ■ Magnetic induction ICPT. ■ Used in cell-phone, music players, and Bluetooth earpieces, etc.
PMA Power matters alliance	PMA standard	Founded by Procter & Gamble and Powermat. 2012	277 kHz – 357 kHz. Up to 5-10W.	■ Magnetic induction. ■ Mobile device ecosystem. ■ A4WP and PMA merge to form industry-leading organization.

Additional standards for EMC, immunity Tests and measurements.
- Radiated and Conducted Emissions — CISPR 11 or CISPR 22
- Radiated Immunity — EN 61000-4-3
- Magnetic Immunity — EN 61000-4-8
- Compliance testing of Wireless Power Transfer products C63.30.
- Radiated EM immunity — ISO 11451-2

TABLE II
WPT SYSTEMS: RECENT AND UNDER STUDY FREQUENCY AND POWER RANGES FOR WPT IN ASIA

WPT System	Country	Frequency Range under consideration	Power Range under considerations	Application
ICPT Inductive coupling, low power	Already in Japan, Korea	Japan: 110kHz–205kHz. Korea: 100kHz–205kHz	-	Mobile devices, portable devices, and industrial Fields.
ICPT Inductive coupling, high power	Japan	Japanese 20.05 kHz - 38 kHz, 42 kHz - 58 kHz, 62 kHz –\\ 100 kHz	Japan: Several W – 1.5kW	Home appliances (operating with high power), and office equipment.
Magnetically coupling resonant MCR WPT	Japan, Korea	Both Japan and Korea : 6.765 MHz -6.795 MHz	Japan: Several W – up to 100W Korea: Unlimited in-band emission limit	Mobile devices, tablets, note-PCs, home appliances (operating with low power)
Capacitive coupling CPT	Japan	Japan: 425 kHz-524 kHz	Japan: Up to 100 W	Portable devices, tablets, home appliances, and office equipment.

I. CONCLUSION

In this paper, the EMI regulations and standards from different countries and international organization have been classified and compared based on different criteria. The comparison provides an easy way to understand the newly released standards and regulations. As an outcome, more efforts have to be done in this field to cover every new application. In addition, in order to compute accurately the human exposure to EMF, a reliable human model is required.

REFERENCE

[1] Miller, J.M., Onar, O.C., Chinthavali, M.: "Primary-side power flow control of wireless power transfer for electric vehicle charging", *IEEE J. Emerg. Sel. Top. Power Electron.*, vol. 3, no.1, pp. 147–162, 2015.

[2] Li, Weihan, et al. "Integrated {LCC} Compensation Topology for Wireless Charger in Electric and Plug-in Electric Vehicles." *IEEE Transactions on Industrial Electronics,* vol. 62, no.7, pp. 4215-4225, 2015.

[3] Li, Weihan, et al. "Comparison study on SS and double-sided LCC compensation topologies for EV/PHEV wireless chargers." *IEEE Transactions on Vehicular Technology,* vol. 65, no. 6, pp. 4429-4439, 2016.

[4] Li, Xing, Chi-Ying Tsui, and Wing-Hung Ki. "A 13.56 MHz wireless power transfer system with reconfigurable resonant regulating rectifier and wireless power control for implantable medical devices." *IEEE Journal of Solid-State Circuits,* vol. 50, no. 4, pp. 978-989, 2015.

[5] Das, Rupam, and Hyoungsuk Yoo. "A Multiband Antenna Associating Wireless Monitoring and Nonleaky Wireless Power Transfer System for Biomedical Implants." *IEEE Transactions on Microwave Theory and Techniques,* vol. 65, no. 7, pp. 2485-2495, 2017.

[6] Fu, Minfan, et al. "Efficiency and optimal loads analysis for multiple-receiver wireless power transfer systems." *IEEE Transactions on Microwave Theory and Techniques,* vol. 63, no. 3, pp. 801-812. 2015.

[7] McDonough, Matthew. "Integration of inductively coupled power transfer and hybrid energy storage system: A multiport power electronics interface for battery-powered electric vehicles." *IEEE Transactions on Power Electronics,* vol. 30, no. 11, pp. 6423-6433. 2015.

[8] Villa, Juan L., et al. "High-misalignment tolerant compensation topology for ICPT systems." *IEEE Transactions on Industrial Electronics,* vol. 59, no. 2, pp. 945-951, 2012.

[9] Lu, Fei, Hua Zhang, Heath Hofmann, and Chunting Chris Mi. "An inductive and capacitive combined wireless power transfer system with LC-compensated topology." *IEEE Transactions on Power Electronics,* vol. 31, no. 12, pp. 8471-8482. 2016.

[10] Sample, Alanson P., David T. Meyer, and Joshua R. Smith. "Analysis, experimental results, and range adaptation of magnetically coupled resonators for wireless power transfer." *IEEE Transactions on Industrial Electronics,* vol. 58, no. 2, pp. 544-554, 2011.

[11] J. W. Kim, H. C. Son, K. H. Kim, and Y. J. Park, "Efficiency analysis of magnetic resonance wireless power transfer with intermediate resonant coil," *IEEE Antennas Wireless Propag. Lett.,* vol. 10, pp. 389–392, May 2011.

[12] Park, Jaehyoung, et al. "A Resonant Reactive Shielding for Planar Wireless Power Transfer System in Smartphone Application." *IEEE Transactions on Electromagnetic Compatibility,* vol. 59, no. 2, pp. 695-703, 2017.

[13] Chen, Xi Lin, et al. "Human exposure to close-range resonant wireless power transfer systems as a function of design parameters." *IEEE transactions on Electromagnetic Compatibility,* vol. 56, no. 5, pp. 1027-1034, 2014.

[14] Song, Chiuk, et al. "Low EMF and EMI Design of a Tightly Coupled Handheld Resonant Magnetic Field (HH-RMF) Charger for Automotive Battery Charging." *IEEE Transactions on Electromagnetic Compatibility,* vol. 58, no. 4, 1194-1206. 2016.

[15] Kim, Minho, et al. "A three-phase wireless-power-transfer system for online electric vehicles with reduction of leakage magnetic fields." *IEEE Transactions on Microwave Theory and Techniques,* vol. 63, no. 11, pp. 3806-3813. 2015.

[16] Lin, James C. "A new IEEE standard for safety levels with respect to human exposure to radio-frequency radiation." *IEEE Antennas and Propagation Magazine,* vol. 48, no. 1, pp. 157-159, 2006.

[17] Online, Available: http://standards.ieee.org/findstds/interps/C95.1-2005.html. *IEEE Standards Interpretations for IEEE Std* C95.1™-2005.

[18] Lin, Jiali, et al. "ICNIRP Guidelines for limiting exposure to time-varying electric and magnetic fields (1 Hz to 100 kHz)." *Health Physics,* vol.99, pp. 818-836, 2010.

[19] Yamanaka, Yukio; SUGIURA, Akira. "Possible EMC regulations for wireless power transmission equipment." In: *Microwave Workshop Series on Innovative Wireless Power Transmission: Technologies, Systems, and Applications (IMWS), 2011 IEEE MTT-S International.* IEEE, 2011, pp. 97-100.

The 2018 International Power Electronics Conference

Comparative Study of Single-Phase Fundamental Component Frequency Estimation Schemes under Time-varying Harmonic Distortion Operation

E. B. Kapisch[1,2*], J. L. Duarte[1], C. A. Duque[2]

1 Electromechanics and Power Electronics Department, Eindhoven University of Technology, Eindhoven, The Netherlands
2 Electrical Circuits Department, Federal University of Juiz de Fora, Juiz de Fora, Brazil
*E-mail: eder.kapisch@engenharia.ufjf.br

Abstract—**This paper presents an adaptation of a fundamental frequency estimation scheme based on a fast-tracking PLL-based structure. The approach is based on a modification of the Adaptive Quadratic Signals Generator Phase-Locked Loop, adding a pre-filtering stage in order to attenuate the influence of the time-varying harmonics. The time-domain discretization of the proposed scheme is also presented. A comparison between the proposed scheme and a highly accurate Zero-Crossing scheme is performed under time-varying harmonic operation conditions through experimental measurements. While the later is able to provide only the frequency value, the proposed scheme is able to provide both the frequency value, the fundamental component wave form, and its quadratic companion signal, keeping the same fast-tracking characteristic and presenting a smoother behavior.**

Keywords—Frequency-Locked Loop, Fundamental Component Estimation, Time-varying Harmonics, Zero-Crossing.

I. INTRODUCTION

The growth in electricity demand has been, evidently, one of the main reasons for the expansion of Electric Power Systems (EPSs) around the world [1]–[5]. Throughout the years, EPSs have become increasingly complex, more and more branched, feeding multiple types of loads, which drain power from the grid in more peculiar ways [6]. Some of the characteristics often present in many of the current loads are the time-varying behavior and high nonlinearity, which are responsible for most causes of deterioration in Power Quality (PQ). Among these several issues, the presence of harmonics in voltage and current waveforms is a major fact [7].

Some examples that one can mention of loads regarded as the main contributors of waveform distortion are the Flexible Alternating Current Transmission Systems (FACTs) devices, saturated magnetic core, arc furnaces, and Variable Frequency Drives (VFDs) [8]. In the current state of the EPSs, renewable energy sources have become increasingly important, and the Integration of these energy sources to the main grid is done through power electronic interfaces. This is another major cause of harmonic distortions [9]. Indeed, grid interfacing

equipment, like voltage source inverters, can bring about harmonic distortion into the voltage and current grid signals due to the non-ideal characteristics of them [10].

Adverse effects can result from harmonic distortion. Among them one can cite the reduction in devices lifespan, erroneous operation of control circuits, interference with communication lines, and increased losses. These effects become even more important within the Smart Grids (SGs) scenario, where the operation must be performed in an integrated way through a real-time communication between control and protection elements [11], [12]. Therefore, accurate and fast-tracking parameter estimation and reliable communication are key aspects for an SG [13], [14].

The efficiency of harmonic mitigation techniques depends on an accurate estimation of the fundamental component of the waveform signals [9]. These techniques are becoming indispensable, especially in the new context of the SGs and its applications, such as the control of power converters.

A crucial aspect in the control of power converters connected to the network is the detection of the fundamental component of the voltage and either the fundamental frequency value or the phase angle thereof [15]. The frequency estimation information is used for the synchronization of the output variables of the converters, power flow calculation, among other purposes. Frequency information is also essential for the control of distributed generation systems, energy storage systems, FACTS [16], power line conditioners, Uninterruptible Power Supplies (UPS) [17], inverters frequency and other power conditioning equipment.

Fig. 1 shows this multiple use of frequency estimation in various applications of the EPS.

In all of these applications, the frequency estimation must be done as quickly and accurately as possible. This becomes quite challenging when the EPS operates under the harmonic distortion condition. Considering more severe adverse conditions, such as the presence of sub-harmonics or also a marked variation in frequency, they

1067

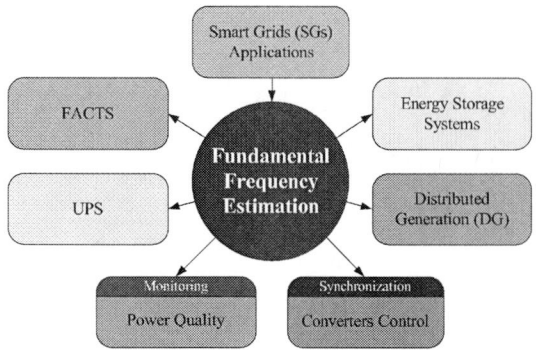

Fig. 1. Multiple usages of EPS frequency estimation information in the context of SGs.

may interfere significantly on the operation of conventional estimators. Therefore, a robust technique to these and other types of PQ disturbances is important.

Many of the algorithms used to estimate the frequency value of the EPSs are based on the assumption that the fundamental component frequency is practically constant, or that it experiences small variations, especially those based on the Fast Fourier Transform (FFT) [18]. This premise has become increasingly far from reality as new sources with intermittent features are incorporated into the grid. The frequency of a distribution network, for instance, can extremely quickly vary during transient events, so it can be very challenging to track it with enough accuracy [19]. Thus, the constant or nearly constant frequency model is no longer appropriate for the current state.

Having said that, this paper presents the comparison of two fundamental frequency estimation schemes. The proposed method is based on Adaptive Quadratic Signals Generator - Phase Locked Loop (AQSG-PLL) [15]. A low-pass pre-filtering stage is added, similar to the approach presented in [20], where zero-crossing with linear interpolation correction is applied to improve precision. The new scheme, so-called PF-AQSG-PLL, can deliver the fundamental and the quadrature components of a measured quantity (voltage or current) in addition to the frequency value, while [20] delivers only the frequency value information. The comparison is performed through practical measurements using a dSPACE®.

This paper is organized as follows. Section II presents the description of the fundamental frequency estimators used as the basis to the proposed scheme. Firstly, the pre-filtered zero-crossing with linear interpolation correction is shown. Then, the AQSG-PLL scheme is presented. In Section III, the proposed scheme as well as its time-domain discretization presented. The approach is based on a combination of the AQSG-PLL and the pre-filtering stage, similar to the one present in the zero-crossing modeling description. Section IV presents the test set-up and some results from these tests. Finally, Section V contains some concluding remarks.

II. MODELING DESCRIPTION OF THE FREQUENCY ESTIMATORS

A. Zero-Crossing with linear interpolation correction

This scheme [20] is composed of a simple zero-crossing detector followed by an interpolation correction stage performed to improve the accuracy of the estimation (Fig. 2). This stage estimates the exact moment of the signal crossing between the two samples, $x_c[n-1]$ immediately before and $x_c[n]$ after the zero-crossing. From the sampling period T_s is it possible to deduct the periods N_a and N_b of (1) and (2) using basic trigonometry.

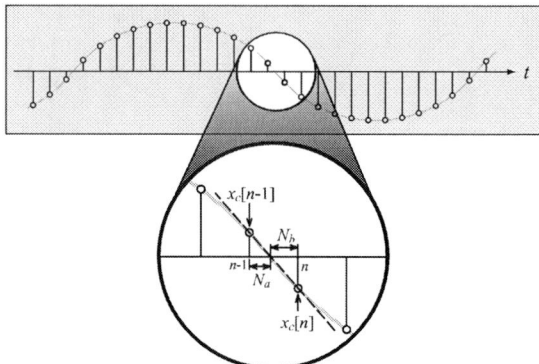

Fig. 2. Samples at zero crossing: linear interpolation.

$$N_a = \frac{x_c[n-1]}{x_c[n-1] - x[n]}; \tag{1}$$

$$N_b = \frac{x_c[n]}{x[n] - x_c[n-1]}; \tag{2}$$

B. Pre-filtering Stage

In order to improve the the reliability of the zero-crossing scheme and avoid double crossing due to noise and other high frequency components, a low-pass filtering stage is applied to the scheme. A 6-order IIR Chebyshev type II was used. Type II Chebyshev (or inverse Chebyshev) filters are known to have a flat passband and equi-ripple in the stopband. In this specific case, this filter was used due its low order and easy implementation characteristics.

Its characteristic curves are shown in Fig. 3, and its transfer function relating its output $Y[z]$ and its input $X[z]$ can be observed through Eq. 3.

$$
\begin{aligned}
H(z) &= \frac{Y(z)}{X(z)} \\
&= \frac{0.0284 - 0.1677z^{-1} + 0.4153z^{-2} - 0.5519z^{-3} +}{1 - 5.7374z^{-1} + 13.7211z^{-2} - 17.5082z^{-3} +} \cdots \\
&\cdots \frac{+0.4153z^{-4} - 0.1677z^{-5} + 0.0284z^{-6}}{+12.5714z^{-4} - 4.8160z^{-5} + 0.7690z^{-6}}
\end{aligned}
\tag{3}
$$

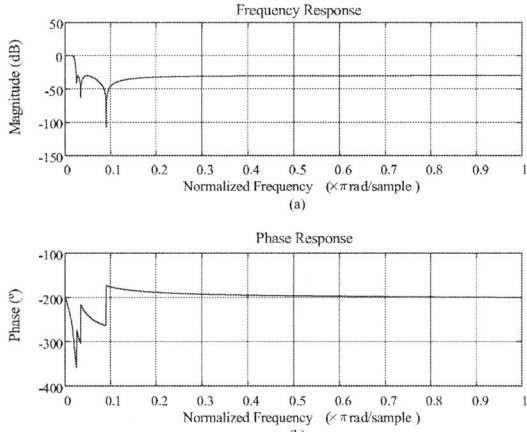

Fig. 3. Low-pass filter characteristic curves. (a) Frequency response. (b) Phase response.

C. AQSG-PLL: Continuous-time modeling

The proposed single-phase frequency estimation scheme is based on AQSG-PLL [15]. The AQSG-PLL is composed of three main blocks, the Adaptive Quadrature Signals Generator (AQSG), the Fundamental Frequency Estimator (FFE) and the Quadrature Companion Generator (QCG). These blocks are described by the set of equations (4) and they are shown in Fig. 4.

$$
\text{AQSG:} \quad
\begin{cases}
\dot{\hat{v}}_{\alpha,1} = \hat{\Omega}_0 \hat{\psi}_1 + \gamma_1 \tilde{v}_{\alpha,1} \\
\dot{\hat{\psi}}_1 = -\varpi_0 \hat{v}_{\alpha,1}
\end{cases}
$$

$$
\text{FFE:} \quad \dot{\hat{\Omega}}_0 = \lambda \hat{\psi}_1 \tilde{v}_\alpha \qquad (4)
$$

$$
\text{QCG:} \quad \hat{v}_{\beta,1} = \hat{\omega}_0 \frac{\hat{\psi}_1}{\varpi_0}
$$

Where the notation $\dot{x} = \frac{dx(t)}{dt}$ means the time derivative of the variable $x(t)$, and $\hat{x}$ is the estimation of $x(t)$. Also, $\psi = \varpi_0 \frac{v_\beta}{\omega_0}$, where ϖ_0 represents the nominal value of the fundamental angular frequency, that is, a positive constant, which can be a simple approximation of the angular frequency ω_0, and v_β is the quadrature companion signal of v_α. The variable $\Omega_0 = \frac{\omega_0^2}{\varpi_0}$ is the square of the real frequency ω_0 scaled by ϖ_0.

In (4), $\hat{\Omega}_0$ is the estimation of Ω_0; $\hat{\psi}_1$ and $\hat{v}_{\alpha,1}$ are fundamental component estimations v_α e ψ, respectively; $\dot{\hat{v}}_{\alpha,1}$ is the fundamental component estimation time derivative of the input v_α; $\tilde{v}_{\alpha,1} \triangleq v_\alpha - \hat{v}_{\alpha,1}$ is the difference between the input signal v_α, from which the frequency estimation is desired, and the fundamental component estimation $\hat{v}_{\alpha,1}$ of this signal, that is, $\tilde{v}_{\alpha,1}$ is the error signal; $\dot{\hat{\psi}}_1$ is the fundamental component estimation time derivative of ψ; $\hat{v}_{\beta,1}$ is the estimation of the fundamental component of the quadrature companion signal v_β; $\gamma_1 > 0$ is a design parameter used to introduce a damping required by the model, and λ is a design parameter known as adaption gain.

III. PROPOSED FUNDAMENTAL COMPONENT FREQUENCY ESTIMATION SCHEME

In order to deal with harmonic distortion scenario, the author originally proposes a Harmonic Compensation Mechanism (HCM) for each individual harmonic present in the system [15]. However, in practical terms, it is difficult to forecast which harmonic will occur and when it will appear. Besides, the complexity of the system will increase expressively as the number of harmonic compensations increase. For this reason, this work proposes the use of the pre-filtering stage presented in Section II applied to the signal before the AQSG-PLL, resulting in the PF-AQSG-PLL, as it is shown in Fig. 5.

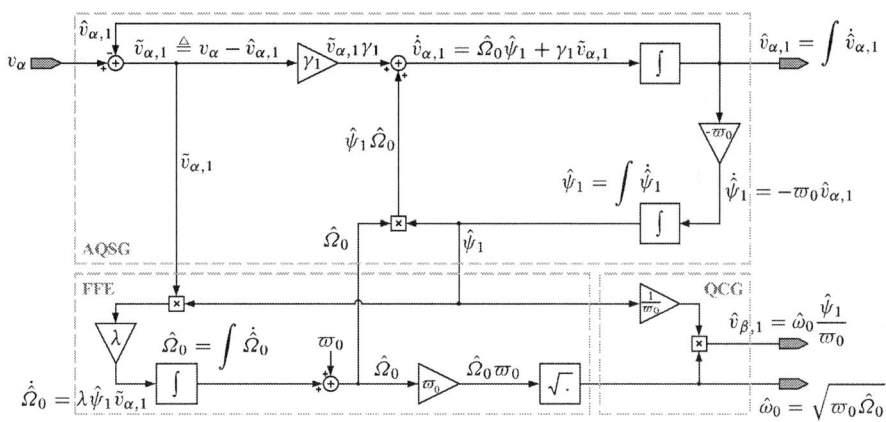

Fig. 4. Estimator AQSG-PLL of the angular frequency ω_0.

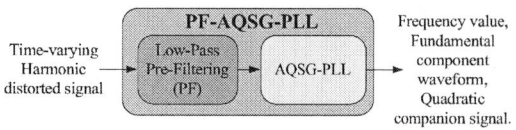

Fig. 5. Proposed pre-filtering stage for the AQSG-PLL.

A. Discretization

In order to the system can be implemented on a digital platform, like either DSPs or FPGAs, transformations from the continuous time domain (s) to the discrete time domain (z) are needed. The three integrators ($\frac{1}{s}$) must be transformed. The trapezoidal integration rule (5) was adopted to the two most inferior integrators and the Adams-Bashforth rule (6) was applied to the superior one. This strategy was planned in order to avoid the "algebraic loops" in the discrete system through the application of an explicit integration method.

$$H_{tr}(z) = \frac{T_s}{2}\left(\frac{z+1}{z-1}\right) \tag{5}$$

$$H_{ab}(z) = \frac{T_s}{2}\left(\frac{3z-1}{z^2-z}\right) \tag{6}$$

IV. EXPERIMENTAL TESTS

A. The set-up

The dSPACE® system based on the DS1104 R&D Controller Board was used to implement the algorithms. This platform was connected to a PC to be programmed and to exchange data. To generate the signals with disturbances and distortions, the CSW5550 from California Instruments® power generator was used. A four-channel RIGOL® DS1104B digital oscilloscope was used.

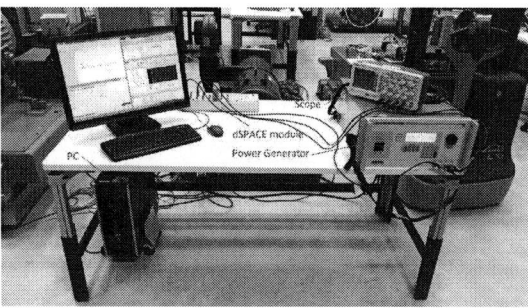

Fig. 6. Set up for experimental tests.

B. Results

For the tests a sampling frequency of 7.68 kHz was used, wich corresponds to 128 samples/cycle in a nominal 60 Hz fundamental signal. The power generator has two three-phase outputs, one on the back, which allows supply power devices with real voltage levels up to 312 V

and 288 A, and other on the front, which has a level attenuation and can be used for the first tests. One phase of this front output was used as input signal for the tests. When an amplitude of 150 V is set to the power output, the front output shows 6.8 V of amplitude. The PF-AQSG-PLL parameters γ_1 and λ were set to 300 and 2.22 respectively, according to [15].

To output the result from the frequency estimation two DAC channel from dSPACE were used, one for the zero-crossing estimation (the blue one on the next figures) and other for the PF-AQSG-PLL estimation (the purple one). As the output of the DAC can reach only ±10 V, a remapping of the values was don in order to use this range to output frequency estimations from 40 Hz to 70 Hz, according to (7):

$$f_{real} = 1.5(f_{scope} + 10) + 40 \tag{7}$$

where f_{scope} is the value shown on the scope screen and f_{real} is the correspondent real value.

1) Case 1: For this test a signal with a step change in frequency from 50 Hz to 60 Hz in a single component signal was done. The amplitude remains the same as 150 V. The result of the estimation can be seen through Fig. 7

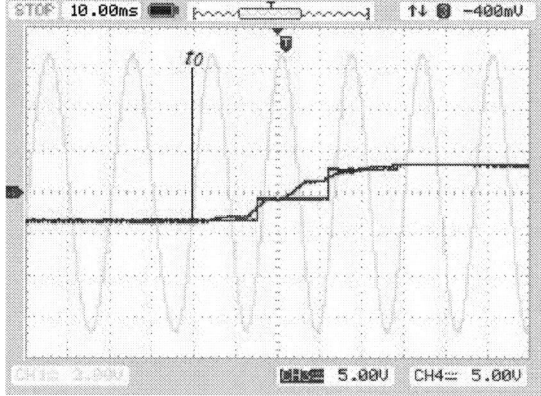

Fig. 7. Change in the fundamental frequency from 50 Hz to 60 Hz.

One can note that from the instant of change t_0 the estimation tracks the new value of frequency in about two and a half cycle of 60 Hz. Before the change the value of both estimations were about -3.36 V, which corresponds to approximately 49.96 Hz. After the change, both reached about 60.04 Hz. However, it must taken be into account the precision of the DAC and the error during the reading on the screen scope. Despite these facts, the estimation can be considered accurate and fast.

2) Case 2: For the second test, a signal with high harmonic distortion content was generated. The composition is: 100% of fundamental (150 V), 10% of 3° harmonic with a phase shift of 13°, 5% of 5° and 5° harmonics with no phase shift, and a 3% no phase shift of the 2° harmonic. The distortion begins in t_0. The frequency remains the same. The result of the estimation can be seen from the Fig. 8.

1070

The 2018 International Power Electronics Conference

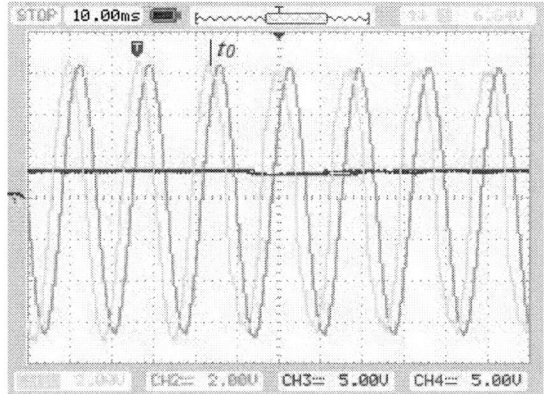

Fig. 8. Tyme-varying harmonic distortion.

One can observe that after the distortion begins, there is a small oscillation in both estimation. However, there is a new channel depicted on the picture. The signal in light blue is the Quadratic Companion Signal related to the input in yellow. It can be observed that this other output from the PF-AQSG-PLL expounds an almos pure sinusoid even during the distortion period. So, this is an advantage of the PF-AQSG-PLL.

3) Case 3: For this case, a harmonic distortion with the same composition of Case 2 is inserted in the input signal. However, this time the amplitude value of all components, including fundamental, drops 20%, so does the frequency, which drops from 60 Hz to 50 Hz. The result can be seen through Fig. 9

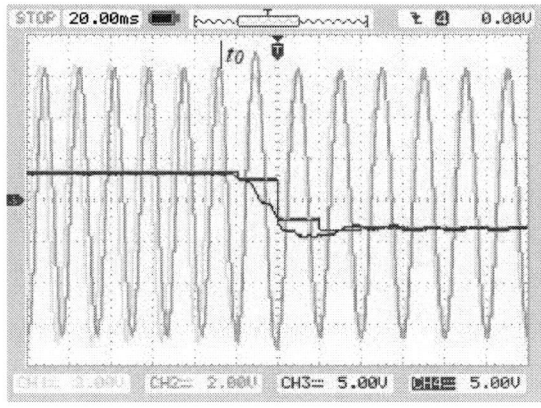

Fig. 9. Time-varying harmonic and amplitude disturbance.

V. CONCLUSION

The improvement resulted from the pre-filtering stage introduced in this work has been experimentally shown to provide very fast and accurate response to the fundamental component frequency estimation. In addition to the frequency value, the PF-AQSG-PLL can deliver an estimation of both fundamental component itself and the quadrature companion signal of the input signal

while keeping the low complexity compared to the zero-crossing which can deliver only the frequency value. Once the natural behavior of the zero-crossing estimation is to update the values estimation every cycle of the signal, it leads to a waveform shape in which one can observe discontinuities in step-form. This does not happen with PF-AQSG-PLL, which presents a smooth and fast-tracking behavior.

ACKNOWLEDGEMENT

Authors would like to thank the Federal University of Juiz de Fora, CAPES and Eindhoven University of Technology for all financial, technical and scientific support on this work.

REFERENCES

[1] K. Qureshi, "Role of advanced nuclear reactor technologies in meeting the growing energy demands," in *2015 Power Generation System and Renewable Energy Technologies (PGSRET)*, pp. 1–5, jun. 2015.

[2] A. P. Putra, R. Sarno, and E. Suryani, "Dynamics simulation model of demand and supply electricity energy public facilities and social sector case study east java," in *2016 International Conference on Information Communication Technology and Systems (ICTS)*, pp. 26–33, out. 2016.

[3] R. Hejeejo, J. Qiu, T. S. Brinsmead, and L. J. Reedman, "Sustainable energy system planning for the management of mgs: a case study in New South Wales, Australia," *IET Renewable Power Generation*, vol. 11, pp. 228–238, abr. 2017.

[4] A. U. Mahin, M. A. Sakib, M. A. Zaman, M. S. Chowdhury, and S. A. Shanto, "Developing demand side management program for residential electricity consumers of Dhaka city," in *2017 International Conference on Electrical, Computer and Communication Engineering (ECCE)*, pp. 743–747, fev. 2017.

[5] D. V. de Sousa Stilpen and V. Cheng, "Solar photovoltaics in Brazil: A promising renewable energy market," in *2015 3rd International Renewable and Sustainable Energy Conference (IRSEC)*, pp. 1–5, dez. 2015.

[6] M. G. Moreira, D. D. Ferreira, and C. A. Duque, "Interharmonic detection and identification based on higher-order statistics," in *2016 17th International Conference on Harmonics and Quality of Power (ICHQP)*, pp. 679–684, out. 2016.

[7] H. C. Lin, C. H. Chen, and L. Y. Liu, "Harmonics and interharmonics measurement using group-harmonic power minimizing algorithm," in *Proceedings of the World Congress on Engineering*, vol. 2, 2011.

[8] G. Rana and A. Mittal, "A review & analysis of harmonics in Variable Frequency Drives (VFDs)," *International Journal of Enhanced Research in Science Technology & Engineering*, vol. 3, pp. 149–153, Feb. 2014.

[9] P. Keerthy, P. Maya, and M. Sindhu, "An adaptive transient tracking harmonic detection method for power quality improvement," in *IEEE Region 10 Symposium (TENSYMP), 2017*, pp. 1–6, IEEE, 2017.

[10] J. Xia, W. Sun, Y. Yin, Z. Xing, and X. Yuan, "Fpga based direct measurement of pwm voltage and inverter disturbance," in *Electrical Machines and Systems (ICEMS), 2016 19th International Conference on*, pp. 1–4, IEEE, 2016.

[11] T. Ilamparithi, S. Abourdia, and T. Kirk, "On the use of real time simulators tor the lest and validation of protection and control systems of micro grids and smart grids," in *2016 Saudi Arabia Smart Grid (SASG)*, pp. 1–5, dez. 2016.

[12] M. M. Rana, L. Li, and S. W. Su, "Microgrid protection and control through reliable smart grid communication systems," in *2016 14th International Conference on Control, Automation, Robotics and Vision (ICARCV)*, pp. 1–6, nov. 2016.

1071

[13] W. Xiaorong and W. Ying, "Study on hierarchical protection & control system in smart grid," in *2014 International Conference on Power System Technology*, pp. 2433–2440, out. 2014.

[14] H. Laaksonen and F. Suomi, "New functionalities and features of ieds to realize active control and protection of smart grids," in *22nd International Conference and Exhibition on Electricity Distribution (CIRED 2013)*, pp. 1–4, jun. 2013.

[15] F. Vasca and L. Iannelli, "Dynamics and control of switched electronic systems," *Advanced Perspectives for Modeling, Simulation and Control of Power Converters*, 2012.

[16] N. Hingorani and L. Gyuyi, "Understanding facts:concepts and technology of flexible ac transmission systems," 2000.

[17] M. Cichowlas, M. Malinowski, M. P. Kazmierkowski, D. L. Sobczuk, P. Rodríguez, and J. Pou, "Active filtering function of three-phase pwm boost rectifier under different line voltage conditions," *IEEE transactions on industrial electronics*, vol. 52, no. 2, pp. 410–419, 2005.

[18] D. Agrez, "Frequency estimation of the non-stationary signals using interpolated dft," in *Instrumentation and Measurement Technology Conference, 2002. IMTC/2002. Proceedings of the 19th IEEE*, vol. 2, pp. 925–930, IEEE, 2002.

[19] M. D. Kusljevic, J. J. Tomic, and L. D. Jovanovic, "Frequency estimation of three-phase power system using weighted-least-square algorithm and adaptive fir filtering," *IEEE Transactions on Instrumentation and Measurement*, vol. 59, no. 2, pp. 322–329, 2010.

[20] P. F. Ribeiro, C. A. Duque, P. M. Ribeiro, and A. S. Cerqueira, *Power systems signal processing for smart grids*. John Wiley & Sons, 2013.

The 2018 International Power Electronics Conference

A Comprehensive Dead-Time Compensation Method for a Three-Phase Dual-Active Bridge Converter with Hybrid Modulation Schemes

Jingxin Hu, Zhiqing Yang and Rik W. De Doncker
Institute for Power Generation and Storage Systems
E.ON Energy Research Center, FEN Research Campus
RWTH Aachen University, Aachen, Germany
Email: post_pgs@eonerc.rwth-aachen.de

Abstract—To achieve a highly efficient operation of a three-phase dual-active bridge (DAB3) converter in a wide load range, modulation schemes such as asymmetrical duty-cycle control (ADCC) are applied to extend the soft-switching range and reduce ac currents. However, the dead time in the DAB3 converter might affect the soft-switching operation, which cannot be compensated by the closed-loop control effectively. In this paper, a comprehensive dead-time compensation method for a DAB3 converter is proposed for different modulation schemes. The proposed method takes different dead-time values for the primary bridge and the secondary bridge into consideration. The switching sequences of the ADCC modulation are adapted accordingly by inserting zero-current intervals to maintain the soft-switching operation. Hence, the deviation of the transmitted power is corrected and soft-switching can be guaranteed. The effectiveness of this method is validated by both simulations and experimental results.

Keywords—*three-phase dual-active bridge, asymmetrical duty-cycle control, soft switching, dead-time compensation.*

I. INTRODUCTION

Among numerous dc-dc converter topologies, the three-phase dual-active bridge (DAB3) converter features inherent soft-switching capability, galvanic isolation, smaller filter size, and bidirectional power conversion capability [1]. Therefore, it is considered in literatures [2]–[5] as a promising topology to integrate renewable energy sources and storage systems into a medium-voltage dc (MVDC) grid.

Due to the fact that renewable energy sources and storage systems often work in partial load conditions, the partial load efficiency of the DAB3 converter is crucial. In order to improve the efficiency of the DAB3 converter, different modulation strategies are studied. With the asymmetrical duty-cycle control (ADCC) [6] [7], zero-current switching (ZCS) or zero-voltage switching (ZVS) can be realized under light load conditions. However, due to the existence of the dead time which is implemented to prevent the shoot-through of one phase leg during the commutation process, the voltage and current waveforms are distorted which affect the steady-state operation and the dynamic performance as addressed in [8] [9]. Compensation methods are also provided to correct the power deviation of the single phase-shift (SPS) modulation by modifying the phase shift in an open-loop manner [10] [11]. However, since the phase shift is the only control variable for the SPS modulation, the compensation of the power deviation can be achieved by a closed-loop control as well. For the ADCC modulation schemes with ZCS characteristics, three

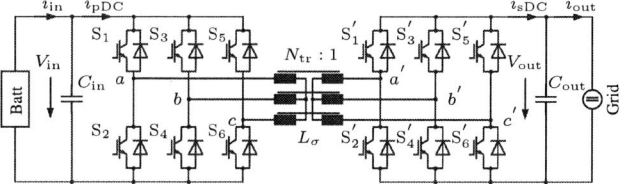

Fig. 1: Topology of the DAB3 converter

control variables i.e. phase shift, duty cycle of the upper device in the primary bridge and duty cycle of the upper device in the secondary bridge are utilized to define the transferred power and ensure a soft-switching operation. In this case, with a normal closed-loop control of the dc voltage, only the power deviation induced by the dead time can be compensated. However, the soft-switching operation cannot be guaranteed with the closed-loop control. Therefore, a dedicated dead-time compensation method is needed for the ADCC modulation. The dead-time compensation for single-phase dual-active bridge (DAB1) converters with ZCS operation is discussed in [12][13]. However, there are a few researches related to DAB3 converters with ZCS modulation such as ADCC in [6]. Therefore, it is essential to investigate the dead-time compensation method to ensure the soft-switching operation of DAB3 converters.

In this paper, the dead-time effect for different modulation schemes of the DAB3 converter are analyzed firstly. It then illustrates how the power deviation could be compensated for both the SPS modulation and the ADCC modulation, and the remaining issue with the soft-switching operation which is fixed by further adaptations of the switching sequence of the ADCC modulation. Another contribution of this paper is that different values of dead time for the primary bridge and the secondary bridge are considered in the compensation method since different devices might be adopted. Thereafter, simulation and experimental results are provided to validate the proposed method.

II. SINGLE PHASE-SHIFT (SPS) CONTROL OF DAB3

A. Power Transmission

DAB3 converter is composed of two three-phase active bridges, which are connected through a medium-frequency transformer as illustrated in Fig. 1. With the conventional SPS

1073

The 2018 International Power Electronics Conference

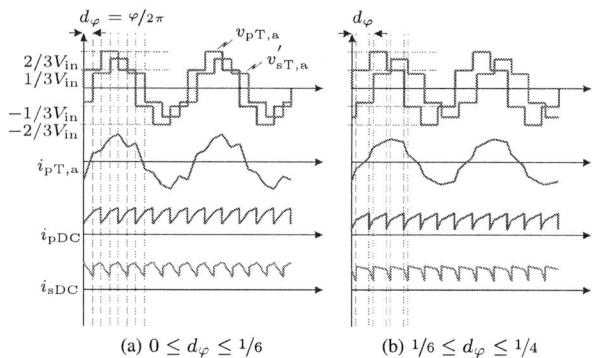

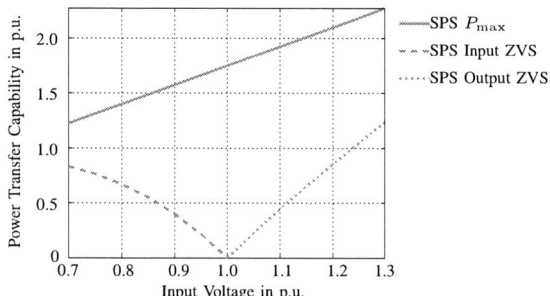

(a) $0 \leq d_\varphi \leq 1/6$ (b) $1/6 \leq d_\varphi \leq 1/4$

Fig. 2: SPS modulation schemes

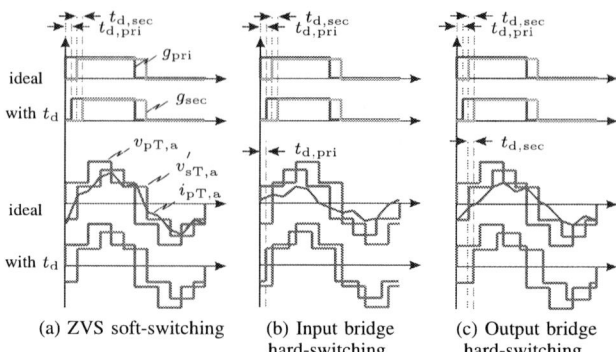

(a) ZVS soft-switching   (b) Input bridge hard-switching   (c) Output bridge hard-switching

Fig. 4: Dead-time influence in SPS modulation

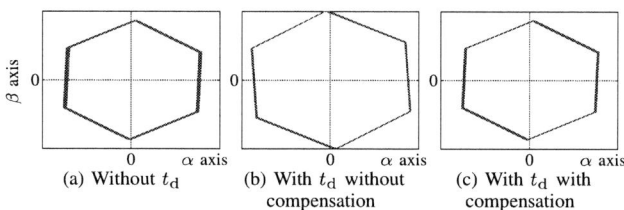

Fig. 3: Soft-switching range with SPS modulation

(a) Without t_d (b) With t_d without compensation   (c) With t_d with compensation

Fig. 5: Current traces in $\alpha - \beta$ diagram with SPS modulation in output hard-switching range

control, both the input and output bridges operate with a fixed duty cycle of 50 %, which results in a six-step waveform of the phase voltage as shown in Fig. 2. The transferred power can be controlled only by the phase shift φ between two bridges as described in (1), where $d_\varphi = \varphi/2\pi$, $\omega = 2\pi f$, f is the switching frequency, L_σ is the total series inductance, $d = N_\text{tr} \cdot V_\text{out}/V_\text{in}$ represents the output-to-input voltage ratio, and N_tr is the turns ratio of the transformer [1]. It is assumed that the output voltage is constant.

$$P_\text{SPS} = \frac{V_\text{in}^2 d}{\omega L_\sigma} \cdot \begin{cases} \frac{4\pi}{3} d_\varphi - 2\pi d_\varphi^2 & , \ 0 \leq d_\varphi \leq 1/6 \\ 2\pi d_\varphi - 4\pi d_\varphi^2 - \frac{\pi}{18} & , \ 1/6 \leq d_\varphi \leq 1/4 \end{cases} \quad (1)$$

It is known that ZVS can be realized with the SPS modulation if currents flow through the anti-parallel diodes during the turn-on process. Therefore, the ZVS boundaries can be determined according to $i_\text{L}(0) \leq 0$ for the input bridge, and $i_\text{L}(\varphi) \geq 0$ for the output bridge respectively. Usually the phase shift φ is limited below $\pi/3$ i.e. $d_\varphi \leq 1/6$ for an efficient operation [6]. Hence only the first mode in Fig. 2a with $0 \leq d_\varphi \leq 1/6$ is considered in this paper. The maximum transferred power, the ZVS boundary of the input bridge and the output bridge are plotted in Fig. 3. The soft-switching capability cannot be guaranteed for the operation range below the dash lines. Depending on the voltage variation, a hard-switching would occur in either the input or the output bridge.

B. Dead-Time Compensation

In a practical implementation, a dead time t_d is added between turn-off and turn-on of two switches in the same phase

leg to prevent the phase leg from shoot-through. During the dead-time interval, the output voltage of a phase leg could be distorted which influences the current waveform consequently. The voltage distortion also depends on the current polarity. If the current flows through the switch when it is turned off, the anti-parallel diode of the complementary switch will take over the current immediately. Consequently, the output voltage of the phase leg changes. With a reverse current polarity, the current flows through the anti-parallel diode when the switch is turned off and the output voltage stays the same after turning off. The complementary switch will turn on after the dead time. Due to these reasons, the voltage polarity reversal phenomenon and phase drift phenomenon are observed and reported in [8].

For the DAB3 converter, the dead-time effect would not appear if the soft-switching could be achieved, especially for ZVS. For a relatively small dead time, since the anti-parallel diode would be turned on firstly, a turn-on delay of the switch caused by the dead-time effect would not influence the real turn-on time, which happens at the current zero-crossing point after a certain time period. Once the dead time is larger than the time difference between the diode turn-on instant and current zero-crossing point, the inductor current would cross zero in the dead-time period, which would generate a dead zone. The width of the dead-zone depends on the length of the dead time. The compensation method is given in [9], however a steady-state error still exists in the dead zone. The analytical solution of the dead-time compensation including the dead zone is introduced in [10] and [11] for DAB1 converters. For DAB3 converters, the dead-time compensation including the dead zone would be much more complicated than DAB1 converters due to the increasing number of operation sub-states. However,

1074

the power deviation caused by the dead zone can be always compensated by a close-loop controller without influencing the ZVS capability. Therefore, the dead zone is not considered in this paper.

The gate signals, current and voltage waveforms are compared under the situations with and without dead-time as shown in Fig. 4. Neglecting the aforementioned dead zone, the voltage waveforms remain the same if the ZVS can be realized. It is noticed that the voltage waveforms would be shifted if a hard-switching occurs. For the hard-switching operation of the input bridge, the voltage waveform of the input bridge would be shifted behind for a period that is equal to the primary side dead time $t_{d,pri}$. For the hard-switching operation of the output bridge, the voltage waveform of the output bridge would be shifted behind for a period that is equal to the secondary side dead-time $t_{d,sec}$. The dead-time effect can be directly observed with the help of the voltage waveforms. Hence, the compensation can be implemented by moving the voltage waveform backward to the desired positions as calculated in (2). It is noticed that the dead-time compensation can be decoupled between the input and output bridge. For DAB3 converter with a high transformer turns ratio, different semiconductor devices will be implemented due to the different voltage and current requirements. Therefore, different dead times are considered here.

$$d_{\varphi,comp} = \begin{cases} d_\varphi + 0 & \text{soft switching} \\ d_\varphi + f \cdot t_{d,pri} & \text{input hard switching} \\ d_\varphi - f \cdot t_{d,sec} & \text{output hard switching} \end{cases} \quad (2)$$

With the help of the space vector, the three-phase ac current traces can be observed in $\alpha - \beta$ diagram. A hexagon is formed in the conventional SPS modulation. Taking an operation point in the output bridge hard-switching range as an example, the ideal current traces are show in Fig. 5a. The area of the hexagon (proportional to the transferred power P_{SPS}) is a second-order function of the contour (proportional to the phase shift d_φ), which can be also reflected through (1). The rotation angle of the hexagon is related to the mismatch of the input and output voltages. The effective phase-shift angle increases once the dead time is included. Hence, the size of the hexagon also increases as illustrated in Fig. 5b. After the dead-time compensation, the size of the hexagon restores to the desired value.

III. ASYMMETRICAL DUTY-CYCLE CONTROL

A. Power Transmission

The ADCC modulation is proposed in [7] to realize the ZCS operation and reduce ac currents in the light load and wide input/output voltage range conditions. According to the transformer current and voltage waveforms, three different modes can be derived, which are the triangular buck (tri-buck) mode, triangular boost (tri-boost) mode and trapezoidal (trap) mode. With the variable duty cycles ($d_1, d_2 \leq 1/3$) in the primary and secondary bridge, the ZCS can be realized as shown in Fig. 6. The power transmission with the ADCC can be calculated by following (3), where $k_d = d^2 + d$. Hybrid modulation is implemented by combining SPS and ADCC

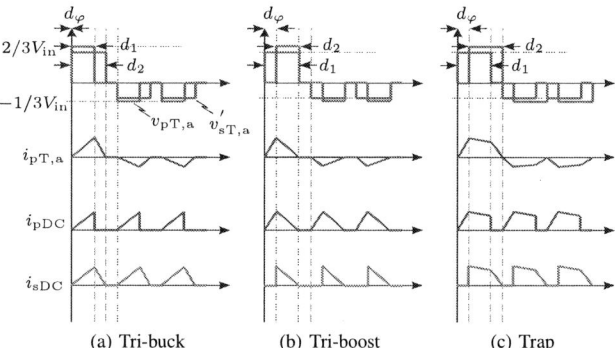

Fig. 6: ADCC modulation schemes

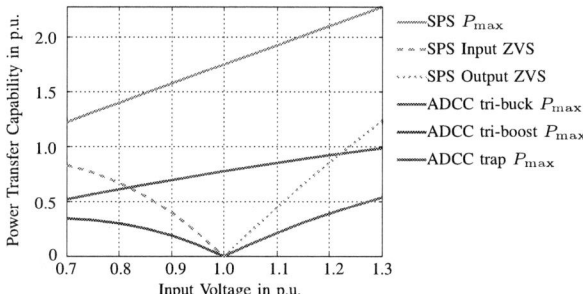

Fig. 7: Soft-switching range with hybrid modulation (SPS + ADCC)

together. In the light load range, ADCC is adopted to achieve ZCS, while SPS is considered for the normal and high power operations. With the hybrid modulation strategy, soft-switching range can be significantly extended as illustrated in Fig. 7. Theoretically, the ZCS can be achieved for all the operation range below the red line. The maximum transferred power of ADCC modulation for different modes can be calculated according to (4).

$$P_{ADCC} = \begin{cases} \frac{2\pi V_{in}^2 (1-d)d_1^2}{\omega L_\sigma} & \text{tri} - \text{buck} \\ \frac{2\pi V_{in}^2 (d-1)d_1^2}{\omega L_\sigma d} & \text{tri} - \text{boost} \\ \frac{2\pi V_{in}^2 (-9(k_d+1)d_1^2 + 6k_d d_1 - d^2)}{9\omega L_\sigma d} & \text{trap} \end{cases} \quad (3)$$

$$P_{ADCC,max} = \begin{cases} \frac{2\pi V_{in}^2 d^2(1-d)}{9\omega L_\sigma} & \text{at } d_2 = \frac{1}{3} & \text{tri} - \text{buck} \\ \frac{2\pi V_{in}^2 d^2(1-d)}{9\omega L_\sigma} & \text{at } d_2 = \frac{1}{3} & \text{tri} - \text{boost} \\ \frac{2\pi V_{in}^2 d^2}{\omega L_\sigma (k_d+1)} & \text{at } d_1 = \frac{k_d}{(k_d+1)} & \text{trap} \end{cases}$$

$$(4)$$

B. Dead-Time Compensation

The dead-time compensation of the ADCC can be analyzed in a similar way as for SPS. The voltage waveforms would be shifted if the dead-time is implemented. However, with the ADCC strategy, not only the phase shift, but also the duty cycles would be influenced as presented in Fig. 8. Therefore, both the phase shift and duty cycles are required to be compensated as calculated in (5) and (6). Fig. 6 shows that

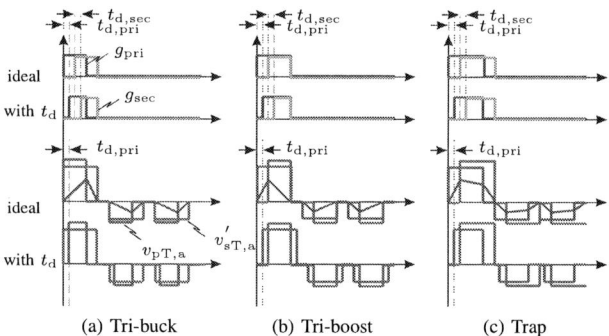

(a) Tri-buck   (b) Tri-boost   (c) Trap

Fig. 8: Dead-time influence in ADCC modulation

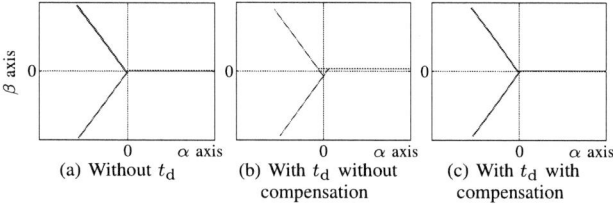

(a) Without t_d (b) With t_d without compensation   (c) With t_d with compensation

Fig. 9: Current traces in $\alpha - \beta$ diagram with tri-buck mode

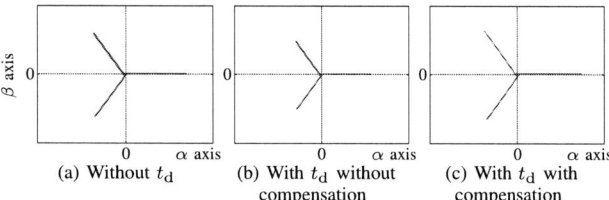

(a) Without t_d (b) With t_d without compensation   (c) With t_d with compensation

Fig. 10: Current traces in $\alpha - \beta$ diagram with tri-boost mode

the positive part of the ac voltage takes no longer than $1/3$ of the switch period, which indicates that only one upper switch of three phases is conducting at each time.

$$
d_{\varphi,\text{comp}} = \begin{cases} d_\varphi + f \cdot t_{d,\text{pri}} & \text{tri} - \text{buck} \\ d_\varphi + f \cdot t_{d,\text{pri}} & \text{tri} - \text{boost} \\ d_\varphi + f \cdot t_{d,\text{pri}} & \text{trap} \end{cases} \quad (5)
$$

$$
d_{1,\text{comp}} = \begin{cases} d_1 + f \cdot t_{d,\text{pri}} & \text{tri} - \text{buck} \\ d_1 & \text{tri} - \text{boost} \\ d_1 + f \cdot t_{d,\text{pri}} & \text{trap} \end{cases} \quad (6)
$$

The current traces are also observed in the $\alpha - \beta$ diagram under the ADCC modulation as shown in Fig. 9. A "Y" form trace that starts and ends at the origin is obtained in the ideal situation. Therefore, the ZCS is realized as presented in Fig. 9a and Fig. 10a for the tri-buck and tri-boost mode, respectively. A small deviation at the origin can be observed due to the implementation of the winding resistance in the transformer. With the dead time, the ZCS capability will be lost for the tri-buck mode as illustrated in Fig. 9b. However, there is no

big influence on the power transmission as the size of the "Y" trace remains almost the same.

For the tri-boost mode, the ZCS can still be guaranteed after implementing a dead-time. However, the power transmission will decrease as shown in Fig. 10b. With the proposed dead-time compensation method, the current traces get back to the desired form. The proposed dead-time compensation method does not consider a waveform overlap which often occurs near the maximum power boundary for the tri-buck and tri-boost mode, and always occurs for the trap mode operation. Therefore, a new modulation scheme based on ADCC with adaptive dead-time compensation will be introduced in the next section.

C. Adaptive Dead-Time Compensation

For the trap mode operation, the positive current waveform of each phase takes exactly $1/3$ of the period in the ideal situation. However, an overlap of the conduction period among the upper switches would be generated after the compensation because the compensated phase shift and duty cycles might exceed the prescribed limitations of the ADCC modulation. Similar problems would also occur for the triangular operations where the operation points are close to the maximum transmission boundaries. As a result, ZCS cannot be achieved and the system efficiency would decrease as illustrated in Fig. 13a. Such a distortion could also occur at the operation points that are close to the maximum transmission boundary for tri-buck and tri-boost mode.

In order to cope with the aforementioned problem, an adaptive dead-time compensation (ADTC) method is proposed. An additional zero-current interval d_{zc} is considered during the modulation. For the triangular mode operations, a zero-current interval would definitely occur at the end of every $1/3$ switching period, which could be used for the dead-time compensation. However, a minimum zero-current interval shall be considered to make sure that the zero-current interval is longer than the dead-time. Otherwise, a waveform overlap between different phases would occur. For the trapezoidal mode operation, an additional zero-current interval is inserted at the end of every $1/3$ switching period as shown in Fig. 11.

$$
P_{\text{ADTC}} = \begin{cases} \dfrac{2\pi V_{\text{in}}^2 (1-d)d_1^2}{\omega L_\sigma} & \text{tri} - \text{buck} \\[2mm] \dfrac{2\pi V_{\text{in}}^2 (d-1)d_1^2}{\omega L_\sigma} & \text{tri} - \text{boost} \\[2mm] \dfrac{2\pi V_{\text{in}}^2 (-9(k_d+1)d_1^2 k_{zc} + 6k_d d_1 - d^2 k_{zc}^2)}{9\omega L_\sigma d} & \text{trap} \end{cases} \quad (7)
$$

$$
P_{\text{ADTC,max}} = \begin{cases} \dfrac{2\pi V_{\text{in}}^2 d^2 (1-d) k_{zc}^2}{9\omega L_\sigma} & \text{at } d_2 = \frac{1}{3} - d_{zc} & \text{tri} - \text{buck} \\[2mm] \dfrac{2\pi V_{\text{in}}^2 (d-1) k_{zc}^2}{9\omega L_\sigma d} & \text{at } d_1 = \frac{1}{3} - d_{zc} & \text{tri} - \text{boost} \\[2mm] \dfrac{2\pi V_{\text{in}}^2 d^2 k_{zc}^2}{\omega L_\sigma (k_d+1)} & \text{at } d_1 = \frac{k_d k_{zc}}{3(k_d+1)} & \text{trap} \end{cases} \quad (8)
$$

The power transmission with ADTC are calculated according to (7). The power boundaries are calculated in (8), where $k_{zc} = 1 - 3 \cdot d_{zc}$, and $d_{zc} = f \cdot t_{zc}$. It is noticed that, for the tri-buck and tri-boost modes the transmission power with ADCT modulation is exactly the same as the one with

The 2018 International Power Electronics Conference

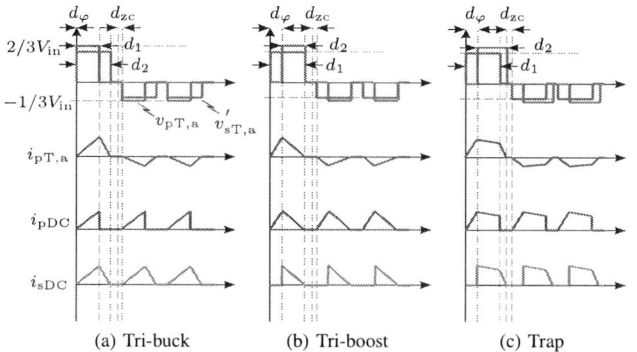

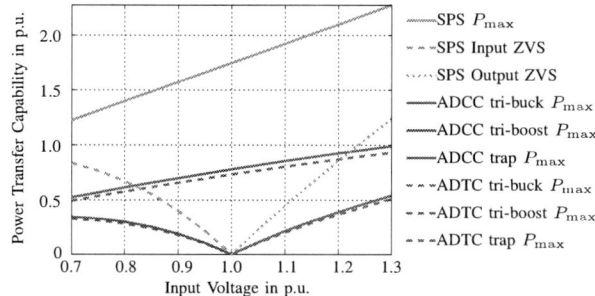

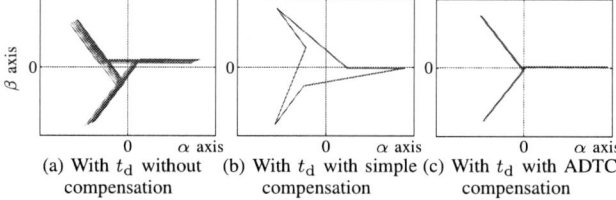

(a) Tri-buck   (b) Tri-boost   (c) Trap

Fig. 11: ADTC modulation schemes

Fig. 12: Soft-switching range with hybrid modulation
(SPS + ADCC + ADTC)

(a) With t_d without compensation   (b) With t_d with simple compensation   (c) With t_d with ADTC compensation

Fig. 13: Current traces in $\alpha - \beta$ diagram with trap mode

ADCC modulation. Only the maximum power boundary is different. An additional zero-current factor k_{zc} is introduced into the ADTC modulation (8) through the zero-current interval compared with the ADCC (4). Usually the t_{zc} is settled longer than the dead-time. In order to have a sufficient margin, $t_{zc} = 2 \cdot \max\{t_{d,pri}, t_{d,sec}\}$ is considered here. The power boundaries with ADTC are plotted in Fig. 12. As the trade-off, the maximum transmission power would slightly decrease if the ADTC is implemented, but ZCS could be guaranteed.

The three-phase ac currents for trap mode operation are observed in $\alpha - \beta$ diagram. With the ADCC modulation, ZCS would not be realized if a dead time is considered since the current traces cannot reach the origin as shown in Fig. 13a. After compensation, the traces under ADCC would be distorted and the ZCS operation is still not realized. By implementing the proposed ATDC modulation, the current traces returns to the origin every ⅓ switching period, which indicates that ZCS is achieved.

The power deviation (defined as the difference between power transmission with and without t_d divided by the power

transmission without t_d) under the different conditions are simulated for the whole operation range. Comparing Fig. 14a and Fig. 14b, the dead zone would be reduced with a decreased dead-time, thus the power deviation would also be reduced especially near the boundaries. However, the power deviation still exists in the light-load conditions and cannot be compensated in the trap mode. By implementing the ADTC method, the deviation could be significantly minimized as shown in Fig. 14c. Theoretically, the power deviation could also be compensated by the close-loop controller, but ZCS cannot be guaranteed. Therefore, not only the smaller power deviation, but also the higher efficiency and the better dynamic performance can be achieved with the ADTC method.

All the previous compensations are calculated based on the assumption that the dead time occurs at the starting point of every switching period, and are only valid for a positive power transmission. In order to provide an overview, operations with the implementation of t_d at the end of the switching period and a negative power transmission are also considered and summarized in the TABLE. I.

IV. EXPERIMENTAL RESULTS

The proposed dead-time compensation method has been tested in a scaled-down converter prototype with an input voltage of 12 V, an output voltage of 96 V, and a switching frequency of 20 kHz.

The decoupled dead-time compensation with SPS modulation is firstly validated. Different dead-times are settled in the primary and secondary bridge with $t_{d,pri} = 1.0\,\mu s$ and $t_{d,sec} = 1.5\,\mu s$. The reference phase shift is $d_\varphi = 0.016$, which corresponds to $t_\varphi = 0.8\,\mu s$. Without the compensation, the phase shift increases due to the hard-switching in the secondary side. The effective phase shift corresponds to a period of $t_\varphi = 2.3\,\mu s$, which is exactly the sum of the ideal phase shift and the dead time in the secondary bridge as shown in Fig. 15. After the compensation, the phase shift is restored to the desired value.

Experimental result with the asymmetrical dead time is also observed for the ADCC modulation i.e for the tri-buck mode as shown in Fig. 16. The implementation of the dead time delays the voltage waveform slightly and also reduces the duty cycle for the primary side. Therefore, ZCS would not occur as shown in Fig. 16. After compensation, the voltage waveform goes back to the desired shape and ZCS can be realized.

The proposed ADTC modulation is tested and compared with the conventional operations for the trap mode. Without the compensation, the ZCS cannot be realized and the current waveform is significantly distorted. With the conventional compensation method, the control parameters would excess the prescribed limitations, hence the ZCS cannot be achieved as shown in Fig. 17a. By implementing the ADTC method, the ZCS can be guaranteed as illustrated in Fig. 17b.

ADTC is also tested with negative power flow. The dead time is implemented at the starting point of a switching period. Compensated phase shift and duty cycles are calculated according to TABLE. I. The experimental results for tri-buck and tri-boost modes are presented in Fig. 18. ZCS could be realized with the proposed ADTC as illustrated.

1077

The 2018 International Power Electronics Conference

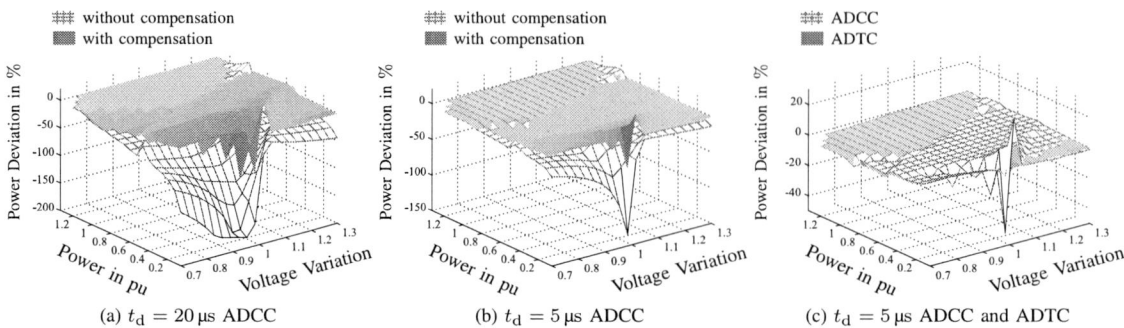

(a) $t_{\mathrm{d}} = 20\,\mu s$ ADCC   (b) $t_{\mathrm{d}} = 5\,\mu s$ ADCC   (c) $t_{\mathrm{d}} = 5\,\mu s$ ADCC and ADTC

Fig. 14: Power deviations of hybrid modulations

TABLE I: Summary of dead-time compensation

Modulation	Positive Power		Negative Power	
	t_{d} **at Start**	t_{d} **at End**	t_{d} **at Start**	t_{d} **at End**
SPS Soft-switching	none	$d_{\varphi,\mathrm{comp}} = d_{\varphi} + (t_{\mathrm{d,pri}} - t_{\mathrm{d,sec}}) \cdot f$	none	$d_{\varphi,\mathrm{comp}} = d_{\varphi} - (t_{\mathrm{d,pri}} - t_{\mathrm{d,sec}}) \cdot f$
SPS Input Hard-switching	$d_{\varphi,\mathrm{comp}} = d_{\varphi} + t_{\mathrm{d,pri}} \cdot f$	$d_{\varphi,\mathrm{comp}} = d_{\varphi} + t_{\mathrm{d,sec}} \cdot f$	$d_{\varphi,\mathrm{comp}} = d_{\varphi} - t_{\mathrm{d,sec}} \cdot f$	$d_{\varphi,\mathrm{comp}} = d_{\varphi} - t_{\mathrm{d,pri}} \cdot f$
SPS Output Hard-switching	$d_{\varphi,\mathrm{comp}} = d_{\varphi} - t_{\mathrm{d,sec}} \cdot f$	$d_{\varphi,\mathrm{comp}} = d_{\varphi} - t_{\mathrm{d,pri}} \cdot f$	$d_{\varphi,\mathrm{comp}} = d_{\varphi} + t_{\mathrm{d,pri}} \cdot f$	$d_{\varphi,\mathrm{comp}} = d_{\varphi} + t_{\mathrm{d,sec}} \cdot f$
ADCC (ADTC) Tri-buck Mode	$d_{\varphi,\mathrm{comp}} = d_{\varphi} + t_{\mathrm{d,pri}} \cdot f$ $d_{1,\mathrm{comp}} = d_1 + t_{\mathrm{d,pri}} \cdot f$	$d_{\varphi,\mathrm{comp}} = d_{\varphi} + t_{\mathrm{d,sec}} \cdot f$ $d_{1,\mathrm{comp}} = d_1 + t_{\mathrm{d,pri}} \cdot f$	$d_{\varphi} = d_{\varphi} - t_{\mathrm{d,sec}} \cdot f$	$d_{\varphi,\mathrm{comp}} = d_{\varphi} + t_{\mathrm{d,pri}} \cdot f$
ADCC (ADTC) Tri-boost Mode	$d_{\varphi,\mathrm{comp}} = d_{\varphi} + t_{\mathrm{d,pri}} \cdot f$	$d_{\varphi,\mathrm{comp}} = d_{\varphi} + t_{\mathrm{d,sec}} \cdot f$	$d_{\varphi,\mathrm{comp}} = d_{\varphi} - t_{\mathrm{d,sec}} \cdot f$ $d_{2,\mathrm{comp}} = d_2 + t_{\mathrm{d,sec}} \cdot f$	$d_{\varphi,\mathrm{comp}} = d_{\varphi} - t_{\mathrm{d,pri}} \cdot f$ $d_{2,\mathrm{comp}} = d_2 + t_{\mathrm{d,sec}} \cdot f$
ADCC (ADTC) Trap Mode	$d_{\varphi,\mathrm{comp}} = d_{\varphi} + t_{\mathrm{d,pri}} \cdot f$ $d_{1,\mathrm{comp}} = d_1 + t_{\mathrm{d,pri}} \cdot f$	$d_{\varphi,\mathrm{comp}} = d_{\varphi} + t_{\mathrm{d,sec}} \cdot f$ $d_{1,\mathrm{comp}} = d_1 + t_{\mathrm{d,pri}} \cdot f$	$d_{\varphi,\mathrm{comp}} = d_{\varphi} - t_{\mathrm{d,sec}} \cdot f$	$d_{\varphi} = d_{\varphi} - t_{\mathrm{d,pri}} \cdot f$

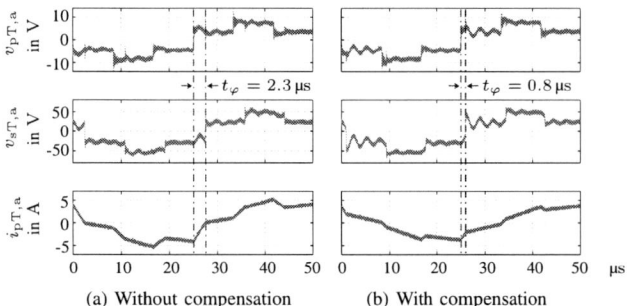

(a) Without compensation   (b) With compensation

Fig. 15: Test of SPS with asymmetrical dead time

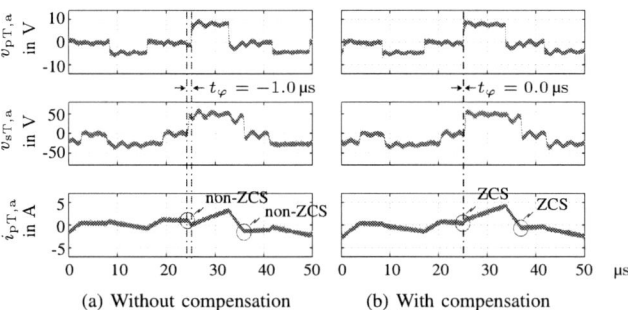

(a) Without compensation   (b) With compensation

Fig. 16: Test of tri-buck mode with asymmetrical dead time

V. CONCLUSION

In this paper, the dead-time compensation method of the DAB3 converter is investigated for different modulation strategies with decoupled effects of the dead time in the primary and secondary bridges. The dead time distorts the voltage and current waveforms of the SPS and ADCC modulation schemes, which brings two main issues: transmitted power deviation and violated ZCS behavior. The deviated power can be compensated by modifying the phase-shift angle and/or duty-cycles in an open-loop manner, or it can also be compensated by a closed-loop controller. However, the ZCS operation cannot be guaranteed by a closed-loop controller. It is achieved in the proposed method by inserting additional zero-current intervals in the ADCC modulation. The proposed compensation method is verified with the experimental measurements, which coincides very well with the theoretical analysis.

ACKNOWLEDGEMENT

Funded by the Federal Ministry of Education and Science (BMBF, FKZ03SF0490), Flexible Electrical Networks (FEN) Research Campus.

The 2018 International Power Electronics Conference

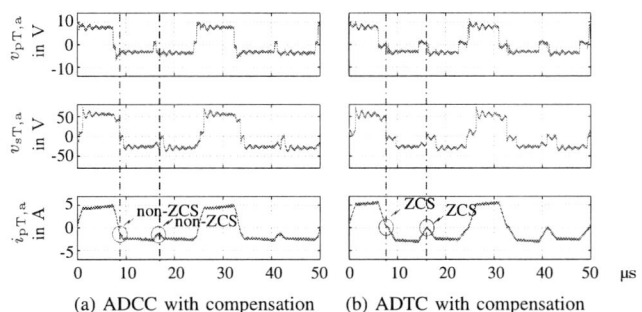

(a) ADCC with compensation   (b) ADTC with compensation

Fig. 17: Test of trap mode

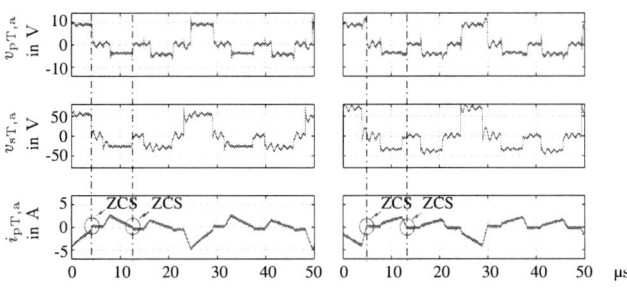

(a) Tri-buck with negative power (b) Tri-boost with negative power

Fig. 18: Test of ADTC with negative power flow

REFERENCES

[1] R. W. De Doncker, D. M. Divan, and M. H. Kheraluwala, "A three-phase soft-switched high-power-density dc/dc converter for high-power applications," *IEEE transactions on industry applications*, vol. 27, no. 1, pp. 63–73, 1991.

[2] R. U. Lenke, *A contribution to the design of isolated dc-dc converters for utility applications.* E. ON Energy Research Center, RWTH Aachen, Univ., 2012.

[3] J. Hu, P. Joebges, and R. W. De Doncker, "Maximum power point tracking control of a high power dc-dc converter for pv integration in mvdc distribution grids," in *Applied Power Electronics Conference and Exposition (APEC), 2017 IEEE.* IEEE, 2017, pp. 1259–1266.

[4] M. Stieneker, B. J. Mortimer, N. R. Averous, H. Stagge, and R. W. De Doncker, "Optimum design of medium-voltage dc collector grids depending on the offshore-wind-park power," in *Power Electronics and Machines for Wind and Water Applications (PEMWA), 2014 IEEE Symposium.* IEEE, 2014, pp. 1–8.

[5] A. Tripathi, K. Mainali, S. Madhusoodhanan, D. Patel, A. Kadavelugu, S. Hazra, S. Bhattacharya, and K. Hatua, "Mvdc microgrids enabled by 15kv sic igbt based flexible three phase dual active bridge isolated dc-dc converter," in *2015 IEEE Energy Conversion Congress and Exposition (ECCE)*, Sept 2015, pp. 5708–5715.

[6] J. Hu, N. Soltau, and R. W. De Doncker, "Asymmetrical duty-cycle control of three-phase dual-active bridge converter for soft-switching range extension," in *Energy Conversion Congress and Exposition (ECCE), 2016 IEEE.* IEEE, 2016, pp. 1–8.

[7] J. Hu, Z. Yang, N. Soltau, and R. W. De Doncker, "A duty-cycle control method to ensure soft-switching operation of a high-power three-phase dual-active bridge converter," in *Future Energy Electronics Conference and ECCE Asia (IFEEC 2017-ECCE Asia), 2017 IEEE 3rd International.* IEEE, 2017, pp. 866–871.

[8] B. Zhao, Q. Song, W. Liu, and Y. Sun, "Dead-time effect of the high-frequency isolated bidirectional full-bridge dc–dc converter: Com-

prehensive theoretical analysis and experimental verification," *IEEE Transactions on Power Electronics*, vol. 29, no. 4, pp. 1667–1680, 2014.

[9] K. Takagi and H. Fujita, "Dynamic control and dead-time compensation method of an isolated dual-active-bridge dc-dc converter," in *Power Electronics and Applications (EPE'15 ECCE-Europe), 2015 17th European Conference on.* IEEE, 2015, pp. 1–10.

[10] D. Segaran, D. G. Holmes, and B. P. McGrath, "Enhanced load step response for a bidirectional dc–dc converter," *IEEE Transactions on Power Electronics*, vol. 28, no. 1, pp. 371–379, 2013.

[11] H. Bai, Z. Nie, and C. Chunting Mi, "A model-based dead-band compensation for the dual-active-bridge isolated bidirectional dc–dc converter," *IEEJ Transactions on Electrical and Electronic Engineering*, vol. 6, no. 6, pp. 517–524, 2011.

[12] N. Schibli, "Symmetrical multilevel converters with two quadrant dc-dc feeding," *EPFL, PhD Thesis*, no. 2220, 2000.

[13] H. van Hoek, J. A. Ferreira, and R. W. De Doncker, *Design and operation considerations of three-phase dual active bridge converters for low-power applications with wide voltage ranges.* Lehrstuhl und Institut für Stromrichtertechnik und Elektrische Antriebe, 2017, no. RWTH-2017-02955.

The 2018 International Power Electronics Conference

Evaluation of a High-Frequency Reactor with a New Wire Guide for a Toroidal Core

Hideki Ayano[1*], Akira Fujimura[1], Yoshihiro Matsui[1]

1 Department of Electrical Engineering, National Institute of Technology, Tokyo College, Tokyo, Japan

*E-mail: ayano@tokyo-ct.ac.jp

Abstract—**This paper proposes a new high-frequency reactor with a wire guide for small-size toroidal core. In theory, reactor size can be reduced in proportion to switching frequency. However, practically, as the switching frequency is increased, the reactor is affected by the parasitic capacitance existing between the windings and between the winding and the core, and the inductance characteristic of the reactor is deteriorated. As a result, there is a limit to increase the switching frequency. This paper proposes two kinds of wire guides with two-layer winding and three-layer winding to reduce the parasitic capacitance. The effects of the wire arrangement and the material of the guide are also evaluated. With the proposed guides, the parasitic capacitance decreases by 0.73 times and 0.66 times compared with the conventional reactor. Moreover, the linear regions of the impedance which can be regarded as a pure inductance are enlarged.**

Keywords—*reactor, parasitic capacitance, high frequency, wire guide*

I. INTRODUCTION

Recently, semiconductor devices such as SiC and GaN have been put into practical use, and low loss and high frequency drive has been realized[1]. Reactors which are used in DC-DC converters can reduce current ripples in proportion to a switching frequency in theory. In other words, the higher the switching frequency is, the smaller the inductance can be used if the current ripple is the same. Especially, since the size and weight of the reactors are large, the reduction in size and weight can greatly contribute to the reduction in size and weight of the converter system as a whole. Furthermore, by increasing the switching frequency and reducing the current ripples, the vibrations of the reactors and the motor system can be reduced, so the acoustic noises can be reduced. Moreover, since the control cycle of the converter can be shortened, the response speed of the current control can be increased. However, practically, as the switching frequency is increased, the reactors are affected by the parasitic capacitances existing between the windings and between the winding and the core, and the inductance characteristic of the reactor is deteriorated[2]. As a result, there is a limit to increase the switching frequency[3]. This problem has been pointed out about not only reactors but also EMI filters and high frequency transformers, and their modeling, impedance evaluation and winding method in the high frequency region have been reported[4]-[9]. The authors had reported the wire guide for a relatively big-size core.[10] The wire guide to reduce the parasitic capacitances for a relatively large-size

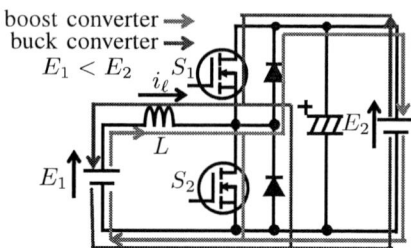

Fig. 1. DC-DC converter system.

core had been reported.

This paper proposes a new reactor, which can be used in high frequency conditions, with a wire guide for a small-size toroidal core. The wire guide is manufactured by using a 3D printer and it is attached at the center hole of the toroidal core. In the case of a small-size core, the winding arrangement and the realization of the wire guide are restricted. This wire guide enables to reduce the parasitic capacitances between the windings and between the winding and the core. The impedance characteristics of the reactor using the proposed wire guides are evaluated, and the reduction effect of the parasitic capacitances and the expanding effect of the frequency range which can be regarded as a pure inductance are shown. Furthermore, the impedance characteristics are evaluated for the cases where the shape of the guide and the material of the guide are changed.

II. PROBLEM WHEN A REACTOR IS USED AT HIGH FREQUENCY

Fig. 1 shows a main circuit of a bidirectional DC-DC converter ($E_1 < E_2$). The system of Fig. 1 is composed of two switches(S_1, S_2) and a reactor. The red line is the current path of the boost converter and the voltage E_1 is boosted to E_2 via the switching devices. The blue line is the current path of the buck converter and the voltage E_2 is stepped down to E_1 via the switching device. This circuit is used in electric vehicles with rechargeable batteries and is operated as the boost converter and the buck converter in the power running and the regenerative modes, respectively.

Fig. 2 shows the simulation results of the current i_ℓ which flows in the inductance L when the switching frequency is changed in the power running operation. In i_ℓs, the ripple currents are superimposed on the direct

1080

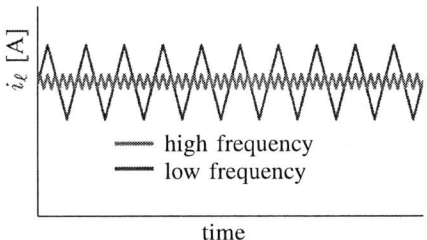

Fig. 2. Currents of the DC-DC converter when the switching frequency is changed.

current. When the ripple currents are large, adverse effects such as vibration and noise of the reactor may occur. The magnitude of the ripple can be calculated by

$$I_{ripple} = \frac{E_1}{L} \times T_{s2}, \tag{1}$$

where, T_{s2} is the conduction time of S_2. From (1), the ripple current can be reduced by (i) increasing the inductance of L or (ii) increasing the switching frequency (decreasing T_{s2}). The method (ii) is used in many cases, in general, because the weight and volume of the reactor becomes large in method (i). From Fig. 2, the higher the switching frequency is, the smaller the ripple current becomes under the ideal condition.

Fig. 3 shows a conventional reactor with a toroidal-core and its equivalent circuit under the high switching frequency condition. Fig. 3(a) is the schematic of a reactor made by winding a copper wire coil around a toroidal core. Since the toroidal core has no gap, the leakage flux is small. However, when the number of windings of the coil is increased, the wires are densely gathered together at the center hole of the toroidal core. In this case, the parasitic capacitances between the copper wires and between the copper wire and the core may affect the impedance characteristic of the reactor.

Fig. 3(b) is the equivalent circuit of the reactor when the switching frequency is high[5]. It is assumed that the coil windings (t_1, t_2, t_3) whose diameters are r are wound at the interval d and the distances between the coil windings and the core are h. In this case, the parasitic capacitances C_d and C_h per unit length, which exists between the windings and between the winding and the core, respectively, can be expressed as follows:

$$C_d = \frac{\pi \varepsilon_0}{\ln\left\{ \frac{d}{2r} + \sqrt{\left(\frac{d}{2r}\right)^2 - 1} \right\}}, \tag{2}$$

$$C_h = \frac{2\pi \varepsilon_0}{\ln\left\{ \frac{h}{r} + \sqrt{\left(\frac{h}{r}\right)^2 - 1} \right\}}. \tag{3}$$

From (2) and (3), if d and h can be increased, C_d and C_h can be reduced, and the impedance characteristic of the reactor can be improved under the high switching frequency condition.

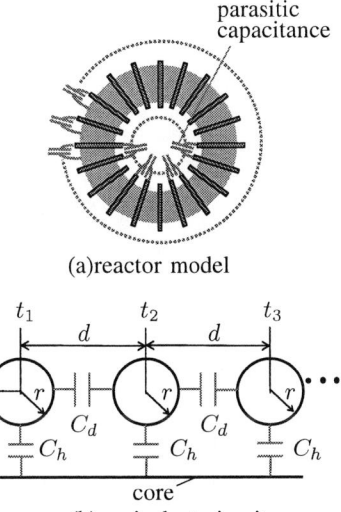

Fig. 3. A conventional reactor with a toroidal-core and its equivalent circuit.

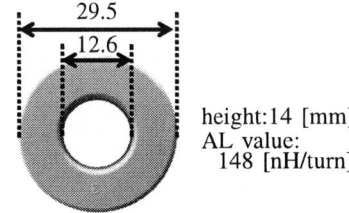

Fig. 4. The core used in the evaluation.

III. THE PROPOSAL OF NEW WIRE GUIDE FOR A TOROIDAL CORE

In the proposed high-frequency reactor, the distances between the windings and between the coil winding and the core are increased by using a resin wire guide. Fig. 4 shows the dimensions of the toroidal core (dust core: HK - 14D (TOHO ZINC CO., LTD.)) used for the evaluation. In the proposed high-frequency reactor, the wire guide is attached at the center hole of the toroidal core. This wire guide enables to reduce the parasitic capacitances between the windings and between the winding and the core. The wire guide is made by 3D printer. In this paper, (i) a reactor with two-layer winding and (ii) a reactor with three-layer winding are proposed.

Fig. 5 shows the proposed wire guides for the reactor with two-layer winding. Fig. 5(a) is the picture of the wire guides. The transparent one and the white one are made of the ultraviolet curing resin and the thermoplastic resin, respectively. Fig. 5(b) shows their dimensions. The thickness of the guide is 1 mm. Figs. 6 and 7 show the pictures and winding order of the proposed high-frequency reactor with two-layer winding, respectively. The enamelled copper wire with a diameter of 0.75 mm is used for the coil winding, and the number of turns is

1081

The 2018 International Power Electronics Conference

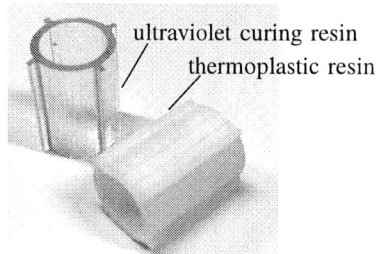

(a) picture of wire guides

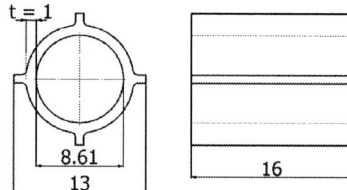

(b) dimensions of the wire guide

Fig. 5.   Proposed wire guides (two-layer winding)

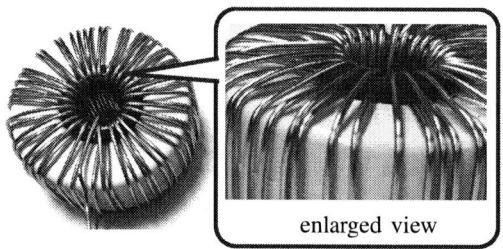

enlarged view

Fig. 6.   Pictures of proposed reactor (two-layer winding).

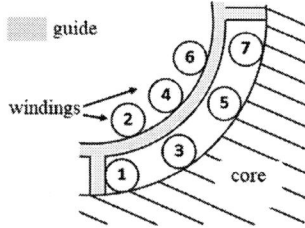

Fig. 7.   Winding order of the proposed reactor (two-layer winding).

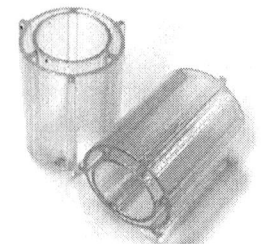

(a) picture of wire guides

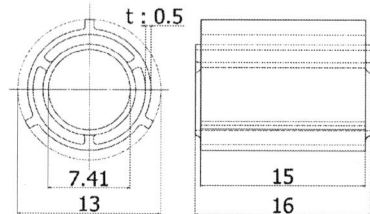

(b) dimensions of the wire guide

Fig. 8.   Proposed wire guides (three-layer winding)

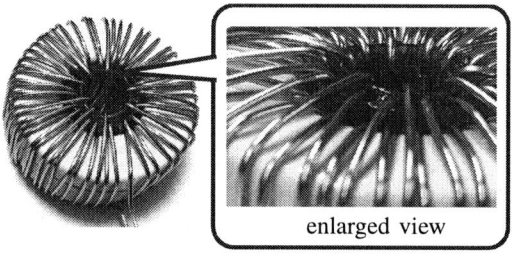

enlarged view

Fig. 9.   Pictures of proposed reactor (three-layer winding).

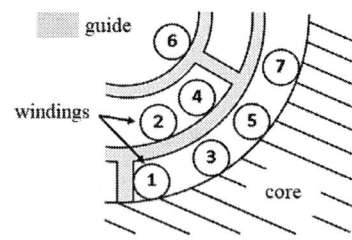

Fig. 10.   Winding order of the proposed reactor (three-layer winding).

56. As shown in the enlarged view of Fig. 6, the length of the guide in the axial direction is slightly longer than that of the core. And, as shown in Fig. 7, the windings are alternately wound around the inner frame of the wire guide and the clearance between the core and the wire guide. The wire guide keeps the distances between the wires and between the wires and the core as far as possible. Especially, the part of the wire guide protruding from the core in its axial direction prevents overlapping of the windings. As a result, the proposed wire guide has the effect of reducing the parasitic capacitances C_d and

C_h in Fig. 3.

Fig. 8 shows the proposed wire guides for the reactor with three-layer winding. Fig. 8(a) is the picture of the wire guides and Fig. 8(b) shows their dimensions. The thickness of the guide is 0.5 mm, so that the windings are not crowded at the inner frame of the wire guide. Since the guide is extremely thin, the guide must be made of the ultraviolet curing resin which can be manufactured with high accuracy. Figs. 9 and 10 show the pictures and winding order of the proposed high-frequency reactor with three-layer winding, respectively. The enamelled

1082

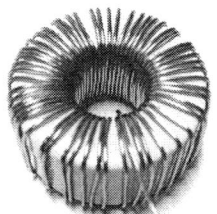 

(a) conventional reactor   (b) concentrated winding reactor

Fig. 11.   Pictures of a conventional reactor and a concentrated winding reactor.

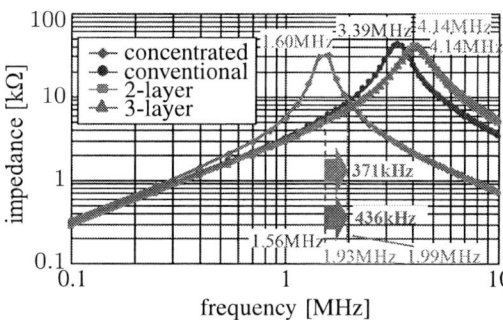

Fig. 12.   Impedance characteristics of proposed and conventional reactors.

copper wire with a diameter of 0.75 mm is used for the coil winding and the number of turns is 56, which are the same condition in Fig. 6. By winding in the order shown in Fig. 10, the ratios of the number of windings at the outer part, the middle part, and the center part can be 3: 2: 1. The wire guide keeps the distances between the wires and between the wires and the core as far as possible, which is the same as Fig. 7. The wire guide of Fig. 8 can use the inner space of the center hole of the toroidal core than that of Fig. 5. The length of the outer frame of the guide in the axial direction is slightly longer than that of the core, and that of the inner frame is slightly longer than that of the outer frame as shown in Fig. 8(b) and Fig. 9. The differences in the length prevent overlapping of the windings. As a result, the proposed wire guide has the effect of reducing the parasitic capacitances C_d and C_h in Fig. 3.

IV.   EVALUATIONS OF THE IMPEDANCE CHARACTERISTICS

A. The effect of the proposed wire guides

In this section, the impedance characteristics of the proposed reactors shown in Figs. 6 and 9 are evaluated. In addition, the conventional reactor and the concentrated winding reactor in which the coil winding is densely wound around a part of the core are also evaluated for comparison. Fig. 11 shows the pictures of a conventional reactor and a concentrated winding reactor, respectively. The toroidal core (HK-14D) shown in Fig. 4 is used for all reactors, and the copper wire diameter and the turns of the coil winding are fixed to be 0.75 mm and 56, respectively. The material of the guides for the two-layer winding and the three-layer winding are the ultraviolet curing resin. For the measurement, the impedance analyzer (E4990A, KEYSIGHT Technology) is used.

Fig. 12 shows the impedance characteristics of each reactor. The resonance frequencies of the conventional reactor, the concentrated winding reactor, the proposed reactors with two-layer winding and with three-layer winding are 3.39 MHz, 1.60 MHz, 4.14 MHz and 4.14 MHz, respectively. Both resonance frequencies of the proposed reactors can be increased by 750 kHz than that of the conventional reactor. The resonance frequency of the concentrated winding reactor is 1.87 MHz lower than that of the conventional reactor, and the characteristics at high frequencies are remarkably deteriorated.

TABLE I.   INDUCTANCES AND CAPACITANCES OF PROPOSED AND CONVENTIONAL REACTORS.

	resonant frequency f_r [MHz]	inductance L [μH]	capacitance C [pF]
concentrated	1.60	542	18.4
conventional	3.39	496	4.45
2-layer	4.14	458	3.23
3-layer	4.14	506	2.93

Table I shows the resonance frequencies f_r, inductances L and parasitic capacitances C of each reactor. In the measurement range, the resistance of each reactor is sufficiently small and can be neglected, and each reactor can be regarded as an LC parallel circuit. However, the influence of resistance appears near each resonance point. Here, L are derived assuming that the reactor can be approximated to be only the inductance L at 100 kHz. C is obtained by substituting f_r and L into

$$C = \frac{1}{(2\pi f_r)^2 L}. \qquad (4)$$

The parasitic capacitance of the concentrated winding reactor is about 4.1 times larger than that of the conventional reactor. This is because the parasitic capacitances C_d between windings in Fig. 3 are increased by densely winding. That is, even when making a conventional reactor, it is important that the coil winding is wound evenly in order to reduce the parasitic capacitance. The parasitic capacitances of the proposed reactors can be reduced by about 0.73 and 0.66 times in the two-layer winding and the three-layer winding, respectively, as compared with that of the conventional reactor. This is because the proposed wire guides reduce the parasitic capacitances C_d and C_h in Fig. 3.

The reactors can be regarded as ideal coils at low frequency where their impedances increase linearly with the frequency as shown in Fig. 12. The inductance of each reactor at 100 kHz obtained from the data in Fig. 12 is defined as L, and the frequency range where the error between L and actual inductance in Fig. 12 is within $\pm 20\%$ is defined as the linear region. The upper limit frequencies of the linear regions for the conventional reactor, the concentrated winding reactor, the reactors with proposed two-layer and three-layer guides are 1.56 MHz, 661 kHz, 1.93 MHz, and 1.99 MHz, respectively.

1083

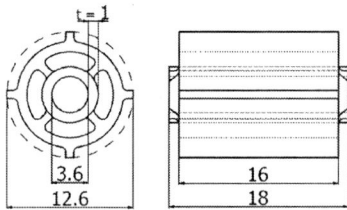

Fig. 13. Dimensions of the wire guide for a comparative experiment

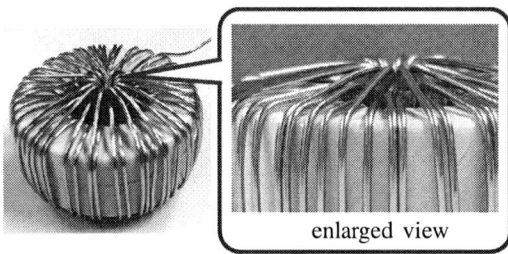

Fig. 14. Pictures of the reactor for the comparative experiment (three-layer winding).

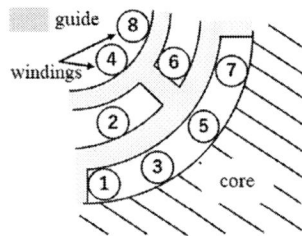

Fig. 15. Winding order of the reactor for the comparative experiment.

Therefore, by using the proposed guides, the frequencies which can be regarded as a pure inductance in the proposed reactors with two-layer and three-layer windings can be increased to 371 kHz and 436 kHz, respectively, than that in the conventional reactor.

B. The effect of the winding arrangement

In this section, the influence of the winding arrangement in the center hole of the toroidal core is evaluated on the reactor with three-layer winding. Fig. 13 shows the dimensions of the guide for comparative evaluation. The thickness of the wire guide in Fig. 13 is thicker than that in Fig. 8. Furthermore, the diameter of the inner frame of the guide of Fig. 13 is smaller than that of Fig. 8. Figs. 14 and 15 show pictures and winding order of the reactor with the guide of Fig. 13. The toroidal core (HK-14D) is used for the reactor and the copper wire diameter and the turns of the coil winding are 0.75 mm and 56, respectively, which are the same as those of the proposed reactor with three-layer winding in Fig. 9. The distance between the windings can be sufficiently obtained between the inner frame and the outer frame.

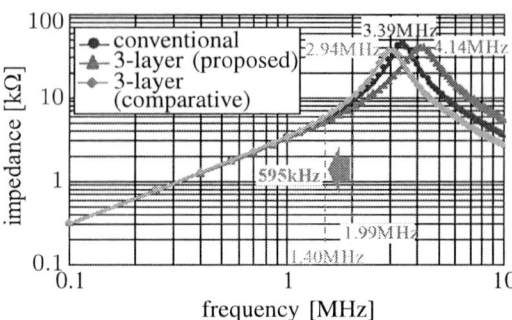

Fig. 16. Impedance characteristics of the reactor for the comparative experiment.

TABLE II. INDUCTANCES AND CAPACITANCES OF THE REACTOR FOR THE COMPARATIVE EXPERIMENT.

	resonant frequency f_r [MHz]	inductance L [μH]	capacitance C [pF]
conventional	3.39	496	4.45
3-layer (proposed)	4.14	506	2.93
3-layer (comparative)	2.94	508	5.77

However, the windings are dense in the inner frame, since the thickness of the wire guide is large and the diameter of the inner frame is small.

Fig. 16 shows the impedance characteristics of the conventional reactor, the proposed reactor with three-layer winding and the reactor for the comparative experiment. Here, the ultraviolet curing resin is used as the material of the guides for the proposed reactor with three-layer winding and the reactor for the comparative experiment. The resonance frequencies of the conventional reactor, the proposed reactor with three-layer winding and the reactor for the comparative experiment are 3.39 MHz, 4.14 MHz and 2.94 MHz, respectively. The resonance frequency of the reactor for the comparative experiment is 1.2 MHz lower than that of the proposed reactor with three-layer winding, and is lower than that of the conventional reactor.

Table II shows the resonance frequencies f_r, inductances L and parasitic capacitances C of each reactor. In the measurement range, the resistance of each reactor is sufficiently small and can be neglected, and each reactor can be regarded as an LC parallel circuit. However, the influence of resistance appears near each resonance point. Here, L are derived assuming that the reactor can be approximated to be only the inductance L at 100 kHz. C is obtained by eq. (4). The parasitic capacitance of the reactor for the comparative experiment is about 2 times larger than that of the proposed reactor with three-layer winding. Furthermore, from Fig. 16, the upper limit frequencies of the linear regions for the conventional reactor, the proposed reactor with three-layer winding and the reactor for the comparative experiment are 1.56 MHz, 1.99 MHz, and 1.40 MHz, respectively. The upper limit frequency of the reactor for the comparative experiment is 595 kHz lower than that of the proposed reactor with three-layer winding. Despite using the guides for

The 2018 International Power Electronics Conference

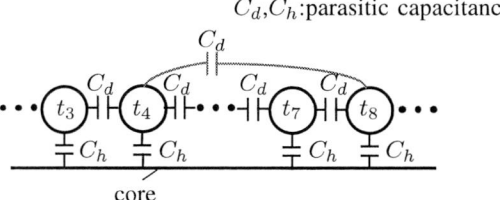

Fig. 17. Equivalent circuit of the reactor for the comparative experiment.

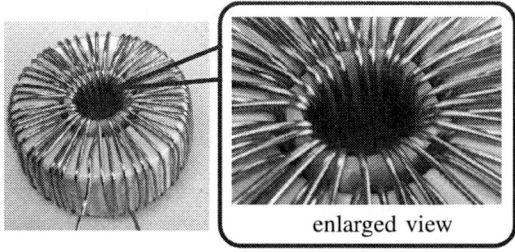

Fig. 18. Pictures of the reactor with a thermoplastic resin guide.

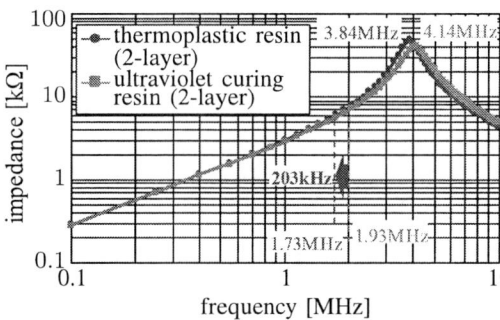

Fig. 19. Impedance characteristics when the material of the guides are different.

TABLE III. INDUCTANCES AND CAPACITANCES WHEN THE WINDING IS WOUND TO THE CENTER OF THE HOLE.

	resonant frequency f_r [MHz]	inductance L [μH]	capacitance C [pF]
thermoplastic resin	3.84	465	3.70
ultraviolet curing resin	4.14	458	3.23

three-layer winding, there is a significant difference in frequency characteristics. This is because, in the reactor for the comparative experiment, the parasitic capacity is arisen at the windings in the inside of the inner frame where the windings are dense as shown in Fig. 17. In other words, due to the parasitic capacitance between Turn 4 and Turn 8 in Figs. 15 and 17, the parasitic capacitances connected in series are decreased, so the combined capacitance of the reactor for the comparative experiment becomes larger than that of the proposed reactor with three-layer winding. On the other hand, in the proposed reactor with three-layer winding, the wire guide is extremely thin and the diameter of the inner frame is large, so that the windings are evenly separated from each other in the inside of the inner frame. Therefore, the consideration of the diameter of the innermost frame and the winding arrangement, so that the distance between the windings is increased, is important to design the wire guide for the high-frequency reactor.

C. The effect of the material of proposed wire guides

In this section, the effect of the material of the wire guide is evaluated on the impedance characteristics of the reactor. As the material of the guide, ultraviolet curing resin and thermoplastic resin (ABS resin) are used. Generally, the tensile strength and the bending strength of the ultraviolet curing resin are larger than those of the thermoplastic resin. In addition, the hardness of the ultraviolet curing resin is higher than that of the thermoplastic resin.

Fig. 18 shows a picture of the reactor with a thermoplastic resin guide for two-layer winding which is shown in Fig. 5. The toroidal core (HK-14D) is used for this reactor, and the copper wire diameter and the turns of the coil winding are 0.75 mm and 56, respectively,

which is the same as the proposed reactor with two-layer winding in Fig. 6. The dimension of the wire guide and the winding order are the same as Figs. 5 and 7, respectively.

Fig. 19 shows the impedance characteristics of the reactors with the ultraviolet curing resin and thermoplastic resin guide for two-layer winding. The resonance frequencies are 4.14 MHz and 3.84 MHz, respectively. The resonance frequency of the reactor with the ultraviolet curing resin guide is 300 kHz higher than that of the reactor with thermoplastic resin guide. This is probably because thermoplastic resin is so soft that the wire guide is deformed when the windings are wound, and the distance between the windings becomes narrower as shown in Fig. 18.

Table III shows the resonance frequencies f_r, inductances L and parasitic capacitances C of each reactor. In the measurement range, the resistance of each reactor is sufficiently small and can be neglected, and each reactor can be regarded as an LC parallel circuit. However, the influence of resistance appears near each resonance point. Here, L are derived assuming that the reactor can be approximated to be only the inductance L at 100 kHz. The parasitic capacitance of the reactor with the ultraviolet curing resin guide can be reduced about 0.87 times than that of the reactor with thermoplastic resin guide. From Fig. 19, the frequencies of linear regions for the reactors with the ultraviolet curing resin guide and the thermoplastic resin guide are 1.93 MHz and 1.73 MHz, respectively. That is, the ultraviolet curing resin is more suitable material of the wire guide for the high frequency reactor than the thermoplastic resin. Furthermore, the difference of material does not affect the reactor characteristics as much as the shape of the wire guides.

1085

V. CONCLUSIONS

This paper has proposed the high-frequency reactor with the wire guide for a small-size toroidal core. This wire guide has enabled to reduce the parasitic capacitances between the windings and between the winding and the core. The wire guide has been manufactured by using a 3D printer. Furthermore, the impedance characteristics have been evaluated for the cases where the shape and the material of the wire guide are changed. The following results were obtained.

- In the conventional reactor, since the coil windings are densely gathered together at the center hole of the toroidal core, the parasitic capacitances between the copper wires and between the copper wire and the core increases. When the switching frequency is increased, the parasitic capacitances become the factor of deteriorating the impedance characteristics of the reactor.

- Two kinds of wire guides with two-layer winding and three-layer winding have been proposed to reduce the parasitic capacitances. These wire guides can increase the distances between the windings and between the winding and the core, and reduce the parasitic capacitances.

- The parasitic capacitances of the proposed reactors can be reduced by about 0.73 and 0.66 times in the two-layer and the three-layer windings, respectively, as compared with that of the conventional reactor. Furthermore, the upper limit frequencies of the linear regions for the proposed reactors with two-layer and three-layer windings can be increased to 371 kHz and 436 kHz, respectively, than that in the conventional reactor. On the other hand, the parasitic capacitance of the concentrated winding reactor is about 4.1 times larger than that of the conventional reactor.

- The parasitic capacitance of the reactor for the comparative experiment is about 2 times larger than that of the proposed reactor with three-layer winding. The upper limit frequency of the linear region for the reactor which is for the comparative experiment is 595 kHz lower than that of the proposed reactor with three-layer winding. This is because the parasitic capacity is arisen at the windings in the inside of the inner frame in the reactor for the comparative experiment. Therefore, the consideration of the diameter of the innermost frame and the winding arrangement so that the distance between the windings is increased is important to design the wire guide for the high-frequency reactor.

- The parasitic capacitance of the reactor with the ultraviolet curing resin guide can be reduced about 0.87 times than that of the reactor with thermoplastic resin guide. This is because the ultraviolet curing resin has high strength and hardness, and there are few problems due to deformation.

This research is supported by JSPS KAKENHI (17K06328)

REFERENCES

[1] R. Khazaka, L. Mendizabal, D. Henry and R. Hanna, "Survey of High-Temperature Reliability of Power Electronics Packaging Components", *IEEE Trans. on Power Electronics*, vol. 30, no. 5, pp. 2456-2464, 2015.

[2] K. Shirakawa, T. Shimizu ,and K. Ishii, "Discussion on Winding Structure of an AC Filter and the Frequency Characteristics", , Proc. of the 2005 JIASC, No.1-17, pp. 111-112 (2005) (in Japanese)

[3] L. Dalessandro, F. S. Cavalcante ,and J. W. Kolar, "Self-Capacitance of High-Voltage Transformers", *IEEE Trans. on Power Electronics*, Vol. 22, No. 5, pp. 2081-2092, 2007.

[4] A. Schroedermeier and D. C. Ludois, "An Integrated Inductor and Capacitor with Co-Located Electric and Magnetic Fields", *IEEE Trans. on Industry Applications*, Vol.53, No.1, p. 380-390, 2017.

[5] L. Middelstädt, S. Skibin, R. Döbbelin and A. Lindemann, "Analytical Determination of the First Resonant Frequency of Differential Mode Chokes by Detailed Analysis of Parasitic Capacitances", IEEE 16th European Conference on Power Electronics and Applications (EPE'14-ECCE Europe), 2014.

[6] I. F. Kovačević, T. Friedli, A. M. Müsing and J. W. Kolar, "3-D Electromagnetic Modeling of Parasitics and Mutual Coupling in EMI Filters", *IEEE Trans. on Power Electronics*, Vol. 29, No. 1, pp. 135-149, 2014.

[7] W. Tan, X. Margueron and N. Idir, "Analytical Modeling of Parasitic Capacitances for a Planar Common Mode Inductor in EMI Filters", 15th International Power Electronics and Motion Control Conference (EPE/PEMC), 2012.

[8] S. Weber, M. Schinkel, S. Guttowski, W. john and H.Reichl, "Calculating Parasitic Capacitance of Three-Phase Common-mode Chokes", PCIM Conference, 2005

[9] I. F. Kovacevic, A. M. Müsing and J. W. Kolar , "PEEC Modelling of Toroidal Magnetic Inductor in Frequency Domain", The 2010 International Power Electronics Conference (IPEC 2010) -ECCE ASIA-, pp. 3158-3165, 2010.

[10] H. Ayano, A. Fujimura, Y. Matsui, T. Hayashi, "A Wire Guide of a Toroidal Core for a High-Frequency Reactor", 2017 Annual Meeting Record I.E.E Japan, 4-021, 2017 (in Japanese).

The 2018 International Power Electronics Conference

Core Loss Evaluation in Powder Cores: A Comparative Comparison between Electrical and Calorimetric Methods

Yuki Ishikura[1*], Jun Imaoka[2], Mostafa Noah[1] and Masayoshi Yamamoto[1]

1 Institute of Materials and Systems for Sustainability, Nagoya University, Furo-cho, Chikusakaku, Nagoya, Aichi 464-8601, Japan

2 Department of Electrical and Electronic Engineering, Kyushu University, 744 Motooka, Nishiku, Fukuoka, 819-0395, Japan

*E-mail: y_ishikura@murata.com

Abstract— **Utilizing powder cores in many power electronics applications have gained interest due to the attractive magnetic properties of powder core materials. For instance, powder cores have soft saturation characteristics and high saturation flux density compared to gapped ferrites. However, due to their low relative-permeability and the low loss characteristics, a precise measurement of the core losses is very difficult to be obtained. One of the common approaches to measure core losses is by measuring the electrical signals of the winding terminals, we hereafter refer it as "electrical methods". Another alternative approach is calorimetric methods, which rely on measuring the core temperature rise. Nonetheless, these methods may have huge measurement error in case of powder cores. This paper evaluates the accuracy of measuring core losses in powder cores, and a comparison between the electrical and calorimetrical methods is presented. Along with the theoretical discussion, simulation and experimental tests are conducted.**

Keywords— *powder core, core loss, electrical methods, calorimetric methods, relative permeability .*

I. INTRODUCTION

The importance of manufacturing efficient and high-power density converters is growing along with the growth of the market in many industrial applications. Powder cores, which have low relative permeability with high dc bias current are very attractive to be utilized in many power electronics applications, since they possess soft saturation characteristic and high saturation magnetic flux density. In addition, magnetic powder cores exhibit lower eddy current loss compared to discrete air gap ferrites, since powder cores have tiny air gaps distributed in material which eliminates fringing effects, as shown in Fig. 1. These attractive features have gained much attention in designing ac and dc inductors [1]-[5].

However, powder cores have some drawbacks such as: (1) Having dc bias characteristic which leads to a large inductance variation at different load levels [4]. (2) Having low relative permeability. This leads to increasing the leakage flux around the core, which forces eddy current to flow nearby conductors, and limit printed

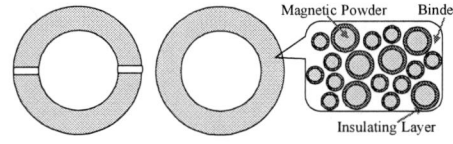

(a) Discrete air gap   (b) Distributed air gap
Fig. 1. Magnetic core with air gap.

circuited board (PCB) layout [6]. (3) The difficulty of measuring accurate core losses. Measuring the actual value of the core loss in these materials is essential, to ensure an optimum thermal design for the passive component [7]-[10].

In general, core loss measurement methods are categorized into two techniques: (1) electrical method [7]-[15], (2) calorimetric method [16]-[20].

1. Electrical methods: electrical methods basically rely on measuring the voltage and current of the winding terminals. By processing these measurements, the core loss to calculate B-H loop can be obtained. Since these methods can be performed within a short time, it is highly reproducible. The accuracy of core loss measurement has been discussed in previous studies [7]-[15]. Among others, in case of powder cores, huge measurement errors are often encountered, which occurs as a result of the phase shift between voltage and current, due to the low relative permeability and the low core loss [11]-[12]. However, in those studies, the B-H analyzer was utilized to measure the core losses, and the error involved with the B-H analyzer was not considered. In addition, since the data which is considered the true value and the data including errors were compared and verified only by the electrical method, which both methods potentially contain errors. Removing the phase-shift error in the core is important while obtaining the core losses in a magnetic material with low relative permeability, and it has been proposed before in [14]-[15]. In this approach, capacitive or inductive components are used to cancel the reactive voltage of the core under test (CUT), and the voltage and current waveforms will be 90° phase-shifted. Implementing this approach will make the measured value of core losses less vulnerable to phase shift errors. However, this method is difficult to implement, since it requires fine tuning of the cancellation element which can be very time consuming.

2. Calorimetric methods: calorimetric methods have been widely used in many power circuits to measure power losses of magnetic components [16]-[20]. The total power losses dissipate as heat, and it leads to a temperature rise of the core, measuring the temperature rise using the calorimetric methods gives a proper measurement of the core losses. Consequently, calorimetric methods are considered to be one of the most promising methods to attain accurate power loss measurements [17]. Calorimetric method can measure core losses under any desired excitations or bias conditions and does not suffer from instrumentation phase shift or time delay errors. However, it has the following disadvantages:
1. The measurement process is very time consuming.
2. The experimental setup is complicated.
3. Inaccurate measurements obtained at low core loss values.

These drawbacks hold back the implementation of powder cores in real industrial applications.

As previously mentioned, measurement methods of core loss are categorized into two techniques; namely, electrical methods and calorimetrical methods. Previous studies reported in the literature only discussed either one of the two methods, in terms of evaluating the accurate measurements of the core losses. Furthermore, B-H analyzer was utilized while evaluating the accuracy of the electrical method [11]-[12]. As a result, the measured value is influenced by the error introduced by the B-H analyzer, therefore, the measured value of the core losses deviates from the actual value.

This paper discusses the measurement accuracy of the core loss in powder cores using the electrical measurement method. Furthermore, a comparison is conducted between the electrical method and a simple calorimetric measurement method. This purpose of this discussion is to reevaluate the measurement accuracy of the electrical measurement in powder cores.

This paper is divided into five sections. Sections II discusses the electrical measurement using the B-H analyzer and the measurement accuracy of the core losses. In addition, this paper proposes implementation method for core loss calculation into circuit simulator. In Section III, the simple calorimetric measurement method is described, and the measurement accuracy of the core loss measurement is evaluated. In section VI, to validate the accuracy of the electrical measurement method in core losses, both electrical measurement and calorimetric measurement are compared through the experimental test. Finally, conclusions and future works are presented in section V.

TABLE I
CHARACTERISTICS OF POWDER CORES USED IN THE TEST.

-	-	Core A	Core B
Product model	-	NPF	NPH-L
Core material	-	Fe-Si	Fe-Si-Al
Relative permeability	μ_r	60	60
Outer Diameter [mm]	OD	26.92	26.92
Inner Diameter [mm]	ID	14.73	14.73
Height [mm]	H	11.18	11.18
Effective Volume [mm³]	V_e	4154	4154
Effective Area [mm²]	A_e	65.4	65.4
Average Path length [mm]	L_e	6.35	6.35
Saturation Flux density [T]	B_{sat}	1.5	1.1
Curie temperature [°C]	T	700	600

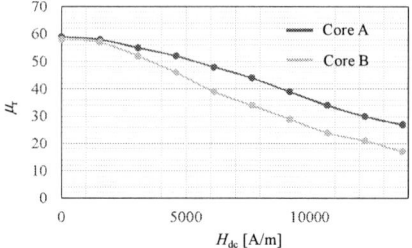

Fig. 2. Relative permeability versus magnetic field intensity.

II. CORE LOSSES OF THE ELECTRICAL METHOD

A. Powder cores under evaluation

The schematic diagram of powder cores is shown Fig. 1 (b). At a microscopic level, magnetic alloy powder particles are separated from one another by binder insulation or by insulation coating. Distributed air gaps in the powder core have three main advantages: (1) eliminating the disadvantages arise from the discrete air gap structure (Fig 1 (a)), which are sharp saturation, fringing loss, and EMI (Electromagnetic interference), (2) despite comparatively low resistivity in the alloy, controlling eddy current losses so that higher B_{sat} alloys may be used at relatively high frequencies. (3) Availability of high range of saturation flux density B_{sat}, and less dependency of temperature.

On the other hand, power electronics system designer must meet inductance of specifications and requirements, such as: total loss, space, cost, EMI, fault-tolerance, thermal performance, and reliability. As previously mentioned, many engineers consider powder cores, among other core materials, as an attractive material to be implemented in various industrial applications. Nonetheless, due to low relative permeability, powder cores suffers from having a non-uniform magnetic flux density within the core structure [21].

In this discussion, two powder cores from POCO Magnetic Co., Ltd, were utilized to evaluate the measurement accuracy of core losses. The specifications of the powder cores are tabulated in Table 1. Core A is composed of iron – silicon (Fe–Si) powder material, and it has high saturation flux density. While, core B is composed of iron – silicon – aluminum (Fe – Si – Al) powder material, and it has low core losses characteristics. The measured values of the relative permeability in each powder core are depicted in Fig. 2. It can be deduced

The 2018 International Power Electronics Conference

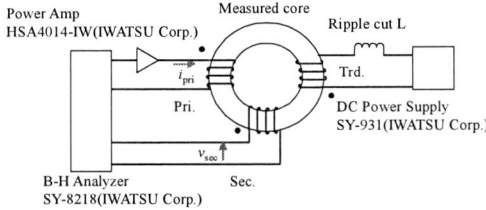

Fig. 3. Overview of electrical measurement system.

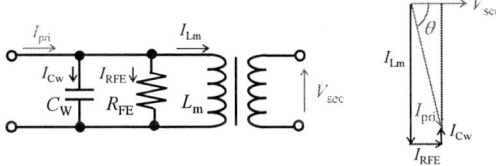

(a) Equivalent circuit   (b) Vector diagram
Fig. 4. Equivalent circuit and vector diagram of CUT.

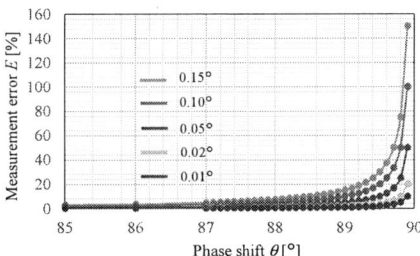

Fig. 5. Measurement error properties in electrical measurements.

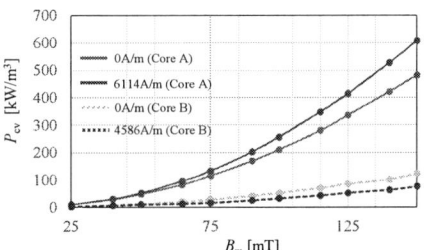

Fig. 6. The core loss versus flux density (20 kHz, 25°C).

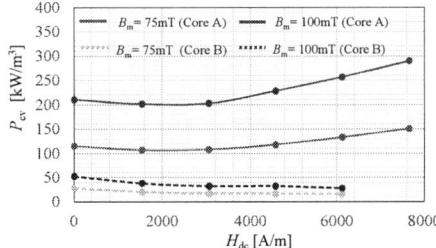

Fig. 7. The core loss characteristic under dc bias conditions
(20 kHz, 25°C).

from Fig.2 that for a certain inductance value, using core A can achieve less number of winding turns than core B, in case of a high magnetic field strength. Hence, it can be said that the core A is superior to the core B in the dc bias characteristic.

B. Test setup of electrical method

Within various electrical methods, using the B-H analyzer is considered as one of the most attractive methods, thanks to repeatability and simplicity [11]-[12]. Fig. 3 shows the electrical measurement system used in this study. The test system consist of a B-H analyzer, a power Amplifier, a dc power supply, and a ripple cut reactor as illustrated in Fig. 3.

The basic concept of measuring the core losses is as follows: three windings are placed around the CUT, and ac current flow in the excitation winding (primary windings), which is proportional to an ac magnetic field strength H_{ac}. The excitation voltage waveform of the primary winding is sinusoidal. The sense winding (secondary winding) voltage is measured and integrated to calculate the flux density B. The core loss per unit volume (P_{cv}) is the enclosed area of B-H loop, multiplied by an excited frequency. In addition, the dc current in the third winding is proportional to the dc magnetic field strength H_{dc}. Core losses can be measured under any dc bias conditions by injecting a dc current into the third winding.

The cores under test (CUT) in this evaluation are core A and core B previously mentioned in section II.A. The primary windings and the secondary windings used in CUT are Litz wire with the diameter of 0.22 mm × 7, to realize low AC resistances. The third winding is single wire with the diameter of 10 mm, with the purpose of achieving low a dc resistance.

C. Accuracy of electrical measurement

Due to the low relative permeability and the low core loss of powder cores, the core loss measurement is very sensitive to phase shift error, and it is very difficult to obtain accurate measurements.

The electrical equivalent circuit of CUT is shown in Fig. 4(a) [11]-[12]. The magnetizing inductance is represented as L_m and the resistance R_{fe} represents the loss in the magnetic core ($R_{fe} = V_{core}^2 / P_{loss}$). The winding capacitances of the primary windings and the secondary windings are denoted by C_w. The vector diagram of current and voltage in the CUT is shown in Fig. 4(b). In this test, since the capacitive current i_{cw} is much lower than the inductive current i_{Lm}, the parasitic winding capacitance can be neglected ($f < 1MHz$) [11]-[12]. In the powder core, the low relative permeability and the low core loss allows for high values of magnetizing inductance current. Therefore, the phase angle between the primary current and the secondary voltage of the CUT approaches 90°. The electrical method is high sensitive to the phase shift error. The phase shift error is caused by delay of voltage and current sensing. The core loss measurement error can be expressed as follows [7]-[15]:

$$E = \frac{\cos(\theta + \phi) - \cos(\theta)}{\cos(\theta)} \cdot 100\% \qquad (1)$$

Where E is the percentage error in the core loss, θ is the actual phase shift between the sensing voltage and the sensing current, and ϕ is the phase shift error. Fig. 5 shows percentage of errors in case of 0.15°, 0.1°, 0.05°, 0.02° and 0.01°. It is noticeable that the when the phase difference get closer to 90°, it produces more than a

1089

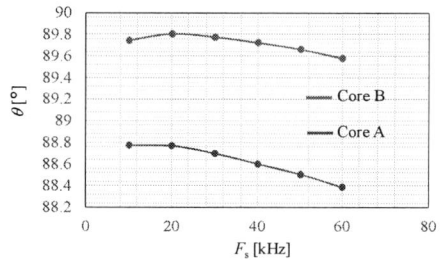

Fig. 8. Phase angle of the voltage and the current (B_m = 75mT).

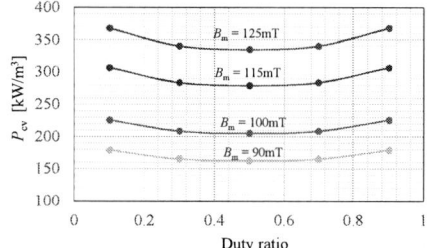

Fig. 9. Core loss in duty cycle dependence
(Core A, f = 20 kHz).

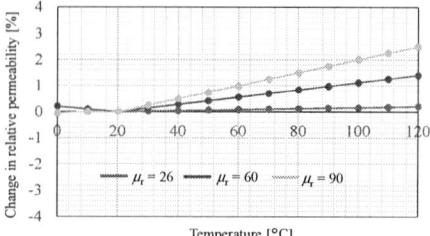

Fig. 10. The temperature dependence of the core loss comparing
25°C and 75°C.

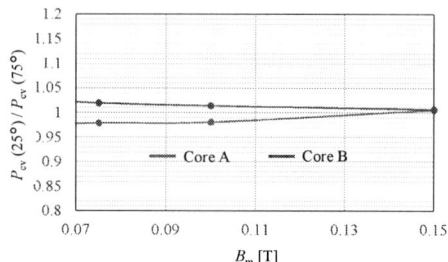

Fig. 11. The temperature dependency of the relative
permeability in cores

100% error. In order to reduce measurement errors, the test is performed using the phase error correction function in B-H analyzer, which can achieve a measurement accuracy tolerance of ±0.15°.

D. Core loss measuremnt

In this section, the measurement results are presented. The measurements were obtained under the following conditions: Excitation voltage is sinusoidal waveform, frequency range is from 20 kHz to 60 kHz, maximum flux density B_m range is from 0.025 T to 0.15 T and the dc magnetic field intensity H_{dc} range is from 0 A/m to 8000 A/m.

Fig. 6 shows the core loss per unit volume $P_{cv} - B_m$ characteristic. It is clear that the core loss for each core increases with an increase in B_m because the core loss depends on the area of the dynamic minor loop, which increases as B_m increases. In addition, the core losses of core A is almost three times of core B. Since core B material consists of Fe – Si –Al, therefore, it exhibits low core loss characteristics [3].

Fig. 7 shows dc bias characteristics of core losses. It is shown that core losses change depending on dc bias magnetic fields, and the two cores can be characterized as follows:
1. The core losses of core A gradually increase with higher values of dc magnetic field intensity.
2. The core losses of core B gradually decrease with higher values of dc magnetic field intensity.
These characteristics are highly dependent on materials.

Fig. 8 shows the phase shift angles for the cores under test. These results had been averaged, after repeatedly conducting this test for five times. The phase-shift angle values of core A lie between 88.4° to 88.8° and in core B is around 89.5° to 89.8°. The tolerance of these results is within ±0.01° during five measurements outputs. Since these cores have phase shift tolerance of ±0.01°, it is

possible to measure materials up to θ =89.8∘ (at f = 20 kHz) when the expected measurement accuracy is under 5%. Although there are few variations between measurements in the measurement error, the measurement accuracy of the B-H analyzer is 0.15°. This measurement accuracy in the B-H analyzer may lead measurement errors which is 10 ~15% in core A and 37.5 ~ 75% in core B. To verify these possibilities, in section IV, measurement results using the electrical method are compared with results using a simple calorimetric method.

In electrical methods, the measured core losses are obtained while the excitation waveform is sinusoidal. In order to use these data in power electronics applications, the common approach for estimating core loss is improved Generalized Steinmetz Equation (iGSE), this formula is used to calculate core losses under square-wave excitation at the respective fundamental frequency [7].

$$ P_{cv} = \frac{1}{T}\int_0^T k_i \left|\frac{dB}{dt}\right|^\alpha (\Delta B)^{\beta-\alpha}\, dt \qquad (2) $$

$$ k_i = \frac{k}{(2\pi)^{\alpha-1}\int_0^{2\pi} k_i \left|\cos\theta\right|^\alpha 2^{\beta-\alpha}\, d\theta} \qquad (3) $$

Where ΔB is peak-to-peak flux density ($\Delta B = 2B_m$) and k_i, α β are material parameters. The iGSE is capable of calculating core loss of any flux waveform, without requiring extra material parameters beyond the steinmetz parameters gained from sinusoidal excitations.

In Fig. 9, the calculated core losses as a function of duty cycles are plotted by performed based on the iGSE (2). In addition, the Steinmetz Premagnetization Graph (SPG) is used for the core loss calculation under dc bias conditions [9]. The SPG is a way of considering the change of Steinmetz parameters of IGSE. By applying the SPG to the measured core loss data, it is possible to

1090

calculate the core loss under any excitation waveform and dc magnetic field bias conditions.

E. Temperature characteristic of core loss

In order to realize accurate core losses measurements, the temperature shall be carefully examined, as it influences the core losses.

Fig 10 shows the dependence of core losses on the temperature, at 25°C and 75°C. It can be clearly understood that the change of core losses is under 2% between 25°C and 75°C. These results indicate that core loss of the two powder cores under the test is not sensitive to a temperature rise from 25°C to 75°C.

Fig. 11 shows the temperature dependence of relative permeability in core A. This figure indicates that the lower the permeability, the less dependence of relative permeability on temperature. Since the relative permeability is almost not affected by temperature variation, the core losses dependence on the temperature can be neglected. It can be concluded from these results that there is no need to consider the core loss dependence on temperature variations up to 75°C.

F. Core loss calculation method using simulation

If the core loss calculation of powder cores can be modeled into time-domain circuit simulation, the core loss of powder cores with respect to various circuits can be sufficiently evaluated before constructing a prototype, so that number of hardware iterations can be reduced. Therefore, this paper proposes implementation method for core loss calculation into circuit simulator, in this section. Using the core loss data obtained by the electrical method, core loss calculation is implemented in the simulation as follows:

Firstly, non-linearity of the relative permeability is modeled. The model based on permeance magnetic circuit is built up in the system-level simulation platform PLECS (Plexim GmbH), by utilizing the magnetic components listed in the library and c-script [22]. The permeance-capacitance based magnetic circuits representing the core can be directly parameterized using the geometrical and material information [23]-[24], as follows:

$$P(H_{dc}) = \mu_0 \cdot \mu_r (H_{dc}) \cdot \frac{A}{l} \qquad (4)$$

Where the cross section area A and magnetic path length l are geometry-related constant parameters once the core is selected, and μ_0 is the permeability of vacuum. The relative permeability $\mu_r (H_{dc})$ accounts for material characteristics which continuously change depending on the magnetic field strength H_{dc}. While, the characteristic of the relative permeability $\mu_r (H_{dc})$ can be expressed by (5) using the fitting model proposed in [25].

$$\mu_r (H_{dc}) = (1 + \frac{p}{1 + (\frac{|H_{dc}|}{q})^r}) \qquad (5)$$

Where, the parameters p, q and r are material-related constant parameters. By solving (4) and (5), the permeance $P(H_{dc})$ then can be represented as follows:

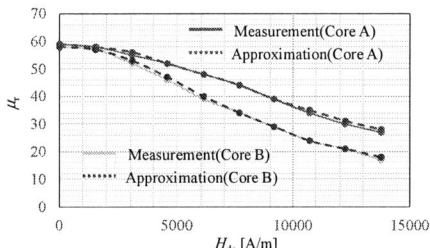

Fig. 12. The comparison between the measured values and approximated results using (5).

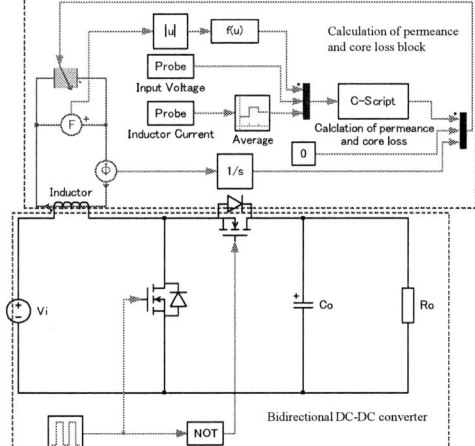

Fig. 13. Core loss calculation model in Bidirectional DC/DC converter with permeance capacitance magnetic circuit.

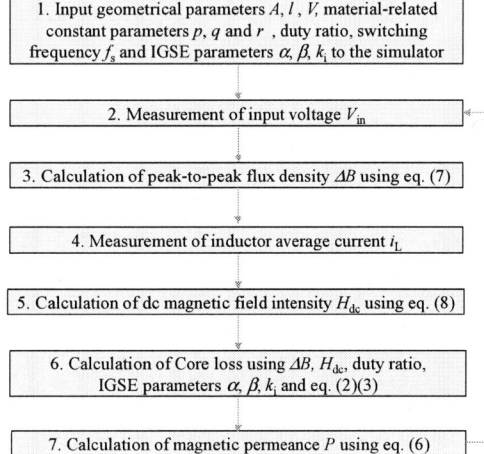

1. Input geometrical parameters A, l, V, material-related constant parameters p, q and r, duty ratio, switching frequency f_s and IGSE parameters α, β, k_i to the simulator

2. Measurement of input voltage V_{in}

3. Calculation of peak-to-peak flux density ΔB using eq. (7)

4. Measurement of inductor average current i_L

5. Calculation of dc magnetic field intensity H_{dc} using eq. (8)

6. Calculation of Core loss using ΔB, H_{dc}, duty ratio, IGSE parameters α, β, k_i and eq. (2)(3)

7. Calculation of magnetic permeance P using eq. (6)

Fig. 14. The flowchart of the simulation algorithm of the core loss calculation including dc bias conditions.

$$P(H_{dc}) = (1 + \frac{p}{1 + (\frac{|H_{dc}|}{q \cdot l})^r}) \frac{\mu_0 \cdot A}{l} \qquad (6)$$

Fig. 12 shows the comparison result between approximated results obtained from (5) and measured values. It is clear that modeling using (5) is consistent with the experimental results. Furthermore, the core loss characteristics calculated using iGSE are implemented in the simulation model.

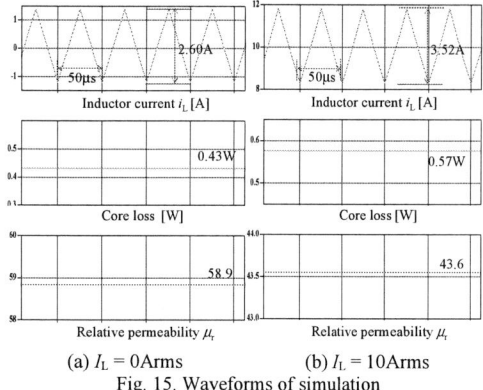

(a) I_L = 0Arms   (b) I_L = 10Arms

Fig. 15. Waveforms of simulation

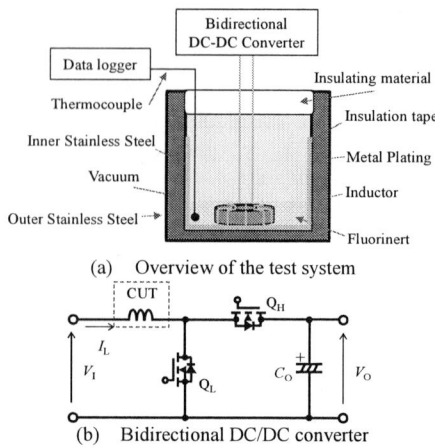

(a)   Overview of the test system

(b)   Bidirectional DC/DC converter

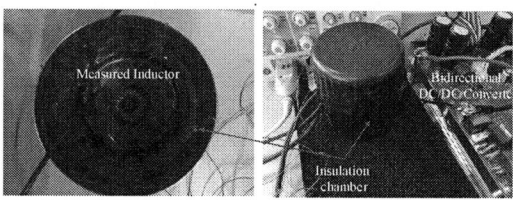

(c)   The picture of test system

Fig. 16. Overview of the test system and picture.

TABLE II

SPECIFICATIONS OF BIDIRECTIONAL Dc-Dc CONVERTER

Input voltage [V]	V_i	10~140
Inductor average current [A]	I_L	0, 8
Switching frequency [kHz]	F_S	20
Duty ratio	D	0.1, 0.5, 0.9

In order to evaluate of core loss calculation method using simulation, we conduct the core loss simulation with the bidirectional dc-dc converter. Fig. 13 shows the circuit configuration of the bidirectional dc-dc converter with the inductor core loss calculation model. In the bidirectional dc-dc converter, the flux density ripple ΔB and the dc bias magnetic field strength H_{dc} of the magnetic core can be calculated as follow

$$\Delta B = \frac{V_i \cdot d}{f_s \cdot N \cdot A} \tag{7}$$

Where, V_i is input voltage, d is duty ratio, f_s is switching frequency, N is number of turn.

$$H_{dc} = \frac{N \cdot I_L}{l} \tag{8}$$

Where, I_L is inductor average current.

Fig. 14 shows the simulation algorithm flowchart to calculate core losses including dc bias conditions in the c-script. For example, the simulation is conducted considering the following two conditions

1. V_i = 20 V, duty = 0.5, f_s = 20 kHz, I_L = 0 A.
2. V_i = 20 V, duty = 0.5, f_s = 20 kHz, I_L = 10 A.

Fig. 15 (a) and (b) show the simulation results in I_L, P_{CV}, μ_r. As shown Fig. 15, when inductor average current I_L change from 0A to 10A, inductor ripple current increases from 2.6A to 3.52A, as a result of a drop in the relative permeability. Since these results almost agree with the dc magnetic field bias characteristic of the relative magnetic permeability shown in Fig. 2, it can be said that the dc magnetic field bias characteristic of the relative permeability can be accurately modeled. Furthermore, when inductor average current I_L change from 0A to 10A, core loss increases 0.43W to 0.57W, as a result of a drop in the relative permeability. The simulation results of the core losses are discussed in section IV.

III.   CORE LOSS OF THE CALORIMETRIC MEASUREMENT

A.   Experimental setup of calorimetrical method

In order to evaluate the measurement accuracy of the electrical method, we examine the accuracy of the calorimetric method under the same powder cores. This had been done to enable us to conduct a fair comparison between the two core loss measurement methods. Fig. 16 (a) and (c) show the test system and a picture of the calorimeter built in this paper to measure core losses. The thermal pot (JBI-273: THERMOS K.K.) has a simple double jacket structure with the vacuum between inner and outer insulating enclosures. This structure allows to separate the inner hot air from the ambient air. During the experiment, the calorimeter is covered with another thermal insulation sheet to minimize the heat leakage. There is certain amount of coolant in the thermal pot. In this test, Fluorinert ™ (FC-43: 3M Corp.) is used as the coolant, which has high electrical insulation and high heat conduction capabilities. Three thermocouples are used for temperature measurements. All of these are attached to the bottom of the thermal pot to measure temperature of the coolant. All wires are passing through the hole in the thermoelectric pot cap. The hole is covered with insulation tape. Fig. 16(b) shows the simplified schematic of its specifications are listed in Table II.

Before measuring inductor losses, the relationship between the loss and temperature rise is measured with a reference heat generator. In this test, the winding resistance of the CUT as the heating element is selected. By using the winding resistance of CUT, it is possible to measure the relationship between losses and temperature rise under the same condition as the actual core loss measurement test. The winding resistor is heated by dc currents. Stirring is necessary for the uniform temperature of the coolant. After a fixed period of time, around five minutes, the injecting current shall be stopped, and the value of temperature rise in the pot shall be taken. The power dissipated by the CUT can be calculated from measured values of the temperature rise ΔT [°C] of the coolant as follows

$$Q = m \cdot c \cdot \Delta T \qquad (9)$$

Where Q [J] is heat energy, m is mass of the coolant, and c [j / (g K)] is the specific heat capacity of the coolant. Since the heat energy Q [J] is proportional to temperature rise ΔT [°C], the power loss [W] can be calculated form the relation of the time and the temperature rise. After conducting the experiments, the CUT shall be cooled down to reach the ambient temperature. By repeating the measurement process under different power loss conditions, the relationship between the power loss and temperature rise can be plotted. Experimental results are shown in Fig. 17. The least square method of the linear regression calculations is utilized to linearize the temperature rise of the calorimeter, and a slope (°C/W: heat resistance) in (9) for each loss was calculated. For example, in Fig. 17, when the CUT is excited at the power level from 0.5W to 2W, the relationship between power level and temperature rise can be expressed as

$$y = 2.7106x + 0.0445 \qquad (10)$$

Where y is the temperature rise of the calorimeter, x is the power, and the second term of equation (10) is the offset temperature of the calorimeter. Microsoft Excel is used to perform the linear regression calculation. Once the reference line equation is obtained, losses of the inductor in the calorimeter box can be calculated.

B. Separation of core loss and copper loss

In calorimetrical methods, since the total loss of the CUT includes the copper loss, it is necessary to separate the core loss and the copper loss in order to calculate the core loss. The copper loss consists of the dc resistance loss and the ac resistance loss. It is extremely difficult to accurately calculate the ac resistance loss caused by the skin effect and the proximity effect. In this study, a single wire that can reduce the effect of the proximity effect is used to calculate the copper loss. Since the proximity effect occurs in case of multi layers windings, the influence of the proximity effect can be eliminated by using a single layer winding. Table III shows the measured ac resistance by using LCR meter at 20 kHz. Using the measured ac resistance and dc resistance, the copper loss can be calculated as follows:

$$P_{cu} = R_{dc}I_{dc}^2 + R_{ac}I_{ac}^2 \qquad (11)$$

Where, P_{cu} is the total copper loss, I_{dc} is the ac current and I_{ac} is the ac current. The core loss can be calculated by subtracting the copper loss from the total loss which is the result of the simple calorimetrical method.

IV. COMPARISON BETWEEN ELECTRIC AND CALORIMETRIC METHODS

In this section, to evaluate the measurement accuracy of electrical method, the experimental and simulation results are compared to the outputs of calorimetrical method. Fig. 18 (a) and (b) show the dependence of the core

TABLE III
INDUCTOR PARAMETERS

-	-	Core A	Core B
Number of turn [Turn]	N	50	50
Wire Diameter [mm]	ϕ	1.0	1.0
DC resistance [mΩ]	R_{dc}	47	50
AC resistance [mΩ]	R_{ac}	53	56

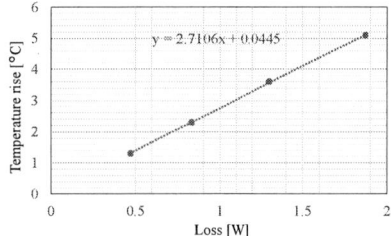

Fig. 17. Reference test result.

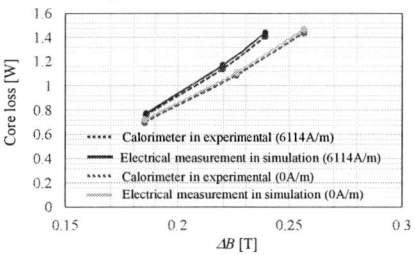

(a)　$P_{CV} - \Delta B$ characteristic in the core A

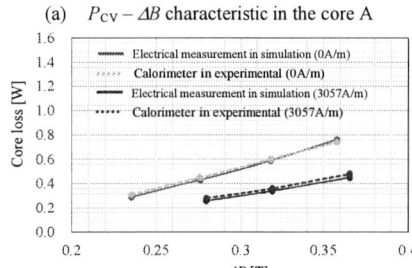

(b)　$P_{CV} - \Delta B$ characteristic in the core B

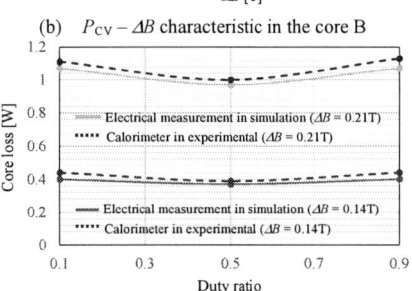

(c)　$P_{cv} - Duty\ cycle$ characteristic in the core A

Fig. 18. Results of comparison between the electrical measurement using the simulation and the calorimetric measurement.

losses on ΔB, under different dc bias conditions. The simple calorimetrical method is compared to electrical method using simulation and experimental results. Shown that core losses change depending on dc bias magnetic fields, and the two cores can be characterized as follow

1. The core losses of the core A gradually increase with higher values of dc magnetic field intensity.
2. The core losses of core B gradually decrease with higher values of dc magnetic field intensity.

This behavior is the same for the electrical and

calorimetrical method. In addition, the measurement tolerance of both methods is below 0.03 W. Since core losses measurement range of the core A is from 0.6 W to 1.4 W, the core loss measurement accuracy range is from 2% to 5%. On the other hand, the core losses measurement range of the core B is from 0.3 W to 0.8 W, and the core loss measurement accuracy range is from 4% to 10%. On the other hand, Fig. 18 (c) shows the characteristics of the duty cycle in the core loss. The measurement tolerance of both methods is below 0.04 W. Since the core loss measurement range of core A is 0.4 W to 1.1 W, the accuracy of core loss measurement is 2 to 10%. As a result, the actual error is sufficiently less than the assumed error mentioned in section II. These results indicate that the accuracy of the electrical method is sufficiently high within these measurement range.

When measuring a smaller core loss range using the calorimetrical method, the measurement accuracy decreases because the loss becomes lower. Therefore, another verification such as adjusting the amount of coolant is necessary. In the electrical method, the frequency increases, or when measuring the core loss of the powder core having less core loss characteristics, the measurement accuracy decreases. These problems shall be addressed in the future.

V. CONCLUSIONS

In this paper, a comparative study of two core loss measurement methods for powder cores has been performed considering dc bias conditions and duty cycles. The first method is the electrical method by the B-H analyzer, the second one is the simple calorimetrical method. In addition, core loss calculation method of powder cores using time-domain circuit simulation is proposed and compared the simple calorimetric method. As a result of the evaluation, it is confirmed that the electrical method is very practical in terms of accuracy and reproducibility since the measurement accuracy is almost the same as the measurement method of simple calorimetry.

REFERENCES

[1] T. Saito, S. Takemoto, and T. Iriyama, "Resistivity and core size dependencies of eddy current loss for Fe-Si compressed cores," *IEEE Trans. on Magnetics*, vol. 41, no. 10, pp. 3301-3303, 2005.

[2] C. Appino, O. Bottauscio, O. de la Barri`ere, F. Fiorillo, A. Manzin, and C. Ragusa, "Computation of eddy current losses in soft magnetic composites," *IEEE Trans. on Magnetics*, vol. 48, no. 11, pp. 3470-3473, 2012.

[3] T. Ishimine, A. Watanabe, T. Ueno, T. Maeda, and T. Tokuoka, "Development of low-iron-loss powder magnetic core material for high-frequency applications," *SEI Tech. Rev*, no. 72, pp. 117-123, 2011.

[4] B. G. You, J. S. Kim, B. K. Lee, G. B. Choi, and D. W. Yoo, "Optimization of powder core inductors of buck-boost converters for hybrid electric vehicles," *J. Electr. Eng. Technol.*, vol. 6, no. 4, pp. 527-534, 2011.

[5] Yiren Wang, Gerardo Calderon-Lopez, Andrew J. Forsyth, "High-Frequency Gap Losses in Nanocrystalline Cores," *IEEE Trans. on Power Electronics*, vol. 32, no. 6, pp. 4683-4690, 2017.

[6] [Online]Available: http://elnamagnetics.com/wp-content/uploads/library/Magnetics-Documents/Leakage_Flux_Considerations_on_KOOL_Mu_E-Cores.pdf

[7] K. Venkatachalam, C. R. Sullivan, T. Abdallah, and H. Tacca, "Accurate prediction of ferrite core loss with nonsinusoidal waveforms using only Steinmetz parameters," in *Proc. IEEE Workshop Comput. Power Electron. ,IEEE Workshop on*, pp. 36-41, 2002.

[8] T. Shimizu and S. Iyasu, "A practical iron loss calculation for ac filter inductors used in PWM inverters," *IEEE Transactions on Industrial Electronics*, vol. 56, no. 7, pp. 2600-2609, 2009.

[9] J. Muhlethaler, J. Biela, J. Kolar, and A. Ecklebe, "Core Losses Under the DC Bias Condition Based on Steinmetz Parameters," *IEEE Trans. on Power Electronics*, vol. 27, no. 2, pp. 953-963, 2012.

[10] R. Beres, X. Wang, F. Blaabjerg, C. Leth Bak, H. Matsumori and T. Shimizu, "Evaluation of core loss in magnetic materials employed in utility grid AC filters," in Proc. of *Applied Power Electronics Conference and Exposition (APEC)*, pp. 3051 – 3057, 2016.

[11] F. Farideh Javidi and Morten Nymand, "Error analysis of high frequency core loss measurement for low-permeability low-loss magnetic cores," *IEEE 2nd Annual Southern Power Electronics Conference (SPEC)*, pp. 1 – 6, 2016.

[12] N. Farideh Javidi, Morten Nymand and Andrew J. Forsyth, "New method for error compensation in high frequency loss measurement of powder cores," in Proc. of *Applied Power Electronics Conference and Exposition (APEC)*, pp. 876 – 881, 2017.

[13] V. Joseph Thottuvelil, Thomas G. Wilson and Harry A. Owen, JR. "High-frequency measurement techniques for magnetic cores," *IEEE Transactions on Power electronics*, vol. 5, no. 1, pp. 41-53, 1990.

[14] D. Hou; M. Mu; F. Lee; Q. Li, "New core loss measurement method for high frequency magnetic materials," *IEEE Tran. on Power Electronics*, Vol.29, No.8, pp. 4374 – 4381, 2014.

[15] D. Hou; M. Mu; F. Lee; Q. Li, "New high frequency core loss measurement method with partial cancellation concept," in *IEEE Transactions on Power Electronics*, vol. 32, no. 4, pp.2987-2994, 2016.

[16] R. Linkous, A.W. Kelley, K.C. Armstrong, "An improved calorimeter for measuring the core loss of magnetic materials," in Proc. of. *15th Annual IEEE Applied Power Electronics Conference & Exposition (APEC)*, vol. 2, pp. 633 – 639, 2000.

[17] Chucheng Xiao, Gang Chen and Willem G. H. Odendaal, "Overview of Power Loss Measurement Techniques in Power Electronics Systems," *IEEE Trans. on Industry Applications*, vol. 43, no. 3, pp. 657-664, 2007.

[18] N S. D. J. Weier, M. A. Shafi and R. A. McMahon, "Precision Calorimetry for the Accurate Measurement of Losses in Power Electronic Devices," in Proc. of. *IEEE Industry Applications Society Annual Meeting*, pp. 1 – 7, 2008.

[19] P. D. Malliband, N. P. van der Duijn Schouten and R. A. McMahon, "Precision calorimetry for the accurate measurement of inverter losses", *The 5th International Conference on Power Electronics and Drive Systems (PEDS)*, vol. 1, pp. 321-326, 2003.

[20] F. Blaabjerg, J. K. Pedersen and E. Ritchie, "Calorimetric measuring systems for characterizing high frequency power electronic components and systems", in Proc. of. *Industry Applications Conference., 37th IAS Annual Meeting*, Vol.2, pp. 1368-1376, 2002.

[21] Han. Cui, Khai D. T. Ngo, Jim Moss, Michele Lim and Ernesto Rey: "Inductor Geometry with Improved Energy Density", *IEEE Transaction on Power Electronics*, vol. 29, no.10, pp.5446-5453, 2014.

[22] [Online]Available: https://www.plexim.com/plecs

[23] J. Allmeling, W. Hammer, J. Schonberger, "Transient Simulation of Magnetic Circuits Using the Permeance-Capacitance Analogy," *in Proc of IEEE 13th Workshop on Control and Modeling for Power Electronics (COMPEL)*, pp. 1-6, 2012.

[24] M. Luo and D. Dujic. "Permeance based modelling of the core corners considering magnetic materialnonlinearity". In: *Annual Conference of the IEEE Industrial Electronics Society (IECON)*, pp. 950–955, 2015.

[25] K. Okamoto, J. Imaoka, M. Shoyama, "Empirical Evaluation of Modeling and Design Method of DC Bias Characteristic in Dust Core Material", *IEICE Technical Report*, vol. 116, no. 429, EE2016-54, pp. 33-38, 2017 (in Japanese)

The 2018 International Power Electronics Conference

Modeling, Magnetic Design, and Simulation Methods Considering DC Superimposition Characteristic of Powder Cores Used in Power Converters

Jun Imaoka[1*], Kenkichiro Okamoto[1], Masahito Shoyama[1], Yuki Ishikura[2], Mostafa Noah[2], and Masayoshi Yamamoto[2]

1 Department of Electrical Engineering, Kyushu University, 744 Motooka, Nishi-Ku, Fukuoka, 819-0395, Japan
2 Institute of Materials and Systems for Sustainability, Nagoya University, Furo-Cho, Chikusaka-Ku, Nagoya, 464-8601, Japan
*Email: imaoka@ees.kyushu-u.ac.jp

Abstract— Powder cores have been gained much attention as one of the attractive magnetic cores used in power converters due to their superior features such as high saturation flux density or distributed air gaps. However, powder cores have a unique feature that the relative permeability of the magnetic core varies depending on the magnetic field intensity. However, modeling the variable relative permeability, the design method for powder cores, and computer simulation methods of non-linear inductance are well not discussed in the related literature. Therefore, this paper proposes a novel modeling, magnetic design methods, and simulation technique considering dc superimposition characteristics of powder cores. The modeling method uses a novel model equation representing the behavior of relative permeability under the dc current superimposition condition, which is helpful to evaluate the performance of powder cores and to properly design various magnetic components. Theoretical analysis has been presented and the effectiveness and the validity of the proposed methods are evaluated through simulation and experimental tests.

Keywords— *Powder core, dc superimposition characteristic, non-linear inductance, relative permeability*

I. INTRODUCTION

Recently, high-power-density and high-efficiency of power converters have been required with expanding the markets such as eco-friendly automotive and aircraft applications. Passive components such as inductors and capacitors increase the power converter volume, and they occupy significant space and weight. In particular, magnetic components such as inductors and transformers, are the main contributors to increase the converter volume and cost.

One of the ways to downsize the magnetic components is driving the power semiconductor devices such as MOSFETs and IGBTs with high-frequency. However, operating at high-frequency has practical boundaries, as it often leads to EMI noises produced by

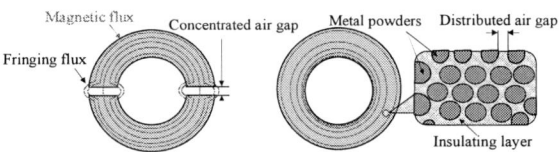

(a)   Concentrated air gap   (b) Distributed air gap (powder core)
Fig. 1. Magnetic core with concentrated or distributed air gaps.

switching devices. This imposes certain limits and restrictions with increasing the switching frequency since additional components shall be installed on the power circuit to tackle the noise, which increases the power circuit volume and size [1]-[2].

An alternative to downsize magnetic components of power converters is utilizing magnetic cores with attractive magnetic properties. Among the magnetic cores used in power converters [3]-[7], powder cores have been received much attention as magnetic cores used for high power application due to the following reasons:

1.   High saturation flux density: in general, the magnetic core is used to obtain the desired inductance with a smaller number of turns. A magnetic core with high saturation flux property, allows the windings to carry higher current levels, and hence a smaller magnetic core can handle higher power. Since powder core usually consists of iron powders, the saturation flux density is usually high on the basis of the Slater-Pauling Curve [8].

2.   Soft-saturation property: soft-saturation means that the inductance value gradually drops as the inductor current increases. Powder cores have this property. In general, ferrite cores have concentrated air-gaps, and the inductance value drops rapidly when the maximum flux density exceeds the core saturation limit, determined by the core material. The soft saturation property of the powder core can be considered as an attractive feature because it has a better reliability while encountering a sudden overcurrent.

The 2018 International Power Electronics Conference

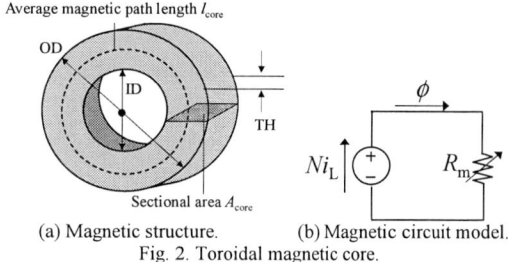

(a) Magnetic structure.　　(b) Magnetic circuit model.

Fig. 2. Toroidal magnetic core.

3.　Eliminating the fringing flux: high permeability magnetic cores such as Mn-Zn ferrite cores require inserting concentrated air gaps to the core when dc current flows through the winding. Although magnetic saturation can be avoided by inserting the air gap as shown in Fig. 1 (a), the concentrated air gap produces fringing flux, which is one of the causes of high-frequency core and winding losses near the concentrated air gaps. For example, in case of nanocrystalline ribbon cores with a concentrated air gap, the eddy current loss on the core surface produced by the fringing flux dramatically increase as reported in [9]-[10]. In [11]-[13], the fringing loss is restrained by edging the corner of the core at air gap or by distributing a concentrated air gap into several air gaps because the expansion range of the fringing flux can be suppressed. However, this process increases the production cost. Conversely, powder cores are manufactured from very fine particles of magnetic materials and are not necessary to insert the concentrated air gaps in general.

Although powder cores have the attractive magnetic properties, the relative permeability of the core is usually very low compared with other magnetic materials such as Mn-Zn ferrite cores, because of the distributed air gaps between metal powders in cores. In addition, powder core has dc superposition characteristics in which the inductance value changes non-linearly depending on the dc current value [14]-[19].

The reason why the inductance value changes non-linearly according to the dc current value can be interpreted as follows: For example, in a toroidal metal powder structure shown in Fig. 1 (b), magnetic flux density on the inner magnetic path of the toroidal core shape becomes higher than that on the outer magnetic path. The reason is that the magnetic reluctance in the inner magnetic path is lower than that in the outer magnetic path because the inner magnetic path length is shorter than the outer one. Therefore, the magnetic flux generated from the magnetomotive force easily flows into the inner side of the toroidal core and magnetic saturation occur from the inner magnetic path. Therefore, powder cores with low-permeability property have the magnetic structural dependency on dc superposition characteristics. As a matter of fact, the structure of insulating layer between the powders shown in Fig. 1 (b) also affects the dc superposition characteristic [20]-[21].

In terms of the designing magnetic components, inductor ripple currents should be designed considering inductance variation caused by dc current values. Designing an inductor considering the maximum value of

ripple current is one of the most important parts in the design process because it contributes to the power losses and temperature rise in power semiconductor devices and other passive components. Furthermore, since the relative permeability of powder cores changes depending on the magnetic field intensity, the dc superposition characteristic of the powder core affects the winding turns to obtain the required inductance value.

A prediction method of inductance variation in powder cores under the dc excitations is proposed by using magnetic circuit theories in [15]. In this study, the dc superimposition characteristic is modeled relying on measured experimental data or data sheets. Nonetheless, modeling based on the material properties is not discussed. In [16]-[17], a modeling method of relative permeability was proposed. However, this modeling method does not explain in details the meaning of coefficients used in the relative permeability model equation.

Therefore, this paper proposes a novel relative permeability modeling method using a mathematical function, simulation technique, and magnetic design method considering dc superimposition characteristics for powder cores used in power converters. The attractive feature of the proposed approach is that it facilitates the implementation of a non-linear inductance into simulators, using a mathematical equation function of the material relative permeability. This paper is divided into six sections. Section II discusses the model equation, which shows that the relative permeability is dependent on magnetic field intensity. The newly developed equation has three coefficients (p, q, and r) indicating the non-linear change of relative permeability depending on magnetic field intensity. The three coefficients can be extracted from experimental tests. The meaning of the proposed model equation and the three coefficients are described in Section II. In Section III, by using the proposed model equation, a magnetic design method of an inductor using a powder core to obtain the desired inductance value is also proposed. Furthermore, accurate magnetic device models in simulations are helpful to estimates the loss, size, and steady-state and dynamic characteristics before manufacturing the prototype. Therefore, an implementation method for the model equation into a circuit simulator is also proposed in section IV. In section V, the proposed magnetic design method is validated through conducting experimental tests. Furthermore, the experimental waveforms are compared with simulation waveforms. Finally, conclusions are presented in section VI.

II.　MODELING OF RELATIVE PERMEABILITY DEPENDING ON MAGNETIC FIELD INTENSITY

A. Experimental Measurement Method of Relative Permeability Under DC Current Superimposed Condition

The magnetic core structure and magnetic circuit model of an inductor are shown in Fig. 2. where Ni_L is the magnetomotive force shown by the product of the number of turns N and the inductor current i_L. R_m means

1096

The 2018 International Power Electronics Conference

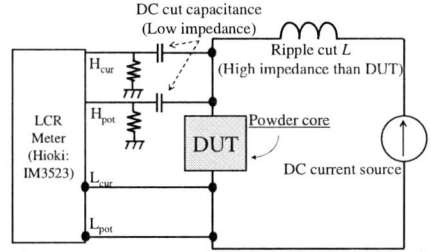

Fig. 3. A measurement system for relative permeability of powder core under DC superimposition condition.

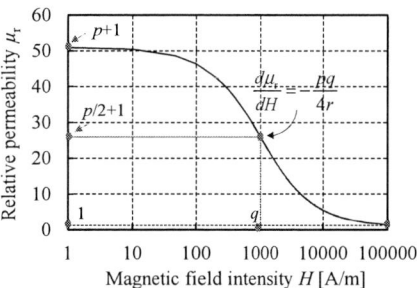

(a) Meaning of the coefficients showing relative permeability. (As an example: p=50, q=1000, r=1)

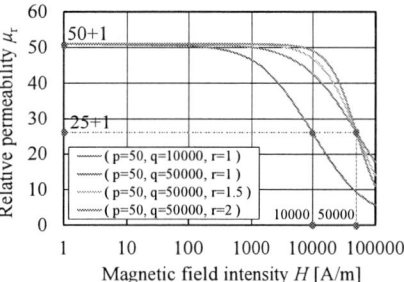

(b) Tendencies toward relative permeability when the coefficients (q, p, r) varies.

Fig. 4. The interpretation of proposed model equation showing relative permeability which depends on magnetic field intensity.

the magnetic reluctance of the closed magnetic path of the toroidal core. l_{core} is the average magnetic path, A_{core} is the sectional area of the core. From this magnetic circuit model, the inductance L_{self} relates to the number of turns N and the structural parameters of the core and given by

$$L_{self} = \frac{N^2}{R_m} = N^2 \cdot P_m = \left(\frac{N^2 \cdot \mu_0 \cdot \mu_r \cdot A_{core}}{l_{core}} \right) \quad (1)$$

where μ_0 is the space permeability, and it equals $4\pi \times 10^{-7}$ H/m. P_m means the permeance of the core. Besides, the inductance L_{self} varies as the function of the inductor current i_L. It can be deduced from (1) that the variable parameter which has the dependency on the inductor current is the relative permeability because N, μ_0, A_{core}, l_{core} are constant parameters, at a fixed value of inductance. Therefore, in order to design an inductor considering dc superimposition characteristic, a precise modeling of relative permeability under dc excitations condition is required.

Fig. 3 shows a measurement system for the relative permeability of powder cores under dc superimposition condition [22]. This system consists of LCR meter (HIOKI, model number: IM3523), two resistors and two capacitors to block the dc component, and two capacitors have low impedance at the measured frequency. These resistors and capacitors are utilized to protect the LCR meter. The ripple cut inductor has a large inductance value to minimize the ac components, and it is connected in series with the dc current source to prevent flowing out of the small signal of the LCR meter to the dc current source side. In order to measure the relative permeability of powder cores, first of all, an optional number of turns are wounded to a powder core. At this time, inductance value is measured while changing dc current value of the dc current source. Then, by using (1), the relative permeability of the cores can be calculated from measured inductance values.

B. Proposed Model Equation Representing Relative Permeability which Changes Depending on Magnetic Field Intensity

Relative permeability of powder cores can be measured by applying the above procedure. However, such many measurement data obtained from the experiments are difficult to process mathematically in designing magnetic components or in implementing dc superimposition characteristics to a circuit simulator. Therefore, we propose a novel model equation showing

the relative permeability obtained from the measured relative permeability.

The proposed model equation is shown by

$$\mu_r(H) = 1 + \frac{p}{1 + (H/q)^r} \quad (2)$$

where p, q, and r are the coefficients to show inductor current dependency on the relative permeability. H is the magnetic field intensity and given by

$$H = \frac{N \cdot i_L}{l_{core}} \quad (3)$$

For the intuitive interpretation of the proposed model equation, the conceptual diagram using (2) is illustrated in Fig. 4 (a). The first term on right side in (2) means the relative permeability of the core when complete magnetic saturation occurs. Therefore, (2) converges to 1 if H becomes infinite. p is a parameter related to initial relative permeability at zero dc current condition. In addition, the initial relative permeability is shown by $1+p$. q means the magnetic field intensity where the initial relative permeability reaches its half value. r is the coefficient related to the negative slope of the relative permeability. The variation of relative permeability when the coefficients (q, p, and r) vary is shown in Fig. 4 (b). As seen in Fig. 4 (b), when r has a high value, the relative permeability tends to rapidly decrease at around the p. In addition, the slope of the relative permeability at q can be expressed as $-pq/4r$. Therefore, the model equation shown in (2) is helpful because the meaning of each coefficient can be intuitively understood. In addition, these coefficients are helpful to evaluate dc superimposition characteristics of various magnetic

1097

The 2018 International Power Electronics Conference

TABLE I
MAGNETIC MATERIALS AND SIZE OF THE TOROIDAL CORES

Type	Magnetic material	Model number	Chemical component	Outer Diameter (OD)	Inner Diameter (ID)	Thickness (TH)	Sectional area A_{core}	Average magnetic path length l_{core}
A	Ferrosilicon powder	BCW155-08150	Fe-Si	27.3mm	14.5mm	11.2mm	71.6mm²	65.7mm
B	Permalloy powder	BCH15-02800	Fe-Ni	27.2mm	14.6mm	11.6mm	73.0mm²	65.7mm
C	Supermalloy powder	BCM15-02800	Ni-Fe-Mo	27.2mm	14.6mm	11.8mm	74.3mm²	65.7mm
D	Sendust powder	BCS15-02750	Fe-Si-Al	27.4mm	14.3mm	11.7mm	76.6mm²	65.5mm
E	Amorphous powder	GLT114D	Fe -P-C-B-Si	27.4mm	14.7mm	11.9mm	75.6mm²	66.1mm

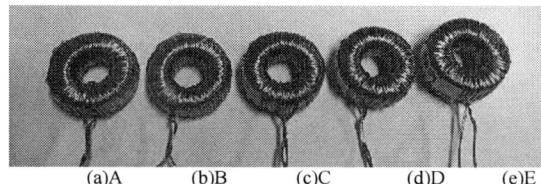

(a)A   (b)B   (c)C   (d)D   (e)E

Fig. 5. Prototypes of the inductors using toroidal core shown in Table II. The number of turns in each core is 65 turns.

materials.

C. Evaluating the Proposed Model Equation

The experimental tests and mathematical analysis are conducted to validate whether the proposed model equation can actually accurately fit the measured values of relative permeability. Five different magnetic materials have been evaluated in this evaluation. Table I shows the magnetic materials and their core size used in this evaluation, and the magnetic cores are shown in Fig. 5. The magnetic cores have almost the same structure to guarantee a fair comparison and to eliminate any dependency of the relative permeability on the core dimensions, as mentioned in section I.

The measurement conditions for this evaluation is as follow: the measured frequency is 50 kHz, the small signal voltage of LCR meter is 1V. In addition, the number of turns in each core is 65 turns and dc current range is 0A to 18A in order to change the dc magnetic field intensity.

The measured results of the relative permeability in each magnetic core are shown in Fig. 6. In Fig. 6, the dots mean the experimentally measured results, and the solid lines show the approximate lines obtained from (2). The coefficients showing relative permeability are automatically parameterized by using computer calculation. The measured coefficients p, q, and r in each core are summarized in Table II.

III. MAGNETIC DESIGN METHOD FOR POWDER CORES

A. Determination of Number of Turns

In this section, a design method of powder core used in power converter is introduced by using the proposed model equation shown in (2). As a case study, the proposed magnetic design method is applied to an inductor used in a boost converter shown in Fig. 7 (a).

Since the model coefficients are already measured by the experiments, and they express the relative

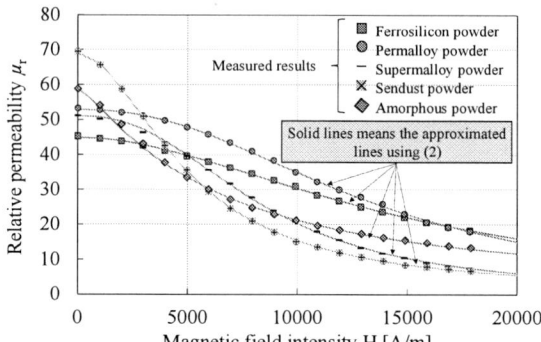

Fig. 6. The comparison results between the measured results and approximated lines using (2).

TABLE II
COEFFICIENTS SHOWING THE RELATIVE PERMEABILITY

Type	p [-]	q [A/m]	r [-]
Ferrosilicon powder (A)	43.9	14300	1.94
Permalloy powder (B)	51.9	13100	2.33
Supermalloy powder (C)	50.1	8260	2.47
Sendust powder (D)	68.0	5020	1.91
Amorphous powder (E)	58.1	6020	1.24

TABLE III
CIRCUIT PARAMETERS

Input voltage	V_i	50V
Output voltage	V_o	100V
Output power	P_o	500W
Duty ratio	d	0.5
Switching frequency	f_{sw}	50kHz
Switching period	T_s	20μs
Inductor average current (maximum power rating)	I_{Lave}	10A
Designed value of inductor ripple current	I_{Lpp}	5A
Inductance at I_{Lave}=10A	L	100μH

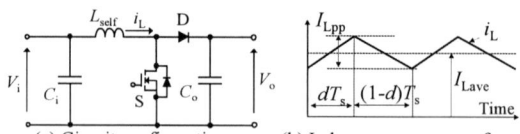

(a) Circuit configuration.   (b) Inductor current waveform.

Fig. 7. Circuit diagram and current waveforms of the boost converter.

permeability of each powder core material. Therefore, these coefficients are used for the magnetic design considering dc superimposition characteristics.

Using (1)-(3), an expression for the self-inductance L_{self} can be obtained, as a function of the variable inductor current:

$$L_{self} = \left(\frac{N^2 \cdot \mu_0 \cdot A_{core}}{l_{core}} \right) \cdot \left(1 + \frac{p}{1 + \left(N \cdot i_L / q \cdot l_{core} \right)^r} \right) \quad (4)$$

1098

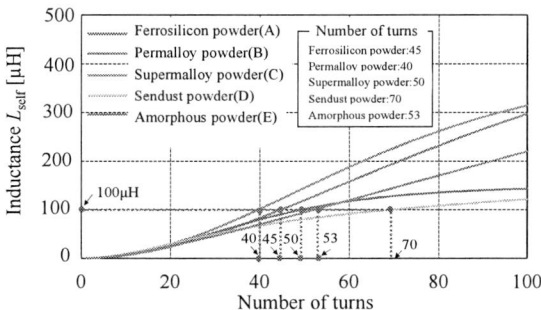

Fig. 8. The relationship between the number of turns and inductance using toroidal core shown in Table II.

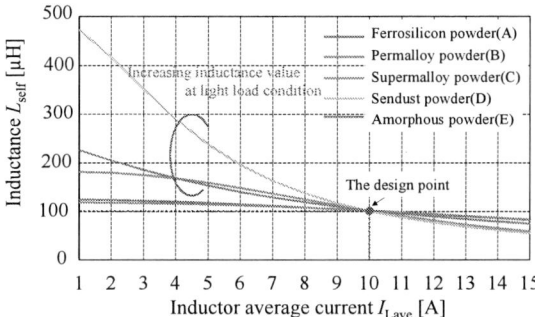

Fig. 9. The relationship between inductance value and inductor average current when using the determined number of turns.

Therefore, once the designed inductance value which fulfills the circuit specification is decided, the number of turns can be derived by substituting the structural parameters of the powder core and the coefficients related to relative permeability into (4). However, equation (4) is difficult to be solved for N, in this case, numerical analysis using computer calculation is used to decide the number of turns N. By conducting numerical analysis, the designed number of turns can be automatically calculated. The circuit parameters for evaluating the proposed magnetic design method are listed in Table III. In general, the inductor ripple current is designed at a maximum power rating of converter because the power losses and temperature rise in active/passive components should be considered in advance. Since most of the high-power converters operate in Continuous Current Mode (CCM) is often used, therefore, in this evaluation, the inductor is designed considering CCM operation.

From Faraday's law, the inductor ripple current I_{Lpp} can be obtained from Fig. 7 (b) and defined as

$$I_{Lpp} = \frac{V_i \cdot d \cdot T_s}{L_{self}} \quad (5)$$

where T_s is the switching period in the main switch S shown in Fig. 7 (a). d is the duty ratio of S.

The inductor ripple current changes non-linearly due to the non-linear behavior of core relative. Nonetheless, the designed self-inductance value can be obtained using (5) and approximated at the operating point of the inductor average current I_{Lave}. In this case, the designed inductance value is 100μH, and as a result, the number of turns can be calculated.

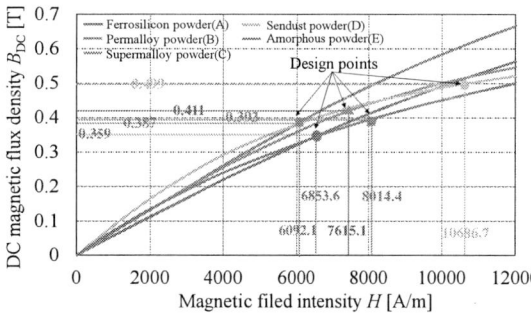

Fig. 10. Single-valued non-linear B-H curves for each core materials.

Using (4), the relationship between inductance value and the number of turns in each magnetic core is shown in Fig. 8. As shown in Fig. 8, the number of turns in each magnetic core differ even though magnetic core sizes are almost same. Therefore, the designed number of turns varies greatly depending on the magnetic properties of each core. In other words, the number of turns depends on the values of coefficients p, q, and r.

It should be noted here that the magnetic materials having high q are effective for reducing the number of turns because these materials can keep high relative permeability under the high current condition. In this case, Core A shown in Table I is designed with 45turns. Similarly, B is 40turns, C is 50turns, D is 70 turns, and E is 53 turns. In order to realize the designed inductance of 00μH at I_{Lave}=10A condition.

B. Inductance variation

The inductance variations while the dc current changes are investigated in this subsection. The inductance variation depending on the dc current value can be investigated according to (4) and considering the calculated number of turns. The inductance variation for each inductor is shown in Fig. 9. As shown in Fig. 9, an inductor utilizing core C, D, E can realize a higher inductance value than that of cores A and B within light load range. A core with high initial permeability property indicated by 1+p tends to realize high inductance value within light load range. Therefore, this property can improve the inductance value in the range of light loads, and it contributes to reducing conduction losses in power devices and passive components because inductor ripple current can be reduced [23]. When using non-linear inductors instead of linear inductors in industrial applications, some advantages have been reported in [24]-[25]. In passive PFC (Power Factor Correction) converters, the non-linear inductors such as powder cores can suppress generation of harmonics in input currents while downsizing magnetic components in comparison with gapped inductors using ferrite cores [24]. In dc/dc converter with maximum power point tracking for photovoltaic, a non-linear inductor is applied to expand the operation range of light load [25]. Using a non-linear inductor, the self-adaption of the circuit to a large range of solar input power can be realized.

1099

C. Dc magnetic flux density in cores

By using the coefficients related to the relative permeability, dc magnetic flux density can be theoretically predicted. Relative permeability μ_r is defined as the rate of a change of a magnetic flux density over the rate of a change of a magnetic field intensity H and divided by the space permeability.

Therefore, the relative permeability at average magnetic field intensity H_{ave} $(=NI_{Lave}/l_{core})$ can be expressed by

$$\mu_r(H) = \frac{1}{\mu_0} \cdot \left. \frac{\partial B}{\partial H} \right|_{H_{ave}} \qquad (6)$$

Therefore, dc magnetic flux density in cores, B_{DC}, can be estimated by integrating magnetic field intensity H as follows.

$$
\begin{aligned}
\left. B_{DC} \right|_{H_{ave}} &= \int_0^{H_{ave}} \mu_r(H) \cdot \mu_0 \cdot dH \\
&= \int_0^{H_{ave}} \left(1 + \frac{p}{1+(H/q)^r} \right) \cdot \mu_0 \cdot dH
\end{aligned}
\qquad (7)
$$

Therefore, by solving (7), single-valued non-linear B-H curves for each core material can be depicted, and the curves are shown in Fig. 10. As a result, the dc flux density caused by the dc average current I_{Lave} can be predicted.

IV. IMPLEMENTATION OF THE PROPOSED MODEL EQUATION TO A CIRCUIT SIMULATOR

This section introduces an implementation method of dc superimposition characteristics of the powder core to a circuit simulator. In general, power electronics engineers use a circuit simulator in order to evaluate the whole circuits including control, power stage, and active/passive components. On the other hand, suppliers of magnetic components use electromagnetic field simulators to evaluate temperature rise, iron losses, and inductance variation of magnetic components. However, since these characteristics affect each other, a comprehensive analysis method on simulators is necessary to evaluate both of power electronics circuits and magnetic components simultaneously. Certainly, there are some software that can conduct a coupled analysis of an electromagnetic field in magnetic components and circuit state, however, a highly functional computer is needed, and much time is required to calculate the electromagnetic field and circuit state simultaneously. One of the effective ways to evaluate magnetic components used in power electronics circuits is to implement magnetic core properties to a circuit simulation. In addition, accurate magnetic device models in simulations are helpful to estimates the loss, size, and steady-state and dynamic characteristics before manufacturing the prototype.

A. Literature Review

In general, there are many implementation methods of dc superimposition characteristics to circuit simulators. For example, using the lookup table which utilizes the measured inductance can be implemented easily and the implementation procedure is simple. However, circuit

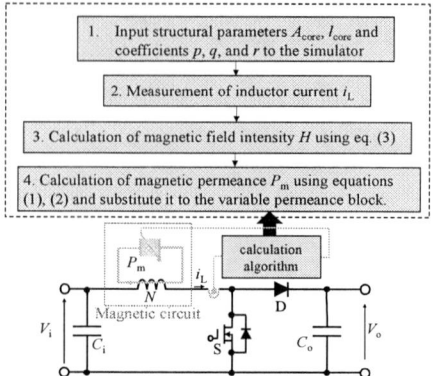

Fig. 11. The flowchart of the simulation algorithm of the coupled analysis of electrical and magnetic circuits.

designers need to carefully type the inductance values into the table in a simulator, and this work may lead to inaccuracy and it wastes a lot of time. In addition, when the number of turns is changed to adjust inductance values, the inductance has to be measured again and retyped into the simulator table.

Another method to express the dc superimposition characteristics is utilizing an equivalent circuit model using ideal passive components and Current-Controlled Voltage Source (CCVS) to reflect the current dependency, this method had been previously proposed in [26]. One of the attractive features of this method is that it allows the dc superimposition characteristics to be implemented in general circuit simulators such as SPICE by using the function of CCVS. However, this method cannot describe the physical meaning of CCVS in the equivalent circuit since inductors do not have CCVS. Furthermore, when the inductor number of turns is changed, the CCVS function needed to be adjusted to coincide with the measured inductance value. Therefore, it is difficult to reflect the dc superimposition characteristics in the circuit simulator while expressing the physical meaning of variable permeability depending on inductor currents using the electrical equivalent circuit model only.

In reference [27]-[28], equivalent circuits using electrical and magnetic circuits were proposed. To implement dc superimposition characteristic to circuit simulators, non-linear magnetic reluctances which have current dependency had been used. In these methods [27]-[28], the physical meaning of inductance reduction can be reflected in the equivalent circuit by using variable magnetic reluctances.

B. An Implementation Method of DC Superimposition Characteristics to a Circuit Simulator

Recently, advanced multi-domain circuit simulators are becoming more popular, which allows the user to simultaneously analyze not only electrical circuit but also thermal and magnetic circuits. In this paper, PLECS (Plexim GmbH) is utilized a software simulation tool, as it can conduct a coupled analysis of both electrical and magnetic circuits [29].

Fig. 11 shows the simulation algorithm flowchart for the electrical and magnetic coupled analysis. The

The 2018 International Power Electronics Conference

TABLE VI
INDUCTOR RIPPLE CURRENT COMPARISON

	Inductor average current 3A			Inductor average current 6A			Inductor average current 10A		
	Experiment	Simulation	Error	Experiment	Simulation	Error	Experiment	Simulation	Error
Ferrosilicon powder (A)	4.04A	4.10A	-1.46%	4.20A	4.37A	-3.89%	4.72A	4.96A	-4.83%
Permalloy powder (B)	4.24A	4.26A	-0.47%	4.52A	4.45A	1.57%	4.92A	4.94A	-0.40%
Supermalloy powder (C)	3.04A	2.88A	5.55%	3.48A	3.40A	2.35%	4.76A	5.00A	-4.80%
Sendust powder (D)	1.60A	1.43A	11.89%	2.80A	2.60A	7.69%	5.08A	5.01A	1.39%
Amorphous powder (E)	2.96A	2.76A	7.24%	3.80A	3.64A	4.40%	4.96A	4.97A	-0.20%

* The error percentages are calculated by (experimental value-simulation value)/simulation value.

simulation algorithm is easy to implement since only changing the variable permeance while corresponding to the magnetic field intensity. Regarding the calculation algorithm inside the simulation, refer to reference [28]. In the next section, we will compare the current waveforms in experiment and simulation to show the validity of our proposed implementation method into the simulator.

V. EXPERIMENTAL EVALUATION

In this section, the experimental and simulation evaluations are conducted with three objectives. The first objective is to show the validity of the magnetic design method using the proposed model equation, which can fulfill the designed inductor ripple current. The second objective is to show the effectiveness of the implementation method of the dc superimposition characteristics of the powder cores into the simulator. The non-linear inductor current waveforms in simulation and experiment are compared especially. The third objective is comparing power conversion efficiency when using powder core with the different magnetic properties such as p, q, and r.

Fig. 12 shows the experimental results when the cores A-E having the designed number of turns are used. These experimental results are measured while setting the dc offset of -10A to show the ripple currents clearly. As shown in Fig. 12, the designed ripple current of 5A and the measured inductor ripple currents are almost identical with each other. In addition, the experimental inductor current waveform when using the sendust cores (D) changes non-linearly rather than linear triangular wave current time by time.

In order to investigate the inductance variation characteristics in each powder core, the inductor average currents are changed from 10A (maximum power rating) to 6A (half load) and 3A (light load). The comparison results of the inductor ripple currents in the experiment and simulation are summarized in Table VI. As shown in Table VI, experimental results are consistent with the simulation results. Therefore, the validity of the modeling method of relative permeability and the implementation method are respectively confirmed.

Finally, power conversion efficiencies when the load changes between light to heavy load were obtained. The measured power conversion efficiencies are shown in Fig. 13. It can be deduced that the Core E inductor has a better power conversion efficiency in the range of light loads condition because of its low-iron loss characteristics. On the other hand, at heavy load condition around 10A, the Core B has the highest power conversion efficiency because it has few winding turns.

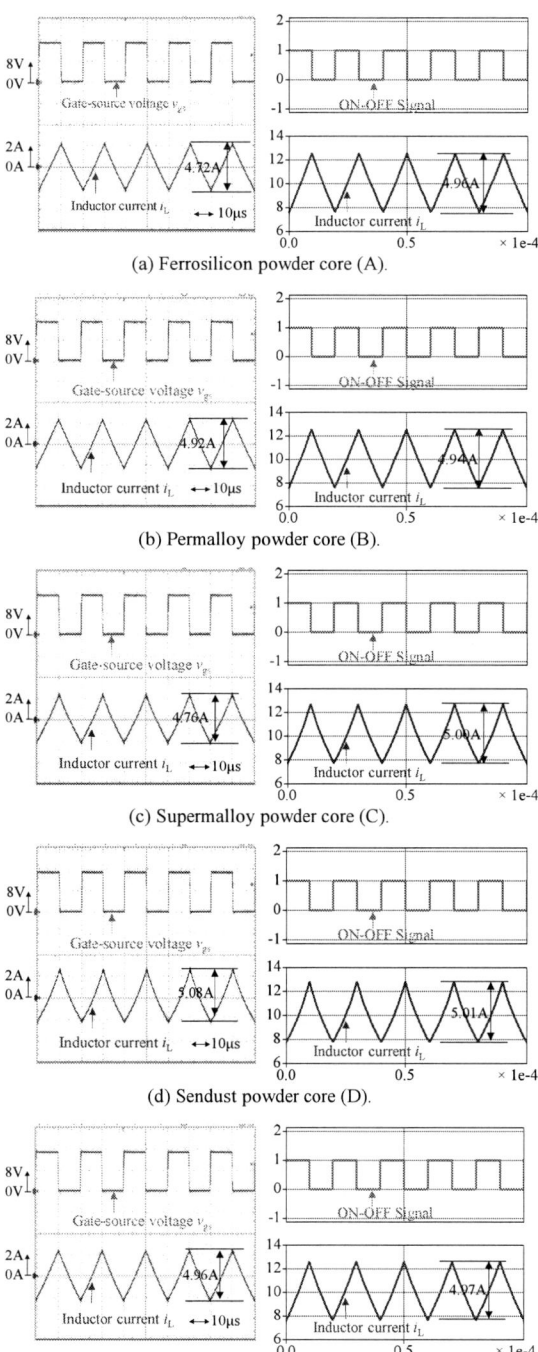

(a) Ferrosilicon powder core (A).

(b) Permalloy powder core (B).

(c) Supermalloy powder core (C).

(d) Sendust powder core (D).

(e) Amorphous powder core (E).

Fig. 12. Comparison of inductor current waveforms in the experiment (left side) and simulation (right side) under the condition that inductor average current is 10A.

1101

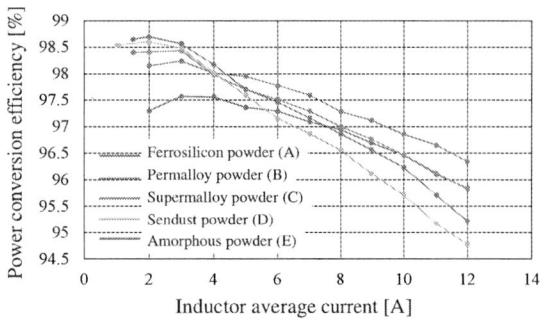

Fig. 13. Power conversion efficiency from light to heavy loads.

VI. CONCLUSIONS

This paper proposed the modeling, design, and simulation methods for powder cores which have non-linear inductance characteristic. A relative permeability which changes depending on magnetic field intensity is modeled for designing inductor using powder cores. The proposed model equation showing the relative permeability has three coefficients and these are parameterized by the experimental results. The design for powder core and implementation methods into simulation circuit simulator were respectively proposed. The validities of analysis were confirmed from both of experiment and simulation.

VII. ACKNOWLEDGEMENT

This work was partially supported by JSPS KAKENHI Grant Number JP16K18059.

REFERENCES

[1] K. Mainali and R. Oruganti, "Conducted EMI Mitigation Techniques for Switch-Mode Power Converters: A Survey," *IEEE Trans. on Power Electron.*, Vol. 25, No. 9, pp. 2344-2356, Sep. 2010.

[2] H. Funato, T. Mori, T. Igarashi, S. Ogasawara, F. Okazaki, and Y. Hirota, "Optimization of Switching Transient Waveform to Reduce Harmonics in Selective Frequency Bands," *IEEJ Journal of Ind. App.*, Vol. 2, No.3, pp. 161-169, May. 2013.

[3] M. Kącki, M. S. Ryłko, J. G. Hayes, and C. R. Sullivan, "Magnetic Material Selection for EMI Filters, " in Proc. IEEE Energy Conversion Cong. & Expo. (ECCE2017), Oct. 2017, pp. 2350-2356.

[4] Y. Itoh, S. Kimura, J. Imaoka and M. Yamamoto, "Inductor loss analysis of various materials in interleaved boost converters", in Proc. IEEE Energy Conversion Cong. & Expos. (ECCE 2014), Sep. 2014, pp.980-987.

[5] M. S. Rylko, K. J. Hartnett, J. G. Hayes, and M. G. Egan, "Magnetic Material Selection for High Power High Frequency Inductors in DC-DC Conversion," in Proc. IEEE Applied Power Electron. Conf. and Expo. (APEC2009), Feb. 2009, pp. 2043-2049.

[6] M. S. Rylko, B. J. Lyons, J. G. Hayes and M. G. Egan, "Revised Magnetics Performance Factors and Experimental Comparison of High-Flux Materials for High-Current DC-DC Inductors," *IEEE Trans. on Power Electron.*, Vol. 26, No. 8, pp. 2112-2126, Aug. 2011.

[7] A. J. Hanson, J. A. Belk, S. Lim, C. R. Sullivan and D. J. Perreault, "Measurements and Performance Factor Comparisons of Magnetic Materials at High Frequency," *IEEE Trans. on Power Electron.*, Vol. 31, No.11, pp. 7909-7925, Nov. 2016.

[8] A. Williams, V. Moruzzi, A. Malozemoff and K. Terakura, "Generalized Slater-Pauling Curve for Transition-Metal Magnets," *IEEE Trans. on Mag.*, Vol. Mag-19, No. 5, Sep. 1983.

[9] Yiren Wang, Gerardo Calderon-Lopez, and Andrew J. Forsyth, "High-Frequency Gap Losses in Nanocrystalline Cores," *IEEE Trans. on Power Electron.*, Vol. 32, No. 6, pp. 2112-2126, Jun. 2017.

[10] S. Nogawa, M. Kuwata, T. Nakau, D. Miyagi, and N. Takahashi, "Study of Modeling Method of Lamination of Reactor Core," *IEEE Trans. on Mag.*, Vol. 42, No. 4, pp. 1455-1458, Apr. 2006.

[11] J. Muhlethaler, J. W. Kolar, and A. Ecklebe, "A Novel Approach for 3D Air Gap Reluctance Calculations,", in Proc. 8th Int. Conf. on Power Electron. and ECCE Asia, May. 2011, pp. 446-452.

[12] R. Jez, "Influence of the Distributed Air Gap on the Parameters of an Industrial Inductor," *IEEE Trans. on Mag.*, Vol. 53, No.11, Article Sequence Number, 8401605, Nov. 2017.

[13] T. Tera, H. Taki, and T. Shimizu, "Loss Reduction of Laminated Core Inductor used in On-board Charger for EVs," *IEEJ Journal of Ind. App.*, Vol. 4, No.5, pp. 626-633, Sep. 2015.

[14] L. Liu, C. Ding, S. Lu, T. Ge, Y. Yan, Y. Mei, K.D.T. Ngo, and G-Q. Lu1, "Design and Additive Manufacturing of Multi-Permeability Magnetic Cores," in Proc. IEEE Energy Conversion Cong. & Expo. (ECCE 2017), Oct. 2017, pp. 881-886.

[15] J. D. Pollock, W. Lundquist and C. R. Sullivan, "Predicting Inductance Roll-Off with DC Excitations," in Proc. IEEE Energy Conversion Cong. & Expo. (ECCE 2011), Sep. 2011, pp. 2139-2145.

[16] G. R. C. Mouli1, J. Schijffelen, P. Bauer and M. Zeman, "Estimation of ripple and inductance roll off when using powdered iron core inductors," in Proc. PCIM Europe 2016, May 2016, pp. 1383-1390.

[17] E. Cardelli, E. Della Torre and E. Pinzaglia, "Identifying the Preisach Function for Soft Magnetic Materials," *IEEE Trans. on Magnetics*, Vol. 39, No. 3, pp. 1341-1334, May. 2003.

[18] J. Imaoka, S. Kimura, Y. Itoh, W. Martinez, M. Yamamoto, M. Suzuki, and K. Kawano, "Feasible Evaluations of Coupled Multilayered Chip Inductor for POL Converters," *IEEJ Journal of Ind. App.*, Vol. 4, No.3, pp. 126-135, May. 2015.

[19] L. Wang, Z. Hu, Y. F. Liu, Y. Pei, X. Yang, and Z. Wang, "A Horizontal-Winding Multipermeability LTCC Inductor for a Low-profile Hybrid DC/DC Converter," *IEEE Trans. on Power Electron.*, Vol. 28, No. 9, pp. 4365-4375, Jun. 2017.

[20] K. Kabeya, S. Yanase, Y. Okazaki and K. Yun, "Magnetic Property of Iron-Dust Cores With Mixture of Ferromagnetic Ferrite Powder and Alumina Powder," *IEEE Trans. on Mag.*, Vol. 50, No. 4, Article Sequence Number: 2800504, Apr. 2014.

[21] K. Shiroki, K. Kawano, H. Matsuura and H. Kishi, "New Type Metal Composite Material for SMD Power Inductor," *Journal of Jpn. Soc. Powder and Powder Metallurgy*, Vol. 61, No. S1. pp, S242-S244, May. 2014.

[22] [Online]Available:https://www.hioki.com/en/products/detail/?product_key=5790

[23] L. Wang, Y. Pei, X. Yang, Z. Wang and Y. Liu, "A Horizontal-winding Multi-permeability Distributed Air-gap Inductor," in Proc of IEEE Applied Power Electron. Conf. and Expo. (APEC), Feb. 2012, pp. 995-1001.

[24] W. H. Wölfle and W. G. Hurley, "Quasi-Active Power Factor Correction With a Variable Inductive Filter: Theory, Design and Practice," *IEEE Trans. on Power Electron.*, Vol. 18, No. 1, pp. 248-255, Jan. 2018.

[25] Longlong Zhang, William Gerard Hurley, and Werner Hugo Wolfle, "A New Approach to Achieve Maximum Power Point Tracking for PV System With a Variable Inductor," *IEEE Trans. on Power Electron.*, Vol. 26, No. 4, pp. 1031-1037, Jan. 2018.

[26] [Online]Available:http://www.murata.com/en-us/about/newsroom/techmag/metamorphosis18/appnote/01-02

[27] O. Ichinokura, K. Sato, and T. Jinzenji, and K. Tajima, "A Spice Model of Porthogonal-Core Transformers," *Journal of Applied Physics*, Vol. 69, No. 8, pp. 4928-4930, Apr. 1991.

[28] J. Allmeling, W. Hammer, and J. Schonberger, "Transient Simulation of Magnetic Circuits Using the Permeance-Capacitance Analogy," in Proc. IEEE 13th Workshop on Control and Modeling for Power Electron. (COMPEL), Jun. 2012, pp. 1-6.

[29] [Online]Available: https://www.plexim.com/plecs

The 2018 International Power Electronics Conference

Modelling and Design of a Medium Frequency Transformer for High Power DC-DC Converters

Miloš Stojadinović*, Jürgen Biela

Laboratory for High Power Electronic Systems, ETH Zürich, Switzerland
*Email: stojadinovic@hpe.ee.ethz.ch

Abstract—Dual Active Bridge (DAB) converters are an interesting solution for battery interfaces in storage systems for traction applications. Due to the environmental conditions and space limitations, the design of the transformer and cooling system is crucial for achieving a high power density. Therefore in this paper a detailed design of a transformer with integrated liquid cooling structure and high isolation voltage is presented. Analytical models for the design are presented and verified with FEM simulations and measurements on a prototype system.

Keywords—Transformer design, modelling, FEM

I. INTRODUCTION

High power DC-DC converters with galvanic isolation are a key element of many applications, as for example medium voltage DC (MVDC) grids [1]–[4], solid-state transformers and power supplies for traction [5]–[8], or more-electric ships [9]. Another application are battery interfaces used in electric vehicles or traction systems [10], [11]. In fig. 1/table I a possible setup and specifications of such a battery interface for a locomotive are given. The system enables to store recuperated energy during braking, what reduces the total energy consumption and enables reuse of the recuperated energy during the acceleration phase. Additionally, energy stored in the batteries can be used to drive the locomotive on non-electrified tracks without a diesel engine, avoiding CO_2 emissions (e.g. in shunt yards). The considered battery interface is based on a dual active bridge (DAB) converter [12]. The series connection of modules at the medium voltage/secondary side allows to use switching devices with lower voltage ratings for all switches. It also makes the transformer isolation requirements more severe. A high efficiency and high power density is required due

Table I. SPECIFICATIONS OF THE BATTERY INTERFACE CONSISTING OF 4 MODULES.

System power (4 modules @ 50 kW)	200 kW
Primary side voltage V_p	530 V..835 V
Nominal primary voltage	710 V
Secondary side voltage V_s	2800 V (4×700 V)
Rated nominal withstand voltage	2.8 kV
Efficiency	>95 %
Power density	>5 kW/dm³
Ambient temperature	75 °C
Cooling medium temperature	60 °C

to the space limitations in the locomotive. The high power density is achieved by optimising the converter design and by pushing the switching frequency of the semiconductor devices to higher values under ZVS condition. Several examples for high power medium voltage DC-DC converters with galvanic isolation can be found in the literature [1], [4]–[8]. Figure 2 gives a comparison of some of these converter systems in the efficiency/power density plane based on data provided in the literature.

For achieving a high power density, the advanced cooling

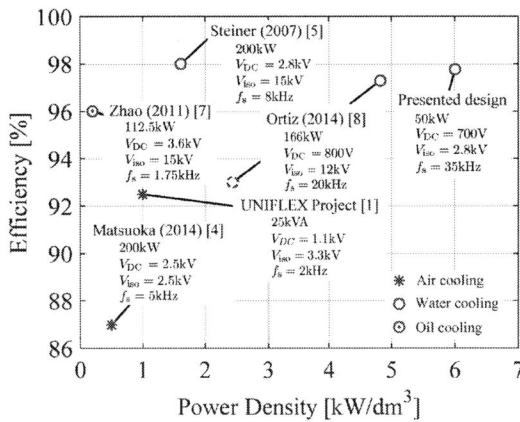

Figure 2. Efficiency / power density comparison of the presented system to the previous state of the art solutions. All the results are given for a single module with full isolation rating of the transformer. For the design presented in [8], both calculated and measured efficiency / power density values are given, where the ⊙ represents the measured values. The presented design only shows the calculated efficiency / power density values.

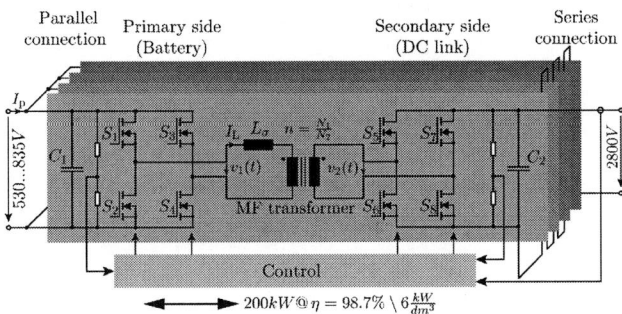

Figure 1. Modular DC-DC converter system based on DAB topology.

The 2018 International Power Electronics Conference

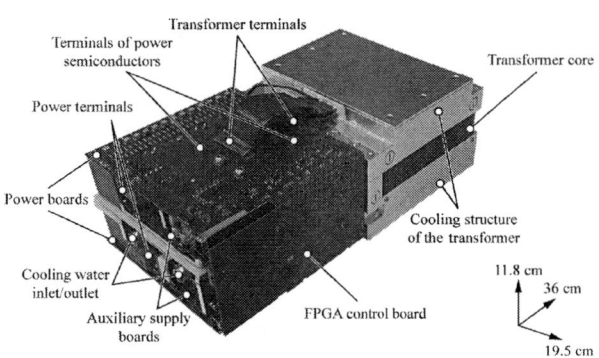

Figure 3. Photo of the single module DAB converter for a modular DC-DC system.

concepts [13] have to be used in transformer. Many designs employ a coaxially wound transformer [5], [14]–[16]. The cooling of such a transformer is usually achieved by using hollow inner conductors [5], [15], [16] through which the de-ionized water is pumped. Another investigated transformer structure is the shell type geometry. The cooling of such a structure can be achieved either with natural or forced convection [17]–[19], using aluminium plates [8] or via heat pipes [20] to conduct the heat from the windings to the core-mounted heat sinks. Due to the switching frequencies in the kHz range, the eddy current losses induced in the structures with aluminium cooling plates lead to higher than predicted temperatures in the transformer. To cope with this issue a thermally conductive coil formers can be used in order to conduct heat from windings to the aluminium heat sink mounted on the transformer core [21]. Unfortunately, the thermal conductivity of such coil formers is considerably lower than for aluminium, which results in larger transformer volumes.

In order to increase the power density, a transformer with integrated liquid cooling is presented in this paper. The cooling system can be operated with tap water, despite the high nominal isolation voltage of $V_{\mathrm{iso}} = 2.8\,\mathrm{kV}$. For designing the integrated cooling structure, analytical thermal models are presented and verified with FEM simulations and measurements. In section II first the short overview of the design procedure is outlined, followed with used models for the transformer design. There, special attention is dedicated to the thermal modelling of the integrated cooling structure with water channels. Results of FEM simulations for different design aspects are given in section III. Finally, experimental measurements on a prototype system are presented in section IV, followed by conclusions.

II. TRANSFORMER DESIGN

The system shown in fig. 1 is used as battery interface in locomotives. Since the secondary side (DC link side) is connected in series, the nominal primary and secondary voltages are almost equal (table I). This enables to use the same

switching devices for both full bridges. For the presented system 1200 V SiC MOSFETs devices are employed. Using SiC MOSFETs under ZVS conditions, enables higher switching frequencies.

In order to determine the parameters for the highest power density, an optimization of the converter modulation scheme and the transformer (fig. 4) is performed for worst case input conditions. The modulation parameters of the DAB converter are the phase shift angle ϕ, clamping interval of the primary side bridge δ_1, the clamping interval of the secondary side bridge δ_2 and frequency f_{s}. Illustrative explanation of the modulation parameters is shown in fig. 5. More details about the used modulation schemes and the description of the implementation are given in [22]. The optimization is multi-objective, i.e. both the system volume and the total system losses are minimized.

Starting with the system specifications (table I) at *step 0* in fig. 4, first the converter electrical model for the initial control parameters is derived in *step 1*, giving at the output the voltage/current waveforms and required leakage inductance (L_σ). The voltage/current waveforms (fig. 5) are used to calculate the losses in the switching devices in *step 2*. In *step 3*, the calculated output power P_{out} (defined by the control parameters) is compared to the required power P_{nom}. In the same step, semiconductor losses are calculated and verified to be below the maximum specified losses. If any of the constraints is not fulfilled, the control parameters are changed and the procedure restarts. In *step 4*, the transformer

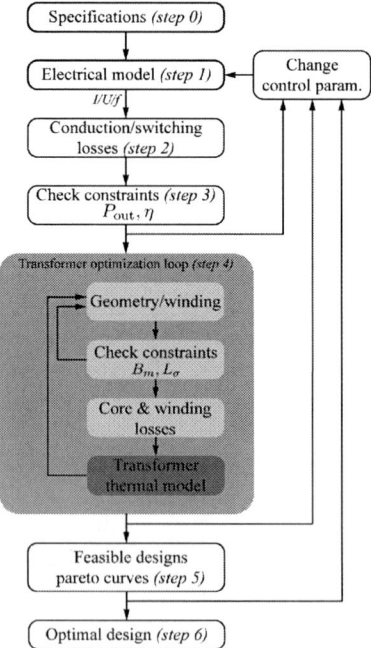

Figure 4. Simplified flow chart for the optimization procedure to find the optimal control variables ϕ, δ_1, δ_2, f_{s} and the optimal transformer design for worst case conditions, by minimizing the DAB converter losses and volume.

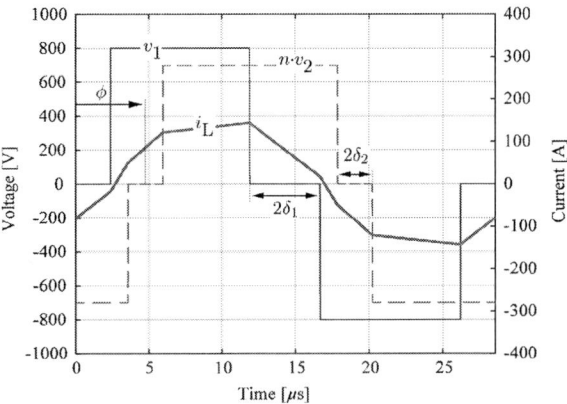

Figure 5. Exemplary voltage and current waveforms at the transformer terminals for a random set of control parameters $[\phi_2, \delta_1, \delta_2]$.

optimization loop is executed which determines a suitable core geometry and winding arrangement that fulfils the following constraints: peak flux density $B_{\mathrm{m}} \leq B_{\mathrm{sat}}$, leakage inductance $L = L_{\mathrm{sigma}}$ and temperature rise $T \leq T_{\mathrm{max}}$. If any of the constraint is not met the calculation restarts with new control parameters. After all the feasible designs are obtained in *step 5*, the optimal design (*step 6*) is chosen at the knee point of the pareto curve.

For the considered system, major challenges are the design of the transformer and the efficient heat removal. In order to improve the heat removal, foil conductors are used for the transformer windings. Using foil only slightly increases the eddy current losses in the windings compared to the litz wire [23], but due to the large copper filling factor and surface area, foil windings are preferred from a thermal and a size point of view. In fig. 6 the CAD drawing of the transformer with integrated cooling system is shown for the specifications given in table I. One of the specifications for the transformer and semiconductor cooling system is to use tap water. Therefore,

the whole liquid cooling structure is at the ground potential. The cooling channels are used for cooling the switches and the transformer. The channel structure can be seen in fig. 10. Because of the high isolation requirement between the transformer windings, the cooling channels are placed only on the outer core legs of the transformer. For removing the heat from the transformer primary winding, first an aluminium bar was considered as a part of the bottom cooling part (fig. 12). The bar is placed between the winding and the transformer middle leg. Due to the induced eddy current losses, the aluminium bar was replaced with an aluminium nitride (AlNi) bar. Aluminium nitride is an electrical isolator which offers high thermal conductivity equivalent to thermal conductivity of aluminium (table II). For cooling the secondary winding and fulfilling the isolation requirement, the transformer is potted using a thermally conductive casting compound *Wepesil VU 4675*, which offers a wide operating temperature range and low hardness.

For isolating the secondary side switches, an aluminium oxide (Al_2O_3) plate is used to separate the secondary side cooling structure, on which the secondary side switches are mounted, and the grounded bottom cooling part. The fixation of the aluminium and Al_2O_3 plate to the main structure is achieved with nylon glass filled bolts. By increasing the outer diameter and rounding the edges of the holes in the aluminium parts (see fig. 7) it is possible to shape the e-field in order to fulfil the given isolation requirements. Table II lists the specifications of the designed medium frequency transformer with important parameters used for thermal modelling.

A. Loss Modelling

In this section, the models used for modelling the transformer core and winding losses are summarized.

Table II. SPECIFICATIONS OF THE MEDIUM FREQUENCY TRANSFORMER.

Element	Material	Thermal Conductivity
Core	N87, N97	$4\,\mathrm{W/(m\,K)}$
Winding isolation	Poly-Pad K10	$0.85\,\mathrm{W/(m\,K)}$
Trafo cold plate	AlNi	$> 150\,\mathrm{W/(m\,K)}$
Switch cold plate	Al_2O_3	$20..30\,\mathrm{W/(m\,K)}$
Potting	Wepesil VU 4675	$1.2\,\mathrm{W/(m\,K)}$

Parameter	Value
Maximum frequency	$35\,\mathrm{kHz}$
Leakage inductance	$26.5\,\mathrm{\mu H}$
Primary turns	20
Secondary turns	20
Primary foil winding thickness	$100\,\mathrm{\mu m}$
Secondary foil winding thickness	$100\,\mathrm{\mu m}$
Primary winding losses	$73\,\mathrm{W}$
Secondary winding losses	$150\,\mathrm{W}$
Core losses N87 (N97)	$45\,\mathrm{W}\ (37\,\mathrm{W})$

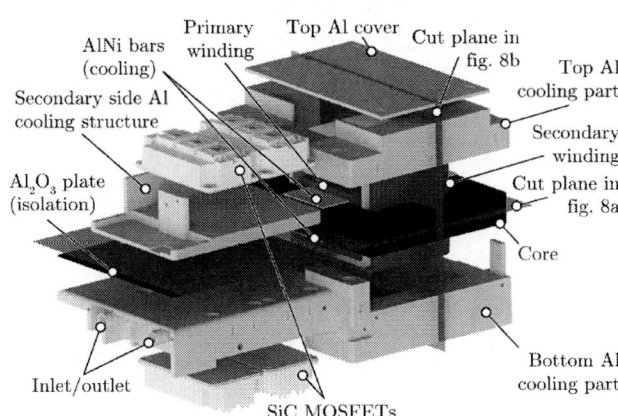

Figure 6. Exploded view CAD drawing of the designed transformer with integrated cooling structure.

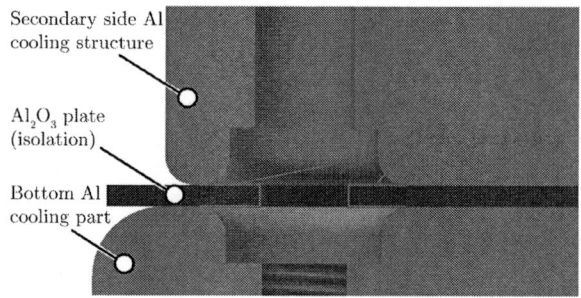

Figure 7. Cut view of the fixation point between bottom cooling part and secondary side cooling structure.

1) Winding Losses: Since foil windings are used for the design of the transformer, the method outlined in [24] is used for calculating the skin and proximity effect losses. The method is based on an 1D approximation of the H-field in the windings, which results in the resistance factor expressed in the term of hyperbolic trigonometric functions of the skin penetration depth.

2) Core Losses: For calculating the transformer core losses the Improved Generalized Steinmetz Equation (iGSE), presented in [25], is used. This procedure takes the derivative of the flux waveform, as well as the peak-to-peak value of the flux into account in order to calculate the core loss. The method offers good precision with low complexity, which is advantageous for optimization problems.

Care must be taken when using manufacturer loss measurements for core materials. These measurements are usually performed on small toroidal cores and the losses might be significantly lower (2-3x) compared to the losses in the cores with different shapes (e.g. lare E-/U-core). It is advisable to compare the losses of the actual used core, usually given just for a single frequency and flux point, to the same point on the loss curves in the material datasheet, and scale the curves accordingly. The core materials which are used and compared during the design procedure are *EPCOS N87* and *EPCOS N97* (table II).

B. Leakage Inductance Modelling

The winding height in transformer design with relatively high isolation requirements is typically smaller than the core window height. In order to calculate the leakage inductance in such design, a formula that employs the *Rogowski* coefficient can be used [26]

$$L_\sigma = \mu_0 N_p^2 \frac{\pi D_{\mathrm{mean}} w_L}{\sqrt{h_{\mathrm{cuP}} \cdot h_{\mathrm{cuS}}}} k_\sigma \qquad (1)$$

where N_p is the number of primary turns, D_{mean} is the mean diameter of the reduced leakage channel, w_L is the width of the reduced leakage channel, h_{cuP} is the primary winding height, h_{cuS} is the secondary winding height, and k_σ is the *Rogowski* coefficient. A simplified illustration of the transformer winding structure is depicted in fig. 8. The expressions for calculating the required parameters in the leakage inductance formula have

been adapted for taking into account the non-circular shape of the windings

$$w_L = \frac{d_P + d_S}{3} + d_L \qquad (2)$$

$$k_\sigma \approx 1 - \frac{d_P + d_S + d_L}{\pi \sqrt{h_{\mathrm{cuP}} \cdot h_{\mathrm{cuS}}}} \qquad (3)$$

$$D_{\mathrm{mean}} = D + d_P + d_L + d_S - \frac{d_S - d_S}{2} \frac{d_P + d_S + 4d_L}{d_P + d_S + 3d_L} \quad (4)$$

where D is the equivalent diameter of the core leg, and can be calculated as

$$D = 2\sqrt{\frac{b_c d_c}{\pi}} \qquad (5)$$

The given formulas for leakage inductance calculation are valid for cases where the heights of the primary and secondary winding are approximately the same.

C. Thermal Modelling

The simplified transformer thermal model used in the optimization procedure (fig. 4) is depicted in fig. 9. The full scale model is more detailed, derived in a matrix form with the size 11×11. It is assumed that the secondary winding bobbin acts as a thermal isolator, and that the primary winding is mainly cooled through the AlNi bar. The picture (fig. 9) is split in two 2D cut planes, each showing a cut view of the transformer with integrated cooling structure (see fig. 6). There also the thermal resistances ($R_{\mathrm{th,r}}$, $R_{\mathrm{th,d}}$) of the water channel are shown. These thermal resistances can be calculated using

$$R_{\mathrm{th,r}} = \frac{0.5 \cdot (w - d)}{bl\lambda_{\mathrm{HS}}} \qquad (6)$$

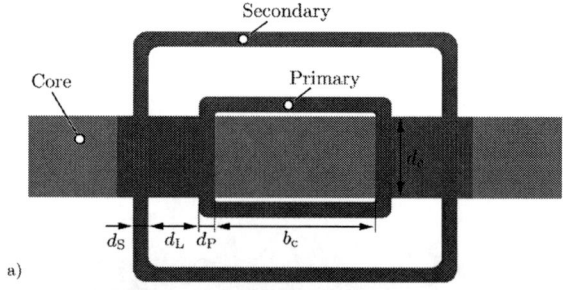

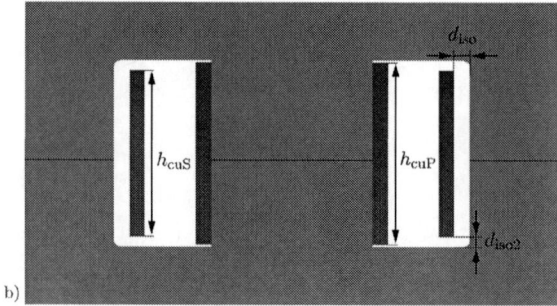

Figure 8. Transformer geometry for calculating leakage inductance.

$$R_{\mathrm{th,d}} = \frac{1}{hl\pi d} \tag{7}$$

where w is the width of the heat sink, b is the height of the heat sink, l is the length of the heat sink, d is the channel diameter, and λ_{HS} is the thermal conductivity of heat sink material. Parameter h is the heat transfer coefficient which can be calculated with

$$h = \frac{N_u \lambda_{\mathrm{fluid}}}{d} \tag{8}$$

For calculating the heat transfer coefficient, it is necessary to know the *Nusselt* number (N_u). The *Nusselt* number is, in general, a function of the average ducted fluid velocity, duct geometry, and the fluids *Prandtl* number (Pr). The author of [27] has derived an analytical model for the generalized *Nusselt* number ($N_{u\sqrt{A}}$) that is suitable for the extruded heat sink model with arbitrary cross-section

$$Nu_{\sqrt{A}} = \left[\left(\left\{ C_2 C_3 \left(\frac{f Re_{\sqrt{A}}}{d} \right)^{\frac{1}{3}} \right\}^5 \right. \right.$$
$$\left. \left. + \left\{ C_1 \left(\frac{f Re_{\sqrt{A}}}{8\sqrt{\pi}} \right) \right\}^5 \right)^{\frac{m}{5}} + \left(C_4 \frac{f(Pr)}{\sqrt{d}} \right)^m \right]^{\frac{1}{m}} \tag{9}$$

with

$$f(Pr) = \frac{0.564}{\left[1 + \left(1.664 Pr^{1/6} \right)^{9/2} \right]^{2/9}} \tag{10}$$

$$f Re_{\sqrt{A}} = \left[\frac{11.8336 \dot{V}}{l \cdot \nu} + (f Re_{\mathrm{fd}})^2 \right]^{1/2} \tag{11}$$

$$f Re_{fd} = \frac{12}{2 \left[1 - \frac{192}{\pi^5} \tanh \frac{\pi}{2} \right]} \tag{12}$$

where $\dot{V}$ is the flow rate in $[m^3/s]$, ν is the kinematic viscosity of the channel fluid, and the parameters C_1, C_2, C_3, C_4 are defined in [27] as

$$C_1 = 3.24 \quad C_2 = \frac{3}{2} \quad C_3 = 0.409 \quad C_4 = 2$$

The blending parameter m is defined by

$$m = 2.27 + 1.65 Pr^{\frac{1}{3}} \tag{13}$$

In order to solve the thermal model of the transformer, additional information on the fluid is required. Due to the multiple parallel pipes, the fluid flow in the different channels is different from the input fluid flow. In order to calculate the velocity of the fluid in the parallel channels, the following expressions [28] are used:

$$\Delta h_{\mathrm{tot}} = \Delta h_1 = \Delta h_2 = \ldots = \Delta h_i \tag{14}$$
$$\dot{V} = \dot{V}_1 + \dot{V}_2 + \ldots + \dot{V}_i \tag{15}$$

where Δh_{tot} is the total head loss of the system, and Δh_i (i $= 1, 2, \ldots$) are the head losses in the individual channels. $\dot{V}$ and $\dot{V}_i$ (i $= 1, 2, \ldots$) designate the input flow rate and flow rate in the individual channels, respectively. For calculating the head losses in the individual channels, the *Moody* relation

$h_{\mathrm{f}} = f\left(L/d\right)\left(V^2/2g\right)$ can be used [28], where V is the fluid velocity, L is the channel length, f is the channel friction factor, and g is the gravitational acceleration. Each channel has a quadratic parallel resistance, and the head loss is related to the total flow rate by

$$h_{\mathrm{f}} = \frac{\dot{V}^2}{\left(\sum \sqrt{K_i / f_i} \right)^2} \tag{16}$$

where

$$\dot{V} = V \cdot \frac{\pi^2 d_i}{4} \qquad K_i = \frac{\pi^2 g \cdot d_i^5}{8 L_i}$$

In the general case, the channel friction factor f_i is a function of the *Reynolds* number and the roughness ratio. Since the *Reynolds* number varies with the fluid velocity, solving of the set of equations must be done iteratively. In the first step, an arbitrary values of f_i are chosen and with them a first estimate of h_{f} is calculated. Then, the resulting flow rate estimate $\dot{V}_i \approx (K_i h_{\mathrm{f}} / f_i)^{1/2}$ is obtained for each channel. Using these results, a new *Reynolds* number and a better estimate of f_i is calculated. Usually, a few iteration steps are sufficient to obtain a satisfactory solution. When the flow in the channel is laminar, a simple expression for f, known as *Darcy* friction

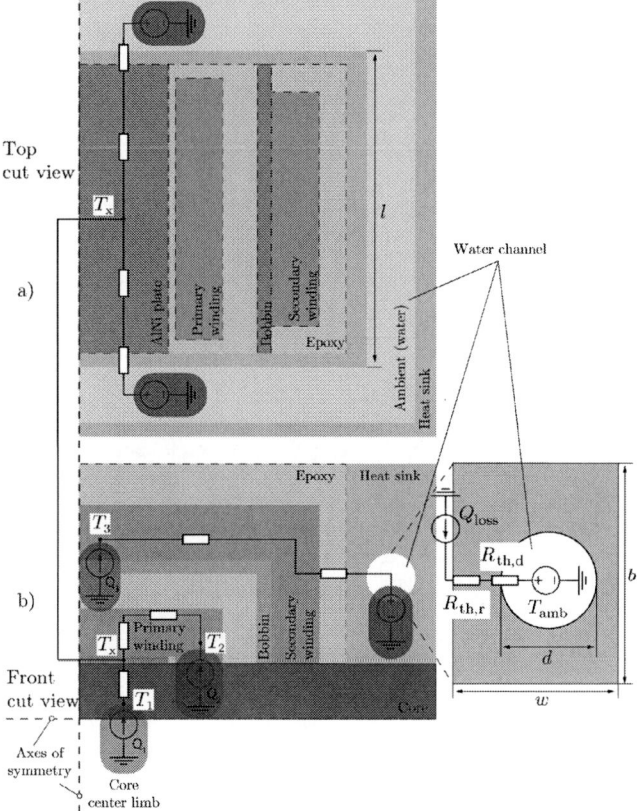

Figure 9. Simplified thermal model of the transformer with model of the channel. The shown thermal network is a reduction of the full scale model having 11 nodes and expressed as 11 × 11 size matrix.

The 2018 International Power Electronics Conference

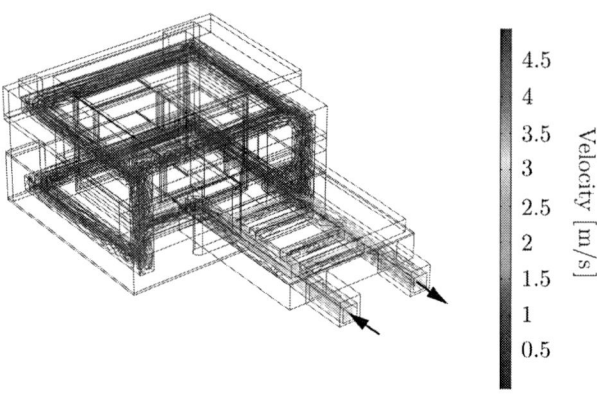

Figure 10. Water velocity field inside the presented converter cooling structure.

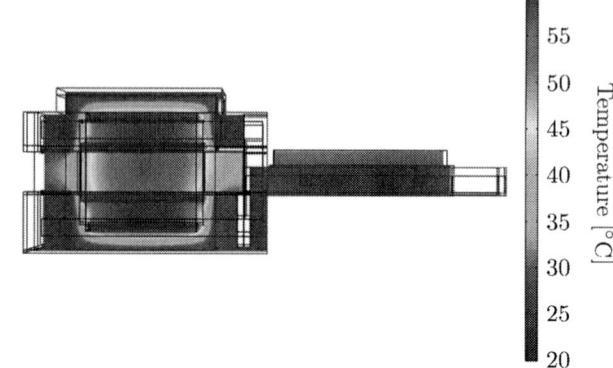

Figure 11. Temperature distribution of the integrated converter structure.

factor, can be used

$$f = \frac{64}{\text{Re}} \qquad (17)$$

where $Re = V \cdot d/\nu$ is the *Reynolds* number. Finally, the obtained fluid flow rate is then used in expression eq. (10) in order to calculate the thermal resistance of the water channel (eqs. (6) and (7)). For the channels which have a turbulent flow, *Haaland's* equation offers good approximation of the turbulent region of the *Moody* chart [28]

$$f = \left[-1.8 \log \left[\frac{6.9}{\text{Re}} + \frac{\epsilon/d}{3.7}^{1.11} \right] \right]^{-2} \qquad (18)$$

where ϵ/d is the channel roughness ratio.

For designs with long channels and/or slow fluid flows, the temperature of the fluid can increase along the axial direction. In order to calculate this temperature rise of the fluid, an 1D energy balance expression [29] is used:

$$q\prime(x) = \dot{V} \rho C_{\text{p}}(T(x) - T_{\text{in}}) \qquad (19)$$

where $q\prime(x) = q/x$ are the power losses per unit of length, ρ is the density of the fluid, C_{p} is the thermal capacity of fluid, and T_{in}, $T(x)$ are the input temperature and temperature at point x in axial direction, respectively.

III. FEM SIMULATIONS

In this section, results of FEM simulations are presented for different design aspects.

A. Lekage Inductance Simulation

For verifying the analytical model of the leakage inductance from section II-B, a 3D magnetic field simulation is performed. For the simulation, an 1 A current excitation of the primary and secondary windings are assumed with opposite directions, and the resulting total magnetic energy is obtained.

The leakage inductance is then calculated from the energy with

$$L_\sigma = \frac{2E_{\text{tot}}}{I^2} \qquad (20)$$

The comparison of the calculated and simulated leakage inductance value is shown in table III.

Table III. COMPARISON BETWEEN THE VALUES OF THE LEAKAGE INDUCTANCE OBTAINED WITH THE ANALYTICAL CALCULATION AND THE 3D FEM SIMULATION.

	Analytical	FEM
Leakage inductance L_σ	26.5 μH	26.2 μH

B. Heat Transfer and CFD Simulations

In order to verify the thermal models presented in section II-C, combined 3D heat transfer and fluid dynamic FEM simulations were performed. The simulated velocity field inside the presented cooling structure is shown in fig. 10, while the temperature distribution is given in fig. 11. The input flow rate at the inlet is assumed to be 27 L/min and the inlet water temperature is equal to 20 °C, which corresponds to the flow rate and water temperature used during experimental measurement. The values of the material parameters for the analytical thermal calculations listed in table II are also used for the 3D FEM simulations. The comparison between the calculated temperatures of the transformer obtained with the presented thermal model from fig. 9 and the results from FEM simulations are given in table IV.

Table IV. CALCULATED TEMPERATURES OF THE TRANSFORMER WITH PRESENTED THERMAL MODEL (FIG. 9).

Temperature	Analytical	FEM
Core mid leg (T_1)	54 °C	56 °C
Primary winding (T_2)	56 °C	59 °C
Secondary winding (T_3)	59 °C	59 °C

1108

The 2018 International Power Electronics Conference

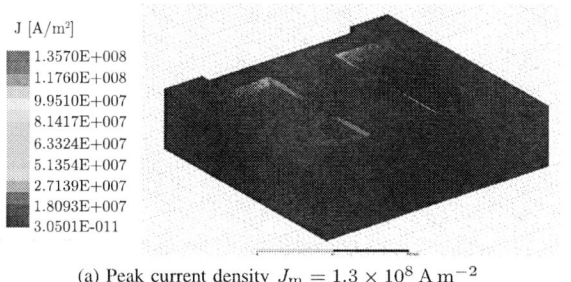
(a) Peak current density $J_{\mathrm{m}} = 1.3 \times 10^8 \,\mathrm{A\,m}^{-2}$

(b) Peak current density $J_{\mathrm{m}} = 4 \times 10^7 \,\mathrm{A\,m}^{-2}$

Figure 12. Eddy currents induced in the bottom cooling structure with a) Al bar and, b) AlNi bar, used for cooling of the primary transformer winding.

C. Eddy Current Simulations

In order to get the eddy current induced losses in the transformer cooling structure 3D FEM simulations were performed. There, two cases have been considered: 1) single Al bar located between the middle transformer leg and the primary winding on the bottom, 2) two AlNi bars located on top and bottom of the middle transformer leg. The surface current densities induced in the bottom cooling parts in case of the Al and AlNi bars are given in fig. 12. The peak current density with the AlNi bars is 3 times lower. The total losses induced in the transformer cooling structure for the two mentioned cases are given in table V.

Table V. INDUCED EDDY CURRENT LOSSES IN THE TRANSFORMER COOLING STRUCTURE.

	Al bar	AlNi bar
Induced losses	75 W	36 W

IV. EXPERIMENTAL RESULTS

A. Leakage Inductance Measurement

The leakage inductance is measured with the impedance analyzer, and the resulting short circuit inductance change with the frequency is depicted in fig. 13. Comparing the measurement results with the results of the analytical calculation and FEM simulation, given in table III, it can be seen that all match very well.

B. Thermal Measurement

In order to measure the temperature of the transformer winding, an NTC temperature probe is attached to the secondary winding prior to potting. For the thermal measurement the primary and secondary winding are supplied with constant DC currents. As a worst case approximation, the value of the supplied current is chosen such that the total transformer losses (core + winding losses), given in table II, are generated in the windings. The water cooling system was operated with constant flow rate of $27 \,\mathrm{L/min}$ and a water temperature of $20\,^\circ\mathrm{C}$. In table VI, the measured secondary winding temperature is given, together with values obtained from analytical calculations and FEM simulations.

Table VI. SECONDARY WINDING TEMPERATURE MEASUREMENT RESULT.

Temperature	Analytical	FEM	Measured
Secondary winding	$59\,^\circ\mathrm{C}$	$59\,^\circ\mathrm{C}$	$57\,^\circ\mathrm{C}$

C. Partial Discharge Measurement

The final measurement which have been performed is the partial discharge measurement. This measurement is performed by a $2.8 \,\mathrm{kV/50\,Hz}$ peak voltage to the secondary side winding while having the primary side winding and core grounded. The results of the partial discharge measurement procedure are depicted in fig. 14. As can be seen, the highest discharge values are lower than $18 \,\mathrm{pC}$, which can, for the given requirements, be regarded as partial discharge free.

V. CONCLUSION

In this paper, the design of a medium frequency transformer with integrated cooling structure is presented. The system integration with advanced cooling design results in a higher power density value (fig. 2). Detailed loss, leakage inductance and thermal models of the medium frequency transformer with integrated cooling structure are used to find the optimal design

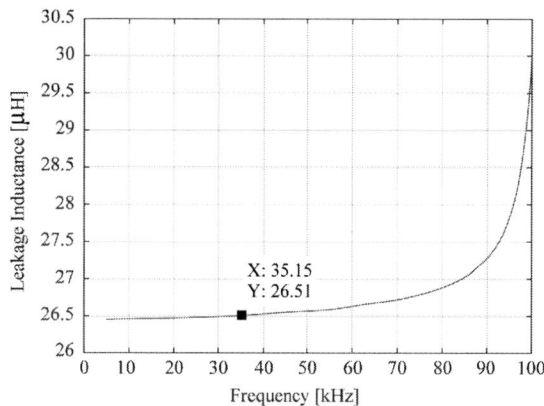

Figure 13. Measured short circuit inductance of the designed transformer. The nominal switching frequency is $35 \,\mathrm{kHz}$

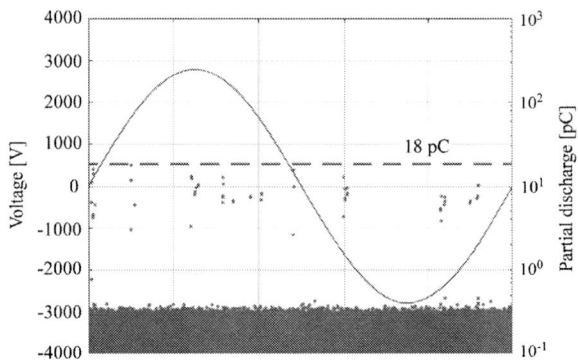

Figure 14.   Results of the partial discharge measurement performed on the designed transformer. The figure shows the cumulative partial discharges during a testing period of 20 min, and their occurrence as a function of the supplied voltage.

parameters using the methodology presented in section II. The resulting design was validated with extensive FEM simulations for various design aspects. The final validation is performed with experimental measurements on the prototype system.

ACKNOWLEDGEMENT

This research is part of the activities of the Swiss Centre for Competence in Energy Research on Efficient Technologies and Systems for Mobility (SCCER Mobility), which is financially supported by the Swiss Innovation Agency (Innosuisse - SCCER program) and *Bombardier Transportation AG Switzerland*. CTI funding grant no.: PFIW-IW 18312.1

REFERENCES

[1] "Uniflex-PM." 2009. [Online]. Available: http://www.eee.nott.ac.uk/uniflex/Documents/Deliverable%20D7_2_FINAL.pdf

[2] G. Reed, G. Kusic, J. Svensson, and Z. Wang, "A case for medium voltage direct current (MVDC) power for distribution applications," in *IEEE-PES Power Systems Conference and Exposition*, 2011.

[3] F. Mura and R. W. De Doncker, "Design aspects of a medium-voltage direct current (MVDC) grid for a university campus," in *IEEE 8th International Conference on Power Electronics and ECCE Asia*, 2011.

[4] Y. Matsuoka, K. Takao, K. Wada, M. Nakahara, K. Sung, H. Ohashi, and S. Nishizawa, "2.5kV, 200kW bi-directional isolated DC/DC converter for medium-voltage applications," in *International Power Electronics Conference*, 2014.

[5] M. Steiner and H. Reinold, "Medium frequency topology in railway applications," in *European Conf. on Power Electronics and Applications*, 2007.

[6] J. Weigel, A. Ag, and H. Hoffmann, "High voltage IGBTs in medium frequency traction power supply," in *European Conference on Power Electronics and Applications*, 2009.

[7] C. Zhao, S. Lewdeni-Schmid, J. Steinke, M. Weiss, T. Chaudhuri, M. Pellerin, J. Duron, and P. Stefanutti, "Design, implementation and performance of a modular power electronic transformer (PET) for railway application," in *European Conference on Power Electronics and Applications*, 2011.

[8] G. Ortiz, "High-power DC-DC converter technologies for smart grid and traction applications," Ph.D. dissertation, ETH Zurich, 2014.

[9] C. Chryssakis and B. J. Vartdal, "Ship electrification and alternative fuels," Motorways of the Seas, 2016.

[10] A. Burke, "Batteries and ultracapacitors for electric, hybrid and fuel cell vehicles," in *Proceedings of the IEEE*, 2007.

[11] S. Vasquez, S. Lukic, E. Galvan, L. Franquelo, and J. Carrasco, "Energy storage systems for transport and grid applications," *IEEE Trans. on Industrial Electronics*, 2010.

[12] R. De Doncker, D. Divan, and M. Kheraluwala, "A three-phase soft-switched high power density DC-DC converter for high power applications," in *IEEE Industry Applications Society Annual Meeting*, 1988.

[13] J. Biela, U. Badstubner, and J. Kolar, "Design of a 5kW, 1-U, 10kW/ltr. resonant DC-DC converter for telecom applications," in *29th International Telecommunications Energy Conference, INTELEC*, 2007.

[14] K. W. Klontz, D. M. Divan, and D. W. Novotny, "An actively cooled 120 kW coaxial winding transformer for fast charging electric vehicles," *IEEE Trans. on Industry Applications*, 1995.

[15] L. Heinemann, "An actively cooled high power, high frequency transformer with high insulation capability," in *IEEE Applied Power Electronics Conference and Exposition*, 2002.

[16] H. Hoffmann and B. Piepenbreier, "Medium frequency transformer for rail application using new materials," in *1st International Electric Drives Production Conference*, 2011.

[17] I. Villar, "Multiphysical characterization of medium-frequency power electronic transformers," Ph.D. dissertation, EPFL, Lausanne, 2010.

[18] P. Shuai and J. Biela, "Design and optimization of medium frequency, medium voltage transformers," in *15th European Conference on Power Electronics and Applications*, 2013.

[19] M. Mogorovic and D. Dujic, "Thermal modeling and experimental verification of an air cooled medium frequency transformer," in *19th European Conference on Power Electronics and Applications (EPE)*, 2017.

[20] M. Pavlovsky, "Electronic dc transformer with high power density," Ph.D. dissertation, Technical University Delft, 2006.

[21] M. A. Bahmani, "Design and optimization considerations of medium-frequency power transformers in high-power DC-DC applications," Ph.D. dissertation, Chalmers University of Technology, 2016.

[22] M. Stojadinovic, E. Kalkounis, F. Jauch, and J. Biela, "Generalized PWM generator with transformer flux balancing for dual active bridge converter," in *Proc. 19th European Conf. Power Electronics and Applications (EPE)*, 2017.

[23] M. H. Kheraluwala, D. Novotny, and D. M. Divan, "Coaxially wound transformers for high-power high-frequency applications," *IEEE Transactions on Power Electronics*, 1992.

[24] E. Bennett and S. C. Larson, "Effective resistance to alternating currents of multilayer windings," *Transactions of the American Institute of Electrical Engineers*, 1940.

[25] K. Venkatachalam, C. Sullivan, T. Abdallah, and H. Tacca, "Accurate prediction of ferrite core loss with nonsinusoidal waveforms using only steinmetz parameters," in *IEEE Workshop on Computers in Power Electronics*, 2002.

[26] V. V. Kantor, "Methods of calculating leakage inductance of transformer windings," *Russian Electrical Engineering*, 2009.

[27] Y. S. Muzychka, "Generalized models for laminar developing flows in heat sinks and heat exchangers," *Heat Transfer Engineering*, 2013.

[28] F. M. White, *Fluid Mechanics*. McGraw-Hill, 1998.

[29] S. Ghiaasiaan, *Convective Heat and Mass Transfer*. Cambridge University Press, 2011.

The 2018 International Power Electronics Conference

Evaluation of Inductor Losses on Z-source Inverter Considering AC and DC Components

Ryuji Iijima, Naoki Kamoshida, Rene Alexander Barrera Cardenas, Takanori Isobe, Hiroshi Tadano
University of Tsukuba

Abstract—This paper discusses ac and dc components of inductor losses on Z-source inverters based on theoretical calculation and experimental measurement. Winding loss with considering dc and ac components of flowing current, and core loss under dc bias application were calculated using Dowell's equation and improved Generalized Steinmetz equation (iGSE) with modification factors for dc bias. The calculated losses were verified by experiments using a 2 kW-class quasi Z-source inverter (QZSI). It is confirmed that the inductor loss is mainly caused by dc winding resistance in the discussed design, and the frequency of the main ac component of the inductor current differs from control frequency. These calculation results have good agreement with the experimental results. Therefore, it is concluded that the discussed loss calculation method can be used to design an optimal inductor for given modulation strategy and control frequency.

Keywords—Z-source inverter, Loss analysis, Dowell's equations, iGSE

I. INTRODUCTION

A Z-source inverter (ZSI) can achieve both boost and inverter operations in one stage by using a short-through mode, in which two switches in inverter leg are turned on simultaneously, and the inductor current in an impedance source is increased [1][2]. Fig. 1 shows a quasi Z-source inverter (QZSI) [2], which is one of the ZSI topologies and is discussed in this paper. Comparison with the conventional voltage-source inverter which is combined with a boost converter, the ZSI topologies can reduce the number of switching devices. On the other hand, two inductors in the impedance source take major volume in the total system. In order to achieve high power density with QZSI topology, the size and the loss reduction of inductors are key factors to be analyzed.

To reduce the inductor size, several modulation methods have been proposed. One method is controlling the distribution of short-through period in every switching cycle to reduce the ripple current in the inductors [3][4][5]. Other methods focus on using high speed switching devices such as SJ-MOSFET [6], SiC-MOSFET [7] and GaN-HENT [8], to achieve the inductor size reduction by increasing operating frequency. Both two approaches can contribute to the size reduction of the inductors; however, details of the inductor loss, which also affect the power density of the inverter, are not reported. The inductor loss depends on its current and voltage waveforms, which are determined by selected modulation technique and operation frequency. Therefore, it can be said that the inductor loss have to be analyzed comprehensively with selected modulation technique and frequencies to evaluate

the impact of the proposed modulation to the power density improvement.

This paper discusses the inductor loss considering ac and dc components in the current flowing into the inductors, with theoretical calculations and experimental evaluations. Generally, the inductor loss consists of a winding loss and an core loss. In this paper, the winding loss can be calculated by using a Dowell's equations, which is considering an ac resistance comes from skin effect and proximity effect [9]. The core loss can be calculated by using an improved Generalized Steinmetz Equation (iGSE) with modification factor which can introduce the dc current bias information to the iGSE [10]. To verify above calculations, a 2-kW class QZSI was fabricated, and the experimentally measured inductor losses were compered with the calculated losses.

This paper is organized as follows. In section II, a space vector modulation performed on the QZSI for selecting switching vectors and calculating time duty ratios is presented. Then in section III, the calculation method for inductor current waveform and inductor design considering the current ripple are discussed. In section IV, the winding loss and the core loss calculation methods are discussed, which can consider dc and ac components. After that, in section V, Steinmetz parameters measurement and the comparison between calculated inductor losses and measured inductor losses are discussed.

II. MODULATION METHOD AND INDUCTOR WAVEFORMS

The QZSI has three kinds of states. An active state outputs dc-link voltage v_{pn} of the inverter to the load through inverter switches. A zero vector state outputs zero voltage to the load. In addition, a short-through state is a distinctive state of ZSI topologies. This state shorts one or more legs to increase the inductor current to make a boosted voltage in capacitors. During the short-through state, zero voltage is applied to the load.

The output ac voltage is modulated from the dc-link voltage, v_{pn}, as usual voltage source inverter (VSI); however, v_{pn} is not constant and can be controlled by short-through period. The modulation method refereed as ZSVM6 [5] is used for discussion in this paper. This method is a kind of space vector modulations, and short-through intervals are inserted between every non-short-through intervals.

The average of v_{pn} in a control cycle, $\overline{v_{pn}}$, becomes equal to the average of C_1 voltage, $\overline{v_{C1}}$, and can be

1111

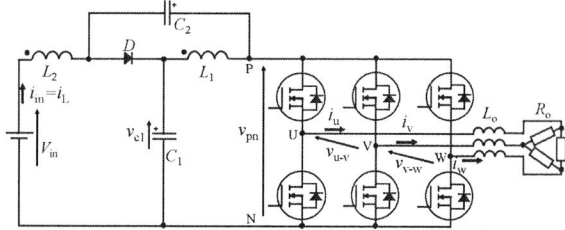

Fig. 1. Circuit diagram of the Quasi Z-source inverter.

expressed as

$$\overline{v_{\text{pn}}} = \overline{v_{\text{C1}}} = \frac{1 - D_{\text{sh}}}{1 - 2D_{\text{sh}}} V_{\text{in}}, \quad (1)$$

where, D_{sh} is the time duty ratio of the short-through state defined as $D_{\text{sh}} = T_{\text{sh}}/T_{\text{control}}$, and T_{sh} is the total duration of the short-through interval in one control cycle, T_{control}. The rest of the interval is non-short-through interval which are assigned for the active state and the zero vector state as same as for the usual VSI.

Fig. 2(a) shows possible switching vectors and a reference space vector. The radius of the inscribed circle in the hexagon shows the maximum output phase voltage amplitude without over-modulation in this method, and it is expressed as $(1/\sqrt{3}) \cdot \overline{v_{\text{pn}}}$. The reference vector v_{ref} is synthesized from two adjacent switching vectors as shown in Fig. 2(b); however, time periods of each switching vectors, t_{a}, t_{b} and t_{z}, and these duty ratio d_{a}, d_{b} and d_{z} should be modified with considering the short-through period as

$$
\begin{aligned}
t_{\text{a}} &= d_{\text{a}} T_{\text{control}} \\
&= m \sin(60° - \theta_{\text{sec}})(1 - D_{\text{sh}}) T_{\text{control}}, \\
t_{\text{b}} &= d_{\text{b}} T_{\text{control}} \\
&= m \sin(\theta_{\text{sec}})(1 - D_{\text{sh}}) T_{\text{control}}, \\
t_{\text{z}} &= d_{\text{z}} T_{\text{control}} \\
&= (1 - D_{\text{sh}} - d_{\text{a}} - d_{\text{b}}) T_{\text{control}}, \quad (2)
\end{aligned}
$$

where, m is the modulation index defined as

$$m = \frac{\sqrt{3}|v_{\text{ref}}|}{\overline{v_{\text{pn}}}}, \quad (3)$$

and can be set as $0 \leq m \leq 1$ from the point of maximum phase-voltage amplitude discussed above.

Fig. 3 shows the switching pattern and the inductor current waveform using the ZSVM6 [5]. Time periods that are calculated by equation 2 can be divided or rearranged during one control cycle. Intervals which are labeled as a, b, Z, and S represent those for the active state switching vectors, V_{a}, V_{b}, the zero state, and the short-through state, respectively. The time periods of the intervals are $t_{\text{a}}/2$, $t_{\text{b}}/2$, $t_{\text{z}}/4$ and $T_{\text{sh}}/6$, respectively.

III. INDUCTOR DESIGN FOR QZSI

The inductor current ripple has various amplitude in every control cycle by changing interval length of A, B

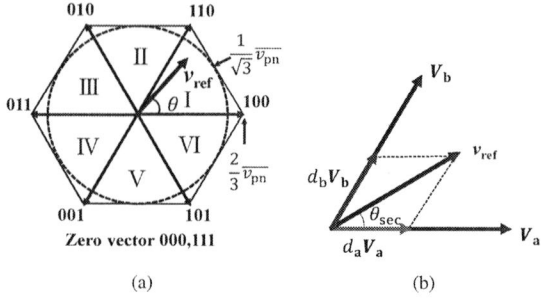

Fig. 2. Schematic of diagram for Space vector Modulation (SVM) method for QZSI. (a) Switching vector hexagon and reference vector v_{ref}. (b) Synthesizing reference voltage vector v_{ref} from two adjacent switching vectors V_{a} V_{b}.

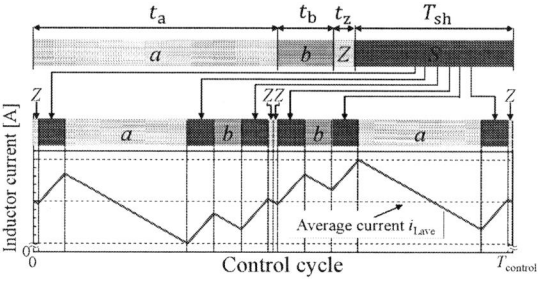

Fig. 3. Switching pattern and the inductor current ripple waveform in one control cycle at $\theta_{\text{sec}} = 10°$, $m = 1.0$ and $D_{\text{sh}} = 0.333$.

and Z as shown in Fig. 3. Differences of inductor current, Δi_{Lsh}, Δi_{La}, Δi_{Lb} and Δi_{Lz} in interval S, A, B and Z, respectively, can be expressed as

$$\Delta i_{\text{Lsh}} = \frac{1}{L}\overline{v_{\text{C1}}}\frac{T_{\text{sh}}}{6}, \quad (4)$$

$$\Delta i_{\text{La}} = -\frac{1}{L}(\overline{v_{\text{C1}}} - V_{\text{in}})\frac{t_{\text{a}}}{2}, \quad (5)$$

$$\Delta i_{\text{Lb}} = -\frac{1}{L}(\overline{v_{\text{C1}}} - V_{\text{in}})\frac{t_{\text{b}}}{2}, \quad (6)$$

$$\Delta i_{\text{Lz}} = -\frac{1}{L}(\overline{v_{\text{C1}}} - V_{\text{in}})\frac{t_{\text{z}}}{4}, \quad (7)$$

where, L is inductance of L_1 and L_2 in the impedance source. The inductor current ripple waveform can be obtained by the current differences in intervals and the order of states as shown in Fig. 3, and those depend on modulation strategy. AC components of the inductor current can be obtained from the waveform by discrete Fourier transform (DFT) analysis.

The maximum current ripple therefore peak current of the inductor is observed at $\theta_{\text{sec}} = 0°$, 60°. The maximum inductor current ripple in a control cycle, Δi_{Lmax}, is

$$\Delta i_{\text{Lmax}} = \frac{1}{L}\left[\overline{v_{\text{C1}}}\frac{2T_{\text{sh}}}{3} - (\overline{v_{\text{C1}}} - V_{\text{in}})\frac{t_{\text{z}}}{2}\right] \quad (\theta_{\text{sec}} = 0°, 60°), \quad (8)$$

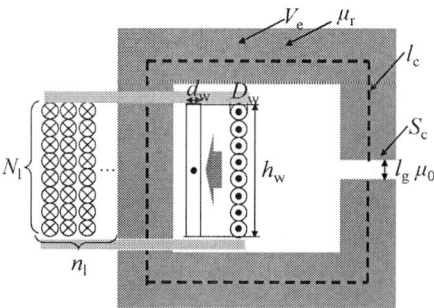

Fig. 4. Schematic view of an inductor for design consideration. Nomenclature of variables and equivalent conductor assumption for the Dowell's equation are shown.

and the peak inductor current, i_{Lpeak}, is

$$i_{\mathrm{Lpeak}} = i_{\mathrm{Lave}} + \frac{\Delta i_{\mathrm{Lmax}}}{2} = \frac{P_{\mathrm{max}}}{V_{\mathrm{in}}} + \frac{\Delta i_{\mathrm{Lmax}}}{2}, \quad (9)$$

where, i_{Lave} is the average inductor current which can be obtained from the maximum power, P_{max}, and the input voltage, V_{in}, of the QZSI. Therefore, the inductance, L to obtain desired the peak inductor current, i_{Lpeak} and the peak inductor current, i_{Lpeak} can be calculated by equation. 8 and 9.

Fig. 4 represents the schematic view of an inductor with core and air-gap for design consideration. A necessary number of turn, N and An air gap, l_{g} which are fundamental parameters to obtain the desired inductance, L at conducting the peak inductor current, i_{Lpeak} can be expressed

$$N = \frac{L i_{\mathrm{Lpeak}}}{S_{\mathrm{c}} B_{\mathrm{max}}}, \quad (10)$$

$$l_{\mathrm{g}} = \frac{\mu_0 N i_{\mathrm{Lpeak}}}{B_{\mathrm{max}}}, \quad (11)$$

where, S_{c} is a cross-sectional area of the core, B_{max} is a maximum flux density of the core, and μ_0 is vacuum permeability.

IV. METHOD FOR INDUCTOR LOSS ANALYSIS

A. Winding Loss Calculation

The winding loss, P_{winding}, can be thought to be generated by a wire dc resistance, R_{dc}, and ac resistance, R_{ac}, which contains loss increase by skin effect and proximity effect, and varies according to frequency and winding configuration. The Dowell's equation can calculate R_{ac} from winding configuration with an approximation of round wires to foil conductors as shown in Fig. 4 [9]. The ac resistance, R_{ac}, in arbitrarily frequency is given as

$$R_{\mathrm{ac}} = \Delta \left[\frac{\sinh 2\Delta + \sin 2\Delta}{\cosh 2\Delta + \cos 2\Delta} + \frac{2(n_1^2 - 1)\eta_{\mathrm{w}}^2}{3} \frac{\sinh \Delta - \sin \Delta}{\cosh \Delta + \cos \Delta} \right] R_{\mathrm{dc}}, \quad (12)$$

where, n_1 is number of stacked layer in the inductor, η_{w} represents porosity factor, and Δ represents the penetration ratio. For round solid wire, Δ is given as

$$\Delta = \sqrt{\eta_{\mathrm{w}}} \frac{d_{\mathrm{w}}}{\delta} = \sqrt{\frac{d_{\mathrm{w}} N_1}{h_{\mathrm{w}}}} \frac{d_{\mathrm{w}}}{\delta} = \sqrt{\frac{N_1}{h_{\mathrm{w}}}} \frac{1}{\delta} \left(\sqrt{\frac{\pi}{4}} D \right)^{\frac{3}{2}}, \quad (13)$$

where, d_{w} represents thickness of the equivalent foil conductor which is converted from the round solid wire diameter, D_{w}. N_1 is the number of turns in one layer, h_{w} is the width of the layer, and δ is the skin depth of the round wire. The ac winding loss at a given frequency can be calculated by rms value of the corresponding frequency component of the inductor current and the ac resistance. For non-sinusoidal current, superimposing losses of all the frequency components can be used.

The dc resistance, R_{dc}, can be calculated by the conductivity of the wire material and its dimension, that is diameter and length. The wire length calculation should include increase of its one-turn length by stacking layers. The dc winding loss can be calculated by multiplying R_{dc} and square of the inductor average current, i_{Lave}. Finally, the total winding loss, P_{winding}, including ac and dc components is the sum of these losses.

B. Core Loss Calculation

Generally, the core loss of the core of inductor, P_{core} can be calculated using the Steinmetz equation [11] as

$$P_{\mathrm{core}} = V_{\mathrm{e}} k f^\alpha \hat{B}^\beta, \quad (14)$$

where, V_{e} is the effective volume of the core, f is the frequency of sinusoidal excitation current, $\hat{B}$ is the peak of the magnetic flux of the core, and k, α, β are referred as Steinmetz parameters which are obtained from a diagram that indicates relationship between core loss per volume and frequency and peak flux density. However the inductor current of the QZSI has dc bias and is nonsinusoidal waveform which varies according to the inverter operation as mentioned in section III. And the core loss cannot be calculated by superposing each frequency components because the core loss is the non linear phenomenon. Therefore, the Steinmetz equation can not be applied the core loss calculation for the QZSI.

This paper introduces an improved Generalized Steinmetz Equation (iGSE) with modification factors considering the dc bias [10]. The iGSE has been proposed for calculating the core loss under non-sinusoidal excitation using the traditional Steinmetz parameters. In reference paper [10], introducing the modification factors to the iGSE have been proposed to consider the dc bias influence. The iGSE with modification factors to calculate the core loss $P_{\mathrm{core\_iGSE}}$ is given as

$$P_{\mathrm{core\_iGSE}} = V_{\mathrm{e}} \frac{1}{T} \int_0^T d_{\mathrm{k}} k_{\mathrm{i}} \left| \frac{dB}{dt} \right|^\alpha (\Delta B)^{d_\beta \beta - \alpha} dt, \quad (15)$$

$$k_{\mathrm{i}} = \frac{k}{(2\pi)^{\alpha-1} \int_0^{2\pi} |\cos \theta|^\alpha 2^{\beta-\alpha} d\theta}, \quad (16)$$

where, $|dB/dt|$ represents slope of the flux density in each interval, ΔB represents the peak to peak of the flux

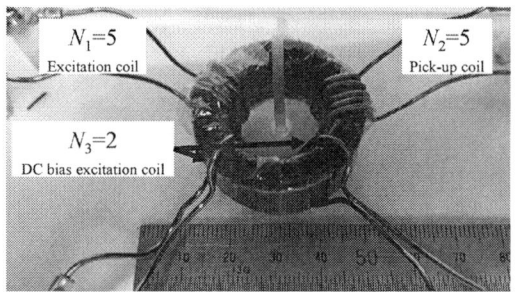

Fig. 5. Experimental setup for the Steinmetz parameters measurement with dc bias.

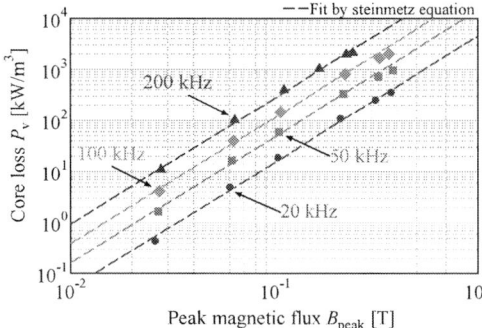

Fig. 6. Measured core loss of ferrite core B64290L0674X087 by sinusoidal excitation without dc bias, and curves by obtained parameters, $k = 0.0166$ (kW/m^3), $\alpha = 1.26$, $\beta = 2.47$.

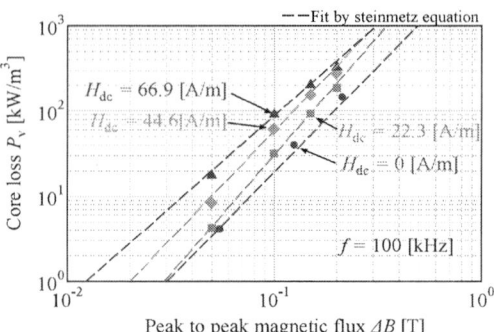

Fig. 7. Core loss of ferrite core B64290L0674X087 with sinusoidal excitation under the dc bias excitation in $f = 100$ (kHz).

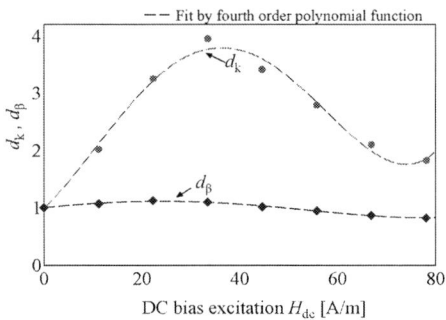

Fig. 8. Modification factor d_k and d_β with H_{dc} for considering the dc bias excitation in iGSE.

density in each interval, d_k, d_β are modification factors which are selected by applied dc bias, and T is the cycle of output fundamental frequency. Values $|dB/dt|$ and ΔB are constant during one switching interval, and those depend on modulation and parameter design. The modification factor, d_k, d_β, introduces variation of the Steinmetz parameters, k and β, caused by the dc bias. These factors can be obtained by measurement of the Steinmetz parameters under the dc bias.

V. EXPERIMENTAL VERIFICATION

A. Measurement of the Steinmetz Parameters and Dc Bias Modification Factors

Fig. 5 shows the experiment setup to measure Steinmetz parameters under the dc bias. This setup has three coils with a core under test (CUT). An excitation coil generates sinusoidal excitation by conducting a sinusoidal current, i_e, at the same time, an induced voltage, v_i, is detected by a pick-up coil. The both coils have same number of turns, and the core loss per unit volume of the CUT, P_{core_cut}, can be calculated as

$$P_{core\_cut} = \frac{1}{V_{e\_cut} T} \int_0^T v_i i_e dt, \quad (17)$$

where, V_{e_cut} is an effective volume of the CUT, T is a cycle of the fundamental frequency of the excitation current, i_e. A dc bias excitation coil generates dc bias

excitation, H_{dc}, by conducting a dc current, I_{dc}, into the coil. H_{dc} can be calculated as

$$H_{dc} = \frac{N_3 I_{dc}}{l_{cut}}, \quad (18)$$

where, l_{cut} is a magnetic path length of the CUT.

Fig. 6 shows a measured loss diagram of a ferrite core, B64290L0674X087 (EPCOS), which is made of N87 as material. The same material was used to fabricate inductors for the QZSI in this paper. The dimension of the CUT is $V_{e\_cut} = 8597$ (mm^3) and $l_{cut} = 89.65$ (mm). From this diagram, the Steinmetz parameters of N87 could be obtained by fitting using equation 14, and the obtained parameters were $k = 0.0166$ (kW/m^3), $\alpha = 1.26$, and $\beta = 2.47$.

Fig. 7 shows the core loss per unit volume, P_{core_cut}, with different peak to peak flux density, ΔB, and dc bias excitation, H_{dc}. This diagram indicates that the core loss can vary with H_{dc}, and measured core losses with various H_{dc} could fit well to equation 14.

Fig. 8 shows the modification factors, d_k, d_β, which are normalized k and β with values obtained with $H_{dc} = 0$, as function of dc excitation. These factors were obtained from core loss diagrams which were measured with $f = 20$, 50, 100 (kHz), $\Delta B = 50$, 100, 150, 200 (mT), and $H_{dc} = 0 \sim 79.1$ (A/m).

1114

TABLE I. SPECIFICATIONS OF FABRICATED 2 kW-CLASS QZSI USING SJ-MOSFET

Input voltage	V_{in}	200 V
Output voltage (RMS)	V_{out}	200 V
Output frequency	f_{out}	50 Hz
Control frequency	$f_{control}$	20 kHz
Short-through ratio	D_{sh}	0.225
Modulation index	m	1
Inverter switches	$Q_{1\sim6}$	IPW60R125CP
		650 V, 18 A
Diode	D	C3D30065D
		650 V, 36 A
Capcitor	C_1, C_2	50 μF

TABLE II. SPECIFICATIONS OF FABRICATED INDUCTOR FOR THE QZSI

Inductance	L_1, L_2	1.6 mH
Maximum ripple	Δi_{Lmax}	1.2 A
Peak inductor current	i_{Lpeak}	11.44 A
Maximum flux density	B_{max}	0.350 T
Relative permeability	μ_r	3600
Maximum dc bias excitation	H_{dc}	69.7 A/m
Core shape		PM 74/59
Core material		N87
Steinmetz parameter	k	0.0166
	α	1.26
	β	2.47
	k_i	0.00112
Modification factor	d_k	1.85
	d_β	0.845
Winding		solid wire
Number of turns	N	66.2
Number of turns in a layer	N_l	16
Number of stacked layers	n_l	5
Width of the layer	h_w	38.3 mm
Wire diameter	D_w	2.0 mm
Air gap	l_g	2.72 mm
Cross-sectional area	S_c	790 mm^2
Effective volume	V_e	101000 mm^3

B. Loss Analysis of QZSI and its Inductor

To verify the inductor loss calculation of the QZSI, a 2 kW-class QZSI using SJ-MOSFET was fabricated with fabricated inductors, and inductor losses were compared between measurement and the calculation. Table I shows the circuit specifications and operating condition of the QZSI, and Table II and Fig. 9 show the fabricated inductor for the QZSI and its design parameters.

Fig. 10 shows operation waveforms at output power P_{out} = 1.98 (kW). From this figure, both inverter and boost operations were confirmed. Measured inductor current ripple Δi_L was 1.5 A, which was 25% larger than the designed value. That is considered due to increasing short-through periods to keep the boosted voltage against losses.

Fig. 11 shows component loss breakdown of the QZSI. The total loss of the circuit was measured by taking the difference between the input power and the output power, which were measured by a digital watt meter, WT-1800. The conduction loss and the switching loss of the inverter part, $Q_1 \sim Q_6$ and the diode, D, were calculated from the device voltage and current waveforms obtained by an oscilloscope, HDO-4034A. The inductor losses were measured by a calorimetric method using two

Fig. 9. The fabricated inductor for the QZSI.

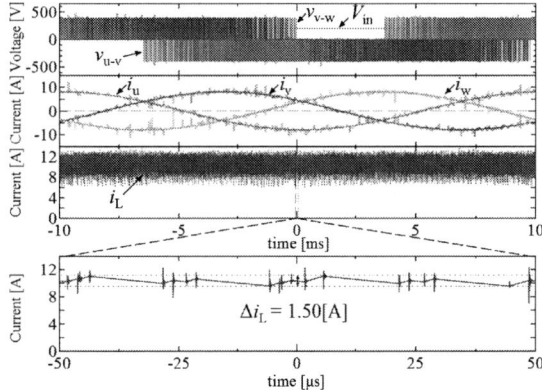

Fig. 10. Operation waveforms of QZSI at P_{out} = 1.98 (kW). The input voltage, V_{in}, the output line-to-line voltage, v_{u-v} and v_{v-w}, the output current, i_u, i_v and i_w, and the inductor current, i_L, are shown.

thermal chambers [12]. It can be seen from Fig. 11 that the inductor loss takes about 16% in the total loss of the QZSI at P_{out} = 1.98 (kW), and it is comparable with the inverter conduction loss and switching loss.

Fig. 12 shows loss breakdown of the two inductors by measurement and calculation at P_{out} = 1.98 (kW). The measured ac winding loss is given by DFT analysis of the current flowing into the inductors and the measured ac resistance as function of frequency.

Fig. 13 shows ac resistance as function of frequency, ac components of the inductor current, the calculated ac winding loss based on those, and the magnitude of the flux density, ΔB in the measurement result and the calculated result at P_{out} = 1.98 (kW).

The dc winding loss occupies most of the total inductor loss in both results as shown in Fig. 12. The calculated dc loss was lower by 14% than the measured loss. This difference came from the measured dc resistance and the calculated dc resistance as shown in Fig. 13.

From Fig. 12, the calculated ac winding resistance was also lower than the measured result. The difference in the ac resistance was larger than the difference in the dc resistance. In comparison between Fig. 13(a) and Fig. 13(b), the calculated ac resistances were lower than the measurement results in all the frequencies. It is con-

The 2018 International Power Electronics Conference

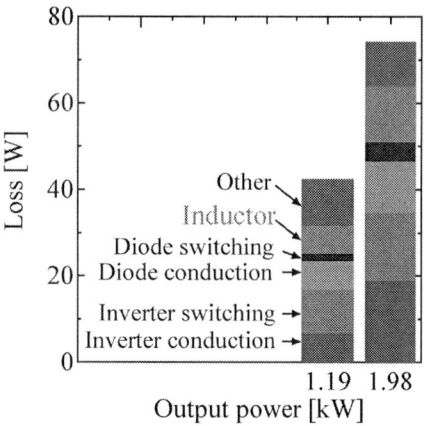

Fig. 11. Loss breakdown of the QZSI at P_{out} = 1.19 (kW) and P_{out} = 1.98 (kW).

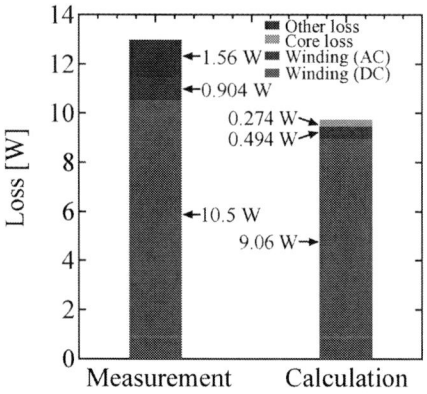

Fig. 12. Loss breakdown of the two inductors by measurement and calculation at P_{out} = 1.98 (kW).

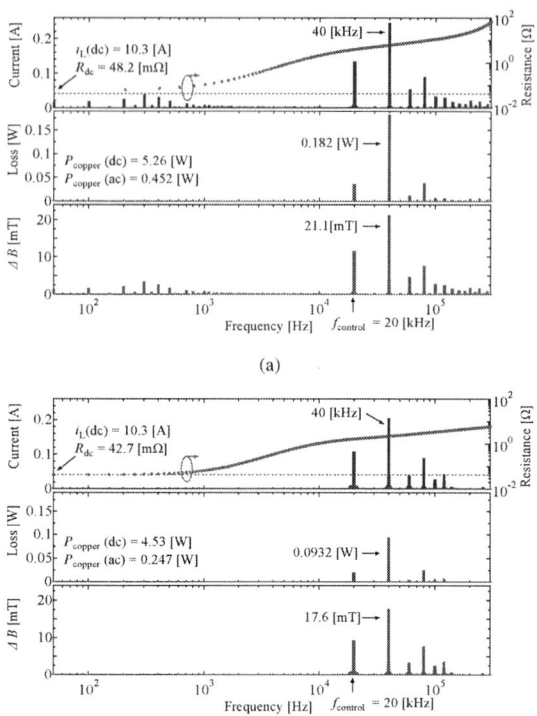

(a)

(b)

Fig. 13. AC components analysis of the inductor current, ac resistance, ac winding loss, and the flux density at P_{out} = 1.98 (kW), based on (a) Experimental measurement, (b) Calculation.

accuracy of the calorimetric loss measurement.

From Fig. 13, the ac components of the inductor current had good agreement between both results, and the main ac component appeared at 40 kHz, which was two times of the control frequency, 20 kHz. From this result, it was confirmed that inductor operation frequency is not equal to the control frequency multiplied by the number of the short-through interval in this modulation.

VI. Conclusion

This paper discussed ac and dc components of the inductor losses on the QZSI from theatrical calculation and experiment with the 2 kW-class fabricated circuit.

The inductor loss was mainly caused by the winding loss, especially dc winding loss, and this agreed with calculated results using the Dowell's equations. To reduce the dc winding loss, reducing wiring length can be thought as effective way by introducing a good core material with higher saturation flux density. It will also reduce the ac winding loss because it can be reduced number of adjacent winding.

The core loss was calculated by the iGSE with modification factors which can introduce dc bias information. The ratio of the calculated core loss was several percents of the total loss since the flux density magnitude was sufficiently low in the discussed inductor design. But the core loss will be increased in case of more higher

sidered that these difference were caused by the winding configuration. The winding configuration is assumed to be lining up to straight in the Dowell's equation [9], but the actual winding configuration doesn't align, and has overlap between adjacent layer wires. It can cause increase of the ac resistance due to proximity effect. Therefore, the ac resistance and corresponding ac winding loss increased in the fabricated inductor than the calculated result.

The calculated core loss takes 2.82% of the total calculated loss and was low-enough in comparison with the winding losses as shown in Fig. 12. Because, from Fig. 13, all the ac components of ΔB were less than 22 mT, which were quite lower than the saturation flux density of the used ferrite core (N87); therefore, the core loss was negligibly low. This calculation does not consider the fringing loss caused by air gap, and this can cause some difference between measurement and calculation. To discuss the more detail of core loss, it is necessary to consider the fringing loss and increase frequency operation or setting higher ripple current by inductor design.

From DFT analysis results, the main frequency component of the inductor current was two times of the control frequency in the discussed modulation strategy. It mean that the inductor operation frequency is not proportional to the number of the short-through intervals. This analysis can be used to evaluate modulation methods and can give suggestion to select core materials and wiring for selected modulation.

As conclusion, discussed evaluation method in this paper is valid for calculating inductor losses of the ZSI based on the core material and dimensions. Obtained results can be used to design an optimal inductor for the ZSI with selected modulation and control frequency.

REFERENCES

[1] F. Z. Peng, "Z-Source Inverter," *IEEE Trans. Ind.* , vol. 39, No. 2, pp. 504-510, Mar./Apr. 2003.

[2] J. Anderson and F. Z. Peng, "Four quasi-Z-Source inverters," *2008 IEEE Power Electronics Specialists Conference*, pp. 2743-2749, 15-19. June. 2008.

[3] F. Z. Peng, M. Shen, and Z. Qian, "Maximum boost control of the Z-source inverter," *IEEE Trans. Ind.* , Vol. 20, No. 4, pp. 833–838, Jul. 2005.

[4] M. Shen, J. Wang, A. Joseph, F. Z. Peng, L. M. Tolbert, and D. J. Adams, "Constant boost control of the Z-source inverter to minimize current ripple and voltage stress," *IEEE Trans. Ind. Appl.* , Vol. 42, No. 3, pp. 770–778, May. 2006.

[5] Y. Liu, B. Ge, H. Abu-Rub and F. Z. Peng, "Overview of Space Vector Modulations for Three-Phase Z-Source/Quasi-Z-Source Inverters," *in IEEE Transactions on Power Electronics*, vol. 29, no. 4, pp. 2098–2108, April 2014.

[6] J. Kitson and N. McNeill, "A high efficiency voltage-fed quasi Z-source inverter with discontinuous input current using super-junction MOSFETs," *8th IET International Conference on Power Electronics, Machines and Drives (PEMD 2016)*,pp. 1-6, 19-21. April. 2016.

[7] R. Iijima, T. Isobe and H. Tadano, "Loss analysis of Z-source inverter using SiC-MOSFET from the perspective of current path in the short-through mode," *2016 18th European Conference on Power Electronics and Applications*, pp. 1-10, Sep. 2016.

[8] Y. Zhou, H. Li and H. Li, "A Single-Phase PV Quasi-Z-Source Inverter With Reduced Capacitance Using Modified Modulation and Double-Frequency Ripple Suppression Control," *IEEE Transactions on Power Electronics*, vol. 31, no. 3, pp. 2166-2173, March 2016.

[9] W. G. Hurley, E. Gath and J. G. Breslin, "Optimizing the AC resistance of multilayer transformer windings with arbitrary current waveforms," *in IEEE Transactions on Power Electronics*, vol. 15, no. 2, pp. 369-376, Mar 2000.

[10] J. Muhlethaler, J. Biela, J. W. Kolar and A. Ecklebe, "Core Losses Under the DC Bias Condition Based on Steinmetz Parameters," *IEEE Transactions on Power Electronics*, vol. 27, no. 2, pp. 953-963, Feb. 2012.

[11] E. C. Snelling, "Soft Ferrites, Properties and Applications," 2nd Ed. London, U.K.: Butterworth, 1988.

[12] K. Orikawa, A. Nigorikawa and J. i. Itoh, "Evaluation on chamber volume and performance for simple calorimetric power loss measurement by two chambers," *2013 1st International Future Energy Electronics Conference (IFEEC)*, Tainan, 2013, pp. 914-919.

The 2018 International Power Electronics Conference

An Integrating Structure of Output Filter for Grid Connected Inverter Based on FMLF Technique

Jie Ma, Yenan Chen, PingPing Chen, Wenxing Zhong, Dehong Xu
Institute of Power Electronics, College of Electrical Engineering, Zhejiang University, Hangzhou, China
E-mail: trumanmar@zju.edu.cn

Abstract—**Output filter of the grid inverter usually has a large footprint in the total system. In order to increase the power density of the grid inverter, an integrated passive integration structure for output filter with flexible multilayer foil (FMLF) is presented. The design of the integrated output filter is discussed. Then a Distributed Electromagnetic Component (DEMC) model of the integrated filter is used to verify the design. Finally, the proposed integrated output filter is applied in a 3kW single-phase inverter. Experimental results are included to show the validity of this integrated filter design.**

Keywords—filter, grid inveter, passive integration, power density.

I. INTRODUCTION

Several types of filters for grid inverters have been studied by predecessors. The L-type filter has larger size to satisfy the power quality requirement. In order to reduce the footprint of the filter, higher orders filter such as LC filters, LCL filter etc. have been introduced for grid inverter applications. [1] [2].

To further reduce the volume of the output filter, passive integration method for the filter is considered. The idea of passive integration is to let the magnetic field and the electric field share the same space. Thus, power or energy density is enhanced and the volume of the passive component is reduced. In addition, the passive integration is able to reduce drawbacks of the parasitic in traditional discrete component composed filters [3].

The passive integration can be roughly classified into two categories shown in Fig 1. One is FMLF integration technique and the other is Printed Circuit Board (PCB) integration technique. Many applications had been developed by PCB integration technique. It can significantly reduce the height of the integrated passive component. Integration of Common Mode（CM） filter was investigated [4]. Both Differential Mode (DM) and CM filter are integrated with an improved solution in [5] with PCB. However, there were following drawbacks for PCB integration. Firstly, it is not suitable for high power application due to the restriction of the thickness of PCB layer. Secondly, the large portion of leakage flux is perpendicular to the PCB layer and will cause higher eddy current loss in PCB layer. Besides winding length is

relatively longer than FMLF winding, which cause higher conducting loss in the PCB winding.

FMLF integration technique was firstly used to integrate a resonant chamber [6]. Then it is extended to EMI filter [7]. Later, this technique was used to integrate both boost inductor and EMI filter [8] [9].

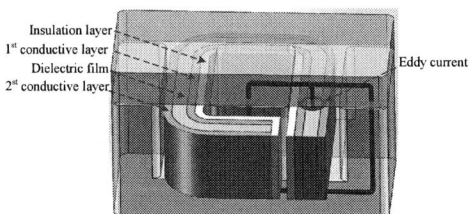

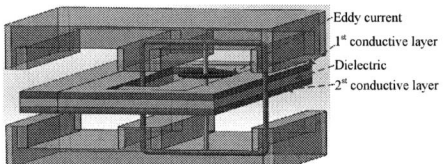

Fig. 1. Integrated PCB and integrated FMLF.

FMLF integration technique has three advantages in comparison to PCB integration technique. Firstly, FMLF integration has lower copper loss due to its shorter length of the FML foil. Secondly, FMLF integration reduces eddy current loss due to its leakage flux aligning with the foil surface. Thirdly, FMLF integration is suitable for high power application since the foil thickness can be custom designed.

In this paper, an integrated passive structure for symmetrical LC filter with flexible multilayer foil (FMLF) is presented. The design of the integrated symmetrical LC filter is discussed. Then a Distributed Electromagnetic Component (DEMC) model of the integrated filter is used to verify the design. Finally, the proposed integrated LC filter is built to apply in a 3kW single-phase inverter. Experimental results and THD analysis are included to show the validity of the integrated filter design.

1118

II. INTEGRATED STRUCTURE DESIGN

A. Integrated structure

Fig. 2 show a LC filter integrated by FMLF winding, four layer FML foil is used. Four layers are insulation layer with white color, 1st conductive layer with bronze color, dielectric film with indigotin color and 2nd conductive layer with bronze color. The two conductive layers are used as winding of the magnetic core to create the inductance while they also operated as the plates of the planar capacitor with the dielectric film. The dielectric film is sandwiched between two FML foils. The integrated component with FMLF can be represented by a low frequency equivalent circuit as shown in Fig 2(b) [10]. It can be looked as an LC filter.

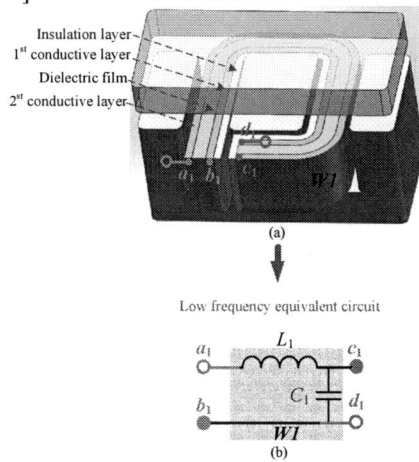

(a)

Low frequency equivalent circuit

(b)

Fig. 2. Integrated LC filter and its low frequency equivalent circuit.

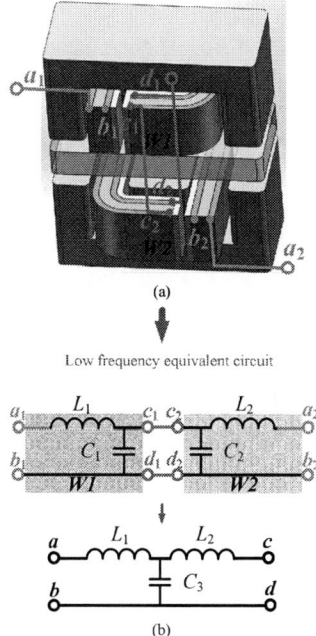

(a)

Low frequency equivalent circuit

(b)

Fig. 3. Integrated LCL filter and its low frequency equivalent circuit.

The integrated LCL filter can be obtained by combining two LC integrated units of Fig 2. The

connection of foil terminals are shown in Fig.3. Two LC filters are back-to-back connected. The low frequency equivalent circuit of the integrated LCL filter can be derived as shown in Fig. 3(b).

According the equivalent circuit in Fig 2, when proper terminals are chosen to connect between different FML foils, different kind of equivalent filter can be obtained. As shown in Fig 4, terminal d_1 connects with terminal c_2, and terminal c_1 connects with d_2. This type of connection can double the capacitance, which is equal to parallel capacitor C_1 and C_2 as shown in low frequency equivalent circuit.

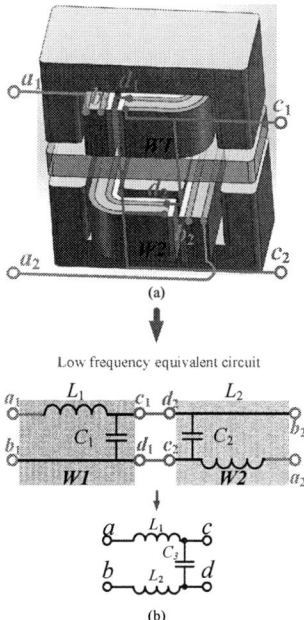

(a)

Low frequency equivalent circuit

(b)

Fig. 4. Integrated LC filter and its low frequency equivalent circuit.

This integrated structure can be used as a symmetrical LC filter. The symmetrical L filter or LC filter are recommended in transformerless PV inverter to reduce the leakage current [11]. It used as output filter for the grid inverter as shown in Fig 5.

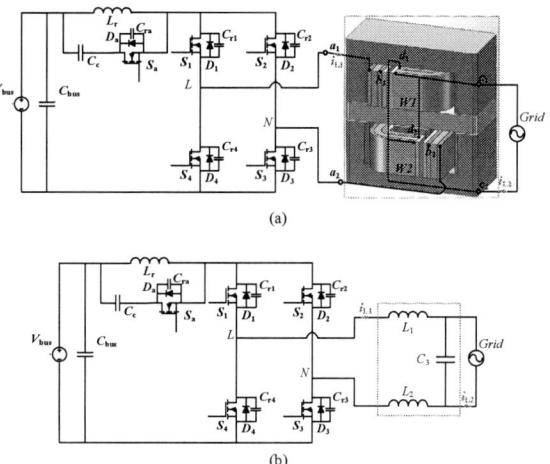

(a)

(b)

Fig. 5. Integrated structure connections with inverter and equivalent circuit.

B. Magnetic circuit analysis

The magnetic circuit of the integrated FMLF W1 is shown in Fig 6. It is assumed that the air gaps of the magnetic core are symmetry. Current from integrated capacitor is ignored because its direction will not form flux. i_{L1} is AC side current of the inverter.

2nd conductive layer of FMLF W1 function as inductor L_1. The inductor current i_{L1} flows into terminal a_1, the flux direction is Φ_{L1}.

The Φ_{L1} can be deduced:

$$\Phi_{L1} = \frac{N_{W1} \cdot i_{L1}}{R_{W1}} \tag{1}$$

Where the N_{W1} is the turns of winding W1; R_{W1} is reluctance of the magnetic loop of flux Φ_{L1}.

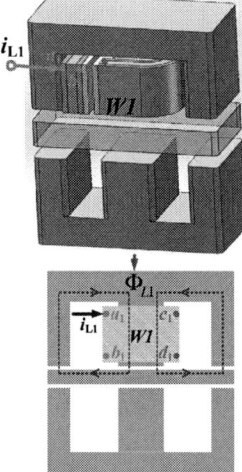

Fig. 6. Magnetic circuit of the integrated FMLF W1.

The magnetic circuit of the integrated FMLF W2 is shown in Fig 7. The grid-side current i_{L2} flows into terminal c_2. The flux direction is Φ_{L2} as shown in the figure.

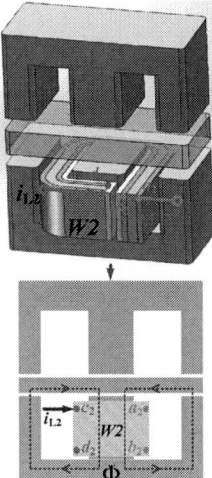

Fig. 7. Magnetic circuit of the integrated FMLF W2.

The Φ_{L2} can be deduced:

$$\Phi_{L2} = \frac{N_{W2} \cdot i_{L2}}{R_{W2}} \tag{2}$$

Fig. 8 gives the whole flux direction of the proposed structure.

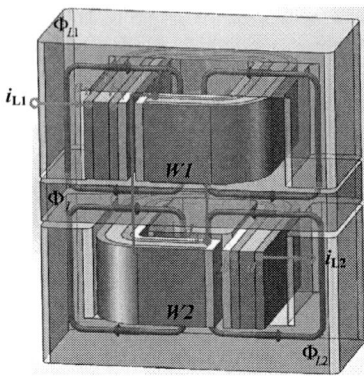

Fig. 8. Magnetic circuit of the integrated FMLF structure.

The flux in the middle I-type core is named Φ_I and can be deduced [12]:

$$\Phi_I = \frac{N_{W1} \cdot i_{L1}}{R_{W1}} - \frac{N_{W2} \cdot i_{L2}}{R_{W2}} = \Phi_{L1} - \Phi_{L2} \tag{3}$$

III. PARAMETER DESIGN AND DEMC MODEL VERIFICATION

A. Parameter design

The parameters of the proposed structure include inductor L_1, L_2 and capacitor C_1, C_2. Both the second conductive layer are used as filter inductor, the parallel conductive layer and the dielectric film form the capacitor [13].

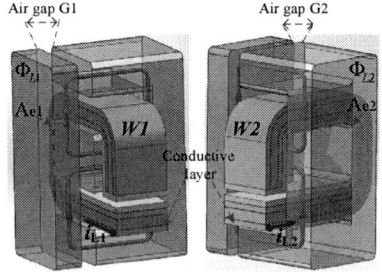

Fig. 9. Parameter of integrated L_1 and L_2.

Output filter L_1 is calculated as

$$L_1 = \frac{N_{W1} \cdot \Phi_{L1}}{i_{L1}} = \frac{N^2_{W1} \cdot \Phi_{L1}}{R_{W1}}$$

$$= N^2_{W1} \cdot \left(\frac{l_{c1}}{\mu_0 \cdot \mu_r \cdot A_{e1}} + \frac{l_{air1}}{\mu_0 \cdot A_{e1}}\right)^{-1} \tag{4}$$

where, μ_0 is the magnetic constant, μ_r is the relative permeability of the ferrite material, l_{c1} and l_{air1} are, the length of the ferrite core and the air gap G1 in R_{W1}

respectively, and A_{e1} is the effective cross-sectional area of R_{W1}.

Output inductor L_2 can be calculated based on the same principle. When l_{c2} and l_{air2} are the length of the ferrite core and the air gap G2 in R_{W2} respectively, and A_{e2} is the effective cross-sectional area of R_{W2}. Output inductor L_2 is expressed as:

$$L_2 = \frac{N_{W2} \cdot \Phi_{L2}}{i_{L2}} = \frac{N^2_{W2} \cdot \Phi_{L2}}{R_{W2}}$$

$$= N^2_{W2} \cdot (\frac{l_{c2}}{\mu_0 \cdot \mu_r \cdot A_{e2}} + \frac{l_{air2}}{\mu_0 \cdot A_{e2}})^{-1} \quad (5)$$

The voltage difference is generated between the adjacent conductive layers of both FMLF W1 and W2. The integrated capacitance C_1 and C_2 can be derived as

$$C_1 = \frac{\varepsilon_0 \cdot \varepsilon_r \cdot \left[N_{W1} \cdot (2 \cdot l_{l1} + 2 \cdot l_{w1}) + \sum_{n=1}^{N_{W1}} 4 \cdot n \cdot t_{ie1} \right] \cdot W_{i1}}{t_{ie1}} \quad (6)$$

$$C_2 = \frac{\varepsilon_0 \cdot \varepsilon_r \cdot \left[N_{W2} \cdot (2 \cdot l_{l2} + 2 \cdot l_{w2}) + \sum_{n=1}^{N_{W2}} 4 \cdot n \cdot t_{ie2} \right] \cdot W_{i2}}{t_{ie2}} \quad (7)$$

where, ε_0 and ε_r are the dielectric constant of the air and the dielectric film respectively, l_{l1}, l_{w1} and t_{ie1} denotes the width, the length and the thickness of the first turn of FMLF W1 respectively, W_{i1} is the height of dielectric film of FMLF W1. l_{l2}, l_{w2} and t_{ie2} denote the width, the length and the thickness of the first turn of FMLF W2respectively, and W_{i2} is the height of dielectric film of FMLF W2.

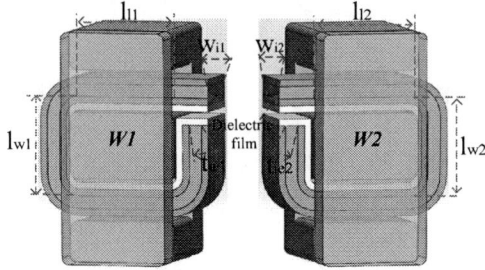

Fig. 10. Parameter of integrated capacitor.

The specific parameters are given in table I.

TABLE I
Parameters of FMLF W1 and W2

FMLF parameters		W1	W2
1st conductive layer	length	2.2m	2.18m
	width	17mm	17mm
	thickness	50μm	50μm
2st conductive layer	length	2.2m	2.18m
	width	17mm	17mm
	thickness	250μm	250μm
Material of the dielectric film		PP	PP
Dielectric constant		3.0	3.0
Thickness of dielectric film		15μm	15μm

Since the integrated capacitor C_1 and C_2 are paralleled, the total capacitance C equals to C_1 plus C_2.

A same core is chosen to produce W1 and W2, the maximum flux for both of them could be the same value [14]. To avoid flux saturation at maximum power, the inequality below should be satisfied:

$$B_{sat} \geq \frac{N_{W1(W2)} \cdot I_{MAX\_L1(MAX\_L2)}}{R_{W1(W2)} \cdot A_{e1(e2)}} \quad (8)$$

where B_{sat} means the saturation flux density of the core.

EE56 core is chosen to make the integrated filter. The material of the core is DMR95, and the AL value is 9000 at 25°C. The other parameters and figure of EE56 core are shown in Fig. 11 and TABLE II.

Finally, FMLF W1 has 22 turns, giving L_1 as 186μH, and FMLF W2 has 22 turns too, giving L_2 as 179μH. Integrated capacitor formed by W1 and W2 is close to 207nF.

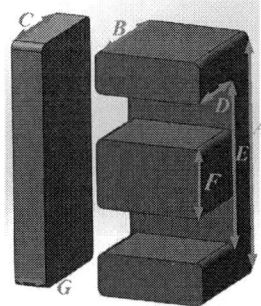

Fig. 11. Parameter of EE56 core and the common I-type core.

TABLE II
Parameters of EE56 core

parameter	value	parameter	value
A	56mm	E	38.2mm
B	28mm	F	17.1mm
C	25mm	G	8.0mm
D	19mm		

B. DEMC model verification

In Fig. 12, W1 is winded on the center leg of the magnetic core. Since the insulation layer is sandwiched into the 1st conductive layer of the first turn and 2nd conductive layer of the 2nd turn, the flying (parasitic) capacitance is generated when the winding is extended to N_{W1} turns, the flying capacitance is named $C_{1(P-1)}$. Dielectric films of each two adjacent turns form capacitance $C_{1(N-1)}$ and C_{1N}. The inductance of each turn of W1 is denoted with L_{1N}. M_{1ij} denotes the mutual inductance of any two turn of W1. The equivalent circuit is also established in Fig. 12. The analysis of integrated capacitor and parasitic capacitor in the model has been investigated in [15] [16].

The analysis of DEMC model of W2 is the same as W1, the only difference is position of the flying (parasitic) capacitor.

The 2018 International Power Electronics Conference

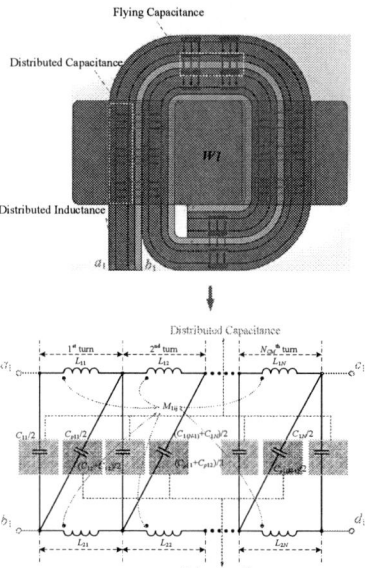

Fig. 12. Equivalent circuit of FMLF W1 with DEMC model.

According to the DEMC model, the proposed integrated structure's equivalent circuit can be given as in Fig 13. M_{W1W2} denotes the mutual inductance between the different turns in same conductive layers and in different layers. L_{1N}, L_{2N} and L_{3N}, L_{4N} denote inductance of 1st conductive layer and 2nd conductive layer of W1 and W2 respectively.

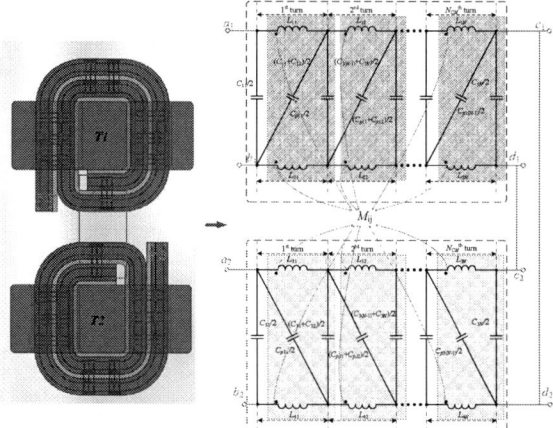

Fig. 13. Equivalent circuit of integrated structure with DEMC model.

As for the proposed structure and its equivalent circuit, the voltage of each node of the structure can be obtained by node voltage analysis. The impedance matrix is Z_{W1W2}.

Here, Z_{ij} (i,j=1, 2, W1+W2) denotes the impedance of each branch of the structure, M_{ij} (i,j=1, 2, W1+W2-1) denotes the mutual inductance. Expression of the impedance matrix is shown in (9), thus, the impedance characteristics of any two terminals can be deduced by (10).

$$V_{ij} = \frac{U_i - U_j}{i_k} \tag{10}$$

Where U_i and U_j are the voltage at any two terminals, the i_k represents the current at the branch between the two terminals, V_{ij} denotes the impedance of this two terminals in the integrated structure.

According to this method, not only the amplitude frequency characteristics could be predicted, but the proper terminals could be chosen to connect between different FML foils to satisfy the design requirements. To verify the proposed model, the theoretical curve of two chosen terminals based on DEMC model is compared with the test curve.

Spectrum analyzer MS4630B is used to measure the impedance characteristic of the integrated structure. First, terminals a_1 and c_1 are connected to the analyzer as shown in Fig. 14.

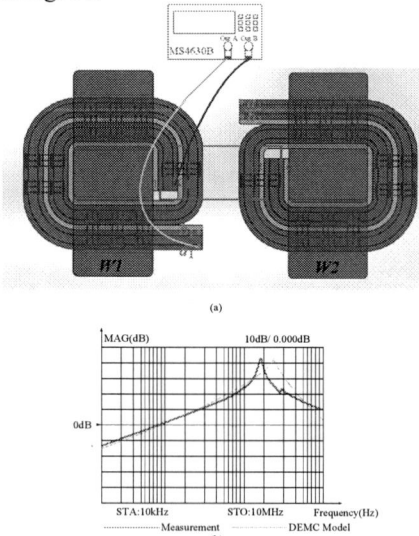

Fig. 14. Test circuit (a) and curves (b) of terminals a_1 and c_1.

$$Z_{W1W2} = \begin{bmatrix} Z_1 & M_{12} & \cdots & M_{1(N_{W1}+N_{W2}-1)} & M_{1(N_{W1}+N_{W2})} \\ M_{21} & Z_2 & \cdots & M_{2(N_{W1}+N_{W2}-1)} & M_{2(N_{W1}+N_{W2})} \\ \cdots & \cdots & \cdots & \cdots & \cdots \\ M_{(N_{W1}+N_{W2}-1)1} & M_{(N_{W1}+N_{W2}-1)2} & \cdots & Z_{(N_{W1}+N_{W2}-1)(N_{W1}+N_{W2}-1)} & M_{(N_{W1}+N_{W2}-1)(N_{W1}+N_{W2})} \\ M_{(N_{W1}+N_{W2})1} & M_{(N_{W1}+N_{W2})2} & \cdots & M_{(N_{W1}+N_{W2})(N_{W1}+N_{W2}-1)} & Z_{(N_{W1}+N_{W2})(N_{W1}+N_{W2})} \end{bmatrix} \tag{9}$$

According to the test curve, the impedance can be obtained with formula (11), and the calculation method has been proposed in [10].

$$L = \frac{100 \cdot 10^{\left(\frac{Z_L}{20}\right)}}{2 \cdot \pi \cdot f} \qquad (11)$$

The curve indicates that when terminals a_1 and c_1 are connected in a circuit, the performance of the integrated structure is almost the same as an inductor in low frequency band. The calculation result is $211\mu H$, which is close to the value of $186\mu H$ measured with a LCR meter. When terminals a_2 and c_2 are connected in the circuit, the performance of the integrated structure is almost the same as in Fig. 14.

When terminals d_1c_2 and c_1d_2 are connected to the analyzer, the test curve and the calculated curve derived with DEMC model are given in Fig. 15. According to the figure, the integrated structure is almost same as a capacitor in low frequency. The capacitance is $211nF$ according to the test curve, which is also close to the value obtained by LCR meter ($207nF$). From Fig. 14 and Fig. 15, it can be seen that the DEMC model can maintain relatively high prediction accuracy in low frequency.

The formula is given in (12).

$$C = \frac{1}{100 \cdot 10^{\left(\frac{Z_L}{20}\right)} \cdot 2 \cdot \pi \cdot f} \qquad (12)$$

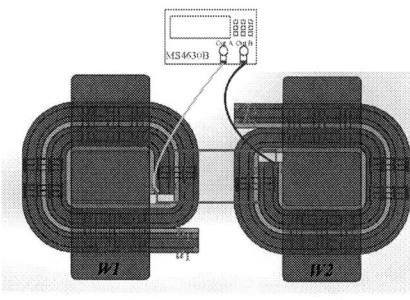

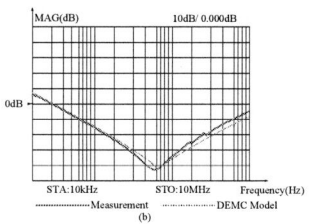

Fig. 15. Test circuit (a) and curves (b) for terminals d_1 and d_2.

When terminals a_1 and a_2 are connected to the analyzer, the test curve and calculated curve are given in Fig. 16. The characteristic of the integrated structure is almost the same as a LC series circuit in low frequency. The calculation result is the same as the result derived above.

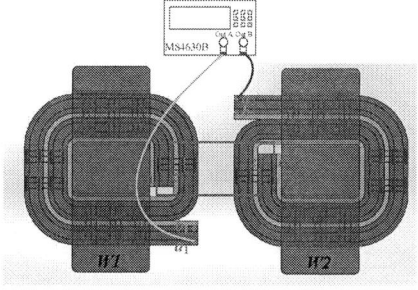

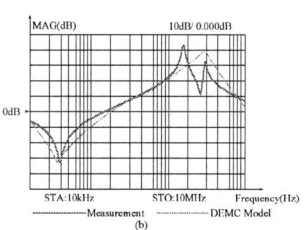

Fig. 16. Test circuit and curve for terminals c_1c_2 and d_1d_2 and DEMC curve.

The tests indicate that when terminals d_1c_2, c_1d_2 terminals a1 and a2 are connected to an inverter as shown in Fig. 5, the proposed design will have the same structure and performance as a symmetrical LC output filter as in Fig. 5(b). In addition, the DEMC model is able to predict the characteristic of the integrated structure with high accuracy.

IV. EXPERIMENTAL VERIFICATION

The performance of the integrated LC output filter is tested on a 3kW single-phase grid inverter prototype is built. The input voltage of the inverter is 360V, and the RMS voltage of the grid is 230V, switching frequency is 50kHz with ZVS double-frequency SPWM modulation, and the current ripple frequency of the symmetrical LC filter is 100kHz [17]. The integrated output filter is used as shown in Fig. 17. It is also compared with a discrete output filter. The parameters of the discrete and the integrated filter are given in TABLE III.

The size comparison between integrated and the original discrete filter are given in Fig. 17. The volume of the integrated filter is reduced by about 30%.

(a)   (b)

Fig. 17. Prototype of the integrated output filter, compared with a distributed one.

The experimental waveform and THD analysis with integrated output filter and the discrete filter are shown in Fig. 18, and 19.

According to the experimental results in Fig. 18 and 19, the performance of the integrated output filter is almost the same as that of the discrete output filter. The total THD at rated power with integrated and discrete output filter are almost the same, which is equal to 3.0%.

TABLE III
Parameters of discrete filter and integrated filter

discrete	value	integrated	value
L_1	192μH	L_1	186μH
L_2	199μH	L_2	179μH
C	220nF	C	207nF

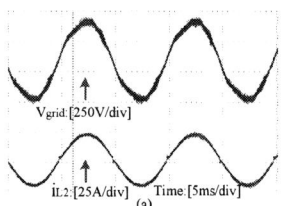

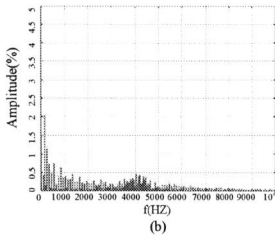

Fig. 18. Experimental results with the integrated filter . (a) current waveforms (b)current THD data.

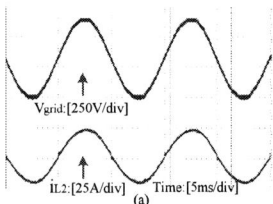

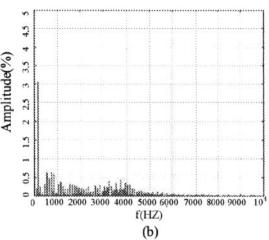

Fig. 19. Experimental results with the distributed filter . (a) current waveforms (b)current THD data.

V. CONCLUSION

Based on FMLF technique, an integrated output filter is designed for the grid inverter. The integrated filter is made by an E-I-E core with FMLF windings. The low frequency equivalent circuit is also derived. Based on the DEMC model, the impedance characteristic of the proposed design is estimated, which is further verified by the experiment. The design procedure of the integrated

filter is presented. The performance of the FMLF integrated filter is able to reach the same performance as the discrete filter. In addition, it can reduce the volume by 30%.

ACKNOWLEDGMENT

This work was supported by the National Natural Science Foundation of China (51337009).

REFERENCES

[1] A. K. S. Bhat, "Analysis and design of LCL-type series resonant converter," in *IEEE Transactions on Industrial Electronics*, vol. 41, no. 1, pp. 118-124, Feb 1994.

[2] Shaoann Shon, "LC filter design, test, and manufacturing," in *IEEE Transactions on Acoustics, Speech, and Signal Processing*, vol. 33, no. 1, pp. 337-337, February 1985.

[3] Deng.Cheng, "Integrated Technology of Flexible Multi-layer Foil for Electromagnetic Components and its Applications." ZheJiang University doctoral dissertation, 2014

[4] J. T. Strydom, J. D. van Wyk and M. A. de Rooij, "Integration of a 1 MHz converter with active and passive stages," *APEC 2001. Sixteenth Annual IEEE Applied Power Electronics Conference and Exposition (Cat. No.01CH37181)*, Anaheim, CA, 2001, pp. 1045-1050 vol.2.

[5] M. Ali, E. Laboure, .. Francois Costa, B. Revol and C. Gautier, "Hybrid Integrated EMC filter for CM and DM EMC Suppression in a DC-DC Power converter," *2012 7th International Conference on Integrated Power Electronics Systems (CIPS)*, Nuremberg, 2012, pp. 1-6.

[6] R. Reeves, "Inductor-capacitor hybrid," in *Electrical Engineers, Proceedings of the Institution of*, vol. 122, no. 11, pp. 1323-1326, November 1975.

[7] Xiaofeng Wu, Dehong Xu, Zhiwei Weng, Y. Okuma and K. Mino, "Modeling of integrated EMI filter with flexible multi-layer (FML) foils," *2009 IEEE 6th International Power Electronics and Motion Control Conference*, Wuhan, 2009, pp. 749-755.

[8] C. Deng, M. Chen, P. Chen, C. Hu, W. Zhang and D. Xu, "A PFC Converter With Novel Integration of Both the EMI Filter and Boost Inductor," in *IEEE Transactions on Power Electronics*, vol. 29, no. 9, pp. 4485-4489, Sept. 2014.

[9] Xiaofeng Wu, X. Dehong, Yanjun Zhang, Yi Chen, Yasuhiro Okuma and Kazuaki Mino, "Integrated EMI filter design with flexible PCB structure," *2008 IEEE Power Electronics Specialists Conference*, Rhodes, 2008, pp. 1613-1617.

[10] Xiaofeng Wu, "Integrated EMI Filter with Flexible Multi-layer Foils." ZheJiang University doctoral dissertation, 2010

[11] R. Gonzalez, E. Gubia, J. Lopez and L. Marroyo, "Transformerless Single-Phase Multilevel-Based Photovoltaic Inverter," in *IEEE Transactions on Industrial Electronics*, vol. 55, no. 7, pp. 2694-2702, July 2008.

[12] D. Pan, X. Ruan, C. Bao, W. Li and X. Wang, "Magnetic Integration of the LCL Filter in Grid-Connected Inverters," in *IEEE Transactions on Power Electronics*, vol. 29, no. 4, pp. 1573-1578, April 2014.

[13] X. Wu, D. Xu, Z. Wen, Y. Okuma and K. Mino, "Design, Modeling, and Improvement of Integrated EMI Filter With Flexible Multilayer Foils," in *IEEE Transactions on Power Electronics*, vol. 26, no. 5, pp. 1344-1354, May 2011.

[14] C. Deng, Z. Wen, C. Hu and D. Xu, "Integration of both EMI filter and Boost inductor for 1 kW PFC converter," *2012 IEEE Energy Conversion Congress and Exposition (ECCE)*, Raleigh, NC, 2012, pp. 4600-4607.

[15] C. Deng *et al.*, "Integration of Both EMI Filter and Boost Inductor for 1-kW PFC Converter," in *IEEE Transactions on Power Electronics*, vol. 29, no. 11, pp. 5823-5834, Nov. 2014.

[16] C. Deng, D. Xu, C. Hu and Z. Wen, "PFC converter with novel integration of both EMI filter and Boost inductors," *2013 IEEE Energy Conversion Congress and Exposition*, Denver, CO, 2013, pp. 3390-3397.

[17] Y. Chen, D. Xu and J. Xi, "Common-Mode Filter Design for a Transformerless ZVS Full-Bridge Inverter," in *IEEE Journal of Emerging and Selected Topics in Power Electronics*, vol. 4, no. 2, pp.405-413, June 2016.

The 2018 International Power Electronics Conference

New Screening Method for Improving Transient Current sharing of Paralleled SiC MOSFETs

Junji Ke, Zhibin Zhao, Peng Sun, Huazhen Huang, James Abuogo, Xiang Cui
State Key Laboratory of Alternate Electrical Power System with Renewable Energy Source
North China Electric Power University
Beijing, China
kejunji@ncepu.edu.cn

Abstract-In this paper the effect of the spread of SiC MOSFET device parameters on transient current sharing is investigated. To aid in this investigation, the coefficient of variation is proposed firstly and then used as the evaluating specification for the spread of device parameters. Two main factors which affect transient current sharing, threshold voltage and trans-conductance, are analyzed based on transfer characteristics of MOSFET. Experimental studies show that threshold voltage has larger spread than trans-conductance. Moreover, there is mutual compensation of the effects of threshold voltage and trans-conductance on SiC MOSFET transient current sharing. Finally, to improve transient current sharing of paralleled devices, this paper proposes the transfer curve screening method. Compared to the traditional single parameter screening method, experimental results show that the current imbalance rate with proposed method is only 3.5%, whereas it reaches 26% with traditional method.

keywords - SiC MOSFET, spread of parameters, the coefficient of variation, current sharing, transfer curve screening.

I. INTRODUCTION

Power devices based on wide band-gap(WBG) materials are starting a new revolution in the field of power electronics. Among the possible alternative WBG semiconductors, silicon carbide (SiC) devices are believed to replace silicon (Si) devices in high power application. SiC devices have higher blocking voltage, faster switching speed and better electro-thermal performance under harsh conditions. This is due to its material feature of high critical field, wide bandgap and high thermal conductivity. As one of WBG devices with more stable structure, comparatively lower cost and relatively mature technology, SiC MOSFET has promising prospect for future applications in electric vehicles [1], high-power-density motor drives [2], converters for photovoltaic power generation [3], solid-state transformers for distribution network, and HVDC transmission [4]. Advanced features of SiC MOSFET, such as high-voltage, high frequency and high temperature capabilities, contribute significantly to the reduction of losses in power device as well as reduction in volume of passive components in the converter circuitry. This greatly improve power density and system efficiency. However, there are also many challenges associated with SiC MOSFET devices, for example, cost, reliability and current capacity. Consequently, with ever-increasing demand for high-power applications, parallel connection is required due to the limitation of current level of single SiC MOSFET chip. Although the circuit layouts of power modules are generally symmetrical, there is still unavoidable variation of device parameters. This poses a great challenge to parallel application of SiC MOSFET due to reduction of device reliability and life-time caused by current distribution imbalance among paralleled units.

To solve the problem of current sharing imbalance for SiC MOSFET, some existing literatures only discuss the consistency of circuit parameters [5-6]. However, the consistency of device parameters is required as a precondition for adopting the symmetric circuit layout to realize current equilibrium. Although the fluctuation of electrical parameters is often limited by strict fabrication process for commercial semiconductors, there is still a certain degree of inconsistency in device parameters attributed to SiC MOSFET's immature production process [7]. As for the effect of device parameters, reference [8] discusses the static and dynamic current sharing characteristics under different threshold voltage and different on-resistance of SiC MOSFET devices. Literature [9] also investigates the influence of threshold voltage and on-resistance by selecting certain number of devices drawn from the same batch and assessing the distribution range of the two parameters. In addition, reference [10-12] all study the spread of SiC JFET devices parameters. Nevertheless, they lack evaluation specification and quantitative analysis for the degree of spread of different parameters. Moreover, the only screening method for improving transient current sharing is single devices parameters screening method like threshold voltage screening for Si MOSFET as reported in [13]. This method, however, takes no consideration of the effect of tans-conductance.

To address the problems pointed out above, this paper samples 30 SiC MOSFET devices manufactured in the same batches by the same company. The coefficient of variation is used to evaluate the spread of main device parameters affecting transient current distribution characteristic for the first time. Thereafter, the relationship between the transient current difference and device parameters is derived based on the transfer characteristics of SiC MOSFET in saturation region. In addition, two pairs of different devices are selected from the samples for switching operation experiment in parallel connection, and the influence of parameters

1125

mismatch on current sharing is analyzed. Finally, the comparison is made between the transfer curve screening method proposed in this paper and traditional threshold voltage screening method.

II. PARAMETERS SPREAD

The transient current distribution of paralleling devices depends mainly on the device transfer characteristics according to reference [14]. Therefore, a sample of 30 devices from the same batch of the same manufacturer are measured under room temperature 25 ℃ and the same test conditions with V_{DS}= 20V. The transfer characteristic curves are shown in Fig.1. As can be observed, even from the same batch of devices, the transfer characteristics of devices still have a large spread.

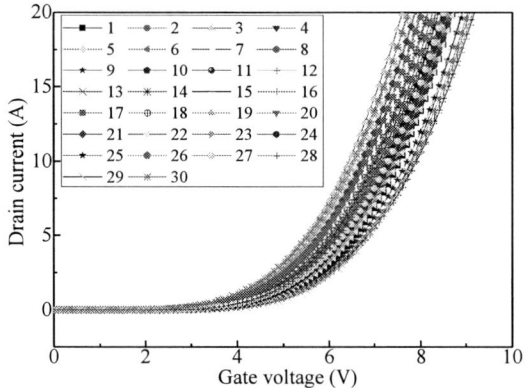

Fig. 1. Transfer curve of SiC MOSFET devices.

The transfer characteristic can also be regarded as a function of threshold voltage and trans-conductance coefficient. During the current rising or falling stage, SiC MOSFET is operating in the saturation region, and the relationship between drain current and gate-source voltage is shown in Eq.1.

$$i_D = g_{fs}\left(v_{GS} - V_{th}\right) \qquad (1)$$

where g_{fs} and V_{th} are trans-conductance coefficient and threshold voltage of SiC MOSFET respectively.

Generally, Eq.1 is used to describe the relationship between channel current and gate-source voltage. However, the drain-source voltage change is very small during this stage. Therefore, the displacement current through drain-source capacitance, induced by dv/dt, can be ignored in Eq.1.

For power MOSFET, the expression of its threshold voltage is shown as follows.

$$V_{th} = \frac{\sqrt{4\varepsilon_s kTN_A \ln\left(N_A/n_i\right)}}{C_{OX}} + \frac{2kT}{q}\ln\left(\frac{N_A}{n_i}\right) - \frac{Q_{OX}}{C_{OX}} \qquad (2)$$

where ε_s, k, T, q, N_A, n_i, C_{OX}, Q_{OX} represent dielectric constant of silicon carbide, Boltzmann constant, kelvin temperature, electron charge, acceptor impurity concentration, intrinsic carrier concentration, gate oxide capacitance, and total effective oxide charge respectively. Q_{OX} includes mobile ion charge, oxide trap charge, fixed oxide charge and interface charge.

Even though the current SiC material technology has been greatly improved, it is still not as good as the mature silicon technology. There are still some problems in its defects control. Slight inconsistencies of these defects may bring about differences in interface and oxide layer quality, thereby resulting in differences in total effective charge Q_{OX} in the oxide layer. According to Eq.2, the Q_{OX} difference can lead to the large spread of V_{th}. In addition, the C_{OX} difference, caused by technological immaturity in the production process such as inconsistent doping concentrations and thickness of the gate oxide layers, can also lead to the fluctuation of threshold voltage [15]. In the parallel application, the spread will increase the imbalance of current sharing, leading to local overcurrent and over-temperature, bringing problems of life-time and reliability of paralleled devices.

The constant-current (CC) method is used to measure the V_{th} in this paper. This method is commonly used in industry due to its simplicity. It defines the value of the gate voltage V_{GS}, corresponding to a given constant drain current, as the V_{th}. The distribution of threshold voltage for 30 SiC MOSFET devices is shown in Fig.2. To quantitatively evaluate the spread of different parameters of SiC devices, this paper proposes to use the coefficient of variation δ, which is defined as shown in the Eq.3, as specification. As shown in Fig.3, V_{th} of device No.29 is the largest and that of device No.30 is the smallest. The mean μ of V_{th} among the 30 devices is 2.09V, the standard deviation σ is 0.29V, and the δ of V_{th} for the 30 devices is 13.8% computed by Eq.3.

$$\delta = \frac{\sigma}{\mu}\times 100\% \qquad (3)$$

where σ and μ are the sample standard deviation and mean respectively for devices parameters.

Generally, according to the statistical analysis theory, the data may be abnormal and should be removed when δ is greater than 15%. Therefore, it shows clearly that the spread of device threshold voltage is large.

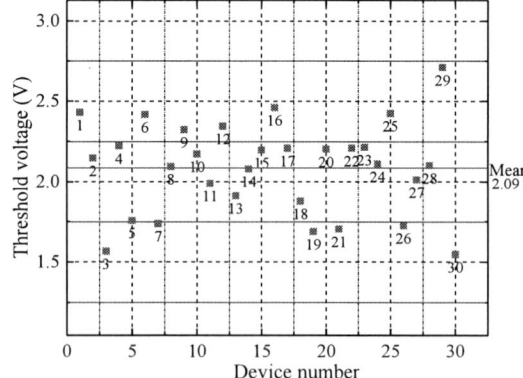

Fig. 2. The spread of threshold voltage for 30 devices

According to Eq.4, trans-conductance coefficient g_{fs} is controlled by the gate bias voltage. Hence the spread of V_{th} also has effects on g_{fs}. Besides, the spread of g_{fs} is also affected by the variation of channel length and channel electron mobility. In this paper, g_{fs} refers to the maximum trans-conductance, corresponding to maximum first derivative point of transfer curve. As shown in Fig.3, the spread of g_{fs} is much smaller than that of V_{th}, and the coefficient of variation is only 3.73%.

1126

$$g_{fs} = \frac{\mu_{ni} C_{OX} Z}{L_{CH}} \left(v_{GS} - V_{th} \right) \qquad (4)$$

where μ_{ni}, C_{OX}, Z, L_{CH} represent the inversion layer electron mobility, characteristic capacitance of the gate oxide layer, length of a cell perpendicular to the cross section and the channel length respectively.

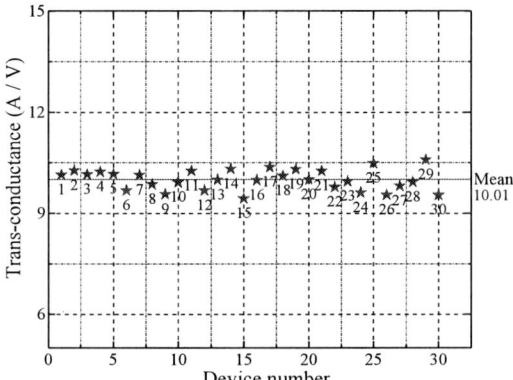

Fig. 3. The spread of trans-conductance for 30 devices.

III. THE EFFECT OF PARAMETERS MISMATCH ON CURRENT SHARING

During the transient process, when gate bias voltage exceeds the V_{th}, the current will start to increase. As shown in Eq.1, drain current of saturation region mainly depends on gate bias voltage, V_{th} and g_{fs}. Assuming that same driver is adopted for paralleled SiC MOSFETs, so that their gate bias voltage are the same, current imbalance can be expressed in Eq.5. Obviously, for paralleled devices, the current sharing imbalance will be more serious with the greater mismatch of either of the two parameters, V_{th} or g_{fs}.

$$i_{D1} - i_{D2} = \frac{\left(g_{fs1} - g_{fs2}\right) i_{DT} + g_{fs1} g_{fs2} \left(V_{th2} - V_{th1}\right)}{g_{fs1} + g_{fs2}} \qquad (5)$$

where i_{D1} and i_{D2} are drain currents; g_{fs1} and g_{fs2} are tranconductance coefficients; and V_{th1} and V_{th2} are threshold voltage for two paralleled devices. i_{DT} is the total current.

When g_{fs} variation is very small, it can be approximated that $g_{fs1}=g_{fs2}=g_{fs}$, Eq.5 can be simplified as:

$$\frac{i_{D1} - i_{D2}}{V_{th1} - V_{th2}} = \frac{\Delta i_D}{\Delta V_{th}} = -\frac{g_{fs1} g_{fs2}}{g_{fs1} + g_{fs2}} < 0 \qquad (6)$$

When V_{th} variation is very small, it can also be approximated that $V_{th1}=V_{th2}$, Eq.5 can be simplified as:

$$\frac{i_{D1} - i_{D2}}{g_{fs1} - g_{fs2}} = \frac{\Delta i_D}{\Delta g_{fs}} = \frac{i_{DT}}{g_{fs1} + g_{fs2}} > 0 \qquad (7)$$

According to Eq.6 and Eq.7, the current imbalance has a negative correlation with V_{th}, and positive correlation with g_{fs}. Therefore, the two parameters have an offsetting contribution to current sharing.

IV. EXPERIMENTAL VERIFICATIONS

A. Test setup description

In order to evaluate the influence of device spread on

current sharing of paralleled SiC MOSFETs, a double pulse test (DPT) platform is set up with two paralleled devices of To-247 package. Two 1200V/36A SiC MOSFET C2M0080120D are used as test device and 1200V/33A SiC SBD C4D20120D is used as freewheeling diode. As shown in Fig.4, experimental platform mainly includes five parts: digital signal processor, DC bus, driver circuit, charging and discharging circuit and an auxiliary power supply. The position of the paralleled devices and the PCB trace layout are nearly symmetrical to avoid the influence of circuit parameters on current sharing.

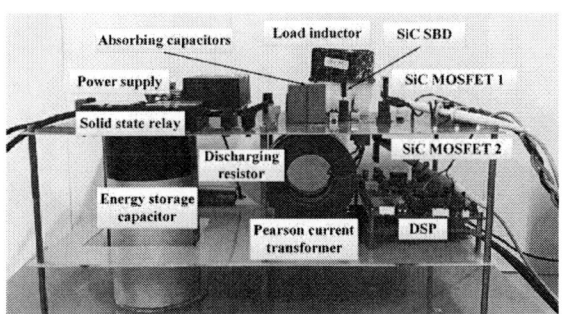

Fig. 4. The test bench of two-paralleled SiC MOSFET devices.

B. Consistency Validation

In order to accurately obtain the current difference between two paralleled devices, it is necessary to reasonably choose the measurement equipments that meet the frequency requirements. For high-speed SiC MOSFETs, the rapid rising and falling transients of current and voltage during switching process have many high-frequency components. This means that the equivalent frequency of the switching transient is far greater than the switching frequency. The maximum equivalent bandwidth of the measured signal can be approximately calculated as

$$BW = \frac{0.35}{\min[t_r, t_f]} \qquad (8)$$

Generally, the rising or falling times range of SiC MOSFET current or voltage is about 20ns-100ns. Actually, in order to ensure the reliability of measurement results, we usually select the bandwidth margin in 3-5 times, that is, in the range of 52.5MHz-87.5MHz. In this paper, the Pearson 6600 is used as current measuring component and the bandwidth is 120MHz. The selected oscilloscope is YOKOGAWA DLM2054 with 500MHz bandwidth. Hence, measuring equipments fully meet the bandwidth requirements of the signal to be measured.

Although two Pearson 6600 with same type are used for two paralleled branches, there may be still slightly difference between them due to the production process tolerance. In addition, there are very small differences among the different channels of the oscilloscope. To exclude the imbalanced current resulted from the difference of measuring component, two Pearson 6600 are simultaneously used to measure one of paralleled device current waveform before making experiments of current sharing. The experimental result is shown in Fig.5. It can be seen that the two current waveforms, meassured by two Pearson 6600, are almost consistent. Therefore, the two current probes can meet the test requirements in this paper.

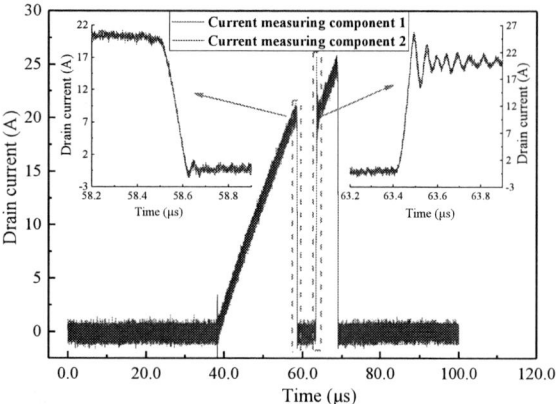

Fig. 5. Consistency validation of current measuring components

In order to individually analyze the influence of device parameter mismatch and avoid the current distribution imbalance caused by inconsistency of circuit parameters, the circuit layout should be guaranteed to be as symmetrical as possible. To verify or calibrate the symmetry of circuit layout, a cross transposition method is proposed as shown in Fig.6. Two devices are selected and inserted into different sockets for switching characteristics test, and then exchange position of the two devices to repeat testing. As shown in Fig.7, current waveforms of the same device in different positions almost the same, and current distribution difference caused by device parameters remains unchanged after exchanging their positions. Therefore, it is verified that the circuit layout is almost symmetrical.

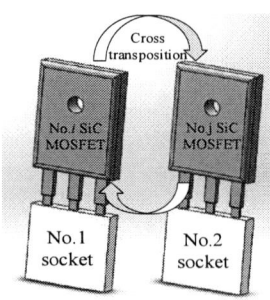

Fig. 6. Cross transposition method

C. Experimental analysis

The circuit layout is proved to be pretty symmetrical for the two parallel branches to assure circuit parameters match. In this case, two pairs of SiC MOSFET devices are selected from the sample of 30 devices, that is, (No.12, No.19) and (No.4, No.30). The parameters of two pairs of devices are shown in TABLE.I. For the first pair, device No.12 has larger V_{th} but lower g_{fs} than those of device No.19. For the second pair, V_{th} and g_{fs} of device No.4 are larger than those of device No.30. In order to

quantitatively describe transient current sharing of paralleled devices, the maximum transient current imbalance α is defined as the ratio of the absolute of the maximum difference of transient current to the base value as shown in Eq.9.

$$\alpha = \frac{|\Delta i_D|_{max}}{I_B} \times 100\% \qquad (9)$$

where $|\Delta i_D|_{max}$ represents the absolute value of the maximum difference of the device transient current. I_B represents the current value at first turn-off.

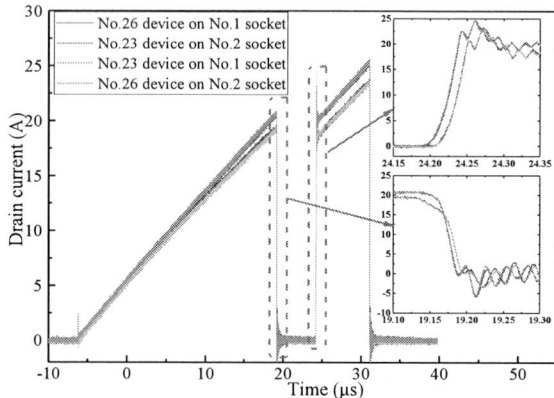

Fig. 7. Current waveform before and after rearrangement

TABLE.I
COMPARISON OF DEVICE PARAMETERS

No.	V_{th} /V	g_{fs} A/V	No.	V_{th} /V	g_{fs} A/V
12	2.36	9.58	4	2.20	10.19
19	1.69	10.02	30	1.55	9.30
Difference	0.67	-0.44	Difference	0.65	0.89

As shown in Fig.8, device No.19 carries larger current in both turn-on and turn-off transients, which is consistent with Eq.5 above. It can also be seen from Fig.8

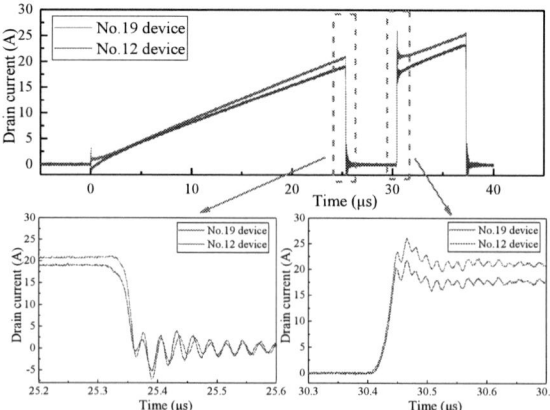

Fig. 8. Current sharing of paralleled devices No.12 and No.19.

that the turn-on delay of device No.19 is smaller while its turn-off delay is larger. Hence, device No.19 turns on faster and turns off slower. In addition, for turn-on transient, the current rising slope of No.19 is steeper. This indicates that the effects of threshold voltage mismatch and trans-conductance mismatch have a positive synthetical effect, leading to more uneven current sharing. Turn-on transient current imbalance degree reached 23.7% calculated by Eq.9. However, for the second pair, Fig.9 presents that turn-on transient current imbalance is only 10.3%. This is because negative synthetical effect between V_{th} and g_{fs} makes the turn-on and turn-off current distribution of the second pair of devices significantly better than that of the first pair of devices.

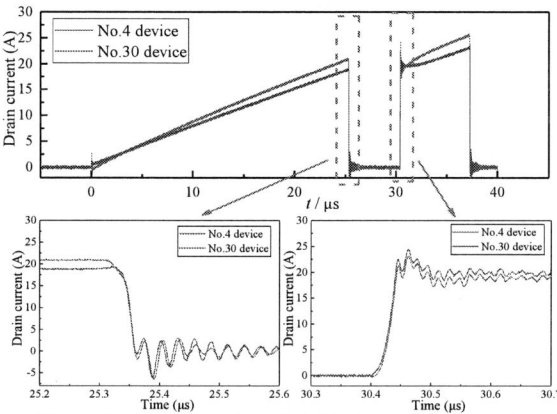

Fig. 9. Current sharing of paralleled devices No.4 and No.30.

V. New Screening Method

A. Process of Transfer Curve Screening

Whether it is Si or SiC device, the existing method is single parameter screening, for example threshold voltage screening. However, even if the threshold voltage difference of SiC MOSFET is very small, there may be still a large difference in the transfer curve due to the mismatch of trans-conductance coefficients between paralleled devices. Consequently, it is not enough to screen threshold voltage before connecting in parallel only. Therefore, this paper proposes a transfer curve screening method to improve transient current sharing of paralleled devices. The procedure of transfer curve screening method is shown in Fig.10. Firstly, two arbitrarily selected transfer curves are discretized. Then the difference of V_{GS} can be obtained under the same drain current by comparing the corresponding points of two curves. The relative difference ε, used in error analysis and data processing, is used to measure the distance of corresponding points of two curves, as shown in Eq.10. Finally, the average of relative difference μ_ε is used for measuring the proximity of curve, as shown in Eq.11. The closest pair of devices, No.20 and No.27, is obtained by comparing transfer characteristic pairwise to each other and calculating the μ_ε in MATLAB. As can be seen in Fig.11, one pair with the smallest difference of V_{th} is device No.15 and device No.17, and the other pair with the smallest difference of transfer curve is device No.20 and device No.27. Obviously, the transfer curves of device No.20 and device No.27 almost coincide.

Although the difference of V_{th} between No.15 and No.17 is very small, the transfer curves are quite different.

$$\varepsilon = \frac{|x_1 - x_2|}{(x_1 + x_2)/2} \tag{10}$$

where x_1 and x_2 respectively represent the abscissa values of the two curves under the same drain current.

$$\mu_\varepsilon = \frac{1}{n} \sum_{k=1}^{n} \varepsilon_k \tag{11}$$

where n and k are respectively the total number of sample points and the serial number.

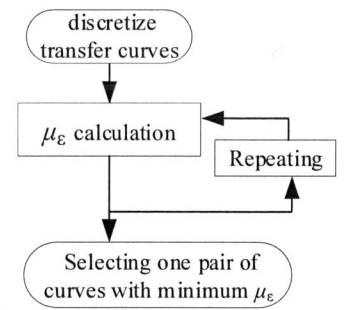

Fig. 10. The procedure of transfer curve screening method

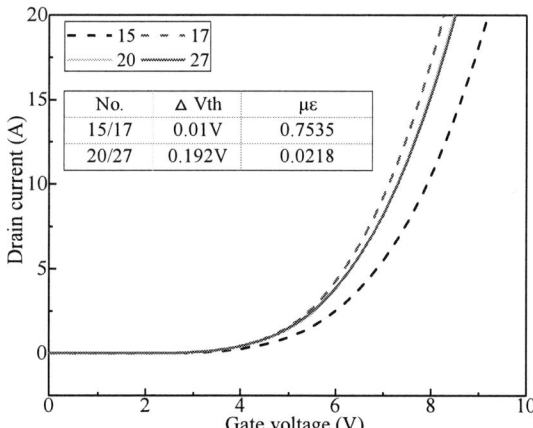

Fig. 11. Transfer curves of devices No.15, No.17, No.20 and No.27.

B. Experimental Verifications

Double pulse tests were performed on these two pairs of devices under clamped inductive load, and the respective experimental results shown in Fig.12 and Fig.13. Although the difference of V_{th} between devices No.15 and No.17 is only 0.01 V, the turn-on transient current imbalance rate between them has reached 26% due to that the μ_ε between the two transfer curves reaches 0.7535. In contrast, the turn-on transient current imbalance rate between device No.20 and device No.27 is only 3.5%. Although the difference of threshold voltage between them is significant, 0.192V, the average of relative difference μ_ε between two transfer curves is just 0.0218. Therefore, there is no doubt that screening threshold voltage alone cannot meet the requirement of current balance. In addition, transfer curve screening method is obviously superior to threshold voltage screening method.

The 2018 International Power Electronics Conference

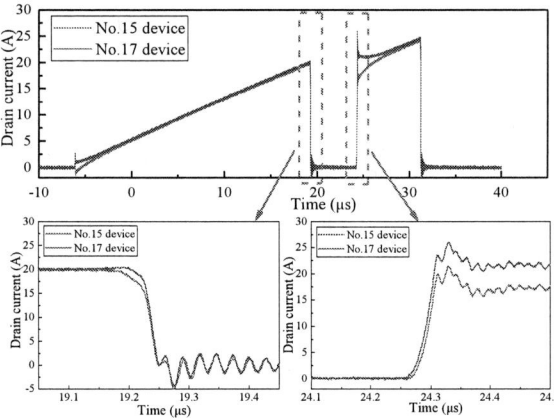

Fig. 12. Current sharing of paralleled devices No.15 and No.17.

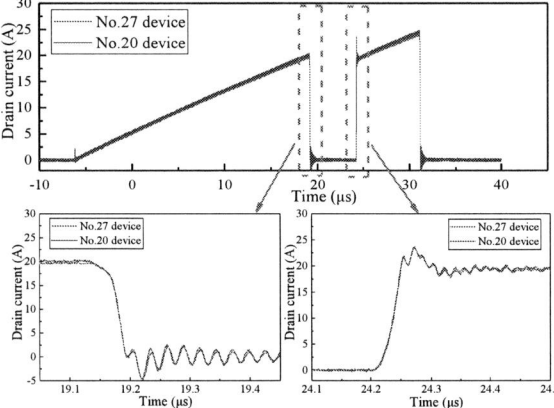

Fig. 13. Current sharing of paralleled devices No.27 and No.20.

VI. CONCLUSIONS

Efforts are made to minimize the difference of device parameters and improve transient current sharing of paralleled SiC MOSFETs. In this paper, evaluation and specification of device parameters spread is first proposed. Thereafter the main factors affecting transient current sharing are investigated by theoretical and experimental analysis. Studies carried out here show that V_{th} spread is much larger than g_{fs} spread. Moreover, the device with larger V_{th} will carry smaller current during turn-on transient, while g_{fs} has an opposite effect. Therefore, the two factors have an synthetical effect for transient current sharing. Finally, a new screening method is proposed to take consideration of the effects of V_{th} and g_{fs} together. Compared to the traditional single parameters screening method, the new transfer curve screening method has better transient current sharing performance, with current imbalance rate as low as 3.5% achievably. In contrast, current imbalance rate may reach 26% with traditional method.

ACKNOWLEDGMENT

The work presented in this paper has been supported by the National Key R&D Program of China (2016YFB0400503).

REFERENCES

[1] K. Hamada, M. Nagao, M. Ajioka, and F. Kawai, "SiC-Emerging Power Device Technology for Next-Generation Electrically Powered Environmentally Friendly Vehicles," *IEEE Transactions on Electron Devices*, vol. 62, no. 2, pp. 278-285, 2015.

[2] "Toyota to trial new SiC Power Semiconductor Technology," Toyota, Japan, http://newsroom.toyota.co.jp/en/detail/5692153, 2015

[3] M. H. Todorovic, F. Carastro, T. Schuetz, R. Roesner, L. Stevanovic, G. Mandrusiak, B. Rowden, F. Tao, P. Cioffi, J. Nasadoski, and R. Datta, "SiC MW solar inverter," in *Proc. PCIM Europe 2016*, pp. 645-652, 2016.

[4] Das M.K, Capell C, Grider D.E, Raju R, Schutten M and Nasadoski J, "10 kV, 120 A SiC Half H-bridge Power MOSFET Modules Suitable for High Frequency, Medium Voltage Applications," *IEEE Energy Conversion Congress and Exposition*, Phoenix, USA, pp. 2689-2692, 2011.

[5] Li H, Munk-Nielsen S, Pham C, and Bęczkowski S, "Circuit mismatch influence on performance of paralleling silicon carbide MOSFETs," in *Conf. EPE* 2014, pp.1-8, 2014.

[6] Li H, Munk-Nielsen S, Wang X, Maheshwari R, Beczkowski S, Uhrenfeldt C, and Franke W.T, "Influences of Device and Circuit Mismatches on Paralleling Silicon Carbide MOSFETs," *IEEE Transactions on Power Electronics*, vol. 31, no. 1, pp.621-634, 2016.

[7] J. Hu, O. Alatise, J. A. O. González, R. Bonyadi, L. Ran and P. A. Mawby, "The Effect of Electrothermal Nonuniformities on Parallel Connected SiC Power Devices Under Unclamped and Clamped Inductive Switching," *IEEE Transaction on Power Electronics*, vol. 31, no. 6, pp. 4526-4535, 2016.

[8] Sadik D P, Colmenares J, Peftitsis D, Lim J.K, Rabkowski J and Nee H.P, "Experimental investigations of static and transient current sharing of parallel-connected silicon carbide MOSFETs," *European Conference on Power Electronics and Applications*, pp.1-10, 2013.

[9] G. Wang, J. Mookken, J. Rice, and M. Schupbach, "Dynamic and static behavior of packaged silicon carbide MOSFETs in paralleled applications," *IEEE Applied Power Electronics Conference and Exposition*, pp. 1478-1483, 2014.

[10] Peftitsis D, Baburske R, Rabkowski J, Lutz J, Tolstoy G and Nee H.P, "Challenges Regarding Parallel Connection of SiC JFETs," *IEEE Transactions on Power Electronics*, vol. 28, no. 3, pp. 1449-1463, 2013.

[11] Kokosis S G, Andreadis I E, Kampitsis G E, Pachos P and Manias S, "Forced Current Balancing of Parallel-Connected SiC JFETs During Forward and Reverse Conduction Mode," *IEEE Transactions on Power Electronics*, vol. 32, no. 2, pp. 1400-1410, 2017.

[12] Lim J K, Peftitsis D, Rabkowski J, Bakowski M and Nee H.P, "Analysis and Experimental Verification of the Influence of Fabrication Process Tolerances and Circuit Parasitics on Transient Current Sharing of Parallel-Connected SiC JFETs," *IEEE Transactions on Power Electronics*, vol. 29, no. 5, pp. 2180-2191, 2017.

[13] "Fuji Power MOSFET," Application Note FA5504N, Fuji Electric.http://pdf.datasheet.directory/datasheets0/fuji_electric/FA 5504.pdf, 2014

[14] Wintrich A, Nascimento J and Leipenat M, "Influence of parameter distribution and mechanical construction on switching behaviour of parallel IGBT," in *Conf. PCIM Europe 2006*. 2006.

[15] Castellazzi A, Johnson M, Piton M, et al. "Experimental analysis and modeling of multi-chip IGBT modules short-circuit behavior," *IEEE Power Electronics and Motion Control Conference*, pp.285-290, 2009.

The 2018 International Power Electronics Conference

PSpice Modeling and Application for SiC Power MOSFET to Evaluate the Power Loss in Full-Bridge Converter

Juan Wei[1*], Fei Lin[1], Zhongping Yang[1], Xianjin Huang[1], Chanjuan Xiao[2], Hao Zhang[2] and Wencai Liang[2]
1 School of Electrical Engineering, Beijing Jiaotong University, Beijing, China
2 CRRC QINGDAO SIFANG CO., LTD, CRRC, Qingdao, China
*E-mail: 16121542@bjtu.edu.cn

Abstract—At present, SiC MOSFET has been widely concerned with its superior characteristics. In order to take advantage of silicon carbide devices and ensure its reliable operation, this paper presents a SiC MOSFET modeling approach based on PSpice simulation software. The model is built based on the parameters in the datasheet and is very easy. The simulation and experimental results show that the model can accurately reflect the static and dynamic characteristics of SiC MOSFET. At the same time, the simulation circuit of the phase-shifted full-bridge converter is built based on the SiC MOSFET model, and the power loss is analyzed and calculated, which consistent with the experimental results. It is shown that the power loss obtained by the simulation circuit is accurate and effective, which provides an important basis for the practical application of SiC MOSFET.

Keywords— SiC MOSFET, modeling, power loss

I. INTRODUCTION

The SiC based power semiconductor devices have characteristics of high operating voltage, high operating frequency and low loss, while they can adapt to more demanding conditions (such as high temperature and high radiation) and have good environmental stability in different high power applications [1]. In addition, compared with Si IGBT and BJT, SiC MOSFET has higher switching speed and lower loss due to MOSFET is unipolar device and there is no tail current. At present, the commercial SiC semiconductor devices has nowadays reached 600V~1700V blocking voltage and 100A~800A current capability [2].

In order to take full advantage of the superiorities of SiC MOSFET and ensure reliable operation in converters, the static characteristics, switching performance and power loss of the devices should be adequately evaluated. Hence, it is necessary to establish an accurate and reliable SiC MOSFET model for engineering analysis and efficiency evaluation. The literatures present a number of models to predict the static and dynamic characteristics of silicon carbide power devices [3]-[9]. However, the existing models have the following shortcomings: parameter extraction requires very expensive test instruments in some literature it is difficult to be widely

used; in addition, many of the literature did not give detailed parameters of the SiC MOSFET model, so it is not conducive to reference; the expressions which describe the device are too complex, and it is not conducive to the entire system simulation.

A PSpice model is proposed in this paper based on the the application of SiC MOSFET in phase-shifted full-bridge converter, which is developed and verified for high power SiC MOSFET modules with voltage and current rating of 1700V and 225A (Cree Semiconductor CAS300M17BM2). The developed model is based on the parameters in the datasheet, thus it is easy to implement. The simulation results of the model are in good agreement with the datasheet, indicating that the model is accurate and reliable. On this basis, the simulation model of phase-shifted full-bridge converter is built and the loss of the devices is analyzed and calculated [11-14]. Finally, the double pulse test platform and prototype of phase-shifted full-bridge converter is set up, some experiments are carried out in the two platforms.

II. MODELING OF SiC MOSFET

The silicon carbide module CAS300M17BM2 (SiC MOSFET + SiC Diode) manufactured by Cree is used in this paper and the power rating is 1700V / 225A. The switching circuit model of the SiC MOSFET is similar to that of the Si MOSFET, as shown in Fig.1. The parasitic parameters C_{GS}/C_{GD}/C_{DS} and the internal gate resistance R_G in the figure are related to the turn-on and turn-off processes of SiC MOSFET.

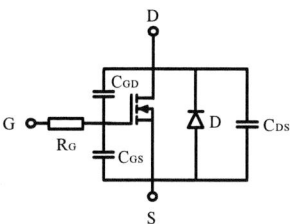

Fig. 1. Internal equivalent model of SiC MOSFET.

A. Static Characteristics Test of SiC MOSFET

In order to obtain more practical and more accurate

parameters of static characteristics, we use professional equipment to test the static characteristics of full silicon carbide modules, including saturation voltage drop, diode forward voltage drop and leakage current. All tests were conducted at room temperature (25 ℃) and elevated temperatures (90 ℃ and 125 ℃), respectively. And the results are shown in Fig.2.

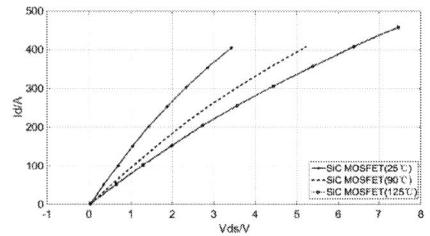

(a) Saturation voltage drop of MOSFET

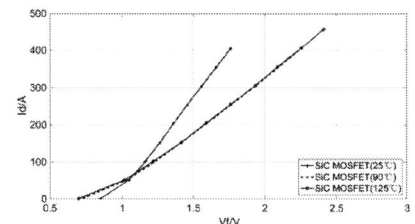

(b) Diode forward voltage drop of MOSFET

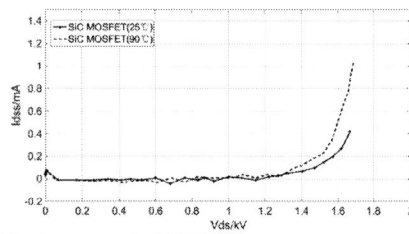

(b) Leakage current of MOSFET

Fig. 2. Test curves of static characteristic

As can be seen from the figure, the device's saturation voltage drop, diode forward voltage drop and leakage current go up as the temperature rises, indicating that the on-state loss and blocking loss will increase when the temperature is rising and this is consistent with the practical application. All the static test results are available for the extraction of modeling parameters below.

B. Extraction of Static Characteristics for SiC MOSFET

The static characteristics of SiC MOSFET mainly include transfer and output characteristics. The modeling method of SiC MOSFET proposed in this paper is based on a set of analytical equations that that represent the characteristics of MOSFET. The current of SiC MOSFET I_D is expressed by drain-source voltage V_{DS} and gate-source voltage V_{GS}, the function expression is as shown in equation (1):

$$\begin{cases} I_D = 0 & V_{GS} < V_T \\ I_D = K_p \left[2\left(V_{GS} - V_T\right)V_{DS} - V_{DS}^2 \right] & V_{DS} < V_{GS} - V_T \\ I_D = K_p \left(V_{GS} - V_T\right)^2 & V_{DS} \geq V_{GS} - V_T \end{cases} \quad (1)$$

As shown in the formula, the working area of MOSFET can be divided into three regions, namely the cut-off region, variable-resistance region and saturation region. In the cut-off area, the current is zero owing to the conductive channel has not yet formed, that is cut-off working condition. In the formula (1), V_T is the threshold voltage and the value of SiC MOSFET used in this paper is typically 2.3V; K_p is conductivity constant and it can be expressed by the formula (2):

$$K_p = \frac{K_p^{'}}{2} \cdot \frac{W}{L} = \frac{\mu_N C_{ox}}{2}\left(\frac{W}{L}\right) \quad (2)$$

Where the intrinsic conductivity factor $K_p^{'} = \mu_N C_{ox}$ is usually constant, μ_N is the electron mobility, C_{ox} is the capacitance per unit area of the oxide layer. W is the channel width, and L is the channel length. Table I shows the parameters of I_{DSS} and V_{GS} in the datasheet.

TABLE I
THE PARAMETERS OF I_{DSS} AND V_{GS} IN THE DATASHEET

Symbol	I_{DSS}	$V_{GS(th)}$
Parameter	Zero Gate Voltage Drain Current	Gate Threshold Voltage
Minimum	/	1.8
Typical	500	2.3
Maximum	1000	/
Unit	µA	V
Condition	V_{DS}=1.7kV V_{GS}=0V	V_{DS}=10V I_D=15mA

It can be seen that V_T=2.3V from the TABLE I, and when V_{GS}=0V, K_p is calculated as

$$I_D = K_p \left(V_{GS} - V_T\right)^2 \quad (3)$$

In the formula, I_D=I_{DSS}=500µA, V_T=2.3V, K_p is obtained by calculation. Then W/L is calculated as

$$\frac{W}{L} = \frac{2K_p}{\mu_n C_{ox}} = 9 \quad (4)$$

TABLE II
MODELING PARAMETERS OF STATIC CHARACTERISTICS

Parameter	Value	Parameter	Value
K_p/A/V^2	9.46e-5	R_G/Ω	3.7
V_T/V	2.3	R_D/Ω	0.001
L/m	2e-6	R_S/Ω	0.01
L/m	1.8e-6		

In summary, the modeling of static characteristic for SiC MOSFET is completed. The parameters of static model which are extracted through the datasheet are shown in TABLE II.

I_S/A	2.543	TT/s	1e-16
R_S/Ω	0.001		

C. Extraction of Dynamic Characteristics for SiC MOSFET

MOSFET has three parasitic capacitances in which the gate-source capacitance C_{GS} and gate-drain capacitance C_{GD} are related to the geometry of the device. The drain-source capacitor C_{DS} is the junction capacitance of the diode, so it is not analyzed here. Normally, the parasitic capacitance appears in the form of input capacitance C_{iss}, output capacitance C_{oss} and transfer capacitance C_{rss} in the datasheet, as shown in equation (5):

$$
\begin{aligned}
C_{rss} &= C_{GD} \\
C_{iss} &= C_{GD} + C_{GS} \\
C_{oss} &= C_{GD} + C_{DS}
\end{aligned} \tag{5}
$$

According to the datasheet, when V_{DS}=1kV, C_{rss}=C_{GD}=0.08nF, C_{iss}=20nF. The values of C_{GS} and C_{GD} will change with V_{GS} and V_{DS} to keep stable, depending on the thickness and type of oxide of insulating layer. The curve between the capacitance and the drain-source voltage has a slight change, but it is common to assume that the capacitances are constant. Consequently, C_{GS}=19.92nF can be obtained according to the formula (5).

The most important parameters are calculated firstly when the parameters of SiC MOSFET model are extracted, then they can be added according to the actual tests.

D. Parameter Extraction of Body Diode Model

(1) Reverse breakdown characteristic

The power level of full silicon carbide module CAS300M17BM2 is 1700V/225A, so the reverse breakdown voltage and reverse breakdown current of body diode are 1700V and 225A, that is BV=1700V and I_{BV}=225A.

(2) Forward characteristic

The forward characteristic data of the diode can be obtained from the datasheet and we can extract the reverse saturation current I_S, the parasitic resistance R_S, the emission coefficient N and the knee current I_{KF} that characterizes the decrease of the forward current amplification factor at high current through inputting data into the PSpice simulation software.

(3) Junction capacitance C_{JO}

What can be obtained according to the reverse voltage and junction capacitance curves in the datasheet are: the value of junction capacitance is 36nF when the pn junction voltage is zero, namely C_{JO}=36nF. In the meanwhile, due reverse recovery current of the module's body diode is zero, then the default value of transit time TT is the minimum value.

TABLE III
MODELING PARAMETERS OF BODY DIODE

Parameter	Value	Parameter	Value
BV/V	1700	N	1.2326
I_{BV}/A	325	C_{JO}/nF	36

III. SIMULATION AND VERIFICATION OF SiC MOSFET MODELING

A. Static Characteristic

In order to verify the accuracy of the above model, it is essential that we carry out some experiments about static characteristics. Among them, M1 is the PSpice model of SiC MOSFET, V1 is the voltage source applied to gate and source, and V2 is the voltage source applied to drain and source. The transfer characteristic curve of the model can be obtained by using the DC Sweep function of PSpice, and compared with the curve in the datasheet, as shown in Fig.3.

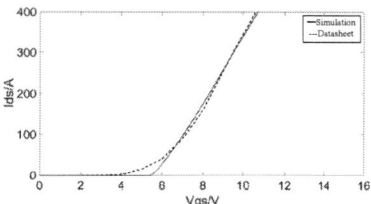

Fig. 3. Comparison of transfer characteristics between simulation and datasheet.

It can be seen from Fig.4 that the curve of simulation is consistent with ones in the datasheet, which verifies the accuracy of established static model of SiC MOSFET.

B. Rise and Fall Time

Fig.4 shows the switching waveform of SiC MOSFET model, where the drain-source voltage is 500V. According to the figure, the rise time of model is 109ns and the fall time is 78ns, at the same time they are 72ns and 56ns respectively in the datasheet. It can be found that the data of simulation is essentially in agreement with ones in the datasheet by comparison, indicating that the SiC MOSFET model is accurate and can be used for system simulation.

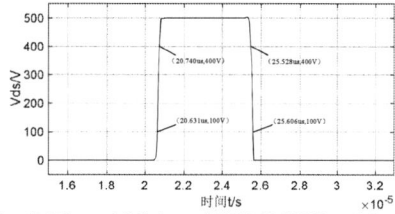

Fig. 4. Rise and fall time of SiC MOSFET model.

C. Dynamic Characteristic

a. The principle of double pulse test (DPT)

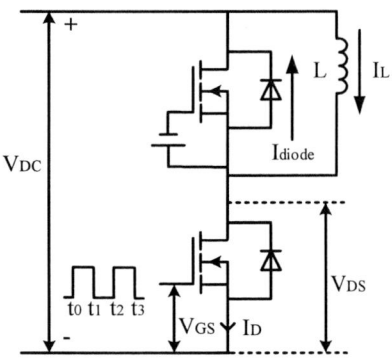

Fig. 5. Double pulse test simulation circuit.

The dynamic characteristics of SiC MOSFET mainly include the turn-on and turn-off characteristics, which are completed in the DPT platform. As shown in Fig.5, the DPT platform consists of a DC voltage source, a load inductance, SiC MOSFET module and drive circuit.

b. The simulation results of SiC MOSFET model

As the switching speed of SiC MOSFET is fairly fast, the parasitic inductance in the power cable will have an impact in it, it is necessary to add parasitic inductance in the simulation circuit. Fig.6 shows the waveforms of DPT simulation of SiC MOSFET model.

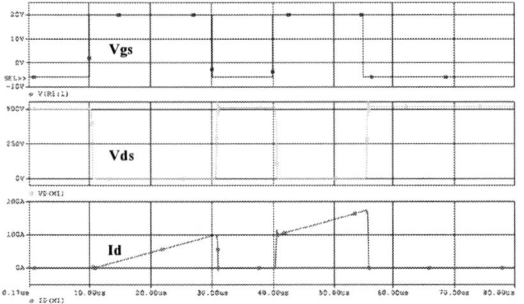

Fig. 6. Turn-on and turn-off simulation waveforms of SiC MOSFET model.

It can be seen that the SiC MOSFET model can accurately realize the functions of turn-on and turn-off processes, which further illustrates the correctness of the model. The comparison between simulation and actual test will be described in detail in Chapter IV.

IV. THEORETICAL CALCULATION AND SIMULATION ANALYSIS OF LOSS FOR SiC MOSFET

The total losses of switching device include static losses and dynamic losses, in which static losses consist of conduction loss and cut-off loss and dynamic losses include turn-on loss and turn-off loss. We can get that the leakage current of device is typically 500μA from the datasheet which almost can be ignored, so the total losses of switching device is:

$$P_{total} = P_{con} + P_{turn\_on} + P_{turn\_off} + P_{diode} \quad (6)$$

A. Calculation formulas of SiC MOSFET losses

a. Conduction loss

The conduction loss of SiC MOSFET can be calculated by the formula (7):

$$P_{con} = I_{rms}^2 R_{DS(on)} t_{on} f_s \quad (7)$$

Where I_{rms} is the current effective value when the device is in conducting state, $R_{DS(on)}$ is the on-resistance of device.

b. Switching losses

The switching losses of SiC MOSFET can be calculated by the formula (8):

$$\begin{cases} P_{turn\_on} = E_{turn\_on} f_s = f_s \int_0^{t_{turn\_on}} v_{ds}(t) \, i_d(t) dt \\ P_{turn\_off} = E_{turn\_off} f_s = f_s \int_0^{t_{turn\_off}} v_{ds}(t) \, i_d(t) dt \end{cases} \quad (8)$$

Where t_{turn_on} and t_{turn_off} are the crossing time of voltage and current during turn-on and turn-off processes respectively, v_{ds} and i_d are drain-source voltage and drain current in the turn-on and turn-off processes, and f_s is the switching frequency.

c. Diode losses

Diode losses contain on-state loss P_{on} and reverse recovery loss P_{rr}. Since the reverse recovery current of SiC power module used in this paper is zero, there is no reverse recovery loss of diode. Therefore, the diode only contains the on-state loss, and the formula is as follows:

$$P_{diode} = P_{on} = V_f I_f f_s \quad (9)$$

Where V_f and If are the forward voltage and current of diode respectively, and f_s is the switching frequency.

B. Simulation of phase-shifted full-bridge converter circuit

Fig.7 shows the simulation of phase-shifted full-bridge converter system. The switches in the circuit are all silicon carbide devices including the diodes in the rectifier circuit, so the frequency in the inverter will increase to 40 kHz.

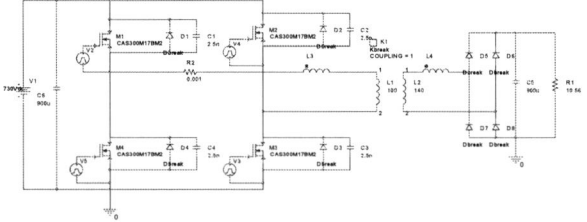

Fig. 7. Simulation circuit of phase-shifted full-bridge converter.

Fig.8 shows the simulation waveforms of the phase-shifted full-bridge converter, figure (a) and (b) display the waveforms of full-bridge inverter and diode rectifier circuit respectively.

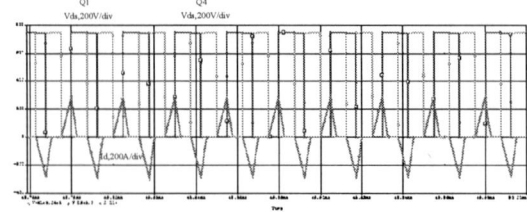

(a) Full-bridge inverter circuit

1134

The 2018 International Power Electronics Conference

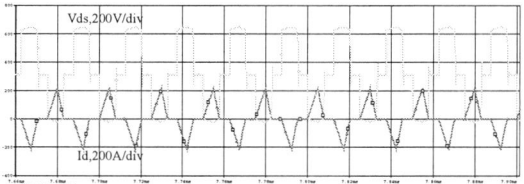

(b) Diode rectifier circuit

Fig. 8. Simulation waveforms of phase-shifted full-bridge circuit.

TABLE IV
THE SIMULATION VALUES OF POWER LOSS

Circuit	Power loss/W
Full-bridge inverter circuit	2343
Diode rectifier circuit	□□□□□

The power loss of the inverter circuit and the diode rectifier circuit can be obtained by the formulas and simulation waveforms as mentioned above, as shown in Table III.

V. EXPERIMENTAL VERIFICATION

A. DPT platform

In order to study the high voltage characteristics of the SiC module CAS300M17BM2 (1700V/225A) and verify the accuracy of dynamic characteristics of the model, a DPT experimental platform is proposed, as shown in Fig.9. The drive board used in the experiments is PT62SCMD17 produced by Cree, which is designed to drive the SiC module CAS300M17BM2.

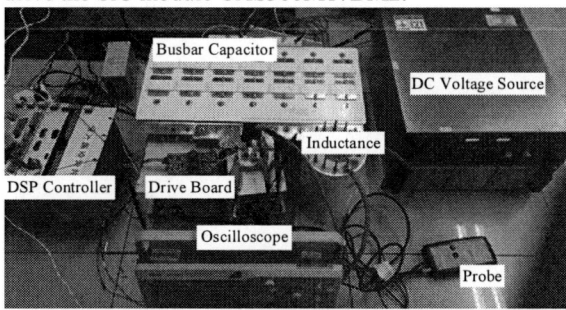

Fig. 9. Double pulse experiment platform.

Fig.10 shows the experimental waveforms of turn-on and turn-off process when the drive resistance is 15Ω. Compared with the waveforms of voltage and current in Fig.6 and Fig.10, it is found that the simulation and experimental waveforms are in good agreement with each other. However, the oscillation of the experiment is more serious, which is caused by the more complicated parasitic parameters in the actual test circuit.

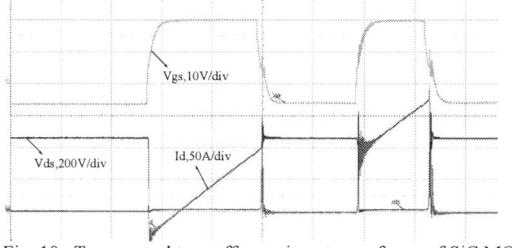

Fig. 10. Turn-on and turn-off experiment waveforms of SiC MOSFET.

B. Prototype platform of SiC MOSFET phase-shifted full-bridge converter

Fig. 11. Prototype platform of phase-shifted full-bridge converter.

The prototype platform of SiC MOSFET phase-shifted full-bridge converter is shown below, and the switches used in inverter circuit and diode rectifier circuit are SiC module CAS300M17BM2. The frequency of full-bridge inverter is 40 kHz, so the transformer applied in the prototype is high-frequency transformer, greatly reducing the volume and improve the power density.

The voltage and current waveforms of inverter and diode rectifier are shown in Fig.12.

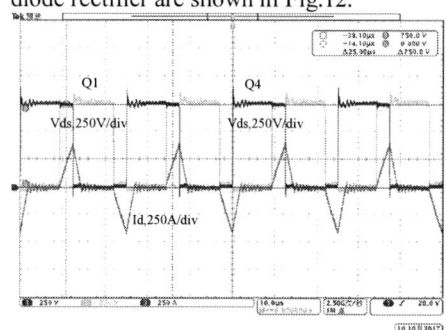

(a) Inverter circuit

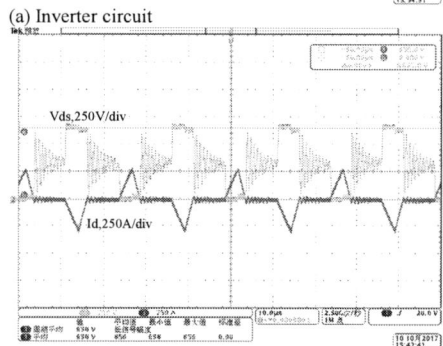

(b) Diode rectifier circuit

Fig. 12. Experimental waveforms of phase-shifted full-bridge circuit.

The power losses of inverter circuit and diode rectifier circuit can be obtained by the formulas in Chapter III, as shown in Table IV. It can be seen from the data in the table that the values of power losses of experiment are generally larger than the ones in simulation. It is because that there are more parasitic parameters in the actual test conditions which will change the voltage and current of switches and then has an impact on the power losses of the circuit.

1135

TABLE V
THE EXPERIMENT VALUES OF POWER LOSS

Circuit	Power loss/W	Error
Full-bridge inverter circuit	2654.32	11.73%
Diode rectifier circuit	□□□□□□	□□□□□

The data on the far right of Table VI show the errors between experimental value and simulation value of power losses. It is found that the errors of inverter circuit and diode rectifier circuit are 10% or so and it can be acceptable, indicating that the values of power losses obtained by simulations are effective. This experiment further validates the accuracy of SiC MOSFET model and provides an important basis for the application of silicon carbide devices in practice.

VI. CONCLUSION

The full-silicon power module CAS300M17BM2 of 1700kV/325A introduced by Cree is used in this paper. The following tasks are completed based on the application of SiC MOSFET in the phase-shifted full-bridge converter SiC MOSFET in the phase-shifted full-bridge converter:

(1) Based on the datasheet, the process of SiC MOSFET modeling is completed. The static and dynamic parameters of the MOSFET are extracted through mathematical equations and data, which makes the parameter extraction process easier and the model is verified by static characteristics simulation and double pulse test.

(2) Based on the phase-shifted full-bridge DC-DC circuit, the theoretical calculation and analysis of the device losses are completed. The phase-shifted full-bridge converter simulation circuit is built in the PSpice simulation software by using the SiC MOSFET model, and the power loss of silicon carbide device in the circuit is obtained through the waveforms of voltage and current.

(3) Double-pulse test platform and phase-shifted full-bridge prototype platform are set up to further verify the accuracy of the model of SiC MOSFET and loss calculation in the phase-shifted full-bridge simulation circuit.

REFERENCES

[1] Sheng Kuang, et al. "Development and Prospect of SiC Power Devices in Power Grid." *Journal of Chinese Electrical Engineering Science* 32.30(2012):1-7. (In Chinese)

[2] Ceccarelli, Lorenzo, F. Iannuzzo, and M. Nawaz. "PSpice modeling platform for SiC power MOSFET modules with extensive experimental validation." *Energy Conversion Congress and Exposition* IEEE, 2017.

[3] Johannesson, Daniel, and M. Nawaz. "Analytical PSpice model for SiC MOSFET based high power modules." *Microelectronics Journal* 53(2016):167-176.

[4] Mudholkar, Mihir, et al. "Datasheet Driven Silicon Carbide Power MOSFET Model." *IEEE Transactions on Power Electronics* 29.5(2014):2220-2228.

[5] Peng Yonglong, Li Rongrong, and Li Yabin. "SiC MOSFET Modeling Based on Key Parameters of PSpice." *Power Electronics* 49.4(2015):54-56. (In Chinese)

[6] Tsolaridis, Georgios, et al. "Development of Simulink-based SiC MOSFET modeling platform for series connected devices." *Energy Conversion Congress and Exposition* IEEE, 2016:1-8.

[7] Phankong, Nathabhat, T. Funaki, and T. Hikihara. "A static and dynamic model for a silicon carbide power MOSFET." *European Conference on Power Electronics and Applications* IEEE, 2009:1-10.

[8] Cui, Yutian, M. Chinthavali, and L. M. Tolbert. "Temperature dependent Pspice model of silicon carbide power MOSFET." *Twenty-Seventh IEEE Applied Power Electronics Conference and Exposition* IEEE, 2012:1698-1704.

[9] Lu, Juejing, et al. "Modeling of SiC MOSFET with temperature dependent parameters and its applications." *Applied Power Electronics Conference and Exposition* IEEE, 2013:540-544.

[10] Ruan Xinbo. "Soft-Switching Technology of Pulse Width Modulated DC/DC Full-Bridge Converter." *Science Press* 2013. (In Chinese)

[11] Xie Jiaji. "Application Research of Auxiliary Inverter Based on SiC MOSFET." Beijing Jiaotong University 2017. (In Chinese)

[12] Chen, Hsin Ju. "Power Losses of Silicon Carbide MOSFET in HVDC Application." (2012).

[13] Liang Mei, et al. "Analytical Model of SiC MOSFET for Accurately Predicting the Switching Performance." *Transactions of China Electrotechnical Society* 32.1(2017):148-158. (In Chinese)

[14] Sintamarean, C., F. Blaabjerg, and H. Wang. "Comprehensive evaluation on efficiency and thermal loading of associated Si and SiC based PV inverter applications." *Industrial Electronics Society, IECON 2013 -, Conference of the IEEE* IEEE, 2013:555-560.

All-SiC Module Packaging Technology

Kento Shirata[1], Norihiro Nashida[1], Hideyo Nakamura[1] and Yoshitaka Nishimura[1*]
1 Fuji electric Co., Ltd., 4-18-1, Tsukama, Matsumoto, Nagano, Japan
*E-mail: nishimura-yoshitaka-m@fujielectric.com

Abstract- **We developed the package advanced technology of the All-SiC module which is applicable to energy saving products of one such as power conditioning sub-system (PCS). Key technologies are 3 dimensional wiring using Cu pins with power board instead of conventional Al wiring and full-mold structure using the thermosetting epoxy resin and transfer molding technology. These technologies lead to small package size, low inductance, and high reliability.**

Keywords-All-SiC module, 3-dimentional wiring, Full-mold structure

I. INTRODUCTION

Public interest in environmental issues such as global warming is increasing year by year, and worldwide society demands less greenhouse gas emissions representative of CO_2. Meeting such a need requires the active utilization of renewable energies and greater energy saving of power electronics equipment. In general, power semiconductors play a key role in the power conversion system of power electronics products. Silicon (Si) devices, which are the conventional mainstream, have undergone various breakthroughs and gradually approach its physical limit. Regarding this background, silicon carbide (SiC) devices, which are the next-generation semiconductors enabling even less power dissipation, are raising expectations for their contribution toward energy saving.

Fuji Electric has developed an All-SiC module using SiC metal-oxide-semiconductor field-effect transistors (SiC-MOSFETs) and SiC Schottky barrier diodes (SiC-SBDs) and using it to a power conditioning subsystem (PCS) for mega solar power plants. This paper describes the packaging and process technology of the All-SiC module.

II. FEATURES OF NEW PACKAGE

In order to achieve more efficient power conversion of high-capacity photovoltaic power generation as mega solar power plants, Fuji Electric started mass production of mega solar PCSs that employ All-SiC modules in 2014. Figure 1 shows the external appearance of the mega solar PCS, its built-in power unit and the All-SiC module. It uses the All-SiC module in voltage boosting circuit and achieves a high efficiency of 98.8% to save energy. Further efforts have been made to reduce the size and weight of the equipment [1].

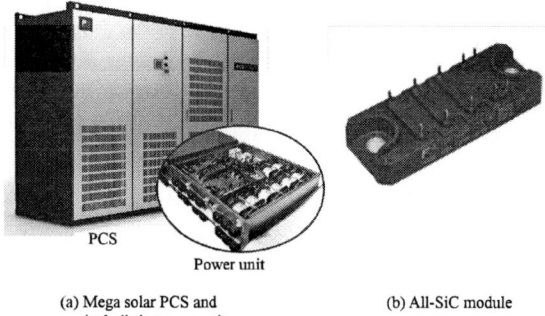

(a) Mega solar PCS and its built-in power unit (b) All-SiC module

Fig. 1 Mega solar PCS and All-SiC module

Figure 2 shows cross-sectional structures of the new package used in the All-SiC module and the conventional package used in a silicon insulated-gate bipolar transistor (Si-IGBT) module. Table I shows the comparison of the typical characteristics of these modules. The new package contains small size SiC chips connected in parallel. In order to flow large current through wiring on the chip, we applied 3-dimensional wiring with Cu pins and a power substrates instead of aluminum wires. For the size advantage, the footprint is reduced to approximately 40% compared with the conventional module. This miniaturization achieved by 3-dimensional wiring effectively reduces the inductance less than a quarter of the conventional one. Additionally, thermal resistance

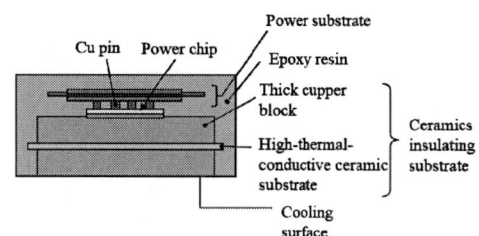

(a) New package (All-SiC module)

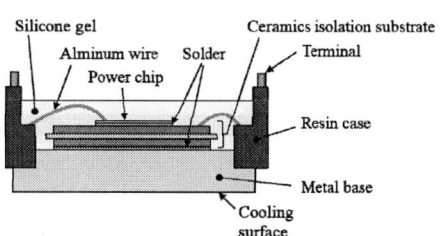

(b) Conventional package (Si-IGBT module)

Fig. 2 Cross-sectional package structure

TABLE I

Comparison of typical characteristics between the new and conventional packages (Relative comparison at 1,200 V/100 A rating)

Package characteristics	Conventional package	New package
Thermal resistance (K/W)	0.469	0.209
Inductance (nH)	52	12
Footprint	1	0.42

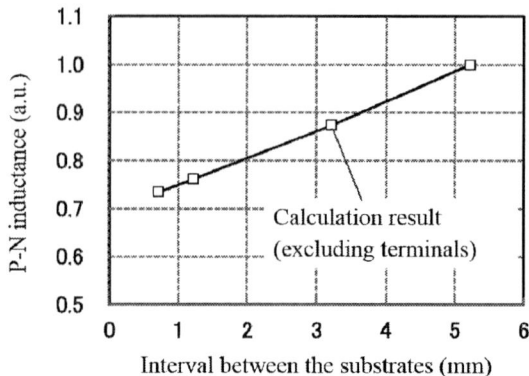

Fig. 3 Relationship between the substrate interval and inductance

is decreased approximately a half of conventional module by using a ceramics insulating substrate consisting of a high-thermal-conductive ceramics substrate (Si_3N_4) bonded with thick copper plates and by adopting a structure without metal base [2], [3]. Furthermore, the use of epoxy resin for the molding resin improves reliability. Molding technology ensures the isolation of the chips and ceramics substrate and also suppresses the distortion of joint area between the chips and Cu pins. Adopting transfer mold forming for this epoxy resin molding eliminates the need for a conventional resin case, leading to miniaturization and productivity improvement. In the new package structure, the epoxy resin is the key component that determines the performance of the module.

III. PACKAGE STRUCTURE DESIGN

A) Internal wiring structure

The SiC-MOSFET allows faster switching compared with the conventional Si-IGBT. To bring out advantage of the ability, it is necessary to reduce surge voltage that increases in proportion to the switching speed and this make it crucial to reduce the inductance of internal wiring.

The new package has achieved miniaturization by adopting 3-dimensional wiring that uses Cu pins and a power substrate as shown in Fig. 2(a). This decreases the wiring distance and reduces self-inductance. Furthermore, we attempted to reduce the inductance further by arranging the power substrate and ceramics insulating substrate in parallel and by connecting the wiring to make the change in the current (di /dt) occur in the opposite directions [4]. In this structure, the closer the 2 substrates are positioned, the more the inductance decreases as shown in Figure 3. Consequently, we set the interval as narrow as possible on the condition that it does not affect the insulation performance or assembly work. As a result, the P-N inductance of the new package is less than 25% (approx. 12nH) compared with the conventional package.

Since a mega solar PCS handles large current, multiple All-SiC modules are connected in parallel. In this case, the inductance is inversely proportional to the number of modules. This is more advantageous for high-speed switching than the conventional case where a smaller number of large-capacity modules were used.

B) Molding structure

A full-mold structure with thermosetting epoxy resin can relief the stress inside the module. In the view of reliability, stress is occurred at joint areas of the chip. Molding can cover surrounding of the chip and other joint areas, and distortion is eased [5].

However, the issue with the full-mold structure is caused by the fact that materials with different linear expansion coefficients are molded together. When the package cured under elevated temperature is returned to normal temperature, internal stress is generated and warpage occurs in the entire module. It is necessary to keep the warpage as low as possible because it may increase stress or thermal resistance when the product is mounted on a cooling fin or cause a pump-out phenomenon of the compound due to the temperature change during operation.

Figure 4 shows the results of the finite element method (FEM) analysis and actual measurement to examine the

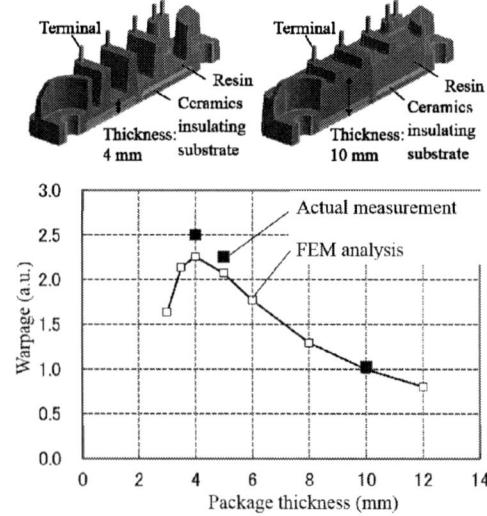

Fig. 4 FEM model of the new package and results of analysis and actual measurement of warpage

relationship between the warpage and the thickness of the main body (thickness of resin) of the new package excluding protrusions. The results indicate that the warpage becomes smaller when the package is either thicker or thinner than a certain thickness. This is probably caused by the fact that, in the region where the package is thin and dominated by the rigidity of the ceramics insulating substrate, the stress on the ceramics insulating substrate decreases with the decrease of the resin thickness, resulting in smaller warpage. On the other hand, in the region where the package is thick and is dominated by the rigidity of the resin, the rigidity of the resin increases further with the increase in the resin thickness so that the package is less affected by the ceramics insulating substrate, resulting in smaller warpage. In real-world situations, however, when the thickness of the ceramics insulating substrate, chip thickness and power substrate lamination are considered, the module would be manufactured in the region where the rigidity of the resin dominates. Consequently, it is effective to produce the All-SiC module with thicker epoxy resin to obtain low warpage module. In this case, it is important to optimize the resin thickness and inductance value because a thicker resin part requires a longer terminal to be extended outside, causing an increase in inductance. The insulation distance between the terminal and ground (creepage/clearance distance) must also be considered.

As a result, the internal structure of the module is designed to arrange most components on the cooling surface side. This is done to reduce inductance by arranging the power substrate and ceramics insulating substrate at a narrower interval and to suppress warpage by making the module thicker. When transfer mold forming is used for this high concentrated structure, the resin hardly flows smoothly and evenly inside the module. Therefore, it requires a mold process design based on an accurate recognition of the resin flow.

IV. MOLD PROCESS DESIGN

A) Simulation techniques and mold design

One of the concerns of the transfer mold forming of the new package is scattered voids (trapped air) and welds (sections where resin flows meet together) inside the module. These are caused by the deterrence of the flow. The chips, Cu pins and other internal components are rationale. The volatile filling speed at the narrow space and that at the other space can also generate voids and welds (resin flow junction part).

To understand this phenomenon, simulations of the resin and air flow were done. Figure 5 shows the simulation model and results of conventional resin flow and a liquid-gas two-phase (resin and air) flow. As a result of this, two-phase flow simulation could clarify welding position, air trap and/or bubbles in the flowing resin. This result is reflected to the mold design.

(a) Flow simulation model

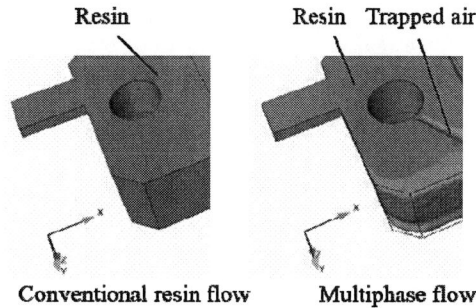

Conventional resin flow   Multiphase flow

(b) Comparison of flow simulations

Fig. 5 Flow simulation model and results

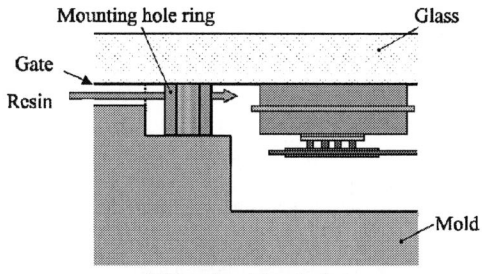

(a) Experimental method

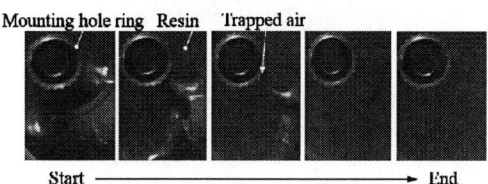

(b) Visualized sample of the area around the mounting hole

Fig. 6 Experiment of resin flow visualization

B) Resin flow visualization

In order to improve mass production quality and productivity, it needs to optimize the mold design and process conditions. This modulation needs to consider factors of the air vent operation and the flow and elimination behavior of bubbles.

Therefore, molding visualization is done by using

experimental glass plate as shown in Fig. 6. As a result of this, it is successful to avoid the air trap and welds by arranging an air vent, optimizing the gate shape and modulating the flow conditions including speed, temperature and pressure. For example, Fig. 6(b) illustrates the process in which the void generated by the air trapped near the mounting hole is pushed down to the inside of the ring and disappears. This corresponds with a result of two-phase flow simulation and we can confirm the behavior of air.

As described above, liquid-gas two-phase flow simulation is important for accurately understanding the mechanism of the void. These technologies are used to develop mold process design techniques that are applied to the mold design and molding conditions. At the result, it is able to achieve the All-SiC module in a full-mold structure.

V. HIGH RELIABILITY

In a power module, thermal stress is generated by temperature rise during device operation, which may cause breakage of a chip junction, etc. ΔT_j power cycling test is a reliability test for evaluating the lifespan of a power module by repeating this device operation.

Figure 7 shows a comparison of ΔT_j power cycle test lifetime. With the test starting temperature at 25°C, the figure plots the temperature amplitude ΔT_j along the horizontal axis and the number of cycles with a cumulative failure rate of 1% [F(t)=1%] along the vertical axis. The solid line shows the power cycling lifetime of the conventional package equipped with Si devices and the plot (○) is the lifetime verified with the new package equipped with Si devices. This result shows that, with the test condition ΔT_j=150 °C, the new package is expected to offer a lifetime more than ten times longer than that of the conventional package.

Accordingly, implantation and epitaxial metal oxide semiconductor (IEMOS), which is the SiC-MOSFET jointly developed with the National Institute of Advanced Industrial Science and Technology, has been mounted in the new package to conduct a power cycling test with ΔT_j=150 °C. The test result confirmed that 50,000 cycles with F(t) = 1%, a lifetime improved by more than 20 times from the conventional package equipped with Si devices, has been achieved [6] (see plot (●) in Figure 7). With the conventional package, breakage such as separation of the chip electrode from wire bonding occurs as the operation temperature increases, and this reduces the lifetime [7]. Meanwhile, the new package is encapsulated in epoxy resin with high heat resistance and breakage of junction between the chip electrode and copper pin is restrained by mitigating the thermal stress generated during operation. In addition, an epoxy resin with a glass transition temperature T_g of 200 °C or higher has been developed [4], which limits variations in mechanical and physical properties such as linear expansion coefficient and elastic modulus to a certain extent within the range of operating temperature and achieves high reliability.

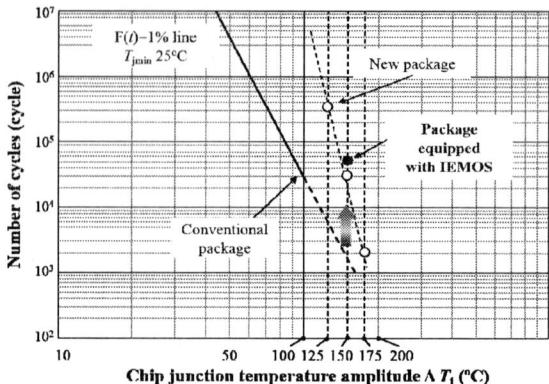

Fig. 7 ΔT_j power cycling test lifetime

VI. CONCLUSION

This paper describes the packaging technology of the All-SiC module. For the design of the All-SiC package, not only structure design but also mold process design using simulation and practical visualization of resin flow are important.

We continue to contribute to developing the power electronics technologies and realizing a low carbon society through developing small- to large-capacity modules and broadening their application to various power electronics.

REFFERENCES

[1] Nashida, N. et al. All-SiC Module for Mega-Solar Power Conditioner. FUJI ELECTRIC REVIEW. 2014, vol.60, no.4, p.214-218.

[2] Horio, M. et al. "New Power Module Structure with Low Thermal Impedance and High Reliability for SiC Devices". Proceedings of PCIM, 2011, p.229-234.

[3] Ikeda, Y. et al. "Investigation on Wirebond-less Power Module Structure with High-density Packaging and High Reliability". Proceedings of ISPSD. 2011, p.272-275.

[4] Horio, M. et al. "Ultra Compact and High Reliable SiC MOSFET Power Module with 200°C Operating Capability". Proceedings of ISPSD. 2012, p.81-84.

[5] Hinata, Y. et al. "Full SiC Power Module with Advanced Structure and its Solar Inverter Application". Proceedings of APEC. 2013, p.604-607

[6] Nashida, N. et al. "All-SiC Power Module for Photovoltaic Power Conditioner System". Proceedings of ISPSD 2014. p.342-345.

[7] Momose, F. et al. New Assembly Technologies for Tj max =175°C Continuous Operation Guaranty of IGBT Module. FUJI ELECTRIC REVIEW. 2013, vol.59, no.4, p.226-229.

A New Smallest 1200V Intelligent Power Module for Three Phase Motor Drives

Minsub Lee, Miran Baek, Junbae Lee, Daewoong Chung

R&D, Infineon Power Semitech, Seoul, Korea
*E-mail: Minsub.Lee@infineon.com, Miran.Baek@infineon.com,
Junbae.Lee@infineon.com, Daewoong.Chung@infineon.com

Abstract— **This paper presents the new smallest 1200V Intelligent Power Module (IPM) of 5A and 10A rating an excellent solution for three phase AC motors and permanent magnet motors in variable speed drives applications such as fan drives, active filter for HVAC and low power motor drives of GPI and Servo drives. This IPM integrates six IGBTs and six diodes with 6 channel Silicon On Insulator (SOI) gate driver in Dual in line (DIP) package with Direct Bond Copper (DBC). This paper provides an overall description regarding electrical characteristics, device performance and package.**

Keywords— *Infineon, CIPOS™, IPM, Three phase Motor drives, Inverter*

I. INTRODUCTION

The importance of energy saving with a regard to environmental issues has been grown bigger and the inverters are increasingly used for applications of a wide range. The new 1200V IPM has been designed and developed in a compact size package of Infineon Technologies with newly developed SOI single gate driver IC. This IPM is called Control Integrated Power System (CIPOS™) Maxi and it is optimized to improve efficiency, reliability and controllability within industrial motor drive applications. This paper describes the features of the internal components as well as the package structure and thermal performance.

II. FEATURES OF DESIGN AND ELECTRICAL PERFORMANCE

A. Overview and circuit configuration

Figures 1 and 2 show the package drawing and the internal block diagram of the CIPOS™ Maxi respectively. It has been designed for the smallest package size as 1200V IPM without dummy pin by optimized internal PCB for gate driver IC and optimized DBC for power devices. CIPOS™ Maxi is composed of 6 IGBTs, 6 diodes in a three phase inverter structure together with a new 1200V SOI single gate driver IC. Especially this single gate driver IC with integrated bootstrap circuit and several protection functions paves the way for minimization. The package development targets the smallest possible size which still meets international industrial standards for insulation distance such as

clearance and creepage. CIPOS™ Maxi devices are available with rated current of 5A and 10A, and each device names are listed in the Table I.

Fig. 1. Package drawing (size: 36mm x 22.7mm x 3.1mm)

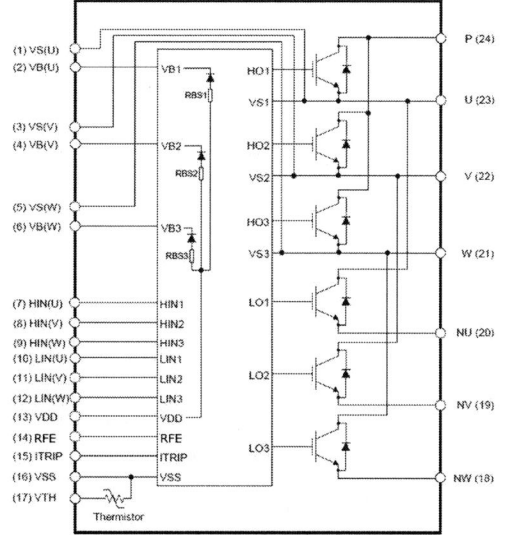

Fig. 2. Internal block diagram

TABLE I

Current rating	Device name	Package
5A	IM818-SCC	24pin, Dual in line
10A	IM818-MCC	

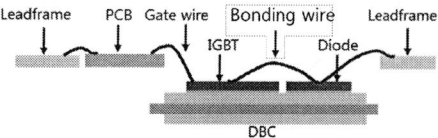

Fig. 3. Internal structure of CIPOS™ Maxi

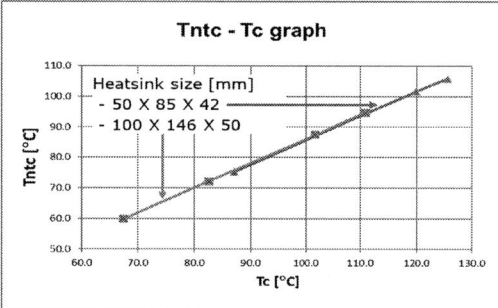

Fig. 4. Relationship between case and thermistor temperature

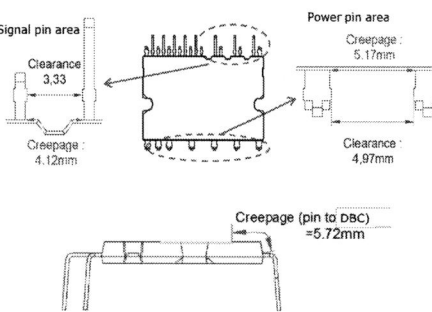

Fig. 5. Clearance and Creepage distance

B. Package

The new package for CIPOS™ Maxi is designed with smallest package size (36mm x 22.7mm x 3.1mm) without dummy pin by optimized internal PCB and DBC structure as shown in Figure 3. Gate driver IC and thermistor are placed on an internal PCB. The IGBTs and diodes are placed on a DBC for higher thermal performance. It adopts Al wire bonding technology for electrical connections between PCB and DBC as well as DBC and Lead-frame. The CIPOS™ Maxi has the independent V_{TH} pin and it is connected to thermistor inside the package, which offers the temperature monitoring function. The thermistor is a NTC type and the case temperature can be tracked by NTC temperature as shown in Figure 4. It is noted that this temperature relationship between NTC and case should be measured on the experimental environment because its relationship can be different according to user's heat dissipation conditions. Also, it is designed with transfer molding technology encapsulation of the internal body and meets all international industrial standards such as clearance and creepage for insulation distance. Figure 5 depicts the insulation distance of pin to pin and pin to DBC.

C. Features of new 1200V SOI single gate driver IC

The new 1200V SOI single gate driver with integrated bootstrap circuit is used to achieve higher levels of integration, reliability and performance in the CIPOS™ Maxi. This SOI gate driver IC disables leakage or latch-up current between adjacent devices structurally. It prevents the latch-up effect even in case of high dV/dt switching and surge under elevated temperature [1]. This gate driver provides several protection functions such as cross-conduction prevention, under voltage lock out, over current detection and enable input. Figure 6 shows the characteristics of the integrated bootstrap circuit.

CIPOS™ Maxi provides an integrated fault output with sleep function and adjustable fault clear time by the RFE pin. There are two situations that would cause the driver IC to report a fault via the RFE pin. The first is an under voltage condition of V_{DD} and the second is if the ITRIP function recognizes a fault. Once the fault condition occurs, the RFE pin is internally pulled to V_{SS} and the fault clear timer is activated; all outputs of this driver are shut down. The RFE output stays in the low state until the fault condition has been removed and the built-in fault clear timer expires which is set to a minimum of 100µs. Once the built-in fault clear timer expires, the voltage on the RFE pin will return to its external pull-up voltage. The output will remain disabled and the fault condition maintained until the voltage on the RFE pin charges up to enable threshold voltage. The charging characteristics is dictated by RC time constant attached to the RFE pin. Figure 7 shows that R_{ext} is connected between the external supply and the RFE pin, while C_{ext} is placed between the RFE and VSS pins. The fault clear time is determined by the charging characteristics of the capacitor where the time constant is set by R_{ext} and C_{ext}.

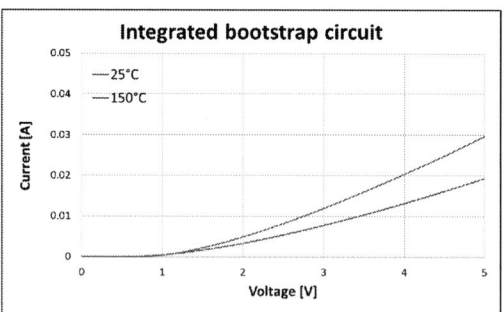

Fig. 6. Bootstrap circuit characteristics

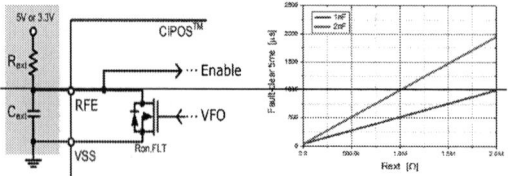

Fig. 7. Circuit for fault-clear time and Adjustable fault-clear time

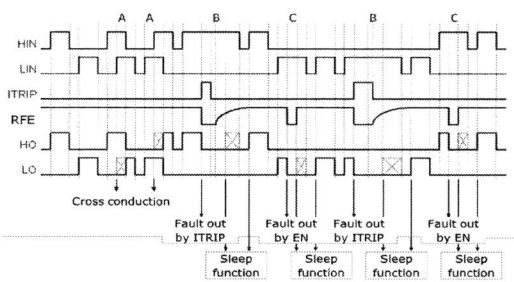

Fig. 8. Timing chart

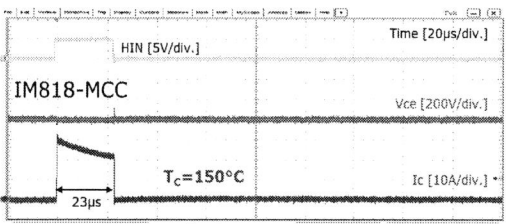

Fig. 9. Short circuit test result of IM818-MCC

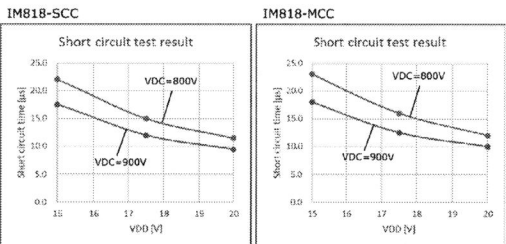

Fig. 10. Short circuit withstand time results under worse condition

Figure 8 illustrates that how the gate driver IC operates to protect device from several abnormal situations by the timing chart regarding protection function such as cross-conduction prevention and sleep function. During interval A, the gate driver received the command to turn-on both the high side and low side switches at the same time, the shoot-through protection has prevented this condition. It is keeping on output channel that is already on ignoring the 2nd input signal. Interval B and C shows the sleep function to protect destruction of IGBTs from unexpected and incorrect operation of the micro controller. [2] After the fault clear time, the driver IC is waiting for a new input signal on LIN/HIN before activating the output stage (LO/HO).

D. Performance

CIPOS™ Maxi offers enhanced operating conditions compared to existing 1200V IPMs such as shortest dead time and shortest input pulse width. Recommend minimum dead time is 0.5µs and minimum pulse width is 1µs.

Short circuit capability is guaranteed up to 10µs under condition of V_{DC}=800V, V_{DD}=15V and $T_J \leq 150$°C. Figure 9 is measured test waveform of an IM818-MCC under condition of V_{DC}=800V, V_{DD}=15V and T_C=150°C. Figure 10 records actual test results regarding short circuit withstand time without any damage under worse conditions such as higher V_{DC} and V_{DD}.

Figure 11 shows allowable maximum operating current by case temperature (T_C) and Figure 12 shows allowable maximum operating current as a function of PWM carrier frequency (f_{SW}) under condition of V_{DC}=600V, V_{DD}=15V, Power Factor (PF)=0.8, Modulation Index(MI)=0.8, Output frequency(FO)=60Hz and SVPWM.

These are simulated by CIPOSIM, a simulation software to estimate losses, junction temperature and allowable maximum operating conditions, provided by Infineon Technologies.

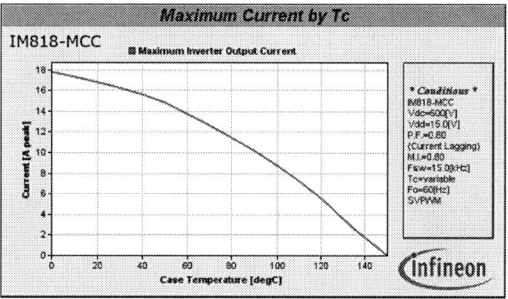

Fig. 11. Allowable maximum current by T_C (V_{DC}=600V, V_{DD}=15V, PF=0.8, MI=0.8, f_{SW}=15 kHz, FO=60Hz, SVPWM)

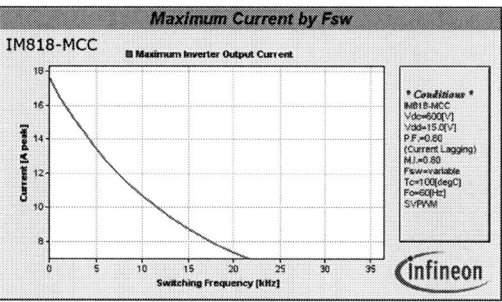

Fig. 12. Allowable maximum current by f_{SW} (V_{DC}=600V, V_{DD}=15V, PF=0.8, MI=0.8, T_C=80°C, FO=60Hz, SVPWM)

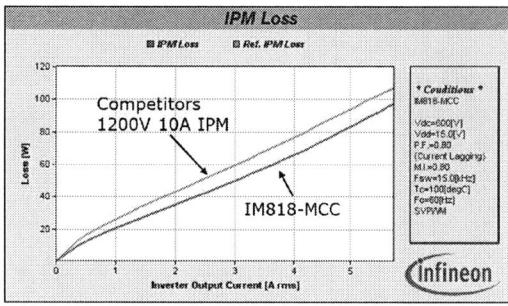

Fig. 13. Inverter loss (V_{DC}=600V, V_{DD}=15V, PF=0.8, MI=0.8, T_C=100°C, f_{SW}=15kHz, FO=60Hz, SVPWM)

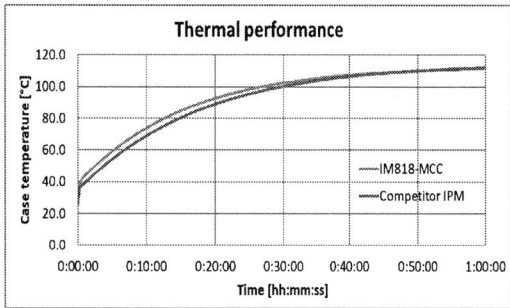

Fig. 14. Thermal performance results (V_{DC}=600V, V_{DD}=15V, PF=1, MI=0.46, Ta=25°C, f_{SW}=15kHz, FO=60Hz, SVPWM)

Figure 13 shows the inverter losses of IM818-MCC compared with competitors 1200V 10A rating IPM under condition of V_{DC}=600V, V_{DD}=15V, PF=0.8, MI=0.8, T_C=100°C, f_{SW}=15kHz, FO=60Hz and SVPWM. IM818-MCC has slightly higher performance at 5Apeak. Figure 14 records evaluation results at using the same heatsink and IM818-MCC shows almost similar thermal performance at the same operating conditions in spite of the package has about half size of the compared 10A IPM.

III. CONCLUSIONS

The new CIPOS™ Maxi is developed with smallest package compared with existing 1200V IPMs. It is offers the chance for costs saving and miniaturization with improved safety and simplified board design for the inverter system.

REFERENCES

[1] R. Keggenhoff, Z.Liang, A. Arens, P. Kanschat, R. Rudolf. "Novel SOI Driver for Low Power Drive Applications", *Power Systems Design Europe, Nov. 2005.*

[2] Taehyun Kim, Minsub Lee, Junbae Lee "Protection Features of Intelligent Power Module against Transient State" *PCIM Europe, 2016*

The 2018 International Power Electronics Conference

Design and Enhancement of ESD Reliability in Circular UHV 300-V nLDMOS Power Components

Shen-Li Chen[1*], Yi-Hao Chao[1], Chih-Ying Yen[1], Jen-Hao Lo[2], Chun-Ting Kuo[2], Yu-Lin Lin[1], Yi-Hao Chiu[1], Pei-Lin Wu[1] and Yu-Lin[1] Jhou

1 Dept. of Electronic Engineering, National United University, Miaoli City, Taiwan
2 Dept. of Integrated Circuits Design and Engineering, Peking University, Wuxi City, China
*E-mail: jackchen@nuu.edu.tw

Abstract—In this paper, parasitic silicon controlled rectifiers (SCRs) in both the drift and drain regions of 0.5-µm 300-V circular ultra-high-voltage n-channel laterally-diffused MOSFETs (UHV-nLDMOSs) were modulated to improve the resistance of the LDMOSs to electrostatic discharge (ESD). Results from a semiconductor curve tracer and transmission-line pulse system behaved that the LDMOSs exhibited normal characteristic curves and the breakdown voltage of the devices increased after their drift-region width was adjusted. This indicates improvements in the ESD resistance of the LDMOSs. Moreover, the secondary breakdown voltage (I_{t2}) of nLDMOSs, after the addition of a central parasitic SCR to their drain center, rose from 2.884 to 3.24 A (12.34%) relative to the reference nLDMOS. The I_{t2} of nLDMOSs with a halved SCR—which reduces the on-resistance path and on-resistance—increased by 62.5% to 4.686 A. The I_{t2} of nLDMOSs with a halved-discrete SCR increased by 58.7% to a respectable 4.577 A, even though the devices had smaller oxide-defined (OD) regions and fewer contact holes. Notably, the I_{t2} of nLDMOSs with a purely central SCR obtained by removing n^+ from their drain regions increased by 57.94% to 4.555 A, comparing favorably to those with parasitic SCRs. The trigger voltage (V_{t1}) of nLDMOSs with parasitic SCRs, which had low on-resistance, was slightly lower than that of the reference nLDMOS. The V_{t1} of LDMOSs with a purely halved SCR was the lowest. The holding voltage (V_h) of nLDMOSs with parasitic SCRs in their drain regions declined by a mere 4.5% compared with the reference nLDMOS.

Keywords—*Drift region, Electrostatic Discharge (ESD), Holding voltage (V_h), Laterally-diffused metal oxide semiconductor (LDMOS), Secondary breakdown current (I_{t2}), Silicon Controlled Rectifier(SCR), Trigger voltage (V_{t1}), Ultra high-voltage(UHV).*

I. INTRODUCTION

The ultra-high-voltage laterally-diffused MOSFET (UHV-LDMOS) has been widely applied in power management circuits, automotive electronics, and integrated circuits for power electronics [1]–[3]. Even this transistor may be capable of operating under high drain voltage, however it provides lower resistance to electrostatic discharge (ESD) than low- and medium-voltage complementary metal–oxide–semiconductor

circuits do. Therefore, the ESD resistance of this UHV device must be improved.

Generally, n-channel LDMOSs (nLDMOSs) are used in input/output ports, and because of their large size, they may function as ESD protection devices in the same time. However, with their low ESD resistance, UHV-LDMOSs are limited in some aspects. For example, if the protection device is not actuated when the V_{t1} becomes excessively high, their protected or protection circuits are damaged; or if the protection device is actuated, it is promptly damaged by an excess concentration of electric currents. In addition, the ESD resistance of UHV-LDMOSs remains underexplored [4]–[8]. Therefore, this paper investigated UHV-nLDMOSs and proposes a method for improving their ESD resistance. The method involves altering the implanted profile in the drain end of the LDMOSs to construct a parasitic silicon controlled rectifier (SCR) and measuring the ESD resistance of the devices in relation to different layout profiles.

II. LAYOUTS IN CIRCULAR UHV 300-V NLDMOS SAMPLES

A. Reference sample of the circular 300-V nLDMOS

As the illustrations of the structure and the layout of circular 300-V UHV-nLDMOSs suggest [Fig. 1 (a) and (b)], because the devices were operated under high voltage, the upper area of the drain end was covered by a poly-2 layer and the lower area by expansive layers of HV N-Well and HV Deep PW, which allowed the breakdown voltage to increase. After the characteristic curve of the devices returned to normality, the gate end was connected to a ground electrode, thus constructing gate-grounded nMOSFETs. Then, such devices discharge external surge currents by dint of parasitic bipolar junction transistors. All tested devices were fabricated using a 0.5 µm manufacturing process developed by the Taiwan Semiconductor Manufacturing Company and had a channel length of 2.5 µm, a channel width of 394.3 µm, and a drift-region (DCG) width of 29 µm.

1145

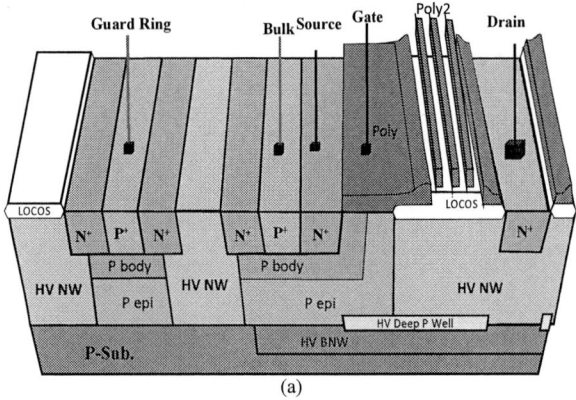

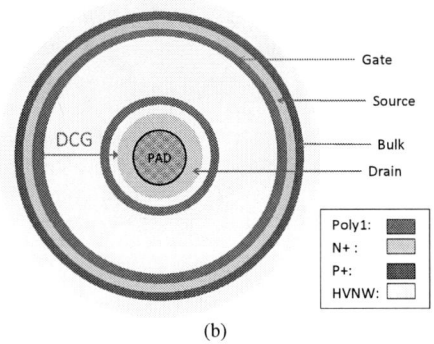

Fig. 1. (a) Device structure, and (b) layout diagram of a 300 V circular nLDMOS.

B. Drift-region modulation in the circular 300-V nLDMOS

As the illustrations of the structure and layout of the modulated DCG of the UHV-nLDMOS indicate [Fig. 1 (a) and (b)], the on-resistance (R_{on}) of the device increased when the DCG expanded. This improved the impedance of the device, enabling it to withstand higher voltages and increasing its breakdown voltage. With an understanding of the relationship between the breakdown and operating voltages, LDMOSs can be reengineered so that they can function under the required operating voltages.

C. Embedded SCR modulation in the circular 300-V nLDMOS

Fig. 2 depicts the layout of a parasitic SCR in the drain end of the UHV-nLDMOS. For the embedded SCR (Fig. 2), the left-hand diagram displays a P^+ region that constitutes one- to two-thirds of the radius of the drain, and the center and outer area of the drain are N^+. Viewed from the center of the drain region, therefore it defined as the "halved SCR," as shown in the left-hand panel of Fig. 2. In the right-hand diagram of the figure is a halved SCR with the P^+ OD region discretely distributed (this SCR is accordingly referred to as the "halved-discrete SCR"), with each OD block encapsulated by a contact hole in accordance with the smallest rule; each discrete OD region measures 1 μm × 1 μm and each contact hole 0.5

μm × 0.5 μm. Three alignments of discrete P^+ OD regions appeared along the outer edge of the radius of the drain center. Although the parasitic SCR structure facilitates the release of ESD from LDMOSs, it caused the holding voltage to decrease, easily leading to the latch-up effect. To address this limitation, in the center of the drain end of the LDMOS, a P^+ region was added whose area accounts for one-third of the oxide-defined (OD) radius of the drain. As such, the parasitic SCR in the device was the "central SCR". Moreover, N^+ was removed from the left-hand panels of Figs. 2 and 3 to yield the "purely central-SCR" and the "purely halved-SCR", as shown respectively in the left- and right-hand panels of Fig. 4. Table I lists all of the testing devices, whose ESD resistance was determined through a transmission-line pulse (TLP) system and subsequently compared.

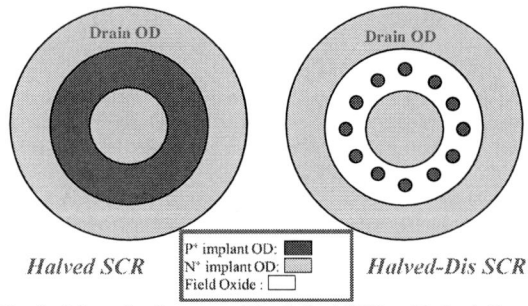

Fig. 2. Schematic diagrams of the halved SCR and halved-discrete SCR in the drain end of a UHV-nLDMOS.

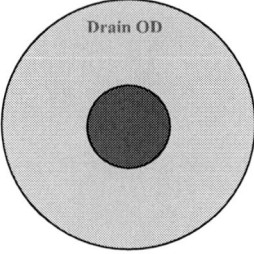

Fig. 3. Layout diagram of the central parasitic SCR in the drain end of a UHV-nLDMOS.

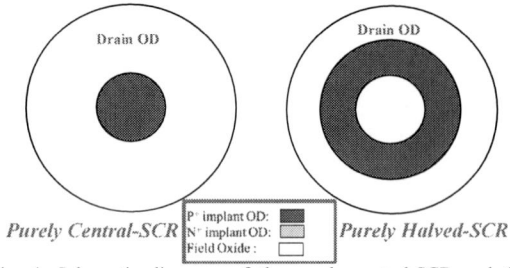

Fig. 4. Schematic diagrams of the purely central-SCR and the purely halved-SCR.

TABLE I
ACRONYMS OF THE UHV 300 V NLDMOS DEVICES

Acronyms	Cell name
Halved	Parasitic Halved-SCR
H_Dis	Parasitic Halved-Discrete SCR

1146

SCR	Central Parasitic SCR
Pure	Purely Central-SCR
P. Halved	Purely Halved-SCR

III. EXPERIMENTAL RESULTS AND DISCUSSION

A. Drift region modulation in the circular 300-V nLDMOS

Fig. 5 and Table II respectively present the results of the breakdown voltage and HBM ESD tests on UHV-nLDMOS samples whose DCG width was changed. LDMOSs with independent gates (parameters also shown in Fig. 5 and Table II) had a normal characteristic curve. Moreover, the breakdown voltage of the gate-grounded nMOSFETs increased as expected after the DCG width of the sample was expanded, because this expansion allowed the breakdown voltage to increase, thereby improving the breakdown voltage. Table II shows the HBM capability of UHV-nLDMOSs with DCG widths of 17 μm, 29 μm, and 42 μm. The average HBM (+) capability of the reference sample (Ref. DUT) was 5250 V, and its average HBM (−) capability was >8000 V because it was an equivalent forward diode.

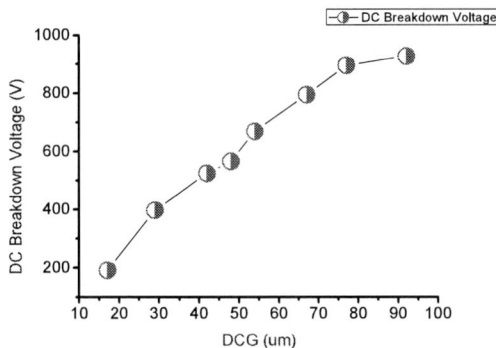

Fig. 5. DC breakdown values of circular 300 V nLDMOS DUTs as DCG changed.

TABLE II
ESD MEASUREMENT VALUES OF CIRCULAR 300 V NLDMOS DUTs AS DCG CHANGED

DCG	HBM1(+)(V)	HBM2(+)(V)	HBM(-)(V)	V_{BK}(V)
D-17	> 8000	> 8000	> 8000	191.1
D-29 (ref.)	5000	5500	> 8000	398.28
D-42	7000	7000	> 8000	524.42

B. Embedded SCR modulation in the circular 300-V nLDMOS

Fig. 6 depicts the snapback voltage–current curve of the modulated parasitic SCR in the nLDMOS drain region, Table III states all crucial measurement parameters. And, Fig. 7 presents changes in the secondary breakdown current (I_{t2}). The nLDMOS samples with a halved SCR had a shortened SCR path and reduced on-resistance, and their I_{t2} increased to 4.686A (a 62.5% increase over the reference nLDMOS

DUT). The capability of nLDMOS samples with a halved-discrete SCR (I_{t2} = 4.577 A) to discharge ESD currents did not exhibit any significant improvement (as compared with the Halved SCR DUT) because their total OD area and the number of contact holes were smaller than those of the nLDMOSs with a halved SCR. Next, the I_{t2} of nLDMOSs, after the addition of a parasitic SCR to their drain center, rose from 2.884 to 3.24 A in comparison to the reference nLDMOS. By contrast, the I_{t2} of the purely halved SCR was a mere 3.317 A, because its short current path caused series resistance to decline.

Fig. 8 shows changes in the trigger and holding voltages of SCRs. Because SCRs have low on-resistance, their trigger voltage (V_{t1}) is slightly lower than that of the reference nLDMOS [9]. Purely halved and central and SCRs, without an LDMOS, had low levels of V_{t1}; in particular, a halved SCR had a lower V_{t1}. A purely central SCR had higher anode series resistance and, therefore, higher V_h. No noticeable changes were found in the V_h of the other SCRs. Overall, the lowest reduction in V_h among the nLDMOSs with parasitic SCRs in the drain region was 4.5%, in relation to the reference nLDMOS. Fig. 9 presents changes the breakdown voltage of the nLDMOSs. Notably, the breakdown voltage of all nLDMOS samples, including the reference nLDMOS, exceeded 392 V, 1.31 times the safe operating voltage. Thus, all the nLDMOSs with parasitic SCRs operated safely.

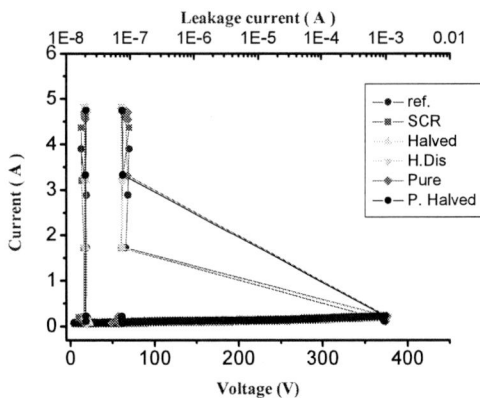

Fig. 6. Leakage and snapback I-V curves of circular 300 V nLDMOS related DUTs.

TABLE III
ELECTRICAL MEASUREMENT VALUES OF CIRCULAR UHV 300 V NLDMOS RELATED DUTs

Circular 300 V		V_{t1} (V)	V_h (V)	I_{t2} (A)	V_{BK} (V)
Ref._nLDMOS		374.751	62.831	2.884±1.506	398.284
SCR	Central SCR	372.735	63.071	3.24±1.535	400.201
	Halved	369.785	60.467	4.686±0.202	399.063
	Halved_Dis	364.738	60.204	4.577±0.247	398.080
	Pure	362.827	71.094	4.555±0.194	397.600
	Pure Halved	351.266	60.000	3.317±0.012	392.340

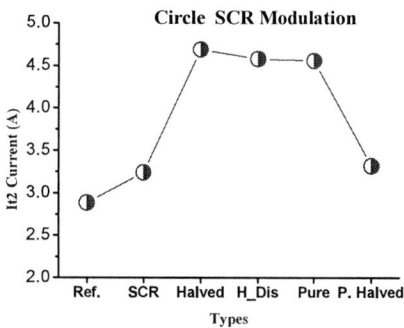

Fig. 7. I_{t2} values of circular 300 V nLDMOS related DUTs.

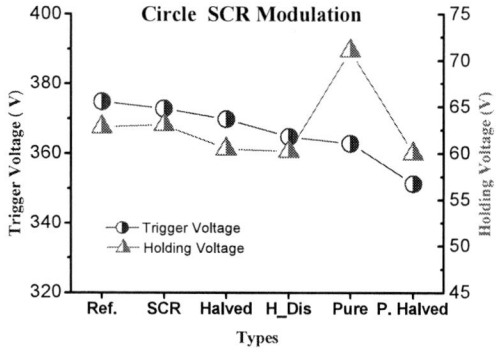

Fig. 8. V_{t1} & V_h values of circular 300 V nLDMOS related DUTs.

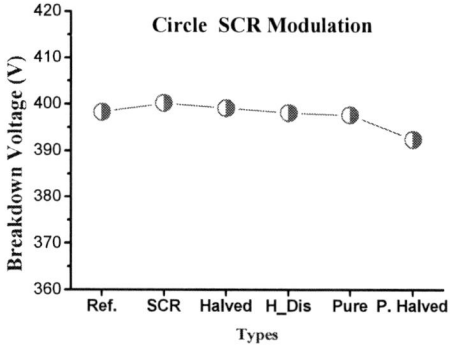

Fig. 9. V_{bk} values of circular 300 V nLDMOS related DUTs.

IV. CONCLUSIONS

After the DCG widths of the circular UHV-nLDMOSs were adjusted, their R_{on} increased, allowing their breakdown voltage to rise proportionally. Thus, the analysis of the relationship between the breakdown and operating voltages suggested that circular UHV-nLDMOSs may operate under safe operating voltages when their DCG width undergoes adjustments.

When a parasitic SCR added to the drain end of an UHV-nLDMOS was converted into a halved parasitic one, the I_{t2} of the LDMOS increased from 3.24 to 4.686 A, a 62.5% improvement over the reference nLDMOS. Even though the same modification was applied to an UHV-nLDMOS with a halved-discrete SCR, the I_{t2} of the

LDMOS rose to a respectable 4.577 A because its total OD area was smaller and it had fewer contact holes. Moreover, the I_{t2} of an UHV-nLDMOS with a purely central SCR was higher than that of LDMOSs with central parasitic SCRs. However, it declined after the purely central SCR was halved. The V_{t1} of LDMOSs dropped slightly below that of the reference LDMOS when they were retrofitted with SCRs that had lower on-resistance. The V_{t1} of LDMOSs with a purely halved SCR was the lowest. When the n^+ OD region was removed from the drain end of nLDMOSs, leaving the center of the purely central SCR, the SCR had higher anode series resistance and higher V_h, and the V_h of the other SCRs exhibited no significant change. The breakdown voltage of all the nLDMOSs tested exceeded 392 V, which is 1.31 times the safe operating voltage, and suggests an adequately safe UHV manufacturing process.

ACKNOWLEDGMENT

In this work, authors would like to thank the National Chip Implementation Center in Taiwan for providing the process information and fabrication platform. And, authors would like to acknowledge the financial support of the Ministry of Science & Technology of Taiwan, through grant number MOST 106-2221-E-239-018.

REFERENCES

[1] H. Chang, J.-J. Jang, M.-H. Kim, E.-K. Lee, D.-E. Jang, J.-S. Park, J.-H. Jung, C.-J. Yoon, S.-R. Bae, C.-H. Park, "Advanced 0.13um smart power technology from 7V to 70V," in *24th International Symposium on Power Semiconductor Devices and ICs*, 2012, pp. 217 - 220.

[2] K. Ko, J. Park, J. Eum, K. Lee, S. Lee, J. Lee, "Proposal of 0.13um new structure LDMOS for automotive PMIC," in *73rd Annual Device Research Conference (DRC)*, 2015, pp. 119-120.

[3] S. Zhou, Y. Song, K. Chien, C. Chen, "Study of safe operating area and improvement for power management integrated circuit," in *China Semiconductor Technology International Conference (CSTIC)*, 2017, pp. 1-3.

[4] J.-H. Lee, T.-C. Kao, C.-L. Chan, J.-L. Su, H.-D. Su, K.-C. Chang, "The ESD failure mechanism of ultra-HV 700V LDMOS," in *IEEE 23rd International Symposium on Power Semiconductor Devices and ICs*, 2011, pp. 188-191.

[5] J.-R. Tsai, Y.-M. Lee, M.-C. Tsai, G. Sheu, S.-M. Yang, "Development of ESD robustness enhancement of a novel 800V LDMOS multiple RESURF with linear P-top rings," in *2011 IEEE Region 10 Conference*, 2011, pp. 760 - 763.

[6] C.-H. Wu, J.-H. Lee, C. Lien, "A Novel Drain Design for ESD Improvement of UHV-LDMOS," *IEEE Trans. on Electron Devices*, vol. 52, no.12, pp. 4135 - 4138, 2015.

[7] Z.Chen, A. Salman, G. Mathur, G. Boselli, "Design and optimization on ESD self-protection schemes for 700VLDMOS in high voltage power IC," in *37th Electrical Overstress/Electrostatic Discharge Symposium (EOS/ESD)*, 2015, pp. 1-6.

[8] S.Kim, D. LaFonteese, D. Zhu, D.S. Sridhar, S. Pendharkar, H. Endoh, K. Boku, "A new ESD self-protection structure for 700V high side gate drive IC," in *29th International Symposium on Power Semiconductor Devices and IC's (ISPSD)*, 2017, pp. 467 - 470.

[9] S.-L. Chen, C.-Y. Yen, C.-H. Yang, Y.-C. Wu, K.-J. Chen, Y.-L. Lin, Y.-H. Chiu, Y.-H. Chao, H.-W. Chen, D. Chen, M. Lo, J.-M. Lin, C.-T. Kuo, J.-H. Lo, "ESD-Immunity Evaluations of a 40 V nLDMOS with Embedded SCRs in the Drain Side," in *IEEE International Symposium on Next-Generation Electronics*, Keelung, Taiwan, May 2017, pp. 1-2.

The 2018 International Power Electronics Conference

A Technology Analysis of Voltage Sharing in Series Connected Power Devices

Z Davletzhanova[1*], O Alatise[1], R Bonyadi[1], J Ortiz-Gonzalez[1], T Dai[1], M Jennings[1], L Ran[1] and P Mawby[1]

1 School of Engineering, University of Warwick, Coventry, UK

*E-mail: z.davletzhanova@warwick.ac.uk

Abstract— Series connected power devices are required for voltage sharing in high voltage applications like grid connected converters. With SiC considered as a strong contender for grid applications, the performance and reliability issues associated with voltage sharing compared to contemporary silicon bipolar devices is important to consider. In applications where series power devices may be at different junction temperatures as a result of the physical architecture of the converter cooling system or differential degradation of the packaging, the zero-temperature coefficient of the power devices determines the voltage sharing and loss distribution in the ON-state while the leakage current and switching synchronization is critical in the OFF-state. In the ON-state, the lower zero-temperature-coefficient (ZTC) point in SiC devices contributes to increasing voltage divergence with the higher thermal resistance device increasingly dissipating more power. In this case, the higher ZTC point in silicon bipolar devices is an advantage although it is a disadvantage for paralleling. Due to the absence of tail currents in SiC devices, they exhibit less voltage divergence during OFF-state transient. The different operating conditions between the series connected devices also determines the voltage sharing during the transient switching of the devices. Using finite-element and experimental measurements, this paper analyzes the technology dependence of voltage sharing in series devices during on-state, off-state and switching.

Keywords— *series connected, SiC/Si power devices, switching mismatch, temperature impact*

I. INTRODUCTION

The increasing role of power electronics in the transmission and distribution system in the near future is an important consideration for converter designers. Power electronics first entered the transmission system as series connected thyristor valve stacks in line commutated current source converter systems. When self-commutated voltage source converters emerged as an alternative for HVDC transmission, series connected IGBTs where used for voltage sharing in the OFF-state with snubbers deployed for voltage sharing and controlling voltage imbalance during switching [1]. However, the emergence of multi-level converters, particularly, modular multilevel converters, obviated the need for series connected of power devices since the IGBTs are required for blocking only the sub-module voltage. Since MMC submodule voltages are lower than the maximum rating of commercially available IGBTs

(6.5 kV IGBTs are commercially available), then the need for series connection of power devices diminished for HVDC systems. However, in the distribution system where power electronics is expected to implement medium and low voltage DC (MVDC and LVDC) networks as well as distribution system FACTS technologies [2], there is a lot of flexibility in converter design. In converters designed for this application, series connected power devices are still being used in tandem with multilevel topologies. For example, IXYS have commercialized 3.3 kV, 6.6 kV and 10 kV 3 level NPC converters implemented in series connected press-pack IGBTs. In this converter, each switching unit is comprised of 2 or 3 series connected press-pack IGBTs [3]. Furthermore, the alternate arm converter, a variant of the MMC, requires series connected power devices. Voltage sharing in these applications is important for analysis of loss distribution and more accurate lifetime estimation.

The blocking voltage and conduction current capability of SiC power devices are increasing. In parallel connected devices, current sharing is critical and this has been investigated for SiC power devices [4-6]. In applications where they are series connected, it is important to understand the voltage sharing characteristics in the ON-state, the OFF-state and during switching [8, 9, 10]. Voltage sharing in the ON-state is important for loss distribution and lifetime estimation while voltage sharing in the OFF state is important for switching loss evaluation. This paper investigates voltage sharing in series connected devices for SiC unipolar devices and silicon bipolar devices through experiments, finite element modelling. Section II discusses the importance of zero temperature coefficient (ZTC) in the ON-state voltage sharing. Section III discusses OFF-state voltage balancing and the impact of leakage current while section IV shows the voltage sharing during the switching transient for different technologies and Section V concludes the paper.

II. ON-STATE VOLTAGE SHARING OF SERIES CONNECTED POWER DEVICES

The ZTC point in the device forward characteristics is the load current at which the forward voltage of the device is temperature invariant. At lower currents, the forward voltage reduces with temperature while at higher currents the forward voltage increases with temperature.

1149

The ZTC arises because there are two competing mechanisms with opposing thermal effects on the power device. As the temperature is increased, the intrinsic carrier concentration increases which tends to lower diode junction voltages and transistor threshold voltages. However, the effective mobility reduces with increased temperature due to increased phonon scattering, hence, the resistivity of the semiconductor increases. At low currents, the junction/threshold voltage dominates whereas at high currents, the voltage drops across parasitic resistances dominates. The ZTC point is the load current at which the two effects are equal and opposite i.e. counter balanced. In transistors, increasing junction temperature will reduce the threshold voltage through a reduction of the surface potential. However, as the load current is increased, the ohmic drop across the series parasitic resistances (drift layer resistance, contact resistance and substrate resistances) increases. Hence, at higher current levels, the forward voltage increases with temperature. SiC unipolar devices generally have lower ZTC points compared to silicon bipolar devices. There are two reasons behind this. First, bipolar devices rely on minority carrier diffusion, hence, minority carrier lifetime is an important factor in ON-state characteristics. Since minority carrier lifetime increases with temperature, carrier density increases, and ON-state conductivity increases with temperature. Secondly, SiC has a wider bandgap than silicon (3.24 eV vs 1.12 eV), hence, at any given temperature the intrinsic carrier concentration of SiC will be lower than silicon since it is inversely related to the energy bandgap. As a result, SiC is more immune to temperature dependent changes in the ON-state resistance below the ZTC point. One of the main objectives of this paper is to investigate how these material property differences impacts voltage sharing in series connected silicon bipolar and SiC unipolar devices.

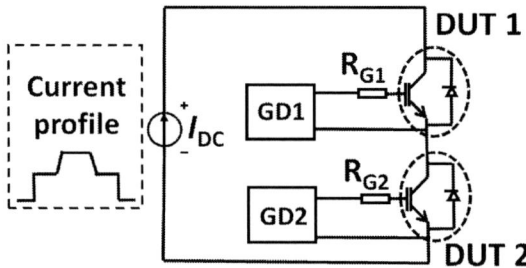

Fig.1. Circuit schematic

Measurements of series connected power devices have been performed using the circuit schematic is shown in Fig. 1. The devices under investigation are 600 V/20A Infineon Field Stop IGBTs with datasheet reference IKW20N60H3 and 650 V/39A ROHM Trench SiC power MOSFET with datasheet reference SCT3060AL.

A. *Experimental and Simulation Results of ON-state voltage sharing*

In order to investigate the impact of temperature and the operating points below and above the ZTC point during on-state, an electro-thermal model was developed and validated through experiment.

During the experiments, two power devices were connected in series and a trapezoidal current was passed through them for duration of 150 seconds using a current source. The voltage across each device was measured on an oscilloscope. Thermal variation between the devices was introduced by varying the sizes of the heatsink connected to both devices. This way, both devices were self-heated through conduction losses, however, the device with the smaller heatsink would exhibit higher junction/case temperatures. By varying the current level fed to the series connected devices, the voltage sharing between the devices below and above ZTC was investigated. The circuit schematic is shown in Fig. 1.

Electro-thermal variation is introduced between the series connected devices by using different size heatsinks. This is valid as the unbalanced electro-thermal degradation due to unbalanced voltage sharing leads to different junction temperatures between the two devices. The DUT with the smaller heatsink will therefore have a higher thermal resistance and a higher case temperature. During the experiment, the devices were thermally isolated. This was deliberately introduced to decouple the impact of the lateral heat transfer and emphasize only on the voltage unbalanced purely caused by the device degradation. The trapezoidal current profile used in the experiment is selected to investigate the impact of operation below and above the ZTC point on the voltage sharing during the ON-state operation. Under normal operation of an inverter, the current waveform is usually sinusoidal which is achieved by applying a PWM signal to the gates of the switches. The switches carry a certain DC current during each ON-state period which can be either below or above the ZTC point. Hence, the voltage sharing between the series connected devices can vary based on the DC value of current. Fig. 2 shows the experimental (Fig. 2(a) and 2(c)) and the simulation results (Fig. 2(b) and 2(d)) of the ON-state voltage sharing between the two series connected IGBT and MOSFETs respectively. When the current is below the ZTC point in IGBTs, the hotter device has a lower V_{ce} and when the device is operating above the ZTC, this is inversed. This was not evident in case of SiC MOSFET due to the very low ZTC of this device.

B. *Electrothermal model development*

Furthermore, a compact electrothermal model of two series Si IGBT and SiC MOSFET was developed in Matlab/Simulink. The model consists of 2 series connected devices with variable junction to case thermal impedances (Z_{TH}). The model comprises of a current source which supplies a load current mission profile. The temperature-dependent forward characteristics of the devices were obtained from static temperature measurements and validated against the datasheets.

Next, the conduction power losses of the two series devices were fed to the Cauer-thermal network of each of the devices. The junction temperature obtained from the Cauer-thermal network was fed back to the forward characteristic lookup table to obtain the correct temperature dependent on-state losses for the next step of the simulation.

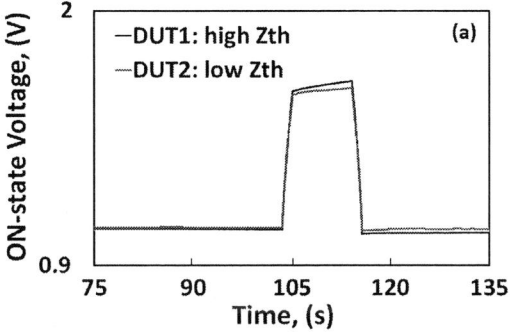

Fig. 2(a). Experimental result of ON-state voltage sharing above/below ZTC for Si IGBT

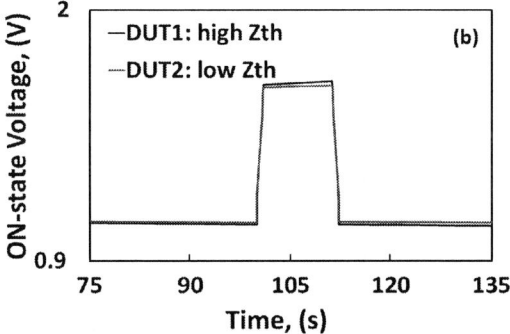

Fig. 2(b). Compact model-based simulation result of ON-state voltage sharing above/below ZTC for Si IGBT

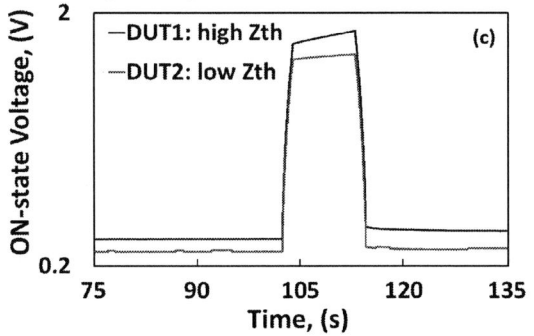

Fig. 2(c). Experimental result of ON-state voltage sharing above/below ZTC for SiC MOSFET

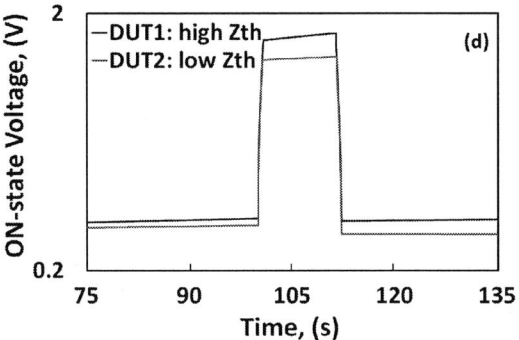

Fig. 2(d). Compact model-based simulation result of ON-state voltage sharing above/below ZTC for SiC MOSFET

In order to obtain the Cauer-thermal network, the transient thermal impedance of the TO-247 (for transistors) packages of devices were obtained through the datasheet and a finite number of elements were chosen on the curve. A rational curve fitting tool was used with specified boundaries based on the device geometry to obtain the coefficients of the rational equation shown below. Equation (1) presents the reconstruction of the Z_{TH} using rational curve fitting. In [7], details are given of how the coefficients of the transfer function are calculated.

$$Z_{th} = \frac{p_1 s^4 + p_2 s^3 + p_3 s^2 + p_4 s + p_5}{s^5 + q_1 s^4 + q_2 s^3 + q_3 s^2 + q_4 s^1 + q_5} \quad (1)$$

Next, the transpose of the equation was used to calculate the thermal resistance and thermal capacitances of each layer of Cauer-thermal network. Table 1 below shows the values of the thermal resistances and capacitances used in the Cauer Network. The values were validated through experimental results to have a match between the case temperature of the devices when a trapezoidal current with a certain duration was applied to the device and the case temperature was measured using a thermocouple attached to the back plate of the device.

TABLE I

CALCULATED THERMAL RESISTANCES AND THERMAL CAPACITANCES

	Si IGBT	SiC MOSFET
R_1	0.0101	0.1164
R_2	0.1832	0.0832
R_3	0.2853	0.2708
R_4	0.1655	0.2845
R_5	0.0121	0.2137
C_1	2.49×10^{-2}	3.34×10^{-4}
C_2	6.07×10^{-4}	9.08×10^{-4}
C_3	1.04×10^{-2}	7.88×10^{-4}
C_4	16.55×10^{-2}	5.28×10^{-3}
C_5	0.01198	0.05874

III. OFF-STATE VOLTAGE SHARING

The differences in electrothermal properties of devices may cause variation in OFF-state voltage blocking capability of series connected devices. The physical architecture of cooling systems in series connected power devices may cause some inevitable temperature variation between series connected power modules, hence, differences in temperature induced leakage current may cause OFF-state voltage divergence. The leakage current is primarily due to carriers generated in the depletion region of the voltage blocking reverse biased PN junction.

For series connected IGBTs with different levels of leakage current, the device with the higher leakage current sets the overall leakage current, hence, the device with the lower leakage current needs to adjust its internal electric field distribution to supply the carriers required to maintain the overall level of leakage. This means that the device with the lower leakage current supplies a greater than normal number of carriers to maintain the flow of leakage current, hence, the device becomes more depleted thereby blocking a higher magnitude of OFF-state voltage.

The temperature dependent intrinsic carrier concentration, space-charge generation lifetime and the gain of the PNP BJT within the IGBT determines the blocking voltage given a defined leakage current. The IGBT with the higher junction temperature has a higher intrinsic carrier concentration and thus, a lower blocking voltage. Fig. 3(a) shows the OFF-state voltage sharing between two series connected IGBTs with two different junction temperatures (85 °C and 135 °C). As can be seen, the device with higher junction temperature has lower blocking voltage. Fig. 3(b) compares the voltage sharing of the series connected SiC MOSFETs at two different junction temperatures. As can be seen, SiC MOSFET is less sensitive to the temperature variation and hence, it shows a more balanced voltage sharing. This is due to the wide bandgap of the device which reduces the impact ionization of the carriers which is the main mechanism determining the voltage sharing during the reverse blocking of the device.

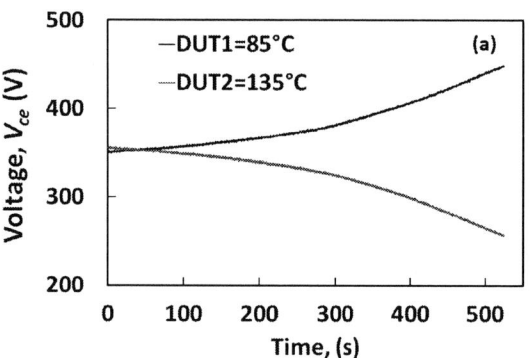

Fig. 3(a). OFF-state voltage sharing of two series connected silicon power IGBTs with temperature mismatch

In designing the power stage with series connected devices, a snubber resistor is used to ensure the voltage sharing during the off-state. The worst-case snubber resistor size is dependent on the maximum leakage current of the devices used in series [11]:

$$\hat{R} \le \frac{nV_D - V_S}{(n-1)\hat{I}_b} \tag{2}$$

In the datasheet of power devices, the values of leakage current are usually shown as a minimum, typical and maximum values, and their temperature dependencies are not illustrated. This makes the optimization of the snubber sizing more challenging. In addition, as the power converter undergoes numbers of cycles, the tolerances of the snubbers degrades and this degradation

on tolerances needs to be considered at the design phase which is out of the scope of this paper.

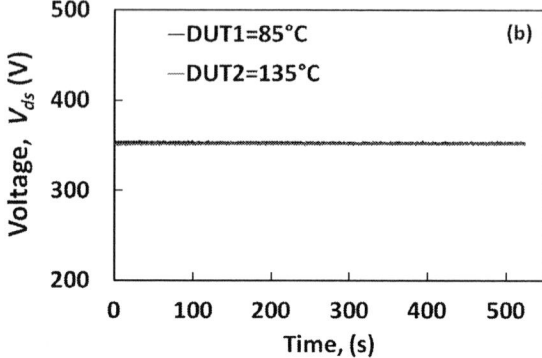

Fig. 3(b). OFF-state voltage sharing of two series connected SiC power MOSFETs with temperature mismatch

However, it can be seen from the equation that the series connected SiC MOSFET would inherently require larger snubber resistors for the OFF-state voltage balance which makes them more desirable due to the lower OFF-state snubber losses. The same devices used on ON-state voltage sharing analysis (silicon IGBT and SiC MOSFET) have been characterized in terms of the temperature dependencies of the leakage currents. As shown in Fig. 4(a), with the rise of temperature, the leakage current increases throughout the whole range of voltage for Si IGBT. Moreover, the breakdown voltage of the device increases with the temperature. The qualitative explanation of this phenomena is that the hot carriers passing through the depletion layer under a high electric field lose part of their energy to optical phonons via scattering thereby resulting in a smaller ionization rate. This effect increases with temperature. In case of SiC MOSFET as shown in Fig. 4(b), the breakdown voltage is shifted to higher voltage with increase of temperature, however, the leakage current is less affected by the temperature and this is the main reason that the series connected SiC MOSFETs show a better voltage sharing during the OFF-state.

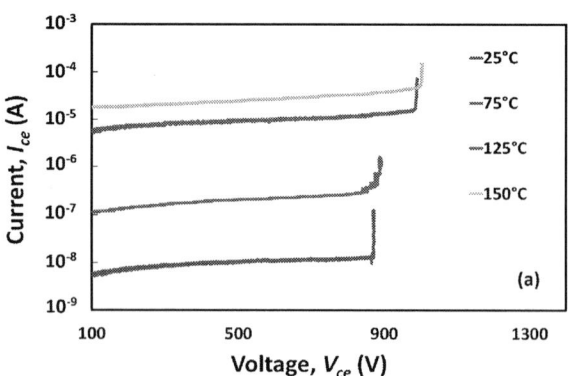

Fig. 4(a). Measured temperature dependency of the leakage current for the silicon IGBT

The 2018 International Power Electronics Conference

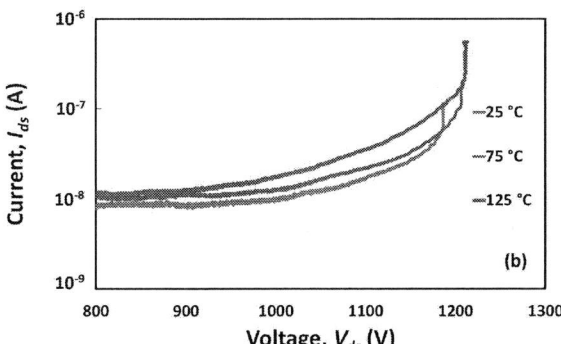

Fig. 4(b). Measured temperature dependency of the leakage current for the SiC MOSFET

IV. DYNAMIC VOLTAGE SHARING

A. Experimental Measurements

To further investigate the voltage imbalance between series connected SiC MOSFETs and silicon IGBTs under dynamic conditions with different operating conditions (such as different temperature and switching rates), the series devices were switched under clamped inductive switching.

As can be seen from circuit schematic Fig. 5 the setup includes a 1mH inductor L, 490 μF dc-link capacitor C_{DC} and a 600V SiC Schottky diode with datasheet reference C3D04060E from Cree/Wolfspeed. The measurements were taken at a dc-link voltage V_{dc}=400 V, unipolar gate drive voltage V_{GG}= 0/18V and external gate resistances R_G=100/120 Ω. The devices under investigation are the same devices as named in the ON-state voltage sharing section.

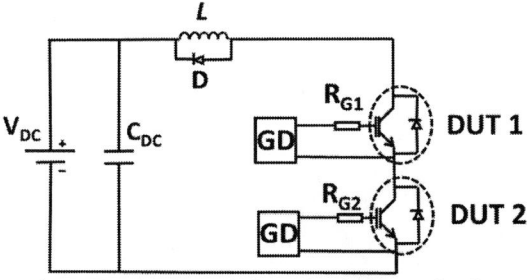

Fig. 5. Circuit schematic for series devices under clamped inductive switching

Similar to the off-state, the differences in the junction temperature affects the voltage balance during switching. Fig. 6(a) and 6(b) show the turn-OFF measurements of the series connected IGBTs and SiC Trench MOSFETs respectively at equal junction temperatures. As can be seen, the voltage divergence during switching is smaller for the SiC MOSFET but still within acceptable limits for the silicon IGBT.

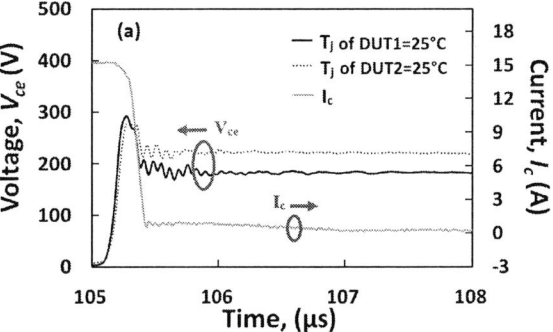

Fig. 6(a). Voltage and current waveforms of series connected silicon IGBTs during turn-OFF

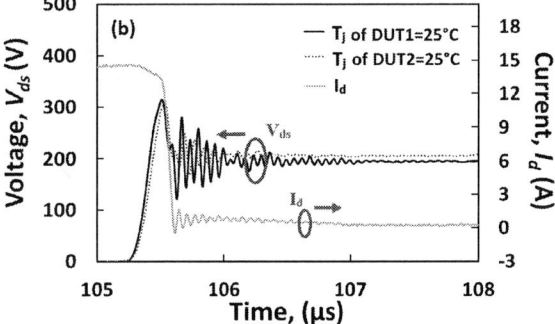

Fig. 6(b). Voltage and current waveforms of series connected SiC MOSFETs during turn-OFF

Fig. 7(a) and 7(b) show the turn-OFF voltage and current transient waveforms of the series connected devices when the junction temperatures have a 60°C temperature difference. As can be seen from the figures, the hotter device switches slower and blocks a smaller voltage.

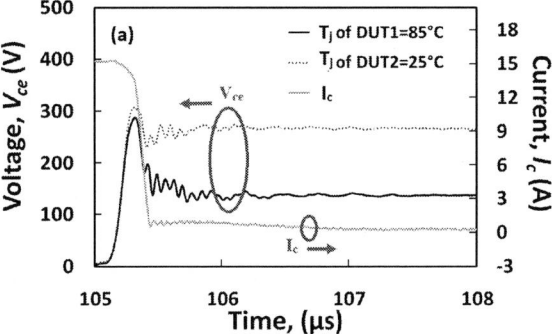

Fig. 7(a). Voltage and current waveforms of series connected silicon IGBTs during turn-OFF with different junction temperatures

1153

The 2018 International Power Electronics Conference

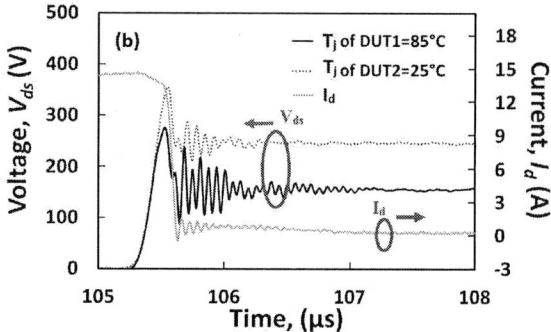

Fig. 7(b). Voltage and current waveforms of series connected SiC MOSFETs during turn-OFF with different junction temperatures

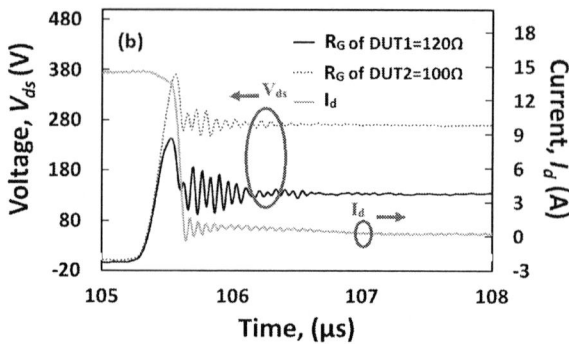

Fig. 8(b). Voltage and current waveforms of series connected SiC MOSFETs during turn-OFF with different gate resistances

The impact of variation in switching speed between series connected devices was also investigated by driving the series devices with slightly different gate resistances under clamped inductive switching conditions. As explained earlier, differences in the switching speed causes dynamic voltage imbalance during the turn-off transient. The switching rate differences can be due to non-uniform degradation of series connected devices, non-uniform operating temperature of the devices, differences in the gate resistance due to the device layout or non-uniform gate wire-bond degradation. Differences in the physical parameters of the device such as threshold voltage, carrier lifetime, carrier concentration and doping of the drift region may also contribute to non-uniform switching speeds.

Fig. 8(a) shows the turn-OFF current and voltage measurements of the series connected IGBTs with DUT2 switched with a smaller gate resistance ($R_G=100 \ \Omega$) than DUT1 ($R_G=120 \ \Omega$). As can be seen that the device with higher switching speed blocks a higher voltage than the slower device which causes a significant divergence in the V_{ce} characteristics. Fig. 8(b) shows similar measurements for the series connected SiC MOSFETs where it can be seen that the voltage divergence is not as significant as for the silicon IGBTs. However, the voltage divergence is higher than that caused by junction temperature variation.

B. Finite Element Simulations

Finite-element models have been developed to describe the physics behind series-connected silicon IGBTs and SiC MOSFETs under clamped inductive switching. The circuit shown in Fig. 5 has been simulated in ATLAS from SILVACO using the mixed mode circuit application to solve the switching transients with the finite-element model. Simulations have been performed to investigate the impact of different switching rates on series connected DUTs under CIS.

The SiC device in the simulation was optimized to a breakdown voltage of 1200 V using a 12 μm depletion layer with a doping of 2×10^{16} cm^{-3}. As shown in Fig. 4 (b) where the leakage current of a 650V rated SiC MOSFET was measured, the breakdown voltage of the device was 1.2 kV. The silicon IGBT is simulated with a drift layer doping of 1.1×10^{14} cm^{-3}, a p-body doping of 2.3×10^{17} cm^{-3}, and a voltage blocking drift layer thickness of 70 μm. Similarly, the IGBT was designed to show a breakdown voltage of 900V as shown in Fig. 4 (a). The circuit in the simulator was identical to the one used in the experiment described in Fig. 5.

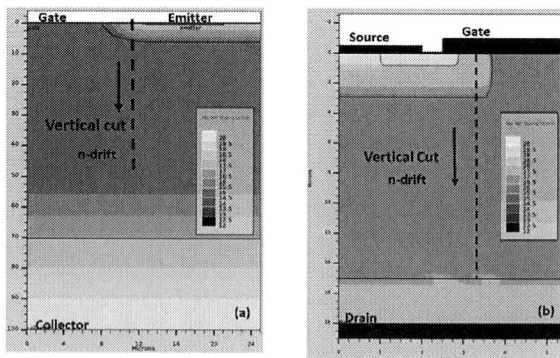

Fig. 9(a). Finite element model of the silicon IGBT and (b) Finite element model of the SiC MOSFET

Fig. 9(a) shows the cross section view of Si IGBT, whereas Fig. 9(b) shows the cross section view of the SiC MOSFET which was modelled in Silvaco and the vertical cut-line through the channel under the gate to the p-pillar down to the drain of the device.

Fig. 10 (a) shows the simulated voltage transients for

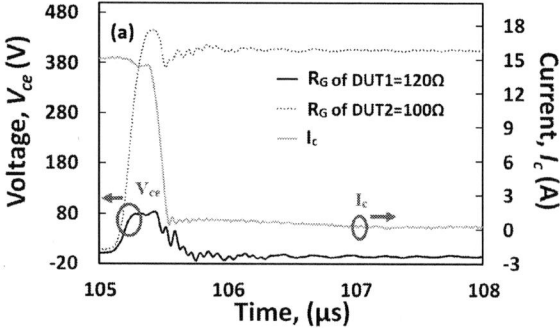

Fig. 8(a). Voltage and current waveforms of series connected silicon IGBTs during turn-OFF with different gate resistances

1154

the series connected IGBTs switched at marginally different rates (R_G=100/120 Ω) whereas Fig. 10(b) shows the simulations performed for the SiC MOSFET. It can be seen that that the simulated voltage divergence rates between the series devices replicate the experimental measurements shown in Fig. 8(a) and 8(b). Seven points in the transient waveform have been identified as can be seen below. The internal electric field along the cross-section of the device (shown along the cut-line in Fig. 9 (a) and 9(b)) is extracted at these time intervals in order to show the internal fields during turn-OFF.

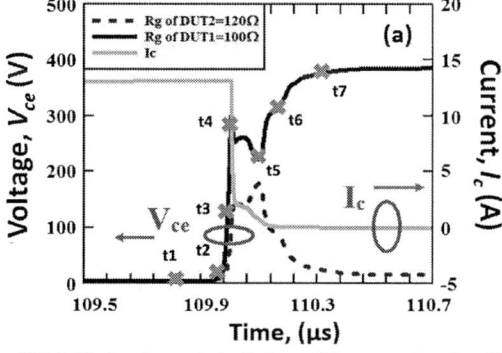

Fig. 10(a). Finite element similation of the current and voltage transients waveforms during turn-OFF of series connected silicon IGBTs under clamped inductive switching

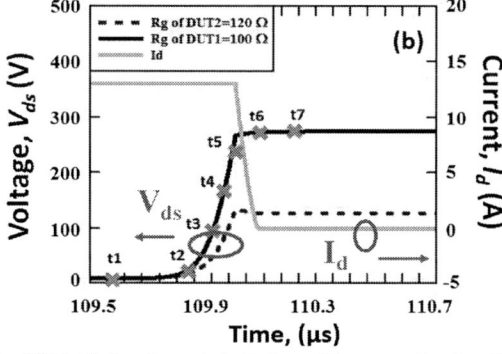

Fig. 10(b). Finite element similation of the current and voltage transients waveforms during turn-OFF of series connected SiC MOSFETs under clamped inductive switching

It can be seen in the silicon IGBT turn-OFF waveform shown in Fig. 10(a) that the voltage divergence does not begin until the start of the tail current phase during the turn-OFF transient. The tail current is due to minority carrier recombination during turn-OFF and when the fast IGBT enters this phase before the slower IGBT, the divergence between the drain voltage characteristics emerges. This tail current is absent in SiC MOSFETs, hence, the divergences emerges immediately as the current is switched.

Fig. 11(a) and 11(b) show the simulated internal electric field across the fast and slow IGBT at different stages of the turn-OFF transient, as it blocks voltage across the reverse biased PN junction. The faster switching device has a higher internal electric field

during the tail current. As the series IGBTs turn-OFF, the faster device completes its recombination phase of the tail current ahead of the slower device. As a result, it is forced to deplete further in order to maintain the current through both devices. This further depletion causes a higher blocking voltage as shown in the finite element results of Fig. 11(a) and Fig. 11(b).

Fig. 12(a) and 12(b) show similar simulations for the series connected SiC MOSFET switched with different gate resistances. The difference between the SiC MOSFET characteristic and the silicon IGBT characteristic is (i) the much higher electric fields in SiC due to the higher critical field and (ii) the lack of a tail current characteristic in the turn-OFF due to its unipolar nature. Similar to the IGBT internal electric field plots, the faster switching SiC MOSFET has a higher internal electric field and therefore blocks a higher voltage since the depletion width is wider.

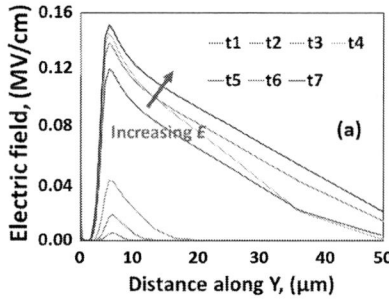

Fig. 11(a). Internal Electric field simulation of the fast IGBT

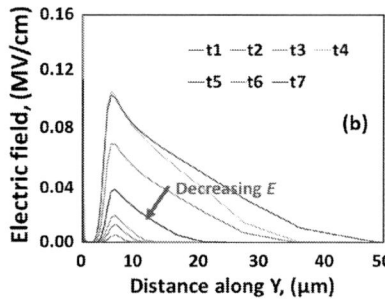

Fig. 11(b). Internal Electric field simulation of the slow IGBT

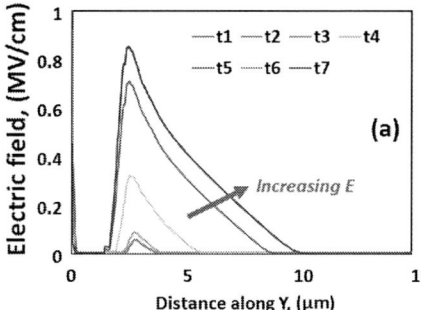

Fig. 12(a). Internal Electric field simulation of the fast SiC MOSFET

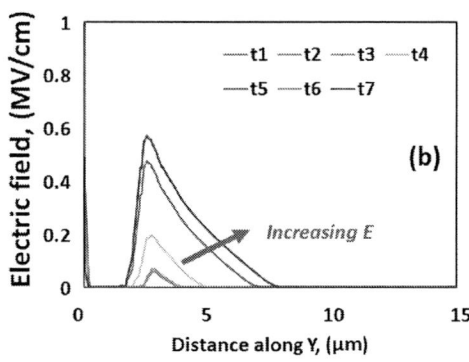

Fig. 12(b). Internal Electric field simulation of the slow SiC MOSFET

V CONCLUSION

The paper investigated the ON-state, OFF-state and switching transient behaviour of series connected IGBT and SiC MOSFETs at different operating conditions (temperature and switching rate) and compared the two technologies through experimental results justified by finite element models as well as compact electro-thermal models. The importance of ZTC point was discussed for voltage sharing during the ON-state operation. It was shown that in case of Si IGBT if the series connected devices are operating at different temperatures and the current is below ZTC point, the device with higher temperature shows a smaller on-state voltage drop while if the current is above ZTC it is vice versa. In case of SiC MOSFET, the ZTC point is very low and the device operates above the ZTC. Consequently, series connected SiC MOSFET which is operating at a higher temperature shows a higher on-state drain-source voltage drop and voltage sharing between the series connected devices is more sensitive to electro-thermal imbalance. In addition, using the FEM of these power devices, the transients of voltage sharing were compared under unbalance switching rate and it was discussed that the voltage imbalance for Si IGBT is highly dependent on the carrier concentration in the drift region during switching while for SiC MOSFET it depends on the switching time constant of the gate voltage and the rate that the MOS-channel cuts the current. It was discussed that SiC MOSFET shows a better performance during the OFF-state and switching transients in comparison to the conventional Si IGBTs, however, during the ON-state due to the very low ZTC point of SiC MOSFET, the voltage sharing becomes highly sensitive to the electro-thermal imbalance.

REFERENCES

[1] F. V. Robinson and V. Hamidi, "Series connecting devices for high-voltage power conversion," 2007 42nd International Universities Power Engineering Conference, Brighton, 2007, pp. 1134-1139.

[2] M. Bocovich, et al, "Overview of series connected flexible AC transmission systems (FACTS)," 2013 North American Power Symposium (NAPS), Manhattan, KS, 2013, pp. 1-6.

[3] "3-level inverter" brochure, IUK-TSM-2014-001 Issue 1, IXYS, UK Westcode, Feb 2014.

[4] J. Hu, O. Alatise, J. A. O. González, R. Bonyadi, L. Ran and P. A. Mawby, "The Effect of Electrothermal Nonuniformities on Parallel Connected SiC Power Devices Under Unclamped and Clamped Inductive Switching," in *IEEE Transactions on Power Electronics*, vol. 31, no. 6, pp. 4526-4535, June 2016.

[5] J. Hu, O. Alatise, J. A. O. Gonzalez, R. Bonyadi, L. Ran and P. Mawby, "Comparative electrothermal analysis between SiC Schottky and silicon PiN diodes: Paralleling and thermal considerations," *2016 18th European Conference on Power Electronics and Applications (EPE'16 ECCE Europe)*, Karlsruhe, 2016, pp. 1-8.

[6] J. O. Gonzalez, O. Alatise, P. Mawby, A. M. Aliyu and A. Castellazzi, "Pressure contact multi-chip packaging of SiC Schottky diodes," *2017 29th International Symposium on Power Semiconductor Devices and IC's (ISPSD)*, Sapporo, 2017, pp. 435-438.

[7] R. Bonyadi *et al.*, "Compact Electrothermal Reliability Modeling and Experimental Characterization of Bipolar Latchup in SiC and CoolMOS Power MOSFETs," in *IEEE Transactions on Power Electronics*, vol. 30, no. 12, pp. 6978-6992, Dec. 2015.

[8] P. R. Palmer *et al.*, "SiC MOSFETs connected in series with active voltage control," 2015 IEEE 3rd Workshop on Wide Bandgap Power Devices and Applications (WiPDA), Blacksburg, VA, 2015, pp. 60-65.

[9] G. Tsolaridis *et al.*, "Development of Simulink-based SiC MOSFET modeling platform for series connected devices," presented at the ECCE conference, Milwaukee, 2016, pp. 1-8.

[10] V. d'Alessandro *et al.*, "SPICE modelling and dynamic electrothermal simulation of SiC power MOSFETs," presented at the ISPSD symposyum, Waikoloa, 2014, pp. 285-288.

[11] B. W. Williams, Power Electronics: Devices, Drivers and applications, Macmillan, 1986.

The 2018 International Power Electronics Conference

Failure Mechanism Analysis and Physics-of-Failure Lifetime Prediction Method for Press-Pack Thyristor of Converter Valve

Ning Liang[1]

Zhigang Zhang[2], Yating Gou[2], Cuicui Liu[2], Zebin Yang[2], Jiangnan Chen[2], Fang Zhuo[2], Feng Wang[2]

1 Maintenance&Test(M&T)Center of EHV Power Transmission Company, Guangzhou, China

2 School of Electrical Engineering, Xi'an Jiaotong University, Xi'an, China

liangning@ehv.csg.cn

Abstract- **The converter valve is critical in HVDC transmission system, meanwhile its reliability facing Special electromagnetic environment special electromagnetic environment threat is significate to be studied from a strategic prospective supported by USA government. Unfortunately, existing reliability evaluation methods at component-level are not suitable in such an extreme environment. In this paper, main failure mechanisms of the most vital component under high special electromagnetic environment press-pack thyristor, are analyzed, and a new physics-of-failure lifetime prediction method is proposed based on loading condition and operating conditions considering concrete internal cooling system of the valve. With the abovementioned method as the basis, reliability evaluation for the whole converter valve system could be derived. The precise analysis process and analytical results are presented to indicate this method and the main damage effects of special electromagnetic environment.**

Keywords- converter valve, failure mechanism, HVDC, lifetime

I. INTRODUCTION

The reliability of converter valve directly affects the reliability of the entire HVDC transmission system. In HVDC system, especially for the project with voltage level equal to or higher than ±800kV, press-pack thyristor is widely adopted due to its high power capacity. Recently USA government has attached great important to the threat assessment of high special electromagnetic environment attack for electric system [1], while the reliability evaluation of converter valve is listed among the most critical and difficult aspects considering its complicated structure. The evaluation method such as Marcov model is well established for normal working conditions at system-level, but the method like failure rate in handbook is no longer feasible at component-level under this extreme environment. In this paper, a new physics-of-failure lifetime prediction method of press-pack thyristor is proposed which is the fundamental reliability evaluation for the whole converter valve system.

_____

This work was supported by the EHV Power Transmission Company of the Southern Power Grid project: Research on Lifetime Evaluation of HVDC Converter Valve (XGYKJXM00000027)

II. GLOBAL APPROACH

The global approach proposed is shown in Fig. 1.

The reliability evaluation for the whole converter valve system under special electromagnetic environment environment is a considerable project highly relying on practical measurement and electromagnetic simulation, in which an urgent problem is lifetime evaluation of component in this specific condition. Therefore the first step is to identify the key component. According to the statistical data of HVDC system faults in recent 10 years, the majority of faults caused by valve main circuit is due to the failure of switching device, therefore press-pack thyristor is studied in this paper. The second section is loading profile generation, which aims to analyze the electrical and thermal stress of thyristor in special electromagnetic environment, and the target is to establish rational thermal model for the transmission of stresses. The last section analyzes the thermal stress and then predicts the lifetime under corresponding special electromagnetic environment condition as a factor of measurement for reliability.

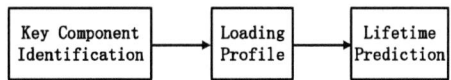

Fig. 1. Global approach flowchart

III. PRESS-PACK THYRISTOR

The press-pack thyristor is characterized by a silicon whole-wafer clamped under pressure to ensure uniform thermal and electrical contacts [2]. The general cross-sectional view of this structure is shown in Fig. 2 [3]. In the middle, silicon wafer is pressed between two molybdenum (Mo) discs, which could allow good pressure distribution at the surface of silicon wafer. The pieces outside of the molybdenum discs are usually nickel-plated copper [4]. On the outside a grooved ceramic housing protects the silicon wafer mechanically. In addition, ceramic insulators are used on both sides creating a hermetic casing as the package, and the valve plate and shell is filled with nitrogen.

1157

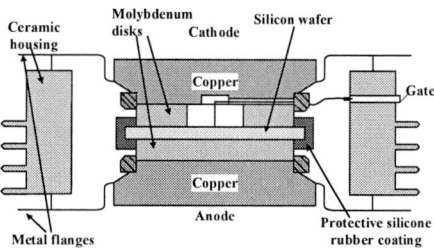

Fig. 2. Press-pack thyristor structure

IV. LOADIND PROFILE

The approach in this section is shown in Fig. 3. It can be seen the loss and thermal characteristics of the thyristor are all included in order to map the special electromagnetic environment profile into the thermal loading. The characteristic of special electromagnetic environment is represented as follows:

$$I(t) = I_0 k(e^{-\alpha t} - e^{-\beta t}) \tag{1}$$

where I0 is current peak value, k is correction factor, α and β are parameters determining the edge of the pulse.

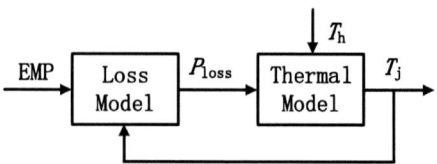

Fig. 3. Loading profile flowchart

The power loss of thyristor is required for the model. For the case of large current (>10kA), the voltage drop is related to both current and junction temperature, modeled by means of an exponential function as [5,6]:

$$V_f = \ln(\frac{I}{I_s} + 1) \times \frac{2kT}{q} + R_s(\frac{T}{T_{ref}})^{krs} \times I \tag{2}$$

where I is the forward current, I_s is the saturation current, k is the Boltzmann's constant, T is the absolute junction temperature, q is the electron charge, R_s is the virtual thyristor resistance at high-injection levels, T_{ref} is the reference temperature (usually 300 K), k_{rs} is the exponent coefficient of the temperature for the series resistance, which equals 1.17 in [7]. The value of saturation current can be expressed as [6]:

$$I_s = AT^3 e^{-\frac{qV_{gT}}{kT}} \tag{3}$$

$$V_{gT} = 1.17 - 0.049 \times \frac{T}{T_{ref}} \times [1 + 0.567\ln(\frac{T}{T_{ref}})] \tag{4}$$

where V_{gT} is the bandgap voltage under different temperature conditions.

By means of fitting curve according to the V-I characteristic curve of device datasheet [8] under different temperature conditions, the unknown parameters R_s and A could be calculated, and the concrete voltage drop model is obtained by combining equations (2)-(4). The target of this section is to converter electrical stress into thermal loading through the proper electro-thermal

model. Generally, thermal models for power semiconductor devices are Foster model and Cauer model, formed by a series of thermal resistance and capacitance. Model parameters are decided by measurement or the transient thermal impedance curve of device datasheet provided by manufactures. These two models are proved effective used in the researches of single device thermal characteristics. However, thyristors in converter valve are working in a complicated environment where the temperature of heat sink T_h is not a constant, affected by the heavy power loss in the valve. Recent researches [9,10] have proved that these two networks should be only connected to a temperature reference and cannot be extended with any other thermal RC networks like heat sink. Therefore, existing thermal models are not applicable and a new thermal model is required fitting this demand, which is shown in Fig. 4[9].

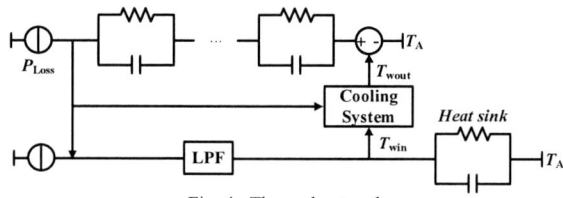

Fig. 4. Thermal network

Two paths are combined together. The first path is a conventional Foster network used for temperature estimation inside the device and parameters are decided by datasheet or experiment as the traditional way. At the end of this path, the heatsink temperature from the other path is connected. The second path is used for the temperature estimation outside. In this path a LPF (low pass filter) from the Foster thermal network is used to model the loss behaviors flowing out of the device, and the filtered loss can create correct temperature behavior of heat sink outside the devices. Besides, heatsink temperature cooling the thyristors is that of water pipe outlet T_{wout}, while its relationship with water pipe inlet T_{win} could be modelled as:

$$T_{wout} = T_{win} + P_{tloss}\frac{60}{W_f k_W} \tag{5}$$

where P_{tloss} is the heat absorbed by internal cooling system within unit time, W_f represents water flow inside the pipe, k_W is specific heat capacity of cooling water. P_{tloss} is total valve power loss and calculated as empirical model [11]:

$$P_{tloss} = \frac{n \times P_T}{0.85} \tag{6}$$

where n is the thyristor quantity in a valve tower, PT is the power loss of a single thyristor. Considering the special press-pack structure and the heat transferred through nitrogen is negligible, the 3-dimension heat flow of thyristor could be simplified along the axial direction as a one dimension model [5]. In this paper, the thyristor model selected is KPc3400-52 produced by Zhuzhou CSR times. For the calculation of heatsink temperature, Yunnan-Guangdong UHVDC transmission project is selected in which thyristor quantity in a 12-pulse converter valve is 720, and cooling system parameters could be found in [12]. Thus, the thermal model could be settled. For the first path, a Foster network containing four RC lumps is used, and the parameters are fitted to well describe the transient thermal impedance curve [6]. For the second path, a three cascaded LPF is used according to the frequency-domain thermal model theory [10]. Based on the model and analysis above, the thermal characteristics of press-pack thyristor under special electromagnetic environment environment could be well described.

The junction temperature of the thyristor under an special electromagnetic environment whose peak value equals 50kA and pulse width equals 0.4ms is shown in Fig. 5. It could be seen that the temperature rise is around 20℃, and T_j attains more than 70℃.

The relationship between junction temperature rise and special electromagnetic environment peak value is shown in Fig. 6. It could be seen that the growth rate of T_j is augmenting with the special electromagnetic environment current peak value getting larger. Also, T_j attains the temperature limit 125℃ of silicon device, which means the device risks the failure directly at this current stress.

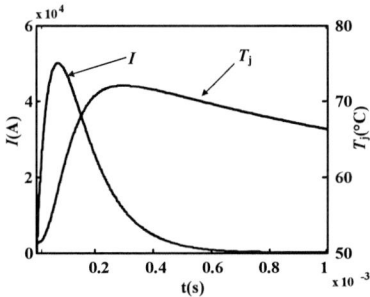

Fig. 5. special electromagnetic environment current and junction temperature

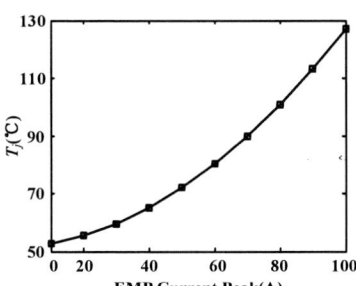

Fig. 6. Current peak and junction temperature

V. LIFETIME PREDICTION

Before choosing the lifetime model, it is supposed to understand different types of electronic packaging of power devices, which is aimed to improve not only electrical but also thermos-mechanical performances in high-stress environment. Basically packaging could be classified into two categories: one is capsule, mainly press-pack devices in which device wafer or chips

clamped under great pressure to ensure uniform contacts; the other is power module using bond wire, solder layer or joint, substrate, base plate, etc. An example of classical thyristor power module is shown in Fig. 7.

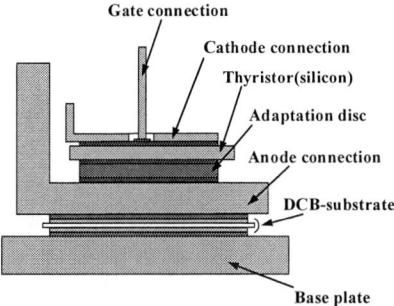

Fig. 7. A classical thyristor power module

The principle of the lifetime modelling is based on the correlation of the number of cycles to failure, Nf, directly gained in the power cycling tests to the corresponding temperature profiles. Two lifetime modelling approaches can be distinguished: empirical and physics-of-failure lifetime models.

Empirical models are purely deduced from statistical analysis of experience and databases of test results for different module technologies. The most widely used empirical lifetime models are the Coffin-Manson model, Coffin-Manson Arrhenius model, Norris-Landzberg model, Bayerer model, etc. [4, 13]. The Bayerer model is shown in equation (7), which considerates the influence of many factors, including the temperature variation, maximum temperature value, influence of heating time, current of wire bond stitch, diameter of bond wire, blocking voltage.

$$N_f = K \cdot (\Delta T_j)^{\beta_1} \cdot e^{\frac{\beta_2}{T_{j,max}}} \cdot t_{on}^{\beta_3} \cdot I^{\beta_4} \cdot V^{\beta_5} \cdot D^{\beta_6} \quad (7)$$

However, overwhelming majority of empirical lifetime model researches are based on power modules, especially IGBT modules, instead of press-pack package device. Also, dominant failure mechanisms of power module are bond wire fatigue, solder fatigue, aluminum metallization, corrosion of interconnections, etc[13, 14]. Nevertheless, the press-pack structure eliminates bond wires and solders layers or solder joints, which could get rid of most of the common reliability problems for power modules. Additionally, the device allows dual sided cooling. For these reasons, empirical models are not suitable for highly reliable device press-pack thyristor, while the main failure mechanisms of thyristor should be recognized first and physics-based lifetime prediction models will be adopted.

The physical models (physics-of-failure, PoF) aim to identify the physical mechanisms of the potential failures, and set up links between failure mechanisms and the lifetime under specified operation conditions through modelling the stress-strain deformations due to thermal-mechanical stresses, which facilitates a more meaningful life prediction approach. In the various physical models, energy-based methods are considered to give better results because they capture test conditions with more accuracy [15]. The basic strategy of energy based models is the assumption that a device fails once the deformation energy accumulated in the device reaches a critical value [15].

For a press-pack thyristor, the main failure mechanism most frequently observed is the layer fatigue and crack caused by the thermal-mechanical stress in adjacent layers under time–temperature exposure because of the mismatch in the coefficient of thermal expansion (CTE) of the different materials. Once a tiny damage is caused in the silicon wafer, a sharp stress concentration at pre-existing damages could lead to the rupture failure under mechanical stresses [14]. In fact, ultimate brittle fracture can occur suddenly without any plastic deformation. Even if the initial crack does not reach the critical length, it can develop by fatigue crack propagation under the influence of the applied stresses until the threshold for brittle fracture is exceeded.

Another second failure mechanism is the delamination at the interface. Layers are connected by tremendous pressure instead of metallurgical joints, so layers can be offset against each other and can glide on each other potentially. Under thermal variations, the CTE mismatch leads to incompatible thermal strains along materials interfaces and initiates delamination [16], therefore interfacial delamination is the other major reliability issue.

In conclusion, the great power losses induced by high stress-mechanical operating condition like special electromagnetic environment could cause great damage to the functionality and finally failure of the device.

As presented before, a special electromagnetic environment generates accumulated deformation energy remained in the device, and the device fails when the deformation reaches the critical value, W_{tot}. The model defined lifetime could be represented as:

$$N_f = C \cdot (\Delta w)^{-n} \qquad (8)$$

Where C and n are the model constants depending on the device characteristics, and Δw is the accumulated energy per cycle or inelastic strain energy density, which is calculated by

$$\Delta w = \int \tau dy \qquad (9)$$

where τ is the stress along strain in a cycle. The calculation of the stress is generally finished with the help of simulation of finite element method, and the relationship between accumulated stress and thermal variations could be simplified as equation (10) according to [17]:

$$\ln(\Delta w) = -12.47 - 1.22\Delta T - 8.02 \times 10^{-3} T_m + \\ 0.11\Delta T^2 + 4.35 \times 10^{-3} T_m \Delta T \qquad (10)$$

where Tm is the mean junction temperature.

The lifetime model for press-back thyristor under special electromagnetic environment can be predicted combining equation (8) and (10), and the damage caused by each pulse could be regarded as consumed lifetime CL according to Miner's rule [12].

$$CL_i = \frac{1}{N_{f\_i}} \qquad (11)$$

$$CL_t = \sum_i \frac{1}{N_{f\_i}} \qquad (12)$$

Where CL_i and CL_t represents respectively the consumed lifetime of a certain pulse and total pulses.

The analytical lifetime prediction of press-pack thyristor caused by different special electromagnetic environment peak value is shown in table I, and a relative simple conclusion could be obtained that the damage on the device is much more severe with the current value increasing.

TABLE I
LIFETIME PREDICTION

Current peak	N_F
20kA	1.78×10^6
50kA	6.21×10^4
100kA	1.27×10^3

VI. CONCLUSIONS AND FUTURE WORK

The method proposed in this paper provides a new feasible method for studying the lifetime prediction and reliability of press-pack thyristor under special electromagnetic environment environment as the basis for the reliability evaluation of whole converter valve. However, the main failure mechanism---crack initiation and propagation is complicated—nearly all models and equations require distributed parameters, such as stress which is dependent on geometry, material and loading levels. Therefore, the further more retailed research concerning thermo-mechanical stress analysis and experimental verification will be continued.

ACKNOWLEDGMENT

This work was supported by the EHV Power Transmission Company of the Southern Power Grid under project XGYKJXM00000027.

REFERENCES

[1] special electromagnetic environment Commission. Report of the commission to assess the threat to the United States from electromagnetic pulse (special electromagnetic environment) attack[J]. Washington DC, 2008, 27..

[2] Yang S, Xiang D, Bryant A, et al. Condition monitoring for device reliability in power electronic converters: A review[J]. IEEE Transactions on Power Electronics, 2010, 25(11): 2734-2752.

[3] Advanced Solutions in Power Systems: HVDC, FACTS, and Artificial Intelligence[M]. John Wiley & Sons, 2016.

[4] Sharifabadi K, Harnefors L, Nee H P, et al. Design, Control, and Application of Modular Multilevel Converters for HVDC Transmission Systems[M]. John Wiley & Sons, 2016.

[5] Profumo F, Tenconi A, Facelli S, et al. Instantaneous junction temperature evaluation of high-power diodes (thyristors) during current transients[J]. IEEE Transactions on Power electronics, 1999, 14(2): 292-299.

[6] Chen X. Research on the thermal characteristic of High Voltage Pulse Thyristor[D]. Huazhong University of Science and Technology, 2015.

[7] Profumo F, Tenconi A, Facelli B, et al. A new approach to the instantaneous junction temperature evaluation of high power diodes (thyristors) during past current transients[C]//Power Electronics Specialists Conference, 1997. PESC'97 Record., 28th Annual IEEE. IEEE, 1997, 1: 154-158.

[8] Zhuzhou CSR times, KPc 3400-46~52 datasheet.

[9] Ma K. Electro-thermal model of power semiconductors dedicated for both case and junction temperature estimation[M]//Power Electronics for the Next Generation Wind Turbine System. Springer International Publishing, 2015: 139-143.

[10] Ma K, He N, Liserre M, et al. Frequency-domain thermal modeling and characterization of power semiconductor devices[J]. IEEE Transactions on Power Electronics, 2016, 31(10): 7183-7193.

[11] W.J.Zhao. Technology of HVDC transmission [J]. 2004.

[12] G.Y.Yang, et al., "Overload Capability Calculation Methods of HVDC Converter Valve," High Voltage Apparatus., 51(12): pp.33-37,2015.

[13] Chung H S, Wang H, Blaabjerg F, et al. Reliability of power electronic converter systems[M]. The Institution of Engineering and Technology, 2016.

[14] Ciappa M. Selected failure mechanisms of modern power modules[J]. Microelectronics reliability, 2002, 42(4): 653-667.

[15] C. Busca, R. Teodorescu, F. Blaabjerg, S. Munk-Nielsen, L. Helle, T. Abeyasekera, P. Rodriguez, An overview of the reliability prediction related aspects of high power IGBTs in wind power applications, Microelectronics Reliability, Volume 51, Issue 9, 2011, Pages 1903-1907, ISSN 0026-2714, http://dx.doi.org/10.1016/j.microrel.2011.06.053.

[16] Durand C, Klingler M, Coutellier D, et al. Power Cycling Reliability of Power Module: A Survey[J]. IEEE Transactions on Device & Materials Reliability, 2016, 16(1):80-97.

[17] Chen Y X. Study on reliability of switched reluctance motor system[D]. China University of Mining and Technology, 2014.

The 2018 International Power Electronics Conference

Surge Voltage Absorption by a Silicon Carbide Avalanche-Diode with P-N structure

K. Koseki[*] and Y. Tanaka

Advanced Power Electronics Research Center, National Institute of Advanced Industrial Science and Technology
(AIST), Tsukuba, Japan
*kunio.koseki@aist.go.jp

Abstract-A SiC avalanche-diode with P-N structure has been developed to utilize it as a surge voltage absorber in power converter circuit. To evaluate the static characteristics, correlation between clamping voltage and avalanche current has been measured. The temperature dependence of the clamping voltage was also measured. The effectiveness in suppressing surge voltage by SiC avalanche-diode has been demonstrated in step-down DC/DC converter. By comparative studies with the RC and RCD snubbers, sufficient performance has been accomplished with avalanche-diode. Moreover, by measuring power efficiency, it is confirmed that the avalanche-diode effectively suppress the surge voltage without introducing additional power dissipation.

Keywords— Avalanche-diode, SiC, snubber, surge voltage.

I. INTRODUCTION

Emerging device technologies with silicon carbide (SiC), such as Metal Oxide Semiconductor Field Effect Transistor (MOSFET), Schottky Barrier Diode (SBD) and Insulated Gate Bipolar Transistor (IGBT) [1, 2] will result in high-power converters with reduced size and enhanced efficiency due to superior material characteristics of SiC than those of silicon. One of the most attractive characteristics of the SiC devices is the faster switching speed, with that switching loss is drastically reduced [3]. On contrary, surge voltage during switching transient is a serious concern because the minimization of parasitic inductance in circuitries and semiconductor packages has a certain limitation. Therefore, a realistic design solution with SiC devices is to limit the switching speed by gate resistor and/or gate-source capacitance or to use devices with higher rated voltage, that has a higher on-state resistance. In those situations, the characteristic superiority of SiC devices is compromised.

A solution to overcome the difficulty and to fully utilize the superiority of SiC-devices is to adopt a surge voltage absorber. Various surge absorbing circuits (e.g., RC and RCD snubber), have been developed to suppress disturbing effects during switching transient. Most of those circuit utilizes capacitors to store surge energy and to minimize the voltage overshoot for the switching element. Ideally, by increasing the snubber capacitance, higher surge energy can be absorbed. The avalanche-diode, on the other hand, acts as a voltage clamper. No energy storage capacitor nor damping resistor are required. Therefore, the simplest snubber circuit can be integrated into semiconductor package or power module without increasing its size.

Silicon avalanche-diode [4, 5] has been widely used as a constant-voltage regulator or a transient voltage suppressor, although a power density of silicon device is limited. On contrary, a SiC avalanche-diode with large current density of 60 kA/cm^2 and clamping voltage of less than 100 V was developed and demonstrated by K. Vassilevski, et. Al. [6], in 1993. To further investigate the potential of the SiC avalanche-diode as a surge absorber, a test sample of a SiC avalanche-diode with higher clamping voltage of about 150 V has been developed at Advanced Power Electronics Research Center (ADPERC) within National Institute of Advanced Industrial Science and Technology (AIST).

In this paper, evaluation of newly developed high-voltage SiC avalanche-diode has been performed and reported.

II. STATIC CHARACTERISTICS

A. Voltage-current correlation

An important characteristic of avalanche-diode as a surge voltage absorber, correlation between anode-cathode voltage and avalanche current, has been evaluated experimentally. Circuit schematic of the setup is depicted in Fig. 2. By turning-on MOSFET for about 400 nsec, stored energy in a capacitor bank (Cb) is applied to both the shunt resistor (Rs) and the avalanche-diode. To avoid enormous in-rash current after a destruction of avalanche-diode, resistance of 470 Ω is chosen for the shunt resistor. During the experimentation, DC charging voltage of the capacitor bank is increased pulse-by-pulse to get the voltage-current correlation of the avalanche-diode. Parameters employed in the experiment is summarized in Table. I. Anode-cathode voltages and avalanche currents are measured by high voltage probe (Tektronics Inc., TPP0850) and wideband current monitor (Pearson Inc., Model 2887), respectively. Monitored voltages and currents are digitized by a digital phosphor oscilloscope (Tektronics Inc., DPO5204B). The device under test (DUT) is the avalanche-diode with P-N structure that has been developed at ADPERC in AIST (see Fig. 1). The diameter of the DUT is 500 μm. The photograph of the assembled experimental setup is shown in Fig. 3.

1162

The 2018 International Power Electronics Conference

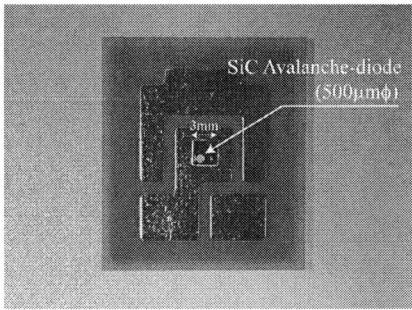

Fig. 1. Photograph of the SiC avalanche-diode.

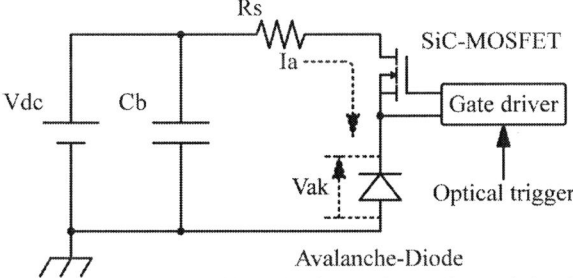

Fig. 2. Circuit schematic of the evaluation test for V-I characteristics of the SiC avalanche-diode.

TABLE I
PARAMETERS EMPLOYED IN THE EXPERIMENTS

Symbol	Meaning	Value
V_{dc}	DC charging voltage	0 – 1200 V
C_b	Bank capacitor	660 μF
R_s	Shunt resistor	470 Ω
τw	Pulse width	400 nsec

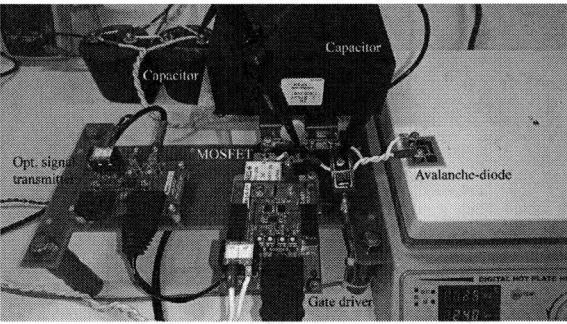

Fig. 3. Photograph of the experimental setup.

Measured anode-cathode voltage and avalanche current with DC charging voltage of 500 V are shown in Fig. 4. An avalanche clamping voltage of about 200 V with avalanche current of about 600 mA has been measured at plat-top period of pulsed voltage and current. By the measured transient waveforms with various charging voltages, correlation between avalanche clamping voltages and currents are summarized in Fig. 5. The increase of clamping voltage is gentle with avalanche current of less than 600 mA in the figure. It is estimated within this current range that the sufficient performance in absorbing surge voltage is accomplished. Further relaxation of increase of clamping voltage at higher current is required to adopt avalanche-diode to various

power converters. It should also be noted that the avalanche-diode tested in the experiment has no relaxation structure for electric field at around electrodes for the evaluation in the simplest structure. Therefore, it is estimated that higher electric field concentrated especially around the periphery reduces maximum avalanche current allowable in the device.

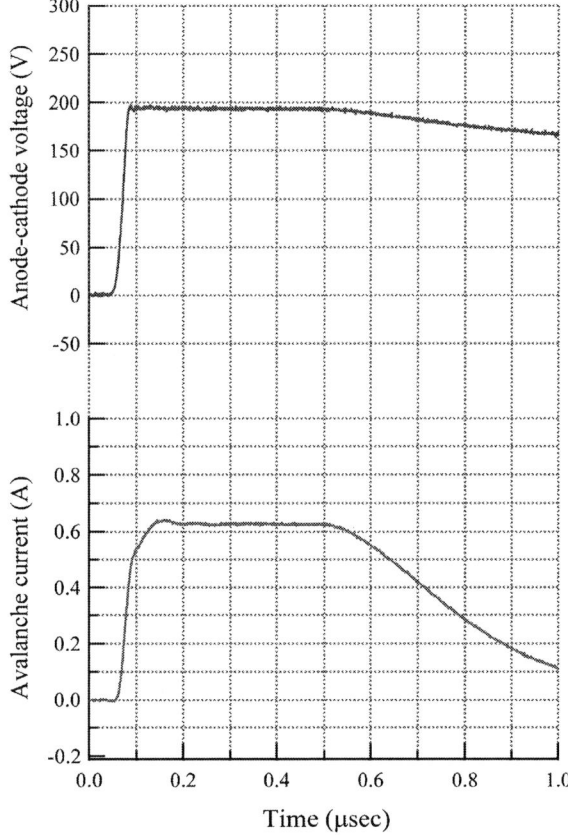

Fig. 4. Measured anode-cathode voltage and reverse current during avalanche breakdown with DC charging voltage of 500 V.

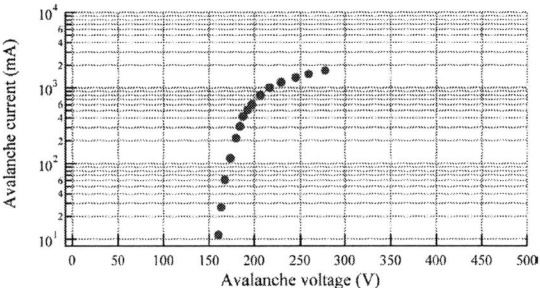

Fig. 5. Measured V-I characteristics of the SiC avalanche-diode.

B. Temperature dependence

Recently, development of the power module with a new bonding technology, with which SiC-MOSFET can be operated at higher temperature up to 250 deg. C, is intensively performed worldwide. To connect the surge absorber directly to the SiC-MOSFET in the power module to maximize the operational performance, it

1163

should also be operated at higher temperature in stable manner. From this view point, avalanche-diode by silicon carbide is an attractive solution.

To evaluate the performance of the avalanche-diode at higher temperature, dependence of the clamping voltage with various temperatures has been measured with the experimental setup as shown in Fig. 3. The temperature of avalanche-diode was controlled by a digital hot-plate (AS ONE, HP-1SA) in the experiment. The result is shown in Fig. 6. In the figure, measured clamping voltages with various current densities are normalized by the measured avalanche current to indicate the impedance of the device in reverse direction. An increase of the clamping voltage of about 10 % with the temperature increase of 25 to 210 deg. C has been measure. Thus, a stable clamping performance in higher temperature has been experimentally confirmed.

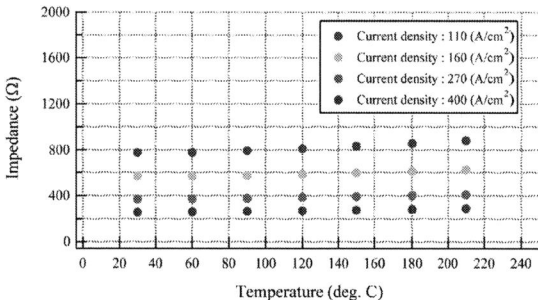

Fig. 6. Measured temperature dependence of the impedance for various current densities.

III. DYNAMIC CHARACTERISTICS

A. Comparative study

To evaluate the dynamic characteristics of the SiC avalanche-diode as a surge absorber, a step-down DC/DC converter has been assembled as shown in Fig. 7. For a comparative study, three types of circuit configurations, (1) without snubber, (2) with RCD snubber (see Fig. 8) and (3) with SiC avalanche-diode, were installed as DUT in the figure and evaluated. As the rated current of the SiC avalanche-diode developed as a test sample is less than 2 A, an inductor (Ls) of 2.5 µH which models a parasitic inductance is added between DC link capacitor (Cin) and switching element (MOSFET) to enlarge the surge voltage. Parameters employed in the experimentation are summarized in Table II. The photograph of the assembled DC/DC converter with SiC avalanche-diode is shown in Fig. 9. In the experiment, the electronic DC load (Takasago, FK1000L2) was used in a constant-power mode.

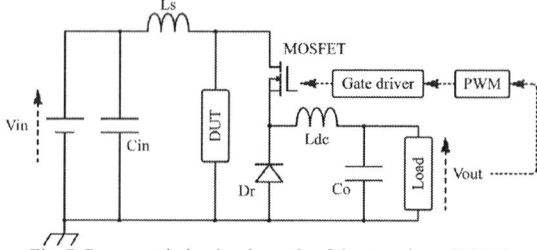

Fig. 7 Conceptual circuit schematic of the step-down DC/DC converter.

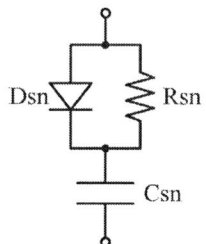

Fig. 8. Circuit schematic of the RCD snubber.

TABLE II
PARAMETERS EMPLOYED IN THE EXPERIMENTS

Symbol	Meaning	Value
V_{in}	DC charging voltage	150 V
V_{out}	Output voltage	50 V
F_s	Switching frequency	90 kHz
Cin	Bank capacitance	220 µF
Ls	Inductor	2.5 µH
	MOSFET	C3M0280090D
Dr	Diode	C3D02060E
Ldc	DC reactor	3.2 mH
Co	Filter capacitance	224 µF
Csn	Capacitance of RCD snubber	1 nF
Rsn	Resistance of RCD snubber	100 Ω

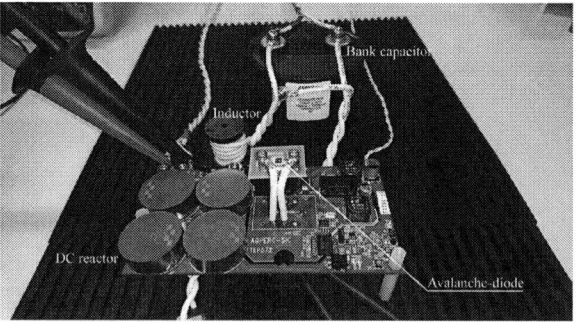

Fig. 9. Photograph of the experimental setup.

Drain-source voltages and DC reactor current measured at the output power of 90 W from DC/DC converter for three types of DUTs are depicted in Fig. 10. Enlarged view of voltages are also shown in Fig. 11. By adoption of the snubber circuits (RCD snubber and avalanche-diode), peak amplitude of voltage overshoot is reduced from about 420 V in snubber-less configuration to about 270 V. Although, oscillation of drain-source voltage has been measured with snubber circuit by

The 2018 International Power Electronics Conference

avalanche-diode, operational performance of the avalanche-diode in suppressing the surge voltage has been confirmed with the actual switching operation in a power converter circuit.

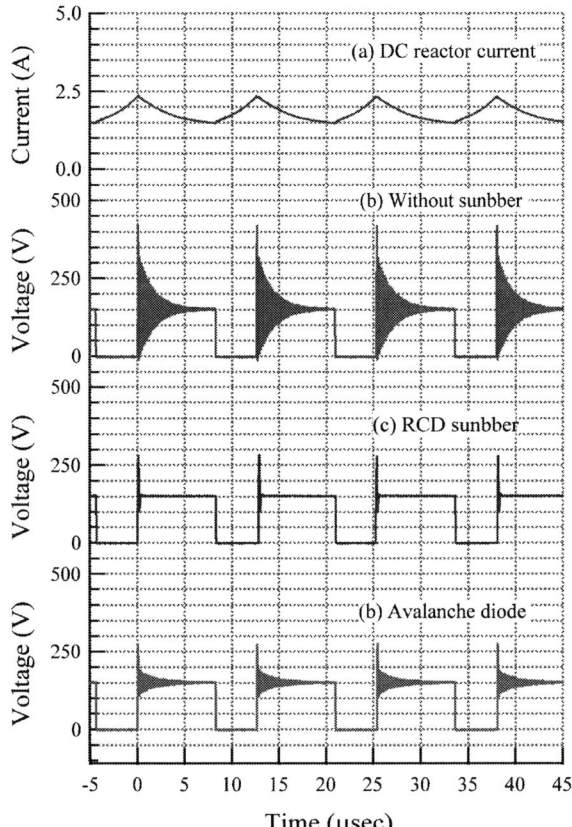

Fig. 10. Measured voltages and current in the comparative study.

Peak amplitude of drain-source voltages with three types of snubber configurations with various output power from DC/DC converter have been measured and shown in Fig. 12. Linear increases of surge voltages with respect to the output power have been measured for both the snubber-less and RCD snubber. For the RCD snubber with a storage capacitance of 1 nF, the increase of the surge voltage is effectively relaxed. With the avalanche-diode, the slope is almost identical with that of snubber-less configuration at lower current and gradually relaxed for higher current.

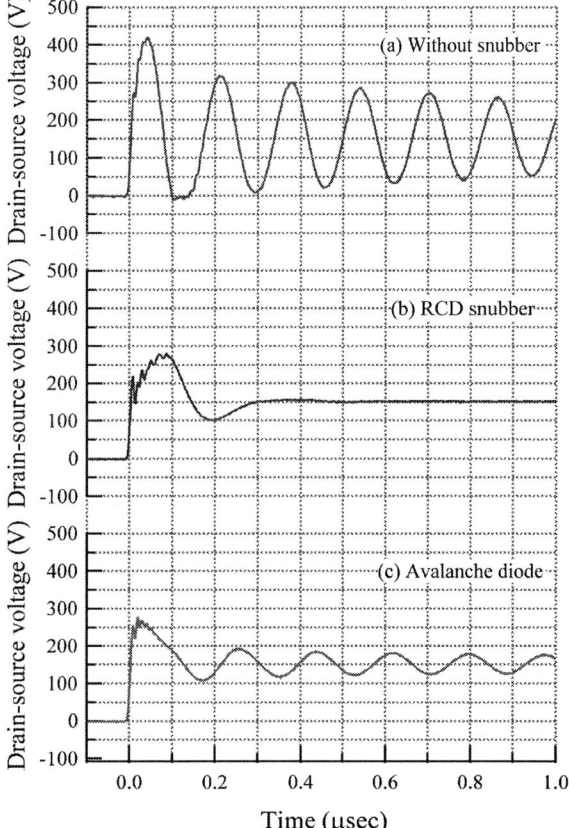

Fig. 11. Enlarged view of drain-source voltages (a) without snubber, (b) with RCD snubber and (c) avalanche diode.

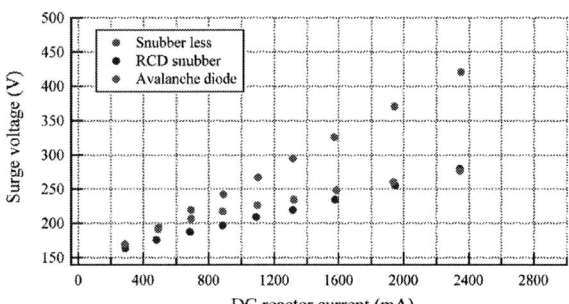

Fig. 12. Dependence of the peak amplitude of surge voltages on switched current for various snubber configurations.

B. Pulsed operation

One of the most attractive circuit configuration with avalanche-diode as a surge absorber is the direct connection to the switching element such as SiC-MOSFET, at which pulsed voltages are applied continuously. It is estimated that only a portion of surge energy is dissipated, and no additional energy deposition happens because the avalanche-diode acts as a voltage clamper. To confirm the estimation, drain-source voltage of MOSFET and avalanche current (reverse current at avalanche-diode) have been measured with an experimental setup shown in Fig. 13. The results are shown in Fig. 14. From the measured drain-source

1165

voltage, surge voltage is effectively absorbed by the avalanche-diode during switching transient. The waveform of the avalanche current also indicates that the power dissipation is localized during switching transient.

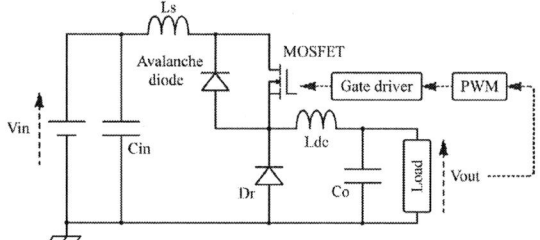

Fig. 13. Conceptual circuit schematic of the experimental setup.

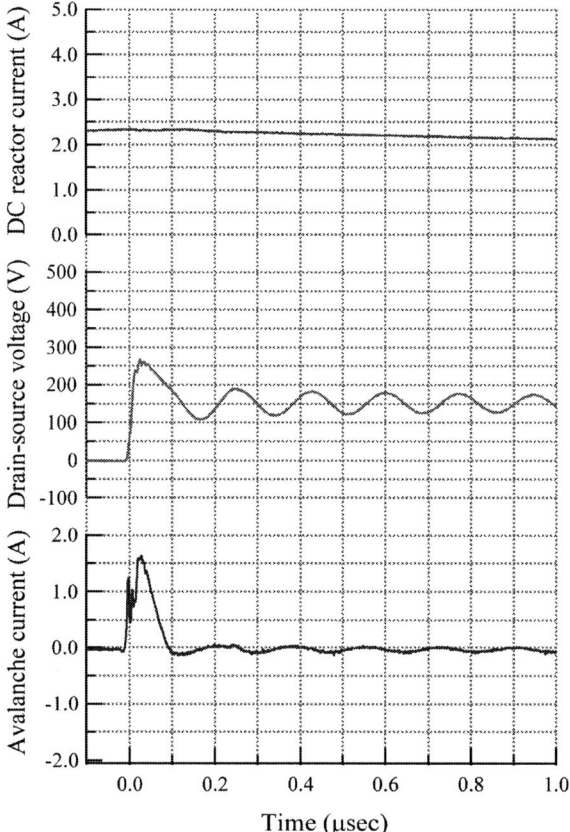

Fig. 14. Voltage and current waveform during experimentation.

To confirm the effectiveness of the SiC avalanche-diode to suppress voltage overshoot without introducing additional power dissipation, comparison of the power efficiency has been performed. Power efficiencies for various output power with the avalanche-diode and snubber-less configuration have been compared. The input and output power were measured by DC Power analyzer (Yokogawa, PZ4000). The results are shown in Fig. 15. It is confirmed that no additional power dissipation is introduced by avalanche-diode as power efficiency for both snubber configuration is almost identical.

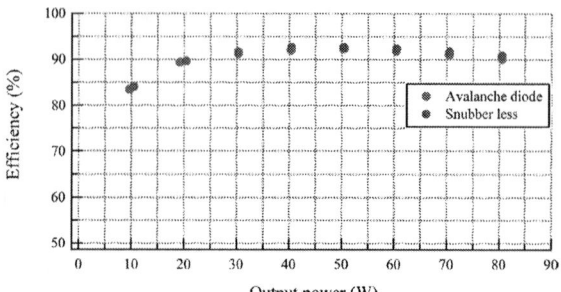

Fig. 15. Comparison of power efficiency with and without avalanche-diode.

IV. SUMMARY

SiC avalanche-diode has been developed at ADPERC in AIST to utilize it as a surge voltage absorber in power converter circuit. To evaluate the static characteristics, correlation between clamping voltage and avalanche current has been measured. The temperature dependence of the clamping voltage was also measured.

The effectiveness in suppressing surge voltage by SiC avalanche-diode has been demonstrated in step-down DC/DC converter. By comparative studies with the RC and RCD snubber, sufficient performance has been accomplished with avalanche-diode. Moreover, by measuring power efficiency, it is confirmed that the avalanche-diode effectively suppress the surge voltage without introducing additional power dissipation.

REFERENCES

[1] Y. Yonezawa, T. Mizushima, K. Takenaka, H. Fujisawa, T. Kato, S. Harada, Y. Tanaka, M. Okamoto, M. Sometani, D. Okamoto, N, Kumagai, S. Matsunakga, T. Deguchi, M. Arai, T. Hatakeyama, Y. Makifuchi, T. Araoka, N. Oose, T. Tsutsumi, M. Yoshikawa, K. Tatera, M. Harashima, Y. Sano, E. Morisaki, M. Takei, M. Miyajima, H. Kimura, A. Otsuki, K. Fukuda, H. Okumura and T. Kimoto, "Low Vf and Highly Reliable 16 kV Ultrahigh Voltage SiC Flip-Type n-channel Implantation and Epitaxial IGBT", in Proc of International Electron Device Meeting (IEDM), 2013 IEEE International, 2013, pp. 6.6.1-6.6.,

[2] Y. Yonezawa, T. Mizushima, K. Takenaka, H. Fujisawa, T. Deguchi, T. Kato, S. Harada, Y. Tanaka, D. Okamoto, M. Sometani, M. Okamoto, M. Yoshikawa, T. Tsutsumi, Y. Sakai, N. Kumagai, S. Matsunaga, M. Takei, M. Arai, T. Hatakeyama, K. Takao, T. Shinohe, T. Izumi, T. Hayashi, K. Nakayama, K. Asano, M. Miyajima, H. Kimura, A. Otsuki, K. Fukuda, H. Okumura and T. Kimoto, "Device Performance and Switching Characteristics of 16 kV Ultrahigh-Voltage SiC Flip-Type n-channel IE-IGBTs". Mater. Sci. Forum, 821-823, (2015), pp. 842-846.

[3] K. Koseki, Y. Yonezawa, T. Mizushima, S. Matsunaga, Y. Iizuka and H. Yamaguchi, "Dynamic Behavior of a Medium-Voltage N-channel SiC-IGBT with Ultra-Fast Switching performance of 300 kV/μs," 2016 19th International Conference on Electrical Machines and Systems (ICEMS), Nov. 2016.

[4] B. Bazin, "Silicon avalanche diode behavior for 10 us surge current in reverse direction", Solid State Electronics, Vol.10, pp509-512, 1967

[5] Y. Hasegawa, T. Kimura and K. Mizuno, "CHARACTERISTICS OF DIODE AFTER AVALANCHE BREAKDOWN", Origin No.45 pp.42-48, 1982.

[6] K. Vassilevski V. A. Dmitriev and A. V. Zorenko, "Silicon Carbide diode operating at avalanche breakdown current density of 60kA/cm2", J. Appl. Phys., vol. 74, no. 12, pp. 7612-7614, 1993.

The 2018 International Power Electronics Conference

Calculation of Thyristor Reliability Parameter of UHVDC Converter Valve in HEMP Environment

Zhigang Zhang[1], Yating Gou[1], Cuicui Liu[1], Zebin Yang[1], Xiaotong Du[1], Jiangnan Chen[1], Fang Zhuo[1], Feng Wang[1]
Yuanliang Lan[2], Caiwang Sheng[2]
1 School of Electrical Engineering, Xi'an Jiaotong University, Xi'an, China
2 Global Energy Interconnection Research Institute co.ltd, Beijing, China
jesuisnicolas02@163.com

*Abstract-*The threat of HEMP (high altitude nuclear electromagnetic pulse) in power system has emerged significant to be assessed. The thyristor converter valve is the core equipment in UHVDC (ultra-high voltage direct current) transmission system, and to evaluate its reliability in HEMP environment is quite important. The failure rate of thyristor is the critical parameter for the overall reliability evaluation, while its value varies depending on the electrical and thermal stress conditions. This paper proposed a modeling method for the calculation of failure rate of thyristor. First, the wideband model of key components as well as stray capacitances in the converter valve is built based on impedance measurement. Second, thermal impedance model of thyristor is built, and the model parameters are calculated according to the transient thermal impedance curve of the device. Based on the models above, the electrical and thermal stresses of thyristor are obtained and analyzed. Finally, the reliability model of thyristor is established, and the parameter failure rate is calculated which could be used in the further study of converter valve reliability. The quantitative analysis is presented in the paper.

Keywords- reliability, HEMP, thermal impedance, wideband model

I. INTRODUCTION

To optimize the actuality of the long-distance allocation between power generation resources and load centers in China, the strategy of UHVDC technology is regarded as a critical method to be developed due to its high transmission efficiency and capacity[1]. Converter valve is the key equipment of transmission system, and its reliability is of great significance to the performance of the entire system.

Nowadays the threat of nuclear explosion for power system components are attached great importance. HEMP is the main destructive effect of nuclear explosion, and it could cause great damage to power equipment, especially converter valve which is relatively vulnerable. The US congress has assessd several times the threat for US infrastructure from EMP attack[2].

The failure rate of thyristor is the core parameter to evaluate the reliability of converter valve, even the entire transmission system. Therefore, it is critical to calculate

and analyze the failure rate of thyristor when the converter valve is in HEMP environment. In this paper, based on previous work, the wideband model of thyristor converter valve is built, and the electrical stress characteristic of thyristors in the valve under HEMP environment is precisely analyzed. Also thermal impedance model is introduced to estimate the thermal stress of thyristor, then the reliability model is introduced to calculate the failure rate on the basis of above results. The quantitative analysis is also presented.

II. CONVERTER VALVE STRUCTURE

The research project converter valve of this paper is the ±800kV DC converter valve of Yunnan-Guangzhou UHVDC transmission project in China. To eliminate the effect of harmonics, the 2×12-pulse valve group structure is used of one pole in converter station, which means four double-valves are connected in series for each phase. The actual structure of a double-valve tower is shown in Fig.1, which includes two single valves from functional perspective. The valve tower is a 6-layer symmetrical structure, in which the intermediate layers are four valve modules, two big shield plates are laid out on the top and bottom, 40 short shields are fixed up on the both sides of transverse bus bars of valve module, and plastic water pipes are arranged inside the layers.

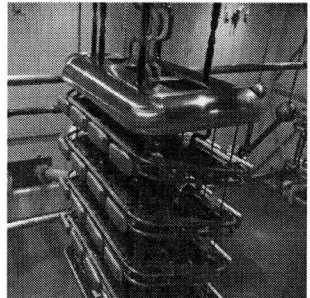

Fig. 1. ±800kV HVDC converter valve tower

The valve module is consisted of two valve sections, and each valve section includes 13 thyristor levels, 2 saturable reactors and compensating capacitor. A thyristor level is composed of a thyristor, a RC snubber circuit, a DC grading resistor and the control unit TE(Thyristor Electronics). The layout of valve module and thyristor level is shown in Fig.2 and Fig.3.

This work was supported by the State Grid Corporation headquarter project: Research on Mechanism and Evaluation Method of Reliability for Converter Valve in HVDC System(5455ZS160015)

The 2018 International Power Electronics Conference

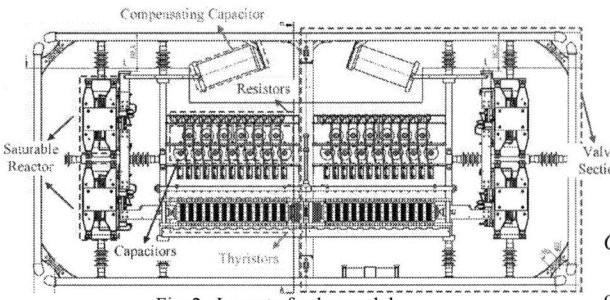

Fig. 2. Layout of valve module

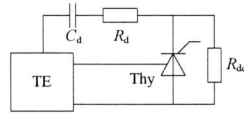

Fig. 3. Layout of thyristor level

III. WIDEBAND MODEL OF CONVERTER VALVE

To conduct the thorough research of the electrical characteristics of converter valve under HEMP, the wideband models of main components and stray capacitances are introduced as follows. The thyristor type studied in this paper is KPE4500-72[3], and the parameters of model are calculated based on impedance measurement of which the method is proposed in [4].

A. Wideband model of thyristor level

For a single thyristor device, a capacitor is usually used as the thyristor wideband model. However, EMP is a wide frequency range excitation source, and therefore a more accurate model which could simulate high frequency characteristics is required[1], as shown in Fig.4.

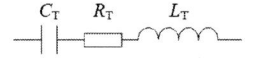

Fig. 4. Wideband model of thyristor

Accordingly, the wideband model of thyristor level is shown in Fig.5, in which R_{SC} and C_{SC} form the snubber circuit, R_G is the grading resistor.

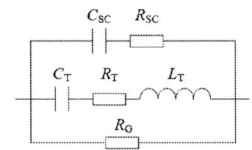

Fig. 5. Wideband model of thyristor level

B. Wideband model of saturable reactor

The saturable reactor is a non-linear device, whose characteristics is related to its current. The wideband model of saturable reactor is shown in Fig.6[5], where L is the main inductance, R_{Cu} is the ohmic losses, R_D is the damping resistance, L_{Leak} is the leakage inductance, R_{Ed} is the eddy current losses, and C_{SR} is the stray capacitance inside the reactor.

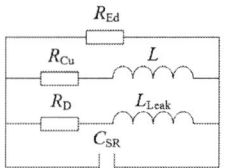

Fig. 6. Wideband model of saturable reactor

C. Stray capacitances of valve module

For a valve module, the stray capacitances are the capacitances between heat sinks, C_{Hs-Hs}, as well as the ground capacitances, C_{Hs-Gd}, as shown in Fig.7 [5-6].

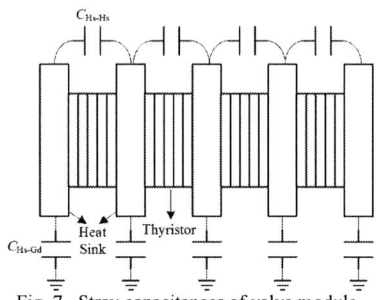

Fig. 7. Stray capacitances of valve module

D. Stray capacitances of valve tower

As a multiple-layer structure, the stray capacitance distribution in the valve tower is shown in Fig.8. The parameters considered are as follows: C_{Ly-Ly} represents the capacitance between one valve section and the section above or below; C_{Ly-Gd} represents stray capacitance between valve module and the ground; C_{Ly-Sh} is the valve module capacitance to the shielding plate; C_{Sh-Gd} is the shielding plate capacitance to ground; C_{VS-VS} is capacitance between valve sections in the same layer. The stray parameters are obtained by measurement in the valve hall.

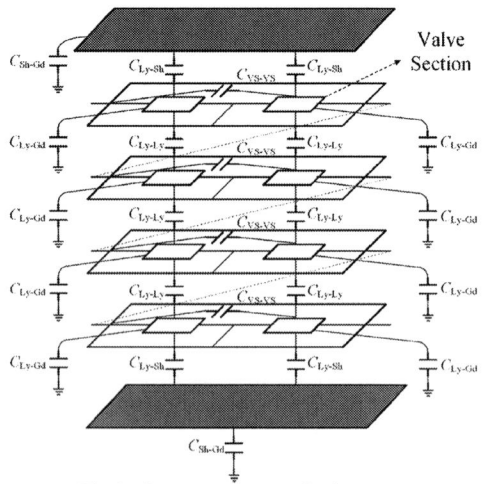

Fig. 8. Stray capacitances of valve tower

IV. THYRISTOR STRESS EVALUATION

A. HEMP waveform and equivalent equation

The functional process of HEMP could be divided into

1168

three sections[7], in which the early stage contains broad frequency signals and could perform destructive effects to converter valve. The waveform of HEMP could be simplified and simulated by the following equation[8]:

$$v(t) = V_0 A(e^{-\alpha t} - e^{-\beta t}) \qquad (1)$$

where V_0 represents the peak value of pulse, A is the correction factor, α and β are the time constants.

Currently many academic and national organizations have proposed some different HEMP standards, including several United States Department of Defense, Bell Laboratory and IEC(International Electrotechnical Commission)[9]. The standard of Bell Laboratory is widely used and therefore the following analysis is based on this standard, whose parameters is: α=4.0×10^{6}s^{-1}, β=4.76×10^{8}s^{-1}, A=1.052, V_0=50kV/m.

B. Thyristor thermal impedance model

The junction temperature of thyristor could be obtained by measurement and calculation. The method of measurement could lead to big error, especially for converter valve thyristor which is extremely difficult to be measured in the circuit. Therefore the calculation method based on thermal impedance is employed in this paper. The thermal impedance model most widely used is a RC network of Foster-type and Cauer-type[10]. Considering the existence of heat sink to be modelled, Foster model is not able to be applied, and Cauer model based on the physical structure of the device is considered to be a relatively correct model to describe the thermal behaviors[9], which is shown in Fig.9, and the meaning of parameters is shown in Tabel I.

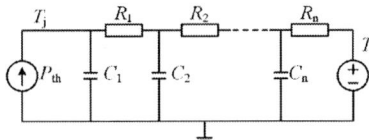

Fig. 9. Topology of Cauer model

TABLE I
PHYSICAL MEANING OF CAUER MODEL PARAMETERS

Symbol	Meaning
P_{th}	Power loss of thyristor
R_i	Thermal resistance
C_i	Thermal capacitance
T_j	Thyristor junction temperature
T_c	Thyristor case plate temperature

C. Electrical stress calculation of valves

To evaluate the electrical stress distribution of the entire converter valve system under HEMP, the four double-valves of single phase are all taken into account and modelled. The simulation diagram is shown in Fig.10. The 3 last valve towers are presented with simplified figure.

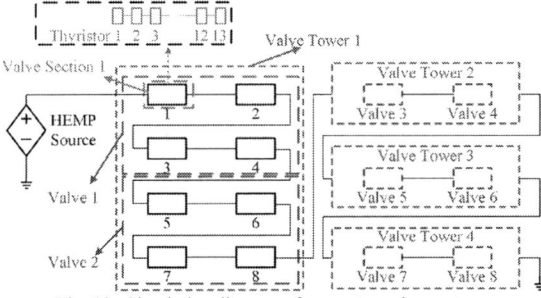

Fig. 10. Simulation diagram of converter valve system

The voltage distribution of different valves is shown in Fig.11. It is noted that the voltage is expressed as per-unit value. The result indicates that the voltage distribution is not uniform, also the degree of unbalance is extreme severe while most of the voltage is assumed by the first valve. Valve 2 and valve 3 withstand nearly 10% of surge pulse, while the rest could be neglected. Therefore the conclusion could be obtained that HEMP mainly affects the devices in the first three valves. The following analysis is concentrated on valve 1. Moreover the voltage of last two valves is too low and not noted in the figure.

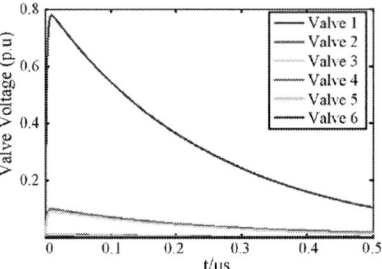

Fig. 11. Voltage distribution of different valves

D. Electrical stress calculation of valve sections and thyristors in valve 1

The voltage distribution in the four valve sections of valve 1 is shown in Fig.12. For one thing, the voltage is not identical of different valve sections due to the stray capacitances. For another thing, it should be noted that the voltage value order is different from valve section order in tower, due to the extremely short time constant of HEMP. Compared with the case in which the valve is exposed to lighting pulse in [5], following figure indicates that the time constant of excitation pulse is another important factor affecting the voltage distribution.

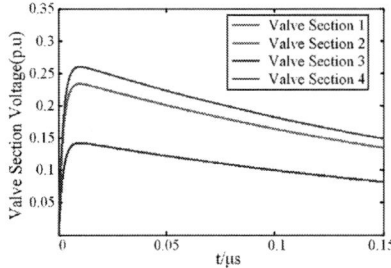

Fig. 12. Voltage distribution of valve sections in valve 1

Choose thyristor 1, 3, 6, 10 and 13 in valve section 1 as example, the voltage and current of these thyristors is shown in Fig.13. The almost overlapping curves indicate that the stray capacitances in valve section has almost little impact. In addition, the conclusion is reached that the electrical stress conditions of thyristors in the same valve section could be regarded as identical.

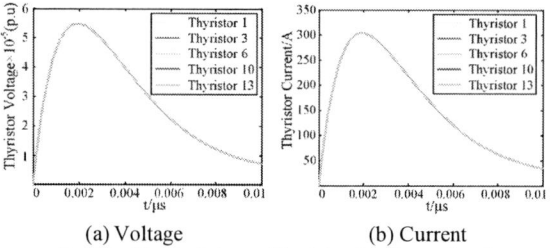

(a) Voltage   (b) Current
Fig. 13. Electrical stress of thyristors in valve section 1

E. Thermal stress calculation of thyristors in valve 1

As mentioned above, the junction temperature of each thyristor could be calculated on the basis of its electrical stress. A four-layer Cauer model is generally considered accurate enough to simulate the transient thermal impedance of thyristor shown in Fig.14, and its parameters are calculated to well fit the original curve.

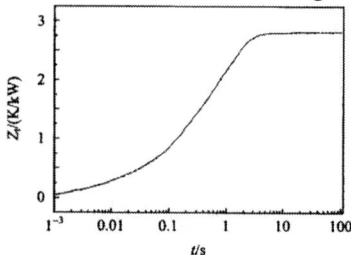

Fig. 14. Transient thermal impedance curve of thyristor

According to the electrical stress calculated above, the thermal characteristics of thyristors could be obtained. Based on the conclusion that electrical stress condition is identical for thyristor in the same valve section, the junction temperature variation of thyristor in each valve section could also be considered identical which is shown in Fig.15.

From the figure, the maximum temperature rise is of thyristors in valve section 4 which is around 5.2°C. This is relatively not much, because the rated capacity is 7.2kV/4500A, which is much more than the electrical stress withstood by the thyristor, and this could explain the small amplitude of temperature rise. It should be noted that the temperature rise order is also different from valve section order in tower.

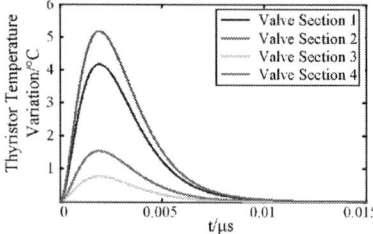

Fig. 15. Junction temperature variation of thyristors in valve 1

V. RELIABILITY PARAMETER CALCULATION

A. Reliability model

In order to evaluate the reliability parameter failure rate, in general it is acquired by the statistics of actual engineering applications and projects under ordinary conditions. However, the stress conditions for thyristor in HEMP environment is much more different, hence the failure rate of thyristor needs to be recalculated.

The reliability model of thyristor is necessary to be established for the calculation of failure rate and device lifetime. Physics-based reliability model is more accurate because its estimation is based on specific failure mechanism[10], whereas this method is exceedingly difficult to be applied for converter valve which contains considerable thyristors of different stress condition and different failure mechanism.

Thus, the reliability model based on reliability data handbook is used. RDF-2000 reliability handbook[11] proposed by Union technique de l'électricité of France takes many important factors into account, including application environment and external stress, and this method is applicable to evaluate the reliability parameters of thyristor under HEMP. The model for failure rate is:

$$\lambda = \left[\underbrace{(\pi_U \times \lambda_0 \times \pi_t)}_{\lambda_{die}} + \underbrace{(\pi_I \times \lambda_{EOS})}_{\lambda_{overstress}} + \underbrace{(2.75 \times 10^{-3} \times \pi_n \times (\Delta T_j)^{0.68} \times \lambda_B)}_{\lambda_{package}} \right] \times 10^{-9} / h \quad (2)$$

This failure rate model includes three parts, related to the normal use, over-stress condition and package technology respectively, and the meaning of these parameters is shown in Table II:

TABLE II
PHYSICAL MEANING OF RELIABILITY MODEL PARAMETERS

Symbol	Meaning
λ_0	Base failure rate
π_U	Use factor
π_t	Temperature mission profile factor
π_I	Influence factor related to the use
λ_{EOS}	Failure rate related to electrical overstress
π_n	Influence factor related to cycles number
ΔT_j	Junction temperature variation
λ_B	Base failure rate of package

The base failure rate is 3×10^{-9}/h, and the rest of parameters are precisely explained in [11]. Based on the analysis above, the calculated failure rate of thyristors in valve 1 is shown in Table III. It is noted that the distinction of thyristors is their location is the valve.

TABLE III
THYRISTOR FAILURE RATE

Thyristor Location	Value/year
Valve section 1	0.0044
Valve section 2	0.0031
Valve section 3	0.0025
Valve section 4	0.0049

According to equation (2), the failure rate increases with larger temperature variation, and the result in Table □ could indicate this phenomon. In fact, the relationship between failure rate and temperature variation could be simplified as Fig.16 shows, which illustrates the high dependency of failure rate on device temperature.

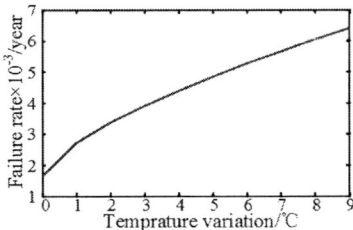

Fig. 16. Failure rate of thyristor

VI. CONCLUSION

In this paper, the structure of the ±800kV converter valve tower and its internal circuit is precisely introduced, the wideband model of the main components in the valve, such as thyristor and saturable reactor, as well as tower stray capacitances are fully discussed. On the basis of these models, the electrical stress of thyristors in the valve tower are calculated and analyzed under HEMP environment. Then the method to build thermal impedance model is introduced, and the thermal stress of valve thyristor is analyzed. Based on the electrical and thermal stresses calculated above, the reliability model of thyristor is presented, and the failure rate of thyristors is clarified. The quantitative evaluation results are presented and analyzed.

The proposed method can be directly used for transmission projects of different voltage scale, also under different HEMP environment, which could be of engineering value for the further study of converter valve reliability.

ACKNOWLEDGMENT

This work was supported by the State Grid Corporation headquarter project: Research on Mechanism and Evaluation Method of Reliability for Converter Valve in HVDC System (5455ZS160015)

REFERENCES

[1] Liu C L, Shuai Q, Qi L, et al. Quantitative analysis of voltage distribution within ±1100kv HVDC converter valve tower under various triasient over-voltage conditions[C]// Lightning Protection. IEEE, 2014:1558-1564.

[2] Foster Jr J S, Gjelde E, Graham W R, et al. Report of the commission to assess the threat to the united states from electromagnetic pulse (emp) attack: Critical national infrastructures[R]. ELECTROMAGNETIC PULSE (EMP) COMMISSION MCLEAN VA, 2008.

[3] Zhuzhou CSR Times Electric Co., Ltd. KPE4500-72 General Thyrisor[R]. Zhuzhou: CSR Times Electric Co., Ltd., 2007

[4] C Fang. "Research on the Method of Wideband Measurement and Circuit Modeling of High Voltage DC Thyristor Modules [D]". North China Electric Power University, 2013

[5] Sun H, Cui X, Qi L, et al. Calculation of overvoltage distribution in HVDC thyristor valves[C]// Electromagnetic Compatibility. IEEE, 2010:540-543.

[6] Qi L, Shuai Q, Cui X, et al. Parameters Extraction and Wideband Modeling of ?1100kV Converter Valve[J]. IEEE Transactions on Power Delivery, 2016, PP(99):1-1.

[7] Ianoz M, Nicoara B I C, Radasky W A. Modeling of an EMP conducted environment[J]. IEEE Transactions on Electromagnetic Compatibility, 2002, 38(3):400-413.

[8] Wik M W. International standardization of immunity to high altitude nuclear electromagnetic pulse (HEMP)[C]// International Symposium on Electromagnetic Compatibility. IEEE, 1992:1-2-4/1-2.

[9] Xie Y Z, Wang Z J, Wang Q S, et al. High altitude nuclear electromagnetic pulse waveform standards: a review[J]. High Power Laser And Particl E Beams, 2003, 15(8):781-787.

[10] Yang J, Tang G, Cao J, et al. Study on Equivalent Circuit Model for HVDC Valve Thyristor Junction Temperature Calculation[J]. Proceedings of the Csee, 2013, 33(15):156-163.

[11] de l'Electricité U T, Telecom F. RDF 2000: Reliability data handbook[J]. UTE C 80, 2000, 810.

The 2018 International Power Electronics Conference

Generalized Stackelberg Game-theoretic Approach for Jointed Energy and Reserve Coordination of Electric Vehicles

Tianyang Zhao[1], Xuewei Pan[2], Lei Li[2], Fei Zhao[2], and Can Wang[2]

[1] Energy research institute @NTU, Nanyang Technology University, Singapore

[2] Harbin Institute of Technology, Shenzhen, China

Abstract—To realize the jointed energy and reserve coordination between the smart grid and electric vehicles (EVs), a Stackelberg game based scheme is proposed. In the game, the smart grid acts as the leader and EVs act as followers. The decision model of EVs is formulated based on network calculus theory, where quality-of-service constraints are considered. In the decision model of smart grid, the marginal utilities of energy and reserve for EVs are considered, resulting in the coupling among EVs and the smart grid, formulating a generalized Stackelberg game (GSG). The existence and uniqueness of the Stackelberg equilibrium are proved, considering the mathematical characteristics of proposed GSG. Introducing the optimal conditions for EVs' decision model, the GSG is reformulated as an optimization problem, solved by commercial software packages. Simulations have been carried out, and the results verify the effectiveness of the proposed method.

Keywords—*Electric vehicle, energy and reserve coordination, Stackelberg game, quality of service*

I. INTRODUCTION

As a powerful demand response sources, electric vehicles (EVs) are playing an important role in the coming energy internet. After connecting to the charging facilities, EVs can not only provide load shifting and other energy-related services for the power systems, but also provide spinning reserve, frequency regulation and other ancillary services [1]. When EVs provide ancillary services for the power systems, the reliability of the power systems and the efficiency of the market can be improved [2]. This papers aims to realize the flexibility collection from EVs, i.e., energy and reserve coordination among EVs, without the participant of aggregators.

As a hot research topic, the jointed energy and reserve management of EVs have been analyzed by some researches [1, 3]. In [3], the arrival process of EVs has been modeled as a Poisson process using the queuing theory, and EVs provide frequency regulation reserve for the power systems during their charging processes. In [1], it has been shown that EVs can provide not only frequency regulation reserve, but also a variety of ancillary services, e.g., spinning reserve.

When providing ancillary services, EVs should adjust the charging processes, e.g., charging time or charge-discharge rates. These adjustments may result in losses, e.g., energy losses, or windfalls, e.g., extra energy, to EV users. When the up reserve is called by the power systems, EVs need to reduce their charging rates, which may cause energy loss. If down reserve is called by the power systems, EVs should increase their charging rates, obtaining extra energy. For windfalls, the power systems may not need to compensate for the down reserve. However, for possible energy losses, the power systems need to compensate for EVs, to stimulate EVs for provision of up reserve.

How to compensate for the reserve provision of EVs is still under discussion. Based on the utility of energy for EV users, a price coordination method for a cluster of charging stations is proposed to realize the coordination among EVs in [4]. In [5], the price competition among charging stations is formulated as a Super-module game. These studies focus on the energy coordination, and the price mechanism for jointed energy and reserve coordination has not been taken into account. In [6], a most recent work has provided a jointed energy and reserve pricing scheme within one charging station. However, this method can only be applied in real-time management, and the quality of service (QoS) of charging services for EVs has not been taken into consideration.

Oriented from network calculus theory, the jointed energy and reserve coordination decision model for each EV is proposed to quantify the energy and reserve utility for EVs. One step further, the jointed energy and reserve price decision model of the smart grid is proposed, where the marginal utility of energy and reserve for each EV are embedded in the decision model to encourage EV users to provide up reserve. The interactive decision making between the smart grid and EVs is formulated as a generalized Stackelberg game (GSG). The smart grid acts as the leader and EVs act as followers. The existence and uniqueness of generalized Stackelberg equilibrium(GSE) are proved. The decision model of each EV is embedded into the decision model of the smart grid as the equilibrium constraints, formulating a mathematical programming with equilibrium constraints (MPEC). The MPEC is solved by commercial software, e.g., Gurobi. The main contributions of this paper are summarized as follows.

(1)A jointed energy and reserve coordination model is proposed for each EV user based on the network calculus theory, where the QoS of charging services are considered.

(2) The bidirectional price and energy/reserve interactively

decision making between the smart grid and EVs is formulated as a GSG.

(3) Considering the mathematical characteristics of proposed GSG, the existence and uniqueness of GSE are proved based on QVI.

The rest of this paper is organized as follows. The jointed energy and reserve scheduling decision model is proposed for EVs in Section II, together with the QoS constraints. In Section III, the energy and reserve coordination between EVs and the smart grid is formulated as a GSG. In Section IV, the mathematical property of the proposed GSG is analyzed. The solution method is given in Section V. Case studies are performed in Section VI. Conclusion of this work is drawn in Section VII.

II. JOINTED ENERGY AND RESERVE DECISION MODEL FOR ELECTRIC VEHICLES WITH QUALITY OF SERVICES CONSTRAINTS

The scale of EVs in this paper is N, and each EV needs to make decisions on its own charging plan and reserve plans. For the i-th EV, it should pay for the energy provided by the smart grid. At the same time, EV can benefit from providing spinning reserves for the smart grid. During the charging and reserve provision process, the utility function U_i of EV i can be expressed as follows

$$
\max_{P_i(t),R_{U,i}(t),R_{D,i}(t)} U_i = \sum_{t=t_{ar,i}}^{t_{dep,i}} [c_i P_i(t) - \frac{d_i}{2} P_i(t)^2
$$
$$
- \lambda_i(t) P_i(t) + \lambda_{U,i}(t) R_{U,i}(t) + \qquad (1)
$$
$$
\lambda_{D,i}(t) R_{D,i}(t)] \Delta t, \forall i \in \mathcal{N}
$$

where $P_i(t)$, $R_{U,i}(t)$ and $R_{D,i}(t)$ are charging rate, up reserve and down reserve of EV i in time slot t, respectively. $t_{ar,i}$ and $t_{dep,i}$ are the time slots when EV i arrives and leaves the charging facility. c_i represents willing price of per unit of energy paid by the EV user i for the charging service. d_i represents the relationship between the willing price and the charging energy, $d_i>0$. $\lambda_i(t)$, $\lambda_{U,i}(t)$ and $\lambda_{D,i}(t)$ are energy price, up reserve price and down reserve price for EV i in time slot t, which are set by the smart grid. Δt is the time step; $\mathcal{N}$ is the set of EVs.

Based on the network calculus theory [7], the energy arrival curve $A_i(t)$, the leaving curve $D_i(t)$ and the minimum energy arrival curve $D_{min,i}(t)$ can be expressed as follows

$$
A_i(t) = \min\{P_{max,i}(t-t_{ar,i})\Delta t, E_{re,i}\}, \forall i \in \mathcal{N} \qquad (2)
$$

$$
D_i(t) = \sum_{t_{ar,i}}^{t} P_i(t)\Delta t, \forall i \in \mathcal{N} \qquad (3)
$$

$$
D_{min,i}(t) = \min\{P_{max,i}(t_{dep,i}-t)\Delta t,
$$
$$
\max[E_{re,i}-P_{max,i}(t_{dep,i}-t)\Delta t, 0]\}, \forall i \in \mathcal{N} \qquad (4)
$$

where $E_{re,i}$ is the required energy of EV i. $P_{max,i}$ is the capacity of the charging facility accessed by EV i.

When EV provides reserve for the smart grid, it needs to meet the following power and energy constraints.

$$
P_i(t)+R_{D,i}(t) \leq P_{max,i}, \forall i \in \mathcal{N}, t \in [t_{ar,i}, t_{dep,i}] \qquad (5)
$$

$$
P_i(t)-R_{U,i}(t) \geq 0, \forall i \in \mathcal{N}, t \in [t_{ar,i}, t_{dep,i}] \qquad (6)
$$

$$
D_i(t)+R_{D,i}(t)\alpha_s \leq A_i(t), \forall i \in \mathcal{N}, t \in [t_{ar,i}, t_{dep,i}] \qquad (7)
$$

$$
D_i(t)-R_{U,i}(t)\alpha_s \geq D_{min,i}(t), \forall i \in \mathcal{N}, t \in [t_{ar,i}, t_{dep,i}] \qquad (8)
$$

$$
R_{D,i}(t), R_{U,i}(t) \geq 0, \forall i \in \mathcal{N}, t \in [t_{ar,i}, t_{dep,i}] \qquad (9)
$$

where $P_{max,i}$ is the maximum charging power of EV i; α_s is the duration maintain time required by spin reserve [8].

When EV is charging, the QoS constraints can be expressed as follows

$$
D_{min,i}(t) \leq D_i(t) \leq A_i(t), \forall i \in \mathcal{N}, t \in [t_{ar,i}, t_{dep,i}] \qquad (10)
$$

For the ease of presentation, the decision model of each EV i can be expressed as the following compact format

$$
\max_{x_i} U_i = -x_i^T \frac{Q_i}{2} x_i + C_i^T x_i
$$
$$
s.t. E_i x_i \leq h_i \qquad (11)
$$

where: $E_i x_i \leq h_i$ represents the constraints(2)-(10) that EV i needs to meet. C_i is a vector composed of c_i、$\lambda_i(t)$、$\lambda_{U,i}(t)$ and $\lambda_{D,i}(t)$. Q_i is a positive definite diagonal matrix consisting of d_i.

III. ENERGY AND RESERVE COORDINATION BASED ON GENERALIZED STACKELBERG GAME

A. Generalized Stackelberg game

For the energy and reserve coordination between the smart grid and EVs, a GSG is proposed, which is described in the following regular expression form, $\Gamma=\{\mathcal{N}\cup\mathcal{G}, \mathcal{S}, \mathcal{U}, g(x)\}$.

$\mathcal{S}$ is a $N+1$ dimensional tuple set. $\mathcal{G}$ is the smart grid. $\mathcal{U}$ is the N dimensional tuple equations, which represent the utility functions of EVs. $g(x)$ is the coupling constraints among the strategy spaces of smart grid and EVs. The GSG game Γ consists of the following two types of decision models and coupling constraint set.

(1) The decision model of the smart grid (refer to section III.B). The smart grid, as a leader, sets the energy prices and reserve prices for EVs, which are $\lambda_i(t)$, $\lambda_{U,i}(t)$ and $\lambda_{D,i}(t)$. The prices should meet constraints (13) - (14). The utility function of the smart grid is (12).

(2) Decision models for each EV (refer to section II). According to the energy and reserve prices released by the smart grid, EVs, as the followers, optimize their own energy and reserve plans, i.e., $P_i(t)$, $R_{U,i}(t)$ and $R_{D,i}(t)$, to maximize their utilities. The energy and reserve plans should meet the constraints (2)-(10). The utility function of EV i is (1).

(3) The coupling constraints g(x) among EVs and smart grid are constraints (16) and (17). These equations are proposed to stimulate EVs to provide up reserve and limit aggregated charging profile of EVs.

B. Decision model of the smart grid

For the smart grid, it can maximize its benefit through optimizing the price signals for each EV. The utility function

U_G can be expressed as follows

$$\max_{\lambda_i(t),\lambda_{U,i}(t),\lambda_{D,i}(t)} U_G = \sum_{t=i}^{T} [C_{ru}(t)\sum_{i\in\mathcal{N}} R_{U,i}(t)$$
$$+ C_{rd}(t)\sum_{i\in\mathcal{N}} R_{D,i}(t) - C_e(t)\sum_{i\in\mathcal{N}} P_i(t) \qquad (12)$$
$$- \sum_{i\in\mathcal{N}} \lambda_{U,i}(t) - \sum_{i\in\mathcal{N}} \lambda_{D,i}(t)]$$

where $C_e(t)$, $C_{ru}(t)$ and $C_{rd}(t)$ are energy price, up reserve price and down reserve price in the power market, respectively. Based on the constraint (9), EV can only provide spinning reserves, whose capacities are positive. It indicates that EV cannot consume reserves. The last two items in (12) show that the smart grid tends to reduce the reserve prices paid to EVs.

When the smart grid is setting the prices, it should be within feasible ranges, meeting the following constraints

$$\lambda_i(t) \geq 0, \lambda_i(t) \leq C_{emax}, \forall i \in \mathcal{N}, t \in [t_{ar,i}, t_{dep,i}] \qquad (13)$$

$$\lambda_{U,i}(t) \geq 0, \lambda_{D,i}(t) \geq 0, \forall i \in \mathcal{N}, t \in [t_{ar,i}, t_{dep,i}] \qquad (14)$$

where C_{emax} is the maximum energy price that the smart grid can set, e.g., the price cap.

When the smart grid is setting the prices, the marginal utilities obtained by EV i through providing energy and obtaining electricity should be considered [6]. EV i is willing to provide the up reserve for the grid only on the condition that the marginal utilities obtained through providing up reserve are no less than the marginal utilities of obtaining electricity. The energy price and up reserve price should meet the following constraints

$$\frac{\partial U_i}{\partial P_i(t)} \leq \frac{\partial U_i}{\partial R_{U,i}(t)}, \forall i \in \mathcal{N}, t \in [t_{ar,i}, t_{dep,i}] \qquad (15)$$

Substituting (1) into (15), we can obtain the following equation

$$c_i - d_i P_i(t) - \lambda_i(t) \leq \lambda_{U,i}(t), \forall i \in \mathcal{N}, t \in [t_{ar,i}, t_{dep,i}] \qquad (16)$$

At the same time, to limit the aggregated charging power of EVs, the decision of the smart grid should also consider the following constraint

$$\sum_{i\in\mathcal{N}} P_i(t) + \sum_{i\in\mathcal{N}} R_{D,i}(t) \leq P_{ev,max}(t), \forall t \in \mathcal{T} \qquad (17)$$

where $P_{ev,max}(t)$ is the charging power limit of EVs, set by the smart grid in time slot period t. $\mathcal{T}=[1,2\ldots,,T]$.

C. Generalized Starkerberg equilibrium

Definition 1: For the GSG Γ, there exists a strategy profile x^*, satisfying the following conditions

$$U_i(x_i^*, x_{-i}^*) \geq U_i(x_i, x_{-i}^*), \forall x_i \in \mathcal{S}_i, i \in \mathcal{N} \cup \mathcal{G} \qquad (18)$$

$$g(x^*) \leq 0 \qquad (19)$$

x^* is the generalized Stackelberg equilibrium (GSE) for GSG Γ. It can be interpreted as follows, according to the price

signals released by the smart grid, EVs optimize their own charging plan and reserve plans to maximize their own utilities. Considering the corresponding optimized response of EVs, the smart grid optimizes energy price and reserve prices for each EV to maximize its own benefits. At the same time, the charging plan and reserve capacity of EVs should meet the constraint (17), and the optimal price of the grid meets the constraint (16).

IV. EXISTENCE AND UNIQUENESS OF GENERALIZED STARKERBERG EQUILIBRIUM

A. Existence of Generalized Starkerberg Equilibrium

For the GSG problem Γ in section III.A, due to the existence of coupling constraints (16) and (17), the pure strategy may not exist, and the existence of GSE is the cornerstone to verify the proposed GSG. The existence of GSE is analyzed via quasi-variational inequality (QVI).

For the GSG problem Γ, the corresponding GSE should meet the following QVI

$$F(x)^T(y-x) \geq 0, \forall y \in \mathcal{S}, g(y) \leq 0 \qquad (20)$$

where x is the solution of the QVI. $F(x)$ is the partial derivative of each player's utility function on its own actions, as follows

$$F(x) = \begin{bmatrix} \nabla_{x_1} U_1(x) \\ \ldots \\ \nabla_{x_N} U_N(x) \\ \nabla_{x_G} U_G(x) \end{bmatrix} \qquad (21)$$

Theorem 1: There exits at least one GSE for the GSG problem Γ.

Proof: Since strategy space $\mathcal{S}_i$ is linear, the decision variables are bounded and continuous, the decision space of each player is tight and convex. At the same time, the utility function U_i of each player is continuously derivable and strictly concave. According to theorem 2.1 in [9], the solution (x, F) of QVI is the equilibrium solution of GNG problem $\mathcal{G}$. At the same time, as $\mathcal{S}$ is tight, there is at least one solution (x, F) for QVI (20). Then at least one GSE exists for Γ.

B. Uniqueness of Generalized Starkerberg Equilibrium

The uniqueness of GSE is important to ensure the stability of the proposed mechanism. For the proposed GSG problem Γ, this section proves that the GSG problem Γ has a unique GSE based on the characteristics of coupling constraint $g(x)$ and QVI (20).

Theorem 2: There is a unique GSE for GSG Γ.

Proof: From theorem 1, it is shown that GSG Γ has at least one GSE. For any GSE x^*,

$$x^{*T} JF(x^*)x^* = -\sum_{i\in\mathcal{N}} \sum_{t=t_{ar,i}}^{t_{dep,i}} [d_i P_i^*(t)^2] \qquad (22)$$

where $JF(x)$ is the first derivative of $F(x)$ with respect to x. As

x^* needs to meet users' QoS constraint (10), $P_i^*(t)$ are not all zeros when at least one EV needs to be charged. Furthermore, when $d_i>0$, (22) is less than 0, indicating $JF(x)$ is a strictly negative matrix. Then QVI (20) has a unique solution(x^*, $F(x^*)$). Combing with theorem 1, it has been shown that the GSG problem Γ has only one unique solution GSE x^* [10].

V. SOLUTION METHOD OF GSE

Considering the characteristics of GSG Γ, an MPEC reformulation is adopted to obtain the unique GSE. The optimal decision model (1)-(10) of each EV is reformulated as equilibrium constraints for the decision model of smart grid. The optimal conditions of each EV are as follows

$$
\begin{aligned}
& Q_i x_i - C_i + E_i^{\mathrm{T}} \mu_i = 0 \\
& E_i x_i \leq h_i \\
& \mu_i \geq 0 \\
& diag(\mu_i)(E_i x_i - h_i) = 0
\end{aligned} \tag{23}
$$

where μ_i is the Lagrangian multiplier corresponding to constraints $E_i x_i \leq F_i$. The fourth equation in (23) represents the relaxation condition between the Lagrangian multiplier and the corresponding constraint. Substituting (23) into the decision model of the smart grid, and the interactively decision model of each EV and the smart grid is thus converted into an MPEC. The existence of bilinear equation in (23) leads to the nonconvex of the constraint condition set of the strategy space, which makes it difficult to solve the MPEC. The complementary conditions in (23) are reformulated into the following constraints based on the big-M method:

$$
\begin{aligned}
& 0 \leq \mu_i \leq B_i M \\
& 0 \leq h_i - E_i x_i \leq (1 - B_i)M
\end{aligned} \tag{24}
$$

where M is a larger scalar, setting to 1 000 000 in this paper. B_i is a binary variable vector corresponding to each constraint.

Based on (23) and (24), the GSG Γ is reformulated to the following optimization problem

$$
\max_x U_G
$$

$$
s.t.\begin{cases}
Mx_N + Nx_{-N} \leq m \\
x_N \in \mathcal{S}_N \\
Q_i x_i - c_i + E_i^{\mathrm{T}} \mu_i = 0, \forall i \in \mathcal{N} \\
E_i x_i \leq h_i, \forall i \in \mathcal{N} \\
\mu_i \geq 0, \forall i \in \mathcal{N} \\
0 \leq \mu_i \leq B_i M, \forall i \in \mathcal{N} \\
0 \leq h_i - E_i x_i \leq (1 - B_i)M, \forall i \in \mathcal{N}
\end{cases} \tag{25}
$$

where $Mx_N + Nx_{-N} \leq m$ expresses constraints (16) and (17). As shown in (25), the optimization problem is a mixed-integer linear programming (MILP) problem. In this paper, Gurobi is adopted to solve this MILP to obtain the equilibrium solution of GSG.

VI. NUMERICAL RESULT

A. Case description

To verify the effectiveness of the proposed method, the driving behaviour and energy requirement data of Beijing is adopted. The objective of this section is to study the variation of utility for each EV and social welfare under different interactive modes. The wholesale market price is obtained from [11]. The reserve price for each period is 0.062¥/kWh [11]. The price cap C_{emax} of is set to 0.67¥/kWh. Δt is set to 1h. The maximum price c_i that each EV is willing to pay is 0.5004¥/kWh and d_i is 0.01¥/kWh². The maximum charging power of each charging facility is set to 7 kW; the maximum charge power of all EVs in the GSG is 25 kW.

The following simulation scenarios are proposed to reveal the effectiveness of the proposed interactive method, as shown in table I.

TABLE I
SIMULATION SCENARIOS

Scenario	Charging strategy	Reserve provision
I	Uncontrolled charging	No
II	Generalized Nash game	No
III	Generalized Stackelberg game	Yes

In table I, the uncontrolled charging refers to the scenario that each EV charges according to its own energy arrival curve $A_i(t)$ and does not provide reserve service for the grid. Scenario I is the benchmark case, which demonstrates the charging characteristic of EV under natural condition. For the generalized Nash game scenario, each EV will manage its charging plan to influence the charging price to achieve the maximized benefits based on linear supply and demand theory. In this case, EV will not provide the reserve for the grid. Scenario II and III are compared to show the variation of social welfare and EVs' own benefits while providing the reserve service.

B. Result analyses

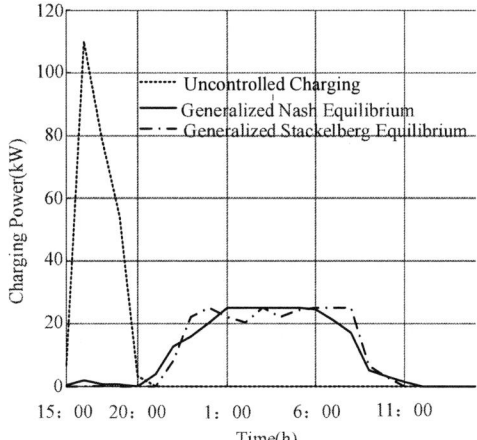

Fig.1 Charging profile of EVs under different strategies

The charging curves of uncontrolled charging, generalized

Nash game, and GSG are shown in Fig.1. It can be observed that, for the uncontrolled charging, the aggregated charging rate reaches 109.93 kW, which exceeds the set maximum EVs charging power (25 kW). On the contrary, the charging profiles of EVs can be limited to the specified limit, which either bases on Generalized Nash equalization or GSG. Furthermore, for the scenario of GSG, EVs will provide reserves for the smart grid and thus are not charging at full rate in the low price periods. For instance, between 1: 00-5: 00, the charging rate of EVs is not 25kW under the GSE.

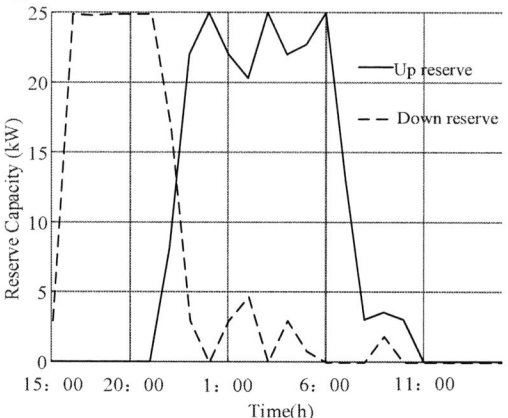

Fig.2 Reserve capacities under generalized Stackelberg equilibrium

In Fig.2, the reserve capacities provided by EVs to the grid are illustrated. Due to the constraint (16), EVs are ensured to obtain sufficient compensation for up reserve provision during 22:00 to 10:00. Compared with Fig.1, Fig.2 shows that during the high energy price periods (, e.g., 15: 00-20: 00), EVs will not provide up reserve. Because energy price is comparatively high during these periods, EVs expect to obtain potential benefit through providing down reserve. The down reverse provided by EV is close to the peak of power rating.

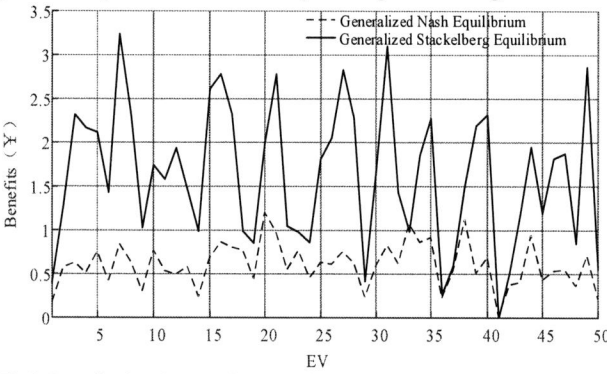

Fig.3. Benefit of each EV under generalized Nash equilibrium and generalized Stackelberg equilibrium

Fig. 3 gives the benefit of each EV under generalized Nash equilibrium and generalized Stackelberg equilibrium. As shown by Fig. 3, when EVs provide reserves to the grid, the benefits of each EV are improved in varying degrees. Compared to GNE, the overall revenue of EVs obtained by GSE is increased by 169.06% and reaches the value of 81.47¥ (For GNE, the overall revenue of EVs is 20.28¥). In

addition, social benefits are increased to 85.44 ¥, which are increased by 182.17%. This is also caused by that EVs are providing up and down reserve to the grid under GSE regulation.

VII. CONCLUSION

In this paper, a GSG method is proposed to realize the jointed energy and reserve management of EVs and the smart grid. In the formulated game, the smart grid acts as the leader to maximize its utility by optimizing the energy and reserve prices for the EVs. Responding to the price signals, EVs act as follower to maximize their own utilities by optimizing the charging and reserve provision plans. The mathematical characteristic of the proposed game has been analyzed. The MPEC reformulation technique is adopted to obtain the equilibrium.

Simulation results show that, the proposed method can, 1) prevent high charging peaks, 2) stimulate EVs to provide reserve, and 3) increase the utility of EVs and society simultaneously.

VIII. ACKNOWLEDGMENT

This work was supported by the National Natural Science Foundation of China (Grant No. 51707047), the Basic Research Plan in Shenzhen City (Grant No. JCYJ 20160531192413576) and Harbin Institute Technology fund (CE29100005).

REFERENCES

[1] R. Li, Q. Wu, S. S. Oren. "Distribution locational marginal pricing for optimal electric vehicle charging management," *IEEE Trans. on Power Systems*, vol.29, no.1, pp: 203-211, 2014.

[2] M. Shafie-Khah, E. Heydarian-Forushani, M.E.H. Golshan, et al. "Optimal trading of plug-in electric vehicle aggregation agents in a market environment for sustainability," *Applied Energy*, vol.162, pp: 601-612, 2016.

[3] A. Y. Lam, K. C. Leung, V. O. Li. "Capacity management of vehicle-to-grid system for power regulation services," In *2012 IEEE Third International Conferences on Smart Grid Communications*, pp: 442-447, IEEE, 2012.

[4] S. Bahrami, P. Mostafa. "Game theoretic based charging strategy for plug-in hybrid electric vehicles," *IEEE Tran. on Smart Grid* , vol. 5, no. 5, pp: 2368-2375, 2014.

[5] W. Lee, X. Lin, S. Robert, and W.W. Vincent. "Electric vehicle charging stations with renewable power generators: A game theoretical analysis," *IEEE Trans. on Smart Grid*, vol.6, no. 2, pp: 608-617, 2015.

[6] T. Zhao, Y. Li, X. Pan, et al. "Real-time Optimal Energy and Reserve Management of Electric Vehicle Fast Charging Station: Hierarchical Game Approach," *IEEE Trans. on Smart Grid*, in press, 2017.

[7] M.A. Zafer, M. Eytan. "A calculus approach to energy-efficient data transmission with quality-of-service constraints," *IEEE/ACM Trans. on Networking*. vol.17, no. 3, pp: 898-911, 2009.

[8] N. Li, U. Canan, M.C. Emil, et al. "Flexible operation of batteries in power system scheduling with renewable energy," *IEEE Trans. on Sustainable Energy*. vol.7, no. 2 pp:685-696, 2016.

[9] F. Facchinei, J. Pang. "Finite-dimensional variational inequalities and complementarity problems," *Springer Science & Business Media*, 2007.

[10] P.T. Harker. "Generalized Nash games and quasi-variational inequalities," *European journal of Operational research*, vol. 54, no. 1 pp: 81-94, 1991.

[11] California ISO - Price Maps[Online]. Available: http : //www. caiso. com/pages/pricemaps

Impedance Influence Analysis of Phase-Locked Loops on Three-Phase Grid-Connected Inverters

Yuncheng Wang, Xin Chen, Yang Zhang, Jie Chen, Chunying Gong
Nanjing University of Aeronautics and Astronautics, Nanjing, P.R.China
E-mail: wychengcc@126.com

Abstract—For grid-connected inverters, phase-locked loop (PLL) is usually adopted for the injected grid current to achieve phase tracking of the grid voltages, which means the PLL will have an inevitable effect on the output impedance characteristics and the impedance stability of grid-connected inverters. Hence, the harmonic linearization method was firstly used to derive the positive- and negative-sequence output impedance models of the three-phase grid-connected inverter with and without PLLs. Second, two kinds of the commonly used PLL circuits that are synchronous reference frame phase locked loop (SRF-PLL) and dual second order generalized integrator frequency locked loop (DSOGI-FLL) were employed to compare and analyze their frequency characteristics. Finally, the simulation and experimental results were given to verify the impedance models of the grid-connected inverter with different PLLs.

Keywords—phase-locked loop, grid-connected inverter, impedance models, impedance stability.

I. INTRODUCTION

With the distributed generation (DG) widely used in recent years, more and more renewable energies, such as PV and wind power, are connected with grid by power electronic equipments[1]. Among these devices, grid-connected inverters are used as the dominant interactive devices that undertake the important mission of transmitting the electricity created by the renewable energies to grid[2].

In a grid-connected inverter, PLL is usually adopted to lock the phase and frequency of grid voltages so as to control the energy transmission effectively between a grid-connected inverter and grid. Therefore, PLL is an indispensable component of a grid-connected inverter.

When PLL is analyzed in the last few years, many literatures have studied how to improve the PLLs under the unbalanced and distorted grid conditions for the stable and reliable operation of grid-connected inverters. Ref. [3] concluded and summarized the PLL synchronization methods under the unbalanced and distorted grid conditions, and introduced the open loop and closed loop phase-locked methods for three-phase grid-connected inverters. When PLL was developed in Ref. [4], the PLL structure was improved and internal regulator parameters were optimized based on synchronous reference frame and digital filter to enhance the robustness in the unbalanced and distorted grid.

However, these literatures didn't study the influences of the PLL on the inverter stability.

When studying the system stability of the grid-connected inverters, the commonly used state space method pays attention to the system controller stability in the inverter. However, when the influences of PLL and grid impedance are considered, the stability analysis will be more complicated. Hence, the impedance-based analysis method was proposed[5], which pays attention to interaction impedance stability between a grid-connected inverter and a grid by modeling them as two individual subsystems with the output impedance characteristics. After that, the impedance stability of the grid-connected inverter is studied by the impedance-based stability criterion[6]. To derive the output impedance models of a grid-connected inverter, different kinds of impedance model methods are proposed. Among these, it is preferred to use the harmonic linearization method, which injects the positive- and negative-sequence small signal voltage perturbations at the output terminal of the inverter. Then the current responses at the perturbation frequencies are measured, and the positive- and negative-sequence impedance models of the inverter are finally derived

After obtaining the impedance models of the grid-connected inverters, the literatures studied the impact of PLL frequency characteristics on the impedance stability of a grid-connected inverter. Ref. [7] modeled the positive- and negative-sequence impedance models of the three-phase grid-connected inverter with the PLLs. However, it only considered the case of the single L filter, but didn't conclude the case of LCL filter. Therefore, it is very essential to analyze the impedance models of a three-phase LCL-type grid-connected inverter.

In this paper, a typical three-phase LCL-type grid-connected inverter was set up to compare the frequency characteristics of different kinds of PLLs and model the output impedances of the inverter for studying the effects of PLLs on the inverter impedances. First, the positive- and negative-sequence output impedance models of the three-phase LCL-type grid-connected inverter with and without PLL were derived by the harmonic linearization method. Second, the frequency characteristics of SRF-PLL and DSOGI-FLL were derived and studied, and the output impedance models of the inverter with the two PLLs were compared. Then, the frequency characteristics of the PLLs and the impedance models of the inverter with different PLLs were verified by simulation results. Finally, a 10kVA three-phase LCL-type grid-connected inverter prototype was set up to make the experimental verification.

II. THE IMPEDANCE MODELS OF THREE-PHASE GRID-CONNECTED INVERTER

The typical block diagram of the three-phase LCL-type grid-connected inverter is shown in Fig. 1.

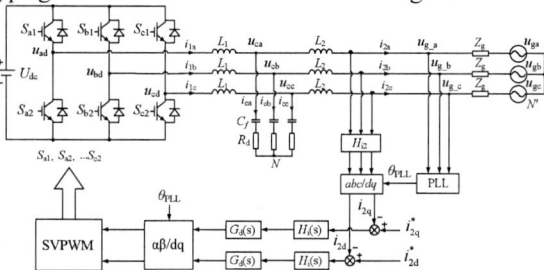

Fig. 1. A typical three-phase grid-connected inverter.

In Fig.1, U_{dc} is the dc input voltage. The LCL filter is made up of L_1, C_f, and L_2. To damp the filter, the passive resistor R_d is used to connect with C_f in series. u_{ca}, u_{cb} and u_{cc} are the capacitor branch voltages. i_{2a}, i_{2b} and i_{2c} are the injected grid currents. The grid impedance Z_g is connected with the three-phase ideal grid voltages, u_{ga}, u_{gb} and u_{gc}, which leads to the voltages at the point of common coupling (PCC) expressing of u_{g_a}, u_{g_b} and u_{g_c}. H_{i2} represents the sampling gain of the injected grid current sensor, and $H_i(s)$ is the current loop regulator function. θ_{PLL} denotes the PLL phase angle. $G_d(s)$ indicates the equivalent transfer function of the system sampling delay loop. In the lab, a three-phase LCL-type grid-connected inverter is set up as a prototype, whose key parameters are written in Table I.

TABLE I
PARAMETERS OF THE THREE-PHASE GRID-CONNECTED INVERTER

Symbol	Meaning	Value
U_{dc}	DC Input voltage	670V
U_g	Grid voltage(RMS)	235V
P_n	Output power	10kVA
f_1	Fundamental frequency	50Hz
f_{sw}	Switching frequency	20kHz
L_1	Inverter-side inductor	1.5mH
C_f	Filter capacitor	6.8μF
R_d	Damping resistor	1.7Ω
L_2	Grid-side inductor	0.2mH

A. The Inverter Ouput Impedances without PLL

Firstly the effect of a PLL is ignored, and the phase angle θ_{PLL} is totally desired. The harmonic linearization method and the symmetrical component method are employed to decompose the three-phase grid-connected inverter into positive- and negative-sequence subsystems. When the injected grid currents are sampled, they are transformed to dq-domain components indicated as $i_{2d}(t)$ and $i_{2q}(t)$, whose equations in frequency domain are written by the Fourier transforming:

$$\mathbf{I}_{2d}[f]=\begin{cases} I_1\cos\varphi_{i1}, & dc \\ \mathbf{I}_p, & f=\pm(f_p-f_1) \\ \mathbf{I}_n, & f=\pm(f_n+f_1) \end{cases} \quad (1)$$

$$\mathbf{I}_{2q}[f]=\begin{cases} I_1\sin\varphi_{i1}, & dc \\ \mp j\mathbf{I}_p, & f=\pm(f_p-f_1) \\ \pm j\mathbf{I}_n, & f=\pm(f_n+f_1) \end{cases} \quad (2)$$

where $\mathbf{I}_p$ and $\mathbf{I}_n$ represent the frequency domain expressions of the positive- and negative-sequence current perturbations respectively, shown as $(I_p/2)e^{\wedge}(\pm j\varphi_{ip})$ and $(I_n/2)e^{\wedge}(\pm j\varphi_{in})$ respectively.

After that the dq-domain current components are controlled by the current control loop, whose outputs are used to generate the duty cycle of the inverter. Finally, the positive- and negative-sequence output impedance models of the inverter without PLL are expressed as:

$$\begin{cases} Z_{id\_p}(s)=\dfrac{L_1s+L_2s+L_1L_2C_fs^3\big/(C_fR_ds+1)+K_{pwm}H_i(s-j\omega_1)G_d(s)}{1+L_1C_fs^2/(C_fR_ds+1)} \\[4mm] Z_{id\_n}(s)=\dfrac{L_1s+L_2s+L_1L_2C_fs^3\big/(C_fR_ds+1)+K_{pwm}H_i(s+j\omega_1)G_d(s)}{1+L_1C_fs^2/(C_fR_ds+1)} \end{cases}$$
(3)

As shown in (3), when the influence of PLL is ignored, the inverter is no longer affected by the PCC voltages, and the current regulator and the filter will have an influence on the impedance characteristics of the grid-connected inverter together. So the inverter is equivalent as the parallel circuit of the ideal current source and the ideal output impedance. Fig. 2 shows the positive-sequence and negative-sequence small signal representations without PLL.

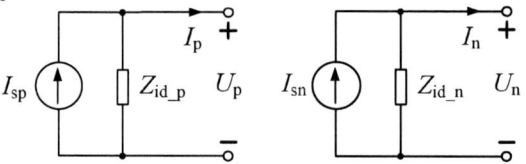

(a) positive-sequence　　(b) negative-sequence
Fig. 2. The small-signal representations of the three-phase grid-connected inverter without PLL.

In Fig. 2, $I_{sp}(s)$ and $I_{sn}(s)$ are the positive- and negative-sequence ideal current sources respectively. $Z_{id\_p}(s)$ and $Z_{id\_n}(s)$ are positive- and negative-sequence ideal output impedances respectively. $I_p(s)$ and $I_n(s)$ are the positive- and negative-sequence injected grid currents respectively. $U_p(s)$ and $U_n(s)$ are the positive-sequence and negative-sequence PCC voltages respectively.

B. The Inverter Ouput Impedances with PLL

When the effect of PLL is involved, the phase angle of the PLL will be affected by the small signal voltage perturbation in the PCC voltages. Then the injected grid currents in dq-domain are obtained as $i_{2d}(t)$ and $i_{2q}(t)$, whose expression in frequency domain are written as:

$$\mathbf{I}_{2d}[f]=\begin{cases} I_1\cos\varphi_{i1}, & dc \\ \mp 2jI_1\sin\varphi_{i1}T_p(\pm j2\pi f_p)U_p+\mathbf{I}_p, & f=\pm(f_p-f_1) \\ \pm 2jI_1\sin\varphi_{i1}T_n(\pm j2\pi f_n)U_n+\mathbf{I}_n, & f=\pm(f_n+f_1) \end{cases} \quad (4)$$

$$\mathbf{I}_{2q}[f]=\begin{cases} I_1\sin\varphi_{i1}, & dc \\ \pm 2jI_1\cos\varphi_{i1}T_p(\pm j2\pi f_p)U_p\mp j\mathbf{I}_p, & f=\pm(f_p-f_1) \\ \mp 2jI_1\cos\varphi_{i1}T_n(\pm j2\pi f_n)U_n\pm j\mathbf{I}_n, & f=\pm(f_n+f_1) \end{cases} \quad (5)$$

where $T_p(s)$ and $T_n(s)$ are the positive- and negative-sequence frequency characteristic functions of the PLL respectively. $\mathbf{U}_p$ and $\mathbf{U}_n$ are the frequency domain expressions of the positive- and negative-sequence voltage perturbations respectively, expressed as $(U_p/2)e^{\wedge}(\pm j\varphi_{up})$ and $(U_n/2)e^{\wedge}(\pm j\varphi_{un})$ respectively.

Finally, the positive- and negative-sequence output impedance models of the inverter are shown as:

$$
\begin{cases}
Z_p(s) = \dfrac{L_1 s + L_2 s + \dfrac{L_1 L_2 C_f s^3}{C_f R_d s + 1} + K_{pwm} H_i(s - j\omega_1) G_d(s)}{1 + \dfrac{L_1 C_f s^2}{C_f R_d s + 1} - K_{pwm}[H_i(s - j\omega_1) + \dfrac{U_1}{K_{pwm} I_1 e^{j\varphi_{i1}}}] T_p(s) I_1 e^{j\varphi_{i1}}} \\[2em]
Z_n(s) = \dfrac{L_1 s + L_2 s + \dfrac{L_1 L_2 C_f s^3}{C_f R_d s + 1} + K_{pwm} H_i(s + j\omega_1) G_d(s)}{1 + \dfrac{L_1 C_f s^2}{C_f R_d s + 1} - K_{pwm}[H_i(s + j\omega_1) + \dfrac{U_1}{K_{pwm} I_1 e^{-j\varphi_{i1}}}] T_n(s) I_1 e^{-j\varphi_{i1}}}
\end{cases} \tag{6}
$$

By comparing (3) and (6), $Z_{PLL\_p}(s)$ and $Z_{PLL\_n}(s)$ are defined as the positive- and negative-sequence equivalent impedances of the PLL respectively. So $Z_p(s)$ and $Z_n(s)$ can be expressed as the parallel forms of $Z_{PLL\_p}(s)$ and $Z_{PLL\_n}(s)$, the equivalent impedances of the PLL, and $Z_{id\_p}(s)$ and $Z_{id\_n}(s)$, the ideal output impedances without the PLL, shown as:

$$
\begin{cases}
Z_p(s) = \dfrac{Z_{PLL\_p}(s) Z_{id\_p}(s)}{Z_{PLL\_p}(s) + Z_{id\_p}(s)} \\[1.5em]
Z_n(s) = \dfrac{Z_{PLL\_n}(s) Z_{id\_n}(s)}{Z_{PLL\_n}(s) + Z_{id\_n}(s)}
\end{cases} \tag{7}
$$

As shown in (3), when the PLL is introduced, the inverter output impedances can be equivalent as the ideal output impedances in parallel with the additional impedances. When concluding the PLL, among the PCC voltages to the current control loop there will exist another feedback loop in parallel.

The another feedback loop of the PLL not only makes the grid voltages affect the characteristics of the current control loop, but also gets the PLL included in the current control loop. Therefore, the positive- and negative-sequence small signal representations with PLL are shown in Fig. 3.

In Fig. 3, $[-U_p(s)]/Z_{PLL\_p}(s)$ and $[-U_n(s)]/Z_{PLL\_n}(s)$ are the positive- and negative-sequence voltage controlled current sources respectively. $Z_{PLL\_p}(s)$ and $Z_{PLL\_n}(s)$ are positive- and negative-sequence equivalent impedances of the PLL respectively.

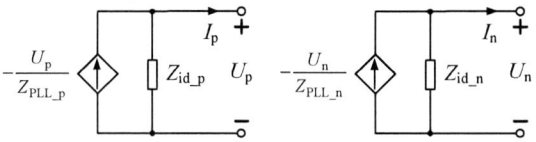

(a) positive-sequence   (b) negative-sequence

Fig. 3. The positive- and negative-sequence small-signal representations of the three-phase grid-connected inverter with PLL.

III. THE FREQUENCY CHARACTERISTICS OF THE PLLs

According to the grid conditions and requests of the grid-connected inverters, there are different kinds of PLLs to choose in the inverter. By now, the most commonly used PLL circuits are SRF-PLL and DSOGI-FLL respectively, whose frequency characteristics are derived and studied below.

A. The Frequency Charactertistic of SRF-PLL

In the three-phase grid-connected inverter, the block diagram of SRF-PLL is shown in Fig. 4.

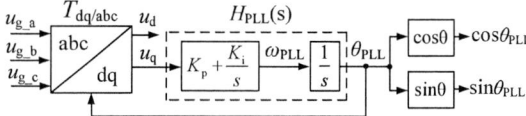

Fig. 4. Block diagram of SRF-PLL.

In Fig. 4, the PCC voltages are sampled and transformed by $T_{dq/abc}$. The PLL regulator comprises the proportional-integral (PI) controller and the integrator, therefore the total transfer function is indicated as $H_{PLL}(s)$, which is expressed as $(K_p + K_i/s)(1/s)$. In order to study the frequency characteristics of SRF-PLL, the harmonic linearization method is adopted, which injects the small-signal voltage perturbations of positive- and negative-sequence into the output terminal of the inverter at the PCC. Finally, $T_{p\_PLL}(s)$, the positive-sequence frequency characteristic function from $\mathbf{U}_p$ to $\cos(\theta_{PLL})$, is written as:

$$
T_{p\_PLL}(s) = \frac{1}{2} \cdot \frac{H_{PLL}(s - j\omega_1)}{1 + U_1 H_{PLL}(s - j\omega_1)} \cdot G_d(s) \tag{8}
$$

Then the negative-sequence frequency characteristic function $T_{n\_PLL}(s)$ from $\mathbf{U}_n$ to $\cos(\theta_{PLL})$ is shown as:

$$
T_{n\_PLL}(s) = \frac{1}{2} \cdot \frac{H_{PLL}(s + j\omega_1)}{1 + U_1 H_{PLL}(s + j\omega_1)} \cdot G_d(s) \tag{9}
$$

Make the comparison between $T_{p\_PLL}(s)$ and $T_{n\_PLL}(s)$, and it is found that the dominant diversity is $H_{PLL}(s)$ in the PLL, which is written as the leftward and rightward phase shift by $j\omega_1$ for positive- and negative-sequence frequency characteristics of the PLL. And the reason for that is $u_q(t)$ has diverse transformation form for the positive- and negative-sequence voltages perturbations in the coordinate transformation.

B. The Frequency Charactertistic of DSOGI-FLL

Fig. 5 shows the block diagram of DSOGI-FLL.

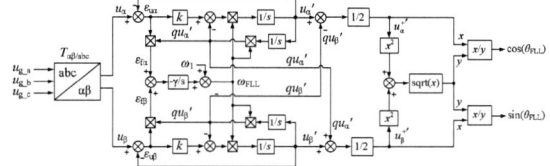

Fig. 5. Block diagram of DSOGI-FLL.

In Fig. 5, the PCC voltages are sampled and transformed by $T_{\alpha\beta/abc}$. k is the damping factor and $q = e^{j(-\pi/2)}$ represents a $90°$ phase-shift operator. ω_{FLL} is the angular frequency and γ is the integration gain. In order to study the frequency characteristic of DSOGI-FLL, the harmonic linearization method is employed here to obtain the positive-sequence frequency characteristic function $T_{p\_FLL}(s)$ from $\mathbf{U}_p$ to $\cos(\theta_{PLL})$:

$$
T_{p\_FLL}(s) = \frac{k\omega_{FLL}s + jk\omega_{FLL}^2}{4U_1(s^2 + k\omega_{FLL}s + \omega_{FLL}^2)} \cdot G_d(s) \tag{10}
$$

Then the negative-sequence frequency characteristic function $T_{n\_FLL}(s)$ from $\mathbf{U}_n$ to $\cos(\theta_{PLL})$ is shown as:

$$T_{n\_FLL}(s) = \frac{k\omega_{FLL}s - jk\omega_{FLL}^2}{4U_1(s^2 + k\omega_{FLL}s + \omega_{FLL}^2)} \cdot G_d(s) \quad (11)$$

When analyzing the frequency characteristics of DSOGI-FLL, it is important to design k and γ to make the dynamic and stability meet the requests. By comparing $T_{p\_FLL}(s)$ and $T_{n\_FLL}(s)$ it is found that the difference between the positive- and negative-sequence voltage perturbations lead to the variable expressions of $T_{p\_FLL}(s)$ and $T_{n\_FLL}(s)$. At the fundamental frequency f_1, $T_{p\_FLL}(s)$ is a band-pass filter with no phase shift, and $T_{n\_FLL}(s)$ is a trap filter of negative infinity magnitude with no phase shift, which makes the negative-sequence voltage perturbation have no influence on the phase angle.

IV. THE FREQUENCY CHARACTERISTICS AND OUTPUT IMPEDANCE VERIFICATION

A. Simulation Verification of SRF-PLL and DSOGI-FLL

For the sake of making the simulation confirmation of the positive- and negative-sequence frequency characteristics of SRF-PLL and DSOGI-FLL, the PLL circuits are both built in the software. The theoretical plots and simulation results of the positive- and negative-sequence frequency characteristics of SRF-PLL and DSOGI-FLL are shown in Fig. 6.

As shown in Fig. 6, $T_{p\_PLL}(s)$ and $T_{n\_PLL}(s)$ are the positive- and negative-sequence frequency characteristic functions of SRF-PLL respectively. $T_{p\_FLL}(s)$ and $T_{n\_FLL}(s)$ are the positive- and negative-sequence frequency characteristic functions of DSOGI-FLL respectively. It can be easily found that the simulation results are in accordance with the theoretical curves, which verifies the theoretical analysis of the frequency characteristics of the PLLs.

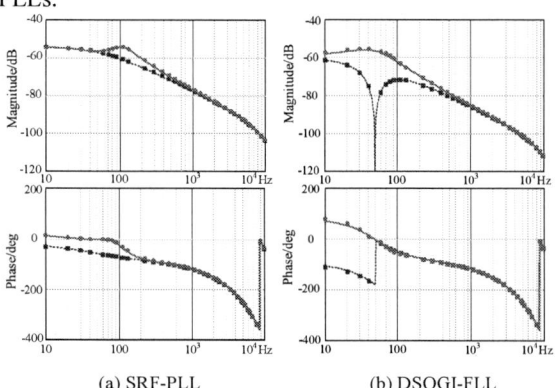

(a) SRF-PLL   (b) DSOGI-FLL

Fig. 6. Plots of the frequency characteristics of the PLLs: (a) Theoretical curve: $T_{p\_PLL}(s)$ red solid line, $T_{n\_PLL}(s)$ blue dash line; (b) Theoretical curve: $T_{p\_FLL}(s)$ red solid line, $T_{n\_FLL}(s)$ blue dash line; Dots: simulation results

It is also found that the frequency characteristics of $T_{p\_PLL}(s)$ and $T_{p\_FLL}(s)$ are similar and both behave as the band-pass filters, whose phases are both zero at the fundamental frequency and hence they can make the phase track of the grid voltages without phase shifting. The differences between $T_{n\_PLL}(s)$ and $T_{n\_FLL}(s)$ behave as that $T_{n\_PLL}(s)$ is a band-pass filter while $T_{n\_FLL}(s)$ is a trap filter, which will make the negative-sequence voltage perturbation of DSOGI-FLL differ from SRF-PLL.

In order to specify the controller parameters of SRF-PLL, the transfer function $T_{PLL}(s)$ is defined as the uniform expression of the PLL without considering the sampling delay function written as:

$$T_{PLL}(s) = \frac{H_{PLL}(s)}{1 + U_1 H_{PLL}(s)} \quad (12)$$

Then, $T_{PLL}(s)$ can be equivalent transformed as:

$$T_{PLL}(s) = \frac{U_1 K_p s + U_1 K_i}{s^2 + U_1 K_p s + U_1 K_i} = \frac{2\xi\omega_n s + \omega_n^2}{s^2 + 2\xi\omega_n s + \omega_n^2} \quad (13)$$

where ω_n is the natural angular frequency and ξ is the damping ratio.

When making a compromise of the steady and dynamic state characteristics of the inverter, ξ is designed as 0.707. Then f_{BW} is defined as the bandwidth of the SRF-PLL, so ω_{BW}, which equals to $2\pi f_{BW}$, is expressed as the angular frequency. Finally, the parameters of the SRF-PLL regulator are obtained as:

$$K_i = \frac{-2\xi^2\omega_{BW}^2 + \omega_{BW}^2\sqrt{1 + 4\xi^4}}{V_1} \quad (14)$$

$$K_p = \sqrt{\frac{4\xi^2 K_i}{V_1}} \quad (15)$$

When choosing the bandwidth f_{BW}, the crucial need is to ensure that SRF-PLL attenuates the low frequency harmonics of the grid voltages and the power factor of the injected grid currents meet the requirements with considering the effects of the grid impedance. Here, the compromise of the bandwidth is made to set f_{BW} as 100Hz.

As for DSOGI-FLL, when designing the damping factor k and the integration gain γ, it is found that the settling time of DSOGI response t_{sDSOGI} can be approximated as:

$$t_{sDSOGI} = \frac{10}{k\omega_{FLL}} \quad (16)$$

It can be found that, the value of k affects the stability and dynamic of DSOGI. The lower the value of k, the bigger the damping and the stronger the filtering effect, but the response will be slower and the settling time will be longer. However, on the other hand, a very high value of k will lead to the faster response and the shorter settling time, and also make the damping smaller and filtering effect weaker. So the tradeoff between stability and dynamic can be achieved with $k = \sqrt{2}$. Similarly, as the integration gain of FLL, γ will also affect the stability and dynamic. The settling time of FLL t_{sFLL} can expressed as:

$$t_{sFLL} = \frac{5k\omega_{FLL}}{\gamma U_1^2} \quad (17)$$

When k is chosen as $\sqrt{2}$, the compromise is made between stability and dynamic to have γ as 0.2.

B. Simulation Verification of Inverter Impedances with PLLs

When making the simulation confirmation of the positive- and negative-sequence output impedance models of the inverter with different kinds of PLLs, the inverter circuit is set up in Saber software. The theoretical plots and simulation results of the positive- and negative-sequence output impedance models of the inverter with SRF-PLL and DSOGI-FLL are shown in Fig. 7 respectively. $Z_{p\_PLL}(s)$ and $Z_{n\_PLL}(s)$ are the positive- and negative-sequence output impedances of the inverter with SRF-PLL respectively. $Z_{p\_FLL}(s)$ and $Z_{n\_FLL}(s)$ are the positive- and negative-sequence output impedances of the inverter with DSOGI-FLL respectively.

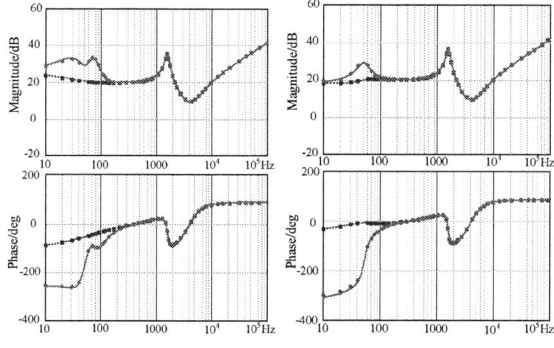

(a) with SRF-PLL   (b) with DSOGI-FLL

Fig. 7. Plots of inverter output impedances using PLLs: (a) Theoretical curve: $Z_{p\_PLL}(s)$ red solid line, $Z_{n\_PLL}(s)$ blue dash line; (b) Theoretical curve: $Z_{p\_FLL}(s)$ red solid line, $Z_{n\_FLL}(s)$ blue dash line; Dots: simulation results

As shown in Fig. 7, the simulation results are same as the theoretical curves, which testify the validity of the impedance modeling of the inverter with different PLLs. It is found that each of the inverter output impedances, which are $Z_{p\_PLL}(s)$ and $Z_{n\_PLL}(s)$, $Z_{p\_FLL}(s)$ and $Z_{n\_FLL}(s)$, differs from the other below the frequency of 100Hz. It is because that the PLL has the dominant effect on the output impedances of the inverter during this frequency range, which shows the importance of considering the PLL influence during the output impedance modeling of the inverter.

C. Experimental Verification of Inverter Impedances with PLLs

It is necessary and important to build a three-phase LCL-type grid-connected inverter in the lab and make the experimental validation of the positive- and negative-sequence output impedance models of the inverter, whose key parameters are listed in Table I.

In the impedance measurement process, the linear amplifier model (LAM) and the Venable 350C, which is known as frequency response analyzer (FRA), are adopted to measure the impedance models of the inverter with different PLLs. The theoretical plots and experimental results of the inverter output impedances with SRF-PLL and DSOGI-FLL are shown in Fig. 8 respectively.

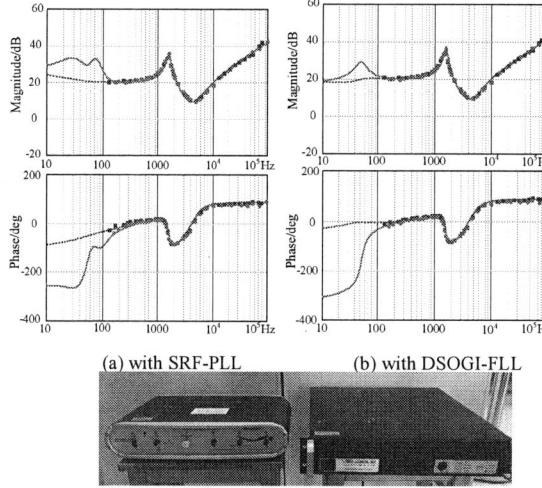

(a) with SRF-PLL   (b) with DSOGI-FLL

(c) FRA and LAM

Fig. 8. Plots of inverter output impedances using PLLs: (a) Theoretical curve: $Z_{p\_PLL}(s)$ red solid line, $Z_{n\_PLL}(s)$ blue dash line; (b) Theoretical curve: $Z_{p\_FLL}(s)$ red solid line, $Z_{n\_FLL}(s)$ blue dash line; Dots: experimental results; (c) picture of FRA and LAM.

As shown in Fig. 8, in the experimental process of measuring impedances, it begins from 100 Hz because the FRA has the lowest frequency limitation and the influences of the fundamental signal are still existing. Consequently, the main differences of the inverter with SRF-PLL and DSOGI-FLL can't be shown clearly unfortunately. However, in the valid frequency range of the impedance measurement, the experimental results are basically consistent with the theoretical plots, which verify the correctness of the positive- and negative-sequence output impedance models of the inverter with different PLLs.

V. CONCLUSION

The harmonic linearization method is employed to model and analyze different kinds of PLLs and the output impedance models of the three-phase LCL-type grid-connected inverter. The drawn conclusions are as follows:

(1) The differences of the impedance characteristics of the three-phase grid-connected inverter with and without the PLL are analyzed in detail, and it is indicated that because of the PLL the small signal representation of the grid-connected inverter should be altered from the ideal current source to the voltage controlled current source, and the additional equivalent impedance would be added in parallel with the ideal output impedance. Therefore, the PLL is the very important part of the grid-connected inverter.

(2) The frequency characteristics of SRF-PLL and DSOGI-FLL are compared and analyzed, and their differences lead to a diversity of the output impedance models of the inverter in the low frequency band. Then the frequency characteristics of the PLLs are verified in simulation, and the eclectic parameters designs of SRF-PLL and DSOGI-FLL are given as a reference.

(3) Finally, the output impedance models of the inverter with different PLLs are verified by the simulation and experimental results. The stability

analysis of the grid-connected inverter with different PLLs interacted with the grid will be given in the future works.

ACKNOWLEDGMENT

This work is financially supported in part by the Science and Technology Project of State Grid Corporation of China (Impedance Analysis and Control Technique Research of Stability and Power Quality of Distributed Generation System).

REFERENCES

[1] K. Macken, M. Bollen, R. Tagawa, "Mitigation of Voltage Dips Through Distributed Generation Systems," *IEEE Trans. Ind. Appli.*, vol. 40, no. 6, pp. 1886-1893, Nov./Dec. 2004.

[2] F. Blaabjerg, R. Teodorescu, M. Liserre, et al, "Overview of control and grid synchronization for distributed power generation systems," *IEEE Trans. Power Electron.*, vol. 53, no. 5, pp. 1398-1409, Oct. 2006.

[3] M. Boyra, J. L. Thomas, "A review on synchronization methods for grid-connected three-phase VSC under unbalanced and distorted conditions," in *Proc. European Conf. on Power Electron. and Appli.*, 2011, pp. 1-10.

[4] F. D. Freijedo, J. Doval-Gandoy, O. Lopez, et al, "Grid-synchronization methods for power converters," in *Proc. Annual Conf. of IEEE Ind. Electron.*, 2009, pp. 522-529.

[5] J. Sun, "Impedance-Based Stability Criterion for Grid-Connected Inverters," *IEEE Trans. Power Electron.*, vol. 26, no. 11, pp. 3075-3078, Nov. 2011.

[6] J. Sun, "Small-Signal Methods for AC Distributed Power Systems-A Review," *IEEE Trans. Power Electron.*, vol. 24, no. 11, pp. 2545-2554, Nov. 2009.

[7] M. Cespedes, J. Sun, "Impedance Modeling and Analysis of Grid-Connected Voltage Source Converters," *IEEE Trans. Power Electron.*, vol. 29, no. 3, pp. 1254-1261, Mar. 2014.

[8] P. Rodriguez, A. Luna, R. Muñoz-Aguilar, et al, "A stationary reference frame grid synchronization system for three-phase grid-connected power converters under adverse grid conditions," *IEEE Trans. on Power Electron.*, vol.27, No.1, Jan. 2012, pp.99-112.

Pulse-Injection-Based Sensorless Control Method with Improved Dynamic Current Response for PMSM

Hechao Wang*, Kaiyuan Lu, Dong Wang, Frede Blaabjerg
Energy technology, Aalborg University, Aalborg, Denmark
*Email hec@et.aau.dk

Abstract- **Sensorless control methods based on pulse injection are widely used in the standstill to low speed machine operation range. Generally, to achieve pulse-injection-based position estimation one or more switching periods for injection are dedicatedly needed, which results in reduced equivalent switching frequency, and consequently poor dynamic current response. In this paper, an improved pulse-injection-based method is proposed for sensorless drives of Permanent Magnet Synchronous Machine (PMSM) operating at standstill to low speed range. The proposed method can effectively increase the current dynamic response without increasing the control complexity. Meanwhile the current ripple could be reduced leading to lower torque ripple and acoustic noise. Experimental results are presented to validate the improved performance of the proposed sensorless method.**

Keywords— SPMSM, sensorless, dynamic response.

I. INTRODUCTION

For position, velocity or torque control in PMSM drives using Field Oriented Control (FOC) principle, the rotor position information is essential. To overcome the disadvantages of using a position sensor, many efforts have been dedicated to estimate the rotor position from the machines voltages and currents [1].

Sensorless control methods could be categorized as: fundamental model based or high frequency model based [2]. For medium-high speed range, the machine fundamental model is often used to estimate the rotor position contained in the machine back-EMF information [3]. Full-order [4] or reduced-order [5] state observers are proposed during last decades. At low speed, the back-EMF is low, resulting in low signal-to-noise ratio and failing in identifying the rotor positon. Thus high frequency (HF) signal injection methods are employed to detect the rotor saliency [6]. Categorized by the types of HF signals used for position estimation, there are steady state HF injection [6-9] and transient voltage pulse(s) injection methods [10, 11].

For steady state HF injection methods, the steady state response of the injected HF signal is utilized for position detection. Some methods inject signals into the stationary reference frame [6]. Then a demodulation process will be employed to extract the negative sequence current which contains the position information. Another kind of steady state HF injection method is implemented in the estimated d-axis, which will reduce the current ripple in

q-axis and hence reduce the torque ripple [9]. However the steady state HF injection methods need filters to demodulate the responding current signals which will degrade the system dynamic performance.

Unlike the steady state HF injection with filters, the pulse-injection-based methods utilize the derivative of the current transient information to estimate the rotor position. However, it often requires one or more switching periods dedicatedly to realize such pulse-injection-based position estimation algorithms [10, 11]. The equivalent control frequency to implement FOC is significantly reduced, and the dynamic response of the current controllers would be sacrificed. Meanwhile, with this kind of methods, the rotor position information is obtained every three [11] or four [10] switching periods, which will degrade the performance of the overall sensorless drive systems. On the other hand, not only the control frequency is reduced but also the injection frequency is limited.

To overcome the noise problems, random frequency signal injection method is proposed [12,13]. But the experimental results show little improvement: the acoustic noise decreases from 55.6 dB to 55.2 dB [12]. Then another way to reduce the acoustic noise is to increase the injection frequency, which will also have good dynamic response [14]. The theoretical maximum injection frequency is equal to the PWM switching frequency. So if the switching frequency is higher than the audible frequency, the acoustic noise could be significantly reduced. Thus the sensorless methods utilizing PWM current ripples to obtain the position information are proposed [15-17]. However, in these methods traditional Space Vector PWM (SVPWM) pattern need to be modified because the current ripple created by SVPWM is too small to be utilized. Meanwhile the current derivative information is hard to be obtained from the sampled current due to the switching transients [18]. Additional hardware may be employed to measure the current derivative directly which is difficult to implement in industrial applications [18]. Therefore, extra effort is needed to keep the injection frequency equal to switching frequency without additional hardware to measure the current derivative.

In this paper, two vectors with opposite directions are injected into the estimated reference frame for position estimation by following the principle that was briefly discussed in [11]. Then, the method is improved by achieving the implementation of the sensorless FOC within one switching period only, where the dynamic response of the current controllers can be maintained as that of sensored FOC. Meanwhile, only one extra current sampling is needed in the middle of the switching period, which will avoid the high frequency oscillation caused by the switching transients. Finally, the proposed method is verified experimentally.

II. FUNDAMENTALS OF SPMSM CONTROL

The basic model of PMSM in dq-reference frame could be described as (1)

$$
\begin{bmatrix} u_d \\ u_q \end{bmatrix} = R \begin{bmatrix} i_d \\ i_q \end{bmatrix} + \omega_r \begin{bmatrix} 0 & -L_q \\ L_d & 0 \end{bmatrix} \begin{bmatrix} i_d \\ i_q \end{bmatrix} + \omega_r \lambda_{mpm} \begin{bmatrix} 0 \\ 1 \end{bmatrix} + \frac{d}{dt} \begin{bmatrix} L_d & 0 \\ 0 & L_q \end{bmatrix} \begin{bmatrix} i_d \\ i_q \end{bmatrix} \tag{1}
$$

where u_d, u_q, i_d, i_q are the stator d- and q-axes voltages and currents respectively; L_d, L_q are the d- and q-axes inductances respectively; R is the stator resistance; ω_r is rotor electrical speed; and λ_{mpm} is the magnitude of the rotor PM flux linkage.

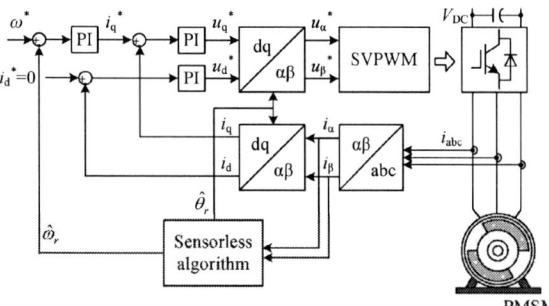

Fig.1 Sendorless FOC drive system for SPMSM.

A sensorless FOC of PMSM drive system is shown in Fig. 1. The rotor position information is essential to achieve FOC. For sensorless control, the rotor position is obtained from position estimation algorithms instead of position sensors. In position estimation algorithms, motor terminal voltages, currents and motor parameters may be involved in the calculation. In this paper, to avoid the uncertainties of motor terminal voltages and motor parameters, only motor line currents are involved in the calculation.

III. IMPROVED SENSORLESS CONTROL ALGORITHM

A. Principle of the method

The stator currents and rotor speed are assumed to be constants during one switching period, which also suggests that the resistance voltage drop, inverter voltage error and back EMF voltage are not changing in one switching period. These constant voltage disturbances may be removed by considering the subtraction of the voltage equations for the two neighboring switching

periods. This gives:

$$
\begin{bmatrix} \Delta u_d \\ \Delta u_q \end{bmatrix} = \begin{bmatrix} L_d & 0 \\ 0 & L_q \end{bmatrix} \frac{d}{dt} \begin{bmatrix} \Delta i_d \\ \Delta i_q \end{bmatrix} \tag{2}
$$

where Δ means the difference between two neighboring switching periods. This equation could be transformed into the αβ-reference frame as:

$$
\begin{bmatrix} \Delta u_\alpha \\ \Delta u_\beta \end{bmatrix} = \begin{bmatrix} L_0 + L_1 \cos 2\theta_r & L_1 \sin 2\theta_r \\ L_1 \sin 2\theta_r & L_0 - L_1 \cos 2\theta_r \end{bmatrix} \frac{d}{dt} \begin{bmatrix} \Delta i_\alpha \\ \Delta i_\beta \end{bmatrix} \tag{3}
$$

where θ_r is the rotor position and $L_0 = (L_d + L_q)/2$, $L_1 = (L_d - L_q)/2$. It could be seen from (3) that the rotor position could be obtained directly by using the motor parameters and motor voltages and currents. However due to the possible variations of motor parameters, a better way is to transfer (3) to the estimated dq-reference frame as described by (4) and the position error contained in $\sin 2\tilde{\theta}_r$ is to be estimated.

$$
\begin{bmatrix} \Delta u_{\hat{d}} \\ \Delta u_{\hat{q}} \end{bmatrix} = \begin{bmatrix} L_0 + L_1 \cos 2\tilde{\theta}_r & L_1 \sin 2\tilde{\theta}_r \\ L_1 \sin 2\tilde{\theta}_r & L_0 - L_1 \cos 2\tilde{\theta}_r \end{bmatrix} \frac{d}{dt} \begin{bmatrix} \Delta i_{\hat{d}} \\ \Delta i_{\hat{q}} \end{bmatrix} \tag{4}
$$

where $\Delta u_{\hat{d}}$, $\Delta u_{\hat{q}}$, $\Delta i_{\hat{d}}$, $\Delta i_{\hat{q}}$ are the stator d- and q-axes voltages and currents respectively in the estimated reference frame; $\tilde{\theta}_r = \theta_r - \hat{\theta}_r$, $\hat{\theta}_r$ is the estimated rotor position. In this form, to reduce the current ripple on the q-axis (thus the torque), it is preferred to inject voltage vectors on the estimated d-axis. Applying the chosen injected voltage vectors to (4), it may be derived that:

$$
\frac{\Delta(\Delta i_{\hat{q}})}{T_s} \approx \frac{d\Delta i_{\hat{q}}}{dt} = \frac{-L_1 \Delta u_{\hat{d}}}{L_0^2 - L_1^2} \sin 2\tilde{\theta}_r \tag{5}
$$

where T_s is the switching period. In one switching period, the differential $d\Delta i_{\hat{q}}/dt$ may be approximated by $\Delta(\Delta i_{\hat{q}})/T_s$. Then the estimated rotor position $\hat{\theta}_r$ could be obtained from $\sin 2\tilde{\theta}_r$ by a PLL.

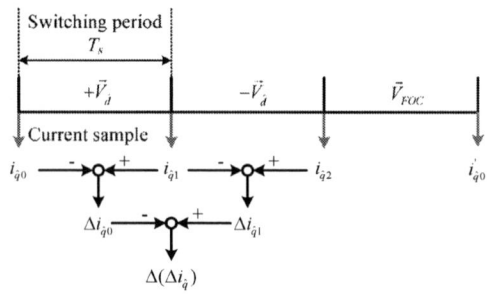

Fig. 2 Implementation of the traditional method.

B. Improved method

It should be noticed that in the traditional method

shown in Fig. 2, in order to obtain the $\Delta i_{\hat{q}}$ caused by the injection voltage vector only one injection voltage vector is applied during each switching period. However, due to the subtraction action, there is no need to separate the injection voltage vector from the FOC voltage vector. In the proposed method, injection voltage vectors are added to the FOC voltage vectors and furthermore, both pulse voltage vectors can be injected within only one switching period as illustrated in Fig. 3. The voltage vector generated by FOC is held to be the same for these two half switching periods. From the FOC implementation point of view, the FOC control loop is updated in every switching period which is the same as sensored FOC. After subtraction of the total voltage vectors applied in these two half switching periods, the FOC voltage component is ideally eliminated, and the voltage variation satisfies (4). The position error may then be obtained in a similar way as in [11]. In this improved method, one switching period is enough to implement both position estimation and FOC algorithms. The cost is to add one extra current sampling in the middle of each switching period which can be easily achieved. It should be noticed here that the middle point of the switching period is far away from the switch ON/OFF point when the voltage command is low. So the current sampling in the middle of the switching period can provide reliable result.

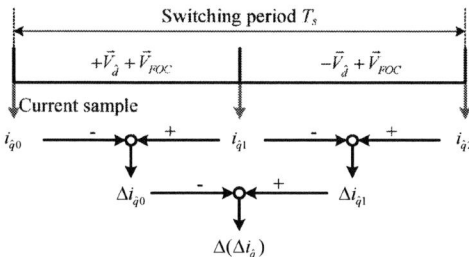

Fig . 3 Improved implementation of injection scheme.

It can be observed from Fig. 2 and 3 that, comparing with the method in [11], the control frequency and the injection frequency of the proposed method are tripled. Therefore the time delay caused be the digital controller will be decreased and the dynamic response of the current loop will be improved, which will potentially allow setting a higher bandwidth for the speed loop. The discretized block diagram of the current control loop is shown in Fig. 4, taking d-axis current as an example. The time delay is $t_d = 1.5 * T_s$ for the improved method (1.5 times the switching period) and $t_d = 4.5 * T_s$ for the traditional method.

The step response of the traditional and proposed methods is shown in Fig. 5. It can be observed that the settling time is significantly reduced with the improved method. Meanwhile with triple injection frequency, the high frequency current ripple will be reduced for the same voltage injection, so that the torque ripple will be reduced to achieve lower acoustic noise level.

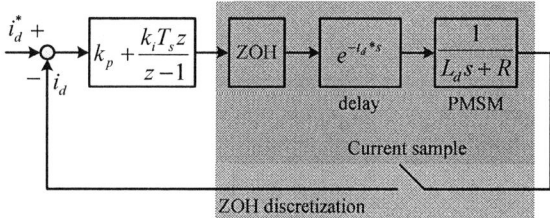

Fig. 4 Current loop with delay.

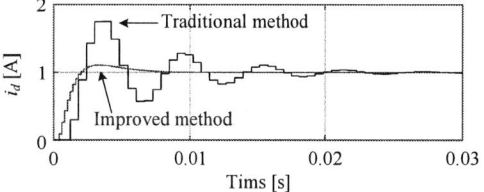

Fig. 5 Step response of different methods.

C. Inverter nonlinearity and cross-saturation effect.

It is well known that the inverter nonlinearity will cause voltage error, which is a function of current [20]. To reduce the inverter voltage error, two opposite voltage vectors are injected in [11]. However in this paper, it should be noticed that the voltage error caused by the inverter nonlinearity is different between the first and second half switching period. So the inverter voltage error could not be well compensated by the subtraction operation. Moreover, it should be noted that the only factor influencing the inverter voltage error is the motor current which is the same factor causing cross-saturation effect. Therefore, position estimation error caused by these two factors could be compensated together.

The cross-saturation effect caused by the stator currents will affect the orientation of the magnetic saliency and introduce positon estimation error to the sensorless drive systems. If taking the cross-saturation effect into account, equation (2) becomes

$$\begin{bmatrix} \Delta u_d \\ \Delta u_q \end{bmatrix} = \begin{bmatrix} L_d & L_{dq} \\ L_{qd} & L_q \end{bmatrix} \frac{d}{dt} \begin{bmatrix} \Delta i_d \\ \Delta i_q \end{bmatrix} \qquad (5)$$

where L_{dq} and L_{qd} are the mutual inductances caused by the cross-saturation effect. To simplify the analysis and calculation, L_{dq} and L_{qd} are set to be equal in this paper. Equation (5) could be simplified if the equation is transformed to the d'q'-reference frame where d'-axis is aligned with the magnetic saliency. The relationship between two reference frames is shown in Fig. 6.

$$\begin{bmatrix} \Delta u_{d'} \\ \Delta u_{q'} \end{bmatrix} = \frac{d}{dt} \begin{bmatrix} L_{d'} & 0 \\ 0 & L_{q'} \end{bmatrix} \begin{bmatrix} \Delta i_{d'} \\ \Delta i_{q'} \end{bmatrix}$$

where $L_{d'} = \dfrac{L_d + L_q}{2} + \dfrac{\sqrt{(L_d - L_q)^2 + 4L_{dq}^2}}{4}$ $\qquad (6)$

$\qquad\quad L_{q'} = \dfrac{L_d + L_q}{2} - \dfrac{\sqrt{(L_d - L_q)^2 + 4L_{dq}^2}}{4}$

where $u_{d'}, u_{q'}, i_{d'}, i_{q'}$ are the stator voltages and currents respectively in the d'q'-reference frame; ε is the position error between the d'-axis and the real machine d-axis. An equation similar with (5) could be obtained as below.

$$\frac{\Delta(\Delta i_{\hat{q}})}{T_s} \approx \frac{d\Delta i_{\hat{q}}}{dt} = \frac{-L_1' \Delta u_{\hat{d}}}{L_0'^2 - L_1'^2} \sin(2\tilde{\theta}_r + 2\varepsilon) \qquad (7)$$

where $L_0' = (L_{d'} + L_{q'})/2$, $L_1' = (L_{d'} - L_{q'})/2$ and $\varepsilon = 0.5 \cdot \arctan 2L_{dq}/(L_d - L_q)$.

In this paper, the position estimation error caused by the cross-saturation effect is compensated straightforward with the sensored experimental results same as [19].

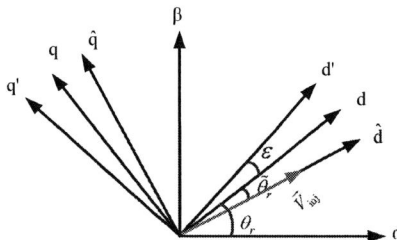

Fig. 6 Relationship among different reference frame.

IV. EXPERIMENT

To verify the improved sensorless method, a SPMSM drive system is built-up with a DC motor as load. The parameters of the SPMSM are shown in Table I. The switching frequency of the inverter was chosen to be 5 kHz.

Table I

Rated power [W]	400	Stator resistance [Ω]	2.3
Max. phase voltage [V]	380	d-axis inductance [mH]	10
Rated current [A]	2.9	q-axis inductance [mH]	13
Rated speed [rpm]	2850	PM flux linkage [Wb]	0.12
Rated frequency [Hz]	95	Pole pairs	2

To compare the current loop dynamic response, PI parameters are set to be identical in both sensorless methods with the rotor locked. The d-axis reference current is set to step from 0A to 4A (which is almost the rated motor current). It can be seen clearly in Fig. 7 that the settling time is much shorter with the proposed method (2ms) than the settling time with the traditional method (5ms), which principally verify the analysis in section II.B.

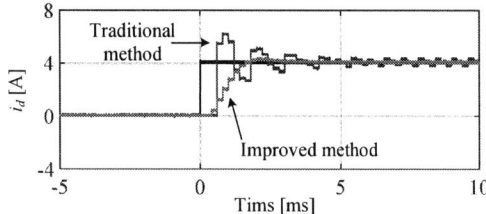

Fig. 7 Measured dynamic response of d-axis current

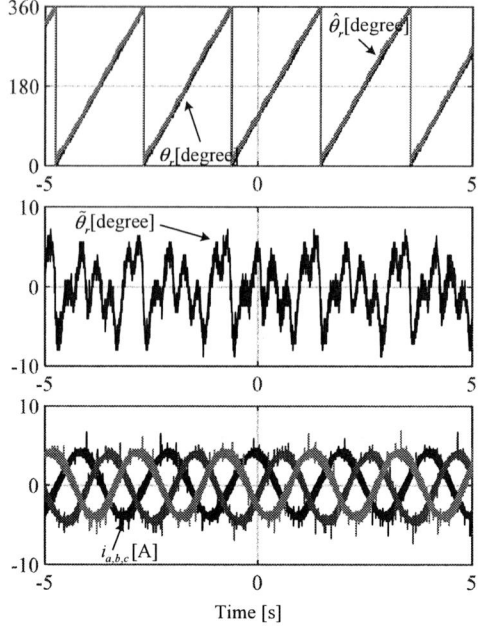

Fig. 8 Steady state wave with the improved method at 15rpm full load (a) Real and estimated position; (b) position estimation error; (c) motor phase currents.

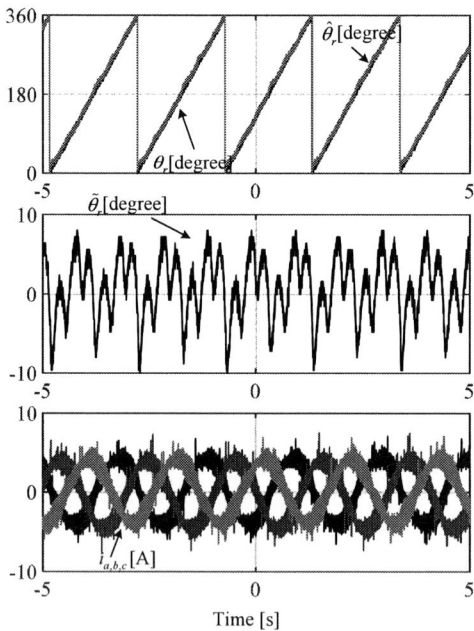

Fig. 9 Steady state wave with the traditional method at 15rpm full load (a) Real and estimated position; (b) position estimation error; (c) motor phase currents.

As discussed in section II.B with triple injection frequency in the improved method, the current ripple will be reduced, as can be observed from Fig 8 and 9. The motor is operated at 15rpm with full load. Comparing with the THD of phase current in traditional method, the THD of phase current in the improved method significantly reduces from 22.7% to 15.5% resulting in

The 2018 International Power Electronics Conference

lower acoustic noise. If the switching frequency could rise to more than 20 kHz, the acoustic noise introduced by voltage injection may become inaudible for the proposed method. While for the traditional method, three times higher switching frequency is required in order to make the acoustic noise inaudible.

When the injection voltage is low (50V) in the traditional method, the drive system will lose control with a full load step. Therefore to keep the traditional drive system stable, a higher voltage (100V) is injected as shown in Fig. 11. However in the improved method, 50V is enough to keep the drive system stable with a full load step. As shown in Fig. 10, the peak-peak value is 11.5 degree with no load and rise to 16.6 degree after full load step which is similar to the result of traditional method with doubled injection voltage vector magnitude. Meanwhile due to the low injection voltage in the proposed method, the phase current ripple is small.

In the speed reversal experiment, during the transient, the position error is well maintained below 10 degrees in the improved method, as it may be observed from Fig. 12. But in the traditional method, shown in Fig. 13, the position error during the transient is 12.2 degree.

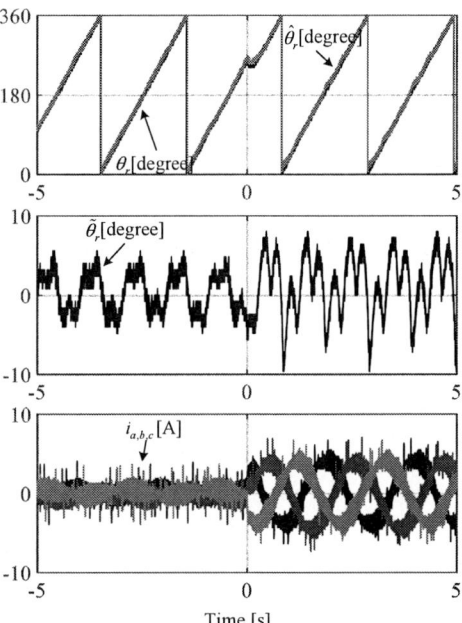

Fig. 11 Step full load response with the traditional method at 15rpm (a) Real and estimated position; (b) position estimation error; (c) motor phase currents.

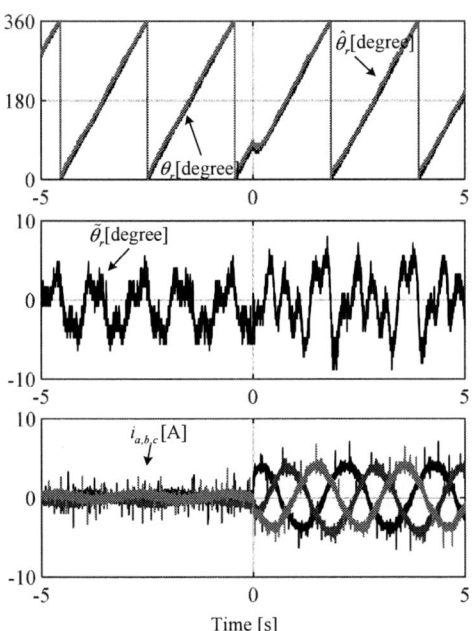

Fig. 10 Step full load response with the improved method at 15rpm (a) Real and estimated position; (b) position estimation error; (c) motor phase currents.

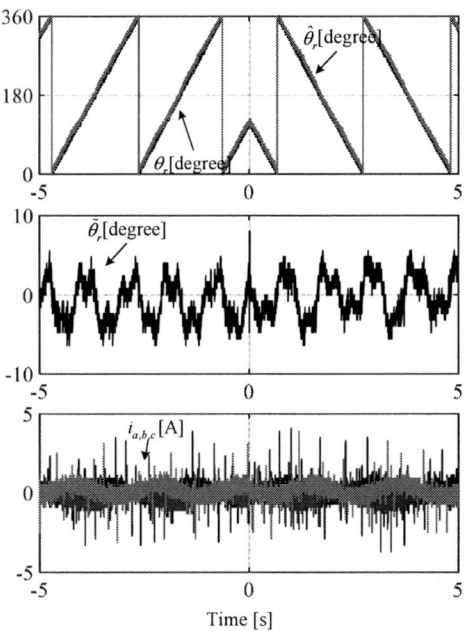

Fig. 12 Speed step change response with the improved method from 15rpm to -15rpm at no load (a) Real and estimated position; (b) position estimation error; (c) motor phase currents.

1187

The 2018 International Power Electronics Conference

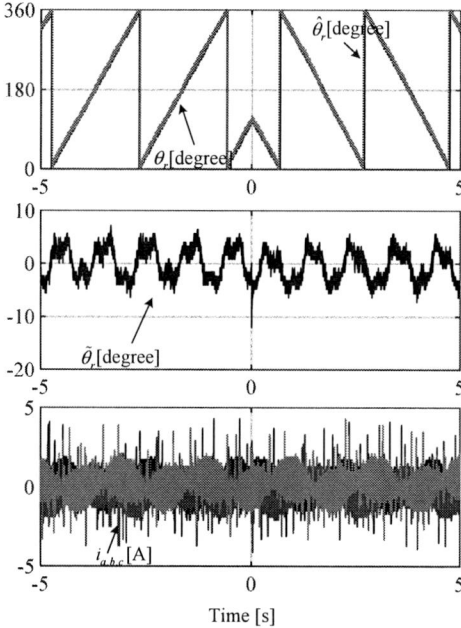

Fig. 13 Speed step change response with the traditional method from 15rpm to -15rpm at no load (a) Real and estimated position; (b) position estimation error; (c) motor phase currents.

V. CONCLUSION

In this paper, an improved pulse-injection-based sensorless method by injecting two opposite voltage vectors in one single switching period is proposed. The position estimation error caused by different factors is analyzed and the compensation method is roughly illustrated. The dynamic response of the current loop is improved with the proposed method. Meanwhile the THD of phase currents caused by the pulse-injection is significantly reduced resulting in lower acoustic noise. In the load and speed step experiments, the improved sensorless method shows a similar performance in position estimation with only half injection voltage comparing with the traditional sensorless method.

REFERENCES

[1] P. P. Acarnley and J. F. Watson "Review of position-sensorless operation of brushless permanent-magnet machines," *IEEE Trans. on Ind. Electron.*, vol. 53, no. 2, pp. 352-362, 2006.

[2] Y. Zhao, C. Wei, Z. Zhang, and W. Qiao, "A Review on Position/Speed Sensorless Control for Permanent-Magnet Synchronous Machine-Based Wind Energy Conversion Systems," *IEEE J.Emerg. Sel. Topics Power Electron.*, vol. 1, no. 4, pp. 203-216, 2013.

[3] I. Boldea, M. C. Paicu, and G. D. Andreescu, "Active Flux Concept for Motion-Sensorless Unified AC Drives," *IEEE Trans. Power Electron.*, vol. 23, no. 5, pp. 2612-2618, 2008.

[4] J. Solsona, M. I. Valla, and C. Muravchik, "Nonlinear control of a permanent magnet synchronous motor with disturbance torque estimation," *IEEE Trans. Energy Convers.*, vol. 15, no. 2, pp. 163–168, 2000.

[5] J. S. Kim, S. L. Sul, "High Performance PMSM Drives without Rotational Position Sensors Using Reduced Order Observer," *IEEE Thirtieth IAS Annual Meeting Conference on Industry Applications*, vol. 1, pp. 75-82, 1995.

[6] M. W. Degner, R. D. Lorenz, "Using Multiple Saliencies for the Estimation of Flux, Position, and Velocity in AC Machines," *IEEE Trans. Ind. Appl.*, vol. 34, no. 5, pp. 1097-1104, 1998.

[7] N. Teske, G. M. Asher, M. Sumner, and K. J. Bradley, "Suppression of Saturation Saliency Effects for the Sensorless Position Control of Induction Motor Drives Under Loaded Conditions," *IEEE Trans. Ind. Electron.*, vol. 47, no. 5, pp. 1142-1150, 2000.

[8] Y. Young-Doo, S. Seung-Ki, S. Morimoto, and K. Ide, "Highbandwidth sensorless algorithm for ac machines based on square-wave-type voltage injection," *IEEE Trans. Ind. Appl.*, vol. 47, no. 3, pp. 1361-1370, 2011

[9] M. J. Corley, R. D. Lorenz, "Rotor Position and Velocity Estimation for a Salient-Pole Permanent Magnet Synchronous Machine at Standstill and High Speeds," *IEEE Trans. Ind. Appl*, vol. 34, no. 4, pp. 784-789, 1998.

[10] M. Schroedl, "Sensorless Control of AC Machines at Low Speed and Standstill Based on the "INFORM" Method," *IEEE IAS Annual Meeting*, pp. 270-277, Oct 1996.

[11] G. Xie, K. Lu, S. K. Dwivedi, J. R. Rosholm and F. Blaabjeg. "Minimum Voltage Vector Injection Method for Sensorless Control of PMSM for Low-Speed Operations," *IEEE Trans.Power Electron.*, vol. 31, no. 2, pp. 1785-1794, 2016.

[12] H. Jiang, M. Summer. "Sensorless torque control of a PM motor using modified HF injection method for audible noise reduction," Power Electronics and Applications (EPE 2011), Sep 2011.

[13] G. Wang, L. Yang, G. Zhang, X. Zhang, and D. Xu, "Comparative investigation of pseudorandom high-frequency signal injection schemes for sensorless IPMSM drives," *IEEE Trans. Power Electron.*, vol. 32, no. 3, pp. 2123-2132, 2017.

[14] S. Kim, Y.-C. Kwon, S.-K. Sul, J. Park, and S.-M. Kim, "Position sensorless operation of IPMSM with near PWM switching frequency signal injection," *Power Electronics and ECCE Asia*, pp. 1660-1665, June, 2011.

[15] S. Kim, J. I. Ha, and S. K. Sul, "PWM switching frequency signal injection sensorless method in IPMSM," *IEEE Trans. Ind. Appl.*, vol. 48, no. 5, pp. 1576–1587, Sep./Oct. 2012.

[16] V. Petrovic, A. M. Stankovic, and V. Blasko, "Position estimation in salient PM synchronous motors based on PWM excitation transients," *IEEE Trans. Ind. Appl.*, vol. 39, no. 3, pp. 835-843, May/Jun. 2003.

[17] Q. Gao, G. Asher, M. Sumner, and P. Makys, "Position estimation of ac machines over a wide frequency range based on space vector pwm excitation," *IEEE Trans. Ind. Appl.*, vol. 43, no. 4, pp. 1001–1011, Jul./Aug. 2007.

[18] J. Holtz, J. Juliet. "Sensorless acquisition of the rotor position angle of induction motors with arbitrary stator windings," *IEEE Trans. Ind. Appl.*, vol 41, no. 6, pp. 1675-1682, 2005.

[19] J. M. Liu, Z. Q. Zhu. "Novel Sensorless Control Strategy with Injection of High-Frequency Pulsating Carrier Signal into Stationary Reference Frame," *IEEE Trans. Ind. Appl..* vol. 50, no. 4, pp. 2574-2583, 2014.

[20] A. R. Munoz, T. A. Lipo. "On-line dead-time compensation technique for open-loop PWM-VSI drives," *IEEE Trans. Power Electron.*, vol. 14, no. 4, pp. 683-689, 1999.

The 2018 International Power Electronics Conference

Influence of Parameter Variations on Operating Characteristics of MTPF Control for DTC-based PMSM Drive System

Keisuke Fujii, Yukinori Inoue, Shigeo Morimoto and Masayuki Sanada
Osaka Prefecture University, Sakai Osaka, Japan
*E-mail: sxb01160@edu.osakafu-u.ac.jp

Abstract- **This study investigated the influence of motor parameter variations on the operating characteristics of maximum torque per flux (MTPF) control of a permanent magnet synchronous motor (PMSM) drive system based on direct torque control. MTPF control, which is used for wide-speed-range operation, is essential for downsizing PMSMs and maximizing their output power at high speed. Stable operation requires accurate motor parameters, including d- and q–axis inductances, permanent magnet flux, and armature resistance, which vary due to motor operating conditions. This paper describes the operating characteristics of MTPF control under motor parameter variations and discusses conditions for stable operation.**

I. INTRODUCTION

Permanent magnet synchronous motors (PMSMs) are used in a wide range of applications, including electric vehicles, air conditioners, and industrial servo drives. The PMSM is capable of higher efficiency compared to an induction motor. Generally, in PMSM drives, an appropriate control method is applied for maximizing output power within the maximum voltage of the inverter and the current limitation of the motor. At low speed, maximum torque per ampere (MTPA) control is adopted for high efficiency, and the current limitation sets the maximum output power. At high speed, flux-weakening (FW) control is applied as the armature reaches its limiting voltage, which corresponds to the maximum voltage available from the inverter. Moreover, if the magnet flux of motor is relatively small, maximum torque per flux (MTPF) control is applied for maximizing the motor's torque and output power [1]–[4].

In the d- and q-axis current control which is often used conventionally, two variables (e.g., the d- and q-axis currents) have to be controlled simultaneously. A control method using only one variable was proposed [1],[2], but it is complicated to calculate the reference current. Also, methods using direct flux vector control or direct torque control (DTC) have been reported. In [3], the reference torque and flux were determined by an approximate equation. In [4], the MTPF control was based on torque angle control, and a method to estimate torque angle was proposed. In [5] and [6], the torque angle was calculated from the stator flux linkage vector and rotor position, and a proportional–integral (PI) controller was used for torque

This work was partly supported by 2016 and 2017 Nagamori Foundation Research Grants in Japan.

angle control. In [7], MTPF control was achieved using the maximum torque angle determined from the stator flux linkage. In this method, only the reference torque variable is used for the MTPF control and the reference flux is separately calculated by the FW control law.

For stable operation, accurate motor parameters are required because they are used to calculate the reference values. However, these parameters vary due to motor nonlinearity [8],[9]. The permanent magnet flux under no load varies according to magnet temperature [8]. Moreover, when the motor is under load, the harmonics in the magnet flux become even higher than those under no load [10],[11]. The d- and q-axis inductances also vary nonlinearly with respect to the load conditions due to magnetic saturation [9], and the armature resistance changes with temperature. These parameter variations and the instabilities that come with them, are inevitable during motor operation. Thus, it is necessary to investigate how motor parameter variations influence operating characteristics in the MTPF region.

This study used simulations to investigate how motor parameter variations affect the operating characteristics of MTPF control of a PMSM drive system based on DTC through simulation. The d- and q-axis inductances, the permanent magnet flux, and the armature resistance were evaluated. In DTC, MTPF control is sensitive to motor parameter variations because the torque of MTPF operation is the maximum torque at a given value for stator flux linkage, and excess torque destabilizes operation. This paper discusses an appropriate margin for the reference torque for MTPF control when other motor parameters are varied. The necessity of this margin for stable operation is confirmed.

II. MAXIMUM OUTPUT POWER CONTROL

In PMSM drives, the motor is controlled to obtain the maximum output power that is feasible with the maximum available inverter voltage and the motor current limit. In DTC, high efficiency and a wide speed range are achieved using the reference torque and flux. Figs. 1 and 2 show the relation between torque and rotor speed and the relation between power and rotor speed, respectively, when the motor parameter values in Table I are used. As shown in Figs. 1 and 2, the PMSM can obtain maximum torque and output power by switching control methods. The reference torque and flux are

1189

calculated as shown in Fig. 3. Fig. 4 is a block diagram of the DTC PMSM drive system.

A. Current Limitation

To meet the current limitation, the torque is controlled so as not to exceed T_{lim} with the following equation [7]:

$$T_{lim} = P_n \hat{\Psi}_s \sqrt{I_{am}^2 - i_M^2}, \qquad (1)$$

where P_n is the number of pole pairs, $\hat{\Psi}_s$ is the amplitude of the estimated stator flux linkage, I_{am} is the limiting armature current, and i_M is the M-axis current.

B. Maximum Torque per Ampere Control

MTPA control is applied in the low-speed region in which the armature voltage V_a does not exceed the limiting voltage V_{am}. The reference flux is obtained from a look-up table. In the MTPA control region, the armature current I_a is equal to the current limit I_{am} and the armature voltage V_a is less than the voltage limit V_{am}.

C. Flux-weakening Control

The armature voltage V_a increases with increasing rotor speed. When the rotor speed reaches the base speed, the armature voltage V_a reaches the limiting voltage V_{am}. Near the voltage limitation, FW control is applied, and the reference flux is given by the following equation [7]:

$$\Psi_{s-FW} = \frac{1}{\omega}\left\{ -R_a i_T + \sqrt{V_{am}^2 - (R_a i_M)^2} \right\}, \qquad (2)$$

where ω is the electrical rotor angular velocity, R_a is the armature resistance, and i_T is the T-axis current.

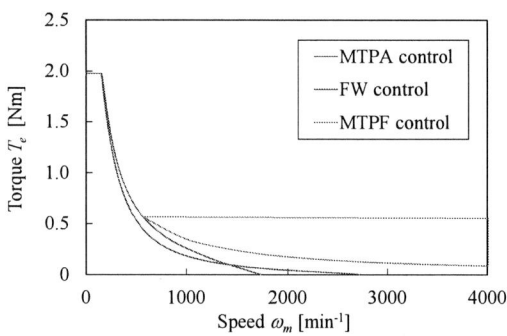

Fig. 1. Speed versus torque by motor control method.

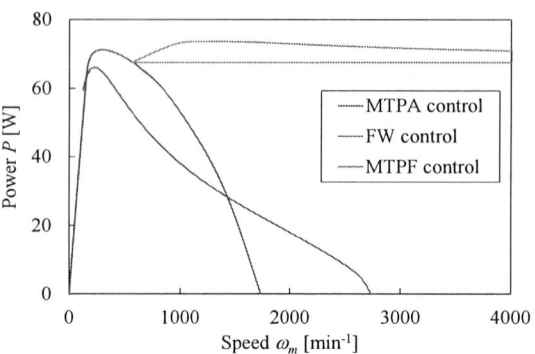

Fig. 2. Speed versus power by motor control method.

TABLE I. PARAMETERS OF TESTED PMSM

Number of pole pairs P_n	2
d-axis inductance L_d	72.2 mH
q-axis inductance L_q	368 mH
Armature resistance R_a	12.04 Ω
Magnet flux linkage Ψ_a	0.0785 Wb
Limiting armature voltage V_{am}	45 V
Limiting armature current I_{am}	2.4 A

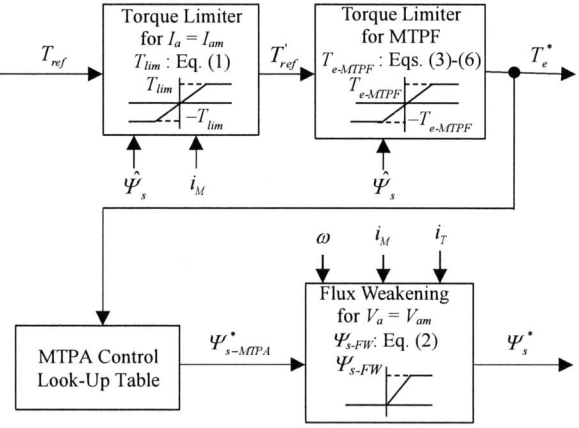

Fig. 3. Calculation of reference torque and flux.

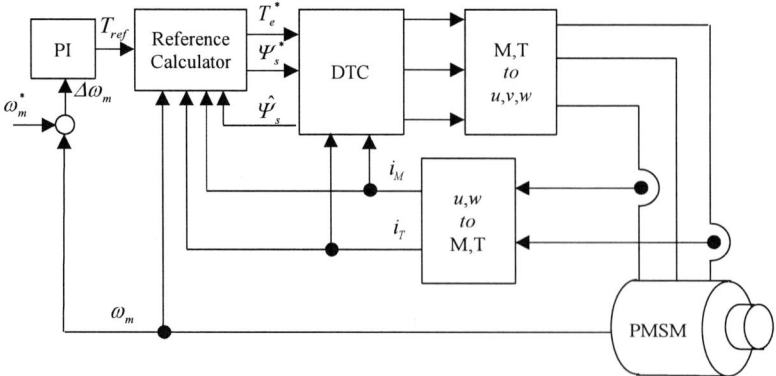

Fig. 4. DTC-based PMSM drive system.

1190

D. Maximum Torque per Flux Control

If $\Psi_{dmin}(=\Psi_a - L_d I_{am}) < 0$, MTPF control can be applied to expand the high-speed operating range. The electromagnetic torque of the PMSM is determined by the stator flux linkage Ψ_s and the torque angle δ. The maximum torque T_{em-dm} is calculated from the maximum torque angle δ_m as the following equations [7]:

$$T_{em-dm} = \frac{P_n \hat{\Psi}_s}{2 L_d L_q} \left\{ 2 \Psi_a L_q \sin \delta_m + \hat{\Psi}_s (L_d - L_q) \sin 2\delta_m \right\}, \quad (3)$$

$$\delta_m = \cos^{-1}\left[\frac{1}{4} \left\{ \frac{a}{\hat{\Psi}_s} - \sqrt{\left(\frac{a}{\hat{\Psi}_s}\right)^2 + 8} \right\} \right], \quad (4)$$

$$a = \frac{L_q}{L_q - L_d} \Psi_a, \quad (5)$$

where Ψ_a is the magnet flux linkage and L_d and L_q are the d- and q-axis inductances, respectively.

The torque versus torque angle characteristic according to (3)-(5) is shown in Fig. 5. The torque of the PMSM is at maximum when the torque angle δ is also at maximum angle δ_m, as is seen in Fig. 5. When the torque angle exceeds that value, operation will become unstable. In DTC, the torque control has to be operated in the region of $\partial T_e/\partial\delta > 0$. In MTPF control, the operating point corresponds to the condition in which the torque is maximum ($\partial T_e/\partial\delta = 0$). However, the torque control is likely to be unstable because the operating point is the boundary of the operating region. Hence, δ has to be below δ_m during both steady and transient states. In this study, the reference torque is given by

$$T_{e-MTPF} = k \cdot T_{em-dm}, \quad (6)$$

where k is a constant ($0 < k \leq 1$) and is given as a margin of the reference torque to the maximum torque at the MTPF operating point [7].

E. Direct Torque Control

Fig. 6 is a block diagram of the direct torque controller with an M-T frame. Estimated torque $\hat{T}_e$ is calculated by

$$\hat{T}_e = P_n \hat{\Psi}_s i_T. \quad (7)$$

The estimated stator flux linkage $\hat{\Psi}_s$ is given by integrating the flux linkage error $\Delta\Psi_s$. The reference induced voltages v_{oM}^*, v_{oT}^* are calculated by [12]

$$\begin{bmatrix} v_{oM}^* \\ v_{oT}^* \end{bmatrix} = \frac{1}{T_s} \begin{bmatrix} (\hat{\Psi}_s + \Delta\Psi_s)\cos\Delta\theta_s^* - \hat{\Psi}_s \\ (\hat{\Psi}_s + \Delta\Psi_s)\sin\Delta\theta_s^* \end{bmatrix}, \quad (8)$$

where T_s is a sampling period and $\Delta\theta_s^*$ is a reference flux linkage variation.

The voltage limiter limits the reference-induced voltage below limit value V_{lim} by considering the voltage saturation of the inverter. The flux compensator compensates the flux linkage error depending on the voltage saturation.

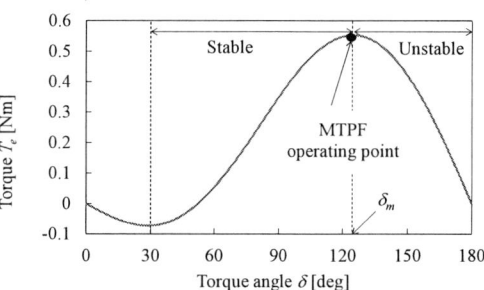

Fig. 5. Torque versus torque angle for $\Psi_s = 0.16$ Wb.

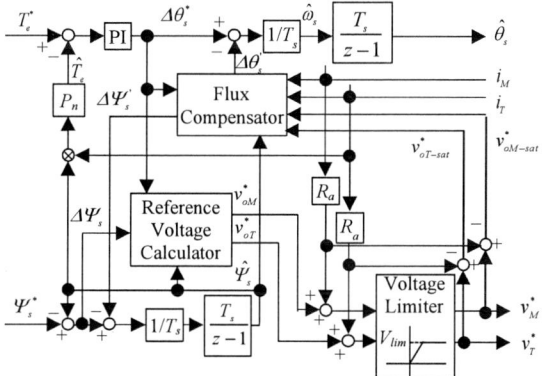

Fig. 6. Direct torque controller with M-T frame.

III. INFLUENCE OF TORQUE MARGIN ON OPERATING CHARACTERISTICS

Generally, the d- and q-axis inductances vary due to magnetic saturation. The permanent magnet flux and the armature resistance vary according to magnet temperature. In the MTPF control based on (3)-(6), accurate motor parameters are required for stable motor operation, and it is necessary to investigate how these parameter variations affect the operating characteristics at high speed.

Simulations were performed using the parameters listed in Table I. The q-axis inductance variation was modeled using the following equation:

$$L_q = 368 + dL_q \cdot |i_q| \text{ (mH)}, \quad (9)$$

where i_q is the q-axis current and dL_q is the coefficient of the q-axis inductance variation.

In this study, the coefficient of the q-axis inductance variation dL_q was determined as −41.8 mH/A when i_q reaches its rated value of 1.76 A and L_q decreases by 20% against 368 mH. The q-axis inductance of motor L_{q_motor} is given by (9) and that of controller $L_{q\_control}$ is constant. Hereafter, the suffixes "_motor" and "_control" mean parameters for the motor and controller, respectively.

Fig. 7 shows the acceleration characteristics as the reference speed was changed from 0 to 4,000 min^{-1} under no load. In Figs. 7(a) and 7(b), when L_{q_motor} is equal to $L_{q\_control}$, the rotor speed reaches the reference speed and the torque angle δ does not exceed the maximum torque angle δ_m. Even if L_{q_motor} is equal to $L_{q\_control}$, k has to be below 1 because δ exceeds δ_m in transient states.

When the inductance variation of L_{q_motor} was 20% (i.e., dL_q is –41.8 mH/A), the rotor speed did not reach the reference speed under k of 0.9800 in (6). Fig. 7(c) shows that δ exceeds δ_m and the torque control becomes unstable as a result.

When the coefficient k was 0.9795, the rotor speed reached the reference speed even when the inductance variation of L_{q_motor} was 20%. Fig. 7(d) shows that the torque angle δ does not exceed the maximum torque angle δ_m and the PMSM is stable. Torque and stator flux decreased at high speed, as seen in Figs. 7(e) and 7(f).

The stable operation of MTPF control is also confirmed by Fig. 8(a), which shows that the armature current decreased from the limiting armature current I_{am} and the armature voltage V_a maintained the limiting armature

voltage V_{am}, while V_a was limited below limit V_{lim} by the voltage limiter at low speed and steady state. Fig. 8(b) compares the power versus speed characteristics. The operating characteristics at dL_q of –41.8 mH/A and k of 0.9795 were approximately equal to the characteristics at dL_q of 0 mH/A and k of 0.9800. In the MTPF control region, i_q was less than 0.36 A and much smaller than its rating of 1.76 A. Therefore, it seems that the effect of the q-axis inductance variation is not significant in MTPF. It also shows that the maximum output power P is obtained in the MTPF control region, as seen in Fig. 2, while P increased to nearly 110 W at low speed, likely because V_a exceeded the limiting armature voltage V_{am} during a transient.

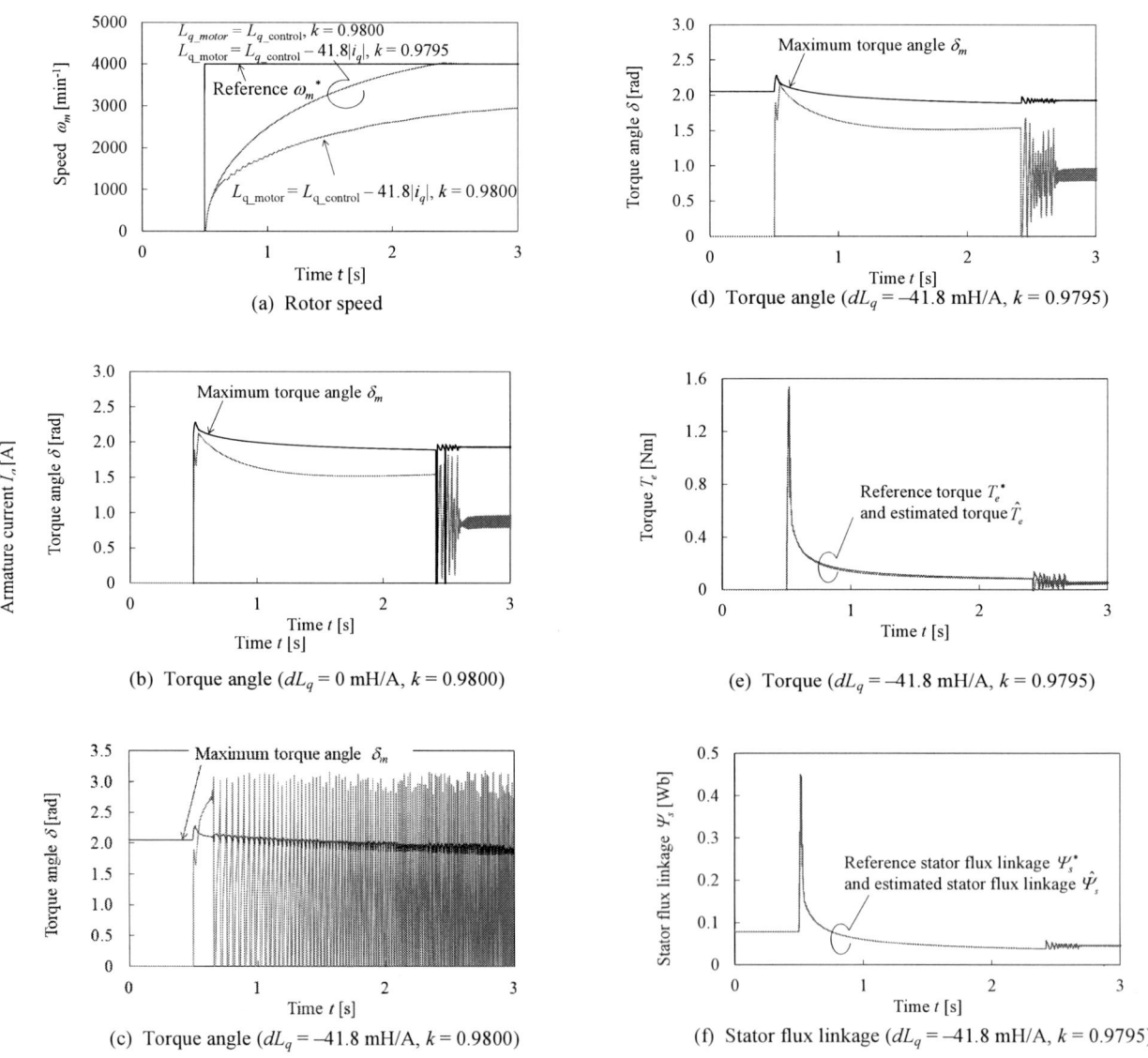

(a) Rotor speed

(b) Torque angle ($dL_q = 0$ mH/A, $k = 0.9800$)

(c) Torque angle ($dL_q = -41.8$ mH/A, $k = 0.9800$)

(d) Torque angle ($dL_q = -41.8$ mH/A, $k = 0.9795$)

(e) Torque ($dL_q = -41.8$ mH/A, $k = 0.9795$)

(f) Stator flux linkage ($dL_q = -41.8$ mH/A, $k = 0.9795$)

Fig. 7. Simulation result for MTPF control (0 min^{-1} to 4,000 min^{-1}, no load).

The effects of varying parameters L_{q_motor}, L_{d_motor}, R_{a_motor}, and Ψ_{a_motor} were evaluated within ±40 %, ±40 %, ±5 %, and ±40 %, respectively. Fig. 9 shows the maximum value of k (k_{max}) that allows stable operation with parameter variations when the reference speed was changed from 0 to 4,000 min⁻¹ under no load. In Fig. 9, when $L_{q\_motor} < L_{q\_control}$, $L_{d\_motor} > L_{d\_control}$, $R_{a\_motor} > R_{a\_control}$, or $\Psi_{a\_motor} < \Psi_{a\_control}$, k_{max} decreased as the parameter variation increased. When $R_{a\_motor} < R_{a\_control}$, the DTC was unstable even if k decreased. This is because there is a positive feedback loop between the armature voltage and current due to the resistance error. Consequently, L_d variation had the greatest effect on stability, and its variation needs to be minimized to achieve MTPF control. The coefficient k was determined by trial and error and regarded as difficult to model.

IV. MATHEMATICAL ANALYSIS OF PARAMETER VARIATIONS

In [13], a robust FW algorithm for DTC-based synchronous reluctance motor drives was reported. This algorithm modified the stator flux reference based on the continuous analysis of duty ratio of the active voltage vector. The stator flux adjustment was derived by differentiating expressions of reference stator flux and the torque–flux relationship with respect to motor speed and d-axis inductance. In this paper, the following equation is presented by differentiating (3) partially with respect to L_d in attempting to calculate variation in torque ΔT_{em-dm}

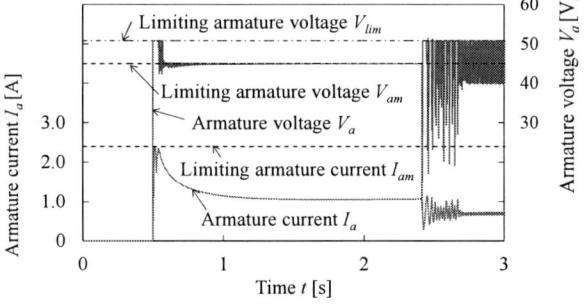

(a) Armature current and armature voltage
($dL_q = -41.8$ mH/A, $k = 0.9795$)

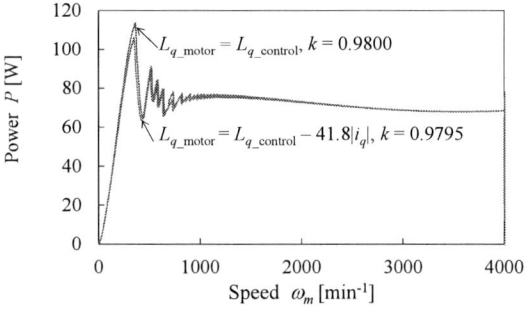

(b) Power versus speed characteristics
Fig. 8. Armature current, voltage, and power characteristics
(0 min⁻¹ to 4,000 min⁻¹, no load).

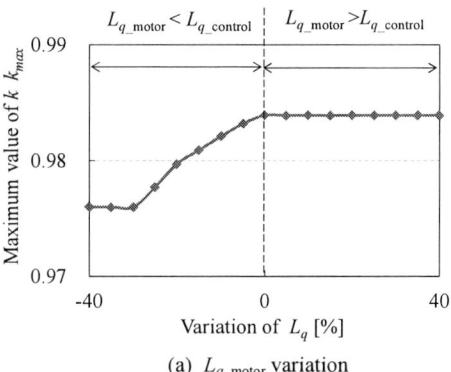

(a) L_{q_motor} variation

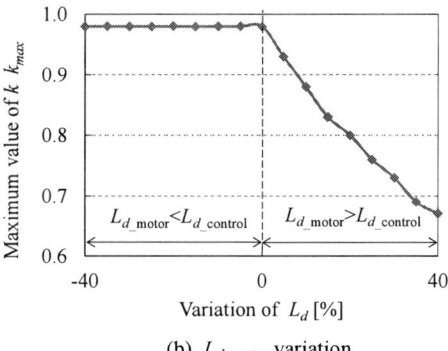

(b) L_{d_motor} variation

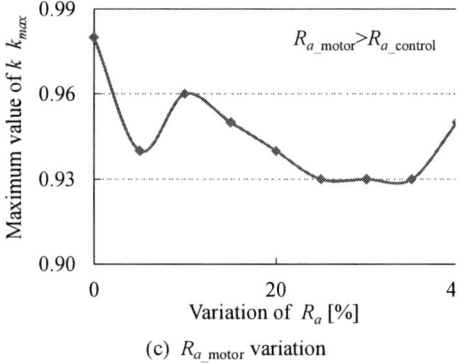

(c) R_{a_motor} variation

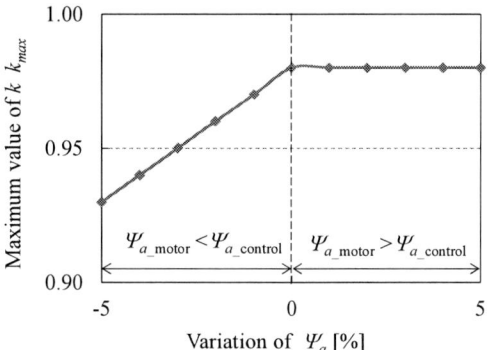

(d) Ψ_{a_motor} variation
Fig. 9. Analysis of maximum coefficient k_{max}
(0 min⁻¹ to 4,000 min⁻¹, no load).

when L_d varies and δ_m and Ψ_s remain constant.

$$\frac{\partial T_{em-dm}}{\partial L_d} = \frac{2L_q P_n \Psi_s}{(2L_d L_q)^2}[\{2\Psi_a L_q \sin\delta_m \\ + \Psi_s(L_d - L_q)\sin2\delta_m\} - L_d \Psi_s \sin2\delta_m] \quad (10)$$

If the control period is short enough, the variation in torque when L_d varies by ΔL_d can be expressed by

$$\Delta T_{em-dm} \cong \frac{\partial T_{em-dm}}{\partial L_d} \times \Delta L_d . \quad (11)$$

The same is true of L_q. The variation in torque with variation of L_q, ΔL_q, can be obtained by

$$\frac{\partial T_{em-dm}}{\partial L_q} = \frac{2L_d P_n \Psi_s}{(2L_d L_q)^2}[\{2\Psi_a L_q \sin\delta_m + \Psi_s(L_d - L_q)\sin2\delta_m\} \\ - L_q\{2\Psi_a L_q \sin\delta_m - \Psi_s(L_d - L_q)\sin2\delta_m\}] \quad (12)$$

$$\Delta T_{em-dm} \cong \frac{\partial T_{em-dm}}{\partial L_q} \times \Delta L_q . \quad (13)$$

Then, Ψ_s can be expressed by

$$\Psi_s = \sqrt{(\Psi_a + L_d i_d)^2 + (L_q i_d)^2} \quad (14)$$

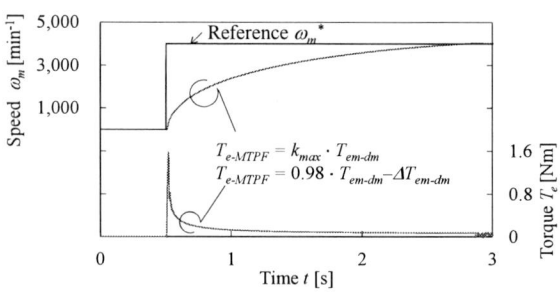

(a) L_{d_motor} is 20% larger than $L_{d\_control}$

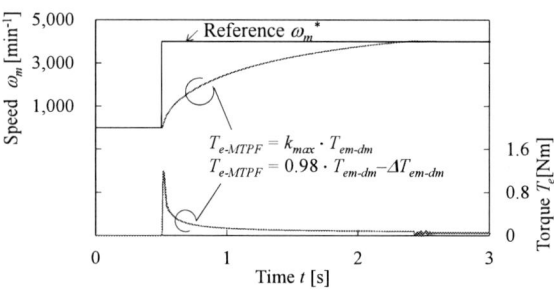

(b) L_{q_motor} is 40% smaller than $L_{q\_control}$

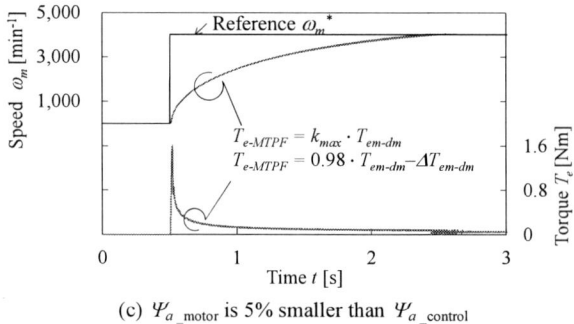

(c) Ψ_{a_motor} is 5% smaller than $\Psi_{a\_control}$

Fig. 10. Comparison of speed and torque characteristics for various combinations of inductance and flux variation.

where i_d is the d-axis current.

The following equation is obtained by differentiating (14) partially with respect to Ψ_a:

$$\frac{\partial \Psi_s}{\partial \Psi_a} = \frac{\Psi_a + L_d i_d}{\sqrt{(\Psi_a + L_d i_d)^2 + (L_q i_d)^2}} . \quad (15)$$

The variation in Ψ_s with variation in Ψ_a ($\Delta\Psi_a$) can be obtained by

$$\Delta\Psi_s \cong \frac{\partial\Psi_s}{\partial\Psi_a}\Delta\Psi_a . \quad (16)$$

Torque is expressed by (17) after Ψ_a varies, and the variation in torque with $\Delta\Psi_a$ can be obtained by

$$T'_{em-dm} = \frac{P_n(\Psi_s + \Delta\Psi_s)}{2L_d L_q}\{2\Psi_a L_q \sin\delta_m \\ + (\Psi_s + \Delta\Psi_s)(L_d - L_q)\sin2\delta_m\} \quad (17)$$

$$\Delta T_{em-dm} = T'_{em-dm} - T_{em-dm}$$
$$= l\Delta\Psi_a\left[m\frac{d\Psi_s}{d\Psi_a} + \left(m + 2n\frac{d\Psi_s}{d\Psi_a}\right)\Psi_s \\ + \left(m\frac{d\Psi_s}{d\Psi_a} + n\left(\frac{d\Psi_s}{d\Psi_a}\right)^2\right)\Delta\Psi_a\right] \quad (18)$$

$$l = \frac{P_n}{2L_q L_d} , \quad (19)$$

$$m = 2L_q \sin\delta_m , \quad (20)$$

$$n = (L_d - L_q)\sin\delta_m . \quad (21)$$

Fig. 10 compares speed and torque characteristics between the reference torque given by (6) and the following equation under the conditions that L_{d_motor} is 20% larger than $L_{d\_control}$, L_{q_motor} is 40% smaller than $L_{q\_control}$, and Ψ_{a_motor} is 5% smaller than $\Psi_{a\_control}$:

$$T_{e-MTPF} = 0.98 \times T_{em-dem} - \Delta T_{em-dm} . \quad (22)$$

These characteristics are almost the same as those seen in Fig. 10. Furthermore, to compare (22) with (6), the following equation is substituted for (22) with coefficient k':

$$k' \cdot T_{em-dm} = 0.98 \times T_{em-dem} - \Delta T_{em-dm} , \quad (23)$$

where k' is expressed by

$$k' = \frac{0.98 \times T_{em-dem} - \Delta T_{em-dm}}{T_{em-dm}} . \quad (24)$$

Fig. 11 shows characteristics of k' as L_{q_motor}, L_{d_motor}, and Ψ_{a_motor} were varied as 0–40%, 0–40%, and 0–5%, respectively, and the reference speed was changed from 0 to 4,000 min⁻¹ under no load. Coefficient k' is defined as the value when $\delta_m - \delta$ is minimum, and, similar to what was seen in Fig. 9, it decreased as the parameter variation increased and the variation of L_d had the greatest effect on stability.

V. Conclusions

This paper discussed the influences of motor parameter variation for MTPF control in a PMSM drive system based on DTC. The necessity of lower reference torque in MTPF control to achieve stable operation was confirmed. In addition, the greatest effect on stability by variation of d-axis inductance L_d was found. However, the coefficient *k* for a lower reference torque is difficult to model. Therefore, a control method with lower sensitivity to parameter variation should be considered.

REFERENCES

[1] Y. Zhang, L. Xu, M. K. Guven, S. Chi, and M. S. Illindala, "Experimental verification of deep flux-weakening operation of a 50 kW IPM machine by using single current regulator," Proceedings of the 1st IEEE Energy Congress Conference and Exposition (ECCE 2009), pp. 3647-3652, 2009.

[2] Y. Wang, X. Wen, and F. Zhao, "A proposed control strategy of PMSM for deep field-weakening and square-wave mode," Proceedings of the 15th International Conference on Electrical Machines and Systems (ICEMS 2012), 2012.

[3] J. Faiz, and S. H. M.-Zonoozi, "A novel technique for estimation and control of stator flux of a salient-pole PMSM in DTC method based on MTPF," IEEE Trans. Industrial Electronics, vol. 50, no. 2, pp. 262-271, 2003.

[4] J. Luukko, O. Pyrhönen, M. Nimela, and J. Pyrhönen, "Limitation of the load angle in a direct-torque-controlled synchronous machine drive," IEEE Trans. Industrial Electronics, vol. 51, no. 4, pp. 793-798, 2004.

[5] G. Pellegrino, R. I. Bojoi, and P. Guglielmi, "Unified direct-flux vector control for AC motor drives," IEEE Trans. Industry Applications, vol. 47, no. 5, pp. 2093-2102, 2011.

[6] G. Pellegrino, E. Armando, and P. Guglielmi, "Direct-flux vector control of IPM motor drives in the maximum torque per voltage speed range," IEEE Trans. Industrial Electronics, vol. 59, no. 10, pp. 3780-3788, 2012.

[7] Y. Inoue, S. Morimoto, and M. Sanada, "Wide-Speed-Range Operation of DTC-Based PMSM Drive System Using MTPF Control," Proceedings of the 2014 International Power Electronics Conference (IPEC2014), 2014

[8] G. Feng, C. Lai, and N. Kar, "Expectation maximization particle filter and Kalman filter based permanent magnet temperature estimation for PMSM condition monitoring using high-frequency signal injection," IEEE Trans. Industrial Informatics, vol. 13, no. 3, pp. 1261–1270, 2016.

[9] X. Liu, H. Chen, J. Zhao, and A. Belahcen, "Research on the performances and parameters of interior PMSM used for electric vehicles," IEEE Trans. Industrial Electronics., vol. 63, no. 6, pp. 3533–3545, 2016.

[10] Z. Azar, Z. Zhu, and G. Ombach, "Influence of electric loading and magnetic saturation on cogging torque, back-EMF and torque ripple of PM machines," IEEE Trans. Magnetics., vol. 48, no. 10, pp. 2650–2658, 2012.

[11] Z. Zhu, D.Wu, and W. Chu, "On-load performance in IPM machines having different slot/pole number combinations considering local magnetic saturation," IEEE Vehicle Power and Propulsion Conference., pp. 1–6, 2016.

[12] T. Seki, Y. Inoue, S. Morimoto, and M. Sanada, "Estimated flux compensation for direct torque control in M–T frame synchronized with stator flux-linkage vector," Proceedings of the 5th IEEE Energy Conversion Congress and Exposition Asia DownUnder 2013 (ECCE Asia DownUnder 2013), pp. 759–764, 2013.

[13] G. H. B. Foo, X. Zhang, and K. W. Lim, "Robust Direct Torque Control of Synchronous Reluctance Motor Drives in the Field-Weakening Region", IEEE Trans. Power Electronics, vol. 32, no. 2, pp. 1289-1297, 2017.

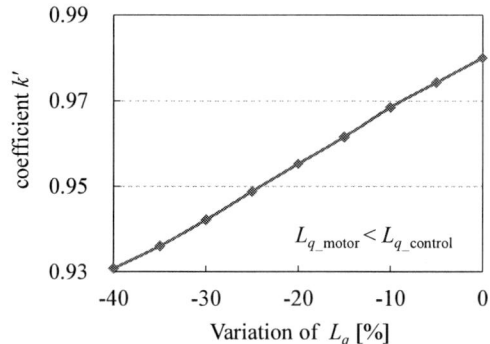

(a) L_{q_motor} variation

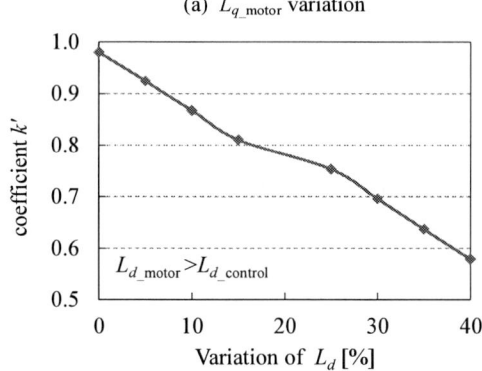

(b) L_{d_motor} variation

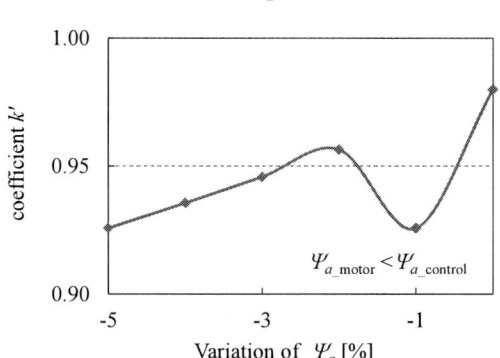

(c) Ψ_{a_motor} variation

Fig. 11. Analysis of coefficient k'
(0 min^{-1} to 4,000 min^{-1}, no load).

A quiet position sensorless control for an IPMSM based on Extended EMF and Voltage injection Synchronized with PWM Carrier

Yuki Ishii[*], Hiroki Yamashita and Hisao Kubota
Graduate School of Electrical Engineering
Meiji University
Kawasaki, Japan
*E-mail: ce171005@meiji.ac.jp

Abstract- This paper presents a new sensorless control method that does not generate acoustic noises for interior permanent magnet synchronous motors (IPMSMs). The magnetic pole position estimation is based on the extended EMF and the voltage injection synchronized with a PWM carrier. The carrier frequency is set high at 15 kHz, where we have difficulty to hear in order to remove the acoustic noise. The noise reduction strategies are proposed in two ways: One is a high carrier frequency sensorless control with a SiC inverter having a small dead-time (0.2 μs) switching devices, the other is a high carrier frequency sensorless control with a new dead-time compensation in addition to the voltage command value based on a polarity of the phase current and whether the PWM carrier is top or bottom. This compensates the effects of the dead-time without deteriorating symmetry of PWM pulse voltages. Effectiveness of high carrier frequency operations are verified by experimental results.

Keywords—IPMSM, Position sensorless control system, Dead-time, High frequency superimposed, Voltage injection

I. INTRODUCTION

Permanent magnet synchronous motors are smaller, lighter and more efficient than induction motors because power supply to the rotor is not required. However, one disadvantage is necessary to control the current in response to the magnetic pole position, so the information about magnetic pole position is required in some form. Generally, the magnetic pole position is measured by using a position sensor such as mechanical encoder or resolver, but there are problems of increased space and cost. Therefore, control methods without position sensors have been researched.

A position sensorless drive system for an IPMSM, which is based on extended EMF (EEMF) and voltage injection synchronized with a PWM carrier, was proposed [1]. In this method, in order to estimate the magnetic pole position using the EEMF, the estimation algorithm is not need to switch with speed, even at low speeds. Further, the voltage injection method in [1] is more efficient and quieter than conventional methods [2].

In this paper, further acoustic noise reduction is carried out in sensorless control of IPMSM proposed in [1]. The

carrier frequency is set much higher (15 kHz) than the method in [1] (5 kHz). The high carrier frequency enables to reduce the noise. But when the carrier frequency increases, the accuracy of sensorless control is degraded. The dead-time influence degrades the accuracy of current differential value which is used for calculation of position estimation. The asymmetry of phase current waveforms by dead-time influence causes this degradation of current differential value. In the sensorless control method proposed in [1], the difference of current value is used as the current differential value.

The current is sampled at the tops and bottoms of carrier wave. The degradation of the current differential value cannot be removed if the general dead-time compensation method is used. Then the dead-time compensation method considering asymmetry is employed. In the proposed method, dead-time compensation voltage is added to voltage command value according to not only the polarity of the phase current but also whether the PWM carrier is top or bottom.

II. PRINCIPLE OF ESTIMATION

A. Estimation method for magnetic pole position based on the EEMF with only fundamental wave

The voltage equation in the rotating coordinate system of IPMSM is shown in (1). p is a differential operator.

$$\begin{bmatrix} v_d \\ v_q \end{bmatrix} = \begin{bmatrix} R_a + pL_d & -\omega L_q \\ \omega L_d & R_a + pL_q \end{bmatrix} \begin{bmatrix} i_d \\ i_q \end{bmatrix} + \begin{bmatrix} 0 \\ \omega \psi_a \end{bmatrix} \quad (1)$$

Using EEMF, (1) can be expressed as (2) [3].

$$\begin{bmatrix} v_d \\ v_q \end{bmatrix} = \begin{bmatrix} R_a + pL_d & -\omega L_q \\ \omega L_q & R_a + pL_d \end{bmatrix} \begin{bmatrix} i_d \\ i_q \end{bmatrix} + \begin{bmatrix} 0 \\ e_{ext} \end{bmatrix} \quad (2)$$

where

$$e_{ext} = (L_d - L_q)(\omega i_d - pi_q) + \omega \psi_a \quad (3)$$

For a sensorless control, the magnetic pole position error $\Delta\theta$ is required. Figure1 shows the relation between d-q axis and γ-δ axis. The γ-δ axis are estimated d-q axis. Replacing the estimated d-axis with the γ-axis, and the estimated q-axis with the δ-axis, (2) becomes (4).

$$\begin{bmatrix} v_\gamma \\ v_\delta \end{bmatrix} \approx \begin{bmatrix} R_a + pL_d & -\omega L_q \\ \omega L_q & R_a + pL_d \end{bmatrix} \begin{bmatrix} i_\gamma \\ i_\delta \end{bmatrix} + \begin{bmatrix} e_\gamma \\ e_\delta \end{bmatrix} \quad (4)$$

where

$$e_{ext}\begin{bmatrix} -\sin\Delta\theta \\ \cos\Delta\theta \end{bmatrix} = \begin{bmatrix} e_\gamma \\ e_\delta \end{bmatrix} \quad (5)$$

Rearranging (4) for EEMF derives (6).

$$\begin{bmatrix} e_\gamma \\ e_\delta \end{bmatrix} = \begin{bmatrix} v_\gamma \\ v_\delta \end{bmatrix} - \begin{bmatrix} R_a + pL_d & -\omega L_q \\ \omega L_q & R_a + pL_d \end{bmatrix} \begin{bmatrix} i_\gamma \\ i_\delta \end{bmatrix} \quad (6)$$

From (5) and (6), the magnetic pole position error $\Delta\theta$ is obtained as (7).

$$\Delta\theta = \tan^{-1}\left(-\frac{e_\gamma}{e_\delta}\right) \quad (7)$$

The estimated speeds are given by a PI compensator, and the estimated position given by integrator from the estimated speed; this is PLL configuration shown as Fig. 1.

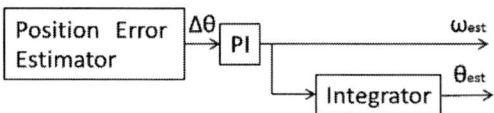

Fig. 1. Estimation of speed and position.

B. Estimation method for magnetic pole position based on EEMF and voltage injection synchronized with PWM carrier

In the above method, it is possible to estimate the magnetic pole position because sufficient EMF is obtained from the motor in high speed region. However, at standstill or low speeds, EEMF becomes low, and it is difficult to estimate the pole position because the decline of the denominator e_δ in (7) increases $\Delta\theta$ and deteriorates the accuracy of position estimation.

In order to increase the EEMF, we use a method of voltage injection to the δ-axis voltage command that is synchronized with PWM carrier as Fig. 2 [4].

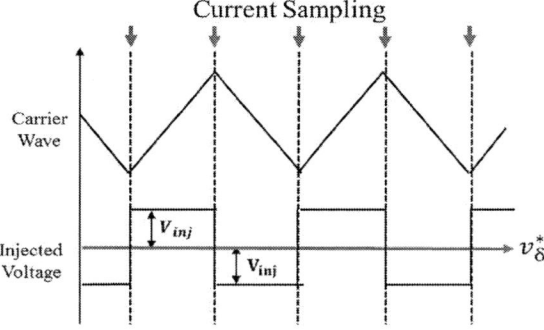

Fig.2. Voltage injection waveforms.

Equation (6) can be written as (9) with injected voltage v_{inj}.

$$\begin{bmatrix} e_\gamma \\ e_\delta \end{bmatrix} = \begin{bmatrix} v_\gamma^* \\ v_\delta^* + v_{inj} \end{bmatrix} - \begin{bmatrix} R_a + pL_d & -\omega L_q \\ \omega L_q & R_a + pL_d \end{bmatrix} \begin{bmatrix} i_\gamma \\ i_\delta \end{bmatrix} \quad (9)$$

The current value is sampled at the tops and bottoms of the triangular carrier waves, and current differential values are the difference between the current sampling value and the previous sampling value.

III. DEAD-TIME COMPENSATION STRATEGY

A. Voltage error by dead-time

Influence of the dead-time is shown in Fig. 3. The gate signal is delayed when the switching devices turn on. The average output voltage error v_E is given by (10).

$$v_E = T_d \times F_s \times V_{dc} \quad (10)$$

where T_d is dead-time [s], F_s is carrier frequency [Hz] and V_{dc} is DC link voltage [V].

B. General dead-time compensation method

In this section, the method of general voltage dead-time compensation is described as shown in Fig. 4. Figure4 shows the carrier wave(a), up and down gate signals(b) and (c), ideal output voltage(d), and real output voltages without/with general voltage dead-time compensation method (e)/(f) when the phase of positive current.

The average output voltage error is added to the voltage references (a)[5]. The absolute value of real output voltage with general voltage dead-time compensation (f) is same as that of ideal output voltage (d). But there is delay time ($T_d/2$) between (f) and (d).

The delay causes asymmetry of current waveform. It causes degrade of position estimation accuracy. When the output voltage pulse is delaying, sampled current value and current difference value is degraded because sampling point is different from desired current component. In the calculation of position estimation, the current difference value is used as current differential value as shown in (6) or (9).

C. Proposed method

In this section, a dead-time compensation method considering asymmetry is described as shown in Fig. 5. Figure5 shows the carrier wave(a), up and down gate signals(b) and (c), ideal output voltage(d), and real output voltages without/with the dead-time compensation method considering asymmetry (e)/(f) when the phase of positive current.

The output voltage error v_E is generated under two conditions. One is when the carrier wave is descending and phase current is positive. The other is when the carrier wave is uprising and phase current is negative. These are shown in Fig. 3. Thus, in order to consider asymmetry, dead-time compensation voltage is added into the voltage command value at only the phase that is being generated output voltage error.

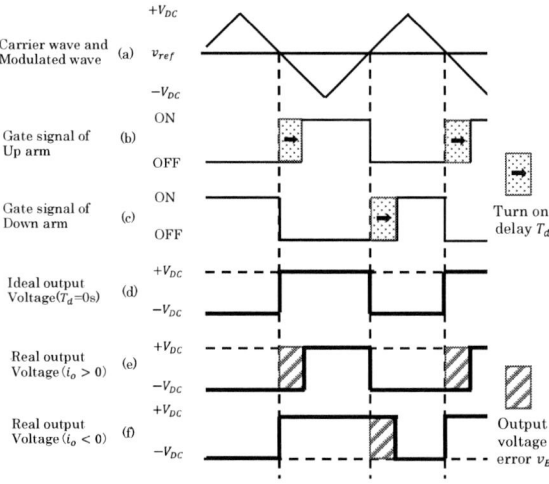

Fig.3. Influence of Dead-time.

A dead-time compensation voltage is added to a voltage command value according to not only the polarity of the phase current but also whether the PWM carrier is top or bottom. A voltage compensation value $2v_E$ is added to the voltage command value at the top of carrier wave when the phase current is positive. A voltage compensation value $2v_E$ is subtracted from the voltage command value at the bottom of carrier wave when the phase current is negative. Figure 5 shows dead-time compensation model waveforms considering asymmetry. This compensation method can decrease asymmetry of output voltage pulse and current waveform.

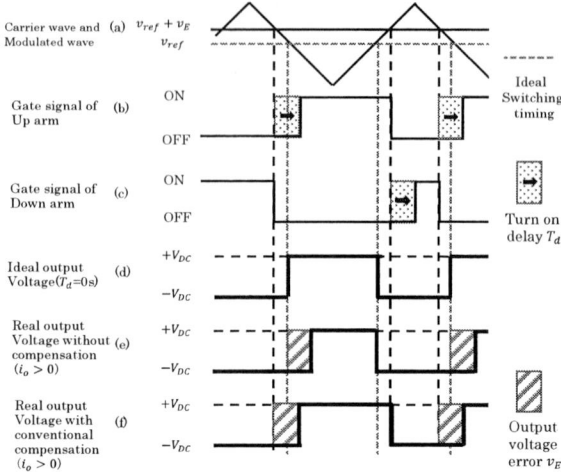

Fig.4. Dead-time compensation model waveforms (general voltage method).

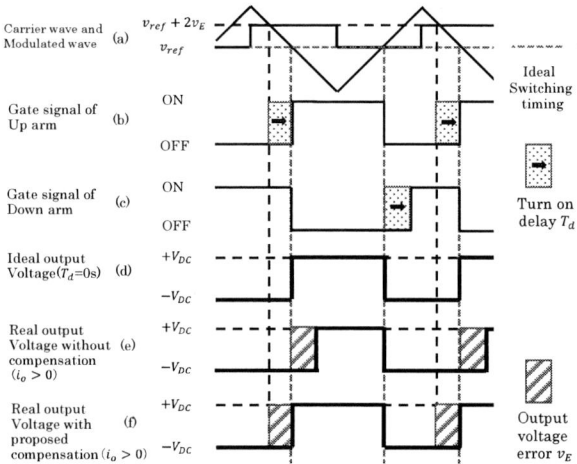

Fig.5. Dead-time compensation model waveforms (Considering asymmetry).

IV. EXPERIMENTAL RESULT

We verify the effectiveness of the proposed methods by actual experiments of IPMSM sensorless drive. Experimental conditions are five patterns. One is using the sensorless position estimation method and carrier frequency is set at 5 kHz. In the other four patterns, carrier frequencies are set at 15 kHz. The dead-time compensation strategies in the four patterns are bellow; the method without dead-time compensation at small dead-time (0.2 µs), the method with dead-time compensation considering asymmetry at normal dead-time (3.0 µs), the method with general dead-time compensation at normal dead-time (3.0 µs) and the method without dead-time compensation at normal dead-time (3.0 µs).

Reference speeds are changed (-40 rad/s→ -20 rad/s → -10 rad/s → 0 rad/s → +10 rad/s → +20 rad/s→ +40 rad/s) in step every 8 sec, the load is -7.95 Nm (rating).

The 2018 International Power Electronics Conference

When the rotor speed is minus, IPMSM is in powering mode. When the rotor speed is plus, IPMSM is in regenerating mode.

Figure6 shows the result of the 5 kHz carrier frequency (carrier frequency F_s =5 kHz). Figure7 and Figure8 show the results of proposed methods. Figure7 is small dead-time (The dead-time was set at T_d =0.2 µs.) method and Fig. 8 is considering asymmetry dead-time compensation method when the dead-time was set at T_d =3.0 µs. Figure9 shows the result of general dead-time compensation at normal dead-time (3.0 µs) and Fig. 10 shows the result of the method without dead-time compensation at normal dead-time (3.0 µs). The experimental conditions are shown in Table 1.

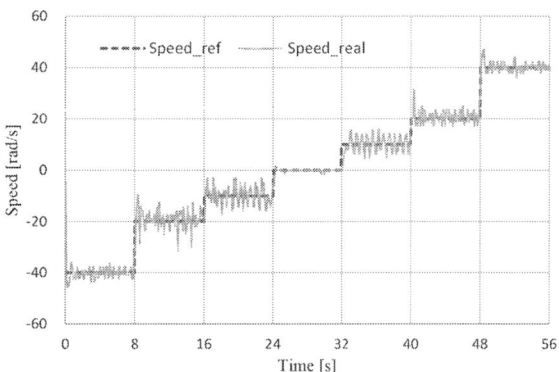

Fig.8. Pole position estimation results with dead-time compensation considering asymmetry (T_d =3.0 µs, F_s =15 kHz).

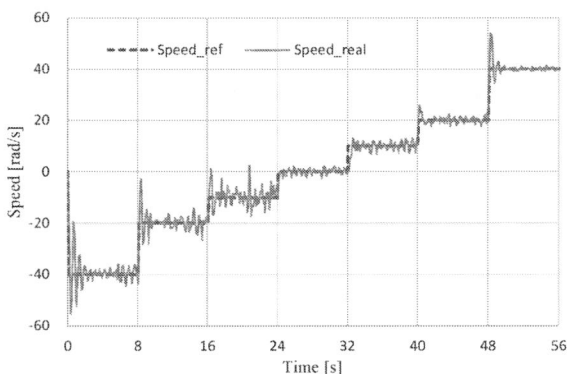

Fig.6. Pole position estimation results without dead-time compensation (T_d =3.0 µs, F_s =5 kHz).

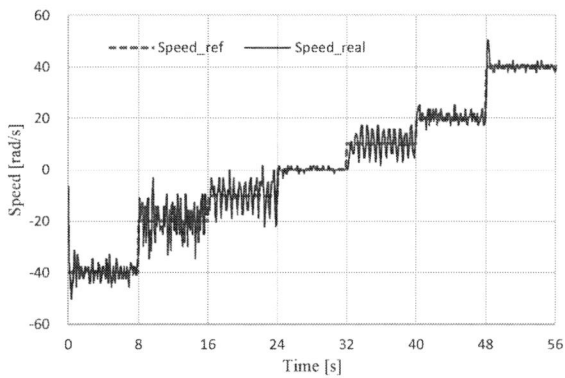

Fig.9. Pole position estimation results with general voltage dead-time compensation (T_d =3.0 µs, F_s =15 kHz).

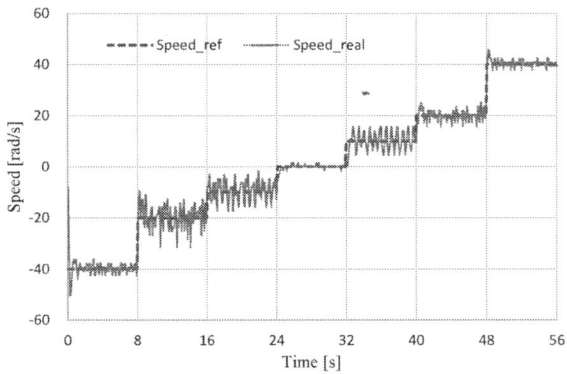

Fig.7. Pole position estimation results without dead-time compensation (T_d =0.2 µs F_s =15 kHz).

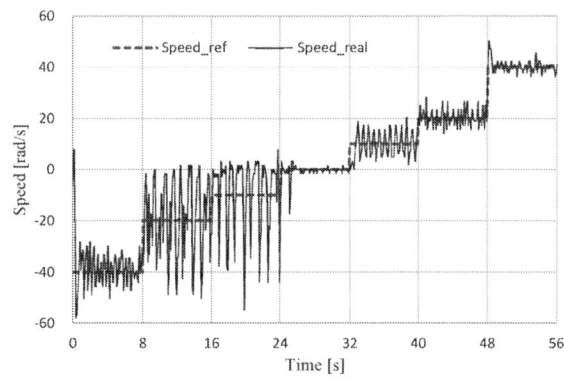

Fig.10. Pole position estimation results without dead-time compensation (T_d =3.0 µs, F_s =15 kHz).

Table 1.
EXPERIMENTAL CONDITIONS

	carrier frequency [kHz]	dead-time [µs]	dead-time compensation		carrier frequency [kHz]	dead-time [µs]	dead-time compensation
Fig.6	5	3.0	no	Fig.8	15	3.0	Considering asymmetry
Fig.7	15	0.2	no	Fig.9	15	3.0	general compensation voltage
				Fig.10	15	3.0	no

1199

The real speed followed the reference speed in first three methods (Fig. 6 ~ Fig. 8). In other words, the carrier frequency increased without degradation of the accuracy of the speed control. Mean while, the real speed had bigger ripple in latter two methods (Fig. 9 and Fig. 10) than proposed methods in Fig. 7 or Fig. 8. Comparing the waveform in Fig. 8 to Fig. 10, the method with dead-time compensation considering assymetry in Fig. 8 is better response than the method with general dead-time compensation in Fig. 9 or the method without dead-time compensation in Fig. 10. Focusing on the carrier frequency, the speed control is normally performed in Fig. 6, on the other hand, in Fig. 10, speed control becomes difficult for the high carrier frequency without dead-time compensation.

The effectiveness of the proposed sensorless control system for acoustic noise reduction is verified experimentally. The noise characteristics are compared with the 5 kHz carrier frequency and two proposed methods using 15 kHz carrier wave. One is small dead-time (0.2 μs) method and another is considering asymmetry dead-time compensation method. The acoustic noise meter is placed at 100 cm from IPMSM. The measuring instrument is LA-3560 (Ono Sokki). The measurement range are 27~140 dB and 10 Hz~20 kHz. Figure11 shows the FFT of the acoustic noise. In Fig. 11, (a) is the FFT of the 5 kHz carrier frequency at no load where dead-time was set at $T_d = 3.0$ μs and carrier frequency was set at F_s =5 kHz without dead-time compensation. (b) is the FFT of proposed small dead-time method at no load where dead-time was set at T_d =0.2 μs and carrier frequency was set at F_s =15 kHz without dead-time compensation. (c) is the FFT of proposed dead-time compensation method at no load where dead-time was set at T_d =3.0 μs and carrier frequency was set at F_s =15 kHz with dead-time compensation considering asymmetry.

In the 5 kHz carrier frequency, the acoustic noise occurred at frequencies of multiples of 5 kHz. Those noise components are very noisy because 5-10 kHz region are easy to hear for human. In Fig. 11 (b) and (c), large part of acoustic noise is nearby 15 kHz. It is difficult for human to hear the acoustic noise at 15 kHz. The 5 kHz and 10 kHz noise seen in Fig. 11 (a) was removed by proposed method. During the experiment, it was noisy in the 5 kHz carrier frequency, but it was quiet in the proposed two methods.

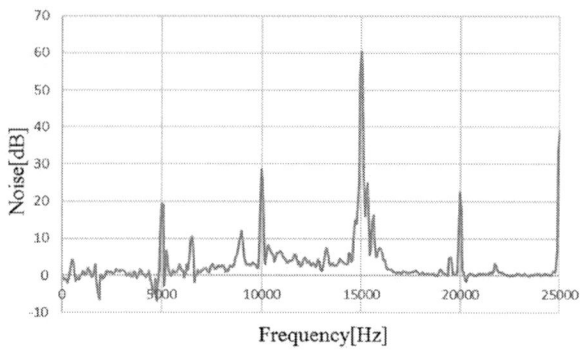

(a) The 5 kHz carrier frequency in without dead-time compensation (T_d =3.0 μs F_s =5 kHz).

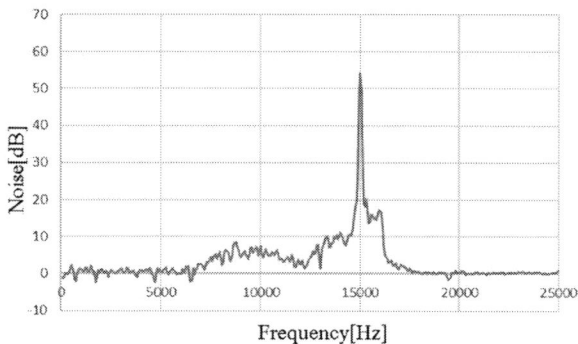

(b) without dead-time compensation (T_d =0.2 μs F_s =15 kHz).

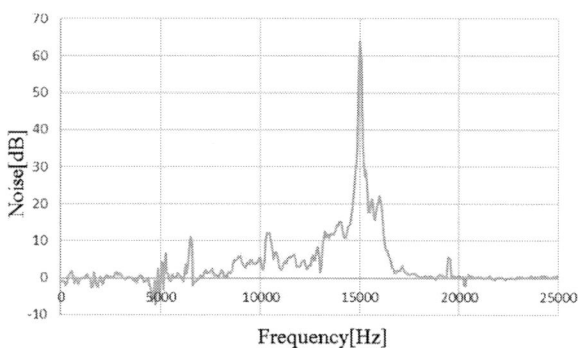

(c) with dead-time compensation considering asymmetry (T_d =3.0 μs F_s =15 kHz).

Fig. 11. Acoustic noise FFT of motor driving.

V. CONCLUSIONS

We proposed a quiet position sensorless control system of an IPMSM. That is based on EEMF and voltage injection synchronized with PWM carrier without switching the estimation algorithm by speed. It is difficult to increase the carrier frequency without dead-time compensation. So we proposed two dead-time countermeasure methods.

One is the method with small dead-time. It is possible to increase the carrier frequency by using the SiC inverter.

It is because high speed switching is possible for the SiC inverter, the dead-time can be set at almost zero and the effect of turn-off current and output voltage delay is small.

The other is the method with a dead-time compensation considering asymmetry. This method compensates not only the output voltage average but also the timing of voltage output. It is possible to increase the carrier frequency with normal dead-time.

By experiment, it is shown that both of the two proposed methods can reduce the acoustic noise and drive the motor quiet.

REFERENCES

[1] Tomoya Yokoyama, Kousuke Uchida and Hisao Kubota, "Position Sensorless Control for IPMSM Based on Extended EMF and Voltage Injection Synchronized with PWM Carrier", IEEJ Trans. IA, Vol. 136, No.3, pp.212-221,2016(in Japanese)

[2] Tomoya Yokoyama and Hisao Kubota, "Acoustic noise characteristics in position sensorless control for IPMSM based on EEMF and voltage injection synchronized with PWM carrier," 2015 IEEE International Conference on Industrial Technology (ICIT), Seville, 2015, pp. 604-609.

[3] Shinji Ichikawa, Zhiqian Chen, Mutuwo Tomita, Shinji Doki, and Shigeru Okuma, "Sensorless Controls of Salient-Pole Permanent Magnet Synchronous Motors Using Extended Electromotive Force Models," IEEJ Transactions on Industrial Applications, vol. 123, no. 12, pp. 1088-1096, 2002 (in Japanese).

[4] Kousuke Uchida and Hisao Kubota, "Position Sensorless Control at Low Speed for IPMSM Based on Extended EMF Using Voltage injection Synchronized with PWM Carrier," Annual Convention Record IEE Japan, 4-110, pp. 187-188, 2012 (in Japanese).

[5] IEEJ research committee on sorting sensorless vector control: "Sensorless vector Control of AC drive system", *Ohmusha,Ltd.* 2016 (in Japanese).

Study of Torque ripple reduction and Torque boost by Modified Trapezoidal Modulation

Satoshi Joryo [1*], Kazuto Tatsumi [1], Toshimitsu Morizane [1], Katsunori Taniguchi [1]

Noriyuki Kimura [1] and Hideki Omori [1]

1 Electrical engineering, Osaka Institute of Technology, Osaka, Japan

*E-mail: satoshi.joryo@outlook.jp

Abstract— **Permanent Magnet Synchronous Motor (PMSM) is compact and highly efficient, so it is widely used as a motor for Electric Vehicles (EVs) drive. However, it does not mean there is no problem at all. Applying high voltage to obtain high torque increases torque ripple. We propose a new modulation for the inverter to drive the PMSM, Modified-Trapezoidal-Modulation (MTM). The greatest advantage of MTM is that it can realize the torque increase and the torque ripple reduction at the same time. In this paper, experiments were conducted to obtain the merits of the proposed method. Results show that the proposed method generates increased torque comparing the result with other modulation techniques and confirmed that MTM reduces torque ripple by 9.27%.**

Keywords— *Modified-Trapezoidal-Modulation (MTM), Permanent Magnet Synchronous Motor (PMSM), Torque Boost, Torque Ripple Reduction*

I. INTRODUCTION

In recent years, electric vehicles (EVs) have attracted a great deal of attention as a new means to prevent emission of exhaust gas and exhaustion of resources in the automobile industry. The demand for EV is expected to increase in the future [1]. EVs often have a permanent magnet synchronous motor (PMSM) as a high torque density and high efficiency traction motor [2][3]. However, in the high speed area, PMSM can not generate higher torque due to its back electromotive force (EMF) and the maximum output voltage of the inverter. Although the over modulation and the one-pulse drive can increase the output voltage on the inverter, it causes the torque ripple.

Therefore, we propose Modified-Trapezoidal-Modulation (MTM) as a modulation that can realize high torque and low torque ripple. In the past paper, the torque ripple reduction and the torque increase is confirmed, but the emphasis of that study is the torque increase on the same DC input voltage [4]. In this paper, MTM is compared on the constant torque output to other modulation techniques for verifying the torque ripple reduction purely.

II. MODIFIED TRAPEZOIDAL MODULATION

Fig. 1 (a) shows the modified trapezoidal waveform. It can be obtained by superimposing a 120 degree rectangular wave on a trapezoidal wave which has a 120 degree flat part. γ is superposition ratio of rectangular

amplitude to the maximum value. When γ = 0, the waveform appears as a trapezoidal wave. When γ = 1, the waveform is a 120 degree rectangular wave. Fig.1 (b)-(e) shows the PWM output waveform when modified trapezoidal wave as a modulating wave on three phase voltage source inverter. Where E_d, v_u, v_v, v_w, and v_{uv} are inverter DC voltage, U-phase voltage, V-phase voltage, W-phase voltage and line voltage of U-phase to V-phase, respectively.

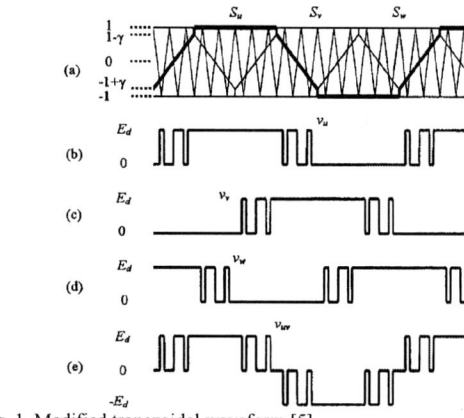

Fig. 1. Modified trapezoidal waveform [5]

Further, the ratio of harmonics can be arbitrarily determined by γ. The relation of the major current harmonic components and γ is shown in Fig. 2.

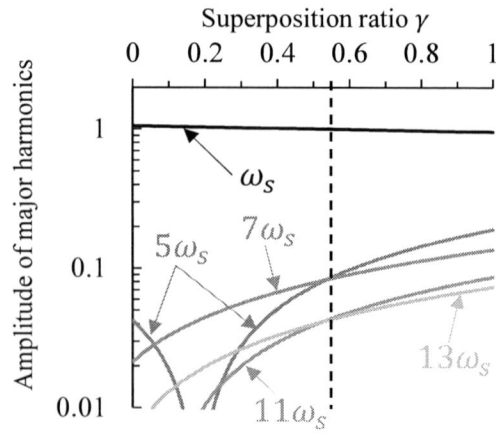

Fig. 2. Major harmonic components with γ

The amplitude of the fundamental wave component by γ is shown in (1), and the amplitude of the nth harmonic component by γ is shown in (2) [6].

$$V_1 = ME_d \left\{ \frac{3\gamma}{\pi} + \frac{6\sqrt{3}(1-\gamma)}{\pi^2} \right\} \tag{1}$$

$$V_n = ME_d \left\{ \frac{4\gamma}{n\pi} \cos\left(\frac{n\pi}{6}\right) + \frac{24(1-\gamma)}{n^2\pi^2} \sin\left(\frac{n\pi}{6}\right) \right\} \cos\left(\frac{n\pi}{6}\right) \tag{2}$$

Where M is a modulation rate on an inverter.

A. Advantages

MTM has three effective characteristics.

Firstly, the switching loss on the inverter can be reduced by MTM. The modified trapezoidal wave has two 120 degree flat part in one fundamental cycle, so the PWM switching time is one-third from a sinusoidal modulation when the inverter is operated in full modulation rate (M=1).

Secondly, the output torque can be increased from sinusoidal modulation owing to the enhancement of the fundamental amplitude. The comparison of the MTM fundamental amplitude and that of SM is shown in Fig. 3.

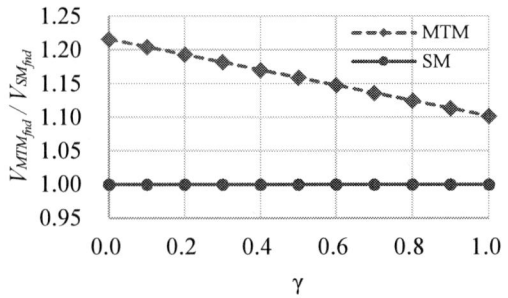

Fig. 3. Comparison of Fundamental Amplitude

Lastly, the torque ripple can be reduced by MTM. The 6th harmonic torque is a large torque ripple component on PMSM. The 6th harmonic torque on the PMSM which has a non-sinusoidal magnetic flux density is approximately defined with (3) [5].

$$\tau_6 \cong \frac{3}{2}\frac{1}{\omega_m}\left\{ E_1(I_7 - I_5) + \cdots \right\} \tag{3}$$

Here, ω_m is the rotor angular velocity, E_1 is the fundamental voltage of EMF, I_5 and I_7 are the harmonic component on the motor line current. According to Fig. 2 and (3), τ_6 reaches minimum when γ=0.54. Actually, the past experiment shows that τ_6 reaches minimum when γ around 0.54 as Fig. 4.

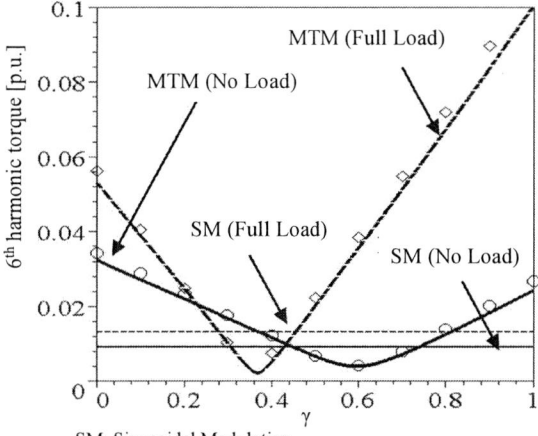

SM: Sinusoidal Modulation
MTM: Modified Trapezoidal Modulation

Fig. 4. 6th harmonic torque vs. γ on non-sinusoidal PMSM [5]

III. SIMULATION

To compare about the torque ripple on the same torque with one-pulse, sinusoidal modulation, sinusoidal over-modulation and MTM, the following simulation was conducted.

A. Simulation configuration

Fig. 5 shows the simulation circuit. The simulation was conducted by PSIM ver11. The PMSM is connected to constant speed load (1800 rpm). The PMSM has a non-sinusoidal magnetic flux density, it is shown as a back EMF wave in Fig. 6. The parameters of the PMSM are shown as Table I. The inverter was operated on constant d-axis voltage, q-axis voltage and full modulation mode (M=1). The DC input voltage of the inverter was varied to maintain the constant torque T=2 N·m. The modulation method could be changed in this simulation. Here, γ was set to 0.54. Then, the modulation rate M in sinusoidal over-modulation was set to M=1.4 to maintain the output torque to MTM in the same DC input voltage (V_{DC}).

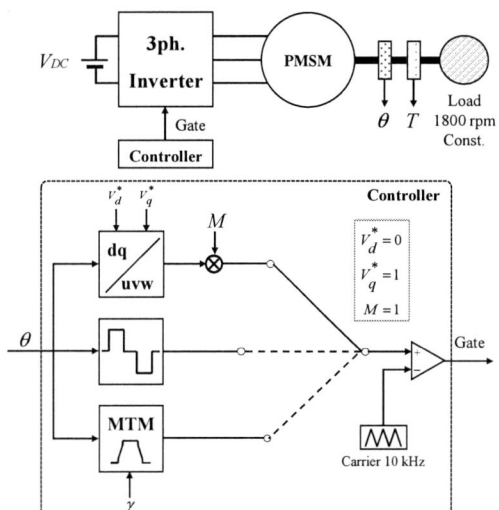

Fig. 5. Configuration of Simulation

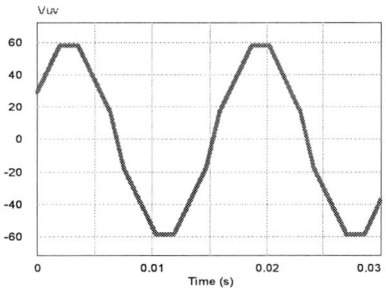

Fig. 6. Back EMF waveform of PMSM on Simulation

TABLE I. PARAMETERS OF PMSM ON SIMULATION

Parameter	Value
Phase Number	3
Number of Poles	4
Flux linkage (mWb)	22.8
L(mH), [L_d=L_q]	2.07

B. Simulation result

The inverter DC bus voltage and the 6th harmonic torque ripple ratio on various modulation is shown in Fig. 7. That is obtained by the FFT analysis of the torque waveform. The torque ripple of MTM is lowest from another modulation method. Although the same torque can be output at the same DC bus voltage both over-modulation and MTM, the torque ripple is 5.97% lower at MTM.

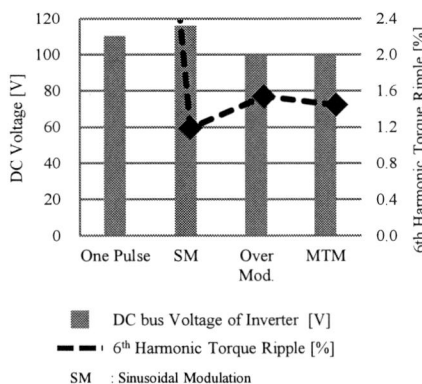

DC bus Voltage of Inverter [V]

6th Harmonic Torque Ripple [%]

SM : Sinusoidal Modulation
MTM : Modified Trapezoidal Modulation

Fig. 7. DC bus Voltage and 6th harmonic torque from simulated result

IV. EXPERIMENT

A. Experimental configuration

Fig. 8 shows the experimental platform. The control program was the same as the simulation. The PMSM was connected to the DC Generator (DCG) as a mechanical load. Table II shows the parameters of the DCG. The DCG armature was connected to the constant resistance load and the DCG field winding was connected to the constant voltage DC power supply (30V). The back EMF waveform of the PMSM is shown in Fig. 9, and the parameters of the PMSM is shown as Table III. In this time, the DC input

voltage of the inverter (V_{DC}) was varied to maintain the constant torque T=5 N·m. Further, γ was set to 0.54 from Fig. 2.

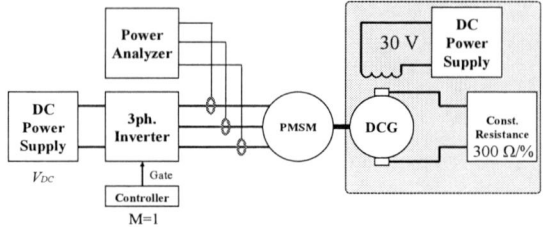

Fig. 8. Experimental Platform

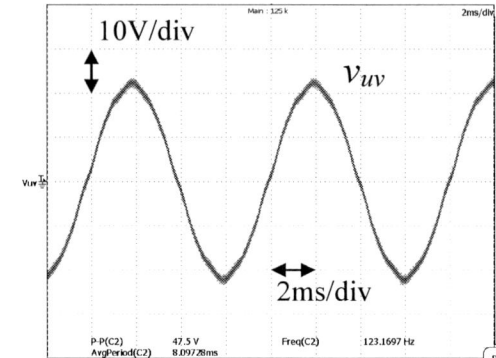

Fig. 9. Back EMF waveform of PMSM used in Experiment

TABLE II. PARAMETERS OF DCG ON EXPERIMENTAL PLATFORM

Parameter	Value
Number of Poles	2
Rated Power (kW)	2
Rated Voltage (V)	100
Rated Current (A)	25
Rated Revolutions (rpm)	1800

TABLE III. PARAMETERS OF PMSM ON EXPERIMENTAL PLATFORM

Parameter	Value
Phase Number	3
Number of Poles	8
Flux linkage (mWb)	80.9
L_d(mH)	0.559
L_q(mH)	1.53
Rated Power (kW)	0.59
Rated Voltage (V)	72
Maximum Torque (N·m) / (rpm)	12.75 / 3750

B. Experimental result

Fig. 10 shows the inverter DC bus voltage and the 6th harmonic torque ripple ratio on various modulation by the

1204

FFT analysis of the torque waveform. The torque ripple with MTM is lowest from another modulation method. The same torque can be output on the same DC voltage by both MTM and over-modulation, but the torque ripple is 9.27% decreased from that of over-modulation.

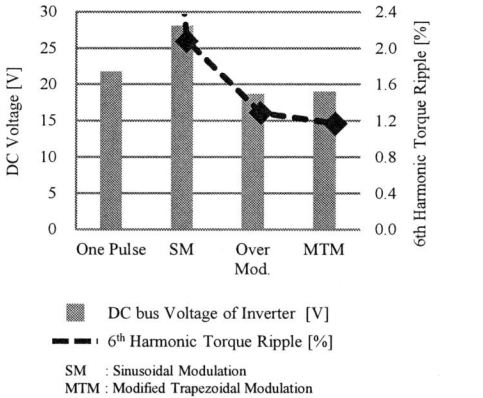

Fig. 10. DC bus Voltage and 6th harmonic torque from experimental result

Furthermore, the comparison of the 6th harmonic torque ratio by changing γ is shown in Fig. 11. The 6th harmonic torque ratio at γ=0.54 is lower than another modulation.

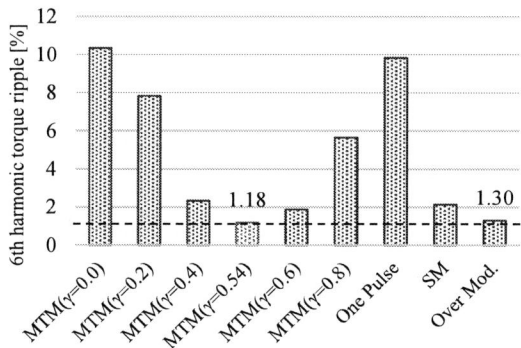

Fig. 11. Experimental comparison of 6th harmonic torque ratio by γ

V. CONCLUSION

In this paper, the advantages of the proposed modified-trapezoidal-modulation (MTM) were confirmed by simulation and experiment. The 6th harmonic torque ripple ratio and the input voltage were compared at the same torque output. As a result, the 6th harmonic torque ripple ratio on MTM is lower than that of other modulations. In addition, the DC bus voltage of the inverter is lower than sinusoidal modulation. It means the higher torque can be output with MTM on the same DC voltage. These facts suggest that the torque increase and the torque ripple reduction can be realized at the same time on the PMSM by simply changing the modulation to MTM.

The next step of this study is to experiment with focusing on the improvement of efficiency.

REFERENCES

[1] International Energy Agency, "Gloal EV Outlook 2017," 2017. Available: https://www.iea.org/publications/freepublications/publication/GlobalEVOutlook2017.pdf. [Accessed: 30- Oct- 2017], p. 6.

[2] T. A. Burress, S. L. Campbell, C. L. Coomer, C. W. Ayers, A. A. Wereszczak, J. P. Cunningham, L. D. Marlino, L. E. Seiber, and H. T. Lin, "Evaluation of the 2010 Toyota Prius hybrid synergy drive system," Oak Ridge Nat. Lab., Oak Ridge, TN, USA, Rep. ORNL/TM-2010/253, 2010.

[3] Y. Sato, S. Ishikawa, T. Okubo, M. Abe, and K. Tamai, "Development of high response motor and inverter system for the Nissan LEAF electric vehicle," presented at the SAE World Congr., Detroit, MI, USA, 2011, Paper 2011-01-0350.

[4] K. Tatsumi, T. Morizane and K. Taniguchi, "Torque Boost and Torque ripple Reduction of PMSM driven by Modified trapezoidal Modulation Inverter," in *IEE-Japan Industry Applications Society Conference*, 2017. (in Japanese)

[5] H. Yonezawa, K. Taniguchi, T. Morizane and N. Kimura, "Modified trapezoidal Modulating Signal suitable for PM Synchronous Motor Drives," *IEEJ Trans. IA*, vol. 125, no. 1, pp. 46-53, 2005. (in Japanese)

[6] K. Taniguchi, "PWM Power Converter System", Kyoritsu Shuppan, 2007, p. 112. (in Japanese)

[7] K. Taniguchi and T. Morizane, "Characteristics of PAM Inverter System for Electric Vehicle," in *IEE-Japan Industry Applications Society Conference*, 2016. (in Japanese)

Fault Diagnosis Method of Current Sensor for Permanent Magnet Synchronous Motor Drives

Guoqiang Zhang[1], Guoxin Wang[1], Gaolin Wang[1]*, Junya Huo[1,2], Lianghong Zhu[2] and Dianguo Xu[1]

1 School of Electrical Engineering & Automation, Harbin Institute of Technology, Harbin, China
2 GD Midea Air-Conditioning Equipment Co., Ltd., Foshan, China
*E-mail: WGL818@hit.edu.cn

Abstract— **The application of permanent magnet synchronous motor (PMSM) is becoming more and more popular. The reliability of PMSM has become a research hotspot. The fault diagnosis method of current sensor for PMSM drive system is studied in this paper. The fault type of the current sensor is analyzed and the fault model is built. Using the method of changing the stationary coordinate frame, the current reference value is compared with the sampled value from the sensor. Then the fault diagnosis method based on the residual error is used to diagnose the fault of the current sensor. Experimental results show that the proposed method could realize the fast and accurate positioning of the fault phase current sensor.**

Keywords— *Permanent magnet synchronous motor, current sensor, fault diagnosis.*

I. INTRODUCTION

Due to the simple structure, reliable operation and great speed performance, permanent magnet synchronous motor (PMSM) has been widely used in high performance driving system and other industrial fields [1]. Typically, in a PMSM drive system, a position sensor and at least two current sensors are used to ensure closed-loop control of the motor [2]. Any sensor failure will cause the motor to run abnormally, or even cause the system to crash. Therefore, the current sensor fault diagnosis of PMSM drive system is one of the hotspots.

The current sensor fault diagnosis methods are mainly divided into three categories, namely, model-based, signal-based and knowledge-based fault diagnosis [3]. The model-based fault diagnosis method compares the measured values from the sensor with those obtained by the observer and then diagnoses the phase current fault. In [2], the estimated value of the speed and phase current was obtained by Extended Kalman filter, which was compared with the measured value to obtain the residual error, then the fault diagnosis of the position and current sensor was realized. A fault diagnosis method based on three adaptive full-order observers was proposed in [4]. However, this method could fail when two current sensors broke down continuously. In [5], the vector control of single current sensor was realized, which can

be controlled by another phase current when a phase current sensor fails. In general, the fault diagnosis methods based on state observer rely on accurate motor parameters. The fault diagnosis of current sensor was completed by coordinate transformation [3]. Artificial neural network method was used to diagnose fault, which is difficult to implement in application [6,7].

In this paper, the fault type of current sensor is analyzed, and the fault of current sensor is modeled. Then the fault diagnose of the current sensor is realized by changing the stationary coordinate system, and the fault tolerance of the motor is performed after the fault is diagnosed. The fault diagnosis and fault tolerant control method of current sensor has been verified by experiment.

II. ANALYSIS OF FAULT DIAGNOSIS PRINCIPLE OF CURRENT SENSOR

A. Current Sensor Failure Analysis

The PMSM current sensor is affected by external environment and its own factors, such as mechanical vibration, corrosion, electromagnetic and noise interference, contact, etc., which can easily malfunction. The specific performance is the low reliability of the sensor output value. The fault type of current sensor can be roughly divided into deviation fault, drift fault, shock failure, short circuit fault, open circuit fault, periodic interference and non-linear dead zone fault, etc.

The fault of current sensor could be simplified as broken line fault, stuck fault, deviation fault and gain fault. At the time of t_0, the sensor has failed, and the sensor failure of the current sensor at t_1 has been relieved. The four fault models are expressed as follows:

Modeling of broken line fault of current sensor:

$$y = \begin{cases} y_r, & 0 \leq t < t_0 \\ 0, & t_0 \leq t < t_1 \\ y_r & t \geq t_1 \end{cases} \tag{1}$$

Modeling of stuck fault of current sensor:

$$y = \begin{cases} y_r, & 0 \leq t < t_0 \\ A, & t_0 \leq t < t_1 \\ y_r & t \geq t_1 \end{cases} \tag{2}$$

Modeling of deviation fault of current sensor:

Supported by the National Key R&D Program of China (2016YFE0102800) and the Natural Science Foundation of Heilongjiang Province (E2016028)

$$y = \begin{cases} y_r, & 0 \le t < t_0 \\ y_r + C, & t_0 \le t < t_1 \\ y_r & t \ge t_1 \end{cases} \quad (3)$$

Modeling of gain fault of current sensor:

$$y = \begin{cases} y_r, & 0 \le t < t_0 \\ By_r, & t_0 \le t < t_1 \\ y_r & t \ge t_1 \end{cases} \quad (4)$$

where y is the output value of the sensor, y_r is the output value of the sensor when working normally, and A, B, and C are constants.

B. Current Sensor Fault Diagnosis Method

The corresponding Clarke transformation and Park transformation are shown as (5) and (6).

$$\begin{bmatrix} i_\alpha \\ i_\beta \end{bmatrix} = \begin{bmatrix} 1 & 0 \\ \dfrac{\sqrt{3}}{3} & \dfrac{2\sqrt{3}}{3} \end{bmatrix} \begin{bmatrix} i_a \\ i_b \end{bmatrix} \quad (5)$$

$$\begin{bmatrix} i_d \\ i_q \end{bmatrix} = \begin{bmatrix} \cos\theta & \sin\theta \\ -\sin\theta & \cos\theta \end{bmatrix} \begin{bmatrix} i_\alpha \\ i_\beta \end{bmatrix} \quad (6)$$

where i_a and i_b are the current values in the three-phase stationary coordinate system, i_α and i_β are the current values in the two-phase stationary frame, and i_d and i_q are the current values in the synchronous rotating coordinate system.

It is not hard to observe from (5) that when the current sensor of phase A fails, i_α and i_β are both fault signals. When the current sensor of phase B fails, i_α is still accurate. The fault diagnosis method of current sensor based on coordinate transformation makes use of the above difference. Therefore, designing a matrix T that meets the following requirements is considered. For (7) and (8), $i_{\alpha'}$ and $i_{\beta'}$ are both fault signals when the current sensor of phase B fails. When the current sensor of phase A fails, $i_{\alpha'}$ is still accurate.

$$\begin{bmatrix} i_{\alpha'} \\ i_{\beta'} \end{bmatrix} = T \begin{bmatrix} 1 & 0 \\ \dfrac{\sqrt{3}}{3} & \dfrac{2\sqrt{3}}{3} \end{bmatrix} \begin{bmatrix} i_a \\ i_b \end{bmatrix} \quad (7)$$

$$\begin{bmatrix} i_d \\ i_q \end{bmatrix} = \begin{bmatrix} \cos\theta & \sin\theta \\ -\sin\theta & \cos\theta \end{bmatrix} T^{-1} \begin{bmatrix} i_{\alpha'} \\ i_{\beta'} \end{bmatrix} \quad (8)$$

where $i_{\alpha'}$ and $i_{\beta'}$ are the current values of the constructed stationary coordinate system respectively, and T and T^{-1} are transformation matrice and the inverse matrice of transformation matrice.

As shown in Fig. 1, the above requirements can be met when the α' axis of the structure is rejoined to the B axis.

The transformation matrix of i_α and i_β to $i_{\alpha'}$ and $i_{\beta'}$ is the Park transformation matrix when electric angle is equal to 120°.

$$T = \begin{bmatrix} \cos 120° & \sin 120° \\ -\sin 120° & \cos 120° \end{bmatrix} \quad (9)$$

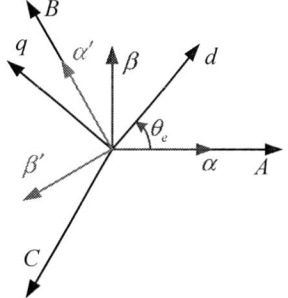

Fig. 1. Relationship of coordinate transformation.

$$\begin{bmatrix} i_{\alpha'} \\ i_{\beta'} \end{bmatrix} = \begin{bmatrix} 0 & 1 \\ -\dfrac{2\sqrt{3}}{3} & -\dfrac{\sqrt{3}}{3} \end{bmatrix} \begin{bmatrix} i_a \\ i_b \end{bmatrix} \quad (10)$$

$$\begin{bmatrix} i_d \\ i_q \end{bmatrix} = \begin{bmatrix} \cos(\theta-120°) & \sin(\theta-120°) \\ -\sin(\theta-120°) & \cos(\theta-120°) \end{bmatrix} \begin{bmatrix} i_{\alpha'} \\ i_{\beta'} \end{bmatrix} \quad (11)$$

Model-based fault diagnosis can be based on residual error or parameter estimation. The fault diagnosis based on coordinate transformation is obviously unable to predict the current. It is a fault diagnosis based on the residual errors. The current given value can be obtained by the inverse Park according to the given value of the PI regulator output.

When the α axis and A axis are combined, the formula for inverse Park transformation can be derived as

$$\begin{bmatrix} i_\alpha^* \\ i_\beta^* \end{bmatrix} = \begin{bmatrix} \cos\theta & -\sin\theta \\ \sin\theta & \cos\theta \end{bmatrix} \begin{bmatrix} i_d^* \\ i_q^* \end{bmatrix} \quad (12)$$

where i_d^* and i_q^* are the given values of the current regulator, and i_α^* and i_β^* are the values of the current given values in the two-phase stationary frame.

When the α' axis and B axis are recombined, the formula for the inverse Park transformation is performed according to the given value of the PI regulator output:

$$\begin{bmatrix} i_{\alpha'}^* \\ i_{\beta'}^* \end{bmatrix} = \begin{bmatrix} \cos(\theta-120°) & -\sin(\theta-120°) \\ \sin(\theta-120°) & \cos(\theta-120°) \end{bmatrix} \begin{bmatrix} i_d^* \\ i_q^* \end{bmatrix} \quad (13)$$

where $i_{\alpha'}^*$ and $i_{\beta'}^*$ are the current given values in the modified two phases of the stationary coordinate system.

C. Current Sensor Fault Diagnosis Algorithm

According to the analysis results, the fault diagnosis and fault phase current location of the current sensor can be realized by analyzing the residual difference between i_α and i_α^*, $i_{\alpha'}$ and $i_{\alpha'}^*$. The fault diagnosis logic diagram is shown in Fig. 2. The structure block diagram of PMSM vector control system based on this fault diagnosis method is shown in Fig. 3.

The specific diagnosis process is described as follows. When $|i_\alpha^*-i_\alpha|<\varepsilon_1$ and $|i_{\alpha'}^*-i_{\alpha'}|<\varepsilon_2$ are both satisfied, the motor current sensor has no fault. When $|i_\alpha^*-i_\alpha|<\varepsilon_1$ and $|i_{\alpha'}^*-i_{\alpha'}|<\varepsilon_2$ are not satisfied, both the A axis and the B axis current sensor fail. When $|i_\alpha^*-i_\alpha|<\varepsilon_1$ satisfies and $|i_{\alpha'}^*-i_{\alpha'}|<\varepsilon_2$ is not satisfied, the B axis current sensor fails. similarly, the A axis current sensor fails.

The 2018 International Power Electronics Conference

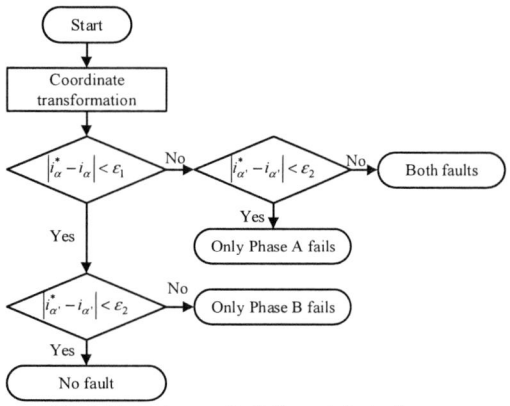

Fig. 2. Current sensor fault diagnosis logic diagram.

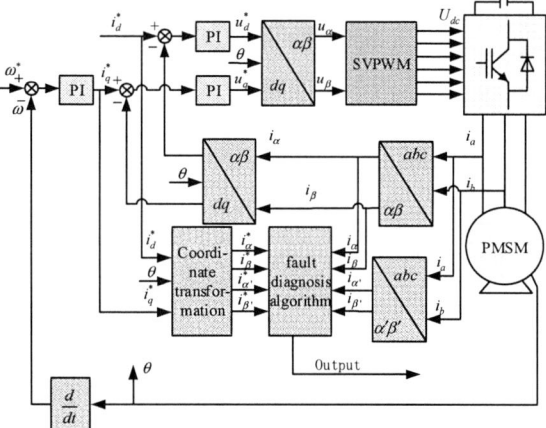

Fig. 3. Block diagram of PMSM based on coordinate transform fault diagnosis.

D. Fault Tolerant Control Method

The fault-tolerant control principle diagram is shown in Fig. 4. When the fault of the current sensor is not diagnosed, the measurement value of phase *A* and phase *B* current sensor are adopted to control the motor. After the fault of phase *A* current sensor is diagnosed, the fault tolerance of the motor is controlled by the measurement of the phase *B* and phase *C* current sensor. After the fault of the phase *B* current sensor is diagnosed, the fault tolerant control of the motor is carried out by the measurement value of phase *A* and phase *C* current sensor.

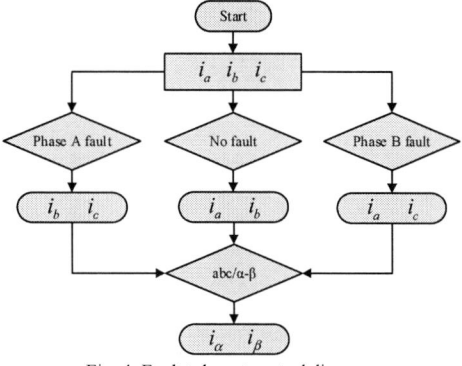

Fig. 4. Fault tolerant control diagram.

III. EXPERIMENTAL RESULTS

A. Experimental Platform

The fault diagnosis algorithm was verified by the experimental platform of permanent magnet synchronous motor drive. The experimental platform includes a 1-kW PMSM, another PMSM is coupled with it to provide the load torque. Two inverters are connected by a common dc bus line. The overall platform is shown in Fig. 5. The vector control and fault diagnosis algorithm is realized by STM32F103VB. The inverter uses PM25RLA120 IPM power module, and the stator current is tested by the PHA20VB15 hall current sensor.

(a) PMSMs

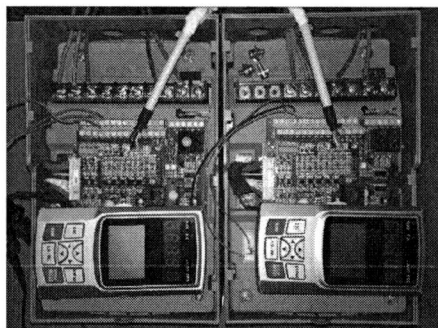

(b) Inverter based on ARM chip

Fig. 5. 1kW PMSM experimental platform.

The motor parameters of the experimental platform are shown in table I.

TABLE I
PARAMETERS OF TESTED PMSM

Parameter type	Value
Rated power	1kW
Rated voltage	220V
Rated speed	3000rpm
Stator resistance	2.75Ω
d axis inductance	7.9mH
q axis inductance	11.7mH
Pole pairs	3

B. Diagnostic Test of Current Sensor Gain Fault

The parameters and thresholds of the PMSM sensor fault diagnosis system need to be set before the current sensor fault diagnosis. The current residuals required for the fault diagnosis system are tested and the thresholds can be set according to the difference.

The motor is operating at 400rpm in no-load condition. When the load is added, the residual difference of the motor *α* axis is shown in Fig. 6. In order to diagnose the motor current sensor, the threshold value of

1208

the motor current fault diagnosis comparator was set by observing the difference of motor current residuals. The specific setting parameters are shown in table II. As a result, the gain and fault diagnosis of motor current sensor can be verified. In order to ensure the normal operation of the motor and prevent accidents, the third phase current sensor is applied to replace the fault phase current sensor for closed-loop control when fault tolerance control is carried out.

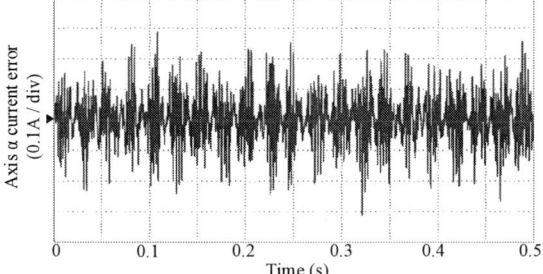

Fig. 6. Residual error of motor α axis current signal.

TABLE II
CURRENT SENSOR FAULT DIAGNOSIS COMPARATOR THRESHOLD

Name	Value
$\lvert i^*_\alpha - i_\alpha \rvert$ Comparator threshold ε_1	$\varepsilon_1 = 0.4$
$\lvert i^*_\alpha - i_\alpha \rvert$ Comparator threshold ε_2	$\varepsilon_2 = 0.4$

Firstly, the motor accelerates to 400rpm and keeps running at a constant speed. Then the rated load is added after the motor stabilizes, and the output of the motor phase A current sensor is scaled up to 1.4 times. Sensor fault diagnosis system can realize the current fault detection, output fault signs and switch to the tolerant control. Phase B and C current sensor outputs replace the fault phase current sensor output and access the closed-loop speed regulating system, the motor steady running. The experimental result is shown as follows.

Fig. 7 is the output waveform of phase A and phase B current sensor before and after the gain failure of phase A current sensor. It can be seen that at 0.25s, the magnitude of phase A current is 1.4 times the non-fault current. At this point, the fault diagnosis system can detect the fault of current sensor. Then the fault tolerance control is carried out by replacing phase A current with phase C current.

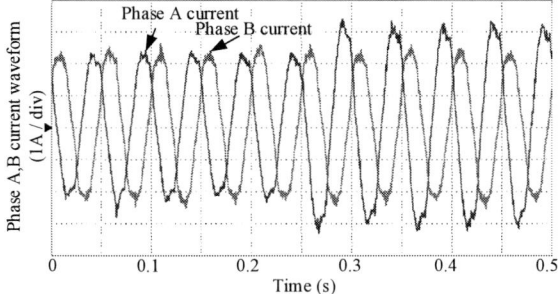

Fig. 7. Phase A and phase B current sensor output waveform.

In order to detect the state change of the motor fault diagnosis system, the signal of the current fault signal should be detected. Fig. 8 shows the current sensor fault signal, the current error of the α axis sensor and the given value. The motor at t_0 will start to accelerate to 400rpm. The load is added at t_1. The current error of the motor α axis is less than the threshold, and the fault mark bit is 0. Phase A current sensor gain failure occurred at t_2. It can be seen from the enlarged diagram that the current signal error is mutated, beyond the threshold range, and the fault diagnosis mark becomes 1. The fault is detected and fault-tolerant control is carried out. The motor continued to operate, unloaded at t_3, and then stopped at t_4.

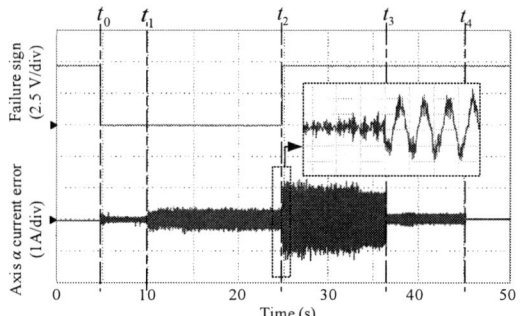

Fig. 8. Fault signal of current sensor and error of α axis.

In order to reflect the diagnosis process of the fault diagnosis system, the change of the α axis current of the motor is observed. Fig. 9 is the comparison diagram of the α axis current and the fault tolerant control current accessing motor system. It can be seen that at 0.25s, the output of the motor sensor has a gain failure and the current amplitude increases. At this point, the fault diagnosis system detects the fault of the current sensor and makes fault tolerance control. The fault current sensor is quickly removed, and the current sampling is implemented by the other two phase current sensors. The motor could operate normally.

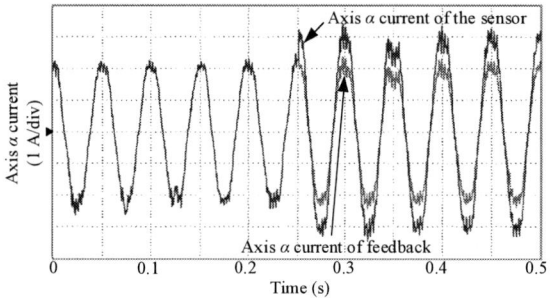

Fig. 9. α axis current of sensor output and feedback.

Fig. 10 shows the fault signal of current sensor and the feedback current waveform of the α axis. At t_0, motor starts up and accelerates to 400rpm. After stable operation, the motor is loaded at t_1. At t_2, phase A current sensor has a gain failure, and then the motor starts to fault tolerant control. The detailed diagram shows that the feedback current signal does not change significantly, the motor continues to operate steadily.

The 2018 International Power Electronics Conference

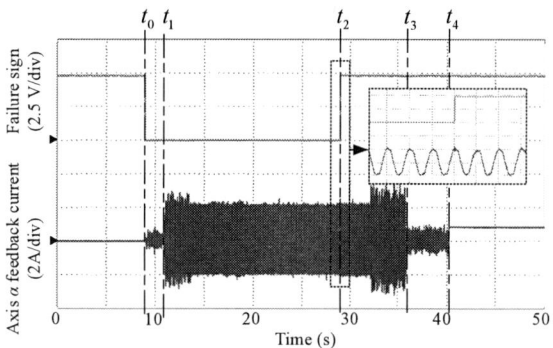

Fig. 10. Fault signal and α axis feedback current.

According to the experimental and analytical results, the fault diagnosis system can accurately diagnose the fault and switch to the fault-tolerant control state in time when the motor current sensor has a gain failure. It could ensure the normal operation of the motor.

C. Diagnostic Test of Current Sensor Broken Line Fault

At t_0 moment, motor starts up. The motor accelerates to 400rpm and then runs at a constant speed. After the motor stabilizes, it is loaded at t_1, then at the t_2, the phase current sensor of the motor A phase is broken. The output of phase A current signal becomes 0. The fault diagnosis system can detect the fault of phase A current sensor and output fault marks for fault tolerance control. The specific experimental results are shown as follows.

Fig. 11 is the output waveform of phase A and phase B current sensor before and after the fault of phase A current sensor. It can be seen that at 0.25s, the output waveform of the motor phase A current sensor becomes 0, and the motor current sensor has a broken line fault.

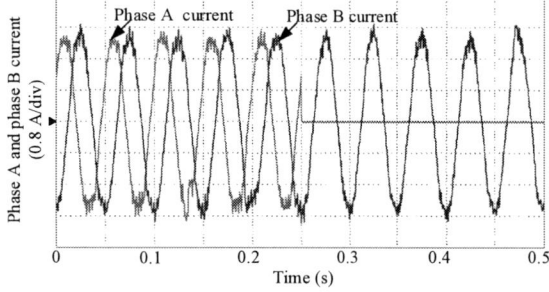

Fig. 11. Phase A and phase B current sensor output waveform.

In order to detect the state change of the motor fault diagnosis system, the signal of the current fault signal should be tested.

Fig. 12 is the fault signal and the current observation error of the α axis. The motor on t_0 will start to accelerate to 400rpm. At time t_1, the current error of the motor α axis is less than the threshold, and the fault mark bit is 0. At the time of t_2, phase A current sensor is broken. It can be seen from the enlarged diagram that the current signal error increases to the threshold range, and the fault diagnosis mark becomes 1. The fault can be detected and then the fault tolerance control is performed.

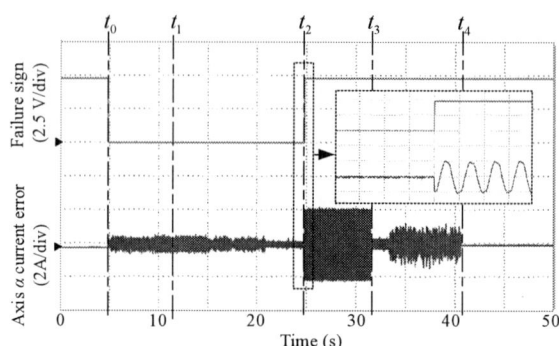

Fig. 12. Fault signal of current sensor and error of α axis.

Fig. 13 is the comparison diagram of the α axis current and the α axis current of the motor closed-loop speed control system obtained by the sensor output. It can be observed that at 0.25s, the output waveform of the α axis current sensor obtained by the motor sensor is broken and the current value becomes 0. At this point, the fault diagnosis system detects the fault of the current sensor and makes fault tolerance control, and the current loop of the motor current closed-loop feedback keeps normal and the motor runs.

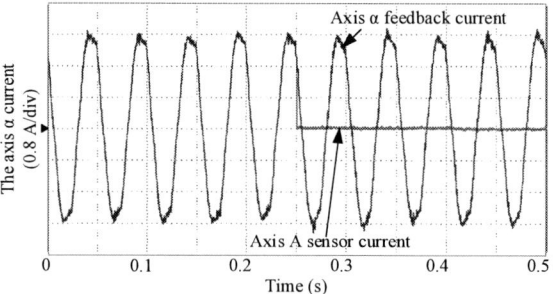

Fig. 13. α axis current of sensor output and feedback.

Fig. 14 is the signal of current sensor fault signal and the feedback current waveform of the α axis. The motor starts to accelerate to 400rpm at t_0. The stable operation is loaded at time t_1. At the time of t_2, phase A current sensor of the motor is broken, then the fault tolerance control is performed. The amplification diagram shows that the feedback current signal has not changed significantly. The motor could operate normally by switching the current sensors.

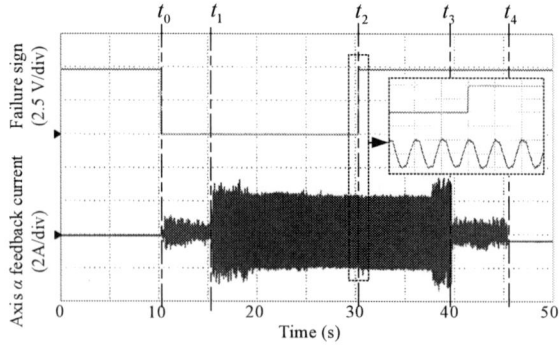

Fig. 14. Fault signal and α axis feedback current.

1210

IV. CONCLUSIONS

In this paper, a method of coordinate transformation of a stationary coordinate system is studied, and the fault diagnosis of the current sensor of PMSM variable frequency drive system is realized. The output current of the sensor obtained by the coordinate transformation is reduced by a given value of the current. By comparing the current residuals to the threshold, the current sensor of the motor can get the fault state of the motor current sensor. The experimental results show that this method can realize the fast and accurate diagnosis of PMSM current sensor fault, and provide the basis for fault tolerance control and ensure the stable operation of the motor.

REFERENCES

[1] Y. Jeong, S. Sul, S. E. Schulz, and N. R. Patel, "Fault Detection and Fault-Tolerant Control of Interior Permanent-Magnet Motor Drive System for Electric Vehicle," *IEEE Trans on Industry Applications*, vol. 41, no. 1, pp. 46-51, 2005.

[2] G. H. Beng Foo, X. Zhang, and D. M. Vilathgamuwa, "A Sensor Fault Detection and Isolation Method in Interior Permanent-Magnet Synchronous Motor Drives Based on an Extended Kalman Filter," *IEEE Trans Industrial Electronics*, vol. 60, no. 8, pp. 3485-3495, 2013.

[3] C. Chakraborty, and V. Verma, "Speed and Current Sensor Fault Detection and Isolation Technique for Induction Motor Drive Using Axes Transformation," *IEEE Trans Industrial Electronics*, vol. 62, no. 3, pp. 1943-1954, 2015.

[4] Y. Yu, Z. Y. Wang, and D. G. Xu, "Speed and current sensors fault detection and isolation based on adaptive observers for induction motor drivers," *Journal of Power Electronics*, vol. 14, no. 5, pp. 967-979, 2014.

[5] V. Verma, C. Chakraborty, S. Maiti, and Y. Hori, "Speed Sensorless Vector Controlled Induction Motor Drive Using Single Current Sensor," *IEEE Trans Energy Conversion*, vol. 28, no. 4, pp. 938-950, 2013.

[6] G. Betta, C. Liguori, and A. Pietrosanto, "An Advanced Neural-Network-Based Instrument Fault Detection and Isolation Scheme," *IEEE Trans Instrumentation and Measurement*, vol. 47, no. 2, pp. 507-512, 1998.

[7] A. Bernieri, G. Betta, A. Pietrosanto, and C. Sansone, "A Neural Network Approach to Instrument Fault Detection and Isolation," *IEEE Trans Instrumentation and Measurement*, vol. 44, no. 3, pp. 747-750, 1995.

The 2018 International Power Electronics Conference

Sensorless Speed Control of Diesel-Generator Systems Based on Multiple SOGI-FLLs

Ngoc Dat Dao[1], Dong-Choon Lee[1], and Dae-Sik Lim[2]

1 Department of Electrical Engineering, Yeungnam University, Korea

2 Bokuk Electric Industrial Company

daongocdat@gmail.com, dclee@yu.ac.kr

Abstract— **This paper proposes a novel method of speed estimation for variable-speed diesel-generator (genset) systems, where the speed of the genset is calculated from the back-EMF frequency of the generator. The back-EMF frequency is extracted from an output phase current. The proposed method is based on multiple and cascade second-order generalized integrators (MC-SOGIs) and separated frequency-locked loops (FLLs). As a result, although the phase currents of the generator connected with a diode rectifier contain high-order harmonics components, the proposed method can estimate the back-EMF frequency accurately with fast response and high stability. Simulation and experimental results are shown to verify the validity of the proposed method.**

Keywords— Diesel engine, frequency-locked loop, genset, second-order generalized integrators, sensorless control.

I. Introduction

Speed information is indispensable for the control of any variable-speed systems. In genset systems, the speed of the engine can be acquired through mechanical sensors such as camshaft or crankshaft speed sensor. However, the use of these sensors increases the cost and hardware complexity of the system. A lot of research efforts have been devoted to eliminate the use of mechanical sensors by a variety of position/speed sensorless control schemes [1]. Most of the methods are applied to the systems with a PWM converter, where the currents of the PMSG are controlled to be sinusoidal by the converter, of which speed and position are estimated based on the generator model. However, in some applications where the output of the generator is connected with a diode rectifier, the output current of the generator is distorted due to high-order harmonic components [2]. Therefore, it is necessary to develop a method to accurately estimate the back-EMF frequency from the output current with a strong immunity against the high-order harmonic components.

For frequency estimation, the SOGI-FLL has been utilized widely in grid synchronization applications [3], [4]. Furthermore, since the SOGI-FLL is an adaptive frequency filter, it has a great potential to apply to variable speed systems. However, some errors may exist in the estimated frequency if the input signals of the SOGI-FLL contain DC [5] or other harmonic components. To improve the filtering capability of the FLLs, a multiple SOGI (MSOGI) FLL has been proposed [6], where each SOGI is responsible for extracting particular harmonic components. The MSOGI-FLL is

effective in eliminating harmonic frequencies that are of integer multiples of the fundamental component, but the method has high sensitivity with unknown frequencies in the input signal such as inter-harmonics, sub-harmonics and DC components [6]. Another method uses a cascade connection of several SOGIs as pre-filters [7]. This method has shown a higher filtering capability than the standard SOGI-FLL. Similar to [7], the improved SOGI-FLL suggested in [8] with a fourth-order transfer function provides an excellent harmonic filtering characteristics.

In this study, in order to extract the back-EMF frequency accurately from a phase current of the generator connected with a diode rectifier, a novel method is proposed which is a combination of multiple and cascade SOGI-FLL method. The method uses multiple SOGI-FLLs as pre-filters to eliminate the fifth-order harmonic since it is the dominant harmonic component in the current signal. Then, the output of the multiple SOGI-FLL is fed to a single standard SOGI-FLL so that other unknown harmonics can be attenuated. The proposed method utilizes two separate FLL blocks, which is different from the existing methods [7], [8] where only one FLL block is used for all SOGIs. That brings an advantage of a degree of freedom for tuning FLLs to achieve faster dynamic response without reducing the immunity to harmonics components. In addition, by using two FLLs, the system is more stable and robust to the frequency disturbance than the case of only one FLL. The operating principle and the dynamic model of the proposed MC-SOGI-FLL are investigated for the stability analysis and the parameter design. Finally, the proposed method is verified by simulation and experimental results.

II. System Description

A. Variable-Speed Genset System

Fig. 1 illustrates the block diagram of the genset system. The mechanical power from the diesel engine is transformed into the AC power by the PMSG. Then, the variable-frequency variable-voltage output of the generator is converted to a DC voltage by the diode rectifier. The DC/DC converter is used to provide a constant DC bus voltage. The DC bus can be connected to the DC grid or to the DC/AC inverter with an AC grid.

A classical genset is driven at a constant speed to generate a fixed electrical frequency and output voltage. Although such a genset has a low initial cost, it gives

1212

The 2018 International Power Electronics Conference

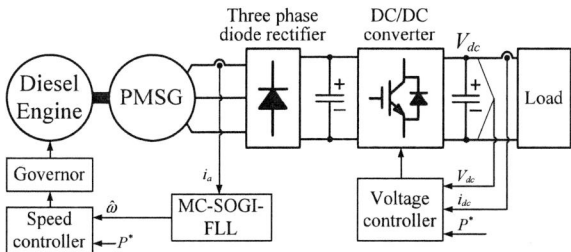

Fig. 1. Block diagram of a genset system.

usually a low efficiency at light-load conditions and the voltage and frequency is fluctuated when the load varies. Many research results have shown that the fuel consumption and emission can be reduced by a variable-speed operation according to the output power demand based on the fuel map of the engine [9], [10]. For a variable-speed operation, the speed information is indispensable, which is acquired by extracting the back-EMF frequency from one phase output current of the generator.

B. Three-Phase Diode Bridge Rectifier

Three-phase diode bridge rectifiers are widely used to convert an AC power from a utility grid or a generator into a DC power source in industry applications. The configuration of the diode rectifier is simple and there is no need to control. The main drawback of the diode rectifier is that it makes the output currents of the generator be distorted. The output currents contain the 5th, 7th, 11th, 13th and other higher harmonic components. The 5th and 7th components are the dominant harmonics which have much higher magnitude than those of others [2]. The percentage of current harmonics increases when the fundamental current component decreases.

III. REVIEW OF EXISTING SOGI-FLL METHODS

A. Standard SOGI-FLL

For a single SOGI as shown in Fig. 2, with the input signal v, the output signals v' and qv' can be expressed in the s -domain by

$$D(s) = \frac{v'}{v}(s) = \frac{k\omega's}{s^2 + k\omega's + \omega'^2} \tag{1}$$

$$Q(s) = \frac{qv'}{v}(s) = \frac{k\omega'^2}{s^2 + k\omega's + \omega'^2}. \tag{2}$$

The transfer functions in (1) and (2) represent the characteristics of a band-pass filter and a low-pass filter, respectively, where the bandwidth of these filters depends only on the gain k, not the estimated frequency ω' [6]. The response of the SOGI is faster by increasing k, but to enhance the filtering capability, the value of k should be decreased. The settling time in the SOGI response can be approximated to

$$t_s = \frac{10}{k\omega'}. \tag{3}$$

B. Multiple SOGI-FLL

The multiple-SOGI-FLL (MSOGI-FLL) is able to

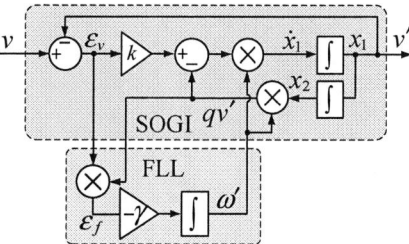

Fig. 2. Block diagram of standard SOGI-FLL.

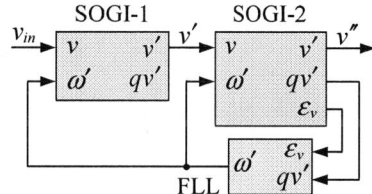

Fig. 3. Block diagram of CSOGI-FLL with a single FLL [7].

obtain the fundamental component without harmonic distortion by using multiple SOGI filters connected in parallel. The harmonics in the inputs are cancelled by the cross-feedback network with the extracted signals of each SOGI. The MSOGI-FLL has shown excellent capability for eliminating harmonic components at certain frequencies. However, the accuracy of the structure can be affected by unknown harmonic components or even by a DC offset in the input signal. The bandwidth of multiple SOGIs can be set to be narrow to enhance the filtering capability, but that would make the system response too slow [6].

C. Cascaded SOGI-FLL

Another method to achieve better harmonic attenuation is to increase the filter-order of the system by connecting the single-SOGI filters in cascade (CSOGI-FLL) [7]. A simple structure of the CSOGI-FLL is shown in Fig. 3 where two SOGIs are connected in series. The first one operates as a band-pass filter and the second one cooperates with a FLL to obtain the frequency of the input signal. Two SOGI filters are adaptively tuned by the same estimated frequency.

The transfer function of this system is

$$D'(s) = \frac{v''}{v} = \frac{k_1 k_2 \omega'^2 s^2}{\left(s^2 + k_1\omega's + \omega'^2\right)\left(s^2 + k_2\omega's + \omega'^2\right)} \tag{4}$$

$$Q'(s) = \frac{qv''}{v} = \frac{k_1 k_2 \omega'^3 s}{\left(s^2 + k_1\omega's + \omega'^2\right)\left(s^2 + k_2\omega's + \omega'^2\right)} \tag{5}$$

IV. PROPOSED MC-SOGI-FLL

The proposed method combines both MSOGI-FLL and CSOGI-FLL structures to obtain better filtering capability with lower computational burden compared with the case that either MSOGI-FLL or CSGOGI-FLL is only used. In addition, the proposed structure utilizes a separated FLL block for each SOGI block connected in series to achieve fast response of frequency estimation and high robustness against frequency variations.

1213

The 2018 International Power Electronics Conference

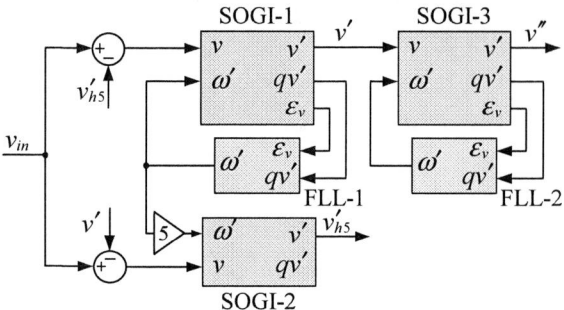

Fig. 4. Block diagram of proposed MC-SOGI-FLL.

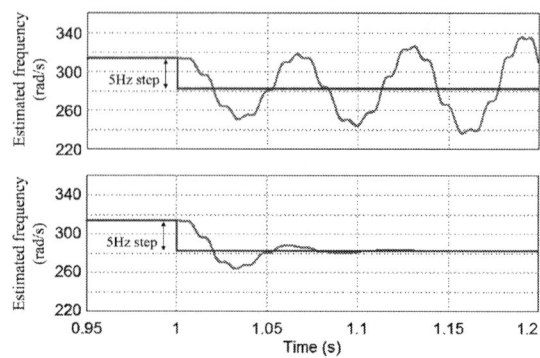

Fig. 5. Estimated frequency from CSOGI-FLL with a single FLL (top) and with two FLLs (bottom) for a frequency step disturbance from 50 Hz to 45 Hz.

Fig. 4 shows the proposed MC-SOGI-FLL used for extracting the back-EMF frequency from the output current of the generator in genset systems. The proposed structure consists of three blocks of SOGIs and two FLL blocks. Two SOGIs are connected in parallel to extract the fundamental and fifth-order harmonic components. Since the fifth-order harmonic component is dominant in the output currents, only one SOGI block is tuned at the fifth-order harmonic frequency.

If the input signal contains high-order harmonics, the state variables of the first SOGI-FLL have the relationship as

$$\dot{\bar{x}}_1 = -\omega^2 \bar{x}_{21} - \omega^2 \sum h^2 \bar{x}_{2h} \qquad (6)$$

where h is an odd integer number (5, 7, 11, 13), which means the order number of harmonic components in the input signal. The bar over the variables represents for steady-state values.

The steady-state error signal can be expressed as

$$\bar{\varepsilon}_v = \left(v - \bar{x}_1\right) = \frac{1}{k\omega'}\left(\dot{\bar{x}}_1 + \omega'^2 \bar{x}_2\right). \qquad (7)$$

Substituting (6) into (7), the steady-state frequency error is obtained as

$$\bar{\varepsilon}_f = \omega' \bar{x}_2 \bar{\varepsilon}_v = \frac{\bar{x}_2^2}{k}\left(\omega'^2 - \omega^2\right) - \underbrace{\frac{\bar{x}_2}{k}\omega^2 \sum (h^2 - 1)\bar{x}_{2h}}_{\Delta \bar{\varepsilon}_{fh}}. \qquad (8)$$

The dynamics of the FLL is given by

$$\dot{\omega}' = -\gamma\bar{\varepsilon}_f = -\frac{\gamma}{k}\bar{x}_2^2\left(\omega'^2 - \omega^2\right) + \gamma\Delta\bar{\varepsilon}_{fh}. \qquad (9)$$

If the value of γ is normalized according to

$$\gamma = \frac{k\omega'}{v'^2 + qv'^2}\Gamma, \qquad (10)$$

the averaged dynamics of the FLL can be expressed as a first-order transfer function as

$$\frac{\bar{\omega}'}{\omega}(s) = \frac{\Gamma}{s + \Gamma}. \qquad (11)$$

As seen in (8), $\bar{\varepsilon}_f$ has the information about an error in frequency estimation which is used as the control signal of the FLL. However, if the input signal contains harmonic components, the frequency estimation will have oscillations due to the second term on the right-hand side in (8).

To reduce the computational time, only one SOGI is used to cancel the 5th order harmonic component. Thus, there are still 7th, 11th and 13th harmonics in the output signal of the first SOGI, which generate harmonic ripples in the frequency estimation. From (9), a high value of γ (or Γ) makes the FLL performance fast, but the harmonic ripples in the estimation become high according to the second term on the right-hand side in (9).

To cancel the ripples in the frequency estimation, a low-pass filter can be applied, but it causes a delay in the estimation. Therefore, a third SOGI is connected in series with the first SOGI to further eliminate the remaining harmonics. A second FLL cooperates with the third SOGI and the estimated frequency has almost no harmonic distortion. Hence, the value of Γ in the second FLL can be set to a high value to achieve fast estimation.

The proposed method uses two separated FLLs which is different from the CSOGI-FLL method proposed in [7] where only one FLL is used for all SOGIs. The advantage of using two separated FLLs is that the gain Γ can be set differently for two FLLs. In this case, Γ_2 is set to much higher than Γ_1 to make the speed estimation faster. If only one FLL is used at the third SOGI, then the gain Γ is limited by the bandwidth of the second-order SOGI (a series of two SOGIs) which is narrower than that of the first-order SOGI (a single SOGI).

Fig. 5 shows the behavior of the CSOGI-FLL with a single FLL [7] and the proposed method with two separated FLLs. For the single FLL, the gain $\Gamma = 200$ and for the two separated FLLs $\Gamma_1 = 50$ and $\Gamma_2 = 200$. As seen from Fig. 5, with the same disturbance of 5 Hz in the frequency of the input signal, the estimated frequency settles to 45 Hz after 0.1 s when using two FLLs, but the output of the single FLL oscillates and it is unstable.

V. SIMULATION RESULTS

To verify the proposed method for speed estimation, a simulation is carried out using the PSIM for a 100 kW variable-speed genset system, where the block diagram of the system is illustrated in Fig. 1 and its parameters are listed in Table I. In the MC-SOGI-FLL (Fig. 4), the first SOGI is used to extract the fundamental component of the input signal, which is the output current of the generator. The second SOGI is tuned at the 5th order

1214

TABLE I
GENSET SYSTEM PARAMETERS

Engine		Generator	
Rated power	110 kW	Rated power	100 kW
Rated speed	1800 rpm	Rated speed	1800 rpm
No. cylinders	6	No. poles	12
Genset inertia	1 kgm^2	Rated voltage	400 V
		Rated current	140 A

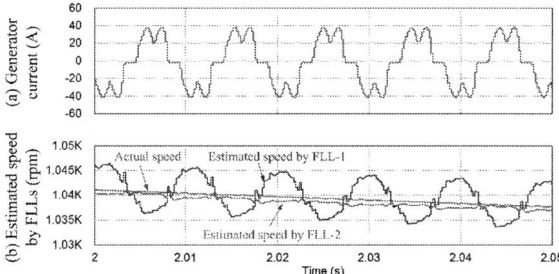

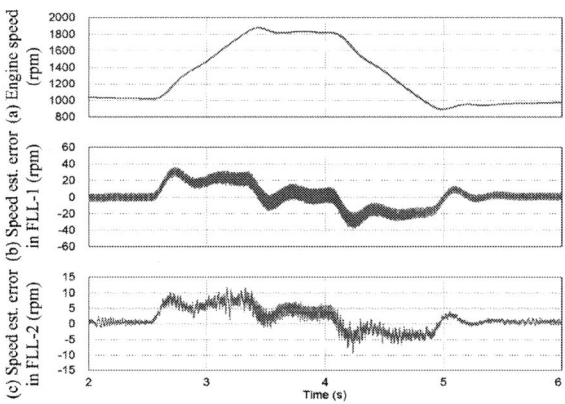

Fig. 6. Performance of MC-SOGI-FLL. (a) Generator current. (b) Speed estimation by FLLs.

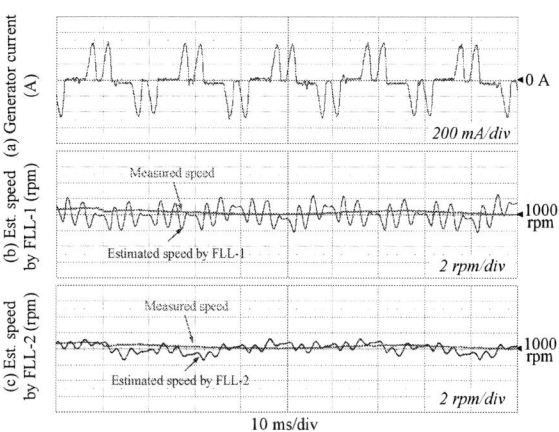

Fig. 8. Experimental performance of MC-SOGI-FLL. (a) Generator current. (b)Measured and estimated speeds by FLL-1. (c) Measured and estimated speeds by FLL-2.

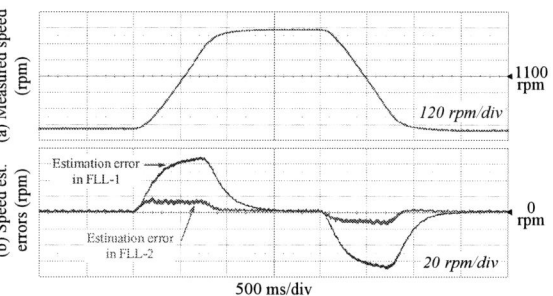

Fig. 9. Experimental performance of MC-SOGI-FLL. (a) Measured speed. (b) Speed estimation errors in FLL-1 and FLL-2.

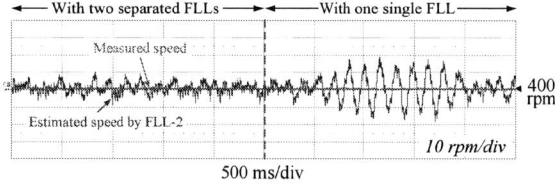

Fig. 10. Performance comparison of speed estimations with two separated FLLs and one single FLL.

Fig. 7. Performance of MC-SOGI-FLL. (a) Engine speed. (b) Speed estimation error in FLL-1. (c) Speed estimation error in FLL-2.

harmonic to be eliminated. Finally, the third SOGI is utilized to enhance the filter capability against the 7th, 11th, 13th order harmonics and DC components. The gains for SOGI-1, SOGI-2 and SOGI-3 are $k_1 = 1$, $k_2 = 0.2$ and $k_3 = 1$, respectively. The gains for FLL-1 and FLL-2 are $\Gamma_1 = 50$ and $\Gamma_2 = 200$, respectively.

Fig. 6 shows the generator current as an input signal of MC-SOGI-FLL and the estimated speed by FLL-1 and FLL-2. The sampling frequency for current measurement and algorithm calculation is 5 kHz. The average value of the current is 27 A and the current THD is 32.6%. A DC offset of -2 A is added to the measured current. As seen from Fig. 6, although the current are much distorted, the speed estimated by the proposed method is highly accurate.

Fig. 7 shows the speed estimation performance when the engine speed varies. The estimated speed is utilized for control of engine speed. During the change of speed, the tracking errors of the speed estimator by FLL-1 and FLL-2 are about 20 rpm and 5 rpm, respectively. The error in the case of FLL-2 is a quarter that of FLL-1 since the gain Γ_2 is four time higher than Γ_1. The speed value

obtained by FLL-1 has ripples from 10 rpm to 20 rpm due to the harmonic components of the generator current.

VI. EXPERIMENT RESULTS

The proposed estimation method has also been verified by experimental results for a system consisting of an induction motor (IM) coupled with a PMSG. The IM emulates the characteristics of a diesel engine which is operated at variable speeds. The output voltage of the PMSG is rectified by a three-phase diode bridge with a DC capacitor. A resistor is connected to the DC capacitor as a load. One current sensor is utilized to measure a phase current of the PMSG. In the experiment, the gains for FLL-1 and FLL-2 are set as $\Gamma_1 = 10$ and $\Gamma_2 = 40$, respectively. These values have been reduced compared with those in the simulation since there is much noise in the experiment.

Fig. 8 shows the performance of the proposed method

at 1000 rpm. The input for the speed estimation is the phase current of the generator, which is distorted and contains high-order harmonic components as shown in Fig. 8(a). However, the speed estimated by the proposed method is accurate with about 2 rpm (0.2%) error compared with the measured value. The error of speed estimation by FLL-1 is 1.5 times higher than that of by FLL-2. Therefore, by utilizing the third SOGI with a separated FLL, the speed estimation is more accurate than that of without the third SOGI.

When the speed of the motor varies, the estimated speed contains errors as shown in Fig. 9. The error is decreased if the gain for the FLL is increased. The gain of FLL-2 can be set higher than that of FLL-1 as $\Gamma_2 = 4\Gamma_2$. As a result, the error with FLL-2 is four times lower than by FLL-1.

Fig. 10 shows the comparison between the performances of the speed estimation in two cases. In the first case, two separated FLLs are used, where the estimated value is stable with a low error of 2.5% (less than 10 rpm). In the second case, only one single FLL is used for all three SOGI blocks. Then, the estimated value is fluctuated unstably where the errors can reach 20 rpm. Thus, it has been proved experimentally that the proposed method is more stable than other existing methods [7], [8].

VII. CONCLUSIONS

In this paper, a novel speed estimation scheme has been proposed for variable-speed diesel-engine generator systems. In the proposed MC-SOGI-FLL method, the speed is estimated by extracting the frequency of a generator output current, which is highly accurate even though the distorted currents exist due to diode rectifier connection. The proposed method can effectively eliminate not only harmonic components but also DC components. Furthermore, the proposed method is more robust to frequency disturbances than other existing methods. The effectiveness of the proposed method has been verified by simulation and experimental results.

REFERENCES

[1] Y. Zhao, C. Wei, Z. Zhang, and W. Qiao, "A review on position / speed sensorless control for permanent-magnet synchronous machine-based wind energy conversion systems," *IEEE J. Emerg. Sel. Top. Power Electron.*, vol. 1, no. 4, pp. 203–216, 2013.

[2] B. Wu, *High Power Converters and AC Drives*, 1st ed. Wiley-IEEE Press, 2006.

[3] P. Rodriguez, A. Luna, M. Ciobotaru, R. Teodorescu, and F. Blaabjerg, "Advanced grid synchronization system for power converters under unbalanced and distorted operating conditions," *IECON Proc. (Industrial Electron. Conf.*, no. 2, pp. 5173–5178, 2006.

[4] P. Rodríguez, A. Luna, R. S. Muñoz-Aguilar, I. Etxeberria-Otadui, R. Teodorescu, and F. Blaabjerg, "A stationary reference frame grid synchronization system for three-phase grid-connected power converters under adverse grid conditions," *IEEE Trans. Power Electron.*, vol. 27, no. 1, pp. 99–112, 2012.

[5] M. Karimi-Ghartemani, S. A. Khajehoddin, P. K. Jain, A. Bakhshai, and M. Mojiri, "Addressing DC component in pll and notch filter algorithms," *IEEE Trans. Power Electron.*, vol. 27, no. 1, pp. 78–86, 2012.

[6] P. Rodríguez, A. Luna, I. Candela, R. Mujal, R. Teodorescu, and F. Blaabjerg, "Multiresonant frequency-locked loop for grid synchronization of power converters under distorted grid conditions," *IEEE Trans. Ind. Electron.*, vol. 58, no. 1, pp. 127–138, 2011.

[7] J. Matas, M. Castilla, J. Miret, L. Garcia De Vicuna, and R. Guzman, "An adaptive prefiltering method to improve the speed/accuracy tradeoff of voltage sequence detection methods under adverse grid conditions," *IEEE Trans. Ind. Electron.*, vol. 61, no. 5, pp. 2139–2151, 2014.

[8] Z. Xin, X. Wang, Z. Qin, M. Lu, P. C. Loh, and F. Blaabjerg, "An improved second-order generalized integrator based quadrature signal generator," *IEEE Trans. Power Electron.*, vol. 31, no. 12, pp. 8068–8073, 2016.

[9] J. Leuchter, P. Bauer, V. Řeřucha, and V. Hájek, "Dynamic behavior modeling and verification of advanced electrical-generator set concept," *IEEE Trans. Ind. Electron.*, vol. 56, no. 1, pp. 266–279, 2009.

[10] S. Choe, Y. K. Son, and S. K. Sul, "Control and analysis of engine governor for improved stability of dc microgrid against load disturbance," *IEEE J. Emerg. Sel. Top. Power Electron.*, vol. 4, no. 4, pp. 1247–1258, 2016.

Robustness of Simplified Speed-Sensorless Vector Control for Induction Motor

Naoki Akao[1], Mineo Tsuji[2], Shin-ichi Hamasaki[3]
1-4 Graduate School of Engineering, Nagasaki University, Nagasaki, Japan
E-mail:mineo@nagasaki-u.ac.jp

Abstract- Many papers have been proposed speed-sensorless vector control systems without encoders from the viewpoint of cost and environmental problems. We have proposed a simplified method of estimating the rotor position from the *d*-axis voltage and controlling the motor speed from the information of *d*-axis voltage too. In order to reveal the robustness of the proposed method, we have performed stability analysis and experiment when the stator resistances changed. To improve the stability of the system, stator resistance compensation is proposed. The simulation and experimental results are shown and discussed.

I. INTRODUCTION

By obtaining the direction of the magnetic flux and controlling the current based on its direction, vector control changes the torque instantaneously. Detection of magnetic flux is indispensable for this control method. For this reason and speed control, a speed sensor is required. However, for cost reduction and miniaturization, sensorless vector control is proposed in many papers. In this case, in order to estimate the magnetic flux, many methods using state observers have been proposed [1]-[5] [10]. Some simplified schemes are also proposed [6]-[9]. However, stability analysis is not described clearly [6]-[8]. We have proposed a simple speed-sensorless vector control method which estimates the magnetic flux without using an observer [11][12].

In order to reveal the robustness of the proposed control method, a linear model is derived. A stability analysis is performed when the stator resistance is changed. Transient responses are observed by simulation and experiment when the stator resistance is changed. Furthermore, we propose a simple identification method of the stator resistance and confirm its usefulness by experiment.

II. SIMPLIFIED INDUCTION MOTOR SENSORLESS VECTOR CONTROL

The rotor flux linkage of the induction motor in the voltage model $(\psi_{rd}^v, \psi_{rq}^v)$ and the current model $(\psi_{rd}^*, \psi_{rq}^*)$ can be obtained by the following models.

Voltage model:

$$e_{sd} = \left(R_s + \sigma L_s p\right) i_{sd} - \dot{\omega}^* \sigma L_s i_{sq} + \frac{M p}{L_r}\psi_{rd}^v - \frac{\dot{\omega}^* M}{L_r}\psi_{rq}^v \quad (1)$$

$$e_{sq} = \dot{\omega}^* \sigma L_s i_{sd} + \left(R_s + \sigma L_s p\right) i_{sq} + \frac{\dot{\omega}^* M}{L_r}\psi_{rd}^v + \frac{M p}{L_r}\psi_{rq}^v \quad (2)$$

Current model:

$$0 = -\frac{M}{\tau_r}i_{sd} + \left(\frac{1}{\tau_r}+p\right)\psi_{rd}^* - \left(\dot{\omega}^* - \omega_r\right)\psi_{rq}^* \quad (3)$$

$$0 = -\frac{M}{\tau_r}i_{sq} + \left(\dot{\omega}^* - \omega_r\right)\psi_{rd}^* + \left(\frac{1}{\tau_r}+p\right)\psi_{rq}^* \quad (4)$$

where, $p = d/dt, \sigma = 1 - M^2/(L_s L_r), \tau_r = L_r/R_r$

We choose the *d-q* axis synchronized with the flux direction θ^* of the current model. Then following equations are derived from (3) and (4)

$$0 = -\frac{M}{\tau_r}i_{sd} + \left(\frac{1}{\tau_r}+p\right)\psi_{rd}^*, \ \psi_{rq}^* = 0 \quad (5)$$

$$\omega^* = \omega_r + \frac{M\,i_{sq}}{\tau_r \psi_{rd}^*}, \ \theta^* = \int_0^t \omega^* dt + \theta^*(0) \quad (6)$$

Figure 1 shows the direction of *d-q* axis. The *d-q* axis is a stationary reference frame. Here,

$$\dot{\psi}_r^v = \psi_{rd}^v + j\psi_{rq}^v \quad (7)$$

$$\dot{\psi}_r^* = \psi_{rd}^* + j\psi_{rq}^* \quad (8)$$

We assume that the rotor flux $\dot{\psi}_r^v = \psi_{r\alpha}^v + j\psi_{r\beta}^v$ of voltage model is true value. By transforming $\dot{\psi}_r^v$ to *d-q* axis which is defined by current model, (1), (2), and (7) are obtained. In *d-q* axis synchronized with the flux $\dot{\psi}_r^*$, it is possible to synchronize the $\dot{\psi}_r^*$ on flux $\dot{\psi}_r^v$ by increasing $\dot{\omega}^*$ when $\psi_{rq}^v > 0$ and decreasing $\dot{\omega}^*$ when $\psi_{rq}^v < 0$. As the result, the $\dot{\omega}^*$ is obtained to satisfy $\psi_{rq}^v = 0$.

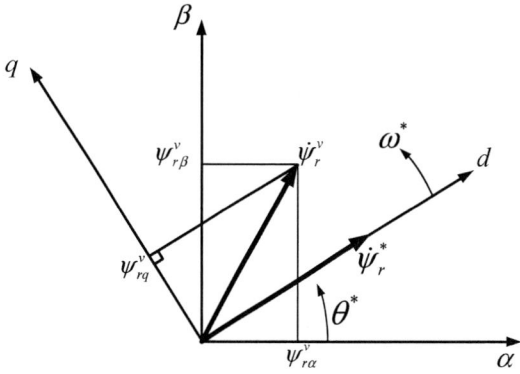

Fig.1 The definition of *d-q* axis.

From (6), the rotor speed is computed by the following equation:

$$\hat{\omega}_r = \omega^* - \frac{M \, i_{sq}}{\tau_r \psi_{rd}^*} \tag{9}$$

By assuming that the d-axis current i_{sd} is controlled its constant reference i_{sd}^* exactly and as the result ψ_{rd}^y is constant, (1) becomes to

$$e_{sd} = R_s i_{sd} - \omega^* \sigma L_s i_{sq} - \frac{\omega^* M}{L_r} \psi_{rq}^y \tag{10}$$

Figure 2 shows the proposed sensorless vector control system [11][12]. The e_{sd}^* is computed as

$$e_{sd}^* = e_d^* + R_s^* i_{sd}^* - \omega^* \sigma L_s i_{sq} \tag{11}$$

where,

$$e_d^* = K_p (i_{sd}^* - i_{sd}) \tag{12}$$

Therefore, following equation is expected from (10) and (11):

$$e_d^* = -\frac{\omega^* M}{L_r} \psi_{rq}^v \tag{13}$$

Equation (13) means that the voltage e_d^* which is the output of d-axis current controller is proportional to the q-axis flux ψ_{rq}^v. Adjusting the ω^* through K_ω shown in Fig. 2, ψ_{rq}^v is controlled as mentioned above. By adding the feed forward compensation, we compute ω^* as

$$\omega^* = \omega_r^* + \frac{i_{sq}}{\tau_r^* i_{sd}^*} - K_\omega e_d^* \tag{14}$$

where, $K_\omega = \mathrm{sign}(\omega^*)|K_\omega|$

From (9) and (14), we can expect

$$K_\omega e_d^* = \omega_r^* - \hat{\omega}_r \tag{15}$$

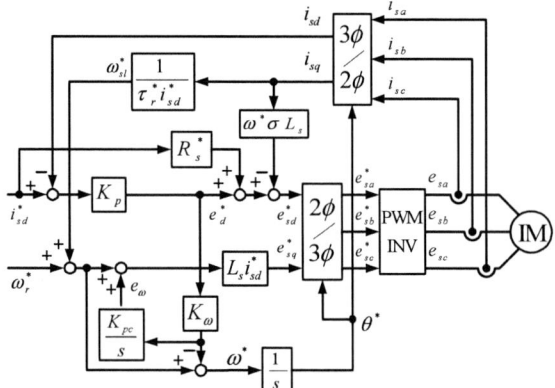

Fig.2 Proposed induction motor speed sensorless vector control system.

Thus, the e_d^* is also proportional to the speed error. By assuming steady state and $\psi_{rd}^y = M i_{sd}^*$, (2) becomes to

$$e_{sq} = \omega^* L_s i_{sd}^* + R_s i_{sq} \tag{16}$$

From (13) and (14), we compute the q-axis voltage as

$$e_{sq}^* = L_s i_{sd}^* (\omega_r^* + \frac{i_{sq}}{\tau_r^* i_{sd}^*} + K_{pc} \frac{K_\omega}{s} e_d^*) \tag{17}$$

This equation is considered as a feedforward plus feedback speed controller. By the integral control of e_d^*, $\psi_{rq}^v = 0$ is expected in steady state. The proposed system of Fig.2 is much simpler than conventional observer-based sensorless systems.

III. STABILITY ANALYSIS

The steady-state and stability analysis of the proposed system is done by choosing d-q axis which rotors with θ^*. The model of IM is expressed by (1) ~ (4) by changing $\psi_{rd}^v \to \psi_{rd}$, $\psi_{rq}^v \to \psi_{rq}$, $\psi_{rd}^* \to \psi_{rd}$, and $\psi_{rq}^* \to \psi_{rq}$. The steady-state values are obtained by setting $p = 0$. Therefore, from (12) and (17)

$$e_d^* = 0 \ , \ i_{sd}^* = i_{sd} \tag{18}$$

The q-axis flux is not zero when stator resistance changes as follows:

$$\psi_{rq} = \frac{L_r i_{sd}}{\omega^* M} (R_s - R_s^*) \tag{19}$$

A linear model is obtained by considering the small perturbations about a steady-state operating taking point of the nonlinear differential equations. The linear model of the system shown in Fig.2 is derived as follows:

$$p\Delta x = A\Delta x + B\Delta r + B_r \Delta T_L \tag{20}$$

where, $\Delta x = \begin{bmatrix} \Delta i_{sd} & \Delta i_{sq} & \Delta \psi_{rd} & \Delta \psi_{rq} & \Delta \omega_r & \Delta e_\omega \end{bmatrix}^T$,

$\Delta r = \begin{bmatrix} \Delta \omega_r^* \end{bmatrix}$, ΔT_L : Load torque

In (20), the stator and rotor resistances variations are considered.

Stability analysis is done by the eigenvalues of system matrix A which is obtained by the linear model (20). The machine parameters are as follows; number of poles $P = 4$, $J = 0.0126 \mathrm{kgm}^2$ (including load), $R_s = 1.54\Omega$, $R_r = 0.787\Omega$, $L_s = L_r = 0.115\mathrm{H}$, $M = 0.11\mathrm{H}$. These constants are measured by experimental induction machine.

In this paper, we discuss the influence on the control system when the resistances value change. We define the nominal value (at 75°C) of the stator resistance as R_{s0}, and k as follows:

$$k = \frac{R_s^*}{R_{s0}} \tag{21}$$

1218

The k represents the ratio when the stator resistance changes from the nominal value. In the experiment, it is difficult to change and measure the actual resistance value, so we change R_s^* in the controller.

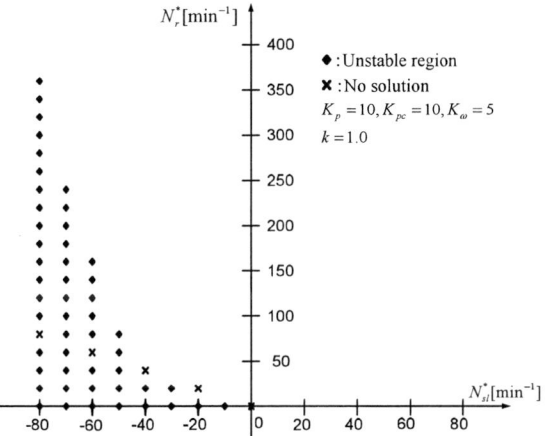

Fig.3 Unstable region ($k = 1.0$).

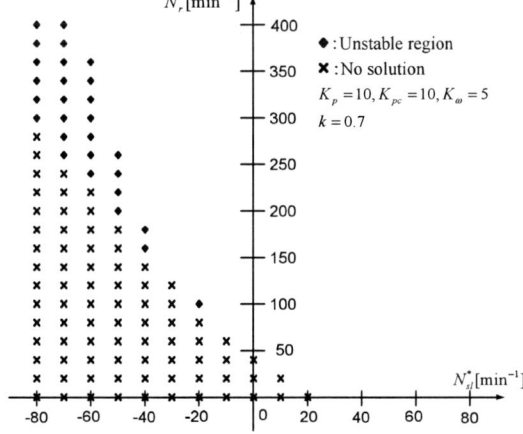

Fig.4 Unstable region ($k = 0.7$).

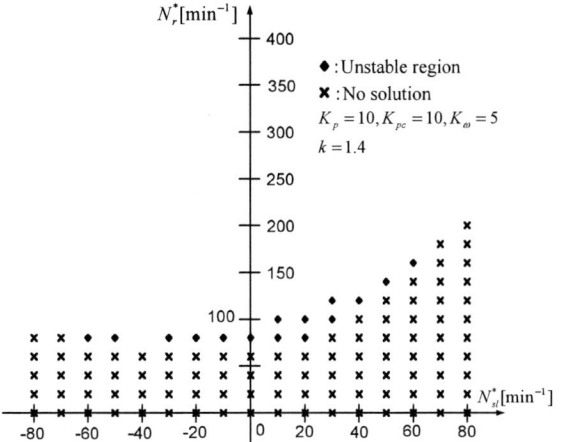

Fig.5 Unstable region ($k = 1.4$).

Figures 3, 4 and 5 show the unstable and no steady-state solution regions for various operating conditions by computing the eigenvalues using the linear model.

Figure 3 shows an unstable region when $k = 1.0$. The horizontal axis is the slip speed N_{sl} and the vertical axis is the speed command N_r^*. In the case of no change in the stator resistance value, the system is stable for motoring operation and an unstable region is seen in the low speed and heavy load regenerating region and plugging region. When synchronous speed $\omega^* = 0$, there is no solution.

Figure 4 shows an unstable region when $k = 0.7$. When the value of R_s^* is smaller than the actual value $R_s (= R_{s0})$, it is observed that both the unstable and no solution regions increase in regenerating and plugging operations.

Figure 5 shows an unstable region when $k = 1.4$. When the value of R_s^* is larger than the actual value $R_s (= R_{s0})$, unstable and no solution regions are seen even in the motoring operation.

IV. IDENTIFICATION OF STATOR RESISTANCE

By considering the steady state in equation (2) and setting $p = 0$, we have

$$e_{sq}^* = \omega^* \sigma L_s i_{sd} + R_s i_{sq} + \frac{\omega^* M}{L_r} \psi_{rd} \tag{22}$$

Comparing (17) and (22), we obtain

$$e_\omega = \frac{R_s i_{sq}}{L_s i_{sd}^*} \tag{23}$$

In Eq. (23), we assume R_s is the true value. And we define

$$R_{sf}^* = e_\omega L_s i_{sd}^* / i_{sq} \tag{24}$$

$$\Delta R_s = R_s^* - R_{sf}^* \tag{25}$$

By using (24) and (25), we identify the stator resistance.

Figure 6 shows flowchart of identification of stator resistance. At first, we calculate ΔR_s and divide the flow into three cases according to the value of ΔR_s. When the value of ΔR_s is positive, subtract 0.01 from the value of R_s^*. When the resistance of value of ΔR_s is negative, add 0.01 to the value of R_s^*. We compute this process every 10ms to identify the stator resistance.

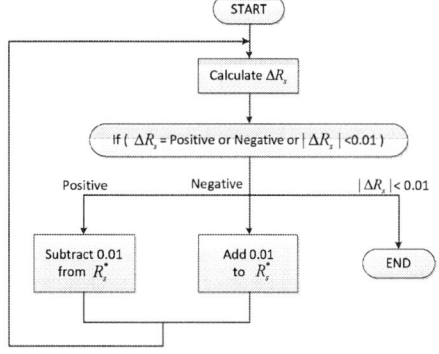

Fig.6 Flowchart of identification of stator resistance.

V. SIMULATION AND EXPERIMENTAL RESULT

In the simulation, a nonlinear model of Fig.2 is solved. In the experiment, PWM inverter fed induction motor drive system is tested by using a digital signal processor (DSP).

Figures 7, 8 and 9 show the simulation and experimental results when the stator resistance R_s^* of the controller is changed. In the simulation and experiment, the step responses from $500\,\text{min}^{-1}$ to $550\,\text{min}^{-1}$ are tested. The load torque T_L is 4Nm , and the control gains are designed as $K_p = 10, K_{pc} = 10, K_\omega = 5$. Actual motor speed N_r, the d-axis voltage e_d^*, the d-axis current i_{sd} , and the q-axis current i_{sq} are shown. Actual stator resistance R_s is not clear. However, in the simulation, $R_s = R_{s0}$ is assumed. Figure 7, 8 and 9 correspond to $k = 1.0$, $k = 0.7$ and $k = 1.4$ respectively. The ratio k is defined in (21). Quick responses of speed are obtained in all cases. Good corelation between the simulation results and experimental ones are observed.

Figure 10 shows the transient responses at low speed ($100\,\text{min}^{-1}$). The load torque T_L is 4Nm. At first, k is set as 1.0, then k is changed to 1.4 at 0.3s or 0.6s. When $k = 1.4$, the steady state solution does not exist as shown in Fig.5. The responses become unstable. In the experimental system, the operation changes to the mode with speed sensor at 2.7s for protection.

Figure 11, 12, 13 and 14 show the simulation and experimental results when identification of stator resistance is excuted. Actual motor speed N_r, the output of integrator e_ω , stator resistance in the controller R_s^*, and R_{sf}^* defined in (24) are shown.

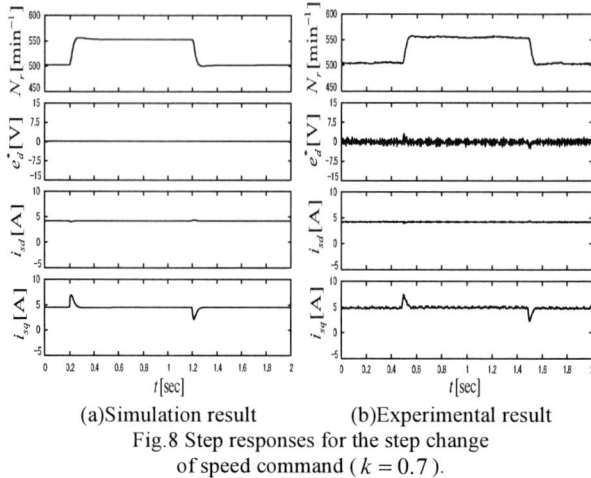

(a)Simulation result   (b)Experimental result

Fig.8 Step responses for the step change of speed command ($k = 0.7$).

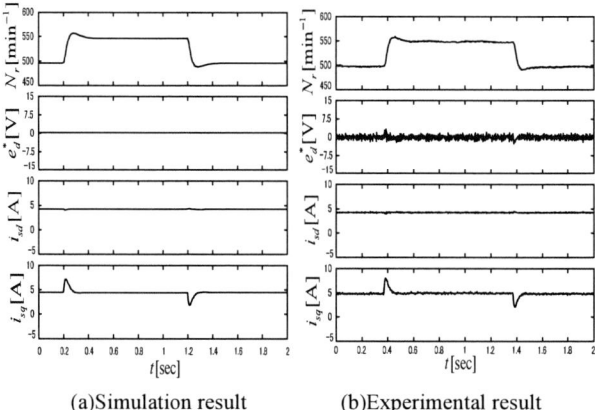

(a)Simulation result   (b)Experimental result

Fig.9 Step responses for the step change of speed command ($k = 1.4$).

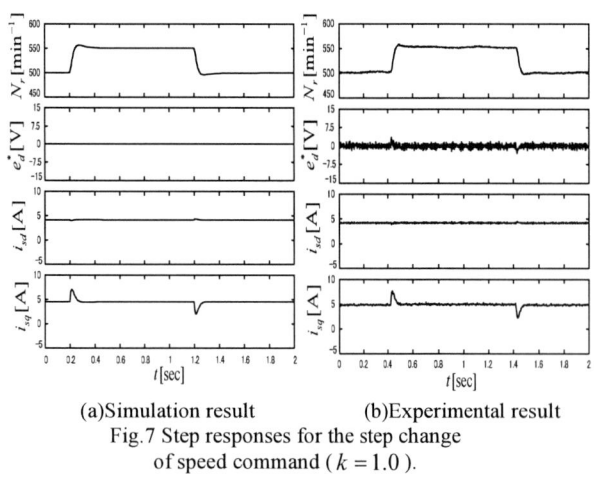

(a)Simulation result   (b)Experimental result

Fig.7 Step responses for the step change of speed command ($k = 1.0$).

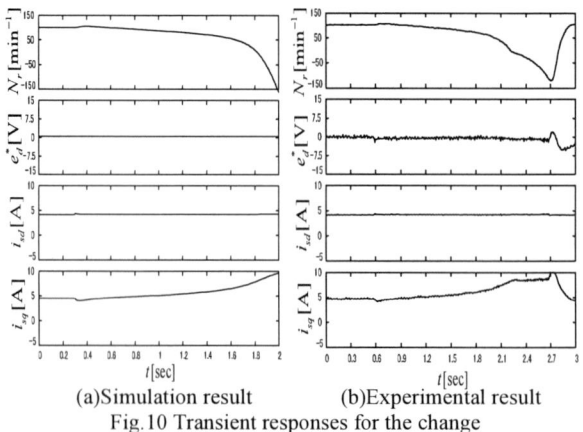

(a)Simulation result   (b)Experimental result

Fig.10 Transient responses for the change of k from 1.0 to 1.4.

The 2018 International Power Electronics Conference

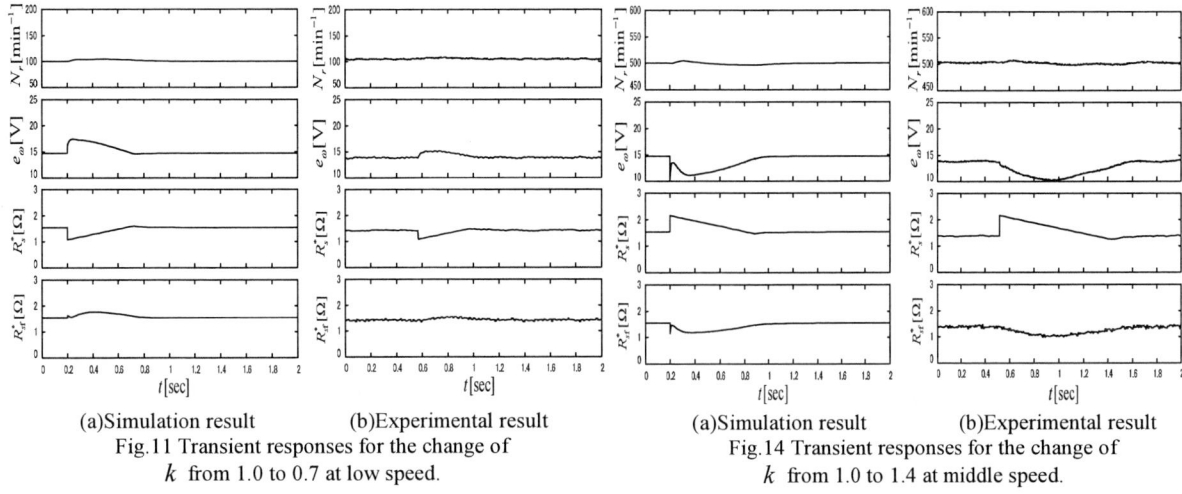

(a)Simulation result   (b)Experimental result

Fig.11 Transient responses for the change of k from 1.0 to 0.7 at low speed.

(a)Simulation result   (b)Experimental result

Fig.14 Transient responses for the change of k from 1.0 to 1.4 at middle speed.

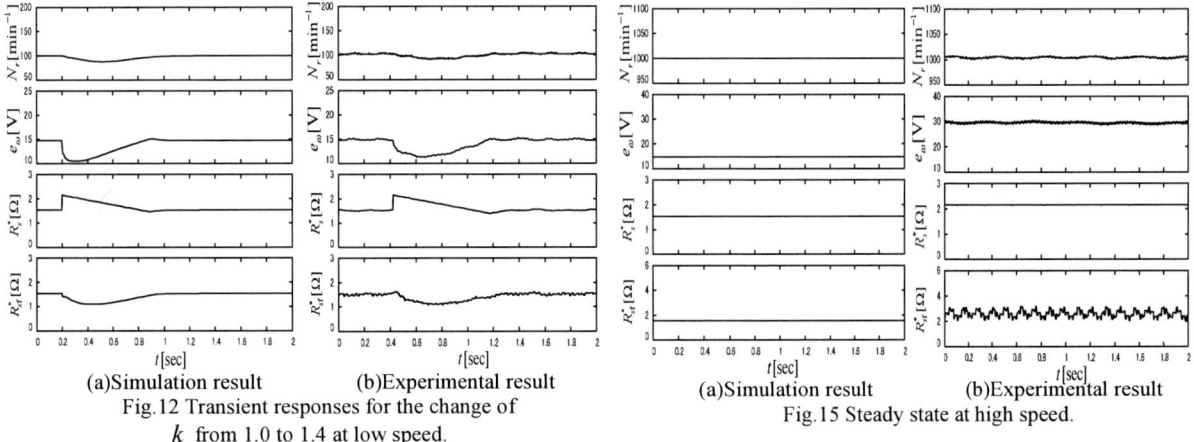

(a)Simulation result   (b)Experimental result

Fig.12 Transient responses for the change of k from 1.0 to 1.4 at low speed.

(a)Simulation result   (b)Experimental result

Fig.15 Steady state at high speed.

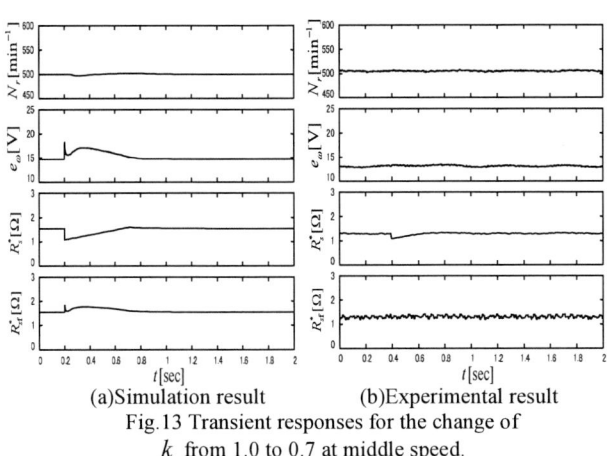

(a)Simulation result   (b)Experimental result

Fig.13 Transient responses for the change of k from 1.0 to 0.7 at middle speed.

Figures 11 and 12 show the results when the ratio $k(R_s^*)$ is changed from 1.0 to 0.7 and from 1.0 to 1.4 at low speed respectively. After the value of R_s^* is changed, it returned to the original steady-state value and the usefulness of identification could be confirmed. In the simulation, the steady-state value of R_s^* converges to R_{sf}^* which is equal to nominal value R_{s0}. However, in the experiment, the steady-state value of R_s^* converges to R_{sf}^* which an estimated value of actual stator resistance.

Figure 13, 14 show the results when the ratio k is changed from 1.0 to 0.7 and from 1.0 to 1.4 at middle speed. Similar waveforms are observed in the simulation and experimental results, and the stator resistance can be compensated.

Figure 15 shows the steady state results at high speed. In the experiment, R_s^* is limited to its maximum value because of large value of e_ω. It is difficult to identify the stator resistance by this method at high speed. However, the identification of stator resistance is important at low speed as shown in Figs. (4) and (5).

VI. CONCLUSION

In this paper, we have studied a simplified speed-sensorless vector control of induction motor proposed in early paper about the influence on the control system when the stator resistance value is changed in simulation and experiment. By taking into account the parameter changes, a linear model is derived. Instability caused by stator resistance change is observed especially at very low speed region. The derived linear model is useful to evaluate the robustness of stator resistances. In addition, we have proposed an identification method of stator resistance and confirmed its usefulness. Good correlation between simulation results and experimental results are obtained.

REFERENCES

[1] C. Schauder : "Adaptive speed identification for vector control of induction motors without rotational transducers", IEEE Trans. Ind. Appl., vol.28, No.5, pp. 1054–1061 (1992)

[2] H. Kubota, K. Matsuse, and T. Nakano : "DSP-based speed adaptive flux observer of induction motor", IEEE Trans. Ind. Appl., vol.29, No.2, pp. 344–348 (1993)

[3] S. Suwankawin and S. Sangwongwanich : "A speed-sensorless IM drive with decoupling control and stability analysis of speed estimation", IEEE Trans. Ind. Electron., vol.49, No.2, pp. 444 – 455 (2002)

[4] M. Hinkkanen : "Analysis and design of full-order flux observers for sensorless induction motors", IEEE Trans. Ind. Electron., vol.51, No.5, pp. 1033-1040 (2004)

[5] M. Hinkkanen, L. Harnefors, and J. Luomi : "Reduced-order flux observer with stator-resistance adaptation for speed sensorless induction motor drives", IEEE Trans. Power Electron., vol.25, No.5, pp. 1173–1183 (2010)

[6] T. Okuyama, N. Fujimoto, and H. Fujii : "Simplified vector control system without speed and voltage sensors – effects of setting errors in control parameters and their compensation –", Trans. IEE Japan, vol.110-D, No.5, pp. 477 – 486 (1990) (in Japanese)

[7] H. Tajima, Y. Matsumoto, H. Umida, and M. Kawano : "Speed sensorless vector control method for an industrial drive system", Proc. of International Power Electronics Conference (IPEC-Yokohama) , pp.1034–1039 (1995)

[8] K. Kondo and K. Yuki : "Study on an application of induction motor speed sensorless vector control to railway vehicle traction", Trans. IEE Japan, vol.125-D, pp. 1–8 (2005) (in Japanese)

[9] N. Yamamura, K. Aiba, and Y. Tsunehiro : "Primary flux control for induction motor drive", Trans. IEE Japan, vol.113-D, No.7, pp.859 – 864 (1993) (in Japanese)

[10] M. Tsuji, S. Chen, K. Izumi, and E. Yamada : "A sensorless vector control system for induction motors using q-axis flux with stator resistance identification", IEEE Trans. Ind. Electron., vol.48, No.1, pp. 185–194 (2001)

[11] M.Tsuji and G.M.Ch.Mangindaan and Y.Kunizaki and S.Hamasaki : " Simplified Speed-Sensorless Vector Control for Induction Motors and Stability Analysis" , IEEJ Journal of Industry Applications, Vol.3 No.2 pp.138-145 (2015)

[12] M.Tsuji, K.Iwamoto and S.Hamasaki : "A Simplified Speed- Sensorless Vector Control for Induction Motor", Proceeding of International Symposium on Power Electronics, Electrical Drives, Automation and Motion (SPEEDAM), Vol.1, pp.522-527 (2016.6)

The 2018 International Power Electronics Conference

Maximum Torque Control Reference Frame Based on a Torque Map for IPMSMs with Large Inductance Variation

Kazuki Ohta[1], Takumi Ohnuma[1*], Shinji Doki[2]

1 Multidisciplinary Eng. Course, National Inst. of Tech., Numazu College, Numazu-city, Shizuoka, Japan
2 Dept. of Info. and Communication Eng., Nagoya University, Nagoya-city, Aichi, Japan
*E-mail: ohnuma@numazu-ct.ac.jp

Abstract—In this paper, we summarize a method of applying a maximum torque control reference frame to IPMSMs with large influence of magnetic saturation. Conventionally, a formula based on the torque equation with motor parameters of IPMSMs has been used to calculate the phase angle of the maximum torque control reference frame. However, the equation is greatly affected by inductance variation due to magnetic saturation. Therefore, we propose the phase angle determination method based on a torque map. The method is not affected by inductance variation because it does not need motor parameters directly. Performance improvement is confirmed by simulation and actual machine experiments.

Keywords—IPMSM, maximum torque control reference frame, magnetic saturation, torque map

I. INTRODUCTION

Interior permanent magnet synchronous motors (IPMSMs) can use not only the magnet torque by the permanent magnet but also the reluctance torque from characteristics of saliency[1]. IPMSMs are applied to various fields, such as EV and HEV traction, where high torque density and efficiency are needed because of their space limitation. However, high-density-desired motors cause large inductance variation due to magnetic saturation.

Rotor position information is needed to control IPMSMs. The position information can be obtained from a position sensor such as a resolver or an encoder attached to the shaft of the motor. The position sensorless control is also highly expected from various viewpoints such as installation space, environmental resistance, and fail safe.

There is a method that uses extended electromotive force model as one effective means of performing position sensorless control[2]. The extended electromotive force model is a model in which both the back electromotive force used in the middle and high speed range and the saliency characteristic used at standstill and low speed range. The model makes it possible to perform position sensorless control in the entire speed range[3]. In this case, additional AC signal current is necessary to induce the salient polarity component used at standstill and low speed range. Because the signal conditions must be determined in consideration of various practical restrictions, influence on noise and drive characteristics occurs

depending on the setting of amplitude and frequency of signal current.

In order to observe extended electromotive force directly by observers, the authors have evaluated the driving characteristics by signal injection with frequencies in the observation band[3]. Although torque ripples are generally generated due to the frequency relatively close to the current for driving the motor, it was already confirmed that signal injection by the maximum torque control reference frame (f-t axes) does not affect the driving characteristics. The f-t axes are a coordinate system rotated by the angle ϕ from the d-q axes, and the phase of ϕ can be calculated based on the torque equation including the motor parameters. However, in Ref.[3], since the tasted motor has a small inductance variation due to magnetic saturation, the performance of the f-t axes with large inductance variation has not been sufficiently discussed.

In recent years, motor parameters for each drive area can be obtained from the design stage with progress of electromagnetic field analysis technology. Furthermore, a torque map for current amplitude and phase can also be obtained. Returning to the definition of the f-t axes, ϕ can be determined if the torque information for the current is known. In this paper, we propose a method determining the phase angle based on a torque map for a motor with large inductance variation due to magnetic saturation.

II. MAXIMUM TORQUE CONTROL REFERENCE FRAME

A. Definition and properties of maximum torque control reference frame

The maximum torque control reference frame (f-t axes) is defined by rotating the d-q axes with an angle ϕ, which is the tangent angle of the constant torque curve as shown in Fig.1. The currents in the f-t axes(i_f, i_t) are the components of the current vector projected on the f-axis and the t-axis, respectively.

The maximum torque control reference frame has the following characteristics.

I) The MTPA control is realized by setting the f-axis current to zero

1223

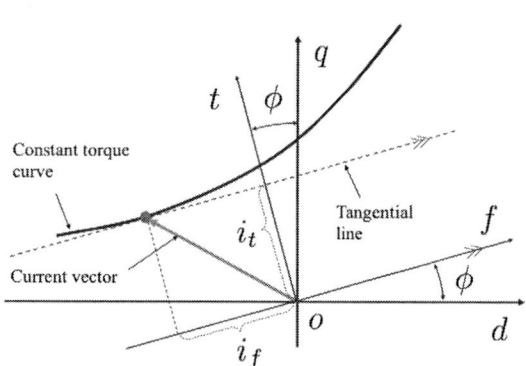

Fig. 1.   Maximum torque control frame(f-t axes)

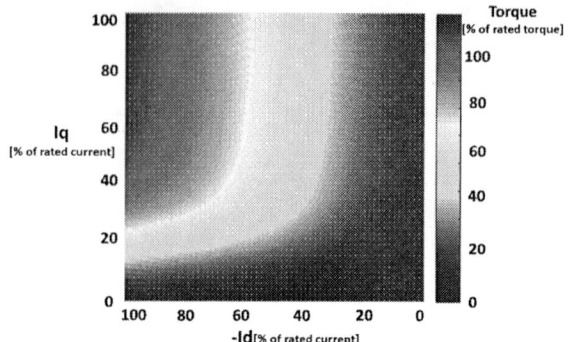

Fig. 2.   Target motor torque map

II)   A small change in current to the f-axis direction does not generate torque ripple

III)   A minute change in current to the t-axis maximizes the torque response (no voltage saturation)

IV)   A voltage limiter that maximizes the torque response[5] can be set by ϕ when voltage saturation is present[6]

The characteristic I means that the MTPA(Maximum Torque Per Ampere) control, which is widely used for high efficiency control, is achieved simply by a rotate transformation. The characteristic II means that a signal injection in the f-axis current does not generate a torque ripple. For this reason, a signal injection to the f-axis is effective way to the position estimation methods with disturbance observer based on the extended electromotive force model at low speeds[3]. The characteristic III means that the maximum torque response control is achieved when a current change is directed to the t-axis under the condition of non voltage saturation[6]. The characteristic IV means that the maximum torque response control is achieved when a voltage limiter is determined based on ϕ under the condition of voltage saturation[5][6]. An accurate estimate of ϕ is needed to realize these controls.

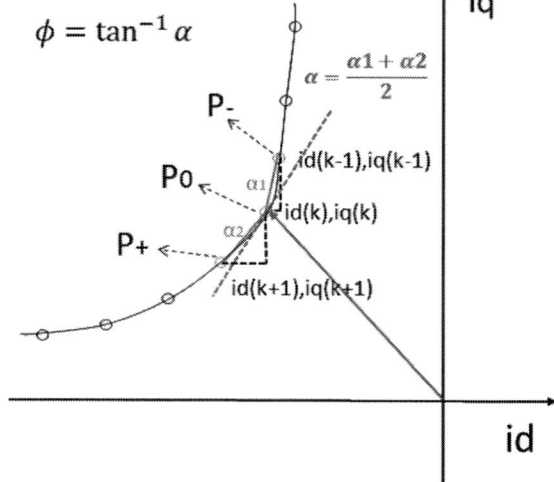

Fig. 3.   How to find an arbitrary constant torque curve and ϕ

$$\alpha_1 = \frac{i_q(k-1) - i_q(k)}{i_d(k-1) - i_d(k)}$$
$$\alpha_2 = \frac{i_q(k) - i_q(k+1)}{i_d(k) - i_d(k+1)}$$
$$\alpha = \frac{\alpha_1 + \alpha_2}{2}$$

Here, i_d and i_q are d- and q-axis currents.

Since α is the slope of the tangent of the constant torque curve at P_0, the angle ϕ is given by equation (1).

$$\phi = tan^{-1}\alpha \tag{1}$$

We get the map of ϕ for all regions corresponding to the torque map.

B. ϕ determination method based on torque map

The method for determining the f-t axes phase angle ϕ is shown in this section, when the torque map for the current amplitude, advance angle, and other parameters are known. The torque map of the motor used during the simulation and actual experiments is shown in Fig.2. Figure 2 is the torque map for the current in the d-q axes obtained by electromagnetic field analysis software. As shown in Fig.3, a constant torque curve is extracted and the slope between the data points is obtained from the central difference. For example, the calculation procedure for calculating ϕ at the P_0 point in Fig.3 is shown below. k represents the data number.

C. ϕ calculation formula based on torque equation

The general torque equation of IPMSM is shown in the equation (2).

$$T = P_n(K_E - (L_q - L_d)i_d)i_q \tag{2}$$

Here, P_n is the number of pole pairs, K_E is the EMF constant, L_d and L_q are d- and q-axis inductances.

Solving the equation (2) for i_q yields the equation (3).

$$i_q = \frac{T}{P_n(K_E - (L_q - L_d)i_d)} \qquad (3)$$

The equation (4) is obtained by differentiating the equation (3) with i_d. Here, torque and inductance are treated as constant values.

$$\frac{di_q}{di_d} = \frac{(L_q - L_d)i_q}{K_E - (L_q - L_d)i_d} \qquad (4)$$

Since the equation (4) is the slope of the tangent of the constant torque curve, ϕ is obtained as the equation (5).

$$\phi = tan^{-1}\frac{(L_q - L_d)i_q}{K_E - (L_q - L_d)i_d} \qquad (5)$$

The equation (5) is a calculation formula that has been conventionally used in Ref.[3]. Motor parameters are used in the equation (5), and the calculation is affected by the sensitivity against parameter variations. In fact, looking at the d- and q-axis inductances (Fig.4,5) obtained by the electromagnetic field analysis, inductance variation occurs significantly.

D. ϕ calculation formula based on torque equation considering magnetic saturation[4]

There are various ways of thinking about how to consider magnetic saturation. One typical method is to regard inductance as a function of current. Then, the d- and q-axis inductances in the equation (2) are expressed as $L_d(i_d, i_q)$, $L_q(i_d, i_q)$, and ϕ is derived similarly as the equation (3)~(5).

$$\phi' = \tan^{-1}\frac{(L_q - L_d)i_q - (\frac{dL_d}{di_d})i_di_q}{K_E - (L_q - L_d)i_d - (\frac{dL_q}{di_q})i_di_q} \qquad (6)$$

The equation (6) is compared with the ϕ calculation formula which does not consider conventional magnetic saturation, and terms of differential of L_d and L_q with respect to i_d and i_q are added. However, if accurate inductance information can be acquired experimentally or in advance from electromagnetic field analysis, an accurate torque map could be obtained. Therefore, it is theoretically possible to calculate ϕ as in the equation (6), in the simulation and actual machine experiments in this paper, the equation (1) is used.

III. SIMULATION

A. Simulation system configuration

The system configuration of the simulation is shown in Fig.6. Simulation conditions are shown in table I. Using the torque as the command value, the t-axis current is calculated in the torque controller. A sine wave signal of arbitrary current amplitude and frequency is given in the f-axis current command. Current control was carried out by a general vector controller. JMAG-RT obtained from electromagnetic field analysis was used for the motor data. In the simulation, to compare the torque ripple

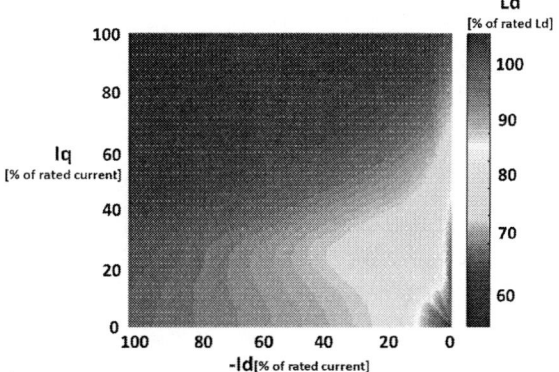

Fig. 4. d axis inductance map of target motor

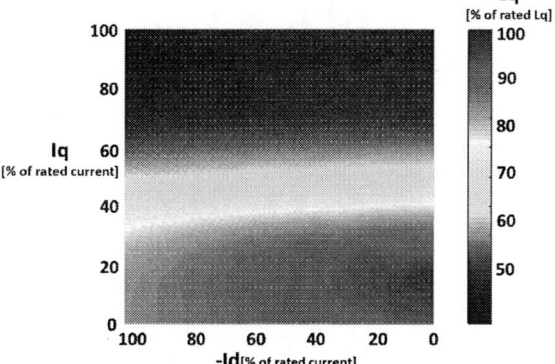

Fig. 5. q axis inductance map of target motor

caused by the signal injection only, the rotation speed was set to 0 rpm and the torque command was given at 50% of the rated value.

B. Torque ripple caused by signal injection

Three cases were set, and torque waveforms were compared by simulation.

Case1 : Using ϕ in the equation (1), without signal injection
Case2 : Using ϕ in the equation (1), with signal injection
Case3 : Using ϕ in the equation (5), with signal injection

Nominal values were used for the d- and q-axis inductances and the EMF constant of the calculation

TABLE I.   FUNDAMENTAL PHYSICAL CONSTANTS

DC link voltage	48[V]
Maximum output	10[kW]
EMF constant	0.00074[Vrms/rpm]
Stator resistance	4.6[mΩ]
L_q/L_d	8
Carrier frequency	8 kHz
Dead time	5 μs
Control period	125 μs
Current controller bandwidth	100 rad/s

The 2018 International Power Electronics Conference

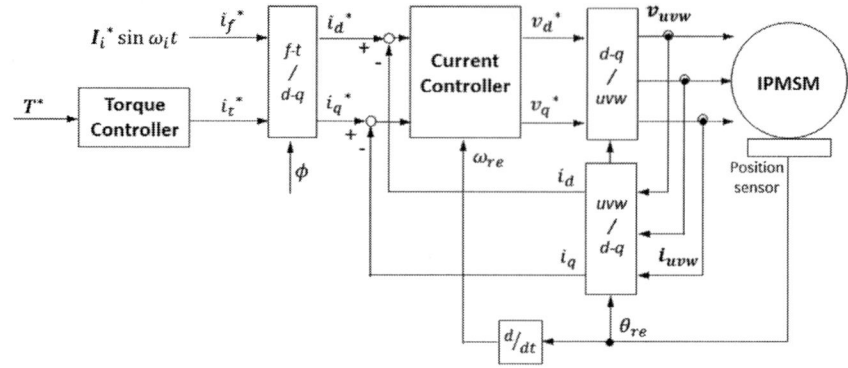

Fig. 6.   Simulation system configuration

formula was used in Case3. Signal current was injected in the f-axis current i_f at about 3% of the rated current and the frequency was set to 100 Hz. Simulation results are shown in Fig.7. Two facts can be found out in comparison with simulation results. First, the average torque at steady state in Cases1 and 2 decreased only about 2% from the command value, while the average torque in Case3 decreased about 12% of the command value. From the results, the formula for calculating ϕ based on the motor parameters, which was conventionally used, can not correctly estimate the MTPA phase in the equation (5), when the parameter variation is remarkable depending on the operating point. Second, the torque ripple caused by signal current is about 0.3%, 2.5% of the average torque in Case2 and Case3, respectively. From this, the tangent direction of the constant torque curve can not be accurately estimated by the conventional equation based on the torque equation (5).

From the above results, in the case of the motor with large parameter variation, it was found that the characteristics I, II of the maximum torque control reference frame listed in Section II.A can not make sufficient performance by the conventional calculation formula. However, even in the case of the motor with large parameter variation, the characteristics I, II were confirmed by simulation in this section.

IV.   EXPERIMENT RESULTS

A. Experiment system configuration

As with the simulation, the experiment system was constructed as shown in the Fig.6. For experimental conditions, the rotation speed was kept constant at 100 rpm on account of a limitation of the load machine, and other conditions were same as the simulation conditions.

B. Average torque comparison by determination methods of ϕ

Figure 8 show torque waveforms with signal injection. Figure 8 on the left is the torque waveform using ϕ based on torque equation. Average torque of the torque waveform is 44.1% because of decreasing torque against command torque. Figure 8 on the right is the torque

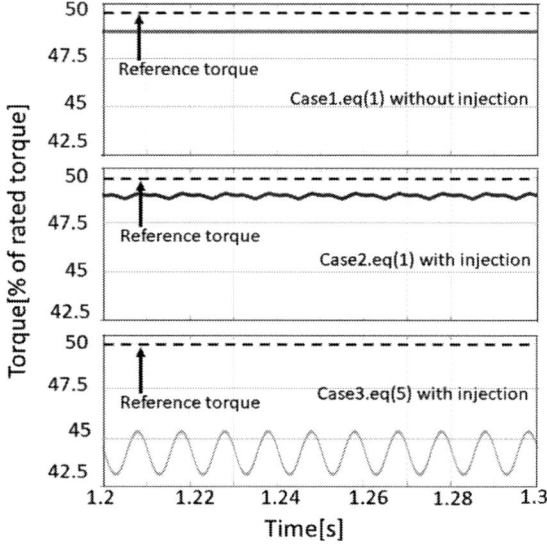

Fig. 7.   Torque waveform comparison (simulation)

waveform using ϕ based on the torque map. Average torque of the torque waveform is 49.7% which follows well with the command torque.

C. Torque ripple comparison by signal injection

Figure 9 show the results of FFT analysis of the torque waveforms with signal injection. The torque ripple caused by the injected signal at 100 Hz is recognized in Fig.9 on the left, which is the case of ϕ calculation based on torque equation. On the other hand, little torque ripple appears at 100 Hz in the Fig.9 on the right, which is the case of ϕ calculation based on the torque map. Even in actual machine experiments, the effectiveness of ϕ calculated by torque map was confirmed.

V.   CONCLUSION

In this paper, we proposed a method to directly decide ϕ from the torque map without using the formula of f-t axes phase angle ϕ obtained from the conventional torque equation. By returning to the definition of the f-t axes, ϕ can be determined from the torque map without going

The 2018 International Power Electronics Conference

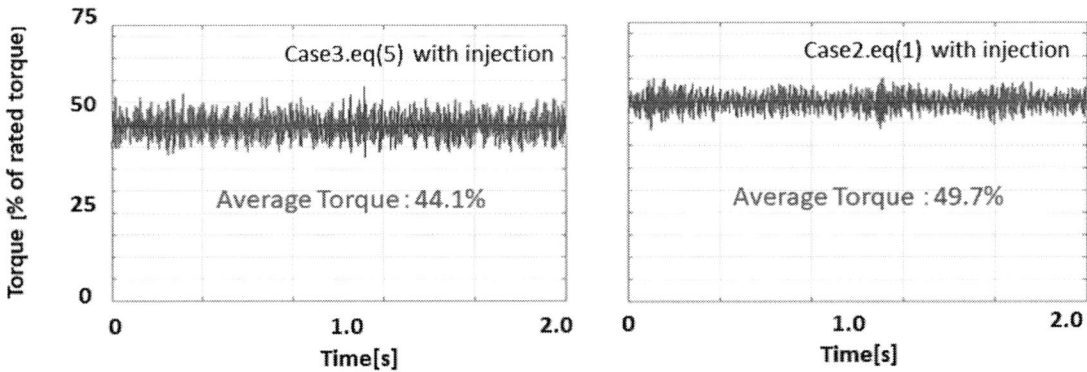

Fig. 8. Torque waveform comparison (actual experiments)

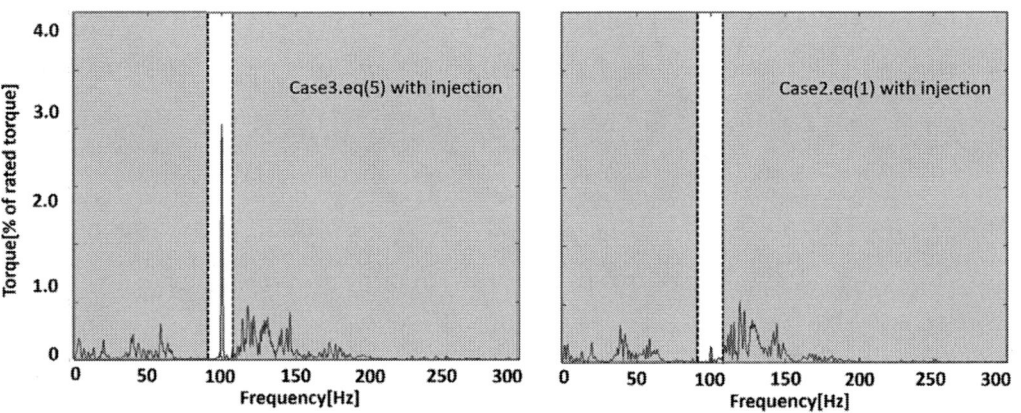

Fig. 9. Signal injection torque FFT using ϕ obtained from torque equation and torque map

through the motor parameters. Even if the parameter variation is large, the f-t axes can be used. In particular, it was confirmed the MTPA control can be realized by setting the f-axis current to zero, and that signal injection without torque ripple can be realized.

REFERENCES

[1] Hyung-Woo Lee, Ki-Doek Lee, Won-Ho Kim, Ik-Sang Jang, Mi-Jung Kim, Jae-Jun Lee, and Ju Lee, "Parameter Design of IPMSM With Concentrated Winding Considering Partial Magnetic Saturation", IEEE Transactions on Magnetics, Vol. 47, no. 10, pp. 3653-3656, 2011

[2] Z.Chen, M.Tomita, S.Doki, S.Okuma: "An extended Electromotive Force Model for Sensorless Control of Interior Permanent-Magnet Synchronous Motors", IEEE,Trans.Ind.Elec, 50-2, pp.288-295 (2003)

[3] T.Ohnuma, S.Doki, S.Okuma, "Extend EMF Observer for Wide Speed Range Sensorless Control of Salient-pole Synchronous Motor Drives." *Proceeding of ICEM10*, 2010.

[4] K.Ohta, H.Iwata, T.Ohnuma, "An Investigation for Optimization of Phase Angle of Maximum Torque Control Reference Frame Considering Magnetic Saturation", YPC1-59, 2016 (in Japanese)

[5] S.Lerdudomsak, S.Doki, S.Okuma, "Novel Voltage Limiter for Fast Torque Responce of IPMSM in Voltage Saturation Region" , IEEJ Trans IA, Vol. 128 No.12, pp. 1346-1356, 2008 (in Japanese)

[6] Yuki Makaino, Tatsuya Iba, Takumi Ohnuma: "Maximum Torque Response Control Based on Maximum Torque Control Frame", *IEEJ Journal of Industry Applications*, Vol.133 No.1 pp.1-6, 2013

PMSM Model Discretization in Consideration of Park Transformation for Current Control System

Masamichi Inoue, Shinji Doki

Department of Information and Communication Engineering, Nagoya University
Furo-cho, Chikusa-ku, Nagoya, Aichi 464-8603, Japan
E-mail: masa.inoue@nagoya-u.jp, doki@nagoya-u.jp

Abstract—This paper describes a new discrete-time PMSM model for discrete design of current vector control system. Discrete design of current control system is a via discrete-time PMSM model. This controller minimizes discretization error and ensure perfect control performance at sampling point. Conventional discretization method presume park transformation runs continuously. When sampling frequency is too lower than fundamental electric frequency, this assumption is invalid and control performance will degrade by modeling error. In this paper, authors propose the method of PMSM model discretization in consideration of discrete-time park transformation. Simulation results show the effectiveness of proposed model.

Keywords—modeling, discretization, high speed motor drive system, Vector control system

I. INTRODUCTION

Permanent Magnet Synchronous Motor (PMSM) has various advantages such as compactness and high efficiency. Thus PMSM has been applied to several fields.[1],[2] Especially in the field of automobile, high power density is desired. High-speed motor drive is a effective measure for this subject. Several designs and control methods of high-speed PMSM are studied. It is assumed that high-speed drive in around kHz will progress in future.[3]

On the other hand, in power electronics, the appearance of SiC or GaN devices raise the switching frequency of power inverters, but the frequency will be up to about 40kHz in such application.[4] The ratio f_d/f_{cc} (f_d : fundamental electric frequency, f_{cc} : control frequency) is about 0.01 in general. However, it is necessary to operate motor by low sampling frequency in contrast with high fundamental electric frequency such as the ratio f_d/f_{cc} is 0.1 because switching frequency restricts control period. In this situation, there is a risk that the influence of discretization error remarkably appears if it is applied to continuous design that is a method of control system design that discretize continuous-time controller and implement computer.

Discrete design is another method of making control system that construct the controller based on discrete motor model directly in discrete-time. This method is not affected by discretization error, hence, various discrete design method have been proposed.[5] Discrete-time motor model is a important factor of discrete design system. So, several PMSM model discretization methods have been also proposed.[6] However, these papers suppose

this transformation runs ideally, in other words, it is assumed that sampling frequency to detect rotor position is sufficiently high. As mentioned before, when the ratio f_d/f_{cc} is 0.1, this assumption is denied because sampling interval is not short enough. Therefore, the modeling error becomes large and thre is a possibility that control performance is deteriorated. This paper propose a new discrete PMSM model while taking discretely running park transformation into account and design current vector control system by proposed model in discrete time.

II. DISCRETE DESIGN OF CURRENT CONTROL SYSTEM

A. Continuous Design and Discrete Design

Continuous design is a general control system design method. Controller is designed in continuous-time based on the continuous-time model and then it is discretized. This design method occurs discretization error greatly. On the other hand, discrete design is another control system design method. In this method, continuous-time model is discretized firstly, and then controller is designed in discrete-time based on discrete-time model. This method minimizes discretization error and ensure control performance at sampling point. The relationship between the design methods is shown in Fig1.

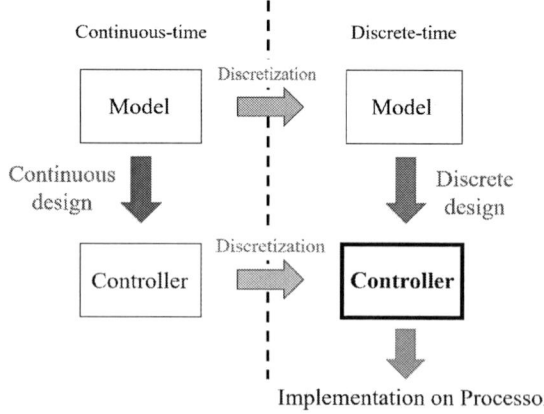

Fig. 1. Relationship between continuous design and discrete design

B. Discrete-time PMSM Model (Conventional Model)

This chapter describes conventional design method of current control system. Current vector control system is shown in Fig2.

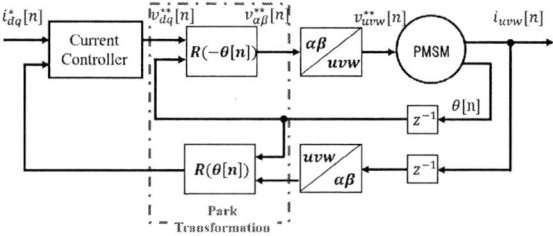

Fig. 2.   Block diagram of current vector control system

Current vector control system contains park transformation shown in (1). It converts AC amount into DC amount or DC amount into AC amount. The current system controls the DC amount.

$$R(\theta) = \begin{bmatrix} \cos\theta & \sin\theta \\ -\sin\theta & \cos\theta \end{bmatrix} \tag{1}$$

Therefore, dq-motor model is commonly used. The mathematical model of PMSM is given as follows:

$$\boldsymbol{v_{dq}}(t) = \begin{bmatrix} R+pL_d & -\omega_{re}L_q \\ -\omega_{re}L_d & R+pL_q \end{bmatrix} \boldsymbol{i_{dq}}(t) + \boldsymbol{e_{dq}}(t) \tag{2}$$

where $\boldsymbol{v_{dq}}$ is dq-axis voltage vector, $\boldsymbol{i_{dq}}$ is dq-axis current vector, $\boldsymbol{e_{dq}}$ is dq-axis electric force vector, R is resistance, L_d, L_q is dq-axis inductance, ω_{re} is electrical angular velocity, K_E is EMF constant and p is differential operator. When (2) is written as a state equation, it becomes (3).

$$\frac{d}{dt}\boldsymbol{i_{dq}}(t) = \boldsymbol{A}\boldsymbol{i_{dq}}(t) + \boldsymbol{B}(\boldsymbol{v_{dq}}(t) - \boldsymbol{e_{dq}}(t)) \tag{3}$$

$$\boldsymbol{A} = \begin{bmatrix} -\frac{R}{L_d} & \frac{\omega L_q}{L_d} \\ -\frac{\omega L_d}{L_q} & -\frac{R}{L_q} \end{bmatrix} \qquad \boldsymbol{B} = \begin{bmatrix} \frac{1}{L_d} & 0 \\ 0 & \frac{1}{L_q} \end{bmatrix} \tag{4}$$

(5) is derived by solving (3). T is control cycle, nT is initial time.

$$\boldsymbol{i_{dq}}((n+1)T) = \exp(\boldsymbol{A}T)\boldsymbol{i_{dq}}(nT)$$
$$+ \int_0^T \exp(\boldsymbol{A}\tau)\boldsymbol{B}(\boldsymbol{v_{dq}}((n+1)T-\tau)$$
$$- \boldsymbol{e_{dq}}((n+1)T-\tau))d\tau \tag{5}$$

In this situation, harmonic components of voltage and current which occurred by switching in the inverter also poses problems, but we assume that the inverter is an ideal linear power amplifier and voltage command is held at zero order because this paper aims at the control error accompanying the discrete implementation. Assuming that the park transformation is ideally running, $\boldsymbol{v_{dq}}$ is constant between the control period. Compute the integral of (5), introduce (6) and discretize it in (7).

$$i[k] = i(kT) \tag{6}$$

$$\boldsymbol{i_{dq}}[n+1] = \exp(\boldsymbol{A}T)\boldsymbol{i_{dq}}[n]$$
$$+ (\exp(\boldsymbol{A}T) - \boldsymbol{I})\boldsymbol{A}^{-1}\boldsymbol{B}(\boldsymbol{v_{dq}}[n] - \boldsymbol{e_{dq}}[n]) \tag{7}$$

The exponential function nomally approximates up to the first order.

$$\boldsymbol{i_{dq}}[n+1] = \boldsymbol{A_d}\boldsymbol{i_{dq}}[n] + \boldsymbol{B_d}(\boldsymbol{v_{dq}}[n] - \boldsymbol{e_{dq}}[n]) \tag{8}$$

$$\boldsymbol{A_d} = \boldsymbol{I} + \boldsymbol{A}T \qquad \boldsymbol{B_d} = \boldsymbol{B}T \tag{9}$$

C. Design of Current Control System in Discrete Time

Design the current control system in discrete time from conventional discrete PMSM model. Cross-coupling exist in the A_d matrix of (8). This coupling can be canceled by inputting the votage as shown below:

$$\begin{bmatrix} v_d^{**}[n] \\ v_q^{**}[n] \end{bmatrix} = \begin{bmatrix} v_d^*[n] + a_{12}i_q[n] \\ v_q^*[n] + a_{21}i_d[n] + b_{22}e_q[n] \end{bmatrix} \tag{10}$$

a_{ij} is respectively component of A_d, b_{ij} is respectively component of B_d. This decoupling control make it possible to control each axis independently. Design the PI control system so that the transfer function from current reference to current has a time lag of first order and time constant is $1/\omega_{cc}$. The PI gains are (11), (12), (13) and (14).

$$K_{pd} = a_{11}(1 - \exp(-\omega_{cc}T)) \tag{11}$$
$$K_{pq} = a_{22}(1 - \exp(-\omega_{cc}T)) \tag{12}$$
$$K_{id} = (1 - a_{11})(1 - \exp(-\omega_{cc}T)) \tag{13}$$
$$K_{iq} = (1 - a_{11})(1 - \exp(-\omega_{cc}T)) \tag{14}$$

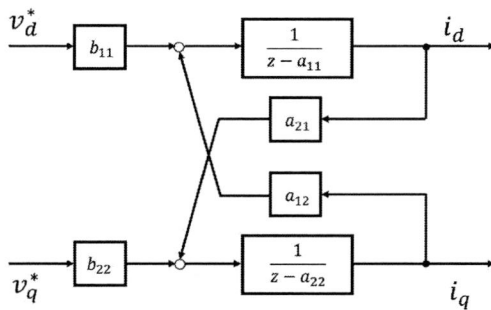

Fig. 3.   block diagram of conventional PMSM model

III.   DISCRETE DESIGN OF CURRENT CONTROL SYSTEM IN CONSIDERATION OF PARK TRANSFORMATION

A. Problem of Conventional PMSM Model

When converting between amounts on dq-axis and on $\alpha\beta$-axis as shown in (1) at section II, B, it is assumed that sampling period of rotor position is sufficiently short and park transformation works continuously and the applied voltage to the real plant is held at zero order as shown in Fig.4. However, the sampling cycle is not short enough when the ratio of f_d to f_{cc} is 0.1. Then, it is necessary to consider the influence of the discretization and implementation of park transformation.

The 2018 International Power Electronics Conference

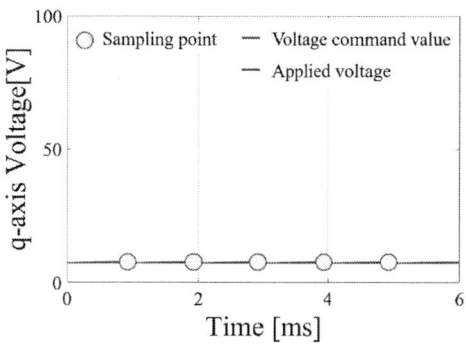

Fig. 4.　The relation between voltage command value and applied voltage (f_d/f_{cc} = 0.01)

One of the influences is control delay. One control cycle is required from sampling the current and rotor position until outputting the voltage command value. Since the rotor is rotating during this period, rotor position at the time of sampling differs from rotor position when outputting the voltage as shown in Fig.5. It has been proposed to calculate the angle $\Delta\theta$ that rotates during one control cycle and advance the phase of the park transformation by $\Delta\theta$ in order to compensate for this control delay as shown in Fig.6. $\Delta\theta$ can be obtained by the product $\omega_{re}T_{cc}$ of the control period and angular velocity, and the park transformation is taken as $R(\theta + \Delta\theta)$.[7]

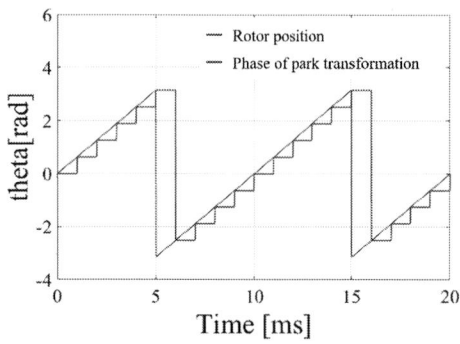

Fig. 5.　The relation between rotor position and the phase of park transformation without compensation

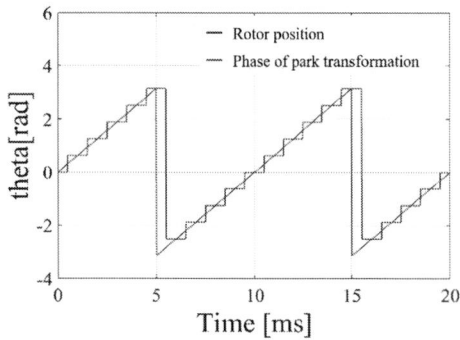

Fig. 6.　The relation between rotor position and the phase of park transformation with phase compensation

However, this method is insufficient when the ratio of f_d/f_{cc} is 0.1. In the conventional compensation method, only rotor position is compensated on average and harmonic components still remain. Therefore, the error of rotor position increases when f_d/f_{cc} is large, so that the error between the voltage command value and the applied voltage to the real plant becomes large as shown in Fig.7 and Fig.8. Modeling error occurs because the assumption of the conventional method does not hold in the case where the transient errors remain. Therefore, the author propose a new discrete PMSM model that takes into consideration of the discrete operation of park transformation explicitly.

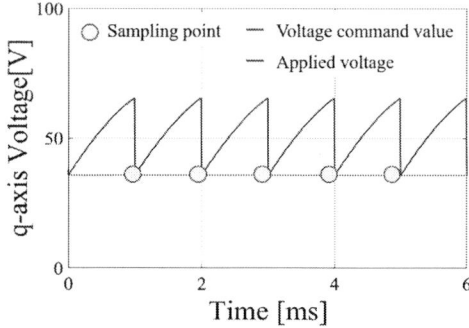

Fig. 7.　The relation between voltage command value and applied voltage (f_d/f_{cc} = 0.1)

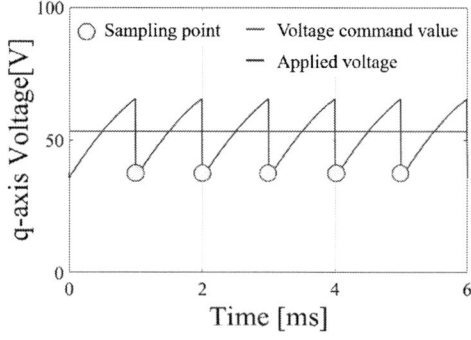

Fig. 8.　The relation between voltage command value and applied voltage with phase compensation(f_d/f_{cc} = 0.1)

B. Proposed Discrete PMSM Model

We discuss proposed PMSM model assuming that ω_{re} is constant between sampling points. The phase of the voltage reference output by the controller and rotor position are matched at time nT shown in Fig.9. However, at time $nT + \Delta t$, the voltage reference is fixed to the same phase as the time nT by zero order hold, whereas rotor position changes as the motor rotates. Applied voltage to the motor changes between sampling points due to phase mismatch between rotor position and voltage reference. The applied voltage considering the phase mismatch is as shown in the following equation.

1230

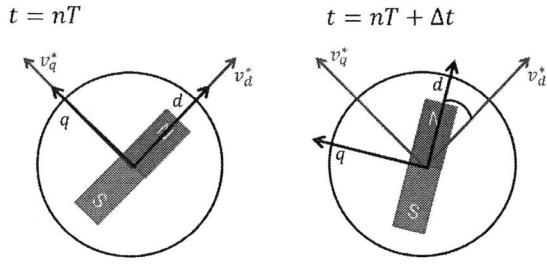

Fig. 9. The relation between rotor position and the phase of voltage reference

$$\begin{bmatrix} v_d(nT+\Delta t) \\ v_q(nT+\Delta t) \end{bmatrix} = \boldsymbol{R}(\omega_{re}\Delta t)\begin{bmatrix} v_d[n] \\ v_q[n] \end{bmatrix} \quad (15)$$

$$\boldsymbol{R}(\theta) = \begin{bmatrix} \cos\theta & \sin\theta \\ -\sin\theta & \cos\theta \end{bmatrix} \quad (16)$$

(17) can be obtained by substituting (15) into (5) and performing integration and discretization. The exponential function is approximated to the third order.

$$\boldsymbol{i_{dq}}[n+1] = \boldsymbol{A_d i_{dq}}[n] + \boldsymbol{B_d v_{dq}}[n] - \boldsymbol{K_d e_{dq}}[n] \quad (17)$$

$$A_d = I + \boldsymbol{A}T + \frac{1}{2}\boldsymbol{A}^2 T^2 + \frac{1}{6}\boldsymbol{A}^3 T^3 \quad (18)$$

$$B_d = \boldsymbol{B}\boldsymbol{V_0} + \boldsymbol{A}\boldsymbol{B}\boldsymbol{V_1} + \frac{1}{2}\boldsymbol{A}^2\boldsymbol{B}\boldsymbol{V_2} + \frac{1}{6}\boldsymbol{A}^3\boldsymbol{B}\boldsymbol{V_3} \quad (19)$$

$$K_d = \boldsymbol{B}T + \frac{1}{2}\boldsymbol{A}\boldsymbol{B}T^2 + \frac{1}{6}\boldsymbol{A}^2\boldsymbol{B}T^3 \quad (20)$$

$$V_0 = \begin{bmatrix} v_{0i} & v_{0j} \\ -v_{0j} & v_{0i} \end{bmatrix} \quad (21)$$

$$V_1 = \begin{bmatrix} v_{1i} & v_{1j} \\ -v_{1j} & v_{1i} \end{bmatrix} \quad (22)$$

$$V_2 = \begin{bmatrix} v_{2i} & v_{2j} \\ -v_{2j} & v_{2i} \end{bmatrix} \quad (23)$$

$$V_3 = \begin{bmatrix} v_{3i} & v_{3j} \\ -v_{3j} & v_{3i} \end{bmatrix} \quad (24)$$

$$v_{0i} = \frac{1}{\omega_{re}}\sin\omega_{re}T \quad (25)$$

$$v_{0j} = \frac{1}{\omega_{re}}(1-\cos\omega_{re}T) \quad (26)$$

$$v_{1i} = \frac{1}{\omega_{re}^2}(1-\cos\omega_{re}T) \quad (27)$$

$$v_{1j} = \frac{T}{\omega_{re}} - \frac{1}{\omega_{re}^2}\sin\omega_{re}T \quad (28)$$

$$v_{2i} = \frac{2T}{\omega_{re}^2} - \frac{2}{\omega_{re}^3}\sin\omega_{re}T \quad (29)$$

$$v_{2j} = \frac{T^2}{\omega_{re}} - \frac{2}{\omega_{re}^3} + \frac{2}{\omega_{re}^3}\cos\omega_{re}T \quad (30)$$

$$v_{3i} = \frac{3T^2}{\omega_{re}^2} - \frac{6}{\omega_{re}^4} + \frac{6}{\omega_{re}^4}\cos\omega_{re}T \quad (31)$$

$$v_{3j} = \frac{T^3}{\omega_{re}} - \frac{6T}{\omega_{re}^3} + \frac{6}{\omega_{re}^4}\sin\omega_{re}T \quad (32)$$

C. Design of Current Control System from Proposed Model

The block diagram of the proposed model is shown in Fig.10.

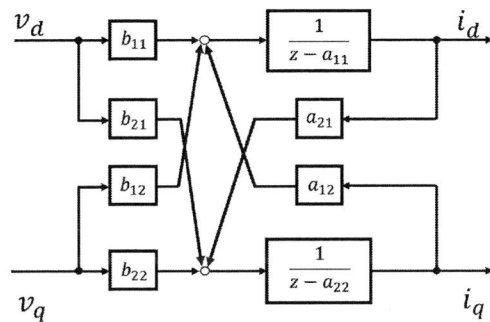

Fig. 10. block diagram of proposed PMSM model

It is necessary to perform voltage type decoupling control since cross-coupling also exist in the B_d in the proposed model. The cross-coupling can be canceled by inputting the voltage as shown in (33).

$$\begin{bmatrix} v_d^{***}[n] \\ v_q^{***}[n] \end{bmatrix} = \begin{bmatrix} C_{11} & C_{12} \\ C_{21} & C_{22} \end{bmatrix}\begin{bmatrix} v_d^{**}[n] \\ v_q^{**}[n] \end{bmatrix}$$
$$= \frac{1}{b_{11}b_{22}-b_{12}b_{21}}\begin{bmatrix} b_{22} & -b_{12} \\ -b_{21} & b_{11} \end{bmatrix}\begin{bmatrix} v_d^{**}[n] \\ v_q^{**}[n] \end{bmatrix} \quad (33)$$

$v_d^{**}[n]$, $v_q^{**}[n]$ and PI gain are the same in (10) - (14).

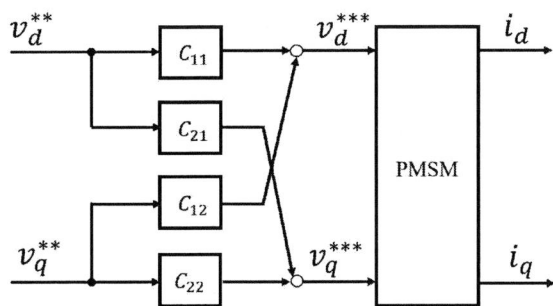

Fig. 11. Cross-coupling compensation

D. Examination of Proposed Model

The exponential function was approximated up to the third order in (17). It is desirable that the approximate order be as small as possible for reasons of calculation load. The optimal approximation order is determined based on the simulation result. Simulation condition is shown in TABLE I Motor parameters are indicated in TABLE II and parameters of the controller are shown in TABLE III. In the proposed model, it is assumed that the voltage reference is held at zero order and the inverter is not taken into consideration. In order to reduce the influence of modeling error by PWM modulation, the carrier frequency is set to 5 kHz and the control cycle is set to 1 ms.

1231

Current control error and calculation load are discussed in this section. Current control error is defined as the error between ideal current response when control with the primary delay and real current response. Control performance is compared by current control error measured when current command step is given in each approximation order. Besides, calculation load is defined as the ratio of the calculation time for conventional model calculation to the time for proposed model calculation. Calculation time is measured by GNU profiler.

TABLE I. SIMULATION CONDTION

CPU	Intel Corei7-4770 @ 3.40GHz
OS	windows10 ver.1709
Simulation Software	matlab 2017a / simulink
Compiler	MinGW-w64 Compiler ver.17.2.0.0

TABLE II. MOTOR PARAMETERS

Symbol	Meaning	Value
P_n	Number of Pole	2
R	Resistance	0.53 Ω
L_d	d-axis Inductance	4.15 mH
L_q	q-axis Inductance	16.74 mH
K_E	EMF constant	0.0916 V/(rad/s)

TABLE III. SETTING CONDITION

Symbol	Meaning	Value
V_{dc}	DC Link Voltage	120 V
f_d	Fundamental electric frequency	100Hz
T	Control period	1 ms
f_c	Carrier frequency	5 kHz
f_{cc}	Bandwidth of current controller	500 rad/s

The relation between the approximate order of the proposed model and the current control error is shown in Fig.12. As the approximation order is increased, the current control error decreases. Overmore, the error becomes almost zero by approximating to the third order. The relation between the ratio of fundamental electric frequency to control frequency and current control error is shown in Fig.13. The current control error is almost 0 irrespective of the ratio f_d/f_{cc} by approximating the exponential function up to the third order. Thus, the approximate order of the proposed model is appropriate for the third order from these results. Meanwhile, the rate of calculation load is shown in Fig.14. Calculation load increased as approximation order is increased. The load when the approximation order is three is about 17 times as conventional current control system. Thus, it is possible that calculation time exceeds control period depending on the situation.

IV. EXPERIMENTATION

A. Experimentation Condition

The conditions of experimentation are same as the foregoing paragraph's simulation (TABLE II-III). The A_d and B_d matrices need to be recalculated for each sampled velocity. However, we assume that the load motor can maintain a constant velocity and use the constant A_d and B_d values calculated in off-line condition in this experimentation for calculation load.

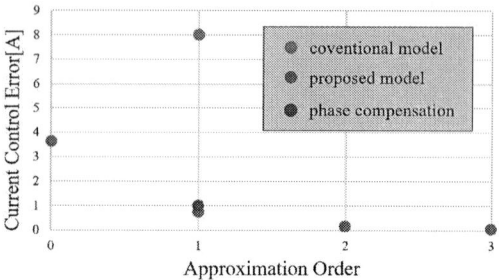

Fig. 12. The relation between approximation order and current control error

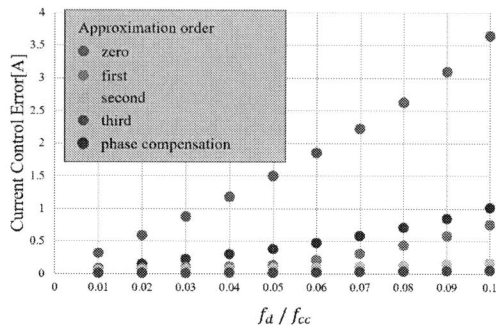

Fig. 13. The relation between the ratio of f_d/f_{cc} and current control error

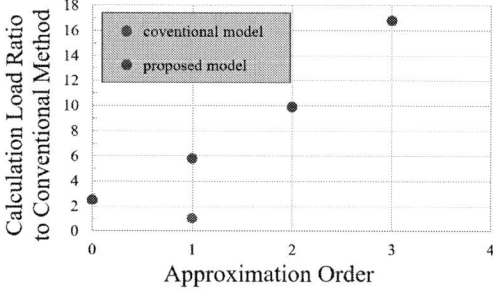

Fig. 14. Calculation Load

B. Experimentation Results

Experimentation results of continuous design of current control system are shown in Fig.15 and Fig.16 and discrete design based on conventional model without compensation, with phase compensation and proposed model are illustrated in Fig.17, Fig.18 and Fig.19 In the proposed model, it can be confirmed that the current pulsation that occurs in the conventional model and phase compensation method is suppressed and the response behaves as designed. Large d-axis current ripple occurs in continuous design and conventional discrete model. The current ripple is suppressed and the response behaves as designed in proposed model. However, some pulsation remains. This is considered to be due to the inability to model in consideration of the inverter and the mismatch of the parameters due to the change in the speed of the motor.

The 2018 International Power Electronics Conference

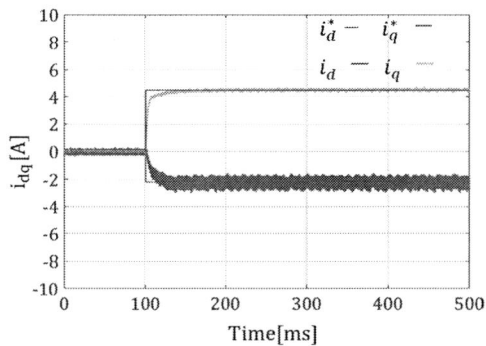

Fig. 15. The experimentation result
(continuous design ($f_d/f_{cc} = 0.01$))

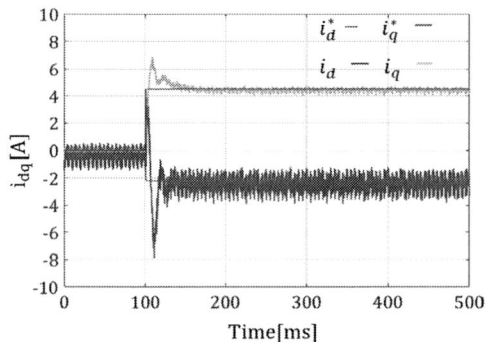

Fig. 16. The experimentation result
(continuous design ($f_d/f_{cc} = 0.1$))

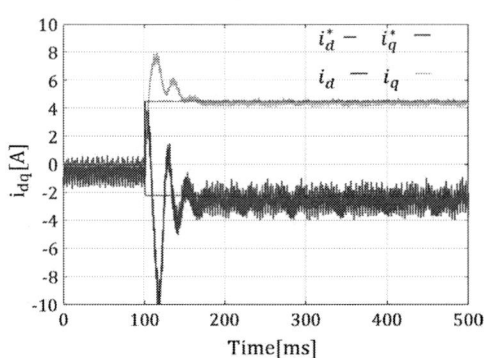

Fig. 17. The experimentation result
(discrete design (conventional model $f_d/f_{cc} = 0.1$))

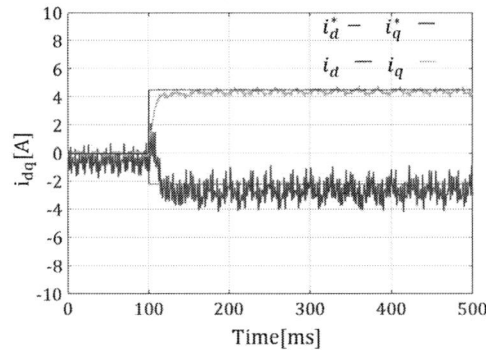

Fig. 18. The experimentation result
(discrete design with phase compensation($f_d/f_{cc} = 0.1$))

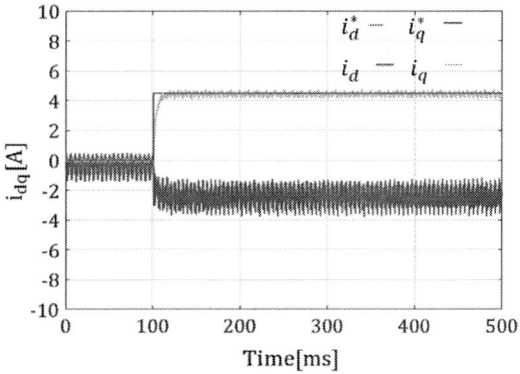

Fig. 19. The experimentation result
(discrete design (proposed model $f_d/f_{cc} = 0.1$))

V. CONCLUSIONS

In this paper, we aim to suppress control error due to modeling error of discrete-time PMSM model and propose a new discrete motor model considering park transformation. we compare the current responses of the conventional model and the proposed model in experimentation and verify the effectiveness of the proposed model.

REFERENCES

[1] K. Kondo and H. Kubota, "Innovative Application Technologies of AC Motor Drive Systems," IFFJ Journal of Industry Applications, Vol.1, No.3, pp.132-140 2012

[2] S. Sui and S. Kim, "Sensorless Control of IPMSM Past, Present, and Future," IEEJ Journal of Industry Applications, Vol.1, No.1, pp.15-23 2012

[3] T. Noguchi, T. Wada, M. Kano and T. Komori, "Low-Voltage-Power-Supply-Fed 1.5-kW 150,000-r/min Ultra-High-Speed PM Motor Designed to Maximize Efficiency and Power Density," IEEJ Trans. IA, Vol.134, No.6, pp.641-648 2014

[4] Z. Yang, T. Niu and Q. Gao, "High switching frequency control scheme for dual- three-phase PMSM and simulation analysis," 2017 IEEE Transportation Electrification Conference and Expo, Asia-Pacific (ITEC Asia-Pacific), Harbin, China, 2017, pp. 1-6.

[5] T. Miyajima, H. Fujimoto and M. Fujituna, "Control method for IPMSM based on perfect tracking control and PWM hold model in overmodulation range," The 2010 International Power Electronics Conference, pp. 593-598, 2010.

[6] A. Navarrete, J. Rivera, J. J. Raygoza and S. Ortega, "Discrete-Time Modeling and Control of PMSM," 2011 IEEE Electronics, Robotics and Automotive Mechanics Conference, Cuernavaca, Morelos, 2011, pp. 258-263.

[7] J. Kudo, T. Noguchi, M. Kawakami and K. Sato, "Methematical Model Errors and Their Compensations of IPM Motor Control System," IEE of Japan Technical Meeting Record, IEE Japan SPC-08-25, 2008, pp. 25-31.(in Japanese)

The 2018 International Power Electronics Conference

Pseudo-Random High-Frequency Sinusoidal Voltage Injection Based Sensorless Control for IPMSM Drives

Guoqiang Zhang[1], Huiying Wang[1], Gaolin Wang[1]*, Junya Huo[1,2], Lianghong Zhu[2] and Dianguo Xu[1]
1 School of Electrical Engineering & Automation, Harbin Institute of Technology, Harbin, China
2 GD Midea Air-Conditioning Equipment Co., Ltd., Foshan, China
*E-mail: WGL818@hit.edu.cn

Abstract— **A sensorless control method based on pseudo-random high-frequency(HF) sinusoidal voltage injection is studied to solve the noise problem caused by the conventional HF injection strategy in this paper. In the low-speed and zero-speed range, HF injection strategy is widely used as a sensorless control strategy for IPMSM drives. However, the harsh noise caused by the HF signals is unexpected. By injecting two HF signals with different frequencies in a random manner, instead of traditional fixed frequency signal, noise problems can be attenuated while sustaining the estimation accuracy. The signal processing is analyzed to select the frequency and amplitude of the pseudo-random high-frequency (PRHF) sinusoidal signals. This method is verified by simulation and experiment on a 2.2-kW IPMSM drive platform.**

Keywords— *Acoustic noise, high-frequency injection, pseudo-random signal, sensorless control.*

I. INTRODUCTION

Recently, permanent magnet synchronic motors (PMSMs) are widely used in lots of high performance drive systems such as household appliances, elevators, electric cars and so on, because of their better characteristics of high efficiency, strong reliability, and compact structure [1],[2]. Vector control is one of the commonly used control strategies of PMSM drives, where the information of rotor position and speed is necessary. The traditional systems adopt mechanical position sensors, however the applications of these sensors bring many problems, such as cost increasing, reliability decreasing, poor adaptability and so on [3]. In the range of low and zero speed, in order to take the place of position sensors, high-frequency signal injection methods on the basis of rotor saliency-tracking technology are highly applied in sensorless interior permanent magnet synchronous motor (IPMSM) drives. Nevertheless, the disadvantages of these methods that the additional torque ripple and high frequency noise caused by the injected signal cannot be ignored. The additional

HF noise causes serious noise pollution [4]. Therefore, how to reduce such noise is an issue needed to study.

There are some researches on noise reduction schemes [5-10]. One is the amplitude adjustment strategy. In [5], the relationship between estimation error of position and the amplitude of high-frequency signal was analyzed, and the minimum amplitude in given error range was chosen to reduce high-frequency noise. Another is frequency adjustment strategy, which increased the signal frequency beyond human auditory sense or adopted low frequency signal to make the noise softer and easier to accept [6],[7]. Besides, by comparison, pseudo-random HF signal injection strategies are easy to implement and have less application restrictions. The noise reduction method in [8] was equipped with injected frequency controller and injected amplitude controller to generate pseudo-random sinusoidal injection signals. The pseudo-random HF square-wave signal injection strategy was proposed in [9]. Furthermore, the frequencies and switching conditions of the HF square-wave signal with the better noise reduction effect has been demonstration in [10].

The ideal noise reduction scheme is to reduce the influence the noise caused by the injection of high frequency signals on the premise of ensuring the accuracy of estimation. In this paper, a pseudo-random high-frequency sinusoidal voltage injection based sensorless control strategy for IPMSM drives is proposed. Two different fluctuating high frequency sinusoidal voltage signals are injected into stator winding asynchronously. The frequency and amplitude limits of the injected signal, as well as switching conditions and probabilities are analyzed. The method is verified by simulations and experiments on a 2.2-kW IPMSM drive platform.

II. PSEUDO-RANDOM HIGH-FREQUENCY VOLTAGE INJECTION STRATEGY

A. Mathematical Model of IPMSM

The mathematical model of IPMSM in *d-q* synchronous reference frame can be expressed as (1).

Supported by the National Key R&D Program of China (2016YFE0102800) and the Natural Science Foundation of Heilongjiang Province (E2016028)

$$\begin{bmatrix} u_d \\ u_q \end{bmatrix} = \begin{bmatrix} R_d + sL_d & -\omega_e L_q \\ \omega_e L_d & R_q + sL_q \end{bmatrix} \begin{bmatrix} i_d \\ i_q \end{bmatrix} + \begin{bmatrix} 0 \\ \omega_e \psi_f \end{bmatrix} \quad (1)$$

where u_d, u_q, i_d, i_q, R_d, R_q, L_d, and L_q are the stator voltages, the stator currents, the stator resistances and the stator inductances respectively, ω_e is the rotor electrical angular speed, ψ_f is the permanent magnet flux linkage, and s is Laplacian operator.

At zero and low speed, the terms related to speed, $\omega_e \psi_f$, $\omega_e L_d$ and $\omega_e L_q$ can be ignored as (2). When the HF voltage is injected, the voltage drops on the stator resistances can be ignored. In summary, the mathematical model becomes as (3), which expresses the relationship between the HF components of stator voltages and currents.

$$\begin{bmatrix} u_d \\ u_q \end{bmatrix} = \begin{bmatrix} R_d + sL_d & 0 \\ 0 & R_q + sL_q \end{bmatrix} \begin{bmatrix} i_d \\ i_q \end{bmatrix} \quad (2)$$

$$\begin{bmatrix} u_{dh} \\ u_{qh} \end{bmatrix} = \begin{bmatrix} sL_{dh} & 0 \\ 0 & sL_{qh} \end{bmatrix} \begin{bmatrix} i_{dh} \\ i_{qh} \end{bmatrix} \quad (3)$$

where u_{dh}, u_{qh}, i_{dh}, i_{qh}, L_{dh}, and L_{qh} are the stator voltages, the stator currents, and the stator inductances in the condition of high-frequency injection respectively.

B. Signal Processing of Traditional Fluctuating High Frequency Sinusoidal Signal Injection Method

The rotor position information obtained from HF voltage signal injection is the estimated rotor position θ_r' rather than the actual position θ_r. Therefore, the new two-phase rotating reference frame established according to the estimated position is different from the actual d-q coordinate, which is called d'-q' coordinate. The relationship among usual reference frames is shown as Fig. 1, where $\Delta \theta_r = \theta_r - \theta_r'$ is the rotor position estimation error.

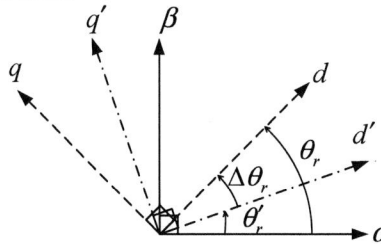

Fig. 1. The relationship among usual reference frames.

The coordinate transforming relations are as follows. $u_{d'h}$, $u_{q'h}$, $i_{d'h}$, $i_{q'h}$ are the stator voltages and the stator currents in d'-q' coordinate in a condition of high-frequency injection respectively, $i_{\alpha h}$ and $i_{\beta h}$ are the stator currents in α-β coordinate in a condition of high-frequency injection, $L_0 = (L_{dh} + L_{qh})/2$, $L_1 = (L_{dh} - L_{qh})/2$, and $K(\theta) = \begin{bmatrix} \cos\theta & \sin\theta \\ -\sin\theta & \cos\theta \end{bmatrix}$ is the transformation matrix.

$$\begin{bmatrix} u_{dh} \\ u_{qh} \end{bmatrix} = K(\Delta\theta_r) \begin{bmatrix} u_{d'h} \\ u_{q'h} \end{bmatrix} \quad (4)$$

$$\begin{bmatrix} i_{\alpha h} \\ i_{\beta h} \end{bmatrix} = K(\theta_r)^{-1} \begin{bmatrix} sL_{dh} & 0 \\ 0 & sL_{qh} \end{bmatrix}^{-1} K(\Delta\theta_r) \begin{bmatrix} u_{d'h} \\ u_{q'h} \end{bmatrix} \quad (5)$$

$$\begin{bmatrix} i_{d'h} \\ i_{q'h} \end{bmatrix} = K(\Delta\theta_r)^{-1} \begin{bmatrix} sL_{dh} & 0 \\ 0 & sL_{qh} \end{bmatrix}^{-1} K(\Delta\theta_r) \begin{bmatrix} u_{d'h} \\ u_{q'h} \end{bmatrix}$$

$$= \frac{1}{s(L_0^2 - L_1^2)} \cdot \quad (6)$$

$$\begin{bmatrix} L_0 - L_1 \cos 2\Delta\theta_r & -L_1 \sin 2\Delta\theta_r \\ -L_1 \sin 2\Delta\theta_r & L_0 + L_1 \cos 2\Delta\theta_r \end{bmatrix} \begin{bmatrix} u_{d'h} \\ u_{q'h} \end{bmatrix}$$

When the high frequency voltage signal u_{inj} is injected into d'-axis, the current response HF components caused by the HF signal are as (7) and (8).

$$\begin{bmatrix} i_{\alpha h} \\ i_{\beta h} \end{bmatrix} = \frac{u_{inj}}{s} \begin{bmatrix} \dfrac{\cos\theta_r \cos\Delta\theta_r}{L_{dh}} + \dfrac{\sin\theta_r \sin\Delta\theta_r}{L_{qh}} \\ \dfrac{\sin\theta_r \cos\Delta\theta_r}{L_{dh}} - \dfrac{\cos\theta_r \sin\Delta\theta_r}{L_{qh}} \end{bmatrix} \quad (7)$$

$$\begin{bmatrix} i_{d'h} \\ i_{q'h} \end{bmatrix} = \frac{u_{inj}}{s(L_0^2 - L_1^2)} \begin{bmatrix} L_0 - L_1 \cos 2\Delta\theta_r \\ -L_1 \sin 2\Delta\theta_r \end{bmatrix} \quad (8)$$

If the position estimation error $\Delta\theta_r$ is small enough, some assumptions can be obtained that $\sin 2\Delta\theta_r \approx 2\Delta\theta_r$, $\sin\Delta\theta_r \approx 0$ and $\cos\Delta\theta_r \approx 1$. Therefore, (9) and (10) can be derived.

$$\begin{bmatrix} i_{\alpha h} \\ i_{\beta h} \end{bmatrix} = \frac{1}{L_{dh}} \begin{bmatrix} \cos\theta_r \\ \sin\theta_r \end{bmatrix} \int u_{inj} \quad (9)$$

$$i_{q'h} = \frac{-2L_1}{(L_0^2 - L_1^2)} \Delta\theta_r \int u_{inj} \quad (10)$$

In order to obtain the position information, a new demodulation coefficient k_d related to the injection signal is needed. In the fluctuating high-frequency voltage signal injection method, the injection signal can be set as $u_{inj} = V_{in} \sin \omega_{in} t$, and the demodulation coefficient is supposed to be $k_d = \sin(\omega_0 t + b_0)$. Taking α-axis as an example, $i_{\alpha h}$, which is processed with high-pass filter (HPF), is multiplied with the demodulated coefficient, as shown in (11).

$$i_{\alpha h} \times k_d = \frac{1}{L_{dh}} \cos\theta_r \left(\int V_{in} \sin \omega_{in} t \, dt \right) \sin(\omega_0 t + b_0)$$

$$= \frac{V_{in}}{\omega_{in} L_{dh}} \left[-\cos \omega_{in} t \cdot \sin(\omega_0 t + b_0) \right] \cos\theta_r$$

$$= \frac{V_{in} \cos\theta_r}{2\omega_{in} L_{dh}} \left[\sin(\omega_{in} t - \omega_0 t - b_0) \right. \quad (11)$$

$$\left. -\sin(\omega_{in} t + \omega_0 t + b_0) \right]$$

$$= K_d \cos\theta_r$$

where K_d is a variable independent of rotor position.

To eliminate the influence of the motor parameters, normalization is usually performed, which means the

variable K_d should be always bigger than zero. From (11), it can be drawn that there must be direct component in K_d, that is $\sin(\omega_{in}t - \omega_0 t - b_0) = 1$ or $\sin(\omega_{in}t + \omega_0 t + b_0) = -1$, and the constraint conditions can be expressed as (12).

$$\begin{cases} \omega_0 = \omega_{in} + 2k_1\pi \\ b_0 = 2k_2\pi - \dfrac{1}{2}\pi \end{cases} \text{or} \begin{cases} \omega_0 = -\omega_{in} + 2k_3\pi \\ b_0 = 2k_4\pi - \dfrac{1}{2}\pi \end{cases} \quad (12)$$

where k_i is an integer, and i equals 1, 2, 3 and 4.

If $k_d = -\cos\omega_{in}t$, then $K_d = \dfrac{V_{in}}{2\omega_{in}L_{dh}}(1 - \cos 2\omega_{in}t)$.

After passing the low pass filter, the demodulated signal becomes as (13)

$$i_{\alpha\theta} = LPF(i_{\alpha h} \times k_d) = K_\theta \cos\theta_r \quad (13)$$

where K_θ is bigger than zero and is independent of rotor position, and $i_{\alpha\theta}$ is the demodulated signal after LPF.

The normalization process is shown in (14) and it is converted to the d'-q' coordinate in (15). When the position error is small enough, it can be inferred that $i_{q\theta\_PU} \approx \Delta\theta_r$, and the rotor position of the motor can be obtained by means of the observer.

$$\begin{bmatrix} i_{\alpha\theta\_PU} \\ i_{\beta\theta\_PU} \end{bmatrix} = \frac{1}{\sqrt{i_{\alpha\theta}{}^2 + i_{\beta\theta}{}^2}} \begin{bmatrix} i_{\alpha\theta} \\ i_{\beta\theta} \end{bmatrix} = \begin{bmatrix} \cos\theta_r \\ \sin\theta_r \end{bmatrix}. \quad (14)$$

$$i_{q\theta\_PU} = -\sin\theta_r' i_{\alpha\theta\_PU} + \cos\theta_r' i_{\beta\theta\_PU} = \sin\Delta\theta_r \quad (15)$$

where $i_{\alpha\theta\_PU}$, $i_{\beta\theta\_PU}$ and $i_{q\theta\_PU}$ are normalized signals.

III. PSEUDO-RANDOM HIGH-FREQUENCY SINUSOIDAL VOLTAGE INJECTION STRATEGY

A. Pseudo-Random High-Frequency Fluctuating Voltage Signal Injection Method

In PRHF method, by injecting two fluctuating HF sinusoidal voltage signals with different frequencies asynchronously in a random manner instead of traditional fixed-frequency signal, noise can be reduced and it can keep the estimation accuracy in the meantime. The block diagram of signal injection, extraction and analysis of PRHF method is shown in Fig. 2.

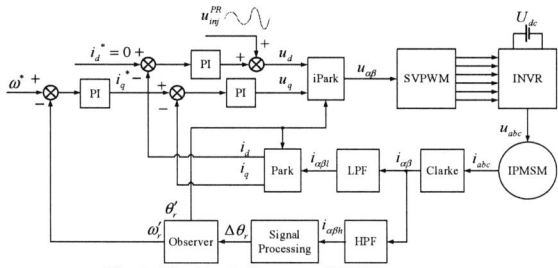

Fig. 2. The block diagram of PRHF strategy.

B. Selection of Pseudo-Random High-Frequency Signal

1) Frequency of PRHF signal

First of all, since the high frequency voltage signal is injected into the motor winding through the inverter, the period of the injected signal must be an integer multiple of the PWM cycle T_s. The PWM carrier frequency of the inverter is $f_s = 6\text{kHz}$. Selecting two sinusoidal signals, $u_1 = V_1 \sin\omega_1 t$ .and $u_2 = V_2 \sin\omega_2 t$, whose frequencies are $f_1 = f_s/n_1$ and $f_2 = f_s/n_2$ respectively, where $n_1 < n_2$, and they are positive integers.

In addition, in order to ensure the estimation precision, the lower frequency f_2 cannot be too small. The probability of occurrence of u_1 is P, then the probability of occurrence of u_2 is $1-P$. When the two signals are determined, the power spectrum is adjusted to be flat by changing P.

2) Amplitude of PRHF signal

The amplitude of the high-frequency signals need to vary with their frequency to keep the amplitude of the HF current component unchanged. It can reduce the harmonic wave when the injected signal is changed. From expression (11), to make the direct component in K_d constant, which means $V_{in}/2\omega_{in}L_{dh}$ should be a constant, the amplitude of the high frequency signal V_{in} must be proportional to its angular frequency ω_{in}. In other words, $V_1 f_1 = V_2 f_2$.

3) Generation method of PRHF signal.

The flow-process diagram of the generation method of PRHF signals is shown in Fig.3. After a complete cycle of sine wave is generated, whether the signal is switched is judged by comparing random number R and the probability P.

C. Demodulation Process of PRHF Method

As mentioned above, the demodulation coefficient $k_d = -\cos\omega_{in}t$. In PRHF method, as ω_{in} changes randomly, the demodulation coefficient needs to be adjusted accordingly, $k_{d1} = -\cos\omega_1 t$ is corresponding to u_1, and $k_{d2} = -\cos\omega_2 t$ is corresponding to u_2. Besides, in the actual experimental system, the generation process and demodulation process of signals are not in the same control cycle, so the effective demodulation coefficients have certain phase delay related to the control cycle.

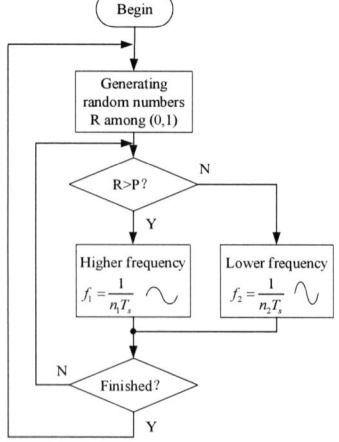

Fig. 3. Diagram of the generation method of PRHF signals.

IV. EXPERIMTENTAL RESULTS

A. Simulation Results

The control strategy is simulated through MATLAB/ Simulink, where the parameters of the 2.2-kw IPMSM are as shown in Table I.

TABLE I
PARAMETERS OF 2.2-KW IPMSM

Parameter	Value	Parameter	Value
Rated Power	2.2kW	Rated Torque	21N·m
Rated Speed	1000rpm	Resistance	2.75Ω
Rated Frequency	50Hz	Flux Linkage	0.56Wb
Rated Voltage	380V	d-axis Inductance	48mH
Rated Current	5.6A	q-axis Inductance	59mH

On the premise of guarantee the integrity of the sine wave and the effect of noise reduction, the injected signals are set as 1kHz/150V and 600Hz/90V. The injected signals and its demodulation coefficient are shown in Fig. 4.

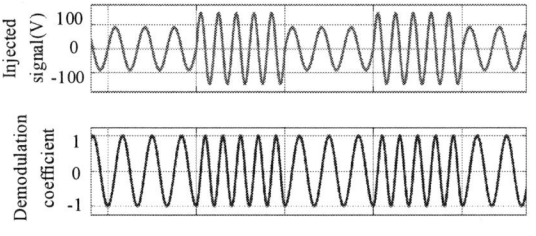

Fig. 4. The injected RPHF signal and demodulation coefficients.

The main purpose of the pseudo-random high-frequency voltage injection method is to reduce the noise caused by the injection of high-frequency signals, which can be reflected in the power spectrum analysis of the phase current. The Simulation PSD results for different methods are as shown in Fig. 5.

Compared with the traditional fixed-frequency injection method, the power spectral density of phase current has been distributed to different frequencies in RPHF method, and the peak value has been weakened in a frequency, which means the noise reduction strategy comes into effect.

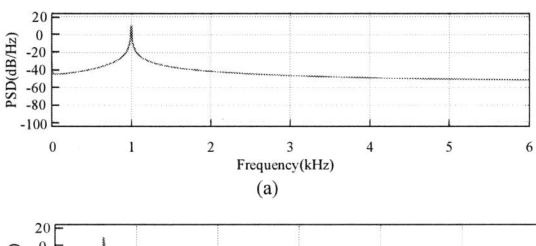

(a)

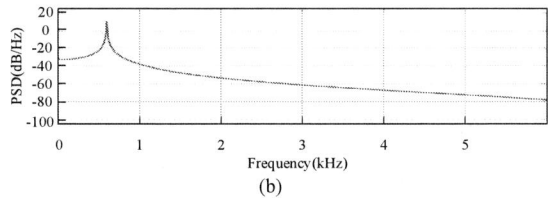

(b)

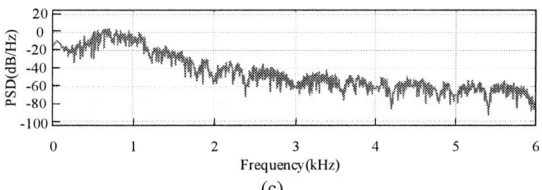

(c)

Fig. 5. Simulation PSD results for different methods. (a)1kHz signal injection. (b)600Hz signal injection. (c)Pseudo-Random signal injection (1kHz and 600Hz).

B. Experimental Results

The pseudo-random high-frequency voltage injection based sensorless control strategy for IPMSM drives is verified at the platform with a 2.2-kW IPMSM shown in Fig. 6. The parameters of the experimental IPMSM are the same as the those of simulation, as shown in Table I. The control algorithm is implemented through STM32F103VB. The PWM carrier frequency is 6kHz, the DC bus voltage is 510V, the control frequency of the inner current loop is 6kHz and the control frequency of the outer speed loop is 1kHz.

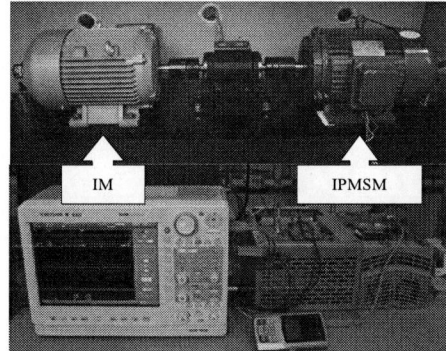

Fig. 6. The experimental platform with a 2.2 kw-IPMSM.

The injected sinusoidal signals are 1kHz/150V and 600Hz/90V, and both of the probabilities are 50%. Experiments were carried out at a given speed of 100rpm with no load. The HF current signals and their enlarged drawing are shown in Fig. 7, where it can be seen that the frequency of the HF component of the current changed. And the demodulated HF corresponding current signals after LPF are shown in Fig.8.

Experiments on steady state performance testing are at speed of 100rpm with no load. The experimental results of RPHF method are shown in Fig. 9.

The performance comparisons between the RPHF injection method and the traditional fixed frequency injection method are shown in Fig. 10.

As can be seen from Fig. 10, the position estimation errors are within 8 degrees, and there is no significant difference between RPHF method and fixed-frequency method. It is proved that the RPHF injection method can guarantee the accuracy of rotor position estimation.

The 2018 International Power Electronics Conference

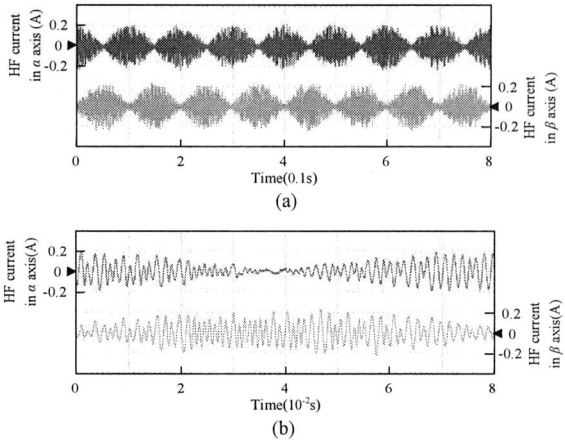

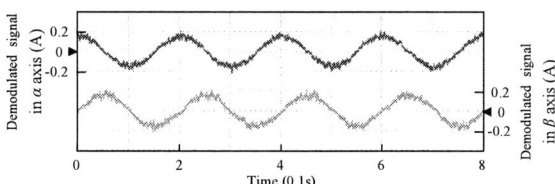

Fig. 7. Experimental results of the HF current. (a)HF current in stationary reference frame. (b) The enlarged drawing of HF current in stationary reference frame.

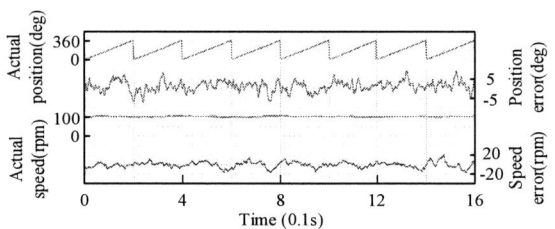

Fig. 8. Experimental results of demodulated current signals in stationary reference frame.

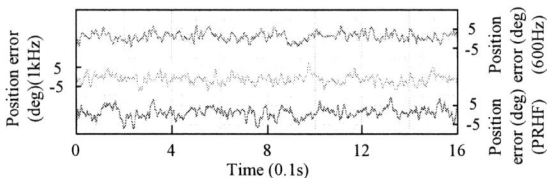

Fig. 9. Experimental results of steady state performance testing at speed of 100rpm with no load.

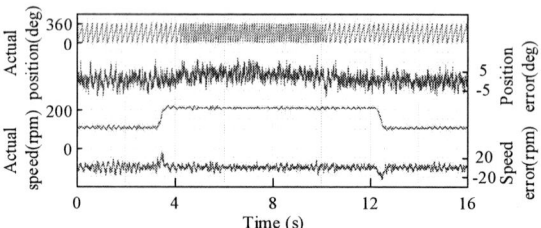

Fig. 10. Comparisons of estimation accuracy between PRHF injection and conventional method at speed of 100rpm with no load.

The experimental results of dynamic characteristic testing at speed from 100rpm to 200rpm with no load by PRHF method are shown as Fig. 11. The position estimation error increases when the speed changes, but still within 10 degrees. It is proved that the PRHF injection strategy has good dynamic performance.

Experimental PSD results of phase current based on different injection methods are shown in Fig. 12.

In traditional fixed-frequency injection method, there is an obvious peak in the power spectrum density of the injected signal frequency (1kHz or 600Hz), which causes

harsh noise. However, in RPHF injection method, the peaks in 1kHz and 600Hz are disappeared, and the PSD of each frequency spectrum changes more gently, which verifies that pseudo-random high-frequency voltage injection based sensorless control strategy has obvious effect in noise reduction.

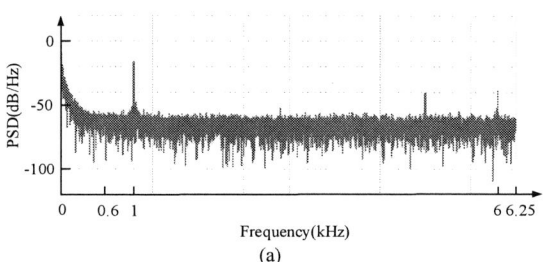

Fig. 11. Experimental results of dynamic characteristic testing at speed from 100rpm to 200rpm with no load.

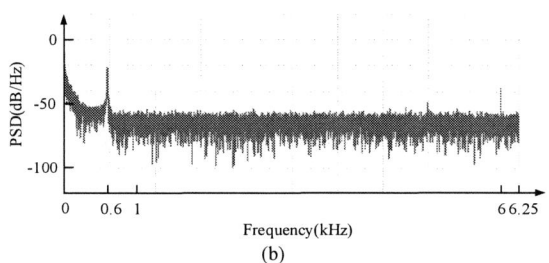

(a)

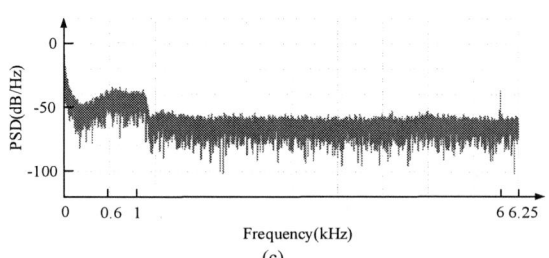

(b)

(c)

Fig. 12. Experimental current PSD results for different injection methods. (a)Fixed-frequency signal (1kHz) injection. (b) Fixed-frequency signal (600Hz) injection. (c)Pseudo-Random signal injection (1kHz and 600Hz).

V. CONCLUSIONS

The sensorless control strategy of IPMSM based on pseudo-random high-frequency voltage signal injection has good steady-state and dynamic performance. By changing the fixed-frequency signal of the traditional method into pseudo-random high-frequency signal, the effect of noise reduction can be realized. Moreover, by the analysis of mathematical model of IPMSM and

1238

demodulation process, selection conditions of the frequency and amplitude of the HF sinusoidal signals are derived to guarantee the accuracy of estimation. Finally, the validity of the control strategy is verified on the 2.2kW-IPMSM platform.

REFERENCES

[1] S.-K. Sul, Y.-C. Kwon, and Y. Lee, "Sensorless Control of IPMSM for Last 10 Years and Next 5 Years," *CES Trans. Electr. Mach. Syst.*, vol. 1, no. 2, pp. 91-99, Jun. 2017.

[2] G. Zhang, G. Wang, and D. Xu, "Saliency-based position sensorless control methods for PMSM drives - A review," *Chin. J. Elect. Eng.*, vol. 3, no. 2, pp. 14-23, Sep. 2017.

[3] G. Wang, R. Yang, and D. Xu, "DSP-based control of sensorless IPMSM drives for wide-speed-range operation," *IEEE Trans. Ind. Electron.*, vol. 60, no. 2, pp. 720-727, Feb. 2013.

[4] D. Kim, Y. Kwon, S. Sul, J. Kim, and R. Yu, "Suppression of injection voltage disturbance for high-frequency square-wave injection sensorless drive with regulation of induced high-frequency current ripple," *IEEE Trans. Ind. Appl.*, vol. 52, no. 1, pp. 302-312, Jan./Feb. 2016.

[5] S. Medjmadj, D. Diallo, M. Mostefai, C. Delpha, and A. Arias, "PMSM drive position estimation: Contribution to the high-

frequency injection voltage selection issue," *IEEE Trans. Energy Convers.*, vol. 30, no. 1, pp. 349-358, Mar. 2015.

[6] S. Kim, J.-I. Ha, and S.-K. Sul, "PWM switching frequency signal injection sensorless method in IPMSM," *IEEE Trans. Ind. Appl.*, vol. 48, no. 5, pp. 1576-1587, Sep./Oct. 2012.

[7] G. Wang, D. Xiao, N. Zhao, X. Zhang, W. Wang, and D. Xu, "Low-Frequency Pulse Voltage Injection Scheme-Based Sensorless Control of IPMSM Drives for Audible Noise Reduction," *IEEE Trans. Ind. Electron.*, vol. 64, no. 11, pp. 8415-8426, Nov. 2017.

[8] S. Taniguchi, K. Yasui, and K. Yuki, "Noise reduction method by injected frequency control for position sensorless control of permanent magnet synchronous motor," in *2014 International Power Electronics Conference (IPEC-Hiroshima 2014 - ECCE ASIA)*, 2014, pp. 2465-2469.

[9] G. Wang, L. Yang, B. Yuan, B. Wang, G. Zhang, and D. Xu, "Pseudo-Random High-Frequency Square-Wave Voltage Injection Based Sensorless Control of IPMSM Drives for Audible Noise Reduction," *IEEE Trans. Ind. Electron.*, vol. 63, no. 12, pp. 7423-7433, Dec. 2016.

[10] G. Wang, L. Yang, G. Zhang, X. Zhang, and D. Xu, "Comparative Investigation of Pseudorandom High-Frequency Signal Injection Schemes for Sensorless IPMSM Drives," *IEEE Trans. Power Electron.*, vol. 32, no. 3, pp. 2123-2132, Mar. 2017.

The 2018 International Power Electronics Conference

AT-NPC 3-Level Inverter-Fed Induction Motor Vector Control With Neutral Point Voltage Control

K. Sudo[1], M. Tsuji[1], S. Hamasaki[1], T. Fukuoka[1], and H. Ichinose[2]

1 Nagasaki University, 1-14 Bunkyo-machi, Nagasaki, 852-8521, Japan

2 Mitsubishi Electric Engineering Co.,Ltd, 6-14 Maruo-machi, Nagasaki, 852-8004, Japan

E-mail : mineo@nagasaki-u.ac.jp

Abstract-In recent years, studies and developments on 3-level inverters have been conducted for the purpose of increasing the capacity of the motor drive system and reducing the harmonic current. In this research, we introduce a control method to keep the neutral point voltage of the smoothing capacitor, and compare and discuss the simulation results and experiment results when the induction motor is vector controlled by the 3-level inverter.

Keywords— 3-level inverter, Induction Motor, Vector control, Neutral point voltage control.

I. INTRODUCTION

In recent years, 3-level inverter-fed induction motor drive system have been studied for the purpose of increasing the capacity of the system and reducing harmonic current. Since the 3-level inverter can output a 5-values stepwise voltages between the lines of the load, the harmonic current can be greatly reduced [1]-[7]. In addition, because only half the DC voltage is applied to the switching element, it is suitable for a large capacity load. Advanced T-type neutral-point-clamped (AT-NPC) 3-level inverter is attracting attention as compared with NPC 3-level inverter that is generally used because AT-NPC inverter has smaller number of elements when currents pass and has smaller conduction loss [8]-[10].

When the induction motor is driven by vector control using the 3-level inverter, the voltage value of the smoothing capacitor diverges because of the current control. Therefore, in this paper, we propose a control method to keep the voltage value of the smoothing capacitor constant, and show its effectiveness by simulation and experimental results of the vector controlled induction motor drive system. The verification is also performed in speed sensorless vector control using a flux observer.

II. 3-LEVEL INVERTER SYSTEM

Figure 1 shows the circuit of AT-NPC 3-level inverter. The 2-level inverter outputs two values for the virtual neutral point of the DC power supply. On the other hand, the 3-level inverter outputs three values for the virtual neutral point of the DC power supply. That is, the line voltage outputs five values. Furthermore, in the 2-level inverter, the switching frequency of the transistor matches the carrier frequency, whereas in the 3-level

inverter the average switching frequency per transistor is half the carrier frequency. Therefore, when the average switching frequency per transistor is equal, the harmonic content contained in the output voltage of the 3-level inverter is equivalent to one quarter of the harmonic component included in the output voltage of the 2-level inverter. So the 3-level inverter can greatly reduce harmonic current. In addition, because only half the DC voltage is applied to the switching element, it is suitable for a high-voltage large-capacity load.

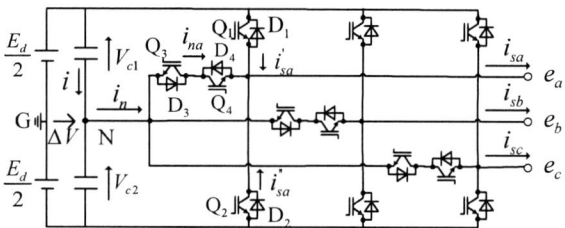

Fig. 1. AT-NPC 3-level inverter.

Figure 2 shows its waveform of the PWM control in *a* phase. PWM control of 3-level inverter is realized by comparing the modulation rate with positive side carrier wave and negative side carrier wave. In positive side, the upper arm of each phase is controlled. In negative side, the lower arm of each phase is controlled. As the result, three values can be outputted, and the fundamental component of the voltage output waveform is proportional to the modulation rate used for the PWM control.

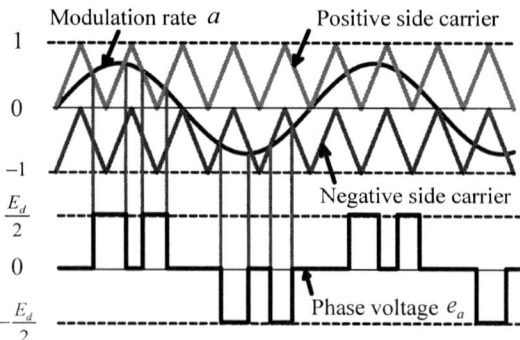

Fig. 2. PWM control of 3-level inverter.

1240

Figure 3 shows a sensorless vector control system using a flux observer in a rotating axis [11]. The flux observer performs the calculations (1) and (2), but for simplicity, the voltage sensor is eliminated and command voltage is used. K_c is the observer gain and $\psi_{rd}^* = M i_{sd}^*$, $\psi_{rq}^* = 0$.

$$p\psi_{rd}^v = \frac{L_r}{M}(e_{sd}^* - R_s^* i_{sd} - \sigma L_s p i_{sd} + \omega^* \sigma L_s i_{sq}) \\ + \omega^* \psi_{rq}^v + K_c(\psi_{rd}^* - \psi_{rd}^v) \tag{1}$$

$$p\psi_{rq}^v = \frac{L_r}{M}(e_{sq}^* - R_s^* i_{sq} - \sigma L_s p i_{sq} - \omega^* \sigma L_s i_{sd}) \\ - \omega^* \psi_{rd}^v + K_c(\psi_{rq}^* - \psi_{rq}^v) \tag{2}$$

The synchronous speed is calculated using the q-axis flux as follows:

$$\omega^* = (K_{wp} + \frac{K_{wi}}{s})\psi_{rq}^v \tag{3}$$

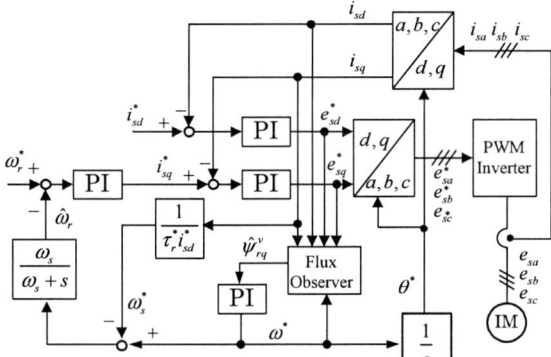

Fig. 3. Sensorless vector control system.

III. NEUTRAL POINT VOLTAGE CONTROL

We explain a-phase as an example. In Fig. 1, ΔV is defined as the neutral point voltage. From Fig. 1, we have

$$\frac{E_d}{2} = \Delta V + V_{c1} \tag{4}$$

$$\frac{E_d}{2} = \Delta V + V_{c2} \tag{5}$$

From (4) and (5),

$$\Delta V = \frac{V_{c2} - V_{c1}}{2} \tag{6}$$

Figure 4 shows a PWM output waveform. From Fig. 4, when the modulation rate a is positive, the output phase voltage considering ΔV is

$$e_a = \frac{T_1}{T}\frac{E_d}{2} + \frac{T_3}{T}\Delta V \\ = a\frac{E_d}{2} + (1-a)\Delta V$$

When the voltage command e_a^* is given, the modulation rate a is computed as follows:

$$a = \frac{e_a^* - \Delta V}{\frac{E_d}{2} - \Delta V} \tag{7}$$

Assuming that ΔV is small

$$a \approx \frac{2e_a^*}{E_d} - \frac{2}{E_d}(1 - \frac{2e_a^*}{E_d})\Delta V \\ = a^* - \frac{2}{E_d}(1 - a^*)\Delta V \tag{8}$$

Also when the modulation rate a is negative, the modulation rate a is computed as follows:

$$a = \frac{e_a^* - \Delta V}{\frac{E_d}{2} + \Delta V} \tag{9}$$

Assuming that ΔV is small, we have

$$a \approx \frac{2e_a^*}{E_d} - \frac{2}{E_d}(1 + \frac{2e_a^*}{E_d})\Delta V \\ = a^* - \frac{2}{E_d}(1 + a^*)\Delta V \tag{10}$$

From Fig. 1, the current equation of each capacitor is described as

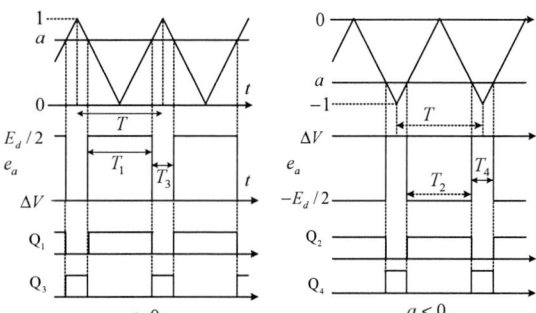

Fig. 4. PWM output waveform.

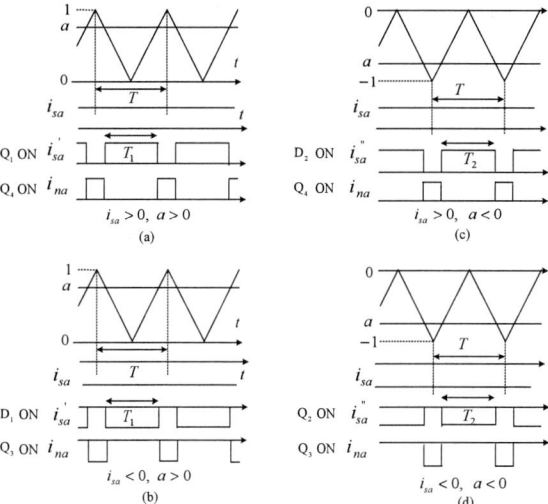

Fig. 5. Current waveforms in each case.

$$i = C\frac{dV_{c1}}{dt} \tag{11}$$

$$i - i_n = C\frac{dV_{c2}}{dt} \tag{12}$$

Therefore, the voltage ΔV satisfies

$$2C\frac{d\Delta V}{dt} = -i_n \tag{13}$$

From (13), when $i_n > 0$, ΔV decreases, and when $i_n < 0$, ΔV increases. Therefore, ΔV can be controlled by controlling i_n as follows:

< Policy 1>

When ΔV is positive, we increase V_{c1} and decrease V_{c2}. From (13), when i_n is positive, we increase i_n further. When i_n is negative, we increase i_n and make it positive.

< Policy 2>

When ΔV is negative, we decrease V_{c1} and increase V_{c2}. From (13), when i_n is negative, we decrease i_n further. When i_n is positive, we decrease i_n and make it negative.

By applying these policies, we control the neutral point voltage as follows:

(1) Case of Modulation factor $a^* > 0$

  (a) Case of $\Delta V > 0$

    From (8), $0 < a < a^*$. In Fig. 4, i_{sa} and a must be positive in order to decrease the modulation factor and increase the average value of i_{na}. So, the control of (8) is applied only to the phase with positive phase current.

    In other words, i_n is increased by shortening T_1 and lengthening T_3 and ΔV is decreased and approaches zero.

  (b) Case of $\Delta V < 0$

    From (8), $0 < a^* < a$. In Fig. 4, i_{sa} and a must be positive in order to increase the modulation factor and decrease the average value of i_{na}. So, the control of (8) is applied only to the phase with positive phase current.

    In other words, i_n is decreased by lengthening T_1 and shortening T_3 and ΔV is increased and approaches zero.

(2) Case of Modulation factor $a^* < 0$

  (a) Case of $\Delta V > 0$

    From (10), $a < a^* < 0$. In Fig. 4, i_{sa} and a must be negative in order to decrease the modulation factor and increase the average value of i_{na}. So, the control of (10) is applied only to the phase with negative phase current.

    In other words, i_n is increased by lengthening T_2 and shortening T_4 and ΔV is decreased and approaches zero.

  (b) Case of $\Delta V < 0$

    From (10), $a^* < a < 0$. In Fig. 4, i_{sa} and a must be negative in order to increase the modulation factor and decrease the average value of i_{na}. So, the control of (10) is applied only to the phase with negative phase current.

    In other words, i_n is decreased by shortening T_2 and lengthening T_4 and ΔV is increased and approaches zero.

IV. SIMULATION AND EXPERIMENTAL RESULTS

In this research, we analyzed by using simulation software named PSIM. Figure 6 shows a block chart of the PSIM. By reading the necessary informations that are capacitor voltages, phase currents, speed from the sensor and speed command, the DLL calculates and outputs the modulation factor. Gates of IGBTs are controlled by comparing the modulation factor with the carrier. In this experiment of this research, each sensors and gate drive circuits are made by ourselves

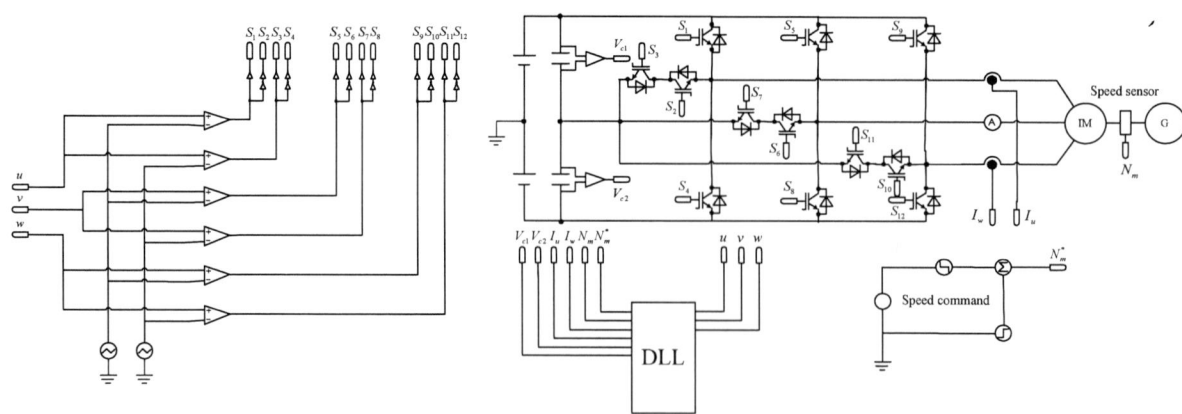

Fig. 6. Block chart of simulation software PSIM

Figures 7 and 8 show simulation results of vector control operation at the speed reference 200min⁻¹, the load torque 4.0Nm, and the DC power supply voltage 200V. Figure 7 shows the case without capacitor voltage control. As the time passes, the capacitor voltage diverges and it turns out that the system is not operated as 3-level inverter. Figure 8 shows the case with capacitor voltage control after starting from 0min⁻¹. There is sustained oscillations of capacitor voltage. But the system is stable at almost 100V.

Figures 9 and 10 show the experimental results of the capacitor voltage with or without of capacitor voltage control. Figure 9 shows the waveform of the capacitor voltage when intentional switching from "with capacitor voltage control" to "without control" is tested. It is observed that the capacitor voltage gradually diverges after the time of switching. Figure 10 shows the waveform of the capacitor voltage when switching from "without capacitor voltage control" to "with control" is executed. It is observed that the diverging capacitor voltage has converged after the capacitor voltage control.

We could show the effectiveness of proposed capacitor voltage control by simulation results and experimental results.

Figures 11, 12 and 13 show the speed and the capacitor voltage responses for the step change of speed command at low speed (50min⁻¹), middle speed (500min⁻¹) and high speed (1000min⁻¹) in vector control with speed sensor. The capacitor voltage control is used in all cases. Each load torque is 4.0Nm (50% of rated torque) and DC power supply voltage is 300V. In the simulation result of Fig. 11, it is confirmed that the actual speed follows quickly its command value. In the experimental results, it seems to be a little pulsated in the low speed region, but it is confirmed that the speed almost follows its command value. The variation in the capacitor voltage is controlled within about ±2 V in the steady state. Therefore, the neutral point voltage control is considered to be accomplished. Also in the medium speed and high speed regions of Fig. 12 and Fig. 13, it is confirmed that the speed follows quickly its command. The capacitor voltage is also stable.

Figures 14, 15 and 16 show the speed step response in the low speed (50min⁻¹), middle speed (500min⁻¹) and high speed (1000min⁻¹) in vector control with speed sensor. The capacitor voltage control is used in all cases. Each load torque is almost no load and DC power supply voltage is 300V. A quick response is obtained for any region in the speed step response. It is confirmed that the sum of the both capacitor voltages apparently becomes larger than the input DC voltage 300V when step-down in the experiment results at the medium speed region and the high speed region. When the speed is stepped down, the mechanical energy of induction generator is stored to the capacitors. It is conceivable that such a result is obtained only by experimental results because the DC sources used in the experiment and simulation are constituted by the rectified power supply and ideal DC source respectively.

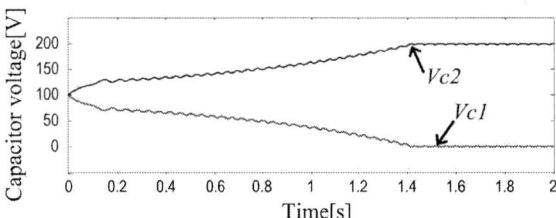

Fig. 7. Simulation result (without capacitor voltage control).

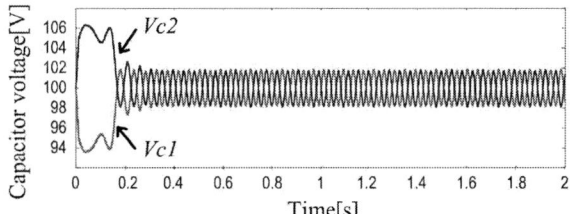

Fig. 8. Simulation result (with capacitor voltage control).

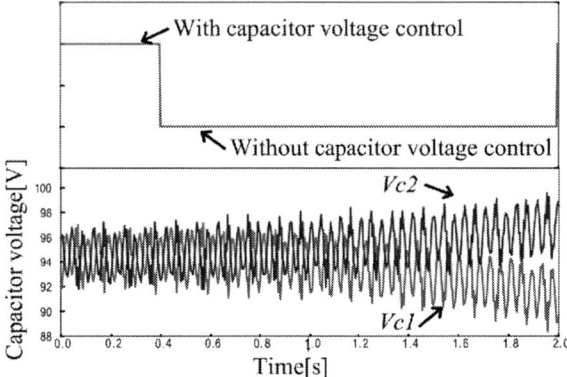

Fig. 9. Experimental result of vector controlled induction motor.

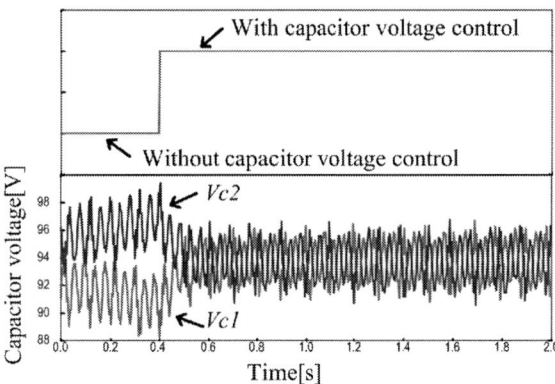

Fig.10. Experimental result of vector controlled induction motor.

The 2018 International Power Electronics Conference

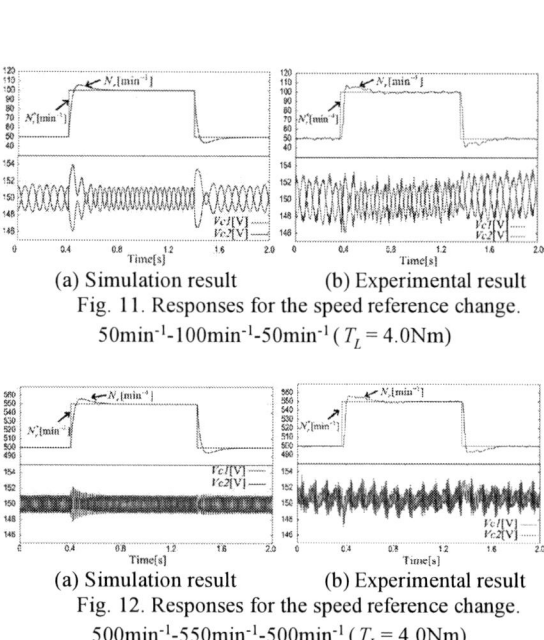

(a) Simulation result   (b) Experimental result

Fig. 11. Responses for the speed reference change.
50min⁻¹-100min⁻¹-50min⁻¹ (T_L = 4.0Nm)

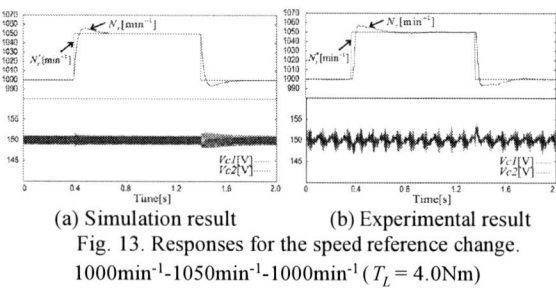

(a) Simulation result   (b) Experimental result

Fig. 12. Responses for the speed reference change.
500min⁻¹-550min⁻¹-500min⁻¹ (T_L = 4.0Nm)

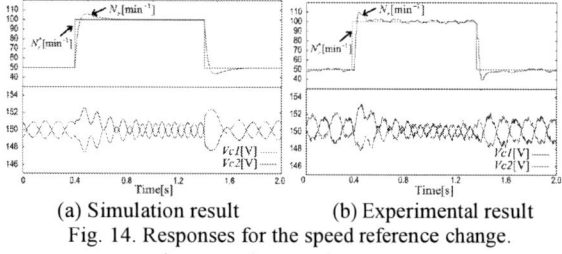

(a) Simulation result   (b) Experimental result

Fig. 13. Responses for the speed reference change.
1000min⁻¹-1050min⁻¹-1000min⁻¹ (T_L = 4.0Nm)

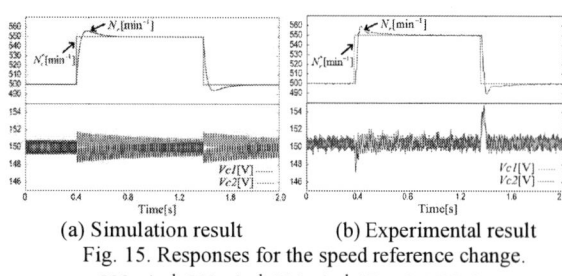

(a) Simulation result   (b) Experimental result

Fig. 14. Responses for the speed reference change.
50min⁻¹-100min⁻¹-50min⁻¹ (T_L = 0.0Nm)

(a) Simulation result   (b) Experimental result

Fig. 15. Responses for the speed reference change.
500min⁻¹-550min⁻¹-500min⁻¹ (T_L = 0.50Nm)

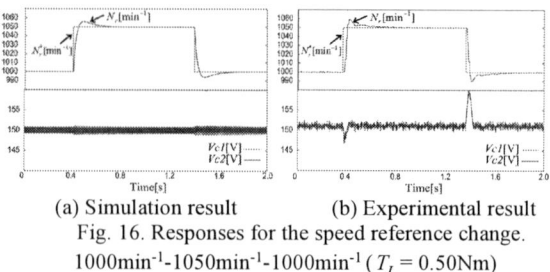

(a) Simulation result   (b) Experimental result

Fig. 16. Responses for the speed reference change.
1000min⁻¹-1050min⁻¹-1000min⁻¹ (T_L = 0.50Nm)

Figures 17, 18 and 19 show the speed and the capacitor voltage responses for the step change of command at 50min⁻¹, 500min⁻¹ and 1000min⁻¹ in speed sensorless vector control. The capacitor voltage control is valid. Each load torque is 4.0Nm and DC power supply voltage is 300V. In the experimental results, it is confirmed that the actual speed follows quickly its command value in each region, but the actual speed increases about 10min⁻¹ with respect to the command value. This may be due to the error of the primary resistance and the effect of PWM inverter voltage control. However, in the simulation results that are ideal, it is confirmed that the speed almost follows the command value in each region. Fluctuations in the capacitor voltage are controlled within about ±2 V in steady state, both experimental and simulation results. Therefore, the neutral point voltage control is effective in the case of sensorless vector control.

Figures 20, 21 and 22 show the speed step response in 50min⁻¹, 500min⁻¹ and 1000min⁻¹ in sensorless vector control. The capacitor voltage control is valid. Each load torque is almost no load and DC power supply voltage is 300V. In the very low speed region, the error between actual speed and its command becomes large in the experimental result. This may also be due to the error of the primary resistance and the effect of voltage control. However, the fluctuation of the capacitor voltage is controlled within about ±4 V in steady state. Therefore, neutral point voltage control is considered to have been achieved. It is confirmed that the capacitor voltage was controlled stably in the middle speed region and the high speed region.

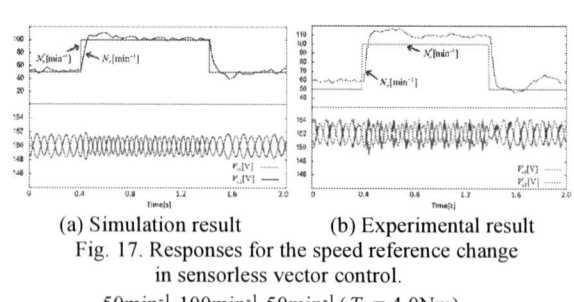

(a) Simulation result   (b) Experimental result

Fig. 17. Responses for the speed reference change
in sensorless vector control.
50min⁻¹-100min⁻¹-50min⁻¹ (T_L = 4.0Nm)

1244

The 2018 International Power Electronics Conference

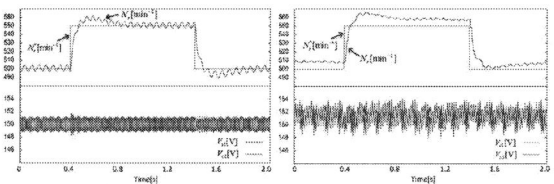

(a) Simulation result (b) Experimental result
Fig. 18. Responses for the speed reference change in sensorless vector control.
500min⁻¹-550min⁻¹-500min⁻¹ (T_L = 4.0Nm)

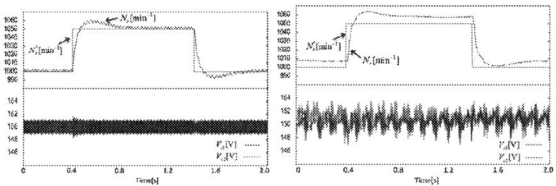

(a) Simulation result (b) Experimental result
Fig. 19. Responses for the speed reference change in sensorless vector control.
1000min⁻¹-1050min⁻¹-1000min⁻¹ (T_L = 4.0Nm)

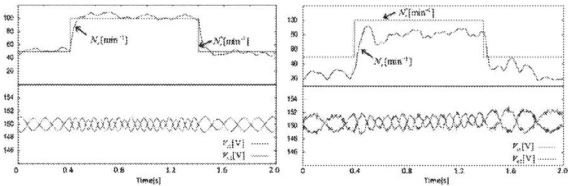

(a) Simulation result (b) Experimental result
Fig. 20. Responses for the speed reference change in sensorless vector control.
50min⁻¹-100min⁻¹-50min⁻¹ (T_L = 0.0Nm)

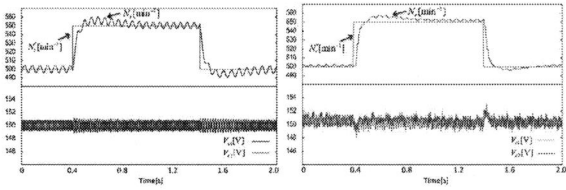

(a) Simulation result (b) Experimental result
Fig. 21. Responses for the speed reference change in sensorless vector control.
500min⁻¹-550min⁻¹-500min⁻¹ (T_L = 0.50Nm)

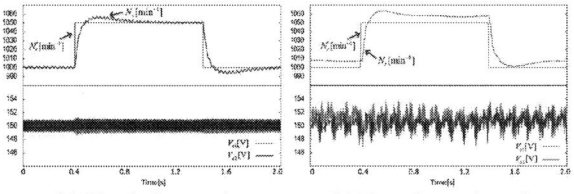

(a) Simulation result (b) Experimental result
Fig. 22. Responses for the speed reference change in sensorless vector control.
1000min⁻¹-1050min⁻¹-1000min⁻¹ (T_L = 0.50Nm)

V. CONCLUSIONS

It is confirmed that the capacitor neutral voltage of AT-NPC 3-level inverter becomes unstable when the vector control is used for induction motor drive. A control method of the capacitor neutral voltage based on the neutral point voltage is proposed and the effectiveness has been confirmed by the simulation and the experimental results. Therefore, vector control with or without speed sensor for the induction motor using AT-NPC 3-level inverter has been accomplished.

REFERENCES

[1] A. Nabae, I. Takahashi and H. Akagi, "A new neutral-point-clamped PWM inverter", *IEEE Trans. Ind. Appl.*, Vol. IA-17, No. 5, pp. 518-523, 1981

[2] J. K. Steinke, "Control strategy for a three phase ac traction drive with three-level GTO PWM inverter", *IEEE Conf.* Rec. PESC 88, pp.431-438, 1988

[3] S. Ogasawara, T. Sawada, K. Abe and H. Akagi, "A vector control system using a neutral-point-clamped voltage source PWM inverter", (in Japanese), *IEEJ Trans.* Vol.111-D, No.11 pp.930-936, 1991

[4] S. Ogasawara, T. Sawada and H. Akagi, "Analysis of the neutral point potential variation of neutral-point-clamped voltage source PWM inverters", (in Japanese), *IEEJ Trans.* Vol.113-D, No.1 pp.42-48, 1993

[5] I. Miyashita, B. Kaku and S, Sone, "A new PWM method for 3-level inverter based on voltage space vector suppressing the neutral point potential variation", *Proc. of International Power Electronics Conference (IPEC)*, pp.506-511, 1995

[6] M. Matsui, "Static var compensator using neutral-point-clamped PWM inverter and its control scheme", *Proc. of International Power Electronics Conference (IPEC)*, pp.488-493, 1995

[7] R. Rojas, T. Ohnishi and T. Suzuki, "PWM control method for NPC inverters with very small DC-link capacitors", *Proc. of International Power Electronics Conference (IPEC)*, pp.494-499, 1995

[8] K. Komatsu, M. Yatsu, S. Miyashita, S. Okita, H. Nakazawa, S. Igarashi, Y. Takahashi, Y. Okuma, Y. Seki and T. Fujihira, "New IGBT modules for advanced neutral-point-clamped 3-level power converters", *Proc. of International Power Electronics Conference (IPEC)*, pp.523-527, 2010

[9] M. Schweizer and Johann W. Kolar, "Design and implementation of a highly efficient three-level t-type converter for low-voltage applications", *IEEE Trans. Power Electron .*, Vol.28, No.2, pp.899-907, 2013

[10] E. AVCI and M. UCAR, "Analysis and design of grid-connected 3-phase 3-level AT-NPC inverter for low-voltage applications", *Turk JElec Eng and Comp Sci*, 25 : pp.2464-2478, 2017

[11] M. Tsuji, S. Chen, K. Izumi, E. Yamada, "A sensorless vector control system for induction motors using q-axis flux with stator resistance identification", *IEEE Trans. Ind. Electron.* Vol.48, No.1, pp.185-194, 2001

Investigation of Various Position Estimation Accuracy Issues in Pulse-Injection-based Sensorless Drives

Hechao Wang*, Kaiyuan Lu, Dong Wang, Frede Blaabjerg
Energy technology, Aalborg University, Aalborg, Denmark
*Email hec@et.aau.dk

Abstract— **For low speed sensorless drives of Surface mounted Permanent Magnet Synchronous Machine (SPMSM), pulse vectors are often injected into the drive to estimate the rotor position. There are many factors that could affect the accuracy of the estimated position such as inverter voltage error, cross-saturation effects and the presence of the speed dependent back electromotive force (EMF). This paper analyzes and evaluates the position estimation error caused by aforementioned factors with experimental results. Detailed quantitative results are given. Meanwhile the interaction between these factors is briefly studied.**

Keywords— *SPMSM, sensorless, position estimation accuracy.*

I. INTRODUCTION

In Field Oriented Control (FOC) of PMSM drives, the rotor position information is essential. Position sensors such as encoder and resolver are preferred to be avoided for reducing the cost and improving the system reliability. A lot of efforts have been dedicated to the control of drives without mechanical sensors, known as sensorless drives [1].

The PMSM sensorless control methods are usually based on either the fundamental model for medium- to high- speed operation range, or the high frequency machine model for standstill to low-speed operation range [2,3]. In medium-high speed operation range, back electromotive force (EMF) voltage in the fundamental model is often utilized to estimate the rotor position. Satisfactory results could be obtained by minimizing the error between the fundamental model and the practical system. However, at low speed, due to the low back EMF voltage, the noise signal accounts for a larger proportion resulting in relative large position estimation error. Thus, the sensorless methods based on the rotor saliency detection, which are normally achieved by injecting steady-state high frequency signals [4-6] or simple voltages pulses [7,8], are generally adopted to obtain the rotor position at low speed.

For the pulse-injection-based sensorless methods, there are many factors that could affect the accuracy of the estimated position, e.g. back-EMF, phase resistance voltage drop, inverter voltage error and machine cross-saturation effects [9]. Some of them can be ignored when performing sensorless control at low speed, such as back-EMF (with the reason of rotor speed $\omega_r \approx 0$) and phase

resistance voltage drop (with the reason of selecting a higher injection voltage) [8]. On the other hand, the inverter voltage error distorts not only the fundamental output voltage but also the injection pulse voltage vector. Both the amplitude and direction of the injection voltage vector are changed due to inverter voltage error leading a position estimation error [10]. Some compensation methods are proposed based on the inverter parameters [11-13]. Furthermore, cross-saturation effects, which will lead to large position estimation errors at load in the low speed range, are studied [14] and compensation methods based on simulation and experimental results are proposed [15-18].

The influence of the aforementioned factors and their compensation techniques are studied separately in the above references. In [8], the position error was simply attributed to the inverter voltage error which is not proper. Detailed quantitative comparison and analysis is not reported in the existing literatures. Meanwhile these influences on position estimation interact with each other and are not separated. Thus the aforementioned influences on position estimation error should be carefully examined and understood synthetically; the weights of the influences of different factors should be compared and the interaction of these factors should be investigated. This will be helpful in developing necessary and effective position error compensation methods for improving the sensorless drive performance.

In this paper, a thorough investigation regarding the influences of different factors on position estimation error is carried out. The position estimation errors caused by these factors are investigated and compared. Thorough experimental results are provided to support the conclusions. The influence of inverter voltage error with different PWM pattern is simply researched. The interaction between these factors is briefly studied. Effective techniques are investigated and implemented to compensate or reduce the influences caused by these factors.

II. SENSORLESS FUNDAMENTALS

The basic model of PMSM in dq-reference frame could be described as:

$$\begin{bmatrix} u_d \\ u_q \end{bmatrix} = R \begin{bmatrix} i_d \\ i_q \end{bmatrix} + \omega_r \begin{bmatrix} 0 & -L_q \\ L_d & 0 \end{bmatrix} \begin{bmatrix} i_d \\ i_q \end{bmatrix}$$
$$+ \omega_r \lambda_{mpm} \begin{bmatrix} 0 \\ 1 \end{bmatrix} + \frac{d}{dt} \begin{bmatrix} L_d & 0 \\ 0 & L_q \end{bmatrix} \begin{bmatrix} i_d \\ i_q \end{bmatrix} \tag{1}$$

where u_d, u_q, i_d, i_q are the stator d- and q-axes voltages and currents respectively; L_d, L_q are the d- and q-axes inductances respectively; R is the stator resistance; ω_r is the rotor electrical speed; and λ_{mpm} is the magnitude of the rotor PM flux linkage.

In pulse-injection-based sensorless methods, in order to simplify the analysis, resistance voltage drop and back-EMF are often ignored at low speed [7]. Then the high frequency model of PMSM in the $\alpha\beta$-reference frame is shown as:

$$\begin{bmatrix} u_{\alpha h} \\ u_{\beta h} \end{bmatrix} = \begin{bmatrix} L_0 + L_1 \cos 2\theta_r & L_1 \sin 2\theta_r \\ L_1 \sin 2\theta_r & L_0 - L_1 \cos 2\theta_r \end{bmatrix} \frac{d}{dt} \begin{bmatrix} i_{\alpha h} \\ i_{\beta h} \end{bmatrix} \tag{2}$$

where θ_r is the rotor position; $u_{\alpha h}$, $u_{\beta h}$, $i_{\alpha h}$, $i_{\beta h}$ are the stator α- and β-axes voltages and currents respectively and $L_0 = (L_d + L_q)/2$, $L_1 = (L_d - L_q)/2$. The rotor position θ_r could be obtained from (2) directly by using machine parameters and injected high frequency terminal voltages and currents. However, due to the uncertainties (inverter voltage error) in estimating the $u_{\alpha h}$, $u_{\beta h}$ voltages and possible variations of machine parameters, (2) is preferred to be transformed to the estimated dq-reference frame for estimating the position error $\tilde{\theta}_r$ instead, giving

$$\begin{bmatrix} u_{\hat{d}h} \\ u_{\hat{q}h} \end{bmatrix} = \begin{bmatrix} L_0 + L_1 \cos 2\tilde{\theta}_r & L_1 \sin 2\tilde{\theta}_r \\ L_1 \sin 2\tilde{\theta}_r & L_0 - L_1 \cos 2\tilde{\theta}_r \end{bmatrix} \frac{d}{dt} \begin{bmatrix} i_{\hat{d}h} \\ i_{\hat{q}h} \end{bmatrix} \tag{3}$$

where $\tilde{\theta}_r = \theta_r - \hat{\theta}_r$; $\hat{\theta}_r$ is the estimated rotor position; $u_{\hat{d}}$, $u_{\hat{q}}$, $i_{\hat{d}}$, $i_{\hat{q}}$ are the stator d- and q-axes voltages and currents respectively in the estimated dq-reference frame. In one switching period, the current differentiation di/dt may be approximated by $\Delta i / T_s$. Therefore, the rotor position error could be obtained by e.g. injecting a voltage vector into the estimated d-axis (giving (4)) or the estimated q-axis (giving (5)) [8]:

$$\frac{\Delta i_{\hat{q}h}}{T_s} \approx \frac{di_{\hat{q}h}}{dt} = \frac{-L_1 u_{\hat{d}h}}{L_0^2 - L_1^2} \sin(2\tilde{\theta}_r) \tag{4}$$

$$\frac{\Delta i_{\hat{d}h}}{T_s} \approx \frac{di_{\hat{d}h}}{dt} = \frac{-L_1 u_{\hat{q}h}}{L_0^2 - L_1^2} \sin(2\tilde{\theta}_r) \tag{5}$$

The current variation due to the injected voltage is then used to estimate the position error [8].

However, as stated above, many items have been neglected in obtaining (2). The influence of such simplification is worthy to be investigated.

If taken (4) as an example, as shown in Fig. 1, the injection vector command $\vec{V}_{inj}^*$ is located on the estimated d-axis. But due to the influence of resistance voltage drop,

back-EMF voltage and inverter voltage error, the injection vector becomes $\vec{V}_{inj}$, of which both the magnitude and direction are changed, and leads to a position estimation error. Meanwhile it should be noticed that all kinds of injection based sensorless methods detect the rotor magnetic saliency (d_{mag}-axis) instead of mechanical saliency (d-axis)[15]. Thus, there is also an error ε between magnetic saliency and mechanical saliency, which is caused by the cross-saturation effect and needs to be compensated.

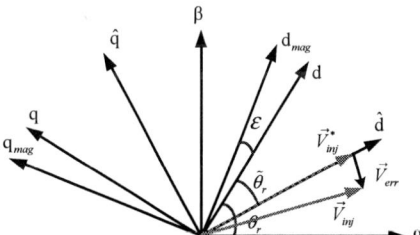

Fig. 1 Relationship among different reference frame.

III. Various Issues on Position Estimation Accuracy

A. Back-EMF

To obtain the high frequency model, e.g. (2), (3), the back-EMF is often ignored for simplification. In low-speed operation, this is acceptable, since the mean position error is around 2.4 degrees only at 4.5 rpm as can be observed in Fig. 2(a), for the test motor reported in the appendix. It is worth to note that in the experiment given in Fig. 2, the q-axis current is kept to be zero (no-load) to avoid any possible cross-saturation effect, while the inverter voltage error is also small due to the low current. However, when the speed increases, neglecting the back-EMF in the high frequency machine model will introduce additional position estimation error, e.g. it can be observed from Fig. 2(b) that the mean position error is increased to 4.1 degrees when operated at 45 rpm. Back-EMF will introduce a positive position error and it may be observed that the introduced position error is proportional to the machine operating speed (Table I). The difficulty occurring here is that it is not obviously clear which speed may be regarded as sufficiently low speed to neglect the back-EMF effects. On the other hand, with back-EMF disturbance presented, the low-speed operation range, where the position estimation error can be limited into an acceptable range, is narrowed.

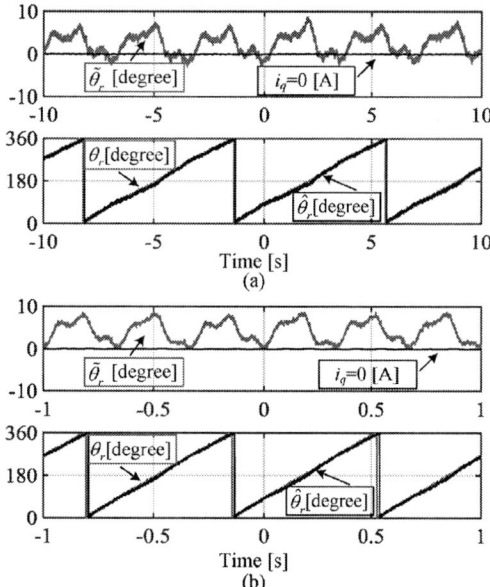

Fig. 2 q-axis current, estimated position error, real rotor position and estimated rotor position with only one voltage vector injecting on the estimated d-axis (a) at 4.5 rpm (b) at 45 rpm.

TABLE I $\tilde{\theta}_r$ AT DIFFERENT SPEEDS WHILE INJECTING AS (4)

Speed [rpm]	4.5	15	30	45	60	120
$\tilde{\theta}_r$ [degree]	2.4	2.8	3.6	4.1	4.8	7.2

To overcome this problem, two opposite injection voltage vectors may be introduced to suppress the influence of the back-EMF. The variation of the current changes detected in the two consequent switching periods with opposite voltage injection vectors may be obtained as [8]:

$$\begin{bmatrix} \Delta u_{\alpha h} \\ \Delta u_{\beta h} \end{bmatrix} = \begin{bmatrix} L_0 + L_1 \cos 2\theta_r & L_1 \sin 2\theta_r \\ L_1 \sin 2\theta_r & L_0 - L_1 \cos 2\theta_r \end{bmatrix} \frac{d}{dt} \begin{bmatrix} \Delta i_{\alpha h} \\ \Delta i_{\beta h} \end{bmatrix} \quad (6)$$

$$\frac{\Delta(\Delta i_{\hat{q}h})}{T_s} \approx \frac{d\Delta i_{\hat{q}h}}{dt} = \frac{-L_1 \Delta u_{\hat{d}h}}{L_0^2 - L_1^2} \sin(2\tilde{\theta}_r) \quad (7)$$

Equation (7) is similar to (4), and the position error may be estimated in a similar way. The implementation of the suggested method is illustrated in Fig. 3.

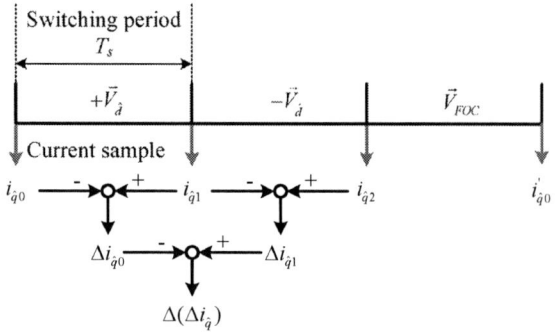

Fig. 3 Implementation of two opposite voltage vectors injection

To illustrate the effectiveness of this two opposite voltage vectors injection method, the obtained position error by injecting on the positive and negative estimated d-axis is shown in Fig. 4. It can be observed that the estimated position errors exhibit a constant small error of around 2.1 degrees at 4.5 rpm and 1.9 degrees at 45 rpm. Table II shows the position errors at different speeds. By comparing Table II with Table I, it is clear that the speed dependent position error term is effectively removed by the adopted opposite voltage vector injection method.

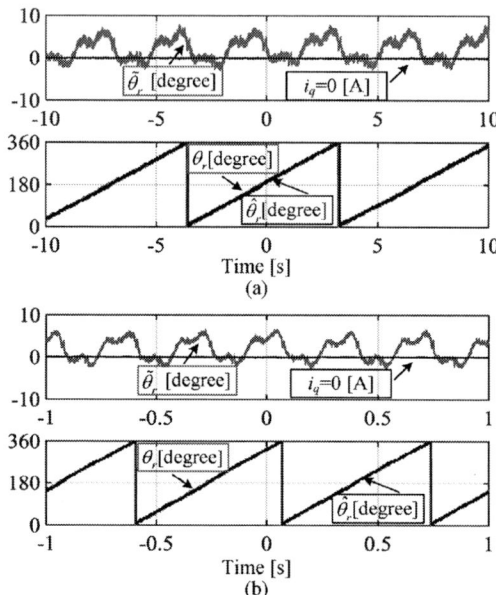

Fig. 4 q-axis current, estimated position error, real rotor position and estimated rotor position with two opposite voltage vectors injecting on the estimated d-axis (a) at 4.5 rpm (b) at 45 rpm.

TABLE II $\tilde{\theta}_r$ AT DIFFERENT SPEEDS WHILE INJECTING AS (7)

Speed [rpm]	4.5	15	30	45	60	120
$\tilde{\theta}_r$ [degree]	2.1	2.2	2.2	1.9	1.7	1.3

B. Inverter voltage error

When supplying voltage to the machine by an inverter, the dead-time, snubber capacitors and some other nonlinear switching characteristics will distort the output voltage, generating voltage error between the real machine terminal voltage and the commanded voltage. This error will affect the rotor position estimation accuracy when the commanded voltage is used in position estimation. Many efforts have been reported to compensate this inverter voltage error. The previously mentioned two opposite voltage vector injection method is also effective in minimizing the influence of inverter voltage error on position estimation. The principle is briefly illustrated below.

The additional inverter voltage errors may be added to (2), giving:

$$\begin{bmatrix} u^i_{\alpha h} \\ u^i_{\beta h} \end{bmatrix} = \begin{bmatrix} L_0 + L_1 \cos 2\theta_r & L_1 \sin 2\theta_r \\ L_1 \sin 2\theta_r & L_0 - L_1 \cos 2\theta_r \end{bmatrix} \frac{d}{dt} \begin{bmatrix} i^i_{\alpha h} \\ i^i_{\beta h} \end{bmatrix} \quad (8)$$
$$+ \begin{bmatrix} u_{\alpha\_inverter} \\ u_{\beta\_inverter} \end{bmatrix}$$

where $u_{\alpha\_inverter}$, $u_{\beta\_inverter}$ are the α- and β-axes inverter voltage error respectively; $u^i_{\alpha h}, u^i_{\beta h}, i^i_{\alpha h}, i^i_{\beta h}$ are the stator α- and β-axes high frequency voltage commands and currents respectively. The inverter voltage error is current dependent; usually the current loop bandwidth is much lower than the switching frequency. Therefore it may be assumed that the inverter voltage errors are approximately the same during two neighboring switching periods [8]. The voltage difference between two neighboring switching periods could then be obtained by subtraction of the voltage equation (8) applied to the two neighboring switching periods giving:

$$\begin{bmatrix} \Delta u^i_{\alpha h} \\ \Delta u^i_{\beta h} \end{bmatrix} = \begin{bmatrix} L_0 + L_1 \cos 2\theta_r & L_1 \sin 2\theta_r \\ L_1 \sin 2\theta_r & L_0 - L_1 \cos 2\theta_r \end{bmatrix} \frac{d}{dt} \begin{bmatrix} \Delta i^i_{\alpha h} \\ \Delta i^i_{\beta h} \end{bmatrix} \quad (9)$$

Equation (9) is very similar to (6), except that the voltage and current variations are now used in (9). The voltage terms representing the inverter voltage errors in (8) are now removed and the position error may be estimated by:

$$\frac{\Delta(\Delta i^i_{\hat{q}h})}{T_s} \approx \frac{d\Delta i^i_{\hat{q}h}}{dt} = \frac{-L_1 \Delta u^i_{\hat{d}h}}{L_0^2 - L_1^2} \sin(2\tilde{\theta}_r) \quad (10)$$

It may be noticed in (1) that if d- and q-axes currents are kept zero, the back-EMF contains only q-axis component. If the estimated position error is small, the back-EMF component could be ignored. Thus, when injecting a single test voltage on the estimated q-axis, the position error is estimated from the resultant current on the estimated d-axis (as shown in (5)), which is not affected by the back-EMF.

Fig. 5 shows the obtained position error by injecting a single test voltage on the estimated q-axis. The estimated position error from (5) changes a little from 4.5 degrees to 4.2 degrees, which suggests that in this injection method, the estimated position is mostly influenced by inverter voltage error but not the back-EMF.

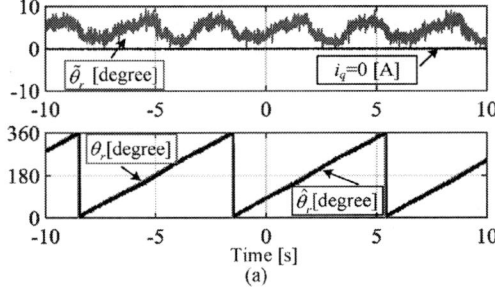

(a)

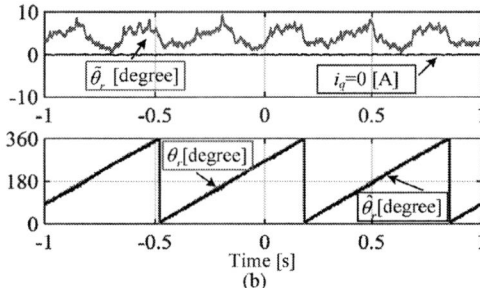

(b)

Fig. 5 q-axis current, estimated position error, real rotor position and estimated rotor position with only one voltage injecting on the estimated q-axis (a) at 4.5 rpm (b) at 45 rpm.

The injection method is continued to include one more opposite injection voltage vector on the estimated q-axis. According to (9)-(10), the inverter voltage error is minimized and the obtained position error is shown in Fig. 6. The estimated position error is reduced to 2.0 degrees comparing with the 4.5 degrees error shown in Fig.5 (a). It should be noticed that in the above experiment the inverter voltage error is quite low because the d- and q-axes currents are kept to zero to avoid the cross-saturation effects.

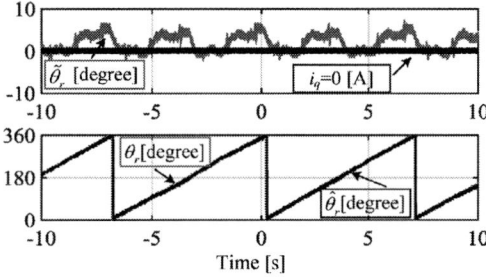

Fig. 6 q-axis current, estimated position error, real rotor position and estimated rotor position with two opposite voltage vectors injecting on the estimated q-axis at 4.5 rpm

To investigate the influence of invert voltage error with different motor currents, meanwhile suppress the influence of the cross-saturation effects, the high frequency pulses are injected only on the positive estimated d-axis (as shown in (5)) while keeping the d-axis current a negative value to avoid saturation. The experiment results are shown in Fig. 7 and Table III. It should be observed that the mean value of the estimated position error decrease from 3.5 degrees to 2 degrees when the d-axis current decrease from 0A to -4A.

TABLE III $\tilde{\theta}_r$ AT DIFFERENT MOTOR CURRENT WHILE INJECTING AS (5)

i_d [A]	0	-1	-2	-3	-4
$\hat{\theta}_r$ [degree]	3.5	2.5	2.4	2.2	2.0

The 2018 International Power Electronics Conference

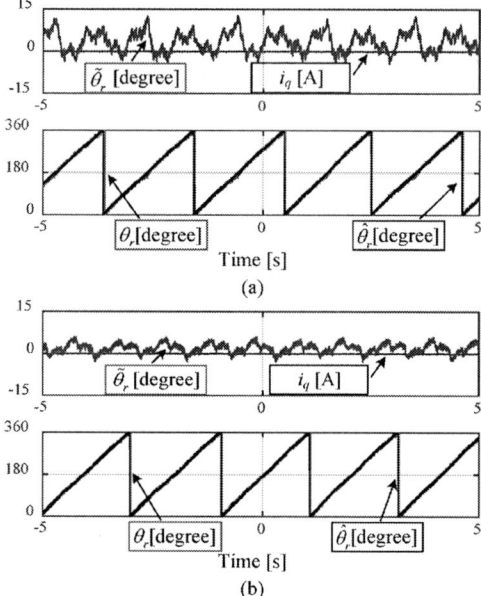

Fig. 7 q-axis current, estimated position error, real rotor position and estimated rotor position with only one voltage injecting on the estimated d-axis (a) with i_d=0A (b) with i_d=-4A.

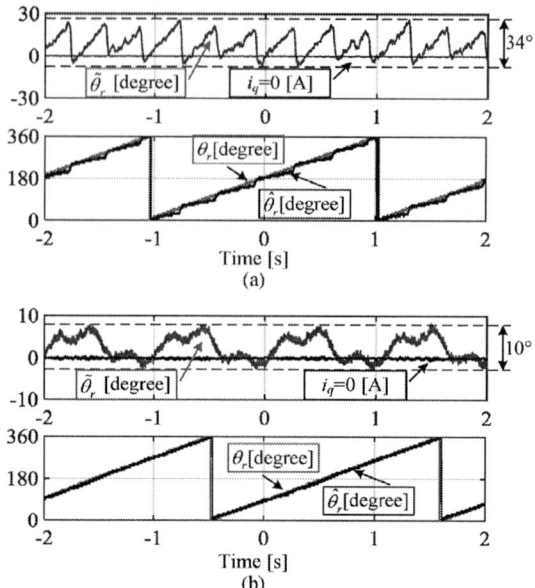

Fig. 8 q-axis current, estimated position error, real rotor position and estimated rotor position with two opposite voltage vectors injecting on the estimated d-axis at 15 rpm (a) injecting 12.5V (b) injecting 200V.

It should be noticed here that, in this experiment the snesorless method is based on traditional PWM pattern. If the PWM pattern is changed, the influence of inverter error will be different. Meanwhile the estimated position error caused by inverter voltage error is related to motor current similar to the influence of cross-saturation. Thus the inverter voltage error influence may strengthen or offset the influence of cross-saturation. For example in [19], the position estimation error caused by these two factors is about 13 degrees with full load. However in this paper, with same setup, the position error is about 23 degrees due to the different PWM patterns.

C. Amplitude of injection vector

Instead of using two opposite injection voltage vectors to suppress the back-EMF and inverter voltage error effects, another effective way is to simply increase the magnitude of the injected voltage. Thus, the back-EMF and inverter voltage error become relatively small, and their influences are reduced as well. Meanwhile, a higher amplitude of the injected voltage vector will cause higher current response, which will increase the measured current signal to noise ratio. The influence of the amplitude of injection vector on the estimated position error is shown in Fig. 8 and Table IV. It may be observed that the position error becomes less dependent on the magnitude of the injected voltage vector as the magnitude increases. There is an injection voltage magnitude threshold value, below which, the estimated position error is large (e.g. 7.8 degrees with 12.5 V injection voltage magnitude) and this is believed to be due to the poor signal-to-noise ratio.

TABLE IV $\tilde{\theta}_r$ WITH DIFFERENT INJECTION VOLTAGES

Amplitude [V]	12.5	25	50	100	200
$\tilde{\theta}_r$ [degree]	7.8	3.0	2.4	2.2	2.1

D. Cross-saturation effects

Almost all kinds of pulse injection sensorless methods are based on detecting the rotor magnetic saliency. However, cross-saturation effects caused by load current will shift the location of the magnetic d-axis away from the mechanical d-axis. This will introduce position error when the mechanical d-axis is taken as the reference d-axis.

When considering cross-saturation effect, (1) becomes

$$\begin{bmatrix} u_d \\ u_q \end{bmatrix} = R \begin{bmatrix} i_d \\ i_q \end{bmatrix} + \omega_r \begin{bmatrix} -L_{dq} & -L_q \\ L_d & L_{dq} \end{bmatrix} \begin{bmatrix} i_d \\ i_q \end{bmatrix} \\ + \omega_r \lambda_{mpm} \begin{bmatrix} 0 \\ 1 \end{bmatrix} + \frac{d}{dt} \begin{bmatrix} L_d & L_{dq} \\ L_{dq} & L_q \end{bmatrix} \begin{bmatrix} i_d \\ i_q \end{bmatrix} \quad (11)$$

where L_{dq} is the cross coupling inductance. Then the following equation could be obtained:

$$\begin{bmatrix} u_{\alpha h}^s \\ u_{\beta h}^s \end{bmatrix} = \begin{bmatrix} L_0^{'} + L_1^{'} \cos 2\theta_r^s & L_1^{'} \sin 2\theta_r^s \\ L_1^{'} \sin 2\theta_r^s & L_0^{'} - L_1^{'} \cos 2\theta_r^s \end{bmatrix} \frac{d}{dt} \begin{bmatrix} i_{\alpha h}^s \\ i_{\beta h}^s \end{bmatrix} \quad (12)$$

where $u_{\alpha h}^s$, $u_{\beta h}^s$, $i_{\alpha h}^s$, $i_{\beta h}^s$ are the stator α- and β-axes voltages and currents respectively while considering the cross-saturation effects, $\theta_r^s = \theta_r + \varepsilon$,

$$L_{d'} = (L_d + L_q)/2 + \sqrt{(L_d - L_q)^2 + 4L_{dq}^2}/4 \quad ,$$

$$L_{q'} = (L_d + L_q)/2 - \sqrt{(L_d - L_q)^2 + 4L_{dq}^2}/4 \quad ,$$

1250

$\varepsilon = 0.5 * \arctan 2L_{dq}/(L_d - L_q)$. From (12) it seems that the position error ε may be calculated directly. However in practice, it is difficult to obtain the information of L_{dq}, which may possibly vary under different load conditions. By using a mechanical encoder, the influence of the cross-saturation effect on the position error could be obtained. Fig. 9 shows the result, where it may be observed that the cross-saturation effect at nearly rated load brings a large position estimation error of around -23.6 degrees. When i_q is reduced to zero to reduce the cross-saturation effect, the position estimation error is around 2.2 degrees only (Table V).

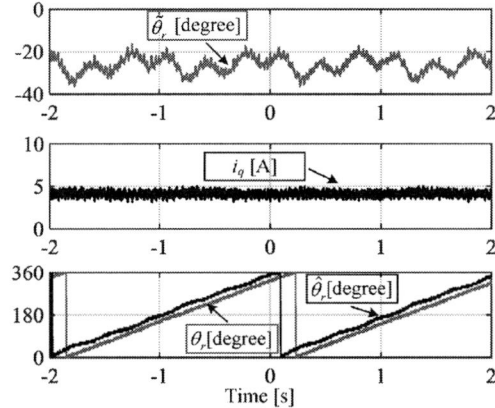

Fig. 9 q-axis current, estimated position error, real rotor position and estimated rotor position with two opposite voltage vectors injecting on the estimated d-axis at 15 rpm at full load condition.

TABLE V $\tilde{\theta}_r$ AT DIFFERENT LOAD CONDITIONS

i_q [A]	0	1	2	3	4
$\tilde{\theta}_r$ [degree]	2.2	-4.4	-11.2	-17.8	-23.6

IV. Conclusion

This paper investigated the factors that could affect the accuracy of the estimated position in pulse-injection-based sensorless methods. It has been observed that at low speed (lower than 60 rpm) the position estimation error caused by speed dependent back-EMF and inverter voltage error is quite small when the stator currents are low. This could be further minimized by injecting two opposite vectors on the estimated d-axis or q-axis. Meanwhile with different PWM patterns the inverter voltage error is different and thus the position estimation is different too. When choosing the amplitude of injection voltage, a compromise should be made between position estimation accuracy and current ripples. The estimated position error caused by cross-saturation effect is much larger than other factors and it is the main position error to be compensated. Appropriate compensation method is required in order to obtain the correct rotor position. If the chosen sensorless method itself could not reduce the influence of inverter voltage error, then the inverter voltage error and cross-saturation effects could be considered and compensated together.

It should be noticed that all the experiments in this paper are operated in steady state. The influences of these factors on position estimation accuracy in transient state require further investigations.

Appendix

The parameters of PMSM involved in these experiments are shown in Table VI.

Table VI

Rated power [W]	400	Stator resistance [Ω]	2.3
Max. phase voltage [V]	380	d-axis inductance [mH]	10
Rated current [A]	2.9	q-axis inductance [mH]	13
Rated speed [rpm]	2850	PM flux linkage [Wb]	0.12
Rated frequency [Hz]	95	Pole pairs	2

References

[1] A. Consoli, G. Scarcella and A. Testa "Industry application of zero-speed sensorless control techniques for PM synchronous motors," *IEEE Trans on Trans. Ind. Appl*, vol. 37, no. 2, pp. 513-521, 2001.

[2] P.P. Acarnley and J.F. Watson "Review of position-sensorless operation of brushless permanent-magnet machines," *IEEE Trans on Industral Electronics*, vol. 53, no. 2, pp. 352-362, 2006.

[3] Y. Zhao, C. Wei, Z. Zhang and W. Qiao, "A Review on Position/Speed Sensorless Control for Permanent-Magnet Synchronous Machine-Based Wind Energy Conversion Systems," *IEEE Journal of Emerging and Selected Topics in Power Electronics*, vol. 1, no. 4, pp. 203-216, 2013.

[4] J. M. Kim, S.J. Kang, S.K. Sul, "Vector control of interior permanent magnet synchronous motor without a shaft sensor," *Twelfth Annual Conference on Applied Power Electronics Conference and Exposition (APEC)*, vol. 2, pp. 743-748, 1997.

[5] M. W. Degner, R. D. Lorenz, "Using Multiple Saliencies for the Estimation of Flux, Position, and Velocity in AC Machines," *IEEE Transaction on Industry Applications*, vol. 34, no. 5, pp. 1097-1104, Sep/Oct 1998.

[6] M. J. Corley, R. D. Lorenz, "Rotor Position and Velocity Estimation for a Salient-Pole Permanent Magnet Synchronous Machine at Standstill and High Speeds," *IEEE Transaction on Industry Applications*, vol. 34, no. 4, pp. 784-789, July/Aug. 1998.

[7] M. Schroedl, "Sensorless Control of AC Machines at Low Speed and Standstill Based on the "INFORM" Method," *IEEE IAS Annual Meeting*, pp. 270-277, Oct 1996.

[8] Ge Xie, Kaiyuan Lu, Sanjeet Kumar Dwivedi et al. "Minimum Voltage Vector Injection Method for Sensorless Control of PMSM for Low-Speed Operations," *IEEE Transaction on Power Electronics*, vol. 31, no. 2, pp. 1785-1794, 2016.

[9] Piippo, A. and Luomi, J. "Inductance harmonics in permanent magnet synchronous motors and reduction of their effects in sensorless control." The XVII International Conference on Electric Machines (ICEM), no. 138, 2006.

[10] J. M. Guerrero, M. Leetmaa, F. Briz, A. Zamarron and R. D. Lorenz. "Inverter Nonlinearity Effects in High-Frequency Signal-Injection-Based Sensorless Control Methods," *IEEE Trans on Trans. Ind. Appl*, vol. 41, no. 2, pp. 618-626, 2005.

[11] C. Silva, G. M. Asher, and M. Sumner, "Influence of dead-time compensation on rotor position estimation in surface mounted PM machines using HF voltage injection," in *Proc. IEEE Power Conversion Conf.*, pp. 1279–1284, 2002.

[12] J.-W. Choi, "A new compensation strategy reducing voltage/current distortion in PWM VSI systems operating with low output voltages," *IEEE Trans. Ind. Appl.*, vol. 31, no. 5, pp. 1001–1008, 1995.

[13] Y. Inoue, K. Yamada, S. Morimoto, M. Sanada, "Effectiveness of Voltage Error Compensation and Parameter Identification for Model-Based Sensorless Control of IPMSM," *IEEE Trans on Trans. Ind. Appl*, vol. 45, no. 1, pp. 213-221, 2009.

[14] E. Capecchi, P. Guglielmi, M. Pastorelli, and A. Vagati, "Positionsensorless control of the transverse-laminated synchronous reluctance motor," *IEEE Trans. Ind. Appl.*, vol. 37, no. 6, pp. 1768–1776, 2001.

[15] Z. Q. Zhu, Y. Li, D. Howe, C. M. Bingham. "Compensation for Rotor Position Estimation Error due to Cross-Coupling Magnetic Saturation in Signal Injection Based Sensorless Control of PM Brushless AC Motors," *Proc. IEEE Electric Machines & Drives Conf.*, pp.208-213, 2007.

[16] S. Ebersberger, B. Piepenbreier "Identification of Differential Inductances of Permancet Magnet Synchronous Machines Using Test Current Signal Injection," *Proc. Symp. Power Electronics, Electrical Drives, Automation and Motion*, pp. 1342-1347, 2012.

[17] J.M.Liu, Z.Q.Zhu. "Novel Sensorless Control Strategy with Injection of High-Frequency Pulsating Carrier Signal into Stationary Reference Frame," *IEEE Transaction on Industry Applications*. vol. 50, no. 4, pp. 2574-2583, 2014.

[18] N. Teske, G. M. Asher, M. Sumner. K.J Bradley. "Suppression of saturation saliency effects for the sensorless position control of induction motor drives under loaded conditions," *IEEE Trans on Ind Electron*. vol. 47, no. 5, 2000.

[19] H. Wang, K. Lu, D. Wang and F. Blaabjerg. "Pulse-Injection-Based Sensorless Control Method with Improved Dynamic Current Response for PMSM," unpublished.

The 2018 International Power Electronics Conference

Position Sensorless Control of Switched Reluctance Motor using Estimated PWM Phase Voltage

Y. Nakazawa[1*], K. Ohyama [2], H. Fujii [3], H. Uehara [3] and Y. Hyakutake [3]

1 Department of Electrical and Information Engineering, National Institute of Technology, Akita College, Akita, Japan
2 Department of Electrical Engineering, Fukuoka Institute of Technology, Fukuoka, Japan
3 Meiwa Manufacturing Co., Ltd, Fukuoka, Japan
*E-mail: nakazawa@akita-nct.ac.jp

Abstract— **This paper verifies a position sensorles control using the estimated pwm phase voltage. pwm phase voltage estimator estimates pwm phase voltage from the voltage drop of DC link voltage, switching device and diode. In the position sensorless control, the position estimation table based on magnetization curves is used. This study omits phase voltage sensors and offers the low cost position sensorless control system.**

Keywords— Switched reluctance motor, phase voltage estimation, position sensorless, voltage pwm control.

I. INTRODUCTION

In recent years, with the increasing necessity for energy conservation, the demand for efficient motors which use a rare earth permanent magnet is increasing. However, there are extremely few deposits of rare earth elements, and the problems of price hike and stability of supply are matters of concern. As a solution of this problem, the switched reluctance motor (SRM) which does not use rare earth elements attracts attention. Since there is no winding in the rotor, SRM has the advantages of robustness, low cost, and high speed rotation. Moreover, since there is no permanent magnet, there is neither heat demagnetization nor the problem that permanent magnet is broken.

A rotor position information is needed to decide the excitation timing of SRM. Therefore, it will be necessary to attach rotor position sensors to a motor shaft, and may be disadvantageous about cost, reliability, and physical envelope.

The fundamental principle of sensorless control is that the rotor position can be estimated from phase voltage and phase current or flux linkage, inductance. Until now, the various sensorless control methods using flux linkage are proposed [1-7].

The position sensorless control method based on flux linkage needs the information of phase voltage. The phase voltage can be measured directly from each phase using voltage sensors, or can be calculated from the switching state of the switching element in an inverter. Since the phase voltage of SRM is pulse voltage, the voltage sensor with sufficient frequency characteristic is required. And such the voltage sensor is very expensive. Therefore, we verified the position sensorles control using the estimated phase voltage in single pulse control [8].

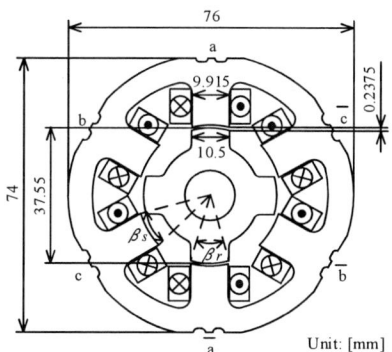

Fig. 1. Cross section of test SRM.

TABLE I
SPECIFICATIONS OF TEST SRM

Parameters	Values
Rated output power P_n	180 [W]
Rated voltage V_n	24 [V]
Number of coil turns	40 [turns/pole]
Winding resistance R	0.088 [Ω] @20°C
Core length	50 [mm]
Stator pole arc β_s	30.25 [°(mech.)]
Rotor pole arc β_r	32.43 [°(mech.)]

This paper estimates the pwm phase voltage from DC link voltage, the voltage drop of switching device and diode, and verify the Less & Less control which is the method applied to the position sensorles control. The Less & Less control needs one DC link voltage sensor and three phase current sensors, and can omits three phase voltage sensors. In the position sensorless control, the position estimation table based on magnetization curves is used. This study verifies its availability by coupled simulation of MATLAB Simulink and PSIM.

II. SPECIFICATIONS OF TEST SRM

Fig. 1 shows the cross section of test SRM and Table 1 shows the specifications. The test SRM is the double

1253

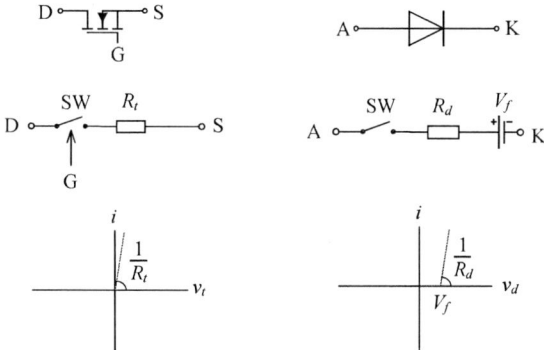

Fig. 2. MOS-FET and diode model used in pwm phase voltage estimator.

salient pole structure of stator 6 poles and rotor 4 poles. The winding wire which made 3 parallel conducting wire of 0.75 mm in diameter is coiled 20 T around a stator pole. The winding wire of pole U and pole   is connected in series. The rated output is 180 W and the rated voltage is 24 V. The origin position of the motor makes unaligned position 0°(mech.). The direction of rotation defines clockwise rotation as positive rotation.

III.   PHASE VOLTAGE ESTIMATION

A. MOS-FET and Diode Model

Fig. 2 shows MOS-FET and diode model which are used for the phase voltage estimator. The MOS-FET model consists of only switch SW and on-resistance R_t. The diode model consists of switch SW, on-resistance R_d, and forward voltage V_f.

B. Phase Voltage v_0 in Excitation Mode

Fig. 3 shows the asymmetrical half bridge inverter of only one phase in each modes. Table 2 shows the switching state corresponding to Fig. 3.

As shown in Fig. 3, when two switching devices is turned on, DC link voltage V_{dc} is applied to winding and the phase current i is flow. The phase voltage v_0 in the excitation mode is expressed by the following equation.

$$v_0 = V_{dc} - 2 \cdot v_t = V_{dc} - 2 \cdot R_t i \qquad (1)$$

where, V_{dc} is DC link voltage , v_t is the voltage drop of switching device, R_t is on resistance of switching device. The phase voltage v_0 in the excitation mode is smaller than the DC link voltage V_{dc} by the voltage drop $2R_t i$ of two switching elements of a high side and a low side.

C. Phase Voltage v_f in Freewheeling Mode

PWM control has soft switching and hard switching. Since hard switching turns on and off the switching devices of both a high side and a low side, the phase

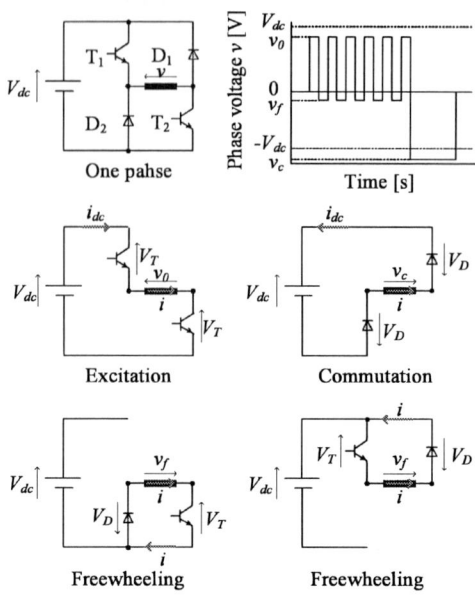

Fig. 3. Pwm phase voltage in each conduction modes.

TABLE II
SWITCHING STATES

Mode	T_1	D_1	T_2	D_2	v
Excitation	on	off	on	off	v_0
Freewheeling	off	off	on	on	v_f
Commutation	off	on	off	on	v_c

voltage in off state is the same as the phase voltage v_c of commutation mode. Soft switching turns on and off the one of switching devices. Generally, a high side switching device is switched as shown in Fig. 3. The phase voltage v_f in the freewheeling mode is expressed by the following equation.

$$v_f = -(v_t + v_d) = -\left\{R_t i + \left(R_d i + V_f\right)\right\} \qquad (2)$$

where, v_d is the voltage drop of diode, R_d is on-resistance, V_f is forward voltage.

D. Phase Voltage v_c in Commutation Mode

When two switching devices is turned off, since the direction of induced electromotive force is reversed, current flows to power supply through diode. The phase voltage v_c in commutation mode is expressed by the following equation.

$$v_c = -(V_{dc} + 2 \cdot v_d) = -\left\{V_{dc} + 2 \cdot \left(R_d i + V_f\right)\right\} \qquad (3)$$

The phase voltage v_c in the commutation mode is smaller than negative DC link voltage $-V_{dc}$ by the voltage drop $2v_d$ of two diodes of a high side and a low side.

The 2018 International Power Electronics Conference

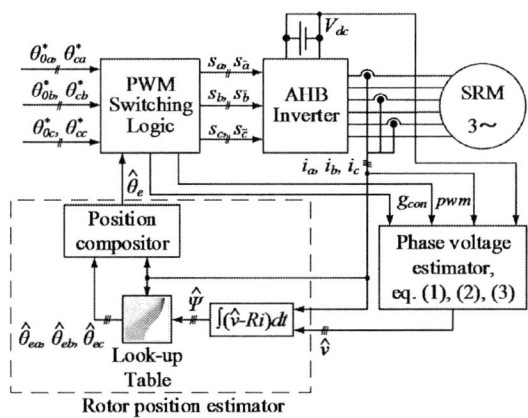

Fig. 4. Verified position sensorless system for pwm phase voltage estimator.

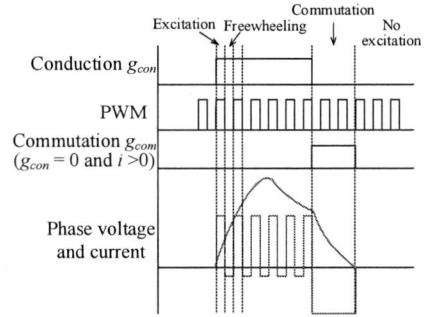

Fig. 5. Excitation signals on PWM control

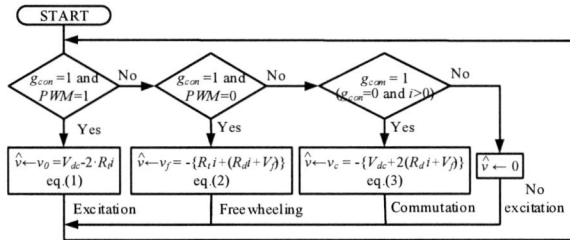

Fig. 6. The flowchart of phase voltage estimator.

E. Phase Voltage Estimator

Fig. 4 shows the position sensorles control system based on magnetization curves for verifying the phase voltage estimation method. The phase voltage estimator estimates the phase voltage in each excitation mode using equations (1) to (3).

Fig. 5 shows the conduction g_{con}, PWM, commutation signals g_{com}, phase voltage and current waveforms generated by the PWM switching logic in Fig. 4. When conduction $g_{con} = 1$ and PWM=1, the mode is excitation mode, when conduction $g_{con} = 1$ and PWM=0, the mode is freewheeling mode. And, when commutation $g_{com} = 1$, the mode is commutation mode. The commutation signal

TABLE III
PARAMETER FOR MOS-FET

Parameters	Inverter model	Phase voltage estimator
On resistance R_t	0.015 Ω	0.015 Ω
Internal diode inductance L_{on}	0 H	-
Internal diode resistance R_{td}	0.01 Ω	-
Internal diode forward voltage V_{tf}	1.6 V	-

TABLE IV
PARAMETER FOR DIODE

Parameters	Inverter model	Phase voltage estimator
On resistance R_d	0.18 Ω	0.18 Ω
Inductance L_d	0 H	-
Forward voltage V_f	1.8 V	1.8 V

TABLE V
SIMULATION CONDITION

Software	
AHB Inverter	PSIM
Other models	MATLAB Simulink
Calculation step	10μs
SR motor	
Power supply voltage V_{dc}	24V
Reference turn-on angle θ_0^*	0°(elec.)
Reference commutation angle θ_c^*	120°(elec.)
Voltage PWM control	High side soft switching
Carrier frequency	10kHz
Duty ratio	0.5
Drive circuit	AHB Inverter
Servo motor for load	
Speed control	1000min⁻¹, 4000min⁻¹

g_{com} is generated by the phase voltage estimator so that it becomes $g_{com} = 1$ when the conduction signal $g_{con} = 0$ and the phase current $i > 0$.

From this relationship, as shown in the flowchart of Fig. 6, the estimated phase voltage in each excitation mode v_0, v_f, v_c is selected using these signals and output from the phase voltage estimator as the estimated phase voltage $\hat{v}$.

F. Simulation Result of Estimated Phase Voltage $\hat{v}$

Tables 3 and 4 show the parameter of MOS-FET and diode used for the inverter model and the phase voltage estimator. The value of the parameter is the value of the data sheet of each element. Inductance which is not described to a data sheet was set to 0 H.

Table 5 shows simulation conditions. As excitation timing is shown in Fig. 4, reference turn-on angle θ_0^* is

1255

The 2018 International Power Electronics Conference

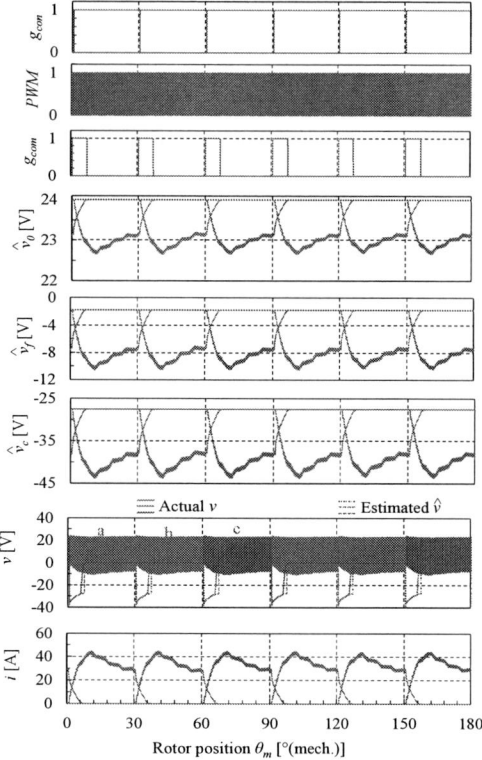

Fig. 7. The estimated pwm phase voltage $\hat{v}$ in 1,000 min^{-1}, 1.15 N·m.

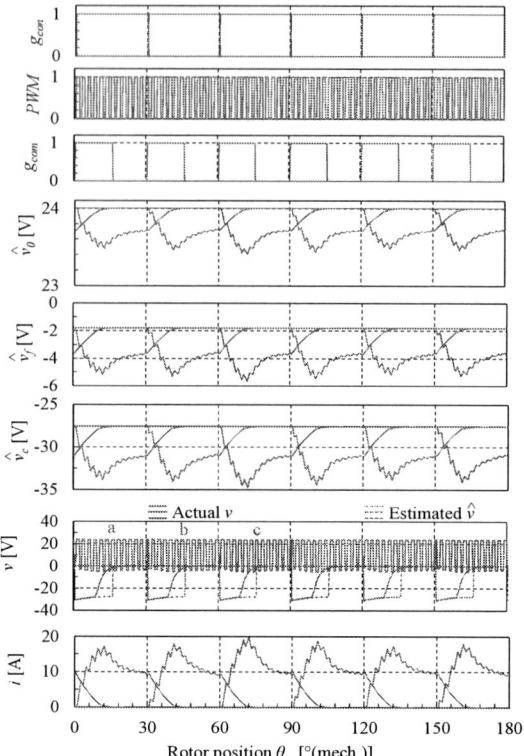

Fig. 8. The estimated pwm phase voltage $\hat{v}$ in 4,000 min^{-1}, 0.18 N·m.

0°. And, reference commutation angle θ_c^* is 120°. The carrier frequency of PWM generator is 10 kHz, duty ratio is 0.5, and high side soft switching by PWM voltage control is carried out. DC link voltage V_{dc} is the rated voltage 24V. The result of speed control by the servo motor which is load is shown.

Fig. 7 shows actual phase voltage v and estimated phase voltage $\hat{v}$ at the 1,000 min^{-1} of rotational speed and 1.15 N·m of load torque. The estimated phase voltage $\hat{v}$ in excitation period is well in agreement with actual phase voltage v. The estimated phase voltage $\hat{v}$ in commutation period is well in agreement with actual phase voltage v to the middle.

Fig. 8 shows actual phase voltage v and estimated phase voltage $\hat{v}$ at 4,000 min^{-1} of rotational speed and 0.18 N·m of load torque. The estimated phase voltage $\hat{v}$ in excitation period are well in agreement with the actual phase voltage v as well as the case of 1,000 min^{-1}. The position estimation error is larger than the case of 1,000 min^{-1} in commutation period. Since the rotational speed is high, large induced electromotive force is produced in SR motor coil. Therefore, large position estimation error is caused by the long time of commutation period.

IV. ROTOR POSITION ESTIMATION

A. Rotor Position Estimaor

The estimated rotor position is calculated by referring to the rotor position estimation table from phase current detected value and flux linkage calculation value.

The flux linkage $\hat{\psi}$ used by table reference is calculated by the following equation using estimated phase voltage $\hat{v}$ and phase current i.

$$\hat{\psi}(\theta, i) = \int (\hat{v} - R \cdot i)\, dt \qquad (4)$$

where, R is the average of DC resistance of winding in each phase. Since the rotor position estimation method cannot be estimated rotor position when phase current does not flow, the "Position compositor" block of Fig. 4 generates the rotor position estimation waveform up to 360° by changing the phase used for rotor position estimation.

B. Rotor Position Estimation Table

The rotor position estimation method uses the rotor position obtained using the two-dimensional static magnetic field analysis of the finite-element-method software ANSYS and the flux linkage table which outputs the value of flux linkage to phase current. The influence of the magnetic saturation of an iron core and a shaft is taken

1256

The 2018 International Power Electronics Conference

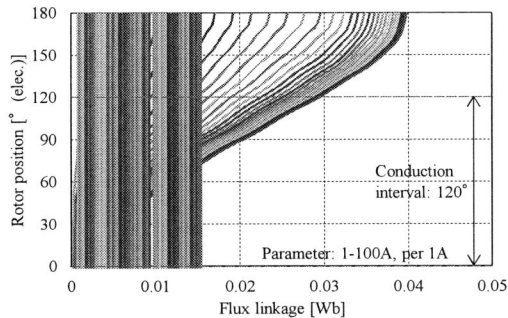

Fig. 9. Rotor position estimate table.

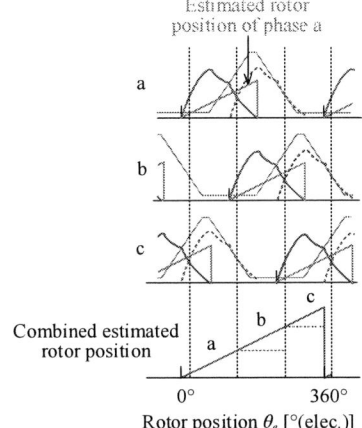

Fig. 10 Position compositor.

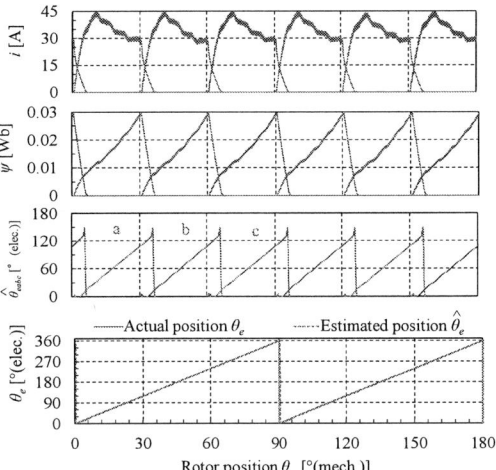

Fig. 11. The estimated rotor position $\hat{\theta}_e$
in 1,000 min⁻¹, 1.15 N·m.

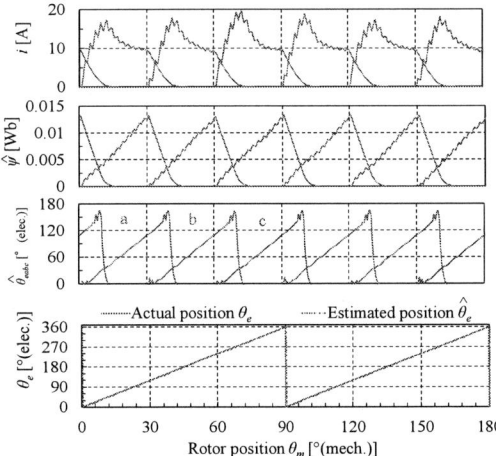

Fig. 12. The estimated rotor position $\hat{\theta}_e$
in 4,000 min⁻¹, 0.18 N·m.

into consideration in the two-dimensional static magnetic field analysis.

The approximate expression for obtaining the rotor position from the flux linkage to each phase current is derived from the flux linkage table by least-squares method. The rotor position estimation table which outputs the rotor position is obtained from phase current and flux linkage obtained by integrating induced electromotive force with respect to time using the approximate expression. Fig. 9 shows the rotor position estimation table. The sensorles control can be carried out by using the estimated rotor position outputted from the rotor position estimation table.

C. Rotor Position Compositor

Fig. 10 shows the composition method of estimated rotor positions of each phase in the "Position compositor" block in Fig. 4.

Here, when reference rotor position is phase a, each phase difference of phase b and phase c is 120, 240° to phase a. Therefore, 120° and 240° are added to the estimated rotor position of phase b and phase c to obtain the combined position estimation waveform. Since the value of the flux linkage to each current is close in from an unaligned angle to overlap angle, the position estimation error is large. So, the "Position compositor" block chooses

the estimated rotor position of the phase into which large current is flowing.

D. Simulation Result of Estimated Rotor Position $\hat{\theta}_e$

Fig. 11 shows the actual rotor position θ_e and the estimated rotor position $\hat{\theta}_e$ when SRM is driven using the rotor position estimator in 1,000 min⁻¹. Although the ripple is looked at the combined estimated rotor position $\hat{\theta}_e$ at the change of the each phase estimated rotor position $\hat{\theta}_{ea}$, $\hat{\theta}_{eb}$, $\hat{\theta}_{ec}$, the estimated rotor position $\hat{\theta}_e$ is the almost same profile as the actual rotor position $\hat{\theta}_e$.

Fig. 12 shows the actual rotor position θ_e and the estimated rotor position $\hat{\theta}_e$ when SRM is driven using the rotor position estimator in 4,000 min⁻¹. As well as 1,000 min⁻¹, although the ripple is looked at the combined estimated rotor position $\hat{\theta}_e$ at the change of the each phase

1257

estimated rotor position $\hat{\theta}_{ea}$, $\hat{\theta}_{eb}$, $\hat{\theta}_{ec}$, there is no influence by operating points.

V. CONCLUSIONS

This paper verified the position sensorles control using the estimated pwm phase voltage. The position sensorles control using the estimated pwm phase voltage enables sensorles drive on various operating points.

The phase voltage v_f in freewheeling only require the information of the on-resistance R_t of MOS-FET, the on-resistance R_d and forward voltage V_f of diode, and it can be easily estimated from phase current i .

In the position sensorless control using magnetization curves, the information of phase voltage v , phase current i , and flux linkage ψ is mainly needed in the excitation period. The error of the estimated phase voltage in the latter half of commutation period does not influence the estimated rotor position. Therefore, the Less & Less control offers lower cost SRM drive system.

REFERENCES

[1] J. P. Lyons, S. R. MacMinn, and M. A. Preston, "Flux-current methods for SRM rotor position estimation," in Industry Applications Society Annual Meeting, 1991, vol.1, pp. 482-487.

[2] W. F. Ray and I. H. Al-Bahadly, "A sensorless method for determining rotor position for switched reluctance motors," in Power Electronics and Variable-Speed Drives, 1994, pp. 13-17.

[3] A. Cheok and N. Ertugrul, "A model free fuzzy logic based rotor position sensorless switched reluctance motor drives," in Industry Applications Conference, Thirty-First IAS Annual Meeting, IAS '96., 1996, vol.1, pp. 76-83.

[4] X. Longya and B. Jianrong, "Position transducerless control of a switched reluctance motor using minimum magnetizing input," in Industry Applications Conference, Thirty-Second IAS Annual Meeting, IAS '97., 1997, vol.1, pp. 533-539.

[5] G. Gallegos-Lopez, P. C. Kjaer, and T. J. E. Miller, "High-grade position estimation for SRM drives using flux linkage/current correction model," in Industry Applications Conference, Thirty-Third IAS Annual Meeting, 1998, vol.1, pp. 731-738.

[6] G. Gallegos-Lopez, P. C. Kjaer, and T. J. E. Miller, "High-grade position estimation for SRM drives using flux linkage/current correction model," Industry Applications, IEEE Transactions on, vol. 35, pp. 859-869, 1999.

[7] Z. Wenyu, L. Chuang, Z. Qiang, C. Jun, and Z. Lei, "A new flux/current method for SRM rotor position estimation," in Electrical Machines and Systems, ICEMS, 2009, pp. 1-6.

[8] Y. Nakazawa, K. Ohyama, H. Fujii, H. Uehara and Y. Hyakutake, "Phase Voltage Estimation for Position Sensorless Control of Switched Reluctance Motor," Electrical Machines and Systems (ICEMS), 2016 19th International Conference on, 2016, pp. 1-4.

The 2018 International Power Electronics Conference

Experimental Confirmation of Thrust and Attractive Force Control of Linear Induction Motor by Two Different Frequency Components

Kenta Sannomiya[1*], Toshimitsu Morizane[1], Noriyuki Kimura[1] and Hideki Omori[1]
1 Electric engineering, Osaka Institute of Technology, Osaka, Japan
*E-mail: m1m17318@st.oit.ac.jp

Abstract— This paper presents a new Maglev control system whereby the thrust and attractive force are controlled simultaneously and independently of the linear induction motor (LIM). It is proposed that two different frequency components be used to achieve the simultaneous and independent control of the thrust and attractive force of the LIM. One of the frequency components drives the LIM with a slip (f_d) and generates a thrust and attractive force. The other frequency component synchronizes the motor speed (f_m) and generates only attractive force. Therefore, this proposed control method simplifies the controller. The controller for thrust and attractive force can be designed separately and independently. In previous research, each controller has PI control and we verify the control performance. Especially the control of attractive force is not sufficient because the gap between the secondary and primary side is constant. In this paper we propose the control coefficient is tuned according to the gap. The mutual relation between the current, gap and the attractive force was confirmed by experiments.

Keywords— Maglev, Motor control, Linear indution motor

I. INTRODUCTION

Linear motors and Electro Magnetic Suspension (EMS) are used in the maglev system. Linear Induction motors (LIMs) have simple structure, flexible mechanism, and direct drive [1][2]. High Speed Surface Transport (HSST) system is an example of a maglev system. In this system, traction is controlled by LIM, and levitation is controlled by EMS as shown in Fig.1 (a) [4]-[7]. On the other hand, LIM generates also attractive force. Therefore disadvantage of this system is that the attractive force generated by LIM disturb the levitation control. In addition, the EMS generates eddy current brake while driving. Therefore, we have proposed the maglev system using only LIMs as shown in Fig.1(b) [3]. In this system, LIMs generates not only the thrust force but also the attractive force. Attractive force generated from LIM is effectively used as levitation force. Therefore, the proposed system is expected to have a low cost, a compact structure and easy maintenance since levitation magnets are unnecessary.

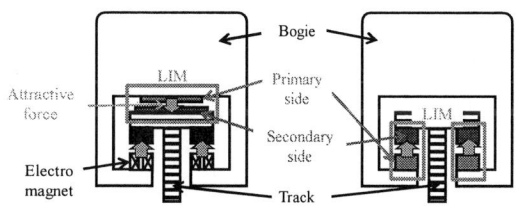

(a) Typical maglev system   (b) Proposed system
Fig1. Typical system construction and proposed system construction

II. CONTROL METHOD OF SUPERIMPOSED FREQUENCY COMPONENT

Our proposed method is that control the thrust force and the attractive force generated from the LIM with the superimposed frequency component simultaneously and independently. Fig.2 shows the sum of thrust force and attractive force generated from two different frequency components. F_{th} and F_{ah} are generated by the higher frequency component f_h. F_{tl} and F_{al} are generated by the lower frequency component. F_a and F_t are the total forces respectively.

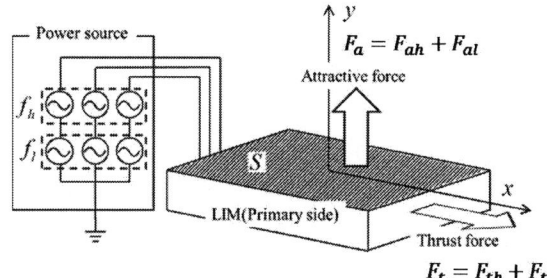

Fig2. Control System of the LIM with two different frequency

The thrust force F_t and attractive force F_a are expressed by the Maxwell stress tensor in (1), (2).

$$F_t = \frac{1}{\mu}\int_S B_x B_y \, ds \tag{1}$$

$$F_a = \frac{1}{2\mu}\int_S (B_y{}^2 - B_x{}^2)\,ds$$
$$= \frac{1}{2\mu}\int_S B_y{}^2\,ds - \frac{1}{2\mu}\int_S B_x{}^2\,ds \tag{2}$$

where, B_x, B_y are the flux densities of x and y-axis respectively, μ is magnetic permeability of the iron core of secondary side, and S is the top surface of primary side. From (2), the total attractive force F_a can be separated into F_{ax} and F_{ax} as described in (3), (4) respectively.

$$F_{ax} = \frac{1}{2\mu} \int_S B_y^2 ds \tag{3}$$

$$F_{ay} = \frac{1}{2\mu} \int_S B_x^2 ds \tag{4}$$

Each flux density of x-axis and y-axis consist of the two frequency components. The flux density of y-axis $B_{yl}(t)$ with low frequency component (f_l) is defined as (5), and the flux density of y-axis $B_{yh}(t)$ with high frequency component (f_h) is defined as (6).

$$B_{yl}(t) = B_{yl} \cdot \sin(2\pi f_l \cdot t + \varphi_{yl}) \tag{5}$$

$$B_{yh}(t) = B_{yh} \cdot \sin(2\pi f_h \cdot t + \varphi_{yh}) \tag{6}$$

Where, B_{yl} is the amplitude of $B_{yl}(t)$, B_{yh} is the amplitude of $B_{yh}(t)$. φ_{yl}, is the phase angular of $B_{yl}(t)$ and φ_{yh} is the phase angular of $B_{yh}(t)$.

Two magnetic flux densities can be added. The total magnetic flux density of the y-axis is expressed as (7).

$$B_y(t) = B_{yl}(t) + B_{yh}(t) \tag{7}$$

The time average of the attractive force $\overline{F_{ay}}$ generated by B_{yl} and B_{yh} is calculated as in (8)

$$
\begin{aligned}
\overline{F_{ay}} &= \frac{1}{T} \int_0^T \frac{1}{2\mu} B_y^2(t) dt \\
&= \frac{1}{T} \int_0^T \frac{1}{2\mu} \left\{ B_{yl}(t) + B_{yh}(t) \right\}^2 dt \\
&= \frac{1}{4\mu} \left\{ B_{yl}^2 + B_{yh}^2 \right\} \\
&= \overline{F_{ayl}} + \overline{F_{ayh}}
\end{aligned} \tag{8}
$$

where F_{ayl} is the time average force of the lower frequency component f_l, F_{ayh} is the time average force of the higher frequency component f_h, and T is time to calculate the average.

In case of T it is assumed to be long enough to calculate the time average, the time average value of "$2B_{yl}(t) \times B_{yh}(t)$" is almost zero. Therefore it can be added the force of LIM generated by superimposed frequency components.

The proposed controller is shown in Fig.3. It has two different frequency components. f_m is synchronous with motor rotational speed, f_m corresponds to f_l. f_d is drive frequency component, f_d corresponds to f_h. When the LIM is driven by the frequency component f_d, LIM generates the thrust and attractive force. When the LIM is driven by the frequency component f_m, the LIM generates only attractive force because slip is zero.

In the proposed control method, it is possible to adjust the total thrust force by controlling only f_d. However, the attractive force generated by the drive frequency

component f_d can't control total attractive force. Therefore, the frequency component f_m adjusts the total attractive force F_a.

Controllers that control each force can be designed independently. Therefore, ordinary vector control can be applied. In addition, this system has other merits. In the typical levitation system, EMS generates eddy current brake.

On the other hand, in the proposed control system, the frequency component synchronous with motor speed f_m doesn't generate the brake force at all. The thrust force of LIM can be effectively used for propulsion.

Fig. 4 shows the control diagram designed by our proposed method. In thrust force controller, ordinary vector control is applied. Attractive force controller includes d-q axis currents, adjust d-axis current to control attractive force. PI control is used for thrust force controller, attractive force controller and summed current control.

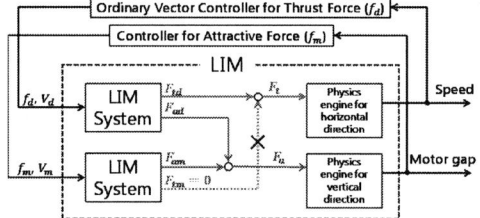

Fig3. Control system of LIM with two different frequency components

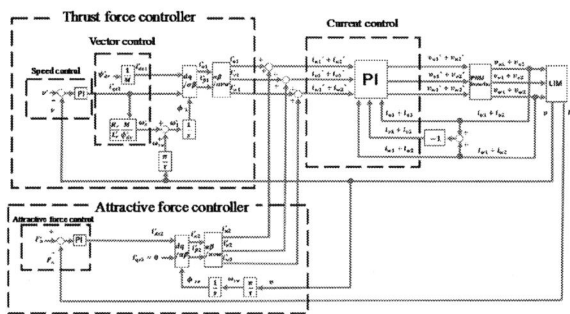

Fig.4 Control diagram of speed and attractive force of LIM

III. EXPERIMENTAL SYSTEM FOR VARIFICATION

We use a LIM with a disk-shaped secondary side to verify the dynamic characteristics. It is possible to measure characteristics of the LIM over a longer time span by using a disk-shaped secondary side. The whole experimental system is shown in Fig. 5. We provide the torque to the secondary side via the timing belt with an additional rotary assist induction motor. It can control the rotational speed of the secondary side to a synchronous speed for the no-load test to identify the LIM parameters. In addition, it is possible to control the load torque with the secondary side in order to measure the power running and regenerative characteristics of the LIM. The load cells are implemented on the primary side as shown in Fig. 5 and measure the thrust and attractive forces directly.

TABLE 1 shows the parameters of primary side and secondary side. The primary side has 4 poles and it is used as a stator. The disk-shaped secondary side consists of an iron core and an aluminum reaction plate. It is used as a rotor. The secondary side rotates and the number of revolutions is measured by the Rotary encoder (OMRON. E6C2-CWZ6C) on the shaft under the secondary disc as shown Fig.6.

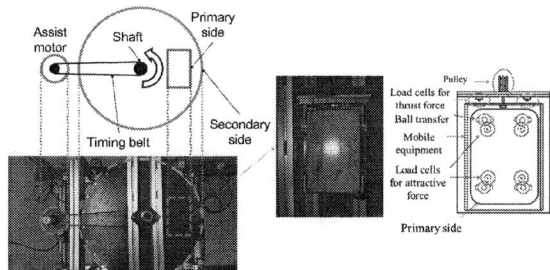

Fig5. Whole experimental system and detail of primary side

Fig. 6. Rotary encoder

When the LIM parameters are identified under the no-load test, the rotational speed of secondary side has to be controlled to the synchronous speed.

TABLE 1 Parameters of the LIM

Primary side		
Size[mm]	230[L]×150[W]×45[H]	
Weight[kg]	7.4	
Pole pitch[mm]	45	
Pole number	4	
Nominal voltage[V]	125	
Nominal current[A]	125VAC	24VDC
	5	3
Secondary side		
Diameter[mm]	700	
Thickness[mm]	Alminum	Iron
	2.0	2.3
Gap[mm]	Mechanical	Magnetic
	9.0	11.0

IV. EXPERIMENTAL RESULTS AND DISCUDDION

The control characteristics of speed and attractive force

of LIM have been confirmed by simulations. In this paper, we examined the control performance of the proposed control method where the two different frequency components are controlled under the condition of air gap 9mm. Fig. 7 is the result of the signal from the load cell for the attractive force under the condition that the load torque is controlled to 0.0214 Nm by the assist motor in a counter direction to the secondary side rotation. It corresponds to 0.102 N of the thrust force of the LIM. The speed of the LIM is controlled to 1.0 m/s in constant. The frequency and voltage of the inverter are 60Hz and 200V respectively. It is measured with an oscilloscope (DL6054, Yokogawa electric machinery Corp). The sampling rate of the oscilloscope is measured at 0.8ms. As shown in Fig.7, the attractive force was vibrated. The vibration results are shown in TABLE 1. According to these results, it was found that the large vibration is synchronous with the shaking of the secondary side, and the small vibration is caused by the slip frequency. In order to confirm the control performance, we have to measure the forces for a longer duration. It is hard to measure the forces for a long duration with an oscilloscope. We use the data logger installed in the PE-EXPART III. Unfortunately, the sampling rate of the PE-EXPART III is measured every 0.5s. Fig.8 shows the same experimental result measured with the data logger of the PE-EXPART III and as the data demonstrates the aliasing occurs. Consequently, the amplitude of the attractive force appears to not be constant. Therefore, we confirm the control performance by the average value of the forces in this paper.

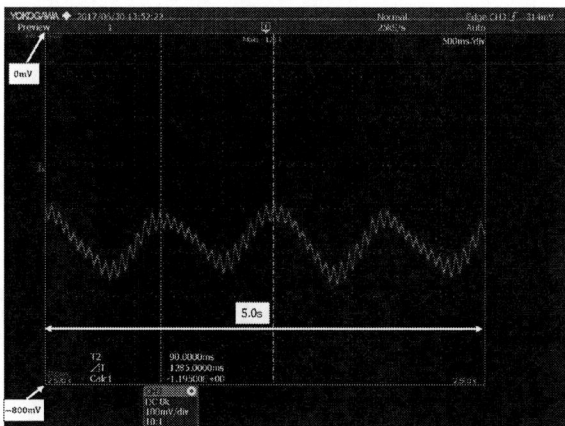

Fig. 7. Reaction of attractive force in constant speed with the oscilloscope (horizontal: 500ms/div, vertical: 100mV/div)

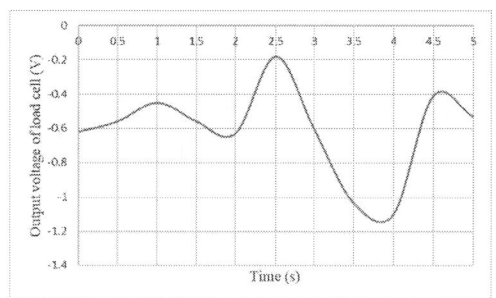

Fig.8. Reaction of attractive force in constant speed with data logger

TABLE 2. Comparison of vibration

Large vibration [Hz]	Shaking of secondary side [Hz]
0.778	0.76
Small vibration[Hz]	Slip frequency [Hz]
11.6	10.5

We examined the control characteristics in two patterns. Experiment1 is the control by only the frequency component f_d. This experiment is the verification of the speed control. Experiment2 is the control by the frequency components f_d and f_m. This experiment is the verification of the speed and attractive force of LIM. These experiments have a same speed profile of LIM as shown in Fig.9. In the two cases, from 0s to 25s, the reference of the velocity is 0 m/s. From 25s to 50s, it accelerates to 1.0 (m/s)/s at a constant rate. From 50s to 70s, it has a constant speed. From 70s to 95s, it decelerates to 1.0 (m/s)/s by the regenerative brake at a constant rate. The load torque is kept to 0.0214 Nm in a counter direction to the secondary side rotation from 0s to 120s.

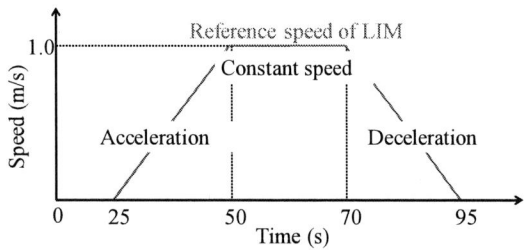

Fig.9. Reference speed profile

Fig.10 and Fig.11 shows the results of speed and thrust force under the experiment1 and experiment2. Fig.10 indicates that the speed control is achieved. As a result, attractive force controller doesn't affect the thrust force controller because both waveforms are almost same.

Fig.12 shows the results of attractive force under the experiment1 and experiment2. In the experiment1, the controller is not able to control the attractive force because this controller has only one frequency component f_d. But in the experiment2, appropriate attractive force can be generated even when the speed is 0.

In Fig.11, the currents of the LIM remain even when the speed is 0m/s from 95s to 120s, because the load torque continues to be provided. Additionally, the attractive force and thrust force generated by the LIM are effectively controlled. It is clear that the controller for each force are able to be controlled simultaneously and independently. However, the controllers could not control the thrust force adequately from 0s to 25s because of the transient phenomena.

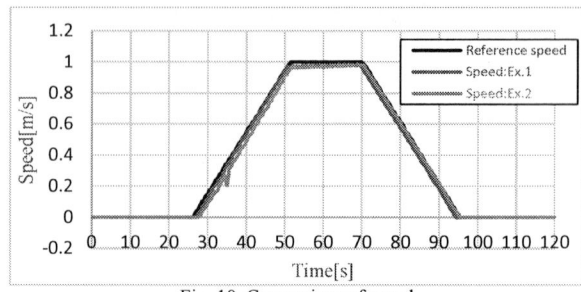

Fig. 10. Comparison of speed

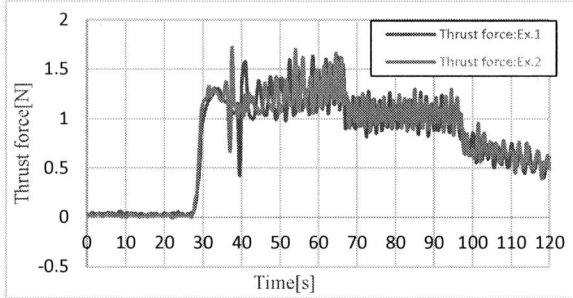

Fig. 11. Comparison of thrust force

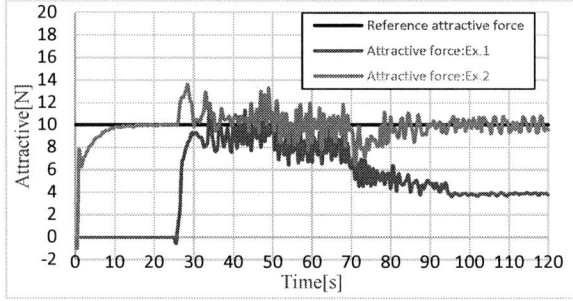

Fig. 12. Comparison of attractive force

Fig.13 shows the mutual relation between the current and the attractive force obtained the air gap was changed. The required current is changed by the length of the gap. We have to tune the control coefficient according to the gap.

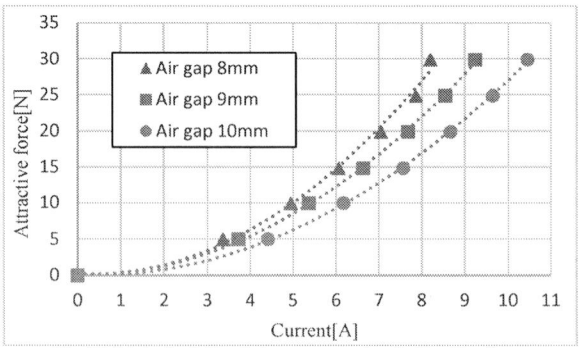

Fig.13. Mutual relation between current and attractive force

V. CONCLISION

We implemented proposed control system by superimposed frequency component to control thrust force and attractive force simultaneously and independently in maglev system. One of the frequency is the drive frequency component f_d and the other is synchronous with the motor speed frequency component f_m. A disk-shaped secondary side was used to verify the control performance of the experimental equipment. The results present study show that speed control and attractive force control can be controlled simultaneously by superimposed frequency.

In this paper, levitation is not controlled and only the attractive force is controlled under the condition that the gap is constant. Further studies are needed in order to levitate the primary side.

In the future, we will install gap sensor on the primary side of LIM and control levitation and speed. We have to tune the control coefficient according to the gap.

REFERENCES

[1] E.Masada, T.Kitano, T.Mizuma and S.Fujiwara.: Recent development in practical application of linear motor cars, T.IEE Japan, Vol.110-D, No.1, 1990, pp.2-13

[2] N.Fujii, T.Harada, Y.Sakamoto, and T.Kayasuga, Compensation method for end effect of linear induction motor, Trans. IEE Japan, Vol.122-D, No.4, 2002, pp.330-337.S. Jacobs and C.P. Bean, "Fine particles, thin films and exchange anisotropy," *in Magnetism*, vol. III, G.T. Rado and H. Suhl, Eds. New York: Academic, 1963, pp. 271-350.

[3] H. Nagano, "Electromagnetic suspension system, HSST" Railway Electrical engineering Association of Japan, Vol.18, No.7, 2007, pp.37-40.

[4] E.Masada, Linear drive technology and application, Ohmsha, 1991, pp.146-148.

[5] The Magnetic Actuator Technical Committee of The Institute of Electrical Engineers of Japan, eds. Linear Motor and Their Applications, IEE Japan, 1991, pp.74-78.

[6] M. Morishita and H. Itoh, "The Self-gap-detecting Zero Power Controlled Electromagnetic Suspension System"; IEEJ Trans. IA, Vol.126, No.12, 2006, pp.1667-1677.

[7] M. Morishita and M. Akashi, "Guide-effective Levitation Control for Electromagnetic Suspension System", Trans. IEE Japan, Vol.119-D, No.10, 1999, pp.1259-1268.

[8] T. Morizane, K. Taniguchi, and N. Kimura, "Characteristics of attractive force of linear induction motor in a novel maglev system driven by the source including high frequency component", in Proc. LDIA 2003, ML07 Birmingham, 2003.

[9] I. Takahashi and Y. Ide: "Decoupling Control of Thrust and Attractive Force of a LIM Using a Space Vector Control Inverter", IEEE Trans. IA, Vol.29, No.1, pp.161-167 (1993)

[10] Kaoru Iwaki, Toshimitsu Morizane, Noriyuki Kimura, and Katsunori Taniguchi "Characteristics of forces of Linear Induction Motor driven by power source including frequency component synchronous with the motor speed", in Proc. ICEMS 2009, 2009, DS1G6-1.

[11] Toshimitsu Morizane, Kosuke Tsujikawa, and Noriyuki Kimura "The measurement of the dynamic characteristics of LIM with experimental equipment using disc-shaped secondary side" in Proc. LDIA2011, LIM-II.4, 2011

[12] Yasuhiro Kotani, Toshimitsu Morizane, Kosuke Tsuzikawa, Noriyuki Kimura, Hideki Omori, "Simulation population and levitation control of linear induction motor in maglev system driven by power source with frequency component synchronous with motor speed," EPE 2013

GA Based Optimized Trajectories of Rotating Speed and *d-q* Axis Currents for an IPMSM

Shuta Kumagai, Kaoru Inoue*, and Toshiji Kato
Department of Electrical Engineering, Doshisha University
Kyotanabe, Kyoto 6100321, JAPAN
*E-mail: kaoinoue@mail.doshisha.ac.jp

Abstract— Interior Permanent Magnet Synchronous Motors (IPMSMs) used in the electrical machineries and apparatuses are often driven under given drive conditions such as operation time, rotating speed, and rotational angle. Energy loss during the variable speed motor operation satisfying the given drive conditions depends on the waveforms of motor currents in *d-q* axis and rotating speed (called as trajectories in this paper). In order to reduce the loss energy of IPMSMs during their variable speed operation satisfying given drive conditions, all loss of IPMSMs should be taken into account. This paper proposes a numerical design methodology of the optimal trajectories of IPMSMs drive systems based on a 3-dimensional loss map in order to reduce loss energy when the operation time, rotating speed and rotational angle are given as drive conditions. The evaluation functions are defined from the drive conditions and the loss map, then the optimal trajectories are derived by using a Genetic Algorism (GA).

Keywords— *IPMSM, Optimial Trajectory, Genetic Algorizm (GA), Energy Saving.*

I. INTRODUCTION

Interior Permanent Magnet Synchronous Motors (IPMSMs) are widely used in the electrical machineries and apparatuses. They are often driven under given drive conditions such as operation time, rotating speed, and rotational angle of motor. Energy loss during the variable speed motor operation satisfying given drive conditions depends on the waveforms of motor currents and rotating speed (called as trajectories in this paper). In order to reduce the loss energy of IPMSMs during its variable speed operation satisfying given drive conditions, all loss of IPMSMs should be taken into account. Various control methods of currents in *d-q* axis and current phase have been reported for efficient drive [1, 2, 3, 4]. These methods focus on the electrical losses. Mechanical and other un-modeled loss should be taken into account.

For squirrel cage induction motors (IMs), the optimal torque and rotating speed trajectories for energy saving have been proposed by using variational method under two drive conditions with respect to operation time and rotating speed [5, 6]. In the reported analytical methods, the electrical copper loss and mechanical rotational damping loss are taken into account. However, iron and un-modeled loss are neglected. A numerical design methodology has been proposed in order to derive optimal trajectories

minimizing energy loss based on a loss map [7]. The loss map will be measured experimentally from the electrical input and the mechanical output power for various motor torque and rotating speed. Hence, all loss of the motor are included in the map. The reported optimal trajectories assure that the rotating speed reaches its desired objective speed at the objective time. Analytical and numerical design approach to derive the optimal trajectories satisfying given drive conditions are important issue for IPMSMs energy saving.

On these stand points, this paper will derive optimal trajectories, such as motor currents in *d-q* axis and rotational speed, of IPMSMs based on a 3-dimensional loss map in order to reduce loss energy under given operation time, objective rotating speed and objective rotational angle. The 3-dimensional loss map is composed by motor currents in *d-q* axis and rotating speed. In this study, the loss map is obtained from the mathematical equations of IPMSMs because this paper proposes a methodology. Experimental measurement of loss map is under preparation. The evaluation functions are defined from the drive conditions and the loss map, then the optimal trajectories are derived by using a Genetic Algorism (GA).

II. DESIGN METHODOLOGY OF OPTIMAL TRAJECTORIES

A. Loss of IPMSM

A mathematical rotor dynamics of IPMSM is shown by

$$J_m \dot{\omega} + \xi \omega + T_L = T \tag{1}$$

where J_m, ξ, ω, T_L, and T are the inertia of the rotor, rotational damping coefficient, rotating speed, load torque, and motor torque induced by the electrical part of IPMSM, respectively.

Figure 1 illustrates the equivalent circuits transformed from three-phase to *d-q* plane [2]. Nomenclatures R_a, R_c, i_d, i_q, i_{od}, i_{oq}, i_{cd}, i_{cq}, Ψ_a, L_d, and L_q represent armature resistance, equivalent iron loss resistance, *d* and *q* axis

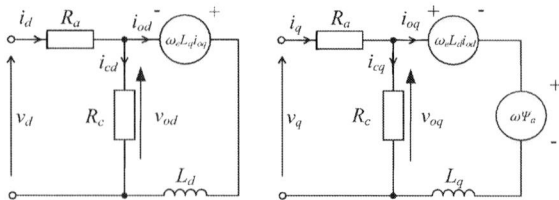

Figure 1 Equivalent circuits of IPMSM in *d-q* plane.

currents, d and q axis currents subtracted iron loss currents i_{cd}, i_{cq}, d and q axis iron loss currents flowing into R_c, flux linkage by permanent magnet, and d, q axis inductance. The d axis current is controlled in negative value to use the reluctance torque. When the number of pole pairs is represented by P_n, electrical rotating speed ω_e is given as $\omega_e = P_n\omega$, and motor torque T is given by

$$T = p\{\Psi_a i_{oq} + (L_d - L_q)i_{od}i_{oq}\} \quad (2)$$

The analytical total loss W is represented by the sum of mechanical rotational damping loss W_m, copper loss W_c, and iron loss W_i and is given by

$$W_{loss} = W_c + W_i + W_m \quad (3)$$

where W_c, W_i, and W_m are shown as follows.

$$W_c = R_a\left(i_d^2 + i_q^2\right) \quad (4)$$

$$W_i = \frac{\omega_e^2\{(L_d i_{od} + \Psi_a)^2 + (L_q i_{oq})^2\}}{R_c} \quad (5)$$

$$W_m = D\omega^2 \quad (6)$$

B. Theoretically calculated loss map

Table I shows the parameters of IPMSM used in this study. Figure 2 shows the theoretically calculated loss map by using Eq. (3). This map has 68921 (41^3) loss data points and are obtained by the loss calculation in rotating speed axis from -240 to 240 rad/s with 12 rad/s division, motor current i_{od} from -10 to 10 A with 0.5 A division, and motor current i_{oq} from -15 to 15 A with 0.75 A division. The motor loss increases depends on the amplitude of currents, phase of currents and rotating speed because the electrical and mechanical loss increase, correspondingly.

C. Discretized system and drive conditions

In order to derive optimal trajectories numerically, the objective motor dynamics Eq. (1) is discretized with a sampling period Δt_s as follows;

$$\begin{aligned}\omega[i + 1] &= A_d\omega[i] + B_d T[i]\\ y[i] &= C_d\omega[i] + D_d T[i]\end{aligned} \quad (7)$$

where i=0, 1, 2, …, $A_d = e^{-\frac{\xi}{Jm}\Delta t_s}$, $B_d = \frac{1}{Jm}\int_0^{\Delta t_s} e^{-\frac{\xi}{Jm}\tau}d\tau$, $C_d = 1$, $D_d = 0$, and $T_L = 0$ for simplicity.

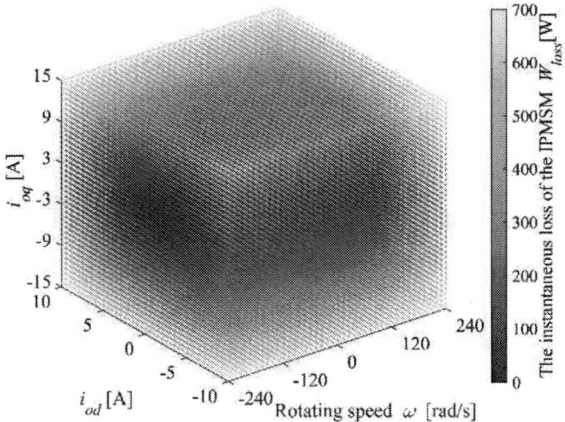

Figure 2 3-dimentional loss map.

TABLE I
MOTOR PARAMETER; MITSUBISHI ELECTRIC MM-EFS71M

Symbol	Meaning	Value
P_n	Number of pole pairs	3
R_a	Armature resistance	1.89 Ω
R_c	Equivalent iron-loss resistance	300 Ω
L_d	d-axis inductance	8.34×10^{-3} H
L_q	q-axis inductance	13.4×10^{-3} H
Ψ_a	Magnet flux-linkage	0.065 Wb
J_m	Inertia	2×10^{-3} kgm^2
ξ	Damping coefficient	4×10^{-4} Nms/rad

Let us define the motor currents in d-q axis and rotating speed trajectories in vector forms as

$$\boldsymbol{i_{od}} = [i_{od}[0] \quad i_{od}[1] \quad \cdots \quad i_{od}[n-1]]' \quad (8)$$

$$\boldsymbol{i_{oq}} = [i_{oq}[0] \quad i_{oq}[1] \quad \cdots \quad i_{oq}[n-1]]' \quad (9)$$

$$\boldsymbol{T} = [T[0] \quad T[1] \quad \cdots \quad T[n-1]]' \quad (10)$$

$$\boldsymbol{\omega} = [\omega[1] \quad \omega[2] \quad \cdots \quad \omega[n]]' \quad (11)$$

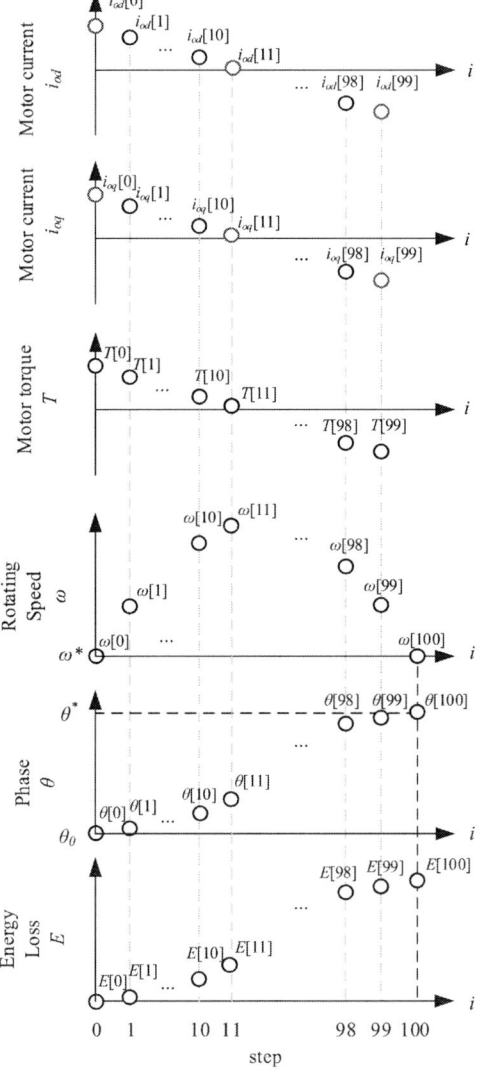

Figure 3 Trajectories to be optimized by GA.

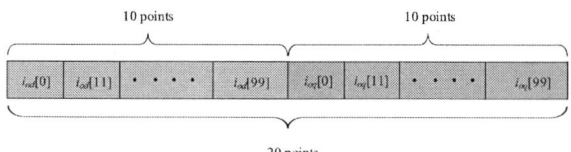

Figure 4 Gene of one individual.

where $[\cdots]'$ represents the transpose of vectors or matrices. Rotational angle is calculated as follows.

$$\theta[n] = \Delta t_s \sum_{i=0}^{n-1} \frac{\omega[i] + \omega[i+1]}{2} \qquad (12)$$

Then, we obtain

$$\boldsymbol{\omega} = \boldsymbol{\alpha}\omega[0] + \boldsymbol{\beta T} \qquad (13)$$

where

$$\boldsymbol{\alpha} = [A_d \quad A_d{}^2 \quad \cdots \quad A_d{}^n]' \qquad (14)$$

$$\boldsymbol{\beta} = \begin{bmatrix} B_d & 0 & \cdots & 0 \\ A_d B_d & B_d & \ddots & \vdots \\ \vdots & \vdots & \ddots & 0 \\ A_d{}^{n-1}B_d & A_d{}^{n-2}B_d & \cdots & B_d \end{bmatrix}. \qquad (15)$$

Drive conditions are given as; initial rotating speed $\omega_0(=\omega(t_0)=\omega[0])$ and rotational angle $\theta_0(=\theta(t_0)= \theta[0])$ at initial time $t=t_0$ (step $i=0$), and objective rotating speed ω^* and objective rotational angel θ^* at objective time $t=t_1$ (step $i=n$) as shown in Fig. 3. Final step n is now set 100, so each trajectories have 100 operation points. The motor is driven from the initial state to the objective one. There are no values of motor currents and torque at t_1 because the trajectories are derived based on a discretized equation numerically.

TABLE II
GA OPERATOR

Meaning	Value
Population size per island	500
Number of island	4
Number of elite per island	20
Selection method	size 2 tournament
Ratio generated by selection	76%
Migration interval	every 500 gen
Migration ratio of population	20%
Ratio generated by mutation	20%
Mutation method	adaptfeasible
Exit condition	100 gens

D. GA based optimization of trajectories

The aim of this study is to derive the optimal trajectories of (8), (9) and (11) satisfying the drive conditions and minimizing the energy loss.

The evaluation function J for optimization is composed by the drive conditions J_1 and J_2, and the loss from the loss map J_3 as

$$J = J_1 + J_2 + J_3. \qquad (16)$$

J should be minimized to obtain the optimal trajectories satisfying the given drive conditions and minimizing the energy loss during operation.

The evaluation function J_1 and J_2 show the given drive conditions and they are given by

$$J_1 = W_1(\omega^* - \omega[n])^2 \qquad (17)$$
$$J_2 = W_2(\theta^* - \theta[n])^2 \qquad (18)$$

respectively, where W_1, $W_2 > 0$ are weights. J_1 means that the rotating speed at the objective time t_1 approaches to its given condition ω^*. Also, J_2 represents that the rotational angle at the objective time t_1 arrives at its given condition θ^*. J_3 represents total loss energy along with the trajectories calculated from the 3-dimensional loss map where $W_3 > 0$ is the weight for optimization.

$$J_3 = W_3 E[n] \qquad (19)$$

$$E[n] = \Delta t_s \sum_{i=0}^{n-1} W_{loss} \qquad (20)$$

In order to solve this optimization problem minimizing the evaluation function J, a Genetic Algorism (GA) is adopted in this paper. GA is effective and efficient to solve

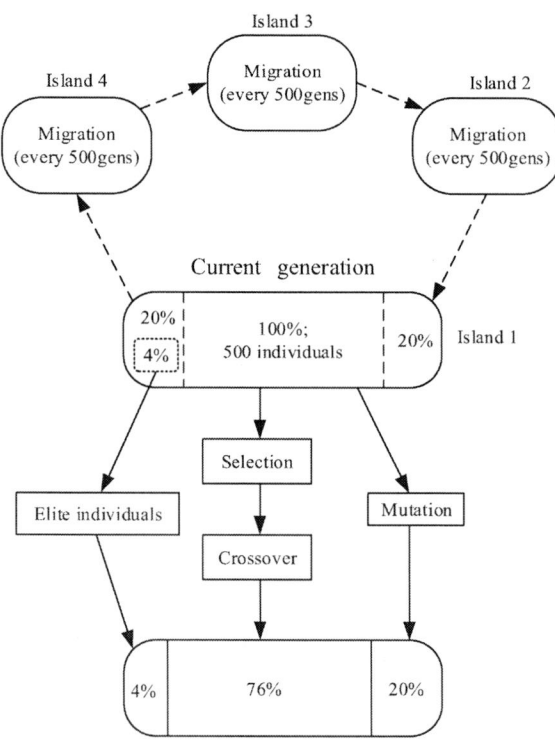

Figure 5 Evolution and migration between 4 islands.

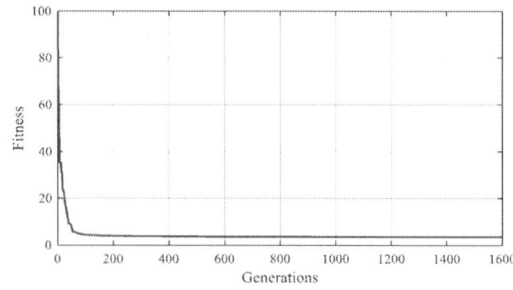

Figure 6 Convergence of Fitness (Evaluation function of the best individual) in the case of $\theta^* = 30$.

The 2018 International Power Electronics Conference

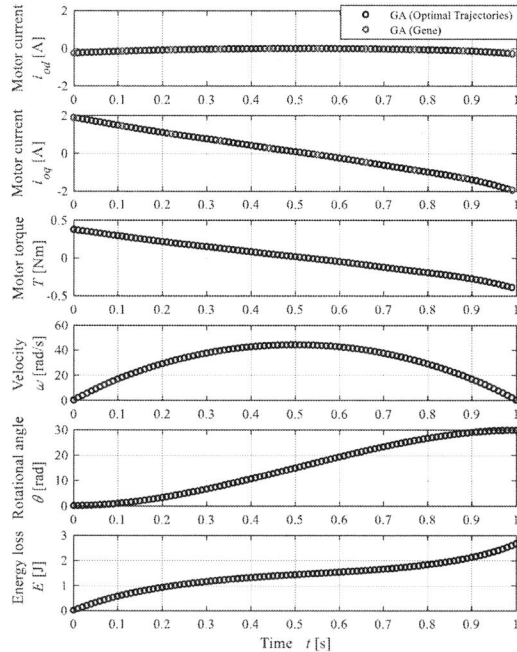

Figure 7 Numerically derived optimal trajectories by using GA ($\theta^* = 30$).

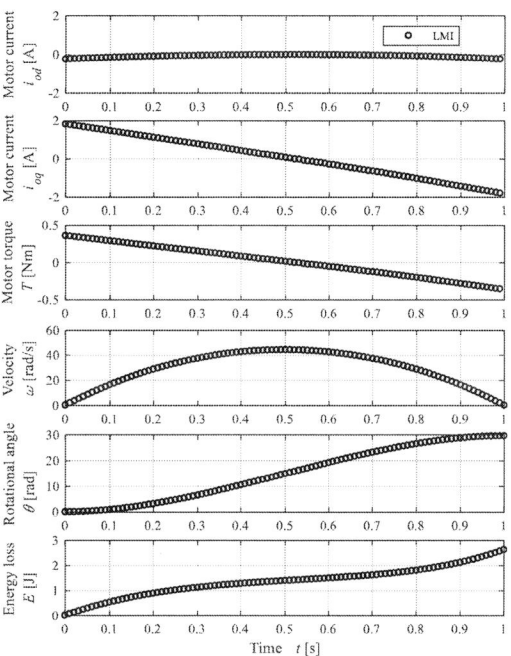

Figure 8 Numerically derived optimal trajectories by using LMI ($\theta^* = 30$).

3-dimensional optimization problem. Figure 3 shows trajectories to be derived. Final step n is set 100, so each trajectories have 100 operation points. In order to reduce calculation time, calculation points of i_{od} and i_{oq} are decimated to one-tenth. When torque, rotating speed, rotational angle and energy loss are calculated, spline interpolation will be used to motor currents. In Fig. 3, black solid circles show the operation points to be calculated and red circles are the gene as shown in Fig. 4. Figure 4 shows the gene of an individual which consists of decimated motor currents i_{od} and i_{oq}. Figure 5 shows the evolution of 500 individuals in one island and migration between four islands. Details of GA operator are shown in Table II.

III. DERIVED OPTIMAL TRAJECTORIES

A. Parameters Setup

Table I and II show the motor and GA parameters used in this study for optimization of trajectories. Drive conditions are set that IPMSM is driven from initial states $\omega_0 = 0$ rad/s, $\theta_0 = 0$ rad at $t_0 = 0$ s to objective states $\omega^* = 0$ rad/s $\theta^* = 30$ rad at $t_1 = 1.0$ s. Weights of evaluation functions are set as W_1 and $W_2=1.0$, $W_3=0.5$ because $E[n]$ in Eq. (19) becomes larger than error evaluations in Eqs. (17) and (18).

B. Derived Optimal Trajectories

LMI optimized trajectories are also derived to compare to GA optimized trajectories. In the LMI optimization, loss minimization control [3] is used as the relationship

between torque and motor currents. Because this study uses theoretically calculated 3-dimensional loss map as shown in Fig. 2, both trajectories derived by GA and LMI will agree. However, when parameter errors and/or un-

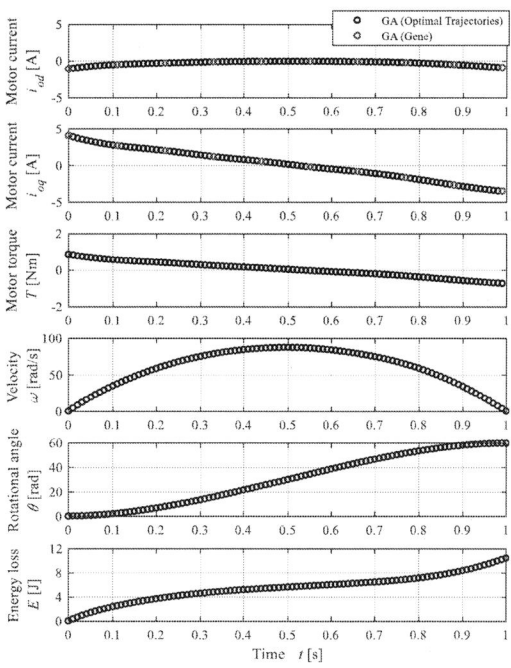

Figure 9 Numerically derived optimal trajectories by using GA ($\theta^* = 60$).

1267

The 2018 International Power Electronics Conference

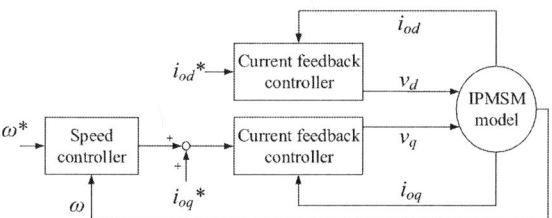

Figure 10 Block diagram of IPMSM drive simulation.

modeled dynamics exist and loss map is obtained experimentally, trajectories between GA and LMI will be different.

Figure 6 shows the convergence of GA fitness (evaluation function of the best individual for each generation). Fitness converges to a constant value, hence the GA optimization is done. Figure 7 shows numerically derived optimal trajectories by using the proposed method (GA). In the figure, black solid circles show the derived optimal trajectories and red circles in motor currents show the gene shown in Fig. 4. The obtained optimal trajectories satisfy the given drive conditions. Figure 8 shows numerically derived trajectories by using LMI. The obtained trajectories also satisfy the given drive conditions and they agree with GA (Fig. 7).

Calculation time to derive trajectories are quite different. Calculation time of proposed GA case is reduced to 76.5% comparing to the LMI case. Hence, proposed GA based optimization method is effective and efficient for optimal trajectories calculation of rotating speed and d-q Axis currents, simultaneously.

Figure 9 shows numerically derived optimal trajectories

by using the proposed method (GA) in the case of $\theta^* = 60$. The obtained optimal trajectories satisfy the given drive conditions. Comparing to the case $\theta^* = 30$ shown in Fig. 7, maximum motor torque and rotating speed are large in order to satisfy the drive condition $\theta^* = 60$ for the same operation time 1 s. As a result, energy loss becomes around 4 times.

C. Simulation Results

In order to evaluate the effectiveness of derived optimal trajectories, simulations are carried out. Matlab and Simulink are used. Figure 10 shows the block diagram of IPMSM drive simulation to let rotating speed and d-axis motor current follow obtained optimal rotating speed and d-axis motor current trajectories as shown in Figs. 7 and 9, respectively. In Fig. 10, ω^* and $i_{od}{}^*$ represent the reference inputs of derived optimal trajectories. Optimal q-axis motor current $i_{oq}{}^*$ trajectory is given as feed-forward input for quick torque response. IPMSM model is composed by Eqs. (1), (2), and Fig. 1.

Figure 11 shows the simulation results in the case of $\theta^* = 30$. In the figure, black lines show the simulation result when the derived optimal trajectories are used as the reference and feed-forward inputs as shown in Fig. 10, blue lines do the constant acceleration and deceleration case for comparison. In the obtained optimal trajectory case, rotating speed, rotational angle satisfy given drive conditions. The amplitude of optimal torque is large at the beginning with slight overshoot, then it follows optimal torque trajectory. Maximum velocity of constant acceleration and deceleration case is larger than that of the proposed optimal trajectory case. Energy loss of proposed

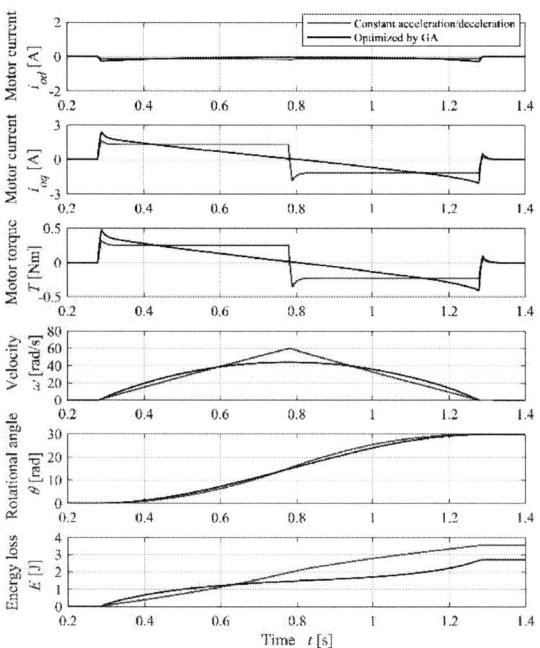

Figure 11 Simulation results when IPMSM is driven according to optimal trajectories and constant acceleration/deceleration ($\theta^* = 30$).

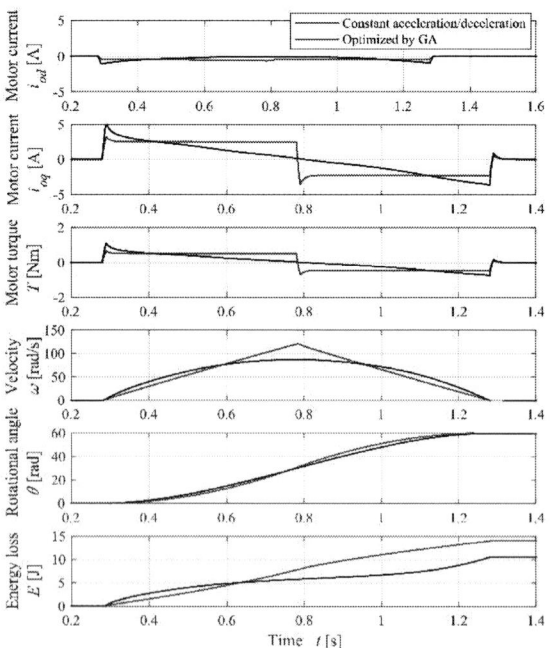

Figure 12 Simulation results when IPMSM is driven according to optimal trajectories and constant acceleration/deceleration ($\theta^* = 60$).

1268

optimal trajectory case is lower than that of constant acceleration and deceleration case. Figure 12 shows the simulation results in the case of $\theta^* = 60$. Each trajectories satisfy given drive conditions. Proposed optimal trajectories reduce energy loss.

These results show the proposed optimal trajectories can satisfy given drive conditions and reduce energy loss of IPMSM comparing to the constant acceleration and deceleration cases.

IV. CONCLUSIONS

This paper has presented a numerical design methodology of the optimal trajectories for IPMSMs by using 3-dimensional loss map. The evaluation functions are defined from the drive conditions and the loss map, then the optimal trajectories are derived by using genetic algorism (GA). Derived trajectories satisfy the given drive conditions and well agreed with trajectories derived by Linea Matrix Inequality (LMI). Calculation time becomes shorter comparing to the case of LMI.

In order to evaluate the effectiveness of derived optimal trajectories, simulations are carried out. IPMSM drive using derived optimal trajectories satisfy given drive conditions and can reduce energy loss comparing to the constant acceleration and deceleration case.

Therefore, proposed GA based optimization method for optimal trajectories calculation of rotating speed and d-q axis currents simultaneously and drive method using derived optimal trajectories are effective and efficient for energy-saving IPMSM drive.

Experimental measurement of the loss map, the optimal trajectories design and experimental verification of its effectiveness are now under preparation.

REFERENCES

[1] T. M. Jahns, G. B. Kliman, and T. W. Newmann, "Interior Permanent-Magnet Synchronous Motors for Adjustable-Speed Drives," *IEEE Transactions on Industry Applications*, Vol. IA-22, No. 4, pp. 738-747 (1986).

[2] S. Morimoto, K. Hatanaka, et al., "Servo Drive System and Control Characteristics of Salient Pole Permanent Magmnet Synchronous Motor," *IEEE Transactions on Industry Applications*, Vol. IA-29, No. 2, pp. 338-343 (1993).

[3] S. Morimoto, Y. Tong, et al., "Loss Minimization Control of Permanent Magnet Synchronous Motor Drives," *IEEE Transactions on Industrial Electronics*, Vol. IE-41, No. 5, pp. 511-517 (1994).

[4] Ronggang Ni, Dianguo Xu, et al., "Maximum Efficiency Per Ampere Control of Permanent-Magnet Synchronous Machines," *IEEE Transactions on Industrial Electronics*, Vol. 62, No. 4, pp. 2135-2143 (2015).

[5] K. Inoue, K. Ogata, and T. Kato, "An Efficient Power Regeneration and Drive Method of an Induction Motor by means of an Optimal Torque Derived by Variational Method" *IEEJ Transactions on Industrial Applications*, Vol. 128, No. 9 (2008).

[6] K. Inoue, M. Minamiyama, and T. Kato, "A Design Methodology of an Optimal Torque Minimizing Energy Loss under Torque Limit for an Induction Motor," *Proceedings of IEEE Energy Conversion Congress & Exposition*, pp. 163-167 (2009).

[7] K. Inoue, N. Okada, et al., "Numerical Design Methodology of Optimal Trajectories for Efficient Induction Motor Drive based on a Loss Map," *Proceedings of IEEE Energy Conversion Congress & Exposition*, pp. 3154-3158 (2013).

The 2018 International Power Electronics Conference

2-degree-of-freedom deadbeat control with disturbance compensation for PMSM drive system using FPGA

Arata Takahashi , Shotaro Takakura , Tomoki Yokoyama
Tokyo Denki University, 5, Asahicho, Senju, Adachiku, Tokyo, Japan
E-mail: yoko@fr.dendai.ac.jp

Abstract—In recent years, permanent magnet synchronous motor is widely used due to its high efficiency and maintenance free features in the industry. high accuracy position control and speed control performance in the servo system is required, so the high response performance in the current control system is required. In this paper, the moter drive system applying the 2-degree-of-freedom deadbeat control with disturbance compensation method is propsed to realize high tracking accuracy and robustness to the system parameters variations. The advantages of the proposed method was compared with the conventional deadbeat control method through simulations and experiments.

Keywords—PMSM, FPGA, deadbeat control, 2-degree-of-freedom control, disturbance compensation control

I. INTRODUCTION

Recently, permanent magnet synchronous motor (PMSM: Permanent Magnet Synchronous Motor) are widely used for industry applications and the high responsibility and the robustness for parameters variations are required[9][10][11]. In recent years, new control methods and approaches for PMSMs were presented. In [1][2][3], the perfect tracking control method based on the exact model of PMSM to control the overmodulation region has been proposed. To control the torque from the linear region of the inverter to the overmodulation region by the same perfect tracking controller, which realizes superior performance compared with the method which switches the plurality of the controllers for each mode. Perfect tracking control method used for controlling linear region of inverter is expanded to be applicable also to the overmodulation region. In addition, a method of realizing a current control system which has high responsible characteristics by applying the deadbeat control has been proposed. However, the conventional deadbeat control system was hard to respond to the load disturbance while the carrier interval. In order to solve this problem, a method of modifying the switching pulse width by 1 MHz high-speed sampling and high-speed control calculation has been proposed in the previous research, and its effectiveness has been confirmed in [6][7][8]. In this paper, 2-degree-of-freedom deadbeat control with disturbance compensation method was proposed and compared with the conventional deadbeat control method and the conventional 2-degree-of-freedom deadbeat control method through simulations and experiments, the advantage of the proposed method was verified.

II. MOTOR DRIVE SYSTEM

Fig.1 shows the block diagram of the motor drive system, which is composed by the FPGA based hardware controller, PWM inverter, IPMSM and load motor. Phase current I_u and I_w are measured by the AD board through the interface board. The encoder pulse is acquired directly by the FPGA board. The FPGA outputs the voltage reference value to the inverter as a PWM pulse. The inverter outputs three-phase current to the PMSM.

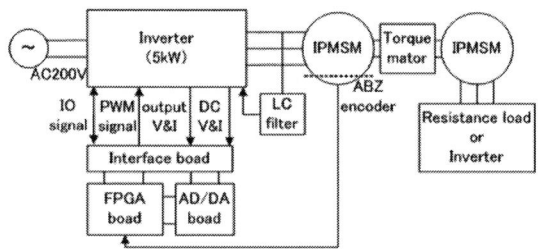

Fig. 1. Diagram of proposed control system

III. DESIGN OF CONTROL METHOD FOR PMSM MOTOR

A. Modeling

Fig.2 shows the d-q frame model of PMSM. (1) shows the voltage model of the PMSM in the d-q frame.

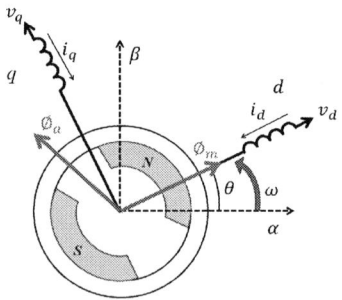

Fig. 2. The d-q frame model of PMSM

$$\begin{bmatrix} V_d \\ V_q \end{bmatrix} = \begin{bmatrix} R & -\omega_e L_q \\ \omega_e L_d & R \end{bmatrix} \begin{bmatrix} I_d \\ I_q \end{bmatrix} + \begin{bmatrix} L_d & 0 \\ 0 & L_q \end{bmatrix} \frac{d}{dt} \begin{bmatrix} I_d \\ I_q \end{bmatrix} + \begin{bmatrix} 0 \\ \omega\psi \end{bmatrix}. \quad (1)$$

1270

$$\psi = \sqrt{3/2}\phi_f$$

ϕ_f, L_d, L_q, ω_e, ψ, R, V_d, V_q, i_d and i_q show the leakage flux, the d-q axis inductance, the rotor speed, the permanent magnet flux, the stator resistance, the d-q-axis stator voltage components and the d-q-axis stator current components respectively.

B. Deadbeat control(DB)

(2) shows the state equation for IPMSM in d-q coordination.

$$\dot{x} = Ax + Bu \qquad (2)$$

, where

$$x = \begin{bmatrix} i_d \\ i_q \end{bmatrix}, u = \begin{bmatrix} V_d \\ V_q - \omega\psi \end{bmatrix},$$

$$A = \begin{bmatrix} -\frac{R}{L_d} & \frac{\omega L_q}{L_d} \\ -\frac{\omega L_d}{L_q} & -\frac{R}{L_q} \end{bmatrix}, B = \begin{bmatrix} \frac{1}{L_d} & 0 \\ 0 & \frac{1}{L_q} \end{bmatrix}. \qquad (3)$$

It is assumed that the PWM pulse is outputted in the center of the carrier period as shown in Fig. 3 and the input voltage V_q is constant during the sampling interval, the state equation for the discrete time model can be described as fllows[7].

$$x(k+1) = Fx(k) + Gu(k) \qquad (4)$$

, where

$$F = e^{AT}, G = e^{\frac{AT}{2}}BE. \qquad (5)$$

In deadbeat control method, the output voltage is calculated by the controller as shown in (6).

$$u_{ref}(k) = G^{-1}\{x(k+1) - Fx(k)\} \qquad (6)$$

, where

$$\begin{bmatrix} V_{dref} \\ V'_{qref} \end{bmatrix} = G^{-1}\left\{ \begin{bmatrix} i_{dref}(k+1) \\ i_{qref}(k+1) \end{bmatrix} - F \begin{bmatrix} i_d(k) \\ i_q(k) \end{bmatrix} \right\}. \qquad (7)$$

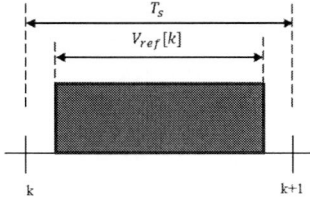

Fig. 3.   PWM pulse model for the deadbeat control

C. Deadbeat control with disturbance compensation(DBDC)

To improve the robustness for the motor parameter variations, the disturbance compensation method is applied to the deadbeat control. The disturbance voltage D_d, and D_q defined as (8) and (9) as follows [10].

$$D_d = R(I_{dref} - I_d) - \omega_e L_d(I_{qref} - I_q) \qquad (8)$$

$$D_q = R(I_{qref} - I_q) - \omega_e L_q(I_{dref} - I_d) \qquad (9)$$

(10) shows the state equation with the disturbance voltage D_d, D_q in d-q coordination.

$$\dot{x} = Ax + Bu \qquad (10)$$

,where

$$x = \begin{bmatrix} I_d \\ I_q \\ D_d \\ D_q \end{bmatrix}, u = \begin{bmatrix} V_d \\ V_q - \omega_e \psi \\ 0 \\ 0 \end{bmatrix},$$

$$A = \begin{bmatrix} -\frac{R}{L_d} & \frac{\omega_e L_q}{L_d} & \frac{1}{L_d} & 0 \\ -\frac{\omega_e L_d}{L_q} & -\frac{R}{L_q} & 0 & \frac{1}{L_q} \\ 0 & 0 & 0 & 0 \\ 0 & 0 & 0 & 0 \end{bmatrix}, B = \begin{bmatrix} \frac{1}{L_d} & 0 & 0 & 0 \\ 0 & \frac{1}{L_q} & 0 & 0 \\ 0 & 0 & 0 & 0 \\ 0 & 0 & 0 & 0 \end{bmatrix}.$$

The discrete time model of equation (10) can be described as follows,

$$x(k+1) = Fx(k) + Gu(k) \qquad (11)$$

,where

$$F = e^{AT}, G = e^{AT}BE. \qquad (12)$$

In deadbeat control with disturbance compensation method, the output voltage is calculated by the controller as shown in (13).

$$u_{ref}(k) = G^{-1}\{x(k+1) - Fx(k)\} \qquad (13)$$

,where

$$\begin{bmatrix} V_{dref} \\ V'_{qref} \\ 0 \\ 0 \end{bmatrix} = G^{-1}\left\{ \begin{bmatrix} i_{dref}(k+1) \\ i_{qref}(k+1) \\ D_{dref}(k+1) \\ D_{qref}(k+1) \end{bmatrix} - F \begin{bmatrix} i_d(k) \\ i_q(k) \\ D_d(k) \\ D_q(k) \end{bmatrix} \right\}. \qquad (14)$$

D. 2-degree-of-freedom deadbeat control with disturbance compensation (2DOF-DBDC)

In order to realize a high tracking accuracy to the reference and a robust characteristics, combine a conventional feedback deadbeat control system and a feedforward controller to compose a 2-degree-of-freedom deadbeat control system. Fig.4 shows the block diagram of the 2-degree-of-freedom deadbeat control system for IPMSM.

The 2018 International Power Electronics Conference

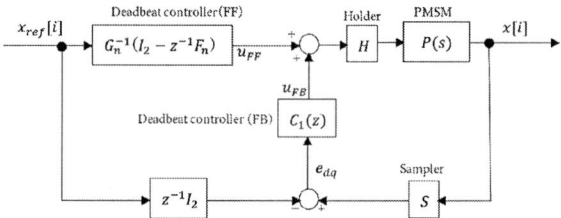

Fig. 4.   Diagram of 2-degree-of-freedom control system

The output voltage of the feedforward controller is calculated as shown in (16) using only the refference current value.

$$u_{FF} = -G_n^{-1}(I_2 - z^{-1}F_n)x_{ref}(k) \qquad (15)$$

The feedback controller calculates the output voltage so that the current error becomes zero by the deadbeat control method. The output voltage of the feedback controller is obtained as shown in equation (17).

$$u_{FB} = -G_n^{-1}F_n x(k) \qquad (16)$$

In 2-degree-of-freedom deadbeat control method, the output voltage is calculated by the controller as shown in (18).

$$u(k) = G_n^{-1}(x_{ref}(k) - F_n x_{ref}(k-1) - F_n x(k)) \quad (17)$$

,where

$$
\begin{bmatrix} V_{dref} \\ V'_{qref} \\ 0 \\ 0 \end{bmatrix} =
$$

$$
G_n^{-1} \left\{ \begin{bmatrix} i_{dref}(k) \\ i_{qref}(k) \\ D_d(k) \\ D_q(k) \end{bmatrix} - F_n \begin{bmatrix} i_{dref}(k-1) \\ i_{qref}(k-1) \\ D_d(k-1) \\ D_q(k-1) \end{bmatrix} - F_n \begin{bmatrix} i_d(k) \\ i_q(k) \\ 0 \\ 0 \end{bmatrix} \right\}. \quad (18)
$$

IV.   SIMULATION

The simulation conditions is shown in Table. I and the parameters of IPMSM is shown in Table. II.

TABLE I.   SIMULATION CONDITION

Parameter	Value
Sampring period	$50\mu s$
Carrier frequency	20kHz
Voltage of inverter	70V
Speed load	200rpm
D-axis reference current	0A
Q-axis reference current	$1.5 \rightarrow 2.0$A
Simulation tools	MATLAB/Simulink

TABLE II.   IPMSM PARAMETER

Parameter	Value
Armature resistance(R)	1.44Ω
D-axis inductance(L_d)	6.9454mH
Q-axis inductance(L_q)	8.8403mH
D axis magnetic flux(ψ)	0.1563wd
Moment of interia(J)	20.0×10^{-3}kgm^2
Pole pairs(P)	8

A. Nominal condition

In this condition, the motor speed reference is settled to 200[rpm] and Q-axis reference current is varied from 1.5[A] to 2.0[A] is evaluated. Simulation conditions with the nominal motor parameters are shown from Fig. 5 to Fig. 8. The torque current characteristics are summarized in Table. III.

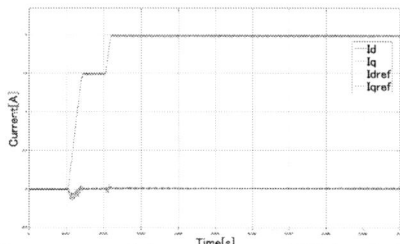

Fig. 5.   Step response of the torque current(DB) (nominal condition)

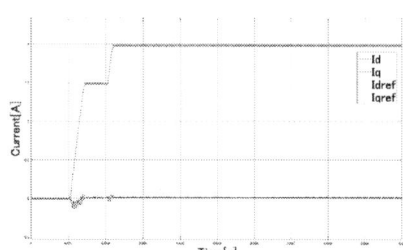

Fig. 6.   Step response of the torque current(DBDC) (nominal condition)

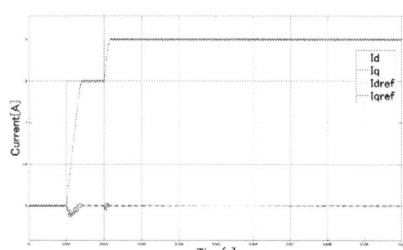

Fig. 7.   Step response of the torque current(2DOF-DB) (nominal condition)

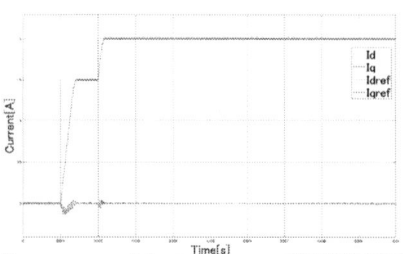

Fig. 8.   Step response of the torque current(2DOF-DBDC) (nominal condition)

1272

The 2018 International Power Electronics Conference

TABLE III.   TORQUE CURRENT CHARACTERISTICS (NOMINAL CONDITION)

Control method	Step response(μs)	Steady state error(mA)
DB	215.8	17.0
DBDC	216.0	17.0
2DOF-DB	146.0	0.0
2DOF-DBDC	146.4	0.0

In the case of the 2DOF-DB and 2DOF-DBDC, the steady-state error is much suppressed and the current response was much improved compared with that of the DB and the DBDC. By applying 2-degree-of-freedom control method, the current response was much improved and the steady-state error is much suppressed.

B. Unnominal condition

Simulations were carried out in the condition that the motor paramaters are varied from the nominal value as the unnominal condition. The motor speed reference is settled to 200[rpm] and Q-axis reference current is varied from 1.5[A] to 2.0[A] is evaluated. Simulation conditions with the unnominal motor parameters are shown from Fig. 9 to Fig. 12. The torque current characteristics are summarized in Table.IV.

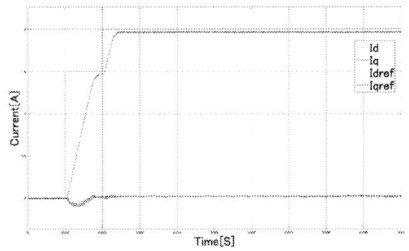

Fig. 9.   Step response of the torque current(DB) (unnominal condition)

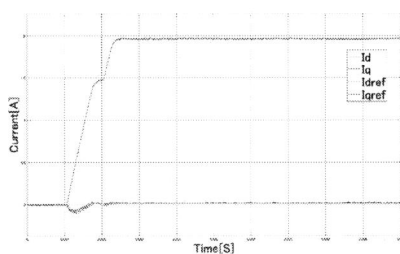

Fig. 10.   Step response of the torque current(DBDC) (unnominal condition)

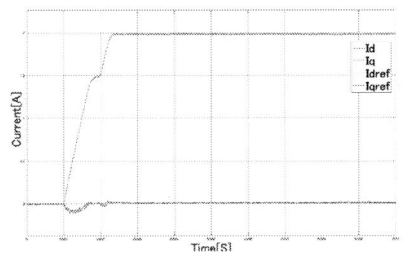

Fig. 11.   Step response of the torque current(2DOF-DB) (unnominal condition)

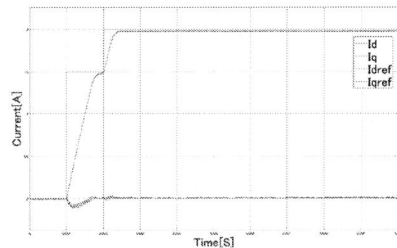

Fig. 12.   Step response of the torque current(2DOF-DBDC) (unnominal condition)

TABLE IV.   TORQUE CURRENT CHARACTERISTICS (UNNOMINAL CONDITION)

Control method	Step response(μs)	Steady state error(mA)
DB	396.0	35
DBDC	416.0	35
2DOF-DB	366.0	19
2DOF-DBDC	367.0	19

In the case of the 2DOF-DB and 2DOF-DBDC, the steady-state error is much suppressed and the current response was much improved compared with that of the DB and the DBDC. By applying 2 degree of freedom control method, the current response was improved and the steady-state error is much suppressed.

V.   EXPERIMENT

The experimental conditions are shown in Table. V, and the parameters of IPMSM are shown in Table. VI

TABLE V.   EXPERIMENTAL CONDITION

Parameter	Value
Sampring period	50μs
Carrier frequency	20kHz
Voltage of inverter	70V
Speed load	200rpm
D-axis reference current	0A
Q-axis reference current	1.5 → 2.0A

TABLE VI.   IPMSM PARAMETER

Parameter	Value
Armature resistance(R)	1.44Ω
D-axis inductance(L_d)	6.9454mH
Q-axis inductance(L_q)	8.8403mH
D axis magnetic flux(ψ)	0.1563wd
Moment of interia(J)	20.0×10^{-3}kgm^2
Pole pairs(P)	8

A. Nominal condition

The condition that the motor speed reference is settled to 200[rpm] and Q-axis reference current is varied from 1.5[A] to 2.0[A] is evaluated. Experimental conditions with the nominal motor parameters are shown from Fig. 13 to Fig. 16. The torque current characteristics are summarized in Table. VII.

The 2018 International Power Electronics Conference

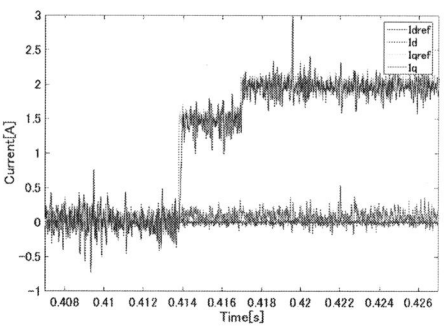

Fig. 13.   Step response of the torque current(DB) (nominal condition)

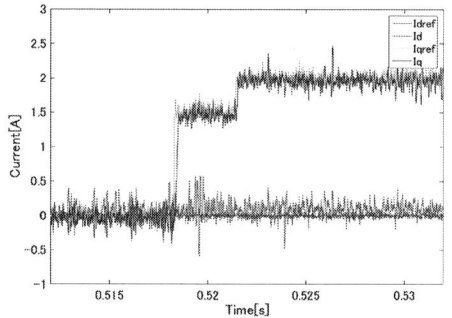

Fig. 14.   Step response of the torque current(DBDC) (nominal condition)

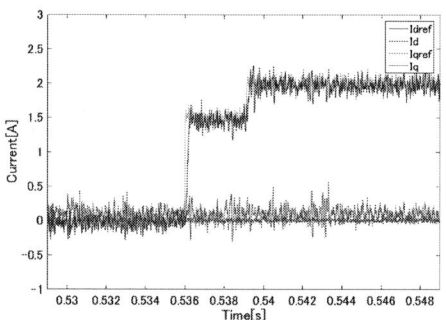

Fig. 15.   Step response of the torque current(2DOF-DB) (nominal condition)

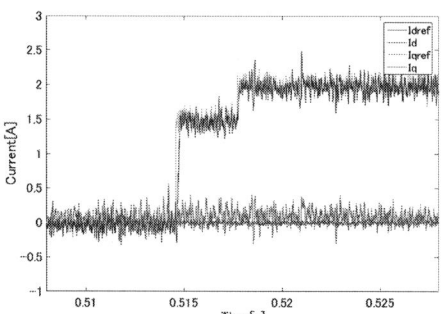

Fig. 16.   Step response of the torque current(2DOF-DBDC) (nominal condition)

TABLE VII.   TORQUE CURRENT CHARACTERISTICS (NOMINAL CONDITION)

Control method	Step response(μs)	Steady state error(mA)
DB	112.5	148
DBDC	98.0	147
2DOF-DB	59.0	148
2DOF-DBDC	55.5	150

In the case of the 2DOF-DB, the current response was much improved compared with that of the DB. In the case of the 2DOF-DBDC, the current response was much improved compared with that of the DBDC. The current response was much improved by applying 2 degree of freedom control method.

B. Unnominal condition

In order to verify the robustness for the motor parameters, the experiments ware carried out that the control gains were settled to 0.5 times of the nominal gains. This condition is equivalent to the situation that the motor parameter fluctuates twice of the nominal value. The motor speed reference is settled to 200[rpm] and Q-axis reference current is varied from 1.5[A] to 2.0[A] is evaluated. Experimental conditions with the unnominal motor parameters are shown from Fig. 17 to Fig. 20. The torque current characteristics are summarized in TableVIII.

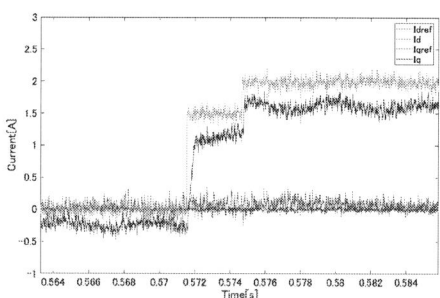

Fig. 17.   Step response of the torque current(DB) (unnominal condition)

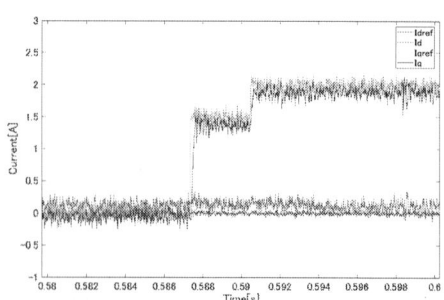

Fig. 18.   Step response of the torque current(DBDC) (unnominal condition)

1274

The 2018 International Power Electronics Conference

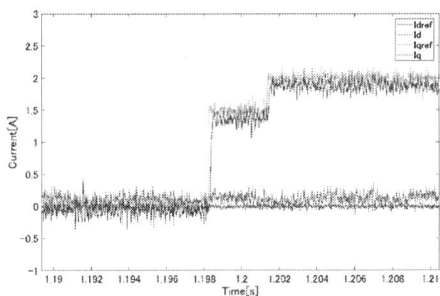

Fig. 19. Step response of the torque current(2DOF-DB) (unnominal condition)

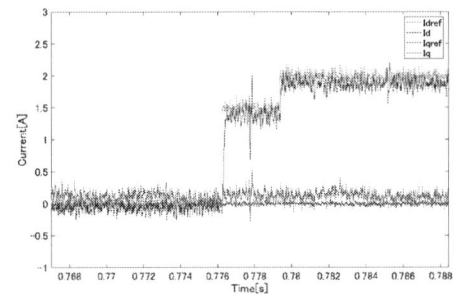

Fig. 20. Step response of the torque current(2DOF-DBDC) (unnominal condition)

TABLE VIII.   TORQUE CURRENT CHARACTERISTICS
(UNNOMINAL CONDITION)

Control method	Step response(μs)	Steady state error(mA)
DB	166.5	379
DBDC	139.5	176
2DOF-DB	82.0	159
2DOF-DBDC	49.5	159

In the case of the DBDC, the steady state error of the current is much suppressed and the current response was much improved compared with the DB. Applying disturbance compensation control method, the disturbance current due to the model error was compensated. In the case of the 2DOF-DB, the steady state error of the current is much suppressed and the current response was much improved compared with the DB. Combining the feedforward controler, the current response was much improved. In the case of the 2DOF-DBDC, the steady state error of the current is much suppressed and the current response was much improved compared with the DBDC and 2DOF-DB. Disturbance compensation control method, and 2 degree of freedom control method improve the current response and suppress the steady state error of the current.

VI. CONCLUSION

The moter drive system applying the 2-degree-of-freedom deadbeat control with disturbance compensation method was propsed. Through simulations and experiments, the current response was much improved in the case of nominal condition and the unnominal condition for the motor parameters was verified.

REFERENCES

[1] Takayuki Miyajima, Hiroshi Fujimoto, Masami Fujitsuna: "Control Method for IPMSM Based on Perfect Tracking Control and PWM Hold Model in Overmodulation Range", inProc. The 2010 International Power Electoronics Conference -ECCE ASIA-IPEC-Sapporo 2010 PROCEEDINGS, Sapporo, Japan, pp.593-598, 2010.

[2] Takayuki Miyajima, Hiroshi Fujimoto, Masami Fujitsuna, "Control Method for IPMSM Based on PTC and PWM Hold Model in Overmodulation Range -Study on Robustness and Comparison with Anti-Windup Control-", inProc. The 2010 IEEE Energy Conversion Congress and Exposition proceedings, Atlanta, USA, pp.2844-2850, 2010.

[3] Takayuki Miyajima, Hiroshi Fujimoto, Masami Fujitsuna: "Control method for IPMSM based on perfect tracking control and PWM hold model in overmodulation range", IEEJ Trans.IA,Vol.130, No.10, p.1153-1160, 2010.

[4] Atsuo Kawamura: "Modern Power electronics", SUURIK-OUGAKUSHA, 2005.

[5] K.P.Gokhale, A.Kawamura, and R.G.Hoft: "Deadbeat microprocessor control of PWM inverter for sinusoidaloutput waveform synthesis" IEEE Trans.Industry Applications, 1987.

[6] Hiroki Uchida, Tomoki Yokoyama: "1MHz Variable Sampling Deadbeat Control of Single Phase PWM Inverter" JIASC, 2012.

[7] Asahi Kitada,Kota Miyata, Kota Tsuchiya, Hiroki Sato, Tomoki Yokoyama, "1MHz Variable Sampling Deadbeat Control for PM motor using FPGA", Power Electronics and ECCE Asia (ICPE-ECCE Asia), Seoul, South Korea, 2015.

[8] Asahi Kitada, Tomoki Yokoyama, "Precise torque control for interior permanent magnet synchronous motor using FPGA based hardware controller", GCC Conference and Exhibition (GCC), 2013.

[9] Ryo Tanabe, Kan Akatsu: "Advanced Torque and Current Control Techniques for PMSMs with a Real-time Simulator Installed Behavior Motor Model" IEEJ Journal of Industry Applications, 2015.

[10] Ryo Tanabe, Kan Akatsu: "Advanced Torque Control of Permanent Magnet Synchronous Motor Using Finite Element Analysis Based Motor Model with a Real-time Simulator" IEEJ Jounal of Industry Applications, 2016.

[11] Kei Matsuura, Yousuke Akama, Kodai Abe, Kiyoshi Ohishi, Hitoshi Haga, Itaru Ando: "Fine Three-Phase Current Reconstruction based on Calculating the Phase-Shifted Voltage Reference Using Only the DC Current Sensor of an Inverter and Its Application to a PM Motor Drive" IEEJ Jounal of Industry Applications, 2016.

Extended EMF-based Simple IPMSM Sensorless Vector Control Using Compensated Current Controller

Takatoshi Inoue[1], Yasumasa Hamabe[2] and Mineo Tsuji[3*] and Shin-ichi Hamasaki[4*]
1-4 Graduate School of Engineering, Nagasaki University, Nagasaki, Japan
3 *E-mail: mineo@nagasaki-u.ac.jp

Abstract— **Vector control of an interior permanent magnet synchronous motor (IPMSM) without the magnetic pole position sensor has been studied. Especially, the extended electromotive force (EMF) based method is one of the famous systems. We proposed a simple method which estimates the magnetic pole position using the output voltage of the *d*-axis PI current controller without using disturbance observer. In this paper, we develop the early proposed method by introducing a new non-interference control. By the results of simulation and experiment, we have improved the magnetic pole position error in the rotor speed ramp responses.**

Keywords— ***Interior Permanent Magnet Synchronous Motor (IPMSM), sensorless vector control, extend electromotive force (EMF), non-interference control***

I. INTRODUCTION

PMSM is applied widely because of high efficiency and small size. Especially, the IPMSM is suitable for high speed drive and possible to produce high torque by utilizing reluctance torque in addition to magnet torque.

In order to realize a vector control of the IPMSM, a resolver or an encoder for detecting rotor position is used because it is necessary to control currents in accordance with the magnetic pole position. However there are demerits that the position sensor makes its size larger and the cost of the device higher. Therefore, many vector control methods without using the position sensor are proposed [1]~[5]. The extended EMF is one of the representative sensorless control schemes.

In general, the extended EMF based sensorless method is estimated by using the disturbance observer. But we have proposed a simple sensorless method based on the extended EMF without using the observer [6]. The rotor speed is estimated by the output voltage of the *d*-axis PI current controller with the non-interference control. We obtained a good response by using early proposed method in the steady state, but we had a problem that the system came to have a large position error in the rotor speed ramp responses.

To improve the performance of the control, we propose a new method that compensates the current controller using the estimated position error. Although a similar current compensation is proposed in earlier papers for ripple of current and speed [7], our proposed method is aimed for the improvement of the performance in the rotor speed ramp responses. We verify utilities of the new proposed method of current controller using the estimated position error by comparing experimental results. The estimated position error is improved by the new compensation of current controller.

II. SENSORLESS SYSTEM

A. Extended EMF Model of IPMSM

Park's equation is known as a typical voltage equation of IPMSM on *d-q* axes synchronized with the magnetic pole position. Modifying Park's equation, a voltage equation using the extended EMF (E_{ex}) is expressed as follows [1]:

$$\begin{bmatrix} v_d \\ v_q \end{bmatrix} = \begin{bmatrix} R_s + pL_d & -\omega_r L_q \\ \omega_r L_q & R_s + pL_d \end{bmatrix} \begin{bmatrix} i_d \\ i_q \end{bmatrix} + \begin{bmatrix} 0 \\ E_{ex} \end{bmatrix} \quad (1)$$

where,

$$E_{ex} = \omega_r \left\{ \left(L_d - L_q \right) i_d + \psi \right\} - \left(L_d - L_q \right) p i_q \quad (2)$$

By converting the reference frame to γ-δ axes which rotate at an estimated speed $\hat{\omega}$ as shown in Fig. 1, the following equation is obtained from (1):

$$\begin{bmatrix} v_\gamma \\ v_\delta \end{bmatrix} = \begin{bmatrix} R_s + pL_d & -\omega_r L_q \\ \omega_r L_q & R_s + pL_d \end{bmatrix} \begin{bmatrix} i_\gamma \\ i_\delta \end{bmatrix} + \begin{bmatrix} e_\gamma \\ e_\delta \end{bmatrix} \quad (3)$$

The second term of the right side is the extended EMF on γ-δ axes. It is expressed as follows:

$$\begin{bmatrix} e_\gamma \\ e_\delta \end{bmatrix} = E_{ex} \begin{bmatrix} \sin \theta_e \\ \cos \theta_e \end{bmatrix} + \left(\hat{\omega} - \omega_r \right) L_d \begin{bmatrix} -i_\delta \\ i_\gamma \end{bmatrix} \quad (4)$$

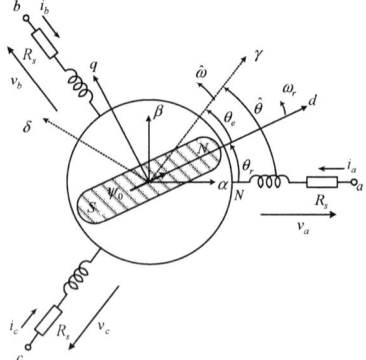

Fig. 1. Model of IPMSM.

B. Early Proposed Method [8]

The early proposed simple sensorless vector control system is shown in Fig. 2. The extended EMF on γ-δ axes are estimated without using disturbance observer shown in Fig. 2.

From (4), when the error of the magnetic pole position is sufficiently small, the following equation is obtained by assuming $\sin\theta_e \simeq \theta_e$ and $\cos\theta_e \simeq 1$.

$$\begin{bmatrix} e_\gamma \\ e_\delta \end{bmatrix} \simeq E_{ex}\begin{bmatrix} \sin\theta_e \\ \cos\theta_e \end{bmatrix} \simeq E_{ex}\begin{bmatrix} \theta_e \\ 1 \end{bmatrix} \tag{5}$$

From (5), the position error θ_e is obtained as

$$\hat{\theta}_e \simeq \frac{e_\gamma^*}{E_{ex}^*} \tag{6}$$

Therefore, we can estimate the position error by using the γ-axes extended EMF e_γ^*.

Fig. 3 shows the part of the compensated current controller. We call this control Method A.

In Fig. 3, d-axis PI current control is given by the following equation.

$$e_\gamma^* = \left(K_{pd} + \frac{K_{id}}{s}\right)\left(i_d^* - i_\gamma\right) \tag{7}$$

We compute the γ-axis voltage reference v_γ^* by the following equation ignoring the differential term of (3).

$$v_\gamma^* = R_s^* i_d^* - L_q^* \hat{\omega}_r i_\delta + e_\gamma^* \tag{8}$$

The non-interference control is applied to the q-axis controller. The voltage v_δ^* is expressed as follows:

$$v_\delta^* = v_{\delta PI}^* + L_d^* \hat{\omega}_r i_\gamma + E_{ex}^* \tag{9}$$

$v_{\delta PI}^*$ is defined as the output voltage of the q-axis PI current controller.

$$v_{\delta PI}^* = \left(K_{pq} + \frac{K_{iq}}{s}\right)\left(i_q^* - i_\delta\right) \tag{10}$$

From (6), the rotor angular speed $\hat{\omega}$ is estimated by the following equation using the output voltage e_γ^* of the d-axis PI current controller.

$$\hat{\omega} = -\left(K_{ep} + \frac{K_{ei}}{s}\right)\frac{e_\gamma^*}{E_{ex}^*} \tag{11}$$

By using damping coefficient and natural angular frequency, the PI speed estimation gains modeled by the block diagram in Fig. 4 are designed as follows:

$$K_{ep} = 2\zeta\omega_n, \; K_{ei} = \omega_n^2 \tag{12}$$

E_{ex}^* is changed as a function of speed by neglecting differential terms of (2) and given by the following equation.

$$E_{ex}^* = \hat{\omega}_r\left\{\left(L_d^* - L_q^*\right)i_d^* + \psi\right\} \tag{13}$$

To reduce the influence of noise, the rotor angular speed is computed by using a following low-pass filter.

$$\hat{\omega}_r = \frac{\omega_c}{s + \omega_c}\hat{\omega} \tag{14}$$

The pole position is obtained by integrating $\hat{\omega}$ as follows:

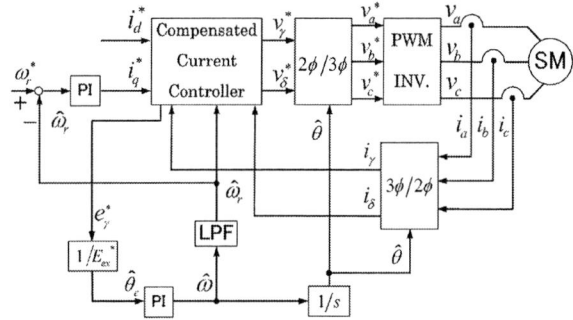

Fig. 2. Sensorless vector control system.

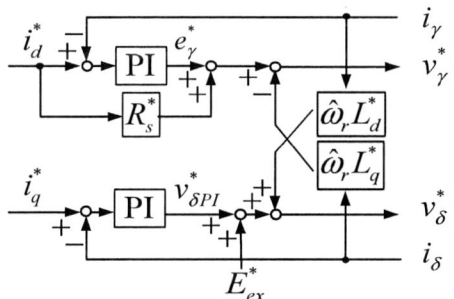

Fig. 3. Compensated Current Controller (Method A).

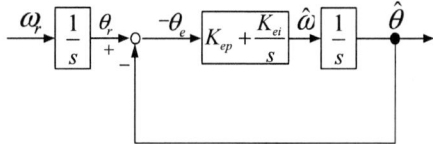

Fig. 4. Block diagram for speed estimation.

$$\hat{\theta} = \frac{1}{s}\hat{\omega} \tag{15}$$

By using the difference between the γ-δ axes and d-q axes, the position error θ_e can be expressed as follows:

$$\theta_e = \hat{\theta} - \theta_r \tag{16}$$

C. Proposed Method

In the case of Method A shown in Fig. 4, we obtained a good response when the speed command change is relatively small. However we had a problem that the system came to have a large position error in the rotor speed ramp responses. To improve the performance of the control, we pay attention to the position error and apply a new compensation of current controller using the estimated position error.

In the steady state, the PI speed estimation controller makes the output voltage e_γ^* of the d-axis PI current controller converge to zero and the position error is converged to zero too. But in the rotor speed ramp response, the position error cannot be controlled to zero. Because the interference terms become the disturbance for current controller of each axis, we apply a compensation by the feed forward control using the estimated position error to the current controller.

When the estimated positon error $\hat{\theta}_e$ occurs between d-q axes and γ-δ axes, the relation of (17) is satisfied.

$$\begin{bmatrix} v_d \\ v_q \end{bmatrix} = \begin{bmatrix} \cos\hat{\theta}_e & -\sin\hat{\theta}_e \\ \sin\hat{\theta}_e & \cos\hat{\theta}_e \end{bmatrix} \begin{bmatrix} v_\gamma \\ v_\delta \end{bmatrix} \qquad (17)$$

We apply the compensation of the voltage command values by using a relation of (17). When the estimated position error is small, the following equation is obtained by assuming $\sin\hat{\theta}_e \simeq \hat{\theta}_e$ and $\cos\hat{\theta}_e \simeq 1$.

By transforming (8) and (9) by $\hat{\theta}_e$ of (17), new γ-δ axes voltages are obtained as follows:

$$\begin{bmatrix} v_\gamma^* \\ v_\delta^* \end{bmatrix} = \begin{bmatrix} e_\gamma^* + R_s^* i_d^* - \hat{\omega}_r L_q^* i_\delta - \hat{\theta}_e v_{\delta PI}^* - \hat{\theta}_e E_{ex}^* - \hat{\theta}_e \hat{\omega}_r L_d^* i_\gamma \\ v_{\delta PI}^* + E_{ex}^* + \hat{\omega}_r L_d^* i_\gamma + \hat{\theta}_e e_\gamma^* + \hat{\theta}_e R_s^* i_d^* - \hat{\theta}_e \hat{\omega}_r L_q^* i_\delta \end{bmatrix} \quad (18)$$

In (18), the interference terms with the position error that cannot be followed in the current controller of each axis are $-\hat{\theta}_e v_{\delta PI}^*$ and $-\hat{\theta}_e E_{ex}^*$ in γ-axis, $\hat{\theta}_e e_\gamma^*$ and $\hat{\theta}_e R_s^* i_d^*$ in δ-axis. A new voltage command values which are used for practical system are shown as follows:

$$\begin{bmatrix} v_\gamma^* \\ v_\delta^* \end{bmatrix} = \begin{bmatrix} e_\gamma^* + R_s^* i_d^* - \hat{\omega}_r L_q^* i_\delta - \hat{\theta}_e v_{\delta PI}^* - \hat{\theta}_e E_{ex}^* \\ v_{\delta PI}^* + E_{ex}^* + \hat{\omega}_r L_d^* i_\gamma + \hat{\theta}_e e_\gamma^* + \hat{\theta}_e R_s^* i_d^* \end{bmatrix} \quad (19)$$

The new compensated current controller (Method B) is shown in Fig. 5.

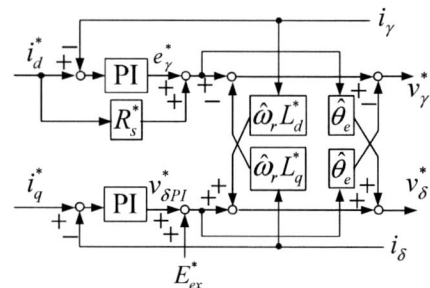

Fig. 5. Compensated Current Controller (Method B).

III. SIMULATION RESULTS

The parameters of controller are designed and shown in TABLE I.

TABLE I
PARAMETERS OF CONTROLLER

Parameter	Value
PI speed controller ω_{sc}	15rad/s
D-axis PI current controller ω_{cd}	1000rad/s
Q-axis PI current controller ω_{cq}	1000rad/s
Low Pass Filter ω_c	300rad/s
Sampling time T	100μs
Carrier frequency	10kHz
Damping coefficient ζ	1.5
Natural angular frequency ω_n	50rad/s

Simulation results are shown in Figs. 6, 7 and 8 in which the each speed references are changed as a ramp function from 700 min^{-1} to 1000 min^{-1}, from 1100 min^{-1} to 1800 min^{-1} and from 1300 min^{-1} to 800 min^{-1} respectively. In these figures, N_r is the actual speed and $\hat{N}_r$ is the estimated speed, i_d is the actual d-axis current and i_d^* is its command, i_q is the actual q-axis current and i_q^* is its command and θ_e is the actual magnetic pole position error. We apply the maximum-torque-per-ampere (MTPA) control to the each axis current command until an induced voltage of a motor reaches the upper limit and switch from MTPA to flux-weakening control when it reaches the limit in Fig. 7.

About simulation results of Method A in Figs.6, 7 and 8, we have the problems that the system has a large θ_e and the large deviation between i_d^* and i_d in the speed ramp responses. By contrast, simulation results of Method B in Figs.6, 7 and 8, θ_e and the deviation between i_d and i_d^* in the speed ramp response are improved.

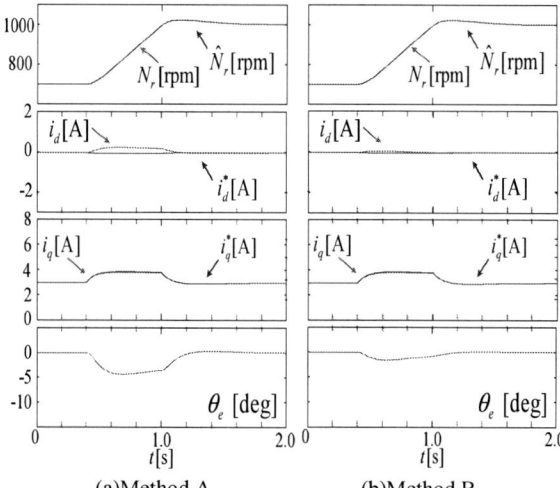

(a)Method A.   (b)Method B.
Fig. 6. Simulation results (from 700 to 1000 min^{-1}).

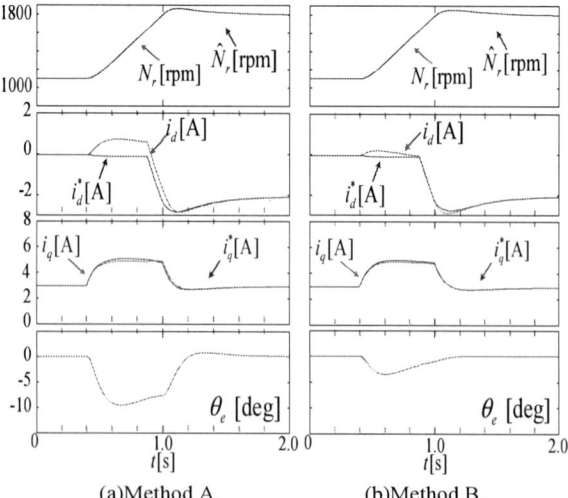

(a)Method A.   (b)Method B.
Fig. 7. Simulation results (from 1100 to 1800 min^{-1}).

The 2018 International Power Electronics Conference

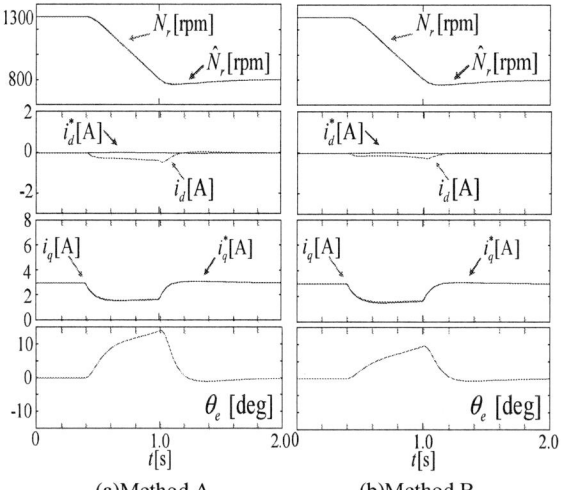

(a)Method A.   (b)Method B.
Fig. 8. Simulation results (from 1300 to 800 min⁻¹).

IV. CONSTITUTION OF THE EXPERIMENTAL SYSTEM

The experimental system is shown in Fig. 9.

This system is constituted by a rectifier and a smoothing capacitor and an IGBT inverter and IPMSM. IPMSM is coupled to a DC motor for load through a torque sensor. A control circuit is composed of DSP (TMS320C33), DC voltage and the AC currents of the motor are detected in DSP through AD converter. PWM gate signal generator sends gate pulse to the PWM inverter lets IGBT have ON or OFF by an amplitude modulation ratio.

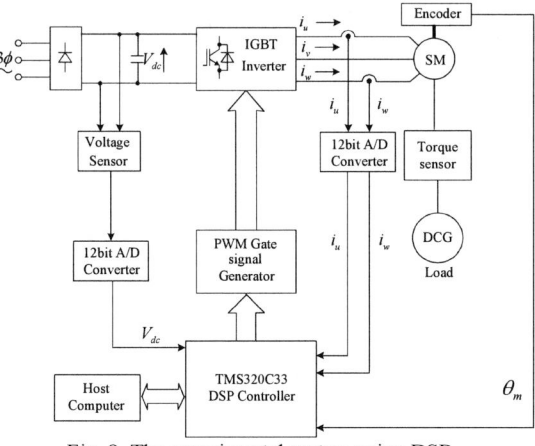

Fig. 9. The experimental system using DSP.

By using a trapezoid approximation of the integral, the digital control equations are expressed as follows:

$$
\begin{aligned}
e_\gamma^*(k) = {} & e_\gamma^*(k-1) \\
& + K_{pd}\left\{\left(i_d^*(k)-i_\gamma(k)\right)-\left(i_d^*(k-1)-i_\gamma(k-1)\right)\right\} \\
& + \frac{K_{id}T}{2}\left\{\left(i_d^*(k)-i_\gamma(k)\right)+\left(i_d^*(k-1)-i_\gamma(k-1)\right)\right\}
\end{aligned}
\tag{20}
$$

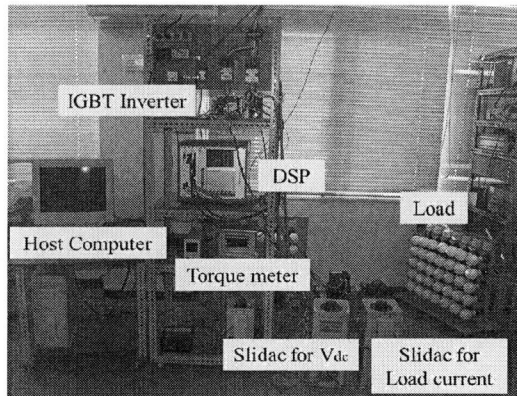

Fig. 10. The experimental equipments.

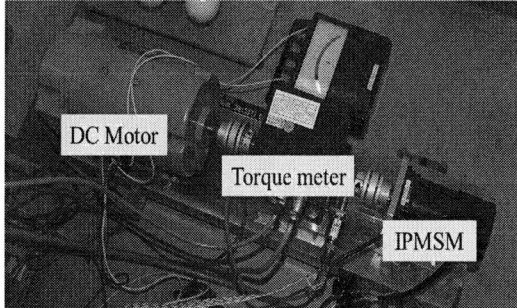

Fig. 11. IPMSM for experiment.

$$
\begin{aligned}
v_{\delta PI}(k) = {} & v_{\delta PI}(k-1) \\
& + K_{pq}\left\{\left(i_q^*(k)-i_\delta(k)\right)-\left(i_q^*(k-1)-i_\delta(k-1)\right)\right\} \\
& + \frac{K_{iq}T}{2}\left\{\left(i_q^*(k)-i_\delta(k)\right)+\left(i_q^*(k-1)-i_\delta(k-1)\right)\right\}
\end{aligned}
\tag{21}
$$

$$
\begin{aligned}
\hat{\omega}(k) = {} & \hat{\omega}(k-1) - K_{ep}\left\{e_\gamma^*(k)-e_\gamma^*(k-1)\right\} \\
& - \frac{K_{ei}T}{2}\left\{e_\gamma^*(k)+e_\gamma^*(k-1)\right\}
\end{aligned}
\tag{22}
$$

$$
\begin{aligned}
\hat{\omega}_r(k) = {} & \hat{\omega}_r(k-1) \\
& + \frac{\omega_c T}{2}\left\{\left(\hat{\omega}_0(k)-\hat{\omega}_r(k)\right)+\left(\hat{\omega}_0(k-1)-\hat{\omega}_r(k-1)\right)\right\}
\end{aligned}
\tag{23}
$$

$$
\hat{\theta}(k) = \hat{\theta}(k-1) + \frac{T}{2}\left\{\hat{\omega}(k)+\hat{\omega}(k-1)\right\}
\tag{24}
$$

The experimental equipments are shown in Fig. 10 and IPMSM used in the experiment is shown in Fig. 11.

V. EXPERIMENTAL RESULTS

The parameters of IPMSM which we used for experiments are shown in TABLE II. Experimental results are shown in Figs. 12, 13 and 14 which correspond to the simulation results shown in Figs. 6, 7 and 8 respectively.

According to the waveforms of experiment of Method A, a good response is obtained in the steady state, but the

1279

system has a large position error and the large deviation between i_d^* and i_d in the speed ramp responses. When the axes of the voltages v_γ^* and v_δ^* do not correspond in the continuous speed changing like a ramp functions, the feedback current also does not follow the current command in Fig.3.

TABLE II
PARAMETERS OF IPMSM

Parameter	Value
Nominal power	800W
Nominal speed	2500min^{-1}
Number of pole P	8
Stator resistance R_s	0.4Ω
D-axis inductance L_d	3.42mH
Q-axis inductance L_q	3.82mH
Armature flux-linkage ψ	8.45mWb

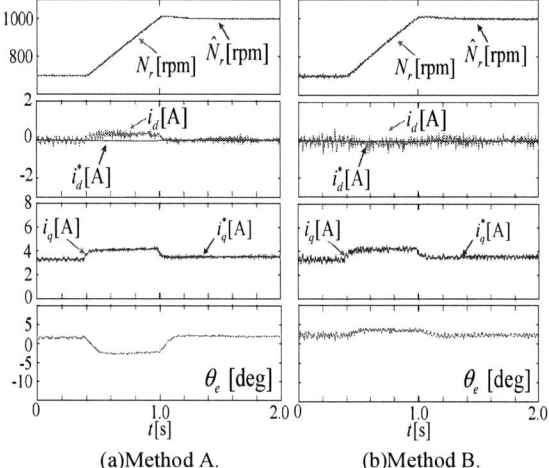

(a)Method A.   (b)Method B.

Fig. 12. Experimental results (from 700 to 1000 min^{-1}).

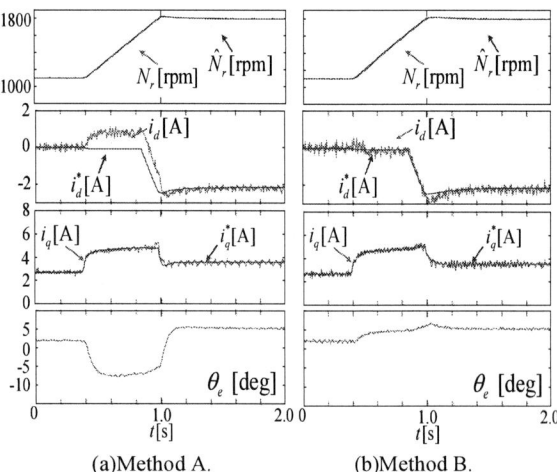

(a)Method A.   (b)Method B.

Fig. 13. Experimental results (from 1100 to 1800 min^{-1})

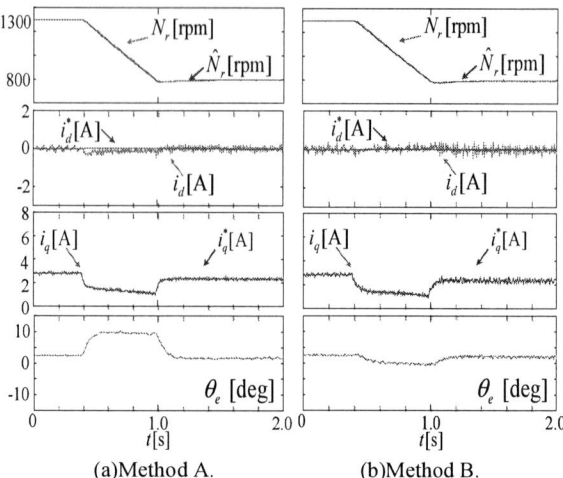

(a)Method A.   (b)Method B.

Fig. 14. Experimental results (from 1300 to 800 min^{-1})

By contrast, concerning the experimental waveforms of the Method B, the positon error θ_e and the deviation between i_d and i_d^* in the speed ramp response are improved. On the other hand, the actual position error θ_e and the estimated speed $\hat{N}_r$ of Method B is more sensible for noise than Method A, because the output voltage e_γ^* of d-axis current controller and the estimated position error $\hat{\theta}_e$ pulsate. We suppose that the influence of the noise becomes remarkable because the interference terms including $\hat{\theta}_e$ is added to the applied voltages v_γ^* and v_δ^* with feed forward control.

As the experimental results, Method B is susceptible to the effect of external noise but reduces the position error and deviation between the actual d-axis current and d-axis current in the speed ramp responses in comparison with Method A.

VI. CONCLUSIONS

For simple IPMSM sensorless vector control using the extended EMF, we have proposed a new system with compensation of current controller using the estimated position error. We have compared the experimental and simulation results obtained by the early proposed method with those of new proposed method. The results of the new compensation of current controller can improve the performance of sensorless vector control in the speed ramp responses. Therefore, it is concluded that the compensation of current controller using the estimated positon error is effective.

REFERENCES

[1] Z. Chen, M. Tomita, S. Doki, and S. Okuma, "An Extended Electromotive Force Model for Sensorless Control of Interior Permanent-Magnet Synchronous Motors," *IEEE Trans. Ind. Electron,* vol.50, no.2, pp.288-295, 2003.

[2] S. Morimoto, K. Kawamoto, M. Sanada, and Y. Takeda, "Sensorless Control Strategy for Salient-Pole

PMSM Based on Extended EMF in Rotating Reference Frame," *IEEE Trans. Ind. Applicat*, vol.38, no4, pp.1054-1061, 2002.

[3]   K. Sakamoto, Y. Iwaji, T. Endo, T. Taniguchi, T. Niki, M. Kawamata, and A. Kawamura, "Position Sensorless Vector Control of Permanent Magnet Synchronous Motors for Electrical Household Appliances," Proc. Of the Power Conversion Conference PCC-Nagoya 2007, LS4-3-3, pp.1119-1125, 2007.

[4]   K. Yamanaka, T. Ohnishi, and M. Hojo, "A Novel Position Sensorless Vector Control of Permanent-Magnet Synchronous Motors," Proc. Of the PCC-Nagoya 2007, DS8-3-1, pp.290-295, 2007.

[5]   J. K. Seok, J. K. Lee, and D.C. Lee, "Sensorless Speed Control of Nonsalient Permanent-Magnet Synchronous Motor Using Rotor-Position-Ttracking PI Controller," *IEEE Trans. Ind. Electron*, vol.53, no2, pp.399-405, 2006.

[6]   M. Tuji, K. Kojima, G. Mangindaan, D. Akafuji, S. Hamasaki, "Stability Study of a Permanent Magnet Synchronous Motor Sensorless Vector Control System Based on Extended EMF Model," *IEEJ Journal of Industry Applications*, vol.1 no.3 pp.148-154, 2012-11.

[7]   K. Takehana, T. Noguchi, "Position-Estimation-Based-Error Compensation in Current Controller of Sensorless IPM Motor," *IEEJ National Convention Record*, no.4-130 pp.1405-1406, 2001.

[8]   Y. Hamabe, M. Tsuji, S. Hamasaki, "Extended EMF-based Simple IPMSM Sensorless Vector Control Considering Cross Coupling Compensation," Proc. of ICEMS, No.DS4G-1-15 pp.1-6, 2016.

Full-band Output Impedance Model of Virtual Synchronous Generator in *dq* framework

Li Wenbing [1], Wang Jianhua [1*], Song Jingyu [2], Luo Fangfang [1], Gao Shang [1], Wu Zaijun [1]

1 Jiangsu Provincial Key Laboratory of Smart Grid Technology and Equipment,
School of Electrical Engineering, Southeast University, Nanjing, China
2 China State Shipbuilding Co., Ltd., Beijing, China
*E-mail: wangjianhua@seu.edu.cn

Abstract-Virtual synchronous generator (VSG), an emerging technique that has been proposed for its possible grid-friendly features in these years. Since VSGs mimic synchronous generators with power electronic devices, they obtain both the high frequency converter's features and low frequency characteristics of the synchronous generator. Full-band output impedance model of the VSG should be investigated for further stability analysis before large scale VSG applications. This paper provides an in-depth investigation of the VSG, whose full-band output impedance model in *dq* framework is proposed, derived and verified, leading to the extraction of information for system stability.

I. INTRODUCTION

Today's power grid is characterized by a high penetration of renewable energy sources, most of which are integrated with power electronic converters. Most of the existing grid-tie converters for renewable sources mimic current source, which are easy to be implemented for grid integration [1]. But their operations depend on power grid as a master. The other type grid-connected converters that feed power to the grid in the form of voltage sources instead of current sources, in a way similar to the conventional power generators, named virtual synchronous generators (VSGs) have been proposed for its possible grid-friendly features in these years. Numerous literatures report the recent developments, such as different VSGs control methods, extend applications as STATCOM and active rectifiers, equivalence between the VSG concepts, droop controllers and synchronous generators, and etc. [2].

However, the massive presence of these controllable converters might lead to undesired interactions between the converters and between the converters and the other elements in the power system, which can jeopardize the system performance and, in the worst case, its stability. For system stability, power quality and safety enhancement concern, many grid codes are issued for these grid-connected converters recently, such as leakage current issue, reactive power compensation, low/high/zero voltage ride through and etc. [3].

Power system stability studies are typically approached by using numerical analysis to determine the actual values of the system's eigenvalues. Although effective in evaluating the stability of the investigated system, one disadvantage of the eigenvalue analysis is that it is bulky and requires the computation of the entire system, which needs to be updated if a different system configuration is to be considered. It is not fit for "plug-and-play" grid-connected converters.

Frequency-domain impedance models offer an alternative method for the investigation of the stability of a system with high penetration power electronic devices. A frequency domain impedance criterion is proposed in [4] to assess the high frequency resonance of the grid-connected single-phase inverter. Its extension of the Nyquist stability criterion, which has been successfully applied for investigation of low frequency sub-synchronous resonance and DC-link stability studies in HVDC systems [5]. As compared to the traditional eigenvalue analysis, the frequency domain impedance approaches present the advantage of facilitating the stability analysis of the system.

Since VSGs mimic synchronous generators with power electronic devices, they obtain both the high frequency converter's features and low frequency characteristics of the synchronous generator. Full-band output impedance model of the VSG should be investigated for further stability analysis before large scale VSG applications. However, most of the existing literatures that mention VSGs inductive impedance, only focus on its phasor model at line frequency for power sharing [6]. Other frequency domain features are omitted.

The aim of this paper is to provide an in-depth investigation of the full-band output impedance model of the VSG in *dq* framework, leading to the extraction of information for system stability. Therefore, this paper is organized as follows. Section II derives output impedance model of VSG in detail. Section III provides the output impedance model verification with software measurement technique, parameters influences are also discussed here. Section IV illustrates experimental results of the VSG prototype for further verification. Section V concludes this paper.

II. INFORMATION

Fig.1 shows the topology of the VSG. A typical three-phase voltage-source inverter is connected to the grid at the point of common coupling(PCC) through an LC filter, which is composed of the inverter-side inductor La,b,c and filter capacitor Ca,b,c. U_{dc} denotes the dc voltage. e_{cabc} and e_{uabc} are the capacitor voltage and output voltage of the inverter, respectively.

The 2018 International Power Electronics Conference

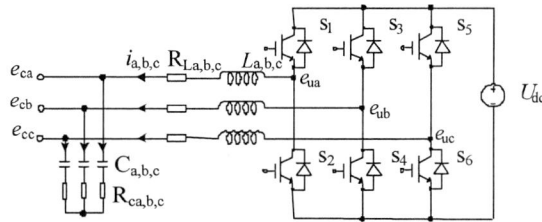

Fig. 1. Topology of the VSG

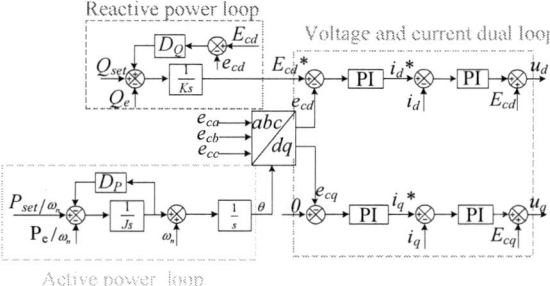

Fig. 2. Control diagram of the VSG

According to Fig. 1, the time-domain mathematical model of three-phase VSG in dq domain can be described as:

$$\begin{bmatrix} e_{ud} \\ e_{uq} \end{bmatrix} = L \frac{d}{dt} \begin{bmatrix} i_d \\ i_q \end{bmatrix} + \begin{bmatrix} R_L & -\omega L \\ \omega L & R_L \end{bmatrix} \begin{bmatrix} i_d \\ i_q \end{bmatrix} + \begin{bmatrix} e_{cd} \\ e_{cq} \end{bmatrix} \quad (1)$$

$$\begin{bmatrix} x_d \\ x_q \end{bmatrix} = \frac{2}{3} \begin{bmatrix} \sin \omega_0 t & \sin(\omega_0 t - \frac{2\pi}{3}) & \sin(\omega_0 t + \frac{2\pi}{3}) \\ \cos \omega_0 t & \cos(\omega_0 t - \frac{2\pi}{3}) & \cos(\omega_0 t + \frac{2\pi}{3}) \end{bmatrix} \cdot \begin{bmatrix} x_a \\ x_b \\ x_c \end{bmatrix} \quad (2)$$

where ω is the angular frequency of the capacitor voltage and Park's transformation of (2) is used to convert the phase domain variables into the dq domain.

For a balanced operation under steady-state, the capacitor voltage and output voltage of the inverter are set as E_{cdq0} and E_{udq0}, respectively. Meanwhile, P_0 and Q_0 denote the active power and reactive power in the DC steady state, so are current in the inductor, I_{dq0} and angular frequency ω_0. Hence, the state equations of VSC at steady DC point are expressed as:

$$\begin{bmatrix} E_{ud0} \\ E_{uq0} \end{bmatrix} = \begin{bmatrix} 0 \\ 0 \end{bmatrix} + \begin{bmatrix} R_L & -\omega_0 L \\ \omega_0 L & R_L \end{bmatrix} \begin{bmatrix} I_{d0} \\ I_{q0} \end{bmatrix} + \begin{bmatrix} E_{cd0} \\ E_{cq0} \end{bmatrix} \quad (3)$$

$$\begin{cases} P_0 = \dfrac{3}{2}(E_{cd0}I_{d0} + E_{cq0}I_{q0}) \\ Q_0 = \dfrac{3}{2}(E_{cq0}I_{d0} - E_{cd0}I_{q0}) \end{cases} \quad (4)$$

The extended summary should be up to It is assumed that any state variable x in (1) is equal to the quiescent value X plus superimposed small ac variation $\hat{x}$. Then, applying the assumption to (1) and only first-order kept, we have:

$$\begin{bmatrix} \hat{e}_{ud} \\ \hat{e}_{uq} \end{bmatrix} = L \frac{d}{dt} \begin{bmatrix} \hat{i}_d \\ \hat{i}_q \end{bmatrix} + \begin{bmatrix} R_L & -\omega_0 L \\ \omega_0 L & R_L \end{bmatrix} \begin{bmatrix} \hat{i}_d \\ \hat{i}_q \end{bmatrix} + \begin{bmatrix} -L \cdot I_{q0} \\ L \cdot I_{d0} \end{bmatrix} \hat{\omega} \quad (5)$$

Making Laplace transformation to (5) gives:

$$\begin{bmatrix} \hat{e}_{ud} \\ \hat{e}_{uq} \end{bmatrix} = \begin{bmatrix} sL + R_L & -\omega_0 L \\ \omega_0 L & sL + R_L \end{bmatrix} \begin{bmatrix} \hat{i}_d(s) \\ \hat{i}_q(s) \end{bmatrix} + \begin{bmatrix} \hat{e}_{cd}(s) \\ \hat{e}_{cq}(s) \end{bmatrix} + \begin{bmatrix} -L \cdot I_{q0} \\ L \cdot I_{d0} \end{bmatrix} \hat{\omega}(s) \quad (6)$$

Then, k_{pwm} is introduced to mimic the PWM function, which is equal to the ratio of the dc voltage and the peak to peak of the delta carrier. Thus, (6) can be further described as :

$$\begin{bmatrix} k_{pwm} \cdot \hat{u}_d \\ k_{pwm} \cdot \hat{u}_q \end{bmatrix} = \begin{bmatrix} sL + R_L & -\omega_0 L \\ \omega_0 L & sL + R_L \end{bmatrix} \begin{bmatrix} \hat{i}_d(s) \\ \hat{i}_q(s) \end{bmatrix} + \begin{bmatrix} \hat{e}_{cd}(s) \\ \hat{e}_{cq}(s) \end{bmatrix} + \begin{bmatrix} -L \cdot I_{q0} \\ L \cdot I_{d0} \end{bmatrix} \hat{\omega}(s) \quad (7)$$

The VSG in this paper adopts the cascaded dual-loop control strategy, which makes the VSG have fast voltage and current response. The control diagram is shown in Fig. 2, where Q_{set} and P_{set} are the set output reactive power and active value, respectively.

According to the voltage and current dual-loop, the modulation signal is:

$$\begin{bmatrix} \hat{u}_d(s) \\ \hat{u}_q(s) \end{bmatrix} = H_c(s) \begin{bmatrix} \hat{i}_d^*(s) - \hat{i}_d(s) \\ \hat{i}_q^*(s) - \hat{i}_q(s) \end{bmatrix} \quad (8)$$

Moreover, the reference current $\hat{i}_{dq}^*$ is expressed as:

$$\begin{bmatrix} \hat{i}_d^*(s) \\ \hat{i}_q^*(s) \end{bmatrix} = H_v(s) \begin{bmatrix} \hat{E}_{cd}^*(s) - \hat{e}_{cd}(s) \\ 0 - \hat{e}_{cq}(s) \end{bmatrix} \quad (9)$$

where $H_c(s)$ and $H_v(s)$ denote the function of current controller and voltage controller.

Aimed at the power-loop control, the following expressions are obtained.

As for the reactive-loop control, we have

$$\begin{cases} \hat{E}_{cd}^*(s) = \dfrac{-\hat{Q}_e(s) - D_Q \hat{e}_{cd}}{Ks} \\ \hat{Q}_e(s) = \dfrac{3}{2}\left(\begin{bmatrix} E_{cq0} & -E_{cq0} \end{bmatrix} \begin{bmatrix} \hat{i}_d(s) \\ \hat{i}_q(s) \end{bmatrix} + \begin{bmatrix} I_{q0} & I_{d0} \end{bmatrix} \begin{bmatrix} \hat{e}_{cd}(s) \\ \hat{e}_{cq}(s) \end{bmatrix} \right) \end{cases} \quad (10)$$

As for the active-loop control, we have

$$\begin{cases} \dfrac{-\hat{P}_e(s)/\omega_n - D_P\hat{\omega}(s)}{Js} = \hat{\omega} \Rightarrow \hat{\omega}(s) = \dfrac{-3\hat{P}_e(s)}{2\omega_n(D_P + Js)} \\ \\ \hat{P}_e(s) = \dfrac{3}{2}\begin{bmatrix} E_{cd0} & E_{cq0} \end{bmatrix}\begin{bmatrix} \hat{i}_d(s) \\ \hat{i}_q(s) \end{bmatrix} + \begin{bmatrix} I_{d0} & I_{q0} \end{bmatrix}\begin{bmatrix} \hat{e}_{cd}(s) \\ \hat{e}_{cq}(s) \end{bmatrix} \end{cases}$$

(11)

In turn, substitute (10) into (9), substitute (9) into (8), and substitute (8) into (7), then we have:

$$\mathbf{B}(s)\big|_{0,0} = 1 - \frac{3k_{pwm} \cdot I_{q0} \cdot H_c(s)H_v(s)}{2Ks} + k_{pwm}H_c(s)H_v(s) + \frac{D_Q \cdot k_{pwm} \cdot H_c(s)H_v(s)}{Ks} + \frac{3L \cdot I_{d0} \cdot I_{q0}}{2\omega_0(D_P + Js)}$$

$$\mathbf{B}(s)\big|_{0,1} = \frac{3k_{pwm} \cdot I_{d0} \cdot H_c(s)H_v(s)}{2Ks} + \frac{3L \cdot I_{q0}^2}{2\omega_0(D_P + Js)}$$

$$\mathbf{B}(s)\big|_{1,0} = \frac{-3L \cdot I_{d0}^2}{2\omega_0(D_P + Js)}$$

$$\mathbf{B}(s)\big|_{1,1} = 1 + k_{pwm}H_c(s)H_v(s) - \frac{3L \cdot I_{d0} \cdot I_{q0}}{2\omega_0(D_P + Js)}$$

$$\mathbf{A}(s)\big|_{0,0} = \frac{3k_{pwm} \cdot E_{cq0} \cdot H_c(s)H_v(s)}{2Ks} + k_{pwm}H_c(s) + sL + \frac{3L \cdot E_{cd0} \cdot I_{q0}}{2\omega_0(D_P + Js)} + R_L$$

$$\mathbf{A}(s)\big|_{0,1} = \frac{-3k_{pwm} \cdot E_{cd0} \cdot H_c(s)H_v(s)}{2Ks} - \omega_0 L + \frac{3L \cdot E_{cq0} \cdot I_{q0}}{2\omega_0(D_P + Js)}$$

$$\mathbf{A}(s)\big|_{1,0} = \omega_0 L - \frac{3L \cdot E_{cd0} \cdot I_{d0}}{2\omega_0(D_P + Js)}$$

$$\mathbf{A}(s)\big|_{1,1} = k_{pwm}H_c(s) + sL - \frac{3L \cdot E_{cq0} \cdot I_{d0}}{2\omega_0(D_P + Js)} + R_L$$

When considering the influence brought by the active input disturbance $\hat{\theta}$, which directly makes perturbations for the dq transform module. Hence, the small perturbations variable, like the variables $\hat{i}_d$, $\hat{i}_q$, $\hat{u}_{cd}$, $\hat{u}_{cq}$ can be corrected as follows.

Considering θ is equal to θ_0 plus $\hat{\theta}$, the transform module can be described as:(zero axis is not considered)

$$T_{dq/abc}(\theta) = \frac{2}{3}\begin{bmatrix} \cos(\theta_0 + \hat{\theta}) & \cos(\theta_0 + \hat{\theta} - \frac{2}{3}\pi) & \cos(\theta_0 + \hat{\theta} + \frac{2}{3}\pi) \\ -\sin(\theta_0 + \hat{\theta}) & -\sin(\theta_0 + \hat{\theta} - \frac{2}{3}\pi) & -\sin(\theta_0 + \hat{\theta} + \frac{2}{3}\pi) \end{bmatrix}$$

The above equation can be further transformed as follows:

$$\begin{cases} T_{dq/abc}(\theta) = T_{dq/abc}(\hat{\theta}) \cdot T_{dq/abc}(\theta_0) \\ \\ T_{dq/abc}(\theta_0) = \frac{2}{3}\begin{bmatrix} \cos(\theta_0) & \cos(\theta_0 - \frac{2}{3}\pi) & \cos(\theta_0 + \frac{2}{3}\pi) \\ -\sin(\theta_0) & -\sin(\theta_0 - \frac{2}{3}\pi) & -\sin(\theta_0 + \frac{2}{3}\pi) \end{bmatrix} \\ \\ T_{dq/abc}(\hat{\theta}) = \begin{bmatrix} \cos\hat{\theta} & \sin\hat{\theta} \\ -\sin\hat{\theta} & \cos\hat{\theta} \end{bmatrix} \end{cases}$$

(13)

Suppose $\hat{i}_{dq} = T_{dq/abc}(\theta_0) * \hat{i}_{abc}$, $I_{dq0} = T_{dq/abc}(\theta_0) * I_{abc0}$, $i_{dq} = T_{dq/abc}(\theta_1) * i_{abc}$ and $\hat{i}_{dq1} = i_{dq} - I_{dq0}$, with (13)

$$-\mathbf{A}(s)\begin{bmatrix} \hat{i}_d(s) \\ \hat{i}_q(s) \end{bmatrix} = \mathbf{B}(s)\begin{bmatrix} \hat{e}_{cd}(s) \\ \hat{e}_{cq}(s) \end{bmatrix}$$

(12)

Thus, $\mathbf{Z1}_{VSG}(s) = \mathbf{B}(s)^{-1}\mathbf{A}(s)$ is the output impedance of VSG, where the filter capacitor is not considered here.

combined, and the corrected current disturbance $\hat{i}_{dq1}$ is given as:

$$\begin{cases} \hat{i}_{d1}(s) = \hat{i}_d(s) + I_{q0} \cdot \hat{\theta}(s) \\ \hat{i}_{q1}(s) = \hat{i}_q(s) - I_{d0} \cdot \hat{\theta}(s) \end{cases}$$

(14)

Similarly, the corrected voltage disturbance $\hat{u}_{dq1}$ is given as :

$$\begin{cases} \hat{u}_{cd1}(s) = \hat{u}_{cd}(s) + E_{uq0} \cdot \hat{\theta}(s) \\ \hat{u}_{cq1}(s) = \hat{u}_{cq}(s) - E_{ud0} \cdot \hat{\theta}(s) \end{cases}$$

(15)

As $\theta(s) = 1/s * \hat{\omega}(s)$, substituting (11) into (14) and (15) gives:

$$\begin{bmatrix} \hat{u}_{cd1}(s) \\ u_{cq1}(s) \end{bmatrix} = \mathbf{A}_u(s)\begin{bmatrix} \hat{i}_d(s) \\ \hat{i}_q(s) \end{bmatrix} + \mathbf{B}_u(s)\begin{bmatrix} \hat{u}_{cd}(s) \\ \hat{u}_{cq}(s) \end{bmatrix}$$

(16)

$$\mathbf{A}_u(s) = \begin{bmatrix} \dfrac{-3E_{cq0}E_{cd0}}{2\omega_n(D_P + Js)s} & \dfrac{-3E_{cq0}E_{cq0}}{2\omega_n(D_P + Js)s} \\ \\ \dfrac{3E_{cd0}E_{cd0}}{2\omega_n(D_P + Js)s} & \dfrac{3E_{cq0}E_{cd0}}{2\omega_n(D_P + Js)s} \end{bmatrix}$$

1284

$$\mathbf{B}_u(s)=\begin{bmatrix} 1-\dfrac{3E_{cq0}I_{d0}}{2\omega_n\left(D_P+Js\right)s} & \dfrac{-3E_{cq0}I_{q0}}{2\omega_n\left(D_P+Js\right)s} \\ \dfrac{3E_{cd0}I_{d0}}{2\omega_n\left(D_P+Js\right)s} & 1+\dfrac{3E_{cd0}I_{q0}}{2\omega_n\left(D_P+Js\right)s} \end{bmatrix}$$

Similar to the form of (16), (14) can be rearranged as:

$$\begin{bmatrix}\hat{i}_{d1}(s)\\ \hat{i}_{q1}(s)\end{bmatrix}=\mathbf{A}_i(s)\begin{bmatrix}\hat{i}_d(s)\\ \hat{i}_q(s)\end{bmatrix}+\mathbf{B}_i(s)\begin{bmatrix}\hat{u}_{cd}(s)\\ \hat{u}_{cq}(s)\end{bmatrix} \quad (17)$$

$$\mathbf{B}_i(s)=\begin{bmatrix} \dfrac{-3I_{q0}I_{d0}}{2\omega_n\left(D_P+Js\right)s} & \dfrac{-3I_{cq0}I_{cq0}}{2\omega_n\left(D_P+Js\right)s} \\ \dfrac{3I_{cd0}I_{cd0}}{2\omega_n\left(D_P+Js\right)s} & \dfrac{3I_{cq0}I_{cd0}}{2\omega_n\left(D_P+Js\right)s} \end{bmatrix}$$

$$\mathbf{A}_i(s)=\begin{bmatrix} 1-\dfrac{3I_{q0}E_{cd0}}{2\omega_n\left(D_P+Js\right)s} & \dfrac{-3I_{q0}E_{cq0}}{2\omega_n\left(D_P+Js\right)s} \\ \dfrac{3I_{d0}E_{cd0}}{2\omega_n\left(D_P+Js\right)s} & 1+\dfrac{3I_{d0}E_{cq0}}{2\omega_n\left(D_P+Js\right)s} \end{bmatrix}$$

Finally, the nominal output impedance of VSG (the filter capacitor is not consider here) $\mathbf{Z}11_{VSG}(s)$ is obtained by substituting (16) and (17) into (12):

$$\mathbf{Z}11_{VSG}=\left(\mathbf{A}_u(s)+\mathbf{B}_u(s)\cdot\mathbf{Z}_{VSG1}\right)\cdot$$
$$\left(\mathbf{A}_i(s)+\mathbf{B}_i(s)\cdot\mathbf{Z}1_{VSG}\right)^{-1} \quad (18)$$

As for the three-phase filter capacitor and series impedance R_C, the impedance $\mathbf{Z}_C(s)$ in the dq axis is expressed as:

$$\mathbf{Z}_C(s)=\begin{bmatrix} \dfrac{1}{sC+1/R_C} & -\omega_0 C \\ \omega_0 C & \dfrac{1}{sC+1/R_C} \end{bmatrix}^{-1} \quad (19)$$

In parallel $\mathbf{Z}_C(s)$ with $\mathbf{Z}1_{VSG}(s)$, the final output impedance of VSG is $\mathbf{Z}_{VSG}(s)=\mathbf{Z}_C(s)//\mathbf{Z}1_{VSG}(s)$.

III. SIMULATION VERIFICATION

The developed output impedance model of VSG are validated in software PSIM 9.0. Fig. 3 describes the impedance measurement in dq coordinate. A series of frequency of small voltage signal in d axis and q axis are injected respectively on the AC side of the converter. Then measure the current in d axis and q axis, and the impedance of the converter is finally obtained.

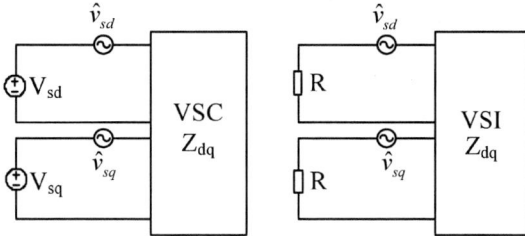

Fig. 3. Impedance measurement diagram in dq coordinates

The specific methods are as follows: A first

perturbation can be made by injecting only voltage $\hat{v}_{sd}(s)$ and setting $\hat{v}_{sq}(s)$ to zero. The response is measured to obtained (13)

$$\begin{bmatrix}\hat{v}_{d1}(s)\\ \hat{v}_{q1}(s)\end{bmatrix}=\mathbf{Z}_{dq}(s)\begin{bmatrix}\hat{i}_{d1}(s)\\ \hat{i}_{q1}(s)\end{bmatrix} \quad (20)$$

The second perturbation is then created by injecting voltage $\hat{v}_{sq}(s)$ and setting $\hat{v}_{sd}(s)$ to zero. Another group of response is measured to obtained (21)

$$\begin{bmatrix}\hat{v}_{d2}(s)\\ \hat{v}_{q2}(s)\end{bmatrix}=\mathbf{Z}_{dq}(s)\begin{bmatrix}\hat{i}_{d2}(s)\\ \hat{i}_{q2}(s)\end{bmatrix} \quad (21)$$

Combine (20) and (21) to get the impedance matrix in dq coordinate:

$$\mathbf{Z}_{dq}(s)=\begin{bmatrix}\hat{v}_{d1}(s) & \hat{v}_{d2}(s)\\ \hat{v}_{q1}(s) & \hat{v}_{q2}(s)\end{bmatrix}\begin{bmatrix}\hat{i}_{d1}(s) & \hat{i}_{d2}(s)\\ \hat{i}_{q1}(s) & \hat{i}_{q2}(s)\end{bmatrix}^{-1} \quad (22)$$

TABLE I
THE PARAMETERS OF MAIN CIRCUIT AND CONTROL LOOP

Symbol	Meaning	Value
U_{dc}	DC voltage	700V
L	Filter inductance	3.2mH
C	Filter capacitor	10uF
f	Switching frequency	100000
D_P	Droop coefficient	10
J	Inertia coefficient	0.05882
D_Q	Droop coefficient	400
K	Inertia coefficient	1.6667
$H_c(s)$	Current controller	20+100/s
$H_v(s)$	Voltage controller	1+20/s

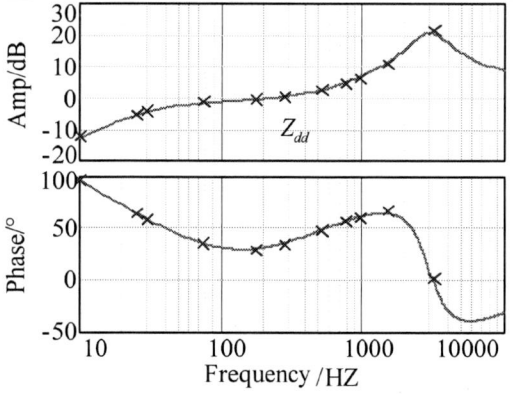

(a)

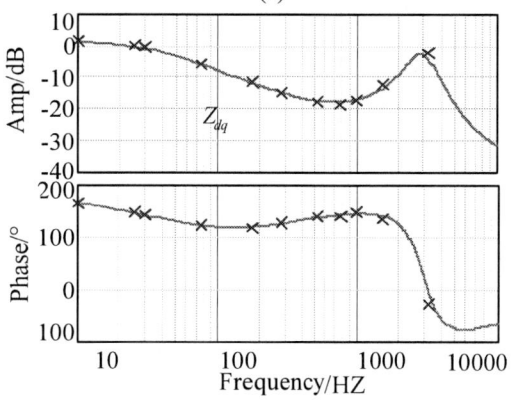

The 2018 International Power Electronics Conference

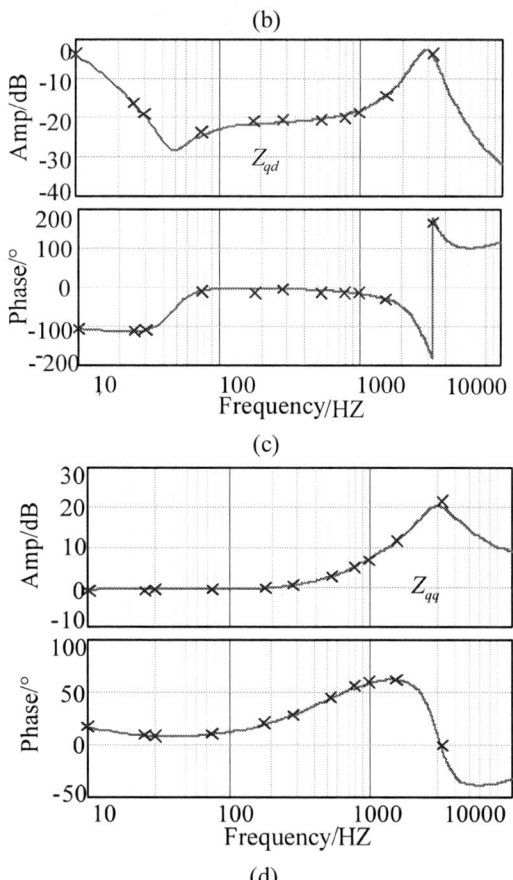

(b)

(c)

(d)

Fig. 4. Impedance validations for VSG(Curve: analytical model. Point: measured impedance)

The main parameters of the VSG in the simulation are shown in Table I. The active power P_{set} is set as 5000W, the reactive power Q_{set} is set 0var, and voltage angular frequency is set 100π. Fig.4 shows the contrast results between the analytical model and the point-by-point results. As shown in Fig. 4, the point-by- point simulation result is in agreement with the analytical model proposed in Section II in the range from 10HZ to 3000HZ, which proves the validity of the proposed impedance model. Moreover, it's found that Z_{dd}, Z_{dq} and Z_{qq} plays the leading role in low-frequency stage, while Z_{dd} and Z_{qq} woks in high-frequency stage. Z_{dq} and Z_{dd} shows inductive and have high phase value in sub-synchronous, which may cause subsynchronous oscillation in high series compensation degree.

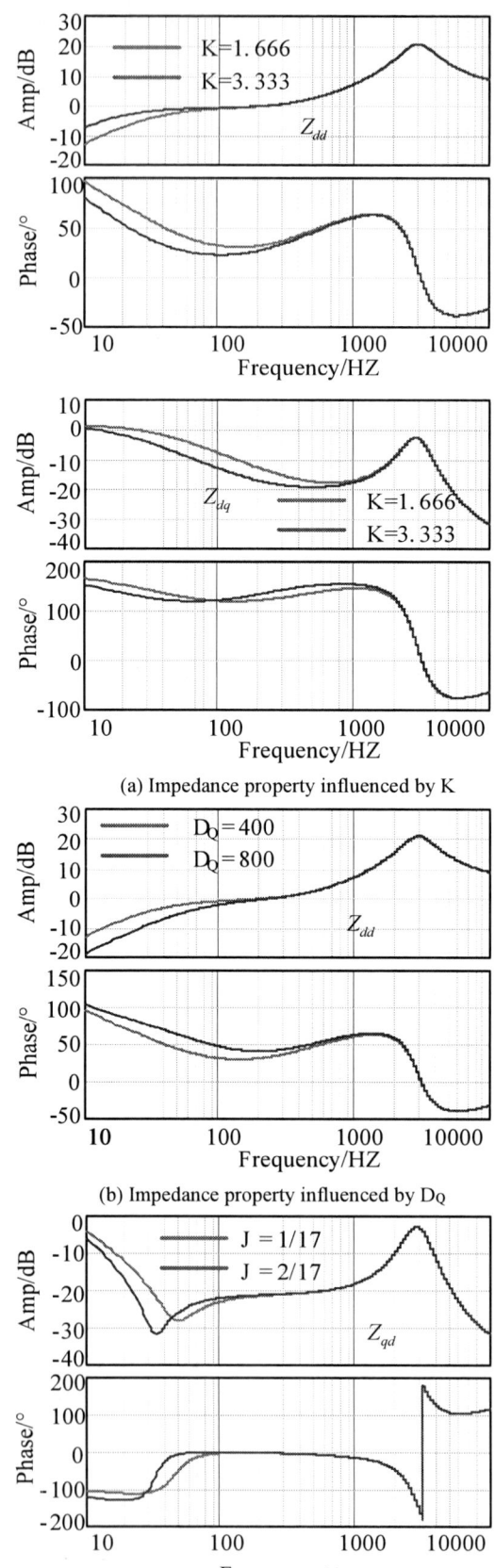

(a) Impedance property influenced by K

(b) Impedance property influenced by D_Q

1286

The 2018 International Power Electronics Conference

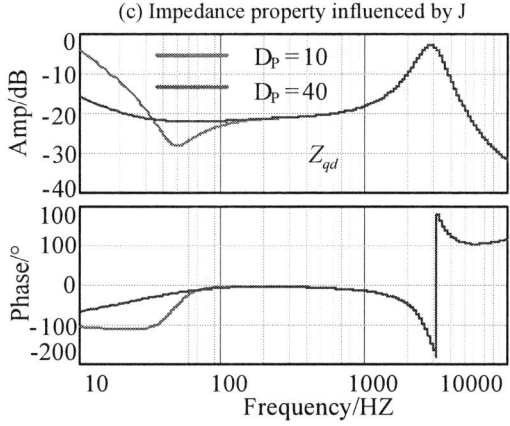

(c) Impedance property influenced by J

(d) Impedance property influenced by D$_P$

Fig. 4. The relationship between *dq* impedance and power loop parameter

Fig. 5 shows the impedance property influenced by power loop parameters. As Fig. 5(a)-(b) shows, the inertia coefficient K and droop coefficient D$_Q$ in reactive loop influence Z_{dd} and Z_{dq}. That is, increasing K can reduce the amplitude and phase in the range from 10-100HZ, while increasing D$_Q$ reduces the amplitude and increases the phase. As Fig. 5(c)-(d) shows, increasing inertia coefficient J will reduce the impedance amplitude and phase value of Z_{qd} around 10HZ, while increasing droop coefficient D$_P$ in active loop can reduce the amplitude and increase the phase of Z_{qd} around 10HZ.

It's seen that proper parameter value plays a leading role to the impedance property of VSG, which will further influence the performance of grid-connected VSG system.

IV. EXPERIMENTAL VERIFICATION

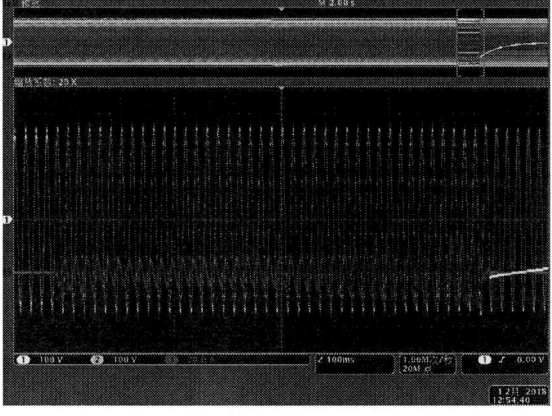

(a) D$_P$ decreases

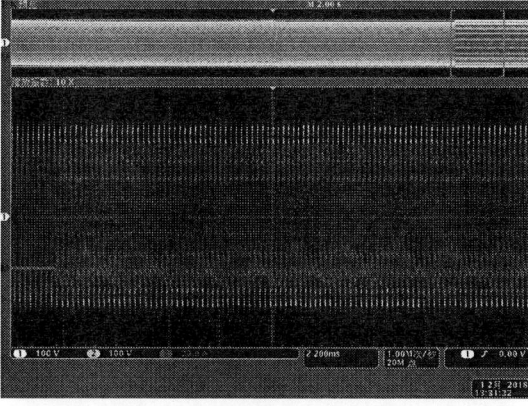

(b) J increases

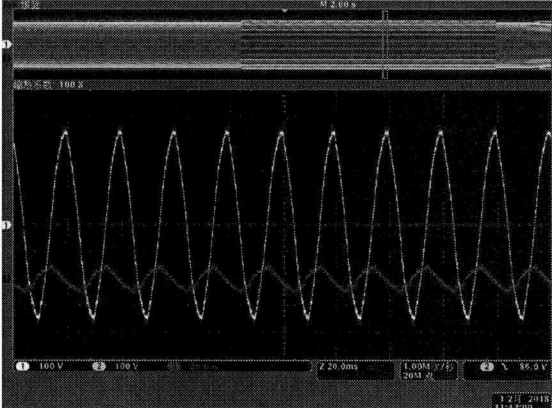

(c) K increases

Fig. 6. VSG unstable waveforms

Fig. 7. VSG stable waveforms

Fig.6 shows VSG unstable waveforms under different situations, which indicates the value of the control parameters will influence the output of VSG. As the experimental results show, increased D$_p$, reduced J and K are beneficial to the stability of the VSG system. Fig.7 shows the output voltage and current when VSG works stable.

V. CONCLUSION

This paper provides an in-depth investigation of the

VSG, whose full-band output impedance model in dq framework is proposed, derived and verified, leading to the extraction of information for system stability.

It is found that Z_{dd}, Z_{dq} and Z_{qq} plays the leading role in low-frequency stage, while Z_{dd} and Z_{qq} woks in high-frequency stage. Z_{dq} and Z_{dd} shows inductive and have high phase value in sub-synchronous, which may cause subsynchronous oscillation in high series compensation degree.

ACKNOWLEDGMENT

This work was supported in part by the National Key Research and Development Program of China (2016YFB0900404), by the Cooperative Innovation Fund of Jiangsu Province-the Prospective and Joint Research Project (BY2015070-18),and Science and Technology Project of STATE GRID Corporation of China (PD71-17-008).

REFERENCES

[1] Chen Z, Xu C, Rygg A, et al. , "Sequence Domain SISO Equivalent Models of a Grid-tied Voltage Source Converter System for Small-Signal Stability Analysis," *IEEE Transactions on Energy Conversion, early access.*

[2] M. Li et al., "A novel virtual synchronous generator control strategy based on improved swing equation emulating and power decoupling method," *2016 IEEE Energy Conversion Congress and Exposition (ECCE)*, Milwaukee, WI, 2016, pp. 1-7.

[3] M. Young, "The PWM strategy on DC-DC converter, " *IEEE Journal of Industry Applications*, vol. 28, no. 15, pp. 123-129, 1989.

[4] R. D. Middlebrook, "Input filter considerations in design and application of switching regulators," *in Rec.1976 IEEE Ind. Appl. Soc. Annu. Meeting,*pp. 366–382.

[5] S. Hiti, V. Vlatkovic, D. Borojevic, and F. C. Lee, "A new control algorithm for three-phase PWM buck rectifier with input displacement factor compensation," *IEEE Trans. Power Electron.*, vol. 9, no. 2, pp. 173–180,Mar. 1994.

[6] J. Alipoor, Y. Miura, and T. Ise, "Power system stabilization using virtual synchronous generator with alternating moment of inertia," *IEEE Journal of Emerging and Selected Topics in Power Electronics*, vol. 3, no.1, pp.451-458, 2015.

The 2018 International Power Electronics Conference

An MTPA Control Method
of a PMSM and a SynRM Based on a DTC
in the Stator Flux Linkage Synchronous Frame

Gimpei Itoh, Yukinori Inoue, Shigeo Morimoto, and Masayuki Sanada
Graduate School of Engineering, Osaka Prefecture University, Sakai, Japan
sxb01016@edu.osakafu-u.ac.jp

Abstract-This paper proposes a method of maximum torque per ampere (MTPA) control for a direct torque control-based (DTC-based) permanent magnet synchronous motor (PMSM) drive system. In the proposed method, the reference flux for MTPA operation is calculated using a mathematical model in a rotating reference frame that is synchronized with the stator flux linkage vector (M-T frame). The mathematical model gives the relationship between the stator flux linkage and the armature current. In this paper, the proposed MTPA control method is applied to three types of motors: an interior PMSM (IPMSM), a surface PMSM (SPMSM), and a synchronous reluctance motor (SynRM). The mathematical models agree with the measured MTPA operating points of the tested motors. Simulation results also demonstrate the validity of the proposed method.

Keywords—Permanent magnet synchronous motor, maximum torque per ampere, mathematical model, direct torque control

I. INTRODUCTION

Permanent magnet synchronous motors are used as high-efficiency motors for industrial applications and household electrical appliances. A DTC method has been reported as a high-performance PMSM drive [1]-[3]. In PMSM control, e.g., MTPA control, a reference frame synchronized with the rotor (*d-q* frame) is generally used [4]-[8]. However, in many DTC-based drive systems, a stationary α-β reference frame is also used for control and estimation of torque and flux. Therefore, MTPA control in DTC requires numerous coordinate transformations and conversions.

The M-T frame is a rotating reference frame based on the stator flux linkage vector. The M-T frame can be applied to various motors [9]. Since DTC requires a reference value of the stator flux linkage for MTPA operation, DTC can easily be applied to a control law in the M-T frame that also requires the stator flux linkage vector.

In previous studies, several mathematical models for MTPA control in the M-T frame have been reported [10],[11]. The previous studies proposed a mathematical model for MTPA in the Ψ_s-i_T plane. In an IPMSM, MTPA control with the mathematical model was reported

[12]. Furthermore, when the motor is an SPMSM, the mathematical model for MTPA control can be approximated as a quadratic function of the T-axis current [11]. On the other hand, when the motor is a SynRM, the mathematical model for MTPA control can be approximated as a power x of the T-axis current. However, the application of DTC to these models has not yet been reported.

This paper proposes a method of MTPA control of a PMSM and a SynRM based on DTC in the M-T frame. By applying mathematical models suitable for the three motors to DTC, a simple MTPA control is available. For determining the parameters of the mathematical models, the operating points of the MTPA condition are measured in the tested motors. The proposed MTPA control is applied to DTC-based PMSM and SynRM drive systems, and the validity of the proposed system is confirmed by the simulation results.

II. MATHEMATICAL MODELS SUITABLE FOR MTPA CONTROL

A. Definition of the M-T Frame

Fig. 1 shows the vector diagram in the M-T frame. In Fig. 1, $\boldsymbol{\psi}_s$ is the stator flux linkage vector, $\boldsymbol{\psi}_a$ is the stator flux linkage vector due to the permanent magnet, $\boldsymbol{i}_a$ is the armature current vector, and i_d, i_q, i_M, and i_T are the *d*-, *q*-, M-, and T-axis components of $\boldsymbol{i}_a$, respectively. The *d-q* frame is based on the permanent magnet stator flux linkage vector $\boldsymbol{\psi}_a$, whereas the M-T frame is based on the

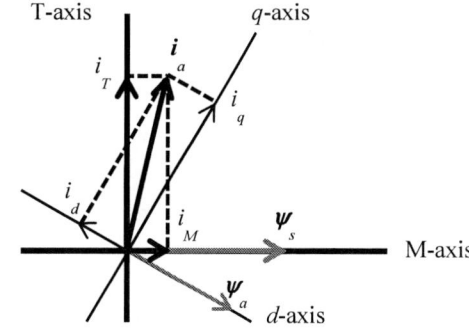

Fig. 1. Vector diagram in the M-T frame.

stator flux linkage vector $\boldsymbol{\psi}_s$ including the armature reaction.

B. Mathematical Model of the MTPA Curve

In the M-T frame, a mathematical model representing the MTPA curve as the relationship between the stator flux linkage Ψ_s and the T-axis current i_T is shown as follows [10]:

$$\Psi_s = \frac{2}{\pi}\left(L_T - b_T i_T\right) i_T \tan^{-1}\left(\frac{L_k}{\Psi_a} i_T\right) + \Psi_a \tag{1}$$

where Ψ_a is the magnet flux linkage, and L_T, L_k, and b_T are the parameters that compose the MTPA curve.

In the SPMSM, the mathematical model representing the MTPA curve can be written as [11]:

$$\Psi_s = k i_T^2 + \Psi_a \tag{2}$$

where k is a coefficient.

In the SynRM, the appropriate model is given by [11]:

$$\Psi_s = k i_T^x \tag{3}$$

where x is an exponent on i_T.

The parameters in (1)-(3) are determined from the measured points under the MTPA condition. The rotor and stator structures of the tested motors, i.e., the IPMSM, the SPMSM, and the SynRM, are shown in Fig. 2, and the parameters of these motors are shown in Table I.

Fig. 3 shows the measured points of the MTPA condition in the tested IPMSM. When the MTPA curve given by (1) passes through points A and B in Fig. 3, L_T and b_T are obtained as follows [10]:

$$\begin{bmatrix} L_T \\ b_T \end{bmatrix} = \begin{bmatrix} \dfrac{i_{TA} i_{TB}}{i_{TB} - i_{TA}}\left(Z_A - Z_B\right) \\[2mm] \dfrac{i_{TA} Z_A - i_{TB} Z_B}{i_{TB} - i_{TA}} \end{bmatrix}$$

$$\begin{bmatrix} Z_A \\ Z_B \end{bmatrix} = \begin{bmatrix} \left(\Psi_{sA} - \Psi_a\right) \Big/ \left(\dfrac{2}{\pi} i_{TA}^2 \tan^{-1}\left(\dfrac{L_k}{\Psi_a} i_{TA}\right)\right) \\[3mm] \left(\Psi_{sB} - \Psi_a\right) \Big/ \left(\dfrac{2}{\pi} i_{TB}^2 \tan^{-1}\left(\dfrac{L_k}{\Psi_a} i_{TB}\right)\right) \end{bmatrix} \tag{4}$$

where i_{TA} and i_{TB} are the T-axis currents at points A and B, and Ψ_{sA} and Ψ_{sB} are the stator flux linkages at points A and B, respectively.

When the value of L_k is given, L_T and b_T are obtained by (4) such that the mean square error of the stator flux linkage at the measured points of the MTPA condition is a minimum. The parameters L_T, L_k, and b_T determined for use in (1) are given as follows:

$L_T = 5.7$ mH, $L_k = 38$ mH, $b_T = -0.13$ mH/A.

Fig. 3 also shows the MTPA curve of (1) using the above parameters. In Fig. 3, (1) is approximately consistent with the measured points of the MTPA condition and is able to express the MTPA curve of the IPMSM.

Fig. 4 shows the measured points of the MTPA condition in the tested SPMSM. When (2) passes through point A in Fig. 4, k is obtained as [11]

$$k = \frac{\Psi_{sA} - \Psi_a}{i_{TA}^2} \tag{5}$$

where i_{TA} is the T-axis current at point A, and Ψ_{sA} is the stator flux linkage at point A.

The parameter k determined for use in (2) is as follows:

$k = 0.15$ mH/A.

Fig. 4 also shows the MTPA curve of (2) using the value for k shown above. In Fig. 4, (2) is approximately consistent with the measured points of the MTPA condition in the range of the T-axis current at point A. Hence, (2) is able to express the MTPA curve of the SPMSM within that range.

Fig. 5 shows the measured points of the MTPA condition in the tested SynRM. When (3) passes through point A in Fig. 5, k and x are obtained by [11]

$$k = \frac{\Psi_{sA}}{i_{TA}^x} \tag{6}$$

where i_{TA} is the T-axis current at point A, and Ψ_{sA} is the stator flux linkage at point A.

When the value of x is given, k is obtained by (6). Specifically, k is determined such that the mean square error of the stator flux linkage at the measured points of the MTPA condition is a minimum. The parameters k and x determined for use in (3) are as follows:

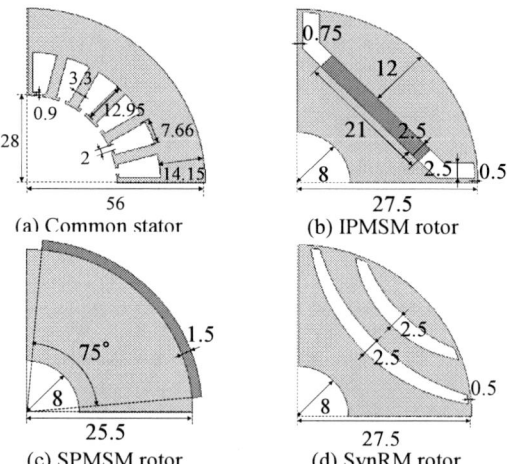

(a) Common stator   (b) IPMSM rotor

(c) SPMSM rotor   (d) SynRM rotor

Fig. 2. Rotor and stator structures of tested motors (Unit: [mm]).

TABLE I. Motor Parameters of Tested Motors

Item [Unit]	Value		
	IPMSM	SPMSM	SynRM
Number of pole pairs P_n		2	
d-axis inductance L_d [mH]	7.26	5.3	5.6
q-axis inductance L_q [mH]	$21.1 - 0.73\lvert i_q \rvert$	5.15	$21.4 - 0.59\lvert i_q \rvert$
Permanent magnet flux linkage Ψ_a [Wb]	0.098	0.128	0
Armature resistance R_a [Ω]		0.814	
Inertia moment J_m [kg m^2]		1.65×10^{-3}	
Coefficient of viscous friction D_r [Nm/(rad/s)]		3.0×10^{-4}	
Rated phase current I_e [A]		3.92	

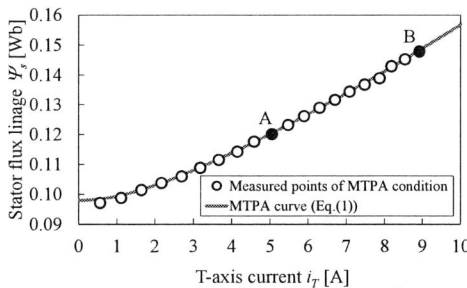

Fig. 3. Measured points of the MTPA condition and the MTPA curve given by (1) in an IPMSM (L_T = 5.7 mH, L_k = 38 mH, b_T = −0.13 mH/A).

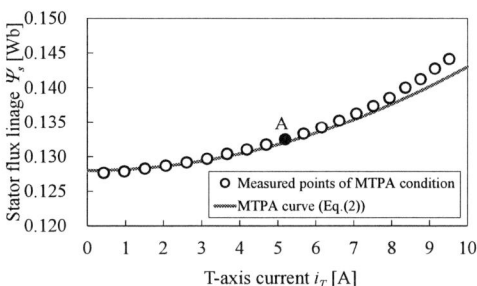

Fig. 4. Measured points of the MTPA condition and the MTPA curve given by (2) in the SPMSM (k = 0.15 mH/A).

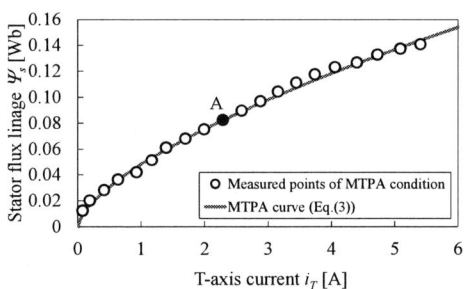

Fig. 5. Measured points of the MTPA condition and the MTPA curve given by (3) in the SynRM (k = 0.045, x = 0.65).

k = 0.048, x = 0.65.

Fig. 5 also shows the MTPA curve of (3) using the parameters shown above. In Fig. 5, (3) is approximately consistent with the measured points of the MTPA condition and can express the MTPA curve of the SynRM.

C. Measurement of Operating Points of the MTPA Condition

The operating points of the MTPA condition shown in Figs. 3-5 can be calculated in terms of the root mean square of a line-to-line voltage V_e, a phase current I_e, and the phase difference φ between V_e and I_e. The T-axis current i_T at the operating points of the MTPA condition is obtained as follows:

$$i_T = \sqrt{3} I_e \cos\left(\varphi - \frac{\pi}{6}\right) \qquad (7)$$

The stator flux linkage Ψ_s at the operating points of the MTPA condition is obtained as

$$\Psi_s = \frac{V_o}{\omega} \qquad (8)$$

where V_o is the induced voltage, and ω is the electric angular velocity. Here, V_o is calculated as follows:

$$V_o = \sqrt{\begin{aligned} &\left(V_e \sin\left(\varphi + \beta - \frac{\pi}{6}\right) - \sqrt{3} R_a I_e \sin\beta\right)^2 \\ &+ \left(V_e \cos\left(\varphi + \beta - \frac{\pi}{6}\right) - \sqrt{3} R_a I_e \cos\beta\right)^2 \end{aligned}} \qquad (9)$$

where β is the current phase.

In this paper, V_e and I_e in (7) and (9) were calculated using the fundamental component. Furthermore, even when V_e and I_e included harmonic components, the validity of the mathematical models was confirmed. However, the total harmonic distortion of the voltage and the current in the measurement is 2% to 20% in the tested motors. Hence, further discussion is necessary when the voltage and current include more harmonic components.

III. DTC-BASED PMSM DRIVE SYSTEM

A. DTC in the M-T Frame

Direct torque control can be applied to motor drive systems regardless of motor type. The DTC-based motor drive system can use the M-T frame because the estimated stator flux linkage vector is used for control. Fig. 6 shows a motor drive system based on DTC. The torque controller in the DTC is based on a proportional and integral (PI) controller. The estimated angular velocity of the stator flux linkage vector ($\hat{\omega}_s$) is obtained from the reference angular variation $\Delta\theta_s^*$. Here, $\hat{\omega}_s$ is integrated by a time integrator, and the estimated position $\hat{\theta}_s$, which is used for coordinate transformation, can be obtained. On the other hand, the estimated stator flux linkage $\hat{\Psi}_s$ is calculated by integrating flux error $\Delta\Psi_s$. From the obtained $\Delta\theta_s^*$, $\hat{\Psi}_s$, and $\Delta\Psi_s$, the reference induced voltages of the M- and T-axis components (v_{oM}^* and v_{oT}^*) are calculated.

B. Reference Flux Calculator With the Mathematical Model of the MTPA

As shown in Fig. 6, the reference stator flux linkage Ψ_s^* is required in DTC. In this paper, Ψ_s^* is calculated by a reference flux calculator using (1)-(3). In the M-T frame, since the stator flux linkage vector has a component only along the M-axis, the electromagnetic torque T_e is calculated using the relationship between the stator flux linkage and the T-axis current, as follows:

$$T_e = P_n \Psi_s i_T \qquad (10)$$

where P_n is the number of pole pairs.

In the conventional system, the reference stator flux linkage of the MTPA condition is generally calculated from the reference torque T_e^* using a look-up table. On the other hand, in the proposed system, the mathematical models of (1)-(3) are used to calculate $\Psi_s^{\prime *}$. However, (1)-(3) are not functions of torque but rather the T-axis current. Therefore, (10) is used to convert the reference torque T_e^* into the T-axis current i_T.

Fig. 7 shows the reference flux calculator for MTPA control [12]. The calculator brings the T-axis current to an operating point of the MTPA. In this method, the average value of i_T and i_{T1} is regarded as a reasonable operating point close to the MTPA condition.

By using the block diagram shown in Fig. 7, MTPA control can be performed. However, in the SynRM, as shown in Fig. 8, when the reference torque greatly varies in the state in which i_T is close to zero, i_{T1} becomes very large. As a result, the average value of i_T and i_{T1} (the value at point A) is far from the desired value of the MTPA operation point (the value at point P). Therefore, a method for bringing point A closer to point P by repeatedly using the calculator in Fig. 7 is used.

Fig. 9 shows a reference flux calculation method that is suitable for MTPA control of the SynRM [13]. In this method, the average value of i_{Tref} and i_{T2} is newly regarded as a reasonable operating point close to the MTPA condition. From Fig. 7, point B is closer to the point P than point A. Therefore, this calculation method is suitable for the SynRM. Fig. 10 shows a reference flux calculator when the reference value calculation is repeated once.

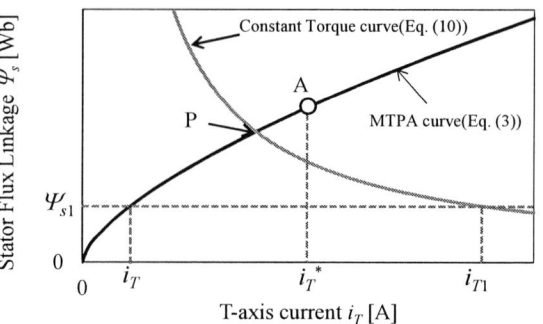

Fig. 8. Reference flux calculation
with the mathematical model in the SynRM.

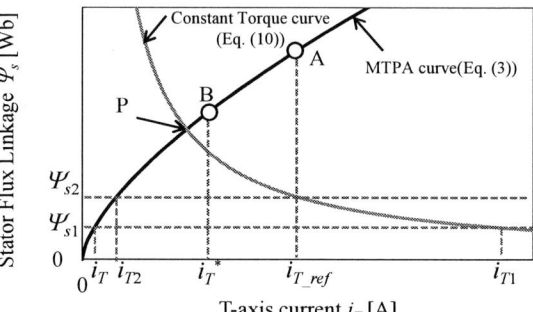

Fig. 9. Reference flux calculation for the SynRM.

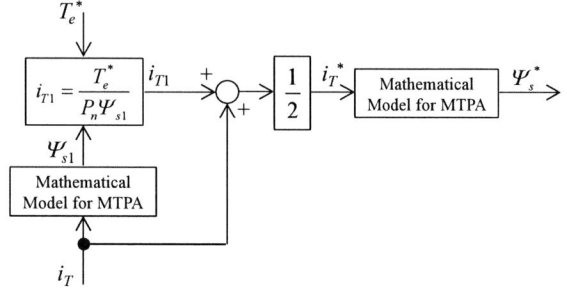

Fig. 7. Reference flux calculator
with the mathematical model of the MTPA.

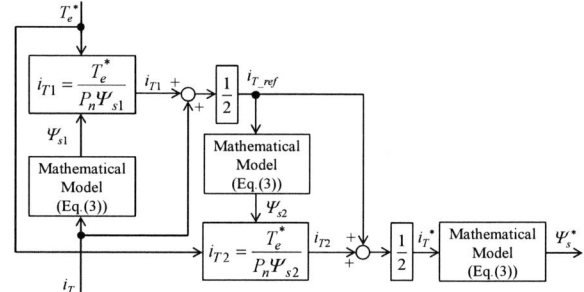

Fig. 10. Reference flux calculator suitable for the SynRM.

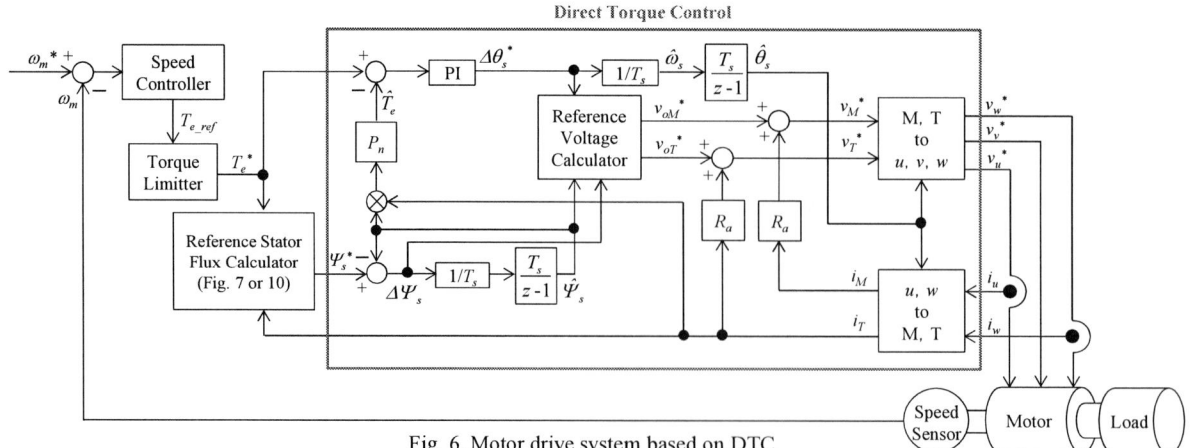

Fig. 6. Motor drive system based on DTC.

The 2018 International Power Electronics Conference

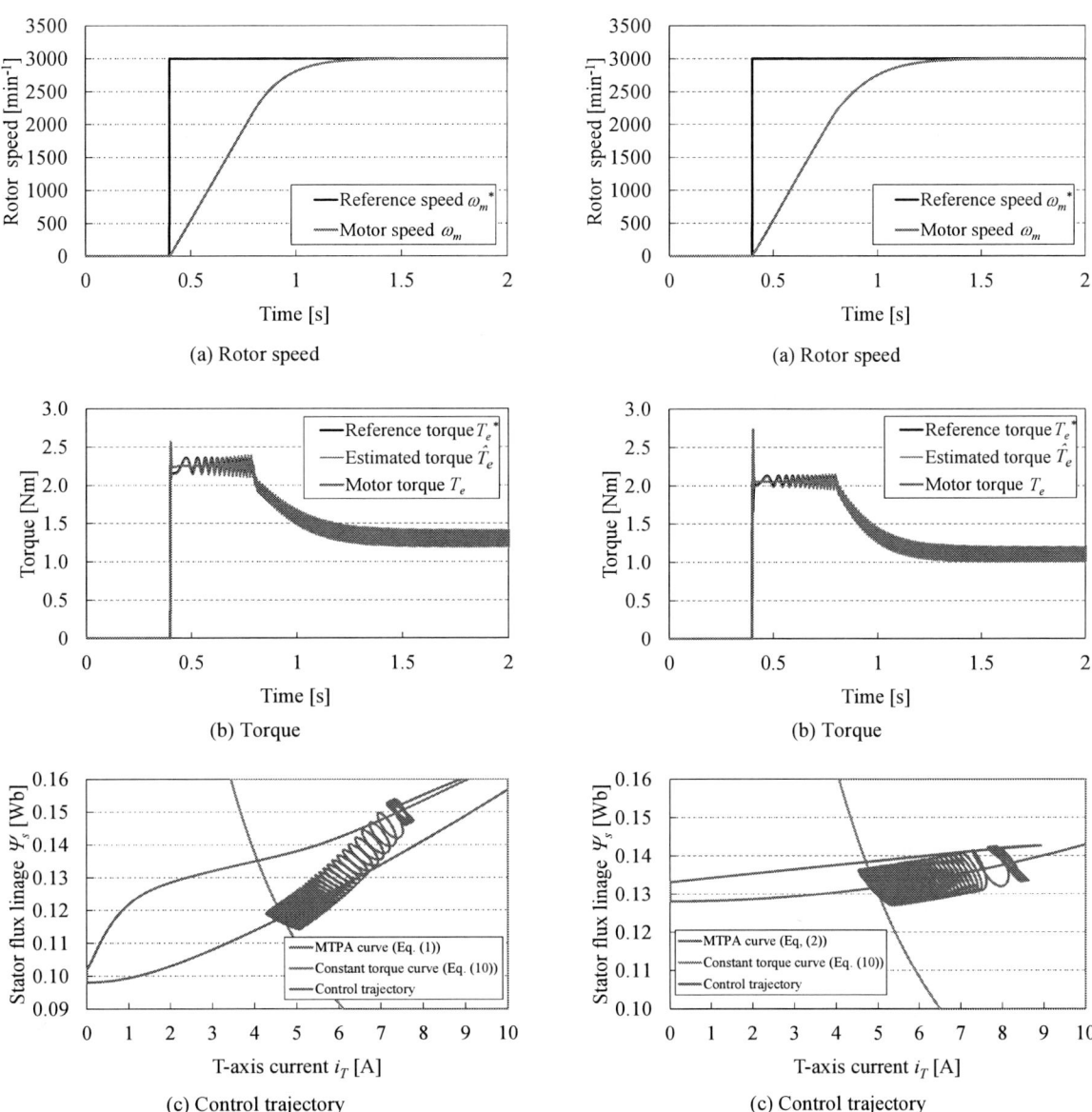

Fig.11. Acceleration characteristic in the IPMSM.

Fig.12. Acceleration characteristic in the SPMSM.

IV. OPERATING CHARACTERISTICS OF THE PROPOSED METHOD

The validity of the proposed system is evaluated by simulation. The parameters of the tested motors are shown in Table I. In this simulation, the reference speed was changed from 0 to 3,000 min⁻¹. Figs. 11-13 show the acceleration characteristics of the proposed system for the three motors. Fig. 11 shows the acceleration characteristic in the IPMSM. In Fig. 11(a), the rotor speed was found to increase stably and reached 3,000 min⁻¹. Fig. 11(b) shows the reference torque, the motor torque, and the estimated torque. In Fig. 11(b), although these values had ripple, the actual and estimated values approximately followed the reference value. Fig. 11(c) shows the control trajectory on the Ψ_s-i_T plane. In Fig.

11(c), the control trajectory did not correspond to the MTPA curve because the reference value increased in very short time. In a steady state, although the control trajectory varied due to the ripple of the stator flux linkage and the T-axis current, the proposed system could control the point of intersection between the MTPA curve and the constant torque curve.

Fig. 12 shows the acceleration characteristic for the SPMSM. In Fig. 12(a), the rotor speed reached 3,000 min⁻¹, and in Fig. 12(b), the actual and estimated values approximately followed the reference value. From Fig. 12(c), MTPA control could be achieved in a steady state.

Fig. 13 shows the acceleration characteristic in the SynRM. In Figs. 13(a) and 13(b), the rotor speed reached 3,000 min⁻¹, and the actual and estimated values approximately followed the reference value. In Fig. 13(c),

1293

The 2018 International Power Electronics Conference

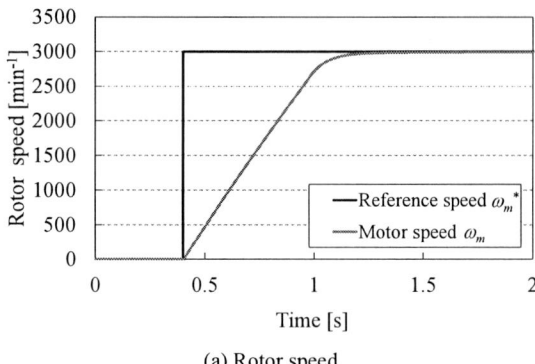

(a) Rotor speed

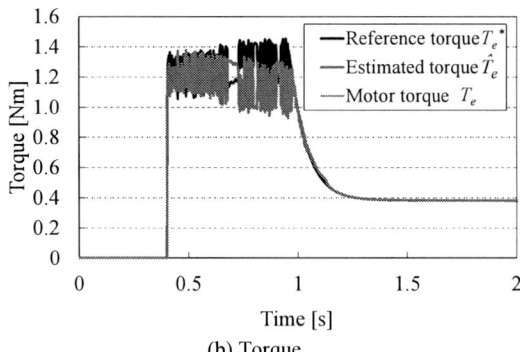

(b) Torque

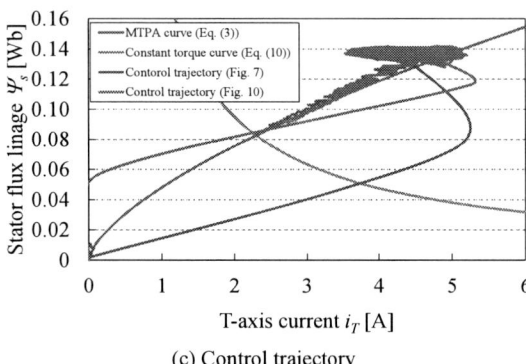

(c) Control trajectory

Fig. 13. Acceleration characteristic in the SynRM.

a control trajectory using the method of Fig. 7 is shown. From Fig. 13(c), the control trajectory of Fig. 10 was closer to the MTPA curve than that of Fig. 7, and more efficient MTPA control could be performed.

V. CONCLUSIONS

In this paper, the mathematical models for MTPA for the IPMSM, SPMSM, and SynRM were applied to a DTC-based PMSM drive system. The validity of the proposed MTPA control was confirmed by simulation. The mathematical models for the MTPA control agrees well with the measured results. The proposed system achieved MTPA control and realized high-efficiency operation.

REFERENCES

[1] M. F. Rahman, L. Zhong, and K. W. Lim, "A Direct Torque-Controlled Interior Permanent Magnet Synchronous Motor Drive Incorporating Field Weakening," *IEEE Transactions on Industrial Applications*, Vol. 34, No. 6, pp. 1246-1253, 1998.

[2] L. Tang, L Zhong, M. F. Rahman, and Y. Hu, "A Novel Direct Torque Controlled Interior Permanent Magnet Synchronous Machine Drive With Low Ripple in Flux and Torque and Fixed Switching Frequency," *IEEE Transactions on Power Electronics*, Vol. 19, No. 2, pp. 346-354, 2004.

[3] C.-H. Choi, J.-K Seok, and R. D. Lorenz, "Wide-Speed Direct Torque and Flux Control for Interior PM Synchronous Motors Operating at Voltage and Current Limits," *IEEE Transactions on Industrial Applications*, Vol. 49, No. 1, pp. 109-117, 2013.

[4] S. Morimoto, M. Sanada, and Y. Takeda, "Wide-Speed Operation of Interior Permanent Magnet Synchronous Motors with High-Performance Current Regulator," *IEEE Transactions on Industry Applications*, Vol. 30, No. 4, pp. 920-926, 1994.

[5] A. Consoli, G. Scarcella, G. Scelba, and A. Testa, "Steady-State and Transient Operation of IPMSMs Under Maximum-Torque-per-Ampere Control," *IEEE Transactions on Industry Applications*, Vol. 46, No. 1, pp. 121-129, 2010.

[6] W. Huang, Y. Zhang, X. Zhang, and G. Sun, "Accurate Torque Control of Interior Permanent Magnet Synchronous Machine," *IEEE Transactions on Energy Conversion*, Vol. 29, No. 1, pp. 29-37, 2014.

[7] B. Cheng, and T. R. Tesch, "Torque Feedforward Control Technique for Permanent Magnet Synchronous Motors," *IEEE Transactions on Industrial Electronics*, Vol. 57, No. 3, pp. 969-974, 2010.

[8] Y. A. -R. I. Mohamed, and T. K. Lee, "Adaptive Self-Tuning MTPA Vector Controller for IPMSM Drive System," *IEEE Transactions on Energy Conversion*, Vol. 21, No. 3, pp. 636-644, 2006.

[9] G. Pellegrino, R. I. Bojoi, and P. Guglielmi, "Unified Direct-Flux Vector Control for AC Motor Drives," *IEEE Transactions on Industrial Applications*, Vol. 47, No. 5, pp. 2093-2102, 2011.

[10] T. Inoue, Y. Inoue, S. Morimoto, and M. Sanada, "Mathematical Model for MTPA Control of Permanent-Magnet Synchronous Motor in Stator Flux Linkage Synchronous Frame," *IEEE Transactions on Industrial Applications*, Vol. 51, No. 5, pp. 3620-3628, 2015.

[11] H. Kamiyama, Y. Inoue, S. Morimoto, and M. Sanada, "Mathematical Model of PMSM and SynRM Under Maximum Torque Per Ampere Condition in a Stator Flux-Linkage Synchronous Frame," *Proc. of the 19th International Conference on Electrical Machines and Systems (ICEMS2016)*, DS3G-4-8, 2016.

[12] T. Inoue, Y. Inoue, S. Morimoto, and M. Sanada, "Maximum Torque Per Ampere Control of a Direct Torque-Controlled PMSM in a Stator Flux Linkage Synchronous Frame," *IEEE Transactions on Industrial Applications*, Vol. 52, No. 3, pp. 2360-2367, 2016.

[13] T. Inoue, Y. Inoue, S. Morimoto, and M. Sanada, "Performance Improvement of SynRM Based on Maximum Torque Per Ampere Control in M-T Frame," *Proc. of the 2014 JIAS Conf.*, Vol.3, pp. 259-262, 2014. (in Japanese)

1294

The 2018 International Power Electronics Conference

EEMFs Excited by Signal Injection for Position Sensorless Control of PMSMs and Their Performance Comparison by Using Imaginary Electromotive Force

Takumi NIMURA, Shota KONDO, Shinji DOKI
Dept. of Information and Communication Engineering, Nagoya University
Furo-cho, Chikusa-ku, Nagoya, Aichi, Japan
E-mail: takumi.nimura@nagoya-u.jp, kondoshota1025@nagoya-u.jp, doki@nagoya-u.jp

Mutuwo TOMITA
Dept. of Electrical and Computer Engineering, National Institute of Technology, Gifu College
2236-2, Kamimakuwa, Motosu-shi, Gifu, Japan
E-mail: mutuwo@nagoya-u.jp

Abstract—In this paper, Extended Electro-motive Force exited by signal injection, which can be utilized at stand-still/Low speed, are discussed. Recently, depending on mathematical expressions for position sensorless control, there are various models of Extended Electromotive Force(EEMF), etc.. By using expression of Imaginary Electromotive Force(IEMF), differences among some models of position sensorless control can easily analyze. In this paper, the IEMF is newly defined including EEMF excited by signal injection, which is evaluated its characteristics. Specifically, we evaluate two types of EEMF excited by signal injection against to the robustness of motor parameters, and finally experiment results are indicated.

Keywords—Electromotive Force, Sensorless, Signal Injection, PMSM

I. Introduction

Permanent Magnet Synchronous Motors (PMSMs) are applied to various fields because of its high power density and high efficiency. In order to ensure vector control of PMSMs, the rotor position information must be detected by some way. However, the position sensor has problems of the cost, space, and a risk of cable disconnection, etc.. Hence, various position sensorless control methods of PMSMs have been proposed[1][2].

Estimating Extended Electromotive Force (EEMF) is one of promising method. The EEMF is well-known to have a high potential in middle/high speed range because that the model enables position estimation by utilizing not only electromotive force but also saliency of inductance[3]. As a matter of fact, the EEMF consists of two components, that one is excited by rotor speed, which is well-known, and another is excited by high frequency signal injection. And it has been reported that drive range of position sensorless control with EEMF is possible to expand to standstill/low speed by using both components[4] although it has some sensivirity to machine parameters.

Recently, not only EEMF, various models for position sensorless control by using EMF have been proposed[5][6][7][8][9]. For compare and evaluate these various models, new concept model, Imaginary Electromotive Force(IEMF) has been proposed[10]. By using IEMF model, the models, EMF, EEMF excited by rotor speed and so on are compared and evaluated under unified manner.

In this paper, the IEMF is newly defined including EEMF excited by signal injection, which is evaluated its characteristics. Specifically, we evaluate two types of EEMF excited by signal injection against to the robustness of motor parameters, and finally experiment results are indicated.

II. Position Sensorless Control Models and Position Estimation System

The circuit equation of PMSMs on the d-q rotating coordinate is given below.

$$\begin{bmatrix} v_d \\ v_q \end{bmatrix} = \begin{bmatrix} R + pL_d & -\omega_{re}L_q \\ \omega_{re}L_d & R + pL_q \end{bmatrix} \begin{bmatrix} i_d \\ i_q \end{bmatrix} + \begin{bmatrix} 0 \\ \omega_{re}K_E \end{bmatrix} \quad (1)$$

$R, L_d, L_q, \omega_{re}, K_E, p$ represent stator resistance, d−axis inductance, q−axis inductance, electrical angular velocity and differential operator respectively. Generally, in the coordinate transformation from equation (1) to d-q coordinate or an optional coordinate, a component $2\theta_{re}$ of rotor position information θ_{re} or the phase component of a double difference between the d-q axis and the optional coordinate is generated. It is difficult to estimate the position information θ_{re} from equation which contains both θ_{re} and $2\theta_{re}$. Therefore, a control models which are not including $2\theta_{re}$ proposed in the case of coordinate transformation from α-β coordinate system to d-q coordinate system or an estimated rotating coordinate[3]. This paper will discuss on the models as mentioned above.

1295

The 2018 International Power Electronics Conference

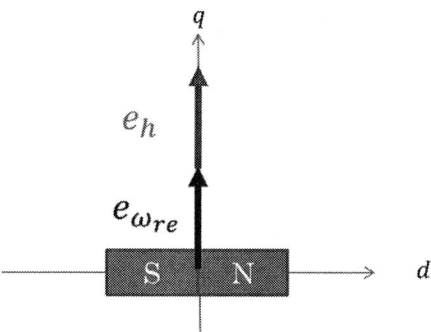

Fig. 1: Phase of EEMF$^{L_d L_q}$

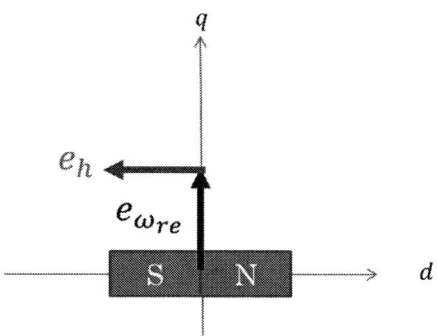

Fig. 2: Phase of EEMF^{L_q}

A. Representation of Imaginary Electromotive Force

Eq.(2), (3) and (4) are calculated by unifying an optional inductances L^* and L^{**} of the diagonal component and the nondiagonal component in the inductance matrix in eq.(1).

$$\begin{bmatrix} v_\alpha \\ v_\beta \end{bmatrix} = \{(R + pL^*)\boldsymbol{I} - \omega_{re}(L^* - L^{**})\boldsymbol{J}\} \begin{bmatrix} i_\alpha \\ i_\beta \end{bmatrix} + e^{\boldsymbol{J}\theta_{re}} \begin{bmatrix} e_d \\ e_q \end{bmatrix}$$

$$(2)$$

$$\begin{bmatrix} e_d \\ e_q \end{bmatrix} = \begin{bmatrix} e_{d_{\omega re}} \\ e_{q_{\omega re}} \end{bmatrix} + \begin{bmatrix} e_{d_h} \\ e_{q_h} \end{bmatrix} \tag{3}$$

$$= \omega_{re} \underbrace{\begin{bmatrix} -(L_q - L^{**})i_q \\ (L_d - L^{**})i_d + K_E \end{bmatrix}}_{e_{\omega re}} + \underbrace{\begin{bmatrix} (L_d - L^*)\dot{i}_d \\ (L_q - L^*)\dot{i}_q \end{bmatrix}}_{e_h} \tag{4}$$

$$\boldsymbol{I} = \begin{bmatrix} 1 & 0 \\ 0 & 1 \end{bmatrix}, \boldsymbol{J} = \begin{bmatrix} 0 & -1 \\ 1 & 0 \end{bmatrix} \tag{5}$$

$$e^{\boldsymbol{J}\theta_{re}} = \begin{bmatrix} \cos\theta_{re} & -\sin\theta_{re} \\ \sin\theta_{re} & \cos\theta_{re} \end{bmatrix} \tag{6}$$

$\boldsymbol{I}, \boldsymbol{J}, e^{\boldsymbol{J}\theta_{re}}$ represent unit matrix, alternating matrix and rotating matrix respectively. $\dot{i}_q$, $\dot{i}_d$ and θ_{re} are represent time derivation by i_q, i_d and the electrical angular position. In this paper, $e_{\omega re}$ and e_h are defined as the components

excited by rotor speed and signal injection in eq.(3) respectively.

Equations (2), (3) and (4) are expression based on Imaginary Electromotive Force (IEMF)[10]. The eq.(3) and (4) are defined as the IEMF. From eq.(4), the phase θ_{emf} of the IEMF is expressed by the following equation.

$$\theta_{emf} = \tan^{-1}\left(\frac{-e_d}{e_q}\right) \tag{7}$$

$$= \tan^{-1}\left\{\frac{-(-\omega_{re}(L_q - L^{**})i_q + (L_d - L^*)\dot{i}_d)}{\omega_{re}\{(L_d - L^{**})i_d + K_E\} + (L_q - L^*)\dot{i}_q}\right\} \tag{8}$$

For example, substituting the inductance $L^* = L_d, L^{**} = L_q$ for IEMF, the model shows EEMF model in ref.[3] and [4]. In this paper, the model is defined as EEMF$^{L_d L_q}$. Substituting the inductance $L^* = L_q, L^{**} = L_q$ for IEMF, the model shows EEMF model in ref.[5]. The model is defined as EEMF^{L_q}. In addition to these, substituting the inductance $L^* = 0, L^{**} = 0$ for IEMF, the model shows the model of estimating armature flux linkage, etc.. It is possible to unifyly consider a model including the EEMF, etc. proposed for position sensorless control by using the IEMF representation.

Eq.(9) and fig.1 show EEMF$^{L_d L_q}$, eq.(10) and fig.2 show EEMF^{L_q}.

$$EEMF^{L_d L_q}:$$
$$\begin{bmatrix} e_d \\ e_q \end{bmatrix} = \omega_{re} \underbrace{\begin{bmatrix} 0 \\ (L_d - L_q)i_d + K_E \end{bmatrix}}_{e_{\omega re}} + \underbrace{\begin{bmatrix} 0 \\ (L_q - L_d)\dot{i}_q \end{bmatrix}}_{e_h}$$

$$(9)$$

$$EEMF^{L_q}:$$
$$\begin{bmatrix} e_d \\ e_q \end{bmatrix} = \omega_{re} \underbrace{\begin{bmatrix} 0 \\ (L_d - L_q)i_d + K_E \end{bmatrix}}_{e_{\omega re}} + \underbrace{\begin{bmatrix} (L_d - L_q)\dot{i}_d \\ 0 \end{bmatrix}}_{e_h}$$

$$(10)$$

From eq.(9) and eq.(10), it is clear that the difference between EEMF$^{L_d L_q}$ and EEMF^{L_q} is component of EEMF excited by signal injection. In order to focus on EEMF

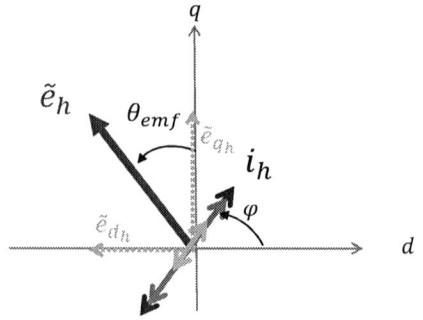

Fig. 3: Phase of IEMF Excited by Signal Injection

1296

The 2018 International Power Electronics Conference

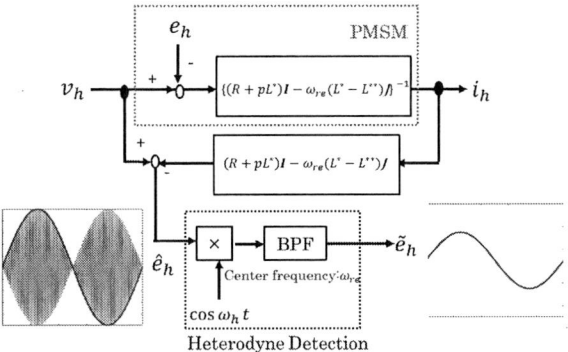

Fig. 4: Estimation System of IEMF Excited by Signal Injection

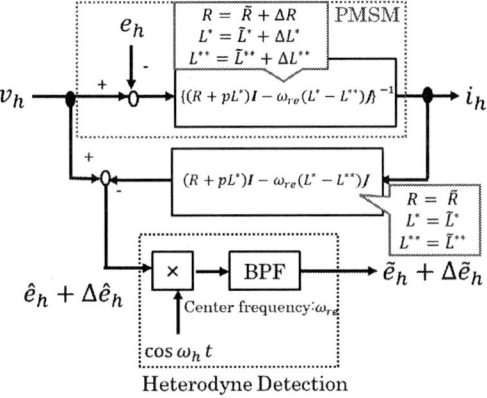

Fig. 5: Estimation System of IEMF Excited by Signal Injection when Parameter Errors Exist

B. Estimation Method of IEMF Excited by Signal Injection

In this section, we describe a method to estimate the EEMF excited by signal injection e_h. First, high frequency signal is injected on the phase which advance φ in angle from d-axis and a linear injection method as a high frequency signal is assumed as shown in fig.3. The high frequency current is shown below.

$$i_{dqh} = (i_h \sin \omega_h t)e^{\boldsymbol{J}\varphi} \tag{15}$$

$$e^{\boldsymbol{J}\varphi} = \begin{bmatrix} \cos \varphi & -\sin \varphi \\ \sin \varphi & \cos \varphi \end{bmatrix} \tag{16}$$

Where, i_h and w_h express the current amplitude of high frequency signal and the angular frequency of high frequency signal. The IEMF excited by signal injection can be calculated by subtracting the voltage drop of the impedance from the voltage v_h as shown in fig.4. Based on eq.(4) and eq.(15), the IEMF excited by signal injection is expressed by following equation.

$$\begin{bmatrix} \hat{e}_{d_h} \\ \hat{e}_{q_h} \end{bmatrix} = \begin{bmatrix} (L_d - L^*)i_d \\ (L_q - L^*)i_q \end{bmatrix} \tag{17}$$

$$= \omega_h i_h \cos \omega_h t \begin{bmatrix} (L_d - L^*)\cos \varphi \\ (L_q - L^*)\sin \varphi \end{bmatrix} \tag{18}$$

It is necessary to extract the rotor position information by detection processing as shown in fig.4, because eq.(18) is the modulated signal with rotation frequency on the carrier wave, high frequency signal. The IEMF excited by signal injection which is extracted by detection processing is given by the following equation.

excited by signal injection, the voltage equation of the component injected high frequency signal of $\text{EEMF}^{L_d L_q}$ and EEMF^{L_q} is shown below.

$\text{EEMF}^{L_d L_q}$:

$$v_h = \{(R + pL_d)\boldsymbol{I} - \omega_{re}(L_d - L_q)\boldsymbol{J}\}i_h + e_h^{L_d L_q} \tag{11}$$

$$e_h^{L_d L_q} = (L_q - L_d)\dot{i}_{qh}e^{\boldsymbol{J}\theta_{re}}\begin{bmatrix} 0 \\ 1 \end{bmatrix} \tag{12}$$

EEMF^{L_q} :

$$v_h = (R + pL_q)\boldsymbol{I}i_h + e_h^{L_q} \tag{13}$$

$$e_h^{L_q} = (L_d - L_q)\dot{i}_{dh}e^{\boldsymbol{J}\theta_{re}}\begin{bmatrix} 1 \\ 0 \end{bmatrix} \tag{14}$$

Note that in EEMF^{L_q}, e_h is rotated to q-axis from d-axis in case of combining with $e_{\omega_{re}}$ which is direction to q-axis.[5]

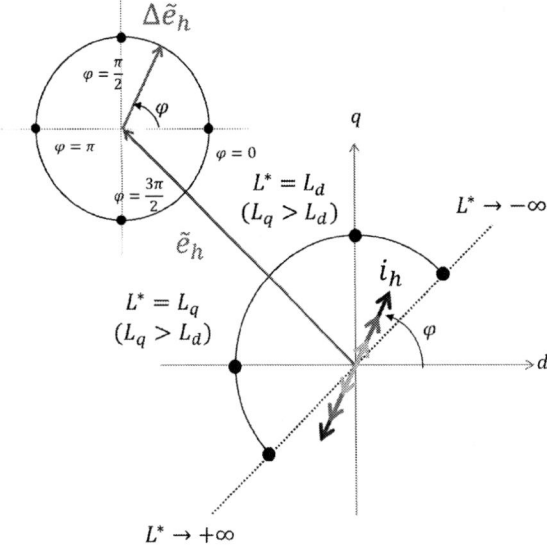

Fig. 6: IEMF Vector when Parameter Errors Exist

1297

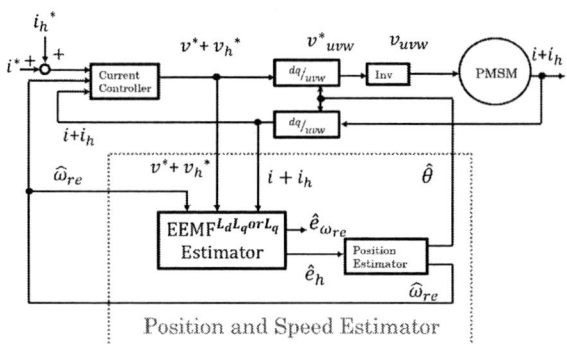

Fig. 7: Position Sensorless Control System

Table1: Parameters of IPMSM

d-axis inductance L_d	4.15m [H]
q-axis inductance L_q	16.74m [H]
EMF constant K_E	0.091 [V · s/rad]
winding resistance R	0.53 [Ω]
pole pairs Pn	2
rated torque T	1.2[Nm]
rated speed ω_{rm}	4000[rpm]
Current limit I_{max}	5[A]

$$
\begin{bmatrix} \tilde{e}_{d_h} \\ \tilde{e}_{q_h} \end{bmatrix} = \frac{\omega_h i_h}{2} \begin{bmatrix} (L_d - L^*)\cos\varphi \\ (L_q - L^*)\sin\varphi \end{bmatrix} \tag{19}
$$

$$
\theta_{emf} = \tan^{-1}\left(\frac{-e_{d_h}}{e_{q_h}}\right) \tag{20}
$$

$$
= \tan^{-1}\left(\frac{-(L_d - L^*)\cos\varphi}{(L_q - L^*)\sin\varphi}\right) \tag{21}
$$

IEMF excited by signal injection is determined by L^* and injected signal phase φ as shown in eq.(19), (20), and (21). Hence, each EEMF model is expressed by substituting L_d, L_q into L^*, L^{**} (EEMF^{L_dL_q} model, eq.(22)), and substituting L_q into L^* (EEMF^{L_q} model, eq.(23)).

$$
EEMF^{L_dL_q} :
$$

$$
\tilde{e}_h^{L_dL_q} = \frac{\omega_h i_h (L_q - L_d)\sin\varphi}{2} e^{\boldsymbol{J}\theta_{re}} \begin{bmatrix} 0 \\ 1 \end{bmatrix} \tag{22}
$$

$$
EEMF^{L_q} :
$$

$$
\tilde{e}_h^{L_q} = \frac{\omega_h i_h (L_d - L_q)\cos\varphi}{2} e^{\boldsymbol{J}\theta_{re}} \begin{bmatrix} 1 \\ 0 \end{bmatrix} \tag{23}
$$

III. Influence of Parameter Error

In this section, we describe the influence of parameter error based on IEMF in the method below. Stator resistance R, optional inductances L^* and L^{**} of plant change from the nominal value $\tilde{R}, \tilde{L}^*$, and $\tilde{L}^{**}$ to the $R = \tilde{R} + \Delta R, L^* = \tilde{L}^* + \Delta L^*$, and $L^{**} = \tilde{L}^{**} + \Delta L^{**}$. Here $\Delta R, \Delta L^*$, and ΔL^{**} are amounts of variation on each

parameter. The IEMF error vector $\Delta\hat{e}_h$ is given as below, which is calculated in fig.5.

$$
\begin{bmatrix} \hat{e}_{\alpha h} + \Delta\hat{e}_{\alpha h} \\ \hat{e}_{\beta h} + \Delta\hat{e}_{\beta h} \end{bmatrix} = \begin{bmatrix} v_{\alpha h} \\ v_{\beta h} \end{bmatrix} - \{(R + pL^*)\boldsymbol{I} - \omega_{re}(L^* - L^{**})\boldsymbol{J}\} \begin{bmatrix} i_{\alpha h} \\ i_{\beta h} \end{bmatrix} \tag{24}
$$

$$
\begin{bmatrix} \Delta\hat{e}_{\alpha h} \\ \Delta\hat{e}_{\beta h} \end{bmatrix} = \{(\Delta R + p\Delta L^*)\boldsymbol{I} - \omega_{re}(\Delta L^* - \Delta L^{**})\boldsymbol{J}\} \begin{bmatrix} i_{\alpha h} \\ i_{\beta h} \end{bmatrix} \tag{25}
$$

$$
= \Delta\hat{e}_R + \Delta\hat{e}_{L^*} + \Delta\hat{e}_{L^{**}} \tag{26}
$$

The IEMF error vector extracted by detection processing $\Delta\tilde{e}_h$ is expressed by the following equation.

$$
\Delta\tilde{e}_{\alpha\beta_h} = \Delta\tilde{e}_R + \Delta\tilde{e}_{L^*} + \Delta\tilde{e}_{L^{**}}
$$

$$
\begin{cases} \Delta\tilde{e}_R = 0 \\ \Delta\tilde{e}_{L^*} = \frac{\omega_h i_h \Delta L^*}{2}\boldsymbol{I}e^{\boldsymbol{J}\theta_{re}}e^{\boldsymbol{J}\varphi}\begin{bmatrix}1\\0\end{bmatrix} \\ \Delta\tilde{e}_{L^{**}} = 0 \end{cases} \tag{27}
$$

From the eq.(27), sensitivity of estimating IEMF excited by signal injection has only L^* in the motor parameters. $\Delta\tilde{e}_R$ and $\Delta\tilde{e}_{L^{**}}$ of parameter error vector R and L^{**} is separated in the heterodyne detection process and its sensitivities have not at all. It is found that the phase of the IEMF excited by signal injection depends on the injected signal phase. The influence on position estimation in IEMF excited by signal injection is minimized by selecting L^* for not affecting parameter variation, or by injecting the HF signal near phase that same direction of IEMF $\varphi = \theta_{emf}$ [11]. Estimating the EEMF error vector, it can be realized by substituting inductance values into L^*, L^{**} in eq.(27). The EEMF error vector excited by signal injection of each EEMF model is expressed by the following equation.

$$
EEMF^{L_dL_q} : \Delta\tilde{e}^{L_dL_q} = \Delta\tilde{e}_R + \Delta\tilde{e}_{L_d} + \Delta\tilde{e}_{L_q}
$$

$$
\begin{cases} \Delta\tilde{e}_R = 0 \\ \Delta\tilde{e}_{L_d} = \frac{\omega_h i_h \Delta L_d}{2}\boldsymbol{I}e^{\boldsymbol{J}\theta_{re}}e^{\boldsymbol{J}\varphi}\begin{bmatrix}1\\0\end{bmatrix} \\ \Delta\tilde{e}_{L_q} = 0 \end{cases} \tag{28}
$$

$$
EEMF^{L_q} : \Delta\tilde{e}^{L_q} = \Delta\tilde{e}_R + \Delta\tilde{e}_{L_q}
$$

$$
\begin{cases} \Delta\tilde{e}_R = 0 \\ \Delta\tilde{e}_{L_q} = \frac{\omega_h i_h \Delta L_q}{2}\boldsymbol{I}e^{\boldsymbol{J}\theta_{re}}e^{\boldsymbol{J}\varphi}\begin{bmatrix}1\\0\end{bmatrix} \end{cases} \tag{29}
$$

It turns out that the EEMF^{L_dL_q} is affected only by ΔL_d from eq.(28), and EEMF^{L_q} is affected only by ΔL_q from eq.(29). Moreover, the influence on position estimation in EEMF^{L_dL_q} model is minimized by injecting the high frequency signal near phase that $\varphi = \frac{\pi}{2}$ (q-axis direction). On the other hand, the influence on position estimation in EEMF^{L_q} model is minimized by injecting the high frequency signal near phase that $\varphi = 0$ (d-axis direction).

The 2018 International Power Electronics Conference

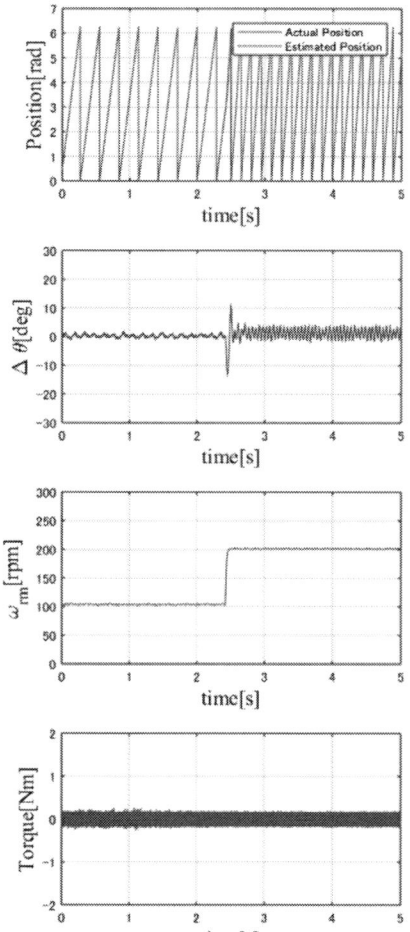

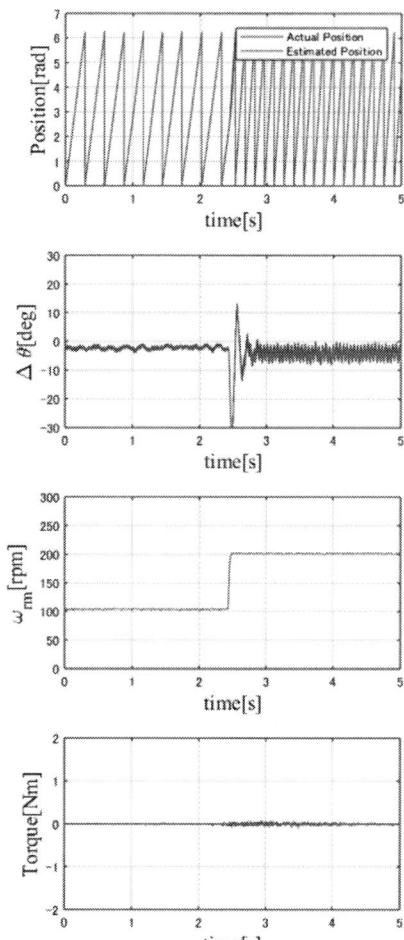

Fig. 8: Result of Position Sensorless Control in EEMF^{L_dL_q}

Fig. 9: Result of Position Sensorless Control in EEMF^{L_q}

IV. EXPERIMENT

The influence of parameter error in the previous section is examined by experiment. Position sensorless control system is shown in fig.7. Table.1 show the conditions of PMSMs parameter and controller. The current controller is general current vector control on d-q coordinate, and the response frequency is set as 4000 [rad/s]. The carrier frequency is set as 10[kHz], and the control period is set as 100 [μs]. The current amplitude of high frequency signal is 1 [A] and the frequency of high frequency signal is 100 [Hz]. The position estimator setting is an error of nominal value 100% under the assumption that rising temperature and

magnetic saturation($\Delta R = \tilde{R}, \Delta L_d = -\tilde{L}_d, \Delta L_q = -\tilde{L}_q$). The phase of the high frequency signal injection is set to q-axis direction in EEMF^{L_dL_q} model, which is set to d-axis direction in EEMF^{L_q} model.

In EEMF, EEMF exited by speed $\hat{e}_{\omega_{re}}$ and excited by signal injection $\hat{e}_h$ are estimated, although this experiment system estimates only EEMF excited by signal injection $\hat{e}_h$. Because this paper clarifies the characteristics of the EEMF excited by signal injection $\hat{e}_h$.

Fig. 8 and 9 show actual position and estimated position, estimated position error, speed, and torque in speed step experience (rated speed 2.5% to 5%). EEMF^{L_dL_q} model and EEMF^{L_q} model estimating EEMF excited by

1299

The 2018 International Power Electronics Conference

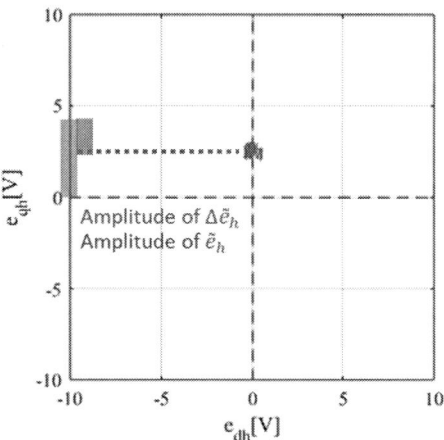

Fig. 10: Result of $\tilde{e}_h + \Delta\tilde{e}_h$ vector in EEMF^{L_dL_q} model

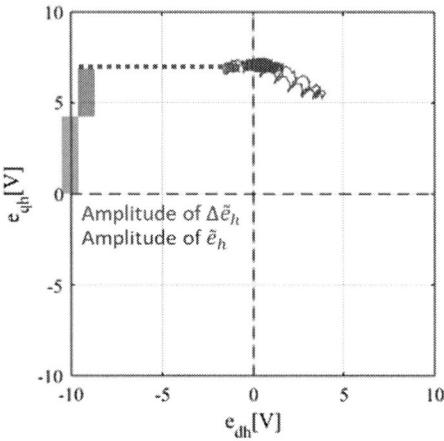

Fig. 11: Result of $\tilde{e}_h + \Delta\tilde{e}_h$ vector in EEMF^{L_q} model (rotated from d-axis to q-axis)

signal injection have robustness against motor parameter in steady state. The EEMF^{L_q} model is superior than the EEMF^{L_dL_q} model regarding torque ripple, because injected signal phase in EEMF^{L_q} model is d-axis. The EEMF^{L_dL_q} model is superior than the EEMF^{L_q} model regarding transient characteristics. The differences in transient characteristics will be examined in the future.

Fig. 10 and Fig. 11 show EEMF vector. In fig. 11, EEMF^{L_q} vector is rotated from d-axis to q-axis in order to compare with EEMF^{L_dL_q} model in fig. 10. The amplitude of EEMF^{L_dL_q} vector decreases because the error vector

$\Delta\tilde{e}_h$ caused by magnetic saturation ΔL_d reduces $\tilde{e}_h$. On the other hand, the amplitude of EEMF^{L_q} vector increases because the error vector caused by magnetic saturation ΔL_q adds to $\tilde{e}_h$. Therefore, EEMF^{L_q} is superior than the EEMF^{L_dL_q} model regarding the amplitude variation of EEMF caused by magnetic saturation.

V. CONCLUSION

This paper showed that an IEMF is newly defined including EEMF excited by signal injection, which is evaluated its characteristics. We evaluated two types of EEMF excited by signal injection against to the robustness of motor parameters, and finally experiment results are indicated.

REFERENCES

[1] Masaru Hasegawa, Shinji Doki : " Trends in Motor Drive Techniques in Japan -Controls for Synchronous Motors with Non-linearity-", IEEJ Trans. Industrial Electronics, IEEJ Journal of Industry Applications Vol. 1 (2012) No. 3 pp.123-131

[2] Seung-Ki Sul, Sungmin Kim : " Sensorless Control of IPMSM: Past, Present, and Future", IEEJ Trans. Industrial Electronics, IEEJ Journal of Industry Applications Vol. 1 (2012) No. 1 pp.15-23

[3] Z.Chen, M.Tomita, S.Doki, and S.Okuma : "An Extended Electromotive Force Model for Sensorless Control of Interior Permanent Magnet Synchronous Motors", IEEE Trans. Industrial Electronics, Vol.50, No.2, pp.288-295(2003)

[4] T.Ohnuma, S.Doki, and S.Okuma : " Extended EMF Observer for Sensorless Control over a Wide Range of Speeds", IEEJ Trans. Industrial Electronics, Vol.131, No.2, pp. 208-218(2011)

[5] R. Saitoh, Y. Makaino, T. Ohnuma : " Adaptive Signal Injection Method Combined with EEMF-based Position Sensorless Control of IPMSM Drives", IEEJ Journal of Industry Applications, Vol.4, No.4, pp.454-459 (2015)]

[6] A. Matsumoto, M. Hasegawa, K. Matsui : " Position Sensorless Control of IPMSMs Based on a Novel Flux Model Suitable for Maximum Torque Control", IEEJ Journal of Industry Applications, Vol.132, No.1, pp.67-77 (2012)

[7] H. Hida, Y. Tomigashi, K. Kishimoto : " Position Sensorless Control for Permanent Magnet Synchronous Motors Based on Maximum Torque Control Frame", IEEJ Journal of Industry Applications, Vol.127, No.12, pp.1190-1196 (2007)

[8] K. Tobari, K. Sakamoto, D. Maeda : " Maximum Torque Control Technique Suitable for Sensorless Permanent Magnet Synchronous Motor Drives", Proc. of the 2006 JIAS Conf., Vol. 1, No. 64, pp. 389-392 (2006)

[9] M. Hasegawa, K. Matsui : " IPMSM Position Sensorless Drives Using Robust Adaptive Observer on Stationary Reference Frame", IEEJ Journal of Industry Applications, Vol.3, No.1, pp.120-127 (2008)

[10] A. Matsumoto, S. Doki, M. Hasegawa : "Discussion about equation representation of mathematical model for control of IPMSM", The Paper of Joint Technical Meeting on Power Engineering, Power Systems Engineering and Semiconductor Power Converter, IEE Japan PE(176)-11-16 (2010)

[11] S. Kondo, S. Doki, M. Tomita : " A discussion about Extended Electromotive Force excited by signal injection and its robustness for position sensorless control of PMSM", The Paper of Joint Technical Meeting on Power Engineering, Motor Drive, IEE Japan MD(70-81)-19-24 (2017)

The 2018 International Power Electronics Conference

Harmonic Current Cancellation Method for PMSM Drive System using Resonant Controllers

Dongsheng Li[1*], Yoshitaka Iwaji[1], Yasuo Notohara[1] and Ken Kishita[2]
1 Research & Development Group, Hitachi, Ltd., Hitachinaka, Japan
2 Hitachi-Johnson Controls Air Conditioning, Inc., Shizuoka, Japan
*E-mail: dongsheng.li.sb@hitachi.com

Abstract— **In a permanent magnet synchronous motor (PMSM) drive system, many nonlinear effects in both the inverter and motor have a negative impact on the motor current and result in current distortion. In this paper, to reduce the current distortion, a resonant controller (RC) is used in parallel with a proportional and integral (PI) regulator in the current controller in the synchronous reference frame. To improve the stability of the RC, a modified transfer function is used. To avoid the interaction between the two types of controllers, a low-pass filter (LPF) is added to suppress the influence of a sudden change of reference. As a result, the harmonic components of the motor current can be reduced greatly, and the dynamic response can be kept the same as that of a PI regulator. Experimental results show that the proposed control method is effective.**

Keywords— *current control; harmonic cancellation; PMSM drive; resonant controller.*

I. INTRODUCTION

In a permanent magnet synchronous motor (PMSM) drive system, many nonlinear effects in both the inverter and motor have a negative impact on the motor current and result in current distortion. These effects include inverter nonlinearity caused by dead-time, turn-on and turn-off delay times, and on-state voltage drops of power devices, the nonlinear characteristics of PMSM, distortions in the electromotive force (EMF) of PMSMs, and DC-link voltage ripple. To suppress the current distortion caused by these nonlinear effects, efforts have been made such as on high-performance methods for compensating for dead time [1], [2] and on methods for compensating for nonlinearity [3], [4]. However, it is difficult to avoid compensation errors if the nonlinear characteristics of power devices and PMSMs are not precisely measured.

For PMSM drive systems, a synchronous reference frame-based control system is commonly used because steady-state error can be reduced to zero easily with the conventional proportional and integral (PI) regulators. However, it is difficult to suppress the harmonic currents caused by the above-mentioned nonlinear effects because of the bandwidth limits of PI regulators.

In recent years, another possible way to control sinusoidal signals has been proposed and applied to current controllers for single- or three-phase inverter systems. It is called the "resonant controller" (RC) [5]-[7]

or "sine transfer function controller" [8]. The main characteristics of the RC is the fact that a very high gain around the resonance frequency can be achieved; thus, the steady-state error of the sinusoidal signal with the resonance frequency can be kept to near zero. As presented in [7], by cascading several RCs tuned to the desired harmonic frequencies, harmonic currents can be selectively compensated.

In this paper, to cancel the harmonic current components caused by the nonlinear effects in a PMSM drive system, an RC is combined in parallel with a PI regulator in the synchronous reference frame. To avoid the interaction between the RC and PI regulator, a low-pass filter (LPF) is added. In the proposed controller, the parameters of the nonlinear characteristics of power devices and PMSMs are not required, only the gain of the RC and the cut-off frequency of the LPF are needed.

II. HARMONIC CANCELLATION METHOD

A. Current Controller for R-L Load

Fig. 1 shows a simple current control system for an *R-L* load. The load consists of a resistor ($R = 0.1\ \Omega$) and an inductor ($L = 2$ mH) connected in series. The current reference is a DC value with an amplitude of 10 A. The current controller consists of a PI regulator only. The inverter is assumed to be an ideal voltage source, but its nonlinear characteristics are replaced by a disturbance ($D(\omega_0)$) with a fixed frequency ($\omega_0 = 200\pi$ rad/s) and fixed amplitude of 5 V. With this disturbance, a harmonic component with the same frequency will appear in the load current.

Fig. 2 shows the waveforms of current reference (i^*) and load current (i). As shown in Fig. 2(a), if there is no disturbance ($D(\omega_0) = 0$), the load current follows the

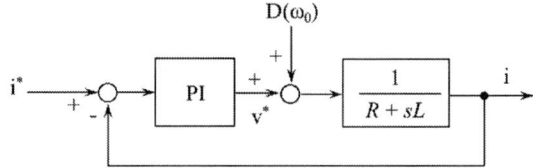

Fig. 1. Current control system for *R-L* load with PI regulator.

1301

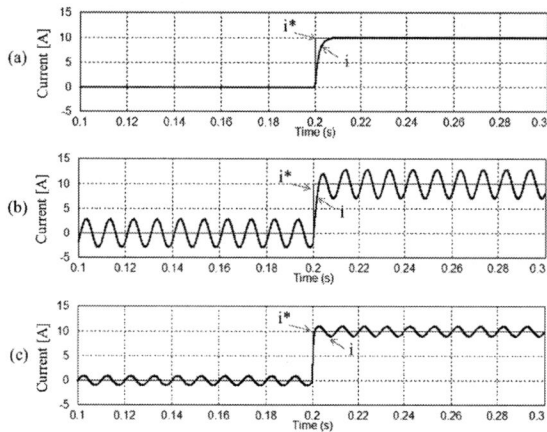

Fig. 2. Waveforms of current reference (i*) and load current (i). (a) Without disturbance ($\omega_{acr} = 200\pi$ rad/s). (b) With disturbance ($\omega_{acr} = 200\pi$ rad/s). (c) With disturbance ($\omega_{acr} = 800\pi$ rad/s).

reference without steady-state error. However, if a disturbance is added to the output voltage, the influence of the disturbance appears in the load current, as shown in Fig. 2(b). The influence can be reduced by increasing the PI regulator response from $\omega_{acr} = 200\pi$ rad/s to $\omega_{acr} = 800\pi$ rad/s, as shown in Fig. 2(c). However, with the higher response, the stability of the current controller decreases, and the suppression effect is not sufficient.

B. Design of Resonant Controller

To suppress the influence of disturbance, an RC is added in parallel with the PI regulator, as shown in Fig. 3. The reference of the RC is set to be zero, and the output of the RC is added to the output of the PI regulator.

The basic transfer function of the RC is a sine transfer function as in (1).

$$G_s(s) = K_s \frac{s}{s^2 + \omega_0^2}$$

(1)

where K_s is the gain of an RC and ω_0 is the resonant frequency.

The open-loop transfer function of the control system includes $G_s(s)$, and an R-L load is (2).

$$G_{CC\_I}(s) = G_s(s) \cdot G_L(s) = \frac{K_s \cdot s}{s^2 + \omega_0^2} \cdot \frac{1}{Ls + R}$$

(2)

where R is the resistance and L is the inductance of an R-L load.

Fig. 4 shows an open-loop Bode diagram of $G_{CC\_I}(s)$. It can be seen that the peak gain was almost infinite at the resonant frequency (100 Hz). However, the phase was near -180° in the region above the resonant frequency. Considering the effect of the computational time of discrete control systems such as a microcontroller unit (MCU), the phase margin of this current control system is not sufficient in practical use.

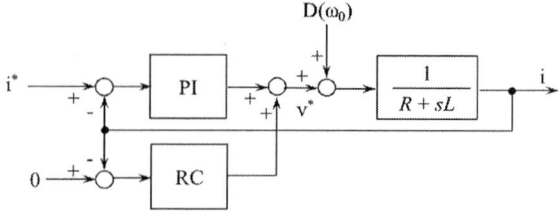

Fig. 3. Current control system for R-L load with PI regulator and RC.

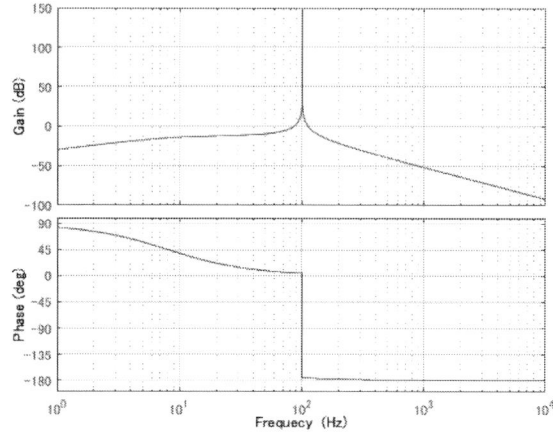

Fig. 4. Bode diagram of open-loop transfer function of $G_{CC\_I}(s)$, $K_s = 200$ V/As.

In this paper, a modified transfer function, shown in (3), is used to improve the stability of the current controller. In (3), a reverse transfer function of R-L load is used to cancel the phase delay of the load. A dumping constant T_a is also added for adjusting the peak gain at the resonant frequency.

$$G_{RC}(s) = \frac{K_{RC} \cdot s}{s^2 + T_a s + \omega_0^2} \cdot (Ls + R)$$

$$= \frac{K_{RC}(Ls^2 + Rs)}{s^2 + T_a s + \omega_0^2}$$

(3)

where K_{RC} is the gain of the resonant controller and T_a is the dumping constant.

Then, the open-loop transfer function becomes (4).

$$G_{CC\_R}(s) = G_{RC}(s) \cdot G_L(s) = \frac{K_{RC} \cdot s}{s^2 + T_a s + \omega_0^2}$$

(4)

Fig. 5 shows a Bode diagram of $G_{CC\_R}(s)$ with different values of T_a. Compared with $G_{CC\_I}(s)$, it is observed that the phase could be kept within -90°, which means that the phase margin is sufficient for this control system. With the dumping constant T_a, the peak gain can be adjusted easily. It should be noted that a smaller peak gain can help to reduce the influence of noises or errors in the detected current or voltage signals.

The 2018 International Power Electronics Conference

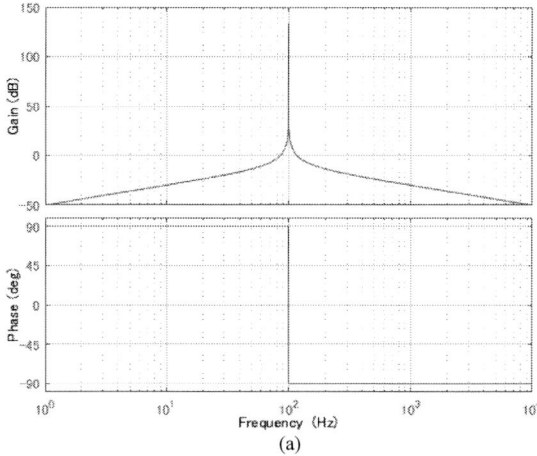

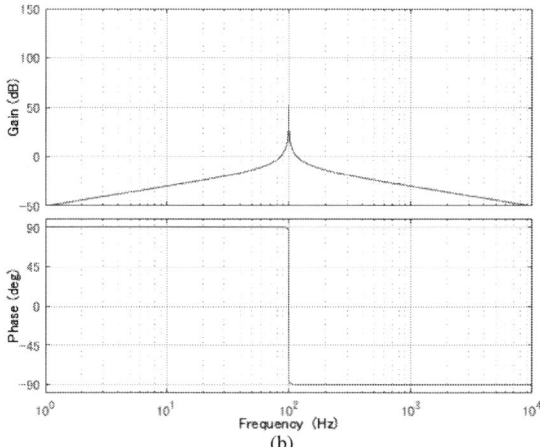

Fig. 5. Bode diagram of open-loop transfer function of $G_{CC\_R}(s)$, K_{RC} = 200 Hz. (a) T_a = 0. (b) T_a = 0.5 Hz.

C. Harmonic Cancellation with RC

With the current control system shown in Fig. 3, the harmonic cancellation was evaluated with K_{RC} = 200 Hz and T_a = 0.5 Hz.

Fig. 6(a) shows the waveforms of current reference (i*) and load current (i), and Fig. 6(b) shows the waveforms of the disturbance and output of the RC. The RC was switched on from 0.4 s. One can observe that the influence of disturbance almost disappeared after the RC was switched on, as shown in Fig. 6(a). Fig. 6(b) shows that the output of the RC was the inverse of the disturbance, which means that the disturbance can be cancelled by the output of the RC.

D. Dynamic Response Improvement with LPF

However, with the controller shown in Fig. 3, when the current reference changed suddenly, there were damped oscillations (ringing) in the load current, as shown in Fig. 7(a), where the current reference changed from 10 to 20 A at 0.5 s and from 20 to 10 A at 0.6 s. The main reason for the oscillations is that the rapid change in current reference lead to an impact component in the input of the RC, as shown in Fig. 7(b).

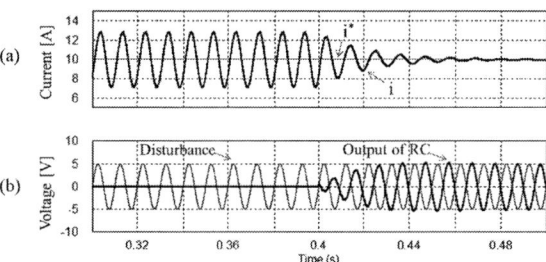

Fig. 6. Waveforms of current and voltage. (a) Current reference (i*) and load current (i). (b) Disturbance and output of RC.

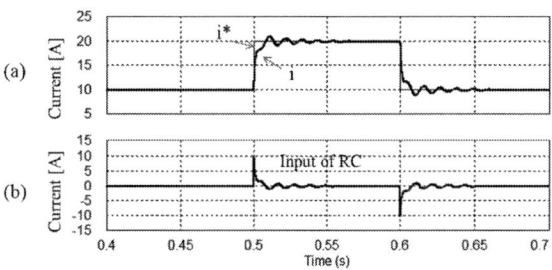

Fig. 7. Waveforms of current and voltage. (a) Current reference (i*) and load current (i). (b) Input signal of RC.

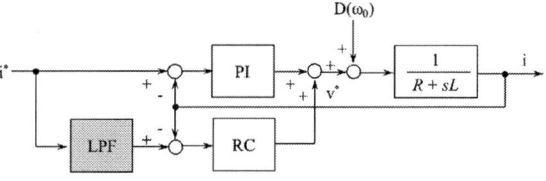

Fig. 8. Current control system for R-L load with added LPF.

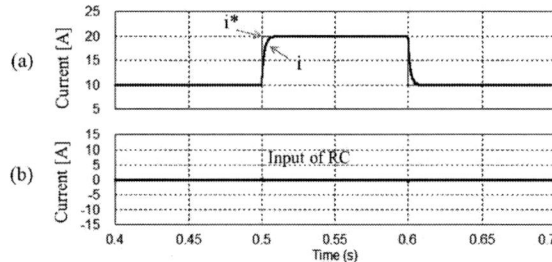

Fig. 9. Waveforms of current and voltage with added LPF. (a) Current reference (i*) and load current (i). (b) Input signal of RC.

To remove the impact component, an LPF was used to generate the reference of the RC from the current reference, as shown in Fig. 8. The cut-off frequency of the LPF was set to be the same as the response frequency of the PI regulator.

Fig. 9 shows the waveforms with the added LPF. Compared with the waveform shown in Fig. 7(b), the impact component in the input of the RC disappeared. The response of the load current was almost the same as the waveform shown in Fig. 2(a).

Fig. 10 shows a comparison of closed-loop Bode diagrams of the current control system with and without the LPF. One can confirm that the transfer characteristics shown in Fig. 10(b) are the same as that of a PI regulator if the LPF is added.

1303

The 2018 International Power Electronics Conference

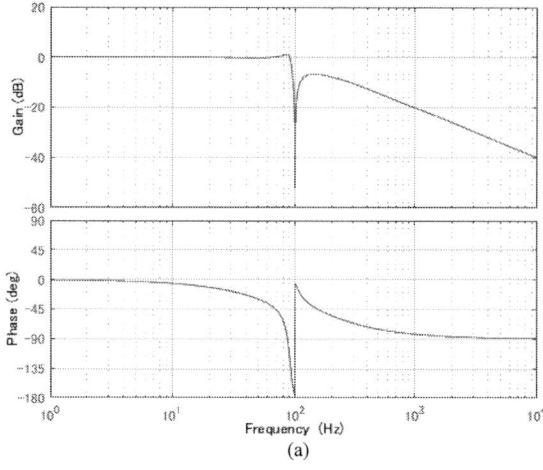

(a)

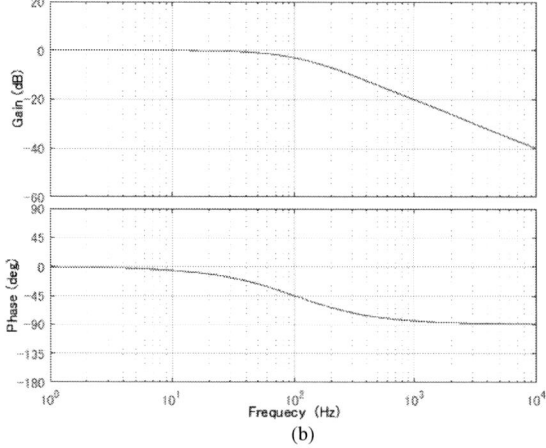

(b)

Fig. 10. Comparison of closed-loop Bode diagrams of current control system with and without LPF. (a) Without LPF. (b) With LPF.

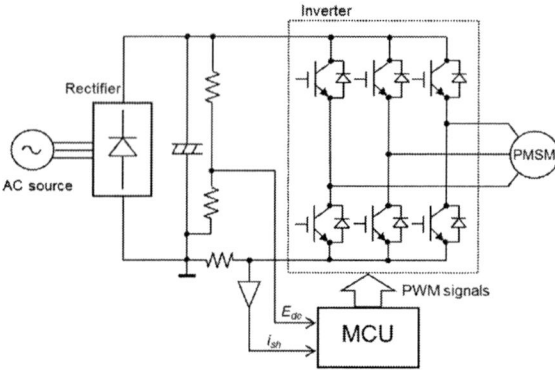

Fig. 11. PMSM drive system.

III. HARMONIC CANCELLATION FOR PMSM DRIVE SYSTEM

A. PMSM Drive System

Fig. 11 shows a low-cost PMSM drive system, where a diode rectifier is used to generate the DC-link voltage and a three-phase inverter is used to drive the motor. With an estimator built in the controller, sensors for the speed or position of a rotor are not needed. The motor currents are reconstructed from the detected shunt current signal (i_{sh}) to reduce the number of current sensors. The drive systems are those that have been commonly applied to compressors, pumps, and fans for energy-saving in recent years.

B. Sensor-less Control

Fig. 12 shows a control block of the PMSM drive system, which is based on the controller without position/speed sensors proposed by [9] [10].

The control is realized by using an approach for directly estimating the position error, where the position error of the rotor can be obtained from (5).

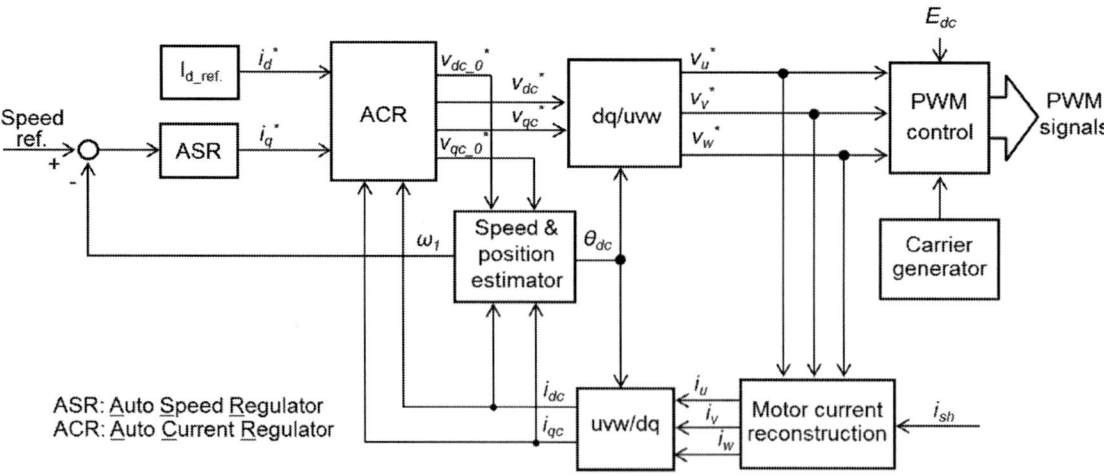

Fig. 12. Speed/position sensor-less control block of PMSM drive system.

1304

The 2018 International Power Electronics Conference

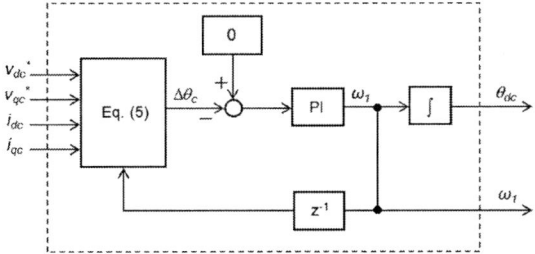

Fig. 13. Control block of speed and position estimator.

$$\Delta\theta_c = tan^{-1}\left(\frac{v_{dc}^* - r \cdot i_{dc} + \omega_1 L_q i_{qc}}{v_{qc}^* - r \cdot i_{qc} - \omega_1 L_q i_{dc}}\right)$$

(5)

where $\Delta\theta_c$ is the estimated position error between the assumed control coordinate (dc-qc axes) and the real coordinate (d-q axes) of the motor, r is the stator winding resistance, L_q is the q-axis inductance of the motor, ω_1 is the estimated rotor speed, v_{dc}^* and v_{qc}^* are the output voltage references in the dc-qc axes, and i_{dc} and i_{qc} are the determined motor currents in the dc-qc axes.

Fig. 13 shows a detailed control block of the speed and position estimator. A PI regulator is employed to adjust the estimated rotor speed (ω_1). Then, the assumed position (θ_{dc}) is generated by using an integrator from ω_1. In other words, the estimator works as a PLL controller to reduce the position error ($\Delta\theta_c$) by adjusting the estimated rotor speed.

It should be noted that this sensor-less controller is not suitable for very low-speed regions (below about 10% of rated speed of PMSM) because (5) is based on the back-electromotive force (EMF) of the motor. The rotor is accelerated by an open-loop speed control with a rated current at start-up in this work.

C. Current Controller with Harmonic Cancellation

Fig. 14 shows the current controller with harmonic current cancellation.

The basic voltage references, $v_{dc\_0}^*$ and $v_{qc\_0}^*$, are calculated by using the cascade vector control method proposed in [11]. Two PI regulators are used to calculate the second current commands, i_d^{**} and i_q^{**}, and basic voltage references are then calculated with the voltage equation of PMSM. The benefit of this vector control method is that the coupling components between the d-q axes can be cancelled theoretically; therefore, the stability of the current controller can be improved in high-speed regions even with a low sampling (calculating) frequency.

To cancel the harmonic components of i_{dc} and i_{qc}, RCs are added in parallel with the vector controller, as shown in Fig. 14. RC_{d1}, RC_{d2}, ... are the RCs for the dc-axis current, and RC_{q1}, RC_{q2}, ... are the RCs for the qc-axis current.

Two LPFs are used to generate reference signals for the RCs from the dc-axis and qc-axis current commands, i_d^* and i_q^*. The sums of each RC output, $v_{dc\_h}^*$ and $v_{qc\_h}^*$, are added to the basic voltage references.

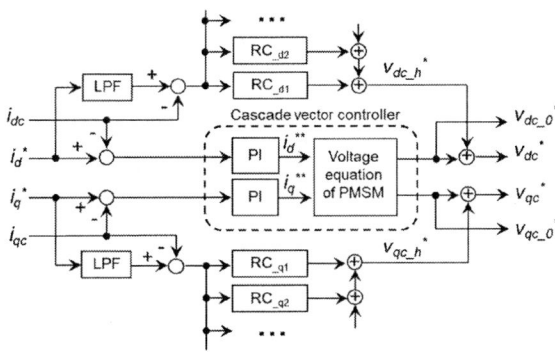

Fig. 14. Current controller for PMSM drive system with harmonic current cancellation.

It should be noted that the outputs of the RCs will lead to ripples in the result of (5) because their outputs are harmonic components. Therefore, the basic voltage references, $v_{dc\_0}^*$ and $v_{qc\_0}^*$, are used in (5) instead of the output voltage references, v_{dc}^* and v_{qc}^*, as shown in Fig. 12 for suppressing the ripples.

D. Selection of Resonant Frequency

As a well-known fact, the orders of harmonic currents caused by nonlinear effects in a three-phase PMSM are mainly $k = 6n$ (n: integer) in the synchronous reference frame, so the resonant frequencies of the RCs are set to be $6\omega_1$, $12\omega_1$, $18\omega_1$, ..., $6n\omega_1$. In fact, the amplitudes of the components above 12th are very small; thus, it is sufficient to compensate for the 6th and 12th components only in practical use.

In some cases, the variation in the characteristics of the power devices and unbalance of the EMF lead to low-order harmonic components such as $2\omega_1$ and $4\omega_1$ in motor currents. In those cases, an RC with $3\omega_1$ is also needed.

IV. EXPERIMENTAL RESULTS

A. Experimental Setup

To verify the effectiveness of the proposed control method, an experiment was carried out with a motor test bench. The control system was implemented by using a 32-bit MCU.

Two types of PMSMs, one a compressor motor (10 kW) and the other a fan motor (0.75 kW), were used. The main parameters are listed in Tables I and II.

B. Experimental Results

Figs. 15(a) and 15(b) show the experimental results of the compressor motor at start-up; two enlarged parts are shown on the bottom of each. As shown in Fig. 15(a), there was a large distortion in the motor current. The main reason for the distortion was the saturation characteristics of the motor because the d-axis current was very large with the open-loop speed control. With the proposed harmonic cancellation method (with 6th RC only), the current

1305

The 2018 International Power Electronics Conference

TABLE I
PARAMETERS USED IN EXPERIMENT OF COMPRESSOR MOTOR

Symbol	Meaning	Value
K_{RC}	Gain of RC	200 Hz
T_a	Dumping constant of RC	0.5 Hz
R_1	Coil resistance of PMSM	0.12 Ω
L_d	d-axis inductance of PMSM	2.1 mH
L_q	q-axis inductance of PMSM	3.0 mH
ω_{acr}	Response of PI regulator	120π rad/s
f_c	Frequency of carrier	6 kHz
V_s	AC voltage	200 V

TABLE II
PARAMETERS USED IN EXPERIMENT OF FAN MOTOR

Symbol	Meaning	Value
K_{RC}	Gain of RC	10 Hz
T_a	Dumping constant of RC	0.5 Hz
R_1	Coil resistance of PMSM	1.5 Ω
L_d	d-axis inductance of PMSM	6.3 mH
L_q	q-axis inductance of PMSM	6.5 mH
ω_{acr}	Response of PI regulator	80π rad/s
f_c	Frequency of carrier	10 kHz
V_s	AC voltage	200 V

distortion could be reduced greatly, as shown in Fig. 15(b).

Fig. 16 shows a comparison of the harmonic spectra of the current waveforms shown in Fig. 15 (the enlarged part on the right). It can be seen that the 5th and 7th older harmonic components were suppressed greatly.

Fig. 17 shows the experimental results of the fan motor in a steady state with a very low rotation speed of 60 min⁻¹. As shown in Fig. 17(a), there was a large distortion in the motor current because of the compensation error of the dead time and variation in the characteristics of the power devices. With the proposed harmonic cancellation method (with 3rd and 6th RCs), the current distortion could be reduced greatly, as shown in Fig. 17(b), and the ripple of the estimated position error, $\Delta\theta_c$, became very small. A small ripple of $\Delta\theta_c$ can reduce the vibration of the motor speed and improve the stability of the sensor-less controller.

Fig. 18 shows a comparison of the harmonic spectra of the current waveforms shown in Fig. 17. It can be seen that the 2nd, 4th and 7th older harmonic components were suppressed greatly.

V. CONCLUSIONS

A control method for cancelling harmonic currents for PMSM drive systems was presented. Resonant controllers are combined with PI regulators to suppress harmonic components. To avoid the interaction between the two types of controllers, an LPF is used to reduce the sudden change in the input of the resonant controller.

Experimental results with a compressor motor (10 kW) and fan motor (0.75 kW) were presented to verify the effectiveness of the proposed control method. The comparisons of the harmonic spectra shows that the harmonic components were suppressed greatly.

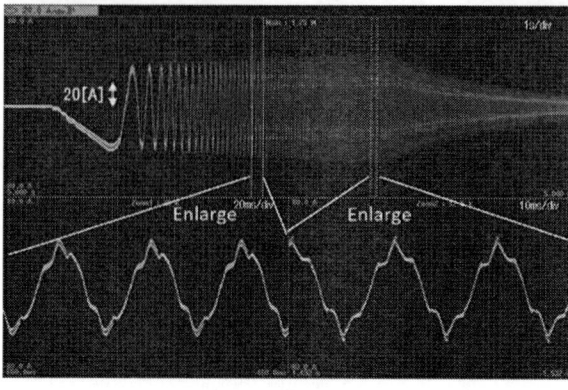

(a)

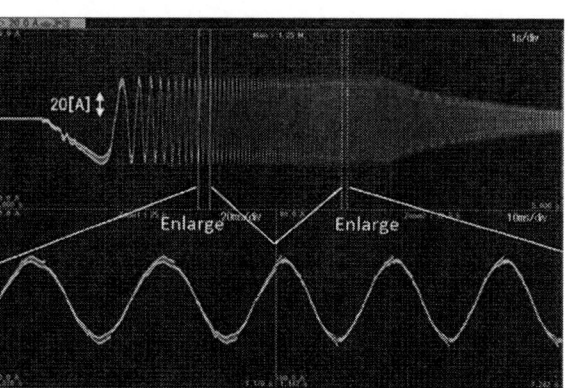

(b)

Fig. 15. Waveforms of motor current at start-up with compressor motor. (a) Without harmonic cancellation. (b) With harmonic cancellation.

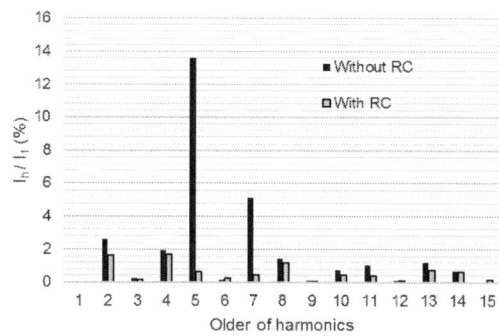

Fig. 16. Comparison of harmonic spectra of motor current at start-up with compressor motor.

The 2018 International Power Electronics Conference

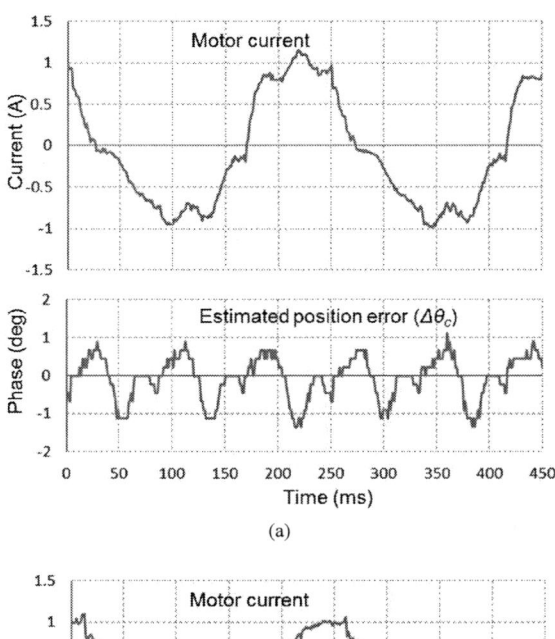

(a)

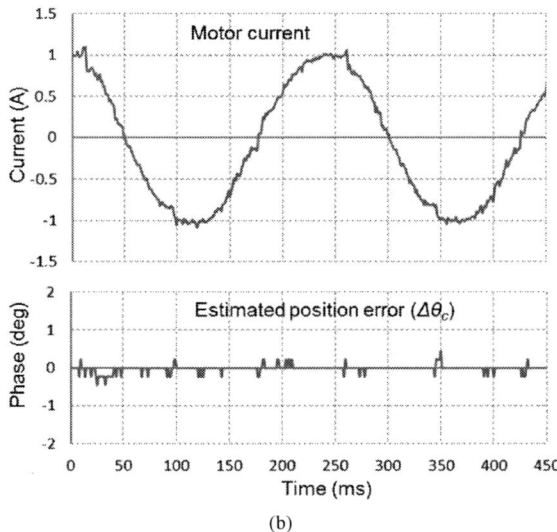

(b)

Fig. 17. Waveforms of motor current and estimated position error with fan motor. (a) Without harmonic cancellation. (b) With harmonic cancellation.

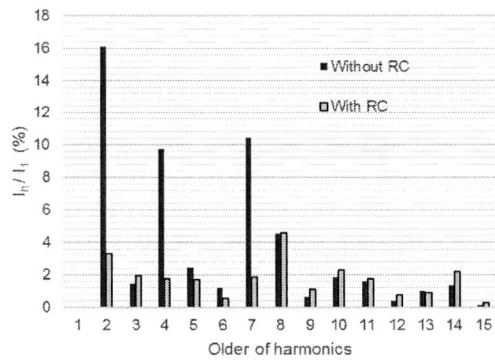

Fig. 18. Comparison of harmonic spectra of motor current at steady state with fan motor.

REFERENCES

[1] N. Urasaki, T. Senjyu, K. Uezato, and T. Funabashi, "An adaptive dead-time compensation strategy for voltage source inverter fed motor drives," *IEEE Trans. on Power Electronics*, vol. 20, no. 5, pp. 1150-1160, 2005.

[2] T. Suzuki, K. Sakamoto, T. Takeuchi, and Y. Notohara, "Embedded-friendly online dead-time compensation using PWM timer," in *Proc. ECCE'2011*, pp. 3906-3912, 2011.

[3] H. Zhao, Q. M. J. Wu, and A. Kawamura, "An accurate approach of nonlinearity compensation for VSI inverter output voltage," *IEEE Trans. on Power Electronics*, vol. 19, no. 4, pp. 1029-1035, 2004.

[4] W. Sun, J. Gao, X. Liu, Y. Yu, G. Wang, and D. Xu, "Inverter nonlinear error compensation using feedback gains and self-tuning estimated current error in adaptive full-order observer," *IEEE Trans. on Industry Applications*, vol. 52, no. 1, pp. 472-482, 2016.

[5] Y. Sato, T. Ishizuka, K. Nezu, and T. Kataoka, "A new control strategy for voltage-type PWM rectifiers to realize zero steady-state control error in input current," *IEEE Trans. on Industry Applications*, vol. 34, no. 3, pp. 480-486, 1998.

[6] D. Zmood and D. G. Holmes, "Stationary frame current regulation of PWM inverters with zero steady-state error," *IEEE Trans. on Power Electronics*, vol. 18, no. 3, pp. 814-822, 2003.

[7] R. Teodorescu, F. Blaabjerg, U. Borup, and M. Liserre, "A new control structure for grid-connected LCL PV inverters with zero steady-state error and selective harmonic compensation," in *Proc. APEC'04*, vol. 1, pp. 580-586, 2004.

[8] S. Fukuda and T. Yoda, "A novel current-tracking method for active filters based on a sinusoidal internal model," *IEEE Trans. on Industry Electronics*, vol. 37, no. 3, pp. 888-895, 2001.

[9] K. Sakamoto, Y. Iwaji, T. Endo, and Y. Takakura, "Position and speed sensorless control for PMSM driver using direct position error estimation," in *IECON'01*, pp. 1680–1685, 2001.

[10] D. Li, T. Suzuki, K. Sakamoto, Y. Notohara, T. Endo, C. Tanaka, and T. Ando, "Sensorless control and drive system of PMSM for compressor applications," in *Proc. IPEMC2006*, vol. 2, 2006.

[11] K. Tobari, T. Endo, Y. Iwaji, and Y. Ito, "Examination of new vector control system of permanent-magnet synchronous motor for high-speed drives," *IEEJ Trans. on Industry Applications*, vol. 129-D, no. 5, pp. 36-45, 2009. (in Japanese)

[12] A. Timbus, M. Liserre, R. Teodorescu, P. Rodriguez, and F. Blaabjerg, "Evaluation of current controllers for distributed power generation systems," *IEEE Trans. on Power Electronics*, vol. 24, no. 3, pp. 654-664, 2009.

[13] S. Han, T. Jo, J. Park, H. Kim, T. Chun, and E. Nho, "Dead time compensation for grid-connected PWM inverter," in *Proc. ECCE-Asia'2011*, pp. 876-881, 2011.

[14] F. Briz, P. García, M. W. Degner, D. Diaz-Reigosa, and J. M. Guerrero, "Dynamic behavior of current controllers for selective harmonic compensation in three-phase active power filters". *IEEE Trans. on Industry Applications*, vol. 49, no. 3, pp. 1411-1420, 2013.

Estimation Error Analysis of Stator Flux Observer for DTC-Based PMSM Drives

Atsushi Shinohara[*] and Kichiro Yamamoto

Department of Electrical and Electronics Engineering, Kagoshima University, Kagoshima, Japan

*E-mail: a-shinohara@eee.kagoshima-u.ac.jp

Abstract— Direct torque control can be applied to any AC motor drive and requires an estimation of stator flux linkage. The stator flux observer was developed for this solution. However, there are few reports about the way of the observer design. This paper analyzes the influence of the armature resistance error in the stator flux observer and proposes the way of the design for PMSM drives.

Keywords— permanent magnet synchronous motor (PMSM), direct torque control (DTC), flux observer.

I. INTRODUCTION

Direct torque control (DTC) was first proposed for induction motor drives [1]. Since DTC can be applied to any AC motor drive [2], there are many researches about applying DTC for permanent magnet synchronous motor (PMSM) drives. In DTC, accurate stator flux linkage is required for high torque response and high performance drives with the conventional knowledge. For example, the maximum torque per ampere (MTPA) control is important for PMSM drives [3]-[5].

For DTC-based PMSM drives, stator flux linkage should be estimated accurately to realize not only the position sensorless drives but also the high-efficiency drives [5].

The stator flux linkage can be estimated by integrating voltage in any AC motor drive. However, there are two major problems in the integrator [6]: (i) saturation by DC drift, and (ii) initial condition. Many researchers have proposed several solutions for these problems [6]-[32].

The existing stator flux observers for PMSM drives are categorized into a few structures depending on their estimation schemes where stator flux linkage is estimated with the integration of voltage or with using motor inductances [6]. To realize wide speed operational range, the combination of the two estimation schemes tend to be used [23]-[32]. Although many stator flux observers have been proposed, there are few researches about the coherent design of the flux observer, and the flux observer has to be designed experimentally.

This paper overviews the design of the stator flux observer for DTC-based PMSM drives and the effect of the armature resistance error. Overviewing the existing flux observers, they are summarized into two structures, which are equivalent in using a conversion of the observer gain. Moreover, the one of the structures can be analyze as the linear controller, and therefore, the bode diagram and the vector diagram analysis can be available. This paper shows the stator flux estimation error due to armature resistance error with vector diagram analysis and with simulations.

II. STATOR FLUX OBSERVER

The voltage model of the PMSM in the α-β stationary reference frame is given by (1).

$$\boldsymbol{\psi}_s = \frac{1}{s}(\boldsymbol{v}_a - R_a \boldsymbol{i}_a) \qquad (1)$$

where $\boldsymbol{v}_a$, $\boldsymbol{i}_a$, and $\boldsymbol{\psi}_s$ are the armature voltage, armature current, and stator flux linkage vectors, R_a is the armature resistance, respectively. s is Laplace operator: it equals to differential operator.

The stator flux linkage vector can be estimated with an integrator based on (1). However, pure integrators have some problems as discussed earlier. One of the solutions for these problems is to estimate the stator flux linkage vector from armature current in the d-q rotating reference frame as following equation:

$$\begin{bmatrix} \psi_d \\ \psi_q \end{bmatrix} = \begin{bmatrix} L_d & 0 \\ 0 & L_q \end{bmatrix} \begin{bmatrix} i_d \\ i_q \end{bmatrix} + \begin{bmatrix} \Psi_a \\ 0 \end{bmatrix} \qquad (2)$$

where i_d and i_q are the d- and q-axis components of the armature current, ψ_d and ψ_q are the d- and q-axis components of the stator flux linkage, L_d and L_q are the d- and q-axis inductances, and Ψ_a is the magnet flux linkage, respectively.

The d- and q-axis components are derived from α- and β-axis components and the rotor position θ as follows:

$$\begin{bmatrix} d \\ q \end{bmatrix} = \begin{bmatrix} \cos\theta & -\sin\theta \\ \sin\theta & \cos\theta \end{bmatrix} \begin{bmatrix} \alpha \\ \beta \end{bmatrix} = \boldsymbol{T} \begin{bmatrix} \alpha \\ \beta \end{bmatrix} \qquad (3)$$

where $\boldsymbol{T}$ is defined as the rotation matrix.

Fig. 1 shows the two types of the flux observers [23]-[32]. In Fig. 1, $L(\theta)$ means the function from the armature current $\boldsymbol{i}_a$ to the stator flux linkage $\boldsymbol{\psi}_s$, using (2) and the

The 2018 International Power Electronics Conference

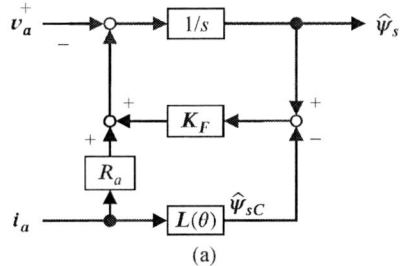 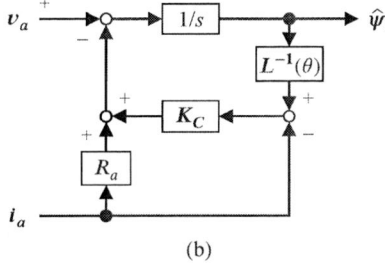

Fig. 1 Existing stator flux observers: (a) Type-F flux observer [7]-[9] and (b) Type-C flux observer [10].

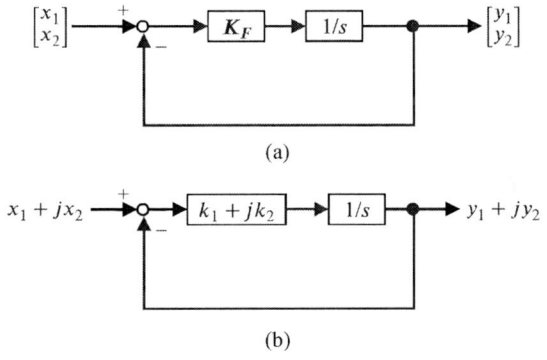

Fig. 2 An example of two-dimensional system whose transfer function is equivalent to $G_c(s)$: (a) Simplified block diagram and (b) equivalent system in complex number when $K_F = k_1 I + k_2 J$.

Fig. 3 Bode diagram of $G_V(s)$ and $G_C(s)$ ($K_F = 10I$ (rad/s))

rotation matrix T. The existing flux observers can be categorized into two types due to the feedback values: one uses the stator flux linkage (see Fig. 1(a): Type-F) and the other uses the armature current (see Fig. 1(b): Type-C).

III. COMPARISON OF STATOR FLUX OBSERVERS

A. Observer Gains

Comparing the two flux observers, their characteristics are equivalent when the each gains K_F and K_C satisfy the following equation:

$$K_C = K_F L(\theta) \qquad (4)$$

Regarding the magnet flux linkage as the constant, the relationship can be described as follows:

$$L(\theta) = \frac{L_d + L_q}{2} I + \frac{L_d - L_q}{2} \begin{bmatrix} \cos 2\theta & \sin 2\theta \\ \sin 2\theta & -\cos 2\theta \end{bmatrix} \qquad (5)$$

where I means the identity matrix.

The equivalent observer gains K_F and K_C rely on the inductances and the rotor position.

B. Observer Analysis as the Linear System

In the Type-F flux observer, the transfer function to the estimated stator flux is analyzed as follows [31]:

$$\hat{\psi}_s = G_V(s)\hat{\psi}_{sV} + G_C(s)\hat{\psi}_{sC} \qquad (6)$$

where

$$\hat{\psi}_{sV} = \frac{1}{s}\left(v_a - \hat{R}_a i_a\right) \qquad (7)$$

$$\hat{\psi}_{sC} = \hat{T}^{-1}\begin{bmatrix} \hat{L}_d & 0 \\ 0 & \hat{L}_q \end{bmatrix}\hat{T}i_a + \hat{T}^{-1}\begin{bmatrix} \hat{\psi}_a \\ 0 \end{bmatrix} \qquad (8)$$

and in (6)~(8), $\hat{X}$ means the estimated value of X.

The transfer functions $G_V(s)$ and $G_C(s)$ are defined by the gain matrix K_F as follows:

$$G_V(s) = \frac{s}{s + K_F} \qquad (9)$$

$$G_C(s) = \frac{K_F}{s + K_F} \qquad (10)$$

Note that $G_V(s)$ and $G_C(s)$ are described as a two-dimensional systems. For example, the system whose transfer function is $G_C(s)$ can be described as a simple feedback system like Fig. 2(a). This system can be also described as a complex filter if K_F is the linear combination of identity matrix I and the $+\pi/2$ rotation matrix J as shown in Fig. 2(b) [33][34].

From the shapes of $G_V(s)$ and $G_C(s)$, the Type-F flux observer combines the advantages of the current model shown in (8) in low speeds and those of the voltage model shown in (7) in high speeds. The border between the low and the high speed is defined by K_F. The bode

1309

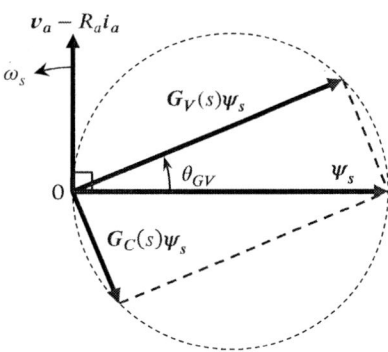

Fig. 4 Phasor diagram of (9) and (10) when $\boldsymbol{K}_F = k_1\boldsymbol{I}$.

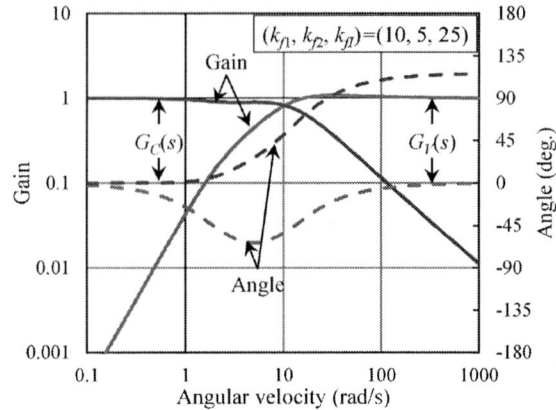

Fig. 5 Bode diagram of $G_V(s)$ and $G_C(s)$ with (12).

diagrams of $\boldsymbol{G_V}(s)$ and $\boldsymbol{G_C}(s)$ are shown in Fig. 3 as an example.

Here, the features of the transfer functions $\boldsymbol{G_V}(s)$ and $\boldsymbol{G_C}(s)$ are shown. Considering the frequency transfer functions of (1), (9), and (10), the relationship among $\boldsymbol{\psi}_s$, $\boldsymbol{G_V}(s)\boldsymbol{\psi}_s$, and $\boldsymbol{G_C}(s)\boldsymbol{\psi}_s$ can be described in phasor diagram as shown in Fig. 4 in steady state when $\boldsymbol{K}_F$ is set as the singular gain like $k_1\boldsymbol{I}$. The three vectors make a rectangular triangle whose angle depends on the ratio of k_1 and the flux vector rotating speed ω_s, as following equations [6][12]:

$$|\boldsymbol{G_V}(s)\boldsymbol{\psi}_s| = \frac{1}{\sqrt{1 + (k_1/\omega_s)^2}}|\boldsymbol{\psi}_s| \qquad (11)$$

$$\theta_{GV} = \arctan\left(\frac{k_1}{\omega_s}\right) \qquad (12)$$

Moreover, the rotating speed of the flux vector is same as that of the induced voltage in steady state [6][12].

When every estimated motor parameter matches the actual values respectively, the following relationship is available.

$$\hat{\boldsymbol{\psi}}_{sV} = \hat{\boldsymbol{\psi}}_{sC} = \hat{\boldsymbol{\psi}}_s = \boldsymbol{\psi}_s \qquad (13)$$

Therefore, the stator flux linkage estimation error can be analyzed by the parameter errors as follows:

$$\widetilde{\boldsymbol{\psi}}_s = \boldsymbol{G_V}(s)\widetilde{\boldsymbol{\psi}}_{sV} + \boldsymbol{G_C}(s)\widetilde{\boldsymbol{\psi}}_{sC} \qquad (14)$$

where $\tilde{X} = \hat{X} - X$ means the estimation error of X.

The both estimation errors $\widetilde{\boldsymbol{\psi}}_{sV}$ and $\widetilde{\boldsymbol{\psi}}_{sC}$ are caused by the parameter errors. Considering that $\boldsymbol{G_V}(s)\boldsymbol{\psi}_s$ and $\boldsymbol{G_C}(s)\boldsymbol{\psi}_s$ are determined as Fig. 4, the influence of the parameter errors in the stator flux linkage can be estimated.

The merit of the linear analysis is that the considerations about the flux observer characteristics and about the effect of parameter errors can be simplified, as

shown in (14). Especially, the analysis is enabled when the observer gain includes PI controller [32].

For example, when $\boldsymbol{K}_F$ is defined as follows:

$$\boldsymbol{K}_F = \left(k_{f1} + \frac{k_{fI}}{s}\right)\boldsymbol{I} + k_{f2}\boldsymbol{J} \qquad (15)$$

(6) can be described as follows:

$$\begin{aligned}\hat{\boldsymbol{\psi}}_s = &\frac{s^2}{s^2 + \left(k_{f1} + jk_{f2}\right)s + k_{fI}}\hat{\boldsymbol{\psi}}_{sV} \\ &+ \frac{\left(k_{f1} + jk_{f2}\right)s + k_{fI}}{s^2 + \left(k_{f1} + jk_{f2}\right)s + k_{fI}}\hat{\boldsymbol{\psi}}_{sC}\end{aligned} \qquad (16)$$

As shown in (16), the transfer function of the flux observer becomes like a second-order filter when $\boldsymbol{K}_F$ includes PI controller [32]. Although $\boldsymbol{K}_F$ includes the imaginary part $(k_{f2} \neq 0)$, the complex filter like Fig. 2(b) can be structured [33][34]. The bode diagram of an example is shown in Fig. 5.

Regardless of k_{f2} and k_{fI} in (15), the bode diagram is available for the observer analysis when $\boldsymbol{K}_F$ is the linear combination of identity matrix $\boldsymbol{I}$ and the $+\pi/2$ rotation matrix $\boldsymbol{J}$, and the vector diagram can be considered from the transfer functions of $\boldsymbol{G_V}(s)$ and $\boldsymbol{G_C}(s)$.

C. Observer Analysis as the Nonlinear System

The linear analysis is difficult for the Type-C flux observer. Then, the nonlinear analysis is applied. For example, when the observer gain $\boldsymbol{K}_C$ is defined as (17), $\boldsymbol{K}_C$ can be determined from the Lyapunov stability as (18) [28]:

$$\boldsymbol{K}_C = k_{c1}\boldsymbol{I} + k_{c2}\boldsymbol{J} \qquad (17)$$

$$\begin{cases} k_{c1} = \frac{R_a}{2}\left(\frac{1}{L_d} + \frac{1}{L_q}\right)L_p + \omega_p L_p \\ k_{c2} = \omega\frac{L_d + L_q}{2} + \text{sgn}(\omega)\frac{L_q - L_d}{2}\sqrt{\omega^2 - \left(\frac{R_a L_p}{L_d L_q}\right)^2} \end{cases} \qquad (18)$$

The 2018 International Power Electronics Conference

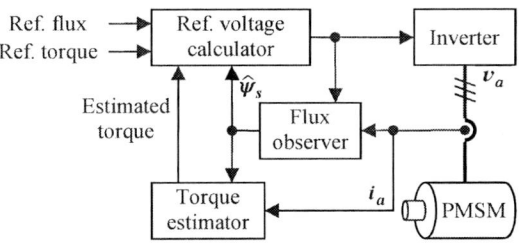

Fig. 6 The block diagram of the DTC PMSM drive.

TABLE I
PARAMETERS OF THE SIMULATED PMSM AND CONTROLLER

Meaning	Value
Number of pole pairs P_n	2
Armature resistance R_a	0.93 Ω
Magnet flux linkage Ψ_a	78.5 mWb
d-axis inductance L_d	9.67 mH
q-axis inductance L_q	24.3 mH
Control period of DTC	100 μs
Rated torque	2.1 Nm
Rated armature current	8.66 A

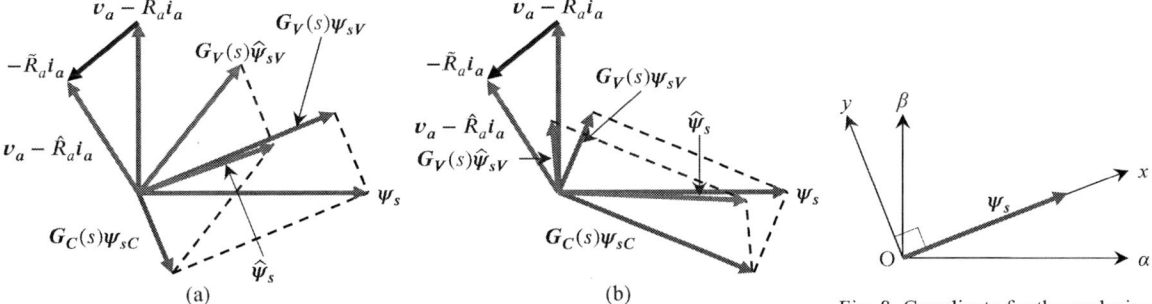

Fig. 7 Analysis of the stator flux linkage estimation error on the flux vector diagram due to the positive armature resistance error: (a) high-speed region and (b) low-speed region.

Fig. 8 Coordinate for the analysis of the stator flux linkage vector.

where ω_p is the distance between the center of the motor poles and that of the observer poles, L_p is the positive inductance value to be coherent the observer dimension.

In Type-C flux observer, the dimension of the observer gains k_{c1} and k_{c2} is equivalent to the resistance (Ω). Thus, in order to decide the gains from the observer pole whose dimension is equivalent to the speed, motor parameters are required. Moreover, the observer gains should be varied depending on speed according to (18). Though similar technique of the variable observer gain was reported in [6][20][22], the variable gain usually makes the observer stability more sensitive. Therefore, the Type-F flux observer is useful rather than the Type-C flux observer.

IV. VALIDATION BY SIMULATION

A. Simulation Setup

Fig. 6 shows the simplified DTC system. Since DTC requires the estimated stator flux linkage, the Type-F flux observer shown in Fig. 1(a) is applied to Fig. 6. In this paper, reference flux vector calculated DTC (RFVC DTC) reported in [6] is applied.

TABLE I lists the parameters of the simulated PMSM and controller.

In order to compare the estimation characteristics, the operation condition was fixed as the reference speed stepped down from 20 rad/s (190 min⁻¹) to 5 rad/s (47.8 min⁻¹) at 0.5 s, the constant load torque of 1 Nm, and the constant reference flux linkage of 90 mWb.

B. Analysis of the Armature Resistance Error

The armature resistance in DTC was set to the larger value than the motor armature resistance.

$$\hat{R}_a = 1.1 R_a \qquad (19)$$

From (7) and (14), the stator flux linkage estimation error can be described as follows:

$$\widetilde{\psi}_s = -\frac{G_V(s)}{s}\widetilde{R}_a i_a = -0.1\frac{G_V(s)}{s}R_a i_a \qquad (20)$$

Fig. 7 shows the flux vector diagram of the Type-F flux observer in the case that both $G_V(s)$ and $G_C(s)$ become first-order filter in steady state. According to the phase characteristics of the first-order filter, the three vectors, ψ_s, $G_V(s)\psi_{sV}$, and $G_C(s)\psi_{sC}$ makes the rectangular triangle whose angle relies on the ratio of the observer gains and the rotation frequency as shown in Fig. 4. The analysis of the vector diagram, as shown in Fig. 7, results in the two phenomena in the estimated stator flux linkage vector: (i) the estimated vector is always smaller than the actual vector, and (ii) the estimated vector leads ahead in the high-speed region and lags behind in the low-speed region.

To analyze the vectors which rotate in steady state, the coordinate which is synchronized to the actual stator flux linkage ψ_s (x-y frame) is used, as shown in Fig. 8. In this frame, due to the estimation error, there is not only the x-axis component of the estimated flux linkage $\hat{\psi}_x$ but also y-axis component of estimated one $\hat{\psi}_y$.

1311

The 2018 International Power Electronics Conference

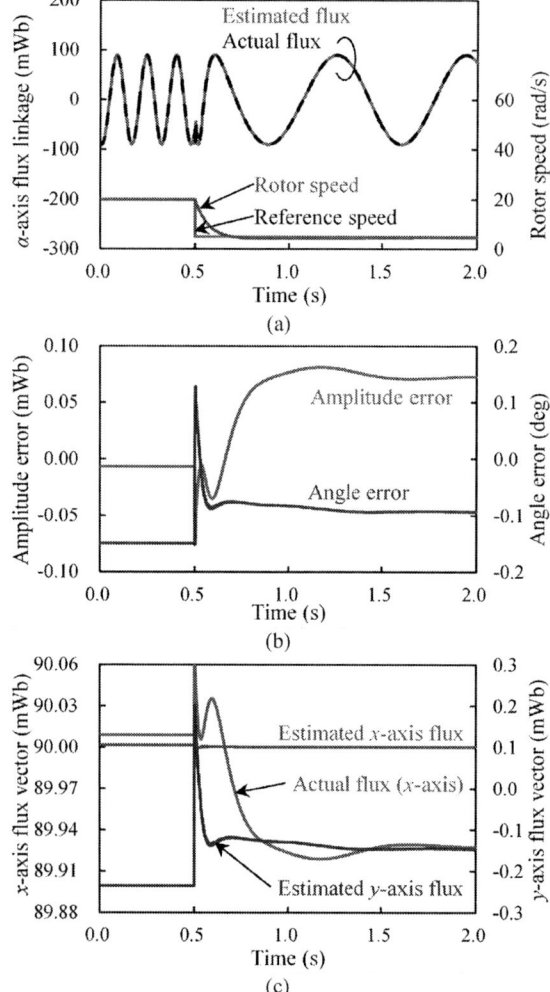

Fig. 9 The estimated stator flux linkage vector characteristics with $K_C = 10I$ under no parameter error : (a) The reference and actual rotor speed and α-axis component of estimated and actual stator flux linkage, (b) the stator flux linkage estimation error, and (c) the estimated and actual flux vector in x-y frame.

C.   Simulation Results With No Parameter Error

First, in order to verify the flux vector estimation error under no parameter error in the flux observer, operating conditions with no parameter error are evaluated.

Fig. 9 shows the estimation characteristics of the stator flux linkage with the Type-F flux observer whose gain K_F was set to $10I$.

DTC controls estimated stator flux linkage so that the amplitude of not actual stator flux but estimated one keeps same as reference value of 90 mWb.

In the case of $K_F = 10I$, the both of cutoff angular frequencies in $G_V(s)$ and $G_C(s)$ are 10 rad/s which is equivalent to 5 rad/s in the rotor speed due to the number of pole pairs P_n of 2. However, because the operating conditions keep the relationship of (13), the estimation error is very small; in particular, the amplitude error is no more than 1 % of the actual value and the angle error is

no more than 0.2 degrees.

Fig. 9(c) shows the analysis result of estimated and actual flux vector in x-y frame defined in Fig. 8. The y-axis component of the estimated flux cannot keep to zero due to angle error shown in Fig. 9(b). This is because of the definition of x-y frame. On the other hand, the x-axis component of actual flux leaves from 90 mWb and decreases. This is because the amplitude of the estimated flux keeps to 90 mWb and there is the amplitude error as shown in Fig. 9(b). As discussed earlier, the difference in x-axis flux is no more than 1 % of the reference value.

D.   Simulation Results About the Flux Estimation Error

Next, in order to verify the influence of the armature resistance error to the estimation characteristics, operating conditions with the armature resistance errors are evaluated. To compare the effects of the observer gains, only the observer gains shown in (15) are different between the two simulations. In this paper, the observer gains are equivalent to ones in Fig. 3 or Fig. 5.

Fig. 10 shows the estimation characteristics of the stator flux linkage with the Type-F flux observer during operations. In each simulation, the estimated stator flux linkage amplitude is controlled to 90 mWb, which is the same as the reference value.

In the case of Fig. 10(a), the cutoff angular frequency in $G_V(s)$ is 10 rad/s which is equivalent to 5 rad/s in the rotor speed due to the number of pole pairs of 2. The convergence of the estimation error is faster because $G_V(s)$ becomes first-order filter. On the other hand, in the case of Fig. 10(b), the estimation error oscillates after the rotor speed decrease. The oscillation is the result of the resonance frequency of $G_V(s)$ because $G_V(s)$ becomes the second-order filter. The resonance frequency may exist when PI controller is applied to the stator flux observer. Therefore, the design of the observer gain k_{fI} in (15) should be careful.

In both simulation results, the estimated stator flux leads ahead the actual one during the rotor speed of 20 rad/s and lags behind the actual one during the rotor speed of 5 rad/s.

In addition, the stator flux linkage estimation error can be analyzed when Type-F flux observer is applied. Fig. 11 shows the detail of flux vector diagram including the armature resistance error. The flux linkage estimation error vector can be described with using the transfer function $G_V(s)$, the armature current vector i_a, and the angle θ_{GV} shown in Fig. 4. Assuming that the flux and the current vectors rotate at the constant speed in steady state, the analyzed flux linkage estimation error e_{cal} can be described as follows:

$$e_{cal} = -K_{GV}T_{GV}\tilde{R}_a i_a \qquad (21)$$

where

1312

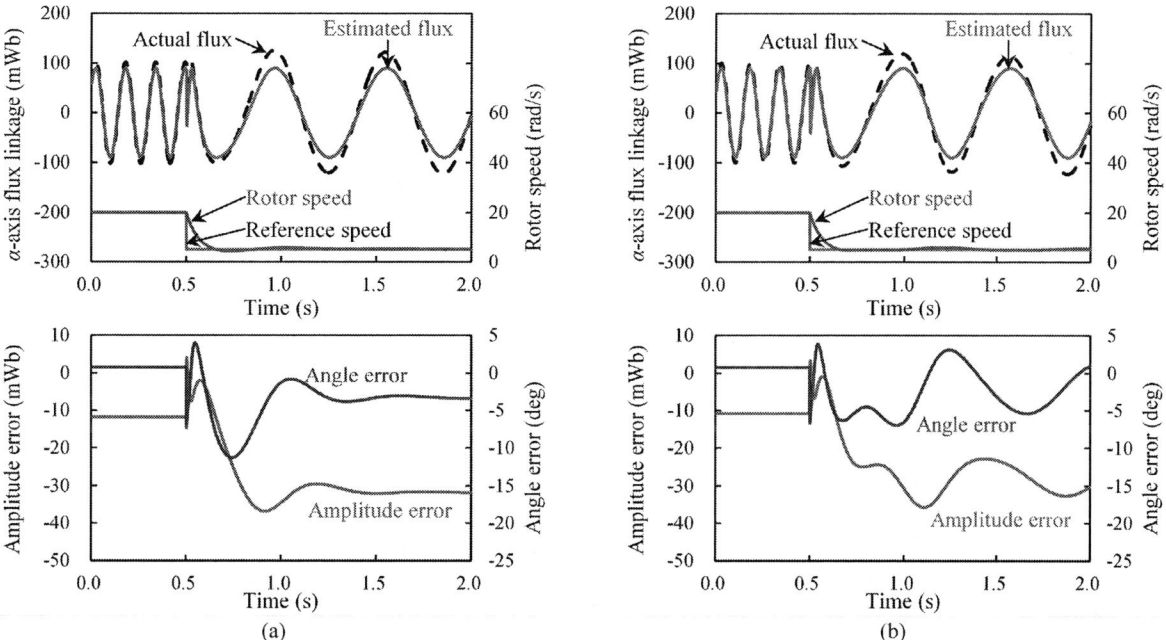

Fig. 10 Operation characteristics of the estimated stator flux linkage with the flux observer under +10% error in Ra. Upper: rotor speed and α-axis component of the estimated and the actual stator flux linkage; lower: the estimation error. (a) $(k_{f1}, k_{f2}, k_{f1}) = (10, 0, 0)$ and (b) $(k_{f1}, k_{f2}, k_{f1}) = (10, 5, 25)$.

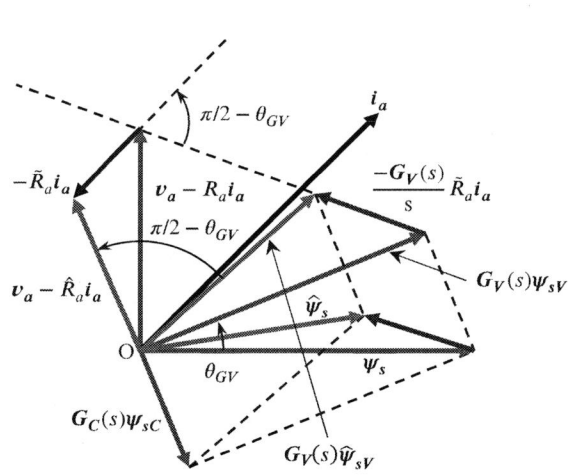

Fig. 11 Analysis of the stator flux linkage estimation error on the flux vector diagram due to the positive armature resistance error in steady state.

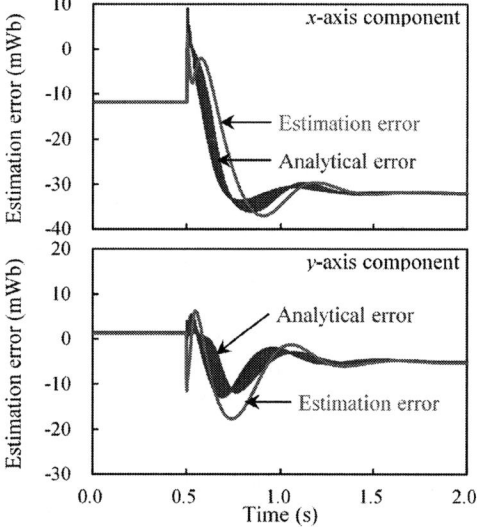

Fig. 12 The stator flux linkage estimation error characteristics due to the armature resistance error in x-y frame under the equivalent conditions to Fig. 10(a), and the analytical errors e_{cal} using (21)-(23).

$$T_{GV} = \begin{bmatrix} \cos(\pi/2 - \theta_{GV}) & -\sin(\pi/2 - \theta_{GV}) \\ \sin(\pi/2 - \theta_{GV}) & \cos(\pi/2 - \theta_{GV}) \end{bmatrix} \quad (22)$$

$$K_{GV} = \left| \frac{G_V(s)}{s} \right| = \frac{1}{\omega_s \sqrt{1 + (k_1/\omega_s)^2}} \quad (23)$$

Fig. 12 shows the stator flux linkage estimation errors in x-y frame under the equivalent conditions to Fig. 10(a). In the steady state where the periods are from 0 s to 0.5 s

and from 1.5 s to 2 s, the analytical error calculated by (21)-(23) matches the "actual" estimation error. From these results, the relationship between the armature resistance error and the stator flux linkage estimation error follows the relationship shown in Fig. 7.

Moreover, the similar analysis may be applied for the inductance error because the inductance error can be treated as the error of $G_C(s)\psi_s$.

1313

V. CONCLUSION

This paper showed the equivalence of the existing stator flux observers and discussed the influence of the armature resistance error with simulation. It was demonstrated that the influence of the error of armature resistance can be considered with using the bode diagram and the vector diagram because the stator flux observer becomes the linear system.

REFERENCES

[1] I. Takahashi and T. Noguchi, "A new quick-response and high-efficiency control strategy of an induction motor," *IEEE Trans. Ind. Appl.*, vol. IA-22, no. 5, pp. 820-827, Sep./Oct. 1986.

[2] G. S. Buja and M. P. Kazmierkowski, "Direct torque control of PWM inverter-fed AC motors –a survey," *IEEE Trans. Ind. Electron.*, vol. 51, no. 4, pp. 744-757, 2004.

[3] S. Morimoto, M. Sanada, and Y. Takeda, "Wide-speed operation of interior permanent magnet synchronous motor with high performance current regulator," *IEEE Trans. Ind. Appl.*, vol. 30, no. 4, pp. 920-926, Jul./Aug. 1994.

[4] A. Shinohara, Y. Inoue, S. Morimoto, and M. Sanada, "Direct calculation method of reference flux linkage for maximum torque per ampere control in DTC-based IPMSM drives," *IEEE Trans, Power Electron.*, vol. 32, no. 3, pp. 2114-2122, Mar. 2017.

[5] A. Shinohara, Y. Inoue, S. Morimoto, and M. Sanada, "Influence of stator flux estimation on reference flux for MTPA operation in PWM-based DTC PMSM drives," *International Journal of Power Electronics*, vol. 8, no. 1, pp. 23-37, 2016.

[6] A. Shinohara, Y. Inoue, S. Morimoto, and M. Sanada, "Comparison of the stator flux linkage estimator for PWM-based direct torque controlled PMSM drives," *in Proc. of IEEE PEDS 2015*, pp. 1035-1040, Jun. 2015.

[7] G. Tan, X. Wu, Z. Ye, Y. Han, and P. Guo, "Dual three-level double-fed induction motor control based on novel stator flux observer," *in Proc. of International Conference on Electrical and Control Engineering (ICECE) 2010*, pp. 3668-3671, 2010.

[8] Y. Wang and Z. Deng, "An integration algorithm for stator flux estimation of a direct-torque-controlled electrical excitation flux switching generator," *IEEE Trans. Energy Convers.*, vol. 27, no. 2, pp. 411-420, Jun. 2012.

[9] Y. Wang and Z. Deng, "Improved stator flux estimation method for direct torque linear control of parallel hybrid excitation switched-flux generator," *IEEE Trans. Energy Convers.*, vol. 27, no. 3, pp. 747-756, Sep. 2012.

[10] J. Hu and B. Wu, "New integration algorithms for estimating motor flux over a wide speed range," *IEEE Trans. Power Electron.*, vol. 13, no. 5, pp. 969-977, Sep. 1998.

[11] M.-H. Shin, D.-S. Hyun, S.-B. Cho, and S.-Y. Choe, "An improved stator flux estimation for speed sensorless stator flux orientation control of induction motors," *IEEE Trans. Power Electron.*, vol. 15, no. 2, pp. 312-318, Mar. 2000.

[12] N. R. N. Idris and A. H. M. Yatim, "An improved stator flux estimation in steady-state operation for direct torque control of induction machines," *IEEE Trans. Ind. Appl.*, vol. 38, no. 1, pp. 110-116 Jan./Feb. 2002

[13] J. Maes and J. A. Melkebeek, "Speed-sensorless direct torque control of induction motors using an adaptive flux observer," *IEEE Trans. Ind. Appl.*, vol. 36, no. 3, pp. 778-785, Mar./Apr. 2000.

[14] J. Holtz and J. Quan, "Drift- and parameter-compensated flux estimator for persistent zero-stator-frequency operation of sensorless-controlled induction motors," *IEEE Trans. Ind. Appl.*, vol. 39, no. 4, pp. 1052-1060, Jul./Aug. 2003.

[15] M. Hinkkanen, "Analysis and design of full-order flux observers for sensorless induction motors," *IEEE Trans. Ind. Electron.*, vol. 51, no. 5, pp. 1033-1040, Oct. 2004.

[16] M. Barut, S. Bogosyan, and M. Gokasan, "Speed-sensorless estimation for induction motors using extended Kalman filters," *IEEE Trans. Ind. Electron.*, vol. 54, No. 1, pp. 272-280, Feb. 2007.

[17] T. O.-Kowalska and M. Dybkowski, "Stator-current-based MRAS estimator for a wide range speed- sensorless induction-motor

drive," *IEEE Trans. Ind. Electron.*, vol. 57, no. 4, pp. 1296-1308, Apr. 2010.

[18] F. C. Dezza, G. Foglia, M. F. Iacchetti, and R. Perini, "An MRAS observer for sensorless DFIM drives with direct estimation of the torque and flux rotor current components," *IEEE Trans. Power Electron.*, Vol. 27, no. 5, pp. 2576-2584, May 2012.

[19] M. Hinkkanen and J. Luomi, "Modified integrator for voltage model flux estimation of induction motors," *IEEE Trans. Ind. Electron.*, vol. 50, no. 4, pp. 818-820, Aug. 2003.

[20] M. Comanescu and L. Xu, "An improved flux observer based on PLL frequency estimator for sensorless vector control of induction motors," *IEEE Trans. Ind. Electron.*, vol. 53, no. 1, pp. 50-56, Feb. 2006.

[21] C. Lascu and G. D. Andreescu, "Sliding-mode observer and improved integrator with DC-offset compensation for flux estimation in sensorless-controlled induction motors," *IEEE Trans. Ind. Electron.*, vol. 53, no. 3, pp. 785-794, Jun. 2006.

[22] D. Stojić, M. Milinković, S. Veinović, and I. Klasnić, "Improved stator flux estimator for speed sensorless induction motor drives," *IEEE Trans., Power Electron.*, vol. 30, no. 4, pp. 2363-2371, Apr. 2015.

[23] M. E. Haque, L. Zhong, and M. F. Rahman, "A sensorless initial rotor position estimation scheme for a direct torque controlled interior permanent magnet synchronous motor drive," *IEEE Trans. Power Electron.*, vol. 18, no. 6, pp. 1376-1383, Jun. 2003.

[24] S. Sayeef, G. Foo, and M. F. Rahman, "Rotor position and speed estimation of a variable structure direct-torque-controlled IPM synchronous motor drive at very low speed including standstill," *IEEE Trans. Ind. Electron.*, vol. 57, no. 11, pp. 3715-3723, Nov. 2010.

[25] Z. Xu and M. F. Rahman, "Comparison of a sliding observer and a Kalman filter for direct-torque-controlled IPM synchronous motor drives," *IEEE Trans. Ind. Electron.*, vol. 59, no. 11, pp. 4179-4188, Nov. 2012.

[26] G. -D. Andreescu, C. I. Pitic, F. Blaabjerg, and I. Boldea, "Combined stator flux observer with signal injection enhancement for wide speed range sensorless direct torque control of IPMSM drives," *IEEE Trans. Energy Convers.*, vol. 23, no. 2, pp. 393-402 Jun. 2008.

[27] G. H. B. Foo and M. F. Rahman, "Sensorless direct torque and flux-controlled IPM synchronous motor drive at very low speed without signal injection," *IEEE Trans. Ind. Electron.*, vol. 57, no. 1, pp. 395-403, Jan. 2010.

[28] G. H. B. Foo and M. F. Rahman, "Direct torque control of an IPM-synchronous motor drive at very low speed using a sliding-mode stator flux observer," *IEEE Trans., Power Electron.*, vol. 25, no. 4, pp. 933-942, Apr. 2010.

[29] W. Xu and R. D. Lorenz, "Reduced parameter sensitivity stator flux linkage observer in deadbeat-direct torque and flux control for IPMSM drives," *IEEE Trans. Ind. Appl.*, vol. 50, no. 4, pp. 2626-2636, Jul./Aug. 2014.

[30] I. Boldea, M. C. Paicu, and G. –D. Andreescu, "Active flux concept for motion-sensorless unified AC drives," *IEEE Trans., Power Electron.*, vol. 23, no. 5, pp. 2612-2618, Sep. 2008.

[31] A. Yousefi-Talouki, P. Pescetto, and G. Pellegrino, "Sensorless direct flux vector control of synchronous reluctance motors including standstill, MTPA, and flux weakening," *IEEE Trans., Ind. Appl.*, vol. 53, no. 4, pp. 3598-3608, Jul./Aug. 2017.

[32] A. Yoo and S. –K. Sul, "Design of flux observer robust to interior permanent-magnet synchronous motor flux variation," *IEEE Trans. Ind. Appl.*, vol.45, no. 5, pp. 1670-1677, Sep./Oct. 2009.

[33] K. Tanaka, M. Hasegawa, and A. Matsumoto, "Extremely precise position estimation in sensorless control of permanent magnet synchronous motors using all-pass filter," *in Proc. of IEEE PEDS 2015*, pp. 652-657, Jun. 2015.

[34] K. Tanaka, M. Hasegawa, and A. Matsumoto, "PMSM sensorless control based on position estimation correction using all-pass filter," *IEEJ Trans. Industry Applications*, vol. 136, no. 3, pp. 238-245, Mar. 2016 (in Japanese).

The 2018 International Power Electronics Conference

Application of Fictitious Reference Iterative Tuning to Controller Design for Various Machines

Hidehiro Ikeda[1*], Kazuya Goto[1], Feili Zhang[1], Kazuya Kayashima[1], Tsuyoshi Hanamoto[2]

1 Department of Electrical Engineering, Nishi-Nippon Institute of Technology, Kanda-machi, Japan
2 Graduate School of Life Science and Systems Engineering, Kyushu Institute of Technology, Kitakyushu, Japan
*E-mail: ikeda@nishitech.ac.jp

Abstract—**This paper presents an application of a fictitious reference iterative tuning method (FRIT), which is data-driven controller design method, to various machines. FRIT can be used to design the controller using only the initial experimental data, without the state equations of the model and the model parameters. The proposed controller is based on a PID control and included multi control loop. The control objects are a two-mass resonance system, a ball screw positioning system, a switched reluctance motor, and a switched reluctance generator. Finally, we confirm the effectiveness of the proposed method through computer simulations and experimental results.**

Keywords—Controller Design, Fictitious Reference Iterative Tuning, and PID Control System.

I. INTRODUCTION

Nowadays, a model-based design (MBD) has been used for controller design in industrial applications. MBD is a method that can reduce time and cost of controller design by imitating an equivalent model of a real object in a computer without using actual equipment, from the design to the testing. Therefore, MBD requires accurate state equations of actual systems and their parameters, which are calculated by the modeling method and the parameter identification. Meanwhile, in the industrial fields in which many motor drive systems are used, experienced technicians often adjust the control system to suit the manufacturing equipment on site. However, engineer shortages in these industrial fields are becoming a serious problem. It is therefore necessary to develop a simple controller design method for the industrial field.

Under these circumstances, to save both the time required and the cost of tuning controllers for motor drive systems, some direct controller tuning methods have been proposed that are based on the transient response data of closed-loop systems without modeling of the plants. A fictitious reference iterative tuning (FRIT) is one of the most promising candidate methods for practical direct parameter tuning [1]~[3]. In FRIT, the controller gains can be designed using only single-shot experimental input-output data without having the model parameters of the object to be controlled. We have proposed a controller gain tuning method for a vibration suppression-type speed controller using FRIT for single-input multi-variable control object [4].

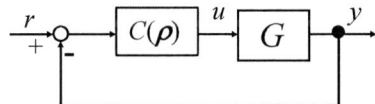

Fig. 1.   Typical closed loop system.

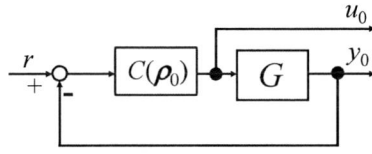

Fig. 2.   Measurement of the initial data.

This paper presents applications of FRIT to various control objects. The proposed controller is based on the PID control and included minor control loops. We confirm the effectiveness of the proposed method, for the two-mass resonance system, the ball-screw positioning system, the switched reluctance motor, and the switched reluctance generator, using computer simulations and experimental results.

II. FICTITIOUS REFERENCE ITERATIVE TUNING (FRIT)

In this paper, the off-line tuning method for several machines when using only single-shot experimental input-output data that is based on the FRIT without use of either the model parameters or the state equation of the model is proposed. While most FRIT designs use only one state variable, this research use especially multiple state variables to design the controller gains when using FRIT.

Fig. 1 shows a typical control system, where G is the transfer function of the object to be controlled, r is the reference signal, ρ represents the controller gain vector, $C(\rho)$ is the controller, u is the control input, and y is the output. In this case, the mathematical model of G is not known in advance and is not required.

To begin with, we performed a single-shot experiment using the initial controller gains ρ_0, and measured the control input u_0 and output y_0, as shown in Fig. 2. Then,

1315

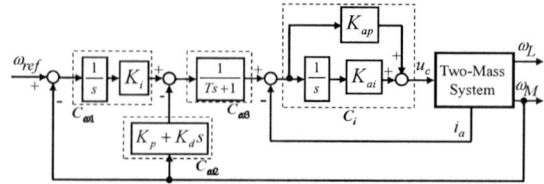

Fig. 3. Reference model showing input of the fictitious reference signal.

we determined the reference model $M(s)$, which satisfies the desired response. The fictitious reference signal $\tilde{\omega}_{ref}$ is then generated using the controller, the control input u_0, and the output y_0, as shown in (1). This means that the initial data u_0 and y_0 can be obtained using any ρ if we input $\tilde{r}(\rho)$ to the closed loop system implemented $C(\rho)$.

$$\tilde{r}(\rho) = C(\rho)^{-1} u_0 + y_0 \qquad (1)$$

Then, the optimal controller gains required to achieve $y_M = y_0$ are determined, as shown in Fig. 3, by an optimization searching method. In this paper, we use the particle swarm optimization or the differential evolution, which are the optimization searching methods, in order to search the FRIT solution rapidly [5], [6]. Finally, these controller gains represent the best answers that enable the desired response of the control system to be obtained. Therefore, the controller design can be performed without any information about the model parameters or the state equations. Examples of applying the proposed design method to various machines in the next chapter.

III. APPICATION OF PROPOSED CONTROLLER DESIGN METHOD USING FRIT

A. Speed Control System for Two-Mass Resonance System

The two-mass resonance system is often used as a control object for suppressing the torsional resonance vibration between the motor and the load in motor drive system [7], [8]. Fig. 4 shows the simplified proposed vibration suppression control system. Where ω_M is the motor angular speed, ω_L is the load angular speed, i_a is the armature current, u_c is the control input, ω_{ref} is the angular speed of the reference signal, $C_{\omega 1}$ is the integral (I) speed controller, $C_{\omega 2}$ is the proportional and derivative (PD) speed controller, $C_{\omega 3}$ is the type of first lag element speed controller, and C_i is the PI current controller.

Though FRIT uses generally one control input and one state variable for the initial experimental data, this proposed method uses one control input u_c and two state variables ω_M and i_a. The controller gain vector is $\rho = [K_p, K_i, K_d, T, K_{ap}, K_{ai}]^T$.

Therefore, the fictitious reference signal $\tilde{\omega}_{ref}$ can be calculated using the following equation without the two-mass resonance model.

$$\tilde{\omega}_{ref}(\rho) = \left(1 + \frac{C_{\omega 2}}{C_{\omega 1}}\right)\omega_{M0} + \frac{i_{a0}}{C_{\omega 1}C_{\omega 3}} + \frac{u_{c0}}{C_{\omega 1}C_{\omega 3}C_{\omega_i}} \qquad (2)$$

Fig. 4. Proposed speed control system for two-mass system.

Fig. 5. Experimental system for two-mass resonance system.

Here, u_{c0}, ω_{M0} and i_{a0} are the initial experimental data. Then, the reference model $M(s)$ as shown in (3), which used depends on the purpose of the system, where the time constant τ is a parameter of the reference model.

$$M(s) = \frac{1}{(\tau s + 1)^n} \qquad (3)$$

We set the $n = 3$ in order to simplify the design, which the τ is calculated using the 99 % response time T_{99} in this paper. The performance index function $J(\rho)$ is then defined in (4) below using ω_{M0}, $M(s)$, and $\tilde{\omega}_{ref}$.

$$J(\rho) = \|M(s)\tilde{\omega}_{ref}(\rho) - \omega_{M0}\|_2 \qquad (4)$$

Fig. 5 shows the experimental system for two-mass resonance system. In this paper, the two-mass resonance model consists of two 300W DC servomotors with a flexible coupling. Where the nominal inertias of the motor and the load are 2.774×10^{-4} kgm^2 and 2.940×10^{-4} kgm^2, respectively, and the nominal stiffness of the flexible coupling is 18.5 Nm/rad. Additionally, in the experiment, the model parameter and state variables are normalized, which 3000 r/min is normalized to 1 pu and 20 A is normalized to 1 pu.

Fig. 6 to 8 show the initial experiment waves ω_{M0}, i_{a0}, and u_{c0}, respectively, that were obtained using the initial controller gains $\rho_0 = [0.1, 30, 0.0001, 0.001, 1, 10]^T$, which is selected so as not to diverge. Where, ω_{ref} is 30 rad/sec. From these figure, the initial rise and the oscillation can be observed.

Fig. 9 shows the comparison of the initial data ω_{M0} and the simulated results for the reference output y_M, where the ρ designed by the proposed method is $\rho_0 = [1.24, 16.7, 6.53 \times 10^{-3}, 8.59 \times 10^{-4}, 2.73, 134.6]^T$. The figure shows that the proposed off-line tuning method

The 2018 International Power Electronics Conference

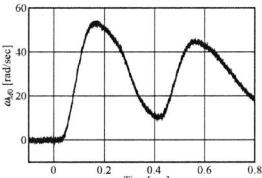

Fig. 6.　Initial ω_{M0} data.

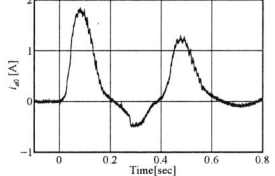

Fig. 7.　Initial i_{a0} data.

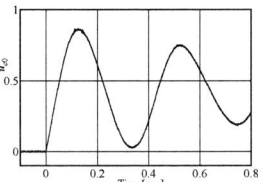

Fig. 8.　Initial u_{c0} data.

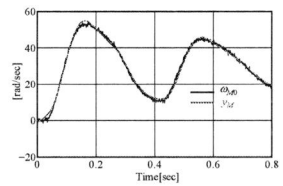

Fig. 9.　ω_{M0} and y_M.

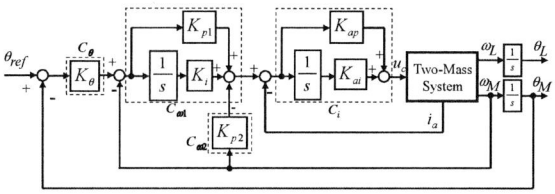

Fig. 11.　Proposed position control system for two-mass system.

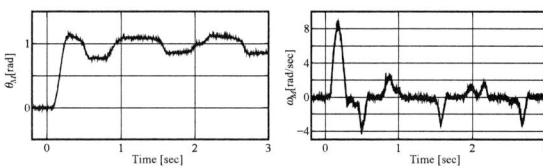

Fig. 12.　Initial θ_{M0} data.　　　Fig. 13.　Initial ω_{M0} data.

works very well, despite the fact that the design was performed using only the initial single-shot experimental data. Fig. 10 shows the experimental results that were obtained form ω_M, ω_L, and i_a, respectively, when using the proposed FRIT method. Where the ω_{ref} is stepped from 30 to 50 rad/sec at 0 sec and the disturbance torque is inputted from 0 to 10 % at 0.5 sec. As shown by this figure, good waves are observed for the reference-following performance and the disturbance response.

B. Position Control System for Two-Mass Resonance System

Next, the proposed method is applied to the position control system of two-mass resonance system. Fig. **??** shows the block diagram of the proposed position control system. Where θ_M is the angle of motor, θ_L is the angle of load, and θ_{ref} is the angle of the reference signal. The controller consists of the P position controller, the PI-P speed controller, and the PI current controller. In this paper, the state variables, which is used for controller, are the θ_M, ω_M, i_a, and u_c. The controller gain vector $\boldsymbol{\rho}$ is $\boldsymbol{\rho} = [K_\theta,\ K_{p1},\ K_{p2},\ K_i,\ K_{ap},\ K_{ai}]^T$. Then, the fictitious reference signal $\tilde{\theta}_{\text{ref}}$ can be calculated using (5). In order to simplify the FRIT design, the reference model

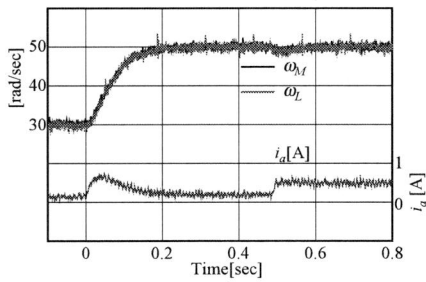

Fig. 10.　Experimental results of speed step response and disturbance response.

is used the (3) and $n = 3$ same as the speed control system for two-mass system. The performance index function $J(\boldsymbol{\rho})$ is defined in (6) below.

$$\tilde{\theta}_{ref}(\boldsymbol{\rho}) = \theta_{M0} + \frac{i_{a0}}{C_\theta C_{\omega 1}}$$
$$+ \frac{u_{c0}}{C_\theta C_{\omega 1} C_i} + \left(1 + \frac{1}{C_{\omega 1} C_{\omega 2}}\right)\frac{\omega_{M0}}{C_\theta} \quad (5)$$

$$J(\boldsymbol{\rho}) = \| M(s)\,\tilde{\theta}_{ref}(\boldsymbol{\rho}) - \theta_{M0} \|_2 \quad (6)$$

Fig. 12 to 15 show the initial experiment waves, where θ_{ref} is 1 rad, respectively, that were obtained using the initial controller gains $\boldsymbol{\rho}_0 = [0.05,\ 0.05,\ 0.05,\ 30,\ 1,\ 10]^T$. In these figures, we can observe both the initial rise and the oscillation. In addition, angle saturation can be observed under the influences of both static friction and Coulomb friction.

Additionally, Fig. 16 and Fig. 17 show the experimental results that were obtained for the motor and load angles (θ_M, θ_L) and the angular speeds (ω_M, ω_L) when using the proposed method ($T_{99} = 0.4$ sec), where $\boldsymbol{\rho} = [6.73 \times 10^{-2},\ 3.20 \times 10^{-5},\ 6.57,\ 145.7,\ 0.743,\ 26.2]^T$. As shown by these figures, good waves are observed for the reference-following performance.

C. Position Control System for Ball Screw Drive

A ball screw drives are widely used in industrial application due to the high accuracy and stiffness. In

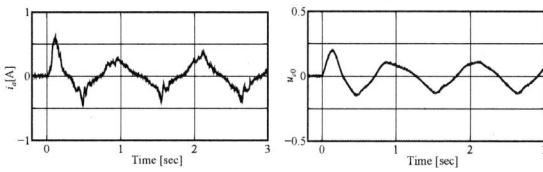

Fig. 14.　Initial i_{a0} data.　　　Fig. 15.　Initial u_{c0} data.

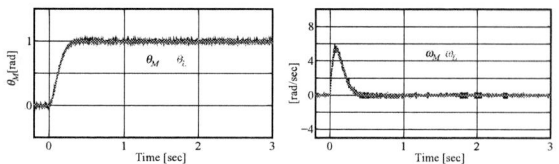

Fig. 16. Experimental results for θ_M and θ_L using the proposed method.

Fig. 17. Experimental results for ω_M and ω_L using the proposed method.

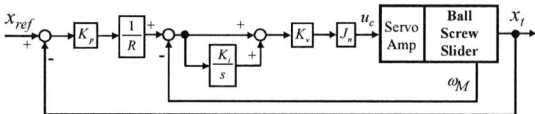

Fig. 18. Position control system of ball screw drive.

order to obtain high positioning accuracy of the ball screw drive, detailed modeling of the object is required. Especially, it is hard to identify the nonlinear frictions. In this paper, we apply the proposed FRIT design method to position control of the ball screw drive so as to design using only initial experimental data without the identification of the ball screw model.

Fig. 18 shows the proposed position control system of the one-axis ball screw drive. In this paper, the ball screw drive system consists of 10 mm lead pitch ball screws, 100 W AC servomotor, servo amplifier controller, and stage, where x_t is the position of the table, ω_M is the motor angular speed, u_c is the control input, x_{ref} is the position of the reference signal, R is the ball screw coefficient, J_n is the nominal value of the total inertia of the system. The position x_t is calculated by the encoder with the motor. Fig. 19 shows the experimental system for ball screw drive.

Position and speed controller is consist of the typical P-PI control, where the controller gain vector ρ is $\rho = [K_p, \ K_v, \ K_i]^T$. The fictitious signal can be calculated by (7). Furthermore, we set the reference model $n = 2$ in consideration for the control object.

$$\tilde{x}_{ref}(\rho) = x_{t0} + \frac{R}{K_p}\omega_{M0} + \frac{Rs}{(s + K_p)K_pK_vJ_n}u_{c0} \quad (7)$$

Fig. 20 to 22 show the position reference x_{ref}, initial experimental data x_{t0}, ω_{M0}, and u_{c0}, respectively, where $\rho_0 = [1, \ 10, \ 100]^T$. The large position error can be observed between x_{ref} and x_{t0}.

Fig. 19. Experimental system for ball screw drive.

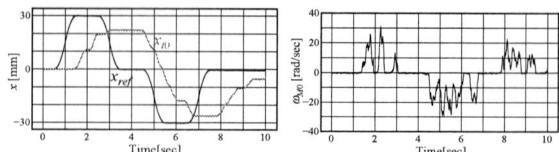

Fig. 20. x_{ref} and x_{t0} data.

Fig. 21. Initial ω_{M0} data.

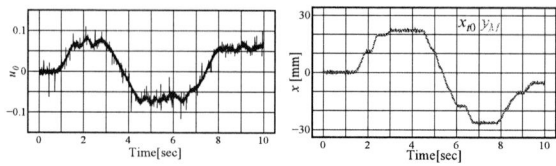

Fig. 22. Initial u_{c0} data.

Fig. 23. x_{t0} and y_M.

Here, the performance index function $J(\rho)$ is defined in (8).

$$J(\rho) = \|M(s)\tilde{x}_{ref}(\rho) - x_{t0}\|_2 \quad (8)$$

Fig. 23 shows the initial experimental result x_{t0} and the reference output y_M using the gains designed by the proposed method, where T_{99} is set to 0.2 sec, where $\rho = [16.7, \ 2800, \ 8.73]^T$. From this figure, the reference output designed by the proposed FRIT method and the initial output x_{t0} is matched very well. Fig. 24 shows the experimental result x_t and x_{ref} using the gains designed by the proposed method. From this figure, we can observe the effectiveness of the proposed method.

Furthermore, Fig. 25 shows the results obtained for variations in the response time, i.e., in the time constant τ of the reference transfer function (3), where $T_{99} = 0.2, 0.3, 0.4, 0.5, 0.6$ sec. As indicated by these figures, based on the reference shape, the response times change satisfactorily, and the proposed method can thus design the response speed arbitrarily.

D. Speed Control for SR Motor

The switched reluctance motor (SRM) is one of the rare-earth-free motors. The SRM simply consists of the stator which has winding and the rotor which is constructed of steel laminations without a permanent magnet. However, the accurate rotor position and the inductance, which is depend on the rotor position and current, should

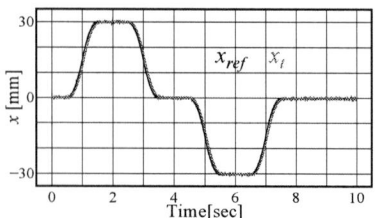

Fig. 24. Experimental results.

1318

The 2018 International Power Electronics Conference

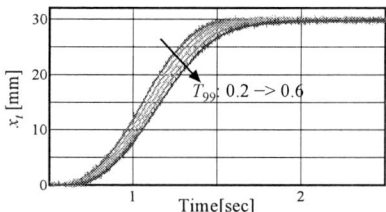

Fig. 25. Experimental results of x_t to various values of T_{99}.

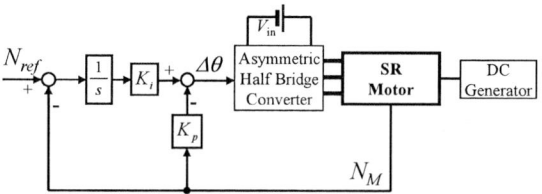

Fig. 26. Speed control system of SRM.

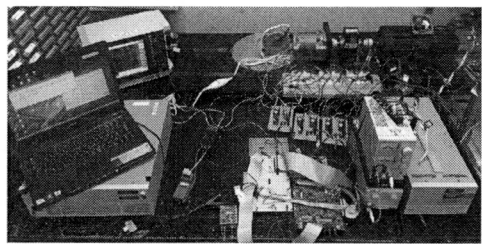

Fig. 27. Experimental system for SRM.

be detected to control accurately. Therefore, the lookup-table of the characteristics of torque and inductance, the shape of the optimal current wave, and the mathematical model of the system, are needed to control the speed of SRM. In this paper, we propose the speed controller using the FRIT design method for the SRM.

Fig. 26 shows the block diagram of the speed control system for the SRM. The speed controller is the simple I-P controller, where the gain vector ρ is $\rho = [K_p, K_i]^T$. In this paper, without using the characteristic look-up table or mathematical model of SRM, the controller gains are selected by FRIT using the control input u_0 and the rotational speed N_M obtained in the initial experiment. Additionally, in order to examine only the applicability of FRIT, the current controller is not used and only the speed controller is utilized.

Fig. 27 shows the experimental system for the SRM. The specification of SRM is 310 W, 6/4 poles, and 5100 rpm. The DC servomotor is coupled as a load. The asymmetric half bride converter (AHBC), which is constructed by two diodes and two IGBTs for each leg, is used as the power-electronics circuit.

The switching signal of the proposed control method is shown in Fig. 28. In this paper, the single pulse excitation is used to turn on/off the IGBTs of each leg for AHBC. The θ_e is the angle using the excitation. If θ_e is 0 deg or 90 deg, the rotor and stator salient pole is located the aligned position, and the phase inductance shows the largest. Meanwhile, if θ_e is 45 deg, each pole is located the unaligned position, and the phase inductance shows the least. The control input is $\Delta\theta$ ($-30\text{deg} \leq \Delta\theta \leq 30\text{deg}$). If $\Delta\theta \geq 0$, θ_{on} is 45 deg and θ_{off} is $\theta_{\text{on}} + \Delta\theta$ else θ_{on} is 0 deg and θ_{off} is $-\Delta\theta$.

The fictitious reference signal $\tilde{N}_M(\rho)$ is defined by (9). The reference model is set to be $n = 3$. The performance index is defined in (10).

$$\tilde{N}_{ref}(\rho) = \frac{s}{K_i}\Delta\theta_0 + \left(1 + \frac{K_p s}{K_i}\right)N_{M0} \qquad (9)$$

$$J(\rho) = \|M(s)\tilde{N}_{ref}(\rho) - N_{M0}\|_2 \qquad (10)$$

Fig. 29 and 31 show the initial experimental waves N_{M0} and $\Delta\theta_0$, where N_{ref} is stepped from 200 to 500 rpm when t is 0 sec, and ρ_0 is $[0.01, 3.0]^T$. In these figures, the overshoot and the oscillation can be observed.

Fig. 30 shows the experimental result using the proposed design method, where T_{99} is 1.0 sec, and the designed ρ is $[0.1065, 0.3201]^T$. From this figure, we can observed good response, and the applicability of the proposed method is verified to the SRM. Fig. 31 and 32 show the experimental results N_M and $\Delta\theta$, where the disturbance torque is increased from 0 to 0.55 Nm when time is 5 sec. As these figures show, good responses were observed in terms of their reference-following performance and disturbance response.

E. Current Control System for SR Generator

Due to the absence of any conductors or magnet on the rotor and the having possibility of wide range drive, the usage of SRM as a generator (switched reluctance

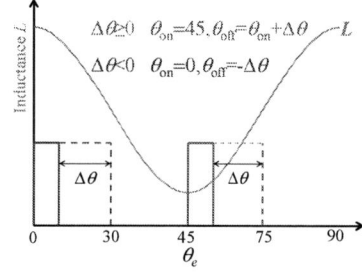

Fig. 28. Switching Signal for SRM Speed Control.

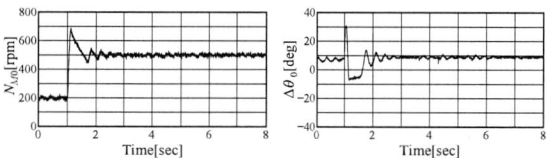

Fig. 29. Initial N_{M0} data.   Fig. 30. Initial $\Delta\theta_0$ data.

1319

The 2018 International Power Electronics Conference

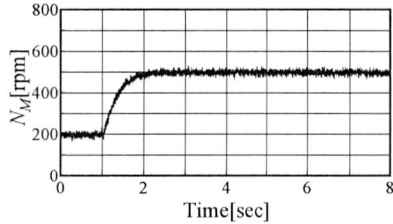

Fig. 31. Experimental result of N_M designed by the proposed method.

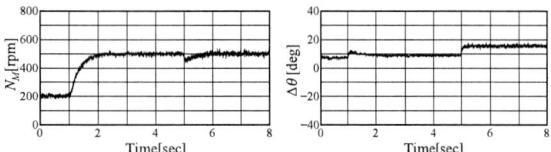

Fig. 32. N_M (load input).   Fig. 33. $\Delta\theta$ (load input).

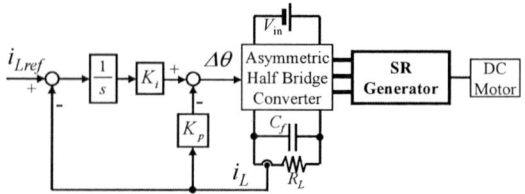

Fig. 34.   Current control system of SRG.

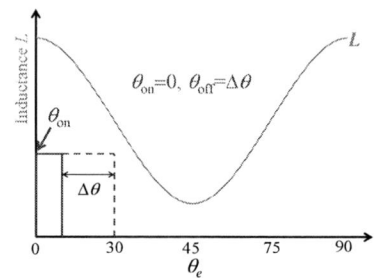

Fig. 35.   Switching Signal for SRG Current Control.

generator; SRG) is considered in recent years. However, the accurate rotor position and the inductance, which is depend on the rotor position and current, should be detected to control accurately. In this subsection, we propose the output current control using the FRIT design method for the SRG. Fig. 34 shows the block diagram of the current control system for the SRG. The each upper diode of the ADHC is connected to the load resistance, and all the regenerative power flowing from the phase inductance is consumed by the load resistance. In addition, a smoothing capacitor is connected in parallel to the load resistance. The controller is the simple I-P controller, where the gain vector ρ is $\rho = [K_p, \ K_i]^T$. The observable state variables are the load current i_L and the control input $\Delta\theta$. The switching signal of the proposed control method is shown in Fig. 35. Where, θ_{on} is fixed 0 deg and the control input $\Delta\theta$ is $0 \leq \Delta\theta \leq 30$ deg.

The specification of SRG is the same as SRM mentioned above. Where, the load resistance R_L is 5 Ω, and the smoothing capacitor C_f is 10 mF. The fictitious reference signal $\tilde{i}_{Lref}(\rho)$ is defined by (11). The performance index is defined in (12).

$$\tilde{i}_{Lref}(\rho) = \frac{s}{K_i}\Delta\theta_0 + \left(1 + \frac{K_p s}{K_i}\right)i_{L0} \quad (11)$$

$$J(\rho) = \|M(s)\tilde{i}_{Lref}(\rho) - i_{L0}\|_2 \quad (12)$$

Fig. 36 and 37 show the initial experimental waves i_{L0} and θ_{off0}, where i_{Lref} is stepped 1 to 2 A when time is 0.1sec, N_M is 1500 rpm, and ρ_0 is $[0.1, \ 500.0]^T$. Fig. 38 shows the experimental result using the proposed design method, where T_{99} is 0.1 sec, n of $M(s)$ is set to 3, and the designed ρ is $[8.458, \ 429.3]^T$. It can be seen that the proposed method is able to apply easily to the SRG.

Additionally, Fig. 39 shows the experimental results (i_L) that were obtained for various values of the speed

reference time parameter, where T_{99}=0.05, 0.1, 0.15, and 0.2. From this figure, the response times change satisfactorily.

IV.   CONCLUSIONS

In this paper, we have proposed the data-driven controller design method using FRIT. The proposed method only uses single-shot initial experimental data. Hence, it was not necessary to use the state equations and the parameters of the control objects as part of the design process. The effectiveness of the proposed method was verified by experiment for the two-mass resonance system, the ball screw system, the switched reluctance motor, and the switched reluctance generator. Our future work is that the stability and robustness will be calculated by the

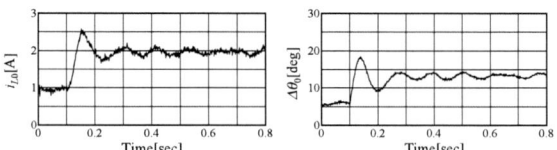

Fig. 36.   Initial i_{L0} data.   Fig. 37.   Initial $\Delta\theta_0$ data.

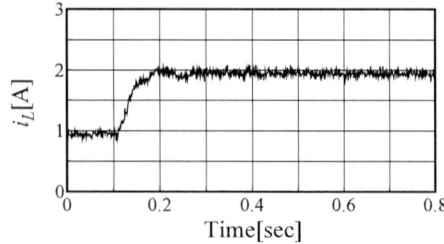

Fig. 38.   Experimental result of i_L designed by the proposed method.

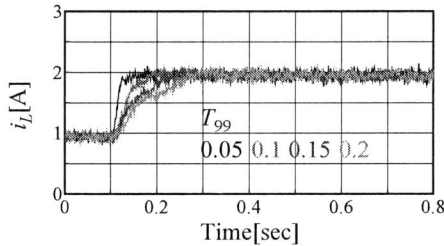

Fig. 39. Experimental responses of i_L to various values of T_{99}.

reference model $M(s)$ using the proposed method, and the influence of the model friction will be verified by the experiment.

REFERENCES

[1] Y. Matsui, H. Ayano, and K. Nakano, "A Controller Tuning for 2-Mass System Using Closed-Loop Transient Data," Proc. of 11th IFAC International Workshop, ThS6T2.1, 2013, pp.564-569.

[2] T. Azuma, and S. Watanabe, "A design of PID controllers using FRIT-PSO," Proc. of 8th ICST, 2014, pp.459-464.

[3] H. T. Nguyen, and O. Kaneko, "Fictitious Reference Iterative Tuning for Cascade PI Controllers of DC Motor Speed Control Systems," IEEJ Trans. on Electronics, Information and Systems, Vol.136, No.5, 2016, pp.710-714.

[4] H. Ikeda, H. Ajishi, and T. Hanamoto, "Application of Fictitious Reference Iterative Tuning to Vibration Suppression Controller for 2-Inertia Resonance System", Proc. of IECON 2015, TS-48, YF-008451, 2015, pp.1825-1830.

[5] H. Ikeda, and T. Hanamoto, "Vibration Suppression Control for Multi-Mass System Using CDM Designed by Group-Based-PSO", Proc. of ICEE 2012, No.P-EM-2, 2012, pp.1160-1165.

[6] H. Ikeda, and T. Hanamoto, "Design of m-IPD Controller of Multi-Inertia System using Differential Evolution", Proc. of IPEC 2014, 21J1-2, 2014, pp.184-189.

[7] Y. Hori, H. Sawada, and Y. Chun, "Slow resonance ratio control for vibration suppression and disturbance rejection in torsional system," IEEE Trans. on Ind. Electron., Vol.46, No.1, 1999, pp.162-168.

[8] K. Szabat, and T. Orlowska-Kowalska, "Vibration Suppression in a Two-Mass Drives System Using PI Speed Controller and Additional Feedbacks - Comparative Study," IEEE Trans. on Ind. Electron., Vol.54, No.2, 2007, pp.1193-1206.

High Efficiency Control for Permanent Magnet Motor Drive System with Fuel Cells Connected in Series with Electric Double-Layer Capacitors

Kichiro Yamamoto*, Fumiya Ohdera and Atsushi Shinohara

Department of Electrical and Electronics Engineering, Kagoshima University, 1-21-40 Korimoto, Kagoshima, Japan

*E-mail: yamamoto@eee.kagoshima-u.ac.jp

Abstract—This paper proposed a strategy to improve efficiency for a permanent magnet motor drive system with fuel cells connected in series with electric double-layer capacitors. The configuration of the proposed system and its operation are explained and equations for losses of the overall system are derived. Moreover, a control method to reduce the losses of overall system is investigated. Finally, experiments of the proposed control are performed and the results demonstrate that the proposed control can improve the efficiency of the overall system.

Keywords— fuel cell, electric double-layer capacitor, voltage booster, permanent magnet synchronous motor

I. INTRODUCTION

Recently, permanent magnet (PM) motors form an increasingly important class of high performance ac motor drives and are used for not only industrial applications but also electric vehicles and other applications [1],[2]. Furthermore, hybrid vehicles and fuel cell (FC) vehicles with PM motors have been developed to reduce CO_2 emission and to adapt depletion of fossil fuels [3]-[5].

Authors have been using a PWM inverter with dc link voltage control circuit to drive a PM motor by a dc voltage source such as battery [6]-[8]. And also, they have combined electric double-layer capacitors (EDLC) with the PM motor drive system and have already proposed a PM motor drive system with regenerating capability augmented by EDLC (Fig. 1) [9]. In the system, however, the EDLC were connected parallel to the battery. Thus, the withstand voltage of the capacitors was required the value equal to or greater than the battery voltage in the system. This disadvantage imposes the necessity of a lot of series connected capacitors on the system and the increase of number of capacitors results in the increase of cost, volume and weight of the system.

To solve the problem, we have proposed the system in which the battery and the EDLC are connected in series (Fig. 2) [10],[11]. In this system, the set of the battery and the EDLC is connected to the dc link of a PWM inverter through a bidirectional voltage booster. In addition, the EDLC itself is also connected to the dc link through the other bidirectional voltage booster. In other words, the system has two bidirectional voltage boosters. One of the boosters is used for both of the battery and the EDLC to control the dc link voltage of the PWM inverter.

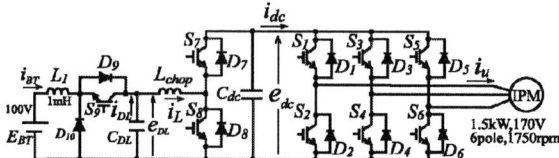

Fig. 1. Circuit configuration with double-layer capacitors in parallel with battery.

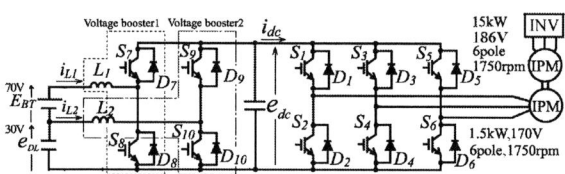

Fig. 2. Circuit configuration with double-layer capacitors in series with battery.

The other is used for only the EDLC to control the current flowing from and to the EDLC.

In this paper, a PM motor drive system with FC connected in series with EDLC is proposed. And also, a strategy to improve efficiency for this system is investigated. First, the configuration of the proposed system and its operation are explained. Moreover, equations for losses of the overall system are derived and the control method to reduce losses of the overall system is investigated. Finally, the effectiveness of the proposed control is validated by experimentl results.

II. SYSTEM CONFIGURATION

The configuration of a PM motor drive system with FC connected in series with EDLC is shown in Fig. 3. This system consists of a PM motor, a PWM inverter, two bidirectional voltage boosters, a FC and an EDLC. In this system, the EDLC is connected in series with the FC to reduce the number of the capacitor cells.

The voltage booster 1 is used for both the FC and the EDLC to control the dc link voltage e_{dc} of the PWM inverter. And the voltage booster 2 is used for only the EDLC to control of the EDLC current i_{L2}. This system operates as a conventional PWM inverter when the dc link voltage reference $e_{dc}*$ is less than the sum of the FC voltage and the EDLC voltage, $e_{FC} + e_{DL}$, and operates as a PWM inverter with variable dc link voltage when $e_{dc}*$ is greater than $e_{FC} + e_{DL}$. The voltage booster 2 controls

The 2018 International Power Electronics Conference

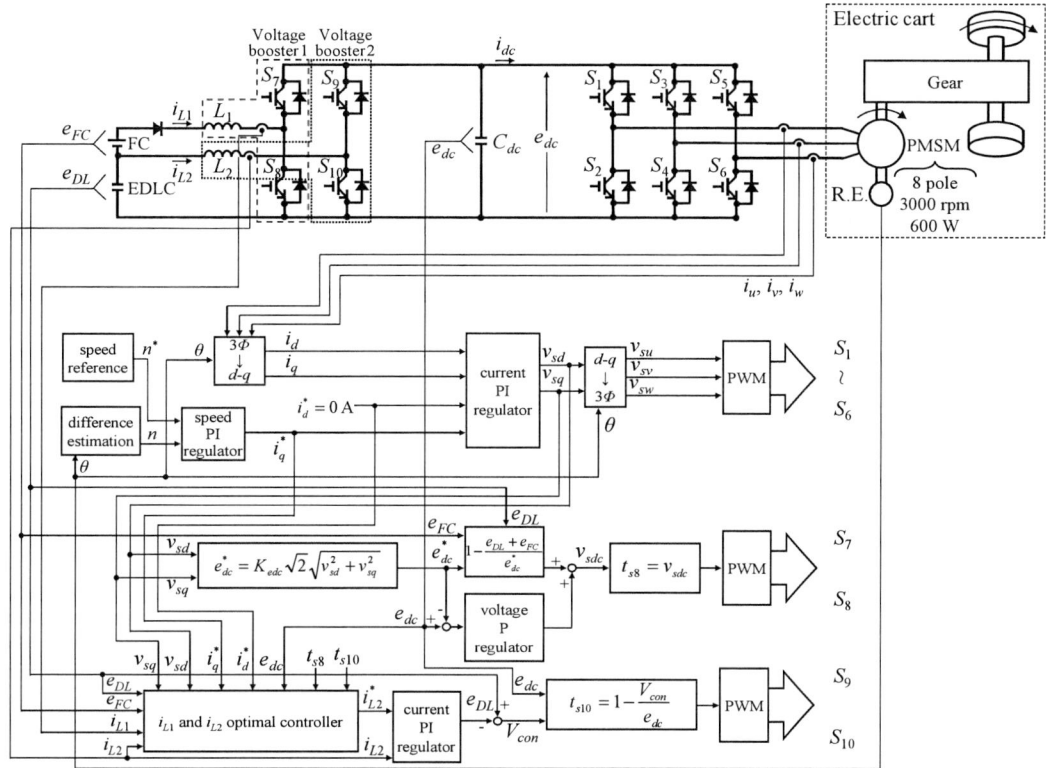

Fig. 3. System configuration of a PM motor drive system with FC connected in series with EDLC.

the EDLC current. As a result, the FC current i_{L1} is indirectly controlled. By the EDLC current control, the FC current can be controlled to any value because the difference between the FC power and the load power is supplied or absorbed by the EDLC.

III. MODELING OF SYSTEM LOSSES

In this section, equations for losses of the proposed system are derived. The overall system losses $P_{\text{system\_loss}}$ include the voltage booster losses $P_{\text{chopper\_loss}}$, the inverter losses $P_{\text{inverter\_loss}}$ and the motor losses $P_{\text{motor\_loss}}$ [12]. Here, we introduce equivalent loss resistances r_{eq1} and r_{eq2}.

$$P_{\text{system\_loss}} = P_{\text{chopper\_loss}} + P_{\text{inverter\_loss}} + P_{\text{motor\_loss}} \quad (1)$$

where

$$P_{\text{chopper\_loss}} = r_{eq1}i_{L1}^2 + r_{eq2}i_{L2}^2,$$

$$P_{\text{inverter\_loss}} = 6\left[\sqrt{\frac{2}{3}}V_{d0}\sqrt{i_d^2+i_q^2}\left(\frac{1}{2\pi}+\frac{m}{8}\cos\varphi\right) + \frac{2}{3}R_{ce}\left(i_d^2+i_q^2\right)\left(\frac{1}{8}+\frac{m}{3\pi}\cos\varphi\right) \right.$$
$$+ \sqrt{\frac{2}{3}}V_{s0}\sqrt{i_d^2+i_q^2}\left(\frac{1}{2\pi}-\frac{m}{8}\cos\varphi\right) + \frac{2}{3}R_{ak}\left(i_d^2+i_q^2\right)\left(\frac{1}{8}-\frac{m}{3\pi}\cos\varphi\right)$$
$$\left. + \frac{E_{on\_nom}+E_{off\_nom}+E_{rec\_nom}}{V_{dc\_nom}\cdot I_{c\_nom}}f_{sw}e_{dc}\sqrt{\frac{2}{3}}\sqrt{i_d^2+i_q^2}\right],$$

$$P_{\text{motor\_loss}} = \frac{v_d^2+v_q^2}{2r_c} + r_a\left(1+\frac{r_a}{r_c}\right)\left(i_d^2+i_q^2\right) - \frac{2r_a}{r_c}\left(v_d i_d + v_q i_q\right).$$

where r_{eq1} and r_{eq2} are equivalent loss resistances for the

voltage booster 1 and 2, V_{d0} and V_{s0} are zero-current conduction voltage drops of IGBTs and freewheeling diodes, R_{ce} and R_{ak} are equivalent series resistances of IGBTs and freewheeling diodes, m is an inverter modulation index, φ is a power factor angle, E_{on_nom} and E_{off_nom} are nominal values of loss energies for IGBT turn-on and turn-off, E_{rec_nom} is nominal value of reverse-recovery loss energy for a freewheeling diode, V_{dc_nom} is nominal value of a collector-emitter voltage, I_{c_nom} is nominal value of a collector current, f_{sw} is an inverter switching frequency, r_a is an armature resistance of PM motor, r_c is an equivalent iron loss resistance.

The voltage booster losses are assumed by using two equivalent loss resistances. Estimation method for r_{eq1} and r_{eq2} is mentioned in detail in Section IV.

The inverter losses are derived from the nominal value described in the datasheet. From a d-axis current i_d and a q-axis current i_q of PM motor, the peak of the collector current I_{cp} can be obtained as

$$I_{cp} = \sqrt{\frac{2}{3}}\sqrt{i_d^2+i_q^2}. \quad (2)$$

The motor losses are derived from d- and q-axis equivalent circuits of PM motor shown in Fig. 4 [12]. In this system, only q-axis current contributes to the electric torque because the motor is a surface magnet motor and it is used with i_d=0 A.

Since the parameters v_d, v_q, i_d, i_q, r_a and r_c in the equation of $P_{\text{motor\_loss}}$ are fixed in accordance with the load conditions, controllable parameters are only e_{dc}, i_{L1}

1323

The 2018 International Power Electronics Conference

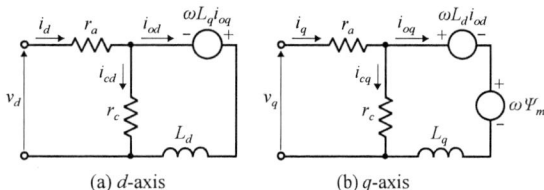

(a) *d*-axis   (b) *q*-axis

Fig. 4. Equivalent circuits of a PM motor [12].

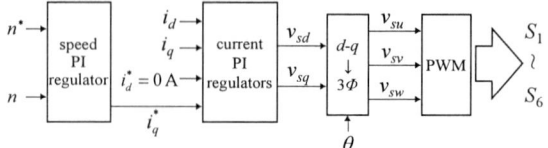

Fig. 5. Block diagram of speed regulator and current regulators.

and i_{L2}. In general, it is well known that the dc link voltage affects the voltage booster losses and the inverter losses. In addition, from equation (1), it is clear that the FC current i_{L1} and the EDLC current i_{L2} affect only the voltage booster losses. In section IV, a dc link voltage control and a FC current control are proposed to improve the efficiency for the overall system.

IV. IMPROVING EFFICIENCY CONTROL

A. Control of DC Link Voltage

As mentioned in Section III, when the dc link voltage e_{dc} decreases, the voltage booster losses and the inverter losses are reduced. However, if the dc link voltage decreases too much, the motor cannot rotate at the desirable rotor speed because of voltage shortage. Therefore, the dc link voltage which can rotate the motor according to the rotor speed reference is needed.

The block diagram of the speed regulator and the current regulators used in our system is shown in Fig. 5. In the figure, the *d*-axis voltage reference v_{sd} and the *q*-axis voltage reference v_{sq} are the *d*- and *q*-axis voltages needed for the motor to rotate at the rotor speed reference n^*. And condition of the dc link voltage that the motor can rotate at n^* is as follows:

$$e_{dc} > \sqrt{2}\sqrt{v_{sd}^2 + v_{sq}^2}. \tag{3}$$

Practically, in order to avoid partial saturation in the PWM voltage, the dc link voltage reference needs the voltage margin factor K_{edc}. Thus, following dc link voltage reference e_{dc}^* is used for the control.

$$e_{dc}^* = K_{edc}\sqrt{2}\sqrt{v_{sd}^2 + v_{sq}^2}. \tag{4}$$

In this paper, experiments were conducted at K_{edc}=1.4. By the dc link voltage control using equation (4), the efficiency of the overall system can be improved while the motor can rotate at the rotor speed reference.

B. Control of FC Current

As explained in Section III, the FC current i_{L1} and the EDLC current i_{L2} affect only the voltage booster losses. Thus, when the FC current and the EDLC current are determined so as to minimize the voltage booster losses, the efficiency of the overall system can be improved. Here, we introduce following equation for power,

$$\left(e_{FC} + e_{DL}\right)i_{L1} + e_{DL}i_{L2} - \left(r_{eq1}i_{L1}^2 + r_{eq2}i_{L2}^2\right) = P_m. \tag{5}$$

In equation (5), we assumed that the input power of the PM motor P_m is $v_{sd}i_d^* + v_{sq}i_q^*$. And P_m includes the loss of inverter. By deriving the EDLC current from equation (5), the following equation is obtained.

$$i_{L2} = \frac{e_{DL} - \sqrt{e_{DL}^2 - 4r_{eq2}\left\{r_{eq1}i_{L1}^2 - \left(e_{FC} + e_{DL}\right)i_{L1} + P_m\right\}}}{2r_{eq2}}. \tag{6}$$

In equation (6), there is a mutual relationship such that when the FC current increases, the EDLC current decreases. Therefore, the ratio between the FC current and the EDLC current is important to reduce the voltage booster losses.

As shown in equation (1), the voltage booster losses $P_{\text{chopper\_loss}}$ are assumed to be $r_{eq1}i_{L1}^2 + r_{eq2}i_{L2}^2$, where r_{eq1} and r_{eq2} are equivalent loss resistances for each voltage booster [13]. These r_{eq1} and r_{eq2} are calculated from the input power and the output power of each voltage booster.

The power flow of two voltage boosters is shown in Fig. 6. From the figure, the input power P_{in1} and the output power P_{out1} of the voltage booster 1 are obtained as

$$P_{in1} = \left(e_{FC} + e_{DL}\right)i_{L1}, \tag{7}$$

$$P_{out1} = e_{dc}i_{L1}\left(1 - d_1\right). \tag{8}$$

Similarly, the input power P_{in2} and the output power P_{out2} of the voltage booster 2 are obtained as

$$P_{in2} = e_{DL}i_{L2}, \tag{9}$$

$$P_{out2} = e_{dc}i_{L2}\left(1 - d_2\right) \tag{10}$$

where d_1 and d_2 are duty ratios of the voltage booster 1 and 2, respectively. And e_{FC}, e_{DL}, e_{dc}, i_{L1} and i_{L2} are values detected using voltage or current sensors.

From equations (7)-(10), the difference between the input power and the output power of the voltage booster 1 is the losses of voltage booster 1, and the difference between the input power and the output power of the voltage booster 2 is the losses of voltage booster 2. By dividing these differences by i_{L1}^2 for the voltage booster 1 and by i_{L2}^2 for the voltage booster 2, following equivalent loss resistances are obtained.

$$r_{eq1} = \frac{\left(e_{FC} + e_{DL}\right) - e_{dc}\left(1 - d_1\right)}{i_{L1}}, \tag{11}$$

$$r_{eq2} = \frac{e_{DL} - e_{dc}\left(1 - d_2\right)}{i_{L2}}. \tag{12}$$

From these equations, it is clear that r_{eq1} and r_{eq2} represent the equivalent resistances for the losses of each

1324

The 2018 International Power Electronics Conference

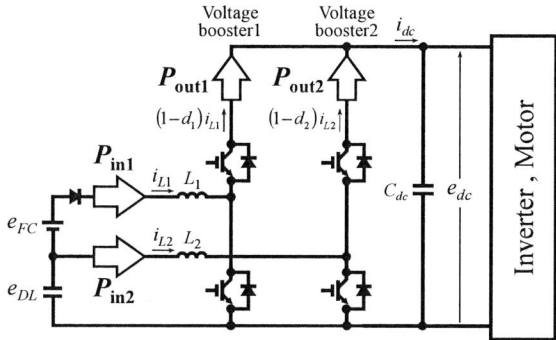

Fig. 6. Power flow of two voltage boosters.

Fig. 7. Electric cart with the motor drive system.

voltage booster, respectively. It is important to note that the equivalent loss resistances include all losses generated by voltage boosters such as switching losses, conduction losses and the losses in inductors L_1 and L_2.

In order to make the equation of voltage booster losses be a function of the FC current, equation (6) is substituted into the equation of voltage booster losses.

$$P_{\text{chopper\_loss}} = r_{eq1}i_{L1} + r_{eq2}\left\{\frac{e_{DL} - \sqrt{e_{DL}^2 - 4r_{eq2}\left\{r_{eq1}i_{L1}^2 - (e_{FC} + e_{DL})i_{L1} + P_m\right\}}}{2r_{eq2}}\right\}^2 . \quad (13)$$

And partial derivative with respect to the FC current makes it as the following:

$$\frac{\partial P_{\text{chopper\_loss}}}{\partial i_{L1}} = (e_{FC} + e_{DL}) + \frac{e_{DL}\left\{2r_{eq1}i_{L1} - (e_{FC} + e_{DL})\right\}}{\sqrt{e_{DL}^2 - 4r_{eq2}\left\{r_{eq1}i_{L1}^2 - (e_{FC} + e_{DL})i_{L1} + P_m\right\}}} . \quad (14)$$

The voltage booster losses can be minimized by setting the FC current such that equation (14) equals to zero. In this case, this special FC current is named as i_{L1_opt}.

In proposed FC current control, the FC current reference i_{L1}^* is set to i_{L1_opt}. By the FC current control using i_{L1_opt}, the efficiency of voltage boosters can be improved. As a result, the efficiency of the overall system can also be improved.

V. EXPERIMENTAL RESULTS

A. Verification of Effectiveness of Proposed DC Link Voltage Control

In order to verify the effectiveness of proposed DC link voltage control and FC current control, an electric cart with the motor drive system shown in Fig. 7 was driven. Before each experiment, the EDLC voltage was adjusted to 20 V in order to equalize initial conditions. The proposed dc link voltage control and the constant dc link voltage control were performed. In the constant dc link voltage control, the dc link voltage references e_{dc}^* were set to 150 V or 200 V. The FC current references i_{L1}^* were set to 1.0 A in both of the two control methods. From these two experimental results, the experimental waveforms and the input powers in the system were compared.

Experimental waveforms for the proposed dc link voltage control and the constant dc link voltage control

(e_{dc}^*=150 V and 200 V) are shown in Fig. 8. From the figure, it is clear that the dc link voltage varies according to the rotor speed reference n^* in the case of the proposed dc link voltage control. Here, the dc link voltage cannot become lower than the sum of FC voltage and EDLC voltage (about 60 V). Therefore, when the dc link voltage reference is greater than the sum of FC voltage and EDLC voltage, the dc link voltage is boosted by voltage booster 1.

The input powers for the proposed dc link voltage control and the constant dc link voltage control (e_{dc}^*=150 V and 200 V) are shown in Fig. 9. In all experiments, the cart was driven under the same load condition. Thus, the efficiency of the overall system can be evaluated by comparing the input powers for the two controls. From Fig. 9, one can find that the input power for the proposed dc link voltage control is the lowest. It means that the efficiency of the overall system is improved by the proposed dc link voltage control. Furthermore, one can also find that the input power increases as the dc link voltage increases. This result is consistent with the equations of the loss derived in Section III.

B. Verification of Effectiveness of Proposed FC Current Control

The proposed FC current control and the constant FC current control were performed. In the constant FC current control, the FC current references i_{L1}^* were set to 0.5 A, 1.0 A and 1.5 A. For both of the two FC current control methods, proposed dc link voltage control was applied. For these FC current control methods, the experimental waveforms, the efficiencies of voltage boosters and the input powers in the system were compared.

Experimental waveforms for the proposed FC current control and the constant FC current control are shown in Fig. 10. From the figure, it is clear that the FC current

1325

The 2018 International Power Electronics Conference

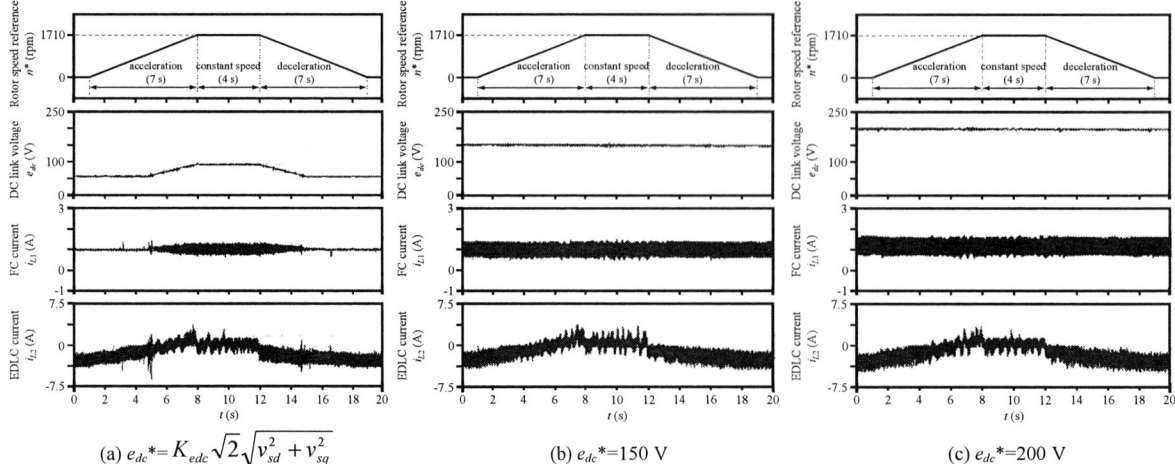

(a) $e_{dc}{}^* = K_{edc}\sqrt{2}\sqrt{v_{sd}^2 + v_{sq}^2}$

(b) $e_{dc}{}^* = 150$ V

(c) $e_{dc}{}^* = 200$ V

Fig. 8. Waveforms for proposed dc link voltage control and constant dc link voltage control ($e_{dc}{}^* = 150$ V and 200 V).

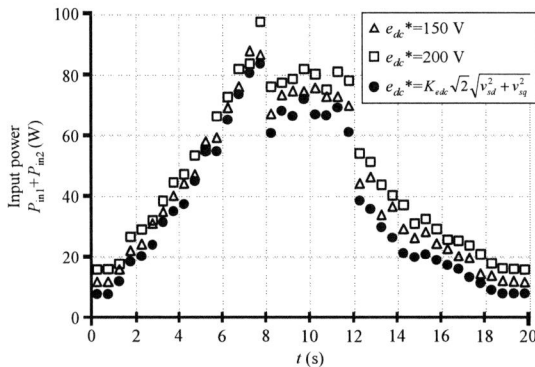

Fig. 9. Input powers for proposed dc link voltage control and constant dc link voltage control ($e_{dc}{}^* = 150$ V and 200 V).

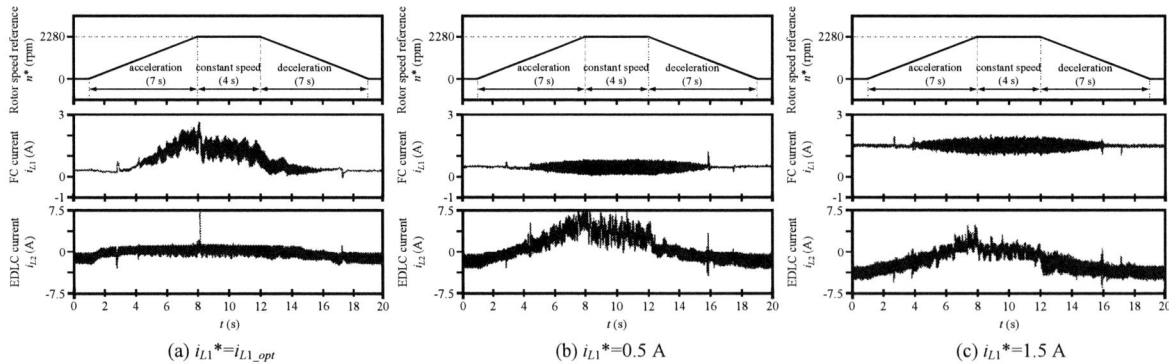

(a) $i_{L1}{}^* = i_{L1\_opt}$

(b) $i_{L1}{}^* = 0.5$ A

(c) $i_{L1}{}^* = 1.5$ A

Fig. 10. Waveforms for proposed FC current control and constant FC current control ($i_{L1}{}^* = 0.5$ A and 1.5 A).

varies according to the rotor speed reference n^* in the case of the proposed FC current control. In the period where the input power of the motor increases (acceleration), the FC current also increases. And, in the period where the input power of the motor hardly changes (constant speed), the FC current is also substantially constant. Moreover, in the period where the input power of the motor decreases (deceleration), the FC current also decreases.

The efficiencies of the voltage boosters with the proposed FC current control and with the constant FC

current control are shown in Fig. 11. The efficiencies were calculated from the measured voltage and current values as the following.

$$\eta_{\text{chopper}} = \frac{-P_{\text{in}1} - P_{\text{in}2} + P_o + |P_{\text{in}1}| + |P_{\text{in}2}| + |P_o|}{P_{\text{in}1} + P_{\text{in}2} - P_o + |P_{\text{in}1}| + |P_{\text{in}2}| + |P_o|} \times 100 \qquad (15)$$

where $P_{\text{in}1} = (e_{FC} + e_{DL})i_{L1}$ is input power of voltage booster 1, $P_{\text{in}2} = e_{DL}i_{L2}$ is input power of the voltage booster 2, $P_o = e_{dc}i_{dc}$ is output power of the voltage boosters.

By using equation (15), it is possible to calculate an

1326

The 2018 International Power Electronics Conference

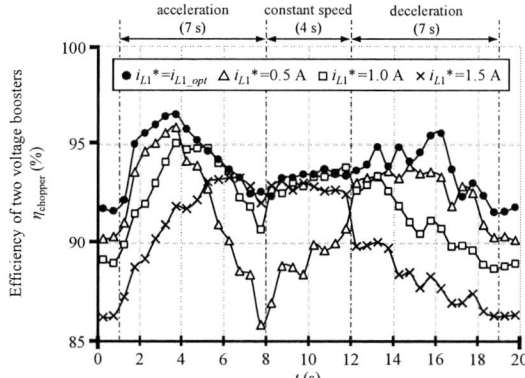

Fig. 11. Efficiencies of voltage boosters with proposed FC current control and with constant FC current control ($i_{L1}*$=0.5 A and 1.5 A).

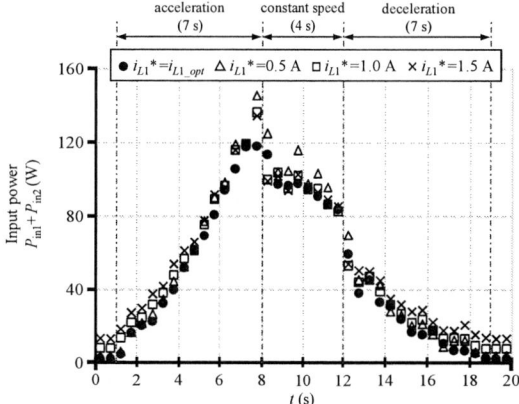

Fig. 12. Input powers for proposed FC current control and constant FC current control ($i_{L1}*$=0.5 A and 1.5 A).

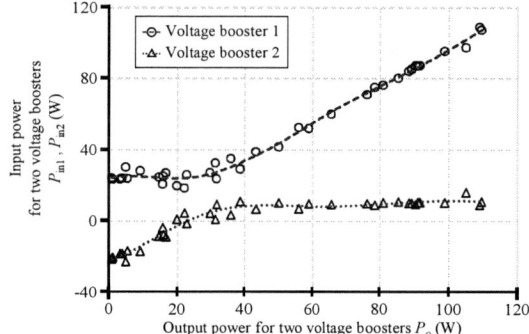

Fig. 13. Input powers vs. output power for two voltage boosters with proposed FC current control.

appropriate efficiency of voltage boosters according to the operating condition of the circuit.

From Fig. 11, one can find that the efficiency of the voltage boosters varies depending on the FC current value. When the FC current reference is 0.5 A, the efficiency of voltage boosters decreases in the period where the input power of the motor is large (acceleration). When the FC current reference is 1.5 A, the efficiency of voltage boosters decreased in the period where the input power of the motor is small (deceleration). Therefore, in order to improve the efficiency of voltage boosters, it is important that when the input power of the motor increases, the FC current also increases, and when the input power of the motor decreases, the FC current also decreases. This is consistent with a behavior of the FC current with the proposed FC current control shown in Fig. 10 (a). For the proposed FC current control, high efficiency is maintained in all period. It means that the efficiency of the voltage boosters is improved by the proposed FC current control.

The input powers for the proposed FC current control and the constant FC current control ($i_{L1}*$=0.5 A and 1.5 A) are shown in Fig. 12. From the figure, it is clear that the input power for proposed FC current control decreases. Therefore, the efficiency of the overall system is also improved by the proposed FC current control.

For the proposed FC current control, each input power vs. sum of the output power for the voltage booster 1 and 2 are shown in Fig. 13. In the range where the output power of two voltage boosters is lower than 20 W, the FC current is limited to the lower limit value (0.25 A) by the controller to prevent deterioration of the FC. Therefore, the EDLC is charged in this period. From Fig. 13, the input power of the voltage booster 2 is kept low and only the input power of the voltage booster 1 is increased, as the load power increases. This result means that the efficiency of the overall system is improved by keeping the EDLC current low.

C. Validation of Equivalent Loss Resistance

The experiment with estimating the equivalent loss resistances and the experiment without estimating loss resistances were conducted.

Experimental waveforms with and without estimation of the equivalent loss resistances are shown in Fig. 14. For the control without estimation, r_{eq1}=0.98 Ω and r_{eq2}=0.18 Ω were used as resistances. These values are measured internal resistance values of the inductor L_1 and L_2. It is clear that the FC current for the control without estimation of the equivalent loss resistances is lower than that for the control with estimation.

The efficiencies of the voltage boosters with and without estimation of the equivalent loss resistances are shown in Fig. 15. From the figure, it is clear that the efficiency of the voltage boosters for the control without estimation are lower than that with estimation and the estimation of the equivalent loss resistances is important to improve the efficiency of the voltage boosters.

VI. CONCLUSION

In this paper, a PM motor drive system with FC connected in series with EDLC was proposed. And also, a strategy to improve efficiency for this system was investigated. Moreover, equations for losses of the overall system were derived and the control method to reduce losses of the overall system was investigated.

In order to verify the effectiveness of the proposed control methods, experiments were conducted by the electric cart with the motor drive system. From the

1327

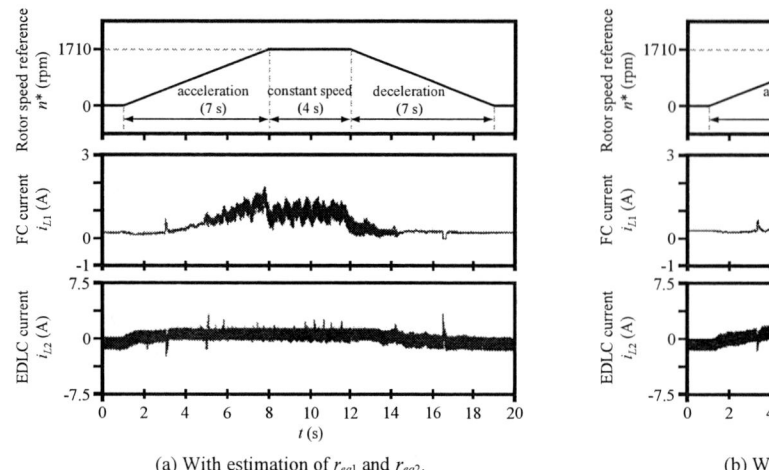

(a) With estimation of r_{eq1} and r_{eq2}.   (b) Without estimation of r_{eq1} and r_{eq2}.

Fig. 14. Waveforms for experiment with and without estimation of r_{eq1} and r_{eq2}

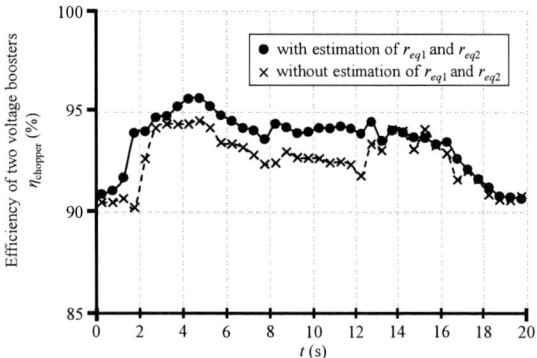

Fig. 15. Efficiencies of voltage boosters with and without estimation of r_{eq1} and r_{eq2}.

results, we conclude the following.

- By the proposed dc link voltage control, the efficiency of the overall system was improved while the motor was rotated at the rotor speed reference.

- By the proposed FC current control, the efficiency of voltage boosters was improved and the efficiency of the overall system was also improved.

- It was confirmed that the efficiency of the overall system was improved by keeping the EDLC current low.

- The efficiency of voltage boosters was improved by estimating the equivalent loss resistances.

- It was shown that proposed dc link voltage control and FC current control are effective for improving the efficiency of the overall system.

REFERENCES

[1] Keiichiro Kondo and Hisao Kubota, "Innovative Application Technologies of AC Motor Drive Systems," IEEJ Journal of Industry Applications, Vol. 1, No. 3, pp. 132-140 (2012)

[2] Daiki Matsuhashi, Keisuke Matsuo, Takashi Okitsu, Tadashi Ashikaga and Takayuki Mizuno, "Comparison Study of Various Motors for EVs and the Potentiality of a Ferrite Magnet Motor," IEEJ Journal of Industry Applications, Vol. 4, No. 3, pp. 174-179 (2014)

[3] Tatsuo Teratani and Shigeru Okuma, "Automotive Technology Evolved by Electrical and Electronic Systems," *IEEJ Trans. IA*, Vol. 125, No. 10, pp. 887-894 (2005) (in Japanese)

[4] Erik Schaltz, Alireza Khaligh, and Peter Omand Rasmussen, "Influence of Battery/Ultracapacitor Energy-Storage Sizing on Battery Lifetime in a Fuel Cell Hybrid Electric Vehicle," *IEEE Trans. on Vehicular Technology*, Vol. 58, No. 8, pp. 3882-3891 (2009)

[5] Li Sun, Kaiwu Feng, Chris Chapman and Nong Zhang, "An Adaptive Power-Split Strategy for Battery-Supercapacitor Powertrain-Design, Simulation, and Experiment," *IEEE Trans. on Power Electronics*, Vol. 32, No. 12, pp. 9364-9375 (2017)

[6] Kichiro Yamamoto, Thomas A. Lipo, Katsuji Shinohara, Yoshihiko Sueyoshi, "Power Loss Reduction and Optimum Modulation Index of PWM Inverter with Voltage Booster for Permanent Magnet Synchronous Motor Drive," IPEC-Tokyo 2000, pp.147-152 (2000)

[7] Kichiro Yamamoto, Katsuji Shinohara, Hitoshi Makishima, " Comparison between Flux Weakening and PWM Inverter with Voltage Booster for Permanent Magnet Synchronous Motor Drive," PCC-Osaka, pp.161-166 (2002)

[8] Kichiro Yamamoto, Katsuji Shinohara, Takahiro Nagahama, "Characteristics of Permanent-Magnet Synchronous Motor Driven by PWM Inverter with Voltage Booster," *IEEE Trans. on Industry Applications*, Vol. 40, No. 4, pp.1145-1152 (2004)

[9] Kichiro Yamamoto, Katsuji Shinohara, Shinya Furukawa, "Permanent Magnet Synchronous Motor Driven by PWM Inverter with Voltage Booster with Regenerating Capability Augmented by Double-Layer Capacitor," IPEC-Niigata 2005, S-65-3, pp.2056-2062 (2005)

[10] Kichiro Yamamoto, Katsuji Shinohara and Akihiro Imakiire, "Steady State Characteristics of PWM Inverter with Voltage Boosters for Permanent Magnet Synchronous Motor Drives," PCC 2007, DS8-3-2, pp.296-301 (2007)

[11] Kichiro Yamamoto, Akihiro Imakiire, Rongyi Lin and Kenichi Iimori, "Comparison of Configurations of Voltage Boosters in PWM Inverter with Voltage Boosters with Regenerating Circuit Augmented by Electric Double-Layer Capacitor," *Proc. of ICEMS*, DSIG2-6, pp.1-6 (2009)

[12] Weiwen Deng, Yang Zhao, Jian Wu, "Energy Efficiency Improvement via Bus Voltage Control of Inverter for Electric Vehicles," *IEEE Trans. on Vehicular Technology*, vol. 66, no. 2, pp. 1063-1073 (2017)

[13] Hugues Renaudineau, Azeddine Houari, Ahmed Shahin, Jean-Philippe Martin, Serge Pierfederici, Farid Meibody-Tabar, Bernard Gerardin, "Efficiency Optimization Through Current-Sharing for Paralleled DC–DC Boost Converters With Parameter Estimation," *IEEE Trans. on Power Electronics*, vol. 29, no. 2, pp. 759-767 (2014)

The 2018 International Power Electronics Conference

Comparative Study of Speed Ripple Reduction by Various Control Methods in PMSM Drive Systems with Pulsating Load

Yuma Komaru, Yukinori Inoue, Shigeo Morimoto and Masayuki Sanada
Osaka Prefecture University, Sakai Osaka, Japan
*E-mail: sxb01080@edu.osakafu-u.ac.jp

Abstract— This paper compared the speed ripple reduction of three different control methods in a permanent magnet synchronous motor when the load torque varied according to rotor position. Generally, speed ripple occurs when the load pulsates. The methods were voltage-current phase difference control, current vector control, and direct torque control. Speed ripple reduction was performed using the angular velocity of the inverter compensation, signal injections to the q-axis current and reference torque, and gain design. This paper describes and compares the effectiveness of each speed ripple reduction method.

Keywords— permanent magnet synchronous motor (PMSM), voltage-current phase difference control, current vector control, direct torque control, speed ripple.

I. INTRODUCTION

Permanent magnet synchronous motors (PMSMs) have excellent characteristics, such as high efficiency and no maintenance, so they are used in various devices, ranging from home appliances to traction motors of electric vehicles. In these applications, the speed ripple, which causes vibration and noise, needs to be minimized. Generally, when the load to a motor drive system pulsates, the load torque fluctuates according to the rotor position [1]. Then, the torque difference between the motor and load increases, and a speed ripple occurs. Therefore, appropriate torque control is needed to reduce speed ripple and vibration of the PMSM.

Typical control methods for PMSM drives include current vector control using a rotor synchronous frame (d-q frame), direct torque control (DTC) using a stator flux linkage synchronous frame (M-T frame), and voltage-current phase difference control, which indirectly controls the current phase using the phase difference between voltage and current.

Since the current vector control uses the d-q frame, its application is facilitated by the well-known voltage equation for PMSMs. In addition, accurate position information is provided by the position sensor, so precise control is possible. Since the control is based on maximum torque per ampere (MTPA), the reference d-axis current $i_d{}^*$ is determined by the reference q-axis current $i_q{}^*$. Therefore, the torque equation depends only on q-axis current i_q. In DTC, a speed sensor is used to obtain the actual rotor speed, which is necessary for rotor speed control. Also, the torque equation is simply

expressed because DTC can always use the M-T frame by estimating the stator flux linkage [2]. The voltage-current phase difference control method, which is based on *V/f* control, is a sensorless rotor position control method [3]. Under MTPA, torque and torque angle have a monotonically increasing relationship, so the torque can be obtained using the torque angle [4].

In the sensorless drive system based on voltage-current phase difference control, the velocity of the inverter is related to variations of the torque and torque angle. Hence, speed ripple reduction by correction of the inverter angular velocity was reported [5]. Speed ripple reduction by correction of the reference armature voltage using the estimated load torque was also investigated [6]. However, since the gain design method for the controller was unclear, it is necessary to change the gain according to the operating conditions, such as speed and average load torque. In contrast, in current vector control and DTC, since the gain design method is established, it is possible to obtain the control characteristics independently of the operating conditions. Furthermore, it is expected that speed ripple reduction can be achieved by appropriately controlling the q-axis current and reference torque [7].

This paper used simulation and experiments to compare the speed ripple reduction when the load torque pulsates according to the rotor position in a PMSM drive system. Three different control methods were evaluated: voltage-current phase difference control, current vector control, and DTC. In the voltage-current phase difference control, speed ripple was controlled by changing the torque according to the load torque using inverter angular velocity compensation. In the current vector control, q-axis current compensation was applied due to the correlation between the q-axis current and torque. In the DTC, reference torque compensation was applied.

II. PMSM DRIVE SYSTEM

In this paper, a common PMSM and load environment was used for all control methods, and the operating characteristics were evaluated for the effectiveness of speed ripple reduction. It was assumed that the load torque T_L pulsates according to the rotor position θ_m as follows:

$$T_L(\theta_m) = \left| \frac{\pi}{2} \bar{T}_L \sin\left(\frac{\theta_m}{2}\right) \right|, \qquad (1)$$

where $\bar{T}_L$ is the average load torque.

A. Voltage-Current Phase Difference Control

Fig. 1 is a block diagram of the voltage-current phase difference control system. In this method, motor control focuses on the correlation between the voltage-current phase difference ϕ and the current phase β, and the voltage-current phase difference ϕ can be controlled by the armature voltage V_a. The reference voltage V_a^* is determined from the error between the reference and estimated values of the phase difference $\Delta\phi$ by a proportional-integral (PI) controller. The product of the multiplication of $\Delta\phi$ and the stabilization gain K_f is subtracted from the reference speed [3].

Based on considerations in [5], as shown in Fig. 1, (2) is given as the inverter output angular velocity compensation $\Delta\omega_1$ to change the torque T_e according to θ_m.

$$\Delta\omega_1 = \frac{\pi}{4} K_\omega \sin\left(\hat{\theta}_m + \theta_k\right), \qquad (2)$$

where K_ω is the speed ripple reduction gain, $\hat{\theta}_m$ is the estimated rotor position, and θ_k is a constant to accommodate the difference between the torque and load torque.

B. Current Vector Control

Fig. 2 is a block diagram of the current vector control method. The reference q-axis current i_q^* is determined from the error between the reference speed ω_m^* and rotor speed ω_m by a PI controller. The reference d-axis current i_d^* is calculated based on MTPA as follows [8]:

$$i_d = \frac{\Psi_a}{2(L_q - L_d)} - \sqrt{\frac{\Psi_a^2}{4(L_q - L_d)^2} + i_q^2}, \qquad (3)$$

where Ψ_a is the magnet flux; L_d and L_q are the d- and q-axis inductances, respectively; and i_d and i_q are the d- and q-axis currents, respectively.

The current control is performed by the PI controllers G_{cd} and G_{cq}, and decoupling control is performed to control the d- and q-axis currents stably in the transient state.

As shown in Fig. 2, (4) is given as the q-axis current compensation Δi_q to change the torque T_e according to θ_m.

$$\Delta i_q = \frac{\pi}{4} K_{iq} \sin\left(\theta_m - \frac{\pi}{2}\right), \qquad (4)$$

where K_{iq} is the speed ripple reduction gain.

In addition, speed ripple reduction is also achieved by adjusting the gain of the speed controller.

C. Direct Torque Control

Fig. 3 is a block diagram of a PMSM drive system based on DTC. A reference calculator restricts the reference torque T_{ref} and calculates the reference stator flux linkage Ψ_s^*, which satisfies the MTPA condition. The limiting torque is determined by the rated current.

As shown in Fig. 3, (5) is given as the reference torque

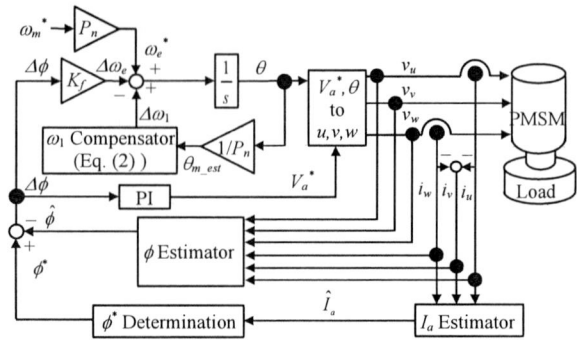

Fig. 1. Voltage-current phase difference control method.

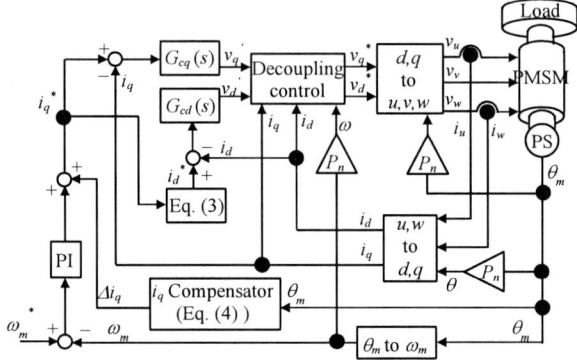

Fig. 2. Current vector control method.

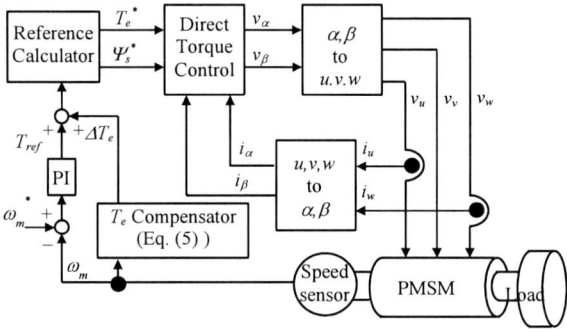

Fig. 3. PMSM drive system with DTC in M-T frame.

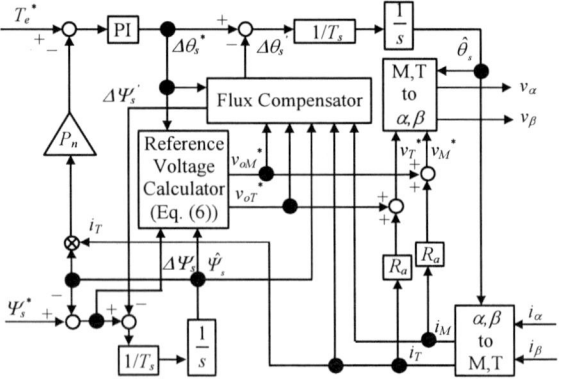

Fig. 4. Direct torque controller with M-T frame.

1330

compensation ΔT_e to change the torque T_e according to θ_m.

$$\Delta T_e = \frac{\pi}{4} K_T \sin(\theta_m + \theta_T), \qquad (5)$$

where K_T is the speed ripple reduction gain and θ_T is a constant to accommodate the difference between the torque and load torque.

Fig. 4 is a block diagram of the DTC in the M-T frame. Using the position variation of the stator flux linkage $\Delta\theta_s^*$, which is determined by a PI controller from the error between the reference torque and estimated torque, the reference induced voltage can be calculated as follows:

$$\begin{bmatrix} v_{oM}^* \\ v_{oT}^* \end{bmatrix} = \frac{1}{T_s} \begin{bmatrix} (\hat{\Psi}_s + \Delta\Psi_s)\cos\Delta\theta_s^* - \hat{\Psi}_s \\ (\hat{\Psi}_s + \Delta\Psi_s)\sin\Delta\theta_s^* \end{bmatrix}, \qquad (6)$$

where T_s is the control period, $\Delta\Psi_s$ is the variation of the stator flux linkage, and $\hat{\Psi}_s$ is the estimated stator flux linkage.

Also, in DTC, the speed ripple is reduced by changing the speed controller K_{sp}, K_{si} and torque controller K_{trq_p}, K_{trq_i}.

III. SIMULATION RESULTS

The characteristics of speed ripple reduction by the three methods were compared by simulation for a control period of 100 μs. The motor parameters are given in Table I, and the controller parameters are in Table II. For cases without and with compensation using (2), (4), and (5), normal gain was used. The damping factor ξ of the torque control system in the DTC was 1.0, and the natural angular frequency ω_n was 800 rad/s. When high gain was applied, the damping factor ξ of the torque control system in the DTC was 5.7, and the natural angular frequency ω_n was 390 rad/s.

TABLE I MOTOR PARAMETERS

d-axis inductance L_d [mH]	8.7
q-axis inductance L_q [mH]	28.3
Number of pole pairs P_n	2
Permanent magnet flux linkage Ψ_a [Wb]	0.108
Inertia moment J_m [kg·m²]	1.65×10^{-3}
Coefficient of viscous friction D_r [N·m·s/rad]	3×10^{-4}
Armature resistance R_a [Ω]	0.64
Rated torque [N·m]	2.5

TABLE II CONTROLLER PARAMETERS

		Normal Gain	High Gain
Voltage-Current Phase Difference Control	K_f	10	-
	K_ω	35	-
	θ_k [°]	0	-
Current vector Control	K_{iq}	1.4	-
	K_{sp}	0.0789	70
	K_{si}	0.66	40
	ω_{cd} [rad/s]	2,000	2,000
	ω_{cq} [rad/s]	2,000	2,000
Direct Torque Control	K_T	0.3	-
	θ_T	-90	-
	K_{sp}	0.0789	5
	K_{si}	0.66	12
	K_{trq_p}	0.358	0.1
	K_{trq_i}	14.3	3.4

Fig. 5 shows the speed and torque characteristics obtained by simulation. In this case, the reference speed ω_m^* was 1,800 min^{-1}, and the average load torque $\overline{T}_L$ was 0.5 N·m. Without compensation, the difference between the torque and load torque increased, and in all control methods, speed ripples of 20 min^{-1} or more occurred due to the difference between the load torque and torque, as shown in Fig. 5(a). In contrast, with compensation, the torque was controlled according to load torque pulsation, and the speed ripples decreased to 3.8 min^{-1} or less, as shown in Fig. 5(b).

Fig. 6 compares the speed ripple reductions of the

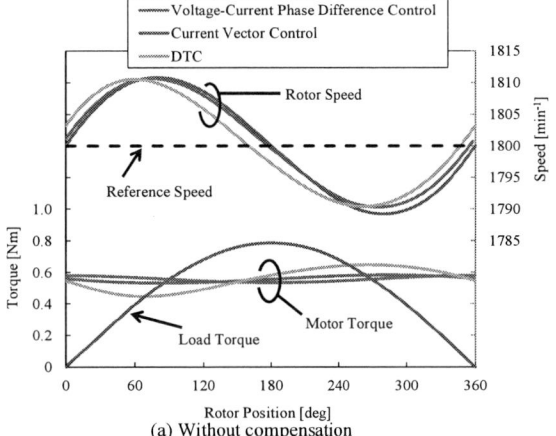

(a) Without compensation

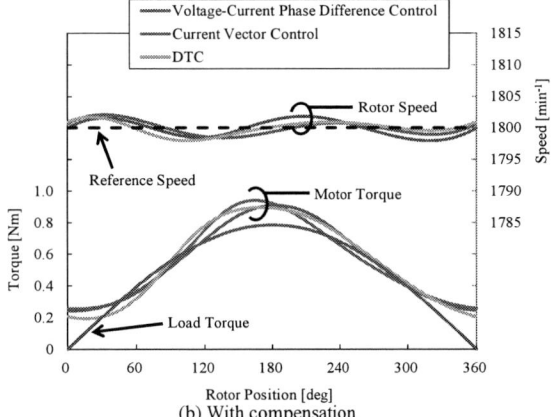

(b) With compensation
Fig. 5. Speed and torque characteristics
(simulation, $\omega_m^* = 1,800$ min^{-1}, $\overline{T}_L = 0.5$ N·m).

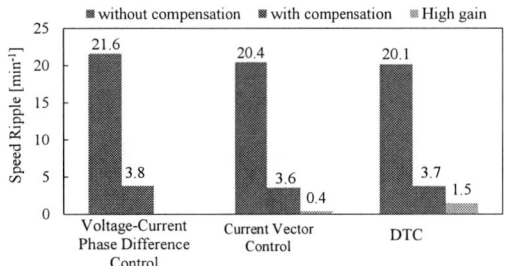

Fig. 6. Comparison of speed ripple reduction
(simulation, $\omega_m^* = 1,800$ min^{-1}, $\overline{T}_L = 0.5$ N·m).

The 2018 International Power Electronics Conference

three control methods. All methods could reduce the speed ripple to 3.8 min⁻¹ or less with compensation. The high gain controller in the PMSM drive system with current vector control produced the largest speed ripple reduction.

Fig. 7 shows the relationship among the speed ripple, variation in the reference speed, and average torque. Fig. 7(a) shows the speed ripple for each control method at an average load torque of 0.5 N·m. In all control methods, the speed ripple could be reduced to 11 min⁻¹ or less in the reference speed range of 900 min⁻¹ to 1,800 min⁻¹ with compensation. In the same speed range, at an average load torque of 0.75 N·m, the speed ripple could be reduced to 12.9 min⁻¹ or less, as shown in Fig. 7(b). As

shown in Fig. 7(c), even when the average load torque was 1.0 N·m, the speed ripple could be reduced to 13.5 min⁻¹ or less in all control methods. It was confirmed that the speed ripple was reduced even when the reference speed and average load torque changed. The methods using high gain in current vector control and DTC could reduce the speed ripple the most. However, these control methods use speed and position sensors. The voltage-current phase difference control method provided adequate control without rotor position or speed sensors.

IV. EXPERIMENTAL RESULTS

Experiments were performed under the same conditions as the simulations using a servomotor as the load. The gate signals of the inverter were generated by pulse width modulation at a carrier frequency of 10 kHz. The DC-link voltage was 150 V, and the control period was 100 μs. In the experiments, the control became unstable and the speed ripple could not be reduced using a high gain. For this reason, experiments were carried out only with the compensation method using (2), (4), and (5). Also, we used the normal gain shown in Table II.

Fig. 8 shows the torque characteristics in voltage-current phase difference control. The reference speed was 1,800 min⁻¹, and the average load torque was 0.5 N·m. As shown in Fig. 8(a), without compensation, since the motor torque is almost constant, the difference between

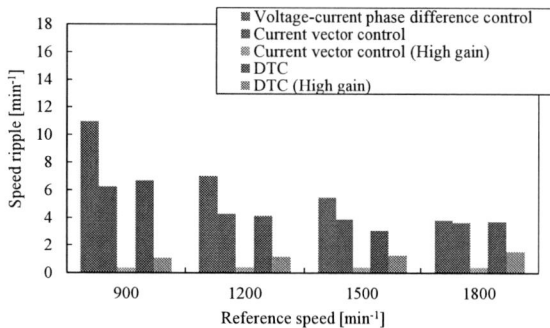

(a) Average load torque of 0.5 N·m

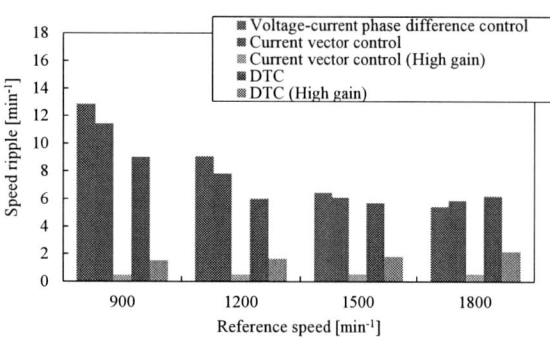

(b) Average load torque of 0.75 N·m

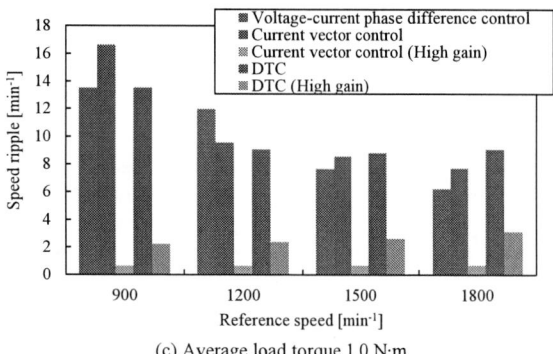

(c) Average load torque 1.0 N·m

Fig. 7. Relationship between reference speed and speed ripple at various load torques (simulation).

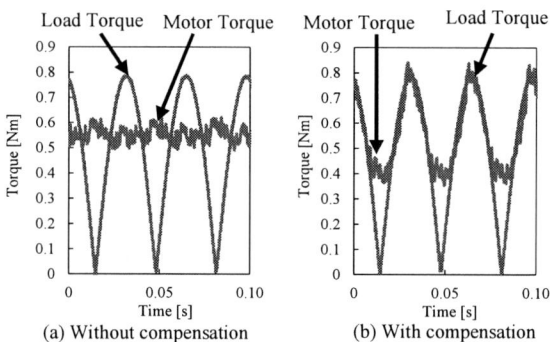

(a) Without compensation   (b) With compensation

Fig. 8. Torque characteristics in voltage-current phase difference control (experiment, $\omega_m^* = 1{,}800$ min⁻¹, $\overline{T}_L = 0.5$ N·m)

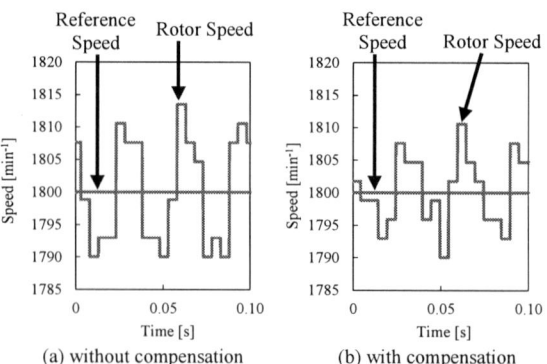

(a) without compensation   (b) with compensation

Fig. 9. Speed characteristics in voltage-current phase difference control (experiment, $\omega_m^* = 1{,}800$ min⁻¹, $\overline{T}_L = 0.5$ N·m).

The 2018 International Power Electronics Conference

the torque and load torque was large. In contrast, with the inverter output angular velocity compensation, the torque was controlled according to the load torque fluctuation, as in the simulation results. Fig. 9 shows the speed characteristics in voltage-current phase difference control by experiment. As shown in Fig. 9(a), without compensation, a speed ripple of 26.4 min^{-1} occurred due to the difference between the load torque and torque. However, with inverter output angular velocity compensation, the speed ripple decreased to 20.5 min^{-1}, as shown in Fig. 9(b). Here, $K_\omega = 47$ and $\theta_k = 25°$.

Fig. 10 shows the q-axis current in current vector control. As shown in Fig. 10(a), without compensation, the q-axis current was almost constant. As shown in Fig. 10(b), with i_q compensation, the q-axis current was sinusoidal. The torque characteristics in current vector control are shown in Fig. 11. As shown in Fig. 11(a), without compensation, the torque was constant, like the q-axis current. As shown in Fig. 11(b), with i_q compensation, the torque was sinusoidal like the q-axis current. These results show that the torque can be controlled using i_q compensation. The speed characteristics in current vector control are shown in Fig. 12. As shown in Fig. 12(a), without compensation, a speed ripple of 26.4 min^{-1} occurred. With i_q compensation, the speed ripple decreased to 20.5 min^{-1}, as shown in Fig. 12(b). Here, $K_{iq} = 1.2$ and $\theta_{iq} = 330°$.

Fig. 13 shows the torque characteristics in DTC. As shown in Fig. 13(a), without compensation, the torque and load torque were in opposite phase. As shown in Fig. 13(b), with T_e compensation, the torque was controlled to be in phase with the load torque. The speed characteristics in DTC are shown in Fig. 14. As shown in Fig. 14(a), without compensation, a speed ripple of 29.3 min^{-1} occurred. With T_e compensation, the speed ripple decreased to 20.5 min^{-1}, as shown in Fig. 14(b). Here, $K_T = 0.35$ and $\theta_T = 330°$.

The speed ripple reductions attained by the three methods are compared in Fig. 15. It can be seen that speed ripples could be reduced using the proposed compensation in all three methods. In the case of no compensation, the speed ripple of the DTC was larger. This is because the motor torque in the DTC was in a phase opposite to the load torque, from Fig. 13(a).

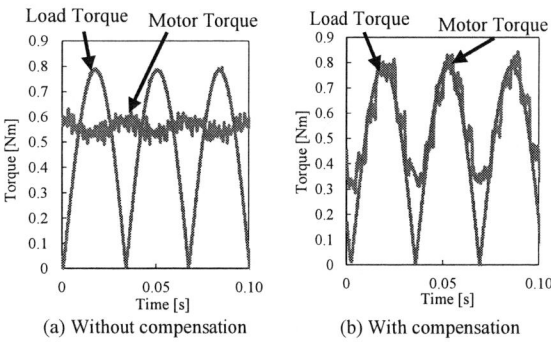

(a) Without compensation   (b) With compensation

Fig. 11. Torque characteristics in current vector control (experiment, $\omega_m^* = 1{,}800$ min^{-1}, $\overline{T}_L = 0.5$ N·m).

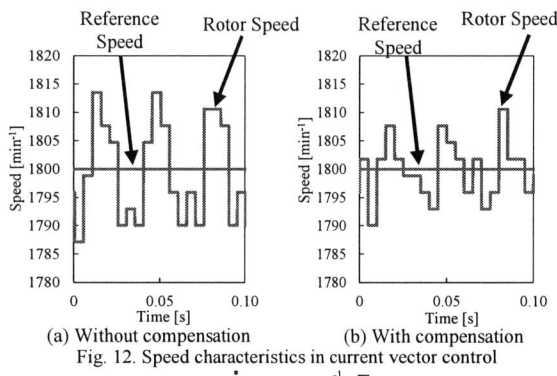

(a) Without compensation   (b) With compensation

Fig. 12. Speed characteristics in current vector control (experiment, $\omega_m^* = 1{,}800$ min^{-1}, $\overline{T}_L = 0.5$ N·m).

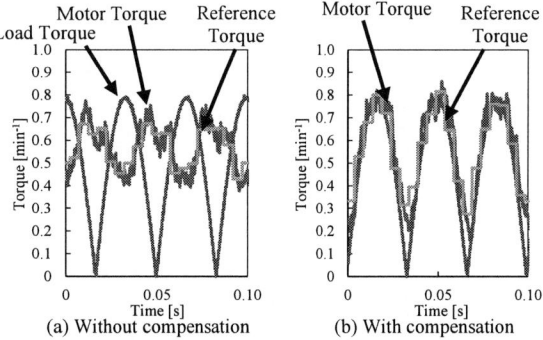

(a) Without compensation   (b) With compensation

Fig. 13. Torque characteristics in DTC (experiment, $\omega_m^* = 1{,}800$ min^{-1}, $\overline{T}_L = 0.5$ N·m).

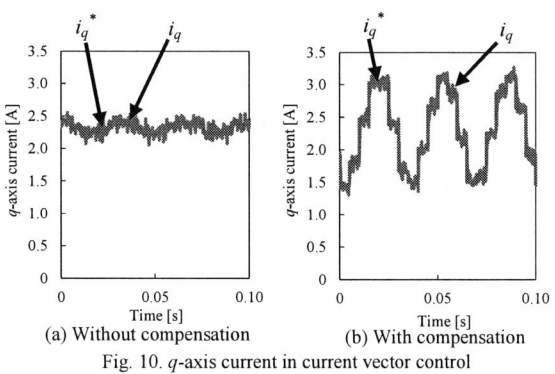

(a) Without compensation   (b) With compensation

Fig. 10. q-axis current in current vector control (experiment, $\omega_m^* = 1{,}800$ min^{-1}, $\overline{T}_L = 0.5$ N·m).

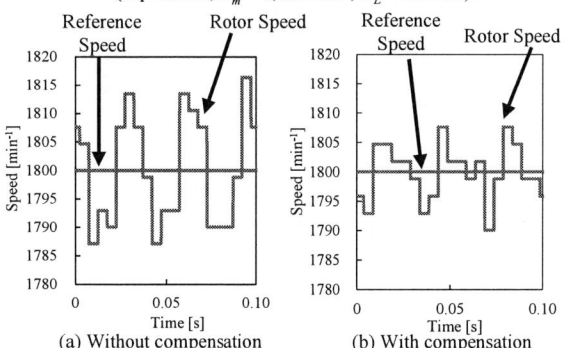

(a) Without compensation   (b) With compensation

Fig. 14. Speed characteristics in DTC (experiment, $\omega_m^* = 1{,}800$ min^{-1}, $\overline{T}_L = 0.5$ N·m).

1333

The 2018 International Power Electronics Conference

Moreover, during compensation, the motor torque had almost the same waveform in all control methods. Therefore, as can be seen from Fig. 15, almost the same rate ripple reduction is seen.

Fig. 16 shows the relationship among the speed ripple, variation in the reference speed, and average torque by experiment. Fig. 16(a) shows the speed ripple for each control method at an average load torque of 0.5 N·m. In all control methods, it can be seen that the speed ripple could be reduced to 29.3 min⁻¹ or less at the reference speed of 900 min⁻¹ to 1,800 min⁻¹ with compensation. In simulation, the ripples at low speed increased, but in the experiments, the speed made almost no difference. In the same speed range, at an average load torque of 0.75 N·m, the speed ripple could be reduced to 35.2 min⁻¹ or less, as shown in Fig. 16(b). As shown in Fig. 16(c), even when the average load torque was 1.0 N·m, the speed ripple could be reduced to 49.8 min⁻¹ or less in all control methods. In these cases, the ripple had a speed dependence similar to that of the simulation. Experiments confirmed that the speed ripple could be reduced even when the reference speed and average load torque were varied.

V. CONCLUSIONS

In this paper compared the speed ripple reduction obtained by voltage-current phase difference control, current vector control, and DTC when the load torque pulsated in a PMSM. It was confirmed that speed ripple can be reduced using the speed ripple reduction methods for various rotor speeds and average load torques in both simulations and experiments. In this paper, only the voltage-current phase difference control operated without rotor position or speed sensors. Therefore, considering the cost and the space saving of a sensorless motor, the voltage-current phase difference control has an advantage over the other methods.

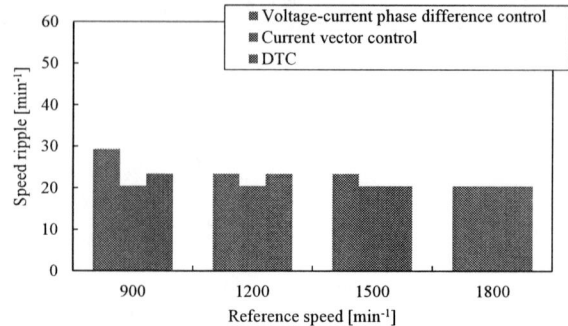

(a) Average load torque of 0.5 N·m

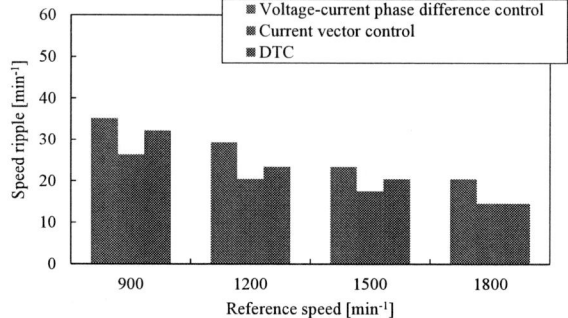

(b) Average load torque of 0.75 N·m

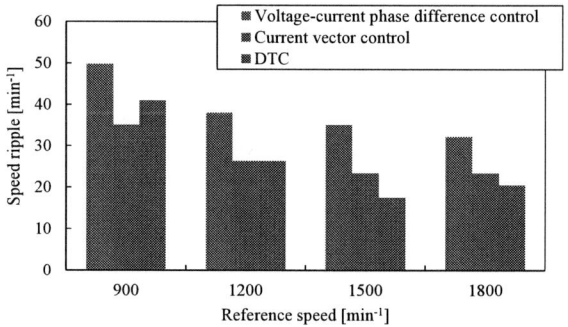

(c) Average load torque of 1.0 N·m

Fig. 16. Relationship between reference speed and speed ripple at various load torques (experiment).

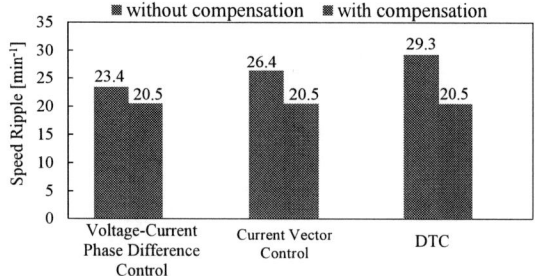

Fig. 15. Comparison of speed ripple reductions
(experiments, $\omega_m^* = 1,800$ min⁻¹, $\overline{T}_L = 0.5$ N·m).

REFERENCES

[1] C. Park, S. Kim, G. Park and J. Seok, "Active Mechanical Vibration Control of Rotary Compressors for Air-conditioning Systems," *Journal of Power Electronics*, Vol. 12, No. 6, pp. 1003-1010, 2012.

[2] Y. Inoue, S. Morimoto, and M. Sanada, "Mechanical Sensorless PMSM Drive System Based on Direct Torque Control in M-T

Frame Synchronized With Stator Flux-Linkage Vector," *Proc. of IEEE 6th International Symposium on Sensorless Control for Electrical Drives (SLED 2015)*, pp. 1-8, 2015.

[3] M. Matsushita, H. Kameyama, Y. Ikeboh, and S. Morimoto, "Sine-Wave Drive for PM Motor Controlling Phase Difference Between Voltage and Current by Detecting Inverter Bus Current," *IEEE Trans. on Industry Applications*, vol. 45, no. 4, pp. 1294-1300, 2009.

[4] Y. Zhang, J. Zhu, W. Xu and Y. Guo, "A Simple Method to Reduce Torque Ripple in Direct Torque-Controlled Permanent-Magnet Variable Amplitude and Angle," *IEEE Transactions on Industrial Electronics*, Vol. 58, No. 7, pp. 2848-2859, 2011.

[5] N. Funamoto, Y. Inoue, S. Morimoto and M. Sanada, "A Speed Ripple Reduction Method with Inverter Angular Velocity Compensation under Pulsatile Load Torque," *Proc. of The 19th International Conference on Electrical Machines and Systems (ICEMS 2016)*, DS3G-1-10 (2016)

[6] G. Sugimori, Y. Inoue, S. Morimoto and M. Sanada, "Speed Ripple Reduction for an Interior Permanent-Magnet Synchronous

1334

Motor Based on Sensorless Voltage-Current Phase Difference Control," *Proc. of IEEE 5th International Symposium on Sensorless Control for Electronical Drives (SLED 2014)*, pp. 56-61, 2014.

[7] F. Genduso, R. Miceli, "Back EMF Sensorless-Control Algorithm for High-Dynamic Performance PMSM," *IEEE Transactions on Industrial Electronics*, Vol. 57, No. 6, pp. 2092-2100, 2010.

[8] S. Morimoto, M. Sanada, Y. Takeda, "Wide-Speed Operation of Interior Permanent Magnet Synchronous Motors with High-Performance Current Regulator," *IEEE Trans. Industry Applications*, vol. 30, no. 4, pp. 920-926, July/Aug. 1994.

Estimation of the parameters of the servo drive system using Particle Swarm Optimization algorithm

Helin Zhu[1], Jae Hyuk Choi[1], Sang Uk Park[1], Jusuk Lee[2], Hyong Gun Lee[3], Hyung Soo Mok[*]

1,* Department of Electrical Engineering, Konkuk University, Seoul, Korea
2 Department of Energy Mechanical Engineering, Gyeonggi College of Science and Technology, Gyeonggi, Korea
3 Research Center, LC-TEK Co. Ltd., Gumi-City, Korea
*E-mail: wngkr3388@naver.com

Abstract— **Most modern industrial or military servo drive application uses Proportion Integral (PI) cascade control structure. The PI gains are highly dependent on the inertia and viscous damping of the drive system, which is impossible to acquire directly. Also, the Modeling and Simulation (M&S) of such system requires exact or approximated value of these parameters. Thus, numbers of approaches are developed to identify these parameters. This paper proposes Particle Swarm Optimization (PSO) based method to estimate the inertia and viscous damping of the servo drive system. The sequence of the method is illustrated, and the detail of the work is presented. In order to find the parameter, experiment has been conducted and position signal and PWM signal data is analyzed. Modeling of the servo drive system is needed to run simulation and compare the results with experiment. The PSO algorithm then automatically finds best approximated parameters with which the simulation gives great agreement with experiment.**

Keywords— *Modeling and Simulation, Parameter estimation, Particle Swarm Optimization*

I. INTRODUCTION

In most modern industrial or military servo drive application, the servo systems are controlled by cascade Proportion Integral (PI) structures [1]. The position control loop is generally the proportion control. However, to meet required high performance, the inner loop controllers which control speed and current are configured to PI structure, which is common method to improve step response and minimize steady state error. The determination of P gains and I gains are highly dependent on electro mechanical constant of the servo motor and parameters of the mechanical system.

Meanwhile, product development process timeline must be shortened due to several reasons. Modeling and Simulation (M&S) is trend to cut down the development cost and time [2].

The challenge to the modelling and control of the servo system is to identify the parameters of the system. Mostly, the servo motors are mass produced with certain specification. And parameters are usually provided with detail experiment. But the mechanical systems are designed to meet specific need and can be assembled by different parts, such as gears, lead crew. These parts can

be made of different material and connected to each other in a complex way. Density of the mechanical parts, connection between each part, gear ratio, they can all contribute to the total inertia and viscous damping of the system. It is extremely hard to directly acquire inertia and viscous damping information if possible. These unknown parameters are basic issues which affect modeling and control greatly.

Several approaches have been worked out to identify important parameters related to modeling and control. Adaptive sliding mode control scheme is proposed [3] by Wei Zhao, Xuemei Ren, Shubo Wang. This approach utilizes TVSM controller to achieve the load output tracking and motors synchronization for the multi-motor driving servo systems with unknown parameters and actuator saturation. A two-step identification method called ARIM for estimating four parameter nonlinear model of a servomechanism is proposed by Ruben Garrido and Antonio Concha [4]. Above methods are using specific controller to estimate parameters. However, by utilizing Particle Swarm Optimization algorithm [5], this paper presented a new parameter estimation approach that doesn't need a controller but experimental data to find best estimated parameters of the system. To implement this method, the system's M&S should be built including electrical and mechanical system. Then, experimental input and output data is needed. By recursively running the simulation whose input is the same input data measured before, PSO algorithm will be continuously looking for the best estimated parameters by comparing simulation and output result every time the simulation is completed until the error between them is less than a given value.

II. PARTICLE SWARM OPTIMIZATION

The Particle Swarm Optimization (PSO) is a computational method which optimizes a problem by trying to improve a candidate solution iteratively until the solutions satisfy given requirement. By having a population of candidate solutions, called particles, the PSO algorithm gather the information of each particle's position and vector to guide swarms to the best-known position. Then, each particle randomly spread to find better solution and this routine continues until the best-known solution is within requirement.

In this paper, the PSO algorithm is applied to find the best estimation of system inertia and viscous damping. The algorithm linked with simulation of the servo drive system automatically finds the best solution, namely, the best approximate of inertia and viscous damping of the system.

Fig. 1 illustrates the PSO based parameter estimation sequence. The method is quite straightforward, and effective in applications. Firstly, the modeling of the servo drive system is necessary and running simulation is one important part of this sequence. This modeling should include every crucial part of servo drive system. The detail of the modeling is presented in later section of this paper. Then, the finite boundary of value from where the best approximated inertia and viscous damping could be found should be determined to minimize calculation time and maximize the accuracy of solutions. Experiment is needed to compare simulation to calculate error which is judgment basis of the best solution.

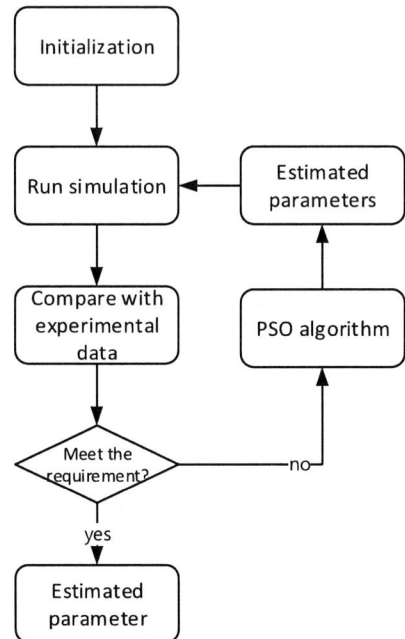

Fig. 1. Parameter estimation sequence based on PSO algorithm.

III. MODELING OF THE SERVO DRIVE SYSTEM

The MATLAB has provided PSO commercial toolbox for convenient use of the algorithm. So, the modeling of the system is done using MATLAB/SIMULINK to better communicate with MATLAB.

The servo drive system tested in this work is position control system to accurately control the angle of the wing. Fig. 2 shows mechanical configuration of the system. Brushless DC motor (BLDC motor) is used to drive the mechanical system. The lead screw connected with motor shaft to translate motor's rotational motion into linear motion. Then the sliding contact translates the linear motion into rotational motion of the wing shaft. The gear

ratio of this mechanical structure increases torque applied the wing shaft and slows down the motion of the wing.

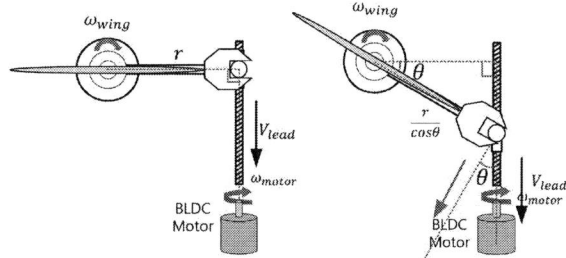

Fig. 2. Motor and mechanical system of the wing angle control system.

Suppose the arm align with wing connected to the sliding contact. The angel between the arm and the line which perpendicular to the lead screw is θ. The length of arm which extends from wing shaft to lead screw is r when $\theta = 0$. It is obvious that as θ changes, the r varies

$$r' = \frac{r}{\cos \theta} \tag{1}$$

The angular velocity of the wing can be expressed as

$$\omega_w = \frac{v_{sc} \cos \theta}{r'} \tag{2}$$

Where, ω_w represents angular velocity of the wing and v_{sc} represents the speed of sliding contact. We define the ratio k so that

$$k = \frac{v_{sc}}{\omega_m} \tag{3}$$

Where, ω_m represents motor shaft angular velocity. Then the angular velocity can be expressed as

$$\omega_w = \frac{k \cos^2 \theta}{r} \omega_m \tag{4}$$

This non-linear mechanical system is modeled in the simulation to mimic its mechanical behavior, and leave the inertia and viscous damping as variable which later becomes the particle of the PSO algorithm. As shown in (4), the non-linearity of total gear ratio between motor and wing makes it difficult to calculate the inertia of the system. However, the PSO algorithm simply finds the best parameters so that the output of the simulation matches the output of the experimental data.

The modeling of the servo drive system takes majority of the time for PSO based parameter estimation approach and it is also the most difficult work to do because if any one part of the modeling gets wrong, the PSO will give deviated or even wrong solution. The model of each part must be carefully tested to makes sure the simulation holds the same characteristic and behavior as its original.

The modeling of BLDC motor is based on motor electrical equation.

$$V_T = 2R_s i_s + 2(L_s - L_M)\frac{d}{dt} i_s + 2e_s \tag{5}$$

Where V_T and i_s represent motor line voltage and phase current, respectively. Although the BLDC motor is commonly driven by three phase full bridge inverter, each moment there are only two phases are being energized and the remaining one phase is disconnected. Two phase resistor R_s is in serial connection and L_s and L_M represent self-inductance and mutual-inductance of each phase coil, respectively. e_s represents back EMF caused by rotating permanent magnet rotor.

In mechanical equation, J and B represent the system inertia and viscous damping. The mechanical equation of BLDC motor and its load is

$$T_e - T_L = J \frac{d\omega_m}{dt} + B\omega_m \qquad (6)$$

Where T_e is the electromagnetic torque. Ii can be expressed by

$$T_e = K_T i_s \qquad (7)$$

Where K_T is torque constant of motor.

In this case, load torque is negligible due to large gear ratio.

$$T_L \approx 0 \qquad (8)$$

In simulation the torque constant connects mechanical and electrical section. The modeling and simulation of the system can be built according to (4), (5), (6) and (7).

Fig. 3 illustrates Modeling & Simulation of the servo drive system. The control circuit of the system is digitized to mimic actual behavior of the controller. The position controller, speed controller, A/D sampling block is triggered by pulse signal which has period equal to the actual control period. In this method the control part can be bypassed, and effective voltage is directly applied to the motor. The BLDC driver chip is analyzed, and its MATLAB/SIMULINK model is built. It basically works as a logical chip and can be configured as continuous block. Inverter and motor block is built with the models provided by Simscape toolbox. The dynamic gear is modeled to mimic the same mechanical behavior as shown in (4).

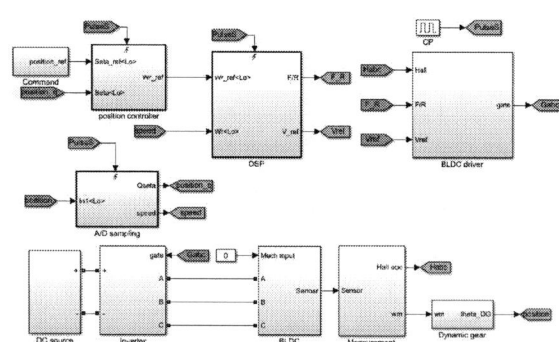

Fig. 3. MATLAB/SIMULINK Modeling and Simulation of the servo drive system.

Experiment is carried out to compare the simulation result with the data. Fig. 4 shows the step response of the wing position (angle). The step response curve contains dynamic characteristic information include control gain, feedback loop and non-linearity of the mechanical system.

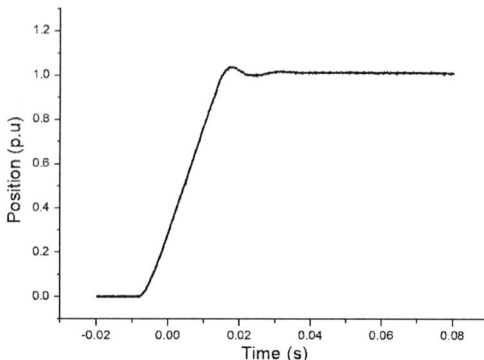

Fig. 4. Step response of wing position (experiment).

Fig. 5 is the output PWM signal, whose duty is proportional to PI control loop's output, namely the reference voltage. By filtering out the high frequency component and scale up to DC link voltage, the PWM signal can be converted to effective DC voltage which is equivalent to PWM chopped pulse voltage.

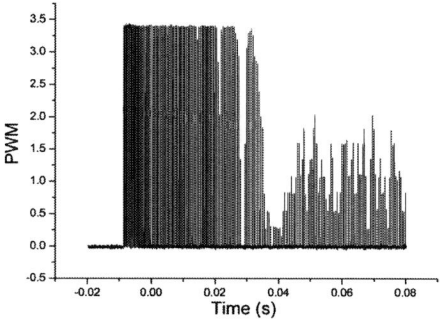

Fig. 5. PWM output signal associated with step response of wing position (experiment).

This equivalence relation is illustrated by Fig. 6 and Fig.7. Fig. 6 shows a typical inverter-motor drive set. In this configuration, V_{dc} is constant and the switching action of the MOSFET switch modulate the average voltage which is applied to the motor, this modulation controls the motor speed and position of the wing. Since the effective voltage is linearly proportional to the PWM duty, we can consider the voltage is as shown in Fig. 7 This is concept of Pulse Amplitude Modulation (PAM) and the BLDC motor is considered operating in six-step mode in this situation.

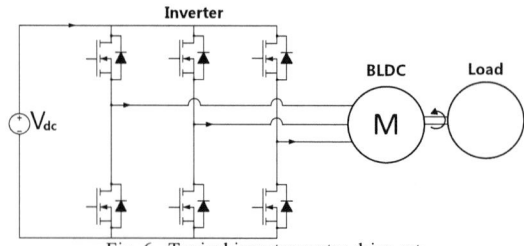

Fig. 6. Typical inverter-motor drive set.

The 2018 International Power Electronics Conference

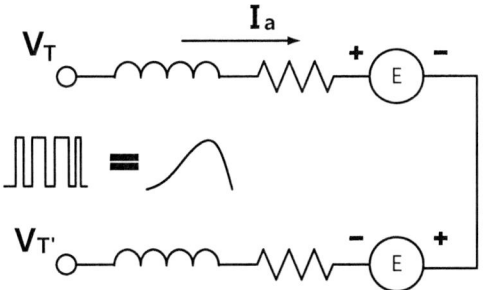

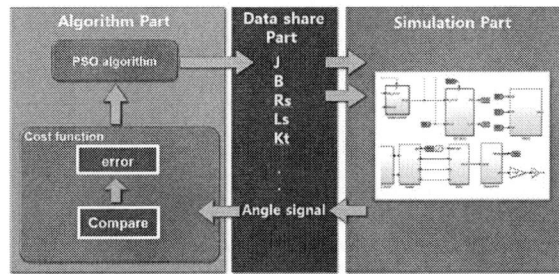

Fig. 9. Parameters estimation process of servo drive system based on PSO algorithm.

Fig. 7. The PWM chopped DC voltage and equivalent average voltage.

Fig. 8 shows the DC link current associated with step response. The motor input current signal is not necessary in estimation algorithm. Yet, the current curve is used to check the consistency with that in the simulation.

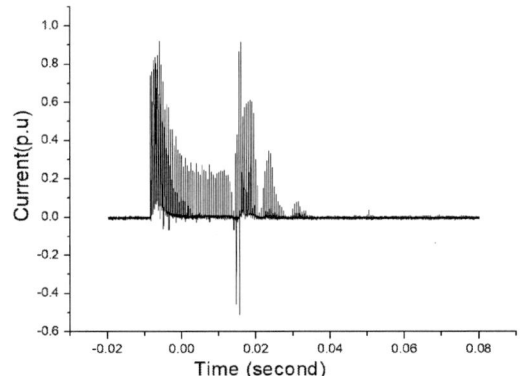

Fig. 8. DC link current associated with step response of wing position (experiment).

IV. PARAMETER ESTIMATION RESULT

After test of each section is completed, the final step is to link run parameter estimation algorithm with simulation. Fig. 9 illustrates the connection of each part. The MATLAB section runs the PSO algorithm and associated objective function.

The objective function is will calculate error between experimental data and simulation result of step response. When error is fed back, PSO algorithm will try to estimate new parameter value and whole process repeats. The parameters are continuously being updated to reach better approximation.

In this work, only the inertia (J) and Viscous damping (B) is estimated by PSO based method since other parameters are already given. But in some cases where the motor parameters are still unknown we can include them as parameters of the PSO algorithm. The MATLB workspace of provide a bridge where PSO algorithm and simulation can exchange result data.

The last estimation attempted by PSO based process gives the best estimated value of inertia and viscous damping. By applying these values to the controller, the step response simulation is carried out. Table I shows numerical comparison between experimental data and simulation. In step response case, experiment shows overshoot of 0.037 (p.u), while simulation shows 0.033 (p.u). The rising time is 17.1ms and 17.3ms, respectively. Fig. 10 illustrates experiment and simulation graph. Comparison of the two curves shows great agreement. Some tiny fraction of error still exists because of unconsidered non-linearities which are difficult to figure out. However, to some extent, the validity of PSO based parameter estimation method is proven.

TABLE I
COMPARISON BETWEEN SIMULATION AND EXPERIMENT

Performance	Simulation	Experiment	Error
Overshoot (p.u)	0.037	0.033	0.004
Rising time (ms)	17.1	17.3	0.1

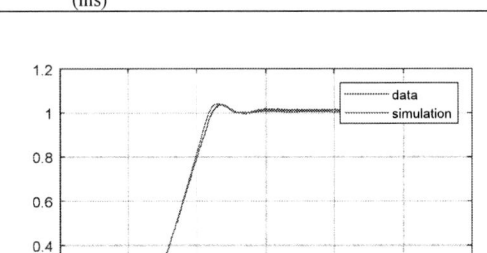

Fig. 10. Comparison of step response between simulation and experiment.

CONCLUSION

In this paper, the problem associate with estimating parameters of servo drive system is issued. Approaches have been developed to find unknown parameters of mechanical system for better control performance and accurate M&S development. This method requires experimental data and accurate modeling and simulation

1339

of the system. By iteratively comparing experiment with simulation result, the PSO algorithm tries to find better parameters that make the error between simulation and experiment minimum. Modeling of electric and mechanical part of the actual dynamic system is necessary step. The process has provided the best estimated value of parameters and simulation has been conducted to compare experiment with simulation result and shows great agreement. Thus, the validity of this method has been proven.

ACKNOWLEDGMENT

This work was supported by "Defense venture support project" of Defense Acquisition Program Administration (DAPA). (No. V160002)

REFERENCES

[1] Ingo Pletschen, Stephan Rohr, Gunther Herrmann and Ralph Kennel "Online Parameter-Estimation of Feedforward Gains in cascaded control structures for Servo Drives" *Power Electronics and Applications (EPE 2011)*, 30 Aug.-1 Sept. 2011.

[2] J.Ramos, "Modeling and Simulation (M&S) issues in operational test and evaluation (OT&E)" *IEEJ Journal of Industry Applications*, vol. 28, no. 15, pp. 123-129, 1989.

[3] Wei Zhao, Xuemei Ren and Shubo Wang, "Parameter Estimation-Based Time-Varying Sliding Mode Control for Multimotor Driving Servo Systems" IEEE/ASME Transactions on Mechatronics, vol. 22, no. 5, pp. 2330-2341, 2017

[4] Ruben Garrido and Antonio Concha, "An Algebraic Recursive Method for Parameter Identification of a Servo Model" IEEE/ASME Transactions on Mechatronics, vol. 18, no. 5, pp. 1572-1580, 2013

[5] Y. Rahmat-Samii, Dennis Gies and Jacob Robinson, "Particle swarm optimization (PSO): A novel paradigm for antenna designs" URSI Radio Science Bulletinm, vol. 76, no. 3, pp. 14-22, 2003

The 2018 International Power Electronics Conference

A Programmable Battery Test System with Energy Recycling Feature Based on Sinusoidal Loading Technique

Chang-Hua Lin[1]*, Guan-Jung Chen[1], Hwa-Dong Liu[1], and Kun-Feng Chen[2]

1 Electrical Department, National Taiwan University of Science & Technology, Taipei, Taiwan

2 The Missile & Rocket Systems Research Division, Chung-Shan Institute of Science and Technology, Longtan, Taiwan

*E-mail: link@mail.ntust.edu.tw

Abstract—A battery testing system using microcontroller chip encompasses low-cost and compact configuration is proposed. The proposed battery test platform consists of a MCU, a signal capturing circuit, a HMI (human machine interface) and a programmable sinusoidal load. To validate the dynamic characteristics of the power battery in practical applications, the wide-range slew-rate and programmable sinusoidal loading feature of implemented sinusoidal load is served as a test load. Furthermore, the extracted energy from the power battery during diagnostic process can be recycled. In addition, the loading current can be programmed by the preset parameters in the user-machine interface. All the analysis of operation modes, simulations, some experimental results, and portable design verify the theoretical predictions and the implemented system.

Keywords— battery test system, energy recycling, sinusoidal load, Class E converter.

I. INTRODUCTION

Recently, owing to the environmental awareness and the energy shortage, the EV [1], including electric scooters, electric vehicles, and hybrid vehicles, etc. have become one of a main stream for development in transportation technology. Nowadays, some researches concentrate on SOC and SOH for power battery is one of important research topic, where the lithium-ion batteries are the most popular voltage source for EVs [2, 3]. Yang et al. [4] has developed a battery testing system to capture the related information of battery by a capturing circuit, and a loading circuit, which is used to release battery energy by constant level current (CC) or other types currents. In fact, the characteristics of the loading circuit of the mentioned above method is totally different from the actual work conditions, which is nonperiodic and irregular shape.

Moreover, it is noteworthy that all of the power consumption during the test process results in thermal issue, accordingly the heat sink and thermal management system are required. Therefore, apart from the above mentioned power waste issue, it also causes some problems such as inflexible, expensive, and complex. Actually, the electrical energy extracted from the power

battery during working condition simulation can be recycled effectively. Currently, two battery test techniques are widely used in the industry. The simplest way, which is essential to add another backup battery [5], the extracted electrical energy from the tested battery is transferred to the designated backup one. Although the next way is more efficient, higher cost and much more complex system are required. The electrical energy extracted from tested battery is converted into ac form and then returns to the power grid by connecting with an inverter [6, 7]. In this paper, the proposed sinusoidal load is utilized to improve the aforementioned drawback. The extracted energy from the test battery can almost be returned to the test battery after ending the test procedure.

II. SCHEME DESCRIPTION OF THE PROPOSED BATTERY TEST SYSTEM

The circuit architecture of the proposed programmable battery test system is depicted in Fig. 1. The proposed system consists of one sinusoidal load, a control core (dsPIC33FJ64GS606), a USB module, an auxiliary power circuit, a signal (voltage and current) capturing circuit, and a dc level offset circuit.

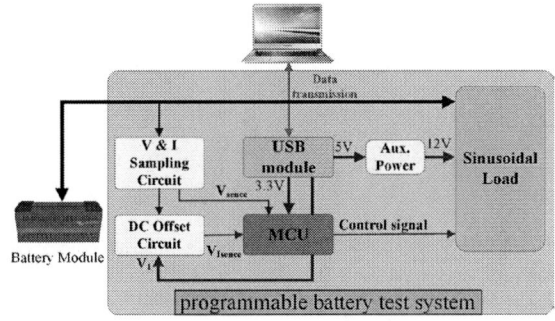

Fig. 1 The circuit architecture of the proposed programmable battery test system.

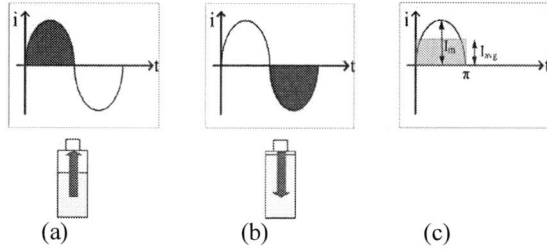

Fig. 2 (a) Extract electrical energy from tested battery. (b) Recycle electrical energy to tested battery. (c) Average value of the current waveform.

In this study, the sinusoidal load is the key part of the proposed battery test system. Actually, the proposed sinusoidal load is used to provide different current conditions for the tested battery. Moreover, Figs. 2(a)-(b) sketch the sinusoidal load is capable of supplying dynamic loading current within the positive half-cycle and transfer the electrical energy within the negative half-cycle. Fig. 2(c) shows the average value of the current in each half-cycle.

III. MODEL ANALYSIS OF THE SINUSOIDAL LOAD

The proposed sinusoidal loading circuit is implemented by the class E topology [8, 9, 10], as depicted in Fig. 3 (a). The original class E topology consists of two inductors (L_1 and L_2), two capacitors (C_1 and C_2), and one MOSFET switch (S).

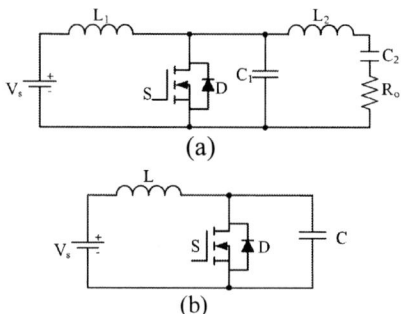

Fig. 3 (a) The original Class E circuit structure. (b) the equivalent circuit of sinusoidal loading circuit.

The former inductor (L_1) of the original class E main circuit is connected with V_S to form a current source. In this study, a tested battery is used as the voltage source V_s. In order to generate the required bidirectional loading behavior, the modified class E circuit structure is employed. First of all, the input inductor is combined with the other components as shown in Fig. 3 (b). And then, the load resistance (R_o) is removed from class E topology to achieve the said circuit characteristics. Moreover, for simplifying the circuit, L_2 and C_2 in the specific frequency range are integrated to a capacitive component C_{eq},

$$C_{eq} = \frac{C_2}{1 - \omega^2 L_2 C_2} \tag{1}$$

Afterward we combine the C_1 and C_{eq} to an equivalent capacitance $C=C_1+C_{eq}$, and then the resonant frequency is

$$f_r = \frac{1}{2\pi\sqrt{LC}} \tag{2}$$

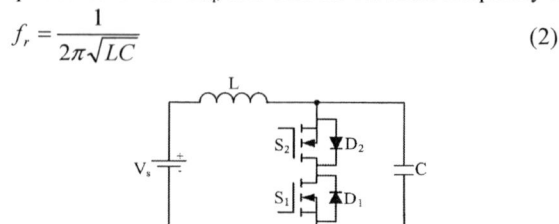

Fig. 4 Circuit topology of the sinusoidal load.

Besides, asymmetrical load current may appear in some operating conditions due to the body diode of power switch. To improve the mentioned above phenomenon, the former power switch in Fig. 3(b) is in series with another power switch S_2 to serve as a sinusoidal load circuit as shown as Fig. 4.

In this study, the switching frequency f_s of the sinusoidal load is set identical with the resonant frequency f_r. Based on the above design, the input inductor current i_L is almost symmetric waveform in each half-cycle. The circuit operation in a switching period is divided into three intervals as follows.

Mode I [t_0-t_1]：

Suppose both the power switches (S_1 and S_2) are off, the initial current $i_{L1}(t_0)=0$ and the voltage of C arrives the maximum value as $V_C(t_0)= -V_{C\_max}$. In this mode, the capacitor C began to resonant with the inductor L, and then the electrical energy is transferred from the capacitor C to the inductor L, as shown in Fig. 5(a). Fig. 5(b) shows the corresponding Laplace transform circuit.

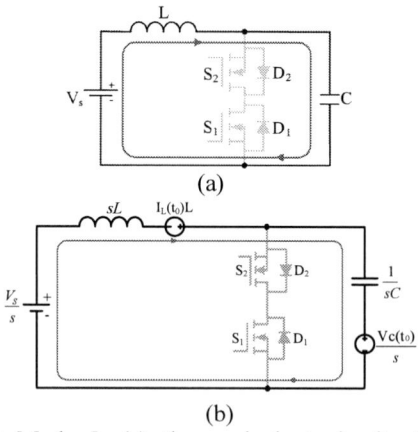

Fig. 5 Mode I. (a) the equivalent circuit. (b) the Laplace transform circuit.

As shown in Fig. 5(b), we can obtain the equation by Kirchhoff's voltage laws (KVL), as follows

$$\frac{V_s}{s} + L \times I_L(t_0) - \frac{V_C(t_0)}{s} = I_L(s)[sL + \frac{1}{sC}] \tag{3}$$

$$V_c(s) = \frac{V_c(t_0)}{s} + I_L(s) \times \frac{1}{sC} \tag{4}$$

1342

By the inverse Laplace transformation, we can get the inductor current $i_L(t)$ and $v_C(t)$ as：

$$I_L(t) = Z[V_s - V_C(t_0)]\sin\omega(t-t_0) + I_L(t_0)\cos\omega(t-t_0) \quad (5)$$

$$V_c(t) = V_c(t_0) + I_L(t) \times \frac{1}{C} \quad (6)$$

where

$$Z = \sqrt{\frac{C}{L}} \quad (7)$$

$$\omega = \sqrt{\frac{1}{LC}} \quad (8)$$

When $t=t_1$, the $v_C(t=t_1)$ resonates to zero voltage, both the power switches are turned on at this moment to achieve zero voltage switching.

Mode II [t_1-t_2]：

At $t=t_1$, the resonant capacitor's voltage is decreased to zero, that is to said, zero voltage switching is achieved when the power switches (S_1 & S_2) are both turned on. And then the resonant inductor L is magnetized by the voltage source V_s. Therefore, the $i_L(t)$ is rising linearly and the electrical energy stored in L. Fig. 6(a)-(b) depicts the equivalent circuit and the Laplace transform circuit, respectively.

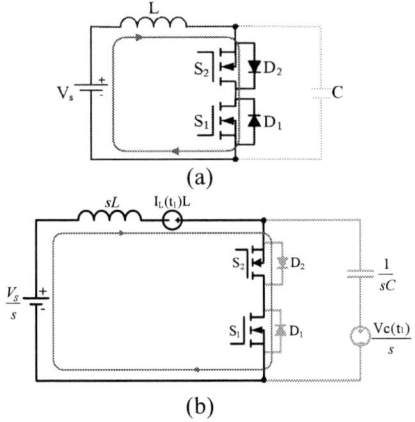

(a)

(b)

Fig. 6. Mode II. (a) the equivalent circuit. (b) the Laplace transform circuit..

By KVL, and $V_C(t_1)=0$, Fig. 6(b) can be obtained as:

$$\frac{V_s}{s} + L \times I_L(t_1) = I_L(s) \times sL \quad (9)$$

And then the inductor current $i_{L1}(t)$ is derived by using the inverse Laplace transformation.

$$I_L(t) = \frac{V_s}{L} \times (t-t_1) + I_L(t_1) \quad (10)$$

Where $i_L(t_1)$ can be indicated as：

$$I_L(t_1) = Z[V_s - V_C(t_0)]\sin\omega(t_1-t_0) + I_L(t_0)\cos\omega(t_1-t_0) \quad (11)$$

When $t=t_2$, the power switches are turned off and this mode is ended.

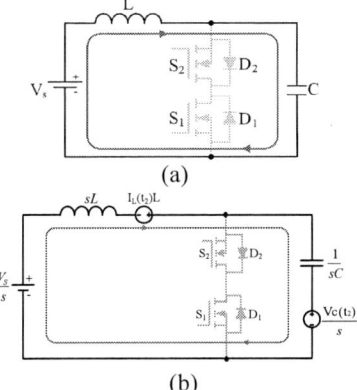

(a)

(b)

Fig. 7. Mode III. (a) the equivalent circuit. (b) the Laplace transform circuit.

Mode III [t_2-t_3]：

Fig. 7(a) depicts the equivalent circuit in Mode III, where the equivalent circuit look exactly the same as the circuit shown in Fig. 5(a). However, the initial current $i_L(t_2)$ and the initial voltage of $V_C(t_2)$ are still different from the initial conditions in Mode I. The resonant inductor L resonates with the resonant capacitor C and transfers the electrical energy stored in L to C. Hence, the Laplace transform circuit can be depicted in Fig. 7(b) and the KVL equation are given as

$$\frac{V_s}{s} + L \times I_L(t_2) - \frac{V_C(t_2)}{s} = I_L(s)[sL + \frac{1}{sC}] \quad (12)$$

$$V_c(s) = \frac{V_C(t_2)}{s} + I_L(s) \times \frac{1}{sC} \quad (13)$$

And then the inductor current $i_L(t)$ and the voltage drop $V_C(t)$ of the resonant capacitor are described as follows by using the inverse Laplace transformation,

$$\begin{aligned}I_L(t) &= Z[V_s - V_C(t_2)]\sin\omega(t-t_2) \\ &+ I_L(t_2)\cos\omega(t-t_2)\end{aligned} \quad (14)$$

$$V_c(t) = V_c(t_2) + I_L(t) \times \frac{1}{C} \quad (15)$$

$$I_L(t_2) = \frac{V_s}{L} \times (t-t_1) + I_L(t_1) \quad (16)$$

When $t=t_3$, the power switches are turned on and this interval is completed.

IV. SIMULATION AND EXPERIMENTAL RESULTS

All the key parameters of the proposed sinusoidal load are listed in Table I. In this study, an 8S1P Li-ion battery module is served as the test target to verify the theoretical feasibility.

TABLE I
Related parameters of the experimentations

Battery Module Voltage	33.6V
Resonant Capacitor	68nF
Resonant Inductor	419.5μH
Power Switch	47N60C3
Operating Frequency	24kHz~30kHz

1343

The 2018 International Power Electronics Conference

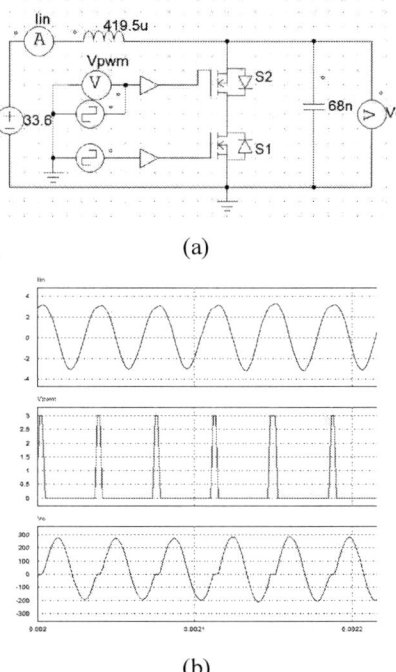

(a)

(b)

Fig. 8. (a)The schematic circuit for simulation by using the simulation software PSIM, (b)The simulation waveform for the input current and capacitor terminal voltage of the proposed resonant load.

One of the simulation result is shown in Fig. 8 according to these parameters shown in Table I. Fig. 8(a) is the schematic circuit for simulation by using the simulation software PSIM. Fig. 8(b) demonstrates the simulation waveform for the input current of the proposed sinusoidal load. The simulation result reveals sinusodial current waveform as the theoretical prediction.

In practical applications of power battery, the loading current is nonperiodical and unpedictable. Therefore, it is not easy to simulate the loading current of the power battery for online test. However, sinusoidal waveform inherently has wide-range slew-rate characteristics, which is suitable for demonstrating the related experimentations.

In the latter experimentations, variable frequency control is used to modulate the current amplitude of the sinusoidal load. Figs. 9-12 show the gating signal and the measured input current of the proposed sinusoidal load under various operating frequency, which ranges from 24kHz to 30kHz. The measured results are really close to the simulation waveform. It is apparent that the amplitudes of the loading current can be modulated by adjusting the operating frequency of the gating signal.

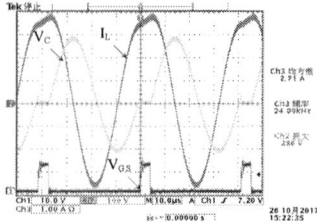

Fig. 9 The gating signal and the measured input current and capacitor terminal voltage of the proposed sinusoidal load at f_s=24kHz.

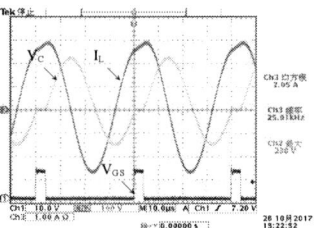

Fig. 10 The gating signal and the measured input current and capacitor terminal voltage of the proposed sinusoidal load at f_s=25kHz.

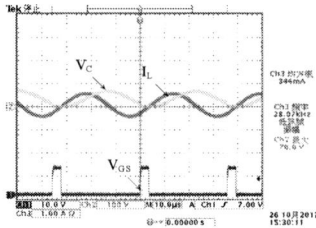

Fig. 11 The gating signal and the measured input current and capacitor terminal voltage of the proposed sinusoidal load at f_s=28kHz.

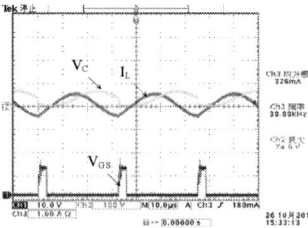

Fig. 12 The gating signal and the measured input current and capacitor terminal voltage of the proposed sinusoidal load at f_s=30kHz.

From (5), (10), and (14) we can see that the lower operating frequency, the higher amplitude of loading current. Actually, from the derived mathematical models in section III, it is clear that the area of the positive half-cycle and the negative half-cycle is not equal, i.e., the energy drawn from the battery during test procedure cannot be recycled completely. However, the proposed

1344

sinusoidal load can recycle most of the drawn energy. It is still superior to the traditional battery tester.

Fig. 13 shows a implemented prototype, which is designed to demonstrate the proposed technical feasibility. Because the extracted electrical energy can be recycled during test procedure and both the MOSFET switches in system achieve ZVS, therefore, it is no need for bulky cooler. Accordingly, the practical volume of the proposed experimental prototype can be reduced significantly and to realize light weight and portable design.

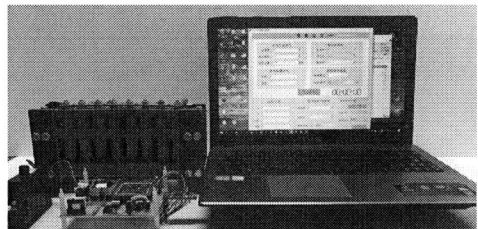

Fig. 13 Prototype of the proposed sinusoidal load.

V.   CONCLUSIONS

A programmed and portable power battery test system is implemented. The proposed sinusoidal loading technique provides wide-range slew-rate discharging conditions for conforming practical work situations, and recycles the extracted energy every cycle without any additional grid-tied equipment or backup battery. Therefore, this system exhibits compact circuit topology, low cost, portable feature. Finally, the recycle rate of the electrical energy during test procedure is verified by the experimental results and up to 89%. The functions of the prototype are verified and close to the theoretical predictions.

ACKNOWLEDGMENT

The financial support of this work was sponsored in part by the Ministry of Science and Technology, Taiwan, R.O.C., Project number: MOST 106-2221-E-011-094-MY3. And the authors would also like to thank National Chung-Shan Institute of Science and Technology for their financial support in part under Project number: NCSIST-104-V409 (107) and invaluable assistance.

REFERENCES

[1]   X. Wang, et al.: "A multi-cell battery pack monitoring chip based on 0.35-μm BCD technology for electric vehicles," IEICE Electron. Express 12 (2015) 20150367. DOI:10.1587/elex.12.20150367

[2]   D. Gharavian, et al.: "ZEBRA battery SOC estimation using PSO-optimized hybrid neural model considering aging effect," IEICE Electron. Express 9 (2012). DOI: 10.1587/elex.9.1115

[3]   D. Gharavian, et al.: "Fast battery charger MCU with adaptive PWM controller using runtime tracking of polarization curve," IEICE Electron. Express 13 (2016) 20160131. DOI:10.1587/elex.13.20160131

[4]   C. H. Yang, H. S. Lo, and H. P. Chui, "Switching-Mode Battery Test System," in Conf. Rec. IEEE IS3C'14, pp. 605-608, 2014.

[5]   K. I. Hwu. and Y. T. Yau, "Active Load for Burn-in Test of Buck-Type DC-DC Converter with Ultra-Low Output Voltage," in Conf. Rec. IEEE APEC'08, pp. 635-638, 2008.

[6]   H. Ma, Q. Guo, X. Han and L. Chen, "Energy Recycling Load System with a High Gain DC-DC Converter for Ultra Low Voltage Power Supplies," in Conf. Rec. IEEE ISIE'13, pp. 1001-1006, 2013.

[7]   M. H. Hung, C. H. Lin, L. C. Lee and C. M. Wang, "State-of-charge and state-of-health estimation for lithium-ion batteries based on dynamic impedance technique," Journal of Power Sources, vol. 268, pp. 861-873, Dec. 2014.

[8]   M. K. Kazimierczuk, "Analysis of Class-E zero-voltage-switching rectifier," IEEE Trans. Circuits Syst., vol. 37, no. 6, pp. 747-755, Jun. 1990.

[9]   Y. Kamito, K. Fukui, and H. Koizumi, "An analysis of the class-E zero-voltage-switching rectifier using the common-grounded multistep-controlled shunt capacitor," IEEE Trans. Power Electron., vol. 29, no. 9, pp. 4807-4819, Sep. 2014.

[10]   L. Roslaniec, A. S. Jurkov, A. Al Bastami, and D. J. Perreault, "Design of single-switch inverters for variable resistance/load modulation operation," IEEE Trans. Power Electron., vol. 30, no. 6, pp. 3200-3214, Jun. 2015.

Development of Large-Capacity Converter for Battery Energy Storage Systems

Hiroyoshi Komatsu[1*], Tatsuji Katayama[1] and Noriko Kawakami[1]

1 Power Electronics Systems Division, Toshiba Mitsubishi-Electric Industrial Systems Corporation (TMEIC), Tokyo, Japan
*KOMATSU.hiroyoshi@tmeic.co.jp

Abstract- Renewable energy sources may cause the power grid instability due to their output fluctuations. One of the countermeasures to such problems is Battery Energy Storage Systems (BESSs). This paper presents the outline and test results of 1.5MW and 2.5MW converters for BESS, suitable for large-scale plants. Both converters are outdoor model and have high environmental durability by applying hybrid cooling method. In terms of 2.5MW converter, adopted the latest 7th-generation IGBT modules, the capacity increases with same dimensions of the 1.5MW converter, increasing the power density up to 1.66 times of the 1.5MW converter. The efficiency of the 1.5 MW converter is achieved over 98% in the wide output range. The efficiency of 2.5MW converter is also more than 98.5% in the wide output range.

Keywords— Battery Energy Storage Systems, Large-Capacity Converter, Neutral point switch tree-level inverter

I. INTRODUCTION

Large scale renewable energy sources are promoted around the world. Regarding photovoltaic (PV) power generation systems, the expected domestic annual installation capacity including replacement of 2030 is 11GW. The installation capacity is increasing in the mid-and-long term. As for the worldwide installation capacity of PV systems, the expected annual installation capacity including replacement of 2030 is 118GW. Regarding the wind turbine generation (WTG) systems, the anticipated annual installation capacity of domestic is 1,500MW in 2030 [1] [2].

However, the output power of renewable energies depends on the weather conditions, which causes large fluctuations in the output power, leading to insufficient frequency adjustment capacity of the power grid.

One of the countermeasures to such problems is the load leveling and system stabilization by the frequency control using Battery Energy Storage Systems (BESSs). In Japan, large-capacity BESSs were introduced in substations of electric power companies to stabilize power systems from fluctuation of the output of renewable energy sources [3] [4]. In the North American market promoting renewable energy positively, the large-scale BESSs are expanding gradually for compensating the fluctuation of large-scale PV systems and the ancillary services.

In this paper, the outline of development and test

TABLE I
SPECIFICATIONS AND RATINGS OF 1.5MW CONVERTER.

Items	Specifications/Rating
Rated capacity	1.5MW /1.5MVA
DC voltage range	750V − 1000V
Rated AC voltage	480Vrms - 50/60Hz
Rated output current	1804Arms
Maximum efficiency	98.7%
Ambient temperature range	-20 ~ 50℃
Cooling	Self-cooling + Forced air cooling
Installation	Outdoor
Dimensions (H*W*D)	2286×5000×1150 mm

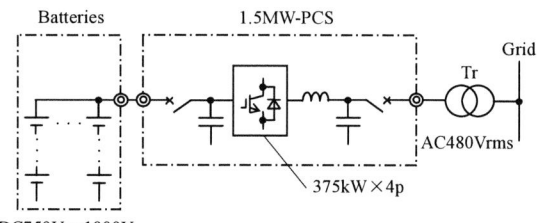

Fig. 1. Schematic diagram of 1.5MW converter.

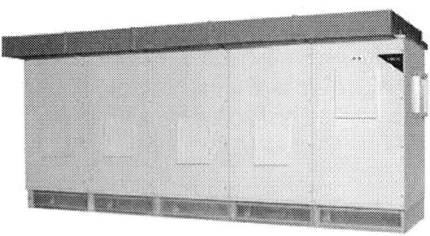

Fig. 2. External view of 1.5MW converter.

results of large-capacity converters for BESSs is described. First, the outdoor type 1.5MW converter which is developed for the North American market is described. The large capacity outdoor model which has

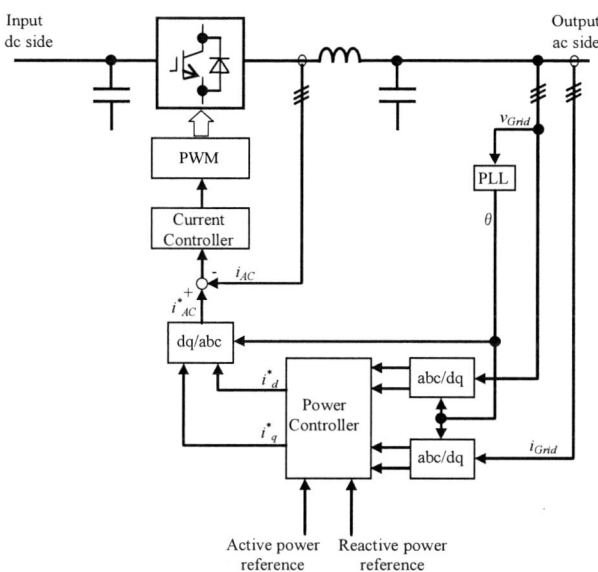

Fig. 3. Control blocks of BESS converter.

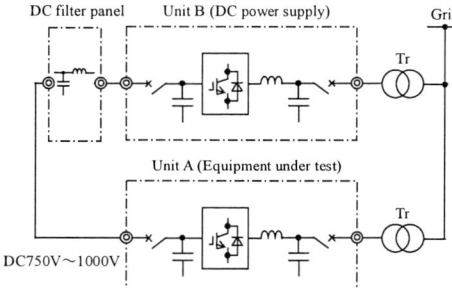

Fig. 4. Circuit configuration for the run-back test.

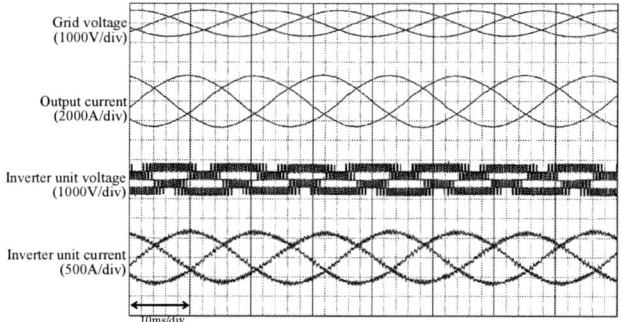

Fig. 5. Output waveforms of 1.5MW converter
at 1.5MW-750Vdc-480Vac

high environmental durability contributes to the reduction of the number of installed units and installation man-hour from the view point of system cost reduction.

Next, outdoor type 2.5MW converter is described. By adopting the latest 7th-generation IGBT modules, the capacity can increase with remaining same dimensions as the 1.5MW converter. In both global and domestic markets, large-scale PV plants with BESSs are expanding gradually. For reduction of the system costs, further increasing of unit capacity of BESS is required. The 2.5MW converter will be suitable for large-scale plants. In addition, 2.5MW converter can be operated with a wide range DC voltage corresponding to the various kinds of secondary batteries.

II. 1.5MW CONVERTER

A. Circuit configuration

Table I shows the specifications of the 1.5MW converter and Figure 1 shows the circuit configuration. Figure 2 shows an external view of the converter. The converter consists of four converter panels and an AC output panel.

In current actual applications, high efficiency converters employ the neutral switch type three-level topology (NPS) [4] [6] [7]. 1.5MW converter has also adopted the NPS topology, achieving high efficiency. In order to achieve high reliability by component reduction, snubber-less configuration is adopted.

B. Environmental durability

Adopting a combination of self-cooling and forced air

cooling, hereafter hybrid cooling, as the cooling method, the converter achieves high environmental durability due to simplicity and better temperature performance.

The advantage of the hybrid cooling in comparison with water cooling is that pumps, valves, flow meters and cooling controller etc. are not needed. The reliability is higher from the view point of reducing the number of parts. In hybrid cooling, fans are used at low speed, so that power loss by fans is lower and maintenance interval is longer.

In addition, the advantage of the hybrid cooling in comparison with the usual forced air cooling is that the converter operates at self-cooling without fans when the output power is low. Therefore the influence of outside air such as humidity and dusts is minimized and high reliability is achieved.

In terms of low temperature specification, adopting high capability heat-pipe, the operation temperature range was expanded. Specification of the minimum operation ambient temperature is –20 degree, however converter cooling capability up to –30 degree was verified at verification test.

C. Control functions

Figure 3 shows the control blocks of BESS converter. The converter usually operates 4 quadrant operations according to active/reactive power references from a

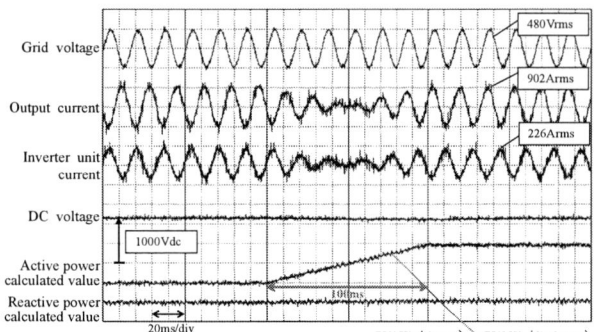

Fig. 6. Response to reference characteristics of 1.5MW converter.

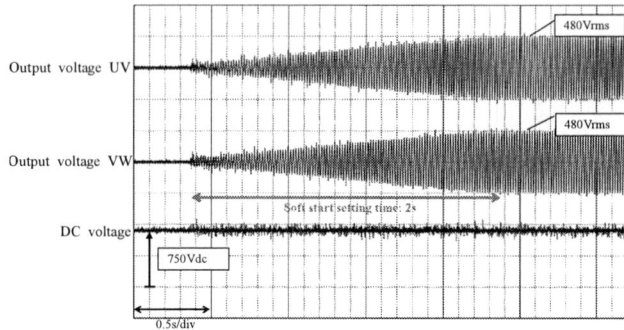

Fig. 7. Stand-alone function characteristics of 1.5MW converter.

system controller.

The converter has the stand-alone operation (also called independent operation) function. If the converter is required to supply the power to important loads when the grid fault occurs, the converter can supply the power independently according to the command from the system controller. Under the stand-alone operation, the converter supplies the power with constant ac voltage and constant frequency to load.

The converter also has the constant current control function (CC), the constant power control function (CP) and the constant voltage control function (CV) for charging batteries.

D. Verification test

Figure 4 shows the circuit configuration of the run-back test. Using this configuration, the verification test was conducted under the full load conditions. In the run-back test, two converters were used. Unit A is the equipment under test (EUT). Unit B is used as a battery simulator. They are connected with transformers and a DC filter panel. The AC side is connected to the power grid. The grid supplies only the loss of unit A and unit B at the load test. With the conditions of 750 - 1000Vdc and 480Vac-60Hz, the converter operation characteristics and efficiency were measured.

Figure 5 shows the output waveforms at the rated power (1.5MW, charge operation) with 480Vac and 750Vdc. The ac current THD is 2.2%. Adopting the three-level topology, a low harmonic distortion was achieved with a small harmonic filter.

Figure 6 shows the waveforms of the reference change. The converter was operated at 1000Vdc, the active power reference was changed from 750 kW (charge) to − 750 kW (discharge) with 100ms ramp rate. The measured active power followed the active power reference very well and the stable operation of the converter was verified.

Figure 7 shows the waveforms of start-up under stand-alone operation mode without load. The soft-start time is set to 2s. With the soft start function, the output voltage of the converter rises smoothly within 2 s to the rated

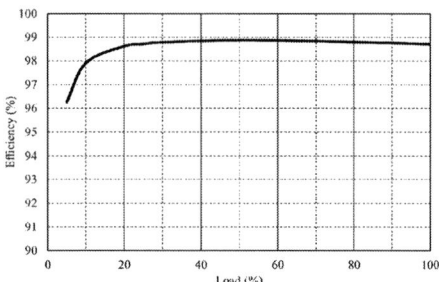

Fig. 8. Efficiency of 1.5MW converter
(Average of charge and discharge).

voltage 480 Vac.

Figure 8 shows the efficiency characteristics of the average of charge and discharge operation at 750Vdc. Maximum efficiency is 98.7%. The efficiency is achieved more than 98% in the wide range of output power.

III. 2.5MW CONVERTER

A. Circuit configuration

Table II shows the specifications and ratings of the 2.5MW converter. Figure 9 shows the circuit configuration and Figure 10 shows an external view of the converter. The converter consists of four converter panels and an AC output panel same as the 1.5MW converter mentioned in chapter II.

The 2.5MW converter also adopted the NPS topology, achieving high efficiency and the converter has snubber-less configuration. In order to increase the capacity with remaining same dimensions as the 1.5MW converter, the latest 7th-generation IGBT modules are adopted. Because the power density of the latest-IGBT modules is larger than the conventional IGBT modules, it is contributed to increase the power density of the 2.5MW converter.

In addition, in accordance with increasing the converter current per panel, the ripple current of DC link capacitors is increased. The capacitors were replaced to high ripple current rating's capacitors.

1348

TABLE II
SPECIFICATIONS AND RATINGS OF 2.5MW CONVERTER.

Items	Specifications/Rating
Rated capacity	2.5MW /2.5MVA
DC voltage range	710V – 1250V
Rated AC voltage	480Vrms - 50/60Hz
Rated output current	3007Arms
Maximum efficiency	98.9%
Ambient temperature range	-20 ~ 50℃
Cooling	Self-cooling + Forced air cooling
Installation	Outdoor
Dimensions (H*W*D)	2286×5000×1150 mm

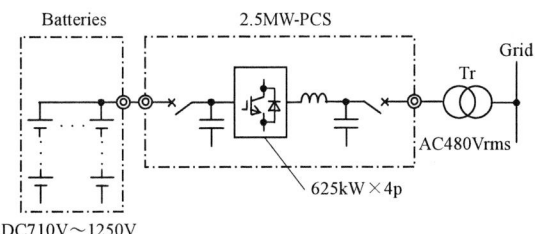

Fig. 9. Circuit configuration of 2.5MW converter.

Fig. 10. External view of 2.5MW converter.

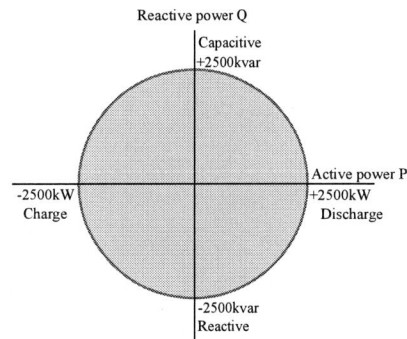

Fig. 11. P-Q characteristics of 2.5MW converter.

Because DC voltage range of 2.5MW converter is wider than 1.5MW converter, the various secondary batteries can be applied. In order to reduce the loss of BESS, battery operating voltage is expected to become higher. The 2.5MW converter is designed to be suitable for the high voltage BESSs.

B. Environmental durability

2.5MW converter which utilized the cooling technology of 1.5MW converter is also outdoor model and has high environmental durability. By adopting high capability heat-pipe, wide converter operating temperature is achieved.

C. Control functions

The 2.5MW converter adopts the same control blocks as the 1.5MW converter. The 2.5MW converter usually operates 4 quadrant operations according to active/reactive power references from a system controller.

Figure 11 shows the operating range of 2.5MW converter at 1000Vdc, 480Vac. Because reactive power operating range is enlarged compared with 1.5MW converter, 2.5MW converter will be suitable for large-scale plants.

D. Verification test

The validity of the design of the developed converter was confirmed by sequence test, operation test, control characteristics test, temperature rise test and so on. Representative results are described below.

Figure 12 shows output waveforms of the rated power operation (2500kW, charge operation) at 480Vac and 710Vdc. The current THD of this operation is 1.5% confirmed to satisfy the IEEE 1547 standard. Adopting the three-level topology, a low-harmonic distortion can be achieved with a small harmonic filter same as the 1.5MW converter.

Figure 13 shows response to the reference change. The converter was operated at dc 1250V, the active power reference was changed from 1250 kW (charge) to – 1250kW (discharge) with 20ms ramp rate. The measured active power followed the active power reference very well and the stable operation of the converter was verified.

Figure 14 shows the efficiency characteristics of the average of charge and discharge at 710Vdc. Maximum efficiency is 98.9%. The efficiency is more than 98.5% in the wide range of output power. By applying the latest 7th generation IGBT, the efficiency was preserved same level as the 1.5MW converter despite the increase in power density.

IV. CONCLUSIONS

This paper presented the outline and the test results of 1.5MW and 2.5MW BESS converters, suitable for large-scale plants. Both converters are outdoor model. High environmental durability and the wide converter operating temperature range were achieved due to high capability heat-pipes. Regarding the efficiency, both converters achieved over 98% in the wide output range by applying NPS topology,

1349

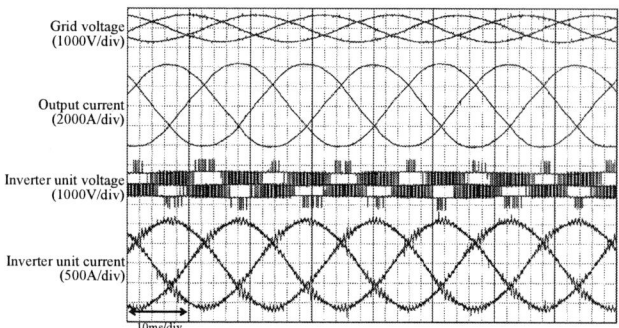

Fig. 12. Output waveforms of 2.5MW converter
at 2.5MW-710Vdc-480Vac

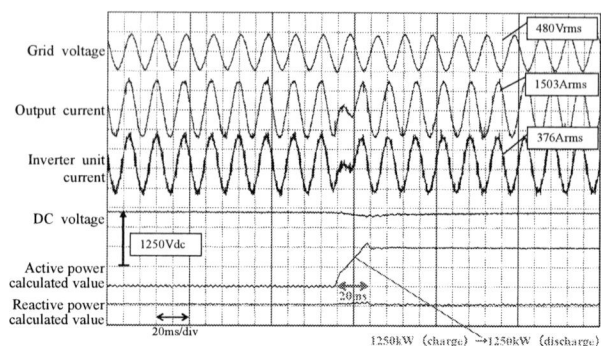

Fig. 13. Response to reference characteristics of 2.5MW converter.

The 2.5MW converter adopting the latest 7th-generation IGBT modules, the output capacity was increased with same dimensions as the 1.5MW converter and the power density is up to 1.66 times of the 1.5MW converter. DC voltage range of 2.5MW converter is wider than 1.5MW converter, so that the various secondary batteries can be applied.

The 1.5MW converter will be verified in a field test in 2018, and the 2.5MW converter will be certificated of the UL standard.

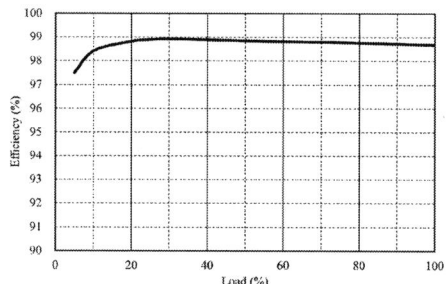

Fig. 14. Efficiency of 2.5MW converter
(Average of charge and discharge).

REFERENCES

[1] Report of PV system market, https://www.fuji-keizai. co.jp/market/17016.html, 24th, Feb. 2017

[2] Report of WTG system market, http://www.newenergy-news.com/, 23th, Jan. 2017

[3] Tohoku Electric Power Co., Inc. Press Release 20th, Feb. 2015, 26th, Feb. 2016

[4] Kyusyu Electric Power Co., Inc. Press Release 3rd, Mar. 2016

[5] T. Tanaka, I. Tominaga, N. Kawakami, "Development of high efficiency PCS for storage batteries," 2011 Japan Industry Application Society Conference Record, pp.1-421 - 1-422, September 2011. (in Japanese)

[6] H. Li, Y. Iijima, N. Kawakami, "Development of Power Conditioning System (PCS) for Battery Energy Storage Systems," the 5th Annual International Energy Conversion Congress and Exhibition for the Asia/Pacific region (ECCE Asia 2013), pp. 1295 -1299, 2013.

[7] K. Yamabe, H.Komatsu, Y. Iijima, "Requirement Analysis and Development of MW-Range PCS for Substation-Scale Battery Energy Storage Systems" The 19th International Conference on Electrical Machines and Systems (ICEMS 2016), pp. 720 -724, 2016

Analysis and Comparison of dc/dc Topologies in Partial Power Processing Configuration for Energy Storage Systems

Maria C. Mira, Zhe Zhang and Michael A. E. Andersen
Department of Electrical Engineering
Technical University of Denmark (DTU)
Kgs. Lyngby, Denmark
mmial@elektro.dtu.dk, zz@elektro.dtu.dk, ma@elektro.dtu.dk

Abstract—This paper presents an analysis and comparison of dc/dc switched-mode power supplies (SMPS) for energy storage systems in partial power processing (PPP) configuration. The advantage of this configuration is that the SMPS only processes the partial power resulting from the voltage difference between the source and the energy storage element, thus allowing for a reduction of the converter power rating. Selection of an appropriate topology for a given system configuration is the key factor in achieving high efficiency power conversion. An analysis and comparison of dc/dc topologies based on component stress factor (CSF) is performed to determine the optimal solution for the evaluated application. Based on the results of the CSF analysis, a dc/dc converter is designed, built and tested. Experimental results prove the feasibility of the PPP configuration with a reduction of the 80% of the power rating compared to the traditional interconnection, which implies a reduction in cost, weight and an increase in efficiency.

Keywords—Converter; dc-dc; partial power converter; partial power processing; electrolyzer cells; energy storage systems.

I. INTRODUCTION

Energy storage elements are indispensable components in renewable energy systems due to the intermittent nature of the energy source [1], [2]. They are as well essential in smart grids in order to balance the energy production and demand [3], and also in stand-alone structures to provide energy in remote locations, where cabling is challenging and expensive. Switched-mode power supplies (SMPS) play an important role in the integration of energy storage elements to provide high efficiency energy conversion [4].

Nowadays, different energy storage technologies are available, such as electrochemical (batteries), thermal (molten salt), mechanical (pumped hydro, compressed air, flywheels), chemical (hydrogen based energy storage), etc. More than 95% of the global energy storage capacity is represented by pumped hydro plants [2], which is a mature technology and allows to store large quantities of energy with high efficiency over a long time. However, it is not suited for distributed generation and have relatively low energy density. Electrochemical energy storage is an emerging technology, which has had significant advances in the last two decades both from technical and cost perspective. It provides high flexibility in terms of energy and power capacity. Nevertheless, batteries lifecycle is limited and there are environmental and safety concerns.

Hydrogen energy storage systems have also attracted research interest in the last years [5]. The advantages of hydrogen is that it can be locally produced, it offers high energy density, long term scalable storage and low enviromental impact. However, the cost of the initial investment is high and there are safety considerations.

II. SYSTEM ANALYSIS AND SPECIFICATIONS

The application under analysis is an energy storage system based on alkaline electrolyzer cells (EC), nevertheless, the system configuration can as well be applied to battery charge systems. The traditional way of interconnecting the elements is shown in Fig. 1 (a), where the load, in this case the EC stack, is connected at the output of a dc/dc converter. In this configuration the converter must be rated at the full power of the EC stack. Figure 1 (b) shows the block diagram of the partial power processing (PPP) configuration, where the EC stack is connected in series with the dc bus (V_{dc}) and the dc/dc converter creating a voltage divider. Therefore, in this arrangement, the input of the SMPS (V_{in}) is set by the voltage difference between the dc bus and the EC stack ($V_{dc} - V_{EC}$), thus, the dc/dc converter only process the differential power between the dc bus and the EC. Figure 2 shows the electrolyzer stack voltage as a function of the current and Table I presents the specifications of the system in PPP configuration. As it can be observed, in PPP arrangement the maximum input power processed by the converter is $P_{in} = 733$ W, compared to the traditional parallel connection, where the converter would have been rated at the maximum EC power, in this case $P_{EC} = 3456$ W. Therefore, a reduction of nearly 80% of the required power of the dc/dc converter is achieved, which considerably helps reducing the cost and weight, as well as increasing the power density and efficiency of the system.

The idea of the series connection of source and load originated in the spacecraft technology for photovoltaic applications [6]. The proposed configuration, called series connected boost unit (SCBU), showed numerous advantages compared to the traditional interconnection. High efficiency and high power density can be achieved because the dc/dc converter only processes a fraction of the total power of the system, which results in small, lightweight and low cost power supplies, which in space implementations are extremely important.

The 2018 International Power Electronics Conference

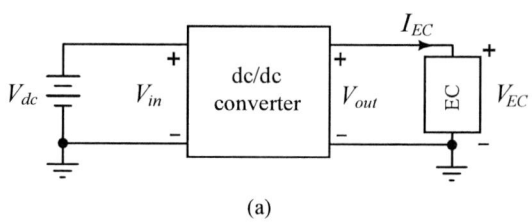

(a)

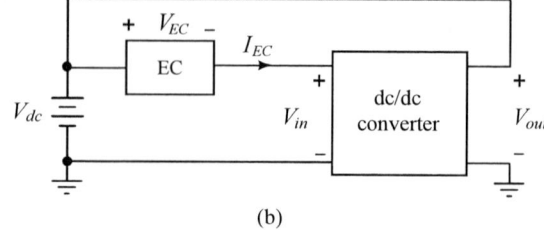

(b)

Fig. 1. Block diagram of the traditional parallel connection (a) and (b) partial power processing configuration (PPP).

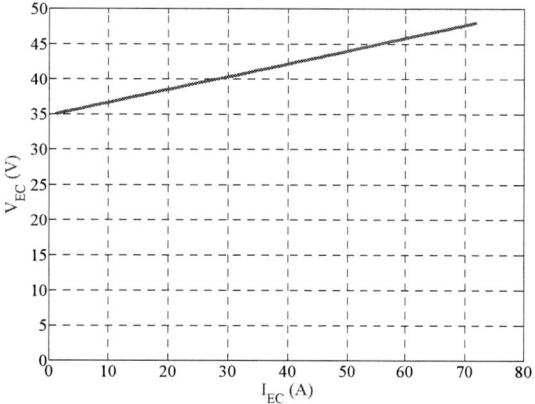

Fig. 2. Electrolyzer stack voltage (V_{EC}) as a function of the current (I_{EC}).

TABLE I
PPP CONFIGURATION SYSTEM SPECIFICATIONS

V_{dc}	50 to 58 V
V_{EC}	35 - 48 V
I_{EC}	1 - 72 A
P_{EC}	3456 W
$V_{in} = V_{dc} - V_{EC}$	2 to 23 V
$P_{in} = V_{in} \cdot I_{EC}$	733 W
V_{out}	50 to 58 V

Depending on the requirements of the application, the series connection of source and load can be performed at the input (input-series-output-parallel ISOP) or at the output of the dc/dc converter (input-parallel-output-series IPOS) [7], where IPOS structure is used in a battery storage system.

Figure 3 and Fig. 4 show the input voltage and input power of the dc/dc converter in PPP configuration, respectively, as a function of the power of the EC, for different values of the output voltage. As it can be observed, the PPP configuration has the disadvantage of presenting a large variation of the converter input voltage, which requires the selected topology to operate efficiently within a wide input to output voltage range. It is important to notice that the point at which the converter power is maximum, will not correspond with the maximum EC, as the characteristic electrolyzer stack curve will make the voltage at the input of the converter decrease as the power increases. It is also important to observe that the system should be designed so the amount of power processed by

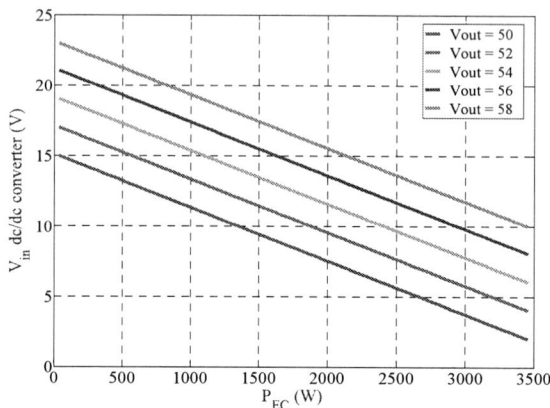

Fig. 3. PPP configuration dc/dc converter input voltage (V_{in}) as a function of the EC power (P_{EC}) for different output voltages.

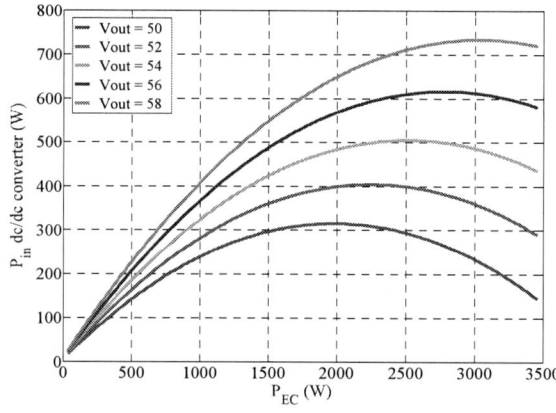

Fig. 4. PPP configuration dc/dc converter input power (P_{in}) as a function of the EC power (P_{EC}) for different output voltages.

the converter tends to zero, thus reducing the losses inserted in the system. Therefore, maximizing the efficiency of the direct power flow, which corresponds to the path to charge the EC, will minimize the power processed by the dc/dc converter and maximize the efficiency of the system.

III. COMPONENT STRESS FACTOR (CSF) ANALYSIS

The topology selection is a key factor in achieving high efficiency, since it will determine the performance of the overall system. From the system specifications it can be observed that the power stage presents a large input/output

1352

The 2018 International Power Electronics Conference

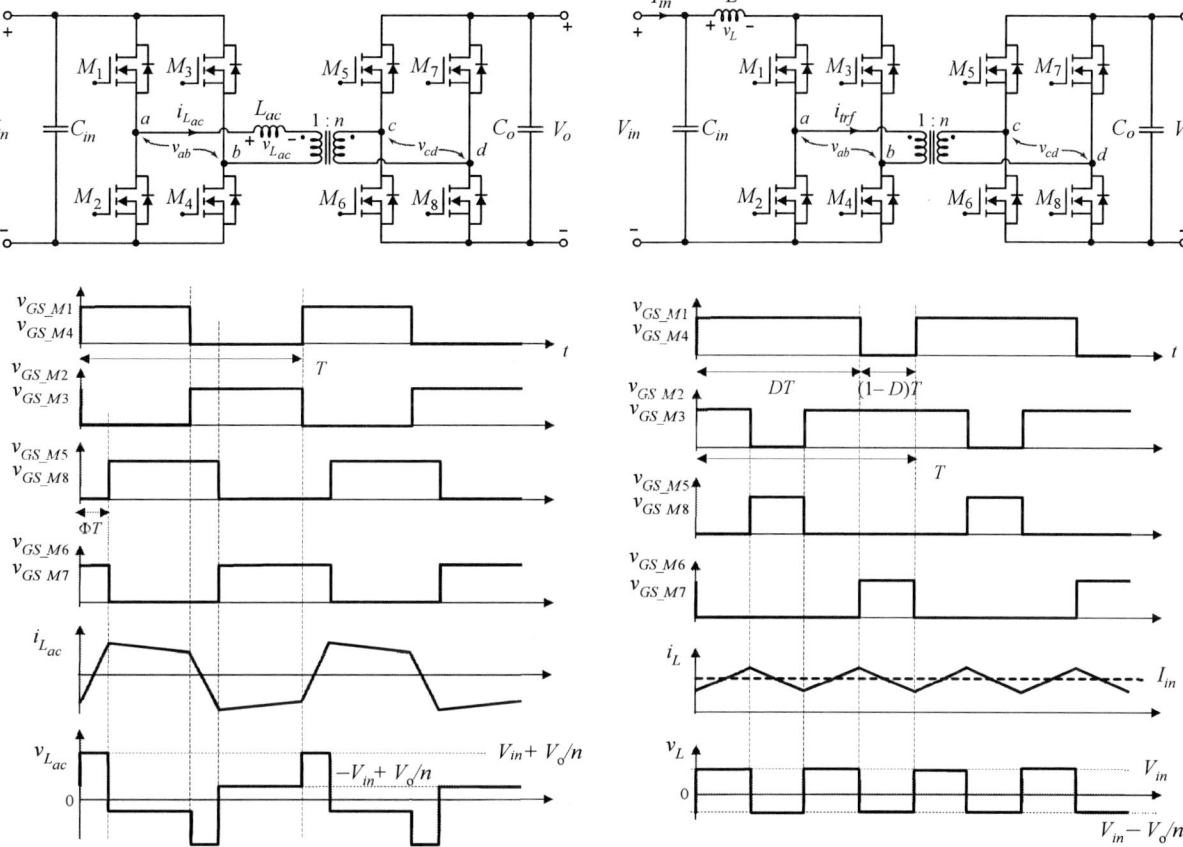

Fig. 5. Dual Active Bridge (DAB) topology and key operating waveforms, top and bottom, respectively.

Fig. 6. Isolated Full Bridge Boost (IFBB) topology and key operating waveforms, top and bottom, respectively.

voltage variation. The challenge is to select a topology that can provide high power conversion efficiency over the whole operating range.

An analysis and review of high efficiency bidirectional dc/dc converters with high voltage gain is performed [8]. Based on the analysis and the system specifications, the topology selection is narrowed down to two candidate topologies: dual active bridge (DAB) and isolated full bridge boost (IFBB) converter. The selection is performed based on complexity in terms of number of active switches, passive components and control. These components will affect the efficiency, cost and reliability of the entire system. Both, DAB and IFBB topologies, have been proved to achieve high efficiency [9], [10], with a reduced component number (low complexity). Figure 5 and Fig. 6 show the schematic of the DAB and IFBB topologies and their operating waveforms, respectively. As it can be observed, both converters present the same number of active switches and passive components. In the DAB the power is delivered to the output through an ac inductor, whose charge and discharge is controlled with the phase-shift angle of the half bridge switching legs. In the IFBB converter, the control parameter is the duty cycle of the primary switches and the input inductor is the component that transfers the energy to the output port.

The analysis of the DAB and IFBB topologies is performed based on component stress factor (CSF) [11]. CSF is a derivation of the component load factors approach (CLF) [12]. CLF is a numerical method, which is calculated based on the components voltage and current stress and normalized to the processed power (volt-amp/watt figure), which makes the calculation dimensionless.

The approach of the CSF analysis is based on the assumption that the evaluated topologies have the same amount of resources: silicon for semiconductors, magnetic material and copper for windings and capacitor volume for energy storage/filter components. A weighing factor is applied to distribute the resources within the topology. The result of the CSF analysis provides an effective way to evaluate the losses in the individual components of the circuit, and consequently, an estimation of the converter performance. Therefore, the analysis gives a quantitative measure to compare the performance of different topologies for a specific application [13], [14].

The CSF method adopts two assumptions in order to simplify the calculations, i. e., the power losses in the converter are neglected (efficiency 100%) and the inductors are large enough to have no ripple current (square waveform).

1353

The stress factor is calculated independently for each component: semiconductors (SCSF), windings (WCSF) and capacitors (CCSF), as shown in (1), (2), (3), respectively. The CSF is related to the power dissipated in the component. In the semiconductors, the conduction losses are calculated with the squared root mean square (rms) current through the device multiplied by the channel on-resistance of the switch. For a given die size, the channel on-resistance is proportional to the voltage rating to the power of 2.5 [15], higher voltage rating will result in a longer channel with smaller cross section. Taking the rated voltage squared gives a good approximation to relate the maximum voltage and the channel on-resistance. Therefore, SCSF is calculated with the breakdown voltage squared, times the rms current squared, and normalized to the square of the processed power, to provide a dimensionless quantity.

Regarding the calculation of the stress factor of magnetic components, to perform a fair comparison, each topology should have the same amount of copper volume, and hence, the same winding area. The windings losses in magnetic components are calculated with the rms current squared, times the winding resistance. The winding resistance is related to the number of turns and the cross-sectional area of the copper. The voltage applied to the windings is proportional to the number of turns. Therefore, the resistance will increase with the square of the number of turns, which is proportional to the voltage squared. The WCSF is then computed as the maximum voltage applied to the windings squared, multiplied by the rms current squared (2). The maximum voltage applied to the winding is calculated as the average voltage applied to the winding over a period, as shown in (4).

The stress factor of capacitors is determined by the resistive losses due to the equivalent series resistance (ESR). The ESR is related to the capacitor volume, and the volume is proportional to the energy storage capacity, thus, the CCSF is calculated with the squared maximum voltage and the rms current as presented in (3).

$$SCSF_i = \frac{\sum_j W_j}{W_i} \cdot \frac{V_{max}^2 \cdot I_{rms}^2}{P_{in}^2} \tag{1}$$

$$WCSF_i = \frac{\sum_j W_j}{W_i} \cdot \frac{V_{max\_avg}^2 \cdot I_{rms}^2}{P_{in}^2} \tag{2}$$

$$CCSF_i = \frac{\sum_j W_j}{W_i} \cdot \frac{V_{max}^2 \cdot I_{rms}^2}{P_{in}^2} \tag{3}$$

$$V_{max\_avg} = \sum_i D_i \cdot |V_i| \tag{4}$$

The distribution of the resources is implemented by the term $\sum_j W_j / W_i$, which represents the weighting factor for component i, where $\sum_j W_j$ is the sum of the individual weights of all components of the same type and W_i is the weight assigned to the component i. In the first iteration, the resources are distributed equally. Based on the results,

the weight distribution can be adjusted. As a result, the component with higher CSF can be assigned with a larger amount of the resource in order to reduce the stress factor.

Once the stress factor for each component is calculated, the total CSF is computed as the sum of component stress factors of the same type as in shown in (5).

$$\begin{cases} SCSF = \sum_i SCSF_i \\ WCSF = \sum_i WCSF_i \\ CCSF = \sum_i SCSF_i \end{cases} \tag{5}$$

From the procedure of the CSF calculation, it can be observed that the analysis accounts for the conduction losses in switches, magnetic components and capacitors. However, it does not consider the switching losses in the semiconductors and the magnetic core losses. From the system specifications presented in Table I, the application under analysis is a low voltage, high current application, therefore, the conduction losses in the semiconductors will dominate over the switching losses. Regarding the magnetic components, the core losses are a function of the magnetic material, the volts-seconds, the peak to peak ac flux density and the switching frequency. As discussed before, the system is characterized by a low input voltage to the dc/dc converter, and the switching frequency is limited to 50 kHz. Therefore, the CSF approach is considered a valid method to compare the topologies for the application under analysis.

Figure 7 (a) to (f) shows the results of the CSF analysis for the DAB and the IFBB topologies for semiconductors (SCSF), windings (WCSF) and capacitors (CCSF), respectively. The graphs show the stress factor values as a function of the power of the EC for different values of the output voltage. The CSFs are normalized to the maximum stress value, which occurs in the DAB topology. At low power levels the DAB converter present very high CSF in semiconductors, winding and capacitors compared to the IFBB topology. This is due to the fact that the DAB topology presents a large rms circulating reactive current. At low power levels the rms current is very large compared to the processed power, which results in large CSF values. As the power of the electrolyzer stack increases, the ratio of the rms current to the processed power is reduced and thus, the CSF. In the IFBB converter the highest CSF occurs at the maximum EC power level (maximum EC current) and minimum output voltage. As it can be observed from the SCSF, the DAB presents a minima for the different output voltages, which corresponds to the point where the converter reactive current is minimized. The IFBB presents higher stress as the input to output voltage transformation ratio is increased, which is an expected result from boost derived topologies. The DAB therefore, shows a reduced SCSF compare to the IFBB, but only in a small range of the operating region. However, the WCSF for the DAB is significantly worse than that of the IFBB in all the operating range due to the increased voltage stress in the magnetic components and the alternating current nature in the resonant inductor.

1354

The 2018 International Power Electronics Conference

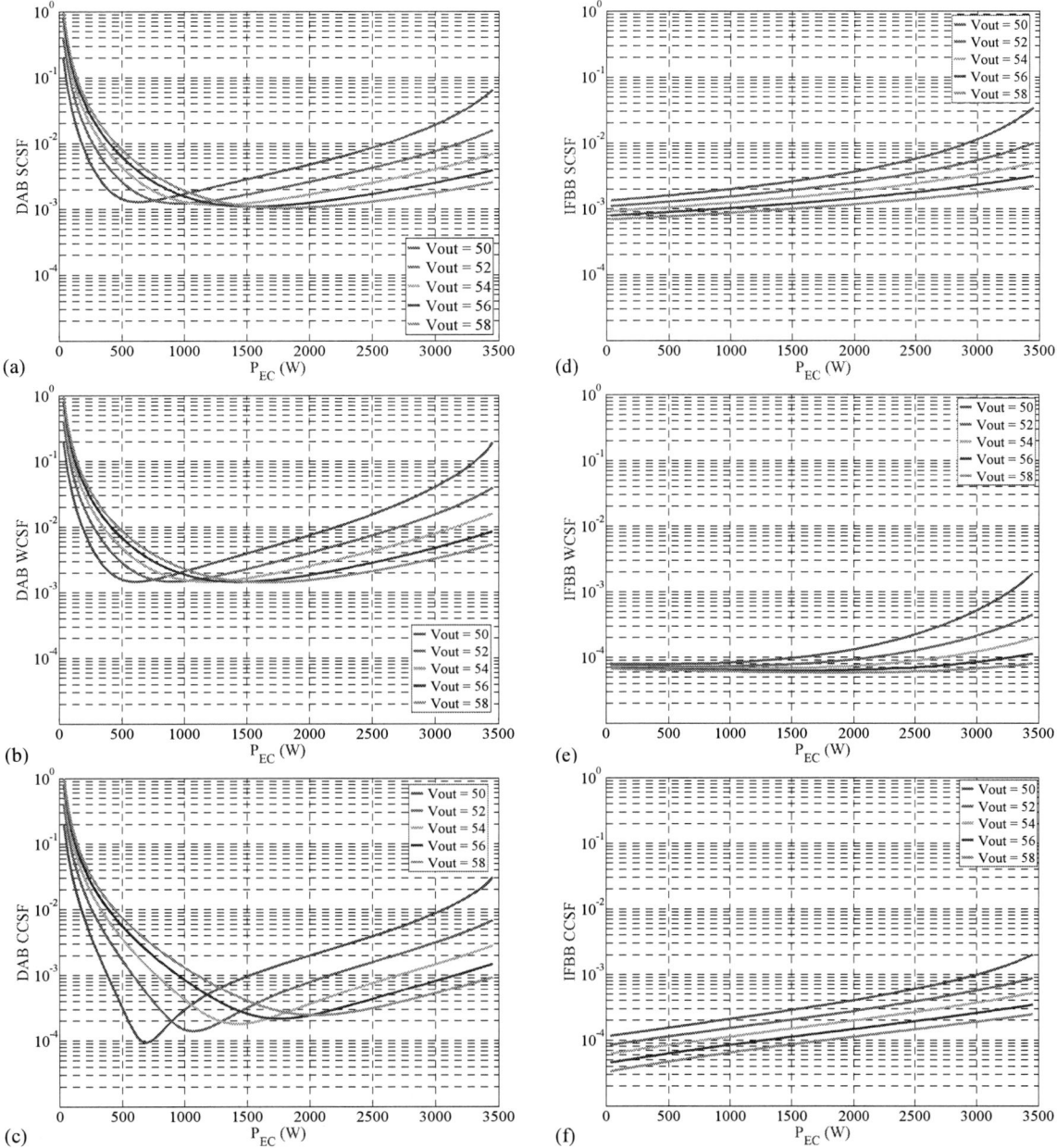

Fig. 7. CSF analysis of DAB: (a) SCSF, (b) WCSF and (c) CCSF; and IFBB converter: (d) SCSF, (e) WCSF and (f) CCSF, as a function of the EC power (P_{EC}) for different output voltages (V_{out}).

Based on the results of the CSF analysis, the IFBB converter is the selected topology to implement the energy storage system in PPP configuration.

IV. IFBB DESIGN & EXPERIMENTAL RESULTS

A prototype of the IFBB is designed and constructed. A printed circuit board (PCB) with 4 layers is designed with special attention to minimize the ac current loops. Due to the high current application, 140 μm (4 oz) PCB copper thickness is used in order to minimize the resistive losses. The main converter components are listed in Table II. The design is based on Silicon (Si) MOSFETs with low R_{DS} on-resistance in order to reduce the conduction losses.

DirectFET technology from Infineon is selected, which provides low package inductance and maximized thermal transfer due to copper drain clip. The high rms current on the IFBB primary side, which is around $I_{rms\_prim\_MOS} = 40$ A, will cause high conduction losses, therefore, primary MOSFETs with very low on-resistance are selected, $R_{DS} = 0.34$ mΩ at $V_{GS} = 10$ V. The maximum rms current on the secondary side is $I_{rms\_sec\_MOS} = 15$ A, which allows to select devices with higher on-resistance than in the primary side. The converter switching frequency is $f_{sw} = 50$ kHz, to limit the switching losses. The magnetic design for the inductor and the transformer is based on planar cores. Planar magnetics offers low profile structure with high surface

1355

TABLE II
PPP IFBB CONVERTER COMPONENTS

$M_1 \sim M_4$	IRL7472L1
$M_5 \sim M_8$	AUIRF7759L2
ISO gate drivers	SI8235AB-D-IS1
Transformer	1:2, EILP43/10/28, Ferrite N87
Inductor	2.3 µH, EILP43/10/28, Ferrite N87
Capacitors C_{in}	20 x 10 µF 50V X7R
Capacitors C_{out}	20 x 10 µF 100V X7S
Current sensor	ACS770LCB-100B
Switching frequency	f_{sw} = 50 kHz
Digital controller	TMS320F2808 DSP

Fig. 8. Experimental prototype of the IFBB converter.

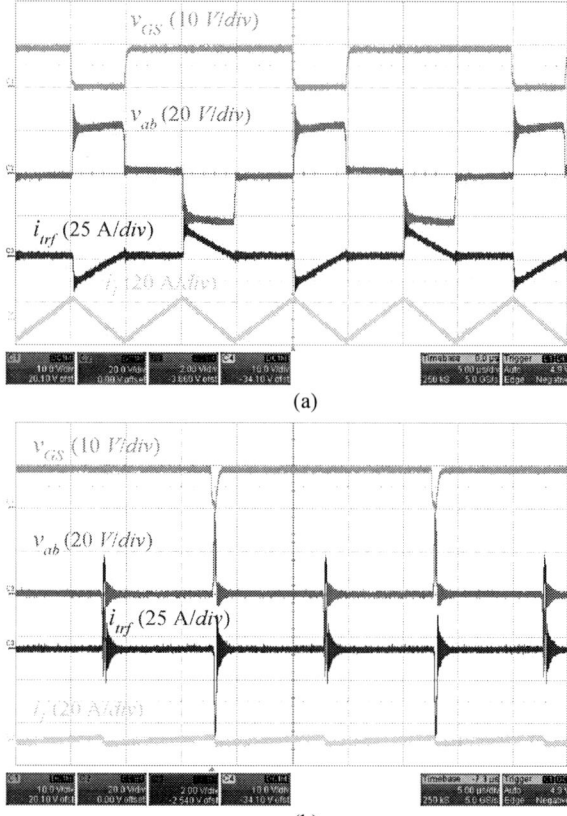

Fig. 9. IFBB operating waveforms, from top to bottom: primary gate-to-source voltage (10 V/div); voltage across the transformer primary side (20 V/div); transformer secondary side ac current (25 A/div) Rogowski coil (20 mV/A); inductor ac current (20 A/div) Rogowski coil (100 mV/A). Time scale 5 µs/div.

area and good thermal characteristics. Moreover, it helps implementing interleaving techniques to achieve low leakage inductance and ac resistance [10]. The inductor is manufactured in a 6 layers PCB with 210 µm (6 oz.) copper thickness. Two PCB stack in parallel in order to reduce the dc resistance. The transformer is manufactured in an 8 layers PCB with 210 µm (6 oz.) copper thickness with a turns ratio of 1:2. The primary winding (P) is formed by two turns in parallel, which are fully interleaved with four turns of the secondary winding (S), in a structure $\frac{P}{2}/S/\frac{P}{2}/S/\frac{P}{2}/S/\frac{P}{2}/S$. Full interleaving winding technique is implemented in order to reduce the transformer leakage inductance. Achieving a low leakage inductance is critical in full bridge boost configurations without snubbers, because the energy stored in the leakage inductance will cause an overshoot in the primary MOSFETs. Full interleaving technique also helps reducing the ac resistance as the magneto motive force (mmf) is always equal to 1[10]. The measured transformer leakage inductance is L_{lk} = 19.4 nH.

The control law is implemented in a digital signal processor (DSP) TMS320F2808 from Texas Instruments. The reading of the inductor current is performed with a high precision Hall effect current sensor, which inserts an extremely low resistance of R_{sens} = 0.1 mΩ.

Figure 8 shows the experimental prototype of the IFBB converter. The primary and secondary MOSFETs are placed on the bottom side of the PCB, in order to transfer the heat to the heat sink through an isolated gap pad material with a thermal conductivity of 4.0 W/mK.

For the experimental test of the IFBB in PPP configuration, the behavior of the EC stack is simulated with an electronic load in series with a power resistor. The electronic load is configured as a constant voltage source and will set the starting point of the characteristic V-I curve shown in Fig. 3. The power resistor (R = 0.185 Ω) connected in series, will provide the slope of the V-I curve as the I_{EC} current increases.

Figure 9 shows the steady state waveforms of the IFBB converter. Figure 9 (a) shows the converter operating at V_{dc} = 50 V, V_{EC} = 38 V, I_{EC} = 13 A, V_{in} = 12 V, P_{EC} = 594 W, P_{conv} = 156 W. In this conditions, the duty cycle D is approx. 75%. Figure 9 (b) shows the IFBB converter waveforms at the system maximum power level, thus, the electrolyzer stack is charging at the maximum current. The

converter operating conditions are V_{dc} = 50 V, V_{EC} = 48 V, I_{EC} = 72 A, V_{in} = 2 V, P_{EC} = 3456 W, P_{conv} = 144 W and the converter's duty cycle is maximum D= 99 %. As it can be observed the inductor current ripple reduces as the input voltage of the converter decreases. However, the peak current on the transformer secondary side increases due to the reduced conducting time of the secondary MOSFETs.

Figure 10 shows an enlarged portion of the steady state waveforms when the converter operates at maximum duty cycle (Fig. 9 (b)). As it can be observed, the voltage across the transformer does not clamp, which indicates that there is no avalanche mode operation of the MOSFETs primary

The 2018 International Power Electronics Conference

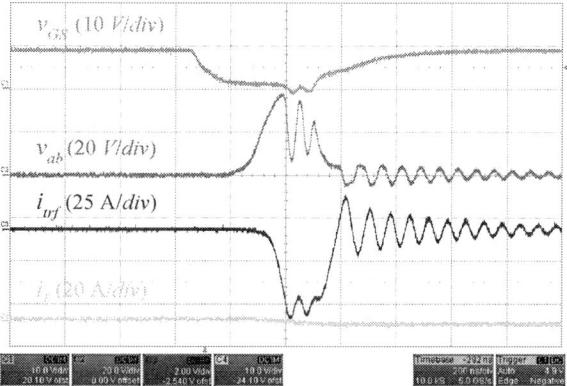

Fig. 10. Enlarged IFBB operating waveforms from top to bottom: primary gate-to-source voltage (10 V/div); voltage across the transformer primary side (20 V/div); transformer secondary side ac current (25 A/div) Rogowski coil (20 mV/A); inductor current ac (20 A/div) Rogowski coil (100 mV/A). Time scale 200 ns/div.

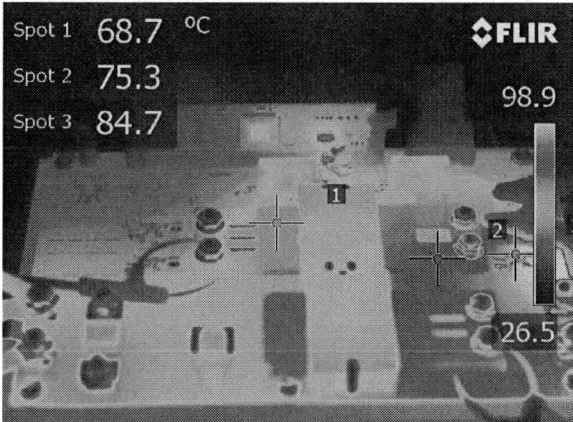

Fig. 11. Thermal image of the IFBB converter in PPP configuration operating at the EC maximum power P_{EC}=3456 W.

side. This proves the low leakage inductance design of the transformer, therefore, no snubber components are required. Fig. 11 presents a thermal image of the IFFB converter operating at the worst conditions, which are minimum input voltage and maximum input current, while the system is working at the maximum power level P_{EC} = 3456 W. The thermal image shows a maximum temperature of 98.9 °C. As the input voltage decreases, the converter current stress on the secondary side reduces, while it increases on the primary side, therefore, the highest temperature appears at the converter primary side.

V. CONCLUSION

This paper presents an analytical comparison of dc/dc topologies in PPP configuration for energy storage systems. Selecting the most appropriate topology for a specific application is crucial in achieving high efficiency. Although the two analyzed converters present the same number of active switches and passive components, the CSF analysis shows a big difference of the converters' performance for the given configuration. Based on the analysis, an IFBB converter is designed and tested. Experimental results shows that in PPP configuration, the converter can handle the maximum system power level,

with a reduction of 80% of the power rating compared to traditional load connection. Therefore, the proposed PPP configuration for energy storage systems achieves a large reduction of the converter size, weight and price.

REFERENCES

[1] J. P. Barton and D. G. Infield, "Energy storage and its use with intermittent renewable energy," *IEEE Trans. Energy Convers.*, vol. 19, no. 2, pp. 441–448, 2004.

[2] D. Stenclik, P. Denholm, and B. Chalamala, "Maintaining Balance: The Increasing Role of Energy Storage for Renewable Integration," *IEEE Power Energy Mag.*, no. december, pp. 31–39, 2017.

[3] H. I. Su and A. El Gamal, "Modeling and analysis of the role of energy storage for renewable integration: Power balancing," *IEEE Trans. Power Syst.*, vol. 28, no. 4, pp. 4109–4117, 2013.

[4] F. Blaabjerg, Z. Chen, and S. B. Kjaer, "Power Electronics as Efficient Interface in Dispersed Power Generation Systems," *Power Electron. IEEE Trans.*, vol. 19, no. 5, pp. 1184–1194, Sep. 2004.

[5] C. Baumann, R. Schuster, and A. Moser, "Economic potential of power-to-gas energy storages," in *International Conference on the European Energy Market, EEM*, 2013.

[6] M. Button, "An Advanced Photovoltaic Regulator Module Array," in *IECEC 96. Proceedings of the 31st Intersociety Energy Conversion Engineering Conference*, 1996, pp. 519–524.

[7] F. Xue, R. Yu, and A. Huang, "Fractional Converter for High Efficiency High Power Battery Energy Storage System," in *2017 IEEE Energy Conversion Congress and Exposition (ECCE)*, 2017, pp. 5144–5150.

[8] K. L. Jørgensen, M. C. Mira, Z. Zhang, and M. A. E. Andersen, "Review of High Efficiency Bidirectional dc-dc Topologies with High Voltage Gain," in *Universities Power Engineering Conference (UPEC), 2017 52nd International*, 2017.

[9] J. Everts, J. Van Den Keybus, F. Krismer, J. Driesen, and J. W. Kolar, "Switching control strategy for full ZVS soft-switching operation of a dual active bridge AC/DC converter," *Conf. Proc. - IEEE Appl. Power Electron. Conf. Expo. - APEC*, pp. 1048–1055, 2012.

[10] M. Nymand and M. A. E. Andersen, "High-efficiency isolated boost DCDC converter for high-power low-voltage fuel-cell applications," *IEEE Trans. Ind. Electron.*, vol. 57, no. 2, pp. 505–514, 2010.

[11] E. Wittenbreder, "Topology Selection by the Numbers," *Technical Witts, Inc.*, pp. 1–19, 2006.

[12] B. Carsten, "Coverter Component Load Factors; A Performance Limitation of Various Topologies," in *International PCI Conference on Power Conversion*, 1988, pp. 31–48.

[13] Z. Zhang, M. C. Mira, and M. A. E. Andersen, "Analytical Comparison of Dual-Input Isolated dc-dc Converter with an ac or dc Inductor for Renewable Energy Systems," in *Future Energy Electronics Conference and ECCE Asia (IFEEC 2017 - ECCE Asia), 2017 IEEE 3rd International*, 2017, pp. 659–664.

[14] M. C. Mira, A. Knott, and M. A. E. Andersen, "A Three-Port Topology Comparison for a Low Power Stand-Alone Photovoltaic System," in *IEEE Power Electronics Conference (IPEC-Hiroshima 2014 - ECCE-ASIA), 2014 International*, 2014, pp. 506–513.

[15] Ned Mohan, *First Course on Power Electronics and Drives.* ed. Minneapolis, USA, 2003.

The 2018 International Power Electronics Conference

Two-Stage Protection for Multi-Channel Power Electronic Converters Fed Large Asynchronous Hydro-Generating Unit

R.R.Semwal, Anto Joseph and Thanga Raj Chelliah, Senior Member IEEE
Hydropower Simulation Laboratory, Dept. of Water Resources Development and Management
Indian Institute of Technology Roorkee, India
E-mail: rrsem.dwt2015@iitr.ac.in, antoj.dwt2014@iitr.ac.in, thangfwt@iitr.ac.in

Abstract- Multi-channeled voltage source inverter (VSI) provided in rotor circuit of doubly fed induction machine (DFIM) ensures control of real and reactive power of the machine alongwith speed variation. VSI is significantly impacted during grid disturbances since DFIM is very sensitive to the grid disturbances as stator/rotor circuit is directly connected to the grid. This paper presents two stage protection circuit for the multi-channeled VSI since it is useful for the continuous operation of the unit in case of marginal grid disturbances. Control system for the dc link chopper circuit protection and crowbar protection is developed based on the dc link voltage of the VSI. In addition, it provides the activation status of the protection circuits during grid disturbances which is economically helpful for the plant authorities. Two-stage protection circuit for a 250 MW DFIM with five channel back-to-back three level VSI is simulated in Matlab/Simulink. A 2.2 kW DFIM is tested in the laboratory to validate the simulation results.

Keywords— doubly fed induction machine, pumped storage power plant, voltage source inverter, two stage protection.

I. INTRODUCTION

Pumped storage power plant has received a renewed attention for bulk energy storage in recent years due to: (i) large energy storage option (in MW range), (ii) ability to provide flexibility to the power system and load balancing. Worldwide, about 140 GW fixed speed PSPP employing synchronous machine has been installed. However, the efficiency of fixed speed system is reduced at partial generation/pumping. Therefore by employing variable speed PSPP, wherein the speed of the turbine is adjusted to get maximum efficiency for an available water head. Large variable speed PSPP (>200 MW) employing synchronous machines is not viable due to: (i) the requirement of power converters of higher rating corresponding to the rating of the machine, (ii) full sized converter is challengeable in terms of cost, size and area requirement in case of underground power house, (iii) rotating over full range of speed a hydro generating unit does not give much difference in efficiency compared to 10-15% speed variation in the unit. Because of these reasons, Variable speed PSPPs with DFIM have gained prominence all over the world (e.g. 400 MW Ohkawachi PSPP, Japan; 250MW Linthal PSPP, Switzerland) since they provide variable speed operation with reduced power converter rating and high dynamic stability [1], [2]. In India, four DFIM units with a unit capacity of 250 MW with the speed variation of -10.73% to +8.33% is under execution stage at Tehri Dam. Voltage source converter with Back-to-Back is preferred in rotor circuit of DFIM since it gives better current waveform and lesser THD. In addition, this converter is used for smooth starting of pump turbine and dynamic/regenerative braking of unit in pumping mode or generation modes [3], [4].

A. Problem Description and Importance of Present Work

From a study, a 250 MW variable speed hydro generating unit provides additional energy storage of 8.95% per year in comparison with synchronous machine based fixed speed PSPP with the speed variation of -10.73% to +8.33% (Tehri PSPP, India). In DFIM, power converters are easily affected by the grid disturbances due to the transformer action principle. Till date, crowbar protection circuit is used for power converters in hydro generators. Once the crowbar protection is enabled, (i) power converters are disconnected from the machine and DFIM acts as the conventional induction machine (squirrel cage induction machine) also it draws more reactive power from the grid and provides instability to the grid if it is weak. (ii) The sudden removal of power converters from the machine leads to the instability of the power network, especially in a sub-synchronous generation. Considering the aforementioned issues the machine need to be shut down immediately when the crowbar protection is enabled [5]-[8]. Even a small disturbance occurrence in the grid may lead to stoppage of the machine. But once the machine is shut down the re-starting the machine will take minimum 20 to 30 minutes, it leads to economic losses and furthermore frequent shutdown of the unit is not accepted by grid operators.

DFIM is also widely used in wind power generating systems (WPGS). The issues on protection of power converters (based on grid disturbances) is also applicable to WPGS [9], [10]. Both active crowbar and dc link chopper protection are available for WPGS. During the grid disturbances active crowbar protection is initially operated and dc link chopper protection is followed after certain time period (based on magnitude of dc link voltage) to reduce the reconnection timing of the unit (i.e. crowbar protection re-enabled) [11]. However, the rating of DFIM available in WPGS (maximum of 8MW) is much smaller in comparison with variable speed PSPP

The 2018 International Power Electronics Conference

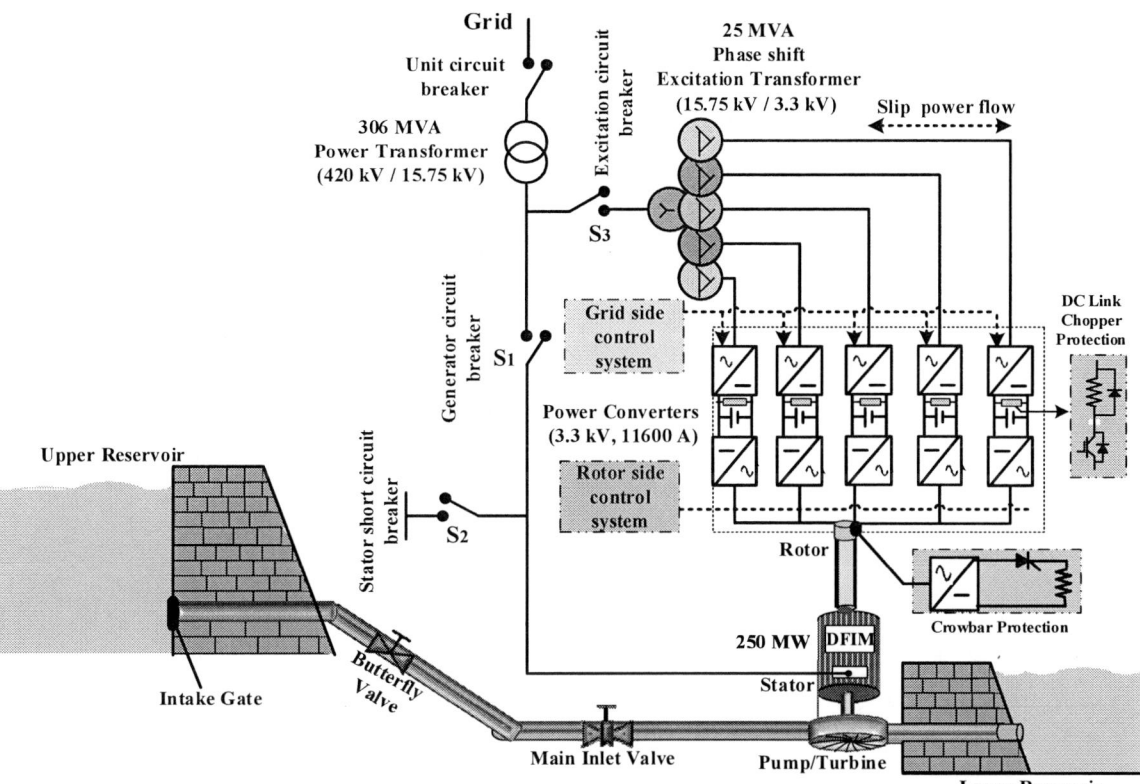

Fig. 1. Hydrological and electrical depiction of a 250 MW variable speed hydrogenerating unit

(e.g. 400MW). In addition, reactive power consumption of one machine during crowbar protection in wind farm does not make any impact on the stability of power grid. Unlike WPGS, DFIM is also operated in motoring mode in variable speed PSPP wherein the speed of the machine is significantly affected during grid disturbances.

Application of DFIM with back-to-back VSI in large capacity has been prominent and with this technology, Linthal variable speed PSPP, Switzerland (4 x 250 MW DFIM with the speed variation of ± 6% synchronous speed (500 rpm)) is recently commissioned (Dec 2015). Problems associated with power converter protection during grid disturbances are yet to be fully solved. This paper analyses the dc link chopper protection in variable speed PSP, as it is initially activated to withstand the dc link voltage (voltage sag/swell) during the grid disturbances so as to keep the machine in continuous operation. But if the fault is very severe then crowbar protection circuit will be enabled.

B. Organization of Paper

The script of paper is organized as follows - design of dc link chopper and crowbar protection is discussed in section II. Section III discusses the DFIM vector control system. To understand the activation of protection circuits, a 250MW DFIM is simulated in MATLAB/Simulink and it is discussed in section IV and concluding remarks are summarized in section V.

II. DESIGN OF DC LINK CHOPPER AND CROWBAR PROTECTION CIRCUIT

The theme of dc link chopper protection is to cut down

excess voltage in the converter during the grid faults. The crowbar protection is not activated as the fault is managed by dc link chopper so as to increase the continuity of machine service. Fig. 1 shows a hydraulic and electrical diagram of a DFIM based pumped storage plant with crowbar and dc link chopper protection.

A. DC Link Chopper Protection Circuit

DC link chopper circuit is connected in parallel to the dc link of the back-to-back power converter. It consists of power semiconductor device (IGBT/IEGT) connected in series with power resistances, an antiparallel diode connected across the resistance. During grid disturbances when the dc-link voltage exceeds a fixed threshold value the dc-link chopper inserts a power resistor in dc-link. Rating of the power resistor is estimated based on maximum power withstanding capacity of the converter and rotor current transients during grid disturbances [12], [13]. Further, maximum permissible voltage via chopper is determined by the lower limit of the resistance which is related to the rotor current transients during grid disturbances. Eqn. (1) and (2) helps the selection of resistance value used in dc link chopper circuit [14].

$$R_{chopper} = \frac{V_{dc}^{2}}{P_{max\_converter}} \qquad (1)$$

$$I_{chopper} \leq \frac{V_{dc\_max}}{R_{chopper}} \qquad (2)$$

The amount of resistance value engaged in chopper circuit is based on the rise of dc link voltage and the

1359

value of resistance engaged is controlled by applying proper PWM signal to power semiconductor device. Further, the power resistor is sized to withstand twice the maximum converter power rating.

B. Crowbar Protection Circuit

Crowbar circuit shortens the rotor windings during severe grid disturbance thereby limiting the rotor voltage and provides an additional path for the rotor current [11]. Crowbar circuit consists of diode rectifier followed by variable resistance power devices connected to make a short circuit. The value of variable resistance is selected depending upon the DC link clamp effect and to limit the short circuit current [15]. The estimation for crowbar resistance is given in eqn. (3).

$$R_{crowbar\_max} = \frac{\sqrt{2}V_{r(max)}\omega_e L'_s}{\sqrt{3.2V_s^2 - 2V_{r(max)}^2}} \quad (3)$$

$$L'_s = \frac{L_s + L_r L_o}{L_r L_o}$$

III. DFIM VECTOR CONTROL AND TWO STAGE PROTECTION SYSTEM

Real (speed) and the reactive power control of DFIM are carried out by the help of grid side converter (GSC) and rotor side converter (RSC) [16] - [19] as shown in Fig.A1. DC link chopper protection and crowbar protection circuit are controlled through the master control system which is based on dc link voltage (V_{dc}) and rotor current transients. During grid disturbances, the rise in dc link voltage is primarily limited by dc link capacitance. If the disturbance is severe then it is limited through dc link chopper circuit and unit is in continuous operation. Further, when the rise in dc link voltage is not limited with dc link chopper circuit then crowbar circuit is operated and unit gets into shutdown.

IV. RESULTS AND DISCUSSION

A. Simulation Results

In order to demonstrate the two stage protection circuit, a 250 MW, DFIM with 5 channel back -to- back IGBT based three level converters (5 MW each) is simulated in MATLAB/Simulink environment with the following settings: (i) RSC switching frequency of 300 Hz, (ii) GSC switching frequency of 500 Hz, (iii) V_{dc} is maintained at 5000 V, (iv) sampling time of $5e^{-4}$s. In the simulation, the machine is controlled to operate at rated power generation (250 MW) and energy storage (250 MW) at generation and pumping mode respectively. The value of dc link capacitance is selected as 18000μF to withstand the rotor currents. The value of power resistor used in dc link chopper and crowbar protection is about 0.2 ohms and 0.005 ohms respectively. DC link voltage is maintained at 5000 volts during normal operation and when grid disturbances occur the dc link capacitances limit the dc link voltage up to 5800 volts. If dc link voltage exceeds the above said voltage rating then dc link

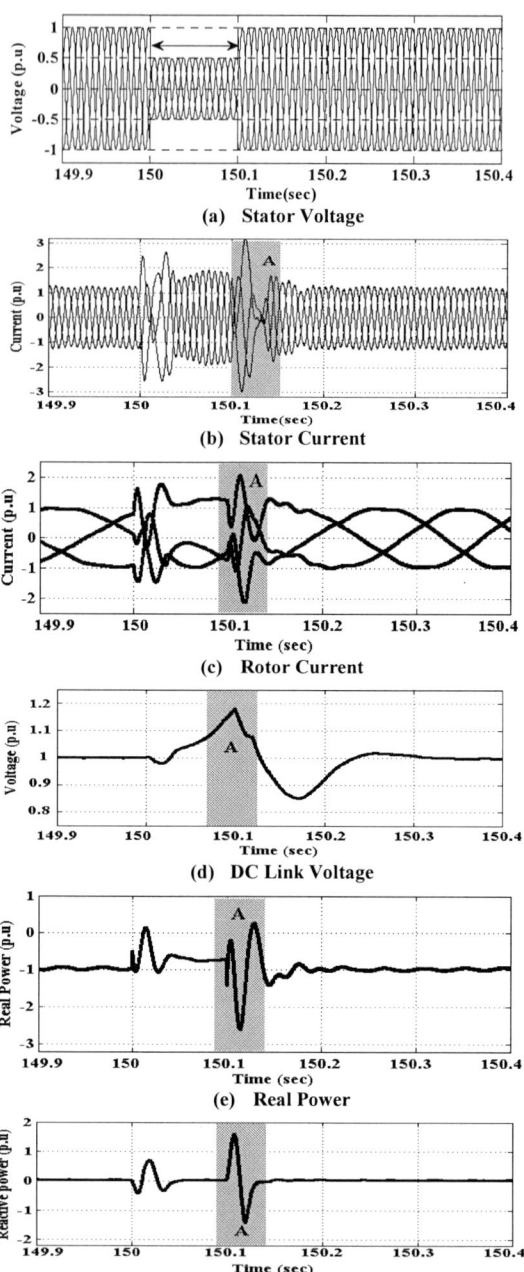

(a) Stator Voltage

(b) Stator Current

(c) Rotor Current

(d) DC Link Voltage

(e) Real Power

(f) Reactive Power

Fig.2. Momentary grid voltage sag fault for a 250 MW DFIM at generation mode (A- DC link voltage protection)

chopper circuit activates and control and limit till voltages reaches about 7200 volts. If the dc link voltage is more than 7200, crowbar protection will get activated and unit gets into shutdown. Momentary grid voltage sag (0.5 p.u depth, time duration 100 ms) is injected in supply grid between 150s and 150.1s at generation mode and results are shown in Fig.2. Further, Table I shows the detailed information about the activation of both dc link chopper and crowbar protection for the various voltage dip profiles. From the table, it is observed that the incorporation of dc link chopper protection leads to increase the continuous operation of the unit in several

The 2018 International Power Electronics Conference

Table 1. Two Stage Protection Response during Grid Disturbances

Mode	Grid Disturbances	Voltage Sag/Swell	DC Link Voltage Variation	DC Link Chopper Protection	Crowbar Protection	Machine Survivability	Economic Gain* (250 MW)
Generation Mode	Momentary Voltage Sag (5 cycles)	0.1 p.u	1.006 p.u	I	I	S	-
		0.2 p.u	1.021 p.u	I	I	S	-
		0.3 p.u	1.043 p.u	I	I	S	-
		0.4 p.u	1.075 p.u	I	I	S	-
		0.5 p.u	1.178 p.u	A	I	S	$ 90750/-
		0.6 p.u	1.236 p.u	A	I	S	$ 90750/-
		0.7 p.u	1.318 p.u	A	I	S	$ 90750/-
		0.8 p.u	1.385 p.u	A	I	S	$ 90750/-
		0.9 p.u	1.692 p.u	A	A	F	-
		1.0 p.u	1.882 p.u	A	A	F	-
	Momentary Voltage Swell (5 cycles)	0.1 p.u	1.027 p.u	I	I	S	-
		0.2 p.u	1.076 p.u	I	I	S	-
Pumping Mode	Momentary Voltage Sag (5 cycles)	0.1 p.u	1.009 p.u	I	I	S	-
		0.2 p.u	1.018 p.u	I	I	S	-
		0.3 p.u	1.034 p.u	I	I	S	-
		0.4 p.u	1.067 p.u	I	I	S	-
		0.5 p.u	1.194 p.u	A	I	S	$ 57750/-
		0.6 p.u	1.258 p.u	A	I	S	$ 57750/-
		0.7 p.u	1.363 p.u	A	I	S	$ 57750/-
		0.8 p.u	1.523 p.u	A	A	F	-
		0.9 p.u	1.621 p.u	A	A	F	-
		1.0 p.u	1.815 p.u	A	A	F	-
	Momentary Voltage Swell (5 cycles)	0.1 p.u	1.015 p.u	I	I	S	-
		0.2 p.u	1.070 p.u	I	I	S	-

DC Link Voltage - 5000 volts (1 p.u)   DC Link Chopper Protection - 5800 volts (1.16 p.u)   Crowbar Protection - 7200 volts (1.44 p.u)

I - Idle Mode   A - Active Mode   S - Survived   F- Failed

* Economic gain with introduction of dc link chopper circuit protection   *Unit Stoppage Time due to Grid Disturbances = 20 minutes

*Financial analysis considered by taking $ 0.11/kWh for generation mode and $ 0.07 for pumping mode

* Grid disturbance occurs 10 times/year

voltage dips and brings more economical benefits. From Fig.2, it is inferred that the stator and rotor phase transient currents increases to 3.1 p.u and 2.1 p.u respectively and leads to saturate the converter. Consequently, dc link voltage of the back-to-back converter increases to 1.17 p.u. This dc link voltage could activate the dc link chopper circuit and unit is in continuous operation during grid disturbance. The amount of resistance variation during the voltage dip is based on the value of dc link voltage. From the test results, it is observed that: (i) instability in both real and reactive power of the machine, (ii) speed of the machine increases during generation mode whereas decreases in pumping mode, (iii) rotor voltage of the machine get disturbed and increases to above the rated value, (iv) during dc link chopper circuit protection operation both GSC and RSC control system are in control.

B. Experimental Validation

A 2.2 kW DFIM with three level neutral point clamped converter (25 A, 425 V) and DS1202 real time controller set-up (shown in Fig. 3) is used to validate simulation results. The control algorithm for controlling grid and rotor side converter is designed in MATLAB/Simulink environment and it is dumped in real time controller. The switching frequency for both GSC and RSC is selected as 1650 Hz. The current, voltage and speed sensor signals are integrated to ADC channel in real time controller for the controlled real and reactive power delivery. Programmable power supply is used to apply the voltage dip profiles to the stator circuit of DFIM. All the rotor phases are connected together through a resistor and contactor for the crowbar circuit operation. The contactor is controlled by the software enabled trigger based on the

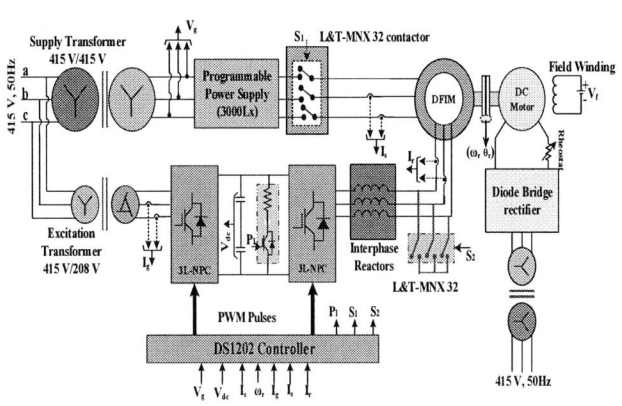

Fig. 3. (a) Experimental Set-up, (b) Experimental block diagram

1361

value of dc link voltage. Sample time is considered as 0.001s. The value of dc link capacitance is selected as 2200µF to withstand the rotor currents. The value of power resistor used in dc link chopper and crowbar protection is about 100 ohms and 14 ohms respectively. DC link voltage is maintained at 325 volts during normal operation and when grid disturbances occur the dc link capacitances limit the dc link voltage up to 361 (1.11 p.u) volts. If dc link voltage exceeds the above said voltage rating then dc link chopper circuit activates and control and limit till voltages reaches about 423 (1.3 p.u) volts. If the dc link voltage value is more than 423 (1.3 p.u) volts, crowbar protection will get activated and the unit gets into shutdown.

DFIM is operated to deliver 0.91 p.u (2 kW) power to grid at 0.9 p.u (1350 rpm) speed. Rotor current (machine side) is rotating at slip frequency of 5 Hz. Momentary grid voltage sag (0.5 p.u depth, time duration 40 ms) is injected in supply grid between 77.9s and 77.98s at generation mode. Fig. 4 shows the test results for the DFIM unit for the without and with dc link chopper protection.

From Fig.4a (i), it is observed that during the momentary voltage dip (0.7 p.u depth for 80 ms), dc link voltage increases up to 1.23 p.u, shown in Fig. 4e (i) and leads to activate the crowbar protection circuit. From the test results, it is observed that the stator, rotor currents and real, reactive power of the DFIM unit gets oscillated as similar to the simulation results. Fig.4a (ii) show the test results for the incorporation of dc link chopper protection in the unit, it shows that the current and power transients are as similar to the case of without dc link chopper protection. However, once the dc link voltage reach more than 1.11 p.u, dc link chopper protection is activated and correspondingly limit the dc link voltage of the back-to-back converter, and the machine is in continuous operation.

V. CONCLUSION

The paper has presented the performance of the DFIM machine during grid disturbances. Also it briefly summarized both the dc link chopper and crowbar protection circuits with the selection of power resistors. Triggering of both dc link chopper and crowbar protection circuit is analyzed during grid disturbances. It makes the machine survivability possible during grid disturbances and result in more economical benefits to the plant operators.

ACKNOWLEDGEMENT

Authors thank Central Power Research Institute, Government of India funding this research work, along with a research fellowship. Data supplied by THDC India Limited for this research is also gratefully acknowledged.

REFERENCES

[1] A. Joseph, T. R. Chelliah, "A Review of Power Electronic Converters for Variable Speed Pumped Storage Plants: Configurations, Operational Challenges and Future Scopes," in IEEE J. Emerg. Sel. Topics Power Electron, vol. 6, no. 1, pp. 103-119, March 2018.

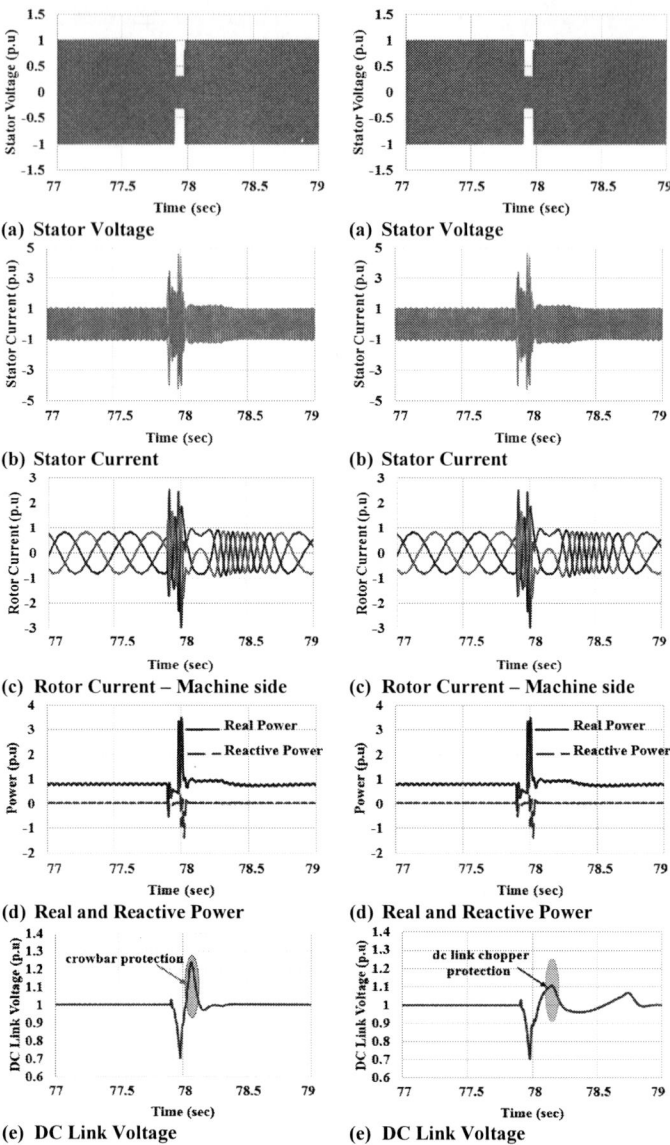

(a) Stator Voltage (a) Stator Voltage

(b) Stator Current (b) Stator Current

(c) Rotor Current – Machine side (c) Rotor Current – Machine side

(d) Real and Reactive Power (d) Real and Reactive Power

(e) DC Link Voltage (e) DC Link Voltage

(i) Wihout dc link protection   (ii) With dc link protection

Fig.4. Momentary grid voltage sag fault (0.7 p.u dip for 80 ms) for a 2.2 kW DFIM at generation mode

[2] T. S. Kuwabara, A. Furuta, H. Kita, and E. Mitsuhashi, "Design and dynamic response characteristics of 400MW adjustable speed pumped storage unit for Ohkawachi power station," IEEE Trans. Energy Convers, vol. 11, no.2, pp. 376-384, June 1996.

[3] J. K. Lung, Y. Lu, W.L. Hung, and W.S. Kao, "Modeling and dynamic simulations of doubly fed adjustable-speed pumped storage units," IEEE Trans. Energy Convers, vol. 22, no.2, pp. 250-258, June 2007.

[4] A. Joseph, K. Desingu, R.R. Semwal, T.R.Chelliah, and D. Khare, "Dynamic performance of pumping mode of 250 MW variable speed hydro-generating unit subjected to power and control circuit faults", IEEE Trans. Energy Convers, vol. 33, no. 1, pp. 430-441, March 2018.

[5] I. Erlich, H. Wrede, and C. Feltes, "Dynamic behaviour of DFIG-based wind turbines during grid faults," presented at IEEE Power Convers. Conf. (PCC 2007), Nagoya, Japan.

[6] T. Kawady, C. Feltes, I. Erlich, and A. I. Taalab, "Protection system behavior of DFIG based wind farms for grid-faults with practical considerations," in Proc. IEEE Power Energy Soc. General Meeting, Minneapolis, Minnesota, USA, Jul. 2010, pp. 1–6.

[7] T. Kawady, H. Shaaban, and A. El-Sherif, "Investigation of grid-support capabilities of doubly fed induction generators during grid faults," in Proc. IET Conf. Renewable Power Generation (RPG 2011), Edinburgh, U.K., Sep., pp. 1–7.

[8] D. J. Atkinson, G. Pannell, W. Cao, B. Zahawi, T. Abeyasekera, and M. Jovanovic, "A doubly-fed induction generator test facility for grid fault ride-through analysis," IEEE Instrum. Meas. Mag., vol. 15, no. 6, pp. 20–27, Dec. 2012.

[9] V. Yaramasu, B. Wu, P.C. Sen, S. Kouro, and M. Narimani, "High-power wind energy conversion Systems: State-of-the-art and emerging technologies," Proc. IEEE, vol. 103, no. 5, pp. 740-788, May 2015.

[10] F. Blaabjerg, and K. Ma, "Future on power electronics for wind turbine systems," IEEE J. Emerg. Sel. Topics Power Electron. vol. 1, no.3, pp. 139-152, September 2013.

[11] G. Pannell, D. J. Atkinson, and B. Zahawi, "Analytical study of grid-fault response of wind turbine doubly fed induction generator," IEEE Trans. Energy Convers., vol. 25, no. 4, pp. 1081–1091, Dec. 2010.

[12] G. Pannell, B. Zahawi, D. J. Atkinson, and P. Missailidis, "Evaluation of the performance of a dc-link brake chopper as a DFIG low-voltage fault-ride-through device," IEEE Trans. Energy convers., vol. 28, no. 3, pp. 535- 542, Sep 2013.

[13] C. Wessels, F. Gebhardt, and F. W. Fuchs, "Fault ride-through of a DFIG wind turbine using a dynamic voltage restorer during symmetrical and asymmetrical grid faults," IEEE Trans. Power Electron., vol. 26, no. 3, pp. 807–815, Mar. 2011.

[14] K. E. Okedu, S. M. Muyeen, R. Takahashi, and J. Tamura, "Wind farms fault ride through using DFIG with new protection scheme," IEEE Trans. Sust. Energy, vol. 3, no. 2, pp. 242–254, Apr. 2012.

[15] G. Pannell, D. Atkinson, and B. Zahawi, "Minimum-threshold crowbar for a fault-ride-through grid-code-compliant DFIG wind turbine," IEEE Trans. Energy Convers, vol.25, pp.750-759, September 2010.

[16] X. Yuan, J. Chai, and Y. Li, "A converter-based starting method and speed control of doubly fed induction machine with centrifugal loads," IEEE Trans. Ind. Appl., vol. 47, no. 3, pp. 1409- 1418, June 2011.

[17] R. Pena, J.C. Clare, and G. M. Asher, "Doubly fed induction generator using back-to-back PWM converters and its application to variable speed wind-energy generation," IEEE Proc. Electric. Power Appl., vol. 143, no. 3, pp. 231- 241, May 1996.

[18] A. Joseph, R. Selvaraj, T.R.Chelliah, and A. Sarma, "Starting and Braking of a Large Variable Speed Hydro-Generating Unit Subjected to Converter and Sensor Faults", IEEE Trans. Ind. Appl., (Accepted).

[19] X. Sun, YS. Lee, and D. Xu, "Modelling, analysis, and implementation of parallel multi-inverter system with instantaneous average-current-sharing scheme," IEEE Trans. Power Electron., vol.18, no. 3, pp.844-856, May 2003.

APPENDIX I: CONTROL DIAGRAM OF MULTI-CHANNEL VSI FED DFIM

APPENDIX II: MACHINE DATA, PLEASE REFER [4]

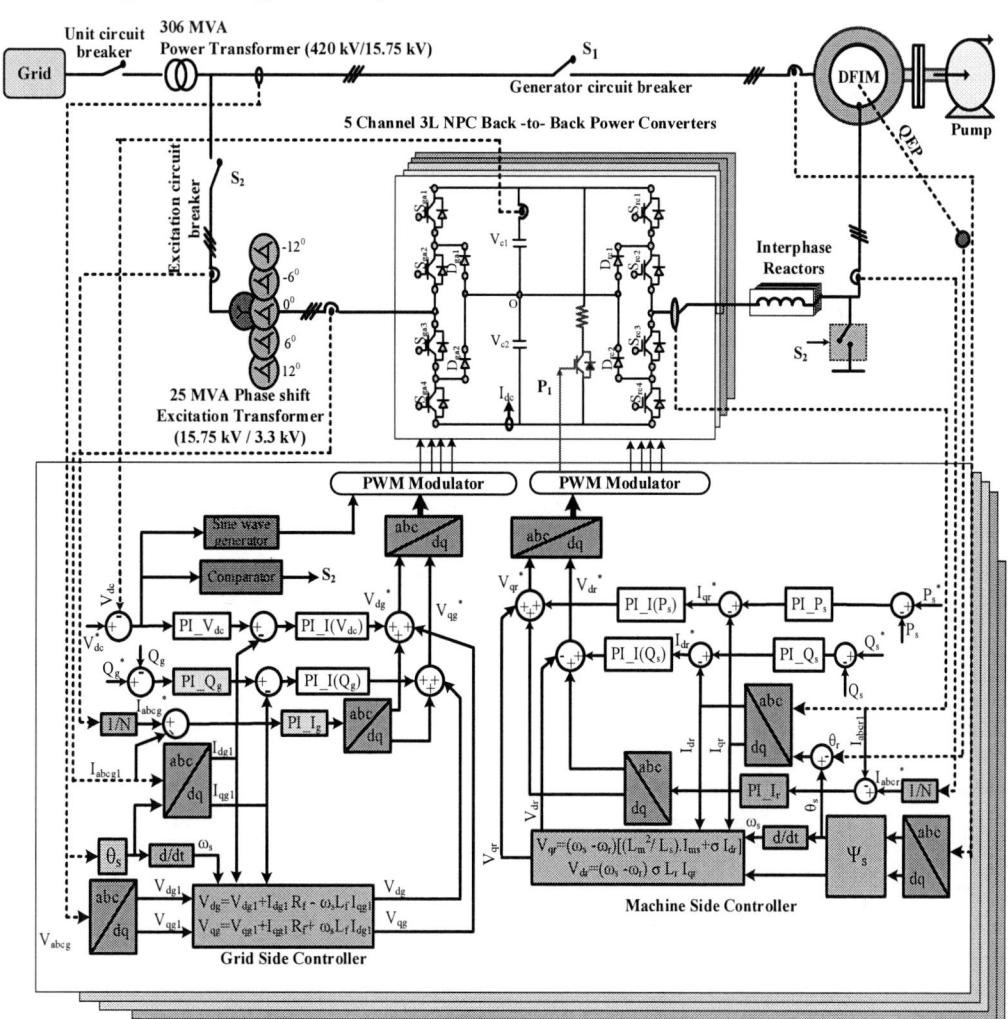

Fig. A1. Control diagram for multi-channel VSI fed DFIM

Current Sharing Control for Series-Parallel Changeover using Battery and Electric Double-Layer Capacitor Bank

Taisei Nishino, Keisaku Isozaki, Naoki Kogai and Kyungmin Sung*
Depart. of Electrical Electronic System Eng., National Institute of Tech.,
Ibaraki College, Hitachinaka, Japan
*E-mail: sung@ee.ibaraki-ct.ac.jp

Abstract— We previously proposed a series-parallel changeover system for connecting an electric double-layer capacitor (EDLC) bank to the main battery, aiming to improve the performance of electric vehicles. In this paper, we propose a control method for output current sharing by connecting the EDLC bank in parallel to the battery. The proposed method can control the ratio of output current in both components to perform the sharing operation. Therefore, the stored energy in the EDLC bank can be used when required by controlling its output current. Experimental results verify the parallel operation with suitable current sharing and suggest the applicability of the method to providing power assistance for light electric vehicles.

Keywords— *electric double-layer capacitor, light electric vehicles, current control.*

I. INTRODUCTION

In recent years, compact light electric vehicles (LEVs) have attracted attention given their low production and running costs, and eco friendly characteristics. In fact, the production of LEVs is rapidly increasing in many countries [1, 2]. Simultaneously, the adoption of a super capacitor known as electric double-layer capacitor (EDLC) used in the energy storage system of LEVs is increasing with the evolution of the electric vehicle [3-6]. Moreover, hybrid energy storage systems, composed of an EDLC bank and a main energy storage unit, which is generally a lead-acid battery, are being developed given the high capacitance, low internal resistance, and rapid charge/discharge characteristics of EDLCs.

In an effort to further improve the performance of LEVs, the regenerative braking energy of the motor can be stored for its subsequent use for locomotion. Hence, the EDLC is convenient for this type of storage given its ability to withstand fast and frequent charge and discharge operations and its high responsiveness when compared to conventional batteries. Nevertheless, EDLCs can provide only 2.5V cell, and hence a bank obtained by connection multiple EDLC cells in series should be used to deliver the voltage level of most electric-vehicle batteries. Furthermore, a bidirectional DC-DC converter is required to control the terminal voltages and power flow of the

EDLC bank, the main battery, and the inverter DC link. Several methods are available to integrate an EDLC bank and a battery using such converters [7-9]. However, these proposed methods only consider the EDLC bank connected in parallel with the battery to mitigate sudden electrical variations that can occur in most applications.

The capacitance of an EDLC bank consisting of several series-connected cells decreases proportionally to the number of cells, and hence, as few as possible cells should be used to guarantee a high capacitance. In addition, when the EDLC bank is at the low voltage side, the DC-DC converter exhibits low efficiency given the high boost ratio, and stored energy below a required threshold to perform voltage boost cannot be used.

To efficiently use EDLC banks, we have previously developed a series-parallel changeover system intended for LEVs containing a battery, an EDLC bank, a bidirectional DC-DC converter, an inverter, and brushless DC (BLDC) motor [10,11]. The system aims to change the power source configuration of the EDLC bank and battery to either parallel or series connection depending on the driving requirements, and appropriately and efficiently use the regenerative braking energy stored in the EDLC bank.

In this paper, we propose a control method for output current sharing that establishes a parallel connection between the EDLC bank and battery in the changeover system, and adjusts a their ratio of output current. Moreover, the EDLC bank voltage determines its output power for saving battery energy. Experiments verify the operation and validate the proposed control method. The proposed combination of parallel and series configurations can provide further assistance in power management and efficiency for LEVs.

II. ENERGY STORAGE AND SUPPLY SYSTEM

A bidirectional DC-DC converter is usually required for a system integrating EDLC bank and battery to provide an interface between the EDLC bank and the inverter DC link. In this section, we first describe the conventional system using a bidirectional DC-DC converter, and then detail the operation of the proposed series-parallel changeover system.

The 2018 International Power Electronics Conference

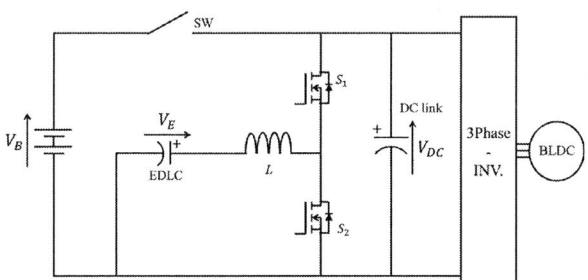

Fig. 1. Energy storage system using bidirectional DC-DC converter.

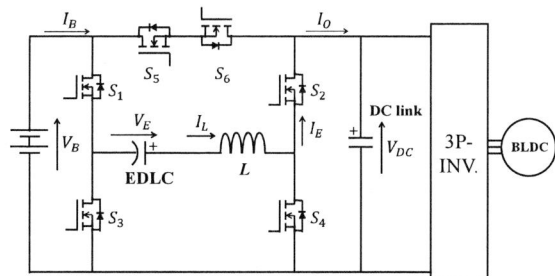

Fig. 3. Proposed series-parallel changeover system.

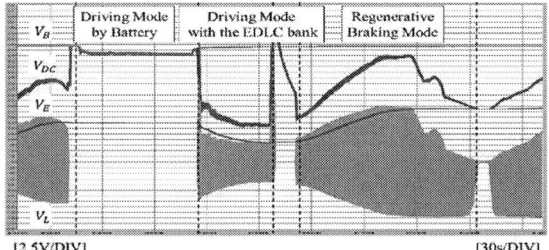

[2.5V/DIV]   [30s/DIV]

Fig. 2. Measurements of operating bidirectional DC-DC converter.

A. Conventional DC-DC Converter System

Fig. 1 shows the basic configuration of a system containing battery and an EDLC bank connected in parallel to a DC-link bus. The EDLC bank is connected to the DC link by the bidirectional DC-DC converter composed of S_1 and S_2, and inductance L. The operation of this system comprises three modes, namely, driving mode using the battery, regenerative braking with the EDLC bank, and driving mode using the EDLC bank.

Fig. 2 shows measurements of the operation using this energy storage and supply system while driving an LEV. In the driving mode using the battery, switch SW is in the ON state, and the bidirectional DC-DC converter does not operate. During regenerative braking with the EDLC bank, SW is in the OFF state, and the bidirectional DC-DC converter operates in buck mode through switch S_1 for the motor to provide energy and charge the EDLC bank. In the driving mode using the EDLC bank, only the bidirectional DC-DC converter operates in boost mode through switch S_2 for the EDLC bank to supply energy to the DC link. However, the controllable voltage range of the EDLC bank is restricted in this mode given the low conversion efficiency, which is caused by the high transfer voltage ratio and saturation of L by the high inductance current. Consequently, limited rations between three and four can be obtained for conversion of the EDLC-bank voltage, but the charged energy that cannot be used given insufficient voltage in boost mode remains in the bank.

B. Series-parallel Changeover System

To improve the operating voltage range of the EDLC bank, we proposed a system to connect the battery and EDLC bank in either parallel or series [10]. Specifically, when the bidirectional DC-DC converter cannot be controlled because a high voltage boost ratio is required, the proposed system switches the connection between the

battery and EDLC bank from parallel to series, such that all the stored energy in the EDLC bank can be supplied to the DC link.

Fig. 3 shows the circuit of the proposed system that extends the conventional configuration by including four additional MOSFETs, two of which replace switch SW in Fig. 1. Specifically, the battery is connected to the DC link via bidirectional switches S_5 and S_6, and the EDLC bank connects to the DC link by the bidirectional DC-DC converter composed of switches S_2 and S_4, and inductance L. In addition, switch S_1 connects the battery and EDLC bank in series, whereas S_3 is used during regenerative braking. The system operation and its different modes are detailed in the sequel.

Mode 1: Battery Driving

In the driving mode using only the battery, the bidirectional switch consisting of S_5 and S_6 is in the ON state. Hence, the battery is directly connected to the DC link whose voltage becomes

$$V_{DC} = V_B. \tag{1}$$

Mode 2: EDLC Bank Driving

This mode can be activated only when sufficient power is available in the EDLC bank, and In this driving mode, the converter operates in boost mode through S_4, whereas the battery is disconnected from the DC link. Then, V_{DC} is given by Eq. (2), where D_{S4} denotes the duty ratio of S_4 in Eq. (3).

$$V_{DC} = \frac{1}{1-D_{S4}} V_E \tag{2}$$

$$D_{S4} = \frac{T_{ON}}{T_{ON}+T_{OFF}} \tag{3}$$

Mode 3: Parallel Battery EDLC Bank Driving

In this mode, the parallel connection between the battery and EDLC bank is achieved with S_5 at the ON state and the converter operating in boost mode through S_4, where voltage of V_{DC} is expressed by Eq. (4), and the current supplied to the DC link is given by Eq. (5). Fig. 4 illustrates the operation during this mode, where the DC-link input current is supplied by both the battery and EDLC bank and a diode voltage drop occurs at S_6. In addition, the stored energy in the EDLC bank contributes to save energy of the battery.

$$V_{DC} = (V_B - V_{D-S6}) \| \frac{V_E}{1-D_{S4}} \quad (4)$$

$$I_O = I_B + I_E \quad (5)$$

Mode 4: Series Battery EDLC Bank Driving

In the driving mode using the series connection between the battery and EDLC bank, only S_1 is at the ON state. Therefore, the battery and EDLC bank contribute to V_{DC} with their respective voltage V_B and V_E, as expressed in Eq. (6), where V_{D-S2} is the voltage drop at S_2. In addition, the series connection makes the same current flow through all the elements indicated in Fig. 5. Likewise, the output voltage of the inverter, which is the terminal voltage for the BLDC motor, increases according to the EDLC bank voltage. Hence, after S_1 turns on, the speed and torque of the BLDC motor increase, thus configuring this mode into a power assistance feature.

$$V_{DC} = V_B + V_E - V_{D-S2} \quad (6)$$

Mode 5: Regenerative Braking for EDLC bank

In this mode, only S_3 is at the ON state and the bidirectional DC-DC converter operates in buck mode through S_2. Consequently, the braking energy flows into the EDLC bank, and V_E is given by Eq. (7), where D_{S2} is the duty ratio of S_2, analogous to that of S_4.

$$V_E = D_{S2} V_{DC} \quad (7)$$

Mode 6: Regenerative Braking for Battery

In this mode, only S_6 is at the ON state, and thus the braking energy flows into the battery for charging.

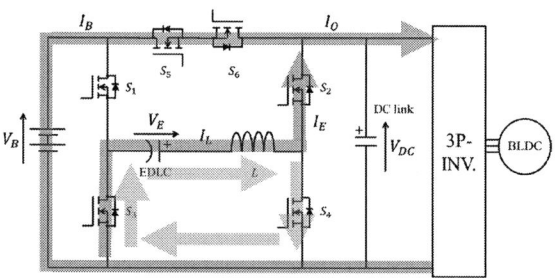

Fig. 4. Current flow during parallel connection of the battery and EDLC bank (Mode 3).

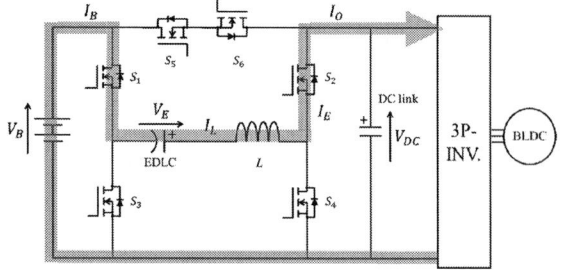

Fig. 5. Current flow during series connection of the battery and EDLC bank (Mode 4).

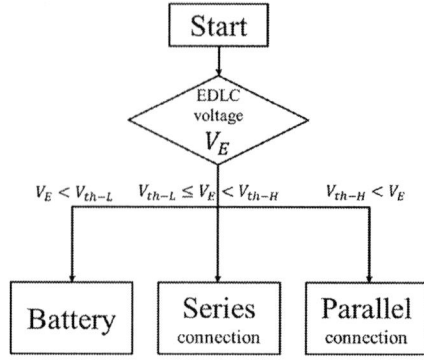

Fig. 6. Series-parallel changeover control.

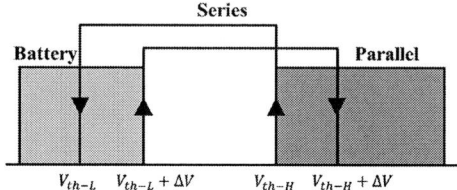

Fig. 7. Change of mode according to voltage of EDLC bank.

III. CONTROL METHOD

A. Selection of Connection Type

In the proposed changeover system, we monitor the EDLC bank voltage, and the operation mode is decided accordingly, as illustrated in Fig. 6 and 7. Threshold voltage V_{th-H} determines the changeover between parallel and series connection, whereas threshold voltage V_{th-L} determines the changeover from series connection to battery mode. Specifically, when EDLC bank voltage V_E is high than V_{th-H}, the control method applies the parallel driving mode. Hence, the output current of the DC-DC converter is supplied from both the battery and EDLC bank, which can share the output current of the battery. Hence, current sharing control is required to regulate the output current in this mode.

Next, when V_E is lower than V_{th-H}, S_1 turns on, thus series connecting the EDLC bank and the battery. Then, this series mode changes to the battery mode when V_E is lower than V_{th-L}, whose value is selected based on the voltage limit during boost conversion. Moreover, we set hysteresis band $\pm\Delta V$ for each threshold voltage to avoid undefined connection mode and jittering. Mode shifting from parallel to series and to can be automatic, but the reverse shifting should be chosen by the driver according to the driving conditions of the LEV.

B. Current Sharing Control

During parallel driving mode, we employ output current sharing control method to regulate the battery and EDLC bank output currents. Fig. 8 shows the block diagram of the proposed control method. The input current to the DC link I_O is the sum of the output currents from the EDLC

The 2018 International Power Electronics Conference

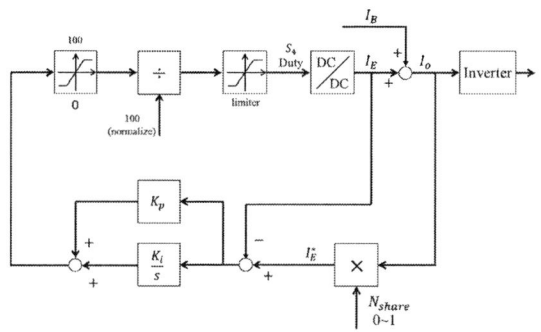

Fig. 8. Diagram of current sharing control.

bank I_E and battery I_B. The current sharing reference, N_{share}, determines ratio between I_E and I_B, and sets output current reference I_E^* of the EDLC bank as $N_{share} \cdot I_O$. Then, I_E is regulated to I_E^* by a PI controller, and the duty ratio of S_4 is controlled accordingly.

IV. EXPERIMENTAL RESULTS AND DISCUSSION

To verify the operation of the proposed system and control method, we performed experiments using an LEV and the experiment specifications listed in Table I. First, we verified the current sharing control during parallel driving mode. The EDLC bank was charged up to a voltage above V_{th-H}, and we varied sharing ratio N_{share} is changed between 0% to several ratio continuously. Fig. 9 shows the results from this experiment, with the BLDC motor speed being approximately constant at 800 rpm with approximate output power of 500 W.

From top to bottom, the graphs in the figure show DC-link voltage V_{DC}, EDLC terminal voltage V_E, inverter input current I_O, EDLC output current I_E, and battery output current I_B. Given that the inverter provides approximately 500 W to the BLDC motor, its input current is close to 13 A. Moreover, the results confirm that the EDLC bank constantly shares current at the intended value N_{share}. Even for step variations of N_{share}, no sudden variations in the DC-link voltage and current occur, and the BLDC motor rotates at an almost constant speed. The output current of the battery varies to compensate the EDLC bank current, with both currents being equal at 50% sharing ratio. In addition, the EDLC bank does not supply current at 0% sharing ratio, it indicating operation in battery mode.

Fig. 10 shows the detailed voltage and current during current sharing at ratio N_{share} of 50%, from where boost operation can be confirmed. These results confirm the appropriate operation of the proposed sharing current control that allows specifying any sharing ratio during driving through the parallel connection between the EDLC bank and battery.

Fig. 11 shows waveforms of the system operation at a 40% sharing ratio with irregular variations of the load on the motor. When the load of the BLDC motor varies, the inverter, EDLC, and battery currents, as well as the speed of the motor vary accordingly.

TABLE I
EXPERIMENT SPECIFICATIONS

Motor	BLDC motor (1 kW, 3 phases, 16 poles)
Battery	48 V - 7.5 Ah (4 series-connected 12 V - 7.5 Ah batteries)
EDLC bank	77.8 F, 22.5 V (9 series-connected 2.5 V – 350 F cells)
DC-link capacitor	5000 µF, 150 V
Inductance L	14 mH or 40 mH
Switching frequency	10 kHz
V_{th-H}	14 V
V_{th-L}	1 V

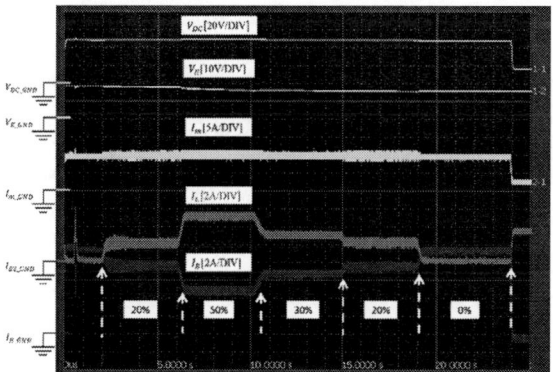

Fig. 9. Voltage and current during current sharing.

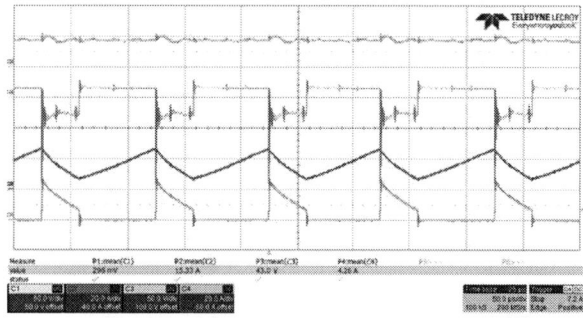

Fig. 10. Voltage and current during current sharing at a 50% ratio. (C1: V_{DC}, C2: V_L, C3: I_L, and C4: I_E ; 50µs/div)

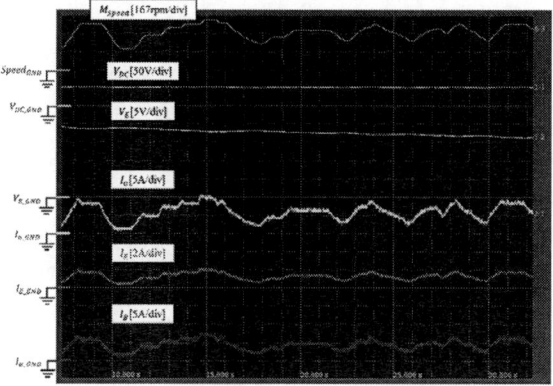

Fig. 11. System behavior under varying load.

1367

The 2018 International Power Electronics Conference

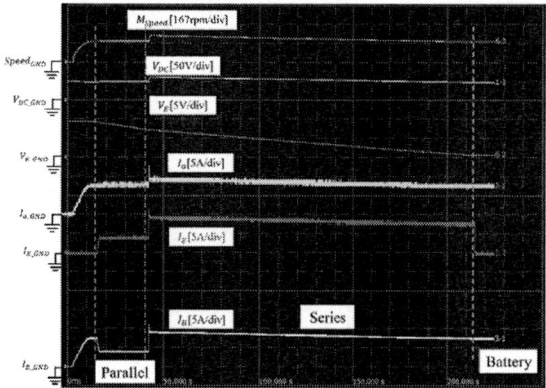

Fig. 12. System behavior during mode switching.

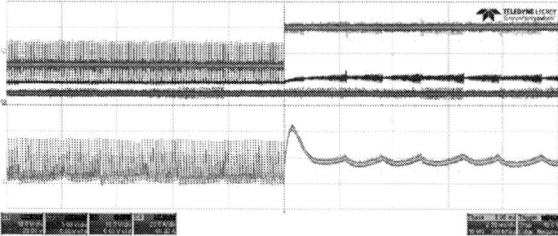

Fig. 13. System waveforms during changeover from parallel to series mode. (C1: gate signal of S_1, C2: V_B, C3: V_{DC} ,and C4: I_O ; 2 ms/div)

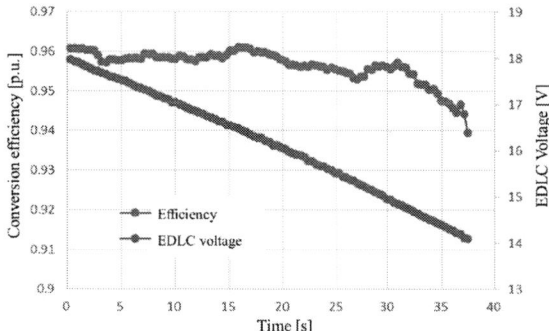

Fig. 14. Conversion efficiency during parallel mode for N_{share} of 30%.

In contrast V_{DC} exhibits an almost constant value. In addition, the average inverter input current is 6.79 A, the average EDLC current is 2.52 A, and the average battery current is 4.39 A, thus verifying that the EDLC bank shares about 40% of the load current and suggesting that the proposed control method stably follows load fluctuations when the LEV drives over rough surfaces.

Fig. 12 shows the results for constant output power and BLDC motor, where the proposed system switches from a battery to parallel and then to series mode. When the energy source changes from the battery to the parallel connection of the battery and EDLC bank, the input current to the inverter increases, and both V_{DC} and the motor speed also increase slightly, whereas V_E gradually decrease as the EDLC bank discharges in during operation in parallel mode. Then, the system changes from parallel

to series mode when the EDLC-bank voltage reaches 14 V. At this point, V_{DC} rapidly increases from 48 V to approximately 62 V along with an increase in the input current and BLDC motor speed, as the available EDLC-bank voltage is added to the battery voltage. Finally, when the EDLC voltage decreases below 1 V, the system switches to battery mode.

During operation in series mode, when the LEV either enters a steep slope or passes over an obstacle, ration N_{share} can reach up to 70% during a short period, and then the system switches to series mode for boosting the output power. Moreover, almost all the energy stored in the EDLC bank can be supplied to the motor. Fig. 13 shows waveforms of the system when it switches from parallel to series mode. Although a sudden inrush current is generated during switching, it can be limited by properly selecting threshold V_{th-L}, inductance L, and the switching of S_2.

The conversion efficiency of the proposed system reaches 96% during battery mode. Given that switches S_5 and S_6 are ON state simultaneously during this mode, the power loss is only related to the conduction of these two MOSFETs. The conversion efficiency during series mode is 88% in average for EDLC-bank voltage from 14 V down to 1 V. Fig. 14 shows the efficiency during parallel mode, where the maximum efficiency of 96% is reached when the EDLC bank has a high voltage level, and as the voltage decreases and the boost ratio increases, the conversion efficiency decreases. The abovementioned efficiency values correspond to the efficiency for the series-parallel changeover proposed in this paper, disregarding the efficiency of the inverter and BLDC motor. Overall, the operation modes retrieved efficiency value around 90%.

V. CONCLUSIONS

In this paper, we propose a series-parallel changeover system with the current sharing control to manage power contribution of battery and EDLC bank in an LEV. By controlling the EDLC-bank output current, the currents of the battery can be adjusted as necessary. In addition, the EDLC bank can be connected in parallel to the battery for energy saving, and when the LEV requires a momentary high power, the proposed control method applies a high current sharing ratio for the EDLC bank to assist the BLDC motor. We verified each operation mode and the changeover in the proposed system through field tests. The results suggest that the power assistance capability and energy saving of the battery energy can improve the performance of LEVs.

REFERENCES

[1] C. Sachs, S. Burandt, S. Mandelj, and R. Mutter, "Assessing the market of light electric vehicles as a potential application for electric In-wheel drives," *2016 6th International Electric drive Production Conference (EDPC)*, pp. 280-285, 2016.

[2] Jonatan J. G. Vilchez, P. Jochem, and W. Fichtner, "EV market development pathways – an application of system dynamics for policy simulation," *EVS27 International Battery, Hybrid and Fuel Cell Electric Vehicle Symposium*, 2013.

[3] S. Chung and O. Trescases, "Hybrid energy storage system with active power-mix control in dual-chemistry battery pack for light

electric vehicles," *IEEE Trans., transportation Electrification*, vol. 3, no. 3, Sep., 2017.

[4] X.D. Xue, K.W.E. Cheng, R. Raman S, J. Chan, J. Mei, and C. D. Xu, "Performance prediction of light electric vehicles powered by body-integrated super-capacitors," *2016 International conference on Electrical systems for Aircraft, Railway, Ship Propulsion and Road Vehicle & International Transportation Electrification Conf.*, pp. 1-6, 2016.

[5] M. Bertoluzzo and G. Buja, "Development of electric propulsion system for light electric vehicle," *IEEE Trans. Industrial Informatics*, vol. 7, no. 3, pp. 428-435, 2011.

[6] O. Laldin, M. Moshirvaziri, and O. Trescases, "Optimal power flow for hybrid ultracapacitor system in light electric vehicles," *IEEE Energy Conversion Congress and Exposition*, pp. 2916-2922, 2011.

[7] M. B. Camara, H. Gualous, F. Gustin, A. Berthon, and B. Dakyo, "DC/DC converter design for supercapacitor and battery power management in hybrid vehicle applications-polynominal contril strategy," *IEEE Trans. Industrial Electron.*, vol. 57, no. 2, pp. 587-597, Feb. 2010.

[8] J. Cao and A. Emadi, "A New Battery/Ultra Capacitor Hybrid Energy Storage System for Electric, Hybrid, and Plug-In Hybrid Electric Vehicles", *IEEE, Trans. Power Electron.*, vol27, NO.1, JAN 2012.

[9] Fangcheng. Liu, Jinjun. Liu, and Linyuan. Zhou; "A Novel Control Strategy for Hybrid Energy Storage System to Relieve Battery Stress", *2010 2nd IEEE International Symposium on Power Electronics for Distributed Generation Systems*, pp.929-934, June 2010

[10] K. Sung, N. Watanabe, and K. Kawamura, "Experimental results of series or parallel changeover system using battery with EDLC for Electric vehicle," *2015 9th International Confeternce on Power Electronics and ECCE Asia* (ICPE-ECCE Asia), 2015.

[11] K. Hata, N. Watanabe, and K. Sung, "A Series or Parallel Changeover System Using Battery with EDLC for EV," *Power Electronics and Applications (EPE), 2013 15th European Conference on*, 2013.

Control Method of Energy Storage System to Improve Output Power of PCS

Mikiya Ishibashi[1*], Hitoshi Haga[1], Kenji Arimatsu[2] and Koji Kato[3]

1 Nagaoka University of Technology, 1603-1 Kamitomioka-cho, Nagaoka, Niigata, Japan
2 Tohoku Electric Power Co., Inc, 7-2-1 Nakayama Aoba-ku, Sendai, Miyagi, Japan
3 Sanken Electric Co., Ltd., 677 Shimoakasaka-Ohnohara, Kawagoe, Saitama, Japan
*E-mail: Mikiya_Ishibashi@stn.nagaokaut.ac.jp

Abstract— This paper proposes an improvement method of a power generation amount of Photovoltaic (PV) system using energy storage system. Power conditioning systems (PCS) of the PV system stops its operation when the generated power of the solar panel is low as in cloudy weather. Even if the solar radiation recovers, the PCS restarts after an interval time. Therefore, the conventional PV system can't fully extract the power of photovoltaic power generation. In the proposed structure, the energy storage system is connected between the PV and the PCS.

This paper discusses conditions for power generation improvement of the PV system using energy storage system. Experiments using solar radiation data on sunny and cloudy days were carried out. On cloudy days PCS output improved by 5.38%. But on sunny days it decreased by 0.58%. The validity of the improvement condition was confirmed by experimental results.

Keywords— *Photovoltaic generation, Power conditioning system, Energy storage system.*

I. INTRODUCTION

In recent years, solar power generation systems such as mega-solar system have rapidly introduced in grid system [1]. Solar power generation system is composed of photovoltaic module (PV) and a power conditioning system (PCS) that outputs generated power of PV to the grid. The PCS may stop when PV power decreases due to a drop in solar radiation. Furthermore, even if the solar radiation recovers, the PCS restarts after an interval of up to 5 minute. Therefore, the conventional PV system can't fully extract the power of photovoltaic power generation [2][3]. For the problem, the authors propose a method using an energy storage system. When the PV power decreases, the PCS powered by the energy storage system. Therefore, the PCS can operate continuous. So, PV power can be acquired continuously.

Generally, when connecting the energy storage system to the photovoltaic power generation system, it is connected to the grid or between the DC/DC converter and the inverter inside the PCS. Often the energy storage system connected to the grid side is used to compensate for fluctuations in PV power. However, this structure can't supply power to the PCS. On the other hand, When the energy storage system is connected between the boost converter and the inverter inside the PCS, it may be possible to supply power to the PCS. However, it is necessary to remodel the PCS circuit because the connection point is inside the PCS. Therefore, when installing to PCS, PCS in operation must be stopped. Moreover, after remodeling the PCS, it is necessary to reacquire authentication about the grid connection. Therefore, between the DC/DC converter and the inverter is not suitable as the connection point of the energy storage system.

This paper proposes a structure that connects the energy storage system in parallel with PV. Power can be supplied to the PCS, no remodeling of PCS is required. Therefore, it is suitable as the connection point of the energy storage system. However, the proposed structure with the energy storage system adds loss because loss occurs in the power converter due to the charge and discharge operation of the energy storage system. Even if the PV power amount increases due to the continuation of the PCS operation, the PCS output power amount may decrease as compared with the conventional configuration when the loss due to the charge and discharge operation of the energy storage system is large.

This paper proposes the system structure and the control method of the energy storage system. In addition, consider the conditions necessary for improving the output power amount of the PCS of the proposed structure. First, the structure and operation of the system are introduced. Next, the control method of the energy storage system used in the proposed structure is explained. After, the proposed structure and the conventional structure are compared by the experimental results. In addition, the improvement conditions of the PCS output power amount in the proposed structure are confirmed by the experimental results.

II. PROPOSED STRUCTURE

Fig. 1(a) shows the structure of the conventional photovoltaic power generation system. It consists of the PV and the PCS. The PCS is composed of a DC/DC converter and an inverter. The DC/DC converter performs MPPT control in order to operate the PV at maximum power. The Inverter converts DC power into

AC power and outputs in to the grid. Fig. 1(b) shows operation of the conventional structure. The PCS stops when solar radiation falls and the PV power decreases. After that the PCS waits even if the solar radiation recovers. Therefore, the conventional PV system can't fully extract the PV power.

For the problem, power is supplied to the PCS when the solar radiation falls by using the energy storage system. Fig. 2 shows photovoltaic generation system structures with energy storage system. Generally, the energy storage system is connected the grid or between the DC/DC converter and the inverter inside the PCS in a photovoltaic generation system. The energy storage system connected to the grid is used to compensate for fluctuations in the PV power [4][5]. However, this structure can't supply power to the PCS. Also, the energy storage system becomes large because not only the DC/DC converter but also the inverter is necessary. On the other hand, when adding the energy storage system between the DC/DC converter and the inverter inside the PCS [6]~[8], the structure has advantages. Because it does not require an inverter. However, remodeling of the PCS is necessary because the connection point is inside the PCS. The PCS must be stopped to connect the energy storage system and the certification on the grid connection of PCS must be gotten again because PCS remodeling is necessary.

Therefore, this paper proposes a structure that connects the energy storage system in parallel with PV. Fig. 3(a) shows the proposed system structure. It doesn't require remodeling of the PCS because the energy storage system is connected to the input terminal of the PCS. Furthermore, the capacity of the energy storage device can be made smaller than that of the power leveling system. Because the energy storage system operates only when the solar radiation falls and supplies power to the PCS. Fig. 3(b) shows the operation of the proposed structure. When PV power drops, the energy storage system supplies power to the PCS. The PCS stop is prevented. And PV power is improved because the PCS can continuously acquire generated power.

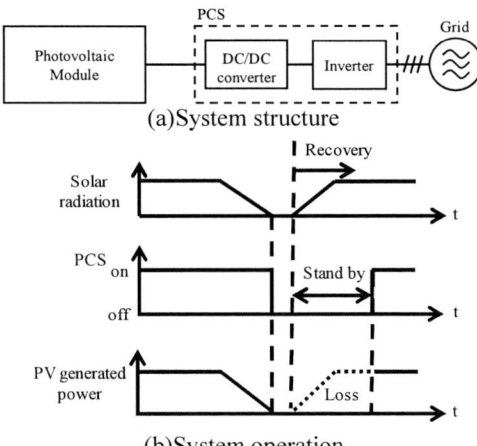

(a)System structure

(b)System operation

Fig. 1. Conventional structure.

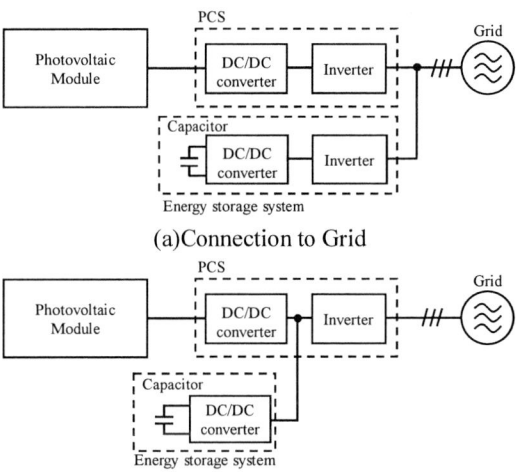

(a)Connection to Grid

(b)Connection between DC/DC converter and inverter

Fig. 2. Connection points of energy storage system.

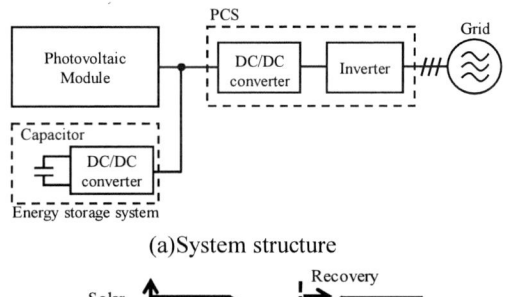

(a)System structure

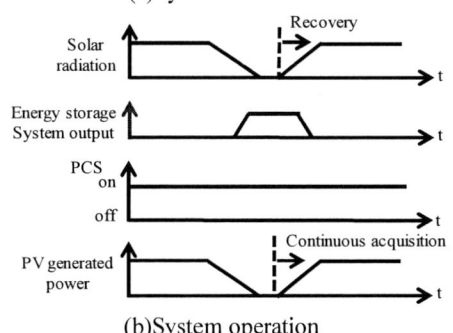

(b)System operation

Fig. 3. Proposed structure.

III. CONTROL METHOD

When the electric power generated by PV decreases, the electric power is supplied from the energy storage system to the PCS. The discharge power of the DC/DC converter is controlled by a power-voltage reference curve in Fig. 4. The power-voltage reference curve is obtained to make limiting the discharge power possible. The PCS has a MPPT control [9]. If the discharge control has the power-voltage relationship of the PV system, it is possible to cause the operating point to follow the peak of power generation using the MPPT control. Therefore, the discharge power can be limited.

When the electric power generated by PV is sufficient, a part of the generated electric power is used for charging the energy storage device. The principle diagram is shown in Fig. 5. The charging power is controlled to be

1371

constant, when charging the energy storage device. This method doesn't disturb the MPPT control of the PCS. The power inputted to the PCS is obtained by subtracting the generated power with the constant charging power set to the storage devices, when the charging power is kept constant. Therefore, the power characteristic of PV matches the maximum power point of the power characteristics of PCS, and the charge can be done without disturbing the MPPT control.

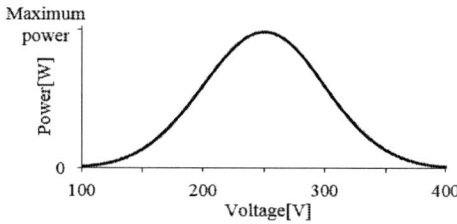

Fig. 4. Power characteristic of discharging power.

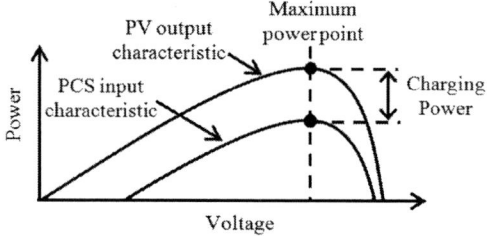

Fig. 5. Power characteristic of charging power.

IV. EXPERIMENTAL STRUCTURE

Fig. 6 shows experimental structure. PV output power, energy storage system output power and PCS output power are measured by a power meter in the experiment. The PV is simulated by programmable power supply with rated 5 kW. Solar radiation data and PV parameters are used to simulate the PV. The PV parameters are shown in Table 1. Solar radiation data for 20 minutes on sunny and cloudy days is used. Rated power of the PCS is 5 kW. When the input voltage of 300 V or more is continued for 10 seconds, it starts up and starts the MPPT control. The PCS stops when the input power becomes less than 500 W. The energy storage system is simulated by bidirectional chopper, DC power supply and electronic load device.

The energy storage system monitors the PV power and performs charge and discharge operations. Fig. 7 shows the control block of charge and discharge operations. Charging is done when the PV power reaches 1500 W or more. The maximum value of charging power is 300 W. The current command value is calculated from the command value of the charging power, and the current is controlled by using the PI controller.

Discharge is carried out when the PV power becomes 500 W or less. The discharge power command value is called from the lookup table using the PV voltage. The current command value is calculated from the discharge power command value of the discharge power, and the current is controlled by using the PI controller. During

charging, bidirectional chopper is used as a step-down chopper that switches S_{2_chop}. During the discharge, bidirectional chopper is used as a step-up chopper that switches S_{1_chop}. The current is monitored by the microcomputer, and the voltage used for capacitor simulation is calculated. The voltage of the DC power supply is set to the calculated capacitor voltage. The capacitor is assumed to be 3 F. The energy storage system performs initial charge until the capacitor reaches 50 V. After, charge and discharge between 50 V and 150 V. The rated power of bidirectional chopper is 500 W. It is 1/10 of the PCS.

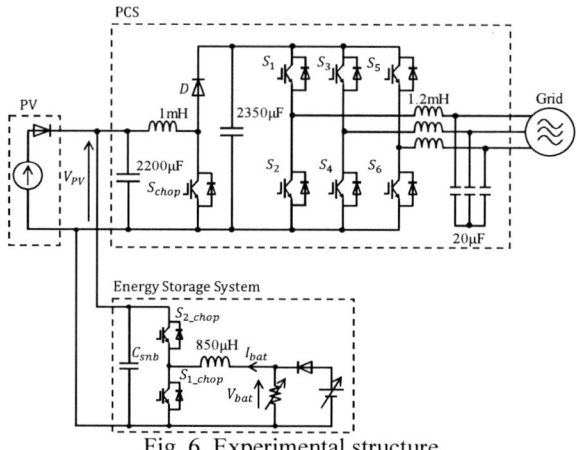

Fig. 6. Experimental structure.

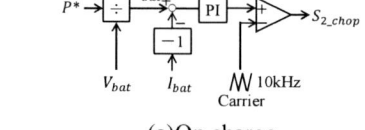

(a)On charge

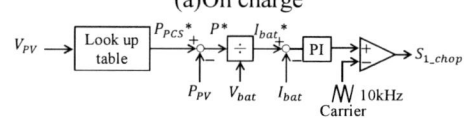

(b)On discharge
Fig. 7. Control brock.

Table 1. PV parameter.

Open circuit voltage	400V
Short circuit current	20A
Maximum power voltage	300V
Maximum power current	16.7A

V. EXPERIMENTAL RESULTS

Fig. 8 shows experimental results on sunny day and cloudy day in the structure of only PV and PCS. On sunny days it is confirmed that PCS doesn't stop until sunset. On the other hand, the stop of the PCS is confirmed on cloudy day. This is because PV power declined when the solar radiation falls. After the PCS

1372

The 2018 International Power Electronics Conference

stops, there is a standby time so PV generated electricity is not acquired.

Fig.9 shows experimental results on sunny day and cloudy day in the proposed structure that energy storage system was added to the PV and the PCS.

On the proposed structure, the PCS is started after the initial charging operation of the energy storage system. On the sunny day, it is confirmed that the charging operation is performed after the PV power reaches 1500 W or more. The voltage of the capacitor rises with the charging operation. Charging stops after the capacitor voltage reaches 150 V. At the sunset, PV output power

will decrease, so the energy storage system discharges. The power storage system discharges until the capacitor voltage decreases to 50 V.

On the cloudy day, it is confirmed that the energy storage system is discharging when PV power decreases after charging. Therefore, the PCS continued operation. And PV power was continuously acquired. As sunset, the capacitor is discharged until the voltage reaches 50 V as well as on the sunny day.

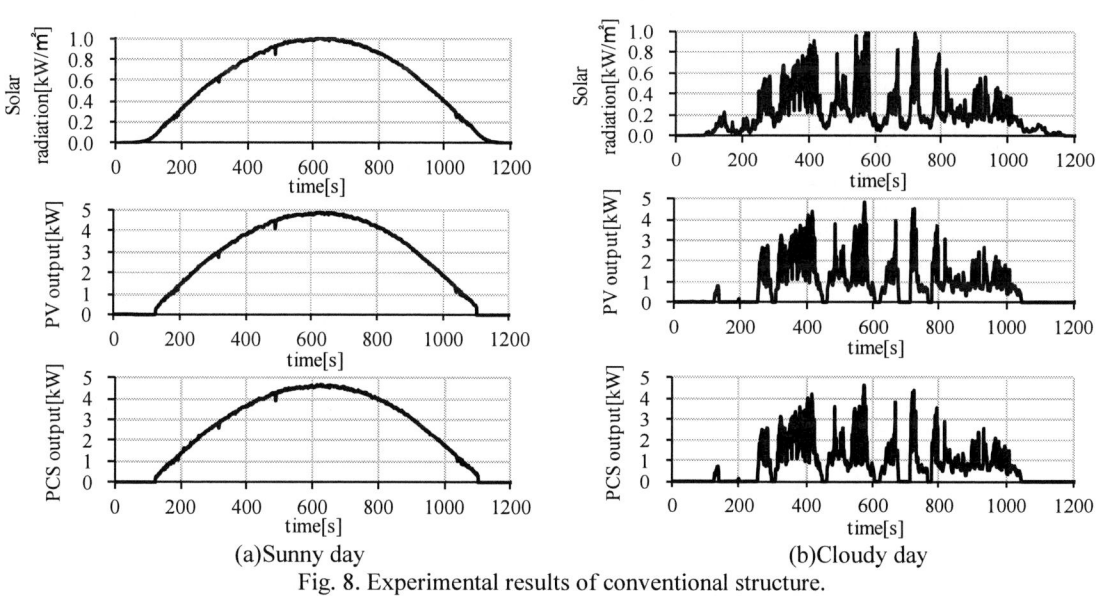

(a)Sunny day   (b)Cloudy day

Fig. 8. Experimental results of conventional structure.

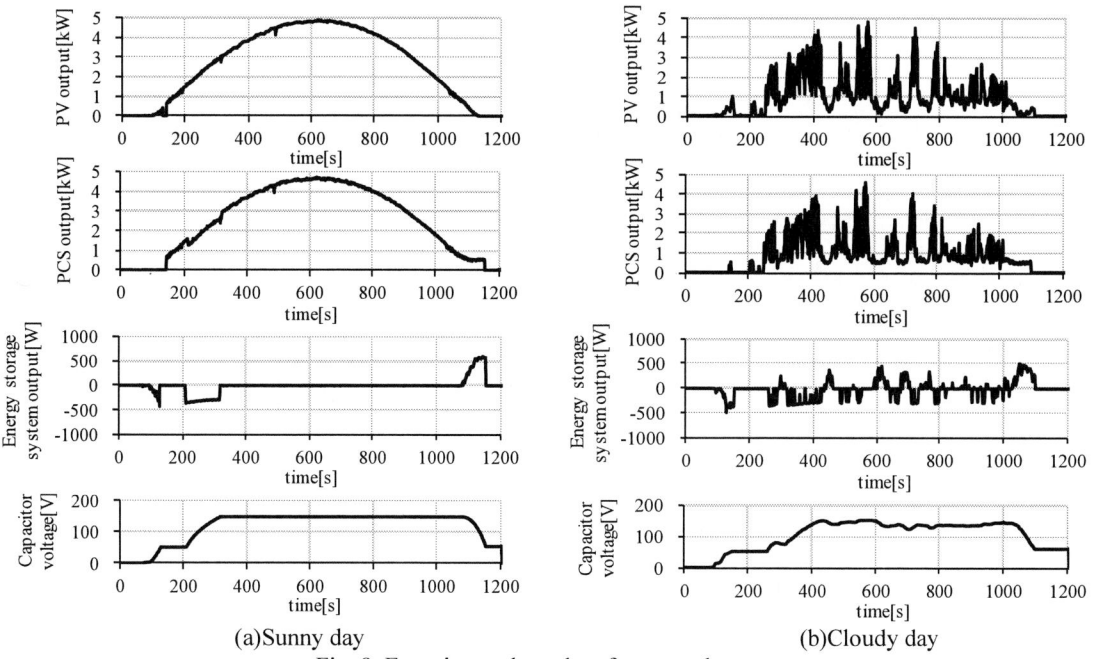

(a)Sunny day   (b)Cloudy day

Fig. 9. Experimental results of proposed structure.

1373

Fig. 10 shows PV generated energy in each weather. PV power generation amount increased by 0.30 % on sunny days. On cloudy days PV power generation amount increased by 10.82 %. On both days, the amount of power acquired from the PV is improved, because the energy storage system performs the initial charging operation. And on sunny days, the amount of power acquired from the PV is improved, because the PCS continues to operate due to the discharging operation when the PV power decreases due to sunset. Furthermore, the amount of power acquired from the PV is improved, because the PCS continues to operate due to the discharge operation when the solar radiations falls.

Fig. 11 shows PCS output power amount in each weather. PCS output power amount decreased by 0.58 % on the sunny day, and improved by 5.38 % on the cloudy day. PCS output power amount improved on cloudy day but decreased on sunny day. This reason is considered.

The condition is considered which the output power amount of the PCS of the proposed structure become larger than that of the conventional structure. In the conventional structure, PV power can't be acquired until the PCS restarts. So loss is generated. On the other hand, PV power can be acquired which is the loss caused by the standby of the PCS in the conventional structure when the operation of the PCS is continued by the discharging operation of the energy storage system on the proposed structure. The power obtained by continuing the operation of the PCS is output to the grid after power conversion by the PCS. Therefore, it is improved by the amount outputted to the grid within the PV generation amount improved by continuous operation of the PCS.

However, a part of PV power is charged and discharged by the energy storage system in the proposed structure. Therefore, loss is generated by the converter of the energy storage system. Fig. 12 shows the energy flow during charging and discharging of the energy storage system. The power input to the energy storage system is converted by the DC/DC converter and charged in the capacitor during charging. At this time, loss occurs in the DC/DC converter. During discharging, the power is output from the capacitor. Then, it is output from energy storage system after power conversion by the DC/DC converter. So, loss is generated by the DC/DC converter. Furthermore, the power output from the energy storage system output to the grid after converted by the PCS. So, also at this time loss is generated by the PCS.

For the above, loss occurs in the DC/DC converter during charging operation and in the DC/DC converter and PCS during discharging operation when part of PV power is charged and discharged by the energy storage system in the proposed structure.

Therefore, if the amount outputted to the grid within PV power generation amount improved by continuous operation of the PCS is larger than loss caused when the charged energy of the energy storage system is output to the grid after the charge and discharge operation and the conversion by the PCS, then the PCS output power amount is improved as compared with the conventional

structure, when the PCS stop is avoided in the proposed structure.

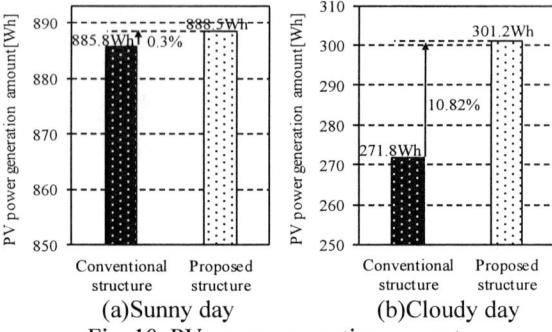

(a)Sunny day   (b)Cloudy day

Fig. 10. PV power generation amount.

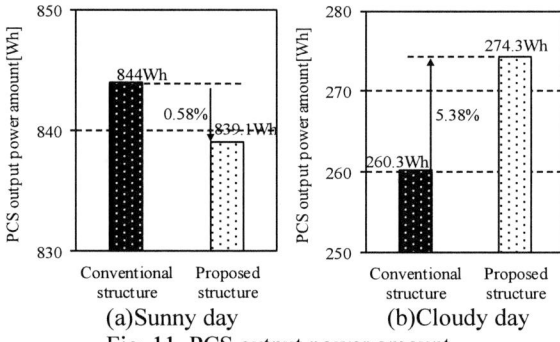

(a)Sunny day   (b)Cloudy day

Fig. 11. PCS output power amount.

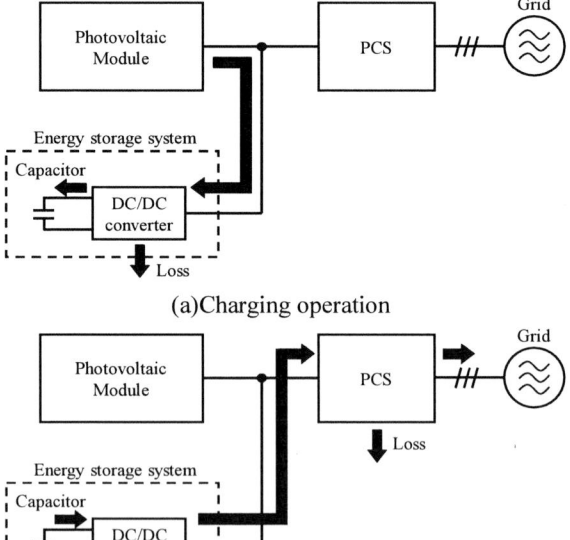

(a)Charging operation

(b)Discharging operation

Fig. 12. Energy flow during charging and discharging of the energy storage system.

Fig. 13 shows the amount outputted to the grid within PV power generation amount improved by continuous operation of the PCS and loss caused by charging and discharging of the energy storage system. The amount is 2.6 Wh that outputted to the grid of the PV generation amount improved by continuous operation of the PCS on the sunny day. And loss of 4.8 Wh is generated when the charged energy of the energy storage system is output from the PCS. Therefore, on the sunny day, the PV output power amount improved but the PCS output power amount decreased of the proposed structure, because the loss due to charge and discharge exceeds.

On the other hand, the amount is 27.5 Wh that outputted to the grid of the PV generation amount improved by continuous operation of the PCS on the cloudy day. And loss of 9.2 Wh is generated when the charged energy of the energy storage system is output from the PCS. Therefore, on the cloudy day, the PV output power amount improved and the PCS output power amount improved too, because the loss due to charge and discharge is less than improved amount.

VI. CONCLUSION

This paper proposes an energy storage system that improves the total PV output power. The main purpose of adding the energy storage system is to continue the operation of the PCS. Since the energy storage system is connected between the PV and the PCS, it has the advantage that it can be added to the existing system. The proposed system consists of DC/DC converter and capacitor. The power rating of the DC/DC converter is sufficiently smaller than that of the PCS. This paper discusses control method of DC/DC converter for improving power output from the PCS. When the solar power generation is large, the DC/DC converter charges the capacitor. When the PV power decreases, the PCS is powered by the energy storage system.

Experiments using solar radiation data on cloudy day and sunny day were carried out. As results of the experiment, it was confirmed that PV output power amount increased by 10.82 % and PCS output power amount increased 5.38 % on cloudy day. On sunny day, PV output power amount increased by 0.3 %, but it was confirmed that PCS output power amount decreased by 0.58 %. Reason about this cause was considered. A part of the PV power is charged and discharged by the energy storage system in the proposed structure, so that loss occurs in the DC / DC converter of the energy storage system. The output power amount of the PCS can't exceed power amount of the conventional structure, when the loss generated by the charge and discharge operation of the energy storage system is larger than the amount output from the PCS of the PV output improvement amount. This was confirmed by comparing the amount outputted to the grid of the PV generation amount improved and the loss generated by the charge and discharge operation.

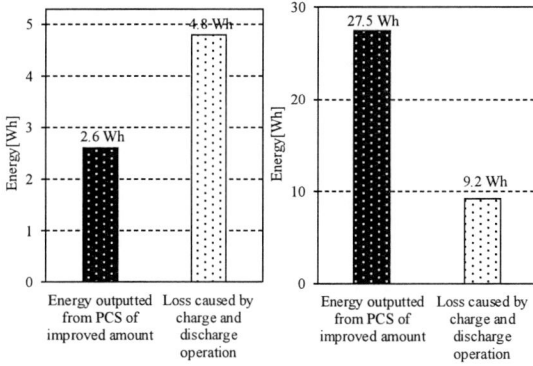

(a)Sunny day　　　　(b)Cloudy day

Fig. 13. Energy about improved amount and loss caused by charge and discharge operation.

REFERENCES

[1] H.Konishi, T.Iwato and M.Kudou: " Development of Large-scale Power Conditioning System in Hokuto Mega-solar Project, " *International Power Electronics Conference* , pp.1975-1979(2010)

[2] S.Machida and T.Tani: "Effect of the Organization of Solar Cell Array and Inverter Components on the System Economical Efficiency," Trans. IEEJ, Vol.120-B, No.5, pp.733-738(2000)

[3] T.Yamada, D.Tokushima, H.Ishikawa, D.Wang and H. Naitoh: " A Novel PV System by using Multiple Inverters, " Trans. IEEJ, Vol.124-B, No.9, pp.1087-1092(2004)

[4] Xiangjun Li , Dong Hui and Xiaokang Lai: " Battery Energy Storage Station (BESS)-Based Smoothing Control of Photovoltaic (PV) and Wind Power Generation Fluctuations, " *IEEE Transactions on Sustainable Energy*, Vol.4, No.2, pp.464-473(2013)

[5] H.Haga, T.Shimao , S.Kondo , K.Kato , Y.Ito , K.Arimatsu and K.Matsuda: " High-efficiency and Cost-minimization Method of Energy Storage System with Multi Storage Devices for Grid Connection, " *International Power Electronics Conference*, pp.415-420(2014)

[6] Y.Riffonneau , S.Bacha , F.Barruel and S.Ploix: "Optimal Power Flow Management for Grid Connected PV systems With Batteries, " *IEEE Transactions on Sustainable Energy*, Vol.2, No.3, pp.309-320(2011)

[7] Wei Jiang, Lei Zhang, Hui Zhao, Huichun Huang and Renjie Hu: " Research on power sharing strategy of hybrid energy storage system in photovoltaic power station based on multi-objective optimization," *IET Renewable Power Generation*, Vol.10, Iss.5, pp.575-583(2016)

[8] Nasif Mahmud , Ahmad Zahedi and Asif Mahmud: "A Cooperative Operation of Novel PV Inverter Control Scheme and Storage Energy Management System Based on ANFIS for Voltage Regulation of Grid-Tied PV System," *IEEE Transactions on Industrial Informatics*, Vol.13, No.5, pp.2657-2668(2017)

[9] T.Esram and P.Chapman: " Comparisn of Photvoltaic Array Maximum Power Point Tracking Techniques," *IEEE Transactions on Energy Conversion*, Vol.22, No.2, pp.439-449(2007)

The 2018 International Power Electronics Conference

A Control Strategy of MMC Battery Energy Storage System Based on Arm Current Control

Liu Danqing[1], Wang Guangzhu[1*], Ou Zhujian[1] and Liu Jiaxing[1]
1 School of Electrical Engineering, Shandong university, Jinan, China
*E-mail: sdwgz@sdu.edu.cn

Abstract— **A control strategy of MMC battery energy storage system(MMC-BESS), which is based on arm current control, is proposed in this paper. Compared with other strategies, there are three technological merits of this strategy. First of all, it could control arm current directly to achieve triplex-control which covers Ac-side current, Dc-bus current and circulating current, without the specific circulating control strategy. In addition, compared with grid side reference current, it could easily give the signal of arm reference current without conversion of coordinates and numerous computations. At last, this strategy realizes a distributed balance control of the capacitance voltage of each sub-module through hierarchical control, and it also can actualize bidirectional flow of active power between any two ends of Dc-side, Ac-side and sub-module battery side. Simulation experiment based on PSIM verifies the effectiveness of this strategy.**

Keywords— *energy storage, MMC, Arm current control, Hierarchical control, capacitor voltage balance.*

I. INTRODUCTION

With the depletion of fossil energy and increasingly serious global climate issues, renewable energy resources, such as photovoltaic generation and wind power generation, are attracting widespread attention around the world. Because renewable energy power generation is heavily depended on climate and other strong volatility and intermittency factors, when they connected to grid on a large scale may result in various problems, like the fluctuation of network voltage or grid frequency and even instable. To resolve these problems, adding energy storage system to new energy generation system is an effective way to ensure supplying stable and sustainable power [1,2].

MMC is an ideal power transform topology for massive energy storage systems. Compared with other conventional two-level or three-level converters used as power conversion system in energy storage, using MMC has several advantages as follow: three-port converter, energy-storage battery put in the sub-module, and power flowed between any two ends of the battery side, Dc side and Ac side. The modularized and multi-level structural features make output voltage with lower contents of harmonic component, and even does not require output filter. Reducing voltage level of the power device is apt to achieve redundancy design and fault ride-through [3-6].

Nearly all the existing energy-storage type MMC current control strategies adopt traditional grid-side current feedback control strategy [3-8]. However, grid-side current is actually the sum of upper arm current and lower arm current, while only controlling the grid-side current will lose the current signals with opposite polarities in upper/lower arm current, the case in point is the circulating current signal inside MMC. Therefore, for traditional grid-side current control strategies, not only current analysis is needed, but also special controller is needed to control the different current components in grid-side current. Meanwhile, a series of transformation of coordinates and scientific calculation are required to get correct grid-side current reference signal. In view of arm current contains all the current information including Ac-side current, Dc-bus current and circulating current, which can be effectively controlled directly by controlling MMC arm current [9-11].

However, the control strategy of MMC-BESS based on arm current has not been discussed. This paper puts forward a control strategy based on arm current that can realize the free flow of power among Dc-side, Ac-side and battery side, with no need for designing additional circulating current controller. Hierarchical control is adopted to get the bridge arm current reference value without conversion of coordinates and numerous computations. Simulation result proves the effectiveness of the strategy, output voltage of the rectifier side or inverter side as well as the sub-module capacitor voltage is well controlled, and circulating current is effectively suppressed.

II. POWER ANALYSIS OF MMC-BESS

The schematic diagram of MMC-BESS adopted in this paper is illustrated in Fig.1. It consists of three parallel-connected phase legs, and each phase leg contains an upper arm and a lower arm. Each arm is constituted by N energy storage sub-modules(ESSM) with an inductor L. Batteries are distributed connected to each SM's capacitor through a bidirectional DC/DC converter, because the SM has substantial second order harmonic current which is harmful to the battery's performance and lifespan. Additionally, with DC/DC converter, the batteries are decoupled from SM, more redundancy for the whole system can be achieved, better charge-discharge of batteries could be obtained. What's more, it could obviously reduce the voltage of battery module and different types of batteries can be connected to same MMC-BESS. In a word, it is actually necessary for energy storage cells to connected with SM via DC/DC converters [4,12].

Where u_{px} and u_{nx} $(x \in \{a,b,c\})$ are respectively output voltages of the upper and the lower arms. i_{px} and i_{nx} are respectively arm currents of the upper and the lower arms of phase x. U_d and I_d are respectively dc bus voltage and current. u_{sx} and i_{sx} are respectively the ac-side voltage and current of phase x.

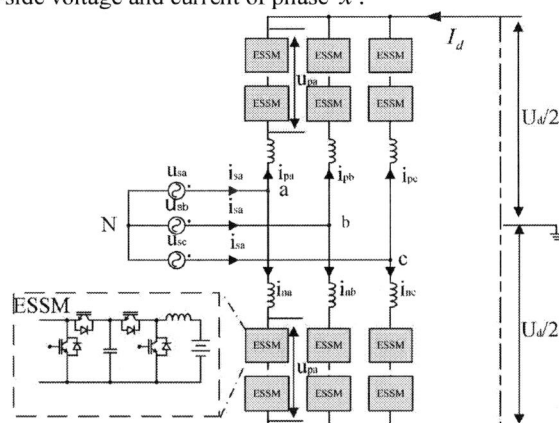

Fig. 1 Basic structure of MMC-BESS

The capacitor of ESSM is the only energy-storage element except inductor in arm and battery energy storage unit in ESSM. When the power absorption or output of ESSM energy storage unit is constant, arm voltage can be controlled by commanding the active power of the arm. In order to choose appropriate control variables for Dc-bus and capacitor voltages, power analysis of MMC-BESS is done as follows.

In steady state, when ignoring the influences of arm inductances, voltages satisfy follow relationship:

$$\begin{cases} u_{px} \approx U_d / 2 - u_{sx} \\ u_{nx} \approx U_d / 2 + u_{sx} \end{cases} \quad (1)$$

The upper/lower arm currents i_{px} and i_{nx} satisfy:

$$\begin{cases} i_{px} = i_{spx} + i_{othx} = i_{spx} + i_{dcx} + i_{cir\_x} \\ i_{nx} = i_{snx} - i_{othx} = i_{snx} - i_{dcx} - i_{cir\_x} \end{cases} \quad (2)$$

Where i_{spx}, i_{snx} are parts of Ac-side current; i_{othx} is the other current in arm current what is containing the circulating current i_{cir_x} and Dc current component i_{dcx}. Relationship formula between Ac-side current i_{sx} and arm currents i_{px} and i_{nx} is as follow:

$$i_{sx} = i_{px} + i_{nx} = i_{spx} + i_{snx} \quad (3)$$

Dc bus current I_d can be expressed by equation:

$$\begin{aligned} I_d &= I_{otha} + I_{othb} + I_{othc} = i_{dca} + i_{dcb} + i_{dcc} \\ &= [(i_{pa} - i_{na}) + (i_{pb} - i_{nb}) + (i_{pc} - i_{nc})] / 2 \end{aligned} \quad (4)$$

Where i_{spx}, i_{snx}, i_{cir_x}, i_{dcx} satisfied the following equations:

$$\begin{cases} \int_0^T U_d i_{spx} dt = \int_0^T U_d i_{snx} dt = \int_0^T U_d i_{cir\_x} dt = 0 \\ \int_0^T U_d i_{dcx} dt = \int_0^T U_{sx} i_{cir\_x} dt = \int_0^T U_{sx} I_d dt = 0 \end{cases} \quad (5)$$

Where T is the period of ac-grid. According to the law of energy conservation, we have (6)

$$P_{dcx} = P_{sx} + P_{px} + P_{nx} \quad (6)$$

Where P_{sx} is the output active power of Ac-side; P_{dcx} is the absorption active power of Dc-bus; P_{px} and P_{nx} are respectively the absorption active power of the upper/lower arms. Each power can be expressed as follow:

$$\begin{cases} P_{sx} = \dfrac{1}{T}\int_0^T u_{sx} i_{sx} dt = \dfrac{1}{T}\int_0^T u_{sx} i_{spx} + \dfrac{1}{T}\int_0^T u_{sx} i_{spx} \\ P_{dcx} = \dfrac{1}{T}\int_0^T U_d i_{dx} dt = \dfrac{1}{T}\int_0^T \dfrac{U_d}{2} i_{px} dt - \dfrac{1}{T}\int_0^T \dfrac{U_d}{2} i_{nx} dt \end{cases} \quad (7)$$

$$\begin{cases} P_{px} = \dfrac{1}{T}\int_0^T u_{px} i_{px} dt = \dfrac{1}{T}\int_0^T \dfrac{U_d}{2} i_{dcx} dt - \dfrac{1}{T}\int_0^T u_{sx} i_{spx} dt \\ P_{nx} = \dfrac{1}{T}\int_0^T u_{nx} i_{nx} dt = \dfrac{1}{T}\int_0^T \dfrac{U_d}{2} i_{dcx} dt - \dfrac{1}{T}\int_0^T u_{sx} i_{snx} dt \end{cases} \quad (8)$$

Based on (7), when Dc-bus voltage U_d keep constant, the absorption active power of Dc-bus P_{dcx} can be controlled by commanding the current i_{dcx}. And control the output active power through adjusting the current i_{spx} and i_{snx}. Based on (8), the active power of upper/lower arm can be controlled by governing i_{spx}, i_{snx} and i_{dcx}, and it also can be used to control the capacitor voltage of upper/lower arm by coordinating with the power control of the SM-energy storage unit. Define Δi_{sx} is as follow:

$$\begin{cases} \Delta i_{sx} = i_{spx} - i_{snx} \\ \Delta P_{sx} = P_{px} - P_{nx} = \dfrac{1}{T}\int_0^T u_{sx} \Delta i_{sx} dt \end{cases} \quad (9)$$

Where ΔP_{sx} represents the active power flow from upper arm to lower arm. According to (9), Δi_{sx} can be used to maintain the voltage balance between the upper and lower arm. Like this, i_{spx} and i_{snx} can also be expressed in the following equations:

$$\begin{cases} i_{spx} = (i_{sx} + \Delta i_{sx}) / 2 \\ i_{snx} = (i_{sx} - \Delta i_{sx}) / 2 \end{cases} \quad (10)$$

III. CONTROL STRATEGY

A. arm current control scheme

The upper/lower arm currents i_{px} and i_{nx} can be controlled directly by corresponding PWM duty cycles. According to equations (3) and (4), Ac-side current and

Dc-bus current could be directly controlled by adjusting i_{px} and i_{nx} without the specific circulating controller.

Due to the three-phase arm current is commanded independently, there is no necessary to consider the voltage balance among the legs. There are some coupling relationship among the three-phase current, and it is necessary to decouple when controlling the three-phase current. The arm current control scheme adopted in this paper is based on the average model in the switching period, and the coupling of three-phase currents has little effect on the control performance.

Arm current control scheme is shown in Fig. 2. In Fig. 2, u_0 is artificially superimposed zero-sequence voltage to improve the utilization of dc voltage; U_{cn} is the rated voltage of SM capacitor; i_{px}^{ref} and i_{nx}^{ref} are respectively the upper/lower arm current references of phase x; d_{px} and d_{nx} are the upper/lower arms' PWM duty cycles; d_{px}^{ctrl} and d_{nx}^{ctrl} are duty cycles output of the upper/lower arm current controller; D_{px} and D_{nx} are steady state PWM duty cycles of the upper/lower arms used as feedforward variables. These duty cycles satisfy:

$$\begin{cases} D_{px} = 0.5 - (u_{sx} + u_0)/(NU_c) \\ D_{nx} = 0.5 + (u_{sx} + u_0)/(NU_c) \end{cases} \quad (11)$$

$$\begin{cases} d_{px} = D_{px} + d_{px}^{ctrl} \\ d_{px} = D_{px} - d_{px}^{ctrl} \end{cases} \quad (12)$$

When working under three-phase balance condition, (3) can also be expressed as:

$$I_d = 3i_{dcx}, x \in \{a,b,c\} \quad (13)$$

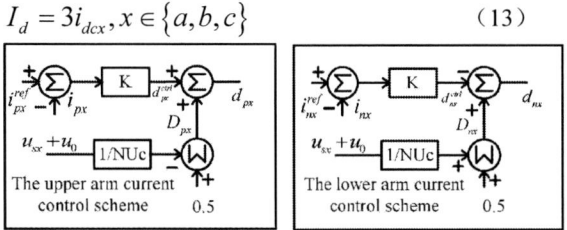

Fig. 2 Arm current control scheme

The duty cycle is realized by PSC-PWM, and the design of controller parameters refers to [13]. It would not be discussed further in paper.

B Dc-bus voltage control scheme

When the MMC-BESS works in a three-phase balanced rectifier condition, the Dc-bus voltage is controlled by I_d. The Dc-bus voltage control scheme is described in Fig. 3, where U_d^{ref} is Dc-bus reference voltage; the output is the Dc-bus current reference component i_{dcx}^{ref}.

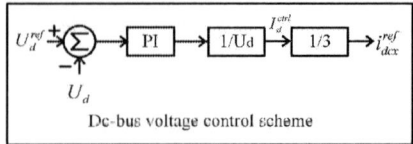

Fig. 3 Dc-bus voltage control scheme

C arm capacitor average voltage control scheme

It was discussed in section II, when i_{dcx} has been determined and the active power of ESSM is constant, the Ac-side active power can be controlled by adjusting i_{sx}, so that the arm capacitor average voltage can also be controlled. Arm capacitor average voltage control scheme is illustrate in Fig. 4, where U_{cx}^{ref} is the capacitor voltage reference; U_{cx_ave} is the average of capacitor voltage; $\sin x$ is the frequency and phase signal of ac-side obtained from phase-locked loop(PLL); The output of the controller is the ac-side current reference i_{sx}^{ctrl}.

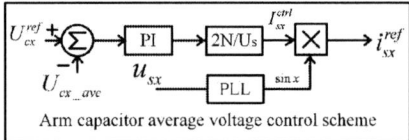

Fig. 4 Arm capacitor average voltage control scheme

D. upper and lower voltage balance control scheme

It was discussed in section II, the power distribution between upper arm and lower arm can be controlled by adjusting Δi_{sx}, so that the capacitor voltage balance between the upper and lower can also be controlled. upper and lower voltage balance control scheme is showed in Fig. 5. The capacitor voltage balance within the individual arm is realized by fine tuning duty cycle of ESSM individually, which is not been discussed in this paper.

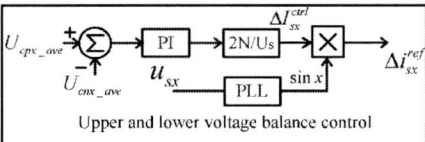

Fig. 5 Upper and lower voltage balance control scheme

E ESSM DC/DC converter control

ESSM DC/DC converter control is described in Fig. 6. The out-loop control of power and inner loop control of inductor current are used to ensure the constant power of the battery energy storage unit through the bidirectional DC/DC converter. Where P_{bat}^{ref} is power reference of battery energy storage unit; i_{p/nx_bat} is inductor current of ESSM-DC/DC converter; d_{p/nx_bat}^{ctrl} is the duty cycle of DC/DC converter.

The 2018 International Power Electronics Conference

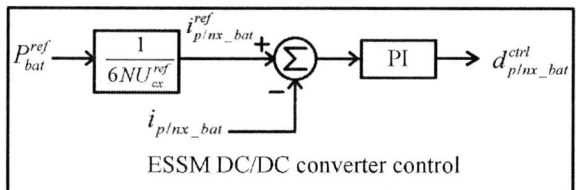

Fig.6 ESSM DC/DC converter control

F control strategy of BESS-MMC

Control strategy of BESS-MMC in rectification condition is showed in Fig. 7. The reference of arm current i_{px}^{ref} and i_{nx}^{ref} are obtained through hierarchical control, which covers Dc-bus voltage control, arm capacitor average voltage control, upper and lower voltage balance control and ESSM DC/DC converter control.

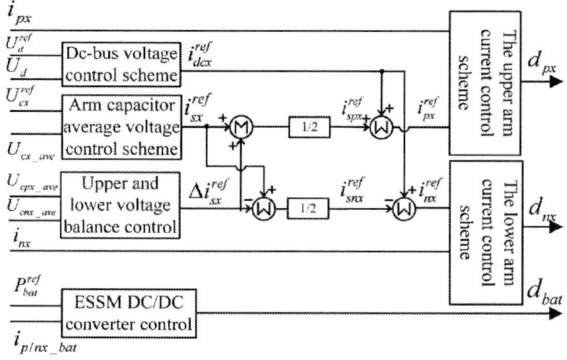

Fig. 7 control strategy of BESS-MMC in rectifier condition

The control strategy of BESS-MMC in inverter condition is similar to the rectifier strategy, which uses Ac-side voltage control scheme instead of Dc-bus voltage control scheme. Details are not given here.

IV. SIMULATION RESULTS

The simulink results in PSIM prove the effectiveness of the control strategy. Simulation experimental parameters are shown in TABLE I. The simulation results of the BESS-MMC operating in the rectifier condition are shown in Fig. 8. The value of P_{bat} is changed from -4kw to -8kw at 0.5 second, and be changed from -8kw to 8kw at 1 second.

TABLE I
SIMULATION EXPERIMENTAL PARAMETERS

Symbol	Meaning	Value
U_d	DC Bus Voltage	800.0 V
U_{ac}	AC Side Voltage	220*sqrt(2) V
U_{bat}	Battery unit voltage	100 V
U_c	ESSM Capacitor Voltage	200 V
R_{dc}	DC Side Resistance	64 Ω
R_{ac}	AC Side Resistance	14.57 Ω
L_{arm}	Arm Inductance	10 mH
L_{essm}	ESSM Inductance	5 mH
C_{essm}	ESSM Capacitance	200 V

f_{ac}	AC Nominal Frequency	50 Hz
f_{car}	Carrier Frequency	5 kHz
f_{essm}	ESSM Carrirt Frequency	20 kHz
N	Number of ESSMs per arm	4

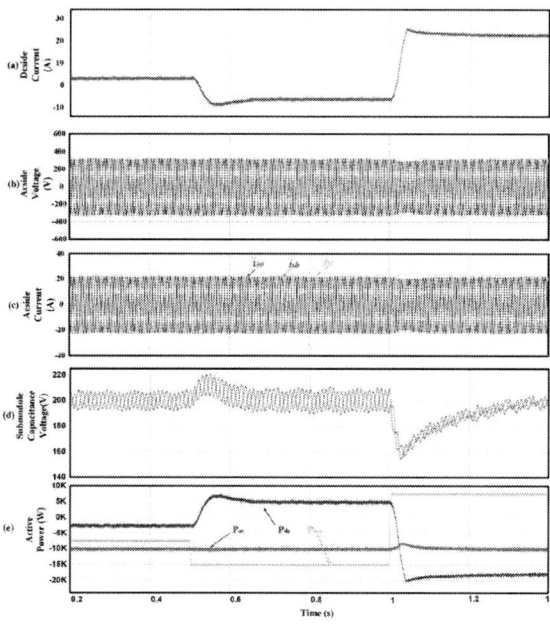

Fig. 8 BESS-MMC operation in rectifier condition.
(a) Dc-bus voltage. (b) Dc-bus current. (c) Ac-side current. (d) The voltage of first capacitor of the upper/lower arms in phase a. (e) The absorbed active power of Dc-bus and Ac-side,The output active power of Battery

The simulation results of the BESS-MMC operating in the inverter condition are shown in Fig. 9. The value of P_{bat} is changed from -7.5kw to -10kw at 0.5 second, and then be changed from -15kw to 7.5kw at 1 second.

Fig. 9 BESS-MMC operation in inverter condition.
(a) Dc-bus current. (b) Ac-side voltage. (c) Ac-side current. (d) The voltage of first capacitor of the upper/lower arms in phase a. (e) The absorbed power of Dc-bus and Ac-side,The output power of Battery

From the simulation results, the capacitor voltage of the ESSM is always controlled well, and the balance can be recovered quickly in the case of power shock. Dc-bus voltage is well controlled in rectifier condition and Ac-side voltage is also well controlled in inverter condition. The active power realizes bidrectional flow between each two of Dc-bus, Ac-side and ESSM side.

V. CONCLUSIONS

A control strategy of MMC-BESS based on arm current control is presents in this paper. Through the hierarchical control strategy to obtain the arm current reference. A power and current double loop controller is used to ensure that the output power of battery storage units are constant. It is verified by simulation experiments under rectifier operating condition and inverter condition that control strategy is effectiveness. The MMC-BESS adopts this strategy can effectively realize the active power bidrectional flow among the Dc-bus, Ac-side and ESSM side, with good dynamic and static performance. When the output power of the ESSM is changed, it will not cause too much disturbance to the grid-side, and the capacitor voltage of the ESSM is well controlled. The research about the battery management and reactive compensation of MMC-BESS adopted this control strategy will be done in near future.

The Project was Supported by Shandong Provincial Natural Science Foundation of China(ZR201709180272).

REFERENCES

[1] Quanyuan Jiang, Haijiao Wang, "Two-Time-Scale Coordination Control for a Battery Energy Storage System to Mitigate Wind Power Fluctuations," *IEEE Trans. on Energy Conversion*, vol.28, March 2013.

[2] Ping Jiang, Huachuan Xiong, "A control scheme design for smoothing wind power fluctuation with hybrid energy storage system," *Aotomation of Electric Power System*, vol.01 pp.122-127, 2013.

[3] Soong, Theodore, and Peter W. Lehn. "Evaluation of emerging modular multilevel converters for BESS application." *IEEE Trans. On Power Delivery*, vol. 29, no. 5, pp. 2086-2094, 2014.

[4] Vasiladiotis, Michail, and Alfred Rufer. "Analysis and control of modular multilevel converters with integrated battery energy storage." *IEEE Trans. on Power Electronics* vol. 30 no. 1 pp. 163-175, 2015.

[5] Zhe Wang, Hua Lin, Yajun Ma and Tao Wang. "A prototype of modular multilevel converter with integrated battery energy storage." *In proc. of IEEE Applied Power Electronics Congerence and Exposition(APEC)*, pp. 434-439, 2017.

[6] J. Xu, P. Zhao and C. Zhao, "Reliability analysis and redundancy configuration of MMC with hybird submodule topologies," *IEEE Trans. on Power Electronics*, vol. 31 no. 4, pp. 2720-2729, 2016.

[7] Ren Bin, Xu Yonghai and Lan Qiaoqian. "A control method for battery energy storage system based on MMC." *In proc. of IEEE 2nd International Future Energy Electronics Conference (IFEEC)*, pp. 1-6, 2015.

[8] Qiang Chen, Rui Li and Xu Cai. "Analysis and Fault Control of Hybrid Modular Multilevel Converter With Intergrated Battery Energy Storage System." *IEEE Trans. on Power Elecronics*, vol.5 no.1 pp. 64-78, 2017

[9] G. Wang, C. Sun, R. Liu, F. Wang, F. Li, "Modular Multilevel Converter control strategy based on arm current control." *Proceeding of the CSEE*, vol. 35, no. 2, pp. 458-464, 2015.(in Chinese)

[10] J. Moon, P. J. Park, D. Kang, Kang, J. "A control method of HVDC-Modular Multilevel Converter based on arm current under unbalanced voltage condition." *IEEE Trans. on Power Delivery*, vol.30 no.2 pp. 529-536, 2015.

[11] Zhujian Ou, Guangzhu Wang. "Modular Multilevel Converter Control Strategy Based on Arm Current Control under Unbalanced Grid Condition," *IEEE Tras. on Power Electronics*, 2017.

[12] Sigurd Byrkjedal wersland; Anirudh Budnar Acharya; Lars Einar Norum. "Integrating battery into MMC submodule using passive technique." *In proc. of IEEE 18th Workshop on Control and Modeling for Power Electronics (COMPEL)*, 2017.

[13] G. Wang, "An arm current direct control scheme for modular multilevel converters." *Automation of Electric Power Systems*, vol. 37, no. 15, pp. 35-39, 46, Aug. 2013

The 2018 International Power Electronics Conference

Equivalent Resistance Control for Maximum Power Transfer Method of Piezoelectric Element in Vibration Power Generation

Kenya Takamura[1], Hiroaki Yamada[1]*,
and Toshihiko Tanaka[1]
1 Graduate School of Sciences and Technology for Innovation,
Yamaguchi University, Yamaguchi, Japan
e-mail: hiro-ymd@yamaguchi-u.ac.jp

Tomoharu Yada[2]* and Hajime Fujiwara[2]
2 New Japan Radio Company, Limited.,
Tokyo, Japan
e-mail: tyada@njr.co.jp

Abstract—This paper proposes a novel equivalent resistance control for maximum power transfer method of piezoelectric element in vibration power generation. In the proposed method, the rectified voltage and current of a diode rectifier are in phase by the boost chopper. Thus, the input side of the boost chopper behaves as a resistor equivalently. For maximizing the generated power of the piezoelectric element, this resistance must be controlled to the internal impedance of the piezoelectric element by the law of the maximum power transfer. The basic principle of the proposed method is discussed, and then confirmed by computer simulation using Piece-wise Linear Electronical Circuit Simulation (PLECS). Simulation results demonstrate that the output power at buck chopper is 7.36 mW by the proposed method. This output power with the proposed method is 3.5 times greater than the output power without the proposed method.

Keywords—*vibration power generation, piezoelectric element, maximum power transfer, equivalent resistance control*

I. INTRODUCTION

The low power electronics for wireless sensor networks (WSNs), have driven numerous researches in the field of energy harvesting [1]- [3]. A wireless sensor node consists of low power microcontroller unit, radio frequency transceiver and microelectromechanical- (MEMS-) based sensor. The task of each wireless sensor node is to collect and transmit data to the another node via a radio link. Distributed wireless sensors are embedded in civil structures, bridges, or in the human body [4]. The batteries for wireless sensor nodes are used as the electrical energy. However, batteries have a limited life span and they are expensive to maintain. The WSNs and embedded systems are required a long-term energy source. In fact, the limited capacity of batteries is one of the main factors constraining the performance and limiting the lifespan of a typical WSN. Energy harvesting is the most promising way of overcoming the challenges currently presented by finite life power sources like batteries [5].

Energy harvesting is the process of extracting small

amount of energy from ambient environment through various sources of energy. The available energy for harvesting is mainly provided by ambient light, ambient radio frequency, thermal sources and mechanical vibration [6]. Among these energies, the vibration energy has large generated power per unit area. Also, compared to ambient light energy, it is energy independent of weather. Piezoelectric element, which is one of the generating element used for vibration power generation, generates AC power. Piezoelectric element has very large internal impedance by depending on vibration frequency [7]- [8]. Therefore, power is small to use as the power supply of the wireless sensor. The buck-boost converter for sensorless power optimization of piezoelectric energy harvester has been proposed [9]. The sensorless power optimization can be achieved under the continuous conduction mode (CCM) operation of the buck-boost converter. However, the large inductor was used for CCM operation. Thus, the power loss of the inductor becomes large. The SSHI (Synchronized Switch Harvesting on Inductor) methods have been proposed [10]- [12]. In the SSHI circuit, the resonant circuit with a switching device are used for increasing the generated power of the piezoelectric element. However, the control circuit of the SSHI is complex because the maximum point of the vibration displacement must be detected for occurrence of the resonance.

This paper proposes a novel equivalent resistance control for maximum power transfer method of piezoelectric element in vibration power generation. In the proposed method, the rectified voltage and current of a diode rectifier are in phase by the boost chopper. Thus, the input side of the boost chopper behaves as a resistor equivalently. For maximizing the generated power of the piezoelectric element, this resistance must be controlled to the internal impedance of the piezoelectric element by the law of the maximum power transfer. Therefore, the proposed control method is very simple as compared to the conventional methods. A computer simulation was implemented to confirm the validity of the proposed method using Piece-wise Linear Electronical Circuit Simulation (PLECS). Simulation results demonstrate that the output power is

7.36 mW by the proposed method.

II. THE PROPOSED EQUIVALENT RESISTANCE CONTROL METHOD

A. Characteristics of the piezoelectric element

Fig.1 shows the piezoelectric element (PPA-2011) and the vibration motor in this paper. The nose-mounted vibration motor is used for generating the vibration of the piezoelectric element. Fig.2 shows a test circuit of the piezoelectric element. As is well know, an equivalent circuit of piezoelectric element is consisted of a voltage source and parallel-connected resistor and capacitor [7]. The diode rectifier, which is used vs-16ctq080 manufactured by VISHAY, is used to convert from AC to DC.

Fig.3 shows the output waveform of the piezoelectric element PPA-2011. From Fig.3, the amplitude of the output voltage is 45 V. The frequency is 17.5 Hz. R_P and C_P are 19 MΩ and 206 nF, respectively. When the vibration frequency is 17.5 Hz, the internal impedance Z_P of the piezoelectric element is given by

$$|Z_P| = \frac{1}{\omega C_P} = 45\,\text{k}\Omega. \tag{1}$$

Therefore, the generated power of the piezoelectric element is maximum when load resistance R of Fig.2 is 45 kΩ by the law of the maximum power transfer.

Fig.4 shows the experimental result of the generated power characteristics in Fig.2. From Fig.4, the generated power depends on the load resistance. The maximum power is about 8.8 mW when R is 45 kΩ.

B. Proposed equivalent resistance control method

Fig.5 shows a power circuit diagram of the proposed vibration power generation system. The proposed vibration power generation system consists of a piezoelectric element, diode rectifier, boost chopper and buck chopper.

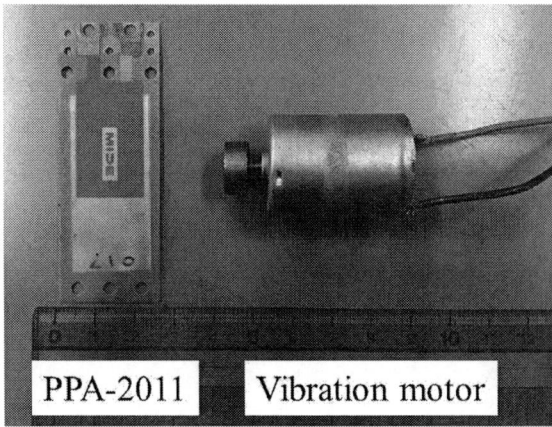

Fig. 1. Piezoelectric element (PPA-2011) and vibration motor used in this paper.

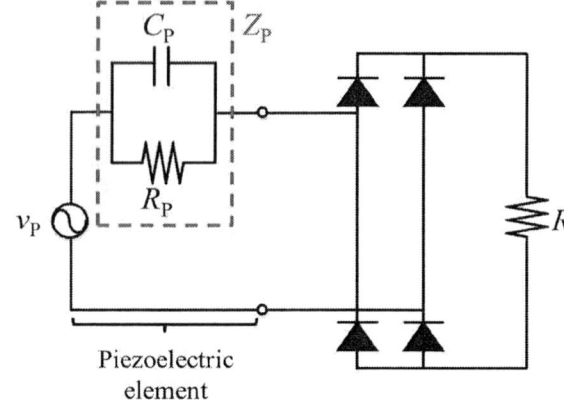

Fig. 2. A test circuit for characterizing the generated power of the piezoelectric element.

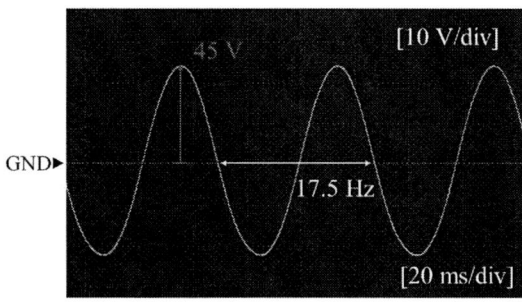

Fig. 3. An output waveform of the piezoelectric element.

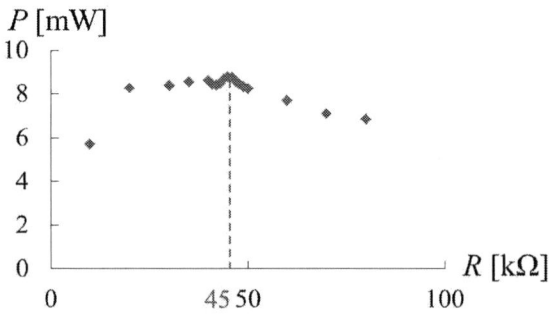

Fig. 4. Experimental result of the generated power characteristics in Fig.2.

In vibration power generation system, the input side of the boost chopper behaves as a resister by the proposed control method. Since it becomes an equivalent circuit to the test circuit of Fig.2, the generated power of the piezoelectric element can be maximized. The proposed equivalent resistance control is performed by boost chopper. The output voltage v_{out} is controlled to be constant at 3.3V.

The 2018 International Power Electronics Conference

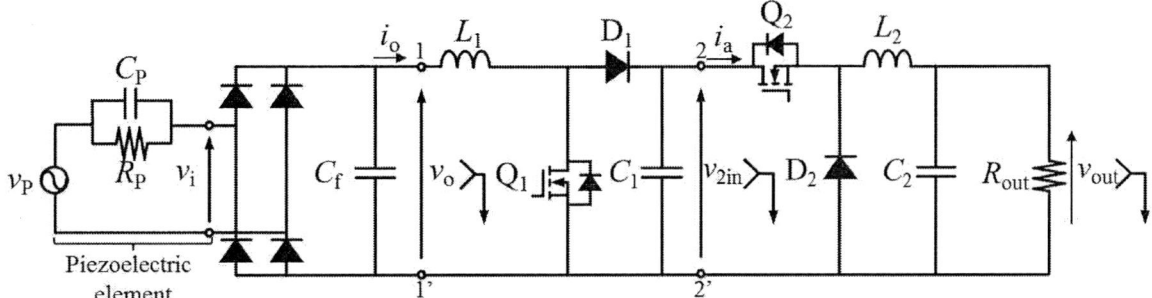

Fig. 5. Power circuit diagram of the proposed vibration power generation system.

Fig.6 shows the control block of the vibration power generation system with the proposed equivalent resistance control. For maximizing the power transfer of the piezoelectric element, the input voltage v_o and current i_o are controlled in phase by using the boost chopper with the proposed equivalent resistance control. The reference input current i_o^* is given by

$$i_o^* = \frac{v_o}{Z_P},\qquad(2)$$

where Z_P is the internal impedance of the piezoelectric element. Therefore, the dc-side of the diode rectifier behaves as a resister. In this case, the generated power of the piezoelectric element is maximized by the law of the maximum power transfer.In Fig.6, the LPFs(Low-Pass Filters) are used for eliminating the switching ripple. The cut-off frequency of the LPFs is 2 kHz.

The over voltage protection control is used in the proposed vibration power generation system. If the generated power is larger than the consumed power, v_{2in} increased by surplus power between the generated and consumed powers. When v_{2in} is larger than 45 V, Q_1 turns off. When v_{2in} is smaller than 40 V, Q_1 is switching. The buck chopper controls the output voltage v_{out} by the PI controller. If v_{2in} is too small, the equivalent resistance control cannot achieve. Thus, Q_2 turns off when v_{2in} is smaller than 20 V.

III. SIMULATION RESULTS

A. Representation of the piezoelectric element model

We confirm the representation of the constructed piezoelectric element model on PLECS. The equivalent circuit of the piezoelectric element in Fig.2 was used for the piezoelectric element model. The maximum value of the piezoelectric element is 45 V and the frequency is 17.5 Hz.

Fig.7 shows the generated power characteristics of the constructed piezoelectric element model on PLECS. From Fig.7, the maximum generated power and the impedance characteristics are the same as Fig.4.

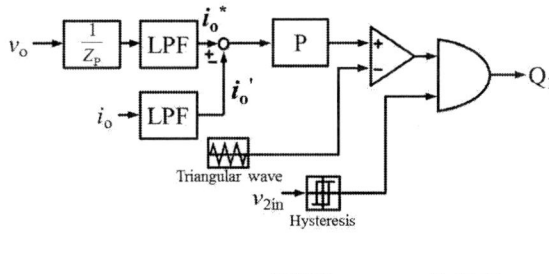

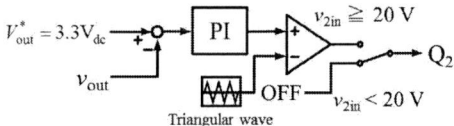

Fig. 6. Control block of the proposed control method.

B. Simulation results for vibration power generation system

A computer simulation using PLECS software is implemented to confirm the validity of the proposed equivalent resistance control method. Table 1 shows the circuit constants of Fig.5. The circuit constants in Fig.8 are the same as Table 1. The on-state resistances of diode and MOSFET are considered on PLECS.

Fig.9 shows the simulated waveforms without the proposed method. The load resistance R_{out} is 5.3 kΩ. v_i and v_o are the input and output voltages of the diode rectifier and boost chopper, i_o and i_o' are the input current of the boost chopper without and with a low-pass filter, v_{out} is the output voltage of the buck chopper, and p_{out} is the output power of the buck chopper. From this simulation results, v_o and i_o' are not in phase. Thus, the input side of the boost chopper does not behaves as an equivalent resistance. The output voltage was controlled by the buck chopper. Thus, the output power p_{out} is 2.10 mW.

1383

The 2018 International Power Electronics Conference

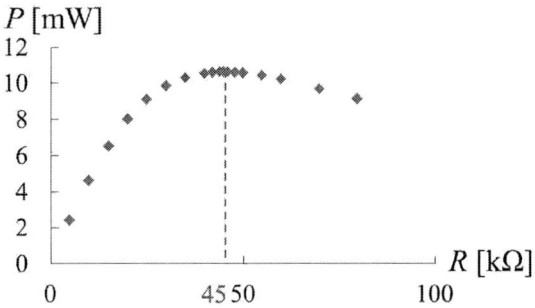

Fig. 7. Generated power characteristics of the constructed piezoelectric element model on PLECS.

TABLE I. CIRCUIT CONSTANTS OF THE PROPOSED SYSTEM

Symbol	Meaning	Value
v_P	Piezoelectric element voltage	45 V
f	Piezoelectric element frequency	17.5 Hz
C_P	Internal capacitance	206 nF
R_P	Internal resistance	19 MΩ
V_f	Forward voltage	0.58 V
R_{on}	on-state resistance of diode	17.5 mΩ
C_f	Filter capacitance	10 pF
L_1	Boost inductance	65 μH
C_1	Boost capacitance	10 μF
f_{SW1}	Switching frequency of boost chopper	20 kHz
$R_{SW_{on}}$	on-state resistance of MOSFET	100 mΩ
L_2	Inductance of buck chopper	633 μH
C_2	Capacitance of buck chopper	10 μF
f_{SW2}	Switching frequency of buck chopper	20 kHz

Fig.10 shows the simulation waveforms with the proposed method. The load resistance R_{out} is 1.5 kΩ. v_{2in} is the input voltage of the buck chopper. From this simulation result, v_o and i'_o were in phase by the equivalent resistance control. Therefore, the input side of the boost chopper behaves as an equivalent resistance. The average values v_o and i_o were 19.7 V and 0.44 mA, respectively. Thus, the equivalent resistance was 45 kΩ. The generated power of the piezoelectric element is maximized. v_{2in} was controlled within 45 V by the over voltage protection control. The output voltage v_{out} can be controlled to be 3.3 V and the output power p_{out} was 7.36 mW. The output power is approximately 70% as compared to maximum power in Fig. 7. The power loss of diode rectifier is responsible for the majority of power loss of the proposed vibration power generation system. The power loss of diode rectifier becomes large because the boost chopper acts in discontinuous conduction mode (DCM).

Fig.11 shows the relationship between the output

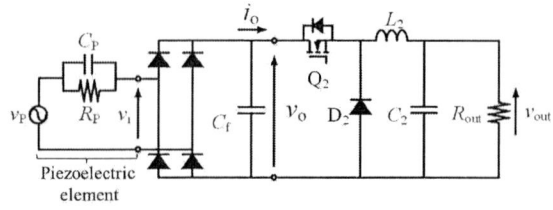

Fig. 8. Power circuit diagram of the vibration power generation system without the proposed control method.

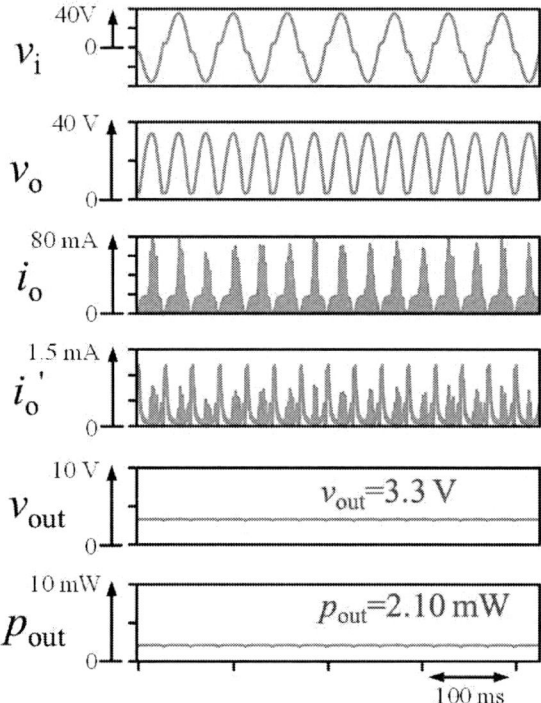

Fig. 9. Simulated waveforms without the proposed method.

power p_{out} and load resistance R_{out} in Fig.5 and Fig.8. In Fig.11, the output voltage v_{out} is at constant 3.3 V. In the case of the vibration power generation system without the proposed method, the minimum load resistance is 5.3 kΩ. However, the inimum load resistance of the vibration power generation system with the proposed method is 1.5 kΩ. As a result, the generated power in Fig.5 can be increased 3.5 times larger than it in Fig.8.

IV. CONCLUSION

This paper proposed a novel equivalent resistance control for maximum power transfer method of piezoelectric element in vibration power generation. In the proposed method, the rectified voltage and current of a diode rectifier are in phase by the boost chopper. Thus, the input side of the boost chopper behaves as a resistor equivalently. For maximizing the generated power of the piezoelectric element, this resistance is controlled to the internal resistance of the piezoelectric element by the law of the maximum power transfer. The proposed control method is very simple as compared to the conventional methods.

A Computer simulation was implemented to demonstrate the validity of the proposed method using PLECS. From the simulation results, we confirmed that the output side of rectifier circuit operates as a resistor when the supply power is maximized. The output power was 7.36 mW with the proposed method. Simulation results demonstrate that this generated power with the proposed method is 3.5 times as large as it without the proposed method.

REFERENCES

[1] G. K. Ottman, H. F. Hofmann, A. C. Bhatt, and G. A. Lesieutre, "Adaptive piezoelectric energy harvesting circuit for wireless remote power supply", *IEEE Transactions on Power Electronics*, Vol.17, No.5, pp. 669-676, 2002.

[2] S. Roundy, P. K. Wright, and J. Rabaey, "A study of low level vibrations as a power source for wireless sensor nodes", *Computer Communications*, Vol.26, No.11, pp.1131-1144, 2003.

[3] D. Guyomar, A. Badel, E. Lefeuvre, and C. Richard, "Toward energy harvesting using active materials and conversion improvement by nonlinear processing", *IEEE Transactions on Ferroelectrics and Frequency Control*, Vol.52, No.4, pp.584-595, 2005

[4] A. Nechibvute, A. Chawanda, and P. Luhanga, "Piezoelectric Energy Harvesting Devices: An Alternative Energy Source for Wireless Sensors", *Hindawi Publishing Corporation Smart Materials Research*, pp.1-3, 2012

[5] S. Boisseau, P. Gasnier, M. Gallardo, G. Despesse, "Self-starting power management circuits for piezoelectric and electret-based electrostatic mechanical energy harvesters", *Journal of Physics*, Vol.476, pp.1-2, 2013

[6] R. Calio, U. B. Rongala, D. Camboni, M. Milazzo, C. Stefanini, G. Petris, and C. M. Oddo, "Piezoelectric Energy Harvesting Solutions", *Sensors*, Vol.14, pp.4755-4757, 2014

[7] K. Takagi, Y. Yamada, and T. Inoue, "Sensorless Parameter Estimation of Piezoelectric Elements with the Consideration of the Internal Resistance Only Based on Impedance Measurement", *The Japan Society of Mechanical Engineers*, Vol.78, No.792, pp.2808-2815, 2012

[8] T. Tanaka, T. Aonuma, K. Natori, and Y. Sato, "A study on Increasing Harvested Energy in Vibration Power Generation Using Piezoelectric Elements" *IEE Japan*, Vol.1, No.93, pp.385-387, 2015

[9] E. Lefeuvre, D. Audigier, C. Richard, and D. Guyomar, "Buck-boost converter for sensorless power optimization of piezoelectric energy harvester," *IEEE Trans. Power Electron.*, Vol.22, No.5, pp.2018-2025, 2007

[10] Y. Li, D. Audigier, P. Combette, J. L. Compeau. E. Lefeuvre, P. Muralt, L. Petit, and C. Richard, "Simple techniques for piezoelectric energy harvesting optimization", *Electronics INASA de Lyon*, Vol.77, pp.74-80, 2014

[11] Y. P. Liu and D. Vasic "Self-Powered Electronics for Piezoelectric Energy Harvesting Devices", *INTECH*, Vol.14, pp.333-346, 2012

[12] J. H. Pedersen, A. Knott, O. C. Thomsen, T. Andersen, T, Sorensen, and P. Spies, "Low Frequency Low Voltage Vibration Energy Harvesting Converter", *DTU Electrical Engineering*, pp.33-35, 2011

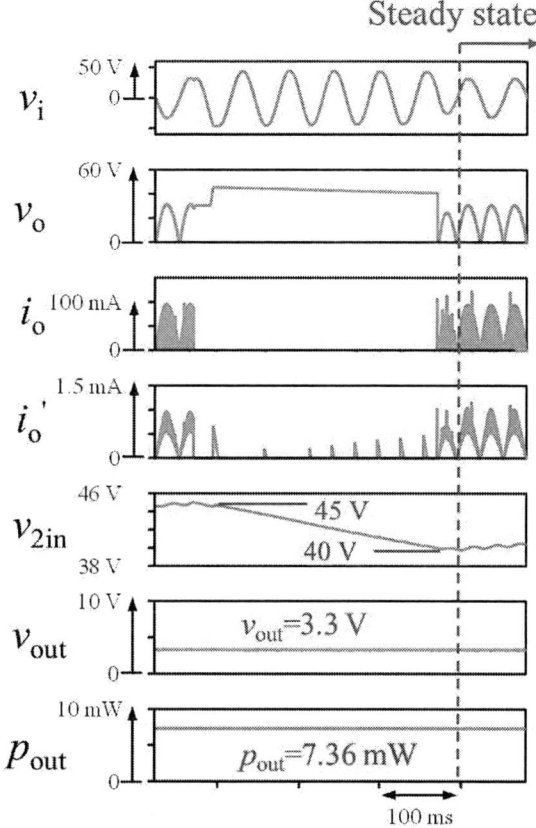

Fig. 10. Simulation waveforms with the proposed method.

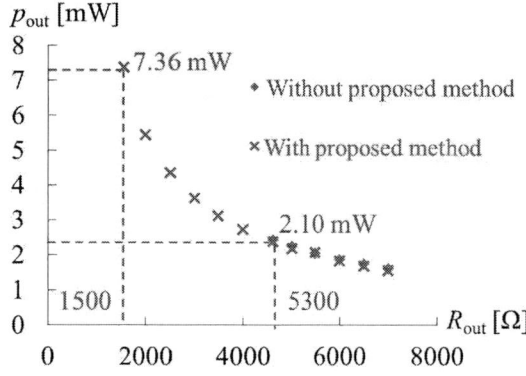

Fig. 11. Relationship between the output power p_{out} and load resistance R_{out}.

The 2018 International Power Electronics Conference

DC Bus Voltage Stabilization for Cascaded Power Converter by Integrating an Extra Port into Load Side PSFB

Jiang You, Weiyan Fan[*] and Mengyan Liao

College of Automation, Harbin Engineering University, Harbin, China

fwy1993@hrbeu.edu.cn

Abstract— **It is well known that the tight regulation of load converter might cause voltage oscillation in DC bus of cascaded power converter system. By sharing the magnetic core of the phase shift full bridge (PSFB), an isolated extra port is proposed to integrate into the PSFB in this paper, which is employed to stabilize the output voltage of PSFB front end DC power source through proper control. Modeling scheme and control strategy are presented. The simulation results demonstrate the validation of the proposed method.**

Keywords— *Cascaded converter, DC bus voltage, Dc bus stabilization, integrated DC bus voltage conditioner.*

I. INTRODUCTION

Cascaded power converter system is composed of different types of converters that have been widely used in a variety of fields due to its flexibility, scalability and redundancy and other advantages. Because of the fact that there is not only with simple series or parallel connection between these converters, the interaction between the source converters and load converters often exists. Furthermore, the high control bandwidth of load converter makes it behave like a constant power load (CPL) which might lead to stability issues in the cascaded system [1-3].

According to Middlebrook criterion the source converter output impedance should be much smaller than the input impedance of load converter, which is required to guarantee the stability of the cascaded power converter system [4]. This requirement can be accomplished by using either passive or active compensation methods. Bulky electrolytic capacitor is the most common used passive method to stabilize the DC bus voltage, but this might lead some issues about power density and reliability [5]. In [6] and [7], the DC bus voltage or current information is used as compensation signal to decrease the source converter output impedance, thus the impedance overlapping can be reduced. While in [8], the DC bus voltage information is integrated into the load converter, in this condition, the phase of load converter input impedance is enhanced to leave -180° in low

This work is supported by the national NSF of China (No. 51479042), the NSF (No. E201238) of Heilongjiang province, China, and the Fundamental Research Funds for the Central Universities (No. GK2040260197)

frequency range, thus the CPL characteristics of load converter is relieved. Though these methods are effective, the control performance of DC bus voltage or the load converter output voltage might be influenced heavily by active compensation methods. For example, the transfer function gain of the DC bus voltage to the output voltage of load converter is increased, the dynamic change of DC bus voltage has more impact on the load converter output voltage. DC bus voltage conditioner is another valid active method to reduce DC bus voltage oscillation [9] [10], which includes a separate converter with extra power switches and energy storage components. This method might increase cost, weight/volume and power loss of the whole system. In [11], by introducing a LC branch, an integrated DC bus voltage conditioner scheme is proposed, in which the DC bus voltage conditioner is implemented by multiplexing utilization of the switches of PSFB, this method can potentially reduce cost and improve power density of the whole system.

Another isolated scheme for integrated DC bus voltage conditioner is proposed in this paper by adding an extra port for the PSFB load converter. In section II, the converter topology modulation and the stability analysis of DC bus voltage is presented firstly. The modeling and control scheme are proposed in section III. Simulation results with brief illustrations are given in section IV. Finally, the conclusion is drawn in section V.

II. SYSTEM DESCRIPTION AND STABILITY ANALYSIS

The topology of the proposed cascaded power converter is shown in Fig. 1. For simplification, the unideal power source is represented by an ideal source v_s and a LC filter. Compared to the conventional PSFB converter, a half bridge isolated port is proposed as an integrated DC bus voltage conditioner. The capacitor C_b is used as energy buffer component, L_b is used to limit the amplitude of high frequency ripple current in the L_bC_b circuit. In this topology, the PSFB is modulated and controlled in the normal way.

As shown in Fig. 1, the output impedance of the power source can be written as (1).

$$Z_o = \frac{L_{dc} + R_{dc}}{L_{dc}C_{dc}s^2 + R_{dc}C_{dc}s + 1} \tag{1}$$

The 2018 International Power Electronics Conference

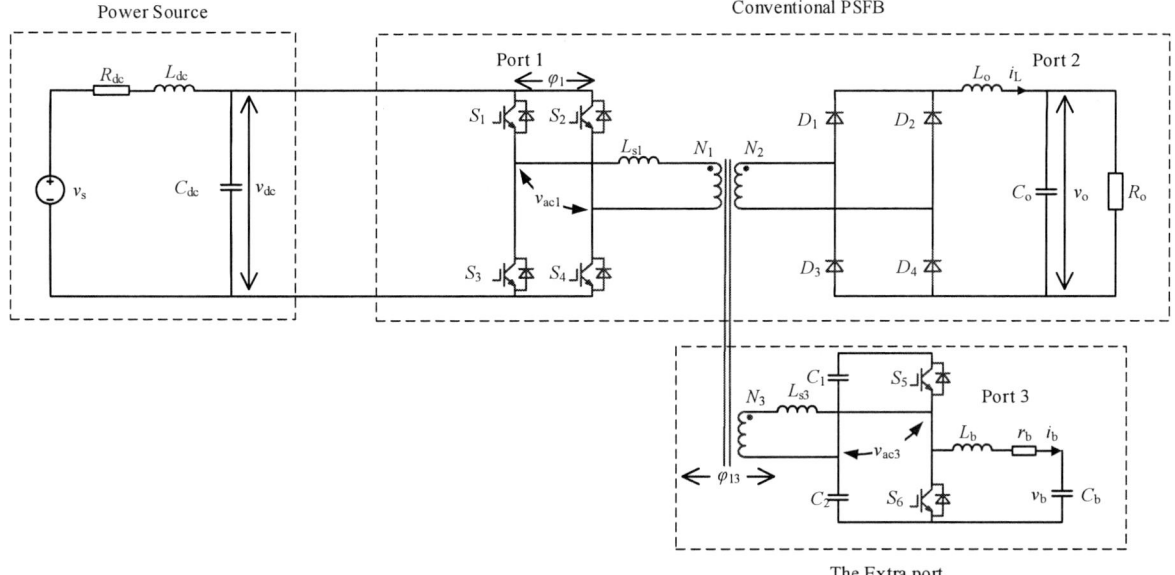

Fig. 1. The topology of the cascaded system with an extra port.

The input impedance of the PSFB with a single voltage control loop can be expressed as (2) if the extra port is not considered.

$$Z_{in} = \frac{1 + G_{vc}F_m G_{vd}}{Y_i(1 + A_v G_{vc}F_m G_{vd}) - A_v G_{vc}F_m G_{id}} \quad (2)$$

In (2), G_{vc} is the voltage controller, F_m is the modulator model, G_{vd} is the transfer function of φ_1-to-v_o, G_{id} is the transfer function of φ_1-to-i_L, A_v is the transfer function of v_{dc} -to- v_o.

Fig. 2 shows the bode diagrams of (1) and (2) by using the parameters listed in TABLE I. It can be seen that there is an impedance overlapping around 100Hz, which might cause stability issue if the DC bus capacitor C_{dc}=900µF. While the DC bus voltage stability can be improved if C_{dc} is increased to 5000µF. The Nyquist diagram given in Fig. 3 also shows that minor loop gains of T_m=Z_o/Z_{in} embrace the point (-1, j0) if C_{dc}=900µF, which indicates the DC bus voltage is unstable.

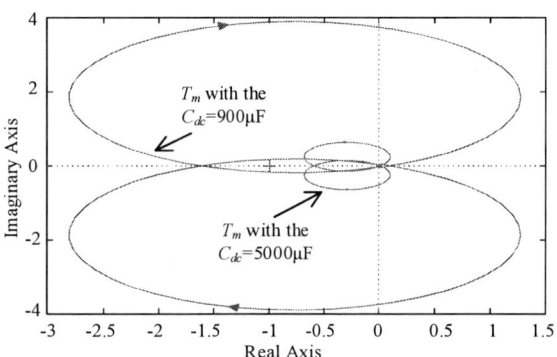

Fig. 3. The Nyquist diagram of T_m with C_{dc}=900µF and C_{dc} =5000µF respectively.

TABLE I
THE SYSTEM PARAMETER

Symbol	Meaning	Value
V_{dc}	DC bus voltage	135 V
V_o	Output voltage	40 V
R_{dc}	DC bus resistance	0.05 Ω
L_{dc}	DC bus inductance	3 mH
C_{dc}	DC bus capacitance	900 µF/5000 µF
R_o	Load	8 Ω/1.7 Ω
L_o	Output inductance	0.15 mH
C_o	Output capacitance	1100 µF
f_s	Switching frequency	20 kHz

III. MODELING AND CONTROL SCHEME

A. Power Flow Calculating

The introduced extra port, port 3 and port 1 are controlled like a dual active bridge (DAB). The modulation scheme is shown in Fig.4. In Fig. 4, the duty cycles of S1-S6 are 50%, the drive signal of S1 is set as reference, φ_1 is the phase shifting between S1 and S2, φ_{13} is the phase shifting between v_{ac1} and v_{ac3}.

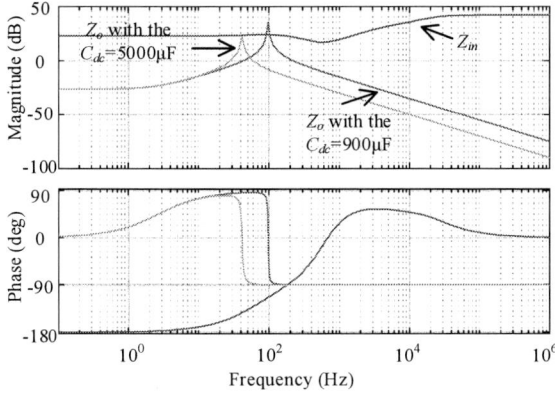

Fig. 2. The bode diagram of Z_{in} and Z_o with C_{dc}=900µF and C_{dc} =5000µF respectively.

1387

The 2018 International Power Electronics Conference

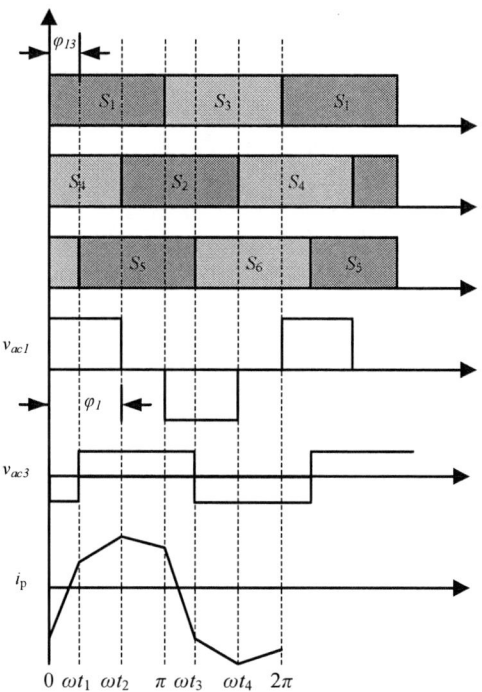

Fig. 4. Modulation scheme with $\varphi_1 > \varphi_{13}$.

Compared with light load condition, the PSFB with heavy load is much easier to lead voltage oscillation in DC bus, and if C_b is properly selected, $\varphi_1 > \varphi_{13}$ can be held in a wide range. As shown in Fig. 4, there are six operation modes in a switching period, the leakage inductor current, i_p is symmetrical in each half cycle, it can be expressed as (3).

$$\begin{cases} i_p(\theta) = \dfrac{V_{dc} + V_b'}{\omega L_s}\theta + i(0) & 0 < \theta \le \varphi_{13} \\[2mm] i_p(\theta) = \dfrac{V_{dc} - V_b'}{\omega L_s}(\theta - \varphi_{13}) + i(\varphi_{13}) & \varphi_{13} < \theta \le \varphi_1 \quad (3) \\[2mm] i_p(\theta) = \dfrac{-V_b'}{\omega L_s}(\theta - \varphi_1) + i(\varphi_1) & \varphi_1 < \theta \le \pi \end{cases}$$

$L_s = L_{s1} + (N_1/N_3)^2 L_{s3}$ is the total primary-referred leakage inductor, V_b' is the primary-referred voltage of the extra port. By using the boundary condition (4) [12].

$$i_p(0) = -i_p(\pi) \qquad (4)$$

The current at the moment of φ_1, φ_{13}, π can be solved in (5).

$$\begin{cases} i_p(\varphi_{13}) = \dfrac{V_b'}{2\omega L_s}\pi - \dfrac{V_{dc}}{2\omega L_s}\varphi_1 + \dfrac{V_{dc}}{\omega L_s}\varphi_{13} \\[2mm] i_p(\varphi_1) = \dfrac{V_b'}{2\omega L_s}\pi + \dfrac{V_{dc} - 2V_b'}{2\omega L_s}\varphi_1 + \dfrac{V_b'}{\omega L_s}\varphi_{13} \quad (5) \\[2mm] i_p(\pi) = -\dfrac{V_b'}{2\omega L_s}\pi + \dfrac{V_{dc}}{2\omega L_s}\varphi_1 + \dfrac{V_b'}{\omega L_s}\varphi_{13} \end{cases}$$

The power delivered from port 1 to port 3 can be calculated as (6).

$$\begin{aligned} P &= \frac{1}{\pi}\int_0^\pi v_{ac1}(\theta)i_p(\theta)d\theta \\ &= \frac{V_{dc}V_b'}{2\pi\omega L_s}\left(-2\varphi_{13}^2 + 2\varphi_1\varphi_{13} - \varphi_1^2 + \pi\varphi_1\right) \end{aligned} \qquad (6)$$

It can be seen from (6), the power is determined by φ_{13} in steady state when φ_1 is constant.

B. Modeling

In Fig. 1, the inductor current i_b, the capacitor voltage v_{c1}, v_{c2} and v_b are selected as the state variables. Firstly, six differential equation groups can be obtained corresponding to the six switching modes in Fig. 4, then (7) can be derived by introducing averaging operator in a switching period.

$$\begin{cases} C_1\dfrac{d\langle v_{C1}\rangle}{dt} = \dfrac{1}{2\pi}\int_{\varphi_{13}}^{\pi+\varphi_{13}} i_p d\theta - \dfrac{1}{2\pi}\int_{\varphi_{13}}^{\pi+\varphi_{13}} i_b d\theta \\[2mm] C_2\dfrac{d\langle v_{C2}\rangle}{dt} = -\dfrac{1}{2\pi}\left[\int_0^{\varphi_{13}} i_p d\theta + \int_{\pi+\varphi_{13}}^{2\pi} i_p d\theta\right] - \dfrac{1}{2\pi}\int_{\varphi_{13}}^{\pi+\varphi_{13}} i_b d\theta \quad (7) \\[2mm] L_b\dfrac{d\langle i_b\rangle}{dt} = -\langle v_b\rangle - \langle i_b\rangle r_b + \dfrac{1}{2\pi}\int_{\varphi_{13}}^{\pi+\varphi_{13}}(v_{C1}+v_{C2})d\theta \\[2mm] C_b\dfrac{d\langle v_b\rangle}{dt} = \langle i_b\rangle \end{cases}$$

Because the duty cycles of S5 and S6 are 50%, if $C_1 = C_2 = C$ in (7) then $v_{c1} = v_{c2}$. A new state variable $v_{c12} = v_{c1} + v_{c2}$ can be introduced to reduce the orders of (7). Besides, i_p is the high frequency AC current in transformer, the average value of i_p is zero in a switch period, so (3) - (5) are used to cancel i_p in (7) [13]. The final small signal model can be obtained as (8).

$$\begin{cases} C\dfrac{d\tilde{v}_{12}}{dt} = A_1\tilde{v}_{dc} + A_2\tilde{\varphi}_{13} - \tilde{i}_b \\[2mm] L_b\dfrac{d\tilde{i}_b}{dt} = -\tilde{v}_b - \tilde{i}_b r_b + \dfrac{1}{2}\tilde{v}_{12} \quad (8) \\[2mm] C_b\dfrac{d\tilde{v}_b}{dt} = \tilde{i}_b \end{cases}$$

In (8)

$$\begin{cases} A_1 = \dfrac{\partial v_{12}}{\partial v_{dc}} = \dfrac{1}{2\pi\omega L_s}\left(\pi\varphi_1 - \varphi_1^2 + 2\Phi_{13}\varphi_1 - 2\Phi_{13}^2\right) \\[2mm] A_2 = \dfrac{\partial v_{12}}{\partial \varphi_{13}} = \dfrac{V_{dc}}{2\pi\omega L_s}\left(2\varphi_1 - 4\Phi_{13}\right) \end{cases} \qquad (9)$$

Φ_{13} is the steady state value of φ_{13}.

The transfer function of φ_{13}-to-v_b can be derived accordingly from (8).

$$G_{v\varphi 13}(s) = \frac{\tilde{v}_b}{\tilde{\varphi}_{13}} = \frac{A_2(1/C_b s)}{2CL_b s^2 + 2Cr_b s + 1 + \dfrac{2C}{C_b}} \qquad (10)$$

1388

The 2018 International Power Electronics Conference

C. Control Scheme

The control scheme of the extra port for DC bus voltage stabilization is presented in Fig. 5.

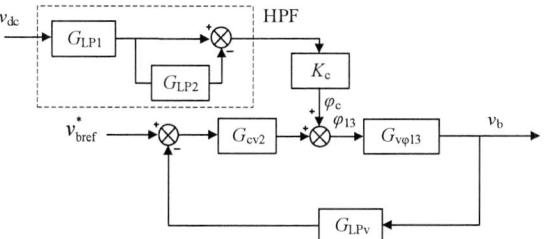

Fig. 5. Control scheme of the extra port.

In this figure, the fluctuation of DC bus voltage is extracted by a high pass filter (HPF) which consists of G_{LP1} and G_{LP2}, the compensation signal φ_c is got by multiplying the output of HPF with an appropriate coefficient K_c, which is used to modify the phase shift angle that is output by the voltage controller G_{vc2}. The output of the controller G_{vc2} is used to regulate the phase shifting, φ_{13} between port 1 and the extra port, which can force C_b to absorb or release power to suppress the change of v_b and attenuate the oscillation of DC bus voltage accordingly.

IV. SIMULATION AND ANALYSIS

A simulation model for the system in Fig.1 is built by using Matlab/Simulink to verify the effectiveness of the proposed method. The parameters of the system are listed in TABLE I. When the DC bus capacitor C_{dc}=900μF, the simulation results of the cascaded PSFB without the proposed extra port is shown in Fig. 6. It can be seen in this figure that continuous oscillation is caused in the DC bus voltage when a sudden load change is added at 0.5s, the output voltage is also degraded heavily.

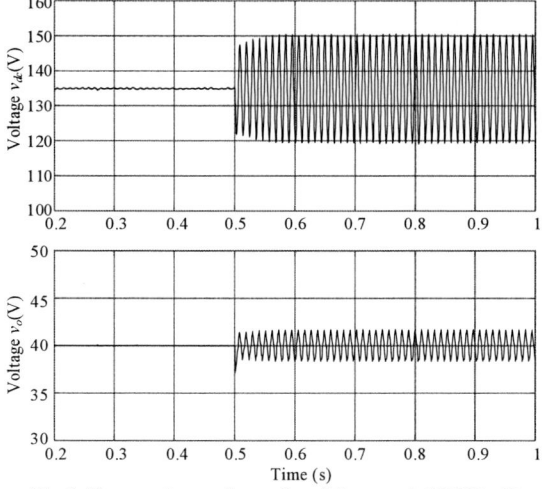

Fig. 6. The waveforms of v_{dc} and v_o of the cascaded PSFB with C_{dc}=900μF (without the proposed extra port).

In Fig. 7, the DC bus capacitor C_{dc} is 5000μF, the DC

bus voltage can be damped gradually to its steady state value, and the dynamic performance of output voltage is also improved. The simulation results shown in Fig.6 and Fig.7 are consistent with the frequency domain analysis presented in section II.

Fig. 8 shows the simulation result with the proposed method and C_{dc}=900μF. When the same load change is exerted in this condition, the oscillation of DC bus voltage is reduced gradually in about 0.1s, and the output voltage also recover to its normal value quickly.

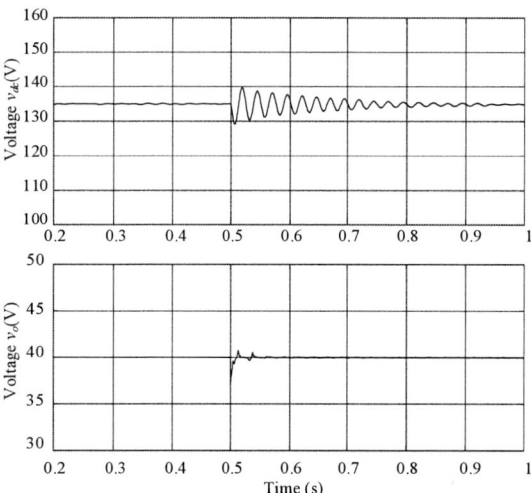

Fig. 7. The waveforms of v_{dc} and v_o of the cascaded PSFB with C_{dc}=5000μF (without the proposed extra port).

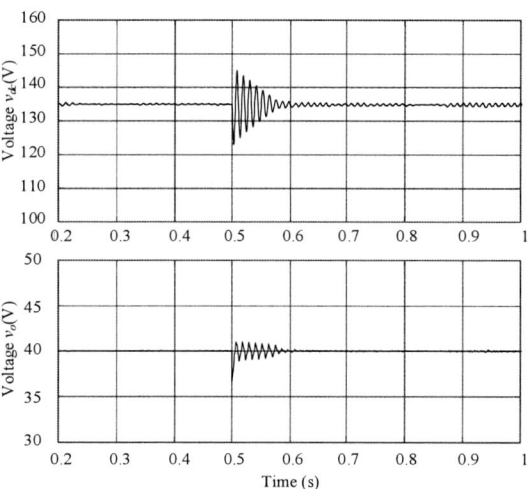

Fig. 8. The waveforms of v_{dc} and v_o with the proposed extra port (with C_{dc}=900μF).

A comparison study of v_{dc} fluctuation is shown in Fig.9. It can be observed that the peak value of the voltage fluctuation is smaller as C_{dc}=5000μF. However, the transient recovery time by using the proposed method is much shorter and it is beneficial to improve power density, reliability and lifetime of the whole power converter system by reducing bulky electrolytic capacitor.

1389

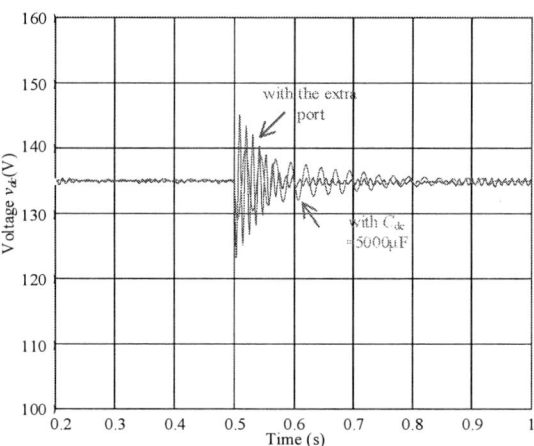

Fig. 9. The waveforms of v_{dc} of the two stable conditions.

V. CONCLUSIONS

In this paper, a novel method is employed to stabilize the DC bus voltage of the cascaded system. in the proposed method, an isolated half bridge circuit is integrated into the conventional PSFB as an extra port. The mathematic analysis and modeling of this extra port is presented. The effectiveness of the proposed method is validated through simulation results.

ACKNOWLEDGMENT

This work is supported by the national NSF of China (No. 51479042), the NSF (No. E201238) of Heilongjiang province, China, and the Fundamental Research Funds for the Central Universities (No. GK2040260197).

REFERENCES

[1] C. M. Wildrick, F. C. Lee, B. H. Cho and B. Choi, "A method of defining the load impedance specification for a stable distributed power system," *IEEE Transactions on Power Electronics*, vol. 10, no. 3, pp. 280-285, May 1995.

[2] X. Li and X. Ruan, "Analysis of an unstable phenomenon related to the different turn-on time between the subsystems in cascaded systems," *2015 IEEE Energy Conversion Congress and Exposition (ECCE)*, Montreal, QC, 2015, pp. 2658-2663.

[3] Kun Xing, Jinghong Guo, Wenkang Huang, Dengming Peng, F. C. Lee and D. Borojevic, "An active bus conditioner for a distributed power system," *30th Annual IEEE Power Electronics Specialists Conference. Record. (Cat. No.99CH36321)*, Charleston, SC, 1999, pp. 895-900 vol.2.

[4] R. D. Middlebrook, "Input filter considerations in design and application of switching regulators," *Conf. Rec. IEEE IAS Annu. Meeting*, pp. 366–382, 1976.

[5] L. Xing, F. Feng and J. Sun, "Optimal Damping of EMI Filter Input Impedance," *IEEE Transactions on Industry Applications*, vol. 47, no. 3, pp. 1432-1440, May-June 2011.

[6] W. Cai, F. Yi, E. Cosoroaba and B. Fahimi, "Stability Optimization Method Based on Virtual Resistor and Nonunity Voltage Feedback Loop for Cascaded DC–DC Converters," *IEEE Transactions on Industry Applications*, vol. 51, no. 6, pp. 4575-4583, Nov.-Dec. 2015.

[7] A. M. Rahimi and A. Emadi, "Active Damping in DC/DC Power Electronic Converters: A Novel Method to Overcome the Problems of Constant Power Loads," *IEEE Transactions on Industrial Electronics*, vol. 56, no. 5, pp. 1428-1439, May 2009.

[8] X. Y. Wang, D. M. Vilathgamuwa and S. S. Choi, "Decoupling Load and Power System Dynamics to Improve System Stability," *2005 International Conference on Power Electronics and Drives Systems*, Kuala Lumpur, 2005, pp. 268-273.

[9] X. Zhang, X. Ruan, H. Kim and C. K. Tse, "Adaptive Active Capacitor Converter for Improving Stability of Cascaded DC Power Supply System," *IEEE Transactions on Power Electronics*, vol. 28, no. 4, pp. 1807-1816, April 2013.

[10] J. You, Z. P. Fan, Y. Hu and M. J. Deng, "Virtual resistor based DBVC and active damping method for DC bus stabilization of cascaded power converters system," *2017 IEEE Transportation Electrification Conference and Expo, Asia-Pacific (ITEC Asia-Pacific)*, Harbin, 2017, pp. 1-6.

[11] Jiang You, D. M. Vilathgamuwa, N. Ghasemi and W. L. Malan, "Analysis and control of integrated DC bus voltage conditioner for cascade power converter system," *2017 IEEE 3rd International Future Energy Electronics Conference and ECCE Asia (IFEEC 2017 - ECCE Asia)*, Kaohsiung, 2017, pp. 522-527.

[12] Hui Li, Fang Zheng Peng and J. S. Lawler, "A natural ZVS medium-power bidirectional DC-DC converter with minimum number of devices," *IEEE Transactions on Industry Applications*, vol. 39, no. 2, pp. 525-535, Mar/Apr 2003.

[13] Hui Li, Fang Zheng Peng and J. Lawler, "Modeling, simulation, and experimental verification of soft-switched bi-directional DC-DC converters," *APEC 2001. Sixteenth Annual IEEE Applied Power Electronics Conference and Exposition (Cat. No.01CH37181)*, Anaheim, CA, 2001, pp. 736-742 vol.2.

Common Mode Current Reduction of Three-phase Cascaded Multilevel Transformerless Inverter for PV System

Wenjie Wang[1], Ke Chen[1], Lijun Hang[1], Anping Tong[2], Yiliang Gan[3]

1: Hangzhou Dianzi University, No 1588, Avenue 2, Hangzhou, China

2: Shanghai Jiao Tong University, 800, Dongchuan Road, Shanghai, China

3: General Office of People's Government of Shuangliu District, Chengdu, China

Abstract — The reduction of common mode leakage current in centralized inverter for grid-connected PV (photovoltaic) system has been well studied. However, the research in cascaded multilevel inverter PV system is not far enough. This paper focuses on analyzing the generation mechanism of common mode leakage current and building equivalent circuit model of cascaded PV system. By clarifying the reason why the conventional cascaded H4 (full-bridge inverter) PV system fails to reduce the common mode leakage current, a new cascaded iH6 inverter PV system is proposed. The related theoretical analysis and simulation results are given, and it is verified that the common mode leakage current in three phase grid-connected system with proposed converters can be greatly reduced compared with the conventional cascaded H4 system.

Keywords—Transformerless; common mode current; common mode voltage; cascaded multi-level inverter; PV grid-connected system.

I. INTRODUCTION

Due to the drawback of low efficiency, large volume and weight, and high cost, the transformer in Grid-connected PV system is removed gradually [1]. The inverter without transformer is becoming popular in low power level (2-10kW) grid-connected domestic distributed PV system due to its high efficiency and low cost [2]. However, the galvanic connection between the grid and PV array, and the parasitic capacitance between the ground and PV panels result in the appearance of common mode leakage current. The leakage current would decrease the conversion efficiency, reduce the grid current quality, and induce the severe conducted and radiated EMI [3].

In order to eliminate the common mode current, its

inducing principle has been well studied [1] and a model of common mode current circuit is given. Basing on this model, several kinds of topologies and modulation method have been proposed in the last few years. However, the application of these topologies and reduction effect of common mode leakage current in the cascaded multilevel inverter have not been studied yet. In this paper, the common mode equivalent model of the cascaded inverter basing on three level inverter common mode model is investigated. The reason why the conventional cascaded H4 PV system fails to reduce the common mode leakage current is clarified and a new cascaded iH6 inverter PV system is proposed. At the end of this paper, the iH6 system's excellent suppression of common-mode leakage current in single-phase and three-phase systems was verified through simulation and experiments.

II. CASCADED H4 INVERTER

Since the three-phase cascaded multilevel converter system is symmetrical and it can be equivalent by three independent single-phase, the common-mode voltage analysis of the three-phase system in this paper is replaced by analyzing the single-phase common-mode voltage.

A single-phase 7-level cascaded H4 inverter for the grid-connected PV system structure is shown as Fig.1(a). C_1, C_2, C_3 are the stay capacitances between PV array and ground, and i_{CM1}, i_{CM2}, i_{CM3} are the common mode leakage current through C_1, C_2, C_3. L is the filter inductor. The inverter's common and differential voltage circuit can be shown as Fig.1(b). U_{CM} is defined as the common mode voltage and U_{DM} is defined as the differential mode voltage.

The 2018 International Power Electronics Conference

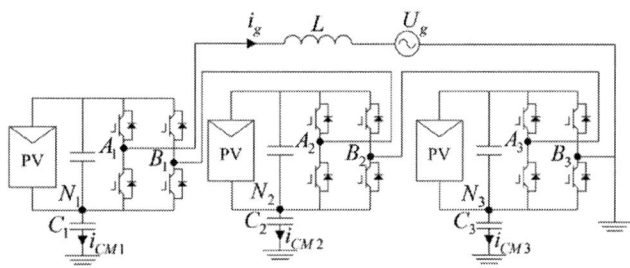

Fig.1 (a) Single phase 7-level cascade H4 inverter PV system

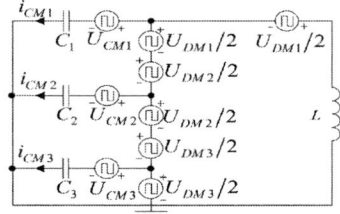

Fig.1 (b) Common mode and differential mode voltage model

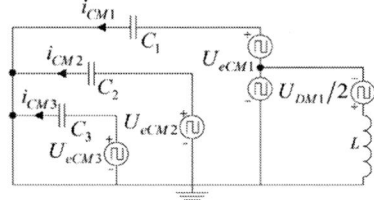

Fig.1 (c) Common mode equivalent circuit

Fig.1(c) shows the simplified common-mode equivalent circuit for single-phase cascaded H4 PV system, where U_{eCM} is defined as the equivalent common mode voltage and it can be derived from Fig.1(b) as following:

$$U_{CM} = (U_{AN} + U_{BN})/2 \tag{1}$$

$$U_{DM} = U_{AN} - U_{BN} = U_{AB} \tag{2}$$

$$U_{eCM1} = U_{CM1} + U_{DM1}/2 + U_{DM2} + U_{DM3} = U_{A_1N_1} + U_{DM2} + U_{DM3} \tag{3}$$

$$U_{eCM2} = U_{CM2} + U_{DM2}/2 + U_{DM3} = (U_{A_2N_2} + U_{B_2N_2})/2 +$$
$$(U_{A_2N_2} - U_{B_2N_2})/2 + U_{DM3} = U_{A_2N_2} + U_{DM3} \tag{4}$$

$$U_{eCM3} = U_{DM3}/2 + U_{CM3} = (U_{A_3N_3} - U_{B_3N_3})/2 +$$
$$(U_{A_3N_3} + U_{B_3N_3})/2 = U_{A_3N_3} \tag{5}$$

The common mode leakage current i_{CM} is excited by U_{eCM} as this equation: $i_{CM} = (C \cdot dU_{eCM})/dt \tag{6}$

Therefore, if the common mode leakage current could be eliminated totally, the equivalent common mode voltage must be kept constant. However, in the cascaded H4 inverter PV system, it is impossible to keep U_{eCM} constant. From equation (3), we can find that U_{eCM1} is not only affected by U_{A1N1}, but also the other inverters differential voltage U_{DM2} and U_{DM3}.It

can be concluded that $U_{A1N1} + U_{DM2} + U_{DM3}$ would not be constant, so the common mode leakage current i_{CM1} could not be zero. This is an inherent drawback of cascaded H4 PV system.

III. PROPOSED CASCADED IH6 INVERTER

In order to solve the above mentioned problem, a new cascaded iH6 inverter is proposed, as shown in Fig.2. The single iH6 inverter was proposed and its topology is shown in Fig.3. Two additional switches T5 and T6 are symmetrically added to the conventional H4 inverter. The main objective of T5 and T6 is to realize the decoupling of dc and ac while UAB is zero. H4 and iH6 all have four working modes under unipolar SPWM strategy.

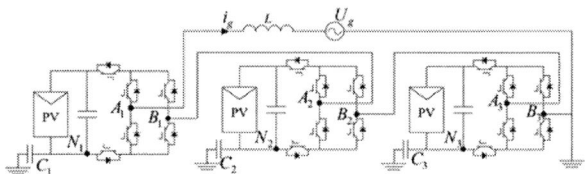

Fig.2   Single-phase 7-level cascaded iH6 inverter

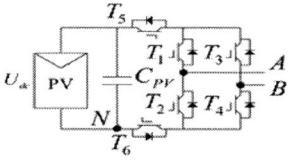

Fig.3   Single-phase three level iH6 inverter

In cascaded H4 system: **Mode1**: $U_{AB} = U_{dc}$, $U_{AN} = U_{dc}$, $U_{DM} = U_{dc}$; **Mode2**: $U_{AB} = 0$, $U_{AN} = U_{dc}$, $U_{DM} = 0$; **Mode3**: $U_{AB} = -U_{dc}$, $U_{AN} = 0$, $U_{DM} = -U_{dc}$; **Mode4**: $U_{AB} = 0$, $U_{AN} = 0$, $U_{DM} = 0$. In cascade iH6 system: **Mode1**: T_1, T_4, T_5 and T_6 are ON, T_2 and T_3 are OFF, $U_{AB} = U_{dc}$, $U_{AN} = U_{dc}$, $U_{DM} = U_{dc}$; **Mode2**: T_1, T_3 and T_6 are ON, T_2, T_4 and T_5 are OFF, $U_{AB} = 0$, $U_{AN} = U_{dc}/2$, , $U_{DM} = 0$; **Mode3**: T_2, T_3, T_5, T_6 are ON, T_1 and T_4 are OFF, $U_{AB} = -U_{dc}$, $U_{AN} = 0$, , $U_{DM} = -U_{dc}$; **Mode4**: T_2, T_4 and T_5 are ON, T_1, T_3 and T_6 are OFF, $U_{AB} = 0$, $U_{AN} = U_{dc}/2$, $U_{DM} = 0$.

In **Mode1** and **Mode3**, U_{AB}, U_{AN} and U_{DM} for iH6 and H4 are totally the same, but in **Mode 2** and **Mode4**, it can be found that U_{AN} of iH6 and H4 system are different while the differential voltage U_{DM} is the same. Taking iH6 system **Mode2** as an example, when T_4, T_5 are turned off, the voltage U_{AN} falls and U_{BN} rises until their values are equal. The grid current flows through T_1 and the antiparallel diode of T_3, or

1392

through T_3 and the antiparallel diode of T_1, if the voltage drop of the devices is ignored, U_{AB} would be zero. Because the former state is **Mode1** or **Mode3**, so $U_{AN} + U_{BN}$ must be equal to U_{dc}. Therefore, in **Mode2**, $U_{AN} = U_{BN} = U_{dc}/2$. In **Mode4**, the same situation happens.

The changing of **Mode2** and **Mode4** is very helpful to suppress the common mode leakage current. As equation (3), (4) and (5) show, the difference of the common mode leakage current in two cascaded inverter PV systems is brought by the difference of U_{AN}. The switching procedure of the four modes in each switching period is:**Mode1<->Mode2**, **Mode2 <->Mode3** or **Mode3<->Mode4**. In the cascaded H4 inverter PV system, the changing of U_{AN} is U_{dc} when **Mode2** switches to **Mode3**. However, in the cascaded iH6 inverter PV system, the change of U_{AN} is $U_{dc}/2$. Thus the equivalent common mode voltage changes less than the cascaded H4 inverter PV system .According to (6),the less change of U_{eCM}, the less value of common mode leakage current is produced.

From the above analysis, the cascaded iH6 PV system can suppress the common mode leakage current more effectively than the cascaded H4 PV system.

IV. COMMON RESONANCE SUPPRESSION

Taking a single-phase 7-level cascaded inverter PV system as an example, there are three iH6-inverters and each inverter has three levels: 1 (U_{dc}), 0 and -1 ($-U_{dc}$). Considering the symmetry of the grid voltage, only the levels 0 and 1 are analyzed. Thus there are **8 States** should be taken into account in the cascaded system: **State0** ([000])~ **State7** ([111]). Only the equivalent resonant circuits of **State2** and **State5** are shown in Fig.4 (a) and Fig.4 (b). More analysis related with the **8 States** will be shown in the final full paper.

From Fig.4, whatever the inverter level is 0 or 1, the stay capacitance always has a resonant circuit loop in the cascaded H4 system. However, in the cascaded iH6 inverter PV system, the stay capacitance is isolated as long as the inverter level is 0. In the unipolar modulation method, the level 0 occupies a certain proportion in an inverter during one line period. In addition, 7 of all **8 States** have the inverter level 0, hence the cascaded iH6 inverter can suppress the high frequency resonant current effectively.

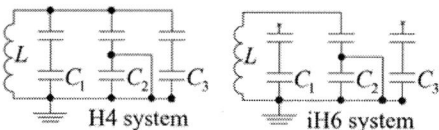

(a) **State2** [010]

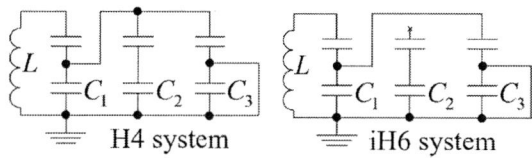

(b) **State5** [101]

Fig.4 The comparing of resonant circuit for **State2** and **State5** in the cascaded H4 and iH6 inverter PV system

Due to the independent equivalence of the cascaded three phase system, it can be concluded that the common mode leakage current suppression by using iH6 in the cascaded single-phase system has the same suppression effect in single-phase system. Simulation and Experimental results Of The Three-Phase 5-Level Cascaded H4 And ih6 Inverter For PV System

Taking the three-phase 5-level system as an example, it is verified that the three-phase ih6 structure also suppresses the common mode leakage current. Fig.5 and Fig.6 show the three-phase 5-level cascaded H4 and iH6 inverter system. The three-phase 5-level cascaded H4 and iH6 inverter systems simulation model were built. C_{1A}, C_{2A}, C_{1B}, C_{1B}, C_{1C}, C_{2C} are the stay capacitances between PV array and ground, and i_{CM1A}, i_{CM2A}, i_{CM1B}, i_{CM1B}, i_{CM1C}, i_{CM2C} are the common mode leakage current through C_{1A}, C_{2A}, C_{1B}, C_{1B}, C_{1C}, C_{2C}.The DC voltage is 100V, the switching frequency is 5 kHz, the inductor is 5mH, three-phase load resistance is 50 ohms each and each PV panel stay capacitance is 1nF. The modulation switching scheme is phase-shifted SPWM. The performance comparison between the two systems is presented.

The 2018 International Power Electronics Conference

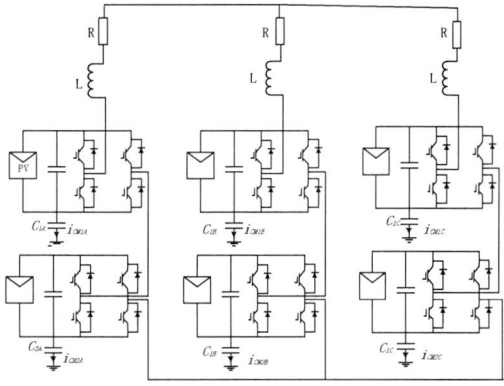

Fig.5   Three-phase 5-level cascaded H4 inverter

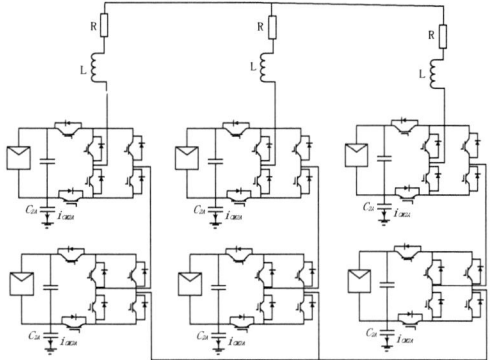

Fig.6   Three-phase 5-level cascaded iH6 inverter

Fig.7 to Fig.9 show the simulation results of three phase 5-level cascaded H4 inverter system. Fig.10 to Fig.12show the simulation results of three-phase 5-level cascade iH6 inverter system. From Fig. 9 and Fig. 12, it can be seen that the common mode current in three-phase iH6 PV system is greatly suppressed.

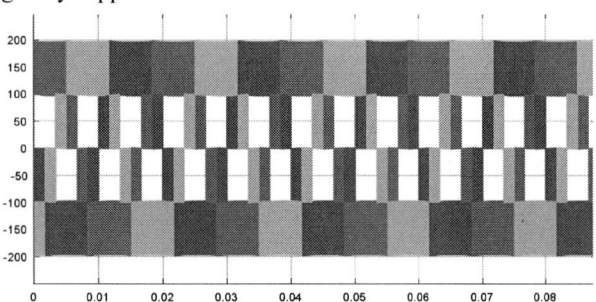

Fig.7 Simulating results of inverter voltage of the cascaded H4 system(50V/div,0.01s/div)

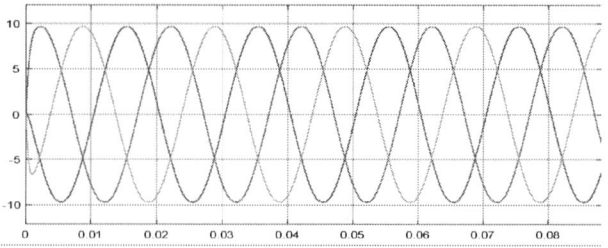

Fig.8 Simulating results of three phase current of the cascaded H4 system(50V/div,0.01s/div)

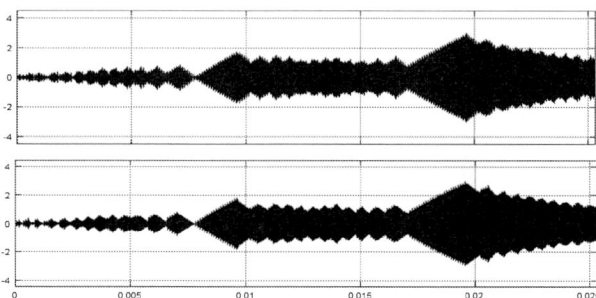

Fig.9 Simulating results of C1A(up) and C2A(down) phase A of the cascaded H4 system(2A/div,0.005s/div)

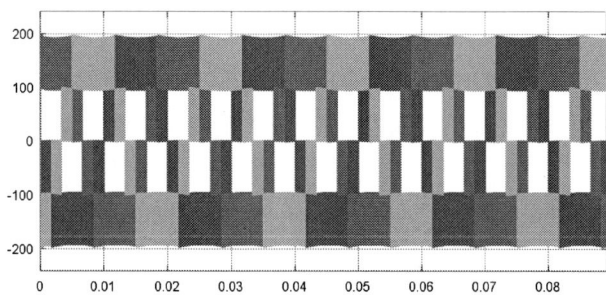

Fig.10 Simulating results of inverter voltage of the cascaded iH6 system(50V/div,0.01s/div)

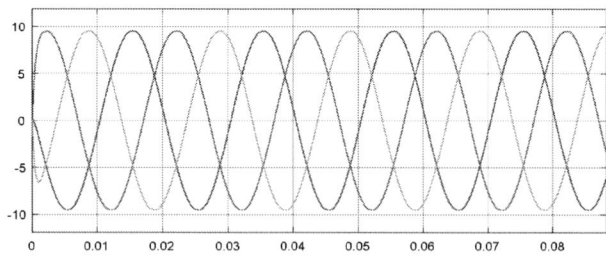

Fig.11 Simulating results of three phase current of the cascaded H4 system(50V/div,0.01s/div)

The 2018 International Power Electronics Conference

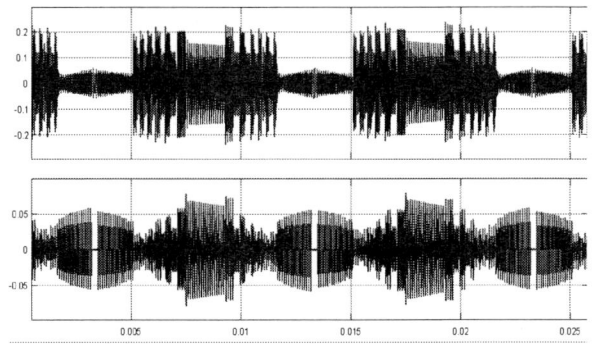

Fig.12 Simulating results of C1A(up) and C2A(down) phase A of the cascaded iH6 system(0.1V/div,0.005s/div)

At the same time, the test bench with cascaded H4 and iH6 bridge is built to verify the theoretical analysis. In the experiments, there are 6 PV simulators, and each PV simulator has an output voltage of 40V. The switching frequency is 2kHz, the inductor is 5mH, three-phase load resistance is 100 ohms.

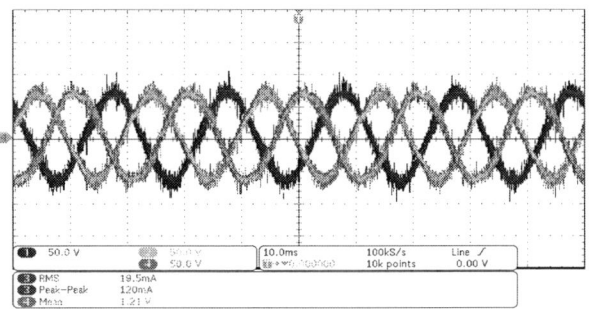

Fig.13 Load phase voltage of three-phase 5-level cascaded H4 cascaded system(50V/div,10ms/div)

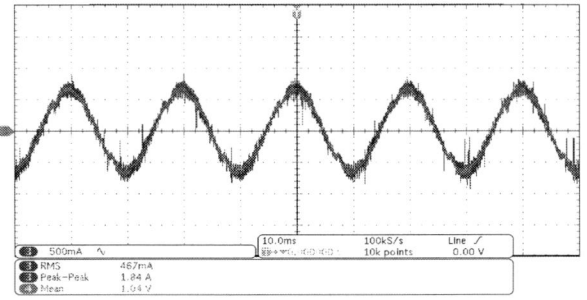

Fig.14. the current of three-phase 5-level cascaded H4 cascaded system in phase A. (500mA/div,10ms/div)

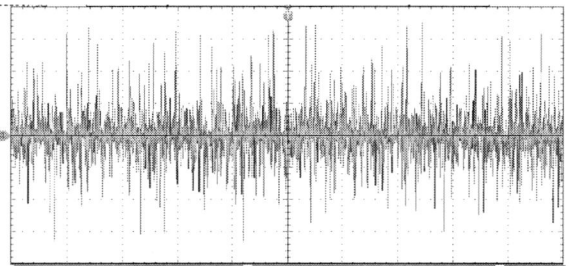

Fig.15 C_{1A} leakage current of three-phase 5-level cascaded H4 cascaded system(100mA/div,10ms/div)

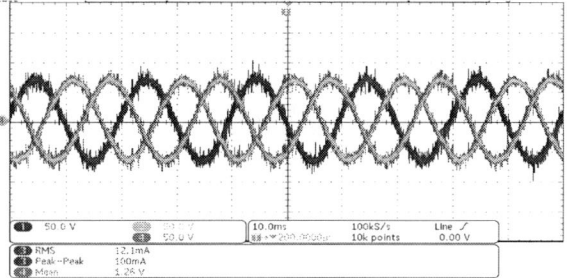

Fig.16 Load phase voltage of three-phase 5-level cascaded iH6 cascaded system(50V/div,10ms/div)

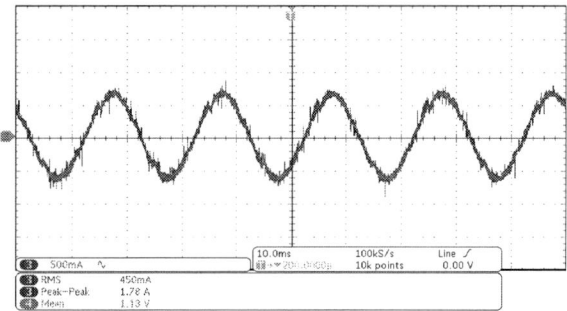

Fig.17 A phase current of three-phase 5-level cascaded iH6 cascaded system(500mV/div,10ms/div)

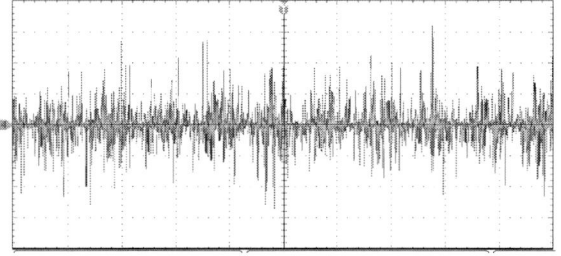

Fig.18 C_{1A} leakage current of three-phase 5-level cascaded iH6 cascaded system(100mA/div,10ms/div)

The experimental results of three phase cascaded multilevel H4 system are shown in Fig. 13 to Fig. 15 respectively. The experimental results of three phase cascaded multilevel iH6 system are shown in Fig. 16 and Fig. 19.. From Fig. 15 and Fig. 18, it can be seen that the common mode current in iH6 PV system is smaller than it in H4. And the load phase voltage and current in iH6 PV system are better than them in H4, the THD is smaller. The experiment verified iH6 system has a good effect to suppress common-mode leakage current and reduce the phase current THD. Fig. 19 shows the test bench of three phase cascaded PV system.

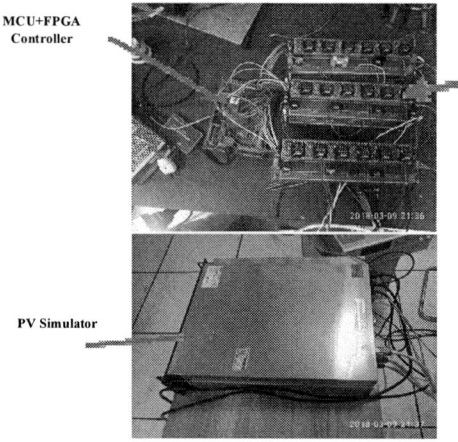

Fig. 19   Photo of test bench

V. CONCLUSIONS

In this paper, the single-phase 7-level system is taken as an example to analyze the theory of cascaded iH6 inverter system which can suppress common-mode leakage current, and by comparing the simulation and experiment of three-phase H4 inverter system and three-phase iH6 system, it is demonstrated that iH6 can suppress common-mode leakage current in the three-phase system. So THE cascaded iH6 inverter PV system can effectively suppress the leakage current without any additional EMI filters.

I. ACKNOWLEDGMENT

This paper is supported by the National Natural Science Foundation of China (NSFC) (Grant No. 51777049 and 51707051).

REFERENCES

[1]   Roberto González, Eugenio Gubía, "Transformerless single-phase multilevel-based photovoltaic inverter", IEEE Trans. Ind. Electron., vol. 55, no. 7, pp. 2694-2702, July,2008.

[2]   H. Xiao and S. Xie, "Leakage current analytical model and application in single-phase transformerless photovoltaic grid-connected inverter," IEEE Trans. Electromagn. Compat., vol. 52, no. 4, pp. 902–913, Nov. 2010.

[3]   Jianhua Wang, Baojian Ji,2, Jianfeng Zhao, Jie Yu, "From H4, H5 to H6 —standardization of full-bridge single phase photovoltaic inverter topologies without ground leakage current issue", IEEE, pp.2419-2425, 2012.

[4]   Bo Yang, Wuhua Li, Yunjie Gu, Wenfeng Cui, Xiangning He, "Improved transformerless inverter with common-mode leakage current elimination for a photovoltaic grid-connected power system", IEEE Trans. Ind. Electron., vol. 27, no. 2, pp.752-762, Feb., 2012.

[5]   E. Gubia, P. Sanchis, A. Ursúa, J. Lopez, and L. Marroyo, "Ground currents in single-phase transformerless photovoltaic systems," Prog. Photovolt.: Res. Appl., vol. 15, no. 7, pp. 629–650, Nov. 2007.

[6]   S. Wang, J. D. van Wyk, and F. C. Lee, "Effects of interactions between filter parasitics and power interconnects on EMI filter performance," IEEE Trans. Ind. Electron., vol. 54, no. 6, pp. 3344–3352, Dec.2007.

[7]   Bailu Xiao, Lijun Hang, Jun Mei, "Modular cascaded H-bridge multilevel PV inverter with distributed MPPT for grid-connected applications", IEEE Trans. Ind. Electron., vol. 51, no. 2, pp.1722-1731, Mar./Apr. 2015.

The 2018 International Power Electronics Conference

Current Sharing/Voltage Sharing Control Strategy for Cascaded DC/DC Converter in Photovoltaic DC Collection System

Bo Chen, Yi Wang, Yanjun Tian and Shilei Wei
State Key Laboratory of Alternate Electrical Power System with Renewable Energy Sources
North China Electric Power University
Baoding, China
Email:812944721@qq.com

Abstract-For Photovoltaic DC collection system, input-parallel-output-series DC/DC converter is widely preferred to realize the voltage conversion between low input voltage and high output voltage. For the seeking of enhancing the stable operation, the output voltage of each module mutual equality must be guaranteed. This paper proposes a current/voltage sharing control strategy for cascaded DC/DC converter in photovoltaic DC collection system. The proposed control method is realized through a double loop contracture, which can realize MPPT of PV module and output voltage sharing between cascaded modules. The effectiveness of the proposed method has been violated by simulation results.

Keyword- photovoltaic; DC/DC converter; input parallel output series; current-sharing / voltage-sharing

I. INTRODUCTION

Photovoltaic DC collection system, as a new networking form of Photovoltaic power can reduce the power conversion stage, the number of cables, reactive circulation, which improves the controllability and conversion efficiency, contributing to the easy integration of the DC power distribution network in the future[1].The high-voltage high-power DC / DC step-up converter is the core equipment in the photovoltaic DC collection system, need to complete the transformation ratio of the high transformation ratio of low-voltage photovoltaic unit access to medium voltage DC distribution system, and to achieve grid-connected photovoltaic power optimization control [2]. In order to reduce the voltage and current stress of the DC/DC module, the multi module input parallel output series (IPOS) structure gains significant attention.

To ensure the stable and efficient operation of the DC collection system, it is necessary to investigate the output voltage sharing control strategy of each module. The current control method for DC-DC converters only need to complete DC voltage conversion, and there is no difference between the modules in series or parallel, can using master-slave control strategy to achieve IPOS current sharing between modules[3].As in [4], An output

Project supported by National Key R&D Program of China(2016YFB0900203)

voltage loop and current loop decoupling control method to achieve output voltage is introduced. Can use separate voltage loop makes the duty cycle of each module the same[5], to achieve the natural voltage division of each module equal. Can use the output voltage loop and pressure ring directly control the output voltage of each module is equal, so can achieve the output voltage between the modules equal[6].But for the Photovoltaic power generation system requires a separate module to achieve Maximum power point tracking(MPPT), which leads to the difference between the control method of this MPPT module and other modules. So need a new control strategy to realize voltage sharing among modules. There is no relevant literature on how PV IPOS systems can simultaneously achieve MPPT and boost control.

This paper studies the control method of PV DC-DC converter based on IPOS structure, which has good dynamic voltage sharing / current sharing while achieving maximum power tracking. Firstly, the topology of IPOS PV DC boost converter and the disadvantages of adopting MPPT control with unified duty ratio are analyzed. Secondly, proposed that the current sharing control of PV module can be used to control the output voltage equal of each module. Then in order to further improve the system start-up and poor dynamic voltage sharing ability in current sharing MPPT control, current sharing and voltage sharing MPPT control strategy is proposed. Finally, through the simulation example of the photovoltaic DC boost system, the current sharing and voltage balancing effects of the three control strategies are analyzed.

II. TOPOLOGY AND MPPT CONTROL OF IPOS PHOTOVOLTAIC DC BOOST CONVERTER

A. Topology of Photovoltaic DC boost converter

IPOS-based photovoltaic DC boost converter topology shown in Figure 1 (a), by the same structure of N DC-DC module input side of the parallel, the output side of the series. The structure of DC-DC module that this text adopts is shown as in Fig. 1 (b). Because the output side is connected in series, the module must adopt the structure of transformer isolation. The primary side of the

1397

transformer DC-AC converter adopts the active clamp full-bridge structure to realize the soft-switching of the switch tubes, by controlling the hysteresis opening and the early switch-off of the clamp switch tube so as to reduce the switching loss[7-8].

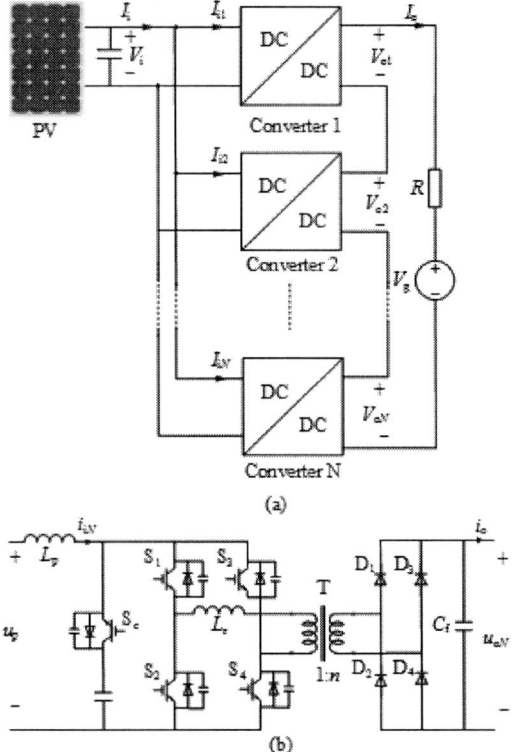

(a)

(b)

Fig. 1   Structure of Photovoltaic IPOS   system.(a) Main circuit structure (b)Active-Clamped Current-Fed Full-Bridge Isolated DC/DC Converter

Due to the unidirectional flow of photovoltaic power generation system, the AC-DC converter on the secondary side of the transformer adopts uncontrolled rectifier structure to reduce the cost. The IPOS structure solves the problem that the current at the input side of the DC boost converter is large, the voltage is low, the voltage at the output side is high, and the current is small, and the modular structure is easy to expand.

B. Unified duty cycle MPPT control

In the ideal case, that is, the parameters of all modules are exactly the same, the drive control and device characteristics are the same, and the losses are the same. Then, the MPPT control can realize the natural voltage sharing of the module. The control block diagram is shown in Fig. 2 .In fact, the parameters of each module capacitor, reactor, high-frequency transformer is difficult to be completely consistent. In addition, the drive control system, once the pulse is not synchronized or pulse loss, can cause imbalance in the power of each module, this will caused overvoltage problem. The problem of circulation and pressure difference between the modules can affect the efficiency of the converter and even the safe operation. Therefore, although the unified duty ratio

MPPT control is simple, it is not suitable for actual engineering application. It is necessary to increase the current sharing control to ensure that the voltage and power of each module in the steady-state and dynamic processes remain equal.

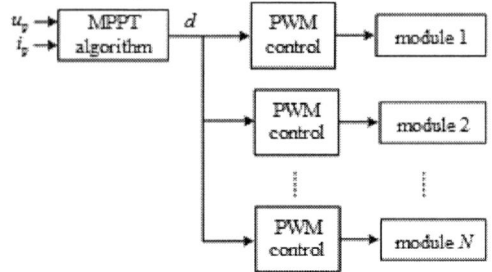

Fig.2 Unified duty cycle MPPT control block diagram

III.   CURRENT-SHARING MPPT CONTROL

This paper surveys the current control strategies for IPOS structures. Voltage sharing and current sharing are necessary for IPOS converter. For the DC-DC boost converter of the Photovoltaic power generation system, MPPT is also required to complete the boost function. When the MPPT control and voltage/current sharing control couple together, the system will deviate from the maximum power point, or even can not be stable. Therefore, a separate module is needed to realize MPPT independently for guaranteeing the power generation efficiency of the PV system, while the other modules are running in voltage/current sharing control. As the MPPT module is passive to achieve voltage/current sharing, so the other components of the voltage/current control need higher dynamic characteristics.

Equation 1 can represent the relationship between the input current and the output voltage of the IPOS structure.

$$u_p \cdot i_{pi} = V_{oi} \cdot I_o \qquad (1)$$

So when the input current of each module is equal, the output voltage can be guaranteed equal.

This paper proposed a control strategy for Photovoltaic DC collection system, as shown in Fig 3.

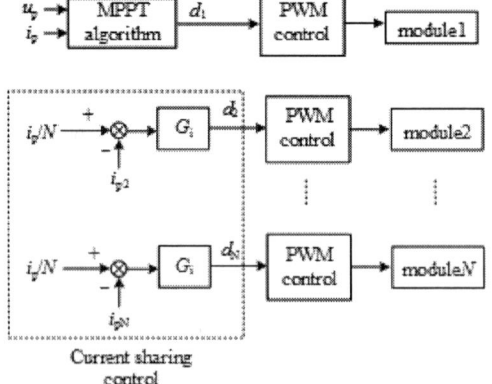

Fig.3 Conventional current sharing control strategy of Photovoltaic IPOS structure

One module is used to realize MPPT, at the same time, other modules use current sharing control. The modules divide the PV output current equally to realize the input current sharing. The value of each module input current is maintained as i_p/N (i_p is the output current for Photovoltaic module), and the MPPT module input current is also i_p/N, so as to realize voltage sharing.

IV. CURRENT-SHARING AND VOLTAGE-SHARING MPPT CONTROL

A. Error Analysis of Current Sharing MPPT Control

As the MPPT module is passive to achieve the current sharing, so under dynamic condition, voltage variations between the MPPT module and the other modules will be engaged, which results in the dynamic unbalancing problem of the current sharing control. Assuming that the current error generated by the current control of the second to N modules is e_i, the input-side current of each module is

$$\begin{cases} i_{p2} = i_{p3} = \cdots = \dfrac{i_p}{N} - e_i \\ i_{p1} = \dfrac{i_p}{N} + (N-1)e_i \end{cases} \quad (2)$$

The output current is

$$i_o \approx \frac{u_p i_p}{V_g} \quad (3)$$

Each module output side voltage is

$$\begin{cases} u_{o1} = \dfrac{u_p i_{p1}}{i_o} \\ u_{o2} = u_{o3} = \cdots = u_{oN} = \dfrac{u_p i_{p2}}{i_o} \end{cases} \quad (4)$$

In the equation, $i_{p1}、 i_{p2}、 \cdots i_{pN}$ is the input current of each module, $u_{o1}、 u_{o2}、 \cdots u_{oN}$ is the output voltage of each module, i_o is the output side current, $u_p、 i_p$ is the voltage and current of the photovoltaic cell, V_g is the output side DC grid voltage.

From equation (2)-(4), we can get the input side current difference ΔI and the output side voltage difference ΔV of module 1 and the remaining modules are

$$\Delta I = i_{p1} - i_{p2} = Ne_i \quad (5)$$

$$\Delta V = u_{o1} - u_{o2} = N\frac{e_i}{i_p}V_g = Ne_u \quad (6)$$

Where, e_u is the error for the second ~ N module output voltage and the reference value of V_g/N. The equation (5)-(6) shows that during the dynamic regulation of other cell currents, if $e_i > 0$, then the output current i_p of photovoltaic cell may change abruptly with the change of light intensity. The input current of module 1 is larger than $N*e_i$ of other modules and the output voltage is larger than that of other modules. Therefore, even if the voltage and current error of each module is not large, the voltage difference between module 1 and other modules

will be enlarged by N times. When there are many modules, module 1 may be damaged due to over-voltage and over-current. Therefore, it is also necessary to introduce the control of differential pressure to improve its dynamic voltage-sharing characteristic.

B. Addition voltage control

To reduce the error in system start-up phase and light intensity changes in the dynamic process, the output voltage loop is added, as in Fig.4. When the MPPT module and the remaining module between the pressure increases, the output voltage loop makes the remaining module duty cycle increases, the output voltage increases, thereby reducing the pressure difference.

The PV module generates the duty cycle of the MPPT algorithm to generate the PWM drive signal for module 1 compared to the sawtooth waveform. The adjustment signal d_{Ni} generated by both current loops is added to the adjustment signal d_{Nu} generated by the equalization loop to generate a PWM drive signal for the module N (N $\neq$ 1) compared with the sawtooth wave.

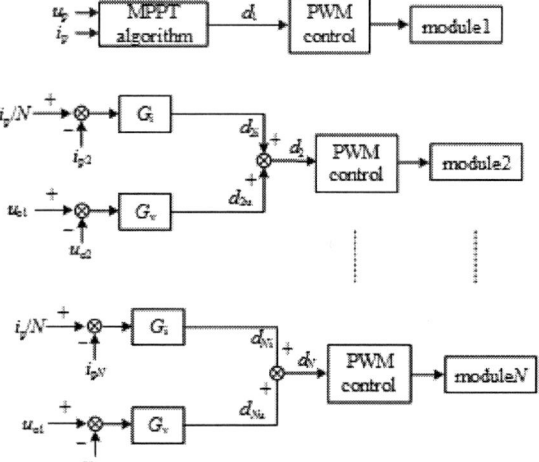

Fig.4 Current sharing and voltage sharing control strategy of Photovoltaic IPOS structure

V. SIMULATION STUDIES

The proposed current sharing and voltage sharing control strategy of cascaded DC/DC converter for Photovoltaic DC collection system is validated by the simulation software MATLAB/Simulink. The performance of current sharing and voltage sharing control strategy of the test system under light intensity change, the parameters of each module are not consistent are presented. The unified duty cycle control strategy and current sharing control strategy have also been simulated under the same conditions to compare the response with current sharing and voltage sharing control strategy.

Table I is the basic parameter values for DC/DC module. Table II is the IPOS structure in the four different simulation parameters.

The 2018 International Power Electronics Conference

TABLE I
BASIC PARAMETER VALUES FOR DC/DC MODULE

Symbol	Meaning	Value
N	Number of modules	3
L_i	Storage inductor of input side	0.3mH
f_s	Switching frequency	5000HZ
C_f	Output filter capacitor	200μF
n_T	Transformer turns ratio	1:5
L_{ri}	Transformer primary leakage inductance	5μH
L_m	Transformer magnetizing inductance	0.286H

TABLE II
PARAMETER VALUES OF THE SYSTEMS UNDER DIFFERENT OPERATING CONDITIONS

	C_{fi} (μF)	L_i (mH)	Light intensity changes(w/ m²)	Line voltage(V)
Condit ion 1	C_{f1}=200 C_{f2}=200 C_{f3}=200	L_1=0.3 L_2=1 L_3=2	0.5s-- (1000— 500); 1s--(500— 750)	15000
Condit ion 2	C_{f1}=200 C_{f2}=150 C_{f3}=250	L_1=0.3 L_2=0.3 L_3=0.3	0.5s-- (1000— 500); 1s--(500— 750)	15000
Condit ion 3	C_{f1}=200 C_{f2}=200 C_{f3}=200	L_1=0.3 L_2=0.3 L_3=0.3	0.5s-- (1000— 500); 1s--(500— 750)	15000

Figure 5 shows the simulation waveform of the MPPT control strategy with unified duty cycle when the input inductance of each module is inconsistent. The simulation conditions are shown in TABLE II condition 1.

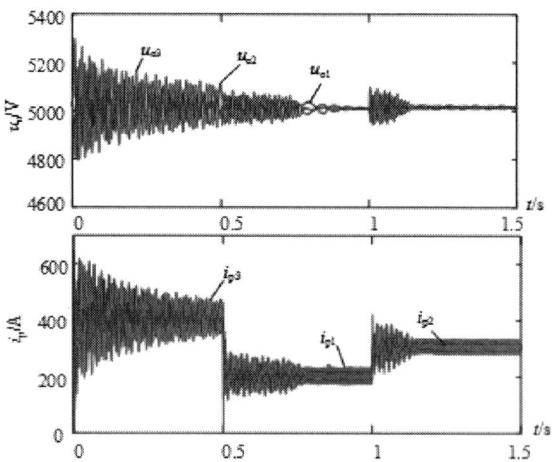

Fig.5 Simulated waveforms of the unified duty cycle MPPT control strategy when the stored energy inductances are not consistent

It can be seen from Figure 5 that during the start up phase of the system and the dynamic process of Light intensity changes, the input current of each module varies inconsistently, resulting in inconsistent changes in the output voltages of the respective modules, thus failing to achieve the output voltage equalization.

Figure 6 shows the simulation waveform of the MPPT control strategy with unified duty cycle when the input inductance of each module is inconsistent. The simulation conditions are shown in TABLE II condition 2

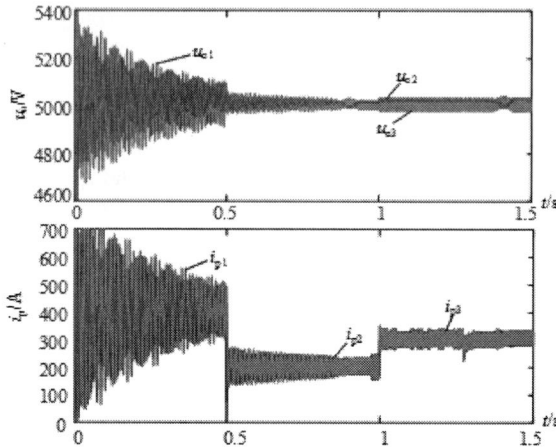

Fig.6 Simulated waveforms of the unified duty cycle MPPT control strategy when the output filter capacitor are not consistent

It can be seen from Figure 6 that during the start up phase of the system, the input current of each module varies greatly and can not be input current sharing,and steady state when the input current of each module and the output voltage is also have some different.

In order to verify the effectiveness of the current sharing MPPT control strategy and the current sharing and voltage sharing control strategy proposed in this paper, the control strategy shown in Figure 3 and Figure 4 is used to simulate the above three conditions.

Fig 7 and Fig 8 show the simulation waveforms when the inductive reactance of the storage inductor is inconsistent with the current sharing MPPT control strategy and the current sharing and voltage sharing MPPT control strategy respectively. The simulation conditions are shown in TABLE II Condition 1.

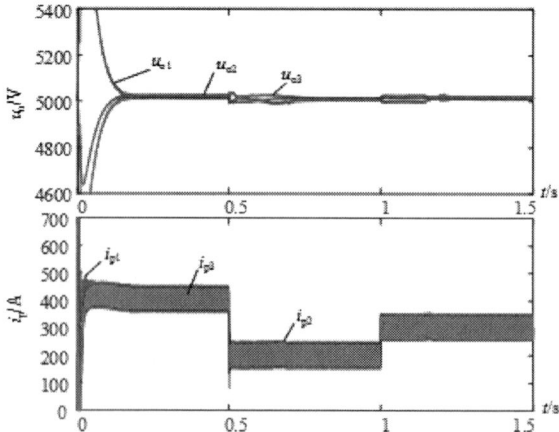

Fig.7 Simulated waveforms of the current sharing MPPT control strategy when the stored energy inductances are not consistent

1400

The 2018 International Power Electronics Conference

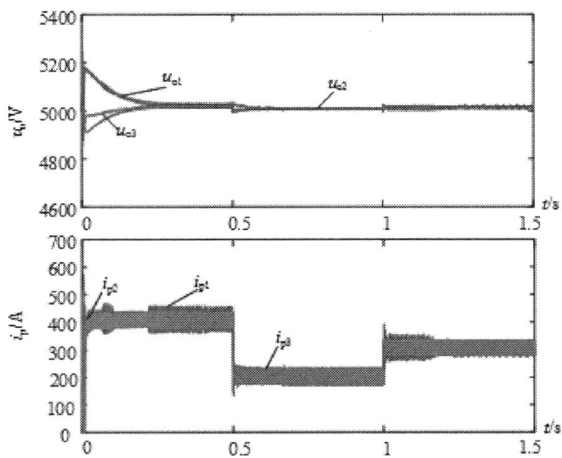

Fig. 8 Simulated waveforms of the current sharing and voltage sharing MPPT control strategy when the stored energy inductances are not consistent

From Fig. 5, Fig. 7 and Fig. 8, it can be seen that, compared with the unified duty cycle control strategy, both current sharing MPPT control strategy and current sharing and voltage sharing MPPT control strategy have better current sharing and voltage sharing effects. However, for current sharing MPPT control strategy, there is a large output voltage difference between module 1 and other modules during the start-up phase and the dynamic change of light intensity, and the voltage difference is amplified by 3 times. However, when the current sharing and voltage sharing MPPT control strategy are used, the output voltage difference between the module 1 and the remaining modules is greatly reduced.

Fig. 9 and Fig. 10 show the simulation waveforms when the output filter capacitors of each module are inconsistent with the current-sharing MPPT control strategy and the current sharing and voltage sharing MPPT control strategy respectively. The simulation conditions are shown in TABLE II Condition 2,

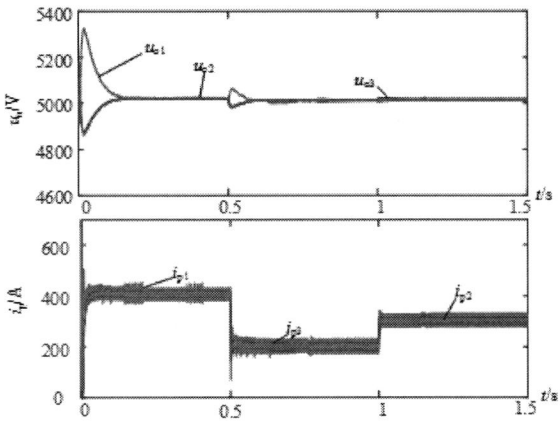

Fig.9 Simulated waveforms of the current sharing MPPT control strategy when the output filter capacitor are not consistent

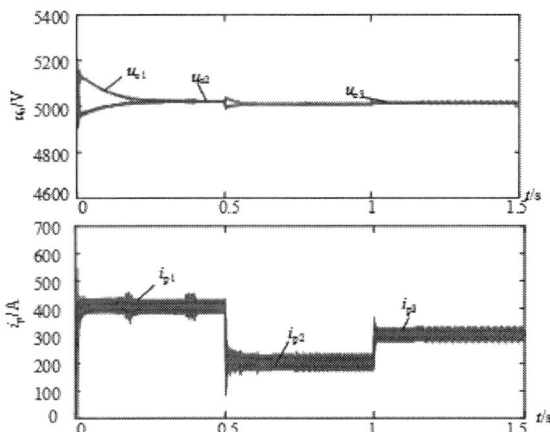

Fig.10 Simulated waveforms of the current sharing and voltage sharing MPPT control strategy when the output filter capacitor are not consistent

It can be seen from Fig. 6, Fig. 9 and Fig. 10 that current-sharing MPPT control strategy and current sharing and voltage-sharing MPPT control strategy, which can be better in the dynamic process in steady-state and light intensity changes of the input current sharing and output voltage sharing, but the use of current-sharing MPPT control strategy, during the start-up phase and light intensity changes in the dynamic process, the module 1 and the remaining modules will have a greater output voltage difference, the voltage difference amplified by 3 times. However, when the current sharing and voltage sharing MPPT control strategy are used, the output voltage difference between the module 1 and the remaining modules is greatly reduced.

VI. CONCLUSION

The unified duty cycle control strategy in PV DC boost system can not achieve the output voltage sharing between modules when the internal parameters of DC boost converter are not matched. In this paper, a current-sharing MPPT control strategy has been proposed, which can guarantee the equalization of the output voltage by controlling the input current of each module. For the current-sharing MPPT control strategy, during system start-up period and solar illumination variations, the MPPT control module and other modules willengage larger voltage deviations, and the proposed current sharing and voltage sharing MPPT control strategy is capable to reduce the output voltage difference. Finally, the simulation results verify the effectiveness of the method.

REFERENCES

[1] J. Echeverría, S. Kouro, M. Pérez and H. Abu-rub, "Multi-modular cascaded DC-DC converter for HVDC grid connection of large-scale photovoltaic power systems," IECON 2013 - 39th Annual Conference of the

1401

IEEE Industrial Electronics Society, Vienna, 2013, pp. 6999-7005.

[2] H. Akagi, "Classification, Terminology, and Application of the Modular Multilevel Cascade Converter (MMCC)," in IEEE Transactions on Power Electronics, vol. 26, no. 11, pp. 3119-3130, Nov. 2011.

[3] I. Federico, E. Jose and F. Luis, "Master–slave DC droop control for paralleling auxiliary DC/DC converters in electric bus applications," in IET Power Electronics, vol. 10, no. 10, pp. 1156-1164, 8 18 2017.

[4] Manias S N, Kostakis G. Modular dc-dc converter for high output voltage applications. Proc. Inst. Elect. Eng.1993, 140 (2): 97-102

[5] Q. Ji, X. Ruan, M. Xu and F. Yang, "Effect of duty cycle on common mode conducted noise of DC-DC converters," 2009 IEEE Energy Conversion Congress and Exposition, San Jose, CA, 2009, pp. 3616-3621.

[6] L. Qu, D. Zhang and B. Zhang, "Active input voltage sharing control scheme for input series output parallel DC/DC converters," IECON 2017 - 43rd Annual Conference of the IEEE Industrial Electronics Society, Beijing, 2017, pp. 744-750.

[7] U. R. Prasanna and A. K. Rathore, "Analysis and design of zero-voltage-switching current-fed isolated full-bridge Dc/Dc converter," 2011 IEEE Ninth International Conference on Power Electronics and Drive Systems, Singapore, 2011, pp. 239-245.

[8] P. U R and A. K. Rathore, "Extended Range ZVS Active-Clamped Current-Fed Full-Bridge Isolated DC/DC Converter for Fuel Cell Applications: Analysis, Design, and Experimental Results," in IEEE Transactions on Industrial Electronics, vol. 60, no. 7, pp. 2661-2672, July 2013.

The 2018 International Power Electronics Conference

PCC Voltage Compensation of PV Inverter with Active Power Decoupling Circuit

Duck-Hwan Hwang[1], Jung-Yong Lee[1] and Younghoon Cho[1*]
Dept. of Electrical Engineering, Konkuk University, Seoul, Republic of Korea
*E-mail: yhcho98@konkuk.ac.kr

Abstract— This paper proposes an algorithm for the point of common coupling(PCC) voltage compensation in single-phase photovoltaic(PV) inverter system. In single-phase power conversion system, the DC link capacitor voltage has a power ripple which is twice of the grid utility frequency. This ripple interferes with the maximum power operation point(MPOP) when tracking the maximum power point(MPP). Also, the power ripple affects on the lifetime of the DC link which consists the electrolytic capacitors. In order to eliminate power ripple, the active power decoupling(APD) circuit has been inserted. And film capacitor has been used instead of electrolytic capacitor to expand the lifetime of PV inverter. In addition, a method for compensating the PCC voltage fluctuation in the PV inverter is introduced. Both the simulations and the experiments for a single-phase PV inverter shows that the proposed method is able to compensate the voltage fluctuation.

Keywords— *Single-phase PV inverter, point of common coupling, acitive-power-decoupling, maximum power point tracking*

I. INTRODUCTION

Recently, the problems of depletion of fossil energy and environmental pollution have promoted researches and developments on renewable energy area [1]. PV power system is relatively easy to be established and to have various capacities, so it is attracting attention as a new renewable energy.

When the power demands rapidly increase, Grid-connected PV inverter supplies generated power to grid. As a result, Stability of the grid is increased. The life time of the PV panel is guaranteed to be 20 years [2-4]. However, Grid-connected PV inverter has a problem due to a short life cycle of electrolytic capacitors [5]. By replacing a electrolytic capacitor with a film capacitor, Life cycle of inverter could have increased over 20 years [6,7].

However, it is difficult to replace the electrolytic capacitors with the film capacitors of same capacitance because film capacitor has lower capacitance density than electrolytic capacitor. Table 1 shows the characteristics depending on the type of capacitor. Replacing the electrolytic capacitor with the film capacitor causes reduction of capacitance and large voltage ripple.

TABLE I
ELECTROLYTIC VERSUS FILM CAPACITORS [8]

Technical parameters	Electrolytic	Film
Capacitance/ unit volume	High	Low
Cost per Joule	Low	High
RMS current rating	20mA/μF	1A/μF
Equivalent series resistance(ESR)	High	Low
Losses	High	Low
Polarity	Yes	No
Operating life cycle	< 20,000h	> 100,000h

In the grid-connected PV system, a fluctuating voltage which has twice frequency of the grid fundamental frequency caused adversely affects in the inverter system.

Many studies have been conducted to mitigate such a problem [9,10]. when an inductor is used to remove ripple using a device with storing energy, this method indicates that the inductor is connected between the neutral node of one leg of the H-bridge inverter and the neutral node of the auxiliary switch leg. At this time, the mode is divided according to the state of the switch to remove the ripple current. Similar to this method, there is a topology in which one of the switches of each leg is changed to a diode. However, when the inductor is used as an energy storage device, the power density of the system is low, leading to an increase the entire volume.

To overcome this problem, a topology using an energy storage device as a capacitor has been proposed. There is a method of reducing the current ripple at the dc terminal by using a topology in which a capacitor is installed at the AC terminal and the neutral terminal between the capacitors and the neutral terminal of the switch are connected to each other.

Another method is to use a compensation circuit in the form of a synchronous buck converter with a separate capacitor except the DC link. In this case, the H-bridge inverter is connected in parallel, and the ripple current is removed by charging and discharging the capacitor controlling the capacitor voltage of the compensation circuit. In this method, there is a method of directly connecting and a method of connecting through a transformer. However, it is difficult to control because of high harmonics of voltage. And also an additional capacitor is needed except the DC link capacitor.

1403

Therefore, it raises the price and size of the system and degrades the efficiency due to the loss of the capacitor. In another topology, two full bridge inverters sharing a single leg topology is proposed, but the power flow is unidirectional. For these reasons, APD research is being conducted which do not use additional capacitor.

In this paper, the electrolytic capacitors of PV inverters are replaced with the film capacitors. In addition, the efficiency of MPPT is increased by using APD circuit that DC link ripple voltage is reduced. Finally, a method for PCC voltage compensation when an APD circuit is applied is presented.

II. SINGLE-PHASE PV SYSTEM

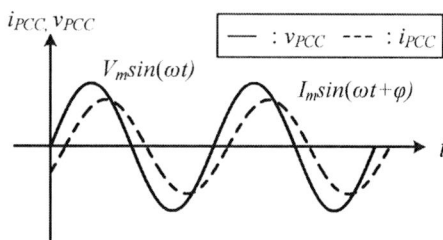

Fig. 1. Single-phase systems

Fig. 1 shows the grid side voltage and current waveforms of single-phase inverter in leading load condition. The inverter output voltage v_{PCC} and current i_{PCC} are expressed as follows:

$$v_{PCC} = V_m \sin(\omega t)$$
$$i_{PCC} = I_m \sin(\omega t - \varphi) \tag{1}$$

V_m and I_m are the maximum value of v_{PCC} and i_{PCC}, ω and φ, ω are the angular frequency and the angle between voltage and current. Equation (2) is obtained by calculating the instantaneous power p_{inv} from Equation (1).

$$p_{inv} = v_{PCC} i_{PCC} = V_m I_m \sin(\omega t) \sin(\omega t - \varphi)$$
$$p_{inv} = \frac{V_m I_m}{2} \cos\varphi - \frac{V_m I_m}{2} \cos(2\omega t - \varphi) \tag{2}$$

Equation (2) consists of the sum of DC component and AC component. The DC component appears in the form of $V_m I_m \cos(\varphi)/2$, and the AC component appears in the form of $V_m I_m \cos(2\omega t - \varphi)/2$. The AC component has the twice frequency of the grid fundamental frequency, and this power ripple exists in the single-phase power system.

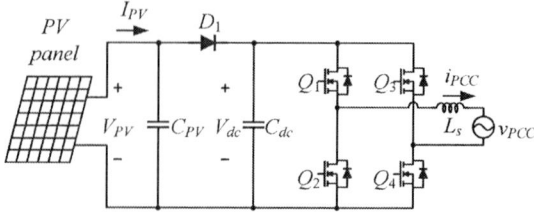

Fig. 2. Power flows in a single-stage PV PCS

Fig.2 shows a single-phase PV inverter system. This system has single-stage conversion state that consisted DC-AC inverter. The power which generated from the PV panel, flows PV side to grid side. DC-link receives active power from the PV panel, and supplying active power to maintain the DC-link voltage, the active power P_{PV} received from the PV panel is sent to the system as a P_{DC}.

$$P_{PV} = P_{DC}$$
$$p_{APD} = p_{AC} \tag{3}$$

Power ripple components P_{AC} is removed using an APD circuit. It is possible to eliminate DC-link voltage ripple. Thus, the trembling of ΔV_{PV} by Low frequency fluctuation can be eliminated, and the performance of the MPPT algorithm can be further improved [11,12].

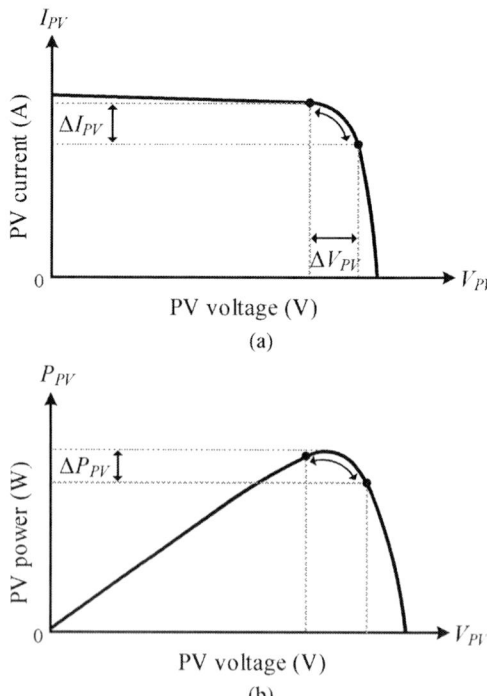

Fig. 3. (a)PV I-V curve, (b)MPP curve

Fig.3 shows the current-voltage characteristic curve and the MPP curve of the PV panel. The above figure shows how DC voltage ripple effects I_{PV} and P_{PV} of the PV panel. A large capacitance electrolytic capacitor reduces the ripple of V_{PV}. In conclusion, the reduction of voltage ripple increases the energy conversion efficiency. However using a large capacitance capacitor increases the size and cost of the system.

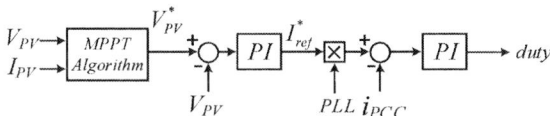

Fig. 4. Single-phase PV inverter control block diagram

Fig.4 shows the single-phase PV inverter control block diagram. In a single-phase PV system, the inverter uses the MPPT algorithm to derive the maximum power point of the PV panel.

III. PCC VOLTAGE COMPENSATION WITH APD CIRCUIT

In Fig.5, the upper and the lower capacitor voltage of the DC-link V_{dc1}, V_{dc2} can be assumed as [13]

$$
\begin{aligned}
V_{dc1} &= \frac{V_{dc}}{2} - V_c \sin(\omega t + \theta) \\
V_{dc2} &= \frac{V_{dc}}{2} + V_c \sin(\omega t + \theta)
\end{aligned}
\tag{4}
$$

V_{dc}, V_c, θ are the DC-link voltage for which the AC component is not considered, the reference voltage to be controlled by the APD circuit, and the phase difference between v_{PCC} and i_{Lf}. The capacitor currents i_{c1} and i_{c2} can be represented as follows:

$$
\begin{aligned}
i_{c1} &= -\omega C_{fl} V_c \cos(\omega t + \theta) \\
i_{c2} &= \omega C_{fl} V_c \cos(\omega t + \theta)
\end{aligned}
\tag{5}
$$

The power of the APD circuit is as follows.

$$
p_{APD} = \omega C_{fl} V_c^2 \sin(2\omega t + 2\theta)
\tag{6}
$$

C_{fl} is the capacitance of C_1 and C_2. If equations (6) and p_{AC} are the same, the ripple power can be canceled.

$$
\omega C_{fl} V_c^2 \sin(2\omega t + 2\theta) = \frac{V_m I_m}{2}\cos(2\omega t - \varphi)
\tag{7}
$$

Equation (7) can be expressed as:

$$
\begin{aligned}
&\omega C_{fl} V_c^2 \left(\sin(2\omega t)\cos(2\theta) + \cos(2\omega t)\sin(2\theta) \right) \\
&= \frac{V_m I_m}{2}\cos(2\omega t - \varphi)
\end{aligned}
\tag{8}
$$

θ, V_c is obtained as:

$$
\theta = \frac{\pi}{4} - \frac{\varphi}{2}
\tag{9}
$$

$$
V_c = \sqrt{\frac{V_m I_m}{2\omega C_{fl}}}
\tag{10}
$$

Fig.5 shows the hardware configuration of PV inverter with APD circuit. The APD circuit is inserted between the boost converter and the full bridge inverter. The APD circuit serves to absorb power at twice the frequency [14].

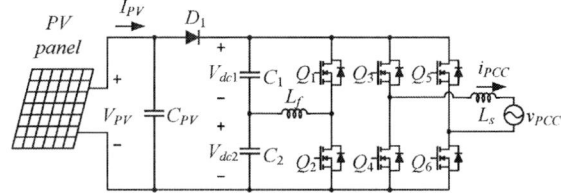

Fig. 5. PCC PV system with APD circuit

The DC-AC inverter performs the MPPT algorithm of the PV panel and maintains the DC link voltage. In accordance with the current reference to maintain the DC link, the inverter generates current.

Fig.6 shows the control configuration of the PV inverter with APD circuit. The controller of APD circuit uses a double loop structure controller. The APD inductor current control has an inner loop and the V_{c2} voltage control has an outer loop.

The APD inductor current control allows the controller to operate stable in transient conditions through current limiting.

V_{c2} voltage control of APD circuit generates V_{c2} voltage reference through V_{PV} and calculated V_c.

The V_{c2} voltage reference is the same low frequency as the grid voltage. therefore, APD voltage control is designed to have a lower bandwidth than the APD inductor control.

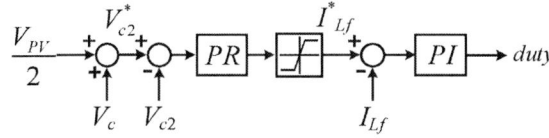

Fig. 6. PV inverter APD circuit control block diagram

TABLE II
SYSTEM PARAMETERS

Symbol	Meaning	Value
L_s	Grid side inductor	2mH
L_f	APD inductor	1mH
C_{PV}	PV capacitor	10uF
C_1, C_2	APD capacitor	155μF
f_{sw}	Switching frequency	10kHz
V_{PV}	PV voltage	350~500V
V_{rms}	Grid voltage in RMS	220V / 60Hz

The simulation was performed by using Power SIM(PSIM) software and the values of the circuit parameters are shown in Table II.

The 2018 International Power Electronics Conference

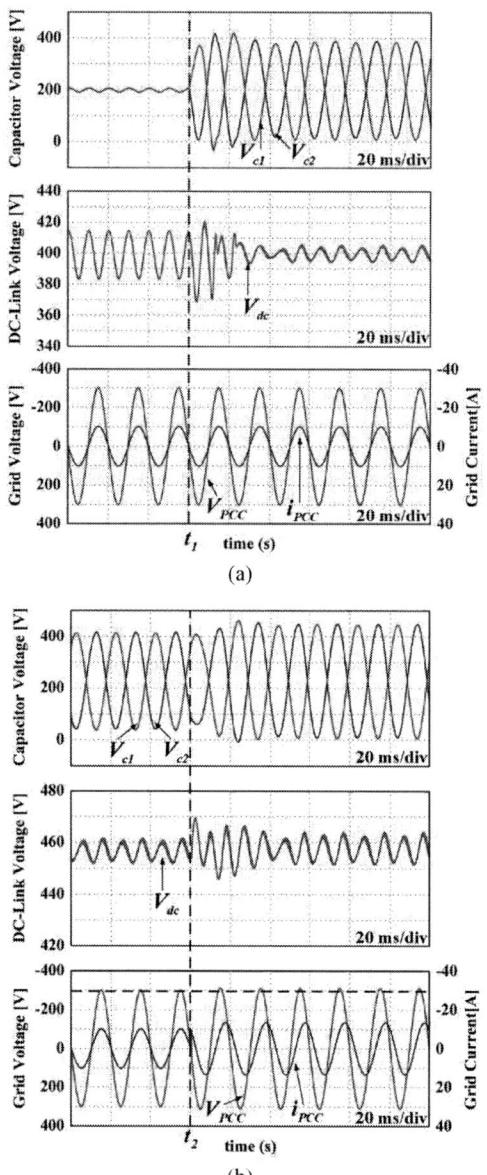

Fig. 7. Simulation results. (a)dynamic characteristic of APD, (b)PCC voltage compensation

Fig. 7(a) shows the simulation using APD circuit control. At first, the APD circuit was not applied and the ripple of V_{DC} was about 30V. The APD circuit is then applied and the ripple of V_{DC} is reduced to 10V

Fig. 7(b) shows the voltage compensated waveform during the control of the APD circuit. When the PCC voltage fluctuates and needs to be compensated out of a certain range, the PCC voltage is operated according to the standard through the reactive power compensation algorithm.

The fluctuation of V_{PV} and I_{PV} fluctuates by P_{PV} before the APD circuit operates. When the APD circuit starts to control, it does not fluctuate at twice the frequency. When voltage compensation control is performed, P_{PV} holds the maximum power point, and the reactive power

increases to compensate for the PCC voltage fluctuation.

When the APD circuit is controlled as shown in Fig.6, it can be confirmed that the phase difference between the V_{c2} voltage and the V_{PCC} voltage is changed.

IV. PCC VOLTAGE COMPENSATION

Photovoltaic inverter provides maximum power to grid through MPPT algorithm. If the output power of the PV inverter is higher than the load, the remaining power is transferred to the grid. On the other hand, power is suppli ed from the grid. Line impedance between PCC and grid causes the voltage drop. Therefore the PCC voltage is higher than grid voltage when the power is transferred to grid from PCC, on the contrary, the PCC voltage is lower than grid voltage when the power is transferred to PCC from grid [15,16].

The Fig.8 shows a grid connected PV inverter system and y_l demonstrates the line admittance between PCC and grid. Equation (11)-(13) are apparent power, active power and reactive power at PCC.

$$\overline{S}_G = \overline{S}_{inv} - \overline{S}_L = \frac{1}{2} \times \overline{V}_{PCC} \times \left\{ \left(\overline{V}_{PCC} - \overline{V}_{grid} \right) \times \overline{y}_l \right\}$$

$$\left(\overline{V_{PCC}} = m V_{PCC}, \overline{V}_{grid} = V_{grid} \angle \delta, \overline{y}_l \angle \gamma \right) \tag{11}$$

$$P_G = P_{inv} - P_L$$
$$= \frac{1}{2} \left\{ m^2 V_{grid}^2 y_l \cos \gamma - m V_{grid}^2 y_l \cos(\delta + \gamma) \right\} \tag{12}$$

$$Q_G = Q_{inv} - Q_L$$
$$= \frac{1}{2} \left\{ -m^2 V_{grid}^2 y_l \sin \gamma - m V_{grid}^2 y_l \sin(\delta + \gamma) \right\} \tag{13}$$

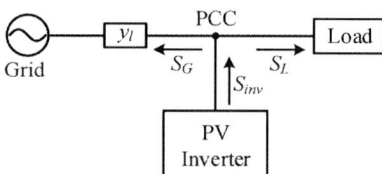

Fig. 8. Distribution system including PV inverter and load

The reference phase of PV inverter is PCC voltage phase as the inverter gets the PCC phase information through PLL algorithm. The m is difference of grid voltage level and PCC voltage level. The equation (14), (15) can be simplified equation (14), (15) by assuming the δ is very small. Thus PCC voltage is compensated by limiting with equation (16).

$$P_{inv} = \frac{1}{2} \left\{ m^2 V_{grid}^2 y_l \cos \gamma - m V_{grid}^2 y_l (\cos \gamma - \delta \sin \gamma) \right\} + P_L \tag{14}$$

$$Q_{inv} = \frac{1}{2} \left\{ -m^2 V_{grid}^2 y_l \sin \gamma + m V_{grid}^2 y_l (\delta \cos \gamma + \sin \gamma) \right\} + Q_L \tag{15}$$

$$|\delta| < 20^{\circ} \tag{16}$$

The reactive power amplitude can be calculated by simplified equation for compensating the PCC voltage. If the PCC voltage is equal with grid voltage by compensating proper reactive power, the $m = 1$. Using equation (17) ~ (19), δ can be rearranged equation (20). Active power and reactive power calculated by δ n equation (20)

$$\begin{aligned} P_{inv} &= p_1\delta + P_L \\ \left(p_1 \right. &= \left. 0.5V_{grid}^{\prime 2}y_l \sin\gamma\right) \end{aligned} \tag{17}$$

$$\begin{aligned} Q_{inv} &= q_1\delta + Q_L \\ \left(q_1 \right. &= \left. 0.5V_{grid}^{\prime 2}y_l \cos\gamma\right) \end{aligned} \tag{18}$$

$$S_{inv}^2 = P_{inv}^2 + Q_{inv}^2 \tag{19}$$

$$\delta = \frac{-b_1 \pm \sqrt{b_1^2 - a_1c_1}}{a_1} \tag{20}$$

$$\begin{pmatrix} a_1 = p_1^2 + q_1^2, \ b_1 = p_1P_L + q_1Q_L, \\ c_1 = P_L^2 + Q_L^2 - S_{inv}^2, \ d_1 = b_1^2 - a_1c_1 \end{pmatrix}$$

Following equation (16), δ have to below 20. d_l have not to be negative due to δ is a real number. Also, according to IEEE 1547, power factor is above 0.9. If any of these conditions were not met, PCC voltage cannot be compensated like grid. Therefore, in this condition, m would be increased or decreased to compensate. Equation (19), (21) ~ (23) is rearranged for obtain δ, when the m is not 1.

$$\begin{aligned} P_{inv} &= mp_1\delta + P_m \\ \left(p_m \right. &= \left. m^2q_1 - mq_1 + P_L\right) \end{aligned} \tag{21}$$

$$\begin{aligned} Q_{inv} &= mq_1\delta + q_m \\ \left(q_m \right. &= \left. -m^2p_1 + mp_1 + Q_L\right) \end{aligned} \tag{22}$$

$$\delta = \frac{-b_m \pm \sqrt{b_m^2 - a_mc_m}}{a_m} \tag{23}$$

$$\begin{pmatrix} a_m = m^2(p_m^2 + q_m^2), \ b_m = m(p_1p_m + q_1q_m), \\ c_m = p_m^2 + q_m^2 - S_D^2, \ d_m = b_m^2 - a_mc_m \end{pmatrix}$$

m would be increased when S_{inv} is larger than S_L, m would be decreased when S_{inv} is smaller than S_L. m is increase or decrease repeatedly until phase and power factor of δ and condition of d_m is met. According to IEEE 1547, m is determined above 0.88 and below 1.1.

V. EXPERIMENTAL RESULTS

This paper presented a PCC voltage compensation of PV inverter with APD circuit on the prototype.

The entire algorithm is implemented using 32bit floating point microcontroller unit of TMS320F28335 manufactured by Texas Instrument

Fig.9 shows the experimental waveform of the PCC voltage compensation algorithm of the PV inverter system with APD circuit.

V_{PCC}, V_g are present the PCC voltage and the grid voltage. V_{c2}, I_{ac} are present the C_2 capacitor voltage of APD circuit and the injected PCC current

Fig.9(a) shows the before the PCC voltage compensation algorithm is applied. Before applying PCC voltage compensation, the grid voltage is 214.8V and the PCC voltage is 214.1V

Fig.9(b) shows the after applying the PCC voltage compensation algorithm. After applying the PCC compensation, the grid voltage is 214.6V and the PCC voltage is 215.3V

In Fig.9(a), phase difference is 0° between V_{PCC} and I_{ac}. in Fig.9(b), phase difference is 25° between V_{PCC} and I_{ac}. Those show that APD circuit inject the reactive power and PCC voltage is compensated.

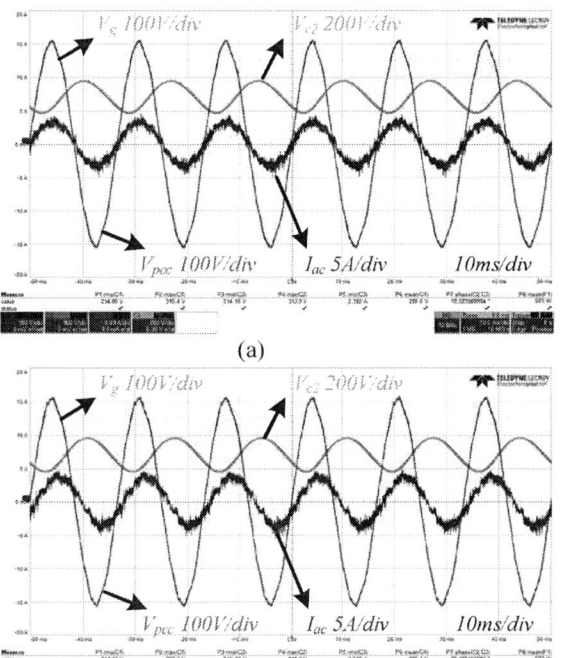

Fig. 9. PCC compensation PV system with APD circuit. (a)before PCC compensation algorithm, (b) after PCC compensation algorithm.

VI. CONCLUSION

In this paper, the APD circuit configuration and the PCC voltage compensation algorithm have been proposed. The film capacitor has been used instead of electrolytic capacitor to increase the lifetime of the DC-link. And the reactive power has been injected into APD circuit to compensate the power ripple. The efficiency of MPPT has been increased by removing the power ripple with APD circuit. Both simulations and experimental results based on the single-phase PV inverter application verify the effectiveness of the proposed method.

ACKNOWLEDGMENT

This work(C0511847) was supported by Business for Cooperative R&D between Industry, Academy, and Research Institute funded Korea Small and Medium Business Administration in 2017.

This work supported by "Human Resources Program in Energy Technology" of the Korea Institute of Energy Technology Evaluation and Planning (KETEP), granted financial resource from the Ministry of Trade Industry & Energy, Republic of Korea. (No.20174030201660).

REFERENCES

[1] O. Ellabban, H. Abu-Rub, and F. Blaabjerg, "Renewable energy resources: Current status, future prospects and their enabling technology," *Renewable and Sustainable Energy Reviews*, vol. 39, pp. 748-764, 2014.

[2] H. Wang and F. Blaabjerg, "Reliability of capacitors for DC-link applications in power electronic converters—An overview," *IEEE Transactions on Industry Applications*, vol. 50, pp. 3569-3578, 2014.

[3] S. Harb and R. S. Balog, "Reliability of candidate photovoltaic module-integrated-inverter (PV-MII) topologies—A usage model approach," *IEEE transactions on power electronics*, vol. 28, pp. 3019-3027, 2013.

[4] C. Rodriguez and G. A. Amaratunga, "Long-lifetime power inverter for photovoltaic AC modules," *IEEE Transactions on Industrial Electronics*, vol. 55, pp. 2593-2601, 2008.

[5] M. A. Ramli, A. Hiendro, K. Sedraoui, and S. Twaha, "Optimal sizing of grid-connected photovoltaic energy system in Saudi Arabia," *Renewable Energy*, vol. 75, pp. 489-495, 2015.

[6] S. B. Kjaer, J. K. Pedersen, and F. Blaabjerg, "A review of single-phase grid-connected inverters for photovoltaic modules," *IEEE transactions on industry applications*, vol. 41, pp. 1292-1306, 2005

[7] H. Hu, S. Harb, X. Fang, D. Zhang, Q. Zhang, Z. J. Shen, *et al.*, "A three-port flyback for PV microinverter applications with power pulsation decoupling capability," *IEEE Transactions on Power Electronics*, vol. 27, pp. 3953-3964, 2012

[8] B. Karanayil, V. G. Agelidis, and J. Pou, "Performance evaluation of three-phase grid-connected photovoltaic inverters using electrolytic or polypropylene film capacitors," *IEEE Transactions on Sustainable Energy*, vol. 5, pp. 1297-1306, 2014.

[9] Y. Sun, Y. Liu, M. Su, W. Xiong, and J. Yang, "Review of active power decoupling topologies in single-phase systems," *IEEE Transactions on Power Electronics*, vol. 31, pp. 4778-4794, 2016.

[10] H. Li, K. Zhang, H. Zhao, S. Fan, and J. Xiong, "Active power decoupling for high-power single-phase PWM rectifiers," *IEEE Transactions on Power Electronics*, vol. 28, pp. 1308-1319, 2013.

[11] D. Verma, S. Nema, A. Shandilya, and S. K. Dash, "Maximum power point tracking (MPPT) techniques: Recapitulation in solar photovoltaic systems," *Renewable and Sustainable Energy Reviews*, vol. 54, pp. 1018-1034, 2016.

[12] J. Ahmed and Z. Salam, "An improved perturb and observe (P&O) maximum power point tracking (MPPT) algorithm for higher efficiency," *Applied Energy*, vol. 150, pp. 97-108, 2015.

[13] Y. Tang and F. Blaabjerg, "A component-minimized single-phase active power decoupling circuit with reduced current stress to semiconductor switches," *IEEE Transactions on Power Electronics*, vol. 30, pp. 2905-2910, 2015.

[14] Y. Tang, F. Blaabjerg, P. C. Loh, C. Jin, and P. Wang, "Decoupling of fluctuating power in single-phase systems through a symmetrical half-bridge circuit," *IEEE Transactions on Power Electronics*, vol. 30, pp. 1855-1865, 2015.

[15] E. Demirok, P. C. González, K. H. B. Frederiksen, D. Sera, P. Rodriguez and R. Teodorescu, "Local Reactive Power Control Methods for Overvoltage Prevention of Distributed Solar Inverters in Low-Voltage Grids," *IEEE Journal of Photovoltaics* 1.2 (2011): 174-182.

[16] A. Cagnano, E. De Tuglie, M. Liserre, and R. A. Mastromauro, "Online optimal reactive power control strategy of PV inverters," *IEEE Transactions on Industrial Electronics*, vol. 58, pp. 4549-4558, 2011.

The 2018 International Power Electronics Conference

A Novel Partial Shading Detection Algorithm Utilizing Power Level Monitoring of Photovoltaic Panels

Thusitha Randima Wellawatta and Sung-Jin Choi

School of Electrical Engineering, University of Ulsan, Ulsan, South Korea

sjchoi@ulsan.ac.kr

Abstract - **Maximum power point tracking (MPPT) under partial shading condition (PSC) is a challenging research topic in the PV array system. As the shaded PV module makes different peak patterns on the power versus voltage curve and misguides the MPPT algorithm, various kinds of global MPP (GMPP) detecting algorithms have been studied. Generally, too frequent execution of the GMPP tracking algorithm reduces the achievable power of PV module due to time spent on the scanning process. Thus, the partial shading detection algorithm is essential for efficient utilization of solar energy source. Based on the theoretical investigation of the characteristic curve patterns under various partial shading conditions, this paper presents a new detection algorithm utilizing power level monitoring. While conventional methods only focus on fast shading patterns, the proposed algorithm always shows superb performance regardless of the partial shading patterns.**

I. INTRODUCTION

Photovoltaic (PV) generation is going to be of immense importance due to its free energy with zero environmental pollution. To maximize the efficiency of the PV array utilization, maximum power point tracking (MPPT) algorithm is essential. The MPPT algorithm always attempts to locate the operating point on the power peak in the characteristic curve.

However, in the real environment, insolation shadows on the PV array are unavoidable and partial shading of PV array makes a considerable energy loss in PV system [1]. This is because multiple peaks can occur on the power vs. voltage curve (P-V curve) and obtainable output power may not be maximized without correct tracking of that peak. In this case, global MPP (GMPPT) algorithm is required to select the highest peak among various local MPPs and give the new control reference to the MPPT algorithm.

There have been a few studies on the GMPPT algorithm [2-9]. To identify the occurrence of the partial shading condition (PSC), the PSC detection algorithm is used. However, conventional methods sometimes cause to the reduction of the system efficiency because they used to waste time on a sequence of repeated scanning of local MPPs, which hinders the optimal operation of PV module [2]. Hence, this study aims to investigate effective PSC detection in order to avoid such an unnecessary GMPPT and improve the system performance further. In this paper, a novel PSC detection algorithm is proposed and demonstrated. Finally, its advantages have been compared with conventional methods.

II. PRINCIPLE OF OPERATION

Basic concept

The PV arrays are engaged with different kind of shadings as shown in Fig.1. The global shading is shown in Fig.1 (a) and can be defined as the shading that can cover the entire PV array temporally. This kind of shading can be occurred due to large moving cloud. Likewise, a part of PV array can be covered by some shadows and it can be defined as PSC as shown in Fig.1 (b). In the real environment, the PSC detection algorithm should be able

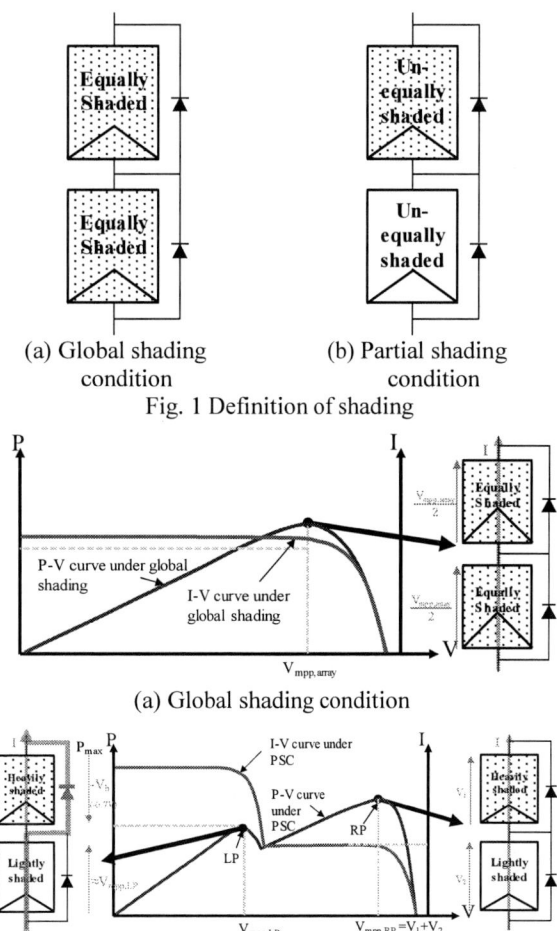

(a) Global shading condition

(b) Partial shading condition

Fig. 1 Definition of shading

(a) Global shading condition

(b) Partial shading condition

Fig 2. P-V and I-V curves two equal panel (Ns = 2)

1409

to distinguish the shading condition.

Figure 2 shows P-V and I-V curves of array with two identical modules which are connected in series configuration. According to Fig.2 (a), maximum MPP current and MPP power are delivered continuously under control of MPPT algorithm. When a global shading occurs, current and power are reduced without activating bypass diodes and new operating point is determined by MPPT. In the PSC, the behavior of characteristic curves is shown in Fig.2 (b). Here, the two peaks are appeared in the P-V curve as a left - side peak (LP) and the right-side peak (RP). The MPPT cannot sense other peaks on P-V curve and keeps the existing peak continuously. Thus, the PSC detection algorithm is important to select the operating point on correct peak by using global peak searching algorithm. The current flow through PV module is

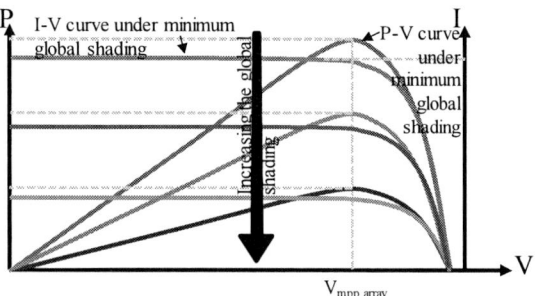

(a) Global shading conditions under 3 different equal insolation levels

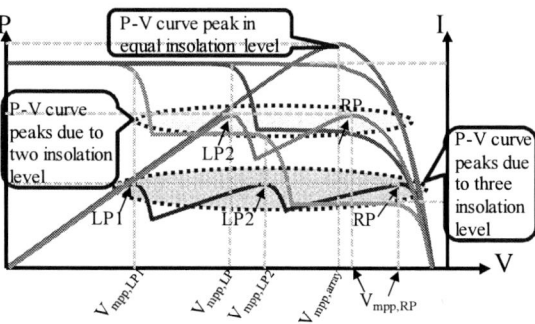

(b) PSC with different no. of peak

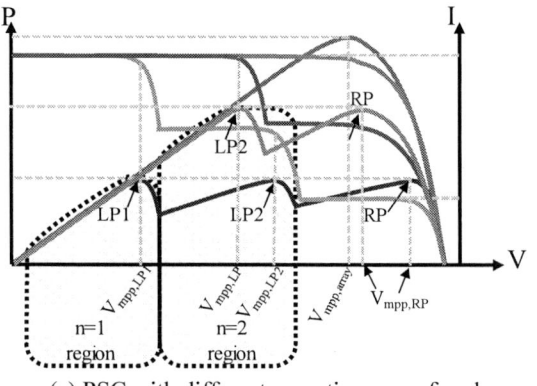

(c) PSC with different operating area of peak

determined by selection of operating peak. When the RP is selected by GMPPT, both modules are contributed to the array power delivery and maximum array current is determined by shaded module. On the contrary, LP is selected, the shaded module is bypassed through the diode due to low current capability and array current is determined by minimally shaded module. Here, power is delivered by minimally shaded module along. Among those two peaks, the highest peak is defined as a global MPP (GMPP). According to the above behavior of characteristic curve, it is also known that the height of the RP is only affected by shading of global insolation [2] whereas the height of LP is always determined by minimally shaded module. The temperature effect due to change of shading is assumed as negligible.

This concept can be extended to multiple numbers of series module array. For an example, three modules array (Ns=3) is explained as shown in Fig.3. The behavior of global shading condition is shown in Fig.3 (a). Here, the current flow and voltage change follow a similar manner as Ns =2 system. However, PSC has two (generally (Ns-1)) possible scenarios as shown in Fig.3 (b). Here, maximally three peaks can be occurred, and the possible number of peaks is determined by the number of modules (bypass diodes) in series connected PV array [3]. In our study, peak occurring regions are defined as shown in Fig.3 (c). LP1 is occurred in "n=1" region and numbers of other regions are incremented according to the peak number. The location of "n=1" can be calculated by using Ns and V_{oc} of the module. When the array covered by one shadow, two peaks are occurred. Here, note that the area of the LP is in "n=2" region. (In Ns=2 system, the region of LP is "n=1"). If two shadows are covered the array, three peaks are appeared and named as LP1, LP2, and RP. The LP1 and LP2 are occurred due to minimally shaded module and module with next shading level, respectively. The RP is always occurred due to maximumly shaded module.

From our observations, the height of the LP can be calculated by using α% of the maximum output power (αP_m) and the ratio, α is determined by (1)

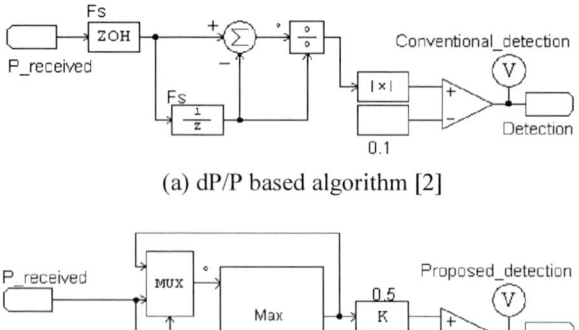

(a) dP/P based algorithm [2]

(b) Proposed ΔP based algorithm
Fig 4: Algorithm implementation in PSIM.

1410

The 2018 International Power Electronics Conference

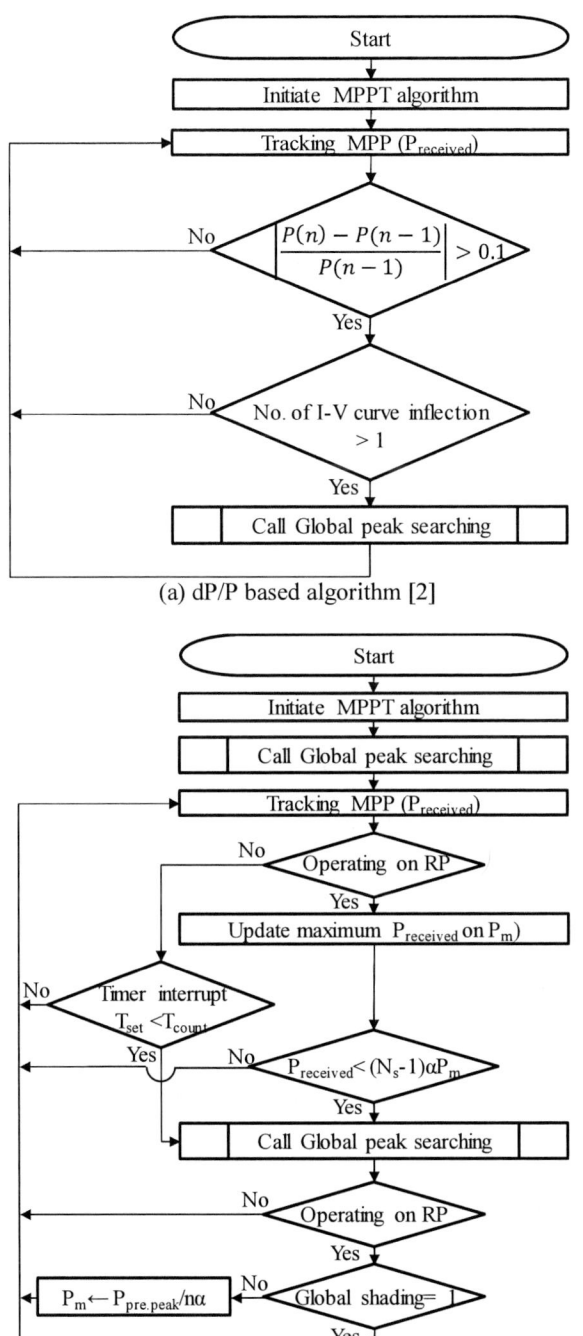

(a) dP/P based algorithm [2]

(b) Proposed ΔP based algorithm
Fig 5: Flow Charts of PSC detection algorithms.

$$\alpha = \frac{1}{N_S}, \qquad (1)$$

where N_S is the number of series connected PV modules and P_m is the stored maximum power value from instantaneous array power. It is found that GMPP searching beyond the αP_m power level is unnecessary and

such a characteristic will be utilized in the proposed algorithm.

Conventional PSC method

Conventional PSC detection algorithm, shown in Fig.4 (a) and Fig.5 (a), monitors the power difference between consecutive two samples [2]. In a sudden change of insolation, dP/P can be considerably higher and if it becomes greater than a threshold which is assumed to be 0.1 in the literature, the algorithm triggers global peak searching. The problems of this method are that it could fail to detect PSC in a smooth change of insolation, or too frequent but unnecessary activation of global peak searching due to PSC detection may occur in a sharp change of insolation (under 0% to 50% shading).

Proposed method

Figure 4 (b) and Fig.5 (b) shows PSIM algorithm chart and the algorithm flow chart of the proposed method, respectively. To check the PSC at the start, the global peak searching subroutine is called and the maximum $P_{received}$ is obtained by MPPT tracking. If the operating peak is not on the RP, the algorithm flows through the timer interrupt loop and operates under the PSC. If the operating peak is on the RP, the maximum power information (P_m) is updated and the instantaneous power ($P_{received}$) is compared with $(N_s-1)\alpha P_m$, continuously. If it gets lower than $(N_s-1)\alpha P_m$, global peak searching subroutine is triggered and it calculates the number of peaks and re-locates the GMPP. If the operating peak is not on the RP, previous procedure was repeated. The number of peak is used to distinguish global shading from partial shading in the proposed algorithm. When the peak searching selects the GMPP as RP, but PSC still exist (no. of peaks > 1), the power of non-shaded array can be predicted by dividing the power of most left peak by $n\alpha$. Here, "n" is the region of the last operated peak as shown in Fig.3 (c). If global peak searching does not identify another peak, it reflects that present operating point is RP and insolation can be changed without PSC (global shading) and the latest

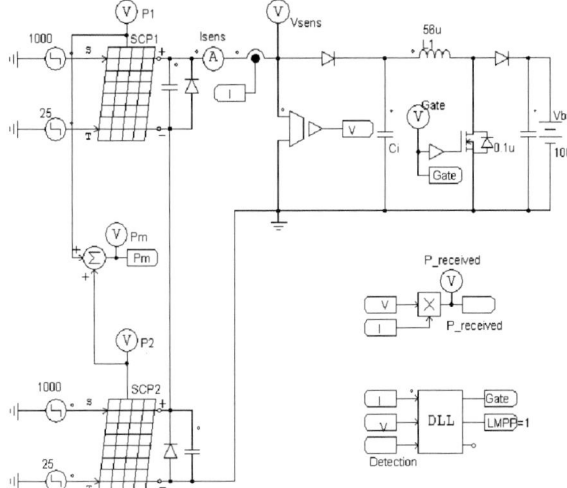

Fig 6: PSIM simulation schematic

1411

The 2018 International Power Electronics Conference

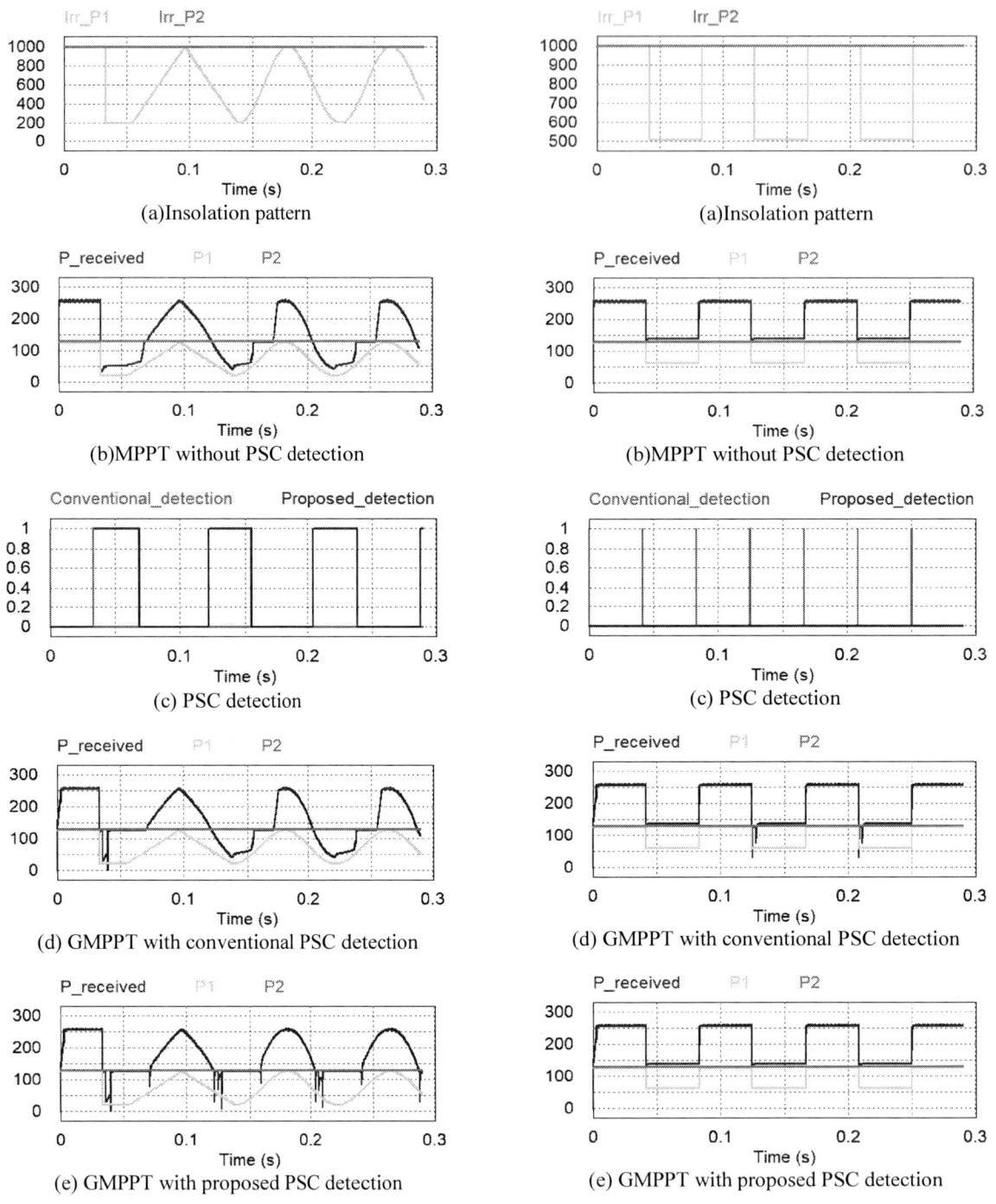

(a)Insolation pattern   (a)Insolation pattern

(b)MPPT without PSC detection   (b)MPPT without PSC detection

(c) PSC detection   (c) PSC detection

(d) GMPPT with conventional PSC detection   (d) GMPPT with conventional PSC detection

(e) GMPPT with proposed PSC detection   (e) GMPPT with proposed PSC detection

Fig 7: Test 1 (Smooth insolation change)   Fig 8: Test 2 (Sharp insolation change)

$P_{received}$ need to be stored in P_m to adapt to the change of insolation.

Comparisons of dP/P algorithm and proposed one are done with simulation. Here, the proposed technique aims at reducing the time taken by a global MPP technique thereby increasing the efficiency. The proposed algorithm successfully overcomes this detection issue in both sharp and smooth insolation changing conditions.

III. ALGORITHM VERIFICATION BY SIMULATION

The existing dP/P PSC algorithm [2] and proposed algorithm are developed in PSIM as Fig.6. Two BP MSX 120 modules are serially connected as an array and a boost converter is used. Power P1 and P2 are the maximum obtainable power of each module. The present output

1412

power is calculated by the product of PV array voltage and current and then is assigned to $P_{received}$. To implement a partial shading, one of the two modules is supplied by the time-varying shading patterns and other modules are fed up with the constant insolation pattern of $1000W/m^2$. P&O (perturb and observation) algorithm with 0.007 ΔD is used as an MPPT algorithm and developed in DLL block. The global peak searching algorithm in [4] is used for implementation. The simulation parameters are shown in Table 1.

A sawtooth and sinusoidal insolation patterns are used to simulate varying insolation. In Test 1, insolation is varied between $1000W/m^2$ and $200 \ W/m^2$ and the detection of PSC is monitored. In Test 2, insolation is varied in the range of $1000W/m^2$ to $500 \ W/m^2$. Under the insolation patterns in Fig. 7, dP/P based algorithm fails to detect smooth insolation changes. In a sharp change of insolation in Fig.8, the conventional algorithm detects PSC. However, such a detection causes a reduction of the overall system because such a small magnitude of insolation change cannot change the highest power peak. In Fig.7 (a) and Fig.8 (a), insolation pattern is shown. The insolation pattern of panel P1 is varied and P2 is kept in constant. The pattern of local MPPT without PSC detection is shown in Fig.7 (b) and Fig.8 (b), where this behavior is generated by P&O algorithm. Conventional and proposed PSC detection, according to the local MPPT are shown in Fig.7 (c) and Fig.8 (c). Here, logically high level, namely "1", indicates PSC detection and low level "0" denotes no PSC detection. When operated with the GMPPT, the behaviors of conventional and proposed PSC detection algorithms are shown in Fig.7 (d), Fig.8 (d) and Fig.7 (e), Fig.8 (e), respectively.

Key observations of the simulation are mentioned in Table 2. An example for the failure of the conventional algorithm is shown here. In such a sample of PV array power, dP/P based algorithm cannot be successfully handled the GMPP, but the proposed method could able to take the correct decision.

Table 1: Simulation parameter

Category	Parameter	Value
PV panel BP MSX 120	P_{max}	120W
	V_{mpp}	33.7V
	I_{mpp}	3.56A
	V_{oc}	42.1V
	I_{sc}	3.87A
	Shunt res.	1000Ω
	Series res.	$0.0015 \ \Omega$
Power circuit	C	22uF
	L	56uH
	V_o	48V
	f_{sw}	100kHz

Table 2: Key observations of Simulation

Power (W)		PSC detection		Status	
1st sample	2nd sample	Conv.	Proposed	Conv.	Proposed
250	225	1	0	Fail	Success
130	120	0	1	Fail	Success

IV. CONCLUSIONS

In this paper, a new PSC detection algorithm is proposed. According to the analysis of the P-V curve, the proposed power level detection algorithm prevents improper global peak searching from occurring, and thus provides a clear distinction between partial shading and global shading. For verification, the algorithm was tested by simulator with complex shading patterns.

REFERENCES

[1] A. Dolara, G. C. Lazaroiu, S. Leva, and G. Manzolini "Experimental investigation of partial shading scenarios on PV (photovoltaic) modules", *Energy*, Vol. 55, pp. 466-475, 2013.

[2] J. Ahmed and Z. Salam, "An Accurate Method for MPPT to Detect the Partial Shading Occurrence in PV System," *IEEE Transactions on Industrial Informatics*, vol. 13, no. 5, pp. 2151 – 2161, 2017.

[3] X. Li, H. Wen, Y. Hu, L. Jiang, and W. Xiao, "Modified Beta Algorithm for GMPPT and Partial Shading Detection in Photovoltaic Systems," *IEEE Transaction on Power Electronics*, DOI 10.1109/TPEL.2017.2697459.

[4] A. Ramyar, H. Iman-Eini and S. Farhangi, "Global Maximum Power Point Tracking Method for Photovoltaic Arrays Under Partial Shading Conditions," in *IEEE Transactions on Industrial Electronics*, vol. 64, no. 4, pp. 2855-2864, April 2017.

[5] M. A. Ghasemi, H. Mohammadian Foroushani and M. Parniani, "Partial Shading Detection and Smooth Maximum Power Point Tracking of PV Arrays Under PSC," in IEEE Transactions on Power Electronics, vol. 31, no. 9, pp. 6281-6292, Sept. 2016.

[6] J. Ma, Tianjiao Zhang, Yu Shi, Xingshuo Li and Huiqin Wen, "Shading pattern detection using electrical characteristics of photovoltaic strings," 2016 IEEE International Conference on Power Electronics, Drives and Energy Systems (PEDES), Trivandrum, pp. 1-4, 2016.

[7] E. Koutroulis and F. Blaabjerg, "A new technique for tracking the global maximum power point of pv arrays operating under partial-shading conditions," *IEEE J. Photovoltaics*, vol. 2, no. 2, pp. 184–190, Apr. 2012

[8] K. S. Tey and S. Mekhilef, "Modified incremental conductance algorithm for photovoltaic system under partial shading conditions and load variation," *IEEE Trans. Ind. Electron*, vol. 61, no. 10, pp. 5384–5392, Oct. 2014.

[9] K. Chen, S. Tian, Y. Cheng, and L. Bai, "An Improved MPPT Controller for Photovoltaic System Under Partial Shading Condition," *IEEE Transactions on Sustainable Energy*, vol. 5, no. 3, pp. 978 - 985, 2014.

The 2018 International Power Electronics Conference

Boost Integrated Three-Phase Solar Inverter using Current Unfolding and Active Damping Methods

Ha Pham N.[1*], Tomoyuki Mannen[2], Keiji Wada[2]

1 School of Electrical and Data Engineering, University of Technology Sydney, Sydney, Australia
2 Department of Electrical and Electronic Engineering, Tokyo Metropolitan University, Tokyo, Japan
*E-mail: phamngocha@ieee.org

Abstract—This paper proposes a three-phase grid connected solar inverter with integrated boost function. The circuit operating principle is based on current unfolding and injection method, which is similar to that of a SWISS rectifier. This approach requires only two high frequency switches operating at only half voltage stress, thus leading to a significant reduction in switching losses. Other switches only operate at line frequency, and therefore can be optimized to reduce conduction losses. The proposed inverter therefore can deliver high efficiency. This paper discusses the basic operating principle as well as control method for the inverter. It is revealed that the output currents of the proposed inverter contains intrinsic oscillation due to current unfolding operation. In order to solve this problem, an active damping method is proposed to stabilize the operation. As a result, stable operation of the proposed method is confirmed by simulation. The feasibility of the proposed inverter is also confirmed using a mini laboratory prototype.

I. INTRODUCTION

Photovoltaic (PV) energy is becoming more and more attractive due to continuous drop in PV panel production cost. Large scale installation of PV energy prefers three-phase grid connection in order to reduce transmission loss. Improving the performance for three-phase inverters is therefore an attractive research topic. For examples, additional boost stage is needed to increase effectiveness of a conventional three-phase PV inverter [1]–[3]. Advanced modulation methods were introduced in [4], [5] to reduce switching loss in three-phase inverters. References [5]–[7] discussed current unfolding and harmonic injection methods to improve the performance of grid-connected three-phase inverters. Other researches applied multilevel approach to reduce voltage stress on high frequency switching devices [8]–[11].

This paper proposes a three-phase inverter with integrated boost function as shown in Fig. 1. Despite having 14 switches, only two high frequency switches are required, the rest are operated at low frequency and thus have negligible switching losses. The low-frequency switches are therefore can be selected to optimise the conduction loss. As a result, the benefit of this inverter is loss reduction and its integrated boost function, resulting in a potential high efficiency. Therefore, the proposed inverter is suitable for large-scale grid connected applications such as solar power plant.

The proposed inverter consists of two parts: a boosting stage which behaves like a voltage-source converter and

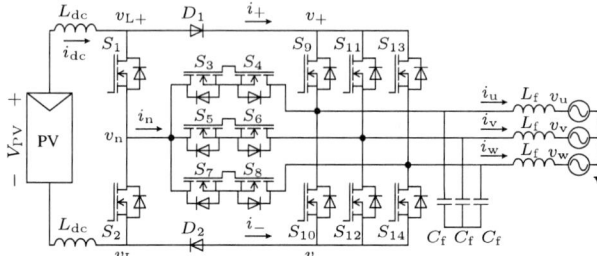

Fig. 1: Circuit configuration of the proposed three-phase solar inverter.

a current unfolding stage which acts like a current-source inverter. This configuration has a similar operating principle compared to a SWISS rectifier [7], but it converts electrical power from dc to three-phase ac.

Similar to other current-source inverters, the switching patterns for the proposed inverter must incorporate overlap time to avoid open circuit conditions. This paper demonstrates that the overlap causes unwanted fluctuation at the output. Therefore, this type of inverter has to deal with oscillation problem unlike the SWISS rectifier where this effect is not as severe thanks to uncontrolled rectifier bridge. In order to suppress oscillation, active damping method using current feedback control was proposed for dc to dc converter [12]. This paper proposes that existing voltage sensors at the output can be used to suppress the oscillation via voltage feedback and changing the duty cycle at the boosting stage.

The operation of the proposed method is studied via a scaled down 20-kVA circuit, connecting a PV to a 400-V three-phase transformer before joining a 6.6-kV grid. Simulation results verify stable operation of the inverter as well as the effectiveness of the active damping method. In addition, a mini laboratory prototype is built to confirm the feasibility of the proposed inverter.

II. CIRCUIT CONFIGURATION

Fig. 1 shows a circuit diagram of the proposed three-phase solar inverter. A PV panel is connected to the positive and negative terminals of the inverter via two dc inductors L_{dc} to reduce common mode noise. The two dc inductors are then connected to two boost converters which control the input current i_{dc} and achieve maximum power point tracking (MPPT) for the PV operation at V_{PV}. The inverter requires only two high frequency switches S_1, S_2 and two diodes D_1, D_2 to control three currents: i_+ at the positive terminal, i_- at the negative terminal, and i_n at the neutral point. Then, it employs 12 switches $S_3 \sim S_{14}$ to unfold those currents to form three-phase currents i_u, i_v, i_w. Those switches only operate at low frequency and thus have negligible switching loss. Since the circuit operates as a current-source inverter with pulsating output currents, a three-phase filter capacitor C_f is needed at the output terminals. Due to the pulsating output currents, there will be small switching ripples remaining at the filter capacitors. Therefore, before connecting to a three-phase grid, a small three-phase filter inductor L_f can be applied to minimize ripple effect on sinusoidal grid.

III. OPERATING PRINCIPLES

A. Low-frequency switching operation

TABLE I: Modulation operation of low-frequency switches in the proposed three-phase inverter.

Switches \ Sectors	I	II	III	IV	V	VI
S_3, S_4	0	1	0	0	1	0
S_5, S_6	1	0	0	1	0	0
S_7, S_8	0	0	1	0	0	1
S_9	1	0	0	0	0	1
S_{10}	0	0	1	1	0	0
S_{11}	0	1	1	0	0	0
S_{12}	0	0	0	0	1	1
S_{13}	0	0	0	1	1	0
S_{14}	1	1	0	0	0	0

Assume that the output three-phase voltages are balanced as:

$$v_u = \sqrt{2}V_o \cos(\omega t), \tag{1}$$

$$v_v = \sqrt{2}V_o \cos(\omega t - 2\pi/3), \tag{2}$$

$$v_w = \sqrt{2}V_o \cos(\omega t + 2\pi/3), \tag{3}$$

where V_o is the rms of phase to neutral voltage and ω is the angular frequency.

The first principle of the proposed three-phase inverter is to fold the three-phase voltages v_u, v_v and v_w at the output to match the terminal voltages v_+, v_-, and v_n from the positive, negative and neutral terminals respectively. Thus, the switching patterns of $S_3 \sim S_{14}$ are determined by the varying relations of the instantaneous values of the phase voltages v_u, v_v, v_w. First, the switches $S_9 \sim S_{14}$ are operated as a synchronous rectifier to enable reversed current flow from the positive and negative terminals to two phases with the highest and lowest voltages. Then, the switches $S_3 \sim S_8$ connect the neutral terminal to the other phase. For this reason, the proposed inverter requires six operating sectors with an equal interval of $\pi/3$ according to its rotating phase angle ωt. These sectors determine the switching sequences for $S_3 \sim S_{14}$ as shown in Table

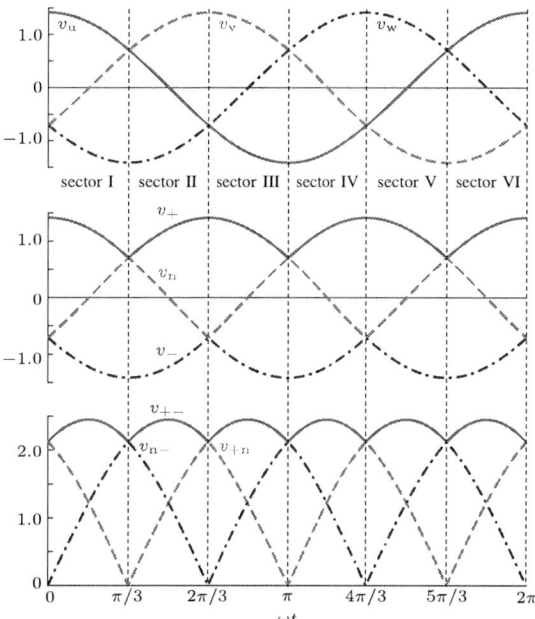

Fig. 2: Voltage folding principle of the proposed three-phase inverter. The voltage scales are displayed as per unit compared to the base value of output rms phase to neutral voltage V_o.

I, where '1' and '0' indicate 'ON' and 'OFF' conditions of the switching devices respectively. It should be noted that the bidirectional switches $S_3 \sim S_8$ are operated at double the line frequency while $S_9 \sim S_{14}$ only switch at line frequency.

For examples, sector I is defined for the case when $v_u > v_v > v_w$. Here, phase u is always positive and thus, is connected to the positive terminal via switch S_9. On the other hand, phase w is always negative and is connected to the negative terminal via switch S_{14}. The remaining phase v can have either positive or negative voltage, so it is connected to the neutral terminal via switches S_5 and S_6 which enable bidirectional output. As a result, the neutral terminal is floating with the middle phase voltage.

Fig. 2 illustrates how the sectors are allocated as well as how the terminal voltages look like. Notice that the voltages are normalized to the base value V_o. It is noted that v_+ at the positive terminal takes positive value while v_- at the negative terminal remains negative all the time

$$v_+ > 0, \tag{4}$$

$$v_- < 0. \tag{5}$$

The neutral terminal is floating, but it always takes a value between the positive and negative terminals as

$$v_+ > v_n > v_-, \tag{6}$$

which is an important condition to implement the two boost converters because the output voltages of those must be positive. Also, in a balanced three-phase system, the sum of those terminal voltages is zero as

$$v_+ + v_n + v_- = 0. \tag{7}$$

The differences in terminal voltages

$$
\begin{aligned}
v_{+-} &= v_+ - v_- \\
v_{+n} &= v_+ - v_n \\
v_{n-} &= v_n - v_-
\end{aligned}
$$

determine the stress level on switching devices and indicate a selection guide for the switches. For examples, the maximum value of v_{+-} applies on $S_9 \sim S_{14}$, while the maximum value of v_{+n} and v_{n-} influences the stress voltages on S_1, D_1 and S_2, D_2 as well as $S_3 \sim S_8$ accordingly. Also, the minimum value of v_{+-} indicates the output of the boosting stages, which must be higher than the maximum input voltage from the PV panel. Therefore, the operating condition of input voltage is given by

$$V_{\mathrm{PV}} < \frac{3\sqrt{2}}{2} V_{\mathrm{o}} \approx 2.12 V_{\mathrm{o}}. \tag{8}$$

B. High-frequency switching operation

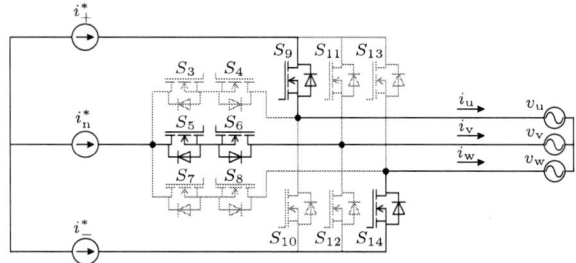

Fig. 3: Equivalent circuit of the proposed three-phase solar inverter for operation in sector I.

The three-phase inverter operates as a current source inverter with an unfolding operation determined by the aforementioned six operating sectors in Table I. Fig. 3 explains how the current unfolding operation is realized with the low frequency switches in section I. Considering the dc side as current sources which actually generate pulse currents. Their average values are determined by the references i_+^*, i_-^*, i_n^*. It can be seen that the phase currents in this sector is connected to the corresponding current sources so that

$$
\begin{aligned}
i_{\mathrm{u}} &= i_+^* \tag{9} \\
i_{\mathrm{v}} &= i_n^* \tag{10} \\
i_{\mathrm{w}} &= i_-^*. \tag{11}
\end{aligned}
$$

The relationship from (9) $\sim$ (11) mean that the desired output currents can be used to extract the reference values for the input current sources. Similarly, applying the unfolding scheme for the other sectors, the full references for terminal currents at the dc side can be achieved.

Fig. 4 shows how a balanced three-phase currents with unity power factor determines the corresponding references for the terminal current sources via the switching

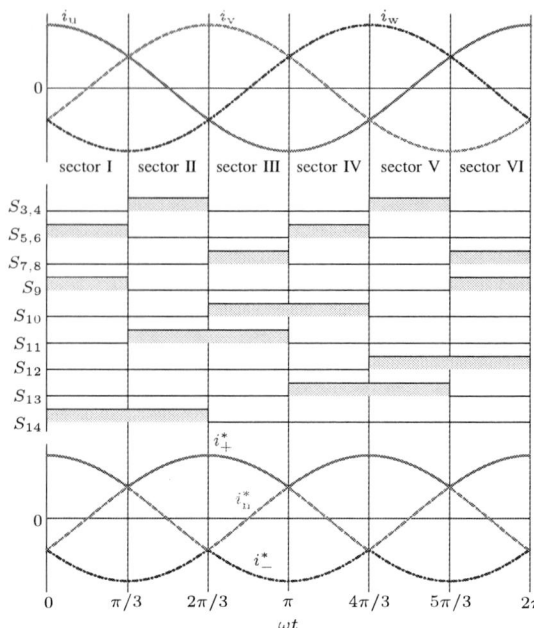

Fig. 4: Unfolding operation of i_+^*, i_n^*, i_-^* to form sinusoidal output three-phase currents i_{u}, i_{v}, i_{w} under unity power factor.

sequences shown in Table I. Notice that i_+^* is always positive and i_-^* always takes negative value so that the boost converters can provide appropriate duty cycles. The remaining neutral current can take either positive or negative value depending on the phase condition. Its value is automatically determined by Kirchhoff's current law as

$$i_n = -i_+ - i_-, \tag{12}$$

and thus, we only need to control i_+ and i_- by adjusting the duty cycles in the two boost converters to achieve the final output three-phase currents.

C. Commutation transition in low-frequency switching operation

Since the low-frequency part of the proposed topology operates as a current source inverter, the output requires filter capacitor and the modulation needs overlap transition instead of blanking time in voltage source inverters. Fig. 5 shows the transition when changing from sector I to sector II. It can be seen that the positive terminal and neutral terminal are shorted because of two paths created by switches S_3, S_4, S_9 and S_5, S_6, S_{11}.

Ideally, this occurs exactly at $\omega t = \pi/3$ when $v_{\mathrm{u}} = v_{\mathrm{w}}$, thus $v_{+n} = 0$ or there should not be any short-circuit problem. However, the overlap time is not zero in practice due to limitation in switching speed as well as to difficulty in synchronization of the ON/OFF timing. In addition, the voltage at the filter capacitor terminals are not the same as the phase voltages due to existing phase currents, e.g.

$$v_{\mathrm{u}} = v_{\mathrm{Cu}} + L_{\mathrm{f}} \frac{di_{\mathrm{u}}}{dt}. \tag{13}$$

The 2018 International Power Electronics Conference

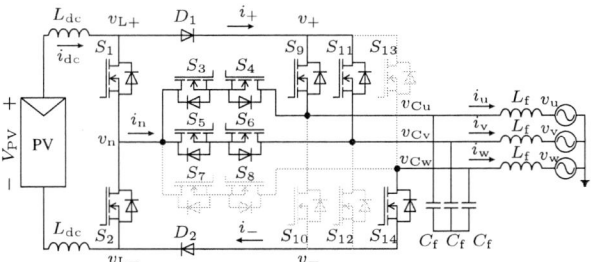

Fig. 5: Commutation transition in switching operation from sector I to sector II.

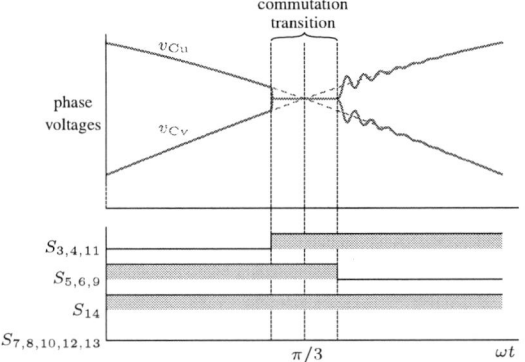

Fig. 6: Voltage waveforms during the commutation transition from sector I to sector II.

Therefore, the transition happens slightly before and ends after $\omega t = \pi/3$ as shown in Fig. 6. When the short circuit occurs, the phase voltage difference is quickly reduced to zero. As a result, the current at the positive terminal i_+ experience a temporary overshoot due to the short circuit. After the transition, the difference will appear again and will cause oscillation in both output voltage and current waveforms because of resonance between output filter capacitors and equivalent circuit inductors. This effect will be demonstrated later by simulation.

D. Active damping method

The aforementioned oscillation at the output due to current unfolding transition will cause distortion as well as increased THD. Thus, this paper proposes an active damping method using voltage feedback at the boost converters to reduce the unwanted effect.

Considering the boost converter at the input, a normal switching operation yields

$$\overline{i_{\text{out}}} = D i_{\text{in}}, \tag{14}$$

where D represents the 'ON' duty cycle of active switch in a boost converter circuit. If the output is connected to a capacitor, resonant oscillation will occur at certain operating conditions.

Fig. 7 illustrates a boost converter equipped with a parallel resistor R at the output. The output current

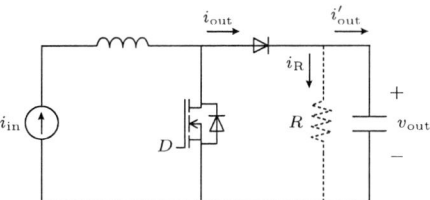

Fig. 7: A boost converter equipped with active damping resistor.

becomes

$$i'_{\text{out}} = i_{\text{out}} - i_{\text{R}}, \tag{15}$$

and thus,

$$\overline{i'_{\text{out}}} = D i_{\text{in}} - \frac{v_{\text{out}}}{R}. \tag{16}$$

This resistor R acts as a damping factor to suppress oscillation in the boost converter. However, real resistor will consumes power and thus increases power loss. Therefore, the duty cycle can be redefined as

$$D' = \frac{\overline{i'_{\text{out}}}}{i_{\text{in}}} = D - \frac{v_{\text{out}}}{i_{\text{in}} R}, \tag{17}$$

to provide active damping without real resistor. Unfortunately, this form will cause unwanted output current reduction which leads to distortion from sinusoidal reference. Thus, we only consider deviation from target value to suppress the unwanted oscillation. Equation (17) is therefore adjusted to be

$$D' = D - \frac{v_{\text{out}} - v^*_{\text{out}}}{i_{\text{in}} R}. \tag{18}$$

This duty cycle provides active damping at the output voltage without affecting the control for large signal.

E. Control method

Applying Kirchoff's voltage law at the dc side of Fig. 1 yields

$$V_{\text{PV}} - 2v_{\text{L}} = v_{\text{L}+} - v_{\text{L}-}, \tag{19}$$

where v_{L} is the voltage across each dc inductor and $v_{\text{L}+}$ and $v_{\text{L}-}$ are the voltages at the output of the upper and lower dc inductors, respectively. The voltage of each input inductor is determined by

$$v_{\text{L}} = L_{\text{dc}} \frac{di_{\text{dc}}}{dt}. \tag{20}$$

Therefore, the total input inductor voltage can be determined using feedback control of i_{dc} as follows

$$2v^*_{\text{L}} = K_{\text{dc}}(i^*_{\text{dc}} - i_{\text{dc}}). \tag{21}$$

At sector I, boosting operation at the positive terminal yields

$$v_{\text{L}+} - v_{\text{n}} = D_+ v_{+\text{n}} \tag{22}$$
$$i_{\text{u}} = i_+ = D_+ i_{\text{dc}}. \tag{23}$$

1417

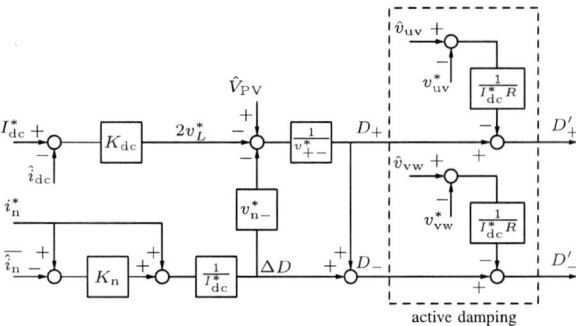

Fig. 8: Control block diagram of the three-phase solar inverter for operation in sector I.

Similarly at the negative terminal, we have

$$v_{L-} - v_n = -D_- v_{n-} \quad (24)$$
$$i_w = i_- = -D_- i_{dc}. \quad (25)$$

The negative sign in the current equation (25) comes from the fact that i_- always takes negative value while i_{dc} is defined to be positive. Considering (12), the sum of (23) and (25) yields

$$i_n = (D_- - D_+) i_{dc} = \Delta D i_{dc}, \quad (26)$$

where

$$\Delta D = D_- - D_+, \quad (27)$$

is the difference in duty cycles of the two boost converters. Thus, the neutral current i_n can be used to determined ΔD as follows

$$\Delta D = \frac{i_n^*}{i_{dc}^*} + K_n \frac{i_n^* - i_n}{i_{dc}^*}, \quad (28)$$

where the secondary part is neutral current feedback used for improving the shape of output currents.

Subtracting (22) to (24) yields

$$
\begin{aligned}
v_{L+} - v_{L-} &= D_+ v_{+n} + D_- v_{n-} \\
&= D_+ v_{+n} + (D_+ + \Delta D) v_{n-} \\
&= D_+ (v_{+n} + v_{n-}) + \Delta D v_{n-} \\
&= D_+ v_{+-} + \Delta D v_{n-}. \quad (29)
\end{aligned}
$$

Taking (19) into consideration, the duty cycle of the upper boost converter can be extracted from (29) as

$$D_+ = \frac{V_{PV} - 2v_L - \Delta D v_{n-}}{v_{+-}}. \quad (30)$$

The proposed inverter can be operated following the relations expressed by (19), (21), (27), (28), and (30). The control method for sector I is summarized in Fig. 8 where active damping expressed by (18) is also included.

TABLE II: Circuit parameters used in simulation

Symbol	Meaning	Value
V_{PV}	MPPT voltage of PV	300 V
V_o	output phase rms voltage	220 V
I_o	output phase rms current	28.2 A
f	line frequency	50 Hz
f_{sw}	switching frequency	20 kHz
L_{dc}	dc inductor	0.5 mH
L_f	ac filter inductor	0.2 mH
C_f	ac filter capacitor	20 μF

IV. SIMULATION RESULTS

Circuit simulation was carried out to test the performance of the proposed inverter. The simulated circuit parameters are shown in Table II. The system is designed to operate at 20 kVA where the MPPT voltage of PV reaches around 300 V. The line frequency is set at 50 Hz, while the switching frequency of S_1, S_2 is set at 20 kHz to avoid acoustic noise. The dc inductors are chosen as 0.5 mH to limit the ripple at the PV input to less than 10% at rated power. The output filters were chosen to limit the current THD to less than 5% at rated condition. The overlap time for commutation in low-frequency switching operation is set to be 100 μs in the simulation.

Fig. 9 shows the operation of the proposed inverter without active damping control. The dc input current I_{dc} is controlled to have a stable value at 62 A. Given that the MPPT voltage is $V_{PV} = 300$ V, the operating power reaches 18.6 kW. It can be seen that the average currents in i_+, i_n, i_- are not ideal due to overlap time implementation in switches $S_3 \sim S_{14}$. Due to the intrinsic glitches, the output currents i_u, i_v, i_w contains oscillating component that comes from resonance between the output filter capacitors and equivalent circuit inductors. Due to this oscillation, it is difficult to achieve a low THD at the output.

Fig. 10 shows the operation of the proposed inverter when active damping control was implemented. The average currents in i_+, i_n, i_- remains problematic, but the duty cycles D_+ and D_- are adjusted to be D'_+ and D'_- to damp the output oscillation. According to this active damping control, the output currents i_u, i_v, i_w contains minimal distortion caused by the overlap time. This proves the effectiveness of the proposed active damping control.

V. EXPERIMENTAL RESULTS

A mini laboratory prototype was set up to demonstrate the proposed inverter. The PV panel was replaced with a dc voltage source. Its input voltage was set to 170 V so that it can provides enough power to run the three-phase load at 2.13 A. For the sake of simplicity in the feasibility test, the grid was replaced by a three-phase Y connected resistive load consisting of 40-Ω resistors as shown in Fig. 11. The circuit parameters used in the experiment is shown in Table III. Notice that the filter capacitors are

The 2018 International Power Electronics Conference

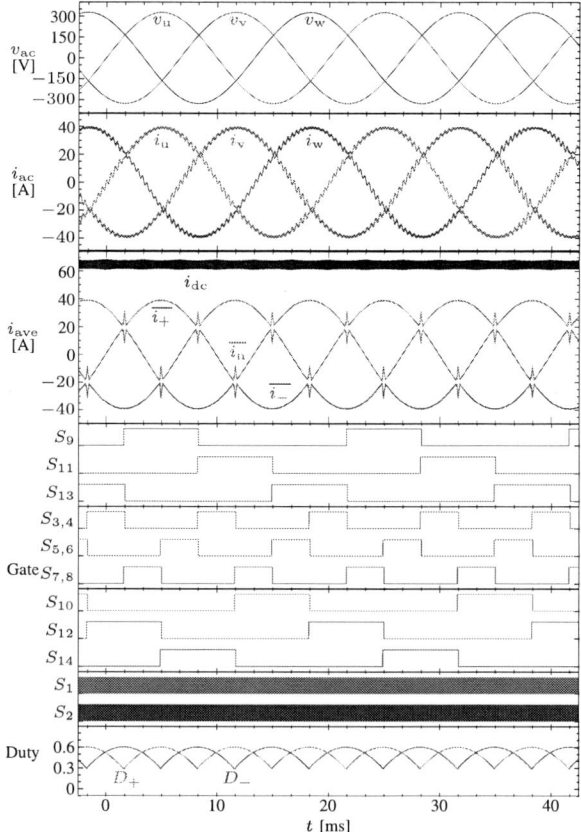

Fig. 9: Simulated waveforms of the proposed inverter without active damping control.

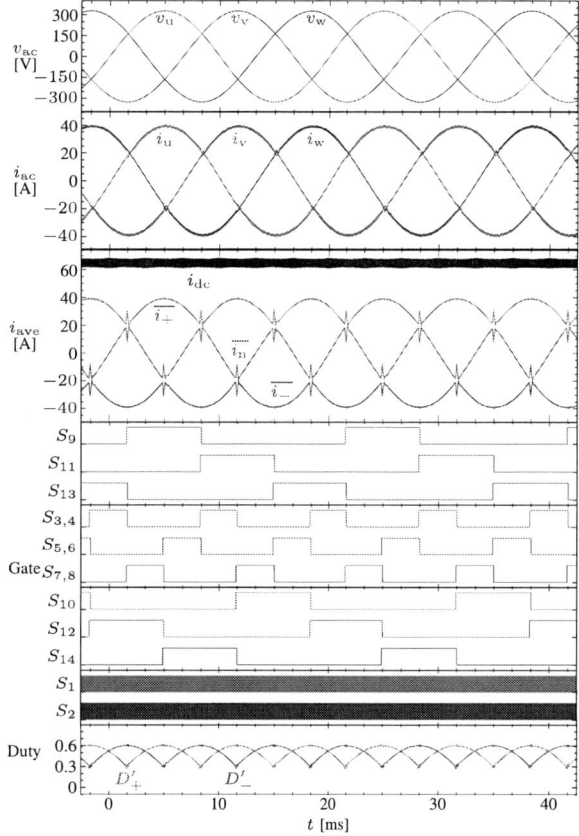

Fig. 10: Simulated waveforms of the proposed inverter when applying active damping control.

TABLE III: Circuit parameters used in experiment

Symbol	Meaning	Value
V_{in}	input voltage	170 V
V_o	output phase rms voltage	85 V
I_o	output phase rms current	2.13 A
f	line frequency	50 Hz
f_{sw}	switching frequency	20 kHz
L_{dc}	dc inductor	0.5 mH
C_f	ac filter capacitor	10 μF
R_l	load resistor	40 Ω

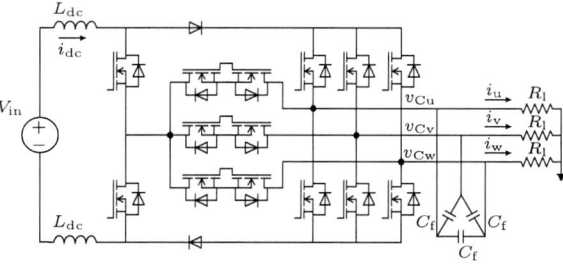

Fig. 11: Experimental setup for feasibility test of the proposed three-phase inverter.

delta connected and the experiment did not employ any filter inductor. The overlap time for commutation in low-frequency switching operation was set to be 25 μs in the experiment.

Fig. 12 shows the measured waveforms on a single phase when operating at 50 Hz. The measured filter capacitor voltage was almost sinusoidal but contained switching ripples due to pulsating output currents in the boost stage. The corresponding output ac current had less switching ripple compared to that of the capacitor voltage due to the presence of parasitic inductance in the resistive load and its connection. Since the parasitic inductance was negligible, the output voltage was proportional with

the output current as $v_{Cu} = R_l i_u$ and thus in phase with the output current, resulting in unity power factor. The input current was controlled to have a constant average value at 3.1 A which is just a little above the maximum required output current of 3 A as a requirement for operating the boost converters.

VI. CONCLUSION

This paper proposed a boost integrated three-phase solar inverter using current unfolding method. The benefits of the proposed topology is that it needs only two high-frequency switches while does not require large output

1419

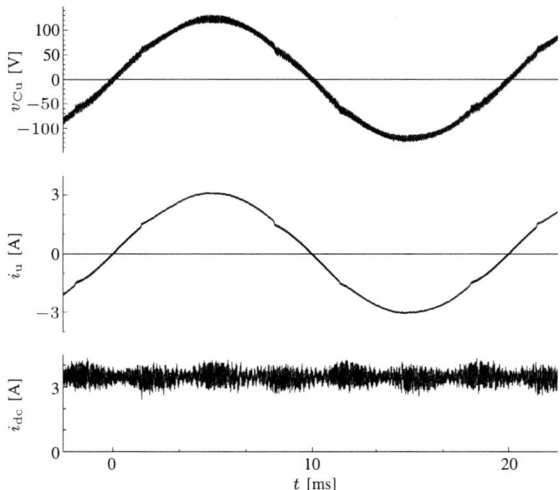

Fig. 12: Experimental verification of the proposed inverter when supporting a balanced three-phase resistive load.

ac inductors. The inverter employs 12 low-frequency switches for unfolding operation, but the conduction loss can be optimized thanks to their negligible switching losses. As a result, high efficiency and/or size reduction in passive components can be achieved.

This paper revealed that the overlap time in switching transition of the unfolding operation causes unwanted oscillation at the output currents. However, it is possible to suppressed the oscillation using active damping method. Simulation study presented stable operation of the proposed inverter and proved the effectiveness of the proposed active damping method. Finally, an experimental verification with resistive load was successfully carried out to confirm the feasibility of the proposed inverter.

REFERENCES

[1] M. de Brito, L. Sampaio, G. Melo and C. A. Canesin, "Three-phase tri-state buck-boost integrated inverter for solar applications," *IET Renewable Power Generation,* vol. 9, no. 6, pp. 557–565, Aug. 2015.

[2] C. Jain and B. Singh, "A Three-Phase Grid Tied SPV System With Adaptive DC Link Voltage for CPI Voltage Variations," *in IEEE Transactions on Sustainable Energy,* vol. 7, no. 1, pp. 337–344, Jan. 2016.

[3] E. Serban, F. Paz and M. Ordonez, "Improved PV Inverter Operating Range Using a Miniboost," *IEEE Transactions on Power Electronics,* vol. 32, no. 11, pp. 8470–8485, Nov. 2017.

[4] D. Yamaguchi and H. Fujita, "A new PV converter for grid connection through a high-leg delta transformer using cooperative control of boost converters and inverters," *2017 IEEE 3rd International Future Energy Electronics Conference and ECCE Asia (IFEEC 2017 - ECCE Asia),* pp. 911–916, 2017.

[5] U. R. Prasanna and A. K. Rathore, "Dual Three-Pulse Modulation-Based High-Frequency Pulsating DC Link Two-Stage Three-Phase Inverter for Electric/Hybrid/Fuel Cell Vehicles Applications," *IEEE Journal of Emerging and Selected Topics in Power Electronics,* vol. 2, no. 3, pp. 477–486, Sept. 2014.

[6] H. Yoo and S. K. Sul, "A new circuit design and control to reduce input harmonic current for a three-phase ac machine drive

system having a very small dc-link capacitor," *2010 Twenty-Fifth Annual IEEE Applied Power Electronics Conference and Exposition (APEC),* pp. 611–618, 2010.

[7] T. B. Soeiro, T. Friedli and J. W. Kolar, "Swiss rectifier – A novel three-phase buck-type PFC topology for Electric Vehicle battery charging," *2012 Twenty-Seventh Annual IEEE Applied Power Electronics Conference and Exposition (APEC),* pp. 2617–2624, 2012.

[8] M. R. Islam, A. M. Mahfuz-Ur-Rahman, M. M. Islam, Y. G. Guo and J. G. Zhu, "Modular Medium-Voltage Grid-Connected Converter With Improved Switching Techniques for Solar Photovoltaic Systems," *IEEE Transactions on Industrial Electronics,* vol. 64, no. 11, pp. 8887–8896, Nov. 2017.

[9] Y. Wang, F. Wang, "Novel Three-Phase Three-Level-Stacked Neutral Point Clamped Grid-Tied Solar Inverter With a Split Phase Controller," *IEEE Trans. Power Electron.,* vol. 28, no. 6, pp. 2856–2866, Jun. 2013.

[10] L. B. G. Campanhol, S. A. O. da Silva, A. A. de Oliveira and V. D. Bacon, "Dynamic Performance Improvement of a Grid-Tied PV System Using a Feed-Forward Control Loop Acting on the NPC Inverter Currents," *IEEE Transactions on Industrial Electronics,* vol. 64, no. 3, pp. 2092–2101, Mar. 2017.

[11] S. K. Chattopadhyay and C. Chakraborty, "A New Asymmetric Multilevel Inverter Topology Suitable for Solar PV Applications With Varying Irradiance," *IEEE Transactions on Sustainable Energy,* vol. 8, no. 4, pp. 1496–1506, Oct. 2017.

[12] A. M. Rahimi and A. Emadi, "Active Damping in DC/DC Power Electronic Converters: A Novel Method to Overcome the Problems of Constant Power Loads," *IEEE Transactions on Industrial Electronics,* vol. 56, no. 5, pp. 1428–1439, May 2009.

Linear Active Disturbance Rejection Control for Isolated Three-Port Converter

Jiang You, Mengyan Liao* and Weiyan Fan

College of Automation, Harbin Engineering University, Harbin, China
liaomengyan417@163.com

Abstract— Because of the coupling and interaction of power delivery among the different ports in isolated three-port converter (TPC) and the small signal linear control model derived from specific steady state point, the parameters of control models will be varied with the change of operation state, this might result in degradation of control performance. In order to solve these problems, linear active disturbance rejection control (LADRC) method is employed for TPC. And an improved control model which is got by introducing virtual resistor is used to design LADRC controller for the TPC in this paper. The simulation results show that the three-port converter can achieve better performance balance in dynamic response and decoupling by using the proposed method.

Keywords— Three-port converter, linear active disturbance rejection control, virtual resistor, decoupling, state observer.

I. INTRODUCTION

Isolated three-port converter gets more and more concerns and studies in the field of renewable energy generation, energy storage system and distribution system [1-2] due to its high power density and flexibility to dispose different energy synchronously through proper control strategy [3-6]. For example, it is studied to improve the fault-tolerant capability of distribution system in multi-electric aircraft [7-8]. Besides, high efficiency soft switching operation can be achieved in this type of converter by proper transformer parameters design and modulation scheme [9-10].

The three windings of the isolation transformer share the same magnetic core in the topology, so interactions of power delivery among the three ports are unavoidable. Traditional decoupling control (TDC) method is usually used for three-port converter, in which, two single input single output (SISO) subsystems can be got by introducing decupling terms [11-12], classical control theory is employed to design controller for each channel. However, since the small-signal models used for controller design are derived by linearizing the original nonlinear model at steady state operation point, the change of operation parameters might lead heavy degradation in dynamic and decoupling control, significant parameters deviation might even cause instability in the control system.

This work is supported by the national NSF of China (No. 51479042), the NSF (No. E201238) of Heilongjiang province, China, and the Fundamental Research Funds for the Central Universities (No. GK2040260197)

LADRC has excellent abilities to tolerate variation of the model parameter and realize natural decoupling [13]. In this method, the impact of model uncertainties and couplings among the different ports are regarded as generalized external disturbances, which can be estimated by the linear extended state observer (LESO). The observed signals are added into the control system to compensate the disturbance. In this condition, the control performance can be improved with the change of model parameters [14].

LADRC based method used for an isolated TPC is studied in the following sections. The topology, modulation scheme and small signal models are presented in section II. The LADRC based TPC control system design method together with the proposed virtual resistor to attenuate the resonant peak for port 1 current control model are given in section III. The simulation and analysis for the proposed control method is shown in section IV. Finally, the conclusion is drawn in section V.

II. MODELING OF THE THREE-PORT CONVERTER

The topology of the isolated three-port converter is shown in Fig. 1(a), in this figure, the port 1 is a power source port (e.g. it can be solar panels) , the power source v_{d1} is series with an inductor L_{d1} with an equivalent series

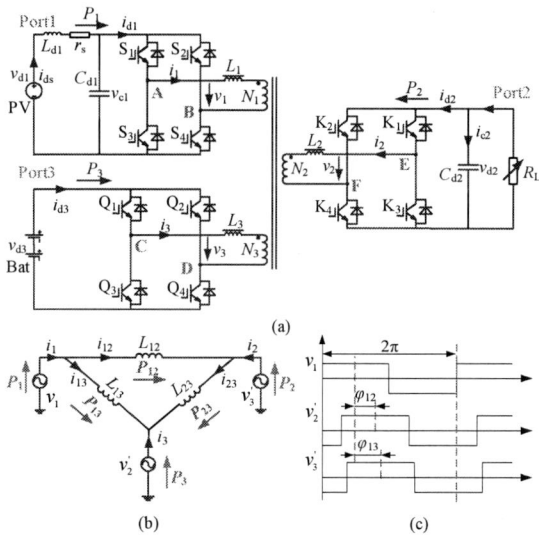

Fig. 1. Topology, equivalent circuit and modulation scheme of the three-port converter. (a) Isolated three-port converter topology. (b) Equivalent Δ-connection circuit. (c) modulation waveform.

resistor r_s. The duty cycles of switching signals for the switches on leg A (S_1 and S_3) and leg B (S_2 and S_4) are fixed at 50%. The phase shifting between leg A and leg B is 180°. The port 2 and port 3 are load port and energy storage port respectively and their switching mode is exactly the same as that of port 1. An equivalent Δ-connection circuit of the TPC is shown in Fig. 1(b) by converting the parameters of port 2 and port 3 to port 1. By setting v_1 as reference, the phase shifting of v_1 and v_2 are φ_{12} and φ_{13} respectively shown in Fig. 1(c).

Since the total power delivered in each switching period is very close to its fundamental component power, the fundamental component analysis method is adopted for mathematic analysis and modeling in this paper. The amplitude of the fundamental component as shown in (1).

$$V_{1f} = \frac{4V_{d1}}{\pi}, \ V_{2f}' = \frac{4N_1 V_{d2}}{\pi N_2}, \ V_{3f}' = \frac{4N_1 V_{d3}}{\pi N_3} \quad (1)$$

where V_{d1}, V_{d2} and V_{d3} are the amplitude of v_1, v_2 and v_3, N_1, N_2 and N_3 are the transformer windings turns for different ports.

According to the Fig. 1(b) the power of port 1, port 2 and port 3 can be written as (2).

$$\begin{cases} P_1 = P_{12} + P_{13} \\ P_2 = -P_{12} + P_{23} \\ P_3 = -P_{13} - P_{23} \end{cases} \quad (2)$$

The simplified equivalent circuit of the power delivery between port 1 and port 2 is shown in Fig. 2.

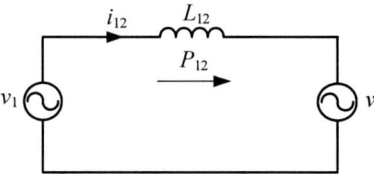

Fig.2 The equivalent circuit with port 3 is opened.

φ_{12} is the phase shift between v_1 and v_2'. If $v_1 = V_{d1} \angle 0$ then $v_2' = V_{2f}' \angle -\varphi_{12}$. The power transmitted between port 1 and port 2 is given in (3).

$$P_{12} = \frac{8N_1}{\pi^2 N_2 \omega_s L_{12}} V_{d1} V_{d2} \sin \varphi_{12} \quad (3)$$

where ω_s is the switching angular frequency. Similarly, P_{12} and P_{13} can be obtained as in (4) and (5) respectively.

$$P_{13} = \frac{8N_1}{\pi^2 N_3 \omega_s L_{13}} V_{d1} V_{d3} \sin \varphi_{13} \quad (4)$$

$$P_{23} = \frac{8N_1^2}{\pi^2 N_2 N_3 \omega_s L_{23}} V_{d2} V_{d3} \sin(\varphi_{13} - \varphi_{12}) \quad (5)$$

According to (2), there always exists a dependent port since $P_1 + P_2 + P_3 = 0$. In this paper, the battery port is considered as a free port for energy buffer, which can absorb and release power according to the power transmission between port 1 and port 2. The power equations of port 1 and port 2 derived from (2)~(5) can be expressed as (6) and (7) respectively.

$$P_1 = \frac{8N_1 V_{d1} V_{d2}}{\pi^2 N_2 \omega_s L_{12}} \sin \varphi_{12} + \frac{8N_1 V_{d1} V_{d3}}{\pi^2 N_3 \omega_s L_{13}} \sin \varphi_{13} \quad (6)$$

$$P_2 = -\frac{8N_1 V_{d1} V_{d2}}{\pi^2 N_2 \omega_s L_{12}} \sin \varphi_{12} + \frac{8N_1^2 V_{d2} V_{d3}}{\pi^2 N_2 N_3 \omega_s L_{23}} \sin(\varphi_{13} - \varphi_{12}) \quad (7)$$

The average current of port 1 and port 2 can be obtained as (8) and (9) respectively.

$$\bar{i}_{d1} = \frac{P_1}{V_{d1}} = \frac{8N_1 V_{d2}}{\pi^2 N_2 \omega_s L_{12}} \sin \varphi_{12} + \frac{8N_1 V_{d3}}{\pi^2 N_3 \omega_s L_{13}} \sin \varphi_{13} \quad (8)$$

$$\bar{i}_{d2} = \frac{P_2}{V_{d2}} = -\frac{8N_1 V_{d1} \sin \varphi_{12}}{\pi^2 N_2 \omega_s L_{12}} + \frac{8N_1^2 V_{d3} \sin(\varphi_{13} - \varphi_{12})}{\pi^2 N_2 N_3 \omega_s L_{23}} \quad (9)$$

(10) can be derived by implementing partial differential operation for (8) and (9) at a steady state operation point A (φ_{120}, φ_{130}) respectively.

$$\begin{cases} G_{11} = \left. \frac{\partial \bar{i}_{d2}}{\partial \varphi_2} \right|_A = -\frac{8N_1 V_{d1} \cos \varphi_{120}}{\pi^2 N_2 \omega_s L_{12}} - \frac{8N_1^2 V_{d3} \cos(\varphi_{130} - \varphi_{120})}{\pi^2 N_2 N_3 \omega_s L_{23}} \\ G_{12} = \left. \frac{\partial \bar{i}_{d2}}{\partial \varphi_3} \right|_A = \frac{8N_1^2 V_{d3}}{\pi^2 N_2 N_3 \omega_s L_{23}} \cos(\varphi_{130} - \varphi_{120}) \\ G_{21} = \left. \frac{\partial \bar{i}_{d1}}{\partial \varphi_2} \right|_A = \frac{8N_1 V_{d2}}{\pi^2 N_2 \omega_s L_{12}} \cos \varphi_{120} \\ G_{22} = \left. \frac{\partial \bar{i}_{d1}}{\partial \varphi_3} \right|_A = \frac{8N_1 V_{d3}}{\pi^2 N_3 \omega_s L_{13}} \cos \varphi_{130} \end{cases} \quad (10)$$

(10) can be rewritten as (11)

$$\begin{bmatrix} \tilde{i}_{d2} \\ \tilde{i}_{d1} \end{bmatrix} = \begin{bmatrix} G_{11} & G_{12} \\ G_{21} & G_{22} \end{bmatrix} \begin{bmatrix} \tilde{\varphi}_{12} \\ \tilde{\varphi}_{13} \end{bmatrix} = G_A \begin{bmatrix} \tilde{\varphi}_{12} \\ \tilde{\varphi}_{13} \end{bmatrix} \quad (11)$$

It can be seen from (10) that the elements in G_A are closely related to the steady state parameters (φ_{120} and φ_{130}) and (11) clearly indicates that $\tilde{i}_{d1}$ and $\tilde{i}_{d2}$ are affected by the cross-coupling terms $G_{21}\tilde{\varphi}_{12}$ and $G_{12}\tilde{\varphi}_{13}$.

By using KCL and KVL laws in Fig. 1(a) and taking advantage of (11), the small signal linearization model of TPC can be obtained in (12).

$$\begin{cases} C_{d2} \frac{d\tilde{v}_{d2}}{dt} = -\frac{\tilde{v}_{d2}}{R_L} - G_{11}\tilde{\varphi}_{12} - G_{12}\tilde{\varphi}_{13} \\ L_{d1} \frac{d\tilde{i}_{ds}}{dt} = \tilde{v}_{d1} - \tilde{v}_{c1} - r_s \tilde{i}_{ds} \\ C_{d1} \frac{d\tilde{v}_{c1}}{dt} = \tilde{i}_{ds} - G_{21}\tilde{\varphi}_{12} - G_{22}\tilde{\varphi}_{13} \end{cases} \quad (12)$$

III. CONTROL STRATEGY OF THREE-PORT CONVERTER

A. Traditional Decoupling Control

By introducing a 2×2 decoupling matrix H, the matrix shown in (13) can be transformed into a diagonal matrix.

$$\begin{bmatrix} G_{11} & G_{12} \\ G_{21} & G_{22} \end{bmatrix} \begin{bmatrix} H_{11} & H_{12} \\ H_{21} & H_{22} \end{bmatrix} = \begin{bmatrix} G_{11} & 0 \\ 0 & G_{22} \end{bmatrix} \quad (13)$$

the decoupling matrix H can be given as (14)

$$H=\begin{bmatrix} H_{11} & H_{12} \\ H_{21} & H_{22} \end{bmatrix}=\begin{bmatrix} \dfrac{G_{11}G_{22}}{G_{11}G_{22}-G_{12}G_{21}} & \dfrac{-G_{12}G_{22}}{G_{11}G_{22}-G_{12}G_{21}} \\ \dfrac{-G_{11}G_{21}}{G_{11}G_{22}-G_{12}G_{21}} & \dfrac{G_{11}G_{22}}{G_{11}G_{22}-G_{12}G_{21}} \end{bmatrix} \quad (14)$$

The TDC structure is shown in Fig. 3. G_v and G_c are the voltage and current controller respectively that can be designed by using classical SISO frequency domain control theory.

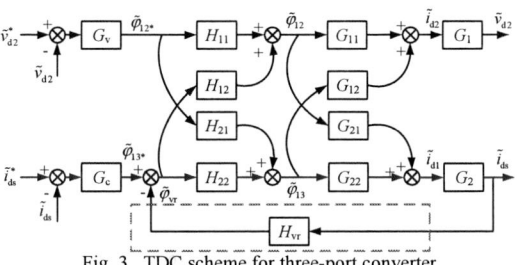

Fig. 3. TDC scheme for three-port converter.

The transfer function of G_1 and G_2 are shown in (15).

$$\begin{cases} G_1(s)=\dfrac{R_L}{R_L C_{d2}s+1} \\ G_2(s)=\dfrac{1}{L_{d1}C_{d1}s^2+r_sC_{d1}s+1} \end{cases} \quad (15)$$

If the equivalent series resistance r_s of L_{d1} in Fig. 1(a) is very small, G_2 will have a very high resonance peak in its bode diagram which is harmful to the performance and stability of the control system. Hence, a virtual resistor compensation network H_{vr} is proposed to solve this problem as shown in the dashed line box of Fig. 4.

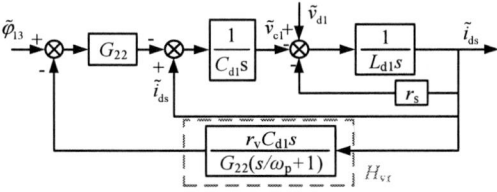

Fig. 4. Schematic diagram of virtual resistor structure.

In Fig. 4, r_v is the designed virtual resistor. In practical application, the sampled i_{ds} is passed through H_{vr} then added to the output of current controller to obtain the final phase shifting between port 1 and port 3. A high frequency pole, ω_p (the value can be between ω_s and $\omega_s/2$) is set in the denominator of H_{vr} to reduce the negative impact of high frequency noise caused by the pure differential operation. Theoretically, if the effect of ω_p is not taken into account, the damping ratio of G_2 will become $(r_s+r_v)/2\times\sqrt{C_{d1}/L_{d1}}$ by introducing virtual resistor r_v.

B. LADRC for Three-Port Converter

In order to realize the LADRC based control system, small signal model shown in (12) is divided into two subsystems. The two differential equations for subsystem 1 and subsystem 2 are shown in (16) and (17) respectively.

$$\ddot{\tilde{i}}_{ds}=-\frac{1}{C_{d1}L_{d1}}\tilde{i}_{ds}-\frac{r_s}{L_{d1}}\dot{\tilde{i}}_{ds}+\frac{G_{22}}{C_{d1}L_{d1}}\tilde{\varphi}_{12}+\frac{1}{L_{d1}}\dot{\tilde{v}}_{d1}$$
$$+(\frac{G_{21}}{C_{d1}L_{d1}}-b_c)\tilde{\varphi}_{13}+b_{01}\dot{\tilde{\varphi}}_{13}=f_c+b_c\tilde{\varphi}_{13} \quad (16)$$

$$\dot{\tilde{v}}_{d2}=-\frac{1}{R_L C_{d2}}\tilde{v}_{d2}-\frac{G_{12}}{C_{d2}}\tilde{\varphi}_{13}$$
$$+(-\frac{G_{11}}{C_{d2}}-b_v)\tilde{\varphi}_{12}+b_{02}\tilde{\varphi}_{12}=f_v+b_v\tilde{\varphi}_{12} \quad (17)$$

where $\tilde{i}_{ds}$ and $\tilde{v}_{d2}$ are the output signals of the two subsystems. $\tilde{\varphi}_{13}$ is the control signal for subsystem 1 and is considered as an external disturbance for subsystem 2, while $\tilde{\varphi}_{12}$ is the control signal for subsystem 2 but regarded as external disturbance for subsystem 1. f_c and f_v are named generalized disturbance that represent both the internal and the external disturbance. In the practical system, the disturbances f_c and f_v are unknown and they are needed to be observed by the measurable signals through LESO.

As for the subsystem 1, $x_{c1}=\tilde{i}_{ds}$, $x_{c2}=\dot{\tilde{i}}_{ds}$ and $x_{c3}=f_c$ are selected as state variables, the augment state space model is shown in (18)

$$\begin{cases} \dot{x}_c=A_c x_c+B_c\tilde{\varphi}_{13}+E_c\dot{f}_c \\ \tilde{i}_{ds}=C_c x_c \end{cases} \quad (18)$$

where

$$A_c=\begin{bmatrix} 0 & 1 & 0 \\ 0 & 0 & 1 \\ 0 & 0 & 0 \end{bmatrix},\ B_c=\begin{bmatrix} 0 \\ b_c \\ 0 \end{bmatrix},\ E_c=\begin{bmatrix} 0 \\ 0 \\ 1 \end{bmatrix},\ C_c=\begin{bmatrix} 1 \\ 0 \\ 0 \end{bmatrix}^T$$

The LESO is constructed as (19)

$$\begin{cases} \dot{z}_c=[A_c-L_cC_c]z_c+[B_c \quad L_c]u_c \\ y_c=z_c \end{cases} \quad (19)$$

where $y_{c1}=z_c=[z_{c1}\ z_{c2}\ z_{c3}]^T$ represent the observer values of $x_c=[\tilde{i}_{ds}\ \dot{\tilde{i}}_{ds}\ f_c]^T$, $u_c=[\tilde{\varphi}_{13}\ \tilde{i}_{ds}]^T$, and L_c is the observer gain that can be designed by using the pole placement method [15] and is shown in (20).

$$L_c=\begin{bmatrix} 3\omega_{oc} & 3\omega_{oc}^2 & \omega_{oc}^3 \end{bmatrix}^T \quad (20)$$

where ω_{o1} is the bandwidth of the observer.

Similarly, the LESO and the corresponding gain vector L_v for subsystem 2 are given in (21) and (22) respectively.

$$\begin{cases} \dot{x}_v=[A_v-L_vC_v]x_v+[B_v \quad L_v]u_v \\ y_v=x_v \end{cases} \quad (21)$$

$$L_v=\begin{bmatrix} 2\omega_{ov} & \omega_{ov}^2 \end{bmatrix}^T \quad (22)$$

where $A_v=\begin{bmatrix} 0 & 1 \\ 0 & 0 \end{bmatrix}$, $B_v=\begin{bmatrix} b_v \\ 0 \end{bmatrix}$, $E_v=\begin{bmatrix} 0 \\ 1 \end{bmatrix}$, $C_v=[1\ 0]$, $u_v=[\tilde{\varphi}_{12}\ \tilde{v}_{d2}]$, $y_v=x_v=[x_{v1}\ x_{v2}]^T\rightarrow[\tilde{v}_{d2}\ f_2]^T$ is the output of observer.

In (16) and (17), if $\tilde{\varphi}_{13}=(u_c-f_c)/b_c$ and $\tilde{\varphi}_{12}=(u_v-f_v)/b_v$ are set, then the two subsystems become cascaded integrators system $\ddot{\tilde{i}}_{ds}=u_c$ and $\dot{\tilde{v}}_{d2}=u_v$. For this type of control system, the control laws of u_c and u_v can be designed as (23) and (24) respectively [15].

$$u_c = k_{pc}(r_c - z_{c1}) - k_{dc}z_{c2} \qquad (23)$$

$$u_v = k_{pv}(r_v - z_{v1}) \qquad (24)$$

It can be seen that (23) and (24) present PD and P type controller respectively, k_{p1}, k_{p2} and k_{d1} are controller coefficients. Under these conditions, the closed loop transfer functions of the subsystem 1 and subsystem 2 can be expressed as (25) and (26) respectively.

$$G_{cL} = \frac{k_{pc}}{s^2 + k_{dc}s + k_{pc}}, \ (k_{pc}=\omega_c^2, \ k_{dc}=2\xi\omega_c) \qquad (25)$$

$$G_{vL} = \frac{k_{pv}}{s + k_{pv}}, \ (k_{pv}=\omega_v) \qquad (26)$$

where ω_c and ω_v are the equivalent control bandwidth, and ξ is the equivalent damping ratio which should be designed to avoid significant oscillations in the transient process. (25) and (26) show that there are no steady state error in the two closed loop control system, and the control performances of the two subsystems are independent with model parameters.

IV. SIMULATION VERIFICATION AND ANALYSIS

The simulation model of the isolated three-port converter is built in Matlab Simulink environment, and the simulation model parameters are listed in TABLE I.

TABLE I
PARAMETERS FOR SIMULATION MODEL

Symbol	Meaning	Value
v_{d1}	DC input voltage	24V
v_{d2}	Output voltage	36V
v_{d3}	Battery voltage	24V
L_{d1}	Input filter inductor	100μH
r_s	Input filter inductor ESR	0.1Ω
C_{d1}	Input filter inductor	1200μF
C_{d2}	Output filter capacitor	1000μF
R_L	Load resistor	45Ω/25Ω
f_s	Switching frequency	20kHz

The steady-state operation point A (0.620, 0.379) is selected, frequency domain controller G_v and G_c are designed firstly by using traditional decoupling control method, the control bandwidths for v_{d2} and i_{ds} subsystems are about 73Hz and 75Hz respectively.

Fig. 5 shows the simulation results of v_{d2} and i_{ds} with a sudden load change at 0.6s, it can be seen from this figure that there is a voltage droop in v_{d2} which causes i_{ds} fluctuation simultaneously, but the dynamic performance of i_{ds} with TDC is much better than the case without decoupling control.

The simulation results by using LADRC method are shown in Fig. 6. In this case the equivalent control bandwidths of LADRC for i_{ds} and v_{d2} subsystems are approximate to that by using TDC respectively. The same load change in Fig. 5 is exerted at 0.6s, it can be seen that the droop of v_{d2} with LADRC is smaller than that with TDC, and the transient recovery time of v_{d2} by using LADRC is much shorter than that of TDC method.

Besides, it seems that the dynamic performance of i_{ds} with LADRC is also better than that with TDC, there is almost no oscillation in i_{ds} in the transient process.

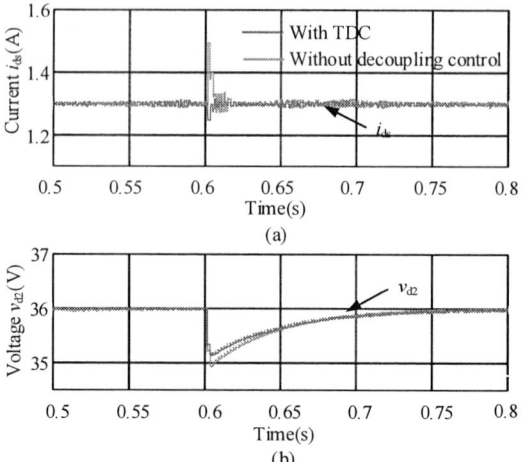

Fig. 5. Waveforms with TDC and without decoupling control (a) i_{ds}. (b) v_{d2}.

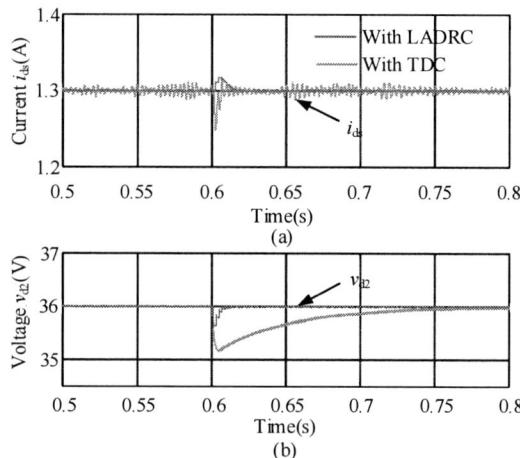

Fig. 6. Simulation results by using LADRC and TDC respectively (a) i_{ds}. (b) v_{d2}.

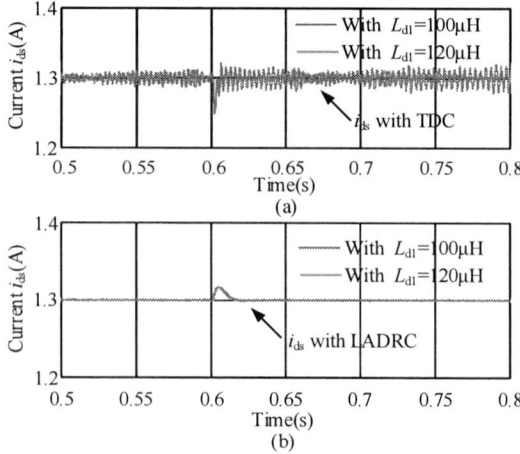

Fig. 7. Simulation results of i_{ds} by using (a) TDC. (b) LADRC.

In Fig. 7, the value of L_{d1} is increased from 100μH to 120μH. The simulation results in Fig. 7 (a) show that the oscillation amplitude is increased under the control of TDC. While in Fig. 7 (b) the waveforms of i_{ds} are almost not changed with LADRC. This indicates that the LADRC method has more stronger performance than traditional decoupling control to tolerate model parameters deviation.

V. CONCLUSION

LADRC based method is employed to control isolated three-port converter in this paper. Mathematic modeling analysis and design procedure are presented. Compared to conventional decoupling control method, the simulation results demonstrate that the proposed method has advantages in decoupling and suppressing the negative impact of model parameters change, so the dynamic performance of the control system can be improved accordingly.

ACKNOWLEDGMENT

This work is supported by the national NSF of China (No. 51479042), the NSF (No. E201238) of Heilongjiang province, China, and the Fundamental Research Funds for the Central Universities(No. GK2040260197).

REFERENCES

[1] L. Wang, Z. Wang and H. Li, "Optimized energy storage system design for a fuel cell vehicle using a novel phase shift and duty cycle control," *2009 IEEE Energy Conversion Congress and Exposition*, San Jose, CA, 2009, pp. 1432-1438.

[2] H. Tao, J. L. Duarte and M. A. M. Hendrix, "Line-Interactive UPS Using a Fuel Cell as the Primary Source," in *IEEE Transactions on Industrial Electronics*, vol. 55, no. 8, pp. 3012-3021, Aug. 2008.

[3] M. Michon, J. L. Duarte, M. Hendrix and M. G. Simoes, "A three-port bi-directional converter for hybrid fuel cell systems," *2004 IEEE 35th Annual Power Electronics Specialists Conference*, 2004, pp. 4736-4742 Vol.6.

[4] H. Tao, A. Kotsopoulos, J. L. Duarte and M. A. M. Hendrix, "Family of multiport bidirectional DC-DC converters," in *IEE Proceedings - Electric Power Applications*, vol. 153, no. 3, pp. 451-458, 1 May 2006.

[5] S. Y. Kim, H. S. Song and K. Nam, "Idling Port Isolation Control of Three-Port Bidirectional Converter for EVs," in *IEEE Transactions on Power Electronics*, vol. 27, no. 5, pp. 2495-2506, May 2012.

[6] Y. Jiang, F. Liu, X. Ruan and L. Wang, "Optimal idling control strategy for three-port full-bridge converter," *2014 International Power Electronics Conference (IPEC-Hiroshima 2014 - ECCE ASIA)*, Hiroshima, 2014, pp. 458-464.

[7] B. Karanayil, M. G. Arregui, V. G. Agelidis and M. Ciobotaru, "Bi-directional isolated multi-port power converter for aircraft HVDC network power transfer," *IECON 2012 - 38th Annual Conference on IEEE Industrial Electronics Society*, Montreal, QC, 2012, pp. 3394-3399.

[8] B. Karanayil, M. Ciobotaru and V. G. Agelidis, "Power Flow Management of Isolated Multiport Converter for More Electric Aircraft," in *IEEE Transactions on Power Electronics*, vol. 32, no. 7, pp. 5850-5861, July 2017.

[9] A. Ajami and P. Asadi Shayan, "Soft switching method for multiport DC/DC converters applicable in grid connected clean energy sources," in *IET Power Electronics*, vol. 8, no. 7, pp. 1246-1254, 2015.

[10] H. Tao, A. Kotsopoulos, J. L. Duarte and M. A. M. Hendrix, "A Soft-Switched Three-Port Bidirectional Converter for Fuel Cell and Supercapacitor Applications," *2005 IEEE 36th Power Electronics Specialists Conference*, Recife, 2005, pp. 2487-2493.

[11] Chuanhong Zhao and J. W. Kolar, "A novel three-phase three-port UPS employing a single high-frequency isolation transformer," *2004 IEEE 35th Annual Power Electronics Specialists Conference*, 2004, pp. 4135-4141 Vol.6.

[12] C. Zhao, S. D. Round and J. W. Kolar, "An Isolated Three-Port Bidirectional DC-DC Converter With Decoupled Power Flow Management," in *IEEE Transactions on Power Electronics*, vol. 23, no. 5, pp. 2443-2453, 2008.

[13] Z. Gao, "Active disturbance rejection control: From an enduring idea to an emerging technology," *2015 10th International Workshop on Robot Motion and Control (RoMoCo)*, Poznan, 2015, pp. 269-282.

[14] Bosheng Sun and Zhiqiang Gao, "A DSP-based active disturbance rejection control design for a 1-kW H-bridge DC-DC power converter," in *IEEE Transactions on Industrial Electronics*, vol. 52, no. 5, pp. 1271-1277, Oct. 2005.

[15] Zhiqiang Gao, "Scaling and bandwidth-parameterization based controller tuning," *Proceedings of the 2003 American Control Conference, 2003.*, 2003, pp. 4989-4996.

Stability Constrained Gain Optimization of Droop Controlled Converters in DC Nanogrids

Soumya Bandyopadhyay*, Laura Ramirez-Elizondo*, and Pavol Bauer*
*Dept. Electrical Sustainable Energy, DCE&S group
TU Delft, Meklweg 04, 2628 CD, Delft, the Netherlands
Email: s.bandyopadhyay@tudelft.nl, l.m.ramirezelizondo@tudelft.nl, and p.bauer@tudelft.nl

Abstract—Autonomous operation of the dc grids with converter interfaced renewable energy sources and energy storage with droop based control can lead to instability. This paper analyzes the stability of droop based closed loop controllers in a dc nanogrid. Linear state space modeling approach is used to model the small signal model of the droop controlled dc-dc converters. The dominant eigenvalues are analyzed, and the effect of closed-loop gains of the converters are investigated. Detailed parametric sensitivity analysis and participation factor of the system parasitics and the controller gains are also presented. Based on that, a segmented droop strategy is proposed to divide the operating ranges into segments with adaptive controller gains to ensure system stability in all of them. A particle swarm optimization algorithm is used to optimize for the converter gains of individual segments.

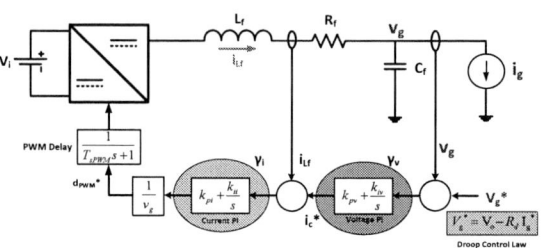

Fig. 1. A droop based feeder interfacing an ESS with a common dc bus in a Microgrid. Droop is the primary control law in this application. Secondary and tertiary control levels can also be augmented on top of this primary control layer.

I. INTRODUCTION

The energy landscape is going through a rapid change with increased penetration of renewable energy sources (RESs) based distributed generation (DG). Due to intermittent nature of the RES, energy storage systems (ESS) are used in conjunction to balance the demand and supply. Since RESs and ESSs are dc based, the recent focus has turned towards DC-based grids as a viable alternative for AC grids. The main advantages of DC grids are the absence of reactive power, no need for synchronization, fewer conversion stages and thus better efficiency etc [1]–[3].

Despite the advantages, dc microgrids have significant challenges with control and stability related to voltage regulation [4]. With the lack of inertia unlike in traditional AC grids, it is important to study the stability analysis of power-sharing based droop control. The bus regulation is handled differently because active power variations are directly related to voltage deviations and not frequency as in ac systems. The inertia of a dc microgrid is provided by the amount of capacitance added to the dc bus. This paper builds a framework to analyze the stability of these droop based converters which form a dc grid. Different approaches are presented in the literature which range from using linearized dc grids at a given equilibrium operating point coupled with eigenanalysis using jacobian linearization [5], [6] or Nyquist stability criteria [7] etc. In this paper, Jacobian linearization techniques will be used to build a small-signal model the non-linear dc grid. A detailed eigenanalysis of the dominant eigenvalues is carried out along with sensitivity analysis and participation factor of

the different system parameters. Based on the results, a piece-wise or segmented droop control strategy is proposed with the goal of ensuring stable operation over a wide range of operating points. To demonstrate the strategy, a particle swarm optimization (PSO) algorithm is used to arrive at an optimized set of controller gains and droop coefficients for different segments. Finally, the dynamic behaviour of the system with optimized gains is presented to demonstrate the validity of the method.

II. SMALL-SIGNAL MODEL OF DROOP CONVERTERS

In this section, the mathematical model of the droop based feeder will be built for further analysis. A droop based feeder (see Figure 1) mainly consist of an ESS interfaced to a dc bus with a converter and a cable. To understand the stability at different operating points, a dynamic model of this feeder is required. Since switching to DC-DC converters make the whole system non-linear, a linearization is needed for stability or eigenanalysis. For that purpose, state-space averaging of the dc-dc converters is used [8]. The details of the modeling procedure are discussed in the following section:

A. Converter modeling

There are many possible ways to model converter behavior. In this study, state space averaging technique will be used. The main reason for this modeling choice is that this technique provides an optimal trade-off between accuracy and computation time. They are well suited for analysis of large electronics systems with switching converters.

The 2018 International Power Electronics Conference

In this study, a 48 V battery with a dc-dc converter is chosen as a case study to supply a grid downstream current of i_g. The dc link nominal voltage is chosen to be 48 V. Since the two voltages are similar, a four switch buck-boost converter is chosen as topology. The equivalent circuit of this converter with device parasitics along with modes of operation is shown in Figure 2. In the first interval ($d_{PWM}T_{sw}$), switches S_1 and S_3 are on, whereas switches S_2 and S_4 are off. During the second interval ($d'_{PWM}T_{sw}$), switches S_2 and S_4 are on and vice versa.

The states of the converter model are the inductor current and the capacitor voltages, i.e, i_{Lc}, v_{Ci} and v_{Co}. The output of the converter are the input and output currents along with the output voltage, i.e, i_i, i_c and v_c. The input of the converter model is the input voltage v_i and duty cycle d_{PWM} which is the output of the control model.. The inputs, states and the outputs of the converter model are shown below in matrix form:

$$U_{conv} = \begin{bmatrix} v_i \\ d_{PWM} \\ i_{Lf} \end{bmatrix} \qquad X_{conv} = \begin{bmatrix} i_{Lc} \\ v_{Ci} \\ v_{Co} \end{bmatrix} \qquad Y_{conv} = \begin{bmatrix} i_{Lc} \\ v_{Co} \\ v_C \\ i_i \end{bmatrix}$$

Interval 1: [0 - $d_{PWM}T_{sw}$]: During this interval, (S_1+S_3) are on while (S_2+S_4) are off. The equivalent circuit is shown in Figure 2. The state equations are shown below:

$$\frac{di_{Lc}(t)}{dt} = \frac{1}{L_c}[v_i - i_{Lc}(t)(r_1 + r_3 + r_L)] \tag{1}$$

$$\frac{dv_{Ci}(t)}{dt} = \frac{1}{r_{Ci}C_i}(v_i - v_{Ci}) \tag{2}$$

$$\frac{dv_{Co}(t)}{dt} = -\frac{i_c}{C_o} \tag{3}$$

The output equations are shown below:

$$i_{Lc} = i_{Lc} \tag{4}$$

$$v_{Co} = v_{Co} \tag{5}$$

$$v_C = v_{Co} + r_{Co}i_{Lf} \tag{6}$$

$$i_i = \frac{v_i - v_{Ci}}{r_{Ci}} + i_{Lc} \tag{7}$$

The above equations are merged into a descriptor state space model:

$$\dot{x}(t) = A_1x(t) + B_1u(t) \tag{8}$$

$$y(t) = C_1x(t) + D_1u(t) \tag{9}$$

Interval 2: [$d_{PWM}T_{sw}$ - T_{sw}]: During this interval, (S_2+S_4) are on while (S_1+S_3) are off. The equivalent circuit is shown in Figure 2. The state equations are shown below:

$$\frac{di_{Lc}(t)}{dt} = \frac{1}{L_c}[r_{Co}i_{Lf} - v_{Co} - i_{Lc}(t)(r_2 + r_4 + r_L + r_{Co})] \tag{10}$$

$$\frac{dv_{Ci}(t)}{dt} = \frac{1}{r_{Ci}C_i}(v_i - v_{Ci}) \tag{11}$$

$$\frac{dv_{Co}(t)}{dt} = \frac{1}{C_o}(i_{Lc} - i_{Lf}) \tag{12}$$

The output equations are shown below:

$$i_{Lc} = i_{Lc} \tag{13}$$

$$v_{Co} = v_{Co} \tag{14}$$

$$v_C = v_{Co} + r_{Co}(i_{Lf} - i_{Lc}) \tag{15}$$

$$i_i = \frac{v_i - v_{Ci}}{r_{Ci}} \tag{16}$$

The above equations are merged into a descriptor state space model:

$$\dot{x}(t) = A_2x(t) + B_2u(t) \tag{17}$$

$$y(t) = C_2x(t) + D_2u(t) \tag{18}$$

State Space Averaging: The average matrices ($A_{conv}, B_{conv}, C_{conv}, D_{conv}$) are averaged over the switching time period:

$$A_{conv} = d_{PWM}A_1 + (1 - d_{PWM})A_2 \tag{19}$$

$$B_{conv} = d_{PWM}B_1 + (1 - d_{PWM})B_2 \tag{20}$$

$$C_{conv} = d_{PWM}C_1 + (1 - d_{PWM})C_2 \tag{21}$$

$$D_{conv} = d_{PWM}D_1 + (1 - d_{PWM})D_2 \tag{22}$$

This concludes the state-space model of the 4-switch buck boost converter. The output filter model is built in the next section.

B. Cable and Filter modeling

The cable and the filter can be combined to form a combined LC filter with a resistance of R_f as shown in Figure 1. The inputs, states and output of the filter model are shown below:

$$U_{cable} = \begin{bmatrix} i_g \\ v_{Co} \\ i_{Lc} \\ d_{PWM} \end{bmatrix} \qquad X_{cable} = \begin{bmatrix} i_{Lf} \\ v_g \end{bmatrix} \qquad Y_{cable} = \begin{bmatrix} i_{Lf} \\ v_g \end{bmatrix}$$

The ouput of the converter current is the input of the cable model. Therefore, the state equations of the cable model are also dependent on the switch status of the converter.

Interval 1: [0 - $d_{PWM}T_{sw}$]: The state equations are shown below:

$$\frac{di_{Lf}(t)}{dt} = \frac{1}{L_f}[v_{Co} - (R_f + r_{Co})i_{Lf} - v_g] \tag{23}$$

$$\frac{dv_{v_g}(t)}{dt} = \frac{1}{C_f}(i_{Lf} - i_g) \tag{24}$$

1427

The 2018 International Power Electronics Conference

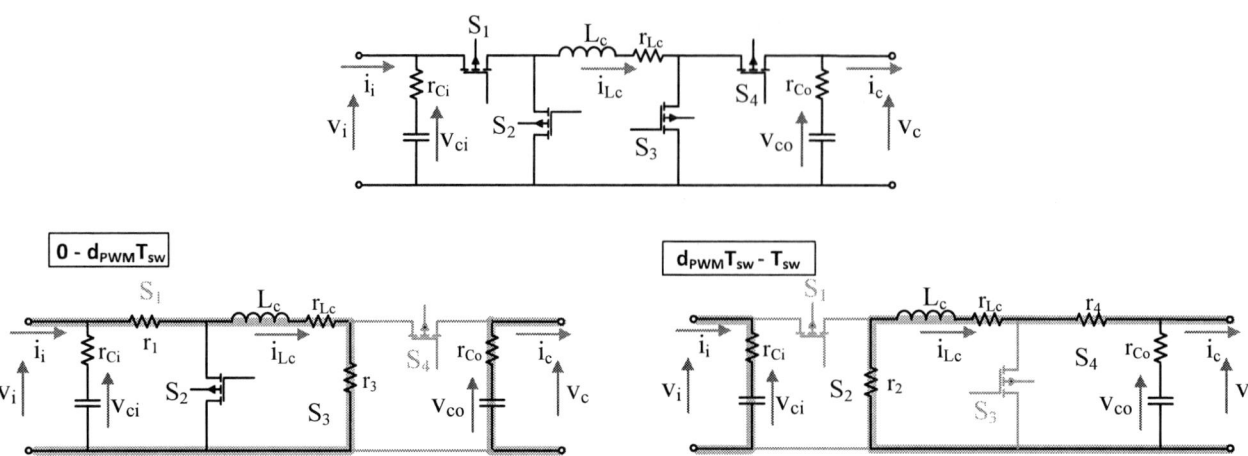

Fig. 2. Four switch buck-boost converter schematic and operating states in buck-boost mode

The above equations are merged into a descriptor state space model:

$$\dot{x}(t) = A_1 x(t) + B_1 u(t) \tag{25}$$

$$y(t) = C_1 x(t) + D_1 u(t) \tag{26}$$

Interval 2: [$d_{PWM}T_{sw}$ - T_{sw}]: The state equations are shown below:

$$\frac{di_{\text{Lf}}(t)}{dt} = \frac{1}{L_f}[v_{\text{Co}} + r_{\text{Co}}i_{\text{Lc}} - (R_f + r_{\text{Co}})i_{\text{Lf}} - v_g] \tag{27}$$

$$\frac{dv_{v_g}(t)}{dt} = \frac{1}{C_f}(i_{\text{Lf}} - i_g) \tag{28}$$

$$\tag{29}$$

The above equations are merged into a descriptor state space model:

$$\dot{x}(t) = A_2 x(t) + B_2 u(t) \tag{30}$$

$$y(t) = C_2 x(t) + D_2 u(t) \tag{31}$$

The average matrices ($A_{\text{cable}}, B_{\text{cable}}, C_{\text{cable}}, D_{\text{cable}}$) are averaged over the switching time period. This concludes the state-space model of the cable cum filter model. The control model is built in the next section.

C. Droop controller modeling

The control model depicted in this study is shown in Figure 1. A droop control strategy is used in this case with an inner current loop and an outer voltage loop. PI controllers are used as voltage and current controllers. A time delay (T_{PWM}) is introduced to take into account the sampling time delay of PWM signal. Based on this, the inputs, states and the output

of the control model are presented below:

$$U_c = \begin{bmatrix} v_{nl} \\ R_d \\ v_g \\ i_{\text{Lf}} \\ i_g \end{bmatrix} \qquad X_c = \begin{bmatrix} g_v \\ g_i \\ d_{\text{PWM}} \end{bmatrix} \qquad Y_c = \begin{bmatrix} d_{\text{PWM}} \end{bmatrix}$$

where g_v, g_i are the outputs of integrators of the voltage and current controllers respectively. v_{nl} and R_d are the droop control parameters. They can be outputs of higher level controllers (secondary control and tertiary control). Since hierarchical control level has dynamics in widely varying time ranges, this study ignores the dynamics of the droop control parameters.

Since there are no inputs which are changing w.r.t switch states there are only one set of state equations:

$$\frac{dg_v}{dt} = v_{nl} - i_g R_d - v_g \tag{32}$$

$$\frac{dg_i}{dt} = k_{\text{pv}}(v_{nl} - i_g R_d - v_g) + k_{\text{iv}}g_v - i_{\text{Lf}} \tag{33}$$

$$\frac{dd_{\text{PWM}}}{dt} = \frac{1}{T_{\text{PWM}}}[-d_{\text{PWM}} + \frac{1}{v_{nl}}[k_{\text{pi}}\{k_{\text{pv}}(v_{nl} \tag{34}$$

$$- i_g R_d - v_g) + k_{\text{iv}}g_v - i_{\text{Lf}}\} + k_{\text{ii}}g_i]]$$

$$\tag{35}$$

From the above state equations, the state matrices of the controller model (A_{control}, B_{control}, C_{control}, D_{control}) can be derived. This section concludes the details of the state space modeling of the converter, controller and the cables. An automated system integration approach based on the modular state space models is presented in the next section.

D. Integrating models to form Global DC grid model

In the previous section, the modular state space models are built. This section is dedicated to building the global

The 2018 International Power Electronics Conference

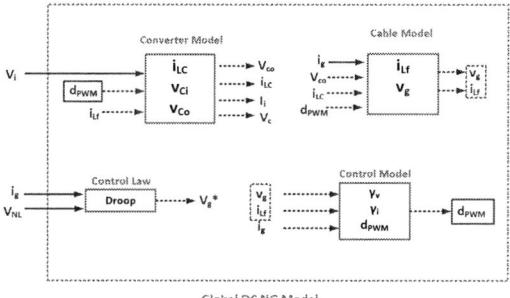

Fig. 3. Building block modeling approach of a DC NG model

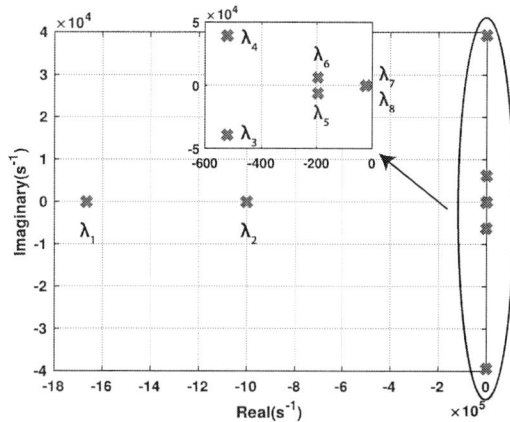

Fig. 5. Plot of the system eigenvalues highlighting the dominant or the critical eigenmodes of the droop system.

DC grid model from the models. As an example, a droop based feeder (see Figure 1) supplying a downstream grid current is chosen as a case study. The individual models are schematically shown in Figure 3. The inter-connections between the models are evident from the figure. Configuring the inputs, outputs of the individual model correctly leads to the global dc feeder model with global inputs, global states and global outputs. This procedure is time-consuming if carried out individually for every feeder with different characteristics. In this section, an automated way of combining the individual models is proposed:

1) In this control problem, the global state vector is :

$$X_{\text{global}} = \begin{bmatrix} i_{\text{Lc}} & v_{\text{Ci}} & v_{\text{Co}} & i_{\text{Lf}} & v_{g} & g_{v} & g_{i} & d_{\text{PWM}} \end{bmatrix} \tag{36}$$

2) The global system input vector (U_{global}) is computed as following:

$$U_{\text{global}} = \begin{bmatrix} V_{i} & i_{g} & v_{nl} \end{bmatrix} \tag{37}$$

This system is comprised of a set of eight non-linear equations that must be linearized over a point of operation x to carry out the analysis of its dominant eigenvalues. In this paper, a simple Jacobian linearization technique is used to linearize the system of equations at an operating point. The final set of equations can be represented in a small signal state space form as shown in:

$$\Delta \dot{x} = A_{\text{sys}}\Delta x + B_{\text{sys}}\Delta u \tag{38}$$

where

$$\Delta x = \begin{bmatrix} \Delta i_{\text{Lc}} \Delta v_{\text{Ci}} \Delta v_{\text{Co}} \Delta i_{\text{Lf}} \Delta v_{g} \Delta g_{v} \Delta g_{i} \Delta d_{\text{PWM}} \end{bmatrix}^{T} \tag{39}$$

The linearized state-space matrix of the system is shown inFigure 4.

III. Frequency Domain Characteristics: Modal Analysis and Sensitivity analysis

The framework of small signal modelling of the droop based feeder is presented in detail in the previous section. The goal of this section is to perform modal analysis on the small signal model to gain insight into the frequency domain behaviour of the model and to identify the relevant modes

of the system. This section is sub-divided into two parts: (a) participation factor analysis, and (b) parametric stability analysis.

A. Participation factor modal analysis

The eigenvalues of any system can be computed if the system matrix is avaliable (A_{sys}) as highlighted in (38). To take control measures based on knowledge of the eigenvalues, the mutual relationships between the eigenmodes and the system states needs to be analysed. Given a system matrix A of order N and eigenvalues (γ_i), the right eigenvector (Φ_i) and left eigenvector (Ψ_i) can be defined as:

$$\mathbf{A}\Phi_i = \gamma_i\Phi_i, \quad \Psi_i\mathbf{A} = \gamma_i\Psi_i \tag{40}$$

The kth element of the eigenvector Φ_i computes the activity of the state variable x_k in the ith mode while that of the eigenvector Ψ_i measures the contribution of this activity to the ith mode [9]. A modified version of participation factor analysis is used in this paper which solves the scaling problem and combines the contribution of the complex conjugate pairs of eigenvalues [10]. Based on that study, the participation factor of γ_i in state k can be computed as:

$$P_{ik} = \Phi_{ki}\Psi_{ik} \quad \text{if} \quad \gamma_i \epsilon \Re \tag{41}$$

$$P_{ik} = 2\text{Re}(\Phi_{ki})\text{Re}(\Psi_{ik}) \quad \text{if} \quad \gamma_i \epsilon C \tag{42}$$

Based on the above analytical framework, participation factors of the different states on system eigenmodes are presented in Table II. It must be noted that the participation factors

TABLE I. Parameter values of chosen system configuration

Subsystem	Parameter	Value	Subsystem	Parameter	Value
Converter	L_c	450 µH	Voltages	v_{nl}	48 V
	$r_{Lc}, r_{Ci}, r_{Co}, r_n$	10 mΩ		v_i	60 V
	C_i, C_o	60 µF		k_{pv}	0.1
	r_f	30 mΩ	Controllers	k_{iv}	10
Cable	C_f	60 µF		k_{pi}	1.2
	L_f	60 µH		k_{ii}	300

1429

The 2018 International Power Electronics Conference

$$\mathbf{A}_{\text{sys}} = \begin{pmatrix} \dfrac{\bar{d}req_3 - req_2}{L_c} & 0 & -\dfrac{\bar{d}'}{L_c} & \dfrac{\bar{d}'r_{Co}}{L_c} & 0 & 0 & 0 & \dfrac{V_{co}+V_i - i_{Lf}r_{Co}+i_{Lc}r_{eq3}}{L_c} \\[2ex] 0 & -\dfrac{1}{r_{Ci}C_i} & 0 & 0 & 0 & 0 & 0 & 0 \\[2ex] \dfrac{\bar{d}'}{L_c} & 0 & 0 & -\dfrac{1}{C_o} & 0 & 0 & 0 & -\dfrac{i_{Lc}}{C_o} \\[2ex] 0 & 0 & 0 & -\dfrac{R_f + r_{Co}}{L_f} & -\dfrac{1}{L_f} & 0 & 0 & 0 \\[2ex] 0 & 0 & 0 & \dfrac{1}{C_f} & 0 & 0 & 0 & 0 \\[2ex] 0 & 0 & 0 & 0 & 0 & 0 & 0 & 0 \\[1ex] 0 & 0 & 0 & -1 & 0 & k_{iv} & 0 & 0 \\[2ex] 0 & 0 & 0 & -\dfrac{k_{pi}}{T_{\text{PWM}}V_g} & 0 & \dfrac{k_{iv}k_{pi}}{T_{\text{PWM}}V_g} & \dfrac{k_{ii}}{T_{\text{PWM}}V_g} & -\dfrac{1}{T_{\text{PWM}}} \end{pmatrix}$$

Fig. 4.   Linearized A matrix of the droop system

TABLE II.   MODES AND PARTICIPATION FACTORS OF DROOP CONTROLLED FEEDER SYSTEM

Eigenvalues	Damping (%)	Frequency (rad/s)	Participation factor (%)
γ_1	100	0	$x_2 = 100$
γ_2	100	0	$x_8 = 100$
γ_3	1.32	39346	$x_3 = 15.3$ $x_4 = 50.6$ $x_5 = 34.7$
γ_4	1.32	39346	$x_3 = 15.3$ $x_4 = 50.6$ $x_5 = 34.7$
γ_5	3.12	6222	$x_1 = 50.7$ $x_3 = 34.7$ $x_5 = 14.8$
γ_6	3.12	6222	$x_1 = 50.7$ $x_3 = 34.7$ $x_5 = 14.8$
γ_7	39.6	52.2	$x_6 = 50.3$ $x_7 = 49.5$
γ_8	39.6	52.2	$x_6 = 50.3$ $x_7 = 49.5$

of different states on different modes are dependent on the operating points since the system has to be linearized. The participation analysis is based on operating conditions and system parameters highlighted in Table I.

B. Parametric stability analysis

The biggest challenge of control design is to tune the regulator parameters to ensure system stability and optimal performance over a wide range of operating points. There are several research in literature tackling this issue [11]–[13]. Stability analysis is verified by looking at the evolution of the dominant eigenvalues or poles of the developed small-signal state-space system for different system parameters. Those parameters are divided into two categories: (a) control parameters: current and voltage controller gains $k_{pv}, k_{iv}, k_{pi}, k_{ii}$ and droop co-efficient R_d, (b) system constants: converter inductance L_c, cable parameters like L_f, C_f, R_f etc. A detailed parametric sensitivity analysis is carried out and the results are shown in Figure 6, 7 and 8. Some observations based on the analysis is presented in the following:

- The variation of operating point (i_g) affects dominant eigenmodes in opposite ways (see Figure 6a). The eigenmode pair $\gamma_{3,4}$ is pushed deep into the left half plane with the increase of i_g. On the contrary, pair $\gamma_{5,6}$ veers into the right half plane which may cause stability problems.

- The sensitivity of dominant eigenvalues with the variation of droop co-efficient R_d depends on the operating current. At high grid current, the positions of the dominant eigenvalues are extremely sensitive to variation of R_d (see Figure 6b). The eigenpair $\gamma_{3,4}$ is adversely affected with increasing R_d. However, at low currents, the effect of R_d on the eigenvalues is comparatively less pronounced (see Figure 6c).

- The proportional gains of the PI regulators of current and voltage loop (k_{pv}, k_{pi}) cause a significant movement of the eigenmodes of the system (see Figure 7a and Figure 7c). This could lead the grid to become unstable in certain conditions. In addition to that, they affect the dominant eigenpairs ($\gamma_{3,4}, \gamma_{5,6}$) in opposite ways. Therefore, it is crucial to determine the values of these gains to ensure stable operation over a wide operating range and also satisfy the required dynamic characteristics of the system.

- The variation of inductances in the circuit (L_c, L_f) has significant effect on the dominant eigenpairs. The cable inductance (L_f) and the cable capacitance (C_f) has very similar effect on the pairs of eigenmodes ($\gamma_{3,4}, \gamma_{5,6}$) as shown Figure 8b and Figure 8c. That's why system designers often propose higher bus capacitance to be added to make the system more stable at the cost of high inrush currents during start-up and higher transient currents during a fault.

- The system becomes more stable with increasing cable resistance (R_f) value as the eigenpair $\gamma_{3,4}$ moves horizontally into the left half plane. However, to minimize the losses in the power cables, it is recommended to decrease the resistance.

This concludes the discussion on the analysis of the evolution of eigenvalues of the droop based dc system. The study identifies the dominant eigenvalues and investigates stability over a wide range of operating points and effect of different system parameters on the overall stability. The results show significant room for improvement of the system dominant eigenvalues sensitive to change in controller gains. To that end, a droop control strategy with optimized controller gains will be proposed in the upcoming section to improve both stability and dynamic performance of the system.

IV. DROOP STRATEGY AND CONTROLLER GAIN OPTIMIZATION

In the previous section, the difficulties concerning the stability of a droop feeder over a wide operating range are discussed. The limitation of linear droop is shown as the system eigenvalues push into the right half plane as the load current increases. In addition to that, the high sensitivity of the dominant eigenmodes of the system on the variation of controller proportional gain values. To that end, this section presents a piecewise droop strategy with adaptive controller gains. To arrive at optimal locations for the system eigenvalues, optimization has been carried out for the controller gains.

A. Segmented Droop Strategy with Adaptive Controller Gains

Segmented droop strategy divides the allowable voltage deviation and the current's range into some segments with different droop coefficients and controller gains (see Figure 9). In this strategy, it is essential to have a high droop co-efficient at low load to improve current-sharing and to have a low droop co-efficient at high load to ensure stable operation. In addition to that, the controller gains will be different in

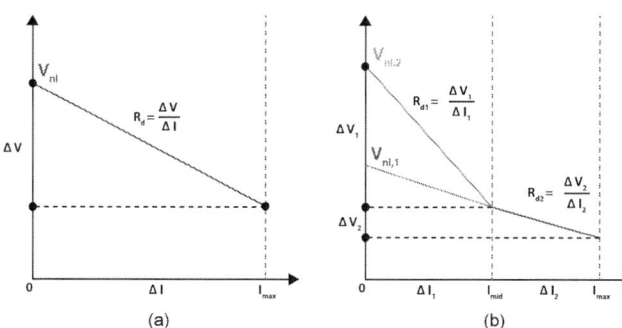

Fig. 9. Droop control strategies: (a) a linear droop control with droop co-efficient R_d, (b) a two segment piece-wise droop curve with two droop co-efficients (R_{d1}, R_{d2}) and reference voltages $(V_{nl,1}, V_{nl,2})$. Also in segmented droop strategy the controller gains will be different.

different segments to ensure system stability and to satisfy dynamic behaviour requirements of the system eigenvalues. In the next part, an optimization algorithm will be used to optimize the controller gains and the droop coefficients in a two-segmented droop strategy.

B. Stability Constrained Gain Optimization

To arrive at optimal locations for the converter system eigenvalues, optimization has been carried out for the controller gains and the droop co-efficient. The droop strategy is designed for the system described in Table I. The optimization algorithm used for this problem is particle swarm optimization (PSO). PSO is an evolutionary gradient free algorithm inspired by the movement of birds or insects in swarm which is gradient free and potentially requires fewer function calls. In this paper, an approach based on placing particles on the border of the search space using a combination of variable clipping and reflecting [14].

The goals of the optimization algorithm will be: (1) Ensure system stability, (2) Make the eigenvalues faster, and (3) Improve the damping of the eigenvalues (by minimizing their complex parts). Since these are multiple objectives, they need to be combined in a single cost or objective function. The resulting objective function is:

$$\xi = k_1 \sum_{i=1}^{N} \sigma_i \ (\forall \sigma_i < 0) + k_2 \sum_{i=1}^{N} \sigma_i \ (\forall \sigma_i > 0) + k_3 \sum_{i=1}^{N} \frac{\sigma_i}{\sqrt{\sigma_i^2 + \omega_i^2}} \tag{43}$$

where

- k_1 = weight assigned to make the eigenvalues faster
- k_2 = penalization of eigenvalues in the right half plane
- k_3 = weight assigned to increasing damping objective

The damping objective weight co-efficient k_3 is chosen to be negative since the optimization algorithm is a minimization algorithm. The chosen weights are:

$$k_1 = 1 \qquad k_2 = 1 \times 10^6 \qquad k_3 = -1 \times 10^3$$

The maximum current ($I_{g,max}$) is chosen to be 12 A and the two droop segments chosen are [1 A -8 A] and [8 A -12 A]. The reference voltage for the second segment $V_{nl,2}$ (see Figure 9b) is chosen to be 48 V. The optimization variables and their ranges are presented in Table III. This concludes the optimization framework setup. In the next section, the optimized controller gains will be presented along with the dynamic performance of the optimized system.

C. Optimized System Simulations

The controller gains and the droop co-efficients are optimized for the two-segmented droop strategy. Their optimized values are presented in Table IV. Figure 10 shows the locations of the optimized dominant eigenvalues in both the droop

TABLE III. OPTIMIZATION VARIABLES AND THEIR RANGE

Variables	Optimization Range	
	Segment 1: (1 A - 8 A)	Segment 2: (8 A - 12 A)
k_{pv}	0.01 - 2	0.01 - 2
k_{pi}	0.01 - 2	0.01 - 2
k_{iv}	1 - 400	1 - 400
k_{ii}	50 -3000	50 -3000
R_d	1 - 5	0.1 - 3

The 2018 International Power Electronics Conference

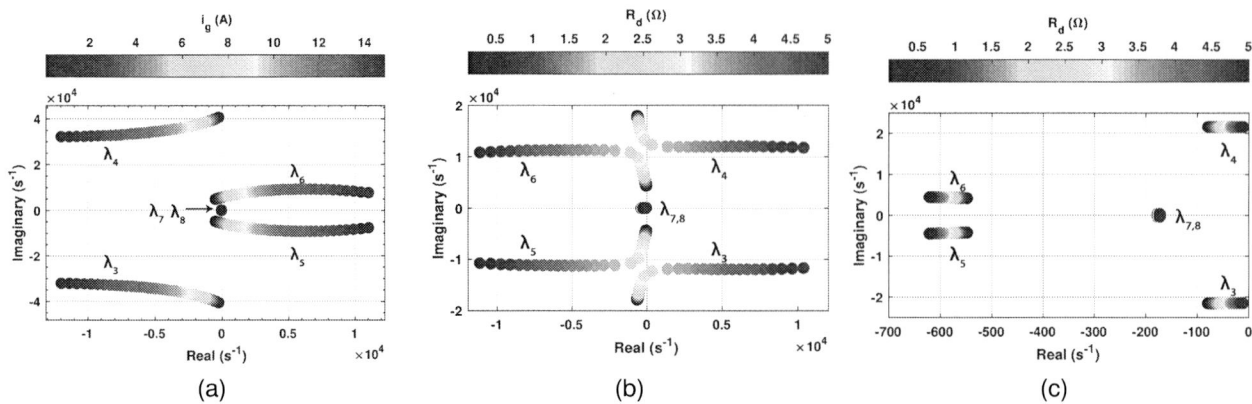

Fig. 6.   Sensitivity analysis of design parameters on dominant eigenvalues of dc droop nanogrid : (a) with operating point downstream load current (i_g), the eigenvalue pair ($\gamma_{5,6}$) moves towards the right half plane and the pair ($\gamma_{3,4}$) is pushed deep into the left half plane as the grid current increases, (b) with the droop co-efficient (R_d) when the grid current is i_g =10 A, and (c) with the droop co-efficient (R_d) when the grid current is i_g =1 A.

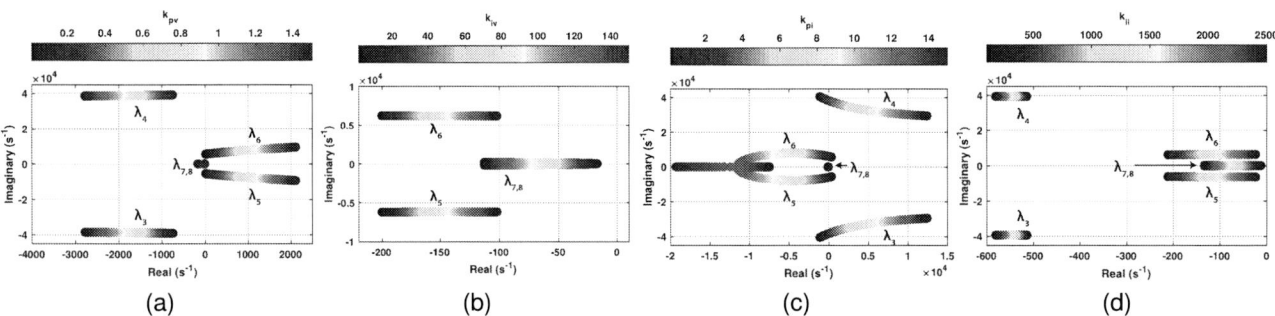

Fig. 7.   Sensitivity analysis of design parameters on dominant eigenvalues of dc nanogrid : (a) with the controller voltage proportional gain (k_{pv}), (b) with the controller voltage integral gain (k_{iv}) and (c) with the controller current proportional gain (k_{pi}), and (d) with the controller current integral gain (k_{ii}).

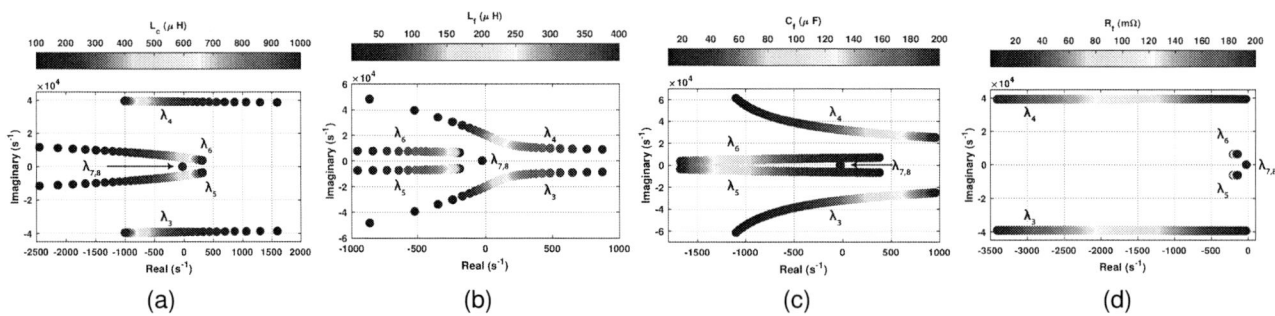

Fig. 8.   Sensitivity analysis of design parameters on dominant eigenvalues of dc nanogrid : (a) with the converter inductance (L_C), (b) with the filter inductor (L_f), (c) with the output filter capacitor (C_f), and (d) with the output filter resistor (R_f)

segments. To visualize the effect of the optimization, some time domain simulations are performed at different operating points at i_g = 1 A, 8 A and 12 A with perturbations of 5 %, 10 % and 20 %. The results of the system simulations are presented in Figure 11. The analysis of the dynamic performance and the dominant eigenvalues are also discussed.

TABLE IV.   OPTIMIZED CONTROL PARAMETERS

Droop Segment	Controller Parameters					
	k_{pv}	k_{pi}	k_{iv}	k_{ii}	R_d	V_{nl}
1-8	0.03	1.36	130	50	3.3	104
8-12	0.01	1.44	200	50	0.1	48

The 2018 International Power Electronics Conference

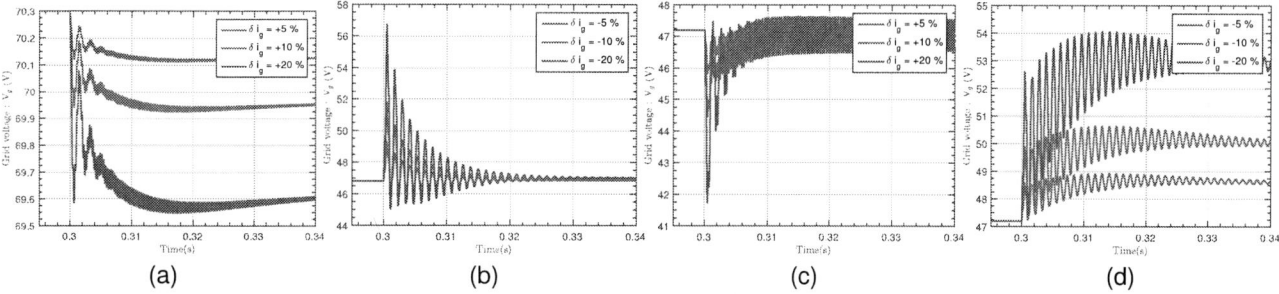

Fig. 11. Effect of perturbations of the load current i_g on the grid voltage v_g at different operating points. (a) $i_g = 1$ A with positive perturbations, where the oscillating frequency is 21,592 rad/s which corresponds with eigenmodes $\gamma_{3,4}$, (b) $i_g = 12$ A with negative perturbations where the oscillating frequency is 5223 rad/s which corresponds with eigenmodes $\gamma_{5,6}$, (c) $i_g = 8$ A with positive perturbations where the oscillating frequency is 18,534 rad/s which corresponds with eigenmodes $\gamma_{3,4}$ and (d) $i_g = 8$ A with negative perturbations where the oscillating frequency is 5575 rad/s which corresponds with eigenmodes $\gamma_{5,6}$.

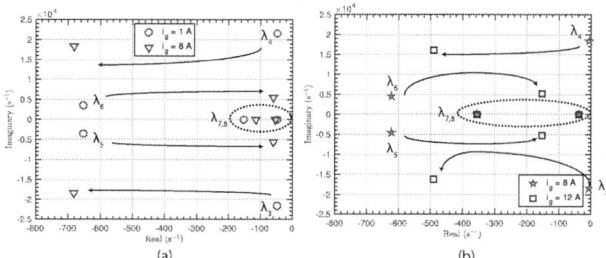

Fig. 10. Plot of the optimized eigenvalues: (a) droop segment 1: [1A -8A] , (b) droop segment 2: [8A -12A]. There are two sets of eigenvalue plots for 8 A for both the segments since it is the boundary between them.

V. CONCLUSION

This paper investigates the transient stability of droop based converter feeders in a DC microgrid over a wide range of operating points. To that end, a small-signal state-space model of the droop controlled system has been developed to evaluate the behaviour of the overall system regarding stability and dynamic performance. The critical or dominant eigenmodes of the system are identified by performing modal analysis on the small signal model. Participation factors of different states on the eigenmodes are measured to identify the states or the system parameters which have a significant influence on the evolution of the dominant eigenvalues.

A parametric sensitivity analysis is performed to reveal the effect of those parameters on the stability of the system. Some notable findings include: (a) the filter inductance, capacitance and resistance have a high impact on the overall stability, (b) the proportional gains of the current and voltage controllers has significant influence on the system stability, (c) the grid current also affects the dominant eigenvalues which can lead to instability at high or low loads, and (d) the droop co-efficient affects system stability at high load currents.Therefore, these values must be optimally chosen not only to meet the dynamic response requirements but also to ensure a stable operation of the system at different operating conditions.

Based on the above results, a segmented droop strategy

is proposed in this paper which allows adaptive controller parameters in different segments of load current ranges to ensure optimal stability and dynamic performance over a wide operating range. Finally, an optimization algorithm is used to optimize the controller gains and the droop coefficients of the individual load segments for optimal operation. The optimized system is tested by perturbations at different operating points by simulations to show the operating of the proposed technique.

REFERENCES

[1] M. Starke, L. M. Tolbert, and B. Ozpineci, "Ac vs. dc distribution: A loss comparison," in *Transmission and Distribution Conference and Exposition, 2008. T&D. IEEE/PES*, pp. 1–7, IEEE, 2008.

[2] J. J. Justo, F. Mwasilu, J. Lee, and J.-W. Jung, "Ac-microgrids versus dc-microgrids with distributed energy resources: A review," *Renewable and Sustainable Energy Reviews*, vol. 24, pp. 387–405, 2013.

[3] L. E. Zubieta, "Are microgrids the future of energy?: Dc microgrids from concept to demonstration to deployment," *IEEE Electrification Magazine*, vol. 4, no. 2, pp. 37–44, 2016.

[4] T. Dragičević, X. Lu, J. C. Vasquez, and J. M. Guerrero, "Dc micro-gridspart i: A review of control strategies and stabilization techniques," *IEEE Transactions on power electronics*, vol. 31, no. 7, pp. 4876–4891, 2016.

[5] N. Bottrell, M. Prodanovic, and T. C. Green, "Dynamic stability of a microgrid with an active load," *IEEE Transactions on Power Electronics*, vol. 28, no. 11, pp. 5107–5119, 2013.

[6] D. Zonetti, R. Ortega, and J. Schiffer, "A tool for power flow analysis of a generalized class of droop controllers for high-voltage direct-current transmission systems," in *Decision and Control (CDC), 2016 IEEE 55th Conference on*, pp. 1564–1569, IEEE, 2016.

[7] A. P. N. Tahim, D. J. Pagano, E. Lenz, and V. Stramosk, "Modeling and stability analysis of islanded dc microgrids under droop control," *IEEE Transactions on Power Electronics*, vol. 30, no. 8, pp. 4597–4607, 2015.

[8] A. Davoudi, J. Jatskevich, and T. De Rybel, "Numerical state-space average-value modeling of pwm dc-dc converters operating in dcm and ccm," *IEEE Transactions on Power Electronics*, vol. 21, no. 4, pp. 1003–1012, 2006.

[9] I. J. Pérez-Arriaga, G. C. Verghese, and F. C. Schweppe, "Selective modal analysis with applications to electric power systems, part i: Heuristic introduction," *ieee transactions on power apparatus and systems*, no. 9, pp. 3117–3125, 1982.

[10] E. H. Abed, M. A. Hassouneh, and W. A. Hashlamoun, "Modal participation factors revisited: One definition replaced by two," in *American Control Conference, 2009. ACC'09.*, pp. 1140–1145, IEEE, 2009.

[11] A. Riccobono and E. Santi, "Comprehensive review of stability criteria for dc power distribution systems," *IEEE Transactions on Industry Applications*, vol. 50, no. 5, pp. 3525–3535, 2014.

[12] S. Anand and B. Fernandes, "Reduced-order model and stability analysis of low-voltage dc microgrid," *IEEE Transactions on Industrial Electronics*, vol. 60, no. 11, pp. 5040–5049, 2013.

[13] A. Alacano, J. J. Valera, G. Abad, and P. Izurza, "Power-electronic-based dc distribution systems for electrically propelled vessels: A multivariable modeling approach for design and analysis," *IEEE Journal of Emerging and Selected Topics in Power Electronics*, vol. 5, no. 4, pp. 1604–1620, 2017.

[14] Y. Del Valle, G. K. Venayagamoorthy, S. Mohagheghi, J.-C. Hernandez, and R. G. Harley, "Particle swarm optimization: basic concepts, variants and applications in power systems," *IEEE Transactions on evolutionary computation*, vol. 12, no. 2, pp. 171–195, 2008.

The 2018 International Power Electronics Conference

SiC Based SSPC for High Voltage Space Applications

D. Marroquí, A. Garrigós, José M. Blanes, R. Gutiérrez
Industrial Electronics Group (IE-g), Universidad Miguel Hernández de Elche.
dmarroqui@umh.es

Abstract- **This paper proposes a novel High Voltage Solid State Power Controller (HVSSPC) implemented with Silicon Carbide (SiC) power semiconductors for satellite high voltage power distribution architectures. It has two different levels of protection. The lowest level (FPGA control) employs digital control techniques using an FPGA device. The FPGA monitors different variables: voltage, current, temperature and eventually sets the required protection setpoints by means of a decision tree. The highest protection level (Hardware control) just uses discrete devices, keeping a predefined level of protection which is preconfigured. The working mode and parameters of the digital control can be selected by telecommand. The HVSSPC control is very robust, in case of critical failure (control power failure, damages in FPGA, Single Event Effects) the hardware level protection remains working. This two level approach of protection provides improved robustness for critical applications in harsh environments.**

Keywords— Aerospace Electronics, Current Limiter, Wide Bandgap Semiconductors, Fault Current Limiters, Surge Protections

I. INTRODUCTION

Nowadays, electrical propulsion technologies are replacing chemical propulsion systems in medium and high power satellite platforms [1], [2]. Its lower mass and volume respect to the chemical propulsion systems allows an increase on the satellite payload as well as more accurate satellite attitude control. This, together with the growing demand of high voltage (HV) power supplies for scientific missions, the increase of Traveling-Wave Tubes (TWTs) and the boost of electric actuators in satellites, results in a renewed interest in distribution architectures and subsequently their protection methods.

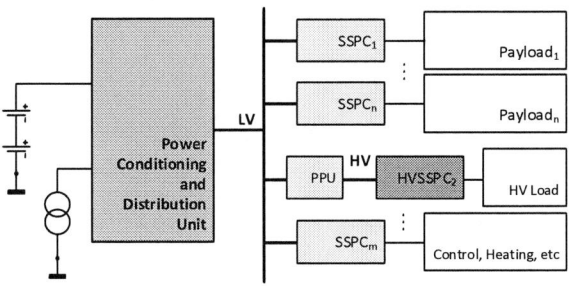

Figure 1: Low voltage centralised PCDU architecture.
Figure 1 shows a classic, centralised bus, consisting of

a central PCU (Power Conditioning Unit) that supplies the bus (traditionally 28V, 50V, 70V, 100V, 120V and 160V) followed by Solid State Power Controllers (SSPCs) or Latching Current Limiters (LCLs) placed at the input of the DC/DC converters which provide voltage conditioning to the load requirements. This low voltage (LV) bus topology is the most widely used today and has the advantage of being used in a large number of missions and be a very robust and reliable technology.

However, several problems are associated to LV bus architectures: connect and disconnect large loads such as thrusters or scientific instruments affects the stability of the bus because they produce large voltage and current peaks [3], [4]. The two conditioning levels reduce the efficiency of the architecture; this is especially critical for high power loads as electrical propulsion.

For these reasons, novel power architectures with low and high voltage distribution (HV) have been recently proposed [5]-[7], please refer to Figure 2.

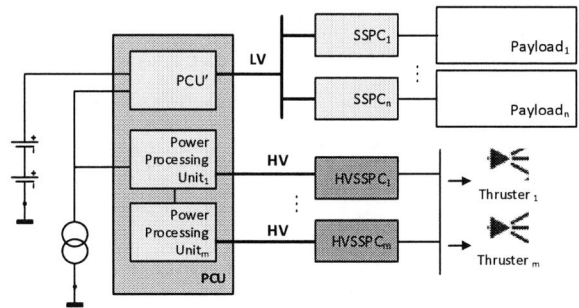

Figure 2: Dual LV and HV PCDU.

In this new scenario, the use of High Voltage SSPC's (HVSSPC) are very interesting for different reasons, like converter parallelization [8] or load protection.

On top of this, with the advent of high voltage silicon carbide power devices, new approaches for solid state protection devices could be devised.

To conclude this section, a new 1kV/2A Solid State Power Controller is proposed in this work.

II. HVSSPC DESCRIPTION

The proposed HVSSPC has two different levels of operation and control. A hardware protection layer (level 2) considers an autonomous protection device without aid of any digital or higher level control subsystem. This is intended for the highest level of protection but with limited capabilities. A software protection layer (level 1) includes a programmable digital device, in this particular case, an FPGA, to expand device capabilities, as shown in Table 1.

TABLE 1:
COMPARATIVE OF OPERATING FEATURES

	LEVEL 1 (Soft)	LEVEL 2 (Hard)
SOA Parametrization	Y	N
Current Limitation	Y	N
Short circuit Protection	Y	Y
Overvoltage Protection	Y	N
Temperature Calibration	Y	N
Remote Control	Y	N
Remote Reset	Y	Y
Aging Device Control	Y	N
Inrush Current	Y	Y
Failure Register	Y	N

A SiC N-MOSFET works as the main element of sectioning and current control. SiC power devices are currently under different investigations by the main space agencies and companies [9],[10], because of its potential in high voltage and high temperature applications. In addition, another critical aspect for space applications, radiation hardness, has been already considered in different studies [11],[12].

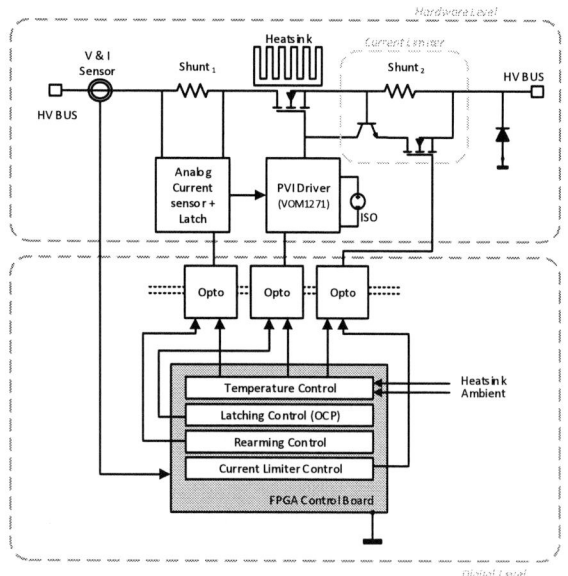

Figure 3: Block diagram of the HVSSPC.

The overall HVSSPC block scheme is shown in Figure 3. Two main blocks compose the system. On one side, the hardware section proposes a fully analog protection system that operates independent from the of the digital system.

The second block is a digital control system. This block drives the direct action circuits. Different sensors are used to control the system behaviour.

The main sensors of the digital control system are: current sensor, supply voltage sensor, ambient temperature sensor and heatsink temperature sensor of the main MOSFET.

The hardware level schematic is shown in Figure 4. Its main parts and the operating equations are also defined. At the end of this section, the working mode is detailed.

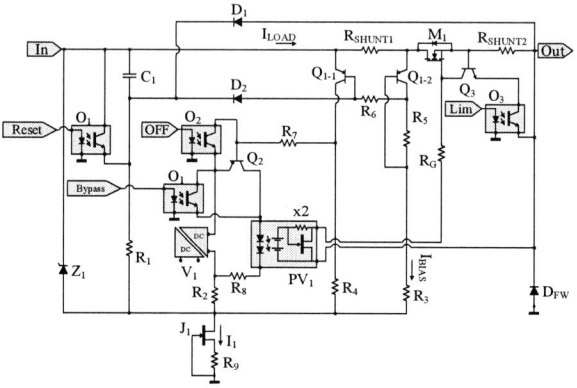

Figure 4: Basic hardware schematic block.

A. Analog current sensor:

A current measuring resistor (R_{SHUNT1}), two paired transistors (Q_{1-1} and Q_{1-2}) and resistors R_3 and R_5 are used for its implementation. This fully analog current sensor allows current measurement without auxiliary power supplies.

In normal operating condition it keeps the bias current I_{BIAS} (1) through transistor Q_{1-2}, while Q_{1-1} remains fully open. When the current value determined by (2) is exceeded, Q_{1-1} is turned ON.

$$I_{BIAS} = \frac{V(Z_1) - I_{LOAD} \cdot R_{SHUNT} - V_{EB}(Q_{1-2})}{R_3} \quad (1)$$

$$I_{LIMIT} = \frac{I_{BIAS} \cdot R_5}{R_{SHUNT}} \quad (2)$$

As is shown in (2) changing R_5 value, the limit and activation value of the HVSSPC can be altered.

The activation/deactivation of Q_{1-1} transistor cause the deactivation/activation of Q_2 transistor. Q_2 provides the energy needed to supply PV_1 driver, allowing the HVSSPC to operate. Deactivation of the Q_2 transistor implies deactivation of the HVSSPC.

In addition, the digital control circuit is equipped with an isolated current sensor for high-level management control.

B. *Reference voltage and bias source:*

A reference voltage is defined for the complete HVSSPC logic circuit. Zener diode (Z_1) provides a stable voltage to implement the analog circuit. Thanks to it, the HVSSPC's design can be enhanced avoiding the need to use components at high Voltages.

For Z_1 biasing, a constant current source is available. It is implemented with a 1200V JFET and a gate resistance. This current source provides stable biasing at a wide range of bus voltages.

For its implementation, the USCi's UJN1205K JFET has been used. A bias current of approx. 1.5mA has been set, sufficient to ensure that the zener, in this case 10V, biases correctly.

This current source loses its given by the following expression (3)

$$W_{J_1} = V_{J_1} \cdot I_{J_1} \approx (V_{In} - V_{Z_1}) \cdot I_{Z_1} \quad (3)$$

C. *Driver:*

A photovoltaic driver (PV_1) is selected to control the M_1 main transistor. Using this driver it's not necessary use auxiliary power supply to provide enough gate voltage to transistor.

Finally, Vishay's VOM1271 driver has been chosen, whose main features are summarized in table 2.

TABLE 2:
VOM1271 MAIN ELECTRICAL CHARACTERISTICS

VOM1271	
Parameter	Value
Typ. Input Current	10mA
Photo-Cell OC Voltage	8.4V @10mA
Photo-Cell SC Current	15µA @10mA
Isolation Voltage	4500V

VOM1271 includes an internal fast discharge circuit. This allows very fast response times and therefore the protection process of the HVSSPC is boosted.

VOM1271 has a typical 10mA operating current required. This current must be drained through the current source and consequently, represents considerable losses. In order to maintain high performance in the HVSSPC, an isolated auxiliary power supply is included to supply the driver. This source can be powered from another lower voltage bus. Two driver circuits have been used in series configuration.

D. *Timing and Latching:*

The HVSSPC provides an analog timing and latching system allowing it to operate without the digital control.

The timing consists of an R-C network (R_1 and C_1). As soon as the current limit is exceeded in the HVSSPC, the D_1 diode is switched off, so that the R_1-C_1 network will start charging up to Z_1 voltage. When the voltage of R_1 allows D_2 to circulate current, the HVSSPC switches off and will be latched until reset.

If the digital control is not working during the timing phase, initially the HVSSPC current is cut suddenly in order to work in the limiting zone until the final latching circuit. The timing is defined by (4), where t_{DELAY} is the driver's time delay and the current limit value is the current limit value given by (2).

$$t_{LIMITING} = R_1 \cdot C_1 \cdot \ln \frac{V(Z_1) - V_F(D_1)}{V_F(D_2)} - t_{DELAY} \quad (4)$$

The M_1 transistor is linear region during the current limiting phase, meeting (5)

$$I_{BIAS} \cdot R_5 = I_{LIMIT} \cdot R_{SHUNT} \quad (5)$$

E. *Direct action circuits:*

Four control signals are available. They enable the HVSSPC to be digitally controlled: Lim, Reset, OFF and Bypass. Please, refer to Figure 4.

Lim: Enables activation of the digitally controlled current limitation. In order to operate this state, the bypass signal must be enabled to avoid analog limitation mode for the HVSSPC.

This current limit is set by the R_{SHUNT2} resistor and transistor Q_3, so the new limit current is set to the value given by (6). V_{BE} usually is 0.7V.

$$I_{Digital_{Lim}} = \frac{V_{BE}(Q_3)}{R_{SHUNT}} \quad (6)$$

Reset: Allows to discharge C_1. C_1 is charged when the HVSSPC is in analog latched state. Discharging it the HVSSPC returns to the initial conditions, allowing the circulation of current through M_1 to the load.

OFF: Allows to force open Q_2 transistor, eliminating the VOM1271 supply and as a result, opening the HVSSPC. OFF command will be used to define the operation of the HVSSPC and define safe operating zones using digital control.

Bypass: This signal provides the HVSSPC the capability to ignore the behaviour of the analog circuit. It enables the PV1 driver to be continuously supplied. As a result, the system can be digitally managed, limiting or opening according to the specific situation.

All direct-action are opto-coupled. The digital control provides a higher level protection. In case of digital control failure, the system keeps working autonomously and safely.

F. Power semiconductors:

Nowadays, different SiC power devices are potential candidates JFET, MOSFET or Cascode. Because of design heritage considerations, MOSFET seems to be the most promising option, but at this moment only N-channel MOSFET are available, then substantial changes are required respect to low voltage P-MOSFET Solid State Power Controllers.

For the selection of the SiC several options have been considered MOSFET as well as different types of semiconductors (Cascode, JFET and MOSFET). Although that there are no SiC MOSFET qualified for space, the N-MOSFET C2M0080120D from Wolfspeed (CREE) has been considered for this study. It is worth to mention that this device has not an equivalent space counterpart but it has been tested under radiation conditions in several projects managed by NASA [10]. Their main characteristics are shown in Table 2.

TABLE 3:
C2M0080120D CHARACTERISTICS

Feature	Value
VDS max	1200V
ID max (@25ºC)	30A
ID max (@100ºC)	24A
RDSon	80mΩ
VGSth	2.6V
Ciss	950pF

In addition, a freewheeling diode (DFW) is included in the HVSSPC. It allows free circulation of current in case of abrupt disconection of heavily inductive loads.

G. Digital control:

The first prototype has been implemented using the Red Pitaya develop card. Red Pitaya includes the necessary elements for actuation and data acquisition. The most interesting features are shown below in Table 4.

TABLE 4:
REDPITAYA FEATURES

Feature	Value
FPGA	Xilinx Zynq 7010 SOC
Processor	Dual Core ARM Cortex A9
RAM	512MB
High Speed ADC (x2)	14bit - 125MS/s
High Speed DAC (x2)	14bit - 125MS/s
Slow ADC	12bit
(x4)	100kS/s

Figure 5 shows the digital control implementation system. The implementation in Vivado (Xilinx) of the different sampling blocks (ADCs) for current, voltage and temperature

Function description:

The HVSSPC has two operating modes. The simplest mode of operation is fully automatic and analog. If all direct action circuits are off, the HVSSPC works completely autonomously.

While the maximum current set in (2) is not reached, Q_{1-1} transistor will not flow current, so Q_2 will provide the current flow to the driver allowing the HVSSPC to be ON.

If the maximum current is exceeded, the system will limit the current for a time set in (4) and the current limit will be given by (2). After this time the HVSSPC will block the current. When the HVSSPC receives the reset or eliminate power supply, current flow restarts.

To operate with digital control, the first step is to turn on the Bypass signal that forces the current flow through PV_1.

This scenario limitation mode, is controlled by Lim signal. It activates the digital limiter and sets the current limit by (6).

Finally, if it is necessary to switch off the HVSSPC, OFF command is available. It can be activated for several reasons, for example: Over voltage, under voltage input, over temperature, over current and limitation time.

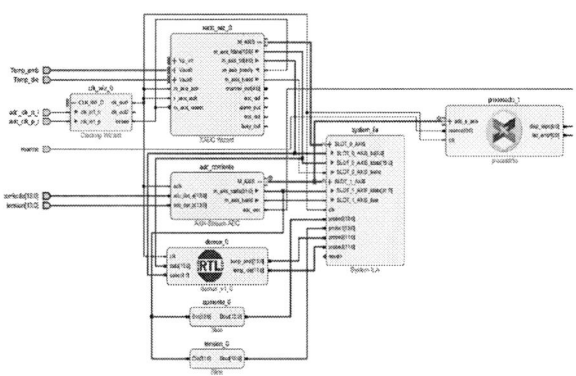

Figure 5: Basic block diagram of the digital control.

III. EXPERIMENTAL RESULTS

To validate the model, a simulation was carried out using LTSpice software using the manufacturer's models.

Figure 6 shows the simulated scheme. Figure 7 shows the resulting output of the differents operating modes. Different operating states are described below.

For the test, a digital operating current limit of 0.5A was set up, and nominal operating current of 0.3A. The timing of analogue limitation has been set to 10ms via the R_1-C_1 network.

t0: Normal operation of the HVSSPC, the digital control is activated and the digital current limiter is also active..

t1: An overload is caused and the digitally controlled limiter actuates, after t1, the digital limiter is turn off so the HVSSPC is turn off also.

t2: The HVSSPC will remain in off state until the RESET signal is activated.

t3: Normal HVSSPC operation, in this case, digital current limiting control is disabled.

t4: HVSSPC is overloaded and the analog protection system starts operating autonomously.

t5: The HVSSPC remains uninterruptedly off until the RESET signal is activated again.

t6: After RESET the HVSSPC runs normally.

t7: The OFF control is enabled and the HVSSPC is forced into an off state.

Figure 8 shows the HVSSPC implemented prototype for experimental validation. The digital control card used is also shown.

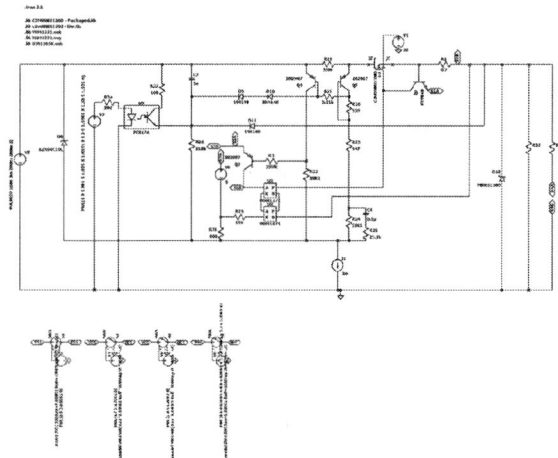

Figure 6: Simulated Circuit in LTSpice.

A Keysight N8937 high-voltage source has been used to carry out the tests. It achieves voltages up to 1500V with a maximum power of 15kW. And wirewound MC14683 resistors from Bourns have been used as load. A bank of resistors has been made with a total of 10 resistances of 100Ω allowing a maximum of 2.25kW to be dissipated

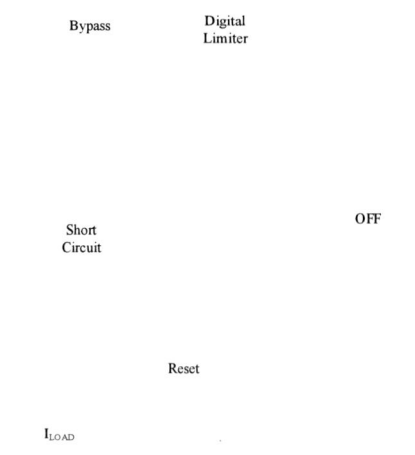

Figure 7: Simulation results.

Three different tests have been included to validate the behaviour of the HVSSPC prototype.

Figure 9 shows the waveforms of a linear bus voltage variation from 750V to 1200V. It can be seen how the HVSSPC remains on until 1200V and 1.4A. Once this point is reached, the switch trips off and MOSFET blocks the main bus voltage. In this kind of test the response speed is not critical because the di/dt is not very high.

The 2018 International Power Electronics Conference

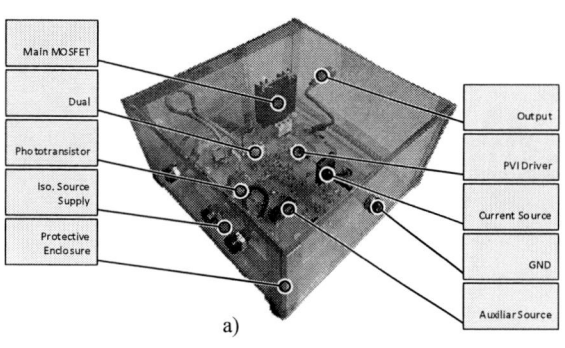

a)

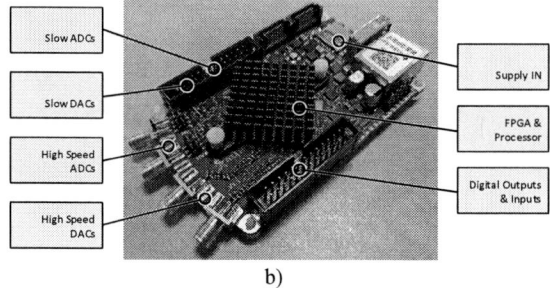

b)

Figure 8: Prototype of the proposed system.
a)Hardware Level b)Digital Level

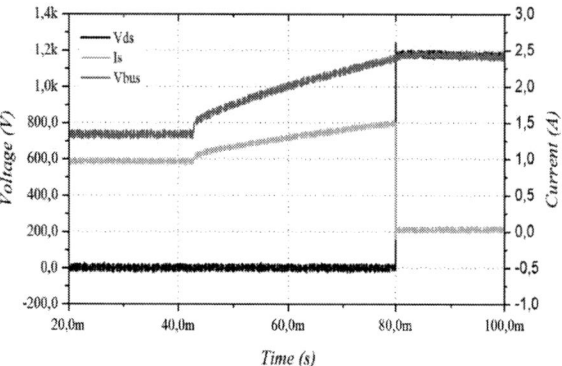

Figure 9: Response to a supply voltage linear increase
from 750V to 1200V with an 800 Ω load.

Figure 10 shows the HVSSPC response to a load step from 1kΩ to 400Ω, with a constant input voltage of 1kV. As it can be seen, the power switch carries the fault current (2.5A) during 25μs, before it finally disconnects.

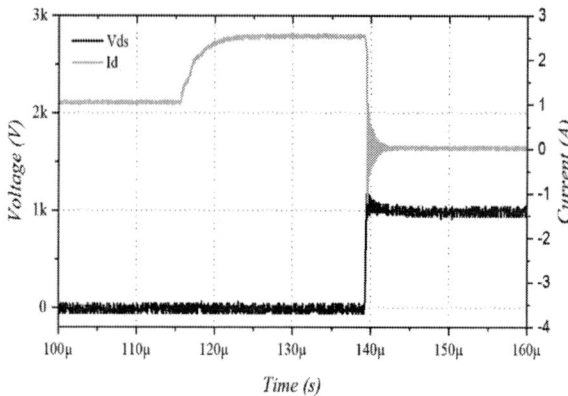

Figure 10: HVSSPC response to load step (1kΩ to 400Ω) with predefined trip-off time.

Figure 11 shows the evaluation of the different direct action circuits. Different operating states are defined below. Here, the R1-C1 network configuration gives a limitation time of approximately 10ms.

t0: HVSSPC operation is normal, the bypass circuit remains deactivated so the analog limiter will operate in fail case.

t1: An overload takes place, after limitation time the HVSSPC blocks the current to the load until the reset signal is produced.

t2: HVSSPC's normal operation, the bypass circuit remains deactivated again.

t3: An Off command is produced. The HVSSPC starts to block the current to the load.

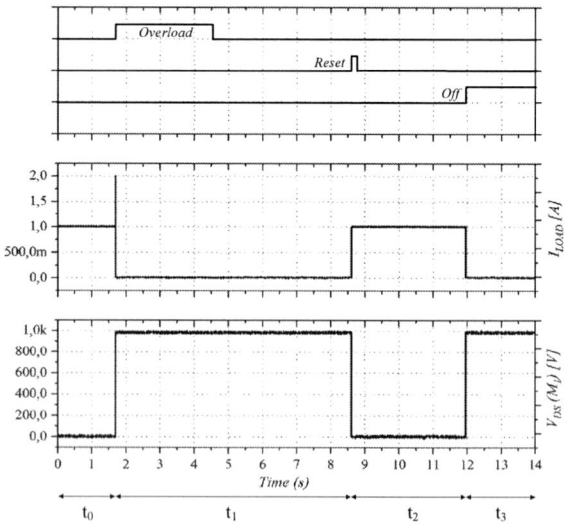

Figure 11: Waveforms of SSPC prototype.

IV. CONCLUSIONS AND FUTURE WORK LINES

The development of new topologies for SSPC for use in high voltage applications with higher intelligence levels that preserves robustness of discrete systems is of special interest for new distribution architectures in different areas [14]-[17].

For this reason, a new topology based in SiC semiconductors, optoisolated driver and a robust hybrid control with two levels controlled by FPGA is proposed. Experimental tests show an excellent behavior of the MOSFET SiC for voltage blocking, obtaining very fast response times at 1000V.

ACKNOWLEDGMENT

This work was supported by the Spanish Ministry of Economy and Competitiveness through the research project ESP2015-68117-C2-2-R.

REFERENCES

[1] M. Gollor, F. Herty, "Characterization of Electrical Propulsion Thrusters as a Load for Electronic Power Supplies", in 2008 European Space Power Conference.

[2] C. R. Spitzer, "The All-Electric Aircraft: A Systems View and Proposed NASA Research Programs," in IEEE Transactions on Aerospace and Electronic Systems, vol. AES-20, no. 3, pp. 261-266, May 1984.

[3] W. Liqiu, W. Chunsheng, N. Zhongxi, L. Weiwei, Z. ChaoHai, and Y. Daren, "Experimental study on the role of a resistor in the filter of hall thrusters," Phys. Plasmas, vol. 18, no. 6, pp. 063508-1-7, 2011.

[4] W. Liqiu, W. Chunsheng, H. Ke, and Y. Daren, "Effect of ionization distribution on the low frequency oscillations mode in hall thrusters," Phys. Plasmas, vol. 19, no. 1, pp. 012107-1-6, 2012.

[5] J. S. Snyder and J. R. Brophy, "Peak power tracking and multi-thruster control in direct drive systems," presented at the 33th Int. Electr. Propul. Conf., The George Washington University, USA, Oct. 6 -10, 2013.

[6] D. Y. Oh, J. S. Snyder, D. M. Goebel, R. R. Hofer, D. F. Landau, T. M, et al., "Solar electric propulsion for discovery class missions," presented at the 33th Int. Electr.

Propul. Conf., The George Washington University, USA, Oct. 6 -10, 2013.

[7] J. A. Hamley, "Direct drive options for electric propulsion systems," IEEE Aerosp. Electron. Sy. Mag., vol. 11, no. 2, pp. 20-24, Feb. 1996.

[8] M. Fu; D. Zhang; T. Li, "New Electrical Power Supply System for All-Electric Propulsion Spacecraft," in IEEE Transactions on Aerospace and Electronic Systems, vol.PP, no.99, pp.1-1

[9] F. Bausier, S. Massetti, F. Tonicello, "Silicon Carbide for space power applications," Proc. of the 10th European Space Power Conference, ESA-ESTEC, April 2014.

[10] Lauenstein, J. M., Casey, M., Samsel, I., LaBel, K., Chen, Y., Ikpe, S. & Topper, A. (2017)." Silicon Carbide Power Devices and Integrated Circuits." NEPP Electronics Technology Workshop; Jun. 2017; Greenbelt.

[11] A. Akturk, R. Wilkins, J. McGarrity and B. Gersey, "Single Event Effects in Si and SiC Power MOSFETs Due to Terrestrial Neutrons," in IEEE Transactions on Nuclear Science, vol. 64, no. 1, pp. 529-535, Jan. 2017.

[12] Lauenstein, J. M., Casey, M., Topper, A., Wilcox, E., Phan, A., Ikpe, S., and LaBel, K. ." Silicon carbide power device performance under heavy-ion irradiation". IEEE Nuclear and Space Radiation Effects Conference (NSREC); Boston;July 2015.

[13] M. Teorörde, F.Grumm, D.Schulz, H.Wattar, J.Lemke, "Implementation of a Solid-State Power Controller for High-Voltage DC Grids in Aircraft": IEEE Power and Energy Student Summit (PESS 2015).

[14] S. Krstic, E. L. Wellner, A. R. Bendre and B. Semenov, "Circuit Breaker Technologies for Advanced Ship Power Systems," in 2007 IEEE Electric Ship Technologies Symposium, Arlington, VA, 2007, pp. 201-208.

[15] D. Izquierdo, A. Barrado, C. Raga, M. Sanz and A. Lazaro, "Protection Devices for Aircraft Electrical Power Distribution Systems: State of the Art," in IEEE Transactions on Aerospace and Electronic Systems, vol. 47, no. 3, pp. 1538-1550, July 2011.

[16] Z. J. Shen, A. M. Roshandeh, Z. Miao and G. Sabui, "Ultrafast autonomous solid state circuit breakers for shipboard DC power distribution," 2015 IEEE Electric Ship Technologies Symposium (ESTS), Alexandria, VA, 2015, pp. 299-305

[17] R. Ouaida, M. Berthou, D. Tournier and J. F. Depalma, "State of art of current and future technologies in current limiting devices," 2015 IEEE First International Conference on DC Microgrids (ICDCM), Atlanta, GA, 2015, pp. 175-180.

The 2018 International Power Electronics Conference

An Improved Voltage-Type Grid-Connected Control Strategy for Compensating Unbalanced Voltage

Liu Hongpeng, Zhou Jiajie and Wang Wei

Department of Electrical Engineering, Harbin Institute of Technology, Harbin, China

Abstract— **During grid faults, the grid-connected inverter (GCI) cannot work well by only using the traditional droop control. Besides, the unbalance factor of voltage/current at Common Coupling Point (PCC) may increase significantly. Therefore, in order to ensure GCI stable operation during grid fault, the GCI should integrate the ability to compensate for the imbalance of the grid. To solve the problem mentioned above, an improved voltage-type grid-connected control strategy is proposed in this paper. On the basis of the positive sequence power droop control, a negative sequence conductance compensation loop is added to keep the PCC voltage balance and reduce the grid current imbalance, thus meeting the PCC power quality requirements. Simulation and experimental results have verified the effectiveness of the improved droop control scheme.**

Keywords— Voltage-type control; Grid-connected inverter; Unbalanced voltage; Droop control.

I. INTRODUCTION

The high capacity single-phase or unbalanced three-phase load connects with the grid resulting in the serious unbalance of the grid voltage [1]-[3]. And yet, the common controller of three-phase grid-connected inverter is mostly designed in the condition of three-phase balance grid. Therefore, the grid imbalance will generate many issues, such as inverter output current distortion, negative sequence components increase of inverter output voltage, low-frequency fluctuation of injection power, etc. Besides, it may affect the safe and stable operation of the local loads connected with PCC [4]-[6].

Some research works have been proposed to achieve the unbalance voltage compensation at PCC [7]. Among them, the current control scheme is widely adopted. Paper [8] proposes a double current loop control scheme, in which the positive and negative sequence components of the grid current are controlled under the positive and negative synchronization reference frames, respectively. However, a low pass filter is needed to extract the dc component of d-q axis current [9]. The delay induced by the filter will affect the control accuracy and stability of the system. A full-feedforward scheme is proposed to reduce the injected grid-current harmonics and unbalanced causer by grid voltage [10]. But the control scheme is implemented in the synchronous d-q frame, which needs the power decoupling control in d-axis and q-axis. In [11], five different strategies are presented to generate inverter reference currents according to the corresponding unbalance control requirements. In the case of the balanced positive-sequence

control (BPSC), although the positive-sequence component of the grid voltage is only used for calculating the grid current references, the currents are sinusoidal and balanced. However, each of control strategies only meet one control objective (i.e. canceling the oscillations in power, obtaining sinusoidal current). A flexible grid voltage support strategy is proposed in [12]. By adjusting the current reference, the output positive sequence voltage increases and the output negative sequence voltage decreases. However, the voltage at PCC is compensated indirectly by adjusting the line current, so that the compensation voltage highly depends on the line impedance of the micro-grid. Thus, a small or unknown line impedance cannot achieve the expected grid voltage compensation effect.

In microgrid, droop control is widely used to achieve seamless transfer between islanded and grid-connected modes. The droop method belongs to voltage-controlled technique and maps the generator terminal parameters with its active and reactive power generations [13]-[16]. However, the grid voltage imbalance compensation is barely concerned in droop-controlled inverters. The *G-H* and *Q-G* droop controls are introduced to share harmonics and unbalanced currents among the distributed generation (DG) units [17]. Paper [18] proposes a control strategy that comprises a voltage control loop, a droop controller and a negative-sequence output impedance controller (NSIC). The voltage loop and droop controller are, respectively, used to regulate the load voltage and to share the average power among the DG units. The NSIC is used to adjust the negative-sequence output impedance of DG such that the negative-sequence currents of power lines will be minimized. The above methods show good performance and improve the power quality of the overall system. However, the proposed method does not apply to the operation in grid-connected mode. A micro-grid hierarchical control strategy considering complex line impedance is proposed in [19]. By using two-layer control structure, microgrids can provide positive sequence and negative sequence power, support the grid voltage and ride-through the voltage dip during the whole fault period. However, the control method requires a high-speed communication line in the micro-network to exchange the voltage imbalance information at the PCC, which increases the system cost and control complexity.

In this paper, the vector relationship of the positive and negative sequence voltage and current is presented and the compensation principle of PCC voltage is analyzed. And then a negative sequence reactive power-conductance (Q-G) loop is introduced to the control system, which improves the unbalanced voltage compensation at PCC. In addition, the small- signal model is developed to evaluate stability of the inverter.

II. IMPROVED VOLTAGE-TYPE UNBALABCE COMPENSATION CONTROL

A. Improved Control Scheme

By using (1) and (2), the positive sequence fundamental active and reactive powers can be controlled under imbalanced grid voltage. But the negative sequence reactive power Q^- is not effectively suppressed. The negative sequence component not only damages the local load, but also influences grid-connected inverter stable operation. Especially when the inverter capacity is large, the unbalanced effects caused by the negative sequence reactive component cannot be ignored.

$$f = f_0 + (K_{pp} + \frac{K_{ip}}{s})(P_0^+ - P^+) \qquad (1)$$

$$U = U_0 + (K_{pq} + \frac{K_{iq}}{s})(Q_0^+ - Q^+) \qquad (2)$$

To solve the above issues, an improved droop control is proposed based on the positive sequence power droop control. By adding a negative sequence reactive power-conductance (Q^--G) loop, the voltage unbalance factor is reduced. Combining with negative sequence extraction, the negative sequence reactive power is obtained as

$$Q^- = u_\alpha^- i_\beta^- - u_\beta^- i_\alpha^- \qquad (3)$$

where u_α^-, u_β^- and i_α^-, i_β^- are negative sequence voltages and currents under two-phase stationary coordinate system, respectively.

The negative sequence reactive conductance G^* is represented as

$$G^* = G_0 - uQ^- \qquad (4)$$

where G_0 is the rated conductance, which is set to zero in this paper. u is droop coefficient, which is determined by the inverter output negative sequence reactive power.

By adding Q^--G loop, the inverter can be controlled as an adjustable negative sequence conductance in the grid-connected mode. The equivalent conductance is proportionate to the negative sequence reactive power produced by the inverter. Then this conductance is multiplied by the negative sequence PCC voltage to generate the compensation current reference. According to the relation between voltage, current and power, the compensating current can be obtained as

$$i_{o\alpha\beta}^- = G^* \cdot u_{o\alpha\beta}^- . \qquad (5)$$

In the grid-connected mode, if the PCC voltage becomes unbalanced, the negative sequence reactive power will increase, thus increasing the equivalent conductance. As a result, the compensation current will increase. The grid current imbalance can be compensated and the PCC voltage imbalance can be suppressed at the same time.

Fig. 1 shows the eventual grid-connected scheme implemented for the three-phase inverter under unbalanced grid. As usual, the voltages and currents in the stationary a-b-c frame is transformed into $i_{o\alpha\beta}$ and $u_{o\alpha\beta}$ in the stationary $\alpha - \beta$ frame. By using the positive and negative sequence extraction scheme, positive and negative sequence voltage and current can be obtained respectively. P^+ and Q^+ and Q^- can then be calculated. The improved droop control loop generates the voltage reference and the negative sequence current compensation reference. Finally, the voltage and current dual-loop controller is used to generate the driving signals of switches.

B. Small-Signal Analysis

Small-signal analysis can next be performed on the voltage and current dual-loop controller for studying the system stability influence of the conductance loop parameter u.

According to the foregoing analysis, the current reference can be obtained as

$$\mathbf{i}^* = -u \cdot \mathbf{u}_o^- \cdot Q^- \qquad (6)$$

where u_o^- is the inverter output negative sequence voltage.

By adding the virtual impedance loop, the output impedance at the fundamental frequency Z_o is approximately equal to $L_v s$. Thus, the negative sequence reactive power at the fundamental frequency can be

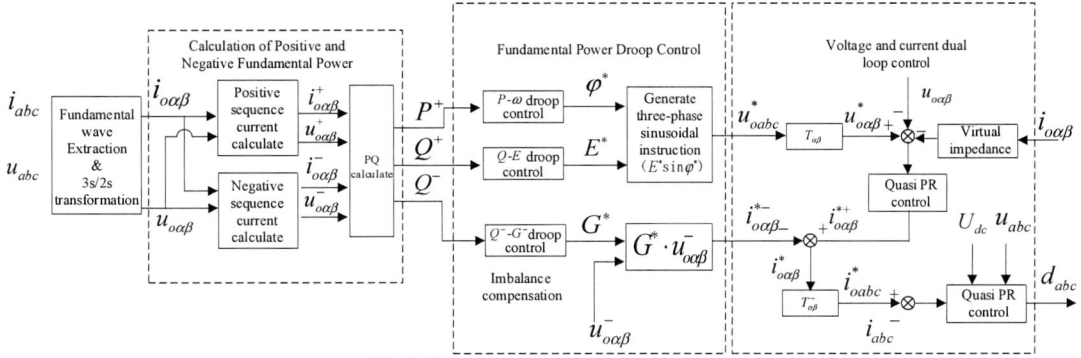

Fig. 1. Overview of improved droop scheme.

The 2018 International Power Electronics Conference

derived as

$$Q^- = 3 \cdot X_v \cdot (I_o^-)^2 \tag{7}$$

where, X_v is the virtual reactance designed by the virtual impedance loop; I_o^- is the fundamental negative sequence component of the inverter output current. Substitute (6) into (7), (8) can be obtained.

$$\mathbf{i}^* = -3 \cdot u \cdot \mathbf{u}_o^- \cdot X_v \cdot (I_o^-)^2 \tag{8}$$

The linearly perturbed expression for the current reference, thus, be derived as

$$\hat{\mathbf{i}}^* = -[3 \cdot \mathbf{u}_o^- \cdot X_v \cdot (\mathbf{I}_o^-)^2 + 6 \cdot \mathbf{u}_o^- \cdot X_v \cdot \mathbf{I}_o^- \cdot \hat{\mathbf{I}}_o^-] \cdot u . \tag{9}$$

In generally, the load impedance of the system is much larger than the line impedance, so negative sequence current can be approximately obtained as

$$\mathbf{I}_o^- \approx \frac{\mathbf{E}^*}{Z_L(j\omega)} \tag{10}$$

where Z_L is the load impedance. Similarly, the negative sequence voltage is given by

$$\mathbf{u}_o^- \approx -Z_o^-(j\omega) \cdot \mathbf{I}_o^- . \tag{11}$$

Substituting (10) into (7), the following expression can be obtained as

$$\hat{\mathbf{i}}^* = -[\frac{3 \cdot X_v \cdot (\mathbf{E}^*)^2}{Z_L^2(j\omega)} \cdot \hat{\mathbf{u}}_o^- + \frac{6 \cdot X_v \cdot (\mathbf{E}^*)^2 \cdot Z_o^-(j\omega)}{Z_L^2(j\omega)} \cdot \hat{\mathbf{I}}_o^-] \cdot u . \tag{12}$$

And then the corresponding small signal perturbation model can be expressed as

$$\hat{\mathbf{u}}_o^- = G_{closed\text{-}loop} \cdot \hat{\mathbf{I}}_o^- \tag{13}$$

where

$$G_{closed\text{-}loop} = \frac{6 \cdot G^-(j\omega) \cdot X_v \cdot Z_o^-(j\omega) \cdot u}{Z_L^2(j\omega) + 3G^-(j\omega) \cdot X_v \cdot u} - \frac{Z_o^-(j\omega) \cdot Z_L^2(j\omega)}{Z_L^2(j\omega) + 3G^-(j\omega) \cdot X_v \cdot u} \tag{14}$$

From the analysis on the small signal perturbation model of negative sequence conductance loop, it can be seen that the bigger the droop coefficient u is, the more obvious the voltage unbalance compensation is. But large value of u may lead to instability of the system. Here, $u = 0.0001, 0.002, 0.02$ and 0.05 are selected to study the root locus. Fig.2 shows the Zero and pole distribution with different droop coefficients u. When $u = 0.05$, there is an unstable pole in the root locus at the right half plane. When $u = 0.001, 0.002$ and 0.02, the zeros and poles are all at the left half plane, showing that the system is stable. Therefore, u is selected as 0.02 in this paper.

III. SIMULATION AND EXPERIMENTAL RESULTS

A. Simulation Results

A three-phase grid-connected inverter system, tied to

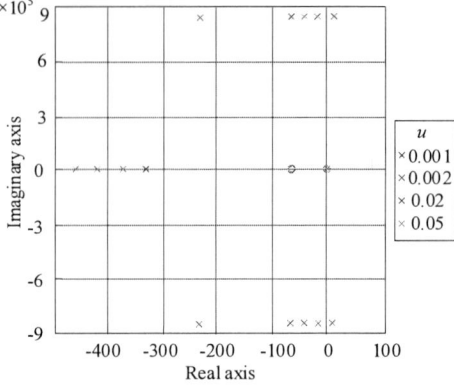

Fig. 2. Zero and pole distribution graphs with different droop coefficients.

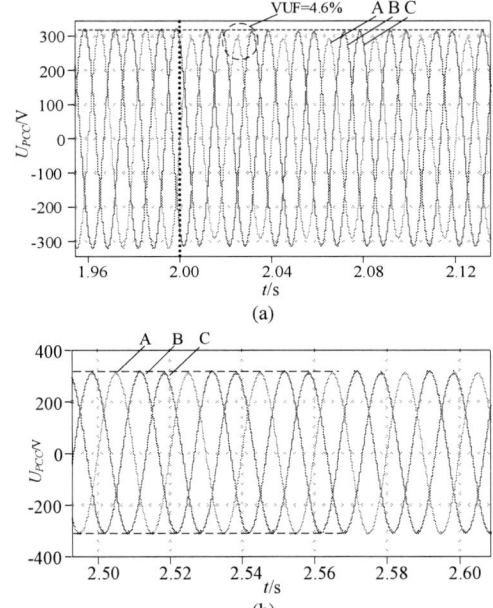

Fig. 3. Grid-connected voltage waveforms (a) before compensation (b) after compensation.

an unbalanced grid, have been simulated in MATLAB/Simulink. Grid voltage is 380V from 0 to 2s and the A-phase voltage is reduced from 220V to 205V after 2s. Fig. 3(a) shows the PCC voltage before using the power compensation. The three-phase voltage imbalance phenomenon occurs at 2s, and the VUF is equal to 4.6%. Fig. 3(b) shows the simulation results when using the proposed control scheme. The VUF reduces to 1% and PCC voltage basically maintain at 220V from 2s to 4s. Hence, the proposed scheme can efficiently compensate the unbalance voltage at PCC.

B. Experimental Results

Experimental testing with a 600 W grid-connected system has been performed for proving the practicality of the proposed scheme. Grid voltage is 110V AC (Peak value), power frequency is 50Hz, and switching frequency is 10 kHz.

Fig. 4 shows the experimental result of the three phase

1444

The 2018 International Power Electronics Conference

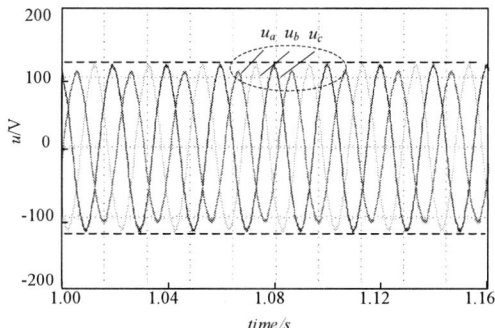

Fig. 4. Waveform of unbalanced grid voltage.

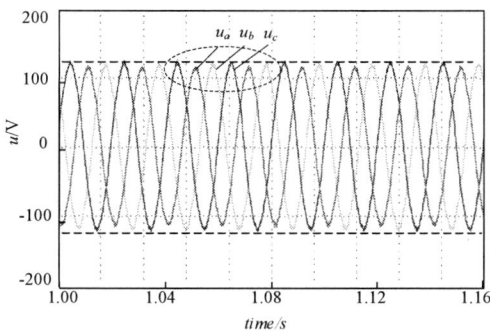

Fig. 5. Experimental results of grid connected voltage when using the positive sequence power droop control.

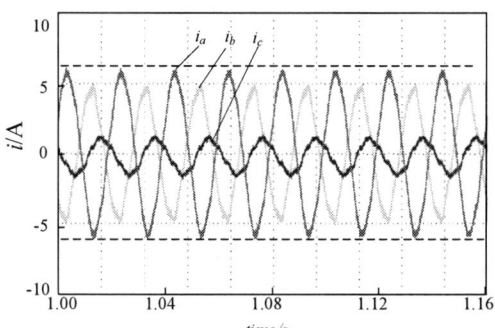

Fig. 6. Experimental results of grid connected current when using the positive sequence power droop control.

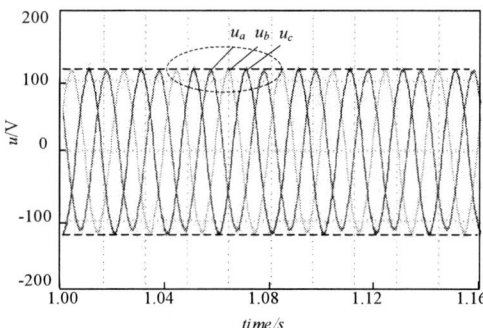

Fig. 7. Experimental results of grid connected voltage when using the negative sequence power droop control.

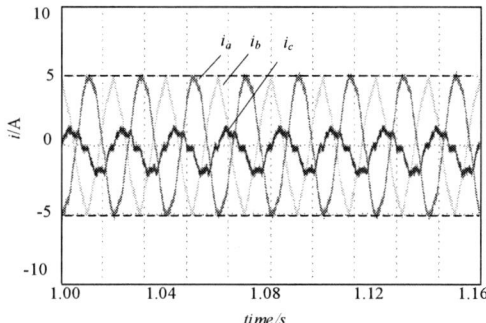

Fig. 8. Experimental results of grid connected current when using the negative sequence power droop control.

current is significantly reduced after compensation. The experimental results verify that the proposed control strategy can compensate the voltage drop and reduce the grid current unbalance factor under unbalanced grid voltage, thus ensuring the safe and reliable operation of grid-connected inverters.

IV. CONCLUSIONS

In this paper, an improved droop control is proposed to suppress voltage imbalance at PCC. The proposed control scheme includes the unbalance compensating loop, active and reactive power droop control loop and voltage and current control loop. The compensating loop regulates the positive and negative sequence output reactive power of the inverter, which ensures the PCC voltage is controlled within the allowable limits. The control system design and stability analysis are discussed. Thus, this proposed control strategy can maintain stable operation of the inverter under unbalanced grid conditions. Simulation and experimental results have verified the performance of the proposed method.

grid voltages. The B and C phase voltages are 23V higher than A phase voltage and the unbalance factor is about 5.3%. Figs. 5 and 6 show the experimental results when using the positive sequence power droop control. U_a is about 4.5%. The amplitude of phase A and B current are 6.0A and 1.1A, respectively. Furthermore, the amplitude of phase C current I_a is bigger than 5A, which exceeds the upper limit of output current for compensating the unbalanced grid voltage.

Figs. 7 and 8 show the three-phase PCC voltage and current waveforms when using improved droop control scheme. The voltage unbalance factor is reduced to 1.5% and the system can operate stably. Due to reducing the negative sequence reactive power, the C phase current, which is seriously unbalanced before compensation, has been greatly increased to 1.5A. At the same time, currents of phase A and B decrease to about 5A, which can avoid overcurrent. The unbalance factor of three phase grid

REFERENCES

[1] M. Savaghebi, A. Jalilian, and J. C. Vasquez, et al, "Secondary control scheme for voltage unbalance compensation in an islanded droop-controlled microgrid," *IEEE Smart Grid*, vol. 3, no. 2, pp. 797–807, 2012.

[2] Y. Han, P. Shen, and X. Zhao, et al, "An enhanced power sharing scheme for voltage unbalance and harmonics compensation in an islanded AC microgrid," *IEEE Trans. Energy Convers.*, vol. 31, no. 3, pp. 1037–1050, 2016.

1445

[3] W. Feng, K. Sun, and Y. J. Guan, et al, "Active power quality improvement strategy for grid-connected microgrid based on hierarchical control," *IEEE Smart Grid*, vol. PP, no. 99, pp. 1–10, 2017.

[4] D. De and V. Ramanarayanan, "Decentralized parallel operation of inverters sharing unbalanced and nonlinear loads," *IEEE Trans. Power Electron.*, vol. 25, no. 12, pp. 3015–3025, 2010.

[5] M. Hamzeh, H. Karimi, and H. Mokhtari, et al, "Harmonic and negative-sequence current control in an islanded multi-bus MV microgrid," *IEEE Smart Grid*, vol. 5, no. 1, pp. 167–176, 2014.

[6] Q. W. Liu, Y. Tao, and X. H. Liu, et al, "Voltage unbalance and harmonics compensation for islanded microgrid inverters," *IET Power Electron.*, vol. 7, no. 5, pp. 1055–1063, 2014.

[7] M. Castilla, J. Miret, and A. Camacho, "Modeling and design of voltage support control schemes for three-phase inverters operating under unbalanced grid conditions," *IEEE Trans. Power Electron.*, vol. 29, no. 11, pp. 6139–6150, 2014.

[8] Y. Suh and T. A. Lipo, "Control scheme in hybrid synchronous stationary frame for PWM AC/DC converter under generalized unbalanced operating conditions," *IEEE Trans. Ind. Appl.*, vol. 42, no. 3, pp. 825–835, 2006.

[9] M. Reyes, P. Rodriguez, and S. Vazquez, et al, "Enhanced decoupled double synchronous reference frame current controller for unbalanced grid-voltage conditions," *IEEE Trans. Power Electron.*, vol. 27, no. 9, pp. 3934–3943, 2012.

[10] X. Wang, X. Ruan, and S. Liu, et al, "Full feed-forward of grid voltage for grid-connected inverter with LCL filter to suppress current distortion due to grid voltage harmonics," *IEEE Trans. Power Electron.*, vol. 25, no. 12, pp. 3119–3127, 2010.

[11] R. Rodriguez, A. V. Timbus, and R. Teodorescu, et al, "Flexible active power control of distributed power generation systems during grid faults," *IEEE Trans. Ind. Electron.*, vol. 54, no. 5, pp. 2583–2592, 2007.

[12] A. Camacho, M. Castilla, and J. Miret, et al, "Reactive power control for distributed generation power plants to comply with voltage limits during grid faults," *IEEE Trans. Power Electron.*, vol. 29, no. 11, pp. 6224–6234, 2014.

[13] J. M. Guerrero, J. C. Vasquez, and J. Matas, et al, "Hierarchical control of droop-controlled AC and DC microgrids—A general approach toward standardization," *IEEE Trans. Ind. Electron.*, vol. 58, no. 1, pp. 158–172, 2011.

[14] H. Bevrani and S. Shokoohi, "An intelligent droop control for simultaneous voltage and frequency regulation in islanded microgrids," *IEEE Smart Grid*, vol. 4, no. 3, pp. 1505–1513, 2013.

[15] Q. C. Zhong and G. C. Konstantopoulos, "Current-limiting droop control of grid-connected invertes," *IEEE Trans. Ind. Electron.*, vol. 64, no. 7, pp. 5963–5973, 2017.

[16] R. R. Kolluri, I. Mareels, and T. Alpcan, et al, "Power sharing in angle droop controlled microgrids," *IEEE Trans. Power Syst.*, vol. 32, no. 6, pp. 4743–4751, 2017.

[17] H. Just and S. Dieckerhoff, "Advanced negative sequence droop control for fault-ride-through operation and system support in weak grids," in *Proc. EPE*, 2017, pp. 1-10.

[18] M. Hamzeh, H. Karimi, and H. Mokhtari, "A new control strategy for a multi-bus MV microgrid under unbalanced conditions," *IEEE Trans. Power Syst.*, vol. 27, no. 4, pp. 2225–2232, 2012.

[19] X. Zhao, J. M. Guerrero, and M. Savaghebi, et al, "Low-voltage ride-through operation of power converters in grid-interactive microgrids by using negative-sequence droop control," *IEEE Trans. Power Electron.*, vol. 32, no. 4, pp. 3128–3142, 2017.

Dual Two-Stage Isolated Bidirectional DC-DC Converter for DC Grid Storage

Gabriel Tibola*, Jorge L. Duarte

Department of Electrical Engineering, Eindhoven University of Technology, Eindhoven, The Netherlands
*E-mail: g.tibola@tue.nl

Abstract—**One of the trends for low-voltage dc grids is the use of a two-line dc distribution in order to facilitate the interface to present-time ac grids, and also to allow the flexible use of diverse loads. Another feature of such dc grids is the distributed storage elements in houses and buildings. This storage system needs an isolated power converter to interface battery and grid. This paper presents the analysis, design and experimental validation of a 30 kW two-stage dual converter, interfacing a ± 350 V_{dc} or ± 380 V_{dc} grid and a battery bank. The first stage consists of a non-isolated synchronous Boost converter with current control. Whereas the second stage is an isolated bidirectional C3LC converter providing same static characteristics in both ways of energy transfer.**

Keywords—*Bidirectional dc-dc converter, dc microgrids, energy storage, LLC converter.*

I. INTRODUCTION

Most of the electrical devices operate with dc currents, many distributed sustainable resources generate energy in dc, and storage components as batteries have dc characteristics. Therefore, dc to ac conversion is usually necessary in order to allow the connection with conventional ac grids. This fact leads to low efficiency and requires frequency synchronization. These drawbacks are reduced by using dc grids, where sources and loads in dc can be directly connected to a dc bus eliminating dc-ac and ac-dc conversion stages. One of the elements commonly found in dc grid system concepts is the distributed local energy storage system (ESS) in houses and buildings. Many converter topologies can be employed to interface ESS and grid. Regarding to the grid, depending on the employed earthing concept, a dc grid for a commercial building can use either one or two dc lines. It is possible to create a mid-point, so a $\pm$ 350 V_{dc} or $\pm$ 380 V_{dc} architecture for the grid can be implemented, offering more flexibility towards different loads [1], [2].

Hence, in view of this kind of grid concept, this paper focuses on the analysis and development of a bidirectional converter for interfacing an ESS to a two-line dc grid. The converter must comprise an internal high-frequency isolation transformer and bidirectional energy flow control. The concept idea have been previously presented in [3] showing the main aspects and simulated results for a proof of concept specification. In addition to the analysis review and extended design criteria, experimental

results in open loop are presented in order to validate the topology principle of operation.

II. TOPOLOGY DEFINITION

Based in the required features, several topologies could be selected. Although, for power levels between 10 to 100 kW, two well known converter families suit better the application: the bidirectional full-bridge converts and the bidirectional resonant converters. For the first group the most promising converter is the dual active bridge (DAB) converters [4]–[6]. On the other side, among the resonant converter, the full-bridge LLC resonant converter [7]–[12] has better performance. Regarding to control, DAB converters operate at fixed frequency and regulate the energy flow using an appropriate ZVS phase-shift method. This would be, in principle, less challenging than the load regulation with variable frequency as in the LLC converters. However, if the LLC operates around the unitary voltage gain, known as load independent point, the design is facilitated. By controlling the energy injected in the primary side using a second converter, the LLC naturally transfers energy from one side to another accordantly to the bridge that is active.

The concept applied to the case study grid is shown in Fig. 1 and consists in dual two stage converter, one in each dc line. The first stage is a synchronous Boost converter [13], [14] with fixed frequency and active current control, while the second stage is cascaded to the first and consists in a bidirectional full-bridge symmetrical LLC-type converter that operates at 50% duty-cycle and fixed frequency tunned at the resonant frequency. By doing so, the LLC operates in the load independent point leading to high efficiency due to ZVS in the full operation range. The approach uses a resonant cell in each side of a transformer in order to create a symmetric unit, which could be named as CLLLC or C3LC [15]. If the cell has identical component values in both sides it works has a convectional LLC converter with the same characteristics in both energy flow directions. Symmetry can be achieved my other means, as in [9], [15]–[21], although the C3LC is chosen due to the feature of blocking dc current in both directions and the simplicity to model and design the converter as an unidirectional LLC.

III. PRINCIPLE OF OPERATION

Along with the separated energy flow control, the use of one converter in each line allows the modularity while analysis and design are performed independently.

This project has received funding in the framework of the joint programming initiative ERA-Net Smart Grids Plus, with support from the European Unions Horizon 2020 research and innovation programme.

The 2018 International Power Electronics Conference

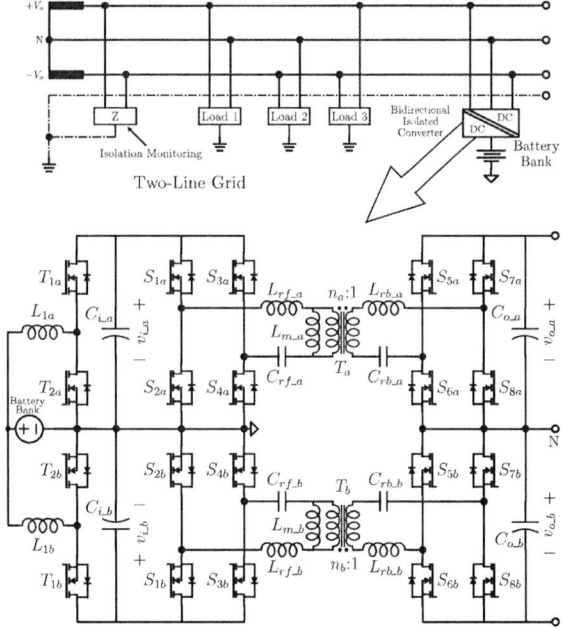

Fig. 1. Dual two-stage isolated bidirectional dc-dc converter interconnecting a two-line dc grid and a battery bank. Letters a and b denote the line.

The converter has two main operational modes: forward (discharging) and backward (charging) modes. For the sake of simplicity, the analysis in this paper considers that one of the LLC full bridges operates in passive way, performing a diode rectification using the switches intrinsic diodes. Synchronous rectification is also possible but it is not discussed in the following.

In forward mode the first stage (synchronous Boost) is set to drain a specific current (positive value) from the battery, according to a desired power. Whereas in the second stage the first bridge is active and the second is passive. The forward resonant inductor L_{rf}, forward resonant capacitor C_{rf} and transformer magnetizing inductor L_m form the LLC resonant tank. As the switching frequency f_s is equal to the forward resonant frequency f_{rf}, the gain of the second stage is unitary and so the grid voltage v_o is reflected to the intermediate buffer capacitor C_i as $v_i = nv_o$, where n is the transformer turns ratio. Hence, as the voltage across C_i is imposed and the current from the Boost tries to increase it, energy flows from battery to grid.

In backward mode the procedure is identical, but the active and passive bridges on the LLC converter are now swapped and the current in the Boost converter is set to charge the battery (negative value). In this mode the LLC resonant tank is formed by the backward elements L_{rb}, C_{rb} and L_m (referred to transformer secondary side). As in the forward mode, v_i is imposed by the grid voltage respecting n and f_s is equal to the backward resonance frequency f_{rb}. Then, when the Boost converter tries to drain current from C_i energy is transferred from grid to battery. Considering that C_i is large enough, allowing

only a small voltage ripple in the intermediate bus, the dynamics of both stages are decoupled and analysis can be done independently, as presented next.

IV. THEORETICAL ANALYSIS

Since the converter is dual and the modules are independent, all analysis and design is performed for one line as illustrates in Fig. 2. Besides, with symmetrical behavior, analysis can be done only in one direction (forward in this case), taking the necessary considerations for backward mode. In order to have the same characteristic in both directions resonant frequencies must be the same ($f_r = f_{rf} = f_{rb}$). In both modes energy delivery occurs twice in a switching cycle.

A. LLC converter

In forward mode, switch pairs (S_1, S_4) and (S_2, S_3) are controlled with complementary pulses at 50% duty-cycle and $f_s = f_r$. Switch pairs (S_6, S_7) and (S_5, S_8) are turned OFF and intrinsic diodes of these pairs (D_{S6}, D_{S7}, D_{S5}, and D_{S8}) perform the rectification. The first cycle happens when the resonant tank is excited with a positive voltage. Hence, the current resonates in the positive direction in the first half of the switching period T_s. The second cycle is the same but in negative direction. Neglecting the dead time (t_d) in the switching legs, there are two equivalent circuits (see Ref. [3]). The main waveforms for the ideal operation are depicted in Fig. 3.

At $t = t_0 = 0$ switch pairs (S_2, S_3) are turned OFF while (S_1, S_4) are turned ON. As the current through the resonant tank is negative, it starts to circulate through the intrinsic diodes of S_1 and S_4 (D_{S1}, D_{S4}). The primary of the transformer acts as a voltage source of nV_o and the magnetizing current i_{Lm} is charging L_m. The difference between the resonant current i_{Lrf} and i_{Lm} passes through the transformer, making diodes D_{S7} and D_{S6} conduct the current $i_D = i_o$, which delivers energy to the load (grid).

When the resonant current cross zero (at $t = t_1$), becoming positive, the current circulating through diodes D_{S1} and D_{S4} is switched to S_1 and S_4 with ZVS. In the rest of the circuit nothing changes and this interval finishes in half of the resonant time (at $t = t_2$) when the switch pair (S_1, S_4) is commanded to turn OFF. In the secondary side the current becomes zero and D_{S7} and D_{S6} stop conducting with ZCS. During the second half

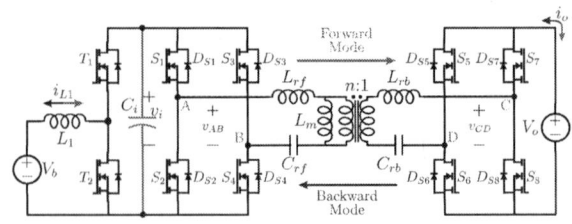

Fig. 2. Proposed converter - one line simplification.

1448

The 2018 International Power Electronics Conference

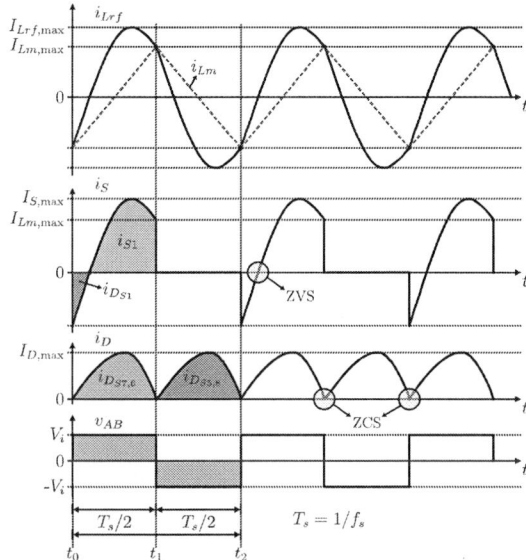

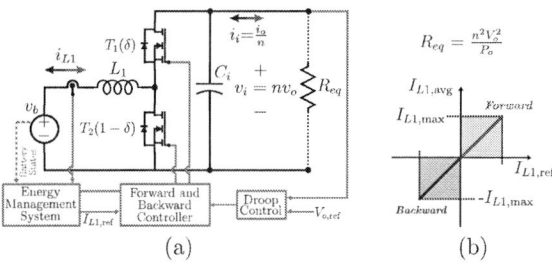

Fig. 4. (a) Synchronous Boost converter with control structure and (b) graphic behavior of the current through L_1 versus reference current.

Fig. 3. Main operational waveforms for $f_s = f_r$.

of the resonant cycle the operation is the same but with the opposite pairs of semiconductors.

The analytical discussion of the operation is rather complex. Though, the first harmonic approximation (FHA) model [22], [23], is used. The FHA is sufficient accurate in this application since the operation is restrict to the load independent point. Therefore, the converter can be ideally seen as a conventional unidirectional LLC with gain relation is expressed by

$$G = \frac{1}{n} \frac{f_n^2 (m-1)}{\sqrt{Q^2 f_n^2 (f_n^2 - 1)^2 (m-1)^2 + (f_n^2 m - 1)^2}} \quad (1)$$

where $m = \frac{L_m + L_r}{L_r}$ is the ratio of total primary inductance L_m to resonant inductance L_{rf}, $f_n = \frac{f_s}{f_r}$ is the parametrized frequency and Q is the load factor, also known as quality factor and it is given by

$$Q = \frac{1}{R_{ac}} \sqrt{\frac{L_{rf}}{C_{rf}}} \quad (2)$$

with R_{ac} in (2) representing the output load reflected to the ac side of the transformer as

$$R_{ac} = \frac{8n^2 V_o^2}{\pi P_o}. \quad (3)$$

B. Synchronous Boost converter

The LLC operating in the load independent point can be seen as a "dc-transformer" with $n : 1$ ratio. Hence, the converter placed between the battery bank and this dc-transformer is actually the responsible to processes and control the energy flow in forward or backward direction. The optimal choice depends on the specification, expected efficiency and desired features such as load independent characteristic. The simplest and most attractive option

is the synchronous dc-dc Boost converter, as shown in Fig. 4(a).

The operation in forward mode is the same as in a conventional Boost converter whereas in backward mode it is like in a Buck converter. The transition between modes is performed by a current mode controller. The reference current $I_{L1,\text{ref}}$ is provided by the energy management system which takes decision of the energy flow direction and level of power according to battery status, current through the inductor and grid voltage. The last one is provided by the smart system or can be obtained directly across the capacitor C_i, not compromising the isolation. Figure 4(b) exemplifies how the average current $I_{L1,\text{avg}}$ through the inductor behaves. If $I_{L1,\text{ref}}$ is positive converter operates in forward mode, while negative values of $I_{L1,\text{ref}}$ indicate backward mode. By doing in this way, charging and discharging processes happen naturally, avoiding severe transients during mode change. The sign of $I_{L1,\text{ref}}$ is also used to determine which bridge of the LLC converter is active, changing the mode accordingly. Considering continuous conduction mode, battery voltage V_b, and the control action δ over T_1, then the synchronous Boost gain (G_{SB}), main voltage and current relations as a function of δ are given by

$$G_{SB} = \frac{V_i}{V_b} = \frac{nV_o}{V_b} = \frac{I_{L1}}{I_i} = \frac{nI_{L1}}{I_o} = \frac{1}{\delta}. \quad (4)$$

V. DESIGN CRITERIA

The design of the LLC converter follows similar procedures found in the literature [7], [11], [12], [19], [21], [24]–[27]. The main aspect of the proposed design is the graphical analysis of the gain, which is a simple and efficient method to determine the LLC parameters. These values are defined using the forward mode and considering the model presented in Fig. 5. The ac resistances R_f and R_b, that combines all the parasitic resistances in each path, are neglected in the analysis. The first step consists in determining the m factor in order to obtain a flat gain curve around the unitary gain for different Q values. The main advantages of smaller m are the possibility for higher gains and the use of narrower switching frequency band. Although, none of this features have relevance in this application. Hence, the increment of m brings the maximum gain of the curves close to the unity and increases the flat area (wider band)

1449

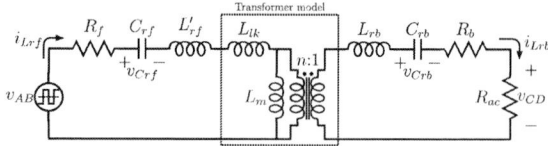

Fig. 5. LLC equivalent circuit model.

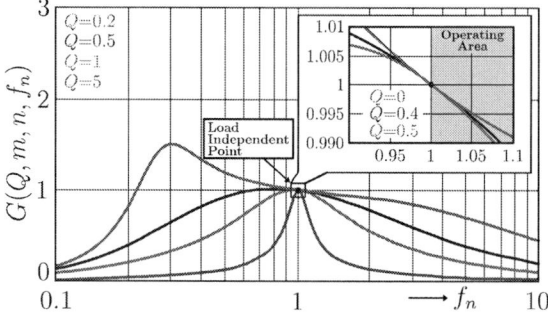

Fig. 6. LLC gain curves in function of f_n for $m=20$ and $n=1$.

around the unitary gain. Figure 6 shows the expected characteristic after selecting a larger m value. In this example $m=20$, and as it can be seen in the zoomed detail around the unitary gain, variations about 10% in f_s causes a minor modification in the gain. This is desired since the resonant frequencies are slightly different in forward and backward directions and also can change with time due the parasitics and variation of component values. Another detail presented in Fig. 6 is the operational area (shaded), which implies that $f_r < f_s \leq 1.1 f_r$ in order to satisfy gain and ZVS features.

The second parameter to be chosen is Q, which by the graphic illustration, must be smaller than 0.5 in order to respect the flat area. Another impact of Q is the capacitor value, that is obtained from equation (2). The smaller is Q the larger is C_{rf}, and despite the desired small value of C_{rf}, it must be chosen in order to also satisfy the maximum rms value of the sinusoidal ac voltage in the resonant frequency. This value is the main constraint of this topology since capacitors have low ac limits for high frequencies and it is given by

$$V_{Crf,\mathrm{rms}} = Z_{rf} I_{Crf,\mathrm{rms}} \qquad (5)$$

where Z_{rf} is the resonant tank impedance $Z_{rf} = \sqrt{\frac{L_{rf}}{C_{rf}}}$. Hence, a good design using film or ceramic capacitors is necessary or the resonant network could use a different symmetric approach to overcome this limitation, such as presented in [9]. The rms value of i_{Crf} is found to become

$$I_{Crf,\mathrm{rms}} = I_{Lrf,\mathrm{rms}} = \frac{2\sqrt{2}V_i}{\pi R_{ac}}\sqrt{\frac{R_{ac}^2}{4\pi^2 L_m^2 f_r^2} + 1}. \quad (6)$$

After defining C_{rf}, the value of L_{rf} is obtained using

$$L_{rf} = \frac{1}{C_{rf}(2\pi f_r)^2} \qquad (7)$$

while L_m is calculated using $L_m = L_{rf}(m-1)$. It is suggested to design L_m at least 10 times larger than L_{rf} in order to easily integrate L_{rf} into the transformer as part of the leakage inductance L_{lk}. The final value of the resonant inductor placed outside the transformer must respect $L'_{rf} = L_{rf} - L_{lk}$. The rms value of i_{Lrf} is calculate using Eq. (6) while the maximum current in the magnetizing inductor is calculate using

$$I_{Lm,\max} = \frac{nV_o}{4L_m f_r}. \qquad (8)$$

This value is important for the maximum turn OFF current stress in the switches, reduction of circulating current, adn also relevant for the definition of the minimum dead time t_d or maximum L_m value according to a predefined t_d as

$$t_{d,\min} = 8C_j L_m f_r = \frac{2nV_o C_j}{I_{Lm,\max}} \qquad (9)$$

where C_j is the sum of the junction capacitance of the switch S and stray capacitances. In order to achieve ZVS, bridge capacitances have to be charged or discharged within the dead time to reverse the bridge voltage.

Once the components in the primary side of the transformer are dimensioned the secondary side capacitor is defined by $C_{rb} = n^2 C_{rf}$. Using the transformer model shown in Fig. 5, the resonant inductor L_{rb} is always placed outside and it is calculated using Eq. (7), by replacing the value of C_{rf} for C_{rb}. This model does not integrated all the resonant magnetics in one component, however, it allows more flexibility in the design and adjustments, specially when the transformer turns ratio is not unitary.

The synchronous boost converter is designed using the same switching frequency as the LLC. The main parameter is the Boost inductor, calculated using

$$L_1 = \frac{V_b(V_i - V_b)}{V_i f_s \Delta I_{L1}} = \frac{V_b(nV_o - V_b)}{nV_o f_s \Delta I_{L1}} \qquad (10)$$

where ΔI_{L1} is the inductor current ripple, chosen to be around 10% to 20% of the maximum average value of the inductor current.

VI. SPECIFICATIONS AND RESULTS

In order to demonstrate the concept and principle of operation a prototype was developed according to the specifications shown in Table I, which also shown the main calculated parameters. One of the converter modules, highlighting the main components, is shown in Fig. 7. The converter was designed to be suitable for two dc grids with nominal voltages of ±350 V and ±380 V. Hence, considering a 20 V fluctuation, the minimum grid voltage is equal to 330 V while the maximum is 400 V.

Taking into account that the Boost converter has an input voltage from the battery bank that varies from 315 V to 567 V, the intermediate bus, which interconnect both stages, was defined to have a minimum voltage of $V_{i,\min} = 600$ V when the grid voltage $V_{o,\min} = 350 - 20 = 330$ V. Transformer relation

The 2018 International Power Electronics Conference

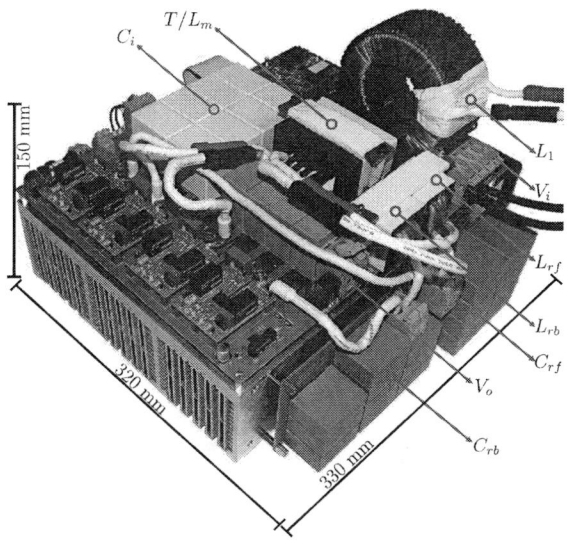

Fig. 7. Photo of the developed prototype (one module) indicating the main passive components. The converter height indicated does take into account the input inductor since this element will be placed in a different location when both modules are connected.

TABLE I. SPECIFICATIONS AND CALCULATED PARAMETERS

Parameter	Value
Rated power (P_o)	15 kW per line
Grid voltage (V_o)	± 350 V and ± 380 V
Grid voltage fluctuation (ΔV_o)	± 20 V
Battery voltage range (V_b)	315 V to 567 V
Stored energy	20 kWh
Switching frequency (f_s)	100 kHz
Boost inductance (L_1)	375 μH
Intermediate capacitor (C_i)	80 μF
Transformer turns ratio (n)	1.78
Resonant capacitor (primary side) (C_{rf})	300 nF
Resonant capacitor (secondary side) (C_{rb})	1000 nF
Resonant inductance (primary side) (L_{rf})	8.44 μH
Resonant inductance (secondary side) (L_{rb})	2.53 μH
Magnetizing inductance (L_m)	167 μH

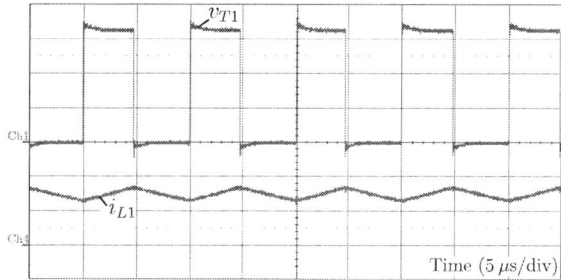

Fig. 8. Experimental results in forward mode for $P_o \approx 5$ kW. Voltage across T_1 (200 V/div), current through L_1 (10 A/div))

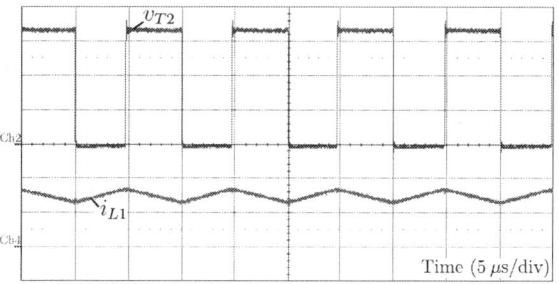

Fig. 9. Experimental results in forward mode for $P_o \approx 5$ kW. Voltage across T_2 (200 V/div), current through L_1 (10 A/div)).

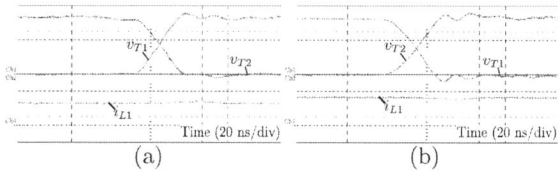

Fig. 10. Experimental results in forward mode for $P_o \approx 5$ kW. Voltages across T_1 and T_2 (200 V/div), and L_1 current (10 A/div) during the (a) turn ON and (b) turn OFF moments of T_2.

is calculated as $n = \frac{V_{i,\min}}{V_{o,\min}}$. According to the design criteria, $m = 20$ and $Q = 0.27$ were selected, then components are defined based in the desired resonant frequency $f_r = f_s = 100$ kHz.

The transformer is designed to have the desired magnetizing inductance, while the resonant inductors are designed and placed outside following the design criteria. The values of L_{rf} and L_{rb} were adjusted changing the air gap in order have the desired resonant frequency. For the semiconductors, both the bidirectional Boost and LLC full-bridges were implemented with 1.2 kV SiC MOSFETS and a maximum dead time of 120 ns.

A. Experimental Results

In order to confirm the working principle, after the resonant cell adjustment, the converter was tested in open loop in both directions of energy transport using resistive loads. At this moment the maximum power achieved was around 5 kW in forward mode and 7 kW in backward mode, limited by the available laboratory supply and resistive loads. However, the results are enough to validate

the topology and further tests will be performed in closed loop using a demonstration grid in order to operate in the maximum power.

1) Forward mode: The switches of the synchronous boost converter are the weakest link of the proposed converter since they operate with hard switching being susceptible to higher losses. Although, by choosing SiC MOSFETs and a making an adequate layout this issue is minimized as can be visualized in the results shown in Fig. 8 and Fig. 9. These results present the voltages across switches T_1 and T_2 together with the current through the inductor L_1. A closed view of the switching moments, considering T_2 as reference, can be seen in Fig. 10. The results are satisfactory presenting a low overvoltage and a relatively short switching time, minimizing losses.

The voltage in the intermediate bus is around twice the value of the input since the duty cycle of the synchronous Boost is 50% for this test. This dc voltage is inverted for the first full bridge and the squared ac voltage is applied to the resonant cell. The voltage between the nodes A and B

1451

The 2018 International Power Electronics Conference

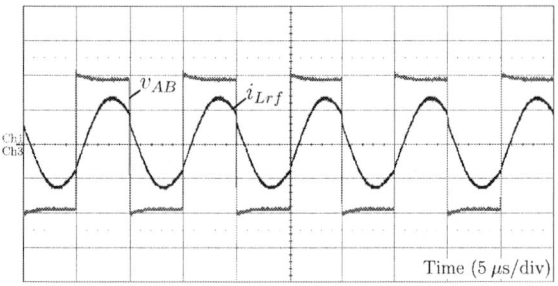

Fig. 11. Experimental results in forward mode for $P_o \approx 5$ kW. Voltage across nodes A and B (350 V/div) and current through the resonant inductor L_{rf} (10 A/div).

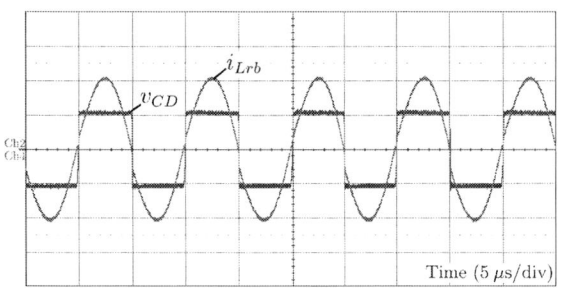

Fig. 12. Experimental results in forward mode for $P_o \approx 5$ kW. Voltage across nodes C and D (350 V/div) and current through the resonant inductor L_{rb} (10 A/div).

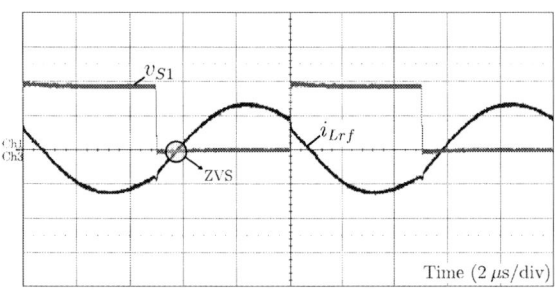

Fig. 13. Experimental results in forward mode for $P_o \approx 5$ kW. Detail of the voltage across S_1 (350 V/div) and current through the resonant inductor L_{rf} (10 A/div).

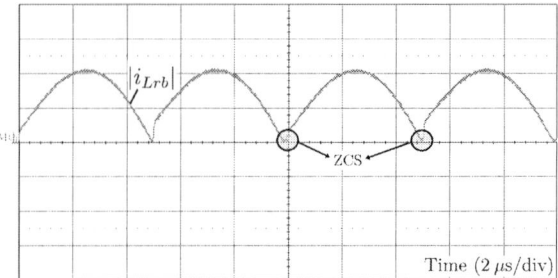

Fig. 14. Experimental results in forward mode for $P_o \approx 5$ kW. Current (10 A/div) through the diodes of the output bridge.

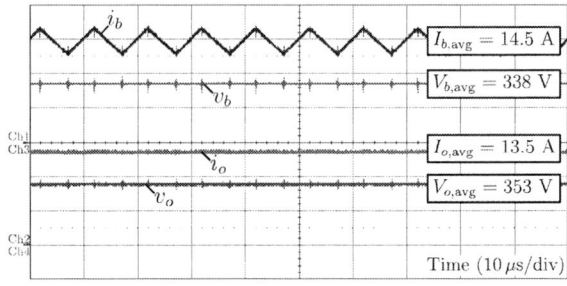

Fig. 15. Experimental results in forward mode for $P_o \approx 5$ kW. Input voltage (200 V/div), input current (5 A/div), output voltage (200 V/div), and output current (5 A/div).

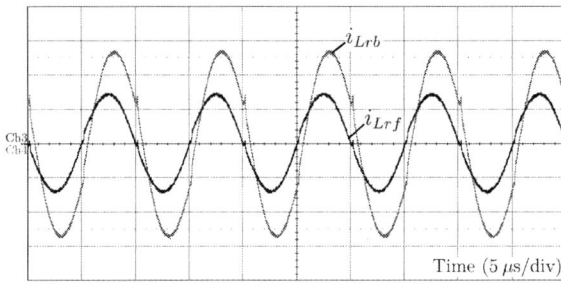

Fig. 16. Experimental results in backward mode for $P_o \approx 5$ kW. Currents through the forward and backward resonant inductors (10 A/div).

(as indicated in Fig. 2) together with the current through the resonant inductor L_{rf} is presented in Fig. 11 while the voltage in the output of the resonant cell together with the current through the secondary side resonant inductor L_{rb} is shown in Fig. 12.

As can be seen, the resonant frequency is very close to the switching frequency, confirming the expected operation in the load independent point reproducing the ideal results presented in Fig. 3. The switches of the first bridge turn ON with ZVS while the intrinsic diodes of the switches in the second bridge turn OFF with ZCS, as expected. These two features are detailed in Fig. 13 and Fig. 14. Figure 15 shows the voltages and currents at the input and output at the same time. The results are

in good agreement and the estimated converter efficiency (η) at this point is around 96%.

2) Backward mode: The results in backward were first obtained in 5 kW and, as expected, the behavior was the same as in forward mode and the waveforms have the same format as can be seen in the currents through the resonant inductors present in Fig. 16. The available load resistors allow this operation to be performed with more power, although the results are basically the same as shown in Fig. 17. It presents the current through the backward inductor L_{rb} and the voltage across the backward capacitor C_{rb}. One particular behavior, noticed in the results in both forward and backward modes, is that the converter is working slightly in discontinuous conduction mode (considering the rectified current) as highlighted in Fig. 14 and Fig. 16. This mainly happens

The 2018 International Power Electronics Conference

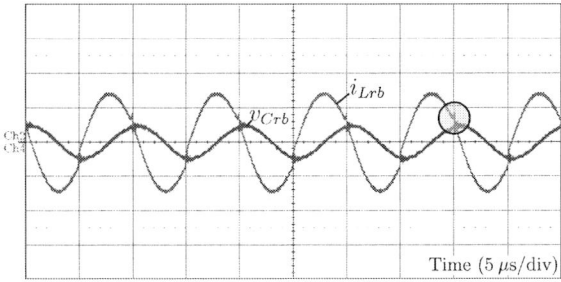

Fig. 17. Experimental results in forward mode for $P_o \approx 7$ kW. Current through L_{rb} (20 A/div) and voltage across C_{rb} (100 V/div).

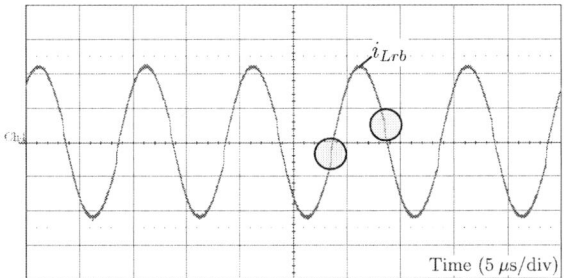

Fig. 18. Experimental results in backward mode to verify the influence of the resistance in the conduction mode. Current through L_{rb} (5 A/div).

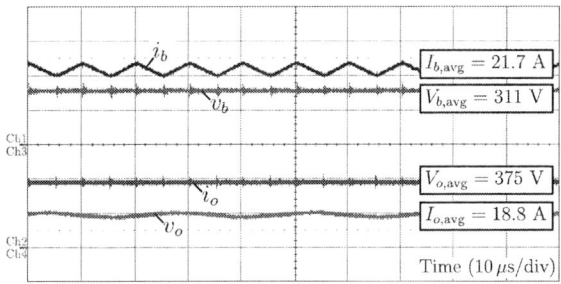

Fig. 19. Experimental results in backward mode for $P_o \approx 7$ kW. Voltage V_b (200 V/div), Current i_b (10 A/div), voltage v_o (200 V/div), and current i_o (20 A/div).

TABLE II. C3LC GAINS IN FORWARD AND BACKWARD MODES FOR DIFFERENT RESISTIVE LOADS

Forward Mode:

R_o (Ω)	V_o (V)	V_i (V)	n
26	357.6	648.9	1.814
52	361.2	651.6	1.804
89.8	364.0	653.8	1.796

Backward Mode:

R_o (Ω)	V_o (V)	V_i (V)	n
14.3	320.1	552.8	1.727
21.2	360.2	628.3	1.744
46.6	380.1	670.1	1.763

because of the load resistance used for the tests and the operation in the load independent point. This result is expect and as far this region is small it does not affect the working principle. Using a smaller value of resistance the converter starts to operate with a slightly inductive behavior, as also expected (see Fig. 6) and demonstrated in the result presented in Fig. 18. The precise model to describe this dependency of the resistive load is not present in this paper since the intention is to use the grid/battery as load/source, where this effect does not take place and the converter operates as described by the model used in this work.

Figure 19 presents the voltages and currents at the input and output at the same time. The results are in good agreement and the estimated converter efficiency in this mode is around 96% as well. A more precise efficiency measurement and efficiency curves will be taken in a future work when operating in closed loop and connected to the grid.

3) Gain: Considering only the LLC converter, expected to operate in the load independent point, the gain related to the intermediate bus voltage V_i and grid voltage V_o should be equal to the transformer gain. Table II shows measurements of the voltages V_i and V_o for different resistive loads and the calculate n for each case. As can be notice, the gains has a minor difference between the measurements in each mode, confirming the operation in the load independent point. Although, there is a contrast between forward and backward mode and the measured gains are different than the turns ration relation which is

1.78. The main explanation for this is the ac values of the impedances in the resonant current path, which are different in each side (see model in Fig. 5). These minor errors are not a problem if the droop control is performed by looking the grid voltage. But in case the intermediate bus is used, then the gain in the control loop must me modified when the converter change mode.

VII. CONCLUSION

This paper proposes the interconnection of a battery energy storage bank to a two-line dc grid network by means of a dual two-stage isolated bidirectional dc-dc converter. The converter is composed of a synchronous Boost cascade with an isolated bidirectional C3LC resonant converter. The C3LC operates with constant duty-cycle and fixed switching frequency equal to the resonant frequency. Doing so, it allows switches to turn ON witch ZVS and rectification diodes to turn OFF with ZCS features in the full power range. In this operation the C3LC naturally reflects the grid voltage to the intermediate buffer capacitor with a gain relation equal to transformers turns ratio. The energy is transferred forward and backwards by controlling the current through the Boost converter inductor. Besides the concept idea, essential analysis, and main design procedure, the paper shows experimental results that demonstrated the operation in good agreement. The developed converter has the flexibility to be applied in two different grid sites (± 350 V_{dc} and ± 380 V_{dc}). Up to this moment, results were obtained in open loop for power levels around 6 kW using resistive loads. Next steps consist in the practicum implementation of the control loop and tests using bidirectional supplies in both sides of the converter and finally the application to a dc microgrid site under development.

1453

ACKNOWLEDGMENT

The DCSMART project has been funded by the ERA-NET Smart Grids Plus Under Horizon 2020, supported by the European Commission Research & Innovation Directorate-General. Moreover, the partner institutions involved in the DCSMART project would like to express their gratitude towards the Netherlands Organization for Scientific Research (NWO) for support and funding the project.

REFERENCES

[1] B. Wunder, L. Ott, J. Kaiser, Y. Han, F. Fersterra, and M. Mrz, "Overview of different topologies and control strategies for dc micro grids," in *Proc. IEEE 1th Int. Conf. on DC Microgrids*, June 2015, pp. 349–354.

[2] B. Wunder, L. Ott, M. Szpek, U. Boeke, and R. Wei, "Energy efficient dc-grids for commercial buildings," in *Proc. IEEE 36th Int. Telecommun. Energy Conf.*, Sept 2014, pp. 1–8.

[3] G. Tibola and J. L. Duarte, "Isolated bidirectional dc-dc converter for interfacing local storage in two-phase dc grids," in *Proc. IEEE 8th Int. Symp. on Power Electron. for Distributed Generation Systems*, April 2017, pp. 1–8.

[4] E. Camacho-Vargas, V. Venegas-Rebollar, and E. L. Moreno-Goytia, "Novel closed loop control of a dab converter for charge/discharge process of ev batteries," in *Proc. IEEE 13th Int. Conf. on Power Electron.*, June 2016, pp. 356–361.

[5] R. W. D. Doncker, D. M. Divan, and M. H. Kheraluwala, "A three-phase soft-switched high power density dc/dc converter for high power applications," in *Proc. IEEE Ind. Appl. Soc. Annu. Meeting*, vol. 1, Oct 1988, pp. 796–805.

[6] B. Hu, X. Zhang, L. Fu, H. Li, C. Yao, Y. Wang, Y. M. Abdullah, and J. Wang, "Comparison study of llc resonant circuit and two quasi dual active bridge circuits," in *Proc. 4th Workshop on Wide Bandgap Power Devices and Appl.*, Nov 2016, pp. 35–41.

[7] A. Hillers, D. Christen, and J. Biela, "Design of a highly efficient bidirectional isolated llc resonant converter," in *Proc. 15th Int. Power Electron. and Motion Control Conf.*, Sept 2012, pp. DS2b.13–1–DS2b.13–8.

[8] S. Hu, J. Deng, C. Mi, and M. Zhang, "Llc resonant converters for phev battery chargers," in *Proc. IEEE 28th Annu. Appl. Power Electron. Conf. and Expo.*, March 2013, pp. 3051–3054.

[9] M. Kim, S. Noh, and S. Choi, "New symmetrical bidirectional l3c resonant dc-dc converter with wide voltage range," in *Proc. IEEE Appl. Power Electron. Conf. and Expo.*, March 2016, pp. 859–863.

[10] J. F. Lazar and R. Martinelli, "Steady-state analysis of the llc series resonant converter," in *Proc. IEEE 16th Appl. Power Electron. Conf. and Expo.*, vol. 2, 2001, pp. 728–735 vol.2.

[11] A. Abramovitz and S. Bronshtein, "A design methodology of resonant llc dc-dc converter," in *Proc. 14th Eur. Conf. on Power Electron. and Appl.*, Aug 2011, pp. 1–10.

[12] H. P. Park and J. H. Jung, "Power stage and feedback loop design for llc resonant converter in high switching frequency operation," *IEEE Trans. on Power Electron.*, no. 99, pp. 1–1, 2016.

[13] J. Font and L. Martinez, "Modelling and analysis of a bidirectional boost converter with output filter," in *Proc. 6th Mediterranean Electrotechnical Conf.*, May 1991, pp. 1380–1383 vol.2.

[14] A. A. Khan, H. Cha, and H. F. Ahmed, "A family of high efficiency bidirectional dc-dc converters using switching cell structure," in *Proc. IEEE 8th Int. Power Electron. and Motion Control Conf.*, May 2016, pp. 1177–1183.

[15] Z. U. Zahid, Z. M. Dalala, R. Chen, B. Chen, and J. S. Lai, "Design of bidirectional dc-dc resonant converter for vehicle-to-grid (v2g) applications," *IEEE Trans. on Transport. Electrific.*, vol. 1, no. 3, pp. 232–244, Oct 2015.

[16] T. Jiang, J. Zhang, X. Wu, K. Sheng, and Y. Wang, "A bidirectional llc resonant converter with automatic forward and backward mode transition," *IEEE Trans. on Power Electron.*, vol. 30, no. 2, pp. 757–770, Feb 2015.

[17] K. Tan, R. Yu, S. Guo, and A. Q. Huang, "Optimal design methodology of bidirectional llc resonant dc/dc converter for solid state transformer application," in *Proc. IEEE 40th Annu. Ind. Electron. Soc. Conf.*, Oct 2014, pp. 1657–1664.

[18] W. Chen, P. Rong, and Z. Lu, "Snubberless bidirectional dc-dc converter with new cllc resonant tank featuring minimized switching loss," *IEEE Trans. on Ind. Electron.*, vol. 57, no. 9, pp. 3075–3086, Sept 2010.

[19] G. Pledl, M. Tauer, and D. Buecherl, "Theory of operation, design procedure and simulation of a bidirectional llc resonant converter for vehicular applications," in *Proc. IEEE Vehicle Power and Propulsion Conf.*, Sept 2010, pp. 1–5.

[20] J. H. Jung, H. S. Kim, M. H. Ryu, and J. W. Baek, "Design methodology of bidirectional cllc resonant converter for high-frequency isolation of dc distribution systems," *IEEE Trans. on Power Electron.*, vol. 28, no. 4, pp. 1741–1755, April 2013.

[21] Z. Lv, X. Yan, Y. Fang, and L. Sun, "Mode analysis and optimum design of bidirectional cllc resonant converter for high-frequency isolation of dc distribution systems," in *Proc. IEEE Energy Convers. Congress and Expo.*, Sept 2015, pp. 1513–1520.

[22] T. Duerbaum, "First harmonic approximation including design constraints," in *Proc. 20th Int. Telecommun. Energy Conf.*, 1998, pp. 321–328.

[23] G. Ivensky, S. Bronshtein, and A. Abramovitz, "Approximate analysis of resonant llc dc-dc converter," *IEEE Trans. on Power Electron.*, vol. 26, no. 11, pp. 3274–3284, Nov 2011.

[24] B. Lu, W. Liu, Y. Liang, F. C. Lee, and J. D. van Wyk, "Optimal design methodology for llc resonant converter," in *Proc. IEEE 21th Annu. Appl. Power Electron. Conf. and Expo.*, March 2006, pp. 6 pp.–.

[25] J. Marquart, S. Nigsch, and K. Schenk, "Design optimization for a high power-density, wide output, high frequency llc resonant converter for lighting applications," in *Proc. Int. Exhibition and Conf. for Power Electron., Intell. Motion, Renewable Energy and Energy Manag.*, May 2016, pp. 1–9.

[26] J. Deng, S. Li, S. Hu, C. C. Mi, and R. Ma, "Design methodology of llc resonant converters for electric vehicle battery chargers," *IEEE Trans. on Veh. Technol.*, vol. 63, no. 4, pp. 1581–1592, May 2014.

[27] X. Zhang, W. You, W. Yao, S. Chen, and Z. Lu, "An improved design method of llc resonant converter," in *Proc. IEEE Int. Symp. on Ind. Electron.*, May 2012, pp. 166–170.

The 2018 International Power Electronics Conference

Modular Multilevel Converter With Capacitor Voltage Self-balancing Using Reduced Number of Voltage Sensors

Taiyuan Yin[1], Yue Wang[1*], Xiaolei Wang[2], Shiyuan Yin[1], Shumin Sun[3] and Guanglei Li[3]

1 State Key Laboratory of Electrical Insulation and Power Equipment, Xi'an Jiaotong University, Xi'an, China
2 School of Electric and Information Engineer, Zhongyuan University of Technology, Zhengzhou, China
3 State Grid Shandong Electric Power Research Institute, Jinan, China
*E-mail: davidfusion@163.com

Abstract—Modular multilevel converter (MMC) has been widely used in high-voltage direct current (HVDC) transmission system due to its advantages of modularity, low switching frequency and excellent output voltage waveforms. Proper operation of MMC is based on capacitor voltage balancing. Conventional MMC capacitor voltage balancing control needs a large number of voltage sensors and sorting algorithm which increases computing burden on the processor. In this paper, a capacitor voltage self-balancing MMC with DC fault ride-through capability is proposed. Capacitor voltage balancing control of this proposed MMC is significantly simple due to its self-balancing auxiliary circuit. Sorting algorithm is not needed, each phase leg only needs one voltage sensor for top SM. So control complexity and computing burden are reduced greatly. Furthermore, DC fault ride-through capability is also reserved. Simulation results validates capacitor voltage of this new MMC topology can be well balanced and DC fault ride-through capability is effective.

Keywords—MMC; capacitor voltage balancing; voltage sensors; DC fault ride-through.

I. INTRODUCTION

MMC-HVDC has attracted a lot of research attention especially after the Trans Bay Cable project in America was accomplished [1]. Due to its advantages of modularity, low switching frequency and excellent output voltage waveforms, MMC has been widely used in high-voltage direct current (HVDC) transmission system [2]. Capacitor voltage balancing is the base of proper operation of MMC. Conventional MMC capacitor voltage balancing control needs voltage information of every SM (sub-module) and sorts them to determine which SM should insert or bypass. MMC used in HVDC always has hundreds of SM to build high voltage. The quantity requirement of voltage sensors is huge and sorting algorithm is very complex due to a large number of SMs in MMC-HVDC [3].

To reduce the computing burden of processor, some researchers improve the capacitor voltage balancing algorithm. Reference [4] divides sub-modules in same

arm into several groups to save sorting time. Reference [5] proposed a grouping-sorting-optimized MPC (GSOMPC) strategy to to reduce the strict requirements of control hardware for sorting and calculation.

Besides improving capacitor voltage balancing algorithm, some researchers proposed several novel MMC topologies which have SM voltage self-balancing capability. These novel MMC topologies don't need voltage information of every SM so that they reduce voltage sensors and computing burden of processor. SM capacitor voltage balancing is realized by auxiliary circuits. Reference [6] proposed a MMC topology with clamped diodes to reduce voltage sensors, however, it needs several transformers at the same time. Reference [7] proposed a new MMC topology with capacitor voltage self-balancing ability, auxiliary voltage-balancing circuit are needed between phases. A MMC topology with clamped diodes was proposed in [8], it only needs six voltage sensors to realize the voltage balancing. However, these topologies don't have DC-fault ride-through ability because of auxiliary diodes and the complex modulation algorithm is not suitable for HVDC-MMC because of too many modules.

This paper proposes a new MMC converter topology with capacitor voltage self-balancing ability, furthermore, this topology and control is very simple with only three voltage sensors. This topology has DC fault ride-through ability compared to the topologies proposed in [6]-[8], which is very important in HVDC transmission system.

This proposed new topology MMC only needs capacitor voltage information of three top SM in each phase so it reduces a large number of voltage sensors. This paper also proposes a quite simple method to keep the sub-module voltage balancing, which reduces the control complexity and computing burden greatly. The effectiveness of this topology and DC fault ride-through ability is validated by simulation.

This work is supported by Science and Technology Project of Shandong Electric Power Company of SGCC (SGSDDK00KJJS1600143)

II. TOPOLOGY ANALYSIS

Similar to conventional MMC, the structure of proposed self-balancing MMC is shown in Fig. 1.

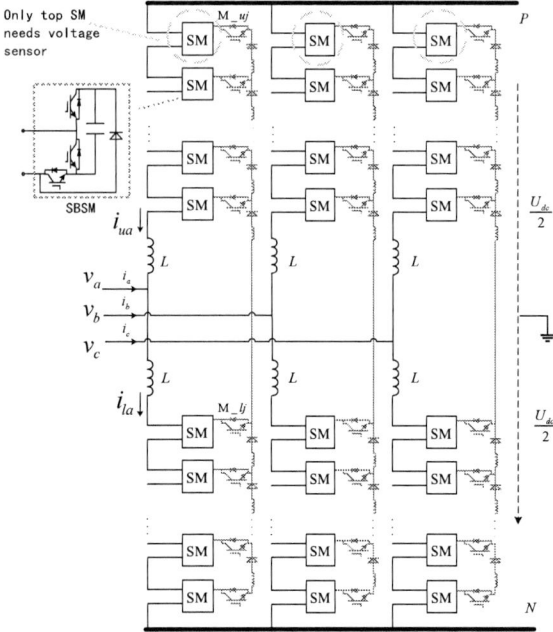

Fig. 1. Structure of proposed self-balancing MMC.

On the grid side, $v_j (j = a,b,c)$ are the alternating-current grid voltage. The proposed converter consists of three phase legs, each containing one upper arm, one lower arm and two arm inductors . Each arm consists of n SMs. The outer terminals of arms are connected to the DC bus. Specially, the adjacent SMs are connected through an auxiliary voltage-balancing circuit including an IGBT, a diode and a small inductor. Design of the small inductor is discussed in [8].

To keep this topology having the DC fault ride-through ability, IGBT is intentionally introduced into auxiliary voltage-balancing circuit between adjacent SMs and SBSM (self-blocking sub-module) [9] are introduced into in this converter. If DC fault ride-through ability is not needed, the IGBT in auxiliary circuit is not needed, either. Auxiliary circuit between adjacent SM can only consist of a diode and a small inductor. At the same time, the SMs which don't have DC fault ride-through ability can also be introduced into this topology. The self-balancing topology is able to simplify as Fig. 5, although simplified topology as Fig. 5 doesn't have DC fault ride-through ability, the SM capacitor voltage self-balancing ability is still reserved.

Detailed connection between SMs and the auxiliary voltage-balancing circuit in Fig.1 is shown in Fig. 2.

When this new MMC converter topology operates in steady state, auxiliary IGBT M _uj (j=1…n) of upper arm (u represents upper arm) and auxiliary IGBT M _lj (l=1…n) of lower arm (l represents lower arm) are always ON. When the voltage of Cuj is larger than the voltage of $Cu(j-1)$ (j=1…n), there will be a current flowing through M _uj to $Cu(j-1)$ when S _$uj2$ is on (neglecting the voltage drop of diode) as shown in Fig. 2, then the voltage of $Cu(j-1)$ will rise and voltage of Cuj will drop until they are equal.

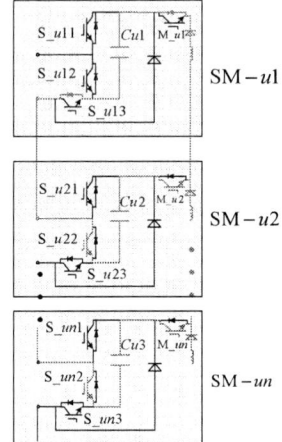

Fig. 2. Self balancing structure of phase leg.

The voltage of SMs in lower arm will have the same relationship as upper arm. So the voltage relation of SMs in upper arm and lower arm is shown as below:

$$\begin{cases} u_{Cu1} \geq u_{Cu2} \geq ... \geq u_{Cun} \\ u_{Cl1} \geq u_{Cl2} \geq ... \geq u_{Cln} \end{cases} \quad (1)$$

In equation (1), u_{Cuj} (j=1…n) represents the voltage of SMj in upper arm, and u_{Cuj} (j=1…n) represents the voltage of SMj in lower arm. As shown in Fig. 1, there is also an auxiliary voltage-balancing circuit between SMn of upper arm and SM1 of lower arm same as other SMs. So $u_{Cun} \geq u_{Cl1}$ when the circulating circuit of MMC is suppressed and the voltage between two inductors of upper arm and lower arm is zero. So equation (2) is obtained as below:

$$u_{Cu1} \geq u_{Cu2} \geq ... \geq u_{Cun} \geq u_{Cl1} \geq u_{Cl2} \geq ... \geq u_{Cln} \quad (2)$$

The DC bus voltage of MMC U_{DC} keeps steady when MMC operates properly. So voltage of SM must meet the relation as equation (3)

$$u_{Cu1} + u_{Cu1} + ... + u_{Cun} + u_{Cl1} + u_{Cl2} + ... + u_{Cln} = 2U_{DC} \quad (3)$$

According to (2) and (3), if u_{Cu1} can be kept as

1456

U_{DC}/n, then the voltage of each SM have to be equal to U_{DC}/n. Thus all SMs capacitor voltage could keep balancing.

To keep all SMs capacitor voltage balancing, only the voltage of the top SM (SM1 of upper arm) of each phase leg are needed to control to U_{DC}/n. Thus only three voltage sensors are needed for the top SM of each phase leg to keep this MMC converter operating well.

III. OPERATION PRINCIPLES

Power control of this new MMC topology uses conventional double loop control as Fig. 3 [10].

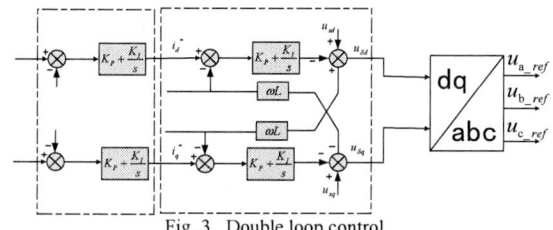

Fig. 3. Double loop control.

u_{j_ref} (j=a,b,c) is the modulation reference signal of each phase which is obtained from the conventional double loop control. To simplify the capacitor voltage control and reduce computing burden of processor, NLM (Nearest Level Modulation) is introduced to this MMC SM voltage balancing control.

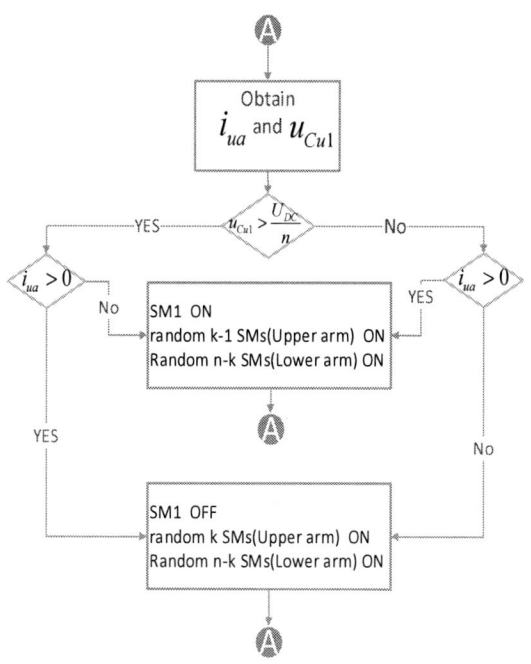

Fig. 4. Voltage balancing control.

Detailed operation principles are shown in Fig. 4. Taking voltage of top SM (SM1) of phase A as an

example to control the capacitor voltage balancing in phase A, the capacitor voltage balancing control of other two phases are the same. The number of needed inserted SM of each arm is obtained from NLM. To keep all SM voltages are equal to U_{DC}/n, firstly SM1 of each phase leg must be controlled as U_{DC}/n. When the number of needed inserted SMs of upper arm is k, whether SM1 is inserted or bypassed is determined by the upper arm current i_{ua} and the voltage of SM1 u_{Cu1}. When $u_{Cu1} > \dfrac{U_{DC}}{n}$ and $i_{ua} > 0$, SM1 is bypassed, k SMs of upper arm and $n-k$ SMs of lower arm are selected arbitrarily to insert into operation; When $u_{Cu1} > \dfrac{U_{DC}}{n}$ and $i_{ua} < 0$, SM1 is putting into operation, $k-1$ SMs of upper arm and $n-k$ SMs of lower arm are selected arbitrarily to put into operation. On the contrary, if $u_{Cu1} < \dfrac{U_{DC}}{n}$, the operation principles are inverse. if $u_{Cu1} < \dfrac{U_{DC}}{n}$ and $i_{ua} < 0$, SM1 is bypassed, k SMs of upper arm and $n-k$ SMs of lower arm are selected arbitrarily to put into operation; When $u_{Cu1} < \dfrac{U_{DC}}{n}$ and $i_{ua} > 0$, SM1 is putting into operation, $k-1$ SMs of upper arm and $n-k$ SMs of lower arm are selected arbitrarily to put into operation. This capacitor voltage control does not need sorting algorithm as a benefit of self-balancing ability of this topology.

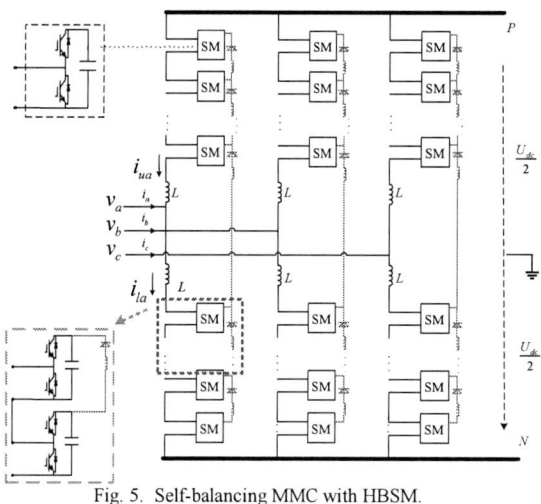

Fig. 5. Self-balancing MMC with HBSM.

As a benefit of IGBTs in the auxiliary circuit, this new MMC topology can transform between conventional MMC and self-balancing MMC topology. So the pre-charge is as simple as conventional MMC ignoring the influence of auxiliary components. When DC fault occurred to this new MMC topology, all the IGBTs in the auxiliary circuit between adjacent SMs should be blocked

1457

to make this topology transform into a conventional MMC topology. At the same time, all SMs should be blocked to ride through the DC fault thanks to the SBSM (self-blocking sub-module) used in this converter.

If DC fault ride-through ability is not needed, the self-balancing topology is able to simplify as Fig. 5.

As shown in Fig. 5, auxiliary IGBT are not needed anymore, only auxiliary diode and small inductor are reserved. Only three voltage sensors are needed for the top SM of each phase leg. And half-bridge sub-module is used in this topology. Voltage balancing operation principle is as same as Fig. 4 discussed. The balancing-branch only consisting of a low power rating diode and an inductor is added to each SM, it will not increase the cost too much.

IV. SIMULATION RESULTS

The simulation is conducted with software MATLAB/Simulink to verify the effectiveness of capacitor voltage balancing ability and DC fault ride-through ability of proposed topology using SBSM. The simulation parameters are as below, DC-link voltage is 2.4kV, total quantity of SMs in each arm is 10, SM capacitor is $3300\,\mu F$, arm inductor is 15mH.

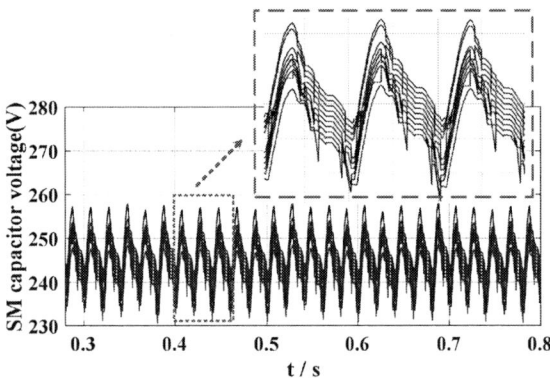

Fig. 6. Proposed new topology MMC capacitor voltages of SMs.

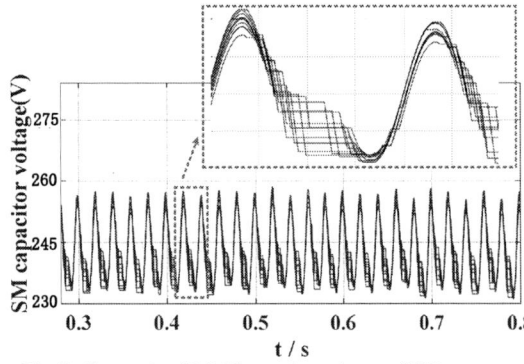

Fig. 7. Conventional MMC capacitor voltages of SMs.

The ten SM capacitor voltages of this proposed MMC

topology are shown in Fig. 6, they have a good balancing performance. Comparing to proposed new MMC topology in this paper, voltage balancing performance of conventional MMC topology which uses capacitor voltage sorting algorithm is also shown in Fig.7.

According to Fig. 6 and Fig. 7, SM capacitor balancing performances of proposed new topology MMC and conventional MMC have a good comparison. All SMs capacitor voltages of proposed new topology MMC could balance around the capacitor voltage reference, which have a similar performance compared to conventional MMC. The voltage differences between neighbouring SMs in new topology MMC is a little larger than which of conventional MMC, it is caused by the voltage drop of auxiliary diodes between neighbouring SMs in new topology MMC. But these two figures still verify that the new topology MMC have the capacitor voltage balancing ability.

To validate the effectiveness of DC fault ride-through ability of this MMC topology. DC dual side short circuit occurred at 0.3s, and 5ms later all the SMs were blocked and all the IGBTs in auxiliary voltage-balancing circuit were also blocked. As shown in Fig. 8, grid AC current operates well before 0.3s, after the DC fault the AC current rose sharply, then short circuit current dropped after 0.305s and short circuit current dropped to zero in 0.31s.

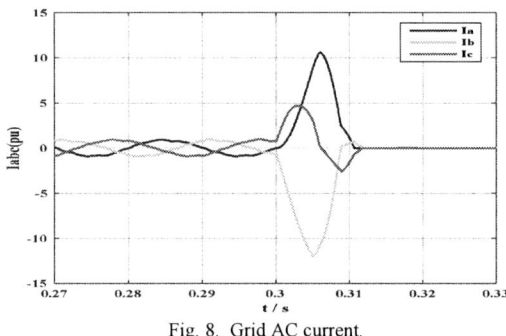

Fig. 8. Grid AC current.

Simulation results as shown in Fig. 6 and Fig. 8 validate the effectiveness of good performance of voltage balancing and DC fault ride-through ability of proposed self-balancing MMC topology.

Due to space limitations, the simulation results figures of new MMC as Fig. 5 are not shown in this paper. New topology MMC as shown in Fig. 5, which only using HBSM, auxiliary diodes and small inductors, also have a good capacitor voltage balancing performance as same as shown in Fig. 6.

V. CONCLUSIONS

This paper proposes a voltage self-balancing MMC topology. Comparing to conventional MMC, this topology only needs three SM voltage sensors and voltage sorting algorithm is not needed, so this topology reduces the computing burden of processor. Compared to other self-balancing topologies in [6]-[8], this proposed topology needs less auxiliary components and voltage sensors, voltage balancing control is more simple especially applicable in MMC-HVDC which contains a large number of modules, the more meaningful advantages are the DC fault ride-through ability and making pre-charging for SM more simple when this new topology MMC starts.

This proposed new topology MMC reduces a large number of voltage sensors for each arm, which only needs three voltage sensors for top SM of phase leg. This novel topology makes the SM voltage balancing quite simple so that reduces the control complexity and computing burden greatly. The effectiveness of this topology and DC fault ride-through ability is validated by simulation.

REFERENCES

[1] Teeuwsen, S. P.. "Modeling the Trans Bay Cable Project as Voltage-Sourced Converter with Modular Multilevel Converter design." *power and energy society general meeting* (2011): 1-8.

[2] Deng, Fujin, and Zhe Chen. "Voltage-Balancing Method for Modular Multilevel Converters Switched at Grid Frequency." *IEEE Transactions on Industrial Electronics* 62.5 (2015): 2835-2847.

[3] Hagiwara, Makoto, and Hirofumi Akagi. "Control and Experiment of Pulsewidth-Modulated Modular Multilevel Converters." *IEEE Transactions on Power Electronics* 24.7 (2009): 1737-1746.

[4] Chu, Zunfang, et al. "An improved voltage balancing method of modular multilevel converters." *conference on industrial electronics and applications* (2014): 360-363.

[5] Liu, Pu, et al. "Grouping-Sorting-Optimized Model Predictive Control for Modular Multilevel Converter With Reduced Computational Load." *IEEE Transactions on Power Electronics* 31.3 (2016): 1896-1907.

[6] Gao, Congzhe, et al. "A DC-Link Voltage Self-Balance Method for a Diode-Clamped Modular Multilevel Converter With Minimum Number of Voltage Sensors." *IEEE Transactions on Power Electronics* 28.5 (2013): 2125-2139.

[7] LIU Hang, ZHAO Chengyong, et al.ones, "Novel Topology of Modular Multilevel Converter With Voltage Self-balancing Ability." *Proceedings of the CSEE* 37.0 (2017): 1-10.

[8] Liu, Xiangdong, et al. "A novel STATCOM based on diode-clamped modular multilevel converters." *IEEE Transactions on Power Electronics* (2016): 1-1.

[9] Qin, Jiangchao, et al. "Hybrid Design of Modular Multilevel Converters for HVDC Systems Based on Various Submodule Circuits." *IEEE Transactions on Power Delivery* 30.1 (2015): 385-394.

[10] Qingrui Tu, Zheng Xu, et al. "Reduced Switching-Frequency Modulation and Circulating Current Suppression for Modular Multilevel Converters." *IEEE Transactions on Power Electronics* 26.3 (2011): 2009-2017.

AUTHOR INDEX

Aapro, Aapo 3156
Abdollahi, Hessamaldin 1719
Abe, Kazuyuki 1567
Abe, Kensho 767
Abe, Kodai 1741, 3890
Abe, Seiya 2360, 2370
Abe, Takashi 2176
Abrishamifar, Adib 2854
Abuogo, James 1125
Acharya, Anirudh Budnar 2630
Acharya, Sayan 3564
Adachi, Masakazu 2237
Afsharian, Jahangir 1537, 3797
Agarwal, Vivek 3471
Agelidis, Vassilios G. 3215
Agostinelli, Matteo 3140
Ahmad, Hamzeh J. 3273
Aiso, Kohei 3186
Akagi, Hirofumi 2352
Akahane, Masashi 2774
Akama, Yousuke 1741
Akao, Naoki 1217
Akatsu, Kan 711, 3186
Alatise, O 1149
Alenius, Henrik 1704, 4205
Ali, Muhammad 528
Ali, Murad 2317
Allmeling, Jost 422, 2199
Almér, Stefan 555
Alsofyani, Ibrahim Mohd 466
Alvarez, S. 4009
Amano, Koki 94
Amei, Kenji 3182
Amin, Mohammad 759
Amrhein, Wolfgang..............................3640
An, Ronghui..................957, 1524, 3251, 3692, 3924
An, Zheng 4001
Andenna, M. 3596
Andersen, A. E. Michael 1351
Andersen, Michael A. E. 607, 4066
Ando, Akinobu 517
Ando, H. 3665
Ando, Masato 1919
Ando, Takashi 3658
Ang, Simon S. 153
Antivachis, Michael............................181
Antonini, Giulio 3588
Antonopoulos, Antonios 2335
Anurag, Anup 3564
Anyapo, Chan 3332
Aoyagi, Kazuki 2237
Aoyama, Masahiro 718, 753
Arai, Takuro 1997
Araumi, Ryunosuke 1877, 3658
Arimatsu, Kenji 1370

Arita, Hideaki 2796, 2820
Arrua, Silvia....................................1719
Asada, Kazunori 3658
Asama, Junichi 4016
Ashizaki, Yusuke 3450
Ashourloo, Mojtaba 2380
Aso, Shinji 3086
Aware, Mohan 1730
Ayano, Hideki 1080
Azad, A N M Wasekul 2416
Azegami, Kazuya 3723
Azuma, S. 3665
Baba, Teppei 2283
Babasaki, Tadatoshi 207
Bach, Hoang Linh 2410
Baek, Jae-Il 108, 2365, 3100, 3533, 3538
Baek, Miran 1141
Bahat-Treidel, Eldad 3607
Bai, Baodong 2638
Baik, Jeong Min 3063
Bak, Yeongsu 1736, 4104
Bakran, Mark-M. 2476
Bandyopadhyay, Soumya 1426
Barrena, Jon Andoni.............................. 759
Barrera-Cardenas, Rene 3431
Bauer, Pavol.............................. 1426, 2630
Bauer, Walter 3640
Bayer, Christoph Friedrich 2410
Bellini, M. 4009
Berg, Matias......................... 963, 4205
Bergveld, H.J. 267
Bertoldi, F. 488
Besselmann, Thomas.............................. 555
Bezha, Minella 3170
Bhattacharya, Subhashish 3564, 3993
Bhowate, Apekshit...............................1730
Bhumkittipich, Krischonme 2430
Biela, J. 1896
Biela, Jürgen 1103, 1509, 2301, 3734
Bilal, Ahmad 2193
Bilsalam, A. 1622
Bin, Zhao 2692
Bixel, Paul 238
Blaabjerg, Frede 439, 746, 1183, 1246, 1711, 1788, 2512, 2604, 2743, 3123, 3164, 3357
Blanes, José M. 1435
Böcker, Jan 3607
Bojoi, R. 732
Bonyadi, R. 1149
Boroyevich, Dushan 790, 3705, 3749, 3985
Bortis, D. 4080
Bortis, Dominik 181
Boynov, K.O......................................161
Braun, Michael 2848, 3074

AUTHOR INDEX

Büdel, Johannes 3034
Bui, M.X. 4174
Bunlaksananusorn, Chanin...............................2490
Burgos, Rolando 790, 3705, 3749, 3985
Cai, Kejun..3965
Cai, Panpan 3495
Cai, Xu 1004, 1491, 2245, 4162, 4220
Canales, F. 4009
Cao, Hu 1816, 3484
Cao, Pengpeng...2973
Cao, Qi 100
Cao, Wu 3002, 3010, 3015
Cardenas, Rene Alexander Barrera 1111
Carvalho, Kelly C. M...3785
Castellazzi, Alberto 130, 2932
Ceballos, Salvador 3117
Celik, Mustafa 1680
Cha, Honnyong....................927, 1046, 2619, 3134
Chae, Beomseok 1977
Chailloux, Thibaut 2153
Chang, Chen-Wei...1617
Chang, Chien-Hsuan..2860
Chang, Liuchen 815, 1472, 1793, 2505
Chang, Yung-Ruei 639, 883
Chanmontree, P. 1622
Chao, Yi-Hao...1145
Charalambous, Apollo 1634
Charoensuksirikul, Supanut 2113
Chattopadhyay, Ritwik.......................................3564
Chazal, Hervé 2158
Chen, Ang-Tung...2102
Chen, Bo 1397
Chen, C. 142
Chen, Ching-Chen 1617
Chen, Ching-Jan 2086
Chen, Chuantong 1598
Chen, Dezhi 2638
Chen, Guan-Jung 1341
Chen, Guo...370
Chen, Hao...3112
Chen, I-Lin 2107
Chen, Jiangnan 1157, 1167
Chen, Jiann-Fuh 2653
Chen, Jie 1015, 1177
Chen, Kai-Hui 3081
Chen, Ke 1391
Chen, Kun-Feng 1341
Chen, Min 878
Chen, Minwu 2547
Chen, Nan 2335
Chen, Pingping 1118
Chen, Shen-Li 1145
Chen, Song...2153
Chen, Tang-Jung 1617
Chen, Tao...1872

Chen, Wan-Jung...3544
Chen, Wenjia 4213
Chen, Wenjie 1062, 2854, 3329
Chen, Wu 1504, 2496
Chen, Xiliang 3329
Chen, Xin 1015, 1177
Chen, Xingxing 1051, 3129, 3439
Chen, Yang 2785
Chen, Yangyang 560
Chen, Yaow-Ming..................................... 639, 883
Chen, Yenan 1118
Chen, Yen-Wen 2576
Chen, Yufeng 3383
Chen, Yu-Jen 275
Chen, Zhe 1758, 2708
Chen, Zhi..2997
Chen, Zhigang 3040
Cheng, Ching-Hsiang 2086
Cheng, Chun-An...2860
Cheng, Hung-Liang 2860
Cheng, Nie 2625
Cheng, Po-Tai.....................503, 1038, 2462, 3549
Cheng, Ran 3877
Cheng, Xiangpeng 2435, 3934
Chengbi, Zeng..2718
Chi, Yongning 1491
Chiba, Akira 3627
Chien, Lin-Hao 2102
Chiu, Huang-Jen 2092, 3151
Chiu, Hui-Lung 123
Chiu, Yi-Hao 1145
Cho, Geum-Bae 2145
Cho, In-Ho 3323
Cho, Shin-Young 1530
Cho, Young Joon 137
Cho, Younghoon 1403
Choe, Chanyang 1598
Choi, Byungcho 1465
Choi, Hyun-Jun 383
Choi, Jae Hyuk 1336
Choi, Jaeho 803
Choi, Joon-Ho 982, 1799
Choi, Seung-Hyun 4049
Choi, Sewan 256
Choi, Sung-Jin 1409
Choi, Youn-Ok...2145
Chou, Shih-Feng 1711
Chou, T.-C. 1912
Choudhury, Abhijit...3401
Chuai, Guoming 3025
Chung, Daewoong..1141
Chung, Henry S. H. 917
Chunkag, V. 1622
Collins, Caspar 1931
Cortes, Camilo 2193

AUTHOR INDEX

Corvasce, C. 3596
Cucala, Asuncion P. 2534
Cui, Shenghui 2250, 2484
Cui, Xiang 1125
Cvetkovic, Igor 790, 3985
Czyz, Piotr 396
D'arco, Salvatore 782, 2003
Da Silva, C. 267
Dahidah, Mohamed S A 3215
Dai, T. 1149
Dai, Wenjing 1015
Daikoku, Akihiro 2796
Danqing, Liu 1376
Dao, Ngoc Dat 1212
Dauphin, Benjamin 3644
Davari, Pooya 746
Davletzhanova, Z 1149
De Doncker, Rik W. 375, 388, 598,
1073, 2250, 2484, 2768, 3729, 3979
Decker, Simon 2848, 3074
Delaforge, Timothé 2158, 3820
Deng, Fujin 1758, 2708
Deng, Jinxin 2992
Deshpande, Prathamesh Pravin.....................4186
Dieckerhoff, Sibylle......................................3607
Dimarino, Christina 3985
Din, Zakiud 2262
Ding, Yong 815
Dinh, Nguyen Duy 363
Diouf, Fatou 2078
Dirksen, Daniel 2410
Divan, Deepak 4001
Doki, Shinji 1032, 1223, 1228, 1295, 1747, 2224
Dong, Hanjing 987
Dong, Mi 1771
Dong, Qinghua 459
Dong, Xiaofeng 4168
Dong, Zhen 459
Dong, Zheng 3768
Driesen, J. 488
Du, Chao 2204
Du, Xiaotong 1167, 2780
Du, Xizhou 1491
Du, Yan 1472, 2877
Du, Zhijiang 84
Duarte, J. L. 946, 1067, 2697
Duarte, Jorge L. 1447, 3840
Dugal, F...3596
Dujic, Drazen..................... 422, 1484, 1498, 2170
Duong, Truong-Duy 982
Duque, C. A. 1067
Eberle, Wilson 927
Ekman, Jonas 3588
Elbaset, Adel A. 3945
Endegnanew, Atsede G. 2003

Endo, Hiroaki 4151
Endres, Tobias Maximilian 2410
Engelmann, Georges 3979
Enomoto, Bruno Yukio 3785
Eto, Haruhi 2097
Faiz, Muhammad Talib 528
Fajri, Poria 3223
Fan, Dongchen 3002, 3010, 3015
Fan, Shengwen 977, 3040
Fan, Weiyan 1386, 1421
Fang, Jingyang 337, 3910
Fang, Ran 4213
Fangfang, Luo 1282
Farkas, Gabor 137
Fayyaz, Asad 130
Felderer, Niklaus 2199
Feng, Chao 2058
Feng, Wei 3678
Ferdowsi, Mehdi 3223
Fernandez, Gabriel 3209
Fernandez-Cardador, Antonio 2534
Fischer, F. 3596
Foo, Gilbert 1724
Formentini, A. 4034
Freijedo, Fracisco D. 1498
Friedrichs, Peter 3584
Fuchs, Simon 2301
Fujii, H. 1253
Fujii, Kansuke 3711
Fujii, Keisuke 1189
Fujii, Toshiyuki 2540, 3578
Fujimoto, Hiroshi 77, 663
Fujimoto, Kazuki 2047
Fujimoto, Yasutaka 571, 681
Fujimura, Akira 1080
Fujita, Atsushi 296
Fujita, Goro 363
Fujita, Hideaki..........................626, 1854, 3813, 3940
Fujiwara, Hajime 1381
Fujiwara, Kazuya 3773
Fukuda, Hiroto 2938
Fukuda, Kenji 2558
Fukui, Tomoya 860
Fukuoka, T. 1240
Fukushima, Kentarou 2176
Fukushima, Takafumi..................................... 3478
Funabiki, Shigeyuki 2449
Funaki, Tsuyoshi 309, 2181, 3092
Funato, Hirohito 94, 2036
Funato, Hiroki 2073
Furukawa, Keita 3349
Furukawa, Kimihisa 3572
Furukawa, Yudai 4193
Furusho, Yasuaki 3711
Gan, Yiliang 1391

AUTHOR INDEX

Ganisetti, V. K. 2907
Gao, Feng.............................2016, 3383, 3965
Gao, Xiaonan 1661
Gao, Zhuo 3455
Garrigós, A. 1435
Gasim, Abdulaziz 2836
Gehlot, Deepak 3471
Geng, Hua 542
Geng, Yiwen 619
Gerada, C. 4034
Gheonjian, Anna 2078
Gietler, Harald 3140
Gohara, Hiromichi 2764
Gondo, Ryota 3490
Gong, Bing 3797
Gong, Chunying 1015, 1177
Gong, Z. 267
Gorodnichev, Anton 375
Goto, Akihisa 2449
Goto, Hiroki 3192
Goto, Kazuya 1315
Goto, Yasuyuki 809
Gou, Yating 1157, 1167
Grimm, Ferdinand 2895
Grossner, Ulrike 3588
Gruber, Wolfgang 3632, 3640, 4028
Gu, Lei 632
Gu, Qing 2963
Guajardo, Cristian Andres Garces 1854
Guan, Bo 1032
Guan, Yajuan 2668, 3678
Guan, Yueshi 614, 3780
Guangzhu, Wang 1376
Guerrero, Josep M. 1498, 2668, 3112
Guerrero, M. Josep 3678
Gui, Yonghao..2668
Guidi, Giuseppe..782, 2003
Guillod, Thomas 396
Gunji, Daisuke 663
Guo, Leilei 904
Guo, Yanjie 3338
Guozhao, Duan 2625
Gupta, K. 267
Gurpinar, Emre 130
Gutiérrez, R. 1435
Ha, Jung-Ik 565, 2500
Ha, Sang-Hyun 3466
Haga, Hitoshi 1370, 3890
Hagiwara, Makoto 3273
Hahashi, Yuji...4059
Haider, M. 4080
Halamicek, Michael 831
Halick, Mohamed 416
Hamabe, Yasumasa 1276
Hamada, Shizunori 227

Hamaguchi, Takumi..3507
Hamasaki, S. 1240
Hamasaki, Shin-Ichi.............1217, 1276, 2938, 3237
Hameyer, Kay 740
Han, Byung-Moon 466
Han, Jung-Kyu 3107, 3533, 4049, 4054
Han, Pengcheng 1027, 2714
Han, Yang 3112
Hanajiri, Kensuke 663
Hanamoto, Tsuyoshi.................................1315, 1698
Hancioglu, Oguz Kaan 1680
Handa, Hiroyuki 3762
Handa, Yuuichi 4059
Hane, Yoshiki 2426
Hang, Lijun 1391, 2866
Hanju, Cha 1985
Hao, Liu 3484
Hao, Xiang 1478
Harnefors, Lennart 3684
Hartmann, S. 3596
Haruna, Junnosuke 94, 2036
Hasegawa, Kazunori 1938
Hasegawa, M. 3665
Hasegawa, Ryuta 2011
Hashempour, Mohammad M. 4198
Hashimoto, Kazuki 3757
Hasler, Jean-Philippe 3684
Hata, Katsuhiro 663
Hata, Ryotaro 2149
Hatakeyama, Tomoyuki 1991
Hataya, Morimasa 410
Hatipoglu, E. 3805
Hatsumi, Takuya 94
Hatta, Yoshiyuki 675
Hattori, Fumiya 2738
Hattori, Keisuke 3286
Haung-Jen, Chiu 645
Hayashi, Nobuo 866
Hayashi, Yuji 356
He, Wangpin 560
He, Xiaokun 1504, 2496
He, Xiaoqiong 1027, 2714
He, Yigang..2317
He, Yingjie 3439
Hendrix, M. A. M. 946, 2697
Heo, Jongwon 726
Hidaka, Yuki 2820
Higuchi, Keiichi 2764
Higuchi, Masato 3952
Higuchi, Shinichi 2216
Hikaru, Naruse 3418
Hikihara, T. 3665
Hikihara, Takashi......................................3654, 3757
Hiller, Marc 3074
Hillers, A. 1896

AUTHOR INDEX

Hillers, André 2301
Hilt, Oliver 3607
Hinz, Arne 598
Hirahara, Hideaki 1960
Hiraki, Eiji 410, 1602, 1610
Hirao, Takashi 2082, 2137
Hirase, Yuko 767
Hirayama, Katsutoshi 4193
Hirayama, Tadashi 3406
Hirokawa, Masahiko 1543, 4133
Hirokawa, Takayuki 296, 410
Hiromoto, Masayuki 3644
Hirose, Keiichi 593, 822
Hirose, Naoki 3791
Hiroshi, Tadano 3431
Hiroshige, Shinichi 3369
Hirota, Takashi 3952
Hoang, Tuan V. 1752
Hoda, Isao 2073
Hofmann, Viktor 2476
Hofmann, Wilfried 3243
Hojo, Masahide 3369
Holenstein, Thomas............................3619
Holmes, D. G. 3670
Hong, Miao 2718
Hongpeng, Liu 1442, 2969
Honjo, Satoshi 2066
Hori, Motohito....................................3396
Hori, Yoichi 77, 663
Horie, Shunsuke 809
Horikoshi, Takahiro 1997
Hoshi, Nobukazu 971, 2660, 3855
Hou, Chung-Chuan 1617
Hou, Lijun 2901
Houran, Mohamad Abou 1062, 2854
Hsieh, Guan-Chyun 123
Hsieh, Hung-I 123
Hsieh, Yao-Ching 3151, 3544
Hsu, Chi-Hsuan 2653
Hu, Jiewen 3985
Hu, Jingxin 1073, 2250, 2484
Hu, Sheng..3052
Hu, Song 370
Hu, Xihong................................614, 3780
Hu, Xing 2262
Huang, Bing-Siang 2092
Huang, Bo-Jia......................................3528
Huang, Chien-Chun 3151
Huang, Huazhen...................................1125
Huang, Jingjin 2980, 4157
Huang, Jingjing 1004, 2688, 2692
Huang, Jun-Xian 1626, 3081
Huang, Lang 1478
Huang, Pin Yu 2165
Huang, Ta-Wei 1626

Huang, Wen-Mei 2576
Huang, Xianjin 1131, 2051
Huang, Xiaoliang 84
Huang, Xuehao 3455
Huang, You-Chun 275
Huemer, Mario 3140
Hui, S. Y. Ron 889, 2552
Hung, Chun-Yao 2576
Hung, Shun-Kang 1575
Huo, Chongcan 987
Huo, Junya 1206, 1234
Hussein, Abdallah........................130, 2932
Huynh, Dang Minh 3086
Hwang, Duck-Hwan 1403
Hwang, Seon-Ik 3323
Hwu, K.I. 851
Hyakutake, Y. 1253
Hyodo, Takashi 2589
Hyunsung, An 1985
Iannuzzi, Diego 2527
Ibuchi, Takaaki 309
Ichinose, H. 1240
Ide, Yuji 3896
Iijima, Ryuji 313, 1111
Iioka, Daisuke 2278
Ikari, Yuki 148
Ikeda, Hidehiro 1315
Ikeda, Yoshinari 3396
Ilves, Kalle 2335
Imai, Kazu 3363
Imai, Makoto 296, 410
Imamori, Satoshi 699
Imaoka, Jun 1087, 1095, 1554, 3773
Imoto, R. 2808
Imtiaz, Abu Saleh 2416
Imura, Takehiro 77, 663
Inaba, Tsuyoshi 4114
Inomata, Kentaro 3952
Inoue, Daisuke 2764
Inoue, Kaoru 1264, 2186, 4151
Inoue, Kent 348
Inoue, Masamichi 1228
Inoue, Takatoshi 1276
Inoue, Y. 704, 2808
Inoue, Yukinori 1189, 1289, 1329, 2802, 2814, 3197
Irino, Yusuke 244
Ise, Toshifumi 775, 2393, 3762, 3902
Ishibashi, Mikiya 1370
Ishibashi, Naoyuki 1543
Ishibashi, Taku 2292
Ishigaki, Shingo 227
Ishiguro, Takahiro 1997, 2011, 3304
Ishihara, Masataka 1610
Ishii, Y. 1834
Ishii, Yuki...1196

AUTHOR INDEX

Ishikawa, Hiroki.................................2176, 3412
Ishikawa, Kohsuke................................2725
Ishikura, Yuki 1087, 1095, 3717
Isobe, Eisuke......................................2042
Isobe, Takanori 313, 1111, 3375, 3431
Isozaki, Keisaku...................................1364
Itaya, Yohei......................................3450
Ito, Kazuhiko.....................................2540
Ito, Yasuaki 1586, 2324
Ito, Yoichi.......................................3086
Ito, Youichi.......................................439
Itoh, Gimpei......................................1289
Itoh, Jun-Ichi...............69, 348, 534, 896, 1567,
 2229, 2237, 2519, 2596, 3349, 3797
Iwabuchi, Akio.....................................439
Iwai, Akinobu.....................................2066
Iwaji, Yoshitaka..................................1301
Iwasaki, Makoto...................................1666
Iwasaki, Tetsuya..................................3490
Iwata, Hiroki 3896
Iyasu, Seiji......................................4059
Iyoda, Isao.......................................2914
Jacobs, Keijo.....................................3292
Jaffar, Hanis Afiqah Binti........................2956
Jain, Prashant....................................3471
Janah, Mounia......................................681
Jang, Duekjin.....................................2619
Jang, Yu-Jin 1655, 3466
Jang, Yun...1736
Jangs, Yujin......................................1562
Jarutus, Neerakorn................................2121
Jehle, Andreas....................................1509
Jennings, M.......................................1149
Jeong, Seog Y.....................................2564
Jeong, Si-Hoon.....................................289
Jeong, Yeonho 838, 2365, 2376
Jhang, Ying-Yi....................................3884
Jhou, Yu-Lin......................................1145
Ji, Guyuan..2921
Jia, Haiyang.......................................998
Jia, Pengyu...................................977, 3040
Jia, Xu...3025
Jiacheng, Wang....................................2986
Jiajie, Zang......................................2986
Jiajie, Zhou......................................1442
Jian, Jun-Min.....................................2653
Jiang, Jinhai......................................84
Jiang, Shuai......................................987
Jiang, Siyue......................................4168
Jiang, Yanfeng....................................2058
Jiang, Yongbin....................................3863
Jianhua, Wang.....................................1282
Jianming, Xu......................................528
Jianqiao, Zhou....................................2986
Jianwen, Zhang....................................2986

Jiaxing, Liu......................................1376
Jikumaru, Takehiro................................177
Jimichi, T..1834
Jimichi, Takushi..................................3729
Jin, Nan..904
Jin, Zheming......................................2668
Jing, Lei...878
Jing, Lyu...2692
Jing, Yang..3383
Jingyu, Song......................................1282
Jing-Yuan, Lin....................................645
Jinjun, Liu.......................................4181
Jinshui, Zhang....................................4181
Jisaki, Jun.......................................3182
Joebges, Philipp..................................375
Jongudomkarn, Jonggrist...........................3902
Jonishi, Akihiro..................................2774
Joryo, Satoshi....................................1202
Joseph, Anto......................................1358
Jumayev, S..161
Jung, Hanul.......................................688
Jung, Hyun-Sam....................................911
Jung, Jae-Jung....................................3557
Jung, Jee-Hoon 289, 383
Jung, Jun-Hyung...................................3323
Jung, Si-Hoon.....................................383
Jungmayr, Gerald..................................3640
Junior, Lourenço Matakas..........................3785
Jynu-Jhe, Jhang...................................645
Kada, Haruya......................................3890
Kadota, Mitsuhiro.................................3572
Kai, Masahiko.....................................1803
Kaicheng, Ding....................................4181
Kaipia, T...2948
Kaishakuji, Hikaru................................2360
Kakigano, Hiroaki 583, 2956
Kamaeguchi, Koki..................................410
Kamakura, Kousuke.................................2756
Kamejima, Takayoshi...............................3286
Kamiya, Naoki.....................................1673
Kamiyama, Naosumi.................................1955
Kamoshida, Naoki..................................1111
Kampeerawar, Warayut..............................3257
Kanai, Naoyuki....................................3396
Kanaya, Kazuhisa..................................2011
Kanazawa, Yasuki..................................2789
Kanchan, R. S.....................................488
Kandula, Prasad...................................4001
Kaneko, Satoshi...................................3396
Kanetani, Kaisei..................................207
Kang, Dong-Hun....................................3030
Kang, Feel-Soon...................................2376
Kang, Kyoung-Suk..................................922
Kang, Tahyun......................................1977
Kang, Yong..2997

AUTHOR INDEX

Kanno, Junya 3299
Kano, Fumihisa 2036
Kanoda, Akihiko 3572
Kanzian, Marc...................................3140
Kapisch, E. B. 1067
Karami, Bagher.................................2854
Karppanen, J. 2948
Kasai, Yuji 2036
Kashihara, Tatsuki 1741
Katayama, Tatsuji 1346
Kato, Hideaki 1580, 1586, 2324
Kato, Hirokazu 3478
Kato, Koji 439, 1370, 3086
Kato, Toshiji...............1264, 2186, 4151
Katoh, Kaoru.....................................233
Katoh, Shinji 2176
Katsuki, Akihiko 1543
Katsura, Seiichiro 669
Katsura, Shogo 767
Katsushi, Terazono 3431
Kawabata, Naoki 2887
Kawabata, Shuma 3406
Kawagoe, Natsuki 3490
Kawaguchi, Hironori 517
Kawaguchi, Jun'ichiro 1828
Kawaguchi, Yuki..............................3572
Kawakami, Masaki 2756
Kawakami, Noriko 1346
Kawamura, Atsuo 318, 1649, 1687, 3916
Kawamura, Itsuo 3396
Kawamura, Kazuki 1567
Kawanishi, Kota 169
Kawashita, Jun 2042
Kayashima, Kazuya 1315
Kaymak, Murat..................................3729
Kazmi, Syed Muhammad Raza 4168
Ke, Junji 1125
Kennel, Ralph...............1661, 2895, 3965
Kezuka, Nobutaka 227
Khan, Ashraf Ali 927
Khan, Faisal......................446, 2416
Khan, Muhammad Mansoor................528
Khan, Usman Ali 927
Khomfoi, Surin.................................1460
Khubchandani, Vasudha....................845
Kiatsookkanatorn, Paiboon................2581
Kida, Masahiro............1586, 2324
Kido, Tatsuya 329
Kikuchi, Ryosuke.............................1877
Kikuchi, Takaaki 2292
Kikuchi, Takeshi 3578
Kikuma, Toshiaki 3299
Kim, Byeongwoo 256
Kim, Chong-Eun 108, 3538
Kim, Dong-Kwan 1655, 3466, 3538

Kim, Gun-Woo 838
Kim, Hansang 1465
Kim, Heung-Geun 927, 1046, 2619, 3134
Kim, Hideaki.................................... 207
Kim, Hyeon-Sik 521
Kim, In-Dong 3229
Kim, Jae-Kuk 3100
Kim, Jang-Mok 3323
Kim, Jin-Hak 1530
Kim, Jin-Young 3229
Kim, Jong-Woo................... 3107, 4049
Kim, Kangsan 256
Kim, Katherine A. 2092, 3063
Kim, Keon Young............................... 4104
Kim, Keon-Woo............108, 1562, 1655, 2365, 2376
Kim, Ki-Mok 2365
Kim, Myong Hwan 2500
Kim, Sanghun 2619
Kim, Sunju 3833
Kim, Yeonjung 1465
Kimura, Hideki 2036
Kimura, Mamoru 1991, 1997
Kimura, Noriyuki 1202, 1259, 2558, 2887, 2914
Kinoshita, Masahiro 3929
Kishimoto, Toshihiko 261
Kishita, Ken 1301
Kitagawa, Wataru 1847, 3507
Kitamura, Akio.................................. 2764
Kitamura, Toshinori 2660
Kiyoshi, Ohishi 1673
Kiyota, Kyohei 3182
Klammer, Bianca 3632
Ko, Chien-Tzu 2107
Kobayashi, Hiroyasu............. 2527, 3490
Kobayashi, Koji 1741
Kobayashi, Marika 2802
Kodaka, Wataru 2589
Kogai, Naoki 1364
Koizumi, Hirotaka 4114
Kolar, J. W. 3805, 4080
Kolar, Johann W. 181, 396, 3619
Kolb, Johannes 2848
Komaru, Yuma 1329
Komatsu, Hiroyoshi 1346
Komatsu, Taiga 2820
Komatsu, Wilson 3785
Komeda, Shohei 3813
Kometani, Haruyuki 711
Kondo, Keiichiro 726, 2047, 2527, 3490
Kondo, Shota 1295
Kondo, Takeshi................................. 4114
Kong, Wei 3460
Kongjeen, Yuttana 2430
Konishi, Akihiro 1602
Konno, Junya 1692

AUTHOR INDEX

Konstantinou, Georgios 3117
Kopta, A. 3596
Kosaka, Takashi 3418
Koseki, K. 1162
Koseki, Takafumi 2042, 2309, 3257
Koshikizawa, Hiroyuki 1567
Kostov, Konstantin 2732
Kouketsu, Masaju 227
Kouno, Yusuke 2176
Kovacevic-Badstübner, Ivana 3588
Kowatari, Hiroki 2660
Koyama, Yushi 2011
Krismer, F. 3805
Krismer, Florian 396
Kubo, Hajime 483
Kubota, Hisao 1196
Kucka, Jakub 1904
Kumada, Keishirou 3396
Kumagai, Shuta 1264
Kumar, Ashish 3993
Kumar, Rajesh 2456
Kumar, S. Gautam 3471
Kumsuwan, Yuttana 2113, 2121
Kunomura, Ken 1803
Kuo, Chun-Ting 1145
Kuraishi, Daigo 3896
Kuraku, Nagendra Vara Prasad 2317
Kuring, Carsten 3607
Kurisaka, Masakatsu 4151
Kurita, Naoyuki 1991
Kurita, Nobuyuki 3640
Kurokawa, Fujio 826, 2097, 2283, 4193
Kurosawa, Nobuhito 1810
Kurumatani, Hiroki 669
Kusaka, Keisuke 69, 348, 2237, 3349
Kusumah, Ferdi Perdana 3870
Kuwata, Gen 177
Kwon, Min-Jun 114
Kyyrä, Jorma 2193, 3870
Lai, Jih-Sheng 3107, 4049
Lai, Jui-Hung 3081
Lan, Yuanliang 1167
Lana, A. 2948
Le, Hanh-Phuc 213
Le, Hoai Nam 2519
Lee, Byoung-Hee 838
Lee, Byung-Kwon 3030
Lee, Chan 688
Lee, Choongin 565
Lee, Dong-Choon 478, 1212
Lee, Hong-Hee 1752
Lee, Hyong Gun 1336
Lee, Il-Oun 1530
Lee, Jae-Bum 3100
Lee, Jia-You 657, 2107

Lee, Joon-Hee 3557
Lee, Junbae 1141
Lee, June-Hee 466
Lee, Jung-Yong 1403
Lee, Jun-Young 3030
Lee, Jusuk 1336
Lee, Kyo-Beum 466, 1736, 4104, 4109
Lee, Kyoung-Won 2145
Lee, Kyung-Hwan 2500
Lee, Min-Su 108
Lee, Minsub 1141
Lee, Nayoung 1562
Lee, Song-Kai 2102
Lee, T. L. 4198
Lee, Tzung-Lin 2576
Lee, Woo-Cheol 114
Lee, Woo-Seok 1530
Lee, Young-Dal 3466, 3538
Lehn, Peter W. 3203
Lei, Qin 2400, 3742
Leng, Darith 1764
Leubner, Martin 3243
Li, Bodong 878
Li, Chi 790, 3705
Li, Dongsheng 1301
Li, Fei 2611
Li, Fujian 2944
Li, Guanglei 1455
Li, Haijin 2270
Li, Haisi 3040
Li, Haoyu 2901
Li, Hong 2058
Li, Hongchang 337, 3910
Li, Jhih-Sian 3081
Li, Jia 2073
Li, Jianfeng 130
Li, Kaiyuan 1517, 1592
Li, Lei 1172
Li, Li 1771
Li, Ming 2973
Li, Mingshen 2668, 3678
Li, Pengcheng 3698
Li, Shufan 3338
Li, Sinan 889, 2552
Li, T.-Y. 1912
Li, Xiaodong 370
Li, Xiaolu Lucia 3768
Li, Xiaoqiang 3910
Li, Xingshuo 453
Li, Xinying 2646
Li, Yan 2245
Li, Yang 795, 1478
Li, Yangman 2901
Li, Yi-Chan 639, 883
Li, Yongdong 1010, 2386

AUTHOR INDEX

Li, Yong-Jyun 275
Li, Yunwei 3958
Li, Yunwei Ryan 1537
Li, Yuze 2997
Li, Zhenjie 84
Li, Zhenwei 998
Li, Zhiqing 100
Liang, Daniel 1943
Liang, Junrui 4122
Liang, Ning 1157
Liang, Wencai 1131
Liao, Chenglin 3338
Liao, Chih-Yi 657
Liao, Hsuan 2653
Liao, Jian-Tang 4233
Liao, Mengyan 1386, 1421
Liaw, C. M. 2907
Lim, Cheon-Yong 1655, 2376, 3533
Lim, Dae-Sik 1212
Lim, Kyungbae 803
Lim, Young-Cheol 982, 1799
Lin, Chang-Hua 1341, 1777
Lin, Cheng-Hung 2092
Lin, Fei 1131, 1816, 2051, 2058, 3484, 3495
Lin, Jin 3460
Lin, Jing-Yuan 3151
Lin, K.-E. 1912
Lin, Min 4133
Lin, Xiang 3460
Lin, Xiaolan 1027
Lin, Xuerui 1537
Lin, Yu-Hsiu 1575
Lin, Yu-Lin 1145
Lisha, Chen 3958
Liske, Andreas 2848
Liu, Baojin 1051, 2944, 3924
Liu, Bi 1872
Liu, Bo 542, 878
Liu, Chao 2245
Liu, Chunhui 3742
Liu, Cuicui 1157, 1167
Liu, Dong 1758, 2708
Liu, Fang 2611, 2992
Liu, Furong 3052
Liu, He 3215
Liu, Hwa-Dong 1341, 1777
Liu, Jia 775, 3902
Liu, Jiaxin 2016
Liu, Jinjun 957, 1051, 1524, 2435, 2646, 2681, 3129, 3176, 3251, 3439, 3692, 3924, 3934
Liu, Junwen 3863
Liu, Kangli 3010, 3015
Liu, Nianzhou 1010
Liu, Ning 2877

Liu, Pang-Jung 2102
Liu, Ruofei 2547
Liu, Shu 3052
Liu, Siqi 1491
Liu, Tao 1478
Liu, Teng 2681, 3176, 3934
Liu, Wei 3164
Liu, Wenzhao 3678
Liu, Xiaosheng 934
Liu, Xicai 1661, 3965
Liu, Xinbo 3455
Liu, Yifu 2400, 3742
Liu, Yu-Chen 2092
Liu, Yuping 1816
Liu, Zeng 957, 1524, 2435, 2681, 3176, 3251, 3692, 3749, 3924
Liu, Zhiyuan 3495
Liu, Zipeng 2681, 3176
Lo, Jen-Hao 1145
Lomonova, E.A. 161
Lopez-Lopez, Alvaro J. 2534
Lotfi, Nima 3223
Lovison, Giorgio 77
Lu, David H. 2404
Lu, David Hongfei 3390
Lu, Kaiyuan 1183, 1246, 2842
Lu, M. Z. 2907
Lu, Shengli 3145
Lu, Shuai 3698
Lu, Y. 267
Luhtala, Roni 547, 2470, 3156
Lunglmayr, Michael 3140
Luo, Min 422, 2199
Luo, Rui 3129, 3439
Luo, Y. 267
Luong, Hoan-Tien 2145
Lyu, Jing 1004, 4162, 4220
Ma, Baohui 2882
Ma, Jie 1118
Ma, Ke 3877
Ma, Shaokang 542
Ma, Tianshu 2703
Ma, Yue 3717
Ma, Zhixun 917, 2688, 2692, 4157, 4162
Mabuchi, Yuichi 3572
Machavolu, Sawanth Krishna 753
Machida, Yuuki 2449
Maharjan, Laxman 1840
Makishima, Shingo 2047
Mannen, Tomoyuki 1414, 1866
Mantooth, H. Alan 153
Mao, Meiqin 815, 1472, 1793, 2505
Mariéthoz, Sébastien 2158, 3820
Marinescu, Radu-Florin 1822
Marroquí, D. 1435

AUTHOR INDEX

Martinez, Wilmar 2193
Maruta, Hidenori 826
Maruyama, Kouji 3396
März, Martin 2410
Masuda, Eisuke 309
Masuda, Mitsuru 88
Masuko, Toshitake 3723
Matsubayashi, Tatsushi 207
Matsuda, Akihiro 2329
Matsuda, Tomohiro 1972
Matsudate, Koki 2022
Matsui, Nobumasa 826, 2283
Matsui, Nobuyuki 3418
Matsui, Teruhisa 1803
Matsui, Yoshihiro 1080
Matsui, Yuto 1847, 3791
Matsuki, Yosuke 2224
Matsumori, Hiroaki 3357
Matsumoto, Satoshi 2360, 2370
Matsumoto, Takashi 2404
Matsumoto, Toshiaki 2011, 3304
Matsumoto, Yasuaki 517
Matsumoto, Yohei 233
Matsumura, Toshiro 809
Matsuo, Keisuke 169
Matsuse, Kouki 169
Mattsson, A. 2948
Mawby, P 1149
Mcgrath, B. P. 3670
Meng, Xin 957, 1549, 3251
Menzi, David 181
Mertens, Axel 1904
Messo, Tuomas 547, 963, 1704, 2470, 3156, 4205
Michihira, Masakazu 992, 3058
Michikoshi, Hisato 2558
Milovanovic, Stefan 1484
Min, Geon-Hong 2500
Minami, Masataka 992, 3058
Mino, Kazuaki 3717
Mira, Maria C. 1351
Mishima, Tomokazu 329, 872
Misra, Mitradatta 3884
Mitsantisuk, Chowarit 3332
Miura, Yushi 775, 2393, 3762
Miwa, Yoshihiro 404
Miyajima, Hiroki 1803
Miyama, Yoshihiro 711
Miyawaki, Satoshi 2738
Miyazaki, Toshimasa 1673
Mizumoto, Yuki 1810
Mizuno, Takayuki 169
Mizuno, Yuji 2283
Mizushima, Takuya 1543
Mocevic, Slavko 3985
Mochidate, Sae 1972

Mogorovic, Marko 2170
Moiannou, Tom 831
Mok, Hyung Soo 1336
Molinas, Marta 759
Moo, Chin-Sien 275, 3544
Moon, Gun-Woo 108, 838, 1562,
　　1655, 2365, 2376, 3100, 3466, 3533, 3538,
　　4049, 4054
Mori, Kazuhisa 233
Morimoto, Hiroaki 2540, 3265
Morimoto, S. 704, 2808
Morimoto, Shigeo ...1189, 1289, 1329, 2802, 2814, 3197
Morimoto, Shinya 2210
Morishima, Naoki 2540, 3450
Moriyama, Hiroyuki 1580, 1586, 2324
Morizane, Toshimitsu 1202, 1259, 2558, 2887, 2914
Mortimer, Benedict J. 598
Motegi, Shin-Ichi 992, 3058
Motohashi, Yuto 753
Motoyama, Hiromasa 356
Mouawad, Bassem 130
Mukaiyama, Naoki 2558
Müller-Hellmann, Adolf 598
Muni, Bishnu Prasad 3471
Murakami, Toshiyuki 575
Nabetani, Yoichi 2404
Nada, Kaho 3578
Nagai, Sakahisa 1687
Nagai, Satoshi 534
Nagao, S. 142
Nagaoka, Naoto 3170
Nagaoka, Shingo 118, 4139
Nagasaka, Kuniaki 1692
Nagashima, Takumi 3490
Nagira, Yoshiki 4016
Naina, Sagar 3046
Nakabayashi, Shigeaki 1692
Nakabayashi, Shigeyuki 517
Nakagawa, Hidehiko 767
Nakahara, Kengo 3237
Nakahara, Mizuki 3572
Nakai, Masanobu 3182
Nakajima, Mizuki 2750
Nakajima, Tatsuhito 1997, 3299
Nakamura, Fuminori 2329
Nakamura, Hideyo 1137
Nakamura, Kenji 2426
Nakamura, Kimikazu 4059
Nakamura, M. 201
Nakamura, Masashi 471
Nakamura, Ritaka 495
Nakano, Hayato 2764
Nakano, Shigeki 2370
Nakao, Hiroshi 196
Nakao, Kazushige 148, 2914

AUTHOR INDEX

Nakao, Yuta 588
Nakashima, Yoshiyasu 196
Nakatsu, Kinya 2082
Nakazawa, Haruo 2404
Nakazawa, Y. 1253
Nakazawa, Yuji 244
Namba, Akihiro 2082
Nanamori, Kimihiro 2789
Naradhipa, Adhistira M. 3833
Narita, Takayoshi 1580, 1586, 2324
Narushima, Hiroki 693
Nashida, Norihiro 1137
Nasr, Miad 2380
Natori, Kenji 588, 1860
Nawaz, Muhammad 2335
Nazib, A. A. 3670
Nee, Hans-Peter 2732, 3292, 3684
Neubert, Markus 3979
Ngamroo, Issarachai 2287
Ngo, Tung 1724
Nguyen, Bang Le-Huy 1046, 3134
Nguyen, Hong-Quan 3426
Nguyen, Minh-Khai 982, 1799, 2145
Nguyen, Tien-The 1046, 3134
Nho, Eui-Cheol 922
Nicolae, Ileana-Diana 1822
Nicolae, Petre-Marian 1822
Nie, Jintong 2963
Niki, Toru 856
Nimura, Takumi 1295
Ninomiya, Tatsuya 2836
Nishikata, Shoji 4227
Nishimura, Yoshitaka 1137
Nishino, Taisei 1364
Nishiyama, Shigeki 2149
Nishizawa, Koroku 2229
Nishizawa, Shin-Ichi 1938
Niu, Haonan 3025
Niyomsatian, K. 4096
Noah, Mostafa 1087, 1095
Noda, Taku 2176
Noda, Yujiro 324
Noguchi, Toshihiko 718, 753
Noh, Seungjun 1598
Nomura, Naofumi 2216
Nomura, Shinichi 2022
Nonogaki, Midori 2292
Noro, Osamu 767
Norrga, Staffan 3292
Norum, Lars 2630
Noto, Yasuyuki 3711
Notohara, Yasuo 1301
Nuchnoi, S. 4096
Nugroho, Dannisworo S. 3855
Nussbaumer, Thomas 3619

Nuutinen, P. 2948
Obara, Hidemine 1649
Oda, Yoshiho 1586, 2324
Ogasawara, Satoshi 2589, 2725, 2796, 3315
Ogawa, Eri 2768
Ogawa, Kazuki 1580
Ogawa, Takuro 866
Ogawa, Tomoyuki 1828, 3265
Ogawa, Toru 2796
Ogino, Hiroshi 517
Oh, Sehoon 688
Ohashi, Hidetomo 2774
Ohdera, Fumiya 1322
Ohguchi, Hideki 699
Ohishi, Kiyoshi 1741, 3332, 3890, 3896
Ohji, Takahisa 3182
Ohnishi, Haruna 3273
Ohno, Takanobu 971
Ohno, Tatsuki 1649
Ohnuma, Naoto 233
Ohnuma, Takumi 1223
Ohnuma, Yoshiya 2738
Ohta, Kazuki 1223
Ohta, Takahiro 517
Ohtake, Asuka 3286
Ohyama, K. 1253
Ohyama, Kazuhiro 2921
Ohyama, Kazunobu 244
Oi, Kazunobu 1890
Oishi, Kazuki 3644
Oiwa, Takaaki 157, 4042
Oka, Toshiomi 2370
Okamoto, Kenkichiro 1095
Okazaki, Yuhei 2335
Okazawa, Toshio 2066
Oki, Yusuke 1828
Okitsu, Takashi 169
Okuda, Takafumi 3654, 3757
Okuno, Kengo 1586, 2324
Okuyama, Ryota 3450
Omori, Hideki 1202, 1259, 2558, 2887
Omori, Shuto 471
Omura, Ichiro 1938
Onishi, Hiroyuki 4139
Onishi, Masami 2082
Ono, Y. 4080
Onozawa, Yuichi 2768
Ooshima, Masahide 3613
Orikawa, Koji 2589, 2725, 3315
Ortiz-Gonzalez, J. 1149
Osawa, Akihiro 2764
Oshima, Takuya 4088
Osman, Ilham 3971
Ota, Ryosuke 3855
Ouaida, Rémy 2153

AUTHOR INDEX

Ouchi, Takayuki 250
Ouyang, Shaodi 1051, 3129
Ouyang, Ziwei 4066
Owaki, Daiki 809
Paiboon, Supakorn 1642
Pairindra, Worapong 1460
Pan, Pengpeng 1504
Pan, Xuewei 1172
Panda, Sanjib Kumar 4186
Pang, Hui 2343
Papadopoulos, C. 3596
Papini, L. 4034
Paramalingam, Jan 2329
Parashar, Sanket 3993
Park, Hwa-Pyeong 289
Park, Jin-Hyuk 4104
Park, Jun H. 2564
Park, Kwon-Sik 922
Park, Moo-Hyun 1562, 3100, 3533
Park, Mu-Hyun 838
Park, Sang Uk 1336
Park, Sanghyeon 282
Partanen, J. 2948
Pasterczyk, Robert 2158
Patel, Prashant 3046
Patel, Utsav 3046
Pathmanathan, M. 488
Patwa, Premal 3046
Pauli, Florian 740
Pecharroman, Ramon R. 2534
Pei, Xuejun 2997
Peltoniemi, P. 2948
Peng, Jinjie 939
Peng, Xu 1027, 2714, 3020
Pengxiang, Zeng 4181
Pham, N. Ha 1414
Pidaparthy, Syam Kumar 1465
Pinomaa, P. 2948
Polmai, Sompob 1764, 2490
Pou, Josep 3117
Prabowo, Yos 3564
Prasanth, Sundararajan 416
Prodic, Aleksandar 831
Promyoo, Adisak 2871
Pueschel, Tilo 190
Pyrhonen, J. 161
Qi, Wenlong 889
Qian, Cheng 1472
Qian, Qinsong 3145
Qiao, Liang 3329
Qin, Zian 1925
Qiu, Maohang 878
Qiu, Zhifeng 939
Rabkowski, Jacek 2129
Radman, Karlo 3632

Radwan, Hamdy 3945
Rahimo, M. 3596, 4009
Rahman, Ahmad Arif Bin Abd 2956
Rahman, Faz 3971
Rahmati, Abdolreza 2854
Ramirez-Elizondo, Laura 1426
Ramos, Niño Christopher 3092
Ran, L 1149
Ran, Li 1931
Rao, Eswar 3471
Rathore, Akshay Kumar 342, 2456
Reinikka, Tommi 1704
Remus, Nico 3243
Ren, Haijun 2714
Ren, Yu 3329
Rencz, Marta 137
Rengarajan, Satish 3564
Riar, Baljit 4074, 4145
Rietmann, Stefan 2301
Rim, Chun T. 2564
Risseh, Arash Edvin 2732
Rivas-Davila, Juan 282, 632, 3848
Robert, Mickaël 2158
Rodriguez-Diaz, Enrique 1498
Roes, M. G. L. 946, 2697
Roinila, Tomi 547, 1704, 1719, 2470, 3156, 4205
Romano, Daniele 3588
Roy, Sourov 446, 2416
Ruan, Liheng 3010, 3015
Rubino, S. 732
Ruf, Andreas 740
Rygg, Atle 759
Sadakata, Hideki 410
Sagawa, Kouhei 2036
Saha, Tarak 4074, 4145
Saito, Tatsuhito 1828, 3265
Saito, Yota 1782
Saitoh, Hiroumi 2278
Sakabe, Tomoki 3058
Sakai, Kazuto 2826
Sakai, Ryosuke 2832
Sakai, Yoshikazu 4114
Sakawaki, Atsushi 244
Sakimoto, Kenichi 767
Sakiyama, Taiki 2186
Sakoda, Kenichi 860
Sakr, Nadim 2078
Sakuma, Kensuke 3522
Sakuraba, Tomokazu 2153
Sakurai, Seiya 3412
Samanta, Suvendu 342
Samermurn, S. 4096
Samizadeh, Mehdi 1062, 2854
Sanada, M. 704, 2808
Sanada, Masayuki ..1189, 1289, 1329, 2802, 2814, 3197

AUTHOR INDEX

Sangwongwanich, Ariya 2512
Sangwongwanich, S. 4096
Sangwongwanich, Somboon 1642, 2581
Sannomiya, Kenta 1259
Sano, Kenichiro 3299
Sano, Toshiki 3896
Santi, Enrico 1719
Sasaki, Masahiro 2774
Sasaki, Masato 3344
Sasongko, Firman 416
Sathik, Mohamed 416
Sato, Fumihiro 250
Sato, Keisuke 3265
Sato, Kenji 3478
Sato, Mitsuru 118
Sato, Motoki 663
Sato, Takashi 3644
Sato, Yasuhiro 2042
Sato, Yukihiko 588, 1860, 1972, 3514, 3522
Satoh, Nobuo 2750
Sayed, Mahmoud A. 3945
Schanen, Jean-Luc 2158
Schletz, Andreas 2410
Schülting, Philipp 388
Schweiker, Daniel 2848
Schweizer, Mario 555
Schwendemann, Rüdiger 3074
See, Kye Yak 2296
Sekiba, Yoichi 2176
Sekimoto, Morimitsu 866
Sekisue, Takayuki 2176
Sekiya, Hiroo 3650, 4127
Semwal, R. R. 1358
Senanayake, Thilak 313
Seng, Tan Chuan 416
Seo, Byuong-Jun 922
Seo, Gab-Su 213
Sera, Dezso 2512
Setiadi, Hadi 626
Settels, Sjef J. 3840
Severson, Eric L. 4020
Sewergin, Alexander 3979
Sha, Yilin 3329
Shabib, G. 3945
Shamseh, Mohammad Bani 3916
Shan, Zhenyu 977
Shang, Gao 1282
Shao, Chi 2866
Shao, Riming 1793
Sharma, Avinash 2456
Sharma, Sohit 1730
Shen, Yanfeng 1788, 1925
Shen, Yatao 815
Shen, Yecheng 2842
Shen, Zhan 1788, 1925

Sheng, Caiwang 1167
Shi, Gang 4220
Shi, Haixu 4168
Shi, Xiangyue 939
Shi, Yong 2877
Shibata, Naoya 3929
Shigeeda, Hidenori 2540
Shigematsu, Koichi 2176
Shigeuchi, Koji 3514, 3522
Shijo, Takuya 324
Shimada, Takae 250
Shimakage, Toyonari 2292
Shimamoto, Keita 2210
Shimao, Tohihiro 439
Shimaoka, Masahiro 1747
Shimizu, Toshihisa 302, 404, 2137, 2165, 3309, 3357
Shimizu, Toshimasa 1803
Shimomura, Shoji 2836
Shimono, Tomoyuki 675
Shimosato, Noboru 261, 3514, 3522
Shimoyama, A. 142
Shin, Sungyong 3418
Shinohara, Atsushi 1308, 1322
Shinohara, Hiroshi 1840
Shinshi, Tadahiko 4016
Shintani, Michihiro 3644
Shirai, Ryo 3309
Shirata, Kento 1137
Shiyuan, Yin 2625
Shoyama, Masahito 1095, 1554, 3773
Shujiang, Duan 2718
Shunsuke, Ohasi 3363
Shuto, Masao 699
Si, Yunpeng 2400, 3742
Sihvo, Jussi 2470
Sih-Yi, Lee 645
Silber, Siegfried 4028
Silventoinen, P. 2948
Simanjorang, Rejeki 416, 2296
Singh, Amit Kumar 4186
Singh, Vijay Kumar 1698
Son, Yung-Deug 3323
Song, Hongyu 3825
Song, Injong 803
Song, Kai 84
Song, Seung-Min 3229
Song, Shuguang 1051, 3129, 3924
Song, Wensheng 1872
Song, Yang 3698
Song, Yipeng 746
Song, Yubo 3877
Soong, Boon-Hee 1517, 1592
Soong, Theodore 3203
Soontorntaweesub, Kittichot 1764
Spiliotis, K. 488

AUTHOR INDEX

Stieneker, Marco 598, 2484
Stock, Alexander 3034
Stojadinovic, Miloš......................................1103
Su, Huiling 795
Su, Jianhui 2877
Su, Yu-Chen 1038, 3549
Sudo, K. 1240
Suetake, A. 142
Suetsugu, Tadashi 4193
Sueuchi, Yuki 1955
Sugahara, Satoshi..2756
Sugahara, T. 142
Suganuma, K. 142
Suganuma, Katsuaki.....................................1598
Sugihara, Yusuke 2789
Sugimoto, Hiroya 3627
Sugimoto, Kazushige......................................767
Sugiyama, Takashi.......................................3578
Suh, Yongsug 1977
Sul, Seung-Ki 521, 911, 3557
Sumida, Hitoshi...2774
Sun, Bainan 607
Sun, Chuan 370
Sun, Haotian 2780
Sun, Jianning 2963
Sun, Kai 3460, 4168
Sun, Lejia 2882
Sun, Peng 1125
Sun, Shumin 1455
Sun, Weifeng 3145
Sun, Xiangdong 2204
Sun, Yongping 560
Sun, Yuchong 3650, 4127
Sung, Kyungmin...1364
Suntio, Teuvo...963
Supanyapong, S..1622
Surakitbovorn, Kawin 632, 3848
Surinkaew, Tossaporn...................................2287
Suul, Jon Are 782, 2003
Suwa, Hiroshi 1997
Suwankawin, S..4096
Suwankawin, Surapong...................................2871
Suzuki, Akio 1840
Suzuki, Dai 157
Suzuki, Hiromitsu 495
Suzuki, Kazuma 1847, 3501, 3507
Suzuki, Kenichiro...511
Suzuki, Toshiki...................................1586, 2324
Suzuki, Yuhei 3390
Suzumori, Hirofumi 2066
Tabata, Yoichiro 329
Tada, Makoto 1580
Tadano, Hiroshi 313, 1111, 3375
Tadano, Yugo 483, 1890
Taguchi, Masashi 826

Taguchi, Yoshiaki...3280
Taiyuan, Yin 2625
Tajima, Katsubumi..2832
Tajyuta, Toshihisa 1840
Takahashi, Akihiko 3896
Takahashi, Akiko 2449
Takahashi, Arata 1270
Takahashi, Isseki 575
Takahashi, Masaki 3186
Takahashi, R. 3665
Takahashi, Shotaro 3315
Takahashi, Tomohira 2796
Takahashi, Toshimichi 227
Takahashi, Yuki 3375
Takakura, Shotaro 1270
Takami, Hiroshi 471
Takamura, Kenya 1381
Takano, Sho 3390
Takasho, Kenta 1890
Takatori, Koji 4139
Takayanagi, Ryohei 3396
Takeda, Kodai 2309
Takemoto, Masatsugu 2589, 2725, 2796, 3315
Takenaka, Hiroshi 3304
Takeno, K. 201
Takenoiri, Shunji 2764
Takeshita, Takaharu 356, 1847, 3501,
3507, 3791, 3945, 4088
Takeuchi, Norikazu 2292
Takeuchi, Yoko..................................1828, 3265
Takiguchi, Masashi 3723
Takimoto, Kazuyasu 3304
Takishima, Kenta 2826
Takubo, Hiromu 3390
Takuma, Shunsuke 2596
Takuno, Tsuguhiro 3578
Tamate, Michio..3315
Tan, Nguyen Anh 478
Tan, Siew-Chong 889
Tanaka, Akira 1960
Tanaka, Takaaki 2604
Tanaka, Takahide 2774
Tanaka, Toshihiko 324, 1381
Tanaka, Tsuguhiro 3929
Tanaka, Y. 1162
Tanemo, Masamichi 2022
Tang, Cheng-Yu 639
Tang, Houjun 528
Tang, Ye 3705
Tang, Yi 337, 428, 434, 3910
Taniguchi, Katsumi 3396
Taniguchi, Katsunori 1202
Taniguchi, Tomoisa 866
Tatsumi, Kazuto 1202
Tatsuta, Fujio 4227

AUTHOR INDEX

Tatte, Yogesh 1730
Tausif, Ali 3833
Tcai, Anatolii 4109
Techama, Pantarote 2490
Teerakawanich, Nithiphat 3332
Teigelkötter, Johannes 3034
Tenconi, A. 732
Teraoka, Kenji 3086
Tey, Kuan-Chung 511
Thai, Van X. 2564
Thummala, Prasanth 4066
Tian, Mofan 998, 2785
Tian, Wei 1661
Tian, Xiaoyu 1771
Tian, Yanjun 1397
Tibola, Gabriel 1447
Tikka, V. 2948
Toba, Akio 1840
Toi, Takato 2229
Tokumaru, Syohei 2938
Tokusaki, Hiroyuki 2589
Tominaga, Isamu 1692
Tomita, Mutuwo 1295
Tong, Anping 1391, 2866
Tran, Hai N. 3833
Tran, Tan-Tai 1799, 2145
Trescases, O. 267
Trescases, Olivier 2380
Tripathi, Ravi Nath 1698
Troppenz, Maria 3607
Trung, Tran Vu 1666
Tsai, Chang-Lin 3151
Tsai, Meng-Jiang 2462
Tsai, Men-Shen 1575
Tsai, Terng-Wei 639, 883
Tsai, Tsung-Lin 3151
Tsai, Yue-Ting 4198
Tse, Chi K. 3768
Tseng, King Jet 1517, 1592
Tseng, Wei-Jing 1626
Tsuchiya, Taichiro 2329
Tsuji, Hitoshi 3717
Tsuji, M. 1240
Tsuji, Mineo 1217, 1276, 2938
Tsukakoshi, Masahiko 238
Tsumura, Akihiko 3490
Tsuno, Masahito 2558
Tsuruta, Ryoji 495
Tsuruta, Yukinori 318
Tsutsumi, Hirohiko 3723
Tu, Yiming 2435, 2681, 3176, 3439, 3934
Tuji, Mineo 3237
Tumerdem, Ugur 1680
Tumurbaatar, Anudari 1972
Uchida, Junichi 1955

Uchida, Yuuki 2750
Uchino, Yuki 324
Uda, Ryosuke 3578
Udagawa, Ikuto 517, 1692
Ueda, Tetsuzo 3762
Uehara, H. 1253
Uematsu, Takeshi 118, 4139
Uemura, Takamasa 860
Ueno, Tsutomu 4151
Uesugi, Yuma 3412
Ueta, Hiroaki 1883
Umeda, Takashi 2814
Umetani, Kazuhiro 410, 1602, 1610
Unamuno, Eneko 759
Uno, Masatoshi 1782, 2030
Unterrieder, Christoph 3140
Ura, A. 704
Urabe, Shinichi 1782
Urata, Kazuki 302
Ute, Ryo 3773
Valente, G. 4034
Van De Ven, B.A.C. 267
Van Duivenbode, Jeroen 3840
Van Lam, Phi 571
Vasquez, C. Juan 3678
Vasquez, Juan C. 1498
Vass-Varnai, Andras 137
Veerachary, M. 845
Vemulapati, U. 3596, 4009
Vobecky, J. 3596
Vukadinovic, Nenad 831
Vyacheslav, Shkodyrev 1966
Wachi, Tsuneshisa 1997
Wada, Haruhisa 3286
Wada, Keiji 1414, 1866, 1919, 2137, 4059
Wakimoto, Hiroki 2404
Wang, Beibei 795
Wang, Bo 459
Wang, Can 1172
Wang, Chao 2386, 2901
Wang, Congling 3112
Wang, Dong 1183, 1246
Wang, Feng 1157, 1167, 2882
Wang, Fusheng 2611, 2992
Wang, Gaolin 1206, 1234
Wang, Guoxin 1206
Wang, Hanyu 2997
Wang, Hao 2270
Wang, Haoyu 100, 3825
Wang, Hechao 1183, 1246
Wang, Hongjie 4074, 4145
Wang, Huai 1021, 1788, 1925, 2604, 2743, 3123
Wang, Huiying 1234
Wang, Jianing 2611
Wang, Jizhe 826, 2097

AUTHOR INDEX

Wang, Jun 3749, 3985
Wang, Kui 1010, 2386
Wang, Laili 2785, 3863
Wang, Liang 3958
Wang, Lifang 3338
Wang, Liwei 927
Wang, Meng 2992
Wang, Naizeng 998, 2785
Wang, Panrui 3383
Wang, Po-Wei 1617
Wang, Qiusheng 2421
Wang, Shike 1524, 3692
Wang, Shinn-Shyong 2086
Wang, Shitao 2866
Wang, Shunyu 3002
Wang, Wei 614, 3780
Wang, Wenjie 1391, 2866
Wang, Xiaolei 1455
Wang, Xiaoqing 878
Wang, Xiaoyang 453
Wang, Xiongfei 1711, 2673, 3164, 3357, 3684
Wang, Yanbo 1758, 2708
Wang, Yangyang 2505
Wang, Yi 1027, 1397, 3495
Wang, Yijie 614, 934, 3780, 3825
Wang, Youyun 2204
Wang, Yu-Chi 657
Wang, Yue 1455, 3863
Wang, Yuncheng 1177
Wang, Zhongxu 2743, 3123
Watanabe, Hiroki 896
Watanabe, Shoichiro 2042
Wei, Baoze 3678
Wei, Feng 1517, 1592
Wei, Jianzhao 2630
Wei, Juan 1131
Wei, Shilei 1397
Wei, Wang 1442, 2969
Wei, Xiaoguang 2343
Wei, Xiuqin 3650, 4127
Wei, Zhang 2969
Wellawatta, Thusitha Randima 1409
Wen, Huiqing 453
Wen, Po-Hsiang 3544
Wenbing, Li 1282
Wickramasinghe, Harith R. 3117
Wijaya, Febry Pandu 3490
Wikström, T. 3596
Winter, Christian 388
Wolf, Mihaela 3607
Wolski, Kornel 2129
Wu, Bin 3797
Wu, Heng 2673
Wu, Hongfei 4168
Wu, Min 3863

Wu, Pei-Lin 1145
Wu, Ping-Heng 503, 3549
Wu, T.-F. 1912
Wu, Tsai-Fu 3884
Wu, Tsung-Hsi 3544
Wu, Xiaojie 619
Wu, Xiaojun 3010, 3015
Wu, Ya'nan 2496
Wu, Zhiqian 1549
Würfl, Joachim 3607
Wyss, Jonas 3734
Xia, Meng 3484
Xia, Yongming 2842
Xiao, Chanjuan 1131
Xiao, Dan 3971
Xiao, Guochun 1549, 2944
Xiao, Jianfang 4157
Xiao, Xi 1966
Xiaoxi, Liu 2969
Xie, Jingwen 3069
Xie, Shaofeng 2547
Xie, Xiaogao 987
Xie, Zhen 2611, 2992
Xiong, Wei 939
Xu, Binci 2270
Xu, Cai 2986
Xu, Dehong 1118, 2270, 2569
Xu, Dewei David 1537, 3797
Xu, Dianguo 459, 560, 614, 934,
            1206, 1234, 3780, 3825, 4213
Xu, Guangzhao 998
Xu, Huadian 2877
Xu, Jin 261, 3514, 3522
Xu, Peng 1478
Xu, Sheng 3002
Xu, Shuang 1793
Xu, Yin-Chi 3884
Xu, Yue 3985
Xuan, Yang 1478
Xuanjie, Gao 2718
Xue, Danhong 2435
Yabuuchi, Tatsushi 233
Yada, Tomoharu 1381
Yamada, Hiroaki 324, 1381
Yamada, Koji 169
Yamaguchi, Daiki 3940
Yamaguchi, Koji 1972
Yamaji, Masaharu 2774
Yamamoto, Aoto 2558
Yamamoto, Hidekazu 2750
Yamamoto, Kichiro 1308, 1322
Yamamoto, Masaya 1782, 2030
Yamamoto, Masayoshi 1087, 1095, 2738, 2789, 3344
Yamamoto, Ryo 4016
Yamamoto, Shu 1949, 1960

AUTHOR INDEX

Yamamoto, Yuuto 3197
Yamanaka, Daisuke 2329
Yamanaka, Kenji 3369
Yamashita, Hiroki.............................1196
Yamashita, Yoshinori 3490
Yamazaki, Katsumi...........................693, 699
Yamazaki, Masahiro..........................207
Yan, Qingzeng 619
Yan, Y.T. 851
Yan, Zhang 4181
Yanagisawa, Yuta 3762
Yang, Chang-Jun 3884
Yang, Cheng-Jhen 639, 883
Yang, Daoshu 1549
Yang, Dongsheng 3357
Yang, Geng 542
Yang, Hong-Tzer...............................4233
Yang, Hui-Chen 2296
Yang, Mei...3958
Yang, Ming 560
Yang, Peng...1966
Yang, Ping 3112
Yang, Renxin 4220
Yang, Sheng-Ming 651, 3426
Yang, Shunfeng 428
Yang, Shuying 2611
Yang, Xu 998, 1062, 1478, 2785, 2854, 3329
Yang, Ying..2973
Yang, Yongheng................ 439, 1021, 1788, 2512, 2743
Yang, Yugang.....................................2703
Yang, Zebin 1157, 1167
Yang, Zhichang..................................2058
Yang, Zhihua 3797
Yang, Zhiqing....................................1073
Yang, Zhongping............. 1131, 1816, 2058, 3484, 3495
Yano, Junya..3723
Yao-Ching, Hsieh 645
Yaoqin, Jia 2441
Yasuda, Takumi..................................992
Yasuda, Yusuke..................................2082
Yaxin, Peng..416
Ye, Han 1504, 2496
Yeh, Shun-Hao 4233
Yelaverthi, Dorai Babu......................4066
Yen, Chih-Ying..................................1145
Yenchamchalit, Kulsomsup................2430
Yi, Hao 2780, 2882
Yijie, Hou...2441
Yin, Shiyuan 1455
Yin, Taiyuan 1455
Yin, Zhijian 1021
Yin, Zhonggang 2204
Yingchun, Xu.....................................2441
Yokokura, Yuki.................. 1673, 1741, 3890, 3896
Yokoyama, T. 3665

Yokoyama, Tomoki 1270, 1877, 1883, 2914, 3363, 3658
Yonezawa, Y. 3603
Yonezawa, Yu 196
Yoon, Bo-Kyung 3063
Yoshida, Souichi 2764
Yoshida, Yukihiro 2832
Yoshihara, Hidemasa 219
Yoshihara, Tohru 1997
Yoshikawa, Gaku 3280
Yoshimi, Daisuke 3952
Yoshimura, Eiji 767
Yoshino, Takuma 3363
Yoshino, Teruo 1692, 3916
Yoshioka, Yusuke 4151
Yoshizawa, Daisuke 238
You, Jiang 1386, 1421
You, Zih-Cing 651
Yu, Yong 459
Yuan, Huawei 889
Yuan, Liqiang 2963
Yuan, Xibo 619, 1634
Yuan, Yiqin 977, 3040
Yue, Wang..2625
Yui, Haiyan 699
Yukita, Kazuto....................................809
Zaijun, Wu 1282
Zaitsu, Toshiyuki 118, 4139
Zaman, Mohammad Shawkat..............2380
Zanchetta, P. 4034
Zane, Regan 4066, 4074, 4145
Zdanowski, Mariusz...........................2129
Zeng, Pengxiang 2646
Zhang, Chen 4220
Zhang, Feili 1315
Zhang, Guoqiang 1206, 1234
Zhang, H. 142
Zhang, Hailong 3863
Zhang, Hao 1131, 1598
Zhang, Hongyang 3684
Zhang, Jianwen 1004
Zhang, Jianzhong 2262
Zhang, Le 3145
Zhang, Lei 3383
Zhang, Lifei 2703
Zhang, Meng 1966
Zhang, Qianfan 3025
Zhang, Runze 1816
Zhang, Shichong 2638
Zhang, Shu 614, 934, 3780
Zhang, Shuai 2944
Zhang, Tengfei 2980
Zhang, Wang 2625
Zhang, Xiaofang 2547

AUTHOR INDEX

Zhang, Xin 917, 953, 1004, 2688,
    2692, 2980, 4157, 4162
Zhang, Xinan 1724
Zhang, Xing 2973, 2992
Zhang, Xueguang 4213
Zhang, Y. 946, 2697
Zhang, Yan 2646
Zhang, Yang 1177
Zhang, Yanping..2204
Zhang, Yaqian 2262
Zhang, Yi 2743, 3123
Zhang, Zhe.............................607, 1351, 3460
Zhang, Zhenbin 1661, 2895, 3965
Zhang, Zhigang 1157, 1167
Zhao, Chongyan 904
Zhao, Fangzhou 1549, 2944
Zhao, Fei 1172
Zhao, Jianfeng.....................3002, 3010, 3015
Zhao, Juan 2051
Zhao, Shengnan...795
Zhao, Tianshu 3020
Zhao, Tianyang 1172
Zhao, Yuanliang..3698
Zhao, Zhengming 2963
Zhao, Zhibin 1125
Zhao, Zhiqing..2714
Zheng, Deyou..2611
Zheng, Xuemei 2901
Zheng, Zedong.......................1010, 2386
Zhong, Wenxing 1118, 2569
Zhou, Dao 1758
Zhou, Dehong 428, 434
Zhou, Fulin 3257
Zhou, Jiuyang..2462
Zhou, Lei..2505
Zhou, Sheng-Zhi...370
Zhou, Victor 1943
Zhou, Yan 934
Zhou, Yimin...2547
Zhu, Cailing 3052
Zhu, Chunbo 84
Zhu, Helin 1336
Zhu, Junjie...3145
Zhu, Lianghong.....................1206, 1234
Zhu, Qingwei 3338
Zhu, Yanlin 2780
Zhu, Ye 2270
Zhujian, Ou 1376
Zhuo, Fang...................1157, 1167, 2780, 2882
Zhuyong, Li 2986
Zischler, Sigrid 2410
Zou, Yaohan 3455